Genomics and Genetics

Edited by
Robert A. Meyers

Related Titles

Meyers, R. A. (ed.)
Encyclopedia of Molecular Cell Biology and Molecular Medicine
16 Volume Set
ISBN-13: 978-3-527-30542-1
ISBN-10: 3-527-30542-4

Dunn, M. J., Jorde, L. B., Little, P. F. R., Subramaniam, S. (eds.)
Encyclopedia of Genetics, Genomics, Proteomics and Bioinformatics
8 Volume Set
ISBN-13: 978-0-470-84974-3
ISBN-10: 0-470-84974-6

Sensen, C. W. (ed.)
Handbook of Genome Research
Genomics, Proteomics, Metabolomics, Bioinformatics, Ethical and Legal Issues
2005
ISBN-13: 978-3-527-31348-8
ISBN-10: 3-527-31348-6

Meksem, K., Kahl, G. (eds.)
The Handbook of Plant Genome Mapping
Genetic and Physical Mapping
2005
ISBN-13: 978-3-527-31116-3
ISBN-10: 3-527-31116-5

Kahl, G.
The Dictionary of Gene Technology
Genomics, Transcriptomics, Proteomics
2004
ISBN-13: 978-3-527-30765-4
ISBN-10: 3-527-30765-6

Hacker, J., Dobrindt, U. (eds.)
Pathogenomics
Genome Analysis of Pathogenic Microbes
2006
ISBN-13: 978-3-527-31265-8
ISBN-10: 3-527-31265-X

Licinio, J., Wong, M.-L. (eds.)
Pharmacogenomics
The Search for Individualized Therapies
2002
ISBN-13: 978-3-527-30380-9
ISBN-10: 3-527-30380-4

Borlak, J. (ed.)
Handbook of Toxicogenomics
Strategies and Applications
2005
ISBN-13: 978-3-527-30342-7
ISBN-10: 3-527-30342-1

Genomics and Genetics

From Molecular Details to Analysis and Techniques

Edited by

Robert A. Meyers

Volume 1

WILEY-VCH Verlag GmbH & Co. KGaA

The Editor

Dr. Robert A. Meyers
RAMTECH LIMITED
122 Escalle Lane
Larkspur, CA 94039
USA

1st Edition 2007
 1st Reprint 2007

■ All books published by Wiley-VCH are carefully produced. Nevertheless, authors, editors, and publisher do not warrant the information contained in these books, including this book, to be free of errors. Readers are advised to keep in mind that statements, data, illustrations, procedural details or other items may inadvertently be inaccurate.

Library of Congress Card No.: applied for

British Library Cataloguing-in-Publication Data: A catalogue record for this book is available from the British Library.

Bibliographic information published by the Deutsche Nationalbibliothek
The Deutsche Nationalbibliothek lists this publication in the Deutsche Nationalbibliografie; detailed bibliographic data are available in the Internet at http://dnb.d-nb.de.

© 2007 WILEY-VCH Verlag GmbH & Co. KGaA, Weinheim

All rights reserved (including those of translation into other languages). No part of this book may be reproduced in any form – by photoprinting, microfilm, or any other means – nor transmitted or translated into a machine language without written permission from the publishers. Registered names, trademarks, etc. used in this book, even when not specifically marked as such, are not to be considered unprotected by law.

Composition: Laserwords Private Ltd, Chennai, India
Printing: betz-druck GmbH, Darmstadt
Bookbinding: Litges & Dopf Buchbinderei GmbH, Heppenheim

ISBN-13: 978-3-527-31609-0

ISBN-10: 3-527-31609-4

Contents

Volume 1

Preface vii

Color Plates ix

1 Molecular Genetics 2
1 Molecular Basis of Genetics 3
  D. Peter Snustad

2 DNA Replication and Transcription 47
  Yusaku Nakabeppu, Hisaji Maki, and Mutsuo Sekiguchi

3 Translation of RNA to Protein 75
  Robert Ashley Cox Henry Rudolph Victor Arnstein

4 Alternatively Spliced Genes 117
  Jane Y. Wu, Liya Yuan, and Necat Havlioglu

5 Repair and Mutagenesis of DNA 171
  Raymond Devoret

6 Molecular Genetics of Genomic Imprinting 189
  Robert Feil, Yuji Goto, and David Umlauf

7 Heterochromatin and Eurochromatin – Organization, Packaging, and Gene Regulation 209
  Boris A. Leibovitch Sarah C. R. Elgin

2 Genomic Organization and Evolution 228
8 Organization of Genes and Genome Domains 229
  Yujing Zeng, Javier Garcia-Frias, and Adam G. Marsh

9 Anthology of Human Repetitive DNA 269
  Vladimir V. Kapitonov, Adam Pavlicek, and Jerzy Jurka

Genomics and Genetics. Edited by Robert A. Meyers.
Copyright © 2007 Wiley-VCH Verlag GmbH & Co. KGaA, Weinheim
ISBN: 978-3-527-31609-0

10	Evolution of Noncoding DNA in Eukaryotes *Josep M. Comeron*	325
11	Horizontal Gene Transfer *Jack A. Heinemann Ralph A. Bungard*	337
12	Molecular Systematics and Evolution *Jeffrey H. Schwartz*	359
13	Genetic Variation and Molecular Evolution *Werner Arber*	385

3 Genomes of Model Organisms — 408

14	E. Coli Genome *Hirotada Mori and Takashi Horiuchi*	409
15	*Drosophila* Genome *Robert D.C. Saunders*	419
16	Malaria Mosquito Genome *Robert A. Holt Frank H. Collins*	441
17	Zebrafish (*Danio rerio*) Genome and Genetics *Ralf Dahm, Robert Geisler, Christiane Nüsslein-Volhard*	469
18	Rat Genome (*Rattus norvegicus*) *Kim C. Worley Preethi Gunaratne*	503
19	Chimpanzee Genome *Ingo Ebersberger*	551

Preface

The *Genomics and Genetics* two volume set was compiled from a selection of key articles from the recently published *Encyclopedia of Molecular Cell Biology and Molecular Medicine* (ISBN 978-3-527-30542-1). The *Genomics and Genetics* set is comprised of 39 detailed articles arranged in six sections covering molecular genetics, genomic organization and evolution, genomes of model organisms, genomic sequencing, genetic engineering, as well as gene medicine and disease. The articles were prepared by eminent researchers from the major research institutions in the United States, Europe and around the globe.

Each article begins with a concise definition of the subject and its importance, followed by the body of the article and extensive references for further reading. The references are divided into secondary references (books and review articles) and primary research papers. Each subject is presented on a first-principle basis, including detailed figures, tables and drawings. Because of the self-contained nature of each article, some overlap among articles on related topics occurs. Extensive cross-referencing is provided to help the reader expand his or her range of inquiry.

The master publication, which is the basis of the *Genomics and Genetics* set, is the *Encyclopedia of Molecular Cell Biology and Molecular Medicine*, which is the successor and second edition of the *Encyclopedia of Molecular Biology and Molecular Medicine*, covers the molecular and cellular basis of life at a university and professional researcher level. This second edition is double the first edition in length and will comprise the most detailed treatment of both molecular and cell biology available today. The Board and I believe that there is a serious need for this publication, even in view of the vast amount of information available on the World Wide Web and in text books and monographs. We feel that there is no substitute for our tightly organized and integrated approach to selection of articles and authors and implementation of peer review standards for providing an authoritative single-source reference for undergraduate and graduate students, faculty, librarians and researchers in industry and government.

Our purpose is to provide a comprehensive foundation for the expanding number of molecular biologists, cell biologists, pharmacologists, biophysicists, biotechnologists, biochemists and physicians as well as for those entering molecular cell biology and molecular medicine from majors or careers in physics, chemistry, mathematics, computer science and engineering. For example there is an unprecedented demand for physicists, chemists and computer scientists who will work with biologists to define the genome, proteome and interactome through experimental and computational biology.

Genomics and Genetics. Edited by Robert A. Meyers.
Copyright © 2007 Wiley-VCH Verlag GmbH & Co. KGaA, Weinheim
ISBN: 978-3-527-31609-0

The Board and I first divided all of molecular cell biology and molecular medicine into primary topical categories and each of these was further defined into subtopics. The following is a summary of the topics and subtopics:

- Nucleic Acids: amplification, disease genetics overview, DNA structure, evolution, general genetics, nucleic acid processes, oligonucleotides, RNA structure, RNA replication and transcription.
- Structure Determination Technologies Applicable to Biomolecules: chromatography, labeling, large structures, mapping, mass spectrometry, microscopy, magnetic resonance, sequencing, spectroscopy, x-ray diffraction.
- Proteins, Peptides and Amino Acids: analysis, enzymes, folding, mechanisms, modeling, peptides, structural genomics (proteomics), structure, types.
- Biomolecular Interactions: cell properties, charge transfer, immunology, recognition, senses.
- Molecular Cell Biology of Specific Organisms: algae, amoeba, birds, fish, insects, mammals, microbes, nematodes, parasites, plants, viruses, yeasts.
- Molecular Cell Biology of Specific Organs or Systems: excretory, lymphatic, muscular, neurobiology, reproductive, skin.
- Molecular Cell Biology of Specific Diseases: cancer, circulatory, endocrine, environmental stress, immune, infectious diseases, neurological, radiation.
- Biotechnology: applications, diagnostics, gene altered animals, bacteria and fungi, laboratory techniques, legal, materials, process engineering, nanotechnology, production of classes or specific molecules, sensors, vaccine production.
- Biochemistry: carbohydrates, chirality, energetics, enzymes, biochemical genetics, inorganics, lipids, mechanisms, metabolism, neurology, vitamins.
- Pharmacology: chemistry, disease therapy, gene therapy, general molecular medicine, synthesis, toxicology.
- Cellular Biology: developmental cell biology, diseases, dynamics, fertilization, immunology, organelles and structures, senses, structural biology, techniques.

We then selected some 340 article titles and author or author teams to cover the above topics. Each article is designed as a self-contained treatment. Each article begins with a key word section, including definitions, to assist the scientist or student who is unfamiliar with the specific subject area. The Encyclopedia includes more than 3000 key words, each defined within the context of the particular scientific field covered by the article. In addition to these definitions, the glossary of basic terms found at the back of each volume, defines the most commonly used terms in molecular and cell biology. These definitions should allow most readers to understand articles in the Encyclopedia without referring to a dictionary, textbook or other reference work.

Larkspur, July 2006

Robert A. Meyers
Editor-in-Chief

Color Plates

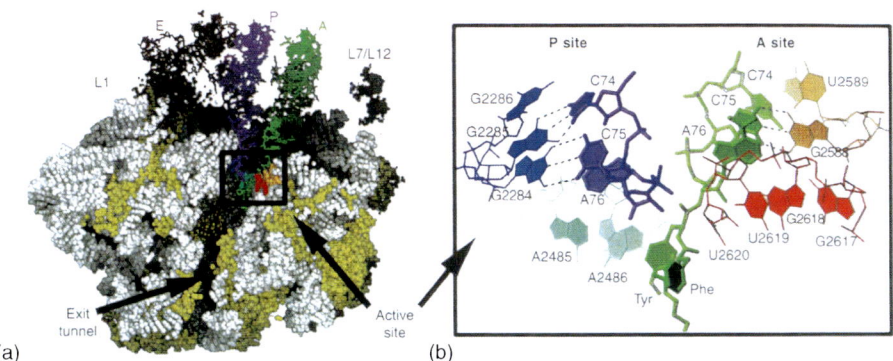

Fig. 11 (p. 98) Fine structure of the peptidyl transferase center interactions of the CCA ends (C74, C75, and A 76) of aminoacyl tRNA and peptidyl-tRNA with the rRNA moiety of the large subunit. Aminoacyl tRNA was represented by a C74 puromycin derivative of tRNA and peptidyl-tRNA was represented by CCA-phenylalanine-caproic acid-biotin. When these model compounds were bound to the A and P sites respectively, the large subunit catalyzed the formation of a peptide bond leading to newly extended peptidyl-tRNA (C74 puromycin–phenylalanine-caproic acid-biotin) in the A site and a deacylated tRNA in the P site. (a) a space filling representation of the 50S subunit (RNA in white and protein in yellow). The tRNAs in the A (green), P (purple), and E (brown) sites are placed according to the structure of the complete ribosome. The subunit has been split though the tunnel and the front section was removed to reveal the peptidyl transferase center (boxed). (b) Base-pairing interactions of CCA sequences of tRNA with 23S rRNA. The analog of deacylated tRNA (purple) is bound to rRNA through base pairing between C74 with G2285) and C75 with G2284. The analog of the newly formed peptidyl-tRNA is bound through base pairing between C75 and G1588. [From Schmeing, et al. (2002) A pretranslocational intermediate in protein synthesis observed in crystals of enzymatically active 50S subunits, *Nat. Struct. Biol.* **9**, 225–230. With permission.]

Genomics and Genetics. Edited by Robert A. Meyers.
Copyright © 2007 Wiley-VCH Verlag GmbH & Co. KGaA, Weinheim
ISBN: 978-3-527-31609-0

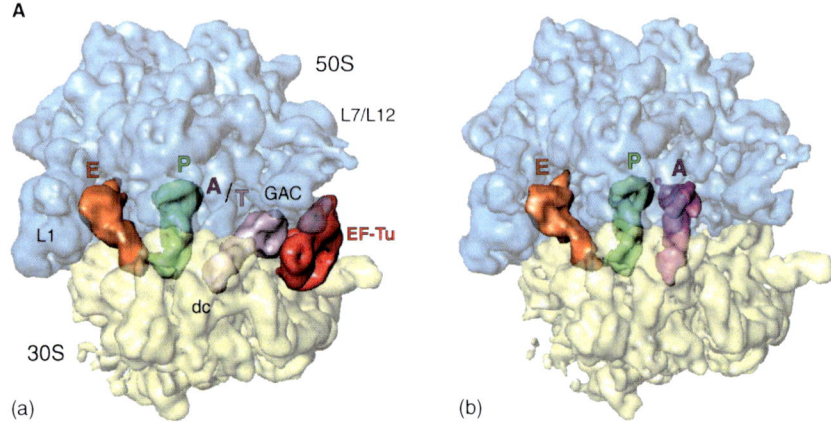

Fig. 13 (p. 102) Cryo-electron microscopy maps showing the incorporation of aminoacyl tRNA into the A site of a bacterial ribosome at a resolution of 0.9 nm and a conformational change in the GTPase-associated center. **A** The location of the binding site for the ternary complex of Phe-tRNA^{Phe}•EF-Tu•GDP-Kir. (a) The antibiotic Kirromycin (Kir) binds to the ternary complex and prevents dissociation of EF-Tu•GDP-Kir from the ribosome after GTP hydrolysis. The stalled complex is located below the L7/L12 stalk of the 50S subunit. EF-Tu is in red and A/T site tRNA is in pale pink. (b) Structure of the ribosome after peptide bond formation. The P site contains tRNA$_f^{Met}$ and the A site contains fMet•Phe-tRNA^{Phe} which is shown in magenta. In both panels, the ribosomal subunits are rendered semitransparent to illustrate the densities for tRNAs in the ribosomal intersubunit space. The sites for the tRNAs are labeled as follows: A/T, A/T site for the aa-tRNA bound to EF-Tu, reaching the decoding site in the A site of the 30S ribosomal subunit; A, aminoacyl site; P, peptidyl site; E, exit site. The P site tRNA is green, the E site tRNA is orange. **B** Conformational change in the GTPase-associated center of the 50S subunit. The cryo-electron microscopy maps show; (a) fMet-tRNA$_f^{Met}$ bound to the P site of the ribosome with a vacant A site; (b) fMet-tRNA$_f^{Met}$ in the P site and Phe-tRNA^{Phe}•EF-Tu•GDP-Kir complex with the A/T site between tRNA colored pink and EF-Tu colored red. The arrow indicates the site at the base of the L7/L12 stalk, which undergoes a conformational change during the binding of the ternary complex; (c) tRNA$_f^{Met}$ bound in the P site and fMet•Phe-tRNA^{Phe} bind in the A site. Insets (b) and (c) focus on the conformational changes of the GTPase-associated center. The view from the intersubunit space allows the movement of the GTPase-associated center to be observed. The E site is shown in orange; (d) Superposition of the 50S ribosomal subunits from the ribosome bound with the stalled ternary complex (semitransparent blue) and from the ribosome bearing a dipeptidyl-tRNA in the A site (solid red). Ribosomal subunits are blue (50S subunit) and yellow (30S subunit). Features of the large ribosomal subunit: L7/L12, stalk of proteins L7/L12 from the 50S ribosomal subunit; GAC, GTPase-associated center; SRC, sarcin-ricin loop; L16, ribosomal protein L16; L1, stalk of ribosomal protein L1. Features of the small ribosomal subunit: h, head of the 30S subunit; dc, decoding center. [From Valle, M., Zavialov, A., Li, W. et al. (2003) Incorporation of aminoacyl tRNA into the ribosome as seen by cryo-electron microscopy, *Nat. Struct. Biol.* **10**, 899–906. With permission.]

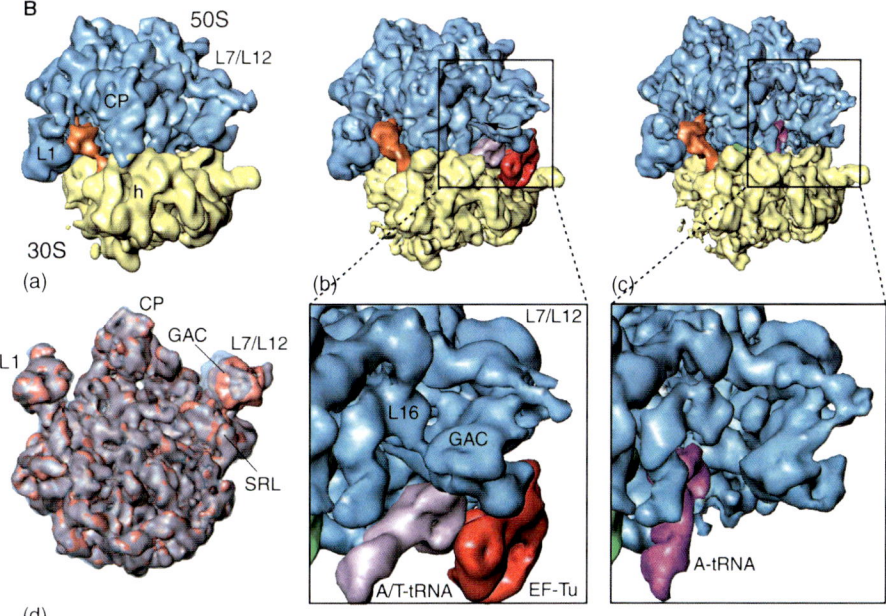

Fig. 13 (Continued)

Fig. 2 (p. 474) Photographs of age-matched adult female (top) and male (bottom) zebrafish. The characteristic horizontal stripes running along the body and fins gave this species its name. Adult zebrafish measure about 3 to 5 cm in length. Females can be recognized by their larger belly accommodating the eggs. Males tend to be more reddish in color than female zebrafish. The scale bar is 1 cm.

A

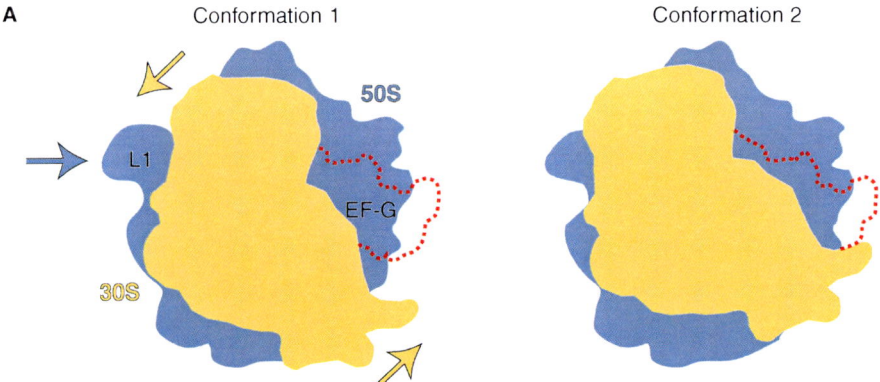

Conformation 1   Conformation 2

Fig. 15 (p. 104) Changes detected by cryo-electron microscopy (1.75–1.84 nm resolution) in the conformation of ribosomes during translocation. (A) After the peptide bond is formed, deacylated tRNA is located in the P site with peptidyl tRNA in the A site (conformation 1), the translocational state. A ratchet-like motion follows when EF-G•GTP is bound to the ribosome leading to conformation 2. The conformational change involves a counterclockwise rotation of the 30S subunit with respect to the 50S subunit with the L1 stalk moving in the direction opposite to the head of the 30S subunit. The outline of EF-G is shown by the dotted line. (B) Conformational changes in the L7/L12 stalk and in EF-G during translocation. The figures show that EF-G (shown in red) links the decoding site with the L7/L12 stalk. The cryo-electron microscopy maps shown in (c) and (d) correspond to conformation 2 in (a) with deacylated tRNA in the P site and peptidyl-tRNA in the A site and EF-G•GTP bound to the ribosome. The L7/L12 stalk (shown on the right of the dotted line) is bifurcated. Part of EF-G is shown close to the A site (marked A in (c)). A third conformation is observed after hydrolysis of GTP to GDP and before EF-G•GDP is released from the ribosome as shown in (a) and (b). In this conformation, the stalk is extended but not bifurcated and part of the EF-G is implicated in an arclike structure. The pretranslocational state is regained (see (a)) when EF-G•GDP is released from the ribosome. At the end of the translocation process, tRNAs in the A and P sites are transferred to the P and E sites, respectively, with the concomitant advance by one codon of mRNA. The insets show the appropriate orientations of the ribosome with 30S subunits and 50S subunits colored yellow and blue respectively. Features of the 30S subunit: b, body; h, head; sp, spur; ch, channel. Features of the 50S subunit: cp, central protuberance; L1, L1 protein; st, L7/L12 stalk. (A was reproduced from, Gas, A., Valle, M., Ehrenberg, M. and Frank, J. (2004). Dynamics of EF-G interaction with the ribosome explored by classification of a heterogenous cryo-EM dataset J. Structural Biol. **147**, 283–290. B was reproduced from Agrawal, R.K., Heagle, A.B., Penczek, P., Grassucci, R.A., Frank, J. (1999) EF-G dependent GTP hydrolysis induces translocation accompanied by large changes in the conformation of the 70S ribosome, Nat. Struct. Biol. **6**, 543–647. With permission).

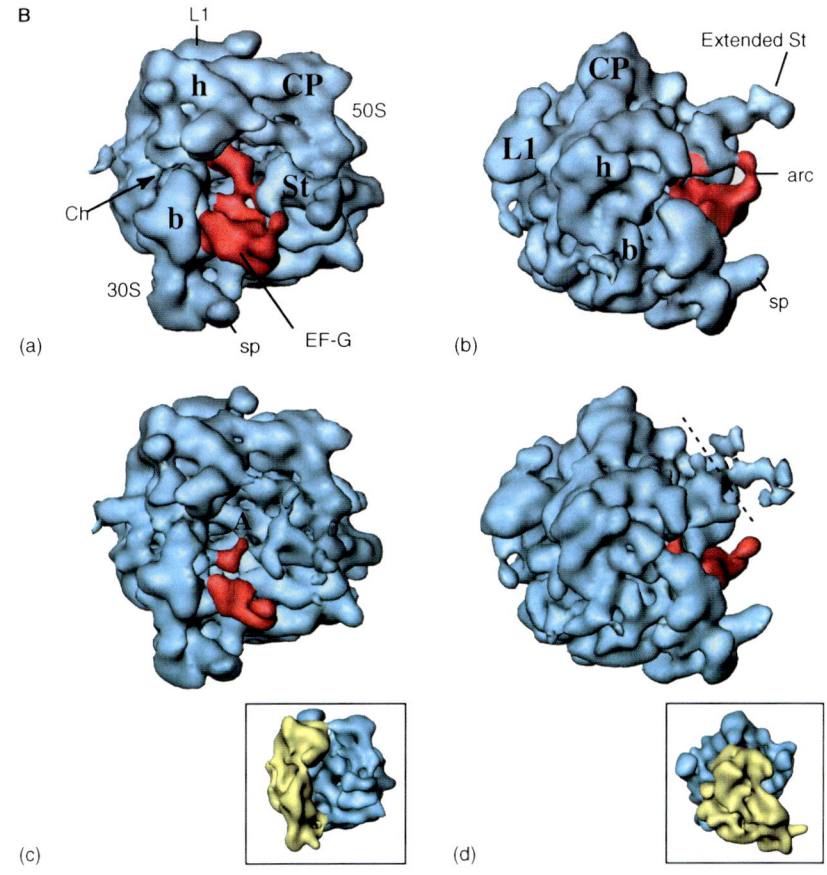

Fig. 15 (*Continued*)

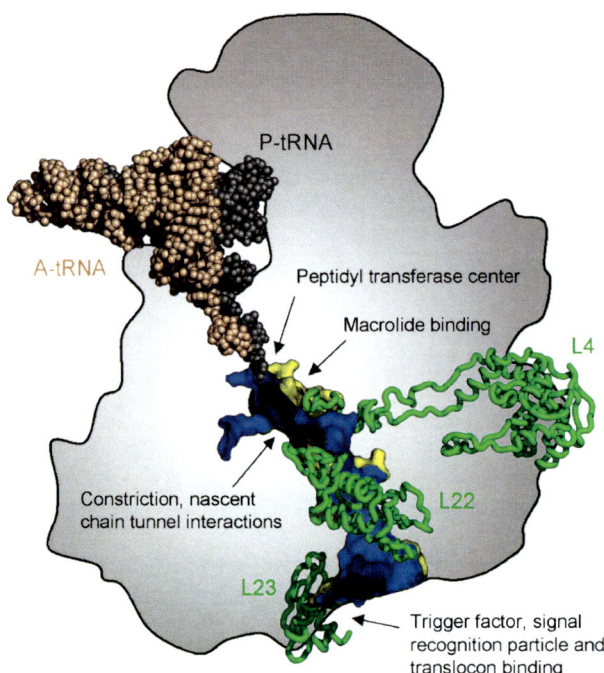

Fig. 16 (p. 106) Protein synthesis and the path of the nascent chain through the ribosome exit tunnel. The contour of the large ribosomal subunit is shown in gray. The aa-tRNA and pept-tRNA are represented by spheres, colored copper, and gray respectively. Ribosomal proteins L4, L22, and L23 are shown as green ribbons. The surface of the ribosomal exit tunnel corresponds to the empty space in the 50S subunit, which is sliced on the side. The surface is colored yellow outside and blue inside. The acceptor stems of aa-tRNA and pept-tRNA point toward the PTC, where peptide bond formation occurs. Nascent polypeptides escape the ribosome through the tunnel. They have to pass the narrow constriction where ribosomal proteins L4 and L22 contact the wall of the tunnel. This region is a target of antibiotics and the site of regulatory interactions between nascent chains and the tunnel. Ribosomal protein L22 mediates the interaction between nascent chains and the tunnel. Ribosomal protein L23 mediates the interaction between nascent chains emerging from the ribosome and cytosolic chaperones such as Trigger Factor (TF) and protein targeting factors such as Signal Recognition Factor (SRFF) and translocons. [From Jenni, S., Ban, N. (2003) The chemistry of protein synthesis and voyage through the ribosomal tunnel, *Curr. Opin. Struct. Biol.* **13**, 212–219. With permission.]

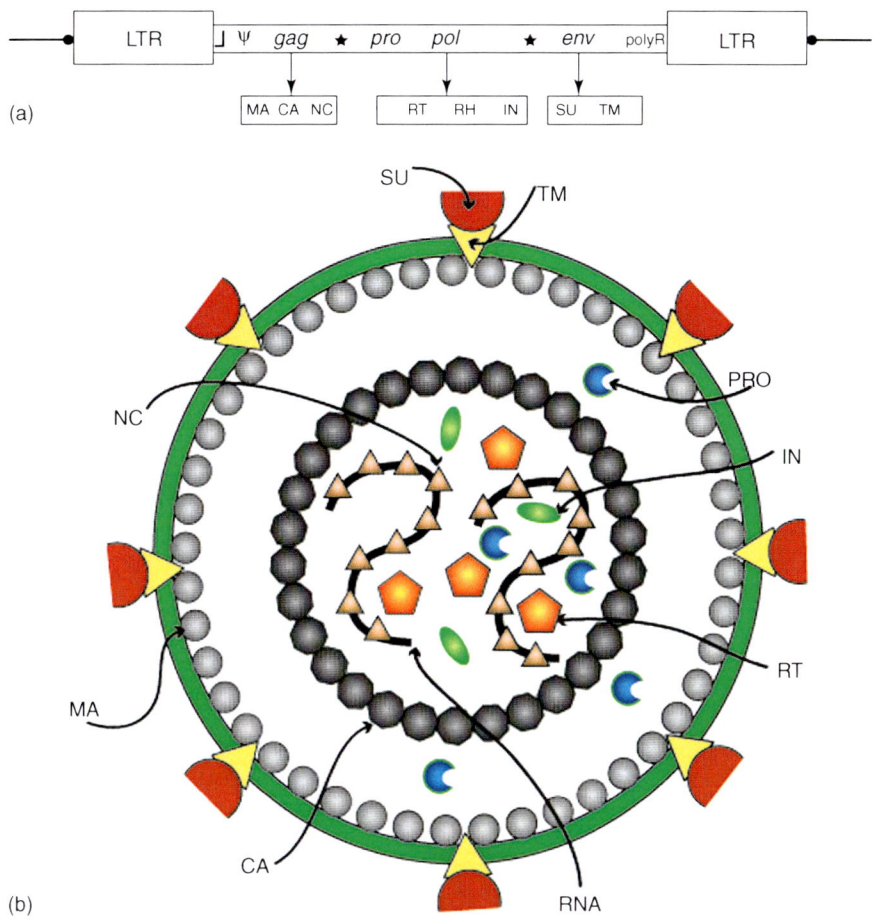

Fig. 8 (p. 292) The general structure of LTR retrotransposons. (a) Structural features of a provirus. Black circles indicate target site duplications; LTR, long terminal repeat; ⌐, primer binding site; Ψ, packaging signal; *gag*, capsid protein gene; *pro*, protease gene; *pol*, polymerase gene; *env*, envelope protein gene; ★, deoxyuridine triphosphatase (dUTPase), which lies between *gag* and *pro* in some Class III retroviruses, and between *pol* and *env* in Class II retroviruses. PPT stands for polypurine tract. The *pro* gene encodes the protease (PR), involved in the processing of the *gag*-, *pol*-, and *env*-encoded proteins. The *pol* gene encodes the reverse transcriptase (RT), ribonuclease H (RH), and integrase (IN). The *env* gene encodes the surface (SU) and transmembrane (TM) proteins. Typically, the *gag* gene encodes the matrix (MA), capsid (CA), and nuclear capsid (NC) structural proteins. (b) Schematic structure of a retroviral particle. The viral envelope (green) is made of a host-cell-produced lipid bilayer. It harbors multiple complexes formed by the *env*-encoded TM and SU components linked together by disulfide bonds. The retroviral capsid formed by the CA surrounds two RNA copies of the retrovirus, tRNAs, RT, IN, and PR.

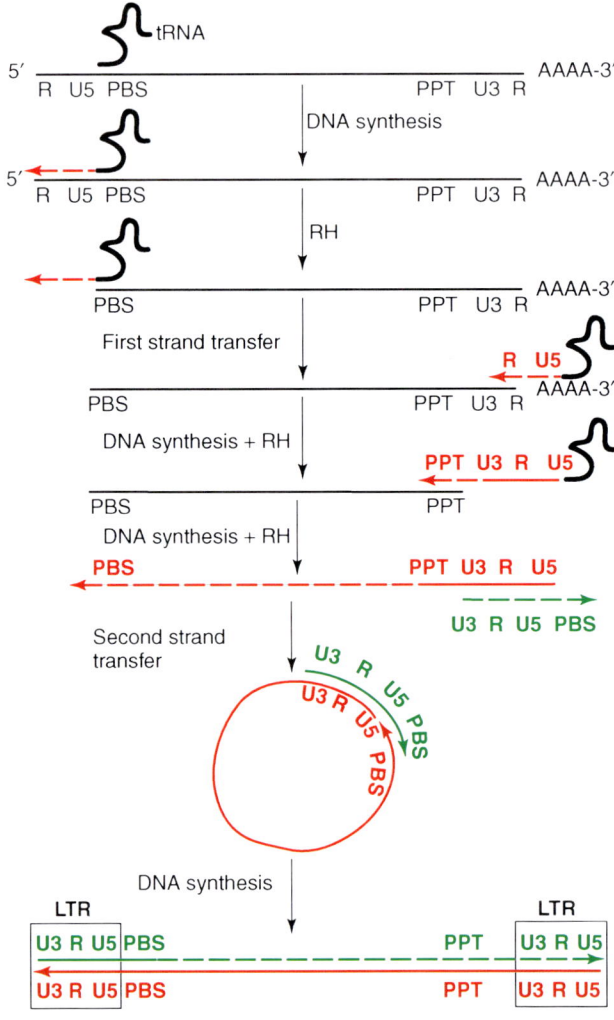

Fig. 9 (p. 294) Reverse transcription of the retroviral RNA. Dotted lines mark newly synthesized DNAs; green, plus-strand DNA; red, minus-strand DNA; RH, RNaseH digestion of RNAs.

Color Plates | xvii

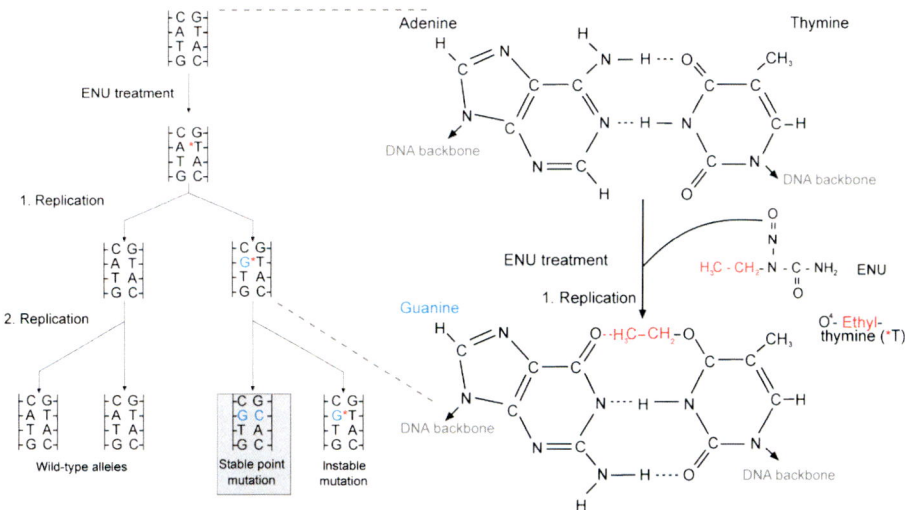

Fig. 4 (p. 479) Mechanism of mutagenesis by ENU leading to an AT → GC transition. The ethyl group in ENU (shown in red or indicated as a red asterisk) is transferred to the O^4-atom of thymine, resulting in an O^4-ethylthymine adduct. Failing an excision of the O^4-ethylthymine adduct by the DNA repair machinery, the O^4-ethylthymine mispairs with a guanine in a subsequent replication of the DNA. In this mispairing, the ENU-derived ethyl group forms a third hydrogen bond with the O^6 of the guanine base in the complementary DNA strand. A successive second replication leads to a complete AT → GC transition in one of the four double helices produced, thus fixing the mutation. Substitutions of bases are indicated in blue. Figure adapted after Noveroske, J.K., Weber, J.S., Justice, M.J. (2000) The mutagenic action of N-ethyl-N-nitrosourea in the mouse, Mamm. Genome 11, 478–483.

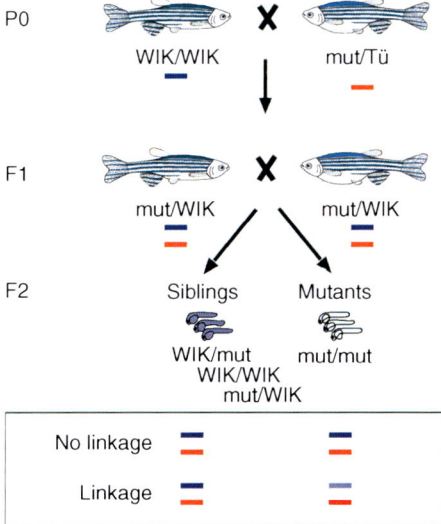

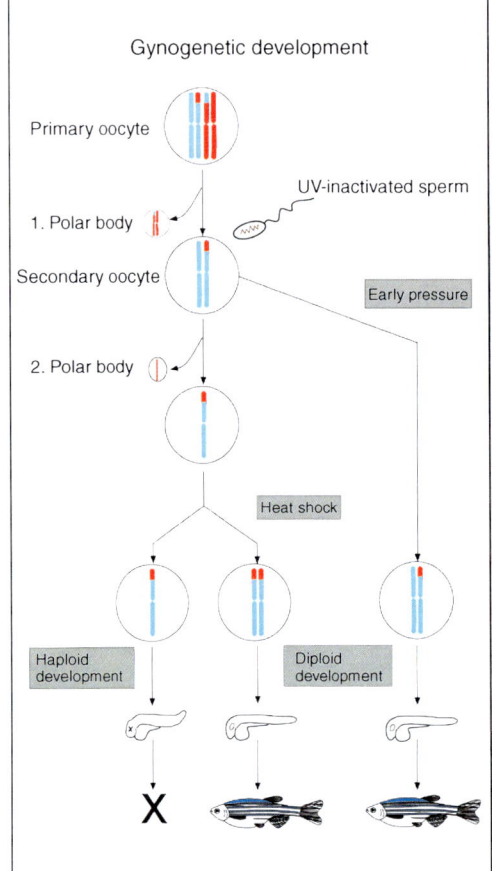
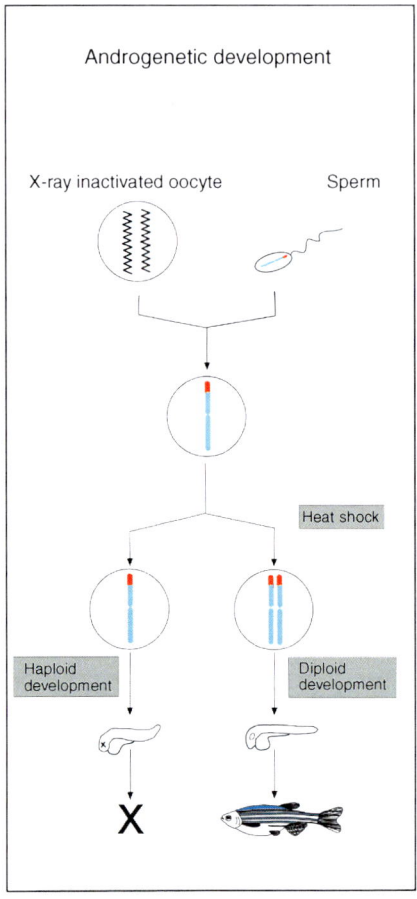

Fig. 8 (p. 489) Schematic representation of the methods used to generate gyno- and androgenetic haploid and diploid zebrafish embryos. For simplicity's sake, only one chromosome is shown. The colors (red and blue) denote the parental origin (maternal and paternal, respectively) of a chromosome/part of a chromosome in the oocyte or the sperm. The presence of two colors in a single chromosome arm indicates a recombination event(s) between the maternal and the paternal chromosome that occurred in the first meiosis leading to the germ cells. The X in the lower row signifies that the respective fish are not viable. For details on the procedures see text.

Fig. 9 (p. 495) Map cross and bulked segregant analysis of a recessive zebrafish mutation. WIK/WIK, reference fish; mut/Tü, fish carrying the mutation in Tü background. Band sizes and intensities of a representative SSLP marker are indicated schematically. In case of no linkage between the mutation and the marker, the intensities in the mutant and sibling pool are the same. In case of a linkage, the Tü band is stronger in the mutant pool and the WIK band in the sibling pool (Figure from Geisler, R. (2002) Mapping and Cloning, in: Nüsslein-Volhard, C., Dahm, R. (Eds.) *Zebrafish, A Practical Approach*, Oxford University Press, Oxford, UK, pp. 175–212.)

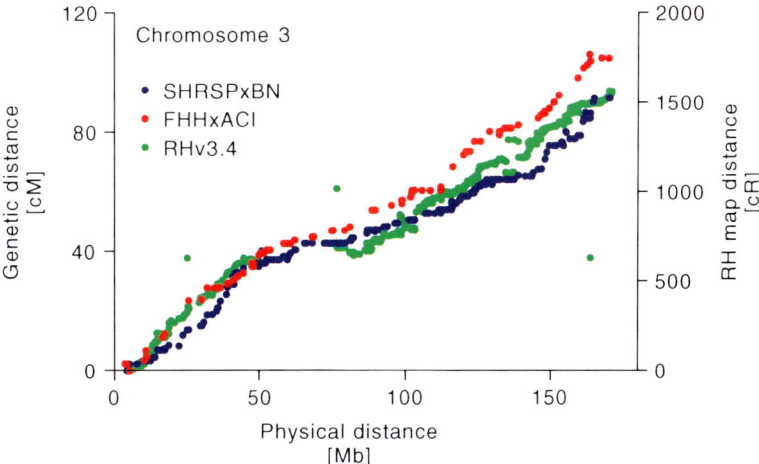

Fig. 2 (p. 511) Map correspondence. Correspondence between positions of markers on two genetic maps of the rat (SHRSPxBN intercross and FHHxACI intercross), on the rat radiation hybrid map, and their position on the rat genome assembly (Rnor3.1). Used with permission from *Nature* **428**, 493–521 (2004).

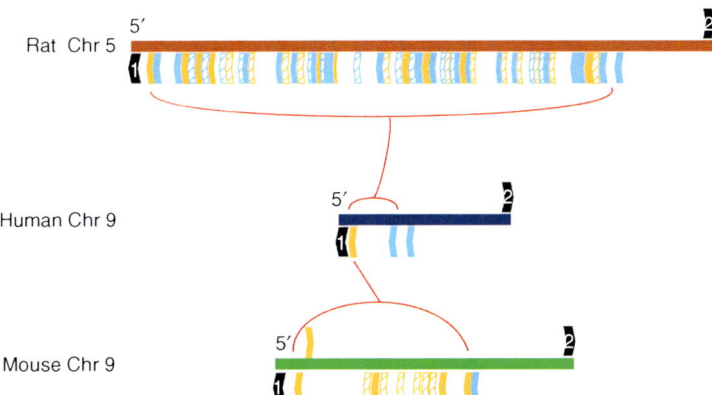

Fig. 13 (p. 538) Adaptive remodeling of genomes and genes. Orthologous regions of rat, human and mouse genomes encoding pheromone-carrier proteins of the lipocalin family (α_{2u} globulins in rat and major urinary proteins in mouse) shown in brown. Zfp37-like zinc finger genes are shown in blue. Filled arrows represent likely genes, whereas striped arrows represent likely pseudogenes. Gene expansions are bracketed. Arrow head orientation represents transcriptional direction. Flanking genes 1 and 2 are TSCOT and CTR1 respectively.

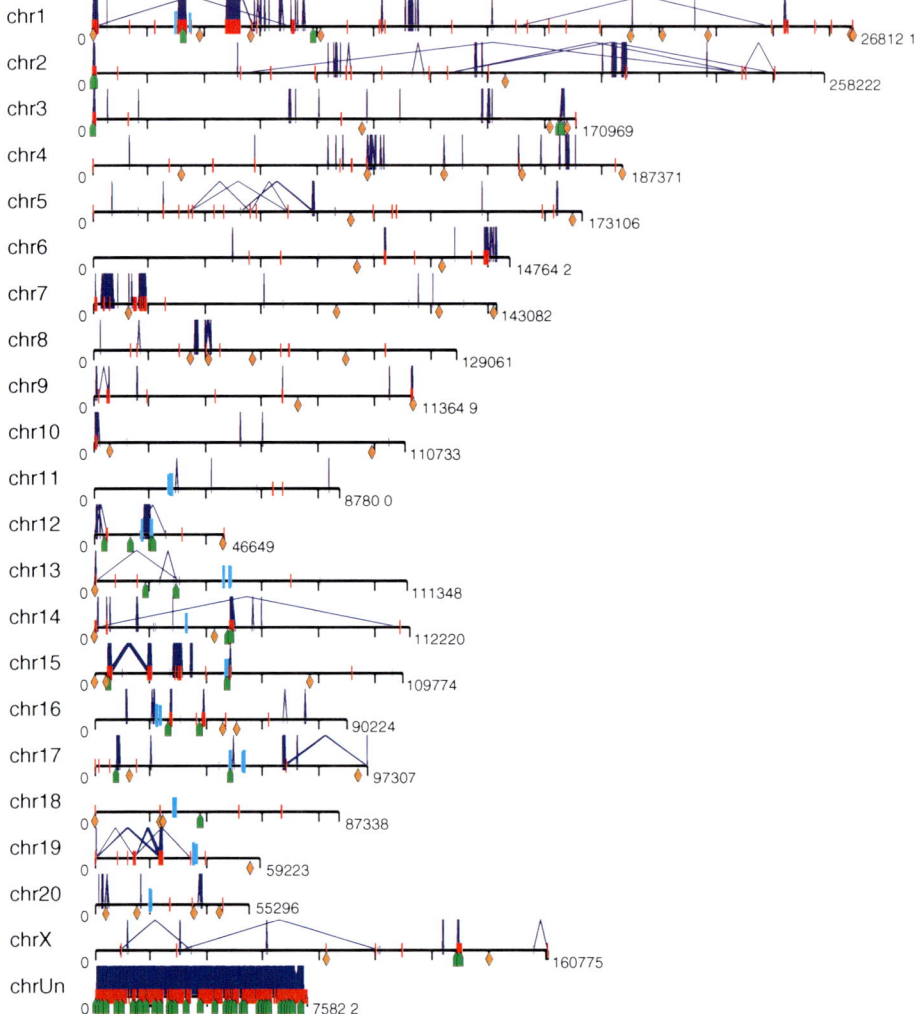

Fig. 3 (p. 512) Distribution of segmental duplications in the rat genome. Interchromosomal duplications (red) and intrachromosomal duplications (blue) are depicted for all duplications with ≥90% sequence identity and ≥20 kb length. The intrachromosomal duplications are drawn with connecting blue line segments; those with no apparent connectors are local duplications very closely spaced on the chromosome (below the resolution limit for the figure). P arms on the left and the q arms on the right. Chromosomes 2, 4–10, and X are telocentric; the assemblies begin with pericentric sequences of the q arms, and no centromeres are indicated. For the remaining chromosomes, the approximate centromere positions were estimated from the most proximal STS/gene marker to the p and q arm as determined by fluorescent in situ hybridization (FISH) (cyan vertical lines; no chromosome 3 data). The chrUn sequence is contigs not incorporated into any chromosomes. Green arrows indicate 1-Mb intervals with more than tenfold enrichment of classic rat satellite repeats within the assembly. Orange diamonds indicate 1-Mb intervals with more than tenfold enrichment of internal (TTAGGG)n-like sequences. For more details, see http://ratparalogy.cwru.edu. Used with permission from *Nature* **428**, 493–521 (2004).

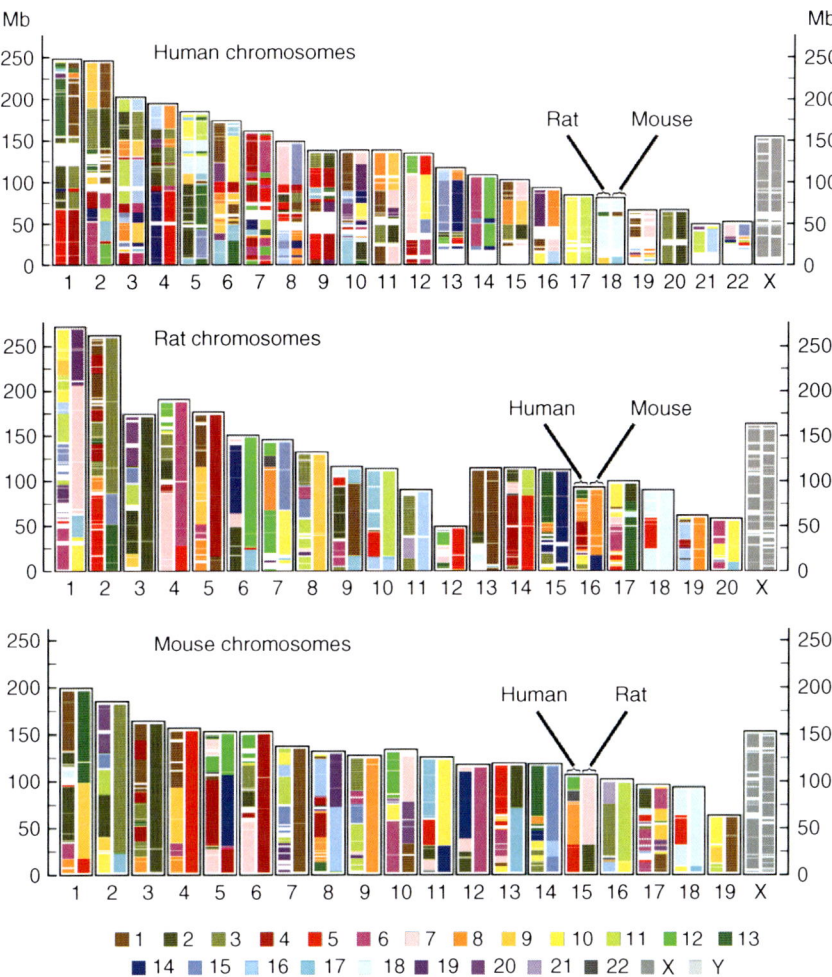

Fig. 4 (p. 514) Map of conserved synteny between the human, mouse, and rat genomes: For each species, each chromosome (x-axis) is a two column boxed pane (p-arm at the bottom) colored according to conserved synteny to chromosomes of the other two species. The same chromosome color code is used for all species (indicated below). For example, the first 30 Mb of mouse chromosome 15 is shown to be similar to part of human chromosome 5 (by the red in left column) and part of rat chromosome 2 (by the olive in right column). An interactive version is accessible (http://www.genboree.org). Used with permission from *Nature* **428**, 493–521 (2004).

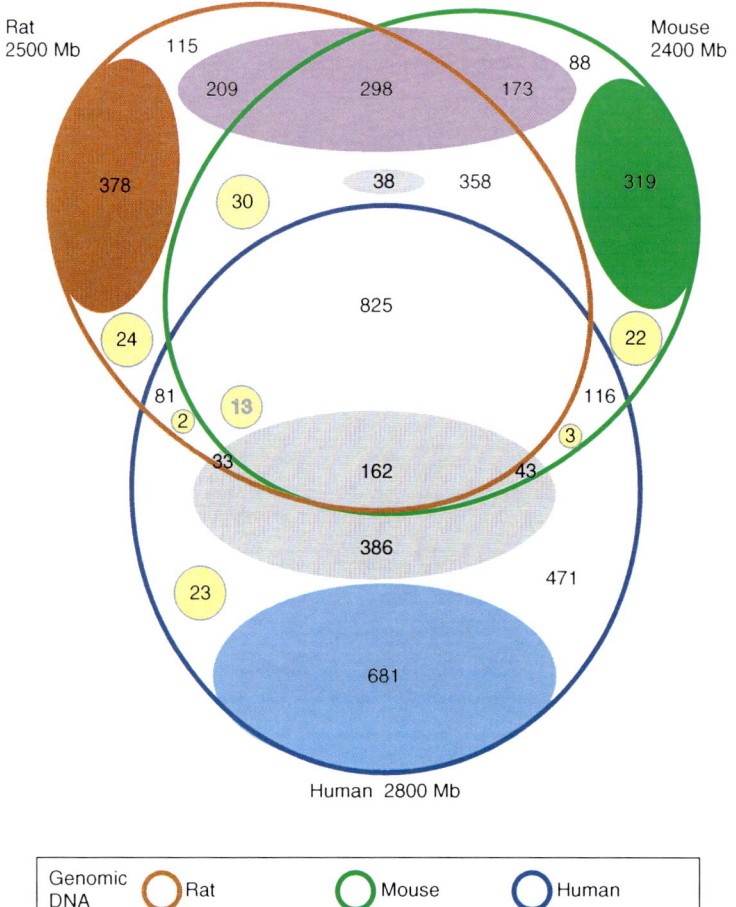
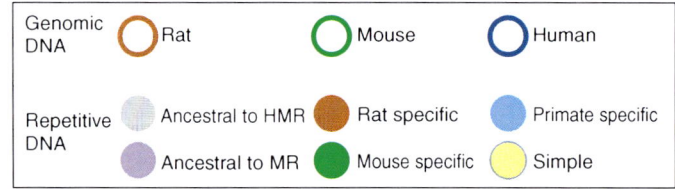

Fig. 7 (p. 517) Aligning portions and origins of sequences in rat, mouse, and human genomes. Each outlined ellipse is a genome, and the overlapping areas indicate the amount of sequence that aligns in all three species (rat, mouse, and human) or in only two species. Nonoverlapping regions represent sequences that do not align. Types of repeats classified by ancestry: those that predate the human–rodent divergence (gray), those that arose on the rodent lineage before the rat–mouse divergence (lavender), species-specific (orange for rat, green for mouse, blue for human), and simple (yellow), placed to illustrate the approximate amount of each type in each alignment category. Uncolored areas are nonrepetitive DNA – the bulk is assumed to be ancestral to the human–rodent divergence. Numbers of nucleotides (in Mb) are given for each sector (type of sequence and alignment category). Used with permission from *Nature* **428**, 493–521 (2004).

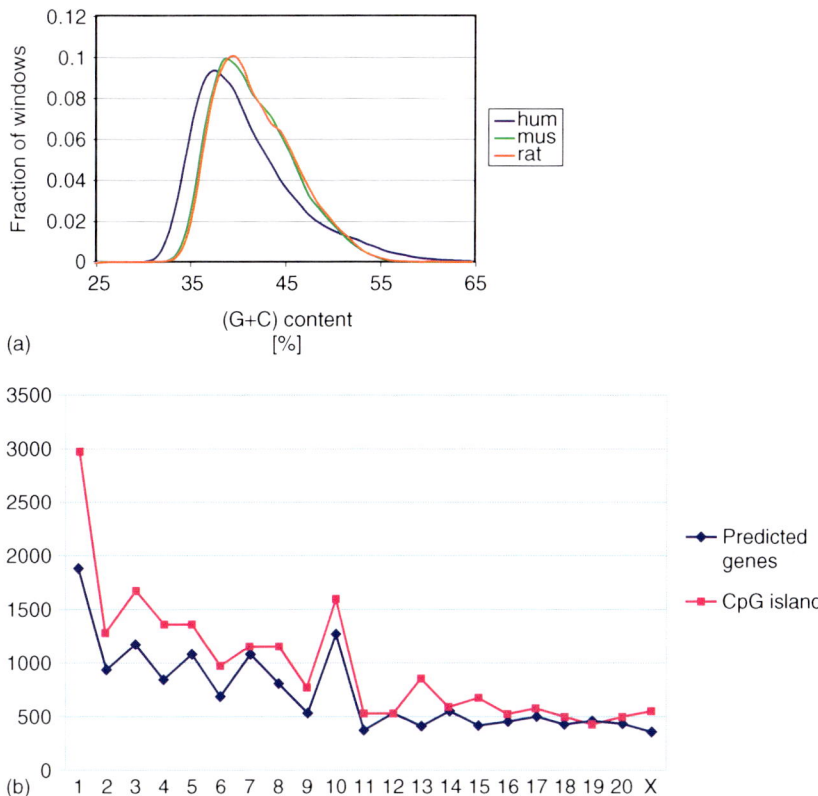

Fig. 8 (p. 520) Base composition distribution analysis. (a) The fraction of 20 kb nonoverlapping windows with a given G + C content is shown for human, mouse, and rat. (b) The number of Ensembl predicted genes per chromosome and the number of CpG islands per chromosome. The density of CpG islands averages 5.9 islands per Mb across chromosomes and 5.7 islands per Mb across the genome. Chromosome 1 has more CpG islands than other chromosomes, yet neither the island density nor ratio to predicted genes exceeds the normal distribution. The number of CpG islands per chromosome and the number of predicted genes are correlated ($R^2 = 0.96$). Used with permission from *Nature* **428**, 493–521 (2004).

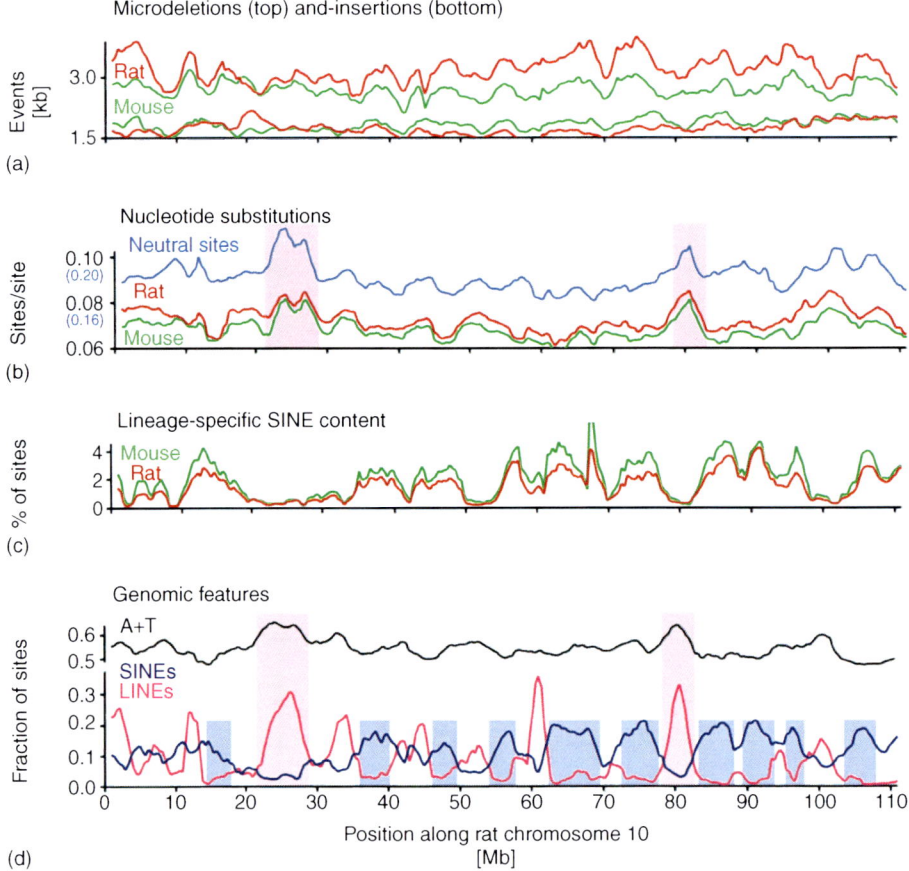

Fig. 9 (p. 521) Variability of several evolutionary and genomic features along rat chromosome 10. (a) Rates of microdeletion and microinsertion events (less than 11 bp) in the mouse and rat lineages since their last common ancestor, revealing regional correlations. (b) Rates of point substitution in the mouse and rat lineages. Red and green lines represent rates of substitution within each lineage estimated from sites common to human, mouse, and rat. Blue represents the neutral distance separating the rodents, as estimated from rodent-specific sites. Note the regional correlation among all three plots, despite being estimated in different lineages (mouse and rat) and from different sites (mammalian vs rodent-specific). (c) Density of SINEs inserted independently into the rat or mouse genomes after their last common ancestor. (d) A + T content of the rat, and density in the rat genome of LINEs and SINEs that originated since the last common ancestor of human, mouse, and rat. Pink boxes highlight regions of the chromosome in which substitution rates, AT content, and LINE density are correlated. Blue boxes highlight regions in which SINE density is high but LINE density is low. Used with permission from *Nature* **428**, 493–521 (2004).

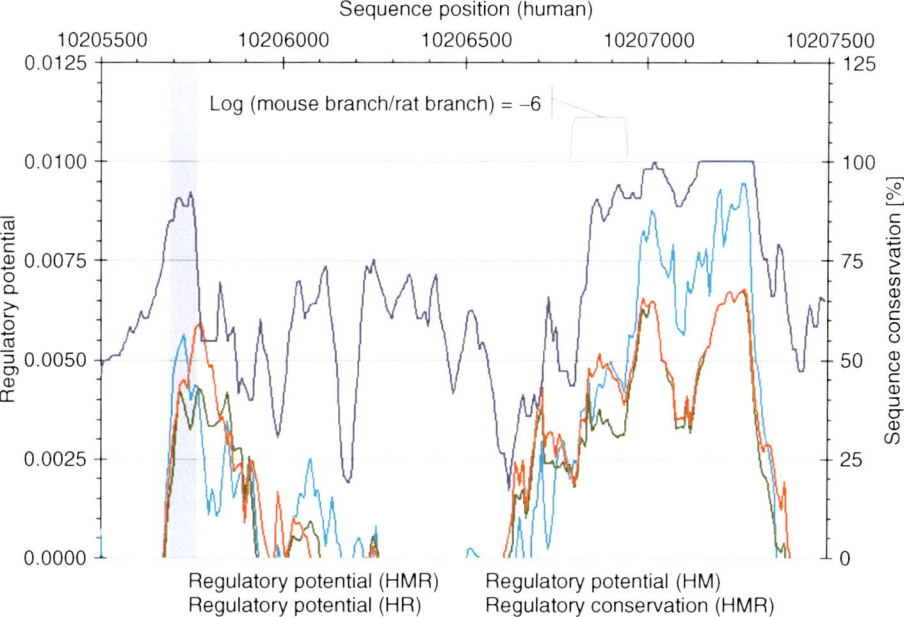

Fig. 11 (p. 530) Close-up of PEX14 (peroxisomal membrane protein) locus on human chromosome 1 (with homologous mouse chromosome 4 and rat chromosome 5). Conservation score computed on 3-way human–mouse–rat alignments presents a clear coding exon peak (gray bar) and very high values in a 504 bp noncoding, intronic segment (right; last 100 bp of alignment are identical in all three organisms). The latter segment showed a striking difference between the inferred mouse and rat branch lengths: the gray bracket corresponds to a phylogenetic tree where the logarithm of mouse to rat branch length ratio is −6. Regulatory potential (RP) scores that discriminate between conserved regulatory elements and neutrally evolving DNA are calculated from 3-way (human–mouse–rat) and 2-way (human-rodent) alignments. Here the 3-way regulatory potential scores are enhanced over the 2-way scores. Used with permission from *Nature* **428**, 493–521 (2004).

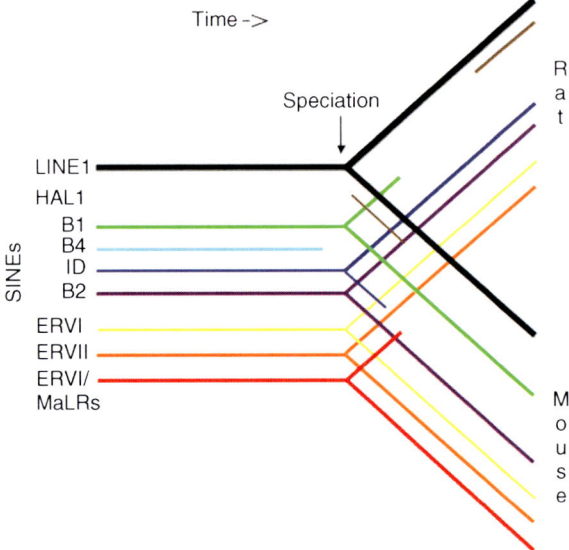

Fig. 12 (p. 535) Historical view of rodent repeated sequences. Relationships of the major families of interspersed repeats (Table 7) are shown for the rat and mouse genomes, indicating losses and gains of repeat families after speciation. The lines indicate activity as a function of time. Note that HAL1-like elements appear to have arisen both in the mouse and rat lineage. Used with permission from *Nature* **428**, 493–521 (2004).

Fig. 14 (p. 540) Evolution of cytochrome P450 (CYP) protein families in rat, mouse, and human. (a) Dendrogram of topology from 234 full-length sequences. The 279 sequences of 300 amino acids; subfamily names and chromosome numbers are shown. Black branches have >70% bootstrap support. Incomplete sequences (they contain Ns) are included in counts of functional genes (84 rat, 87 mouse, and 57 human) and pseudogenes (including fragments not shown; 77 rat, 121 mouse, and 52 human). Thus, 64 rat genes and 12 pseudogenes were in predicted gene sets. Human CYP4F is a null allele due to an in-frame STOP codon in the genome, although a full-length translation exists (SwissProt P98187). Rat CYP27B, missing in the genome, is "incomplete" since there is a RefSeq entry (NP_446215). Grouped subfamilies CYP2A, 2B, 2F, 2G, 2T, and CYP4A, 4B, 4X, 4Z, occur in gene clusters; thus nine loci contain multiple functional genes in a species. One (CYP1A) has fewer rat genes than human, seven have more rodent than human, and all nine have different copy numbers. CYP2AC is a rat-specific subfamily (orthologs are pseudogenes). CYP27C has no rodent counterpart. Rodent-specific expansion, rat CYP2J is illustrated below. (b) The neighbor-joining tree, with the single human gene, contains clear mouse (Mm) and rat (Rn) orthologous pairs (bootstrap values >700/1000 trials shown). Bar indicates 0.1 substitutions per site. (c) All rat genes have a single mouse counterpart except for CYP2J 3, which has further expanded in mouse (mouse CYP2J 3a, 3b, and 3c) by two consecutive single duplications. The genes flanking the CYP2J orthologous regions (rat chromosome 5, 126.9–127.3 Mb; mouse chromosome 4, 94.0–94.6 Mb; human chromosome 1, 54.7–54.8 Mb) are hook1 (HOOK1; pink) and nuclear factor I/A (NFIA; cyan). Genes (solid) and gene fragments (dashed boxes) are shown above (forward strand) and below (reverse strand) the horizontal line. No orthology relation could be concluded for most of these cases. Used with permission from *Nature* **428**, 493–521 (2004).

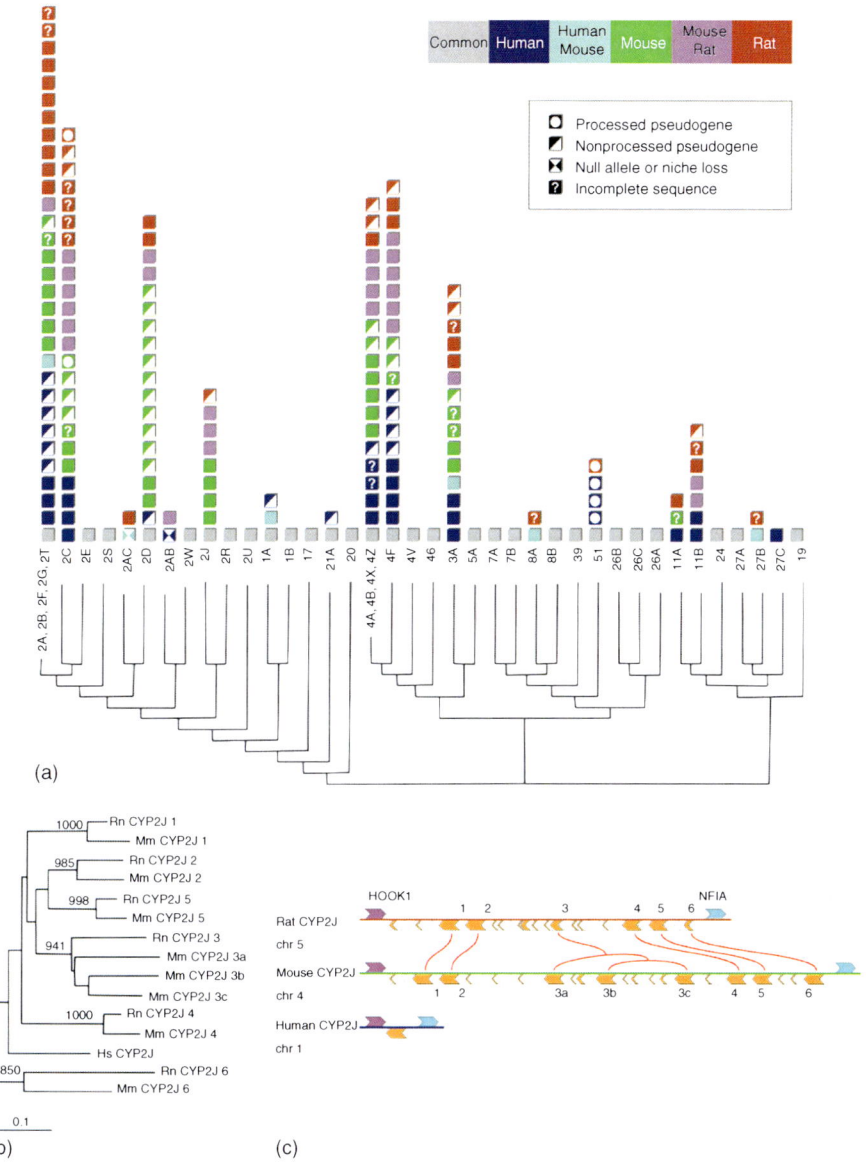

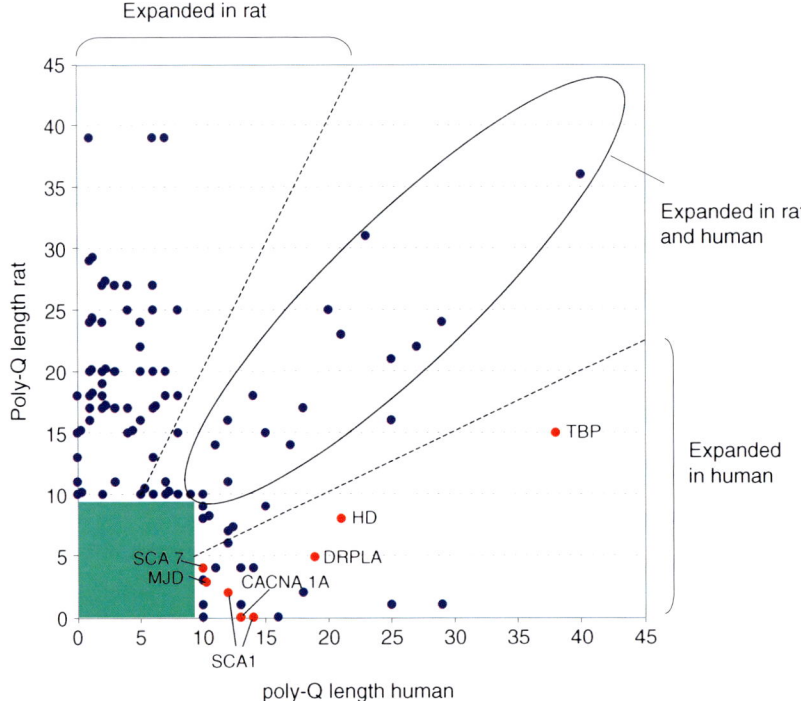

Fig. 16 (p. 543) Polyglutamine repeat length comparison between human and rat. Points represent protein poly-Q length for rat and human. Red points correspond to repeats in genes associated with human disease: SCA1, spinocerebellar ataxia 1 protein, or ataxin1; SCA7, spinocerebellar ataxia 7 protein; MJD, Machado–Joseph disease protein; CACNA1A, spinocerebellar ataxia 6 protein, or calcium channel α-1A subunit isoform 1; DRPLA, dentatorubro–pallidoluysian atrophy protein; HD, Huntington's disease protein, or huntingtin; TBP, TATA binding protein or spinocerebellar ataxia 17 protein. Repeat lengths over 10 were examined; green shading delineates the range not included in the analysis. Also noted are a set that are expanded in rat and human (black circle) and a set where repeats are expanded in the rat. Used with permission from *Nature* **428**, 493–521 (2004).

Color Plates | xxix

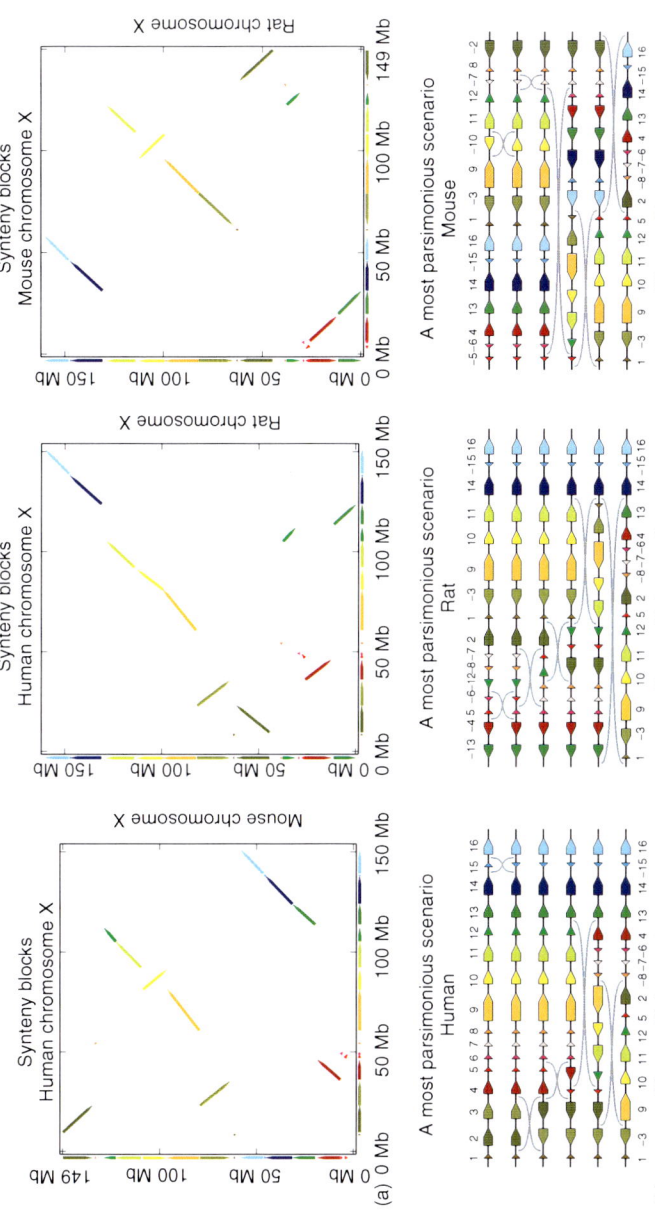

Fig. 6 (p. 516) X chromosome in each pair of species: (a) GRIMM–Synteny computes 16 three-way orthologous segments (≥300 kb) on the X chromosome of human, mouse, and rat, shown for each pair of species, using consistent colors. (b) The arrangement (order and orientation) of the 16 blocks implies that at least 15 rearrangement events occurred during X chromosome evolution of these species. The program MGR determined that evolutionary scenarios with 15 events are achievable and all have the same median ancestor (located at the last common mouse–rat ancestor). Shown is a possible (not unique) most parsimonious inversion scenario from each species to that ancestor. Note the last common ancestor of human, mouse, and rat should be on the evolutionary path between this median ancestor and human. Used with permission from *Nature* **428**, 493–521 (2004).

Part 1
Molecular Genetics

1
Molecular Basis of Genetics

D. Peter Snustad
University of Minnesota, Minnesota, USA

1	**Genetic Information**	**8**
1.1	Four-letter Alphabet	8
1.2	The Gene, the Basic Unit of Function	8
1.3	Genes are Located on Chromosomes	9
1.4	Autosomal and Sex-linked Inheritance	11
1.5	Mendelian versus Quantitative Traits	14
1.6	Genomics: Analyses of Entire Genomes	14
2	**DNA Replication: The Genotypic Function**	**15**
2.1	Semiconservative Replication	15
2.2	Proofreading	15
2.3	The Complex Replisome	16
3	**Gene Expression: The Phenotypic Function**	**19**
3.1	Colinearity between Gene and Polypeptide	20
3.2	Transcription	20
3.3	Introns and RNA Splicing	21
3.4	RNA Editing	22
3.5	Translation	23
3.6	Inteins	25
3.7	Some Complex Gene/Protein Relationships	25
3.8	Pathways of Gene Expression are Often Complex	26
3.9	Pleiotropy and Epistasis	28
3.10	Penetrance and Expressivity	28
3.11	Intragenic Complementation	29
3.12	Intergenic Noncomplementation	29
3.13	Regulation of Gene Expression	29
3.14	RNA-mediated Gene Silencing	31

Genomics and Genetics. Edited by Robert A. Meyers.
Copyright © 2007 Wiley-VCH Verlag GmbH & Co. KGaA, Weinheim
ISBN: 978-3-527-31609-0

4		**Mutation: The Evolutionary Function** 32
4.1		Dominance, Partial Dominance, and Codominance 32
4.2		Transitions, Transversions, and Frameshift Mutations 33
4.3		Gain-of-function and Loss-of-function Mutations 34
4.4		Suppressor and Enhancer Mutations 34
4.5		Chromosome Aberrations 35
4.6		Polyploidy 35
4.7		Nondisjunction and Aneuploidy 35
4.8		Expanding Trinucleotide Repeats in Humans 36

5		**Recombination: New Combinations of Genes to be Acted on by Natural or Artificial Selection** 36
5.1		Segregation 36
5.2		Independent Assortment 37
5.3		Linkage, Crossing-over, and Genetic Maps of Chromosomes 38
5.4		Transposable Genetic Elements 39
5.5		Physical Maps of Chromosomes 40
5.6		Map-position-based Identification of Genes 41
		Acknowledgements 43

Bibliography 43
Books and Reviews 43
Primary Literature 43

Keywords

Aneuploidy
The presence of an extra or missing chromosome (or chromosomes) in a cell or an organism. Any condition in which the chromosomes are not all present in equal number.

Attenuation
A mechanism by which gene expression in prokaryotes is regulated by the premature termination of transcription.

Autosomal Inheritance
The pattern of inheritance observed for genes located on chromosomes other than the sex chromosomes (chromosomes carrying sex-determining genes).

CentiMorgan (cM)
The measure of the distance between markers on genetic maps based on the average number of crossover events that occur during meiosis. A distance of 1 cM indicates

that 1 recombinant chromosome is present among every 100 chromosomes produced during meiosis.

Codon
The unit of three contiguous nucleotides in mRNA specifying the incorporation of one amino acid in the polypeptide produced by translating that mRNA on the polyribosomes.

Complementation Test
The introduction of two recessive mutations into the same cell but on different chromosomes (a *trans* heterozygote) to determine whether the mutations are both in the same gene or are in two different genes. If the mutations are both in the same gene, the $m1+/+m2$ heterozygote will exhibit a mutant phenotype, whereas if they are in two different genes, the *trans* heterozygote will exhibit the wild-type phenotype.

Crossing-over
A recombination process through which new combinations of genes on specific chromosomes are produced by the breakage and reunion of their DNA molecules.

Dominance
The control of the phenotype of an organism by one allele of a gene correlated with the exclusion of any effect of the other allele.

Epistasis
The interaction of nonalleles. Any condition in which an allele of one gene masks the expression of one or more alleles of a different gene (a nonallele).

Exon
A nucleotide sequence of a gene that corresponds to a sequence that is present in the final processed RNA product of the gene.

Frameshift Mutation
A mutation that alters the codon reading frame of a gene, by either inserting or deleting one or more nucleotide pairs in multiples other than three.

Gain-of-function Mutation
A mutation that produces a gene product with a new function.

Genetic Map
A diagram of a chromosome with distances between markers based on recombination frequencies (centiMorgans).

Heritability
The proportion of the total phenotypic variability present for a trait that results from genetic factors rather than environmental effects.

Independent Assortment
During meiosis, each pair of homologous chromosomes lines up at the metaphase plate independently of every other pair. As a result, the alleles of a gene on one chromosome segregate independently of the alleles of a gene on any nonhomologous chromosome.

Intron
A sequence of nucleotide pairs in a gene that is not present in the mature RNA because it is excised from the primary transcript during processing.

Loss-of-function Mutation
A mutation that impairs or abolishes the expression of a gene or renders its product less active or nonfunctional.

Meiosis
The process by which the chromosome number in reproductive cells is reduced to half the number present in other (somatic) cells of the body. Chromosomes duplicate once and cells duplicate twice to produce haploid gametes.

Mitosis
The separation of the daughter chromatids produced by chromosome replication and the division of the cytoplasm to produce two identical progeny cells.

Mutation
A heritable change in the structure of the genetic material of an organism. When used in the broad sense, mutations include both "point mutations," involving changes in the structure of individual genes, and gross changes in chromosome structure (chromosome aberrations). In the narrow sense, mutations include only "point mutations." The term *mutation* is used to refer to (1) the *process* by which the change occurs and (2) the *result* of the process, the *alteration* in the gene or genetic material.

Nondisjunction
The failure of homologous chromosomes or sister chromatids to separate or disjoin from each other during meiosis or mitosis.

Polyploidy
The presence of three or more copies of each chromosome in a cell or an organism.

Recombination
The generation of new combinations of genes in progeny, which were not present together in either of the parents, either (1) by independent assortment of nonhomologous chromosomes during meiosis or (2) by crossing-over (breakage and exchange of parts) of homologous chromosomes during meiosis or mitosis.

Segregation
The separation of the maternal and paternal chromosomes, and thus the alleles of genes in heterozygotes, from each other during the reductional division of meiosis.

Sex-linked Inheritance
A pattern of inheritance that occurs when the gene controlling a trait is located on a chromosome carrying genes that determine the sex phenotype of the organism.

Suppressor Mutation
A mutation that partially or completely eliminates the phenotypic effect of another mutation.

Transition
A mutation resulting from the replacement of one purine with the other purine and/or one pyrimidine with the other pyrimidine.

Transposable Genetic Element
A DNA unit that can move from one location in a genome to another location or even to a different genome.

Transversion
A mutation resulting from the replacement of a purine with a pyrimidine and/or a pyrimidine with a purine.

The phenotype of a living organism is controlled by its genotype, the summation of its genetic information, acting within the constraints imposed by the environment in which the organism exists. Much of the genetic material of an organism is organized into basic functional units called genes, which specify RNA and/or protein products. Some genes encode one primary gene product, either an RNA molecule or polypeptide. Other genes produce two or more related polypeptides by RNA editing, differential transcript splicing, or the assembly of genes from gene segments during development. The genetic information of all living organisms, whether viruses, bacteria, corn plants, or humans, is stored in the sequence of bases (purines and pyrimidines) or base pairs in the deoxyribonucleic acid (DNA) present in their chromosomes. In some viruses, the genetic information is stored in the sequence of bases in ribonucleic acid (RNA). The genetic information is encoded using a four-letter alphabet: the four bases adenine (A), guanine (G), cytosine (C), and thymine (T). In RNA, uracil (U) replaces the thymine present in DNA. In the double-stranded DNA present in most cellular organisms, adenine and thymine form one base pair (A:T) and guanine and cytosine form a second base pair (G:C).

The genetic material of an organism must carry out three essential functions: (1) the genotypic function, transmission of the genetic information from generation to generation; (2) the phenotypic function, directing the growth and development of the offspring into mature, reproductive adults; and (3) the evolutionary function, mutation, allowing organisms to evolve in response to changes in the environment. Mutation produces new genetic variability, which provides the raw material for evolution. Recombination of genetic material occurs by the independent assortment of nonhomologous chromosomes and by crossing-over between homologous chromosomes. This recombination provides new combinations of genes and thus new phenotypes on which natural selection acts during the process of evolution.

1
Genetic Information

The genetic information of living organisms is stored in large macromoleules called *nucleic acids*. These nucleic acids are of two types: *DNA* contains the sugar 2′-deoxyribose and *RNA* contains the sugar ribose. In all eukaryotic organisms, the genetic information is stored in giant DNA molecules located in one to many chromosomes, the number depending on the species. In some viruses that contain no DNA, the genetic information is stored in RNA.

1.1
Four-letter Alphabet

The genetic information is stored in nucleic acids using a four-letter alphabet: the four bases *adenine* (A), *guanine* (G), *cytosine* (C), and *thymine* (T) in DNA or *uracil* (U) in RNA. In DNA, which has a double-stranded structure in cellular organisms, the bases are present in pairs: A with T and G with C (Fig. 1). Although a four-letter alphabet may seem too simple to store enough information to produce the vast phenotypic variability observed in living organisms, recall that the Morse code is based on just two symbols – dots and dashes. Moreover, computers perform their amazing feats using a binary code composed of 0s and 1s. Even with just four letters, a vast amount of genetic information can be stored in the large nucleic acids present in living cells. Consider, for example, that one complete copy of the human genome (all the genetic information in one complete set of human chromosomes) contains three-billion (3×10^9) base pairs of DNA. Since the number of different sequences of 4 letters used n at a time is 4^n, one can see that the human genome has the capacity to store a huge amount of information with $n = 3 \times 10^9$.

1.2
The Gene, the Basic Unit of Function

The basic functional unit of genetic information is the *gene*, defined operationally by the complementation test and most commonly specifying the amino acid sequence of one polypeptide chain or the nucleotide sequence of one RNA molecule. Different forms of a given gene are called alleles. The wild-type alleles of a gene are those that exist at relatively high frequencies in natural populations and yield wild-type or "normal" phenotypes; they are usually symbolized by a + or a symbol with a + superscript (e.g. w^+ for the allele that yields wild-type red eyes in fruit flies). Alleles of a gene that result in abnormal or non-wild-type phenotypes are called

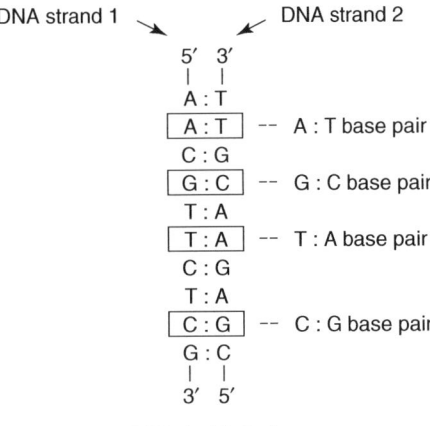

Fig. 1 Two-dimensional view of a segment of double-stranded DNA showing the four base pairs used to store genetic information.

mutant alleles. They are usually symbolized by one to three letters written in italics (e.g. w and w^{ap} for the alleles that cause white and apricot eye color respectively in fruit flies). Many eukaryotes such as corn plants, fruit flies, and humans contain two copies of their genome in most cells, two copies of each of their chromosomes; such eukaryotes are called diploids. Thus, diploid organisms may contain two different alleles of any given gene, in which case they are heterozygous (e.g. w/w^+, w^{ap}/w^+, w/w^{ap}) or two identical copies of a given gene, in which case they are homozygous (e.g. w^+/w^+, w/w, w^{ap}/w^{ap}). A w^{ap}/w^+ heterozygous fruit fly has wild-type red eyes. The w^+ allele is expressed in this heterozygous fly; w^+ is thus called the dominant allele. The w^{ap} allele is not expressed in this heterozygous fly; it is said to be recessive because its effect on the phenotype is masked by the w^+ allele.

The *complementation test* is performed by producing cells or organisms that contain two recessive mutant genes located on two different chromosomes, that is, *trans* heterozygotes, and by determining whether these cells or organisms have mutant or wild-type phenotypes. If the two mutant genes are allelic, that is, the defects or mutations are in the same gene, the *trans* heterozygote will have a mutant phenotype. If the two mutant genes are not allelic, that is, the mutations are in two different genes, the *trans* heterozygote will have the wild-type phenotype. The rationale behind the complementation test is illustrated in Fig. 2.

1.3
Genes are Located on Chromosomes

One of the important discoveries in biology was that the genetic information of organisms is present in structures called chromosomes, because the transmission of these organelles could be followed during cell division and reproduction. This research culminated in the "Chromosome Theory of Inheritance," which is the core of modern genetics. The DNA molecules that carry the genetic information are packaged into chromosomes with the aid of proteins and RNA molecules. In a eukaryotic cell, the chromosomes are present in a membrane-bounded compartment called the nucleus. The prokaryotic equivalent is the nucleoid, which is not surrounded

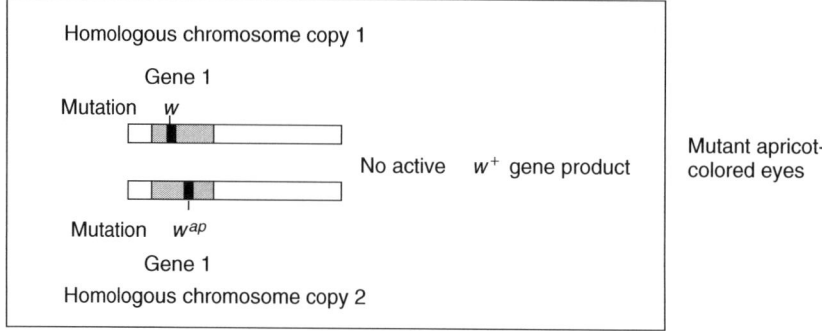

(a) Two mutations in the same gene.

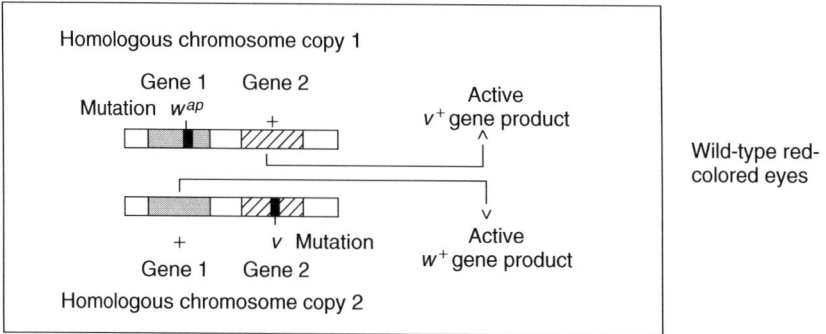

(b) Two mutations in two different genes.

Fig. 2 Illustration of the complementation test used to operationally define the gene, the basic unit of function of genetic material. The operation is to place the two recessive mutations in question in the same cell or cells of a multicellular organism on two separate chromosomes, that is, to construct a *trans* heterozygote, and to determine whether this cell or organism has a mutant or a wild-type phenotype. If the phenotype is mutant, the two mutations are in the same gene; this is illustrated for the w (white eyes) and w^{ap} (apricot eyes) mutations of *Drosophila* in (a). If the phenotype of the *trans* heterozygote is wild type, the two mutations are in two different genes and the two mutations are said to complement each other. Complementation between the w^{ap} and v (vermilion eye color) mutations of *Drosophila* is illustrated in (b); note that active (wild-type) products of both genes (w^+ and v^+) are present in the *trans* heterozygote shown in (b) – thus, the wild-type phenotype.

by a membrane. An important function of the chromosomes is to ensure the proper distribution of the genetic material to daughter cells during cell division.

Each chromosome contains a single large DNA molecule packaged in a matrix of RNA and protein. The DNA is highly compacted by coiling and supercoiling. If the DNA in a single human chromosome were uncoiled so that it was a perfectly linear molecule, it would measure between 2.5 and 8.5 cm in length. During cell division, this DNA molecule is present in a chromosome that is only about 0.5 μm in diameter and 3 to 10 μm in length. Eukaryotic chromosomes have several distinct structural features that are visible under the light microscope (Fig. 3). After replication, each chromosome is composed of

Fig. 3 The structure of a highly condensed replicated eukaryotic chromosome. (Reproduced from Snustad, D.P., Simmons, M.J., *Principles of Genetics*, 3rd edition, Copyright © 2003 by John Wiley & Sons, Inc., Hoboken, NJ, USA. This material is used by permission of John Wiley & Sons, Inc.)

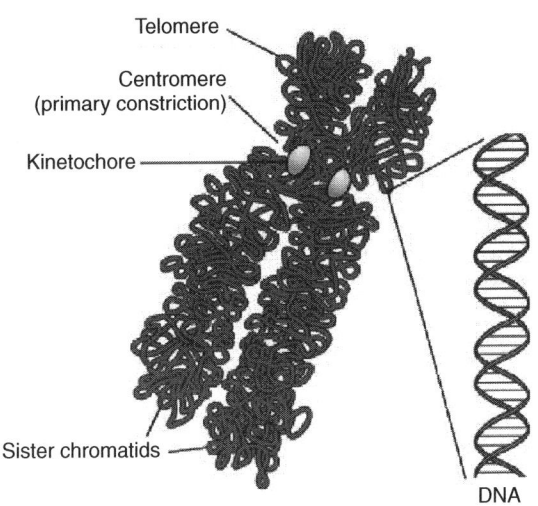

two sister chromatids joined together at a constriction called the centromere The kinetochore, a protein structure present in each centromere, plays an important role in the separation of daughter chromosomes during cell division. The position of the centromere varies from the middle to the end of the chromosome. The ends of the chromosomes are called telomeres; they contain unique molecular structures that enhance chromosome stability.

1.4
Autosomal and Sex-linked Inheritance

In many eukaryotes, sex phenotype is controlled by key regulatory genes that are present on chromosomes that are not present as morphologically identical or nearly identical pairs of homologous chromosomes. In humans, for example, an important male-determining gene, *SRY* (for *s*ex-determining *r*egion of the *Y*), which encodes the testis-determining factor (TDF), is located on a small chromosome called the *Y chromosome*. This chromosome is normally present only in males (there are rare exceptions). During reproductive divisions, it pairs with a much larger chromosome called the *X chromosome*, which is present in two copies in females. The X and Y chromosomes share common terminal regions, allowing them to pair during meiosis. This chromosomal mechanism of sex determination is called the XX-XY mechanism. It occurs in humans and other mammals, the fruit fly *Drosophila melanogaster*, and a number of other species. In some species, the Y chromosome is completely absent and males have one less chromosome than females (the XX-XO mechanism of sex determination). The X and Y chromosomes are referred to as sex chromosomes, and all the other chromosomes (present in morphologically identical pairs) are called *autosomes*. Humans, for example, contain the 2 sex chromosomes and 44 autosomes (22 homologous pairs).

The human Y chromosome plays a major role in sex determination; however, it is small and contains very few genes that affect other traits. In contrast, the X chromosome is large and contains a large number of important genes. The absence of an X chromosome is lethal in humans. Given that females contain two X chromosomes and males contain one X and one Y, whereas both females and males contain two copies of each autosome, traits controlled by genes located on sex chromosomes will exhibit different patterns of inheritance than traits controlled by genes on autosomes

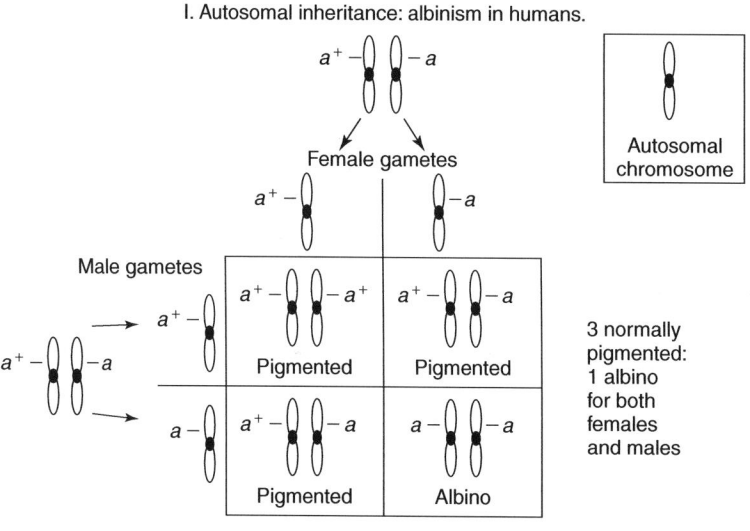

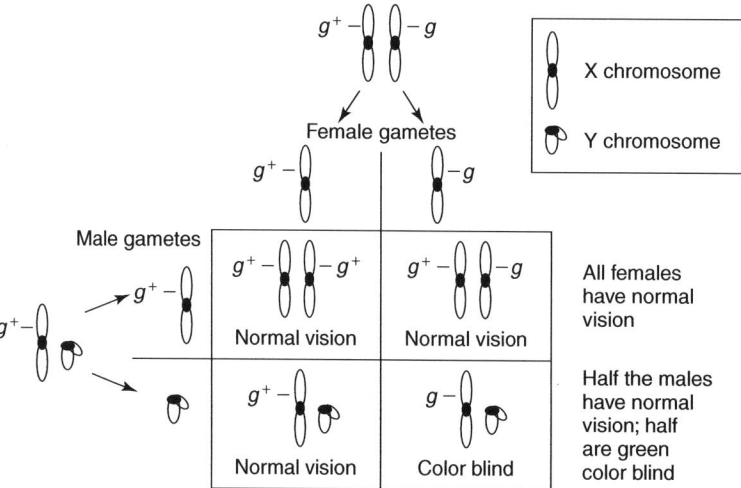

Fig. 4 Autosomal versus sex-linked (X-linked) inheritance. (a) Autosomal inheritance: the expected results are shown for a mating between two individuals who are both heterozygous for a recessive allele causing albinism. (b) X-linked inheritance: the predicted results are shown for a mating between a woman who is heterozygous for a recessive allele causing green color blindness and a male with normal vision.

(Fig. 4). In humans, albinism is caused by a recessive mutation in any of several autosomal genes, whereas green color blindness is caused by a recessive mutant allele of a gene on the X chromosome.

Note that the X chromosome of a male is always passed to his daughters, because the offspring would be male if it received a Y chromosome. Given that the X chromosome contains many essential genes, a son has to get an X chromosome from his mother. Because a male has only one X chromosome, recessive traits such as color blindness and hemophilia are expressed much more frequently in males. Males are *hemizygous* ("half that of a zygote") for X-linked genes; if a recessive X-linked allele is present, it will be expressed. Females contain two X chromosomes; they will have to be homozygous for a recessive X-linked allele to express the trait. Thus, if an X-linked recessive allele is present in a population with a frequency of q, and if there is random mating in the population with respect to this trait, the recessive trait will occur with a frequency of q in males and a frequency of q^2 in females. To express the trait, females will have to obtain one X chromosome carrying the mutant allele from the mother via the egg and a second X chromosome carrying the mutant allele from the father through the sperm. These are independent events, and the probability that two independent events will both occur is equal to the product of the probabilities of the individual events ($q \times q = q^2$). Thus, it is easy to see why males express X-linked traits more frequently than females. In contrast, traits controlled by genes located on autosomes are usually expressed with equal frequency in the two sexes.

Females who receive a recessive X-linked mutant allele from their father often do not express it because they receive the wild-type allele from their mother. However, they will transmit the mutant allele to half of their hemizygous sons, who will express the trait. These heterozygous women are referred to as "carriers" of the trait; they carry the mutation, but do not show any effect of its presence.

Females have two copies of every gene on the X chromosome; males have only one. In humans and other mammals, this difference in gene dosage is corrected by inactivating one of the two X chromosomes in each cell in females. This inactivation occurs at random, so half of a woman's cells express her maternal X chromosome and the other half express her paternal X chromosome. Women are, therefore, mosaics for the expression of genes on their X chromosomes. The inactivation occurs early during development and results in a highly condensed chromosome that is visible under the light microscope as a densely staining "Barr body." The inactivated X chromosome undergoes reactivation during oogenesis so that all eggs contain active X chromosomes.

Sex is determined by many different mechanisms in other species. In birds, males contain two identical sex chromosomes and females contain two distinct sex chromosomes. In ants, bees, wasps, and other Hymenoptera, males have only one copy of each chromosome, whereas females carry two copies. Males are "haploid"; they develop from unfertilized eggs. Females are diploid, developing from fertilized eggs as in most other eukaryotes. In still other species, special sex chromosomes are not present, and environmental factors play key roles in sex determination. Thus, when considering sex-linked inheritance, one must not extrapolate from humans and fruit flies to other species.

1.5
Mendelian versus Quantitative Traits

Mendel studied pea plants that differed in specific phenotypic characteristics such as red flowers versus white flowers, tall versus dwarf, round seeds versus wrinkled seeds, and so on. As a result, he was able to classify the progeny of his crosses into distinct phenotypic classes and calculate the frequency of each class. Such traits are now referred to as *Mendelian traits*, because they yield the predicted monohybrid, dihybrid, and so on, segregation ratios in genetic crosses. Many other traits show continuous variation such that the progeny of crosses cannot be placed into discreet phenotypic classes. In humans, height and weight exhibit continuous variation from the shortest or smallest individual to the tallest or heaviest individual. The inheritance of such traits must be studied using quantitative measurements; therefore, the traits are commonly called *quantitative traits*. The genes that control quantitative traits are no different than the genes that control Mendelian traits. The difference is that quantitative traits are influenced by a large number of genes; they are multifactorial, being influenced by many factors, both genetic and environmental. Geneticists have developed statistical tools that can be used to estimate the number of genes affecting a trait, the proportion of the phenotypic variability that is genetic (caused by genes), and the proportion that is environmental (due to environmental factors). For a given trait, the proportion of the total phenotypic variability that is caused by genetic factors is the trait's *heritability*. Estimates of heritability for traits such as yield in grains and growth rate in domestic livestock have played an important role in the enhanced agricultural productivity realized in developed countries of the world during the last half century.

1.6
Genomics: Analyses of Entire Genomes

The science of genetics began with Mendel's monohybrid, dihybrid, and trihybrid crosses in garden peas at the monastery in Brünn (now Brno in the Czech Republic). Most of the early studies focused on the effects of one or a few genes. As the science matured, slot-blot and dot-blot hybridization technology allowed geneticists to simultaneously examine the expression of many genes. These blot hybridization procedures culminated in the development of microarray technologies that facilitated studies of hundreds to thousands of genes at once. Then, in 1995, the complete nucleotide sequence of the genome of the bacterium *Haemophilus influenzae* was published. The nucleotide sequences of many other bacteria were subsequently reported, along with the sequence of the genome of the yeast *Saccharomyces cerevisiae*. The nearly complete sequences of several model systems – the worm *Caenorhabditis elegans*, the fruit fly *D. melanogaster*, and the plant *Arabidopsis thaliana* – followed. Then, in early 2001, two drafts of the sequence of the human genome were published, followed in 2002 by drafts of the sequences of the genomes of two subspecies of rice.

Sophisticated computer programs were developed that allowed scientists to scan these sequences and identify open reading frames (ORFs), sequences with no "stop" signals that would prevent the synthesis of a protein product in at least one of the three reading frames.

These tools could be used to predict the presence and location of all the genes in a genome with reasonable accuracy. Gene-specific hybridization probes were then synthesized on the basis of the gene sequences and were used to prepare whole-genome microarrays. New technologies were used to array thousands of oligonucleotide hybridization probes on silicon wafers only a few square centimeters in size. These microarrays, commonly called "gene chips," allow geneticists to simultaneously study the expression of all the genes in an organism. All these developments led to a new subdiscipline of genetics – *genomics* – focused on the structure and function of entire genomes.

2
DNA Replication: The Genotypic Function

The genetic information of an organism must be transmitted from cell to cell during development and from generation to generation during reproduction. This transfer of genes from parents to offspring – the *genotypic function* – occurs by the accurate replication of DNA, that is, by the production of two progeny DNA molecules that are identical to the parental DNA molecule.

2.1
Semiconservative Replication

When Watson and Crick worked out the double-helix structure of DNA in 1953, they recognized that the complementary nature of the two strands – A paired with T and G paired with C – might play an important role in its replication. If the two strands of a parental double helix of DNA separated, the base sequence of each parental strand could serve as a template for the synthesis of a new complementary strand, producing two identical progeny double helices. This process is called *semiconservative replication* because the parental double helix is half conserved, each parental single strand remaining intact (Fig. 5). Meselson and Stahl documented the semiconservative replication of DNA in *E. coli* in 1958.

2.2
Proofreading

DNA replication is amazingly accurate with only about one error for every billion bases incorporated. This accuracy is necessary to keep the mutation load at a tolerable level, especially in large genomes such as those of mammals, which contain 3×10^9 nucleotide pairs. On the basis

Fig. 5 Semiconservative DNA replication. The single strands of the parental double helix are separated and each strand serves as a template for the synthesis of a complementary strand of DNA. The process results in the production of two progeny double helices that are identical to the parental double helix.

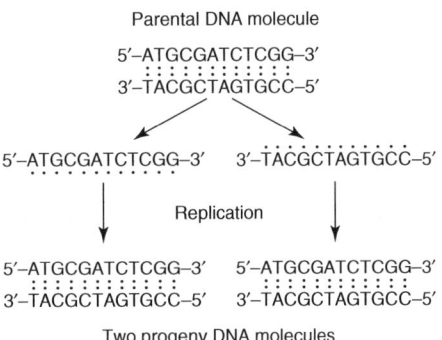

of the dynamic structures of the four nucleotides in DNA, the observed fidelity of DNA replication is much higher than expected. The thermodynamic changes in nucleotides that allow the formation of hydrogen-bonded base pairs other than A:T and G:C predict error rates of 10^{-5} to 10^{-4} or 1 error per 10 000 to 100 000 incorporated nucleotides. The predicted error rate of 10 000 times the observed error rate raises the question of how this high fidelity of DNA replication is achieved. The answer is that a mechanism for *proofreading* the nascent DNA chain as it is being synthesized has evolved in living organisms.

DNA proofreading involves scanning the termini of nascent DNA chains for errors and correcting them before continuing chain extension. This process is carried out by a $3' \rightarrow 5'$ exonuclease activity that is built into DNA polymerases. When a template–primer DNA has a terminal mismatch (an unpaired or incorrectly paired base or sequence of bases at the $3'$ end of the primer), the $3' \rightarrow 5'$ exonuclease activity of the DNA polymerase clips off the unpaired base or bases (Fig. 6). When an appropriately base-paired terminus is produced, the $5' \rightarrow 3'$ polymerase activity of the enzyme begins resynthesis by adding nucleotides to the $3'$ end of the primer strand. In mutant organisms that lack the proofreading activity of polymerases, the mutation rate is orders of magnitude higher than that in organisms with normal proofreading activity. Thus, proofreading is an important component of the semiconservative replication of DNA.

2.3
The Complex Replisome

DNA replication is complex, requiring the participation of a large number of proteins, and only a few of the most

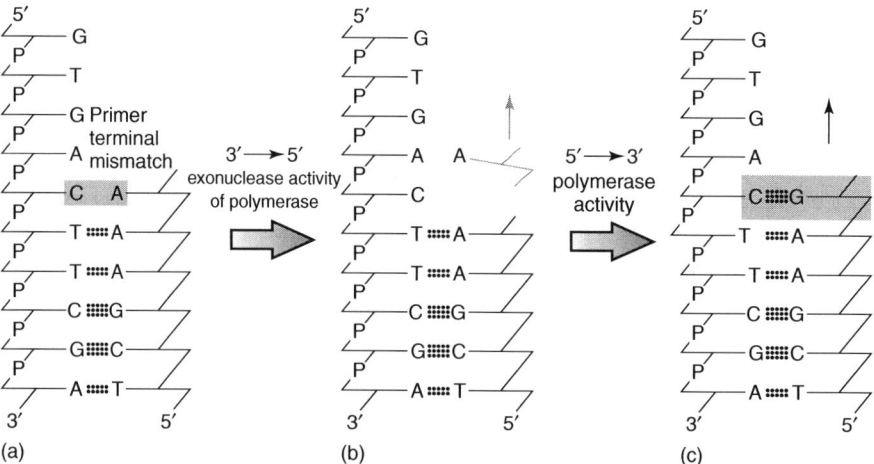

Fig. 6 Proofreading by the $3' \rightarrow 5'$ exonuclease activity of DNA polymerases during DNA replication. If DNA polymerase is presented with a template and primer containing a $3'$ primer terminal mismatch (a), the $3' \rightarrow 5'$ exonuclease activity will cleave off the mismatched terminal nucleotide (b). Then, given a correctly base-paired primer terminus, DNA polymerase will catalyze a $5' \rightarrow 3'$ covalent extension of the primer strand (c). (Reproduced from Snustad, D.P., Simmons, M.J., *Principles of Genetics*, 3rd edition, Copyright © 2003 by John Wiley & Sons, Inc., Hoboken, NJ, USA. This material is used by permission of John Wiley & Sons, Inc.)

important features of the process will be considered here. A nucleic acid chain has a chemical polarity based on the phosphodiester bonds that link the 5′ and 3′ carbons of adjacent nucleotides; that is, each chain will have a 5′ carbon at one end and a 3′ carbon at the other end. The complementary strands of a DNA double helix have opposite chemical polarity, with one strand 5′ → 3′ and the other strand 3′ → 5′, moving unidirectionally along the molecule. Because both nascent strands are extended as a replication fork moves along a parental double helix, one strand is extended at the macromolecular level in the 5′ → 3′ direction and the other strand in the 3′ → 5′ direction. However, DNA polymerases only catalyze 5′ → 3′ synthesis; at the molecular level, all synthesis is 5′ → 3′.

DNA synthesis is continuous on the progeny strand that is being extended in the overall 5′ → 3′ direction but is discontinuous on the strand growing in the overall 3′ → 5′ direction (Fig. 7). Discontinuous replication occurs by the synthesis of short DNA strands (1000 to 2000 nucleotides long in bacteria and 100 to 200 nucleotides long in eukaryotes). The short DNA strands are called "Okazaki fragments" after the scientists who discovered them.

The Okazaki fragments are initiated by short RNA primers synthesized by DNA primase. The RNA primers are subsequently replaced by DNA sequences by the combined 5′ → 3′ exonuclease and polymerase activities of a repair DNA polymerase (DNA polymerase I in *Escherichia coli*), and the Okazaki fragments are then joined by DNA ligase.

DNA replication involves many additional enzymes and other proteins. Replication requires that the two strands of a parental DNA molecule be separated during the synthesis of new complementary strands. Given that each gyre, or turn, of DNA is about 10 nucleotide pairs long, a DNA molecule must be rotated 360° once for every 10 replicated base pairs. In *E. coli*, DNA replicates at a rate of about 30 000 nucleotides per minute. Thus, a replicating DNA molecule must spin at 3000 revolutions per minute to facilitate the unwinding of the parental DNA strands. The unwinding process is catalyzed by enzymes called DNA helicases. The unwound strands are prevented from re-pairing by becoming coated with single-strand DNA-binding proteins (SSB proteins). The binding of SSB proteins to single-stranded DNA is cooperative; that is, the binding of the first SSB monomer stimulates the binding of additional monomers at contiguous sites on the DNA chain. Because of the cooperativity of SSB protein binding, an entire single-stranded region of DNA is rapidly coated with SSB protein.

Bacterial chromosomes contain circular molecules of DNA. With DNA spinning at 3000 revolutions per minute during unwinding of the parental strands, a swivel or axis of rotation is required to prevent tangles of supercoils from forming ahead of the replication fork. The required axes of rotation are provided by enzymes called DNA topoisomerases. The topoisomerases catalyze transient breaks in DNA molecules but use covalent linkages to themselves to hold on to the cleaved molecules, allowing subsequent reformation of the cleaved bonds. The transient single-strand break produced by the activity of topoisomerase I provides an axis of rotation that allows the segments of DNA on opposite sides of the break to spin independently, with the phosphodiester bond in the intact strand serving as

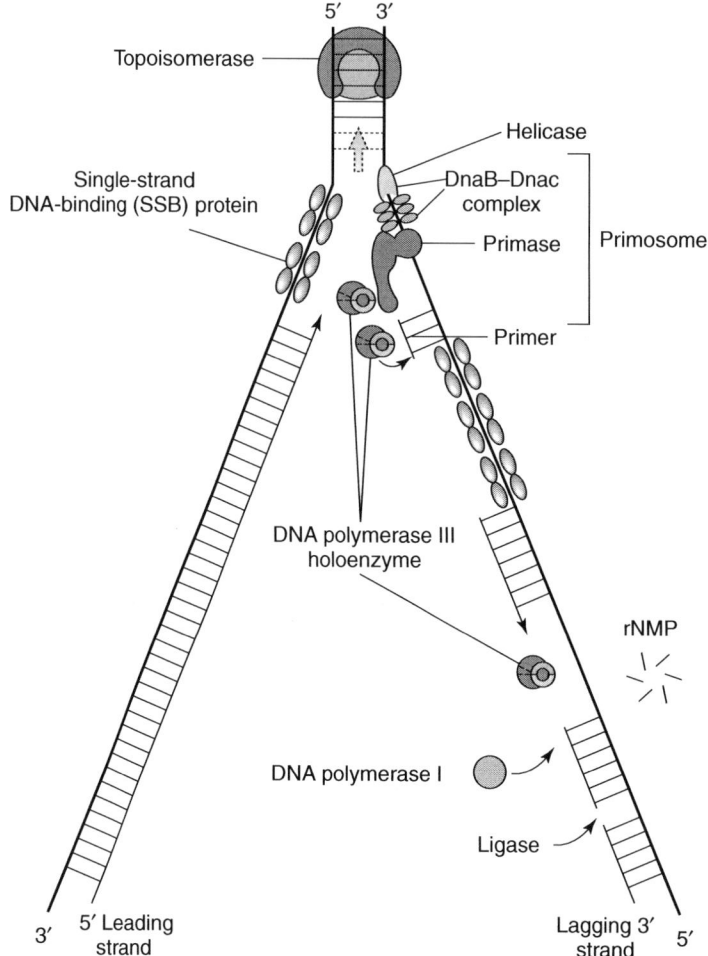

Fig. 7 Diagram of a replication fork in *E. coli* showing the major components of the replisome (rNMP, ribonucleoside monophosphates). (Reproduced from Snustad, D.P., Simmons, M.J., *Principles of Genetics*, 3rd edition, Copyright © 2003 by John Wiley & Sons, Inc., Hoboken, NJ, USA. This material is used by permission of John Wiley & Sons, Inc.)

a swivel. Thus, during DNA replication, only a short segment of DNA in front of the replication fork needs to spin – the segment up to the closest transient nick by topoisomerase I. Another *E. coli* topoisomerase called DNA gyrase is also required for DNA replication. DNA gyrase uses energy from ATP and introduces and removes negative (left-handed) supercoils in DNA.

DNA replication also requires a number of different proteins that are involved in the initiation or priming of synthesis at specific origins of replication. These proteins bind to specific nucleotide sequences at the origin and induce localized strand separation or "bubbles" in which the synthesis of nascent strands begins. All the enzymes and DNA-binding proteins involved in replication assemble into a *replisome* at each replication fork and act in concert as the fork moves along the parental DNA molecule (Fig. 7).

3
Gene Expression: The Phenotypic Function

The genetic information controls the growth and development of the organism, be it a virus, a bacterium, a plant, or an animal. This genetic information must be expressed accurately – the *phenotypic function* – both spatially and temporally to produce the appropriate three-dimensional form of the organism. In multicellular organisms, the genetic information must control the growth and differentiation of the organism from the single-celled zygote to the mature adult. To accomplish this phenotypic function, each gene of an organism must be expressed at the proper time and in the proper cells during development. The initial steps in the pathways of gene expression, transcription and translation, are quite well elucidated; these steps are illustrated for the expression of the human β-globin gene in Fig. 8.

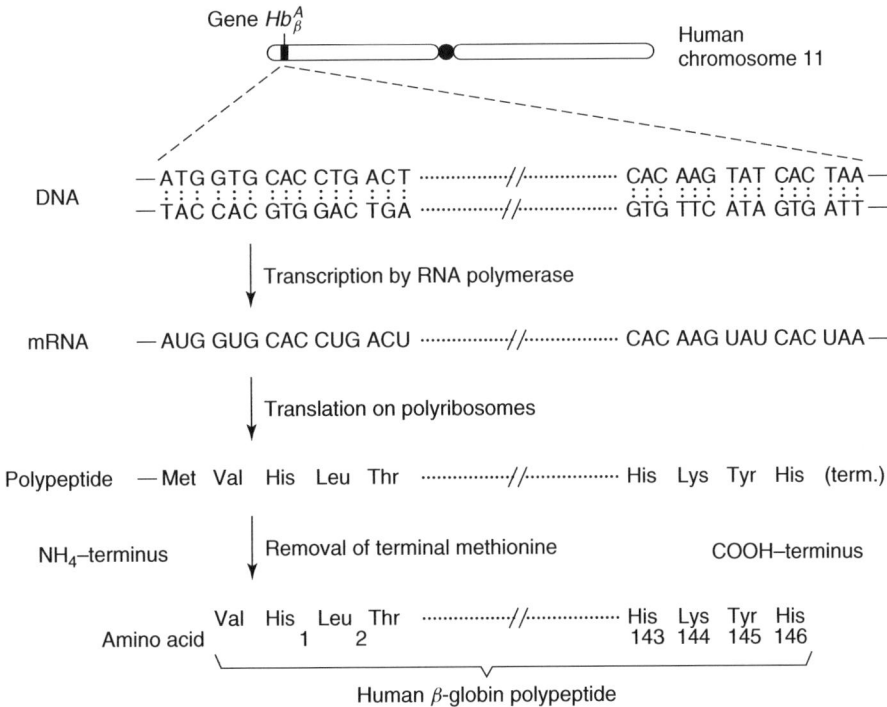

Fig. 8 Schematic diagram showing the first steps in the expression of the human gene (Hb_β^A) encoding β-globin: transcription, translation, and the proteolytic removal of the amino-terminal methionine residue from the primary translation product. For simplicity, only the terminal portions of the coding sequence and the polypeptide product are shown.

3.1
Colinearity between Gene and Polypeptide

The genetic information is stored in linear sequences of nucleotide pairs in DNA (or nucleotides in RNA, in some cases). Transcription and translation convert this genetic information into *colinear* sequences of amino acids in polypeptides, which function as the key intermediaries in the genetic control of the phenotype. The first three base pairs of the coding sequence of a gene specify the first amino acid of the polypeptide, the next three base pairs (four to six) specify the second amino acid, and so on, in a colinear fashion (Fig. 9). Although the coding regions of most of the genes in higher eukaryotes are interrupted by noncoding sequences called introns, their presence does not invalidate the concept of colinearity. The presence of introns in genes simply means that there is no direct correlation in physical distances between the positions of base-pair coding triplets in a gene and the positions of amino acids in the polypeptide specified by that gene. The coding sequences and the polypeptides that they encode are still colinear.

3.2
Transcription

The first step in gene expression, *transcription*, involves converting genetic information stored in the form of base pairs in double-stranded DNA into the sequence of bases in a single-stranded molecule of messenger RNA (mRNA). This process is catalyzed by enzymes called RNA polymerases and occurs when one strand of the DNA is used as a template to synthesize a complementary strand of RNA using the same base-pairing rules as for DNA replication, except that uracil is incorporated into RNA at positions where thymine would be present in DNA (see Fig. 8, top). Like replication, transcription is a complex process involving numerous proteins called transcription factors. Transcription is initiated at specific sequences within regions called promoters located adjacent to the gene. Transcription initiation factors bind to these sequences and induce localized unwinding of the DNA molecule. RNA synthesis occurs within these locally unwound "transcription bubbles" as RNA polymerase moves along the DNA template strand. Chain extension occurs in the $5' \to 3'$ direction by a mechanism very similar to DNA synthesis. The termination of transcription also occurs at specific nucleotide sequences,

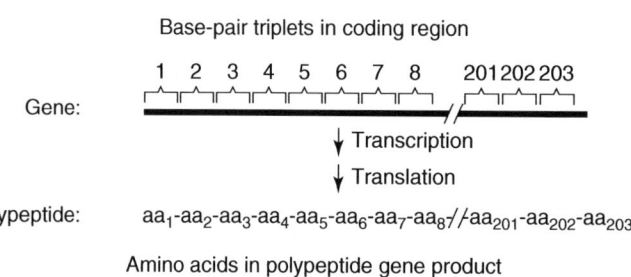

Fig. 9 Colinearity between the base-pair triplets in the coding region of a gene and the amino acid sequence in the polypeptide product of the gene.

sometimes with the aid of other termination proteins.

3.3 Introns and RNA Splicing

Most of the genes of eukaryotes are interrupted by sequences that are not represented in the mature RNA products of these genes. These sequences are called *introns* (for *intervening sequences). The coding sequences and other sequences that are present in the final RNA products of these interrupted genes are called *exons* (for *expressed sequences). The structure of a typical eukaryotic gene is shown in Fig. 10.

Not all eukaryotic genes contain introns, so they are not required for expression. When present, the number of introns per gene varies from one to over 50, and the intron size varies from about 50 nucleotide pairs to thousands of nucleotide pairs. The human *DMD* gene, which is responsible for Duchenne muscular dystrophy when nonfunctional, is one of the largest known. The *DMD* gene contains 78 introns and is over 2.5-million nucleotide pairs in length. Rare genes of Archaea and of a few viruses of prokaryotes also contain introns. In the case of these "split" genes, the primary transcript contains the entire sequence of the gene and the intron sequences are excised during RNA processing.

For genes that encode proteins, the splicing mechanism must be precise; it must join exon sequences with accuracy to the single nucleotide to assure that codons in exons distal to introns are read correctly. Accuracy to this degree would seem to require precise splicing signals, presumably nucleotide sequences within introns and

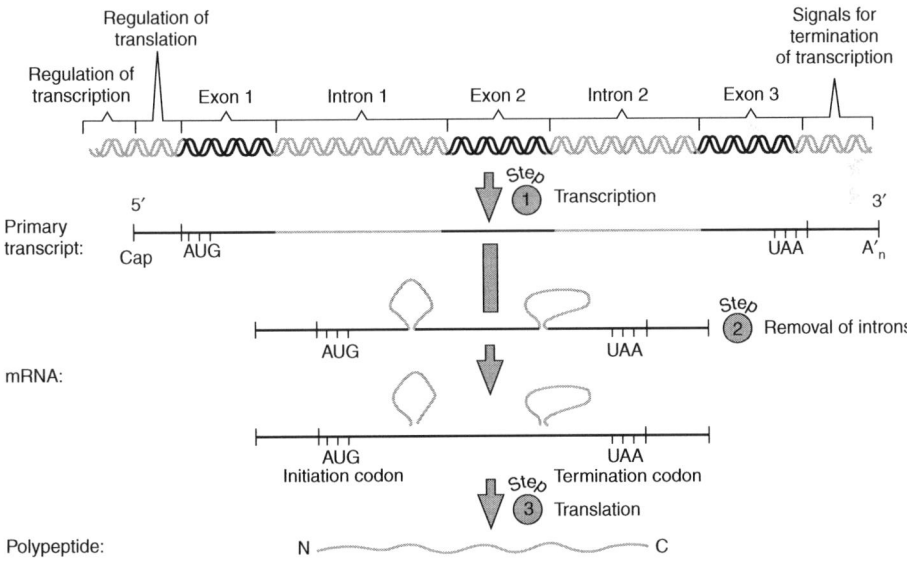

Fig. 10 Structure of a typical eukaryotic gene. Note that the intron sequences are present in the primary transcript but are removed during nuclear processing to produce the mature mRNA prior to its export to the cytoplasm for translation. (Reproduced from Snustad, D.P., Simmons, M.J., *Principles of Genetics*, 3rd edition, Copyright © 2003 by John Wiley & Sons, Inc., Hoboken, NJ, USA. This material is used by permission of John Wiley & Sons, Inc.)

at the exon–intron junctions. However, in the primary transcripts of nuclear genes, the only completely conserved sequences of different introns are the dinucleotide sequences at the ends of introns. In primary transcripts, all introns begin with GU and end with AG. For nuclear genes, there is also one somewhat conserved sequence located about 30 nucleotides from the 3' splice site. The introns of genes of mitochondria and chloroplasts also contain conserved sequences, but they are different from those of nuclear genes.

There are three major classes of intron excision from RNA transcripts. The introns of tRNA precursors are excised by precise endonucleolytic cleavage and ligation reactions catalyzed by special splicing endonuclease and ligase activities. The introns of some rRNA precursors are removed autocatalytically in a unique reaction mediated by the RNA molecule itself. No protein enzymatic activity is involved. The introns of nuclear pre-mRNA transcripts are excised in two-step reactions carried out by complex ribonucleoprotein particles called spliceosomes. For details about the splicing mechanisms, see [Lewin (2000) or Alberts et al. (2002)].

3.4
RNA Editing

According to the central dogma of molecular biology, genetic information flows from DNA to RNA to protein during gene expression. Normally, the genetic information is not altered in the mRNA intermediary. However, the discovery of *RNA editing* has shown that exceptions do occur. RNA editing processes alter the information content of gene transcripts in two ways: (1) by changing the structures of individual bases and (2) by inserting or deleting uridine monophosphate residues.

The first type of RNA editing, which results in the substitution of one base for another base, is rare. This type of editing was discovered in studies of the apolipoprotein-B *(apoB)* genes and mRNAs in rabbits and humans. Apolipoproteins are blood proteins that transport certain types of fat molecules in the circulatory system. In the liver, the *apoB* mRNA encodes a large protein 4563 amino acids long. In the intestine, the *apoB* mRNA directs the synthesis of a protein only 2153 amino acids long. Here, a C residue in the pre-mRNA is converted to a U, generating an internal UAA translation–termination codon, which results in the truncated apolipoprotein. The C → U conversion is catalyzed by a sequence-specific RNA-binding protein with an activity that removes amino groups from cytosine residues. In some transcripts present in plant mitochondria, most of the Cs are converted to U residues.

A second, more complex type of RNA editing occurs in the mitochondria of trypanosomes. In this case, uridine monophosphate residues are inserted (occasionally deleted) into gene transcripts, causing major changes in the polypeptides specified by the mRNA molecules. This RNA editing process is mediated by guide RNAs transcribed from distinct mitochondrial genes. The guide RNAs contain sequences that are partially complementary to the pre-mRNAs to be edited. Pairing between the guide RNAs and the pre-mRNAs results in gaps with unpaired A residues in the guide RNAs. The guide RNAs serve as templates for editing, as Us are inserted in the gaps in pre-mRNA molecules opposite the As in the guide RNAs. In some cases, two or more different guide RNAs participate in the editing of a single pre-mRNA. For some unknown reason, RNA editing plays a major role in the expression of genes in the mitochondria of trypanosomes and plants.

3.5
Translation

During *translation*, the sequence of bases in the mRNA molecule is converted ("translated") into the specified sequence of amino acids in the polypeptide gene product according to the rules of the genetic code (Fig. 8, center). Each amino acid is specified by one or more codons, and each codon contains three nucleotides. Of the 64 possible nucleotide triplets, 61 specify amino acids and 3 specify polypeptide chain termination. Translation occurs on ribosomes, which are complex macromolecular structures located in the cytoplasm (Fig. 11). Translation involves three types of RNA, all of which are transcribed from DNA templates (chromosomal genes). In addition to mRNAs, 3 to 5 RNA molecules (rRNA molecules) are present as part of the structure of each ribosome, and 40 to 60 RNA molecules (tRNA molecules) function as adaptors by mediating the incorporation of the proper amino acids into polypeptides in response to specific nucleotide sequences in mRNAs. The amino acids are attached to the correct tRNA molecules by a set of

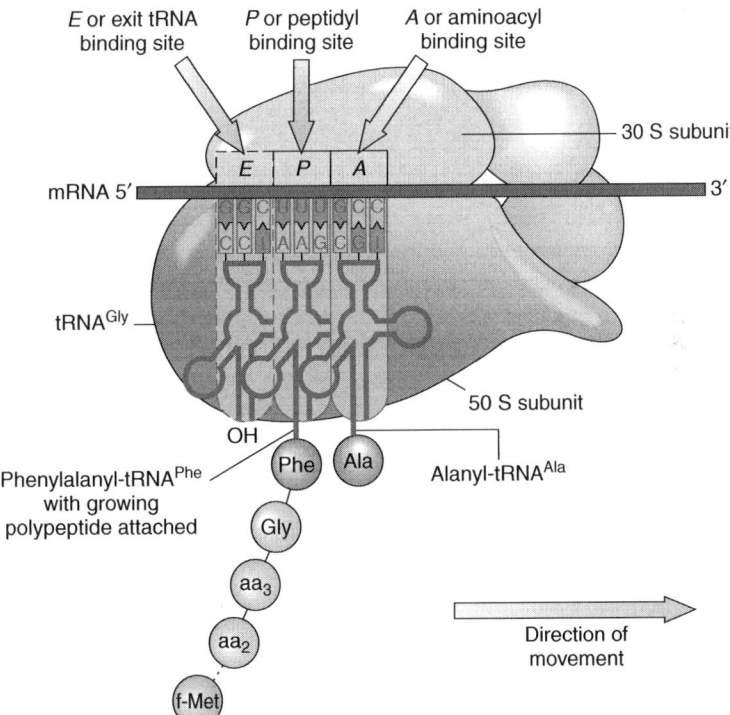

Fig. 11 Diagram of a bacterial 70S ribosome showing the three tRNA binding sites and the base pairing between the codons of the mRNA and the anticodons of the tRNAs. (Reproduced from Snustad, D.P., Simmons, M.J., *Principles of Genetics*, 3rd edition, Copyright © 2003 by John Wiley & Sons, Inc., Hoboken, NJ, USA. This material is used by permission of John Wiley & Sons, Inc.)

activating enzymes called aminoacyl tRNA synthetases. The tRNA molecules contain nucleotide triplets called anticodons, which base pair with the codons in mRNA during the translation process.

The ribosomes may be thought of as workbenches, complete with machines and tools needed to make a polypeptide. They are nonspecific in the sense that they can synthesize any polypeptide (any amino acid sequence) encoded by a particular mRNA molecule, even an mRNA from a different species. Each mRNA molecule is simultaneously translated by several ribosomes, resulting in the formation of polyribosomes. The translation of the sequence of nucleotides in an mRNA molecule into the sequence of amino acids in its polypeptide product can be divided into three stages: (1) polypeptide chain initiation; (2) chain elongation; and (3) chain termination.

In *E. coli*, the translation initiation process involves the 30S subunit of the ribosome, a special initiator tRNA, an mRNA molecule, three soluble protein initiation factors IF-1, IF-2, and IF-3, and one molecule of GTP. In the first stage of the initiation of translation, a free 30S subunit interacts with an mRNA molecule and the initiation factors. The 50S subunit joins the complex to form the 70S ribosome in the final step of the initiation process. The addition of the 50S ribosomal subunit to the complex positions the initiator tRNA, methionyl-tRNA$_f^{Met}$, in the peptidyl (*P*) site of the ribosome with the anticodon of the tRNA aligned with the AUG initiation codon of the mRNA. Methionyl-tRNA$_f^{Met}$ is the only aminoacyl tRNA that can enter the *P* site directly, without first passing through the aminoacyl (*A*) site. With the initiator AUG positioned in the *P* site, the second codon of the mRNA is in register with the *A* site, dictating the aminoacyl tRNA binding specificity at that site and setting the stage for the second phase in polypeptide synthesis, chain elongation.

The process of polypeptide chain elongation is basically the same in both prokaryotes and eukaryotes. The addition of each amino acid to the growing polypeptide occurs in three steps: (1) binding of an aminoacyl tRNA to the *A* site of the ribosome; (2) transfer of the growing polypeptide chain from the tRNA in the *P* site to the tRNA in the *A* site by the formation of a new peptide bond; and (3) translocation of the ribosome along the mRNA to position the next codon in the *A* site. During step 3, the nascent polypeptide-tRNA and the uncharged tRNA are translocated from the *A* and *P* sites to the *P* and *E* sites respectively. These three steps are repeated in a cyclic manner throughout the elongation process. Polypeptide chain elongation proceeds rapidly. In *E. coli*, all three steps required for the addition of one amino acid to the growing polypeptide chain to occur in about 0.05 s. Thus, the synthesis of a polypeptide containing 300 amino acids takes only about 15 s.

Polypeptide chain elongation undergoes termination when any of the three chain-termination codons (UAA, UAG, or UGA) enters the *A* site on the ribosome. These three stop codons are recognized by soluble proteins called release factors (RFs). In *E. coli*, there are two release factors, RF-1 and RF-2. RF-1 recognizes termination codons UAA and UAG; RF-2 recognizes UAA and UGA. In eukaryotes, a single release factor (eRF) recognizes all three termination codons. Termination is completed by the release of the mRNA molecule from the ribosome and the dissociation of the ribosome into its subunits.

3.6
Inteins

Occasionally, the primary translation product of a gene contains one or more short amino acid sequences, called *inteins*, that excise themselves from the nascent polypeptide. Inteins occur in both eukaryotic and prokaryotic polypeptides; one of the first inteins discovered is in the RecA protein (required for recombination) in *Mycobacterium tuberculosis*, the bacterium that causes tuberculosis. The ability to carry out intein excision is a function of the structure of the primary translation product and is thus encoded in the gene just like any other amino acid sequence.

3.7
Some Complex Gene/Protein Relationships

As discussed in Section 3.3, most eukaryotic genes are split into expressed sequences (exons) and intervening sequences (introns). In some cases, transcripts of split genes may undergo several different types of splicing, making the relationships between genes and proteins more complex than the usual one gene–one polypeptide. In other cases, expressed genes are assembled from "gene pieces" during the development of the specialized cells in which they are expressed.

When the transcripts of an interrupted gene undergo alternate pathways of transcript splicing, different exons are joined to produce a related set of mRNAs that encode a family of closely related polypeptides. These interrelated polypeptides are called *protein isoforms*. Alternate splicing pathways are often tissue-specific, producing related proteins that carry out similar, but not necessarily identical, functions in different types of cells. The mammalian tropomyosin genes produce complex families of protein isoforms. Tropomyosins regulate muscle contraction in animals. One mouse tropomyosin gene produces at least 10 different polypeptides by alternate pathways of transcript splicing. Genes of this type do not fit the one gene–one polypeptide concept very well. Such genes can be defined as DNA sequences that are single units of transcription and encode a set of protein isoforms.

Genetic information is not always organized into genes of the type described in Section 1.2. In a few cases, genes are assembled from a storehouse of gene segments during the development of an organism. The immune system of vertebrate animals depends on the synthesis of proteins called antibodies to provide protection against infections by viruses, bacteria, toxins, and other foreign substances. Each antibody contains four polypeptides, two identical heavy chains and two identical light chains. The light chains are of two types: kappa and lambda. Each antibody chain contains a variable region, which exhibits extensive diversity from antibody to antibody, and a constant region, which is largely the same in all antibodies. In germ-line chromosomes, the DNA sequences encoding these antibody chains are present in gene segments, and the gene segments are joined together to produce genes during the differentiation of the antibody-producing B-lymphocytes from progenitor cells. The B-lymphocytes subsequently differentiate into antibody-secreting plasma cells.

This process of *gene assembly during development* is illustrated in Fig. 12. A kappa light chain gene is assembled from three gene segments V_k (V for variable region), J_k (J for joining segment), and

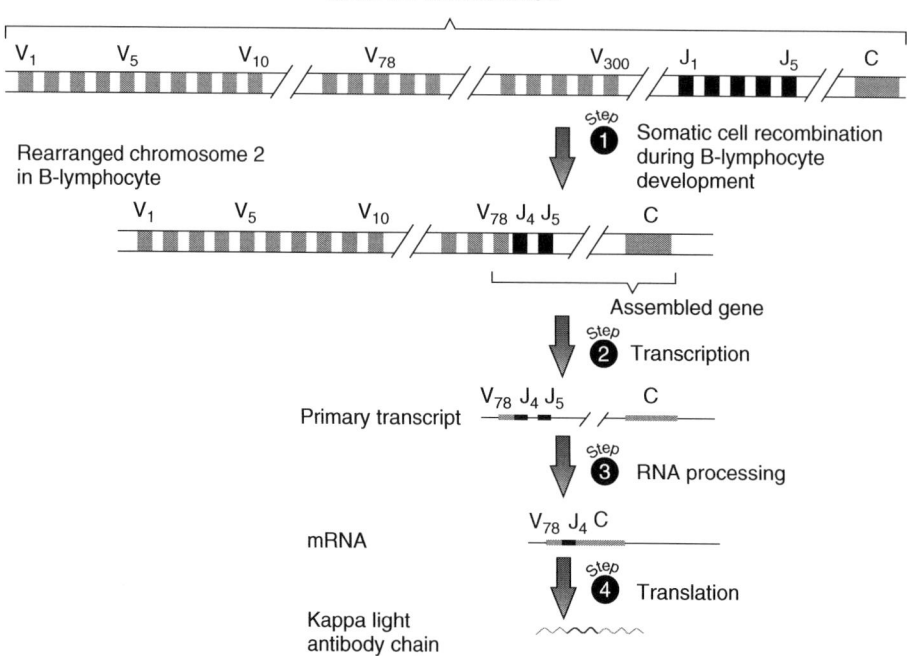

Fig. 12 Assembly of a gene encoding an antibody kappa light chain from gene segments during B-lymphocyte differentiation in humans. (Reproduced from Snustad, D.P., Simmons, M.J., *Principles of Genetics*, 3rd edition, Copyright © 2003 by John Wiley & Sons, Inc., Hoboken, NJ, USA. This material is used by permission of John Wiley & Sons, Inc.)

C_k (C for constant region) during B-lymphocyte development. Together, the V_k and J_k gene segments encode the variable region of the kappa light chain, whereas the C_k gene segment encodes the constant region. No functional V_k–J_k–C_k kappa light chain gene is present in any human germ-line chromosome. Instead, human chromosome 2 contains a cluster of about 300 V_k gene segments, another cluster of 5 J_k gene segments, and a single C_k gene segment (Fig. 12). During the differentiation of each B-lymphocyte, recombination joins one of the V_k gene segments to one of the J_k gene segments. Any J_k segments remaining between the newly formed V_k–J_k exon and the C_k gene segment become part of an intron that is removed during the processing of the primary transcript. Similar somatic recombination events are involved in the assembly of the genes encoding antibody heavy chains, lambda light chains, and T-lymphocyte receptor proteins.

3.8
Pathways of Gene Expression are Often Complex

The pathway through which a gene exerts its effect on the phenotype of the organism is often long and complex, especially in multicellular eukaryotes (Fig. 13). Pathways of gene action frequently involve protein–protein and other macromolecular interactions, cell–cell interactions and

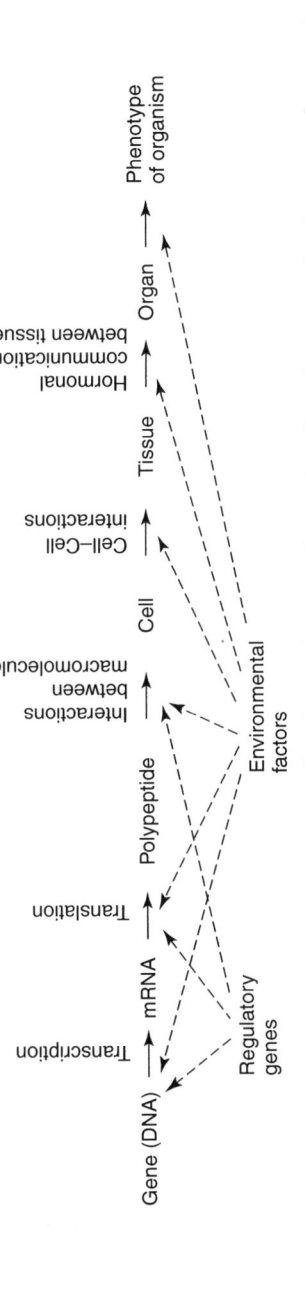

Fig. 13 A typical pathway of gene expression showing some of the factors that can influence the effect of a gene on the phenotype of an organism.

intercellular communication by hormones and other signal molecules, tissue and organ interactions, and restrictions imposed by environmental factors. Note that, although not shown here, each gene also has an effect on the phenotype of the population in which the organism lives (population genetics) and ultimately on the phenotype of the biosphere (ecological genetics).

3.9
Pleiotropy and Epistasis

Sometimes, a single gene influences many aspects of an organism's phenotype. When this occurs, the gene is said to be pleiotropic. The mutant gene that causes sickle-cell anemia in humans provides a classic example of *pleiotropy*. This inherited disorder results from a mutation in the β-globin gene that changes the sixth amino acid from a glutamic acid in the wild-type β-globin to a valine in the sickle-cell polypeptide. This amino acid substitution alters the conformation of the polypeptide, causing aggregation of hemoglobin molecules and the development of grossly deformed red blood cells. These sickle-shaped cells have reduced capacity to transport oxygen and, therefore, cause the anemia. However, the sickle cell allele not only causes hemolytic anemia in the homozygous state but also has many other effects such as enlarged spleen, impaired growth, recurrent pain, and increased susceptibility to microbial and viral infections. Perhaps the allele's most unexpected pleiotropic effect is the enhanced resistance to *Paramecium falciparum* malaria, which it provides when present in the heterozygous state.

When two or more genes influence a trait, an allele of one of them may mask the effects of alleles of the other gene on the phenotype. When an allele has such an overriding effect, it is said to be epistatic to the other genes. The occurrence of such interactions between different genes (nonalleles) is called *epistasis*. There are many examples of epistasis. In *Drosophila*, a recessive mutation in the cinnabar gene causes the eyes of the fly to be bright red. A recessive mutation in another gene results in white eyes. Flies that are homozygous for both these mutations have white eyes. The white mutation is, therefore, epistatic to the cinnabar mutation. In humans and other mammals, recessive mutant alleles that cause albinism are usually epistatic to other genes affecting eye and hair color.

3.10
Penetrance and Expressivity

Sometimes, a mutant gene can be present in an organism without having any effect on its phenotype. Such a gene is said to exhibit *incomplete penetrance*. In humans, polydactyly, the presence of extra fingers and toes, provides an example of incomplete penetrance. This condition is caused by a dominant mutation that results in polydactyly in some, but not all, individuals who are known to carry it. Penetrance is measured by determining what proportion of the individuals who carry a gene exhibit the trait made possible by its presence. Incomplete penetrance can be a serious problem in pedigree analysis, sometimes resulting in the assignment of incorrect genotypes.

In other cases, the presence of a gene may result in a range of phenotypic effects from mild to severe. Such a gene is said to exhibit *variable expressivity*. The dominant Lobe eye mutation in *Drosophila* provides an example of variable expressivity; its phenotypic effect varies from tiny compound eyes, through a full range of intermediates, to large, lobulated eyes.

3.11
Intragenic Complementation

The results of complementation tests are usually unambiguous when recessive mutations that result in the synthesis of no gene product, partial gene products, or totally defective gene products are used – for example, deletions of segments of genes or polypeptide chain–terminating mutations. When mutations that cause amino acid substitutions are used, the results are sometimes ambiguous because of the occurrence of a phenomenon called *intragenic complementation*.

The functional forms of some proteins are dimers or higher multimers consisting of two or more polypeptides. These polypeptides may be either the products of a single gene or the products of two or more different genes. When the active form of the protein contains two or more homologous polypeptides, intragenic complementation sometimes occurs. *Inter*genic complementation (discussed in Section 1.2) and *intra*genic complementation are distinct phenomena.

As a simple example of intragenic complementation, consider an enzyme that functions as a homodimer, that is, a protein containing two copies of a specific gene product. An organism that is heterozygous for two different mutations in the gene will produce some dimers that contain the two different mutant polypeptides. We call these heterodimers. Such heterodimers may have partial or complete (wild-type) function; when this occurs, intragenic complementation has occurred. In such cases, the *trans* heterozygote has a wild-type phenotype or a phenotype intermediate between mutant and wild type. Why such heteromultimers should be active when the two corresponding homomultimers are inactive is not clear. Apparently, the wild-type sequence of amino acids in the nonmutant segment of one mutant polypeptide somehow compensates for the mutant segment of the polypeptide encoded by the second mutant allele, and vice versa. In the case of noncomplementing mutations in a gene encoding a multimeric protein, the heteromultimers are nonfunctional, just like the mutant homomultimers.

3.12
Intergenic Noncomplementation

In some cases, mutations known to be located in two different genes fail to complement each other. This phenomenon is called *intergenic noncomplementation*, and it has become a powerful tool for identifying genes encoding products that interact. If two different polypeptides are part of a macromolecular complex, mutations in the two genes encoding them may not complement each other. The presence of mutant polypeptides in the complex may render the entire complex nonfunctional, even if wild-type products of both genes are present. Thus, the occurrence of intergenic noncomplementation provides evidence that the two gene products interact in some way. Intergenic noncomplementation is allele-specific; not all mutations in the genes involved will exhibit the phenomenon.

3.13
Regulation of Gene Expression

In all organisms, gene expression is highly regulated so that energy is used to synthesize gene products only when those products are needed for the growth and differentiation of the organism. In higher eukaryotes, only a small proportion of the genes in any genome are expressed in a given cell type. Thus, gene expression is highly programmed such that genes needed to make neurons are only turned on in developing nerve cells, genes needed to make red blood cells are only expressed in progenitors of erythrocytes, and so on. Most microorganisms exhibit a striking ability to regulate the expression of specific genes in response to environmental signals. The expression of particular genes is turned on when the products of these genes are needed for growth. Their expression is turned off when the gene products are no longer needed. Gene expression is regulated at several different levels: transcription, mRNA processing, mRNA turnover, translation, and posttranslation (Fig. 14). However, the regulatory mechanisms with the largest effects on phenotype act at the level of transcription.

Most regulatory mechanisms fit into two general categories: (1) the rapid turn-on or turn-off of gene expression in response to environmental changes (especially important in microorganisms) and (2) preprogrammed circuits of gene expression (important in all organisms).

Certain "housekeeping" gene products – tRNA molecules, rRNA molecules, ribosomal proteins, and the like – are essential components of all living cells. These genes are continually being expressed in most cells; they are referred to as *constitutive genes*. Other gene products are needed only in the presence of specific metabolites. They are not expressed in the absence of the metabolite but are turned on when the metabolite is present. This process is called *induction*. Genes whose expression is regulated in this manner are called *inducible genes*; their products, if enzymes, are called *inducible enzymes*. Enzymes that are involved in catabolic pathways are often inducible. Other genes are turned on and they synthesize their gene products unless the metabolite synthesized by those products is present in the environment. Then, they are turned off. This process is called *repression*, and genes that are regulated in this manner are called *repressible genes*. Enzymes that are components of anabolic pathways are often repressible. Both induction and repression occur at the level of transcription.

The regulation of gene expression, induction or repression, can be accomplished by both positive and negative control mechanisms. Both mechanisms involve the participation of *regulator genes* – genes whose products regulate the expression of other genes. In *positive control mechanisms*, the product of the regulator gene functions by turning on the expression of one or more genes, whereas in *negative control mechanisms*, the product of the regulator gene is involved in shutting off the expression of genes. Positive and negative regulation mechanisms can both mediate either inducible or repressible gene expression.

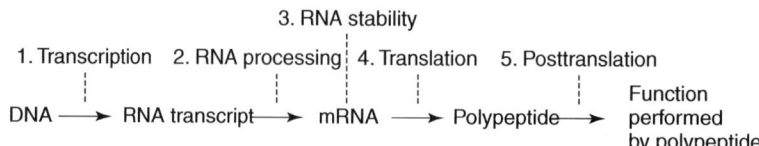

Fig. 14 Pathway of gene expression showing five stages at which gene expression is regulated.

A given gene is expressed when RNA polymerase binds to its promoter and synthesizes an RNA transcript that contains the coding region of the gene. The product of the regulator gene acts by binding to a site called the regulator protein binding site (*RBS*) adjacent to the promoter of the structural gene (or genes). When the product of the regulator gene is bound to *RBS*, transcription of the structural gene(s) is turned on in a positive control system or is turned off in a negative control system. The regulator gene products are called *activators* in positive control systems and *repressors* in negative control systems. Whether a regulator protein can bind to the *RBS* depends on the presence or absence of *effector molecules* in the cell. The effectors are usually small molecules such as amino acids, sugars, and similar metabolites. The effector molecules involved in induction of gene expression are called *inducers*; those involved in repression of gene expression are called *corepressors*.

The effector molecules (inducers and corepressors) bind to regulator gene products (activators and repressors) and cause changes in the three-dimensional structures of these proteins. Such changes in protein conformation are called *allosteric transitions*. In the case of activators and repressors, the allosteric transitions alter their ability to bind to regulator protein binding sites adjacent to the genes that they control.

In microorganisms, *operons* – negatively and coordinately regulated units of gene expression – play important regulatory roles. Each operon contains one to many structural genes, adjacent to promoter and operator sequences. The operator is the binding site for the repressor (inducible system) or the repressor/corepressor complex (repressible system). When the repressor or repressor/corepressor complex is bound to the operator, it prevents RNA polymerase from transcribing the structural genes of the operon. When the repressor is not bound to the operator, RNA polymerase binds to the promoter and transcribes the genes in the operon. Another important regulatory mechanism in microorganisms – attenuation – occurs by the control of premature termination of transcription.

Regulation of gene expression in eukaryotes usually involves the binding of several proteins called *transcription factors* to upstream promoters and to regulatory sequences called *enhancers* and *silencers*, which can be located either upstream of, downstream from, or within the genes that they regulate. As the names suggest, enhancers and silencers increase and decrease, respectively, the levels of gene expression.

3.14
RNA-mediated Gene Silencing

RNA-mediated gene silencing (*RMGS*) was first studied in plants where it was called *posttranscriptional gene silencing* (*PTGS*). Similar RNA-induced gene silencing processes were subsequently discovered in animals and fungi and were called *RNA interference* (*RNAi*) and *quelling*, respectively. RMGS is triggered by the presence of double-stranded RNA (dsRNA). The dsRNA may be composed of two complementary or partially complementary single strands or a self-complementary RNA molecule containing complementary regions such that it folds back on itself and forms a hairpin-like structure. The dsRNAs are degraded to 21- to 22-nucleotide-long dsRNA fragments called *small interfering RNAs* (*siRNAs*) by the ribonuclease "Dicer." The siRNA fragments subsequently become associated with an endonuclease-containing complex called RISC for "RNAi silencing complex." This complex binds to and degrades mRNA molecules that contain sequences complementary to the sequence of the siRNAs. RMGS requires several other gene products, including an RNA-dependent RNA polymerase, an RNA helicase, and the proteins in RISC. Additional components of the RMGS pathways are currently being investigated in both plants and animals.

RMGS is believed to be a defense mechanism protecting the host organism from viral infections, runaway transposons, and other invading DNAs and dsRNAs. In addition, the recent identification of families of naturally occurring *microRNAs* with self-complementary sequences that form "hairpin" structures and give rise to siRNA-like fragments suggests that RMGS may play important roles in regulating growth and development and other important natural processes. Finally, the mechanisms by which PTGS, RNAi, and quelling occur in various species are probably similar but not necessarily identical. Indeed, there are probably multiple pathways by which RMGS can occur, even within a species.

In addition to their normal biological function(s), RMGS has become an important tool for reverse genetic analyses, allowing researchers to shut off genes in a controlled manner. In *C. elegans*, RNAi was used to systematically "knock out" the expression of each of the genes on chromosomes I and III respectively. In plants, chimeric gene cassettes that direct the synthesis of hairpin RNAs and induce PTGS have proven to be effective in producing gene knockouts. Thus, RMGS is an important tool in the field of functional genomics.

4
Mutation: The Evolutionary Function

Although the genetic information must be transmitted from generation to generation with considerable accuracy, it is not static, but it undergoes occasional change or *mutation – the evolutionary function* – to produce new genetic variability that provides the raw material for ongoing evolution. The new variant genes produced by mutation are called mutant alleles and often result in abnormal or mutant phenotypes. When used in the narrow sense, mutation refers only to changes in the structures of individual genes. However, in the broad sense, mutation refers to any heritable change in the genetic material and includes gross changes in chromosome structure or chromosome aberrations. In addition, the word mutation refers not only to the change in the genetic material but also to the process during which the change occurs.

4.1
Dominance, Partial Dominance, and Codominance

When mutations occur producing new mutant alleles, these alleles can exhibit a range of interactions with the original or wild-type allele. The phenomenon of interactions between various alleles of a given gene is called *dominance*. If an organism that is heterozygous for a wild-type allele and a mutant allele has the same phenotype as an organism that is homozygous for the mutant allele, the mutant allele is *dominant* and the wild-type allele is *recessive*. If the heterozygote has the wild-type phenotype, the reverse is true: the mutant allele is recessive and the wild-type allele is dominant. If the phenotype of the heterozygote is intermediate between the phenotypes of the respective homozygotes, the mutant allele exhibits *partial dominance* (sometimes called semidominance) or no dominance (if the phenotype is precisely intermediate to the phenotypes of the two homozygotes). If both alleles produce their phenotypic effects in heterozygotes, the alleles are said to be *codominant*. Humans with the AB blood type (genotype $I^A I^B$), for example, have both type A antigens and type B antigens on their red blood cells. Thus, the I^A and I^B alleles are codominant.

4.2
Transitions, Transversions, and Frameshift Mutations

Point mutations within individual genes may be either base-pair substitutions or the insertion or deletion of one or a few contiguous base pairs. Base-pair substitutions usually result in the substitution of a single amino acid in the mutant polypeptide gene product. Base-pair substitutions are of two types: *transitions* and *transversions*. Transitions occur when one purine is substituted for the other purine and the corresponding pyrimidine for pyrimidine substitutions take place in the complementary strand of DNA. Transversions involve purine for pyrimidine and pyrimidine for purine substitutions. Of the 12 different base-pair substitutions, 4 are transitions and 8 are transversions (Fig. 15).

Sickle-cell anemia in humans is the result of a single base-pair substitution, a transversion, in the adult β-globin gene. This disorder occurs in individuals who are homozygous for the altered β-globin gene. This single base-pair substitution in the Hb_β^S gene changes the sixth amino acid of the β-globin polypeptide from glutamic acid in Hb_β^A homozygotes to valine in Hb_β^S homozygotes (Fig. 16). This one amino acid change in the human β-globin chain results in sickle-shaped red blood cells and in sickle-cell anemia in individuals homozygous for the Hb_β^S allele. Thus, a single base-pair substitution in DNA can have a very large effect on the phenotype of the organism harboring the mutation.

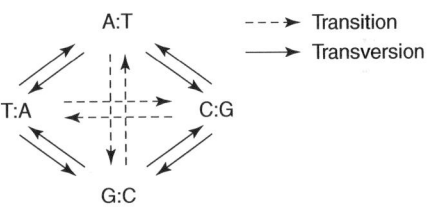

Fig. 15 Base-pair substitutions in DNA.

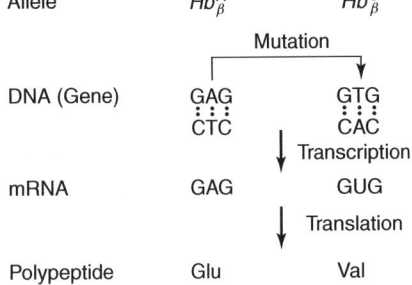

Fig. 16 The mutational origin of sickle-cell anemia in humans.

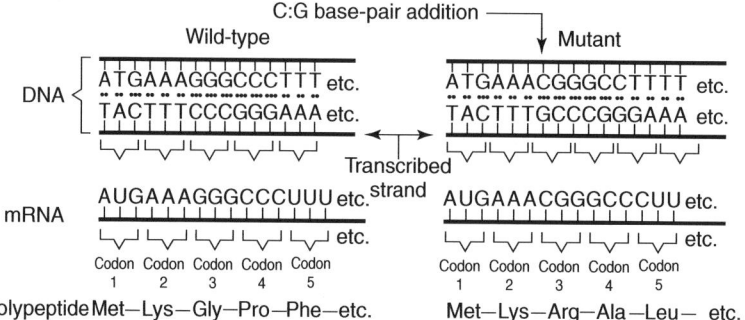

Fig. 17 The effect of a frameshift mutation, in this case a C:G base-pair insertion. (Reproduced from Snustad, D.P., Simmons, M.J., *Principles of Genetics*, 3rd edition, Copyright © 2003 by John Wiley & Sons, Inc., Hoboken, NJ, USA. This material is used by permission of John Wiley & Sons, Inc.)

The insertion or deletion of one or two base pairs within the coding sequence of a gene alters the codon reading frame in the mRNA (Fig. 17); thus, such mutations are referred to as *frameshift mutations*. Frameshift mutations usually result in totally nonfunctional gene products.

produce mutant alleles that encode products with new functions. Some gain-of-function mutations result in the synthesis of altered polypeptides with novel activity. Others result in the synthesis of the gene products in cells or tissues in which they are not normally synthesized.

4.3
Gain-of-function and Loss-of-function Mutations

Mutations may also be classified on the basis of their effect on the function of the altered gene products. Mutations that diminish or eliminate the activity of a gene product are called *loss-of-function* mutations. Most recessive mutations are simple loss-of-function mutations. However, some loss-of-function mutations are dominant (for example, when the mutant product interferes with the activity of the wild-type gene product) or partially dominant (for example, when the threshold level of gene product activity requires two functional copies of the gene). Many dominant mutations are *gain-of-function* mutations; they

4.4
Suppressor and Enhancer Mutations

Two classes of mutations have proven to be especially valuable to researchers investigating the genetic control of a specific biological process. Given an organism with a mutant phenotype caused by a mutation in one gene that affects a trait, researchers can often identify other genes affecting the trait by screening for *suppressor mutations* or *enhancer mutations*. Suppressor mutations partially or completely cancel the effect of the original mutation, whereas enhancer mutations cause the mutant phenotype to be more extreme. In both cases, the mutations usually occur in genes that encode products involved in the same pathway or process, acting either upstream of or downstream from the product of the gene altered by the original mutation.

4.5
Chromosome Aberrations

There are four types of gross chromosome rearrangements or *chromosome aberrations*: duplications, deletions, inversions, and translocations. A *duplication* is the occurrence of a segment of a chromosome in two or more copies per genome. A *deletion* or *deficiency* results from the loss of a segment of a chromosome. An *inversion* occurs when a segment of a chromosome is turned end-for-end relative to its orientation in a normal chromosome. A *translocation* results when a segment of a chromosome is broken off and becomes reattached to another chromosome. All four types of chromosome aberrations can have major effects on the phenotype and fertility of an organism.

4.6
Polyploidy

Numerical changes in chromosome number are referred to as changes in *ploidy*. *Euploidy* involves variations in the number of copies of the genome or complete sets of chromosomes in a cell or an organism. *Aneuploidy* occurs when there are missing or extra copies of one or a few chromosomes.

All the other sections of this article have focussed on the genetics of *diploid* ($2n$) organisms such as humans and fruit flies. Diploid organisms contain two copies of the genome in their germ-line cells and most, but not all, somatic cells. However, not all organisms are diploids; some animals, for example, certain fish and salamanders, and many plants are *polyploid*, that is, contain more than two copies of the genome. Many important crop species are polyploids: seedless watermelons are triploid ($3n$), cotton is a tetraploid ($4n$), bread wheat is a hexaploid ($6n$), and strawberries are octaploids ($8n$). Polyploidy has played an important role in the evolution of plants.

4.7
Nondisjunction and Aneuploidy

During mitosis, each chromosome duplicates and the two daughter chromosomes separate, with one going to each progeny cell. Similarly, during meiosis, the two homologous chromosomes separate during the first division, and then the two daughter chromosomes (previously chromatids) move to separate cells during the second division. The result of a normal mitosis or meiosis is that all the progeny cells receive complete sets of chromosomes. Occasionally, however, homologous chromosomes and daughter chromosomes fail to separate or disjoin properly during mitosis or meiosis. The failure of chromosomes to separate properly during mitosis or meiosis is called *nondisjunction*. Nondisjunction produces cells or gametes with extra or missing chromosomes; it is the primary cause of aneuploidy.

Whereas euploidy is not detrimental *per se*, aneuploidy is, especially in animals. In humans, aneuploidy is almost always lethal. Excluding individuals who are mosaics containing cells with two or more different karyotypes (chromosome sets), only seven types of aneuploidy exist in humans. *Monosomy* occurs when one member of a pair of chromosomes is missing; *trisomy* occurs when three copies of a chromosome are present. In humans, only one monosomic condition, Turner syndrome (X0, the presence of a single X chromosome), is viable. Monosomy for any of the autosomes is lethal. Five trisomic conditions are viable in humans: Klinefelter syndrome (XXY), Down syndrome (trisomy 21), Patau syndrome (trisomy 13), Edward syndrome (trisomy 18), triplo-X (XXX), and XYY. Trisomy for any of the other autosomes is lethal. The message is clear; the genes in our genome have evolved as a coadapted set of instructions. Each of the genes in our genome must be expressed at the proper time and in the proper place during development to produce a normal human being. Anything that disrupts the delicate balance of preprogrammed gene expression required for normal development is highly deleterious.

4.8
Expanding Trinucleotide Repeats in Humans

Another class of mutations – *expanding trinucleotide repeats* – are responsible for several inherited neurological diseases in humans. Tandemly repeated sequences of one to six nucleotide pairs are known as simple tandem repeats. Such repeats are dispersed throughout genomes, including the human genome. Repeats of three nucleotide pairs, trinucleotide repeats, are known to increase in copy number and give rise to mutant alleles. In humans, several trinucleotides undergo such increases in copy number. Expanded CGG trinucleotide repeats on the X chromosome are responsible for fragile X syndrome, the most common form of inherited mental retardation. Normal X chromosomes contain from 6 to about 50 copies of the CGG repeat at the *FRAXA* site. Mutant X chromosomes contain up to 1000 copies of the tandem CGG repeat at this site. CAG and CTG trinucleotide repeats are involved in several inherited neurological diseases,

including Huntington disease, myotonic dystrophy, Kennedy disease, dentatorubral pallidoluysian atrophy, Machado Joseph disease, and spinocerebellar ataxia. In all these neurological disorders, the severity of the disease is correlated with trinucleotide copy number – the higher the copy number, the more severe the disease symptoms. In addition, the expanded trinucleotides associated with these diseases are unstable in somatic cells and between generations. This instability gives rise to the phenomenon of "anticipation," which is the increasing severity of the disease or earlier age of onset that occurs in successive generations as the trinucleotide copy number increases.

5
Recombination: New Combinations of Genes to be Acted on by Natural or Artificial Selection

Mutation produces new genetic variability, but the resulting mutant genes must be placed in new combinations with previously existing genes so that natural selection (or artificial selection in the case of plant and animal breeding) can preserve those combinations that produce the organisms best adapted to specific environments (or desired by the breeder). These new combinations are produced by recombination mechanisms that are essential to the process of evolution.

5.1
Segregation

In an organism that is heterozygous for two different alleles of a gene, such as tall (*D*) and dwarf (*d*) in Mendel's peas, the two alleles segregate from each other during the formation of gametes. The

Fig. 18 Segregation of alleles during the first division of meiosis.

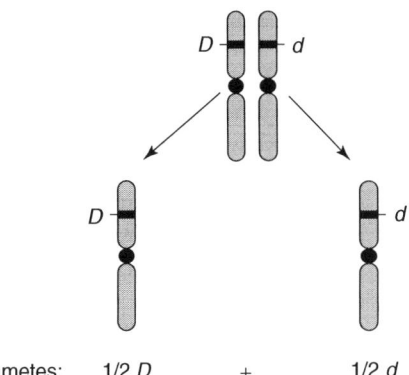

biological basis of this segregation is the pairing and subsequent separation of homologous chromosomes during the first (reductional) division of meiosis (Fig. 18).

5.2
Independent Assortment

New combinations of genes on nonhomologous chromosomes are produced by the independent assortment of chromosomes during the first division of meiosis (Fig. 19). Consider a cross between two double homozygotes, for example, $AA\ BB$ and $aa\ bb$, where genes A and B are on different chromosomes. The F_1 progeny (F_1 for first filial generation) will be double heterozygotes ($Aa\ Bb$). Half of the gametes produced by F_1 progeny will have the same combination of alleles as the

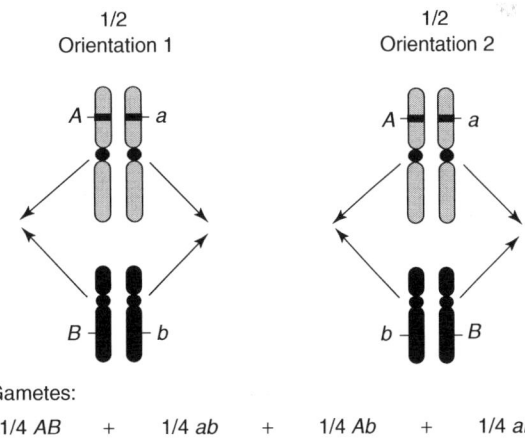

Fig. 19 The generation of new combinations of genes by the independent assortment of nonhomologous chromosomes during meiosis.

parents (1/4 AB + 1/4 ab). The other half will have new (recombinant) combinations of the alleles (1/4 Ab and 1/4 aB). The recombination of pairs of alleles on different chromosomes is the result of the *independent assortment* of the homologous chromosomes during the first division of meiosis. Each pair of homologous chromosomes lines up at the metaphase plate independently of every other pair. Both maternal chromosomes may be on the same side of the metaphase plate, with the two paternal chromosomes on the other side, or one maternal chromosome may be on the same side of the metaphase plate as the paternal member of the other chromosome pair, and vice versa. The result is that the segregation of the alleles of a gene on one chromosome occurs independently of the segregation of the alleles of any gene on a nonhomologous chromosome.

5.3
Linkage, Crossing-over, and Genetic Maps of Chromosomes

Because organisms contain far more genes than chromosomes, not all pairs of alleles can assort independently of one another. Humans, for example, contain 23 pairs of chromosomes and approximately 35 000 genes. Thus, each chromosome contains many genes. Genes that are relatively close together on the same chromosome tend to stay together; if less than 50% of the gametes produced during meiosis in a double heterozygote have a recombinant combination of the alleles (see Section 5.2), the genes are said to be *linked*. New combinations of the alleles of genes on the same chromosome are produced by *crossing-over* (breakage and exchange of parts) between homologous chromosomes during meiosis and mitosis (Fig. 20). During an early stage (prophase) of meiosis, the homologous chromosomes pair (a process called *synapsis*) side by side. This pairing facilitates crossing over, which occurs by cutting DNA strands with enzymes called endonucleases, exchanging single strands of DNA from homologous chromosomes with the aid of recombination proteins, and rejoining the strands of DNA in new combinations with the assistance of DNA polymerases and DNA ligases. DNA ligases seal single-strand breaks in DNA molecules; they play important roles in replication, recombination, and DNA repair processes.

The frequency of crossing over and thus the frequency at which recombinant combinations of linked genes occur are approximately proportional to the distance between genes on the chromosome.

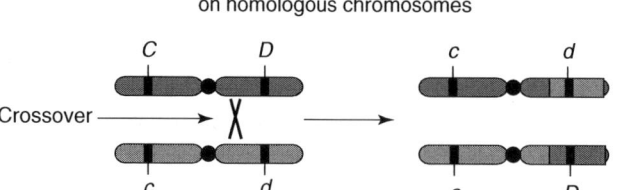

Fig. 20 The generation of new combinations of genes by crossing-over between genes located on homologous chromosomes during meiosis or mitosis.

This relationship is used to prepare *genetic maps* of chromosomes, which show the linear order and the relative positions of genes on a chromosome. Crosses are performed between organisms with different combinations of the alleles of genes on a specific chromosome, and the frequency of gametes carrying recombinant combinations of the alleles of the genes is determined. As an interesting historical sidelight, this genetic mapping procedure was developed by Alfred H. Sturtevant while he was still an undergraduate working in Thomas Hunt Morgan's laboratory at Columbia University in 1911. When genes are relatively far apart on a chromosome, multiple crossovers may occur between them. When two crossovers (or an even number) occur between two genes, the second cancels out the first, and all the resulting gametes will have parental combinations of the alleles of the genes. Thus, map distances are based on the average number of crossovers between genes. One map unit (one centiMorgan, cM) is the genetic distance in which an average of 1 chromatid out of a 100 will have undergone a crossover event. For distances less than 20 map units, the map distance is approximately equal to the proportion of recombinant chromosomes produced in percent, or 1 cM equals the distance yielding 1% recombinant gametes. Figure 21 contains an abbreviated genetic map of the X chromosome in *D. melanogaster*.

5.4
Transposable Genetic Elements

Living organisms contain remarkable DNA elements that can jump from one site in the genome to another site. These mobile elements are called *transposable genetic elements* or *transposons*. The insertion of one of these transposons into a gene will often render the gene nonfunctional, producing a mutant allele. Indeed, the wrinkled allele that Mendel studied in the garden pea was caused by the insertion of a transposable element.

Transposable elements are present in both prokaryotes and eukaryotes and

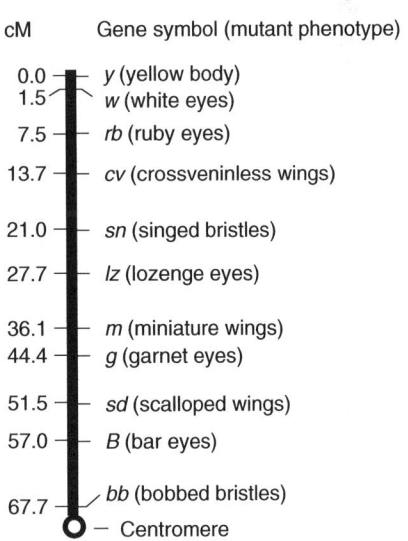

Fig. 21 Abbreviated genetic map of the X chromosome of *Drosophila*. Distances are given in centiMorgans (approximately equal to the percentage of recombinant chromosomes).

exhibit a variety of structures and mechanisms of transposition. However, most of them can be grouped into one of three classes. *Cut-and-paste transposons* move by excision from one site and insertion into another site. *Replicative transposons* move during replication, with the old copy remaining at the initial site and the nascent copy being inserted at a new site. Thus, the copy number increases with each transposition. *Retrotransposons* move via the synthesis of DNA copies of the RNA transcribed from the element ("retro" for reverse flow of genetic information, RNA → DNA, rather than DNA → RNA as in transcription). Some of the retrotransposons are closely related to retroviruses, or RNA tumor viruses, that use this RNA → DNA lifestyle.

Transposable genetic elements are major components of genomes, making up more than 40% of the DNA in the human genome and perhaps as much as 80% of the maize genome. They are responsible for a large number of spontaneously occurring mutations and play important roles in genome evolution. Crossing-over between two copies of a transposon in a chromosome will produce deletions or inversions depending on their orientations with respect to each other. Two transposons can also move the segment of DNA between them from one location to another if they transpose simultaneously. Indeed, transposable elements are in part responsible for the rapid spread of antibiotic and drug resistance in bacteria, a phenomenon of major concern to the medical community.

5.5
Physical Maps of Chromosomes

Whereas genetic maps are based on recombination frequencies, *physical maps* of chromosomes are based on the molecular distances – base pairs (bp), kilobases (kb, 1000 bp), and megabases (Mb, 1-million bp) – separating genes or molecular markers. Physical distances do not correlate directly with genetic map distances because recombination frequencies are not always proportional to molecular distances. However, the two are often reasonably well correlated in euchromatic regions of chromosomes. In humans, 1 cM is equivalent, on average, to about 1 Mb of DNA.

When mutations change the nucleotide sequences in restriction enzyme cleavage sites, the enzymes no longer recognize them. Other mutations may create new restriction sites. These mutations result in variations in the lengths of the DNA fragments produced by digestion with various restriction enzymes. Such *restriction fragment length polymorphisms*, or *RFLPs*, have proven invaluable in constructing detailed genetic and physical maps of chromosomes. The RFLPs can be mapped just like other genetic markers; they segregate in crosses as though they are codominant alleles. Correlations between genetic and physical maps can be established by positioning clones of genetically mapped genes or RFLPs on the physical map. Markers that are mapped both genetically and physically are called *anchor markers*. They anchor the physical map to the genetic map and vice versa. *Sequence-tagged sites* (*STSs*) – unique genomic DNA sequences 200 to 500 base pairs long – and *expressed-sequence tags* (*ESTs*) – DNA copies of mRNAs – are often used as hybridization probes to anchor physical maps to genetic maps.

In humans, the most useful RFLPs involve short sequences that are present as tandem repeats. The number of copies of each sequence present at a given site on a chromosome is highly variable. These sites, called *variable number tandem repeats* (*VNTRs*) are highly polymorphic. VNTRs vary in fragment length because of differences in the number of copies of the repeated sequence between the restriction sites. *Microsatellites* are another class of polymorphisms that have proven extremely valuable in constructing high-density maps of eukaryotic chromosomes. Microsatellites are polymorphic tandem repeats of sequences only two to five nucleotide pairs long. They are called microsatellites because they are a subset of the satellite sequences present in the highly repetitive DNA of eukaryotes. When two or more segments of a chromosome have been identified, mapped physically, and shown to overlap, they are said to form a *contig*, and their combined physical maps form a *contig map*.

5.6
Map-position-based Identification of Genes

The availability of detailed genetic and physical maps of chromosomes permits researchers to isolate genes on the basis of their location in the genome. This approach, called *positional cloning*, can be used to identify any gene regardless of its function. The steps in positional cloning are illustrated in Fig. 22. The gene is first mapped to a specific region of a chromosome by genetic crosses or, in the case of humans, by pedigree analysis. The gene is next localized on the physical map of this region of the chromosome.

Positional cloning involves "walking" or "jumping" along a chromosome until the desired gene is reached. Chromosome walks are initiated by the selection of a molecular marker (e.g. RFLP or gene clone) close to the gene and the use of this clone as a hybridization probe to screen a genomic library for overlapping sequences. Physical maps are constructed for the overlapping clones identified in the library screen, and the restriction fragment farthest from the original probe is used to screen a second genomic library constructed by using a different restriction enzyme or prepared from a partial digest of genomic DNA. Repeating this procedure several times and isolating a series of overlapping genomic clones allow a scientist to walk the required distance along a chromosome to the desired gene.

When the distance from the closest molecular marker to the gene of interest is large, a technique called chromosome jumping can be used to speed up an otherwise laborious walk. Each jump can cover a distance of 100 kb or more. Like a walk, a jump is initiated by using a molecular probe such as an RFLP as a starting point. However, with chromosome jumps, large DNA fragments are prepared by partial digestion of genomic DNA with a restriction endonuclease. The large genomic fragments are then circularized with DNA ligase. A second restriction endonuclease is used to excise the junction fragment from the circular molecule. This junction fragment will contain both ends of the long fragment; it can be identified by hybridizing DNA fragments separated by gel electrophoresis to the initial molecular probe. A restriction map of the junction fragment is prepared, and a restriction fragment that corresponds to the distal end of the long genomic fragment is cloned and used to initiate a chromosome walk or a second chromosome jump. Chromosome jumping has proven especially useful in work with large genomes such as the human genome.

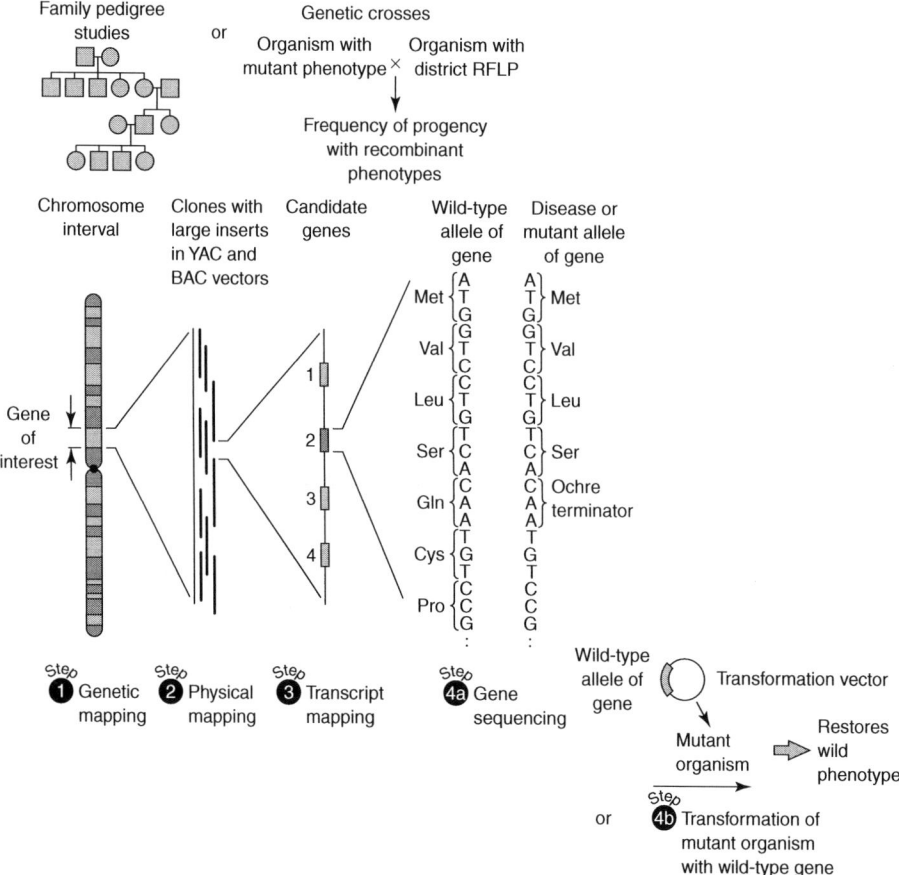

Fig. 22 Steps involved in cloning of genes on the basis of their location in the genome (positional cloning). The gene is first mapped to as small a region of a chromosome as possible. In humans, mapping must be done using family pedigree data, and candidate genes must be screened by comparing the sequences of genes from "normal" individuals with the sequences from individuals with the trait or genetic disorder. In other species, mapping is done using data from genetic crosses, and candidate genes are screened by transforming mutant organisms with the wild-type allele of the gene and seeing whether it restores the wild-type phenotype. (Reproduced from Snustad, D.P., Simmons, M.J., *Principles of Genetics*, 3rd edition, Copyright © 2003 by John Wiley & Sons, Inc., Hoboken, NJ, USA. This material is used by permission of John Wiley & Sons, Inc.)

Verification that a clone of the desired gene has been isolated is accomplished in various ways. In organisms such as *Drosophila*, verification is achieved by introducing the wild-type allele of the gene into a mutant organism and showing that it restores the wild-type phenotype. In humans, verification commonly involves determining the nucleotide sequences of the wild-type gene and several mutant alleles and showing that the coding sequences of the mutant genes are defective and unable to produce functional gene products.

Acknowledgements

Figures 3, 6, 7, 10, 11, 12, 17, and 22 have been reproduced from Snustad, D.P., Simmons, M.J., *Principles of Genetics*, 3rd edition, Copyright © 2003 by John Wiley & Sons, Inc., Hoboken, NJ, USA. This material is used by permission of John Wiley & Sons, Inc.

Bibliography

Books and Reviews

Alberts, B., Johnson, A., Lewis, J., Raff, M., Roberts, K., Walter, P. (2002) *Molecular Biology of the Cell*, 4th edition, Garland Publishing, New York.

Bateson, W. (1909) *Mendel's Principles of Heredity*, Cambridge University Press, Cambridge, UK.

Beckwith, J.R., Zipser, D. (1970) *The Lactose Operon*, Cold Spring Harbor Laboratory Press, Cold Spring Harbor, New York.

Bell, S.P., Dutta, A. (2002) DNA replication in eukaryotic cells, *Annu. Rev. Biochem.* **71**, 333–374.

Benkovic, S.J., Valentine, A.M., Salinas, F. (2001) Replisome-mediated DNA replication, *Annu. Rev. Biochem.* **70**, 181–208.

Carlson, E.A. (1966) *The Gene: A Critical History*, W. B. Saunders, Philadelphia, PA.

Crow, J.F. (2000) Two centuries of genetics: a view from half-time, *Annu. Rev. Genomics Hum. Genet.* **1**, 21–40.

Crow, J.F., Dove, W.F. (Eds.) (2000) *Perspectives on Genetics, Anecdotal, Historical, and Critical Commentaries, 1987–1998*, University of Wisconsin Press, Madison, WI.

Davies, K. (2001) *Cracking the Genome, Inside the Race to Unlock Human DNA*, Free Press, New York.

Drake, J.W. (1970) *The Molecular Basis of Mutation*, Holden-Day, San Francisco, CA.

Echols, H. (2001) *Operators and Promoters: The Story of Molecular Biology and Its Creators*, University of California Press, Berkeley, CA.

Karp, G. (2001) *Cell and Molecular Biology: Concepts and Experiments*, 3rd edition, John Wiley & Sons, New York.

Keegan, L.P., Gallo, A., O'Connell, M.A. (2001) The many roles of an RNA editor, *Nat. Rev. Genet.* **2**, 869–878.

Kornberg, A., Baker, T.A. (1992) *DNA Replication*, 2nd edition, W. H. Freeman, San Francisco, CA.

Lee, T.I., Young, R.A. (2000) Transcription of eukaryotic protein-coding genes, *Annu. Rev. Genet.* **34**, 77–137.

Lewin, B. (2000) *Genes VII*, Oxford University Press, New York.

Lewis, R. (2003) *Human Genetics, Concepts and Applications*, 5th edition, WCB/McGraw-Hill, Dubuque, IA.

Lodish, H., Berk, A., Zipursky, S.L., Matsudaira, P., Baltimore, D., Darnell, J. (2000) *Molecular Cell Biology*, 4th edition, W. H. Freeman, San Francisco, CA.

Mount, D.W. (2001) *Bioinformatics, Sequence and Genome Analysis*, Cold Spring Harbor Laboratory Press, Cold Spring Harbor, New York.

Neidhardt, F.C., Curtis, R. III, Ingraham, J.L., Lin, E.C.C., Low, K.B., Magasanek, B., Reznikoff, W.S., Riley, M., Schaechter, M., Umbarger, H.E. (Eds.) (1996) Escherichia coli and Salmonella typhimurium, *Cellular and Molecular Biology*, 2nd edition, ASM Press, Washington, DC.

Peters, J.A. (Ed.) (1959) *Classic Papers in Genetics*, Prentice Hall, Englewood Cliffs, NJ.

Ptashne, M., Gann, A. (2002) *Genes & Signals*, Cold Spring Harbor Laboratory Press, Cold Spring Harbor, New York.

Snustad, D.P., Simmons, M.J. (2003) *Principles of Genetics*, 3rd edition, John Wiley & Sons, Hoboken, NJ.

Stahl, F.W. (1979) *Genetic Recombination*, W. H. Freeman, San Francisco, CA.

Watson, J.D., Hopkins, N.H., Roberts, J.W., Steitz, J.A., Weiner, A.M. (1987) *Molecular Biology of the Gene*, 4th edition, Vols. I and II, Benjamin/Cummings, Menlo Park, CA.

Primary Literature

Adams, M.D., and 194 co-authors (2000) The genome sequence of *Drosophila melanogaster*, *Science* **287**, 2185–2195.

Avery, O.T., MacLeod, C.M., McCarty, M. (1944) Studies on the chemical nature of the substance inducing transformation in pneumococcal types, *J. Exp. Med.* **79**, 137–148.

Beadle, G.W., Tatum, E.L. (1941) Genetic control of biochemical reactions in neurospora, *Proc. Natl. Acad. Sci. U.S.A.* **27**, 499–506.

Benzer, S. (1961) On the topography of the genetic fine structure, *Proc. Natl. Acad. Sci. U.S.A.* **47**, 403–415.

Brenner, S., Jacob, F., Meselson, M. (1961) An unstable intermediate carrying information from genes to ribosomes for protein synthesis, *Nature* **190**, 576–581.

Brenner, S., Stretton, A.O.W., Kaplan, S. (1965) Genetic code: the 'nonsense' triplets for chain termination and their suppression, *Nature* **206**, 994–998.

Brody, S., Yanofsky, C. (1963) Suppressor gene alteration of protein primary structure, *Proc. Natl. Acad. Sci. U.S.A.* **50**, 9–16.

Cairns, J. (1963) The bacterial chromosome and its manner of replication as seen by autoradiography, *J. Mol. Biol.* **6**, 208–213.

Creighton, H.B., McClintock, B. (1931) A correlation of cytological and genetical crossing over in Zea mays, *Proc. Natl. Acad. Sci. U.S.A.* **17**, 492–497.

Crick, F.H.C., Barnett, L., Brenner, S., Watts-Tobin, R.J. (1962) General nature of the genetic code for proteins, *Nature* **192**, 1227–1232.

Delbrück, M., Bailey, W.T. Jr. (1946) Induced mutations in bacterial viruses, *Cold Spring Harbor Symp. Quant. Biol.* **11**, 33–37.

Fleischmann, R.D., and 39 co-authors (1995) Whole-genome random sequencing and assembly of *Haemophilus influenzae* Rd, *Science* **269**, 496–512.

Fraenkel-Conrat, H., Singer, B. (1957) Virus reconstitution II. Combination of protein and nucleic acid from different strains. *Biochem. Biophys. Acta* **24**, 540–548.

Fraser, A.G., Kamath, R.S., Zipperien, P., Martinez-Campos, M., Sohrmann, M., Ahringer, J. (2000) Functional genomic analysis of *C. elegans* chromosome I by systematic RNA interference, *Nature* **408**, 325–330.

Freese, E. (1959) The specific mutagenic effect of base analogues on phage T4, *J. Mol. Biol.* **1**, 87–105.

Garen, A., Garen, S. (1963) Complementation in vivo between structural mutants of alkaline phosphatase from *E. coli*, *J. Mol. Biol.* **7**, 13–22.

Gönczy, P., Echeverri, C., Oegema, K., Coulson, A., Jones, S.J.M., Copley, R.R., Duperon, J., Oegema, J., Brehm, M., Cassin, E., Hannak, E., Kirkham, M., Pichler, S., Flohrs, K., Goessen, A., Leidel, S., Alleaume, A.-M., Martin, C., Özlu, N., Bork, P., Hyman, A.A. (2000) Functional genomic analysis of cell division in *C. elegans* using RNAi of genes on chromosome III, *Nature* **408**, 331–336.

Hershey, A.D., Chase, M. (1952) Independent functions of viral protein and nucleic acid in growth of bacteriophage, *J. Gen. Physiol.* **36**, 39–56.

Huberman, J.A., Riggs, A.D. (1968) On the mechanism of DNA replication in mammalian chromosomes, *J. Mol. Biol.* **32**, 327–341.

International Human Genome Sequencing Consortium, Lander, E.S., and 248 co-authors (2001) Initial sequencing and analysis of the human genome, *Nature* **409**, 860–921.

Jacob, F., Monod, J. (1961) Genetic regulatory mechanisms in the synthesis of proteins, *J. Mol. Biol.* **3**, 318–356.

Kornberg, A. (1960) Biologic synthesis of deoxyribonucleic acid, *Science* **131**, 1503–1508.

Lederberg, J., Tatum, E.L. (1946) Novel genotypes in mixed culture of biochemical mutants of bacteria, *Cold Spring Harbor Symp. Quant. Biol.* **11**, 113, 114.

Lederberg, J. (1947) Gene recombination and linked segregations in *Escherichia coli*, *Genetics* **32**, 505–525.

McClintock, B. (1956) Controlling elements and the gene, *Cold Spring Harbor Symp. Quant. Biol.* **21**, 197–216.

McManus, M.T., Sharp, P.A. (2002) Gene Silencing in mammals by small interfering RNAs, *Nat. Rev. Genet.* **3**, 737–747.

Mendel, G. (1866) Experiments in Plant Hybridization, in: Peters, J.A. (Ed.) *Classic Papers in Genetics*, pp. 1–20, Prentice Hall, Englewood Cliffs, NJ. (Reprint of William Bateson's translation of Mendel's paper Versuche über Pflanzenhybriden. *Verh. Naturforsch. Ver Brünn* 4, 3–47.)

Meselson, M.S., Stahl, F.W. (1958) The replication of DNA in *Escherichia coli*, *Proc. Natl. Acad. Sci. U.S.A.* **44**, 671–682.

Muller, H.J. (1927) Artificial transmutation of the gene, *Science* **66**, 84–87.

Nirenberg, M.W., Matthaei, J.H. (1961) The dependence of cell-free protein synthesis in *E. coli* upon naturally occurring or synthetic polyribonucleotides, *Proc. Natl. Acad. Sci. U.S.A.* **47**, 1588–1602.

Okazaki, R., Okazaki, T., Sakabe, K., Sugimoto, K., Kainuma, R., Sugino, A., Iwatsuki, N.

(1968) In vivo mechanism of DNA chain growth, *Cold Spring Harbor Symp. Quant. Biol.* **33**, 129–143.

Ptashne, M. (1967) Specific binding of the λ phage repressor to λ DNA, *Nature* **214**, 232–234.

Schlesinger, M.J., Levinthal, C. (1963) Hybrid protein formation of *E. coli* alkaline phosphatase leading to *in vitro* complementation, *J. Mol. Biol.* **7**, 1–12.

Speyer, J.F., Lengyel, P., Basilio, C., Ochoa, S. (1962) Synthetic polynucleotides and the amino acid code, IV, *Proc. Natl. Acad. Sci. U.S.A.* **48**, 441–448.

Sturtevant, A.H. (1913) The linear arrangement of six sex-linked factors in *Drosophila*, as shown by their mode of association, *J. Exp. Zool.* **14**, 43–59.

The Arabidopsis Genome Initiative (2000) Analysis of the genome sequence of the flowering plant *Arabidopsis thaliana*, *Nature* **408**, 796–815.

The *C. elegans* Sequencing Consortium (1998) Genome sequence of the nematode *C. elegans*: a platform for investigating biology, *Science* **282**, 2012–2018.

Taylor, J.H., Woods, P.S., Hughes, W.L. (1957) The organization and duplication of chromosomes as revealed by autoradiographic studies using tritium-labeled thymidine, *Proc. Natl. Acad. Sci. U.S.A.* **43**, 122–128.

Tettelin, H., and 38 co-authors (2001) Complete genome sequence of a virulent isolate of *Streptococcus pneumoniae*, *Science* **293**, 498–506.

Venter, J.C., and 272 co-authors (2001) The sequence of the human genome, *Science* **291**, 1304–1351.

Waterhouse, P.M., Helliwell, C.A. (2003) Exploring plant genomes by RNA-induced gene silencing, *Nat. Rev. Genet.* **4**, 29–38.

Watson, J.D., Crick, F.H.C. (1953) Molecular structure of nucleic acids. A structure for deoxyribose nucleic acid, *Nature* **171**, 737, 738.

Watson, J.D., Crick, F.H.C. (1953) Genetical implications of the structure of deoxyribonucleic acid, *Nature* **171**, 964–967.

Yanofsky, C., Carlton, B.C., Guest, J.R., Helinski, D.R., Henning, U. (1964) On the colinearity of gene structure and protein structure, *Proc. Natl. Acad. Sci. U.S.A.* **51**, 266–272.

2
DNA Replication and Transcription

Yusaku Nakabeppu[1], Hisaji Maki[2], and Mutsuo Sekiguchi[3]
[1] *Kyushu University, Fukuoka, Japan*
[2] *Nara Institute of Science and Technology, Nara, Japan*
[3] *Biomolecular Engineering Research Institute, Osaka, Japan*

1	**DNA Replication** 49	
1.1	Structural Aspects of DNA Replication	49
1.1.1	Replicon: Unit of Replication	49
1.1.2	Replication Fork 50	
1.2	Biochemistry of DNA Replication	51
1.2.1	DNA Polymerase 52	
1.2.2	DNA Helicase and Primase	54
1.2.3	Other Replication Proteins	55
1.3	Regulation of DNA Replication	56
1.3.1	Initiation Mechanisms in Prokaryotes	56
1.3.2	Initiation Mechanisms in Eukaryotes	58
2	**Transcription** 59	
2.1	RNA Polymerase and Transcriptional Apparatus	59
2.1.1	General RNA Polymerase Architecture	59
2.1.2	Transcription Initiation	61
2.1.3	Promoter Escape 63	
2.1.4	Transcription Elongation	63
2.2	Transcription Unit 64	
2.2.1	Bacterial Transcription Unit	65
2.2.2	Eukaryotic Transcription Unit	65
2.3	Regulation of Transcription	67
2.3.1	Transcriptional Regulation in Bacteria	67
2.3.2	Transcriptional Regulation in Eukaryotes	68
	Bibliography 70	
	Books and Reviews 70	
	Primary Literature 70	

Genomics and Genetics. Edited by Robert A. Meyers.
Copyright © 2007 Wiley-VCH Verlag GmbH & Co. KGaA, Weinheim
ISBN: 978-3-527-31609-0

Keywords

Origin (ori)
A unique site on the chromosome from which DNA replication starts in either one or both directions.

Replication Fork
A moving front of DNA replication at which a double-stranded DNA helix is separated into two newly replicated helices.

DNA Polymerase
An enzyme that catalyzes the formation of DNA chain by adding deoxyribonucleotides to the 3′-hydroxyl end of a preexisting DNA strand.

Promoter
A region on a DNA molecule in which an RNA polymerase binds and initiates transcription.

Transcription Factor
A class of regulatory protein that binds to a promoter or to a nearby sequence of DNA to facilitate or prevent initiation of transcription.

RNA Polymerase
An enzyme that transcribes an RNA molecule from one strand (template strand) of a DNA molecule.

■ Genetic information of organisms is kept in DNA in the form of an array of nucleotide sequences, and the cell must precisely replicate its chromosomal DNA. The self-complementary nature of the double-stranded DNA molecule is the basis for accurate replication, and organisms are equipped with DNA polymerase enzyme, which adds deoxyribonucleotides to preexisting DNA chains in the $5' \rightarrow 3'$ direction, along the template strand of DNA. One strand (the leading strand) is synthesized continuously while the other (the lagging strand) is synthesized discontinuously and is then sealed.

To extract information from DNA, a limited region of the DNA, which constitutes the gene or the gene cluster (operon), is transcribed into RNA. This process is catalyzed by RNA polymerase. As is the case with DNA polymerase, this enzyme is composed of multiple subunits and catalyzes the addition of ribonucleotides to the growing RNA chain in the $5' \rightarrow 3'$ direction, along the template DNA. Both the startpoint and the frequency of transcription are determined by the promoter, a specific sequence of DNA to which RNA polymerase and transcription regulatory proteins can bind.

1
DNA Replication

Precise transmission of chromosomal DNA from generation to generation is crucial to cell propagation. This can only be achieved when chromosomal DNA is accurately replicated, providing two copies of the entire genome for faithful distribution into each daughter cell. To this end, the cell has special mechanisms to maintain the high fidelity of DNA replication, to segregate the replicated DNA, and to coordinate DNA replication and cell division tightly within the cell cycle.

Interest and research in DNA replication have been fueled by increasing concerns with mutagenesis, evolution, carcinogenesis, and cancer therapy. Spontaneous mutations, apparently a major force in evolution, arise largely from errors during DNA replication. Cancer cells, having multiple mutations in genes controlling the cell cycle, escape the tight regulation of cell cycle and start their chromosomal DNA replication abnormally. This uncoupling of cell division and DNA synthesis may frequently induce chromosome disorders and additional mutations, thus accelerating the malignancy. Most anticancer treatments involve inhibition of DNA replication in cancer cells.

Many basic concepts and techniques in modern biotechnology are derived from research on DNA replication. The development of cloning vectors has depended on the knowledge of the replication mechanisms of bacterial plasmids and phages. Genetic engineering, DNA sequencing, and the polymerase chain reaction (PCR) utilize replicative enzymes, such as DNA polymerase and DNA ligase.

1.1
Structural Aspects of DNA Replication

The double-stranded nature of DNA provides the basis for its semiconservative replication. The two strands of the parental duplex contain complementary sequence information. DNA synthesis by a polymerase also makes use of this complementarity in generating the daughter duplex. A faithful copy can be made by complementary base pairing between the free nucleotide substrate and the parental DNA strand template. Thus, the daughter duplex consists of one parental strand and one new strand. Complementary base pairing also allows for the correction of replication errors by several repair mechanisms.

1.1.1 Replicon: Unit of Replication

DNA replication does not start at random locations but at particular sites, called the *origins of DNA replication*. A DNA whose replication starts from an origin and proceeds bidirectionally or unidirectionally to terminus site(s) is called a *replicon*, a unit of DNA replication. In bacterial cells, the circular chromosome contains a unique origin and DNA replication proceeds bidirectionally from the origin to the terminus (Fig. 1(a)). Therefore, the whole bacterial genome (4700 kb for *Escherichia coli*) is a single replicon. This genome structure is also common to most bacterial plasmids and some phages, although some of them are replicated unidirectionally.

On the other hand, eukaryotic cells contain multiple replication origins on each chromosome (Fig. 1(b)). In order for the eukaryotic cell to replicate its huge genome (2900 Mb per human haploid genome) in a relatively short period (about 7 h for the S-phase in a cultured animal cell), many replicons function simultaneously. Each replicon is 40 to 100 kb in size

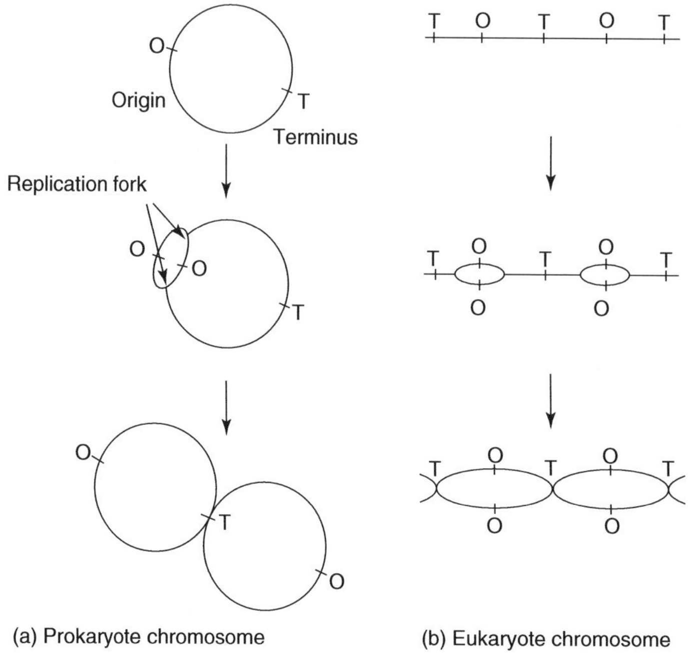

Fig. 1 Unit of DNA replication in prokaryotic and eukaryotic cells. Symbols: O = replication origin, T = replication terminus.

and the number of replicons utilized depends on the growth state of the cell: more rapidly growing cells use more replicons. There are some exceptional cases, such as the early embryonic divisions of *Drosophila* embryos, where the duration of the S-phase is compressed by the simultaneous functioning of a large number of replicons.

1.1.2 Replication Fork

In each replicon, replication is continuous from the origin to the terminus and is accompanied by the movement of the replicating point, called the *replication fork*. Both parental DNA strands are concurrently replicated at the fork. Since DNA polymerase can extend a DNA chain only in the $5' \rightarrow 3'$ direction, replication at a fork is semi-discontinuous (Fig. 2): DNA synthesis is continuous on one strand (leading strand) and discontinuous on the other (lagging strand). Short pieces of DNA, called *Okazaki fragments*, are repeatedly synthesized on the lagging-strand template, with these Okazaki fragments being a few thousand nucleotides in bacterial cells and a few hundred nucleotides in eukaryotic cells.

The velocity of fork movement has been estimated from the size of a replicon and the length of its replication period. In *E. coli*, the replication fork proceeds at 1000 bp/s and one Okazaki fragment is synthesized every 1 to 2 s. In eukaryotic cells, the rate of fork movement is slow, 10 to 100 bp/s. The distinction between prokaryotic and eukaryotic replication may be due to a difference in the replication machinery or in the structure of the chromosome.

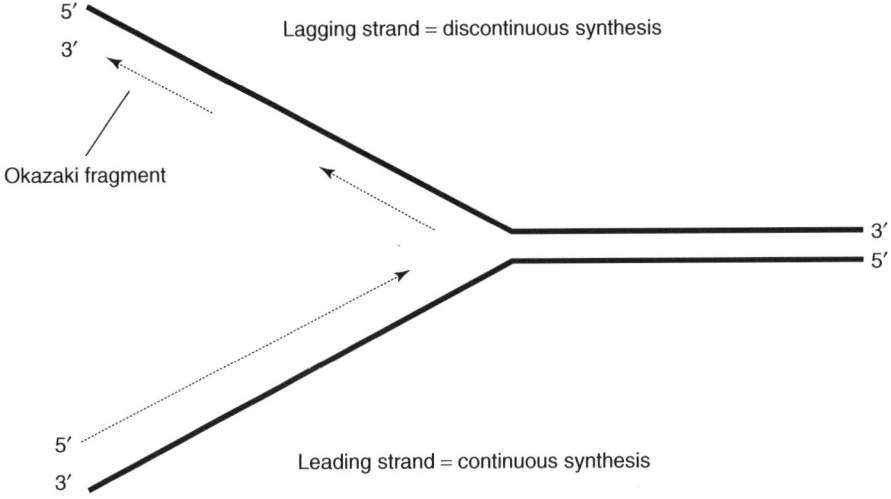

Fig. 2 Semi-discontinuous DNA synthesis at the replication fork.

1.2 Biochemistry of DNA Replication

DNA replication is a complex process involving numerous enzymes at the replication fork. In *E. coli*, more than 20 different proteins participate in DNA replication. These were identified by screening of mutants defective in DNA replication and by purifying enzymes required for *in vitro* DNA synthesis. On the basis of their biochemical roles in different stages of chromosomal DNA replication, these proteins are classified into several categories: (1) proteins involved in the initiation of replication: DnaA, HU, DNA gyrase, single-stranded DNA binding protein (SSB), DnaB, DnaC, RNaseH1, and RNA polymerase; (2) proteins required for chain elongation and connecting Okazaki fragments: DnaG (primase), DNA polymerase III (Pol III) holoenzyme, SSB, DNA polymerase I (Pol I), RNaseH1, and DNA ligase; (3) a swivel to relieve torsional stress in advance of the replication fork: DNA gyrase; (4) proteins for the termination of replication and segregation of daughter molecules: terminus binding protein (Tus), DNA gyrase, and DNA topoisomerase IV; (5) primosome proteins required for reformation and restart of replication fork when it accidentally collapses: PriA, PriB, PriC, DnaT, DnaB, DnaC, and DnaG. Among these proteins, the DnaB helicase, primase (DnaG), and Pol III holoenzyme are the basic components acting at the replication fork, forming a multiprotein complex called a *replisome* (Fig. 3).

Replication of many bacterial plasmids and small phages requires most of the host-encoded replication proteins, although the initiation step of these extra chromosomal replicons differs from that at the host origin. Large phages such as T4 and T7 encode their own replication proteins, whose activities resemble those of the *E. coli* replication proteins. In eukaryotic cells, chain elongation, termination, and segregation appear to be carried out by activities very similar to the *E. coli* replication proteins. Proteins involved in the

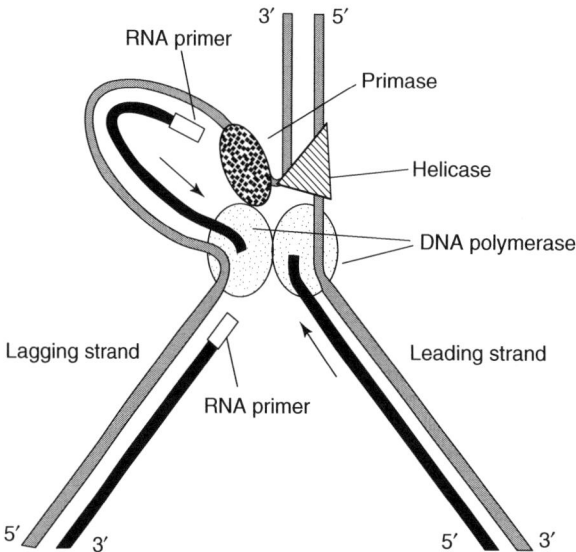

Fig. 3 Hypothetical structure of replication machinery. Primase, DNA helicase, and twin DNA polymerases form a "replisome."

initiation of chromosomal replication in eukaryotic cells have also been identified.

1.2.1 DNA Polymerase

DNA polymerase plays a primary role in replication, catalyzing synthesis of a DNA chain complementary to the template DNA strand as follows:

$$(dNMP)_n + dNTP \longrightarrow (dNMP)_{n+1} + PP_i$$

The reaction absolutely requires three components: (1) deoxynucleosidetriphosphate (dNTP) as a substrate; (2) single-stranded DNA as a template; and (3) a short RNA or DNA as a primer. The primer must be annealed to the template by base pairing and its 3′-terminus must possess a free 3′-OH group. Chain growth is exclusively 5′ → 3′ since the polymerization mechanism is a nucleophilic attack by the 3′-OH group of the primer on the α-phosphate of the incoming dNTP. The products are a new primer, longer by one nucleotide, and an inorganic pyrophosphate. 2′,3′-dideoxy NTP can be incorporated by some DNA polymerases, but blocks chain elongation because it lacks a free 3′-OH group. This chain termination property is utilized in some DNA sequencing methods.

Both prokaryotic and eukaryotic cells possess multiple numbers of distinct DNA polymerases (Table 1), which differ in their structures as well as in their biochemical and biological functions. A 3′ → 5′ exonuclease activity is associated with all replicative DNA polymerases except the eukaryotic Pol α. The fidelity of DNA synthesis is increased by the 3′ → 5′ exonuclease activity, which allows for removal of nucleotides incorrectly inserted by the polymerase.

In *E. coli*, Pol III holoenzyme is the major replicative polymerase for both leading-

Tab. 1 DNA polymerases found in prokaryote and eukaryote.

Organism	DNA polymerase	Processivity	Editing exonuclease	Function	Unique features
E. coli	Pol I	Low	Yes	Repair	$5' \to 3'$ exonuclease associated
	Pol II	Medium	Yes	TLS	Structural homology to eukaryotic Pol α
	Pol III	High	Yes	Replication	Asymmetric dimer with twin active site
	Pol IV	Low	No	TLS	Capable of induction of frameshifts
	Pol V	Low	No	TLS	Activated by RecA protein
Mammal	Pol α	Medium	No	Replication	Primase associated
	Pol β	Low	No	Repair	Small single-polypeptide
	Pol γ	High	Yes	Mitochondrial replication	Preference to ribohomopolymer template
	Pol δ	High	Yes	Replication, repair	PCNA-dependent processivity
	Pol ε	High	Yes	Replication, repair	PCNA-independent processivity
	Pol ζ	Low	No	TLS	Mismatch extender polymerase
	Pol η	Low	No	TLS	Error-free bypass DNA synthesis
	Pol ι			TLS	Most error-prone polymerase
	Pol κ	Medium	No	TLS	Related to E. coli Pol IV
	Pol λ	Low	No	Repair in meiosis?	Homologous to Pol β
	Pol μ	Low	No	Somatic hyper mutation?	Expressed preferentially in lymphoid

Note: TLS: translesion DNA synthesis.

and lagging-strand synthesis. Pol I participates in lagging-strand synthesis by eliminating primer RNAs and also has a role in repair synthesis. The other three polymerases, Pol II, Pol IV, and Pol V, are involved in translesion DNA synthesis and are induced when the chromosome DNA is damaged. The Pol III holoenzyme is a huge multiprotein complex that consists of 10 distinct polypeptides. This enzyme extends the DNA chain with a high processivity (>500 kb of DNA can be continuously synthesized without dissociation of polymerase from the template) and a high catalytic efficiency (the velocity of chain elongation is 1000 nucleotides second at 37 °C). The catalytic core, composed of α-, ε-, θ-subunits, contains the polymerase activity and a $3' \rightarrow 5'$ exonuclease for proofreading. τ-subunit connects two of the catalytic core subassemblies. The remaining six auxiliary subunits enhance the processivity of the core by clamping it onto the template. They also promote the repeated association of the polymerase necessary for discontinuous synthesis of the lagging strand. Among the auxiliary subunits, a dimer of β-subunit shows a ring-shaped structure and acts as a sliding clamp. Loading of the β-clamp onto the DNA is facilitated by a five-subunits subassembly called γ-complex. Structural analysis of the Pol III holoenzyme and studies on a reconstituted replication fork suggest that the holoenzyme is a functionally asymmetric dimer with twin polymerase active sites: one half of the dimer has high processivity and might be a polymerase for continuous synthesis of the leading strand; the other half has recycling capacity needed for the lagging-strand synthesis. Thus, a single molecule of Pol III holoenzyme acts at the replication fork catalyzing concurrently both leading and lagging-strand synthesis.

In eukaryote cells, nuclear DNA replication requires three polymerases, Pol α, Pol δ, and Pol ε. Pol γ is a sole polymerase participating in mitochondrial DNA replication. Pol β, a small single polypeptide, functions in base excision repair. There are many other polymerases that were recently identified. Some of them are involved in translesion DNA synthesis, but biological roles of the remaining polymerases are unknown. Among the replicative polymerases, Pol α is unique in possessing a primase activity, the only such activity so far identified in eukaryotic cells, suggesting that Pol α may play a role in the priming of DNA synthesis. However, DNA synthesis by Pol α, lacking the $3' \rightarrow 5'$ exonuclease activity, is inaccurate and shows a low processivity. These enzymatic characteristics make Pol α a poor candidate for the major replicative polymerase. On the other hand, Pol δ and Pol ε both possess a $3' \rightarrow 5'$ exonuclease activity and are able to carry out highly processive DNA synthesis with the aid of a sliding clamp (PCNA) and a clamp loader (RF-C). Therefore, these polymerases are better suited for chromosome replication. Yeast mutant strains defective in the $3' \rightarrow 5'$ exonuclease activity of either Pol δ or Pol ε increase spontaneous mutation rate, suggesting that both polymerases are responsible for chain elongation during chromosome replication. However, the division of labor between the polymerases remains ambiguous.

1.2.2 DNA Helicase and Primase

DNA polymerase cannot replicate a duplex DNA without assistance. This enzyme requires single-stranded DNA as a template and cannot synthesize a chain without an RNA or DNA primer annealed to the template. Two other enzymes enable polymerase to work on a duplex DNA. One is a DNA helicase, which opens up the duplex at the replication fork to provide a single-stranded template. The other is a primase, which synthesizes a short RNA to prime DNA chain elongation.

Several DNA helicases have been identified from *E. coli* and its phages, including DnaB, T7 gp4, and T4 gp41. Biochemical characterization of these activities *in vitro* DNA replication systems suggests that the primary replicative helicase binds to and moves on the lagging-strand template in the 5′ → 3′ direction unwinding the duplex as it goes. This helicase action requires ATP hydrolysis. In addition, genetic studies indicate that these helicases are essential for DNA replication and growth of the cell or its phages, respectively. Another common property of the replicative helicases is an intimate association with a primase. The T7 gp4 has both primase and helicase activity within the same polypeptide. T4 gp41 greatly enhances the primase activity of T4 gp61. A similar functional interaction has been observed between DnaB protein and *E. coli* primase (DnaG protein).

Although several DNA helicase activities have been found in eukaryotic cells, the replicative helicase has not been clearly identified. Large T-antigen encoded by the SV40 virus is required for the replication of SV40 DNA and has a 3′ → 5′ helicase activity. Moreover, human DNA Pol α, which contains a primase subunit, has a specific affinity to the large T-antigen helicase. Most recently, MCM2-7 proteins that form a ring-shaped complex around DNA and that are required for chain elongation during chromosomal replication were suggested to be a replicative helicase.

Primases can start a new chain when copying a duplex DNA, but, like DNA polymerases, require a single-stranded DNA as the template. Primases can utilize dNTPs in place of rNTPs, although they absolutely require an rNTP (ATP in most cases) to initiate the primer. Under physiological concentrations, rNTPs are preferentially used during primer synthesis. In prokaryotes, primer synthesis is initiated at preferred sequences on the template: 3′-GTC for *E. coli* primase (DnaG), 3′-CTG(G/T) for T7 phage primase (gp4), and 3′-TTG for T4 phage primase (gp61). The sequence preference of the initiation of primer synthesis by eukaryotic primases is uncertain. All primases found in various organisms extend the primer chain in the 5′ → 3′ direction to a particular length: 10 to 12 nucleotides by DnaG protein, 4 nucleotides by T7 gp4, 5 nucleotides by T4 gp61, and 8 to 12 nucleotides by eukaryotic primases.

A functional interaction between primase and replicative helicase has been extensively studied using *E. coli* replication proteins. On most templates, DnaG exhibits a very feeble priming activity, which can be greatly enhanced if DnaB first binds that DNA. This stimulation of primase activity is further increased when the DnaB helicase is activated to its processive form at the replication fork. The activation of DnaB protein is a key step in the initiation of DNA replication and is catalyzed in several ways (see Sect.1.3.2).

1.2.3 Other Replication Proteins

In addition to the components of the replisome, several auxiliary proteins are needed for DNA replication. Single-stranded DNA binding protein (SSB) protects the exposed template and stabilizes the structure of the replication fork. *E. coli* mutants defective in SSB are inviable; DNA synthesis in these conditional mutants stops quickly under nonpermissive conditions. *E. coli* contains about 300 molecules of SSB per cell, at a level sufficient to cover about 1400 nucleotides of the lagging-strand template at each replication fork. Eukaryotic counterpart to the *E. coli* SSB has been identified in the *in vitro* replication system. This protein, RF-A, consists of three distinct subunits, binds tightly to ssDNA, and is required for the helicase action of the virus-encoded T-antigen at the SV40 origin.

Enzymes to remove the primer RNA and fill the gap, such as RNaseH1 and

DNA polymerase I of *E. coli*, are essential for the sealing of Okazaki fragments by DNA ligase. Mutants defective in either DNA polymerase I or DNA ligase show a massive accumulation of short Okazaki fragments under restrictive conditions. In eukaryote, a 5′ → 3′ exonuclease called *FEN1* removes the primer RNA.

DNA topoisomerases function as swivels, to relieve torsional stress produced ahead of the replication fork. From the analysis of mutants, it appears that DNA gyrase (a type II topoisomerase from *E. coli*) acts as the swivel in prokaryotic cells. In eukaryotes, either type I or type II topoisomerases can provide the swivel action needed. Another important role of topoisomerases in replication is the decatenation of daughter molecules. This function is provided by type II topoisomerases in both prokaryotic and eukaryotic cells and allows the separation of the daughter molecules prior to segregation.

1.3 Regulation of DNA Replication

Initiation of DNA replication is a major determinant of the cell cycle, and thus is tightly regulated in all organisms. Although signals that turn on DNA replication in eukaryotic and prokaryotic cells are different, how these signals are transduced to the initiation machinery is one of the most important issues in cell biology. The mechanisms of initiation have been detailed using systems that initiate and replicate a minichromosome *in vitro*. These various mini-chromosomes contain replication origins from bacterial chromosomes (*E. coli oriC*), bacteriophages (λ *ori*, P1 phage *ori*), and plasmids (ColE1 *ori*, F plasmid *oriS*). In contrast to the prokaryotic origins, the sequences required for an origin of replication vary significantly between different eukaryotic organisms. Chromosomal origins have been isolated from yeast as well as from several eukaryotic viruses. However, attempts to identify such autonomously replicating segments from animal or plant chromosomes have been unsuccessful. Instead, a more direct approach, using electron microscope and two-dimensional gel electrophoresis techniques, has been taken to identify replication origins in the higher eukaryotes. Although it has been difficult to develop *in vitro* systems for eukaryotic DNA replication, genetic studies with yeast mutants that affect initiation processes of chromosomal replication have successfully revealed the initiation mechanisms of eukaryotic DNA replication.

1.3.1 Initiation Mechanisms in Prokaryotes

A DNA segment carrying a replication origin can be isolated as a mini-replicon. DNA fragments can be linked to a selectable marker and introduced into the cell. Those fragments that allow for the maintenance of the marker are likely to contain an origin for autonomous replication of the DNA. Analysis of deletion and base-substitution mutants of the mini-replicon for their replication capacity will define a minimal region of the replication origin and essential motifs in that region. Replication origins that have been examined in this way can be classified into two groups from their structural similarities (Fig. 4). One class, to which most prokaryotic replicons belong, consists of 4 to 7 repeats of an initiator-protein binding site and an AT-rich region. The other group (ColE1 plasmid *ori*, T7 primary *ori*, T4 primary *oriA*) carries a transcriptional promoter instead of the initiator-protein binding sites.

The reconstitution of initiation in the test tube has elucidated several events during this process as follows: (1) recognition of the origin; (2) opening of a particular region in the origin; (3) loading of a replicative helicase onto the single-stranded DNA region, and (4) initial priming for leading- and lagging-strand synthesis. Some bacterial plasmids and phages are replicated unidirectionally. A major determinant for uni- or bidirectionality is the number of replicative helicases loaded at the origin. One helicase can form one replication fork. Thus, bidirectional replication involves two replicative helicases.

The initiation mechanism at the origin carrying the initiator-protein binding site has been depicted from studies on *in vitro* replication of *oriC* plasmid DNA (Fig. 4(a)). At least eight proteins are required for the *in vitro* reconstitution of the *oriC* minichromosome. A key event is the formation of the initial complex in which the initiator protein (DnaA protein) tightly binds to its 9 bp recognition sequence in *oriC*. Following this stage, an AT-rich region is opened in an ATP-dependent manner, and the DnaB helicase and SSB are loaded onto the single-stranded region to form the prepriming complex. Primase and DNA polymerase III holoenzyme then start the DNA synthesis. Several plasmids (F-factor and R-factor) and bacteriophages (λ, P1) in this class of replicon encode their own initiator proteins that specifically recognize and bind the origin sequence. Most of these origin-binding proteins can open the duplex DNA by themselves but fail to load the DnaB helicase. To do this, they contain one or two DnaA recognition sequences in their origins and utilize the DnaA protein as a landmark for entrance and activation of the replicative helicase.

Replicons without an initiator protein use transcription for origin recognition and duplex opening (Fig. 4(b)). Initiation of ColE1 plasmid replication requires RNA synthesis started 555 bp upstream from the transition point between RNA and DNA. The transcript (RNA II) forms an RNA–DNA hybrid around the transition point, is processed by RNaseH1, and serves as the initial primer for the leading strand DNA synthesis. DNA pol I extends the leading-strand DNA by about 400 nucleotides. This initial chain elongation, resulting in duplex opening, does not involve any helicase action but

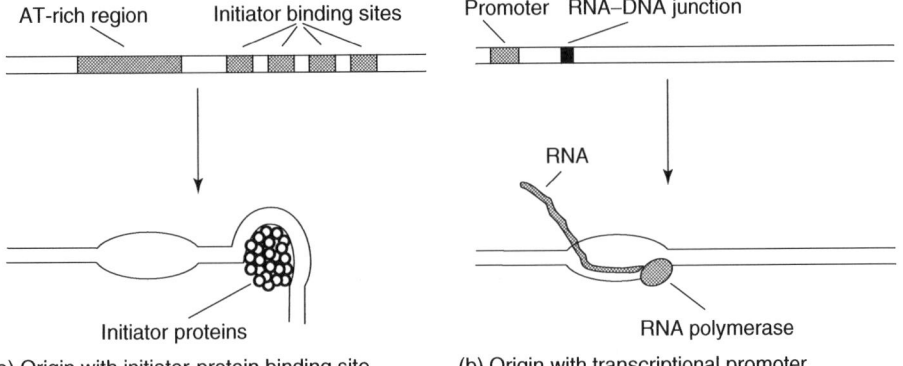

(a) Origin with initiator-protein binding site (b) Origin with transcriptional promoter

Fig. 4 Two types of replication origins and related mechanisms of initiation.

requires that the template be negatively supercoiled. The loading of DnaB helicase is facilitated by primosome proteins, which are assembled into a complex on a specific DNA segment (primosome assembly site) about 150 bp downstream from the origin.

1.3.2 Initiation Mechanisms in Eukaryotes

The yeast *Saccharomyces cerevisiae* is the only eukaryote in which there is a detailed understanding of the sites used for initiation of chromosomal DNA replication. In yeast, these sites are called *ARSs* (autonomously replicating sequences) because they were initially identified by their ability to confer high-frequency transformation and self-replication on plasmids introduced into cells. By these criteria, there are 200 to 400 ARS elements in the yeast genome. ARS elements are relatively small, about 100 to 150 bp, and consist of one or more copies of an essential 11 bp-long AT-rich ARS consensus sequence, as well as several other less conserved elements. In other organisms, the sequences required to direct the initiation of DNA replication are more complex. In the fission yeast, *Schizosaccharomyces pombe*, initiation of replication requires origin sequences spread over at least 800 to 1000 bp. Although these sequences contain several AT-rich sequences of 20 to 50 bp that are important for origin function, they do not exhibit the strong sequence similarity observed for the *S. cerevisiae* ARS elements. Metazoan origins are still less well defined and can extend over thousands of base pairs of DNA. In addition, the sites of initiation are not always tightly linked within these regions.

Eukaryotic origins of replication direct the formation of a number of protein complexes leading to the assembly of two bidirectional DNA replication forks. These events are initiated by the formation of the pre-replicative complex (pre-RC) at the origins of replication during G1 (Fig. 5). Pre-RC formation involves the ordered assembly of a number of replication factors including ORC, Cdc6, Cdt1, and Mcm2–7 proteins. The ARS consensus sequence is the binding site for the multisubunit origin recognition complex (ORC), which binds constitutively throughout the cell cycle and is essential for initiation. Several proteins are recruited to the ARS during the G1 phase of the cell cycle to form a pre-RC, including the multisubunit minichromosome maintenance (MCM) complex, which has ATPase and

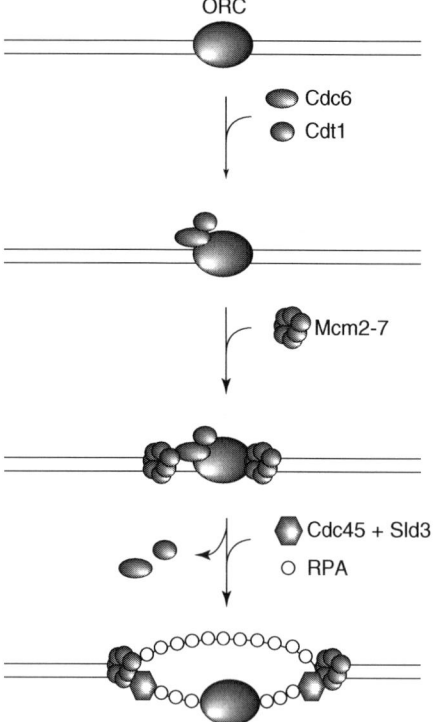

Fig. 5 Initiation mechanisms of eukaryotic DNA replication.

helicase activity and is needed for both initiation and fork progression. The regulation of pre-RC formation is a key element of the mechanisms coordinating DNA replication with the cell cycle. Once formed, this complex awaits activation by at least two kinases that trigger the transition to DNA replication. As with the formation of the pre-RC, the transition to replication involves the ordered assembly of additional replication factors that facilitate unwinding of the DNA at the origin and culminates in the association of the multiple eukaryotic DNA polymerases with the unwound DNA.

2
Transcription

The transfer of information from DNA to protein begins with the synthesis of RNA molecules in a process called *transcription*. In this process of transcription, the genetic information carried in DNA is transferred to several kinds of RNA molecules (mRNA, tRNA, rRNA, and others), so that an mRNA encoding a polypeptide in its turn translates the information in the four-letter language of the nucleic acids to the twenty-letter language of the amino acids with the coordinated actions of tRNA and rRNA as part of the mechanism of protein synthesis.

2.1
RNA Polymerase and Transcriptional Apparatus

DNA-dependent RNA polymerase, which catalyzes the polymerization of RNA from ribonucleotides in a DNA-dependent manner, is the enzyme responsible for the transcription of DNA into RNA. The RNA polymerase binds to a specific DNA sequence, called the *promoter*, and unwinds the duplex DNA for about one turn of the helix to expose a short stretch of single-stranded DNA so that complementary base pairing can be made with the incoming ribonucleotides. The enzyme joins two of the ribonucleoside triphosphate monomers and then moves along the DNA strand, extending the growing RNA chain in the $5' \rightarrow 3'$ direction until it encounters a second special sequence, called the *terminator*, which signals where the RNA synthesis should stop. After the transcription is completed, each RNA chain is released from the DNA template as a free, single-stranded RNA molecule. In addition to RNA polymerase, other protein factors are required for efficient transcription, and they are responsible for determining the transcription efficiency of specific transcription units.

2.1.1 General RNA Polymerase Architecture

RNA polymerases are large multisubunit enzymes, which consist of 5 to 15 subunits and whose sizes are up to 0.6 MDa. In eubacteria and archaebacteria, a single type of RNA polymerase synthesizes all classes of RNA molecules, except for the primer RNA, which is synthesized by primase. The bacterial RNA polymerases consist of 5 subunits (2 α-subunits and one each of β, β', and ω), and a few more transcription factors, such as σ, ρ, and NusA for *E. coli* RNA polymerases, are required to initiate transcription from the proper start site in the promoter of duplexed DNA template (Table 2).

There are three known distinct types of RNA polymerases in eukaryotic cells. RNA polymerase I (pol I) makes ribosomal RNA and pol II mostly makes messenger RNA, while pol III makes small RNA molecules, such as transfer

Tab. 2 Subunit composition of E. coli RNA polymerase and transcription factors.

Subunit	Gene	Number of amino acids	Molecular weight [Da]	Function
α	rpoA	329	36 512	Connecting $\beta\beta'$ subunits
β	rpoB	1342	150 618	Catalyzing RNA synthesis substrate nucleotide binding
β'	rpoC	1407	155 163	Template binding and association with σ-subunit
ω	rpoZ	91	10 230	Regulation of transcription
$\sigma 70$	rpoD	613	70 263	Recognition of general promoters
ρ	rho	419	46 974	Termination
NusA	nusA	494	54 536	Elongation, termination

Tab. 3 Eukaryotic RNA polymerases[a].

Polymerase	Location	Genes transcribed	Polymerase activity in cells [%]
pol I	Nucleolus	Genes for 28S, 5.8S and 18S ribosomal RNAs (rRNAs)	50–70
pol II	Nucleoplasm	Protein-coding genes Most genes (U1, U2, U4, U5) of small nuclear RNAs (snRNAs)	20–40
pol III	Nucleoplasm	Genes for tRNAs, 5S rRNA Small nucleolar RNAs (snoRNAs) Small cytopasmic RNAs (scRNAs)	10

[a] Mitochondria and chloroplast contain distinctive RNA polymerases.

RNA, 5S ribosomal RNA, and a few other small RNA molecules (Table 3). Among eukaryotic RNA polymerases, yeast S. cerevisiae RNA polymerases have so far been the most extensively characterized. Each RNA polymerase molecule consists of two large subunits, 160 ~ 200 kDa and 130 ~ 150 kDa, which share amino acid sequence homology with the β'- and β-subunits of the bacterial enzymes respectively, and a collection of smaller polypeptides. In pol II, Rpb3 and Rpb11 share amino acid sequence homology with the bacterial α-subunit, while Rpb6 shares a homologous sequence with the ω-subunit. Among the three types of eukaryotic RNA polymerases, five subunits (Rpb5, 6, 8, 10, and 12) are shared. pol I and pol III share two additional subunits (AC19 and AC40), both of which share amino acid sequence homology with the bacterial α-subunit.

As shown in Fig. 6, S. cerevisiae pol II and Thermus aquaticus RNA polymerase share a general architecture, which consists of five core subunits, β', β, α/α, and ω or their counterparts in the pol II (Table 4).

Fig. 6 General architecture of RNA polymerases. Schematic presentation for RNA polymerase subunits in bacterial (α, α, β, β', ω) and eukaryotic enzymes (Rbp1–12).

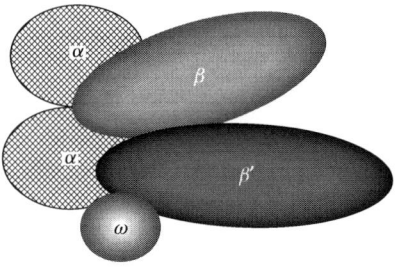

Thermus aquaticus RNA polymerase

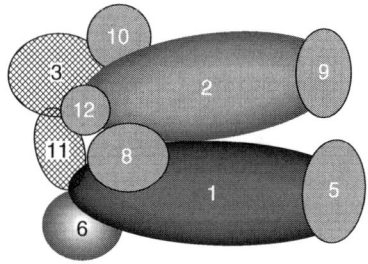

Saccharomyces cerevisiae RNA polymerase II

The bacterial β' and β subunits or the yeast Rpb1 and 2 subunits form the central masses with a deep cleft between them, respectively. α-homodimer in the bacterial enzyme or Rpb3–Rpb11 heterodimer in the yeast pol II anchor the two large subunits, while the fifth subunit ω and Rpb6 support and stabilize either the large subunit β or Rpb1.

2.1.2 Transcription Initiation

None of the eukaryotic RNA polymerases can initiate transcription by themselves; rather, they require a group of general transcription initiation factors for initiation at proper sites. pol II requires several general initiation factors (TFIIA, IIB, IID, IIE, IIF, IIH), and the formation of a functional preinitiation complex on the core promoters, which are defined as the minimal sequences required to recruit the appropriate RNA polymerase and initiate transcription, involves ordered

Tab. 4 Subunit structure of RNA polymerase II.

Subunit	Amino acid residues (% identity to human)			
	Human	S. cerevisiae	A. thaliana	Notes
Rpb1	1970	1733(49)	1860(55)	[a]Bacterial β'
Rpb2	1174	1224(56)	1188(65)	[a]Bacterial β
Rpb3	275	318(39)	319(39)	[a]Bacterial α
Rpb4	142	221(23)	138(37)	
Rpb5	210	215(44)	205(54)	[b]RNA pol I, II, III
Rpb6	127	155(47)	144(47)	[a]Bacterial ω
Rpb7	172	171(43)	176(53)	
Rpb8	150	146(35)	146(47)	[b]RNA pol I, II, III
Rpb9	125	122(38)	114(58)	
Rpb10	67	70(69)	71(75)	[b]RNA pol I, II, III
Rpb11	117	120(46)	116(54)	[a]Bacterial α
Rpb12	58	70(33)	51(53)	[b]RNA pol I, II, III

[a]Homologous subunit in bacterial RNA polymerase.
[b]Common subunits for eukaryotic RNA polymerase I, II, III.

interactions of an array of general initiation factors on the core promoters (Fig. 7).

TFIID, a high molecular weight complex comprising TATA-binding protein (TBP) and more than 10 TBP-associated factors (TAF$_{II}$s), binds to the TATA box of the promoter; then TFIIA and B enter to form the DAB complex. The non-phosphorylated form of RNA polymerase II (pol IIa), together with TFIIF, binds to the DAB complex on the promoter, followed by association of TFIIE and TFIIH to form a closed preinitiation complex. The TFIIH XPB DNA helicase, composed

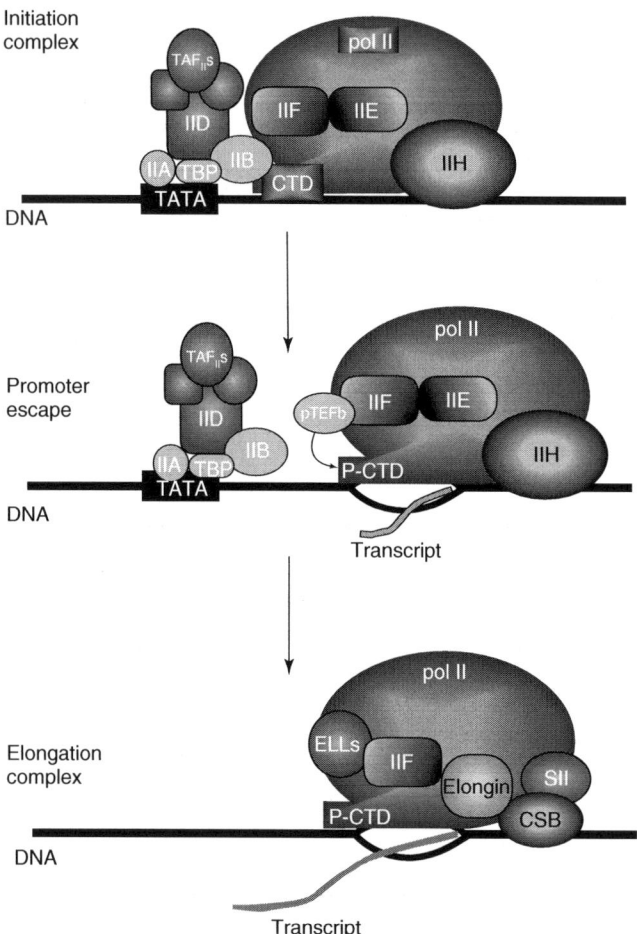

Fig. 7 Formation of a functional initiation complex on a TATA-containing promoter by general initiation factors. After completion of the final initiation complex, pol II initiates transcription and moves away from the promoter and dissociates from the DAB complex after synthesizing 10- to 15-nucleotide-long transcripts (promoter escape). During the promoter escape step, the elongation complex is released from the promoter to start elongation by phosphorylation of the CTD by pTEFb. P-CTD: phosphorylated CTD.

of nine subunits, and two of which, XPB and XPD, are catalytic subunits for the DNA helicase, converts the closed preinitiation complex to both an initiation- and an escape-competent transcriptional intermediate (an open complex) in a single ATP-dependent step, probably by unwinding the DNA downstream from the transcription start site, and the synthesis of the first phoshodiester bond of nascent transcript begins.

RNA polymerase III requires different combinations of transcription initiation factors to form a preinitiation complex on the promoters of 5S RNA, tRNA, and other small RNA genes. Three pol III specific initiation factors, TFIIIA, B, and C, are required for transcription of the 5S RNA genes, TFIIIB and C for the tRNA genes, and TFIIIB and TFIID for the U6 gene. TFIII A and C, as well as TFIID, recognize and bind to specific DNA elements while TFIIIB has by itself no sequence–specific interaction with DNA. RNA polymerase I also requires several transcription factors to form an active initiation complex on the promoter of rDNA.

2.1.3 Promoter Escape

After initiating transcription, pol II moves away from the promoter and dissociates from the DAB complex after synthesizing 10- to 15-nucleotide-long transcripts (promoter escape, Fig. 7).

The largest subunit of pol II (Rpb1) contains an unusual repetitive domain, which is not found in the homologous large subunits of pol I and III or in the bacterial β'-subunit. This carboxyl-terminal domain (CTD), called the *tail* of pol II, consists of multiple heptapeptide repeats of the consensus sequence YSPTSPS. The heptapeptide sequence is found in the largest subunits of most eukaryotic RNA polymerase II. The deletion of most or all of the CTD is lethal for yeast, *Drosophila*, and mouse cells, thereby indicating that the CTD has an essential role in transcription. The CTD of RNA polymerase II in the initiation complex is hypophosphorylated, while pTEFb (positive transcription elongation factor b), a cyclin-dependent CTD kinase complex composed of CDK9 and one of several cyclins such as T1, T2, and K, highly phosphorylates the CTD during the step of promoter escape, and thus promotes a dissociation of the elongation complex of pol II from the DAB complex.

2.1.4 Transcription Elongation

During RNA chain extension, RNA polymerase II forms a stable elongation complex, and in the complex downstream DNA duplex enters the enzyme between a pair of mobile "jaws" formed by Rpb1, Rpb5, and Rpb9 subunits and then extends through the cleft toward the active site for RNA synthesis with Mg^{2+}. The DNA–RNA hybrid extends upward beyond the active site and the template strand is locked near the catalytic site in place by a clamp formed by parts of, Rpb1, Rpb2, and Rpb6. In the active site, the nontemplate strand of DNA is held within the Rpb2 subunit and the RNA transcript; it then extrudes through the complex via a channel formed by parts of both of the two large subunits, and the growing RNA end is located above the pore/secondary channel, from which nucleoside triphosphates (NTPs) may be supplied during RNA synthesis (Fig. 8).

mRNA synthesis is carried out efficiently in eukaryotic cells at rates of 1200 to 1500 nucleotides/min. However, purified mammalian pol II extends the RNA chain *in vitro* at rates of only 300 to 400 nucleotides/min, thus indicating

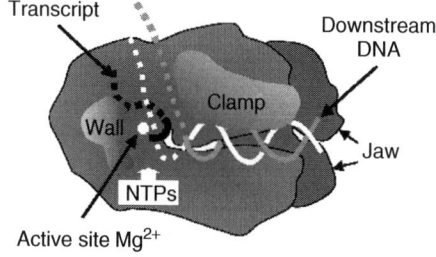

Fig. 8 The RNA polymerase II elongation complex. During transcript (black line) extension, RNA polymerase II forms a stable elongation complex in which a downstream DNA duplex (gray lines) enters the enzyme between a pair of mobile "jaws" and extends through the cleft toward the active site for RNA synthesis. The DNA–RNA hybrid extends upward beyond the active site and the template strand (white line) is locked near the catalytic site. The pore/secondary channel where the nucleoside triphosphates (NTPs) enter is shown.

that the elongation step of RNA synthesis by eukaryotic RNA polymerase II in vivo has to be supported by various transcription factors or elongation factors to achieve higher rates of elongation. Furthermore, unusual template structures, damage on the template, or imbalance in the cellular ribonucleotide pool, which cause either a pause or an arrest of transcription, can be overcome by factors that directly or indirectly interact with the elongation complex. Pausing occurs when RNA polymerase II undergoes modest and reversible backsliding on the DNA template, thus resulting in a misalignment of 3′-OH terminus of the nascent transcript with the catalytic site, while irreversible backsliding of the enzyme causes arrest since the misalignment becomes even more severe.

Elongation factors fall into two broad functional classes based on their abilities to either prevent transient pausing by pol II or reactivate arrested pol II. The majority of elongation factors including ELLs, Elongin, and CSB are capable of suppressing transient pausing by pol II. TFIIF, one of the general transcription factors also has an activity to suppress transient pausing. In addition, the phosphorylation of the pol II CTD by pTEFb is required to support the efficient elongation by pol II, as well as for the promoter escape, as described above (Fig. 7).

Among many elongation factors, only the SII family elongation factor has a capacity to reactivate the arrested pol II by triggering the polymerase-associated endoribonuclease that cleaves the nascent transcript upstream of its 3′-OH terminus, thus resulting in the formation of a new 3′-OH terminus. The newly generated 3′-OH, which is correctly positioned to the catalytic site of the enzyme, can thus be reextended.

During transcript elongation, pol II is hypothesized to face a major obstacle-nucleosome in which the DNA is tightly wrapped into the chromatin. It is likely that a class of chromatin elongation factors, such as an elongator, alters the structure of the nucleosome through histone acetylation, chromatin remodeling, and histone dissociation, thereby facilitating the elongation step by pol II.

2.2 Transcription Unit

Transcription takes place in limited regions of genomic DNA, and only one of the two DNA strands is used as a template. The promoter, an oriented DNA sequence along the template DNA, determines the startpoint of the region to be transcribed and also determines which of the two strands is copied. A transcription unit extends to the terminator, at which the RNA polymerase stops adding nucleotides to the growing RNA chain. The critical feature of the transcription unit is that it constitutes a stretch of DNA expressed via the production of a single RNA molecule. A transcription unit may include only one or several genes. A typical transcription unit of *E. coli* and various transcription units for three eukaryotic RNA polymerases are shown in Fig. 9.

2.2.1 Bacterial Transcription Unit

Bacterial promoters are identified by two short conserved sequences centered at −35 and −10 relative to the startpoint. In *E. coli*, TTGACA and TATAAT are consensus sequences for the −35 and the −10 boxes of the promoters, respectively, which are recognized by the RNA polymerase holoenzyme carrying σ^{70}. Promoters whose −10 and −35 sequences closely approximate the consensus sequences are strong promoters, and the distance separating the two sequences, usually 16 to 18 bp, is also important. Bacterial RNA polymerase terminates transcription at two types of sites; factor-independent sites contain a GC-rich hairpin followed by a run of U residues, while ρ factor–dependent termination requires ρ-specific sequences.

2.2.2 Eukaryotic Transcription Unit

In eukaryotes, the promoters recognized by each type of RNA polymerase are distinct. The promoters for RNA polymerase I are located upstream from the transcription startpoint, and the sole product of transcription by pol I is a large precursor that contains the sequences of the major rRNA (28S, 5.8S, 18S). Termination occurs at a discrete site and apparently requires few factors.

The promoter elements that are necessary and sufficient for the specific initiation by RNA polymerase II with its general factors are referred to as either minimal or core promoter elements, and are located around the transcription startpoint. The most common core elements are the TATA box, which is present about 20 to 30 bases upstream of the transcription start site, the initiator element (Inr) located at the start site, and a downstream promoter element (DPE), which is present about 30 bases downstream of the transcription start site. Specific genes may contain at least one or all of these elements, which may function in conjunction with other cis-elements such as UAS (upstream activating sequence) or enhancers, which positively regulate transcription initiation.

The termination capacity of RNA polymerase II is regulated through the polyadenylation of the transcript. The signal sequence for polyadenylation, AAUAAA, is located 10 to 30 nucleotides upstream of the polyadenylation site, which is often immediately after the dinucleotide 5′-CA-3′ and is followed 10 to 20 nucleotides later by the GU-rich region. When the elongation complex of pol II, to which the cleavage and polyadenylation specificity factor (CPSF) and the cleavage stimulation factor (CstF) bind through the CTD, transcribes across the polyadenylation signal sequence, CPSF binds to the signal sequence and CsTF binds GU-rich sequence on the transcript. Next, the transcript is cleaved to create a new 3′ end, to which the poly (A) tail is added by poly(A) polymerase associated with the bound CPSF and CstF, followed by the termination of transcription.

For RNA polymerase III, there are two types of promoters: the promoters for U6

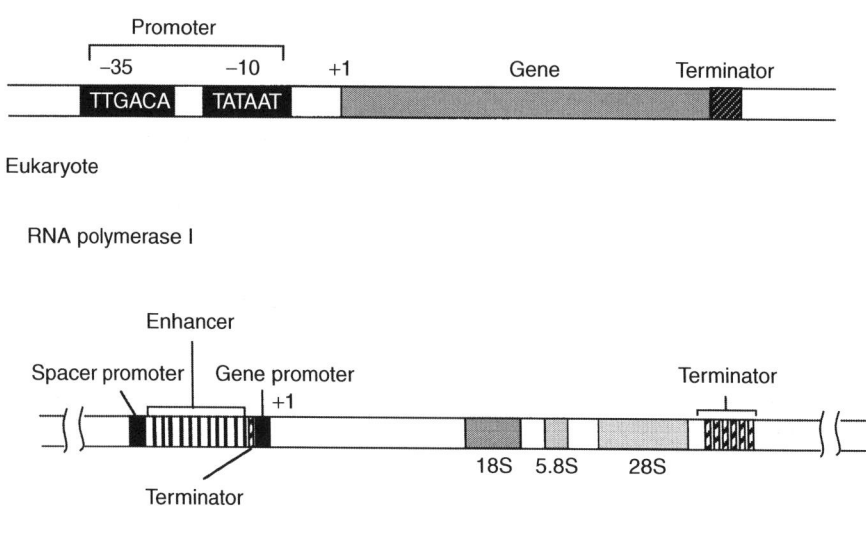

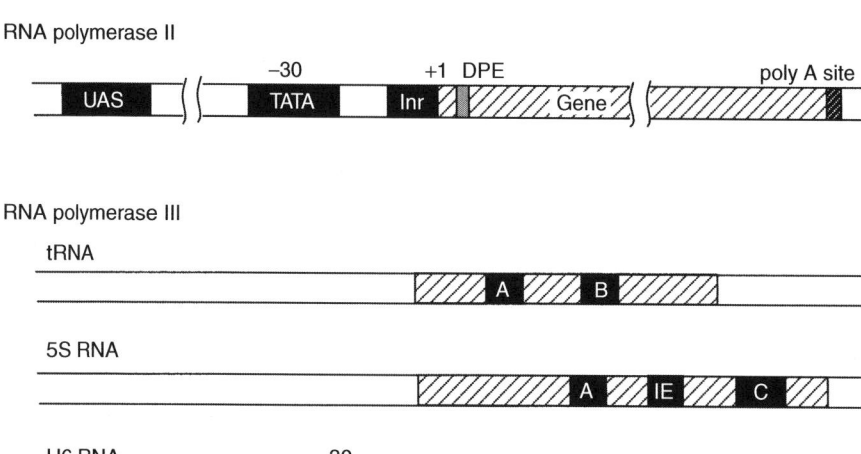

Fig. 9 Various types of transcriptional units in prokaryotic and eukaryotic cells. Inr, the initiator element; UAS, upstream activating sequence; DPE, a downstream promoter element; A, B, IE, C, intragenic cis-elements in genes for tRNA and 5S RNA; DSE, the distal sequence element; PSE, proximal sequence element.

are upstream from the startpoint, while those of the 5S RNA and tRNA genes lie downstream from the startpoint. The external promoter for the U6 gene consists of the TATA box and the distal and proximal sequence elements (DSE and PSE), which enhance promoter activity. Intragenic A and B boxes in the tRNA gene or the B box in the 5S RNA gene can both be recognized by, and bound by, TFIII C, while the IE and C boxes in 5S RNA are involved in interaction with TFIIIA. Transcription by RNA polymerase III terminates within a run of 4 U residues embedded in a GC-rich region, in a manner similar to that observed in the prokaryotic factor-independent terminators.

2.3
Regulation of Transcription

Transcriptional regulation is the primary means of controlling gene expression, and it is essential for the developmental program of multicellular organisms, as well as for the switching on and off of gene expression in response to intra- or extracellular signals. In any organism, RNA synthesis can be regulated by mainly controlling the efficiency of initiation.

2.3.1 Transcriptional Regulation in Bacteria

In bacterial cells, there are several different mechanisms by which the modulation of transcription initiation is achieved. Various σ-factors modify the binding ability of RNA polymerase to specific promoter sequences, thus enabling it to initiate at a different set of promoters.

In a set of genes whose expression is coinduced in response to intra- or extracellular signals, a common repressor protein binds to specific DNA sequences, called an *operator*, in a group of promoters and represses their transcription by inhibiting RNA polymerase binding to the promoters. After responding to a specific signal, such repressor molecules are inactivated, and the RNA polymerase can thereby initiate transcription from the promoters. For example, the LexA protein, which represses the SOS-regulon, is proteolytically inactivated by an activated form of the RecA protein in response to DNA damage. The modulations of genes (an operon) involved in the biosynthesis of a specific metabolite are mediated by the repressors that directly sense the intracellular status of the product. For example, the LacI repressor protein binds to the operator region of the *lac* operon promoter in the absence of lactose or its metabolite, thereby prohibiting transcription initiation on the promoter, and it dissociates from the promoter after binding to a metabolite of lactose, thus allowing for the transcription of the *lac* operon.

Promoters requiring positive activator proteins for efficient expression are characterized by a poor fit to the consensus sequence in the -35 region. The protein activators bind to the region near the -35 box, and stimulate initiation either by improving the affinity of RNA polymerase for the sequence or by increasing the rate of isomerization of the initiation complex. The complex of cAMP with catabolite activator protein (CAP) enhances the affinity of RNA polymerase to the *lac* promoter about 20-fold. Ada-protein-accepted alkyl groups from DNA damaged by alkylating agents as a repair reaction, in turn, strongly bind to Ada boxes in the promoters of Ada regulon, such as *alkA, aidB* and *ada* itself, and enhance the affinity of RNA polymerase to the promoters, thus establishing an adaptive status to the alkylating agents.

In addition to the regulation of initiation, the level of transcription can be regulated by the premature termination of transcription or attenuation. For example, in most bacterial amino acid biosynthesis operons, an early termination of transcription occurs at the attenuator site that lies from 100 to 200 bases downstream of the start site, when an abundance of the charged cognate tRNA signals a sufficiency of the amino acid. Attenuation in these operons is regulated in a complex mechanism involving charged cognate tRNA, ribosome, and ρ-protein.

2.3.2 Transcriptional Regulation in Eukaryotes

In multicellular organisms, each cell selects a set of genes from their genome to be expressed at a given time and under specific stimuli, and this selectivity and specificity of gene expression is a fundamental mechanism for controlling the cell fate such as cell proliferation, differentiation, and cell death as well as cellular functions. In eukaryotic cells, transcription by RNA polymerase II, a process for the synthesis of mRNA encoding polypeptides, which are major determinants of diversity of each cell, is regulated more dynamically than transcription by pol I and III that provide rather common elements, such as rRNA and tRNA for cell function.

The genes in the nuclear genome are usually packaged into chromatin consisting of nucleosomes and other chromatin proteins, and their promoters are transcriptionally inactive because of their limitations for access by the transcription apparatus (Fig. 10). An assortment of regulatory elements for RNA polymerase II transcription can be scattered both upstream and downstream of the transcription start site for a gene. Each gene has a particular combination of positive and negative regulatory cis-elements that are uniquely arranged regarding the number, type, and spatial array. Usually, cis-elements are arrayed within several hundred base pairs of the initiation site, but some elements can exert their control over much greater distances (1 to 30 kb). These elements are binding sites for sequence-specific DNA binding proteins, such as AP-1 (JUN/FOS heterodimer), that either directly or indirectly mediate activation or repression of transcription of the gene. The region near the startpoint, in particular the TATA box (if there is one), is responsible for selecting the exact startpoint and it is called the *promoter*, as described above. The elements further upstream determine the efficiency with which the promoter is used and are known as UAS (*upstream activating sequence*) or enhancers (Fig. 9).

In order to transcribe a particular gene, the chromatin structure has to be remolded, thus enabling various DNA binding proteins and transcription apparatuses to access a particular promoter (Fig. 10). First, inactive genes are localized to condensed chromatin domains or so-called heterochromatin compartments, in which each promoter is assembled in regularly spaced nucleosomes. A sequence-specific DNA binding protein binds to a particular cis-element, and then recruits a specific chromatin-remodeling complex such as SWI/SNF with other proteins through protein–protein interaction. The chromatin-remodeling complex alters the twist and writhe of nucleosomal DNA thereby affecting its accessibility to DNA binding proteins. Following chromatin remodeling, the prebound protein or another DNA binding protein on the second cis-element recruits another class of chromatin modifiers with histone acetyltransferase (HAT) activity, such as PCAF (p300/CBP-associated factor), CBP (CREB binding protein), and p300. HATs acetylate specific lysine residues in histone amino–termini and weaken the interactions between the histone octamer and DNA, thus further facilitating decondensation of nucleosome structure. The retention of SWI/SNF complex on nucleosomal promoter is significantly stabilized by histone acetylation.

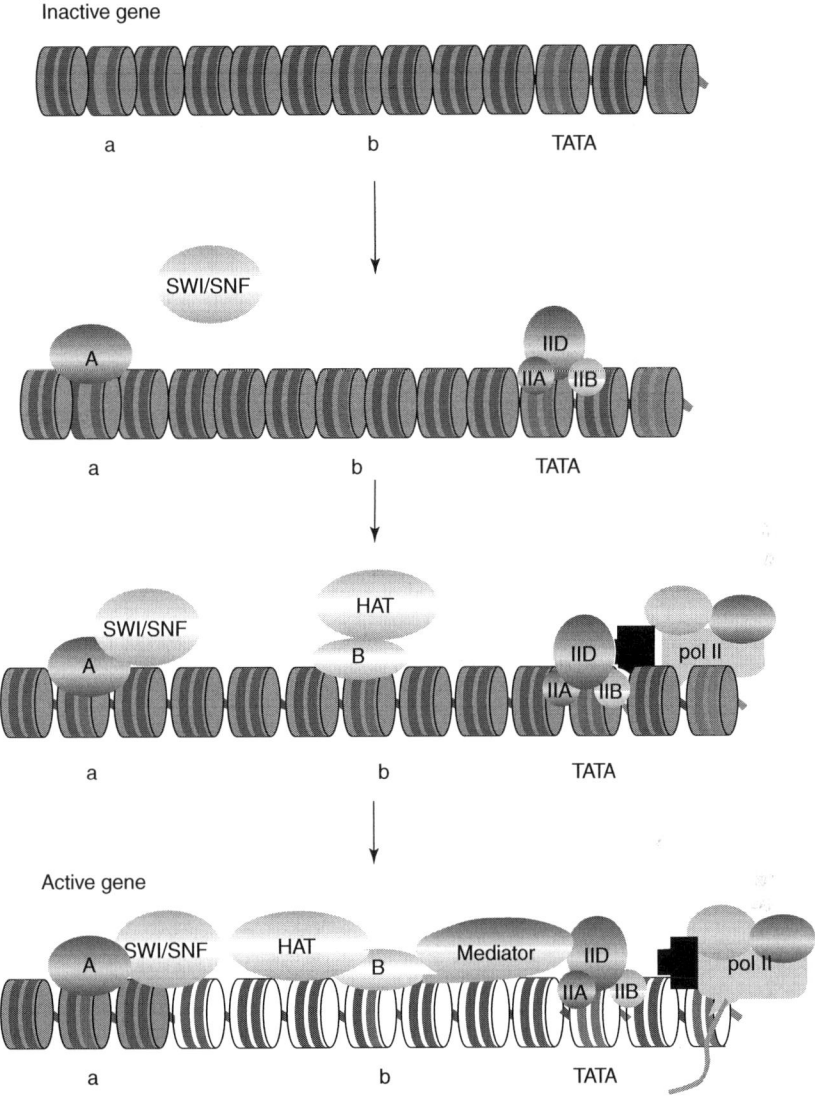

Fig. 10 Transcription activation in eukaryotes. Genes in the nuclear genome are usually packaged into chromatin consisting of nucleosomes and other chromatin proteins. In order to transcribe a particular gene, the chromatin structure has to be remolded, and thus various DNA binding proteins (A and B) can bind cis-elements (a and b) and further promote the formation of the transcription initiation complex on a particular promoter (TATA).

In addition to the acetylation of histones, phosphorylation and methylation of histones or non-histone proteins are likely to play important roles in both chromatin modification and control of gene expression.

Either during or after the remodeling or modification steps of chromatin, a partial initiation complex (TFIIA/B and TBP) or a complete initiation complex (TFIIA/B, TBP/TFIID, and pol II) with other general factors are recruited. Again, the assembly of the initiation complex is facilitated through histone modification by the promoter-targeted HAT. Finally, a mediator complex such as DRIP, ARC, TRAP, and/or coactivator such as p300/CBP or p160/SRC are recruited to the initiation complex through protein–protein interaction with DNA binding proteins such as CREB or nuclear receptors, which specifically bind to one of the cis-elements on the promoter. The mediator or coactivator is likely to be a bridge between a given DNA binding protein or activator and pol II via direct associations with specific subunits of the complex, thus ensuring the formation of a functional preinitiation complex on the core promoter.

The repression of a certain gene expression is also regulated through the modification of chromatin by histone deacetylases (HDACs), which remove acetyl groups at lysine residues in histones, which are transferred by HATs. The deacetylation of histones causes a particular region of chromatin to be condensed as heterochromatin, which thus results in the repression of gene expression.

See also Alternatively Spliced Genes; Molecular Basis of Genetics; Genomic Sequencing.

Bibliography

Books and Reviews

Alberts, B., Johnson, A., Lewis, J., Raff, M., Roberts, K., Walter, P. (2002) *Molecular Biology of the Cell*, 4th edition, Garland Publishing, New York.

Bell, S.P., Dutta, A. (2002) *Annu. Rev. Biochem.* **71**, 333–374.

Brown, T.A. (2002) *Genomes*, 2nd edition, BIOS Scientific Publishers, Oxford.

Cramer, P., Bushnell, D.A., Fu, J., Gnatt, A.L., Maier-Davis, B., Thompson, N.E., Burgess, R.R., Edwards, A.M., David, P.R., Kornberg, R.D. (1999) Architecture of RNA polymerase II and implications for transcription mechanism, *Science* **288**, 640–649.

Cramer, P. (2002) Multisubunit RNA polymerases, *Curr. Opin. Cell Biol.* **12**, 80–97.

Dvir, A., Conaway, J.W., Conaway, R.C. (2001) Mechanism of transcription initiation and promoter escape by RNA polymerase II, *Curr. Opin. Genet. Dev.* **11**, 209–214.

Featherstone, M. (2002) Coactivators in transcription initiation: here are your orders, *Curr. Opin. Genet. Dev.* **12**, 149–155.

Kornberg, A., Baker, T.A. (1992) *DNA Replication*, 2nd edition, Freeman, New York.

Lindahl, T., Sedgwick, B., Sekiguchi, M., Nakabeppu, Y. (1988) Regulation and expression of the adaptive response to alkylating agents, *Annu. Rev. Biochem.* **57**, 133–157.

Marians, K.J. (1992) Prokaryotic DNA replication, *Annu. Rev. Biochem.* **61**, 673–720.

Nakabeppu, Y., Ryder, K., Nathans, D. (1988) DNA binding activities of three murine jun proteins: Stimulation by Fos, *Cell* **55**, 907–915.

Rachez, C., Freedman, L.P. (2001) Mediator complexes and transcription, *Curr. Opin. Cell Biol.* **13**, 274–280.

Waga, S., Stillman, B. (1998) *Annu. Rev. Biochem.* **67**, 721–751.

Zahng, G., Campbell, E.A., Minakhin, L., Richter, C., Severinov, K., Darst, S.A. (1999) Crystal structure of thermus aquaticus core RNA polymearse at 3.3 A resolution, *Cell* **98**, 811–824.

Primary Literature

Andrulis, E.D., Werner, J., Nazarian, A., Erdjument-Bromage, H., Tempst, P., Lis, J.T. (2002) The RNA processing exosome is linked to elongating RNA polymerase II in Drosophila, *Nature* **420**, 837–841.

Arai, K., Arai, N., Shlomai, J., Kornberg, A. (1980) Replication of duplex DNA of phage ϕX174 reconstituted with purified enzymes, *Proc. Natl. Acad. Sci. U.S.A.* **77**, 3322–3326.

Bar-Nahum, G., Nudler, E. (2001) Isolation and characterization of sigma(70)-retaining transcription elongation complexes from Escherichia coli, *Cell* **106**, 443–451.

Bell, S.D., Jackson, S.P. (2001) Mechanism and regulation of transcription in archaea, *Curr. Opin. Microbiol.* **4**, 208–213.

Bell, S.P., Stillman, B. (1992) ATP-dependent recognition of eukaryotic origins of DNA replication by a multiprotein complex, *Nature* **357**, 128–134.

Bhaumik, S.R., Green, M.R. (2002) Differential requirement of SAGA components for recruitment of TATA-box-binding protein to promoters in vivo, *Mol. Cell. Biol.* **22**, 7365–7371.

Bjorklund, S., Buzaite, O., Hallberg, M. (2001) The yeast mediator, *Mol. Cells* **11**, 129–136.

Boube, M., Joulia, L., Cribbs, D.L., Bourbon, H.M. (2002) Evidence for a mediator of RNA polymerase II transcriptional regulation conserved from yeast to man, *Cell* **110**, 143–151.

Buck, S.W., Sandmeier, J.J., Smith, J.S. (2002) RNA polymerase I propagates unidirectional spreading of rDNA silent chromatin, *Cell* **111**, 1003–1014.

Bushnell, D.A., Kornberg, R.D. (2003) Complete, 12-subunit RNA polymerase II at 4.1-A resolution: Implications for the initiation of transcription, *Proc. Natl. Acad. Sci. U.S.A* **100**, 6969–6973.

Butler, J.E., Kadonaga, J.T. (2002) The RNA polymerase II core promoter: a key component in the regulation of gene expression, *Genes Dev.* **16**, 2583–2592.

Cairns, J. (1963) The bacterial chromosome and its manner of replication as seen by autoradiography, *J. Mol. Biol.* **6**, 208–213.

Cashel, M., Hsu, L.M., Hernandez, V.J. (2003) Changes in conserved region 3 of Escherichia coli sigma 70 reduce abortive transcription and enhance promoter escape, *J. Biol. Chem.* **278**, 5539–5547.

Conaway, R.C., Brower, C.S., Conaway, J.W. (2002) Emerging roles of ubiquitin in transcription regulation, *Science* **296**, 1254–1258.

Dallmann, H.G., Kim, S., Pritchard, A.E., Marians, K.J., McHenry, C.S. (2000) Characterization of the unique C terminus of the *Escherichia coli* τ DnaX protein. Monomeric C-τ binds α and DnaB and can partially replace τ in reconstituted replication forks, *J. Biol. Chem.* **275**, 15512–15519.

Davis, J.A., Takagi, Y., Kornberg, R.D., Asturias, F.A. (2002) Structure of the yeast RNA polymerase II holoenzyme: mediator conformation and polymerase interaction, *Mol. Cells* **10**, 409–415.

Dvir, A., Conaway, J.W., Conaway, R.C. (2001) Mechanism of transcription initiation and promoter escape by RNA polymerase II, *Curr. Opin. Genet. Dev.* **11**, 209–214.

Eissenberg, J.C., Ma, J., Gerber, M.A., Christensen, A., Kennison, J.A., Shilatifard, A. (2002) dELL is an essential RNA polymerase II elongation factor with a general role in development, *Proc. Natl. Acad. Sci. U.S.A.* **99**, 9894–9899.

Ellison, V., Stillman, B. (2001) Opening of the clamp: an intimate view of an ATP-driven biological machine, *Cell* **106**, 655–660.

Falkenberg, M., Gaspari, M., Rantanen, A., Trifunovic, A., Larsson, N.G., Gustafsson, C.M. (2002) Mitochondrial transcription factors B1 and B2 activate transcription of human mtDNA, *Nat. Genet.* **31**, 289–294.

Freiman, R.N., Tjian, R. (2003) Regulating the regulators: lysine modifications make their mark, *Cell* **112**, 11–17.

Fuller, R.S., Kaguni, J.M., Kornberg, A. (1981) Enzymatic replication of *Escherichia coli* chromosome, *Proc. Natl. Acad. Sci. U.S.A.* **78**, 7370–7374.

Funnell, B.E., Baker, T.A., Kornberg, A. (1987) In vitro assembly of a prepriming complex at the origin of the *E. coli* chromosome, *J. Biol. Chem.* **262**, 10327–10334.

Gill, G. (2001) Regulation of the initiation of eukaryotic transcription, *Essays Biochem.* **37**, 33–43.

Gonzalez, F., Delahodde, A., Kodadek, T., Johnston, S.A. (2002) Recruitment of a 19S proteasome subcomplex to an activated promoter, *Science* **296**, 548–550.

Grummt, I. (1999) Regulation of mammalian ribosomal gene transcription by RNA polymerase I, *Prog. Nucleic Acid Res. Mol. Biol.* **62**, 109–154.

Hartzog, G.A. (2003) Transcription elongation by RNA polymerase II, *Curr. Opin. Genet. Dev.* **13**, 119–126.

Hiasa, H. Marians, K.J. (1999) Initiation of bidirectional replication at the chromosomal origin is directed by the interaction between helicase and primase, *J. Biol. Chem.* **274**, 27244–27248.

Hilleren, P., Parker, R. (1999) Mechanisms of mRNA surveillance in eukaryotes, *Annu. Rev. Genet.* **33**, 229–260.

Hirota, Y., Ryter, A., Jacob, F. (1968) Thermosensitive mutants of *E. coli* affected in the processes of DNA synthesis and cellular division, *Cold Spring Harbor Symp. Quant. Biol.* **33**, 677–693.

Huberman, J., Riggs, A.D. (1968) On the mechanism of DNA replication in mammalian chromosomes, *J. Mol. Biol.* **32**, 327–341.

Jacob, F., Brenner, S., Cuzin, F. (1963) On the regulation of DNA replication in bacteria, *Cold Spring Harbor Symp. Quant. Biol.* **28**, 329–348.

Katayama, T., Kurokawa, K., Crooke, E., Sekimizu, K. (1998) The initiator function of DnaA protein is negatively regulated by the sliding clamp of the *E. coli* chromosomal replicase, *Cell* **94**, 61–71.

Kim, D.K., Yamaguchi, Y., Wada, T., Handa, H. (2001) The regulation of elongation by eukaryotic RNA polymerase II: a recent view, *Mol. Cells* **11**, 267–274.

Kitagawa, R., Ozaki, T., Moriya, S., Ogawa, T. (1998) Negative control of replication initiation by a novel chromosomal locus exhibiting exceptional affinity for *E. coli* DnaA protein, *Genes Dev.* **12**, 3032–3043.

Kornberg, T., Gefter, M.L. (1971) Purification and DNA synthesis in cell-free extracts: properties of DNA polymerase II, *Proc. Natl. Acad. Sci. U.S.A.* **68**, 761–764.

Kugel, J.F., Goodrich, J.A. (2002) Translocation after synthesis of a four-nucleotide RNA commits RNA polymerase II to promoter escape, *Mol. Cell. Biol.* **22**, 762–773.

Lee, J., Chastain, P.D., Kusakabe, T., Griffith, J.D., Richardson, C.C. (1998) Coordinated leading and lagging strand DNA synthesis on a minicircular template, *Mol. Cell* **1**, 1001–1010.

Lee, S.K., Yu, S.L., Prakash, L., Prakash, S. (2002) Requirement of yeast RAD2, a homolog of human XPG gene, for efficient RNA polymerase II transcription. Implications for Cockayne syndrome, *Cell* **109**, 823–834.

Maki, H., Maki, S., Kornberg, A. (1988) DNA Polymerase III holoenzyme of *Escherichia coli*. IV. The holoenzyme is an asymmetric dimer with twin active sites, *J. Biol. Chem.* **263**, 6570–6578.

Marahrens, Y., Stillman, B. (1992) A yeast chromosomal origin of DNA replication defined by multiple functional elements, *Science* **255**, 817–823.

Masternak, K., Peyraud, N., Krawczyk, M., Barras, E., Reith, W. (2003) Chromatin remodeling and extragenic transcription at the MHC class II locus control region, *Nat. Immunol.* **4**, 132–137.

Masukata, H., Tomizawa, J. (1990) A mechanism of formation of a persistent hybrid between elongating RNA and template DNA, *Cell* **62**, 331–338.

McHenry, C., Kornberg, A. (1977) DNA polymerase III holoenzyme of *Escherichia coli*. purification and resolution into subunits, *J. Biol. Chem.* **252**, 6478–6484.

Meselson, M., Stahl, F.W. (1958) The replication of DNA in *E. coli*, *Proc. Natl. Acad. Sci. U.S.A.* **44**, 671–682.

Myers, L.C., Kornberg, R.D. (2000) Mediator of transcriptional regulation, *Annu. Rev. Biochem.* **69**, 729–749.

Okazaki, T., Kurosawa, Y., Ogawa, T., Seki, T., Shinozaki, K., Hirose, S., Fujiyama, A., Kohara, Y., Machida, Y., Tamanoid, F., Hozumi, T. (1979) Structure and metabolism of the RNA primer in the discontinuous replication of prokaryotic DNA, *Cold Spring Harbor Symp. Quant. Biol.* **43**, 203–219.

Pillai, B., Verma, J., Abraham, A., Francis, P., Kumar, Y., Tatu, U., Brahmachari S.K., Sadhale, P.P. (2003) Whole genome expression profiles of yeast RNA polymerase II core subunit, Rpb4, in stress and nonstress conditions, *J. Biol. Chem.* **278**, 3339–3346.

Reeve, J.N. (2003) Archaeal chromatin and transcription, *Mol. Microbiol.* **48**, 587–598.

Salinas, F., Benkovic, S.J. (2000) Characterization of bacteriophage T4-coordinated leading- and lagging-strand synthesis on a minicircle substrate, *Proc. Natl. Acad. Sci. U.S.A.* **97**, 7196–7201.

Selleck, W., Howley, R., Fang, Q., Podolny, V., Fried, M.G., Buratowski, S., Tan, S. (2001) A histone fold TAF octamer within the yeast TFIID transcriptional coactivator, *Nat. Struct. Biol.* **8**, 695–700.

Shamoo, Y., Steitz, T.A. (1999) Building a replisome from interacting pieces: sliding clamp complexed to a peptide from DNA polymerase and a polymerase editing complex, *Cell* **99**, 155–166.

Shilatifard, A., Conaway, R.C., Conaway, J.W. (2003) The RNA polymerase II elongation complex, *Annu. Rev. Biochem* **72**, 693–715.

Shim, E.Y., Walker, A.K., Shi, Y., Blackwell, T.K. (2002) CDK-9/cyclin T (P-TEFb) is required in two postinitiation pathways for transcription in the C. elegans embryo, *Genes Dev.* **16**, 2135–2146.

Stukenberg, P.T., Studwell-Vaughan, P.S., O'Donnell, M. (1991) Mechanism of the sliding beta-clamp of DNA polymerase III holoenzyme, *J. Biol. Chem.* **266**, 11328–11334.

Svejstrup, J.Q. (2002) Mechanisms of transcription-coupled DNA repair, *Nat. Rev. Mol. Cell. Biol.* **3**, 21–29.

Tan, Q., Prysak, M.H., Woychik, N.A. (2003) Loss of the Rpb4/Rpb7 Subcomplex in a Mutant Form of the Rpb6 Subunit Shared by RNA Polymerases I, II, and III, *Mol. Cell. Biol.* **23**, 3329–3338.

Taube, R., Lin, X., Irwin, D., Fujinaga, K., Peterlin, B.M. (2002) Interaction between P-TEFb and the C-terminal domain of RNA polymerase II activates transcriptional elongation from sites upstream or downstream of target genes, *Mol. Cell. Biol.* **22**, 321–331.

Tekotte, H., Davis, I. (2002) Intracellular mRNA localization: motors move messages, *Trends Genet.* **18**, 636–642.

Tougu, K., Marians, K.J. (1996) The interaction between helicase and primase sets the replication fork clock, *J. Biol. Chem.* **271**, 21398–21405.

Tsurimoto, T., Melendy, T., Stillman, B. (1990) Sequential initiation of lagging and leading strand synthesis by two different polymerase complexes at the SV40 DNA replication origin, *Nature* **346**, 534–539.

Waga, S., Stillman, B. (1994) Anatomy of a DNA replication fork revealed by reconstitution of SV40 DNA replication *in vitro*, *Nature* **369**, 207–212.

Weinreich, M., Liang, C., Stillman, B. (1999) The Cdc6p nucleotide-binding motif is required for loading mcm proteins onto chromatin, *Proc. Natl. Acad. Sci. U.S.A.* **96**, 441–446.

Wyrick, J.J., Aparicio, J.G., Chen, T., Barnett, J.D., Jennings, E.G., Young, R.A., Bell, S.P., Aparicio, O.M. (2001) Genome-wide distribution of ORC and MCM proteins in *S. cerevisiae*: high-resolution mapping of replication origins, *Science* **294**, 2357–2360.

Xiao, T., Hall, H., Kizer, K.O., Shibata, Y., Hall, M.C., Borchers, C.H., Strahl, B.D. (2003) Phosphorylation of RNA polymerase II CTD regulates H3 methylation in yeast, *Genes Dev.* **17**, 654–663.

Yamazaki, K., Aso, T., Ohnishi, Y., Ohno, M., Tamura, K., Shuin, T., Kitajima, S., Nakabeppu, Y. (2003) Mammalian elongin A is not essential for cell viability but required for proper cell cycle progression with limited alteration of gene expression, *J. Biol. Chem.* **278**, 13585–13589.

Young, B.A., Anthony, L.C., Gruber, T.M., Arthur, T.M., Heyduk, E., Lu, C.Z., Sharp, M.M., Heyduk, T., Burgess, R.R., Gross, C.A. (2001) A coiled-coil from the RNA polymerase β' subunit allosterically induces selective nontemplate strand binding by σ^{70}, *Cell* **105**, 935–944.

Zyskind, J.W., Smith, D.W. (1980) Nucleotide sequence of the *S. typhimurium* origin of DNA replication, *Proc. Natl. Acad. Sci. U.S.A.* **77**, 2460–2464.

3
Translation of RNA to Protein

Robert Ashley Cox and Henry Rudolph Victor Arnstein
National Institute for Medical Research, London, England, UK

1	**Introduction** 78	
2	**mRNA Structure and the Genetic Code** 80	
2.1	Structure 80	
2.2	Prokaryotic mRNA 81	
2.3	Eukaryotic mRNA 81	
2.4	The Genetic Code 82	
2.4.1	Deviations from the Standard Genetic Code 82	
3	**Transfer RNA (tRNA)** 83	
4	**Ribosome Structure and Function in Translation** 86	
4.1	Ribosome Structure 86	
4.2	The Ribosome Cycle in Translation 90	
4.2.1	Formation of Preinitiation Complexes 90	
4.2.2	Initiation Complex Formation: Joining of the Large Ribosomal Subunit 94	
4.2.3	Polypeptide Chain Synthesis: The Elongation–Translocation Cycle 95	
4.3	High-resolution Structural Studies of the Ribosome 100	
5	**Translational Control of Gene Expression** 104	
5.1	mRNA Stability 105	
5.2	Control by Interaction of Proteins with mRNA 107	
5.3	Control by mRNA Structure 108	
5.4	Control by Modification of Translation Factor Activity 108	
5.4.1	Initiation Factors 108	
5.4.2	Other Translation Factors 111	
5.5	Effects of Antisense Polynucleotides 111	
5.6	Availability of Amino Acids, tRNA Abundance, and Codon Usage 111	

Genomics and Genetics. Edited by Robert A. Meyers.
Copyright © 2007 Wiley-VCH Verlag GmbH & Co. KGaA, Weinheim
ISBN: 978-3-527-31609-0

5.6.1	Amino Acids 111
5.6.2	Abundance of tRNAs and Codon Usage 111
5.7	Modulation of Ribosome Activity 111
5.8	Ribosome-inactivating Proteins 112
6	**Concluding Remarks** 112
	Bibliography 113
	Books and Reviews 113
	Primary Literature 113

Keywords

Anticodon
Three consecutive bases in tRNA that bind to a specific mRNA codon by complementary antiparallel base pairing.

Antiparallel Base pairing
Pairing through specific hydrogen bonds between base residues of two polynucleotide chains or two segments of a single chain with phosphodiesterbonds running in the $5' \rightarrow 3'$ direction in one chain or segment and in the $3' \rightarrow 5'$ direction in the other. In DNA and RNA, the hydrogen bonds are usually formed between complementary base pairs (adenine (A) with either thymine (T) or uracil (U), and guanine (G) with cytosine (C)).

Codon
Three consecutive bases in mRNA or DNA, which code for an amino acid in protein synthesis.

Elongation
The stepwise addition of amino acids to the carboxyl terminus of a growing polypeptide chain.

Fidelity
The accuracy with which the RNA sequence is translated into the correct amino acid sequence.

Initiation
A multistep reaction between ribosomal subunits, charged initiator transfer RNA and messenger RNA that results in apposition of the ribosome-bound initiator Met-tRNA with an AUG initiator codon in mRNA. In this position, the ribosome is poised to form the first peptide bond.

Polarity
The asymmetry of a polynucleotide or polypeptide. In DNA, the two strands have opposite polarity; that is, they run in opposite directions (5′ → 3′ and 3′ → 5′). The polarity of a polypeptide is defined as running from the N-terminus to the C-terminus.

Reading Frame
One of three possible ways of translating groups of three nucleotides in mRNA. The appropriate reading frame is determined by the initiation codon.

Template Strand
The strand of the DNA double helix that is used as a template for transcription of RNA. It has a base sequence complementary to the RNA transcript.

Termination
The end of polypeptide synthesis, which is signaled by a codon for which there is no corresponding aminoacyl tRNA. When the ribosome reaches a termination codon in the mRNA, the polypeptide is released and the ribosome–mRNA–tRNA complex dissociates.

Translation
The stepwise synthesis of a polypeptide with an amino acid sequence determined by the nucleotide sequence of the mRNA coding region. The genetic code relates each amino acid to a group of three consecutive nucleotides termed a *codon*. Decoding of mRNA takes place in the 5′ → 3′ direction, and the polypeptide is synthesized from the amino to the carboxyl terminus.

Translocation
The stepwise advance of a ribosome along the mRNA, one codon at a time, with simultaneous transfer of peptidyl tRNA from the A site to the P site of the ribosome.

■ Proteins have a linear primary sequence of no more than 20 encoded amino acids and this underlying simplicity is used to express great subtlety in structure and versatility in function. For these reasons, proteins are essential to the structure and function of living cells. The assembly of polypeptide chains from amino acids and their subsequent modifications, leading to the final three-dimensional protein structure, are exceptionally complex processes; many components are involved and much of the cell's energy is utilized. Each peptide bond requires the expenditure of four high-energy phosphate bonds (ATP is converted to AMP and two molecules of GTP are converted to GDP). This value excludes the energy used for initiation and release of the polypeptide chains and the cost of synthesizing and processing mRNA. The linear amino acid sequence of a protein is encoded within the gene as a linear deoxyribonucleotide sequence. Early steps in the biosynthesis of a protein include transcription of the gene and appropriate processing of the transcript leading to the production of mature messenger RNA (mRNA). We describe the mechanisms involved in translating mRNA to produce a polypeptide chain which has the amino acid sequence specified by the gene. Translation takes place after mRNA is bound to small ribonucleoprotein particles called *ribosomes* through the mediation of aminoacyl tRNA which links the linear nucleotide sequence of mRNA with the linear amino acid sequence of the encoded polypeptide. The genetic message is read sequentially as mRNA moves relative to the ribosome. Thus, the polypeptide chain is formed by the sequential incorporation of amino acids into peptide linkage. The process of translation is achieved with high fidelity with less than one error per 10 000 amino acids incorporated into protein. The mechanisms of protein synthesis are essentially the same for all cells irrespective of whether they have a nucleus (eukaryotes) or not (prokaryotes).

1
Introduction

A gene or cistron is defined as the region of DNA that is transcribed into a functional RNA. The transcript functions either as such (e.g. tRNA, rRNA, snRNA) or as a messenger (mRNA), which after processing or editing as required, normally codes for one or more polypeptide chains in the translation process. A polynucleotide such as RNA is an asymmetrical polymer assembled from nucleoside triphosphates by a stepwise mechanism linking the 5′ position of one nucleotide by a phosphate bridge to the 3′ position of the adjacent nucleotide. In the finished polynucleotide chain, the first nucleotide residue has a 5′ position, which is not linked to another nucleotide, whereas the last nucleotide has an unlinked 3′ position. Thus, polynucleotide synthesis proceeds from the 5′ to the 3′ terminus and the polymer is said to have a 5′ to 3′ polarity. Usually, linear RNA sequences are written with the 5′ terminus on the left and the 3′ terminus on the right (Fig. 1a). Within the RNA chain, some bases may form antiparallel base pairs (Fig. 1b, as found in DNA).

The polypeptide chains of proteins are also asymmetrical polymers in which the amino acid residues are linked by peptide bonds between their α-amino and carboxyl groups (Fig. 2), leaving a free α-amino group at one end (the amino terminus) of the polymer and a free α-carboxyl group at the opposite end (the carboxyl terminus). The significance of the polarity of the RNA and proteins will become evident when the process of protein biosynthesis is explained below.

The genetic information stored in DNA is not usable directly for making proteins. Rather, it must be copied into a primary RNA transcript containing the coding sequences by an enzymatic transcription of segments of DNA containing the genes. In prokaryotes, the primary transcript is also the messenger (mRNA); that is, it can be used directly in polypeptide synthesis. In contrast, in eukaryotes the primary transcript (precursor-mRNA) is often much larger in size than the mature mRNA and requires extensive processing involving the excision of intervening and other noncoding sequences.

Messenger RNA serves as the template for protein synthesis; that is, the linear

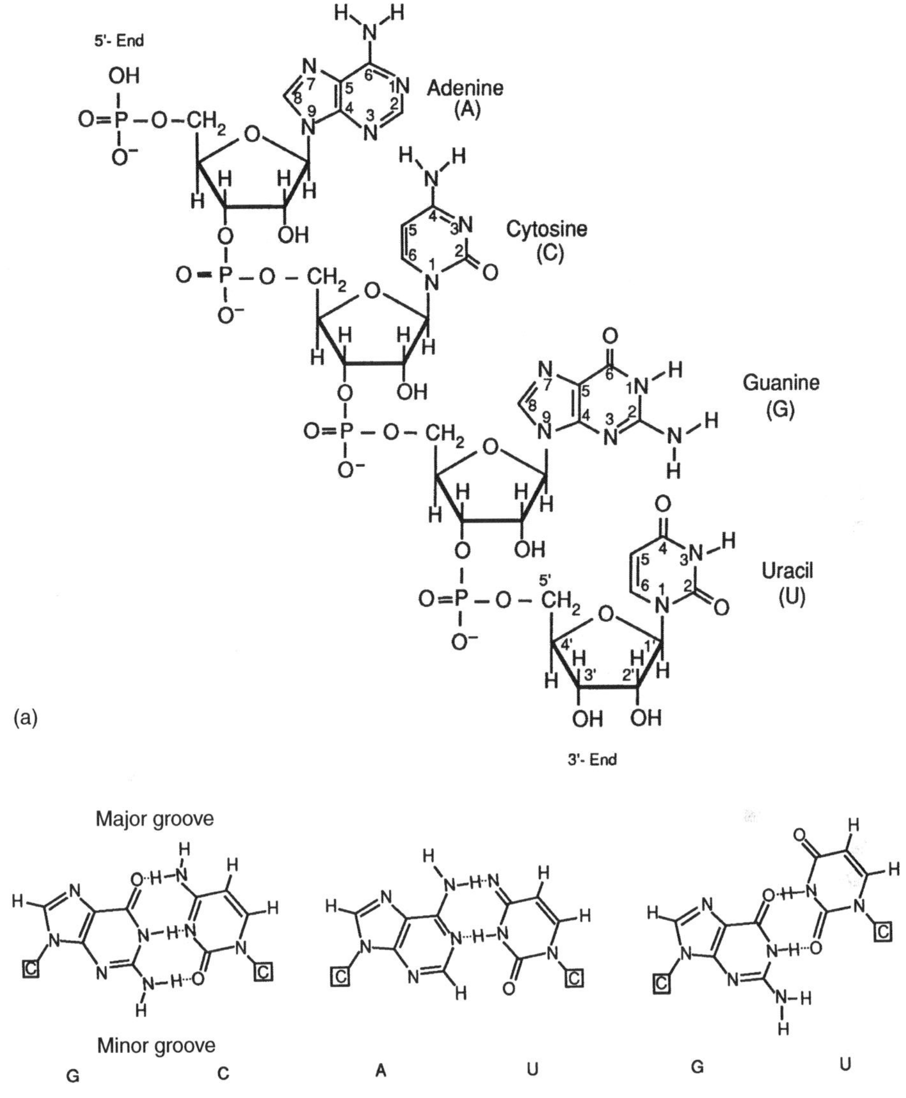

Fig. 1 Structural elements of ribonucleic acids. (a) Primary structure indicating the numbering system for purines, pyrimidines, and ribose. (b) Base-pairing interactions commonly found in RNA: (i) G + C base pair; (ii) A + U base pair; (iii) G + U base pair. Pairs (i) and (ii) are Watson and Crick base pairs; (iii) is a special type of base pairing found in intramolecular bihelical regions. The minor groove is on the side of the base pair with the glycosidic bond of which the carbon atom C 1′ of ribose is boxed. Note that DNA contains base pairs of types i and ii only but with thymine (5-methyluracil) in place of uracil. [From Arnstein, H.R.V., Cox, R.A. (1992) *Protein Biosynthesis*, Oxford University Press, Oxford. With permission.]

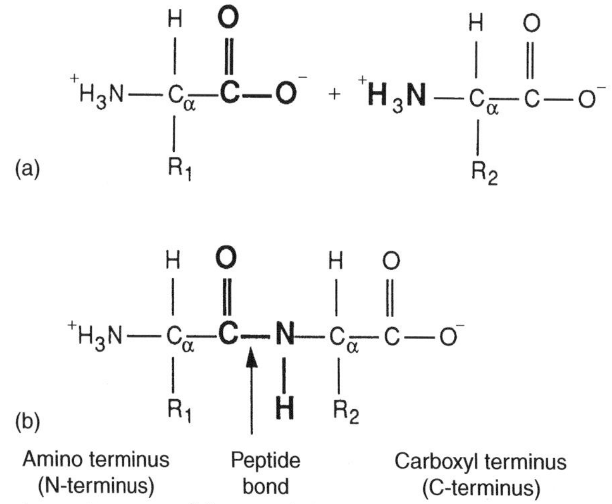

Fig. 2 Structure of the peptide bond. (a) Two L-amino acids with different side chains, R_1 and R_2; (b) a dipeptide formed from the two amino acids shown in (a).

nucleotide sequence of the mRNA dictates the amino acid sequence of the polynucleotide encoded originally by the gene. Conventionally, gene and mRNA nucleotide sequences are written in the 5′ to 3′ direction, which corresponds to the direction in which mRNA is decoded during polypeptide synthesis: The mRNA is read in the 5′ to 3′ direction, and the polypeptide is synthesized from the amino toward the carboxyl terminus.

The mechanism whereby RNA is translated into protein is complex, and the cell devotes considerable resources to the translational machinery. The components include 20 different amino acids, as well as transfer RNAs, aminoacyl tRNA synthetases, ribosomes, and a number of protein factors which cycle on and off the ribosomes and facilitate various steps in initiation of translations, elongation of the nascent polypeptide chain, and termination of synthesis with release of the completed polypeptide from the ribosome. The process depends on a supply of energy provided by ATP and GTP. The rate of protein synthesis is typically in the range of 6 (immature red blood cells of the rabbit) to 20 (*Escherichia coli* growing optimally) peptide bonds per second at 37 °C.

2
mRNA Structure and the Genetic Code

2.1
Structure

The sequence information of a gene that is copied (transcribed) into the nucleotide sequence of RNA from the complementary strand of DNA is called *the template strand*. The primary transcript is a single strand of RNA, which is a faithful copy of the nontemplate strand of DNA with substitution of U residues in place of T residues found in DNA. Sometimes, the primary transcript is altered, as described below, before it functions as mRNA; in these cases, the original unmodified transcript is the precursor or pre-mRNA. Usually, mRNAs have nontranslated sequences at the 5′ and

3' ends in addition to the coding domain. These noncoding sequences sometimes affect the efficiency of translation and the stability of mRNA. The structures of typical mRNAs are shown in Fig. 3. The decoding process involves base pairing between three bases (designated a codon) in the mRNA and the three-base anticodon of a transfer RNA (tRNA). In a separate reaction, each tRNA is first linked to a particular amino acid and thus the pairing of mRNA with tRNA determines the sequence of amino acids in the resulting protein.

2.2
Prokaryotic mRNA

In prokaryotes, pre-mRNA usually undergoes little or no modification so that pre-mRNA and mRNA are very similar if not identical. Because pre-mRNA is colinear with DNA, DNA and proteins are usually colinear in these organisms. Gene expression in prokaryotes usually involves the cotranscription of several adjacent genes, and translation of mRNA sequences into polypeptides may begin at the 5' end of mRNA while transcription is still in progress at the 3' end.

2.3
Eukaryotic mRNA

In eukaryotes, the genetic information is stored mainly in the nucleus and to a minor degree in some organelles (mitochondria and chloroplasts). The description that follows pertains only to nuclear genes. Eukaryotic genes are more complicated than prokaryotic genes because the coding region is often discontinuous:

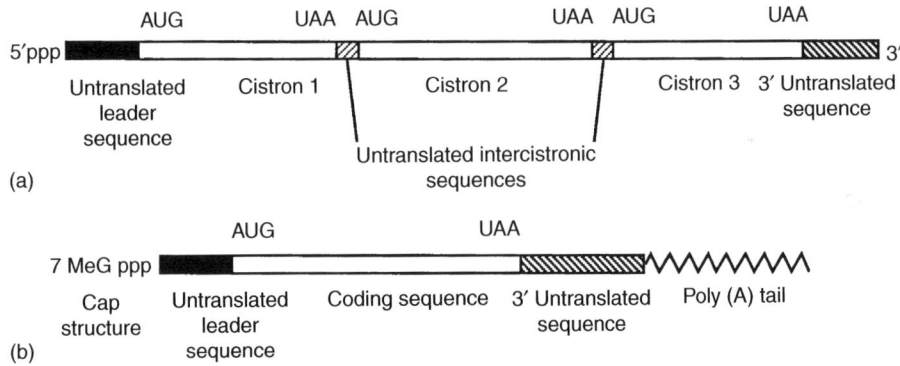

Fig. 3 Structure of typical mRNAs.
(a) *Prokaryotic mRNA*. At the beginning of the transcript, there are two additional phosphate residues linked by pyrophosphate bonds to the 5' phosphate group of the terminal nucleotide (see Fig. 1). The 5' untranslated sequence often contains a ribosome-binding site (the Shine–Dalgarno sequence), which increases the efficiency of translation. AUG and UAA are representative initiation and termination codons, respectively. Cistrons (gene sequences) are separated by short (typically 12 to 24 nucleotides) noncoding intercistronic sequences. (b) *Eukaryotic mRNA*. Typically, the processed transcripts are monocistronic. The 5' end is usually modified by the addition of 7-MeG, known as a cap structure, which is linked by two pyrophosphate groups to the terminal nucleotide of mRNA. The coding sequence is located between 5' and 3' untranslated sequences. In the majority of cases, the 3' end is modified by the addition of 25 to 250 adenylate residues, termed the poly(A) tail. [From Arnstein, H.R.V., Cox, R.A. (1992) *Protein Biosynthesis*, Oxford University Press, Oxford. With permission.]

the coding sequences or exons are interrupted by intervening sequences (introns). Thus, genes and proteins are usually *not* colinear in eukaryotes. In the nucleus, a complicated set of splicing reactions removes all the introns from the pre-mRNA and fuses the exons into a continuous coding sequence. Other processing steps involve adding a "cap" to the 5' end of the mRNA and a poly(A) "tail" to the 3' end (see Fig. 3b). After completion of these nuclear maturation steps, the mRNA is transported to the cytoplasm where it is translated. As is the case with prokaryotic mRNA, the coding region is flanked by 5' and 3' nontranslated sequences.

2.4
The Genetic Code

The genetic code is triplet, comma-less, and nonoverlapping. As a consequence, a nucleotide sequence has three possible reading frames (Fig. 4). Because mRNA is normally translated into a unique polypeptide, an essential step in the translation process is the selection of the appropriate reading frame. This is achieved by starting translation at the initiation codon, usually AUG or less frequently GUG, which ensures that the following codons are read in phase within the required reading frame.

Of the 64 theoretically possible triplets in the genetic code, 61 sense codons correspond to 20 genetically encoded amino acids found in all, or nearly all, proteins. When GUG is used as the initiation codon, it codes for methionine by interaction with the anticodon of the initiator Met-tRNA^{Met}, whereas elsewhere it codes for valine. All other codons specify only one amino acid but many amino acids are specified by two or more (up to six) codons; the code is unambiguous, but degenerate. The remaining three codons, termed *nonsense codons*, usually signify termination of synthesis and release of the finished polypeptide chain.

2.4.1 Deviations from the Standard Genetic Code

One of the nonsense codons, UGA, has an additional function in the synthesis of selenoproteins. The process involves the initial synthesis of selenocysteyl-tRNA from a novel seryl-tRNA and selenium. This tRNA contains an anticodon that is able to decode UGA and insert selenocysteine residues into the growing polypeptide chain, but only at UGA codons in a particular context of neighboring nucleotides. The insertion of selenocysteine residues into the polypeptide also requires a specific elongation factor T, which differs from the factor used for the

```
     5'                              3'
    |AUG|GUA|UUC|AG |...
(a)  fMet   Val   Phe   —

     5'                              3'
     A|UGG|UAU|UCA|G . |...
(b)     Trp   Tyr   Ser   —

     5'                              3'
     AU|GGU|AUU|CAG|...
(c)     Gly   Ile   Gln
```

Fig. 4 Translation of a polynucleotide sequence into three alternative polypeptides using different reading frames.

incorporation of other aminoacyl tRNAs (see Sect. 4.2.3).

Mitochondria and chloroplasts, as well as certain organisms such as mycoplasma and ciliated protozoa, use a few non-standard genetic codons. For example, methionine is usually coded for by AUG (or occasionally GUG), but in human mitochondria this codon is replaced by AUA. Variations in the genetic code are thought to have arisen as a result of the loss of some tRNA genes and mutational pressure on DNA, giving rise to a predominance of either AT- or GC-rich codons.

3
Transfer RNA (tRNA)

Transfer RNA nomenclature: The amino acid linked to a charged tRNA is indicated by a prefix and the specificity of the tRNA in the aminoacylation reaction is shown as a superscript on the right; for example, Phe-tRNA^{Phe} indicates phenylalanine-specific tRNA charged with phenylalanine. The anticodon may be indicated as a right subscript or, alternatively, in the superscript after the amino acid; for example, tRNA$_{UGC}$ or tRNA$^{Ala/UGC}$. The right-hand subscript position is sometimes used to indicate the organism from which the tRNA is derived (e.g. tRNA$^{Val}_{yeast}$). The initiator tRNA, which is specific for methionine, is termed Met-tRNA$^{Met}_f$ or Met-tRNA$^{Met}_i$. Often the superscript ^{Met} is omitted. In prokaryotes and in the mitochondria of eukaryotes, the methionine residue of the charged initiator tRNA is formylated by transformylase using N^{10}-formyltetrahydrofolate as the donor, giving N-formylmet-tRNA$_f$. Commonly, this charged tRNA is termed fMet-tRNA$_f$. The methionine-specific elongator RNA, which inserted methionine into internal positions of the growing peptide chain is termed tRNA$^{Met}_m$ (or tRNA$_m$) when uncharged, and Met-tRNA$^{Met}_m$ (or Met-tRNA$_m$) when charged. By relating individual codons of mRNA to the cognate amino acids, tRNA functions as a key bilingual intermediate in the translation of the genetic code. All tRNAs are single-stranded molecules about 80 nucleotides long with a common 3′ terminal CCA sequence. Most of the bases are standard but some (e.g. pseudoU, dihydroU, and T) are derived by modification after transcription of the transfer RNA genes.

The secondary structure of tRNA is usually presented in two dimensions as a cloverleaf to highlight the regions of base pairing (Fig. 5a). X-ray crystallography reveals that additional hydrogen bonds give rise to an L-shaped tertiary structure (Fig. 5b). The CCA sequence carrying the amino acid is located distal to the anticodon.

The decoding process involves antiparallel base pairing between the three bases of mRNA codons and the complementary anticodons of tRNA during peptide bond formation (Fig. 6). The first and middle bases of the codon form conventional base pairs with the third and middle base of the anticodon, respectively, but the third base of the codon pairs with the first base of the anticodon by a less stringent interaction (e.g. base pairing of G with U as well as with C), giving rise to degeneracy. This so-called "wobble" considerably reduces the number of tRNA species required to decode the 61 sense codons. Thus, the protein synthesis system in the cytosol of eukaryotes contains only a few more than 40 different tRNAs, and in mitochondria, 22 to 24 species are sufficient.

The attachment of amino acids to tRNA involves the formation of an ester bond between the α-carboxyl group of the amino acid and the 3′ hydroxyl group of the terminal adenosine of tRNA. It requires specific enzymes, the aminoacyl tRNA synthetases. There are 20 different synthetases, each specific for one of the 20 amino acids, and each enzyme recognizes something unique in the structure of its cognate tRNA. The structural determinants that ensure accuracy of this

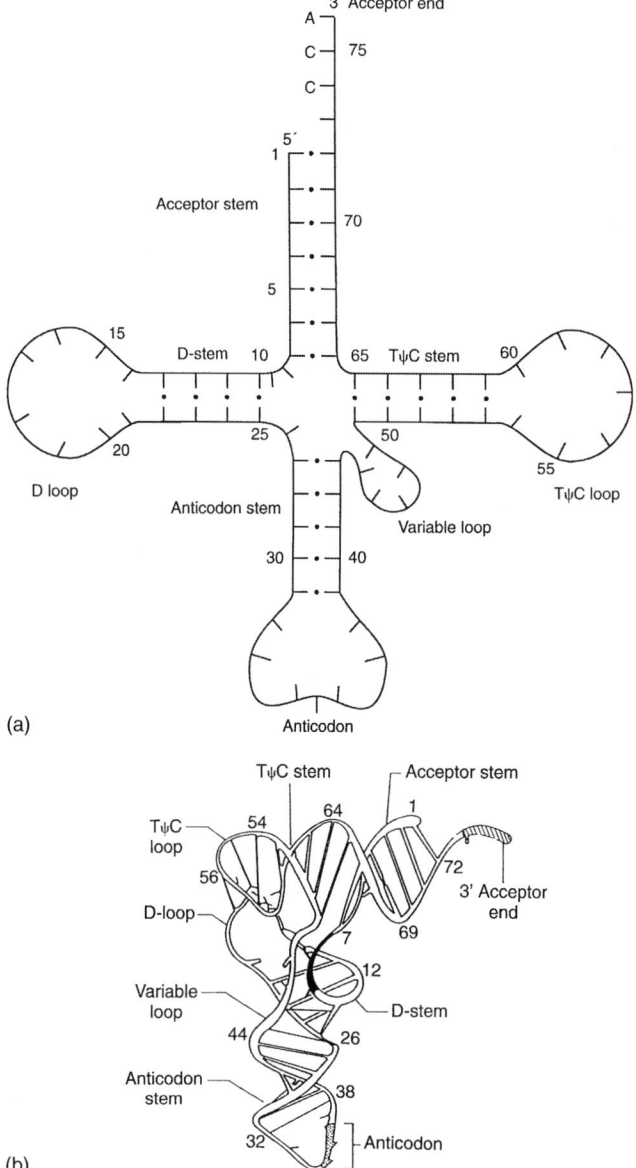

charging reaction vary for different tRNAs. The anticodon may play a part but sometimes even a single base elsewhere is sufficient to determine the specificity of the tRNA-synthetase interaction. The accuracy of the synthetase reaction in attaching an amino acid to its cognate tRNA is critically important to the fidelity of the

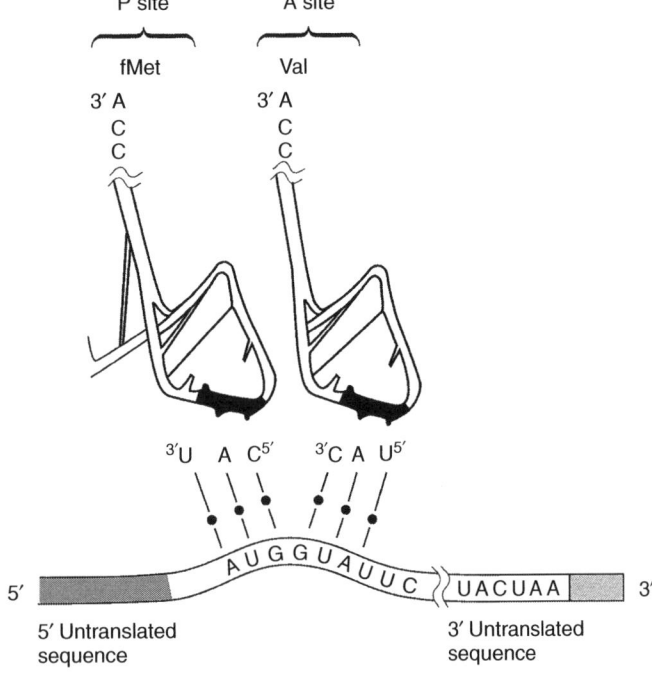

Fig. 6 Schematic illustration of base pairing between a codon and its anticodon. The diagram shows the interaction at the P site of the ribosome (see Fig. 9) between the initiation codon AUG of mRNA and the anticodon CAU of fMet-tRNA$_f^{Met}$ and the interaction at the A site between the codon GUA of mRNA and the anticodon UAC of Val-tRNA^{Val}. The polarity of an RNA species runs from the 5′ end to the 3′ end. The fragment of tRNA is representative of the general structure (see Fig. 5b) placed in the appropriate orientation. The interaction between the codon and the anticodon is antiparallel, and the three base pairs have a bihelical conformation. The shaded regions of mRNA denote untranslated sequences.

Fig. 5 Structure of phenylalanyl-transfer RNA. (a) Secondary cloverleaf structure. The 5′ and 3′ ends of the molecule are marked; the continuous line represents the sugar phosphate backbone. The short lines denote base residues and the dots denote base pairing through standard hydrogen bonds (see Fig. 1). The D loop contains dihydrouracil residues; the TψC loop contains thymine and pseudouridine. (b) Tertiary structure. The abbreviations for unusual bases are defined in (a). The sugar phosphate backbone is represented by double parallel lines. Standard base pairs are represented by short double parallel lines, and nonstandard base pairs by single lines. The anticodon sequence is stippled, and the acceptor end is shaded. [From Arnstein, H.R.V., Cox, R.A. (1992) *Protein Biosynthesis*, Oxford University Press, Oxford. With permission].

translation process. Once the aminoacyl tRNA has been formed, the subsequent incorporation of the amino acid residue into a polypeptide does not depend on the amino acid itself but only on the interaction between the anticodon of the aminoacyl tRNA with the codon of mRNA. Thus, an error in the synthetase reaction would lead to the incorporation of an inappropriate amino acid into the polypeptide.

Synthesis of aminoacyl tRNA (III) from amino acids (I) requires activation of the amino acid carboxyl group with formation of an intermediate enzyme-bound aminoacyladenylate (II) (Scheme 1).

The energy for the reaction (two high-energy phosphate bonds) is provided by ATP and stored in the ester bond of the aminoacyl tRNA to be used subsequently for peptide bond synthesis.

4
Ribosome Structure and Function in Translation

4.1
Ribosome Structure

Ribosomes are high molecular–weight complexes of RNA (rRNA) and proteins (Table 1), and the electron-dense particles are easily visualized within cells by electron microscopy (see Fig. 7). Usually, several

$$E + R\text{-}CH(NH_3^+)\text{-}CO_2^- + ATP \rightarrow E \bullet R\text{-}CH(NH_3^+)\text{-}CO \bullet AMP + PP_i \rightarrow R\text{-}CH(NH_3^+)\text{-}CO \bullet tRNA + AMP + E$$
$$\downarrow 2P_i$$

(I)   (II)   (III)

Where E = aminoacyl tRNA synthetase

Scheme 1

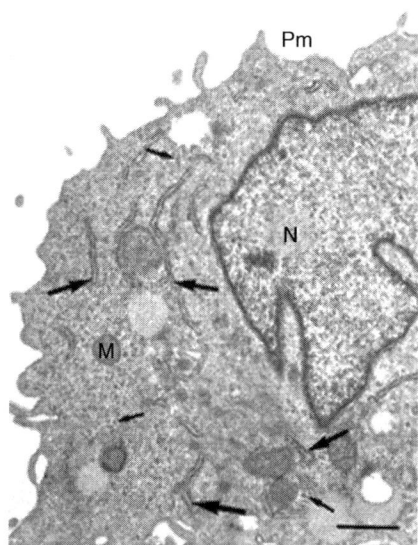

Fig. 7 Electron micrograph of a thin section through part of an epithelial cell showing plasma membrane (Pm), nucleus (N), mitochondria (m), and rough endoplasmic reticulum (large arrows) to which are attached numerous polysomes. Groups of free ribosomes (small arrows) occur in the cytoplasm. Bar represents 1000 nm. [Micrograph by I. D. J. Burdett, National Institute for Medical Research, London, UK.]

ribosomes become attached to one mRNA molecule giving rise to polyribosomes (also called *polysomes*; see Fig. 7). Ribosomes from various sources (prokaryotes, eukaryotic cytoplasm, mitochondria, chloroplasts, and kinetoplasts) vary in size from 20 to 30 nm in diameter, but all are composed of a large and a small subparticle or subunit and perform similar functions in protein synthesis. In general, cytoplasmic ribosomes of eukaryotes are larger than ribosomes of prokaryotes (Table 1). The principal functional domains of the ribosome and associated components are given in Fig. 8. More detailed resolution of the ribosome structure has allowed the

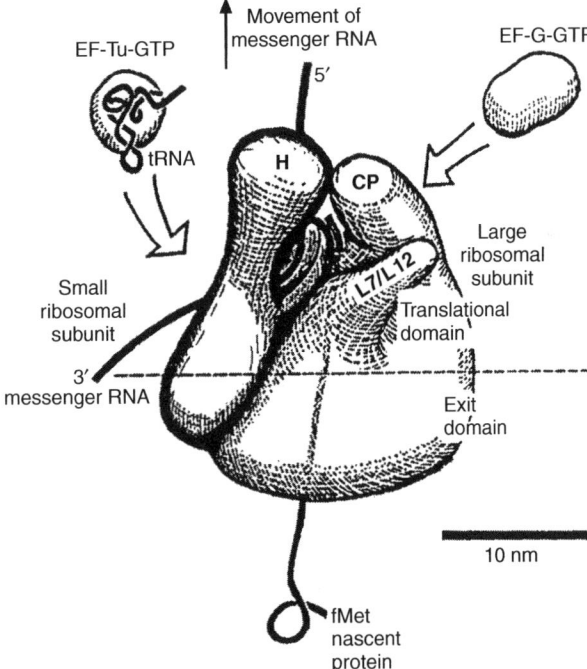

Fig. 8 Model of the *E. coli* ribosome, based on low resolution electron microscopy. The diagram shows the relative orientation of the large and small subunits and other functional components involved in polypeptide chain synthesis. H, head of the small subunit; CP, central protuberance and L7/L12 stalk of the large subunit; EF-Tu, elongation factor required for binding aminoacyl tRNA to the ribosomal A site; EF-G, elongation factor required for translocation of the peptidyl-tRNA from the A to the P site. The broken line indicates the boundary between the translational and exit domains that are involved in peptide bond formation and extrusion of the nascent polypeptide chain, respectively. Further details of ribosome structure and function are given in Fig. 11 for peptide bond formation, in Figs. 12 and 13 for the delivery of aminoacyl tRNA to the A site and Figs. 14 and 15 for translocation and Fig. 16 for the location of the nascent polypeptide chain. [From Arnstein, H.R.V., Cox, R.A. (1992) *Protein Biosynthesis*, Oxford University Press, Oxford. With permission.]

Tab. 1 Properties of ribosomes and ribosomal subunits.

Source	$S_{20,w}$	Size [nm][a]	Mass [Mda]	RNA:protein [w/w]	Axial ratio	rRNA mass [Mda]	$S_{20,w}$ (nominal)
Ribosomes							
Prokaryotes							
E. coli	70	22.5 ± 2.5	2.6–2.9	2:1	–[f]		
Eukaryotes							
Cytoplasm	80	28.0 ± 2.8	3.4–4.5	1:1	–		
Chloroplast	70	22.5 ± 2.5	2.5–3.3	1:1	–		
Mitochondria[b]							
Protists	60–80		ca. 3.25	1:1.38	–		
Animals	55–60		2.7–3.2	1:1.7–1:4	–		
Fungi	67–80		4.2	1:1.13–1:1.6	–		
Higher plants	78		–	–	–		
Small ribosomal subunits							
Prokaryotes							
E. coli	30	22.0 ± 2.2	0.95	1:1	2:1	0.6	16
Eukaryotes							
Cytoplasm	36–41	25.0 ± 2.5	1.4	1:1	2:1	0.7	18
Chloroplast	28–35	22.0 ± 2.2	1.2	1:1	2:1	0.6	16
Mitochondria[b]							
Protists	32–45		–	–	–	0.2–0.35	11–16
Animals	32–40		1.1–1.6	–	–	0.31	12–14
Fungi	32–29		1.6–1.7	–	–	0.45–0.64	15–19
Higher plants	44		–	–	–	0.64	18

Large ribosomal subunits						
Prokaryotes						
E. coli	50	22.5 ± 2.2	1.75	2:1	1:1	1.17 / 0.04	23 / 5
Eukaryotes						
Cytoplasm	60	28.0 ± 2.8	2.1–3.1	1:1	1:1	1.2–1.75 / 0.05 / 0.4 / 1.1 / 0.04 / 0.03[d]	25–28 / 5.8 / 5 / 23 / 5 / 4.55[d]
Chloroplast	46–54	22.5 ± 2.2	2.4	1:1	1:1		
Mitochondria[b]							
Protists	45–60	—	—	—	—	0.4–0.74	12–24
Animals	25–35	1.65–2.0	—	—	—	0.5	16–21
Fungi	50	—	—	—	—	0.77–1.22	21–25
Higher plants	60	—	—	—	—	1.25 / 0.04[e]	26 / 5[e]

[a] Largest dimension.
[b] Mitochondrial ribosomes are preferentially associated with the inner mitochondrial membrane and are more diverse than cytoplasmic or chloroplast ribosomes. Particular features include posttranscriptional oligoadenylation of the 3′ ends of both the smaller and larger rRNA components. Except for mitochondrial ribosomes of higher plants, the essential sequence motifs 5S rRNA and 5.8S rRNA are present in the large rRNA component and are not found as individual species.
[c] Mitochondria of protists such as *Trypanosoma brucei* are also known as kinetoplasts.
[d] 4.5S rRNA corresponding to the 100 nucleotides at the 3′ end of the 23S rRNA in eubacteria occurs as a separate species in chloroplasts of higher plants.
[e] This species is present only in plant mitochondria and is absent from all other mitochondrial ribosomes.
[f] Data not available.

placement of mRNA, aminoacyl tRNA, peptidyl tRNA, and the nascent polypeptide chain (see Sect. 4.3).

The small subunit comprises a single rRNA of $0.3–0.7 \times 10^6$ Da and single copies of 20 to 30 unique proteins. It has a major function in binding initiator tRNA and mRNA in the initiation of protein synthesis and in decoding the genetic message.

The large subunit comprises a high molecular–weight rRNA ($0.6–1.7 \times 10^6$ Da) and often one or two smaller rRNAs ($0.03–0.05 \times 10^6$ Da) and 30 to 50 different proteins are present, with one exception, as single copies. The large subunit binds aminoacyl tRNA at the A site, peptidyl tRNA at the P site, and discharged tRNA at the E (exit) site. The large subunit contains the peptidyl transferase center and, unusually, this enzyme activity resides in the rRNA molecule itself rather than in the associated ribosomal proteins. This subunit is also involved in binding elongation factor G, which is required for translocation (Fig. 8).

4.2
The Ribosome Cycle in Translation

Polypeptide synthesis can be divided into three stages: initiation, elongation, and termination. Initiation involves the binding of a ribosome to mRNA with the initiation codon correctly placed in the P site. Elongation leads to the stepwise increase in the length of the polypeptide chain through the transfer of the growing chain to the amino group of aminoacyl tRNA. Termination of chain elongation and release of the completed polypeptide occurs when a termination codon reaches the A site. All stages require the participation of protein factors. Advances in establishing the structures of translational factors by X-ray crystallography have been rapid in the last decade.

4.2.1 Formation of Preinitiation Complexes

The ribosome cycle starts with the stepwise formation of an initiation complex from mRNA, charged initiator tRNA, and ribosomal subunits. A number of preinitiation complexes are formed as intermediates and the process is facilitated by initiation factors. In outline, prokaryotic and eukaryotic systems are similar, but there are a few differences, particularly as regards the complexity of the initiation factors and details of the mechanisms.

Prokaryotic systems. Three proteins, initiation factors IF-1, IF-2, and IF-3 (see Table 2), are required for the initiation of protein biosynthesis (see Fig. 9a). Ribosomal subunits are released by dissociation of ribosomes following translation of the mRNA. Dissociation is facilitated by the combined action of the initiation factors IF-1 and IF-3; IF-1 increases the rate of dissociation and IF-3 acts as an antiassociation factor when bound to the (30S) ribosomal subunit, thereby displacing the equilibrium in favor of subunit formation. Initiation factor IF-2 is also able to bind to the small 30S subunit and this association is stabilized by IF-1 and GTP, the latter acting as a steric effector without being hydrolyzed at this stage. IF-2 plays a central role in binding fMet-tRNA$_f$ to the 30S preinitiation complex by specific recognition of the N-formylmethionine residue attached to the initiator tRNA, thus restricting this interaction to charged initiator tRNA. All three factors bind to the 30S ribosomal subunit near the 3' end of the 16S ribosomal RNA at adjacent sites that are

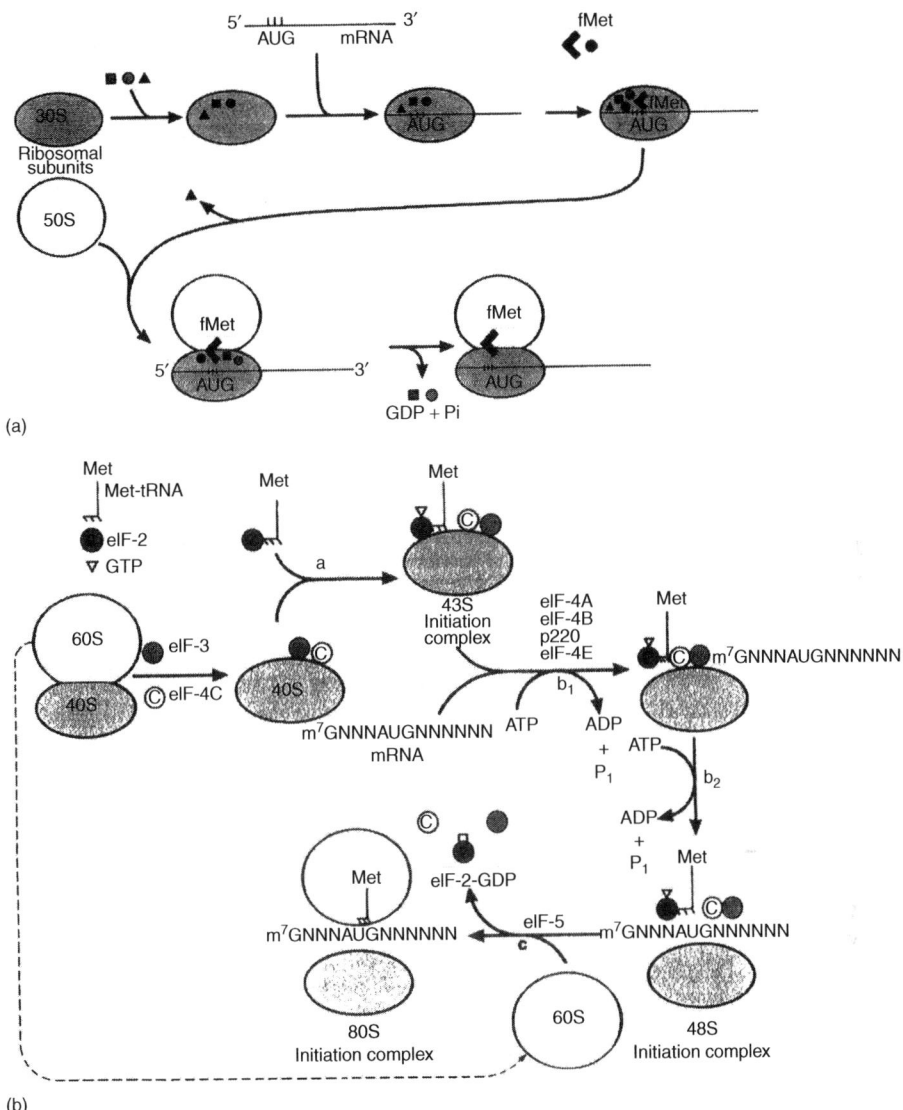

Fig. 9 Schematic diagram illustrating the formation of initiation complexes, (a) prokaryotic initiation. The following symbols are used for initiation factors and other components involved in the formation of the 70S initiation complex: ■, IF-1; ●, IF-2; ▲, IF-3; <, fMet-tRNA$_f$; ▽, GTP; other symbols are defined in the figure. (b) Eukaryotic initiation complexes in mammalian protein biosynthesis from components defined in the figure. The process may be divided into three stages: **a**, formation of a 43S preinitiation complex; **b**, binding of mRNA with formation of a 48S preinitiation complex; and **c**, synthesis of the 80S initiation complex containing the initiator tRNA in the correct position for peptide bond formation. [From Arnstein, H.R.V., Cox, R.A. (1992) *Protein Biosynthesis*, Oxford University Press, Oxford. With permission.]

Tab. 2 Prokaryotic initiation factors from *E. coli*.

Factor	M_r [kDa]	Properties and function
IF-1	9	Stimulates activity of IF-2; accelerates dissociation of unprogrammed ribosomes to subunits.
IF-2	100	Binds fMet-tRNA$_f$ to the ribosomal P site by a GTP-requiring reaction.
IF-3	22	Binds natural mRNAs to the small ribosomal subunit probably by facilitating base pairing between the untranslated leader sequence and the 3' end of 16S rRNA; prevents ribosomal subunit association when bound to the small subunit.

From Arnstein, H.R.V., Cox, R.A. (1992) *Protein Biosynthesis*, Oxford University Press, London. With permission.

located at the interface between the small and large ribosomal subunits.

In the next step, the initiator tRNA and mRNA associate with the 30S–IF-1–IF-2–IF-3 complex with release of IF-3. There is evidence from *in vitro* experiments that the binding of mRNA precedes that of the initiator tRNA.

Messenger RNA binds to the small ribosomal subunit immediately before formation of the final initiation complex with the initiation codon correctly positioned in the P site (see Fig. 9a). In the case of bacterial and bacteriophage messengers, the molecular recognition mechanism proposed by Shine and Dalgarno in 1974 involves base pairing between short nucleotide sequences, most often CUCC, near the 3' end of the 16S ribosomal RNA and a complementary region, usually consisting of 3 to 9 bases on the 5' side of the mRNA initiation codon, which has been found to be present in nearly all of more than 150 bacterial and bacteriophage messengers. Studies with mutants and mRNA fragments indicate that, in addition to the Shine–Dalgarno interaction, outlying upstream sequences in the leader region may also provide recognition signals between mRNAs and ribosomes, possibly by ensuring that the Shine–Dalgarno sequence is in an appropriate conformation.

The Shine–Dalgarno mechanism is also found in chloroplast protein synthesis as judged from sequence analysis of the 16S rRNA and mRNAs, but apparently not in mammalian mitochondria where the initiator codon occurs either directly at, or only a few nucleotides downstream from, the 5' end of mRNA, which excludes the possibility of mRNA–rRNA base pairing in this region.

Eukaryotic systems. At least 12 proteins, the eukaryotic initiation factors (eIF) (see Table 3), are needed for initiation of protein biosynthesis (see Fig. 9b). The dissociation of cytosolic 80S ribosomes is facilitated by a complex initiation factor, eIF-3 (M_r approx. 5–700 000), consisting of 9 to 11 polypeptide chains, which binds to the small (40S) ribosomal subunit and prevents its reassociation to 80S ribosomes. Thus, this factor has antiassociation activity, but low molecular–weight proteins with similar activity have also been reported, and a protein, eIF-4C, of M_r 20 000, seems to function as an accessory factor to eIF-3 in the formation of a 43S ribosomal preinitiation complex. Also, another protein factor, eIF-6, of M_r 24 000, prevents

Tab. 3 Eukaryotic initiation factors.

Initiation factor	Synonym	M_r (kDa) of factors and subunits	Properties and function
eIF-1		15	Stabilizes initiation complexes
eIF-2		α, 38[a]; β, 35[a]; γ, 55	GTP-dependent binding of Met-tRNA$_f$ to the small ribosomal subunit
eIF-2B	GEF	27, 37, 52, 67, 85[a]	Conversion of eIF-2-GDP into eIF-3-GTPs
eIF-3		9–11 subunits 24–170[a]	Associates with 40S subunit to maintain dissociation; binds mRNA to 43S preinitiation complex
eIF-4A	50 kDa component of CBP-II	50	ATP-dependent unwinding of the secondary structure of the mRNA 5' region; stimulates translation of exogenous mRNA in cell-free systems
eIF-4B		80[a]	mRNA binding; stimulates cell-free translation; ATPase activity of eIF-4A and eIF-4F; AUG recognition and recycling eIF-4F
eIF-4C		17	Ribosome dissociation; 60S subunit joining
eIF-4D		17	Formation of first peptide bond
eIF-4E	CBP-I; 24 kDa CBP	24–28[a]	Binds mRNA cap structure
eIF-4F	CBP-II; cap binding protein complex	24 (CBP-I)[a] 50 (eIF-4A), 200[a]	ATPase; unwinds mRNA secondary structure; stimulates cell-free translation
e-IF-4G		220	Stimulates protein synthesis by interacting with eIF-4E and poly(A) protein to circularize polysomes
eIF-5		60[a]	GTPase; release of eIF-2 and eIF-3 from preinitiation complex to allow joining of the 60S subunit
eIF-6		24	Antiassociation activity; binds o the 60S ribosomal subunit.

[a] Denotes subunit can be phosphorylated *in vivo*.
Note: GEF: guanine nucleotide exchange factor; CBP: cap binding protein; From Arnstein, H.R.V., Cox, R.A. (1992) *Protein Biosynthesis*, Oxford University Press, London. With permission.

reassociation by binding to the large (60S) ribosomal subunit.

Initiation factor eIF-2 gives a stable binary complex with GTP which binds the initiator tRNA, Met-tRNA$_f$, forming a ternary complex. Interaction of this ternary complex with the 40S ribosomal subunit containing bound initiation factors eIF-3 and eIF-4C gives rise to the 43S preinitiation complex, which is competent to bind messenger RNA in the presence of three further initiation factors, eIF-4A, eIF-4B, and eIF-4E, together with ATP.

The binding of cytosolic eukaryotic messenger RNAs to the small ribosomal subunit probably does not involve base-paring with the 18S rRNA, as no uninterrupted sequences of the Shine–Dalgarno type have been found. Instead, a "scanning model" has been proposed, in which the preinitiation complex, composed of the 40S ribosomal subunit, Met-tRNA$_f^{Met}$ and associated initiation factors binds at or near the 5′ cap of the mRNA and slides along the messenger until it encounters the first AUG triplet, at which point the 60S ribosomal subunit joins to give rise to the 80S initiation complex. Recognition of the cap is facilitated by cap binding proteins (CBP), which mediate an ATP-dependent melting of the mRNA secondary structure at the 5′ terminal region to allow the mRNA to thread through a channel in the neck of the 40S subunit. The cap structure is required for efficient binding and translation even in cases where the initiating AUG codon occurs hundreds of nucleotides downstream. As a rule, scanning by the 40S subunit stalls at the first AUG codon, which is recognized mainly by interaction with the anticodon of the Met-tRNA$_f^{Met}$. However, this recognition also depends in some way on eIF-2 and may be modulated as a result of deviation of the mRNA structure from the consensus sequence GCCGCCA/G CC$\underline{A}$UGG. Sequence context may also account for the rare cases where initiation occurs downstream of the 5′ proximal AUG codon. Where this sequence context is unfavorable, initiation becomes inefficient, hence most 40S subunits will tend to initiate further along the mRNA at another AUG triplet in a more favorable context. This model also explains rare cases where initiation is not restricted to one particular AUG codon and translation of a single mRNA gives rise to two proteins.

4.2.2 Initiation Complex Formation: Joining of the Large Ribosomal Subunit

The last event in the initiation of protein synthesis involves the joining of the large ribosomal subunit to the preinitiation complex (Fig. 9). In the prokaryotic system, association of the 50S subunit with the 30S preinitiation complex takes place with hydrolysis of GTP by the GTPase activity of IF-2 and release of IF-1, IF-2, GDP, and P$_i$. GTP hydrolysis is essential for release of IF-2 from the initiation complex, which is prerequisite for allowing the fMet-tRNA$_f$ to engage in the formation of the first peptide bond. In eukaryotic protein synthesis, the 80S initiation complex is formed by joining the 60S ribosomal subunit to the 48S preinitiation complex consisting of the 40S ribosomal subunit, eIF-2, eIF-3, GTP, Met-tRNA$_f$, mRNA, and possibly eIF-4C. This coupling reaction requires an additional factor, eIF-5, which mediates the hydrolysis of GTP to GDP with release of eIF-2-GDP, P$_i$, and eIF-3 from the 48S preinitiation complex.

By this stage, all initiation factors have been released and are available for recycling, although the exact steps at which factors are released from intermediate complexes are not known in every case.

There is thus an initiation factor cycle within the ribosome cycle, and regulation of the activity of factors, particularly eIF-2, is an important control mechanism in translation (see Sect. 5.4.1).

4.2.3 Polypeptide Chain Synthesis: The Elongation–Translocation Cycle

In the initiation complex, location of the charged initiator tRNA in the P site of the ribosome (see Fig. 10a) allows transfer of the methionine residue to the amino group of another aminoacyl tRNA in the A site (Fig. 10b) by peptidyl transferase to form dipeptidyl tRNA (see Fig. 10c). Functional insertion of Met-tRNA$_f$ directly into the P site can be demonstrated using the trinucleotide AUG as a synthetic mRNA and another trinucleotide, for example, UUU, to bind an acceptor aminoacyl tRNA (in this case Phe-tRNA).

It is possible to measure the peptidyl transferase activity of the large subunit in the absence of mRNA by using the antibiotic puromycin, which resembles the 3′ terminal region of Phe-tRNA in structure; as an artificial acceptor it forms methionyl puromycin with Met-tRNA$_f$, which can be assayed.

Elongation. The first peptide bond is formed when the aminoacyl tRNA in the ribosomal A site is converted into the corresponding methionyl-aminoacyl-tRNA by transfer of the methionyl (or N-formylmethionyl) residue from the charged initiator tRNA in the P site (Fig. 10c). In artificial cell-free systems, any N-substituted aminoacyl tRNA, such as peptidyl tRNA or N-acetylaminoacyl-tRNA, can function in peptide bond synthesis as a donor in the P site in place of the charged initiator tRNA. The reaction is catalyzed by the peptidyltransferase activity of the large ribosomal subunit (see Fig. 11). No soluble cofactors appear to be involved, but monovalent cations (K^+) at a concentration of 100 mM or more and divalent cations (Mg^{2+}) below 2 mM are required.

Efficient entry of aminoacyl tRNA into the ribosomal A site requires the participation of an elongation factor, termed EF-Tu in prokaryotes (see Table 4) and eEF-1 (EF-1A) in eukaryotes (see Table 5), and GTP. The appropriate elongation factor forms a ternary complex with GTP and all aminoacyl tRNAs except initiator

Tab. 4 Properties of prokaryotic elongation and termination factors from *E. coli*.

	M_r [kDa]	Properties and function
Elongation factors		
EF-Tu	43	N-terminal acetyl-serine; heat labile; binds aminoacyl tRNA to the ribosomal A site
EF-Ts	30	Heat stable; regeneration of EF-Tu-GTP
EF-G	77	GTP-dependent translocation of peptidyl-tRNA and its mRNA codon from the A site to the P site.
Termination (release) factors		
RF1	36	Requires UAA or UAG codons for hydrolysis of peptidyl-tRNA
RF2	38	Requires UAA or UGA codons for hydrolysis of peptidyl-tRNA
RF3	46	Enhances RF1 and RF2 activity

From Arnstein, H.R.V., Cox, R.A. (1992) *Protein Biosynthesis*, Oxford University Press, London. With permission.

tRNA, but not with uncharged tRNA, thus ensuring that only appropriately charged tRNAs are efficiently bound in the A site. A special elongation factor showing extensive homology with both EF-Tu and IF-2 is involved in the synthesis of selenoproteins (see Sect. 2.4.1) from selenocysteyl-tRNA^{UCA} in E. coli.

The above-mentioned aminoacyl tRNA binding reaction catalyzed by EF-Tu is the

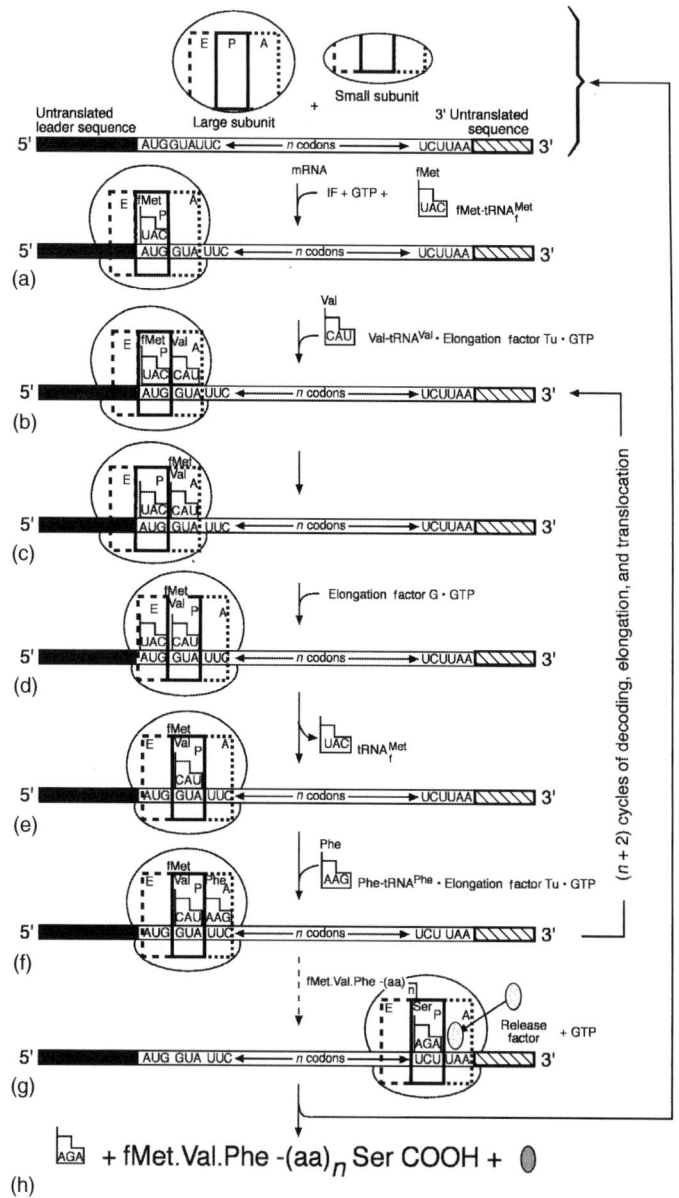

rate-limiting step in the elongation cycle; peptide bond formation and translocation are much faster. The initial binding of the ternary complex to the ribosome is readily reversed, but the interaction is stabilized by the subsequent codon recognition, which induces the GTPase conformation of EF-Tu leading immediately to the hydrolysis of the GTP component of the ternary complex to GDP. Hydrolysis of the GTP moiety causes a further change in the conformation of EF-Tu from the GTP-binding to the GDP-binding form. This conformational change leads to the release of aminoacyl tRNA, allowing its CCA end to align with the peptidyl transferase center of the ribosome and the instantaneous formation of the peptide bond. The elongation factor is later released from the ribosome as a complex with GDP. The operation of EF-Tu is thus similar to that of initiation factor 2 which binds charged initiator tRNA to the small ribosomal subunit. The kinetic changes and the changes in ribosome structure that accompany these events are enumerated below (see Sect. 4.3).

Following dissociation from the ribosome, the EF-Tu–GDP complex interacts with another elongation factor, EF-Ts,

Tab. 5 Eukaryotic elongation and termination factors.

	M_r [kDa]	Properties and function
Elongation factors from various yeast, animal, and plant cells		
eEF-1A (EF-1_L or eEF-Tu)	50–60	Analogous to EF-Tu
eEF-1B (eEF-Ts)	30	Analogous to EF-Ts
eEF-2	105	Contains essential SH groups and one residue of a posttranslationally modified histidine residue; GTP-dependent translocation analogous to EF-G
eEF-3	126	GTPase and ATPase activity; function not fully defined
Termination or release factors		
eRF-1	110	Two 55-kDa subunits; binds to the ribosome A site by a GTP and termination codon dependent reaction; hydrolyzes peptidyl-tRNA in the P site
eEF-2		GTPase; stimulates eRF-1 activity

From Arnstein, H.R.V., Cox, R.A. (1992) *Protein Biosynthesis*, Oxford University Press, London. With permission.

Fig. 10 Schematic view of the biosynthesis of a polypeptide by translation of prokaryotic mRNA. (a) Formation of the initiation complex with fMet-tRNA$_f^{Met}$ in the P site from mRNA, ribosomal subunits, initiation factors (IF), GTP, and fMet-tRNA$_f^{Met}$ (for details, see Fig. 8). The binding sites for aminoacyl tRNA, peptidyl-tRNA, and uncharged tRNA are designated A, P, and E, respectively. (b) Decoding of the codon GUA with Val-tRNA^{Val}. (c) Formation of the peptide bond by transfer of fMet to Val-tRNA^{Val} forming fMet-Val-tRNA^{Val}. (d) Translocation of tRNA$_f^{Met}$ to the E site and of fMet-Val-tRNA^{Val} to the P site with codon UUC aligned with the A site. (e) Ejection of tRNA$_f^{Met}$ from the ribosome. (f) Decoding of UUC with Phe-tRNA^{Phe}. (g) Decoding of UAA with release factor. Steps (b) to (f) constitute the elongation–translocation cycle. The position shown is reached by repeating the cycle $n + 2$ times after formation of fMet.Val-tRNA. (h) Release of the completed polypeptide, fMet.Val.Phe-(aa)$_n$Ser COOH, ribosomal subunits, release factor, mRNA and tRNA^{Ser}. The cycle may be repeated starting at (a).

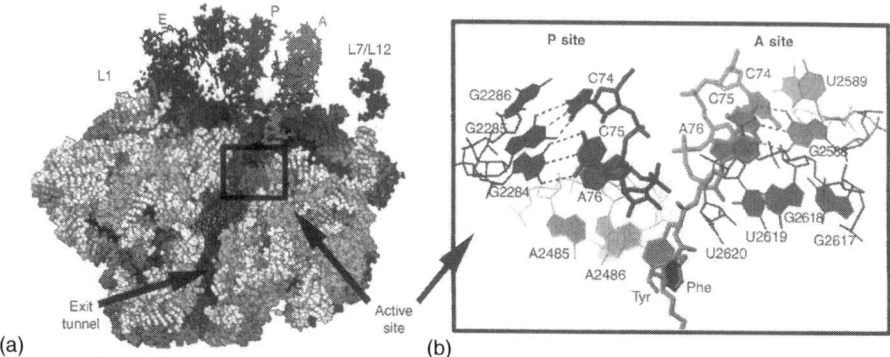

Fig. 11 Fine structure of the peptidyl transferase center interactions of the CCA ends (C74, C75, and A 76) of aminoacyl tRNA and peptidyl-tRNA with the rRNA moiety of the large subunit. Aminoacyl tRNA was represented by a C74 puromycin derivative of tRNA and peptidyl-tRNA was represented by CCA-phenylalanine-caproic acid-biotin. When these model compounds were bound to the A and P sites respectively, the large subunit catalyzed the formation of a peptide bond leading to newly extended peptidyl-tRNA (C74 puromycin–phenylalanine-caproic acid-biotin) in the A site and a deacylated tRNA in the P site. (a) a space filling representation of the 50S subunit (RNA in white and protein in yellow). The tRNAs in the A (green), P (purple), and E (brown) sites are placed according to the structure of the complete ribosome. The subunit has been split though the tunnel and the front section was removed to reveal the peptidyl transferase center (boxed). (b) Base-pairing interactions of CCA sequences of tRNA with 23S rRNA. The analog of deacylated tRNA (purple) is bound to rRNA through base pairing between C74 with G2285) and C75 with G2284. The analog of the newly formed peptidyl-tRNA is bound through base pairing between C75 and G1588. [From Schmeing, et al. (2002) A pretranslocational intermediate in protein synthesis observed in crystals of enzymatically active 50S subunits, *Nat. Struct. Biol.* **9**, 225–230. With permission.] (see color plate p. v).

with formation of an EF-Tu–EF-Ts heterodimer and release of GDP. Reaction of the heterodimer with GTP regenerates the EF-Tu–GTP complex required for binding aminoacyl tRNA. The sequence of events is similar in eukaryotes with eEF-1A (M_r 50 000) corresponding to EF-Tu and eEF-1B (M_r 30 000) to EF-Ts.

Selection of the specific aminoacyl tRNA to be bound at the ribosomal A site is by base pairing between the relevant mRNA codon and the tRNA anticodon. Because this interaction involves only a triplet of bases and hence a maximum of nine hydrogen bonds (see Fig. 1B), it is intrinsically unstable at physiological temperatures and is stabilized by a network of interactions with 16S rRNA in the decoding site where bases of 16S rRNA interact with the codon–anticodon complex. It is inferred that these interactions allow sufficient time for peptide bond synthesis to occur. Also, the codon–anticodon pairing must be monitored for fidelity in order to minimize errors in translation. In *E. coli*, there is genetic and biochemical evidence that one of the proteins of the small ribosomal subunit, S12, is involved in ensuring the fidelity of normal translation and in causing the mistranslation that occurs in the presence of the antibiotic streptomycin due to incorrect codon–anticodon interactions.

Translocation. Translocation involves the movement of the ribosome along the mRNA in the 5' → 3' direction. Immediately after synthesis of the first peptide bond, the ribosomal A site contains dipeptidyl tRNA while uncharged initiator tRNA remains in the P site. Thus, both these sites are occupied, and to allow the next aminoacyl tRNA to enter the A site it is necessary to eject the uncharged tRNA and shift the dipeptidyl tRNA from the A into the P site. This translocation (Fig. 10d) takes place as a concerted process involving movement of both messenger RNA and dipeptidyl RNA together into the P site, leaving the A site occupied by the next mRNA codon and free to accept the cognate aminoacyl tRNA (see Fig. 10e). At the same time, the deacylated tRNA moves first into an E (exit) site with subsequent ejection when the next aminoacyl tRNA enters the A site.

Translocation requires the participation of another elongation factor (EF-G in prokaryotes (Table 4) and EF-2 in eukaryotes (Table 5)) and GTP. It seems that when EF-G and GTP bind to the ribosome, translocation occurs but GTP hydrolysis is required only subsequently to release EF-G and GDP. The location of the EF-G binding site of the ribosome overlaps with that for EF-Tu; thus, EF-G must be released before EF-Tu–aminoacyl-tRNA–GTP complex can enter the A site. Analogous reactions occur in eukaryotic systems. The kinetic steps involved in translocation and the structural changes that take place are described below (Sect. 4.3).

After translocation, the ribosomal P site is occupied by dipeptidyl tRNA and the vacant A site contains the third mRNA codon. Entry of the next aminoacyl tRNA, selected as before by the codon–anticodon interaction into the A site (Fig. 10f) enables peptide bond synthesis to continue and repeated operation of the elongation–translocation cycle gives rise to a stepwise elongation of the nascent polypeptide chain, each complete cycle elongating the chain by one amino acid residue and moving the mRNA by one codon in the 5' to 3' direction. When the end of the coding sequence is reached, one of the termination (or stop) codons has entered the A site, translation stops, and the completed polypeptide chain is released.

Termination. (See Figs. 10g-h.) The presence of one of the three termination codons, UAA, UAG, or UGA, in the A site results in the binding of a release factor (Table 4) instead of an aminoacyl tRNA to the ribosome. In prokaryotes, two release factors have been identified, one (RF1) recognizing UAA and UAG, the other (RF2) functioning with UGA. Ribosomal binding and release of RF1 and RF2 are stimulated by a third factor, RF3, which interacts with GTP and GDP. In eukaryotic cells, such as reticulocytes, one release factor (eRF) has been found to function with all three termination codons, and the binding of this factor to ribosomes is stimulated by GTP but not GDP. Although the details are not entirely clear, GTP hydrolysis appears to be required for the release of the finished polypeptide chain by cleavage of the peptidyl tRNA bond and completion of the termination process leading to dissociation of the release factor from the ribosome.

Thus, at the end of the ribosome cycle, the coding sequence of messenger RNA has been translated to produce a particular polypeptide chain, and all the components involved become available for reuse in another round of the cycle (Fig. 10h). Usually, several ribosomes, each at a different stage in completing the process of translation, are attached to one mRNA molecule, giving rise to polyribosomes (also called *polysomes*; Fig. 7). In eukaryotic cells, the efficiency of protein synthesis is stimulated by factor eIF4-G (Table 3), which interacts with both factor eIF-4E and a poly A-binding protein to link the 5′ and 3′ ends of mRNA. The resultant circularized polysomes show an enhanced ability to reinitiate after release of the ribosomal subunits from the messenger RNA at the end of a round of translation.

4.3 High-resolution Structural Studies of the Ribosome

Early structural studies based on electron microscopy provided sufficient information to allow the construction of static models showing the location of mRNA and the relative orientations of the ribosomal subunits (Fig. 8). The modern technique of high-resolution cryo-electron microscopy (cryo-EM) has provided views of the ribosome at a level of resolution (0.9 nm) approaching that achieved by X-ray diffraction studies. Although prokaryotic ribosomes are the principal subjects of structural studies because they are simpler than eukaryotic ribosomes, the insights into structure and function are thought to apply to all ribosomes.

During the last 10 years, X-ray diffraction studies have led to considerable advances in the elucidation of the structure of the ribosome in more detail, culminating in the determination of the *E. coli* ribosome at 0.78-nm resolution and of the small and large subunits at resolutions of 0.30 and 0.25 nm respectively. The structures have revealed the identity of each amino acid and each nucleotide. The findings provide insights at the atomic level into the reactions leading to the decoding of mRNA and the formation of the peptide bond (Fig. 11). Moreover, the importance of the role of rRNA in ribosome function has become evident from the structures; the functional regions of both small and large subunits are rich in RNA. The peptide bond is formed by nucleophilic attack on the ester carboxyl group of peptidyl tRNA bound to the site by the α-amino group of aminoacyl tRNA in the A site. Peptide bond formation is catalyzed by the peptidyl transferase center of the large subunit of the ribosome. Structural studies confirm that this catalytic site comprises 23S rRNA sequences only. The location of the peptidyl transferase center and the role of the CGA ends of tRNA are illustrated in Fig. 11. The three-dimensional structures also highlight the dynamic aspects of ribosome function leading to the view that the ribosome is a highly sophisticated motor driven by GTP with rRNA playing a leading role.

The first of the two GTP requiring steps is the delivery of aminoacyl tRNA to the A site of the ribosome. The distinct kinetic steps involved as the ternary complex of aminoacyl tRNA, EF-Tu and GTP binds to the ribosome are illustrated in Fig. 12 and the conformational changes that have been found to take place are depicted in Fig. 13. Please note that one molecule of GTP is converted to GDP irrespective of whether peptide bond formation ensues. The peptide bond is formed soon after the selection of the aminoacyl tRNA (the accommodation step in Fig. 12), leading to the newly formed and extended peptidyl tRNA residing in the A site. The second GTP requiring step then follows; namely, translocation of aminoacyl tRNA from the A site to the P site of the ribosome. The kinetically resolved stages of the translocation are shown in Fig. 14 and the conformational changes so far identified are illustrated in Fig. 15.

The ribosome also influences the secondary and tertiary structure (folding)

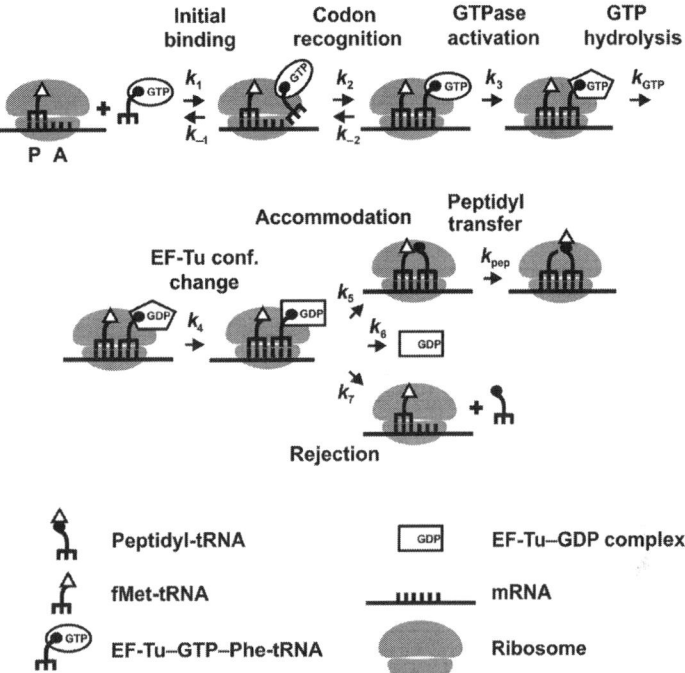

Fig. 12 Kinetically resolved stages in the interaction of ternary complex of aminoacyl tRNA•EF-Tu•GTP with the A site of the ribosome. Individual steps are shown by the numbered rate constants and by k_{GTP} and k_{pep}. The factor EF-Tu is shown in different conformations in GTP and GDP bound states and in the activated GTPase state. The ribosome is able to distinguish between cognate aminoacyl tRNA (for example, with anticodon 3'AAG 5') and near cognate aminoacyl tRNA (for example, with anticodon 3'GAG 5'): this ability is called proof reading. [From Gromadski, K.B., Rodnina, M.V. (2004) Streptomycin interferes with conformational coupling between codon recognition and GTPase activation on the ribosome, *Nat. Struct. Mol. Biol.* **11**, 316–322. With permission.]

of the nascent polypeptide chain during its passage through the 10-nm long × 1 to 2 nm-diameter tunnel that is present in the large ribosomal subunit (Fig. 16). The wall of the tunnel comprises nucleotides of 23S rRNA and nonglobular parts of two ribosomal proteins, L4 and L2; proteins L23 and L29 flank the exit from the tunnel. Thus, the nascent polypeptide chain passes through an environment that influences the way in which it folds. As nascent polypeptide chains emerge from the tunnel, they are free to encounter molecular chaperones that prevent accumulation and aggregation of newly synthesized proteins. For example, the chaperone Trigger factor (TF) has a binding site involving protein L23 close to the exit of the tunnel. Signal recognition particles that convey secretory and transmembrane proteins to their appropriate destinations have bonding sites near to the exit of the tunnel in the regions of proteins L23 and L29.

The higher resolution studies of subunits have also revealed the mode

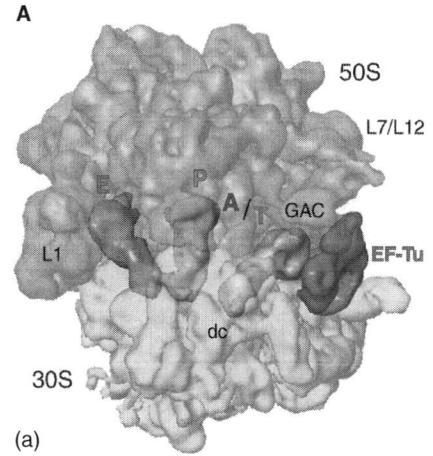
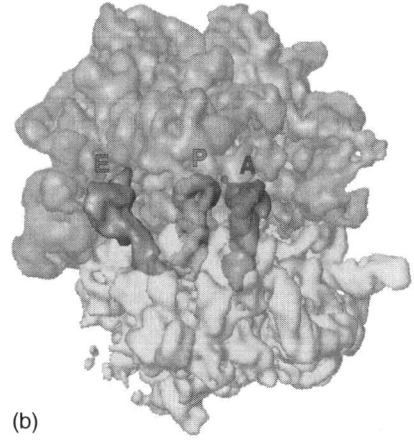

(a)　　　　　　　　　　　　　　　　(b)

Fig. 13 Cryo-electron microscopy maps showing the incorporation of aminoacyl tRNA into the A site of a bacterial ribosome at a resolution of 0.9 nm and a conformational change in the GTPase-associated center. **A** The location of the binding site for the ternary complex of Phe-tRNA^{Phe}•EF-Tu•GDP-Kir.
(a) The antibiotic Kirromycin (Kir) binds to the ternary complex and prevents dissociation of EF-Tu•GDP-Kir from the ribosome after GTP hydrolysis. The stalled complex is located below the L7/L12 stalk of the 50S subunit. EF-Tu is in red and A/T site tRNA is in pale pink.
(b) Structure of the ribosome after peptide bond formation. The P site contains tRNA$_f^{Met}$ and the A site contains fMet•Phe-tRNA^{Phe} which is shown in magenta. In both panels, the ribosomal subunits are rendered semitransparent to illustrate the densities for tRNAs in the ribosomal intersubunit space. The sites for the tRNAs are labeled as follows: A/T, A/T site for the aa-tRNA bound to EF-Tu, reaching the decoding site in the A site of the 30S ribosomal subunit; A, aminoacyl site; P, peptidyl site; E, exit site. The P site tRNA is green, the E site tRNA is orange. **B** Conformational change in the GTPase-associated center of the 50S subunit. The cryo-electron microscopy maps show;
(a) fMet-tRNA$_f^{Met}$ bound to the P site of the ribosome with a vacant A site; (b) fMet-tRNA$_f^{Met}$ in the P site and Phe-tRNA^{Phe}•EF-Tu•GDP-Kir complex with the A/T site between tRNA colored pink and EF-Tu colored red. The arrow indicates the site at the base of the L7/L12 stalk, which undergoes a conformational change during the binding of the ternary complex; (c) tRNA$_f^{Met}$ bound in the P site and fMet•Phe-tRNA^{Phe} bind in the A site. Insets (b) and (c) focus on the conformational changes of the GTPase-associated center. The view from the intersubunit space allows the movement of the GTPase-associated center to be observed. The E site is shown in orange; (d) Superposition of the 50S ribosomal subunits from the ribosome bound with the stalled ternary complex (semitransparent blue) and from the ribosome bearing a dipeptidyl-tRNA in the A site (solid red). Ribosomal subunits are blue (50S subunit) and yellow (30S subunit). Features of the large ribosomal subunit: L7/L12, stalk of proteins L7/L12 from the 50S ribosomal subunit; GAC, GTPase-associated center; SRC, sarcin-ricin loop; L16, ribosomal protein L16; L1, stalk of ribosomal protein L1. Features of the small ribosomal subunit: h, head of the 30S subunit; dc, decoding center. [From Valle, M., Zavialov, A., Li, W. et al. (2003) Incorporation of aminoacyl tRNA into the ribosome as seen by cryo-electron microscopy, *Nat. Struct. Biol.* **10**, 899–906. With permission] (see color plate p. vi).

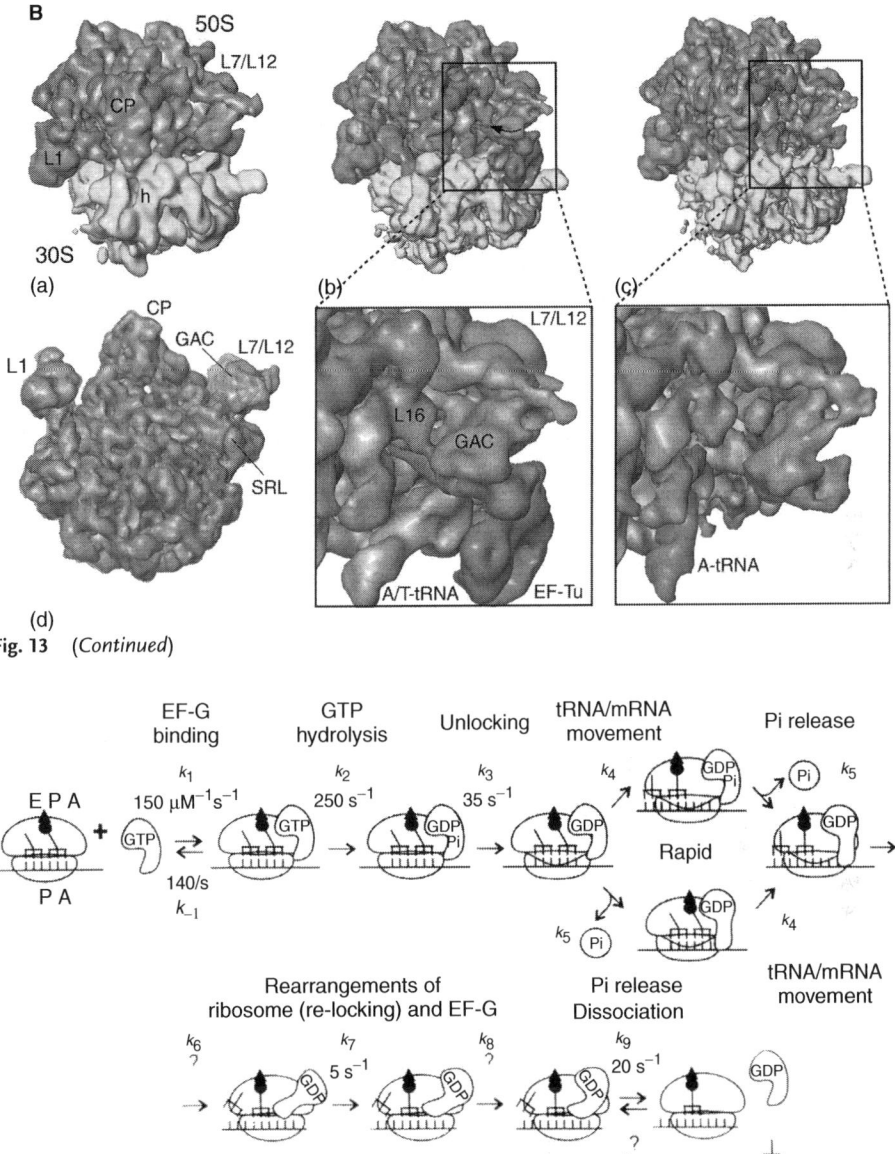

Fig. 13 (Continued)

Fig. 14 A kinetic model of translocation. The diagram indicates several steps identified by several biochemical and rapid kinetic techniques. Ribosomes are shown in two conformations, closed and open. In the closed or locked conformation, bound tRNAs interact extensively with the ribosome. These interactions need to be disrupted to allow translocation (the open or unlocked conformation) and to be reestablished (relocked) after translocation. EF-G is shown in different conformations and orientations based on cryo-electron microscopy (see Fig. 5). [Reproduced from Peske, F., Savelsbergh, A., Katunin, V.I., Rodnina, M.V., Wintermeyer, W. (2004) Conformational changes of the small ribosomal subunit during elongation factor G-dependent tRNA-mRNA translocation, *J. Mol. Biol.* **343**, 1183–1194. With permission].

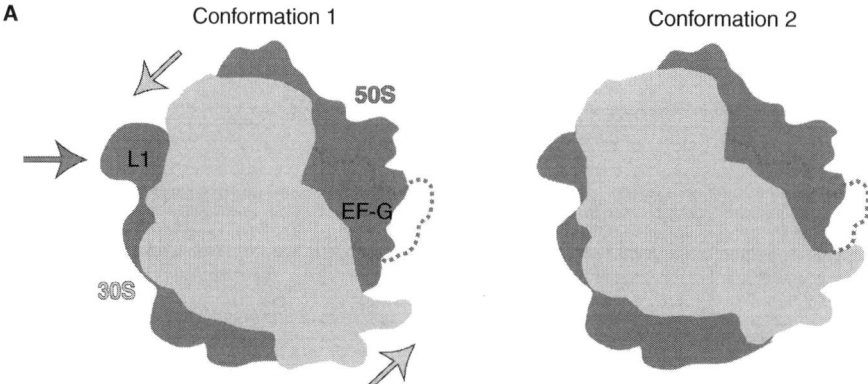

Fig. 15 Changes detected by cryo-electron microscopy (1.75–1.84 nm resolution) in the conformation of ribosomes during translocation. (A) After the peptide bond is formed, deacylated tRNA is located in the P site with peptidyl tRNA in the A site (conformation 1), the translocational state. A ratchet-like motion follows when EF-G•GTP is bound to the ribosome leading to conformation 2. The conformational change involves a counterclockwise rotation of the 30S subunit with respect to the 50S subunit with the L1 stalk moving in the direction opposite to the head of the 30S subunit. The outline of EF-G is shown by the dotted line.
(B) Conformational changes in the L7/L12 stalk and in EF-G during translocation. The figures show that EF-G (shown in red) links the decoding site with the L7/L12 stalk. The cryo-electron microscopy maps shown in (c) and (d) correspond to conformation 2 in (a) with deacylated tRNA in the P site and peptidyl-tRNA in the A site and EF-G•GTP bound to the ribosome. The L7/L12 stalk (shown on the right of the dotted line) is bifurcated. Part of EF-G is shown close to the A site (marked A in (c)). A third conformation is observed after hydrolysis of GTP to GDP and before EF-G•GDP is released from the ribosome as shown in (a) and (b). In this conformation, the stalk is extended but not bifurcated and part of the EF-G is implicated in an arclike structure. The pretranslocational state is regained (see (a)) when EF-G•GDP is released from the ribosome. At the end of the translocation process, tRNAs in the A and P sites are transferred to the P and E sites, respectively, with the concomitant advance by one codon of mRNA. The insets show the appropriate orientations of the ribosome with 30S subunits and 50S subunits colored yellow and blue respectively. Features of the 30S subunit: b, body; h, head; sp, spur; ch, channel. Features of the 50S subunit: cp, central protuberance; L1, L1 protein; st, L7/L12 stalk. (A was reproduced from, Gas, A., Valle, M., Ehrenberg, M. and Frank, J. (2004). Dynamics of EF-G interaction with the ribosome explored by classification of a heterogenous cryo-EM dataset J. Structural Biol. **147**, 283–290. B was reproduced from Agrawal, R.K., Heagle, A.B., Penczek, P., Grassucci, R.A., Frank, J. (1999) EF-G dependent GTP hydrolysis induces translocation accompanied by large changes in the conformation of the 70S ribosome, Nat. Struct. Biol. **6**, 543–647. With permission) (see color plate p. viii).

of action of several antibiotics such as paromomycin, streptomycin, and spectinomycin, which modify the decoding function of the small subunit, and chloramphenicol, puromycin, and vernamycin, which affect peptide bond formation.

5
Translational Control of Gene Expression

Cells need to synthesize specific proteins in the required amounts at particular times and to deliver them to the correct locations.

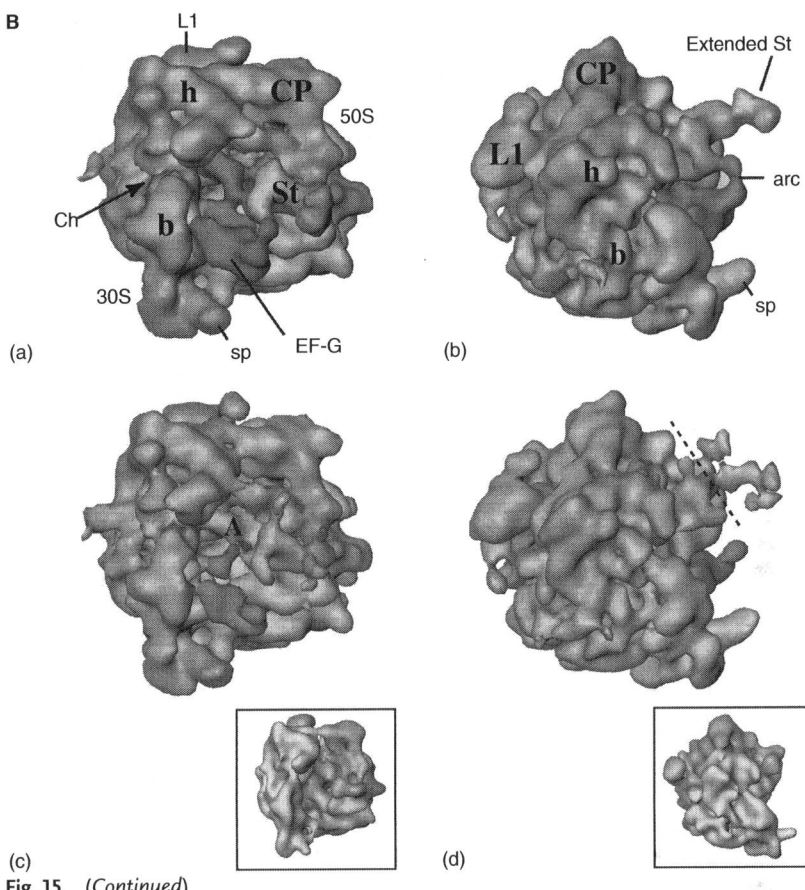

Fig. 15 (Continued)

These processes depend on a great many interactions and numerous mechanisms exist for the translational control of gene expression. Examples of the ways used by cells to control the translation of mRNA are presented in the sections that follow. This material is intended to be illustrative rather than comprehensive, and it is to be expected that novel control mechanisms will continue to be discovered.

5.1
mRNA Stability

Provided all other components of the translational systems are present in optimum amounts, control of translation may be achieved through the availability of the relevant messenger RNAs at the site of protein synthesis. The steady state level of mRNA is determined by the rate of its synthesis and degradation. Control of transcription is of major importance for synthesis in both prokaryotes and eukaryotes. The stability of mRNA depends on its primary and secondary structure as well as on the presence of factors such as stabilizing proteins and nucleases.

The secondary structure of mRNA is determined by its nucleotide sequence. For any mRNA, a number of coding sequences are possible because of the degeneracy

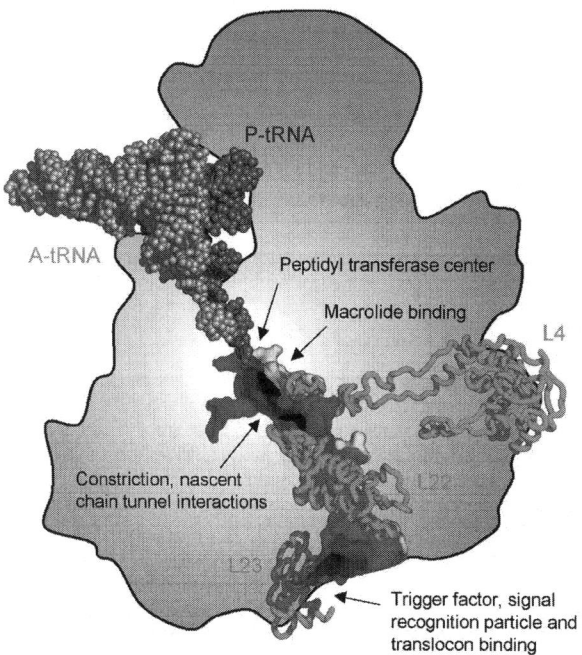

Fig. 16 Protein synthesis and the path of the nascent chain through the ribosome exit tunnel. The contour of the large ribosomal subunit is shown in gray. The aa-tRNA and pept-tRNA are represented by spheres, colored copper, and gray respectively. Ribosomal proteins L4, L22, and L23 are shown as green ribbons. The surface of the ribosomal exit tunnel corresponds to the empty space in the 50S subunit, which is sliced on the side. The surface is colored yellow outside and blue inside. The acceptor stems of aa-tRNA and pept-tRNA point toward the PTC, where peptide bond formation occurs. Nascent polypeptides escape the ribosome through the tunnel. They have to pass the narrow constriction where ribosomal proteins L4 and L22 contact the wall of the tunnel. This region is a target of antibiotics and the site of regulatory interactions between nascent chains and the tunnel. Ribosomal protein L22 mediates the interaction between nascent chains and the tunnel. Ribosomal protein L23 mediates the interaction between nascent chains emerging from the ribosome and cytosolic chaperones such as Trigger Factor (TF) and protein targeting factors such as Signal Recognition Factor (SRFF) and translocons. [From Jenni, S., Ban, N. (2003) The chemistry of protein synthesis and voyage through the ribosomal tunnel, *Curr. Opin. Struct. Biol.* **13**, 212–219. With permission] (see color plate p. x).

of the genetic code. Thus, degeneracy allows for particular features of secondary structure (often termed *cis* factors) to be favored, which may act either to stabilize or destabilize the mRNA, according to the needs of the cell, by determining its susceptibility to degradative enzymes (often termed *trans*-acting factors).

Whereas the structure of mRNA determines its susceptibility to degradative enzymes, the detailed mechanisms are complex. In prokaryotes, the enzymes involved include two endonucleases (RNase E and RNase III) and two exonucleases (polynucleotide phosphorylase and RNase II). Other nucleases may be active in particular cases such as phage infection. In eukaryotes, a major pathway involves the removal of the 3' poly(A) tail (deadenylation), followed by removal of the 5' cap, which renders the mRNA susceptible to rapid endonucleolytic degradation in the 5' → 3' direction.

5.2
Control by Interaction of Proteins with mRNA

Throughout the ribosome cycle, dynamic protein-mRNA interactions are functionally important in the initiation, elongation, and termination of polypeptide synthesis. In addition, more stable associations between proteins and mRNAs have been observed, particularly in eukaryotic cells. These messenger ribonucleoprotein complexes (mRNPs) occur both in polyribosomes and free in the cytosol, some of the latter being either temporarily or permanently unavailable for translation. Thus, protein-mRNA interactions contribute to the efficiency with which mRNAs are translated. Some proteins, such as the poly(A)-binding protein (p78), are present in most if not all mRNPs, whereas others appear to be cell specific and mRNA selective. In unfertilized sea urchin eggs and *Xenopus* oocytes, for example, untranslated messenger is sequestered by association with proteins that prevent translation until later stages of development. Duck reticulocytes contain globin mRNP, which cannot be translated *in vitro*, whereas the mRNA obtained by deproteinizing the complex can be translated, showing that in this case translation is prevented by the mRNP proteins.

Formation of a site-specific mRNA-protein complex is involved in the translational control of the biosynthesis of ferritin, an iron storage protein, which is stimulated in response to the presence of iron. In this instance, a cytoplasmic repressor protein of 85 kDa binds to a highly conserved 28-nucleotide stem-loop structure in the 5' untranslated region of ferritin mRNAs in the absence of iron. In the presence of iron, the protein dissociates from the mRNA, which is then available for translation. A similar loop motif occurs in the 3' untranslated region of transferrin receptor mRNA, which is also subject to translational control by an iron-responsive repressor.

During the cell cycle, histone mRNA is destabilized after completion of DNA replication, resulting in a 30- to 50-fold decrease. This change appears to be due to an increase in the level of free histones, which form a complex with histone mRNA by interaction with a stem-loop structure at the extreme 3' terminus. Formation of this histone–histone mRNA complex is thought to activate a ribosome-associated 3' → 5' exonuclease, which degrades the histone mRNA. During the S phase, newly synthesized DNA binds free histones to form nucleosomes, thus preventing the degradation of histone mRNA at this stage of the cell cycle.

Specific regulation of gene expression at the level of translation also exists in prokaryotes. For example, the synthesis of *E. coli* threonyl-tRNA synthetase is negatively autoregulated by an interaction of the tRNA-like leader sequence of its mRNA with the synthetase that inhibits translation by preventing the binding of ribosomes. The synthetase is displaced from the mRNA by tRNA^{Thr}, which thus acts as a translational antirepressor. This regulatory mechanism allows the cell to maintain a balance between the tRNA synthetase and its cognate tRNA.

Similarly, there is a mechanism used to control the synthesis of proteins encoded by a polycistronic mRNA. In this case, selective binding of the ribosomal protein to the region of the mRNA involved in the initiation of translation leads to the regulatory protein controlling both its own synthesis and that of other ribosomal proteins. A specific example is the role of ribosomal protein S4, which acts as a translational repressor of four ribosomal proteins (S4, S11, S13, and L17). Protein S4 appears to function as a repressor through an unusual "pseudoknot" linking a hairpin loop upstream of the ribosome-binding site with sequences 2 to 10 codons downstream of the initiation codon. (A pseudoknot structure contains intramolecular base pairs between base residues in the loop of a stem-loop structure and distal complementary regions of the RNA.) Stabilization of this structure by S4 would prevent the binding of ribosomes, and this control mechanism may contribute to the coordinated synthesis of the different ribosomal proteins required for ribosome assembly.

to allow the synthesis of a single protein from two or more overlapping genes by suppression of an intervening termination codon. Several retroviruses use this mechanism to move from one reading frame to another in the expression of the viral RNA-dependent DNA polymerase. Other examples of the operation of such a frameshift include the synthesis of the reverse transcriptase enzymes of several retrotransposons, such as the yeast Ty1. The mechanism of changing the reading frame involves "slippery" sequences and a complex folding of the mRNA into a structure termed a *pseudoknot*.

In *E. coli* phages, translational control of the three cistrons of Q$_\beta$, f2, and related bacteriophage RNAs (Fig. 17) accounts for the synthesis of coat protein:replicase:A protein in the approximate ratio of 20 : 5 : 1. These quantitative differences are due to the differential and independent initiation of translation at each cistron as a result of differences in the secondary structure of the mRNA initiation sites. Furthermore, *in vivo* there is a delay in the synthesis of coat protein and this temporal control involves translational repression of the cistron by ribosomal protein S1. In addition, S1 functions as one of the subunits of the f2 RNA replicase; therefore, association of the newly synthesized translation product of the f2 replicase cistron with S1 will favor its dissociation from the phage RNA, thus allowing translation of the coat protein cistron to start.

5.3
Control by mRNA Structure

The secondary structure of some eukaryotic mRNAs regulates translation by a mechanism involving a ribosomal frameshift which gives rise to a directed change of the translational reading frame

5.4
Control by Modification of Translation Factor Activity

5.4.1 Initiation Factors
In eukaryotes, initiation of protein synthesis is inhibited by phosphorylation of the initiation factor eIF-2. In particular cases,

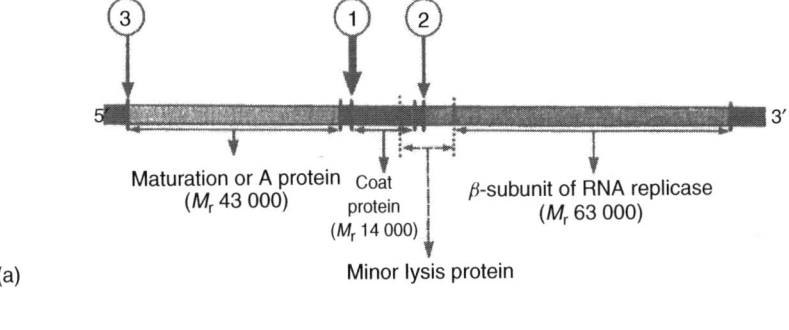

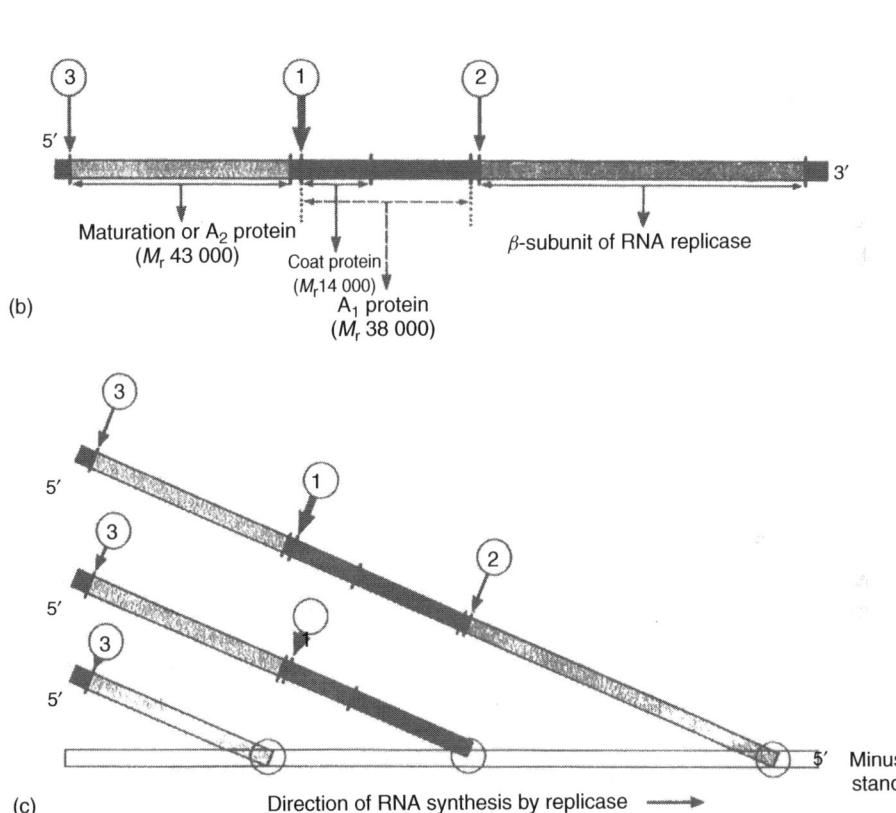

Fig. 17 Translational control of bacteriophage synthesis: (a) arrangements of MS2 and f2 bacteriophage cistrons, (b) arrangement of Qβ bacteriophage cistrons, and (c) replicative intermediate synthesizing new plus strands on the complementary minus-strand copy of the original bacteriophage RNA. The ribosome-binding sites (RBS) are indicated by the numbered arrows. The major RBS (1) binds ribosomes efficiently but can be blocked by ribosomal protein S1. The secondary RBS (2) becomes available only after translation of at least part of the coat protein cistron by ribosomes. RBS 3 is masked by the secondary structure of native bacteriophage RNA but is accessible in nascent RNA (c) or *in vitro* when the secondary structure is destroyed. Noncoding regions are shown in black. [Reproduced from Arnstein, H.R.V., Cox, R.A. (1992) *Protein Biosynthesis*, Oxford University Press, Oxford. With permission].

phosphorylation of eIF-2 is stimulated by a lack of heme or the presence of double-stranded RNA. Two different protein kinases capable of phosphorylating the α-subunit of eIF-2 have been characterized. One substrate, called the heme-controlled repressor (HCR) or heme-regulated inhibitor (HRI), is a cytoplasmic protein (95 000 Da) that is activated by phosphorylation. Double-stranded RNA activates phosphorylation of a 67 000-Da protein, which in turn phosphorylates eIF-2. Thus, a cascade of protein phosphorylation is involved. Phosphorylated eIF-2 is unable to exchange GDP for GTP, which prevents it from functioning in the binding of initiator tRNA to ribosomes.

Conditions other than lack of heme (e.g. heat shock, serum deprivation, or the presence of oxidized glutathione), which are known to inhibit protein synthesis, also stimulate the phosphorylation of eIF-2α. Conversely, the activity of eIF-4F is decreased by dephosphorylation of the 24-Da subunit (see Table 3) rather than by phosphorylation. Thus, a number of different kinases and phosphatases are involved in modulating the activities of different factors.

Small RNAs may also be involved in regulating the translation of mRNA in eukaryotic cells. Of the stimulatory RNAs, the best characterized is a small RNA of about 160 nucleotides, which accumulates in cells after infection with adenovirus. This virus-associated RNA, VA-RNA$_1$ sustains general protein synthesis by inhibiting the phosphorylation of the α-subunit of initiation factor eIF-2.

Two features of mRNA structure are implicated in several regulatory processes involving initiation factors, namely, the 5′ cap structure and a cytoplasmic polyadenylation element (cPE) located within the 3′ untranslated region. For example, embryonic cells have mRNAs that are stored until particular stages in development are reached. During storage, eIF-4E is thought to bind to the 5′ cap structure and a protein cPEB is bound to the cPE element; a third protein binds to both eIF-4E and cPEB. Similar nucleoprotein structures have been proposed to regulate the synthesis of key proteins during progression through the cell division cycle. Furthermore, eIF-2 and eIF-4E are targets for changing the patterns of protein synthesis associated with tumorigenesis. Phosphorylation of the α-subunit of eIF-2 (eIF-2α) regulates its activity. Phosphorylation of a small family of eIF-4E binding proteins (4E-BPs) regulates the binding of eIF-4E to eIF-4G. The activities of the appropriate kinases and phosphatases may, in turn, be controlled by cellular and viral oncogenes and tumor-suppressor genes.

A novel pathway for the initiation of translation has been found which permits initiation in a manner that is independent of the 5′ end of mRNA. Particular RNA sequences are involved, termed *internal RNA entry sites* (IRES). IRES elements have been found in both viral and cellular RNAs. The IRES elements intervene in the initiation of translation during cellular stress and during infection by viruses, for example, poliovirus or hepatitis C virus. The cricket paralysis virus RNA possesses an IRES element, which, in the absence of initiation factors and initiator tRNA, can bind to the P site and assemble 80S ribosomes. In this way, the reading frame is set so that the first codon is in the A site. The first translocation step takes place without a peptide bond being formed. However, aminoacyl tRNA is delivered to the P site so that chain elongation can proceed normally. The cricket paralysis virus IRES bound to the small subunit and to ribosomes was visualized by cryo-electron microscopy.

5.4.2 Other Translation Factors

A Ca^{2+}/calmodulin-dependent protein kinase phosphorylates eEF-2. The phosphorylated factor appears to be inactive. Dephosphorylation of the factor by phosphatase restores its activity.

5.5 Effects of Antisense Polynucleotides

Antisense RNAs, which are polynucleotides with base sequences complementary to messenger RNAs, have been found in both prokaryotes and eukaryotes. Natural antisense RNAs are not common but synthetic RNAs directed at specific targets have been widely studied. It has been demonstrated that they can function as inhibitors of messenger RNA translation. In prokaryotes, the most effective inhibitors appear to have a base sequence complementary to the 5′ leader region, including the Shine–Dalgarno sequence, which is involved in the binding of mRNA to the small ribosomal subunit. In eukaryotes, translation of mRNA is inhibited by polyribonucleotides complementary to the 5′ untranslated region of mRNA, indicating a direct effect on initiation as in prokaryotes. Polynucleotides complementary to the 3′ untranslated region of mRNA also inhibit translation in some cells, and this effect may be due to destabilization of mRNA by ribonucleases specific for double-stranded RNA. The effect of antisense polynucleotides is not restricted to translation of mRNAs, but transcription as well as the processing of transcripts may also be inhibited.

5.6 Availability of Amino Acids, tRNA Abundance, and Codon Usage

5.6.1 Amino Acids

Polypeptide synthesis depends on an adequate supply of tRNAs charged with the 20 protein amino acids and appropriate interactions between their anticodons and the codons of mRNA. Peptide chain elongation is decreased or inhibited by lack of amino acids or other conditions giving rise to an imbalance or deficiency in aminoacyl tRNAs.

5.6.2 Abundance of tRNAs and Codon Usage

Different tRNAs are present in the cytosol in unequal amounts, and elongation rates are slower at codons corresponding to rare tRNA species. The existence of synonymous codons raises the question of preferential use of some codons and its possible significance in relation to translational efficiency and control. In some bacteria (e.g. *Pseudomonas aeruginosa*, which has a high content of $G + C$, 67.2%, in DNA), the most common codons are those with the strongest predicted codon–anticodon interaction – that is $G + C$ base pairs – but this preference is not universal and, for example, does not apply to *E. coli*, which has a lower proportion of $G + C$ (50%). Although codon usage may affect elongation rates, it is probably of less importance in translational control than the secondary structure of mRNA in relation to the rate of initiation of protein synthesis.

5.7 Modulation of Ribosome Activity

Specific ribosomal components have an important function in relation to the fidelity of protein synthesis. Thus, in *E. coli* ribosomal protein, S12 determines the accuracy of codon–anticodon interactions and modulates the translational error frequency in the presence of the antibiotic streptomycin.

To what extent reversible modifications of ribosomal constituents are involved in translational control of protein synthesis is uncertain. Although phosphorylation of ribosomal protein S6 increases with cell proliferation, it is not known whether this change is directly related to the accompanying increase in protein synthesis by an effect on the translation rate.

5.8
Ribosome-inactivating Proteins

Many molds and plants produce toxins, which are protective agents, termed ribosome-inactivating proteins (RIPs), directed at particular cells and their ribosomes. These toxins are classified as either type I or type II RIPs according to the number of polypeptide chains.

Type I RIPs comprise a single polypeptide chain; for example, α-sarcin, an extracellular cytotoxin produced by *Aspergillus giganteus*, consists of a single chain of 150 amino acid residues.

Type II RIPs comprise two polypeptide chains, A and B. The A chain has the ability to inactivate ribosomes, and the B chain is a galactose-specific lectin responsible for the entry of the toxin into the target cell. Ricin, which is isolated from castor oil beans, is representative of type II RIPs.

Ribosomes are inactivated as a result of the RNA N-glycosidase activity of RIPs. These toxins have different specificities for particular cells and ribosomes. However, the target site for all RIPs is an adenylate residue (position 2660 in the *E. coli* 23S rRNA sequence) located within a highly conserved sequence of 12 nucleotides ($5'_{2654}$AGUACGAGAGGA$_{2665}3'$). Cleavage of the GpA$_{2660}$ internucleotide bond or depurination of A$_{2660}$ is sufficient to inactivate the ribosome. The target residue is located in the loop region of a stem–loop element of the secondary structure, which is termed the α-sarcin stem–loop. Thus, the target adenylate is either directly or indirectly essential for ribosome function. The α-sarcin stem–loop is known to be important for binding elongation factors EF-Tu and EF-G to the ribosome. RIPs have attracted interest as active components of reagents directed at particular targets such as cancer cells.

6
Concluding Remarks

The control of protein synthesis, either by regulation of the amount of mRNA available for translation or by the efficiency with which it is translated, is important in cell growth and development as a factor determining the level of cellular and extracellular proteins. Subversion of this control occurs in cells infected by viruses when the viral nucleic acid uses the protein-synthesizing machinery of the host cell and thereby changes normal cell metabolism in favor of the synthesis of viral proteins needed for the production of virus progeny.

The polypeptide chains of all proteins are synthesized by the process described above. This mechanism gives rise to primary polypeptide chains, which are often further modified; for example, by cleavage into smaller peptides, by structural modification of selected amino acid residues, by splicing of the polypeptide chain, or by the formation of covalent bonds between polypeptide chains. Some of these secondary modifications are related to the correct folding of polypeptide chains and to the production of active enzymes or peptide hormones from inactive precursors (e.g. insulin from proinsulin). Also, the transport of proteins within the cell or the secretion of extracellular proteins is often linked to structural changes in polypeptide chains either during or after completion of synthesis.

See also DNA Replication and Transcription.

Bibliography

Books and Reviews

Arnstein, H.R.V., Cox, R.A. (1992) *Protein Biosynthesis*, Oxford University Press, London.

Garrett, R.A., Douthwaite, S.R., Matheson, A.T., Moore, P.B., Noller, H.F. (Eds.) (2000) *The Ribosome: Structure, Function, Antibiotics and Cellular Interactions*, American Society for Microbiology, Washington, DC.

Hershey, J.W.B., Merrick, W.C. (2000) Pathway and Mechanism of Initiation of Protein Synthesis, in: Sonenberg, N., Hershey, J.W.B., Mathews, M.B. (Eds.) *Translational Control of Gene Expression*, Cold Spring Harbor Laboratory Press, Cold Spring Harbor, NY, pp. 33–88.

Hershey, J.W.B., Mathews, M.B., Sonenberg, N. (Eds.) (1996) *The Pathway and Mechanism of Eukaryotic Protein Synthesis*, Cold Spring Harbor Laboratory Press, Cold Spring Harbor, NY.

Hill, W.E., Dahlberg, A., Garrett, R.A., Moore, P.B., Schlessinger, D., Warner, J.R. (1990) *The Ribosome, Structure, Function and Evolution*, American Society for Microbiology, Washington, DC.

Matheson, A.T., Davies, J., Hill, W., Dennis, P. (1996) *Frontiers in Translation*, National Research Council of Canada, Ottawa, Ontario.

Nierhaus, K.H., Subramanian, A.R., Erdmann, V.A., Franceschi, F., Wittmann-Liebold, B. (Eds.) (1994) *Translational Apparatus: Structure, Function, Regulation and Evolution*, Plenum Press, New York.

Nierhaus, K.H., Wilson, D.N. (Eds.) (2004) *Protein Synthesis and Ribosome Structure: Translating the Genome*, Wiley-VCH Verlag, Weinheim.

Primary Literature

Agrawal, R.K., Heagle, A.B., Penczek, P., Grassucci, R.A., Frank, J. (1999) EF-G-dependent GTP hydrolysis induces translocation accompanied by large conformational changes in the 70S ribosome, *Nat. Struct. Biol.* **6**, 643–647.

Agrawal, R.K., Sharma, R.S., Kiel, M.C., Irokawa, G., Booth, T.M., Spahn, C.M.T., Grassucci, R.A., Kaji, A., Frank, J. (2004) Visualization of ribosome-recycling factor on the *Escherichia coli* 70S ribosome: Functional implications, *Proc. Natl. Acad. Sci. U.S.A.* **101**, 8900–8905.

Al-Karadaghi, S., Kristensen, O., Liljas, A. (2000) A decade of progress in understanding the structural basis of protein synthesis, *Prog. Biophys. Mol. Biol.* **73**, 167–193.

Auerbach, T., Bashan, A., Yonath, A. (2004) Ribosomal antibiotics: structural basis for resistance, synergism and selectivity, *Trends Biotechnol.* **22**, 570–576.

Ban, N., Nissen, P., Hansen, J., Moore, P.B., Steitz, T.A. (2000) The complete atomic structure of the large ribosomal subunit at 2.4 Å resolution, *Science* **289**, 905–920.

Bashan, A., Agmon, I., Zarivach, R., Schluenzen, F., Harms, J., Berisio, R., Bartels, H., Franceschi, F., Auerbach, T., Hansen, H.A.S., Kossoy, E., Kessler, M., Yonath, A. (2003) Structural basis of the ribosomal machinery for peptide bond formation, translocation, and nascent chain progression, *Mol. Cell.* **11**, 91–102.

Blanchard, C.S., Gonzalez R.L. Jr., Kim, H.D., Chu, S., Puglisi, J.D. (2004) tRNA selection and kinetic proofreading in translation, *Nat. Struct. Mol. Biol.* **11**, 1008–1014.

Cao, Q., Richter, J.D. (2002) Dissolution of the maskin-eIF4E complex b cytoplasmic polyadenylation and poly(A)-binding protein controls cyclin B1 mRNA translation and oocyte maturation, *EMBO J.* **21**, 3852–3862.

Carter, A.R., Clemons, W.M., Brodersen, D.E., Morgan-Warren, R.J., Wimberly, B.T., Ramakrishnan, V. (2000) Functional insights from the structure of the 30S ribosomal subunit and its interactions with antibiotics, *Nature* **407**, 340–349.

Castagnetti, S., Ephrussi, A. (2003) Orb and a long poly(A) tail are required for efficient oskar translation at the posterior pole of the Drosophila oocyte, *Development* **130**, 835–843.

Charlesworth, A., Cox, L.L., MacNicol, A.M. (2004) Cytoplasmic polyadenylation element (CPE)- and CPE binding protein (CPEB)-independent mechanisms regulate early class maternal mNA translational activation in Xenopus oocytes, *J. Biol. Chem.* **279**, 17650–17659.

Clemens, M.J. (2004) Targets and mechanisms for the regulation of translation in malignant transformation, *Oncogene* **23**, 3180–3188.

Cormier, P., Pyronnet, S., Salaun, P., Mulner-Lorillon, O., Sonenberg, N. (2003) Cap-dependent translation and control of the cell cycle, *Prog. Cell Cycle Res.* **5**, 469–475.

Cropp, T.A., Schultz, P.G. (2004) An expanding genetic code, *Trends Genet.* **20**, 625–630.

Dever, T. (2002) Gene-specific regulation by general translation factors, *Cell* **108**, 545–556.

Diedrich, G., Spahn, C.M.T., Stelzi, U., Schäfer, M.A., Wooten, T., Bochkariov, D.E., Coopeman, B.S., Traut, R.R., Nierhaus, K.H. (2000) Ribosomal protein L2 is involved in the association of the ribosomal subunits, tRNA binding A and P sites and peptidyl transfer, *EMBO J.* **19**, 5241–5250.

Dodson, R.E., Shapiro, D.J. (2002) Regulation of pathways of mRNA destabilization and stabilization, *Prog. Nucleic Acid Res. Mol. Biol.* **72**, 129–164.

Gao, H., Valle, M., Ehrenberg, M., Frank, J. (2004) Dynamics of EF-G interaction with the ribosome explored by classification of a heterogeneous cryo-EM dataset, *J. Struct. Biol.* **147**, 283–290.

Gebauer, F., Hentze, M.W. (2004) Molecular mechanisms of translational control, *Nat. Rev. Mol. Cell Biol.* **5**, 827–835.

Gilbert, R.J.C., Fucini, P., Connell, S., Fuller, S.D., Nierhaus, K.H., Robinson, C.V., Dobson, C.M., Stuart, D.I. (2004) Three-dimensional structures of translating ribosomes by cryo-EM, *Mol. Cell* **14**, 57–66.

Green, R., Lorsch, J.R. (2002) The path to perdition is paved with protons, *Cell* **110**, 665–668.

Green, R., Noller, H.F. (1997) Ribosomes and translation, *Annu. Rev. Biochem.* **66**, 674–716.

Groisman, I., Jung, M.Y., Sarkissian, M., Cao, Q., Richter, J.D. (2002) Translational control of the embryonic cell cycle, *Cell* **109**, 473–483.

Gromadski, K.B., Rodnina, M.V. (2004) Streptomycin interferes with conformational coupling between codon recognition and GTPase activation on the ribosome, *Nat. Struct. Mol. Biol.* **11**, 316–322.

Grunberg-Manago, M. (1999) Messenger RNA stability and its role in control of gene expression in bacteria and phages, *Annu. Rev. Genet.* **33**, 193–227.

Halic, M., Becker, T., Pool, M.R., Spahn, C.M.T., Grassucci, R.A., Frank, J., Beckmann, R. (2004) Structure of the signal recognition particle interacting with the elongation-arrested ribosome, *Nature* **427**, 808–814.

Hansen, J.L., Schmeing, T.M., Moore, P.B., Steitz, T.A. (2002) Structural insights into peptide bond formation, *Proc. Natl. Acad. Sci. U.S.A.* **99**, 11670–11675.

Hardesty, B., Kramer, G. (2001) Folding of a nascent peptide on the ribosome, *Prog. Nucleic Acid Res. Mol. Biol.* **66**, 41–66.

Harms, J., Schluenzen, F., Zarivach, R., Bashan, A., Gat, S., Agmon, I., Bartels, H., Franceschi, F., Yonath, A. (2001) High resolution structure of the large ribosomal subunit from a mesophilic eubacterium, *Cell* **107**, 679–688.

Hazelrigg, T. (2004) Lost in translation gets an oskar, *Dev. Cell* **6**, 611–613.

Hellen, C.U., Sarnow, P. (2001) Internal ribosome entry sites in eukaryotic mRNA molecules, *Genes Dev.* **15**, 1593–1612.

Herr, A.J., Atkins, J.F., Gesteland, R.F. (2000) Coupling of open reading frames by translational bypassing, *Annu. Rev. Biochem.* **69**, 343–372.

Hirokawa, G., Inokuchi, H., Kaji, H., Igarashi, K., Kaji, A. (2004) *In vivo* effect of inactivation of ribosome recycling factor - fate of ribosomes after unscheduled translation downstream of open reading frame, *Mol. Microbiol.* **54**, 1011–1021.

Hoang, L., Frederick, K., Noller, H.F. (2004) Creating ribosomes with an all-RNA 30S subunit P site, *Proc. Natl. Acad. Sci. U.S.A.* **101**, 12439–12443.

Horton, L.E., Bushell, M., Barth-Baus, D., Tilleray, V.J., Clemens, M.J., Hensold, J.O. (2002) p53 activation results in rapid dephosphorylation of the eIF4E-binding protein 4E-BP1, inhibition of ribosomal protein S6 kinase and inhibition of translation initiation, *Oncogene* **21**, 5325–5334.

Ibba, M., Soll, D. (2000) Aminoacyl-tRN Asynthesis, *Annu. Rev. Biochem.* **69**, 617–650.

Jan, E., Hinzy, T.G., Sarnow, P. (2003) Divergent tRNA-like element supports initiation, elongation and termination of protein biosynthesis, *Proc. Natl. Acad. Sci. U.S.A.* **100**, 15410–15415.

Jenni, S., Ban, N. (2003) The chemistry of protein synthesis and voyage through the ribosomal tunnel, *Curr. Opin. Struct. Biol.* **13**, 212–219.

Katunin, V.I., Muth, G.W., Strobel, S.A., Wintermeyer, W., Rodnina, M.V. (2002) Important contribution to catalysis of peptide bond formation by a single ionizing group within the ribosome, *Mol. Cell* **10**, 339–346.

Krab, I.M., Parmeggiani, A. (2002) Mechanisms of EF-Tu, a pioneer GTPase, *Prog. Nucleic Acid Res. Mol. Biol.* **71**, 513–551.

Meyer, S., Temme, C., Wahle, W. (2004) Messenger RNA turnover in eukaryotes: pathways and enzymes, *Crit. Rev. Biochem. Mol. Biol.* **39**, 197–216.

Nakamura, A., Sato, K., Hanyu-Nakamura, K. (2004) Drosophila cup is an eIF4E binding protein that associates with Bruno and regulates oskar mRNA translation in oogenesis, *Dev. Cell* **6**, 69–78.

Nissen, P., Hansen, J., Ban, N., Moore, P.B., Steitz, T.A. (2000) The structural basis of ribosome activity in peptide bond synthesis, *Science* **289**, 920–930.

Noller, H.F. (2004) The driving force for molecular evolution of translation, *RNA* **10**, 1833–1837.

Peske, F., Savelsbergh, A., Katunin, V.I., Rodnina, M.V., Wintermeyer, W. (2004) Conformational changes of the small ribosomal subunit during elongation factor G-dependent tRNA-mRNA translocation, *J. Mol. Biol.* **343**, 1183–1194.

Prats, A.-C., Prats, H. (2002) Translational control of gene expression: role of IREs and consequences for cell transformation and angiogenesis, *Prog. Nucleic Acid Res. Mol. Biol.* **72**, 367–413.

Ramakrishnan, V. (2002) Ribosome structure and the mechanism of translation, *Cell* **108**, 557–572.

Rodnina, M.V., Wintermeyer, W. (2003) Peptide bond formation on the ribosome: structure and mechanism, *Curr. Opin. Struct. Biol.* **13**, 334–340.

Rodnina, M.V., Savelsbergh, A., Wintermeyer, W. (1999) Dynamics of translation on the ribosome: molecular mechanics of translocation, *FEMS Microbiol. Rev.* **23**, 317–333.

Ryabova, L.A., Pooggin, M.M., Hohn, T. (2002) Viral strategies of translation initiation: ribosomal shunt and reinitiation, *Prog. Nucleic Acid Res. Mol. Biol.* **72**, 1–39.

Sachs, A.B. (2000) Cell cycle-dependent translation initiation: IRES elements prevail, *Cell* **101**, 243–245.

Sachs, A.B., Varani, G. (2000) Eukaryotic translation initiation: there are (at least) two sides to every story, *Nat. Struct. Biol.* **7**, 356–361.

Schlüenzen, F., Pyetan, E., Fucini, P., Yonath, A., Harms, J.M. (2004) Inhibition of peptide bond formation by pleuromutilins: the structure of the 50S ribosomal subunit from *Deinococcus radiodurans* in complex with tiamulin, *Mol. Microbiol.* **54**, 1287–1294.

Schmeing, T.M., Seila, A.C., Hansen, J.L., Freeborn B., Soukup, J.K., Scaringe, S.A., Strobel, S.A., Moore, P.B., Steitz, T.A. (2002) A pre-translocational intermediate in protein synthesis observed in crystals of enzymatically active 50S subunits, *Nat. Struct. Biol.* **9**, 225–230.

Schmidt, E.V. (2004) The role of c-myc in regulation of translation initiation, *Oncogene* **23**, 3217–3221.

Sengupta, J., Nilsson, J., Gursky, R., Spahn, C.M.T., Nissen, P., Frank, J. (2004) Identification of the versatile scaffold protein RACK1 on the eukaryotic ribosome by cryo-EM, *Nat. Struct. Mol. Biol.* **11**, 957–962.

Shine, J., Dalgarno, L. (1974) The 3′-terminal sequence of *Escherichia coli* 16S ribosomal RNA: complementarity to nonsense triplets and ribosome binding sites, *Proc. Natl. Acad. Sci. U.S.A.* **71**, 1342–1346.

Shuman, S. (2001) Structure, mechanism and evolution of the mRNA capping apparatus, *Prog. Nucleic Acid Res. Mol. Biol.* **66**, 1–40.

Sonenberg, N., Dever, T.E. (2003) Eukaryotic translation factors and regulators, *Curr. Opin. Struct. Biol.* **13**, 56–63.

Spahn, C.M.T., Jan, E., Mulder, A., Grassucci, R.A., Sarnow, P., Frank, J. (2004) Cryo-EM visualization of a viral internal ribosome entry site bound to human ribosomes: the IRES functions as an RNA-based translation factor, *Cell* **118**, 465–475.

Spremulli, L.L., Coursey, A., Navratil, T., Hunter, S.E. (2004) Initiation and elongation factors in mammalian mitochondrial protein biosynthesis, *Prog. Nucleic Acid Res. Mol. Biol.* **77**, 211–161.

Vagner, S., Galy, B., Pyronnet, S. (2001) Irresistible IRES. Attracting the translation machinery to internal ribosome entry sites, *EMBO Rep.* **2**, 893–898.

Valle, M., Zavialov, A., Li, W., Stagg, S.M., Sengupta, J., Nielsen, R.C., Nissen, P., Harvey, S.C., Ehrenberg, M., Frank, J. (2003) Incorporation of aminoacyl-tRNA into the ribosome as seen by cryo-electron microscopy, *Nat. Struct. Biol.* **10**, 899–906.

Vanzi, F., Vladimirov, S., Knudsen, C.R., Goldman, Y.E., Cooperman, B.S. (2003) Protein synthesis by single ribosomes, *RNA* **9**, 1174–1179.

Vila-Sanjurjo, A., Schuwirth, B.-S., Hau, C.W., Cate, J.H.D. (2004) Structural basis for the control of translation initiation during stress, *Nat. Struct. Mol. Biol.* **11**, 1054–1059.

Wild, K., Halic, M., Sinning, I., Beckmann, R. (2004) SRP meets the ribosome, *Nat. Struct. Mol. Biol.* **11**, 1049–1053.

Wilusz, C.J., Wilusz, J. (2004) Bringing the role of mRNA decay in the control of gene expression into focus, *Trends Genet.* **20**, 491–497.

Wimberly, B.T., Brodersen, D.E., Clemons, W.M. Jr., Morgan-Warren, R.J., Carter, A.P., Vonrhein, C., Hartsch, T., Ramakrishnan, V. (2000) Structure of the 30S ribosomal subunit, *Nature* **407**, 327–339.

Yonath, A. (2002) The search and its outcome: high-resolution structures of ribosomal particles from mesophilic, thermophilic, and halophilic bacteria at various functional states, *Annu. Rev. Biophys. Biomed. Struct.* **31**, 257–273.

Yonath, A., Bashan, A. (2004) Ribosomal crystallography: initiation, peptide bond formation, and amino acid polymerization are hampered by antibiotics, *Annu. Rev. Microbiol.* **58**, 233–251.

Yusupov, M.M., Yusupova, C.Z., Baucom, A., Lieberman, K., Earnest, T.N., Cate, J.H.D., Noller, H.F. (2001) Crystal structure of the ribosome at 5.5 A resolution, *Science* **292**, 883–896.

4
Alternatively Spliced Genes

Jane Y. Wu[1], Liya Yuan[2], and Necat Havlioglu[2]
[1] *Northwestern University Feinberg, School of Medicine Center for Genetic Medicine, Chicago, IL, USA*
[2] *Washington University School of Medicine, St. Louis, MO, USA*

1	**Pre-mRNA Splicing and Splicing Machinery**	**119**
1.1	Splicing Machinery: Spliceosome	119
1.2	Splicing Signals	120
1.3	Spliceosomal UsnRNP Biogenesis	129
1.4	Spliceosome Assembly	130
1.5	Biochemical Mechanisms of pre-mRNA Splicing	133
2	**Alternative pre-mRNA Splicing**	**134**
2.1	Alternative Splicing and its Role in Regulating Gene Activities and Generating Genetic Diversity	134
2.1.1	Different Patterns of Alternative Splicing	135
2.1.2	Alternative Splicing and Genetic Diversity	135
2.2	Mechanisms Underlying Alternative Splicing Regulation	136
2.2.1	Splicing Signals and Splicing Regulatory Elements	137
2.2.2	Trans-acting Splicing Regulators	140
2.3	Tissue-specific and Developmentally Regulated Alternative Splicing	143
2.4	Regulation of Alternative Splicing in Response to Extracellular Stimuli	144
3	**Pre-mRNA Splicing and Human Diseases**	**144**
3.1	Splicing Defects in Human Diseases	145
3.2	Molecular Mechanisms Underlying Splicing Defects Associated with Disease	149
4	**Perspectives on Diagnosis and Treatment of Diseases Caused by pre-mRNA Splicing Defects**	**152**
4.1	Diagnosis of Human Diseases Caused by Splicing Defects	152

Genomics and Genetics. Edited by Robert A. Meyers.
Copyright © 2007 Wiley-VCH Verlag GmbH & Co. KGaA, Weinheim
ISBN: 978-3-527-31609-0

4.2	Potential Therapeutic Approaches	153
4.2.1	Oligonucleotide-based Approaches: Antisense, RNAi, and Chimeric Molecules	153
4.2.2	Ribozymes	153
4.2.3	SMaRT	154
4.2.4	Chemical Compounds	154
5	**Concluding Remarks**	154
	Acknowledgment	155
	Bibliography	155
	Books and Reviews	155
	Primary Literature	158

Keywords

Pre-mRNA
Nascent transcripts that are precursors of mature messenger RNAs.

Exon
Gene regions that are present in mature messenger RNA transcripts.

Intron
Intervening sequences that are removed from pre-mRNA transcripts and not represented in mature mRNA species.

Pre-mRNA Splicing
The process of removing introns from pre-mRNAs and ligating exons to form mature mRNA transcripts.

UsnRNPs
Uridine-rich small nuclear ribonucleoprotein particles.

Spliceosome
The macromolecular machine in which pre-mRNA splicing reactions take place.

5′ and 3′ Splice Sites
Sequences that spliceosomes recognize at 5′ and 3′ ends of an intron. They are also named splice donor and splice acceptor sites respectively.

Alternative Splicing
The process in which cells selectively use different combinations of splice sites to generate two or more mRNA transcripts from a single pre-mRNA species.

Splicing Regulatory Elements
Specific sequences in introns or exons that are recognized by trans-acting factors to allow either enhancement or repression of splicing.

Splicing Regulators
Trans-acting factors that interact with splicing regulatory elements to either stimulate or inhibit splicing.

■ An important step of eukaryotic gene expression is pre-mRNA splicing, the process of removing intervening sequences (introns) from the nascent transcript (messenger RNA precursor, or pre-mRNA). The discovery of split genes in the viral genome and subsequent research in the field of pre-mRNA splicing have greatly advanced our understanding of mammalian gene regulation. Studies on pre-mRNA splicing have also facilitated sequence analyses of the human genome. With the completion of human genome sequencing, it is now further appreciated that pre-mRNA splicing and alternative splicing play critical roles in regulating gene expression and in enhancing genetic diversity.

Evolutionarily, the basic machinery for pre-mRNA splicing appears to be highly conserved among different species of metazoans. In *Saccharomyces cerevisiae*, although only a small percentage of genes undergo splicing, more than 100 genes have been identified that are either dedicated to or involved in pre-mRNA splicing. In mammals, pre-mRNA splicing is a crucial step for gene expression because the vast majority of mammalian transcription units contain one or more introns that must be accurately removed to form mature and functional messenger RNA (mRNA) species. In this chapter, we review the current knowledge about mammalian pre-mRNA splicing, with special emphasis on the aspects related to the pathogenesis or treatment of human diseases.

1
Pre-mRNA Splicing and Splicing Machinery

1.1
Splicing Machinery: Spliceosome

Pre-mRNA splicing is the first step of the major post-transcriptional processes in eukaryotic gene expression (Fig. 1). The biochemical reactions of pre-mRNA splicing occur in a macromolecular machine named the *spliceosome*. This large RNA–protein complex contains, in addition to the pre-mRNA substrate, several uridine-rich small nuclear ribonucleoprotein (UsnRNP) particles as well as a myriad of non-snRNP protein factors. The splicing machinery is similar in complexity and size to that of the protein synthesis machinery, the ribosome. In all cases studied so far, splicing reactions take place in the mature spliceosome inside the nucleus.

For processing the majority of introns (the major class, also called the U2-type

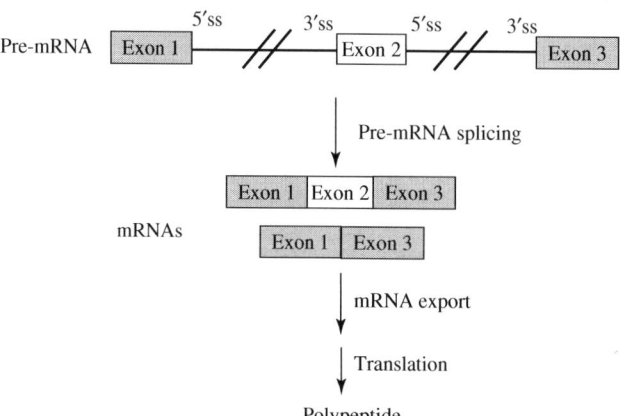

Fig. 1 The schematics for major posttranscriptional processes in eukaryotic gene expression. Different splice isoforms of mRNAs can be produced as a result of alternative selection of splice sites.

intron), the spliceosome contains U1, U2, U4/6, and U5snRNPs. The splicing of the minor class of introns (also called the U12-type) occurs in the spliceosome containing U11 and U12, in addition to U4atac, U6atac, and U5snRNPs (Figs. 2 and 3). This chapter will focus on the splicing of introns in the major class. This is because this class accounts for more than 99% of known introns. Most of our knowledge about pre-mRNA splicing has also come from studies of the major class of introns.

Comparison of genes coding for components of splicing machinery, including both snRNAs and protein factors, reveals a high degree of conservation through evolution. Amazing similarity is found among spliceosomal components from yeast, fruit fly, and human. A recent study using nanoscale microcapillary liquid chromatography tandem mass spectrometry to analyze partially purified human spliceosomes assembled on model splicing substrates revealed 145 distinct spliceosomal proteins. This suggests that the spliceosome is one of the most complex macromolecular machines for mammalian gene expression. Table 1 shows the currently known human proteins identified in functional spliceosomes assembled on model splicing substrates. The role of individual proteins in splicesome assembly and in splicing regulation will be discussed.

1.2
Splicing Signals

The sites for cleavage and ligation in splicing reactions are defined by conserved splicing signals in the pre-mRNA, including the 5′ splice site (5′ss), the branch site, the polypyrimidine tract, and the 3′ splice site (3′ss) consensus sequence. These cis-elements are important sites for RNA–RNA and protein–RNA interactions. Systematic analyses of both yeast and higher eukaryotic splicing signal sequences led to the classification of U2 and U12 types of introns based on the key UsnRNPs involved in the recognition of the branch sites in the corresponding introns (Fig. 2).

In general, the spliceosome for U2-type introns contains U1, U2, and U4/U6,

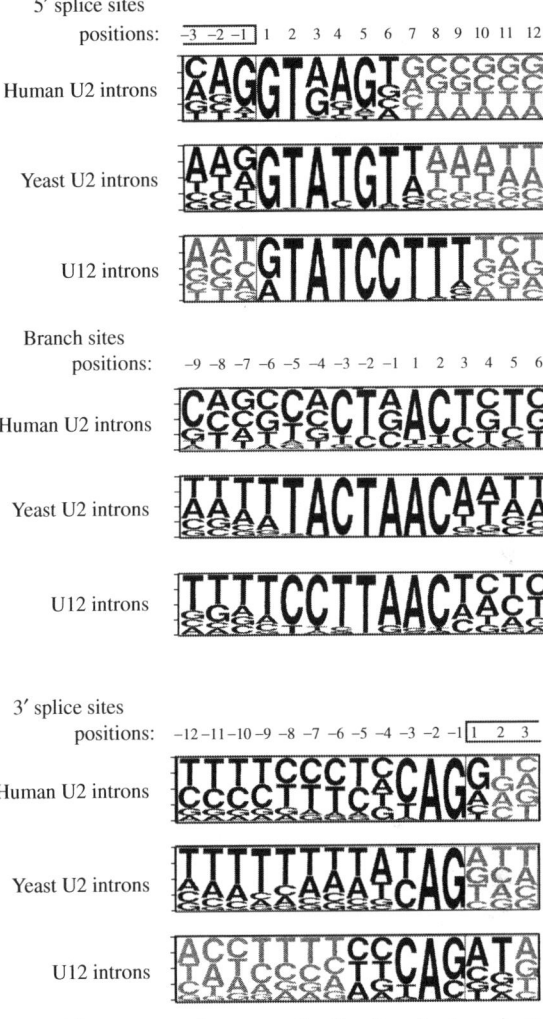

Fig. 2 The sequence features at the 5′ splice site, branch site, and 3′ splice site as derived from representative sets of introns of corresponding types. (Modified with permission from Burge, C.B., Tuschl, T.H., Sharp, P.A. (1999). in: The RNA World, Gesteland, R.F., Cech, T.R., Atkins, J.F. (Eds) *The RNA World*, Cold Spring Harbor Laboratory Press, New York, pp. 525–560). The frequency of each nucleotide (A, G, T, or C) at each sequence position is depicted by the heights of the corresponding letters in the diagram.

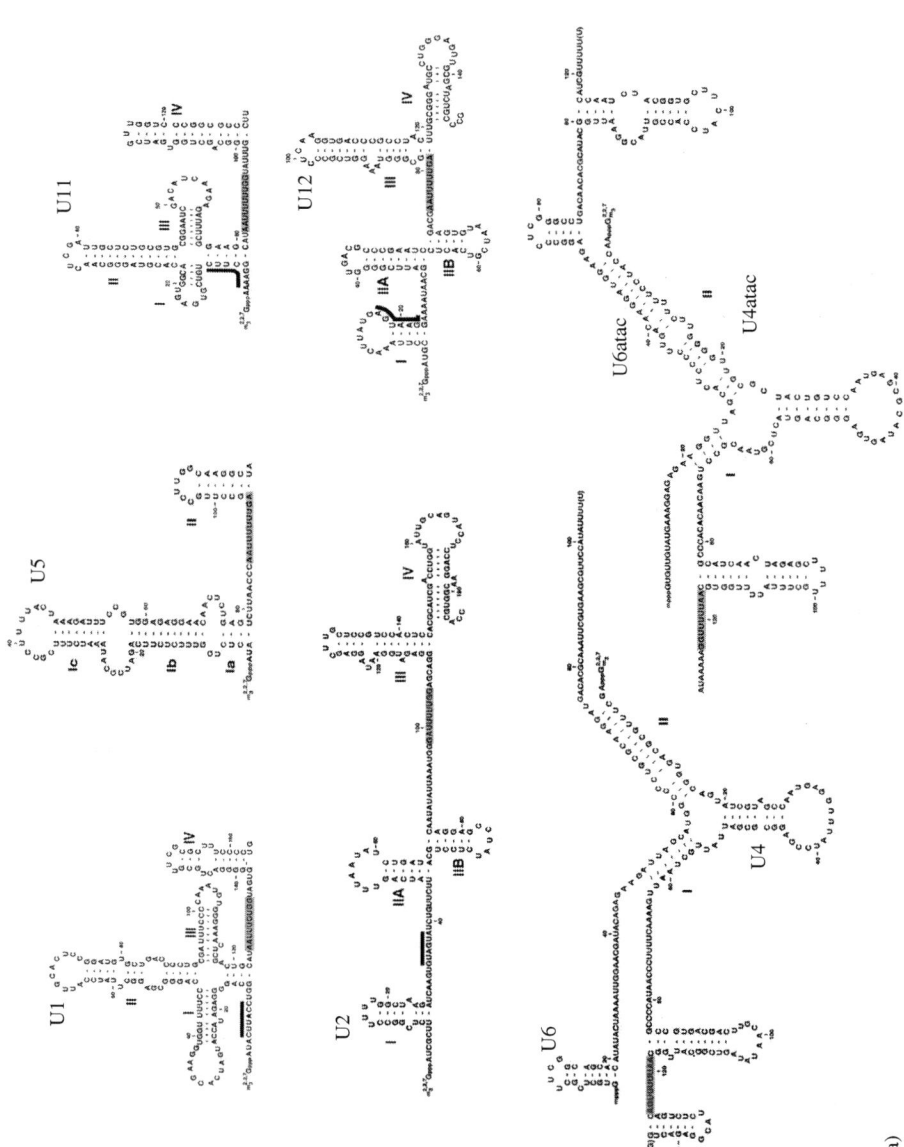

(a)

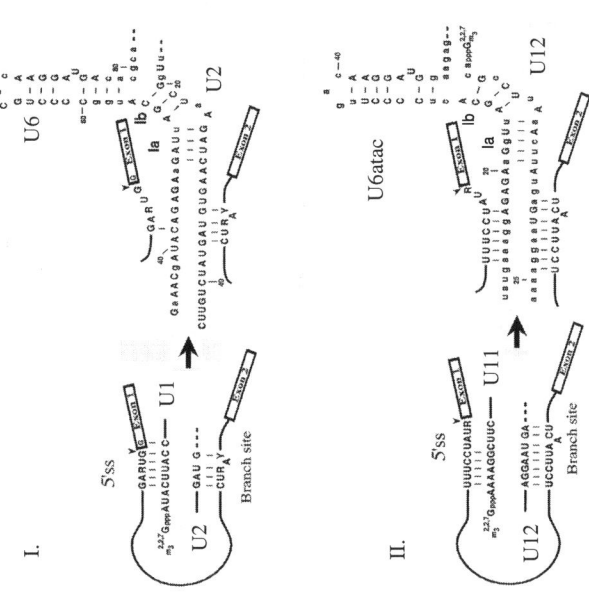

Fig. 3 Sequences of human spliceosomal UsnRNAs and their interactions with splice sites. Panel (a) Sequences and secondary structures of spliceosomal snRNAs: U1, U2, U4/6 (for the U2 type), and U11, U12, U4atac/U6atac snRNA (for the U12 type) as well as U5snRNAs (for both classes). The 5' to 3' orientation of U4 and U4atac is from right to left, and that of U1, U11, U2, U12, and U5snRNAs is from left to right. The RNA helices are indicated by Roman numerals, and nucleotide positions by Arabic numerals. The shaded nucleotides are binding sites for Sm proteins, and regions marked with black lines form base-pairing interactions with the pre-mRNA. (Modified with permission from Burge, C.B., Tuschl, T.H., Sharp, P.A. (1999). in: The RNA World, Gesteland, R.F., Cech, T.R., Atkins, J.F. (Eds) The RNA World, Cold Spring Harbor Laboratory Press, New York, pp. 525–560.) Panel (b) Base-pairing interactions in the U2 and U12 types of spliceosome. The exon-intron boundaries at 5' splice sites are marked by arrowheads. During early stages of spliceosome assembly, U1 or U11snRNP interacts with the 5' splice site, and U2 or U12snRNP with the branch sites. These interactions are displaced in mature spliceosome by U6 or U6atac associated with U2 and U12 respectively. In these drawings, the distances between interacting nucleotides do not reflect corresponding physical distances in the spliceosome. (Adapted from Tarn, W.Y., Steitz, J.A. (1996b) Highly diverged U4 and U6 small nuclear RNAs required for splicing rare AT-AC introns, Science **273** (5283), 1824–1832.)

Tab. 1 Proteins detected in human spliceosomes.

Acce.#	Protein name	Yeast homolog (SGD ORF)	Motifs	Cal. MWt
A. snRNP proteins				
Sm/LSm core proteins				
P14678	Sm B/B'	SMB1 (YER029C)	Sm 1 and 2	24 610
P13641	Sm D1	SMD1 (YGR074W)	Sm 1 and 2	13 282
P43330	Sm D2	SMD2 (YLR275W)	Sm 1 and 2	13 527
P43331	Sm D3	SMD3 (YLR147C)	Sm 1 and 2	13 916
P08578	Sm E	SME1 (YOR159C)	Sm 1 and 2	10 804
Q15356	Sm F	SMX3 (YPR182W)	Sm 1 and 2	9725
Q15357	Sm G	SMX2 (YFL017W-A)	Sm 1 and 2	8496
Q9Y333	hLSm2	LSM2 (YBL026W)	Sm 1 and 2	10 835
Q9Y4Z1	hLSm3	LSM3 (YLR438C-A)	Sm 1 and 2	11 714
Q9Y4Z0	hLSm4	LSM4 (YER112W)	Sm 1 and 2	15 350
Q9Y4Y8	hLSm6	LSM6 (YDR378C)	Sm 1 and 2	9128
Q9UK45	hLSm7	LSM7 (YNL147W)	Sm 1 and 2	11 602
U1snRNP-specific proteins				
P08621	U1-70kD	SNP1 (YIL061C)	RS, 1 RRM	70 081
P09012	U1 A	MUD1 (YBR119W)	2 RRMs	31 280
P09234	U1 C	YHC1 (YLR298C)	ZF	17 394
AB034205	LUC7A	LUC7 (YDL087C)	RS-rich	51 466
U2snRNP-specific proteins				
AF054284	SAP155 (SF3b155)	HSH155 (YMR288W)	PP2A	1 45 815
Q13435	SAP145 (SF3b145)	CUS1 (YMR240C)	Pro-rich	97 657
AJ001443	SAP130 (SF3b130)	RSE1 (YML049C)		1 35 592
Q15459	SAP114 (SF3a120)	PRP21 (YJL203W)	SWAP, UBQ	88 886
Q15428	SAP62 (SF3a66)	PRP11 (YDL043C)	ZF	49 196
A55749	SAP61 (SF3a60)	PRP9 (YDL030W)	ZF	58 849
Q15427	SAP49 (SF3b49)	HSH49 (YOR319W)	2 RRMs	44 386
P09661	U2 A'	LEA1 (YPL213W)	Leu-rich	28 444
P08579	U2 B''	MSL1 (YIR009W)	2 RRMs	25 486
Q9Y3B4	p14	SNU17 (YIR005W)	1 RRM	14 584
U5snRNP-specific proteins				
AB007510	U5-220kD (PRPC8)	PRP8 (YHR165C)		2 73 785
O75643	U5-200kD	BRR2 (YER172C)	DExD, Helicase	1 94 479
BC002360	U5-116kD	SNU114 (YKL173W)	G domain	1 09 478
BC001666	U5-102kD	PRP6 (YBR055C)	HAT	1 06 925
BC002366	U5-100kD	PRP28 (YDR243C)	DExD, Helicase, RS	95 583
BC000495	U5-52kD	SNU40 (YHR156C)	GYF	37 646
AF090988	U5-40kD		WD40s	39 299
O14834	U5-15kD	DIB1 (YPR082C)		16 786
U4/U6snRNP-specific proteins				
T50839	U4/U6-90kD (HPRP3)	PRP3 (YDR473C)	PWI	77 529
BC007424	U4/U6-60kD (HPRP4)	PRP4 (YPR178W)	WD40s	58 321

Tab. 1 (continued)

Acce.#	Protein name	Yeast homolog (SGD ORF)	Motifs	Cal. MWt
AL050369	U4/U6-61kD (PRPF31)	PRP31 (YGR091W)	NOP	55 424
AF036331	U4/U6-20kD	CPH1 (YDR155C)	Cyclophilin	19 207
P55769	U4/U6-15.5kD	SNU13 (YEL026W)	Putative RRM	14 174
U4/U6.U5 tri-snRNP specific proteins				
T00034	Tri-snRNP 110kD	SNU66 (YOR308C)	RS	90 255
AF353989	Tri-snRNP 65kD	SAD1 (YFR005C)	RS, UBQ, ZF	65 415
X76302	Tri-snRNP 27kD		RS	18 859
B. Non-snRNP spliceosomal proteins				
SR proteins				
Q08170	SRp75		2 RRMs, RS	56 792
Q05519	p54/SFRS11		1 RRM, RS	53 542
Q13247	SRp55		2 RRMs, RS	39 568
Q13243	SRp40		2 RRMs, RS	31 264
Q07955	ASF/SF2		2 RRMs, RS	27 613
Q16629	9G8		1 RRM, RS	27 367
Q01130	SC35		1 RRM, RS	25 575
Q13242	SRp30c		2 RRMs, RS	25 542
AF057159	hTra2		1 RRM, RS	21 935
P23152	SRp20		1 RRM, RS	19 330
Other spliceosomal proteins with known motifs				
NP_057417	SRm300		RS-rich	251 965
P26368	U2 AF65	MUD2 (YKL074C)	RS, 3 RRMs	53 501
Q01081	U2AF35		RS	27 872
Y08765	SF1	BBP (Ylr116w)	ZF, KH	68 632
AB018344	hPrp5	PRP5 (YBR237W)	DExD, Helicase, RS	117 461
O60231	hPrp2	PRP2 (YNR011C)	DExD, HELICc	119 172
P38919	IF4N	FAL1 (YDR021W)	DExD, HELICc	46 833
P17844	p68	DBP2 (YNL112W)	DExD, Helicase	69 148
Q92841	p72	DBP2 (YNL112W)	DExD, HELICc	72 371
T00333	fSAP164		DExD	163 986
Q08211	RHA	(YLR419W)	DExD, HELICc, DSRM	140 877
O00571	DDX3	DBP1 (YPL119C)	DExD, HELICc, SR	73 243
Q9UJV9	Abstrakt		DExD, HELICc	69 738
P42285	fSAP118	MTR4 (YJL050W)	DExD	117 790
BC002548	fSAP113		DExD	113 671
NP_055792	fSAP152 (Acinus)		1 RRM	151 887
Q14011	CIRP (cold inducible RNA-binding)		1 RRM, Gly-rich	18 648
BC007871	SPF45		1 RRM, Gly-rich	44 962
NP_008938	CF I-68kD (cleavage factor)		1 RRM	59 209

(continued overleaf)

Tab. 1 (continued)

Acce.#	Protein name	Yeast homolog (SGD ORF)	Motifs	Cal. MWt
AC004858	fSAP94		1 RRM, PWI	94 122
P52298	hCBP20	CBP20 (YPL178W)	1 RRM	18 001
BC003402	fSAP47	ECM2 (YBR065C)	1 RRM, ZF	46 896
BC006474	fSAPa		1 RRM, SWAP	Partial
AAB18823	TAT-SF1	CUS2 (Ynl286w)	2 RRMs	85 759
I55595	fSAP59 (transcription cofactor)		2 RRMs	58 657
U76705	IMP3		2 RRMs, 4 KH	63 720
NP_073605	OTT		3 RRMs	1 02 135
BC008875	PUF60		3 RRMs	58 171
Q15097	PAB2	PAB1 (YER165W)	4 RRMs, PABP	58 518
AF356524	SHARP (transcription cofactor)		4 RRMs	4 02 248
P41223	fSAP17	CWC14 (YCR063W)	ZF	16 844
O43670	ZNF207 (transcription factor)		ZF	50 751
AF044333	PLRG1	PRP46 (YPL151C)	WD40s	57 194
BC008719	hPrp19	PRP19 (YLL036C)	WD40s	55 181
BC006849	hTEX1	TEX1 (YNL253W)	WD40s	38 772
BC002876	fSAP57	PFS2 (YNL317W)	WD40s	57 544
BC003118	fSAP35		WD40s	34 849
AF083383	SPF38		WD40s	34 290
Other spliceosomal proteins with known motifs				
XM_087118	hFBP3		1 WW	24 297
AF049523	hFBP11	PRP40 (YKL012W)	2 WW, 1 FF	Partial
T08599	CA150		3 WW, 6 FF	1 23 960
XP_042023	CyP64	CPR3 (YML078W)	Cyclophilin, WD40	64 222
AF271652	PPIL3	CPR6 (YLR216C)*	Cyclophilin	18 155
S64705	PPIL2	CYP5 (YDR304C)*	Cyclophilin, Ubox	58 823
BC003048	PPIL1	CYP2 (YHR057C)*	Cyclophilin	18 237
Q9UNP9	PPIE	CPH1 (YDR155C)*	Cyclophilin	33 431
AY029347	hPrp4 kinase		TyrKc	1 16 973
Q9NYV4	CrkRS	CTK1 (YKL139W)	TyrKc	1 64 155
T12531	TIP39	(YLR424W)*	Gly-rich	96 820
BC001403	CF I-25kD (cleavage factor)		NUDIX	26 227
Q09161	hCBP80	CBP80 (YMR125W)	MIF4G	91 839
AB046824	fSAPb	CWC22 (YGR278W)	MIF4G	Partial
T02672	fSAPc		ERM	Partial
AF255443	hCrn	CLF1 (YLR117C)	HAT	99 201
BC007208	XAB2 (transcription cofactor)	SYF1 (YDR416w)	HAT	1 00 010

Tab. 1 (continued)

Acce.#	Protein name	Yeast homolog (SGD ORF)	Motifs	Cal. MWt
Q13573	SKIP	PRP45 (YAL032C)	SKIP	61 494
A53545	hHPR1	HPR1 (YDR138W)	DEATH	75 627
AF083385	SPF30**		TUDOR	26 711
NP_055095	SPF31		DnaJ	30 986
Q13123	RED		RED	65 630
Spliceosomal proteins without known motifs				
NP_001244	CDC5L	CEF1 (YMR213W)		92 250
AAK21005	ASR2			100 665
AF441770	hTHO2	THO2 (YNL139C)		169 573
I39463	fSAP79			78 536
AK027098	fSAP24			23 671
BC001621	SNP70			69 998
BC006350	fSAP71			70 521
P55081	MFAP1			51 855
AJ276706	WTAP			44 243
O15355	PP2Cγ	PTC3 (YBL056W)		59 272
AJ279080	fSAP105 (putative transcription factor)			104 804
NP_056311	fSAP121			121 193
BC004442	fSAP33	ISY1 (YJR050W)		32 992
AF161497	CCAP2	CWC15 (YDR163W)		26 610
Q9Y5B6	GCFC (putative transcription factor)			29 010
NP_073210	DGSI			52 567
NP_110517	NAP (nuclear protein inducing cell death)			65 173
NP_056299	fSAP29			28 722
NP_037374	fSAP23 (putative transcription factor)			22 774
BC004122	fSAP11			10 870
BC000216	fSAP18			18 419
AF081788	SPF27			26 131
C. Late-acting spliceosomal proteins				
Catalytic step II and late-acting proteins				
Q92620	hPrp16	PRP16 (YKR086W)	DExD, Helicase, RS	140 473
Q14562	hPrp22	PRP22 (YER013W)	DExD, Helicase, RS	139 315
O43143	hPrp43	PRP43 (YGL120C)	DExD, Helicase	92 829
BC010634	hSlu7	SLU7 (YDR088C)	ZF	68 343
AF038392	hPrp17	PRP17 (YDR364C)	WD40s	65 521
BC000794	hPrp18	PRP18 (YGR006W)		39 860

(continued overleaf)

Tab. 1 (continued)

Acce.#	Protein name	Yeast homolog (SGD ORF)	Motifs	Cal. MWt
Spliced mRNP/EJC proteins				
AF048977	SRm160		PWI, RS	93 519
Q13838	UAP56	SUB2 (YDL084W)	DExD, Helicase	48 991
PJC4525	RNPS1		1 RRM, Ser-rich	34 208
NM_005782	Aly	YRA1 (YDR381W)	1 RRM	26 861
Q9Y5S9	Y14		1 RRM	19 889
P50606	Magoh		Mago_nashi	17 164

Source: (Adapted with permission from Zhou, Z., Licklider, L.J., Gygi, S.P., Reed, R. (2002) Comprehensive proteomic analysis of the human spliceosome, Nature 419(6903), 182–185).
Note: A. snRNP proteins. B. Non-snRNP spliceosomal proteins. C. Late-acting spliceosomal proteins. In each list, spliceosomal proteins are grouped by their structural motifs. Corresponding GenBank accession numbers are shown (Acce#), together with their yeast homologs (as SGD open reading frame, SGD ORF), sequence motifs, and calculated molecular weight (Cal. Mwt).
COLD: cold shock RNA-binding domain; Cyclophilin: cyclophilin type peptidyl–prolyl cis–trans isomerase; DEATH: domain found in proteins involved in cell death; DExD: DExD/H-like helicases superfamily; DSRM: double-stranded RNA-binding motif; ERM: ezrin/radixin/moesin family motif; fSAP: functional spliceosome-associated protein; FF: two conserved F residues; G domain: GTP-binding domain that contains a P-loop motif; GYF: contains conserved G-T-F residues; HAT: Half-A-TPR (tetratrico-peptide repeat); Gly-rich: peptide sequences rich in G, R, S residues; HELICc: helicase superfamily c-terminal domain; KH: hnRNP K homology RNA-binding domain; MIF4G: middle domain of eukaryotic initiation factor 4G; NOP: putative snoRNA binding domain; NUDIX: mutT-like domain; PABP: poly-adenylate binding protein, unique domain; PP2A: protein phosphatase 2A repeat; PWI: domain in splicing factors; RED: protein with extensive stretch of alternating R and E or D; RRM: RNA recognition motif; RS: arginine–serine-rich domains; SKIP: conserved domain found in chromatinic proteins; Sm: snRNP Sm proteins; SWAP: suppressor-of-white-apricot splicing regulator; TUDOR: a domain present in several RNA-binding proteins; TyrKc: tyrosine kinase, catalytic domain; UBQ: ubiquitin homologs; Ubox: modified RING finger domain; WD40s: WD40 repeats, structural repeats of the beta propeller domain; WW: domain with 2 conserved W residues, interacting with proline-rich polypeptides; ZF: Zinc finger domain.

whereas the U12-type of spliceosome involves U11, U12, and U4atac/U6atac, with both sharing U5snRNP. Although nucleotide sequences of U11, U12, and U4atac/U6atac snRNAs are different from their counterparts in the spliceosomes for U2-type introns, the predicted secondary structures of the corresponding snRNAs have striking similarity (Fig. 3a). In addition, the specific interactions between the spliceosomal snRNAs and pre-mRNA substrates also appear to be highly similar, as shown in Fig. 3(b). Both U2- and U12-types of introns can be found in the same genes. There is evidence suggesting that the two types of spliceosomes may interact and share some protein components in addition to the common U5snRNP. The evolutionary origins of these two types of spliceosomes remain largely speculative.

U12-class introns contain more conserved sequences at the 5′ss. The sequences around branch sites of the U12-class introns are also highly conserved with the TCCTTAAC (the underlined A as the branch site) consensus signal located approximately 10 to 20 nucleotides upstream of the 3'ss. A typical U12-class intron lacks a polypyrimidine tract between the branch site and the 3'ss.

In the yeast, *S. cerevisiae*, only a small fraction of genes contain introns, and these are usually short introns (approximately 240 introns, averaging 270 nucleotides in length). Splicing signals in yeast introns are highly conserved and may contain sufficient information for defining splice junctions, especially considering the number of genes devoted to or associated with pre-mRNA splicing in the yeast genome.

In mammals, pre-mRNA transcripts are usually much longer and contain multiple introns of variable sizes. In humans, the average size of exons is 150 nucleotides, and that of introns is approximately 3500 nucleotides. Mammalian introns can be as large as 500 kbp. The basic splicing signals in mammalian pre-mRNAs are degenerate, especially in the case of U2-type of introns. The branch sites for U2-type introns are highly divergent. At both the 5'ss and 3'ss, only two nucleotides (/GT at the 5'ss and AG/ at 3'ss) are highly conserved. As a result, the nucleotide sequences surrounding the splice junctions, the 5'ss and 3'ss, usually contain only a limited amount of information. This is not sufficient for conferring the specificity required to achieve accurate splice site selection. In mammals, the recognition of not only exon–intron junction sequences but also the regulatory elements in intronic and exonic regions is important for defining splice junctions and maintaining splicing fidelity. In addition, multiple networks of interactions among the machineries for transcription, cap formation, splicing, and polyadenylation may also influence splice site selection. This high degree of degeneracy in the splicing signals in mammalian pre-mRNA transcripts provides the flexibility for alternative selection and pairing of different splice sites, a fundamental mechanism for regulating alternative splicing.

1.3
Spliceosomal UsnRNP Biogenesis

As essential subunits of the splicing machinery, spliceosomal UsnRNPs contain not only uridine-rich snRNAs but also a number of polypeptides. U1, U2, U4, and U5snRNAs are transcribed by RNA polymerase II as precursors containing additional 3' nucleotides. After acquiring a monomethylated guanosine (m7G) cap structure, these pre-UsnRNAs are exported to the cytoplasm in a pathway dependent on the m7G cap, the cap-binding complex (CBC), RanGTP, and phosphorylated adaptor for RNA export. In the cytoplasm, pre-UsnRNAs interact with Sm proteins including B/B', D3, D2, D1, E, F, and G (Table 1A) to form the snRNP core structure.

The Sm protein-binding sites in U1, U2, U4, and U5snRNAs are highly conserved, containing two stem-loop structures flanking PuAU4-6GPu sequence (Fig. 3a). A number of proteins interacting with Sm proteins have been identified, including the protein product of SMN (survival of motor neuron) gene, SMN-interacting protein/Gemin2, Gemin3, and Gemin4. Genetic defects in the SMN gene cause spinal muscular atrophy (SMA), possibly by interfering with UsnRNP Sm core assembly and therefore deficiency in UsnRNP biogenesis.

Following UsnRNP Sm core assembly, the m7G cap of their snRNA is converted to the 2,2,7-tri-methylated guanosine (m3G), and the 3' extra nucleotides of pre-UsnRNAs are removed. These core UsnRNPs are imported into the nucleus to be assembled into the spliceosome. This UsnRNP import process requires not only general nuclear import factors such as Importin-β but also specific factors such as Snurportin-1 that recognizes the m3G cap of the UsnRNP and interacts with general factors for snRNP import into the nucleus.

Before the association of UsnRNP particle-specific proteins, the UsnRNAs undergo internal modifications including pseudouridylation and 2'-O-methylation. Such modifications appear to be necessary for the assembly of a functional UsnRNP, as shown for the human U2snRNP. These posttranscriptional modifications are mediated by a small nucleolar (sno)RNA-guided mechanism through the action of snoRNPs. For example, U85 snoRNP directs both 2'-O-methylation and pseudouridylation of U5snRNA. Following these posttranscriptional internal modifications of snRNAs, the assembly of UsnRNPs is completed with the association with individual UsnRNP-specific proteins (Table 1A). Among these UsnRNP-specific proteins, U5-220 kDa (also named Prp8 or PRPC8) is a crucial spliceosomal protein that is most highly conserved through evolution. Prp8 interacts with sequences of all major splicing signals including 5'ss, branch site, and 3'ss. It has been proposed that Prp8 plays a critical role in catalysis by aligning 5'ss, and 3'ss at the catalytic center.

U6snRNP plays an important role at the catalytic center of the spliceosome, and its biogenesis has several unique features. The U6snRNA is transcribed by RNA polymerase III, and its cap structure is a γ-monomethyl group. The U6snRNP does not contain an Sm protein. Instead, U6snRNP contains seven Sm-like proteins (Lsm2, 3, 4, 5, 6, 7, and 8) interacting with the U-rich region at the 3' end of the U6snRNA. The assembly of U6snRNP is believed to occur in the nucleus. The base-pairing interactions between U4snRNA and U6snRNA lead to the formation of the U4/U6snRNP, which associates with U5snRNP as the U4/U6.U5 tri-snRNP complex to join the spliceosome. The formation of U4/U6.U5 tri-snRNP complex and association of tri-snRNP with the spliceosome require PRPF31 (human homolog of yeast Prp31p) in addition to other proteins.

1.4
Spliceosome Assembly

A large number of studies show that mammalian spliceosomes are assembled on the pre-mRNA splicing substrate in an orderly fashion. A recent study, on the other hand, reported a penta-snRNP particle purified from the yeast extract. The finding of this "preassembled" penta-snRNP particle suggests the potential importance of concurrent multisite interactions during spliceosome assembly among spliceosomal components and different intronic as well as exonic regions. Many components of spliceosomes are conserved between yeast and human. In fact, much of our knowledge of spliceosome assembly is based on genetic studies in yeast and on biochemical experiments using both yeast and mammalian systems. The mammalian homologs of key spliceosomal components have been identified (Table 1). Systematic proteomic studies of spliceosomes assembled on model splicing substrates confirm the high degree of conservation between mammalian and yeast spliceosomal proteins (Table 1).

Spliceosome assembly is a highly dynamic process, with multiple RNA–RNA, RNA–protein, and protein–protein interactions. During spliceosomal assembly and activation, there are a series of conformational rearrangements. This process involves not only changes in the RNA conformation but also remodeling of snRNPs and exchanges of protein factors. Studies over the past two decades have only provided a sketch of this multicomponent interactive process. At the catalytic center of the functionally mature spliceosome, the 5′ss and 3′ss must be juxtaposed precisely to ensure accurate cleavage and ligation.

The classical view of spliceosome assembly features stepwise interactions between the pre-mRNA substrate and different spliceosomal UsnRNPs. Immediately after transcription, nascent pre-mRNA transcripts interact with hnRNP (heterogeneous nuclear ribonucleoprotein) proteins (Table 2) to form *H complex*. Spliceosome assembly is initiated with the recognition of the 5′ splice site by U1snRNP in an ATP-independent manner. Efficient interaction between the U1snRNP and 5′ splice site requires U1snRNP-specific proteins (such as U-70 K) and possibly other proteins. Recognition of the branch site and 3′ splice site requires cooperative binding of SF1 (also named the *branch point binding protein*, BBP) to the branch site and binding of U2 auxiliary factor (U2AF65 and U2AF35 as a

Tab. 2 Human H complex proteins.

Acce. #	Protein name	Motifs	Cal. MWt
Q13151	hnRNP A0	2 RRMs	30 841
P09651	hnRNP A1**	2 RRM	38 715
P22626	hnRNP A2/B1	2 RRMs	37 430
P51991	hnRNP A3	2 RRMs	39 686
P07910	hnRNP C1/C2	1 RRM	33 299
Q14103	hnRNP D0	2 RRMs	38 434
P26599	hnRNP I/PTB**	4 RRMs	57 221
Q07244	hnRNP K	KH	50 976
P14866	hnRNP L	3 RRMs	60 187
O43390	hnRNP R	3 RRMs	70 943
AL031668	hnRNP RALY	1 RRM	32 214
P35637	FUS/hnRNP P2	1 RRM	53 426
B54857	NF-AT 90 k	2 DSRMs, ZF	73 339
AJ271745	NFAR-1	DSRM, ZF	76 033
A54857	NF-AT 45 k	DSRM, ZF	44 697
AF037448	GRY-RBP	3 RRMs	69 633
P43243	Matrin3		94 623
O43684	hBUB3	WD40s	37 155
Q15717	HuR	3 RRMs	36 062
Q92804	TAFII68	1 RRM, ZF	61 830
P16991	YB1**	COLD	35 924
P16989	DBPA	COLD	40 060
P08107	HSP70	HSP70	70 052
P11142	HSP71	HSP70	70 898
P11021	GRP78	HSP70	72 116

The H complex proteins that are involved in alternative splicing regulation are marked with "**."

heterodimer) to the intronic sequence at the 3′ splice site. Splicing reactions can occur in the absence of detectable U1snRNP. However, efficient splicing of most introns appears to require U1snRNP. The formation of such an early complex (*E complex*) is considered to be the commitment step that directs the nascent pre-mRNA transcript into the splicing pathway.

A distinct feature of the formation of mammalian splicing commitment complex is the involvement of a family of proteins named SR proteins that are not found in the genome of *S. cerevisiae*. SR proteins share common structural features, with one or two RNA recognition motifs (RRMs) of the RNP-consensus (RNP-cs) type at the amino-terminus and a carboxyl domain rich in arginine and serine residues (RS domain, Table 1). These proteins play important roles in recognizing exons and mediating interactions between splice sites. SR proteins interact with exonic sequences and recruit splicing factors for the formation of "cross-exon" complex, (i.e. exon definition). SR proteins also interact with 5′ss and facilitate the "cross-intron" recognition by mediating protein–protein interactions between U1snRNP and factors associated with the branch site 3′ss including U2snRNP, U2AF, and possibly other proteins.

The binding of U1snRNP and SR proteins promotes the interaction of the 17S U2snRNP with the branch site sequence. The stable association of U2snRNP with the branch site to form the *A complex* is ATP-dependent and requires the U2snRNP-specific proteins SF3a [SF3a60/SAP61 (Prp9), SF3a66/SAP62 (Prp11) and SF3a120/SAP114 (Prp21)]. This process also requires non-snRNP splicing factors including hPrp5 (a putative ATP-dependent RNA helicase). Upon the integration of the U2snRNP complex, a U2snRNP-protein, p14, contacts the branch adenosine residue and interacts with other U2snRNP proteins. The next ATP-dependent step is the association of U4/U6.U5 tri-snRNP complex to form the *B1 complex*, in which the interaction of the 5′ss with U1snRNA is destabilized by U5-100 kDa (hPrp28), a putative RNA helicase. The 5′ss sequence is then engaged in interactions with U6snRNA around the intronic region and with U5snRNA at the exonic region. After escorting the U6snRNP into the B1 complex, the U4snRNP is released to form the *B2 complex*, a transition that requires ATP and possibly U5-200 kDa (human homolog of Brr2 in yeast, another putative RNA helicase). The next ATP-dependent transition is the formation of the *C1 complex*, a process involving the activity of yet another possible RNA helicase hPrp2 (human homolog of Prp2/Ynr011c). This may directly lead to the formation of the catalytic site for the *first step of splicing*, the cleavage at the 5′ss with the formation of a lariat intermediate (see Fig. 4). U2, U6, and U5snRNPs are associated with the catalytically active form of splicing complex in which the first step of splicing occurs. The *second step of splicing* (cleavage at the 3′ss and ligation of exons) requires an additional set of protein factors including hPrp16, hPrp17, hPrp18, and hSlu7 (Table 1). The human homolog of the yeast Prp16, hPrp16, contains an RS domain in addition to DEXD-box helicase and ATPase domains. The transition to form the *C2 complex* is again ATP-dependent. The catalytic center for the second step is also formed by U6snRNA, U2snRNA, and/or associated proteins. The splicing products, ligated exons, and the lariat intron, are then released from the spliceosome by another DEXD-box containing RNA helicase together with other factors. The release of the lariat intron and dissociation of U2, U5, and U6snRNPs from the *I complex* requires another putative RNA helicase, hPrp43. The released U6snRNP is reannealed with U4snRNP and associated with U5snRNP to form the U4/U6.U5 tri-snRNP complex to enter another cycle of spliceosome assembly. The lariat intron is debranched by the enzyme Dbr1. Many spliceosomal components are presumably recycled to form new spliceosomes.

It is clear that spliceosome assembly is a highly dynamic process involving multiple networks of RNA–RNA, protein–RNA, and protein–protein interactions at each step. Our understanding of this complex process, as well as its regulation, remains limited.

1.5
Biochemical Mechanisms of pre-mRNA Splicing

The complex and dynamic process of spliceosome assembly culminates in the formation of the catalytic core in which 5′ss and 3′ss are precisely juxtaposed. The biochemical reactions of pre-mRNA splicing involve two transesterification steps (Fig. 4). The first step is the cleavage at the 5′ss and the formation of the lariat intermediate with a 2′-OH displacing a 3′-OH group. The second step is the cleavage at the 3′ splice site with concomitant ligation of the 5′ and 3′ exons in which a 3′-OH displaces another 3′-OH group. Although it remains controversial, these two steps of splicing reactions may take place in two catalytic sites. In addition to the chemical differences between the

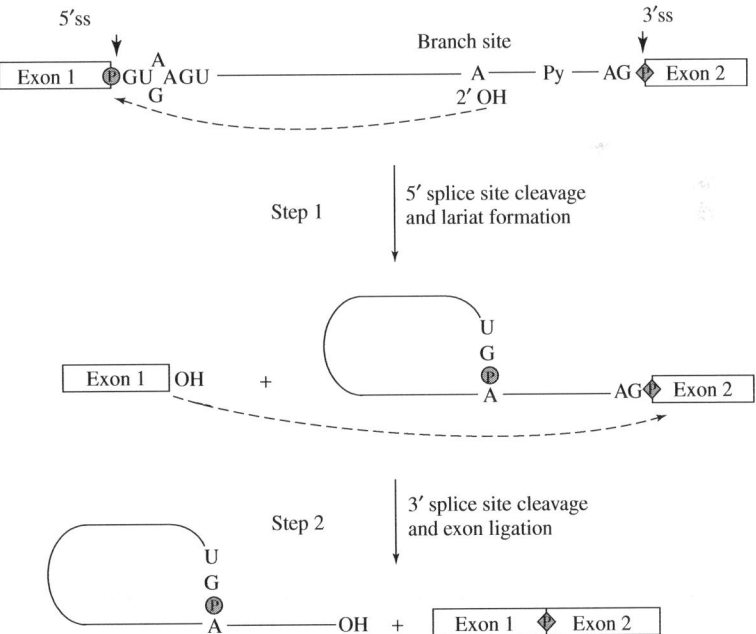

Fig. 4 The biochemical mechanisms of pre-mRNA splicing: two steps of transesterification. The phosphodiester linkages are indicated by the letter "p" inside a circle or a diamond.

two reaction steps, the stereoselectivity of the transesterification reactions and the difference in the metal-ion dependence support the existence of two reaction sites.

Accumulating evidence suggests that pre-mRNA splicing may be RNA-catalyzed with U6snRNA at the center for catalysis and protein factors including PRP8 as components of the catalytic core. U6snRNA plays a critical role in catalysis and in interacting with pre-mRNA and other spliceosomal snRNAs. In yeast, U6snRNA contributes to spliceosomal catalysis by coordinating catalytic metals. Precise catalytic mechanisms of mammalian pre-mRNA splicing remain to be elucidated.

2
Alternative pre-mRNA Splicing

2.1
Alternative Splicing and its Role in Regulating Gene Activities and Generating Genetic Diversity

The vast majority of vertebrate genes contains at least one intron. It is estimated that more than 50% of human genes undergo *alternative splicing*, the process that generates distinct splicing products from the same pre-mRNA transcript by using different splice sites (Fig. 5). Alternative splicing is an important mechanism for regulating gene activities in eukaryotic species. Alternative splicing events can

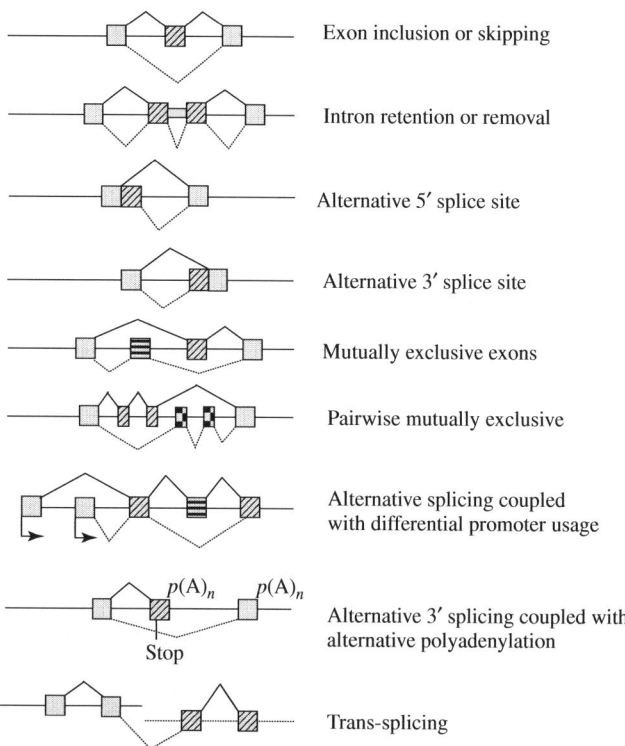

Fig. 5 Diverse patterns of alternative splicing. Alternatively spliced regions or exons are illustrated as shaded boxes.

occur in the protein-coding regions or in the regulatory regions of genes, including both 5′ and 3′ untranslated regions. Alternative splicing can affect peptide coding capacity or influence mRNA stability or translational control of the transcription products. Alternative splicing regulation can also be coupled with transcription, polyadenylation, RNA editing, or mRNA exporting processes. A recent example for the coupling of splicing with other processes of gene regulation is that steroid hormone receptors can simultaneously regulate transcription and alternative splicing by recruiting coregulators involved in both processes. In addition, alternative splicing can affect posttranslational modifications of protein products.

2.1.1 Different Patterns of Alternative Splicing

A number of distinct alternative splicing patterns have been reported (Fig. 5). Most common alternative splicing events include exon inclusion/skipping, intron removal/retention, and alternative selection of competing 5′ or 3′ splice sites. More complex patterns of alternative splicing include mutually exclusive or cassette types of exon inclusion. Alternative selection of terminal exons can be coupled with differential promoter usage or polyadenylation. Furthermore, recent studies have documented possible alternative trans-splicing of mammalian genes, although it may occur only at a low frequency in mammals.

2.1.2 Alternative Splicing and Genetic Diversity

Alternative splicing regulates gene activities involved in every aspect of cell survival and function. It is a major mechanism for generating the complexity of mammalian proteomes. Alternative splicing contributes to proteome expansion by a number of mechanisms, such as the usage of distinct translation start sites, in-frame nucleotide deletion or insertion, changes in peptide sequence, and alternative usage of different translation stop codons. Such changes in peptide sequence or length may lead to the formation of proteins with distinct properties, including biochemical/biophysical characteristics, subcellular localization (secreted versus membrane associated, membrane-tethered versus cytoplasmic, cytoplasmic versus nuclear), posttranslational modifications (glycosylation, phosphorylation, or lipid modification), or interactions with other cellular components.

Alternative splicing can be an excellent mechanism for generating functionally antagonistic products from the same genetic locus and for the fine-tuning of gene activities at the posttranscriptional level. For example, alternative splicing of a number of genes critical for cell death leads to the formation of both cell death–promoting and cell death–preventing splicing isoforms. These include genes encoding for death ligands, death receptors, Bcl-2 superfamily of death regulators, caspases, and other cell-death regulatory genes. Several human caspase genes utilize alternative splicing to produce protein products that either contain or lack their enzyme active sites, resulting in antagonistic activities in cell death.

The nervous system is a good example where alternative splicing is utilized to generate extreme functional diversity. A vast number of genes involved in neural development and function undergo complex alternative splicing. Some genes encoding neural receptors and axon guidance molecules can generate hundreds to even thousands of different splicing isoforms. Many ion channels and neurotransmitter receptors have different alternative splicing isoforms with distinct electrophysiological properties. Alternative splicing regulation has a significant impact on the proper function of the nervous system, including learning, memory, and behavior development.

2.2
Mechanisms Underlying Alternative Splicing Regulation

The fundamental mechanisms underlying alternative splicing are the intricate interactions among trans-acting splicing factors and cis-elements in the pre-mRNA substrates, leading to selection of different splice sites. These cis-elements include splice sites and splicing regulatory sequences (splicing enhancers or silencers). The sizes of the exons or introns, as well as secondary structures of the pre-mRNA also influence splice site selection. The highly degenerate nature of mammalian splicing signals and the existence of large numbers of trans-acting factors allow versatile RNA–protein and RNA–RNA interactions during different stages of spliceosome assembly. Such interactions can be modulated by both exonic and intronic regulatory elements. The recognition of splice sites and cis-acting splicing regulatory sequences is mediated by a network of interactions between pre-mRNA and trans-acting factors including snRNPs and non-snRNP splicing factors. A number of splicing regulators, both positive and negative, play important roles in alternative splicing regulation (Table 3). The final outcome of alternative splicing of a given gene, including the production and the delicate balance of distinct isoforms,

Tab. 3   Pre-mRNA splicing regulators.

Protein name	Motifs	RNA-binding sites
A. Splicing activators		
SRp75	2 RRMs, RS	
SRp55	2 RRMs, RS	ESEs
SRp40	2 RRMs, RS	
ASF/SF2	2 RRMs, RS	
9G8	1 RRM, RS	
SC35	1 RRM, RS	
SRp30c	2 RRMs, RS	
hTra2	1 RRM, RS	
SRp20	1 RRM, RS	
SRm300	RS-rich	
SRm160	PWI, RS	
SF1 (BBP)	ZF, KH, CCHC,	ISE
KSRP	KH	
NOVA1	3 KH	(UCAUY)3
rSLM-2	KH, STAR	
CUG-BP	RRMs	ISE
ETR3	3 RRMs,	ISE
Halfpint (D.m.)	RRM (homolog of PUF60)	

Tab. 3 (continued)

Protein name	Motifs	RNA-binding sites
TIA-1	3 RRMs	U-rich
p72	DExD, Helicase, RGG	
YB1	Cold box	A/C-rich ESE
B. Splicing repressors		
hnRNP A1	2 RRMs, Gly-rich	ESSs or ISSs
hnRNP I/PTB	4 RRMs	U/C-rich
nPTB	4 RRMs	U/C-rich
SmPTB	4 RRMs	U/C-rich
Elav (D.m.)	3 RRMs	
Sxl (D.m.)	3 RRMs	
Mec8 (C.e.)	2 RRMs	
Fox1 (C.e.)	RRM	
PSI (D.m.)	4 KH	ISS
QKI-5	KH	ISS
SRp30c	2 RRMs, RS	
SRrp35	1 RRM, RS	
SRp38	1 RRM, RS	
SRrp40	1 RRM, RS	
SPF45	1 RRM, Gly-rich	
RSF1 (D.m.)	1 RRM, GRS	
SWAP (D.m.)	RS	
C. Bifunctional splicing regulators		
p54/SFRS11	1RRM, RS	
SRrP86	1RRM, RS	
p32		
HnRNP H	3RRMs, Gly	ESE or ISE ESS
Napor1	3RRMs	

Note: A. Splicing activators. B. Splicing repressors. C. Bifunctional splicing regulators.
STAR: signal transduction and activation of RNA.
For other abbreviations, see footnotes for Table 1.

is determined by combinatorial effects of multisite interactions among pre-mRNA, essential spliceosomal components, and regulatory factors.

Genetic and biochemical studies in *Drosophila* genes have tremendously advanced our understanding of alternative splicing regulatory mechanisms, although some mechanisms may not be conserved in corresponding mammalian systems. For example, *Drosophila* sex determination genes are under extensive regulation by alternative splicing, and this regulatory mechanism is not conserved in mammalian sex determination. In general, *Drosophila* genes contain more small introns than mammalian genes. Here, we focus on mammalian alternative splicing regulation.

2.2.1 Splicing Signals and Splicing Regulatory Elements

Unlike the yeast spliceosome, mammalian spliceosomes have the daunting task of searching for small exons in the vast sea of introns. This process of "exon recognition" or "exon definition" is particularly remarkable because sequences at mammalian splice sites are so highly degenerate (see Fig. 2). A large number of sequence elements with similarity to authentic splice sites (pseudosplice sites) can be found in both intronic and exonic regions). Further complicating the issue, some of these elements act as *cryptic splice sites* that are only used by the splicing machinery when the authentic splice sites are altered by mutations. Therefore, specific recognition of authentic splice sites and correct pairing of corresponding 5′ and 3′ splice sites is a central issue for both constitutive splicing and alternative splicing regulation. The intrinsic sequence degeneracy of mammalian splice sites determines that alternative splicing is a rule rather than an exception during mammalian gene expression.

In addition to splice sites, sequence elements in both intronic and exonic regions modulate alternative splicing. Such regulatory elements can either enhance or suppress splicing, and are hence named exonic splicing enhancers (ESEs) and intronic splicing enhancers (ISEs), or exonic splicing silencers (ESSs) and intronic splicing silencers (ISSs).

A number of ESE motifs have been identified using biochemical systematic evolution of ligands by exponential enrichment (SELEX) or bioinformatical approaches. A/G-rich (also called purine-rich) and A/C-rich elements are among ESE motifs characterized by biochemical studies. Proteins containing SR domains play a major role in recognizing A/G-rich ESEs and recruiting other spliceosomal components (including snRNPs and other protein factors), thereby promoting the usage of neighboring splice sites. A cold-box protein, YB-1, has been shown to enhance splicing by interacting with an A/C-rich ESE. Another ESE in HIV-1 tev-specific exon interacts with hnNRP H and SR protein SC35 to enhance splicing. The SELEX method has been used to identify preferred binding sequences for individual SR proteins, and optimal binding sites for individual SR proteins are degenerate. ESE prediction programs based on SELEX and computational analysis of human genes have been developed (http://exon.cshl.org/ESE). These programs are useful in predicting alternative splicing patterns of natural pre-mRNA substrates in cells.

ISE elements have been studied in a number of genes, including c-src, β-tropomyosin, calcitonin/calcitonin gene-related peptide gene (CGRP), fibronectin, nonmuscle myosin heavy chain, cardiac troponin T (cTNT), FGFR-2, $\alpha 2$ subunit of glycine receptor, and other genes. Such ISEs may contain sequences similar to 5′ splice site, U-rich element adjacent to the regulated 5′ splice site, UGCAUG element, (UCAUY)3-containing sequences, or CUG-containing motif.

A number of ESSs have been characterized in different genes such as β-tropomyosin, CD44, and viral genes including HIV Tat, bovine papillomavirus type-1, and Rous sarcoma virus. They do not share any obvious sequence motifs. Their activities are often associated with interactions with proteins of the hnRNP family, including hnRNPA1, hnRNP H, and hnRNP F.

A variety of ISSs have been analyzed in alternatively spliced genes including hnRNP A1, fibroblast growth factor receptor 2 (FGFR2), caspase 2, $GABA_AR\gamma 2$ (γ-aminobutyric acid receptor typeA $\gamma 2$ subunit), NMDA R1 (*N*-methyl-D-aspartate receptor R1 subunit;), clathrin light chain B, and HIV tat. Many of these ISSs contain extended polypyrimidine tracts. Some splicing repressor elements contain sequences similar to authentic splice sites, or decoy splice sites. It has been proposed that such decoy splice sites mediate nonproductive interactions to suppress the usage of upstream 5′ splice site. Again, several hnRNP proteins, including polypyrimidine tract binding protein (PTB, also known as hnRNP I), hnRNP A1, and hnRNP H play important roles in splicing repression by ISSs.

One of the common features of these cis-regulatory elements is that they are not simple sequence elements that act independently of each other. Sequence elements have been identified that can act to stimulate the splicing of one exon but repress another exon (e.g. in FGFR2 gene). Some regulatory elements contain both enhancer and silencer domains. The splicing of IgM exons M1 and M2 is regulated by a sequence containing juxtaposed splicing enhancer and silencer elements. Alternative splicing of protein 4.1R pre-mRNA generates multiple isoforms. This alternative splicing event is critical for red blood cell membrane biogenesis during erythroid differentiation and is regulated by multiple cis-elements and trans-acting regulators.

Splicing regulatory elements usually contain multiple binding sites for splicing regulators and function by recruiting other spliceosomal components to form RNP-like complexes. An evolutionarily conserved 100 bp intronic suppressor element in the caspase 2 (casp-2) gene, In100, specifically inhibits its exon 9 splicing. Alternative splicing of this 61 bp exon 9 leads to the formation of two functionally antagonistic products, casp-2L and casp-2S. Casp-2L product promotes cell death, whereas casp-2S prevents cell death. The In100 element contains a decoy 3′ splice site juxtaposed to a PTB-binding domain, both of which contribute to the full activity of In100 in inhibiting exon 9 inclusion. The upstream portion of In100 contains a sequence with features of an authentic 3′ splice site (including a branch site, a polypyrimidine tract, and AG dinucleotide). However, this site is not used under normal conditions. This sequence is only recognized as a 3′ss when the site is isolated with the downstream PTB-binding domain deleted. Biochemical and cell culture experiments show that this decoy 3′ acceptor site interacts nonproductively with the 5′ splice site of the alternative exon 9, thus repressing the efficient use of the 5′ splice site of exon 9 despite a high level of U1snRNP binding to this 5′ splice site. Downstream of the decoy 3′ splice site resides the second functional domain that interacts with PTB. The binding of PTB to CU-rich motifs within this downstream domain juxtaposed to the decoy 3′ acceptor site correlates well with the repressor activity of this domain. PTB can modulate recognition of the adjacent decoy 3′ acceptor site. In addition to factors interacting with an authentic 3′ splice site (such as U2AF or U2snRNP), PTB as well as other proteins interact with In100 and contribute to recognition of the In100 decoy 3′ splice site by the spliceosome as an intronic repressor element, rather than as an authentic 3′ splice site. The regulatory role of PTB in casp-2 alternative splicing and its mechanism of action appear to be distinct from other systems (Sect. 2.2.2). A recent survey of known human genes involved in cell death regulation suggests that In100-like intronic elements (i.e. 3′ splice site juxtaposed to PTB-binding domains) may represent a general intronic splicing repressor motif. Such intronic elements may play a role in regulating alternative splicing of other cell-death genes.

Another example of the complexity of splicing regulatory elements is in the alternative splicing of exon 10 in the human *tau* gene. Alternative splicing of this exon is associated with the pathogenesis of dementias (Sect. 3.1). In this case, both exonic and intronic regulatory elements play important roles in controlling exon 10 inclusion. In addition, in the exonic region, both positive and negative elements are involved and form a multidomain composite regulatory element.

2.2.2 Trans-acting Splicing Regulators

The recognition and selection of splice sites are determined during spliceosome assembly, especially at early steps of spliceosomal formation. Splicing activators facilitate interaction of U1snRNP with 5′ splice site and of U2snRNP with 3′ splice site, whereas splicing repressors suppress the recognition of splice sites. A number of proteins involved in spliceosome assembly also play important roles in regulating alternative splicing (see Table 1).

Several families of splicing activators have been reported (Table 3). SR proteins are among the best characterized splicing activators. The interactions between SR proteins and ESEs play a critical role in exon recognition by the spliceosome. ESEs are present in both constitutively and alternatively spliced exons. By mediating protein–protein and RNA–protein interactions during early steps of spliceosome assembly, SR proteins coordinate the communication between 5′ and 3′ splice sites and promote interactions between exonic enhancers and splice sites. Enhancer complexes are usually multicomponent complexes, containing not only SR proteins but also other RS domain–containing proteins such as U2AF.

Some RS domain–containing proteins, such as SRm300 and SRm160, act as coactivators for ESE function. The RS domain in SR proteins can be differentially phosphorylated, providing another level of regulation. The phosphorylation status of SR proteins regulates protein–protein interactions, intracellular distribution, and activities of SR proteins. Differential phosphorylation of SR proteins has been shown to play a role in regulating gene expression during development.

SF1, KSRP, NOVA-1, and rSLM-2 are RNA-binding proteins containing heterogeneous nuclear ribonucleoprotein K-type homology (KH) domain. They can enhance splicing by interacting with ISEs. SF1, a protein important for branch site recognition during spliceosome assembly, binds to GGGGCUG repeats in an ISE to activate the recognition of a 6-bp micro exon in cTNT gene. KSRP interacts with UGCAUG sequence and stimulates the neuronal-specific exon inclusion in c-src. NOVA-1 enhances the splicing of exon E3A in the $\alpha 2$ subunit of the glycine receptor gene (GlyRα2). The rat Sam68-like mammalian protein (rSLM-2) is a member of the STAR (signal transduction and activation of RNA) protein family. It can influence the splicing pattern of the CD44v5, human transformer-2beta, and tau minigenes in transfected cells.

Several members of CUG-BP and ETR-like factors (CELF) family activate the splicing of genes including cTNT, muscle-specific chloride channel, insulin receptor, and NMDA R1. Some of these CELF proteins activate splicing by interacting with ISEs.

TIA-1, a mammalian homolog of yeast NAM8 protein, interacts with the U-rich intronic sequence adjacent to the 5′ splice site of the K-SAM alternative exon in the FGFR2 gene. The activation of this splice site is U1snRNP-dependent, and TIA-1 may function by facilitating U1-snRNP binding to the 5′ splice site.

Human Y-box binding protein (YB-1) was initially identified as a transcription factor interacting with single-stranded DNA in response to cold shock. YB-1 contains a five-stranded ß-barrel known as the "COLD" domain and accessory domains rich in basic amino acids and aromatic groups. This domain arrangement in YB-1 protein with specific interaction domain containing ß-sheets and a basic domain providing generic RNA binding is reminiscent of the modular structure of the SR proteins. YB-1 stimulates the splicing of the human CD44 alternative exon v4 by interacting with the A/C-rich element in the exonic splicing enhancer. Another protein, DEAD-box RNA helicase p72, has been reported to affect the splicing of alternative exons containing AC-rich exon enhancer elements. The mechanism of YB-1 or other proteins interacting with the A/C-rich type of ESEs in splicing activation remains to be elucidated.

HnRNP proteins hnRNPA1 A1 and PTB are among the best characterized splicing repressors. HnRNP A1 can interact with either ESSs or ISSs to prevent exon recognition by SR proteins or U2 assembly. A model has been proposed for interactions between hnRNP A1 and ESS based on studies on HIV Tat exon 3 splicing. In this model, the high-affinity binding of hnRNPA1 to ESS promotes nucleation of multiple A1 along the exon. The formation of this inhibition zone can be blocked by SF2/ASF, but not by another SR protein SC35, providing an explanation for differential antagonism between hnRNP A1 and different SR proteins. Two different models have been proposed for the function of hnRNPA1 in splicing silencing mediated by ISSs. In the case of HIV Tat intron 2 splicing, the interaction of hnRNP A1 with an alternative branch point sequence blocks the recognition of the branch site by U2snRNP. In the second model, the cooperative binding of hnRNP A1 to two intronic elements flanking the alternative exon inhibits exon inclusion by a looping-out mechanism during autoregulation of hnRNP A1 gene splicing.

PTB interacts with U/C-rich elements in ISSs in a number of genes including n-src, FGFR2, $GABA_A R\gamma 2$, tropomyosin, NMDA R1 exon 5, clathrin light chain B, caspase-2, and calcitonin/CGRP genes. Depletion of PTB using an RNA interference approach demonstrates that PTB is a negative regulator of exon definition in cultured cells. PTB-binding sites are frequently located in the intronic regions upstream of the regulated 3′ splice sites, although functionally active PTB-binding sites are also found in intronic elements downstream of the alternatively spliced exons (such as in c-src and caspase-2 genes). Mechanisms by which PTB represses splicing are not clear yet. One model for repression by PTB is via competition with U2AF binding to the polypyrimidine tract to block early spliceosome formation. This model is based on studies of alternative splicing of $GABA_A R\gamma 2$, NMDA R1 exon 5, and clathrin light chain B genes. In these genes, the high-affinity PTB binding to the long polypyrimidine tracts immediately upstream of the neural-specific exons represses the inclusion of these exons in nonneural tissues. A similar repression mechanism may be used in suppressing the inclusion of muscle-specific exons in rat α- or β-tropomyosin. PTB exists in a range of different tissues as an abundant splicing repressor. The repressor activity of PTB may be modulated by other regulatory proteins. Two tissue-specific PTB-related proteins have been found, neuron-enriched (nPTB,) and smooth muscle–enriched (SmPTB). For example, nPTB can compete with PTB to promote inclusion of neuronal-specific N1 exon inclusion in c-src. The expression of SmPTB correlates with the smooth muscle–specific suppression of α-tropomyosin exon 3, which is included in nonsmooth muscle cells. The release of PTB suppression can be also achieved by cell-type specific CELF proteins such as ETR3 and Napor-1.

One exception to the general inhibitory activity of PTB has been reported where PTB stimulates the inclusion of an alternative 3'-terminal exon. In this case, the splicing regulation is coupled with alternative polyadenylation. In addition, PTB has also been implicated in translational control of viral transcripts. More comprehensive understanding of the biological roles of PTB in gene regulation requires further investigation.

Several proteins containing an RS domain and RRM-cs domains have also been reported to act as splicing repressors, including SRrp30c, SRrp35, SRp38, and SRrp40 (see Table 3). SR proteins are generally hyperphosphorylated by the kinase SRPK1. Dephosphorylated SRp38 is required for the splicing repression in mitotic cells. Another RS domain-containing protein, the *Drosophila* suppressor-of-white-apricot protein (SWAP) suppresses its own pre-mRNA splicing.

An RRM-containing spliceosomal protein, SPF45, represents a late-acting splicing regulator. In *Drosophila*, SPF45 blocks splicing at the second step by interacting with Sex-lethal (Sxl) protein, indicating that 3' splice site recognition and splicing regulation can occur at the second catalytic step.

Proteins containing KH domains can also act as splicing repressors. For example, a nuclear isoform of quaking (qk) protein, QKI-5, regulates alternative splicing of myelin-associated glycoprotein gene by interacting with an intronic splicing repressor element. A *Drosophila* splicing repressor, P-element somatic inhibitor (PSI), is required for the soma-specific inhibition of splicing of P-element pre-mRNA both *in vitro* and *in vivo*.

Several splicing regulators can act as either positive or negative splicing regulators, depending on the sequence context of the pre-mRNA substrates. We classify these splicing regulators as bifunctional splicing regulators. Some SR proteins or SR-domain containing proteins can repress splicing of certain pre-mRNAs but activate splicing of other substrates. SRrp86 is an 86-kDa related to SR proteins. It can function as either an activator or a repressor by regulating the activity of other SR proteins. Other proteins can repress splicing by interacting with SR proteins. For example, an ASF/SF2-interacting protein, p32, was shown to act as a splicing regulator by inactivating the function of ASF/SF2.

HnRNP H and the related protein hnRNP F contain three RRMs of RNP-cs type. HnRNP H interacts with G-rich elements in either splicing enhancers or silencers. When binding to a splicing enhancer in the HIV env gene, hnRNP H acts as an activator by promoting the assembly of a complex containing SC35 and U1snRNP. When interacting with ESSs in the β-tropomyosin pre-mRNA or HIV Tat exon 2, or with a negative regulator of splicing (NRS) in the Rous Sarcoma Virus (RSV) genome, hnRNP H serves as a splicing repressor.

Napor1, a splice variant of ETR3, is expressed at a high level in the forebrain but at a low level in the cerebellum. Overexpression of Napor1 exerts opposite effects on two different exons in NMDAR1, inhibiting exon 5 inclusion but stimulating exon 21 inclusion.

With more alternative splicing events characterized in detail, more cases of bifunctional splicing regulators may emerge. It is conceivable that these splicing regulators interact with different sequence elements and different spliceosomal components. When these interactions promote the productive recognition of splicing signals by the splicing machinery, such proteins act as splicing activators. On the other hand, when the same proteins compete with other splice activators or facilitate recognition of splicing silencers leading to nonproductive interactions between the spliceosome and splicing substrates, these proteins behave as splicing repressors.

2.3
Tissue-specific and Developmentally Regulated Alternative Splicing

Tissue-specific alternative splicing has been studied in mammalian systems for more than a decade. However, only a limited number of tissue-specific splicing regulators have been identified so far. These include Nova-1, Nova-2, nPTB, and SmPTB. Cell-type specific splicing regulators have also been found in *Drosophila* and *Caenorhabditis elegans*, such as neuronal-specific Elav (embryonic lethal and abnormal vision), ovary-specific Halfpint, and muscle-specific Mec8. From a large number of biochemical, molecular, and genetic studies, the emerging general theme is that no single factor dictates the tissue specificity of alternative splicing of any genes. Instead, multicomponent interactions among tissue-specific splicing regulators, pre-mRNA, and general splicing factors determine the specific alternative splicing pattern of a given gene in a tissue- or cell-type specific manner. On the other hand, genetic deletion of a single splicing regulator, even those highly tissue-specific ones, leads to defects in splicing of multiple target genes. For example, Nova-1 deletion in mice affects alternative splicing of $GABA_AR\gamma2$, $GlyR\alpha2$ and perhaps other unknown target genes. Similarly, the *Drosophila* neural-specific splicing regulator Elav has at least three known target genes, neuroglian (nrg), erectwing, and armadillo.

In addition to tissue-specific expression of splicing regulators, a number of other mechanisms modulate tissue-specific alternative splicing. Tissue-specific combination of different splicing regulators, relative concentrations of distinct splicing factors, and differential modification of splicing regulators all contribute to the tissue-specific alternative splicing of individual genes in a given tissue or cell type. For example, the level of the splicing repressor PTB in neural tissues is lower than in other tissues, providing an explanation for a more permissive environment for exon inclusion in a number of genes in neural tissues. Although most SR proteins are expressed in a wide range of tissues, their expression patterns vary in different tissues, both in different isoforms and at different levels. Many SR proteins have different isoforms either because of alternative splicing or posttranslational modifications. Kinases or phosphatases that regulate phosphorylation of different splicing factors can also be tissue specific. One example is the SR protein–specific kinase 2 (SRPK2), which is highly expressed in the brain and presumably capable of regulating activities of target SR proteins.

It is not clear yet how the extremely complex alternative splicing pattern of different genes in a given cell or a specific tissue is coordinated. In some cases, the alternative splicing events of different genes appear to be coregulated by the same protein. For example, the splicing regulator PTB can differentially recognize neural and non-neural substrates. The expression pattern of PTB in different regions of the brain at different developmental stages supports a role for PTB to act as an alternative splicing coordinator for different splicing target genes.

2.4
Regulation of Alternative Splicing in Response to Extracellular Stimuli

Extracellular signals can induce changes in alternative splicing of different genes. For example, growth factors or hormones stimulate alternative splicing of intracellular responsive genes. Treatment with growth factors induces changes in alternative splicing of phosphotyrosine phosphatase PTP-1B gene. The alternative splicing pattern of protein kinase C beta is changed by insulin. Activation of calmodulin-dependent kinases (CaM kinases) stimulates changes in alternative splicing of BK potassium channels. In most cases, however, signal transduction pathways involved in splicing regulation have not been characterized.

A number of genes important for the functioning of the nervous system show activity-dependent changes in their splicing. Alternative splicing of syntaxin 3 changes in response to induction of long-term potentiation. NMDAR1 alternative splicing is modulated by both pH and Ca^{2+}. Stress hormones regulate alternative splicing of potassium channels. A cis-acting sequence named *calcium-responsive RNA element* (CaRRE) in the stress-inducible exon of BK potassium channels and in NMDAR1 exon 5 has been identified. This element mediates CaM kinase-dependent repression of the inclusion of the CaRRE-containing exons. The mechanisms by which the CaRRE causes exon skipping remain to be elucidated.

Drastic changes in the cell growth environment are expected to affect RNA metabolism, including alternative splicing. For example, ischemia in mice induces changes in alternative splicing of several genes examined. Accompanying such changes in alternative splicing are changes in the intracellular distribution of several splicing regulators. It is possible that these splicing regulators undergo changes in posttranslational modifications or at other levels induced by ischemia.

Chemical compounds, when administrated to animals or applied to cells in culture, can also cause changes in alternative splicings. Sodium butyrate has been tested in transgenic mice and shown to increase exon 7 inclusion of the SMN gene. Aclarubicin treatment induces changes in SMN gene alternative splicing in cultured cells. Treatment of A549 lung adenocarcinoma cells with cell-permeable ceramide, D-e-C(6) ceramide, downregulates the levels of Bcl-xL and caspase 9b splicing isoforms. The functional responses of nuclear splicing machinery to drug treatment opens the possibility of correcting aberrant or defective splicing using therapeutic agents (Sect. 4.2).

3
Pre-mRNA Splicing and Human Diseases

Aberrant pre-mRNA splicing has been implicated in the pathogenesis of a number of human diseases. Alterations, or dysregulation of either constitutive pre-mRNA splicing or alternative splicing can cause disease phenotypes. Molecular genetic studies of human diseases have revealed a wide range of mutations that cause diseases because of their effects on pre-mRNA splicing. Investigating the molecular nature of these genetic defects and pathogenetic mechanisms will facilitate both diagnosis of such diseases and development of new therapeutic approaches.

3.1
Splicing Defects in Human Diseases

Genetic defects that cause aberrant pre-mRNA splicing can be found in every category of disease, from malignancies affecting a certain tissue or organ to diseases or syndromes involving multiple systems. Splicing mutations have been identified in genes involved in the pathogenesis of diseases in every system. Detailed information about these mutations can be found in databanks including HGMD (the Human Gene Mutation Database, http://www.uwcm.ac.uk/uwcm/mg/) as well as in a large number of publications. As shown in Table 4, we use a few examples to illustrate the diverse range of disease phenotypes and genes involved. The current understanding of these splicing defects will be summarized.

Malignancies are a major cause of mortality. Splicing mutations in genes critical for either cell proliferation or cell death result in malignancies in different systems. These include oncogenes, tumor suppressor genes, and genes involved in cell death. Only a few examples are included here to demonstrate the complexity of the involvement of defective/aberrant splicing in tumorigenesis. Cancer of the same tissue or organ can be caused by splicing mutations in different genes. For example, splicing defects in p53, p51, CD44, and members of epidermal growth factor receptor (EGFR) family lead to lung cancer, the most common cancer in humans. Breast cancer has been associated with splicing mutations in a number of genes, including HER-2/neu, p53, mdm2, and BRCA1 or BRCA2. Glioblastomas are also associated with splicing mutations in different genes. On the other hand, splicing defects in a single gene can lead to tumorigenesis in different tissues. For example, CD44 aberrant splicing has been associated with the development of a range of different tumors, such as breast cancer, prostate cancer, lung cancer, and Wilm's tumor. CD44 is a polymorphic family of cell surface glycoproteins important for cell adhesion and migration. Aberrant splicing of CD44 has been associated with the invasive behavior and metastasis of several types of tumors. Alternative CD44 splicing variants have been correlated with the poor prognosis of tumors. Another gene is Mdm2, whose alternative splicing defects have been implicated in oncogenesis in multiple tissues. Oncogenic splicing variants of Mdm2 that lack a domain important for its function in interacting with the tumor suppressor p53 have been found in breast cancer, glioblastoma, and rhabdomyosarcoma. Alterations in the balance of different FGFR2 splicing isoforms have been correlated with progression of prostate cancer. A significant fraction of neurofibromatosis type 1 (NF1) cases are associated with an aberrantly spliced NF1 gene. Production of tumor antigens as a result of alternative splicing has been associated with neuroblastoma and other tumors. The role of these tumor antigens in the development and metastasis of these tumors remains to be investigated. In pediatric acute myeloid leukemia (AML), splicing isoforms of gene fusion transcripts produced as a result of

Tab. 4 Examples of human diseases associated with aberrant or defective pre-mRNA splicing.

	Diseases	Genes	Aberrant splicing
Malignancies			
	Lung cancer	p53, p51, CD44, EGFR	Aberrant or defective splicing products
	Cancer in different tissues	p53, CD44	Defective or aberrant splicing isoforms
	Wilm's tumor	CD44	Aberrant splicing isoforms
	Breast cancer	HER2/neu	Aberrant splicing and imbalance of splicing isofors
	Breast and ovarian cancer	BRCA1, BRCA2	Exon skipping
	Breast cancer, rhabdomyosarcoma	Mdm2	Oncogenic variants lacking p53-binding domain
	Glioblastomas	Mdm2	Oncogenic variants lacking p53-binding domain
	Neurofibromatosis type 1	NF-1	Exon skipping
	Neuroblastoma	HUD, HUC, NNP-1, α-internexin	Production of tumor antigens
	AML	MLL-SEPTIN6	MLL-SEPTIN6 fusion-splicing variants
Cardiovascular diseases			
	Hypercholesterolemia	LDLR	Cryptic ss usage, exon skipping
	Hypertriglyceridemia	Hepatic lipase	Cryptic ss usage
	Marfan syndrome	Fibrillin-1	Cryptic ss usage, exon skipping
	Cardiomyopathy	cTNT	Aberrant splicing isoforms
	Hypertension	G-protein $\beta 3$	Exon skipping
Metabolic diseases			
	Glycogen storage disease, type II	Lysosomal α-glucosidase	Cryptic ss usage
	Hereditary tyrosinemia, Type I	Fumarylacetoacetate hydrolase	Exon skipping
	Acute intermittent porphyria	Porphobilinogen deaminase	Exon skipping
	Ceruloplasmin deficiency	Ceruloplasmin	Cryptic ss usage
	Fabry's disease	Lysosomal α galactosidase A	Cryptic ss usage
	Tay–Sach's Disease	β-hexosaminidase	Intron retention, exon skipping
	Sandhof disease	β hexosaminidase β-subunit	Cryptic ss usage

Tab. 4 (continued)

	Diseases	Genes	Aberrant splicing
Neurodegenerative diseases			
	FTDP-17	Tau	Imbalance of splicing isoforms
	Alzheimer's disease	Presenilin-1	Exon skipping
	Alzheimer's disease	Presenilin-2	Exon skipping
	Ataxia telangiectasia	ATM	Cryptic ss usage
	Multiple sclerosis	CD45	Imbalance of splicing isoforms
	Spinal muscular atrophy	SMN1, SMN2	Exon skipping
	Retinitis pigmentosa	HPRP3, PRPF31, PRPC8	Unknown
Psychiatric disorders			
	Schizophrenia	$GABA_A R\gamma 2$, NCAM	Imbalance of splicing isoforms
	Schizophrenia	NMDAR1	Aberrant alternative splicing
	ADHD	Nicotinic acetylcholine receptor	Aberrant alternative splicing
Other syndromes or diseases			
	Cystic fibrosis	CFTR	Exon skipping
	IGHD II	GH-1	Imbalance of splicing isoforms
	Frasier syndrome	WT 1	Imbalance of splicing isoforms
	Epilepsy	AMPA receptor	Imbalance of splicing isoforms
	Menkes disease	MNK	Exon skipping
	Beta-thalassemia	β-globin	Cryptic ss usage
	Metachromatic leukodystrophy	Arylsulfatase A	Cryptic ss usage
	Myotonic dystrophy	DMPK	(CUG)n expansion:aberrant splicing

chromosomal translocation have been reported. Such aberrant gene fusion-splicing products may contribute to pathogenesis of tumors caused by chromosomal translocation. Finally, the development of drug resistance has been associated with changes in alternative splicing of members of multidrug resistance genes or genes involved in drug metabolism.

Cardiovascular diseases are among the leading causes of human death. Aberrant splicing of structural genes such as cTNT is associated with cardiomyopathy. Splicing defects resulting in the formation of a defective receptor, such as the low-density lipoprotein receptor (LDLR), lead to familial hypercholesterolemia (FH). A single-nucleotide change altering the function of intracellular signal transduction molecules can cause diseases affecting multiple systems. For example, a single-nucleotide polymorphism leading to the formation of a G-protein beta3 subunit splice variant has been associated with hypertension. Aberrant or defective splicing can disrupt the function of genes encoding a wide range of proteins important for the function of the cardiovascular system, from cell surface receptors to intracellular signaling molecules. Both structural genes and regulatory genes can be affected by splicing mutations. These examples show that splicing mutations can affect development/formation of the cardiovascular system and regulation of cardiovascular system function.

A range of metabolic diseases is caused by splicing mutations, affecting the production of a single metabolic enzyme. For example, the glycogen storage diseases, Sandhof's disease, and Fabry's disease can be caused by the usage of a single cryptic splice site in the corresponding genes. Hereditary tyrosinemia (type I) and acute intermittent porphyria can develop because of improper exon skipping in the genes encoding for respective metabolic enzymes.

The critical role of splicing regulation for the normal function of the nervous system is demonstrated by the large number of neurodegenerative and psychiatric disorders associated with aberrant splicing. Splicing defects in *tau* and presenilin genes have been identified in different types of dementia, including frontotemporal dementia with parkinsonism linked to chromosome 17 (FTDP-17) and Alzheimer's disease. Splicing mutations that cause cryptic splice site usage in the ataxia telangiectasia (ATM) gene have been identified in ATM patients. Multiple sclerosis has been associated with imbalances of different splicing isoforms of CD45. A number of studies have shown that genetic mutations in ubiquitously expressed protein factors that affect spliceosome formation can lead to diseases with specific neurological manifestations, such as spinal muscular atrophy and retinitis pigmentosa. For example, mutations in SMN1 or SMN2 cause spinal muscular atrophy. Defects in genes essential for spliceosomal assembly, including HPRP3, PRPF31, and PRPC8, have been found in autosomal dominant retinitis pigmentosa. The molecular mechanisms underlying such neuronal-specific diseases caused by defects in the general splicing factors remain to be investigated.

Psychiatric disorders, including attention deficit/hyperactivity disorder (ADHD) and schizophrenia, have been associated with aberrant splicing. Imbalance of different splicing isoforms or aberrant splicing products of several genes were detected in the brain tissues of patients with schizophrenia. These genes include $GABA_AR\gamma2$, NMDAR1, neuronal nicotinic acetylcholine receptors, and neural cell adhesion molecule (NCAM). Changes in splicing isoforms of astroglial AMPA receptors have also been reported in human temporal lobe epilepsy. Several other examples of diseases caused by splicing mutations or disregulation of alternative splicing are listed in Table 1, including cystic fibrosis, familial isolated growth hormone deficiency type II (IGHD II), Frasier's syndrome, Menkes disease, β-thalassemia, and metachromatic leukodystrophy. Again, a variety of aberrant splicing events in the corresponding genes cause the formation of defective gene products and consequently the disease phenotypes.

3.2 Molecular Mechanisms Underlying Splicing Defects Associated with Disease

Genetic defects in pre-mRNA splicing that cause human diseases can be classified into two categories, cis-acting mutations that alter expression or function of single genes and trans-acting defects that affect the components of splicing machinery or regulators of alternative splicing. Both cis-acting and trans-acting splicing mutations can result in clinical manifestations, either primarily involving a single tissue/organ or affecting multiple systems. These mutations can cause diseases by their direct effects on single genes or by indirect mechanisms such as disrupting expression or regulation of multiple genes.

Cis-acting splicing mutations can cause diseases by either producing defective/aberrant transcripts or by simply altering the delicate balance of different naturally expressed splicing isoforms. Development of disease phenotypes can be the result of loss of function of the involved genes or gain-of-function toxicities associated with the aberrant gene products. A large number of studies have focused on relationships between cis-acting splicing mutations and specific defects in the formation or function of the affected genes. However, the relationship is much less understood between genetic defects in trans-acting splicing factors (such as spliceosomal components or splicing regulators) and specific disease phenotypes.

At least four types of molecular mechanisms have been described for cis-acting splicing mutations that cause human diseases (Fig. 6). These mutations often lead to the formation of defective protein products or the loss of function of the genes involved as a result of RNA or protein instability. Exon skipping is perhaps the most commonly reported mechanism for the production of defective gene products. Activation of cryptic splice sites has also been

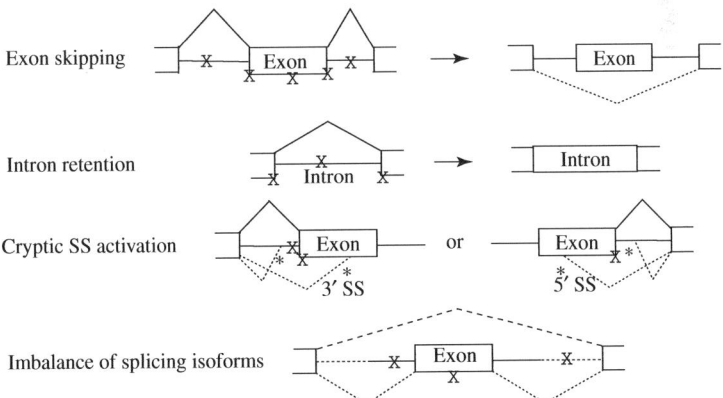

Fig. 6 Mechanisms for cis-acting splicing mutations. Positions of mutations are marked by "X," and they can be at the splice sites or within intronic or exonic sequences. The cryptic splice sites are marked by "*." Normal splicing events are illustrated by solid lines, and aberrant splicing events are depicted in dotted lines.

frequently associated with the production of aberrant or defective gene products. The failure to remove introns or intron retention is another mechanism for the formation of truncated protein products or the complete loss of gene function, because the inclusion of intronic sequences often introduces premature stop codons in gene transcripts. Finally, disturbance in the balances among naturally occurring splicing isoforms can lead to human disease. These mechanisms are not exclusive of each other. Multiple mutations acting by different mechanisms have been found in the same genes, leading to similar disease phenotypes. For example, both exon skipping and cryptic splice site activation contributing to the formation of defective LDLR lead to hypercholesterolemia. Similarly, exon skipping or cryptic splice site usage in Fibrillin-1 gene have been associated with Marfan syndrome.

Exon skipping can be caused by mutations at the splice sites or in splicing enhancers (either ESEs or ISEs). Translationally silent mutations can be functionally significant in splicing. For example, third codon changes that do not affect peptide sequences have often been overlooked as disease-causing mutations. Recently, such mutations have been examined at the splicing level. Such "silent" mutations may cause significant disruption in splicing, because defective function of ESEs leads to either improper exon skipping or imbalance of natural splicing isoforms.

Activation of cryptic splice sites in a large number of genes has been associated with human pathogenesis (see examples in Table 4). Single-nucleotide changes at the authentic splice sites result in inactivation of authentic splice sites. The consequences can be failures in exon inclusion or intron removal in some genes, or selection/activation of cryptic splice sites in other genes.

Intron retention has been examined in a number of disease genes. The average size of introns in human genes is approximately 3 kbp. The retention of even a small single intron may have catastrophic effects on the function of the genes affected because of the formation of defective or truncated peptides. Alternatively, intron retention can also cause a complete loss of expression of the mutated genes as a result of nonsense-mediated mRNA decay or instability of the gene products. Tay–Sach's disease can be caused by either intron retention or exon skipping in the splicing of the β-hexosaminidase gene.

Disturbance in the delicate balance of different natural splicing isoforms is now being recognized as a common mechanism for splicing diseases. It has been identified in neurodegenerative diseases, psychiatric disorders, and other diseases. These types of splicing defects can be caused by point mutations at splice junctions or in splicing regulatory elements (enhancers or silencers) located both in exon and intronic regions. FTDP-17, caused by mutations in the human *tau* gene, is a good example of such splicing defects. In addition to mutations that affect biophysical or biochemical function of Tau proteins, a large number of mutations that alter the ratio of natural splicing isoforms have been identified in FTDP-17 patients. The *tau* gene undergoes complex alternative splicing during the development of the nervous system. Six splicing isoforms are produced, three containing exon 10 and three lacking exon 10. These isoforms are named Tau4R and Tau3R respectively, because exon 10 encodes one of four microtubule-binding repeats. The balance among different Tau isoforms is important for normal brain function, and disruption of the balance of Tau isoforms leads to the development of FTDP-17. Maintaining the appropriate balance of different Tau isoforms is controlled by complex interactions between the splicing machinery and various splicing regulatory elements in the *tau* gene. At the 5′ splice site of exon 10, a stem-loop type of secondary structure modulates the recognition of this 5′ splice site by U1snRNP. A number of intronic or exonic mutations destabilizing this stem-loop structure leads to an increase in Tau4R production. Other regulatory elements residing in exon 10, either splicing enhancers or silencers, may be disrupted by point mutations or small deletions, leading to the imbalance of different Tau isoforms. These mutations may increase (such as N279K) or decrease (such as L284L, del280K) the activity of the exonic splicing enhancer.

Other examples of diseases caused by disruption of the balance of different alternative splicing isoforms include Frasier syndrome, familial isolated growth hormone deficiency type II (IGHD II) and atypical cystic fibrosis. Such splicing defects have also been implicated in the pathogenesis of other diseases such as multiple sclerosis and schizophrenia. In Frasier syndrome, intronic mutations are found to change the ratio of +KTS to −KTS isoforms of Wilms tumor suppressor (WT1) gene products. The formation of these two splicing isoforms is the result of selection of two competing 5′ splice sites in exon 9. These two splice sites are separated by 9 nucleotides encoding KTS (lysine-threonine-serine). The use of the upstream and the downstream 5′ splice sites produces −KTS and +KTS isoforms, and the balance of these isoforms is important for kidney and gonad development. Mutations that decrease the use of the downstream +KTS 5′ splice site with an increase in the −KTS isoform have been associated with the majority of Frasier syndrome cases. These examples clearly show the significant role of splicing defects in human pathogenesis.

Spinal muscular atrophy (SMA) is an example that demonstrates the complexity of mechanisms underlying diseases caused by defects in trans-acting factors, such as proteins essential for spliceosomal snRNP biogenesis. SMA is a leading cause of infant mortality. Pathologically, it is characterized by degeneration of motor neurons in the anterior horn of the spinal cord leading to muscular atrophy. Genetic defects causing SMA include deletions or mutations in the survival of motor neuron genes (SMN). There are two copies of highly homologous SMN genes in humans, *SMN1* and *SMN2*. The *SMN1* gene is deleted or mutated in the majority of SMA patients. The SMN proteins are detected as distinctive speckles termed *gems* within the nucleus. SMN proteins contain an RNP1 motif and are critical for snRNP biogenesis and therefore, the formation of the spliceosome. Both SMN genes are expressed in a wide range of tissues. Although the predicted peptide sequences of *SMN1* and *SMN2* genes are identical, the gene products produced from the two genes in cells are different. A single translationally silent C to T nucleotide change at position +6 of exon 7 leads to the inefficient inclusion of exon 7 in the *SMN2* gene product, and thus the failure of *SMN2* to replace the function of *SMN1* and to provide protection against SMA. It is not clear why mutations in the *SMN* gene specifically cause motor neuron−specific disease. Although the *SMN2* gene is not usually mutated in SMA patients, studies on stimulating *SMN2* exon 7 alternative splicing suggest a potential therapeutic approach based on the activation of alternative splicing of genes homologous to mutated genes (Sect. 4.2).

Myotonic dystrophy (DM), an autosomal dominant disease, is an example of trans-acting genetic mutations in which remarkable progress has been made. DM is a most common form of muscular dystrophy affecting both skeletal muscle and smooth muscles. Type I myotonic dystrophy (DM1) is caused by a CTG trinucleotide expansion in the 3′ untranslated region (3′UTR) of the DMPK (DM protein kinase) gene on chromosome 19q13.3. Type II myotonic dystrophy (DM2) is associated with a large CCTG repeat expansion in the intron 1 of the ZNF9 gene. An "RNA gain-of-function" hypothesis has been proposed that these CTG or CCTG repeat expansions cause the formation of aberrant RNA transcripts containing large CUG or CCUG repeats. Such RNA transcripts containing long tracks of CUG/CCUG repeats may disrupt the normal function of certain RNA-binding proteins and induce secondary aberrant splicing of other genes. The RNA-binding proteins involved are likely to be the splicing regulators of CUG-BP family. The disruption of CUG-BP protein functions leads to aberrant splicing of genes including cardiac troponin T, insulin receptor, muscle-specific chloride channel, and tau. The aberrant splicing of these downstream genes can explain the cardiac phenotype, insulin resistance and myotonia found in DM patients. These findings demonstrate the complex roles of alternative splicing regulation in human pathogenesis.

Several trans-acting splicing defects have been reported in autosomal dominant retinitis pigmentosa (adRP). Recent genetic studies demonstrate that mutations in genes encoding general splicing factors such as HPRP3, PRPC8, or PRPF31 cause adRP. It remains unclear why these seemingly general defects in the splicing machinery cause such neuron-specific diseases. The underlying pathogenetic mechanisms await further investigation.

4
Perspectives on Diagnosis and Treatment of Diseases Caused by pre-mRNA Splicing Defects

4.1
Diagnosis of Human Diseases Caused by Splicing Defects

In the past three decades, significant advances have been made in the diagnosis of human genetic diseases, including diseases caused by splicing mutations. However, currently available molecular diagnostic tools remain limited for detecting splicing mutations.

Definitive diagnosis of splicing diseases should be based on the following criteria. First, the clinical manifestations should correlate with defects in a given gene. Second, corresponding splicing mutations are detected in genomic DNA samples of the patients. Third, aberrant or defective splicing products should be detected in the affected tissues or cells from patients. Detection of genomic mutations relies on DNA sequence analysis. In most cases, efforts in genomic DNA sequence analysis have been focusing on either exonic regions or near splice junctions. Detection of aberrant or defective splicing products from patient samples can be technically challenging. The most frequently used methods include RT-PCR and analyses of defective protein products. In general, these methods are still limited to research studies. Systematic approaches to detecting disease-causing splicing mutations are yet to be developed for diagnostic applications in clinics.

4.2
Potential Therapeutic Approaches

Molecular studies and mechanistic characterization of splicing mutations that cause human disease have led to the development of a number of potential therapeutic approaches. They can be classified into four types, oligonucleotide-, ribozyme-, trans-splicing- and chemical compound-based approaches. These approaches are still at the stage of research and development.

4.2.1 Oligonucleotide-based Approaches: Antisense, RNAi, and Chimeric Molecules

Several groups have tested oligonucleotide-based methods for designing therapies to treat diseases caused by aberrant pre-mRNA splicing. These approaches include using modified antisense oligonucleotides or RNA interference (RNAi). Antisense oligonucleotides could be used to restore the proper function of mRNAs that are disrupted by splicing mutations. Antisense oligonucleotides designed to block cryptic sites in the mutated globin gene in β-thalassemia were reported to increase wild-type mRNA and decrease aberrant mRNA. A lentiviral vector-based system using a modified U7snRNA containing a sequence blocking the aberrant splice sites, reduced aberrant pre-mRNAs and increased levels of the correctly spliced β-globin mRNA and protein. The production of stress-induced aberrant AChE-R mRNA was reduced by antisense oligonucleotides. Another study using the antisense oligonucleotide approach reported the reversal of the aberrant splicing pattern caused by mutations associated with FTDP-17 in the human *tau* gene. On the other hand, antisense oligonucleotides have been used to increase the inclusion of exon 7 of SMN2 in order to develop therapeutic agents for spinal muscular atrophy. 2'-O-Methylated antisense oligoribonucleotides were used to modify the splicing pattern of the dystrophin pre-mRNA in the mdx mouse model of Duchenne muscular dystrophy. Recently, RNAi was used to selectively degrade alternatively spliced mRNA isoforms in *Drosophila* by treating cultured cells with dsRNA corresponding to an alternatively spliced exon.

One of most common splicing defects associated with human diseases is exon skipping. Because an RS domain can act as a splicing activator, small chimeric molecules containing a minimal synthetic RS domain covalently linked to an antisense moiety have been tested to target defective BRCA1 or SMN2 pre-mRNA transcripts and shown to restore splicing. These oligonucleotide-based approaches have shown feasibility in correcting splicing defects and treating the corresponding splicing diseases.

4.2.2 Ribozymes

Ribozymes are RNAs that catalyze biochemical reactions in cells, especially cleavage of other nucleic acids. Efforts have been made in developing derivatives from small naturally occurring RNAs, including the hammerhead, the hairpin, the tRNA processing ribonuclease P (RNase P), and group I and group II ribozymes as therapeutic agents. Ribozymes can be used to reduce aberrant or defective splicing transcripts. In addition, a trans-splicing group I ribozyme was shown to convert mutant transcripts to normal mRNAs in the beta-globin and p53 genes.

4.2.3 SMaRT

Spliceosome-mediated RNA trans-splicing (SMaRT) was developed utilizing the endogenous trans-splicing activity in mammalian cells to correct aberrant splicing. SMaRT-mediated repair was reported to partially correct splicing defects in cystic fibrosis transmembrane conductance (CFTR) gene in cultured cells and bronchial xenografts.

4.2.4 Chemical Compounds

A number of chemical compounds have been shown to interact with RNA and/or RNA-binding proteins. Using chemical compounds to correct pre-mRNA splicing defects is being actively explored as a new therapeutic approach. Although the underlying mechanisms remain unclear, aclarubicin and sodium butyrate increase the inclusion of exon 7 in SMN2 transcripts in fibroblasts derived from spinal muscular atrophy patients or in transgenic mice, suggesting the therapeutic potential of chemical compounds in treating diseases associated with defective splicing.

Other chemical compounds are being tested that are capable of modifying splicing regulators, for example, kinases or phosphatases, that influence SR protein functions.

5
Concluding Remarks

After more than two decades of studies, a general picture of mammalian pre-mRNA splicing and alternative splicing regulation has begun to emerge. The basic components of the mammalian splicing machinery have been identified. The highly dynamic process of spliceosome assembly involves multiple networks of RNA–RNA, RNA–protein and protein–protein interactions. Recognition of splice sites and splicing regulatory sequences has been investigated in a number of genes, leading to identification and characterization of both cis-acting regulatory elements and trans-acting factors. We have begun to appreciate the contribution of alternative pre-mRNA splicing to creating genetic diversity, especially in mammals. A large number of splicing mutations that cause human diseases are being identified and characterized. We now have a glimpse of the complex picture of the involvement of aberrant or defective splicing in the pathogenesis of human diseases. Furthermore, efforts are being made to improve the diagnosis and treatment of diseases associated with pre-mRNA splicing defects.

Despite the significant progress, we are only at the beginning stage of understanding the molecular mechanisms controlling pre-mRNA splicing and alternative splicing regulation. A number of important questions remain to be addressed.

For the majority of genes, we have little knowledge about their complete expression profiles of different splicing isoforms in different cell types. We do not know how the splicing events of different genes are coordinated during development or in response to environmental changes. How a cell senses the environmental stimuli and responds by producing different splicing products remains largely unknown. Molecular pathways that transduce extracellular signals into the nuclear splicing machinery have not yet been delineated. Further development of new technologies such as using microarray approaches to examine splicing isoforms under different physiological or pathological conditions at the genome level may be important for understanding the role of alternative pre-mRNA splicing in the biology of mammalian cells.

The basic components of mammalian splicing machinery are now characterized at the molecular level. However, the dynamic interactions among different components of the splicing machinery during spliceosome assembly and recognition of splicing regulatory elements are far from being understood. Little is known about the structural basis of such multicomponent interactions. Furthermore, the catalytic mechanisms of pre-mRNA splicing remain to be elucidated.

On the basis of studies of the relatively few model genes, a number of cis-acting elements and trans-acting splicing regulators have been identified. The mechanisms by which the splicing machinery specifically recognizes authentic versus decoy, or pseudosplice sites remain unclear. Some of the splicing regulatory factors are also important players for spliceosome assembly. Spliceosomal proteins also play a role in other processes of gene regulation. Further work needs to be carried out to understand the relationship between pre-mRNA splicing and other processes of gene expression and regulation, including transcription, RNA editing, RNA transport, translational control, and post-translational modification.

Studies of human diseases caused by splicing mutations or aberrant splicing have significantly advanced our understanding of human genetic diseases. Tremendous effort is still required to understand pathogenetic mechanisms underlying diseases caused by splicing defects, to develop efficient diagnostic tools and to design effective therapeutic approaches for these diseases.

Acknowledgment

We apologize to readers and our colleagues for having referred only to reviews instead of original research articles because of the space limit. The work in the authors' laboratory has been generously supported by Leukemia and Lymphoma Society, Muscular Dystrophy Association, Alzheimer's Association, the Society for Progressive Supranuclear Palsy, and NIH (AG17518, GM53945 and EY014576). We thank Drs Chris Burge, Reinhard Lührmann, Robin Reed, Phil Sharp, and Joan Steitz for sharing information and for giving permissions to modify figures from their previous publications. We are grateful to Drs Edward Benz, Doug Black, James Bruzik, Mariano Garcia-Blanco, Warren Gish, David Ornitz, Arnold Strauss, and members of Wu laboratory for their critical reading of the manuscript. We thank German G. Leparc, Joe Gee Seo, and Ying Zhang for the data bank searches and Mrs. Angie Hantak for outstanding assistance.

See also Molecular Basis of Genetics.

Bibliography

Books and Reviews

Adams, M.D., Rudner, D.Z., Rio, D.C. (1996) Biochemistry and regulation of pre-mRNA splicing, *Curr. Opin. Cell. Biol.* **8**(3), 331–339.

Bartel, F., Taubert, H., Harris, L.C. (2002) Alternative and aberrant splicing of MDM2 mRNA in human cancer, *Cancer Cell* **2**(1), 9–15.

Baserga, S.J., Yang, X.D., Steitz, J.A. (1993) The Diverse World of Small Ribonucleoproteins, in: Gesteland, R.F., Atkins, J.F. (Eds.) *The RNA World*, Cold Spring Harbor Laboratory Press, Cold Spring Harbor, New York, pp. 359–381.

Berget, S.M. (1995) Exon recognition in vertebrate splicing, *J. Biol. Chem.* **2**, 2411–2414.

Black, D.L. (2000) Protein diversity from alternative splicing: a challenge for bioinformatics and post-genome biology, *Cell* **103**(3), 367–370.

Black, D.L. (2003) Mechanisms of alternative pre-mRNA splicing, *Annu. Rev. Biochem.* **72**, 291–336.

Black, D.L., Grabowski, P.J. (2003) Alternative pre-mRNA splicing and neuronal function, *Prog. Mol. Subcell Biol.* **31**, 187–216.

Blencowe, B.J. (2000) Exonic splicing enhancers: mechanism of action, diversity and role in human genetic diseases, *Trends Biochem. Sci.* **25**(3), 106–110.

Bilbao, D., Valcarcel, J. (2003) Getting to the heart of a splicing enhancer, *Nat. Struct. Biol.* **10**(1), 6–7.

Brett, D., Pospisil, H., Valcarcel, J., Reich, J., Bork, P. (2003) Alternative splicing and genome complexity, *Nat. Genet.* **30**(1), 29–30.

Brow, D.A. (2002) Allosteric cascade of spliceosome activation, *Annu. Rev. Genet.* **36**, 333–360.

Buchholz, T.A., Wazer, D.E. (2002) Molecular biology and genetics of breast cancer development: a clinical perspective, *Semin. Radiat. Oncol.* **12**(4), 285–295.

Burge, C.B., Tuschl, T.H., Sharp, P.A. (1999) in: Gesteland, R.F., Cech, T.R., Atkins, J.F. (Eds.) *The RNA World*, Cold Spring Harbor Laboratory Press, New York, pp. 525–560.

Caceres, J.F., Kornblihtt, A.R. (2002) Alternative splicing: multiple control mechanisms and involvement in human disease, *Trends Genet.* **18**, 186–193.

Cartegni, L., Chew, S.L., Krainer, A.R. (2002) Listening to silence and understanding nonsense: exonic mutations that affect splicing, *Nat. Rev. Genet.* **3**(4), 285–298.

Chabot, B., LeBel, C., Hutchison, S., Nasim, F.H., Simard, M.J. (2003) Heterogeneous nuclear ribonucleoprotein particle A/B proteins and the control of alternative splicing of the mammalian heterogeneous nuclear ribonucleoprotein particle A1 pre-mRNA, *Prog. Mol. Subcell Biol.* **31**, 59–88.

Cooper, T.A., Mattox, W. (1997) The regulation of splice-site selection, and its role in human disease, *Am. J. Hum. Genet.* **61**(2), 259–266.

Dredge, B.K., Polydorides, A., Darnell, R.B. (2001) The splice of life: alternative splicing and neurological disease, *Nat. Neurosci. Rev.* **2**, 43–50.

Faustino, N.A., Cooper, T.A. (2003) Pre-mRNA splicing and human disease, *Genes Dev.* **17**(4), 419–437.

Forch, P., Valcarcel, J. (2003) Splicing regulation in Drosophila sex determination, *Prog. Mol. Subcell Biol.* **31**, 127–151.

Fu, X.D. (1995) The superfamily of arginine/serine-rich splicing factors, *RNA* **1**(7), 663–680.

Goedert, M., Spillantini, M.G. (2001) Tau gene mutations and neurodegeneration, *Biochem. Soc. Symp.* **67**, 59–71.

Goldstrohm, A.C., Greenleaf, A.L., Garcia-Blanco, M.A. (2001) Co-transcriptional splicing of pre-messenger RNAs: considerations for the mechanism of alternative splicing, *Gene* **277**(1–2), 31–47.

Grabowski, P.J. (1998) Splicing regulation in neurons: tinkering with cell-specific control, *Cell* **92**(6), 709–712.

Grabowski, P.J., Black, D.L. (2001) Alternative RNA splicing in the nervous system, *Prog. Neurobiol.* **65**, 289–308.

Graveley, B.R. (2000) Sorting out the complexity of SR protein functions, *RNA* **6**(9), 1197–1211.

Hastings, M.L., Krainer, A.R. (2001) Pre-mRNA splicing in the new millennium, *Curr. Opin. Cell Biol.* **13**, 302–309.

Hertel, K.J., Lynch, K.W., Maniatis, T. (1997) Common themes in the function of transcription and splicing enhancers, *Curr. Opin. Cell Biol.* **9**(3), 350–357.

Holmila, R., Fouquet, C., Cadranel, J., Zalcman, G., Soussi, T. (2003) Splice mutations in the p53 gene: case report and review of the literature, *Hum. Mutat.* **21**(1), 101–102.

Konarska, M.M. (1998) Recognition of the 5′ splice site by the spliceosome, *Acta Biochim. Pol.* **45**(4), 869–881.

Krämer, A. (1996) The structure and function of proteins involved in mammalian pre-mRNA splicing, *Annu. Rev. Biochem.* **65**, 367–409.

Lopez, A.J. (1998) Alternative splicing of pre-mRNA: developmental consequences and mechanisms of regulation, *Annu. Rev. Genet.* **32**, 279–305.

Maniatis, T., Reed, R. (2002) An extensive network of coupling among gene expression machines, *Nature* **416**(6880), 499–506.

Maniatis, T., Tasic, B. (2002) Alternative pre-mRNA splicing and proteome expansion in metazoans, *Nature* **418**(6894), 236–243.

Manley, J.L. (2002) Nuclear coupling: RNA processing reaches back to transcription, *Nat. Struct. Biol.* **9**(11), 790–791.

Manley, J.L., Tacke, R. (1996) SR proteins and splicing control, *Genes Dev.* **10**(13), 1569–1579.

Maroney, P.A., Romfo, C.M., Nilsen, T.W. (2002) Functional recognition of 5′ splice site by U4/U6.U5 tri-snRNP defines a novel ATP-dependent step in early spliceosome assembly, *Mol. Cell* **6**, 317–328.

Nilsen, T.W. (2002) The spliceosome: no assembly required? *Mol. Cell* **9**(1), 8–9.

Nissim-Rafinia, M., Krem, B. (2002) Splicing regulation as a potential genetic modifier, *Trends Genet.* **18**, 123–127.

Moore, M.J. (2000) Intron recognition comes of AGe, *Nat. Struct. Biol.* **7**(1), 14–16.

Mount, S.M. (2000) Genomic sequence, splicing, and gene annotation, *Am. J. Hum. Genet.* **67**(4), 788–792.

Mount, S.M., Salz, H.K. (2000) Pre-messenger RNA Processing Factors in the *Drosophila* Genome, *J. Cell Biol.* **150**, F37–F43.

Paushkin, S., Gubitz, A.K., Massenet, S., Dreyfuss, G. (2002) The SMN complex, an assemblyosome of ribonucleoproteins, *Curr. Opin. Cell Biol.* **14**(3), 305–312.

Ponta, H., Wainwright, D., Herrlich, P. (1998) The CD44 protein family, *Int. J. Biochem. Cell Biol.* **30**(3), 299–305.

Proudfoot, N.J., Furger, A., Dye, M.J. (2002) Integrating mRNA processing with transcription, *Cell* **108**, 501–512.

Reed, R. (1996) Initial splice-site recognition and pairing during pre-mRNA splicing, *Curr. Opin. Genet. Dev.* **6**(2), 215–220.

Reed, R. (2000) Mechanisms of fidelity in pre-mRNA splicing, *Curr. Opin. Cell Biol.* **12**(3), 340–345.

Reed, R., Hurt, E. (2002) A conserved mRNA export machinery coupled to pre-mRNA splicing, *Cell* **108**, 523–531.

Roberts, G.C., Smith, C.W. (2002) Alternative splicing: combinatorial output from the genome, *Curr. Opin. Chem. Biol.* **6**(3), 375–383.

Rosonina, E., Blencowe, B.J. (2002) Gene expression: the close coupling of transcription and splicing, *Curr. Biol.* **12**(9), R319–R321.

Sharp, P.A. (1994) Split genes and RNA splicing, *Cell* **77**(6), 805–815.

Smith, C.W., Valcarcel, J. (2000) Alternative pre-mRNA splicing: the logic of combinatorial control, *Trends Biochem. Sci.* **25**(8), 381–388.

Soret, J., Tazi, J. (2003) Phosphorylation-dependent control of the pre-mRNA splicing machinery, *Prog. Mol. Subcell Biol.* **31**, 89–126.

Staley, J.P., Guthrie, C. (1998) Mechanical devices of the spliceosome: motors, clocks, springs, and things, *Cell* **92**(3), 315–326.

Steitz, J.A. (1984) Autoantibody probes for mammalian gene expression, *Harvey. Lect.* (1984–85) **80**, 39–47.

Stoilov, P., Meshorer, E., Gencheva, M., Glick, D., Soreq, H., Stamm, S. (2002) Defects in pre-mRNA processing as causes of and predisposition to diseases, *DNA Cell Biol.* **21**(11), 803–818.

Suzuki, T., Nishio, K., Tanabe, S. (2001) The MRP family and anticancer drug metabolism, *Curr. Drug Metab.* **2**(4), 367–377.

Tacke, R., Manley, J.L. (1999) Determinants of SR protein specificity, *Curr. Opin. Cell Biol.* **11**(3), 358–362.

Tarn, W.Y., Steitz, J.A. (1997) Pre-mRNA splicing: the discovery of a new spliceosome doubles the challenge, *Trends Biochem. Sci.* **22**(4), 132–137.

Tupler, R., Perini, G., Green, M.R. (2001) Expressing the human genome, *Nature* **409**(6822), 832–833.

Valadkhan, S., Manley, J.L. (2002) Intrinsic metal binding by a spliceosomal RNA, *Nat. Struct. Biol.* **9**(7), 498–499.

Valcarcel, J., Green, M.R. (1996) The SR protein family: pleiotropic functions in pre-mRNA splicing, *Trends Biochem. Sci.* **21**(8), 296–301.

Villa, T., Pleiss, J.A., Guthrie, C. (2002) Spliceosomal snRNAs: Mg(2+)-dependent chemistry at the catalytic core? *Cell* **109**(2), 149–152.

Wagner, E.J., Garcia-Blanco, M.A. (2001) Polypyrimidine tract binding protein antagonizes exon definition, *Mol. Cell Biol.* **21**(10), 3281–3288.

Wang, J., Manley, J.L. (1997) Regulation of pre-mRNA splicing in metazoa, *Curr. Opin. Genet. Dev.* **7**(2), 205–211.

Will, C.L., Luhrmann, R. (2001) Spliceosomal UsnRNP biogenesis, structure and function, *Curr. Opin. Cell Biol.* **13**(3), 290–301.

Wu, J.Y., Tang, H., Havlioglu, N. (2003) Alternative pre-mRNA splicing and regulation of programmed cell death, *Prog. Mol. Subcell Biol.* **31**, 153–185.

Primary Literature

Abovich, N., Rosbash, M. (1997) Cross-intron bridging interactions in the yeast commitment complex are conserved in mammals, *Cell* **89**, 403–412.

Andreasi, C., Jarecki, J., Zhou, J., Coovert, D.D., Monani, U.R., Chen, X., Whitney, M., Pollok, B., Zhang, M., Androphy, E., Burghes, A.H.M. (2001) Aclarubicin treatment restores SMN levels to cells derived from type I spinal muscular atrophy patients, *Hum. Mol. Genet.* **24**, 2841–2849.

Arinobu, Y., Atamas, S.P., Otsuka, T., Niiro, H., Yamaoka, K., Mitsuyasu, H., Niho, Y., Hamasaki, N., White, B., Izuhara, K. (1999) Antagonistic effects of an alternative splice variant of human IL-4, IL-4 delta2, on IL-4 activities in human monocytes B cells, *Cell Immunol.* **191**, 161–167.

Auboeuf, D., Honig, A., Berget, S.M., O'Malley, B.W. (2002) Coordinate regulation of transcription and splicing by steroid receptor coregulators, *Science* **298**(5592), 416–419.

Barnard, D.C., Li, J., Peng, R., Patton, J.G. (2002) Regulation of alternative splicing by SRrp86 through coactivation and repression of specific SR proteins, *RNA* **8**(4), 526–533.

Berget, S.M., Moore, C., Sharp, P.A. (1977) Spliced segments at the 5' terminus of adenovirus 2 late mRNA, *Proc. Natl. Acad. Sci. U.S.A.* **74**(8), 3171–3175.

Berglund, J.A., Abovich, N., Rosbash, M. (1998) A cooperative interaction between U2AF65 and mBBP/SF1 facilitates branchpoint region recognition, *Genes Dev.* **12**, 858–867.

Berglund, J.A., Chua, K., Abovich, N., Reed, R., Rosbash, M. (1997) The splicing factor BBP interacts specifically with the pre-mRNA branchpoint sequence UACUAAC, *Cell* **89**(5), 781–787.

Biesiadecki, B.J., Elder, B.D., Yu, Z.B., Jin, J.P. (2002) Cardiac troponin T variants produced by aberrant splicing of multiple exons in animals with high instances of dilated cardiomyopathy, *J. Biol. Chem.* **277**(52), 50275–50285.

Bolduc, L., Labrecque, B., Cordeau, M., Blanchette, M., Chabot, B. (2001) Dimethyl sulfoxide affects the selection of splice sites, *J. Biol. Chem.* **276**(20), 17597–17602.

Boucher, L., Ouzounis, C.A., Enright, A.J., Blencowe, B.J. (2001) A genome-wide survey of RS domain proteins, *RNA* **7**(12), 1693–1701.

Brand, K., Dugi, K.A., Brunzell, J.D., Nevin, D.N., Santamarina-Fojo, S. (1996) A novel A→G mutation in intron I of the hepatic lipase gene leads to alternative splicing resulting in enzyme deficiency, *J. Lipid Res.* **37**(6), 1213–1223.

Bruzik, J.P., Van Doren, K., Hirsh, D., Steitz, J.A. (1988) Trans splicing involves a novel form of small nuclear ribonucleoprotein particles, *Nature* **335**(6190), 559–562.

Buckanovich, R.J., Darnell, R.B. (1997) The neuronal RNA binding protein Nova-1 recognizes specific RNA targets in vitro and in vivo, *Mol. Cell Biol.* **17**(6), 3194–3201.

Buratti, E., Baralle, F.E. (2001) Characterization and functional implications of the RNA binding properties of nuclear factor TDP-43, a novel splicing regulator of CFTR exon 9, *J. Biol. Chem.* **276**, 36337–36343.

Buratti, E., Dork, T., Zuccato, E., Pagani, F., Romano, M., Baralle, F.E. (2001) Nuclear factor TDP-43 and SR proteins promote in vitro and in vivo CFTR exon 9 skipping, *EMBO J.* **20**, 1774–1784.

Cáceres, J.F., Stamm, S., Helfman, D.M., Krainer, A.R. (1994) Regulation of alternative splicing in vivo by overexpression of antagonistic splicing factors, *Science* **265**, 1706–1709.

Cao, W., Jamison, S.F., Garcia-Blanco, M.A. (1997) Both phosphorylation and dephosphorylation of ASF/SF2 are required for pre-mRNA splicing in vitro, *RNA* **3**(12), 1456–1467.

Caputi, M., Zahler, A.M. (2001) Determination of the RNA binding specificity of the heterogeneous nuclear ribonucleoprotein (hnRNP) H/H'/F/2H9 family, *J. Biol. Chem.* **276**(47), 43850–43859.

Caputi, M., Zahler, A.M. (2002) SR proteins and hnRNP H regulate the splicing of the HIV-1 tev-specific exon 6D, *EMBO J.* **21**(4), 845–855.

Carlo, T., Sierra, R., Berget, S.M. (2000) A 5' splice site-proximal enhancer binds SF1 and activates exon bridging of a microexon, *Mol. Cell Biol.* **20**(11), 3988–3995.

Carstens, R.P., McKeehan, W.L., Garcia-Blanco, M.A. (1998) An intronic sequence element mediates both activation and repression of rat fibroblast growth factor receptor 2 pre-mRNA splicing, *Mol. Cell Biol.* **18**(4), 2205–2217.

Carstens, R.P., Wagner, E.J., Garcia-Blanco, M.A. (2000) An intronic splicing silencer causes skipping of the IIIb exon of fibroblast growth factor receptor 2 through involvement of polypyrimidine tract binding protein, *Mol. Cell Biol.* **20**(19), 7388–7400.

Cartegni, L., Krainer, A.R. (2002) Disruption of an SF2/ASF-dependent exonic splicing enhancer in SMN2 causes spinal muscular atrophy in the absence of SMN1, *Nat. Genet.* **30**(4), 377–384.

Cartegni, L., Krainer, A.R. (2003) Correction of disease-associated exon skipping by synthetic exon-specific activators, *Nat. Struct. Biol.* **10**(2), 120–125.

Celotto, A.M., Graveley, B.R. (2002) Exon-specific RNAi: a tool for dissecting the functional relevance of alternative splicing, *RNA* **8**(6), 718–724.

Chakarova, C.F., Hims, M.M., Bolz, H., Abu-Safieh, L., Patel, R.J., Papaioannou, M.G., Inglehearn, C.F., Keen, T.J., Willis, C., Moore, A.T., Rosenberg, T., Webster, A.R., Bird, A.C., Gal, A., Hunt, D., Vithana, E.N., Bhattacharya, S.S. (2002) Mutations in HPRP3, a third member of pre-mRNA splicing factor genes, implicated in autosomal dominant retinitis pigmentosa, *Hum. Mol. Genet.* **11**, 87–92.

Chalfant, C.E., Rathman, K., Pinkerman, R.L., Wood, R.E., Obeid, L.M., Ogretmen, B., Hannun, Y.A. (2002) De novo ceramide regulates the alternative splicing of caspase 9 and Bcl-x in A549 lung adenocarcinoma cells. Dependence on protein phosphatase-1, *J. Biol. Chem.* **277**(15), 12587–12595.

Chang, J.G., Hsieh-Li, H.M., Jong, Y.J., Wang, N.M., Tsai, C.H., Li, H. (2001) Treatment of spinal muscular atrophy by sodium butyrate, *PNAS* **98**, 9808–9813.

Charlet, B.N., Logan, P., Singh, G., Cooper, T.A. (2002a) Dynamic antagonism between ETR-3 and PTB regulates cell type-specific alternative splicing, *Mol. Cell* **9**(3), 649–658.

Charlet, B.N., Savkur, R.S., Singh, G., Philips, A.V., Grice, E.A., Cooper, T.A. (2002b) Loss of the muscle-specific chloride channel in type 1 myotonic dystrophy due to misregulated alternative splicing, *Mol. Cell* **10**(1), 45–53.

Chen, C.D., Kobayashi, R., Helfman, D.M. (1999) Binding of hnRNP H to an exonic splicing silencer is involved in the regulation of alternative splicing of the rat b-tropomyosin gene, *Genes Dev.* **13**, 593–606.

Chiba-Falek, O., Parad, R.B., Kerem, E., Kerem, B. (1999) Variable levels of normal RNA in different fetal organs carrying a cystic fibrosis transmembrane conductance regulator splicing mutation, *Am. J. Respir. Crit. Care Med.* **159**, 1998–2002.

Chou, M.Y., Rooke, N., Turck, C.W., Black, D.L. (1999) hnRNP H Is a Component of a Splicing Enhancer Complex That Activates a c-src Alternative Exon in Neuronal CellsMol, *Cell Biol.* **19**, 69–77.

Chow, L.T., Gelinas, R.E., Broker, T.R., Roberts, R.J. (1977) An amazing sequence arrangement at the 5′ ends of adenovirus 2 messenger RNA, *Cell* **12**(1), 1–8.

Chui, D.H., Hardison, R., Riemer, C., et al. (1998) An electronic database of human hemoglobin variants on the world wide web, *Blood* **91**, 2643–2644.

Claes, K., Vandesompele, J., Poppe, B., Dahan, K., Coene, I., De Paepe, A., Messiaen, L. (2002) Pathological splice mutations outside the invariant AG/GT splice sites of BRCA1 exon 5 increase alternative transcript levels in the 5′ end of the BRCA1gene, *Oncogene* **21**(26), 4171–4175.

Cooper, S.R., Taylor, J.K., Miraglia, L.J., Dean, N.M. (1999) Pharmacology of antisense oligonucleotide inhibitors of protein expression, *Pharmacol. Ther.* **82**, 427–435.

Cote, J., Dupuis, S., Jiang, Z., Wu, J.Y. (2001a) Caspase-2 pre-mRNA alternative splicing: Identification of an intronic element containing a decoy 3′acceptor site, *Proc. Natl. Acad. Sci. U.S.A.* **98**(3), 938–943.

Cote, J., Dupuis, S., Wu, J.Y. (2001b) Polypyrimidine track-binding protein binding downstream of caspase-2 alternative exon 9 represses its inclusion, *J. Biol. Chem.* **276**(11), 8535–8543.

Coulter, L.R., Landree, M.A., Cooper, T.A. (1997) Identification of a new class of exonic splicing enhancers by in vivo selection, *Mol. Cell Biol.* **17**(4), 2143–2150.

Cowper, A.E., Caceres, J.F., Mayeda, A., Screaton, G.R. (2001) Serine-arginine (SR) protein-like factors that antagonize authentic SR proteins and regulate alternative splicing, *J. Biol. Chem.* **276**(52), 48908–48914.

Daoud, R., Da Penha Berzaghi, M., Siedler, F., Hubener, M., Stamm, S. (1999) Activity-dependent regulation of alternative splicing patterns in the rat brain, *Eur. J. Neurosci.* **11**(3), 788–802.

Daoud, R., Mies, G., Smialowska, A., Olah, L., Hossmann, K.A., Stamm, S. (2002) Ischemia induces a translocation of the splicing factor tra2-beta 1 and changes alternative splicing patterns in the brain, *J. Neurosci.* **22**(14), 5889–5899.

Das, S., Levinson, B., Whitney, S., Vulpe, C., Packman, S., Gitschier, J. (1994) Diverse mutations in patients with Menkes disease often lead to exon skipping, *Am. J. Hum. Genet.* **55**(5), 883–889.

Deguillien, M., Huang, S.C., Moriniere, M., Dreumont, N., Benz, E.J. Jr., Baklouti, F. (2001) Multiple cis elements regulate an alternative splicing event at 4.1R pre-mRNA during erythroid differentiation, *Blood* **98**(13), 3809–3816.

Del Gatto-Konczak, F., Bourgeois, C.F., Le Guiner, C., Kister, L., Gesnel, M.C., Stevenin, J., Breathnach, R. (2000) The RNA-binding protein TIA-1 is a novel mammalian splicing regulator acting through intron sequences adjacent to a 5′ splice site, *Mol. Cell Biol.* **20**(17), 6287–6299.

Eldridge, A.G., Li, Y., Sharp, P.A., Blencowe, B.J. (1999) The SRm160/300 splicing coactivator is required for exon-enhancer function, *Proc. Natl. Acad. Sci. U.S.A.* **96**(11), 6125–6130.

Fackenthal, J.D., Cartegni, L., Krainer, A.R., Olopade, O.I. (2002) BRCA2 T2722R is a deleterious allele that causes exon skipping, *Am. J. Hum. Genet.* **71**(3), 625–631.

Fairbrother, W.G., Yeh, R.F., Sharp, P.A., Burge, C.B. (2002) Predictive identification of exonic splicing enhancers in human genes, *Science* **297**(5583), 1007–1013.

Feltes, C.M., Kudo, A., Blaschuk, O., Byers, S.W. (2002) An alternatively spliced cadherin-11 enhances human breast cancer cell invasion, *Cancer Res.* **62**(22), 6688–6697.

Fischer, U., Liu, Q., Dreyfuss, G. (1997) The SMN-SIP1 complex has an essential role in spliceosomal snRNP biogenesis, *Cell* **90**(6), 1023–1029.

Fogel, B.L., McNally, M.T. (2000) A Cellular Protein, hnRNP H, Binds to the Negative Regulator of Splicing Element from Rous Sarcoma Virus, *J. Biol. Chem.* **275**, 32371–32378.

Forch, P., Puig, O., Kedersha, N., Martinez, C., Granneman, S., Seraphin, B., Anderson, P., Valcarcel, J. (2000) The apoptosis-promoting factor TIA-1 is a regulator of alternative pre-mRNA splicing, *Mol. Cell* **6**(5), 1089–1098.

Forch, P., Puig, O., Martinez, C., Seraphin, B., Valcarcel, J. (2002) The splicing regulator TIA-1 interacts with U1-C to promote U1 snRNP recruitment to 5′ splice sites, *EMBO J.* **21**(24), 6882–6892.

Fouraux, M.A., Kolkman, M.J., Van der Heijden, A., De Jong, A.S., Van Venrooij, W.J., Pruijn, G.J. (2002) The human La (SS-B) autoantigen interacts with DDX15/hPrp43, a putative DEAH-box RNA helicase, *RNA* **8**(11), 1428–1443.

Fu, X.D., Maniatis, T. (1990) Factor required for mammalian spliceosome assembly is localized to discrete regions in the nucleus, *Nature* **343**(6257), 437–441.

Fu, X.D., Maniatis, T. (1992) Isolation of a complementary DNA that encodes the mammalian splicing factor SC35, *Science* **256**(5056), 535–538.

Fujimaru, M., Tanaka, A., Choeh, K., Wakamatsu, N., Sakuraba, H., Isshiki, G. (1998) Two mutations remote from an exon/intron junction in the beta-hexosaminidase beta-subunit gene affect 3′-splice site selection and cause Sandhoff disease, *Hum. Genet.* **103**(4), 462–469.

Garcia-Blanco, M.A., Jamison, S.F., Sharp, P.A. (1989) Identification and purification of a 62,000-dalton protein that binds specifically to the polypyrimidine tract of introns, *Genes Dev.* **3**(12A), 1874–1886.

Garcia-Blanco, M.A., Anderson, G.J., Beggs, J., Sharp, P.A. (1990) A mammalian protein of 220 kDa binds pre-mRNAs in the spliceosome: a potential homologue of the yeast PRP8 protein, *Proc. Natl. Acad. Sci. U.S.A.* **87**(8), 3082–3086.

Ge, H., Zuo, P., Manley, J.L. (1991) Primary structure of the human splicing factor ASF reveals similarities with Drosophila regulators, *Cell* **66**(2), 373–382.

Gebhardt, F., Zanker, K.S., Brandt, B. (1998) Differential expression of alternatively spliced c-erbB-2 mRNA in primary tumors, lymph node metastases, and bone marrow micrometastases from breast cancer patients, *Biochem. Biophys. Res. Commun.* **247**(2), 319–323.

Ghanem, M.A., Van Steenbrugge, G.J., Van Der Kwast, T.H., Sudaryo, M.K., Noordzij, M.A., Nijman, R.J. (2002) Expression and prognostic value Of CD44 isoforms in nephroblastoma (Wilms tumor), *J. Urol.* **168**(2), 681–686.

Gooding, C., Kemp, P., Smith, C.W. (2003) SmPTB: a novel polypyrimidine tract binding protein (PTB) paralog expressed in smooth muscle cells, *J. Biol. Chem.* in press.

Gozani, O., Patton, J.G., Reed, R. (1994) A novel set of spliceosome-associated proteins and the essential splicing factor PSF bind stably to pre-mRNA prior to catalytic step II of the splicing reaction, *EMBO J.* **13**(14), 3356–3367.

Graveley, B.R., Hertel, K.J., Maniatis, T. (1998) A systematic analysis of the factors that determine the strength of pre-mRNA splicing enhancers, *EMBO J.* **17**(22), 6747–6756.

Graveley, B.R., Maniatis, T. (1998) Arginine/serine-rich domains of SR proteins can function as activators of pre-mRNA splicing, *Mol. Cell* **1**(5), 765–771.

Groenen, P.J.T.A., Wansink, D.G., Coerwinkel, M., Van den broek, W., Jansen, G., Wieringa, B. (2000) Constitutive and regulated modes of splicing produce six major myotonic dystrophy protein kinase (DMPK) isoforms with distinct properties, *Hum. Mol. Genet.* **9**, 605–616.

Gui, J.F., Lane, W.S., Fu, X.D. (1994) A serine kinase regulates intracellular localization of splicing factors in the cell cycle, *Nature* **369**(6482), 678–682.

Grover, A., Houlden, H., Baker, M., Adamson, J., Lewis, J., Prihar, G., Pickering-Brown, S., Duff, K. (1999) Hutton M 5′ splice site mutations in tau associated with the inherited dementia FTDP-17 affect a stem-loop structure that regulates alternative splicing of exon 10, *J. Biol. Chem.* **274**, 15134–15143.

Guo, N., Kawamoto, S. (2000) An intronic downstream enhancer promotes 3′ splice site usage of a neural cell-specific exon, *J. Biol. Chem.* **275**(43), 33641–33649.

Hall, S.L., Padgett, R.A. (1996) Requirement of U12 snRNA for in vivo splicing of a minor class of eukaryotic nuclear pre-mRNA introns, *Science* **271**(5256), 1716–1718.

Hartmuth, K., Urlaub, H., Vornlocher, H.P., Will, C.L., Gentzel, M., Wilm, M., Luhrmann, R. (2002) Protein composition of human prespliceosomes isolated by a tobramycin affinity-selection method, *Proc. Natl. Acad. Sci. U.S.A.* **99**(26), 16719–16724.

Hasegawa, Y., Kawame, H., Ida, H., Ohashi, T., Eto, Y. (1994) Single exon mutation in arylsulfatase A gene has two effects: loss of enzyme activity and aberrant splicing, *Hum. Genet.* **93**(4), 415–420.

He, Y., Smith, S.K., Day, K.A., Clark, D.E., Licence, D.R., Charnock-Jones, D.S. (1999) Alternative splicing of vascular endothelial growth factor (VEGF)-R1 (FLT-1) pre-mRNA is important for the regulation of VEGF activity, *Mol. Endocrinol.* **13**, 537–545.

Hechtman, P., Boulay, B., De Braekeleer, M., Andermann, E., Melancon, S., Larochelle, J., Prevost, C., Kaplan, F. (1992) The intron 7 donor splice site transition: a second Tay-Sachs disease mutation in French Canada, *Hum. Genet.* **90**(4), 402–406.

Hibi, K., Trink, B., Patturajan, M., Westra, W.H., Caballero, O.L., Hill, D.E., Ratovitski, E.A., Jen, J., Sidransky, D. (2000) AIS is an oncogene amplified in squamous cell carcinoma, *Proc. Natl. Acad. Sci. U.S.A.* **97**(10), 5462–5467.

Hinek, A., Zhang, S., Smith, A.C., Callahan, J.W. (2000) Impaired elastic-fiber assembly by fibroblasts from patients with either Morquio B disease or infantile GM1-gangliosidosis is linked to deficiency in the 67-KD spliced variant of β galactosidase, *Am. J. Hum. Genet.* **67**, 23–36.

Honig, A., Auboeuf, D., Parker, M.M., O'Malley, B.W., Berget, S.M. (2002) Regulation of alternative splicing by the ATP-dependent DEAD-box RNA helicase p72, *Mol. Cell Biol.* **22**(16), 5698–5707.

Hou, V.C., Lersch, R., Gee, S.L., Ponthier, J.L., Lo, A.J., Wu, M., Turck, C.W., Koury, M., Krainer, A.R., Mayeda, A., Conboy, J.G. (2002) Decrease in hnRNP A/B expression during erythropoiesis mediates a pre-mRNA splicing switch, *EMBO J.* **21**(22), 6195–6204.

Huh, G.S., Hynes, R.O. (1994) Regulation of alternative pre-mRNA splicing by a novel repeated hexanucleotide element, *Genes Dev.* **8**(13), 1561–1574.

Hutton, M., Lendon, C.L., Rizzu, P., Baker, M., Froelich, S., Houlden, H., Pickering-Brown, S., Chakraverty, S., Isaacs, A., Grover, A., Hackett, J., Adamson, J., Lincoln, S., Dickson, D., Davies, P., Petersen, R.C., Stevens, M., de Graaff, E., Wauters, E., van Baren, J., Hillebrand, M., Joosse, M., Kwon, J.M., Nowotny, P., Heutink, P., et al. (1998)

Association of missense and 5′-splice-site mutations in tau with the inherited dementia FTDP-17, *Nature* **393**, 702–705.

Ismaili, N., Sha, M., Gustafson, E.H., Konarska, M.M. (2001) The 100-kda U5 snRNP protein (hPrp28p) contacts the 5′ splice site through its ATPase site, *RNA* **7**(2), 182–193.

Jacquenet, S., Mereau, A., Bilodeau, P.S., Damier, L., Stoltzfus, C.M., Branlant, C. (2001) A second exon splicing silencer within human immunodeficiency virus type 1 tat exon 2 represses splicing of Tat mRNA and binds protein hnRNP H, *J. Biol. Chem.* **276**(44), 40464–40475.

Jensen, K.B., Dredge, B.K., Stefani, G., Zhong, R., Buckanovich, R.J., Okano, H.J., Yang, Y.Y. (2000) Nova-1 regulates neuron-specific alternative splicing and is essential for neuronal viability, *Neuron* **25**(2), 359–371.

Jiang, Z., Cote, J., Kwon, J.M., Goate, A.M., Wu, J.Y. (2000) Aberrant splicing of tau pre-mRNA caused by intronic mutations associated with the inherited dementia frontotemporal dementia with parkinsonism linked to chromosome 17, *Mol. Cell. Biol.* **20**, 4036–4048.

Jiang, Z., Tang, H., Havlioglu, N., Zhang, X., Stamm, S., Yan, R., Wu, J.Y. (2003) Mutations in tau gene exon10 associated with FTDP-17 alter the activity of an exonic splicing enhancer to interact with Tra2beta, *J. Biol. Chem.*

Jiang, Z.H., Zhang, W.J., Rao, Y., Wu, J.Y. (1998) Regulation of Ich-1 pre-mRNA alternative splicing and apoptosis by mammalian splicing factors, *Proc. Natl. Acad. Sci. U.S.A.* **95**, 9155–9160.

Jonghe, C.D., Cruts, M., Rogaeva, E.A., Tysoe, C., Singleton, A., Vanderstichele, H., Meschino, W., Dermant, B., Vanderhoeven, I., Backhovens, H., et al. (1999) Aberrant splicing in the presenilin-1 intron 4 mutation causes presenile Alzheimer's disease by increased Abeta42 secretion, *Hum. Mol. Genet.* **8**, 1529–1540.

Kalbfuss, B., Mabon, S.A., Misteli, T. (2001) Correction of alternative splicing of Tau in frontotemporal dementia and parkinsonism linked to chromosome 17, *J. Biol. Chem.* **276**, 42986–42993.

Kan, J.L., Green, M.R. (1999) Pre-mRNA splicing of IgM exons M1 and M2 is directed by a juxtaposed splicing enhancer and inhibitor, *Genes Dev.* **13**(4), 462–471.

Kanopka, A., Muhlemann, O., Akusjarvi, G. (1996) Inhibition by SR proteins of splicing of a regulated adenovirus pre-mRNA, *Nature* **381**(6582), 535–538.

Kanopka, A., Muhlemann, O., Petersen-Mahrt, S., Estmer, C., Ohrmalm, C., Akusjarvi, G. (1998) Regulation of adenovirus alternative RNA splicing by dephosphorylation of SR proteins, *Nature* **393**(6681), 185–187.

Kennedy, C.F., Kramer, A., Berget, S.M. (1998) A role for SRp54 during intron bridging of small introns with pyrimidine tracts upstream of the branch point, *Mol. Cell. Biol.* **18**(9), 5425–5434.

Klamt, B., Koziell, A., Poulat, F., Wieacker, P., Scamber, P., Berta, P., Gessler, M. (1998) Frasier syndrome is caused by defective alternative splicing of WT1 leading to an altered ratio of WT1 +/− KTS splice isoforms, *Hum. Mol. Genet.* **7**, 709–714.

Kohtz, J.D., Jamison, S.F., Will, C.L., Zuo, P., Luhrmann, R., Garcia-Blanco, M.A., Manley, J.L. (1994) Protein-protein interactions and 5′-splice-site recognition in mammalian mRNA precursors, *Nature* **368**(6467), 119–124.

Konforti, B.B., Koziolkiewicz, M.J., Konarska, M.M. (1993) Disruption of base pairing between the 5′ splice site and the 5′ end of U1 snRNA is required for spliceosome assembly, *Cell* **75**(5), 863–873.

Krainer, A.R., Conway, G.C., Kozak, D. (1990) The essential pre-mRNA splicing factor SF2 influences 5′ splice site selection by activating proximal sites, *Cell* **62**(1), 35–42.

Krainer, A.R., Maniatis, T., Ruskin, B., Green, M.R. (1984) Normal and mutant human beta-globin pre-mRNAs are faithfully and efficiently spliced in vitro, *Cell* **36**(4), 993–1005.

Kraus, A., Neff, F., Behn, M., Schuermann, M., Muenkel, K., Schlegel, J. (1999) Expression of alternatively spliced mdm2 transcripts correlates with stabilized wild-type p53 protein in human glioblastoma cells, *Int. J. Cancer* **80**(6), 930–934.

Labourier, E., Blanchette, M., Feiger, J.W., Adams, M.D., Rio, D.C. (2002) The KH-type RNA-binding protein PSI is required for Drosophila viability, male fertility, and cellular mRNA processing, *Genes Dev.* **16**(1), 72–84.

Labourier, E., Bourbon, H.M., Gallouzi, I.E., Fostier, M., Allemand, E., Tazi, J. (1999) Antagonism between RSF1 and SR proteins for both splice-site recognition in vitro and

Drosophila development, *Genes Dev.* **13**(6), 740–753.

Ladd, A.N., Charlet, N., Cooper, T.A. (2001) The CELF family of RNA binding proteins is implicated in cell-specific and developmentally regulated alternative splicing, *Mol. Cell. Biol.* **21**(4), 1285–1296.

Laggerbauer, B., Achsel, T., Luhrmann, R. (1998) The human U5-200kD DEXH-box protein unwinds U4/U6 RNA duplices in vitro, *Proc. Natl. Acad. Sci. U.S.A.* **95**(8), 4188–4192.

Lallena, M.J., Chalmers, K.J., Llamazares, S., Lamond, A.I., Valcarcel, J. (2002) Splicing regulation at the second catalytic step by Sex-lethal involves 3′ splice site recognition by SPF45, *Cell* **109**(3), 285–296.

Lander, E.S., et al. (2001) Initial sequencing and analysis of the human genome, *Nature* **409**, 860–921.

Larriba, S., Bassa, L., Gimenez, J., Ramos, M.D., Segura, A., Nunes, V., Estivill, X., Casals, T. (1998) Testicular CFTR splice variants in patients with congenital absence of the vas deferens, *Hum. Mol. Genet.* **7**, 1739–1743.

Lavigueur, A., La Branche, H., Kornblihtt, A.R., Chabot, B. (1993) A splicing enhancer in the human fibronectin alternate ED1 exon interacts with SR proteins and stimulates U2 snRNP binding, *Genes Dev.* **7**, 2405–2417.

Le Guiner, C., Lejeune, F., Galiana, D., Kister, L., Breathnach, R., Stevenin, J., Del Gatto-Konczak, F. (2001) TIA-1 and TIAR activate splicing of alternative exons with weak 5′ splice sites followed by a U-rich stretch on their own pre-mRNAs, *J. Biol. Chem.* **276**(44), 40638–40646.

Le Hir, H., Moore, M.J., Maquat, L.E. (2000) Pre-mRNA splicing alters mRNP composition: evidence for stable association of proteins at exon-exon junctions, *Genes Dev.* **14**(9), 1098–1108.

Lim, S.R., Hertel, K.J. (2001) Modulation of survival motor neuron pre-mRNA splicing by inhibition of alternative 3′ splice site pairing, *J. Biol. Chem.* **276**(48), 45476–45483.

Lin, C.G., Bristol, L.A., Jin, L., Dykes-Hoberg, M., Crawford, T., Clawson, L., Rothstein, J.D. (1998) Aberrant RNA processing in a neurodegenerative disease: the cause for absent EAAT2, a glutamate transporter in amyotrophic lateral sclerosis, *Neuron* **20**, 589–602.

Lisbin, M.J., Qiu, J., White, K. (2001) The neuron-specific RNA-binding protein ELAV regulates neuroglian alternative splicing in neurons and binds directly to its pre-mRNA, *Genes Dev.* **15**(19), 2546–2561.

Liu, H.X., Chew, S.L., Cartegni, L., Zhang, M.Q., Krainer, A.R. (2000) Exonic splicing enhancer motif recognized by human SC35 under splicing conditions, *Mol. Cell. Biol.* **20**(3), 1063–1071.

Liu, X., Jiang, Q., Mansfield, S.G., Puttaraju, M., Zhang, Y., Zhou, W., Cohn, J.A., Garcia-Blanco, M.A., Mitchell, L.G., Engelhardt, J.F. (2002) Partial correction of endogenous DeltaF508 CFTR in human cystic fibrosis airway epithelia by spliceosome-mediated RNA trans-splicing, *Nat. Biotechnol.* **20**(1), 47–52.

Liu, W., Qian, C., Francke, U. (1997) Silent mutation induces exon skipping of fibrillin-1 gene in Marfan syndrome, *Nat. Genet.* **16**(4), 328–329.

Llewellyn, D.H., Scobie, G.A., Urquhart, A.J., Whatley, S.D., Roberts, A.G., Harrison, P.R., Elder, G.H. (1996) Acute intermittent porphyria caused by defective splicing of porphobilinogen deaminase RNA: a synonymous codon mutation at -22 bp from the 5′ splice site causes skipping of exon 3, *J. Med. Genet.* **33**(5), 437–438.

Lorson, C.L., Hahnene, E., Androphy, E.J., Wirth, B. (1999) A single nucleotide in the SMN gene regulates splicing and is responsible for spinal muscular atrophy, *PNAS* **96**, 6307–6311.

Lou, H., Helfman, D.M., Gagel, R.F., Berget, S.M. (1999) Polypyrimidine tract-binding protein positively regulates inclusion of an alternative 3′-terminal exon, *Mol. Cell. Biol.* **19**(1), 78–85.

Lou, H., Yang, Y., Cote, G.J., Berget, S.M., Gagel, R.F. (1995) An intron enhancer containing a 5′ splice site sequence in the human calcitonin/calcitonin gene-related peptide gene, *Mol. Cell. Biol.* **15**(12), 7135–7142.

Lukas, J., Gao, D.Q., Keshmeshian, M., Wen, W.H., Tsao-Wei, D., Rosenberg, S., Press, M.F. (2001) Alternative and aberrant messenger RNA splicing of the mdm2 oncogene in invasive breast cancer, *Cancer. Res.* **61**(7), 3212–3219.

Lundquist, E.A., Herman, R.K., Rogalski, T.M., Mullen, G.P., Moerman, D.G., Shaw, J.E. (1996) The mec-8 gene of C. elegans encodes a protein with two RNA recognition motifs and regulates alternative splicing of unc-52 transcripts, *Development* **122**(5), 1601–1610.

Lykke-Andersen, J., Shu, M.D., Steitz, J.A. (2001) Communication of the position of exon-exon junctions to the mRNA surveillance machinery by the protein RNPS1, *Science* **293**(5536), 1836–1839.

Lynch, K.W., Weiss, A. (2001) A CD45 polymorphism associated with multiple sclerosis disrupts an exonic splicing silencer, *J. Biol. Chem.* **276**, 24341–24347.

Makarova, O.V., Makarov, E.M., Liu, S., Vornlocher, H.P., Luhrmann, R. (2002a) Protein 61K, encoded by a gene (PRPF31) linked to autosomal dominant retinitis pigmentosa, is required for U4/U6*U5 tri-snRNP formation and pre-mRNA splicing, *EMBO J.* **21**, 1148–1157.

Makarov, E.M., Makarova, O.V., Urlaub, H., Gentzel, M., Will, C.L., Wilm, M., Luhrmann, R. (2002b) Small nuclear ribonucleoprotein remodeling during catalytic activation of the spliceosome, *Science* **298**, 2205–2208.

Mankodi, A., Takahashi, M.P., Jiang, H., Beck, C.L., Bowers, W.J., Moxley, R.T., Cannon, S.C., Thornton, C.A. (2002) Expanded CUG repeats trigger aberrant splicing of ClC-1 chloride channel pre-mRNA and hyperexcitability of skeletal muscle in myotonic dystrophy, *Mol. Cells* **10**(1), 35–44.

Mann, C.J., Honeyman, K., Cheng, A.J., Ly, T., Lloyd, F., Fletcher, S., Morgan, J.E., Partridge, T.A., Wilton, S.D. (2001) Antisense-induced exon skipping and synthesis of dystrophin in the mdx mouse, *Proc. Natl. Acad. Sci. U.S.A.* **98**(1), 42–47.

Markovtsov, V., Nikolic, J.M., Goldman, J.A., Turck, C.W., Chou, M.Y., Black, D.L. (2000) Cooperative assembly of an hnRNP complex induced by a tissue-specific homolog of polypyrimidine tract binding protein, *Mol. Cell. Biol.* **20**(20), 7463–7479.

Maruyama, T., Miyake, Y., Yamamura, T., Tajima, S., Funahashi, T., Matsuzawa, Y., Yamamoto, A. (1998) A novel point mutation in a splice acceptor site of intron 1 of the human low density lipoprotein receptor gene which causes severe hypercholesterolemia: an unexpected absence of exon skipping, *Hum. Mutat.* **11**(6), 480–481.

McGrory, J., Cole, W.G. (1999) Alternative splicing of exon 37 of FBN1 deletes part of an 'eight-cysteine' domain resulting in the Marfan syndrome, *Clin. Genet.* **55**(2), 118–121.

Matsumura, T., Osaka, H., Sugiyama, N., Kawanishi, C., Maruyama, Y., Suzuki, K., Onishi, H., Yamada, Y., Morita, M., Aoki, M., Kosaka, K. (1998) Novel acceptor splice site mutation in the invariant AG of intron 6 of alpha-galactosidase A gene, causing Fabry disease, *Hum. Mutat.* **11**, 483.

Matter, N., Marx, M., Weg-Remers, S., Ponta, H., Herrlich, P., König, H. (2000) Heterogeneous ribonucleoprotein A1 is part of an exon-specific splice-silencing complex controlled by oncogenic signaling pathways, *J. Biol. Chem.* **275**, 35353–35360.

Mayeda, A., Krainer, A.R. (1992) Regulation of alternative pre-mRNA splicing by hnRNP A1 and splicing factor SF2, *Cell* **68**(2), 365–375.

McCullough, A.J., Berget, S.M. (2000) An intronic splicing enhancer binds U1 snRNPs to enhance splicing and select 5′ splice sites, *Mol. Cell. Biol.* **20**(24), 9225–9235.

McGrory, J., Cole, W.G. (1999) Alternative splicing of exon 37 of FBN1 deletes part of an 'eight-cysteine' domain resulting in the Marfan syndrome, *Clin. Genet.* **55**(2), 118–121.

McKie, A.B., McHale, J.C., Keen, T.J., Tarttelin, E.E., Goliath, R., Van Lith-Verhoeven, J.J., Greenberg, J., Ramesar, R.S., Hoyng, C.B., Cremers, F.P., MacKey, D.A., Bhattacharya, S.S., Bird, A.C., Markham, A.F., Inglehearn, C.F. (2001) Mutations in the pre-mRNA splicing factor gene PRPC8 in autosomal dominant retinitis pigmentosa (RP13), *Hum. Mol. Genet.* **10**, 1555–1562.

Mercatante, D., Kole, R. (2000) Modification of alternative splicing pathways as a potential approach to chemotherapy, *Pharmacol. Ther.* **85**, 237–243.

Merendino, L., Guth, S., Bilbao, D., Martinez, C., Valcarcel, J. (1999) Inhibition of msl-2 splicing by sex-lethal reveals interaction between U2AF35 and the 3′ splice site AG, *Nature* **402**(6763), 838–841.

Min, H., Turck, C.W., Nikolic, J.M., Black, D.L. (1997) A new regulatory protein, KSRP, mediates exon inclusion through an intronic splicing enhancer, *Genes Dev.* **11**(8), 1023–1036.

Min, H., Chan, R.C., Black, D.L. (1995) The generally expressed hnRNP F is involved in a neural-specific pre-mRNA splicing event, *Genes Dev.* **9**(21), 2659–2671.

Modafferi, E.F., Black, D.L. (1997) A complex intronic splicing enhancer from the c-src pre-mRNA activates inclusion of a heterologous exon, *Mol. Cell. Biol.* **17**(11), 6537–6545.

Moore, M.J., Sharp, P.A. (1993) Evidence for two active sites in the spliceosome provided by stereochemistry of pre-mRNA splicing, *Nature* **365**(6444), 364–368.

Nabholtz, J.M., Reese, D.M., Lindsay, M.A., Riva, A. (2002) HER2-positive breast cancer: update on breast cancer international research group trials, *Clin. Breast Cancer* **3**(Suppl. 2), S75–S79.

Nelson, K.K., Green, M.R. (1990) Mechanism for cryptic splice site activation during pre-mRNA splicing, *Proc. Natl. Acad. Sci. U.S.A.* **87**(16), 6253–6257.

Nesic, D., Kramer, A. (2001) Domains in human splicing factors SF3a60 and SF3a66 required for binding to SF3a120, assembly of the 17S U2 snRNP, and prespliceosome formation, *Mol. Cell. Biol.* (19), 6406–6417.

Nissim-Rafinia, M., Krem, B. (2002) Splicing regulation as a potential genetic modifier, *Trends Genet.* **18**, 123–127.

Nicoll, M., Akerib, C.C., Meyer, B.J. (1997) X-chromosome-counting mechanisms that determine nematode sex, *Nature* **388**(6638), 200–204.

Nguyen, V.N., Mirejovsky, T., Melinova, L., Mandys, V. (2000) CD44 and its v6 spliced variant in lung carcinomas: relation to NCAM, CEA, EMA and UP1 and prognostic significance, *Neoplasma* **47**(6), 400–408.

Ono, Y., Ohno, M., Shimura, Y. (1994) Identification of a putative RNA helicase (HRH1), a human homolog of yeast Prp22, *Mol. Cell. Biol.* **14**(11), 7611–7620.

Ono, R., Taki, T., Taketani, T., Kawaguchi, H., Taniwaki, M., Okamura, T., Kawa, K., Hanada, R., Kobayashi, M., Hayashi, Y. (2002) SEPTIN6, a human homologue to mouse Septin6, is fused to MLL in infant acute myeloid leukemia with complex chromosomal abnormalities involving 11q23 and Xq24, *Cancer Res.* **62**(2), 333–337.

Oshika, Y., Nakamura, M., Tokunaga, T., Ohnishi, Y., Abe, Y., Tsuchida, T., Tomii, Y., Kijima, H., Yamazaki, H., Ozeki, Y., Tamaoki, N., Ueyama, Y. (2000) Ribozyme approach to downregulate vascular endothelial growth factor (VEGF) 189 expression in non-small cell lung cancer (NSCLC), *Eur. J. Cancer* **36**(18), 2390–2396.

Pagani, F., Buratti, E., Stuani, C., Bendix, R., Dork, T., Baralle, F.E. (2002) A new type of mutation causes a splicing defect in ATM, *Nat. Genet.* **30**, 426–429.

Perez-Tur, J., Froelich, S., Prihar, G., Crook, R., Baker, M., Duff, K., Wragg, M., Busfield, F., Lendon, C., Clark, R.F., et al. (1995) A mutation in Alzheimer's disease destroying a splice acceptor site in the presenilin-1 gene, *NeuroReport* **7**, 297–301.

Petersen-Mahrt, S.K., Estmer, C., Ohrmalm, C., Matthews, D.A., Russell, W.C., Akusjarvi, G. (1999) The splicing factor-associated protein, p32, regulates RNA splicing by inhibiting ASF/SF2 RNA binding and phosphorylation, *EMBO J.* **18**(4), 1014–1024.

Pinto, A.L., Steitz, J.A. (1989) The mammalian analogue of the yeast PRP8 splicing protein is present in the U4/5/6 small nuclear ribonucleoprotein particle and the spliceosome, *Proc. Natl. Acad. Sci. U.S.A.* **86**, 8742–8746.

Ploos Van Amstel, J.K., Bergman, A.J., Van Beurden, E.A., Roijers, J.F., Peelen, T., Van Den Berg, I.E., Poll-The, B.T., Kvittingen, E.A., Berger, R. (1996) Hereditary tyrosinemia type 1: novel missense, nonsense and splice consensus mutations in the human fumarylacetoacetate hydrolase gene; variability of the genotype-phenotype relationship, *Hum. Genet.* **97**(1), 51–59.

Poola, I., Speirs, V. (2001) Expression of alternatively spliced estrogen receptor alpha mRNAs is increased in breast cancer tissues, *J. Steroid Biochem. Mol. Biol.* **78**, 459–469.

Poltora, Z., Cohen, T., Neufeld, G. (2000) The VEGF splice variants: properties, receptors, and usage for the treatment of ischemic diseases, *Herz* **25**, 126–129.

Puig, O., Gottschalk, A., Fabrizio, P., Seraphin, B. (1999) Interaction of the U1 snRNP with nonconserved intronic sequences affects 5' splice site selection, *Genes Dev.* **13**(5), 569–580.

Puttaraju, M., Dipasquale, J., Baker, C.C., Mitchell, L.G., Garcia-Blanco, M.A. (2001) Messenger RNA repair and restoration of protein function by spliceosome-mediated RNA trans-splicing, *Mol. Ther.* **4**(2), 105–114.

Query, C.C., Bentley, R.C., Keene, J.D. (1989) A common RNA recognition motif identified within a defined U1 RNA binding domain of the 70K U1 snRNP protein, *Cell* **57**(1), 89–101.

Rappsilber, J., Ajuh, P., Lamond, A.I., Mann, M. (2001) SPF30 is an essential human splicing factor required for assembly of the U4/U5/U6 tri-small nuclear ribonucleoprotein into the spliceosome, *J. Biol. Chem.* **276**(33), 31142–31150.

Rappsilber, J., Ryder, U., Lamond, A.I., Mann, M. (2002) Large-scale proteomic analysis of the human spliceosome, *Genome Res.* **12**(8), 1231–1245.

Richard, M.M., Erenberg, G., Triggs-Raine, B.L. (1995) An A-to-G mutation at the +3 position of intron 8 of the HEXA gene is associated with exon 8 skipping and Tay-Sachs disease, *Biochem. Mol. Med.* **55**(1), 74–76.

Rogan, P.K., Faux, B.M., Schneider, T.D. (1998) Information analysis of human splice site mutations, *Hum. Mutat.* **12**, 153–171.

Ruskin, B., Zamore, P.D., Green, M.R. (1988) A factor, U2AF, is required for U2 snRNP binding and splicing complex assembly, *Cell* **52**(2), 207–219.

Reyes, J.L., Gustafson, E.H., Luo, H.R., Moore, M.J., Konarska, M.M. (1999) The C-terminal region of hPrp8 interacts with the conserved GU dinucleotide at the 5′ splice site.å, *RNA* **5**(2), 167–179.

Sanford, J.R., Bruzik, J.P. (1999) Developmental regulation of SR protein phosphorylation and activity, *Genes Dev.* **13**(12), 1513–1518.

Sato, N., Imaizumi, K., Manabe, T., Tanigushi, M., Hitomi, J., Katayama, T., Takunari, Y., Morihara, T., Yasuda, Y., Takagi, T., Kudo, T., Tsudo, T., Itoyama, Y., Makifuchi, T., Fraser, P.E., St George-Hyslop, P., Tohyama, M. (2001) Increased production of beta amyloid and vulnerability to endoplasmic reticulum stress by an aberrant spliced form of presenilin 2, *J. Biol. Chem.* **276**, 2108–2114.

Savkur, R.S., Philips, A.V., Cooper, T.A. (2001) Aberrant regulation of insulin receptor alternative splicing is associated with insulin resistance in myotonic dystrophy, *Nat. Genet.* **29**(1), 40–47.

Seifert, G., Schroder, W., Hinterkeuser, S., Schumacher, T., Schramm, J., Steinhauser, C. (2002) Changes in flip/flop splicing of astroglial AMPA receptors in human temporal lobe epilepsy, *Epilepsia* **43**(Suppl. 5), 162–167.

Shin, C., Manley, J.L. (2002) The SR protein SRp38 represses splicing in M phase cells, *Cell* **111**(3), 407–417.

Siatecka, M., Reyes, J.L., Konarska, M.M. (1999) Functional interactions of Prp8 with both splice sites at the spliceosomal catalytic center, *Genes Dev.* **13**(15), 1983–1993.

Siebel, C.W., Kanaar, R., Rio, D.C. (1994) Regulation of tissue-specific P-element pre-mRNA splicing requires the RNA-binding protein PSI, *Genes Dev.* **8**(14), 1713–1725.

Siffert, W., Rosskopf, D., Siffert, G., Busch, S., Moritz, A., Erbel, R., Sharma, A.M., Ritz, E., Wichmann, H.E., Jakobs, K.H. (1998) Horsthemke B association of a human G-protein beta3 subunit variant with hypertension, *Nat. Genet.* **18**(1), 45–48.

Sigalas, I., Calvert, A.H., Anderson, J.J., Neal, D.E., Lunec, J. (1996) Alternatively spliced mdm2 transcripts with loss of p53 binding domain sequences: transforming ability and frequent detection in human cancer, *Nat. Med.* **2**(8), 912–917.

Simard, M.J., Chabot, B. (2002) SRp30c is a repressor of 3′ splice site utilization, *Mol. Cell. Biol.* **22**(12), 4001–4010.

Singh, R., Valcarcel, J., Green, M.R. (1995) Distinct binding specificities and functions of higher eukaryotic polypyrimidine tract-binding proteins, *Science* **268**(5214), 1173–1176.

Sontheimer, E.J., Steitz, J.A. (1993) The U5 and U6 small nuclear RNAs as active site components of the spliceosome, *Science* **262**(5142), 1989–1996.

Sontheimer, E.J., Sun, S., Piccirilli, J.A. (1997) Metal ion catalysis during splicing of premessenger RNA, *Nature* **388**(6644), 801–805.

Spike, C.A., Davies, A.G., Shaw, J.E., Herman, R.K. (2002) MEC-8 regulates alternative splicing of unc-52 transcripts in C. elegans hypodermal cells, *Development* **129**(21), 4999–5008.

Stevens, S.W., Ryan, D.E., Ge, H.Y., Moore, R.E., Young, M.K., Lee, T.D., Abelson, J. (2002) Composition and functional characterization of the yeast spliceosomal penta-snRNP, *Mol. Cell.* **9**(1), 31–44.

Sun, H., Chasin, L.A. (2000) Multiple splicing defects in an intronic false exon, *Mol. Cell. Biol.* **20**, 6414–6425.

Staley, J.P., Guthrie, C. (1998) Mechanical devices of the spliceosome: motors, clocks, springs, and things, *Cell* **92**(3), 315–326.

Staley, J.P., Guthrie, C. (1999) An RNA switch at the 5′ splice site requires ATP and the DEAD box protein Prp28p, *Mol. Cell.* **3**(1), 55–64.

Stamm, S., Zhu, J., Nakai, K., Stoilov, P., Stoss, O., Zhang, M.Q. (2000) An alternative-exon database and its statistical analysis, *DNA Cell. Biol.* **19**(12), 739–756.

Stark, J.M., Cooper, T.A., Roth, M.B. (1999) The relative strengths of SR protein-mediated associations of alternative and constitutive exons can influence alternative splicing, *J. Biol. Chem.* **274**(42), 29838–29842.

Stevens, S.W., Ryan, D.E., Ge, H.Y., Moore, R.E., Young, M.K., Lee, T.D., Abelson, J. (2002) Composition and functional characterization of the yeast spliceosomal penta-snRNP, *Mol. Cell.* **9**(1), 31–44.

Stickeler, E., Fraser, S.D., Honig, A., Chen, A.L., Berget, S.M., Cooper, T.A. (2001) The RNA binding protein YB-1 binds A/C-rich exon enhancers and stimulates splicing of the CD44 alternative exon v4, *EMBO J.* **20**(14), 3821–3830.

Stickeler, E., Kittrell, F., Medina, D., Berget, S.M. (1999) Stage-specific changes in SR splicing factors and alternative splicing in mammary tumorigenesis, *Oncogene* **18**(24), 3574–3582.

Stoss, O., Olbrich, M., Hartmann, A.M., Konig, H., Memmott, J., Andreadis, A., Stamm, S. (2001) The STAR/GSG family protein rSLM-2 regulates the selection of alternative splice sites, *J. Biol. Chem.* **276**(12), 8665–8673.

Sun, H., Chasin, L.A. (2000) Multiple splicing defects in an intronic false exon, *Mol. Cell. Biol.* **20**(17), 6414–6425.

Tange, T.O., Damgaard, C.K., Guth, S., Valcarcel, J., Kjems, J. (2001) The hnRNP A1 protein regulates HIV-1 tat splicing via a novel intron silencer element, *EMBO J.* **20**(20), 5748–5758.

Tarn, W.Y., Steitz, J.A. (1996a) A novel spliceosome containing U11, U12, and U5 snRNPs excises a minor class (AT-AC) intron in vitro, *Cell* **84**(5), 801–811.

Tarn, W.Y., Steitz, J.A. (1996b) Highly diverged U4 and U6 small nuclear RNAs required for splicing rare AT-AC introns, *Science* **273**(5283), 1824–1832.

Teigelkamp, S., Mundt, C., Achsel, T., Will, C.L., Luhrmann, R. (1997) The human U5 snRNP-specific 100-kD protein is an RS domain-containing, putative RNA helicase with significant homology to the yeast splicing factor Prp28p, *RNA* **3**(11), 1313–1326.

Todd, R.D., Lobos, E.A., Sun, L.W., Neuman, R.J. (2003) Mutational analysis of the nicotinic acetylcholine receptor alpha 4 subunit gene in attention deficit/hyperactivity disorder: evidence for association of an intronic polymorphism with attention problems, *Mol. Psychiatry* **8**(1), 103–108.

Tysoe, C., Whittaker, J., Xuereb, J., Cairns, N.J., Cruts, M., Vanbroeckhoven, C., Wilcock, G., Rubinsztein, D.C. (1998) A presenilin-1 truncating mutation is present in two cases with autopsy-confirmed early-onset Alzheimer disease, *Am. J. Hum. Genet.* **62**, 70–76.

Vacek, M.M., Ma, H., Gemignani, F., Lacerra, G., Kafri, T., Kole, R. (2003) High-level expression of hemoglobin A in human thalassemic erythroid progenitor cells following lentiviral vector delivery of an antisense snRNA, *Blood* **101**(1), 104–111.

Valadkhan, S., Manley, J.L. (2001) Splicing-related catalysis by protein-free snRNAs, *Nature* **413**(6857), 701–707.

Valcarcel, J., Singh, R., Zamore, P.D., Green, M.R. (1993) The protein Sex-lethal antagonizes the splicing factor U2AF to regulate alternative splicing of transformer pre-mRNA, *Nature* **362**(6416), 171–175.

Van Buskirk, C., Schupbach, T. (2002) Half pint regulates alternative splice site selection in Drosophila, *Dev. Cell.* **2**(3), 343–353.

Van Lith-Verhoeven, J.J., Van Der Velde-Visser, S.D., Sohocki, M.M., Deutman, A.F., Brink, H.M., Cremers, F.P., Hoyng, C.B. (2002) Clinical characterization, linkage analysis, and PRPC8 mutation analysis of a family with autosomal dominant retinitis pigmentosa type 13 (RP13), *Ophthalmic Genet.* **23**, 1–12.

Vithana, E., Al-Maghtheh, M., Bhattacharya, S.S., Inglehearn, C.F. (1998) RP11 is the second most common locus for dominant retinitis pigmentosa, *J. Med. Genet.* **35**, 174–175.

Vithana, E.N., Abu-Safieh, L., Allen, M.J., Carey, A., Papaioannou, M., Chakarova, C., Al-Maghtheh, M., Ebenezer, N.D., Willis, C., Moore, A.T., Bird, A.C., Hunt, D.M., Bhattacharya, S.S. (2001) A human homolog of yeast pre-mRNA splicing gene, PRP31, underlies autosomal dominant retinitis pigmentosa on chromosome 19q13.4 (RP11), *Mol. Cell.* **8**, 375–381.

Vorgerd, M., Burwinkel, B., Reichmann, H., Malin, J.P., Kilimann, M.W. (1998) Adult-onset glycogen disease type II: phenotypic and allelic heterogeneity in German patients, *Neurogenetics* **1**, 205–211.

Wagner, E.J., Garcia-Blanco, M.A. (2002) RNAi-mediated PTB depletion leads to enhanced exon definition, *Mol. Cells* **10**(4), 943–949.

Watanabe, T., Sullenger, B.A. (2000) Induction of wild-type p53 activity in human cancer cells by ribozymes that repair mutant p53 transcripts, *Proc. Natl. Acad. Sci. U.S.A.* **97**(15), 8490–8494.

Wang, H.Y., Lin, W., Dyck, J.A., Yeakley, J.M., Songyang, Z., Cantley, L.C., Fu, X.D. (1998) SRPK2: a differentially expressed SR protein-specific kinase involved in mediating the interaction and localization of pre-mRNA splicing factors in mammalian cells, *J. Cell. Biol.* **140**(4), 737–750.

Wang, J., Takagaki, Y., Manley, J.L. (1996) Targeted disruption of an essential vertebrate gene: ASF/SF2 is required for cell viability, *Genes Dev.* **10**(20), 2588–2599.

Watanabe, T., Sullenger, B.A. (2000) Induction of wild-type p53 activity in human cancer cells by ribozymes that repair mutant p53 transcripts, *Proc. Natl. Acad. Sci. U.S.A.* **97**(15), 8490–8494.

Weidenhammer, E.M., Ruiz-Noriega, M., Woolford, J.L. Jr. (1997) Prp31p promotes the association of the U4/U6 × U5 tri-snRNP with prespliceosomes to form spliceosomes in Saccharomyces cerevisiae, *Mol. Cell. Biol.* **17**, 3580–3588.

Weidenhammer, E.M., Singh, M., Ruiz-Noriega, M., Woolford, J.L. Jr. (1996) The PRP31 gene encodes a novel protein required for pre-mRNA splicing in Saccharomyces cerevisiae, *Nucleic Acids Res.* **24**, 164–170.

Will, C.L., Schneider, C., MacMillan, A.M., Katopodis, N.F., Neubauer, G., Wilm, M., Luhrmann, R., Query, C.C. (2001) A novel U2 and U11/U12 snRNP protein that associates with the pre-mRNA branch site, *EMBO J.* **20**(16), 4536–4546.

Will, C.L., Urlaub, H., Achsel, T., Gentzel, M., Wilm, M., Luhrmann, R. (2002) Characterization of novel SF3b and 17S U2 snRNP proteins, including a human Prp5p homologue and an SF3b DEAD-box protein, *EMBO J.* **21**(18), 4978–4988.

Wimmel, A., Schilli, M., Kaiser, U., Havemann, K., Ramaswamy, A., Branscheid, D., Kogan, E., Schuermann, M. (1997) Preferential histiotypic expression of CD44-isoforms in human lung cancer, *Lung Cancer* **16**(2–3), 151–172.

Wu, J.Y., Maniatis, T. (1993) Specific interactions between proteins implicated in splice site selection and regulated alternative splicing, *Cell* **75**, 1061–1070.

Wu, J.A., Manley, J.L. (1991) Base pairing between U2 and U6 snRNAs is necessary for splicing of a mammalian pre-mRNA, *Nature* **352**(6338), 818–821.

Wu, J.I., Reed, R.B., Grabowski, P.J., Artzt, K. (2002) Function of quaking in myelination: regulation of alternative splicing, *Proc. Natl. Acad. Sci. U.S.A.* **99**(7), 4233–4238.

Wu, S., Romfo, C.M., Nilsen, T.W., Green, M.R. (1999) Functional recognition of the 3' splice site AG by the splicing factor U2AF35, *Nature* **402**(6763), 832–835.

Xie, J., Black, D.L. (2001) A CaMK IV responsive RNA element mediates depolarization-induced alternative splicing of ion channels, *Nature* **410**(6831), 936–939.

Xie, J., McCobb, D.P. (1998) Control of alternative splicing of potassium channels by stress hormones, *Science* **280**(5362), 443–446.

Yang, L., Embree, L.J., Hickstein, D.D. (2000) TLS-ERG leukemia fusion protein inhibits RNA splicing mediated by serine-arginine proteins, *Mol. Cell. Biol.* **20**(10), 3345–3354.

Yang, L., Embree, L.J., Tsai, S., Hickstein, D.D. (1998) Oncoprotein TLS interacts with serine-arginine proteins involved in RNA splicing, *J. Biol. Chem.* **273**(43), 27761–27764.

Yang, F., Hanson, N.Q., Schwichtenberg, K., Tsai, M.Y. (2000) Variable number tandem repeat in exon/intron border of the cystathionine beta-synthase gene: a single nucleotide substitution in the second repeat prevents multiple alternate splicing, *Am. J. Med. Genet.* **95**, 385–390.

Yang, X.L., Miura, N., Kawarada, Y., Terada, K., Petrukhin, K., Gilliam, C., Sugiyama, T. (1997) Two forma of Wilson disease protein produced by alternative splicing are localised in distinct cellular compartments, *Biochem. J.* **326**, 897–902.

Yazaki, M., Yoshida, K., Nakamura, A., Furihata, K., Yonekawa, M., Okabe, T., Yamashita, N., Ohta, M., Ikeda, S. (1998) A novel splicing mutation in the ceruloplasmin gene rsponsible for hereditary ceruloplasmin deficiency with hemosiderosis, *J. Neurol. Sci.* **156**, 30–34.

Yeakley, J.M., Fan, J.B., Doucet, D., Luo, L., Wickham, E., Ye, Z., Chee, M.S., Fu, X.D.

(2002) Profiling alternative splicing on fiber-optic arrays, *Nat. Biotechnol.* **20**(4), 353–358.

Yeakley, J.M., Tronchere, H., Olesen, J., Dyck, J.A., Wang, H.Y., Fu, X.D. (1999) Phosphorylation regulates in vivo interaction and molecular targeting of serine/arginine-rich pre-mRNA splicing factors, *J. Cell. Biol.* **145**(3), 447–455.

Yean, S.L., Wuenschell, G., Termini, J., Lin, R.J. (2000) Metal-ion coordination by U6 small nuclear RNA contributes to catalysis in the spliceosome, *Nature* **408**(6814), 881–884.

Yu, L., Heere-Ress, E., Boucher, B., Defesche, J.C., Kastelein, J., Lavoie, M.A., Genest, J. (1999) Familial hypercholesterolemia. Acceptor splice site (G→C) mutation in intron 7 of the LDL-R gene: alternate RNA editing causes exon 8 skipping or a premature stop codon in exon 8, *Atherosclerosis* **146**, 125–131.

Zachar, Z., Chou, T.B., Kramer, J., Mims, I.P., Bingham, P.M. (1994) Analysis of autoregulation at the level of pre-mRNA splicing of the suppressor-of-white-apricot gene in Drosophila, *Genetics* **137**(1), 139–135.

Zahler, A.M., Lane, W.S., Stolk, J.A., Roth, M.B. (1992) SR proteins: a conserved family of pre-mRNA splicing factors, *Genes Dev.* **6**(5), 837–847.

Zahler, A.M., Neugebauer, K.M., Lane, W.S., Roth, M.B. (1993) Distinct functions of SR proteins in alternative pre-mRNA splicing, *Science* **260**(5105), 219–222.

Zamore, P.D., Patton, J.G., Green, M.R. (1992) Cloning and domain structure of the mammalian splicing factor U2AF, *Nature* **355**(6361), 609–614.

Zapp, M.L., Stern, S., Green, M.R. (1993) Small molecules that selectively block RNA binding of HIV-1 Rev protein inhibit Rev function and viral production, *Cell* **74**(6), 969–978.

Zhang, L., Liu, W., Grabowski, P.J. (1999) Coordinate repression of a trio of neuron-specific splicing events by the splicing regulator PTB, *RNA* **5**(1), 117–130.

Zhang, W., Liu, H., Han, K., Grabowski, P.J. (2002) Region-specific alternative splicing in the nervous system: implications for regulation by the RNA-binding protein NAPOR, *RNA* **8**(5), 671–685.

Zhang, W.J., Wu, J.Y. (1996) Functional properties of p54, a novel SR protein active in constitutive and alternative splicing, *Mol. Cell. Biol.* **16**(10), 5400–5408.

Zheng, Z.M., Huynen, M., Baker, C.C. (1998) A pyrimidine-rich exonic splicing suppressor binds multiple RNA splicing factors and inhibits spliceosome assembly, *Proc. Natl. Acad. Sci. U.S.A.* **95**, 14088–14093.

Zhou, Z., Licklider, L.J., Gygi, S.P., Reed, R. (2002) Comprehensive proteomic analysis of the human spliceosome, *Nature* **419**(6903), 182–185.

Zhu, J., Mayeda, A., Krainer, A.R. (2001) Exon identity established through differential antagonism between exonic splicing silencer-bound hnRNP A1 and enhancer-bound SR proteins, *Mol. Cells* **8**(6), 1351–1361.

5
Repair and Mutagenesis of DNA

Raymond Devoret
Institute Curie, Orsay, France

1	**The Structure of DNA Determines its Functions in Repair and Mutagenesis** 173	
1.1	When Donated by Both Parents, the CFTR Mutations are Deleterious 174	
2	**DNA-damaging Agents are Mutagens** 176	
2.1	A DNA Lesion is Different from a Mutation 176	
3	**Characteristic Mutation Types** 176	
4	**Silent Base Substitutions** 179	
4.1	Synonymous Mutations are Neutral 180	
5	**Four Main Mechanisms that Generate Spontaneous Mutations** 181	
6	**Induced Mutations** 182	
7	**Mutagenesis in Bacteria and Carcinogenesis in Mammals** 183	
8	**Carcinogens Leave No Definite Mutational Signatures on DNA** 184	
9	**Humans Use Mutagenesis to Produce their Immunity Repertoire** 184	
	Acknowledgments 185	
	Bibliography 185	
	Books and Reviews 185	
	Primary Literature 185	

Genomics and Genetics. Edited by Robert A. Meyers.
Copyright © 2007 Wiley-VCH Verlag GmbH & Co. KGaA, Weinheim
ISBN: 978-3-527-31609-0

Keywords

DNA-damaging Agents
Physical or chemical agents, for example, such as Xrays or aflatoxin B_1, which affect DNA functions by distorting its structure. The newly formed structure is often termed *DNA lesion*, which often precedes the appearance of a mutation. Note that the transmogrification of a DNA lesion may give rise to a mutation. Yet, note that a lesion is not a mutation.

DNA Repair
A collective term that encompasses at least three dozens of biochemical processes that are involved in restoring the structure of DNA after it has sustained damage.

Gene
A segment of DNA, the expression of which results in the production of a messenger RNA (mRNA) that translates into a protein.

Genetic
Qualifies what pertains to DNA, whereas, *hereditary* specifies a trait passed from a parent to its progeny. The Down syndrome is a genetic disease, resulting from a triplication of human chromosome 21. In contrast, mucoviscidosis is a hereditary disease caused by a mutation located in human chromosome 7.

Genome
Designates the whole set of chromosomes present in a cell.

Mutagens
Physical or chemical agents that alter DNA, increasing the occurrence of mutations.

Mutagenesis
Designates the overall process that results in the induction of a mutation.

Mutation
A DNA change that is recorded durably and passed on to the offspring. *Induced mutations* result from the error-prone repair of DNA lesions or from DNA replication errors. *Spontaneous mutations* correspond to naturally occurring DNA changes.

SOS Mutagenic Repair
In prokaryotes, it is an induced error-prone DNA replication, involving a specialized polymerase that causes the synthesis of a base across a lesion (termed TLS, standing for trans-lesion DNA synthesis). TLS is a major SOS function regulated by the activation of RecA protein.

■ During their lifetime, all the cells of a living organism are exposed constantly to harmful agents such as radiations and chemicals. DNA (deoxyribonucleic acid), a crucial cell constituent, is highly sensitive to deleterious agents. A large majority of cells survive when exposed chronically at low doses since DNA damage induces cell enzymes that repair DNA lesions. Repair acts very efficiently when DNA stops replicating after the occurrence of lesions. After repair has taken place, DNA replication resumes, followed by cell division. While various repair processes concur to restore cell survival, each of them is never completely efficient and lesions remain in the DNA, acting as stepping-stones. Resumption of DNA synthesis on the strand opposite the remaining lesions permits the bypassing of these obstacles. A process, called *trans-lesion DNA synthesis* is one of the most efficient processes generating mutations. Mutations come into view in cells that survive the DNA-damaging treatment.

In short, a mutation is a DNA change that is recorded durably and passed on to the offspring after replication. A number of mutations in *germ* cells have *harmful* effects for they often cause hereditary diseases. Similarly, mutations that accumulate in dividing body cells (*somatic cells*) can generate cancers. Yet, most of the occurring mutations behave as *neutral* in their usual environment. Selection forces the genome to retain the *beneficial* mutations.

1
The Structure of DNA Determines its Functions in Repair and Mutagenesis

DNA is the essential component of chromosomes in all cells. Each chromosome is made of an extremely long DNA molecule with two intertwined *complementary strands* (a double helix). Each strand consists of a poly(phosphate sugar) backbone from which project Adenine (A), Cytosine (C), Guanine (G), and Thymine (T) bases at each sugar. The two strands form a *duplex* maintained by hydrogen bonds between complementary bases: A binds to T and G to C. The four bases A, C, G, or T may combine into a linear chain that constitutes a *genetic sequence* (see Fig. 1).

DNA of the whole set of chromosomes in a cell is called the *genome*. A bacterium like *Escherichia coli*, a natural resident of the colon of mammals, is made of one chromosome, whereas human cells carry 23 chromosome pairs. In general, the cells of complex organisms like mammals have more chromosomes per cell than lower organisms like microbes.

The double-stranded structure of DNA permits its duplication, one strand providing the template for the making of its copy. Not only has such a structure the possibility of replicating but also of repairing itself. If, for example, ultraviolet light hits a strand of the DNA double helix, the damaged piece undergoes: (1) excision of the damaged DNA piece that leaves a gap; and (2) DNA resynthesis to close the gap, using the undamaged strand as a template to make a correct new strand.

Repair processes are many, involving more than fifty identified proteins. The three major processes that repair DNA are (1) *excision*; (2) *recombination*; and (3) *mismatch repair*.

(a) AAGGCAAACCTACTGGTCTTATGT   original sequence
(b) AAGGCAAA**T**CTACTGGTCTTATGT   transition
(c) AAGGCAAACCTACTG**C**TCTTATGT   transversion
     Del(ACCTA)
(d) AAGGCAA^V^CTGGTCTTATGT   deletion of ACCTA
     Ins(AAAGC)
(e) AAGGCAAACCTACT^V^AAAGC^V^GGTCTTATGT   insertion of AAAGC

Fig. 1 Types of mutations: (a) wild type original sequence; (b) transition from C to **T** (underlined bold T); (c) transversion from G to **C** (underlined bold C); (d) deletion (Del) of the sequence **ACCTA**, the ^{V} sign indicates from where it has been taken out; and (e) insertion (Ins) of the sequence **AAAGC**, the two ^{V} signs indicate where the sequence has been put in.

There are two types of excision repair depending on the length of excised DNA: nucleotide excision repair (NER) or base excision repair (BER). Various different types of recombination repair exist depending on whether the recombining DNAs are *homologous, nonhomologous,* or *nonhomologous end joining.* More complicated repair processes use a combination of all of these. Repair, by definition, should lead to the full restoration of the original DNA structure. In fact, almost all repair processes leave a low percentage of errors, some of which give rise to mutations.

The main function of DNA is to code for the hereditary traits that are passed from parents to progeny. Recently, geneticists have identified the long string of bases present in many genomes, from bacteria to man. The mapping of the human genome has resulted in building a compendium of the number and positioning of the 7000 or so genes, whose mutations are responsible for human hereditary diseases.

When a mutation occurs within a gene, the protein encoded by the gene is often altered (see Figs. 1b, c and 2c); this alteration may produce a visible change (*phenotype*) of the organism under study. The actual mutation itself (*genotype*) is invisible to the naked eye, whereas the technique of DNA sequencing will locate it precisely.

Since a mutation is a DNA change, one should define it in relation to a preceding, parental DNA sequence (Fig. 1a). Consequently, geneticists have agreed to define a *standard genetic sequence* for each organism studied. The adopted standard is usually the most widespread type that fits in its environment.

Mutations fall into three classes termed *deleterious, neutral,* or *beneficial* according to their effects on the organism in which they arose. These qualifications are only valid in relation to a given environment.

1.1
When Donated by Both Parents, the CFTR Mutations are Deleterious

Human cells carry 22 pairs of chromosomes called *autosomes* plus a pair of sex chromosomes. The probability for a child to suffer from *cystic fibrosis* is one out of 1600 children. This disease is dramatic; the fluidity of lung secretions disappear, stifling normal pulmonary aeration. The cause of the disease is genetic: the affected child has received from the two

(a) Ile Cys Ile Lys Ala Leu Val Leu Leu Thr amino acid sequence

ATA TGT ATA AAG GCA CTG GTC CTG TTA ACA DNA sequence

ATA TGT ATA AAG GCA CTG GT**A** CTG TTA ACA base substitution

Ile Cys Ile Lys Ala Leu **Val** Leu Leu Thr synonymous amino acid

(b) Ile Cys Ile Lys Ala Asn Val Leu Leu Thr amino acid sequence

ATA TGT ATA AAG GCA AAC GTC CTG TTA ACA DNA sequence

ATA TGT ATA AAG GCA AAC **T**TC CTG TTA ACA base substitution

Ile Cys Ile Lys Ala Asn **Phe** Leu Leu Thr missense amino acid

(c) Ile Cys Ile Lys Ala Asn Val Leu Leu Thr amino acid sequence

ATA TGT ATA AAG GCA AAC GTC CTG TTA ACA DNA sequence

ATA TGT ATA **T**AG GCAAACGTCCTGTTAACA stop codon

Ile Cys Ile **Ter** polypeptide termination

Fig. 2 Types of substitutions in a protein-coding region: (a) synonymous, (b) missense, and (c) nonsense. In each case, the top sequence is wild type while the bottom sequence is the mutated one.

parental chromosomes 7, a mutated cystic fibrosis CFTR gene. In Europe, the frequency of people carrying only one CFTR mutation is 1/40. Such heterozygotes show a decrease in the fluidity of mucus excretion that may pass unnoticed. Geneticists have postulated that the selection of one CFTR mutation results from past cholera epidemics. People, ill with cholera, suffer from an almost 15 L of diarrhea per day; they mostly die of dehydration more than from infection sickness. Individuals with only one mutated cystic fibrosis CFTR mutation leak much less water and can overcome the infection with cholera germs. In countries where cholera is endemic, individuals with only one CFTR mutation seem to be resistant to cholera.

The probability for a child to receive two CFTR mutations is 1/1600. This fraction shows the high frequency of cystic fibrosis in the human population. Resistance to cholera in a human population is proportional to the frequency of occurrence of spontaneous CFTR mutations.

2
DNA-damaging Agents are Mutagens

Naturally, mutations arise at low frequency. To increase the frequency of appearance of mutations, Herman Muller thought of irradiating Drosophila eggs with X rays. He discovered that X rays increased the mutation frequency as a function of the absorbed dose. Moreover, chemicals such as mustard gas (yperite) showed a similar dose response. Agents that produce mutations are denoted *mutagens*.

2.1
A DNA Lesion is Different from a Mutation

The practical knowledge of DNA lesions that we have is that after too much bathing in the sun, our skin becomes red and itching. Solar radiations, named UVA (320–400 nm), produce various *DNA lesions* within our skin cells. Photobiologists have identified more than 20 lesions, depending on their 3D-structure and shapes.

Why is it that with so many diverse lesions produced, we observe that 85% of them are single base pair changes, in just a DNA base? For example, Adenine (A) may change into Cytosine (C), Guanine (G), or Thymine (T). Benzopyrene, a chemical inhaled with tobacco smoke, when metabolized, gives rise to a chemical derivative, benzopyrene dihydrodiol that sticks to DNA. In our body, this compound produces a single base mutation: an A or a T or a C or a G.

What is the relationship between a DNA lesion and a mutation? In the preceding example, a benzopyrene-dihydrodiol-base complex bound to one DNA strand is taken out by excision, which leaves a gap of around 10 bases (in bacteria) or 30 bases (in human cells). DNA resynthesis will fill the gap, sometimes in an error-prone way, giving rise to a mutation. Recall that a DNA lesion is not a mutation even if nowadays one can synthesize a piece of DNA with an altered base *in vitro*.

Biochemical reactions, namely, *translesion DNA-synthesis*, proceed on the DNA to restore a functional sequence at the site where there was a lesion. Most unexcised DNA lesions block DNA replication. SOS-induced-repair polymerases bypass the DNA lesions, allowing the cells to survive. At least 50 repair proteins concur to cell survival. Repair enzymes act all the more efficiently as DNA lesions are not too many.

Interestingly, the lesions that trigger an SOS response in bacteria induce a cascade of repair processes. Structural restoration of a DNA double helix is a primordial step permitting replication, cell division, and cell survival to take place.

3
Characteristic Mutation Types

By the use of various techniques, biologists proved that all living species possess a similar DNA structure. In all genomes, there are four nucleic bases, A, C, G, T, whether in microorganisms or mammals.

1. *Point mutations*: The most frequent are base substitutions. Substitutions

happen between purines, Adenine (A) and Guanine (G), or between pyrimidines, Cytosine (C) and Thymine (T) (Fig. 1b); they are called transitions (A capital letter for each base name is used to stress what base the symbol stands for). An AT base pair may give rise to a GC (AT → GC) and, conversely, a GC base pair to an AT (GC → AT); the two-way reactions are illustrated as AT ↔ GC.

Transversions are substitutions of a purine for a pyrimidine and *vice versa* (Fig. 1c). There are four reciprocal base pair changes: AT ↔ CG; AT ↔ TA; GC ↔ TA; GC ↔ CG. Base substitutions that occur in protein-coding regions affect the structure of the expressed protein with the exception of the third base of a codon when the change does not modify the amino acid (see neutral or synonymous mutation, Fig. 2a).

2. *Deletions and insertions*: Examples of deletions and insertions are seen in Fig. 1(d) and (e). Note that an unequal crossing-over between two chromosomes may result in a deletion of a DNA segment in one chromosome and in an addition in the other.

3. *Frameshifts*: A small deletion or insertion within a gene, whose number of bases is not a multiple of three, causes a shift in the reading frame, denoted frameshift (Fig. 3a). Then, the coding sequence downstream displays a wrong phase. The new phasing changes the nature of the encoded amino acids; or else a new stop codon may be brought upstream into phase, producing a shortened protein. Figure 3(a) shows a protein sequence that may not have normal activity. Figure 3(b) also illustrates the fact that an addition of a base results into a +1 frameshift that may remove a preexisting stop signal giving rise to an elongated protein.

4. *DNA rearrangements*: Deletions of a segment of a gene or of sets of genes may reduce or eliminate protein functions

(a) Lys Ala Leu Val Leu Leu Thr Ile Cys Ile Ter

   AAG GCA CTG GTC CTG TTA ACA ATA TGT ATA TAA TACCATCGCAATAGGG

       Del(G)

   AAG GCA CTG ᵛTCC TGT TAA CAATATGTATATAATACCATCGCAATAGGG

   Lys Ala Leu Phe Cys Ter

(b) Lys Ala Asn Val Leu Leu Thr Ile Cys Ile Ter

   AAG GCA AAC GTC CTG TTA ACA ATA TGT ATA TAA TACCATCGCAATAGGG

       Ins(G)

   AAG GCA AAC ᵛGᵛGT CCT GTT AAC AAT ATG TAT ATA ATA CCA TCG CAA AGT GG

   Lys Ala Asn Gly Pro Val Asn Asn Met Tyr Ile Ile Pro Ser Gln Ter

Fig. 3 Examples of frameshifts caused by deletion or insertion. (a) A deletion of a G causing premature termination. (b) An insertion of a G obliterates a stop codon. Termination codons are underlined. In each case, the top sequence is wild type while the bottom sequence is the mutated one.

(Fig. 3). In addition, insertions of large DNA segments can occur, as can duplications, inversions, and other more complicated DNA rearrangements. All these processes are the likely results of recombination between DNA sequences sharing homology. Agents that break chromosomes like X rays or yperite, facilitate DNA rearrangements.

Effect of a Bcl2 rearrangement: When a piece of human chromosome 14, carrying the *Bcl2* gene, and thus expressing a function preventing cell death, is translocated to chromosome 18, then the *Bcl2* gene gets expressed constantly. This leads to the production of a modified immunoglobulin, normally eliminated from the body by cell death. The modified immunoglobulin is kept alive here because of the *Bcl2* translocation. Thus, a pool of unhealthy cells builds gradually; point mutations accumulate in those unhealthy cells and will eventually give rise to cancer cells.

5. *Insertion of transposable genetic elements*: Transposable genetic elements, also called transposons or jumping genes, are present in most organisms. They are highly mobile around chromosomes. If a transposon inserts into a gene encoding a protein, it may cut the gene into two pieces and the protein may not be expressed as a whole. This is the negative property of transposon insertion. On the contrary, if a transposon with active promoters inserts itself into the regulatory region of a gene, it may strongly increase the expression of downstream genes.

6. *Mutation reversions: true- or pseudoreversion of a mutation*: Sometimes after a few generations, the phenotype of a mutant organism seems to have reversed to the wild type original and a question arises whether the mutation reverted to the original parental type? Or else, has the mutation remained in the DNA while another mutation had occurred elsewhere, counterbalancing the first one?

7. *Mutation hotspots*: Mutations may occur at random along the genome, but more often than not, they crop up at some sites designated hotspots. There is such a hotspot in the *E. coli* chromosome: the dinucleotide CpG, in which one finds the cytosine frequently methylated, changes to TpG, introducing an error.

Another specific mechanism can take place, designated *induced activation of cytosine deaminase* (AID). The deamination of methyl cytosine gives rise to thymine in the absence of replication. This process converts cytosine to uracil and adenine to hypoxanthine. It may be the main cause of immunoglobin hypermutation, which may result from the conversion of cytosine to uracil with the AID enzyme.

The dinucleotide TpT is a mutation hotspot in prokaryotes (but not in eukaryotes). The particular 3-D shape of a DNA region generates some targets for specific enzymes like methylases. In bacteria, DNA regions containing short *palindromes* like 5'GCCGGC3', 5'GGCGCC3', and 5'GGGCCC3' are prone to mutate more frequently than other regions. In eukaryotic genomes, short tandem repeats, resulting from *slipped-strand mispairing* are often hotspots (Fig. 4a,b).

8. *Mutations at regulatory sequences*: DNA encodes sequences that control the expression of proteins, such as promoters

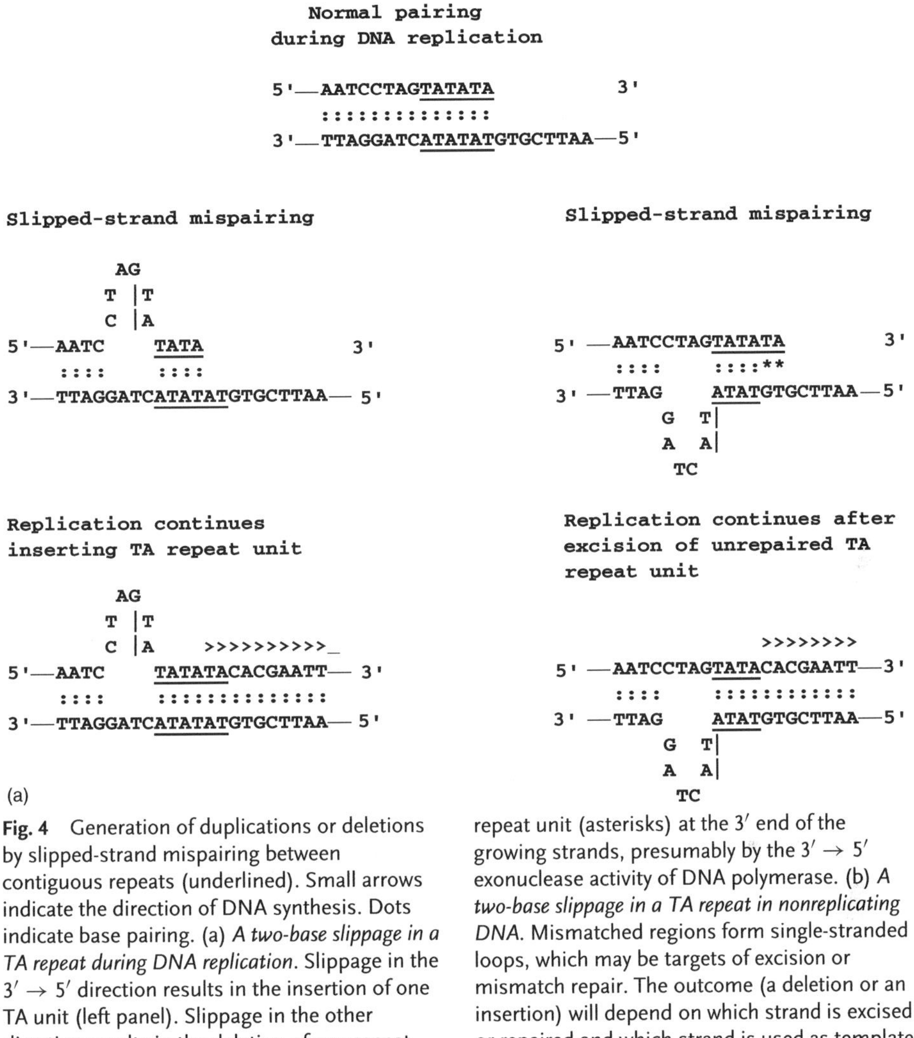

```
              Normal pairing
           during DNA replication

           5'—AATCCTAGTATATA                3'
              ::::::::::::::
           3'—TTAGGATCATATATGTGCTTAA—5'
```

Slipped-strand mispairing   Slipped-strand mispairing

```
          AG
        T |T
        C |A
5'—AATC     TATA              3'         5' —AATCCTAGTATATA               3'
   ::::     ::::                             ::::        ::::**
3'—TTAGGATCATATATGTGCTTAA— 5'            3' —TTAG        ATATGTGCTTAA—5'
                                                      G  T|
                                                      A  A|
                                                      TC
```

Replication continues   Replication continues after
inserting TA repeat unit   excision of unrepaired TA
                                         repeat unit

```
          AG
        T |T
        C |A      >>>>>>>>>>_                         >>>>>>>>
5'—AATC     TATATACACGAATT— 3'           5' —AATCCTAGTATACACGAATT—3'
   ::::     :::::::::::::::                  ::::        :::::::::::::
3'—TTAGGATCATATATGTGCTTAA— 5'            3' —TTAG        ATATGTGCTTAA—5'
                                                      G  T|
                                                      A  A|
                                                      TC
```

(a)

Fig. 4 Generation of duplications or deletions by slipped-strand mispairing between contiguous repeats (underlined). Small arrows indicate the direction of DNA synthesis. Dots indicate base pairing. (a) *A two-base slippage in a TA repeat during DNA replication.* Slippage in the 3' → 5' direction results in the insertion of one TA unit (left panel). Slippage in the other direction results in the deletion of one repeat unit (right panel). The deletion shown in the right panel results from excision of the unpaired repeat unit (asterisks) at the 3' end of the growing strands, presumably by the 3' → 5' exonuclease activity of DNA polymerase. (b) *A two-base slippage in a TA repeat in nonreplicating DNA.* Mismatched regions form single-stranded loops, which may be targets of excision or mismatch repair. The outcome (a deletion or an insertion) will depend on which strand is excised or repaired and which strand is used as template in the DNA repair process. (Modified scheme from Levinson and Gutman).

and operators. Mutations in these sequences may alter the cell concentration of proteins encoded downstream of the regulatory sequence. Overexpression or underexpression of proteins may sharply modify the function of a gene in an organism. Since many proteins work in a concerted action, a sharp increase or decrease in the cell concentration of a protein may disturb the very existence of an organism.

4
Silent Base Substitutions

A mutation that does not cause any amino acid change in the expressed protein is

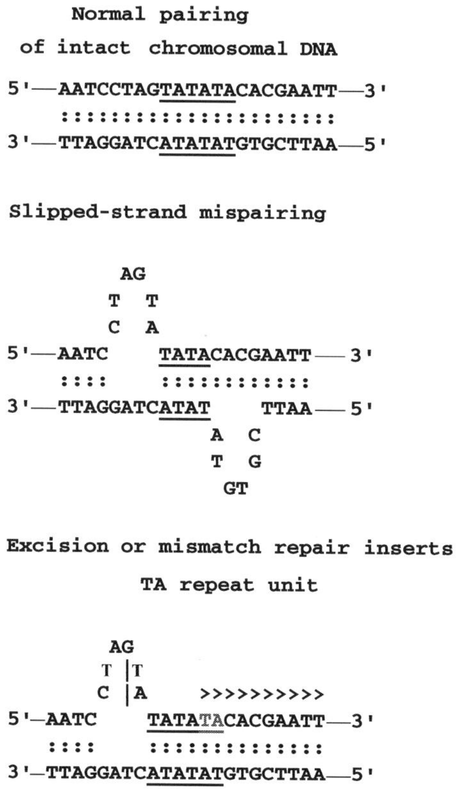

Normal pairing
of intact chromosomal DNA

```
5'—AATCCTAGTATATACACGAATT—3'
    ::::::::::::::::::::::
3'—TTAGGATCATATATGTGCTTAA—5'
```

Slipped-strand mispairing

```
            AG
           T  T
           C  A
5'—AATC       TATACACGAATT —3'
   ::::       ::::::::::::
3'—TTAGGATCATAT       TTAA—5'
               A  C
               T  G
                GT
```

Excision or mismatch repair inserts
TA repeat unit

```
            AG
           T |T
           C |A    >>>>>>>>>
5'—AATC       TATATACACGAATT—3'
   ::::       ::::::::::::::
3'—TTAGGATCATATATGTGCTTAA—5'
```

(b)

Fig. 4 (Continued)

dubbed *synonymous* and may become *silent*. In Fig. 2(a), the protein codes for the valine amino acid but the DNA sequence reveals a base change.

A *nonsynonymous* mutation alters an amino acid, resulting in a *missense* or a *nonsense* mutation. A missense mutation specifies an amino acid different from the one originally encoded, for instance, valine (Val) becomes phenylalanine (Phe) (Fig. 2b). A nonsense mutation changes a codon into one of the three termination codons TAG, TAA, or TGA. Then, a truncated protein is produced since lysine (Lys) becomes Stop (Fig. 2c).

4.1
Synonymous Mutations are Neutral

In a DNA sequence that codes for a protein, each of the sense codons can mutate to nine other codons by means of a single base substitution. For example, the triplet CCU that codes for proline (Pro) can give rise to three *synonymous* substitutions CCC, CCA, or CCG, which also code for proline (Pro). In a different order, the triplet can give rise to six *nonsynonymous* substitutions, which code for different amino acids, UCU for serine (Ser), ACU for threonine (Thr), GCU for alanine (Ala), CUU for leucine (Leu), CAU for histidine (His), or CGU for arginine (Arg).

Since the genetic code defines 61 sense codons, it is reckoned that the possible codon substitutions are $61 \times 9 = 549$. If we assume that base substitutions may occur at random and that all codons are equally frequent in regions coding for proteins, we can compute the expected proportion of the different types of base substitutions. Because of the structure of the genetic code, *synonymous* substitutions occur mainly at the *third* position of codons. Indeed, 69% of all the possible *synonymous* changes in a codon are at the third-base position.

In contrast, substitutions at the *second* position of codons are *nonsynonymous*, and so are the vast majority of base changes at the *first* position (96%).

Synonymous amino acid replacements do not alter either protein structure or function. From an evolutionary viewpoint, these mutations are *neutral* since they do not seem to affect the physiology of the mutated organisms. Such synonymous mutations accumulate with time since there is no counterselection for their emergence. Synonymous mutations indeed generate *change at the DNA level*, as indicated by the use of restriction enzymes and sequencing techniques. Over time, accumulation of synonymous mutations increases DNA divergence between individuals within a species. Then, a new species can emerge by segregation. Chimpanzee and man share 98.5% of the structure of their proteins. DNA sequences in the two species have diverged upon evolution from a common ancestor. The consequence of the divergence is a gradual loss of DNA homology that hinders recombination between the two species and thus the formation of hybrids.

A species is defined by the property of the individuals within a group to mate and give rise to a progeny since the parental DNAs can recombine. They can do so if, and only if, the DNAs are homologous. If genomic DNA accumulates third-base changes with time, the DNA of the mating pair turns out to be too divergent with respect to their respective ancestors. The resulting diverged individuals may attempt to mate, but they cannot give rise to progeny.

Since the act of mating is followed by DNA recombination, the latter is successful only if it arises between two homologous DNAs. Mating is doomed to fail if it occurs between two heterologous DNAs. The first step in recombination is the pairing of two pairs of DNA strands. When two heterologous DNA chains attempt to pair, base-to-base *mismatches* arise between bases that should otherwise complement. Mismatches are removed by the induction of a very efficient repair process called *mismatch repair*. Radman and associates have demonstrated that recombination between two heterologous DNAs aborts through the action of *mismatch repair*.

5
Four Main Mechanisms that Generate Spontaneous Mutations

Spontaneous mutations may result from four main mechanisms: (1) DNA polymerization errors, (2) spontaneous oxidative DNA damage, (3) loss of a purine base, or less frequently, loss of a pyrimidine base, and (4) deamination of cytosine or of adenine.

1. *DNA polymerization errors*: Point mutations, usually transitions, arise from two successive faulty processes: (1) an error in chromosome replication resulting in a mismatch; and (2) followed with inefficient repair.

For example, DNA polymerase III may incorporate *in vitro* an incorrect base at a frequency of 10^{-6} to 10^{-7} per base pair per cell per generation. In *E. coli* bacteria, mismatch repair will remove the incorrect base, reducing the overall rate of replication and postreplication errors down from 10^{-9} to 10^{-10} per base pair per cell per generation.

Replication slippage, also denoted *slipped-strand mispairing*, is the major mechanism that accounts for frameshift mutations occurring at contiguous short repeats (Fig. 4).

2. *Incorporation of an oxidized base*: Exposed to oxygen, guanine is subject to oxidation that produces 8-oxo-G. This abnormal base has a tendency to slip into DNA, generating transversions. Fortunately, an elaborate repair system comes into play that either removes 8-oxo-G or decreases the consequences of its incorporation. Such a repair system is ubiquitous, protecting DNA from oxidative lesions.

Note that the oxygen we breathe is far from harmless. Oxygen is a DNA-damaging agent and may have a role in the aging process. Ames and associates have correlated the number of oxidative hits to DNA with the amount of DNA breakdown products released in rat urine and human urine. They have shown that the number of oxidative hits to DNA per cell per day is around 100 000 in rats and about 10 000 in humans. The 10-fold difference may result in part from the elevated metabolic rate of rodents, at least five times higher than that of mammals.

The mutagenic action of 8-oxo-G implies that there is no clear-cut difference between "natural" and "artificial" DNA-damaging agents. The distinction between spontaneous and induced mutations is also vague since the air we breathe does also insert oxides into DNA.

3. *Loss of a nucleic base*: It is the most frequently arising spontaneous lesion. If the chemical bond between sugar and base breaks, releasing a free base and leaving a gap in the sequence of bases, such an event generates an abasic site (apurinic or apyrimidinic). Loss of a purine base is more frequent than loss of a pyrimidine. A pyrimidine base strengthens DNA, which is in general sturdier than RNA.

Replication over an abasic site may produce a point mutation since the polymerase has no way of determining the identity of the missing base it has to copy. Adenine (A) seems frequently placed opposite the missing base. When T is the missing base, the placement of an A restores the normal coding.

4. *Deamination*: This process may convert cytosine to uracil and adenine to hypoxanthine. The former conversion seems to be involved as the main cause of immunoglobin hypermutation. This process would account for the creation of half a million antibodies, a process permitting our immunity repertoire to be constituted.

6
Induced Mutations

They may result from three different mechanisms: (1) the replacement of a base in the DNA by a base analog that confuses the polymerase; (2) an imperfect fit between a chemically altered base with another base during replication; and (3) the replacement of a base with a gap that lacks coding information.

1. *Base replacement by a base analog*: 2-Aminopurine, a base analog that mimics a normal base, causes frequent mispairing when incorporated into DNA.
2. *Mispairing by alkylation of bases*: Alkylation with ethylmethanesulfonate or with MNNG (N-methyl-N'-nitro-N-nitrosoguanidine) causes mispairing. Although these two chemicals modify different bases at various positions, alkylation at the O^6 position of guanine and at the O^4 position of thymine causes specific mispairing with T and with G, leading to GC $\leftrightarrow$ AT and AT $\leftrightarrow$ GC transitions. *In vivo*, O^6-alkylguanine and O^4-alkylthymine are two chemicals that cause the most mutations through alkylation.
3. *Noncoding lesions*: Ultraviolet light C (256 nm) generates pyrimidine dimers and 6–4 pyrimidine-pyrimidone compounds in DNA. Such lesions sharply bend DNA, distorting the precise pairing of bases upon replication.

7
Mutagenesis in Bacteria and Carcinogenesis in Mammals

Bruce Ames asked an interesting question: do mutagens cause cancer? If they do, it should be possible practically to replace the usual lengthy cancer tests on mice, lasting months and years, with a much shorter test involving bacteria instead of animals. Ames and collaborators devised a mutagenicity test using a *Salmonella* tester strain (his^-) that requires histidine for growth. The his^- tester bacteria treated with various carcinogens produced his^+ mutants that were able to grow on a medium devoid of histidine.

To make the test even more efficient, the bacterial tester strains were engineered in order to: (1) prevent the removal of the DNA lesions produced by the chemical tested; (2) increase the mutagenic rate by introducing a mutator plasmid in the tester strains; and (3) render the tester strains permeable to chemical compounds in changing its cell wall structure.

In the experimental procedure, one mixes the chemical with rat liver enzymes to reproduce the metabolization of the compound by a human liver. In the test, a transient physiological chimera combines rat enzymes with permeable bacteria to reproduce the human situation.

The results of the tests have shown that more than 65% of the 3000 chemical compounds tested, which are rodent carcinogens also cause mutations in bacteria. The converse is true: most mutagens are active carcinogens. Using a bacterial test to detect potential carcinogens has saved time, money, and animal suffering.

Note that even though bacteria are evolutionarily distant from rodents and humans, they provide an excellent tool for establishing that mutagenicity correlates with carcinogenicity. This is because DNA damage is the origin of both processes.

Because the *Salmonella* mutagenicity test is rapid, accurate, and relatively simple, many government agencies and biological laboratories have tested new chemicals for their mutagenicity and their potential carcinogenicity all over the world. In the European Union, the bacterial mutagenicity test has become mandatory. The overall result is that even though bacteria are evolutionarily more than a billion years distant from rodents and humans, they provide an excellent tool for measuring the mutagenic potential of chemical compounds. Government agencies and biological laboratories are testing newly found chemicals for their mutagenicity and potential carcinogenicity all over the world. In the European Union, the *Salmonella* mutagenicity test has become mandatory.

8
Carcinogens Leave No Definite Mutational Signatures on DNA

In human cells, a gene called *p53* expresses a protein that inhibits the occurrence of cancers; wild type protein P53 looks as an *anti-oncogene*. Measurements of the occurrence of p53 gene mutations demonstrated that in more than half of all cancers taken together, 90% of the p53 mutations are point mutations, two-third of them being transitions GC → AT.

In bronchial cancers, one finds that 45% of p53 mutations are transversions GC → TA. Benzopyrene, a mutagen produced by cigarettes or car exhausts, gives rise to a low level of GC → TA transversions.

Is the presence in cancer cells of a GC → TA transversion in the *p53* gene a case for making benzopyrene responsible for the production of the mutation? So many carcinogens and environmental pollutants are causing GC → TA transversions. How is it possible to incriminate benzopyrene itself in causing a tumor? Moreover, a point mutation is so common and so minute a DNA change that it is hard to take it as an indelible mark left over by a carcinogen.

In order to provide a legal proof for carcinogenicity of a mutated DNA sequence, a sequence must be long and varied enough to offer a valuable correlation. Moreover, correlation may indicate a causality link if the correlation index is extremely high.

The probability of finding an identical copy of a given DNA fingerprint is below 10^{-10}. Since the human population has not reached 10^{10} yet, fingerprinting provides a valid legal proof of identification. There is a large difference in the probability of occurrence of a given microsatellite with that of a transversion.

Few chemicals are legally recognized carcinogens. For instance, asbestos, a mineral, causes a unique and almost exclusive type of tumor called *mesothelioma*. Another compound, radioactive iodine, ingested by Ukrainian children in the aftermath of the Chernobyl nuclear accident is instrumental in breaking a specific chromosome in thyroid cancer cells. That type of causal relationship between mutations induced by radioactive iodine and a specific tumor is rather infrequent in practice. The probability for carcinogens to leave a "mutational signature" on DNA is so low that this concept has never been accepted as evidence in court.

9
Humans Use Mutagenesis to Produce their Immunity Repertoire

Mutations in DNA sequences in the immunoglobulin locus are involved in *class switch recombination* (CSR). These mutations are base substitutions, deletions, or duplications. Such mutations are primarily dependent upon the *induced activation of cytosine deaminase* or AID. One finds them during *V gene somatic hypermutation*. A correlation between a pattern of mutations and a mutational complex might generate them. This is due to the observation of the V gene mutation pattern of XP-V patients afflicted with the cancer-prone *Xeroderma pigmentosum variant syndrome*. The patients are deficient in DNA polymerase η (polη), suggesting that this enzyme is responsible for a large part of the mutations occurring at A–T bases. After analysis of switched memory B cells from two XP-V patients, it has been deduced that polymerase polη is also an A–T mutator during CSR, in both the switch region of tandem repeats as well as upstream of it. This suggests that the same error-prone translesional polymerase is involved, together with AID, in CSR.

Acknowledgments

It is a pleasure to acknowledge Li and Graur for their figure models. I also thank Pr. Joann Sweasy (Yale University) and Dr. Evelyne Sage (Institut Curie) for their insightful comments, and also thank Mrs. Melanie Pierre for her kind and indefatigable help.

Bibliography

Books and Reviews

Ames, B.N. (1979) Identifying environmental chemicals causing mutations and cancer, *Science* **204**, 587–593.

Devoret, R. (1979) Bacterial tests for potential carcinogens, *Sci. Am.* **241**, 40–50.

Echols, H., Goodman, M.F. (1991) Fidelity mechanisms in replication, *Annu. Rev. Biochem.* **60**, 477–511.

Friedberg, E.C., Walker, G.C., Siede, W. (1995) *DNA Repair and Mutagenesis*, American Society for Microbiology, Washington, DC.

Griffiths, A.J.F., Miller, J.H., Suzuki, D.T., Lewontin, R.C., Gelbart, W.M. (1993) *An Introduction to Genetic Analysis*, W. H. Freeman and Company, New York.

Li, W.H., Graur, D. (1991) *Fundamentals of Molecular Evolution*, Sinauer Associates, Sunderland.

Radman, M. (1975) SOS Repair Hypothesis, Phenomenology of an Inducible DNA Repair Which is Accompanied by Mutagenesis, in: Hanawalt, P., Setlow, R. (Eds.) *Molecular Mechanisms for Repair of DNA*, Plenum Publishers, New York.

Radman, M., Wagner, R. (1988) The high fidelity of DNA duplication, *Sci. Am.* **259**, 24–30.

Walker, G.C. (1984) Mutagenesis and inducible responses to deoxyribonucleic acid damage in Escherichia coli, *Microbiol. Rev.* **48**, 60–93.

Primary Literature

Bailone, A., Sommer, S., Knezevic, J., Dutreix, M., Devoret, R. (1991) A RecA protein mutant deficient in its interaction with the UmuDC complex. *Biochimie* **73**, 479–484.

Becherel, O.J., Fuchs, R.P.P., Wagner, J. (2002) Pivotal role of the β-clamp in translesion DNA synthesis and mutagenesis in E. coli cells, *DNA Repair* **1**, 703–708.

Blanco, M., Herrera, G., Collado, P., Rebollo, J.E., Botella, L.M. (1982) Influence of RecA protein on induced mutagenesis, *Biochimie* **64**(8–9), 633–636.

Boudsocq, F., Campbell, M., Devoret, R., Bailone, A. (1997) Quantitation of the inhibition of Hfr × F⁻ recombination by the mutagenesis complex UmuD'C, *J. Mol. Biol.* **270**, 201–211.

Bruck, I., Woodgate, R., McEntee, K., Goodman, M.F. (1996) Purification of a soluble UmuD'C complex from Escherichia coli. cooperative binding of UmuD'C to single-stranded DNA, *J. Biol. Chem.* **271**, 10767–10774.

Burnouf, D.Y., Olieric, V., Wagner, J., Fujii, S., Reinbolt, J., Fuchs, R.P., Dumas, P. (2004) Structural and biochemical analysis of sliding clamp/ligand interactions suggest a competition between replicative and translesion DNA polymerases, *J. Mol. Biol.* **335**, 1187–1197.

Cordonnier, A.M., Fuchs, R.P. (1999) Replication of damaged DNA, molecular defect in xeroderma pigmentosum variant cells, *Mutat. Res.* **435**, 111–119.

Dalrymple, B.P., Kongsuwan, K., Wijffels, G., Dixon, N.E., Jennings, P.A. (2001) A universal protein-protein interaction motif in the eubacterial DNA replication and repair systems, *Proc. Natl. Acad. Sci. U.S.A.* **98**, 11627–11632.

Dickerson, S.K., Market, E., Besmer, E., Papavasiliou, F.N. (2003) AID mediates hypermutation by deaminating single stranded DNA, *J. Exp. Med.* **197**, 1291–1296.

Dorner, T., Lipsky, P.E. (2001) Smaller role for pol eta? *Nat. Immunol.* **2**, 982–984.

Dutreix, M., Moreau, P.L., Bailone, A., Galibert, F., Battista, J.R., Walker, G.C., Devoret, R. (1989) New recA mutations that dissociate the various RecA protein activities in Escherichia coli provide evidence for an additional role for RecA protein in UV mutagenesis, *J. Bacteriol.* **171**, 2415–2423.

Faili, A., Aoufouchi, S., Gueranger, Q., Zober, C., Leon, A., Bertocci, B., Weill, J.-C., Reynaud, C.-A. (2002) AID dependent somatic hypermutation occurs as a DNA single-strand event in the BL2 cell line, *Nat. Immunol.* **3**, 815–821.

Faili, A., Aoufouchi, S., Weller, S., Vuillier, F., Stary, A., Sarasin, A., Reynaud, C.-A., Weill, J.-C. (2004) DNA Polymerase eta is involved in hypermutation occurring during immunoglobulin class switch recombination, *J. Exp. Med.* **199**, 265–270.

Friedberg, E.C., Wagner, R., Radman, M. (2004) Specialized DNA polymerases, cellular survival, and the genesis of mutations, *Science* **296**(5573), 1627–1630.

Goodman, M.F. (2002) Error-prone repair DNA polymerases in prokaryotes and eukaryotes, *Annu. Rev. Biochem.* **71**, 17–50.

Goodman, M.F., Tippin, B. (2000) Sloppier copier DNA polymerases involved in genome repair, *Curr. Opin. Genet. Dev.* **10**, 162–168.

Kannouche, P., Fernandez de Henestrosa, A.R., Coull, B., Vidal, A.E., Gray, C., Zicha, D., Woodgate, R., Lehmann, A.R. (2002) Localization of DNA polymerases eta and iota to the replication machinery is tightly co-ordinated in human cells, *EMBO J.* **21**, 6246–6256.

Lang, T., Maitra, M., Starcevic, D., Li, S.X., Sweasy, J.B. (2004) A DNA polymerase beta mutant from colon cancer cells induces mutations, *Proc. Natl. Acad. Sci. U.S.A.* **101**(16), 6074–6079.

Lehmann, A.R., Kirk-Bell, S., Arlett, C.F., Paterson, M.C., Lohman, P.H., de Weerd-Kastelein, E.A., Bootsma, D. (1975) Xeroderma pigmentosum cells with normal levels of excision repair have a defect in DNA synthesis after UV-irradiation, *Proc. Natl. Acad. Sci. U.S.A.* **172**, 219–223.

Livneh, Z. (2001) DNA damage control by novel DNA polymerases, translesion replication and mutagenesis, *J. Biol. Chem.* **276**, 25639–25642.

Maor-Shoshani, A., Livneh, Z. (2002) Analysis of the stimulation of DNA polymerase V of Escherichia coli by processivity proteins, *Biochemistry* **41**, 14438–14446.

Masutani, C., Kusumoto, R., Iwai, S., Hanaoka, F. (2000) Mechanisms of accurate translesion synthesis by human DNA polymerase eta, *EMBO J.* **19**, 3100–3109.

Matic, I., Babic, A., Radman, M. (2003) 2-aminopurine allows interspecies recombination by a reversible inactivation of the Escherichia coli mismatch repair system, *J. Bacteriol.* **185**(4), 1459–1461.

Matic, I., Taddei, F., Radman, M. (2000) No genetic barriers between Salmonella enterica serovar typhimurium and Escherichia coli in SOS-induced mismatch repair-deficient cells, *J. Bacteriol.* **182**(20), 5922–5924.

Nagaoka, H., Muramatsu, M., Yamamura, N., Kinoshita, K., Honjo, T. (2002) Activation-induced deaminase (AID)-directed hypermutation in the immunoglobulin Smu region, implication of AID involvement in a common step of class switch recombination and somatic hypermutation, *J. Exp. Med.* **195**, 529–534.

Napolitano, R., Janel-Bintz, R., Wagner, J., Fuchs, R.P. (2000) All three SOS-inducible DNA polymerases (Pol II, Pol IV and Pol V) are involved in induced mutagenesis, *EMBO J.* **19**, 6259–6265.

Ohmori, H., Friedberg, E.C., Fuchs, R.P., Goodman, M.F., Hanaoka, F., Hinkle, D. et al (2001) The Y-family of DNA polymerases, *Mol. Cell* **8**, 7–8.

Pages, V., Fuchs, R.P. (2002) How DNA lesions are turned into mutations within cells? *Oncogene* **21**, 8957–8966.

Pham, P., Bransteitter, R., Petruska, J., Goodman, M.F. (2003) Processive AID-catalysed cytosine deamination on single stranded DNA simulates somatic hypermutation, *Nature* **424**, 103–107.

Pham, P., Bertram, J.G., O'Donnell, M., Woodgate, R., Goodman, M.F. (2001) A model for SOS-lesion targeted mutations in Escherichia coli, *Nature* **409**, 366–370.

Reina-San-Martin, B., Difilippantonio, S., Hanitsch, L., Masilamani, R.F., Nussenzweig, A., Nussenzweig, M.C. (2003) H2AX is required for recombination between immunoglobulin switch regions but not for intra-switch region recombination or somatic hypermutation, *J. Exp. Med.* **197**, 1767–1778.

Reuven, N.B., Arad, G., Maor-Shoshani, A., Livneh, Z. (1999) The mutagenesis protein UmuC is a DNA polymerase activated by UmuD', RecA, SSB and is specialized for translesion replication, *J. Biol. Chem.* **274**, 31763–31766.

Reuven, N.B., Arad, G., Stasiak, A.Z., Stasiak, A., Livneh, Z. (2001) Lesion bypass by the Escherichia coli DNA polymerase V requires assembly of a RecA nucleoprotein filament, *J. Biol. Chem.* **276**, 5511–5517.

Reynaud, C.-A., Aoufouchi, S., Faili, A., Weill, J.-C. (2003) What role for AID, mutator, or assembler of the immunoglobulin mutasome? *Nat. Immunol.* **4**, 631–638.

Roca, A.I., Cox, M.M. (1997) RecA protein, structure, function, and role in recombinational DNA repair, *Prog. Nucleic Acid Res. Mol. Biol.* **56**, 129–223.

Sarasin, A. (2003) An overview of the mechanisms of mutagenesis and carcinogenesis, *Mutat. Res.* **544**(2–3), 99–106.

Sassanfar, M., Roberts, J.W. (1990) Nature of the SOS-inducing signal in Escherichia coli, the involvement of DNA replication, *J. Mol. Biol.* **212**, 79–96.

Sommer, S., Bailone, A., Devoret, R. (1993) The appearance of the UmuD'C protein complex in Escherichia coli switches repair from homologous recombination to SOS mutagenesis, *Mol. Microbiol.* **10**, 963–971.

Sommer, S., Boudsocq, F., Devoret, R., Bailone, A. (1998) Specific RecA amino acid changes affect RecA-UmuD'C interaction, *Mol. Microbiol.* **28**, 281–291.

Starcevic, D., Dalal, S., Sweasy, J.B. (2004) Is there a link between DNA polymerase beta and cancer? *Cell Cycle* **3**, 998–1001.

Sutton, M.D., Walker, G.C. (2001) Managing DNA polymerases, coordinating DNA replication, DNA repair, and DNA recombination, *Proc. Natl. Acad. Sci. U.S.A.* **98**, 8342–8349.

Sutton, M.D., Narumi, I., Walker, G.C. (2002) Posttranslational modification of the umuD-encoded subunit of Escherichia coli DNA polymerase V regulates its interactions with the beta processivity clamp, *Proc. Natl. Acad. Sci. U.S.A.* **99**, 5307–5312.

Tang, M., Shen, X., Frank, E.G., O'Donnell, M., Woodgate, R., Goodman, M.F. (1999) UmuD$_2$C is an error-prone DNA polymerase, Escherichia coli pol V, *Proc. Natl. Acad. Sci. U.S.A.* **96**, 8919–8924.

Tang, M., Pham, P., Shen, X., Taylor, J.S., O'Donnell, M., Woodgate, R., Goodman, M.F. (2000) Roles of E. coli DNA polymerases IV and V in lesion-targeted and untargeted SOS mutagenesis, *Nature* **404**, 1014–1018.

Wagner, J., Etienne, H., Janel-Bintz, R., Fuchs, R.P.P. (2002) Genetics of mutagenesis in E. coli, various combinations of translesion polymerases (Pol II, IV and V) deal with lesion/sequence context diversity, *DNA Repair* **1**, 159–167.

Wagner, J., Fujii, S., Gruz, P., Nohmi, T., Fuchs, R.P. (2000) The beta clamp targets DNA polymerase IV to DNA and strongly increases its processivity, *EMBO Rep.* **1**, 484–488.

Wagner, J., Gruz, P., Kim, S.-R., Yamada, M., Matsui, K., Fuchs, R.P.P., Nohmi, T. (1999) The dinB gene encodes a novel E. coli DNA polymerase, DNA PolIV, involved in mutagenesis, *Mol. Cell* **4**, 281–286.

Yavuz, S., Yavuz, A.S., Kraemer, K.H., Lipsky, P.E. (2002) The role of polymerase eta in somatic hypermutation determined by analysis of mutations in a patient with xeroderma pigmentosum variant, *J. Immunol.* **169**, 3825–3830.

Zeng, X., Winter, D.B., Kasmer, C., Kraemer, K.H., Lehmann, A.R., Gearhart, P.J. (2001) DNA polymerase eta is an A-T mutator in somatic hypermutation of immunoglobulin variable genes, *Nat. Immunol.* **2**, 537–541.

6
Molecular Genetics of Genomic Imprinting

Robert Feil, Yuji Goto, and David Umlauf
Centre National de la Recherche Scientifique, Montpellier, France

1	**Genomic Imprinting** 191	
1.1	Embryological Evidence 192	
1.2	Imprinted Chromosomal Domains 194	
2	**Imprinted Genes** 196	
3	**Molecular Mechanisms** 198	
3.1	Imprinting-control Regions 198	
3.2	Reading the Imprint 200	
4	**Imprinting and Disease** 201	
5	**Evolution of Imprinting** 203	
	Bibliography 204	
	Books and Reviews 204	
	Primary Literature 204	

Keywords

Androgenetic Embryo
An embryo with two paternal genomes, and no maternal genome, produced by nuclear transplantation.

Chromatin
DNA packaged around nucleosomes. The degree of packaging differs between active (euchromatic) and inactive (heterochromatic) chromosomal regions.

Genomics and Genetics. Edited by Robert A. Meyers.
Copyright © 2007 Wiley-VCH Verlag GmbH & Co. KGaA, Weinheim
ISBN: 978-3-527-31609-0

DNA Methylation
Attachment of methyl (CH$_3$) groups to the bases of DNA. In mammals, DNA methylation occurs at cytosines that are followed by guanines (at CpG dinucleotides).

Epigenetic Modification
Any heritable, but reversible, alteration of DNA or associated nucleosomes above the level of the DNA sequence. This additional layer of information may indicate the parental origin of the chromosome.

Genomic Imprinting
A parent-of-origin-dependent mechanism whereby certain gene loci become expressed from only the maternal or only the paternal chromosome.

Histone Modification
The histones in nucleosomes can be altered by covalent modifications. At imprinting-control regions, these modifications are different between the parental alleles.

Imprinting and Behavior
Some imprinted domains are associated with behavioral phenotypes, and genetic disruption of certain imprinted genes gives aberrant behavior.

Imprinting and Cancer
The epigenetic maintenance of imprinting is frequently deregulated in cancer. Since imprinted genes are important in cell proliferation and differentiation, such deregulation is probably involved in the process of tumorigenesis.

Imprinting and Growth
Many imprinted genes influence fetal growth and development. Imprinted genes that enhance growth are mostly expressed from the paternal allele. Several other imprinted genes, which reduce growth, are expressed from the maternal allele.

Imprinting-control Regions
DNA sequence elements that are essential for imprinted gene expression. They are modified by DNA methylation and epigenetic modifications on the chromatin.

Nucleosome
The basic structural unit of chromatin, consisting of ~150 bp of DNA wrapped around an octamer of histone proteins (two each of four different histones).

Nutrient Transfer
Imprinted genes are important for the development of the extraembryonic membranes. These are essential for nutrient transfer to the developing embryo.

Parthenogenesis
The derivation of offspring from eggs only. Parthenogenesis is viable in some animal groups, such as in bird species, but is embryonic lethal in mammals because of the functional nonequivalence of the maternal and the paternal genome.

Uniparental Disomy
Inheritance of a particular chromosome in two copies from one parent, with absence of the chromosome from the other parent.

> Genomic imprinting is a developmental mechanism in mammals and other organisms leading to repression or expression of genes depending on whether they are inherited from the mother or the father. The imprinted expression of genes is regulated by various epigenetic alterations, including DNA methylation and covalent modifications at histones. A large number of imprinted genes have been identified in placental mammals. Mostly clustered in the genome, these play important roles in embryonic and extraembryonic development, and in behavior. In humans, genetic and epigenetic alterations at imprinted genes are involved in different disease syndromes and in cancer.

1
Genomic Imprinting

Gene expression is not determined solely by the DNA code itself. It depends also on different epigenetic features. The term *epigenetic* is used to refer to mechanisms that do not involve changes in the DNA sequence and that are heritable from one cell generation to the next. Unlike heritable changes due to mutation or directed gene rearrangement (such as in the immunoglobulin genes), epigenetic modifications are reversible and can be removed from genes and chromosomes without leaving behind any permanent change to the genetic material. The main epigenetic modifications by which gene expression can be altered are DNA methylation and modifications to the chromatin. A well-known epigenetic mechanism is X-chromosome inactivation. In mammalian X-chromosome inactivation, sequential epigenetic modifications lead to the (random) transcriptional repression of one of the two X-chromosomes in all the somatic cells of females.

In this article, we consider a particular class of epigenetic imprints, those that mark the parental origin of genomes, chromosomes, and genes. Genes regulated by such "genomic imprinting" are expressed depending on whether they are on the maternally or on the paternally derived chromosome. Some imprinted genes are expressed only from the paternal chromosome, whereas others are exclusively expressed from the maternal chromosome. During the last fifteen to twenty years, imprinting has evolved from the initial observations in mouse embryos to a rapidly expanding field with importance for mammalian development and genetics, and human disease. A large number of imprinted genes have now been identified. In addition, molecular studies have unraveled the underlying molecular mechanisms. Imprinting is not unique to mammals but is known to occur in seed plants and invertebrate species as well. This article, however, focuses on the regulation and role of autosomal imprinted genes in mammals.

Following the discovery of genomic imprinting, and the identification of the first imprinted genes in mammals, researchers in the field hypothesized that the epigenetic marks that regulate parent-of-origin-dependent expression are established in either the female or the male germ line and (after fertilization) are maintained throughout development. This epigenetic information needs to be removed upon passage of the imprinted gene through the germ line in the developing fetus, however, so that new imprints can be established. Recent studies on DNA methylation and other epigenetic modifications showed that, indeed, there are three distinct phases in imprinting: establishment of the imprint in the male or female germ line, somatic maintenance of the imprint after fertilization, and its erasure upon (re-)passage through the germ line (Fig. 1).

1.1
Embryological Evidence

Embryological studies in the mouse provided the first evidence that, in mammals, both a maternal and a paternal genome are required for the production of viable offspring. Significantly, it was found that monoparental embryos, carrying either two maternal or two paternal genomes, cannot develop to term. Such monoparental embryos were obtained by nuclear transplantation, immediately following the fertilization of the egg by the sperm. By replacing the female pronucleus (female genome) with a male pronucleus (male genome), for example, it was possible to produce androgenetic embryos (which have two paternal genomes). Conversely, embryos with two maternal genomes (gynogenotes) were made by replacing the male pronucleus with a female pronucleus. Embryos with two maternal genomes were derived by activation of unfertilized eggs (parthenogenesis) as well. Intriguingly, both gynogenetic (parthenogenetic) and androgenetic embryos survived only for a few days after implantation in the uterus and were found to have major developmental abnormalities.

Gynogenetic (parthenogenetic) and androgenetic embryos have different developmental phenotypes (Fig. 2). After

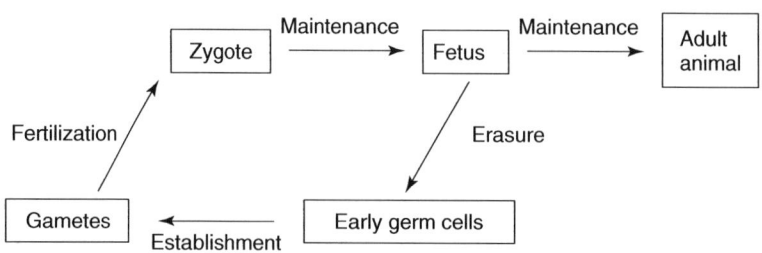

Fig. 1 Ontogeny of genomic imprinting: germ line establishment, somatic maintenance, and erasure.

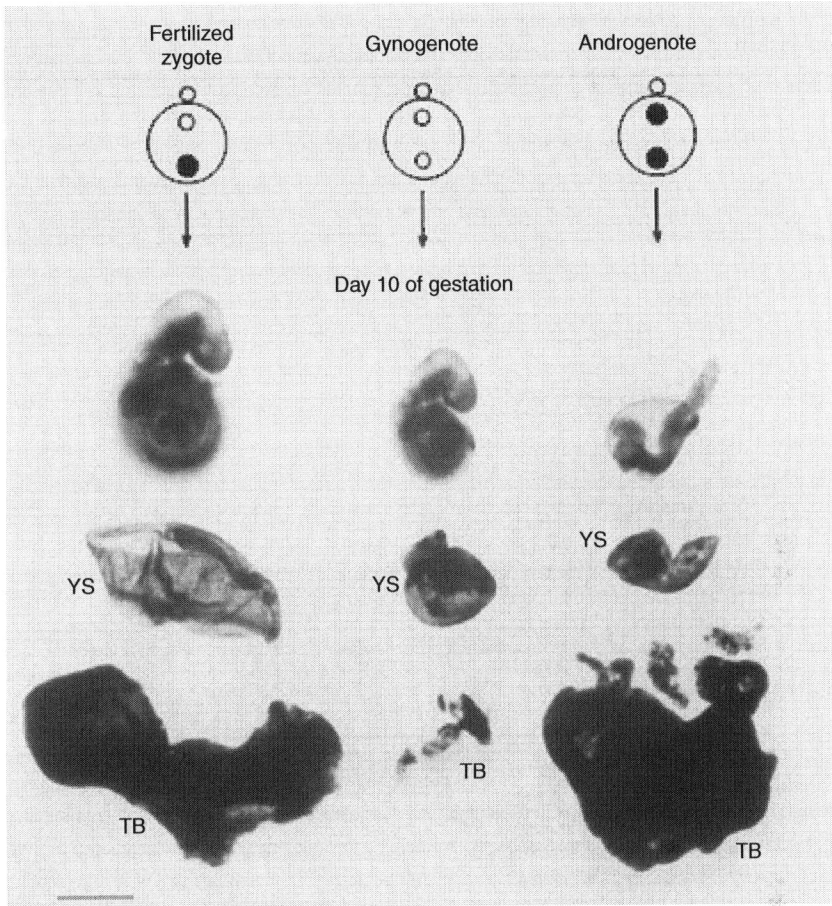

Fig. 2 Normal, androgenetic, and gynogenetic mouse embryos and their extraembryonic membranes at day 10 of gestation. Shown are the embryo, the yolk sac (YS), and the trophoblast (TB).

implantation into recipient females, parthenogenetic conceptuses develop to only about day 10 of gestation, with an apparently normal but small embryo. Development of the extraembryonic membranes (yolk sac and trophoblast), in contrast, is severely deficient, and these are the tissues that are important for nutrient transfer to the embryo. The phenotype of androgenetic conceptuses is opposite to that of parthenogenetic conceptuses. Whereas the extraembryonic tissues are normal in the androgenetic conceptuses, the embryo proper is retarded and progresses rarely beyond the four- to six-somite stage. The investigations on monoparental embryos established that both the parental genomes are required for normal mammalian development. They also provided evidence for the existence of genetic loci at which expression depends on the parental origin of the gene. In parthenogenetic and androgenetic embryos, individual imprinted genes are either expressed from both the gene copies (double gene dose) or are not expressed at all. Cumulatively, the aberrant levels of expression of imprinted genes are responsible for the striking phenotypes of the two types of monoparental embryos.

1.2
Imprinted Chromosomal Domains

The embryological evidence from the monoparental embryos was reinforced by genetic studies demonstrating that specific chromosomal domains are subject to imprinting. Particularly, mice that were heterozygous for chromosomal translocations were intercrossed to obtain embryos and offspring with uniparental disomy for individual chromosomes (or chromosomal regions). Since during meiosis there is sometimes nondisjunction at the chromosome with the translocation, some of the resulting gametes comprise two copies of the translocated chromosome, whereas others contain none. Embryos that arise from two of such opposite gametes will have two copies of all the chromosomes, but for the translocated chromosome, both the copies will be paternal or maternal. By using different translocation lines, such uniparental disomic embryos were generated for almost all autosomal chromosomes. Phenotypic analyses unraveled the role of subsets of imprinted genes that reside in two paternal or two maternal copies in the different uniparental disomies. These studies also revealed that the maternal and paternal copies of individual chromosomal regions have frequently opposite roles in development and after birth (Fig. 3).

One of the imprinted domains is on the distal portion of mouse chromosome 7. When present in two maternal copies (maternal disomy), it leads to reduced growth and fetal death, whereas paternal disomy of this distal region is associated with enhanced growth and embryonic death. Some 10 imprinted genes have been mapped to this region. Several of these are part of the insulin-like growth factor/insulin signaling pathway (IGF/INS pathway). Being key players in the regulation of fetal growth and development, they contribute to the opposite growth phenotypes in the maternal and paternal distal-7 disomies. The corresponding chromosomal region in humans, chromosome 11p15.5, is involved in the Beckwith–Wiedemann syndrome (BWS), a human growth disorder that can be caused by paternal disomy of this imprinted region.

Another chromosomal domain with opposite phenotypes in paternal and maternal disomies is the proximal portion of chromosome 11. Mice with paternal disomy of this region are larger than their normal littermates, whereas maternal disomy mice are smaller. This indicates that there are imprinted genes in this region, of which aberrant levels of expression in the maternal and paternal disomies cause their abnormal growth. So far, two imprinted genes have been identified in this domain, *U2af1-rs1* and *Grb10*. The latter could be responsible for the phenotypes of the maternal and paternal disomies. Its main embryonic transcript is expressed from the maternal allele only, and it encodes a protein with a negative effect on the growth-regulating IGF/INS pathway.

Disomy phenotypes at a few other imprinted domains involve abnormal postnatal behavior. Paternal disomy for distal mouse chromosome 2, for instance, gives offspring that are hyperactive, whereas maternal disomy is associated with reduced activity after birth. Such behavioral phenotypes emphasize that imprinted genes can affect behavior. A small number of

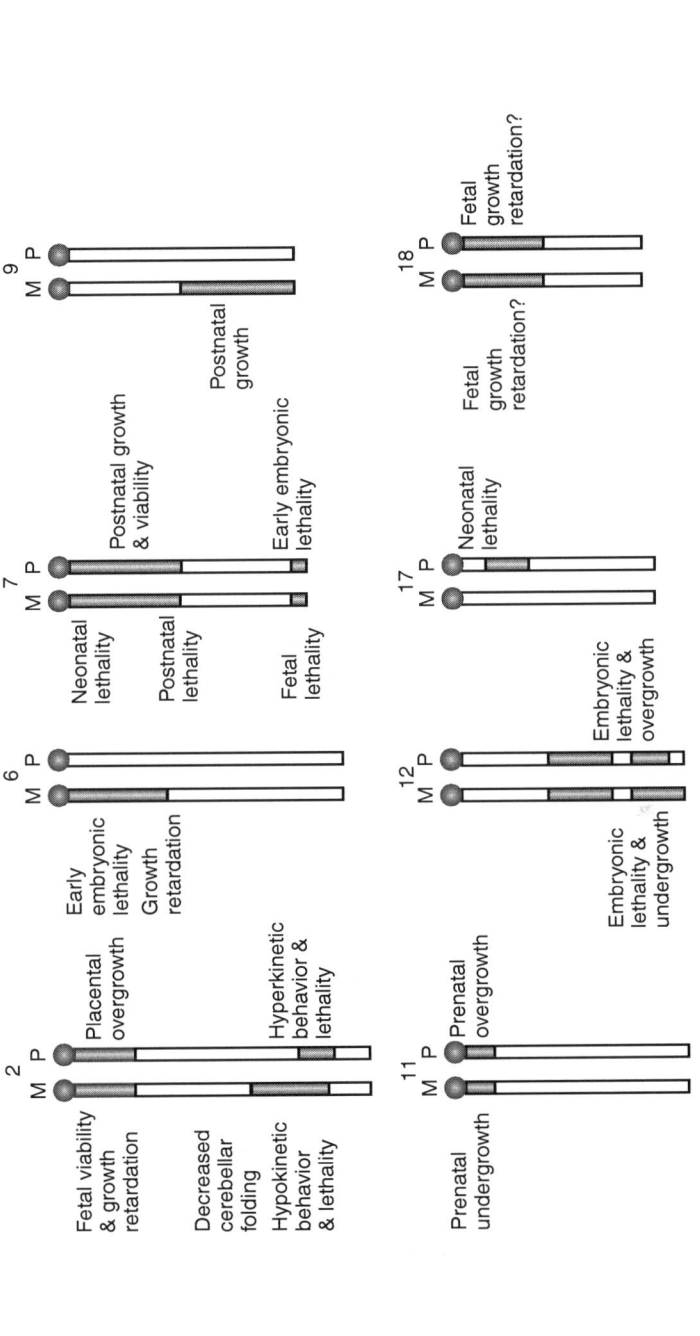

Fig. 3 Imprinted chromosomal domains in the mouse with the associated developmental phenotypes in maternal (indicated to the left) and paternal (to the right) disomies.

imprinted genes were discovered on distal mouse chromosome 2. Two of these have neuroendocrine functions (*Gnas* and *Gnas-xl*) and are involved in the behavioral phenotypes of the maternal and paternal disomy mice.

In total, 12 chromosomal regions with imprinting phenotypes have been identified on 8 different autosomal chromosomes (Fig. 3). The large majority of the known imprinted genes maps to these chromosomal regions. Probably, the remainder of the genome comprises few imprinted genes or contains imprinted genes that give rise to minor phenotypes only when present in two maternal or two paternal copies.

2
Imprinted Genes

It is unknown which proportion of mammalian genes is imprinted and estimates vary between about 100 and a 1000 genes. To date, however, some 70 imprinted genes have been detected in the mouse and most of these are imprinted in humans as well. A consistent feature of imprinted genes is that they are organized in clusters in the genome. These clusters are hundreds to several thousands of kilobases in size and are similarly organized in humans and mice. We selected several imprinted clusters as examples, and we describe their roles in development and behavior. A comprehensive presentation of imprinted genes is given elsewhere.

A well-known imprinted cluster is on distal mouse chromosome 7 (Fig. 4) and on the corresponding chromosome 11p15.5 in humans. This cluster comprises 11 imprinted genes. Several of these genes play key roles in fetal growth and development. The insulin-like growth factor-2 gene (*Igf2*), at the proximal side of the cluster, is expressed from the paternal allele only. Transgenic mice inheriting a null *Igf2* allele from the father are much smaller than their littermates; maternal inheritance of the targeted allele does not alter the phenotype. This strong paternal effect on fetal growth is primarily due to the loss of IGF2 in the extraembryonic membranes, which decreases nutrient transfer to the developing fetus. The neighboring insulin-2 gene (*Ins2*), also of the IGF/INS pathway, is located at about 20 kb from *Igf2*. In the yolk sac, it is the paternal chromosome that expresses *Ins2*, whereas the maternal chromosome is repressed. The paternal expression of *Igf2* and *Ins2* is regulated by an "imprinting-control region" downstream of *Igf2*, close to a maternally expressed imprinted gene (*H19*) that produces a noncoding RNA. At the distal side of the cluster, the *Cdkn1c* gene (also called *p57Kip2*) codes for a cyclin-dependent kinase inhibitor. This imprinted gene is expressed from the maternal allele only. When *Cdkn1c* expression is ablated by gene targeting in the mouse, offspring are enhanced in size and also display other similarities to the Beckwith–Wiedemann syndrome in humans. Interestingly, a similar growth phenotype arises as a consequence of *Igf2* overexpression in mice. One role of CDKN1C could therefore be to inhibit the growth-promoting action of IGF2. Thus, several genes in the imprinted gene cluster are involved in the regulation of fetal growth and seem part of the same signaling pathway.

Four of the other imprinted genes at the distal 7 cluster display allelic expression in the extraembryonic tissues, in which they are expressed from the maternal allele. One of these, *Ascl2* (also named *Mash2*) encodes a transcription factor that is essential for placental development. Genetic studies

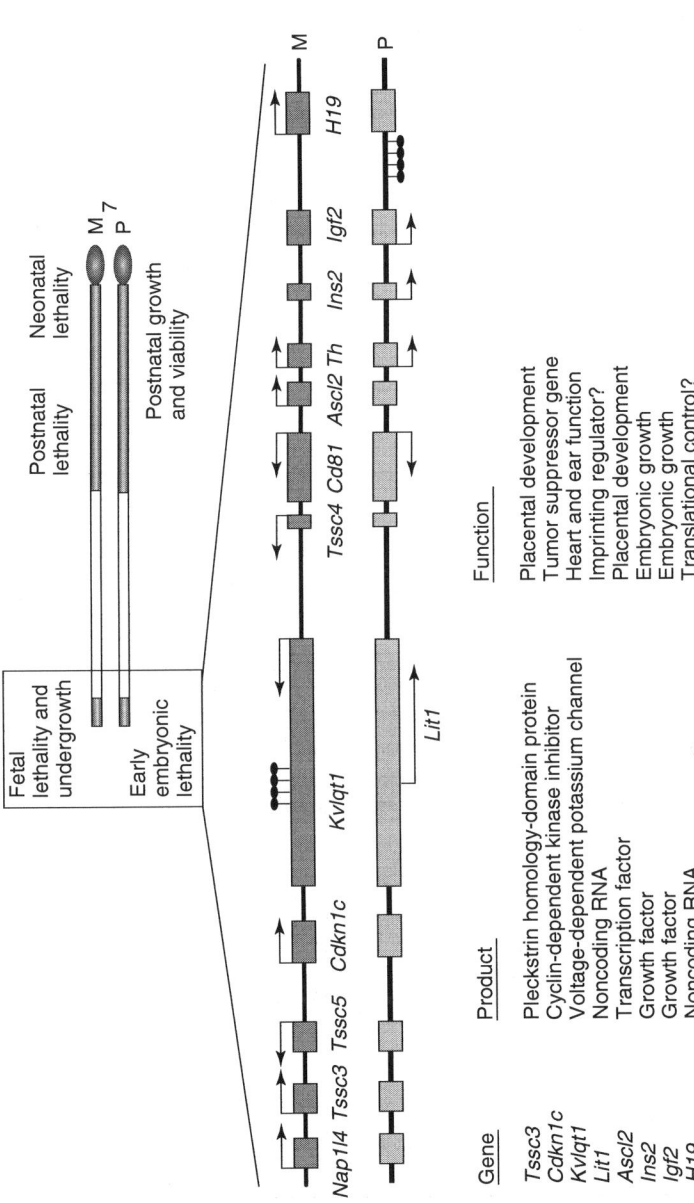

Fig. 4 The imprinted domain on distal mouse chromosome 7 (human chromosome 11q15). Shown are the imprinted genes and their known functions. Lollypops indicate the allele-specific DNA methylation at the two imprinting-control regions.

show that the imprinting of the four extraembryonic genes, and that of *Cdkn1c*, is regulated by a second "imprinting-control region," which is located in the central portion of the cluster.

Amongst the imprinted genes that influence the IGF/INS pathway, there is also the IGF2-receptor gene (*Igf2r*) on mouse chromosome 17. *Igf2r* is expressed exclusively from the maternal allele and exerts a negative effect on growth by reducing the levels of active IGF2. Whereas most imprinted mouse genes are imprinted in humans as well, *Igf2r* is one of the exceptions. In humans, this gene is expressed from both the parental alleles.

Some chromosomal domains comprise imprinted genes that are expressed predominantly in the brain. One of these clusters maps to the central portion of mouse chromosome 7 (and to human chromosome 15q11–q13) and comprises a large number of genes that are all expressed from the paternal chromosome only (Fig. 5c). In humans, loss of expression at these genes (*SNRPN*, *ZNF127*, *NDN*, and others) leads to the Prader–Willi syndrome, a variable disorder that is partly due to a hypothalamic defect (see below). Biallelic expression of the genes and loss of expression of a neighboring imprinted gene (*UBE3A*) is associated with the clinically distinct Angelman syndrome (AS). The regulation of imprinting in this domain is complex and involves at least two distinct genetic elements.

A minority of imprinted genes are not part of an imprinted gene cluster. One of these is the U2af1-related sequence-1 gene (*U2af1-rs1*) on proximal mouse chromosome 11. This intronless gene is repressed on the maternal chromosome and encodes a brain-specific RNA splicing factor homologous to the splicing factor U2AF. The imprinted *U2af1-rs1* gene has arisen via a retrotransposition event in rodents, and in humans there is no equivalent imprinted gene.

3
Molecular Mechanisms

3.1
Imprinting-control Regions

The expression of imprinted genes is regulated by epigenetic modifications that mark the parental alleles to be active or repressed. These epigenetic modifications are put onto key regulatory elements, depending on the parental origin of the allele, and lead to the allelic gene expression.

At all imprinted loci, there are sequence elements at which DNA methylation is present on one of the two parental alleles only. At many of these "differentially methylated regions" (DMRs), the DNA methylation originates from either the egg or the sperm. After fertilization, the allelic methylation is maintained in the somatic cells. Regions with such a germ line methylation mark are essential in the control of imprinting. They are referred to as *imprinting-control regions*. Most imprinting-control regions are rich in CpG dinucleotides and correspond to CpG islands.

At the imprinted *U2af1-rs1* gene, DNA methylation is present exclusively on the repressed maternal allele (Fig. 5a). This differential DNA methylation becomes established during oogenesis along its CpG island, located at the 5′ side of the gene, and spreads throughout the entire maternal gene during early embryonic development.

The maternally expressed *Igf2r* gene has an imprinting-control region within the second intron that is methylated on the

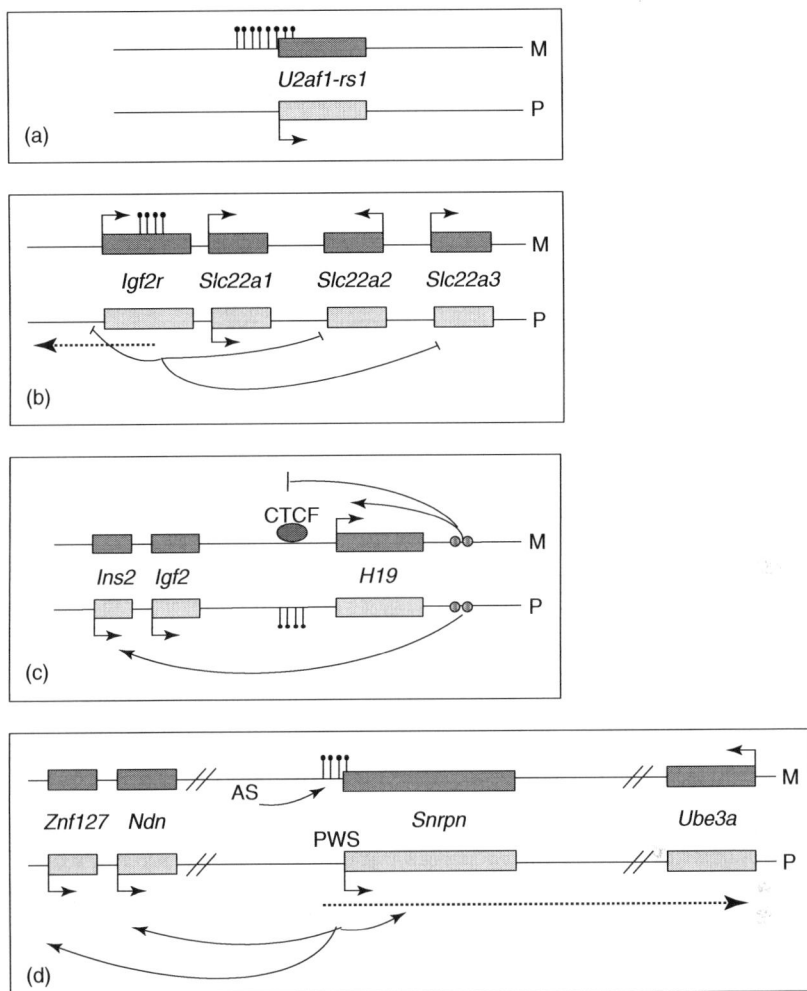

Fig. 5 Reading the imprint. Imprinting-control regions confer allelic gene expression at the (a) *U2af1-rs1* gene, (b) the Igf2r locus, (c) the imprinted cluster containing the *Igf2* gene, and (d) at the PWS/AS region in the mouse. Lollypops indicate the allele-specific DNA methylation at the imprinting-control regions. Antisense transcripts are shown as interrupted lines; circles indicate transcriptional enhancers.

maternal allele (Fig. 5b). This maternal methylation is established during oogenesis and is maintained in all the somatic lineages. The intronic imprinting-control region is essential for the allelic repression at the locus: removal by gene targeting leads to expression from both the parental alleles.

Most imprinting-control regions are methylated on the maternal allele. However, in some, the DNA methylation is found at the paternal allele, and it is the maternal allele that is unmethylated. One of these paternal methylation marks controls the allelic expression of the *Igf2* and *Ins2* genes on distal mouse chromosome 7

(Fig. 5c). This region, a CpG island located upstream of the close-by *H19* gene, acquires its DNA methylation during spermatogenesis. After fertilization, this paternal mark is maintained in all the somatic tissues. Deletion of the control region gives rise to biallelic expression of *Igf2* and *Ins2*.

It is unclear why imprinting-control regions attract DNA methylation in either the female or the male germ line. Several studies suggest, however, that close-by tandemly-repeated sequences might be essential in this choice process.

More is known about the DNA methyltransferases (DNMT) that are involved in the germ line establishment of the methylation marks. The methyltransferases DNMT3A and DNMT3B are essential in this process. In addition, a DNMT-like protein (DNMT3L) is required for the establishment of methylation imprints as well, particularly in the female germ line. Once established, allelic patterns of DNA methylation need to be maintained in the developing embryo. The maintenance methyltransferase, DNMT1, plays an important role in this process and differential chromatin features are likely to be involved as well. At imprinting-control regions, pronounced differences in histone modifications have been detected between the parental alleles. Levels of histone acetylation are low on the allele that comprises methylated DNA, whereas high levels of acetylation are present on the chromatin of the opposite, unmethylated, allele. In addition, there are strong allelic differences in histone methylation at specific lysine residues on histone H3. Whereas methylation of lysine residue 9 of H3 is detected on the parental allele that has DNA methylation, it is on the opposite parental allele (without DNA methylation) that there are high levels of H3 lysine-4 methylation. At several imprinted loci, there is also evidence for allele-specific chromatin compaction, occurring in association with the differential histone modifications.

It is not yet understood how DNA methylation, histone acetylation, and histone methylation are mechanistically linked at imprinting-control regions. However, at several imprinting-control regions, (allelic) DNA methylation was found to be linked to histone deacetylation (the removal of the acetyl group from the histones). This link is brought about by proteins that bind the methylated DNA and attract large protein complexes that comprise histone deacetylases. It is to be explored also to which extent the differential histone modifications are important in the somatic maintenance (and germ line establishment) of the allelic patterns of DNA methylation at imprinted loci. Nonhistone proteins, binding to the unmethylated allele of many imprinting-control regions, are likely to be involved in the maintenance of the allelic DNA methylation as well.

3.2
Reading the Imprint

Imprinting-control regions are comparable in that they all have allele-specific DNA methylation and differential chromatin organization. The way in which this gives rise to imprinted gene expression differs between loci. The simplest scenario, whereby differential methylation and associated chromatin features lead to imprinted gene expression, is that of the *U2af1-rs1* gene on mouse chromosome 11 (Fig. 5a). Here, methylated DNA and compacted chromatin are present across the promoter on the maternal allele. As a consequence, the gene can be transcribed from the paternal allele only.

The imprinting-control region of *Igf2r* regulates allelic expression in a rather different way (Fig. 5b). Here, the maternal methylation covers the promoter of an antisense transcript. As a consequence, this antisense transcript (named *Air*) is produced from the (unmethylated) paternal allele only. Via a yet-unclear mechanism, this paternal antisense transcript represses the paternal *Igf2r* gene and two flanking ion-transporter genes (*Slc22a2* and *Slc22a3*). A similar antisense transcript is produced at the imprinting-control region that regulates the extraembryonic tissue-specific imprinted genes on distal mouse chromosome 7 (Fig. 3).

Another example of how a germ line mark conveys imprinted expression is provided by the *Ins2-Igf2-H19* locus on distal mouse chromosome 7 (Fig. 5c). Here, the imprinting-control region located upstream of the noncoding *H19* gene, is methylated on the paternal allele and acts as a chromatin boundary on the unmethylated maternal allele. In fact, this upstream element has multiple recognition sites for a zinc finger protein called CTCF. The binding of CTCF is prevented by methylation. This chromatin protein is therefore associated with the unmethylated maternal allele only, at which it forms a specialized chromatin structure. This unusual structure insulates the *Igf2* and *Ins2* promoters from their enhancers (located downstream of *H19*). As a consequence, *Igf2* and *Ins2* are repressed on the maternal chromosome. This maternal repression is not exclusively at the transcriptional level but influences posttranscriptional processes as well.

The central portion of mouse chromosome 7 (Fig. 5d) corresponds to the Prader–Willi syndrome (PWS) region and Angelman syndrome (AS) region on human chromosome 15q11–13. The key regulatory element in this domain is the 5′ portion of the *SNRPN* gene and it is methylated on the maternal chromosome. This imprinting-control region is essential for the paternal expression of *SNRPN* and its flanking genes (including *Znf127* and *Ndn*). When the element is deleted on the paternal chromosome, these brain-specific genes are all no longer expressed. Upstream of the *SNRPN* gene, a paternal RNA of several hundreds of kilobases in size is produced as well. This transcript is in antisense orientation to a gene at the far extremity of the imprinted domain. This gene, *UBE3A*, is the only one in the cluster that is repressed on the paternal chromosome. The *SNRPN* imprinting-control region itself is regulated by a second control region, which is located further upstream, and is essential for the acquisition of the allelic DNA methylation at *SNRPN*. Precisely how the allelic expression and repression is brought about along thousands of kilobases remains to be unraveled. It has been observed, however, that there is differential timing of replication in the S-phase between the parental chromosomes along the entire domain. Such a differential replication timing has been detected at other imprinted loci as well. Future work should investigate the role of the differential replication timing and whether it reflects differential chromatin organization along entire imprinted domains.

4
Imprinting and Disease

In many genetic diseases, the clinical manifestations depend on whether the mutation is inherited from the mother or the father. Although imprinting is suspected to be involved, causal genes and molecular mechanisms are yet to be identified for most of these disorders. Clinical phenotypes can be associated with uniparental disomies as well, similarly as in the mouse. Additionally, imprinting can become deregulated during embryonic development, by epigenetic alterations or by somatic mutations, resulting in loss or biallelic expression of imprinted genes. Such somatic loss of imprinting can result in specific disease phenotypes as well.

Beckwith–Wiedemann syndrome (BWS) is a fetal overgrowth syndrome with a high incidence of embryonal tumors, including Wilms' tumor of the kidney and rhabdomyosarcoma. Genetically, the syndrome is linked to the cluster of growth-related imprinted genes comprising *IGF2* (see Fig. 4). Paternal disomy of this domain is responsible for a proportion of cases and leads to a double dose of *IGF2* expression and loss of expression of *CDKN1C* and other maternally expressed genes in the cluster. BWS can also be caused by genetic mutations at *CDKN1C* and by alterations at the *KVLQT1* gene, where there is one of the two imprinting-control centers of the cluster. The finding that the growth syndrome can be caused by mutations at different places in the imprinted domain supports the idea that its genes are coregulated and involved in the same biological functions. The majority of the BWS cases are sporadic, however, and apparently without genetic mutations. These are mostly caused by epigenetic alterations in the developing embryo. In some of the sporadic cases, for instance, there is aberrant, biallelic methylation at the imprinting-control region at the *H19* gene (Fig. 5c). This results in expression of *IGF2* from both the parental chromosomes during development and therefore in fetal overgrowth.

The neurobehavioral Angelman syndrome (AS) includes mental retardation, ataxia, and hyperactivity and arises from maternal deletion or paternal disomy of the imprinted domain on chromosome 15q11–13. Prader–Willi syndrome (PWS), on the other hand, arises from paternal deletion or maternal disomy of this imprinted domain. This opposite syndrome involves mild mental retardation, obesity due to hyperphagia, and hypogonadism. Cases with small genetic deletions have been identified, and analysis of these patients has revealed that the PWS and AS are caused by distinct regions in the large imprinted domain (Fig. 5d). The smallest identified deletions in PWS remove the imprinting element at the 5' portion of *SNRPN*. This gives loss of expression of *SNRPN, NDN, ZNF127*, and several other genes in the cluster. The smallest deletions in AS removes the control region that is essential for the establishment of the epigenetic imprint at *SNRPN*. Consequently, there is expression of *SNRPN, NDN*, and *ZNF127* from both the parental alleles and loss of expression of the *UBE3A* gene located at the end of the cluster. The latter seems to be the main cause of the clinical phenotype of AS.

Amongst other imprinting disorders are Albright Hereditary Osteodystrophy (AHO), linked to a cluster of imprinted neuroendocrinal genes on human chromosome 20q, and transient neonatal diabetes mellitus (TNDM), linked to chromosome 6q24–25. The latter is mostly sporadic and is caused by aberrant expression of the imprinted gene *ZAC*. This zinc finger protein–encoding gene has a CpG island with maternal DNA methylation. It was discovered recently that in cases of TNDM without genetic defects, this imprinting-control region has lost its methylation.

Epigenetic alterations at imprinting-control regions occur frequently in tumors as well. This has been observed in Wilms' tumor of the kidney, but also in lung cancer, breast cancer, and various other cancers. In particular, *IGF2* was found to be expressed from both the parental alleles during tumorigenesis and this could confer a proliferative advantage to the cells. In many cases, the biallelic *IGF2* expression is caused by acquisition of DNA methylation at the imprinting-control region upstream of *H19*, similarly as in BWS. This epigenetic alteration occurs early in tumor formation and could be linked to the pathological tendency of tumorigenic cells to acquire methylation at CpG islands.

When early embryos are taken from their natural environment and put into a culture dish, it can lead to aberrant imprinting as well. This was observed in the mouse and in domestic animals. It is unclear, at present, whether loss of imprinting due to embryo culture is mechanistically comparable to that in human imprinting disorders or in tumors. However, culture of embryos and early embryonic cells can also induce aberrant DNA methylation at imprinting-control regions. This results in biallelic, or loss of, imprinted gene expression and can have pronounced phenotypic consequences at later developmental stages. An important issue to be investigated is whether there are culture conditions that do not affect imprinting and would thus be best suitable for *in vitro* culture and manipulation procedures in animals and humans.

5
Evolution of Imprinting

There is a lot of interest in how broadly imprinting is conserved amongst mammals. Also, in species other than the mouse, both the parental genomes are essential for normal development. Parthenogenesis, for instance, leads to embryonic lethality in humans, pigs, and sheep. In the latter (ruminant) species, parthenogenetic conceptuses die shortly after implantation, due to deficient development and functioning of the extraembryonic membranes. These studies indicate that imprinting is conserved amongst different groups of mammals. Indeed, most of the known imprinted mouse genes are also imprinted in humans and, as far as this has been analyzed, in other placental mammals as well.

Evolutionary biologists have proposed several hypotheses to explain why imprinting has arisen in placental mammals and to account for the different imprinting-related phenotypes. In placental mammals, there is continuous transfer of nutrients from the mother animal to the developing offspring, and this determines their development and growth. Possibly, the most attractive theory of imprinting says that paternally inherited genes tend to increase nutrient transfer and thereby the growth of the developing fetus. This would enhance their chances of being propagated to future generations. Maternally derived genes, however, would be best propagated by limiting the growth of the developing fetus. This is because too high a burden of nutrient transfer compromises the reproductive success of the mother animal and hence of all its offspring. During the evolution of placental mammals, there would therefore have been a 'parental tug-of-war' between these opposing maternal and paternal strategies, leading to balanced combinations of expression levels of maternally and paternally derived genes. As outlined with different examples, imprinted genes such as *Igf2*, *Ins2*, and *Igf2r* indeed play important roles in nutrient transfer and growth, for instance, by promoting or reducing the development of the extraembryonic tissues. Other imprinted genes are important in determining the activity of the newborn animals, which, again, could have an impact on nutrient transfer, but now after birth.

See also Molecular Basis of Genetics.

Bibliography

Books and Reviews

Beechey, C.V., Cattanach, B.M.C., Selley, R.L. (2002) MRC Mammalian Genetics Unit, Harwell, United Kingdom. World Wide Web site – Mouse imprinting data and references (http://www.mgu.har.mrc.ac.uk/imprinting/imptables.html).

Feil, R., Khosla, S. (1999) Genomic imprinting in mammals: an interplay between chromatin and DNA methylation? *Trends Genet.* **15**, 431–435.

Lee, J.T. (2003) Molecular links between X-inactivation and autosomal imprinting: X inactivation as the driving force for the evolution of imprinting? *Curr. Biol.* **13**, R242–R254.

Ohlsson, R., Tycko, B., Sapienza, C. (1998) Mono-allelic expression: 'there can only be one', *Trends Genet.* **14**, 435–438.

Reik, W., Walter, J. (2001) Genomic imprinting: parental influence on the genome, *Nat. Rev. Genet.* **2**, 21–32.

Sleutels, F., Barlow, D.P., Lyle, R. (2000) The uniqueness of the imprinting mechanism, *Curr. Opin. Genet. Dev.* **10**, 229–233.

Surani, M.A. (1998) Imprinting and the initiation of gene silencing in the germ line, *Cell* **93**, 309–312.

Tilghman, S.M. (1999) The sins of the fathers and mothers: genomic imprinting in mammalian development, *Cell* **96**, 185–193.

Wilkins, J.F., Haig, D. (2003) What good is genomic imprinting? *Nat. Rev. Genet.* **4**, 359–368.

Primary Literature

Barlow, D.P., Stöger, R., Herrmann, B.G., Saito, K., Schweifer, N. (1991) The mouse insulin-like growth factor type-2 receptor is imprinted and closely linked to the *Tme* locus, *Nature* **349**, 84–87.

Bartolomei, M.S., Zemel, S., Tilghman, S.M. (1991) Parental imprinting of the mouse *H19* gene, *Nature* **351**, 153–155.

Bourc'his, D., Xu, G.L., Lin, C.S., Bollman, B., Bestor, T.H. (2001). *Dnmt3l* and the establishment of maternal genomic imprints, *Science* **294**, 2536–2539.

Brandeis, M., Kafri, T., Ariel, M., Chaillet, J.R., McCarrey, J., Razin, A., Cedar, H. (1993) The ontogeny of allele-specific methylation associated with imprinted genes in the mouse, *EMBO J.* **12**, 3669–3677.

Buiting, K., Saitoh, S., Gross, S., Dittrich, B., Schwartz, S., Nicholls, R.D., Horsthemke, B. (1995) Inherited microdeletions in the Angelman and Prader-Willi syndromes define an imprinting centre on human chromosome 15, *Nat. Genet.* **9**, 395–400.

Cattanach, B.M., Kirk, M. (1985) Differential activity of maternally and paternal derived chromosome regions in mice, *Nature* **315**, 496–498.

Cavaillé, J., Seitz, H., Paulsen, M., Ferguson-Smith, A.C., Bachellerie, J.P. (2002) Identification of tandemly-repeated C/D snoRNA genes at the imprinted human 14q32 domain reminiscent of those at the Prader-Willi/Angelman syndrome region, *Hum. Mol. Genet.* **11**, 1527–1538.

Chaillet, J.R., Vogt, T.F., Beier, D.R., Leder, P. (1991) Parental-specific methylation of an imprinted transgene is established during gametogenesis and progressively changes during embryogenesis, *Cell* **66**, 77–83.

Charlier, C., Segers, K., Karim, L., Shay, T., Gyapay, G., Cockett, N., Georges, M. (2001) The callipyge mutation enhances the expression of coregulated imprinted genes in cis without affecting their imprinting status, *Nat. Genet.* **27**, 367–369.

Cockett, N.E., Jackson, S.P., Shay, T.L., Farnir, F., Berghmans, S., Showder, G.D., Nielsen, D.M., Georges, M. (1996) Polar overdominance at the ovine callipyge locus, *Science* **273**, 236–238.

Constancia, M., Hemberger, M., Hughes, J., Dean, W., Ferguson-Smith, A., Fundele, R.,

Stewart, F., Kelsey, G., Fowden, A., Sibley, C., Reik, W. (2002) Placental-specific IGF-II is a major modulator of placental and fetal growth, *Nature* **417**, 945–948.

Cui, H., Cruz-Correa, M., Giardello, F.M., Hutcheon, D.F., Kafenok, D.R., Brandenburg, S., Wu, Y., He, X., Powe, N.R., Feinberg, A.P. (2003) Loss of IGF2 imprinting: a potential marker of colorectal cancer risk, *Science* **299**, 1753–1755.

Dean, W., Bowden, L., Aitchison, A., Klose, J., Moore, T., Meneses, J.J., Reik, W., Feil, R. (1998) Altered imprinted gene methylation and expression in completely ES cell-derived mouse fetuses: association with aberrant phenotypes, *Development* **125**, 2273–2282.

Debaun, M.R., Niemitz, E.L., Feinberg, A.P. (2003) Association of in vitro fertilization and Beckwith-Wiedemann syndrome and epigenetic alterations of LIT1 and H19, *Am. J. Hum. Genet.* **72**, 156–160.

DeChiara, T.M., Robertson, E.J., Efstratiadis, A. (1991) Parental imprinting of the mouse insulin-like growth factor II, *Cell* **64**, 849–859.

Doherty, A.S., Mann, M.R., Tremblay, K.D., Bartolomei, M.S., Schultz, R.M. (2000) Differential effects of culture on imprinted H19 expression in the pre-implantation embryo, *Biol. Reprod.* **62**, 1526–1535.

Feil, R., Walter, J., Allen, N.D., Reik, W. (1994) Developmental control of allelic methylation in the imprinted mouse *Igf2* and *H19* genes, *Development* **120**, 2933–2943.

Feil, R., Boyano, M.D., Allen, N.D., Kelsey, G. (1997) Parental chromosome-specific chromatin conformation in the imprinted *U2af1-rs1* gene in the mouse, *J. Biol. Chem.* **272**, 20893–20900.

Ferguson-Smith, A.C., Cattanach, B.M., Barton, S.C., Beechey, C.V., Surani, M.A. (1991) Embryological and molecular investigations of parental imprinting on mouse chromosome 7, *Nature* **352**, 609–610.

Ferguson-Smith, A.C., Sasaki, H., Cattanach, B.M., Surani, M.A. (1993) Parent-origin-specific epigenetic modification of the mouse H19 gene, *Nature* **362**, 751–755.

Fitzpatrick, G.V., Soloway, P.D., Higgins, M.J. (2002) The brain on microarrays, *Nat. Genet.* **32**, 426–431.

Fournier, C., Goto, Y., Ballestar, E., Delaval, K., Hever, A.M., Esteller, M., Feil, R. (2002) Allele-specific histone lysine methylation marks regulatory regions at imprinted mouse genes, *EMBO J.* **23**, 6560–6570.

Gregory, R.I., Randall, T.E., Johnson, C.A., Khosla, S., Hatada, I., O'Neill, L.P., Turner, B.M., Feil, R. (2001) DNA methylation is linked to deacetylation of histone H3, but not H4, on the imprinted genes Snrpn and U2af1-rs1, *Mol. Cell. Biol.* **21**, 5426–5436.

Gunaratne, P.H., Nakao, M., Ledbetter, D.H., Sutcliffe, J.S., Chinault, A.C. (1995) Tissue-specific and allele-specific replication timing control in the imprinted human Prader-Willi syndrome region, *Genes Dev.* **9**, 808–820.

Hajkova, P., Erhardt, S., Lane, N., Haaf, T., El Maarri, O., Reik, W., Walter, J., Surani, A. (2002) Epigenetic reprogramming in mouse primordial germ cells, *Mech. Dev.* **117**, 15–23.

Hata, K., Okano, M., Lei, H., Li, E. (2002) Dnmt-3L cooperates with the Dnmt3 family of de novo DNA methyltransferases to establish maternal imprints in mice, *Development* **129**, 1983–1993.

Hatada, I., Nabetani, A., Arai, Y., Ohishi, S., Suzuki, M., Miyabara, S., Nishimune, Y., Mukai, T. (1997) Aberrant methylation of an imprinted gene U2af1-rs1 (SP2) caused by its own transgene, *J. Biol. Chem.* **272**, 9120–9122.

Howell, C.Y., Bestor, T.H., Ding, F., Latham, K.E., Mertineit, C., Trasler, J.M., Chaillet, J.R. (2001) Genomic imprinting disrupted by a maternal effect mutation in the *Dnmt1* gene, *Cell* **104**, 829–838.

Humpherys, D., Eggan, K., Akutsu, H., Hochedlinger, K., Rideout, W.M., Biniszkiewcz, D., Yanagimachi, R., Jaenisch, R. (2001) Epigenetic instability in ES cells and cloned mice, *Science* **293**, 95–97.

Inoue, K., Kohda, T., Lee, J., Ogonuki, N., Mochida, K., Noguchi, Y., Tanemura, K., Kaneko-Ishino, T., Ishino, F., Ogura, A. (2002) Faithful expression of imprinted genes in cloned mice, *Science* **295**, 297.

Izumikawa, Y., Naritoma, K., Hariyama, K. (1991) Replication asynchrony between homologs 15q11.2: cytogenetic evidence for genomic imprinting, *Hum. Genet.* **87**, 1–5.

Jouvenot, Y., Poirier, F., Jami, J., Paldi, A. (1999) Bi-allelic transcription of *Igf2* and *H19* in individual cells suggests a post-transcriptional contribution to genomic imprinting, *Curr. Biol.* **9**, 1199–1202.

Judson, H., Hayward, B.E., Sheridan, E., Bonthron, D.T. (2002) A global disorder of imprinting in the human female germ line, *Nature* **416**, 539–542.

Khosla, S., Aitchison, A., Gregory, R., Feil, R. (1999) Parental allele-specific chromatin configuration in an insulator/imprinting control element upstream of the mouse *H19* gene, *Mol. Cell. Biol.* **19**, 2556–2566.

Khosla, S., Dean, W., Brown, D., Reik, W., Feil, R. (2001) Culture of pre-implantation mouse embryos affects fetal development and the expression of imprinted genes, *Biol. Reprod.* **64**, 918–926.

Killian, J.K., Byrd, J.C., Jirtle, J.V., Munday, B.L., Stoskopf, M.K., MacDonald, R.G., Jirtle, R.L. (2000) M6P/IGF2R imprinting evolution in mammals, *Mol. Cells* **5**, 707–716.

Kitsberg, D., Selig, S., Brandeis, M., Simon, I., Keshet, I., Driscoll, D.J., Nicholls, R.D., Cedar, H. (1993) Allele-specific replication timing of imprinted gene regions, *Nature* **364**, 459–463.

Knoll, J.H., Cheng, S.D., Lalande, M. (1994) Allele specificity of DNA replication timing in the Angelman/Prader-Willi syndrome imprinted chromosomal region, *Nat. Genet.* **6**, 41–46.

Lee, J., Inoue, K., Ono, R., Ogonuki, N., Kohda, T., Kaneko-Ishino, T., Ogur, S., Ishino, F. (2002) Erasing genomic imprinting memory in mouse clone embryos produced from day 11.5 primordial germ cells, *Development* **129**, 1807–1817.

Li, E., Beard, C., Jaenisch, R. (1993) Role for DNA methylation in genomic imprinting, *Nature* **366**, 362–365.

Mager, J., Montgomery, N.D., de Villena, F.P., Magnuson, T. (2003) Genome imprinting regulated by the mouse Polycomb group protein Eed, *Nat. Genet.* **33**, 502–507.

McGrath, J., Solter, D. (1984) Completion of mouse embryogenesis requires both the maternal and paternal genomes, *Cell* **37**, 179–183.

Milligan, L., Forné, T., Antoine, E., Weber, M., Hémonnot, B., Dandolo, L., Brunel, C., Cathala, G. (2002) Turnover of primary transcripts is a major step in the regulation of mouse *H19* gene expression, *EMBO Rep.* **3**, 774–779.

Muscatelli, F., Abrous, D.N., Massacrier, A., Boccaccio, I., Le Moal, M., Cau, P., Cremer, H. (2000) Disruption of the mouse *Necdin* gene results in hypothalamic and behavioral alterations reminiscent of the human Prader-Willi syndrome, *Hum. Mol. Genet.* **12**, 3101–3110.

Nabetani, A., Hatada, I., Morisaki, H., Oshimura, M., Mukai, T. (1997) Mouse *U2af1-rs1* is a neomorphic imprinted gene, *Mol. Cell. Biol.* **17**, 789–798.

Paulsen, M., El-Maarri, O., Engemann, S., Strodicke, M., Franck, O., Davies, K., Reinhardt, R., Reik, W., Walter, J. (2000) Sequence conservation and variability of imprinting in the Beckwith-Wiedemann syndrome cluster in human and mouse, *Hum. Mol. Genet.* **9**, 1829–1841.

Pant, V., Mariano, P., Kanduri, C., Mattson, A., Lobanenkov, V., Heuchel, R., Ohlsson, R. (2003) The nucleotides responsible for the direct physical contact between the chromatin insulator protein CTCF and the *H19* imprinting control region manifest parent of origin-specific long-distance insulation and methylation-free domains, *Genes Dev.* **17**, 586–590.

Perk, J., Makedonski, K., Lande, L., Cedar, H., Razin, A. (2002) The imprinting mechanism of the Prader-Willi/Angelman regional control center, *EMBO J.* **21**, 5807–5814.

Rainier, S., Johnson, L.A., Dobry, C.J., Ping, A.J., Grundy, P.E., Feinberg, A.P. (1993) Relaxation of imprinted genes in human cancer, *Nature* **362**, 747–749.

Reik, W., Brown, K.W., Schneid, H., Le Bouc, Y., Bickmore, W., Maher, E.R. (1995) Imprinting mutations in the Beckwith-Wiedemann syndrome suggested my altered imprinting patterns in the IGF2-H19 domain, *Hum Mol. Genet.* **4**, 2379–2385.

Reik, W., Collick, A., Norris, M.L., Barton, S.C., Surani, M.A. (1987) Genomic imprinting determines methylation of parental alleles in transgenic mice, *Nature* **328**, 248–251.

Ripoche, M.A., Kress, C., Poirier, F., Dandolo, L. (1997) Deletion of the *H19* transcription unit reveals the existence of a putative imprinting control element, *Genes Dev.* **11**, 1596–1604.

Sasaki, H., Jones, P.A., Chaillet, J.R., Ferguson-Smith, A.C., Barton, S.C., Reik, W., Surani, M.A. (1992) Parental imprinting: potentially active chromatin of the repressed maternal allele of the mouse insulin-like growth factor II (Igf2) gene, *Genes Dev.* **6**, 1843–1856.

Schoenherr, C.J., Levorse, J.M., Tilghman, S.M. (2002). CTCF maintains differential methylation at the *Igf2/H19* locus, *Nat. Genet.* **33**, 66–69.

Simon, I., Tenzen, T., Reubinoff, B.E., Hillman, D., McCarrey, J.R., Cedar, H. (1999) Asynchronous replication of imprinted genes is established in the gametes and maintained during development, *Nature* **401**, 929–932.

Sleutels, F., Zwart, R., Barlow, D.P. (2002) The non-coding Air RNA is required for silencing autosomal imprinted genes, *Nature* **415**, 810–813.

Smith, R.J., Dean, W., Konfortova, G., Kelsey, G. (2003) Identification of novel imprinted genes in a genome-wide screen for maternal methylation, *Genome Res.* **13**, 558–569.

Stöger, R., Kubicka, P., Liu, C.G., Kafri, T., Razin, A., Cedar, A., Barlow, D. (1993) Maternal-specific methylation of the imprinted mouse *Igf2r* locus identifies the expressed locus as carrying the imprinting signal, *Cell* **73**, 61–71.

Strain, L., Warner, J.P., Johnston, T., Bonthron, D.T. (1995) A human parthenogenetic chimaera, *Nat. Genet.* **11**, 111–113.

Surani, M.A.H., Barton, S.C., Norris, M.L. (1984) Development of reconstituted mouse eggs suggests imprinting of the genome during gametogenesis, *Nature* **308**, 548–550.

Szabo, P., Tang, S.H., Rentsendorj, A., Pfeifer, G.P., Mann, J.R. (2000) Maternal-specific footprints at putative CTCF sites in the *H19* imprinting-control region give evidence for insulator function, *Curr. Biol.* **10**, 607–610.

Swart, R., Sleutels, F., Wutz, A., Schinkel, A.H., Barlow, D.P. (2001) Bidirectional action of the *Igf2r* imprint control element on upstream and downstream imprinted genes, *Genes Dev.* **15**, 2361–2366.

Takagi, N., Sasaki, M. (1975) Preferential inactivation of the paternally derived X chromosome in the extraembryonic membranes of the mouse, *Nature* **256**, 640–642.

Tucker K.L., Beard C., Bausmann J., Jackson-Grusby L., Laird, P.W., Lei, H., Li, E., Jaenisch, R. (1996) Germ-line passage is required for establishment of methylation and expression patterns of imprinted but not nonimprinted genes, *Genes Dev.* **10**, 1008–1020.

Varrault, A., Bilanges, B., Mackay, D.J., Basyuk, E., Ahr, B., Fernandez, C., Robinson, D.O., Bockaert, J., Journot, L. (2001) Characterization of the methylation-sensitive promoter of the imprinted ZAC gene supports its role in transient neonatal diabetes mellitus, *J. Biol. Chem.* **276**, 18653–18656.

Webber, A.L., Ingham, R.S., Levorse, J.M., Tilghman, S.M. (1998) Location of enhancers is essential for the imprinting of H19 and Igf2 genes, *Nature* **391**, 711–715.

Weksberg, R., Shen, D.R., Fei, Y.L., Song, Q.L., Squire, J. (1993) Disruption of insulin-like growth factor 2 imprinting in Beckwith-Wiedemann syndrome, *Nat. Genet.* **5**, 143–150.

Xin, Z., Allis, C.D., Wagstaff, J. (2001) Parent-specific complementary patterns of histone H3 lysine 9 and H3 lysine 4 methylation at the Prader-Willi syndrome imprinting center, *Am. J. Hum. Genet.* **69**, 1389–1394.

Yokomine, T., Kuroiwa, A., Tanaka, K., Tsudzuki, M., Matsuda, Y., Sasaki, H. (2001) Sequence polymorphisms, allelic expression status and chromosome localisation of the chicken IGF2 and *MPR1* genes, *Cytogenet. Cell Genet.* **93**, 109–113.

Yoon, B.J., Herman, H., Sikora, A., Smith, L.T., Plass, C., Soloway, P.D. (2002) Regulation of DNA methylation of *Rasgrf1*, *Nat. Genet.* **30**, 92–96.

Young, L.E., Fernandes, K., McEvoy, T.G., Butterwith, S.C., Gutierrez, C.G., Carolan, C., Broadbent, P.J., Robinson, J.J., Wilmut, I., Sinclair, K.D. (2001) Epigenetic change in *IGF2R* is associated with fetal overgrowth after sheep embryo culture, *Nat. Genet.* **27**, 153–154.

7
Heterochromatin and Eurochromatin – Organization, Packaging, and Gene Regulation

Boris A. Leibovitch and Sarah C. R. Elgin
Washington University, St. Louis, MO, USA

1	Introduction	211
2	Cytological Observations	212
3	**Biochemical Characteristics of Heterochromatin**	**215**
3.1	DNA in Heterochromatin	215
3.2	Modification of DNA in Heterochromatin	217
3.3	Histone Modifications and Chromatin Structure	218
3.4	Specific Heterochromatin Proteins	219
4	**Heterochromatin Behavior**	**221**
4.1	Self-association	221
4.2	Late Replication	222
4.3	Recombination	222
5	**Conclusions: Euchromatin/Heterochromatin Relationships**	**222**
	Bibliography	**224**
	Books and Reviews	224
	Primary Literature	224

Keywords

DNA Modification
Enzymatic addition of chemical groups to DNA occurring after DNA synthesis; in eukaryotes, most commonly methylation of cytosine.

Genomics and Genetics. Edited by Robert A. Meyers.
Copyright © 2007 Wiley-VCH Verlag GmbH & Co. KGaA, Weinheim
ISBN: 978-3-527-31609-0

Epigenetics
A mode of gene regulation, based not on the DNA sequence itself but on stable packaging, heritable during cell division.

Heterochromatin
The silenced domains of chromatin, characterized as maintaining a condensed form, defined by a set of distinctive biochemical properties and cytological behaviors.

Heterochromatin Protein 1 (HP1)
A conserved protein predominantly associated with pericentric heterochromatin, capable of interacting both with a specifically modified histone and with the enzyme that generates the modification.

Histones
A family of low molecular weight proteins, very basic and evolutionarily conserved, which form the core structure for nucleosome assembly in chromatin formation.

Nucleosomes
The basic repeating subunit of chromatin consisting of a tetramer of histones H3 and H4 plus two dimers of histones H2A and H2B, with ca. 170 bp of DNA wrapped around in two left-handed turns; each particle connects to the next by a short stretch of linker DNA, which interacts with histone H1.

Posttranslational Modification of Proteins
An enzymatic addition or removal of various chemical groups to proteins (including histones) after synthesis, including acetylation, methylation, phosphorylation, and so on.

Repetitive DNA
The DNA sequences that are represented by multiple copies in the genome; the copies may be in tandem arrays or scattered as individual copies.

Transposable Elements
Various families of moderately repetitive DNA sequences originally capable of transposition; their distribution and numbers vary between different individuals and strains.

> The genomes of eukaryotes are packaged in chromatin, a complex of DNA with histones in a nucleosomal array, and folded further through interactions with other proteins into higher order domains. The different levels of packaging help define whether genes are available for expression, as in euchromatin, or generally silenced, as in heterochromatin. The condensed heterochromatin domains are characterized

by a group of features, including low gene density, high levels of repetitious sequences, location at the nuclear periphery, late replication, and little recombination during meiosis, apparently reflecting a particular pattern of posttranslational modifications of the histones, association of heterochromatin-specific proteins, and regular spacing of nucleosomes.

1
Introduction

In 1928, Heitz observed that the staining of liverwort *Pellia epiphylla* interphase nuclei is not uniform, suggesting the presence of two fractions of chromatin as manifested by weakly stained areas and areas that maintain the intense staining of metaphase chromosomes. Weakly stained areas were designated as *euchromatin*, while the dark positively heteropycnotic areas were labeled *heterochromatin*. The fate of these regions was traced from interphase to metaphase and back to interphase; it was observed that these regions remained differentially stained in almost all phases of the cell cycle, showing a common level of condensation only in metaphase. The most prominent heterochromatic regions are located mainly in or close to the centromeres and telomeres of chromosomes. The development of better microscopes and application of various genetic, biochemical, and molecular approaches, coupled with the examination of different species in similar studies, have provided a much deeper understanding of the underlying mechanisms generating this different packaging. Table 1 describes the major differences between these two chromatin fractions, giving both cytological and biochemical characteristics. The reader should bear in mind that the differences described reflect the sum of many observations made in a wide range of different organisms, cell types, and so on. Heterochromatic regions may manifest all or only some of these features. It seems likely that a similar condensed chromatin organization can be achieved in different ways, using redundant molecular mechanisms. Recent data also indicate that some of the features characteristic of heterochromatin may be utilized to downregulate the expression of genes in euchromatic domains. This implies the existence in the nucleus of a continuum of chromatin states, potentially generated by various combinations of the features listed in Table 1.

Some of the properties of heterochromatin are reversible; several must be transiently altered or re-created following replication, indicating the presence of various mechanisms to accomplish this with fidelity. This, in turn, implies that the formation of heterochromatin (or heterochromatin-like structures) may be used as a mechanism of epigenetic inheritance, maintaining particular states (silent or activatable) of gene expression. The heterochromatin located around centromeres, telomeres, and some other regions maintained in that form in all cell types is usually referred as *constitutive heterochromatin*. Regions that show a cell-type-specific (or otherwise selected) heritable silencing of particular genes, chromosome domains or whole chromosomes are referred to as *facultative heterochromatin*. Many questions remain as to how constitutive and

Tab. 1 Distinctions between euchromatic and constitutive heterochromatic regions of complex genomes.

Feature	Euchromatin	Heterochromatin
Proportion of genome	50–80%	20–50%
Interphase appearance	Decondensed (lightly stained)	Condensed (densely stained – heteropycnotic)
Chromosomal location	Distal arms	Usually pericentromeric, telomeric
Nuclear location	General	Often clumped (nuclear periphery, around nucleolus)
Sequence complexity	High; unique DNA, interspersed middle repetitious sequences	Low; repetitious DNA: satellite DNA, blocks of middle repetitious sequences
Gene density	High	Low
DNA methylation	CpG islands hypomethylated	Fully methylated (in mammals and plants)
Histone modification;	High histone acetylation; specific amino acids are methylated	Low histone acetylation; specific amino acids are methylated
Nucleosome spacing	Irregular	Regular
Nuclease accessibility	Variable	Low
Mechanism of gene silencing	Individual genes	Large domains
Meiotic recombination	Normal	Absent or significantly decreased
Replication timing	Throughout S phase	Late S phase

facultative domains are targeted for packaging in a heterochromatic form, a process sometimes referred to as *heterochromatinization*. This article will summarize our current knowledge concerning the differentiation of chromatin into these two fractions, and the possible role of heterochromatin in genome organization and regulation of gene expression.

2
Cytological Observations

Observations using light microscopy showed that the dark heteropycnotic chromatin visible in interphase nuclei corresponds to the strongly stained domains in metaphase chromosomes surrounding centromeres and telomeres. Small, intensely stained islands within the chromosome arms, referred to as *interstitial* or *intercalary heterochromatin*, have been observed in chromosomes of some organisms. In interphase, the strongly stained domains lie at the periphery of the nucleus, close to the nuclear membrane. These observations were confirmed subsequently by immunostaining with antibodies to proteins specifically enriched in heterochromatin (see Sects. 3.3 and 3.4), and by *in situ* hybridization with DNA probes complementary to the repetitive DNA fractions that are concentrated around centromeres. Analyses using electron microscopy (EM) to examine interphase nuclei have also demonstrated the presence of two chromatin fractions, with the electronically dense, tightly packaged fraction located mainly at the nuclear periphery and around the nucleolus.

The patterns seen on EM photographs, by light microscopy after staining with dyes or immunostaining, or with *in situ* hybridization of repetitious sequences, all indicate that heterochromatin in the interphase nucleus is organized in several large blocks, which tend to localize at the nuclear membrane. This suggests that the heterochromatic regions of different chromosomes in the interphase nucleus interact with each other, as the number of spots observed is smaller than the total number of centromeres and other heterochromatic regions. Careful analysis of the spatial organization of chromosomes within the nucleus of some organisms has shown that the distribution of heterochromatin is not random. The centromeres and surrounding pericentromeric heterochromatin tend to concentrate on one side of the nucleus, while telomeric regions of the same chromosomes are clustered on the opposite side. This phenomenon is often referred to as *Rabl orientation*, honoring Carl Rabl, who described it first in 1885. The positioning of the heterochromatic telomeres and centromeres implies that the euchromatic regions in between also assume defined positions within the nuclei. Direct observations have confirmed the existence of the so-called chromosomal territories with limited fixed positions for different chromosomal segments in specific areas of the nucleus.

While constitutive heterochromatin (e.g. pericentromeric and telomeric regions) is present in an invariant pattern at given sites and in given chromosomes, facultative heterochromatin is represented by chromosomal regions or whole chromosomes that are capable of existing in one state in some cells and in the other state in other cells. In some organisms, such as the mealy bug *Planococcus citri*, the full haploid set of paternal chromosomes is found in the heterochromatic state. Mammals demonstrate the well-known example of X-chromosome inactivation in female nuclei. The inactivated X-chromosome forms the so-called Barr body, usually identified as a condensed mass at the periphery of female nuclei. The number of Barr bodies correlates well with the number of inactivated X-chromosomes in cases in which the individual carries more than two X-chromosomes. Identification of the inactive X-chromosome as the condensed Barr body led to the suggestion that the formation of heterochromatin might be a mechanism for downregulating gene expression. In the normal situation, this inactivation serves to compensate for the differences in dosage of X-chromosomal genes between females, with two X-chromosomes, and males, with single X-chromosome. X-chromosome inactivation serves to diminish the negative effect of additional X-chromosomes, and such aneuploids survive, while most cases of human aneuploidy (extra copies of a chromosome) are lethal. While the decision of which X chromosome to inactivate appears to be random in somatic cells, the decision once made is stably inherited through multiple rounds of mitosis, a critical observation indicating that changes in chromatin structure play a role in epigenetic inheritance. In these cases, gene activity is determined not by the DNA sequence *per se*, but by the particular packaging of the gene in chromatin.

A second phenomenon demonstrating the link between gene inactivation and the location and packaging of a gene in a heterochromatic form is position effect variegation (PEV), described in Drosophila, yeast, and mammals. In these cases, a gene normally located in a euchromatic environment is positioned adjacent to or within heterochromatin by chromosomal rearrangement or transposition (Fig. 1).

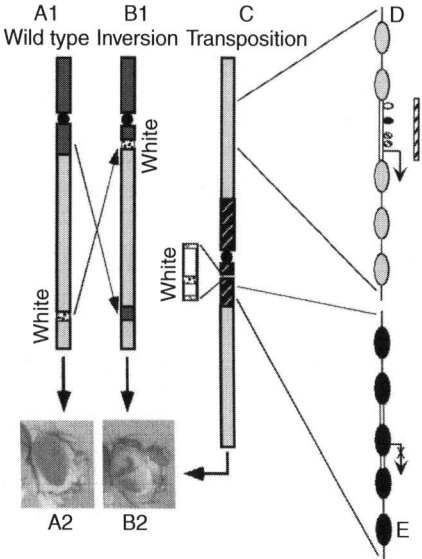

Fig. 1 Position effect variegation in Drosophila. (A1, A2) The wild type *white* gene (spotted rectangle) is required to generate red fly eyes; it is localized in a euchromatic segment (light grey) of the X chromosome. (B1, B2) Inversions of the chromosome segment with one breakpoint in pericentromeric heterochromatin (black) and one breakpoint adjacent to *white* position the gene adjacent to heterochromatin, leading to inactivation of the gene in some eye progenitor cells, resulting in a variegating phenotype. (C, B2) The same result is observed when *white* is transposed into heterochromatin as a part of a transposon. (D) Genes in euchromatic domains are generally packaged in an irregular pattern of nucleosomes (grey ovals); one observes accessible (open, nucleosome free) sites at or near the 5′ end of genes, allowing interaction with transcription factors (small circles) that can regulate initiation of transcription from the start site (bent arrow). Such open sites are often called DH sites (DNase I hypersensitive sites) or HS sites (hypersensitive sites) because they are readily cleaved by nucleases. (E) In heterochromatic domains, the nucleosomes (black ovals) have a characteristic, different pattern of histone modification, and interact with a subset of chromosomal proteins, some specific to these domains. Within heterochromatic regions, the nucleosomes show very regular spacing, and appear to mask promoters and other regulatory sites, inhibiting transcription (crossed bent arrow).

In both cases, one observes that the gene is stochastically inactivated in some cells, but remains active in others, resulting in a variegating phenotype; the active or inactive status, once set, is mitotically inherited in the progeny of the cells, while the phenotype, at least in yeast, is stably maintained over many generations, that is, through meiotic divisions. The variegating gene in such situations is now found located close to the nuclear periphery, with heterochromatin, at a higher frequency. Inactivation of some genes during normal development may occur concomitantly with their spatial relocation to heterochromatin domains. Genetic analysis has shown that a screen can identify mutations in other loci that suppress or enhance the pattern of variegation, observed as a loss in reporter gene silencing or an increase in reporter gene silencing. Such second-site modifiers have frequently been found to be genes that encode chromosomal proteins, or proteins required for chromatin assembly and function. The study of such suppressors and enhancers of variegation [Su(var)s and E(var)s] has greatly facilitated studies of the biochemical properties of heterochromatin and the resultant epigenetic marks.

3
Biochemical Characteristics of Heterochromatin

3.1
DNA in Heterochromatin

Kinetic analysis of the renaturation of genomic DNA has shown that the genomes of higher eukaryotes are made up not only of unique sequences (present in one copy per haploid genome), which renature slowly, but also of repetitious sequences (present in multiple copies per haploid genome), which renature more quickly. Constitutive heterochromatin is enriched in repetitive DNA and has relatively little unique DNA. Most of the genes encoding proteins are present in a single copy. Genetic analysis has identified only a few loci that map within the pericentric heterochromatin. Sequencing of several eukaryotic genomes has confirmed the conclusion that heterochromatin is gene-poor, and provided more precise knowledge about the distribution and features of DNA in both chromatin fractions. Interestingly, analysis of the DNA sequences from Drosophila has predicted the presence of at least ∼100 genes embedded in the pericentric heterochromatin; thus, while it is gene-poor, the region is not devoid of genes.

Repetitive DNA may be roughly divided into two classes based on the copy number – middle repetitive and highly repetitive DNA. Genes encoding the large ribosomal RNAs, 5S RNA, and histones fall into the middle repetitive class in most eukaryotic organisms. However, the majority of the middle repetitive DNA consists of intact and/or damaged transposable elements. In some organisms, including humans, up to ca. 50% of the DNA is derived from transposable elements, representing different families, including pro-retroviruses. These repeats can be scattered as single copies, found at apparently random sites within genomic DNA, or may be grouped in tandem or inverse arrays. Sometimes different transposable elements are found nested together, one inside of another. Just as unique DNA sequences are characteristic of euchromatin, a high frequency of clusters and/or nested copies of moderate repeats/transposable elements is also a feature of heterochromatin. A significant fraction of the transposable elements, particularly those copies in heterochromatin, carry various mutations that render them incapable of transposing.

Several investigations have suggested that the presence of repetitious elements may serve as a signal for targeting formation of heterochromatin. For example, the *Drosophila melanogaster* gene *white* is normally found at a euchromatic site and is fully expressed, giving flies a red eye; however, a tandem array of *white* genes, generated from a transgene, results in inactivation, regardless of the site of insertion within the chromosomes, leading to a variegated eye phenotype. This variegation, similar to PEV, is the result of stochastic inactivation of the cluster of *white* genes in some progenitor cells of the eye in which this gene is normally active. However, the presence of repetitive DNA is not sufficient to trigger the formation of heterochromatin in all cases. For example, the histone genes, repetitive in most eukaryotes, are not packaged in cytologically visible heterochromatin, nor do they manifest other biochemical characteristics of heterochromatin. However, the repetitious rRNA genes, found in the nucleolus, can manifest heterochromatic properties, including the silencing of marker genes inserted into this domain.

The second class of repeats, present almost exclusively in heterochromatin, is highly repetitive DNA, also known as *satellite* DNA. Satellite DNA consists of short sequences (from ca. 10 bp to a few hundred bp) repeated in tandem arrays up to a million times. Some satellite DNA fractions are chromosome-specific, while others occur on all chromosomes, though in varying amounts. In some species, including the kangaroo rat *Dipodomys ordii* and some beetles from the coleopteran family *Tenebrionidae*, satellite DNA accounts for the majority of the genomic DNA. If the base composition of a given repeat is GC- or AT-rich, that DNA may be separated from the bulk of the DNA in a cesium chloride (CsCl) density gradient, giving a separate "satellite" peak, heavier or lighter than the bulk DNA, respectively. Single copies or clusters of transposable elements often interrupt such arrays. Cloning and sequencing of the DNA in the boundary regions between heterochromatin and euchromatin (often called transition zones) has shown that this part of the genome consists of a considerable mix of both types of repeats, with an intermediate density of genes.

The pericentromeric location of the most highly repeated sequences raises the question of their possible participation in the formation of the centromere (the constriction that serves as a platform for assembly of kinetochore proteins) and the kinetochore (the complex of proteins to which the nuclear spindle fibers (microtubules) attach to accomplish chromosome segregation). The matter is controversial, with several competing points of view. Some evidence has suggested that satellite DNA is both necessary and sufficient to form a centromere, while other experiments indicate that there is no requirement for a particular DNA sequence. In the latter view, the centromere is a specialized chromatin structure maintained in an epigenetic pattern. This structure may arise in different ways, for example, with the help of specific proteins such as the centromere-specific histone CENP-A, a histone H3 variant.

In addition to pericentromeric heterochromatin, differentially stained regions that have many characteristics of heterochromatin are found near or at telomeres. In most organisms, the telomere consists of $(TG)_n$ repeats that interact with a group of specialized proteins, creating a specific chromatin structure. The system enables telomeres to support DNA replication without any loss of DNA due to unidirectional copying, to prevent the fusion of chromosome ends, and to direct the chromosome ends to sites close to the nuclear membrane. Localization is thought to involve interactions between telomere proteins and the nuclear lamina, a structure just below the inner nuclear membrane. Drosophila differs in having clusters of particular transposable elements at the ends of the chromosomes, flanked downstream by different satellite-like repeats – telomere-associated sequence (TAS). It is unclear which feature – DNA sequence and/or proteins – makes telomeric regions in all organisms heterochromatic. Studies of genes affecting PEV of visible markers transposed into or close to telomeres have shown that while many of the same genes affect silencing in telomeres and mating-type loci in *Saccharomyces cerevisiae*, different groups of genes influence telomeric and pericentric PEV in Drosophila. This implies a different chromatin structure in telomeric and pericentromeric heterochromatin in this organism.

3.2
Modification of DNA in Heterochromatin

In mammals and plants (but not in Drosophila or yeast), there is a strong correlation between heterochromatic regions of chromosomes and the degree of DNA methylation. Historically, the role of DNA methylation in gene silencing and formation of heterochromatin was first demonstrated by the observation that DNA of the inactive X-chromosome is hypermethylated. Methylation in mammals usually occurs behind the replication fork at position C5 of cytosine bases in CpG dinucleotides. C-residues in other sequences have been found to be modified, but at a lower frequency. This modification does not occur exclusively within heterochromatic domains. Many euchromatic genes have CpG-rich elements (so-called "CpG islands") in their 5′ upstream region, which are important for the regulation of the gene through interaction with the transcriptional regulatory protein Sp1. Methylation of these islands is usually an indicator of gene silencing; presumably, methylation blocks binding by Sp1 to its target sequence. Removal of 5′mC (for example by replicating DNA in the presence of 5-azacytidine, a form of C that cannot be methylated) can reactivate these genes. The same drug can activate silenced genes on the inactive X-chromosome in female mammalian nuclei.

DNA methylation is a reversible process. Only the parental strand of any newly replicated DNA double helix will contain the methylated cytosine. In the usual situation, this remaining 5′mC will interact with 5′mC-binding proteins that direct site-specific enzymes, the maintenance DNA methyltransferases, to methylate the cytosine of the CpG on the opposite DNA strand. If this interaction is blocked, or if this interaction and subsequent modification fails for any reason, the DNA of half of the granddaughter cells will lose 5′mC at this site. However, a second group of DNA methyltransferases has been identified that are capable of *de novo* methylation of cytosines on the nonmodified DNA template. These DNA methylases restore methylation of some genes and in heterochromatin after the massive demethylation of DNA that occurs as part of normal early development in mammals. We do not know, however, how these enzymes are targeted to the sequences that need to be so modified. Some sequences that one might expect to be silenced are not demethylated during this general event, among them proviral sequences of IAP particles (which resemble transposable elements in the mouse genome) and imprinted genes. Imprinted genes are expressed specifically from the chromosome inherited from the mother or the chromosome inherited from the father, with the allele from the other parent being silenced. Not surprisingly, these genes in mammals exhibit differential DNA methylation; the silenced genes show additional features of facultative heterochromatin.

3.3 Histone Modifications and Chromatin Structure

The different cytological appearance of heterochromatin has led to the suggestion that the molecular organization and DNA packaging in the underlying chromatin must be different. Data from the last few years has shed significant light on this subject. Chromatin is a chain of repeating subunits (nucleosomes) of ca. 200 bp, with approximately 146 bp of DNA folded around a histone octamer (core) consisting of a tetramer of $(H3 + H4)_2$ and two dimers of $(H2A + H2B)$. These core particles are joined together by short (20–60 bp) linkers, which are generally associated with histone H1, other linker histones such as H5, or HMG proteins. Nucleosome location can generally be regarded as sequence-independent, although DNA characteristics such as bendability can influence nucleosome positioning. Histones interact with each other to form the octamer and bind DNA primarily through their histone-fold domains, leaving their N-terminal tails available for other interactions. Presumably these interactions stabilize the higher-order folding of the basic chain into 30 nm fibers and higher order packaging arrays. Posttranslational modification of the histones plays a major role in determining their interactions with other chromatin components (DNA, other histones, and non-histone proteins), and hence in stabilizing or destabilizing higher order structures.

The most common *in vivo* modifications are acetylation, methylation, phosphorylation, ubiquitination, or polyADP-ribosylation. Most of the sites of modification are clustered in the N-terminal regions, changing the total charge and hydrophobicity of a given histone tail. In addition, several sites of modifications have been described in the central histone-fold motifs. Histone modifications are catalyzed by enzymes that introduce or remove certain groups at one or more positions. The enzymes are often present as part of large multimeric chromatin-remodeling complexes, targeted by associated gene-specific activators and repressors. The very complex pattern of histone modifications has led to the idea of a histone "code," defining different packaging states. In particular, modification of histone H3 by methylation at lysine 9 (giving H3-mK9) has been associated with packaging into heterochromatin, while acetylation in general, and acetylation of H4 at lysine 5 and lysine 8 specifically, are associated with packaging in euchromatic form.

A second type of complex, the ATP-dependent chromatin remodeling complex, defines nucleosome formation behind the replication fork and/or destabilizes nucleosomes in an already mature nucleosomal array. All remodeling complexes of this type include a subunit possessing a DNA- and/or nucleosome-inducible ATPase activity. Various versions of these complexes are simultaneously present in a given cell, modified by the temporary addition or removal of different subunits. Such complexes are involved in both activation and repression of many genes. The exact composition of the complexes, their targets for action, and their mechanism(s) for localization at the appropriate targets are all currently under investigation. Complexes of both types, capable of histone modification and of ATP-dependent remodeling, are essential for gene activation, as even within euchromatic domains nucleosome arrays render many enhancer and promoter sites inaccessible to regulatory proteins and/or general transcription factors and RNA polymerase. In addition, nucleosomes suppress RNA elongation during transcription.

Nucleosome arrays in heterochromatin are structurally similar but not identical to those in euchromatin. Digestion of chromatin with micrococcal nuclease reveals a more regular spacing in heterochromatic arrays, with a narrower cleavage site in the linker. Heterochromatic regions, overall, are less accessible to both general and site-specific nucleases, and to most other reagents that interact with DNA. These properties suggest that some characteristic features of nucleosome structure, positioning, and mobility are different in heterochromatin and euchromatin, presumably reflecting differences in internucleosomal interactions as well as differences in the associated chromosomal proteins. These changes may be due to the various histone modifications mentioned above. Certainly the average pattern of histone modifications differs between heterochromatin and euchromatin, creating a distinct signature for each domain using the histone code. However, the histone modification pattern is likely to vary for different genes in a given domain (e.g. for euchromatic genes that are actively being transcribed vs. not being transcribed; for inducible vs. uninducible genes, etc.), and for different types of heterochromatin, creating a continuum of chromatin states. The discovery of the patterns of posttranslational histone modifications, with the realization that enzymes responsible for the modifications had in some cases been previously identified as activators or repressors of gene expression, was one of major recent breakthroughs in the chromatin field.

The differences in heterochromatic nucleosome arrays, specifically the more regular spacing, have suggested that specific chromatin-remodeling complexes might be involved in assembling heterochromatin. *In vitro* assays have demonstrated that two Drosophila chromatin remodeling complexes, CHRAC and ACF, can generate arrays with heterochromatin-like regular spacing. The human WSTF–ISWI and ACF complexes appear to be the mammalian homologs of CHRAC and ACF, based both on their composition and by their ability to space nucleosomes regularly on DNA. These complexes are concentrated in heterochromatic regions in late S-period, the time when the bulk of the heterochromatin is replicated.

3.4
Specific Heterochromatin Proteins

Heterochromatin is concentrated in the pericentric and telomeric regions, and is rich in satellite DNA and other repetitious DNA. Several different proteins that specifically interact with satellite DNA have been found in vertebrates and in Drosophila. Most of these proteins interact as well with homologous sequences scattered in euchromatic regions, and some also function in the regulation or chromatin organization of euchromatic genes. Most of these proteins interact with other proteins to form multimeric complexes. For example, Drosophila GAGA factor, the protein specifically interacting with $(GA)_n/(CT)_n$ short repeats in the promoter regions of many genes, regulating their expression, is found also to interact with $(AAGAG)_n$ and $(AAGAGAG)_n$ satellite DNA fractions, concentrated in the pericentromeric heterochromatin of several chromosomes. This protein has been found in several multimeric complexes, and interacts with the chromatin-remodeling complex NURF. Interestingly, mutations in GAGA factor enhance PEV of genes transferred to heterochromatin, suggesting that its dominant role is in support of gene expression. Whether GAGA factor association has an impact on the chromatin structure of the satellites where it binds is yet to be demonstrated.

The first protein specifically characterized as a heterochromatin protein, Drosophila Heterochromatin Protein 1 (HP1), binds poorly to DNA, with little specificity. HP1 is thought to play a structural role in heterochromatin, since an antipodal dosage dependence is observed: PEV is suppressed by mutations in the gene for HP1 that result in a loss of functional HP1, and is enhanced by an increase in the amount of HP1, for example from a duplication or a transgene. This type of action is also observed for a few other chromosomal proteins, *SU(VAR)3–9* (an H3 methyl-transferase), *SU(VAR)3–7* (a zinc finger protein), and HP2 (a potential structural protein). Homologs of HP1 have been found in the yeast *Schizosaccharomyces pombe*, worms, other insects, amphibians, mammals, and plants. Full inactivation of HP1 through loss or missense mutations, or by the expression of intracellular antibodies, is lethal. From one to three HP1 homologs have been identified in different organisms, but the proteins are not identical, nor can they substitute for each other in a genetic analysis; all are required. Immunocytological analyses indicate that the homologs have different patterns of nuclear distribution, with only one of them restricted to pericentric heterochromatin. The three proteins in mammals can physically interact with each other. In transgenic Drosophila, expression of the genes for the HP1 human homologs also enhances PEV, suggesting a conserved functional interaction with the Drosophila machinery that establishes and maintains heterochromatin, including the originally identified HP1, encoded by *Su(var)205*.

The HP1's are small proteins, ca. 200 to 250 amino acids, with two conserved domains joined by a more variable hinge region. The N-terminal chromo domain was first recognized by homology with a conserved domain of Polycomb, a protein required for developmentally maintained silencing of homeotic genes in the euchromatin. The chromo domain has since been identified in several other chromosomal proteins. HP1 can dimerize through interactions involving the chromo shadow domain, and this structure appears to form a platform for interactions with many other chromosomal proteins, including *SU(VAR)3–9* and HP2.

The binding of HP1 to nucleosomes is facilitated by methylation of lysine 9 in histone H3, a histone modification characteristically enriched in heterochromatin. The modified histone tail fits nicely within the fold generated by the HP1 chromo domain. Missense mutations that lead to a loss of *Su(var)* activity (loss of silencing) are mutations that result in a loss of this stable interaction. This H3 modification is driven by one (or perhaps more) specific histone methyltransferase encoded by homologous genes found in mammals, yeast (*S. pombe*) and Drosophila. The founding member in Drosophila is *SU(VAR) 3–9*. As the name implies, mutations of the gene encoding this protein suppress the PEV in Drosophila. This enzyme in mammals and flies forms complexes with HP1 and often colocalizes with HP1 in heterochromatic regions of chromosomes. The ability of HP1 to interact with both the modified histone (H3-mK9) and the enzyme responsible for the modification suggests a mechanism for epigenetic inheritance of the modification pattern, and for the spreading of a heterochromatic domain once initiated.

HP1 can also interact with various proteins that are associated with repressor activity in euchromatic domains. The best-studied cases are in mammals. Gene repression is generally accompanied by histone deacetylation and, at least in some cases, by methylation of histone 3 at lysine 9. For example, a complex of mammalian HP1, SUV39H1, and the transcriptional repressor Rb has recently been described. Retinoblastoma protein (Rb) plays an important role in the induction of the G1-S transition during the cell cycle; mutations of Rb have been associated with several tumors. It has been suggested that Rb acts as a negative regulator by binding SUV39H1 (the human homolog of SU(VAR)3–9), resulting in local methylation of lysine 9 of histone H3, thereby recruiting HP1 to "fix" the gene in an inactive state. The process in this case is highly specific, involving only one nucleosome located on the promoter element of the cyclin D gene, repressed by Rb *in vivo*. Rb has also been found in complexes with histone deacetylase. It is unclear whether both silencing complexes are essential for complete inactivation, or whether cells and genes may choose between these modes of inactivation, one (deacetylation) being easily reversible, the other (methylation) being stable through generations. Since SUV39H1 cannot methylate acetylated lysine 9 in histone H3, the order of events seems obvious in this case. However, it is unclear whether the process can stop after the first step (deacetylation) without progressing to the next step (H3 methylation and binding of HP1). These data (and more) suggest that heterochromatin proteins play roles beyond the packaging of heterochromatic domains, being involved in regulating the activity of targeted euchromatic genes. It is unclear, however, how these processes are managed to limit the silenced domain to one or a few nucleosomes, whereas heterochromatic domains appear able to spread over large distances (at least 10 kb, observed in *S. pombe*).

4
Heterochromatin Behavior

4.1
Self-association

Many of the proteins enriched in heterochromatin, including HP1, are able to form homomeric and heteromeric complexes. This no doubt contributes to the association of heterochromatic regions in a small number of blocks, easily visible on a cytological level. This property of heterochromatic proteins may be utilized as a mechanism for gene inactivation. The best studied case involves the Ikaros family proteins, recently analyzed in mammalian pre-B cells. Several homologous proteins of this family have two domains, one participating in protein–protein interactions and the other in DNA binding. The Ikaros proteins show sequence-specific binding to DNA and form homo- and heteromeric complexes with each other. Most of the target sequences are multiple short repeats comprising a satellite DNA fraction concentrated in pericentric heterochromatin. However, identical short repeats are present in the regulatory regions of many genes. Ikaros suppresses the activity of the terminal deoxynucleotidyl transferase gene at the appropriate stage of lymphocyte differentiation by competing with a gene-specific activator for a partly overlapping binding site. Reduced chromatin accessibility for restriction enzymes and DNA methylation are observed at the promoter of the repressed gene. Importantly, the inactivated gene is placed within or next to a heterochromatin block, most likely due to the formation of multimeric complexes between the Ikaros protein bound at the promoter and that bound in the pericentric heterochromatin. Whether or not silencing is dependent on this relocalization remains to be seen. Similar observations of relocalization have been made for some other genes in this cell lineage.

4.2
Late Replication

Early observations on incorporation of labeled precursors during DNA synthesis showed that the diffuse weakly staining euchromatin was replicated primarily during the first half of S-phase, while the heteropycnotic heterochromatic regions showed incorporation of label primarily during the second half of S-phase. These observations were confirmed by direct analysis of replication times for different DNA sequences: those enriched in constitutive heterochromatin were replicated at the end of S-phase. The same observations have held true for chromosomes and for separate chromosomal domains that showed the characteristics of facultative heterochromatin. For example, the inactive X-chromosome replicates late in S-phase, while the active X replicates early. It is still unclear what drives this replication schedule. There are many possibilities, ranging from the idea that the densely packaged heterochromatin may be less accessible to the replication machinery, to the suggestion that replication of heterochromatin requires proper amounts of specific chromatin-remodeling complexes such as WSTF–ISWI and ACF.

4.3
Recombination

Careful genetic mapping in several eukaryotes has shown that the frequency of recombination is usually significantly decreased toward centromeres and telomeres. Crossing-over is very low near heterochromatin domains in general. Similar observations have been made for regions located within or close to the silenced mating-type loci in yeast (*S. cerevisiae*), domains that appear heterochromatic by the criteria discussed above (silencing of transposed reporter genes, low accessibility to nuclease digestion, lack of histone acetylation, etc.). We do not know why heterochromatic regions are less susceptible to recombination. The ability of heterochromatic regions to "stick together" should facilitate the exchange of genetic material. In fact, spontaneous or induced chromosomal rearrangements with breakpoints in heterochromatin are seen more frequently then those with breakpoints in euchromatin. Of course, this may be explained simply as a consequence of the more frequent survival of the resulting progeny, due to the low gene density in regions of constitutive heterochromatin. A plausible explanation might lie in the inaccessibility of densely packed chromatin in this region to the recombinational machinery.

5
Conclusions: Euchromatin/Heterochromatin Relationships

The above discussion has provided many examples that point to the differences between heterochromatin and euchromatin, while at the same time emphasizing that no one single feature provides a unique identifier of heterochromatin. A constellation of several characteristics is required to define a region as heterochromatic, and this combination may be different in different heterochromatic domains. For example, HP1 is commonly associated with pericentric heterochromatin, but is not associated with the inactive X-chromosome in mammals. Furthermore, different organisms may employ different strategies to generate a heterochromatic domain – one that has a regular chromatin structure, is inaccessible to a variety of probes, and is lacking in meiotic recombination. For example, while many plants and animals (notably mammals) utilize DNA methylation as a heritable mark of heterochromatin formation and gene silencing, Drosophila, yeast, and many other organisms do not appear to exhibit significant levels of DNA methylation. While organisms from the yeast *S. pombe* to Drosophila and humans utilize a strategy of H3-K9 methylation and HP1 association to form heterochromatin, the yeast *S. cerevisiae* does not. It is clear that higher eukaryotes have a number of histone methyltransferases that can modify H3-K9, and that these play both overlapping and distinct roles. Histone H3 and H4 hypoacetylation does seem to be a common strategy, but one used in different ways in different organisms. There has been some speculation that the range of biochemical mechanisms used by an organism to maintain heterochromatin structure, and concomitant silencing, may reflect the size of its genome. This may be correlated with the load of retroviruses and transposable elements – sequences that need to be silenced – commonly carried in the genomes of organisms with higher C values.

While heterochromatin is gene-poor, it is not devoid of genes. Models of heterochromatin structure must be able to accommodate the observation that the activity of these genes depends on the presence of normal heterochromatin structure; inactivation of HP1 leads to a loss of function of these genes. At the same time that we have noted specific biochemical features associated with heterochromatin formation, we have found that the same features can be found within euchromatic domains. For example, HP1 and H3-mK9 have been found not only in pericentric heterochromatin, but are also utilized in repressive regulation of some normal eukaryotic genes. Thus, one definition for *heterochromatic genes* may be those that are dependent on HP1 for optimal expression, whereas *euchromatic genes* might be defined as those exhibiting a loss of expression on association with HP1. As more genes are discovered that map to heterochromatic domains, it will be of interest to see if this pattern holds for other biochemical marks as well.

A critical question now under intense study is that of targeting heterochromatin formation: how is the pattern of differential chromatin packaging established? Recent data indicate that small interfering RNAs may play a role here, in addition to the regulation achieved by the RNA interference system by targeting transcription products for degradation (posttranscriptional silencing). A similar small RNA fraction could interact either with DNA in heterochromatin, for example, repeated sequences or transposable elements, or with proteins enriched in heterochromatin. Some chromodomain-containing proteins, notably MOF, can interact with RNA. Most compelling, mutations in several known genes participating in the formation of such small RNAs affect heterochromatin silencing and centromeric function in the yeast *S. pombe*. One prerequisite for the formation of such small RNAs is the formation of long double stranded RNA, for example by bidirectional transcription. Interestingly, many transposable elements (commonly overrepresented in heterochromatin domains) are capable of bidirectional transcription. Thus heterochromatin formation may have derived from the need to silence invading transposons and retroviruses, to be adapted later in evolution to help create structures such as the centromere, and to help regulate gene expression throughout the genome. Experiments during the next several years should clarify many of the issues raised above.

See also Gene Mapping and Chromosome Evolution by Fluorescence–Activated Chromosome Sorting; Gene Mapping by Fluorescence In Situ Hybridization; Molecular Genetics of Hemophilia.

Bibliography

Books and Reviews

Dernburg, A.F., Karpen, G.H. (2002) A chromosome RNAissance, *Cell* **111**, 159–162.

Eissenberg, J.C., Elgin, S.C.R. (2000) The HP1 protein family: getting a grip on chromatin, *Curr. Opin. Genet. Dev.* **10**, 204–210.

Georgopoulos, K. (2002) Haematopoietic cell-fate decisions, chromatin regulation and Ikaros, *Nat. Rev. Immunol.* **2**, 162–174.

Grewal, S.I., Elgin, S.C.R. (2002) Heterochromatin: new possibilities for the inheritance of structure, *Curr. Opin. Genet. Dev.* **12**, 178–187.

Huang, Y. (2002) Transcriptional silencing in *Saccharomyces cerevisiae* and *Schizosaccharomyces pombe*, *Nucleic Acids Res.* **30**, 1465–1482.

Jones, P.A., Takai, D. (2001) The role of DNA methylation in mammalian epigenetics, *Science* **293**, 1068–1070.

Lamb, J.C., Birchler, J.A. (2003) The role of DNA sequence in centromere formation, *Genome Biol.* **4**, 214.1–214.4.

Lyon, M.F. (2002) X-chromosome inactivation and human genetic disease, *Acta Paediatr. Suppl.* **91**, 107–112.

Neely, K.E., Workman, J.L. (2002) The complexity of chromatin remodeling and its links to cancer, *Biochim. Biophys. Acta.* **1603**, 19–29.

Parada, L., Misteli, T. (2002) Chromosome positioning in the interphase nucleus, *Trends Cell Biol.* **12**, 425–432.

Richards, E.J., Elgin, S.C.R. (2002) Epigenetic codes for heterochromatin formation and silencing: rounding up the usual suspects, *Cell* **108**, 489–500.

Turner, B.M. (2002) Cellular memory and the histone code, *Cell* **111**, 285–291.

Primary Literature

Akhtar, A., Zink, D., Becker, P.B. (2000) Chromodomains are protein-RNA interaction modules, *Nature* **407**, 405–409.

Bannister, A.J., Zegerman, P., Partridge, J.F., Miska, E.A., Thomas, J.O., Allshire, R.C., Kouzarides, T. (2001) Selective recognition of methylated lysine 9 on histone H3 by the HP1 chromo domain, *Nature* **410**, 120–124.

Bongiorni, S., Mazzuoli, M., Masci, S., Prantera, G. (2001) Facultative heterochromatization in parahaploid male mealybugs: involvement of a heterochromatin-associated protein, *Development* **128**, 3809–3817.

Bozhenok, L., Wade, P.A., Varga-Weisz, P. (2002) WSTF-ISWI chromatin remodeling complex targets heterochromatic replication foci, *EMBO J.* **21**, 2231–2241.

Bruvo, B., Pons, J., Ugarkovic, D., Juan, C., Petitpierre, E., Plohl, M. (2003) Evolution of low-copy number and major satellite DNA sequences coexisting in two Pimelia species-groups (Coleoptera), *Gene* **312**, 85–94.

Cheutin, T., McNairn, A.J., Jenuwein, T., Gilbert, D.M., Singh, P.B., Misteli, T. (2003) Maintenance of stable heterochromatin domains by dynamic HP1 binding, *Science* **299**, 721–725.

Collins, N., Poot, R.A., Kukimoto, I., Garcìa-Jimènez, C., Dellaire, G., Varga-Weisz, P.D. (2002) An ACF1/ISWI chromatin-remodeling complex is required for DNA replication through heterochromatin, *Nat. Genet.* **32**, 627–632.

Cryderman, D.E., Tang, H., Bell, C., Gilmour, D.S., Wallrath, L.L. (1999) Heterochromatic silencing of Drosophila heat shock genes acts at the level of promoter potentiation, *Nucleic Acids Res.* **27**, 3364–3370.

Czermin, B., Schotta, G., Hulsmann, B.B., Brehm, A., Becker, P.B., Reuter, G., Imhof, A. (2001) Physical and functional association of SU(VAR)3–9 and HDAC1 in Drosophila, *EMBO Rep.* **2**, 915–919.

Farkas, G., Gausz, J., Galloni, M., Reuter, G., Gyurkovics, H., Karch, F. (1994) The *Trithorax-like* gene encodes the Drosophila GAGA factor, *Nature* **371**, 806–808.

Filesi, I., Cardinale, A., van der Sar, S., Cowell, I.G., Singh, P.B., Biocca, S. (2002) Loss of heterochromatin protein 1 (HP1) chromodomain function in mammalian cells by intracellular antibodies causes cell death, *J. Cell Sci.* **115**, 1803–1813.

Hatch, F.T., Mazrimas, J.A. (1974) Fractionation and characterization of satellite DNAs of the kangaroo rat (*Dipodomys ordii*), *Nucleic Acids Res.* **1**, 559–575.

Hediger, F., Neumann, F.R., Van Houwe, G., Dubrana, K., Gasser, S.M. (2002) Live imaging of telomeres. yKu and Sir proteins define redundant telomere-anchoring pathways in yeast, *Curr. Biol.* **12**, 2076–2089.

Heitz, E. (1928) Das heterochromatin der Moose, *Jehrb. Wiss. Botanik* **69**, 762–818.

Hilliker, A.J. (1985) Assaying chromosome arrangement in embryonic interphase nuclei of *Drosophila melanogaster* by radiation induced interchanges, *Genet. Res.* **47**, 13–18.

Hoskins, R.A., Smith, C.D., Carlson, J.W., Carvalho, A.B., Halpern, A., Kaminker, J.S., Kennedy, C., Mungall, C.J., Sullivan, B.A., Sutton, G.G., Yasuhara, J.C., Wakimoto, B.T., Myers, E.W., Celniker, S.E., Rubin, G.M., Karpen, G.H. (2002) Heterochromatic sequences in a Drosophila whole-genome shotgun assembly, *Genome Biol.* **3**, 85.1–85.16.

Hwang, K.K., Eissenberg, J.C., Worman, H.J. (2001) Transcriptional repression of euchromatic genes by Drosophila heterochromatin protein 1 and histone modifiers, *Proc. Natl. Acad. Sci. U.S.A.* **98**, 11 423–11 427.

Ishii, K., Arib, G., Lin, C., Van Houwe, G., Laemmli, U.K. (2002) Chromatin boundaries in budding yeast: the nuclear pore connection, *Cell* **109**, 551–562.

James, T.C., Elgin, S.C.R. (1986) Identification of a nonhistone chromosomal protein associated with heterochromatin in *Drosophila melanogaster* and its gene, *Mol. Cell Biol.* **6**, 3862–3872.

James, T.C., Eissenberg, J.C., Craig, C., Dietrich, V., Hobson, A., Elgin, S.C.R. (1989) Distribution patterns of HP1, a heterochromatin-associated nonhistone chromosomal protein of Drosophila, *Eur. J. Cell Biol.* **50**, 170–180.

Lachner, M., O'Carroll, D., Rea, S., Mechtler, K., Jenuwein, T. (2001) Methylation of histone H3 lysine 9 creates a binding site for HP1 proteins, *Nature* **410**, 116–120.

Lane, N., Dean, W., Erhardt, S., Hajkova, P., Surani, A., Walter, J., Reik, W. (2003) Resistance of IAPs to methylation reprogramming may provide a mechanism for epigenetic inheritance in the mouse, *Genesis* **35**, 88–93.

Lu, B.Y., Emtage, P.C., Duyf, B.J., Hilliker, A.J., Eissenberg, J.C. (2000) Heterochromatin protein 1 is required for the normal expression of two heterochromatin genes in Drosophila, *Genetics* **155**, 699–708.

Nakayama, J., Rice, J.C., Strahl, B.D., Allis, C.D., Grewal, S.I.S. (2001) Role of histone H3 lysine 9 methylation in epigenetic control of heterochromatin assembly, *Science* **292**, 110–113.

Ng, H.H., Ciccone, D.N., Morshead, K.B., Oettinger, M.A., Struhl, K. (2003) Lysine-79 of histone H3 is hypomethylated at silenced loci in yeast and mammalian cells: a potential mechanism for position-effect variegation, *Proc. Natl. Acad. Sci. U.S.A.* **100**, 1820–1825.

Nicolas, E., Roumillac, C., Trouche, D. (2003) Balance between acetylation and methylation of histone H3 lysine 9 on the E2F-responsive dihydrofolate reductase promoter, *Mol. Cell Biol.* **23**, 1614–1622.

Nielsen, A.L., Oulad-Abdelghani, M., Ortiz, J.A., Remboutsika, E., Chambon, P., Losson, R. (2001) Heterochromatin formation in mammalian cells: interaction between histones and HP1 proteins, *Mol. Cell* **7**, 729–739.

Noma, K., Allis, C.D., Grewal, S.I.S. (2001) Transitions in distinct histone H3 methylation patterns at the heterochromatin domain boundaries, *Science* **293**, 1150–1155.

Rabl, C. (1885) Uber Zelltheilung, *Morphol. Jahrb.* **10**, 214–330.

Saccani, S., Natoli, G. (2002) Dynamic changes in histone H3 Lys 9 methylation occurring at tightly regulated inducible inflammatory genes, *Genes Dev.* **16**, 2219–2224.

Schotta, G., Ebert, A., Krauss, V., Fischer, A., Hoffmann, J., Rea, S., Jenuwein, T., Dorn, R., Reuter, G. (2002) Central role of Drosophila SU(VAR)3–9 in histone H3-K9 methylation and heterochromatic gene silencing, *EMBO J.* **21**, 1121–1131.

Shaffer, C.D., Stephens, G.E., Thompson, B.A., Funches, L., Bernat, J.A., Craig, C.A., Elgin, S.C.R. (2002) Heterochromatin protein 2 (HP2), a partner of HP1 in Drosophila heterochromatin, *Proc. Natl. Acad. Sci. U.S.A.* **99**, 14 332–14 337.

Smothers, J.F., Henikoff, S. (2001) The hinge and chromo shadow domain impart distinct targeting of HP1-like proteins, *Mol. Cell Biol.* **21**, 2555–2569.

Sun, F.L., Cuaycong, M.H., Elgin, S.C.R. (2001) Long-range nucleosome ordering is associated with gene silencing in *Drosophila melanogaster* pericentric heterochromatin, *Mol. Cell Biol.* **21**, 2867–2879.

Sun, X., Le, H.D., Wahlstrom, J.M., Karpen, G.H. (2003) Sequence analysis of a functional Drosophila centromere, *Genome Res.* **13**, 182–194.

Tamaru, H., Selker, E.U. (2001) A histone H3 methyltransferase controls DNA methylation in *Neurospora crassa*, *Nature* **414**, 277–283.

Volpe, T.A., Kidner, C., Hall, I.M., Teng, G., Grewal, S.I., Martienssen, R.A. (2002) Regulation of heterochromatic silencing and histone H3 lysine-9 methylation by RNAi, *Science* **297**, 1833–1837.

Xiao, H., Sandaltzopoulos, R., Wang, H.M., Hamiche, A., Ranallo, R., Lee, K.M., Fu, D., Wu, C. (2001) Dual functions of largest NURF subunit NURF301 in nucleosome sliding and transcription factor interactions, *Mol. Cell* **8**, 531–543.

Ye, Q., Callebaut, I., Pezhman, A., Courvalin, J.C., Worman, H.J. (1997) Domain-specific interactions of human HP1-type chromodomain proteins and inner nuclear membrane protein LBR, *J. Biol. Chem.* **272**, 14 983–14 989.

Zhao, T., Heyduk, T., Allis, C.D., Eissenberg, J.C. (2000) Heterochromatin protein 1 binds to nucleosomes and DNA in vitro, *J. Biol. Chem.* **275**, 28 332–28 338.

Part 2
Genomic Organization and Evolution

8
Organization of Genes and Genome Domains

Yujing Zeng[1], Javier Garcia-Frias[1], and Adam G. Marsh[2]
[1] *Department of Electrical and Computer Engineering, University of Delaware, Newark, DE, USA*
[2] *Department of Marine Studies, University of Delaware, Lewes, DE, USA*

1	**Macroscale Distribution**	231
1.1	Gene Number Estimates	232
1.2	Gene Prediction	233
1.2.1	Public Consortium	234
1.2.2	Celera Effort	235
1.3	Gene Distribution	235
2	**Meso-scale Organization**	237
2.1	Gene Distribution versus Chromosomal Banding	238
2.2	Gene Distribution versus Sequence Composition	239
2.2.1	Uneven Distribution of GC Content Over the Genome	239
2.2.2	Isochores	240
2.2.3	Gene Properties versus GC Content	241
2.2.4	Evolutionary Hypothesis About Isochores	243
2.2.5	Recent Discussions About the Isochore Concept	244
3	**Microscale Structure**	245
3.1	Gene Model	245
3.2	CpG Islands	248
3.3	GGG Trinucleotide	251
3.4	Pseudogene Distributions	252
3.5	Transposable Elements	253
4	**Metascale Integration**	253
4.1	Associative Expression Networks	254
4.2	Functional Coordination and Physical Mapping	255

Genomics and Genetics. Edited by Robert A. Meyers.
Copyright © 2007 Wiley-VCH Verlag GmbH & Co. KGaA, Weinheim
ISBN: 978-3-527-31609-0

4.3	Physical Colocation and Dislocation	257
4.4	Dynamic Expression Linkages	259
4.5	Importance of Expression Controls	261

Bibliography 262
Books and Reviews 262
Primary Literature 262

Keywords

Chromosomal Bands
Transverse light and dark bands on the chromosomes produced by different staining techniques.

CpG Islands
Dinucleotide motif of a cytidine and guanidine (5′-C-p-G-3′), commonly unmethylated, within a GC-rich chromosomal domain.

EST (Expressed-Sequence Tag)
Short (300–500 nucleotide) single read DNA sequence derived from a randomly selected cDNA clone.

Exon
A coding DNA segment within a gene.

GC Content
The distribution of Guanine and Cytosine nucleotides among the genome, which is correlated with various structural and functional properties of the genome fragments.

Intron
A noncoding DNA segment within a gene.

Isochores
Long DNA segments, usually greater than 300 kb, which are characterized by a fair compositional homogeneity.

Promoter
DNA segment composed of the first few hundred nucleotides located "upstream" (on the 5′ side) of a gene, and which controls the transcription of such a gene.

Transcript
mRNA sequence produced by transcription from DNA.

Splicing
The process that removes all the introns to produce mature mRNA.

■ The human genome encodes the information about the development, physiology, and evolution of our own species. Among the whole DNA stretch with more than 3000 million nucleotides, the most interesting portions are those composed of genes, the DNA fragments that code for a defined biochemical function, such as the production of a particular protein. It is believed that genes are responsible for the major biological functions of the genome, and knowledge about them and their encoded proteins is crucial for basic biology, biomedical research, biotechnology, and health care.

In this contribution, we provide an overview of the ongoing research on the gene distribution in the human genome. It has been known for a long time that the distribution of protein-coding genes among human chromosomes is extremely uneven. DNA fragments with different gene densities show different compositional properties, which are believed to be correlated with gene composition, function, and evolution. Around 98% of the human genome does not code directly for proteins and therefore was once dismissed as "junk DNA." However, many studies have shown that these "junk" sections contain many important features essential for the biological function of a genome. The challenge to understanding how genomes "work" involves more than just the identification of protein-coding segments. We must understand the mechanisms by which all the components of a genome interact to produce a functional cellular system.

1
Macroscale Distribution

In order to perfectly define the human genome, it would be necessary to precisely locate each and every single gene. At present, this is not possible because gene identification from the human genome drafts is a complicated process given the multiple sequence elements that combine to ultimately produce (express) an operational gene. A gene is a locus of co-transcribed exons, which in the human genome tend to be short. Exons are separated by introns, which are sequences of long lengths (some of them exceeding 10 kb) that do not code for protein sequences. Transcription of a gene domain produces heteronuclear RNA (hnRNA), which is a linear copy of the coding DNA strand (except with complementary nucleotides). This hnRNA transcript contains all the exon and intron sequences. Post-transcriptional RNA processing splices out the intronic sequences and prepares the RNA transcript for translation as a mature messenger RNA molecule (mRNA). What is surprising is that these mature coding segments of mRNA (which are responsible for producing all the proteins in our bodies) represent only ~1.5% of our total genome DNA sequence. Thus, looking for genes in our genome is like looking for needles in a haystack. In addition, a single gene may give rise to multiple transcripts by alternative splicing, and alternative transcription initiation and termination sites. Given the current limited knowledge about gene structure and function, none of the existing algorithms for gene prediction and annotation are perfect, and therefore we can only estimate the gene distribution based on our current understanding of how genes are structured and expressed.

1.1
Gene Number Estimates

Before the publication of the human genome drafts, estimates of the total number of genes in humans were proposed in the literature. However, different approaches produced different estimates, with some as high as 120 000, even when the methods utilized for the estimation were similar. One of the reasons for this variability was the necessity to use strong assumptions in order to obtain the estimate, which made it very difficult to assess the reliability of the predicted number. For instance, Ewing and Green estimated the number of genes by comparing a database of ESTs (expressed-sequence tags) with chromosome 22 and with a nonredundant set of mRNA sequences. The idea, explained in more detail in their work, is to estimate the total number of genes by counting the amount of overlap between two sets of genes obtained independently. Both comparisons (ESTs with chromosome 22 and ESTs with the mRNA sequences) produced an estimate in the order of 35 000 genes. This estimate is very different from the 60 000–70 000 genes predicted by Fields et al., who used a similar method to compare a set of ESTs with mRNA sequences. The difference is that Ewing and Green eliminate single unconfirmed ESTs in the counting process, thereby resulting in a more conservative estimate. Other methods predicted 80 000 genes by estimating the number of CpG islands in the genome and assuming that all islands are associated with genes and that 56% of the genes have an island. These assumptions were challenged by Ewing and Green, who suggest an estimate below 30 000. Other gene number estimates based on reassociation kinetics and genome comparisons with other species also produce very different results.

With the publication of the two drafts of the human genome sequence (The human genome sequence is available at http://genome.ucsc.edu/) in 2001, more comprehensive studies about gene distributions in the human genome have become possible. Although the genome sequence is still being updated, both the public and the private sequencing efforts have obtained similar estimates of around 30 000 genes for humans. These studies, however, are far from being complete and conclusive. There are several difficulties: first, the draft sequences do not provide a perfect coverage of the human genome and there are still errors in the sequence data. Therefore, even the best estimate from the sequence might not be close to the real one. Second, and more importantly, gene extraction from the genome is not a trivial task, and even the best gene prediction tools are not likely to be 100% reliable. Therefore, it is fair to say that our knowledge about the gene distribution in the human genome, although much better than before the publication of the draft sequences, is still far from complete. For instance, the work of Das et al., published after the human genome draft, suggests, by using experimental evidence, a higher estimate in the range between

41 000 and 45 000 genes is obtained. As we will see in the following sections, these estimates are still based on some assumptions. However, as an advantage over other methods that do not use direct sequence data, the availability of the genome draft allows the refinement of these assumptions. Notice also that the methods that are applied directly over the genome sequence to identify genes obtain not only estimates about the gene number but also about the gene distribution.

1.2
Gene Prediction

There are different methods to locate genes in the human genome. The simplest one is to perform alignments between known genes contained in databases such as RefSeq and the genome sequence. Although this approach is very powerful to locate genes that are already known, it does not allow the discovery of any new genes. Another related possibility is to make gene predictions by performing sequence comparison with previous known genes and proteins in other organisms. As before, the problem with this approach is that it cannot extract new genes, since, obviously, they are not included in the comparison set. A third method is based on homology studies between the genome sequence and libraries of ESTs. Obviously, the success of this method depends on the completeness and quality of the libraries, which should consist of relatively large ESTs matching the genes. Finally, *de novo* (*ab initio*) techniques try to recognize groups of exons directly from the genome sequence. In order to do so, these methods define gene models using *a priori* information about the gene structure (sequence composition, intron/exon structure and length, splice sites, etc.), and perform gene extraction by identifying regions in the genome sequence that correspond to the gene model.

Several models have been used in different *ab initio* methods, including neural networks, decision trees, and, more frequently, hidden Markov models (HMMs), which are the base of well-known tools such as Genscan and Genie. If our understanding of the cell mechanisms to identify genes were complete (or in other words, if *a priori* information about the gene structure were perfectly available), and provided that the DNA sequence were perfectly known, *ab initio* techniques would be able to extract and locate all genes contained in the genome. Unfortunately, this *a priori* knowledge is not available, and the lack of knowledge leads to several types of errors in gene prediction:

- False positives, resulting when sections of the genome sequence that do not correspond to genes are identified as such by the prediction method. This type of prediction error can occur for several reasons such as spurious predictions due to alignments produced by chance or by the presence of pseudogenes.
- Missing genes, resulting when sections of the genome corresponding to genes are not identified as such by the prediction method. This type of prediction error can occur for several reasons, including the inability of the method to either detect all gene components or to properly connect correctly detected components (fragmentation).
- Inaccurate genes, resulting when a gene is identified but the prediction is either incomplete or partially wrong. This type of prediction error occurs when the method misses some exons or incorporates wrongly predicted exons in the gene prediction.

The gene prediction methods just mentioned are not mutually exclusive. In fact, they are usually applied together with the objective to develop complex prediction tools capable of achieving the best possible sensitivity and specificity – that is, joint optimization to miss as few genes as possible with the least number of false positives.

1.2.1 Public Consortium

The first step in the effort from the public consortium to predict genes in the human genome draft is to utilize the Ensembl system. This system uses the *ab initio* tool Genscan to generate gene predictions that are confirmed by similarities to ESTs, mRNAs, and protein motifs. Each gene component (exon/intron) is confirmed independently, which increases the likelihood of fragmentation. In order to reduce this fragmentation, Genie, another *ab initio* technique, is utilized jointly with Ensembl. These Ensembl/Genie results are compared with known genes contained in RefSeq, SWISSPROT, and TrEMBL databases. At the end of this process, the total number of gene predictions is almost 32 000. From this prediction, approximately 15 000 of them correspond to known genes, 4000 come from the combination of Ensemble and Genie, and 13 000 from Ensembl alone.

In order to validate the quality of these predictions, different approaches are utilized to estimate the sensitivity and specificity of the prediction stage. The first approach used in the landmark paper of the public consortium is to compare these predictions with new independently discovered genes not used in the prediction process. Thirty-one genes newly discovered in chromosome X were considered in the aforementioned paper. From these genes, 28 are included in the draft, and 19 were successfully predicted, which gives an estimate for the sensitivity of the gene prediction process of 68%. Moreover, on average 79% of each gene was detected and the fragmentation can be estimated as 1.4 gene predictions per true gene. However, this approach should be taken with caution because the number of genes limits the accuracy of the statistical analysis. A higher sensitivity of 85% is obtained in the second approach, when the predicted genes are compared to a set of 15 294 mouse cDNAs generated by the RIKEN Genomic Sciences Center. Specifically, 81% of the genes in the mouse showed similarity to the human genome sequence and 69% were similar to the predicted human genes, resulting in a sensitivity of $69/81 = 85\%$. The final approach compares the predicted genes with the gene annotations in chromosome 22, which consisted of 539 confirmed genes and 133 pseudogenes. This comparison suggests a rate of false-positive or spurious predictions (including pseudogenes predictions) at around 20%.

The final estimate of the number of genes results from the development showed above: 17 000 gene predictions of unknown genes with a 20% false-positive rate and a fragmentation factor of 1.4 lead to an estimate of 9500 new unknown genes. If these gene predictions are assumed to contain 60% of all previously unknown human genes (following the validation results in the previous paragraph), the number of previously unknown human genes can be estimated as 16 000. Considering also the 15 000 known genes contained in the human genome draft, the total number of genes in the human genome can be estimated to be around 31 000. Notice that this number is highly dependent on the estimates of specificity, sensitivity, and fragmentation of the gene prediction method utilized for the analysis.

1.2.2 Celera Effort

In order to perform gene prediction, Celera Corporation, led by Craig Venter, developed an approach called *Otto*. Otto considers different lines of evidence to locate genes. The first step in the process is to identity gene boundaries by examining matches of the genome sequence to EST and protein databases. These matches are combined into bins in such a way that a single gene is expected for each one of the bins.

Then, different procedures are applied to extract genes. A first set of genes is extracted from the genome by comparing the transcript database RefSeq with these bins. A total of ~6500 genes resulted from this process, which identified genes when some transcript in the RefSeq database matched the genome sequence for at least 50% of its length with more than 92% identity. These ~6500 genes are the ones in which Otto has a higher confidence. For the rest of the genome regions that have sequence similarity but do not present a clear match to known genes, Otto considers four lines of evidence (similarity to known proteins, similarity to human ESTs and cDNAs, similarity to rodent ESTs and cDNA, and conservation between mouse and human DNA) and tries to develop a gene model using Genscan. If the model is reasonable, and predicted exons are supported by at least one of the lines of evidence, a gene is considered identified. The number of additional genes obtained in this way by Otto was ~11 000. Considering the ~6500 genes identified before, the number of genes predicted by Otto is ~17 500 if only one line of evidence is required. The number decreases to ~17 000, ~15 500, and ~12 500 if two, three, or four lines of evidence are utilized.

It is interesting to note that Otto does not try to directly identify genes with gene prediction tools such as Genscan. Instead, the gene boundaries are predicted by first looking at homology evidence. The reason for this approach is that most gene prediction methods have problems performing boundary prediction. The specificity, sensitivity, and fragmentation rate of this method are superior to those of Genscan. However, predictions from Otto are very conservative, since they depend on ESTs and some gene transcripts are not included in the existing databases. Therefore, Otto's predictions are complemented with additional ones produced by three *de novo* techniques (GRAIL, Genscan, and FgenesH). The number of *de novo* predictions that did not overlap with known genes or Otto results was around 58 000. From this number ~21 500 are supported by one line of evidence, ~8600 by two, ~5000 by three, and ~1900 by four. Therefore, the total number of predicted genes obtained by combining these predictions with the ones proceeding from Otto, would be ~39 000, ~26 500, ~23 000, and ~20 000 depending if one, two, three, or four lines of evidence are required for the *de novo* methods (and assuming that one line of evidence is enough for the Otto predictions). Around 1000 additional genes can be predicted by identifying regions outside the original bins where there was a match between an EST and the genome sequence across a splice junction, and at least one line of evidence was present.

1.3 Gene Distribution

As the sequencing of larger and larger regions of the human genome was completed, there was a series of increasingly comprehensive gene maps constructed and cross-referenced to the human genetic map. The gene map published in 1996 by Schuler et al. confirmed an uneven gene distribution among different chromosome bands. Two years later, a new gene map with higher coverage and accuracy was published, showing that the distribution of genes across individual chromosomes presents striking fluctuations. Significantly, higher than average gene densities are found in chromosomes 1, 11, 17, 19, and 22, while lower than average densities are found in chromosomes 4, 5, 8, 13, 18, and X.

The publication of the two draft sequences in 2001 represents a milestone in the study on human genomics. We are now closer than ever to understanding how our own genomes are structured in terms of the distribution and organization of genes within and between chromosomes. The first, full, genome structural analysis was completed by the International Human Genome Sequencing Consortium (IHGSC), which aimed at creating an initial integrated gene index (IGI) and an associated integrated protein index (IPI) for the human genome. Then the chromosomal distribution of the IGI gene set was examined. On average, the gene density in the human genome is 11.1 genes per Mb. However, this number presents significant variations among chromosomes, from 26.8 genes per Mb for chromosome 19 to 6.4 genes per Mb for chromosome Y. The number for chromosome Y could be even smaller, because the high number of pseudogenes contained in chromosome Y is likely to have generated numerous false-positives from the gene prediction studies.

Several months after the publication of the human genome, a draft physical map with annotations for a majority of the human transcripts was published by Zhuo et al. The map was generated by assembling the sequence clusters in the UNIGENE database into nonredundant sequence contigs, and then aligning these clusters to the human genome draft. As shown in Fig. 1, the estimated transcript densities show a significant difference among all the human chromosomes. The

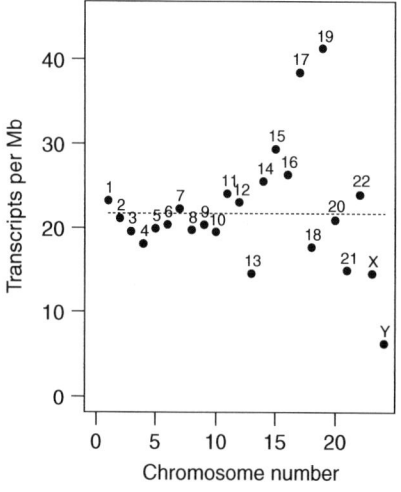

Fig. 1 Transcript densities on the different human chromosomes. The broken line indicates the average transcript density for the whole genome (from Zhuo, D. et al. (2001) Assembly, annotation, and integration of UNIGENE clusters into the human genome draft, *Genome Res.* **11**, 904–918).

highest density, found in chromosome 19, is above 40 transcripts per Mb, which is almost twice the value of the average density over the whole genome. The lowest gene densities were found in chromosomes X, Y, 21, and 13, which is consistent with the previous gene density estimates.

According to the record updated on January 25, 2003, there are 18 761 human genes reported in RefSeq that have their function and coding sequences known. Figure 2 shows the chromosomal distribution of these genes, excluding 225 genes that have not been located in the draft genome. The uneven gene distribution indicates that the human genome is organized in discrete regions, and that functional characteristics may get distributed in a related pattern.

2
Meso-scale Organization

With an estimate of around 30 000 genes and over 3 Gb nucleotides in the human genome sequence, the inevitable question that springs to mind is "where exactly are the genes located?" In other words, how are our 30 000 genes distributed over our 3 000 000 000 bp genome? Many contributions in the literature have tried to address this question during the last 30 years. As described in Sect. 1.3, genes are distributed unevenly among chromosomes. Moreover, different lines of evidence, which will be reviewed in this section, have led to the same conclusion: the gene distribution along each chromosome is conspicuously nonuniform. In this section, we move our observation from the

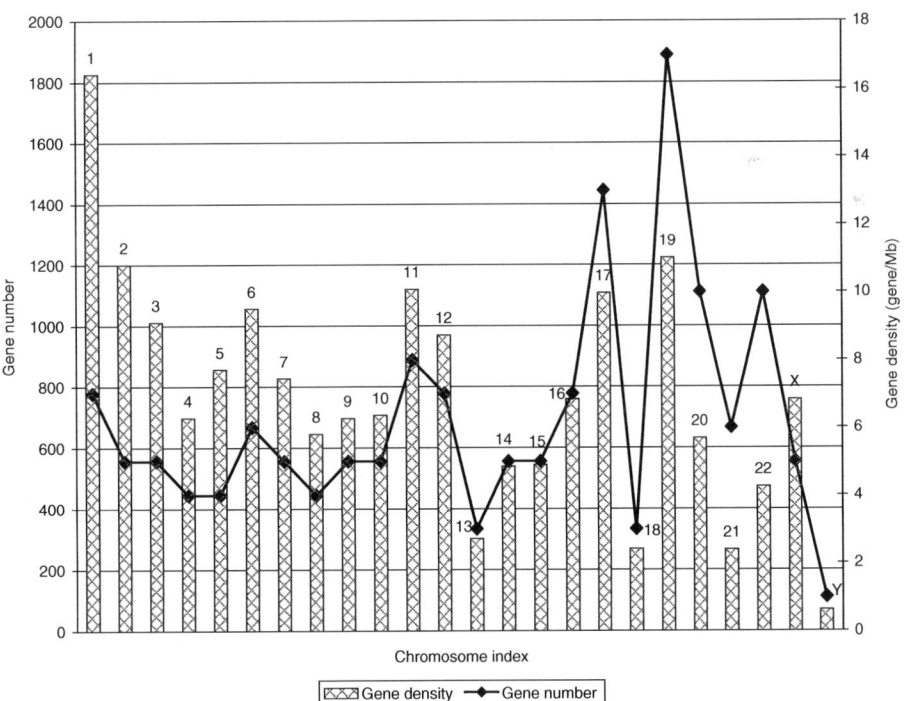

Fig. 2 Chromosomal distribution of the 18 761 genes reported in RefSeq as of January 25, 2003.

chromosome to a finer scale, and describe the gene distribution according to several biochemical features in the human genome.

2.1
Gene Distribution versus Chromosomal Banding

The uneven gene distribution in the human genome was hinted at long before the gene map on the whole genome was estimated, by the observation of the banding patterns produced when chromosomes are stained. More than 30 years ago, it was discovered that certain dyes selectively stain some regions of metaphase chromosomes more intensely than other regions, and that the resulting transverse banding patterns are specific for individual chromosomes.

Banding characteristics were readily applied for the identification of individual chromosomes. Shortly afterwards, other applications were discovered, with several studies showing that chromosomal bands might have functional and structural significance. Giemsa, for example, is a permanent DNA dye that has affinity for A+T rich DNA regions. After a pretreatment that denatures the DNA sequence with barium hydroxide, Giemsa darkens the regions of constitutive heterochromatin, DNA fragments that are always retained in a compact organization and contain almost no active genes. These dark bands are called C-bands and cover around 17 to 20% of the human chromosome complement. They are mostly distributed in pericentromeric regions. One exception is the Y chromosome that has a relatively large heterochromatic segment in its long arm, suggesting a low gene density on that chromosome.

The remaining 80% of the genome, the euchromatic component, is divided into G-, R-, and T-bands. G-bands are produced by staining the chromosomes with Giemsa, after subjecting them to a brief proteolytic treatment. R-banding (referred to as reverse Giemsa, since the resulting bands are roughly the reverse of G-bands) requires pretreatment of the chromosomes, prior to Giemsa staining, in a hot physiological alkaline solution. Under such conditions, the AT-richer DNA sections are denatured faster, reducing their affinity for Giemsa staining. Finally, T-bands are an extremely heat-resistant subset of R-bands, obtained by repeating the heat pretreatment. The distribution of all these bands along the human chromosomes is shown in the ideograms (Fig. 3), which are used broadly in cytogenetic analyses (e.g. for the detection of chromosome aberrations causing disease).

Chromosomal bands correspond to chromosome domains with different properties. For instance, it has been shown that early and late phases of mammalian DNA replication correlate with the banding pattern: R-bands replicate early, while G-bands are associated with a late replication time. This result has been confirmed by analysis of replication times across R/G boundaries. Moreover, the work of Matassi et al. has revealed that genes located closely on a chromosome have similar evolutionary rates, which may imply a relationship between evolutionary rates and chromosomal bands. One of the most interesting properties is the relationship between gene density and chromosome bands.

Chromosomal bands exhibit a clear correlation with gene density: Generally speaking, the gene density of R-bands is higher than that of G-bands.

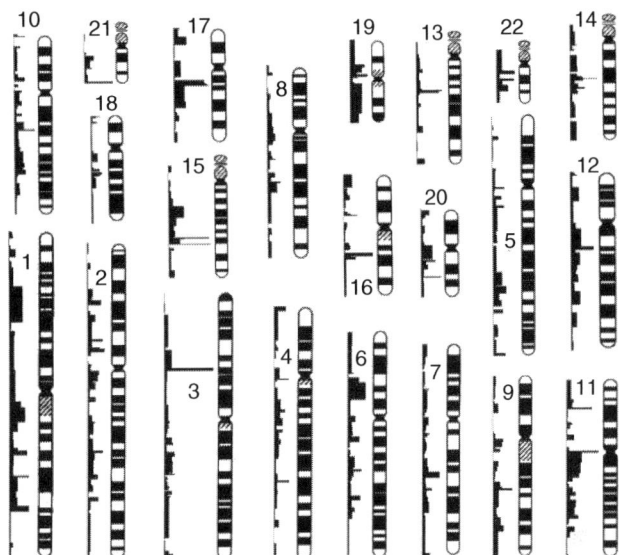

Fig. 3 Gene distributions on chromosomes 1–22. A plot of the average number of estimated gene *loci* across a moving window of 100 Kb is aligned to the ideograms of each chromosome. Genes are heterogeneously distributed both within and among chromosomes (see Caron, H., et al. (2001) The human transcriptome map: Clustering of highly expressed genes in chromosomal domains, *Science* **291**, 1289–1292).

Furthermore, over half of the genes are contained in T-bands, which represent only 15% of all bands. The combination of this result with the ideograms makes it is possible to obtain a rough picture of the gene distribution over each chromosome.

The base compositions of G-, R-, and T-bands are very different from each other: R-bands are richer in GC content than G-bands, and the 27% GC-richest R-bands correspond almost perfectly to the T-bands. In the context of the previous discussion, this fact suggests that the gene density may be related to the local compositional pattern of the DNA sequence, and establishes a connection between cytogenetic observations and molecular analysis.

2.2 Gene Distribution versus Sequence Composition

2.2.1 Uneven Distribution of GC Content Over the Genome

More than 30 years ago, ultracentrifugation experiments based on the use of sequence-specific ligands, such as silver ions, were performed to achieve high-resolution DNA fractionation. These experiments led to the discovery of the striking compositional heterogeneity in the nucleotide composition of human DNA. Specifically, it was found that the distribution of the GC content over the whole genome is far from uniform, since there is a substantial variation in average GC content among large DNA fragments.

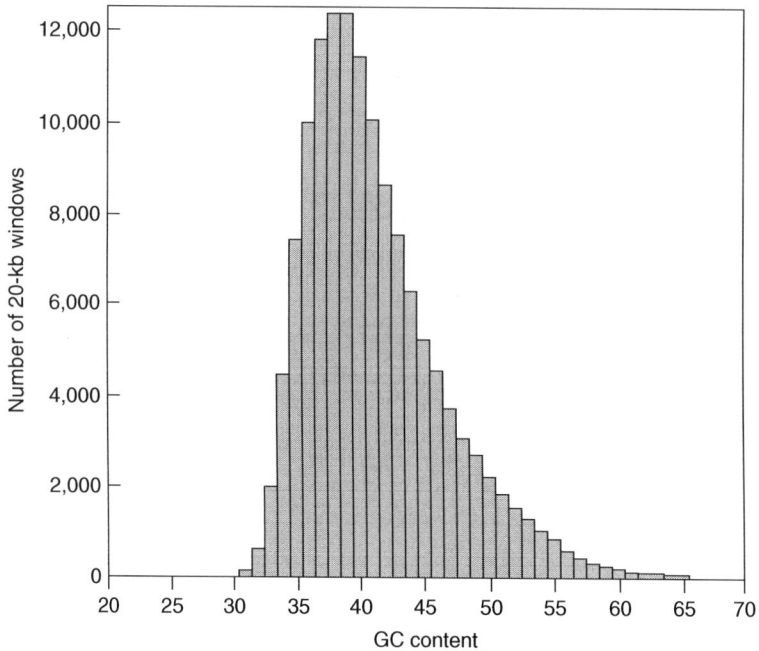

Fig. 4 GC content histogram obtained from the human genome draft (from IHGSC, *Nature* **409**, 2001).

The recent publication of the human genome confirms the uneven GC content distribution in a direct and global manner. Figure 4 shows the GC content histogram obtained from the human genome drafts using 20-kb windows. The figure confirms that local GC content undergoes long-range excursions from its genome-wide average of 41%. Recently, Pavlicek et al. obtained a compositional map of the human chromosomes by scanning the human genome sequence with a 100 Kb moving window. The map shows the large proportion of the genome characterized by long-stretch fragments with low GC content. The regions of GC content greater than 46% constitute only 12.5% of the whole-genome sequence. It is also obvious from the map that the GC content distribution is chromosome specific.

2.2.2 Isochores

Bernardi et al. proposed that the long-range variation in GC content reflects that the human genome is a mosaic of isochores, defined as long DNA segments (greater than 300 kb on average) that are characterized by a fair compositional homogeneity (above a 3-kb size level). Isochores in the human genome have been classified into five different families, depending on their percentage in GC compositions. Families L1 and L2 have a GC content lower than the human genome average, specifically in the range <37% for family L1 and 37 to 42% for family L2. The combination of both families makes up 62.5% of the human genome. A higher GC content than the average is observed in families H1, H2, and H3, which cover the ranges 42 to 46%, 46 to 52% and >52%, and constitute 25, 8.3, and

4.2% of the human genome respectively. Implicit in the isochore definition is the idea that the large-scale GC compositional heterogeneity is discrete or discontinuous, with abrupt changes from higher to lower (or vice versa) GC levels.

Different methods have been applied to determine the isochore families to which human genes belong. All these studies suggest a very nonuniform distribution of genes throughout isochore families: Almost 30% of all human genes reside in the roughly 4% of the genome that belongs to the GC-richest family, H3; in comparison, DNA fragments in families L1 and L2 have much lower gene density, which is less than one-sixth of that of H3.

The relationship between GC content and gene richness has been addressed quantitatively by Mouchiroud et al., who reported a direct positive linear relationship between the GC_3 values (GC levels of third codon positions) of coding sequences and the GC levels of the DNA regions in which they are embedded. This correlation permits the positioning of the distribution profile of coding sequences relative to that of DNA fragments. Consequently, it makes it possible to estimate the relative gene density with respect to different GC contents. In the work of Bernardi et al., this relative gene concentration was calculated by dividing the percentage of genes located in a given GC_3 interval by the percentage of DNA located in the corresponding GC range. With the availability of the draft human genome sequence, the gene density can be estimated directly. Figure 5 presents the gene density as a function of the GC content level, estimated by aligning 9315 known genes on the draft human genome. The results are similar to the earlier estimates, and show the high nonuniformity of the gene distribution in the human genome. Notice that the gene density increases from a very low average level in GC-poor regions to a much higher level in GC-rich regions.

2.2.3 Gene Properties versus GC Content

According to the slope change in gene concentration at the boundary between isochore families H1 and H2 (shown in Fig. 5), Bernardi et al. defined two "gene spaces" in the human genome: the "genome core" and the "empty space." The "genome core" consists of the DNA fragments belonging to isochore families H2 and H3. Although it only represents around 12% of the genome, it embeds more than half of the human genes. The other 88% of the genome has a very low gene density (one gene per 50–150 kb) and therefore is referred to as the "empty space" or the "empty quarter" (from the classical name of the Arabian desert). This division into core and empty spaces is also validated by other observations, such as (1) the similarity of the gene concentrations in each space; (2) the similarity of the heptanucleotide comprising the AUG initiation codon of human genes observed in L/H1 and H2/H3 isochores respectively, and (3) the identical chromosomal distribution of H2 and H3 isochores.

As indicated before, the compositional pattern of coding sequences is directly related to that of their embedding isochores. Therefore, it is expected that genes in the genome core should have a relative high GC content, while this content should be low for those genes in the empty space. Several studies have confirmed these expectations. Besides the compositional difference, the gene structures in the two spaces are significantly different. Interestingly, long genes are scarce in the core genome and more frequent in the empty quarter. A primary reason for this

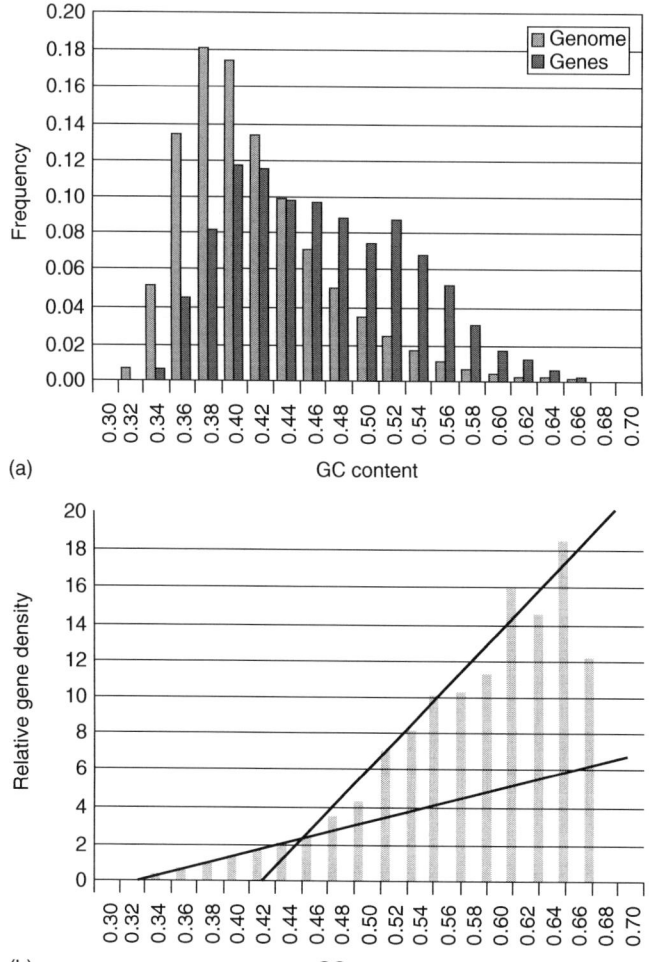

Fig. 5 (a) GC content distribution in 9315 known genes and in the whole genome. (b) Gene density as a function of GC content, obtained by taking the ratio of the data in (a). Values are less accurate at higher GC levels because the denominator is small. The two slopes represented here are associated to the gene core and empty space (from IHGSC, *Nature* **409**, 2001 and Bernardi et al, *Gene* **276**, 2001, see original papers for details).

length difference is the fluctuation in intron sizes, which, in the empty quarter are on average three times longer than in the genome core. In contrast, coding properties such as exon length or number of exons in a gene, are not so dependent on GC content. This is corroborated in Fig. 6, which shows the dependence of the mean values of exon and intron lengths with respect to the GC content. As shown in the figure, the intron size decreases as a function of local GC content, with a

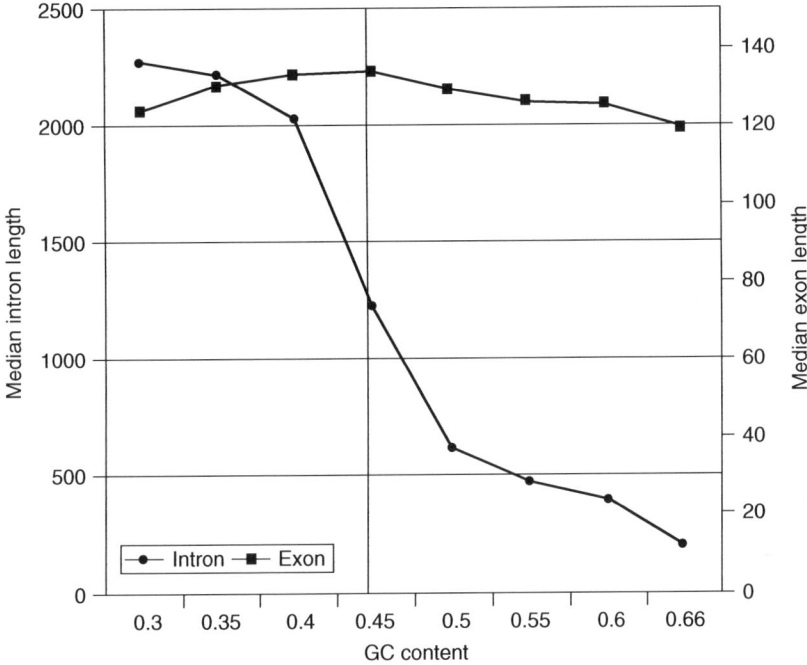

Fig. 6 Dependence of the exon and intron average length on the GC content (from IHGSC, *Nature* **409**, 860–921, 2001, see original paper for details).

sharp transition occurring at ~45% GC, in agreement with the boundary between the genome core and the empty quarter.

Different efforts have been carried out to localize DNA fragments with different GC-richness on human metaphase chromosomes. The results support that the two gene spaces are characterized by different structural and functional properties. GC-poor isochores are located in G-bands and in H3⁻ R-bands (in which H3 isochores are not detected), which generally exhibit a closed chromatin structure. On the other hand, the chromosomal localization of the GC-richest isochore family (H3) corresponds to T-bands. Moreover, since isochore families are unevenly distributed over the chromosomes, some chromosomes (17, 19, 22) have a very high percentage of genome core fragments, whereas other ones (4, 13, 15, 18, X, and Y) have a very small percentage or no genome core segments at all. The explanation for such an uneven distribution is still unclear.

2.2.4 Evolutionary Hypothesis About Isochores

As explained in previous sections, a correlation is evident between %GC content and the gene density of a chromosomal domain. One of the questions raised by this distribution is which one is under the greatest selective pressure: does the %GC content of a chromosomal domain control the number of genes located within that domain or does the clustering of genes into local domains determine the %GC content? The basic question being asked is at what level does natural selection operate to produce such clearly demonstrable isochore domains in the human genome. Although this is a simple question, its answer is buried under the complexity of separating gene function from genome structure. We know that in coding domains there is a codon bias toward GC dinucleotides in warm-blooded vertebrates that result in a higher proportion of hydrophobic and amphipathic amino acid residues in proteins. One of the consequences of maintaining protein structure at higher body temperatures (37 °C) may be an increased need for hydrophobic interactions, thus resulting in a strong selective pressure for greater %GC representation in coding domains. However, the regulation of gene expression events appears to be just as critically controlled by %GC composition. Large-scale alterations in chromosome structure related to GC hydrogen bonding in double-stranded DNA can alter the accessibility of expression regulators to promoter domains.

It is difficult to isolate the impact that nucleotide sequence has on the selective fitness of an organism because a gene sequence is operational at two levels: (1) gene level function, and (2) structural conformation of local chromosomal domains. The difficulty of dealing with the concept of isochores is that this phenomenological observation crosses these two functional boundaries. Identifying the most proximal cause of such sequence anomalies (% frequencies showing such large departures from a random frequency distribution) is difficult because of the interdependence of structure and function.

2.2.5 Recent Discussions About the Isochore Concept

In this chapter, we have presented the standard definition of the isochore concept. However, the plausibility of the "isochore" concept has been questioned in several recent papers. Most of the concerns focus on the definition of "compositional homogeneity" – which some recent papers interpret as relative to random sequences – on the capability of ultracentrifugation experiments to define intrasequence heterogeneity, and on the fraction of the human genome to which the isochore concept applies. Some of these concerns have been addressed by Bernardi and Clay. Other authors point to the fact that isochore families were not originally defined from sequence data, but from ultracentrifugation experiments that resulted in 5 (after a somewhat arbitrary reduction from 13) major overlapping Gaussian distributions of absorbance, each one associated with a distribution of GC content. Therefore, because of the overlapping, a DNA fragment with a given GC content could correspond to different isochore families, which is a drawback to the definition of isochores based only on the sequence data. In order to overcome this problem, the boundaries (in CG content) among isochore families can be defined in such a way that they do not overlap, as we directly did when we introduced isochores. However, this definition is different from the original one.

The concepts of homogeneity and heterogeneity are largely subjective, and therefore much more research is necessary to develop better methods to perform segmentation of the genome into regions of different GC content. However, the existence of GC-rich and GC-poor regions in the human genome has now been fully corroborated by the draft sequences. Moreover, many studies have indicated an interesting correlation between GC content domains and various biological properties, including transcriptional regulation, local DNA replication, patterns of codon usage, and the posttranscriptional processing of mRNA. Although much more work is still necessary before reaching a clear conclusion, the study of the distribution of guanidine and cytidine nucleotides will be a primary route to a better understanding of the gene distribution in the human genome.

3 Microscale Structure

In Sect. 1, we saw that the most recent estimates predict that the human genome contains a number of genes in a 30 000 to 40 000 range, a surprisingly small amount if we consider that much less complex organisms such as *Arabidopsis* and *Caenorhabditis* are also supposed to have a similar number of genes. As a first attempt to explain this fact, it seems intuitive to think that we may need to look outside of the coding sequences for the mechanisms that generate the complexities inherent in the human development. This intuition has been confirmed by an increasing amount of recent work in the literature, which suggests that the explanation is buried in introns and intergenetic regions, once referred to as "junk sequence." An important piece of supporting evidence is the obvious correlation between the GC levels of coding sequences and their context DNA fragments in which they are embedded. The study of the patterns in these noncoding regions may provide hints on the gene distribution over the genome.

3.1 Gene Model

Human genes are characterized by long introns, which on average constitute more than 90% of a gene. The median of the intron sizes (~1000 bp) is about 8 times the median of internal exon sizes. Moreover, the distributions of gene size, intron size, and the number of introns in a gene present significant fluctuations, much higher than those of exons. This is shown in Fig. 7, where the exon and intron size distributions are shown for the human, the worm, and the fly. It is interesting to note that the exon distribution is very similar in the three species, while the intron distributions (and therefore the overall gene size distribution) presents a much higher difference among the three species, with the human distribution being the one with greater variability. Another interesting observation is the dependence between intron sizes and the local GC-level: on average, GC-rich regions show a scarcity of long introns, while the median intron size in GC-poor regions is about twice that encountered in the whole genome. This special pattern is believed to be related to the splicing mechanism. For example, Bernardi suggests that long introns in the gene-poor fragment may facilitate alternative splicing in tissue dependent genes.

Despite the controversy over the exact nature of isochores (whether phenomenological or functional), it is clear that the most prominent feature of protein-coding genes is their higher than average GC nucleotide (guanidine and cytidine) compositions. Majewski has compiled sequence analysis data for 10 858 identified human genes into a "model" gene composed of

- 2000 bp promoter region
- first exon (250 bp, mostly nontranslated)
- first intron (just 500 bp on the 5' and 3' ends)

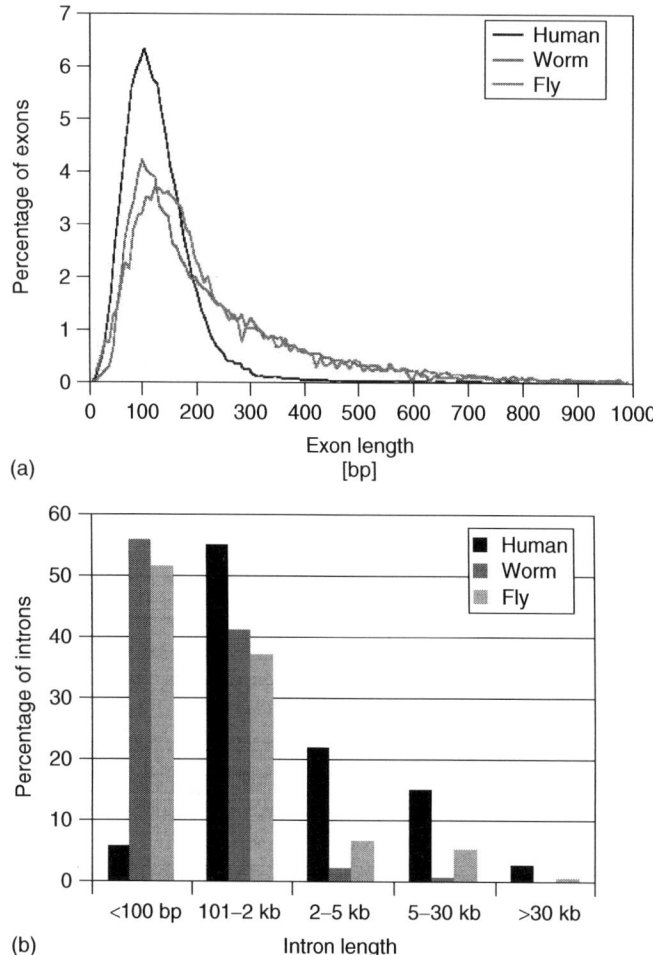

Fig. 7 Size distributions of exons, introns, and short introns for worm, fly, and human genomes (from IHGSC, *Nature* **409**, 860–921, 2001, see original paper for details).

- internal exon (500 bp)
- internal intron (just 500 bp on the 5′ and 3′ ends)
- terminal exon (500 bp)
- termination domain

In most vertebrates, exons and introns can be much larger than 1000 bp total. For this sequence analysis, the calculation of frequency distributions for each of these domains was executed by examining nucleotides at the same sequence positions relative to the splice sites and/or boundaries known to exist in the 10 000 + genes used for this study. Each of the essential gene elements categorized above have been condensed into a grand average in terms of nucleotide composition (Fig. 8).

Although genes have higher GC compositions, the distribution of G and C

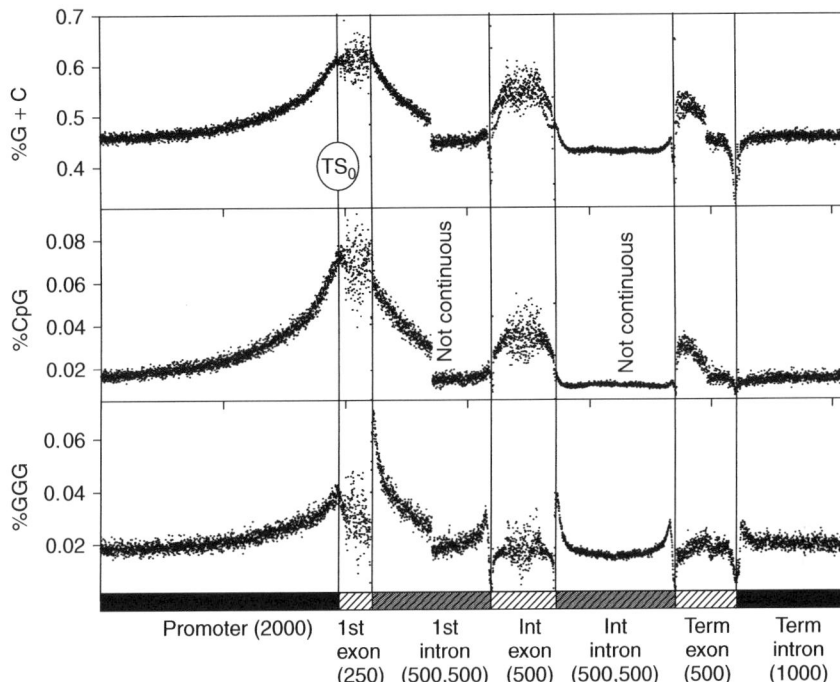

Fig. 8 The distribution of sequence elements within an archetypal model gene are shown for a composite of 10 858 currently known human genes (Majewski and Ott 2002). Gene positions were aligned relative to the transcriptional start site (TS$_0$) and nucleotide frequencies were then averaged by relative position to determine the frequency of (a) % composition of cytidine and guanidine nucleotides, (b) 'CpG' dinucleotide motifs, and (c) "GGG" trinucleotide repeats. The sequence composition evidences several striking patterns in terms of the distribution of G and C nucleotides across all described genes (int = internal; term = terminal). [These distributions are replotted from the original Fig. 5 of Majewski, J. and Ott, J. (2002) Distribution and characterization of regulatory elements in the human genome, *Genome Res.* **12**, 1827–1836 using the original data kindly provided by J. Majewski.]

nucleotide frequencies is heterogeneous, with distinct intragenic locales. Outside of a gene domain, the GC content is between 42 and 45%. As one moves (5' to 3') into the promoter domain (2000 bp from the first exon) the GC content steadily increases to upwards of 60% at the beginning of the first exon (transcription start site, TS$_0$). The first exon has an average GC composition of 55 to 60% without any spatial differentiation. Most interestingly, at the splice junction between the first exon and the first intron there is a sudden increase in the GC composition. The 5' end of the first intron (the splice junction with the first exon) has a GC distribution of nearly 65%. This frequency then rapidly declines further into the intron, but exhibits another rise as one approaches the 3' splice junction with the first internal exon, as shown in Fig. 8.

The internal exons exhibit an interesting feature with a peak in GC content between 100 to 150 bp away from the splice junction on either end, and then a decline in GC frequency as the splice site is approached. The internal introns evidence an opposite distribution with peaks in GC frequencies right at the splice junctions, and then declines to average levels roughly 200 bp away from the junction. The presence of such significant spatial shifts in nucleotide frequencies (especially at boundary junctions) is suggestive of functional mechanisms for regulating gene transcription and RNA posttranscriptional processing. However, it is unlikely that the mere sequence composition of a gene element provides a specific enough mechanism for exerting fine-scale regulatory controls. Rather, spatial shifts in GC distributions are more likely to reflect changes in the distribution of specific regulatory motifs that have higher proportions of GC nucleotides.

The archetypical gene model provides a unique look at how human genes are structured. The most significant feature in terms of the potential regulation of gene expression rates is the distance, both upstream and downstream from the TS_0 site, for which significant biases for the presence of GC motifs can be found. There are several regulatory motifs in this wide domain whose mechanistic actions to control gene transcription events are determined by cytosine and guanidine nucleotides, such as CpG and GGG motif. The potential regulatory domain for direct transcriptional control over a single gene locus comprises a ~2000 bp window around the TS_0. This domain results in a large number of potential regulatory interactions among the special motif distributions; however, regulatory regions can exist much farther upstream. This range of potential control states is thus likely to establish the high recombinatorial repertoire in potential expression patterns that is essential for the evolution of complex network interactions.

3.2
CpG Islands

"CpG islands" (CGIs) are one of the most prominent regulatory motifs in the human genome. The name refers to a simple dinucleotide motif of a cytidine and guanidine (5'-C-p-G-3') within a GC-rich chromosomal domain. The frequency of CpG dinucleotides in the human genome is much less than expected, as a result of a methylation/mutation process. Generally, about 60 to 90% of the CpG sequences within a genome are methylated. DNA methylation is a reaction, specific to cytosines in CpG dinucleotides, that transfers a methyl group from S-adenosylmethionine to the C5 position of cytosine. Methylated cytosines have a high mutation rate to thymines (CpG to TpG), which has been confirmed by many studies on DNA polymorphism or genetic diseases. This process, known as GC suppression, leads to an overall reduction in the frequency of GC content to about 41% of all nucleotides and a further reduction in the frequency of CpG dinucleotides to about a quarter of their expected frequency. CpG islands (those C–G motifs in GC-rich domains) are primarily unmethylated and are the exception to CpG underrepresentation in the genome.

CpG islands make up 1 to 2% of the DNA sequence. They are of great importance because the cytidine base can be methylated such that the methyl group extends upwards into the major groove of the DNA helix (Fig. 9). The intrusion of these polar methyl groups essentially disrupts the potential interactions between the local nucleotide bases and transcriptional regulators in this domain by altering the

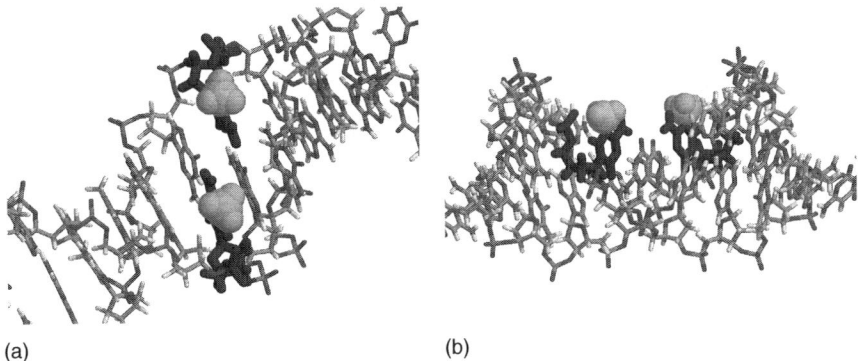

Fig. 9 Three-dimensional structure models for the methylation site in a CpG motif; (a) top view perpendicular to the major groove, (b) side view parallel to the major groove. The 5-methyl-cytidine residues are drawn in the dark wireframe representation, the methyl group atoms are shown with Van der Walls atomic radii as gray spheres. The primary impact of the methylation appears to be a redistribution of charges around the C–G hydrogen bonding. The nonpolar methyl group is adjacent to the amino (–NH$_2$) group that would establish one of three hydrogen bonds with the complementary G-residue hydroxyl group (coordinate data from PDB ref# 1IG4).

local charge distribution. Usually, both the (+) strand motif and the complementary (−) strand motif cytidines are methylated (m5C) to increase the density of methyl groups within a small area (see Fig. 9):

$$5'-C5m-G-3'$$
$$3'-G-m5C-5'$$

This provides an effective blocking mechanism to "silence" regions of DNA from being transcribed and is one of the most active areas of current research into epigenetic mechanisms of gene regulation.

CGIs commonly occur in the promoter domains of functional genes. Overall, this pattern suggests that the majority of the methylated CpG motifs play a role in gene silencing (blocking transcription of noncoding regions, repetitive DNA elements, transposable elements, pseudogenes, etc.). But in active promoter domains, the methylation states of CpG motifs are not fixed such that changes in methylation can be used for up-and-down regulation of transcriptional activities at a specific gene locus.

An attractive property of CGIs is their significant structural difference from the rest of the genome. In the work of Gardiner–Garden and Frommer, CGIs were formally defined as sequences greater than 200 bp in length, with a GC content greater than 0.5, and with an observed to expected GC content ratio (CpG$_{obs}$/CpG$_{exp}$), greater than 0.6. This definition has been widely used in later studies. Recent work has suggested more specific structures for CGIs, especially for those potentially corresponding to promoter regions. This makes it practical to search for CGIs using computing tools. In the work from the International Human Genome Sequencing Consortium, a computer program was implemented to identify CpG islands and 28 890 CGIs were found in the draft human genome sequence with repeat regions masked. The count of 28 890 CGIs is reasonably close to previous estimates of about 35 000, which were obtained by experimental methods. The convenience of CGIs searching has led to effective approaches for gene estimation. In 2000, Grunau et al. determined features that discriminate the promoter-associated and promoter-nonassociated CpG islands and used them for large-scale human promoter mapping.

As shown in Fig. 10, the density of CpG islands shows important variations in different chromosomes and correlates positively with estimates of the gene density. Chromosome Y is the one presenting the lowest density, with 2.9 islands per Mb. On the opposite extreme, chromosome 19 presents a density of 43 islands per Mb. The average for all the genomes is 10.5 islands per Mb, with most of the chromosomes having 5 to 15 islands per Mb. Similar trends are observed when the percentage of nucleotides contained in CpG islands is considered.

The coherent distribution of CpG islands within the model gene demonstrates the general feature that these motif elements comprise for all human genes (Fig. 8). There is a fourfold increase in CpG frequency in the promoter domain as one moves 5′ to 3′ from −2000 bp down to the first exon (0 bp at transcriptional start site). The high frequency occurrence of CpG islands is maintained through the first exon (+250 bp) and into the first intron. This distribution on either side of the first exon boundary suggests a potential regulatory influence of CpG methylation patterns extending from −1000 bp upstream of the TS_0, across the first exon (0–250), and then 500 bp into the first intron.

The position of CpG islands at the TS_0 appears to be a crucial mechanism for temporal and spatial regulation of gene expression rates. In Fig. 11, we have summarized gene expression data for 24 tissue types from both adults and embryos (orthologs in mouse embryos) on the basis of the presence/absence of a CpG island at the TS_0 site. In human adult tissues, there is a significant trend for genes exhibiting tissue-specific expression patterns to have a CpG island at the

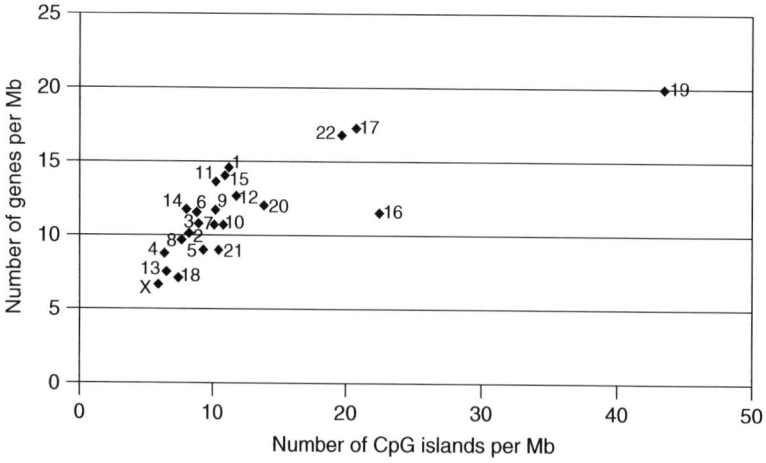

Fig. 10 Number of CpG islands per Mb versus number of genes per Mb (from IHGSC, *Nature* **409**, 860–921, 2001, see original paper for details).

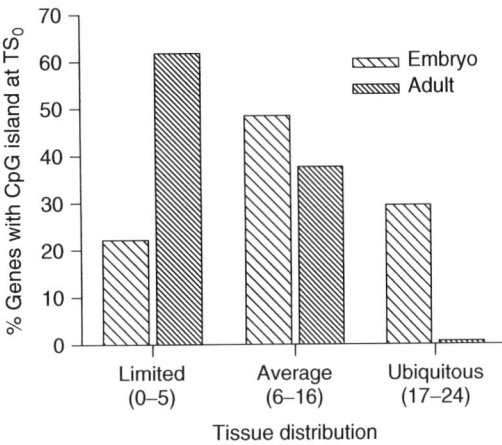

Fig. 11 Frequency of genes with CpG-island regulatory elements in their transcriptional start domains. Genes are separated into three categories depending on the number of tissues that express those genes (collated expression data from 24 tissue types). Expression categories range from "limited" (expression in less than 5 tissue types) to ubiquitous (expression in more than 17 tissue types). Regulation of gene expression in adult tissues clearly demonstrates a frequency relationship between tissue-specificity and the occurrence of CpG islands at the TS_0 (data replotted from Ponger, L., et al. (2001) Determinants of CpG islands: Expression in early embryo and isochore structure, Genome Res. 11, 1854–1860).

TS_0 (62%), in contrast to genes that are ubiquitously expressed (housekeeping genes) where only 0.7% exhibited a TS_0 CpG motif (Fig. 11). This pattern strongly indicates that CpG methylation may be one of the primary mechanisms by which tissue-specific expression patterns in adults are regulated. The pattern of CpG distributions in embryonic gene regulation does not appear to be correlated with tissue-specific expression.

3.3
GGG Trinucleotide

Another important regulatory motif is the "GGG" trinucleotide, which will obviously be more prevalent in GC-rich (isochore) domains. These motifs have been shown to be overrepresented within introns and have been implicated to function in regulating RNA splicing during posttranscriptional processing. In the archetypal model gene, the distribution of GGG trinucleotides shows a predominant location at the 5′ ends of intronic (noncoding) domains. This distribution supports the hypothesis that these motifs are critical control components of RNA posttranscriptional processing by the spliceosome protein complex. Also, it is quite significant that the GGG frequency is much lower than expected near the TS_0 and across the first exon. This deficiency suggests that the GGG motif does not commonly play a role in the regulation of gene expression rates; however, a few examples of GGG-mediated gene regulation do exist. It is important to keep in mind that the archetypal model just represents an overall average nucleotide distribution.

A related motif with less known function is the "CCC" trinucleotide. Splicing control elements (SCEs) can only process nucleotide motifs on the coding (transcribed) strand so that "GGG" and "CCC" distributions are not functionally reflexive (simple reverse complements of each other). The frequency of CCC motifs on the coding strand must present a distinct regulatory structure for SCE recognition, and although the representation of CCC motifs is not as significant in the archetypal model gene as are the GGG motifs, they evidence the same prominence in intron borders and scarcity in the promoter and first exon domains. Consequently, the observed increase in %GC content flanking the TS_0 can primarily be attributed to an increase in the frequency of CpG islands, and not the distribution of trinucleotide repeats of cytidine or guanidine.

3.4
Pseudogene Distributions

Pseudogenes are a subclass of genes in which the coding domain is no longer subjected to selection forces acting on cellular fitness. There are two primary routes of pseudogene formation: (1) duplication of a chromosomal region, and (2) reverse transcription of an mRNA molecule into cDNA and consequent reintegration of that processed transcript back into the genome. In either case, the end result is that two copies of an operative gene now exist in the genome, one of which will continue to function under the selective pressures it has been adapted for, and the other is now independent of any selection pressures (i.e. it does not impact the cellular fitness of an organism if the copy is expressed or not expressed, because the original operative gene is continuing its function). The two cases are easily distinguished because with a gene domain duplication, all the noncoding regulatory regions are preserved in the copy (promoter, introns, terminal domain), while in the reverse transcription of mRNA (retropseudogene), the mRNA transcript has been processed such that the promoter, introns, and much of the terminal domain have been removed, or modified. For reverse-transcribed pseudogenes, they must become integrated back into the genome downstream from an active promoter for any possibility of transcriptional activity.

Pseudogenes can be actively transcribed and in that respect are truly "genes" other than the fact that there is no selective pressure on them to actually do anything useful. On chromosomes 21 and 22 (the most thoroughly studied to date), a total of 454 pseudogenes have been identified, and estimates based on the draft human genome to date suggest that there will be between 23 000 and 33 000 pseudogenes in total. (http://bioinfo.mbb.yale.edu/genome/pseudogene/) Thus, pseudogenes are likely to play an important role in evolution by allowing operative genes to obtain new functions without immediate detrimental effects to an organism. However, most pseudogenes appear to be destined for genomic oblivion once they are formed. Without a strong selection on function, random drift in the nucleotide sequence rapidly (on an evolutionary timescale) can introduce premature termination signals or frameshifts that result in nonfunctional mRNA transcripts. There are very few documented cases of pseudogenes evolving into new operative genes.

The documentation of pseudogenes in a genome is not easy. It is likely that there are numerous operative genes in the human genome that have a pseudogene origin. Because most genes themselves are located in GC isochore domains, they have higher than average %GC compositions. Once a pseudogene is formed, random mutational drift will tend to reduce the %GC composition toward the genome average. If the pseudogene adapts to provide a new function, its phylogenetic sequence affinity can be obscured by the shift in nucleotide composition away from its ancestral operative gene. Because only ~1.5% of the human genome is directly involved in coding for protein sequences, there is a large background of DNA for many recombinatorial possibilities to arise from rearranging DNA. A recent study has been able to identify the incorporation of several mitochondrial genes into the human genome, suggesting a new route for pseudogene incorporation when eggs are fertilized by sperm. As our understanding of these processes increases, it is likely that we will be able to develop the computational tools to more fully identify operative-pseudogene relationships.

3.5
Transposable Elements

There are sequence elements in the genome that are very efficient at duplicating and moving their local DNA domains. These transposable elements (TEs) were originally described as parasitic DNA, because of their ability to self-replicate unchecked within a genome environment. However, our understanding of their role in genome structuring is just beginning to identify them as absolutely necessary or at least mutualistic elements. At present we know little about their molecular mechanism of operation, and are left to only speculate on the potential importance of their role in genome evolution. As discussed in the previous section on pseudogenes, the ability to create copies of genes is an important mechanism for either deriving novel functions from existing proteins, or to give existing proteins different patterns of expression. TEs appear to be the primary vehicle by which genomic material is shuffled around into alternative recombinations as a natural experiment resulting in new genes and expression patterns. "NEW" is not necessarily better and these alterations may or may not be selectively advantageous. However, such rearrangements can provide a greater molecular diversity within an organism for selection to then act upon. A large body of literature on TEs exists, particularly in terms of structure and function in *Drosophila*, and the reader should consult these resources.

4
Metascale Integration

The first pivotal study of functional genomics lead to Jacob & Monod's seminal description of the *Lac* operon in *Escherichia coli*. Their summary description of the "one gene equals one protein" organization of DNA set the stage for the next three decades of genomic research focusing on linear, direct models of gene interactions. This approach worked well for the single-target cloning strategies that were implemented in the 1970s to late 1980s. However, in the 1990s, as high-throughput technologies were developed, the collection of gigabytes of gene expression data opened up new avenues for approaching the functional organization of a genome as a three-dimensional structure, that is, one in which the genetic code was not just operative in a long, one-dimensional array of four nucleotides. It is now clearly evident that gene location within a chromosomal domain has a significant impact on transcriptional regulation.

The interaction between genes, their protein products and their chromosomal locations are now routinely considered in network organizations of "connectivity." Long gone (in terms of the volume of literature that has appeared in the last five years) is the focus of identifying linked companion genes in operon-like, downstream–upstream units. Even the paradigm of "one gene equals one protein" needs qualification as alternative splicing, bidirectional coding and post-transcriptional mRNA processing has lead us to realize that single coding domains may produce several proteins, given the multiple mechanisms that are involved between the initiation of gene transcription and the folding of a final protein product.

Given the importance of describing how genes are "connected," there is considerable attention being focused on identifying the mechanistic relationships that may exist between coordinated patterns of gene expression (i.e. network associations) and the physical mapping of those transcriptionally linked genes. With the current estimate of only ~35 000 genes in the human genome, there is strong support for the idea that interactive mechanisms between chromosomal locations and mRNA processing are just as important in determining the physiological "fitness" of genes (e.g. their contribution to organismal survival) as are primary nucleotide sequences.

In this section, we will cover the essential relationships between expression connectivity and physical location, with the goal of describing the potential links between gene location, function, and evolution.

4.1
Associative Expression Networks

The field of proteomics has established the clear necessity of describing the coordination between gene expression rates and functional protein activities. This includes a focus on describing the multiple interactions between genes and/or proteins in an environmental context. Correlating single gene expression events (mRNA transcription) with protein-level functioning (specific activities) has been relatively unfruitful, with many variations in gene expression levels not reflected by corresponding changes in their protein pools. We now know that even for such "housekeeping" proteins as lactate dehydrogenase and Na^+- and K^+-ATPase, levels of gene expression and levels of functional protein activity are usually not directly correlated. The regulation of gene expression and protein activities appears to be coordinated at a higher level of physiological organization. We call this organizational structure a "network" to reflect the many parallel connections between gene expression events and protein activity levels.

Gene expression experiments, such as microarrays and gene chips, are widely used to identify genes that appear to function together because of their coordinated expression behavior. However, the best strategy for identifying such "coordination" in expression data is open for debate. The most general approach adopted has been one of paired correlation analysis to produce gene clusters associated by expression patterns. To extract finer-scale interactions between genetic components in terms of induction, repression, modulation, and coordination requires a more concerted analytical scheme to deduce network connectivities between gene or protein members. This level of analysis involves establishing an interactive structure or network around which the data can be quantitatively assessed. Both Boolean and Bayesian network models are currently employed for resolving finer features of interactions and dependencies in multivariate expression data.

One of the key components to describing complex network interactions is designing experimental manipulations of the expression system that will provide a wide range of expression data. The strategy is described as "extreme pathway" modeling or "perturbed expression" profiling. The idea is to generate as many potential variants of expression of different targets due to treatment stressors or transgenic manipulations (knockouts, overexpression).

Gene and protein interactions are then assessed across this diversity of expression patterns. In general, genetic systems now appear to be organized into "small-world" networks, which means they have discrete subnets or local communities that respond coordinately. This observation can simplify an interactive network analysis by limiting the size of the subnet units to be considered. Complete network datasets are cumbersome and suffer from statistical insensitivity because of the large numbers of component members. Isolating smaller subnet units in perturbed expression profiles can lead to the identification of statistically robust, local-community interactions.

An awareness of the adaptive versatility of subnet structuring is rapidly developing in environmental genomics. Simple connections between densely integrated subnets allow for (1) a large degree of flexibility in coordinating the regulation of biochemical and cellular activities, (2) provide a large recombinatorial mechanism for the *de novo* appearance of new functionalities, and (3) are robust in terms of their internal buffering capacity against external inputs. Subnet rewiring is potentially one of the dominant mechanisms of epigenetic interactions impacting species' environmental adaptations and evolution. Using high-order structure analyses to relate complex expression data (microarrays) to cellular function has a large potential to provide new insights into how genomic-level interactions are structured in adaptive network associations.

4.2
Functional Coordination and Physical Mapping

Unfortunately, one of the fundamental, guiding principles of expression network analyses is that coordinately transcribed genes will be functionally related. This assumption is based on the assumption that gene transcription rates are independent and that equivalent patterns of expression between two genes are thus indicative of parallel promoter activities. However, recent studies that have mapped transcriptional activities (from microarray and serial analysis of gene expression (SAGE) analyses) to the corresponding chromosomal gene locations have demonstrated that transcriptionally active domains are heterogeneously distributed throughout the genomes of humans, flies, nematodes, and yeasts. These domains are a very significant feature of mRNA expression dynamics because genes that are closely located on chromosomes are likely to evidence "correlated" expression patterns simply because of their physical location, rather than any functional association.

A Human Transcriptome Map (HTM) has been assembled by the integration of several independent databases from SAGE profile data, the draft human genome, and chromosomal maps. This HTM is accessible through a web portal based at the Amsterdam Medical Center, University of Amsterdam. (http://bioinfo.amc.uva.nl/HTM-bin/index.cgi/) Although the statistical determination of regions of increased gene expression (RIDGE) domains needs further exploration, the basic observation of heterogeneous distributions of transcriptional activity for 2.5×10^6 SAGE transcript tags across 12 tissues types is visually evident in the excellent array of web tools the HTM site provides for exploring specific expression locations. On an average, RIDGEs contain between 6 and 30 mapped genes per centiray (genes cR^{-1}) in contrast to the 1 to 2 genes cR^{-1} in domains that are low in transcriptional activity. In addition, RIDGE segment genes evidence average expression levels that are sevenfold higher than the overall genome average. A rough calculation using gene targets and gene transcription rates would suggest that total transcriptase activity in RIDGEs is between 20 and 200-fold higher than in weakly expressed domains.

The selective mechanism (i.e. the advantage in molecular and biochemical "fitness") for the reason why genes are heterogeneously distributed across chromosomal domains and the reason why transcriptional activities evidence such large spatial differences remains unknown, but certainly not for a lack of speculation. Although there are many logical ideas on the table, the stumbling block is that biological systems are not logical; they do not arise from optimized design principles. Instead, chromosome structure and function have been determined purely by operational efficiency. Logically, the human genome could be as small as ~50 Mb, but operationally, functional competency has required 3.3 Gb, at least at this point in time in the ongoing evolution of the hominid genus (*Homo*).

With this in mind, there appeared to be preliminary support for a hypothesis that tissue-specific patterns of gene expression were coordinated by clustering genes into functional domains in order to unify transcriptional controls. This direct explanation could account for a large component of the variable distribution of genes into distinct, tissue-specific cassettes. However, in a summary comparison of the expression levels of over 11 000 genes from 14 different tissues, Lercher et al. demonstrated that those genes that are tissue-specific in their expression do not cluster together into local domains. To the contrary, their analysis shows that it is the ubiquitous housekeeping genes that do show strong positional clustering within a chromosome. This distribution suggests that the apparent clustering of genes that share high expression rates is primarily a consequence of the local clustering of housekeeping genes, or more generally, genes whose products are necessary in most human cell types.

Although the polytene banding patterns on chromosomes are direct evidence of large-scale spatial heterogeneity in chromosome structure, there does not appear to be any correlation between specific chromosomal structures and transcriptional activity. The expression domains in the fly genome reveal ~200 groups of adjacent genes that have similar patterns of expression across 80 experimental conditions. These groups have 10 to 30 nonrelated genes (in terms of cellular function) covering an average domain size of 100 Kb (range of 20–200 Kb). However, these domains do not evidence any spatial correlations to polytene structure.

The intrachromosome domain model of nuclear organization suggests that genes are more likely to be located at the surface of chromosome territories. In humans and mice, there is substantial evidence to indicate such a differential location of active genes at the surface of chromosomal territories. But the locations of both housekeeping and tissue-specific genes do not appear to be confined to just the peripheral domains of chromosomal territories, indicating that the tight and loose banding of chromatin may not restrict the transcriptional machinery from accessing different locations. There is clearly a near-neighbor effect on the transcription rate of most genes and it is likely that the intercorrelation of gene density and local transcriptional activity influences the gene organization of chromosomes, rather than the activity of individual promoter domains functioning in isolation.

4.3
Physical Colocation and Dislocation

For a better understanding of the distribution of functionally related genes, we have summarized the chromosomal distributions of the genes coding for enzymes involved in the glycolysis and the tricarboxylic acid pathways (TCA) using the Kyoto Encylopedia of Genes and Genomes (Table 1; http://www.genome.ad.jp/kegg/kegg2.html). All human cells are dependent upon the oxidation of glucose via glycolysis and the TCA cycle for generating metabolic energy (ATP and reduced nicotinamide adenine dinucleotide (NADH,H$^+$)). Although these pathways are highly integrated in terms of the production (glycolysis) and utilization (TCA) of acetyl-coenzyme A, the constituent enzymes have very different intracellular locations: glycolysis = cytoplasm; TCA = mitochondria. Thus, one might expect that the transcriptional requirements for maintaining the protein pools in these different cellular locations could serve as a selective force to organize the genes into local domains (i.e. mitchondrial group, etc.). The physical map locations for these genes were obtained from the LocusLink database at the U.S. National Center for Biotechnology Information. (http://www.ncbi.nlm.nih.gov/LocusLink/)

Using the normalized chromosomal ideograms, Fig. 12 traces the sequential order of gene loci that are involved in the catabolism of glucose. Although these

Tab. 1 A serial list of the enzymes involved in glycolysis and the TCA cycle. Chromosomal locations were obtained from the LocusLink database and the loci codes used in Fig. 12 are indicated in the "map" column.

EC#	Enzyme Name	Location	map
2.7.1.1	Hexokinase	10q22	a
5.3.1.9	Glucose-P-isomerase	19q13.1	b
2.7.1.11	6-phosphofructokinase	21q22.3	c
4.1.2.13	Fructose-bis-P aldolase	16q22-q24	d
1.2.1.12	GAP-dehydrogenase	12p13	e
2.7.2.3	Phosphoglycerate kinase	xq13	f
5.4.2.1	Phosphoglycerate mutase	7q31-q34	g
4.2.1.11	Enolase	1p36.3-p36.2	h
2.7.1.40	Pyruvate kinase	1q21	i
1.1.1.27	Lactate dehydrogenase	11p15.4	j
1.2.1.51	Pyruvate dehydrogenase	xp22.2-p22.1	k
4.1.3.7	Citrate synthase	12p11-qter	l
4.2.1.3	Aconitase	9p22-p13	m
1.1.1.41	Isocitrate dehydrogenase	15q25.1	n
1.2.4.2	a-keto dehydrogenase	7p14-p13	o
2.3.1.61	S-succinyltransferase	14q24.3	p
6.2.1.4	Succinyl-CoA synthetase	2p11.2	q
1.3.5.1	Succinic dehydrogenase	5p15	r
4.2.1.2	Fumarase	1q42.1	s
1.1.1.37	Malate dehydrogenase	2p13.3	t
6.4.1.1	Pyruvate carboxylase	11q13.4	u
3.1.3.11	Fructose-bisphosphatase	9q22.3	v

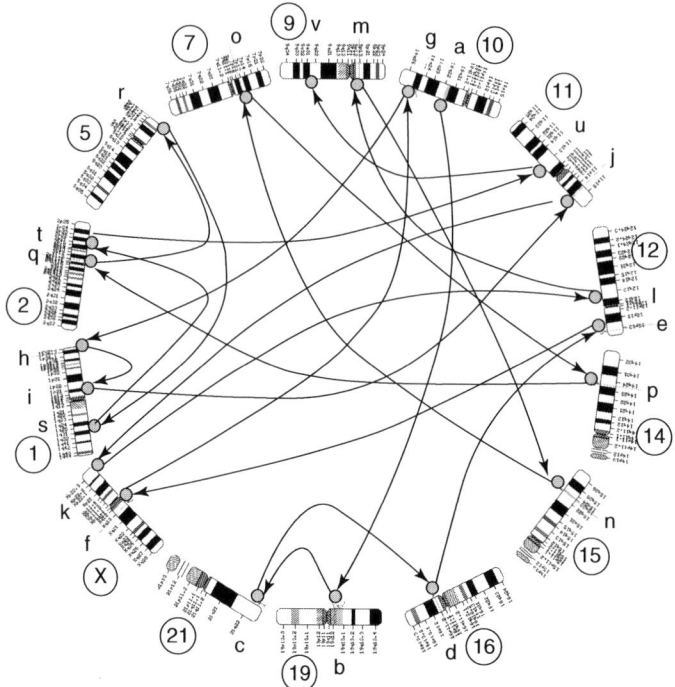

Fig. 12 The chromosomal locations of the gene loci involved in glycolysis and the TCA cycle. Although these metabolic pathways are dominant features in all human cells and are organized into two discrete cellular compartments (cytoplasm and mitochondria), there is no apparent colocalization of these genes within common chromosomal domains.

genes are as fundamental as one can get in terms of housekeeping genes in humans, they are not localized into operational groups of "glucose metabolism" or "mitochondrial locations" or "TCA cycle" cassettes in which their transcription rates could be coordinately controlled.

This result may appear to contradict the positional analysis of Lercher et al. where they conclude that it is the housekeeping genes that do show strong positional clustering within a chromosome. The above analysis of the glucose oxidation housekeeping enzymes (Fig. 12) is meant to illustrate that even though "housekeeping" genes appear to be coordinately located, "housekeeping" is only an operational definition that could easily apply to more than 80% of the genes expressed in any tissue. Consequently, with so many "housekeeping" genes in a genome, and those genes concentrated into domains, there is a high probability that neighboring genes in a location will be "housekeeping" even though they may share no functional relationship in terms of their biochemical activities.

A first glance at Fig. 12 begs the question "is this an organized pathway?" This is because we think of pathways as linear, chain reactions and anticipate a linear organization in the underlying genetic structure (such as in a *Lac* operon type model). However, in most "housekeeping" pathways, component metabolites are often shared with multiple other metabolic pathways, creating an interactive network association of biochemical activities. Thus, there is no realistic mechanism for all the protein-coding loci of a pathway to be colocated if each gene member of that pathway is also a component in several other biochemical pathways. So the startling pattern evident in Fig. 12 is that the 22 primary genes involved in the glucose oxidation machinery are dislocated across 14 chromosomes. And in no instance are any two sequential protein genes colocated.

4.4
Dynamic Expression Linkages

In looking at the dislocation of genes coding for glycolytic proteins in Fig. 12, we struggle with an inability to understand the design by asking the question "why?" Why should functionally related genes be distributed randomly (without apparent pattern) across half of our chromosomes? However, a more direct route for us in terms of deciphering a design mechanism would be to start by trying to answer the question "why not?" From this perspective, a list of observations favoring a random distribution can be summarized as follows:

1. Natural selection is operative at the level of phenotype dysfunction. As long as any process (molecular, biochemical, cellular) does not diminish the overall fitness of the organism in its entirety, that trait is selectively neutral (the mechanism of natural selection eliminates individual genotypes with deleterious traits). In terms of gene organization, it makes no difference in selective fitness for an organism if metabolic pathway genes are colocated or not. As long as their expression levels are appropriately regulated, the way in which they are distributed across the chromosomes does not impact metabolic performance.

2. No chromosome is an island. Distributing functionally related genes across many chromosomes establishes an interdependency on chromosome composition. This could serve to maintain genome integrity over evolutionary timescales because the loss or gain of a chromosome is then catastrophic (i.e. severely compromises the fitness of the individual). Note that the glucose oxidation pathway alone "links" 14 of 24 chromosomes.

3. If two genes need to be coexpressed simultaneously, they do not have to reside next to each other. The only necessary constraint is that they need to be located downstream from promoter domains that respond to the same regulatory controls. The implication here is a very important point because it suggests that although gene location may not show distributional patterns, it is likely that promoter and regulatory sequences do.

4. Genomes are incredibly successful at adapting to the necessity of generating multiple expression profiles (development, tissue-specific, stress-response, environmental adaptation). This genotype to phenotype expression requires a great potential for recombinatorial diversity. Keeping pathway components separated allows for a large biological potential in expression diversity because sequential pathway reactions are not "hard-wired" (directly linked) to one another.

Of these four points, the first and second fall within the larger realm of organismal evolution and are beyond the scope of this article. The third and fourth points represent the current frontiers of exploration in understanding eukaryotic genome organization.

The physical colocation of genes is not a necessary requirement of coordinated expression. In fact, it would appear to be the exception rather than the rule in terms of how genes are distributed across chromosomes. Given that genes are unequally distributed across chromosomes and that local transcriptional activities vary as a function of local gene densities (Sect. 4.2), a first hypothesis for a mechanism to establish parallel transcriptional controls would be for genes requiring similar expression patterns to be located in similar genetic neighborhoods (i.e. similar gene densities and transcriptional activities). To assess this potential for expression linkage, we have examined the local domains of the glycolytic enzymes (first nine enzymes in Table 1) looking at the number of sequence tags within a ±1 cR window surrounding a locus and the expression levels of those tagged transcripts. These cR measurements have been assessed using the GB-94 panel (GeneMap'99) where, 1 cR corresponds to ~280 kb of DNA. In Fig. 13, the number of SAGE tags are used as an index of the transcriptional activity surrounding each glycolytic gene locus, with the domain total equivalent to the total amount of transcription in the 2 cR window and the gene total being the

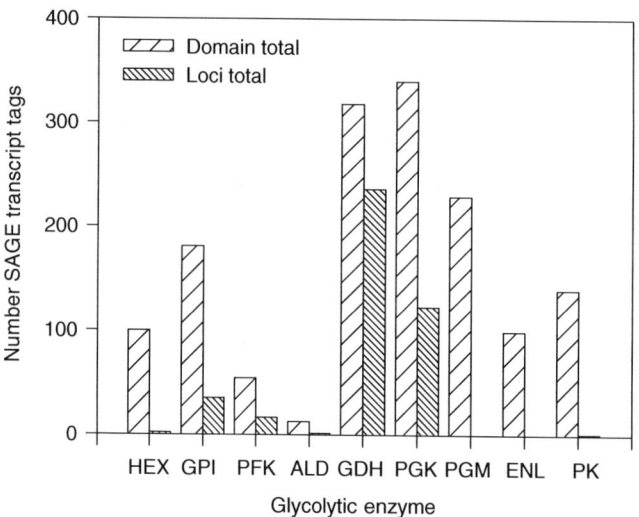

Fig. 13 Transcriptional activity in the loci domains for the glycolytic enzymes presented in Table 1. SAGE transcript tags for a ±1 cR region (GeneMap'99-GB4) around each gene loci were enumerated using the Human Transcriptome Map resources for average expression levels across 12 tissue types total. The histogram bars plot: (1) the total transcriptional activity over a 2 cR region (number of SAGE marker hits), and (2) the transcriptional activity of each gene locus (see chromosomal location data in Table 1). Despite the coordinated, biochemical activity of these genes for glycolysis, there appears to be no direct common transcriptional features regarding the chromosomal domains where these genes reside.

transcriptional activity of the individual locus. Here, there is no apparent relationship between the transcriptional domains of sequential glycolytic gene loci. The only two genes that come close to having similar domains are GDH and PGK, each with high domain totals and high individual levels of expression.

Much significance has been placed on the domain organization of genes into regions of differential expression activity; however, this appears to be a simple tautological observation, that is, that areas dense with genes show higher expression levels because they have more genes. This was one of the fundamental hypotheses of Bernardi in describing the phenomenological nature of isochore organization, which we have seen substantiated. Thus, future efforts to correlate gene expression levels across wide domains do not appear to be a fruitful avenue of research. The genetic complexity of a 500-Mb window is just too great to summarize with a single, average metric. Although it is a conceptually enticing approach in terms of design simplification, measurements of SAGE hits, %GC, GC_3 frequency, CpG islands, isochore boundaries, and so on, cannot describe the molecular mechanism underlying the dynamic expression linkages that are evident.

Intriguing new insights into the distribution of cis-regulatory elements across chromosomes is beginning to show distinctive spatial patterns that could potentially lead to an understanding of why genes are distributed the way they are. Characterizing sequence structures over small windows (~100 bp) has revealed a remarkable degree of consistent patterns in upstream regulatory sequences and exon/intron splicing boundaries. These patterns indicate a consistent spatial organization in regulatory domains that is independent of chromosomal location. Two big challenges to overcome in clearly identifying such cis-regulatory mechanisms is the difficulty of finding statistically significant sequence motifs in such short motifs and the development of quantitative models of regulatory interactions that can explain sequence structural organization.

4.5
Importance of Expression Controls

Another new frontier in assessing genomic organization is describing the mechanistic basis for the recombinatorial diversity of gene expression activities. In eukaryotes, gene promoter domains are rarely triggered by single activators or repressors. Instead, multiple interactions between several target sites within a promoter domain, and between several activating or repressing protein factors at each of those sites, results in the regulation of transcription rates. Consequently, the expression of a gene within a cell is more accurately described as a probabilistic function, rather than a deterministic one, that is, all cells within a tissue do not have the same expression profile at any instant in time. There is the potential for significant variation in expression levels, which is one of the key features of the successful organization of a genome in perpetuating a biological system.

We are now just beginning to see data quantifying levels of variance in gene expression rates. In a comprehensive study of expression patterns in yeast cells, only 7% of the statistically significant changes in gene expression levels following experimental stimuli were actually functional, that is, directly related to altering the cellular physiology of the yeast to accommodate the stimuli. The remaining 93% of "significant" gene expression activity was essentially "random." In vertebrates, the significance of accounting for individual variance in gene expression levels has been documented in fish, and this approach is now being applied to screening human microarray expression data for disease detection by establishing a critical threshold of detection relative to differences in expression levels between different individuals.

Thirty years ago, scientists were amazed at how similar in composition the chromosomes of chimpanzees and humans appeared to be. This lead to the hypothesis that one of the primary forces in species evolution was not mutations in coding gene domains, but subtle alterations in the timing and expression levels of those genes. When it became clear that humans essentially had only three times the number of genes as in the fly *Drosophila*, we realized that coding gene domains were not the only evolutionary game in play. The large difference in biological system organization between humans and flies is likely impacted by the timing, magnitude, and coordination of gene expression events just as significantly as gene exon domains. Overall, the probabilistic (not deterministic) nature of gene expression controls will likely become one of the primary mechanisms determining how and where genes are distributed within the human genome.

See also Gene Mapping and Chromosome Evolution by Fluorescence–Activated Chromosome Sorting; Gene Mapping by Fluorescence In Situ Hybridization; Molecular Basis of Genetics; Construction and Applications of Genomic DNA Libraries.

Bibliography

Books and Reviews

Dale, J., von Schantz, M. (2002) *From Genes to Genomes: Concepts and Applications of DNA Technology*, Wiley, Chichester.

Lander, E.S. et al. (2001) Initial sequencing and analysis of the human genome, *Nature* **409**, 860–921.

Malcolm, S., Goodship, J. (Eds.) (2001) *Genotype to Phenotype*, BIOS Scientific, Oxford; Academic Press, San Diego, CA..

Marks, J. (2002) *What it means to be 98% Chimpanzee: Apes, People, and their Genes*, University of California Press, Berkeley.

Maroni, G. (2001) *Molecular and Genetic Analysis of Human Traits*, Blackwell Science, Malden, Mass.

Miesfeld, R.L. (1999) *Applied Molecular Genetics*, John Wiley, New York.

Saccone, C., Pesole, G. (2003) *Handbook of Comparative Genomics: Modern Methodology*, Wiley-Liss, New York.

Venter, J.C. et al. (2001) The sequence of the human genome, *Science* **291**, 1304–1351.

Weissmann, G. (2002) *The Year of the Genome: A Diary of the Biological Revolution*, Times Books, New York.

Primary Literature

Adachi, N., Lieber, M.R. (2002) Bidirectional gene organization: a common architectural feature of the human genome, *Cell* **109**, 807–809.

Adams, M.D. et al. (1991) Complementary DNA sequencing: expressed sequence tags and human genome project, *Science* **252**, 1651–1656.

Albert, R., Barabasi, A.L. (2000) Dynamics of complex systems: scaling laws for the period of Boolean networks, *Phys. Rev. Lett.* **84**, 5660–5663.

Albert, R. et al. (2000) Error and attack tolerance of complex networks, *Nature* **406**, 378–382.

Alon, U. et al. (1999) Broad patterns of gene expression revealed by clustering analysis of tumor and normal colon tissues probed by oligonucleotide arrays, *Proc. Natl. Acad. Sci. U.S.A.* **96**, 6745–6750.

Alvarez-Valin, F. et al. (2002) Isochores, GC(3) and mutation biases in the human genome, *Gene* **300**, 161–168.

Amaral, L.A.N. et al. (2000) Classes of small-world networks, *Proc. Natl. Acad. Sci. U.S.A.* **97**, 11149–11152.

Antequera, F., Bird, A. (1993) Number of CpG islands and genes in human and mouse, *Proc. Natl. Acad. Sci. U.S.A.* **90**, 11995–11999.

Antequera, F., Bird, A. (1999) CpG islands as genomic footprints of promoters that are associated with replication origins, *Curr. Biol.* **9**, 661–667.

Arkhipova, I.R. (2001) Transposable elements in the animal kingdom, *Mol. Biol.* **35**, 157–167.

Bailey, J.A. et al. (2002) Recent segmental duplications in the human genome, *Science* **297**, 1003–1007.

Barthelemy, M., Amaral, L.A.N. (1999) Small-world networks: evidence for a crossover picture, *Phys. Rev. Lett.* **82**, 5180–5180.

Bernardi, G. (1993) The isochore organization of the human genome and its evolutionary history – a review, *Gene* **135**, 57–66.

Bernardi, G. (1993) The vertebrate genome: isochores and evolution, *Mol. Biol. Evol.* **10**, 186–204.

Bernardi, G. (1995) The human genome: organization and evolutionary history, *Annu. Rev. Genet.* **29**, 445–476.

Bernardi, G. (2000) Isochores and the evolutionary genomics of vertebrates, *Gene* **241**, 3–17.

Bernardi, G. (2001) Isochores: myth or reality? *Med. Sci.* **17**, 1097.

Bernardi, G. (2001) Misunderstandings about isochores. Part 1, *Gene* **276**, 3–13.

Bernardi, G. et al. (1973) The specificity of deoxyribonucleases and their use in nucleotide sequence studies, *Nat. N. Biol.* **246**, 36–40.

Bernardi, G. et al. (1985) The mosaic genome of warm-blooded vertebrates, *Science* **228**, 953–958.

Bird, A.P. (1986) CpG-rich islands and the function of DNA methylation *Nature* **321**, 209–213.

Bolouri, H., Davidson, E.H. (2002) Modeling DNA sequence-based cis-regulatory gene networks, *Dev. Biol.* **246**, 2–13.

Bortoluzzi, S. et al. (1998) A comprehensive, high-resolution genomic transcript map of human skeletal muscle, *Genome Res.* **8**, 817–825.

Brazma, A., Vilo, J. (2000) Gene expression data analysis, *FEBS Lett.* **480**, 17–24.

Brudno, M. et al. (2001) Computational analysis of candidate intron regulatory elements for tissue-specific alternative pre-mRNA splicing, *Nucleic Acids Res.* **29**, 2338–2348.

Burton, R.S. et al. (1999) Genetic architecture of physiological phenotypes: Empirical evidence for coadapted gene complexes, *Am. Zool.* **39**, 451–462.

Bustamante, C.D. et al. (2002) A maximum likelihood method for analyzing pseudogene evolution: Implications for silent site evolution in humans and rodents, *Mol. Biol. Evol.* **19**, 110–117.

Cargill, M. et al. (1999) Characterization of single-nucleotide polymorphisms in coding regions of human genes, *Nat. Genet.* **22**, 231–238.

Caron, H. et al. (2001) The human transcriptome map: Clustering of highly expressed genes in chromosomal domains, *Science* **291**, 1289–1292.

Clay, O., Bernardi, G. (2001) Compositional heterogeneity within and among isochores in mammalian genomes. II. Some general comments, *Gene* **276**, 25–31.

Cohen, B.A. et al. (2000) A computational analysis of whole-genome expression data reveals chromosomal domains of gene expression, *Nat. Genet.* **26**, 183–186.

Corneo, G. et al. (1968) Isolation and characterization of mouse and guinea pig satellite deoxyribonucleic acids, *Biochemistry* **7**, 4373–4379.

Corthals, G.L. et al. (2000) The dynamic range of protein expression: a challenge for proteomic research, *Electrophoresis* **21**, 1104–1115.

Coulondre, C. et al. (1978) Molecular basis of base substitution hotspots in Escherichia coli, *Nature* **274**, 775–780.

Crawford, D.L. et al. (1999) Evolutionary physiology of closely related taxa: analyses of enzyme expression, *Am. Zool.* **39**, 389–400.

Crawford, D.L. et al. (1999) Evolutionary analysis of TATA-less proximal promoter function, *Mol. Biol. Evol.* **16**, 194–207.

Cross, S.H. et al. (2000) CpG island libraries from human chromosomes 18 and 22: landmarks for novel genes, *Mamm. Genome* **11**, 373–383.

Das, M. et al. (2001) Assessment of the total number of human transcription units, *Genomics* **77**, 71–78.

Davidson, E.H. et al. (2002) A genomic regulatory network for development, *Science* **295**, 1669–1678.

de Krom, M. et al. (2002) Stochastic patterns in globin gene expression are established prior to transcriptional activation and are clonally inherited, *Mol. Cell* **9**, 1319–1326.

de la Fuente, A. et al. (2002) Linking the genes: inferring quantitative gene networks from microarray data, *Trends Genet.* **18**, 395–398.

Deloukas, P. et al. (1998) A physical map of 30 000 human genes, *Science* **282**, 744–746.

DeRisi, J.L. et al. (1997) Exploring the metabolic and genetic control of gene expression on a genomic scale, *Science* **278**, 680–686.

D'Haeseleer, P. et al. (2000) Genetic network inference: from co-expression clustering to reverse engineering, *Bioinformatics* **16**, 707–726.

Dolstra, H. et al. (2002) Bi-directional allelic recognition of the human minor histocompatibility antigen HB-1 by cytotoxic T lymphocytes, *Eur. J. Immunol.* **32**, 2748–2758.

D'Onofrio, G. (2002) Expression patterns and gene distribution in the human genome, *Gene* **300**, 155–160.

D'Onofrio, G. et al. (1999) Evolutionary genomics of vertebrates and its implications, *Ann. N. Y. Acad. Sci.* **870**, 81–94.

Drouin, R. et al. (1994) High-resolution replication bands compared with morphologic G- and R-bands, *Adv. Hum. Genet.* **22**, 47–115.

Duret, L. et al. (1995) Statistical analysis of vertebrate sequences reveals that long genes are scarce in GC-rich isochores, *J. Mol. Evol.* **40**, 308–317.

Dutrillaux, B. (1977) A theoretical model of structural chromosome rearrangement induction, *Ann. Genet.* **20**, 221–226.

Dutrillaux, B., Covic, M. (1974) Factors influencing the heat denaturation of chromosomes, *Exp. Cell Res.* **85**, 143–153.

Echols, N. et al. (2002) Comprehensive analysis of amino acid and nucleotide composition in eukaryotic genomes, comparing genes and pseudogenes, *Nucleic Acids Res.* **30**, 2515–2523.

Eisen, M.B. et al. (1998) Cluster analysis and display of genome-wide expression patterns, *Proc. Natl. Acad. Sci. U.S.A.* **95**, 14863–14868.

Eisen, M.B. et al. (1999) Cluster analysis and display of genome-wide expression patterns, *Proc. Natl. Acad. Sci. U.S.A.* **96**, 10943.

Epstein, C., Butow, R. (2000) Microarray technology-enhanced versatility, persistent challenge, *Curr. Opin. Biotechnol.* **11**, 36–41.

Ewing, B., Green, P. (2000) Analysis of expressed sequence tags indicates 35 000 human genes, *Nat. Genet.* **25**, 232–234.

Fields, C. et al. (1994) How many genes in the human genome? *Nat. Genet.* **7**, 345–346.

Finnegan, E.J. (2002) Epialleles – a source of random variation in times of stress, *Curr. Opin. Plant Biol.* **5**, 101–106.

Forst, C. (2002) Network genomics – a novel approach for the analysis of biological systems in the post-genomic era, *Mol. Biol. Rep.* **29**, 265–280.

Forst, C.V., Schulten, K. (2001) Phylogenetic analysis of metabolic pathways, *J. Mol. Evol.* **52**, 471–489.

Friedman, N. et al. (2000) Using Bayesian networks to analyze expression data, *J. Comput. Biol.* **7**, 601–620.

Frith, M.C. et al. (2002) Statistical significance of clusters of motifs represented by position specific scoring matrices in nucleotide sequences, *Nucleic Acids Res.* **30**, 3214–3224.

Fujiyama, A. et al. (2002) Construction and analysis of a human-chimpanzee comparative clone map, *Science* **295**, 131–134.

Gallo, A. et al. (2002) Micro-processing events in mRNAs identified by DHPLC analysis, *Nucleic Acids Res.* **30**, 3945–3953.

Gardiner-Garden, M., Frommer, M. (1987) CpG islands in vertebrate genomes, *J. Mol. Biol.* **196**, 261–282.

Gasch, A.P. et al. (2000) Genomic expression programs in the response of yeast cells to environmental changes, *Mol. Biol. Cell* **11**, 4241–4257.

Ge, H. et al. (2001) Correlation between transcriptome and interactome mapping data from Saccharomyces cerevisiae, *Nat. Genet.* **29**, 482–486.

Gerstein, M., Jansen, R. (2000) The current excitement in bioinformatics, analysis of whole-genome expression data: how does it relate to protein structure and function, *Curr. Opin. Struct. Biol.* **10**, 574–584.

Giaever, G., et al. (2002) Functional profiling of the Saccharomyces cerevisiae genome, *Nature* **418**, 387–391.

Gibson, G. (2002) Developmental evolution: getting robust about robustness, *Curr. Biol.* **12**, R347–R349.

Gibson, G. (2002) Microarrays in ecology and evolution: a preview, *Mol. Ecol.* **11**, 17–24.

Girvan, M., Newman, M.E.J. (2002) Community structure in social and biological networks, *Proc. Natl. Acad. Sci. U.S.A.* **99**, 7821–7826.

Goncalves, I. et al. (2000) Nature and structure of human genes that generate retropseudogenes, *Genome Res.* **10**, 672–678.

Greenbaum, D. et al. (2002) Analysis of mRNA expression and protein abundance data: an approach for the comparison of the enrichment of features in the cellular population of proteins and transcripts, *Bioinformatics* **18**, 585–596.

Grunau, C. et al. (2000) Large-scale methylation analysis of human genomic DNA reveals tissue-specific differences between the methylation profiles of genes and pseudogenes, *Hum. Mol. Genet.* **9**, 2651–2663.

Halushka, M.K. et al. (1999) Patterns of single-nucleotide polymorphisms in candidate genes for blood-pressure homeostasis, *Nat. Genet.* **22**, 239–247.

Harrison, P.M., Gerstein, M. (2002) Studying genomes through the aeons: protein families, pseudogenes and proteome evolution, *J. Mol. Biol.* **318**, 1155–1174.

Harrison, P.M. et al. (2002) Molecular fossils in the human genome: identification and analysis of the pseudogenes in chromosomes 21 and 22, *Genome Res.* **12**, 272–280.

Hartemink, A.J. et al. (2001) Using graphical models and genomic expression data to statistically validate models of genetic regulatory networks, *Pac. Symp. Biocomput.* 422–433.

Hatzimanikatis, V. et al. (1999) Proteomics: theoretical and experimental considerations, *Biotechnol. Prog.* **15**, 312–318.

Hochachka, P.W. et al. (1998) Integrating metabolic pathway fluxes with gene-to-enzyme expression rates, *Comp. Biochem. Physiol.* **120B**, 17–26.

Holmquist, G.P. (1992) Chromosome bands, their chromatin flavors, and their functional features, *Am. J. Hum. Genet.* **51**, 17–37.

Holter, N.S. et al. (2000) Fundamental patterns underlying gene expression profiles: simplicity from complexity, *Proc. Natl. Acad. Sci. U.S.A.* **97**, 8409–8414.

Huang, S. (1999) Gene expression profiling, genetic networks, and cellular states: an integrating concept for tumorigenesis and drug discovery, *J. Mol. Med.* **77**, 469–480.

Hubbard, T., Birney, E. (2000) Open annotation offers a democratic solution to genome sequencing, *Nature* **403**, 825.

Hurst, L.D., Eyre-Walker, A. (2000) Evolutionary genomics: reading the bands, *BioEssays* **22**, 105–107.

Ideker, T. et al. (2001) Integrated genomic and proteomic analyses of a systematically perturbed metabolic network, *Science* **292**, 929–934.

Ioshikhes, I.P., Zhang, M.Q. (2000) Large-scale human promoter mapping using CpG islands, *Nat. Genet.* **26**, 61–63.

Jablonka, E. et al. (1998) 'Lamarckian' mechanisms in Darwinian evolution, *Trends Ecol. Evol.* **13**, 206–210.

Jackson, R.B. et al. (2002) Linking molecular insight and ecological research, *Trends Ecol. Evol.* **17**, 409–414.

Jacob, F., Monod, J. (1961) Genetic regulatory mechanisms in the synthesis of proteins, *J. Mol. Biol.* **3**, 318–356.

Jones, P.A., Takai, D. (2001) The role of DNA methylation in mammalian epigenetics, *Science* **293**, 1068–1070.

Kerr, M.K., Churchill, G.A. (2001) Bootstrapping cluster analysis: assessing the reliability of conclusions from microarray experiments, *Proc. Natl. Acad. Sci. U.S.A.* **98**, 8961–8965.

Kholodenko, B.N. et al. (2002) Untangling the wires: a strategy to trace functional interactions in signaling and gene networks, *Proc. Natl. Acad. Sci. U.S.A.* **99**, 12841; **99**, 15245.

Kidd, D. (2001) Profiling serine hydrolase activities in complex proteomes, *Biochemistry* **40**, 4005–4015.

Kidwell, M.G., Lisch, D.R. (2001) Perspective: transposable elements, parasitic DNA, and genome evolution, *Evolution* **55**, 1–24.

Kikukawa, Y. et al. (2002) The 26S proteasome Rpn10 gene encoding splicing isoforms: Evolutional conservation of the genomic organization in vertebrates, *Biol. Chem.* **383**, 1257–1261.

King, M.C., Wilson, A.C. (1975) Evolution at two levels: molecular similarities and biological differences between humans and chimpanzees, *Science* **188**, 107–116.

Kohler, C., Grossniklaus, U. (2002) Epigenetics: the flowers that come in from the cold, *Curr. Biol.* **12**, R129–R131.

Kulp, D. et al. (1996) A generalized hidden Markov model for the recognition of human genes in DNA, *Proc. Int. Conf. Intell. Syst. Mol. Biol.* **4**, 134–142.

Lander, E.S. et al. (2001) Initial sequencing and analysis of the human genome, *Nature* **409**, 860–921.

Larsen, F., et al. (1992) CpG islands as gene markers in the human genome, *Genomics* **13**, 1095–1107.

Leibovitch, B.A. (2002) Chromatin structure, heterochromatin, and transposable genetic elements: Are these teammates? *Mol. Biol.* **36**, 189–195.

Lercher, M.J. et al. (2002) Clustering of housekeeping genes provides a unified model of gene order in the human genome, *Nat. Genet.* **31**, 180–183.

Li, W. (2001) Delineating relative homogeneous G + C domains in DNA sequences, *Gene* **276**, 57–72.

Liang, F. et al. (2000) Gene index analysis of the human genome estimates approximately 120 000 genes, *Nat. Genet.* **25**, 239, 240.

Long, M., Langely, C.H. (1993) Natural selection and the origin of *jingwei*, a chimeric processed functional gene in Drosophila, *Science* **260**, 91–95.

Lukashin, A.V., Fuchs, R. (2001) Analysis of temporal gene expression profiles: clustering by simulated annealing and determining the optimal number of clusters, *Bioinformatics* **17**, 405–414.

MacKenzie, S. et al. (2002) Post-transcriptional regulation of TNF-alpha during in vitro differentiation of human monocytes/macrophages in primary culture, *J. Leukocyte Biol.* **71**, 1026–1032.

Mahy, N.L. et al. (2002) Gene density and transcription influence the localization of chromatin outside of chromosome territories detectable by FISH, *J. Cell Biol.* **159**, 753–763.

Mahy, N.L. et al. (2002) Spatial organization of active and inactive genes and noncoding DNA within chromosome territories, *J. Cell Biol.* **157**, 579–589.

Majewski, J., Ott, J. (2002) Distribution and characterization of regulatory elements in the human genome, *Genome Res.* **12**, 1827–1836.

Makalowski, W. (2000) Genomic scrap yard: how genomes utilize all that junk, *Gene* **259**, 61–67.

Marcotte, E.M. (2001) Measuring the dynamics of the proteome, *Genome Res.* **11**, 191–193.

Marsh, A.G. et al. (2000) Gene expression and enzyme activities of the sodium pump during sea urchin development: implications for indices of physiological state, *Biol. Bull.* **199**, 100–107.

Mattick, J.S., Gagen, M.J. (2001) The evolution of controlled multitasked gene networks: The role of introns and other noncoding RNAs in the development of complex organisms, *Mol. Biol. Evol.* **18**, 1611–1630.

McAdams, H.H., Arkin, A. (1999) It's a noisy business! Genetic regulation at the nanomolar scale, *Trends Genet.* **15**, 65–69.

McCullough, A.J., Berget, S.M. (1997) G triplets located throughout a class of small vertebrate introns enforce intron borders and regulate splice site selection, *Mol. Cell. Biol.* **17**, 4562–4571.

Miklos, G.L., John, B. (1979) Heterochromatin and satellite DNA in man: properties and prospects, *Am. J. Hum. Genet.* **31**, 264–280.

Mira, A., Ochman, H. (2002) Gene location and bacterial sequence divergence, *Mol. Biol. Evol.* **19**, 1350–1358.

Mouchiroud, D. et al. (1991) The distribution of genes in the human genome, *Gene* **100**, 181–187.

Musio, A. et al. (2002) Heterogeneous gene distribution reflects human genome complexity as detected at the cytogenetic level, *Can. Genet. Cytogenet.* **134**, 168–171.

Nakao, M. (2001) Epigenetics: interaction of DNA methylation and chromatin, *Gene* **278**, 25–31.

Newman, S.A., Muller, G.B. (2000) Epigenetic mechanisms of character origination, *J. Exp. Zool.* **288**, 304–317.

Nijhout, H.F. (2002) The nature of robustness in development, *BioEssays* **24**, 553–563.

Oleksiak, M.F. et al. (2001) Utility of natural populations for microarray analyses: isolation of genes necessary for functional genomic studies, *Marine Biotech.* **3**, S203–S211.

Oleksiak, M.F. et al. (2002) Variation in gene expression within and among natural populations, *Nat. Genet.* **32**, 261–266.

Oliver, J.L. et al. (2002) Isochore chromosome maps of the human genome, *Gene* **300**, 117–127.

Ozbudak, E.M. et al. (2002) Regulation of noise in the expression of a single gene, *Nat. Genet.* **31**, 69–73.

Pal, C., Miklos, I. (1999) Epigenetic inheritance, genetic assimilation and speciation, *J. Theor. Biol.* **200**, 19–37.

Papin, J.A. et al. (2002) The genome-scale metabolic extreme pathway structure in

Haemophilus influenzae shows significant network redundancy, *J. Theor. Biol.* **215**, 67–82.

Pavlicek, A. et al. (2002) A compact view of isochores in the draft human genome sequence, *FEBS Lett.* **511**, 165–169.

Pe'er, D. et al. (2001) Inferring subnetworks from perturbed expression profiles, *Bioinformatics* **17**, S215–S224.

Pesole, G., et al. (1999) Isochore specificity of AUG initiator context of human genes, *FEBS Lett.* **464**, 60–62.

Petrov, D.A., Hartl, D.L. (2000) Pseudogene evolution and natural selection for a compact genome, *J. Hered.* **91**, 221–227.

Pierce, V.A., Crawford, D.L. (1997) Phylogenetic analysis of thermal acclimation of the glycolytic enzymes in the genus Fundulus, *Physiol. Zool.* **70**, 597–609.

Plass, C. (2002) Cancer epigenomics, *Hum. Mol. Genet.* **11**, 2479–2488.

Ponger, L. et al. (2001) Determinants of CpG islands: expression in early embryo and isochore structure, *Genome Res.* **11**, 1854–1860.

Price, N.D. et al. (2002) Determination of redundancy and systems properties of the metabolic network of Helicobacter pylori using genome-scale extreme pathway analysis, *Genome Res.* **12**, 760–769.

Pruitt, K.D., Maglott, D.R. (2001) RefSeq and LocusLink: NCBI gene-centered resources, *Nucleic Acids Res.* **29**, 137–140.

Roest Crollius, H. et al. (2000) Estimate of human gene number provided by genome-wide analysis using Tetraodon nigroviridis DNA sequence, *Nat. Genet.* **25**, 235–238.

Roy, P.J. et al. (2002) Chromosomal clustering of muscle-expressed genes in Caenorhabditis elegans, *Nature* **418**, 975–979.

Saccone, S. et al. (1992) The highest gene concentrations in the human genome are in telomeric bands of metaphase chromosomes, *Proc. Natl. Acad. Sci. U.S.A.* **89**, 4913–4917.

Saccone, S. et al. (1996) Identification of the gene-richest bands in human chromosomes, *Gene* **174**, 85–94.

Saccone, S. et al. (1999) Identification of the gene-richest bands in human prometaphase chromosomes, *Chromosome Res.* **7**, 379–386.

Saccone, C. et al. (2002) Molecular strategies in Metazoan genomic evolution, *Gene* **300**, 195–201.

Schlosser, G. (2002) Modularity and the units of evolution, *Theory Biosci.* **121**, 1–80.

Schuler, G.D. et al. (1996) A gene map of the human genome, *Science* **274**, 540–546.

Shapiro, J.A. (2002) A 21(st) century view of evolution, *J. Biol. Phys.* **28**, 745–764.

Sherlock, G. (2000) Analysis of large-scale gene expression data, *Curr. Opin. Immun.* **12**, 201–205.

Smolke, C.D., Keasling, J.D. (2002) Effect of gene location, mRNA secondary structures, and RNase sites on expression of two genes in an engineered operon, *Biotechnol. Bioeng.* **80**, 762–776.

Somogyi, R., et al. (1997) The gene expression matrix: Towards the extraction of genetic network architectures, *Nonlinear Anal. Theory Methods Appl.* **30**, 1815–1824.

Spellman, P.T., Rubin, G. (2002) Evidence for large domains of similarly expressed genes in the Drosophila genome, *J. Biol.* **1**, 5.(article)

Strehl, S. et al. (1997) High-resolution analysis of DNA replication domain organization across an R/G-band boundary, *Mol. Cell Biol.* **17**, 6157–6166.

Sumner, A.T. (1982) The nature and mechanisms of chromosome banding, *Can. Genet. Cytogenet.* **6**, 59–87.

Thieffry, D., Romero, D. (1999) The modularity of biological regulatory networks, *Biosystems* **50**, 49–59.

Toh, H., Horimoto, K. (2002) Inference of a genetic network by a combined approach of cluster analysis and graphical Gaussian modeling, *Bioinformatics* **18**, 287–297.

Venter, J.C. et al. (2001) The sequence of the human genome, *Science* **291**, 1304–1351.

Wagner, A. (2002) Estimating coarse gene network structure from large-scale gene perturbation data, *Genome Res.* **12**, 309–315.

Wagner, A., Fell, D.A. (2001) The small world inside large metabolic networks, *Proc. R. Soc. London, Ser. B-Biol. Sci.* **268**, 1803–1810.

Waterston, R. et al. (1992) A survey of expressed genes in Caenorhabditis elegans, *Nat. Genet.* **1**, 114–123.

Watts, D.J., Strogatz, S.H. (1998) Collective dynamics of 'small-world' networks, *Nature* **393**, 440–442.

Wessels, L.F. et al. (2001) A comparison of genetic network models, *Pac. Symp. Biocomput.* 508–519.

Williams, E.J.B., Hurst, L.D. (2002) Clustering of tissue-specific genes underlies much of the similarity in rates of protein evolution of linked genes, *J. Mol. Evol.* **54**, 511–518.

Woischnik, M., Moraes, C.T. (2002) Pattern of organization of human mitochondrial pseudogenes in the nuclear genome, *Genome Res.* **12**, 885–893.

Wolf, Y.I. et al. (2002) Scale-free networks in biology: new insights into the fundamentals of evolution? *BioEssays* **24**, 105–109.

Wray, G.A. (2002) Evolution of the gene network underlying wing polyphenism in ants, *Science* **297**, 249–252.

Yang, T.H., Somero, G.N. (1996) Activity of lactate dehydrogenase but not its concentration of messenger RNA increases with body size in barred sand bass, Paralabrax nebulifer (Teleostei), *Biol. Bull.* **191**, 155–158.

Yang, T.H., Somero, G.N. (1996) Fasting reduces protein and messenger RNA concentrations for lactate dehydrogenase but not for actin in white muscle of scorpion fish (Scorpaena guttata, Teleostei), *Mol. Marine Biol. Biotechnol.* **5**, 153–161.

Yeung, K.Y. et al. (2001) Validating clustering for gene expression data, *Bioinformatics* **17**, 309–318.

Zhuo, D. et al. (2001) Assembly, annotation, and integration of UNIGENE clusters into the human genome draft, *Genome Res.* **11**, 904–918.

Zoubak, S. et al. (1996) The gene distribution of the human genome, *Gene* **174**, 95–102.

9
Anthology of Human Repetitive DNA

Vladimir V. Kapitonov, Adam Pavlicek, and Jerzy Jurka
Genetic Information Research Institute, 2081 Landings Drive, Mountain View, CA, USA

1	Introduction	273
2	**Tandemly Arrayed Repeats**	**273**
2.1	Microsatellites and Minisatellites	274
2.2	Satellites and Telomeric Repeats	275
2.2.1	Centromeric Satellites	276
2.2.2	Telomeric and Subtelomeric Repeats	278
3	**Interspersed Repeats**	**279**
3.1	Non-LTR Retrotransposons	281
3.1.1	L1 Non-LTR Retrotransposons	281
3.1.2	L1-dependent Nonautonomous Non-LTR Retrotransposons	283
3.1.3	CR1-like Non-LTR Retrotransposons	286
3.2	LTR Retrotransposons	287
3.3	DNA Transposons	297
3.3.1	hAT-like Transposons	298
3.3.2	Mariner/Tc1-like Transposons	300
3.3.3	piggyBac-like Transposons	300
3.3.4	MuDR-like Transposons	300
4	**Repetitive Elements and Human Diseases**	**301**
5	**Potential Contribution of Repetitive DNA to Cellular Functions**	**313**
6	**Databases and Programs for Analysis of Repetitive DNA**	**316**
6.1	Databases of Repetitive Elements	316
6.2	Programs for Detection and Analysis of Repetitive DNA	317

Genomics and Genetics. Edited by Robert A. Meyers.
Copyright © 2007 Wiley-VCH Verlag GmbH & Co. KGaA, Weinheim
ISBN: 978-3-527-31609-0

Bibliography 319
Books and Reviews 319
Primary Literature 319
Seminal Early References 320
Recent Key Primary Literature 321

Keywords

AP Endonuclease
An enzyme that cleaves apyrimidinic or apurinic sites in DNA.

Autonomous Transposable Element
A mobile element encoding enzyme(s) necessary for its proliferation in the host genome. All other mobile elements are called nonautonomous.

Consensus Sequence
A DNA or protein sequence made of nucleotides or amino acid residues prevalent at each position in a set of multiply aligned homologous sequences.

cDNA
A DNA copy of an RNA molecule created during reverse transcription.

DNA Transposon
A transposable element that moves from one genomic site to another in the form of DNA only.

Family of Repetitive Elements
Multiple DNA copies derived from a number of related transposable elements called source, founder, or master genes. A repetitive family can be divided into subfamilies, each corresponding to a separate founder gene.

Interspersed Repeats
Randomly inserted copies of currently or historically active transposable elements.

Long Interspersed Repeat (LINE) Element
A non-LTR retrotransposon encoding reverse transcriptase and endonuclease catalyzing nuclear reverse transcription and integration of cDNA in the host genome.

Long Terminal Repeats (LTRs)
200 to 3000 bp long identical DNA sequences flanking a provirus sequence.

LTR Retrotransposons
A class of retrotransposons that are reverse transcribed in the cytoplasm. The resulting cDNA is integrated into the host genome, forming a provirus flanked by long terminal repeats.

Microsatellites
DNA sequences, typically shorter than 1000 bp, composed of tandemly repeated units 1 to 10 bp long (1–6 bp or 1–13 bp ranges are also used). Microsatellites are also known as short tandem repeats (STR) or simple sequence repeats (SSRs).

Minisatellites
DNA sequences composed of tandemly repeated units longer than in microsatellites. Minisatellites span from 1 kb to 100 kb and are also known as variable number of tandem repeats (VNTRs).

Non-LTR Retrotransposons
A class of retrotransposons that are transposed via target-primed reverse transcription. Autonomous non-LTR retrotransposons encode reverse transcriptase.

Processed Pseudogene (Retropseudogene)
A pseudogene derived from usually spliced and polyandenylated host mRNA by LINE-mediated reverse transcription.

Provirus
A DNA copy of a retrovirus inserted into the genome.

Pseudogene
A nonfunctional gene copy.

Recombination
The exchange of DNA or RNA molecules.

Repetitive Elements (Repeats)
DNA fragments present in multiple copies in the genome without clearly assigned biological function.

Ribonuclease H (RNase H)
An enzyme that degrades the RNA strand of RNA–DNA hybrid molecules.

Retrotransposition (Retroposition)
A process involving transcription, processing of mRNA, translation, reverse transcription of the transcribed mRNA, and integration of the resulting cDNA into the host genome.

Retrotransposon (Retroposon)
A transposable element that is reproduced via reverse transcription of an RNA intermediate. They can be split into two classes: LTR and non-LTR retrotransposons.

Retrovirus
An RNA virus, whose life cycle includes cytoplasmic reverse transcription of the RNA transcribed from its provirus integrated into the host genome.

Reverse Transcriptase
An RNA-dependent DNA polymerase.

Satellite DNA
An array of tandemly repeated units spanning up to 10^6 bp and present mainly in centromeric and paracentromeric heterochromatin. Satellite DNA usually forms a prominent "satellite band" separable from the rest of genomic DNA by density gradient centrifugation.

Short Interspersed Repeat (SINE) Element
A nonautonomous non-LTR retrotransposon derived mostly from host genes containing a pol III internal promoter. SINE elements are retrotransposed by reverse transcriptase encoded by LINE elements.

Superfamily of TEs
A group of related families of transposable elements.

Tandem Repeats
DNA units arranged in a head-to-tail manner.

Target-primed Reverse Transcription (TPRT)
A mechanism of integration of non-LTR retrotransposons into the genome.

Transposable Elements (TEs, Transposons, Mobile Elements)
DNA sequences that can move to different positions in the genome either by excision and reinsertion, or by retrotransposition.

Transposase
An enzyme encoded by a DNA transposon and involved in its insertion and excision from the host DNA.

Transposition
A relocation of a DNA sequence from one genomic site to another.

■ Human repetitive DNA includes tandemly arrayed and interspersed repeats. Tandemly arrayed microsatellites, minisatellites, telomeric repeats, and centromeric satellites constitute a significant portion of the genomic DNA, particularly of heterochromatin. Furthermore, around 45% of the genome is represented by so-called interspersed repeats that are mostly remnants of retrovirus-like LTR retrotransposons, non-LTR retrotransposons, and DNA transposons inserted in the genomic DNA over millions of years. Currently, the most active are L1 and Alu families of non-LTR retrotransposons, and they are known to cause genetic diseases through insertion into genes. Furthermore, repetitive sequences, particularly Alu elements, can stimulate illegitimate recombinations producing chromosomal instabilities leading to genetic disorders. Analysis of repetitive DNA is an important part of genome studies, and it is based on specialized databases and computer programs.

1
Introduction

The generally synonymous terms *repetitive elements*, *repetitive (reiterated) sequences*, and *repeats* describe a broad variety of DNA sequences with copies present in multiple locations throughout the genome. They can be divided into families of similar or sometimes identical elements. This definition also covers families of genes with known functions, such as RNA genes and other multigene families. However, the subject of this chapter is limited to repetitive elements that do not have a clearly assigned biological function.

Historically, the first categorization of repetitive DNA into "highly repetitive" and "middle repetitive" was based on DNA denaturation/renaturation kinetics. This categorization refers to the physical properties of DNA that cannot be translated satisfactorily into a classification of repeats based on cloning and detailed sequence studies. In general, highly repetitive DNA primarily includes the tandemly repeated sequences described in Sect. 2, whereas the middle-repetitive DNA includes repetitive elements interspersed with single-copy DNA (Sect. 3). However, the middle-repetitive DNA may contain sequences with a very high copy number such as the Alu family described in Sect. 3.1.2.2.

On the basis of chromosomal distribution, repetitive elements are divided into *tandem repeats*, consisting of direct repetitions of the same motif, and *interspersed repeats*. Interspersed repeats are mostly inactive copies of transposable elements (TEs). Together, these elements represent around 50% of the sequenced human genome. Some abundant interspersed repeats are relatively old and diverse and cannot be clearly distinguished from "unique" DNA. This is due to the acquisition of random mutations and sequence fragmentation over time. Therefore, it is believed that the real contribution of repetitive DNA to the human genome is underestimated, and that the cited proportion of 50% only reflects the current limits of detectability.

2
Tandemly Arrayed Repeats

Tandemly arrayed repeats are composed of multiple head-to-tail repetitions of simple

(a) SSR

(b) Satellite 1

(c) Satellite α

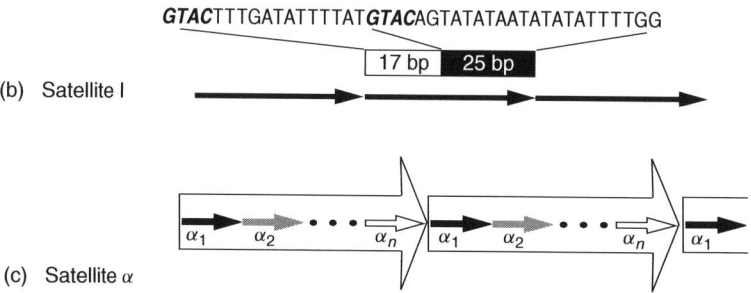

Fig. 1 Structure of human tandem repeats. (a) Schematic structure of simple sequence repeats. Arrows represent units ordered in head-to-tail fashion. (b) Structure of human classical satellite 1. Satellite 1 was originally described as a 17mer (A) and a 25mer (B) arranged in an alternating pattern A-B-A-B. This structure was inferred from the cutting of human DNA by the *RsaI* restrictase (the *RsaI* cutting sites are highlighted). Further sequence analyses revealed that satellite 1 is in fact composed of a single 42 bp-long basic unit. (c) Diagram of the hierarchical structure of human alpha satellites, showing the organization of a typical alpha satellite DNA region. Monomeric 171 bp units are represented by the small labeled arrows. The hierarchical structure of the alphoid DNA is shown by the large arrows.

sequence motifs (Fig. 1a). Human tandem repeats are composed of units ranging in size from 1 to ~1000 bp. On the basis of unit length and genomic distribution, they are somewhat arbitrarily divided into microsatellites, minisatellites, and satellites. Microsatellites are tandem arrays of short units, while minisatellites are composed of longer patterns. Micro- and minisatellites tend to be scattered all over chromosomal DNA, with some bias toward telomeric ends. The moderate degree of tandem repetition and high degree of dispersion throughout the chromosomes are the major features distinguishing micro- and minisatellites from satellites. Satellites are long, tandemly arrayed sequences located in well-defined chromosomal areas such as pericentromeric, subtelomeric, and telomeric regions. Satellites may consist of simple micro- or minisatellite-like units, but these are often organized into higher-order structures.

2.1
Microsatellites and Minisatellites

Originally, microsatellites were referred to as simple sequence repeats (SSRs) and minisatellites as variable number of tandem repeats (VNTR). Currently, the term simple sequence repeats is applied to both micro- and minisatellites. Micro- and minisatellite repeats are defined by three parameters: the *pattern* (or sequence of the unit), unit *length*, and the *number* of units.

The boundary separating micro- and minisatellites differs considerably in the scientific literature. Microsatellites are usually defined as repetition of 1–6 base pair (bp) long units, but the upper boundary is often extended to 10–13 bp. Microsatellites are much more abundant than minisatellites.

The distribution of microsatellites in the human genome appears to be more uniform than that of minisatellites, with approximately one microsatellite for every 6 kb of DNA. Almost one-third of all potential microsatellites are simple runs of A. Such runs are often introduced as poly(A) tails of non-LTR retroposons, in particular, of Alu repeats (Sect. 3.1.2.2). Therefore, the abundance of $(A)_n$ may not be directly comparable to the abundance of other, spontaneously expanding, microsatellites. Other highly abundant microsatellites ranked in descending order are $(AC)_n$, $(AAAN)_n$, $(AAAAN)_n$, and $(AAN)_n$, including the complementary sequences. Many of these A-rich microsatellites are also spin-offs of poly(A) tails of non-LTR retroposons. Microsatellites composed of $(CAG/CTG)_n$ and $(CGG/CCG)_n$ trinucleotide repeats are less abundant, but nevertheless very important due to their connection with human diseases (Sect. 4).

The variation patterns of microsatellites and minisatellites appear to be very similar. In each case, the majority of variations are due to differences in the number of tandemly repeated units, not from mutations of the individual units. Despite the similarities between micro- and minisatellites, there are several different underlying mechanisms involved in their evolution. Indeed, the traditional division of tandem repeats into microsatellites and minisatellites is based on early hypotheses about mechanisms driving their expansion. For microsatellites, "replication slippage" during DNA replication was suggested, while minisatellite expansion was thought to result from unequal crossing-over. Nevertheless, our current knowledge suggests that processes driving SSR expansion are more complex and range from polymerase slippage and gap repair to DNA recombination and gene conversion. Another pathway of SSR expansion involves an initial comobilization of the microsatellite or minisatellite with transposition of another repeat, as in the case of poly(A) tails of non-LTR retrotransposons (Sect. 3.1).

Loci containing micro- or minisatellites produce a very large number of variant alleles, and the sequence length polymorphism makes micro- and minisatellites some of the most informative genomic markers. The most widely used microsatellites are the $(CA/GT)_n$ dinucleotide repeats. The variation at microsatellite loci has been used in genetic linkage analysis to establish the locations of disease genes as well as in forensics, human population genetics, and cancer genetics.

Many other kinds of simple repetitive sequences represent variable permutations of a limited number of motifs, sometimes interspersed with unique sequences. Such "composed" or "cryptically simple repeats" occur up to ten times more often in the genomic DNA than in any equivalent random sequence. They result from a systematic reshuffling of various DNA fragments differing in length and sequence composition. Therefore, they are of interest from the standpoint of genetic variability. These repeats have not yet been systematically classified or explored.

2.2
Satellites and Telomeric Repeats

Typically, satellites are arrays of 10^3–10^7 tandemly repeated units, or monomers, predominantly located in the well-defined chromosomal regions, such as centromeres and heterochromatin. Traditionally, the term *satellites* was restricted to centromere-specific repeats; however, telomeric repeats can also be viewed as satellite-like sequences. Because of their distinct base composition, many satellites were originally discovered as a result of their separation from bulk DNA by centrifugation in buoyant density gradients. Satellites with GC content similar to that of the unique genomic DNA, which could not be separated by centrifugation, have been called *cryptic satellites*. The function of satellite DNA remains unclear. Satellites represent up to 20% of the genomic DNA, but they are underrepresented in the sequenced portion of the genome.

Table 1 lists representative human satellite sequences collected in Repbase Update (RU), the reference collection of repetitive sequences (Sect. 6.1). The most extensively studied centromere- and telomere-specific repeats are reviewed in the next two sections. In addition, there are several chromosome-specific tandem repeats, some of which are listed in Table 1.

2.2.1 Centromeric Satellites

Human centromeric DNA contains a variety of different satellite repeats. The major noncryptic satellite fractions were

Tab. 1 Human satellites and satellite-like sequences in repbase update.

Repbase	Description
Centromeric	
SAR	Satellite I
GAATG	Satellites II and III, both contain 5-ATTCC basic unit
HSATII	Satellite II (example, diverged copy)
ALR	Alpha satellite
ALR1	Alpha satellite subfamily 1
ALR2	Alpha satellite subfamily 2
BSR	Beta satellite
GSAT	Gamma satellite
GSATX	Gamma X satellite, X chromosome specific
CER	48-bp satellite
SN5	SN5 satellite
LSAU	Heterochromatic satellite
HSAT4	35-bp centromeric repeat
HSAT5	Centromeric repeat, 14-bp unit
GSATII	Centromere-associated repeat
D20S16	Centromere-associated repeat
Telomeric	
TAR1	Telomere-associated repeat sequence, TAR1 or SUBTEL_sat in RepeatMasker
REP522	Repetitive subtelomeric-like sequence
Others	
SATR1	Primate satellite-like sequence
SATR2	Primate satellite-like sequence
MSR1	37 bp minisatellite repeats, specific to chromosome 19, but detected in chromosome 19-derived segmental duplications located in other chromosomes
MER22	SSTI moderately repetitive family, strictly tandem repeat, appearing at 2 chromosomal locations only
ACRO1	Satellite/acromeric in all short arms of acromeric chromosomes
HSATI	Chromosome 22 satellite

defined as classical satellites I, II, III, and IV. Classical satellite I was found to be dominated by a simple component called *satellite 1*. Satellite 1 contains a 42-bp basic unit, originally described as a 17-mer (A) and a 25-mer (B), arranged in an alternating pattern: A-B-A-B (Fig. 1b). The major simple components of classical satellites II and III are named *satellites 2 and 3*, both containing the GAATG unit (sometimes described by the complementary ATTCC). Repetitive sequences from the satellite DNA fraction IV are nearly identical to those in fraction III.

Among other well-defined and extensively studied components of classical satellites II and III are the so-called *alpha satellites*, first discovered in African green monkeys and subsequently in other primates. Alpha satellite DNA is composed of a 171-bp unit (Fig. 1c). Interchromosomal sequence identities suggest that all human alpha satellites arose from two or three different ancestral monomers, which in turn diverged into 12 distinct types of alphoid monomers. They can differ by as much as 20 to 40% from one another by sequence identity.

Human chromosomes contain alpha satellites organized hierarchically into higher-order repeat arrays (Fig. 1c), which can differ between different chromosomal locations. On the basis of this organization, the alpha satellites from the human genome have been divided into five suprachromosomal families: the first family consists of alpha satellites located on chromosomes 1, 3, 5, 6, 7, 10, 12, 16, and 19; the second comprises chromosomes 2, 4, 8, 9, 13, 14, 15, 18, 20, 21, and 22; the third is located on chromosomes 1, 11, 17, and X; the fourth family is found on chromosomes 13, 14, 15, 21, 22, and Y; and, finally, the fifth family was detected on chromosomes 5, 6, 13, 14, 19, and 21.

Members of each suprachromosomal family share a common alpha satellite repeat unit. The size, conservation, and organization of these higher-order structures depend on the chromosome and on the physical distance from the centromere. Apart from the five classical suprachromosomal families, other higher-order structures have also been described. For example, the central region of the chromosome X centromere contains a specific alpha satellite array arranged into 2-kb long highly identical blocks.

The evolution of alpha satellites is complex, and is driven by intra- and interchromosomal recombination and gene conversion. The recombination occurs during chromosomal pairing and leads to an unequal crossing-over between chromosomes. As a result, stretches of alpha satellite DNA expand and contract at a very rapid rate. In addition, the alpha satellite regions of chromosomes within the same suprachromosomal families may undergo exchanges between nonhomologous chromosomes, thus increasing the mode and tempo at which these sequences evolve. However, the intrachromosomal recombination within alpha satellites appears to be the predominant evolutionary driving force. In general, the average intrafamily identity of alpha satellite monomers (~90%) is much higher than the interchromosomal identity (~75%). In addition, the average intrafamily identity within a species (99% in human and gorilla) is generally greater than that between species (90% between the human and gorilla satellite monomers). Humans and gorillas share the basic monomer as well as a higher-order repeat unit for suprachromosomal family 3, suggesting that this alpha satellite structure predates the human/great ape divergence. Thus, the alpha satellites within each species have evolved in a "concerted" manner, giving rise to a significantly lower nucleotide divergence within a species than between species.

Alpha satellites have been found only within primate genomes, usually located around the centromere, and can bind key centromere proteins. Occasionally, however, ectopic centromeres are formed in noncentromeric segments of chromosomes, producing cytogenetic aberrations including visible karyotype disorders. These new centromeres were termed *neocentromeres*. However, not all neocentromeres detected in humans are derived from alpha satellites. About 60 reported neocentromeres completely lack alpha satellites but maintain their ability to bind main centromeric proteins. These neocentromeric sequences are promising material for the construction of artificial chromosomes. However, these results also suggest that centromeric repeats, including alpha satellites, may not have any essential role in the genome.

Table 1 lists other reported human centromeric satellites. The human *beta* (Sau3A) satellite family is defined by the 68-bp GC-rich monomer. This satellite is present in pericentromeric regions and in the short arms of human acrocentric chromosomes. *Gamma* satellite DNA was initially detected as a 220-bp repeat located on chromosome 8. A second subfamily of gamma satellite, *gamma X*, was identified in the centromeric region of the X chromosome. Both families share 220-bp units 62% identical to each other. The CER satellite family, characterized by the 48-bp unit, is present on chromosomes 13, 14, 15, 21, and Y.

Among poorly defined satellite families are the Sau (*Lsau*) and Sn5 families characterized by the 498- and 455-bp units, respectively. Many centromeric satellites listed in Table 1 were defined by sequence analyses. Classification of centromeric satellites continues, and additional families may be detected during the course of sequencing centromeric and heterochromatin regions.

2.2.2 Telomeric and Subtelomeric Repeats

In contrast to the highly variable monomers in centromeric satellites, the *TTAGGG* telomeric monomers are conserved in all vertebrates. They form tandem arrays 3 to 20 kb long and their conservation is indicative of the importance of telomeres for normal chromosomal maintenance and viability. Historically, telomeres have been found to be associated with the nuclear envelope, and they have been implicated in chromosomal positioning within the cell. Telomeres are implicated in aging, immortalization of cells, and chromosomal rearrangements.

The maintenance of telomeric repeats presents a unique challenge to the cell, since standard DNA replication would result in the progressive shortening of the telomere over time. The replication of human chromosomes is catalyzed by DNA polymerases. However, DNA polymerases require short template primers for the initiation of replication. Thus, the ends of chromosomes or telomeres would become gradually shorter over time through the exclusive use of this mode of replication. The length of telomeres is maintained by a special replication mechanism for the ends of chromosomes involving a telomere-specific terminal transferase. This enzyme, called *telomerase*, is a ribonucleoprotein (RNP), and it uses the RNA component as a template to initiate the replication of the telomeres. The enzymatic addition of bases is catalyzed by the protein component. Faithful replication of telomeres protects chromosome ends from degradation, fusion, and recombination.

The length of telomeres appears to depend on the number of cell divisions with a general tendency toward shortening. This has been implicated in chromosomal aberrations, involving translocations in differentiated and abnormal cells. The length of telomeres has been used as a measure of the age of individual cells and is implicated indirectly in aging. It is still unclear whether changes in the telomere length have a direct impact on the aging process of a cell or whether they are the consequence of aging, possibly through a decrease in the telomerase fidelity.

The 100 to 300-kb long sequences adjacent to telomeric repeats are called *subtelomeric repeats*. They represent the boundary between the telomere and the rest of the chromosome. Most subtelomeric repeats contain motifs similar to TTAGGG, but there are exceptions such as *REP522*. Unlike telomeric repeats, most subtelomeric repeats are not substantially conserved in vertebrates. In fact, they are highly polymorphic in humans and even in different chromosomes. The distal ends of human chromosomes are characterized by an enhanced rate of recombination and aberrations in subtelomeric regions have been shown to be associated with several congenital abnormalities. The high divergence of subtelomeric repeats may prevent recombinations between nonhomologous chromosomes and increase their stability.

3
Interspersed Repeats

Usually, interspersed repeats are scattered remnants of transposable elements that were transpositionally active either in the human genome or genomes of our evolutionary ancestors. Active TEs have produced families and subfamilies of interspersed repeats of different age and size. There are over 600 families of human interspersed repeats represented in Repbase (Sect. 6.1). Most of them originated and became extinct during the last 30 to 150 million years but some of the oldest ones ceased to proliferate as long as 200 million years ago and are barely identifiable by sequence analysis. Detectable interspersed repeats represent ~45% of the human genome. Given the limits of detectability, it is very likely that the bulk of the remaining nonprotein coding DNA is also composed of very old, unrecognizable remnants of TEs.

The consensus sequence of a TE can be reconstructed by applying the so-called *simple majority rule* to its multiply aligned interspersed copies (Fig. 2). The average similarity of diverged repeats to their consensus sequence (y) is higher than the average similarity between individual repeats (x). The approximate relation between the two variables is given by the following simple formula:

$$y = \frac{1 + \sqrt{12x - 3}}{4}.$$

In many cases, consensus sequence is a quite accurate approximation of the original TE that generated a particular family of repeats. It can be further improved by correcting stop codons in open reading frames (ORFs) or restoring rapidly mutating CpG dinucleotides, which sometimes erroneously converge to CpA or TpG in the consensus sequence. At least one TE reconstructed this way, named *Sleeping Beauty*, has been shown to transpose *in vivo*.

Transposable elements harbored by the human genome can be divided into three

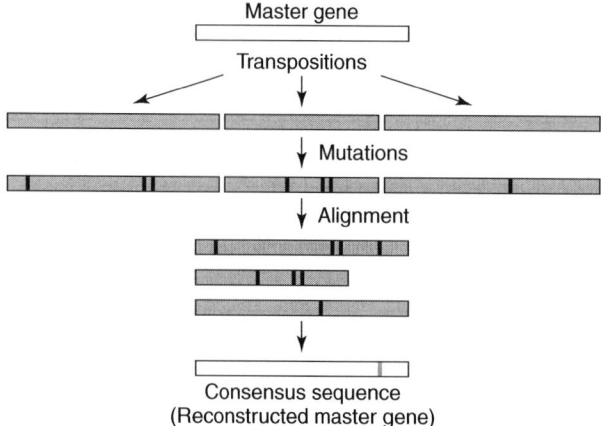

Fig. 2 Molecular paleontology of transposable elements. The top white rectangle represents an ancient active transposon that multiplied into interspersed repeats shown as gray rectangles subsequently marked by black vertical bars indicating mutations. Most of these mutations can be "undone" after multiple alignments and construction of a consensus sequence representing the reconstructed transposon (bottom white rectangle). The gray bar in the consensus sequence indicates an unresolved mutation.

groups. Two of these groups belong to a class of retroelements that include non-LTR retrotransposons and endogenous retrovirus-like elements also known as *long terminal repeat* (LTR) retrotransposons. Retroelements proliferate via reverse transcription of their RNA expressed in the host cell. An RNA copy of the transcribed retroelement is used as a template to produce its DNA copy, the so-called cDNA, which can be inserted in the genome. In non-LTR retrotransposons, a nick in the host DNA primes cDNA formation, whereas in retroviral-like elements, the DNA nicking and cDNA integration occur after reverse transcription. The nicking and cDNA formation are catalyzed by the endonuclease/integrase (IN) and reverse transcriptase (RT), respectively, encoded by the transposable element itself or by one of its autonomous relatives present in the genome.

Unlike DNA transposons, active retroelements do not relocate from their original position.

The third group of TEs is represented by the *cut-and-paste* DNA transposons, called so because they are excised (cut) from the original genomic location and inserted (pasted) into a new site. Typically, a single enzyme called *transposase* mediates the process.

All groups of TEs are composed of autonomous and nonautonomous elements. An autonomous element encodes a complete set of enzymes characteristic of its family. Nonautonomous families use the protein machinery of their autonomous relatives. Therefore, the nonautonomous TEs can be viewed as parasites competing for enzymes encoded by the autonomous elements. Both groups also require additional cellular factors for amplification.

3.1
Non-LTR Retrotransposons

Detectable copies of non-LTR retrotransposons contribute ~34% of the human genome compared to the ~11% representing other detectable interspersed repeats. Non-LTR retroelements appear to be strictly intracellular parasites, inherited for millions of years rather than passed horizontally. For historical reasons, some autonomous non-LTR retrotransposons are referred to as long interspersed repeats (LINEs) due to the fact that they are several kilobases long. Analogously, their short nonautonomous associates are referred to as short interspersed repeats (SINEs). Autonomous non-LTR retrotransposons in eukaryotic species are currently divided into 13 superfamilies or clades: L1, CRE, NeSL, R4, R2, RTE, Tad1, R1, LOA, I, Ingi, Jockey, CR1, and Penelope. Only two of them, L1 and CR1 superfamilies are represented in the human genome. The human population continues to carry active elements of L1 non-LTR retrotransposons whereas CR1-like retroelements are extinct. L1-dependent Alu and SINE, VNTRs and ALU (SVA) nonautonomous retroelements also continue to proliferate in the human population.

3.1.1 L1 Non-LTR Retrotransposons

The typical L1 (or LINE1) autonomous retrotransposon is ~6 kb long, and contains two open reading frames referred to as ORF1 and ORF2, encoding the ~500-aa and ~1000-aa proteins, respectively (Fig. 3). The function of the L1-ORF1p protein is not well understood. Since it has a common leucine zipper (LZ) motif present in its sequence, it presumably binds RNA/DNA molecules. The L1-ORF2 protein includes the apurinic–apyrimidinic endonuclease (APE), reverse transcriptase (RT), and conserved cysteine-rich (C) domains. The endonuclease domain is involved in nicking the host DNA, followed by priming of an RNA-directed DNA synthesis that is catalyzed by the RT domain.

Out of several hundred thousand L1 copies present in the human genome, about 5000 are full-length elements and only around 100 contain intact open reading frames. In addition to ORF1 and ORF2, the full-length L1 elements contain 5' and 3' untranslated regions (UTRs), each several hundred bp long. The 5' UTR includes a poorly characterized internal pol II transcription promoter. Unlike most human genes that contain pol II promoters upstream of the transcription start site, the L1 promoter is located downstream of the start site. Occasionally, the 5' L1 UTR may also promote transcription that starts in the strand complementary to the 5' UTR and extends to the genomic region flanking the L1 5' end. The 3' L1 UTR includes an AATAAA polyadenylation signal and a polyA tail.

In addition to the transcription, the L1 life cycle includes export of the transcribed

Fig. 3 Schematic structure of the L1 non-LTR retrotransposon. It consists of a 5' untranslated region (5' UTR) with pol II internal promoter, two open reading frames (ORF1 and ORF2), a 3' untranslated region, a polyA signal (AATAAA), and a polyA tail. L1 elements are flanked by target site duplications (TSDs). The ORF1 protein includes a conserved leucine zipper motif (LZ). The ORF2 protein includes domains for the apurinic/apyrimidinic-like endonuclease (APE), RT, and a conserved cysteine-rich motif (C).

RNA from the nucleus to the cytoplasm, translation and formation of ribonucleoprotein particles, which are transported back into the nucleus, reverse transcription, and integration into the genome. L1 elements lack splice signals and therefore, L1 mRNA is bi-cistronic with two open reading frames separated by a short noncoding spacer. Translation of ORF1 is likely to be initiated at the 5′ UTR by ribosomal scanning. ORF2 translation may be initiated by an internal ribosomal entry site (IRES) or by the so-called *ribosomal shunting* or *jumping*. Also, it is unclear how the L1-encoded proteins enter the nucleus. Usually, nuclear export of proteins as large as those encoded by ORF1 and ORF2 is driven by specific amino acid sequences called *nuclear localization signals*. As there are no known nuclear localization signals encoded by either ORF1 or ORF2 proteins, the mechanism involved in their entry into the nucleus remains a mystery. It is even unclear if the ORF1 translation products ever enter the nuclear compartment.

The model of reverse transcription and integration of L1 into the genome invokes a coupled process known as *target-primed reverse transcription* (TPRT). According to this model, the APE domain of the ORF2 protein recognizes a 6-bp target site described by the TTAAAA consensus sequence, and nicks the second DNA strand between the bases complementary to TT and AAAA (Fig. 4). The free 3′-OH end of the nick is then used for priming reverse transcription of L1 RNA by the ORF2 reverse transcriptase, starting from the 3′ polyA tail. Eventually, a second nick is created on the first strand, preferably ~15 bp downstream of the first nicking site, leading to a staggered break and, after integration, to the formation of target site duplications (TSDs) flanking the integrated sequence. The typical size of TSDs generated by L1 retrotransposition is ~15 bp, but the size may vary and sometimes even short deletions of target sequences are observed. The final integration steps, including

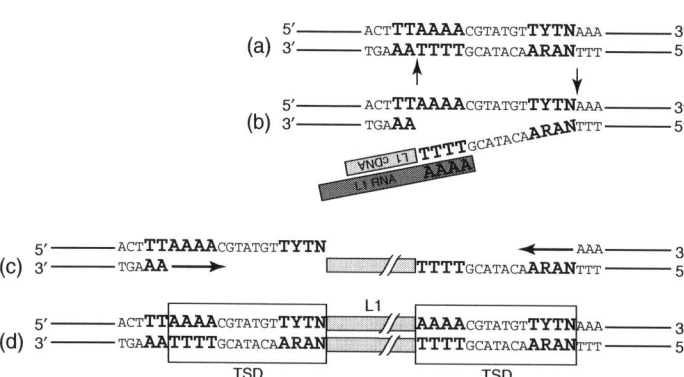

Fig. 4 The target-primed reverse transcription (TPRT) mechanism. (a) The L1 endonuclease cleaves the antiparallel DNA strand between 3′ AA and 3′ TTTT, and exposes a free 3′ hydroxyl group at the nick, serving as a primer for reverse transcription of L1 mRNA. (b) The second nicking occurs ~15 nucleotides from the first nicking site, preceded by a TYTN consensus sequence. (c) The remaining steps include elimination of L1 RNA, DNA synthesis, and repair, probably mediated by the host enzymes. (d) Integrated L1 retroelement and its flanking sites.

synthesis of the second strand, ligation, and elimination of the L1 RNA, are probably mediated by host enzymes.

The majority of L1 elements undergo 5′ truncations and inversions during retroposition and, as a result, they are represented mostly by 3′ terminal portions that are shorter than 1 kb in length. Possible reasons for the truncation may be digestion of the L1 RNA by cellular ribonucleases, or premature termination of the reverse transcription, or both.

Occasionally, transcription of L1 elements does not stop at the L1 polyadenylation signal but continues until another transcription termination signal encoded by the host DNA is encountered. As a result, host sequences adjacent to the 3′ L1 terminus and the new polyA sites may be transduced by the L1 to a new genomic location. Over 10% of the analyzed L1 insertions in the human genome contain 3′ transduced sequences that range from 30 bp to several kb in length. This represents an attractive mechanism that could be involved in exon shuffling in the host genome. However, no real exons have yet been detected in L1-transduced sequences. Therefore, it is possible that the L1-mediated 3′ transduction is suppressed in the vicinity of human genes. Some L1 retrotransposons can also be accidentally transcribed from upstream promoters leading to transduction of genomic DNA flanking the 5′ ends of L1 elements. Also, new L1 subfamilies can be generated by the addition of both 5′ and 3′ flanking DNA.

3.1.2 L1-dependent Nonautonomous Non-LTR Retrotransposons

There are two major groups of L1-dependent non-LTR retrotransposons present in the human genome: processed pseudogenes, and SINEs that include Alu and SVA elements. Elements from both groups are flanked by ∼15-bp TSDs, have the 3′ polyA tails, and are inserted preferentially into the TTAAAA target sites. It has also been demonstrated experimentally that the propagation of processed pseudogenes and Alu elements is mediated by the L1-encoded RT. Therefore, both processed pseudogenes and Alu repeats are classified as nonautonomous non-LTR retroelements mobilized by the L1 retroposition machinery. The L1 reverse transcriptase, in a cis-acting manner, preferentially binds the same mRNA molecule it was translated from. Therefore, Alu and other nonautonomous retropseudogenes probably intercept the L1 RT while interacting with ribosomal components.

3.1.2.1 **Processed pseudogenes** Processed pseudogenes, or retropseudogenes, are products of occasional retrotransposition of spliced mRNA formed during regular transcription of the host genes (Fig. 5). Since transcription of these genes is usually driven by external pol II promoters, processed pseudogenes do not include the promoters. Therefore, it is extremely unlikely that they can be repeatedly transcribed and retrotransposed. Nevertheless, the human genome is estimated to contain at least 30 000 retropseudogenes formed from a variety of host genes. The most abundant retropseudogenes appear to be derived from highly expressed genes ranging from those coding for small nuclear, nucleolar, and cytoplasmic RNA, to those coding for ribosomal proteins.

3.1.2.2 **Alu repeats** Alu elements were first discovered using DNA renaturation–reassociation experiments that were performed during the early era of repeat studies. They are represented by more

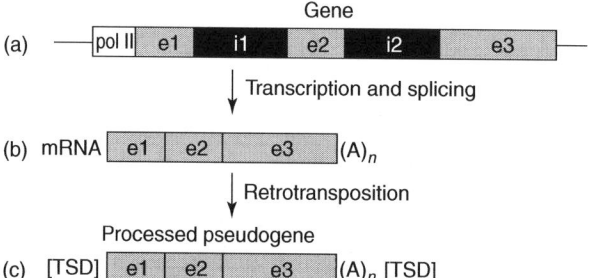

Fig. 5 Formation of processed pseudogenes (retropseudogenes). (a) The original gene that is transcribed into RNA, excluding pol II promoter. Exons e1–e3 and introns i1, i2 are indicated. (b) mRNA afer splicing and polyadenylation. (c) L1-retrotransposed, processed retropseudogene with ~15 bp TSDs.

than one million copies in the human genome, which makes them numerically the most abundant of all human interspersed repeats. The name "Alu" is derived from an *AluI* restriction site present in these elements.

Full-length Alu elements are ~300-bp long (Fig. 6) dimers composed of two homologous 120- and 148-bp left and right monomers respectively, derived from the *7SL RNA* gene ~100 million years (Myr) ago. The 7SL RNA molecule is an important component of the signal recognition particle (SRP) that binds to nascent signal sequences and transiently arrests translation. The *7SL RNA* gene contains the pol III internal promoter composed of the conserved ~15-bp A and B boxes. The right monomer contains an additional 31-bp 7SL-derived sequence

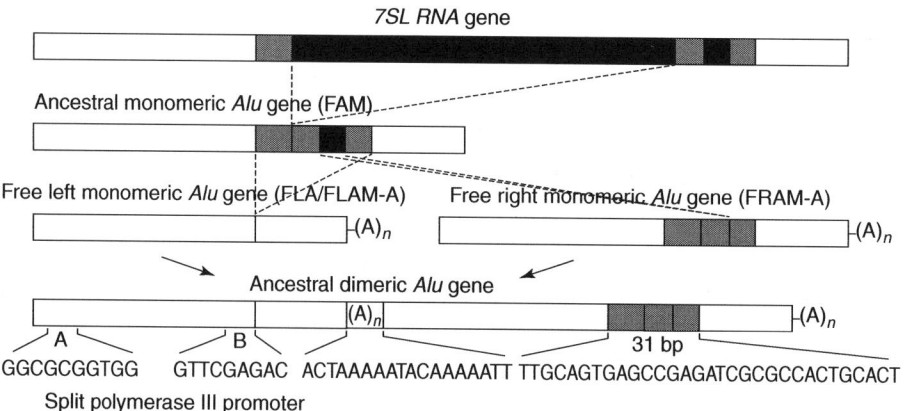

Fig. 6 A scheme for the origin and early evolution of human Alu master (source) genes. Dimeric Alu was derived from the *7SL RNA* gene after a two-step deletion process leading first to an ancestral family of Alu-like monomers (FAM) and then to other monomers, of which two, FLA/FLAM-A, and FRAM, merged into a precursor of dimeric Alu characteristic of primates.

relative to the left monomer, and the monomers are separated by the A-rich central region. The right monomer also contains a 3'-oligo-dA-rich tail (or polyA tail), usually 20 to 30 bp long. Free Alu monomers such as FLA, FLAM, and FRAM were very active during the early era of Alu evolution, and at least one of them, BC200, has been recruited as a regulated gene expressed in the brains of higher primates.

Probably around one hundred Alu elements out of one million appear to be currently participating in retrotransposition. It is unclear why Alu retroposition is confined to this limited number of "master" or "source" genes. One often invoked explanation is that most retroposed elements remain untranscribed. For example, pol III transcription depends on flanking sequences and most Alu elements may be inserted into genomic sites that do not complement their pol III promoters. Furthermore, most Alu sequences are heavily methylated at CpG positions that are also present in the promoter region. The methylation occurs at position 5 of cytosine residues. Deamination of 5-methyl cytosine produces thymine – a mutation that is likely to escape the repair mechanism. As a result, CpG dinucleotides mutate to TpG and CpA at a rate approximately ten times higher than that of the non-CpG dinucleotides. Other regions of rapid mutations include A-rich linker and tail regions often associated with microsatellite expansion. This rapid rate of mutations represents another potential way to eliminate or mitigate the ability of any particular Alu repeat to mobilize. Thus, the majority of Alu repeats appear to be fossil relics that integrate in the genome and accumulate mutations in a random manner, characteristic of pseudogenes.

The human Alu family has been divided into nine major subfamilies: AluJo, Jb, Sz, Sx, Sg, Sq, Sp, Sc, and Y, ordered approximately from the oldest to the youngest (Table 2). AluJo and Jb form a separate branch distinct from the AluS branch that includes all the other seven major subfamilies. Each Alu subfamily represents a set of elements derived from a small group of closely related master genes. As a result, members of the same family share, in principle, the same consensus sequence, which is different from

Tab. 2 Average chromosomal density and estimated age of major Alu subfamilies.

Alu subfamily[a]	AluJo	AluJb	AluSz	AluSx	AluSg	AluSq	AluSp	AluSc	AluY
Number[b]	216 465	130 297	192 932	155 182	88 697	106 344	61 465	47 071	146 176
Proportion[c]	18.91	11.3	16.9	13.6	7.8	9.3	5.4	4.1	12.8
Genomic density (%)	1.60	1.1	1.9	1.5	0.9	1.0	0.6	0.5	1.4
Age (Myr)[d]	81	81	48	37	31	44	37	35	19

[a] Alu subfamilies were classified by censor. Alu classification may slightly differ between different programs and parameters used for Alu detection.
[b] Numbers of Alu elements from major subfamilies present in the sequenced part of the human genome (2805 Mb).
[c] Proportions of elements from the corresponding subfamilies relative to all Alu sequences.
[d] Age (in million years), based on Kimura's distance.

consensus sequences for other subfamilies. AluJo and AluJb are the two oldest subfamilies that proliferated ~65 million years ago. AluY is the youngest among major AluS subfamilies as it was retrotransposed 5–30 million years ago. AluY gave rise to a number of still younger sub(sub)families of which the following 22 have been reported so far: AluYa1, Ya4, Ya5, Ya8, Yb3a1, Yb3a2, Yb8, Yb9, Ybc3a, Yc1, Yc2, Yd2, Yd3, Yd3a1, Yd8, Ye2, Ye5, Yf1, Yf2, Yg6, Yh9, Yi6. At least some of these sub(sub)families still contain elements actively proliferating in human populations, as indicated by their insertion polymorphism. Among the largest and best-studied active sub(sub)families are AluYa5 and Yb8. However, some of the smaller ones such as AluYb9, Yc2, Yg6 and Yi6 are also of growing interest in human population genetics studies.

Densities of recently retroposed Alu elements are about three times higher on chromosome Y than on chromosome X, and about two times higher than on autosomes. This indicates that Alu elements are primarily retroposed in paternal germlines. However, these original chromosomal proportions change very rapidly with time. The most striking long-term trend is the accumulation of Alu elements in GC-rich and gene-rich regions. At the same time L1 elements are predominant in AT-rich and gene-poor regions. Given that newly retrotransposed Alu and L1 elements are inserted into regions of similar GC content, the evolution of the human genome is characterized by massive subsequent redistributions of Alu repeats. The most likely factor involved in the redistributions may be related to the different patterns of postinsertion duplications and deletions.

3.1.3 CR1-like Non-LTR Retrotransposons

CR1 is one of the most abundant and widely distributed clades of non-LTR retrotransposons that are present in the genomes of birds, amphibians, fishes, reptiles, invertebrates, and mammals. Typically, autonomous CR1-like elements are several kb long and contain two open reading frames, ORF1 and ORF2, coding for multidomain proteins. The ORF1p may be involved in DNA/RNA binding, but other interesting functions are also likely (see below). The ORF2 protein includes the APE and RT domains, which are distantly similar to those identified in the L1 ORF2 protein. CR1-like elements do not generate target site duplications.

3.1.3.1 L2 and L3 retroelements
Two human families of long interspersed repeats called L2 and L3 were most likely amplified from active autonomous CR1-like retroelements (Fig. 7). Given the high sequence diversity, their retrotransposition probably occurred 200 to 300 million years ago, long before mammalian radiation. Together with the nonautonomous elements described below, L2 and L3 represent ~5% of the human genome. Most of the preserved L2 and L3 copies are 5′ truncated, which, in addition to their very old age, impedes reconstruction of their source genes. For example, the ~3300-bp L2 consensus sequence does not seem to contain the ORF1 protein that is present in other CR1-like retroelements. However, it is not clear whether this is due to incomplete reconstruction or because the active L2 source gene lacked such a protein. The ORF1 protein is present in the L3 consensus sequence, and it contains the esterase domain that is also present in the ORF1 proteins encoded by CR1-like elements in birds, reptiles, and fishes. Unlike L1,

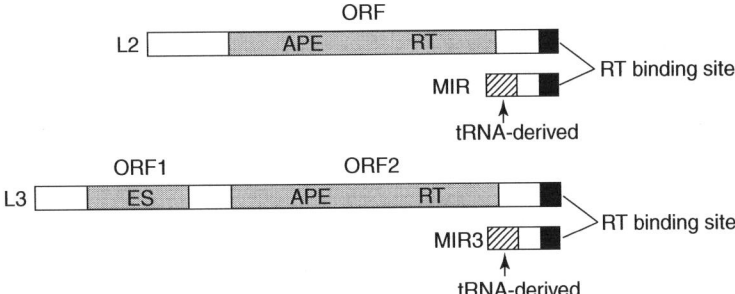

Fig. 7 L2 and L3 autonomous CR1-like retrotransposons and their nonautonomous companions MIR and MIR3. L2 and L3 encode enzymes containing endonuclease (APE) and reverse transcriptase (RT) domains. L3-ORF1 is homologous to esterase (ES).

neither L2 nor L3 elements are flanked by target site duplications. Furthermore, instead of the 3' polyA tail, L2 and L3 carry the $(TGAA)_3$ and $(GATTCTAT)_2$ microsatellites at their 3' termini respectively.

3.1.3.2 **Nonautonomous MIR and MIR3 elements** MIR and MIR3 are considered to be nonautonomous elements amplified by L2- and L3-encoded proteins, respectively. MIR is a classical 262-bp SINE element derived from a tRNA-like sequence and a 50 bp sequence fragment homologous to the 3' terminus of the L2 element (Fig. 7). This terminus was probably recognized by the L2-encoded RT that led to amplification of MIRs. Approximately 4% of the human genome is made up of L2 and MIR copies.

MIR3 is viewed as a nonautonomous element that parasitized the L3 retroposition machinery (Fig. 7). Analogously to MIR and L2, MIR3 and L3 also share the ∼50-bp 3' termini, which are likely to serve as binding sites of the L3-encoded RT. In addition, the ∼140-bp 5' terminal portions of the MIR3 and MIR consensus sequences are 81% identical to each other. This portion is homologous to tRNA and probably includes the pol III internal promoter essential for proliferation of MIR3 and MIR retroelements. Approximately 0.5% of the human genome is derived from L3 and MIR3 elements.

3.2
LTR Retrotransposons

Human long terminal repeat retrotransposons are retrovirus-like elements retrotransposed in the genome and inherited by the host from generation to generation. It is believed that human LTR retrotransposons are descendants of exogenous retroviruses that repeatedly infected germlines of human ancestors and became endogenous. Endogenous retroviruses retain their ability to retrotranspose within cells for some time, but cannot be transmitted efficiently between cells. Their subsequent evolution reflects adaptation to an intracellular environment, simplification, and formation of nonautonomous copies.

Most human endogenous retroviruses (HERVs) are extinct and are represented only by mutated interspersed copies, particularly by LTRs. The interspersed copies of LTR retrotransposons constitute around 9% of the human genome. Although LTR retrotransposons are not as abundant as the non-LTR retrotransposons, they are by far the most diverse and complex TEs harbored by our genome. Approximately 200 families of LTR retrotransposons have been identified in the human genome. A vast majority of them were identified and reconstructed using computer-assisted sequence analysis.

On the basis of the similarities to known animal exogenous retroviruses, all LTR retrotransposons present in the human genome can be divided into three classes (Table 3).

Class I non-LTR retrotransposons are similar to gamma retroviruses including mammalian type C retroviruses such as murine leukemia virus (MLV). All elements from this class are flanked by 4- and 5-bp TSDs that are generated upon their integration in the genome (Table 4).

Class II non-LTR retroelements are similar to beta retroviruses represented by mammalian type B and D retroviruses such as the mouse mammary tumor virus (MMTV), Mason-Pfizer monkey virus, and primate lentiviruses. The latter include human immunodeficiency virus (HIV). All Class II retroelements are characterized by 6-bp target site duplications (Table 5). They were probably active in hominoids including *Homo sapiens*.

Class III families represent the oldest LTR retrotransposons detected in the human genome. Some of them were active in the common ancestor of primates and other mammals. Class III elements amplified to a larger copy number than Class I and Class II elements together. Class III families are related to spumaviruses such as human foamy virus, and they are characterized by 5-bp target site duplications (Table 6).

Structural features of LTR retrotransposons
A typical retrovirus is inserted into the genome in the form of a provirus containing two copies of 200 to 3000-bp long LTRs, which are identical to each other at the time of insertion (Fig. 8). Most LTRs have conserved 5′-TG and CA-3′ termini, but in some families such as HERVK11 and HERVK11D, LTRs contain TA-3′ termini instead of CA-3′. The internal portion of a provirus is defined as a sequence flanked

Tab. 3 Classification of human LTR retrotransposons.

Human LTR retrotransposons	Percentage of the genome [%]	Target site duplications [bp]	Related exogenous retroviruses	Age [Myr]
Class I	2.5	4–5	Gamma retroviruses (C type[a]): Murine leukemia virus	20–80
Class II	0.5	6	Beta retroviruses (B type[b], D type[c]): mouse mammary tumor virus; Mason–Pfizer monkey virus; lentiviruses	1–35
Class III	6	5	Spumaviruses: human foamy virus	20–150

Distinctive features seen in transmission electron micrographs:
[a] Central and spherical viral core, "C particles".
[b] Accentric and spherical viral core, "B particles".
[c] Cylindrical core, "D particles".

Tab. 4 Class I endogenous retroviruses.

Family name	Names in repbase update		Alternative names[a]	Coding status	tRNA[b]
	Internal portion	LTR			
HERV-I	HERVI	LTR10B, LTR10A, LTR10C, LTR10E	RTVL-I	Gag, PR, RT, RH, IN, ENV	I
HERVIK10D	HERVIP10D	LTR10D		Gag, PR, RT, RH, IN, ENV	K
HERVIP10F	HERVIP10F	LTR10F		Gag, PR, RT, RH, IN, ENV	I/P
HERVIP10FH	HERVIP10FH	LTR10F		Nonautonomous	I/P
HERV-E	HERVE	LTR2		Gag, PR, RT, RH, IN, ENV	E
Harlequin	HARLEQUIN	LTR2B, LTR2C		Nonautonomous	E
HERV3	HERV3	LTR4	BaEV, HERV-R, ERV3	Gag, PR, RT, RH, IN, ENV	R/L
HERV15	HERV15I	LTR15	RRHERV-I, HSRIRT	Gag, PR, RT, RH, IN, ENV	I
HERV17	HERV17	LTR17	HERV-W	Gag, PR, RT, RH, IN, ENV	R/W
HERVS71	HERVS71	LTR6A, LTR6B	S71	Gag, PR, RT, RH, IN, ENV	T
HERV9	HERV9	LTR12, LTR12B, LTR12C, LTR12D, LTR12E	HERV-pHE1, ERV9, PTR5, HRES-1/1	Gag, PR, RT, RH, IN, ENV	R
MER52AI	MER52AI	MER52A MER52B, MER52C, MER52D		Gag, PR, RT, RH, IN	P
HERVG25	HERVG25	LTR25		Gag, PR, RT, RH, IN, Env	G
HUERS-P3	HUERS-P3	LTR9	HuRRS-P, HERV-P	Gag, PR, RT, RH, IN, Env	P
HUERS-P3B	HUERS-P3B	LTR9B		Nonautonomous Env	P
HUERS-P2	HUERS-P2	LTR1, LTR1B, LTR1C, LTR1D		Nonautonomous Gag	P
HUERS-P1	HUERS-P1	LTR8, LTR8A		?	P
HERV-H	HERVH	LTR7A, LTR7B	RTVL-H, RTVL-H2, RGH	Gag, PR, RT, RH, IN, Env	H
HERV19	HERV19I	LTR19A		Nonautonomous	F
HERVFH19	HERVFH19I	LTR19B, LTR19C,	HERV-F	Gag, PR, RT, RH, IN, Env	F

(continued overleaf)

Tab. 4 (continued)

Family name	Names in repbase update		Alternative names[a]	Coding status	tRNA[b]
	Internal portion	LTR			
HERVH48	HERVH48I	MER48	HERV-Fb	Gag, PR, RT, RH, IN, Env	F
HERVFH21	HERVFH21I	LTR21A, LTR21B	HERV-XA34, HERV-F(XA34)	Gag, PR, RT, RH, IN, Env	F
HERV46	HERV46I	LTR46		Gag, PR, RT, RH, IN, Env	F
PABL_A	PABL_AI	PABL_A	HERV-Rb	Nonautonomous	R/L
PABL_B	PABL_BI	PABL_B		Gag, PR, RT, RH, IN, Env	R/L
LOR1	LOR1_I	LOR1		Nonautonomous	E
MER51	MER51I	MER51A-E MER4		Nonautonomous	E
MER110	MER110I	MER110, MER110A		?	?
HERV23	HERV23I	LTR23		Nonautonomous	?
HERV30	HERV30I	LTR30		Gag, PR, RT, RH, IN, Env	R/L
MER34B	MER34B_I	MER34B		Nonautonomous	?
HERV35	HERV35I	LTR35		Nonautonomous	?
HERV38	HERV38I	LTR38, LTR38B-C		Nonautonomous	M
HERV39	HERV39	LTR39		Nonautonomous	R
HERV43	HERV43I	LTR43		Nonautonomous (Env only)	H
HERV45	HERV45I	LTR45, LTR45, LTR45B-C		? PR, RT, RH, IN	?
HERV49	HERV49I	LTR49		Nonautonomous	?
HERV70	HERV70_I	LTR70		Nonautonomous	?
PRIMA4	PRIMA4_I	PRIMA4_LTR		Gag, PR, RT, RH, IN, Env	R/W
MER4I	MER4I	MER4		Nonautonomous	R/W
MER4BI	MER4BI	MER4B		Nonautonomous (+remnants of class III-derived gag and pol)	R/F
MER41	MER41I	MER41A-B, D-G		Nonautonomous	R/W
PRIMA41	PRIMA41	MER41C		Gag, PR, RT, RH, IN, Env	R/W?
MER57	MER57I	MER57A, MER57B	HERV-HS49C23	Nonautonomous	S

Notes: Internal sequences are not reconstructed yet for LTR24, LTR24B, LTR24C, LTR26, LTR26E, LTR27, LTR27B, LTR28, LTR29, LTR31, LTR34, LTR36, LTR37A-B, LTR44, LTR48, LTR48B, LTR51, LTR54, LTR54B, LTR56, LTR58, LTR59, LTR60, LTR61, LTR65, LTR68, LTR72, LTR72B, LTR73, LTR76, LTR77, MER4C, MER4D, MER4E, MER72, MER72B, MER90.
[a]Names given to partially sequenced retroviruses, and secondary names given to individual sequences of retroviruses whose consensus sequences were deposited in repbase update much earlier.
[b]tRNAa are designated by the corresponding one-letter amino acid code.

Tab. 5 HERVK-like endogenous retroviruses (class II).

Family	Internal portion	LTRs	Alternative names
HERVK10	HERVK	LTR5	HERV-K(HML-2)
HERVK3	HERVK3I	LTR3, LTR3B	HERV-K(HML-6)
HERVK11	HERVK11I	MER11A, MER11B, MER11C	HML-8
HERVK11D	HERVK11DI	MER11D	HML-7
HERVK9	HERVK9I	MER9	HML-3
HERVK22	HERVK22I	LTR22, LTR22A, LTR22B	HML-5
HERVK13	HERVK13I	LTR13	HML-4
HERVK14	HERVK14I	LTR14A, LTR14B	HML-1
HERVK(C4)	HERVKC4	LTR14	HML-10
HERVK14C	HERVK14CI	LTR14C	
HERVK(HML9)	HERVK(HML9)I	HERVK(HML9)_LTR	HML-9

Tab. 6 HERV-L-like endogenous retroviruses (class III).

Family	Repbase update		ENV
	Internal portion	LTRs	
HERV-L	HERVL	MLT2A1, MLT2A2	
ERVL	ERVL	MLT2B1, MLT2B2	
HERVL68	HERVL68I	MER68A, MER68B	
HERVL74	HERVL74	MER74, MER74A-C	
HERV16	HERV16	LTR16, LTR16A, LTR16A1, LTR16C-D	
HERV18[a]	HERV18	LTR18A, LTR18B	+
HERV32	HERV32	LTR32	
HERVL40	HERVL_40	LTR40A, LTR40B, LTR40C	
HERV47	HERV47	LTR47A, LTR47B	
HERVL66	HERVL66I	LTR66	+
HERV52	HERV52I	LTR52	
HERV57	HERV57I	LTR57	
HERVL70	MER70I	MER70A, MER70B	
THE1B	THE1BR	THE1B	
MSTA	MSTAR	MSTA	
MLT1	MLT1R	MLT1	
MLT1C	MLT1CR	MLT1C	
Unknown	Unknown	LTR33, LTR33A, LTR41, LTR42, LTR50, LTR53, LTR67, LTR69, LTR75, MER54, MER54A, MER54B, MER73, MER74B, MER88, MaLRs, MLT2-like, MLT1-like	

[a]This family is also known as HERV-S.

by LTRs. Usually it is several kb long, and encodes a set of proteins necessary for the retroviral life cycle. These proteins are encoded by the *gag* (group specific antigen), *pro* (protease), *pol* (polymerase), and *env* (envelope) genes (Fig. 8). Also, some Class II and Class III retroviruses encode deoxyuridine triphosphatase (dUTPase).

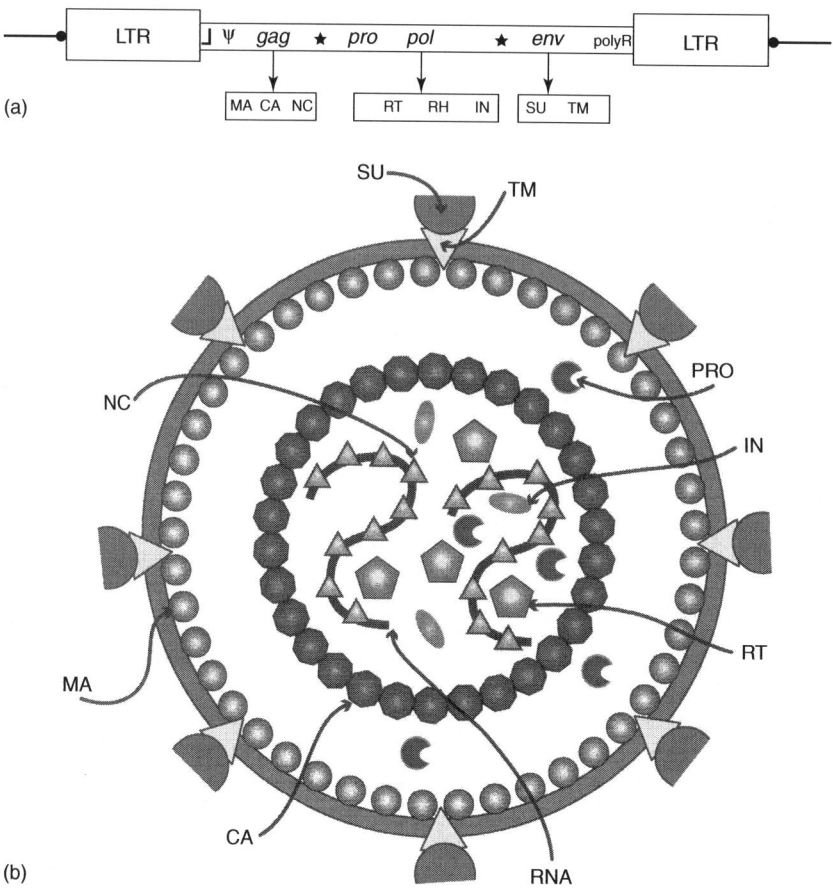

Fig. 8 The general structure of LTR retrotransposons. (a) Structural features of a provirus. Black circles indicate target site duplications; LTR, long terminal repeat; ⌐, primer binding site; Ψ, packaging signal; *gag*, capsid protein gene; *pro*, protease gene; *pol*, polymerase gene; *env*, envelope protein gene; ★, deoxyuridine triphosphatase (dUTPase), which lies between *gag* and *pro* in some Class III retroviruses, and between *pol* and *env* in Class II retroviruses. PPT stands for polypurine tract. The *pro* gene encodes the protease (PR), involved in the processing of the *gag*-, *pol*-, and *env*-encoded proteins. The *pol* gene encodes the reverse transcriptase (RT), ribonuclease H (RH), and integrase (IN). The *env* gene encodes the surface (SU) and transmembrane (TM) proteins. Typically, the *gag* gene encodes the matrix (MA), capsid (CA), and nuclear capsid (NC) structural proteins. (b) Schematic structure of a retroviral particle. The viral envelope (green) is made of a host-cell-produced lipid bilayer. It harbors multiple complexes formed by the *env*-encoded TM and SU components linked together by disulfide bonds. The retroviral capsid formed by the CA surrounds two RNA copies of the retrovirus, tRNAs, RT, IN, and PR. (See color plate p. xi).

Historically, LTR retrotransposons were viewed as endogenous retrovirus-like elements that did not encode the env protein. Currently, they are considered as nonautonomous retrovirus-derived elements, and the originally introduced distinction between retroviruses and LTR retrotransposons is becoming obsolete.

Gag is a multifunctional structural polyprotein separated into matrix, capsid, and nucleocapsid structural proteins by the *pro*-encoded protease (PR), which is related to cellular aspartic proteases.

The *pol* gene encodes a polyprotein precursor processed by the protease to reverse transcriptase and integrase. RT contains a ribonuclease H domain (RH), which digests RNA strands in RNA–DNA duplexes. RT encoded by LTR retrotransposons is capable of both RNA- and DNA-dependent DNA syntheses. It is the most conserved endoretroviral protein. The retroviral IN may be distantly related to $DD_{35}E$ transposases (see Sect. 3.3.2).

The *env*-encoded protein precursor is cleaved by the protease into the envelope, surface and transmembrane glycoproteins. It is thought that viral particles bind host cells through interactions between the surface glycoproteins and specific cell surface receptors (Fig. 8). Such interactions trigger fusion of the cell and virus membranes, followed by intrusion of the virus into the cell cytoplasm.

Transcription Transcription of the provirus starts at the pol II promoter present in the long terminal repeat (see LTR; Fig. 8a; Fig. 9). LTR is composed of three regions, called U3, R, and U5. The U3 region contains the pol II promoter. Additional signals affecting transcription, including enhancers, were also detected in the U3 region. The start of the LTR transcription corresponds to the U3-R boundary. Conversely, the R-U5 boundary is defined by the transcription stop-site. The transcription termination is mediated by the AATAAA or ATTAAA polyadenylation signals usually present in the U3 region. Therefore, the transcription is terminated in the second LTR only. In some retroviruses, including HIV, the polyadenylation signal is placed downstream of the U3-R boundary. However, the first LTR is still bypassed during transcription, presumably due to suppression of the transcription termination by a small distance between the transcription start-site and the polyadenylation signal.

Reverse transcription Reverse transcription of retroviral RNAs occurs in the cytoplasm of the infected host cell. It begins either when the viral particles enter the cytoplasm or when nuclear transcription of the already integrated provirus produces its functional RNA. The cytoplasmic localization of the retroviral reverse transcription is in sharp contrast to nucleus-based reverse transcription of non-LTR retrotransposons (Fig. 4). The retroviral reverse transcription is believed to follow a multistep process outlined schematically in Fig. 9 and described below.

1. The reverse transcription starts with the minus-strand DNA synthesis primed by the 3′ end of a tRNA molecule annealed to the primer binding site (PBS). Usually, the PBS is complementary to the ∼18-bp 3′ end of the virus-specific tRNA that starts just a few nucleotides downstream of the U5 region.
2. Minus-strand DNA synthesis proceeds to the 5′ end of the RNA, producing a 100 to 150 bp DNA molecule called *minus-strand strong-stop DNA*, which is released from the original RNA–DNA duplex by the RNaseH mediated digestion of the complementary RNA.
3. Following the digestion, the minus-strand strong-stop DNA performs the first strand transfer, annealing to the R-region at the 3′ end of the plus-strand RNA.
4. After the strand transfer, minus-strand DNA synthesis continues, and it is

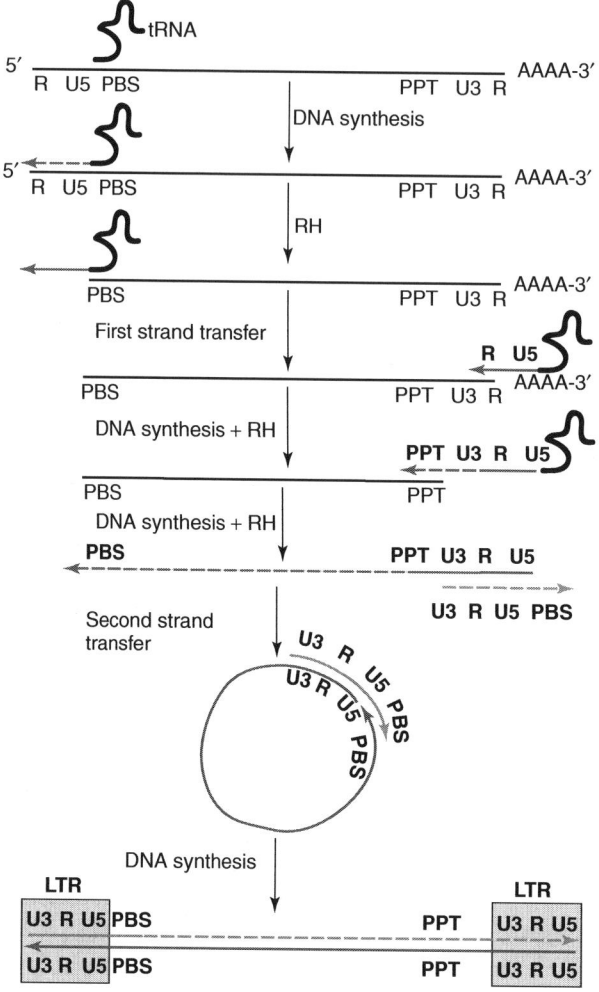

Fig. 9 Reverse transcription of the retroviral RNA. Dotted lines mark newly synthesized DNAs; green, plus-strand DNA; red, minus-strand DNA; RH, RNaseH digestion of RNAs. (See color plate p. xii).

accompanied by digestion of the RNA template, except for a short RNaseH-resistant polypurine tract (PPT).

5. The 3' end of the PPT primes plus-strand DNA synthesis, which continues until the PBS-like portion of the tRNA molecule is reverse transcribed. As a result, the so-called *plus-strand strong-stop DNA* is generated.

6. RNaseH digests the PBS-annealed tRNA molecule.

7. The second strand transfer is performed by annealing the complementary PBS segments present at the termini of the plus-strand and minus-strand DNAs.

8. The plus and minus DNA strands serve as templates for each other.

The replication is completed when RT synthesizes the entire plus and minus DNA strands. The final product is a linear duplex DNA.

Life cycle of LTR retrotransposons The life cycle of an LTR retrotransposon starts with the transcription of its provirus copy integrated in the genome. The resulting mRNA is transported from the nucleus to the cytoplasm, where it is translated. Usually, reverse transcription of the proviral mRNA occurs in the viral particles in the cytoplasm. The resulting cDNA is then transported back to the nucleus, together with the retroviral integrase where the integrase inserts the cDNA into the host genome, producing a new provirus flanked by identical LTRs and short TSDs.

Subsequent homologous recombination between the LTRs may result in elimination of the provirus with the exception of a solo LTR (Fig. 10). The number of endogenous retroviruses that retain their internal portions flanked by both LTRs is about ten times smaller than the number of solo LTRs. Therefore, recombination between the flanking LTRs is a significant factor in elimination of LTR retrotransposons. Gene conversion between LTRs can also occur at a very low rate. Therefore, the average divergence between proviral LTRs does not adequately reflect the age of the corresponding family of LTR retrotransposons.

RNA recombination of LTR retrotransposons LTR retrotransposons copackage two copies of their RNAs, which can recombine with each other, in the same viral particle. Two RNAs representing essentially distinct LTR retrotransposons can also be copackaged together if they share similar ψ packaging signals. Retroviral recombination is usually mediated by the RNA template switch that occurs during minus-strand DNA synthesis, when RT is paused by a break in the reverse transcribed RNA. The RT pausing can also be caused by a specific secondary structure of the RNA template and by internal repeats.

Characteristics of human LTR retrotransposons The most abundant and ancient group of LTR retrotransposons preserved in the human genome are class III retrotransposons (Table 6). For example, the HERVL16 and HERVL33 families are more than 200 million years old. Despite their high abundance, they have remained undetected for a long time due to their extensive modification by mutations. Most

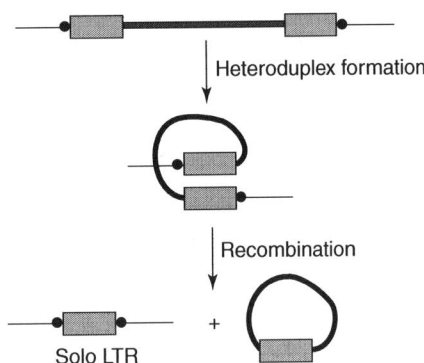

Fig. 10 Formation of a solo LTR through the illegitimate recombination between proviral LTRs.

Class III families of retroelements reconstructed so far are nonautonomous, including HERV-L, which was discovered as the first representative of class III elements. For example, the HERV-L family does not contain *env*, HERVL66 misses *gag* and protease, and HERVL68 does not encode any proteins at all. The most abundant Class III families are MaLRs (Mammalian apparent LTR retrotransposons) and THE1 elements, which probably did not encode any functional proteins at the time of their activity. MaLRs (THE1) were recently classified as members of Class III based on the observation that some of their noncoding sequences were derived from HERV-L-like elements. HERV18 was the first Class III element found to contain *env*. Many different families of LTR retrotransposons are still represented by their LTRs only. As shown in Tables 4 and 6, these families were reliably assigned to their proper classes based on similarities to LTRs from classified retroelements.

Class I LTR retrotransposons (Table 4) include dozens of nonautonomous families that are members of the same group, also called the MER4 or MER4I group after the MER4 element that was discovered over a decade ago and later identified as an LTR flanking the internal portion of a retrovirus-like element (MER4I). Other Class I families of LTR retrotransposons were discovered and classified on the basis of their sequence similarity to MER4I. All these families are characterized by 4-bp target site duplications. Analogously to MaLR and THE1 elements, the originally discovered MER4I-like retrotransposons, do not code for any proteins. Furthermore, the human genome also harbors autonomous Class I families whose members were ancestral to nonautonomous MER4I-like elements. For example, the internal portion of the PRIMA4 autonomous retrovirus is similar to the entire MER4I sequence; even their LTRs are similar. Analogously, the autonomous PRIMA41 retrovirus is an ancestor of the MER41 family of MER4I-like nonautonomous elements. Class I elements were involved in vigorous multiple recombinations between RNA copies of different retroviruses. As a result, many Class I families are chimeras composed of parts that came from retroviruses dissimilar to each other, for example, *Harlequin*.

Class III and Class I elements were probably engaged in extensive complementation, meaning that various families mutually assisted each other by sharing necessary enzymatic and packaging activities. For instance, the RNA of nonautonomous proviruses can be copackaged into functional viral cores produced by autonomous LTR retroelements and exogenous retroviruses. These viral cores can be released from the host cell and can infect other cells. Given the low fidelity of reverse transcription ($\sim 10^{-5}$ per nucleotide per cDNA molecule), the original nonautonomous retrovirus can be dramatically transformed after multiple rounds of reverse transcription and intercellular migrations. As a result of such transformation, the similarity between the original provirus and the transformed retrotransposon can wane rather fast.

Human Class II elements are represented by 11 families (Table 5) with ages ranging from ~ 35 million to less than 5 million years. The oldest Class II family is HERVK22. On the other hand, some HERVK10 elements are polymorphic in the human population, and contain nearly intact ORFs coding for the standard retroviral proteins. In addition to the standard proteins, some HERVK elements encode dUTPase and K-Rev. The latter is a 105-aa protein, also called cORF, which is a functional homolog of the HIV Rev protein. K-Rev mediates nuclear export of the unspliced HERVK RNA by binding to both the host Crm1 nuclear export factor and to a cis-acting retroviral RNA target.

Mechanisms of retroviral evolution are poorly understood since very few intermediate forms of exogenous retroviruses are available for analysis. Therefore, LTR retroelements deposited in mammalian genomes provide us with a unique opportunity to study retrovirus-like elements, which became extinct millions of years ago.

3.3
DNA Transposons

All DNA transposons detected in the human genome represent the so-called *cut-and-paste* category since they use excision from host DNA (cut) followed by reinsertion (paste) as two basic steps in their life cycle. Both steps are mediated by the transposon-encoded enzyme called *transposase*, which recognizes terminal inverted repeats (TIRs; Fig. 11). The transposon insertion is accompanied by target site duplication.

The cut-and-paste process is thought to be associated with the active replication process. As shown in Fig. 12, the excision is more likely in already replicated DNA segments, whereas the insertion occurs preferentially in nonreplicated segments. On the basis of this model, the predicted multiplication rate of cut-and-paste DNA transposons is 1.5 per replication cycle; but the real rate is expected to vary between 1 and 1.5.

To date, no active DNA transposons have been found in the human genome. All

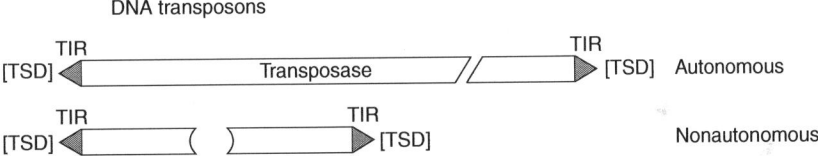

Fig. 11 Schematic structure of autonomous and nonautonomous DNA transposons.

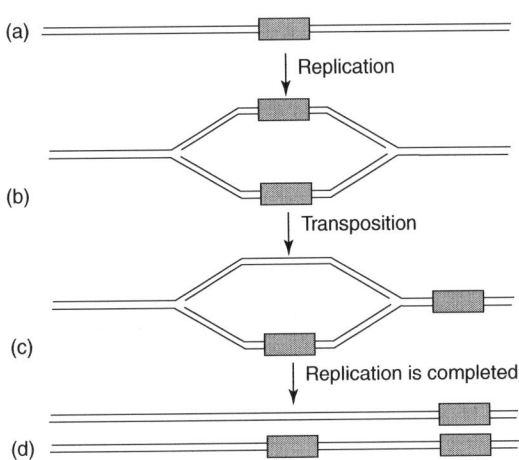

Fig. 12 Relationship between DNA transposition and host replication.

basic properties of human DNA transposons were deduced from sequence analysis of human repetitive DNA, followed by reconstruction of DNA and protein sequences encoded by once-active elements. On the basis of their structural commonalties, eukaryotic DNA transposons can be divided into eight different superfamilies, of which six are represented in the human genome (Table 7). DNA transposons that belong to the same superfamily, even those from different vertebrates, plants, and insects, are characterized by common structural hallmarks such as similar TIRs, transposases, and the same characteristic sizes of TSDs. Additional superfamily characteristics include conserved termini of TIRs, usually 2 to 3 bp long. Like other classes of TEs, DNA transposons are represented by autonomous and nonautonomous elements.

The human genome harbors families derived from over 29 autonomous transposons. Of these, over 25 belong to the hAT and mariner/Tc1 superfamilies (Table 8). However, only 13 of them have been reconstructed and characterized in more detail. Some of the most abundant families of nonautonomous hAT elements, such as MER5A and Cheshire_A, are represented by more than 10^4 copies each. Overall, the transposon-derived interspersed repeats account for ~5% of the human genomic DNA.

3.3.1 hAT-like Transposons

The superfamily of hAT-like elements was named after the initials of *hobo*, *Activator* and *Tam* transposons identified in the fruit fly and maize. They are flanked by ~15-bp terminal inverted repeats that include characteristic 5'-CA and TG-3' termini, and generate 8-bp target site duplications upon integration in the genome. Autonomous hAT-like transposons are ~3000-bp long and encode ~500-aa transposase similar to transposases encoded by all known hAT-like transposons detected in different eukaryotic species. Analogously, the ORFs in the human hAT-like transposons are also expected to encode the transposases.

The most abundant hAT elements are the nonautonomous deletion products of the corresponding autonomous transposons. They retain TIRs, but the

Tab. 7 Superfamilies of cut-and-paste DNA transposons represented in humans.

Superfamily	Related bacterial transposases	Termini	TSDs
Mariner/Tc1	IS630	CA-TG	2 (TA)
hAT	–	CA-TG	8
piggyBac	–	CCC-GGG	4 (TTAA)
MuDr[a]	IS256	GGG-CCC	9–10
Harbinger[b]	IS5	GGG-CCC	3
P[c]	–	CA-TG	7–8

[a] Only nonautonomous *Ricksha* has *MuDr* hallmarks: 9-bp TSDs, 70-bp TIR, 5'-GGG and CCC'- 3' termini.
[b] *Harbinger*- and
[c] *P* – like transposases are present as single-copy genes conserved in mammals including humans.

Tab. 8 Families of DNA transposons identified in the human genome.

Superfamily	Autonomous elements	Nonautonomous elements	%
hAT	Blackjack	MER81, MER94	0.06
	Charlie1	Charlie1A, Charlie1B	0.21
	Charlie2	Charlie2A, Charlie2B	0.08
	Charlie3	MER1A, MER1B	0.11
	Charlie4	MER80	0.05
	Charlie5	MER3, MER33	0.40
	Charlie6		$0.04 \cdot 10^{-1}$
	Charlie7		0.04
	Charlie8	Charlie8A, MER102	0.08
	Charlie9	MER112	0.03
	Charlie10	MER5A, MER5B, MER5C	0.42
	Cheshire	Cheshire_A, Cheshire_B (MER58, MER58A-C)	0.31
	Zaphod	Zaphod2 (Fordprefect)	0.03
	Unknown	MER20, MER20B, MER30, MER30B, MER45, MER45A-C, MER45R, MER63A-D, MER69A-C, MER91A-C, MER96, MER96B, MER97A-C, MER99, MER106, MER106B, MER107, MER117, MER119, ORSL	0.54
			2.4
Mariner/Tc1	Golem (Tigger3)	Golem_A, Golem_B, Golem_C (Tigger3)	0.14
	HSMAR1	MADE1	0.04
	HSMAR2		0.04
	HSTC2	MER104, MER104A-C, Kanga1, Kanga2	0.06
	Tigger1		0.32
	Tigger2	MER28 (Tigger2a)	0.15
	Tigger5	TiggerA-B	0.01
	Tigger6	Tigger6A-B	0.03
	Tigger7	MER44A-D	0.09
	Zombi (Tigger4)	Zombi_A-C	0.03
	Unknown	MARNA, Tigger8, Tigger9, MER2, MER2B, MER6, MER6A-C, MER8, MER53, MER82	1.95
			2.8
piggyBac	Looper	MER75, MER85	0.02
MuDR	Unknown	Ricksha, Ricksha_0	0.03
	Merlin1_HS		$0.01 \cdot 10^{-3}$
Unclassified	Unknown	MER81, MER94, MER121, MER122, MER103, MER105, MER113, MER116	0.13

transposase is either partially or completely deleted. Typically, the number of nonautonomous copies representing a particular hAT family is at least an order of magnitude higher than the copy number of the corresponding autonomous elements. Sometimes, nonautonomous elements differ from their autonomous companions by possessing additional internal fragments (for example, MER5A-B and Charlie10). It is possible that these additional fragments represent genomic DNA captured by the nonautonomous transposons (analogously to MuDR transposons in the *Arabidopsis thaliana* genome).

The most abundant human hAT families are *Charlie10*, *Charlie5*, *Cheshire* and their nonautonomous companions represented by ~60 000, ~50 000 and ~40 000 copies respectively (see Table 8). As indicated above, there are no active DNA transposons in the human genome. The youngest hAT elements are *Charlie3* repeats that are ~8% diverged from their family consensus sequence. Given the standard neutral mutation rate in the human genome, this corresponds to ~35 million years. Most of the human hAT-like families are much older and they might have been active in a common ancestor of primates and rodents.

3.3.2 Mariner/Tc1-like Transposons

The Mariner/Tc1 superfamily reflects the names of the two founding DNA transposons *mariner* and *Tc1*, first discovered in the *Drosophila melanogaster* and *Caenorhabditis elegans* genomes. Members of this superfamily are characterized by 2-bp target site duplications (usually TA), by ~28-bp TIRs, and by the 5'-CA...TG-3' termini. However, some elements may have different termini; for example, HSMAR1 and HSMAR2 are characterized by the 5'-TT...AA-3' and 5'-CG...CG-3', respectively. Mariner/Tc1 transposases are distantly related to a broad set of bacterial transposases similar to that encoded by IS630 insertion element, and are characterized by the conserved $DD_{35}E$ motif.

3.3.3 piggyBac-like Transposons

The piggyBac superfamily includes DNA transposons with TTAA TSDs, the ~15-bp terminal inverted repeats, the 5'-CCC and GGG-3' termini, and the unique piggyBac transposase. The piggyBac transposon was first discovered as an active mobile element in the cabbage looper, *Trichoplusia ni*. The human transposon called *Looper* was the first piggyBac-like element identified outside cabbage looper. It was reconstructed from copies transposed ~100 million years ago. Since then, piggyBac-like elements were identified in nonmammalian vertebrates, insects, and fishes. In addition to Looper, the human genome harbors MER75 and MER85 transposons that were active 10 to 50 million years ago. Only ~2000 copies of the piggyBac-like elements are present in the human genome.

3.3.4 MuDR-like Transposons

MuDR/Mu transposons, also known as *Mutators*, were first discovered in maize. They are characterized by 9 to 10-bp TSDs and by ~70 to 400-bp TIRs, typically ending with 5'-GGG and CCC-3' conserved termini. However, some families of MuDR transposons in the *A. thaliana* genome can transpose without any discernible TIRs. Different autonomous MuDR transposons encode similar MuDR-like transposases, which share common conserved motifs with bacterial IS256-like transposases. Currently, only two families of MuDR-like transposons are known to be harbored by the human genome. One of them is represented by the nonautonomous transposons Ricksha and Ricksha_0, and characterized by 9 to 10-bp TSDs, ~70-bp TIRs, and 5'-GGG and CCC-3' termini. Ricksha and Ricksha_0 were active 40 to 100 million years ago and are present in only ~1000 copies in the human genome. Ricksha is a composite transposon that differs from Ricksha_0 by having a ~700-bp insertion of the HERVL endogenous retrovirus including its long terminal repeat (MLT2B). Presumably, the HERVL provirus was inserted into an active Ricksha_0 copy, which amplified into the Ricksha subfamily. A second family of human MuDR-like transposons was reported recently as the Merlin1_HS autonomous transposon, which encodes a transposase that is the most similar to the IS1016 bacterial transposase. Since the IS1016 and IS256 transposases are distantly related, Merlin1_HS can be considered to be a member of the MuDR superfamily, quite different from standard MuDR transposons. Merlin1_HS is characterized by the 8-bp TSDs and 21-bp TIRs with 5′-GGA and TCC-3′ termini. The human genome harbors fewer than 100 copies of Merlin1_HS. They are ~20% divergent from the consensus sequence, which indicates their activity some 100 million years ago.

4
Repetitive Elements and Human Diseases

The potential pathogenic impact of repetitive elements can manifest itself through a broad spectrum of pathways. The first category of disorders is linked to the expression of active TEs, particularly to the production of proteins. Such proteins can interfere with cellular processes and, most importantly, they can provoke host immune reactions against cells containing TE antigens. The second group of pathogenic effects linked to repetitive DNA includes mutations and genetic rearrangements such as satellite expansion, repeat insertions, deletions, illegitimate recombination, or nucleolytic activity of proteins encoded by TEs. These mutations can occur both in germline or somatic cells.

The pathogenic potential of TE-derived protein products is far from being understood. Relevant research was done mostly on human endogenous retroviruses (HERVs; Sect. 3.1.2). HERVs have been proposed as causative agents of autoimmune diseases such as rheumatoid arthritis, systemic lupus erytrematosus, insulin-dependent diabetes mellitus, and multiple sclerosis. Other potentially HERV-linked disorders include schizophrenia and several types of cancer. In many such cases, retroviral mRNAs or even entire particles were detected in malignant tissues. Retroviral antigens such as the *gag* gene product may indeed trigger immune reaction against affected tissues leading to autoimmune abnormalities. However, it is still unclear whether expression of retroviral mRNAs and proteins causes or merely correlates with the pathological processes. Increased expression of TEs is typical of many cellular alterations including cell transformations. This indicates that rather than inducing the pathogenesis, the expression of TEs may simply reflect relaxed transcription control in malignant cells.

Repeat-induced DNA rearrangements represent a wide range of mutation defects related to the amplification of repetitive elements and genetic recombination. Genetic diseases can be caused by both simple and interspersed repeats. The unstable expansions of simple sequence repeats, namely, trinucleotides, are involved in a number of genetic diseases. The fragile X syndromes (known as FRAXA and FRAXE) involve CGG and GGC triplet expansions. Myotonic dystrophy can be caused by the CTG/CAG repeat expansion. Other examples of diseases associated with moderate expansion of CAG repeats include neural diseases such as Kennedy's disease, Huntington's disease, spinocerebellar ataxia 1 (SCA1) and dentatorubropallidoluysian atrophy (DRPLA). The disease-related triplet expansions are much more frequent than contractions, and occur primarily in GC-rich genes.

Centromeric and telomeric repeats can also trigger chromosomal instabilities in the human genome. Recombinations between both centromeric and telomeric satellites were reported in cases of chromosome rearrangements, chromosome fragility, or jumping chromosomal translocations.

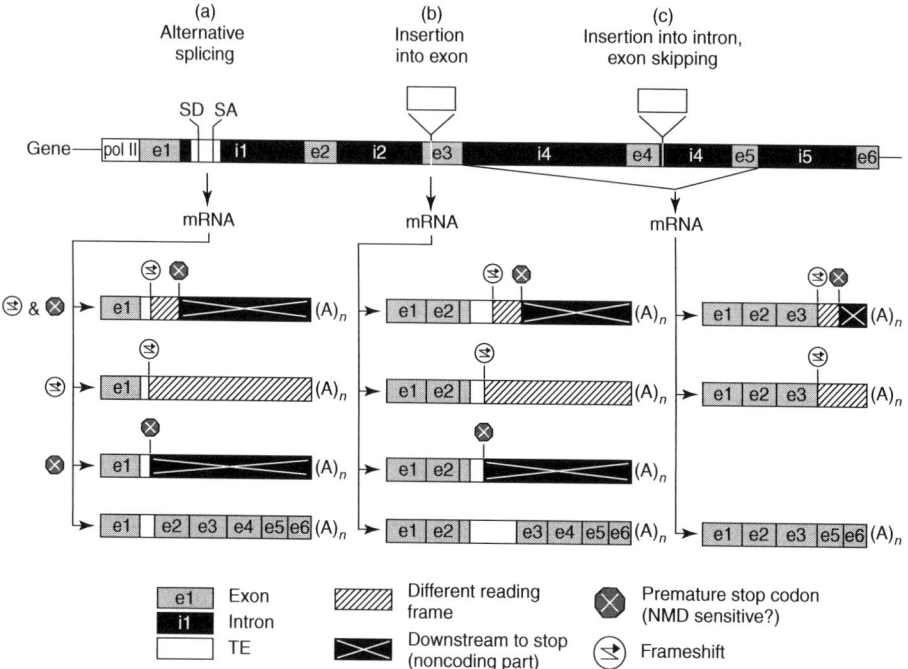

Fig. 13 Transposable elements and human disease. Three different mechanisms of TE-mediated disruptions affecting the gene functionality: (a) intronic repeats that contain splice donor (SD) and splice acceptor (SA) sites, can provide a new exon. (b) Insertion of transposable element into exon. (c) Insertion of transposable element into intron, disrupting proper splicing can lead to exon skipping. All three disruptions lead to similar changes on the mRNA level. Insertion or deletion in coding sequences may change the reading frame, if the indel length is not a multiple of three. Long frameshifts are likely to induce premature stop codon(s) into mRNA (case 1). mRNAs with premature stop codon (i.e. located more than 50–55 bp upstream of the last exon–exon junction) are likely to be destroyed by the nonsense-mediated RNA decay (NMD), and the corresponding protein is not produced. In some cases, particularly for short frameshifts, the random stop codon is found close enough to the last exon–exon boundary and the protein can be synthesized, albeit with an aberrant C-terminal domain (case 2). Note that mRNAs completely lacking stop codons are also destroyed by the cellular RNA control, although by a different pathway than the nonsense-mediated RNA decay. Insertions containing stop codon without change in the reading frame (case 3). Again, if the stop is more than 50 to 55 bp 5′ from the last exon–exon junction, the RNA is likely to be NMD-sensitive and the protein is not produced. Finally, if neither frameshifts nor new premature stop codons are induced, insertions lead to creation of a protein with a new repeat derived domain. Exon-skippings induce deletions of some domains. These new proteins may or may not be functional.

Interspersed repeats can affect alternative splicing and cause inclusion of intronic repeats into cellular mRNAs. This can alter or destroy the mRNA or protein functionality and be manifested as a disease phenotype. Notably, Alu repeats contain cryptic splice sites in the antisense orientation that, after acquiring specific mutations, can give rise to new Alu-derived exons (Fig. 13a). The majority of such Alu cassettes are alternatively spliced and produce premature termination (stop) codons. The affected mRNAs are probably destroyed by cellular mRNA quality control machinery, particularly by the nonsense-mediated RNA decay (NMD) pathway, which degrades mRNAs with premature stop codons at the time of protein translation. However, if the new exon becomes constitutively spliced in, the NMD control prevents efficient translation of the affected mRNA, and thus causes a deleterious phenotype at the protein level. So far, there are two known cases of disease-related Alu splicing into cellular mRNAs. One, reported in a case of ornithine-delta aminotransferase deficiency, was caused by alternative Alu splicing, creating a premature stop codon. In a second instance, a splice-mediated insertion of an Alu sequence in the precursor of alpha 3 type IV collagen mRNA caused autosomal recessive Alport syndrome in the affected patient. The affected proteins were not expressed in either case, despite the fact that the insertion preserved the open reading frame in the second case.

Insertion into genomic DNA is an integral part of the amplification of interspersed repeats. An integration into coding sequences is likely to cause an insertional inactivation of the targeted gene, which subsequently manifests itself as a genetic abnormality. To date, there are 41 published repeat insertions associated with human diseases (Table 9). All the insertions are either L1 elements (14 cases), L1-dependent Alu (23 cases), or SVA elements (4 cases). The Alu and L1 insertions come nearly exclusively from the two active subfamilies AluY and L1-Ta respectively (Table 9). No recent insertions of endogenous retroviruses or DNA transposons have been reported.

From Table 9 it is clear that some genes are overrepresented among insertion targets, particularly coagulation factors VIII and XI and the dystrophin gene. Furthermore, 42% (19 out of 41) of all documented insertions are on chromosome X, which represents only 5% of the human genome. The bias is particularly strong for L1 insertions 11/14 (79%), and less so for L1-dependent nonautonomous Alu and SVA elements, which represent 26% (6/23) and 25% (1/4) of all documented insertions respectively. Chromosome X is present in only one functional copy in both males and females, due to female imprinting (inactivation) of one X copy. Therefore, the X-linked genes are more vulnerable to mutations. The excess of detected insertions on the X chromosome may reflect the detection bias for dominant phenotypes rather than any preferential insertion patterns on this chromosome.

The majority of insertions listed in Table 9 occurred in exons and the remaining intronic insertions appear to have disrupted splicing or other essential signals located within introns. The deleterious effect of intraexonic insertions is mainly due to the incorporation of premature stop codons into the reading frame and mRNA degradation by NMD (Fig. 13b). Other defects can be caused by intron inclusion or exon skipping (Fig. 13a, c), leading to either truncation of the protein or induction of premature stop codons and NMD sensitivity.

Tab. 9 Insertions of interspersed repeats and human diseases.

Repeat	Locus	Chrom	Gene	Symbol	Phenotype
Alu/Ya5	–	–	Human homolog of moloney murine virus leukemia (MMLV) retrovirus insertion locus	MLVI-2	Associated with leukemia
1x Alu/Yb8 1x Alu/Ya5	NM_000141	10q26.13	Fibroblast growth factor receptor 2 isoform 1	FGFR2	Apert syndrome
Alu/Y	NM_000062	11q12.1	Serine/cysteine proteinase inhibitor	C1NH	C1 inhibitor deficiency
Alu/Ya5	NM_000190	11q23.3	Porphobilinogen deaminase (hydroxymethylbilane synthase)	PBGD/HMBS	Acute intermittent porphyria
Alu/Y	NM_000059	13q13.1	Breast cancer 2	BRCA2	Breast cancer
Alu/Ya5	NM_000267	17q11.2	Neurofibromin 1	NF1	Neurofibromatosis
Alu/Ya5	NM_000789	17q23.3	Angiotensin I converting enzyme isoform 1	ACE	Protection against myocardial infarction and age-related macular generenation?
Alu/n.a.	NM_000092	2q36.3	Alpha 4 type IV collagen precursor	COL4A3	Alport syndrome
Alu/Ya4	NM_000388	3q21.1	Calcium-sensing receptor	CASR	Familial hypocalciuric hypercalcemia
Alu/Yb8	NM_000055	3q26.1	Butyrylcholinesterase	BCHE	Acholinesterasemia
Alu/n.a.	NM_001119	4p16.3	Adducin 1, transcript variant 1	ADD1	Huntington disease
Alu/Yb1	NM_006264	4q21.3	Protein tyrosine phosphatase, nonreceptor type 13	PTPN13 (APO-1/CD95)	Autoimmune lymphoproliferative syndrome
Alu/Yb8	NM_000038	5q22.2	Adenomatosis polyposis coli	APC	Hereditary desmoid disease
Alu/Ya5	NM_000503	8q13.3	Eyes absent homolog 1 (Drosophila)	EYA1	Branchio-oto-renal syndrome
Alu/Ya5	NM_000167	Xp21.2	Glycerol kinase	GK	Glycerol kinase deficiency

TE	Accession	Locus	Gene	Symbol	Disease
Alu/Ya5	NM_000206	Xq13.1	Interleukin 2 receptor, gamma	IL2RG	X-linked severe combined immunodeficiency
Alu/Y	NM_000061	Xq22.1	Bruton agammaglobulinemia tyrosine kinase	BTK	X-linked agammaglobulinemia
Alu/Ya5	NM_000133	Xq27.1	Coagulation factor IX	F9	Hemophilia B
Alu/Ya3a1	NM_000133	Xq27.1	Coagulation factor IX	F9	Hemophilia B
Alu/n.a.	NM_000133	Xq27.1	Coagulation factor IX	F9	Hemophilia B
Alu/Yb8	NM_000132	Xq28	Coagulation factor VIII	F8	Hemophilia A
Alu/Yb8	AF074480	6p22.3	CMP-N-acetylneuraminic acid (CMP-Neu5Ac) hydroxylase	CMAH	Inactivated in humans, fixed
L1/Ta	NM_000518	11p15.4	b-Globin	HBB	Beta thalassemia
L1/Ta	NM_000038	5q22.2	Adenomatosis polyposis coli	APC	Colon cancer
L1/Ta	NM_006731	9q31.2	Fukutin	FCMD	Fukuyama-type congenital muscular dystrophy
L1/Ta	NM_006915	Xp11.3	Retinitis pigmentosa 2 gene	RP2	X-linked retinitis pigmentosa 2
2x L1/Ta	NM_000397	Xp11.4-Xp21.1	Cytochrome b-245, beta polypeptide	CYBB	Chronic granulomatous disease
4x L1/Ta	NM_000109	Xp21.1–2	Dystrophin	DMD	3x muscular dystrophy, 1x X-linked dilated cardiomyopathy
L1/Ta	NM_000133	Xq27.1	Coagulation factor IX	F9	Hemophilia B
2x L1/Ta 1x L1/pre-Ta	NM_000132	Xq28	Coagulation factor VIII	F8	Hemophilia A
SVA	NM_003126	1q23.1	Alpha spectrin	SPTA1	Hereditary elliptocytosis
SVA	NM_006731	9q31.2	Fukutin	FCMD	Fukuyama-type congenital muscular dystrophy
SVA	NM_000061	Xq22.1	Bruton agammaglobulinemia tyrosine kinase	BTK	X-linked agammaglobulinemia
SVA	NM_000527	19p13.2	Low density lipoprotein receptor	LDLR	Familial hypercholesterolemia

Source: The table is based on the updated online database of TE insertions into the human genome (http://www.med.upenn.edu/genetics/labs/kazazian/human.html).

Table 9 compiles both *de novo* insertions, such as in the case of L1-induced Fukuyama muscular dystrophy, and inherited insertions causing familial disorders, such as an Alu insertion into the adenomatosis polyposis coli (APC) gene, causing hereditary desmoid disease. The table also contains one somatic L1 insertion that inactivates APC, indicating that the L1 retrotransposition can take place in both germline and somatic tissues. Most phenotypes caused by insertional inactivations are negatively selected in the population and subsequently eliminated. Nevertheless, in cases of nonessential genes with mild defects, the insertions may eventually become fixed in the population. For example, an Alu insertion seems to have inactivated the CMP-*N*-acetylneuraminic acid hydroxylase (CMAH) in the human lineage after the evolutionary separation of human and chimp, since all primates except humans contain a functional copy of CMAH. In rare cases, insertions may even produce a positive phenotype. For instance, AluYa5 insertion into intron 16 of angiogenesin-converting enzyme (ACE) seems to provide protection against myocardial infarction and age-related macular degeneration, since genotypes without the insertion have a higher incidence of the aforementioned diseases in comparison to individuals having Alu insertion in both alleles.

Table 10 shows a list of the genetic defects attributed to recombination between interspersed repeats. Preliminary indications for linkage between Alu polymorphisms and genetic diseases, based solely on polymorphism studies, are not included. While for some genes only a single case of recombination between interspersed repeats is known, other loci including LDLR, C1NH, alpha-globin, APC, HEXB or MLL represent hotspots for recombinations. Many independent cases of such recombination have been reported. As in the case of insertions, there is a detection bias for recombinations in X-linked genes. Twelve out of 54 (22%) deletion/duplication loci are located on chromosome X.

In mammalian cells, two major recombination pathways exist: *homologous recombination* and nonhomologous end joining, also known as *nonhomologous recombination*. In the great majority of genes listed in Table 10, homologous recombination between TEs was detected, but nonhomologous recombination between one or more TEs was also found in many genes.

The majority of known rearrangements were reported for direct repeats located within or near one gene. While some rearrangements listed in Table 10 are inherited (i.e. passed through germline cells), others occur primarily during mitosis in somatic tissues. This includes many leukemia- or other cancer-related recombinations. Figure 14 shows a model of mitotic intrachromosomal recombination between direct repeats. Unequal sister chromatid exchange produces one duplicated and one deleted copy. Reciprocal intrachromatid exchange, on the other hand, results in one deletion and one circular episomal DNA, which is frequently lost. The mechanism is the same as for the creation of solo LTRs (see Fig. 10). Mispairing and crossing-over between repeats, leading to deletion, have been reported in 54 genes, compared to duplications in only five.

Internal indels can result in altered lengths of the protein, if the reading frame is preserved, or in premature stop codons within the reading frame. Typical outcomes of internal exon/gene deletions and duplications are shown in Fig. 14. RNAs with premature stop codons

Tab. 10 Recombination of interspersed repeats and human diseases.

Repeat	Mutation	Locus	Chrom	Gene	Symbol	Phenotype
Alu	Deletion	NM_000511	19q13.33	Alpha(1,2)fucosyl-transferase	FUT2	ABO-Bombay phenotype
Alu	Deletion	NM_000033	Xq28	ATP-binding cassette, subfamily D, member 1	ABCD1	Adrenoleukodystrophy
Alu	Deletion	NM_000061	Xq22.1	Bruton's tyrosine kinase	BTK	Agammaglobulinemia, X-linked
Alu	Deletion	NM_000273	Xp22.22	Albinism ocular type 1 gene	OA1	Albinism ocular type 1
L1	Deletion	NM_001847	Xq23	Type IV alpha 6 collagen	COL4A6	Alport syndrome and diffuse leiomyomatosis
Alu [2]	Deletion	NM_000021	14q24.2	Presenilin-1	PSEN1	Alzheimer's disease, early onset
Alu	Deletion	NM_000062	11q12.1	Complement component 1 inhibitor precursor	C1NH/SERPING1	Angioedema, hereditary
Alu	Deletion	NM_003190	6p21.32	Tapasin	TAPBP	Bare lymphocyte syndrome, type I
Alu/L1 [3]	Translocation	t(9;11)(p24;q23)	–			Bipolar affective disorder (manic depression)
Alu	Deletion	NM_007295	17q21.31	Breast cancer 1, early onset	BRCA1	Breast and/or ovarian cancer
Alu [4]	Deletion	NM_000059	13q13.1	Breast cancer 2	BRCA2	Breast cancer
Alu	Deletion	NM_005228	7p11.2	Epidermal growth factor receptor	EGFR	Cancer, glioblastomas
Alu	Deletion	NM_000321	13q14.2	Retinoblastoma	RB1	Cancer predisposition

(continued overleaf)

Tab. 10 (continued)

Repeat	Mutation	Locus	Chrom	Gene	Symbol	Phenotype
Alu	Deletion	NM_000038	5q22.2	Adenomatous polyposis coli	APC	Colorectal cancer, adenomatous polyposis
Alu	Deletion	NM_000251	2p21	Mismatch repair protein	MSH2	Colorectal cancer, nonpolyposis
Alu	Deletion	NM_000249	3p22.3	MutL homolog 1	MLH1	Colorectal cancer, nonpolyposis
Alu	Translocation	t(1;22)(q23;q11)	–	–	–	–
Alu	Deletion	NM_000208	19p13.2	Insulin-receptor	INSR	Diabetes, insulin-resistant
Alu	Duplication	NM_000302	1p36.22	Lysyl hydroxylase 1/procollagen-lysine, 2-oxoglutarate 5-dioxygenase	PLOD	Ehlers Danlos syndrome
Alu	Translocation	t(11;22)	–	–	–	Ewing's sarcoma
Alu	Deletion	NM_005243	22q12.2	Ewing sarcoma breakpoint region 1	EWSR1	Protection against Ewing's sarcoma
Alu	Deletion	NM_000169	Xq22.1	Alpha-galactosidase A	GLA	Fabry disease
Alu	Deletion	NM_000135	16q24.3	Fanconi anemia, complementation group A	FANCA	Fanconi anemia
Alu	Deletion	NM_000419	17q21.31	Integrin alpha 2b precursor, platelet glycoprotein IIb, CD41B	ITGA2B	Glanzmann's thrombasthenia
L1	Deletion	NM_000293	16q12.1	Phosphorylase kinase, beta subunit	PHKB	Glycogen storage disease
Alu	Deletion	NM_000152	17q25.3	Alpha-glucosidase	GAA	Glycogen storage disease type II
Alu	Deletion	NM_002133	22q12.3	Heme oxygenase-1	HMOX1	Heme oxygenase-1 deficiency
Alu	Deletion	NM_000132	Xq28	Coagulation factor VIII	F8	Hemophilia A

Alu	Deletion	NM_004402	1p36.32	Caspase-activated DNase, beta polypeptide	CAD	Hepatoma
L1	Deletion	NM_000249	3p22.3	MutL homolog 1	MLH1	Hereditary nonpolyposis colorectal cancer
Alu	Transposition/Deletion	NM_000202	Xq28	Iduronate-2-sulfatase	IDS	Hunter disease
Alu	Deletion/Duplication	NM_000527	19p13.2	Low density lipoprotein receptor	LDLR	Hypercholesterolemia, familial
Alu	Deletion	NM_000064	19p13.3	Alpha chain of complement component 3	C3	Hyperlipidemia, familial combined
Alu	Deletion	NM_000384	2p24.1	Apolipoprotein B precursor	APOB	Hypobetalipoproteinemia
Alu	Deletion/Duplication	NM_000194	Xq26.2	Hypoxanthine phosphoribosyl-transferase	HPRT	Lesch–Nyhan syndrome
Alu	Deletion/Duplication/Translocation	Numerous translocations [1]	–	Several, including MLL, MOZ and CPB	–	Leukemia, various acute/chronic forms
Alu	Deletion	NM_000546	17p13.1	Tumor protein p53	TP53	Li-Fraumeni syndrome

(continued overleaf)

Tab. 10 (continued)

Repeat	Mutation	Locus	Chrom	Gene	Symbol	Phenotype
Alu	Deletion	NM_003982	14q11.2	Solute carrier family 7	SLC7A7	Lysinuric protein intolerance
HERV15 [5]	Duplication	AZFa locus	Yq11.21	DEAD/H (Asp-Glu-Ala-Asp/His) box polypeptide/*Drosophila* fat facets related	DBY/USP9Y	Male infertility
Alu	Deletion	NM_002608	22q13.1	Platelet-derived growth factor beta	PDGFB	Meningioma
Alu	Translocation	t(1;19)(q21.3;q13.2)	–	–	PAFAH1B3-CLK2 fusion	Mental retardation, ataxia and atrophy of the brain
Alu [6]	Deletion	NM_000512	16q24.3	N-acetylgalactosamine-6-sulfatase precursor	GALNS	Mucopolysaccharidosis type IVA
Alu/L1 [3], Alu	Deletion	NM_000109	Xp21.1–2	Dystrophin	DMD	Muscular dystrophy
Alu	Deletion	NM_000533	Xq22.2	Proteolipid protein gene	PLP1	Pelizaeus–Merzbacher disease
Alu	Deletion	NM_001171	16p13.11	ATP-binding cassette, subfamily C, member 6	ABCC6	Pseudoxanthoma elasticum
Alu	Deletion	NM_000330	Xp22.13	Retinoschisis factor	RS1	Retinoschisis
Alu	Deletion	NM_000521	5q13.3	Hexosaminidase B, beta polypeptide	HEXB	Sandhoff disease
Alu	Deletion	NM_000022	20q13.12	Adenosine deaminase	ADA	Severe combined imminodeficiency (SCID)
Alu	Deletion	del(5)(q12.3)	–	–		Small-cell lung cancers

Alu	Translocation	t(X;18)			Synovial sarcoma
Alu	Deletion	NM_005502	9q31.1	ATP-binding cassette transporter, subfamily A member 1	ABCA1	Tangier disease
Alu	Deletion	NM_000520	15q24.1	Hexosaminidase A, alpha polypeptide	HEXA	Tay Sachs disease
Alu	Deletion	del(16)(p)	—	Alpha-globin gene cluster	—	Thalassemia, alpha
Alu	Deletion	del(11)(p)	—	Beta-globin gene cluster	—	Thalassemia, delta, gamma
Alu	Deletion	NM_000488	1q25.1	Serine (or cysteine) proteinase inhibitor, clade C (antithrombin)	SERPINC1	Thrombophilia
Alu	Deletion	NM_000548	16p13.3	Tuberous sclerosis 2	TSC2	Tuberous sclerosis
Alu	Deletion	NM_025237	17q21.31	Sclerosteosis	SOST	van Buchem disease
Alu [2]	Deletion	NM_000552	12p13.31	von Willebrand factor A	VWF	von Willebrand disease
Alu	Deletion	NM_000377	Xp11.23	Wiskott-Aldrich syndrome protein	WAS	Wiskott-Aldrich syndrome
Alu	Translocation	DXYS5 locus	—	—		XX maleness

Notes: The table represents a significant update of previously published reviews. Each row corresponds to one or several rearrangements affecting the same gene. [1]: There are dozens of cases of Alu involvement in recombination leading to various forms of both acute and chronic leukemias, and related carcinomas. The disorders known to be, at least partially, caused by Alu recombinations are: t(8;16)(p11;p13), t(11;18)(q21;q21), t(2;8), t(14;19), t(9;11), t(14;18), t(11;19)(q23;p13.3), t(10;16)(q22;p13), t(4;11). Philadelphia translocation t(9;22) and variants, 7q22-q34 inversion, and dup(11)(q23). [2]: Nonhomologous Alu recombination. [3]: Nonhomologous recombination between Alu and L1 elements. [4]: Nonhomologous recombination induced by a polyA tail of an Alu element. [5]: Homologous recombination between two HERV15 elements produced duplication of a 780-kb segment on chromosome Y harboring two genes DBY and USP9Y. [6]: Specific deletion of a single Alu monomer in an intronic Alu element, probably by nonhomologous recombination.

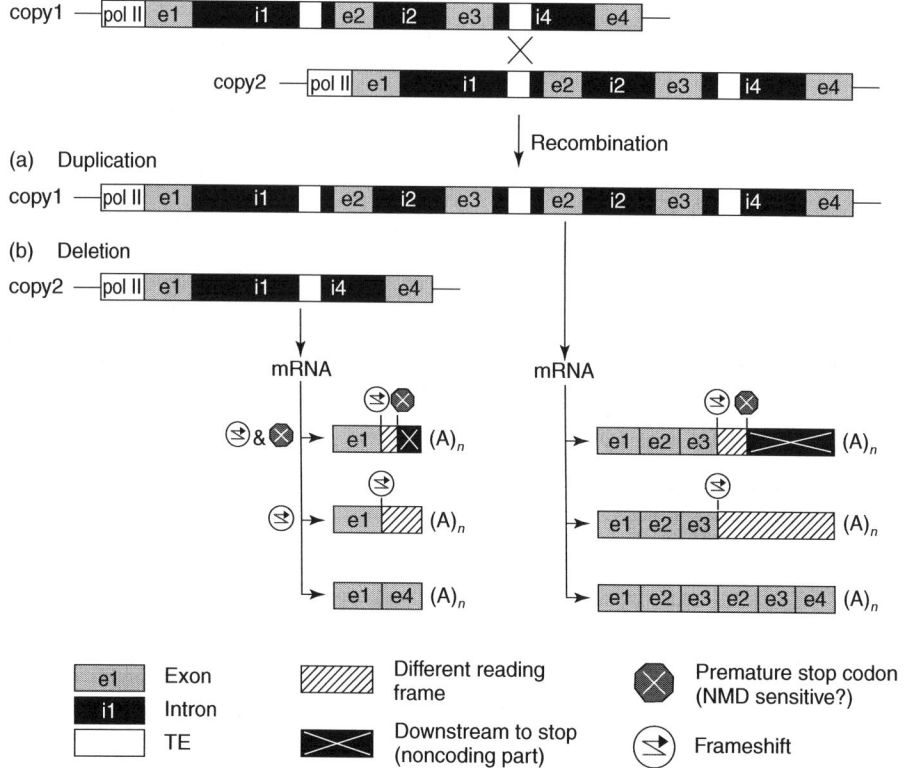

Fig. 14 Recombinations between repetitive elements and human disease. Figure shows classical recombination between two direct intragenic repeats, producing (a) exons duplication; and (b) exon deletion. The effects are similar to those described in Fig. 13.

are likely to be NMD-sensitive, and the resulting phenotype will be manifested by lack of the corresponding protein.

The prevalence of Alu-mediated rearrangements is overwhelming: 55 out of the 61 genes in Table 10 are affected by Alu-induced recombinations. Only three genes with L1 homologous recombinations have been reported so far, and only one case is known for endogenous retroviruses (Table 10). One deletion and one translocation were induced by nonhomologous Alu-L1 recombinations. The excess of detected Alu recombinations is probably related to the Alu genomic distribution and structure. A typical Alu is about 300 bp long, and this size seems to be sufficient for efficient homologous recombination in mammalian cells, where the minimal length requirement is around 150 bp. Alus are by far the most frequent repeats found in human genes. L1 and endogenous retroviruses are underrepresented in GC- and gene-rich regions and are often truncated or rearranged. The rearrangements disrupt the long sequence similarity necessary for homologous recombination. In summary, Alu elements frequently provide closely spaced, highly identical copies prone to recombination in GC-rich and gene-rich regions, and thus represent the main risk for human genetic disorders caused by recombination between repetitive elements.

Alu sequences contain several regions with elevated occurrence of recombination breakpoints. These include promoters and the AT-rich linker between the left and right Alu monomer. Moreover, Alu elements harbor a 26-bp motif (positions 22–47 at the Alu consensus), possibly associated with increased frequency of recombination. The region contains the pentanucleotide motif core CCAGC, which is also part of chi, and an 8-bp sequence known to stimulate recBC-mediated recombination in *Escherichia coli*.

Alu repeats have also been implicated in the etiology of some recurrent somatic chromosome rearrangements detected in neoplastic tissues. The large-scale rearrangements lead to visible karyotype changes such as Philadelphia translocation, which is associated with chronic myelogenous leukemia. Chromosomal aberrations found in cancers result in loss of heterozygosity and in cancer progression. Apart from somatic translocation, nonlethal cytogenetically balanced chromosomal translocations in germline cells can lead to constitutional congenital disorders. An Alu-mediated translocation, t(9;11)(p23;q23), was shown to cause a familial form of the bipolar affective disorder (manic depression). Also, a patient with translocation t(1;19)(q21.3;q13.2) induced by nonhomologous Alu-L1 recombination was diagnosed with brain atrophy, ataxia, and mental retardation.

Some rearrangements may protect against deleterious changes. For example, the reciprocal translocation t(11;22)(q24;q12) fusing the Ewing's sarcoma (EWS) gene on 22q.12 with the *FLI1* gene on 11q.24, is associated with about 80% of primitive neuroectodermal tumors, including Ewing's sarcoma. However, a 2.4 kb Alu-mediated deletion in ESW intron 6, located near the translocation hotspot, seems to stabilize the region, since the carriers of this allele have a tenfold decreased incidence of EWS compared to populations without the 2.4 kb deletion.

In summary, repetitive elements contribute to human diseases through a variety of mechanisms. While the pathogenetic effects of repeat-encoded mRNAs and proteins are still debatable, repeat insertions and recombinations are significant contributors to genetic variability and abnormalities. The overall frequency of such events is unknown. Typical PCR-based detection of human mutations fails to amplify many large-scale variations, and the resulting defects are likely to be underrepresented in mutation databases. Also, germline recombination producing large rearrangements can terminate embryonic development in early stages, thus providing no opportunity for prenatal diagnostics.

The impact of retrotransposon insertions is small compared with the effect of deletions. Insertions correspond to about 0.1% of all human diseases. Alu-mediated recombinations are estimated to contribute to only 0.3% of human diseases. However, since large rearrangements (about 7.4% of all human mutations) often lack precise determination of breakpoints, it is possible that the impact of Alu recombination is higher.

5 Potential Contribution of Repetitive DNA to Cellular Functions

Satellites and TEs are thought to be involved in a broad variety of processes directly related to the host cell. They can affect host replication, imprinting, gene silencing, transcriptional and posttranscriptional regulation, alternative splicing, and evolution of genes and proteins. Below are critically reviewed examples of such involvements.

It is believed that alpha satellites are functional elements in human centromeres, and they bind key proteins involved in host replication (see Sect. 2.2.1). However, the complete lack of alpha satellites in *de novo*-created (functional) neocentromeres indicates that they may represent well-adapted selfish DNA.

It has been proposed that L1s may be involved in inactivating chromosome X in female somatic cells. This view is based on the observed overrepresentation of L1 elements on human chromosome X, particularly near the X inactivation center. However, the lack of L1 clusters in the mouse X inactivation center indicates that L1 accumulation near the inactivation start may not be functionally related to the imprinting of the X chromosome.

As mentioned in Sect. 3, TEs contain different transcription regulation sites including promoters, enhancers, and polyadenylation signals. Accordingly, insertions of TEs into regulatory regions may alter the gene expression, and thus they can be negatively selected. For example, the majority of L1 and HERV elements present in introns are in the opposite orientation to the transcription direction of the corresponding genes. The most likely explanation is that polyA signals of these retrotransposons prematurely terminate the transcription if inserted in the direct orientation. Nevertheless, at least some TE-derived polyA signals may be recruited by cellular genes. Pol II transcription signals encoded by endogenous retroviruses and DNA transposons may serve as mobile "ready-to-use" blocks involved in the evolution of host genes. In several cases, TE-derived promoters and enhancers are incorporated into human genes. However, repetitive elements are found in many genomic regions, including promoters. Therefore, repeats present in these regions may be tolerated rather than being functional.

Potentially important regulation of genes by TEs may also occur at the RNA level. Antisense transcripts from promoter regions of TE inserted into introns can inhibit gene expression. For instance, a HERV-K element inserted into intron 9 of complement C4 gene in the orientation opposite to the gene transcription produces an antisense mRNA modulating the gene expression. Transposable elements are frequently found in untranslated regions of many mRNAs. Therefore, these mRNAs can also be regulated by transcripts derived from other similar copies of genomic TEs. Expressed SINE sequences can be acquired by the host as novel RNA genes. Indeed, *BC200*, a small RNA gene expressed in the primate brain, is the best-studied example of an Alu element recruited by the genome.

Protein-coding TEs can also be recruited as host protein-coding genes. Table 11 lists examples of TE-derived genes present in the human genome. Well-known examples of such genes are V(D)J recombinase and syncytin. V(D)J recombinase, which catalyzes rearrangements of immunoglobulin and T-cell receptor genes in lymphocytes, shares enzymatic characteristics with mariner/Tc1-like DNA transposases. Syncytin, a cell-fusion protein probably participating in placental morphogenesis and antiviral defense, was derived from the *env* gene encoded by the HERV17 (HERV-W) retrovirus. While the V(D)J

Tab. 11 Known human protein-coding genes derived from transposable elements.

GenBank ID	Gene name	Related transposon family	Type
pID 7959287	KIAA1513	Tc1/Pogo/Tc2	DNA transposon
pID 7662294	KIAA0766	hAT	DNA transposon
pID 7513096	JRKL	Tc1/Pogo	DNA transposon
pID 7513011	KIAA0543	hAT/Tip100/Tip100	DNA transposon
pID 7243087	KIAA1353	hAT	DNA transposon
pID 7243087	Buster2	hAT	DNA transposon
pID 7021900		hAT/Charlie3	DNA transposon
pID 6581097	Buster3	hAT	DNA transposon
pID 6581095	Buster1	hAT	DNA transposon
pID 6453533		piggyBac/MER85	DNA transposon
pID 4758872	DAP4,pP52^{rIPK}	hAT/Tip100/Zaphod	DNA transposon
pID 4504807	Jerky	Tc1/Pogo/Tigger4/Zombi	DNA transposon
EST 6986275	Sancho	Tc1/Pogo	DNA transposon
pID 4263748		hAT	DNA transposon
pID 4160548	Tramp	hAT	DNA transposon
pID 3413884	KIAA0461	Tc1/Pogo/Tc2	DNA transposon
pID 3327088	KIAA0637	hAT	DNA transposon
pID 29863	CENP-B	Tc1/Pogo	DNA transposon
pID 2231380		Tc1/Mariner/Hsmar1	DNA transposon
pID 131827	V(D)J recombinase (RAG1/2)	Tc1/Mariner	DNA transposon
pID 10439762		hAT/Tip100	DNA transposon
pID 10439744		hAT/Tip100	DNA transposon
pID 10439678		hAT	DNA transposon
pID 10047247	KIAA1586	hAT/Tip100	DNA transposon
BAC 4309921		piggyBac/MER85	DNA transposon
BAC 3522927		hAT/Tam3	DNA transposon
NP_776172		Harbinger	DNA transposon
NT_006413		P	DNA transposon
pID 4773880	Syncytin	HERV-W	LTR retrotransposon
pID 1196425	HERV-3 env	HERV-R	LTR retrotransposon
NP_055883	PEG10	Gypsy	LTR retrotransposon
AL117190	Rtl1	Gypsy	LTR retrotransposon
XP_291322		Gypsy	LTR retrotransposon
NP_689907		Gypsy	LTR retrotransposon
XP_293405		Gypsy	LTR retrotransposon
NP_065820		Gypsy	LTR retrotransposon
NP_036449		Gypsy	LTR retrotransposon
Z75407		Gypsy	LTR retrotransposon

recombinase is an ancient gene, possibly derived from the DNA transposase ~400 million years ago, syncytin is less than 5 million years old. Some host proteins listed in Table 11, like syncytin and Jerky, are entirely derived from proteins encoded originally by TEs. The remaining protein-coding genes were formed by the fusion of the preexisting host proteins with transposon-encoded proteins. Such fusions might have been initiated by alternative splicing.

DNA transposons are most abundant among recruited genes, despite their relatively low abundance in the genome compared to other TEs (Table 11). Rtl1 is the only clear example of RT recruited in the host gene. Genes derived from DNA transposase are probably involved in specific DNA binding and cleavage (e.g. the V(D)J recombinase). Surprisingly, eight genes are derived from Gypsy-like LTR retrotransposons, whose remnants are otherwise not detectable in the human genome. Two genes are derived from *env* genes encoded by HERV-W and HERV-R elements. Proteins from *env*-derived genes may prevent retroviral infections by saturating the cellular receptors for exogenous retroviruses.

The extent of the contribution of TEs to the host protein-coding capacity is still not well known. The number of well-supported examples of TE-derived host proteins is relatively small (Table 11). On the other hand, analyses of human mRNA sequences disclosed several hundreds of mRNAs with one or several exons derived from TEs. However, some of these TE-containing sequences can represent common contaminations in EST libraries. In addition, many of the transcripts contain short reading frames, indicating that they do not code for functional proteins. Indeed, it should be noted that the majority of cellular polymerase II transcripts do not code for any proteins and never reach the cytoplasm. Cellular mRNA control mechanisms such as the nonsense-mediated RNA decay protect cells from potentially pathogenic truncated or aberrant proteins. Indeed, products of TE-containing transcripts are seldom detected at the protein level.

In conclusion, while there are a few well-documented examples of novel functional genes derived from TEs, the extent of this phenomenon is unknown. Many proteins deposited in public databases are hypothetical ORFs predicted by statistical methods based on computational analysis of genomic DNA and/or cDNA sequences. These methods are less than 90% accurate and, as a result, they may include TEs in erroneously predicted proteins. Analogously, databases of sites involved in transcription regulation may include regions whose "regulatory" effects are shown *in vitro* only.

6
Databases and Programs for Analysis of Repetitive DNA

About half of the human genome is composed of various repetitive elements. Detection of repetitive elements is a basic step in many biologically important analyses, including, but not limited to, sequence assembly during genome sequencing, genome annotation, similarity searches, and gene and coding sequence prediction. Therefore, specialized tools and databases have been developed for the detection and masking of repetitive elements.

6.1
Databases of Repetitive Elements

The first database of human repetitive elements, Repbase, was published in 1992. Currently it contains over 620 families and subfamilies of human repeats. Repbase became a regularly updated general database of eukaryotic repetitive elements, renamed as Repbase Update (RU). RU is accompanied by the monthly online journal Repbase Reports, which publishes newly discovered eukaryotic repeats.

Currently, RU is released in the EMBL and Fasta formats. Only the EMBL format contains a detailed description and annotation of every RU consensus sequence. In addition, a version of RU manually tailored for RepeatMasker is available. RU contains several sections dedicated to separate eukaryotic species. Human TEs are listed in the section "humrep.ref." Most of the abundant processed pseudogenes can be obtained from the "pseudo.ref" library. The "simple.ref" library contains the most common simple repeat motifs.

Apart from Repbase Update, there are other useful online resources for studies of human repeats. Table 12 lists selected databases as well as related resources devoted to human repetitive elements.

6.2
Programs for Detection and Analysis of Repetitive DNA

There are two basic approaches for detection of repetitive elements in biological sequences. The first is based on the detection of similarity with previously known repeats. The second *de novo* approach is based on a

Tab. 12 Databases of human repeats, and related sources.

Database	URL description
Repbase Update	http://www.girinst.org/Repbase_Update.html
	De facto the only specialized database of eukaryotic repetitive elements. To date, Repbase Update contains consensus sequences and annotation for more than 4000 different eukaryotic repetitive elements. The content can be searched through sequence similarity search on the Censor web server, or through the text search using keywords.
HERVd	http://herv.img.cas.cz/
	The Database of Human Endogenous Retroviruses (HERVs) maps all known endogenous retroviruses in the human genome. Retroviruses can be searched using several criteria such as family, length, chromosome etc., or by fasta similarity search. Several characteristics including the GC content, CpGs are calculated.
Human TE Insertion	http://www.med.upenn.edu/genetics/labs/kazazian/human.html
	The database lists known insertions of repetitive elements into the human genome associated with human genetic diseases.
Exapted retrogenes	http://www-ifi.uni-muenster.de/exapted-retrogenes/tables3.html
	Examples of vertebrate regulatory elements, parts of coding regions or genes generated by retroelements. Vertebrate genes generated by retrosequences.
UCSC GoldenPath	http://genome.ucsc.edu/
	The UCSC human genome assembly contains extensive annotation files including tables with maps of repetitive elements in the human genome.
Ensembl	http://www.ensembl.org/
	The ensembl genome assembly contains extensive annotation files including tables with maps of repetitive elements in the human genome.

search for characteristic properties of particular repeats. Table 13 lists some of the most popular programs developed for the detection of repetitive elements.

Similarity-based methods typically compare a library of known repetitive sequences against genomic DNA using general programs for sequence similarity search such as blast, wublast or cross_match. The repetitive library can be any custom set of repeats; however, specialized programs nearly exclusively rely on consensus sequences from Repbase. *De novo* detection methods do not require prior knowledge of repeats and the guiding principle of this approach is to search for the presence of multiple copies in the genome.

Standard detection programs for interspersed repeats use similarity-based detection of repeats. Censor and MaskerAid require wublast, while RepeatMasker uses the Smith-Waterman algorithm implemented in cross_match. For large quantities of data or for quick searches, it is better to use Censor or

Tab. 13 Programs for detection of repetitive sequences in biological sequences.

Program (package)	URL description
Censor	http://www.girinst.org/Censor_Server.html
	General package for similarity-based identification of interspersed and tandem repeats in both nucleic and amino acid sequences. Requires wublast and Repbase Update repeat libraries. Both online and local versions are available.
RepeatMasker	http://ftp.genome.washington.edu/cgi-bin/RepeatMasker
	General package for similarity-based identification of interspersed and tandem repeats in nucleic acid sequences. Requires cross_match and Repbase Update repeat libraries. Both online and local versions are available.
MakerAid	http://sapiens.wustl.edu/maskeraid/
	Program performs a similarity-based detection of interspersed and tandem repeats in nucleic acid sequences. Can be used with RepeatMasker for fast repeat identification. Requires wublast and external repeat libraries. Local versions only.
Tandem repeat finder	http://tandem.biomath.mssm.edu/trf/trf.html
	An excellent program to for *de novo* detection of tandem repeats in DNA sequences. Both online and local versions available.
Satellites	http://bioweb.pasteur.fr/seqanal/interfaces/satellites.html
	Program to for *de novo* detection of tandem repeats in DNA sequences. Online version only.
Sputnik	http://espressosoftware.com:8080/esd/pages/sputnik.jsp
	Scans for 2–5 long tandem repeats using *de novo* search.
EMBOSS	http://www.hgmp.mrc.ac.uk/Software/EMBOSS/index.html
	A general open source package for analysis of biological sequences. Two included programs etandem and equicktandem detect tandem repeats using *de novo* search. Local versions only.

MaskerAid, since they are much faster than cross-match-based RepeatMasker. RepeatMasker and MaskerAid are restricted to nucleic acid sequences only. Censor, on the other hand, can perform virtually any similarity search implemented in the wublast package, including translated searches, and can even detect repetitive elements in protein sequences. All these programs produce maps of repetitive elements in query sequences, with the exact position of each repeat. These three programs also "mask" sequences, that is, they replace detected repeats by special characters, usually by "N" for nucleic and "X" for protein sequences. Such masking is essential for many database searches or gene predictions. Moreover, all these programs detect the most frequent simple repeats.

Unlike in the case of interspersed repeats, *de novo* detection is the primary approach in the analysis of simple repeats. There are several packages for the detection of microsatellites with short unit sizes, but for minisatellites with large units there are only a few suitable programs. In such situations, Tandem Repeat Finder can be applied.

In summary, for routine detection of known repeats it is best to use Censor, RepeatMasker, or MaskerAid with appropriate repeat libraries (Repbase Update libraries). These programs also detect the majority of simple sequence repeats, particularly the most abundant microsatellites. Specialized analysis of simple sequence repeats may involve specialized *de novo* detection programs. The majority of programs listed in Table 13 can be installed only on Linux/Unix-related systems. Users of other operating systems may use several programs that have online submission forms, as specified in Table 13.

See also Molecular Basis of Genetics; Genomic Sequencing.

Bibliography

Books and Reviews

Batzer, M.A., Deininger, P.L. (2002) Alu repeats and human genomic diversity, *Nat. Rev. Genet.* **3**, 370–379.

Berg, D.E., Howe, M.M. (Eds.) (1989) *Mobile DNA*, American Society for Microbiology, Washington.

Coffin, J.M., Hughes, S.H., Varmus, H.E. (Eds.) (1997) *Retroviruses*, Cold Spring Harbor Laboratory Press, New York.

Cooper, D.N. (2003) *Encyclopedia of The Human Genome*, Nature Publishing Group, London.

Jurka, J. (1998) Repeats in genomic DNA: mining and meaning, *Curr. Opin. Struct. Biol.* **8**, 333–337.

Lower, R., Lower, J., Kurth, R. (1996) The viruses in all of us: characteristics and biological significance of human endogenous retrovirus sequences, *Proc. Natl. Acad. Sci. U.S.A.* **93**, 5177–5184.

Maraia, R.J. (Ed.) (1995) *The Impact of Short Interspersed Elements (SINEs) on the Host Genome*, Springer, New York.

Ostertag, E.M., Kazazian, H.H. Jr. (2001) Biology of mammalian L1 retrotransposons, *Annu. Rev. Genet.* **35**, 501–538.

Smit, A.F. (1996) The origin of interspersed repeats in the human genome, *Curr. Opin. Genet. Dev.* **6**, 743–748.

Weiner, A.M., Deininger, P.L., Efstratiadis, A. (1986) Nonviral retroposons: genes, pseudogenes, and transposable elements generated by the reverse flow of genetic information, *Annu. Rev. Biochem.* **55**, 631–661.

Wilkinson, D., Mager, D., Leong, J. (1994) Endogenous Human Retroviruses, in: Levy, J. (Ed.) *The Retroviridae*, Plenum Press, New York, pp. 465–535.

Primary Literature

Seminal Early References

Britten, R.J., Davidson, E.H. (1971) Repetitive and non-repetitive DNA sequences and a speculation on the origins of evolutionary novelty, *Q. Rev. Biol.* **46**, 111–137.

Britten, R.J., Baron, W.F., Stout, D.B., Davidson, E.H. (1988) Sources and evolution of human Alu repeated sequences, *Proc. Natl. Acad. Sci. U.S.A.* **85**, 4770–4774.

Dombroski, B.A., Mathias, S.L., Nanthakumar, E., Scott, A.F., Kazazian, H.H. Jr. (1991) Isolation of an active human transposable element, *Science* **254**, 1805–1808.

Corneo, G., Ginelli, E., Polli, E. (1967) A satellite DNA isolated from human tissues, *J. Mol. Biol.* **23**, 619–622.

Esnault, C., Maestre, J., Heidmann, T. (2000) Human LINE retrotransposons generate processed pseudogenes, *Nat. Genet.* **24**, 363–367.

Feng, Q., Moran, J.V., Kazazian, H.H. Jr., Boeke, J.D. (1996) Human L1 retrotransposon encodes a conserved endonuclease required for retrotransposition, *Cell* **87**, 905–916.

Hattori, M., Kuhara, S., Takenaka, O., Sakaki, Y. (1986) L1 family of repetitive DNA sequences in primates may be derived from a sequence encoding a reverse transcriptase-related protein, *Nature* **321**, 625–628.

Jeffreys, A.J., Wilson, V., Thein, S.L. (1985) Hypervariable 'minisatellite' regions in human DNA, *Nature* **314**, 67–73.

Jelinek, W.R., Toomey, T.P., Leinwand, L., Duncan, C.H., Biro, P.A., Choudary, P.V., Weissman, S.M., Rubin, C.M., Houck, C.M., Deininger, P.L., Schmid, C.W. (1980) Ubiquitous, interspersed repeated sequences in mammalian genomes, *Proc. Natl. Acad. Sci. U.S.A.* **77**, 1398–1402.

Jurka, J. (1997) Sequence patterns indicate an enzymatic involvement in integration of mammalian retroposons, *Proc. Natl. Acad. Sci. U.S.A.* **94**, 1872–1877.

Jurka, J., Smith, T. (1988) A fundamental division in the Alu family of repeated sequences, *Proc. Natl. Acad. Sci. U.S.A.* **85**, 4775–4778.

Martin, F., Olivares, M., Lopez, M.C., Alonso, C. (1996) Do non-long terminal repeat retrotransposons have nuclease activity? *Trends Biochem. Sci.* **8**, 283–285.

Moos, M., Galwitz, D. (1983) Structure of two human beta-actin-related processed genes one of which is located next to simple repetitive sequence, *EMBO J.* **2**, 757–761.

Moran, J.V., Holmes, S.E., Naas, T.P., DeBerardinis, R.J., Boeke, J.D., Kazazian, H.H. Jr. (1996) High frequency retrotransposition in cultured mammalian cells, *Cell* **87**, 917–927.

Ono, M., Yasunaga, T., Miyata, T., Ushikubo, H. (1986) Nucleotide sequence of human endogenous retrovirus genome related to the mouse mammary tumor virus genome, *J. Virol.* **60**, 589–598.

Prosser, J., Frommer, M., Paul, C., Vincent, P.C. (1986) Sequence relationships of three human satellite DNAs, *J. Mol. Biol.* **187**, 145–155.

Quentin, Y. (1988) The Alu family developed through successive waves of fixation closely connected with primate lineage history, *J. Mol. Evol.* **27**, 194–202.

Repaske, R., O'Neill, R.R., Steele, P.E., Martin, M.A. (1983) Characterization and partial nucleotide sequence of endogenous type C retrovirus segments in human chromosomal DNA, *Proc. Natl. Acad. Sci. U.S.A.* **80**, 678–682.

Schmid, C.W., Deininger, P.L. (1975) Sequence organization of the human genome, *Cell* **6**, 345–358.

Schmid, C.W., Jelinek, W.R. (1982) The Alu family of dispersed repetitive sequences, *Science* **216**, 1065–1070.

Shafit-Zagardo, B., Maio, J.J., Brown, F.L. (1982) KpnI families of long, interspersed repetitive DNAs in human and other primate genomes, *Nucleic Acids Res.* **10**, 3175–3193.

Smit, A.F., Riggs, A.D. (1996) Tiggers and DNA transposon fossils in the human genome, *Proc. Natl. Acad. Sci. U.S.A.* **93**, 1443–1448.

Soriano, P., Meunier-Rotival, M., Bernardi, G. (1983) The distribution of interspersed repeats is nonuniform and conserved in the mouse and human genomes, *Proc. Natl. Acad. Sci. U.S.A.* **80**, 1816–1820.

Ullu, E., Tschudi, C. (1984) Alu sequences are processed 7SL RNA genes, *Nature* **312**, 171–172.

Van Arsdell, S.W., Dension, R.A., Bernstein, L.B., Weiner, A.M., Manser, T., Gesteland, R.F. (1981) Direct repeats flank three small nuclear RNA pseudogenes in the human genome, *Cell* **26**, 11–17.

Vanin, E.F. (1985) Processed pseudogenes: characteristics and evolution, *Annu. Rev. Genet.* **19**, 253–272.

Weber, J.L., May, P.E. (1989) Abundant class of human DNA polymorphisms which can be typed using the polymerase chain reaction, *Am. J. Hum. Genet.* **44**, 388–396.

Weiner, A.M. (1980) An abundant cytoplasmic 7S RNA is complementary to the dominant interspersed middle repetitive DNA sequence family in the human genome, *Cell* **22**, 209–218.

Recent Key Primary Literature

Amor, D.J., Choo, K.H. (2002) Neocentromeres: role in human disease, evolution, and centromere study, *Am. J. Hum. Genet.* **71**, 695–714.

Andersson, M.-L., Lindeskog, M., Medstrand, P., Westley, B., May, F., Blomberg, J. (1999) Diversity of human endogenous retrovirus class II-like sequences, *J. Gen. Virol.* **80**, 255–260.

Antonarakis, S.E. (1998) Recommendations for a nomenclature system for human gene mutations, *Hum. Mutat.* **11**, 1–3.

Batzer, M.A., Deininger, P.L., Hellmann-Blumberg, U., Jurka, J., Labuda, D., Rubin, C.M., Schmid, C.W., Zietkiewicz, E., Zuckerkandl, E. (1996) Standardized nomenclature for Alu repeats, *J. Mol. Evol.* **42**, 3–6.

Bedell, J.A., Korf, I., Gish, W. (2000) MaskerAid: a performance enhancement to repeat masker, *Bioinformatics* **16**, 1040–1041.

Berkhout, B., Jebbink, M., Zsiros, J. (1999) Identification of an active reverse transcriptase enzyme encoded by a human endogenous HERV-K retrovirus, *J. Virol.* **73**, 2365–2375.

Brosius, J. (1999) RNAs from all categories generate retrosequences that may be exapted as novel genes or regulatory elements, *Gene* **238**, 115–134.

Brouha, B., Schustak, J., Badge, R.M., Lutz-Prigge, S., Farley, A.H., Moran, J.V., Kazazian, H.H. Jr. (2003) Hot L1s account for the bulk of retrotransposition in the human population, *Proc. Natl. Acad. Sci. U.S.A.* **100**, 5280–5285.

Chou, H.H., Hayakawa, T., Diaz, S., Krings, M., Indriati, E., Leakey, M., Paabo, S., Satta, Y., Takahata, N., Varki, A. (2002) Inactivation of CMP-N-acetylneuraminic acid hydroxylase occurred prior to brain expansion during human evolution, *Proc. Natl. Acad. Sci. U.S.A.* **99**, 11736–11741.

Chureau, C., Prissette, M., Bourdet, A., Barbe, V., Cattolico, L., Jones, L., Eggen, A., Avner, P., Duret, L. (2002) Comparative sequence analysis of the X-inactivation center region in mouse, human, and bovine, *Genome Res.* **12**, 894–908.

Cordonnier, A., Casella, J.F., Heidmann, T. (1995) Isolation of novel human endogenous retrovirus-like elements with foamy virus-related pol sequence, *J. Virol.* **69**, 5890–5897.

Deininger, P.L., Batzer, M.A. (1999) Alu repeats and human disease, *Mol. Genet. Metab.* **67**, 183–193.

Deininger, P.L., Batzer, M.A., Hutchison III, C.A., Edgell, M.H. (1992) Master genes in mammalian repetitive DNA amplification, *Trends Genet.* **8**, 307–311.

Dewannieux, M., Esnault, C., Heidmann, T. (2003) LINE-mediated retrotransposition of marked Alu sequences, *Nat. Genet.* **35**, 41–48.

Duncan, C., Biro, P.A., Choudary, P.V., Elder, J.T., Wang, R.R., Forget, B.G., de Riel, J.K., Weissman, S.M. (1979) RNA polymerase III transcriptional units are interspersed among human non-alpha-globin genes, *Proc. Natl. Acad. Sci. U.S.A.* **76**, 5095–5099.

Goncalves, I., Duret, L., Mouchiroud, D. (2000) Nature and structure of human genes that generate retropseudogenes, *Genome Res.* **10**, 672–678.

Grimaldi, G., Skowronski, J., Singer, M.F. (1984) Defining the beginning and the end of KpnI family segments, *EMBO J.* **3**, 1753–1759.

Hohjoh, H., Singer, M.F. (1997) Sequence-specific single-strand RNA binding protein encoded by the human LINE-1 retrotransposon, *EMBO J.* **16**, 6034–6043.

Jeffreys, A.J., Tamaki, K., MacLeod, A., Monckton, D.G., Neil, D.L., Armour, J.A. (1994) Complex gene conversion events in germline mutation at human minisatellites, *Nat. Genet.* **6**, 136–145.

Jurka, J., Klonowski, P., Dagman, V., Pelton, P. (1996) CENSOR – a program for identification and elimination of repetitive elements from DNA sequences, *Comput. Chem.* **20**, 119–121.

Jurka, J., Kapitonov, V.V. (1999) Sectorial mutagenesis by transposable elements, *Genetica* **107**, 239–248.

Jurka, J. (2000) Repbase update: a database and an electronic journal of repetitive elements, *Trends Genet.* **16**, 418–420.

Jurka, J., Krnjajic, M., Kapitonov, V.V., Stenger, J.E., Kokhanyy, O. (2002) Active Alu elements are passed primarily through paternal germlines, *Theor. Popul. Biol.* **61**, 519–530.

Kapitonov, V., Jurka, J. (1996) The age of Alu subfamilies, *J. Mol. Evol.* **42**, 59–65.

Kapitonov, V.V., Jurka, J. (2003) The esterase and PHD domains in CR1-like non-LTR retrotransposon, *Mol. Biol. Evol.* **20**, 38–46.

Kazazian, H.H. Jr. (1998) Mobile elements and disease, *Curr. Opin. Genet. Dev.* **8**, 343–350.

Kazazian, H.H. Jr., Goodier, J.L. (2002) LINE drive retrotransposition and genome instability, *Cell* **110**, 277–280.

Kolosha, V.O., Martin, S.L. (2003) High-affinity, non-sequence-specific RNA binding by the open reading frame 1 (ORF1) protein from long interspersed nuclear element 1 (LINE-1), *J. Biol. Chem.* **278**, 8112–8117.

Kolomietz, E., Meyn, M.S., Pandita, A., Squire, J.A. (2002) The role of Alu repeat clusters as mediators of recurrent chromosomal aberrations in tumors, *Genes Chromosomes Cancer* **35**, 97–112.

Korenberg, J.R., Rykowski, M.C. (1988) Human genome organization: Alu, lines, and the molecular structure of metaphase chromosome bands, *Cell* **53**, 391–400.

Lander, E.S., Linton, L.M., Birren, B., Nusbaum, C., Zody, M.C., Baldwin, J., Devon, K., Dewar, K., Doyle, M., FitzHugh, W. et al. (2001) Initial sequencing and analysis of the human genome, *Nature* **409**, 860–921.

Lee, C., Wevrick, R., Fisher, R.B., Ferguson-Smith, M.A., Lin, C.C. (1997) Human centromeric DNAs, *Hum. Genet.* **100**, 291–304.

Lehrman, M.A., Schneider, W.J., Sudhof, T.C., Brown, M.S., Goldstein, J.L., Russel, D.W. (1985) Mutation on LDL receptor: Alu-Alu recombination deletes exons encoding transmembrane and cytoplasmic domains, *Science* **227**, 140–146.

Lerman, M.I., Thayer, R.E., Singer, M.F. (1983) KpnI family of long interspersed repeated DNA sequences in primates: polymorphism of family members and evidence for transcription, *Proc. Natl. Acad. Sci. U.S.A.* **80**, 3966–3970.

Li, T.H., Schmid, C.W. (2001) Differential stress induction of individual Alu loci: implications for transcription and retrotransposition, *Gene* **276**, 135–141.

Lower, R. (1999) The pathogenic potential of endogenous retroviruses: facts and fantasies, *Trends Microbiol.* **7**, 350–356.

Lyon, M.F. (1998) X-chromosome inactivation: a repeat hypothesis, *Cytogenet. Cell. Genet.* **80**, 133–137.

Mager, D.L. (1999) Human endogenous retroviruses and pathogenicity: genomic considerations, *Trends Microbiol.* **7**, 431–432.

Martignetti, J.A., Brosius, J. (1993) BC200 RNA: a neural RNA polymerase III product encoded by a monomeric Alu element, *Proc. Natl. Acad. Sci. U.S.A.* **90**, 11563–11567.

Medstrand, P., Mager, D.L. (1998) Human-specific integrations of the HERV-K endogenous retrovirus family, *J. Virol.* **72**, 9782–9787.

Mi, S., Lee, X., Li, X., Veldman, G.M., Finnerty, H., Racie, L., LaVallie, E., Tang, X.Y., Edouard, P., Howes, S., Keith, J.C. Jr., McCoy, J.M. (2000) Syncytin is a captive retroviral envelope protein involved in human placental morphogenesis. *Nature* **403**, 785–789.

Moran, J.V., DeBerardinis, R.J., Kazazian, H.H. Jr. (1999) Exon shuffling by L1 retrotransposition, *Science* **283**, 1530–1534.

Morrish, T.A., Gilbert, N., Myers, J.S., Vincent, B.J., Stamato, T.D., Taccioli, G.E., Batzer, M.A., Moran, J.V. (2002) DNA repair mediated by endonuclease-independent LINE-1 retrotransposition, *Nat. Genet.* **31**, 159–165.

Paces, J., Pavlicek, A., Paces, V. (2002) HERVd: database of human endogenous retroviruses, *Nucleic Acids Res.* **30**, 205–206.

Okada, N., Hamada, M., Ogiwara, I., Ohshima, K. (1997) SINEs and LINEs share common 3′ sequences: a review, *Gene* **205**, 229–243.

Prak, E.T., Kazazian, H.H. Jr. (2000) Mobile elements and the human genome, *Nat. Rev. Genet.* **1**, 134–144.

Schlotterer, C., Tautz, D. (1992) Slippage synthesis of simple sequence DNA, *Nucleic Acids Res.* **20**, 211–215.

Schueler, M.G., Higgins, A.W., Rudd, M.K., Gustashaw, K., Willard, H.F. (2001) Genomic and genetic definition of a functional human centromere, *Science* **294**, 109–115.

Scott, A.F., Schmeckpeper, B.J., Abdelrazik, M., Comey, C.T., O'Hara, B., Rossiter, J.P., Cooley, T., Heath, P., Smith, K.D., Margolet, L. (1987) Origin of the human L1 elements: proposed progenitor genes deduced from a consensus DNA sequence, *Genomics* **1**, 113–125.

Shafit-Zagardo, B., Brown, F.L., Zavodny, P.J., Maio, J.J. (1983) Transcription of the KpnI families of long interspersed DNAs in human cells, *Nature* **304**, 277–280.

Singer, M.F. (1982) SINEs and LINEs: highly repeated short and long interspersed sequences in mammalian genomes, *Cell* **28**, 433–434.

Smit, A.F. (1993) Identification of a new, abundant superfamily of mammalian LTR-transposons, *Nucleic Acids Res.* **21**, 1863–1872.

Smit, A.F., Toth, G., Riggs, A.D., Jurka, J. (1995) Ancestral, mammalian-wide subfamilies of LINE-1 repetitive sequences, *J. Mol. Biol.* **246**, 401–417.

Smit, A.F., Riggs, A.D. (1995) MIRs are classic, tRNA-derived SINEs that amplified before the mammalian radiation, *Nucleic Acids Res.* **23**, 98–102.

Smit, A.F. (1999) Interspersed repeats and other mementos of transposable elements in mammalian genomes, *Curr. Opin. Genet. Dev.* **9**, 657–663.

Speek, M. (2001) Antisense promoter of human L1 retrotransposon drives transcription of adjacent cellular genes, *Mol. Cell. Biol.* **21**, 1973–1985.

Swergold, G.D. (1990) Identification, characterization, and cell specificity of a human LINE-1 promoter, *Mol. Cell. Biol.* **10**, 6718–6729.

Toth, G., Gaspari, Z., Jurka, J. (2000) Microsatellites in different eukaryotic genomes: survey and analysis, *Genome Res.* **10**, 967–981.

Voliva, C.F., Jahn, C.L., Comer, M.B., Hutchison, C.A., Edgell, M.H. (1983) The L1Md long interspersed repeat family in the mouse: almost all examples are truncated at one end, *Nucleic Acids Res.* **11**, 8847–8859.

Wallace, M.R., Andersen, L.B., Saulino, A.M., Gregory, P.E., Glover, T.W., Collins, F.S. (1991) A de novo Alu insertion results in neurofibromatosis type 1, *Nature* **353**, 864–866.

Wellinger, R.J., Sen, D. (1997) The DNA structures at the ends of eukaryotic chromosomes, *Eur. J. Cancer* **33**, 735–749.

10
Evolution of Noncoding DNA in Eukaryotes

Josep M. Comeron
The University of Iowa, Iowa City, IA

1	The C-value Paradox 326	
2	Noncoding Genome Size: Passive versus Selective Trait 327	
3	Deletion Bias (DB) Based on Small Indels 328	
3.1	DB in Drosophila 328	
3.2	What is the True DB in Drosophila? 328	
3.3	DB in Other Eukaryotes 329	
3.4	Overall Rate of DNA Loss due to Small Indels and Genome Size 329	
4	Transposable Elements (TE) and DNA Gain 330	
5	Variation across and among Genomes: The Influence of Recombination 330	
6	Noncoding DNA is not Free of Genetic Information or Evolutionary Effect 331	
7	Summary 332	
	Acknowledgment 333	
	Bibliography 333	
	Book and Reviews 333	
	Primary Literature 333	

Genomics and Genetics. Edited by Robert A. Meyers.
Copyright © 2007 Wiley-VCH Verlag GmbH & Co. KGaA, Weinheim
ISBN: 978-3-527-31609-0

Keywords

Deletion Bias
The study of small insertions and deletions (indels) in most eukaryotes shows an excess of deletions over insertions, forecasting DNA loss.

Effective Population Size (Ne)
A theoretical number of individuals that would explain the amount of genetic variation observed within a species in the absence of selection. Ne is somewhat related to, but always smaller than, the actual number of individuals. Ne is positively associated with the effectiveness of selection and varies among species as well as across genomes.

> The amount of haploid nuclear DNA (or C-value) among eukaryotes varies more than 4 orders of magnitude, with brewer's yeast *Saccharomyces cerevisiae* (1.2×10^7 bp) or the parasitic microsporidium *Encephalitozoon intestinalis* (3×10^6 bp) as prototypical examples of the smallest genomes and several amoebae mentioned as the eukaryotes with largest genomes with $>6 \times 10^{11}$ bp (e.g. *Amoeba dubia* and *Chaos chaos*). Differences in the amount of noncoding DNA are also observed within a given taxonomic group, with algae varying more than 5000-fold, more than 1000-fold among flowering plants or invertebrates, more than 300-fold among vertebrates, and 100-fold among amphibians and insects. An apparent "paradox" arises when the size of the nuclear genome is compared to organismal complexity (morphological or developmental) if we assume that the amount of genetic material (genome size) should somehow correspond to the amount of genetic information. The so-called C-value paradox or enigma was coined to describe this lack of correspondence. Later findings revealed that coding sequences account for only a small proportion of the genomic DNA in most eukaryotes. This observation does not solve the C-value paradox, but shifts the debate from the number of genes responsible for a given degree of complexity (the G-value paradox) to the amount of noncoding DNA and the causes for its tremendous variation. In fact, recent studies based on fully sequenced genomes and expression data in model eukaryotes, including humans, have shown that complex gene regulation and the number of different transcripts – with a high fraction of genes having multiple transcripts – and not the number of genes, would better represent biological "complexity," hence explaining the G-value paradox.

1
The C-value Paradox

Many cause/consequence relationships have been proposed on the basis of phenotypic correlations. The variation in noncoding genome size among eukaryotes has been associated with biological factors, both cellular and organismal: cell and nucleus size, cell division rate, metabolic rate, developmental rate, developmental complexity, and overall gene expression.

Indeed, eukaryotes with large genomes tend to have larger nucleus and cells, and species with high developmental rate to developmental complexity ratio tend to have reduced genomes; other associations are only observed in particular lineages.

However, an aspect not always considered is that many biological factors are associated with species or population parameters such as generation time, population size, and effective population size (*Ne*). This indirect association between cellular or developmental factors and *Ne* is pertinent in the analysis of any genomic characteristic (or any biological feature for that matter) because *Ne* is a main parameter influencing natural selection. Population genetic theory predicts and evolutionary analyses confirm that *Ne* is positively associated with the effectiveness of selection. (Increased effectiveness of selection will favor the fixation of advantageous mutations and the removal of deleterious mutations.) As a consequence, it is not direct to distinguish between cause, effect, and indirect or coincidental association; many proposed causal associations illustrate this difficulty. For instance, a causal association between fast developmental rate (and short generation time) and small genomes can be proposed. However, many species with short generation time also have large population sizes and *Ne*, and therefore the observed small genome is likely the result of more efficient selective constraints favoring small genomes while the negative relationship between developmental rate and genome size would be secondary.

Moreover, several cases of closely related species – with equivalent cellular and developmental parameters – exemplify that the size of noncoding DNA can change very rapidly in evolutionary terms independent of biological factors. For instance, closely related Drosophila species show severalfold differences in noncoding DNA and there is significant variation even at intraspecific level. Another piece of information comes from genomic analyses that have shown that several features of noncoding DNA such as the amount of intergenic DNA (or gene density), and intron presence and size also show a heterogeneous distribution across genomes. Thus, an integral explanation to the presence and size of noncoding DNA should be able to explain differences between distant groups or taxa, differences between closely related species with equivalent cellular/developmental features, as well as differences across a given genome.

2
Noncoding Genome Size: Passive versus Selective Trait

One possible explanation to the size of noncoding DNA and genome size in eukaryotes is that the amount of noncoding DNA is the passive result of mutational tendencies only; the manifestation of a mutation-drift equilibrium between insertions and deletions. Under this mutational-only scenario, variation in genome size is explained by differences in the patterns of insertion and deletion of "selfish" DNA or due to variation in the intrinsic rates of DNA loss through small insertions and deletions (indels). Special attention has been paid recently to the latter process, with several eukaryotes showing an excess of small deletions over insertions (the so-called deletion bias; DB). In particular, it has been proposed that differences in the rate of DNA loss caused by small indels are a major factor influencing genome size.

Under a selective scenario, genome size represents the equilibrium of mutational tendencies to either increase or decrease the amount of DNA counterbalanced by selective tendencies: a mutational selection–drift equilibrium. Mutational tendencies can certainly vary among species, with differences in the amount of repetitive DNA and its tendency to expand in the genome (mostly associated with transposable elements; TE), as well as variation in DB. It follows, however, that a mutational-only hypothesis – which assumes that the presence of noncoding DNA has no fitness costs – would predict either the collapse of noncoding DNA or the unlimited expansion of noncoding DNA. A more likely proposal assumes that both mutational tendencies (DNA expansion by TE amplification and DNA loss by DB) are balanced as a result of selective constraints.

The selective causes acting on genome size can be varied and be particular to groups or taxa according to specific cellular or developmental requirements (e.g. fast cell division), targeting replication costs, or biological effects of bulk DNA. As indicated earlier, the effectiveness of selection acting on noncoding DNA would also depend on parameters such as N_e and recombination (see the following) and so the outcome of the equilibrium would be species specific. Moreover, both N_e and recombination vary across genomes. Thus, a mutation selection–drift equilibrium is expected to be highly variable among taxa, between closely related species, and also vary across a given genome.

3
Deletion Bias (DB) Based on Small Indels

3.1
DB in Drosophila

The struggle to determine the mutational DB on the basis of small indels is best exemplified in Drosophila, in which three different approaches give three different results. The first significant attempt to investigate DB in Drosophila, studied nonallelic copies of non long–terminal repeat retrotransposable elements, so-called "dead-on-arrival" (DOA) elements. The study of *Helena* DOA elements gives a DB of 8.7, suggesting a very high rate of DNA loss. The second approach investigated defective TEs, generating a DB of 3.6. The third approach is the study of indels at polymorphic level in DNA sequences likely free of selective constraints (intergenic regions and introns), with a DB that ranges between 1.4 and 1.9, evidencing a significant but moderate rate of DNA loss.

3.2
What is the True DB in Drosophila?

As indicated by B. Charlesworth, the study of nonallelic DOA elements (that are either fixed or segregate at high frequencies within a species) could bias the estimates of DB if selection favors deletions in recently inserted elements, hence causing an overestimation of DB. The rational for arguing selection against inserted repetitive elements (favoring deletions and shorter versions of the elements) is based on the notion that shorter elements would reduce the probability of ectopic recombination and its deleterious effects. On the other hand, population genetic theory predicts that polymorphic studies will estimate mutational parameters that can only be closer to the true values compared to analyses that use mutations that are either fixed or segregate at high frequencies within a species. If indels are neutral, all approaches should give equivalent results, but if indels are under weak/moderate selection, the study of indels at polymorphic level is least influenced by selection. Strong selective constraints on the sequences used to investigate polymorphic indels, however, could also reduce the number of deletions observed at polymorphic level, hence causing underestimates in DB.

Several observations suggest that the estimates of DB based on different polymorphic studies is close to the true DB. First, an *ad hoc* theoretical model presuming that long deletions in introns are under strong selection would forecast a lower DB, but even in this case it cannot explain the observed data. Second, the study of polymorphic indels in intergenic regions also produces a very low DB (1.7) and, therefore, the argument that the observed DB is biased because of selection is not likely to be correct. And third, a recent study on the frequency of several families of TEs in populations has generated a key result to understand the observed discrepancy in DB based on long repetitive elements. This study shows that there is a negative correlation between the length of TEs within a family and the frequency in the population, in agreement with the ectopic recombination model that claims that deletions would be selectively favored in long repetitive elements. The consequence of this observation is that estimates of DB based on long repetitive elements will be biased, showing an excess of deletions than those expected by mutational tendencies only.

3.3
DB in Other Eukaryotes

Measures of DB have been obtained in other eukaryotes also on the basis of different approaches. One of the first studies to detect an excess of small deletions over insertions was applied to mammals by analyzing processed pseudogenes. Large-scale analyses of pseudogenes in humans and murids show an average DB of 2.74. The study of pseudogenes in the insect *Podisma pedestris* and in *Caenorhabditis elegans* shows DB of 2.7 and 3.8, respectively. The study of DOA *Lau1* elements in the Hawaiian crickets of the genus *Laupala* and *Maui* elements in two pufferfish species suggests a DB of 2.7 and 1.3–2.0, respectively.

3.4
Overall Rate of DNA Loss due to Small Indels and Genome Size

The rate of DNA loss depends on DB as well as the size of indels, and in most species deletions are longer than insertions. In order of increasing genome size, the overall rate of DNA loss observed in the different eukaryotes is the following: 1.8 in *Drosophila melanogaster* (based on polymorphic indels in intergenic regions and introns), 9 to 15 in Pufferfish, 3.8 in Laupala, 2.6 in rat/mouse, 1.7 in humans, and 3.6 in Podisma. Conversely, *C. elegans* shows a net gain of DNA based on small indels. As T.R. Gregory concluded recently, the use of updated estimates of DNA loss for mammals and Drosophila raise considerable doubts about the strength of any correlation between DNA loss rate by small indel bias and genome size. Therefore, although the mutational rate of DNA loss caused by small indels can vary evolutionarily, and might itself be the result of selection, it cannot explain the observed differences in noncoding DNA content and genome size between species. Moreover, this mutational explanation cannot account also for the observed differences across genomes.

4
Transposable Elements (TE) and DNA Gain

TE elements are widespread in eukaryotic genomes and comprise in many cases a substantial portion of euchromatic genome size, including humans where TEs explain 45% of intergenic and intronic sequences. The recent invasion/amplification of particular TE families causes rapid increases in genome size, temporarily altering any specific equilibrium among selective and long-term mutational tendencies. Two examples of such effect are maize, in which genome size has recently doubled due to TE invasion, and Drosophila, with substantial differences in TE presence between closely related species, as well as among populations of the same species. The opposite trend is also detected, with species that have recently lost repetitive DNA showing a sudden reduction in genome size albeit having similar mutational tendencies associated with small indels. For instance, tetraodontid (smooth) and diodontid (spiny) pufferfish have similar DB but smooth pufferfish shows a twofold reduction in genome size because of a recent reduction in repetitive elements. This latter case also illustrates that any study of mutational tendencies toward DNA gain or loss should include small indels (i.e. DB), as well as the rate of large insertions and TE amplification.

It is possible to argue that a generic mutational tendency of TE to replicate and increase genome size might have favored a mutational mechanism associated with DNA repair of small indels with a tendency toward DNA loss, as observed in most eukaryotes. In this regard, small differences in DB among taxa would reflect long-term tendencies associated with TE amplification in the genome.

The additional influence of selective forces that can be species specific and/or vary across genomes would generate the final outcome.

5
Variation across and among Genomes: The Influence of Recombination

Many genomic analyses have revealed heterogeneity of noncoding DNA presence across eukaryotic genomes. In Drosophila, the amount of intergenic DNA is higher in genomic regions with low recombination (and crossing-over) rates. Introns are also longer and more frequent in genes located in regions with severely reduced rates of crossing-over. An equivalent trend is observed in humans, with longer intergenic sequences, and introns in isochores are associated with reduced rates of crossing-over. *S. cerevisiae* also shows an influence of recombination on the presence of noncoding DNA. A single exception to this tendency is observed in *C. elegans,* in which an atypical pattern of crossing-over along chromosomes has been reported.

Interestingly, a negative relationship between recombination rates and the amount of noncoding DNA is observed among species. The study of nine distant eukaryotes with well-characterized genetic and physical maps (Fig. 1) reveals a significant negative relationship between recombination rate and a measure of the amount of noncoding DNA once the number of genes is taken into account. (Albeit the study is based on a small number of species, the observed relationship is not caused by any single species.) This observation suggests that similar, or at least overlapping, forces are shaping the size of noncoding DNA among species and

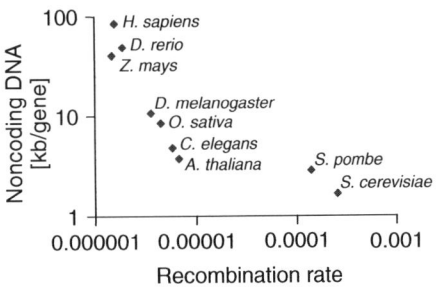

Fig. 1 Relationship between recombination rate and gene density among evolutionary distant model eukaryotes. The nine eukaryotes with well-characterized genetic and physical maps are: Homo sapiens, Danio rerio (zebra fish), Zea mays (maize), Oryza sativa (rice), D. melanogaster, C. elegans, Arabidopsis thaliana, S. cerevisiae, and Schizosaccharomyces pombe (fission yeast). Recombination rate indicates the average percentage of meiotic recombination (cM) between adjacent base pairs (total genetic map/total physical map), taking also into account the number of chromosomes to appraise the recombination caused by the random segregation of chromosomes. Although A. thaliana is a predominantly selfing plant, this feature has only recently been acquired, with all closely related species being outcrossing and, therefore, the recombination rate used for A. thaliana reflects the recent ancestral outcrossing condition. Nonparametric Spearman's rank correlation $R = -0.95$ ($p = 0.00009$). The least significant relationship after removing any one species is $R = -0.929$ ($p = 0.0009$). The removal of both distantly related yeasts (S. cerevisiae and S. pombe) from the analysis does not influence the outcome either ($R = -0.893$, $p = 0.007$).

across genomes. These equivalent findings across and among genomes could be explained if the mechanism of mismatch repair associated with recombination instigates DNA loss. Although this possibility needs to be investigated thoroughly in different species, in D. melanogaster there are similar patterns of indels in regions with normal and reduced rates of crossing-over.

An alternative explanation, invoking natural selection, emerges when we consider the influence that recombination has on the effectiveness of selection. Indeed, theoretical and experimental investigations predict that recombination will have an impact on the power or effectiveness of selection. In particular, most models of selection forecast that recombination will raise the effectiveness of selection. The total length of genetic map in a species as well as the number of chromosomes will play a role in the total meiotic recombination rate. Recombination rates and Ne (mentioned earlier) are two parameters that have been usually limited to population genetics studies but they should be included in any general explanation to genome architecture, the amount of noncoding DNA and genome size due to their influence to the effectiveness of selection. Moreover, both recombination and Ne are influenced by species-specific factors as well as vary across genomes.

6
Noncoding DNA is not Free of Genetic Information or Evolutionary Effect

Noncoding DNA in eukaryotes has been often labeled as "junk" DNA, useless (with no genetic information) as well as with no evolutionary effect (neutral to selection). In fact, however, noncoding DNA can be evolutionarily relevant by containing genetic information as well by its only presence. The number of known noncoding genes and especially noncoding RNA genes (ncRNA) has recently increased rapidly. Many of these ncRNA function directly as RNA while others play an essential role in gene regulation, with antisense RNAs being an important class of these genes. Moreover, noncoding DNA, intergenic or intronic, harbors most of the information on gene regulation.

On the other hand, the presence of DNA with no genetic information between genes or between exons should not be assumed to be devoid of evolutionary effect. For instance, the presence of introns and their length can have a direct fitness effect on transcription costs and be a trait under selection. However, the presence of DNA sequences with no direct fitness effects can also be a selective trait. Note that the significant parameter associated with recombination is the total recombination between genes or exons (i.e. clusters of sites under selection) and this is influenced by the rate per physical unit as well as by the physical distance. Therefore, under realistic conditions for most eukaryotes, an organism might increase the effectiveness of selection by increasing the amount of noncoding DNA, intergenic or intronic. Under this scenario, the presence of "junk" DNA can be viewed as a modifier of recombination with a definite evolutionary long-term effect and can itself be a selective trait. The long-term fitness effects of increasing the physical distance between clusters of sites under selection (genes, exons, or regulatory regions) and recombination, is expected to differ among species (according to the overall recombination rate) and across genomes (according to variation in recombination rates along chromosomes).

7
Summary

The study of noncoding DNA is no longer restricted to phenotypic correlations and all recent observations support the notion that noncoding genome size in eukaryotes is the result of a highly dynamic balance between mutational and selective forces. Population parameters that influence levels and patterns of polymorphisms also influence the effectiveness of selection and have the potential to influence genomic features. Therefore, the evolution of noncoding DNA (intergenic and intron presence, and size) ought to be investigated from a population genetics perspective (i.e. the so-called *population genomics* approach). The presence of regulatory sequences and noncoding genes (e.g. ncRNA genes) would add genetic information required for organismal "complexity," but it can explain only a fraction of noncoding DNA.

Mutational tendencies are strongly influenced by TE invasion and amplification, which can be species specific and cause sudden changes in genome size. This "one-way ticket to genome obesity" is somewhat counteracted by a moderate tendency toward DNA loss due to small indels observed in most eukaryotes. The widespread presence of TEs in eukaryotes might have caused a selective pressure on mechanisms of DNA repair favoring DNA loss. The study of small indels only cannot produce an accurate picture of mutational tendencies. On the other hand, selective forces on the presence of noncoding DNA can vary in different species and/or taxa and be associated with replication costs, nuclear architecture, or other biological effects of bulk DNA. Importantly, the actual impact of these selective tendencies is determined by recombination rates and Ne parameters that can vary across genomes, as well as among species and should be included in any general explanation to genome evolution. Future studies on variation of recombination rates, indel evolution, and Ne in many eukaryotes will give us a broader and more precise picture on the selective and mutational forces influencing the evolution of noncoding DNA.

Acknowledgment

The author thanks Dr. A. Llopart for her critical review of the manuscript. Josep M. Comeron is funded by National Science Foundation (NSF), DEB03-44209.

Bibliography

Book and Reviews

Cavalier-Smith, T. (1982) Skeletal DNA and the evolution of genome size, *Annu. Rev. Biophys. Bioeng.* **11**, 273–302.

Cavalier-Smith, T. (1985) *The Evolution of Genome Size*, John Wiley & Sons, New York.

Comeron, J.M. (2001) What controls the length of noncoding DNA? *Curr. Opin. Genet. Dev.* **11**, 652–659.

Gregory, T.R. (2002) Genome size and developmental complexity, *Genetica* **115**, 131–146.

Hartl, D.L. (2000) Molecular melodies in high and low C, *Nat. Rev. Genet.* **1**, 145–149.

Ohno, S. (1972) So much 'junk' DNA in Our Genome, in: Smith, H.H. (Ed.) *Evolution of Genetic Systems*, Gordon and Breach, New York, pp. 366–370.

Primary Literature

Awadalla, P. (2003) The evolutionary genomics of pathogen recombination, *Nat. Rev. Genet.* **4**, 50–60.

Bennett, M.D. (1971) The duration of meiosis, *Proc. R. Soc. Lond., B, Biol. Sci.* **178**, 259–275.

Bensasson, D., Petrov, D.A., Zhang, D.X., Hartl, D.L., Hewitt, G.M. (2001) Genomic gigantism: DNA loss is slow in mountain grasshoppers, *Mol. Biol. Evol.* **18**, 246–253.

Biemont, C. (1999) Distribution of transposable elements in Drosophila species, *Genetica* **105**, 43–62.

Blumenstiel, J.P., Hartl, D.L., Lozovsky, E.R. (2002) Patterns of insertion and deletion in contrasting chromatin domains, *Mol. Biol. Evol.* **19**, 2211–2225.

Carvalho, A.B., Clark, A.G. (1999) Intron size and natural selection, *Nature* **401**, 344.

Castillo-Davis, C.I., Mekhedov, S.L., Hartl, D.L., Koonin, E.V., Kondrashov, F.A. (2002) Selection for short introns in highly expressed genes, *Nat. Genet.* **31**, 415–418.

Cavalier-Smith, T. (1978) Nuclear volume control by nucleoskeletal DNA, selection for cell volume and cell growth rate, and the solution of the DNA C-value paradox, *J. Cell Sci.* **34**, 247–278.

Charlesworth, B. (1996) The changing sizes of genes, *Nature* **384**, 315–316.

Claverie, J.M. (2000) What if there are only 30,000 human genes? *Science* **291**, 1255–1257.

Comeron, J.M. (2004) Selective and mutational patterns associated with gene expression in humans: influences on synonymous composition and intron presence, *Genetics* **167**, 1293–1304.

Comeron, J.M., Kreitman, M. (2000) The correlation between intron length and recombination in Drosophila. Dynamic equilibrium between mutational and selective forces, *Genetics* **156**, 1175–1190.

Comeron, J.M., Kreitman, M. (2002) Population, evolutionary and genomic consequences of interference selection, *Genetics* **161**, 389–410.

Comeron, J.M., Kreitman, M., Aguade, M. (1999) Natural selection on synonymous sites is correlated with gene length and recombination in Drosophila, *Genetics* **151**, 239–249.

Crow, J.F., Kimura, K. (1970) *An Introduction to Population Genetics Theory*, Harper and Row, New York.

Day, T., Gloor, G.B. (1997) Homology requirements for targeting heterologous sequences during P-induced gap repair in Drosophila melanogaster, *Genetics* **147**, 689–699.

Devos, K.M., Brown, J.K.M., Bennetzen, J.L. (2002) Genome size reduction through illegitimate recombination counteracts genome expansion in Arabidopsis, *Genome Res.* **12**, 1075–1079.

Doolittle, W.F., Sapienza, C. (1980) Selfish genes, the phenotype paradigm and genome evolution, *Nature* **284**, 601–603.

Duret, L., Mouchiroud, D., Gautier, C. (1995) Statistical analysis of vertebrate sequences reveals that long genes are scarce in GC-rich isochores, *J. Mol. Evol.* **40**, 308–317.

Evgen'ev, M., Zelentsova, H., Mnjoian, L., Loluectova, H., Kidwell, M.G. (2000) Invasion of Drosophila virilis by the Penelope transposable element, *Chromosoma* **109**, 350–357.

Graur, D., Shuali, Y., Li, W.-H. (1989) Deletions in processed pseudogenes accumulate faster in rodents than in humans, *J. Mol. Evol.* **28**, 279–285.

Gregory, T.R. (2004) Insertion-deletion biases and the evolution of genome size, *Gene* **324**, 15–34.

Hey, J. (1998) Selfish genes, pleiotropy and the origin of recombination, *Genetics* **149**, 2089–2097.

Hey, J., Kliman, R.M. (2002) Interactions between natural selection, recombination and gene density in the genes of Drosophila, *Genetics* **160**, 595–608.

Hill, W.G., Robertson, A. (1966) The effect of linkage on limits to artificial selection, *Genet. Res.* **8**, 269–294.

Kliman, R.M., Irving, N., Santiago, M. (2003) Selection conflicts, gene expression, and codon usage trends in yeast, *J. Mol. Evol.* **57**, 98–109.

Lander, E.S., Linton, L.M., Birren, B., Nusbaum, C., Zody, M.C., et al. (2001) Initial sequencing and analysis of the human genome, *Nature* **409**, 860–921.

Mirsky, A.E., Ris, H. (1951) The desoxyribonucleic acid content of animal cells and its evolutionary significance, *J. Gen. Physiol.* **34**, 451–462.

Neafsey, D.E., Palumbi, S.R. (2003) Genome size evolution in pufferfish: a comparative analysis of diodontid and tetraodontid pufferfish genomes, *Genome Res.* **13**, 821–830.

Parsch, J. (2003) Selective constraints on intron evolution in Drosophila, *Genetics* **165**, 1843–1851.

Petrov, D.A. (2002a) DNA loss and evolution of genome size in Drosophila, *Genetica* **115**, 81–91.

Petrov, D.A. (2002b) Mutational equilibrium model of genome size evolution, *Theor. Popul. Biol.* **61**, 531–544.

Petrov, D.A., Hartl, D.L. (1998) High rate of DNA loss in the Drosophila melanogaster and Drosophila virilis species groups, *Mol. Biol. Evol.* **15**, 293–302.

Petrov, D.A., Hartl, D.L. (2000) Pseudogene evolution and natural selection for a compact genome, *J. Hered.* **91**, 221–227.

Petrov, D.A., Lozovskaya, E.R., Hartl, D.L. (1996) High intrinsic rate of DNA loss in Drosophila, *Nature* **384**, 346–349.

Petrov, D.A., Chao, Y.C., Stephenson, E.C., Hartl, D.L. (1998) Pseudogene evolution in Drosophila suggests a high rate of DNA loss, *Mol. Biol. Evol.* **15**, 1562–1567.

Petrov, D.A., Sangster, T.A., Johnston, J.S., Hartl, D.L., Shaw, K.L. (2000) Evidence for DNA loss as a determinant of genome size, *Science* **287**, 1060–1062.

Petrov, D.A., Aminetzach, Y.T., Davis, J.C., Bensasson, D., Hirsh, A.E. (2003) Size matters: non-LTR retrotransposable elements and ectopic recombination in Drosophila, *Mol. Biol. Evol.* **20**, 880–892.

Powell, J.R. (1997) *Progress and Prospects in Evolutionary Biology. The Drosophila Model*, Oxford University Press, New York.

Ptak, S.E., Petrov, D.A. (2002) How intron splicing affects the deletion and insertion profile in Drosophila melanogaster, *Genetics* **162**, 1233–1244.

Robertson, H.M. (2000) The large srh family of chemoreceptor genes in Caenorhabditis nematodes reveals processes of genome evolution involving large duplications and deletions and intron gains and losses, *Genome Res.* **10**, 192–203.

San Miguel, P., Gaut, B.S., Tikhonov, A., Nakajima, Y., Bennetzen, J.L. (1998) The paleontology of intergene retrotransposons of maize, *Nat. Genet.* **20**, 43–45.

Schaeffer, S.W. (2002) Molecular population genetics of sequence length diversity in the Adh region of Drosophila pseudoobscura, *Genet. Res.* **80**, 163–175.

Smit, A.F. (1999) Interspersed repeats and other mementos of transposable elements in mammalian genomes, *Curr. Opin. Genet. Dev.* **9**, 657–663.

Thomas, C.A. (1971) The genetic organization of chromosomes, *Annu. Rev. Genet.* **5**, 237–256.

Vinogradov, A.E. (1995) Nucleotypic effect in homeotherms: body-mass-corrected basal metabolic rate of mammals is related to genome size, *Evolution* **49**, 1249–1259.

Wright, S. (1931) Evolution in Mendelian populations, *Genetics* **16**, 97–159.

Zuckerkandl, E. (1976) Gene control in eukaryotes and the c-value paradox "excess" DNA as an impediment to transcription of coding sequences, *J. Mol. Evol.* **9**, 73–104.

Zuckerkandl, E. (2002) Why so many noncoding nucleotides? The eukaryote genome as an epigenetic machine, *Genetica* **115**, 105–129.

11
Horizontal Gene Transfer

Jack A. Heinemann and Ralph A. Bungard
University of Canterbury, Christchurch, New Zealand

1	**Horizontal Gene Transfer Versus Introgression**	339
2	**Vectors and Pathways of HGT** 341	
2.1	Transformation and Pathogenic Bacteria	341
2.2	Conjugation 342	
2.3	Parasites, Symbionts and Pathogens	343
2.3.1	Viruses and Transduction 343	
2.3.2	Parasitic Plants 344	
2.3.3	Parasitic and Pathogenic Animals	344
2.4	Hybridization 345	
3	**Different Ways to Observe HGT**	345
3.1	Descriptive Bioinformatics 347	
3.2	Experimentation 349	
3.2.1	Seminal Experimentalist Approaches	349
3.2.2	Modern Process Approaches 349	
4	**Barriers to HGT** 351	
4.1	Transfer and Inheritance 352	
4.2	Introgression 352	
5	**Final Thoughts** 354	
	Acknowledgment 354	
	Bibliography 354	
	Books and Reviews 354	
	Primary Literature 355	

Genomics and Genetics. Edited by Robert A. Meyers.
Copyright © 2007 Wiley-VCH Verlag GmbH & Co. KGaA, Weinheim
ISBN: 978-3-527-31609-0

Keywords

Bacteriophage (phage)
Viruses specific to bacteria.

Competition Model
Derives from an experimental demonstration of the evolutionary forces that determine when genes reproduce by HGT. On the basis of the hypothesis that different genes reproduce faster by HGT in different environments and the change in environment determines when genes become mobile. Stability of the environment then influences which genes introgress. Produces unique predictions of how genes are recruited to HGT vectors.

Complexity Hypothesis
Posits that genes least likely to introgress are of a type that are involved in highly interdependent large biochemical complexes. The interactions lead to constraints that would preclude replacement by genes that have diverged.

Conjugation
A form of DNA transfer mediated by infectious elements (called *plasmids*) normally found in bacteria. Plasmids transfer from bacteria to all known forms of life.

Hybridization (to form hybrids)
The initial product of a sexual cross between individuals that are considered to be of different varieties, subspecies, species, or genera and that might often be revealed by low fecundity or infertile offspring.

Introgression
"The permanent incorporation of genes from one set of differentiated populations (species, subspecies, races, and so on) into another" (quote from Stewart Jr. et al.).

Phylogeny
The evolutionary connections between organisms inferred from assumed descent via a common ancestor.

Transduction
The transfer of genes by viruses to different organisms within the infectious range of the virus.

Transformation
The change in the phenotype of organisms due to the inheritance of a gene. In molecular biology jargon, it often is a term that also refers to the uptake of DNA regardless of whether it alters the phenotype.

Vectors
Viruses, gametes, organisms, plasmids, or other agents that can deliver genes by HGT.

Vertical Gene Transfer
Reproduction of genes in the context of whole genome replication and reproduction of the organism or entire cell of an organism.

> Horizontal gene transfer (HGT) may be defined as any occurrence of heritable material passing between organisms, asynchronous with reproduction of the organisms. It represents replication of heritable material outside the context of parent to offspring (i.e. vertical) reproduction. Three types of evidence traditionally lead to claims of HGT. Firstly, biochemical and genetic observations of real-time (usually laboratory recreations of) gene flow between two genotypically distinguishable individuals. Secondly, DNA sequences that are common within a species but inconsistently found in that genus, and finally, direct evidence of processes that can create genes that reproduce horizontally. HGT is now confirmed to occur between all biological kingdoms and has consequences even when genes fail to introgress.

1
Horizontal Gene Transfer Versus Introgression

Awareness of HGT (horizontal gene transfer) has risen in recent times in part due to some spectacular scientific debates surrounding, particularly, the release of genetically modified organisms (GMOs) as food crops and the spread of antibiotic-resistant bacteria that cause disease. The recent controversies belie the long history of research in HGT, sometimes treating it as if it had been discovered only since genome sequences began to be completed. On the contrary, HGT is a general and pervasive biological phenomenon, but it has been packaged into separate boxes that go by other names. We attempt here to identify what is fundamental to HGT so that it may be recognized in the various other boxes it appears in, and to place it center stage in the modern scientific discourses that depend on understanding it.

There is genuine confusion about what HGT is and what it is not. Part of this confusion results from semantics, with some authors simply preferring the term lateral gene transfer (LGT) to HGT. To resolve this issue, C.I. Kado suggested that the term LGT be reserved for gene transfers between closely related partners and HGT for more distant. This proposal has some merit, but it presupposes that the important defining characteristic of HGT resides in the organisms that are donors and recipients of genes rather than the genes themselves. Besides the obvious problem of knowing when partners are close enough to call exchanges between them (LGT), it is entirely possible for two partners to swap a gene that is phylogenetically more distant to both the donor and recipient than either is to one another.

The most important source of confusion is the tendency to substitute an outcome of HGT with a description of HGT. Two outcomes are most commonly associated with HGT. One outcome is that HGT can generate genotypic diversity in a species or offer a selective advantage to recombinants within a species. This has tempted many to consider HGT a process by which organisms, especially the asexual bacteria, recover some of the benefits of sex. There is no denying that HGT can provide genes that benefit the recipient and introduce genetic diversity into populations, but these outcomes are based on the biology of the recipient organism and are only one of the consequences of HGT.

A second outcome is introgression. Genes, and sometimes entire chromosomes, can be introduced into a separate, differentiated population. For a gene to become permanent, it almost certainly must have provided an adaptive function, or survived a fortunate bottleneck. Moreover, the recombinant genome must be found in many if not all individuals of a species. For that to occur, the recombinant genome must outcompete most or all genomes that lacked the new gene.

The concept of introgression is normally reserved for describing vertical gene transfer over the barriers erected by meiosis. However, aspects of the concept also apply to HGT. Whether an alien gene is introduced during (e.g. hybridization) or asynchronously with reproduction (i.e. HGT), there are barriers to it becoming a stable and common part of the genome of the species. Alien genes likewise must be present in a significant proportion of a population in order to be detected. Overcoming these barriers can lead to alien genes and chromosomes becoming a characteristic feature of a species.

The process by which genes introgress, however, is only part of the way genes evolve by HGT (Fig. 1). Genome sequencing reveals past gene transfers by finding alien genes in genomes. The number of alien genes found by such methods only reflects the number of genes that have introgressed, not the number or frequency of transfer. This one outcome of HGT is certainly important evidence for HGT, but it is ill suited to measuring its impact and frequency. Measurements of HGT that rest with genomic DNA sequences are essentially a measure of the range and frequencies in which genomes retain genes, not the range and frequencies of gene transfers that make the genes available in the first place (Fig. 1).

Making HGT the same as an outcome of HGT, namely, the "acquisition of 'alien genes' by a particular genome" (quote from Koonin et al. in Syvanen and Kado) substitutes historic description for understanding how HGT happens biochemically and how it influences organisms physiologically. Quite rightly, this focus leads to strong criticisms about proper identification of DNA sequences that are indeed alien.

Understanding how genes introgress, therefore, is not the same as understanding HGT. HGT is not limited to transfers that change organismal genotypes. For example, some genomes may evolve almost exclusively by HGT. Viruses and transposable elements like *mariner* are arguably the product of evolution that has little to do with organismal genomes and their vertical reproduction.

The essence of HGT as an evolutionary process, and not just an outcome, must be preserved in any definition. HGT may

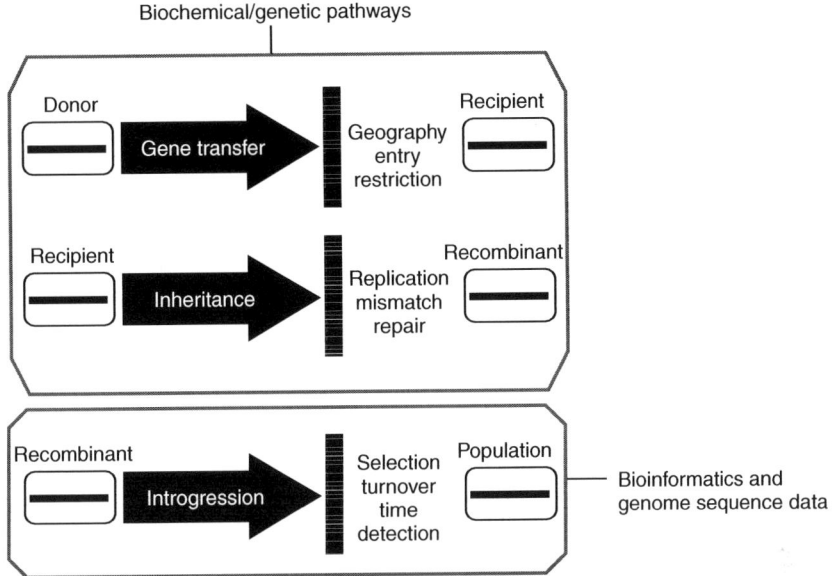

Fig. 1 Different levels of analysis used to study horizontal gene transfer. Symbols: rounded rectangles = cells or organisms, solid rectangles = mobile gene, brick wall = listed barriers. Only process experiments as discussed in Sect. 3 can cover all levels.

be defined as any occurrence of heritable material passing between organisms, asynchronous with reproduction of the organisms. It represents replication of heritable material outside the context of parent to offspring (i.e. vertical) reproduction. This definition permits different means of transfer – such as viruses or transduction, conjugation, and transformation – and different outcomes of transfer – such as serial infection or recombination – to be united in legitimate study about how genes evolve through HGT.

2
Vectors and Pathways of HGT

2.1
Transformation and Pathogenic Bacteria

Bacteria can at times become competent for the uptake of DNA from their surrounding environment. Should part or all of that DNA become stably incorporated into the genome, they are called *transformants*. Competence may be expressed constitutively in a proportion of the population or be induced. It is usually associated with the expression of a set of proteins, called *com* for competence, that include or are in addition to structures (e.g. vesicles and pili in some gram-negative bacteria) associated with the ability to bind extracellular DNA and translocate it into the cytoplasm.

The most widely studied examples of naturally competent bacteria are *Streptococcus pneumoniae* and *Bacillus subtilis* (of the gram-positives) and *Haemophilus influenzae* and *Neisseria gonorrhoeae* (of the gram-negatives). Although the molecular details of transformation varies between organisms, in each case DNA is taken into the cell in single-stranded form where it is thought to serve as a substrate for single-strand specific DNA binding recombinases (e.g. analogs of Escherichia coli's RecA protein) that initiate a search for similar sequences of DNA in the recipient genome. This recombination process is aborted or reversed, depending on the activities of the endogenous mismatch-repair (MMR) pathways, if the incoming strand creates too many mismatched base pairs. Depending on the efficiency of MMR at the time, the incoming DNA may survive through a subsequent round of DNA replication and become immortalized in the recipient genome.

The uptake process has been characterized to the molecular level for a small number of bacteria and a small number of environmental conditions that induce competence. Many bacteria that are not known to be competent have homologs of *com* genes, which might indicate that they do have the potential to take up DNA from the extracellular environment. It remains unknown how many bacteria may take up DNA and all the environments in which they may be induced to take up DNA naturally. Even the icon of noncompetent bacteria, *E. coli*, was recently shown to develop a natural competence in laboratory studies.

In contrast, no special competence seems to be required for eukaryotic cells to take up DNA by transformation. DNA can pass into animal cells *in situ* simply through food and has been reported to pass through the placenta and recombine into the mammalian genome. Animal cells can also be transformed *in vivo* and *in vitro* using bacteria that cross into the cytoplasm. Viruses also can produce proteins that serve to transfer genes within multicellular organisms, bypassing the need to reinfect from the outside.

2.2
Conjugation

Bacterial conjugation in its broadest sense includes the conjugative plasmid groups discovered in gram-negative bacteria, the T-DNA of the Ti plasmid, pheromone-responsive plasmids and conjugative transposons. The focus in this overview is on the model conjugative systems defined by the IncP and IncF groups of plasmids of gram-negative bacteria.

Plasmids are similar to viruses but they lack any known extracellular form. Though loosely defined as extrachromosomal, like some viruses they can be found integrated into chromosomes at times. They have also been defined as accessory elements or parasites because hosts cured of plasmids can survive and reproduce in at least one environment. Conjugative plasmids got their name because for a time it was thought that they replaced the need for sex in prokaryotes. This idea derived from the use of genes with adaptive value, such as antibiotic resistance genes, to monitor the transfer of the plasmids. None of the above-mentioned characteristics is universal, so we have opted to define plasmids by their ability to reproduce in at least one environment in which the host cannot. That niche may not be physical but temporal, such as provided by HGT.

Conjugation mediated by the IncF and IncP plasmids requires, at a minimum, a cis-acting DNA sequence called *the origin of transfer* (*oriT*). All other functions (called *tra*) act in trans thus allowing plasmids with all trans-acting functions to also transfer plasmids with no or a few trans-acting functions. The trans-acting gene products are divided further into those involved in DNA metabolism (and usually specific to a particular *oriT*) and those involved in DNA transport and cell–cell interactions (and thus most likely to interact with a greater range of other plasmids). The conjugative genes that are specific to DNA metabolism introduce a "nick" at *oriT* and initiate the unwinding and concomitant transfer of DNA to a recipient cell. Both strands are used as templates for the synthesis of a complementary strand, one in the donor cell and one in the recipient.

The tumor-inducing (Ti) plasmid from *Agrobacterium tumefaciens* is a plasmid with two conjugative systems. One system mobilizes the entire plasmid; the second system mobilizes a single region of the plasmid, called *the T-DNA*. It is the T-DNA that is recovered from crown gall tissue of plants infected by *A. tumefaciens*, a discovery that unambiguously demonstrated that conjugation was a means to transfer DNA across biological kingdoms.

It has become clear over the past decade that the DNA transport apparatus of conjugation is the ancestor, or at least a sibling, of other macromolecular transport systems such as the type IV protein secretion pathways. Bacteria that secrete proteins into human cells during infection may also secrete plasmids using the same apparatus. Consistent with this speculation is the ability of conjugative plasmids to transfer from bacteria to eukaryotic cells, and the ability of bacteria to support conjugation *in situ* and inside human cells.

2.3
Parasites, Symbionts and Pathogens

Parasites, symbionts, and pathogens are excellent gene vectors. Viruses, like many pathogens, can have much broader transfer ranges than symptom ranges. Some viruses transfer between plants and animals, and bacteriophage can penetrate the human central nervous system. These vectors also transmit a wide range of molecules in addition to DNA, such as RNA and proteins. These molecules are relevant here because prions and dsRNA are horizontally transferred epigenes capable of transforming the germline.

2.3.1 Viruses and Transduction

Transduction is formally viral-mediated transfer of nonvirus genes between hosts. While it is not uncommon for viruses to acquire genes from their hosts, and thus acquire new virulence or host range characteristics, transduction is distinctive because, like transformation, it delivers genes most recently of chromosomal origin that become stably integrated into the recipient genome.

Generalized transducing bacteriophage, the viruses specific to bacteria, package host DNA instead of viral DNA at some low frequency during infection (on the order of 0.001%). This phenomenon is called *generalized transduction* because any region of the chromosome may be packaged. The size of the phage determines how much DNA may be transduced by a single phage, which is normally only 1 to 2% of the host genome. The mature phage particle retains the ability to attach and initiate an infection, but delivers cellular rather than viral DNA to the recipient. Since the phage particle does not contain the viral genome, the infection does not result in the production of new viruses.

All temperate phages that integrate into a chromosome as a part of their life cycle have the potential to be *specialized* transducing phage. During the process of excising from the chromosome to begin an infectious cycle, DNA flanking the phage genome can be packaged at some low frequency. This phenomenon is known as *specialized* because only a small number of genes are ever transduced by this phage. Phage integrating into the chromosome usually does not enter at random, but instead enters at special DNA sequences in the host chromosome. Thus, it is only the genes flanking integration sites that are transduced.

Bacterial genomes are littered with the skeletons of phage, many of which no longer have an infectious life cycle. These skeletons are revealed when they are the source of genes that promote bacterial virulence (pathogenicity islands) or symbiosis (symbiosis islands). They may also provide sequences into which still active temperate phages could integrate by homologous recombination or by a site-specific mechanism. These phage remnants also make it difficult to say what a host gene is and what a viral gene is not. This ambiguity is part of the reason why we consider viral infection itself a demonstration of HGT rather than simply when an arbitrary marker is acquired in the course of a phage infection.

2.3.2 Parasitic Plants

Around 1% of all flowering plants (angiosperms) are parasitic, attaching to host plant species from which they extract some, if not all, of the carbon, nutrients, and water they require for growth. Some of these parasites, like the Rafflesiaceae, have evolved to such an extent that the only outward signs of the parasite are when they exit the host to flower. It has long been known that parasitic plants not only exchange carbon and nutrients with their host but that they also exchange complex molecules like alkaloids and pathogens including viruses, bacteria, and phytoplasmas. This intimate relationship may also facilitate HGT between plant species. A multigene phylogenetic analysis of the parasitic Rafflesiaceae has shown that some genes from both the nuclear and the mitochondrial genomes are distinct to the order Malpighiales while other mitochondrial genes are closely associated with their obligate host *Tetrastigma*. The Rafflesiaceae are obligate host parasites, meaning that they only parasitize one host species. However, other parasitic plant species like the *Cuscuta* can attach to a broad range of hosts and can even simultaneously parasitize multiple hosts. This raises the possibility that any one host species may be continually exposed to DNA (either naked or within vectors like viruses and bacteria) from multiple, unrelated species. It will not be surprising if introgressions of many genes are discovered because of the long and continuing history of plant–plant parasitism.

2.3.3 Parasitic and Pathogenic Animals

The massive transfer of DNA to eukaryotes is widely believed to have happened when the different kinds of symbionts or parasites, such as those that were the free-living relatives of the mitochondria and chloroplast, fused with the common ancestor of eukaryotes. The transfer of DNA from organelles to the nucleus may have been a passive event, such as happens when invasive bacteria die inside animal cells, or an active process as indicated in laboratory studies following transgene transfer from yeast mitochondria to the nucleus.

Other intracellular parasites may transfer DNA to the host nucleus. Human, rabbit, and bird cells acquire the DNA from the parasite *Trypanosoma cruzi*, a protozoan that can cross the human placenta and transform the germline of animals.

The DNA donor *T. cruzi* is an example of a parasite/pathogen delivered to its human host by a biting insect vector. Other biting insects may be direct vectors of genes. The egg predatory mite *Proctolaelaps regalis* is the most likely vector of P transposable elements to the Drosophilids. The mite does not have the element in its genome, so it may have introduced it into the germline of Drosophila after feeding on a carrier.

2.4
Hybridization

While hybridization is the outcome of sexual reproduction, it can also double as a vector for HGT. Even infertile hybrids can serve to promote HGT. For example, some hybrid plants convert to apomixis, an asexual mode of reproduction, increasing the chances of introgression. Infertile asexual hybrids or hybrids with low fertility may also produce offspring in time. When hybridization reduces fertility by creating two or more unmatched chromosomes, in time these chromosomes can duplicate, and thus pair. Once all the chromosomes again are partnered in meiosis, the hybrid can return to sexual modes of reproduction. Meanwhile, alien genes and transposable elements carried by alien chromosomes, whether or not the entire chromosomes should ever introgress, can transfer into the original genome.

The impact that mixing of previously separated populations has on the likelihood of interspecies gene transfer and introgression, should not be underestimated. Although there is an immediate increase in the chance of hybridization from breaking geographical boundaries between, say, a native species and an introduced one, the chances of more widespread gene transfer appear to be markedly increased if the introduced alien also acts as a bridging species, that is, a species that can link, through hybridization, two or more species that previously were genetically, geographically, morphologically, or physiologically separated to an extent that prevented hybridization.

These "hybrid-bridge" species have the potential to effectively act as multipliers of introgression, as they not only permit gene transfer between themselves and a resident native but also between multiple native and/or introduced species that were previously separated so much that the hybridization event was unlikely (Fig. 2). Simple examples of "hybrid-bridge" species can be envisaged: for example, a species with a flowering time that is intermediate between two resident species could allow gene flow between species previously separated by flowering morphology; likewise, a species with a broad habitat tolerance could link species from habitats that were previously geographically separated. In the face of current discoveries of gene introgressions through hybridization, it seems only a matter of time before an example of long-distance HGT driven by hybrid-bridge species is recognized.

3
Different Ways to Observe HGT

The science of horizontal gene transfer is well represented by researchers using two different but compatible approaches. The newest approach is to use bioinformatic descriptions of putative alien genes or tracts of alien DNA sequence to infer how and how often HGT happens, and what the consequences are to the recombinant genomes. This approach has the strength of looking over many different species simultaneously because of the accumulation of complete genome sequences and the rapid nature of *in silico* analysis. The bioinformatics approach of

1.

Separation prevents HGT through hybridization

2.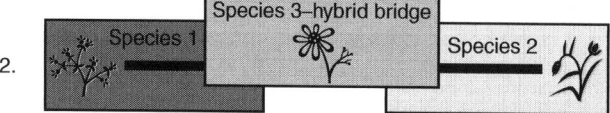

Introduction of an intermediate "bridge" species

3.

Hybridization creates a mechanism for HGT between previously separate gene pools

Fig. 2 Hybridization provides a mechanism for HGT between differentiated and separate multicellular eukaryotes. Symbols: rectangles represent populations of species in which HGT is confined; arrows represent the potential flow if genetic material in HGT. Removal of historical barriers to HGT, such as geographical separation, may lead to even greater potential for HGT. In this example, (1) two species may traditionally be unable to hybridize because of barriers such as habitat separation or differences in flowering morphology or physiology. (2) The introduction of a species that can hybridize with both Species 1 and 2 could bridge this separation. (3) The consequences of this "hybrid bridge" would mean that HGT, which previously could not occur between Species 1 and 2, would now be possible.

HGT, however, is limited by how well its techniques identify true HGT events, how accurate its predictions of gene function are, and by the very necessary fact that it is concentrating exclusively on DNA that has transferred, converted into vertically inherited material, and introgressed, that is, outcomes of HGT. Moreover, each event in the trillions of cells of a single large multicellular organism, or among the recipients of species that number 10^{20} individuals or more, is extremely easy to overlook. Below, the various bioinformatics tools for identifying HGT will be discussed.

The second approach is to observe HGT. Those observations can include biochemical descriptions of gene transfer pathways to eco-evolutionary experiments that measure how genes sort between horizontal and vertical mechanisms of inheritance. The experimental approach has the strength of using empirical biology to discover phenomena not immediately apparent from DNA sequences. It can in addition discover epigenes and other important molecules that may also transfer horizontally, and, importantly, it has the comparative power required to test hypotheses about HGT processes. This approach is limited by the generation time of experimental organisms, the infrastructure needed to maintain different kinds of organisms, the realism of laboratory recreations of ecosystems in which gene transfers may occur, and the very limited capacity to physically sample enough biological material in order to detect HGT. Below, some novel process experiments and their relevance to understanding the ecology of mobile genes will be presented.

3.1
Descriptive Bioinformatics

Bioinformatics has produced compelling evidence from sequenced genomes of past gene transfers. Those transfer events that are detectable using bioinformatics are so because they are either

- recent;
- large fragments of DNA; or
- small but extremely different from the genome as a whole and out of place in the organismal phylogeny, in particular.

The introduction of new genes into a genome can create an anomalous phylogenetic tree, incongruent with other gene-based trees or trees built from morphological features. For example, the genes for the various aminoacyl tRNA synthetases, aaRS, create different phylogenetic trees in bacteria. This appears to result from different aaRS genes having been acquired by bacteria from various eukaryotes at different times. A particularly striking example is provided by ProRS, where a eukaryotic version is found in *Mycobacterium leprae* but not in the close relative *Mycobacterium tuberculosis*. In this case, the corresponding prokaryotic ProRS was lost from *M. leprae*;

in some examples of aaRS transfer, both the prokaryotic and eukaryotic versions are maintained.

Evidence for HGT based on analyses that identify candidate genes that violate a phylogenetic order is prone to false-positive errors due, for example, to long-branch attraction. The most pervasive source of false-negative and false-positive identifications of alien genes comes from the initial construction of the tree based on putative orthologous genes. These genes diverged when the species diverged and represent the divergence of the two organisms from a common ancestor. However, they can be difficult to distinguish from paralogous genes created when a gene duplicates creating two alleles, which can then change independently over time. These reflect only the history of the genes within, instead of between, lineages.

Identifying which of multiple paralogous genes is a true orthologous gene is sometimes done by assuming that the most similar of the paralogous genes in the different species are the orthologous genes. However, accepting this *a priori* assumption can render the comparison redundant. Moreover, that test can never be independently verified because, no matter how close the sequences match, the possibility that one species lost the true orthologous gene since the speciation event can never be excluded. Indeed, the genes may be orthologous but because the gene was transferred horizontally, the comparison misrepresents the ancestry of the organism. Mistakes have been made by the inadvertent comparisons of paralogous genes.

Recent transfers may be indicated by a significant divergence in the composition of the DNA from the parameters expected on the basis of average characteristics of the genome. A horizontally transferred gene preserved in a vertically reproducing lineage may be identified by a significant deviation from the average G + C content of the genome or from the normal codon bias of the host. Both of these indicators are quantitative. Thus, they are useful if large tracts of sequences with one or more large deviations from accepted norms are being analyzed.

The origin of sequences becomes increasingly difficult to determine the shorter they become or the closer to accepted ranges they appear. Shorter sequences may be the usual consequence of HGT that involves integration of alien genes with less sequence similarity to the recipient genome.

The outcome of many described HGT events is insertion of difficult-to-spot subgene domains. These nucleotide insertions, deletions, or replacements create "mosaic" genes that may have altered biochemical properties, for example, genes making bacteria resistant to antibiotics. Mosaics are caused by biochemical barriers, primarily MMR, that incompletely remove DNA during recombination.

MMR can saturate under stress or falter through mutation, promoting interspecies HGT ~10 000-fold and creating "mutators." Depending on the proficiency and mechanism of MMR in an organism, strands of DNA from different species may initially be paired, with the invading strand subsequently degraded. Should some mispairs escape repair, or if the MMR machinery became saturated, stretches of the recipient genome could be "massaged" into a closer match with the donor DNA over short intervals.

Recombination frequencies can vary significantly depending on the activity of MMR, resulting in mosaic genes composed of sequences from two highly divergent genes. In fact, a mutation in bacteria is 50 times more likely to have occurred by recombination than by DNA damage or replication errors. Since mutations are often revealed by very small changes in DNA sequence, it is normally impossible to distinguish between their origin as a polymerase or a mismatch-repair error. Mosaic genes can serve as molecular versions of "hybrid-bridges" (Sect. 2.4), facilitating higher frequency recombination between species that find sequence similarity in some parts of the mosaic gene.

The added difficulty of identifying genes that transfer frequently is that they may not mutate at the same rate as genes that reproduce vertically. The genomes of horizontally mobile vectors tend to be fluid. Over time, certain vectors may be replaced by related vectors carrying different genes, such as plasmids with an expanding repertoire of antibiotic resistance genes. Tracing their ancestry by sequence structure allows relationships to be determined only over very short periods of time. For example, the structure of infectious retroviruses can change at 10^4 to 10^6 times the rate of other genes and defective retroviruses that are reproducing in synchrony with the host. The phylogeny of these viruses is confined to tracing the residues of defective viruses trapped in chromosomes of organisms or monitoring divergences only on decade timescales.

Determining ancestry of organisms without relying upon nucleic acid or protein sequences is possible, but much more difficult. In the cases of ribonuclease H and the A8 subunit of mitochondria, ancestry was inferred from three-dimensional structures or other biophysical characteristics because the primary nucleic or amino acid sequences had lost such information.

3.2 Experimentation

Within the experimentalist approach are two mutually supportive schools. The first school is characterized by a quest to understand the biochemistry and pathways behind observable gene transfers. The work of that group is well represented in the literature on conjugation, transduction, and transformation. The second school elevates HGT to an evolutionary process – like sex, mutagenesis, or symbiosis – that can be characterized by how it evolved and to what selective forces it responds to.

3.2.1 Seminal Experimentalist Approaches

The origin of the experimentalist approach for studying HGT can be traced to the first description of mechanisms, including the pioneering work of the 1940s and 1950s with phage, transformation, and conjugation. Those studies were overtly focused on the biology of the organism or the vector, rather than on how HGT shaped them. For example, Luria and Delbrück's famous experiment measuring phage resistance had little to do with understanding how viruses themselves evolved, but it required a comprehensive understanding of the life cycle of the phage in order to demonstrate an evolutionary truth about whether mutations follow or precede selection. Avery et al. used natural competence in *pneumoccocus* to all but prove that DNA was the source of the genes behind the virulence traits they studied, but they used HGT to identify the transforming principle rather than using the transforming principle in a study of HGT. Hayes' use of the antibiotic streptomycin revealed the unidirectional transfer of genes from a donor to recipient by conjugation, but it would take almost 40 years before his trick was resurrected to begin to understand why horizontally mobile plasmids would come to carry antibiotic resistance genes.

In the late 1960s, research groups with a specific focus on HGT emerged. Their pioneering work provided the first hints that both proteins and DNA and RNA could physically transfer between species. The team of Maurice Stroun and Philippe Anker were arguably the first to detect hints that gene transfer from the plant pathogen *A. tumefaciens* was behind the crown gall tumors it caused. This work would take another 10 years, and a suite of new techniques, to be confirmed. Stroun and Anker also produced compelling evidence with the techniques of their time for the horizontal transfer of proteins, a theme that has gained prominence with the demonstration of the interspecies infectivity of prions and the description of the Type IV DNA and protein secretion systems that bacteria use for both conjugation and the transfer of proteins to eukaryotic hosts. What made the work of Stroun and Anker different from that of their contemporaries was that while they were inspired by the biology of the organisms they chose, their experiments were process-oriented, designed to detect gene transfer rather than to focus on the biology of particular organisms.

3.2.2 Modern Process Approaches

Research that focuses on the process of HGT is becoming more prominent, especially as the threat of gene transfer from transgenic crops and antibiotic-resistant bacteria command the attention of the scientific community. Past descriptions of gene transfers and full descriptions of biochemical mechanisms, when available, have proved of limited use in predicting when, where, and how often gene transfer might create an unwelcome biological outcome. In this section, two contemporary models that promise predictive power are contrasted.

Why and when are some genes reproducing horizontally while other genes, or at other times these same genes, are reproducing vertically? Two different models have been constructed in answer to this question. The first model, called *the complexity hypothesis*, argues that some genes are more mobile than others simply because of the complexity of the processes in which they are involved. Those genes whose products are closely reliant on, or associated with, many other molecular partners will simply not be functional if removed from their current context. This model is supported by bioinformatic analysis of HGT events in sequenced genomes. The second model, termed *the competition model*, argues instead that the two forms of reproduction (HGT and VGT) favor different genes in different environments. This model suggests that genes transit to horizontal reproduction when that mode of reproduction outpaces alleles of the same gene reproducing vertically. This model is supported by laboratory evolution experiments.

The complexity hypothesis posits that operational genes are more mobile than informational genes. Informational genes are those with products that are associated within large complexes and may be involved in such biochemical activities as transcription, translation, and replication. Operational genes, in contrast, may work alone or in very simple assemblies and thus more easily function after transfer. From six complete prokaryotic genomes that were analyzed for 132 orthologous genes, the dual contentions that gene transfer was a continuous process and that those genes most likely to introgress were operational were supported. Again, this analysis does not say how frequently genes are transferred, but how frequently different types of transferred genes transit into stable alien genes in new species.

The competition model differs from the complexity hypothesis for predicting the attributes of genes that make them transit to horizontal reproduction. It is based on the view that the current function of a protein is not sufficient to predict future roles in different environments. These roles in different environments, rather than in their cellular context alone, determine which genes transit from vertical to horizontal and back to vertical reproductive strategies.

The competition model is not at odds with the complexity hypothesis because it would accept that operational genes may more frequently, but not exclusively, enjoy a differential reproduction as mobile genes. Consistent with this expectation, there is bioinformatics evidence of informational gene transfers and even contemporary discoveries of plasmid vectors of informational genes. Interestingly, the mobility of these genes correlates with the relatively recent introduction of modern antibiotics by humans. In the case of the ribosomal S14 protein, there is compelling evidence that transfer and introgression may be facilitated by antibiotic pressure selecting for alleles of the *rps14* gene that could confer resistance. Likewise, the eukaryotic-type IleRS has been discovered on plasmids that deliver it to bacteria susceptible to the antibiotic mupirocin. Within the expectations of the competition model, these informational genes can have a role in some environments that makes them transit from a vertical to a horizontal reproductive strategy. The competition model goes further to explain how the genes get mobilized, not just how transferred genes introgress.

A new study comparing the complete genomes of 116 prokaryotes supports the operational–informational division that is central to the complexity hypothesis. The huge amount of sequence information that makes up this study also revealed a second class of operational genes that are as immobile as the information genes. It also showed that among the genes most commonly found to be mobile were predominantly those associated with plasmids and phage, cell surface proteins, DNA binding (particularly restriction-modification), and pathogenicity. These categories include genes that are frequently classed as PSK, or postsegregational killing genes.

The distinction of PSK genes in this new study also provides tacit support for the competition model: PSK genes are defined as any combination of genes that simultaneously produce and suppress a toxic effect on the host. PSK genes attack other infectious elements, for example, competing plasmids and viruses, but do not necessarily increase the absolute rate of reproduction of the vector or the host. PSK genes have been found to be a significant disadvantage to vectors that are forced to reproduce at the rate of cell division. Only when PSK is coupled to HGT do the vectors bearing PSK genes demonstrate competitiveness with vectors that did not have such genes.

The PSK method of attack is to kill cellular offspring that lose the genes. If a bacterium is infected by an invading virus or an incompatible plasmid, the PSK genes kill the bacterium along with the invading horizontally mobile element. In doing so, the PSK-bearing vector does not benefit with an immediate increase in reproduction, but it does eliminate a competitor.

Antibiotic resistance, novel virulence determinants, and restriction-modification genes are likely to be PSK genes and thus have evolved in the same way. The structural and functional attributes of PSK are independent of the benefits such genes might provide the organismal host in the presence of antibiotics or other lethal agents. The advantages PSK genes confer upon the HGT vector are even more important than the cost to organismal hosts for carrying the genes.

PSK genes evolve by HGT, at least some of the time. However, these phenotypes are only indirectly related to how PSK genes, like antibiotic resistance, converted from chromosomal to horizontally transferred genes. Organisms with these genes may in time have found some use for them, explaining introgression, but that is not an explanation for how the genes came to evolve on horizontally mobile elements.

The competition model predicts that genes capable of causing cell death will, under the right circumstances, transit to horizontal reproduction. It is thus noted with some irony that classic PSK systems, also called *suicide genes*, have been recommended by a prominent panel of the US National Research Council as promising bioconfinement tools for application to genetically engineered microbes.

4
Barriers to HGT

If HGT were so common and frequent, why is there not more evidence of it? The definitive answer to this question is some ways off, because, as discussed above, the preponderance of searches for it so far have been designed to detect introgression into organismal genomes. Toward the end of the last century, however, several groups independently started asking this question. Their motivation was not to discount HGT, but to yield biologically interesting hypotheses about the biological barriers to either transfer *or* introgression (Fig. 1).

There is no doubt that there is a raft of situations, circumstances, and processes that prevent or at least reduce the rate of HGT. These are referred to here as *transfer, inheritance, and introgression barriers*.

4.1
Transfer and Inheritance

Barriers to gene transfer and inheritance include those that prevent geographical access to genes, entry of genes into the cell or organism, stability of DNA (for example, restriction enzymes) and stability of integration (recombination and mismatch repair) or an inherent compatibility between important cis-acting replication sequences and the enzymes of replication.

Geographical barriers spatially separate species to an extent that prevents exposure to their respective genes. Entry barriers include impenetrable cell walls that prevent transformation or transduction, and differences in form (like flower shape) that prevent hybridization. Virus infectious ranges may be determined by the presence or absence of a specific receptor. The genetic material, whether it is protein, RNA or DNA, may be enzymatically degraded upon entry into a cell, or specially sequestered and denatured or depolymerized. Immediately following transfer and stabilization, the gene must replicate or, in the case of nucleic acids, physically recombine with a chromosome or other cellular replicon.

4.2
Introgression

Another layer of barriers may prevent introgression. Developmental barriers specifically deny incoming genes access to the germline. Genetic barriers prevent recombination (Sect. 2.1) or they prevent chromosome pairing during meiosis (Sect. 2.4). Selection barriers block the expansion of the recombinant genotype within the population or species. This can be due to the gene having no or a negative effect on host fitness or due to expression barriers.

Expression barriers arise from incompatibility of an alien gene or polypeptide product with the host expression biochemistry, roughly that referred to as the *central dogma reactions*. The central dogma reactions are well understood as individual processes and are summarized but are treated in depth elsewhere.

Failure of cis-acting sequences of alien genes or their transcripts or products, to be properly recognized or processed by any of the reactions between transcription and the proper placement of a mature protein may prevent the gene from providing a function that could make its new host differentially competitive. Regulatory sequences on alien genes may not effectively recruit specific transcription factors to bind to promoters, terminators, and enhancers. The alien gene may produce an mRNA that is spliced differently or not at all, or which requires special editing (Fig. 3). Translated polypeptides may not fold properly in the recipient or may not sort to the correct subcellular location.

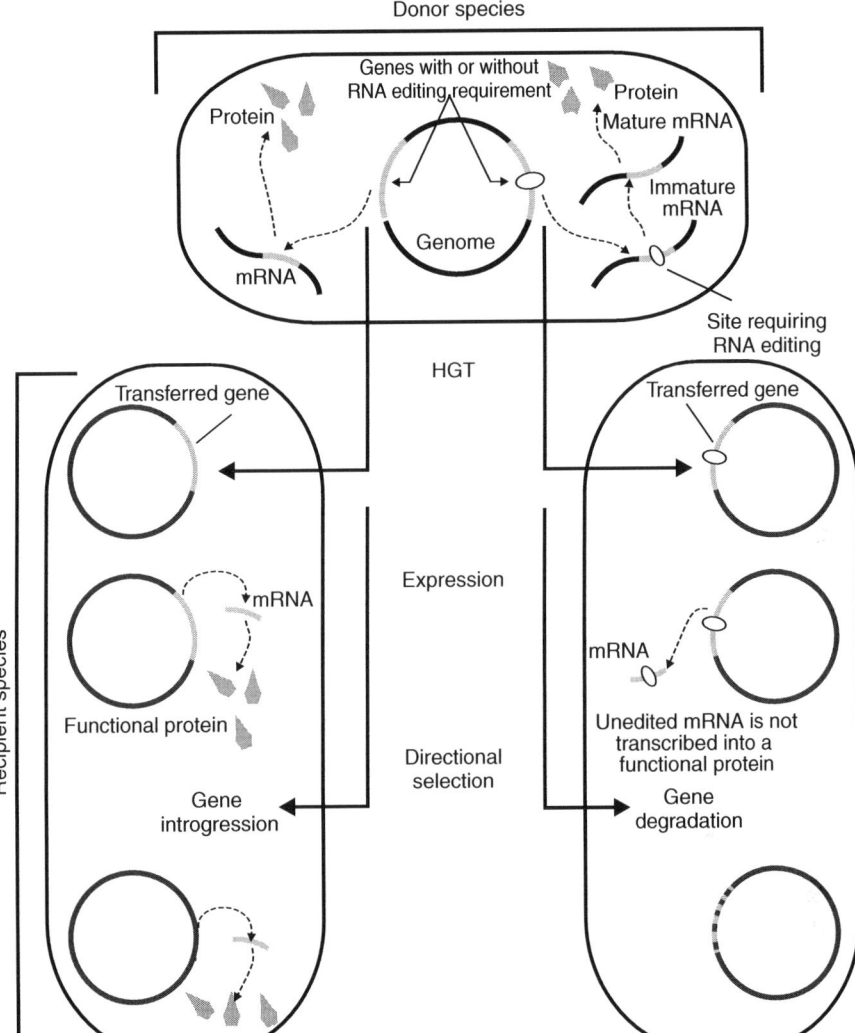

Fig. 3 RNA editing as an example of a molecular barrier to introgression. Symbols: rounded rectangles represent distinct donor or recipient species; circles represent genomes; arrows represent the linear sequence of events from HGT from donor to recipient genomes to the processes of gene expression and finally either intergression or gene degradation; protein mRNA strands and proteins are labeled as such. Some genes require donor-specific posttranscriptional RNA editing. Failure to edit transcripts could uncouple HGT from introgression if recipients do not have a similar RNA editing pathway. Genes without an RNA editing requirement are more likely to properly translate and may therefore introgress through directional selection.

5
Final Thoughts

The ultimate limitation of HGT is not transfer, but our ability to detect it. Gene transfers between closely related organisms may not be detected in genome sequences. While even small changes in existing genes may be important for function and introgression, they may be too small to distinguish from the noise of mutation through vertical descent. Small changes may arise by gene transfer and a combination of mutational amelioration and recombination, but these are almost always credited to repaired DNA damage and polymerase errors.

There are examples of genes that have overcome all the barriers to HGT described in Sect. 4. Of these barriers, only those that prevent transfer are fundamental to the evolution of all mobile genes. Barriers to inheritance and introgression only affect genes that tie their fates once again to the fate of an organism.

Acknowledgment

Jack A. Heinemann thanks the University of Canterbury for support on U6570. Ralph A. Bungard acknowledges support from the Marsden Fund of New Zealand.

See also Molecular Basis of Genetics.

Bibliography

Books and Reviews

Brochier, C., Philippe, H., Moreira, D. (2000) The evolutionary history of ribosomal protein RpS14: horizontal gene transfer at the heart of the ribosome, *Trends Genet.* **16**, 529–533.

Bungard, R.A. (2004) Photosynthetic evolution in parasitic plants: insight from the chloroplast genome, *Bioessays* **26**, 235–247.

Christie, P.J., Vogel, J.P. (2000) Bacterial type IV secretion: conjugation systems adapted to deliver effector molecules to host cells, *Trends Microbiol.* **8**, 354–360.

Cogoni, C., Macino, G. (2000) Post-transcriptional gene silencing across kingdoms, *Curr. Opin. Genet. Dev.* **10**, 638–643.

Doolittle, W.F. (1999) Lateral genomics, *Trends Biochem. Sci.* **24**, M5–M8.

Eberhard, W.G. (1989) Why do bacterial plasmids carry some genes and not others? *Plasmid* **21**, 167–174.

Eberhard, W.G. (1990) Evolution in bacterial plasmids and levels of selection, *Quart. Rev. Biol.* **65**, 3–18.

Ellstrand, N.C. (2003) Current knowledge of gene flow in plants: implications for transgene flow, *Phil. Tran. Roy. Soc. Lon. B* **358**, 1163–1170.

Gelvin, S.B. (2003) Agrobacterium-mediated plant transformation: the biology behind the "gene-jockeying" tool, *Microbiol. Mol. Biol. Rev.* **67**, 16–37.

Grohmann, E., Muth, G., Espinosa, M. (2003) Conjugative plasmid transfer in gram-positive bacteria, *Microbiol. Mol. Biol. Rev.* **67**, 277–301.

Heinemann, J.A. (1991) Genetics of gene transfer between species, *Trends Genet.* **7**, 181–185.

Heinemann, J.A. (1999) How antibiotics cause antibiotic resistance, *Drug Discov. Today* **4**, 72–79.

Heinemann, J.A., Roughan, P.D. (2000) New hypotheses on the material nature of horizontally transferred genes, *Ann. N. Y. Acad. Sci.* **906**, 169–186.

Heinemann, J.A., Silby, M.W. (2003) Horizontal Gene Transfer and the Selection of Antibiotic Resistance, in: Amábile-Cuevas, C.F. (Ed.) *Multiple Drug Resistant Bacteria*, Horizon Scientific Press, Wymondham, Norfolk, pp. 161–178.

Kurland, C.G., Canback, B., Berg, O.G. (2003) Horizontal gene transfer: A critical view, *Proc. Natl. Acad. Sci. U.S.A.* **100**, 9658–9662.

Lawrence, J.G., Ochman, H. (2002) Reconciling the many faces of lateral gene transfer, *Trends Microbiol.* **10**, 1–4.

Martin, W. (2003) Gene transfer from organelles to the nucleus: frequent and in big chunks, *Proc. Natl. Acad. Sci. U.S.A.* **100**, 8612–8614.

Ragan, M.A. (2002) Reconciling the many faces of lateral gene transfer. Response from Ragan, *Trends Microbiol.* **10**, 4.

Schaecter, M. (Ed.) (2004) *The Desk Encyclopedia of Microbiology*, Elsevier Academic Press, San Diego, CA and London.

Souza, V., Eguiarte, L.E. (1997) Bacteria gone native vs. bacteria gone awry?: Plasmid transfer and bacterial evolution, *Proc. Natl. Acad. Sci. U.S.A.* **94**, 5501–5503.

Stewart Jr., C.N., Halfhill, M.D., Warwick, S.I. (2003) Transgene introgression from genetically modified crops to their wild relatives, *Nat. Rev. Genet.* **4**, 806–817.

Syvanen, M. Kado, C.I. (Eds.) (2002) *Horizontal Gene Transfer*, Academic Press, San Diego, CA.

Woolhouse, M.E.J., Taylor, L.H., Haydon, D.T. (2001) Population biology of multihost pathogens, *Science* **292**, 1109–1112.

Zhaxybayeva, O., Lapierre, P., Gogarten, J.P. (2004) Genome mosaicism and organismal lineages, *Trends Genet.* **20**, 254–260.

Primary Literature

Adler, L.S. (2000) Alkaloid uptake increases fitness in a hemiparasitic plant via reduced herbivory and increased Pollination, *Am. Nat.* **156**, 92–99.

Arber, W. (2003) Elements for a theory of molecular evolution, *Gene* **317**, 3–11.

Avery, O.T., MacLeod, C.M., McCarty, M. (1944) Studies on the chemical nature of the substance inducing transformation of Pneumococcal types, *J. Exp. Med.* **79**, 137–158.

Balsalobre, L., Ferrandiz, M.J., Linares, J., Tubau, F., de la Campa, A.G. (2003) Viridans group Streptococci are donors in horizontal transfer of topoisomerase IV genes to Streptococcus pneumoniae, *Antimicrob. Agents Chemother.* **47**, 2072–2081.

Bernstein, H., Byerly, H.C., Hopf, F.A., Michod, R.E. (1985) Genetic damage, mutation, and the evolution of sex, *Science* **229**, 1277–1281.

Carrera, M.R.A., Kaufmann, G.F., Mee, J.M., Meijler, M.M., Koob, G.F., Janda, K.D. (2004) Treating cocaine addiction with viruses, *Proc. Natl. Acad. Sci. U.S.A.* **101**, 10416–10421.

Citovsky, V., Zambryski, P. (1993) Transport of nucleic acids through membrane channels: Snaking through small holes, *Annu. Rev. Microbiol.* **47**, 167–197.

Claverys, J.-P., Martin, B. (2003) Bacterial 'competence' genes: signatures of active transformation, or only remnants? *Trends Microbiol.* **11**, 161–165.

Claverys, J.-P., Prudhomme, M., Mortier-Barriere, I., Martin, B. (2000) Adaptation to the environment: Streptococcus pneumoniae, a paradigm for recombination-mediated genetic plasticity? *Mol. Microbiol.* **35**, 251–259.

Condit, R. (1990) The evolution of transposable elements: conditions for establishment in bacterial populations, *Evolution* **44**, 347–359.

Cooper, T.F., Heinemann, J.A. (2000) Postsegregational killing does not increase plasmid stability but acts to mediate the exclusion of competing plasmids, *Proc. Natl. Acad. Sci. U.S.A.* **97**, 12543–12648.

Cupples, C.G., Cabrera, M., Cruz, C., Miller, J.H. (1990) A set of lacZ mutations in Escherichia coli that allow rapid detection of specific frameshift mutations, *Genetics* **125**, 275–280.

Davis, C.C., Wurdack, K.J. (2004) Host-to-parasite gene transfer in flowering plants: phylogenetic evidence from Malpighiales, *Science* **305**, 676–678.

Denamur, E., Lecointre, G., Darlu, P., Tenaillon, O., Acquviva, C., Sayada, C., Sunjevaric, I., Rothstein, R., Elion, J., Taddei, F., Radman, M., Matic, I. (2000) Evolutionary implications of the frequent horizontal transfer of mismatch repair genes, *Cell* **103**, 711–721.

Doerfler, W., Schubbert, R., Heller, H., Kammer, C., Hilger-Eversheim, K., Knoblauch, M., Remus, R. (1997) Integration of foreign DNA and its consequences in mammalian systems, *Trends Biotechnol.* **15**, 297–301.

Doucet-Populaire, F., Trieu-Cuot, P., Andremont, A., Courvalin, P. (1992) Conjugal transfer of plasmid DNA from Enterococcus faecalis to Escherichia coli in digestive tracts of gnotobiotic mice, *Antimicrob. Agents Chemother.* **36**, 502–504.

Doucet-Populaire, F., Trieu-Cuot, P., Dosbaa, I., Andremont, A., Courvalin, P. (1991) Inducible transfer of conjugative transposon Tn1545 from *Enterococcus faecalis* to *Listeria monocytogenes* in the digestive tracts of gnotobiotic mice, *Antimicrob. Agents Chemother.* **35**, 185–187.

Eckardt, N.A. (2001) A sense of self: the role of DNA sequence elimination in allopolyploidization, *Plant Cell* **13**, 1699–1704.

Evans, E., Alani, E. (2000) Roles for mismatch repair factors in regulating genetic recombination, *Mol. Cell. Biol.* **20**, 7839–7844.

Feil, E.J., Maynard Smith, J., Enright, M.C., Spratt, B.G. (2000) Estimating recombinational parameters in Streptococcus pneumoniae from multilocus sequence typing data, *Genetics* **154**, 1439–1450.

Ferguson, G.C., Heinemann, J.A., Kennedy, M.A. (2002) Gene Transfer between Salmonella enterica Serovar Typhimurium inside epithelial Cells, *J. Bacteriol.* **184**, 2235–2242.

Filee, J., Forterre, P., Laurent, J. (2003) The role played by viruses in the evolution of their hosts: a view based on informational protein phylogenies, *Res. Microbiol.* **154**, 237–243.

Funchain, P., Yeung, A., Stewart, J., Clendenin, W.M., Miller, J.H. (2001) Amplification of mutator cells in a population as a result of horizontal gene transfer, *J. Bacteriol.* **183**, 3737–3741.

Gerdes, K., Rasmussen, P.B., Molin, S. (1986) Unique type of plasmid maintenance function: postsegregational killing of plasmid-free cells, *Proc. Natl. Acad. Sci. U.S.A.* **83**, 3116–3120.

Gibbs, M.J., Weiller, G.F. (1999) Evidence that a plant virus switched hosts to infect a vertebrate and then recombined with a vertebrate-infecting virus, *Proc. Natl. Acad. Sci. U.S.A.* **96**, 8022–8027.

Grillot-Courvalin, C., Goussard, S., Huetz, F., Ojcius, D.M., Courvalin, P. (1998) Functional gene transfer from intracellular bacteria to mammalian cells, *Nat. Biotechnol.* **16**, 862–866.

Hayes, W. (1952) Recombination in *Bact. coli* K 12: Unidirectional transfer of genetic material, *Nature* **169**, 118–119.

Haygood, R., Ives, A.R., Andow, D.A. (2003) Consequences of recurrent gene flow from crops to wild relatives, *Phil. Tran. Roy. Soc. Lon. B* **270**, 1879–1886.

Heinemann, J.A. (1999) Genetic evidence for protein transfer during bacterial conjugation, *Plasmid* **41**, 240–247.

Heinemann, J.A., Sprague Jr., G.F. (1989) Bacterial conjugative plasmids mobilize DNA transfer between bacteria and yeast, *Nature* **340**, 205–209.

Heinemann, J.A., Traavik, T. (2004) Problems in monitoring horizontal gene transfer in field trials of transgenic plants, *Nat. Biotechnol.* **22**, 1105–1109.

Houck, M.A., Clark, J.B., Peterson, K.R., Kidwell, M.G. (1991) Possible horizontal transfer of Drosophila genes by the mite Proctolaelaps regalis, *Science* **253**, 1125–1128.

Hughes, A.L., Friedman, R. (2003) Genome-wide survey for genes horizontally transferred from cellular organisms to baculoviruses, *Mol. Biol. Evol.*, **Advance Access April 25, 2003**, http://mbe.oupjournals.org/cgi/reprint/msg107v101.

Humayun, M.Z. (1998) SOS and Mayday: multiple inducible mutagenic pathways in Escherichia coli, *Mol. Microbiol.* **30**, 905–910.

Jacobs, H.T. (1991) Structural similarities between a mitochondrially encoded polypeptide and a family of prokaryotic respiratory toxins involved in plasmid maintenance suggest a novel mechanism for the evolutionary maintenance of mitochondrial DNA, *J. Mol. Biol.* **32**, 333–339.

Jain, R., Rivera, M.C., Lake, J.A. (1999) Horizontal gene transfer among genomes: the complexity hypothesis, *Proc. Natl. Acad. Sci. U.S.A.* **96**, 3801–3806.

Ke, C., Tsai, J.H. (1985) Transmission of huanglungbin agent from citrus to periwinkle by dodder, *Annu. Rev. Phytopathol.* **75**, 1324–1324.

Keeling, P.J., Palmer, J.D. (2001) Lateral transfer at the gene and subgenic levels in the evolution of eukaryotic enolase, *Proc. Natl. Acad. Sci. U.S.A.* **98**, 10745–10750.

Kunik, T., Tzfira, T., Kapulnik, Y., Gafni, Y., Dingwall, C., Citovsky, V. (2001) Genetic transformation of HeLa cells by Agrobacterium, *Proc. Natl. Acad. Sci. U.S.A.* **98**, 1871–1876.

Luria, S.E., Delbrück, M. (1943) Mutations of bacteria from virus sensitivity to virus resistance, *Genetics* **28**, 491–511.

Matic, I., Rayssiguier, C., Radman, M. (1995) Interspecies gene exchange in bacteria: the role of SOS and mismatch repair systems in evolution of species, *Cell* **80**, 507–515.

McClure, M.A. (2001) Evolution of the DUT gene: horizontal transfer between host and pathogen in all three domains of life, *Curr. Protein Pept. Sci.* **2**, 313–324.

Mihaylova, V.T., Bindra, R.S., Yuan, J., Campisi, D., Narayanan, L., Jensen, R., Giordano, F., Johnson, R.S., Rockwell, S., Glazer, P.M. (2003) Decreased expression of the DNA mismatch repair Gene Mlh1 under hypoxic stress in mammalian cells, *Mol. Cell. Biol.* **23**, 3265–3273.

Naito, T., Kusano, K., Kobayashi, I. (1995) Selfish behavior of restriction-modification systems, *Science* **267**, 897–899.

Nakamura, Y., Itoh, T., Matsuda, H., Gojobori, T. (2004) Biased biological functions of horizontally transferred genes in prokaryotic genomes, *Nat. Genet.* **36**, 760–766.

Nicolaisen, M. (2004) Transmission of branch-inducing phytoplasma PoiBI from poinsettia (Euphorbia pulcherrima) to crown-of-thorns (E-milii), *HortScience* **38**, 551–552.

Nitz, N., Gomes, C., de Cássia Rosa, A., D'Souza-Ault, M.R., Moreno, F., Lauria-Pires, L., Nascimento, R.J., Teixeira, A.R.L. (2004) Heritable integration of kDNA minicircle sequences from Trypanosoma cruzi into the avian genome: insights into human Chagas Disease, *Cell* **118**, 175–186.

O'Callaghan, D., Cazevieille, C., Allardet-Servent, A., Boschiroli, M.L., Bourg, G., Foulongne, V., Frutos, P., Kulakav, U., Ramuz, M. (1999) A homologue of the Agrobacterium tumefaciens VirB and Bordetella pertussis Ptl type IV secretion systems is essential for intracellular survival of Brucella suis, *Mol. Microbiol.* **33**, 1210–1220.

Rayssiguier, C., Thaler, D.S., Radman, M. (1989) The barrier to recombination between *Escherichia coli* and *Salmonella typhimurium* is disrupted in mismatch-repair mutants, *Nature* **342**, 396–401.

Resources, Bo.Aa.N., Agriculture, Bo., Sciences, Bo.L. (2004) *Biological Confinement of Genetically Engineered Organisms*, The National Academies Press, Washington, DC.

Robertson, H.M., Soto-Adames, F.N., Walden, K.K.O., Avancini, R.M.P., Lampe, D.J. (2002) The Mariner Transposons of Animals: Horizontally Mobile Jumping Genes, in: Syvanen, M., Kado, C.I. (Eds.) *Horizontal Gene Transfer*, Academic Press, San Diego, CA, pp. 173–185.

Shafer, K.S., Hanekamp, T., White, K.H., Thorsness, P.E. (1999) Mechanisms of mitochondrial DNA escape to the nucleus in the yeast Saccharomyces cerevisiae, *Curr. Genet.* **36**, 183–194.

Snow, A.A. (2002) Transgenic crops – why gene flow matters, *Nat. Biotechnol.* **20**, 542.

Stroun, M., Anker, P. (1971) Bacterial nucleic acid synthesis in plants following bacterial contact, *Mol. Gen. Genet.* **113**, 92–98.

Stroun, M., Anker, P. (1973) Transcription of spontaneously released bacterial deoxyribonucleic acid in frog auricles, *J. Bacteriol.* **114**, 114–120.

Stroun, M., Anker, P., Auderset, G. (1970) Natural release of nucleic acids from bacteria into plant cells, *Nature* **227**, 607–608.

Stroun, M., Anker, P., Maurice, P., Gahan, P. (1977) Circulating nucleic acids in higher organisms, *Int. Rev. Cytol.* **51**, 1–48.

Thomashow, M.F., Nutter, R., Postle, K., Chilton, M.-D., Blattner, F.R., Powell, A., Gordon, M.P., Nester, E.W. (1980) Recombination between higher plant DNA and the Ti plasmid of Agrobacterium tumefaciens, *Proc. Natl. Acad. Sci. U.S.A.* **77**, 6448–6452.

Tsen, S.-D., Fang, S.-S., Chen, M.-J., Chien, J.-Y., Lee, C.-C., Tsen, D.H.-L. (2002) Natural plasmid transformation in Escherichia coli, *J. Biomed. Sci.* **9**, 246–252.

Waters, V.L. (2001) Conjugation between bacterial and mammalian cells, *Nat. Genet.* **29**, 375–376.

Whitman, W.B., Coleman, D.C., Wiebe, W.J. (1998) Prokaryotes: the unseen majority, *Proc. Natl. Acad. Sci. U.S.A.* **95**, 6578–6583.

Yadav, N.S., Postle, K., Saiki, R.K., Thomashow, M.F., Chilton, M.-D. (1980) T-DNA of a crown gall teratoma is covalently joined to host plant DNA, *Nature* **287**, 458–461.

12
Molecular Systematics and Evolution

Jeffrey H. Schwartz
University of Pittsburgh, Pittsburgh, PA, USA

1	The Beginning of Molecular Systematics	362
2	The Molecular Assumption	363
3	DNA Hybridization	366
4	Mitochondrial DNA	370
5	DNA Sequences	371
6	"Evo-devo"	375
7	Positional Information and Shape	377
8	"Mutation"	378
9	Toward a Theory of Evolutionary Change	379
10	Molecules and Systematics: Looking Toward the Future	380

Bibliography   382
  Books and Reviews   382
  Primary Sources   382

Keywords

African Apes
Chimpanzees, bonobos, and gorillas.

Anthropoidea
The so-called higher primates; the subordinal rank that subsumes New and Old World monkeys plus the hominoids.

Clade
A group of related organisms; a group of organisms united by common ancestry.

Derived Feature
A relative character state determined by how restrictively shared a feature under study is.

Hominids
Humans and their fossil relatives.

Hominoids
The group of anthropoid primates that includes gibbons and siamangs, chimpanzees, gorillas, orangutans, and humans.

Homology
Similarity due to common ancestry.

Large-bodied Hominoids
Orangutans, humans, and African apes.

Lesser (Small-bodied) Hominoids
The gibbons and siamangs.

Monophyletic Group
A clade.

Outgroup
A taxon that is outside of (a sister taxon of) the focal group of phylogenetic reconstruction.

Phylogeny Reconstruction
The generation of a theories of evolutionary relationships among taxa.

Primitive Feature
A relative character state determined by how broadly shared a feature under study is.

Primitive Retention
A feature that is shared by the members of a group by inheritance from a common ancestor.

Sister Taxon
A taxon that is the closest relative of another.

Systematics
The study of the pattern and relationships of organisms, including the identification of species and delineation of groups of taxa.

Taxic
Relating to taxa.

Taxon (Plural: Taxa)
Any taxonomic rank, for example, species, genus, family, order, and class.

Studies that rely on genetic and molecular information to address evolutionary questions fall roughly into two different categories: the reconstruction of the evolutionary relationships of organisms (including the times of divergence of groups or lineages), and the formation and emergence of morphological novelties that distinguish or characterize different organisms. The former endeavor is sometimes referred to as *molecular systematics*, and when it is applied to primates and especially the relatedness of humans and apes, "molecular anthropology." The debate over the regularity of molecular change, resulting in a molecular clock, lies within the realm of molecular systematics, and especially that of molecular anthropology. In the past decade, in particular, the primary focus of molecular systematists has been DNA sequences, both nuclear and mitochondrial. Throughout, the assumption has been that the degree of similarity reflects the degree of evolutionary relatedness, because differences accrue once lineages have diverged from a common ancestor. The popularity of molecular systematics in recent years is also predicated on the notion of "a law of large numbers," that is, the thousands of bases that produce DNA sequences. The question is whether these assumptions are sustainable.

The rather newly defined discipline of evolution and development, or "evo-devo" as it has been nicknamed, is less involved in the interpretation of molecular data for purposes of reconstructing evolutionary relationships than it is in trying to identify molecular elements – whether they be transcription factors or other kinds or classes of proteins, as well as "genes" themselves – that are relevant to, or participate in, the processes of development of structure and form. Here, what is important is not a particular RNA, DNA, or amino acid sequence, but the signal transduction pathways or sequences of communication between different molecules in the regulation of development and the origin of structure. Sometimes, hypotheses of when and at what level within groups of organisms specific features emerged are generated as a result of overlaying this information on a presumed theory of relationships of the organisms under consideration. It appears that insights from developmental genetics will prove to be more useful than sequence data alone for systematic and phylogenetic inquiry.

This entry will attempt to summarize the main aspects of these fields of inquiry, including their underlying assumptions, and suggest possible avenues for future research that might bring these disciplines together in light of new ways of thinking about evolution.

1
The Beginning of Molecular Systematics

The first inquiries into systematics and phylogenetic reconstruction by investigation of the "blood relationship" of organisms can be traced to a few individuals, of whom the best known is George H. F. Nuttall. In the research that preceded his monograph of 1904, *Blood Immunity and Blood Relationship: A Demonstration of Certain Blood-Relationships Amongst Animals by Means of the Precipitin Test for Blood*, Nuttall sought to demonstrate that the degree of similarity between animals in their blood serum proteins was a reflection of their evolutionary closeness. The idea was simple: produce an antiserum (antibodies or antigens) to blood serum proteins of one animal, mix the antiserum with blood serum proteins of another animal, and measure the amount of precipitin that settled out. The more profound the precipitation, the greater the similarity (because of a greater number of antibody–antigen binding sites), and, consequently, the closer the evolutionary relationship between the two organisms being compared.

The strength of the antiserum played an important role in determining just how many species might belong to the same group. A weak antiserum would provoke reactivity in the sera of only the most closely related of organisms, while a stronger solution would produce a precipitate when combined with the sera of a greater number of related animals. The more distantly related the groups being compared, the less the reactivity between serum and antiserum. In addition to strength of reactivity, Nuttall thought that the reaction rate was also an indicator of closeness of relatedness. In the end, he decided that "the zoological relationships between animals are best demonstrated by means of powerful antisera." Consequently, he concluded, "if we accept *the degree of blood reaction as an index of the degree of blood relationship* within the Anthropoidea [the so-called higher primates, New and Old World monkeys plus the hominoids, which includes gibbons, chimpanzees, gorillas, orangutans, and humans], then we find that the Old World apes are more closely allied to man than are the New World apes, and this is exactly in accordance with the opinion expressed by Darwin."

Although Nuttall believed that his approach accurately revealed the evolutionary relationships of animals, with the exception of reference in 1922 by the paleontologist, W. K. Gregory, to Nuttall having demonstrated "the anthropoid heritage of man," his work went largely unacknowledged until the 1960s, when various publications claimed to have demonstrated the evolutionary relationships of organisms through study of elements of their biochemistry.

2
The Molecular Assumption

Perhaps the most influential of these publications was by Emile Zuckerkandl and Linus Pauling on fetal and adult hemoglobin, which included a comparison of sequences between human, gorilla, horse, and fish. Although one might question the comparison of gill- with lung-bearing animals, the endeavor demonstrated that the fish was more dissimilar to humans than to the horse, while the gorilla was the most similar to humans in hemoglobin sequence. Since this pattern of taxic distance mirrored the morphologically accepted scheme of evolutionary relatedness of these animals, Zuckerkandl and Pauling suggested that their observations "can be understood at once if it is *assumed* [emphasis added] that in the course of time the hemoglobin-chain genes duplicate, [and] that the descendants of the duplicate genes 'mutate away' from each other." From this assumption, they felt justified in proposing the following: "[O]ver-all similarity must be an expression of evolutionary history," with descendants "mutating away" and becoming "gradually more different from each other." In other words, the more time that elapses after a lineage's divergence, the greater the molecular difference its succession of descendants would accumulate. Consequently, evolutionary closeness became synonymous with molecular difference. Also assumed in this nexus of assumption is the notion of constant and gradual unilinear change.

There is, however, nothing in Zuckerkandl and Pauling's work that justifies the notion that "overall similarity" is "an expression of evolutionary history." They merely assumed such a correspondence, just as they assumed that difference also meant change and the same kind of change, which, in turn, was achieved at a constant and gradual rate. However, the discovery of a hundred nucleotide differences between two taxa does not translate into the rate at which presumed substitutions occurred. One needs additional information, such as a suggestion of the time at which these differences began to accumulate. And then one has to assume that the rate of substitution over this period of time was gradual and that differences did not arise during a few concentrated phases of replacement.

Nevertheless, Zuckerkandl and Pauling's effort resulted in the "molecular assumption" – continual molecular change and its continual accumulation over evolutionary time – which would thereafter become the foundation of molecular systematics. As Adalgisa Caccone and Jeffrey Powell would write in 1989: "Virtually all molecular phylogenetic studies...have a major underlying assumption: the genetic similarity or difference among taxa is an indication of phylogenetic relatedness. Lineages that diverged more recently will be genetically more similar to one another than will be lineages with more ancient splits." If one embraces the assumption, the rest may follow logically, but not necessarily because of biological demonstration. One of the major extensions of the "molecular assumption" is the notion of a "molecular clock," the existence of which was promoted initially by Vincent Sarich and Allan Wilson.

Beginning with their first major paper in 1966, Sarich and Wilson sought to elucidate the evolutionary relationships of primates, and especially of humans and the large-bodied apes, using the technique of microcomplement fixation (MC'F), which requires only minute amounts of serum and antiserum for study of immunological reactivity. Their molecule of choice was the larger blood serum protein, albumin. The degree of reactivity achieved between albumin and anti-albumin was translated into an "index of dissimilarity," which was subsequently referred to as *immunological distance* (ID). The closer an ID value was to 1.0, the greater was the overall molecular similarity and thus, given an assumed equivalence between ID and evolutionary relationship, the presumption of closeness of relatedness between the organisms under study. As an apparent check on the validity of this approach, Sarich and Wilson conceived of the "test of reciprocity." For example, if chimpanzee serum was cross-reacted with antihuman serum (produced by injecting another animal, for instance, a rabbit or chicken, with human serum) the first time around, human serum was cross-reacted with antichimpanzee serum the next time to see if the resultant ID values were reasonably similar. Predictably, the test of reciprocity usually confirmed the initial immunological finding.

The supposed evolutionary arrangement of the primates that Sarich and Wilson achieved was more consistent with the general pattern of primate relationships based on comparative morphology than it was in disagreement. Because of this concordance, they concluded that "the MC'F data are in qualitative agreement with the anatomical evidence, on the basis of which the apes, Old World monkeys, New World monkeys, prosimians, and nonprimates are placed in taxa which form a series of decreasing genetic relationship to man." In 1967, in their second article on this topic, Sarich and Wilson argued that molecular change must have occurred at a constant rate among all major groups of primates. In the third and last article of this series, which was also published in 1967, they concluded that the small amount of difference they detected between hominoids in their albumin reflected little molecular change and that this, in turn, implied that little time had elapsed since the separation of the hominoid lineages. Since, as they claimed, molecular change ticked away at a constant rate, IDs represented a "molecular clock" that could reveal the times at which the various hominoids – actually any species – diverged and went off on their own evolutionary paths.

After calibrating the molecular clock on the basis of paleontologists' interpretation, based on fossils, of when the Old World monkey and hominoid lineages might have separated, Sarich and Wilson calculated divergence dates of about 10 million years ago for the gibbon lineage, eight million years ago for the orangutan lineage, and a mere five million years ago for the human and African ape lineages. The major implication of this calculation was that fossils older than five million years, such as *Ramapithecus* (now referred to the genus *Sivapithecus*) from c. 12- to 14-million-year old deposits in Indo-Pakistan, could not, as paleontologists concluded from study of morphology, be hominid.

However, while molecular systematists were in general agreement on the premise of the molecular assumption – that degrees of similarity reflect degrees of evolutionary closeness – not all of them embraced the notion of a molecular clock that "ticked" at a constant rate. Among the most vocal dissenters was Morris Goodman, who in the early 1960s was the leading proponent of Nuttall's work. Although Goodman went beyond claiming that molecular similarity indicated a very close relationship between humans and the African apes to advocating that these three hominoids should also be placed in the same taxonomic family, Hominidae (a position that for years caused many comparative primate anatomists and paleontologists to be skeptical of molecular systematics in general), he nonetheless tried to accommodate the paleontologists' identification of pre-five-million-year old fossils as being hominid.

In contrast to Sarich (who in 1971 was the first molecular systematist to reject morphology as being phylogenetically informative), Goodman accepted the paleontologically derived date of at least 14 million years as the time of divergence between hominids and their potential ape relatives. As a consequence, though, Goodman then had to interpret the immunological and biochemical data in a way that would make them compatible with a deep timescale of hominoid diversification. He did so by constructing a selectionist argument to explain the apparent acceleration or deceleration in rates of molecular change that *de facto* must have occurred, given the paleontologically established dates for the earliest appearances of each of the large-bodied hominoids. Thus, for instance, if the various hominoid lineages had indeed originated between 18 and 14 million years ago, the extreme similarity between the various extant hominoids in their albumin had to have resulted from a slowdown in the rate of molecular change of that particular blood serum protein.

Goodman proposed that the inferred deceleration in rate of molecular change in large-bodied hominoid albumin was related to the fact that, although they are similar to other anthropoid primates in developing hemochorial placentation (the most intimate of placental modes in the approximation of maternal and fetal blood systems), large-bodied hominoids differ from other anthropoids in having very long gestation periods. He argued that in animals with hemochorial placentation but fairly short gestation periods, there would be enough slack in the system to allow unimpeded molecular change to occur and accrue. The fetus would be born before the mother's immune system could produce antibodies to it. A large-bodied hominoid's gestation period, however, would be long enough not only for maternal-fetal immunological incompatibility to build up but also for maternally produced antibodies to diffuse through the placenta, with deleterious consequences to the fetus.

The dilemma which then had to be confronted was: How could an animal have both hemochorial placentation and a long gestation period? Goodman's answer was that natural selection must have acted to reduce the possibility of immune responses from the mother toward her fetus's proteins. Consequently, there had to have been a slowdown of molecular change that was commensurate with a prolongation of the gestation period. Although albumin had been the basis of Sarich and Wilson's constant-rate molecular clock, Goodman used the same molecule as the exemplar of how the molecular clock could run at different rates.

Concurrent with the various approaches – immunological reactivity as well as gel-column electrophoretic separation – that were brought to bear during the 1960s and 1970s on the determination of evolutionary relationships was an increased interest in protein sequences. The first major study was Zuckerkandl and Pauling's 1962 analysis of hemoglobin, and, for some years thereafter, hemoglobin was the most intensely studied molecule. Beginning in 1973, however, a research group headed by A. E. Romero-Herrera analyzed myoglobin sequences across a number of primarily mammalian taxa. As more taxa were added to the study, the resultant, most "parsimonious" arrangement of the hominoids placed the gibbon rather than the orangutan closer to a human-African clade. Only by arguing for a more complicated scheme of myoglobin "evolution" in hominoids – including, for instance, the condition that there had to have been various "back-mutations" to a state similar to the presumed unchanged state – could one arrive at the more commonly accepted arrangement of divergences: gibbon first, then the orangutan, and then a human-African ape group.

But while protein sequencing was seen by some, especially Zuckerkandl as late as 1987, as the molecular level on which to focus for purposes of determining evolutionary relationships (primarily because amino acid sequences are less subject to the problems with DNA, such as insertions, deletions, and back-mutations), there was a growing consensus that the best systematic information lay at an even deeper "genetic" level. Since base differences at the third position of a codon (a nucleotide triplet) did not yield different resultant amino acids, the concern of DNA sequence advocates was that demonstration of similarity in protein sequences could produce "false" phylogenies because the underlying nucleotide sequences themselves may be different. Consequently, it became imperative to get to the level of DNA – nuclear DNA – itself, especially because the general belief or at least expectation (which was based on bacteria and inferred for metazoans) was the existence of a direct correlation between specific DNA sequences and specific genes. And, as genes were conceptualized, there was also supposed to be correspondence between one or perhaps a few genes and a specific feature or structure. In addition, because DNA sequences were composed of thousands upon thousands of bases, the apparent massive scale of the comparison had the appeal of providing an overwhelming amount of phylogenetically relevant information.

3
DNA Hybridization

Although DNA sequencing was possible by the late 1970s, it was an expensive and laborious procedure, which made the endeavor itself, much less the comparison of DNA sequences for phylogenetic purposes, prohibitive. There was, however, another way in which DNA sequence information could be achieved: DNA–DNA hybridization.

The theory behind using DNA–DNA hybridization (or just DNA hybridization) as an approximation of overall similarity in DNA sequences between different taxa lay in the fact that if the two strands that form the helical structure of nuclear DNA were dissociated (which could be accomplished by subjecting it to heat), their complementary bases would rebond and the two strands would reform their original helical organization. However, it is not only two cleaved but also originally helically arranged DNA strands from the same individual that will reassociate (reanneal). Analogous to the geneticist Theodosius Dobzhansky's discovery earlier in the twentieth century with chromosomes of different lengths from different species of fruit fly, any two strands of DNA, from different individuals of even different species, will attempt to anneal.

As in Dobzhansky's fruit-fly experiments, in which the larger chromosome would loop and fold so that whichever of its loci were also present in the shorter one would match up, single strands of DNA (derived from heat splitting or melting of a double helix) from different organisms would attempt to recombine or hybridize at complementary base positions. The more complementarity there was between hybridized DNA strands, the greater was the intensity of heat needed to break down the bonds holding them together. Consequently, it seemed logical to conclude that the heat (ΔT) it took to disassociate a DNA–DNA hybrid (that is, the more thermally stable the hybrid was), the more similar the annealed strands were in their nucleotide sequences. From this apparent demonstration, one could then invoke the molecular assumption to explain how higher ΔT's, reflecting greater molecular similarity, were also a reflection of closer evolutionary relatedness. As logical, though, as this thought experiment might appear to be, it relies on an assessment of overall similarity, which does not identify the bases or sequence positions that underlie differences in ΔT's. Conversely, similar ΔT's in different taxa might not be the result of the same regions hybridizing.

One of the first applications of the technique of DNA hybridization to evolutionary questions was published in 1968 by R. J. Britten and D. E. Kohne, who were primarily concerned with learning more about the genome. They discovered that the genomes of higher organisms – but not of bacteria or viruses – contained "hundreds of thousands of copies of DNA sequences." However, not only did repeated DNA sequences represent a considerable portion of a genome, Britten and Kohne also found that they were "trivial and permanently inert." Only a small fraction of a genome was composed of unrepeated, unique, single-copy DNA sequences, which apparently constituted the active or functional elements of that genome.

Britten and Kohne also speculated on pathways or mechanisms that might produce genomic change, which they argued must be considered on two different levels: change in nucleotide sequence, and the origin of new families of nucleotide sequences. As Zuckerkandl and Pauling had assumed for differences in hemoglobin, Britten and Kohne proposed that changes in nucleotide sequences – which would be identified as point mutations – occur slowly over time. However, this model would explain only the "divergence of pre-existing families" of nucleotide sequences, not the introduction of new families of sequences, as was assumed would eventually happen in the gradual-accumulation-of-change model that dominated molecular systematics. The introduction of new families of sequences, Britten and Kohne alternatively suggested, must "result from relatively sudden events," called *saltatory replications*. Accordingly, "saltatory replications of genes or gene fragments occurring at infrequent intervals during geologic history might have profound and perhaps delayed results on the course of evolution." Although Britten and Kohne's distinctions between types of molecular change were not at the time embraced by molecular systematists, their suggestion of "saltatory replication" would certainly appear to have been borne out with the subsequent identification of vertebrate regulatory genes involved in segmentation – *Hox-a*, *Hox-b*, *Hox-c*, and *Hox-d* – as replicates of the orthologous *Antennapedia* gene identified in fruit flies.

One of the early attempts at using DNA hybridization to reconstruct relationships among the primates was published in 1976 by Raoul Benveniste and George Todaro, whose primary focus was actually on the distribution among mammals of the type C viral gene, the presence of which, they discovered, was particularly characteristic of animals found in Asia, including the orangutan and the gibbon, but which, unexpectedly, was also present in humans. After reviewing these data, Benveniste and Todaro turned to DNA hybridization for sorting out the relationships of various primates. Relying on previous studies, they estimated "the effect of mismatched base-pairs on thermal stability" as being "between 0.7 and 1.7 °C per 1% altered pairs." Since, depending on the animal, nuclear DNA consists of anywhere from 10 to 100 million nucleotide pairs, every 1% of difference in nucleotide sequence between DNA hybrids would result in a drop of about 1.0 °C of the heat necessary to melt the hybrid molecule.

Although Benveniste and Todaro calculated a 1.1% sequence difference between humans and both African apes and a 2.4% difference between humans and the orangutan, their published ranges of ΔT's actually demonstrate overlap between all large-bodied hominoids (although one does not know the identity of nucleotide difference and similarity). Taking into consideration the fact that presumed-to-be-homologous DNA sequences are not the same length (that is, they do not consist of the same number of nucleotides) in all animals, including the large-bodied hominoids (see below for human and orangutan), one also does not know from DNA hybridization experiments the details of how a short sequence from one species aligns with a longer sequence of another species.

Probably because DNA hybridization seemingly provided a way of getting closer to the supposedly most "informative" genetic level, questions about what, exactly, similarity in ΔT's meant were not explored. But other questions were.

In the 1980s, Charles Sibley and Jon Ahlquist reviewed earlier DNA hybridization experiments and concluded that a major problem with them all was that only a limited number of cross hybridizations had been tested. This, Sibley and Ahlquist claimed, was the reason previous studies had such difficulty in refining certain evolutionary relationships, particularly those between humans and the African apes. Although there was not particular morphological support for this theory of relationship, molecular systematists often grouped these three hominoids together. However, the level of resolution in the majority of the molecular work was never fine enough to determine which two of the three hominoids were the most similar (and thus presumably the most closely related).

Sibley and Ahlquist felt another source of error in these earlier studies resulted from using all of an animal's DNA, repeated as well as single-copy, to form hybrids. They argued that it was important to remove all repeated DNA – since it was redundant – and to work only with single-copy DNA.

Before Sibley and Ahlquist decided to tackle the evolutionary relationships of the large-bodied hominoids, they had spent years applying their DNA hybridization technique to the phylogeny of birds, through which they also developed the notion of a uniform average rate of genomic change (UAR). UARs, they believed, characterized the nature of molecular, and thus evolutionary, change in all taxa. The notion of a UAR also seemed to negate the need for debates over constant versus irregularly running molecular clocks, which were directed at individual molecules. As they saw it, the entire genome could change at a uniform average rate, even though the rates at which individual molecules change may be quite different.

Sibley and Ahlquist argued that the power of DNA hybridization in "discovering" the phylogenetic relationships of organisms came from sampling the entire genome of an animal. They claimed that, since an organism's genome is composed of millions of nucleotides, the "law of large numbers" provides the checks and balances necessary to rule out "false" similarities or parallelisms. Accordingly, since DNA hybrid strands will only link up along those stretches of sequences that are complementary, there should be no question about homology. Conversely, sequences that did not line up were not homologous. At least that was the argument.

Sibley and Ahlquist's melting temperatures for their DNA hybrids produced numbers, which they then converted into phylogenetic distances using a procedure called *average linkage*. "Average linkage" begins "by clustering the closest pair or pairs of taxa," after which "one links the taxa which have the smallest average distance to any existing cluster," and on it goes until "all taxa are linked." The underlying assumption, which permits the linking of the most similar pairs, is that the DNA hybridization technique "measures the net divergence between the homologous nucleotide sequences of the species being compared."

Many aspects of Sibley and Ahlquist's phylogenetic reconstructions of birds were consistent with theories of relatedness derived from study of morphology. However, there were some significant differences, for instance, with regard not only to the broader phylogenetic relationships of, but also to the details of relatedness among, the flycatchers. By the 1980s, however, it was common practice, when molecular and the morphological phylogenies were in discord, to opt for the former. As Caccone and Powel argued in 1989, once the molecular assumption is accepted, overall similarity becomes the yardstick for determining closeness of relatedness. Indeed, whether it is a presumed albumin clock or a UAR, the assumption of ongoing molecular change validates the molecular assumption, which, in turn, demands that a molecular phylogeny is correct in its entirety, even if other sources of information contradict it.

Sibley and Ahlquist's "law of large numbers" and UAR appealed to most molecular systematists. This appeal was probably the primary reason paleontologists finally succumbed to the notion that molecular phylogenies had greater authority in deciphering evolutionary relationships when morphologically based theories were in conflict with them. But not all molecular systematists were convinced. Among them was Alan Templeton, who objected to the use of DNA hybridization because it did not allow one to determine the polarity of the similarity, which is necessary to address the question: "Is similarity due to distant or recent common ancestry?" For, only by studying the actual sequences of nucleotides can one determine the identity and from this attempt to infer the significance of similarity and dissimilarity. Sibley and Ahlquist countered, however, that there was "no reason to expect data derived from base sequences to improve on those from amino acid sequences, which have produced contradictory results." In their support, they also quoted A. E. Friday (who had collaborated with Romero–Herrera on myoglobin sequencing): "Phylogenetic conclusions derived from a study of nucleotide sequences will be subject to the same suspicions as those derived from amino acid sequences."

4
Mitochondrial DNA

As the field of molecular systematics was expanding its sphere of investigation during the 1970s, DNA located outside the nucleus, in a cell's mitochondria, was attracting attention. In a study published in 1979, W. Brown, M. George, and A. Wilson noted that there was more difference between humans and a sampling of Old World monkeys in their mitochondrial (mt) DNA than there was among these same primates in their nuclear DNA. By assuming that the divergence between the human and Old World monkey lineages occurred over 20 million years ago (based on morphological studies of the primate fossil record), Brown et al. concluded that the differences between these humans and Old World monkeys could be explained if nucleotide substitutions occurred at a slower pace —5 to 10 times slower – in nuclear than in mtDNA (in other words, if mtDNA "evolved" 5 to 10 times faster than nuclear DNA). On the basis of this interpretation, Brown et al. calculated that mtDNA data are the most accurate for studying evolutionary events that occurred "within the past 3–10 million years."

In subsequent publications, they proceeded to analyze the mtDNA of the large-bodied hominoids, which, largely because of Sarich and Wilson's molecular clock, were assumed to have diverged within this time period and which, because of this assumption, were thus amenable to such analysis. Although their data showed the fewest differences (interpreted as substitutions) between humans and chimps, Brown et al. concluded that humans were related to an African ape group. Given the molecular assumption, however, the data should have yielded a human–chimpanzee sister group, as Maryellen Ruvolo and a number of collaborators concluded in 1991 from their analysis of mtDNA.

In addition to the fact that mtDNA is single-, not double-stranded, the allure of using mtDNA was and continues to be twofold. One assumption is that, unlike nuclear DNA, mtDNA is supposedly inherited only through the maternal line and thus not subject to the complexities that occur through recombination of maternal and paternal DNA. The second assumption is the existence of a "hotspot" in the D-looped configuration of mtDNA and that this is the primary site of molecular activity. This region is called the *hypervariable zone* and its study has been interpreted as providing evidence of evolutionary change. Although crucial to the use of mtDNA for purposes of phylogenetic reconstruction, these assumptions are probably incorrect.

The introduction of paternal mtDNA into fertilized eggs has been reported in the literature from time to time. For example, in 1992, Allan Wilson and coworkers demonstrated that this happens in mice, as part of the sperm's tail, which does contain mtDNA, penetrates the ovum's membrane. More recently, John Maynard-Smith and coworkers calculated that there had to have been a recombination of maternal and paternal mtDNA during human evolution, and concluded that this possibility should cause systematists to reconsider the seemingly inviolable "fact" that humans and chimpanzees are closely related.

As Erika Hagelberg reviewed at great length, there is increasing indication that mtDNA is not as exempt from paternal inheritance and recombination as was initially believed. With regard to so-called hypervariable sites being regions of preferentially high mutation rates, which therefore lend themselves to phylogenetic analysis,

Hagelberg pointed out that "there is no direct evidence of hypervariability," although "most researchers believe that anomalous patterns of DNA substitution are best explained by mutation." Indeed, she writes, "because the notion of hypervariability fits with the received view of mtDNA clonality [maternal inheritance only], anomalies are seldom questioned." Hagelberg gives an interesting example. Depending on which subject's mtDNA is used, one can reconstruct "our most-recent female common ancestor...[as having] lived just 6000 years ago, a date more consistent with Biblical Eve than Mitochondrial Eve." As she cautions: "The picture is far from simple, and it is clear that extreme care must be taken in the interpretation of mtDNA phylogenetic trees in the face of possible recombination...[T]here are enough unexplained patterns in mtDNA data to warrant reassessment of the conclusions of many mtDNA studies."

5
DNA Sequences

In spite of Sibley and Ahlquist's and Friday's warnings about nucleotide sequence data not being any more reliable than protein sequence data for reconstructing phylogenetic relationships, the increasing ease with which DNA could be sequenced could not be ignored by molecular systematists. Nuclear as well as mtDNA were sequenced. Regardless of the genetic "level" under scrutiny, the interpretation of similarity or difference in DNA sequences under comparison was still predicated on the molecular assumption. The weight of DNA sequence data as being key to deciphering evolutionary relationships assumed special significance in some areas of molecular systematics because of the supposed information content of nuclear DNA in general. As Elizabeth Bruce and Francisco Ayala wrote as early as 1979 in an article on blood serum proteins: "Information macromolecules – that is, nucleic acids and proteins – document evolutionary history...[Thus] degrees of similarity in such macromolecules reflect, on the whole, degrees of phylogenetic propinquity."

Almost coincident with the rise in popularity of comparing DNA sequences for purposes of inferring evolutionary relationships came the cautionary notes. Of particular importance in this regard is the question of alignment, which, in 1991, J. A. Lake was among the first to address. As mentioned above, not all DNA sequences chosen for comparison, if homologous (e.g. representing the same "gene" or segment) are the same length. Therefore, decisions must be made with regard to how to subdivide the shorter sequence in order to align its nucleotides with those of the longer sequence. Typically, the alignment of compared sequences is presented in the literature without justification of the assumptions and decisions that produced the alignment, which, in turn, was then used as the basis for the phylogenetic analysis. But, as Lake warned in the title of his article, "[t]he order of sequence alignment can bias the selection of tree topology." In addition to assumptions that inform the decision to break up a short sequence so that its bases align with those of a longer sequence is the issue of whether, for one or another sequence, bases may have been added or inserted, removed or deleted, or one base substituted for another.

In *what it means to be 98% chimpanzee*, Jonathan Marks provides an example of these problems with DNA sequences from a human and from an orangutan to show three different ways in which their bases (C,T,G,A) can be aligned and the interpretive consequences. Even beforehand, the assumption must be made that the 40 bases in the human sequence actually have homologous counterparts in the 54 bases in the orangutan sequence.

There may be seven differences or there may be eleven differences, depending on how we decide the bases correspond to each other across the species – and that is, of course, assuming that a one-base gap is

```
HUMAN CCTCCGCCGCGCCG     CTCCGC GCCGCCGGGCA           CGGCC              CCGC
ORANG CC           GTCGCCTCCGCCACGCCGCGCCACCGGGCCGGGCCGGCCCGGCCCGCCCCGC

HUMAN      CCTCCGCCGCGCCGCT         CCGCGCCGCCGGGCACGGCCCCGC
ORANG CCGTCGCCTCCGCCACGCCGCGCCACCGGGCCGGGCCG GCCCGGCCCGCCCCGC

HUMAN CCTCCGCCGCGCCG    CT CCGCGCCGCCGGG CAC   GGCC              CCGC
ORANG           CCGTCGCCTCCGCCACGCCGCGCCACCGGGCCGGGCCGGCCCGGCCCGCCCCGC
```

As Marks comments:

"Tabulate the differences. The top one invokes five gaps and six base substitutions; the middle has only two gaps but nine base substitutions. And the bottom one has five gaps and only three base substitutions. The three pairs of sequences differ in the assumptions about which base in one species corresponds to which base in the other. While we might, by Occam's Razor, choose the alignment that invokes the fewest inferred hypothetical evolutionary events, we still have to decide whether a gap "equals" a substitution. Does the bottom one win because it has a total of only eight differences? Or might the middle one win because a gap should be considered rare and thereby "worth," say, five base substitutions?

The problem is that we cannot know which is "right," and the one we choose will contain implicit information about what evolutionary events have occurred, which will in turn affect the amount of similarity we tally. How similar is this stretch of DNA between human and orangutan? also equivalent to a five-base gap and to a base substitution.

In a more general sense, however, the problem of taking *quantitative* estimates of difference between entities that differ in *quality* is prevalent throughout the genetic comparison of human and ape. The comparison of DNA sequences presupposes that there are corresponding, homologous sequences in both species, which of course there must be if such a comparison is actually being undertaken. But other measurements have shown that a chimpanzee cell has 10% more DNA than a human cell. (this doesn't mean anything functionally, since most DNA is functionless.) But how do you work that information into the comparison, or into the 99.44% similarity [between human and chimp]?" [comment added].

These concerns have not, however, been widely appreciated by molecular systematists, especially molecular anthropologists, who not only portray the analysis of DNA sequences as neutral and objective but also use the assumption of relatedness to inform the way in which they analyze the sequences they have aligned according to certain assumptions. Exemplary in this regard is the multiple DNA sequence analysis Maryellen Ruvolo and collaborators published in 1997, in which they sought to resolve the supposed dilemma of to which African ape humans are more closely related. In order to pursue this question, they assumed first that the orangutan was the sister taxon of a clade or evolutionary group consisting of humans, the chimpanzee, and the gorilla. Consequently, the differences in the orangutan had to be considered primitive relative to any similarities that were delineated between humans and one or the other of the African apes.

With the ever-growing popularity of the parsimony-based phylogenetic computer program PAUP (phylogenetic analysis using parsimony), it is common practice to "root" a phylogenetic analysis in a taxon that is chosen as the primitive outgroup – that is, the taxon that diverged earlier than the others – prior to the analysis taking place. Rooting parsimony or any of the other available clustering analyses (for example, nearest-neighbor joining or maximum likelihood, which are essentially similar to Sibley and Ahlquist's linking technique) in a particular taxon may be necessary for the algorithm to "work." Nevertheless, this procedure artificially determines character polarity since, by definition, the outgroup (the taxon in which the tree is rooted) is defined from the outset as being primitive in its entirety. In turn, the taxa to which the outgroup is the supposed primitive sister taxon are predetermined as being derived in whatever ways they differ from it.

The widespread use of this algorithm-based approach to analyzing nuclear and mtDNA as well as protein sequences presents its own set of problems and assumptions. Consider the molecular assumption: Since molecular change is supposedly continually occurring and being accumulated as a lineage proceeds through time, the degree of molecular similarity reflects the antiquity or recency of lineage divergence. Accordingly, each lineage accumulates it own unique array of molecular changes, which should make a lineage more distinctive (that is, different) the longer it is in existence. Although tautological, this assumption explains why more recently diverged taxa are more similar than more anciently divergent lineages. On the other hand, in order to root an algorithm for purposes of generating presumed phylogenetic relationships, one must assume that the taxon chosen as the earlier-divergent outgroup is totally primitive relative to the taxa to which it is supposed to be the sister taxon. Yet, it is the molecular assumption that validates the use of overall similarity as the key to resolving phylogenetic relationships by contrasting it with the unique differences that earlier-divergent lineages accumulated along their own, unique evolutionary trajectories. Clearly, both assumptions cannot be correct at the same time. Either the earlier divergent-taxa or lineages did not change, but remained primitive (which is the logical extension of identifying a taxon as the outgroup in which to root a computer analysis), or they did change by accumulating their own suites of molecular difference (the basis of the molecular assumption), in which case they are at least in some aspects derived (and uniquely so, for that matter, because of their unique molecular histories) and not primitive relative to the taxa to which they are being compared.

In the realm of morphological systematics, according to Hennigian or cladistic principles, overall similarity is not *de facto* a clue to evolutionary relatedness. Similarity must be sorted out into features that reflect a hierarchy of inheritance: primitive features from ancient ancestors, and derived features from recent ancestors.

Since the pattern of life is one of a hierarchy of nested sets of smaller and smaller clades (groups of related taxa), that which is considered primitive versus that which is considered derived depends on the level in the hierarchy of nested clades one is investigating. Primitive features – features retained in descendants – do not elucidate the relationships of these taxa. Only derived features can. It is also important conceptually to recognize that a derived feature at one level of the hierarchy is a primitive retention at another. There is no theoretical reason why this approach to systematics cannot be applied to molecular data. The major difficulty is that molecular systematists would have to sample and compare a wide range of taxa. This is the only way in which relative primitiveness and derivedness can be determined. It cannot be justified by *a priori* assumptions of directionality, as underlies the molecular assumption, or by choosing an outgroup on the basis of its presumed evolutionary relationship to other taxa. However, even from the beginning, it is also crucial to realize that shared similarity does not translate directly into a demonstration of relatedness. Taxa may be similar, not because they inherited changes that distinguished a recent common ancestor, but because they share primitive retentions, that is, features that have not changed in a succession of ancestors.

Nevertheless, it is becoming increasingly popular in the literature for molecular studies on the relatedness of taxa to be identified as being "cladistic." One argument is that nucleotide bases – C, G, T, A – represent alternative character states. On one level this may appear logical, but it is actually misleading since none of them represents a character. Phylogenetically relevant alternative molecular character states would be better represented by comparison, for example, of arrangements of "gene" sequences with regard to *cis* and *trans* elements, patterns of introns and exons and of methylation of transposons or other elements, and pathways of molecular communication. Another argument in support of molecular studies being cladistic derives from the claim that molecular similarity is equivalent to synapomorphy; that is, shared similarity represents shared derived character states. This conclusion is, of course, only a restatement of the molecular assumption: the most recently diverged taxa share more recently accumulated (equate with derived) molecular states. Thus, the supposedly shared derived molecular states are delineated *a posteriori*; in other words, after the algorithm of choice has clustered taxa on the basis of their greater or lesser degrees of similarity (depending on the algorithm), typically after rooting the tree in a particular taxon (which, as pointed out above, at once defines it as being primitive and the taxa being compared to it as derived in their shared similarities). This, however, is not how a cladistic analysis proceeds. The endeavor of hypothesizing primitive versus derived character states occurs prior to hypothesizing relationships – which is the only way in which such a methodology can actually be employed.

The assumption of continual molecular change – whether through point mutations affecting nuclear or mtDNA, or altering amino acids in protein sequences – is also of interest. Recall that this idea was initially framed by Zuckerkandl and Pauling as a way of explaining their data: "overall similarity *must* [emphasis added] be an expression of evolutionary history," with descendants "mutating away" from each other, becoming "gradually more different from each other." It is this assumption that proposes that earlier-divergent taxa accumulate their own molecular differences, while the most recently divergent taxa are similar because of the longer shared history of accumulated molecular changes and shorter time of independent molecular change. The existence of molecular clocks and UAR is predicated on this notion. Nevertheless, it must be recognized that this is an extrapolation – an explanation of how something might come to be. It has not been demonstrated.

The contradiction is that while constant molecular change is predicted through the molecular assumption as occurring during gametogenesis, or in some way as to be passed on to offspring, in molecular biology, it is well known that the only source of constant molecular change is ultraviolet radiation, which produces a mutation rate of 10^{-8}–10^{-9}. But the other element of UV-derived mutation is that it is random, with the potential of affecting either somatic or sex cells and also with regard to the molecule that is affected. Thus, while there might appear to be concordance between the reality of the physical world in which there is a relatively constant UV-provoked mutation and the concept of a constantly "evolving" molecular world, this is an illusion.

The notion of constant and accumulative mutation affecting sex cells is of further interest because it also contradicts the basic tendency of cells to remain in homeostasis. As seen, for example, in the roles of heat shock proteins (HSP) – maintaining cell membrane physical states through lipid transport, eliminating reading errors that occur during transcription or translation, DNA repair, chaperoning other proteins, and ensuring proper folding of proteins as they emerge from the ribosomes – the basic propensity of a cell is to resist change. Intuitively, this should make sense inasmuch as unabated molecular change would undermine the integrity of cell function, as would also be the case with a constant accumulation of point mutations, and more probably lead to the death of organisms than to change.

6
"Evo-devo"

In 1975, Mary-Claire King and Allan Wilson surveyed all available data on blood serum proteins, as well as the results of DNA hybridization, with regard to humans and chimpanzees. Although their publication is cited as having demonstrated the relatedness of these two hominoids, this was not their intention. As they stated: "the only two species which have been compared by all of these methods are chimpanzees...and humans," and thus "a good opportunity is...presented for finding out whether the molecular and organismal estimates of distance agree." The result was that humans and chimpanzees differed in their genetic makeup only by about one percent. King and Wilson concluded that "all the biochemical methods agree in showing that the genetic distance between humans and the chimpanzee is probably too small to account for their substantial organismal differences." In order to explain how humans and chimpanzees could be virtually identical in their genes but markedly different animals, King and Wilson suggested that humans and chimpanzees must be different in those genes that regulate development.

Since then, studies on the regulation of development have expanded exponentially, not only with regard to distinguishing between regulatory and structural genes, but especially with regard to the array of molecules that induce gene transcription and communicate via signal transduction pathways to produce structure. Interestingly, those animals that have been studied in depth – such as the fruit fly, zebra fish, frog, chick, mouse, human – demonstrate a commonality of "homeotic genes" (which contribute to segmental patterning and segment identify). In turn, through their protein products (transcription factors), homeotic genes control or at least affect gene expression. Time and time again during development, the same proteins (e.g. various growth factors, trans-inducing and bone-modifying proteins) and regulatory genes (and their products) are coopted to produce what in adult organisms are different morphologies.

In 1994, Lewis Wolpert summarized the situation: "During development, differences are generated between cells in the embryo that then lead to spatial organization (pattern formation), changes in form, and the generation of different cell types. Genes control development by controlling cell behavior." But one should not be too gene-centric in envisioning the emergence of form from genes and gene products alone. For, while there might be genetic regulation of some cells' activities, the results (e.g. cellular asymmetry, cell membrane elasticity or rigidity, compressive forces) might produce physical or mechanical responses, which may not themselves be genetically based, but which, nonetheless, greatly affect cell geometry and ultimately organismal form.

"Cell behavior," to return to Wolpert, has many different levels of meaning.

In the 1970s, Søren Løvtrup argued that one must recognize the importance of epigenesis in development: especially that changes in properties of the fertilized egg can alter the chronology and spatial organization of patterns of cellular diversification. Since the larger clades of multicellular organisms possess the same kinds of cells, as well as the same chemical substances that form the immediate environment of the cells, variation in the spatial and chronological organization of cellular differentiation must be at least one of the keys to the emergence of evolutionary novelty. For instance, whether a cell divides symmetrically or asymmetrically (which can be affected even by the positions of the chromosomes relative to the center or periphery of the cell) can greatly impact the spatial relationships of cells and, therefore, eventually have an effect on organismal shape. As Pere Alberch has emphasized, the development of organismal form and structure is also a function of the physical and mechanical properties of cells' sizes, shapes, and spatial relationships.

The application to evolutionary questions of discoveries in the regulation of development has given rise to the field of "evo-devo" (evolution and development), in which the interrelationship between the "genetics" and the "epigenetics" of development has become a primary focus. Indeed, Løvtrup's concern with the influence on metazoan development of differences in cellular differentiation during gastrulation appears to be even more germane to an understanding of the emergence of form and its conservation across taxa as well as of the emergence of differences in form, whether their expression constitutes variation (individual differences) or diversity (species differences).

7
Positional Information and Shape

One of the ongoing questions in developmental biology is how cells acquire positional information, not only in terms of entire structures themselves (e.g. where limbs will grow) but also with regard to how cells acquire information to contribute to regional shapes of a structure (e.g. the different segments of a limb).

In invertebrates, wing (e.g. as in a fruit fly) and limb (e.g. as in the brine shrimp-like crustacean, *Artemia*) positioning involves activation of the regulatory genes *nubbin* and *apterous*. In vertebrates, various regulatory genes, especially *Hox* genes, and also *dlx* (distal-less), are known to be involved in segmentation and limb positioning. In fish and tetrapods, the *Hoxd11-13* genes are expressed along the posterior margin of the enlarging limb bud; however, in tetrapods, this homeodomain expands anteriorly across the distal (lower) end of the limb bud. Additionally, in vertebrates, *Hox* genes are not only recruited in the formation of a segmented trunk but also through regional activation, in the segmentation of the hindbrain. Regional activation of *sonic hedgehog*, however, in part, underlies forebrain segmentation.

Eye development is also of interest in this regard. For although what used to be thought of as "master-control genes," such as *Pax-6*, were found to participate in signal transduction pathways leading to eye formation in invertebrates (typically multilensed, rigid) as well as in vertebrates (single lensed, deformable), there is at least one element that vertebrates have in common: Even though there are differences among vertebrates so far studied with regard to when in ontogeny and how often and in how many different regions the *Rx* gene (which is also recruited in fruit-fly eye development) is activated, it is always expressed in the vertebrate forebrain. Thus, at one level, one can hypothesize that the last common ancestor of vertebrates was characterized by early activation of the *Rx* gene in the presumptive forebrain and that differences between taxa are the result of differences in other aspects of *Rx* gene expression: for example, the different proteins in the lenses of amphibians and mammals may be due in part to down- or upstream effects of the *Rx* gene being expressed later on in development in the amphibian (frogs) retina, whereas in mammals (mice), the *Rx* gene is expressed early on in the presumptive eye itself.

In addition to considering morphological differences in light of differences in regional (as well as in overlapping) domains of regulatory gene expression, it is becoming increasingly clear that differences in fields of molecular gradients (morphogenetic fields) also play a role. As C. Owen Lovejoy, M. J. Cohn, and T. D. White hypothesized in 1999 in their discussion of the evolution of human pelvic form, "if a particular PI [positional information] gradient were to span n cell diameters, and those cells defined the ultimate anteroposterior dimension of the presumptive ilium (superoinferior in the adult human), then a slight increase in the steepness of its slope would cause that signal to span fewer cells, 'distorting' the presumptive anlagen and substantially altering downstream adult morphology." In other words, although it would seem to be a process involving myriad steps, "the transformation of the common ancestral pelvis [in its entirety] into that of early hominids may have been as 'simple' as a slight modification of a gradient" [comment added]. Thus, in addition to differences in gene expression and pathways of molecular communication, as well as to the physical and mechanical consequences of cellular organization, morphological novelty in metazoans (and presumably plants as well) may also be affected by altering the domains of morphogenetic fields.

But what is the source of differences in gene or molecular gradient expression? Lovejoy et al. suggest that one need not seek the answer in mutation, which is a position that Sean Carroll has recently also strongly argued.

8
"Mutation"

The concept of "mutation" is about as slippery as that of a "gene." It means different things to different researchers, and, interestingly, the differing concepts seem to "work" in their disparate intellectual contexts. With regard to mutation, the "textbook" notions of preceding decades included point mutation, gene duplication, and chromosomal rearrangement. The latter was basic to the earlier experimental studies and theoretical considerations of the fruit-fly geneticist, Theodosius Dobzhansky. Dobzhansky's emphasis on chromosomal rearrangement as a potential source of evolutionary novelty was subsequently adopted by the developmental biologist, Richard Goldschmidt, in his theory of systemic mutation, which he argued would lead to the abrupt appearance of novel form. Unfortunately, Goldschmidt is best remembered, and consequently criticized, for suggesting that such novelties would emerge in individuals he identified as "hopeful monsters."

One of Goldschmidt's major theoretical thrusts, however, was distinguishing between what he identified as micromutation (leading to variation and microevolution) and macromutation (leading to the origin of species or evolution). The small mutations that fruit-fly population geneticists, such as Thomas Hunt Morgan, inferred lay behind small phenotypic changes, Goldschmidt identified as micromutations, which, he argued, led only to the survival of species, not to their origin. The latter required a larger source of genetic disturbance, and for that he turned to chromosomal rearrangement. The logic is understandable: If the chromosome theory was correct (that, indeed, units of heredity or genes were contained on chromosomes – as Morgan presented it, like beads in a necklace), then manipulating them on a grand scale (producing a systemic mutation) might yield evolutionarily significant novelty, that is, new species. A major problem with Goldschmidt's theory, however, was that he did not provide a mechanism by which more than one individual would be the bearer of the novelty and, thus, of the systemic mutation underlying it.

Point mutations, commonly the result of UV radiation, are random with regard to affecting somatic or sex cells. In addition, if they do not interfere with cell function, such point mutations are not a reliable source of potential genetic and subsequently morphological novelty. Indeed, it appears that point mutations do not often cause any noticeable effect, either genetically or phenotypically. Gene duplication – as seen, for instance, in the emergence of *Hoxa-d* – does sometimes occur, but knockout experiments have demonstrated that duplication typically reflects redundancy, not a source of phenotypic novelty.

It may be true that manipulation of levels of thyroxin or retinoic acid during ontogeny can affect the size of an organism, or the shapes of some of its features, just as a mother's diet can affect methylation in its fetus or fetuses, and thus aspects of her progeny's postnatal growth. However, these disturbances only affect an individual during its lifetime, and should not be expected to be repeated across generations and under the influence of fluctuating environmental stimuli (e.g. diet, temperature, amount of daylight). The problems, then, that still must be addressed are as follows: How does a genetic or cellular change remain "fixed" or constant, and how do many individuals come to bear it?

It is important to realize that there is a difference between change at the genetic level and what is perceived as phenotypic change. Common in the literature on the genetics of evolution is the mistake of conflating the two as constituting macromutation – a misconception that derives from the confusion Dobzhansky introduced with regard to the terms micromutation and macromutation when, in 1941, he sought to discredit Goldschmidt. Nevertheless, especially with the increasing awareness from molecular biology that there are not "genes for" features, we must be vigilant in making a distinction between what appears morphologically to have been the result of a macromutation (e.g. developing feathers instead of scales) and the underlying genetic–epigenetic interactive pathway.

9
Toward a Theory of Evolutionary Change

The question at hand, then, is the articulation of a mechanism that can first provide the potential for genetic novelty. Building on my original theory for the sudden appearance of morphological novelty (through the silent spread of recessive "mutations"), the molecular biologist, Bruno Maresca, and I have proposed that the opportunity for genetic novelty may lie in overstressing cells to the extent that their HSPs can no longer maintain genetic homeostasis; that is, they cannot fulfill their roles as chaperones, respond to the needs of the cell membrane, and, perhaps most importantly, for this discussion, properly fold other proteins and assist in DNA repair. Although first identified in heat shock experiments, HSPs can be affected by a variety of stresses, including diet (saturated vs unsaturated fatty acids), wind, aridity, and cold. Since most multicellular plants and animals possess HSPs, the theory is more widely applicable than metazoan-centric Darwinian and neo-Darwinians models of evolutionary change, the latter of which relies on unwarranted extrapolations from fruit-fly population genetics.

Since most multicellular organisms have a window of heat shock response, they can "adapt" to normal fluctuations in their environmental circumstances. If environmental change exceeds this window (as when seasons change), most organisms can "reset" it, often in less than two months. Until this window is reset, the stress induces an increase in HSP production. If, however, there is a spike in environmental stress that exceeds an organism's ability to reset its HSP response, HSP function will fail, and opportunities for introducing genetic novelty will emerge – especially as a result of improper protein folding and inefficient DNA repair. In the former situation, improperly folded proteins may, for example, no longer recognize (or be recognized by) promoter or enhancer regions to which they would normally bind, but they may now be capable of binding to different sites. An obvious result could be the activation or deactivation of a "gene" or "genes," and, thus, the creation of one or more new developmentally significant signal transduction pathways. With regard to inefficient DNA repair, genetic novelty of a different sort can be introduced, with obvious potential consequences. In both cases, however, the fact that the environmental stress will be at least regional (if not global), the circumstances exist for more, perhaps many more, than one individual of a species to be affected (not, however, necessarily in the same way).

However, while it might be tempting to extrapolate immediately from these possible sources of genetic novelty to the emergence of evolutionarily relevant morphological novelty, one must be cautious. First, the effects of extreme environmental spikes on HSPs must be actualized during gametogenesis. If they are not, offspring will not be affected. Second, if the effects do not kill the individuals that inherit any of these genetically based novelties, they will probably not be expressed; that is, these genetic novelties will be in the "recessive" state. Consequently, there will not be an immediate phenotypic reflection of these genetic changes. In the recessive state, however, they can spread "silently" throughout the population, until it becomes sufficiently saturated with heterozygotes that homozygotes for the genetic novelty will be produced. If the resultant phenotypic expression – cellular or greater – does not kill its bearers, they may continue to reproduce themselves, as heterozygotes will also contribute to the numbers of individuals bearing the phenotypic novelty. Thus, the spread of a genetic basis for potential phenotypic novelty may take numerous generations before there is any statistical possibility of the phenotype being expressed. In other words, there will be a temporal disjunction between the disruption of cellular and genetic homeostasis, and what will be seen as the abrupt or sudden appearance of phenotypic novelty, and in some number of individuals. In addition, one must bear in mind that, during periods of "silently spreading" genetic novelty, there could be other environmental spikes that would contribute to the pool of potential for genetic novelty and also then phenotypic novelty (however defined). Superficially, this process – or at least the sudden appearance of phenotypic novelty – may seem macromutational, but, clearly, it is not, at least in the original sense of Goldschmidt or even that of Dobzhansky. Indeed, something as simple as slight changes in protein folding could have major cascading morphological effects.

10
Molecules and Systematics: Looking Toward the Future

It may be widely believed, and even true at some level, that, as Sean Carroll has recently reiterated, "genomes diverge as a function of time." However, the observation that genomes may be different (in whatever ways difference, and similarity, may be identified and defined) does not in and of itself provide clues to how this difference was achieved. No doubt, some difference is due to the rare and random effects of UV radiation. In addition, genomic difference may be due to failures in DNA repair. There may be something intuitively appealing about Sibley and Ahlquist's the "law of large numbers" – the idea that organisms are closely related because they share "lots" of their genome. However, as Jonathan Marks points out, humans share about 25% of their genome with bananas. Essentially, there is nothing in an observation of genomic difference or similarity that directly translates into the "molecular assumption" and, consequently, a theory of evolutionary relatedness.

Can, then, molecular information be useful in systematics and phylogenetic reconstruction?

The answer is yes, but it will have to be at the level of cell biology and pathways of molecular communication. As King and Wilson came close to predicting many decades ago, it is not through the study of molecular or genomic similarity of organisms that we will come to understand their biology, but through the investigation of those elements that underlie the development of their biology. This makes sense. For, if something as simple as the inactivation or deletion of a transcriptional enhancer can result in a more caudal repositioning of the sacrum, or if the expansion of a morphogenetic gradient can transform in its entirety a narrow pelvic girdle with tall, thin ilial blades into a broad, deep, and squat structure, then it is by seeking to identify the similarity or difference in these molecular events that we may more profitably explore the molecular basis of morphology and, consequently, the evolutionary relationships of complex organisms.

This is, perhaps, a timely occasion to both question and expand our perceptions of what is or will be evolutionarily revealing at the molecular level. There has been a steady increase in the number of studies that demonstrate virtual molecular identity between taxa that are morphologically very different and then express amazement at this apparent contradiction. As Sean Carroll pointed out with regard to the importance placed on the human and chimpanzee-genome projects – especially since so much money has been poured into them in the hope that forthcoming comparisons will instantaneously provide answers to any questions – demonstrating molecular similarity does not translate into deciphering the pathways that make these organisms so different in hard- and soft-tissue anatomy, physiology, reproductive biology, cognitive abilities, and behavior. Here, the "law of large numbers" fails to be enlightening. For, in contrast to the bacterial world, in the metazoan world, a one-to-one correspondence between a "gene" (a sequence of nucleotides bound by start and stop codons) and a "gene product" (a protein or amino acid sequence) is not there. In multicellular animals, RNA essentially directs the "show," for example, in reading select bases and splicing specific introns, as it composes different proteins from the same stretches of DNA. The surprise "The International Chimpanzee Chromosome 22 Consortium" had at finding upon comparing human chromosome 21 with its apparent orthologue in the chimpanzee, chromosome 22 – not only that these hominoids differ by 83% in their amino acid sequences but also that this large difference is produced from very similar DNA sequences – should serve as a lesson: While there may be appeal to the "law of large numbers" that comparison of chromosomes and especially of entire genomes purportedly represents, in the end, this molecular level may not be the evolutionarily informative hotspot everyone has been seeking.

See also Gene Mapping and Chromosome Evolution by Fluorescence–Activated Chromosome Sorting; Genetic Variation and Molecular Evolution.

Bibliography

Books and Reviews

Frontiers in biology: development, *Science* (1994) 266, 561–614.

Gerhart, J.C., Kirschner, M.W. (1997) *Cells, Embryos, and Evolution: Toward a Cellular and Developmental Understanding of Phenotypic Variation and Evolutionary Adaptability*, Blackwell Science, New York.

Marks, J. (2003) *What it Means to be 98% Chimpanzee*, (revised paperback edition). University of California Press, Berkeley, CA.

Müller, G.B., Newman, S.A. (Eds.) (2000) *Origination of Organismal Form*, MIT Press, Cambridge, MA.

Raff, R.A. (1996) *The Shape of Life: Genes, Development, and the Evolution of Animal Form*, University of Chicago Press, Chicago, IL.

Schwartz, J.H. (1987) *The Red Ape: Orangutans and Humans Origins*, Basic Books, New York; in press.

Schwartz, J.H. (1999) *Sudden Origins: Fossils, Genes, and the Emergence of Species*, John Wiley, New York.

Primary Sources

Averoff, M., Cohen, S.M. (1997) Evolutionary origin of insect wings from ancestral gills, *Nature* 385, 627–630.

Awadella, P., Eyre-Walker, A., Maynard Smith, J. (1999) Linkage disequilibrium and recombination in hominid mitochondrial DNA, *Science* 286, 2524–2525.

Baker, R.H., Xiaobo, Y., DeSalle, R. (1998) Assessing the relative contribution of molecular and morphological characters in simultaneous analysis trees, *Mol. Phylogenet. Evol.* 9, 427–436.

Benveniste, R.E., Todaro, G.J. (1976) Evolution of type C viral genes: evidence for an Asian origin of man, *Nature* 261, 101–8.

Britten, R. (2002) Divergence between samples of chimpanzee and human DNA sequences is 5%, counting indels, *Proc. Natl. Acad. Sci. U.S.A.* 99, 13633–13635.

Britten, R.J., Kohne, D.E. (1968) Repeated sequences in DNA, *Science* 161, 529–40.

Brown, W.M., George, M. Jr., Wilson, A.C. (1979) Rapid evolution of mitochondrial DNA, *Proc. Natl. Acad. Sci. U.S.A.* 76, 1967–71.

Brown, W.M., Prager, E.M., Wang, A., Wilson, A.C. (1982) Mitochondrial DNA sequences of primates: tempo and mode of evolution, *J. Mol. Evol.* 18, 225–39.

Bruce, E.J., Ayala, F.J. (1979) Phylogenetic relationships between man and the apes: electrophoretic evidence, *Evolution* 33, 1040–1056.

Caccone, A., Powell, J.R. (1989) DNA divergence among hominoids, *Evolution* 43, 925–942.

Carroll, S.R. (2003) Genetics and the making of Homo sapiens, *Nature* 422, 849–857.

Cohn, M.J., Patel, K., Krumlauf, R., Wilkinson, D.G., Clarke, J.D.W., Tickle, C. (1997) Hox9 genes and vertebrate limb specification, *Nature* 387, 97–101.

Crockford, S.J. (2003) Thyroid rhythm phenotypes and hominid evolution: a new paradigm implicates pulsatile hormone selection in speciation and adaptation changes, *Comp. Biochem. Physiol., A* 135, 105–129.

Cronin, J.E., Sarich, V.M. (1980) Tupaiid and Archonta phylogeny: the Macromolecular Evidence, in: Luckett, W.P. (Ed.) *Comparative Biology and Evolutionary Relationships of Tree Shrews*, Plenum Press, New York, pp. 293–312.

Czelusniak, J., Goodman, M., Moncrief, N.D., Kehoe, S.M. (1990) Maximum parsimony approach to construction of evolutionary trees from aligned homologous sequences, *Methods Enzymol.* 183, 601–615.

Dobzhansky, T. (1935) *Drosophila miranda*, a new species, *Genetics* 20, 377–391.

Dobzhansky, T. (1941) *Genetics and the Origin of Species*, 2nd edition, Columbia University Press, New York.

Felsenstein, J. (1988) Phylogenies from molecular sequences: inference and reliability, *Annu. Rev. Genet.* 22, 521–565.

Ferris, S.D., Wilson, A.C., Brown, W.M. (1981) Evolutionary tree for apes and humans based on cleavage maps of mitochondrial DNA, *Proc. Natl. Acad. Sci. U.S.A.* 78, 2431–2436.

Figdor, M., Stern, C. (1993) Segmental organization of embryonic diencephalon, *Nature* 363, 630–634.

Goldschmidt, R. (1940) *The Material Basis for Evolution*, Yale University Press, New Haven, CT, (reprinted 1982).

Goodman, M. (1962) Immunochemistry of the primates and primate evolution, *Ann. N.Y. Acad. Sci.* 102, 219–234.

Goodman, M., Tashian, R.E. (Eds.) (1976) *Molecular Anthropology*, Plenum Press, New York.

Goodman, M., Braunitzer, G., Stangl, A., Schrank, B. (1983) Evidence on human origins from haemoglobins of African apes, *Nature* **303**, 546–48.

Goodman, M., Olson, C.B., Beeber, J.E., Czelusniak, J. (1982) New perspectives in the molecular biological analysis of mammalian phylogeny, *Acta Zoolo. Fennica* **169**, 1–73.

Graur, D., Martin, W. (2004) Reading the entrails of chickens: molecular timescales of evolution and the illusion of precision, *Trends Genet.* **20**, 80–86.

Gregory, W.K. (1922) *The Origin and Evolution of the Human Dentition*, Williams and Wilkins, Baltimore, MD.

Hagelberg, E. (2003) Recombination or mutation rate heterogeneity? Implications for mitochondrial Eve, *Trends Genet.* **19**, 84–90.

Hasegawa, M., Yano, T. (1984) Maximum likelihood method of phylogenetic inference from DNA sequence data, *Bull. Biometri. Soc. Jpn.* **5**, 1–7.

Hasegawa, M., Kishino, H., Yano, T. (1985) Dating of the human-ape splitting by a molecular clock of mitochondrial DNA, *J. Mol. Evol.* **22**, 160–174.

Hedges, S.B., Kumar, S. (2002) Vertebrate genomes compared, *Science* **297**, 1283–1285.

Hedges, S.B., Kumar, S. (2003) Genomic clocks and evolutionary timescales, *Trends Genet.* **19**, 200–206.

Hedges, S.B., Kumar, S., Tamura, K., Stoneking, M. (1992) Human origins and analysis of mitochondrial DNA sequences, *Science* **255**, 737–739.

Horai, S., Hayasaka, K., Kondo, R., Tsugane, K., Takahata, N. (1995) Recent African origin of modern humans revealed by complete sequence of hominid mitochondrial DNAs, *Proc. Natl. Acad. Sci. U.S.A.* **92**, 532–536.

Hunt, P., Gulisano, M., Cook, M., Sham, M.H., Faiella, A., Wilkinson, D., Boncinelli, E., Krumlauf, R. (1991) A distinct Hox code for the branchial region of the vertebrae head, *Nature* **353**, 861–864.

King, M.-C., Wilson, A.C. (1975) Evolution at two levels in humans and chimpanzees, *Science* **188**, 107–88.

Lake, J.A. (1991) The order of sequence alignment can bias the selection of tree topology, *Mol. Biol. Evol.* **8**, 378–385.

Lovejoy, C.O., Cohn, M.J., White, T.D. (1999) Morphological analysis of the mammalian postcranium: a developmental perspective, *Proc. Natl. Acad. Sci. U.S.A.* **96**, 13247–13252.

Lovejoy, C.O., McCollum, M.A., Reno, P.I., Rosenman, B.A. (2003) Developmental biology and human evolution, *Annu. Rev. Anthropol.* **32**, 85–109.

Løvtrup, S. (1974) *Epigenetics: A Treatise on Theoretical Biology*, Wiley, New York.

Lowenstein, J.M., Molleson, T., Washburn, S.L. (1982) Piltdown jaw confirmed as orang, *Nature* **299**, 294.

Lowenstein, J.M., Sarich, V.M., Richardson, B.J. (1981) Albumin systematics of the extinct mammoth and Tasmanian wolf, *Nature* **291**, 409–411.

Lumsden, A., Krumlauf, R. (1996) Patterning the vertebrate neuraxis, *Science* **274**, 1109–1115.

Maresca, B., Schwartz, J.H. Environmental change and stress protein concentration as a source of morphological novelty: sudden origins, a general theory on a mechanism of evolution, in manuscript.

Marshall, C.R. (1991) Statistical tests and bootstrapping: assessing the reliability of phylogenetics based on distance data, *Mol. Biol. Evol.* **8**, 386–391.

Mathers, P., Grinberg, A., Mahon, K., Jamrich, M. (1997) The Rx homeobox gene is essential for vertebrate eye development, *Nature* **387**, 604–607.

Morgan, T.H. (1916) *A Critique of the Theory of Evolution*, Princeton University press, Princeton, NJ.

Muragaki, Y., Mundlos, S., Upton, J., Olsen, B.R. (1996) Altered growth and branching patterns in synpolydactyly caused by mutations in HOXD13, *Science* **272**, 548–551.

Nuttall, G.H.F. (1904) *Blood Immunity and Blood Relationship*, Cambridge University Press, Cambridge, MA.

O'hUigin, C.O., Satta, Y., Takahata, N., Klein, J. (2002) Contribution of homoplasy and of ancestral polymorphism to the evolution of genes in anthropoid primates, *Mol. Biol. Evol.* **19**, 1501–1513.

Osborn, J.F. (1978) Morphogenetic Gradients: Fields Versus Clones, in: Butler, P.M., Joysey, K.A. (Eds.) *Development, Function, and Evolution of Teeth*, Academic Press, New York, pp. 171–201.

Oster, G., Alberch, P. (1982) Evolution and bifurcation of developmental programs, *Evolution* **36**, 444–459.

Pennisi, E. (1999) Genetic study shakes up out of Africa theory, *Science* **283**, 1828.

Romero-Herrera, A.E., Lehmann, H., Castillo, O., Joysey, K.A., Friday, A.E. (1976) Myoglobin of the orangutan as a phylogenetic enigma, *Nature* **261**, 162–64.

Romero-Herrera, A.E., Lehmann, H., Joysey, K.A., Friday, A.E. (1978) On the evolution of myoglobin, *Philos. Trans. R. Soc. Lond. B Biol. Sci.* **283**, 61–183.

Rubenstein, J.L.R., Martinex, S., Shimamura, K., Puelles, L. (1994) The embryonic vertebrate forebrain: the prosomeric model, *Science* **266**, 578–580.

Ruvolo, M. (1997) Molecular phylogeny of the hominoids: inferences from multiple independent DNA sequence data sets, *Mol. Biol. Evol.* **14**, 248–265.

Ruvolo, M., Disotell, T.R., Allard, M.W., Brown, W.M., Honeycutt, R.L. (1991) Resolution of the African hominoid trichotomy by use of a mitochondrial gene sequence, *Proc. Natl. Acad. Sci. U.S.A.* **88**, 1570–1574.

Samollow, P.B., Cherry, L.M., Whitte, S.M., Rogers, J. (1996) Interspecific variation at the Y-linked *RPS4Y* locus in hominoids: implications for phylogeny, *Am. J. Phys. Anthropol.* **101**, 333–343.

Sarich, V.M. (1971) A Molecular Approach to the Question of Human Origins, in: Dolhinow, P., Sarich, V.M. (Eds.) *Background for Man*, Little, Brown, Boston, MA, pp. 60–81.

Sarich, V.M., Wilson, A.C. (1966) Quantitative immunochemistry and the evolution of primate albumins: micro-complement fixation, *Science* **154**, 1563–66.

Sarich, V.M., Wilson, A.C. (1967a) Rates of albumin evolution in primates, *Proc. Natl. Acad. Sci. U.S.A.* **58**, 142–148.

Sarich, V.M., Wilson, A.C. (1967b) Immunological time scale for hominid evolution, *Science* **158**, 1200–1203.

Schwartz, J.H. (2001) A review of the systematics and taxonomy of Hominoidea: history, morphology, molecules, and fossils, *Ludus Vitalis* **IX**, 15–45.

Shubin, N., Alberch, P. (1986) A morphogenetic approach to the origin and basic organization of the tetrapod limb, *Evol. Biol.* **20**, 319–387.

Shubin, N., Tabin, C., Carroll, S. (1997) Fossils, genes and the evolution of animal limbs, *Nature* **388**, 639–648.

Sibley, C.G., Ahlquist, J.E. (1983) Phylogeny and Classification of Birds Based on the Data of DNA-DNA Hybridization, in: Johnston, R.F. (Ed.) *Current Ornithology*, Vol. 1, Plenum Press, New York, pp. 245–92.

Sibley, C.G., Ahlquist, J.E. (1984) The phylogeny of the hominoid primates, as indicated by DNA-DNA hybridization, *J. Mol. Evol.* **20**, 2–15.

Sordino, P., van der Hoeven, F., Duboule, D. (1995) Hox gene expression in teleost fins and the origin of vertebrate digits, *Nature* **375**, 678–681.

Stauffer, R.L., Walker, A., Ryder, O.A., Lyons-Weiler, M., Hedges, S.B. (2001) Human and ape molecular clocks and constraints on paleontological hypotheses, *J. Hered.* **92**, 469–474.

Summerbell, D. (1981) Evidence for regulation of growth, size and pattern in the developing chick limb bud, *J. Embryol. Exp. Morphol.* **65**, 129–150.

The International Chimpanzee Chromosome 22 Consortium. (2004) DNA sequence and comparative analysis of chimpanzee chromosome 22, *Nature* **429**, 382–388.

Tickle, C. (1992) A tool for transgenesis, *Nature* **358**, 188–189.

Wolpert, L. (1980) Positional Information and Pattern Formation in Limb Development, in: Pratt, R.M., Christiansen, R.L. (Eds.) *Current Research Trends in Prenatal Craniofacial Development*, Elsevier/North Holland, New York, pp. 89–101.

Zuckerkandl, E. (1987) On the molecular evolutionary clock, *J. Mol. Evol.* **26**, 34–46.

Zuckerkandl, E., Pauling, L. (1962) Molecular Disease, Evolution, and Genic Heterogeneity, in: Kasha, M., Pullman, B. (Eds.) *Horizons in Biochemistry*, Academic Press, New York, pp. 189–225.

13
Genetic Variation and Molecular Evolution

Werner Arber
Biozentrum, University of Basel, Klingelbergstrasse 70, Basel, Switzerland

1	**Introduction** 387	
2	**Principles of Molecular Evolution** 388	
2.1	Evolutionary Roles of Genetic Variation, Natural Selection, and Isolation 388	
2.2	Molecular Mechanisms of the Generation of Genetic Variation 390	
3	**Genetic Variation in Bacteria** 391	
4	**Local Changes in the DNA Sequences** 392	
5	**Intragenomic DNA Rearrangements** 393	
5.1	Site-specific DNA Inversion at Secondary Crossover Sites 394	
5.2	Transposition of Mobile Genetic Elements 395	
6	**DNA Acquisition** 396	
7	**The Three Natural Strategies Generating Genetic Variations Contribute Differently to the Evolutionary Process** 397	
8	**Evolution Genes and Their Own Second-order Selection** 399	
9	**Arguments for a General Relevance of the Theory of Molecular Evolution for All Living Organisms** 400	
10	**Conceptual Aspects of the Theory of Molecular Evolution** 401	
10.1	Pertinent Scientific Questions 401	

Genomics and Genetics. Edited by Robert A. Meyers.
Copyright © 2007 Wiley-VCH Verlag GmbH & Co. KGaA, Weinheim
ISBN: 978-3-527-31609-0

10.2 Philosophical Values of the Knowledge on Molecular Evolution 402
10.3 Aspects Relating to Practical Applications of Scientific Knowledge on Molecular Evolution 404

Bibliography 405
Books and Reviews 405
Primary Literature 405

Keywords

Biological Evolution
A nondirected, dynamic process of diversification resulting from the steady interplay between spontaneous mutagenesis and natural selection.

DNA Rearrangement
Results from mostly enzyme-mediated recombination processes, which can be intra- or intermolecular.

Evolution Gene
Its protein product acts as a generator of genetic variations and/or as a modulator of the frequency of genetic variations.

Gene Acquisition
Results from horizontal transfer of genetic information from a donor cell to a receptor cell. With bacteria, this can occur in transformation, conjugation, or phage-mediated transduction.

Natural Selection
Results from the capacity of living organisms to cope with the encountered physicochemical and biological environments. Largely depending on its genetic setup and physiological phenotype, each organism may have either a selective advantage or a selective disadvantage as compared to the other organisms present in the same ecosystem.

Spontaneous Mutation
Defined here as any alteration of nucleotide sequences occurring to DNA without the intended intervention of an investigator. The term mutation is used here as a synonym of genetic variation.

Transposition
DNA rearrangement mediated by a mobile genetic element such as a bacterial insertion sequence (IS) element or a transposon.

Variation Generator
An enzyme or enzyme system whose mutagenic activity in the generation of genetic variation has been documented.

The comparison of DNA sequences of genes and entire genomes offers interesting insights into the possible evolutionary relatedness of genetic information of living organisms. Together with a relatively rich database from experimental microbial genetics, conclusions can be drawn on the molecular mechanisms by which genetic variations are spontaneously generated. A number of different specific mechanisms contribute to the overall mutagenesis. These mechanisms are here grouped into three natural strategies of the spontaneous generation of genetic variations: local changes of DNA sequences, intragenomic rearrangement of DNA segments, and acquisition of foreign DNA by horizontal gene transfer. These three strategies have different qualities with regard to their contributions to the evolutionary process. As a general rule, none of the known mechanisms producing genetic variants is clearly directed. Rather, the resulting alterations in the inherited genomes are more random. In addition, usually only a minority of resulting variants provide a selective advantage. Interestingly, in most of the molecular mechanisms involved, the products of so-called evolution genes are involved as generators of genetic variation and/or as modulators of the frequencies of genetic variation. Products of evolution genes work in tight collaboration with nongenetic factors such as structural flexibilities and chemical instabilities of molecules, chemical and physical mutagens, and random encounter. All of these aspects contributing to the spontaneous generation of genetic variations together form the core of the theory of molecular evolution. This theory brings neo-Darwinism to the molecular level. In view of the increasing evidence coming particularly from microbial genetics, knowledge of molecular evolution can be seen as a confirmation of Darwinism at the level of biologically active molecules, in particular, nucleic acids and proteins. Philosophical and practical implications of this knowledge will be briefly discussed.

1
Introduction

Evolutionary biology has traditionally devoted its major attention to the comparison of phenotypical traits of higher organisms, both of those actually living and of those that have been extinct (paleontological fossils). The resulting theory of descent is, together with other criteria, at the basis of the systematic classification of living organisms. Darwin's theory of natural selection brought a new element into the understanding of the long-term development of forms of life. Natural selection is the result of organisms coping with the encountered living conditions that are dependent on both the environmental physicochemical conditions and the activities of all living forms in a particular ecological niche. The Darwinian theory of evolution also postulated that intrinsic properties of life are not entirely stable and principally identical for all organisms of a given species. In the so-called modern synthesis, in which evolutionary biology and genetics became integrated, transmissible phenotypic variations representing the substrate of natural selection were explained as due to genetic variations (or mutations). Shortly thereafter, deoxyribonucleic acid (DNA) was identified as the carrier of genetic information. DNA is thus also the target for mutagenesis. Within the last few decades a rapid development of molecular genetics with novel research strategies leading to genomics, sequencing of entire genomes, and functional studies of genes and their products paved the way for hitherto inaccessible knowledge on the basis of life and its multiple manifestations. This also relates to the process of molecular evolution. A synoptical insight into the various molecular mechanisms contributing to the generation of genetic variations represents a molecular synthesis between the neo-Darwinian theory and molecular genetics. This synthesis can confirm the Darwinian evolution at the molecular level.

2
Principles of Molecular Evolution

The principles of molecular evolution to be outlined here are founded on

1. the neo-Darwinian theory of evolution with its three pillars of genetic variation, natural selection, and isolation;
2. the solidly established microbial genetics database;
3. DNA sequence comparisons with bioinformatic tools;
4. physicochemical knowledge on the reactivity, conformational flexibility, and chemical stability of biologically active molecules.

2.1
Evolutionary Roles of Genetic Variation, Natural Selection, and Isolation

The long-term maintenance of any form of life requires a relatively high stability of its genetic information. However, rare occasional genetic variations occur in all organisms. This gives rise to mixed populations of organisms with the parental genome and organisms having one or more alterations in their genome. These populations are steadily submitted to natural selection. The experience shows that, in general, favorable genetic variations are considerably less frequent than unfavorable ones. The latter provide a selective disadvantage. Indeed, genetic variants with unfavorable genetic alterations will sooner or later get eliminated from propagating populations, which will become enriched for organisms carrying favorable genetic variations. It should be noted that by far not all alterations in the nucleotide sequences of a genome will lead to a change in the phenotype of the organism. However, such silent and neutral mutations may later become physiologically relevant in conjunction with still other, upcoming DNA sequence changes.

Natural selection is by no means a constant element. It varies both in time and in space. This is due to variations in the physicochemical environmental conditions as well as to variations of the life activities of all the different organisms present in a particular ecological niche and forming an ecosystem. Since a genetic variation may also affect the influence that the organism exerts on the other organisms present in the same ecosystem (e.g. think of weeds and pathogenicity effects, but also of beneficial, synergistic effects), any novel mutation may not only influence the life of the concerned organism itself but also the lives of other cohabitants of the same ecological niche.

The third pillar of biological evolution – besides genetic variation and natural selection – is isolation. Evolutionary biologists define two different aspects of isolation. One of these is *geographic isolation*, which may seriously reduce the number of potential habitats for an organism. The other type of isolation is called *reproductive isolation*. For example, two distantly related diploid organisms may not be fertile in sexual reproduction. But reproductive isolation can also be seen in a wider definition to seriously limit the possibility of horizontal transfer of segments of genetic information between two different kinds of organisms.

As summarized in Fig. 1, genetic variation drives biological evolution. Complete genetic stability would render any

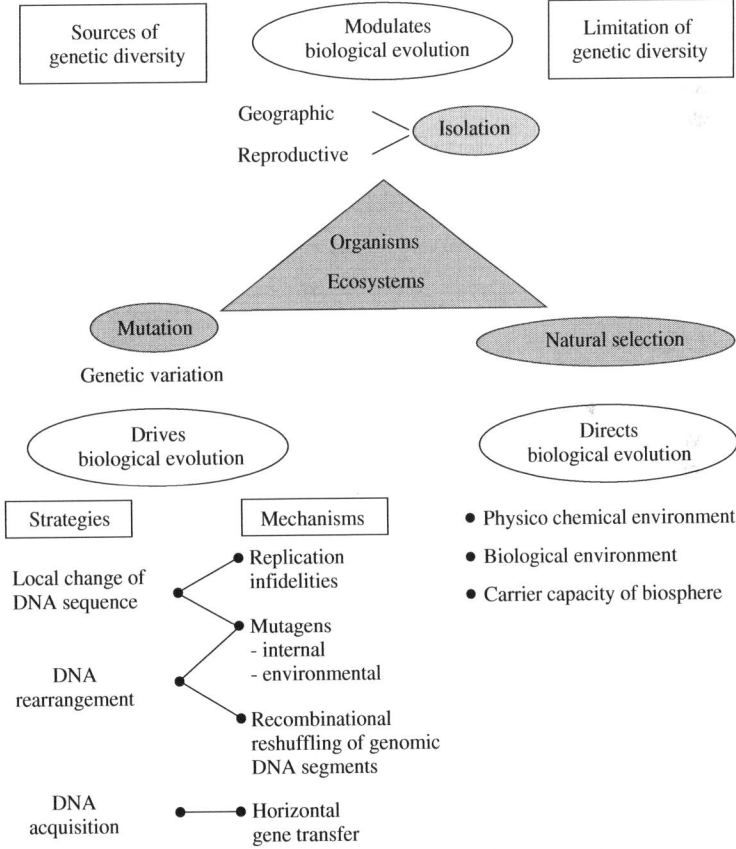

Fig. 1 Synoptical presentation of major elements of the theory of molecular evolution. A number of specific mechanisms, each with its own characteristics, contribute to the four groups of mechanisms of genetic variation listed. Each of the specific mechanisms follows one and sometimes more than one of the three principal, qualitatively different strategies of genetic variation.

evolutionary process impossible. A very high frequency of genetic variation would rapidly lead to the extinction of the concerned organisms because of the stated prevalence of unfavorable mutations in the spontaneous generation of genetic variations. It is natural selection together with the available sets of genetic variants that determines the direction of biological evolution, or in other words, the directions in which the branches of the tree of evolution grow. Finally, the geographic and reproductive isolations modulate the evolutionary process.

2.2
Molecular Mechanisms of the Generation of Genetic Variation

The concept to be presented here requires the reader to question some long-established textbook knowledge, such as the claim that spontaneous mutations would largely result from errors, accidents, and illegitimate processes. We defend here the alternative view that living nature actively cares for biological evolution

1. by making use of intrinsic properties of matter such as a certain degree of chemical instability and of structural flexibility of molecules, and
2. by having developed genetically encoded systems, the products of which are involved in the generation of genetic variations and in modulating the frequencies of genetic variation.

It might be relevant to mention here that the term mutation is differently defined in classical genetics and in molecular genetics. In classical genetics, a mutant is any variant of a parental form, showing in its phenotypic properties some alteration that becomes transmitted to the progeny.

In contrast, it has become a habit in molecular genetics to call any alteration in the parental nucleotide sequence of the genome a mutation, whether it has phenotypic consequences or not. There is good reason to assume that in most spontaneously occurring mutagenesis events, the specific mechanism involved will not pay attention to whether the sequence alteration at the involved target site will cause a phenotypic alteration or not. Therefore, for studies on mechanisms and on the statistics of their occurrence, it is indicated to follow the molecular genetics definition of the term mutation, which we use here as a synonym of genetic variation. We will use the term *spontaneous mutation* to label any type of DNA sequence alteration unintended by the investigator. This definition says nothing about whether a mutation relates to a phenotypic change.

At present, the research on genetic variation mainly follows two strategies. First, the increasing availability of entire and partial genome sequences offers excellent means to compare regulatory sequence elements, specific domains of genes, entire genes, groups of genes, and entire genomes with regard to DNA sequence homologies and genome organization. Within a given species, this can reveal a genetic polymorphism. Between more or less related species, it can give a reliable measure for evolutionary relatedness. For example, the molecular clock – an indicator of evolutionary relatedness – is based on single nucleotide alterations. Sequence comparisons can often suggest how sequence alterations could have occurred in the course of past evolution. Secondly, a more reliable insight into the generation of genetic variations can be gained by the observation of individual processes of mutagenesis. In view of the large size of genomes and of the rare and random occurrence of spontaneous DNA sequence alterations, this approach is relatively difficult. However, a rich database is already available from microbial genetics, particularly from bacteria and their viruses and plasmids. Their relatively small genomes are haploid so that phenotypic effects caused by genetic variation normally become rapidly manifested. With appropriate selection and screening techniques, this allows one to identify occasional, functionally relevant mutations in populations. On the other hand, investigations on structural alterations in the genomes of individual bacterial colonies, for example, by the study of restriction fragment length polymorphism, can reveal when and where on the genome a DNA rearrangement must have occurred.

A quite solid, general result of this type of experimental investigations reveals that in the spontaneous generation of genetic variations, it is not just a single mechanism at work. Rather, a number of mechanistically different processes contribute to the overall mutagenesis. We will discuss selected examples below. Interestingly, a critical evaluation of the situation shows that the specific mechanisms of mutagenesis often depend both on nongenetic elements and on activities of specific enzymes, the products of the so-called evolution genes. The multitude of thus identified, distinct mechanisms contributing to the formation of genetic variations can be grouped into three qualitatively different natural strategies (Fig. 1). These are

1. local changes in the DNA sequences,
2. rearrangement of DNA segments within the genome,
3. the acquisition by horizontal transfer of a DNA segment originating in another kind of organism.

Selected examples for each of these strategies will be discussed in Sects. 4, 5, and 6.

3
Genetic Variation in Bacteria

Several seminal discoveries, largely based on work carried out with microorganisms between 1943 and 1953, were essential for the later development of molecular genetics.

1. It was realized that bacteria and bacteriophages have genes that can mutate, and that spontaneous mutations in microorganisms normally arise independently of the presence of selective agents. It was also learned that the genetic information of bacteria and of some bacteriophages is carried in DNA molecules rather than in other biological macromolecules such as proteins.
2. The newly discovered phenomena of DNA transformation, bacterial conjugation, and bacteriophage-mediated transduction revealed natural means of horizontal gene transfer between different bacterial cells.
3. It was seen that horizontal gene transfer has natural limits, including systems of host-controlled modification, which are today known as *DNA restriction–modification systems*.
4. Mobile genetic elements were identified as sources of genetic instability and were seen to represent mediators of genetic rearrangements. While such rearrangements are often caused by transposition, they can also result from the integration of a bacteriophage genome into the genome of its bacterial host strain, which is thereby rendered lysogenic.

It is at the end of this fruitful period of discoveries, in 1953, that structural analysis of DNA molecules led to the double-helix model. The filamentous structure of DNA molecules made it clear as to how genetic information could be contained in the linear sequences of nucleotides. The double helical nature of the model also offered an understanding of semiconservative DNA replication at the molecular level and thus of information transfer into progeny.

Many classical microbial genetic investigations were carried out with *Escherichia coli* K-12. Its genome is a single circular DNA molecule (chromosome) of about 4.6×10^6 bp. In periods of growth, the rate of spontaneous mutagenesis is about 10^{-9} per bp and per generation. This represents one new mutation in every few hundred cells in each generation. *E. coli* has several well-studied bacteriophages and plasmids. This material facilitates investigations on life processes in these bacteria.

Under good growth conditions, the generation time of *E. coli*, measured between one cell division and the next, is very short, on the order of 30 min. Upon exponential growth, this leads to a multiplication factor of 1000 every 5 h. Thus, a population of 10^9 cells representing 30 generations is reached from an inoculum of a single cell in only 15 h. This rapid growth rate greatly facilitates population genetic studies and thus facilitates investigations on the evolutionary process.

On the filiform DNA molecules of *E. coli* and its bacteriophages and plasmids, the genetic information is relatively densely stored as linear sequences of nucleotides or base pairs. Genes depend on the presence of continuous sequences of base pairs (reading frames) that encode specific gene products, usually proteins, and of expression control signals that ensure the occurrence of gene expression at the relevant time with the needed efficiency. Mutations can affect reading frames as well as control signals, both of which represent specific DNA sequences. In addition, some specific DNA sequences relate to the control of the metabolism of the DNA molecules, in particular, their replication.

In the following Sects. 4, 5, and 6, we will describe selected examples of mechanisms contributing each in its specific manner to the formation of genetic variants. We will group them into the already mentioned three major natural strategies (Fig. 1), although, as we will later see, some of the specific mechanisms involve more than one of these strategies.

4
Local Changes in the DNA Sequences

The process of DNA replication is one of the important sources of genetic variation, which may depend both on structural features of the substrate DNA and on functional characteristics of the replication fork.

Some of the "infidelities" of DNA replication are likely to depend on tautomeric forms of nucleotides, that is, a structural flexibility inherent in these organic compounds. Base pairing depends on specific structural forms. Conformational changes of nucleotides can result in a mispairing if short-living, unstable tautomeric forms are "correctly" used in the synthesis of the new complementary strand upon DNA replication. For this reason, we do not consider mutations resulting in this process as errors and call them *infidelities*. DNA replication is indeed one of the sources of nucleotide substitution, and this plays an important role in the evolutionary development of biological functions.

An inherent low degree of chemical instability of nucleotides represents another source of mutagenesis. For example, cytosine can undergo oxidative deamination to become uracil. Upon replication, this gives rise to an altered base pairing and results also in nucleotide substitution.

Other replication infidelities that may relate to slippage in the replication fork can result in either the deletion or the insertion of one or a few nucleotides in the newly synthesized DNA strand. If such mutations occur within reading frames for protein synthesis, the phenotypic effect may be drastic. This is the case when, in the protein synthesized from a gene affected by a frameshift mutation, the amino acid sequence downstream of the site of mutation strongly differs from that of the nonmutated product. In addition, the size of such a mutated protein is usually altered depending on the chance occurrence of an appropriate stop codon in the new reading frame.

Proofreading devices and other enzymatic repair systems prevent replication infidelities from producing mutations at intolerably high rates. Generally, they act rapidly after replication by screening for imperfect base pairing in the double helix. Successful repair thereby requires specific means to distinguish the newly replicated DNA strand from its template, the complementary parental strand. Because of these correction activities, many primary mispairings are removed before they have the opportunity to become fixed as mutations. DNA repair systems modulate the frequencies of mutagenesis.

Genetic information of some viruses, and sometimes also segments of genetic information of chromosomal origin, may pass through RNA molecules, which may later become retrotranscribed into DNA. No efficient repair systems are known to act at the level of RNA. Indeed, RNA replication shows a higher degree of infidelity than DNA replication. In consequence, genetic information that becomes replicated as RNA molecules generally suffers increased mutation rates.

A relatively large number of internal and environmental chemicals exert mutagenic effects by means of molecular mechanisms that in many cases are well understood and often cause local DNA sequence changes. Some intermediate products of the normal metabolic activities of a cell may be mutagenic and may thus contribute to spontaneous local mutagenesis. The mutagens of the environment include not only a multitude of chemical compounds but also ultraviolet radiation and some physicochemical constraints such as elevated temperature, which influence the chemical stability of nucleotides. Each of these mutagens and mutagenic conditions contributes in a specific way to the generation of genetic variations.

Again, many of the potential sequence alterations brought about by internal and environmental mutagens are efficiently repaired by enzymatic systems. However, since the efficiency of such repair is rarely 100%, evolutionarily relevant mutations persist.

5
Intragenomic DNA Rearrangements

Various recombination processes are well known to mediate DNA rearrangements, which often result in new nucleotide sequences.

While in haploid organisms general recombination is not essential for propagation, it influences genetic stability at the population level in various ways as a generator of new sequence varieties. For example, it can bring about sequence duplications and deletions by acting at segments of homology that are carried at different locations in a genome. General recombination is also known to act in the reparation of damage caused to DNA by ionizing radiation. In this case, an intact genome can become assembled from undamaged segments of sister copies of the chromosome by homologous recombination. The best-known contribution of general recombination to genetic diversity is meiotic recombination bringing about random recombinants between the paternal and the maternal chromosomes in diploid organisms.

Two other widely spread types of recombination systems are dealt with in more detail in Sects. 5.1 and 5.2: site-specific recombination and transposition. Both are known to contribute to genetic variation. Still other recombination processes, such as the one mediated by DNA gyrase can, for the time being, perhaps best be grouped as illegitimate recombinations. This group may contain several different molecular mechanisms that act with very low efficiency and have remained at least in part unexplained.

5.1
Site-specific DNA Inversion at Secondary Crossover Sites

Genetic fusions represent the results of joining together segments of two genes (gene fusions) or of two operons (operon fusions) that are not normally together. An *operon* is a set of often functionally related genes that are copied into messenger RNA (i.e. transcribed) as a single unit. As a result of this organization, those genes are coordinately regulated; that is, they are turned on or off at the same time. Therefore, in an operon fusion, one or more genes are put under a different transcription control, but the genes *per se* remain unchanged. In contrast, gene fusion results in a hybrid gene composed of sequence motifs and often of functional domains originating in different genes.

In site-specific DNA inversion, a DNA segment bordered by specific DNA sequences acting as sites of crossing-over becomes periodically inverted by the action of the enzyme DNA invertase. Depending on the location of the crossover sites, DNA inversion can give rise to gene fusion or to operon fusion. The underlying flip-flop system can result in microbial populations composed of organisms with different phenotypic appearances: if, for example, the DNA inversion affects the specificity of phage tail fibers, as is the case with phages P1 and μ of *E. coli*, phage populations with two different host ranges will result.

Occasionally, a DNA sequence that deviates considerably from the efficiently used crossover site, a so-called *secondary crossover site*, can serve in DNA inversion, which thus involves a normal crossover site and a secondary crossover site. This process results in novel DNA arrangements, many of which may not be maintained because of lethal consequences or reduced fitness; but if a few new sequences are beneficial to the life of the organism, these may be selectively favored. This DNA rearrangement activity can thus be looked at as evolutionarily important. Since many different DNA sequences can serve in this process as secondary crossover sites, although at quite low frequencies, site-specific DNA inversion systems act as variation generators in large populations of microorganisms. I have thus postulated that this evolutionary role of DNA inversion systems may be more important than their much more efficient flip-flop mechanism, which can at most help a microbial population to more readily adapt to two different, frequently encountered environmental conditions. As a matter of fact, other strategies could be used as well for the latter purpose.

Computer-aided comparison of DNA sequences quite often reveals that independent genes may consist of a particular domain with high homology and of other DNA sequences showing no significant signs of relatedness. DNA inversion using secondary sites of crossing-over is a potential source of such mosaic genes. DNA inversion can span over relatively large distances in DNA molecules and has the advantage of not loosing any DNA sequences located between the two sites of crossing-over. Deletion formation represents another source for gene fusion, but it has the disadvantage that the DNA sequences between the sites of crossing-over are usually eliminated.

5.2
Transposition of Mobile Genetic Elements

Nine different mobile genetic elements have been found to reside, often in several copies, in the chromosome of *E. coli* K-12 derivatives. This adds up to occupation of about 1% of the chromosomal length by such insertion sequences, also called *IS elements*. At rates on the order of 10^{-6} per individual IS element and per cell generation, these mobile genetic elements undergo transpositional DNA rearrangements. These include simple transposition of an element and more complex DNA rearrangements such as DNA inversion, deletion formation, and the cointegration of two DNA molecules. Because of different degrees of specificity in the target selection upon transposition, the IS-mediated DNA rearrangements are neither strictly reproducible nor fully random. Transposition activities thus also act as variation generators. In addition to DNA rearrangements mediated by the enzyme transposase, which is usually encoded by the mobile DNA element itself, other DNA rearrangements just take advantage of extended segments of DNA homologies at the sites of residence of identical IS elements at which general recombination can act. Altogether, IS elements represent a major source of genetic plasticity of microorganisms.

Transposition occurs not only in growing populations of bacteria but also in prolonged phases of rest. This is readily seen with bacterial cultures stored at room temperature in stabs (little vials containing a small volume of growth medium in agar). Stabs are inoculated with a drop of a bacterial culture taken up with a platinum loop, which is inserted ("stabbed") from the top to the bottom of the agar. After overnight incubation, the stab is tightly sealed and stored at room temperature. Most strains of *E. coli* are viable in stabs during several decades of storage. That IS elements exert transpositional activities under these storage conditions is easily seen as follows.

A stab can be opened at any time, a small portion of the bacterial culture removed, and the bacteria well suspended in liquid medium. After appropriate dilution, bacteria are spread on solid medium. Individual colonies grown upon overnight incubation are then isolated. DNA from such subclones is extracted and fragmented with a restriction enzyme. The DNA fragments are separated by gel electrophoresis. Southern hybridization with appropriate hybridization probes can then show whether different subclones reveal restriction fragment length polymorphisms

(RFLPs), which are indicative of the occurrence of mutations during storage.

If this method is applied to subclones isolated from old stab cultures, and if DNA sequences from residential IS elements serve as hybridization probes, an extensive polymorphism is revealed. None or only little polymorphism is seen with hybridization probes from unique chromosomal genes. Good evidence is available that transposition represents a major source of this genomic plasticity observed in stabs, which at most allow for a very residual growth at the expense of dead cells. One can conclude that the enzymes promoting transposition are steadily present in the stored stabs. Indeed, the IS-related polymorphism increases linearly with time of storage for periods as long as 30 years. In a culture of *E. coli* strain W3110 analyzed after 30 years of storage, each surviving subclone had suffered on the average about a dozen RFLP changes as identified with hybridization probes from eight different residential IS elements, of which IS5 was the most active. Lethal mutations could of course not be identified in this study.

Lethal mutations that affect essential genes for bacteriophage reproduction can be accumulated in the prophage state of the phage genome in its lysogenic host. Such mutants can be screened for their inability to produce infective phage particles upon induction of phage reproduction. Experiments were carried out with *E. coli* lysogenic for a phage P1 derivative grown in batch cultures at 30 °C for about 100 generations allowing for alternative periods of growth and rest. Most of the independent lethal mutants could thereby be identified to be caused by the transposition of an IS element from the host chromosome into the P1 prophage that is maintained in its host as a plasmid. In these experiments, IS2 was the most active element and it mainly inserted into a few preferred regions of the P1 genome, but each time at another site. The used insertion targets did not show any detectable homology or similarity with each other. This is another good example for an enzymatically mediated variation generator. The experiment as such identifies IS transposition as a major source of lethal mutagenesis.

There is no evidence available that bacterial mobile genetic elements would play an essential role in the bacterial life span extending from one cell division to the next. However, these elements are major players in the evolution of bacterial populations. As we have seen here, they contribute to intragenomic DNA rearrangements. Depending on the target sequences involved, the resulting mutations may often be lethal by interrupting essential reading frames or expression control regions. Favorable mutations may be relatively rare, but these can contribute to evolutionarily advantageous developments of the genome. In Sect. 6, we will see that mobile genetic elements also play important roles in the natural strategy of DNA acquisition.

6
DNA Acquisition

While the mutagenesis mechanisms belonging to the strategies of local changes in the DNA sequences (Sect. 4) and of intragenomic DNA rearrangements (Sect. 5) are exerted within the genome and can affect any part of the genome, an additional strategy of spontaneous sequence alterations depends on an external source of genetic information. In DNA or gene acquisition, genetic information indeed originates from an organism other than the one undergoing mutagenesis. In bacteria, DNA acquisition can occur by means of transformation, conjugation, or virus-mediated transduction. In the latter two strategies of horizontal gene transfer, a plasmid or a virus, respectively, acts as natural gene vector.

The association and dissociation of chromosomal genes with natural gene vectors often arises from transpositional activities and from general recombination acting at IS elements that are at different chromosomal locations. These mechanisms have been well studied with conjugative plasmids and with bacteriophage genomes serving in specialized transduction. For example, composite transposons, which are defined as two identical IS elements flanking a segment of genomic DNA (often with more than one gene unrelated to the transposition process), are known to occasionally transpose into a natural gene vector and, after their transfer into a receptor cell, to transpose again into the receptor chromosome. Hence, together with other mechanisms, such as site-specific and general recombination, transposition represents an important promoter of horizontal gene transfer.

Several natural factors seriously limit gene acquisition. Transformation, conjugation, and transduction depend on surface compatibilities of the bacteria involved. Furthermore, upon penetration of donor DNA into receptor cells, the DNA is very often confronted with restriction endonucleases. These enzymes cause a fragmentation of the invading foreign DNA, which is subsequently completely degraded. Before fragments become degraded, however, they are recombinogenic and may find a chance to incorporate all or part of their genetic information into the host genome. Therefore, we interpret the role of restriction systems as follows: they keep the rate of DNA acquisition low, and at the same time they stimulate the fixation of relatively small segments of acquired DNA to the receptor genome. This strategy of acquisition in small steps can best offer microbial populations the chance to occasionally extend their biological capacities without extensive risk of disturbing the functional harmony of the receptor cell by acquiring too many different functions at once. These considerations have their relevance at the level of selection exerted on the hybrids resulting from horizontal DNA transfer. This selection represents one of the last steps in the acquisition process.

DNA acquisition by horizontal gene transfer is a particularly interesting source of new genetic information for the receptor bacterium because the chance that the acquired DNA exerts useful biological functions is quite high – most likely, it has already assumed the same functions in the donor bacterium.

An interesting hypothesis links the universality of the genetic code with the important role played by horizontal gene transfer in the evolutionary development of the living world. According to this view, those organisms using the most common genetic language would, in the long term, be able to profit best from the increasing worldwide pool of genetic functions under the pressure of adapting to changing living conditions.

7
The Three Natural Strategies Generating Genetic Variations Contribute Differently to the Evolutionary Process

Biological evolution is a systemic process. As outlined above, many different specific mechanisms contribute to generate genetic diversity that represents at any time the substrate for natural selection. The building up of functional complexity is a stepwise process, in which many random attempts of genetic alterations become rapidly rejected, while relatively few novel sequences are approved as favorable by natural selection and are maintained and amplified. The genome can thus be seen as a cabinet in which key information from favorable historical developments is stored. Stepwise, additional favorable information is added. In the context of changing selective conditions, stored information, having lost its functionally beneficial relevance, may be deleted or favorably altered. As we have seen, a multitude of different mechanisms are behind this dynamic process. For a better understanding of the events, we have grouped the identified mechanisms into three major natural strategies of genetic variation: local changes in the DNA sequences, intragenomic DNA rearrangements, and DNA acquisition. These three strategies have different qualities with regard to their contributions to biological evolution.

The local DNA sequence change is probably the most frequently involved strategy of genetic variation. Indeed, its frequency, which depends primarily on intrinsic properties of matter, chemical instability and conformational flexibility, would be intolerably high if it would not be modulated by efficient enzymatic systems of DNA repair. Local sequence changes bring about nucleotide substitution, the deletion and the insertion of one or a few base pairs, or a local scrambling of a few base pairs. These sequence changes can contribute to a stepwise improvement of a biological function. It must be kept in mind that the functional test for such improvement is carried out by natural selection. In principle, a long series of stepwise local sequence changes could also be expected to bring about a novel biological function. However, this kind of long-term process can gain efficiency only once natural selection starts to be exerted on such upcoming function.

In contrast, the reshuffling of DNA segments within the genome can be considered as a tinkering with existing elements, whereby favorable gene fusions and operon fusions may occasionally result. DNA rearrangement can also be the source of gene duplication and higher amplification, which are widely recognized contributions to the evolutionary progress.

In Sect. 6, we have already pointed to the evolutionarily high efficiency of horizontal gene transfer. As a matter of fact, DNA acquisition allows the recipient organism to share in the success of evolutionary developments made by others. In drawing the evolutionary tree of bacteria, DNA acquisition should be accounted for by more or less randomly adding temporal horizontal shunts between individual branches. It must be kept in mind that usually only small DNA segments flow through such shunts in horizontal gene transfer.

Several of the specific mechanisms of genetic variation employ, strictly speaking, more than one of the three strategies shown in Fig. 1. In transposition of IS elements, for example, a chromosomal DNA segment consisting of the mobile genetic element can undergo a translocation and thereby become inserted at a new target site. As a rule, the target sequence thereby gets duplicated, which usually involves a few nucleotides. Thus, this transposition event will consist of both a DNA rearrangement and a local sequence change.

As far as we know, most of the well-studied microbial strains use in parallel each of the three natural strategies for the generation of genetic variations. In addition, bacteria very often use not only one, but several different specific mechanisms for mutagenesis by each of the strategies. Dissimilar specific mechanisms often work with different efficiencies as reflected by their contribution to the overall mutation rate. For any given strategy, it might be less relevant which specific mechanism is at work than the fact that the particular strategy finds its application with an evolutionarily useful efficiency. In other words, specific mechanisms may substitute for each other within a strategy, at least to some degree. This rule does not apply between the strategies because of the difference in the qualities of their evolutionary contributions.

The efficiency displayed by a given specific mechanism of spontaneous mutagenesis may depend on both internal (e.g. availability of enzymes that mediate mutagenesis) and external factors (e.g. environmental stress). It is also to be noted that some mechanisms may act more or less randomly along a DNA molecule, while other mechanisms may show regional or site preferences for their activities. In view of these considerations, we tend to assume that an evolutionarily fit (or well prepared) organism should best be able to use a few specific mutagenesis mechanisms for each of the three strategies to generate genetic variations. In Sect. 8, we will explain what we mean by evolutionary fitness.

8
Evolution Genes and Their Own Second-order Selection

The attentive reader will have seen in the description of some specific mechanisms contributing to the spontaneous mutagenesis that besides a number of nongenetic factors, specific products of genes are very often at work. These gene products can belong to systems for repair of DNA damage and will, in this case, modulate the frequency of mutagenesis. Similarly, restriction enzymes seriously reduce both the chance of DNA acquisition and the size of a DNA segment that may eventually be acquired by the recipient cell. Other gene products such as transposases and other mediators of DNA recombination act as generators of genetic variations. Since variation generators and modulators of the frequencies of genetic variation are evolutionary functions, we call the underlying genetic information *evolution genes*. In the microbial world, these evolution genes generally play no essential role in the physiology of individual lives going by cell division from one generation to the next. Under laboratory conditions, neither restriction–modification systems nor enzymes for DNA rearrangements are needed for the propagation of bacteria. The role of such enzyme systems is primarily evolutionary and becomes manifest at the level of populations.

We assume that evolutionary genes are themselves submitted to selective pressure. However, such selection cannot follow the rules of direct selection for improvements of essential functions such as those of housekeeping genes. Rather, the selection for the presence and improvement of a variation generator will be exerted at the level of populations. Clearly, it will also be an individual that may one day undergo a mutation, which improves an evolutionary function. This function will also be exerted in its progeny in which appropriate genetic variants of genes for directly selected products will be either more or less abundant. Any genetic alteration that affects an evolution gene and proves in the long term to be of higher evolutionary value will be maintained and will provide an evolutionary benefit to the carrier of the involved gene. In the long run, this will lead to fine-tuning of the evolutionary functions of both variation generators and of modulators of the frequency of genetic variation. The underlying indirect selection based on the cells ability to provide genetic variants at a well-balanced level is called *second-order selection*.

We must be aware that some gene products may exert their essential functions for the benefit of both the life of the individual and the evolutionary progress of the population. In these cases, we assume that evolutionary selection is exerted for both kinds of functions and will eventually bring the gene to a fine-tuned state to optionally carry out its functions for each of the different purposes. However, as we have already mentioned, a number of gene products involved in genetic variation are inessential for the lives of individual bacterial cells. Similarly, the products of many housekeeping genes are inessential for biological evolution.

9
Arguments for a General Relevance of the Theory of Molecular Evolution for All Living Organisms

Largely on the basis of evidence from microbial genetics, we have so far postulated that the products of a number of evolution genes contribute, each in its specific way, to the generation of genetic variants at evolutionarily useful frequencies. Thereby, the sources of mutagenesis may relate either to the activity of the evolution gene product itself (e.g. a transposase) or to a nongenetic factor (e.g. a chemical mutagen or an intrinsic structural flexibility of a nucleotide). In many cases, nongenetic factors and products of evolution genes cooperate in the formation of genetic variants at physiologically tolerable and evolutionarily beneficial levels. This is, for example, the case in spontaneous mutagenesis by an environmental mutagen when some of the primary damage on the DNA gets successfully repaired while some other damage leads to a fixed mutation.

The theory of evolution postulates that life on Earth started almost four billion years ago with primitive, unicellular microorganisms. It is in the first two billion years that microbes must have developed the basis for the actual setup of evolutionary strategies and the underlying evolution genes. One can postulate that the acquired evolutionary capacities could have allowed some microbial populations to undergo a division of labor in more and more stable associations of cells. This development might later have led to multicellular organisms. In this kind of development, the evolutionary fitness of the involved organisms might have been an important precondition. At still later stages of further evolutionary development, the three natural strategies for generating genetic variations (Sects. 4 to 7) must have continued exerting their evolutionary influence together with some other factors such as the formation of symbiotic associations. As a matter of fact, we attach considerable evolutionary relevance to endosymbiosis of higher organisms with bacteria. Such situations of cohabitation may form an ideal condition for occasional horizontal gene transfer between the close associates.

A scientifically justified quest for further experimental proof of the postulates of the theory of molecular evolution remains quite difficult to be answered. Clearly, there is a need for research on the spontaneous generation of genetic variation in higher organisms, ideally at the level of the genomes. While this is already quite difficult in microorganisms, it is of increased perplexity with the much larger genomes of higher organisms. However, sequence comparisons now offer fruitful ways to search for sequence homologies, sequence similarities, single nucleotide polymorphisms, as well as for traces of intragenomic DNA rearrangements and of horizontal transfer of genetic information. Data so far available are in support of the principles of molecular evolution outlined in Sect. 2, which are likely to be valid for any kind of living organism.

Some of the general evolutionary strategies developed in microorganisms must have also turned out to be useful for the developmental and physiological processes at somatic levels of higher organisms. The generation of antibody diversity in the immune systems of vertebrates by genetic rearrangements and so-called *somatic mutagenesis* is a good example. Another example is the enzymatic repair of DNA damage caused in somatic cells by external mutagens such as UV irradiation.

These considerations illustrate that whatever gene function may prove to be useful for whatever particular purpose, it may be evolutionarily retained and in the course of time further finetuned. We have already encountered this principle in the microbial world, where we have postulated multifunctional enzymes (such as working both for the physiology of the cells and for an evolutionary task) to become evolutionarily improved both by direct and by secondorder selection for the various functions. This may also be the case in higher organisms.

10
Conceptual Aspects of the Theory of Molecular Evolution

With reference to Sect. 2.2, it is fair to again explicitly state that for the time being, evolution genes and evolution functions are a concept rather than a fully proven fact. This concept is based on a particular way to interpret numerous available experimental data. We will briefly analyze the difficulties to clarify the situation in a scientific debate. This will be followed by pointing to philosophical and more practical values of a deeper understanding of the molecular processes that drive biological evolution.

10.1
Pertinent Scientific Questions

In the history of scientific investigations, biologists have often searched for evidence that living organisms would be able to specifically modify, or adapt, their genetic information in order to better cope with upcoming changes in the living conditions. Most of these attempts have failed to give the expected response. In other cases, where a certain degree of specific adaptation could be observed, specific causal explanations have sometimes been found upon deeper investigation. However, there is at present no good scientific evidence for a general rule that genetic alterations would always be directed toward a specific goal. This situation favors the view that spontaneous mutations affect DNA more or less randomly, which is in line with the general observation that only a minority of spontaneous mutants prove to be favorable under the encountered living conditions.

The postulate of evolution genes that act as generators of genetic variations may be a surprise in this context. This has to do with a widely followed concept of genetic information as a strict program for the fulfillment of a specific task. This definition does apply to many housekeeping genes, the products of which efficiently catalyze a reaction that reliably always yields the same product. This is not how a variation generator that does not work efficiently and that yields a different product from case to case functions. A good example is the transposition of mobile genetic elements. However, not all scientists see the primary function of a mobile genetic element in genetic variation. Rather, some colleagues interpret IS translocation, which often goes along with the replication of the element, as a selfish activity. This interpretation considers mobile genetic elements as parasites with the argument that their activity would often harm their host cell.

This discussion shows that the concept of evolution genes cannot easily be defended by referring to scientific evidence. Rather, the concept reflects an attitude of the observer of natural events. According to the view defended in this article, nature actively cares for biological evolution. The products of evolution genes are actively involved in generating different kinds of genetic variants at frequencies insuring both a certain genetic stability required for maintaining the concerned form of life and a low frequency of genetic variations as the driving force of evolution. This interpretation recognizes biological evolution as an essential principle of self-organization of life on Earth.

Another pertinent question that cannot find an easy scientific answer relates to the evolutionary function of viruses. At least some viruses are clearly identified to sometimes act as gene vectors in horizontal gene transfer. Some of them also temporarily integrate their genome into the host genome. This relates to the lysogenic state of bacteria as well as to endogenous viruses such as retroviruses that reside in many higher organisms. Again, one may wonder if these viruses primarily fulfill evolutionary functions for the evolutionary development of their hosts or whether they should rather be looked at as parasites that may carry out some evolutionary function by accident.

While prokaryotic organisms have genomes that are relatively densely packed with functional genes, many higher organisms have extended segments of intergenic DNA sequences, some of which are highly repetitive. Some of these noncoding sequences are highly homogeneous with regard to their nucleotide composition. While the biological roles played by noncoding regions are still not well understood, it has been postulated that compositional constraints may influence natural selection. These aspects have not been covered in this article and they may more specifically relate to the molecular evolution of higher organisms, in addition to the principles outlined here largely on the basis of evidence from microbial genetics.

10.2
Philosophical Values of the Knowledge on Molecular Evolution

One of the central questions of human curiosity is to know where life – and more specifically human life – comes from. The Darwinian theory opposed to the idea of a specific act of creation for each particular form of life, the alternative explanation of a steady evolutionary development implying the descent of the actual species from common ancestors. The directions of evolution are thereby given by natural selection acting steadily on all available forms of life including all present variants. Until recently, the sciences could not specifically explain how genetic variants are generated. With recently developed research strategies, molecular genetics can now fill this gap. The branch of science called *molecular evolution* explains that there is not just a single source for genetic variants. Rather, many different specific mechanisms contribute to the generation of genetic variants at low frequencies. These mechanisms follow one and sometimes more than one basic strategies of evolutionary development. These are local changes in the DNA sequences, rearrangements of DNA segments within the genome, and acquisition of segments of foreign DNA by horizontal gene transfer. As a general rule, spontaneous mutagenesis is not specifically directed; it is at least to some extent random, so that only a minor fraction of spontaneous genetic variants turn out to be favorable for the concerned organism and thus provide it with a selective advantage. Nevertheless, new knowledge on the precise molecular mechanisms of the generation of genetic variations provides strong evidence that, in many cases, specific enzymes are involved – the products of evolution genes. These products work in tight collaboration with nongenetic factors that can be intrinsic properties of matter or environmental conditions. This view of the evolutionary process represents the core of a theory of molecular evolution and it can be seen as an extension of neo-Darwinism at the molecular level.

It should be clearly stated that the theory of molecular evolution does not explain the origin of life. It can, however, explain that biological evolution exerted in all living beings is a steady, dynamic process that is actively promoted not only by intrinsic properties of matter but also by the intervention of products of evolution genes or more generally of evolutionary functions of many different gene products. A recently published book devoted to these exciting insights is entitled *Darwin in the Genome*.

The high philosophical value of this extension of our worldview is obvious and merits to be widely discussed and evaluated in its various cultural dimensions. One interesting aspect is the implied duality of the genome. Indeed, evolution genes are located together with all the other genes on the genome and on accessory DNA molecules such as plasmids and viral genomes. While probably a major part of gene products carries out functions to benefit the cell and the individual, often a multicellular organism, probably a minority of gene products works for the biological evolution of the concerned population. Note, however, that the generation of a novel genetic variant obviously also occurs in an individual cell. But this act of creation has only a small chance to bring to the concerned cell and to its descendants a selective advantage. More often, the mutation is unfavorable and renders the life of the concerned organism more troublesome. As long as the spontaneous rate of mutagenesis remains low (mostly thanks to the intervention of evolution genes), unfavorable mutations are tolerable at the level of propagating populations.

Incidentally, the situation described here offers a possible explanation to the quite difficult theodicean question: why does God, despite His love for the human creature, admit that physically evil events such as a mutation causing an inheritable disease can occur to individuals? As a matter of fact, Genesis describes creation as a stepwise process, which implies the permanent evolutionary expansion of the diversity of life forms. Genesis also states that God evaluated this system as good.

Hence, biological evolution occurs according to God's intention to amplify diversity of life on our planet. Occasional unfavorable mutations affecting rare individuals in populations is a sacrifice brought to the creative force residing in the system of molecular evolution.

In brief, the genetic information contained in each genome – of bacteria as well as of all higher organisms – represents an internal duality. It serves individuals for the fulfillment of their individual lives and it serves populations for a slow but steady expansion of life forms and thus of biodiversity.

10.3
Aspects Relating to Practical Applications of Scientific Knowledge on Molecular Evolution

Living organisms today occupy an amazing variety of ecological niches on our planet Earth. These niches include extreme physicochemical conditions such as elevated temperatures, high pressure, and quite unusual compositions of chemical elements. However, despite the intrinsic potential of the living world to evolutionarily expand, one can estimate the carrying capacity of the planet for life to be in the order of 10^{30} living cells. Although this is a very large number, it seriously limits free expansion of life in its various forms.

The following reflections should help illustrate this statement. An adult human being carries in the order of 10^{13} cells. The human population today, thus occupies a share of about 10^{23} cells of the available 10^{30}. Incidentally, this happens to be close to the average available for each of the estimated 10^7 different species of organisms on the planet. Bacteria propagate by cell division as outlined in Sect. 3. Extending the reflections made there, one can conclude that from an inoculum with one single bacterial cell, one can theoretically expect to obtain 10^{30} cells within only 50 h. In reality, growth will be stopped much earlier by lack of nutrition, but this reflection illustrates well the enormous internal forces for expansion of a given form of life.

Similarly, a high potentiality for evolutionary expansion toward more diversity resides in the mechanisms of molecular evolution that are described in this article. These mechanisms can serve us as a basis to better understand both the origin and the steady replenishment of biodiversity as well as the internal limits set to evolutionary expansion. This knowledge can and should increasingly be used as a background for any measures taken toward the protection of biodiversity and of habitats for diverse forms of life. Last but not the least, a better understanding of the evolutionary process can be of help to render the development of agricultural and related practices more sustainable.

Genetic engineering offers ample new possibilities for the sustainable production of medical drugs, to obtain food of higher quality, and to reduce the nocent impact of the human civilization on the environment. The serious reservations made by large parts of the human population impede many of the proposed biotechnological applications. A part of these concerns refer to unpredictable long-term effects of genetically modified organisms (GMO) that are released into the environment, such as in agricultural applications. Scientific assessments of long-term and, in particular, evolutionary effects of such applications are thus required. Knowledge of the natural strategies of molecular evolution can provide a good basis for such studies. As a matter of fact, in genetic engineering, DNA sequence alterations are brought about within the genome by site-directed mutagenesis in studies of the biological functions of specific genes. In addition, well-defined segments of DNA are introduced into other organisms either in view of amplifying the particular DNA segment or in view of harvesting a specific gene product. GMO can also directly serve in applications as GM food and for bioremediation by microorganisms. A candid comparison of these practices involving genetic engineering with the natural strategies of generation of genetic variations reveals a high degree of similarities. The amounts of nucleotides or lengths of DNA sequences involved in these genetic modifications, both in genetic engineering and in the natural genetic variation, are of the same order of magnitude. Depending on the strategy involved, they may concern one to a few base pairs, or in other instances, a DNA segment containing a sequence domain or one to a few genes, both in intragenomic DNA rearrangements and in the horizontal transfer of DNA between two different organisms. Thus, one can principally expect that long-term evolutionary risks of GMO compare with the biohazard intrinsic to the natural process of biological evolution. Similar risks may also be inherent in classical breeding techniques.

These considerations ask for a more integral, holistic, and critical evaluation of the impact of past and present human activities on the natural process of biological evolution. Such assessments should address any human impact on genetic variation, natural selection, and isolation. As far as we know from the long-term history, the foundations of life and its evolutionary development on Earth are relatively robust. This is good news for us human beings, but it should not exempt us from a responsible and well-reflected use of our scientific knowledge in any attempt to render our own lives more easy and comfortable.

See also Molecular Basis of Genetics.

Bibliography

Books and Reviews

Caporale, L.H. (Ed.) (1999) *Molecular Strategies in Biological Evolution*, Vol. 870, Annals of the New York Academy of Sciences, New York, NY.

Caporale, L.H. (2003) *Darwin in the Genome: Molecular Strategies in Biological Evolution*, McGraw-Hill, New York.

Kucherlapati, R., Smith, G.R. (Eds.) (1988) *Genetic Recombination*, American Society for Microbiology, Washington, DC.

Moses, R.E., Summers, W.C. (Eds.) (1988) *DNA Replication and Mutagenesis*, American Society for Microbiology, Washington, DC.

Shapiro, J.A. (1983) *Mobile Genetic Elements*, Academic Press, New York.

Primary Literature

Arber, W. (1991) Elements in microbial evolution, *J. Mol. Evol.* **33**, 4–12.

Arber, W. (1993) Evolution of prokaryotic genomes, *Gene* **135**, 49–56.

Arber, W. (1995) The generation of variation in bacterial genomes, *J. Mol. Evol.* **40**, 7–12.

Arber, W. (2000) Genetic variation: molecular mechanisms and impact on microbial evolution, *FEMS Microbiol. Rev.* **24**, 1–7.

Arber, W. (2002) Evolution of prokaryotic genomes, *Curr. Top. Microbiol. Immunol.* **264/I**, 1–14.

Arber, W. (2002) Molecular evolution: comparison of natural and engineered variations, *The Pontifical Academy of Sciences. Scripta Varia* **103**, 90–101.

Arber, W. (2003) Elements for a theory of molecular evolution, *Gene.* **317**, 3–11.

Arber, W. (2003) Traditional Wisdom and Recently Acquired Knowledge in Biological Evolution, *Proceedings of UNESCO Conference* on "Science and the Quest for Meaning"; in press.

Arber, W., Hümbelin, P., Caspers, P., Reif, H.J., Iida, S., Meyer, J. (1981) Spontaneous mutations in the *Escherichia coli* prophage P1 and IS-mediated processes, *Cold Spring Harbor Symp. Quant. Biol.* **45**, 38–40.

Arber, W., Naas, T., Blot, M. (1994) Generation of genetic diversity by DNA rearrangements in resting bacteria, *FEMS Microbiol. Ecol.* **15**, 5–14.

Bernardi, G. (2000) Isochores and the evolutionary genomics of vertebrates, *Gene* **241**, 3–17.

Bernardi, G. (2000) The compositional evolution of vertebrate genomes, *Gene* **259**, 31–43.

Drake, J.W. (1991) Spontaneous mutation, *Annu. Rev. Genet.* **25**, 125–146.

Echols, H., Goodman, M.F. (1991) Fidelity mechanisms in DNA replication, *Annu. Rev. Biochem.* **60**, 477–511.

Galas, D.J., Chandler, M. (1989) Bacterial Insertion Sequences, in: Berg, D.E., Howe, M.M. (Eds.) *Mobile DNA*, American Society for Microbiology, Washington, DC, pp. 109–162.

Glasgow, A.C., Hughes, K.T., Simon, M.I. (1989) Bacterial DNA Inversion Systems, in: Berg, D.E., Howe, M.M. (Eds.) *Mobile DNA*, American Society for Microbiology, Washington, DC, pp. 637–659.

Iida, S., Hiestand-Nauer, R. (1987) Role of the central dinucleotide at the crossover sites for the selection of quasi sites in DNA inversion mediated by the site-specific Cin recombinase of phage P1, *Mol. Gen. Genet.* **208**, 464–468.

Lorenz, M.G., Wackernagel, W. (1994) Bacterial gene transfer by natural genetic transformation in the environment, *Microbiol. Rev.* **58**, 563–602.

Naas, T., Blot, M., Fitch, W.M., Arber, W. (1994) Insertion sequence-related genetic variation in resting *Escherichia coli* K-12, *Genetics* **136**, 721–730.

Naas, T., Blot, M., Fitch, W.M., Arber, W. (1995) Dynamics of IS-related genetic rearrangements in resting *Escherichia coli* K-12, *Mol. Biol. Evol.* **12**, 198–207.

Sandmeier, H. (1994) Acquisition and rearrangement of sequence motifs in the evolution of bacteriophage tail fibers, *Mol. Microbiol.* **12**, 343–350.

Sengstag, C., Arber, W. (1983) IS2 insertion is a major cause of spontaneous mutagenesis of the bacteriophage P1: non-random distribution of target sites, *EMBO J.* **2**, 67–71.

Weber, M. (1996) Evolutionary plasticity in prokaryotes: a panglossian view, *Biol. Philos.* **11**, 67–88.

West, S.C. (1992) Enzymes and molecular mechanisms of genetic recombination, *Annu. Rev. Biochem.* **61**, 603–640.

Woese, C.R. (1987) Bacterial evolution, *Microbiol. Rev.* **51**, 221–271.

Part 3
Genomes of Model Organisms

Genomics and Genetics. Edited by Robert A. Meyers.
Copyright © 2007 Wiley-VCH Verlag GmbH & Co. KGaA, Weinheim
ISBN: 978-3-527-31609-0

14
E. Coli Genome

Hirotada Mori[1] and Takashi Horiuchi[2]
[1] *Faculty of Pharmaceutical Sciences, Kyushu University, Fukuoka, Japan*
[2] *Research and Education Center for Genetic Information,*
Nara Institute of Science and Technology, Ikoma, Japan

1	The *E. coli* Genome Sequence	410
2	ORF Predicted from the Complete Genome Sequence of *E. coli*	410
3	Repetitive Sequences, Sites, RNA Genes, etc.	411
4	Post Sequence Genome Project	413
4.1	Transcriptome Analysis	414
4.2	Proteome and Interactome Analysis	414
5	*E. coli* Genome as a Model for Systems Biology	414
6	Useful *E. coli* Websites	415
	Bibliography	416
	Books and Reviews	416
	Primary Literature	416

Keywords

Escherichia coli
A member of γ protepbacteria Enterobacteriaceae normally found in the human gastrointestinal tract among numerous strains. K-12 has been used for most experiments in molecular biology.

Genome
The total complement of genetic material in a cell or individual.

Genomics and Genetics. Edited by Robert A. Meyers.
Copyright © 2007 Wiley-VCH Verlag GmbH & Co. KGaA, Weinheim
ISBN: 978-3-527-31609-0

Functional Genomics
A field of global functional analysis of genome based on complete sequence data.

■ Since Wollman & Bronfenbrenner introduced this organism in the 1930s, *Escherichia coli* has become one of the most thoroughly studied organisms in biology. The discovery of conjugation by Lederberg and Tatum in the 1940s using sets of defined mutants triggered a revolution in biology through clarification of the concept of "gene." It is a clear testament to the importance of this organism in modern biology that nobody can escape from using *E. coli* to study genes or genomes. For over 50 years, extensive and sustained efforts of many microbiologists, geneticists, biochemists and bioinformatists around the world contributed to the detailed analysis of the functions of many genes and gene products in these bacteria.

1
The *E. coli* Genome Sequence

The *E. coli* genome sequence was determined through independent efforts of both the United States and Japanese groups by using two different strains of *E. coli* – MG1655 and W3110 respectively. These strains diverged from the same ancestral strain about 50 years ago causing slight differences between them, including the large inversion involving the ribosomal RNA genes (see below). The complete genome sequence analysis revealed precise differences between the two strains. Comparison of the genomes of strains MG1655 and W3110 revealed relatively infrequent base substitution mutations, in the order of 10^{-5} per bases in the protein-coding region. On the other hand, dynamic genome rearrangements such as insertions and deletions are not uncommon. In summary, the *E. coli* genome can tolerate large rearrangements, such as insertion, deletion and recombination rather than microscale changes like base substitution. In 2001, pathogenic *E. coli* O157, which is very closely related to *E. coli* K-12, was subjected to genomic sequencing by two groups (in the US and Japan). The results obtained with this pathogenic strain agreed well with the above findings with strain K-12. The *E.coli* O157 acquired its pathogenicity mainly by horizontal gene transfer mediated by a temperate bacteriophage.

2
ORF Predicted from the Complete Genome Sequence of *E. coli*

Extensive studies of a few model organisms including *E. coli* and *Saccharomyces cerevisiae* elucidated the functions of many genes and gene products particularly by using techniques of genetics, biochemistry and physiology, and so on. Complete genome sequence of *E. coli* revealed the presence of about 4400 predicted ORFs. More than 2000 genes have so far been characterized and classified by their function, whereas the rest remain functionally uncharacterized (Fig. 1). Network analysis

Functional category	Number of ORFs	%
Unknown	2001	45.5
Transport/Binding proteins	377	8.6
Energy metabolism	363	8.3
Other categories	300	6.8
Cell envelope	204	4.6
Translation	163	3.7
Central intermediary metabolism	156	3.6
Amino acid biosynthesis	138	3.1
Biosynthesis of cofactors, prosthetic group, carriers	129	2.9
Cellular processes	127	2.9
Regulatory functions	124	2.8
Purines, pyrimidines, nucleosides, and nucleotides	107	2.4
Replication	92	2.1
Fatty acid/Phospholipid metabolism	60	1.4
Transcription	52	1.2

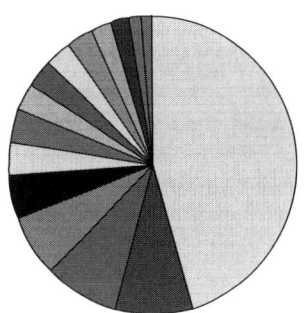

Fig. 1 Functional classification of *E. coli* ORFs.

of mutual functional relationships between genes or gene products has just started.

Accumulation of numerous genomic sequence data accelerated comparative approach, providing the basis for distinction between two ways of genes to be homologous: orthologous genes are homologues between species, and paralogues are those that descend directly from copies of a gene that duplicated within an ancestral genome. Orthologues are therefore good instruments for building phylogenetic trees of organisms, and paralogues are useful for understanding the course of protein evolution. Comparative analysis using the full set of genes in *E. coli* revealed two classes of genes: paralogues and nonparalogues. Sets of paralogous genes could be considered to have descended from ancestral genes by duplication and divergence. The concepts of "protein family" and "module" were discussed by Riley and her colleagues: "module" defined as a long conserved segment, usually larger than 100 amino acid residues, within a protein and an element of sequence that has independent evolutionary history. It may also correspond to a functional unit and contain several domains that are defined as structural units. These functional units are capable of independent existence but in some cases, joined to another module of independent function in a multimodule protein. For example, proteins involved in transport or binding proteins form a large protein family in *E. coli*, and some of them (enzyme, transport protein and transcriptional regulator protein) share modules for binding certain ligands.

3
Repetitive Sequences, Sites, RNA Genes, etc.

Many kinds of repetitive sequences are found within the genome and some of them have important physiological function. Distribution of repeats on the chromosome is not random and seems to be related to some feature of DNA replication. Repetitive sequences in *E. coli* are encountered in many contexts. Various classes repeats are present in diverse prokaryotes, including *E. coli*. Coding sequences such as ribosomal RNA genes, transfer RNA genes, and insertion sequences are present usually in multiple copies but their copy numbers are relatively low. Other interspersed repetitive DNA sequences are relatively short but abundant and located within intercistronic noncoding sequences. The latter families of repetitive sequences are listed in Table 1.

E. coli has 7 copies of rRNA coding genes (*rrn*) and an additional copy of 5S rRNA coding gene. Other rRNA related DNA sequences named TRIP, showing significant similarity to 5S rRNA, were recently found, which may have important roles related to 5S rRNAs. The rRNA genes are located within half of the chromosome that contains the origin of DNA replication (*oriC*), whereas many of the TRIP sequences are located in the other half of the chromosome. Moreover, *rrn* and some of the related TRIP sequences are distributed symmetrically on the leading strands on both sides of *oriC*. Consistent with this location, transcription of *rrn* operons generally proceeds away from the replication origin (Fig. 2).

Repetitive sequences such as rRNA genes provide a driving force for genome rearrangement. A large inversion of the genome between rRNA genes has been documented in *E. coli*, and one such inversion was found in the strain W3110. Although the direction of *rrn* gene transcription is strictly restricted away from the replication origin, the inversion in *E. coli* appears to be stable because the geometric relationship between *rrn* operons and the replication origin are preserved.

Insertion Sequences (IS) represent another family of repetitive elements and cause genetic variation among different strains of *E. coli*: both the abundance and distribution of insertion sequences can vary in different strains. Strain W3110, whose sequence was determined by the Japanese group, contains about 60 ISs

Tab. 1 Extragenic highly repetitive sequence families.

Sequence	Size [bp]	Copy number
BIME	40 ~ 400	~800
Ter	30	63
BoxC	56	36
Iap	29	23
IRU	127	19
RSA	152	6

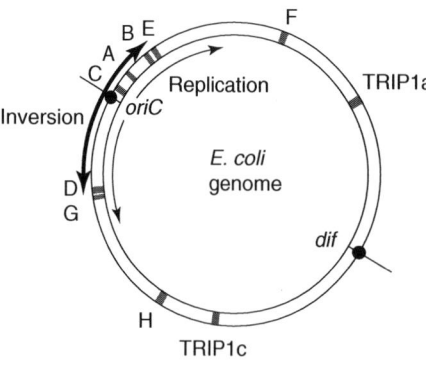

Fig. 2 Ribosomal RNA genes (*rrnA-H*) and related sequences. The arrow represents the inversion loci of W3110 between D (*rrnD*) and E (*rrnE*).

that belong to at least 10 distinct families and at least 10 ISs differ in their location when compared to another strain MG1655, whose sequence was determined by the US group. These results testify to a great variability in both number and family of ISs in closely related strains. Although it is not possible to decide whether these elements were primarily selected on the basis of their beneficial character, or whether they act as genomic parasites, there are a number of examples in which direct selection for certain mutants resulted from transcriptional activation mediated by IS.

Interspersed repetitive sequences represent relatively short (usually less than 500 bp), noncoding, intercistronic and dispersed elements found in bacterial genomes. Six classes of highly repetitive sequences, BIME, IRU, Box C, RSA, iap, and Ter Sequences have been identified (Table 1). Although they are highly repeated, they do not constitute more than 2% of the total E. coli DNA and this explains the failure of identification by the classical experimental methods. These sequences were mostly discovered by computer analysis of sequence data.

None of these sequences encode proteins, and are dispersed throughout the chromosome. BIME is the best documented case, originally identified as REP (Repetitive Extragenic Palindromic), which consists of a 38 b consensus palindromic sequence and can form a stable stem-loop structure with a 5 bp variable loop in the middle. Clusters of REP element with several sequences form a complex mosaic structure called BIME (Bacterial Interspersed Mosaic Elements). Approximately 800 BIMEs appear to be homogeneously distributed over the entire chromosome. BIMEs are found in extragenic locations at the 3′ ends of operon or between two genes of an operon, but rarely upstream of the first gene of an operon. Specific interactions of DNA gyrase, IHF and DNA polymerase I with BIMEs were reported, although the functional importance is unclear. The BIME-Pol I interaction suggests possible amplification of BIME regions.

4
Post Sequence Genome Project

Upon completion of genome sequencing, efforts were made to construct various resources including a complete set of ORF clones and of deletion mutants. These resources should provide basic tools for systematic functional analysis of the genomes that would eventually develop into systems biology.

Furthermore, genome-wide analysis of transcriptional regulation (transcriptome), protein dynamics (proteome), flow of metabolites (metabolome) have rapidly developed to make the best use of DNA sequence information ("-ome" is a Greek suffix for "whole").

Although E. coli is among the most thoroughly studied genetic systems, half of the genes had not been characterized experimentally. One can now identify all of their genetic components, such as protein-coding genes, RNA genes, and regulatory elements. It is, in other words, limited number of genes to be analyzed, although almost half of them remain to be functionally analyzed. Moreover, global analysis of regulatory network of transcription, protein–protein interaction and so on, has just started.

DNA sequence analysis identified about 4400 protein-coding genes in E. coli. The power of genetically tractable model organisms is that they can facilitate the analysis of physiological gene function *in vivo*. Precise genetic manipulation is particularly important for functional genomics. Genome sequence data permitted to design oligo DNA primers for precise amplification of ORFs and a complete set of histidine tagged ORF clones, fused with and without GFP gene, has been obtained and is available upon request for *E. coli* (reviewed by Mori et al.).

One approach for systematic functional analysis is gene deletion (replacement) by homologous recombination. Targeted gene replacement, once thought to be difficult in *E. coli*, can now be dealt with by using techniques developed by Wanner and colleagues. These systematic approaches to construct resources for functional genomics would rapidly raise *E. coli* to one of the leading organisms in the field of functional genomics.

4.1
Transcriptome Analysis

With the advance of genome research, many novel techniques have been developed including DNA microarray or DNA chip technology that are extremely useful for analysis of global gene expression. Generally, DNA microarray is defined as an orderly arrangement of tens to hundreds of thousands of unique DNA molecules of known sequence usually on a glass slide. Each unique DNA molecule is individually synthesized on a rigid silicon plate (generally called *DNA chip* and developed by Affymetrix Co.) or presynthesized DNA molecules (synthetic oligonucleotides or polymerase chain reaction (PCR) products) are spotted and immobilized on the slide glass. The use of DNA microarray to study *E. coli* gene regulation was first illustrated by Blattner and colleagues, and has rapidly expanded to analyze various aspects of transcriptional regulation.

4.2
Proteome and Interactome Analysis

Technology for global analysis of cellular proteins has been established since 1975 as two-dimensional polyacrylamide gel electrophoresis developed by O'Farrell. And *E. coli* has a long history for protein cataloging in 2D gel. However, new technology called *matrix-assisted laser desorption ionization-mass spectrometry* (MALDI-MS) accelerated protein identification dramatically. Currently, the most widely used method for proteome analysis is 2DE, followed by identification by mass spectrometry in high-throughput. Proteomics has now developed on global scales not only for identification and cataloging of all the proteins expressed in a cell but for global analysis of protein–protein interaction, called interactome, using the yeast two-hybrid system or related methods. As an alternative method for global interaction analysis, His-tagged-pulldown technique widely used as a fusion tag in expression of recombinant proteins has been used to allow affinity purification. A complete set of His-tagged ORF clones has already been established, and the comprehensive protein–protein interaction analysis is now under way. Mass spectrometry is also used for identification of copurified interacting candidate proteins.

5
***E. coli* Genome as a Model for Systems Biology**

Biology itself is now standing at the turning point from the past descriptive biology to the modern quantitative systems biology. Many systematic approaches such as DNA microarray or proteome analysis have been adopted that should help to understand the cells comprehensively as a system. Among model organisms, *E. coli* is especially important in this field because of enormous accumulation of biological knowledge and experimental methodologies. Scientists can now describe in exquisite detail many of the molecular parts and processes in a cell. Complete genetic information is now available for many organisms in all life kingdoms. Yet, we still do not have a comprehensive computational model of a living cell.

The International *E. coli* Alliance (IECA) was formed to tackle this fundamental biological problem in November 2002. IECA's mission is to consolidate global efforts to understand a living bacterial cell. Scientists around the world are working together to create a complex computer model, integrating all of the dynamic molecular interactions required for the life of a simple, self-replicating cell. An *E. coli* cell model will have immediate practical benefits and will significantly advance the field of computational systems biology. Generation of a computerized *E. coli* cell will add powerful new tools to our existing arsenal of discovery, including virtual experimentation and mathematical simulation. Ultimately, these biological and computational tools could be useful in both drug discovery and in the design of bioenhanced nanomachines. Furthermore, development of a virtual system for experimentation on the *E. coli* cell will be extremely useful for understanding more complex cells and contribute to the development and validation of *in silico* models of human cells. Biology is now changing to become a big science and *E. coli* might become one of the leading organisms in this new biology as was the case in the field of molecular genetics.

6
Useful *E. coli* Websites

As Internet technology advanced, the importance of websites rapidly increased.

Tab. 2 Useful websites of *Escherichia coli*.

URL	Institution
http://ecoli.aist-nara.ac.jp	Nara Institute of Science and Technology, Japan
http://www.genome.ad.jp/	Kyoto University, Japan
http://www.shigen.nig.ac.jp/ecoli/pec/index.jsp	The Institute of National Genetics, Japan
http://gib.genes.nig.ac.jp/	The Institute of National Genetics, Japan
http://genome.gen-info.osaka-u.ac.jp/bacteria/o157/	The University of Osaka, Japan
http://www.cifn.unam.mx/Computational_Genomics/regulondb/	Universidad Nacional Autonoma de Mexico, Mexico
http://redpoll.pharmacy.ualberta.ca/CCDB/	University of Alberta, Canada
http://www.uni-giessen.de/~gx1052/ECDC/ecdc.htm	Justus-Liebig-University, Germany
http://genolist.pasteur.fr/Colibri/	Institute Pasteur, France
http://web.bham.ac.uk/bcm4ght6/res.html	The University of Birmingham, UK
http://colibase.bham.ac.uk/	The University of Birmingham, UK
http://biocyc.org/ecocyc/	SRI International, USA
http://www.genome.wisc.edu/	University of Wisconsin – Madison, USA
http://bmb.med.miami.edu/EcoGene/EcoWeb/	University of Miami School of Medicine, USA
http://www.vetsci.psu.edu/ecoli.cfm	Penn State University, USA
http://www.ncbi.nlm.nih.gov/	National Center for Biotechnology Information, USA

The useful websites of *E. coli* are listed in Table 2.

See also Molecular Basis of Genetics; Genomic Sequencing.

Bibliography

Books and Reviews

Koonin, E.V., Galperin, M.Y. (Eds.) *Sequence – Evolution – Function Computational Approaches in Comparative Genomics*, Kluwer Academic Publishers.

Mahillon, J., Leonard, C., Chandler, M. (1999) *Res. Microbiol.* **150**, 675–687.

Mori, H., Isono, K., Horiuchi, T., Miki, T. (2000) *Res. Microbiol.* **151**, 121–128.

Neidhardt, F.C. (Ed.) *Escherichia coli and Salmonella*, ASM Press.

Tucker, C.L., Gera, J.F., Uetz, P. (2001) *Trends Cell Biol.* **11**, 102–106.

Primary Literature

Allen, T.E., Herrgard, M.J., Liu, M., Qiu, Y., Glasner, J.D., Blattner, F.R., Palsson, B.O. (2003) *J. Bacteriol.* **185**, 6392–6399.

Allen, T.E., Palsson, B.O. (2003) *J. Theor. Biol.* **220**, 1–18.

Blattner, F.R., Plunkett, G. III, Bloch, C.A., Perna, N.T., Burland, V., Riley, M., Collado-Vides, J., Glasner, J.D., Rode, C.K., Mayhew, G.F., et al. (1997) *Science* **277**, 1453–1474.

Chen, M., Hofestaedt, R. (2003) *In Silico Biol.* **3**, 0030.

Chiang, D.Y., Brown, P.O., Eisen, M.B. (2001) *Bioinformatics* **17**(Suppl. 1), S49–S55.

Corbin, R.W., Paliy, O., Yang, F., Shabanowitz, J., Platt, M., Lyons, C.E. Jr., Root, K., McAuliffe, J., Jordan, M.I., Kustu, S., et al. (2003) *Proc. Natl. Acad. Sci. U.S.A.* **100**, 9232–9237.

Covert, M.W., Schilling, C.H., Famili, I., Edwards, J.S., Goryanin, II, Selkov, E., Palsson, B.O. (2001) *Trends Biochem. Sci.* **26**, 179–186.

Datsenko, K.A., Wanner, B.L. (2000) *Proc. Natl. Acad. Sci. U.S.A.* **97**, 6640–6645.

Deane, C.M., Salwinski, L., Xenarios, I., Eisenberg, D. (2002) *Mol. Cell. Proteomics* **1**, 349–356.

Edwards, J.S., Palsson, B.O. (1998) *Biotechnol. Bioeng.* **58**, 162–169.

Edwards, J.S., Palsson, B.O. (2000) *Biotechnol. Prog.* **16**, 927–939.

Edwards, J.S., Palsson, B.O. (2000) *BMC Bioinf.* **1**, 1.

Edwards, J.S., Palsson, B.O. (2000) *Proc. Natl. Acad. Sci. U.S.A.* **97**, 5528–5533.

Edwards, J.S., Ibarra, R.U., Palsson, B.O. (2001) *Nat. Biotechnol.* **19**, 125–130.

Edwards, J.S., Ramakrishna, R., Palsson, B.O. (2002) *Biotechnol. Bioeng.* **77**, 27–36.

Ehrenberg, M., Elf, J., Aurell, E., Sandberg, R., Tegner, J. (2003) *Genome Res.* **13**, 2377–2380.

Fiehn, O. (2002) *Plant Mol. Biol.* **48**, 155–171.

Figeys, D., Gygi, S.P., Zhang, Y., Watts, J., Gu, M., Aebersold, R. (1998) *Electrophoresis* **19**, 1811–1818.

Galperin, M.Y., Koonin, E.V. (1999) *Genetica* **106**, 159–170.

Gerdes, S.Y., Scholle, M.D., Campbell, J.W., Balazsi, G., Ravasz, E., Daugherty, M.D., Somera, A.L., Kyrpides, N.C., Anderson, I., Gelfand, M.S., et al. (2003) *J. Bacteriol.* **185**, 5673–5684.

Gilson, E., Saurin, W., Perrin, D., Bachellier, S., Hofnung, M. (1991) *Nucleic Acids Res.* **19**, 1375–1383.

Goryanin, I., Hodgman, T.C., Selkov, E. (1999) *Bioinformatics* **15**, 749–758.

Gutierrez-Rios, R.M., Rosenblueth, D.A., Loza, J.A., Huerta, A.M., Glasner, J.D., Blattner, F.R., Collado-Vides, J. (2003) *Genome Res.* **13**, 2435–2443.

Gygi, S.P., Rist, B., Aebersold, R. (2000) *Curr. Opin. Biotechnol.* **11**, 396–401.

Hayashi, T., Makino, K., Ohnishi, M., Kurokawa, K., Ishii, K., Yokoyama, K., Han, C.G., Ohtsubo, E., Nakayama, K., Murata, T., et al. (2001) *DNA Res.* **8**, 11–22.

Herrgard, M.J., Covert, M.W., Palsson, B.O. (2003) *Genome Res.* **13**, 2423–2434.

Hill, C.W., Harnish, B.W. (1981) *Proc. Natl. Acad. Sci. U.S.A.* **78**, 7069–7072.

Hutchison, C.A., Peterson, S.N., Gill, S.R., Cline, R.T., White, O., Fraser, C.M., Smith, H.O., Venter, J.C. (1999) *Science* **286**, 2165–2169.

Itoh, T., Okayama, T., Hashimoto, H., Takeda, J., Davis, R.W., Mori, H., Gojobori, T. (1999) *FEBS Lett.* **450**, 72–76.

Jiao, Z., Baba, T., Mori, H., Shimizu, K. (2003) *FEMS Microbiol. Lett.* **220**, 295–301.

Kolisnychenko, V., Plunkett, G. III, Herring, C.D., Feher, T., Posfai, J., Blattner, F.R., Posfai, G. (2002) *Genome Res.* **12**, 640–647.

Koonin, E.V., Tatusov, R.L., Rudd, K.E. (1995) *Proc. Natl. Acad. Sci. U.S.A.* **92**, 11921–11925.

Labedan, B., Riley, M. (1995) *Mol. Biol. Evol.* **12**, 980–987.

Labedan, B., Riley, M. (1995) *J. Bacteriol.* **177**, 1585–1588.

Liang, P., Labedan, B., Riley, M. (2002) *Physiol. Genomics* **9**, 15–26.

Lockhart, D.J., Dong, H., Byrne, M.C., Follettie, M.T., Gallo, M.V., Chee, M.S., Mittmann, M., Wang, C., Kobayashi, M., Horton, H., Brown, E.L. (1996) *Nat. Biotechnol.* **14**, 1675–1680.

Maharjan, R.P., Ferenci, T. (2003) *Anal. Biochem.* **313**, 145–154.

Masuda, N., Church, G.M. (2003) *Mol. Microbiol.* **48**, 699–712.

Milo, R., Shen-Orr, S., Itzkovitz, S., Kashtan, N., Chklovskii, D., Alon, U. (2002) *Science* **298**, 824–827.

Mrowka, R., Liebermeister, W., Holste, D. (2003) *Nat. Genet.* **33**, 15–16; author reply 16–17.

Natale, D.A., Galperin, M.Y., Tatusov, R.L., Koonin, E.V. (2000) *Genetica* **108**, 9–17.

O'Farrell, P.H. (1975) *J. Biol. Chem.* **250**, 4007–4021.

Oshima, T., Aiba, H., Masuda, Y., Kanaya, S., Sugiura, M., Wanner, B.L., Mori, H., Mizuno, T. (2002) *Mol. Microbiol.* **46**, 281–291.

Overbeek, R., Fonstein, M., D'Souza, M., Pusch, G.D., Maltsev, N. (1999) *Proc. Natl. Acad. Sci. U.S.A.* **96**, 2896–2901.

Overbeek, R., Larsen, N., Pusch, G.D., D'Souza, M., Selkov, E. Jr., Kyrpides, N., Fonstein, M., Maltsev, N., Selkov, E. (2000) *Nucleic Acids Res.* **28**, 123–125.

Perna, N.T., Plunkett, G. III, Burland, V., Mau, B., Glasner, J.D., Rose, D.J., Mayhew, G.F., Evans, P.S., Gregor, J., Kirkpatrick, H.A., et al. (2001) *Nature* **409**, 529–533.

Picataggio, S.K., Templeton, L.J., Smulski, D.R., LaRossa, R.A. (2002) *Methods Enzymol.* **358**, 177–188.

Posfai, G., Kolisnychenko, V., Bereczki, Z., Blattner, F.R. (1999) *Nucleic Acids Res.* **27**, 4409–4415.

Posfai, G., Koob, M.D., Kirkpatrick, H.A., Blattner, F.R. (1997) *J. Bacteriol.* **179**, 4426–4428.

Price, N.D., Reed, J.L., Papin, J.A., Famili, I., Palsson, B.O. (2003) *Biophys. J.* **84**, 794–804.

Ramakrishna, R., Edwards, J.S., McCulloch, A., Palsson, B.O. (2001) *Am. J. Physiol. Regul. Integr. Comp. Physiol.* **280**, R695–R704.

Reed, J.L., Palsson, B.O. (2003) *J. Bacteriol.* **185**, 2692–2699.

Reed, J.L., Vo, T.D., Schilling, C.H., Palsson, B.O. (2003) *Genome Biol.* **4**, R54.

Rhodius, V.A., LaRossa, R.A. (2003) *Curr. Opin. Microbiol.* **6**, 114–119.

Rhodius, V., Van Dyk, T.K., Gross, C., LaRossa, R.A. (2002) *Annu. Rev. Microbiol.* **56**, 599–624.

Richmond, C.S., Glasner, J.D., Mau, R., Jin, H., Blattner, F.R. (1999) *Nucleic Acids Res.* **27**, 3821–3835.

Riley, M., Labedan, B. (1997) *J. Mol. Biol.* **268**, 857–868.

Rudd, K.E. (1998) *Microbiol. Mol. Biol. Rev.* **62**, 985–1019.

Sabatti, C., Rohlin, L., Oh, M.K., Liao, J.C. (2002) *Nucleic Acids Res.* **30**, 2886–2893.

Sabina, J., Dover, N., Templeton, L.J., Smulski, D.R., Soll, D., LaRossa, R.A. (2003) *J. Bacteriol.* **185**, 6158–6170.

Schilling, C.H., Edwards, J.S., Letscher, D., Palsson, B.O. (2000) *Biotechnol. Bioeng.* **71**, 286–306.

Selinger, D.W., Cheung, K.J., Mei, R., Johansson, E.M., Richmond, C.S., Blattner, F.R., Lockhart, D.J., Church, G.M. (2000) *Nat. Biotechnol.* **18**, 1262–1268.

Sharples, G.J., Lloyd, R.G. (1990) *Nucleic Acids Res.* **18**, 6503–6508.

Shen-Orr, S.S., Milo, R., Mangan, S., Alon, U. (2002) *Nat. Genet.* **31**, 64–68.

Soga, T., Ohashi, Y., Ueno, Y., Naraoka, H., Tomita, M., Nishioka, T. (2003) *J. Proteome Res.* **2**, 488–494.

Soga, T., Ueno, Y., Naraoka, H., Matsuda, K., Tomita, M., Nishioka, T. (2002) *Anal. Chem.* **74**, 6224–6229.

Soga, T., Ueno, Y., Naraoka, H., Ohashi, Y., Tomita, M., Nishioka, T. (2002) *Anal. Chem.* **74**, 2233–2239.

Storz, G. (2002) *Science* **296**, 1260–1263.

Tao, H., Bausch, C., Richmond, C., Blattner, F.R., Conway, T. (1999) *J. Bacteriol.* **181**, 6425–6440.

Tatusov, R.L., Galperin, M.Y., Natale, D.A., Koonin, E.V. (2000) *Nucleic Acids Res.* **28**, 33–36.

Tatusov, R.L., Koonin, E.V., Lipman, D.J. (1997) *Science* **278**, 631–637.

Tatusov, R.L., Natale, D.A., Garkavtsev, I.V., Tatusova, T.A., Shankavaram, U.T., Rao, B.S., Kiryutin, B., Galperin, M.Y., Fedorova, N.D., Koonin, E.V. (2001) *Nucleic Acids Res.* **29**, 22–28.

Thieffry, D., Salgado, H., Huerta, A.M., Collado-Vides, J. (1998) *Bioinformatics* **14**, 391–400.

Tomita, M., Hashimoto, K., Takahashi, K., Shimizu, T.S., Matsuzaki, Y., Miyoshi, F., Saito, K., Tanida, S., Yugi, K., Venter, J.C., Hutchison, C.A. III (1999) *Bioinformatics* **15**, 72–84.

Tomita, M. (2001) *Trends Biotechnol.* **19**, 205–210.

Tweeddale, H., Notley-McRobb, L., Ferenci, T. (1998) *J. Bacteriol.* **180**, 5109–5116.

Uchiyama, I. (2003) *Nucleic Acids Res.* **31**, 58–62.

Vanbogelen, R.A. (1996).

Vazquez, A., Flammini, A., Maritan, A., Vespignani, A. (2003) *Nat. Biotechnol.* **21**, 697–700.

Wei, Y., Lee, J.M., Richmond, C., Blattner, F.R., Rafalski, J.A., LaRossa, R.A. (2001) *J. Bacteriol.* **183**, 545–556.

Yang, C., Hua, Q., Baba, T., Mori, H., Shimizu, K. (2003) *Biotechnol. Bioeng.* **84**, 129–144.

Yu, B.J., Sung, B.H., Koob, M.D., Lee, C.H., Lee, J.H., Lee, W.S., Kim, M.S., Kim, S.C. (2002) *Nat. Biotechnol.* **20**, 1018–1023.

Zheng, M., Wang, X., Templeton, L.J., Smulski, D.R., LaRossa, R.A., Storz, G. (2001) *J. Bacteriol.* **183**, 4562–4570.

Zipkas, D., Riley, M. (1975) *Proc. Natl. Acad. Sci. U.S.A.* **72**, 1354–1358.

15
Drosophila Genome

Robert D.C. Saunders
Department of Biological Sciences, The Open University, Milton Keynes, UK

1	**Introduction** 421	
2	**An Overview of the *Drosophila* Genome** 422	
3	**The Chromosomes of *Drosophila*** 422	
3.1	Mitotic Chromosomes 422	
3.2	Polytene Chromosomes 423	
3.3	Chromosome Aberrations 425	
3.3.1	Using Polytene Chromosomes to Map Genes 426	
4	**The *Drosophila* Genome Project** 427	
4.1	Euchromatin and the Genes 427	
4.1.1	Duplicated Genes, Pseudogenes 427	
4.2	Transposable Elements 428	
4.3	Heterochromatin 430	
4.4	Long-range Genome Structure 430	
5	**Genetic Analysis in *Drosophila*** 431	
5.1	Mutagenesis 431	
5.1.1	Reverse Genetics 432	
5.1.2	Classes of Mutant Allele 432	
5.1.3	Position Effect Variegation 433	
5.1.4	Interfering RNA 433	
5.1.5	Targeted Mutagenesis 433	
5.2	Genetic Mapping 433	
5.2.1	Microsatellite and RFLP Markers 434	
6	**Prospects** 435	

Genomics and Genetics. Edited by Robert A. Meyers.
Copyright © 2007 Wiley-VCH Verlag GmbH & Co. KGaA, Weinheim
ISBN: 978-3-527-31609-0

Bibliography 435
Books and Reviews 435
Primary Literature 435

Keywords

Polytene Chromosome
Highly replicated chromosomes in an interphase-like chromatin conformation, with each chromatid precisely synapsed with its sisters. This results in a banded structure, permitting high-resolution mapping by light microscopy.

Heterochromatin
A term describing those regions of the chromosome that are highly condensed in metaphase chromosomes and undergo late replication in S-phase. Heterochromatin generally corresponds with those regions of the genome containing simple sequence repetitive, or satellite, DNA. There are few genes in heterochromatic regions.

Euchromatin
Euchromatin corresponds to the single copy and middle repetitive DNA of the genome, in which most of the genes are located. Unlike heterochromatin, euchromatin does polytenize.

***In situ* Hybridization**
Cloned sections of DNA can be accurately mapped to polytene chromosomes by *in situ* hybridization, a process in which chemically or radioactively labeled nucleic acid is hybridized to denatured chromosomes on a microscope slide. Signals are detected immunochemically or by application of a radioactive emulsion.

Balanced Lethal Stocks
Recessive lethal stocks are maintained as balanced stocks, in which the population consists of flies heterozygous for the mutant chromosome and for a corresponding balancer chromosome. Balancer chromosomes possess multiple inversions, which suppress recombination, and several genetic markers, including a dominant visible, and are themselves recessive lethal or sterile, so that balancer homozygotes are either inviable or sterile.

Drosophila melanogaster is arguably one of the most important model organisms available to the modern researcher. This chapter seeks to review the structure of the *Drosophila* genome and the main genetic and molecular techniques currently in use. The *Drosophila* genome has recently been sequenced, and this chapter sets the biology of *Drosophila* in the context of recent updates to the sequence annotation.

1
Introduction

Drosophila melanogaster is arguably the genetically best-characterized metazoan organism. Since its adoption for genetic analysis, its many advantages as a laboratory animal have led to its use as a model system for a wide variety of biomedical researches. The *Drosophila* genome is relatively small, approximately 170×10^6 bp, and is distributed between four pairs of chromosomes. The polytene chromosomes of *Drosophila* have provided a physical map for genetic studies since they were mapped in the 1930s, and continue to be a powerful tool in molecular genetic analysis.

This article describes the features of the *Drosophila* genome relevant to genetic and molecular studies in *Drosophila*.

In the early years of this century, *D. melanogaster* (then known as *Drosophila ampelophila*) was adopted by Thomas Hunt Morgan as an experimental organism in the then new science of genetics. *Drosophila* proved to be an ideal organism for genetic research: it has a short life cycle, 10 to 14 days at 25 °C, and very modest requirements in terms of husbandry. In fact, a modern *Drosophila* laboratory may maintain a thousand or more genetically distinct stocks. Many of the basic concepts of genetics were established using the *Drosophila* species, including important advances in population genetics and evolutionary theory, particularly with *Drosophila pseudoobscura* and related species. However, it is *D. melanogaster* that has made the biggest impact on modern molecular genetic research. With the advent of molecular cloning technology, *Drosophila* found itself ideally placed to benefit from integrated genetics and molecular biology.

As a model organism, *Drosophila* has been an essential tool for the elucidation of basic genetic processes, common to many eukaryotes.

1. For many areas of biomedical research: Examples include aging, neurobiology, embryology, pattern formation, signal transduction, and cell cycle regulation. In particular, research in these areas is facilitated by the sophisticated genetic analysis possible with *Drosophila*.
2. For genome mapping and analysis: *D. melanogaster* has a moderately sized genome, 170×10^6 bp, in comparison with those of yeasts, nematode worms, the mouse, humans, and other species in which large-scale molecular mapping is carried out. Table 1 shows the haploid genome sizes of these species. Thus, the *Drosophila* genome is ideal for use as a model for genome mapping in higher eukaryotes, and additionally has features such as polytene chromosomes, which simplify mapping.
3. Finally, many medically and economically important pests and disease vectors are members of the order Diptera. Examples are the anopheline mosquitoes (malaria vectors), aedine mosquitoes (vectors of yellow and dengue fevers), and the tsetse flies (vector of trypanosomiasis, or sleeping

Tab. 1 The haploid genome sizes of some species.

Organism	Genome size $[\times 10^6$ bp]
Escherichia coli	4
Saccharomyces cerevisiae	16
Schizosaccharomyces pombe	15
Caenorhabditis elegans	100
Drosophila melanogaster	170
Homo sapiens, Mus musculus	3,000

sickness) to name but a few. Several features of the biology and life cycle of these insects mean they are not good research organisms, from the point of view of laboratory culture and genetics. *Drosophila* is a model system for the molecular genetic analysis of these insects.

In this article, I shall review the genetics and molecular biology of *D. melanogaster*, with respect to its use as a model system.

2
An Overview of the *Drosophila* Genome

The haploid genome of *D. melanogaster* is relatively small, 170×10^6 bp, compared with the typical mammalian genome of approximately 3×10^9 bp. *Drosophila* genomic DNA is unmethylated, in contrast to most other higher eukaryotes. The composition of the *Drosophila* genome in terms of single copy and repetitive DNA fractions is presented in Table 2.

About 21% of the genome comprises highly repetitive simple sequence DNA, known as *satellite DNA*. Satellite DNA is present in large blocks of repeated short (5 to 10 bp) units principally located close to centromeric regions. There are several classes of satellite, distinguishable by the repeat unit sequence, and these classes often have chromosome-specific distribution. One satellite, the 1.688 or 359 bp satellite, is atypical, comprising arrays of repeated 359 bp units confined to the X chromosome. In general, satellite DNA is concentrated in heterochromatin, which will be discussed below.

A substantial proportion, 18%, of the genome is moderately repetitive. Some of this repetition is due to the ribosomal RNA and histone gene families, but much is accounted for by mobile DNA, or transposable elements. Transposable elements are found in variable, dispersed locations within the genome. Several transposable elements have proven to be very useful in the experimental manipulation and analysis of *Drosophila*, particularly the *P* element, which is described in detail in Sect. 4.2.

The composition of the *Drosophila* genome has, of course, been largely redefined by the successes of the *Drosophila* genome projects, and in particular that of the Berkeley Drosophila Genome Project (BDGP). This will be discussed in some detail below.

3
The Chromosomes of *Drosophila*

3.1
Mitotic Chromosomes

The mitotic chromosome complement of *D. melanogaster* is illustrated diagrammatically in Fig. 1. There are two metacentric autosomes, chromosomes 2 and 3; one tiny dotlike autosome, chromosome 4; and the heteromorphic sex chromosomes, the acrocentric X chromosome and the submetacentric Y chromosome. Sex determination in *D. melanogaster* is based on the ratio of X chromosomes to autosomes. XX individuals have a ratio of 1, and develop as

Tab. 2 The composition of the *Drosophila melanogaster* genome.

Single copy sequences	61%	103.7×10^6
Satellite DNA	21%	35.7×10^6
rDNA, histones, etc.	3%	5.1×10^6
Transposable elements	9%	15.3×10^6
Foldback DNA	6%	10.2×10^6
Total genome size	–	170×10^6

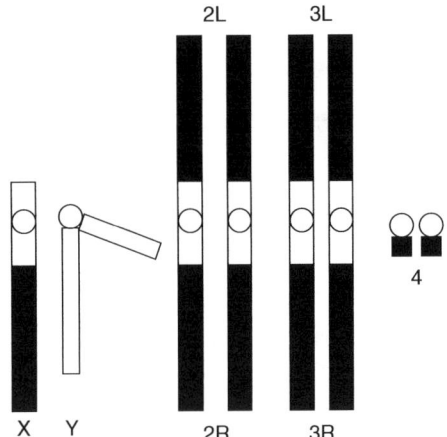

Fig. 1 Diagrammatic representation of the diploid complement of mitotic chromosomes of *D. melanogaster*. The karyotype shown is XY (male). Circles and white boxes represent centromeres and heterochromatin respectively.

females, while XY individuals have a ratio of 0.5, and develop as males. Thus, male *Drosophila* are hemizygous for X-linked loci. XO individuals are male, but sterile, since the Y chromosome carries factors essential for male fertility.

Chromatin can be subdivided on microscopic criteria into heterochromatin and euchromatin, with heterochromatin being characterized by having a highly condensed state and late replication in S-phase. Modern molecular genetic research has shown that, in general, the genes lie in the euchromatin, while the heterochromatin is composed principally of simple sequence repetitive DNA. Indeed, the term heterochromatin is often used as a synonym for satellite DNA. There are exceptions to this rule, including the male fertility factors of the Y chromosome mentioned above, and a few autosomal loci.

3.2
Polytene Chromosomes

In common with many species of Diptera, *Drosophila* possesses polytene chromosomes in several tissues, most notably the larval salivary glands. The salivary glands of third instar larvae are easy to identify and dissect, and their chromosomes are simple to prepare for analysis. Polytene chromosomes are very large, highly replicated chromosomes that are invaluable in genetic research using *Drosophila* (and many other Diptera) as a model.

Polytene chromosomes are found in many species of Diptera including *Drosophila*, where the easiest to work with are found in the salivary glands of the third instar larva. The polytene chromosomes of *Drosophila* represent the euchromatic portion of the genome, in a highly amplified condition, in which each chromatid is replicated approximately 1000-fold, and each copy is precisely aligned, resulting in a highly reproducible transverse banding pattern along the chromosomes (Fig. 2). The pericentromeric heterochromatin, and the heterochromatic Y chromosome remain unpolytenized and are located together with the centromeres in the chromocenter, to which each of the chromosome arms are attached. Thus, the polytene chromosome complement is roughly star shaped, with each major chromosome arm of approximately equal length radiating from the chromocenter.

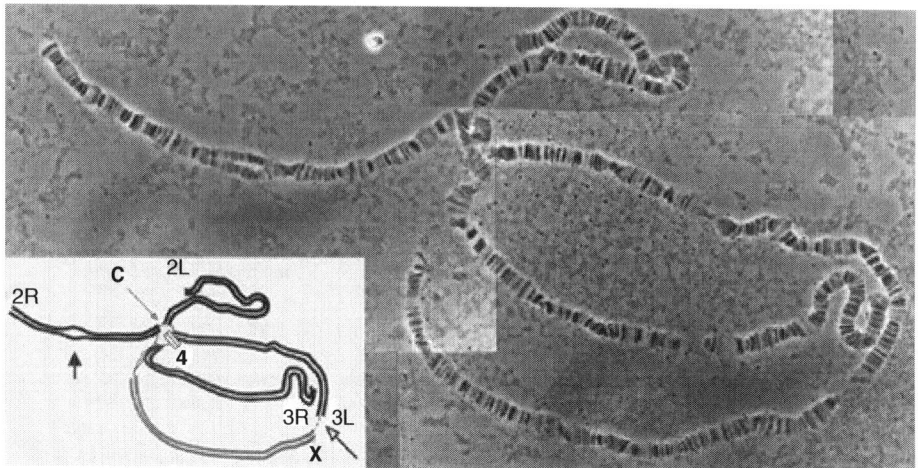

Fig. 2 The polytene chromosomes of *D. melanogaster*. Phase contrast micrograph of a preparation of *D. melanogaster* polytene chromosomes. The specimen has been fixed in 17% lactic acid, 50% acetic acid, and is unstained. Each chromosome arm extends from the chromocenter. Inset: diagrammatic representation of the photograph. Each chromosome arm is labeled. The X chromosome is shaded in light gray, chromosome 2 in dark gray, and chromosome 3 in black, with chromosome 4 in white. **C**: chromocenter, **X**: X chromosome, **2L**, **2R**: left and right arms of chromosome 2, **3L**, **3R**: left and right arms of chromosome 3, **4**: chromosome 4. The solid arrow points to a region of asynapsis in arm 2L. The open arrow points to ectopic pairing between the telomeres of 3L and X.

The arm corresponding to chromosome 4 is about 5% of the length of the others. This arrangement is shown in Fig. 2, with a diagrammatic interpretation in the inset.

In the 1930s, the importance of polytene chromosomes for genetic research was realized and highly detailed cytogenetic maps were created on the basis of the banding pattern of the chromosomes. The map devised by Calvin Bridges in the mid-1930s is basically that used to this day, and provides a reference system of map locations understandable by all researchers possessing a copy of the map. Each major arm is divided into 20 divisions numbered 1 to 20 (X chromosome), 21 to 60 (second chromosome), and 61 to 100 (third chromosome). The tiny fourth chromosome is allocated two divisions, 101 and 102. The telomeric divisions are 1, 21, 60, 61, 100, and 102. The numbered divisions are each further broken down into six subdivisions, lettered A to F. This basic map was further refined by Bridges and Bridges to include a number for each band on the map. Thus, each band of the polytene chromosome complement can be unambiguously identified (at least in principle).

Genes identified by mutation can be accurately mapped to the polytene chromosomes using a huge collection of chromosome rearrangements, such as deficiencies, inversions, duplications, transpositions, and translocations. At the simplest level, this involves determining whether a mutant allele is located within a given deficiency (see Sect. 3.3.1).

The resolution afforded by the use of polytene chromosomes is very high;

features of 5 to 100 kb can be recognized on the map by light microscopic analysis, which can rapidly provide positional information at a precision useful by molecular cloning standards. *In situ* hybridization can map cloned segments of DNA to single band resolution, several orders of magnitude better than that easily possible with metaphase chromosomes, and this technique is consequently invaluable for many gene cloning strategies and physical genome mapping projects.

3.3
Chromosome Aberrations

A great many chromosome aberrations, affecting all chromosomes have been recovered and analyzed. The gene and chromosomal aberration data sets of the *Drosophila* database FlyBase contain details of many thousands of chromosomal aberrations. These include the following:

a. *Deficiencies* (Fig. 3). Deficiency is the terminology used in *Drosophila* research for a chromosomal deletion. In a deficiency chromosome, a segment of the chromosome is deleted. In practice, the largest deficiency of practical utility is about one polytene map division. Flies carrying deficiencies larger than this have severely reduced fitness and viability. Furthermore, there are a number of haploinsufficient loci in the genome, which are inviable when hemizygous (e.g. when uncovered by a deficiency). Telomeric deficiencies can be unstable, unless the specialized telomeric sequences have been replaced.

b. *Inversions* (Fig. 3). Inversion chromosomes are those in which a section has been inverted with respect to the wild-type gene order. Inversion and normal chromosomes synapse to form inversion loops, in both meiotic and polytene chromosomes. Chromosomes bearing inversions suppress recombination within the inverted region when heterozygous with wild-type sequence chromosomes. *Balancer chromosomes* are chromosomes carrying several inversions, together with

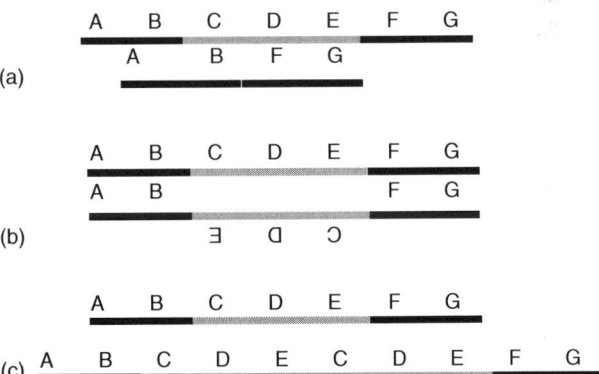

Fig. 3 Chromosome rearrangements. In each panel, a standard order chromosome bearing loci A–G is shown above the rearranged chromosome. The region involved in the rearrangement is shaded. (a) Deficiency. Loci C, D, and E have been deleted. (b) Inversion. Loci C, D, and E have been inverted. (c) Duplication. Loci D, D, and E have been duplicated in tandem.

suitable dominant and recessive markers, and are used to effectively suppress recombination along the whole chromosome. The use of balancer chromosomes is described in Sect. 5.1. Chromosome inversions that contain the centromere are referred to as *pericentric inversions* and those that do not are *paracentric inversions*.

c. *Duplications.* (Fig. 3). These chromosomes carry an additional segment derived from a homologous or other chromosome.

d. *Translocations.* Translocations involve at least two chromosome breaks, and involve the reciprocal interchange of chromosome sections between two or more nonhomologous chromosomes. More than two chromosomes can be rearranged.

e. *Transpositions.* Essentially, a transposition is where a chromosome section is inserted in another chromosome, involving three breaks.

f. *Compound chromosomes.* These chromosomes are derived from the normal chromosome complement by breakage and fusion of chromosome arms. Compound X chromosomes (two X chromosomes sharing a single centromere) affect the pattern of inheritance of normal X and Y chromosomes: in flies of normal karyotype, Y chromosomes are patroclinous (inherited from the father), while in a typical compound X stock, they are matroclinous (inherited from the mother). Compound chromosomes involving the attachment of X and Y chromosomes, whole autosome fusions, and whole autosome arm fusions can be constructed.

The genetic technologies for constructing novel aberrations and compound chromosomes are well established for *D. melanogaster*. Translocations can be combined to create segmental aneuploids, synthetic deficiencies or duplications, when necessitated by the lack of conventional deficiencies for a particular chromosome region. Recombination between inversion chromosomes can also be used to synthesize deficiencies and duplications.

3.3.1 Using Polytene Chromosomes to Map Genes

The principle of deficiency mapping a recessive mutant allele is to test whether a deficiency chromosome "uncovers" the mutation by examining the phenotype of a stock carrying both the deficiency chromosome and the recessive allele-bearing chromosome. If the mutant phenotype is expressed, the locus must lie within the region deleted in the deficiency chromosome (Fig. 4). In addition to simple

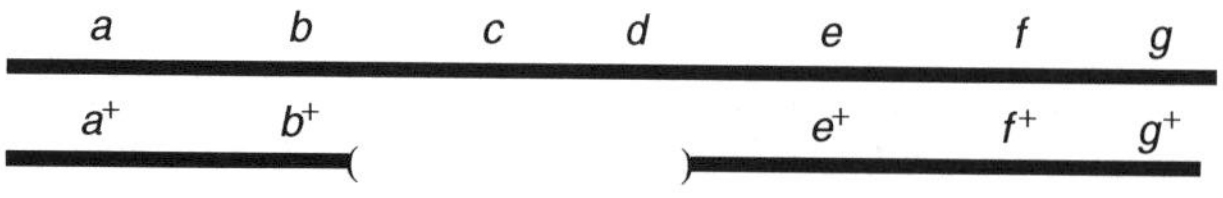

Fig. 4 Deficiency mapping. This figure illustrates the principle of deficiency mapping. A chromosome bearing recessive mutation at several loci (*a* to *g*) is heterozygous with a deficiency chromosome. The region deleted is indicated by parentheses. In this case mutants at loci *c* and *d* will be uncovered by the deficiency, and will display a mutant phenotype. Conversely, loci *a*, *b*, *e*, *f*, and *g* are heterozygous with wild-type alleles on the inversion chromosome and do not display mutant phenotype.

deficiency chromosomes, the *Drosophila* researcher can create "synthetic deficiencies," by combining other, often more complex, rearrangements. If a recessive allele's homozygous phenotype is identical to its hemizygous phenotype, it is likely to be an amorphic (null) allele, while if the hemizygous phenotype is stronger, it is likely to be a hypomorphic allele (see Sect. 5.1.2).

In some cases, a chromosome rearrangement inactivates or otherwise affects the expression of a gene, and can be used to pinpoint that gene's location both cytologically and molecularly.

The polytene chromosome map location of a gene is often a vital prelude to both further genetic analysis and its molecular cloning. This is discussed further in Sect. 3.2.

4
The *Drosophila* Genome Project

The *Drosophila* genome was sequenced by a collaboration of the Berkeley Drosophila Genome Project (BDGP), Celera Genomics, and the European Drosophila Genome Project (EDGP). Since the publication of the original draft sequence in 2000, two further revisions have been published, the most recent at the time of writing being Release 3.0. In Release 3.0, the euchromatin is represented by high quality sequence of 116.8 Mbp, with mere X gaps – in fact chromosome arm 3R is represented by a single contiguous sequence.

Identification of genes in the *Drosophila* genome has been achieved by combined strategy, in which data gleaned by conventional genetic and molecular analyses have been combined with purely computer-based techniques of *in silico* gene prediction. The input of data from conventional sources has been of particular utility in identifying alternatively spliced gene transcripts, and defining the 5' and 3' ends of genes.

The total of predicted genes in the *Drosophila* genome is 13 676. This is, perhaps, a rather small number; the nematode worm *C. elegans* appears to have around 19 000 genes, and morphologically at least appears a simpler organism.

4.1
Euchromatin and the Genes

98% of identified genes are located in the euchromatin, with the remainder lying in heterochromatic sequences. The total number of euchromatic genes is 13 379. The average number of exons per gene is 4.6, very similar to the genes of *C. elegans* and the plant *Arabidopsis thaliana*, but considerably fewer than the genes of the human genome (8.9).

Some interesting observations concerning the structure of *Drosophila* genes can be made. Firstly, nesting of genes is common: about 7.5% of genes lie within introns of other genes. Generally, such genes lie in opposite orientation to the gene in which they are located. Similarly, 15% of identified genes overlap another gene. There are 26 cases of genes that are interleaved with another gene. Finally, 48 examples of dicistronic genes – in which two genes with distinct protein products produce a single transcript – have been identified.

4.1.1 Duplicated Genes, Pseudogenes

There are comparatively few examples of duplicated genes and pseudogenes in the *Drosophila* genome. There are a mere 17 examples of pseudogenes identified by the

BDGP, of which 15 are evidently generated by a recombinational mechanism, as they contain introns, and of the remaining examples, one appears to be the result of retrotransposition (it has a poly-A tail and no introns), while the other is so scrambled that its origin cannot be determined.

4.2
Transposable Elements

Transposable elements, or transposons, are segments of DNA that are able to excise from the genome, or replicate, and insert at a second location. These elements fall into the middle repetitive faction of the genome, since they are generally present in copy numbers varying from a few tens to several hundred. *Drosophila* possesses a great many transposable elements (there are 1572 in Release 3 of the *Drosophila* genome sequence), classified into several families on the basis of structure and transposition mechanism. Thus, different strains of *D. melanogaster* have different chromosomal locations for each type of transposable element. The transposition activity of transposable elements can result in a genetic syndrome known as *hybrid dysgenesis*: high mutation rates, high frequency of chromosome rearrangements, sterility, and male recombination (male recombination does not normally occur in *Drosophila*). In general, each type of transposable element displays particular preferred sites of insertion, determined by DNA sequence and other less well-understood factors.

The *P* element is the best-characterized transposable element. Genetically, the activity of this element is seen as hybrid dysgenesis resulting from the cross of *P* element containing (P strain) males with females lacking *P* elements (M strain). Crossing M males with P females does not cause hybrid dysgenesis. Molecular analysis revealed the *P* element to be a 2.9-kb segment of DNA that encodes two proteins: a transposase, and a transposase repressor, depending on alternative splicing of four exons. Transposase is normally only expressed in the germ line. Additionally, the 31-bp terminal repeats of the element and other factors (not encoded by the element) are also required for transposition.

Cloned *P* elements may be injected into *Drosophila* embryos, and will subsequently integrate into the genome. By *in vitro* genetic manipulation, a rich variety of artificial *P* elements have been constructed for use in inserting exogenous DNA into the *Drosophila* germ line. These elements do not themselves encode transposase (so that they will be stable once integrated into the genome) and therefore require a source of transposase, which can be supplied by a defective *P* element encoding transposase, but lacking one or both of the terminal repeats. A more elegant method is to inject embryos of a *Drosophila* strain carrying a stably integrated element expressing transposase. Examples of *P* element derivatives are described below.

1. *Transformation vectors.* One use of germ-line transformation is to provide direct experimental evidence of the identity of a cloned gene. This can be done by determining whether or not a given segment of DNA identified in a cloning project can rescue a mutant phenotype when introduced into the genome by *P* element mediated germ-line transformation. Transformation vectors are defective elements that lack transposase, but possess the terminal repeats. In addition, they have a

selectable marker gene so that genetically transformed flies can be selected or recognized. Typical marker genes include the *Drosophila* gene $white^+$, and the bacterial antibiotic resistance gene neo^R. $white^+$ restores wild-type eye color in individuals of $white^-$ background, and neo^R confers resistance to the antibiotic G418, to which *Drosophila* are normally sensitive.

2. *Mutagenesis.* Insertion of a P element, or another transposable element often affects that gene's pattern of expression. It is therefore possible to make use of transposon mobilization as a mutagen. Mutants induced in this way present a considerable advantage when the mutated gene is to be cloned. Targeted mutagenesis will be discussed in Sect. 5.1.1.

3. *Enhancer trapping.* P elements that contain the *E. coli LacZ* gene lacking a functional promoter sequence have been constructed. Upon insertion, these elements express *LacZ* only if there is a neighboring gene with an enhancer able to direct expression of the *LacZ* transgene. *LacZ* expression is conveniently detected with a histochemical or an immunological assay. Owing in part to the high gene density of the *Drosophila* genome, a large proportion of enhancer trap elements do express *LacZ*, often in an intricate temporal and spatial pattern that is frequently informative about the nearby gene's identity or function. Enhancer trapping is therefore an efficient means of detecting genes.

4. *The Gal4-UAS system.* This system comprises two distinct element insertions. The first carries the *S. cerevisiae Gal4* gene, which encodes a transcription factor that activates transcription from an upstream activating sequence (UAS_{Gal4}). The second element carries the transgene of interest, under the control of the UAS_{Gal4}, and which is therefore expressed only in cells expressing Gal4. The Gal4 expression may be controlled by a specific promoter, or by neighboring genomic enhancers (in which case the Gal4 element functions as an enhancer trap). This bipartite system is extremely flexible, since a large collection of lines expressing Gal4 in distinct spatial and temporal patterns is available. These can be combined with any UAS element carrying a transgene, which may be a reporter gene such as *LacZ*, or a mutant gene predicted to have a dominant phenotype. One exciting possibility is to express toxin genes (such as ricin A chain) in specific cell types, to investigate the developmental consequences of ablation of particular cells.

5. *FLP–FRT.* This is also a two-component system, in which one element carries the gene encoding the *S. cerevisiae* 2 µm plasmid fragment length polymorphism (FLP) recombinase, and the other the target sequences for this enzyme (FRT). The expression of FLP can be controlled in several ways as can the Gal4 described above. Cells in which FLP is expressed will rearrange sequences flanked by the FRT sites. Depending on the orientation of the two FRT sequences, this can be an inversion or an excision, enabling a gene between the sequences to be turned on or off.

6. Elements bearing visible marker genes can be used as dominant markers for the purpose of generating chromosomal deficiencies. For example, a w^- stock (w mutants have white eyes) bearing a $P[w^+]$ element inserted at, for example, 96A will have normal reddish-brown eyes. Following X irradiation,

flies with this chromosome region deleted will have lost the P[w^+] element, and will have white eyes. These w^- stocks can be analyzed by polytene cytogenetics to reveal which, if any, carry a chromosomal deletion.

4.3 Heterochromatin

Heterochromatin is characterized by particular staining patterns that reflect differential condensation in comparison to euchromatin. Typical characteristics include late replication and a higher level of condensation than seen for euchromatin. Typically, heterochromatin is located in pericentromeric regions, and near telomeres. Furthermore, the Y chromosome is entirely heterochromatic. Heterochromatin does not polytenize, and is not represented in the polytene chromosome maps; it forms the chromocentre, from which the polytenized chromosome arms radiate. Detailed cytological analyses of mitotic chromosomes have permitted banding maps of the heterochromatin to be constructed on the basis of differential staining patterns; such maps are of low resolution, consisting of 61 defined bands.

Sequencing the DNA corresponding to heterochromatin presents considerable difficulty because of its highly repetitive nature. For this reason, the BDGP's heterochromatic sequence remains draft quality – of the estimated 59 Mb heterochromatin (to which must be added the 41 Mb Y chromosome, which is entirely heterochromatic), an unfinished draft-quality assembly of 20.7 Mb has been produced and distributed over 2597 scaffolds. Nevertheless, several conclusions may be drawn.

Some 297 protein-coding genes have been identified within heterochromatin. This total includes several loci known from genetic analyses to map within heterochromatin. Several genes that do not encode proteins have been identified; these include 5.8S, 2S, 18S, and 28S rRNA genes.

The general organization of heterochromatic DNA is in blocks of approximately 20 kb to 1 Mb consisting of satellite DNA, separated by sections of between 5 and 50 kb of more complex sequences that contain a high density of transposable elements.

4.4 Long-range Genome Structure

That the structure of the *Drosophila* genome involves long-range interactions is in no doubt, as clearly indicated by genetic phenomena, such as transvection, in which regulatory interactions between homologous chromosomes are revealed. One of the outcomes of the *Drosophila* genome sequence was the ability to examine the transcriptional activity of the whole genome with reference to specific treatments or biological processes, by microarray studies. Analyses of several microarray experiments encompassing over 80 experimental conditions has revealed that there appears to be considerable clustering of genes which share expression profiles.

Over 20% of the *Drosophila* genes surveyed appeared to be clustered in groups of 10 to 30 genes with similar expression patterns. This is likely to reflect a regulation at the level of chromatin structure. However, there appeared to be no correlation with the banding pattern of polytene chromosomes, or with sites of known attachment to the nuclear scaffold.

5 Genetic Analysis in *Drosophila*

Despite the availability of the complete genome sequence, genetic analytical techniques remain very important. Many of the genes identified in the genome have no known mutant alleles. Examination of mutant phenotype, often in combination with mutation of other genes, has the potential to reveal their biological function.

5.1 Mutagenesis

A typical genetic screen for mutations, in this case autosomal recessive lethals, is shown in Fig. 5. A large variety of mutagens are available, which can differ widely in the nature of the lesions produced. For example, ethylmethanesulphonate (EMS), administered in the food, causes predominantly single base pair changes, whereas X rays induce chromosomal deletions and other rearrangements. The use of transposable elements as mutagens is described in Sect. 4.1.

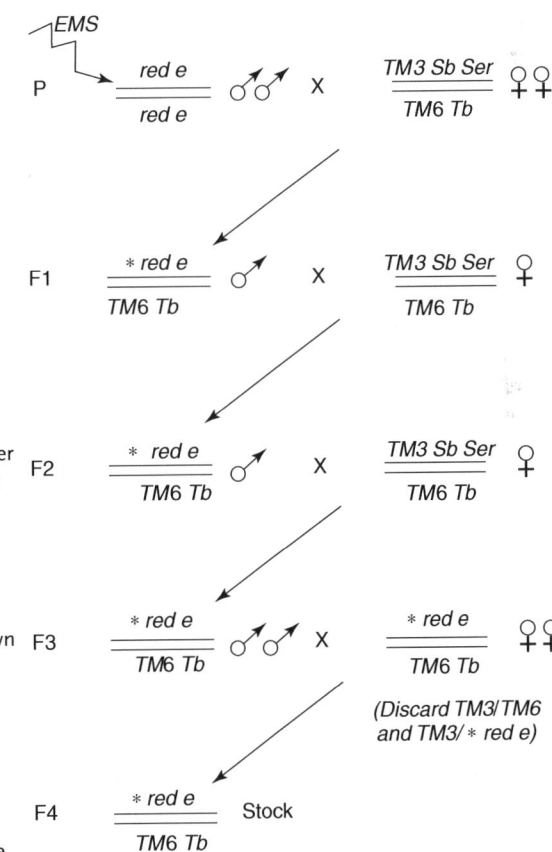

Fig. 5 Mutagenesis scheme. This scheme is designed to recover autosomal recessive lethal mutations. With minor modification, however, other types of phenotype may be screened for. Full descriptions of marker mutations and balancer chromosomes can be found in *The Genome of Drosophila melanogaster*. All females in crosses are virgin. The markers in this figure are: *red*: red malpighian tubules (wild-type tubules are whitish, this recessive mutation also results in brown eyes in the adult); *ebony (e)*: adult has black body, recessive; *Stubble (Sb)*: dominant mutation affecting adult bristle morphology; *Serrate (Ser)*: dominant mutation causing serrated wing margins; *Tubby (Tb)*: dominant mutation causing tubby larvae and adults, but more easily scored in larvae. The mutagenized chromosome is indicated by an asterisk.

Male flies are treated with mutagen, and mated *en masse* to virgin females heterozygous for two balancer chromosomes. All subsequent matings are single-pair matings, again with virgin females. The target chromosome is marked with recessive markers, to facilitate later genetic manipulations (the mutagenized chromosome is labeled with an asterisk in subsequent crosses in Fig. 5). Note that the F1 and F2 crosses are identical – this is necessary when the mutagen, such as EMS, causes lesions requiring one or more cell divisions to become fixed. The F3-cross yields a balanced stock in which flies homozygous for the mutagenized chromosome (easily recognized because of the recessive markers) can be examined for the phenotype of interest. In the case of recessive lethals, this class would be absent. Balancer chromosomes have multiple inversions, which prevent recombination from occurring, and carry a recessive lethal mutation to render balancer chromosome homozygotes inviable. In general, balancer chromosomes carry a dominant marker, so they can be followed in genetic crosses. Because balancer chromosomes suppress recombination, these features will not be separated by recombination. Some X chromosome balancer chromosomes carry a female sterile mutation rather than a lethal, so that the balancer chromosome will be male viable.

Many classes of mutant phenotype can be discovered, using appropriate genetic systems, from dominant temperature-sensitive lethals, to simple visible mutations, from female sterile mutants to behavioral mutations.

5.1.1 Reverse Genetics

Often, the *Drosophila* researcher is in the position of having cloned a gene, but is without a corresponding mutant. Unlike other model organisms, there is no efficient means of targeted mutagenesis in *Drosophila*. A useful technique in this regard is to use P element mutagenesis as described in Sect. 4.2 and detect insertions in the gene of interest using PCR with one gene-specific primer and one P element specific primer.

As an alternative to PCR screening P element insertions, saturation mutagenesis of a chromosomal region known to include the gene of interest, and spanned by a small chromosome deficiency can be undertaken. Mutants displaying an appropriate phenotype could then be examined molecularly, to investigate whether the gene in question is mutated.

A very powerful technique of targeted mutagenesis has been developed by Golic and colleagues. It is clear that it is of general utility, but it is uncertain at the time of writing whether it will be widely adopted by the *Drosophila* research community.

5.1.2 Classes of Mutant Allele

Several classes of mutant allele may be recovered following mutagenesis. The simplest is the loss-of-function mutation, either complete (null mutation or amorph) or partial (hypomorph). Amorphs and hypomorphs are generally recessive, and can be formally distinguished by comparing their phenotypes in combination with a deficiency (see Sect. 3.3.1). Other allelic states are hypermorphs, in which there is a gain of gene function, and neomorphs, in which a novel function is acquired. Neomorphs are generally dominant alleles. Mutants of any or all classes can be recovered for a given gene, using appropriate mutagenesis protocols.

An additional complication is that an allele's phenotypic effect may be temperature-sensitive. At its extreme, this can be seen as a true conditional mutant, in which environmental conditions can entirely suppress the mutant phenotype.

5.1.3 Position Effect Variegation

As described earlier, most genes are located within euchromatin. In comparison with euchromatin, heterochromatin has unusual properties, connected with differences in chromatin organization and composition. These properties influence gene expression. When a gene, for example *white*, is relocated by chromosome rearrangement so that it is located close to heterochromatin, its expression is affected by the heterochromation. The heterochromatin appears to "heterochromatinize" its neighboring euchromatin. This occurs to different extents in different clonal cell lineages within the developing organism. In the case of a *white* allele mutated by an inversion bringing it into close proximity of the centric heterochromatin, it may be active in some ommatidia of the compound eye, and inactive in others, resulting in a mottled eye with red and white facets (*white* mutants have white eyes rather than the wild-type reddish color). The *white* allele w^{m4} is an example. There are loci that display the opposite effect; an example is *light*, which maps to the pericentric heterochromatin of chromosome arm 2L: rearrangements that relocate *light* to euchromatic regions result in reduced levels of expression. Mutants that enhance or suppress PEV are known, and characterization of these genes is illuminating with respect to chromatin structure.

5.1.4 Interfering RNA

Perhaps the single most important molecular biological technique to have been developed in recent years is that of RNA interference. *Drosophila* researchers have been able to apply this new technology in several ways – firstly by injecting embryos or transfecting cultured cells with double-stranded RNA corresponding to a gene under investigation, and subsequently examining the phenotypic consequences, and secondly by generating transgenic strains, which encode a transcript suitable for genetic interference.

5.1.5 Targeted Mutagenesis

Rong and Golic published a method for targeted mutagenesis, in which *in vitro* mutated genes are inserted in the *Drosophila* genome by P-element techniques and then substituted for the wild type allele by a recombination based procedure. A detailed description of this procedure is out of the scope of this article, but can be found in the cited references. If widely adopted, this technique has great potential.

5.2 Genetic Mapping

Genetic, or recombination mapping, in *D. melanogaster* has a long history. The theoretical basis of recombination mapping was devised and implemented using *Drosophila* in 1913. The map unit, or centimorgan (cM) corresponds to one percent recombination. Over short map distances, these values are additive, but over longer distances, complications arise from multiple crossovers. Correlating the cytogenetic and recombination map positions of a number of loci has demonstrated that the recombination rate varies along the chromosome: recombination is effectively suppressed close to telomeres and centromeres. Thus, 1 cM represents a larger stretch of DNA near the telomere or centromere than in the middle of a chromosome arm.

Under normal circumstances, there is no recombination in D. melanogaster males. The special circumstances in which male recombination is observed include hybrid dysgenesis. Additionally, recombination in the tiny fourth chromosome does not occur, except in triploid females.

Recombination mapping is now generally undertaken as a starting point to cytological mapping, the localization of a gene on the polytene chromosome map, and its subsequent molecular cloning. The correlation between the polytene chromosome map and the recombination map permits the approximate cytogenetic map position to be determined from recombination map data. The recombination map order of loci can be important in cloning projects.

A typical scheme for mapping a recessive lethal mutation is shown in Fig. 6. The first two crosses in this scheme are carried out *en masse*, while subsequent crosses are single-pair matings. Each recombinant stock generated in this scheme can be classified on the basis of which markers are present on the recombinant chromosome, and whether the lethal is present. This permits easy assignment of the lethal to one interval between markers. Further recombination mapping may be undertaken, concentrating on that interval, to yield a more precise map position.

5.2.1 Microsatellite and RFLP Markers

Genetic mapping in *Drosophila* has rarely used microsatellites and restriction fragment length polymorphism (RFLP) markers, due to the availability of many visible markers and chromosome rearrangements, permitting the easy localization of genes to a molecularly

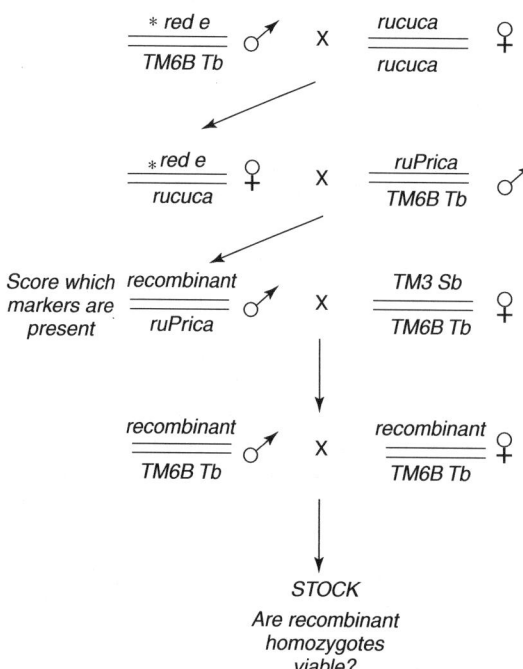

Fig. 6 Recombination mapping scheme. Markers are as described in the legend to Fig. 5, and are as follows. The chromosome *rucuca* is multiply marked with the following recessives: *ru* (*roughoid*: eyes are rough in appearance); *h* (*hairy*: extra hairs on wing veins); *th* (*thread*: aristae are threadlike, instead of branched); *st* (*scarlet*: adult eye is bright red); *cu* (*curled*: wings curled upwards); *sr* (*stripe*: stripe on dorsal surface of thorax); *e*; *ca* (*claret*: adult eyes are ruby red). Flies homozygous for both *st* and *ca* have orange eyes. The chromosome *ruPrica* is as *rucuca*, but in addition contains the dominant marker *Prickly* (*Pr*), which causes the adult bristles to be shortened, with twisted tips. As in Fig. 5, all females are virgin.

relevant precision. However, microsatellites and RFLPs are of central importance for recombination mapping in genetically less tractable insects, such as mosquitoes.

6 Prospects

To some extent, *Drosophila* is a late starter with respect to the development of molecular physical maps. This is undoubtedly due to having had an excellent physical map dating from the 1930s, the polytene chromosome map. It is an enduring legacy to the mapping talents of Calvin Bridges that his polytene chromosome map has continued to be relevant to researchers undertaking detailed molecular genetic studies undreamt of by the early geneticists.

See also Molecular Basis of Genetics; Genomic Sequencing

Bibliography

Books and Reviews

Adams, M.D., Sekelsky, J.J. (2002) From sequence to phenotype: reverse genetics in *Drosophila melanogaster*, *Nat. Rev. Genet.* 3, 189–198.

Ashburner, M. (1989) *Drosophila, A Laboratory Handbook*, Cold Spring Harbor Laboratory Press, Cold Spring Harbor, New York.

Duffy, J.B. (2002) GAL4 system in Drosophila: a fly geneticist's Swiss army knife, *Genesis* 34(1–2), 1–15.

Lefevre, G. (1976) A Photographic Representation and Interpretation of the Polytene Chromosomes of *Drosophila Melanogaster* Salivary Glands, in: Ashburner, M., Novitski, E. (Eds.) *The Genetics and Biology of Drosophila*, Vol. 1a, Academic Press, New York, pp. 31–66.

Sorsa, V. (1988) *Chromosome Maps of Drosophila*, CRC Press, Boca Raton, Florida.

St Johnston, D. (2002) The art and design of genetic screens: *Drosophila melanogaster*, *Nat. Rev. Genet.* 3, 176–188.

The FlyBase Consortium (2003). The FlyBase database of the Drosophila genome projects and community literature, *Nucleic Acids Res.* 31, 172–175, http://flybase.org/

Primary Literature

Adams, M.D., Celniker, S.E., et al. (2000) The genome sequence of *Drosophila melanogaster*, *Science* 287, 2185–2195.

Agard, D.A., Sedat, J.W. (1983) Three-dimensional architecture of a polytene nucleus, *Nature* 302, 676–681.

Ajioka, J.W., Garza, D., Johnson, D., Carulli, J.P., Jones, R.W., Hartl, D.L. (1990) Genome Evolution Analyzed by Cloning Large Fragments of *Drosophila* DNA in Yeast Artificial Chromosomes, *Molecular Evolution*, A. R. Liss.

Ajioka, J.W., Smoller, D.A., Jones, R.W., Carulli, J.P., Vellek, A.E.C., Garza, D., Link, A.J., Duncan, I.W., Hartl, D.L. (1991) *Drosophila* genome project: one-hit coverage in yeast artificial chromosomes, *Chromosoma* 100, 495–509.

Arbeitman, M.N., et al. (2002) Gene expression during the life cycle of Drosophila melanogaster, *Science* 297(5590), 2270–2275.

Ashburner, M., et al. (1999) An exploration of the sequence of a 2.9-Mb region of the genome of *Drosophila melanogaster*: The *Adh* region, *Genetics* 153, 179–219.

Ballinger, D., Benzer, S. (1989) Targeted gene mutations in *Drosophila*, *Proc. Natl. Acad. Sci. U.S.A.* 86, 9402–9406.

Benos, P.V., et al. (2001) From first base: the sequence of the tip of the X chromosome of *Drosophila melanogaster*, a comparison of two sequencing strategies, *Genome Res.* 11, 710–730.

Bentley, A., MacLennan, B., Calvo, J., Dearolf, C.R. (2000) Targeted recovery of mutations in Drosophila, *Genetics* 156, 1169–1173.

Bergman, C.M., et al. (2002) Assessing the impact of comparative genomic sequence data on the functional annotation of the Drosophila genome, *Genome Biol.* 3(12), RESEARCH0086-6.

Bibikova, M., Golic, M., Golic, K.G., Carroll, D. (2002) Targeted chromosomal cleavage and mutagenesis in Drosophila using zinc-finger nucleases, *Genetics* **161**(3), 1169–1175.

Bier, E., et al. (1989) Searching for pattern and mutation in the *Drosophila* genome with a P-laxZ vector, *Genes Dev.* **3**, 1273–1287.

Bieschke, E.T., Wheeler, J.C., Tower, J. (1998) Doxycycline-induced transgene expression during *Drosophila* development and aging, *Mol. Gen. Genet.* **258**, 571–579.

Boutanaev, A.M., Kalmykova, A.I., Shevelyov, Y.Y., Nurminsky, D.I. (2002) Large clusters of co-expressed genes in the Drosophila genome, *Nature* **420**(6916), 666–669.

Bridges, C.B. (1935) Salivary chromosomes. With a key to the banding of the chromosomes of *Drosophila melanogaster*, *J. Hered.* **26**, 60–64.

Bridges, C.B., Bridges, P.N. (1939) A new map of the second chromosome: a revised map of the right limb of the second chromosome of *Drosophila melanogaster*, *J. Hered.* **30**, 475–476.

Bridges, P.N. (1941) A revision of the salivary gland 3R-chromosome map of *Drosophila melanogaster*, *J. Hered.* **32**, 299–300.

Celniker, S.E., Wheeler, D.A. et al. (2002) Finishing a whole-genome shotgun: release 3 of the Drosophila melanogaster euchromatic genome sequence, *Genome Biol.* **3**(12), RESEARCH0079-9.

Clemens, J.C., Worby, C.A., Simonson-Leff, N., Muda, M., Maehama, T., Hemmings, B.A., Dixon, J.E. (2000) Use of double-stranded RNA interference in *Drosophila* cell lines to dissect signal transduction pathways, *Proc. Natl. Acad. Sci. U.S.A.* **97**, 6499–6503.

Cooley, L., Kelley, R., Spradling, A.C. (1988) Insertional mutagenesis of the *Drosophila* genome with a single P element, *Science* **239**, 1121–1128.

Danilevskaya, O.N., Petrov, D.A., Pavlova, M.N., Koga, A., Kurenova, E.V., Hartl, D.L. (1992) A repetitive DNA element, associated with telomeric sequences in *Drosophila melanogaster*, contains open reading frames, *Chromosoma* **102**, 32–40.

Deak, P., Omar, M.M., et al. (1997) P element insertion alleles of essential genes on the third chromosome of *Drosophila melanogaster*: Correlation of physical and cytogenetic maps in chromosomal region 86E-87F, *Genetics* **147**, 1697–1722.

Elbashir, S.M., Martinez, J., Patkaniowska, A., Lendeckel, W., Tuschl, T. (2001) Functional anatomy of siRNAs for mediating efficient RNAi in Drosophila melanogaster embryo lysate, *EMBO J.* **20**(23), 6877–6888.

Engels, W.R., Preston, C.R. (1984) Formation of chromosome rearrangements by P factors in *Drosophila*, *Genetics* **107**, 657–678.

Garza, D., Ajioka, J.W., Burke, D.T., Hartl, D.L. (1989) Mapping the *Drosophila* genome with yeast artificial chromosomes, *Science* **246**, 641–646.

Garza, D., Ajioka, J.W., Carulli, J.P., Jones, R.W., Johnson, D.H., Hartl, D.L. (1989) Physical mapping of complex genomes, *Nature (Product review)* **340**, 577–578.

Gavrila, L., Ecovoiu, A.A., et al. (2001) Localizing genes in *Drosophila melanogaster* polytene chromosomes by fluorescence in situ hybridization, *J. Cell Mol. Med.* **5**(1).

Giordano, E., Rendina, R., Peluso, I., Furia, M. (2002) RNAi triggered by symmetrically transcribed genes in *Drosophila melanogaster*, *Genetics* **160**, 637–648.

Gloor, G.B., Nassif, N.A., Johnson-Schlitz, D.M., Preston, C.R., Engels, W.R. (1991) Targeted gene replacement in Drosophila via P element-induced gap repair, *Science* **253**(5024), 1110–1117.

Gong, W.J., Golic, K.G. (2003) Ends-out, or replacement, gene targeting in Drosophila, *Proc. Natl. Acad. Sci. U.S.A.* **100**(5), 2556–2561.

Hartl, D.L. (1992) Genome Map of *Drosophila Melanogaster* Based on Yeast Artificial Chromosomes, in: Tilghman, Davies (Eds.) *Genome Analysis*, Vol. 4, Cold Spring Harbor Laboratory Press, Cold spring Harbor, pp. 39–69.

Hartl, D.L., Ajioka, J.W., Cai, H., Lohe, A., Lozovskaya, E.R., Smoller, D.A., Duncan, I.A. (1992) Towards a *Drosophila* genome map, *Trends Genet.* **8**(2), 70–75.

Hartl, D.L., Nurminsky, D.I., Jones, R.W., Lozovskaya, E.R. (1994) Genome structure and evolution in *Drosophila*: applications of the framework P1 map, *Proc. Natl. Acad. Sci. U.S.A.* **91**, 6824–6829.

Hochstrasser, M., Mathog, D., Gruenbaum, Y., Saumweber, H., Sedat, J.W. (1986) Spatial organization of chromosomes in the salivary gland nuclei of *Drosophila melanogaster*, *J. Cell Biol.* **102**, 112–123.

Hoheisel, J.D., Lennon, G.G., et al. (1991) Use of high coverage reference libraries of *Drosophila melanogaster* for relational data analysis. A step towards mapping and sequencing of the genome, *J. Mol. Biol.* **221**, 903–914.

Hoskins, R.A., et al. (2002) Heterochromatic sequences in a Drosophila whole-genome shotgun assembly, *Genome Biol.* **3**(12), RESEARCH0085-5.

Huang, A.M., Rehn, E.J., Rubin, G.M. (2000) Recovery of DNA Sequences Flanking P-element Insertions: Inverse PCR and Plasmid Rescue, in: Sullivan, W., Ashburner, M., Hawley, R.S. (Eds.) *Drosophila Protocols*, CSHL Press, Cold Spring Harbor, pp. 429–437.

Jack, J.W., Judd, B.H. (1979) Allelic pairing and gene regulation: a model for the *zeste-white* interaction in Drosophila, *Proc. Natl. Acad. Sci. U.S.A.* **76**, 1368–1372.

Judd, B.H. (1988) Transvection: allelic crosstalk, *Cell* **53**, 841–843.

Kafatos, F.C., Louis, C., Savakis, C., Glover, D.M., Ashburner, M., Link, A.J., Siden-Kiamos, I., Saunders, R.D.C. (1991) Integrated maps of the *Drosophila* genome: progress and prospects, *Trends Genet.* **7**, 155–161.

Kaiser, K., Goodwin, S.F. (1990) "Site-directed" transposon mutagenesis of Drosophila, *Proc. Natl. Acad. Sci. U.S.A.* **87**, 1686–1690.

Kaminker, J.S., et al. (2002) The transposable elements of the Drosophila melanogaster euchromatin: a genomics perspective, *Genome Biol.* **3**(12), RESEARCH0084-4.

Kennerdell, J.R., Carthew, R.W. (1998) Use of dsRNA-mediated genetic interference to demonstrate that *frizzled* and *frizzled 2* act in the Wingless pathway, *Cell* **95**, 1017–1026.

Kennerdell, J.R., Carthew, R.W. (2000) Heritable gene silencing in *Drosophila* using double-stranded RNA, *Nat. Biotechnol.* **18**, 896–898.

Kornberg, T. (2002) Another arrow in the Drosophila quiver, *Proc. Natl. Acad. Sci. U.S.A.* **99**(15), 9607–9608.

Laird, C.D., McCarthy, B.J. (1969) Molecular characterization of the *Drosophila* genome, *Genetics* **63**, 865–882.

Leach, T.J., Chotkowski, H.L., Wotring, M.G., Dilwith, R.L., Glaser, R.L. (2000) Replication of heterochromatin and structure of polytene chromosomes, *Mol. Cell Biol.* **20**(17), 6308–6316.

Lindsley, D.L., Zimm, G.G. (1992) *The Genome of Drosophila Melanogaster*, Academic Press, San Diego.

Lozovskaya, E.R., Petrov, D.A., Hartl, D.A. (1993) A combined molecular and cytogenetic approach to genome evolution in *Drosophila* using large-fragment DNA cloning, *Chromosoma* **102**, 253–266.

Madueno, E., et al. (1995) A physical map of the X chromosome of *Drosophila melanogaster*: cosmid contigs and sequence tagged sites, *Genetics* **139**, 1631–1647.

Mathog, D., Sedat, J.W. (1989) The three-dimensional organization of polytene nuclei in male *Drosophila melanogaster* with compound XY or ring X chromosomes, *Genetics* **121**, 293–311.

Mathog, D., Hochstrasser, M., Gruenbaum, Y., Saumweber, H., Sedat, J.W. (1984) Characteristic folding pattern of polytene chromosomes in *Drosophila* salivary gland nuclei, *Nature* **308**, 414–421.

Miklos, G.L., Rubin, G.M. (1996) The role of the genome project in determining gene function: insights from model organisms, *Cell* **86**(4), 521–529.

Miklos, G.L.G., Healy, M.J., Pain, P., Howells, A.J., Russell, R.J. (1984) Molecular and genetic studies on the euchromatin-heterochromatin transition region of the X chromosome of *Drosophila melanogaster*, *Chromosoma* **89**, 218–227.

Miklos, G.L.G., Yamamoto, M.-T., Davies, J., Pirrotta, V. (1988) Microcloning reveals a high frequency of repetitive sequences characteristic of chromosome 4 and the β-heterochromatin of *Drosophila melanogaster*, *Proc. Natl. Acad. Sci. U.S.A.* **85**, 2051–2055.

Misra, S., et al. (2002) Annotation of the Drosophila melanogaster euchromatic genome: a systematic review, *Genome Biol.* **3**(12), RESEARCH0083-RESEARCH0083.

Myers, E.W., et al. (2000) A whole-genome assembly of *Drosophila*, *Science* **287**, 2196–2204.

Ohler, U., Liao, G.C., Niemann, H., Rubin, G.M. (2002) Computational analysis of core promoters in the Drosophila genome, *Genome Biol.* **3**(12), RESEARCH0087-7.

Painter, T.S. (1933) A new method for the study of chromosome rearrangements and the plotting of chromosome maps, *Science* **78**, 585–586.

Painter, T.S. (1934) A new method for the study of chromosome aberrations and the plotting of chromosome maps in *Drosophila melanogaster*, *Genetics* **19**, 175–188.

Peter, A., et al. (2002) Mapping and identification of essential gene functions on the X chromosome of Drosophila, *EMBO Rep.* **3**, 34–38.

Potter, C.J., Turenchalk, G.S., Xu, T. (2000) Drosophila in cancer research, *Trends Genet.* **16**(1), 33–39.

Roman, G., Endo, K., Zong, L., Davis, R.L. (2001) P[Switch], a system for spatial and temporal control of gene expression in Drosophila melanogaster, *Proc. Natl. Acad. Sci. U.S.A.* **98**(22), 12602–12607.

Rong, Y.S., Golic, K.G. (2000) Gene targeting by homologous recombination in Drosophila, *Science* **288**, 2013–2018.

Rong, Y.S., Golic, K.G. (2001) A targeted gene knockout in Drosophila, *Genetics* **157**(3), 1307–1312.

Rong, Y.S., Titen, S.W. et al. (2002) Targeted mutagenesis by homologous recombination in D. melanogaster, *Genes Dev.* **16**(12), 1568–1581.

Rørth, P. (1996) A modular misexpression screen in Drosophila detecting tissue specific phenotypes, *Proc. Natl. Acad. Sci. U.S.A.* **93**, 12418–12422.

Rørth, P., Szabo, K. et al. (1998) Systematic gain-of-function genetics in Drosophila, *Development* **125**, 1049–1057.

Scalenghe, F., Turco, E., Edstrom, J.E., Pirrotta, V., Melli, M. (1981) Microdissection and cloning of DNA from a specific region of Drosophila melanogaster polytene chromosomes, *Chromosoma* **82**, 205–216.

Schubeler, D., Scalzo, D., Kooperberg, C., van Steensel, B., Delrow, J., Groudine, M. (2002) Genome-wide DNA replication profile for Drosophila melanogaster: a link between transcription and replication timing, *Nat. Genet.* **32**(3), 438–442.

Sharakhov, I.V., Fillippova, M.A., et al. (1993) beta-Heterochromatin in Drosophila – molecular organization and functioning – characterization of the dna sequence from the d-melanogaster proximal beta-heterochromatin associated with the nuclear envelope, *Genetika* **29**(3), 393–402.

Siden-Kiamos, I., et al. (1990) Towards a physical map of the Drosophila melanogaster genome: mapping of cosmid clones within defined genomic divisions, *Nucelic Acids Res.* **18**, 6261–6270.

Smoller, D.A., Petrov, D., Hartl, D.L. (1991) Characterization of bacteriophage P1 library containing inserts of Drosophila DNA of 75–100 kilobase pairs, *Chromosoma* **100**, 487–494.

Spellman, P.T., Rubin, G.M. (2002) Evidence for large domains of similarly expressed genes in the Drosophila genome, *J. Biol.* **1**, 5.

Spierer, A., Spierer, P. (1984) Similar level of polyteny in bands and interbands of Drosophila giant chromosomes, *Nature* **307**, 176–178.

Spradling, A.C., Stern, D., Beaton, A., Rhem, E.J., Laverty, T., Mozden, N., Misra, S., Rubin, G.M. (1999) The Berkeley Drosophila genome project gene disruption project: single P-element insertions mutating 25% of vital Drosophila genes, *Genetics* **153**(1), 135–77.

Stapleton, M., et al. (2002) The Drosophila gene collection: identification of putative full-length cDNAs for 70% of D. melanogaster genes, *Genome Res.* **12**(8), 1294–1300.

Stapleton, M., et al. (2002) A Drosophila full-length cDNA resource, *Genome Biol.* **3**(12), RESEARCH0080-0.

Stark, A., Brennecke, J., Russell, R.B., Cohen, S.M. (2003) Identification of Drosophila microRNA targets, *Public Library Sci. Biol.* **1**(3), 1–13.

Stebbins, M.J., Yin, J.C.P. (2001) Adaptable doxycycline-regulated gene expression systems for Drosophila, *Gene* **270**, 103–111.

Stebbins, M.J., Urlinger, S., Byrne, G., Bello, B., Hillen, W., Yin, J.C.P. (2001) Tetracycline-inducible systems for Drosophila, *Proc. Natl. Acad. Sci. U.S.A.* **98**, 10775–10780.

Toba, G., Ohsako, T., Miyata, N., Ohtsuka, T., Seong, K.-H., Aigaki, T. (1999) The gene search system: a method for efficient detection and rapid molecular identification of genes in Drosophila melanogaster, *Genetics* **151**, 725–737.

Tomancak, P., et al. (2002) Systematic determination of patterns of gene expression during Drosophila embryogenesis, *Genome Biol.* **3**(12), RESEARCH0088-8.

Tower, J., Karpen, G.H., Craig, N., Spradling, A.C. (1993) Preferential transposition of Drosophila P elements to nearby chromosomal sites, *Genetics* **133**, 347–359.

van Roessel, P., Brand, A.H. (2000) GAL4-mediated ectopic gene expression in Drosophila, in: Sulliva, W., Ashburner, M., Hawley, R.S. (Eds.) *Drosophila Protocols*, Cold Spring Harbor Laboratory Press, Cold Spring Harbor, pp. 439–447.

White, K.P., Rifkin, S.A., Hurban, P., Hogness, D.S. (1999) Microarray analysis of *Drosophila* development during metamorphosis, *Science* **286**, 2179–2184.

Yamamoto, M.-T., Mitchelson, A., Tudor, M., O'Hare, K., Davies, J.A., Miklos, G.L.G. (1990) Molecular and cytogenetic analysis of the heterochromatin-euchromatin junction region of the *Drosophila melanogaster* X chromosome using cloned DNA sequences, *Genetics* **125**, 821–832.

16
Malaria Mosquito Genome

Robert A. Holt[1] and Frank H. Collins[2]
[1] *Canada's Michael Smith Genome Science Centre, Vancouver, BC, Canada*
[2] *University of Notre Dame, Notre Dame, IN, USA*

1	**The Mosquito and Infectious Disease** 443	
1.1	Scope and Impact of Mosquito-borne Disease 443	
1.2	Strategies for Malaria Control 444	
2	***A. gambiae*, the Principal Vector of Malaria** 445	
2.1	The Mosquito and the Malaria Parasite 445	
2.2	*A. gambiae* is a Polymorphic Species 446	
3	**The *A. gambiae* Genome** 447	
3.1	The PEST Strain 447	
3.2	The Draft Genome Sequence 448	
3.3	Genome Assembly, Genome size, and Physical Mapping 449	
3.4	Genetic Variation 450	
3.5	Gene Content 452	
3.6	Comparison of the *Anopheles* and *Drosophila* Proteomes 453	
3.7	Genes Regulated by Blood Feeding 455	
4	**Utility of the *A. gambiae* Genome** 456	
4.1	Odorant Receptors and Novel Repellents/Attractants 456	
4.2	Insecticide Resistance Genes 457	
4.3	Insecticide Targets 459	
4.4	Immune-response Genes and Genetic Vector Control Strategies 460	
	Bibliography 461	
	Books and Reviews 461	
	Primary Literature 463	

Genomics and Genetics. Edited by Robert A. Meyers.
Copyright © 2007 Wiley-VCH Verlag GmbH & Co. KGaA, Weinheim
ISBN: 978-3-527-31609-0

Keywords

BAC
Bacterial artificial chromosome. A single copy cloning vector for large DNA fragments (on the order of 100 to 200 kbp).

Contig
A segment of contiguous sequence, uninterrupted by gaps, assembled from a number of overlapping sequence reads.

Cytogenetic Form
Subspecies of *Anopheles gambiae*, as classified by paracentric chromosomal inversions.

DDT
The pesticide dichlorodiphenyltrichloroethane.

Mate Pair
The two sequence reads derived from either end of a DNA fragment.

Orthologs
Genes in different species arising from a single ancestral gene (e.g. human α-globin and mouse α-globin).

Paralogs
Genes in the same species, arising from local gene duplication (e.g. human α-globin and human β-globin).

PEST
The pink eye standard laboratory strain of *Anopheles gambiae*.

Read
A DNA sequence, approximately 500 bp in length, that is derived from a template DNA molecule. The terms "read," "trace," "lane," and "electropherogram" are often used interchangeably.

SNP
Single nucleotide polymorphism.

Scaffold
A set of contigs separated by sequence gaps but of defined order and orientation.

Vector
An organism that carries disease-causing pathogens from one host to another.

WHO
World Health Organization.

Shotgun
A method of whole-genome sequencing where high molecular weight DNA from an organism is randomly fragmented and sequence reads are obtained from fragment ends. Sequencing is done at some level of redundancy (coverage) such that each nucleotide in the genome is represented in multiple reads. The consensus genome sequence is reconstructed computationally on the basis of overlaps among sequence reads.

Sporogeny
The stages of the *Plasmodium* development cycle that take place within the mosquito vector.

Mosquito-borne illnesses exact an enormous toll on human health; malaria parasites alone currently infect approximately 500 million people, more than 1 million of whom will die this year. *A. gambiae* is the principal malaria vector in sub-Saharan Africa, where 90% of the world's deaths due to malaria occur. Historically, the greatest successes in prevention of malaria, yellow fever, and other mosquito-borne disease have come from controlling the mosquito vector. Current work on vector control holds much promise and can be addressed more efficiently by genomic approaches. The reference genome sequence of *A. gambiae sensu stricto* (a representative member of the cryptic *A. gambiae* species complex) has yielded novel insecticide targets, has facilitated the identification of genes involved in insecticide resistance, and has led to the identification of genes that underlie the strong preference of this vector for human blood. Germ-line transformation of the mosquito has been achieved and the reference genome sequence will facilitate efforts to develop a tractable genetic control strategy for mosquitoes. The publication of the *A. gambiae* genome is a landmark in medical entomology, both because this was the first vector genome to be fully sequenced and because of the impact of this mosquito on public health.

1
The Mosquito and Infectious Disease

1.1
Scope and Impact of Mosquito-borne Disease

Mosquitoes, which are members of the insect family Culicidae of the order Diptera, the flies, are medically the most important group of arthropod vectors, or transmitters, of human pathogens. Mosquitoes transmit a large number of human viral pathogens, including the viruses that cause dengue, yellow fever, and West Nile encephalitis, as well as several nematodes species that cause lymphatic filariasis. But, by far, the most important mosquito-borne disease is malaria, which is caused by protozoan parasites of the genus *Plasmodium*. Malaria shares with HIV/AIDS and tuberculosis the distinction of being one of the three most important disease-specific causes of human mortality in the world today. Estimates of the public health impact of malaria are imprecise, but it is widely assumed that approximately half a billion people are currently infected with malaria parasites and that more than 1 million die each year, principally infants and very young children. Although four different *Plasmodium* species infect humans, nearly all of the mortality is caused by the parasite *Plasmodium falciparum*.

1.2
Strategies for Malaria Control

The etiologic agent of human malaria and the pathogen's mode of transmission by mosquitoes were both determined in the latter years of the nineteenth century by Charles Laveran (who discovered the parasite) and Ronald Ross and Battista Grassi (they independently identified the vector), but malaria has been recognized as an important human disease for centuries. By the fifth century, Hippocrates had clearly described the clinical symptoms associated with three different species of malaria parasites, and it was widely recognized in those times that the disease was caused by association with marshy environments with bad air (thus the name malaria).

Antimalarial drugs and interventions targeted at the vector remain the primary public health measures for dealing with malaria. Two very important discoveries emerged during World War II: the antimalarial drug chloroquine, and the insecticide DDT (dichlorodiphenyltrichloroethane). Both were extremely effective, very inexpensive to produce, unencumbered by patents, and had remarkably low levels of toxicity. (The adverse environmental impacts of DDT, which is a highly stable molecule that can accumulate to toxic levels in species at or near the top of the food chain, were entirely due to its widespread use for control of insects of agricultural importance.) In the decades of the 1950s and 1960s, chloroquine and DDT (and related types of antimalarials and insecticides) formed the basis for very effective, countrywide malaria-control programs that were implemented in the framework of a World Health Organization-directed worldwide Malaria Eradication Campaign. Malaria was permanently eradicated from most developed countries in the temperate-climate regions of the world and prevalence was significantly reduced in many other malaria-endemic countries. Unfortunately, worldwide eradication was an impossible goal, and by the 1970s, the responsibility for managing and funding malaria-control programs was returned to national health programs and budgets.

Several important lessons emerged from the WHO (World Health Organization) Malaria Eradication Campaign. Vector control using insecticides like DDT sprayed on the walls of houses was an extremely effective malaria-control strategy. However, it was costly and required well-managed teams of technically skilled people. Moreover, it was effective only in countries with sufficient infrastructures like roads and bridges for vector-control teams to reach the population that was at risk. Control programs based exclusively on antimalarial drugs like chloroquine were even more costly because of the requirement for active case detection. But as active control programs lapsed, effective and inexpensive antimalarial drugs like chloroquine found their way into malaria-endemic communities through either the national primary health systems or through the marketplace.

For most of sub-Saharan Africa, with the notable exceptions of some of the countries in southern Africa and around some of the major cities, the organized malaria-control programs developed during the WHO Malaria Eradication Campaign were not even attempted, particularly in rural areas. Nonetheless, inexpensive antimalarial drugs like chloroquine became widely available, even in relatively remote rural communities in most of Africa. Unfortunately, chloroquine-resistant *P. falciparum* appeared in coastal Kenya in 1978, and in the subsequent two decades, resistant parasites became the prevalent form almost throughout sub-Saharan Africa. Although the health impact of the publicly available and affordable antimalarials like chloroquine was never assessed, the emergence of parasite resistance to this drug has been coincident with an equally poorly documented perception that malaria-specific mortality in Africa has been increasing. Today, chloroquine is an effective antimalarial only in parts of west Africa, and unfortunately, replacement drugs like Fansidar (sulfadoxine/pyramethamine), mefloquine, and – more recently – artemisinin-based drugs (based on the ingredients in *Artemisia annua*) are either very costly or are themselves provoking the emergence of resistant parasite strains. As resistance to existing antimalarial drugs becomes more entrenched, there is increased need for novel and effective vector-control strategies.

2
A. gambiae, the Principal Vector of Malaria

2.1
The Mosquito and the Malaria Parasite

Malaria parasites are intracellular human parasites that undergo cycles of invasion, growth, and mitotic replication in erythrocytes. The parasites are transmitted among human hosts by *Anopheles* mosquitoes, which ingest parasites when they feed on the blood of an infected human host. The malaria parasites undergo a complex developmental cycle, referred to as the *sporogenic developmental cycle*, in the mosquito vector. A small fraction of the *Plasmodium*-infected erythrocytes are differentiated forms called *gametocytes* that are triggered by hypoxanthine in the mosquito's midgut to develop into male and female gametes. These fuse to form diploid zygotes (the only diploid stage in the *Plasmodium* life cycle), which traverse the mosquito's midgut cells and then undergo a meiotic division and differentiate further into growth stages called the *oocysts*. After approximately one week of growth and mitotic divisions, a single oocyst will produce several thousand specialized, haploid forms called *sporozoites*. On maturation of the oocysts, the sporozoites are released into the mosquito's open circulatory system through which they are transported passively to the salivary glands. These parasite stages invade the salivary gland cells and accumulate in the salivary ducts and acini of these cells, from where they are injected into a host capillary when the mosquito next takes a blood meal. In the human host, the sporozoites are transferred passively through the circulatory system to the liver, where they invade hepatocytes. In these cells, the parasite undergoes approximately a dozen rounds of mitotic replication that result in the production of several thousand *merozoites*. On completion of this cycle of replication, the infected hepatocytes rupture and release the merozoites into the circulatory system, where they invade erythrocytes and initiate the blood cell cycle of infection that causes disease.

Plasmodium falciparum malaria is widely distributed throughout the widely distributed throughout the world, especially in tropical regions bounded by approximately 30° north and south latitude. Over this broad distribution, parasite prevalence in human populations with endemic *P. falciparum* ranges from low levels of a few percent to peak endemicity levels, approaching 100% in rainy seasons when transmission is maximum. Of the approximately 500 or more *Anopheles* species found worldwide, several dozen are involved in the transmission of malaria parasites. The *Anopheles* vectors are the major determinant of local levels of disease endemicity. The majority of *Anopheles* species are probably physiologically competent to support sporogonic development of the parasite. In order for malaria parasites to remain endemic in a particular human community, the vector population must effectively propagate each infected human case to – on average – more than one new human host. With an effective case reproductive rate of less than 1, the parasite cannot be stably maintained in a human population.

Four important features of mosquito biology determine whether a particular *Anopheles* species will be effective as a malaria vector: (1) It must be a physiologically suitable host. (2) It must also be a moderately abundant species in order to maintain a density sufficient to achieve transmission. (3) The mosquito must also take blood meals from people. Because the vector must both acquire the infection in one blood meal and then take a second meal on another human host after the sporogonic cycle is complete, blood-meal host choice is an important determinant of vectorial effectiveness. (4) Finally, and most important, the age structure of the vector population must be such that a sufficient portion survives long enough to allow sporogonic development to be completed.

2.2
A. gambiae is a Polymorphic Species

Approximately 90% of the world's malaria-specific mortality occurs in sub-Saharan Africa, where the primary vector is *A. gambiae*. A second important African vector is the mosquito *Anopheles arabiensis*, which is a close relative of *A. gambiae* in a cluster of seven related African species known as the *A. gambiae sibling-species complex*. This extraordinary concentration of malaria in Africa is largely a consequence of the close ecological association of *A. gambiae* and *A. arabiensis* with people. *A. gambiae*, in particular, takes blood meals almost exclusively from people; its larval stages develop in temporary pools of water, like tire ruts, animal hoof-prints, and irrigated agricultural fields that are produced by human activity; and when not laying eggs or mating, the adult females spend most of their time in human dwellings either blood feeding or resting while eggs develop.

Perhaps, because of the close association this mosquito has with people, it is also a highly polymorphic species. Like a number of other insects, especially in the order Diptera, *A. gambiae* and its sibling species have giant polytene chromosomes in many of their tissues, and these have facilitated population analysis. During the past three decades, careful studies of the polytene chromosomes of field-collected *A. gambiae* from countries in west Africa have revealed patterns of distribution of a number of paracentric chromosome inversions that suggest the presence of at least five different reproductively isolated chromosome forms designated as Bissau, Bamako, Mopti, Forest, and Savanna. (All *Anopheles* genomes are based on three pairs of chromosomes, two autosomes, and an X and Y sex-determining pair, and in most *Anopheles* species, cells of either the larval salivary glands or nurse cells of developing ovaries contain polytene chromosomes that, when properly prepared, will show a reproducible, species-specific banding pattern. In addition to their value in species identification and population studies, polytene chromosomes are excellent for physical mapping of cloned DNA.) Most of the inversions that distinguish chromosomal forms are restricted to the right arm of chromosome 2. In geographic regions where two or more of the chromosome forms are present, inversion heterozygotes expected from interbreeding of forms are either totally absent or are present at levels significantly below Hardy–Weinberg expectations. Three of these forms, Bamako, Mopti, and Savanna, have been carefully studied in Mali, where they exhibit clear differences in ecology. In spite of considerable effort by several groups to identify molecular markers diagnostic of these chromosome forms, the only molecule found thus far that distinguishes different *A. gambiae* populations is the rDNA, where sequences in both the IGS and ITS regions can be used to identify what are referred to as *S and M ribosomal DNA (rDNA) types*. In Mali, the M rDNA type is associated exclusively with Mopti cytogenetic specimens and the S rDNA type is found only in Savanna and Bamako specimens. In other parts of west Africa, however, the association between karyotype and rDNA type is less predictable. Both S and M rDNA types are found in Forest, Bissau, and Savanna cytogenetic forms.

The frequencies of polymorphic paracentric inversions are clearly important characters in *A. gambiae* as well as in its sibling species, *A. arabiensis*. In both species, inversion frequencies have been shown to vary along clines associated with climatic and ecological parameters, and the data from Mali, supporting the presence of chromosomally distinct *A. gambiae* forms that are both ecologically and genetically differentiated is quite robust. Unfortunately, karyotyping of wild specimens is both difficult and possibly biased, and the lack of congruence between the chromosomal inversion and rDNA molecular markers means that the population structure of this important African vector remains unclear.

3
The *A. gambiae* Genome

3.1
The PEST Strain

The *A. gambiae* PEST (Pink Eye STandard) strain was chosen for genome sequencing because it had a fixed chromosomal arrangement, a sex-linked pink-eye mutation that can readily be used as an indicator of cross-colony contamination, and because clones from two different PEST-strain BAC (Bacterial Artificial Chromosome) libraries had already been end sequenced and physically mapped. The pink-eye mutation originated in a colony called *A. gambiae LPE*, established in 1951 at the London School of Hygiene and Tropical Medicine from mosquitoes collected in Lagos, Nigeria. In 1986, this mutation was introduced into a colony of *A. gambiae* from western Kenya by crossing males of the LPE strain with female offspring of wild caught Kenyan *A. gambiae* (the Savanna form), selecting males from the F2 of this cross and then crossing them again with additional female offspring of wild caught Kenyan *A. gambiae*. From the F2 offspring of this second outcross to Kenyan mosquitoes, a strain was selected that was fixed for pink eye. This outcrossing scheme was repeated once more in 1987, producing a pink-eye strain with a genetic composition largely constituted of the western Kenya Savanna cytogenetic form. In each of these crosses, several hundred female offspring of at least 20 wild caught mosquitoes were used in the cross. This strain, designated *A. gambiae* PE, was polymorphic for the inversions 2La (32%) and 2Rbc (19%). The 2Rbc inversion is characteristic of Mopti, indicating that the original LPE strain from Nigeria was the Mopti form. This inversion was apparently balanced by the uninverted form because no 2Rbc/bc individuals were detected in the colony. From this PE strain, Mukabayire and Besansky selected a set of nine families whose female parent and at least 20 female offspring were fixed for the standard chromosome karyotype. The progeny of these nine families were pooled to form the *A. gambiae* PEST strain. This strain may clearly have some Mopti-derived DNA as the standard karyotype is shared by Mopti and Savanna and the original PE strain did have the 2Rbc inversion rather than the 2Rb that is typical of Savanna. When tested, this colony was fully susceptible to *P. falciparum* from western Kenya.

3.2
The Draft Genome Sequence

Obtaining the draft genome sequence of *A. gambiae* was a collaborative effort, with sequence data provided by Celera, The Institute for Genomics Research, and the French National Sequencing Centre, Genoscope. The whole-genome shotgun method was adopted for this project, given its speed and efficiency and in consideration of its previously successful use for the *Drosophila* genome project. Paired end reads were derived from 2 254 546 plasmid clones with 2-, 10-, or 50-kb inserts, (50 kbp plasmid libraries had the bulk of their insert sequence excised during the initial steps of library construction, and only the insert ends were propagated in *Escherichia coli*.) and approximately 40 000 large-insert BAC clones, to give a total of 10.2-fold sequence coverage of the genome, estimated from CoT analysis to be about 260 Mbp. All plasmid and BAC libraries were constructed using DNA extracted from several hundred adult PEST-strain mosquitoes. Libraries from male and female mosquitoes were constructed separately and sequenced in equal proportion, giving approximately equal coverage of the autosomes and accordingly less coverage of the X and Y chromosomes. All electropherograms are downloadable from the NCBI trace archive (http://www.ncbi.nlm.nih.gov/Traces/trace.cgi?) or Ensembl trace server (http://trace.ensembl.org).

The WGS data set was assembled using Celera's assembly software, where algorithms are executed in a stepwise manner. An initial Blastn comparison of each sequence read to every other sequence read is used to detect overlaps (the overlap criteria being 40 or more base pairs with 6% or less mismatch). Reads within each cluster of overlapping sequences are then assembled into a contiguous consensus sequence (contig). Contigs that appear from read depth to be single copy in the genome are ordered and oriented into longer scaffolds, using mate pairs (two sequence reads derived from either end of a clone insert). Mate pairs are also used to guide computational intrascaffold gap filling, whereby one mate is located in the sequence flanking either side of a gap and its partner is placed within the gap. The resulting assembly is thus a set of typically very long scaffolds that are ordered and oriented sets of contigs whose intervening gaps are of known mean length and whose standard deviations are based on estimated library insert sizes.

The published draft of the *A. gambiae* whole-genome assembly was comprised of 8987 scaffolds, spanning 278 million base pairs (Mbp) of sequence. It is important to note that a small number of very large scaffolds cover most of the genome (91% of the sequence is represented by 303 scaffolds greater than 30 kb), and that many small scaffolds are expected to fit within intrascaffold gaps. In the entire assembly, there are 8986 gaps between scaffolds and 9964 gaps within scaffolds. Gaps within scaffolds are largely due to repetitive sequences that could not be uniquely placed on the basis of sequence overlap or mate-pair information and may also occasionally be due to a simple lack of coverage. Genbank accession numbers for the 8987 genome scaffolds are AAAB01000001 through AAAB01008987 and the scaffold sequences are downloadable as a set from ftp://ftp.ncbi.nih.gov/genbank/genomes/ Anopheles_gambiae/Assembly_scaffolds.

While use of the whole-genome shotgun approach for this project was well justified by time and cost savings, it is important to understand that the present form of the assembly is that of a draft genome sequence and, as with all draft genomes, there are numerous gaps that will require focused finishing efforts. Further, because of the fact that the plasmid libraries were derived from whole adult mosquitoes, some contaminating sequence from commensal gut bacteria was expected and were in fact observed. Scaffolds with hits to bacteria were flagged in their Genbank record as containing putative contaminating sequence of possible bacterial origin. In total, 213 short scaffolds were flagged, which together comprised 36.5 kb or 0.013% of the genome. These scaffolds were not culled from the assembly because of the remote possibility that some may represent real horizontal transfer events.

3.3
Genome Assembly, Genome size, and Physical Mapping

The sum of all the assembled *Anopheles* genome scaffolds is 278 Mb, which is significantly larger than the genome size of 260 Mb predicted by CoT analysis. There are several reasons why the assembly is close to, but slightly larger than the genome size predicted by CoT analysis. First, many small scaffolds may fit into gaps within other larger scaffolds. Second, because the X chromosome accounts for close to 20% of the genome but is underrepresented in males relative to the autosomes, CoT analysis using a mixed-sex pool of mosquitoes is expected to underestimate genome size. The third likely cause of discrepancy is the unanticipated assembly of highly variant alleles into separate contigs, as discussed further below (Sect. 3.4 Genetic variation).

The final step in defining the reference genome for *A. gambiae* was to order and

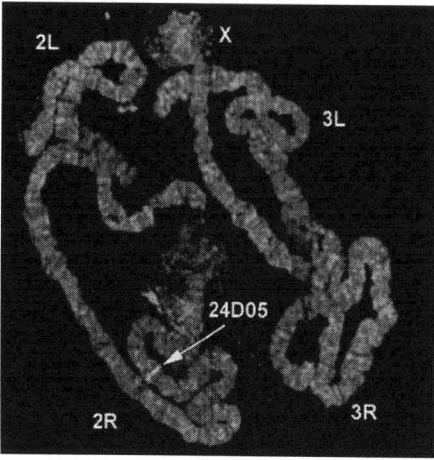

Fig. 1 Illustration of fluorescent *in situ* hybridization using a BAC clone probe. The *A. gambiae* polytene chromosome complement was isolated from the ovarian nurse cells of half-gravid mosquitoes. The probe, 24D08, is a BAC clone from the PEST library and is labeled with Cy3-dUTP red; the polytene chromosome is stained with Yoyo-1 iodide green (Molecular Probes).

orient the scaffolds on the three chromosome pairs, using physical mapping data. The chromosomal locations of assembled scaffolds were determined with sequence-tagged genomic DNA BAC clones that were mapped by *in situ* hybridization to ovarian nurse-cell polytene chromosomes (Fig. 1). For the initial assembly, approximately 2000 BAC clones were mapped and these data enabled the chromosomal location and orientation of scaffolds, constituting about 227 Mbp. Photo archive data showing the results of physical mapping are viewable at http://www.anobase.org. Since the initial assembly, the order and orientation of some of the smaller scaffolds have been confirmed or corrected, and ~7 Mbp more scaffolded sequence has been placed and oriented on the physical genome by *in situ* mapping of selected BAC and cDNA clones. Currently, 172 of the largest scaffolds have been assigned to chromosomal locations (a total of ~233 Mbp), and gaps between almost one-third of these scaffolds have been crossed by BAC clones or short BAC tiling paths (these BACs remain to be sequenced). Several dozen scaffolds (<10 Mbp) can be assigned only to nonspecific centromeric heterochromatic regions and about 1 Mbp of scaffolded sequence has been identified as possible Y-chromosome sequence; thus, about 243 Mbp of the 278 Mbp assembly has been assigned to locations with some level of confidence. Most of the remaining ~8700 unassigned scaffolds (~35 Mbp) are less than 20 kb. The majority of these appear to be repetitive heterochromatic sequences that, in *A. gambiae*, are clustered around the centromeres, the telomeres, and in a few islands in the euchromatic chromosome arms. Most of these scaffolds do not appear to contain genes and many are unlikely to be assigned to specific chromosomal locations by currently available approaches.

3.4
Genetic Variation

Analysis of high-quality mismatches within contig multiple sequence alignments revealed approximately half-a-million single nucleotide polymorphisms (SNPs) in the *A. gambiae* PEST-strain genome. The distribution of SNPs along the chromosomes is bimodal, with some regions showing roughly one SNP every 10 kb and others showing approximately one SNP every 200 bp. Given that the genome of the PEST strain appears to have resulted from a complex introgression of divergent Mopti and Savanna chromosomal forms, we expect that some genomic regions may be derived only from one or the other form, yielding a low density of SNPs, while other genomic regions may continue to segregate both divergent forms, yielding a high SNP density. Interestingly, the X chromosome has a markedly lower average level of polymorphism and does not show this bimodal pattern, perhaps due to male hemizygosity and hence, a lower rate of introgression is seen on this chromosome. In addition, heterozygosity of the X is expected to be depressed because of the selection for homozygosity of the X-linked pink-eye mutation.

As expected, the overwhelming majority of SNPs lie in intergenic regions, but there is still an abundance of SNPs within functional genes. Introns and intergenic regions have virtually identical heterozygosities, but the silent coding positions appear to have more than twofold enrichment of variability. Generally, silent coding sites are considered as having more stringent constraints than introns or intergenic regions because of biased codon usage, and this is reflected in a lower diversity of silent sites in most organisms. The reason for elevated silent variation in *A. gambiae* is at present unknown. Nucleotides with strong functional constraints, such as splice donors, splice acceptors, and stop codons have the lowest heterozygosity, and nonsynonymous (missense) positions are also evidently low in heterozygosity.

The bimodal SNP distribution in the PEST strain, and hence, the existence of two approximately equally frequent haplotypes in a substantial fraction of the genome, presented a unique challenge for assembly of the sequence data set, and it is important to note that some assembly artifacts have resulted. Where the sequence variation between the two haplotypes is small, they have been assembled together into a single contig, with the most frequent nucleotide at each position contributing to the consensus sequence. This is the manner in which all current whole-genome assembly algorithms handle genetic variation, even though the resulting consensus sequence does not precisely represent any native haploid genome. In the case of *A. gambiae*, where the haplotypes are more divergent, they have occasionally been split into two separate contigs. This is problematic because the assembler only allows a single consensus sequence to represent any given position in the genome. As a result, in the draft assembly, there are situations in which the second redundant contig (1) has been placed in the nearest gap, causing an artificial tandem duplication within the scaffold; (2) has been left out of the scaffold as singleton; or (3) where the problem is at a scaffold end, the duplicate has been placed at the beginning of the neighboring scaffold. The phenomenon described here was initially revealed as a larger separation of mate pairs in parts of the assembly than what the library insert sizes had predicted. Comprehensive screening of the 1 644 078 total mate pairs in the assembly revealed a total of 726 regions, with this diagnostic mate-pair signature covering a total of approximately 7.7% of the assembly. The amount of sequence that is doubly represented in the genome assembly because of this artifact is estimated at about 23.5 Mbp.

While the bimodal pattern of SNP variation is a unique feature of the PEST strain and is not representative of any wild mosquito population, individual variants represent real polymorphisms in *A. gambiae* and will provide valuable marker sets for genetic studies. The approximately 500 000 PEST-strain SNPs are downloadable from ftp://ftp.ncbi.nih.gov/genbank/genomes/Anopheles_gambiae or www.ensembl.org/anopheles_gambiae. A further 75 000 SNPs have been identified by comparing light sequence coverage (1.2×) from a lab strain of the Mopti cytogenetic form to the assembled PEST-strain sequence, and these SNP data are also available at www.ensembl.org/anopheles_gambiae.

Understanding the genetic diversity of anopheline mosquitoes and the molecular basis of cytogenetic form–associated traits will be accelerated by genomics and will likely point the way to more efficient strategies of vector-targeted malaria control. Efforts are presently underway to analyze large regions of the A. gambiae Y chromosome as population genetic markers and as indicators of the evolutionary relationships of the different A. gambiae forms and other species closely related to A. gambiae. Efforts to use other parts of the genome to define and identify such species and forms have been complicated by presumed interspecific and interform recombination over many regions of the different genomes. As a presumably nonrecombining part of the genome, the Y-chromosome sequence will be an invaluable source of material for teasing apart population structure and for developing rapid diagnostics for field studies. Studies are also underway to obtain wild isolates of A. gambiae from several different regions of Africa and sequence segments of DNA at regular intervals across the genomes of multiple individuals from each isolate. This survey of genetic diversity among different isolates of A. gambiae will facilitate the teasing apart of anopheline mosquito population structure and will play a crucial role in understanding mosquito biology, particularly those biological traits that influence vectorial capacity.

3.5
Gene Content

The initial automated annotation of the draft A. gambiae genome sequence was accomplished using established annotation pipelines at Celera and the Ensembl group at the European Bioinformatics Institute/Sanger Institute. Both pipelines use a combination of homology-based evidence and *ab initio* gene-finding algorithms to model gene structure. At the inception of the project, there were only approximately 6000 EST sequences available in public databases. However, given the evolutionary distance between *Drosophila* and *Anopheles* (~250 million years), obtaining wide representation from *Anopheles* EST libraries was essential for identifying the open reading frames of A. gambiae genes, and an additional approximately 80 000 EST sequences were derived from whole mosquitoes, as an integral part of the *Anopheles* genome project.

For proteome analysis, a nonredundant set of predicted protein sequences from the two pipelines (Celera and Ensembl) was constructed by selecting, for each locus, the annotation containing the largest number of exons. A total of 15 189 automated predictions were derived in this manner and, after removal of putative transposable elements and other contaminants, served as the set for initial analysis of the A. gambiae proteome. As is typical with automated genome annotations, it is likely that some false-positive predictions (pseudogenes, bacterial contaminants, and transposons) and false-negative predictions (*Anopheles* genes that were not computationally predicted) remain, and there are undoubtedly numerous errors in defining the precise boundaries of these putative gene structures. Automated annotation, while being a cost-effective method for an initial high-level view of the genome, does not substitute for the careful manual curation that will be undertaken by the community over a period of time. Since the initial publication of the *Anopheles* genome, the annotation has been under continuous development by the Ensembl group in coordination with the International *A. gambiae* Genome Consortium. The present version of the annotated genome is available at www.ensembl.org/Anopheles_gambiae. A tutorial on the Ensembl Anopheles site is available at www.ensembl.org/Anopheles-gambiae/documents/mosquito_doc.pdf.

3.6
Comparison of the *Anopheles* and *Drosophila* Proteomes

Analysis of the annotated gene set has revealed many interesting features of the proteome of *A. gambiae*, particularly in comparison with *Drosophila melanogaster*, the closest relative of this vector (*A. gambiae* and *D. melanogaster* shared a common Dipteran ancestor approximately 250 million years ago.) to have a completely sequenced genome. In total, 47.2% of the *Anopheles* protein set is composed of 1:1 orthologs (defined as *reciprocal best Blast matches*), and 13.8% is composed of "many-to-many" orthologs, where paralogous gene families have expanded/contracted unevenly. A further 17.9% of the *Anopheles* protein set is represented by proteins that have a homologous relationship to *Drosophila* proteins, but the match cannot be easily defined as orthologous or paralogous. For example, proteins in this group might share one or more protein domains or might be divergent members of large and diverse protein families. 10.0% of *Anopheles* proteins had their best match to a noninsect species and 11.1% of *Anopheles* proteins do not show homology to proteins from any other species. Of these 1437 *Anopheles*-specific genes, 575 are supported by EST evidence and 522 show homology to other *A. gambiae* proteins. While some of these proteins may represent *de novo Anopheles* genes or rapidly evolving ancestral genes that are now mutated beyond recognition, it is likely that homologous sequences for these proteins will be discovered as the genomes of organisms with closer evolutionary relationships to *A. gambiae* are sequenced. Interestingly, in comparing the *A. gambiae* and *D. melanogaster* proteomes, sequence similarity is highest among structural molecules, transporters, and enzymes and is lowest among genes involved in defense and immunity. This is consistent with the results from other recent whole-genome projects that suggest that the most plastic gene families are those that mediate the direct interaction of the organism with its environment, such as genes involved in olfaction, detoxification, and immune response. In comparison, developmental genes are highly conserved between *D. melanogaster* and *A. gambiae*, with 85% having single 1:1 orthologs (best reciprocal matches) between the two species.

Given the evolutionary distance between *A. gambiae* and *D. melanogaster* conserved gene order (synteny) is recognizable only over short distances. Approximately 34% of *A. gambiae* genes colocalize in small microsyntenic clusters, defined as chromosomal regions where at least two orthologs or orthologous groups are separated by no more than five unrelated genes in the intervening DNA. In total, there are 948 such microsyntenic blocks in the *A. gambiae* genome, when compared to *D. melanogaster*, the largest block containing 31 orthologous genes.

Most clusters contain substantially fewer genes and show evidence of local duplications, inversions, and translocations. In contrast, homology between *A. gambiae* and *D. melanogaster* on the level of whole-chromosome arms is easily recognizable upon mapping of 1:1 orthologs. The most conserved pair of chromosomal arms are *D. melanogaster*2L (Dm2L) and *A. gambiae*3R (Ag3R), with 67% of Ag3R orthologs and 83% of microsyntenic clusters being significantly similar to Dm2L. *A. gambiae* chromosomal arm 2L appears to be largely homologous to Dm3L and Dm2R, with approximately 42% of *A. gambiae* orthologs found on each of these two *D. melanogaster* arms. Similarly, Ag2R shares homology with Dm3R. Ortholog content of Ag3L indicates that this chromosomal arm has modest homology to both Dm2R (30%) and Dm3L (22%). Homology between the X chromosomes of the two species is limited, with significant shuffling of chromosomal segments to and from autosomes.

The predicted *Anopheles* protein set has been classified on the basis of protein domains, and their functional categories, using InterPro and the Gene Ontology (GO). The relative abundance of the majority of proteins containing InterPro domains is similar between the mosquito and fly. As expected, insect-specific cuticle and chitin-binding domains and the insect-specific olfactory receptors are overrepresented relative to noninsects. However, comparing *A. gambiae* and *D. melanogaster*, only 6 of the 200 most abundant protein domains differ significantly in abundance. However, some of these differences are dramatic, such as the striking expansion of fibrinogen domain–containing proteins observed in *A. gambiae* relative to *D. melanogaster*. There are 58 *A. gambiae* and 13 *D. melanogaster* fibrinogen domain–containing proteins and only a single pair is a 1:1 ortholog. It is tempting to speculate that fibrinogen domains, which were originally found in proteins that mediate blood coagulation in mammals, could be conferring some function to mosquitoes that enables blood feeding, such as inhibition of blood-meal coagulation. Further, the serine proteases (central effectors of proteolytic processes) are well represented in both insect genomes, but *Anopheles* has nearly 100 additional members, perhaps again reflecting the hematophagous adaptation of this Dipteran.

Major differences in gene number are seen in the peroxidase system, which mediates the oxidation of diverse substrates. A total of 18 peroxidases were observed in *A. gambiae* and only 10 were observed in *D. melanogaster*. The *Anopheles* peroxidases have high homology to the salivary peroxidases of the mosquito *A. albimanus*, suggesting the possibility that this gene family has expanded to facilitate the blood-feeding process.

In a separate comparative analysis using the LEK algorithm to identify protein families with different representation in *A. gambiae* and *D. melanogaster*, a prominent expansion of a 19-member family of odorant receptors (ORs) was detected in *A. gambiae*. The members of this family do not show substantial sequence similarity to ORs from *D. melanogaster* or any other organism. While between *A. gambiae* and *D. melanogaster*, there is significant conservation of the G protein-coupled receptor superfamily as a whole, this *A. gambiae* OR subfamily appears to be unique and may play an important role in mosquito-specific behavior that includes host seeking, as discussed further in Sect. 4.1. In addition to the conspicuous protein family expansions discussed above, many additional expansions were seen in diverse families. These include critical components of the visual system, structural components of the cell adhesion and contractile machinery, energy-generating glycolytic enzymes, anabolic and catabolic enzymes involved in protein and lipid metabolism, transporters, and detoxification enzymes.

Identification of gene family contractions have not been a focus of analysis of the *A. gambiae* genome, even though gene loss is considered to be an important mediator of evolution. It is ambiguous when doing a binary comparison whether there is expansion in one species or contraction in the other, relative to the state of the common ancestor. This can only be revealed by analysis of orthology across several species. Further, it is difficult to distinguish real gene losses from simple false-negative annotation results. Given that the overall gene count in *A. gambiae* and *D. melanogaster* is not significantly different, with each species having roughly 14 000 annotated gene loci, the overall number of expansions must be balanced by an equal number of reductions in each species, as they evolve to fill their ecological niches.

3.7
Genes Regulated by Blood Feeding

Newly emerged female mosquitoes are sustained by sugar meals as they seek the blood meal that is necessary to support egg development. Active host seeking is facilitated by olfactory, temperature, and visual cues. After locating a host and extracting several times their weight in blood, the female mosquito appears to undergo profound metabolic reprogramming, commensurate with the task of processing the massive blood bolus.

The *A. gambiae* genome project included an EST-sequencing program utilizing two source libraries. One library was constructed using DNA from whole adult female mosquitoes before blood feeding, when in a relatively quiescent metabolic state, and the second library was constructed 24 h after a blood meal, when the mosquitoes were under the metabolic burden of converting the blood meal into eggs. This approach facilitated identification of genes activated or inactivated by blood feeding in a hematophagous insect. Approximately 40 000 ESTs were sequenced from each library and assembled into approximately 7000 consensus sequences, each representing a distinct gene locus. 435 genes showed significantly altered transcription following blood-meal ingestion, as determined by Chi-squared analysis. Analysis of gene expression pre- and post blood feeding has revealed both expected and unexpected changes (Table 1). As expected, there is dramatic upregulation (up to 144 fold) of transcripts encoding serine proteases and other proteases, transcripts involved in transcriptional control and nuclear regulation, and transcripts representing the protein synthesis machinery that are involved in converting the digested blood proteins into new functional mosquito protein products. Further, transcripts involved in gluconeogenesis, lipid metabolism, and purine metabolism are upregulated, as are a small number of transcripts, such as ubiquitin-conjugating enzymes and cytosolic aminopeptidases, whose protein products mediate intracellular degradation of protein. Finally, there is upregulation of oogenesis-associated proteins and lipids 24 h post blood meal, including several vitellogenins (yolk constituents) and members of the phenoloxidase pathway that mediates eggshell melanization. In terms of decreased expression, genes involved in sugar metabolism (for example, salivary and midgut maltases, glycolysis, and oxidative phosphorylation) are downregulated, which is not unexpected, given the presumed decreased reliance on sugar metabolism with the high intake of a protein-rich blood meal. More intriguing is the marked downregulation post blood meal of striated muscle proteins, including actin, myosin, tafazzin, and troponin, as well as ATP-dependent cation transporters and Ca^{2+} binding proteins involved in motor signaling. Taken together with the observed decrease in transcription of opsins, photoreceptor associated proteins, cuticular proteins associated with the eye lens, and pheromone-binding proteins, the post blood-meal expression pattern suggests that after a blood meal, mosquitoes become detached from their environment and sacrifice mobility in order to devote metabolic attention to the monumental task of blood-meal processing and egg development. The set of genes upregulated by blood-meal digestion presents a rich source of novel insecticide targets and will allow identification of promoters that can activate transgenes in a blood-meal responsive manner.

4
Utility of the A. gambiae Genome

4.1
Odorant Receptors and Novel Repellents/Attractants

The distinct preference of A. gambiae for human blood meals contributes substantially to the menace of this vector. Very

Tab. 1 Representative messages significantly more transcribed or less transcribed in female A. gambiae mosquitoes 24 h after human blood-meal ingestion.

Category	↕	Gene description	Gene symbol	EST count (no blood)	EST count (blood)
Extracellular protein digestion	↑	Trypsin-2 precursor	AgCP10888	1	144
Intracellular protein degradation	↑	Ubiquitin	AgCP12328	0	47
Nuclear regulation	↑	Histone	AgCP10826	2	40
Protein synthesis	↑	Elongation factor 1a	AgCP3905	163	278
Oogenesis	↑	Vitellogenin	AgCP2518	0	878
Gluconeogenesis	↑	Aspartate amino transferase	AgCP8491	1	10
Glycolysis	↓	Succinate dehydrogenase	AgCP3730	24	7
Oxidative phosphorylation	↓	Glycerol-3-phosphate dehydrogenase	AgCP3306	51	11
Extracellular sugar digestion	↓	Salivary maltase	AgCP12790	10	0
Striated muscle	↓	Flightin (striatal muscle protein 27)	AgCP3724	71	2
Vision	↓	Opsin	AgCP12420	276	184
Chemosensation	↓	Pheromone-binding protein	AgCP11481	7	0

little is known about the molecular basis of this selective behavior, although it almost certainly involves specific human odors and mosquito pathways for detecting and responding to these odors. Continuing to explore the diversity of ORs and odorant binding proteins encoded in the A. gambiae genome will lead to a better understanding of the mechanisms of host preference and will uncover potential targets for a new generation of specific attractants/repellents. Preliminary analysis of the draft genome sequence identified a total of 276 G protein-coupled receptors, including 79 candidate ORs. The majority of these ORs (64 of 79) show expression only in olfactory tissues, as determined by RT-PCR experiments. Comparison to D. melanogaster revealed a similar total number of ORs, but only a single unequivocal orthologous pair of ORs between the two species. The extensive lineage-specific expansion of ORs likely reflects the ecological and physiological relevance of these receptors and the importance of each species being able to detect relevant chemicals – for example, rotting fruit odors for D. melanogaster and human host odors for Anopheles. AgOr1, a putative A. gambiae OR, has recently been identified as a possible mediator of human host finding. Initial clues that AgOr1 may be important in host finding include its expression exclusively in the olfactory tissue of female mosquitoes, and the downregulation of its expression in these tissues after a blood meal. Hallem EA et al. provided further evidence by engineering D. melanogaster neurons lacking endogenous ORs and by transfecting these with AgOr1. Electrophysiological measurements of the transfected cells showed that a component of human sweat, 4-methylphenol, elicited a strong response.

4.2
Insecticide Resistance Genes

During the 1950s and early 1960s, the WHO malaria eradication campaign succeeded in eradicating malaria from Europe and dramatically reduced its prevalence in many other parts of the world, primarily through programs that combined mosquito control agents like DDT with antimalarial drugs like chloroquine. However, malaria and anopheline mosquitoes remain entrenched in sub-Saharan Africa and the appearance of chloroquine-resistant parasites and insecticide-resistant mosquitoes is contributing to the current rise of malaria in Africa. Even control programs based on insecticide-impregnated bed nets, which are now advocated by WHO and are being widely implemented in Africa are threatened by the development of resistance to pyrethroids, the insecticide class of choice for this application. Insecticide resistance is often metabolic, where the insecticide will induce the expression of an enzyme that degrades it. Metabolic resistance is mediated largely by one of the three classes of enzymes, depending on the particular insecticidal agent in question. These are carboxylesterases (CEs), glutathione-S-transferases (GSTs) and cytochrome P450s. Ranson et al. cataloged members of these three major insecticide resistance–related families in the A. gambiae genome. Comparison to D. melanogaster revealed that a considerable expansion of these families has occurred in the mosquito. For example, there are 51 versus 31 CEs and 111 versus 90 cytochrome P450 enzymes in A. gambiae and D. melanogaster respectively. Secure orthologs between D. melanogaster and A. gambiae, identified by rigorous analysis of phylogenetic trees comprise less than

15% of these supergene families, indicating that gene families have radiated independently from common ancestral genes, presumably driven by the different metabolic requirements of the two species for natural compounds. Over the past decade, several resistance loci for different agents have been mapped in the mosquito genome, using classical genetic methods, and inspection of gene content within the boundaries of mapped resistance loci has been revealing. For example, it has been observed that several large clusters of cytochrome P450 genes colocalize with a known pyrethroid resistance locus involved in oxidative metabolism of the insecticide. Further, a major DDT resistance locus colocalizes with a cluster of eight epsilon (insect-specific) GST genes on chromosome 3R. Biochemical characterization of the members of this GST cluster has shown that only one of these eight enzymes, GSTE2-2, is able to metabolize DDT. Western blots using antibodies raised against this GST indicated that its expression is elevated in a DDT-resistant strain (ZAN/U) of *A. gambiae*, relative to an insecticide-sensitive strain (Kisumu). Subsequent and more comprehensive quantitative PCR studies in these two mosquito strains have indicated that not just GSTE2-2, but in total, five of the eight GSTs genes in this cluster are significantly overexpressed in the DDT-resistant strain. Gene dosage analysis showed no evidence for gene amplification, suggesting perhaps that coexpression of genes in the epsilon cluster is controlled by a common regulatory element.

In addition to metabolic resistance, insensitivity to an insecticide can arise through mutations that disrupt binding of the insecticide molecule to its target protein. A well-characterized example is the insensitivity to two major insecticide families, organophosphates and carbamates, that arises through mutation in their target protein Acetylcholinesterase (AChE). Only one AChE gene is present in *D. melanogaster* (called ace-2) and it is well established that resistance in *D. melanogaster* is caused by mutations at this locus. Two AChE genes, ace-1 and ace-2, have been identified in *A. gambiae*, through sequence searches of the reference genome, using human and *D. melanogaster* AChEs as queries. The *A. gambiae* ace-1 gene shows 52% amino acid identity and the *A. gambiae* ace-2 gene shows 83% amino acid identity to the single *D. melanogaster* AChE, and both *Anopheles* genes contain the conserved FGESAG active site motif. Phylogenetic analysis suggests that the presence of two AChE genes is the ancestral state, and one of these genes has been lost in *D. melanogaster*. It is well established that mutation of the target site of the single *D. melanogaster* AChE gene causes resistance to organophosphates and carbamates in this species, Interestingly, however, in organophosphate and carbamate resistant *Culex pipiens* and *A. gambiae* strains, insensitivity results from a substitution (G119S) near the catalytic site of the ace-1 gene, not a mutation of the ace-2 gene, which is presumably the direct *D. melanogaster* ortholog.

Understanding the common mechanisms of insecticide resistance in *A. gambiae* is of practical importance. The possibility now exists of screening new compounds for tolerance liability by determining whether they cause mutation in known hotspots in genes previously shown to confer resistance or induction of genes with a detoxifying function. Further, specific mutations in resistance-conferring genes can be used as markers to monitor the spread of resistance to insecticides in use in Africa, and can provide important information to help guide ongoing chemical-based vector-control programs.

4.3
Insecticide Targets

While most chemical agents used in the battle against mosquito-borne disease are broad-spectrum agents developed for the purpose of agricultural pest control, the possibility now exists of finding and exploiting physiological and biochemical systems that are unique to the mosquito, in order to produce new and highly selective agents for mosquito control. The reference mosquito genome sequence has helped reveal a variety of systems that may be good candidates for the development of more selective agents for mosquito control. For example, regulatory peptides act as neurochemicals and hormones that guide mosquito reproduction and development. A total of 35 genes encoding putative regulatory peptides have been annotated in the *A. gambiae* genome and agents that mimic or block the action of these peptides may be effective insecticides. Ecdysteroid peptide hormones, which govern gene expression during female reproduction and larval development are of particular interest. Ovary ecdysteroids released after a blood meal stimulate the fat body to begin secreting yolk proteins. The presence of an ortholog of *A. gambiae* ovary ecdysteroid hormone in the mosquito *Aedes aegypti* but the absence of an ortholog in *D. melanogaster* suggests that there is a unique role for this peptide in hematophagous insects and, thus, an opportunity for development of specific antimosquito agents. Similarly, diuresis is a process that has a critical function in the mosquito that may be exploited. After blood feeding, female mosquitoes engage in rapid diuresis to decrease the volume of the blood meal and allow flight. Genes for all four diuretic hormones known to regulate fluid secretion in insects have been identified in the *A. gambiae* genome and present additional targets for incapacitating the mosquito. Also of interest, the receptor for FRMF amide, a cardioexcitatory tetrapeptide first identified in clam, has been identified in the mosquito genome. In functional studies, the tetrapeptide significantly increased the frequency of spontaneous contractions of the heart in mosquito larvae, suggesting the possibility of developing agonists for this receptor as a larvicide.

Mining of reference genome sequences has revealed that enzyme glutathione reductase is absent from Diptera and is functionally substituted by the thioredoxin system. Thioredoxins are small ubiquitous thiol proteins that efficiently cleave disulfide bonds in a number of other proteins and take part in redox control of numerous cellular processes such as protein folding, signaling, and transcription. The key enzyme in this system is thioredoxin reductase, a single copy gene located on the *A. gambiae* X chromosome and shares 52% sequence identity with its human ortholog. The sequence of the catalytic redox center is different between the *A. gambiae* (Thr-Cys-Cys-SerOH) and human (Gly-Cys-selenocysteine-GlyOH) proteins and this difference provides an attractive avenue for the development of a selective inhibitor.

Differences between mammalian and insect metabotropic glutamate receptors have also been revealed by whole-genome comparison. The *D. melanogaster* metabotropic glutamate receptor (mGluR) shares a very similar pharmacological profile with its mammalian orthologs, being activated or inhibited by the same natural or synthetic ligands. However, a new receptor in mGluR subclass has been found in the mosquito, the honeybee *Apis mellifera*, and *D. melanogaster*, called AmXR, HBmXR, and *DmXR* respectively. No direct orthologs of these novel insect receptors are detectable in *C. elegans* or in human genome. The native ligand for these receptors is unknown, although it has been shown that extract from *D. melanogaster* and *A. gambiae*, but not *C. elegans* or mouse brain, caused dose-dependent receptor activation. Sensitivity to formaldehyde (which masks amino groups) and insensitivity to HCl (which disrupts peptide bonds) suggests that the native ligand is an amino acid–like molecule, but the native ligand is clearly not glutamate, given that residues contacting γ-carboxyl group of glutamate are not conserved with the mGluR, and heterologous expression showed no stimulation by glutamate or mGluR agonists AMPA, kainate, and NMDA quisqualate. Because the XR receptors appear to be insect specific, compounds that disrupt their function may, if lethal, form a promising new class of specific insecticide.

4.4
Immune-response Genes and Genetic Vector Control Strategies

To successfully complete its complex life cycle, the malaria parasite must evade both the human and mosquito immune systems. Parasites ingested with the blood meal develop within the mosquito gut into ookinetes, which subsequently burrow into the gut epithelium and form oocysts. When the oocysts burst, sporozoites are released and these travel to the salivary glands to await transfer to a new host. However, large losses in parasite number occur during this invasion because of mosquito defenses such as melanotic parasite encapsulation. These defenses do not comprise an adaptive immune system, such as that found in mammals, but rather belong to the ancient innate immune system that most metazoans rely on for dealing with invading microorganisms. Facilitating these innate mosquito defenses offers a new and promising strategy for malaria control.

In the past decade, genetic selection in the laboratory has yielded several interesting *A. gambiae* strains, including some that are completely refractory to *Plasmodium*. These strains have been the subject of intensive study in a number of different laboratories, and the *A. gambiae* genome is facilitating the identification of candidate genes for *Plasmodium* resistance. More recently, Christophides et al. screened the *A. gambiae* genome for genes implicated in the innate immune response of the mosquito and identified 242 players from 18 different gene families including, for example, pattern-recognition receptors that bind pathogen-specific molecules, second messengers that can amplify or dampen immune signals, and effector molecules such as the prophenoloxidases that mediate melanization. There is marked diversification of these genes relative to their *Drosophila* homologs, likely reflecting the adaptation of *Anopheles* and *Drosophila* to different immune challenges represented by different ecological and physiological conditions. Recent functional screening (gene silencing using dsRNA) of a subset of these candidate immune-response genes in *A. gambiae* has yielded compelling findings for several pattern-recognition receptors. Functional knockout of the *A. gambiae* leucine-rich repeat protein (LRIM1) led to a substantial (3.6-fold) increase in oocyst numbers, indicating that this gene has a strong protective effect against the invading parasite. Similarly, functional knockout of the pattern-recognition receptor thioester-containing protein 1 (TEP1) in *A. gambiae* led to a fivefold increase in the number of oocysts developing on the midgut epithelium and a complete abolishment of parasite melanization. Conversely, functional knockout of two different *A. gambiae* C-type lectines (CTL4 and CTLMA2) resulted in a paradoxical enhancement of immune response, with massive melanization of invading ookinetes. It is not clear how these mosquito CTL genes act to protect the developing malaria parasite within the mosquito and no obvious homologs of these genes have been detected in any other organism. LRIM1, TEP1, CTL4, and CTLMA2 are the first mosquito genes to be identified as having an important role in the immune response of the mosquito to the invading malaria parasite. While it is unlikely that gene silencing using dsRNA will be practical as a malaria-control measure, these exciting observations are nonetheless the first steps along the road to the development of a genetic-based transmission-blocking strategy for malaria. Germ-line transformation of the mosquitoes *A. stephensi* and *A. gambiae* has recently been achieved, bringing the possibility of malaria control through the introduction of transgenic mosquitoes another step closer to reality. While genetic control strategies for malaria vectors are complicated by several factors, including the diversity of naturally occurring cytogenetic forms that are reproductively isolated, the selective pressure applied to the *Plasmodium* population, plus all of the usual risks associated with the release of transgenic insects, there is much promise that transgenics will play a key role in malaria control in the coming decades.

See also Molecular Basis of Genetics.

Bibliography

Books and Reviews

Adams, M.D., Celniker, S.E., Holt, R.A., Evans, C.A., Gocayne, J.D., Amanatides, P.G., Scherer, S.E., Li, P.W., Hoskins, R.A., Galle, R.F., George, R.A., Lewis, S.E., Richards, S., Ashburner, M., Henderson, S.N., Sutton, G.G., Wortman, J.R., Yandell, M.D., Zhang, Q., Chen, L.X., Brandon, R.C., Rogers, Y.H., Blazej, R.G., Champe, M., Pfeiffer, B.D., Wan, K.H., Doyle, C., Baxter, E.G., Helt, G., Nelson, C.R., Gabor, G.L., Abril, J.F., Agbayani, A., An, H.J., Andrews-Pfannkoch, C., Baldwin, D., Ballew, R.M., Basu, A., Baxendale, J., Bayraktaroglu, L., Beasley, E.M., Beeson, K.Y., Benos, P.V., Berman, B.P., Bhandari, D., Bolshakov, S., Borkova, D., Botchan, M.R., Bouck, J., Brokstein, P., Brottier, P., Burtis, K.C., Busam, D.A., Butler, H., Cadieu, E., Center, A., Chandra, I., Cherry, J.M., Cawley, S., Dahlke, C., Davenport, L.B., Davies, P., de Pablos, B., Delcher, A., Deng, Z., Mays, A.D., Dew, I., Dietz, S.M., Dodson, K., Doup, L.E., Downes, M., Dugan-Rocha, S., Dunkov, B.C., Dunn, P., Durbin, K.J., Evangelista, C.C., Ferraz, C., Ferriera, S., Fleischmann, W., Fosler, C., Gabrielian, A.E., Garg, N.S., Gelbart, W.M., Glasser, K., Glodek, A., Gong, F., Gorrell, J.H., Gu, Z., Guan, P., Harris, M., Harris, N.L., Harvey, D., Heiman, T.J., Hernandez, J.R., Houck, J., Hostin, D., Houston, K.A., Howland, T.J., Wei, M.H., Ibegwam, C., Jalali, M., Kalush, F., Karpen, G.H., Ke, Z., Kennison, J.A., Ketchum, K.A., Kimmel, B.E., Kodira, C.D., Kraft, C., Kravitz, S., Kulp, D., Lai, Z., Lasko, P., Lei, Y., Levitsky, A.A., Li, J., Li, Z., Liang, Y., Lin, X., Liu, X., Mattei, B., McIntosh, T.C., McLeod, M.P., McPherson, D., Merkulov, G., Milshina, N.V., Mobarry, C., Morris, J.,

Moshrefi, A., Mount, S.M., Moy, M., Murphy, B., Murphy, L., Muzny, D.M., Nelson, D.L., Nelson, D.R., Nelson, K.A., Nixon, K., Nusskern, D.R., Pacleb, J.M., Palazzolo, M., Pittman, G.S., Pan, S., Pollard, J., Puri, V., Reese, M.G., Reinert, K., Remington, K., Saunders, R.D., Scheeler, F., Shen, H., Shue, B.C., Siden-Kiamos, I., Simpson, M., Skupski, M.P., Smith, T., Spier, E., Spradling, A.C., Stapleton, M., Strong, R., Sun, E., Svirskas, R., Tector, C., Turner, R., Venter, E., Wang, A.H., Wang, X., Wang, Z.Y., Wassarman, D.A., Weinstock, G.M., Weissenbach, J., Williams, S.M., Woodage, T., Worley, K.C., Wu, D., Yang, S., Yao, Q.A., Ye, J., Yeh, R.F., Zaveri, J.S., Zhan, M., Zhang, G., Zhao, Q., Zheng, L., Zheng, X.H., Zhong, F.N., Zhong, W., Zhou, X., Zhu, S., Zhu, X., Smith, H.O., Gibbs, R.A., Myers, E.W., Rubin, G.M., Venter, J.C. (2000) The genome sequence of *D. melanogaster*, *Science* **287**, 2185–2195.

Besansky, N.J., Powell, J.R. (1992) Reassociation kinetics of *Anopheles gambiae* (Diptera: Culicidae) DNA, *J. Med. Entomol.* **29**, 125–128.

Catteruccia, F., Nolan, T., Loukeris, T.G., Blass, C., Savakis, C., Kafatos, F.C., Crisanti, A. (2000) Stable germline transformation of the malaria mosquito *Anopheles stephensi*, *Nature* **405**, 959–962.

Christophides, G.K., Zdobnov, E., Barillas-Mury, C., Birney, E., Blandin, S., Blass, C., Brey, P.T., Collins, F.H., Danielli, A., Dimopoulos, G., Hetru, C., Hoa, N.T., Hoffmann, J.A., Kanzok, S.M., Letunic, I., Levashina, E.A., Loukeris, T.G., Lycett, G., Meister, S., Michel, K., Moita, L.F., Muller, H.M., Osta, M.A., Paskewitz, S.M., Reichhart, J.M., Rzhetsky, A., Troxler, L., Vernick, K.D., Vlachou, D., Volz, J., von Mering, C., Xu, J., Zheng, L., Bork, P., Kafatos, F.C. (2002) Immunity-related genes and gene families in Anopheles gambiae, *Science* **298**, 159–165.

Coluzzi, M., Sabatini, A., Petrarca, V., Di Deco, M.A. (1979) Chromosomal differentiation and adaptation to human environments in the *Anopheles gambiae* complex, *Trans. R. Soc. Trop. Med. Hyg.* **73**, 483–497.

Hill, C.A., Fox, A.N., Pitts, R.J., Kent, L.B., Tan, P.L., Chrystal, M.A., Cravchik, A., Collins, F.H., Robertson, H.M., Zwiebel, L.J. (2002) G protein-coupled receptors in *Anopheles gambiae*, *Science* **298**, 176–178.

Holt, R.A., Subramanian, G.M., Halpern, A., Sutton, G.G., Charlab, R., Nusskern, D.R., Wincker, P., Clark, A.G., Ribeiro, J.M., Wides, R., Salzberg, S.L., Loftus, B., Yandell, M., Majoros, W.H., Rusch, D.B., Lai, Z., Kraft, C.L., Abril, J.F., Anthouard, V., Arensburger, P., Atkinson, P.W., Baden, H., de Berardinis, V., Baldwin, D., Benes, V., Biedler, J., Blass, C., Bolanos, R., Boscus, D., Barnstead, M., Cai, S., Center, A., Chaturverdi, K., Christophides, G.K., Chrystal, M.A., Clamp, M., Cravchik, A., Curwen, V., Dana, A., Delcher, A., Dew, I., Evans, C.A., Flanigan, M., Grundschober-Freimoser, A., Friedli, L., Gu, Z., Guan, P., Guigo, R., Hillenmeyer, M.E., Hladun, S.L., Hogan, J.R., Hong, Y.S., Hoover, J., Jaillon, O., Ke, Z., Kodira, C., Kokoza, E., Koutsos, A., Letunic, I., Levitsky, A., Liang, Y., Lin, J.J., Lobo, N.F., Lopez, J.R., Malek, J.A., McIntosh, T.C., Meister, S., Miller, J., Mobarry, C., Mongin, E., Murphy, S.D., O'Brochta, D.A., Pfannkoch, C., Qi, R., Regier, M.A., Remington, K., Shao, H., Sharakhova, M.V., Sitter, C.D., Shetty, J., Smith, T.J., Strong, R., Sun, J., Thomasova, D., Ton, L.Q., Topalis, P., Tu, Z., Unger, M.F., Walenz, B., Wang, A., Wang, J., Wang, M., Wang, X., Woodford, K.J., Wortman, J.R., Wu, M., Yao, A., Zdobnov, E.M., Zhang, H., Zhao, Q., Zhao, S., Zhu, S.C., Zhimulev, I., Coluzzi, M., della Torre, A., Roth, C.W., Louis, C., Kalush, F., Mural, R.J., Myers, E.W., Adams, M.D., Smith, H.O., Broder, S., Gardner, M.J., Fraser, C.M., Birney, E., Bork, P., Brey, P.T., Venter, J.C., Weissenbach, J., Kafatos, F.C., Collins, F.H., Hoffman, S.L. (2002) The genome sequence of the malaria mosquito *Anopheles gambiae*, *Science* **298**, 129–149.

Mongin, E., Louis, C., Holt, R.A., Birney, E., Collins, F.H. (2004) The *Anopheles gambiae* genome: an update, *Trends Parasitol.* **20**, 49–52.

Ranson, H., Claudianos, C., Ortelli, F., Abgrall, C., Hemingway, J., Sharakhova, M.V., Unger, M.F., Collins, F.H., Feyereisen, R. (2002) Evolution of supergene families associated with insecticide resistance, *Science* **298**, 179–181.

Zdobnov, E.M., von Mering, C., Letunic, I., Torrents, D., Suyama, M., Copley, R.R., Christophides, G.K., Thomasova, D., Holt, R.A.,

Subramanian, G.M., Mueller, H.M., Dimopoulos, G., Law, J.H., Wells, M.A., Birney, E., Charlab, R., Halpern, A.L., Kokoza, E., Kraft, C.L., Lai, Z., Lewis, S., Louis, C., Barillas-Mury, C., Nusskern, D., Rubin, G.M., Salzberg, S.L., Sutton, G.G., Topalis, P., Wides, R., Wincker, P., Yandell, M., Collins, F.H., Ribeiro, J., Gelbart, W.M., Kafatos, F.C., Bork, P. (2002) Comparative genome and proteome analysis of *Anopheles gambiae* and *Drosophila melanogaster*, *Science* **298**, 149–159.

Primary Literature

Altschul, S.F., Gish, W., Miller, W., Myers, E.W., Lipman, D.J. (1990) Basic local alignment search tool, *J. Mol. Biol.* **215**, 403–410.

Ashburner, M., Ball, C.A., Blake, J.A., Botstein, D., Butler, H., Cherry, J.M., Davis, A.P., Dolinski, K., Dwight, S.S., Eppig, J.T., Harris, M.A., Hill, D.P., Issel-Tarver, L., Kasarskis, A., Lewis, S., Matese, J.C., Richardson, J.E., Ringwald, M., Rubin, G.M., Sherlock, G. (2000) Gene ontology: tool for the unification of biology. The gene ontology consortium, *Nat. Genet.* **25**, 25–29.

Bauer, H., Gromer, S., Urbani, A., Schnolzer, M., Schirmer, R.H., Muller, H.M. (2003) Thioredoxin reductase from the malaria mosquito *Anopheles gambiae*, *Eur. J. Biochem.* **270**, 4272–4281.

Blandin, S., Shiao, S-H., Moita, L.F., Janse, C.J., Waters, A.P., Kafatos, F.C., Levashina, E.A. (2004) Complement-like protein TEP1 is a determinant of vectorial capacity in the malaria vector *Anopheles gambiae*, *Cell* **116**, 661–670.

Bryan, J.H., Di Deco, M.A., Petrarca, V., Coluzzi, M. (1982) Inversion polymorphism and incipient speciation in *Anopheles gambiae* s.str. in The Gambia, West Africa, *Genetica* **59**, 167–176.

Collins, F.H., Sakai, R.K., Vernick, K.D., Paskewitz, S., Seeley, D.C., Miller, L.H., Collins, W.E., Campbell, C.C., Gwadz, R.W. (1986) Genetic selection of a Plasmodium-refractory strain of the malaria vector *Anopheles gambiae*, *Science* **234**, 607–610.

Coluzzi, M., Petrarca, V., Di Deco, M.A. (1985) Chromosomal inversion intergradation and incipient speciation in *Anopheles gambiae*, *Boll. Zool.* **52**, 45–63.

Coluzzi, M., Sabatini, A., della Torre, A., Di Deco, M.A., Petrarca, V. (2002) A polytene chromosome analysis of the *Anopheles gambiae* species complex, *Science* **298**, 1415–1418.

Ding, Y., Ortelli, F., Rossiter, L.C., Hemingway, J., Ranson, H. (2003) The *Anopheles gambiae* glutathione transferase supergene family: annotation, phylogeny and expression profiles, *BMC Genomics* **4**, 35.

Duttlinger, A., Mispelon, M., Nichols, R. (2003) The structure of the FMRFamide receptor and activity of the cardioexcitatory neuropeptide are conserved in mosquito, *Neuropeptides* **37**, 120–126.

Favia, G., Lanfrancotti, A., Spanos, L., Siden-Kiamos, I., Louis, C. (2001) Molecular characterization of ribosomal DNA polymorphisms discriminating among chromosomal forms of *Anopheles gambiae* s.s, *Insect Mol. Biol.* **10**, 19–23.

Favia, G., Dimopoulos, G., della Torre, A., Toure, Y.T., Coluzzi, M., Louis, C. (1994) Polymorphisms detected by random PCR distinguish between different chromosomal forms of *Anopheles gambiae*, *Proc. Natl. Acad. Sci. U.S.A.* **91**, 10315–10319.

Favia, G., della Torre, A., Bagayoko, M., Lanfrancotti, A., Sagnon, N.F., Toure, Y.T., Coluzzi, M. (1997) Molecular identification of sympatric chromosomal forms of Anopheles gambiae and further evidence of their reproductive isolation, *Insect Mol. Biol.* **6**, 377–383.

Fournier, D., Mutero, A., Pralavorio, M., Bride, J.M. (1993) *D. melanogaster* acetylcholinesterase: mechanisms of resistance to organophosphates, *Chem. Biol. Interact.* **87**, 233–238.

Gentile, G., Slotman, M., Ketmaier, V., Powell, J.R., Caccone, A. (2001) Attempts to molecularly distinguish cryptic taxa in *Anopheles gambiae* s.s, *Insect Mol. Biol.* **10**, 25–32.

Gibbs, R.A., Weinstock, G.M., Metzker, M.L., Muzny, D.M., Sodergren, E.J., Scherer, S., Scott, G., Steffen, D., Worley, K.C., Burch, P.E., Okwuonu, G., Hines, S., Lewis, L., DeRamo, C., Delgado, O., Dugan-Rocha, S., Miner, G., Morgan, M., Hawes, A., Gill, R., Holt, R.A., Adams, M.D., Amanatides, P.G., Baden-Tillson, H., Barnstead, M., Chin, S., Evans, C.A., Ferriera, S., Fosler, C., Glodek, A., Gu, Z., Jennings, D., Kraft, C.L., Nguyen, T., Pfannkoch, C.M., Sitter, C., Sutton, G.G., Venter, J.C., Woodage, T., Smith, D., Lee, H.M., Gustafson, E.,

Cahill, P., Kana, A., Doucette-Stamm, L., Weinstock, K., Fechtel, K., Weiss, R.B., Dunn, D.M., Green, E.D., Blakesley, R.W., Bouffard, G.G., De Jong, P.J., Osoegawa, K., Zhu, B., Marra, M., Schein, J., Bosdet, I., Fjell, C., Jones, S., Krzywinski, M., Mathewson, C., Siddiqui, A., Wye, N., McPherson, J., Zhao, S., Fraser, C.M., Shetty, J., Shatsman, S., Geer, K., Chen, Y., Abramzon, S., Nierman, W.C., Havlak, P.H., Chen, R., Durbin, K.J., Egan, A., Ren, Y., Song, X.Z., Li, B., Liu, Y., Qin, X., Cawley, S., Cooney, A.J., D'Souza, L.M., Martin, K., Wu, J.Q., Gonzalez-Garay, M.L., Jackson, A.R., Kalafus, K.J., McLeod, M.P., Milosavljevic, A., Virk, D., Volkov, A., Wheeler, D.A., Zhang, Z., Bailey, J.A., Eichler, E.E., Tuzun, E., Birney, E., Mongin, E., Ureta-Vidal, A., Woodwark, C., Zdobnov, E., Bork, P., Suyama, M., Torrents, D., Alexandersson, M., Trask, B.J., Young, J.M., Huang, H., Wang, H., Xing, H., Daniels, S., Gietzen, D., Schmidt, J., Stevens, K., Vitt, U., Wingrove, J., Camara, F., Mar Alba, M., Abril, J.F., Guigo, R., Smit, A., Dubchak, I., Rubin, E.M., Couronne, O., Poliakov, A., Hubner, N., Ganten, D., Goesele, C., Hummel, O., Kreitler, T., Lee, Y.A., Monti, J., Schulz, H., Zimdahl, H., Himmelbauer, H., Lehrach, H., Jacob, H.J., Bromberg, S., Gullings-Handley, J., Jensen-Seaman, M.I., Kwitek, A.E., Lazar, J., Pasko, D., Tonellato, P.J., Twigger, S., Ponting, C.P., Duarte, J.M., Rice, S., Goodstadt, L., Beatson, S.A., Emes, R.D., Winter, E.E., Webber, C., Brandt, P., Nyakatura, G., Adetobi, M., Chiaromonte, F., Elnitski, L., Eswara, P., Hardison, R.C., Hou, M., Kolbe, D., Makova, K., Miller, W., Nekrutenko, A., Riemer, C., Schwartz, S., Taylor, J., Yang, S., Zhang, Y., Lindpaintner, K., Andrews, T.D., Caccamo, M., Clamp, M., Clarke, L., Curwen, V., Durbin, R., Eyras, E., Searle, S.M., Cooper, G.M., Batzoglou, S., Brudno, M., Sidow, A., Stone, E.A., Payseur, B.A., Bourque, G., Lopez-Otin, C., Puente, X.S., Chakrabarti, K., Chatterji, S., Dewey, C., Pachter, L., Bray, N., Yap, V.B., Caspi, A., Tesler, G., Pevzner, P.A., Haussler, D., Roskin, K.M., Baertsch, R., Clawson, H., Furey, T.S., Hinrichs, A.S., Karolchik, D., Kent, W.J., Rosenbloom, K.R., Trumbower, H., Weirauch, M., Cooper, D.N., Stenson, P.D., Ma, B., Brent, M., Arumugam, M., Shteynberg, D., Copley, R.R., Taylor, M.S., Riethman, H., Mudunuri, U., Peterson, J., Guyer, M., Felsenfeld, A., Old, S., Mockrin, S. and Collins, F. (2004) Genome sequence of the Brown Norway rat yields insights into mammalian evolution, *Nature* **428**, 493–521.

Githeko, A.K., Brandling-Bennett, A.D., Beier, M., Atieli, F., Owaga, M., Collins, F.H. (1992) The reservoir of Plasmodium falciparum malaria in a holoendemic area of western Kenya, *Trans. R. Soc. Trop. Med. Hyg.* **86**, 355–358.

Grossman, G.L., Rafferty, C.S., Clayton, J.R., Stevens, T.K., Mukabayire, O., Benedict, M.Q. (2001) Germline transformation of the malaria vector, Anopheles gambiae, with the piggyBac transposable element, *Insect Mol. Biol.* **10**, 597–604.

Hallem, E.A., Nicole Fox, A., Zwiebel, L.J., Carlson, J.R. (2004) Olfaction: mosquito receptor for human-sweat odorant, *Nature* **427**, 212–213.

Kanzok, S.M., Fechner, A., Bauer, H., Ulschmid, J.K., Muller, H.M., Botella-Munoz, J., Schneuwly, S., Schirmer, R., Becker, K. (2001) Substitution of the thioredoxin system for glutathione reductase in *D. melanogaster*, *Science* **291**, 643–646.

Knols, B.G., de Jong, R., Takken, W. (1995) Differential attractiveness of isolated humans to mosquitoes in Tanzania, *Trans. R. Soc. Trop. Med. Hyg.* **89**, 604–606.

Mason, G.F. (1967) Genetic studies on mutations in species A and B of the *Anopheles gambiae* complex, *Genet. Res.* **10**, 205–217.

Mitri, C., Parmentier, M.L., Pin, J.P., Bockaert, J., Grau, Y. (2004) Divergent evolution in metabotropic glutamate receptors. A new receptor activated by an endogenous ligand different from glutamate in insects, *J. Biol. Chem.* **279**, 9313–9320.

Mukabayire, O., Besansky, N.J. (1996) Distribution of T1, Q, Pegasus and mariner transposable elements on the polytene chromosomes of PEST, a standard strain of *Anopheles gambiae*, *Chromosoma* **104**, 585–595.

Mukabayire, O., Caridi, J., Wang, X., Toure, Y.T., Coluzzi, M., Besansky, N.J. (2001) Patterns of DNA sequence variation in chromosomally recognized taxa of *Anopheles gambiae*: evidence from rDNA and single-copy loci, *Insect Mol. Biol.* **10**, 33–46.

Myers, E.W., Sutton, G.G., Delcher, A.L., Dew, I.M., Fasulo, D.P., Flanigan, M.J., Kravitz, S.A., Mobarry, C.M., Reinert, K.H., Remington, K.A., Anson, E.L., Bolanos, R.A.,

Chou, H.H., Jordan, C.M., Halpern, A.L., Lonardi, S., Beasley, E.M., Brandon, R.C., Chen, L., Dunn, P.J., Lai, Z., Liang, Y., Nusskern, D.R., Zhan, M., Zhang, Q., Zheng, X., Rubin, G.M., Adams, M.D., Venter, J.C. (2000) A whole-genome assembly of *D. melanogaster*, *Science* **287**, 2196–2204.

Olson, M.V. (1999) When less is more: gene loss as an engine of evolutionary change, *Am. J. Hum. Genet.* **64**, 18–23.

Osta, M.A., Christophides, G.K., Kafatos, F.C. (2004) Effects of mosquito genes on plasmodium development, *Science* **303**, 2030–2032.

Ortelli, F., Rossiter, L.C., Vontas, J., Ranson, H., Hemingway, J. (2003) Heterologous expression of four glutathione transferase genes genetically linked to a major insecticide-resistance locus from the malaria vector *Anopheles gambiae*, *Biochem. J.* **373**, 957–963.

Price, D.A., Greenberg, M.J. (1977) Structure of a molluscan cardioexcitatory neuropeptide, *Science* **197**, 670–671.

Ribeiro, J.M. (2003) A catalogue of *Anopheles gambiae* transcripts significantly more or less expressed following a blood meal, *Insect Biochem. Mol. Biol.* **33**, 865–882.

Riehle, M.A., Garczynski, S.F., Crim, J.W., Hill, C.A., Brown, M.R. (2002) Neuropeptides and peptide hormones in *Anopheles gambiae*, *Science* **298**, 172–175.

Rubin, G.M., Yandell, M.D., Wortman, J.R., Gabor Miklos, G.L., Nelson, C.R., Hariharan, I.K., Fortini, M.E., Li, P.W., Apweiler, R., Fleischmann, W., Cherry, J.M., Henikoff, S., Skupski, M.P., Misra, S., Ashburner, M., Birney, E., Boguski, M.S., Brody, T., Brokstein, P., Celniker, S.E., Chervitz, S.A., Coates, D., Cravchik, A., Gabrielian, A., Galle, R.F., Gelbart, W.M., George, R.A., Goldstein, L.S., Gong, F., Guan, P., Harris, N.L., Hay, B.A., Hoskins, R.A., Li, J., Li, Z., Hynes, R.O., Jones, S.J., Kuehl, P.M., Lemaitre, B., Littleton, J.T., Morrison, D.K., Mungall, C., O'Farrell, P.H., Pickeral, O.K., Shue, C., Vosshall, L.B., Zhang, J., Zhao, Q., Zheng, X.H., Lewis, S. (2000) Comparative genomics of the eukaryotes, *Science* **287**, 2204–2215.

Shahabuddin, M., Pimenta, P.F. (1998) Plasmodium gallinaceum preferentially invades vesicular ATPase-expressing cells in Aedes aegypti midgut, *Proc. Natl. Acad. Sci. U.S.A.* **95**, 3385–3389.

Takken, W., Knols, B.G. (1999) Odor-mediated behavior of Afrotropical malaria mosquitoes, *Annu. Rev. Entomol.* **44**, 131–157.

Toure, Y.T., Petrarca, V., Traore, S.F., Coulibaly, A., Maiga, H.M., Sankare, O., Sow, M., Di Deco, M.A., Coluzzi, M. (1994) Ecological genetic studies in the chromosomal form Mopti of *Anopheles gambiae* s.str. in Mali, West Africa, *Genetica* **94**, 213–223.

Toure, Y.T., Petrarca, V., Traore, S.F., Coulibaly, A., Maiga, H.M., Sankare, O., Sow, M., Di Deco, M.A., Coluzzi, M. (1998) The distribution and inversion polymorphism of chromosomally recognized taxa of the *Anopheles gambiae* complex in Mali, West Africa, *Parassitologia* **40**, 477–511.

Venter, J.C., Adams, M.D., Myers, E.W., Li, P.W., Mural, R.J., Sutton, G.G., Smith, H.O., Yandell, M., Evans, C.A., Holt, R.A., Gocayne, J.D., Amanatides, P., Ballew, R.M., Huson, D.H., Wortman, J.R., Zhang, Q., Kodira, C.D., Zheng, X.H., Chen, L., Skupski, M., Subramanian, G., Thomas, P.D., Zhang, J., Gabor Miklos, G.L., Nelson, C., Broder, S., Clark, A.G., Nadeau, J., McKusick, V.A., Zinder, N., Levine, A.J., Roberts, R.J., Simon, M., Slayman, C., Hunkapiller, M., Bolanos, R., Delcher, A., Dew, I., Fasulo, D., Flanigan, M., Florea, L., Halpern, A., Hannenhalli, S., Kravitz, S., Levy, S., Mobarry, C., Reinert, K., Remington, K., Abu-Threideh, J., Beasley, E., Biddick, K., Bonazzi, V., Brandon, R., Cargill, M., Chandramouliswaran, I., Charlab, R., Chaturvedi, K., Deng, Z., Di Francesco, V., Dunn, P., Eilbeck, K., Evangelista, C., Gabrielian, A.E., Gan, W., Ge, W., Gong, F., Gu, Z., Guan, P., Heiman, T.J., Higgins, M.E., Ji, R.R., Ke, Z., Ketchum, K.A., Lai, Z., Lei, Y., Li, Z., Li, J., Liang, Y., Lin, X., Lu, F., Merkulov, G.V., Milshina, N., Moore, H.M., Naik, A.K., Narayan, V.A., Neelam, B., Nusskern, D., Rusch, D.B., Salzberg, S., Shao, W., Shue, B., Sun, J., Wang, Z., Wang, A., Wang, X., Wang, J., Wei, M., Wides, R., Xiao, C., Yan, C., Yao, A., Ye, J., Zhan, M., Zhang, W., Zhang, H., Zhao, Q., Zheng, L., Zhong, F., Zhong, W., Zhu, S., Zhao, S., Gilbert, D., Baumhueter, S., Spier, G., Carter, C., Cravchik, A., Woodage, T., Ali, F., An, H., Awe, A., Baldwin, D., Baden, H., Barnstead, M., Barrow, I., Beeson, K., Busam, D., Carver, A., Center, A., Cheng, M.L., Curry, L., Danaher, S., Davenport, L., Desilets, R., Dietz, S., Dodson, K.,

Doup, L., Ferriera, S., Garg, N., Gluecksmann, A., Hart, B., Haynes, J., Haynes, C., Heiner, C., Hladun, S., Hostin, D., Houck, J., Howland, T., Ibegwam, C., Johnson, J., Kalush, F., Kline, L., Koduru, S., Love, A., Mann, F., May, D., McCawley, S., McIntosh, T., McMullen, I., Moy, M., Moy, L., Murphy, B., Nelson, K., Pfannkoch, C., Pratts, E., Puri, V., Qureshi, H., Reardon, M., Rodriguez, R., Rogers, Y.H., Romblad, D., Ruhfel, B., Scott, R., Sitter, C., Smallwood, M., Stewart, E., Strong, R., Suh, E., Thomas, R., Tint, N.N., Tse, S., Vech, C., Wang, G., Wetter, J., Williams, S., Williams, M., Windsor, S., Winn-Deen, E., Wolfe, K., Zaveri, J., Zaveri, K., Abril, J.F., Guigo, R., Campbell, M.J., Sjolander, K.V., Karlak, B., Kejariwal, A., Mi, H., Lazareva, B., Hatton, T., Narechania, A., Diemer, K., Muruganujan, A., Guo, N., Sato, S., Bafna, V., Istrail, S., Lippert, R., Schwartz, R., Walenz, B., Yooseph, S., Allen, D., Basu, A., Baxendale, J., Blick, L., Caminha, M., Carnes-Stine, J., Caulk, P., Chiang, Y.H., Coyne, M., Dahlke, C., Mays, A., Dombroski, M., Donnelly, M., Ely, D., Esparham, S., Fosler, C., Gire, H., Glanowski, S., Glasser, K., Glodek, A., Gorokhov, M., Graham, K., Gropman, B., Harris, M., Heil, J., Henderson, S., Hoover, J., Jennings, D., Jordan, C., Jordan, J., Kasha, J., Kagan, L., Kraft, C., Levitsky, A., Lewis, M., Liu, X., Lopez, J., Ma, D., Majoros, W., McDaniel, J., Murphy, S., Newman, M., Nguyen, T., Nguyen, N., Nodell, M., Pan, S., Peck, J., Peterson, M., Rowe, W., Sanders, R., Scott, J., Simpson, M., Smith, T., Sprague, A., Stockwell, T., Turner, R., Venter, E., Wang, M., Wen, M., Wu, D., Wu, M., Xia, A., Zandieh, A., Zhu, X. (2001) The sequence of the human genome, *Science* **291**, 1304–1351.

Vernick, K.D., Fujioka, H., Seeley, D.C., Tandler, B., Aikawa, M., Miller, L.H. (1995) Plasmodium gallinaceum: a refractory mechanism of ookinete killing in the mosquito, *Anopheles gambiae*, *Exp. Parasitol.* **80**, 583–595.

Waterston, R.H., Lindblad-Toh, K., Birney, E., Rogers, J., Abril, J.F., Agarwal, P., Agarwala, R., Ainscough, R., Alexandersson, M., An, P., Antonarakis, S.E., Attwood, J., Baertsch, R., Bailey, J., Barlow, K., Beck, S., Berry, E., Birren, B., Bloom, T., Bork, P., Botcherby, M., Bray, N., Brent, M.R., Brown, D.G., Brown, S.D., Bult, C., Burton, J., Butler, J., Campbell, R.D., Carninci, P., Cawley, S., Chiaromonte, F., Chinwalla, A.T., Church, D.M., Clamp, M., Clee, C., Collins, F.S., Cook, L.L., Copley, R.R., Coulson, A., Couronne, O., Cuff, J., Curwen, V., Cutts, T., Daly, M., David, R., Davies, J., Delehaunty, K.D., Deri, J., Dermitzakis, E.T., Dewey, C., Dickens, N.J., Diekhans, M., Dodge, S., Dubchak, I., Dunn, D.M., Eddy, S.R., Elnitski, L., Emes, R.D., Eswara, P., Eyras, E., Felsenfeld, A., Fewell, G.A., Flicek, P., Foley, K., Frankel, W.N., Fulton, L.A., Fulton, R.S., Furey, T.S., Gage, D., Gibbs, R.A., Glusman, G., Gnerre, S., Goldman, N., Goodstadt, L., Grafham, D., Graves, T.A., Green, E.D., Gregory, S., Guigo, R., Guyer, M., Hardison, R.C., Haussler, D., Hayashizaki, Y., Hillier, L.W., Hinrichs, A., Hlavina, W., Holzer, T., Hsu, F., Hua, A., Hubbard, T., Hunt, A., Jackson, I., Jaffe, D.B., Johnson, L.S., Jones, M., Jones, T.A., Joy, A., Kamal, M., Karlsson, E.K., Karolchik, D., Kasprzyk, A., Kawai, J., Keibler, E., Kells, C., Kent, W.J., Kirby, A., Kolbe, D.L., Korf, I., Kucherlapati, R.S., Kulbokas, E.J., Kulp, D., Landers, T., Leger, J.P., Leonard, S., Letunic, I., Levine, R., Li, J., Li, M., Lloyd, C., Lucas, S., Ma, B., Maglott, D.R., Mardis, E.R., Matthews, L., Mauceli, E., Mayer, J.H., McCarthy, M., McCombie, W.R., McLaren, S., McLay, K., McPherson, J.D., Meldrim, J., Meredith, B., Mesirov, J.P., Miller, W., Miner, T.L., Mongin, E., Montgomery, K.T., Morgan, M., Mott, R., Mullikin, J.C., Muzny, D.M., Nash, W.E., Nelson, J.O., Nhan, M.N., Nicol, R., Ning, Z., Nusbaum, C., O'Connor, M.J., Okazaki, Y., Oliver, K., Overton-Larty, E., Pachter, L., Parra, G., Pepin, K.H., Peterson, J., Pevzner, P., Plumb, R., Pohl, C.S., Poliakov, A., Ponce, T.C., Ponting, C.P., Potter, S., Quail, M., Reymond, A., Roe, B.A., Roskin, K.M., Rubin, E.M., Rust, A.G., Santos, R., Sapojnikov, V., Schultz, B., Schultz, J., Schwartz, M.S., Schwartz, S., Scott, C., Seaman, S., Searle, S., Sharpe, T., Sheridan, A., Shownkeen, R., Sims, S., Singer, J.B., Slater, G., Smit, A., Smith, D.R., Spencer, B., Stabenau, A., Stange-Thomann, N., Sugnet, C., Suyama, M., Tesler, G., Thompson, J., Torrents, D., Trevaskis, E., Tromp, J., Ucla, C., Ureta-Vidal, A., Vinson, J.P., Von Niederhausern, A.C., Wade, C.M., Wall, M., Weber, R.J., Weiss, R.B., Wendl, M.C., West,

A.P., Wetterstrand, K., Wheeler, R., Whelan, S., Wierzbowski, J., Willey, D., Williams, S., Wilson, R.K., Winter, E., Worley, K.C., Wyman, D., Yang, S., Yang, S.P., Zdobnov, E.M., Zody, M.C., Lander, E.S. (2002) Initial sequencing and comparative analysis of the mouse genome, *Nature* **420**, 520–562.

Weill, M., Lutfalla, G., Mogensen, K., Chandre, F., Berthomieu, A., Berticat, C., Pasteur, N., Philips, A., Fort, P., Raymond, M. (2003) Comparative genomics: insecticide resistance in mosquito vectors, *Nature* **423**, 136–137.

Yeates, D.K., Wiegmann, B.M. (1999) Congruence and controversy: toward a higher-level phylogeny of Diptera, *Annu. Rev. Entomol.* **44**, 397–428.

Zdobnov, E.M., Apweiler, R. (2001) InterProScan: an integration platform for the signature-recognition methods in InterPro, *Bioinformatics* **17**, 847–848.

Zheng, L., Cornel, A.J., Wang, R., Erfle, H., Voss, H., Ansorge, W., Kafatos, F.C., Collins, F.H. (1997) Quantitative trait loci for refractoriness of *Anopheles gambiae* to Plasmodium cynomolgi B, *Science* **276**, 425–428.

Zieler, H., Nawrocki, J.P., Shahabuddin, M. (1999) Plasmodium gallinaceum ookinetes adhere specifically to the midgut epithelium of Aedes aegypti by interaction with a carbohydrate ligand, *J. Exp. Biol.* **202**(Pt 5), 485–495.

17
Zebrafish (*Danio rerio*) Genome and Genetics

Ralf Dahm[1], *Robert Geisler*[2], *and Christiane Nüsslein-Volhard*[2]
[1] *Medical University of Vienna, Vienna Austria*
[2] *Max Planck Institute for Developmental Biology, Tübingen, Germany*

1	**The Zebrafish as a Model Organism for Developmental Biology**	**472**
1.1	General Remarks	472
1.2	Advantages of the Zebrafish as a Model Organism	472
1.3	Zebrafish Development	472
1.4	Phylogeny	475
1.5	Laboratory Strains	475
2	**Genetic Tools and Methods Available to Study Development in the Zebrafish**	**478**
2.1	General Remarks	478
2.2	Genetic Screens (Forward Genetics)	478
2.2.1	Chemical Mutagenesis	478
2.2.2	Insertional Mutagenesis	480
2.2.3	Radiation Mutagenesis	483
2.3	Reverse Genetics	483
2.3.1	Morpholino Knockdown	484
2.3.2	TILLING	486
2.3.3	RNA Interference	488
2.4	Parthenogenesis	488
2.5	Transgenesis	492
2.6	Microarrays	493
3	**Mapping of the Zebrafish Genome**	**493**
3.1	Genetic Mapping	493
3.2	Radiation Hybrid Mapping	495

Genomics and Genetics. Edited by Robert A. Meyers.
Copyright © 2007 Wiley-VCH Verlag GmbH & Co. KGaA, Weinheim
ISBN: 978-3-527-31609-0

| 3.3 | Physical Mapping 496 |
| 3.4 | Genome Sequencing 497 |

4	**Resources on the Internet 498**
4.1	General 498
4.2	Mapping 498
4.3	Sequencing 498
4.4	Resource Centers 498

Bibliography 498
Books and Reviews 498
Primary Literature 499

Keywords

Bacterial Artificial Chromosomes (BACs)
Genomic clones with 100 to 200 kb insert size used for physical mapping and genome sequencing.

Expressed Sequence Tags (ESTs)
Single-pass sequences of cDNA clones that provide rapid access to genes as well as information on expression and polymorphism.

Forward Genetics
Methods that identify an unknown gene by its mutant phenotype, usually by mutagenesis screening with a chemical mutagen or retrovirus, and cloning the affected locus.

Knockdown
Abolishment of activity of a known gene by injection of morpholino oligonucleotides complementary to its transcript into the fertilized egg, mainly feasible for genes required early in development.

Microsatellite Markers
Genetic markers consisting of a sequence of short repeats of variable length (in particular $(CA)_n$) flanked by a pair of primer sequences, useful for genetic mapping because they are widespread, codominant, and easily scored by PCR amplification and gel electrophoresis.

Positional Cloning
Molecular cloning of mutant loci for which no obvious candidate gene exists, by genetically mapping the mutation to a specific contig or genomic clone, and sequencing the transcripts of this critical region.

Radiation Hybrid Mapping
Mapping method for sequence-tagged site (STS) markers, based on fusion cell lines generated from irradiated cells of the species of interest and rodent cells. If two markers are found by PCR to be present or absent in the same radiation hybrid cell lines, they are likely to be close to each other.

Restriction Fingerprinting
Physical mapping method in which genomic clones are digested with a restriction enzyme, overlaps between the clones are identified by their similar restriction patterns, and contigs are constructed from the overlapping clones.

Reverse Genetics
Methods that identify the mutant genotype of a known gene for further investigation of its function, for example by knockdown or TILLING.

Single Nucleotide Polymorphisms (SNPs)
Changes of a single nucleotide between strains, useful for genetic mapping, for example by sequencing.

Targeting Induced Local Lesions In Genomes (TILLING)
Identification of mutations in a known gene by screening of a library of mutagenized animals, either by sequencing or with CEL-I heteronuclease; also referred to as targeted knockout.

Teleost Genome Duplication
Hypothesized genome duplication (tetraploidization) event in the evolution of ray-finned fish, followed by functional specialization or loss of duplicate genes; as a result, some mammalian genes have two zebrafish orthologs with distinct functions and expression domains.

> Owing to its transparent, easily accessible embryos, simple breeding, and short generation time, the zebrafish has become one of the most important model organisms to study the genetic control of embryonic development. Zebrafish research initially focused on forward genetics (mutagenesis screening), but reverse genetic methods, transgenesis, and microarrays are now equally available to characterize known genes. The zebrafish genome consists of 25 chromosome pairs and has an estimated haploid size of 1.5 Gb. Several genome maps exist for the zebrafish. The meiotic MGH map is generally used for placement of mutant loci and, in combination with the T51 radiation hybrid map and BAC fingerprinting, forms the basis of genome sequencing. A genome sequence is being generated by a combination of whole-genome shotgun and BAC-based sequencing, and is expected to be finished in 2005.

1
The Zebrafish as a Model Organism for Developmental Biology

1.1
General Remarks

In recent years, the zebrafish has become a favorite model organism for biologists studying developmental as well as cell biological processes. It is uniquely suited to identifying novel regulatory genes in vertebrate development and elucidating their function in the developing animal. No other vertebrate model organism offers the same combination of transparent and easily accessible embryos, efficient mutagenesis screening that has led to several hundred characterized mutations, and more recently, a sequenced genome as well as microarray and knockout technologies.

In addition, the zebrafish is of interest to the pharmaceutical industry because the embryos can be used for screening small molecule libraries in high-throughput pipelines. This approach can be applied to toxicology and, with mutants that are models for human diseases, to drug discovery and development.

The potential of the zebrafish system has been only partially realized so far. A main reason is that prior to genome sequencing, cloning of zebrafish mutations relied mainly on candidate genes and was therefore biased to finding mutations in genes already known from other organisms. However, this is changing rapidly, and novel regulatory genes are increasingly being discovered and characterized in the zebrafish.

1.2
Advantages of the Zebrafish as a Model Organism

The zebrafish has many properties that make it ideally suited as a model organism for experimental biologists:

- Zebrafish are easy to maintain and breed, minimizing the cost and effort to keep large populations of animals.
- They develop externally (outside of the female), allowing easy access to the embryos at all stages of development.
- The embryos are sturdy and large enough for experimental manipulations, such as microinjections or transplantation of cells. However, the embryos are small enough to fit into standard 96-well plates, facilitating automated analyses.
- Sperm samples can easily be obtained (even from live fish), stored, and later used for *in vitro* fertilizations.
- Embryonic development of the zebrafish is synchronous, that is, at any given time, all the embryos in a clutch will be at the same developmental stage.
- Compared with other vertebrate models for development, such as the mouse or *Xenopus*, the organs in zebrafish larvae are composed of fewer cells, making them easier to study. However, the organs still function in much the same way as those in larger animals.
- Zebrafish have large numbers of offspring.

1.3
Zebrafish Development

In addition to these advantages, there are two main embryological advantages that set the zebrafish apart from other vertebrate model organisms: its very rapid early development and the fact that its development can easily be observed in great detail.

For a vertebrate, embryonic development in the zebrafish is extremely rapid (see Fig. 1). Embryogenesis takes approximately 2 days, the larvae hatch on the

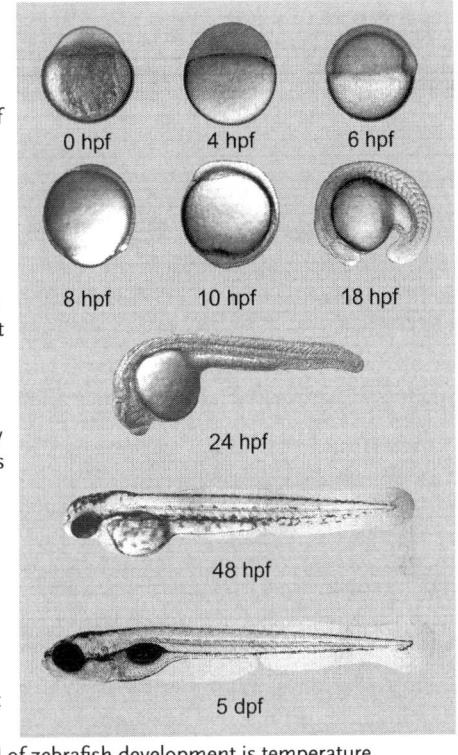

Fig. 1 A photographic overview of key aspects of early zebrafish development. hpf: hours postfertilization; dpf: days postfertilization. 0 hpf: At the one-cell stage immediately after fertilization, the zygote sits on top of a large ball of yolk. 4 hpf: Four hours following fertilization, the cells have undergone several rounds of cleavages to form a mass of cells located on top of the yolk. Please note that in contrast to species such as *Xenopus*, the yolk cell does not undergo cleavage divisions. 6 hpf: Gastrulation begins with epiboly of the cells. At this stage, the antero-posterior and dorso-ventral axes can be determined. 8 hpf: Onset of thickening in the region of the future head. 10 hpf: First somite forms (right) and the optic lobes evaginate from the diencephalon. 18 hpf: Body plan becomes recognizable, and the first voluntary muscle contractions are observable as movements of the tail. 24 hpf: Most vertebrate-specific body features and organs are already recognizable, including a compartmentalized brain, eyes, ears, and all internal organs. However, the majority of the organs are not yet functional. Notably though, the heart starts beating just before the end of the first day of development. Over the next 24 hpf, numerous cell types differentiate, and the organs gradually begin to function. 48 hpf: Zebrafish start to hatch and start swimming. 5 dpf: Larvae swim well and search for food independently. The speed of zebrafish development is temperature dependent. Zebrafish develop at temperatures ranging from approximately 18 to 32 °C. The higher the temperature, the faster the development. Normally, zebrafish are kept at temperatures ranging between 25 and 28.5 °C. In this text, all statements relating to the development of any given structure at any time point always refer to development having occurred at 28.5 °C. (Adapted from Haffter, P., Granato, M., Brand, M., Mullins, M.C., Hammerschmidt, M., Kane, D.A., Odenthal, J., van Eeden, F.J., Jiang, Y.J., Heisenberg, C.P., Kelsh, R.N., Furutani-Seiki, M., Vogelsang, E., Beuchle, D., Schach, U., Fabian, C., Nusslein-Volhard, C. (1996) The identification of genes with unique and essential functions in the development of the zebrafish, *Danio rerio*, Development **123**, 1–36.)

third day of development and as of day five, the vast majority of cell types have differentiated and the organs have taken up their function. This rapid development allows the observation of developmental processes and the completion of experiments generally within a few hours to days.

Despite the rapidity of embryonic and early larval development in zebrafish, the time needed to reach sexual maturity (generation time) is between two and four months (depending on the conditions under which they are raised; see Fig. 2).

The development of the zebrafish is similar to the embryogenesis in higher-ordered vertebrates, including humans and other mammals. But unlike in mammals, the embryonic development of zebrafish occurs outside the female. This allows easy access to the embryos without the need to sacrifice the female. Moreover, unlike in birds and reptiles, the shell of the zebrafish eggs (the chorion) is transparent, allowing the continuous observation of the developing embryo in its "natural" environment without the need to open the

Fig. 2 Photographs of age-matched adult female (top) and male (bottom) zebrafish. The characteristic horizontal stripes running along the body and fins gave this species its name. Adult zebrafish measure about 3 to 5 cm in length. Females can be recognized by their larger belly accommodating the eggs. Males tend to be more reddish in color than female zebrafish. The scale bar is 1 cm. (See color plate p. vii).

shell. This property is shared with, for example, the African claw-toed frog *Xenopus*, another model for early development.

The most important advantage of the zebrafish, however, is that the embryos themselves are completely transparent during early development. This is due both to their relatively small size and the absence of pigment throughout the first 36 h of development (the time when pigmentation in the melanocytes and the cells of the retinal pigment epithelium begins). Yet, even after the onset of pigmentation, most parts of the body can still be observed between the relatively few melanophores. Also, in contrast to *Xenopus*, in zebrafish, the yolk does not participate in the cleavage divisions of the early embryo but remains distinct from the embryo throughout development.

This transparency of the zebrafish embryos and larvae allows the visualization of many processes that occur deep inside the developing animal. These include, the development of internal organs, such as the brain, the cardiovascular system or the digestive tract, the migration of cells as well as the path finding of axons. In the zebrafish, these developmental processes can be visualized in real time without the need for invasive intervention (e.g. dissection or histological sectioning) or sophisticated imaging techniques (e.g. MRI).

The external development and the transparency of the zebrafish embryos facilitate numerous experimental approaches. These include, for example, laser manipulations, such as cell ablation or uncaging experiments as well as the generation of extensive fate maps. Moreover, in techniques, such as mRNA *in situ* hybridization and antibody staining to test for the expression patterns of genes and proteins, the translucency of the embryos in many cases obliterates the need for sectioning as the staining can be performed and imaged on whole embyros.

In contrast to mammals, zebrafish can generate large numbers of offspring. When kept under optimal conditions, a single female can lay up to 200 eggs per week. This greatly facilitates high-throughput approaches and statistical analyses, including the reliable identification of mutant phenotypes, for example in a genetic screen. This fact, together with the small size and easy maintenance of zebrafish embryos as well as their rapid embryonic development, allows for forward genetic approaches to be taken that are often too labor intensive and too costly to be carried out with other vertebrate models, such as the mouse. The zebrafish thus combines being a vertebrate with many of the advantages of invertebrate model organisms.

For a summary of key properties of commonly used animal model organisms, please refer to Table 1.

1.4
Phylogeny

The zebrafish (*Danio rerio*, formerly also known as *Brachydanio rerio*) is a small tropical freshwater fish. Its natural habitats are rivers of South Asia, mainly northern India as well as northern Pakistan, Nepal, and Bhutan. The zebrafish belongs to the family of the cyprinids (*Cyprinidae*) in the class of ray-finned fishes (*Actinopterygii*) and within this class to the bony fishes (teleosts or *Teleostei*) to which most extant ray-finned fishes belong (Fig. 3). The teleost fishes comprise other model organisms:

- Medaka (*Oryzias latipes*), a species, which is mainly used to study developmental biology.
- The three-spined stickleback (*Gasterosteus aculeatus*) and the guppy (*Poecilia reticulata*), the two species with a long tradition in behavioral studies and more recently, in population genetics also.
- The two pufferfish species *Takifugu rubripes* (also referred to as *Fugu rubripes*) and *Tetraodon nigroviridis* (recently also referred to as *Monotreta nigroviridis*), which with their relatively small and simple genomes are important organisms to understand genome architecture, organization, and function. *Takifugu rubripes* was the first fish and second vertebrate species whose genome was fully sequenced (in 2002). The genome of *T. nigroviridis* is currently also being sequenced.

These five species belong to the teleost crown group, a large radiation whose last common ancestor with the zebrafish lived about 100 million years ago (a similar evolutionary time span as that between humans and mice). The lineages leading to the cyprinids and mammals split about 400 to 450 million years ago.

It is assumed that during the evolution of the ray-finned fishes, there has been a genome duplication event, which lead to a tetraploidization of the genome. This duplication was followed by a functional specialization of some of the duplicated genes on the one hand and the loss of other genes on the other hand. As a result, an estimated 20% of mammalian genes have two zebrafish orthologs with distinct functions and expression domains. This subfunction partitioning can help to understand different aspects of the mammalian gene function as they can be studied in isolation.

1.5
Laboratory Strains

Several wild-type laboratory strains of the zebrafish have been established by the major research laboratories. While these strains are inbred to some extent, none of them is completely isogenic, and all of them have somewhat overlapping distributions of marker alleles. The choice of a mapping strain is thus often a matter of personal preference. Only the most common strains will be mentioned here.

The primary strains used for mutagenesis experiments are AB and Tübingen (Tü). The AB strain was generated by G. Streisinger in Eugene, Oregon, by crossing the strains obtained at a pet shop. It has been used for a major mutagenesis screen by the laboratory of W. Driever, and for several smaller

Tab. 1 Comparison of key properties of commonly used model organisms.

Organism	Size egg (mm)	Length adult	Number of offspring	Duration of embryogenesis	Generation time	Genome size [bp]	Number of genes
Caenorhabditis elegans	0.05	1 mm	300–350/life time	20 h	50 h	0.097×10^9	20 000
Drosophila melanogaster	0.5	3 mm	50/day	24 h	9 days	0.18×10^9	13 600
Zebrafish (Danio rerio)	0.7	5 cm	200/week	2 days	2–3 months	1.5×10^9	est. 30 000
Xenopus laevis	1–1.3	10–14 cm	300–1000/2–3 months	2–2.5 days	1–2 years	3.1×10^9	Unknown
Xenopus tropicalis	0.6–0.8	4–5 cm	1000–3000/2–3 months	2 days	3–4 months	1.7×10^9	Unknown
Chicken (Gallus gallus)	60–80	30 cm	1/1–2 days	6 days	9 months	1.1×10^9	Unknown
Mouse (Mus musculus)	0.1	6–10 cm	4–8/4–6 weeks	14 days	2 months	2.6×10^9	30 000

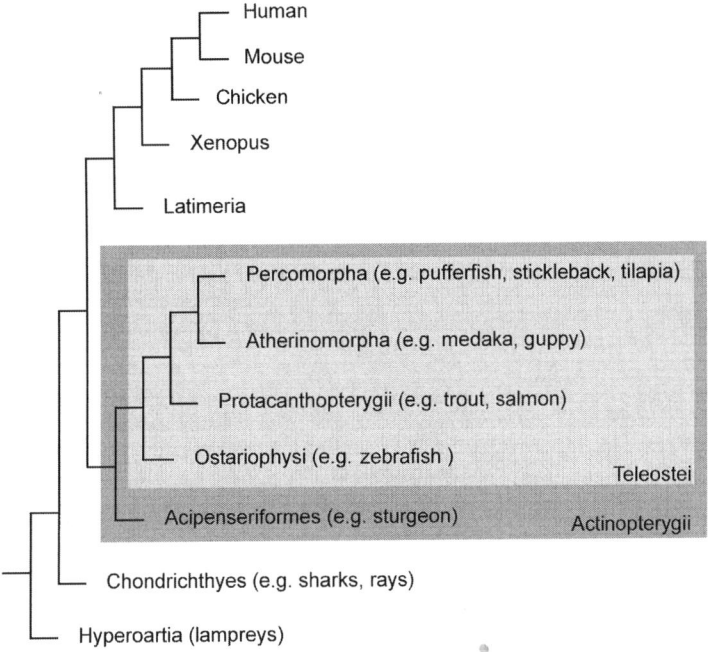

Fig. 3 Schematic representation of the major vertebrate lineages.

screens. The Tübingen strain, likewise originating from a pet shop, has been used as the genetic background of most mutagenesis screens including the two largest ones, performed in the laboratory of C. Nüsslein-Volhard. For this reason, it was also chosen as the strain to be sequenced in the zebrafish genome project.

Wild-type India Calcutta (WIK) was generated from a wild catch in India as a highly polymorphic reference strain for mapcrosses with mutations in the Tü background. In crosses of Tü × WIK, 68% of microsatellite markers give a scorable polymorphism. India (IN) was used for the MGH meiotic mapping panel in combination with AB; however, it is more difficult to breed than WIK and may contain mutations that affect viability. The strain SJD was isogenized over several generations by the laboratory of S. L. Johnson, and it is more polymorphic in crosses to Tü (75% scorable polymorphism). It would thus be better suited as a reference strain, but it is more difficult to breed because of a skewed sex ratio (a common problem with inbred zebrafish strains).

In the zebrafish, visible markers are not as useful for the maintenance of mutations as, for example, in *Drosophila* because it has 25 chromosomes and lacks balancer chromosomes (which would prevent recombination between the marker and the mutant locus). A common strain with visible markers is Tübingen long fin (TL, formerly named Tup long fin), which carries a recessive mutation (leo^{t1}) that causes spotting, and a dominant mutation (lof^{dt2}) causing long fins.

Wild-type strains can be obtained either from the Zebrafish International Resource Center (ZIRC) or from the laboratories in which they originated.

2
Genetic Tools and Methods Available to Study Development in the Zebrafish

2.1
General Remarks

When compared to other model organisms, such as the nematode *Caenorhabditis elegans*, the fruit fly *Drosophila melanogaster*, or the mouse *Mus musculus*, the number of tools and methods that are available when working with the zebrafish is still relatively limited. This is not surprising when considering that the zebrafish is a comparatively recent addition to the list of model organisms. It was established largely by George Streisinger in the late 1970s and early 1980s. Since then, the zebrafish community has grown steadily, the tools and methods available to zebrafish researchers have increased considerably over the past two decades, and there is now a substantial number of protocols, reagents, as well as transgenic and mutant zebrafish lines.

2.2
Genetic Screens (Forward Genetics)

To date, genetic screens are the most successful approach for identifying genes with an essential function in the development of an organism. Genetic screens rely on the random induction of mutations in the genome and the subsequent search for individuals with a mutant phenotype. They have the advantage that they are relatively unbiased with respect to prior knowledge, and thus also allow the identification of previously unknown genes. However, being able to identify mutants for a given process, morphological structure or behavior relies on the capability to develop an assay for the detection of the corresponding phenotypes. Furthermore, as genetic screening relies on the identification of a phenotype, it selects genes with unique or at least partially nonredundant functions. Genes with redundant functions, however, cannot be detected in genetic screens.

The possibility to perform large-scale genetic screens in the zebrafish is a powerful feature of this model organism. When compared to other vertebrate model organisms (mouse, chicken, frog), the generation time of the zebrafish is similar (2–4 months). But the large number of offspring greatly facilitates phenotypic and genetic analyses, such as the collecting of sufficient numbers of mutant individuals for different assays and phenotypic analyses as well as of recombinant individuals for the subsequent mapping of mutations (see below). Moreover, the moderate space required to breed large numbers of zebrafish significantly reduces the cost of such undertakings. Finally, the speed of zebrafish development and the ease with which it can be observed greatly facilitate the identification of mutant phenotypes for virtually any organ, tissue, or cell type.

Several mutagenesis protocols have been established for the zebrafish. The mutagens used include ENU, radiation (mainly γ-rays), as well as insertional mutagenesis with retroviruses or transposable elements.

2.2.1 Chemical Mutagenesis
The generation of mutants by chemical mutagenesis has been one of the most successful approaches in elucidating the function of genes and the genetic control of biological processes, particularly those of embryonic development. One of the most widely used substances to induce mutations is the chemical *N*-ethyl-*N*-nitrosourea (ENU). ENU is a potent mutagen, which has been used to induce mutations in various organisms. For

example, in the mouse, it has the highest mutation rate of all germ cell mutagens examined, and also for the zebrafish, ENU has been the mutagen of choice in the majority of genetic screens carried out to date.

ENU is a synthetic alkylating agent, which induces mutations by transferring its ethyl group to oxygen or nitrogen radicals in individual bases of DNA, including the N1, N3, and N7 groups of adenine, the O2 and N3 groups of cytosine, the O6, N3, and N7 groups of guanine, and the O2, O4, and N3 groups of thymine. Generally, ENU induces point mutations: The ethylation of a base can lead to the modified base being misread during DNA replication, resulting in the integration of a noncomplementary base in the second DNA strand (see Fig. 4). Following a second round of replication, one of the four double helices derived from the initially modified double helix carries a complete base-pair substitution. ENU-induced mutations chiefly involve the base thymidine (AT to TA transversions and AT to GC transitions as shown in Fig. 3 are the most common mutations induced by ENU). However, chromosomal rearrangements (translocations, inversions, deletions) have also been described for post-meiotically mutagenized male germ cells.

When animals are exposed to ENU (in the case of zebrafish by bathing them in a solution containing ENU), it induces mutations in all cells of the body, including in the precursors of the germ cells. It has been shown for mice that the rate of mutagenesis by ENU in testes is highest in spermatogonial stem cells

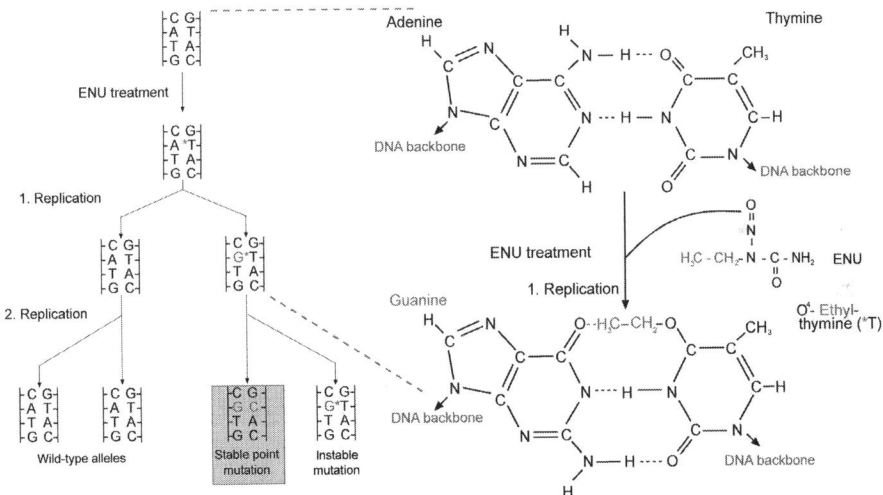

Fig. 4 Mechanism of mutagenesis by ENU leading to an AT → GC transition. The ethyl group in ENU (shown in red or indicated as a red asterisk) is transferred to the O^4-atom of thymine, resulting in an O^4-ethylthymine adduct. Failing an excision of the O^4-ethylthymine adduct by the DNA repair machinery, the O^4-ethylthymine mispairs with a guanine in a subsequent replication of the DNA. In this mispairing, the ENU-derived ethyl group forms a third hydrogen bond with the O^6 of the guanine base in the complementary DNA strand. A successive second replication leads to a complete AT → GC transition in one of the four double helices produced, thus fixing the mutation. Substitutions of bases are indicated in blue. Figure adapted after Noveroske, J.K., Weber, J.S., Justice, M.J. (2000) The mutagenic action of N-ethyl-N-nitrosourea in the mouse, *Mamm. Genome* **11**, 478–483. (See color plate p. xiii).

where optimal doses of ENU result in one mutation per gene in every 175 to 655 gametes (corresponding to a specific locus rate of $1.5–5.7 \times 10^{-3}$). In the zebrafish, the specific locus rate ranges from 0.6 to 4.8×10^{-3}, indicating that either the mutagenesis rates that can be achieved are slightly lower than those for the mouse or that the optimum conditions for mutagenesis in the zebrafish have not yet been found.

As the mutations induced by ENU primarily occur only in one of the DNA strands (see Fig. 3), there is a difference in the offspring resulting from mutagenized germ cells, depending on whether the exposure to ENU occurred before or after meiosis: mutations in premeiotic spermatogonia lead to sperm, which carry the mutation in both strands (due to DNA replication during meiosis), whereas mutations in postmeiotic germ cells result in sperm with only one of the DNA strands mutagenized. As a consequence, animals arising from spermatogonia carry the mutation in every cell. However, the progeny from germ cells that were mutagenized postmeiotically are mosaic with respect to the mutation.

Figure 5 depicts a typical breeding scheme for a two generation ENU mutagenesis screen in the zebrafish. Genetic screens using the zebrafish have resulted in the isolation of nearly 2000 mutant lines up until today. The vast majority of these mutations are recessive and affect embryonic or larval development (most are embryonic or larval lethal). However, a few dominant mutations as well as some mutations with adult phenotypes have been isolated. For a list of published mutants with brief phenotypic and – where available – molecular characterizations, refer to the searchable database at http://www.zfin.org.

2.2.2 Insertional Mutagenesis

The most significant disadvantage of working with mutants generated by chemical mutagenesis is the difficulty in identifying the mutation. The localization of chemically mutated genes in the genome requires a laborious genetic mapping of the mutated locus followed by positional cloning and/or a candidate gene approach (see below). This drawback can be circumvented if the mutagenesis is achieved by the insertion of a known sequence (e.g. a transposon or a retroviral vector), which can be used as a starting point for the sequencing of the regions flanking the insert. This strategy has been successfully applied in mutagenesis screens in other model organisms. Famously, the P-element transposon has been used to generate numerous mutants in *Drosophila*, and retroviral vectors were employed to mutate, for example, mice.

In the zebrafish, insertional mutants have been generated using one of three methods: injection of plasmid DNA, transposons, and pseudotyped retroviral vectors. However, the only efficient protocols for insertional mutagenesis in the zebrafish have been established using retroviruses.

Briefly, retroviral vectors are injected into zebrafish embryos at the blastula stage, resulting in mosaic founder fish carrying insertions also in the precursors of the germ cells (up to 25–30 insertions in the germ line per individual with a given insertion occurring in 3–20% of the gametes). Incrossing of these founders generates a nonmosaic F_1 generation. Typically, approximately 20% of these fish have 5 to 10 insertions per individual. The F_1 fish carrying multiple insertions are incrossed to raise F_2 families with at least 10 independently segregating inserts,

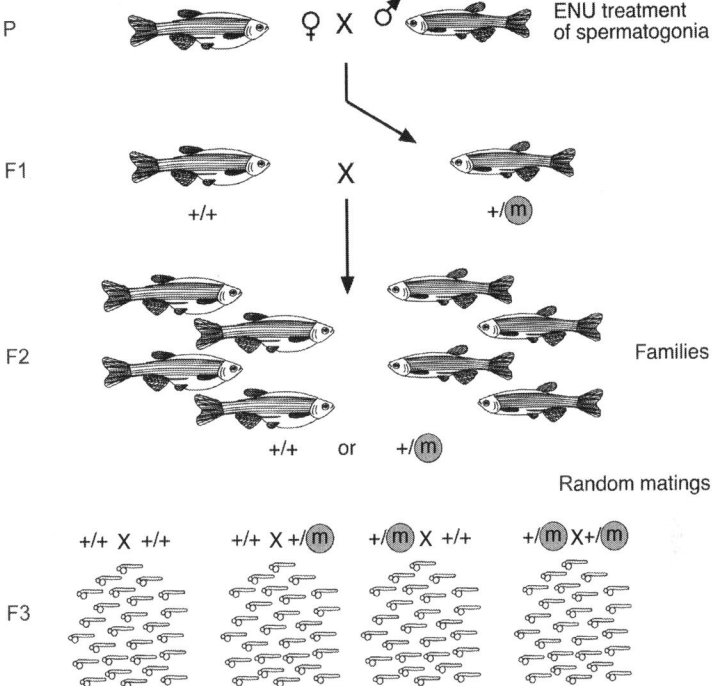

Fig. 5 Depiction of a mutagenesis and crossing scheme. Male fish are mutagenized with ENU and subsequently mated with wild-type female fish. The fish of the resulting F_1 generation are heterozygous for individual mutations (depicted as "m" in this example). These F_1 fish are again mated with wild-type fish in single-pair matings. The offspring from each of these matings are commonly referred to as "F_2 families." In these F_2 families, approx. 50% of the individuals are heterozygous for a particular mutation (+/m); the other 50% carry two wild-type alleles of this locus (+/+). Random single-pair matings between the siblings of an F_2 family give rise to the F_3 generation. This is the first generation in which a particular mutation can occur in a homozygous state and which hence first show a mutant phenotype for recessive mutations (the vast majority of ENU-induced mutations are recessive; rarely dominant phenotypes can already be observed in the F_1 and F_2 generations). On the level of the mutated locus, four different cases can be distinguished: Two homozygous F_2 individuals are mated (+/+ X +/+), a genotypically wild-type F_2 female is mated with a heterozygous F_2 male (+/+ X +/m), a heterozygous F_2 female is mated with a genotypically wild-type F_2 male (+/m X +/+), and two heterozygous F_2 individuals are mated (+/m X +/m). In the case of a recessive mutation, the latter cross results in phenotypically mutant F_3 offspring (approximately 25% of the individuals). (Figure adapted from Haffter, P., Granato, M., Brand, M., Mullins, M.C., Hammerschmidt, M., Kane, D.A., Odenthal, J., van Eeden, F.J., Jiang, Y.J., Heisenberg, C.P., Kelsh, R.N., Furutani-Seiki, M., Vogelsang, E., Beuchle, D., Schach, U., Fabian, C., Nusslein-Volhard, C. (1996) The identification of genes with unique and essential functions in the development of the zebrafish, *Danio rerio*, *Development* **123**, 1–36.)

which in turn are incrossed and their offspring (F3) screened for phenotypes.

In a zebrafish mutagenesis screen using a pseudotyped retroviral vector, approximately one in every 80 insertions results in an embryonic/early larval phenotype that can be identified on the basis of morphologically observable criteria. With an average number of 12 insertions that can be screened per F_2 family, the mutagenesis rate in zebrafish insertional mutagenesis screens using retroviruses is an estimated seven- to tenfold lower than in equivalent genetic screens using ENU. This means that significantly larger numbers of insertion lines have to be screened in order to obtain the same number of mutant phenotypes as when using an ENU protocol. However, once an interesting phenotype resulting from an insertion has been identified, the affected gene can be cloned much faster than is normally the case for ENU-induced mutations (approximately 2 weeks for insertions compared to several months or more for ENU mutations).

To date, over 500 zebrafish mutants with retroviral insertions have been identified. The range of phenotypes, which can be obtained with either chemical or insertional mutagenesis, is comparable. As observed in genetic screens based on chemical mutagenesis, roughly one-third of these mutants show defects in the development of one or very few parts of the body, organs, or cell types, while two-thirds display general or pleiotropic phenotypes, such as necrosis, heart edema, or small heads and eyes.

A major advantage of mutagenesis approaches relying on the insertion of retroviral or transposon sequences is that they allow a very easy cloning of the disrupted genes via the inserted sequences. The experience with cloning the mutated genes in the zebrafish insertional mutants has shown that in 80% of the cases, it was sufficient to sequence up to a few kilobases of the genomic DNA flanking the inserted sequence to identify the gene. This has led to the cloning of over 280 disrupted genes in the zebrafish to date – including some whose function *in vivo* had not previously been described. In comparison, approximately 100 chemically mutated zebrafish genes were cloned to date.

The differences in the efficiencies of mutagenesis and identification of the mutated genes lead to a difference in the kinds of genes, which tend to be identified using the two methods. Owing to the very large numbers of mutants generated in chemical mutagenesis screens, generally only those displaying specific developmental phenotypes are kept. The comparatively lower numbers generated in insertional mutagenesis screens leave more capacity to also keep mutant lines with less specific phenotypes. The need to focus on specific phenotypes in chemical mutagenesis screens is increased by the fact that the effort required to clone a chemically mutated gene is considerably bigger when compared to the identification of a gene disrupted by an insertion. This results in mutations in genes with more pleiotropic effects during development being underrepresented in collections of mutants resulting from chemical mutagenesis screens. Additionally, the cloning of the majority of ENU-mutated zebrafish genes at least partially relied on a candidate gene approach (favoring the identification of known genes), while the cloning of insertion mutations is less biased. As a consequence, the majority of ENU mutations identified to date are in transcription factors as well as signaling molecules and their receptors. The insertion mutants, on the other hand, also identified a number of genes involved in more general cellular processes, such as the control of the cell cycle, vesicle trafficking, protein sorting between organelles or protein glycosylation.

2.2.3 Radiation Mutagenesis

In the past, ionising radiation has been a commonly used means for inducing mutations in a wide range of organisms. Ionising radiation carries enough energy to cause double-strand breaks in DNA helices and thus leads to chromosomal rearrangements, such as deletions, inversions, and translocations. As the majority of mutations induced by ionising radiation tend to be large chromosomal deficiencies generally comprising more than one gene, the usefulness of such mutants in the study of gene function is limited. Moreover, many radiation-induced mutations tend to be dominant lethal, and therefore mutants carrying such mutations cannot be maintained. For these reasons, ionising radiation has only been used infrequently and on a comparatively small scale to generate zebrafish mutants, and, compared to chemical and insertional mutagenesis, radiation plays a marginal role in the zebrafish as a means to generate mutants.

There are however studies in which γ-radiation has been used to induce mutations in the zebrafish, and several mutants with developmental defects have been isolated. γ-rays were shown to efficiently induce mutations, for example, in mature sperm cells, fertilized eggs, and blastula embryos of the zebrafish with specific locus mutation rates of up to 10^{-2}. With the current protocols, the survival rate to adulthood is approximately 25% for zebrafish raised using irradiated sperm in *in vitro* fertilizations and between 25 and 50% for irradiated zygotes (http://zfin.org/zf_info/zfbook/chapt7/7.9.html).

2.3 Reverse Genetics

In recent years, genome sequencing projects have yielded an enormous wealth of information, including numerous predicted genes with, as of yet, unidentified functions. The elucidation of the functions of these genes is one of the major challenges in the years to come. Reverse genetics approaches offer efficient ways to address this challenge. They allow to specifically knock out or knock down genes of interest either by generating mutants in these genes or by preventing the translation of the corresponding mRNA(s) into proteins respectively. The function of the targeted genes is subsequently elucidated via analyses of the resulting phenotypes.

For two decades, the mouse was the only vertebrate model organism for which reverse genetics methods could be established. These methods are based on homologous recombination in embryonic stem (ES) cells: an endogenous locus is replaced by a homologous extra-genomic construct engineered to carry a desired mutation. Following a selection step, ES cells carrying the mutation are introduced into mouse blastocysts where they participate in the normal development and contribute progeny to all organs including the germ line. These chimeric mice (always males) are mated with wild-type females to generate heterozygous offspring, which is generally outcrossed to generate a larger number of heterozygous F_2 individuals. The F_2 is then incrossed to produce the first homozygous embryos.

Despite efforts to establish zebrafish ES cell lines, there is currently no protocol for ES cell–based knockout or targeted gene expression by homologous recombination in the zebrafish. However, the advent of reverse genetics technology in the form of morpholino knockdown and, more recently, the TILLING (Targeting Induced Local Lesions In Genomes) approach had a significant impact on zebrafish research. Particularly, in combination with the zebrafish genome-sequencing project, which is nearing its completion, these two techniques will increasingly contribute to elucidate the function of novel genes. Taken together, the morpholino and TILLING techniques will allow reverse genetics approaches to be performed routinely and on a large scale in the zebrafish.

2.3.1 Morpholino Knockdown

Morpholinos provide a very rapid means to obtain a phenotype, and thus assay the *in vivo* function of a specific gene. The technique is based on injecting modified oligonucleotides that prevent the expression of the targeted gene by blocking the translation of the corresponding mRNA(s). As zebrafish produce large numbers of offspring, which develop externally, they are ideally suited to large-scale knockdown approaches. Importantly, their eggs are large enough for sufficient quantities of morpholinos to be injected to block the expression of target genes over the full duration of embryonic development. Since the establishment of protocols for morpholino knockdown in the zebrafish, the technique has rapidly spread through the zebrafish community and is now widely used to rapidly confirm or exclude candidate genes for mutants obtained in mutagenesis screens as well as to obtain phenotypes for novel genes. Particularly for the latter, the zebrafish will play an increasingly important role in the future.

When performing a targeted mutagenesis of a locus by homologous recombination in mice, it typically takes one year to obtain the first homozygous offspring and another six months to generate the first 100 homozygous embryos to work with. Importantly, one heterozygous female has to be sacrificed each time to obtain (on an average) two homozygous embryos. This makes targeted mutagenesis in the mouse a time-consuming, laborious, and costly process. In comparison, the knockdown of genes by morpholino injection in zebrafish is faster, as there is no need for time-consuming manipulations of ES cells or for the breeding and crossing of several generations of animals to obtain homozygous individuals. Instead, the desired phenotype can be observed in the "first" generation from within a few hours to days. Also, in a single experiment, hundreds of embryos can be injected in a couple of hours.

Morpholinos cannot be used in mice, mainly because the morpholinos are rapidly diluted during mouse development, and their effect is thus lost before a number of interesting processes occur. This is due to two features characteristic for mammalian development: Firstly, the eggs are small such that only relatively small quantities of morpholino can be injected. Secondly, the embryos rapidly grow following implantation into the uterus.

Morpholinos are short oligonucleotides (18- to 25-bases long). In contrast to naturally occurring nucleic acids, the backbone of morpholinos is chemically modified: Each of the four bases A, C, G, and T is linked to a six-membered morpholine ring (hence the name) instead of the sugar moieties normally occurring in nucleic acids. These rings are connected with each other via nonionic phosphorodiamidate units (see Fig. 6). This chemical modification renders morpholinos resistant to degradation by nucleases. Morpholinos are generally designed to be complementary (antisense) to the region including the START

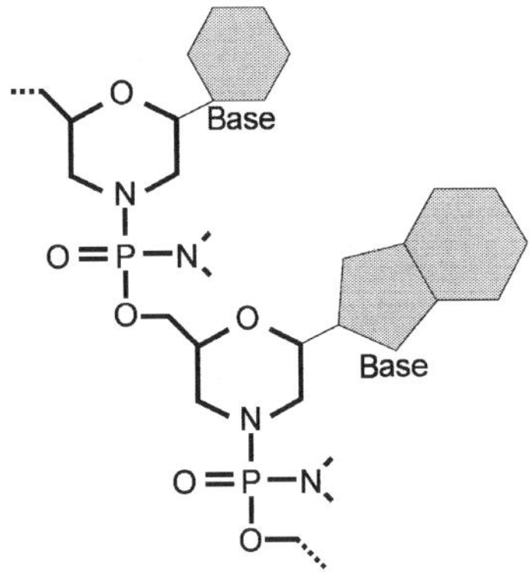

Fig. 6 Schematic representation of the backbone of a morpholino oligonucleotide (morpholino or MO). Instead of being linked to a ribose molecule (as is the case in RNA) or deoxyribose molecule (as in DNA), the bases in morpholinos are linked to a six-membered morpholine ring. These rings are connected via nonionic phosphorodiamidate groups (instead of an ionic phosphate group as in RNA and DNA). The bases linked to the morpholine ring are the same as in DNA: adenine, cytosine, guanine, and thymine.

codon of translation (ATG) or a crucial splice site in the mRNA of a targeted gene.

When injected into a cell, the morpholinos bind to their target sequence in the endogenous mRNA transcripts and block their translation into protein. The resulting effect of a morpholino injection is a phenotype that closely resembles or is identical to a loss-of-function mutant. In principle, any gene can be knocked down using morpholinos, including those that act early during development as maternally deposited mRNAs (a class for which it is relatively difficult to obtain homozygous mutants).

In zebrafish, morpholinos are generally injected into the egg yolk shortly after fertilization. They exert their effect almost immediately and effectively block the translation of the targeted mRNA for between three and four days of development (which in the zebrafish comprises the entire period of embryonic development). After that time, the effect of the morpholinos gradually diminishes as they are progressively diluted (and degraded) in the growing animal. As a consequence, only phenotypes that manifest themselves during this period can be observed, and genes acting later than three to four days after fertilization can generally not be targeted by morpholino injection.

Unlike nucleic acids used to knock down genes in other model organisms ("interfering RNA," RNAi; see below), morpholinos are not charged and not toxic even at high concentrations, allowing large amounts to be injected. Varying the amount of morpholino injected makes it possible to attain different dose-dependent strengths of a phenotype – not unlike the situation observable in allelic series.

Morpholinos are also very well suited to overcome the problem of redundancy of gene functions – a weakness of mutagenesis screens and homologous recombination experiments. By simultaneously injecting two (or more) morpholinos targeted to related genes, these genes can be inactivated simultaneously. Similarly, morpholinos can be used to knock down additional genes in a mutant background. This way, the knockdown of gene functions through morpholinos is likely to contribute considerably to elucidating the genetic networks acting in development.

Despite their advantages, the knockdown with morpholinos has significant limitations. Morpholino injections can only lead to lack-of-function phenotypes and never serve to generate the variety of sophisticated mutations achievable with homologous recombination in mice (e.g. conditional knockouts and knockins). Moreover, the dilution of the morpholino as the embryo grows restricts the time when phenotypes can be reliably identified and characterized to the first, approximately, three days of development. Also, a morpholino knockdown phenotype is not as reliable and reproducible as a genetic mutant. In many instances, the knockdown phenotypes will faithfully phenocopy the mutant phenotype, but the injection of morpholinos can result in nonspecific phenotypes, such as widespread cell death and neuronal degeneration. These have to be separated from genuine phenotypes by including a control morpholino in each experiment (e.g. a morpholino with the same sequence except for three mismatched bases). Furthermore, a morpholino targeted at a particular gene may in some instances "cross react" and block the production of other gene products. Finally, the knockdown by morpholino injection is transient and not passed on to subsequent generations. This necessitates new injections for each experiment that is being undertaken with knockdown fish. However, many of these drawbacks are overcome by the TILLING technique to generate knockout mutants.

2.3.2 TILLING

TILLING is an effective method to isolate mutants with defined mutations, including the knockout of genes. It constitutes a random mutagenesis and the subsequent screening for mutations in target genes. The method was initially developed for *Arabidopsis thaliana* but has since been used in other species, including the zebrafish.

In TILLING, mutations are randomly induced in the genome via chemical mutagenesis (e.g. with ENU as described above). The mutagenized fish are mated with wild-type animals to produce F_1 fish, which carry individual mutations in a heterozygous state (Fig. 7). However, instead of using these F_1 fish to raise F_2 families and subsequently screening their offspring for phenotypes as would be done in a genetic screen, the F_1 fish are directly examined for the presence of mutations. This is done by taking biopsies (fin clips) from which genomic DNA is isolated. This DNA is then screened for mutations in (known) genes of interest. Screening can be done either by sequencing parts of the genes with specific primer pairs flanking a region of interest or with an enzyme-mediated assay based on the property of the plant endonuclease CEL-I to recognize and cleave heteroduplexes. The possibility to automate certain steps of the TILLING procedure allows the screening of large numbers of mutagenized genomes for mutations in genes of interest. Once a mutation of interest has been identified in the genome of a particular fish, outcrossing this individual to a wild-type fish generates more offspring carrying the mutation. Incrossing of this F_2 generation results in homozygous F_3 individuals, which can be phenotypically analyzed.

The TILLING procedure has recently been successfully tested in the zebrafish and is now used on a large scale to generate zebrafish knockout mutants in a broad range of genes (see www.zf-models.org). In a test study, a library of 4608 male and female F_1 zebrafish derived from ENU-mutagenized founder fish was screened for mutations in 16 genes. The overall mutation frequency in this experiment was one mutation in approximately 235 000 bp. For the majority of target genes, only one PCR amplification product (amplicon) was screened for

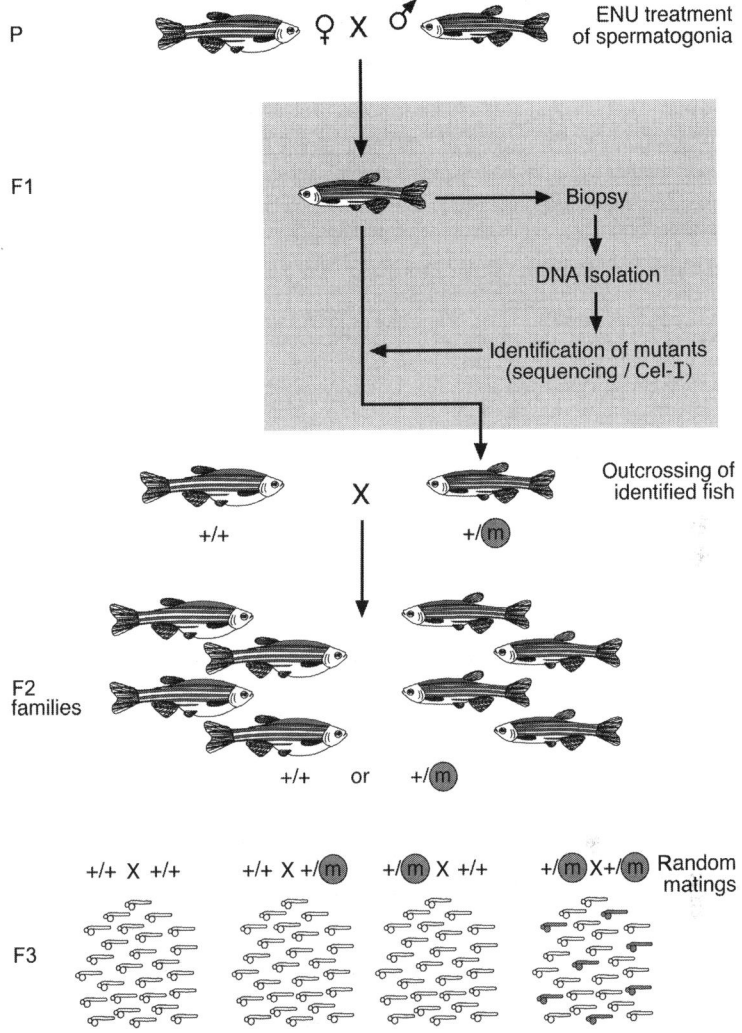

Fig. 7 Schematic illustration of the procedure used to generate, identify, and recover TILLING mutants. For details on the procedure see text.

mutations. The size of the amplicon was such that, on average, one-quarter of the coding sequence was screened. In addition to coding sequence, these amplicons comprised an average of 26% noncoding sequences, such as untranslated regions and intronic sequences.

This approach resulted in the identification of 255 mutations. The majority (80%) of all mutations found were located in coding sequences, whereas 20% were found in noncoding regions. Of the mutations identified in coding regions, most were missense mutations resulting in the exchange of an amino acid, roughly one-third were silent, and approximately one in ten were nonsense, resulting in a premature STOP codon (14 in total). Of the mutations identified in noncoding regions, approximately 10% (seven in total) altered splice donor or acceptor consensus sequences and are presumed to cause a change in the splicing of the affected genes. A detailed characterization of the mutants arising from this TILLING project has not yet been done. However, it is presumed that the 14 nonsense mutations and the seven mutations affecting splice donor/acceptor sites will most likely give rise to complete loss-of-function or knockout phenotypes for these genes. This would signify that, in total, 21 null mutations (roughly 10%) could be identified in 13 of the 16 target genes in a period of a few months.

These pilot studies suggest that TILLING can be performed in a high-throughput setup in the zebrafish. Importantly, the study has shown that the TILLING technique is sufficiently sensitive to detect mutations against a background of naturally occurring single nucleotide polymorphisms (SNPs) – a finding of particular importance for species such as the zebrafish for which no extensively inbred lines exist. It must, however, be noted that in contrast to methods relying on homologous recombination in mice, TILLING cannot be used to introduce designed mutations. Instead, it relies on the spectrum of mutations inducible randomly by chemical mutagenesis.

2.3.3 RNA Interference

Similar to the knockdown with morpholinos, RNA interference (RNAi) is a method to transiently knockdown specific gene products. RNAi relies on a conserved cellular response to double-stranded RNA (dsRNA). When dsRNA is introduced into a cell, it leads to a series of events resulting in the sequence-specific degradation and translational inhibition of the homologous endogenous mRNA and thus the posttranscriptional silencing of the targeted gene. Since its discovery in *C. elegans* in 1998, RNAi as a mechanism to silence genes has been employed in a variety of fungi, plant and animal species, including model organisms such as *Neurospora, Arabidopsis, C. elegans, Drosophila,* and mice.

There have also been attempts to establish RNAi-mediated knockdown protocols for the zebrafish. However, the ratio of specific phenotypes was generally low and variable, and there have been studies reporting significant non-specific effects of RNAi on the development of zebrafish. Although dsRNA-mediated posttranscriptional silencing appears to be an almost universal cellular mechanism and it should be possible to develop protocols to overcome these problems in principle, this has not yet been achieved, and morpholino injection is currently the method of choice to generate knockdown phenotypes in the zebrafish.

2.4 Parthenogenesis

As the fertilization of the eggs in zebrafish occurs outside the body of the female, the process can be manipulated experimentally, for instance, to induce the development of gynogenetic and androgenetic haploid or diploid embryos. The generation of individuals with a genome derived from only one parent (uniparental inheritance) can be useful for F_1 mutant screens as such screens require significantly fewer progeny to be raised in order to identify recessive mutations of interest than conventional diploid screens, which necessitate the production of an F_3 generation (see above). Moreover, gynogenetic and androgenetic individuals can be used to determine the frequencies of meiotic recombination events during gametogenesis in either females or males, as well as to study genomic imprinting and sex determination.

For an overview of the procedures employed to generate gyno- and androgenetic haploid and diploid embryos, see Fig. 8.

Generation and development of gynogenetic haploid embryos. Haploid embryos are generated by fertilizing eggs with sperm that has previously been inactivated by irradiation with UV light. UV irradiation destroys the sperms' DNA but does not affect their capability to activate the eggs. The developing embryos accordingly

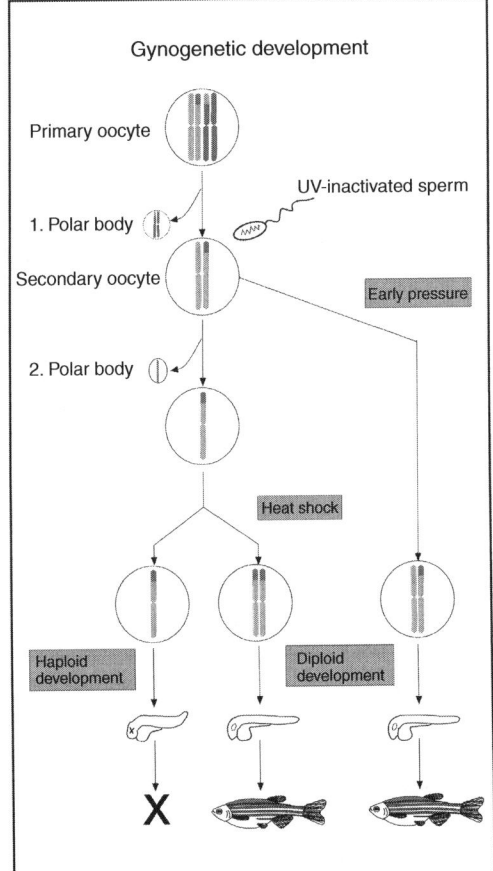
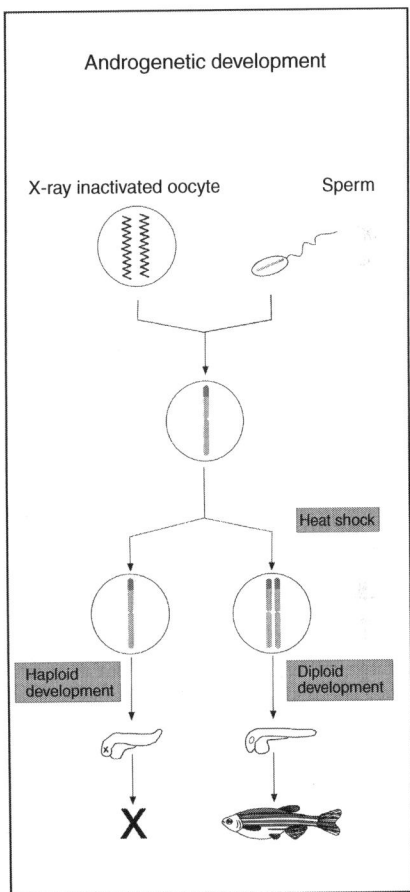

Fig. 8 Schematic representation of the methods used to generate gyno- and androgenetic haploid and diploid zebrafish embryos. For simplicity's sake, only one chromosome is shown. The colors (red and blue) denote the parental origin (maternal and paternal, respectively) of a chromosome/part of a chromosome in the oocyte or the sperm. The presence of two colors in a single chromosome arm indicates a recombination event(s) between the maternal and the paternal chromosome that occurred in the first meiosis leading to the germ cells. The X in the lower row signifies that the respective fish are not viable. For details on the procedures see text. (See color plate p. xiv).

are haploid, lacking the paternal genetic complement. In consequence, any mutation(s) present in the maternal genome will manifest themselves in the developing haploid progeny. This can be exploited, for example, to directly screen the offspring of mutagenized females or of heterozygous F_1 females arising from the mating of a mutagenized male. If a mutation is present in both DNA strands of a locus (e.g. all insertions and all ENU mutations in the F_1), 50% of the progeny of these females will show the mutant phenotype. The phenotypes resulting from the induced mutations in the maternal genome are frequently similar to the phenotype in diploid individual homozygous for the mutation, although the haploid phenotypes tend to be stronger.

However, these advantages of haploid screens are counterbalanced by significant drawbacks. The viability of haploid embryos is significantly reduced when compared to diploid individuals. They complete embryogenesis but generally arrest quite early during larval development. Haploid embryos usually die by day four or five of development (none live to maturity). As a consequence, haploid screens are only really practical to uncover mutations that produce a phenotype within, approximately, the first three days of development. Moreover, haploid embryos typically display a set of background phenotypes, such as smaller cells, a shorter, thicker, and occasionally bent body axis, as well as defects in organogenesis (e.g. less distinct brain morphology, variable number of ears, defects in the cardiovascular system: swollen pericardial cavity, reduced circulation of blood) against which mutant phenotypes have to be identified. However, as these background phenotypes are relatively consistent, an experienced screener can often distinguish them from genuine phenotypes resulting from a mutation. Additionally, as with all genetic screens, the ratio of individuals showing a particular phenotype to those not showing the phenotype is a useful measure to decide. Nonetheless, the haploid background phenotypes may impede the observation of some phenotypes. Owing to these disadvantages, haploid screens are of only minor significance. However, they may prove valuable to identify mutations that affect early development with comparatively less effort. Once a mutation has been identified, the carrier fish can be bred to produce individuals diploid for the mutation.

Generation and development of gynogenetic diploid embryos. In contrast to haploid embryos, parthenogenetic diploid embryos display less severe developmental abnormalities, and can even survive to become fertile adults. This considerably facilitates the identification of mutant phenotypes and removes the restriction to day three of development up until which phenotypes can be identified in a haploid background. The generation and screening of gynogenetic diploid fish is particularly useful for the identification of maternal-effect mutations as well as of recessive mutations affecting adult traits.

Gynogenetic diploid embryos are generated by diploidizing haploid embryos produced through the fertilization of eggs with UV-inactivated sperm. Following the activation of an oocyte with UV-inactivated sperm, there are two phases during which there are two complements of the maternal genome in the zygote: prior to the completion of the second meiotic division (i.e. before the second polar body is extruded from the zygote) and later during the first mitosis (after the DNA has been replicated, but before zygote undergoes the first cleavage division). If either the extrusion of the second polar body or the first cleavage is prevented, the embryo will possess a diploid maternal complement (see Fig. 8).

Exposure of eggs in the metaphase of the second meiotic division to a high hydrostatic pressure causes the spindle to disintegrate and the two complements of chromosomes are not separated. As a consequence, there is no extrusion of the polar body, and the egg carries a diploid female complement, which is replicated and distributed to the daughter cells as a normal diploid zygotic complement would. This method to prevent the formation of the second polar body is known as *early pressure* (*EP*). The survival rate following a diploidization with EP typically ranges between 50 and 80%. The gynogenetic diploid fish are viable, develop normally, and become fertile but are not as robust generally as diploid fish with a paternal and a maternal complement.

Despite the fact that the genome of EP-induced gynogenetic diploid fish derives from a single complement, not all loci are identical (i.e. homozygous). The heterozygosity is due to the fact that the sister chromatids present in the oocyte in the second meiosis have already undergone recombination during the first meiosis. The probability of a given locus to be homozygous in an embryo depends on the distance of this locus from the centromere. Loci closely linked to the centromere will be homozygous in nearly all embryos (i.e. 50% of embryos are homozygous for one allele and the other 50% are homozygous for the other allele). The greater the distance of a locus from the centromere, however, the fewer embryos will be homozygous for this locus. As a consequence, EP-based mutagenesis screens show a bias toward the identification of loci nearer the centromeres. Moreover, the fact that the ratio of phenotypically wild-type to mutant embryos is not constant owing to the unpredictable occurrence of recombination events, which hampers the reliable distinction of mutant phenotypes from background phenotypes.

An alternative method for generating gynogenetic diploid embryos is through a heat shock. The separation of the chromosomes and the ensuing cytokinesis of the first mitotic division can be prevented by exposing the egg to a heat shock (generally for 2 min at a temperature of 41 °C starting at 13 min following fertilization of the eggs with UV-inactivated sperm). The efficiency of this procedure to produce diploid embryos is not as high as that usually obtained with EP. Generally, only between 10 and 20% of the individuals survive the procedure and grow up to become fertile adults. Following the heat shock, development continues normally with cleavage divisions. What would have been the second such cleavage is now the first and produces the first two daughter cells.

The outcome of a diploidization induced by heat shock is different from that resulting from exposure of the egg to EP. In embryos created through heat shock, the meioses of the oocyte have been completed, and one set of sister chromatids of the remaining female complement of chromosomes extruded in a polar body. As the diploid state of these embryos is achieved through the replication of a single set of chromatids (i.e. one double helix), the embryos are homozygous for all loci, and 50% of the progeny will carry any given locus in a homozygous state. This advantage of (heat shock) HS-induced gynogenetic diploidy is, however, counterbalanced by the lower survival rates when compared to EP-based methods.

Generation and development of androgenetic haploid and diploid embryos. Similar to the production of gynogenetic diploid embryos carrying only a maternal genetic contribution, androgenetic haploid and diploid embryos can be produced. In these cases, the DNA of the oocytes is destroyed by exposing them to X-rays or UV light, and they are subsequently fertilized with normal sperm. The efficiency of androgenetic haploid production ranges from 10 to 50%. These embryos can either be left to develop as haploid androgenetic embryos or they can be diploidized by preventing the first mitotic cleavage division by heat shock as described above.

The androgenetic haploid and the diploid embryos also show distinct phenotypes. Like gynogenetic haploid embryos, androgenetic haploid embryos display a characteristic set of phenotypes, including a shorter, broader body as well as smaller melanocytes, and their development arrests around day four. The fact that both androgenetic and gynogenetic haploid embryos display very similar phenotypes indicates that these are due to the haploid state of these fish and independent of the parental origin of the genome. Androgenetic diploid embryos exhibit a retarded embryonic and larval development compared to normal diploid embryos, but later catch up in size and develop into fertile adult fish. However, the irradiation of the eggs necessary for androgenesis frequently causes sufficient cellular damage to make the phenotypic screening of the resulting embryos very difficult.

2.5
Transgenesis

The expression of exogenous genes or endogenous genes under the control of modified enhancer/promoter elements is a commonly used way of studying the functions of a gene *in vivo*. There are techniques to achieve both transient and stable expressions of transgenes in the zebrafish. For transient expression studies, the DNA constructs are generally injected into fertilized eggs or early embryos. However, these transient expressions generally result in only half of the injected embryos showing an expression in the expected cell types (e.g. as compared to *in situ* hybridization patterns obtained for the gene). Moreover, the expression is frequently not found in all the cells of the expected cell type. In order to achieve an expression of a transgene that faithfully mimics the expression of an endogenous promoter in the zebrafish, a stable transgenic line has to be created.

To achieve a stable transgenic zebrafish line, several techniques have been developed. These include the infection with retroviral vectors, penetration with DNA-coated microprojectiles, and electroporation. However, the introduction of transgenes is commonly achieved by microinjection of plasmid DNA into the embryo at the one-cell stage. This technique is fast (over 100 eggs can be injected in an hour) and comparatively gentle on the embryos. Furthermore, it results in high rates of integration of up to 25% for transient transfections and between 5 and 30% for stable transgenic lines (with germ lines mosaic for the transgene; between 1 and 50% of the F_1 offspring of these transgenic fish are hemizygous for the transgene).

In recent years, several genetic tools have become available that allow a sophisticated control and analysis of the expression of transgenes in zebrafish. These include:

– Reporter genes, such as the green fluorescent protein (GFP) as well as its derivative cyan (CFP) and yellow (YFP) versions under the control of

tissue- or cell type–specific promoters and enhancers for promoter/enhancer studies or cell lineage experiments.
- Promoters (e.g. the hsp70 promoter) that can be activated by heat shock for analyses of the effect of a transgene at specific times during development.
- The Gal4-UAS system to efficiently drive the expression of transgenes.

Owing to the transparency of its embryos, the spatiotemporal expression of fluorescent transgenes, such as GFP, in the various organs and tissues can be particularly well observed and imaged. This possibility has been exploited in numerous studies and is currently being used in a GFP-based enhancer trap screen performed by the laboratory of T. Becker at the University of Bergen in Norway.

2.6
Microarrays

Microarray hybridization allows to study the expression of thousands of genes in parallel. Probes representing individual transcripts are spotted or synthesized on a glass slide or silicon chip, and hybridized with fluorescently labeled target cDNA from a mutant, knockdown, or other conditions of interest. Microarrays are either based on cDNA clones or oligonucleotides. For a species like zebrafish where a large amount of genome sequence has already been generated, oligonucleotide arrays are advantageous as cross-hybridization can be ruled out by selecting unique oligonucleotide sequences for the probes. Two sets of 16 399 65-mer and 14 240 50-mer zebrafish oligonucleotides, as well as a photolithographically synthesized array representing 14 900 transcripts, are commercially available for spotting. Microarrays have only recently been introduced in the zebrafish field but hold great promise to unravel regulatory pathways in development and to discover genes of interest that are not accessible by reverse genetics methods because of redundant functions.

3
Mapping of the Zebrafish Genome

3.1
Genetic Mapping

Genetic mapping of mutant loci in the zebrafish can be carried out either by half-tetrad analysis or by genome scanning, usually with microsatellite markers. The latter is generally the method of choice because it is technically less demanding and more accurate.

Four genetic maps have been generated for the zebrafish (HS, MGH, MOP, and GAT; cf. Table 2). The established reference map for the zebrafish is the meiotic MGH map produced by the laboratory of M. Fishman at the Massachusetts General Hospital, Boston. It is based on microsatellite markers (CA repeats) scored on a panel of 44 diploid F_2 fish from a cross between the strains AB and India. When the MGH map was published in 1998, it contained 705 markers covering 2350 centimorgans (cM) and was already complete, that is it had one linkage group for each of the 25 zebrafish chromosomes. Production of the map continued until 2001 when a total of 3845 markers was scored.

In order to place a zebrafish mutation on the MGH map, mutant carriers are crossed with a polymorphic reference line. F_1 brothers and sisters are mated, and phenotypically mutant F_2 fish and their wild-type siblings are collected. A genome scan is then carried out by amplifying a subset of the MGH markers both from

Tab. 2 Mapping panels for the zebrafish.

Name	Type	Markers placed as of February 2004	Laboratory
An integrated map of the zebrafish genome (ZMAP)	Computed	30 566 (1)	J. Postlethwait
Goodfellow T51 (T51)	Radiation Hybrid	14 989 (2)	R. Geisler, Y. Zhou
Loeb/NIH/5000/4000 (LN54)	Radiation Hybrid	4192 (1)	I. Dawid
Heat shock (HS)	Meiotic	3888 (1)	J. Postlethwait, W. Talbot
Boston MGH Cross (MGH)	Meiotic	3881 (1)	M. Fishman
Mother of Pearl (MOP)	Meiotic	676 (1)	J. Postlethwait
Gates et al (GAT)	Meiotic	383 (1)	W. Talbot

Sources: (1) http://zfin.org
(2) http://wwmap.tuebingen.mpg.de

a mutant and from a sibling pool and comparing the band intensities of the marker alleles between the pools (batched segregant analysis, Fig. 9). If a significant difference is detected, the marker in question is likely to be linked to the mutation. This potential linkage can be confirmed by repeating the PCR with the individual mutant and wild-type F_2 fish that were used for the pooling, and counting the recombinations. An unambiguous map position can be obtained by identifying linkage to additional markers. To date, over 400 zebrafish mutations have been mapped at a resolution of approximately 1 cM (600 kb) by genome scanning with a set of 192 microsatellite markers, mainly by the laboratory of R. Geisler. An alternative to this approach is the typing of single nucleotide polymorphism (SNP) markers by microarray hybridization, which has been demonstrated by W. Talbot. Microarrays offer the prospect of performing batched segregant analysis for a very large number of markers in parallel once sufficiently accurate detection methods are available.

Results of such rough mapping can be sufficient for cloning a mutation of interest if a suitable candidate gene can be identified, and if evidence such as a matching expression pattern or a morpholino-induced knockdown phenotype can be obtained for the gene. In the absence of good candidate genes or supporting evidence, however, the map position needs to be refined. This involves identifying the closest polymorphic MGH markers and collecting recombinants for these markers from a large number of individuals (typically 1000–2000 embryos, corresponding to 2000–4000 meioses). The recombinants are then used for mostly SNP-based mapping of genomic clones or expressed sequence tags (ESTs) from the region of the mutation. SNP are typically identified and typed by sequencing. Genome sequencing showed that SNPs within the Tü strain occur at an average frequency of 1/200 bp. In the vicinity of the mutation, recombinations should be lost. The region in which no recombinations are detected on either side is referred to as the critical region; when starting out with a sufficient number of recombinants, it can be reduced to a single contig of the zebrafish genome sequence, or even to a single genomic clone. Transcripts of the critical region are

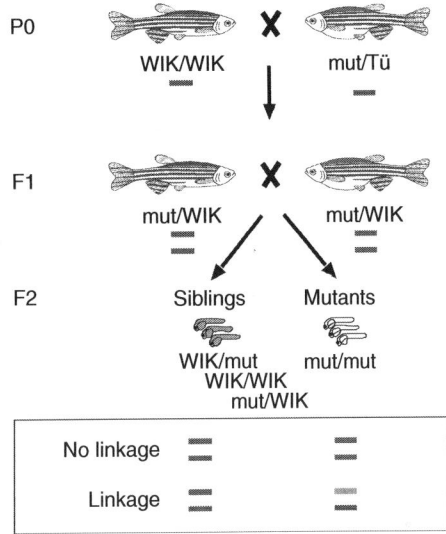

Fig. 9 Map cross and bulked segregant analysis of a recessive zebrafish mutation. WIK/WIK, reference fish; mut/Tü, fish carrying the mutation in Tü background. Band sizes and intensities of a representative SSLP marker are indicated schematically. In case of no linkage between the mutation and the marker, the intensities in the mutant and sibling pool are the same. In case of a linkage, the Tü band is stronger in the mutant pool and the WIK band in the sibling pool (Figure from Geisler, R. (2002) Mapping and Cloning, in: Nüsslein-Volhard, C., Dahm, R. (Eds.) *Zebrafish, A Practical Approach*, Oxford University Press, Oxford, UK, pp. 175–212.) (See color plate p. xiii).

identified and, if necessary, cloned based on annotation available at the ENSEMBL website, and then sequenced in order to find the mutation.

3.2
Radiation Hybrid Mapping

Mapping panels for radiation hybrid (RH) mapping are generated by fusing lethally irradiated cells of the species of interest with mouse or hamster cells. Hybrid cell lines are then grown, each of which retains a large number of random fragments of the species of interest, either inserted into the host chromosomes or as separate minichromosomes. Markers to be mapped are tested for their presence or absence in DNA of the cell lines by PCR. If two markers are present in many of the same RH lines (i.e. they have a similar RH score vector), they are likely to be close to each other in the genome. By ordering markers such that the ones with the most similar score vectors are next to each other, maps can be generated. Radiation hybrid mapping is thus similar to genetic mapping, with radiation-induced breaks in place of genetic recombination sites. Distances are given in centirays (1 cR indicates 1% breakage). The main advantages over genetic mapping are a higher resolution and the ability to map almost any sequence-tagged site (STS) marker without the need to identify a polymorphism (since RH lines are interspecies hybrids). For these reasons, RH mapping has become the method of choice for mapping of genes and genomic clones in the zebrafish.

The T51 radiation hybrid panel for the zebrafish was produced by the laboratory of P. Goodfellow. A map of the zebrafish genome based on the T51 panel containing 1451 STSs was generated by the laboratory of R. Geisler in 1999, demonstrating the feasibility of radiation hybrid mapping in a nonmammalian vertebrate. Since then the laboratories of R. Geisler and Y. Zhou have added several thousand STS markers to the map, mainly ESTs and BAC ends. As of January 2004, 14 989 markers could be placed on the T51 map, corresponding to an average marker spacing of 100 kb. The map is anchored on the meiotic MGH map by microsatellite markers scored on both panels.

The LN54 radiation hybrid panel for the zebrafish was produced by a consortium led by I. Dawid. It has a lower resolution than the T51 panel and was used for the mapping of 4192 markers. Mapping services are available for the RH maps; markers are either mapped on request or RH DNA is provided to interested researchers by the mapping labs, and the resulting data can be submitted online.

Data from these mapping efforts have been used for studies of the conserved synteny between zebrafish and human, which revealed that 80% of the genes belong to conserved synteny groups. Although a genome duplication is believed to have taken place in the teleost lineage, most human genes have only one ortholog in zebrafish, perhaps indicating an extensive loss of duplicated genes (and those paralog pairs that do exist in zebrafish are sufficiently diverged so that they do not pose a problem for mapping).

3.3
Physical Mapping

A physical map of even higher resolution than a radiation hybrid map can be obtained by fingerprinting of genomic clones with large inserts such as Bacterial Artificial Chromosomes (BACs, 100–200 kb insert size). This method identifies contigs made up of overlapping genomic clones. Overlaps are detected either by hybridization with oligonucleotide probes (oligonucleotide fingerprinting) or by restriction digest (restriction fingerprinting). In the former case, clones are deemed to overlap if they hybridize with the same probes, in the latter, if they share a sufficient number of restriction fragments of similar size. In the zebrafish, the oligonucleotide fingerprinting method was used only on chromosome 20, while restriction fingerprinting was carried out on the entire genome by the laboratories of R. Geisler, S. Humphray, and R. Plasterk. Four BAC libraries were constructed from the Tü line for the purposes of fingerprinting and sequencing. A total of 213 000 BACs from these four libraries were fingerprinted by restriction digest with the enzyme HindIII, resulting in 3775 contigs with a combined 21-fold genome coverage. Contig data are available online for use in the positional cloning of zebrafish mutations, and are also used by the Sanger Institute to select BACs for the clone-based sequencing of the zebrafish genome. Furthermore, the contigs are being anchored on the T51 radiation hybrid map by mapping of BAC ends in a collaboration of R. Geisler and Y. Zhou. Most of the contigs are expected to be anchored by the end of 2004. The resulting physical map of the zebrafish genome is used by the ENSEMBL project for assembling of the genome sequence.

Cytogenetics in the zebrafish is lagging behind since the 25 chromosome pairs are relatively small and similar in size and are therefore difficult to identify visually. No sex chromosomes are known. An alternative is offered by fluorescent *in situ* hybridization (FISH), but because of the relative ease of genetic mapping, this is not commonly carried out in zebrafish, and FISH data are available only for some chromosomes. As a consequence of this, the zebrafish community agreed in 2003 to number zebrafish chromosomes following the numbering of the genetic linkage groups (LGs), in contrast to other organisms where chromosomes are numbered by size.

3.4 Genome Sequencing

The sequencing of zebrafish transcripts is largely based on the EST approach. ESTs are obtained by single-pass sequencing of cDNA clones and offer rapid access to transcribed sequences as well as information on genetic polymorphism. In the zebrafish, EST sequencing has been carried out mostly by the WashU-Zebrafish Genome Resources Project led by S. L. Johnson, which has generated over 200 000 ESTs, while smaller numbers have been produced by the laboratories of C. C. Liew and J. Peng. The WashU ESTs have been and assembled in over 25 000 "WZ" clusters, which represent the majority of all zebrafish genes, while those of J. Peng have been assembled in "IMBC" clusters. In a separate effort, zebrafish "UniGene" clusters are being assembled by the NCBI. Complementing these efforts, the Zebrafish Gene Collection is a recent NIH-sponsored effort to generate full-length cDNA libraries and sequences. Curated information on zebrafish genes is available from NCBI's LocusLink service as well as from the ZFIN and ENSEMBL websites.

In February 2001, the Sanger Institute began a zebrafish genome project, which follows a combination of two strategies: sequencing of BAC clones selected from the physical map of the genome and whole-genome shotgun sequencing. The BACs are selected to represent a minimum tiling path across the fingerprint contigs. The genome size is estimated to be 1.5 to 1.6 Gb. As of February 2004, 1.0 Gb of BAC clone sequences (0.6-fold coverage) and 9.1 Gb of whole-genome shotgun sequences (5.7-fold coverage) have been generated. In addition, sequencing of BAC ends has been carried out both by the Sanger Institute and by the laboratory of S. Schuster. A finished genome sequence (with approximately 10-fold coverage) is expected to be available by the end of 2005.

Assembly of the zebrafish genome sequence has proved more difficult than in other organisms because of two factors:

1. There is a high degree of polymorphism that is visible in genome sequencing because both the BACs and the whole-genome shotgun library used for sequencing were made from sperm or embryos derived from several adult fish (due to constraints of the DNA isolation method employed).
2. Additionally, the zebrafish genome contains a high number of repeats. A repeat library was compiled by R. Waterman that contains repeats found more than 50 times and which covers approximately 40% of the genome. Masking of repeats is therefore essential when performing similarity searches or designing PCR primers for zebrafish sequences.

As of February 2004, the sequence was assembled into 62 895 supercontigs (which still contain short stretches of unknown sequence). These supercontigs are mostly based on FPC contigs, which are in turn anchored on the T51 RH map by BAC ends, thus providing rough chromosomal positions. The genome sequence is subsequently annotated by the ENSEMBL project of the Sanger Institute and the European Molecular Biology Laboratory and is accessible with annotations through a genome browser on the ENSEMBL website.

4 Resources on the Internet

4.1 General

The Zebrafish Information Network ⟨http://zfin.org/⟩

Community database and repository of genome data, which provides information on genes, maps, stocks, expression patterns, publications, researchers, and suppliers.

4.2 Mapping

MGH Zebrafish WWW Server ⟨http://zebrafish.mgh.harvard.edu/⟩

Homepage for the MGH genetic map.

Tübingen Map of the Zebrafish Genome ⟨http://wwwmap.tuebingen.mpg.de/⟩

Homepage for the T51 RH map (calculated with RH TSP Map software), mutant mapping, data submission to the T51 map.

The Children's Hospital Zebrafish Genome Project Initiatives

⟨http://zfrhmaps.tch.harvard.edu/ZonRHmapper/⟩

Alternate homepage for the T51 RH map (calculated with SAMapper), instant mapping service.

4.3 Sequencing

ENSEMBL Zebrafish Genome Browser ⟨http://www.ensembl.org/Danio_rerio/⟩

Assembled and annotated zebrafish genome sequence.

The Sanger Institute: Danio rerio Sequencing Project ⟨http://www.sanger.ac.uk/Projects/D_rerio/⟩

Information and links on genome sequencing, database of fingerprinting contigs, online RepeatMasker.

WashU-Zebrafish Genome Resources Project ⟨http://zfish.wustl.edu/⟩

Database of ESTs and EST clusters, BLAST server for EST clusters.

NCBI Zebrafish Genome Resources ⟨http://www.ncbi.nlm.nih.gov/genome/guide/zebrafish/index.html⟩

Homepage for NCBI resources and services including sequence databases, BLAST, and a map viewer.

Zebrafish Gene Collection ⟨http://zgc.nci.nih.gov/⟩

Full-length cDNA libraries and sequences.

4.4 Resource Centers

Zebrafish International Resource Center ⟨http://zfin.org/zf_info/stckctr/stckctr.html⟩

Wild-type and mutant zebrafish stocks, miscellaneous clones, antibodies and materials, zebrafish pathology, and health services.

RZPD German Resource Center for Genome Research ⟨http://www.rzpd.de/⟩

cDNA and genomic libraries, library pools and filters, and individual clones.

See also Molecular Basis of Genetics; Genomic Sequencing; Shotgun Sequencing (SGS); Transgenic Fish; Whole Genome Human Chromosome Physical Mapping.

Bibliography

Books and Reviews

Amemiya, C.T., Zhong, T.P., Silverman, G.A., Fishman, M.C., Zon, L.I. (1999) Zebrafish

YAC, BAC, and PAC genomic libraries, *Methods Cell Biol.* **60**, 235–258.

Amsterdam, A. (2003) Insertional mutagenesis in zebrafish, *Dev. Dyn.* **228**, 523–534.

Clack, J.A. (2002) *Gaining Ground – The Origin and Evolution of Tetrapods*, Indiana University Press, Indiana, IN, p. 369

Detrich, H.W., Westerfield, M., Zon, L.I. (Eds.) (1999a) *The Zebrafish: Biology*, Methods in Cell Biology Series, Vol. 59, Academic Press, San Diego, CA.

Detrich, H.W., Westerfield, M., Zon, L.I. (Eds.) (1999b) *The Zebrafish: Genetics and Genomics*, Methods in Cell Biology Series, Vol. 60, Academic Press, San Diego, CA.

Frohnhöfer, H.G. (2002) Table of Zebrafish Mutations, in: Nüsslein-Volhard, C., Dahm, R. (Eds.) *Zebrafish, A Practical Approach*, Oxford University Press, Oxford, UK, pp. 237–292.

Geisler, R. (2002) Mapping and Cloning, in: Nüsslein-Volhard, C., Dahm, R. (Eds.) *Zebrafish, A Practical Approach*, Oxford University Press, Oxford, UK, pp. 175–212.

Gilmour, D.T., Jessen, J.R., Lin, S. (2002) Manipulating Gene Expression in the Zebrafish, in: Nüsslein-Volhard, C., Dahm, R. (Eds.) *Zebrafish, A Practical Approach*, Oxford University Press, Oxford, UK, pp. 121–143.

Gong, Z. (1999) Zebrafish expressed sequence tags and their applications, *Methods Cell Biol.* **60**, 213–233.

Hannon, G.J. (2002) RNA interference, *Nature* **418**, 244–251.

Johnson, S.L., Zon, L.I. (1999) Genetic backgrounds and some standard stocks and strains used in zebrafish developmental biology and genetics, *Methods Cell Biol.* **60**, 357–359.

Kimmel, C.B. (1989) Genetics and early development of zebrafish, *Trends Genet.* **5**, 283–288.

Kwok, C., Critcher, R., Schmitt, K. (1999) Construction and characterization of zebrafish whole genome radiation hybrids, *Methods Cell Biol.* **60**, 287–302.

Nüsslein-Volhard, C., Dahm, R. (Eds.) (2002) *Zebrafish, A Practical Approach*, Practical Approach Series, Oxford University Press, Oxford, UK, p. 303.

Pelegri, F. (2002) Mutagenesis, in: Nüsslein-Volhard, C., Dahm, R. (Eds.) *Zebrafish, A Practical Approach*, Oxford University Press, Oxford, UK, pp. 145–174.

Pelegri, F., Schulte-Merker, S. (1999) A Gynogenesis-Based Screen for Maternal-Effect Genes in the Zebrafish, *Danio rerio*, in: Detrich, H.W., Westerfield, M., Zon, L.I. (Eds.) *The Zebrafish: Genetics and Genomics*, Methods in Cell Biology Series, Vol. 60, Academic Press, San Diego, CA, p. 1–20.

Sola, L., Gornung, E. (2001) Classical and molecular cytogenetics of the zebrafish, Danio rerio (Cyprinidae, Cypriniformes): an overview, *Genetica* **111**, 397–412.

Talbot, W.S., Schier, A.F. (1999) Positional cloning of mutated zebrafish genes, *Methods Cell Biol.* **60**, 259–286.

Till, B.J., Colbert, T., Tompa, R., Enns, L.C., Codomo, C.A., Johnson, J.E., Reynolds, S.H., Henikoff, J.G., Greene, E.A., Steine, M.N., Comai, L., Henikoff, S. (2003) High-throughput TILLING for functional genomics, *Methods Mol. Biol.* **236**, 205–220.

Walker, C. (1999) Haploid Screens and Gamma-Ray Mutagenesis, in: Detrich, H.W., Westerfield, M., Zon, L.I. (Eds.) *The Zebrafish: Genetics and Genomics*, Methods in Cell Biology Series, Vol. 60, Academic Press, San Diego, CA, pp. 44–70.

Primary Literature

Amsterdam, A., Burgess, S., Golling, G., Chen, W., Sun, Z., Townsend, K., Farrington, S., Haldi, M., Hopkins, N. (1999) A large-scale insertional mutagenesis screen in zebrafish, *Genes Dev.* **13**, 2713–2724.

Barbazuk, W.B., Korf, I., Kadavi, C., Heyen, J., Tate, S., Wun, E., Bedell, J.A., McPherson, J.D., Johnson, S.L. (2000) The syntenic relationship of the zebrafish and human genomes, *Genome Res.* **10**, 1351–1358.

Chen, W., Burgess, S., Golling, G., Amsterdam, A., Hopkins, N. (2002) High-throughput selection of retrovirus producer cell lines leads to markedly improved efficiency of germ line-transmissible insertions in zebra fish, *J. Virol.* **76**, 2192–2198.

Colbert, T., Till, B.J., Tompa, R., Reynolds, S., Steine, M.N., Yeung, A.T., McCallum, C.M., Comai, L., Henikoff, S. (2001) High-throughput screening for induced point mutations, *Plant Physiol.* **126**, 480–484.

Corley-Smith, G.E., Lim, C.J., Brandhorst, B.P. (1996) Production of androgenetic zebrafish (*Danio rerio*), *Genetics* **142**, 1265–1276.

Cresko, W.A., Yan, Y.L., Baltrus, D.A., Amores, A., Singer, A., Rodriguez-Mari, A., Postlethwait, J.H. (2003) Genome duplication, subfunction partitioning, and lineage divergence: Sox9 in stickleback and zebrafish, *Dev. Dyn.* **228**, 480–489.

Davidson, A.E., Balciunas, D., Mohn, D., Shaffer, J., Hermanson, S., Sivasubbu, S., Cliff, M.P., Hackett, P.B., Ekker, S.C. (2003) Efficient gene delivery and gene expression in zebrafish using the sleeping beauty transposon, *Dev. Biol.* **263**, 191–202.

Driever, W., Solnica-Krezel, L., Schier, A.F., Neuhauss, S.C., Malicki, J., Stemple, D.L., Stainier, D.Y., Zwartkruis, F., Abdelilah, S., Rangini, Z., Belak, J., Boggs, C. (1999) A genetic screen for mutations affecting embryogenesis in zebrafish, *Development* **123**, 37–46.

Ekker, M., Ye, F., Joly, L., Tellis, P., Chevrette, M. (1999) Zebrafish/mouse somatic cell hybrids for the characterization of the zebrafish genome, *Methods Cell Biol.* **60**, 303–321.

Felsenfeld, A.L., Walker, C., Westerfield, M., Kimmel, C., Streisinger, G. (1990) Mutations affecting skeletal muscle myofibril structure in the zebrafish, *Development* **108**, 443–459.

Fire, A., Xu, S., Montgomery, M.K., Kostas, S.A., Driver, S.E., Mello, C.C. (1998) Potent and specific genetic interference by double-stranded RNA in *Caenorhabditis elegans*, *Nature* **391**, 806–811.

Fürnzler, D., Her, H., Knapik, E.W., Clark, M., Lehrach, H., Postlethwait, J.H., Zon, L.I., Beier, D.R. (1998) Gene mapping in zebrafish using single-strand conformation polymorphism analysis, *Genomics* **51**, 216–222.

Gaiano, N., Amsterdam, A., Kawakami, K., Allende, M., Becker, T., Hopkins, N. (1996) Insertional mutagenesis and rapid cloning of essential genes in zebrafish, *Nature* **383**, 829–832.

Gates, M.A., Kim, L., Egan, E.S., Cardozo, T., Sirotkin, H.I., Dougan, S.T., Lashkari, D., Abagyan, R., Schier, A.F., Talbot, W.S. (1999) A genetic linkage map for zebrafish: comparative analysis and localization of genes and expressed sequences, *Genome Res.* **9**, 334–347.

Geisler, R., Rauch, G.-J., Baier, H., van Bebber, F., Broß, L., Davis, R.W., Dekens, M., Finger, K., Fricke, C., Gates, M.A., Geiger, H., Geiger-Rudolph, S., Gilmour, D., Glaser, S., Gnügge, L., Habeck, H., Hingst, K., Holley, S., Keenan, J., Kirn, A., Knaut, H., Lashkari, D., Maderspacher, F., Martyn, U., Neuhauss, S., Neumann, C., Nicolson, T., Pelegri, F., Ray, R., Rick, J., Roehl, H., Roeser, T., Schauerte, H.E., Schier, A.F., Schönberger, U., Schönthaler, H.-B., Schulte-Merker, S., Seydler, C., Talbot, W.S., Weiler, C., Nüsslein-Volhard, C., Haffter, P. (1999) A radiation hybrid map of the zebrafish genome, *Nat. Genet.* **23**, 86–89.

Goff, D.J., Galvin, K., Katz, H., Westerfield, M., Lander, E.S., Tabin, C.J. (1992) Identification of polymorphic simple sequence repeats in the genome of the zebrafish, *Genomics* **14**, 200–202.

Golling, G., Amsterdam, A., Sun, Z., Antonelli, M., Maldonado, E., Chen, W., Burgess, S., Haldi, M., Artzt, K., Farrington, S., Lin, S., Nissen, R., Hopkins, N. (2002) Insertional mutagenesis in zebrafish rapidly identifies genes essential for early vertebrate development, *Nat. Genet.* **31**, 135–140.

Gong, Z., Yan, T., Liao, J., Lee, S.E., He, J., Hew, C.L. (1997) Rapid identification and isolation of zebrafish cDNA clones, *Gene* **201**, 87–98.

Haffter, P., Granato, M., Brand, M., Mullins, M.C., Hammerschmidt, M., Kane, D.A., Odenthal, J., van Eeden, F.J., Jiang, Y.J., Heisenberg, C.P., Kelsh, R.N., Furutani-Seiki, M., Vogelsang, E., Beuchle, D., Schach, U., Fabian, C., Nüsslein-Volhard, C. (1996) The identification of genes with unique and essential functions in the development of the zebrafish, *Danio rerio*, *Development* **123**, 1–36.

Halpern, M.E., Ho, R.K., Walker, C., Kimmel, C.B. (1993) Induction of muscle pioneers and floor plate is distinguished by the zebrafish no tail mutation, *Cell* **75**, 99–111.

Hatta, K., Kimmel, C.B., Ho, R.K., Walker, C. (1991) The cyclops mutation blocks specification of the floor plate of the zebrafish central nervous system, *Nature* **350**, 339–341.

Henion, P.D., Raible, D.W., Beattie, C.E., Stoesser, K.L., Weston, J.A., Eisen, J.S. (1996) Screen for mutations affecting development of zebrafish neural crest, *Dev. Genet.* **18**, 11–17.

Johnson, S.L., Africa, D., Horne, S., Postlethwait, J.H. (1995) Half-tetrad analysis in zebrafish: mapping the *ros* mutation and the centromere of linkage group I, *Genetics* **139**, 1727–1735.

Johnson, S.L., Gates, M.A., Johnson, M., Talbot, W.S., Horne, S., Baik, K., Rude, S., Wong, J.R.,

Postlethwait, J.H. (1996) Centromere-linkage analysis and consolidation of the zebrafish genetic map, *Genetics* **142**, 1277–1288.

Kauffman, E.J., Gestl, E.E., Kim, D.J., Walker, C., Hite, J.M., Yan, G., Rogan, P.K., Johnson, S.L., Cheng, K.C. (1995) Microsatellite-centromere mapping in the zebrafish (Danio rerio), *Genomics* **30**, 337–341.

Kawakami, K., Amsterdam, A., Shimoda, N., Becker, T., Mugg, J., Shima, A., Hopkins, N. (2000) Proviral insertions in the zebrafish hagoromo gene, encoding an F-box/WD40-repeat protein, cause stripe pattern anomalies, *Curr. Biol.* **10**, 463–466.

Kelly, P.D., Chu, F., Woods, I.G., Ngo-Hazelett, P., Cardozo, T., Huang, H., Kimm, F., Liao, L., Yan, Y.L., Zhou, Y., Johnson, S.L., Abagyan, R., Schier, A.F., Koster, R.W., Fraser, S.E. (2001) Tracing transgene expression in living zebrafish embryos, *Dev. Biol.* **233**, 329–346.

Kimmel, C.B., Ballard, W.W., Kimmel, S.R., Ullmann, B., Schilling, T.F. (1995) Stages of embryonic development of the zebrafish, *Dev. Dyn.* **203**, 253–310.

Knapik, E.W., Goodman, A., Ekker, M., Chevrette, M., Delgado, J., Neuhauss, S., Shimoda, N., Driever, W., Fishman, M.C., Jacob, H.J. (1998) A microsatellite genetic linkage map for zebrafish (Danio rerio), *Nat. Genet.* **18**, 338–343.

Knapik, E.W., Goodman, A., Atkinson, O.S., Roberts, C.T., Shiozawa, M., Sim, C.U., Weksler-Zangen, S., Trolliet, M.R., Futrell, C., Innes, B.A., Koike, G., McLaughlin, M.G., Pierre, L., Simon, J.S., Vilallonga, E., Roy, M., Chiang, P.W., Fishman, M.C., Driever, W., Jacob, H.J. (1996) A reference cross DNA panel for zebrafish (Danio rerio) anchored with simple sequence length polymorphisms, *Development* **123**, 451–460.

Kurita, K., Burgess, S.M., Sakai, N. (2004) Transgenic zebrafish produced by retroviral infection of in vitro-cultured sperm, *Proc. Natl. Acad. Sci. U.S.A.* **101**, 1263–1267.

Kwok, C., Korn, R.M., Davis, M.E., Burt, D.W., Critcher, R., McCarthy, L., Paw, B.H., Zon, L.I., Goodfellow, P.N., Schmitt, K. (1998) Characterization of whole genome radiation hybrid mapping resources for non-mammalian vertebrates, *Nucleic Acids Res.* **26**, 3562–3566.

Li, Y.X., Farrell, M.J., Liu, R., Mohanty, N., Kirby, M.L. (2000) Double-stranded RNA injection produces null phenotypes in zebrafish, *Dev. Biol.* **217**, 394–405.

Lin, S., Gaiano, N., Culp, P., Burns, J.C., Friedmann, T., Yee, J.K., Hopkins, N. (1994) Integration and germ-line transmission of a pseudotyped retroviral vector in zebrafish, *Science* **265**, 666–669.

Lo, J., Lee, S., Xu, M., Liu, F., Ruan, H., Eun, A., He, Y., Ma, W., Wang, W., Wen, Z., Peng, J. (2003) 15,000 unique zebrafish EST clusters and their use in microarray for profiling gene expression patterns during embryogenesis, *Genome Res.* **13**, 455–466.

McPherson, J.D., Chevrette, M., Dawid, I.B., Johnson, S.L., Ekker, M. (1999) Radiation hybrid mapping of the zebrafish genome, *Proc. Natl. Acad. Sci. U.S.A.* **96**, 9745–9750.

Moens, C.B., Yan, Y.L., Appel, B., Force, A.G., Kimmel, C.B. (1996) valentino: a zebrafish gene required for normal hindbrain segmentation, *Development* **122**, 3981–3990.

Mohideen, M.A., Moore, J.L., Cheng, K.C. (2000) Centromere-linked microsatellite markers for linkage groups 3, 4, 6, 7, 13, and 20 of zebrafish (Danio rerio), *Genomics* **67**, 102–106.

Molina, A., Di Martino, E., Martial, J.A., Muller, M. (2001) Heat shock stimulation of a tilapia heat shock protein 70 promoter is mediated by a distal element, *Biochem. J.* **356**, 353–359.

Nasevicius, A., Ekker, S.C. (2000) Effective targeted gene 'knockdown' in zebrafish, *Nat. Genet.* **26**, 216–220.

Nechiporuk, A., Finney, J.E., Keating, M.T., Johnson, S.L. (1999) Assessment of polymorphism in zebrafish mapping strains, *Genome Res.* **9**, 1231–1238.

Oates, A.C., Bruce, A.E., Ho, R.K. (2000) Too much interference: injection of double-stranded RNA has nonspecific effects in the zebrafish embryo, *Dev. Biol.* **224**, 20–28.

Postlethwait, J.H., Talbot, W.S. (2000) Genetic linkage mapping of zebrafish genes and ESTs, *Genome Res.* **10**, 558–567.

Postlethwait, J.H., Woods, I.G., Ngo-Hazelett, P., Yan, Y.-L., Kelly, P.D., Chu, F., Huang, H., Hill-Force, A., William, S.T. (2000) Zebrafish comparative genomics and the origins of vertebrate chromosomes, *Genome Res.* **10**, 1890–1902.

Postlethwait, J.H., Yan, Y.L., Gates, M.A., Horne, S., Amores, A., Brownlie, A., Donovan, A., Egan, E.S., Force, A., Gong, Z., Goutel, C., Fritz, A., Kelsh, R., Knapik, E., Liao, E., Paw, B., Ransom, D., Singer, A., Thomson, M., Abduljabbar, T.S., Yelick, P., Beier, D., Joly, J.S., Larhammar, D., Rosa, F., Westerfield, M., Zon, L., Johnson, S., Talbot, W.S. (1998) Vertebrate genome evolution and the zebrafish gene map, *Nat. Genet.* **18**, 345–349.

Postlethwait, J., Johnson, S., Midson, C.N., Talbot, W.S., Gates, M., Ballinger, E.W., Africa, D., Andrews, R., Carl, T., Eisen, J.S., Horne, S., Kimmel, C.B., Hutchinson, M., Johnson, M., Rodriguez, A. (1994) A genetic linkage map for the zebrafish, *Science* **264**, 699–703.

Rawls, J.F., Frieda, M.R., McAdow, A.R., Gross, J.P., Clayton, C.M., Heyen, C.K., Johnson, S.L. (2003) Coupled mutagenesis screens and genetic mapping in zebrafish, *Genetics* **163**, 997–1009.

Raz, E., van Luenen, H.G., Schaerringer, B., Plasterk, R.H., Driever, W. (1998) Transposition of the nematode *Caenorhabditis elegans* Tc3 element in the zebrafish *Danio rerio*, *Curr. Biol.* **8**, 82–88.

Scheer, N., Camnos-Ortega, J.A. (1999) Use of the Gal4-UAS technique for targeted gene expression in the zebrafish, *Mech. Dev.* **80**, 153–158.

Stickney, H.L., Schmutz, J., Woods, I.G., Holtzer, C.C., Dickson, M.C., Kelly, P.D., Myers, R.M., Talbot, W.S. (2002) Rapid mapping of zebrafish mutations with SNPs and oligonucleotide microarrays, *Genome Res.* **12**, 1929–1934.

Streisinger, G., Walker, C., Dower, N., Knauber, D., Singer, F. (1981) Production of clones of homozygous diploid zebra fish (*Brachydanio rerio*), *Nature* **291**, 293–296.

Stuart, G.W., McMurray, J.V., Westerfield, M. (1988) Replication, integration and stable germ-line transmission of foreign sequences injected into early zebrafish embryos, *Development* **103**, 403–412.

Tawk, M., Tuil, D., Torrente, Y., Vriz, S., Paulin, D. (2002) High-efficiency gene transfer into adult fish: a new tool to study fin regeneration, *Genesis* **32**, 27–31.

Ton, C., Hwang, D.M., Dempsey, A.A., Tang, H.C., Yoon, J., Lim, M., Mably, J.D., Fishman, M.C., Liew, C.C. (2000) Identification, characterization, and mapping of expressed sequence tags from an embryonic zebrafish heart cDNA library, *Genome Res.* **10**, 1915–1927.

Ungar, A.R., Helde, K.A., Moon, R.T. (1998) Production of androgenetic haploids in zebrafish with ultraviolet light, *Mol. Mar. Biol. Biotech.* **7**, 320–326.

Wargelius, A., Ellingsen, S., Fjose, A. (1999) Double-stranded RNA induces specific developmental defects in zebrafish embryos, *Biochem. Biophys. Res. Commun.* **263**, 156–161.

Westerfield, M., Liu, D.W., Kimmel, C.B., Walker, C. (1990) Pathfinding and synapse formation in a zebrafish mutant lacking functional acetylcholine receptors, *Neuron* **4**, 867–874.

Wienholds, E., Schulte-Merker, S., Walderich, B., Plasterk, R.H. (2002) Target-selected inactivation of the zebrafish rag1 gene, *Science* **297**, 99–102.

Wienholds, E., van Eeden, F., Kosters, M., Mudde, J., Plasterk, R.H., Cuppen, E. (2003) Efficient target-selected mutagenesis in zebrafish, *Genome Res.* **13**, 2700–2707.

Woods, I.G., Kelly, P.D., Chu, F., Ngo-Hazelett, P., Yan, Y.L., Huang, H., Postlethwait, J.H., Talbot, W.S. (2000) A comparative map of the zebrafish genome, *Genome Res.* **10**, 1903–1914.

Xiao, T., Shoji, W., Zhou, W., Su, F., Kuwada, J.Y. (2003) Transmembrane sema4E guides branchiomotor axons to their targets in zebrafish, *J. Neurosci.* **23**, 4190–4198.

Yan, Y.L., Talbot, W.S., Egan, E.S., Postlethwait, J.H. (1998) Mutant rescue by BAC clone injection in zebrafish, *Genomics* **50**, 287–289.

Zardoya, R., Meyer, A. (1996) Evolutionary relationships of the coelacanth, lungfishes, and tetrapods based on the 28S ribosomal RNA gene, *Proc. Natl. Acad. Sci. U.S.A.* **93**, 5449–5454.

Zhao, Z., Cao, Y., Li, M., Meng, A. (2001) Double-stranded RNA injection produces nonspecific defects in zebrafish, *Dev. Biol.* **229**, 215–223.

18
Rat Genome (*Rattus norvegicus*)

Kim C. Worley and Preethi Gunaratne
Human Genome Sequencing Center, Baylor College of Medicine, Houston, TX, USA

1	**The Rat and the Rat Genome** 506	
2	**The Assembly Strategy and Results** 507	
3	**Features of the Rat Genome** 511	
3.1	Genome Size 511	
3.2	Telomeres and Centromeres 511	
3.3	Orthologous Chromosomal Segments and Large-scale Rearrangements 513	
3.4	Segmental Duplications 513	
3.5	Gains and Losses of DNA 515	
3.6	Substitution Rates 518	
3.7	G + C Content and CpG Islands 519	
3.8	Shift in Substitution Spectra between Mouse and Rat 519	
3.9	Evolutionary Hotspots 520	
3.10	Covariation of Evolutionary and Genomic Features 522	
4	**Evolution of Genes** 522	
4.1	Construction of Gene Set and Determination of Orthology 522	
4.2	Properties of Orthologous Genes 523	
4.3	Indels and Repeats in Protein-coding Sequences 524	
4.4	Transcription-associated Substitution and Asymmetry 525	
4.5	Conservation of Intronic Splice Signals 526	
4.6	Gene Duplications 526	
4.7	Conservation of Gene-regulatory Regions 529	
4.8	Pseudogenes and Gene Loss 529	
4.9	*In Situ* Loss of Rat Genes 530	
4.10	Noncoding RNA Genes 532	

Genomics and Genetics. Edited by Robert A. Meyers.
Copyright © 2007 Wiley-VCH Verlag GmbH & Co. KGaA, Weinheim
ISBN: 978-3-527-31609-0

5	**Evolution of Transposable Elements** 532	
5.1	LINE-1 Activity in the Rat Lineage 532	
5.2	Different Activity of SINEs in the Rat and Mouse Lineage 534	
5.3	Colocalization of SINEs in Rat and Mouse 535	
5.4	Endogenous Retroviruses and Derivatives 536	
5.5	Simple Repeats 536	
5.6	Prevalent, Medium-length Duplications in Rodents 537	
6	**Rat-specific Biology** 537	
6.1	Chemosensation 537	
6.2	α_{2u} Globulin Pheromones 538	
6.3	Detoxification 539	
6.4	Proteolysis 539	
7	**Human Disease Gene Orthologs in the Rat Genome** 539	
8	**Summary** 544	
	Acknowledgment 545	
	Bibliography 545	
	Books and Reviews 545	
	Primary Literature 545	

Keywords

BAC
A bacterial artificial chromosome, a large-insert (~200 kb) cloning vector for genomic sequences.

Contig
A contiguous set of overlapping segments of DNA.

Finished sequence
Complete, contiguous sequence generated with an accuracy of 1 error per 10 000 bp.

Genome
The entire complement of nuclear DNA in an individual or the representative sequence composite from several individuals.

Nonsynonymous
Nucleotide changes in the coding region of a gene that change the amino acid sequence of the translated protein. Some amino acid changes are more deleterious to

the protein function than others. The nonsynonymous substitution rate is the number of nonsynonymous substitutions per nonsynonymous site K_A.

ORF
Open reading frame – a sequence that translates without internal stop codons into a protein sequence.

Orthologous
Sequence regions in different organisms that originated from the same sequence in the last common ancestor of the organisms.

Scaffold
A set of contigs with sequence gaps between them that are linked by mate-pair information or marker information that may or may not give an estimated gap size.

Synonymous
Nucleotide changes in the coding region of a gene that do not change the amino acid sequence of the translated protein. The synonymous substitution rate is the number of synonymous substitutions per synonymous site K_S.

Whole-genome Shotgun (WGS)
Sequence generated randomly from a genome. Usually sheared genomic DNA is subcloned into a library in plasmid vectors, the library is then sampled and clones sequenced from both ends of the inserts.

The Brown Norway rat was the third mammalian genome to be sequenced. The three-way comparison of the human, rat, and mouse sequences resolves details of mammalian evolution and allows divergence events to be placed on different branches of the evolutionary tree. The comparison of the human to invertebrate and rodent genomes highlights the consequences of evolution over 1000 million years and 75 million years, while the comparison of the two rodent genomes describes changes that occurred in the 12 to 24 million years since the common ancestor of the rat and mouse.

A number of insights came from these comparisons:

The rat genome is 2.75 gigabases (Gb), smaller than human (2.9 Gb), and slightly larger than mouse (2.6 Gb). The three genomes encode similar numbers of genes. The majority of the genes have persisted without deletion or duplication and with well-conserved intron–exon structures. The exceptions are members of gene families that have expanded through gene duplication. Genes found in rat, but not mouse included genes producing pheromones, or involved in immunity, chemosensation, detoxification, or proteolysis.

Human genes known to be associated with disease have orthologs in the rat, but their rates of synonymous substitutions are significantly different from other genes.

Three percent of the rat genome is in large segmental duplication, primarily located near the centromeres. Expansions of major gene families are due to these genomic duplications.

About 40% of the rat genome aligns orthologously to mouse and human; this eutherian core of the genome contains the vast majority of exons and known regulatory elements (which comprise 1–2% of the genome). Only a portion of this core (5–6%) appears to be under selective constraint in rodents and primates, while the remainder appears to be evolving neutrally. Outside the eutherian core, the majority of the 30% of the rat genome that aligns only with mouse is rodent-specific repeats. More than half of the nonaligning sequence is rat-specific repeats.

More genomic changes have occurred in the rodent lineages than in the primate. Large rearrangements include approximately 250 rearrangements between a murid ancestor and human (the majority between the eutherian ancestor and the murid ancestor, and approximately 50 each from the murid ancestor to the rat and mouse). There is a threefold higher base substitution rate in neutral DNA along the rodent lineage than along the human lineage, with the rate on the rat branch 5 to 10% higher than along the mouse branch. Microdeletions are more frequent than microinsertions in both the rat and mouse branches. Most interestingly, there is a strong correlation between local rates of microinsertions and microdeletions, nucleotide substitutions, and transposable element insertions in the rat and mouse lineages, although the events occurred independently since the divergence of the two branches.

1
The Rat and the Rat Genome

The rat, although revered as the first sign in the Chinese zodiac and bearer of the Hindu god Ganesh, is a known carrier of over 70 diseases. Rats are involved in the transmission of several infectious diseases to man, including cholera, bubonic plague, typhus, leptospirosis, cowpox, and hantavirus infections. A major agricultural pest, rats and other rodents consume approximately one-fifth of the annual food harvest.

The laboratory rat (*Rattus norvegicus*) has contributed to human health by testing new drugs, and improving the understanding of essential nutrients and the pathobiology of human disease. It was the first mammal domesticated for scientific research (1828). The rat has been the model of choice for physiologists and nutritionists, and there are over 234 inbred strains developed to study genetic diseases.

The rat genome-sequencing project produced a draft sequence that, unlike mouse and human, would not ultimately be finished to remove all sequence gaps and produce a high base accuracy. For this reason, the quality of the draft was important. Although gaps remained, the overall sequence quality supported detailed analyses.

The sequence was generated from two inbred females, (BN/SsNHsd/Mcwi) by a network of centers led by the Human Genome Sequencing Center, Baylor College of Medicine. An international team representing over 20 groups contributed to the analysis reported and summarized here.

2
The Assembly Strategy and Results

The rat genome project used a combined WGS (whole-genome shotgun) with BAC clone strategy (Fig. 1). The project benefited from the logistically simpler WGS sequence generation and the local sequence assembly to resolve duplications afforded by BAC clones. The Atlas assembly suite, designed to combine these data sets, provided a BAC fisher to present localized combined BAC + WGS data to the public prior to the availability of the complete assembly.

Over 44 million DNA sequence reads were generated (Table 1). After removal of low-quality reads and vector contaminants, 36 million reads were used in the assembly where 34 million reads were retained. This was a sevenfold sequence coverage, with 60% of the reads from the WGS. Coverage estimates range from 7.3x (when estimated from the entire "trimmed" length of the sequence data) to 6.9x (when estimated from the sequence with quality of Phred20 or higher).

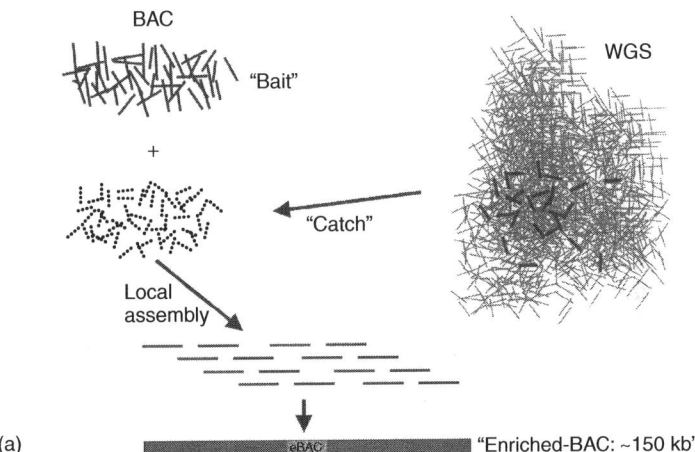

Fig. 1 The new "combined" sequence strategy and Atlas software.
(a) Formation of "eBACs." The strategy combined the advantages of both BAC and WGS sequence data. Modest sequence coverage (~1.8-fold) from a BAC is used as "bait" to "catch" WGS reads from the same region of the genome. These reads, and their mate pairs, are assembled using Phrap to form an enriched BAC or "eBAC." This stringent local assembly retains 95% of the "catch." (b) Creation of higher-order structures. Multiple eBACs are assembled into bactigs on the basis of sequence overlaps. The bactigs are joined into superbactigs by large clone mate-pair information (at least 2 links), extended into ultrabactigs using additional information (single links, FPC contigs, synteny, markers), and ultimately aligned to genome mapping data (RH and physical maps) to form the complete assembly. Used with permission from *Nature* **428**, 493–521 (2004).

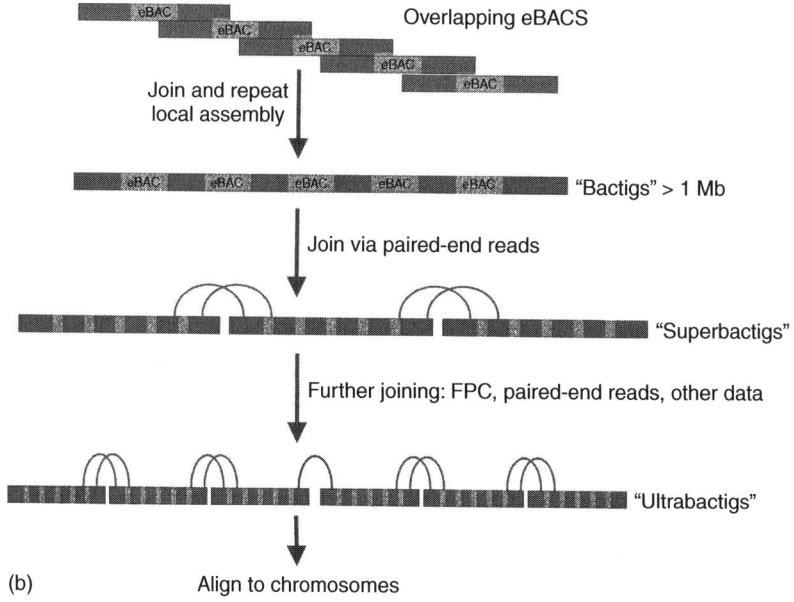

Fig. 1 (Continued)

Simultaneous to the sequencing, a "fingerprint contig" (FPC) map was developed, which was used in combination with the ongoing sequencing to identify BACs for sequence skimming. The parallel development of mapping and sequencing resources permitted the data-gathering phase of the project to be completed in less than two years.

The statistics of the rat draft assembly (v. 3.1) are given in Table 2. The current assembly (v. 3.4) splices in a number of finished BAC sequences to the v. 3.1 assembly. Much of the gene and protein feature analysis was developed using earlier assemblies (v. 2.0 and 2.1), while the genome description is based on v. 3.1.

The majority of the genome is in contigs larger than the expected mammalian gene (N50 = 38 kb). These contigs are linked into 783 larger scaffolds; those anchored to the radiation hybrid map had an N50 of 5.4 Mb, while the smaller scaffolds that could not be anchored had an N50 of 1.2 Mb.

The quality of the assembled sequence was assessed using sequence from finished BACs, the comparison showed that the bases within contigs were of high quality (1.32 mismatches per 10 kb), similar to the finished sequence. The majority of mismatches occurred at the ends of contigs in regions that average 750 bp and total <0.9% of the genome. Only six mismatches (insertions or deletions) were found within contigs when compared to 13 Mb of finished sequence or one per 2.2 Mb.

The assembly accuracy was judged in comparison to linkage and radiation hybrid maps. The majority of the genetic markers (13/3824) and sequence-tagged sites (96.9%) had consistent chromosome placement (Fig. 2). The maps are congruent with the assembly except for possible mismapped markers.

Tab. 1 Clones and reads used in the RGSP.

Insert size [kb][a]	Source or vector	Reads [millions]				Bases [billions]		Sequence coverage[b]		Clone coverage[c]
		All[d]	Used	Paired	Assembled	Trimmed	≥Phred20	Trimmed	≥Phred20	
2–4	Plasmid	9.6	8.6	7.4	7.9	4.8	4.5	1.8	1.6	3.70
4.5–7.5	Plasmid	4.5	4.3	3.6	3.6	2.4	2.3	0.87	0.82	2.96
10	Plasmid	8.4	7.2	6.4	6.4	4.1	3.8	1.5	1.4	11.63
50	Plasmid	1.7	1.3	1.0	1.1	0.69	0.65	0.25	0.24	9.47
150–250	BAC	0.32	0.31	0.26	0.26	0.18	0.16	0.07	0.06	9.26
Total WGS		24.5	21.7	18.7	19.2	12.1	11.3	4.4	4.1	37.0
2–5	BAC skims	19.6	14.6	13.2	14.5	8.0	7.7	2.9	2.8	4.8[e]
Total		44.1	36.3	31.9	33.7	20.2	19.0	7.3	6.9	41.8

[a] Grouped in ranges of sizes for individual libraries tracked to specific multiples of 0.5 kb.
[b] Total bases in used reads divided by sampled genome size including all cloned and sequenced euchromatic or heterochromatic regions.
[c] Estimated as sum of insert sizes divided by sampled genome size.
[d] WGS reads available on the NCBI Trace Archive as of March 21, 2003; BAC skim reads attempted at BCM–HGSC as of May 12, 2003; BAC end reads obtained directly from TIGR.
[e] Refers to coverage from 2–5 kb subclones from BACs. The BACs that were skimmed amounted to 1.58x clone coverage.

Tab. 2 Statistics of the RGSP draft sequence assembly.

Features[a]	Number	N50 Length [kb]	Bases [Gb]	Bases plus gaps [Gb][b]	Percentage of genome[c]			
					Sampled [2.78 Gb]		Assembled [2.75 Gb]	
					Bases	Bases + Gaps	Bases	Bases + Gaps
Anchored contigs	127 810	38	2.476	2.481	89.1	89.2	90.0	90.2
Anchored superbactig scaffolds	783	5 402	2.476	2.509	89.1	90.3	90.0	91.2
Anchored ultrabactigs	291	18 985	2.476	2.687	89.1	96.6	90.0	97.7
Unanchored superbactigs, main scaffolds	134	1210	0.056	0.062	2.0	2.2	2.0	2.3
Unanchored ultrabactigs	128	1529	0.056	0.069	2.0	2.5	2.0	2.5
All superbactigs, main scaffolds	917	5301	2.533	2.571	91.1	92.5	92.1	93.5
Minor scaffolds	4345	8	0.033	0.038	1.2	1.4	1.2	1.4

[a] Anchored sequences are those that can be placed on chromosomes because they contain known markers. The main scaffold for each superbactig is the largest set of contigs (in terms of total contig sequence), which can be ordered and oriented using mate-pair links and ordering of BACs. Scaffolds, which cannot be ordered and oriented with respect to the main scaffold, are termed *minor scaffolds*.
[b] Ambiguous bases (Ns) are counted in the gap sizes, and excluded in the base counts.
[c] Computed as bases plus gaps divided by estimated genome size. Sampled genome size is based on oligonucleotide frequency statistics of unassembled WGS reads. Assembled genome size is based on cumulative contig sequence following assembly.

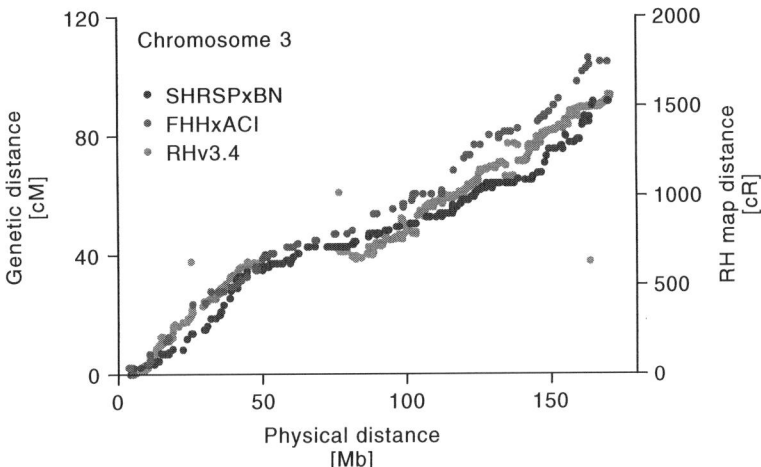

Fig. 2 Map correspondence. Correspondence between positions of markers on two genetic maps of the rat (SHRSPxBN intercross and FHHxACI intercross), on the rat radiation hybrid map, and their position on the rat genome assembly (Rnor3.1). Used with permission from *Nature* **428**, 493–521 (2004). (See color plate p. xv.)

3
Features of the Rat Genome

3.1
Genome Size

Genome assemblies are usually smaller than the actual genome size owing to underrepresentation of sequences due to cloning bias and sequencing and assembly difficulties. However, equating assembled genome size with the euchromatic, clonable portion of the genome (CpG) does not take into account the heterochromatic sequence included in the assembly. The rat genome size was estimated by two methods: scaling the assembled genome size by the fraction of features found in the assembly, and measuring the clonable or sampled genome size on the basis of the distribution of short oligomers in the WGS reads before assembly and in the assembly. Both estimates gave a relatively consistent measure of estimated genome size of 2.75 Gb. This conservative estimate was still considerably higher than the size estimated for the mouse draft genome sequence, which has different repeat content and appears to have underrepresented segmental duplications because of technical reasons.

3.2
Telomeres and Centromeres

The rat has both metacentric and telocentric chromosomes, unlike the wholly telocentric mouse chromosomes. As expected, the draft sequence does not contain complete telomere and centromere sequences. The approximate physical location of the centromeres relative to the genomic sequence is shown in Fig. 3. Several of the putative centromere positions coincide with segmental duplication blocks and classical satellite repeat clusters, consistent with enrichment of these sequence features in rat pericentromeric DNA. Human subtelomere regions are characterized by an abundance of segmentally duplicated DNA and an enrichment of internal (TTAGGG)n-like sequence islands.

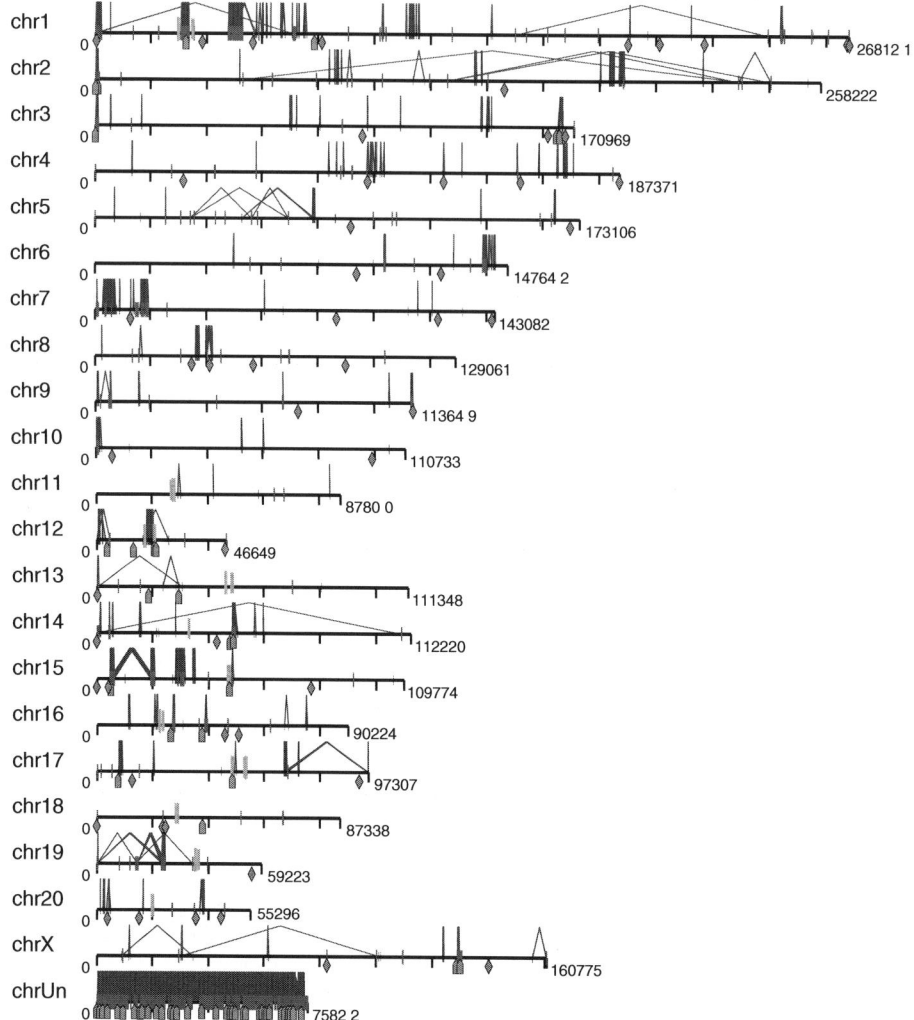

Fig. 3 Distribution of segmental duplications in the rat genome. Interchromosomal duplications (red) and intrachromosomal duplications (blue) are depicted for all duplications with ≥90% sequence identity and ≥20 kb length. The intrachromosomal duplications are drawn with connecting blue line segments; those with no apparent connectors are local duplications very closely spaced on the chromosome (below the resolution limit for the figure). P arms on the left and the q arms on the right. Chromosomes 2, 4–10, and X are telocentric; the assemblies begin with pericentric sequences of the q arms, and no centromeres are indicated. For the remaining chromosomes, the approximate centromere positions were estimated from the most proximal STS/gene marker to the p and q arm as determined by fluorescent *in situ* hybridization (FISH) (cyan vertical lines; no chromosome 3 data). The chrUn sequence is contigs not incorporated into any chromosomes. Green arrows indicate 1-Mb intervals with more than tenfold enrichment of classic rat satellite repeats within the assembly. Orange diamonds indicate 1-Mb intervals with more than tenfold enrichment of internal (TTAGGG)n-like sequences. For more details, see http://ratparalogy.cwru.edu. Used with permission from *Nature* **428**, 493–521 (2004). (See color plate p. xvi.)

Approximately one-third of the euchromatic rat subtelomeric regions contain similar features, suggesting that Rnor3.1 might extend very close to the chromosome ends.

3.3
Orthologous Chromosomal Segments and Large-scale Rearrangements

Multimegabase regions of chromosomes have been passed from the primate-rodent ancestor to human and murid rodent descendants with minimal rearrangements of gene order. These regions, bounded by the breaks that occurred during ancient large-scale chromosomal rearrangements are referred to as *orthologous chromosomal segments*. The sequence of these rearrangements was tentatively reconstructed using the human genome and other outgroup data (see Fig. 4). Inspection shows events that preceded and follow the rat–mouse divergence are interleaved. At 1-Mb resolution, multiple methods report virtually indistinguishable sets of orthologous chromosomes segments: 278 between human and rat, 280 between human and mouse, and 105 between rat and mouse. The larger numbers of breaks in orthologous segments between the human and the rodents is expected because of the greater evolutionary distance.

Understanding the number and timing of rearrangement events that have occurred in each of the three individual lineages (see tree in Fig. 5a) since the common primate-rodent ancestor required a more detailed analysis. The X chromosome is presented here as an example; its history is easier to trace completely since rearrangements between the X and the autosomes are rare. There are 16 human–mouse–rat orthologous segments of at least 300 kb in size (Fig. 6a). The most parsimonious scenario requires 15 inversions in the descent from the primate-rodent ancestor. Outgroup data from cat, cow, and dog resolve the timing of these rearrangements more precisely. Most of these events occurred in the rodent lineage: five (or four) before the divergence of rat and mouse, five in the rat lineage and five in the mouse lineage. At most one rearrangement occurred in the human lineage since divergence from the common ancestor with rodents. The analysis of the whole genome showed similar results. The assignment of the considerable rearrangement activity to the rodent branch following primate-rodent divergence is consistent with previous, lower resolution studies.

3.4
Segmental Duplications

Segmental duplications are defined here as genomic regions of at least 5 kb in length that are repeated with >90% identity remaining between the copies. The rat has approximately 2.9% of its bases in these duplicated regions, whereas the human has 5 to 6%, and the mouse has 1.0 to 2.0%. These duplicated structures are particularly challenging to assemble, so some of the mouse–rat difference is attributable to the BAC-based approach used in the rat assembly compared to the WGS-mouse approach. Most of these regions have less than 99.5% identity and are, therefore, not simply overlapping sequences that were not joined by the assembly program. Nearly 44% of these blocks of segmental duplications are mapped to the "unplaced" chromosome in Rnor3.1 indicating the difficulty of anchoring these elements to the genome.

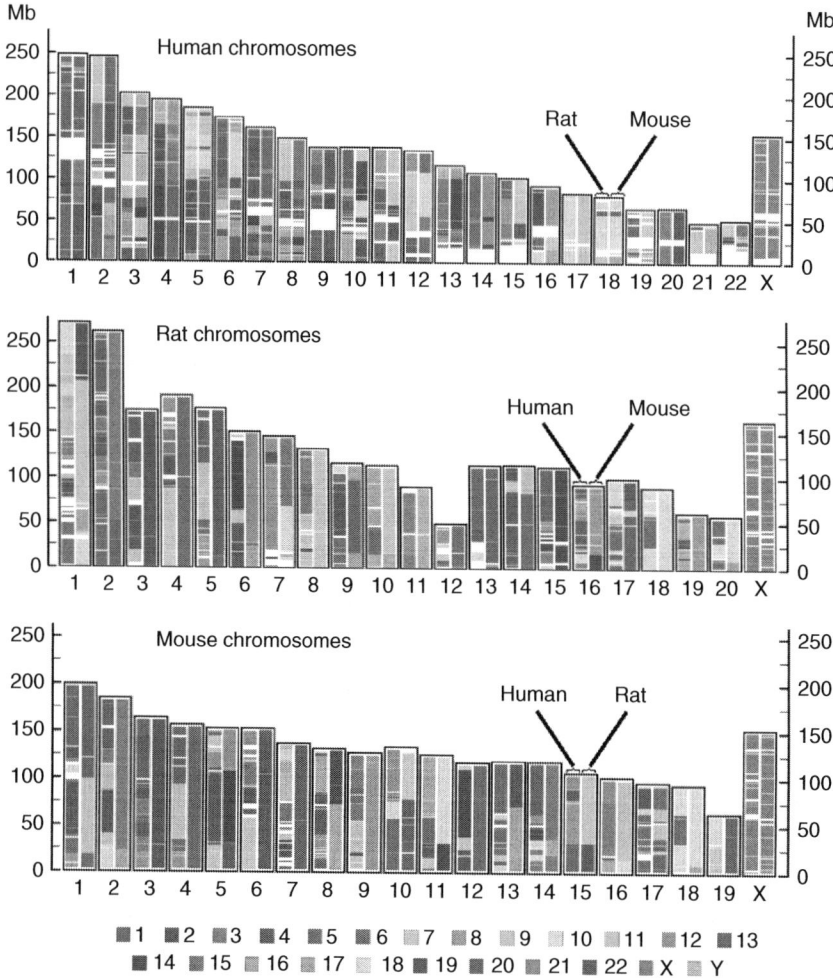

Fig. 4 Map of conserved synteny between the human, mouse, and rat genomes: For each species, each chromosome (x-axis) is a two column boxed pane (p-arm at the bottom) colored according to conserved synteny to chromosomes of the other two species. The same chromosome color code is used for all species (indicated below). For example, the first 30 Mb of mouse chromosome 15 is shown to be similar to part of human chromosome 5 (by the red in left column) and part of rat chromosome 2 (by the olive in right column). An interactive version is accessible (http://www.genboree.org). Used with permission from Nature **428**, 493–521 (2004). (See color plate p. xvii.)

Intrachromosomal duplications are three times more common than interchromosomal duplications and are significantly enriched near the telomeric and centromeric regions (Fig. 3). The pericentromeric accumulation of segmental duplications in the rat seems to be a general property of mammalian chromosome architecture.

There is considerable clustering of segmental duplications. For many of the largest clusters, the underlying sequence

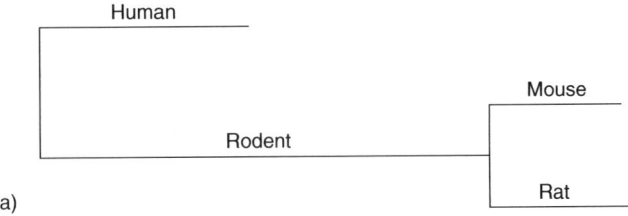

Fig. 5 Substitutions and microindels (1–10 bp) in the evolution of the human, mouse, and rat genomes: (a) The lengths of the labeled branches in the tree (top panel) are proportional to the number of substitutions per site inferred from all sites with aligned bases in all three genomes. (b) The table shows the midpoint and variation in these branch length estimates when estimated from different sequence alignment programs and different neutral sites, including sites from ancestral repeats, fourfold degenerate sites in codons, and rodent-specific sites ("in neutral sites only" row). Other rows give midpoints and variation for microindels on each branch of the tree. Used with permission from Nature **428**, 493–521 (2004).

alignments show a wide range of sequence identity, suggesting that duplication events have occurred continuously over millions of years. In contrast, an analysis of all duplicated regions showed a bimodal distribution consistent with bursts of segmental duplication (particularly intrachromosomal duplication) that occurred approximately 5 and 8 Myr ago.

The segmental duplications of the rat genome were of considerable interest because they represent an important mechanism for the generation of new genes. Sixty-three (of 4532 total) NCBI reference sequence genes were completely or partially located within duplicated regions. As discussed in the following, many of these genes present in multiple copies, belong to recently duplicated gene families, and contribute to distinctive elements of rat biology.

3.5 Gains and Losses of DNA

In addition to large rearrangements and segmental duplications, genome architecture is strongly influenced by insertion and deletion events that add and remove DNA over evolutionary time. To characterize the origins and losses of sequence elements in the human, mouse, and rat genomes, the nucleotide bases were categorized using alignment data and annotations of the insertions of repetitive elements (Fig. 7).

18 Rat Genome (Rattus norvegicus)

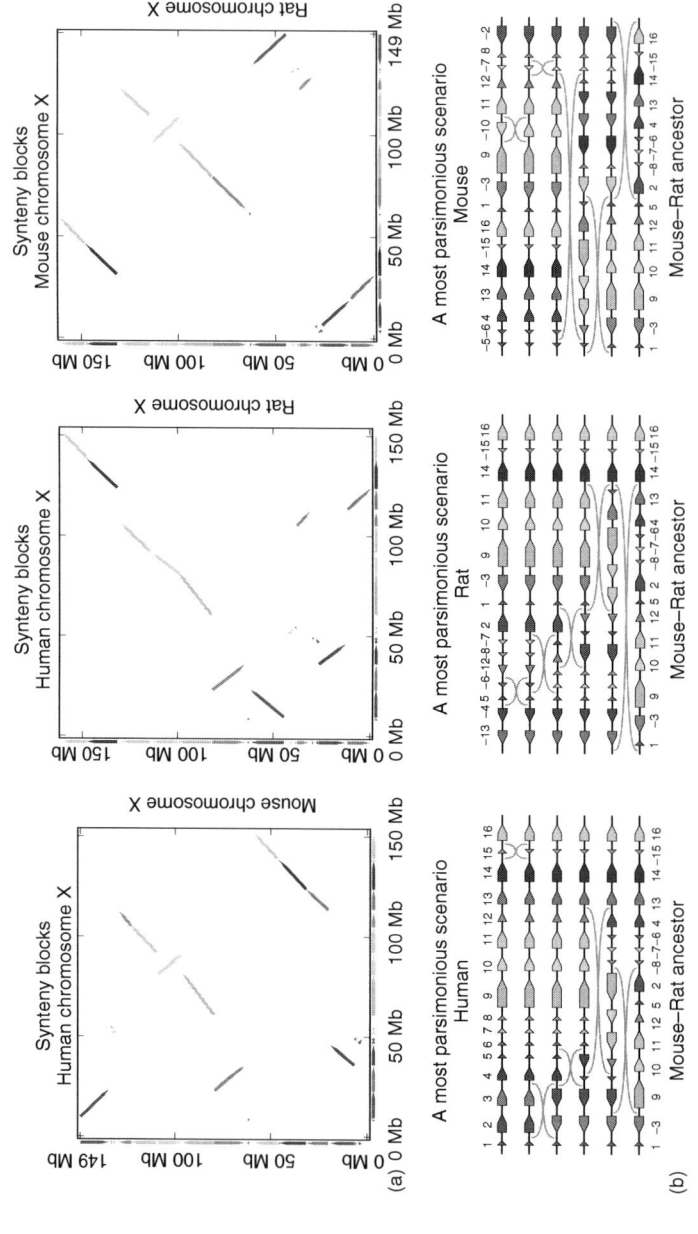

Fig. 6 X chromosome in each pair of species: (a) GRIMM–Synteny computes 16 three-way orthologous segments (≥300 kb) on the X chromosome of human, mouse, and rat, shown for each pair of species, using consistent colors. (b) The arrangement (order and orientation) of the 16 blocks implies that at least 15 rearrangement events occurred during X chromosome evolution of these species. The program MGR determined that evolutionary scenarios with 15 events are achievable and all have the same median ancestor (located at the last common mouse–rat ancestor). Shown is a possible (not unique) most parsimonious inversion scenario from each species to that ancestor. Note the last common ancestor of human, mouse, and rat should be on the evolutionary path between this median ancestor and human. Used with permission from Nature **428**, 493–521 (2004). (See color plate p. xxv.)

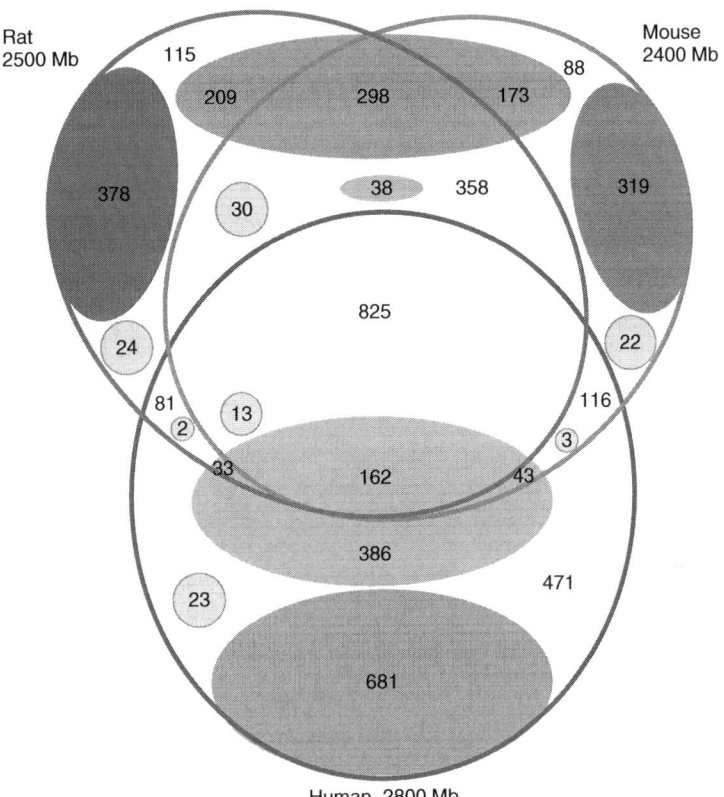

Fig. 7 Aligning portions and origins of sequences in rat, mouse, and human genomes. Each outlined ellipse is a genome, and the overlapping areas indicate the amount of sequence that aligns in all three species (rat, mouse, and human) or in only two species. Nonoverlapping regions represent sequences that do not align. Types of repeats classified by ancestry: those that predate the human–rodent divergence (gray), those that arose on the rodent lineage before the rat–mouse divergence (lavender), species-specific (orange for rat, green for mouse, blue for human), and simple (yellow), placed to illustrate the approximate amount of each type in each alignment category. Uncolored areas are nonrepetitive DNA – the bulk is assumed to be ancestral to the human–rodent divergence. Numbers of nucleotides (in Mb) are given for each sector (type of sequence and alignment category). Used with permission from *Nature* **428**, 493–521 (2004). (See color plate p. xviii.)

The estimates of the amount of repeats represent lower bounds because some repeats, especially the older ones, are not recognized even though the rodent repeat database was greatly expanded by analyzing the rat and mouse genomes.

About a billion nucleotides (39% of the euchromatic rat genome) align in all three species, constituting an "ancestral core," which is retained in these genomes. The ancestral core contains 94 to 95% of the known coding exons and regulatory regions. The levels of three-way conservation of "mammalian ancestral repeats" (transposon relics retained in all three species) confirm estimates that 5 to 6% of the human genome is accumulating substitutions more slowly than the neutral rate in the three lineages, and hence may be under purifying selection. In this constrained fraction, noncoding regions outnumber coding regions regardless of the constrain, an observation that supports comparative analyses on limited subsets of the genome. The preponderance of noncoding elements in the most constrained fraction of the genome underscores the likelihood that they play critical roles in mammalian biology.

About 700 Mb (28%) of the rat euchromatic genome aligns only with the mouse. At least 40% of this is rodent-specific repeats that are inserted on the branch from the primate-rodent ancestor to the murid ancestor. Some of the remainder is ancestral mammalian repeats or single copy DNA deleted in the human lineage but retained in rodents (Fig. 7). Although this 700 Mb of rodent-specific DNA is primarily neutral, it may also contain some functional elements lost in the human lineage in addition to sequences representing gains of rodent-specific functions, including some coding exons.

3.6
Substitution Rates

The alignment data allow relatively precise estimates of the rates of neutral substitutions and microindel events ($<=10$ bp). As in previous studies comparing human and mouse, both synonymous fourfold degenerate ("4D") sites in protein-coding regions and sites in mammalian ancestral repeats were used in this analysis. In addition, the primarily neutral rodent-specific sites that did not align to human discussed earlier were used.

Estimates for the neutral substitution level between the two rodents range from 0.15 to 0.20 substitutions per site, while estimates for the entire tree of human, mouse, and rat range from 0.52 to 0.65 substitutions per site (Fig. 5). This difference was predictable because of the evolutionary closeness of the two rodents. Less predictable was the finding that for all classes of neutral sites analyzed, the branch connecting the rat to the common rodent ancestor is 5 to 10% longer than the mouse branch. Thus, for as yet unknown reasons, the rat lineage has accumulated substantially more point substitutions than the mouse lineage since their last common ancestor.

Four-way alignments with sequence from orthologous ancestral repeats in the three genomes and the repeat consensus sequences (to approximate the progenitor repeat sequence) were used to distinguish substitutions on the branches from the primate-rodent ancestor to either the human or rodent ancestor. This revealed a three-to-one speed-up in rodent substitution rates relative to human, larger than estimated in the mouse analysis but consistent with more recent studies, which also use multiple sequence alignments.

Estimates for rates of microdeletion events are, for all branches, approximately twofold higher than rates of microinsertion (Fig. 5b), suggesting a fundamental difference in the mechanisms that generate these mutations. Furthermore, there are substantial lineage specific rate differences. The rat lineage has accumulated microdeletions more rapidly than the mouse, while the opposite holds true for microinsertions. As with substitutions, both microinsertions and microdeletion rates are substantially slower in the human lineage. The size distributions of microindels (1–10 bp) were heavily weighted toward the smallest indels and similar for insertions, deletions, and both the mouse and rat branches.

3.7
G + C Content and CpG Islands

The G + C content of the rat varies significantly across the genome (Fig. 8a) in a distribution that more closely resembles that of mouse than human. The variation of G + C content is coupled with differences in the distribution of CpG islands – short regions that are associated with the 5′ ends of genes and gene regulation and that escape the depletion of CpG dinucleotides owing to deamination of methylated cytosine. The 2.6-Gb rat genome assembly contains 15 975 CpG islands in nonrepetitive sequences of the genome. This is similar to the 15 500 CpG islands reported in the 2.5-Gb mouse genome, but far fewer than the 27 000 reported in the human genome. The distribution of CpG islands by chromosome is shown in Fig. 8b.

The changes in CpG island content predate the rat-mouse split and are consistent with the accelerated loss of CpG dinucleotides in rodents compared with humans. It remains possible, however, that the greater number of human regions with extremely high G + C content are due to distributional changes occurring in the primate, rather than the rodent lineage.

3.8
Shift in Substitution Spectra between Mouse and Rat

The nonrepetitive fraction of the rat genome is enriched for G + C content relative to the mouse genome by ∼0.35% over 1.3 billion nucleotides. This is a subtle but substantial difference that may be explained in part by differences in the spectra of mutation events that have accumulated in the mouse and rat lineages. The small minorities of substitutions where events can be assigned to either the mouse or the rat lineage by virtue of a nucleotide match in the alignment column between human and only one rodent were studied. Of the ∼117 million alignment columns meeting this criteria, ∼60 million involve a change in the rat lineage versus ∼57 million in the mouse, reflecting the increase in the rates of point substitution in the rat lineage (Fig. 5b). Fifty percent of these events in rat involve a substitution from an A/T to a G/C compared to only 47% of all mouse changes. The complementary change, G/C to A/T, is more common in the mouse lineage (38% vs. 35% in rat). There is not a similar substantial difference between changes that do not alter G + C content. In addition, this bias is not confined to particular transition or transversion events, nor can it be explained simply as a result of divergent substitution rates of CpG dinucleotides. Thus, this shift appears to be a general

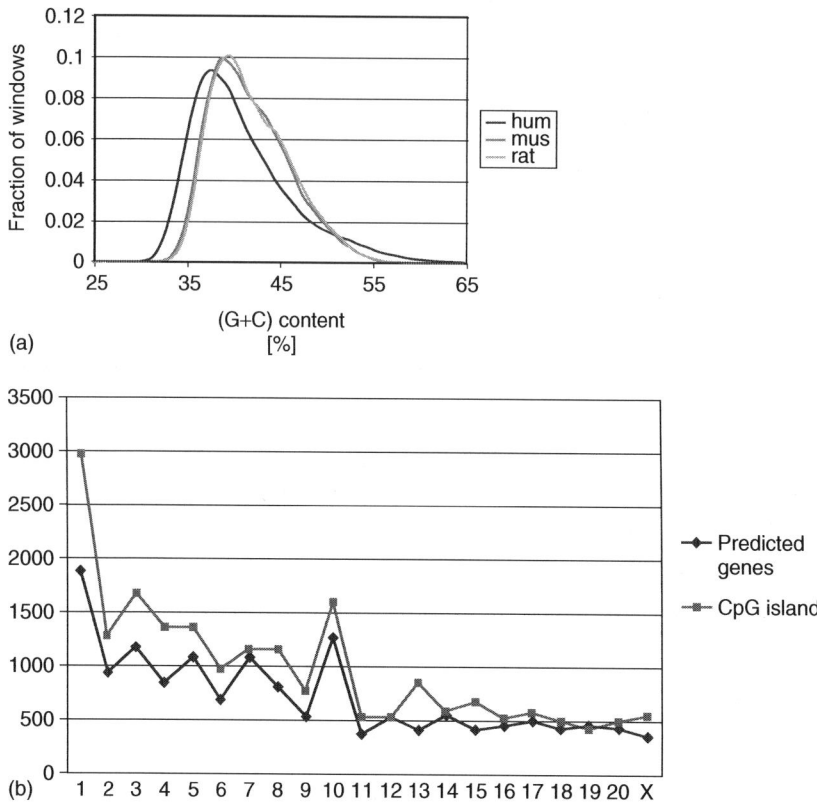

Fig. 8 Base composition distribution analysis. (a) The fraction of 20 kb nonoverlapping windows with a given G + C content is shown for human, mouse, and rat. (b) The number of Ensembl predicted genes per chromosome and the number of CpG islands per chromosome. The density of CpG islands averages 5.9 islands per Mb across chromosomes and 5.7 islands per Mb across the genome. Chromosome 1 has more CpG islands than other chromosomes, yet neither the island density nor ratio to predicted genes exceeds the normal distribution. The number of CpG islands per chromosome and the number of predicted genes are correlated ($R^2 = 0.96$). Used with permission from *Nature* **428**, 493–521 (2004). (See color plate p. xix.)

change that results in an increase in G + C content in the rat genome. Biochemical changes in the repair or replication enzymes might be responsible, and the observation that recombination rates are slightly higher in rat than in mouse may suggest a role for G + C biased mismatch repair. However, population genetic factors, such as selection, cannot be ruled out.

3.9 Evolutionary Hotspots

Comparison of the two rodent genomes, using human as outgroup, reveals regions that are conserved yet under different levels of constraint in mouse and rat. These regions may have distinct functional roles and contribute to species-specific differences. In 5055 regions >=100 bp, with at least a tenfold difference in the estimated number of substitutions per site on the mouse and rat branches and <0.25 substitutions per site on the human branch (to avoid fast-evolving regions), analysis found them enriched twofold in transcribed regions. Thirty-nine percent of mouse hot spots were found in 18% of the

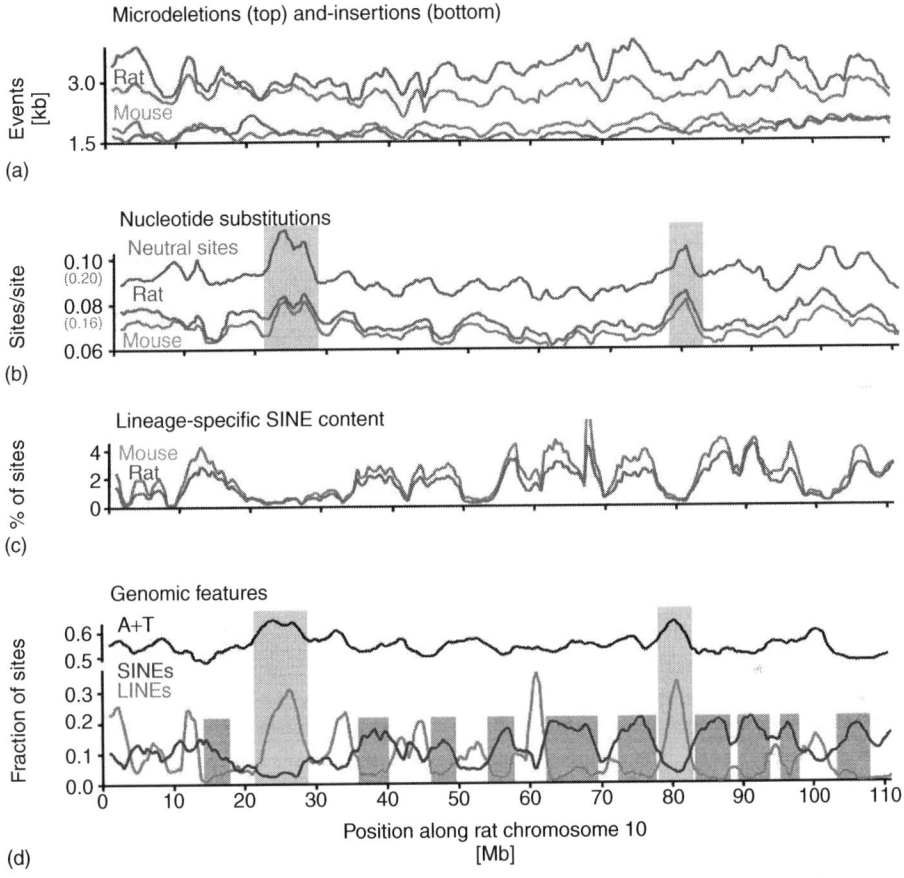

Fig. 9 Variability of several evolutionary and genomic features along rat chromosome 10. (a) Rates of microdeletion and microinsertion events (less than 11 bp) in the mouse and rat lineages since their last common ancestor, revealing regional correlations. (b) Rates of point substitution in the mouse and rat lineages. Red and green lines represent rates of substitution within each lineage estimated from sites common to human, mouse, and rat. Blue represents the neutral distance separating the rodents, as estimated from rodent-specific sites. Note the regional correlation among all three plots, despite being estimated in different lineages (mouse and rat) and from different sites (mammalian vs rodent-specific). (c) Density of SINEs inserted independently into the rat or mouse genomes after their last common ancestor. (d) A + T content of the rat, and density in the rat genome of LINEs and SINEs that originated since the last common ancestor of human, mouse, and rat. Pink boxes highlight regions of the chromosome in which substitution rates, AT content, and LINE density are correlated. Blue boxes highlight regions in which SINE density is high but LINE density is low. Used with permission from *Nature* **428**, 493–521 (2004). (See color plate p. xx.)

mouse genome covered by RefSeq genes; and 17% of rat hotspots were found in 8% of the rat genome covered by RefSeq genes. Similar numbers were also observed with coding exons and EST regions. Half of all hotspots in the mouse genome lie totally in noncoding regions.

3.10
Covariation of Evolutionary and Genomic Features

A high-resolution analysis of the genomic and evolutionary landscape of rat chromosome 10 uncovered strong correlations between certain microevolutionary features. Strongly correlated, in particular, are the local rates of microdeletion ($R^2 = 0.71$; Fig. 9a), microinsertion ($R^2 = 0.56$; Fig. 9a), and point substitution ($R^2 = 0.86$; Fig. 9b) between the two independent lineages of mouse and rat. In addition, microinsertion rates are correlated with microdeletion rates ($R^2 = 0.55$; Fig. 9a). These strong correlations are also observed in an independent genome-wide analysis, both on the original data and after factoring out the effects of $G + C$ content. Perhaps surprisingly, substantially less correlation is seen between microindel and point substitution rates (compare Figs. 9a and b).

The local point substitution rate in sites common to human, mouse, and rat strongly correlates with that in rodent-specific sites ($R^2 = 0.57$; Fig. 9b, blue line versus red/green). These two classes of sites, while interdigitated at the level of tens to thousands of bases, constitute sites that are otherwise evolutionarily independent. This result confirms that local rate variation is not solely determined by stochastic effects and extends, at high resolution, the previously documented regional correlation in rate between 4D sites and ancestral repeat sites.

4
Evolution of Genes

A substantial motivation for sequencing the rat genome was to study protein-coding genes. Besides being the first step in accurately defining the rat proteome, this fundamental data set yields insights into differences between the rat and other mammalian species with a complete genome sequence. Estimation of the rat gene content is possible because of relatively mature gene-prediction programs and rodent-transcript data. Mouse and human genome sequences also allow characterization of mutational events in proteins such as amino acid repeats and codon insertions and deletions. The quality of the rat sequence also allows us to distinguish between the functional genes and pseudogenes.

An estimated 90% of rat genes possess strict orthologs in both mouse and human genomes. Studies also identified genes arising from recent duplication events occurring only in rat, and not in mouse or human. These genes contribute characteristic features to rat-specific biology, including aspects of reproduction, immunity, and toxin metabolism. In contrast, almost all human "disease genes" have rat orthologs. This underscores the importance of the rat as a model organism in experimental science.

4.1
Construction of Gene Set and Determination of Orthology

The Ensembl gene-prediction pipeline has predicted 20 973 genes with 28 516 transcripts, and 205 623 exons in rat. These genes contain an average of 9.7 exons, with a median exon number of 6.0. At least 20% of the genes are alternatively spliced, with an average of 1.3 transcripts predicted per gene. Of the 17% single exon transcripts, 1355 contain frameshifts relative to the predicted protein and 1176 are likely processed pseudogenes. The majority of transcripts are supported by rodent (61%) or vertebrate (72%) transcript evidence, and have at least one untranslated region predicted (60%). The coding densities ranged from 1.2 to 2.2% and the coding fraction of RefSeq genes covered by these predictions ranged from 82 to 98%. The number of coding exons per gene and average exon length were similar in the three species. Differences were observed in intron length, with an average of 5338 bp in human, 4212 bp in mouse, and 5002 bp in rat.

4.2
Properties of Orthologous Genes

Orthology relationships were conservatively predicted and 12 440 rat genes showed clear, unambiguous 1:1 correspondence with a gene in the mouse genome. Accounting for potential errors 86 to 94% of the rat genes have 1:1 mouse orthologs. The remaining genes were associated with lineage-specific gene family expansions or contractions. Surprisingly, a similar proportion (89–90%) of rat genes possessed a single ortholog in the human genome, perhaps because of less resolution in the draft genomes. The majority of nucleotide changes within protein-coding regions that reflected synonymous or nonsynonymous substitutions yielded K_A/K_S ratios of less than 0.25 indicating purifying selection. Values of 1 suggest neutral evolution, and values greater than 1 indicate positive selection. Examination of ortholog pairs in orthologous genomic segments (Table 3) showed a slight increase in median K_S values, indicating that the rat lineage has more neutral substitutions in gene coding regions than the mouse lineage.

Rat genes shared with mouse, but with no counterparts in human are expected to reflect either a rapidly evolving gene set, or genes that may have arisen from noncoding DNA, or been converted to pseudogenes in the human lineage. Thirty-one ENSEMBL rat genes were collected that have no nonrodent homolog in current databases. These are twofold overrepresented among genes in paralogous gene clusters, and threefold overrepresented among genes whose proteins are likely to be secreted. This is consistent with observations that clusters of paralogous genes, and secreted proteins, evolve relatively rapidly.

Tab. 3 1:1 orthologous genes in human, mouse, and rat genomes[b].

	Human/Mouse	Human/Rat	Mouse/Rat
1:1 ortholog relationships	11 084	10 066	11 503
Median K_S values[a]	0.56 (0.39–0.80)	0.57 (0.40–0.82)	0.19 (0.13–0.26)
Median K_A/K_S values[a]	0.10 (0.03–0.24)	0.09 (0.03–0.21)	0.11 (0.03–0.28)
Median % amino acid identity[a]	88.0% (74.4–96.3%)	88.3% (75.9–96.4%)	95.0% (88.0–98.7%)
Median % nucleotide identity[a]	85.1% (77.4–90.0%)	85.1% (77.8–89.9%)	93.4% (89.2–—95.7%)

[a] Numbers in parentheses represent the 16th and 83rd percentiles.
[b] Data obtained from Ensembl, *Homo sapiens* version 11.31 (24 841 genes), *Mus musculus* version 10.3 (22 345 genes), *Rattus norvegicus* version 11.2 (21 022 genes).

The paucity of rodent-specific genes indicates that *de novo* invention of complete genes in rodents is rare. This is not unexpected, since the majority of eukaryotic protein-coding genes are modular structures containing coding and noncoding exons, splicing signals, and regulatory sequences; and the chances of independent evolution and successful assembly of these elements into a functional gene are small, given the relatively short evolutionary time available since the mouse–rat split. However, individual rodent-specific exons may arise more frequently, particularly if the exon is alternatively spliced. Of the 2302 potential novel rodent-specific exons, with transcript support, none matched human transcripts but approximately half (1116) appear to be present in alternative splice forms found in rodents. These exons are speculated to contain the few successful lineage-specific survivors of the constant process of gene evolution, by birth and death of individual exons.

4.3
Indels and Repeats in Protein-coding Sequences

In contrast to small indels occurring in the bulk of the genome, indels within protein-coding regions are likely lethal, or deleterious, and rapidly removed from the population by purifying selection. Indel rates within rat coding sequences were 50-fold lower than in bulk genomic DNA. The whole genome excess of deletions compared to insertions (Fig. 5b) was also evident in coding sequences. Deletions are ∼16% more likely than insertions to be removed from coding sequences by selection.

Owing to the triplet nature of the genetic code, indels of multiples of 3 nucleotides in length (3n indels) are less likely to be deleterious. Direct comparison of 3n indel rates between bulk DNA (0.77 indel/kb for mouse, 0.83 indel/kb for rat) and coding sequence (0.087 indel/kb for mouse and 0.084 indel/kb for rat) showed that 3n indels were ninefold underrepresented in coding sequences. At least 44% of indels were duplicative insertion or deletion of a tandemly duplicated sequence, collectively termed *sequence slippage*. Sequence slippage contributed approximately equally to observed insertions and deletions, the overall excess of deletions could be attributed specifically to an excess of nonslippage deletion over nonslippage insertion in both mouse and rat lineages. Of slippage indels, 13% were trinucleotide repeats known to be particularly prone to sequence slippage and encode homopolymeric amino acid tracts.

Other characteristics of amino acid repeat variations were searched to gain better understanding of dynamic changes in length of homopolymeric amino acid tracts on gene evolution and disease susceptibility. Most species-specific amino acid repeats (80–90%) were found in indel regions and regions encoding species-specific repeats were more likely to contain tandem trinucleotide repeats than those encoding conserved repeats. This was consistent with involvement of slippage in the generation of novel repeats in proteins and extended previous observations for glutamine repeats in a more limited human–mouse dataset.

The percentage of proteins containing amino acid repeats was 13.7% in rat, 14.9% in mouse, and 17.6% in human. The most frequently occurring tandem amino acid repeats were glutamic acid, proline, alanine, leucine, serine, glycine, glutamine, and lysine. Tandem trinucleotide repeats were significantly more abundant in human than in rodent coding sequences, in striking contrast to the frequencies observed in bulk genomic sequences (29 trinucleotide repeats/Mb in rat, 32 repeats/Mb in mouse and 13 repeats/Mb in human, Sect. 5.5). The conservation of human repeats was higher in mouse (52%) than in rat (46.5%), suggesting a higher rate of repeat loss in the rat lineage than in the mouse lineage.

Functional consequences of these inframe changes in rat, mouse, and human were investigated through clustering of proteins based on annotation of function and cellular localization, and mapping indels onto protein structural and sequence features. The rate that indels accumulated in secreted (3.9×10^{-4} indel/aa) and nuclear (4.0×10^{-4}) proteins is approximately twice that of cytoplasmic (2.4×10^{-4}) and mitochondrial (1.4×10^{-4}) proteins. Likewise, ligand-binding proteins acquire indels (3.1×10^{-4}) at a higher rate than enzymes (2.1×10^{-4}). These trends exactly mirror those observed for amino acid substitution rates, suggesting tight coupling of selective constraints between indels and substitutions. Transcription regulators showed the highest rate of indels (4.3×10^{-4}), a finding that may relate to the overrepresentation of homopolymorphic amino acid tracts in these proteins.

Known protein domains exhibited 3.3-fold fewer indels than expected by chance, again paralleling nucleotide substitution rate differences between domains and nondomain sequences. Transmembrane regions were refractory to accumulating indels, exhibiting a sixfold reduction compared to that expected by chance. Low-complexity regions were 3.1-fold enriched, reflecting their relatively unstructured nature and enrichment for indel prone trinucleotide repeats. Mapping of indels onto groups of known structures revealed indels are 21% more likely to be tolerated in loop regions than the structural core of the protein.

An interesting observation was indel frequency and amino acid repeat occurrence both correlated positively with G + C coding sequence content of the local sequence environment. This may in part be explained by the correlation of polymerase slippage-prone trinucleotide repeat sequences and G + C content. There is also a positive correlation between CpG dinucleotide frequency and coding sequence insertions, but not deletions. This effect diminishes rapidly with increasing distance from the site of the insertion.

4.4
Transcription-associated Substitution and Asymmetry

A significant strand asymmetry for neutral substitutions in transcribed regions has been reported. Within an intron the higher rate of A → G substitutions over that of T → C substitutions, together with a smaller excess of G → A over C → T substitutions, leads to an excess of G + T over C + A on the coding strand. The asymmetries are hypothesized to be a by-product of transcription-coupled repair in germline cells. Examining the 3-way alignments of rat, mouse, and human verified that the strand asymmetries for neutral substitutions exist in introns across the rat genome (Table 4). These asymmetries are also seen if the study is limited to ancestral repeat sites, excludes ancestral repeat sites, excludes CpG dinucleotides, is limited to positions flanked by sites that are identical in the aligned sequences (in the case of observations 2 and 3 in Table 4), or considers introns of RefSeq genes for human or mouse. Thus, it appears that strand asymmetry of substitution events

Tab. 4 Strand asymmetry of substitutions in introns of rat genes.

1	Base frequencies on coding strand (G + T)/(C + A)	Rat genome 1.060	
2	Ratio of purine transitions to pyrimidine transitions rate(A ↔ G)/rate(C ↔ T)	Rat–Mouse 1.036	Rat–Human 1.036
3	Rate of transitions rate(A → G)/rate(T → C) rate(G → A)/rate(C → T)	Rat 1.058 1.017	Mouse 1.091 1.00*

Data in (1) were computed from the rat genome, those in (2) were computed from pairwise alignments, and data in (3) were computed from 3-way alignments. All values except * were highly significant (p-values $<10^{-4}$).

within transcribed regions of the genome is a robust genome-wide phenomenon.

4.5 Conservation of Intronic Splice Signals

The dynamics of evolution of consensus splice signals in mammalian genes was examined using 6352 human–mouse–rat orthologous introns from 976 genes. Intron class is extremely well conserved: no conversion between U2 and U12 introns, nor switching within U12 introns between the major AT–AC and GT–AG subtypes was observed, although U12 switching has been documented at larger evolutionary distances. In contrast, conversions between canonical GT–AG and noncanonical GC–AG subtypes of U2 introns are not uncommon. Only ~70% of GC–AG introns are conserved between human and mouse/rat, and only 90% are conserved between mouse and rat. Using human as outgroup, we detected 9 GT to GC conversions after divergence of mouse and rat (from 6282 introns likely to be GT–AG prior to human and rodents split), and 2 GC to GT conversions (from 34 GC–AG introns likely to predate human and rodent split). Given the higher rate of conversion from GT to GC than the reverse, these results give some indication of the degree to which mutation from T to C is tolerated in donor sites. This substitution appears to be better tolerated in introns with very strong donor sites, since in these introns the proportion of GC donor sites is ~11%, which is much higher than the 0.7% overall frequency of GC donor sites in U2 introns. Very few other noncanonical configurations in U2 introns are conserved, suggesting that most correspond to transient, evolutionarily unstable states, pseudogenes, or misannotations.

4.6 Gene Duplications

Duplication of genomic segments represents a frequent and robust mechanism for generating new genes. Since there were no compelling data showing rat-specific genes arising directly from noncoding sequences, gene duplications were examined to measure their potential contribution to rat-specific biology. A previous study showed such gene clusters in mouse without counterparts in human are subject to rapid, adaptive evolution. Using methods that directly identified paralogous clusters found 784 rat paralog clusters containing 3089 genes. This was lower than in mouse (910 clusters per 3784 genes), but the difference probably reflects the larger number of gene predictions from the mouse assembly. Using methods that analyzed genomic segmental duplications, it appears that the timing of expansion of these individual families, is reflected in local gene duplication and retention within clusters. Neutral substitution rate varies among orthologs by approximately twofold (Fig. 10). Rates of change among ancestral gene duplications (those that predate the mouse/rat split) were relatively constant. Mouse-specific and rat-specific duplications occurred at similar rates, except for those with $K_S < 0.04$ that are reduced in mouse-specific duplications (Fig. 10), though this may be accounted for by different assembly methods.

The rat paralog pairs that probably arose after the rat/mouse split (12–24 Myr ago) have a K_S value of ≥ 0.2 (Table 3). Six hundred and forty nine $K_S < 0.2$ gene duplication events were found in rat, a lower number than found in mouse (755). For both rodents, this represents a likelihood of a gene duplicating between 1.3×10^{-3} and 2.6×10^{-3} every Myr. This is consistent with a previous estimate for Drosophila genes, and an order of magnitude lower than an estimate for *Caenorhabditis elegans* genes.

Immunoglobulin, T-cell receptor α-chain, and α_{2u} globulin genes appear to be duplicating at the fastest rates in the rat genome (Table 5). Since divergence with mouse, these rat clusters have increased gene content severalfold. This recapitulates previous observations that rapidly evolving and duplicating genes are overrepresented in olfaction and odorant-detection, antigen

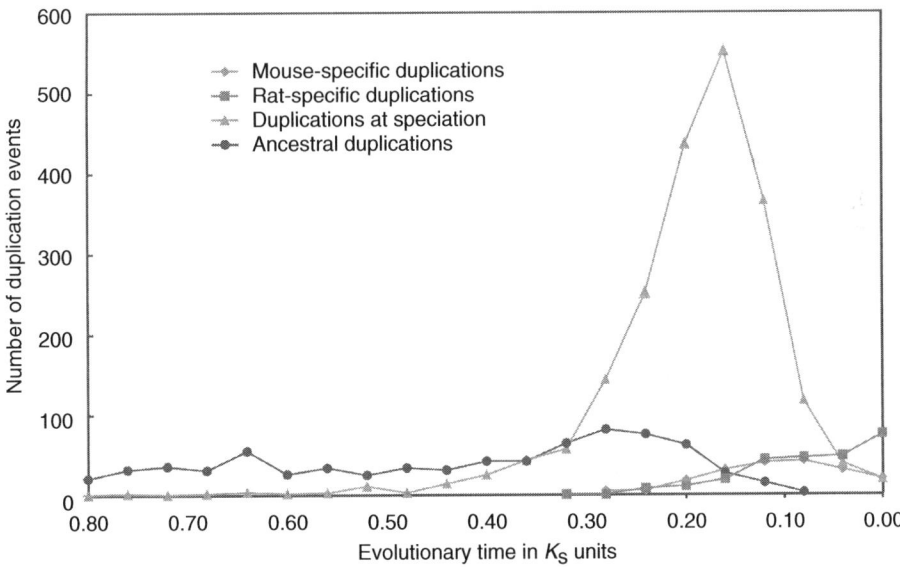

Fig. 10 Variation in the frequency of gene duplications during the evolutionary histories of the rat and mouse. The sequence of gene duplication events was inferred from phylogenetic trees determined from pairwise estimates of genetic divergence under neutral selection (K_S). The median K_S value for mouse:rat 1:1 orthologs is 0.19; this value corresponds to the divergence time of the mouse and rat lineages. Used with permission from *Nature* **428**, 493–521 (2004).

Tab. 5 Recent gene duplications ($K_S < 0.2$) in the rat lineage. Duplications involving retroviral genes, fragmented genes with internal repeats, and likely pseudogene clusters were removed from this list. Only gene clusters exhibiting at least 3 duplications are shown.

Cluster ID/ Chrom.	Recent duplication events	Numbers of genes involved	Extant/ Ancestral cluster size	Annotation	Process
249/4	38	53	60/22	Immunoglobulin κ-chain V	Immunity
640/15	38	47	53/15	TCR α-chain V	Immunity
346/6	25	35	44/15	Immunoglobulin heavy chain V	Immunity
190/3	22	42	168/146	Olfactory receptor	Chemosensation
578/13	16	28	59/43	Olfactory receptor	Chemosensation
400/8	15	26	82/67	Olfactory receptor	Chemosensation
743/20	15	21	37/22	Olfactory receptor	Chemosensation
72/1	12	22	102/90	Olfactory receptor	Chemosensation
500/10	12	18	32/20	Olfactory receptor	Chemosensation
51/1	6	7	16/10	Glandular kallikrein	Reproduction?
256/4	6	8	10/4	Vomeronasal receptor V1R	Chemosensation
488/10	6	10	11/5	Olfactory receptor	Chemosensation
644/15	6	10	14/8	Granzyme serine protease	Immunity
4/1	5	6	9/4	Trace amine receptor, GPCR	Neuropeptide receptors?
248/4	5	9	15/10	Vomeronasal receptor V1R	Chemosensation
393/8	5	10	31/26	Olfactory receptor	Chemosensation
522/10	5	8	19/14	Keratin-associated protein	Epithelial cell function
550/11	5	8	17/12	Olfactory receptor	Chemosensation
635/15	5	9	20/15	Olfactory receptor	Chemosensation
79/1	4	8	38/34	Olfactory receptor	Chemosensation
88/1	4	6	11/7	Olfactory receptor	Chemosensation
109/1	4	7	43/39	Olfactory receptor	Chemosensation
294/5	4	5	5/1	α_{2u} globulin	Chemosensation
310/5	4	5	11/7	Olfactory receptor	Chemosensation
353/7	4	7	13/9	Olfactory receptor	Chemosensation
399/8	4	5	6/2	Ly6-like urinary protein	Chemosensation?
638/15	4	6	6/2	Ribonuclease A	Immunity
690/17	4	6	21/17	Prolactin paralog	Reproduction
239/4	3	6	6/3	Prolactin-induced protein	Reproduction
253/4	3	4	5/2	Camello-like N-acetyltransferase	Developmental regulator
274/4	3	6	20/17	Ly-49 lectin natural killer cell protein	Immunity
297/5	3	4	5/2	Interferon-α	Immunity
523/10	3	4	6/3	Keratin-associated protein	Epithelial cell function
746/20	3	5	6/3	MHC class 1b (M10)	Chemosensation

recognition, and reproduction. An examination of duplicated genomic segments showed this enrichment for most of the same genes and also elements involved in foreign compound detoxification (cytochrome P450 and carboxylesterase genes). Together these are exciting findings because each of these categories can easily be associated with a familiar feature of rat-specific biology, and further investigation could explain some differences between rats and their evolutionary neighbors.

4.7
Conservation of Gene-regulatory Regions

As the third mammal to be fully sequenced, the rat can add significantly to the utility of nucleotide alignments for identifying conserved noncoding sequences. This power increases roughly as a function of total amount of neutral substitution represented in the alignment, and rat adds about 15% to the human-mouse comparison (Fig. 5). Many conserved mammalian noncoding sequences are expected to have regulatory function, and can be predicted using further analyses based upon these alignments.

Typical genome-wide human–mouse–rat alignments show strong conservation for a coding exon, as well as for several noncoding regions (Fig. 11). For example, the intronic region, in Fig. 11 contains 504 bp that are highly conserved in human, mouse, and rat. The last 100 bp of this alignment block are identical in all three species. Peaks in regulatory potential score are correlated with conservation score, and in the highly conserved intronic segment, they are higher for the three-way regulatory potential score than for the two-way scores using human and just one rodent. These data are illustrative, but form the foundation of ongoing efforts to identify genome sequences involved in gene regulation.

Requiring conservation among mammalian genomes greatly increases specificity of predictions of transcription factor binding sites. Transcription factor databases such as TRANSFAC contain known transcription factor binding sites and some knowledge of their distribution, but simply searching a sequence with these motifs provides little discriminatory power. Using a set of 164 weight matrices for 109 transcription factors extracted from TRANSFAC finds 186 792 933 matches in the April 2003 reference human genome sequence, but this was reduced to only 4 188 229 by demanding conservation in the human–mouse–rat three-way alignments. This is a 44-fold increase in specificity.

4.8
Pseudogenes and Gene Loss

Pseudogenes found in rat were classified according to whether they arose from retrotransposition, in which case they integrated into the genome randomly, or whether they arose from tandem duplication and neutral sequence substitution. Using human–rat synteny, 80% of pseudogenes exhibited no significant similarity to the corresponding human orthologous region, and therefore were considered to be retrotransposed, processed pseudogenes. The total pseudogene count and processed pseudogene proportion are consistent with those found for human. Pseudogenes are normally not subjected to selective constraint and therefore accumulate sequence modifications neutrally. Indeed, nearly all of our identified pseudogenes ($97 \pm 3\%$) evolved under neutrality according to a

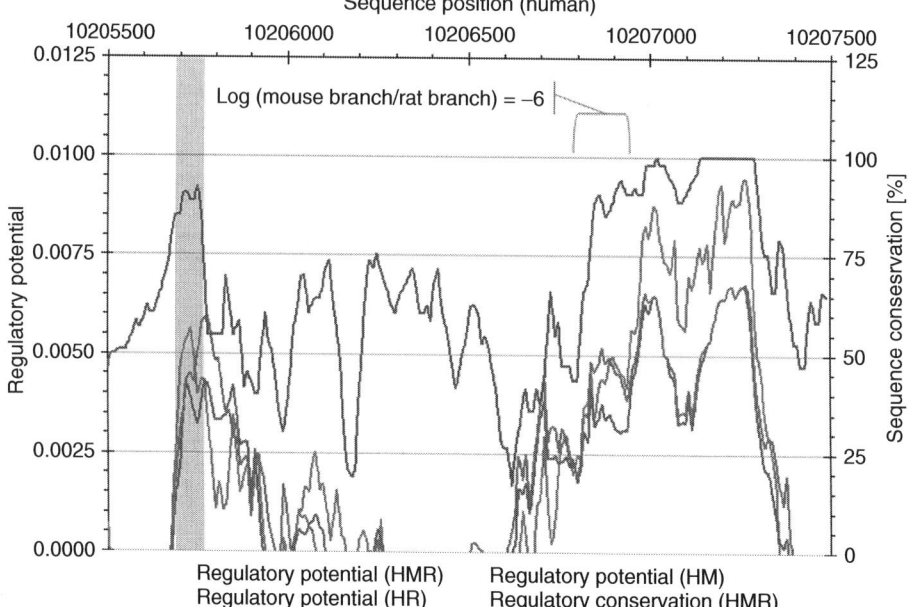

Fig. 11 Close-up of PEX14 (peroxisomal membrane protein) locus on human chromosome 1 (with homologous mouse chromosome 4 and rat chromosome 5). Conservation score computed on 3-way human–mouse–rat alignments presents a clear coding exon peak (gray bar) and very high values in a 504 bp noncoding, intronic segment (right; last 100 bp of alignment are identical in all three organisms). The latter segment showed a striking difference between the inferred mouse and rat branch lengths: the gray bracket corresponds to a phylogenetic tree where the logarithm of mouse to rat branch length ratio is −6. Regulatory potential (RP) scores that discriminate between conserved regulatory elements and neutrally evolving DNA are calculated from 3-way (human–mouse–rat) and 2-way (human-rodent) alignments. Here the 3-way regulatory potential scores are enhanced over the 2-way scores. Used with permission from *Nature* **428**, 493–521 (2004). (See color plate p. xxi.)

K_A/K_S test, and therefore, are consistent with being pseudogenic.

As with the human genome, the largest group of rat pseudogenes (totaling 2188) consists of ribosomal protein genes. Other large rat pseudogene families arose from olfactory receptors (552, see Sect. 6.1), glyceraldehyde 3-phosphate dehydrogenase (251), protein kinases (177), and RNA binding RNP-1 proteins (174). Pseudogenes homologous to a meiotic spindle-associated protein spindlin are particularly numerous in rat (at least 53 copies), compared to mouse (approximately 3 copies). This suggests that spindlin pseudogenes may have distributed rapidly by a recently active transposable element.

4.9
In Situ Loss of Rat Genes

As an organism evolves, its need for certain genes may be reduced, or lost, owing to changes in its ecological niche. Loss of selective constraints leads to accumulation of nonsense and/or frameshift mutations without retrotransposition or duplication. These nonprocessed pseudogenes are

Tab. 6 Candidate rat pseudogenes, orthologous to mouse and human functional genes.

Mouse gene	Human gene	Strand	Rat genome Coordinates[a]	Frameshifts stops[b]	Annotation
ENSMUSG00000013611	ENSG00000174226	+	7:92752590–92807556	1/0	Sorting nexin
ENSMUSG00000024364	ENSG00000158402	+	18:62742414–62770427	2/0	Dual specificity phosphatase CDC25c
ENSMUSG00000026293	ENSG00000077044	+	9:95634847–95692601	1/0	Diacylglycerol kinase δ
ENSMUSG00000026785	ENSG00000160447	+	3:9210762–9229984	5/0	Protein kinase PKNβ
ENSMUSG00000026829	ENSG00000148288	+	3:7662414–7664521	2/2	Forssman glycolipid synthetase
ENSMUSG00000027426	ENSG00000125846	+	3:125918806–125924149	1/1	Zinc finger protein 133
ENSMUSG00000028000	ENSG00000138799	–	2:221272797–221304350	1/0	Complement factor I
ENSMUSG00000029203	ENSG00000078140	–	14:44385206–44441888	1/0	Ubiquitin-protein ligase E2 (HIP2)
ENSMUSG00000030270	ENSG00000144550	–	20:8332585–8362331	3/0	Copine (membrane trafficking)
ENSMUSG00000035449	ENSG00000167646	+	1:67374986–67381472	1/0	Cardiac troponin I
ENSMUSG00000037029	ENSG00000105261	+	1:82728049–82730272	1/0	Zinc finger protein 146
ENSMUSG00000037432	ENSG00000158142	+	9:42465695–42498651	1/1	Dysferlin-like protein
ENSMUSG00000039660	ENSG00000167137	–	3:9320401–9326997	4/0	Similar to yeast YMR310c RNA binding protein
ENSMUSG00000042653	ENSG00000137634	+	8:4938446–4939091	1/0	Brush border 61.9 kDa-like protein

[a] Coordinates from rat v2.0.
[b] Mouse genes were used as templates for predicting rat pseudogenes.

interesting since they link environmental changes to genomic mutation events. However, predicted pseudogenes with disrupted reading frames might also be indicative of errors in genome sequence or assembly. By constraining the search to orthologous genomic regions, we identified 14 rat putative nonprocessed pseudogenes (Table 6) with apparently functional, single human, and mouse orthologs.

4.10
Noncoding RNA Genes

The abundance and distribution of noncoding RNAs was investigated in rat. Cytoplasmic tRNA gene identification in rodents is complicated by tRNA derived ID SINE elements (B2 and ID). TRNAscan–SE predicted 175 943 tRNAs (genes and pseudogenes); however, the majority (175 285) were SINEs identified by RepeatMasker. This is far greater than found in mouse (24 402/25 078) or human (25/636). Of the remaining 666 predictions, 163 were annotated as tRNA pseudogenes and 4 were annotated as undetermined by tRNAscan–SE. An additional 68 predictions were removed because their best database match in either human, mouse, or rat tRNA databases matched tRNAs with either a different amino acid or anticodon (violating the wobble rules that specify the distinct anticodons expected). The total of 431 tRNAs (including a single selenocysteine tRNA) identified in the rat genome is comparable to that for mouse (435) and human (492). These three species share a core set of approximately 300 tRNAs using a cutoff of $\geq 95\%$ sequence identity and $\geq 95\%$ sequence length.

A total of 454 noncodingRNAs (other than tRNAs) were identified by sequence comparison to known noncoding RNAs. These include 113 micro (mi) RNAs, 5 ribosomal RNAs, 287 small nucleolar (sno) RNAs, and small nuclear (sn) RNAs; 49 various other ncRNAs such as signal recognition particle (SRP) RNA, 7SK RNA, telomerase RNA, Rnase P RNA, brain-specific repetitive (Bsr) RNA, noncoding transcript abundantly expressed in brain (Ntab) RNA, small cytoplasmic (sc) RNA, and 626 pseudogenes. Complete 18S and 28S rRNA genes and more rRNAs were not identified presumably owing to assembly issues.

5
Evolution of Transposable Elements

Most interspersed repeats are immobilized copies of transposable elements that have accrued substitutions in proportion to their time spent fixed in the genome (for introduction see references in Nature paper). About 40% of the rat genome draft is identified as interspersed repetitive DNA derived from transposable elements, similar to mouse (Table 7) and lower than for the human (almost 50%). The latter difference is mainly because of the lower substitution rate in the human lineage, which allows us to recognize much older (Mesozoic) sequences as interspersed repeats. Almost all repeats are derived from retroposons, elements that procreate via reverse transcription of their transcripts. As in mouse, there is no evidence for activity of DNA transposons since the rat–mouse split. Many aspects of the rat and the mouse genomes' repeat structure are shared; here we focus on the differences.

5.1
LINE-1 Activity in the Rat Lineage

The long interspersed nucleotide element (LINE-1 or L1) is an autonomous

Tab. 7 Composition of interspersed repeats in the rat genome.

	Rat				Mouse	
	Copies [×10³]	Total length [Mb]	Fraction of genome [%]	Lineage specific [%]	Fraction of genome [%]	Lineage specific [%]
LINEs:	657	594.0	23.11	11.70	20.10	9.74
LINE-1	597	584.2	22.73	11.70	19.65	9.74
LINE-2	48	8.4	0.33	–	0.38	–
L3/CR1	11	1.4	0.06	–	0.06	–
SINEs:	1360	181.3	7.05	1.52	7.78	1.80
B1(Alu)	384	42.3	1.65	0.16	2.53	0.92
B4(ID_B1)	359	55.4	2.15	0.00	2.25	0.00
ID	225	19.6	0.76	0.54	0.20	0.00
B2	328	55.2	2.15	0.68	2.29	0.74
MIR	109	13.0	0.51	–	0.56	–
LTR elements:	556	232.4	9.04	1.84	10.28	2.85
ERV_classI	40	24.9	0.97	0.56	0.79	0.36
ERV_classII	141	83.4	3.24	1.02	4.13	1.73
ERV-L (III)	74	21.6	0.84	0.04	1.08	0.23
MaLRs	302	102.5	3.99	0.22	4.27	0.53
DNA elements:	108	20.9	0.81	–	0.86	–
Charlie(hAT)	80	14.8	0.58	–	0.60	–
Tigger(Tc1)	18	4.0	0.16	–	0.17	–
Unclassified:	14	7.3	0.28	–	0.37	–
Total	2690	1036	40.31	14.90	39.45	14.26
Small RNAs:	8	0.6	0.03	0.01	0.03	0.01
Satellites:	14	6.4	0.25	?	0.31	?
Simple repeats:	897	61.1	2.38	?	2.41	?

Data for Rnor3.1 and October 2003 mouse (MM4), excluding Y chromosome, using the December 17 2003 version of RepeatMasker. To highlight the differences between rat and mouse repeat content, column 5 and 6 show the fraction of the genomes comprised of lineage specific repeats. The LINE-1 numbers include all HAL1 copies, while all BC1 scRNA and >10% diverged tRNA–Ala matches, far more common than other small RNA pseudogenes and closely related to ID, have been counted as ID matches.

retroelement, with an internal RNA polymerase II promoter and two open reading frames (ORFs). ORF1 is an RNA binding protein with chaperone-like activity suggesting a role in mediating nucleic acid strand transfer steps during L1 reverse transcription. ORF2 encodes a protein with both reverse transcriptase and DNA endonuclease activity. LINEs are usually 5′ truncated so that only a small subset extends to include the promoter region and can function as a source for more copies.

Many classes of LINE-like elements exist, but only L1 has been active in rodents. Over half a million copies, in variable stages of decay, comprise 22% of the rat genome. Although over 10% of the human genome is L1 copies introduced before the rodent–primate split, only 2% of the rat genome could be recognized as such because of the fast substitution rate in the rodent lineage. Thus, probably well over one quarter of all rat DNA is derived directly from the L1 gene.

Following the mouse/rat split, L1 activity appears to have increased in rat. The 6 rat-specific L1 subfamilies, represented by 150 000 copies, cover 12% of the rat genome. Mouse-specific L1 copies accumulated over the same period cover only 10% of the genome (Table 7). This greater accumulation of L1 copies could explain some of the size difference of the rat and mouse genome.

In addition to the traditional L1 elements, there are 7500 copies (10 Mb) of a nonautonomous element that is derived from L1 by deletion of most of its ORF2. On the basis of their low divergence, the presently identified HAL1-like (HAL1 for Half-a-LINE) elements operated only a few million years ago in the mouse lineage (MusHAL1) and still propagate in the rat genome (RNHAL1). RNHAL1 only contains an ORF1, while MusHAL1 encoded an endonuclease as well, though no reverse transcriptase. The repeated origin and high copy number of HAL1s suggests that the ORF1 product, which binds strongly to its mRNA, may render this transcript a superior target for L1 mediated reverse transcription. In this way, HAL1 resembles the nonautonomous, endogenous retrovirus-derived MaLR elements (in the following), which, for over 100 million years, retained only the retroviral gag ORF that encodes an RNA binding protein. A potential advantage of HAL1 over L1 is its shorter length, which, considering the usual 5′ truncation of copies, increases the chance that a copy includes the internal promoter elements and can become a source gene.

5.2
Different Activity of SINEs in the Rat and Mouse Lineage

The most successful usurpers of the L1 retrotransposition machinery are SINEs. These are small RNA-derived sequences with an internal RNA polymerase III promoter. The human Alu SINE was shown to be transposed by L1. L1 lacks sequence specificity and rodent and primate SINE sequences are unrelated to L1. Though any transcript can be retroposed, as evidenced by the numerous mammalian processed pseudogenes, L1-dependent SINEs probably have features that make them especially efficient targets of the L1 reverse transcriptase.

While before the radiation of most mammalian orders L1 was at least as active as L2, the L2 dependent MIR was the only known (and very abundant) SINE of that time. All of the currently active SINEs in different mammalian orders appear to have arisen after the demise of L2 (and consequently MIR), as if an opportunity (or necessity) arose for the creation and expansion of other SINEs.

Four different SINEs are distinguished in rat and mouse. The B1 element and the primate Alu originated from a 7SLRNA gene, probably just before the rodent–primate split and after the speciation from most other eutherians, where Alu/B1 elements are not known. The other SINEs (B4, B2, ID) are rodent-specific and have tRNA-like internal promoter regions.

The fortunes of these SINEs during mouse and rat evolution have been different (Fig. 12). B4 probably became extinct before the mouse–rat speciation, while B2 has remained productive in both lineages, scattering >100 000 copies in each genome after this time. Interestingly, the fate of the B1 and ID SINEs has been opposite in rat and mouse. While B1 is still active in mouse, having left over 200 000 mouse-specific copies in its trail, the youngest of the 40 000 rat-specific B1 copies are 6 to 7% diverged from their source, indicating a relatively early extinction in the rat lineage.

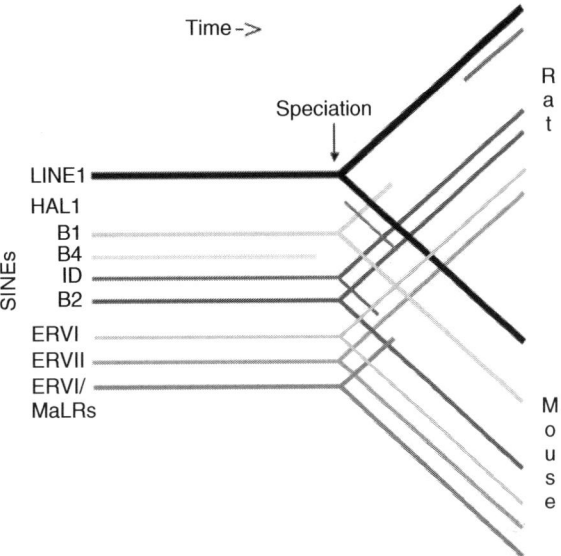

Fig. 12 Historical view of rodent repeated sequences. Relationships of the major families of interspersed repeats (Table 7) are shown for the rat and mouse genomes, indicating losses and gains of repeat families after speciation. The lines indicate activity as a function of time. Note that HAL1-like elements appear to have arisen both in the mouse and rat lineage. Used with permission from *Nature* **428**, 493–521 (2004). (See color plate p. xxii.)

On the other hand, after the mouse–rat split only a few hundred ID copies may have inserted in mouse, while this heretofore minor SINE (∼60 000 copies predate the speciation) picked up activity in rat to produce 160 000 ID copies.

5.3
Colocalization of SINEs in Rat and Mouse

Although the fates of SINE families differ, the number of SINEs inserted after speciation in each lineage is remarkably similar, ∼300 000 copies. As MIR was replaced by L1 driven SINEs, it appears that the demise of B1 in rat allowed the expansion of IDs. Moreover, these independently inserted and unrelated SINEs (ID and B1 only share a mechanism of retroposition) accumulated at orthologous sites: the density of rat-specific SINEs in 14 243 ∼100 kb windows in the rat genome is highly correlated ($R^2 = 0.83$) with the density of mouse-specific SINEs in orthologous regions in mouse (Fig. 5). These data corroborate and refine the observation of a strong correlation between the location of primate and rodent-specific SINEs in 1-Mb windows. At 100 kb, no correlation is seen for interspersed repeats other than SINEs.

Insertions of SINEs at the same location have been reported, and the correlation could reflect the existence of conserved hotspots for SINE insertions. However, primate data does not support this. Likewise, gene conversions do not significantly contribute to the observed correlation.

Figure 9(c) displays the lineage-specific SINE densities on rat chromosome 10 and in the mouse orthologous blocks, showing a stronger correlation than any other feature. The cause of the unusual distribution patterns of SINEs, accumulating in gene-rich regions where other interspersed repeats are scarce, apparently is a conserved feature independent of the primary sequence of the SINE and effective over smaller than isochore-sized regions.

In the human genome, the most recent (unfixed) Alus are distributed similar to L1, whereas older copies gradually take on the opposite distribution of SINEs. However, this temporal shift in SINE distribution pattern is not observed in mouse or rat.

Some regions of high LINE content coincide with regions that exhibit both higher AT content and an elevated rate of point substitution (Fig. 9, pink rectangles). In a genome-wide analysis, LINE content correlates strongly with substitution rates, and about 80% of this correlation is explained by higher rates in AT-rich regions. SINE density shows the opposite correlation both on chromosome 10 (Fig. 9) and genome-wide.

These phenomena, in conjunction with an overall trend in substitution rates toward AT-richness, suggest a model by which quickly evolving regions accumulate a higher-than-average AT content, which attracts LINE elements. While distinct cause–effect relationships such as this remain largely speculative, these results reinforce the idea that local genomic context strongly shapes local genomic features and rates of evolution.

5.4
Endogenous Retroviruses and Derivatives

Retrovirus-like elements are the other major contributors to interspersed repeats in the rodent genome. These have several 100-bp long terminal repeats (LTRs) with transcriptional regulatory sequences that flank an internal sequence that, in autonomous elements, encodes all proteins necessary for retrotransposition. All mammalian LTR elements are endogenous retroviruses (ERVs; classes I, II, or III) or their nonautonomous derivatives.

The most productive retrovirus in mammals has been the class III element ERV-L, primarily through its ancient nonautonomous derivatives called "MaLRs" with 350 000 copies occupying ~5% of the rat genome (Table 7). Human ERV-L and MaLR copies are >6% diverged from their reconstructed source genes and must have died out around the time of our speciation from new world monkeys. In mouse, several thousand almost identical MaLR and ERV-L copies suggest sustained activity. In contrast, rat ERV-L activity was silenced a few million years ago given that the least diverged MaLR and ERV-L copies differ by >4% from each other. Other class III ERVs were active earlier in rodent evolution, before the mouse–rat speciation. Class I and class II elements still thrive in rat, where 2 of 4 class I and 4 of 9 class II autonomous ERVs appear to be active.

5.5
Simple Repeats

Interspersed simple sequence repeats (SSRs), regions of tandemly repeated (1–6 bp) units that probably arise from slippage during DNA replication and can expand and compress by unequal crossing over. Remarkable differences were noted between the human and mouse genomes' SSR content with threefold to fivefold more base pairs contained in near (>90%) perfect SSRs in mouse. SSRs are both more frequent and on average longer in mouse. Polypurine (or polypyrimidine) repeats are especially (10-fold) overrepresented in the mouse genome. As discussed in Sect. 4.3, this contrasts sharply with the greater frequency of triplet repeats coding for amino acids in human relative to the rodents.

Rat and mouse SSR content are more similar, with genome coverage of ~1.4% (for >90% perfect elements compared to 0.45% in human) and are of similar average length. For example, the most common mammalian SSR, the (CA)n repeat, averages 42 bp long in mouse and 44 bp in rat. Some potentially significant differences are that polypurine SSRs are of similar average length but 1.2-fold more common in mouse, while the rare SSRs containing CG dimers are 1.5-fold more frequently observed in rat.

5.6
Prevalent, Medium-length Duplications in Rodents

In addition to the transpositionally derived and simple repeats, the rat and mouse genomes contain a substantial amount of medium-length unclassified duplications (typically 100–5000 bp). These are readily seen in self-comparisons and in intrarodent comparisons after masking the known repeats, but they are much less prevalent in comparisons with the human genome. A substantial fraction of the rodent genomes consists of currently unexplained repeats, which may include (1) novel families of low-copy rodent interspersed repeats; (2) extensions of known but not fully characterized rodent repeats; and (3) duplications generated by a mechanism different from transposition.

6
Rat-specific Biology

The rat genome sequence and predicted gene set reveals genetic differences between rats and mice that might specify their differences in physiology and behavior. In particular, recently duplicated genes are enriched in elements involved in chemosensation and functional aspects of reproduction (Table 5). The differences in the gene complements of rat and mouse are illustrated by in-depth analyses of olfactory receptors (ORs), pheromones, cytochromes P450, proteases, and protease inhibitors.

6.1
Chemosensation

The ability to emit and sense specific smells is a key feature of survival for most animals in the wild. Rat and mouse pheromones, vomeronasal receptors, and ORs genes were duplicated frequently during the time since the common ancestor of rats and mice (Table 5). The rapid evolution of these genes is attributed to conspecific competition, in particular sexual selection.

The rat genome has 1866 ORs in 113 locations: 69 multigene clusters, and 49 single genes. Extrapolating for the 9.8% of the genome not in the assembly, there are ~2070 OR genes and pseudogenes. The rat therefore has ~37% more OR genes and pseudogenes than the ~1510 ORs of the mouse, assuming similar representation of recently duplicated sequences in the two genome assemblies used. Of the 1774 OR sequences that are not interrupted by assembly gaps, 1216 (69%) encode intact proteins, while the remaining 547 (31%) sequences are likely pseudogenes with in-frame stop codons, frameshifts, and/or interspersed repeat elements. Fewer mouse OR homologs are pseudogenes (~20%), but the larger family size in rat still leaves it with substantially more intact ORs than the mouse (~1430 vs ~1210). Striking rat-specific expansions of two ancestral clusters account for much of the difference in OR family size and pseudogene content between rat and mouse, although many other clusters exhibit more subtle changes. Significant differences between human and mouse in OR families have also been reported, but the functional implications of OR repertoire size on the ability of different species to detect and discriminate odorants are not yet known.

6.2
α_{2u} Globulin Pheromones

The α_{2u} globulin genes are odorant-binding proteins that also contribute to essential survival functions in animals. α_{2u} globulin homologs are likely to be highly heterogeneous among murid species. Distinct homologs and genomic arrangement differences have been observed among mouse strains. The evolution of α_{2u} globulin genes on rat chromosome 5 has remodeled this genomic region (Fig. 13). The orthologous human genomic region contains a single homolog, suggesting that the common ancestor of rodents and human possessed one gene. The genome of C57BL/6J mice contains 4 homologous genes, and 7 pseudogenes, while the rat genome contains 10 α_{2u} globulin genes and 12 pseudogenes in a single region.

Phylogenetic trees constructed using amino acid, and noncoding DNA, sequences show that, surprisingly, the rat α_{2u} globulin gene clusters appear to have arisen recently via a rapid burst of gene duplication since the rat/mouse split (Table 5). This is consistent with the Rfp37-like zinc finger-like pseudogene having accompanied virtually all of the rat-specific α_{2u} globulin gene duplications (Fig. 13). The sequences of these genes are

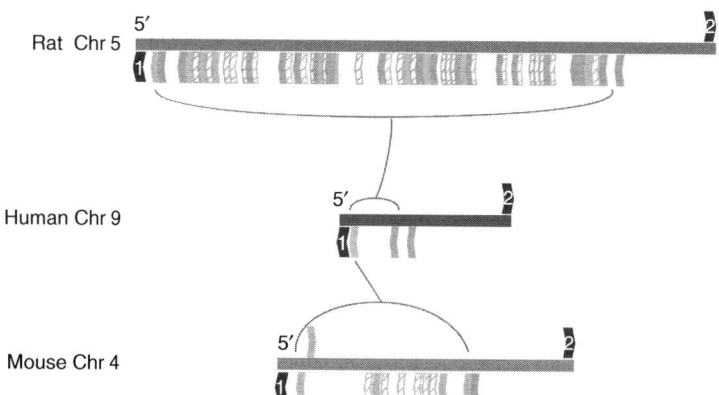

Fig. 13 Adaptive remodeling of genomes and genes. Orthologous regions of rat, human and mouse genomes encoding pheromone-carrier proteins of the lipocalin family (α_{2u} globulins in rat and major urinary proteins in mouse) shown in brown. Zfp37-like zinc finger genes are shown in blue. Filled arrows represent likely genes, whereas striped arrows represent likely pseudogenes. Gene expansions are bracketed. Arrow head orientation represents transcriptional direction. Flanking genes 1 and 2 are TSCOT and CTR1 respectively. (See color plate p. xv.)

also evolving rapidly with median K_A/K_S values of 0.77 and 1.06, for rat and mouse genes respectively. Amino acid sites that appear to have been subject to adaptive evolution are situated both within the ligand-binding cavity, and on the solvent-exposed periphery of the α_{2u} globulin structure. This demonstrates how genome analysis can reveal the imprint of adaptive evolution from megabase to single base levels.

The rapid evolution of these genes, and the remodeling of their genomic regions, can be attributed to the known roles of rat α_{2u} globulins and mouse homologs in conspecific competition and sexual selection. These proteins are pheromones and pheromone carriers that are present in large quantities in rodent urine, and act as scent markers indicating dominance and subspecies identity.

6.3
Detoxification

Cytochrome P450 is a well-recognized participant in metabolic detoxification and rapid evolution is observed within this family. These enzymes are particularly relevant to clinical and pharmacological studies in humans as they metabolize many toxic and endogenous compounds. Rodents are important model organisms for understanding human drug metabolism, so it is important to identify 1:1 orthologs and species-specific expansions and losses. Compared to human genes, there are clear expansions of several rodent P450 subfamilies, but there are also significant differences between rat and mouse subfamilies (Fig. 14a). The fastest evolving subfamily seems to be CYP2J containing a single gene in human, but at least 4 in rat and 8 in mouse (Fig. 14b,c). CYP2J enzymes catalyze the NADPH-dependent oxidation of arachidonic acid to various eicosanoids, which in turn possess numerous biological activities including modulation of ion transport, control of bronchial and vascular smooth muscle tone, and stimulation of peptide hormone secretion. The genomic ordering of genes and their phylogenetic tree indicate an ongoing expansion in the rodents (Fig. 14b,c). This suggests that adaptive evolution has been involved in diversifying their functions.

6.4
Proteolysis

Protease genes also represent an example of rapid evolution in the rat genome. The rat contains 626 protease genes (~1.7% of the genes), similar to mouse (641) and more than human (561). One hundred and two rat protease genes are not found in human, and 42 are absent from mouse. Several rat gene families have expanded; most are involved in reproductive or immunological functions, and have evolved independently in the rat and mouse lineages.

These gene family expansions dramatically illustrate how large-scale genomic changes have accompanied species-specific innovation. Positive selection of duplicated genes has afforded the rat an enhanced repertoire of precisely those genes that allow reproductive success despite severe competition from both within its own, and with other species. This serves as a general illustration of the importance of chemosensation, detoxification, and proteolysis in innovation and adaptation.

7
Human Disease Gene Orthologs in the Rat Genome

The rat is recognized as the premier model for studying the physiological aspects of

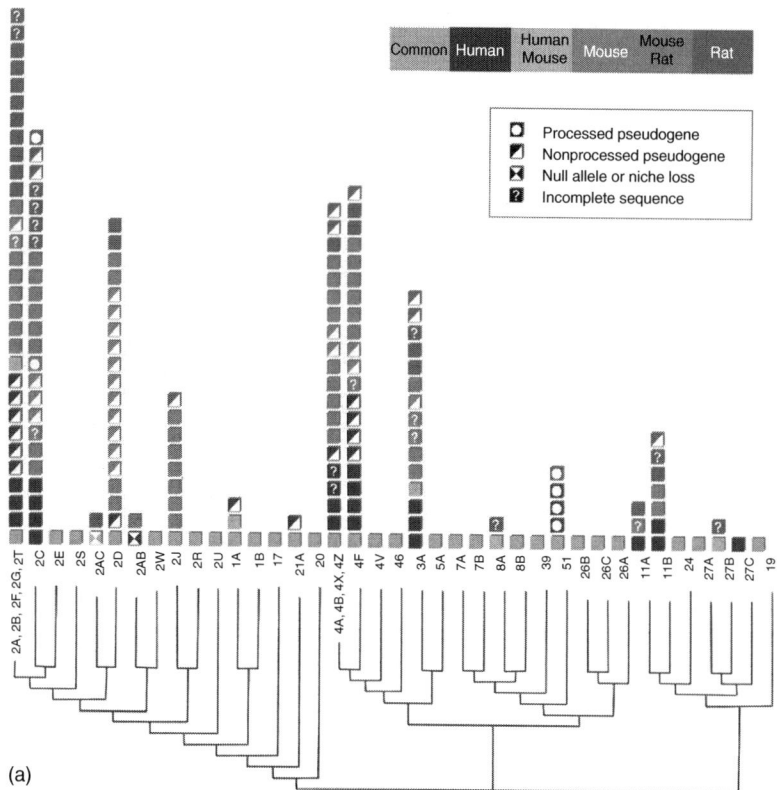

Fig. 14 Evolution of cytochrome P450 (CYP) protein families in rat, mouse, and human. (a) Dendrogram of topology from 234 full-length sequences. The 279 sequences of 300 amino acids; subfamily names and chromosome numbers are shown. Black branches have >70% bootstrap support. Incomplete sequences (they contain Ns) are included in counts of functional genes (84 rat, 87 mouse, and 57 human) and pseudogenes (including fragments not shown; 77 rat, 121 mouse, and 52 human). Thus, 64 rat genes and 12 pseudogenes were in predicted gene sets. Human CYP4F is a null allele due to an in-frame STOP codon in the genome, although a full-length translation exists (SwissProt P98187). Rat CYP27B, missing in the genome, is "incomplete" since there is a RefSeq entry (NP_446215). Grouped subfamilies CYP2A, 2B, 2F, 2G, 2T, and CYP4A, 4B, 4X, 4Z, occur in gene clusters; thus nine loci contain multiple functional genes in a species. One (CYP1A) has fewer rat genes than human, seven have more rodent than human, and all nine have different copy numbers. CYP2AC is a rat-specific subfamily (orthologs are pseudogenes). CYP27C has no rodent counterpart. Rodent-specific expansion, rat CYP2J is illustrated below. (b) The neighbor-joining tree, with the single human gene, contains clear mouse (Mm) and rat (Rn) orthologous pairs (bootstrap values >700/1000 trials shown). Bar indicates 0.1 substitutions per site. (c) All rat genes have a single mouse counterpart except for CYP2J 3, which has further expanded in mouse (mouse CYP2J 3a, 3b, and 3c) by two consecutive single duplications. The genes flanking the CYP2J orthologous regions (rat chromosome 5, 126.9–127.3 Mb; mouse chromosome 4, 94.0–94.6 Mb; human chromosome 1, 54.7–54.8 Mb) are hook1 (HOOK1; pink) and nuclear factor I/A (NFIA; cyan). Genes (solid) and gene fragments (dashed boxes) are shown above (forward strand) and below (reverse strand) the horizontal line. No orthology relation could be concluded for most of these cases. Used with permission from *Nature* **428**, 493–521 (2004). (See color plate p. xxiii.)

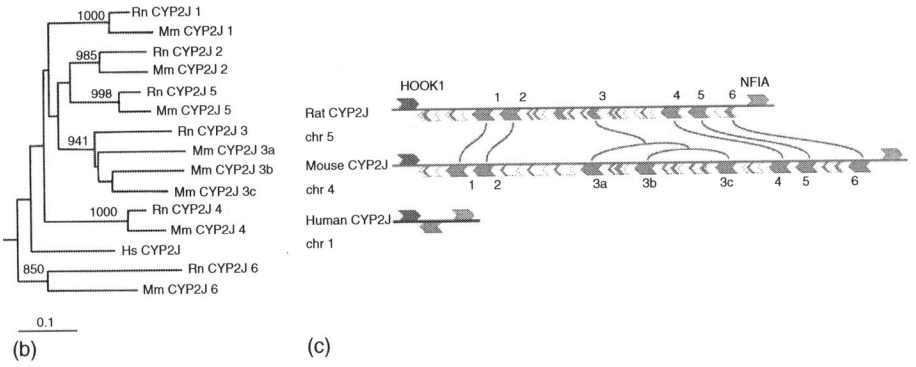

Fig. 14 (Continued)

many human diseases, but it has not had as prominent a role in the study of simple genetic disease traits. The rat genome provides an opportunity to use the more than 1000 human Mendelian disorders that have associated loci and alleles to link the rat to human disease examples. The precise identification of the rat orthologs of human genes that are mutated in disease creates further opportunities to discover and develop rat models for biomedical research.

Predicted rat genes were compared with 1112 well-characterized human disease genes that were verified and classified on the basis of pathophysiology. So, 844 (76%) have 1 : 1 orthologs in the rat as predicted by Ensembl. These predictions are likely to be of high quality since 97.4% of the 11 422 rat:human 1 : 1 orthologs predicted by Ensembl were found in orthologous genomic regions.

The proportion of human disease genes with single orthologs in the rat (76%) was higher than the proportion for all human genes (46%). Careful analysis of the remaining 268 human genes not predicted to show 1 : 1 orthology indicated that only six of the human disease genes lack likely rat orthologs among genome, cDNA, EST, and protein sequences. Thus, it appears that, in general, genes involved in human disease are unlikely to have diverged, or to have become duplicated, deleted, or lost as pseudogenes, between rat and human divergence.

Comparisons of K_S, K_A, and the K_A/K_S ratio values of human disease orthologs with those of all remaining ortholog pairs found only the K_S distributions differed significantly indicating that coding regions of human disease genes and their rat counterparts have mutated more rapidly than the nondisease genes. Factors influencing the specific loci could cause this result, or the disease genes may characteristically reside in genomic regions that exhibit higher mutation rates.

The disease gene set was next grouped into 16 disease-system categories and analyzed (Fig. 15). Only five disease systems exhibited significant K_A/K_S differences with respect to the remaining samples. Neurological and malformation-syndrome disease categories manifested K_A/K_S ratios consistent with purifying selection acting on these gene sets. In contrast, the pulmonary, hematological, and immune categories manifested the highest median K_A/K_S ratios, consistent with a role for more positive selection, or reduced selective constraints, among these genes.

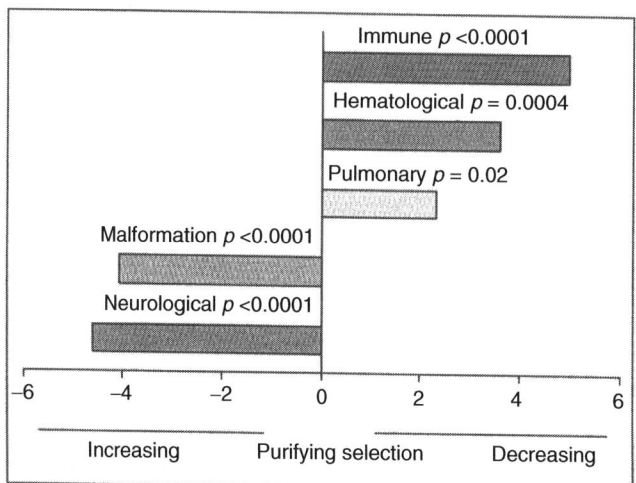

Fig. 15 Selective constraints differ for human disease systems in the rat genome. Human disease-system categories showing significant differences ($p < 0.05$) in a nonparametric test [Mann–Whitney–Wilcoxon] comparing K_A/K_S (human:rat) ratios. P-values from two-level tests between genes from one disease system and the remaining genes. (Mean–Mean0)/Std0 values from multilevel tests from sixteen categorized disease systems. Negative values for neurological (−4.63) and malformation-syndrome (−4.04) categories were observed to be consistent with K_A/K_S ranges where purifying selection predominates. Immune, hematological, and pulmonary categories show positive values of 4.98, 3.59, and 2.34, respectively. Used with permission from *Nature* **428**, 493–521 (2004).

Orthologs of more diverse phyla demonstrated consistent results.

These results demonstrate that various disease systems exhibit significantly different average evolutionary rates. The higher rates noted for the immune system disease genes are consistent with rapid diversification of the functions of the immune systems of rodents and humans. This is expected for genes involved in controlling species-restricted infectious agents if strong adaptive pressure acts during host–pathogen coevolution. Thus, results of studies of these rodent genes may be less directly relevant to our understanding of human immune system diseases than results obtained for other pathophysiology disease systems where conservation is greater and purifying selection is stronger.

A number of genes were examined that harbor triplet nucleotide repeats, and are involved in human neurological disorders, such as Huntington disease, a condition caused by CAG triplet repeat expansion producing abnormally long polyglutamine tracts in an otherwise normal protein. Analysis of the rat:human orthologs of these disease genes indicated that repeat length is substantially shorter in all cases in the rat than in the normal human gene (Fig. 16). To date, there are no naturally occurring rat strains described that exhibit neurological disease associated with repeat expansion mechanisms. The rat may lack repeat

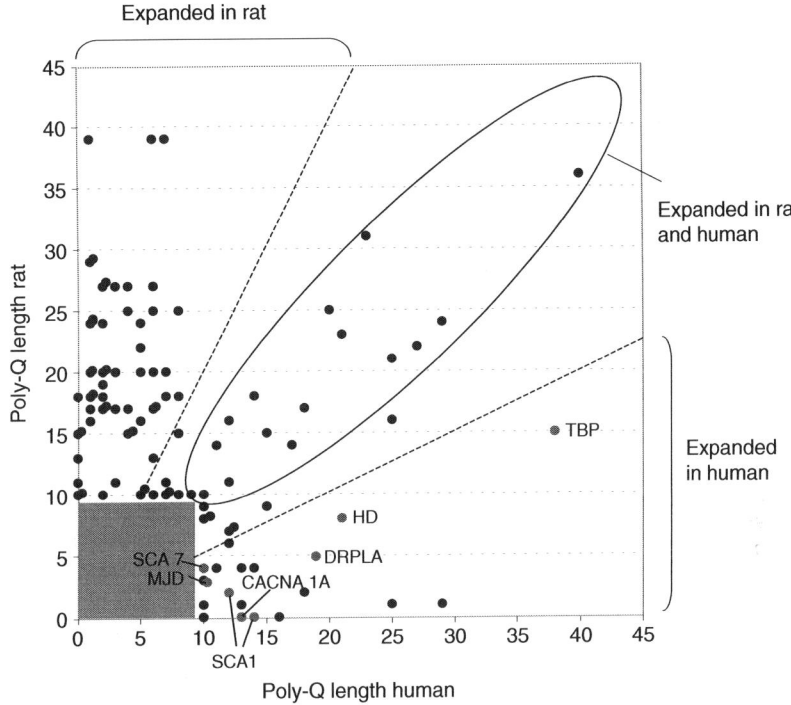

Fig. 16 Polyglutamine repeat length comparison between human and rat. Points represent protein poly-Q length for rat and human. Red points correspond to repeats in genes associated with human disease: SCA1, spinocerebellar ataxia 1 protein, or ataxin1; SCA7, spinocerebellar ataxia 7 protein; MJD, Machado–Joseph disease protein; CACNA1A, spinocerebellar ataxia 6 protein, or calcium channel α-1A subunit isoform 1; DRPLA, dentatorubro–pallidoluysian atrophy protein; HD, Huntington's disease protein, or huntingtin; TBP, TATA binding protein or spinocerebellar ataxia 17 protein. Repeat lengths over 10 were examined; green shading delineates the range not included in the analysis. Also noted are a set that are expanded in rat and human (black circle) and a set where repeats are expanded in the rat. Used with permission from *Nature* **428**, 493–521 (2004). (See color plate p. xxiv.)

expansion mutational mechanisms or these orthologs may fail to reach a "critical repeat length" susceptible to such mutational mechanisms. Other human genes without known disease association also contain glutamine repeats that are much shorter in the rat orthologs, and thus may be susceptible to mutations that arise through repeat expansion mechanisms. In Fig. 16, it may also be observed that a relatively high proportion of repeats are significantly longer in the rat than in their corresponding human ortholog.

In addition to enabling the direct comparison of rat/human disease orthologs, the rat genome sequence itself is an invaluable aid for discovery of other rat genes to study as disease models. First, genes underlying disease phenotypes with simple inheritance that have been mapped to chromosomal regions can be more easily pursued in both species. Indeed, the rearrangements of conserved segments between the two species have significant value since they tighten the boundaries of the mapped disease regions and reduce the number of genes potentially associated with the disease phenotype. Second, the identification of multiple alleles contributing to quantitative and complex trait differences that are involved in disease processes can be pursued with more accuracy, both in the initial association phases, and in subsequent efforts to detect causative alleles.

8
Summary

As the third mammalian genome to be sequenced, the rat genome has provided both predictable and surprising information about mammalian species. Although clear at the outset that ongoing rat research would benefit from the resource of a genome sequence, there was uncertainty about how many new insights would be found, especially considering the superficial similarities between the rat and the already sequenced mouse. Instead, the results of the sequencing and analysis have generated some deep insights into the evolutionary processes that have given rise to these different species. In addition, the project was invaluable for further developing the methods for the generation and analysis of large genome sequence data sets.

The generation of the rat draft tested the new "combined approach" for large genome sequencing. The high quality of the overall attests of this overall strategy and the supporting software, provide a suitable approach to this problem. With a BAC "skimming" component in the underlying data set, the assembly recovered a fraction of the genome that was expected, by analogy to the mouse project to be difficult to assemble from pure WGS data. The BAC skimming component also allowed progressive generation of high quality local assemblies that were of use to the rat community as the project developed.

The rat genome data have improved the utility of the rat model enormously. The rat gene content provides a parts list that can be explored with the high degree of confidence and precision that is appropriate for biomedical research. The resources for physical and genetic mapping have also been improved, since the relative position of individual markers is now known with confidence and there are computational resources to bridge the process of genetic association with gene modeling and experimental investigation.

The expected benefit of a third mammalian sequence providing an outgroup by which to discriminate the timing of events that had already been noted between mouse and human was fully realized. Using the three sequences and other partial data sets from additional organisms, it was possible to measure some of the overall faster rate of evolutionary change in the rodent lineage shared by mice and rats, as well as the peculiar acceleration of some aspects of rat-specific evolution. The observation of specific expanded gene families in the rat should provide material for targeted studies for some time.

The sequence of the rat genome represents the beginning of the full analysis of the mammalian genome and its complex evolutionary history. Much of the additional data required to complete this story will be from other genomes, distantly related to rat. Nevertheless, the prospect of a finished rat genome, polymorphism data from analysis of other rat strains, and targeted cDNA clone collections will provide rat-specific reagents for routine use in research. Together with the ongoing efforts to fully develop methods to genetically manipulate whole rats and provide effective "gene knockouts" the current and future rat genome resources will ensure a place for this organism in genomic and biomedical research for some time.

Acknowledgment

This chapter is based largely on the publication of the rat genome sequence by the Rat Genome Sequencing Project Consortium (*Nature* **428**, 493–521 (2004)). We thank E. Eichler, K. Fetchel, R. Hardison, P. Havlak, K. Kalafus, M. Krzywinski, A. Milosavljevic, L. Pachter, P. Pevzner, C. Ponting, K. Roskin, A. Sidow, A. Smit, G. Tessler, and D. Torrents, for the figures. The rat genome-sequencing project was funded primarily by the NHGRI and NHLBI of the NIH.

See also Molecular Basis of Genetics; Genomic Sequencing.

Bibliography

Books and Reviews

Chiaromonte, F. et al. (2003) The share of human genomic DNA under selection estimated from human–mouse genomic alignments, *Cold Spring Harb. Symp. Quant. Biol.* **68**, 245–254.

Danielson, P.B. (2002) The cytochrome P450 superfamily: biochemistry, evolution and drug metabolism in humans, *Curr. Drug Metab.* **3**, 561–597.

Hedrich, H.J. (2000) in: Krinke, G.J. (Ed.) *History, Strains, and Models in the Laboratory Rat*, Academic Press, San Diego, CA, pp. 3–16.

Lopez-Otin, C., Overall, C.M. (2002) Protease degradomics: a new challenge for proteomics, *Nat. Rev. Mol. Cell Biol.* **3**, 509–519.

Lynch, M., Conery, J.S. (2000) The evolutionary fate and consequences of duplicate genes, *Science* **290**, 1151–1155.

Montoya-Burgos, J.I., Boursot, P., Galtier, N. (2003) Recombination explains isochores in mammalian genomes, *Trends Genet.* **19**, 128–130.

Ohno, S. (1970) *Evolution by Gene Duplication*, Springer, Berlin, Germany.

Prak, E.T., Kazazian, H.H. Jr. (2000) Mobile elements and the human genome, *Nat. Rev. Genet.* **1**, 134–144.

Reddy, P.S., Housman, D.E. (1997) The complex pathology of trinucleotide repeats, *Curr. Opin. Cell Biol.* **9**, 364–372.

Scarborough, P.E., Ma, J., Qu, W., Zeldin, D.C. (1999) P450 subfamily CYP2J and their role in the bioactivation of arachidonic acid in extrahepatic tissues, *Drug Metab. Rev.* **31**, 205–234.

Smit, A.F. (1999) Interspersed repeats and other mementos of transposable elements in mammalian genomes, *Curr. Opin. Genet. Dev.* **9**, 657–663.

Weiner, A.M. (2002) SINEs and LINEs: the art of biting the hand that feeds you, *Curr. Opin. Cell Biol.* **14**, 343–350.

Willson, T.M., Kliewer, S.A. (2002) PXR, CAR and drug metabolism, *Nat. Rev. Drug Discov.* **1**, 259–266.

Primary Literature

Adams, M.D. et al. (2000) The genome sequence of *Drosophila melanogaster*, *Science* **287**, 2185–2195.

Adkins, R.M., Gelke, E.L., Rowe, D., Honeycutt, R.L. (2001) Molecular phylogeny and divergence time estimates for major rodent groups: evidence from multiple genes, *Mol. Biol. Evol.* **18**, 777–791.

Alba, M.M., Guigo, R. (2004) Comparative analysis of amino acid repeats in rodents and humans, *Genome Res.* **14**, 549–554.

Alba, M.M., Santibanez-Koref, M.F., Hancock, J.M. (1999) Conservation of polyglutamine tract size between mice and humans depends on codon interruption, *Mol. Biol. Evol.* **16**, 1641–1644.

Altschul, S.F., Lipman, D.J. (1990) Protein database searches for multiple alignments, *Proc. Natl. Acad. Sci. U.S.A.* **87**, 5509–5513.

Antequera, F., Bird, A. (1993) Number of CpG islands and genes in human and mouse, *Proc. Natl. Acad. Sci. U.S.A.* **90**, 11995–11999.

Aparicio, S., et al. (2002) Whole-genome shotgun assembly and analysis of the genome of *Fugu rubripes*, *Science* **297**, 1301–1310.

Bailey, J.A., et al. (2002) Recent segmental duplications in the human genome, *Science* **297**, 1003–1007.

Batzoglou, S., et al. (2002) ARACHNE: a whole-genome shotgun assembler, *Genome Res.* **12**, 177–189.

Benit, L., et al. (1997) Cloning of a new murine endogenous retrovirus, MuERV-L, with strong similarity to the human HERV-L element and with a gag coding sequence closely related to the Fv1 restriction gene, *J. Virol.* **71**, 5652–5657.

Bird, A.P. (1980) DNA methylation and the frequency of CpG in animal DNA, *Nucleic Acids Res.* **8**, 1499–1504.

Birdsell, J.A. (2002) Integrating genomics, bioinformatics, and classical genetics to study the effects of recombination on genome evolution, *Mol. Biol. Evol.* **19**, 1181–1197.

Boffelli, D., et al. (2003) Phylogenetic shadowing of primate sequences to find functional regions of the human genome, *Science* **299**, 1391–1394.

Bourque, G., Pevzner, P.A. (2002) Genome-scale evolution: reconstructing gene orders in the ancestral species, *Genome Res.* **12**, 26–36.

Bourque, G., Pevzner, P.A., Tesler, G. (2004) Reconstructing the genomic architecture of ancestral mammals: lessons from human, mouse, and rat genomes, *Genome Res.* **14**, 507–516.

Bray, N., Pachter, L. (2004) MAVID Constrained ancestral alignment of multiple sequence, *Genome Res.* **14**, 693–699.

Burge, C.B., Padgett, R.A., Sharp, P.A. (1998) Evolutionary fates and origins of U12-type introns, *Mol. Cell* **2**, 773–785.

Cai, L., et al. (1997) Construction and characterization of a 10-genome equivalent yeast artificial chromosome library for the laboratory rat, *Rattus norvegicus*, *Genomics* **39**, 385–392.

Cantrell, M.A., et al. (2001) An ancient retrovirus-like element contains hot spots for SINE insertion, *Genetics* **158**, 769–777.

Cavaggioni, A., Mucignat-Caretta, C. (2000) Major urinary proteins, 2U-globulins and aphrodisin, *Biochim. Biophys. Acta* **1482**, 218–228.

Chakrabarti, K., Pachter, L. (2004) Visualization of multiple genome annotations and alignments with the K-BROWSER, *Genome Res.* **14**, 716–720.

Chen, R., Sodergren, E., Gibbs, R., Weinstock, G.M. (2004) Dynamic building of a BAC clone tiling path for genome sequencing project, *Genome Res.* **14**, 679–684.

Cheung, J., et al. (2003) Recent segmental and gene duplications in the mouse genome, *Genome Biol.* **4**, R47 [online].

Clark, A.J., Hickman, J., Bishop, J. (1984) A 45-kb DNA domain with two divergently orientated genes is the unit of organisation of the murine major urinary protein genes, *EMBO J.* **3**, 2055–2064.

Cooper, G.M., et al. (2004) Characterization of evolutionary rates and constraints in three mammalian genomes, *Genome Res.* **14**, 539–548.

Cooper, G.M., Brudno, M., Green, E.D., Batzoglou, S., Sidow, A. (2003) Quantitative estimates of sequence divergence for comparative analyses of mammalian genomes, *Genome Res.* **13**, 813–820.

Costas, J. (2003) Molecular characterization of the recent intragenomic spread of the murine endogenous retrovirus MuERV-L, *J. Mol. Evol.* **56**, 181–186.

Dehal, P., et al. (2002) The draft genome of *Ciona intestinalis*: insights into chordate and vertebrate origins, *Science* **298**, 2157–2167.

Dermitzakis, E.T., et al. (2002) Numerous potentially functional but non-genic conserved sequences on human chromosome 21, *Nature* **420**, 578–582.

Dermitzakis, E.T., et al. (2003) Evolutionary discrimination of mammalian conserved non-genic sequences (CNGs), *Science* **302**, 1033–1035.

Dewannieux, M., Esnault, C., Heidmann, T. (2003) LINE-mediated retrotransposition of marked Alu sequences, *Nat. Genet.* **35**, 41–48.

Duret, L., Mouchiroud, D. (2000) Determinants of substitution rates in mammalian genes: expression pattern affects selection intensity but not mutation rate, *Mol. Biol. Evol.* **17**, 68–74.

Eichler, E.E. (1998) Masquerading repeats: paralogous pitfalls of the human genome, *Genome Res.* **8**, 758–762.

Eichler, E.E. (2001) Segmental duplications: what's missing, misassigned, and misassembled – and should we care? *Genome Res.* **11**, 653–656.

Elnitski, L., et al. (2003) Distinguishing regulatory DNA from neutral sites, *Genome Res.* **13**, 64–72.

Emes, R.D., Goodstadt, L., Winter, E.E., Ponting, C.P. (2003) Comparison of the genomes of human and mouse lays the foundation of genome zoology, *Hum. Mol. Genet.* **12**, 701–709.

Emes, R.D., Beatson, S.A., Ponting, C.P., Goodstadt, L. (2004) Evolution and comparative genomics of odorant- and pheromone-associated genes in rodents, *Genome Res.* **14**, 591–602.

Felsenfeld, A., Peterson, J., Schloss, J., Guyer, M. (1999) Assessing the quality of the DNA sequence from the human genome project, *Genome Res.* **9**, 1–4.

Goff, S.A., et al. (2002) A draft sequence of the rice genome (*Oryza sativa* L. ssp. *japonica*), *Science* **296**, 92–100.

Graves, J.A., Gecz, J., Hameister, H. (2002) Evolution of the human X – a smart and sexy chromosome that controls speciation and development, *Cytogenet. Genome Res.* **99**, 141–145.

Green, P. (2002) Whole-genome disassembly, *Proc. Natl. Acad. Sci. U.S.A.* **99**, 4143–4144.

Green, H., Wang, N. (1994) Codon reiteration and the evolution of proteins, *Proc. Natl. Acad. Sci. U.S.A.* **91**, 4298–4302.

Gumucio, D.L., et al. (1992) Phylogenetic footprinting reveals a nuclear protein which binds to silencer sequences in the human and globin genes, *Mol. Cell. Biol.* **12**, 4919–4929.

Gurates, B., et al. (2002) WT1 and DAX-1 inhibit aromatase P450 expression in human endometrial and endometriotic stromal cells, *J. Clin. Endocrinol. Metab.* **87**, 4369–4377.

Guy, J., et al. (2003) Genomic sequence and transcriptional profile of the boundary between pericentromeric satellites and genes on human chromosome arm 10p, *Genome Res.* **13**, 159–172.

Haldi, M.L., et al. (1997) Construction of a large-insert yeast artificial chromosome library of the rat genome, *Mamm. Genome* **8**, 284.

Hardison, R., et al. (1993) Comparative analysis of the locus control region of the rabbit-like gene cluster: HS3 increases transient expression of an embryonic-globin gene, *Nucleic Acids Res.* **21**, 1265–1272.

Hardison, R.C., et al. (2003) Covariation in frequencies of substitution, deletion, transposition, and recombination during eutherian evolution, *Genome Res.* **13**, 13–26.

Havlak, P., et al. (2004) The Atlas genome assembly system, *Genome Res.* **14**, 721–732.

Hayward, B.E., Zavanelli, M., Furano, A.V. (1997) Recombination creates novel L1 (LINE-1) elements in *Rattus norvegicus*, *Genetics* **146**, 641–654.

Hillier, L.W., et al. (2003) The DNA sequence of human chromosome 7, *Nature* **424**, 157–164.

Horvath, J.E., et al. (2003) Using a pericentromeric interspersed repeat to recapitulate the phylogeny and expansion of human centromeric segmental duplications, *Mol. Biol. Evol.* **20**, 1463–1479.

Huang, H., et al. Evolutionary conservation of human disease gene orthologs in the rat and mouse genomes. *Genome Biol.* (submitted).

Hubbard, T., et al. (2002) The Ensembl genome database project, *Nucleic Acids Res.* **30**, 38–41.

Hurst, L.D. (2002) The Ka/Ks ratio: diagnosing the form of sequence evolution, *Trends Genet.* **18**, 486–487.

Hurst, J.L., et al. (2001) Individual recognition in mice mediated by major urinary proteins, *Nature* **414**, 631–634.

International Human Genome Sequencing Consortium. (2001) Initial sequencing and analysis of the human genome. *Nature* **409**, 860–921.

Jaffe, D.B., et al. (2003) Whole-genome sequence assembly for mammalian genomes: arachne 2, *Genome Res.* **13**, 91–96.

Jensen-Seaman, M.I., et al. (2004) Comparative recombination rates in the rat, mouse, and human genomes, *Genome Res.* **14**, 528–538.

Kalafus, K.J., Jackson, A.R., Milosavljevic, A. (2004) Pash: efficient genome-scale sequence anchoring by positional hashing, *Genome Res.* **14**, 672–678.

Kent, W.J., Baertsch, R., Hinrichs, A., Miller, W., Haussler, D. (2003) Evolution's cauldron: duplication, deletion, and rearrangement in the mouse and human genomes, *Proc. Natl. Acad. Sci. U.S.A.* **100**, 11484–11489.

Kirkness, E.F., et al. (2003) The dog genome: survey sequencing and comparative analysis, *Science* **301**, 1898–1903.

Kolbe, D., et al. (2004) Regulatory potential scores from genome-wide 3-way alignments of human, mouse and rat, *Genome Res.* **14**, 700–707.

Krzywinski, M., et al. (2004) Integrated and sequence-ordered BAC and YAC-based physical maps for the rat genome, *Genome Res.* **14**, 766–779.

Kwitek, A.E., et al. (2004) High density rat radiation hybrid maps containing over 24,000 SSLPs, genes, and ESTs provide a direct link to the rat genome sequence, *Genome Res.* **14**, 750–757.

Levinson, G., Gutman, G.A. (1987) Slipped-strand mispairing: a major mechanism for DNA sequence evolution, *Mol. Biol. Evol.* **4**, 203–221.

Li, X., Waterman, M.S. (2003) Estimating the repeat structure and length of DNA sequences using L-tuples, *Genome Res.* **13**, 1916–1922.

Loots, G.G., Ovcharenko, I., Pachter, L., Dubchak, I., Rubin, E.M. (2002) rVista for comparative sequence-based discovery of functional transcription factor binding sites, *Genome Res.* **12**, 832–839.

Ma, B., Tromp, J., Li, M. (2002) PatternHunter: faster and more sensitive homology search, *Bioinformatics* **18**, 440–445.

Makalowski, W., Boguski, M.S. (1998) Evolutionary parameters of the transcribed mammalian genome: an analysis of 2,820 orthologous rodent and human sequences, *Proc. Natl. Acad. Sci. U.S.A.* **95**, 9407–9412.

Margulies, E.H., Blanchette, M., Haussler, D., Green, E. (2003) Identification and characterization of multi-species conserved sequences, *Genome Res.* **13**, 2507–2518.

Marra, M.A., et al. (1997) High throughput fingerprint analysis of large-insert clones, *Genome Res.* **7**, 1072–1084.

Martin, S.L., Bushman, F.D. (2001) Nucleic acid chaperone activity of the ORF1 protein from the mouse LINE-1 retrotransposon, *Mol. Cell. Biol.* **21**, 467–475.

Misra, S., et al. (2002) Annotation of the *Drosophila melanogaster* euchromatic genome: a systematic review. *Genome Biol.* **3**(12), RESEARCH0083. Epub 2002 Dec. 31.

Modrek, B., Lee, C.J. (2003) Alternative splicing in the human, mouse and rat genomes is associated with an increased frequency of exon creation and/or loss, *Nat. Genet.* **34**, 177–180.

Mouse Genome Sequencing Consortium. (2002) Initial sequencing and comparative analysis of the mouse genome. *Nature* **420**, 520–562.

Mural, R.J., et al. (2002) A comparison of whole-genome shotgun-derived mouse chromosome 16 and the human genome, *Science* **296**, 1661–1671.

Murphy, W.J., Fronicke, L., O'Brien, S.J., Stanyon, R. (2003) The origin of human chromosome 1 and its homologs in placental mammals, *Genome Res.* **13**, 1880–1888.

Murphy, W.J., Sun, S., Chen, Z.Q., Pecon-Slattery, J., O'Brien, S.J. (1999) Extensive conservation of sex chromosome organization between cat and human revealed by parallel radiation hybrid mapping, *Genome Res.* **9**, 1223–1230.

Murphy, W.J., Bourque, G., Tesler, G., Pevzner, P., O'Brien, S.J. (2003) Reconstructing the genomic architecture of mammalian ancestors using multispecies comparative maps, *Hum. Genomics* **1**, 30–40.

Myers, E.W., et al. (2000) A whole-genome assembly of *Drosophila*, *Science* **287**, 2196–2204.

Myers, E.W., Sutton, G.G., Smith, H.O., Adams, M.D., Venter, J.C. (2002) On the sequencing and assembly of the human genome, *Proc. Natl. Acad. Sci. U.S.A.* **99**, 4145–4146.

Nadeau, J.H., Taylor, B.A. (1984) Lengths of chromosomal segments conserved since divergence of man and mouse, *Proc. Natl. Acad. Sci. U.S.A.* **81**, 814–818.

Nekrutenko, A. (2004) Identification of novel exons from rat-mouse comparisons, *J. Mol. Evol.* **59**(5), 703–708.

Nekrutenko, A., Makova, K.D., Li, W.H. (2002) The KA/KS ratio test for assessing the protein-coding potential of genomic regions: an empirical and simulation study, *Genome Res.* **12**, 198–202.

Nekrutenko, A., Chung, W.Y., Li, W.H. (2003) An evolutionary approach reveals a high protein-coding capacity of the human genome, *Trends Genet.* **19**, 306–310.

Nelson, D.R. (1999) Cytochrome P450 and the individuality of species, *Arch. Biochem. Biophys.* **369**, 1–10.

Osoegawa, K., et al. (2004) BAC Resources for the rat genome project, *Genome Res.* **14**, 780–785.

Ostertag, E.M., Kazazian, H.H. Jr. (2001) Biology of mammalian L1 retrotransposons, *Annu. Rev. Genet.* **35**, 501–538.

Pennacchio, L.A., Rubin, E.M. (2001) Genomic strategies to identify mammalian regulatory sequences, *Nat. Rev. Genet.* **2**, 100–109.

Pevzner, P., Tesler, G. (2003a) Human and mouse genomic sequences reveal extensive breakpoint reuse in mammalian evolution, *Proc. Natl. Acad. Sci. U.S.A.* **100**, 7672–7677.

Pevzner, P., Tesler, G. (2003b) Genome rearrangements in mammalian evolution: lessons from human and mouse genomes, *Genome Res.* **13**, 37–45.

Pruitt, K.D., Maglott, D.R. (2001) RefSeq and LocusLink: NCBI gene-centered resources, *Nucleic Acids Res.* **29**, 137–140.

Puente, X.S., Lopez-Otin, C.A. (2004) A genomic analysis of rat proteases and protease inhibitors, *Genome Res.* **14**, 609–622.

Puente, X.S., Sanchez, L.M., Overall, C.M., Lopez-Otin, C. (2003) Human and mouse proteases: a comparative genomic approach, *Nat. Rev. Genet.* **4**, 544–558.

Quentin, Y. (1994) A master sequence related to a free left Alu monomer (FLAM) at the origin of the B1 family in rodent genomes, *Nucleic Acids Res.* **22**, 2222–2227.

Rat Genome Sequencing Project Consortium. (2004) Genome sequence of the Brown Norway rat yields insights into mammalian evolution, *Nature* **428**, 493–521.

Riethman, H., et al. (2004) Mapping and initial analysis of human subtelomeric sequence assemblies, *Genome Res.* **14**, 18–28.

Roskin, K.M., Diekhans, M., Haussler, D. (2003) in: Vingron, M., Istrail, S., Pevzner, P., Waterman, M. (Eds.) *Proceedings of the 7th Annual International Conference on Research in Computational Biology (RECOMB 2003)*, ACM Press, New York, pp. 257–266, doi:10.1145/640075.640109.

Rothenburg, S., Eiben, M., Koch-Nolte, F., Haag, F. (2002) Independent integration of rodent identifier (ID) elements into orthologous sites of some RT6 alleles of *Rattus norvegicus* and *Rattus rattus*, *J. Mol. Evol.* **55**, 251–259.

Rouquier, S., Blancher, A., Giorgi, D. (2000) The olfactory receptor gene repertoire in primates and mouse: evidence for reduction of the functional fraction in primates, *Proc. Natl. Acad. Sci. U.S.A.* **97**, 2870–2874.

Roy-Engel, A.M., et al. (2002) Non-traditional Alu evolution and primate genomic diversity, *J. Mol. Biol.* **316**, 1033–1040.

Saitou, N., Nei, M. (1987) The neighbor-joining method: a new method for reconstructing phylogenetic trees, *Mol. Biol. Evol.* **4**, 406–425.

Salem, A.H., et al. (2003) Alu elements and hominid phylogenetics, *Proc. Natl. Acad. Sci. U.S.A.* **100**, 12787–127891.

Salem, A.H., Kilroy, G.E., Watkins, W.S., Jorde, L.B., Batzer, M.A. (2003) Recently integrated Alu elements and human genomic diversity, *Mol. Biol. Evol.* **20**, 1349–1361.

Schwartz, S., et al. (2003) Human–mouse alignments with BLASTZ, *Genome Res.* **13**, 103–107.

Smit, A.F. (1993) Identification of a new, abundant superfamily of mammalian LTR-transposons, *Nucleic Acids Res.* **21**, 1863–1872.

Springer, M.S., Murphy, W.J., Eizirik, E., O'Brien, S.J. (2003) Placental mammal diversification and the Cretaceous-Tertiary boundary, *Proc. Natl. Acad. Sci. U.S.A.* **100**, 1056–1061.

Stanyon, R., Stone, G., Garcia, M., Froenicke, L. (2003) Reciprocal chromosome painting shows that squirrels, unlike murid rodents, have a highly conserved genome organization, *Genomics* **82**, 245–249.

Steen, R.G., et al. (1999) A high-density integrated genetic linkage and radiation hybrid map of the laboratory rat, *Genome Res.* **9**, (insert), AP1–AP8.

Stenson, P.D., et al. (2003) Human Gene Mutation Database (HGMD): 2003 update, *Hum. Mutat.* **21**, 577–581.

Tagle, D.A., et al. (1988) Embryonic epsilon and globin genes of a prosimian primate (*Galago crassicaudatus*). Nucleotide and amino acid sequences, developmental regulation and phylogenetic footprints, *J. Mol. Biol.* **203**, 439–455.

Taylor, M.S., Ponting, C.P., Copley, R.R. (2004) Occurrence and consequences of coding sequence insertions and deletions in mammalian genomes, *Genome Res.* **14**, 555–566.

The International Human Genome Mapping Consortium. (2001) A physical map of the human genome, *Nature* **409**, 934–941.

Thomas, J.W., et al. (2003a) Pericentromeric duplications in the laboratory mouse, *Genome Res.* **13**, 55–63.

Thomas, J.W., et al. (2003b) Comparative analyses of multi-species sequences from targeted genomic regions, *Nature* **424**, 788–793.

Torrents, D., Suyama, M., Bork, P. (2003) A genome-wide survey of human pseudogenes, *Genome Res.* **13**, 2559–2567.

Trinklein, N.D., Aldred, S.J., Saldanha, A.J., Myers, R.M. (2003) Identification and functional analysis of human transcriptional promoters, *Genome Res.* **13**, 308–312.

Tuzun, E., Bailey, J.A., Eichler, E.E. (2004) Recent segmental duplications in the working draft assembly of the brown Norway rat, *Genome Res.* **14**, 493–506.

Venter, J.C., et al. (2001) The sequence of the human genome, *Science* **291**, 1304–1351.

Ventura, M., Archidiacono, N., Rocchi, M. (2001) Centromere emergence in evolution, *Genome Res.* **11**, 595–599.

Vitt, U., et al. (2004) Identification of candidate disease genes by EST alignments, synteny and expression and verification of Ensembl genes on rat chromosome 1q43-54, *Genome Res.* **14**, 640–650.

Waterston, R.H., Lander, E.S., Sulston, J.E. (2002) On the sequencing of the human genome, *Proc. Natl. Acad. Sci. U.S.A.* **99**, 3712–3716.

Waterston, R.H., Lander, E.S., Sulston, J.E. (2003) More on the sequencing of the human genome. *Proc. Natl. Acad. Sci. U.S.A.* **100**, 3022–3024; author reply (2003) **100**, 3025–3026.

Wingender, E., et al. (2001) The TRANSFAC system on gene expression regulation, *Nucleic Acids Res.* **29**, 281–283.

Wolfe, K.H., Sharp, P.M. (1993) Mammalian gene evolution: nucleotide sequence divergence between mouse and rat, *J. Mol. Evol.* **37**, 441–456.

Yang, S., et al. (2004) Patterns of insertions and their covariation with substitutions in the rat, mouse and human genomes, *Genome Res.* **14**, 517–527.

Yang, Z., Goldman, N., Friday, A. (1994) Comparison of models for nucleotide substitution used in maximum-likelihood phylogenetic estimation, *Mol. Biol. Evol.* **11**, 316–324.

Yap, V.B., Pachter, L. (2004) Identification of evolutionary hotspots in the rodent genomes, *Genome Res.* **14**, 574–579.

Young, J.M., et al. (2002) Different evolutionary processes shaped the mouse and human olfactory receptor gene families, *Hum. Mol. Genet.* **11**, 535–546.

Yu, J., et al. (2002) A draft sequence of the rice genome (*Oryza sativa* L. ssp. *indica*), *Science* **296**, 79–92.

Yunis, J.J., Prakash, O. (1982) The origin of man: a chromosomal pictorial legacy, *Science* **215**, 1525–1530.

Zhang, X., Firestein, S. (2002) The olfactory receptor gene superfamily of the mouse, *Nat. Neurosci.* **5**, 124–133.

Zhang, Z., Harrison, P., Gerstein, M. (2002) Identification and analysis of over 2000 ribosomal protein pseudogenes in the human genome, *Genome Res.* **12**, 1466–1482.

19
Chimpanzee Genome

Ingo Ebersberger
Institute for Bioinformatics, Heinrich-Heine-University Düsseldorf, Germany

1	**Human and Great Ape Phylogeny** 553	
1.1	Phylogenetic Analyses Based on Protein Comparisons 555	
1.1.1	Immunological Comparison of Proteins 555	
1.1.2	Electrophoretic Comparison of Proteins 555	
1.1.3	Amino Acid Sequence Comparison 556	
1.2	Phylogenetic Analyses Based on DNA Sequence Comparisons 556	
1.2.1	Comparative DNA–DNA Hybridization 557	
1.2.2	Mitochondrial DNA 558	
1.2.3	Nuclear DNA 558	
1.3	Genetic Diversity 559	
2	**Comparison of the Chimpanzee and Human Genomes** 561	
2.1	Cytogenetic Differences 561	
2.2	Subchromosomal Rearrangements and Repeat Content 562	
2.3	Substitutional DNA Sequence Differences 563	
2.4	Insertions and Deletions 564	
3	**Chimpanzees in Evolutionary Research** 565	
3.1	DNA Sequence Evolution 565	
3.2	From Genotype to Phenotype 567	
3.3	Comparative Analysis of Gene Expression 568	
4	**Chimpanzees in Biomedical Research** 569	
4.1	Are Chimpanzees Good Model Organisms? 569	
4.2	Biomedical Differences Between Humans and Chimpanzees 570	

Genomics and Genetics. Edited by Robert A. Meyers.
Copyright © 2007 Wiley-VCH Verlag GmbH & Co. KGaA, Weinheim
ISBN: 978-3-527-31609-0

| 4.3 | Chimpanzees and Infectious Diseases | 571 |
| 4.4 | Ethical Considerations | 572 |

5	**The Chimpanzee Genome Project**	**573**
5.1	The Quest for Differences	573
5.2	Sequencing Strategy and Requirements	573
5.3	Human, Chimpanzee, and... Future Perspectives	575

Bibliography 576
Books and Reviews 576
Primary Literature 576

Keywords

Genetic Distance
An estimate of the amount of nucleotide substitutions between species. Genetic distance serves as a measure of the time elapsed since two species last shared a common ancestor.

Great Apes
Collective name for the orangutans (*Pongo pygmaeus*) from Borneo and Sumatra and for the African apes: chimpanzees (*Pan troglodytes*), bonobos (*Pan paniscus*), and gorillas (*Gorilla gorilla*).

Lineage Sorting
The distribution of alleles to the daughter populations upon a split of the ancestral population, for example, during speciation.

Monophyly
A group of species in a phylogenetic tree is monophyletic if all included species are descendants from a common ancestor to which no other species outside the group can be traced. Monophyletic groups are also referred to as clades or branches of a phylogenetic tree.

Phenotype
The complete set of organizational properties of an individual. It resembles the expressed information encoded in the genotype.

Phylogeny
A hypothesis about the evolutionary relationships of species expressed in a tree. All parts of a phylogenetic tree are inferred except the tips, which represent extant species or fossil remains.

On the basis of comparative molecular studies, chimpanzees are the closest living relatives of humans. On the level of their genomes, this is reflected in a nearly identical chromosomal organization and an average DNA sequence difference only about 10 times higher than that between any two humans. This renders the comparison of chimpanzees and humans, and the resulting catalog of their genetic differences, suitable to gain relevant insights into human evolution and human disease. The collection of genetic differences between both species as a whole comprises an ideal data set to analyze how DNA sequences change over time on a high-resolution scale. Of further interest are those differences that are located in the functional regions of the genome. They form the genetic basis of the distinct biological properties of humans and chimpanzees and thus might bear an answer to the question "What makes us human?" More practically, these differences provide access to the molecular factors that account for the different disease spectra in humans and chimpanzees, and might, therefore, open up new therapeutic approaches to fight human disease.

The relevance of chimpanzees as a subject in evolutionary and biomedical research is obvious. However, their close relationship to humans demands ethical guidelines that clearly distinguish chimpanzees from common genetic model organisms in laboratory research.

1
Human and Great Ape Phylogeny

For more than a century, scientists have been busy reconstructing the phylogenetic relationship of humans and the great apes. On the basis of the presence of shared morphological or molecular characters, individual species can be grouped to clades of an evolutionary tree. At the end of the nineteenth century, morphological evidence led to the generally accepted view that the great apes together with gibbons (*Hylobates* sp. and *Symphalangus* sp.) and humans share a common ancestor from which no other living organism descended. In phylogenetic terms, they represent the monophyletic group of hominoids within the primates (Fig. 1).

While this view has remained unchallenged since then, the relative order in which the lineages to the individual hominoid species emerged – often referred to as the "branching order" – provoked intense debate. From the shared consensus that the gibbon lineage diverged first, several competing hypotheses were proposed concerning the branching order of the remaining species. According to one favored scenario, the African apes (chimpanzees, bonobos, and gorillas) branched off second, allying humans with orangutans (Fig. 2a). An alternative view already taken by Darwin and Huxley positions the African apes closer to humans than the orangutans. Obviously, this hypothesis requires a further decision whether the closest relatives of chimpanzees and bonobos are the gorillas – as most morphologists believed – (Fig. 2b) or whether either of the species is closer to humans (Fig. 2c, d). The limited number of phylogenetic informative morphological characters as well as their sometimes

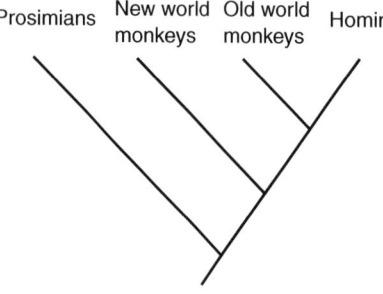

Fig. 1 A phylogenetic tree for the anthropoid primates. The prosimians combine the Lemur species from Madagascar, the nocturnal Lorises from South and Southeast Asia and from Africa, and the Tarsiers from certain islands of Southeast Asia. Generally, prosimians seem to be more ancestral in their morphology compared to the other three groups. The New World monkeys are confined to South and Central America and are, with one exception, the owl monkey, active during the day. They differ from the third group, the Old World monkeys typically in having a flat nose and separate nostrils. Old World monkeys such as rhesus macaques and baboons are found in Africa as well as in certain parts of Asia. The hominoids combine the gibbons (Asia), the orangutans (Asia), the African apes (gorillas, chimpanzees, bonobos), and humans.

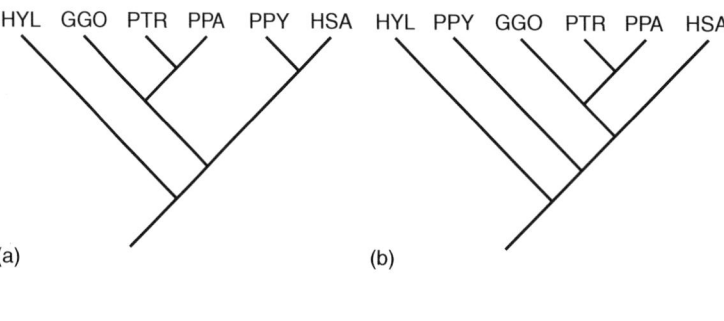

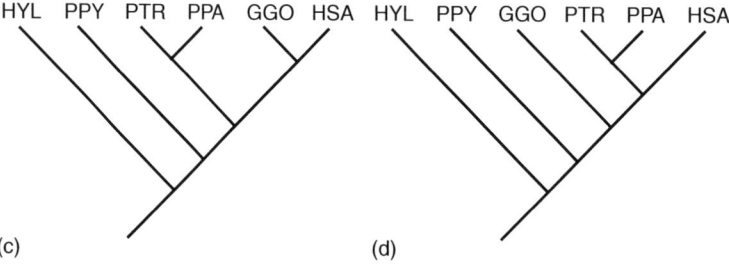

Fig. 2 Alternative phylogenies of the hominoids. HYL: gibbons; PPY: orangutans; GGO: gorillas; PTR: chimpanzees; PPA: bonobos; HSA: humans.

problematic interpretation rendered morphologists unable to conclusively solve the question of hominoid phylogeny.

In the second half of the twentieth century, the expansion of molecular biology provided a new and presumably more powerful approach to phylogenetic studies. The genetic distance between species as reflected in the amount of structural differences between their proteins and nucleic acids was introduced as a more objective measure of their evolutionary relationship. Under the assumption that the accumulation of such differences behaves like a clock that is synchronized in all species under consideration, those species with a smaller genetic distance are more closely related.

1.1
Phylogenetic Analyses Based on Protein Comparisons

1.1.1 Immunological Comparison of Proteins

The first attempt to resolve primate phylogeny using molecular evidence was based on immunological comparisons of their serum proteins. As the underlying principle, an antiserum against a protein from species X is obtained by immunizing, for example, rabbits or chickens. The degree of similarity between the protein of species X and the corresponding protein in a second species Y is then determined by the extent of cross-reactivity of anti-X-serum and Y-protein. A quantitative measure of similarity is achieved since the antiserum comprises a mixture of antibodies, each recognizing a distinct surface structure (epitope) of the protein used for immunization. The more closely two species are related, the more similar are their proteins and thus, the more epitopes are shared among them. As a consequence, the extent of cross-reactivity increases.

Initial immunological comparisons of primate proteins used antisera against whole blood serum, a rather poorly defined protein mixture. Their results supported the rough classification of anthropoid primates into the major groups already suggested by morphologists (Fig. 1). However, no further resolution was possible since the antigenic properties of hominoid blood sera are too similar to each other to be differentiated by these means.

In later years, increasing understanding of the molecular processes of immune response in combination with the ability to generate antisera against individual proteins facilitated the development of more sensitive immunological assays. On the basis of such improved methods, the African apes and not orangutans were assigned as the closest relatives of humans. But subsequent determination of the branching order of the individual African ape species and humans failed for mainly the following reason. For proteins that differ only slightly, the correlation between the observed immunological difference and the genetic distance is not absolute. As an example, protein A from species X differs from the corresponding protein in species Z in three positions. By chance, none of these differences affect the antigenic properties of the respective proteins; therefore, no difference will be detected in an immunological assay with anti-A_X-serum. Protein A from a third species, Y, however, differs only in a single position from A_X which, again by chance, affects an epitope recognized by the anti-A_X-serum. The resulting lower cross-reactivity with anti-A_X-serum erroneously suggests a higher genetic distance between species Y and X than between species Z and X and thus, a grouping of X and Z to the exclusion of Y. This fuzziness in combination with a substantial experimental effort for the individual assays renders immunological comparisons not suitable to infer the evolutionary relationships of closely related species.

1.1.2 Electrophoretic Comparison of Proteins

Proteins contain amino acids with basic or acidic side chains and thus are charged molecules under most pH-conditions. This circumstance is exploited to separate proteins on a matrix according to their size and their net charge by application of an electric field. Additional resolution can be achieved by a preceding separation of the proteins according to their isoelectric point – the pH where the net charge is 0 (2D electrophoresis). Variants of a particular protein that differ in their net charge because of sequence differences involving charged amino acids can be identified by their different mobility on the matrix. The application of protein electrophoresis in phylogenetic analyses is based on the fact that the degree of difference in the mobility pattern between related proteins approximately reflects their extent of amino acid sequence difference. Thus, similar to immunological comparisons, protein electrophoresis provides an estimate of the genetic distance between species that forms the basis of phylogenetic conclusions.

The most extensive study to reconstruct the phylogeny of hominoids by means of protein electrophoresis was based on a 2D electrophoresis of 383 fibroblast proteins. Concordant to conclusions from immunological distances of the hominoids, it supports an early separation of the gibbons followed by the lineage leading to orangutans. Notably, the study suggested a smaller genetic distance between humans and chimpanzees than between either species and gorillas. This lends support to the model that humans are the closest relatives to chimpanzees and bonobos (Fig. 2d).

However, similar to the immunological comparison of proteins, electrophoretic analyses can detect only a certain subset of amino acid differences between proteins. Only about 30% of all amino acid substitutions alter the charge and, therefore, the mobility pattern of a protein during electrophoresis. For closely related species, this poses the same resolution problem as already discussed for the immunological comparison of proteins.

1.1.3 Amino Acid Sequence Comparison

The direct comparison of their amino acid sequences represents the most accurate way to assess the degree of difference between two proteins. Even though experimentally elaborative, it provides the opportunity to also trace those changes in the protein sequence that neither affect the immunological properties nor the charge of a protein.

Despite this better resolution, the phylogenetic insights from comparative sequence analyses of hominoid proteins are limited. A data set of 1271 amino acid positions from 9 different polypeptides revealed an average similarity between human and chimpanzee proteins of 99.6%. A similar analysis with a smaller data set comparing humans, chimpanzees, and gorillas determined the average amino acid sequence similarity between humans and gorillas and between chimpanzees and gorillas to be slightly lower (99.3%). Even though this implies a closer relationship between humans and chimpanzees to the exclusion of gorillas, the number of informative positions was far too low to allow decisive conclusions.

Hindered by the disproportionate experimental effort, this data set has never been substantially enlarged. Rather, the focus of evolutionary biologists shifted from the comparison of proteins to the comparison of DNA sequences between species.

1.2
Phylogenetic Analyses Based on DNA Sequence Comparisons

At least two reasons suggest that comparative analyses of DNA sequences are more suitable for reconstructing the phylogeny of closely related species than the comparison of proteins. First, the molecular clock of DNA sequences ticks faster than that of protein sequences and, therefore, provides a better resolution in time. This is because in coding sequences of genes, three types of positions can be distinguished: nondegenerated sites where any of the three possible nucleotide substitutions is nonsynonymous, that is, it alters the encoded amino acid. Twofold-degenerated sites where only two of the three possible nucleotide substitutions are nonsynonymous. And finally, fourfold-degenerated sites where no nucleotide substitution changes the encoded amino acid. Thus, only a subset of DNA sequence changes translates into changes of the respective protein. Furthermore, a change of the protein sequence generally interferes with protein function and is therefore commonly removed by purifying selection. Therefore, the accumulation of nonsynonymous DNA sequence differences and the consequent accumulation of amino acid sequence differences in proteins is considerably slow. In contrast, nucleotide sequence changes that leave the encoded protein unaltered are less likely subject to purifying selection and thus, accumulate differences at a substantially higher rate. Second, comparative molecular analyses can be extended to those regions in the genome that are not protein-coding, and in the optimal case, not functional at all. The latter sequences should be most informative for phylogenetic analyses of closely related species since they are likely to evolve free of selective constraint.

1.2.1 Comparative DNA–DNA Hybridization

DNA occurs in nature usually as a double helix where the two complementary strands are connected by hydrogen bonds formed between guanine and cytosine, and adenine and thymine respectively. Heating DNA duplexes breaks the hydrogen bonds and the two strands are separated – the DNA is "denatured" or "melted". The progress of this process can be quantified as it gradually increases the UV absorption at 260 nm of the DNA solution. Upon cooling, complementary DNA strands will rehybridize to duplex DNA. Performing the rehybridization in the presence of DNA from a second species, heteroduplexes can be formed where the two complementary strands stem from different species.

The temperature required to separate 50% of the DNA duplexes, also called the *melting temperature* (T_m), increases, depending on a number of factors: (1) The guanine and cytosine content (GC content) of the DNA: guanine and cytosine pair with three hydrogen bonds whereas adenine and thymine pair only with two hydrogen bonds; (2) the length of the analyzed fragments; (3) the degree of sequence similarity between the two strands forming the duplex. Accordingly, controlling for base composition and duplex length, the difference between the melting temperature of DNA heteroduplexes and that of the corresponding homoduplexes – both strands stem from the same species – (ΔT_m) serves as a measure for the genetic difference between species. For closely related species, the correlation between ΔT_m and the number of mismatches between the complementary strands of a DNA duplex is approximately linear.

An application of this method to the problem of hominoid phylogeny confirmed the conclusions from protein comparisons that gibbons, followed by orangutans, are the most diverged species. Furthermore, additional support was given to the view that humans and chimpanzees shared a common ancestor after the separation of the gorilla lineage.

However, conclusions based on hybridization experiments have been treated with caution since ΔT_m is likely to be partially determined by factors of uncertain identity. The unclear influence of experimental conditions and of genomic organization such as the presence of duplications, which facilitates hybridization of nonhomologous DNA sequences, were among the main points of critique.

1.2.2 Mitochondrial DNA

In the pre-PCR era, the use of nucleic acids for phylogenetic analyses was substantially complicated by the need to reproducibly isolate the region of interest – usually a fragment of a few hundred base pairs (bp) – from a genome several million to few billion base pairs in size. To minimize this problem, initial focus was laid on the genome of mitochondria, which has a size of only approximately 16 000 bp in humans, and is generally present in more than 1000 copies per cell.

Isolated mitochondrial genomes from the hominoid species have been compared by two different methods. Restriction enzymes – enzymes that cut DNA at specific recognition sites – can be used to test for the presence and relative position of their respective cleavage site in a DNA sequence (the phylogenetic informative character in this approach). On the basis of restriction maps of their mitochondrial genomes, chimpanzees were placed – in contrast to results from other molecular comparisons – closer to gorillas than to humans. However, little statistical support was provided for that conclusion and alternative phylogenies were almost as likely.

Similar to the situation with comparative analyses of proteins, the most accurate way to determine the exact degree of difference between nucleic acids is to directly compare their sequences. Parts, and later, the sequences of the entire mitochondrial genome of hominoids were determined and compared. In these studies, chimpanzees appear again to be closer related to humans than to gorillas.

1.2.3 Nuclear DNA

Meanwhile, modern techniques of molecular biology provide a rapid and reproducible access to the DNA sequence of almost any region in a genome. As a consequence, a variety of data sets have been collected from the nuclear genome that address the question of hominoid phylogeny with particular regard to the relationship of humans, chimpanzees, and gorillas (Table 1). For practical reasons, early studies compared DNA sequences of coding regions from genes. Later, increasing focus was laid on the comparison of noncoding and potentially nonfunctional DNA sequences such as introns, pseudogenes, or intergenic regions. These DNA sequences appear to be the best possible data source for phylogenetic studies on closely related species. However, as can be clearly seen in Table 1, the supported phylogenetic relationship of humans, chimpanzees, and gorillas varies according to the analyzed genomic region.

Apparently, two factors have prevented the hominoid phylogeny from being determined by means of molecular comparisons: first, the insufficient accuracy and resolution of most comparative methods, and second, ambiguous information content in the molecular data itself. In order to determine which of the alternative phylogenies represents the true order in which the lineages to the individual hominoid species emerged (species tree), it is necessary to understand why phylogenetic trees estimated from DNA

Tab. 1 Comparative DNA sequence data sets to resolve the phylogeny of humans, chimpanzees, and gorillas.

Number of genomic loci	Total compared length [bp]	Numbers of loci supporting phylogeny[a]			
		(H,C)G	(H,G)C	(C,G)H	(H,C,G)
45	46 855	23	8	8	6
53	24 237	31	10	12	0
51	62 530	26	10	12	3

Notes: H: human; C: chimpanzee; G: gorilla.
[a] The grouped species shared a common ancestor after the separation of the third species. (H,C,G) represents those loci that could not resolve the phylogeny (trifurcation).

sequences (gene trees) can differ from the species tree. Mainly two problems need to be considered: first, sampling only a finite number of nucleotides from a locus produces a stochastic sampling error and thus, bears a probability to infer an incorrect gene tree; second, the model where the time point at which corresponding DNA sequences from different species last shared a common ancestor is identical to the time point of speciation is too simplistic. Figure 3 shows an example where the presence of sequence variants of the analyzed locus in the ancestral population in combination with lineage sorting can cause gene tree–species tree incongruities. Especially for species where the amount of genetic polymorphisms in their ancestral population was high and the time between subsequent speciation events was short – both seem to apply to the common ancestor of chimpanzees, humans, and gorillas – lineage sorting renders phylogenetic conclusions based on only a single locus unreliable. Therefore, a joint analysis of multiple independent DNA sequence data sets was required to eventually end the debate concerning the phylogeny of hominoids with sufficient statistical support. Humans are the closest relatives of chimpanzees and based on molecular evidence, both species separated approximately 4.6 to 6.2 million years ago. The likewise estimated date for the speciation of gorillas ranges from 6.2 to 8.4 million years before present.

1.3
Genetic Diversity

DNA sequence evolution is a continual process in time. Therefore, DNA sequence

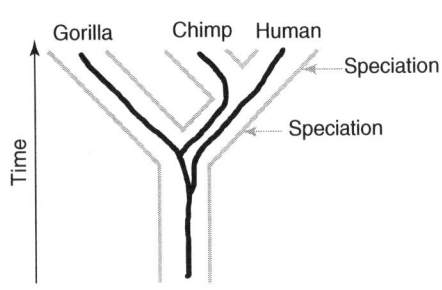

Fig. 3 Gene tree–species tree incongruities due to ancestral polymorphisms and lineage sorting. The gray and black lines represent the species tree (true phylogeny) and the gene tree respectively.

differences exist not only as divergence between species but also as diversity within species. In analogy to the use of divergence to reconstruct a phylogenetic tree that helps infer the evolutionary relationships among species, diversity data can be used to model a within-species tree representing the relationships of sequence variants at a given locus in a population. Such a genealogical tree can help clarify the demographic history of an individual species. Following a genealogical tree backwards in time, the genetic lineages represented by the individual alleles will gradually join until only a single lineage remains. This lineage from which all extant sequence variants are derived is termed the *most recent common ancestor* (MRCA). Making an assumption about the mutation rate and the long-term population size of the species – diversity increases with an increase of either parameter – the estimated time point of the MRCA can be inferred and reflects the amount of diversity present in the extant population.

For chimpanzees and humans, the date of their MRCAs as assessed from single nucleotide polymorphism data is 1 900 000 years and 500 000 years before present respectively. Accordingly, contemporary chimpanzees are two to fourfold more diverse compared to contemporary humans. What can explain the differing amounts of genetic diversity between both species? Since there is clear indication that the mutation rate in chimpanzees and humans is approximately the same, a difference in their population history is proposed as the underlying cause. Contemporary humans are, presumably, descendants of a small African population that subsequently expanded and spread over the entire world. Owing to the small size of the founding population, the majority of DNA sequence variation present in the ancestral human population has been lost, and thus, the genetic diversity in present day humans is low. Chimpanzees, in contrast, are believed to have not undergone such a substantial population size decrease. Therefore, a larger proportion of old genetic lineages could contribute to their present day diversity.

From the analysis of genetic diversity in chimpanzees, a second interesting aspect emerges. According to the location of their habitats in Africa, three groups of chimpanzees are distinguished: western, central, and eastern chimpanzees. From the perspective of DNA sequence variation of the mitochondrial genome and at a Y chromosomal locus, each of the three groups appear to be genetically unique. No shared DNA sequence variants for these loci have been observed, which indicate an independent evolution of the three chimpanzee populations. As a consequence, they were given the status of a subspecies: western chimpanzee (*Pan troglodytes verus*), central chimpanzee (*P. t. troglodytes*) and eastern chimpanzee (*P. t. schweinfurthii*). However, no such genetic uniqueness is seen at autosomal and X chromosomal loci. This was initially taken as an indication that the respective chimpanzee populations do not evolve independently and raised doubts on the biological significance of their classification into different subspecies. Recently, an alternative explanation has been proposed that would be consistent with an independent evolution of the chimpanzee populations. Subsequent to the reproductive isolation of populations, shared ancestral polymorphisms will gradually become extinct by genetic drift, that is, the stochastic loss of alleles caused by the fact that some individuals in a generation have no offspring whereas others have several. Eventually, no ancestral alleles will be retained and the respective populations are genetically unique. It is now crucial for the case of the chimpanzee populations that the time needed to complete this process depends on the effective population size and thus varies among the analyzed loci. Mitochondrial and Y chromosomal loci will have removed ancestral sequence variants first, since their effective population size is the smallest – only one sequence variant is passed on to the next generation. X chromosomal loci will need on an average triple the time since they have a threefold higher effective population size – two copies in females, and one copy in males. And the time for autosomal loci to become genetically unique is about four times that of mitochondrial and Y chromosomal loci since on an average four copies – two copies each in males and females – are passed on to the next generation. As a consequence, the presence of shared sequence variants at biparentally inherited loci in the three chimpanzee populations in contrast to their genetic uniqueness at mitochondrial and Y chromosomal loci could indicate that the subspecies started to evolve independently – in evolutionary scales – only recently. Therefore, the time was not sufficient to establish genetic uniqueness at all loci.

2
Comparison of the Chimpanzee and Human Genomes

Early comparative studies of their proteins and DNA sequences have already provided evidence that chimpanzees and humans are genetically highly similar. However, differences exist that are informative in two ways. They help understand how genomes diverge subsequent to reproductive isolation of two populations. And they help identify those changes that form the genetic basis of the phenotypic properties differentiating chimpanzees from humans.

Four categories of differences between the genomes of chimpanzees and humans will be discussed in the following text.

2.1
Cytogenetic Differences

The microscopic comparison of chimpanzee and human chromosomes revealed a small number of cytogenetic differences, that is, differences in either chromosome number or differences in presence, order, and position of large chromosomal segments. Their detection was greatly simplified by chromosome banding techniques as well as fluorescence *in situ* hybridization (FISH), methods that allow a specific and reproducible labeling of individual chromosomal regions. The most apparent difference between the genomes of chimpanzees and humans is the number of chromosome pairs per cell. While chimpanzees – as well as gorillas and orangutans – have 24 pairs, humans have only 23. A comparison of the chromosome banding patterns between the species revealed that human chromosome 2 is present as two separate chromosomes (12 and 13) in chimpanzees (Fig. 4a). A subsequent molecular characterization revealed that a fusion at the ends (telomeres) of two ancestral chromosomes gave rise to human chromosome 2.

Seven additional chromosomes, human chromosomes 4, 5, 9, 12, 15, 16, and 17 differ between humans and chimpanzees by an inverted arrangement of chromosomal segments. In all cases, the inversion breakpoints are located on either side

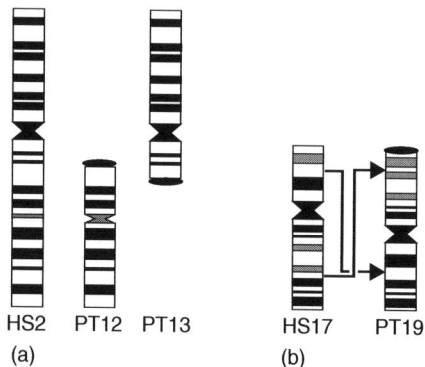

Fig. 4 Examples of cytogenetic differences between humans and chimpanzees. Chromosomes and chromosome bands are not drawn to scale. (a) Human chromosome 2 (HS2) is present as two separate chromosomes in chimpanzees (PT12, PT13). (b) Human chromosome 17 (HS17) and chimpanzee chromosome 19 (PT19) differ by a pericentric inversion. Arrows mark the inversion breakpoints.

of the centromere (pericentric inversion) (Fig. 4(b)). To date, the inversion breakpoints of a single human chromosome, human chromosome 17, have been analyzed on the DNA sequence level, and the breakpoints in three additional chromosomes, human chromosome 4, 9, and 12, have been mapped to intervals of 1 to 2 Mb. So far, no gene was found to be structurally modified by these inversions.

Finally, a number of chromosomes differ between chimpanzees and humans in the amount of constitutive heterochromatin, chromosomal regions that remain condensed throughout the cell cycle and generally do not contain genes.

Cytogenetic differences are discussed to comprise evolutionary important changes that contributed to the initial separation of chimpanzees and humans. Somewhat in contrast to this view, currently no evidence exists that the function of a gene has been affected by any of the observed chromosomal rearrangements. However, pericentric inversions can establish reproductive isolation between populations without the necessity of an altered gene function. In a heterozygous state, an individual carries one normal and one inverted copy of the chromosome, pericentric inversions can substantially interfere with the fertility of the individual. Crossing-over during meiosis in the inverted region generates recombination products that are generally incompatible with life. Thus, it can be speculated that the pericentric inversions differentiating the chromosomes of contemporary chimpanzees and humans comprised early reproductive barriers that facilitated the separation of the two species.

2.2
Subchromosomal Rearrangements and Repeat Content

In contrast to the well-defined extent of cytogenetic differences between humans and chimpanzees, the amount of genomic differences whose size is below the resolution of cytogenetic methods is still unclear. Subtelomeric regions, those genomic segments that are located directly adjacent to the chromosome ends, appear to be dynamic structures that undergo rapid changes over short evolutionary distances. For example, a subtelomeric region formed by a 32-bp repeat unit present at many chromosome ends of chimpanzees and almost all chromosome ends of gorillas is not found in humans. Similarly, a genomic locus that exists only as a single copy in the subtelomeric region of chimpanzee chromosome 17 is found in eight additional chromosomes in humans. The reason for these rapid changes in the subtelomeric region of chromosomes is still unclear.

From the analysis of the human genome draft sequence, it emerged that about 5% of the DNA sequence – preferentially in the regions around the centromeres – represents evolutionary recent duplications of chromosomal segments with a DNA sequence identity of >90%. Interestingly, initial calculations reveal that more than 7000 exons are transcribed from these duplicated segments. Thus, gene duplications due to segmental duplications provide good candidates for species-specific evolution. Comparing the duplication pattern for human chromosome 22 between humans and chimpanzees, differences are seen, which imply that loss and gain of duplicated elements is an ongoing process in human and chimpanzee evolution. However, comparative analyses focusing on segmental duplications have just started. Therefore, the exact amount of differentially duplicated segments in the chimpanzee and human genomes as well as the number and identity of possibly involved genes remains to be determined.

Around 45% of the human genome consists of repetitive DNA sequences due to the expansion of various groups of mobile elements in the genome. As an example, a particular class of repeats termed *Alu-elements* are present in more than one million copies in the human genome comprising ~10% of the entire DNA sequence. Even though the overall mobility of transposable elements has markedly decreased over the past 35 to 50 million years, transposition is still ongoing in the human and, presumably, also in the chimpanzee genome.

Repetitive elements can propagate genetic differences between species by mainly two ways. First, the insertion of individual transposable elements into regulatory or coding regions of genes can alter expression and function of the affected genes. As an example, humans differ from chimpanzees – and from all other primates – by an inactivation of a gene *CMAH* whose product is essential for the generation of a certain glycoprotein. This inactivation is due to a human-specific insertion of an Alu-element that replaced a 92-bp long exon and, in addition, causes a frameshift in the subsequent coding sequence. The gene modified thus is expressed into a truncated and nonfunctional protein and is responsible for the only proven biochemical difference between chimpanzee and human cells to date. Second, intrachromosomal crossing-over mediated by direct or inverted repetitive elements can cause deletions and inversions, respectively. For example, exon 35 of the human tropoelastin gene has been deleted presumably by a recombination event between two flanking Alu-elements.

In summary, several factors can account for genetic differences between humans and chimpanzees on a subchromosomal scale. However, a determination of the extent to which they each contribute to the total amount of differences between both genomes must await more comprehensive studies.

2.3
Substitutional DNA Sequence Differences

The most frequently detected difference between chimpanzee and human DNA sequences is caused by substitutions of individual nucleotides. Initial studies suggested that per 200 compared nucleotide positions, only about 3 differ between humans and chimpanzees, equivalent to a mean sequence difference of around

1.5%. Many of these estimates were based on the comparison of short stretches of DNA from only one or few regions of the genome. However, different regions in the human genome differ in aspects such as gene content, transcription frequency, base composition, and recombination frequency all of which are suspected to influence the evolutionary rate of DNA sequences. Thus, sampling of DNA sequences from a multitude of loci distributed throughout the genome is required to obtain a relevant estimate of the average amount of substitutional DNA sequence differences between two species. On the basis of genome-wide comparisons of DNA sequences between humans and chimpanzees, their average amount of DNA sequence difference was recently determined to be 1.2%. The results furthermore reveal that indeed genomic regions vary substantially in their divergence among species (Fig. 5).

In absolute numbers, a mean DNA sequence difference of 1.2% corresponds to a total of about 40 million substitutional DNA sequence differences between the human and chimpanzee genomes. However, only a tiny proportion of the human genome – about 3% – is believed to be relevant for gene function. This suggests that less than (only) 1.2 million substitutional differences account for the distinct phenotypes of humans and chimpanzees.

2.4
Insertions and Deletions

Processes that are capable of removing nucleotides from DNA or adding nucleotides to DNA (deletions and insertions respectively) comprise a further source of genetic difference between species. Insertions and deletions mediated by transposable elements or by duplications of chromosomal segments have already been

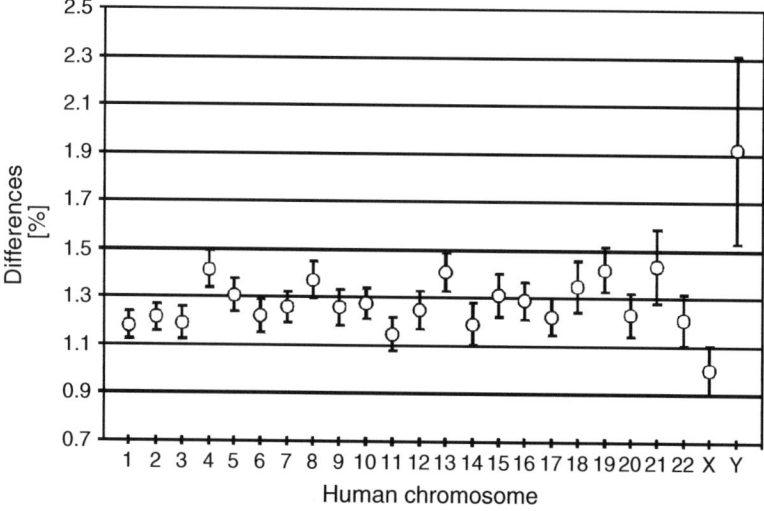

Fig. 5 Mean DNA sequence differences between humans and chimpanzees by human chromosome. Bars indicate 95% confidence intervals. [Reprinted with permission from Ebersberger, I., Metzler, D., Schwarz, C., Paabo, S. (2002) Genomewide comparison of DNA sequences between humans and chimpanzees, *Am. J. Hum. Genet.* **70**, 1490–1497.]

discussed above (Sect. 2.2). Each of these events generally affects a large number of DNA sequence positions; however, they presumably comprise considerably rare occurrences during human and chimpanzee evolution. Insertions and deletions extending only over a few nucleotides apparently occur far more frequently in DNA sequence evolution. A comparison of around only 700 kbp of genomic DNA sequence between humans and chimpanzees has detected already over 1000 individual insertions and deletions. About half of these extend over only 1 to 2 nucleotides and 95% do not exceed 20 bp in length.

Currently, it is estimated that a total of about 4% of DNA sequence in the chimpanzee and human genomes has no counterpart in the respective other species. This indicates that insertions and deletions contribute substantially both in the number of events and in the number of affected positions to the DNA sequence differences between species. Since insertions and deletions also appear to be commonly located in functional regions of the genome, they need to be considered as likely causes for phenotypic differences between humans and chimpanzees.

3
Chimpanzees in Evolutionary Research

Prior to describing the relevance of chimpanzees in evolutionary analyses – I will only focus on biological evolution and omit the complex topic of the evolution of culture – the term *evolution* itself shall be briefly discussed. In the broadest sense, evolution can be regarded as mere change of distinct characters over time. In the more specific view of biologists, evolution represents heritable changes in a population that spread over many generations.

The only changes that can be stably inherited comprise modifications in the DNA sequence. Thus, in a strict sense, biological evolution is, ultimately, synonymous to DNA sequence evolution and evolutionary research aims at identifying the factors that determine emergence and accumulation of DNA sequence changes. In reality, however, it is far more practicable to study biological evolution at two levels: molecular evolution comprised of the change of DNA sequences over time, and the evolution of phenotypes leading to the differentiation of individual species. This distinction appears to be indispensable for the following reasons. The genetic basis of the vast majority of observed phenotypic differences between two species is generally, and in the better case, not well understood. Therefore, it is not feasible to study their evolution on the DNA sequence level. Furthermore, the majority of DNA sequence changes, at least in the genomes of mammals, has no effect on the phenotype since approximately 97% of the genome has no obvious function. As a consequence, the evolution of DNA sequences is in major parts independent of the evolution of phenotypes.

However, in order to arrive at a comprehensive understanding of biological evolution, the interdependence of DNA sequence evolution and phenotypic evolution needs to be determined. Here, the analysis of chimpanzees in comparison to humans is likely to provide relevant new insights.

3.1
DNA Sequence Evolution

The advantage of DNA sequence comparisons between chimpanzees and humans to shed light on the factors that determine DNA sequence evolution is twofold.

First, corresponding DNA sequences in the two species differ only in about 1.2% of the positions. This reduces the risk that individual positions have changed more than once in one species, or parallel in both species since the sequences last shared a common ancestor. Such multiple or parallel substitutions blur the correlation between the true and observed number of changes and comprise a substantial problem in the evolutionary analysis of DNA sequences. Second, the close relationship between chimpanzees and humans justifies the hope that factors determining rate and mode of DNA sequence change do not differ substantially between the two species. The approach to analyze DNA sequence evolution by a DNA sequence comparison between chimpanzees and humans is straightforward. Different regions in the human genome vary in divergence to the chimpanzee. Any genomic feature that is found to covary with the divergence can serve as a candidate to influence DNA sequence evolution. However, in order to correctly interpret the data, one pitfall needs to be regarded that has been already mentioned in Sect. 1.3. Genomic regions can differ in their divergence between humans and chimpanzees even though they evolve at the same rate. This problem arises since the time point at which the human and chimpanzee DNA sequences last shared a common ancestor is not the same throughout the genome. It is required to take this into account in order to identify regions that truly differ in their evolutionary rate.

When DNA sequences from different regions of the chimpanzee genome are compared to the corresponding sequences in humans, it is seen that different chromosomes vary in the amount of DNA sequence differences between humans and chimpanzees (Fig. 5). Interestingly, the extent of divergence for the individual chromosomes is correlated with the evolutionary time they spend in the male germ line. The X chromosome, which spends the least time in the male germ line, displays also the least divergence (1.0%), the Y chromosome, which is confined to the male germ line, displays the highest divergence (1.9%), and the divergence of the autosomes, which spend an intermediate time in the male germ line ranges between that of the sex chromosomes (1.2%). Currently, it is assumed that a mutation rate about threefold higher in the male germ line compared to the female germ line is responsible for this observation. However, it is noteworthy that the evolutionary rates also differ among autosomes that spend the same amount of time in the male and female germ line respectively. This indicates the existence of further factors that influence the accumulation of DNA sequence differences in the human genome. Recently, a second genomic feature has been found to covary with DNA sequence divergence between humans and chimpanzees. Regions with a high recombination rate also display a high divergence between the two species, while regions with a lower recombination rate have diverged less. This can be taken as an indication that the process of recombination is mutagenic and thus, drives DNA sequence evolution.

The amount of comparative DNA sequence data between chimpanzees is still limited. However, more extensive studies are likely to give further insights into how different factors interact to shape the rate and pattern of DNA sequence evolution in the genomes of humans and chimpanzees.

3.2
From Genotype to Phenotype

Maybe the most widely noticed field in evolutionary research is concerned with the emergence of new phenotypes that contribute to the uniqueness of individual species. Mainly three aspects are in the focus of interest: (1) When did the new phenotype arise? (2) Was it chance or selective advantage that caused its fixation in the species? (3) What are the genetic changes that underly the particular phenotype? While the answers to the first two questions are mainly of academic interest, the correlation between genotype and phenotype is – as will be discussed in Sect. 3.3 – of immediate practical use.

When comparing humans and chimpanzees, the impression is that both species are strikingly alike, yet show a certain set of well-defined differences. Meanwhile, an extensive catalog of phenotypic differences between humans and chimpanzees has been collected. Among the most well-known examples are bipedal walking, enlarged brain size, and spoken language, which distinguish humans from chimpanzees and all other primates. This set of phenotypic differences is opposed by a considerably low amount of genetic differences (see Sect. 2), which provides an excellent starting point for the correlation of genotype and phenotype. Two approaches can be chosen for such a venture. In a bottom-to-top strategy, the genomes of humans and chimpanzees are screened for differences that affect gene function. This can be either changes in the coding region of genes resulting in a structural change of the gene product, or in regions that control the expression pattern of a gene. While the identification of differences potentially affecting gene function is considerably simple, the subsequent determination of the likely effect on the phenotype is experimentally extremely demanding. In the first step, the function of the affected gene must be determined, and in the next step, the influence of the genetic change on the respective function of the gene product must be determined. This latter step will be especially problematic for such changes that affect the expression pattern of genes rather than the structure of the gene product. Eventually, inferences of the likely phenotypic consequence of the genetic change can be made. The reverse approach, a top-to-bottom strategy, starts from a recognized phenotypic difference between humans and chimpanzees. Certain genes that are known or suspected to contribute to the phenotype of interest are assigned as candidates whose DNA sequence can be screened for differences between humans and chimpanzees. Once species-specific differences are found for a candidate gene, the subsequent procedure resembles that of the bottom-to-top strategy with the subtle but essential difference that a particular phenotype has been already correlated with the gene.

To date, the study of candidate genes has been successfully applied at least in two studies, one of which originated from the observation of a structural difference between the cell surfaces of humans and great apes. It could be shown that this is due to an inactivation of an enzyme on the human lineage caused by a deletion in the corresponding gene (see Sect. 2.4). Even though this comprises a marked biochemical difference between human and chimpanzee cells with possible implications for the intercellular cross talk, the relevance of this difference for human evolution is still unclear. A second study focused on a gene *FOXP2*, which when mutated, causes a distinct type of speech impairment in humans.

A comparison of this gene between humans, chimpanzees, and other primates revealed that it carries two human-specific nucleotide substitutions, each of which translates into an amino acid substitution in the encoded protein. This becomes especially interesting in the light that this protein is evolutionarily highly conserved. Between humans and mice, only three amino acid changes are observed; two of which occur on the human lineage. Furthermore, there is indication that this gene has been the target of selection during recent human evolution. Both observations in combination can be judged as an indication that the respective gene could have a certain relevance for the evolution of spoken language in humans.

These examples demonstrate that, indeed, the comparison of humans and chimpanzees is capable of detecting genetic changes that can be correlated with phenotypic differences between both changes. This justifies the hope that in a step-by-step process, even the genetic frameworks of traits as complex as (spoken) language can be identified.

3.3
Comparative Analysis of Gene Expression

The phenotypic appearance of a species is not only dependent on the functional integrity of its genes but also on their correct activation with respect to time, place, and extent. Genetic changes that alter this construction plan of a species provide a powerful tool to achieve substantial phenotypic changes over a very short time period. In view of the low genetic difference between humans and chimpanzees, it was considered in early comparative studies that changes in a few "master genes" – genes that control gene expression at certain key points of development – account for the phenotypic differences between both species.

To date, such master changes, if they exist, remain to be identified. However, as a first step, evidence was provided that the pattern of gene expression indeed varies in a species-specific way. When the expression pattern of nearly 18 000 genes is analyzed for several human and chimpanzee individuals, it is found to be more similar among individuals of the same species than between species (Fig. 6). Starting from this observation, follow-up studies have to determine whether the evolution of gene expression is mainly due to regulatory changes in individual genes or whether a cluster of functionally related genes changes their expression in a concerted manner due to a switch in a shared master gene. Furthermore, those individual genes that have substantially changed

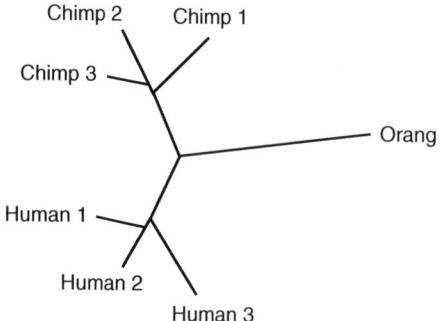

Fig. 6 Relative extent of the differences in the expression pattern within and between species. The expression pattern jointly determined from nearly 18 000 genes is found to be more similar within humans and chimpanzees respectively, than between both species.

their expression pattern in humans or chimpanzees comprise candidates for phenotypic differences between both species and thus, should be the subject of comparative studies as described in the previous section (Sect. 3.2).

4
Chimpanzees in Biomedical Research

The cure or prevention of diseases is one of the major interests in human research. In the study of disease-causing processes on the cellular and molecular level, nonhuman primates in general, and chimpanzees in particular, can be regarded as important model organisms. Their similarity to humans renders them susceptible to a number of pathogens correlated with human diseases. Thus, the infection of primates under controlled laboratory conditions allows the study of the onset and progress of the respective disease with a resolution far beyond what can be achieved by observatory studies in humans. Furthermore, as a result of the growing understanding of disease-causing processes, new therapeutical strategies and medicines will be developed that specifically target individual pathways in the human metabolism. Therefore, the traditional model organisms that are rather distantly related to humans may no longer be suitable to test the efficacy and safety of these medicines. Rather, model organisms such as the chimpanzee are required that resemble the human metabolism as closely as possible. Eventually, certain human diseases and pathogens appear to affect chimpanzees either to a lesser extent than humans, or not at all. The identification of the underlying molecular basis of this reduced susceptibility will provide starting points for new therapeutical approaches to cure the respective disease.

4.1
Are Chimpanzees Good Model Organisms?

The arguments listed above comprise a clear vote for a wide use of chimpanzees in biomedical research. Accordingly, numerous research facilities founded chimpanzee colonies, and in the 1980s, breeding programs were started to insure that enough chimpanzees were available for research needs. However, it appears that the similarity between chimpanzees and a (hypothetical) good model organism for biomedical research is very limited. Such an organism would be rather mouse-sized than chimp-sized. Not only is that desirable due to the limitation of laboratory space and budget for food, it also greatly facilitates handling of the animals during the experiments. Who ever has seen somebody pulling chimps by their tail out of the cage? And this is not because their tails are pretty much nonexisting! Furthermore, a good model organism would have a generation time and a life span substantially shorter than that of chimpanzees. The benefits would be twofold. First, it increases the value of the model organism for research on transgenic animals. The specific inactivation of genes in an organism (knockout) as well as their overexpression or exchange with functional homologs from other species (knockin) gains increasing importance in the study of function and evolution of genes and gene products. A minimal time span between the genetic modification of the individual – this generally happens at a very early stage in development – and the latest possible observation of the corresponding phenotypic effect is desirable for both kinds of studies. The latter time point is obviously determined by the life span of the model organism. Furthermore, once an individual carrying the desired genetic modification has been generated – which is still a costly, time-consuming, and experimentally elaborative venture – its maintenance as a laboratory strain is generally desirable. This, however, is only feasible when the generation time of the respective species is in a reasonable range of at most a year. A slowly developing, late reproducing, and long living species such as chimpanzees is thus, not particularly well suited for standard research on transgenic animals. Second, an early (natural) death of the research animal elegantly solves the problem of what to do with the animals once the experiments are over. Chimpanzees, however, have an average life span of around 40 to 50 years, which is well beyond the duration of most experimental studies. The relevance of this problem becomes immediately obvious when one imagines a group of chimpanzees infected with a human pathogen that need to be taken care of once they are no longer useful for experimental studies.

Clearly, strong arguments exist to use chimpanzees as experimental animals in biomedical research. However, the practical disadvantages are apparent. Together with a number of ethical concerns that I will discuss below (Sect. 4.4), the use of chimpanzees as a standard experimental model organism is clearly voted against. Meanwhile, this has been widely recognized and the number of chimpanzees held as laboratory animals has substantially decreased over the past years. In the case of Europe, invasive experimental studies on chimpanzees have entirely ended.

4.2
Biomedical Differences Between Humans and Chimpanzees

Practical experience has shown that chimpanzees are not suitable as experimental model organisms for studying human diseases, or for testing the efficacy and safety of new pharmaceuticals. Nevertheless, a certain relevance remains for chimpanzees in biomedical research. From long-term observations in primate centers and zoos, differences in susceptibility and severity of certain human diseases between humans and chimpanzees have emerged. A few examples are listed in Table 2.

Tab. 2 Differences in the disease spectra between humans and chimpanzees.

Medical condition	Humans	Chimpanzees	Evidence
HIV progression to AIDS	Common	Very rare	Definite
Late complication in hepatitis B/C	Frequent	Uncommon	Definite
Malaria[a]	Susceptible	Resistant	Definite
Menopause	Universal	Rare	Definite
Influenza A symptomatology	Moderate to severe	Mild	Likely
Alzheimer's disease pathology	Common	Rare	Likely
Myocardial infarction	Common	Uncommon	Likely
Epithelian cancers	Common	Rare	Likely

[a] Caused by *Plasmodium falciparum* infection.

Although the evidence is in parts fragmentary and vague – in certain cases it has mere anecdotal character – the medical relevance of involved diseases such as AIDS, hepatitis, or certain types of cancer is sufficient to draw attention to this issue. Differences in the disease spectrum between two species can be attributed to two different causes: differences in environmental influences such as diet and differing genetic predispositions. In recent years, the environmental conditions for captive chimpanzees have become increasingly similar to those of humans. This, however, had no substantial effect on the differences of their respective disease spectra. Therefore, the existence of crucial genetic differences between the two species must be assumed whose identification comprises the main relevance of biomedical studies on chimpanzees. Conveniently, the identification of candidates for these genetic differences can be achieved by a comparison of the genome sequences of chimpanzees and humans and does not require experimental studies on chimpanzees.

As an outlook, the medical consequences of biomedical studies on chimpanzees are likely to be revolutionary. Once the genetic basis is understood as to why this species is less susceptible to certain human diseases, cure or even prevention of these diseases can be imagined by mimicking the chimpanzee phenotype in well-defined disease-relevant parts of human biology.

4.3
Chimpanzees and Infectious Diseases

Maybe the most widely recognized biomedical difference between chimpanzees and humans is their different susceptibility to AIDS upon infection with HIV-1 (Human Immunodeficiency Virus 1), the most common cause of AIDS in humans. While chimpanzees rarely display AIDS-like syndromes, the progression to AIDS in humans is common. This difference is generally explained by the evolutionary history of HIV-1. It is suggested that HIV-1 originated from the chimpanzee virus SIV (Simian Immunodeficiency Virus) that crossed the species border only about 50 years ago. Thus, the low susceptibility to AIDS in chimpanzees could be due to a long coevolution of virus and host in a way that contemporary chimpanzees get infected by the virus but due to an efficient control of the virus load do not develop the disease. Humans, in contrast, who have become exposed to the virus only recently, lack the immunological ability for virus control and thus, suffer from AIDS. Obviously, the missing development of the disease renders chimpanzees inadequate as a model to study the progress of AIDS. Other primate species, such as the rhesus macaque appear to be more suitable since AIDS-like symptoms have been observed in these species subsequent to infection with SIV. However, the genetic reasons for the adapted immune response that cause the relative resistance against AIDS in chimpanzees are of considerable interest. They might point toward novel therapeutic approaches to prevent the progression to AIDS in HIV-infected humans. Initial insights suggest that differences in the allele combination in the major histocompatibility complex – a collection of genes correlated with immune response – between humans and chimpanzees is responsible for their varying susceptibility to AIDS.

Similar to the case of infection with HIV-1 and progression to AIDS, chimpanzees reduce the risk of long-term complications of hepatitis infections by an effective control of the virus load in the organism. Thus, there appears to be a general capability in chimpanzees to deal more effectively with virus infections than humans. It will be the main task of future studies to determine the exact elements of the virus–host relationship in chimpanzees that account for this better virus control.

In conclusion, although chimpanzees are not suited as experimental model organisms, there appears to be sufficient reason for the comparison of humans and chimpanzees to become a major field in biomedical research. However, one should be aware that the insights from past biomedical studies in chimpanzees have not met the high initial expectations. The risk remains that the biomedical relevance of genetic differences between humans and chimpanzees is too difficult to interpret just by observatory studies. As a consequence, a substantially increased experimental effort might be required whose realization ultimately depends on the establishment of new experimental systems that replace chimpanzees as a model organism. Such systems should combine a maximum of the advantages provided by chimpanzees as a model organism with a minimum of its practical and ethical concerns.

4.4
Ethical Considerations

Chimpanzees, like all other great apes, are an endangered species that, due to human interaction, face the extinction of their wild populations. From this perspective, it seems somewhat odd to discuss their use for experimental research beneficial primarily to humans. However, chimpanzees breed well in captivity, which insures that enough animals are available for research without affecting the wild populations.

In addition, it can be argued that every study, even those with fatal outcome for the participating individual, eventually benefits chimpanzees as a species. For example, vaccines and drugs to cure human diseases developed on the basis of experimental research on chimpanzees will be effective also in chimpanzees. A scenario can be imagined where these pharmaceuticals help save a wild chimpanzee population whose survival is endangered by such a disease. In combination, both arguments would, presumably, be sufficient for a well-controlled use of almost any other species – even one that faces its extinction – as an experimental animal. However, the situation is more complex for chimpanzees. An increasing number of comparative studies reveal that there is no clear-cut qualitative difference distinguishing humans from chimpanzees in a number of areas that have initially been regarded as "specific to humans." Many points of similarities exist in cognition and psychology between chimpanzees and humans. For example, it was repeatedly suggested that chimpanzees can understand and use rudimentary forms of language – even though they cannot speak due to anatomical limitations – and there is certainly tool usage in chimpanzees. In addition, only recently, it was appreciated that chimpanzees have cultural traditions similar to humans. Given these insights and given the high genetic similarity between humans and chimpanzees, it might be worthwhile to reconsider the clear-cut distinction in "human being" and "animal". The resulting implications for the use of chimpanzees in research are obvious and well recognized. In brief, only such studies should be done with chimpanzees that would be done with humans who are considered not to be able to provide informed consent. As a legal consequence, the government of New Zealand has extended three basic human rights to chimpanzees: (1) the right not to be deprived of life; (2) the right not to be subjected to torture and cruel treatment; and (3) the right not to be subjected to medical or scientific research that is not in the best interest of this individual.

Whether these ventures and agreements are sufficient to protect chimpanzees from abuse in human research will be seen in the future. As soon as comparative studies of humans and chimpanzees lead to fundamental questions that can be solved neither by DNA sequence comparisons nor by transgenic experiments in model systems, research proposals will arise that include the generation of transgenic – that will be ultimately "humanized" – chimpanzees. Therefore, it is an ethical imperative to accompany an extensive comparative study on the chimpanzee genome with a program that studies ethical, legal, and social implications of such research.

5
The Chimpanzee Genome Project

Even under careful consideration of all practical and ethical limitations that are correlated with comparative studies between chimpanzees and humans, the insights based on the availability of a chimpanzee genome sequence are likely to be substantial. As a consequence, the US National Human Genome Research Institute (NHGRI) has recently assigned chimpanzees the top priority for whole-genome sequencing, a decision that might be a pioneer for future research on nonhuman primates. The choice of sequencing the genome of chimpanzees rather than that of the rhesus macaque – the most widely used primate in biomedical research – seems to put more weight on the evolutionary aspect of a human–nonhuman primate comparison than on its immediate value for experimental biomedical studies.

5.1
The Quest for Differences

With respect to the list of eukaryotes whose genome has been entirely sequenced, the choice of determining the genome sequence of chimpanzees comprises the first step into the next phase of comparative genomics. Comparing the genome sequences from yeast, worm, fly, mouse, and human led to the identification of shared, and thus, evolutionarily conserved elements. Follow-up studies mainly aimed at determining the function of these conserved elements and generating a rough view on the genetic framework underlying organismal organization. However, adding chimpanzees to that list shifts the focus ultimately from conserved genomic elements toward the differences between two genomes. As a consequence, the initial outcome of a genome comparison between humans and chimpanzees will be a comprehensive map of their DNA sequence differences. As a second step, the functional consequences of these differences need to be analyzed and, eventually, a fine-scale view on the functional organization of the human genome will emerge.

5.2
Sequencing Strategy and Requirements

Sequencing an entire genome with more than three billion bases is still an enterprise that requires careful planning and strategical considerations. Recent sequencing efforts, including the genomes of *Drosophila*,

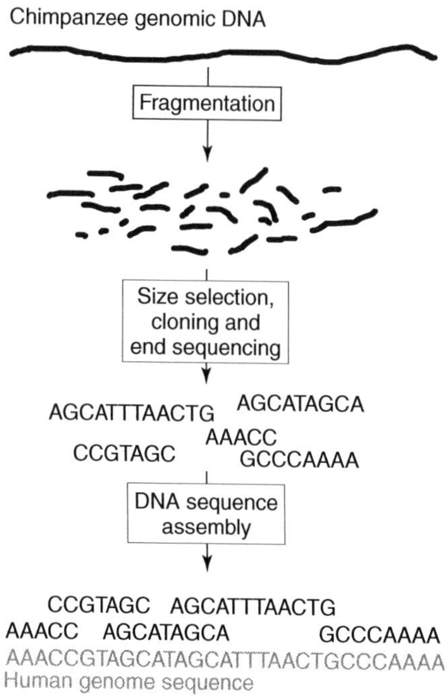

Chimpanzee genomic DNA
↓
Fragmentation
↓
↓
Size selection, cloning and end sequencing
↓
AGCATTTAACTG AGCATAGCA
         AAACC
CCGTAGC   GCCCAAAA
↓
DNA sequence assembly
↓
CCGTAGC AGCATTTAACTG
AAACC AGCATAGCA   GCCCAAAA
AAACCGTAGCATAGCATTTAACTGCCCAAAA
Human genome sequence

Fig. 7 Whole-genome shotgun strategy to sequence the chimpanzee genome. The entire genome of an individual is fragmented at random positions and genomic libraries with a defined insert size are generated. Subsequently, random clones are picked from the genomic libraries and the DNA sequences of their insert ends are determined. In the assembly step, the individual DNA sequences are aligned to reconstruct a contiguous consensus sequence that represents the original genomic sequence of the chimpanzee. This step is greatly facilitated for the chimpanzee genome since the human genome sequence can serve as a guideline for the assembly.

human, and mouse clearly demonstrated that the whole-genome shotgun sequencing strategy is the fastest and most efficient way to determine an initial draft sequence of a genome. The general principle of this strategy is outlined in Fig. 7. While the generation of the shotgun sequence reads is considerably straightforward, their assembly to a comprehensive consensus sequence comprises the crucial and most problematic step. This step, however, will be greatly facilitated for the chimpanzee genome. Given the high similarity between both genomes, it will be in most cases straightforward to align the chimpanzee DNA sequence fragments to the human genome sequence. However, special care must be given to those genomic regions that differ between the two species. Human regions not present in chimpanzees will cause gaps in the alignment, while chimpanzee DNA sequences not present in humans will fail to align against the human genome. In addition, recent duplications as well as inversions of genomic regions comprise further obstacles on the way toward a comprehensive draft version of the chimpanzee genome.

Following the shotgun phase of a genome project, remaining gaps and uncertainties in the draft sequence must be removed in the so-called finishing phase. This sequence-finishing is tedious, time-consuming and thus, expensive. However, it is assumed that the majority of comparative studies between chimpanzees and humans will not substantially benefit from a finished, rather than a draft version of the chimpanzee genome. Thus, it is currently considered to initially leave the chimpanzee genome sequence in a draft status. It is estimated that such a draft sequence will cover more than 90% of the chimpanzee genome.

In conclusion, a first draft of the chimpanzee genome sequence can be generated with considerably low effort and is likely to be available by the end of 2003. However, there is one important issue when working with this draft sequence that played only a very minor role in previous comparative studies of genome sequences. Errors in the draft sequence of the chimpanzee genome, but also in the human genome sequence due to errors in the assembly process or due to sequencing errors suddenly gain relevance. An attempt to identify conserved regions between distantly related species is conservative with respect to the quality of the draft sequence. Regions that have been found conserved between species using early versions of the draft sequence have a high probability of remaining conserved after sequencing errors in the draft sequence have been corrected. It is self-evident that the situation is certainly more difficult when the focus lays on DNA sequence differences between genomes as in the case of the human–chimpanzee comparison. To minimize this problem, every DNA sequence position in the draft version of the chimpanzee genome should get assigned a confidence value to allow a quick assessment of its reliability.

5.3
Human, Chimpanzee, and... Future Perspectives

With the availability of the chimpanzee genome draft sequence, the first step into the next stage of comparative genomics will be completed. A comprehensive catalog of genetic differences between humans and chimpanzees will emerge that ultimately reveals those changes responsible for the distinct phenotypic traits of both species. The identification of those presumably few functional changes and the dating of their presumed occurrence will be the main task of future studies. To accomplish this goal, it will be necessary to exactly determine the extent and pattern of genetic variation both within humans and within chimpanzees. This not only helps in identifying those genuine genetic differences between both species that cannot be attributed to polymorphisms in either species but also helps pinpoint those regions in the genomes of chimpanzees and humans for which fewer than average sequence variants exist. Such a reduction of diversity can comprise signatures of selection during recent human and chimpanzee evolution and provide hints for the likely importance of changes that are close by in functional regions of the genome.

Eventually, the detection of genuine genetic differences between humans and chimpanzees can only then be interpreted in a meaningful way when the lineage on which the corresponding change occurred has been determined. Therefore, the genome of a third, more distantly related species is required as an outgroup that helps infer whether an observed DNA sequence difference between humans and chimpanzees is due to a change on the human or the chimpanzee lineage. Presumably, a second anthropoid primate would be optimal as such an outgroup. In order to make full use of the genomic comparison of humans and chimpanzees, the genome sequence of the outgroup species should be determined immediately following the completion of the chimpanzee genome sequence. Likely candidate species are two Old World monkeys regularly used in biomedical research: the rhesus macaque (*Macaca mulatta*), and the baboon (*Papio hamadryas*), or the orangutan (*P. pygmaeus*). Which of these species to choose depends largely on future plans to use primates as experimental animals.

See also Molecular Basis of Genetics; Genomic Sequencing.

Bibliography

Books and Reviews

Bontrop, R.E. (2001) Non-human primates: essential partners in biomedical research, *Immunol. Rev.* **183**, 5–9.

Gagneux, P., Varki, A. (2001) Genetic differences between humans and great apes, *Mol. Phylogenet. Evol.* **18**, 2–13.

Goodman, M. (1999) The genomic record of Humankind's evolutionary roots, *Am. J. Hum. Genet.* **64**, 31–39.

Hacia, J.G. (2001) Genome of the apes, *Trends Genet.* **17**, 637–645.

Kaessmann, H., Paabo, S. (2002) The genetical history of humans and the great apes, *J. Intern. Med.* **251**, 1–18.

King, M.C., Wilson, A.C. (1975) Evolution at two levels in humans and chimpanzees, *Science* **188**, 107–116.

Olson, M.V., Varki, A. (2003) Sequencing the chimpanzee genome: insights into human evolution and disease, *Nat. Rev. Genet.* **4**, 20–28.

Primary Literature

Arnason, U., Gullberg, A., Janke, A. (1998) Molecular timing of primate divergences as estimated by two nonprimate calibration points, *J. Mol. Evol.* **47**, 718–727.

Arnason, U., Gullberg, A., Janke, A., Xu, X. (1996a) Pattern and timing of evolutionary divergences among hominoids based on analyses of complete mtDNAs, *J. Mol. Evol.* **43**, 650–661.

Arnason, U., Xu, X., Gullberg, A. (1996b) Comparison between the complete mitochondrial DNA sequences of Homo and the common chimpanzee based on nonchimeric sequences, *J. Mol. Evol.* **42**, 145–152.

Bailey, J.A., Gu, Z., Clark, R.A., Reinert, K., Samonte, R.V., Schwartz, S., Adams, M.D., Myers, E.W., Li, P.W., Eichler, E.E. (2002a) Recent segmental duplications in the human genome, *Science* **297**, 1003–1007.

Bailey, J.A., Yavor, A.M., Viggiano, L., Misceo, D., Horvath, J.E., Archidiacono, N., Schwartz, S., Rocchi, M., Eichler, E.E. (2002b) Human-specific duplication and mosaic transcripts: the recent paralogous structure of chromosome 22, *Am. J. Hum. Genet.* **70**, 83–100.

Britten, R.J. (2002) Divergence between samples of chimpanzee and human DNA sequences is 5%, counting indels, *Proc. Natl. Acad. Sci. U.S.A.* **99**, 13633–13635.

Caccone, A., Powell, J.R. (1989) DNA divergence among hominoids, *Evolution* **43**, 925–942.

Casane, D., Boissinot, S., Chang, B.H., Shimmin, L.C., Li, W. (1997) Mutation pattern variation among regions of the primate genome, *J. Mol. Evol.* **45**, 216–226.

Chen, F.C., Li, W.H. (2001) Genomic divergences between humans and other hominoids and the effective population size of the common ancestor of humans and chimpanzees, *Am. J. Hum. Genet.* **68**, 444–456.

Chou, H.H., Hayakawa, T., Diaz, S., Krings, M., Indriati, E., Leakey, M., Paabo, S., Satta, Y., Takahata, N., Varki, A. (2002) Inactivation of CMP-N-acetylneuraminic acid hydroxylase occurred prior to brain expansion during human evolution, *Proc. Natl. Acad. Sci. U.S.A.* **99**, 11736–11741.

Chou, H.H., Takematsu, H., Diaz, S., Iber, J., Nickerson, E., Wright, K.L., Muchmore, E.A., Nelson, D.L., Warren, S.T., Varki, A. (1998) A mutation in human CMP-sialic acid hydroxylase occurred after the Homo-Pan divergence, *Proc. Natl. Acad. Sci. U.S.A.* **95**, 11751–11756.

deGroot, N.G., Otting, N., Doxiadis, G.G., Balla-Jhagjhoorsingh, S.S., Heeney, J.L., van Rood, J.J., Gagneux, P., Bontrop, R.E. (2002) Evidence for an ancient selective sweep in the MHC class I gene repertoire of chimpanzees, *Proc. Natl. Acad. Sci. U.S.A.* **99**, 11748–11753.

Ebersberger, I., Metzler, D., Schwarz, C., Paabo, S. (2002) Genomewide comparison of DNA sequences between humans and chimpanzees, *Am. J. Hum. Genet.* **70**, 1490–1497.

Enard, W., Khaitovich, P., Klose, J., Zollner, S., Heissig, F., Giavalisco, P., Nieselt-Struwe, K.,

Muchmore, E., Varki, A., Ravid, R., Doxiadis, G.M., Bontrop, R.E., Paabo, S. (2002a) Intra- and interspecific variation in primate gene expression patterns, *Science* **296**, 340–343.

Enard, W., Przeworski, M., Fisher, S., Lai, C., Wiebe, V., Kitano, T., Monaco, A., Paabo, S. (2002b) Molecular evolution of FOXP2, a gene involved in speech and language, *Nature* **418**, 869–872.

Felsenstein, J. (1987) Estimation of hominoid phylogeny from a DNA hybridization data set, *J. Mol. Evol.* **26**, 123–131.

Ferris, S.D., Wilson, A.C., Brown, W.M. (1981) Evolutionary tree for apes and humans based on cleavage maps of mitochondrial DNA, *Proc. Natl. Acad. Sci. U.S.A.* **78**, 2432–2436.

Gagneux, P., Wills, C., Gerloff, U., Tautz, D., Morin, P.A., Boesch, C., Fruth, B., Hohmann, G., Ryder, O.A., Woodruff, D.S. (1999) Mitochondrial sequences show diverse evolutionary histories of African hominoids, *Proc. Natl. Acad. Sci. U.S.A.* **96**, 5077–5082.

Gibbons, A. (1998) Which of our genes make us human?, *Science* **281**, 1432–1434.

Goldman, D., Giri, P.R., O'Brien, S.J. (1987) A molecular phylogeny of the hominoid primates as indicated by two-dimensional protein electrophoresis, *Proc. Natl. Acad. Sci. U.S.A.* **84**, 3307–3311.

Goodman, M. (1962) Evolution of the immunologic species specificity of human serum proteins, *Hum. Biol.* **34**, 104–150.

Goodman, M., Koop, B.F., Czelusniak, J., Fitch, D.H., Tagle, D.A., Slightom, J.L. (1989) Molecular phylogeny of the family of apes and humans, *Genome* **31**, 316–335.

Goodman, S. (2001) Europe brings experiments on chimpanzees to an end, *Nature* **411**, 123.

Hasegawa, M., Kishino, H., Yano, T. (1987) Man's place in Hominoidea as inferred from molecular clocks of DNA, *J. Mol. Evol.* **26**, 132–147.

Hayakawa, T., Satta, Y., Gagneux, P., Varki, A., Takahata, N. (2001) Alu-mediated inactivation of the human CMP-N-acetylneuraminic acid hydroxylase gene, *Proc. Natl. Acad. Sci. U.S.A.* **98**, 11399–11404.

Horai, S., Satta, Y., Hayasaka, K., Kondo, R., Inoue, T., Ishida, T., Hayashi, S., Takahata, N. (1992) Man's place in Hominoidea revealed by mitochondrial DNA genealogy, *J. Mol. Evol.* **35**, 32–43.

Ijdo, J.W., Baldini, A., Ward, D.C., Reeders, S.T., Wells, R.A. (1991) Origin of human chromosome 2: an ancestral telomere-telomere fusion, *Proc. Natl. Acad. Sci. U.S.A.* **88**, 9051–9055.

Johnson, M.E., Viggiano, L., Bailey, J.A., Abdul-Rauf, M., Goodwin, G., Rocchi, M., Eichler, E.E. (2001) Positive selection of a gene family during the emergence of humans and African apes, *Nature* **413**, 514–519.

Kaessmann, H., Wiebe, V., Paabo, S. (1999) Extensive nuclear DNA sequence diversity among chimpanzees, *Science* **286**, 1159–1162.

Kehrer-Sawatzki, H., Schreiner, B., Tanzer, S., Platzer, M., Muller, S., Hameister, H. (2002) Molecular characterization of the pericentric inversion that causes differences between chimpanzee chromosome 19 and human chromosome 17, *Am. J. Hum. Genet.* **71**, 2.

McConkey, E.H., Fouts, R., Goodman, M., Nelson, D., Penny, D., Ruvolo, M., Sikela, J., Stewart, C.B., Varki, A., Wise, S. (2000) Proposal for a human genome evolution project, *Mol. Phylogenet. Evol.* **15**, 1–4.

McConkey, E.H., Goodman, M. (1997) A human genome evolution project is needed, *Trends Genet.* **13**, 350–351.

McConkey, E.H., Varki, A. (2000) A primate genome project deserves high priority, *Science* **289**, 1295–1296.

Monfouilloux, S., Avet-Loiseau, H., Amarger, V., Balazs, I., Pourcel, C., Vergnaud, G. (1998) Recent human-specific spreading of a subtelomeric domain, *Genomics* **51**, 165–176.

Muchmore, E.A. (2001) Chimpanzee models for human disease and immunobiology, *Immunol. Rev.* **183**, 86–93.

Muchmore, E.A., Diaz, S., Varki, A. (1998) A structural difference between the cell surfaces of humans and the great apes, *Am. J. Phys. Anthropol.* **107**, 187–198.

Muller, S., Wienberg, J. (2001) "Bar-coding" primate chromosomes: molecular cytogenetic screening for the ancestral hominoid karyotype, *Hum. Genet.* **109**, 85–94.

Nickerson, E., Nelson, D.L. (1998) Molecular definition of pericentric inversion breakpoints occurring during the evolution of humans and chimpanzees, *Genomics* **50**, 368–372.

O'hUigin, C., Satta, Y., Takahata, N., Klein, J. (2002) Contribution of homoplasy and of ancestral polymorphism to the evolution of genes in anthropoid primates, *Mol. Biol. Evol.* **19**, 1501–1513.

Royle, N.J., Baird, D.M., Jeffreys, A.J. (1994) A subterminal satellite located adjacent to telomeres in chimpanzees is absent from the human genome, *Nat. Genet.* **6**, 52–56.

Ruvolo, M. (1997) Molecular phylogeny of the hominoids: inferences from multiple independent DNA sequence data sets, *Mol. Biol. Evol.* **14**, 248–265.

Ruvolo, M., Pan, D., Zehr, S., Goldberg, T., Disotell, T.R., Dornum, M.V. (1994) Gene trees and hominoid phylogeny, *Proc. Natl. Acad. Sci. U.S.A.* **91**, 8900–8904.

Satta, Y., Klein, J., Takahata, N. (2000) DNA archives and our nearest relative: the trichotomy problem revisited, *Mol. Phylogenet. Evol.* **14**, 259–275.

Satta, Y. (2001) Comparison of DNA and protein polymorphisms between humans and chimpanzees, *Genes Genet. Syst.* **76**, 159–168.

Schmid, C.W., Marks, J. (1990) DNA hybridization as a guide to phylogeny: chemical and physical limits, *J. Mol. Evol.* **30**, 237–246.

Schwartz, J.H. (1984) The evolutionary relationships of man and orangutans, *Nature* **308**, 501–505.

Sibley, C.G., Ahlquist, J.E. (1984) The phylogeny of the hominoid primates, as indicated by DNA-DNA hybridization, *J. Mol. Evol.* **20**, 2–15.

Sibley, C.G., Comstock, J.A., Ahlquist, J.E. (1990) DNA hybridization evidence of hominoid phylogeny: a reanalysis of the data, *J. Mol. Evol.* **30**, 202–236.

Stone, A.C., Griffiths, R.C., Zegura, S.L., Hammer, M.F. (2002) High levels of Y-chromosome nucleotide diversity in the genus Pan, *Proc. Natl. Acad. Sci. U.S.A.* **99**, 43–48.

Thomas, J.W., Touchman, J.W. (2002) Vertebrate genome sequencing: building a backbone for comparative genomics, *Trends Genet.* **18**, 104–108.

Toder, R., Grutzner, F., Haaf, T., Bausch, E. (2001) Species-specific evolution of repeated DNA sequences in great apes, *Chromosome Res.* **9**, 431–435.

Varki, A. (2000) A chimpanzee genome project is a biomedical imperative, *Genome Res.* **10**, 1065–1070.

Vigilant, L., Paabo, S. (1999) A chimpanzee millennium, *Biol. Chem.* **380**, 1353–1354.

Whiten, A., Goodall, J., McGrew, W.C., Nishida, T., Reynolds, V., Sugiyama, Y., Tutin, C.E., Wrangham, R.W., Boesch, C. (1999) Cultures in chimpanzees, *Nature* **399**, 682–685.

Winter, H., Langbein, L., Krawczak, M., Cooper, D.N., Jave-Suarez, L.F., Rogers, M.A., Praetzel, S., Heidt, P.J., Schweizer, J. (2001) Human type I hair keratin pseudogene phihHaA has functional orthologs in the chimpanzee and gorilla: evidence for recent inactivation of the human gene after the Pan-Homo divergence, *Hum. Genet.* **108**, 37–42.

Wise, C.A., Sraml, M., Rubinsztein, D.C., Easteal, S. (1997) Comparative nuclear and mitochondrial genome diversity in humans and chimpanzees, *Mol. Biol. Evol.* **14**, 707–716.

Yunis, J.J., Prakash, O. (1982) The origin of man: a chromosomal pictorial legacy, *Science* **215**, 1525–1530.

Printed and bound in the UK by
CPI Antony Rowe, Eastbourne

Genomics and Genetics

Edited by
Robert A. Meyers

Related Titles

Meyers, R. A. (ed.)
Encyclopedia of Molecular Cell Biology and Molecular Medicine
16 Volume Set
ISBN-13: 978-3-527-30542-1
ISBN-10: 3-527-30542-4

Dunn, M. J., Jorde, L. B., Little, P. F. R., Subramaniam, S. (eds.)
Encyclopedia of Genetics, Genomics, Proteomics and Bioinformatics
8 Volume Set
ISBN-13: 978-0-470-84974-3
ISBN-10: 0-470-84974-6

Sensen, C. W. (ed.)
Handbook of Genome Research
Genomics, Proteomics, Metabolomics, Bioinformatics, Ethical and Legal Issues
2005
ISBN-13: 978-3-527-31348-8
ISBN-10: 3-527-31348-6

Meksem, K., Kahl, G. (eds.)
The Handbook of Plant Genome Mapping
Genetic and Physical Mapping
2005
ISBN-13: 978-3-527-31116-3
ISBN-10: 3-527-31116-5

Kahl, G.
The Dictionary of Gene Technology
Genomics, Transcriptomics, Proteomics
2004
ISBN-13: 978-3-527-30765-4
ISBN-10: 3-527-30765-6

Hacker, J., Dobrindt, U. (eds.)
Pathogenomics
Genome Analysis of Pathogenic Microbes
2006
ISBN-13: 978-3-527-31265-8
ISBN-10: 3-527-31265-X

Licinio, J., Wong, M.-L. (eds.)
Pharmacogenomics
The Search for Individualized Therapies
2002
ISBN-13: 978-3-527-30380-9
ISBN-10: 3-527-30380-4

Borlak, J. (ed.)
Handbook of Toxicogenomics
Strategies and Applications
2005
ISBN-13: 978-3-527-30342-7
ISBN-10: 3-527-30342-1

Genomics and Genetics

From Molecular Details to Analysis and Techniques

Edited by

Robert A. Meyers

Volume 2

WILEY-
VCH

WILEY-VCH Verlag GmbH & Co. KGaA

The Editor

Dr. Robert A. Meyers
RAMTECH LIMITED
122 Escalle Lane
Larkspur, CA 94039
USA

1st Edition 2007
1st Reprint 2007

■ All books published by Wiley-VCH are carefully produced. Nevertheless, authors, editors, and publisher do not warrant the information contained in these books, including this book, to be free of errors. Readers are advised to keep in mind that statements, data, illustrations, procedural details or other items may inadvertently be inaccurate.

Library of Congress Card No.: applied for

British Library Cataloguing-in-Publication Data: A catalogue record for this book is available from the British Library.

Bibliographic information published by the Deutsche Nationalbibliothek
The Deutsche Nationalbibliothek lists this publication in the Deutsche Nationalbibliografie; detailed bibliographic data are available in the Internet at http://dnb.d-nb.de.

© 2007 WILEY-VCH Verlag GmbH & Co. KGaA, Weinheim

All rights reserved (including those of translation into other languages). No part of this book may be reproduced in any form – by photoprinting, microfilm, or any other means – nor transmitted or translated into a machine language without written permission from the publishers. Registered names, trademarks, etc. used in this book, even when not specifically marked as such, are not to be considered unprotected by law.

Composition: Laserwords Private Ltd, Chennai, India
Printing: betz-druck GmbH, Darmstadt
Bookbinding: Litges & Dopf Buchbinderei GmbH, Heppenheim

ISBN-13: 978-3-527-31609-0

ISBN-10: 3-527-31609-4

Contents

Volume 2

Preface vii

Color Plates ix

4 Genomic Technologies 580
20 Genomic Sequencing 581
    Todd Charles Wood and Jeffrey P. Tomkins

21 Construction and Applications of Genomic DNA Libraries 605
    Eugene R. Zabarovsky

22 Shotgun Sequencing (SGS) 633
    Jun Yu, Gane Ka-Shu Wong, Jian Wang, Huanming Yang

23 Whole Genome Human Chromosome Physical Mapping 677
    Cassandra L. Smith, Giang H. Nguyen, Denan Wang, Nickolaev Bukanov

24 Serial Analysis of Gene Expression 693
    Jacques Marti Jean-Marc Elalouf

25 Gene Mapping by Fluorescence In Situ Hybridization 713
    Barbara G. Beatty and Henry H.Q. Heng

26 Gene Mapping and Chromosome Evolution by Fluorescence–Activated Chromosome Sorting 749
    Malcolm A. Ferguson-Smith

5 Genetic Engineering 768
27 Gene Targeting 769
    Michael M. Seidman and John H. Wilson

28 Genetic Engineering of Antibody Molecules 807
    Manuel L. Penichet and Sherie L. Morrison

Genomics and Genetics. Edited by Robert A. Meyers.
Copyright © 2007 Wiley-VCH Verlag GmbH & Co. KGaA, Weinheim
ISBN: 978-3-527-31609-0

29 Transgenic Fish — 831
Pinwen Peter Chiou, Jenny Khoo, Chung Zoon Chun, Thomas T. Chen

30 Transgenic Mice in Biomedical Research — 863
J. Willem Voncken Marten Hofker

31 Transgenic Plants for Food Use — 905
Susanne Stirn Horst Lörz

6 Gene Medicine and Disease — 938

32 Human Genetic Variation and Disease — 939
Lynn B. Jorde

33 Alzheimer's Disease — 955
Jun Wang, Silva Hecimovic and Alison Goate

34 Triplet Repeat Diseases — 985
Stephan J. Guyenet Albert R. La Spada

35 Molecular Genetics of Down Syndrome — 1045
Charles J. Epstein

36 Molecular Genetics of Hemophilia — 1057
Francesco Giannelli Peter M. Green

37 Gene Therapy and Cardiovascular Diseases — 1085
Michael E. Rosenfeld and Alan D. Attie

38 Somatic Gene Therapy — 1117
M. Schweizer, E. Flory, C. Münk, Uwe Gottschalk, K. Cichutek

39 Mutagenesis, Malignancy and Genome Instability — 1135
Garth R. Anderson, Daniel L. Stoler, Jeremy D. Bartos

Cumulative List of Contributors — 1153

Subject Index — 1159

Preface

The *Genomics and Genetics* two volume set was compiled from a selection of key articles from the recently published *Encyclopedia of Molecular Cell Biology and Molecular Medicine* (ISBN 978-3-527-30542-1). The *Genomics and Genetics* set is comprised of 39 detailed articles arranged in six sections covering molecular genetics, genomic organization and evolution, genomes of model organisms, genomic sequencing, genetic engineering, as well as gene medicine and disease. The articles were prepared by eminent researchers from the major research institutions in the United States, Europe and around the globe.

Each article begins with a concise definition of the subject and its importance, followed by the body of the article and extensive references for further reading. The references are divided into secondary references (books and review articles) and primary research papers. Each subject is presented on a first-principle basis, including detailed figures, tables and drawings. Because of the self-contained nature of each article, some overlap among articles on related topics occurs. Extensive cross-referencing is provided to help the reader expand his or her range of inquiry.

The master publication, which is the basis of the *Genomics and Genetics* set, is the *Encyclopedia of Molecular Cell Biology and Molecular Medicine*, which is the successor and second edition of the *Encyclopedia of Molecular Biology and Molecular Medicine*, covers the molecular and cellular basis of life at a university and professional researcher level. This second edition is double the first edition in length and will comprise the most detailed treatment of both molecular and cell biology available today. The Board and I believe that there is a serious need for this publication, even in view of the vast amount of information available on the World Wide Web and in text books and monographs. We feel that there is no substitute for our tightly organized and integrated approach to selection of articles and authors and implementation of peer review standards for providing an authoritative single-source reference for undergraduate and graduate students, faculty, librarians and researchers in industry and government.

Our purpose is to provide a comprehensive foundation for the expanding number of molecular biologists, cell biologists, pharmacologists, biophysicists, biotechnologists, biochemists and physicians as well as for those entering molecular cell biology and molecular medicine from majors or careers in physics, chemistry, mathematics, computer science and engineering. For example there is an unprecedented demand for physicists, chemists and computer scientists who will work with biologists to define the genome, proteome and interactome through experimental and computational biology.

Genomics and Genetics. Edited by Robert A. Meyers.
Copyright © 2007 Wiley-VCH Verlag GmbH & Co. KGaA, Weinheim
ISBN: 978-3-527-31609-0

The Board and I first divided all of molecular cell biology and molecular medicine into primary topical categories and each of these was further defined into subtopics. The following is a summary of the topics and subtopics:

- Nucleic Acids: amplification, disease genetics overview, DNA structure, evolution, general genetics, nucleic acid processes, oligonucleotides, RNA structure, RNA replication and transcription.
- Structure Determination Technologies Applicable to Biomolecules: chromatography, labeling, large structures, mapping, mass spectrometry, microscopy, magnetic resonance, sequencing, spectroscopy, x-ray diffraction.
- Proteins, Peptides and Amino Acids: analysis, enzymes, folding, mechanisms, modeling, peptides, structural genomics (proteomics), structure, types.
- Biomolecular Interactions: cell properties, charge transfer, immunology, recognition, senses.
- Molecular Cell Biology of Specific Organisms: algae, amoeba, birds, fish, insects, mammals, microbes, nematodes, parasites, plants, viruses, yeasts.
- Molecular Cell Biology of Specific Organs or Systems: excretory, lymphatic, muscular, neurobiology, reproductive, skin.
- Molecular Cell Biology of Specific Diseases: cancer, circulatory, endocrine, environmental stress, immune, infectious diseases, neurological, radiation.
- Biotechnology: applications, diagnostics, gene altered animals, bacteria and fungi, laboratory techniques, legal, materials, process engineering, nanotechnology, production of classes or specific molecules, sensors, vaccine production.
- Biochemistry: carbohydrates, chirality, energetics, enzymes, biochemical genetics, inorganics, lipids, mechanisms, metabolism, neurology, vitamins.
- Pharmacology: chemistry, disease therapy, gene therapy, general molecular medicine, synthesis, toxicology.
- Cellular Biology: developmental cell biology, diseases, dynamics, fertilization, immunology, organelles and structures, senses, structural biology, techniques.

We then selected some 340 article titles and author or author teams to cover the above topics. Each article is designed as a self-contained treatment. Each article begins with a key word section, including definitions, to assist the scientist or student who is unfamiliar with the specific subject area. The Encyclopedia includes more than 3000 key words, each defined within the context of the particular scientific field covered by the article. In addition to these definitions, the glossary of basic terms found at the back of each volume, defines the most commonly used terms in molecular and cell biology. These definitions should allow most readers to understand articles in the Encyclopedia without referring to a dictionary, textbook or other reference work.

Larkspur, July 2006

Robert A. Meyers
Editor-in-Chief

Color Plates

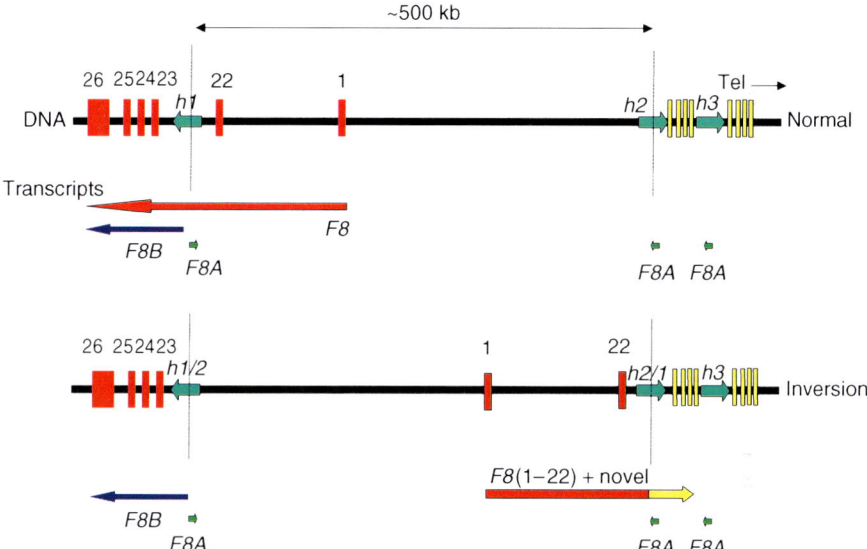

Fig. 5 (p. 1066) Diagram of *int22h*-related inversion, showing its transcriptional consequences. Top: normal structure of DNA region and *F8* gene transcripts (not drawn to scale). *F8* = transcript encoding coagulant factor, *F8A* and *F8B* = transcripts arising from CpG island in *int22h* repeat. Green arrows in genomic DNA represent *int22h* repeats (*h1* = *int22h-1*, *h2* = *int22h-2*, *h* = *int22h-3*). Red bars indicate some numbered exons of the *F8* gene, yellow bars are exons unrelated to the *F8* gene. Vertical dotted lines are sites where intranemic homologous recombination occurs and leads to the inversion. Tel → shows direction to telomere. Arrow point in transcripts indicates 3′ end. Bottom: structure of DNA region with inversion resulting from recombination of *int22h-1* with *int22h-2*. *H1/2* and *h2/1* are the recombined *int22h* repeats. The transcripts show that the *F8* transcript contains only exons 1 to 22 of *F8* attached to the normally unrelated yellow exons and that exons 23 to 26 are present only in the *F8B* transcript. NB inversions involving *int22h-3* yield analogous transcripts.

Genomics and Genetics. Edited by Robert A. Meyers.
Copyright © 2007 Wiley-VCH Verlag GmbH & Co. KGaA, Weinheim
ISBN: 978-3-527-31609-0

x | *Color Plates*

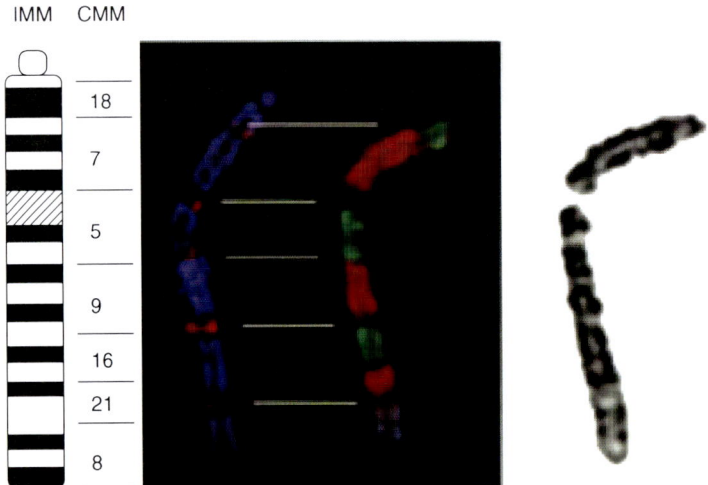

Fig. 5 (p. 762) Indian muntjac (IMM) chromosome 3 to which chromosome-specific paint probes of the Chinese muntjac (CMM) have been hybridized. This reveals that this chromosome has been formed by the tandem fusion of seven entire chromosomes of the Chinese muntjac. A centromere probe (red) indicates that remnants of ancestral centromeres are still present at the sites of fusion.

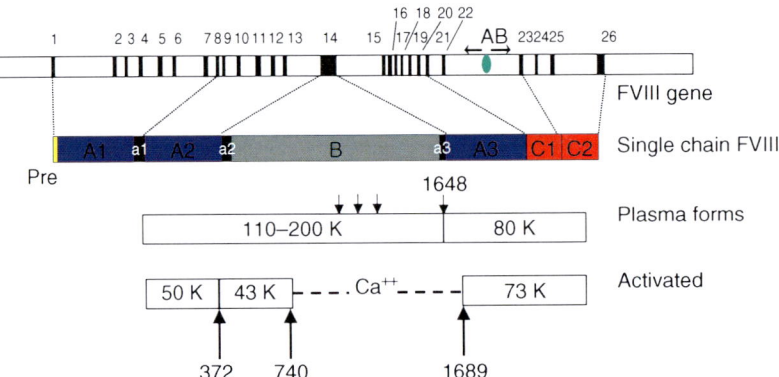

Fig. 2 (p. 1061) Factor VIII gene (*F8*) and protein. Diagram of *F8* gene (top bar) and FVIII (lower bars). Exons are indicated by black boxes and numbered (1–26). Arrows in opposite directions over intron 22 indicate the presence of CpG island (green box) associated with the start of two transcripts of opposite orientation (A, B). Dotted lines indicate exons coding for the different protein domains (see text). Arrows in plasma and activated forms indicate site of proteolytic cleavages (see text) and numbers associated with arrows indicate position of amino acid at the amino terminal end of cleavage sites. Numbers in bars are weight in Daltons of individual protein segments.

Color Plates | xi

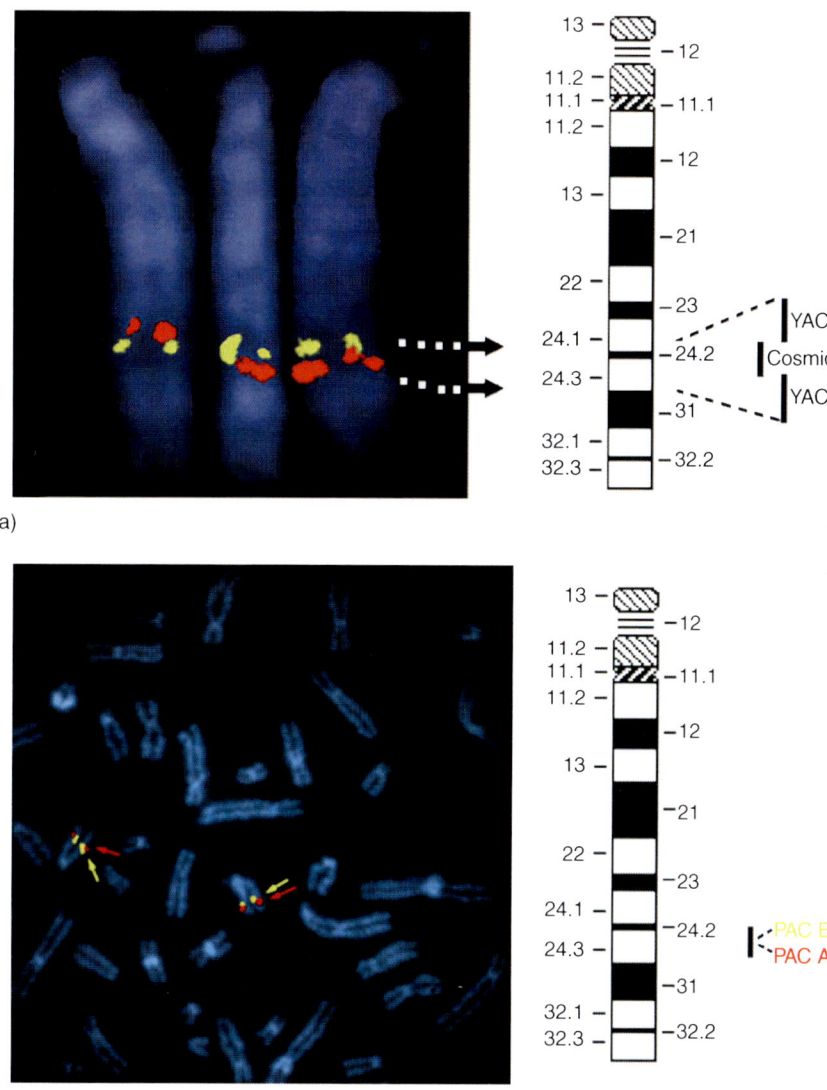

(a)

(b)

Fig. 4 (p. 739)

xii | Color Plates

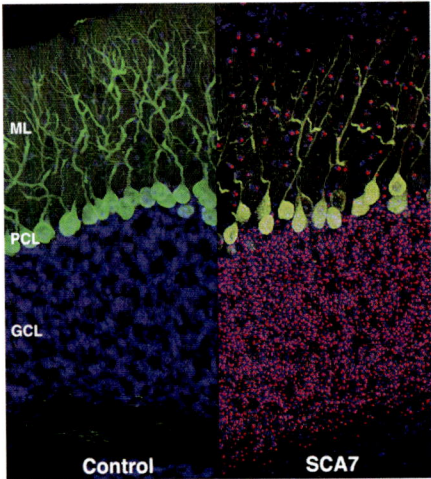

Fig. 5 (p. 1008) Noncell autonomous Purkinje cell degeneration in a mouse model of spinocerebellar ataxia type 7 (SCA7). Confocal microscopy analysis of cerebellar sections from a SCA7 transgenic mouse (SCA7) created with an ataxin-7 CAG-92 containing murine prion protein expression vector and from an age- and sex-matched nontransgenic littermate (Control). Staining with an anti-ataxin-7 antibody (magenta), a calbindin antibody (green), and DAPI (blue) reveals a healthy, normal-appearing cerebellum characterized by properly oriented Purkinje cells with extensive dendritic arborization in the "Control" mice. However, SCA7 transgenic mice display pronounced Purkinje cell degeneration as evidenced by decreased dendritic arborization and displacement of Purkinje cell bodies. Interestingly, although numerous neurons in the granule cell layer (GCL) and the molecular layer (ML) display aggregates of ataxin-7, there is no accumulation of mutant ataxin-7 in the degenerating Purkinje cells due to lack of appreciable expression there. As the Purkinje cells degenerate without expressing the mutant protein, the degeneration is described as noncell autonomous. (Adapted from Garden, G.A., Libby, R.T., Fu, Y.H., Kinoshita, Y., Huang, J., Possin, D.E., Smith, A.C., Martinez, R.A., Fine, G.C., Grote, S.K., et al. (2002) Polyglutamine-expanded ataxin-7 promotes noncell-autonomous Purkinje cell degeneration and displays proteolytic cleavage in ataxic transgenic mice, *J. Neurosci.* **22**, 4897–4905, used with permission of the *Journal of Neuroscience*.)

Fig. 4 (p. 739) Relational mapping on metaphase chromosomes using two fluorochromes. (a) Three probes: Pairwise mapping was used to order two YACs and one cosmid probe within a YAC contig, which mapped to 14q24.2–24.3. The two YACs were labeled with DIG or biotin and the cosmid with biotin. DIG was detected with rhodamine (red) and biotin with FITC (yellow). Left chromosome: DIG-YAC 1 plus cosmid; middle chromosome: DIG-YAC 2 plus cosmid; right chromosome: Biotin-YAC 1 plus DIG-YAC 2. Thus, the order shown is YAC 1 – cosmid – YAC 2. Note: YAC 2 often demonstrated double-hybridization signals in a lateral orientation suggesting two domains of strong hybridization or clustering of Alu sequences within the YAC (see right chromosome). (b) Two probes: Two PAC probes localized to 14q24-31 by FISH were labeled with different haptens (DIG-PAC A and biotin-PAC B) and hybridized simultaneously to normal human metaphase chromosomes counterstained with DAPI. PAC A (detected with rhodamine, red) localized telomeric to PAC B (detected with FITC, yellow). [From Barbara G. Beatty, Stephen W. Scherer. (2002), Fish: A Practical Approach, reprinted by permission of Oxford University Press.]

(a)

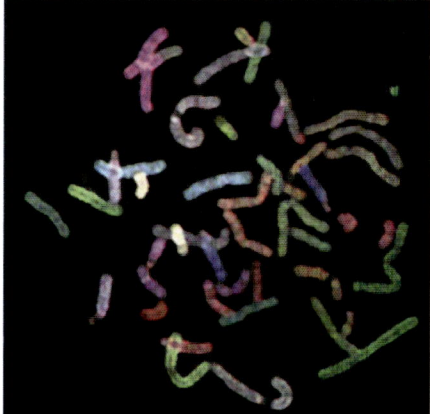

(b)

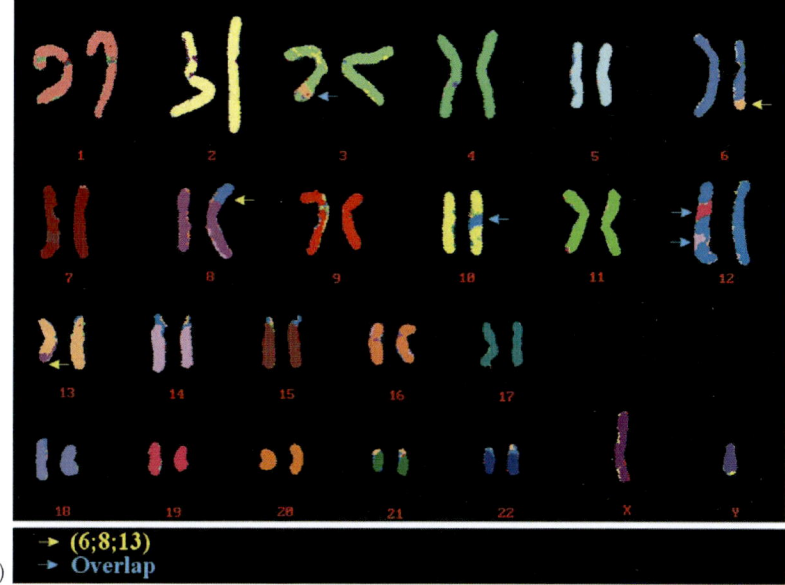

→ (6;8;13)
→ Overlap

Fig. 2 (p. 755) Multicolor FISH. (a) Using a combination of five fluorochromes, each chromosome pair has a distinctive color. (b) This allows the identification of interchromosomal translocations between chromosomes 6, 8, and 13 in the karyotype.

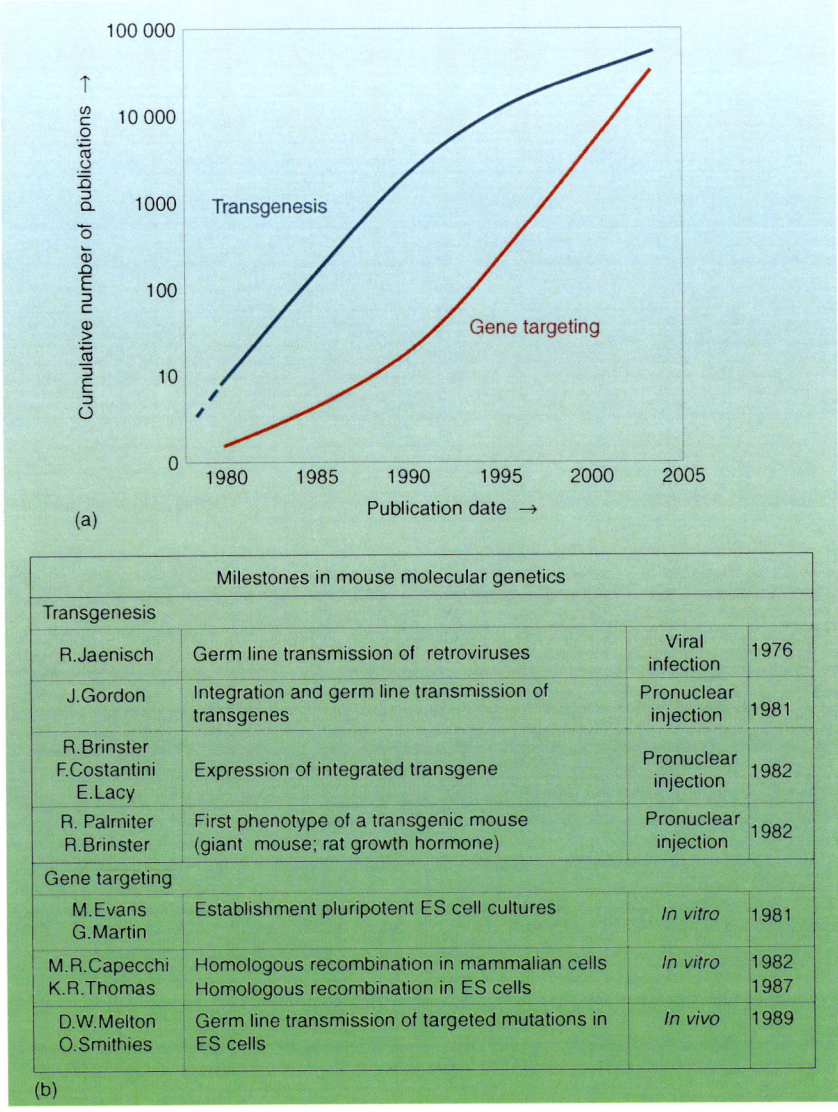

Fig. 1 (p. 868) Genetically modified mouse models in biomedical science. (a) A rough estimation of published scientific output using genetically modified mouse models. The blue line represents reports on transgenic models and the red line on gene targeting. Note: with the advent of conditional gene targeting, which utilizes transgenesis in conjunction with targeted mutagenesis, a strict separation of transgenesis and gene targeting into the twentieth century is difficult. (b) Milestones in mouse molecular genetics. The first report on germ-line transmission of integrated retroviruses (via infection) dates back to the mid-1970s. Subsequently, pronuclear microinjection of fertilized oocytes and subsequent proof of transgene integration, expression, and germ-line transmission, were pioneered in different laboratories. Similarly, breakthroughs in the use of embryonic stem (ES) cell technology are listed. Many other excellent laboratories have since contributed to refinement of technologies and strategies; these are listed throughout this chapter.

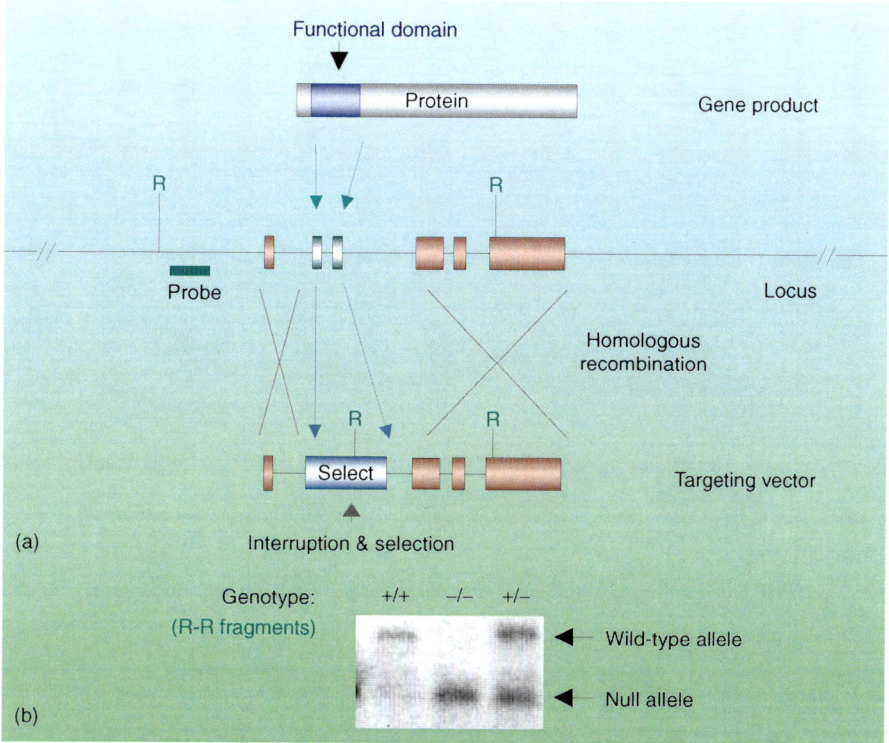

Fig. 8 (p. 881) Basic design of a conventional targeting vector. (a) The targeting vector duplicates part of the locus to be mutated (homologous sequences), but is interrupted by a positive selection marker (*select*), for instance, an antibiotic resistance gene. Many conventional targeting vectors also contain a negative selection marker. The herpes simplex virus Thymidine Kinase gene is often included: the TK gene product converts a harmless compound, for instance, Ganciclovir, into a cytotoxic substance. The positioning of the TK cassette at one of the ends of a targeting vector results in exclusion of TK from the genome upon homologous recombination, whereas TK is cointegrated in case of random integration of a targeting vector. (b) The vector is designed such that by means of an "external probe" (blue-grey box: *probe*; from within the to be targeted locus, but outside the homology regions used in the targeting construct) and strategically chosen restriction endonuclease sites (R) a shift is detected upon homologous recombination. The figure shows the three possible genotypes; two wild-type (+/+) alleles, two knockout (−/−) alleles (homozygous knockout), and a combination of both in the heterozygous (+/−) condition (see also Fig. 10). Mutagenesis vectors may be introduced into ES cells via several means; the most commonly used method is electroporation (electroshock) by which the cellular membrane is damaged, allowing uptake of DNA in cells.

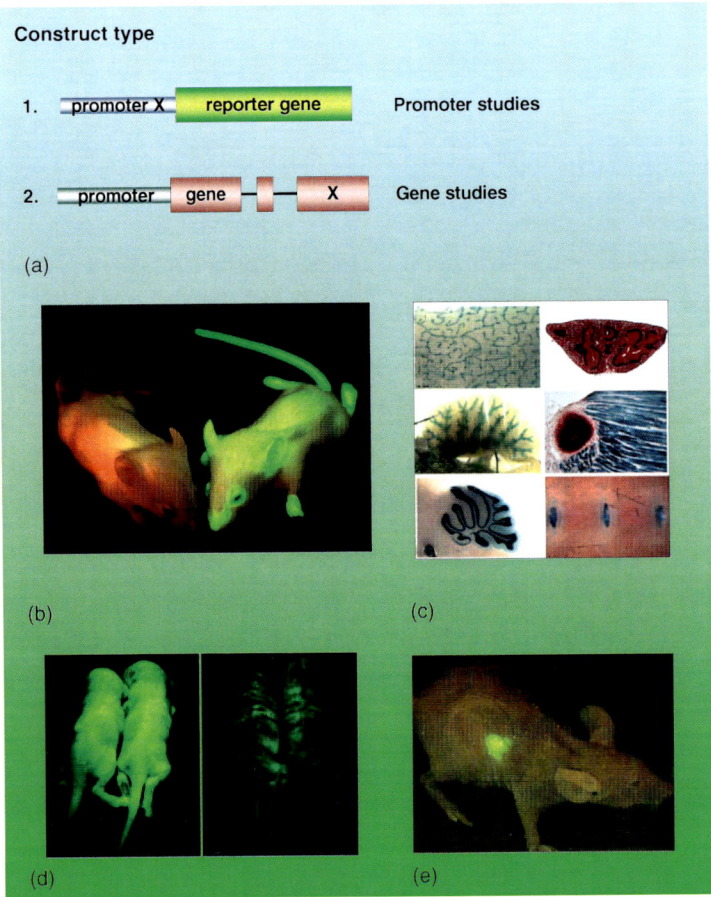

Fig. 2 (p. 872) (legend continued on next page) Transgenesis; concepts. (a) Original scientific applications of transgenesis in the mouse were roughly divided in two areas: (1) studies of gene promoter activity (and/or other regulatory elements) and (2) studies of the effect(s) of overexpression (or a mutated form) of gene X. Promoter studies typically make use of *reporter* genes (green box) that reveal their presence through specific catalytic activity. Often-used reporter genes include the green fluorescent protein jelly fish (b, d), bacterial beta-galactosidase (c), (also see Fig. 3), and fire fly luciferase (d). (b) Transgene expression is controlled by a keratinocyte-specific promoter (source image: http://www.mshri.on.ca/nagy/cre.htm; Maatman, R., Gertsenstein, M., de Meijer, E., Nagy, A. et al. (2003) Aggregation of embryos and embryonic stem cells, *Methods Mol. Biol.* **209**, 201–230). (c) Differential expression of beta-galactosidase (upper left to lower right): Lymphatic vessels (intestinal surface), marginal zones between red and white pulp (spleen), bronchial epithelium (lungs), Large skeletal muscles (limb), granule cell layer (cerebellum), Intervertebral discs (source images: Valenzuela, D.M., Murphy, A.J., Frendewey, D., Gale, N.W. et al. (2003) High-throughput engineering of the mouse genome coupled with high-resolution expression analysis, *Nat. Biotechnol.* **21**, 652–659). (d) Fluorescent images of an X-linked enhanced green fluorescent protein (EGFP) transgene; transgenic male offspring (left) of a mating between an X-linked EGFP transgenic female and a nontransgenic male have ubiquitous green fluorescence in the skin owing to the presence of a single active X chromosome. Transgenic female pups (right) have mosaic (tortoiseshell) green fluorescence in the skin owing to random inactivation of one

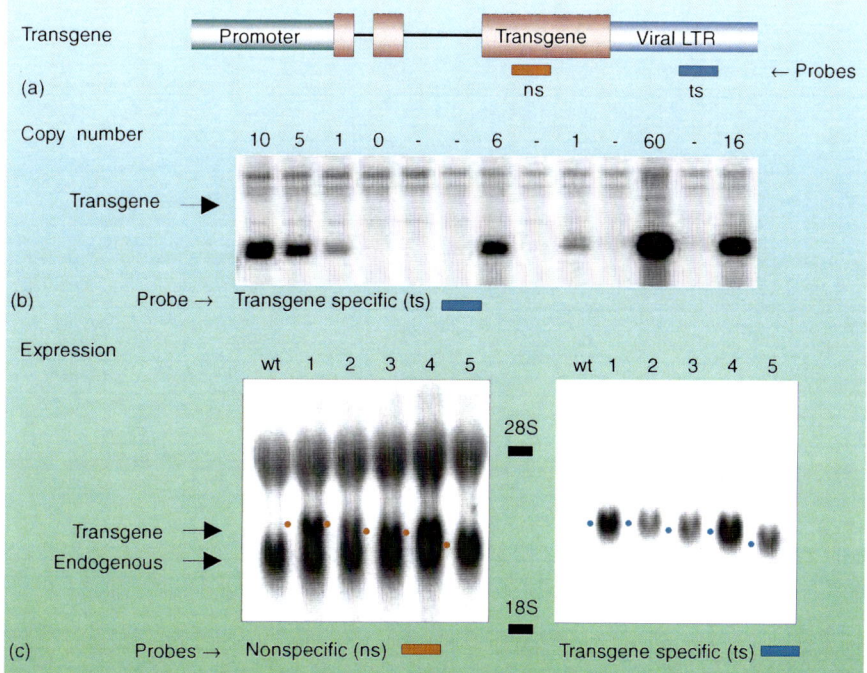

Fig. 3 (p. 874) Transgene construction and detection. (a) Schematic representation of a hypothetical transgene. The transgene is derived from mouse genomic sequences (brown). Transgene expression is under control of a general promoter and terminates in a retroviral long terminal repeat (LTR) element. Probes used for detection of transgene DNA or RNA are depicted as a brown box (nonspecific (ns) probe, which detects both the endogenous gene and the transgene) or a blue box (a transgene-specific (ts) probe, which only detects the transgene). (b) Copy number determination by Southern blot analysis with a ts probe. Bold print above image represent DNA samples with known copy numbers, resp. 10, 5, 1, and no transgene copies inserted; the remaining lanes show four founders with varying copy numbers (image adapted from: (http://www.healthsystem.virginia.edu/internet/transgenic-mouse/southerndesign.cfm)) (c) Detection of transgene expression by Northern blot hybridization clearly demonstrates the need for transgene specific probes (right panel). Left panel: expression signal of endogenous gene masks transgene expression.

Fig. 2 (legend continued from previous page) X chromosome (source image: Hadjantonakis, A.K., Dickinson, M.E., Fraser, S.E., Papaioannou, V.E. (2003) Technicolour transgenics: imaging tools for functional genomics in the mouse, *Nat. Rev. Genet.* **4**, 613–625). (e) An external image of adenovirus-encoded GFP gene expression acquired from a nude mouse in the light box 72 h after portal vein injection.

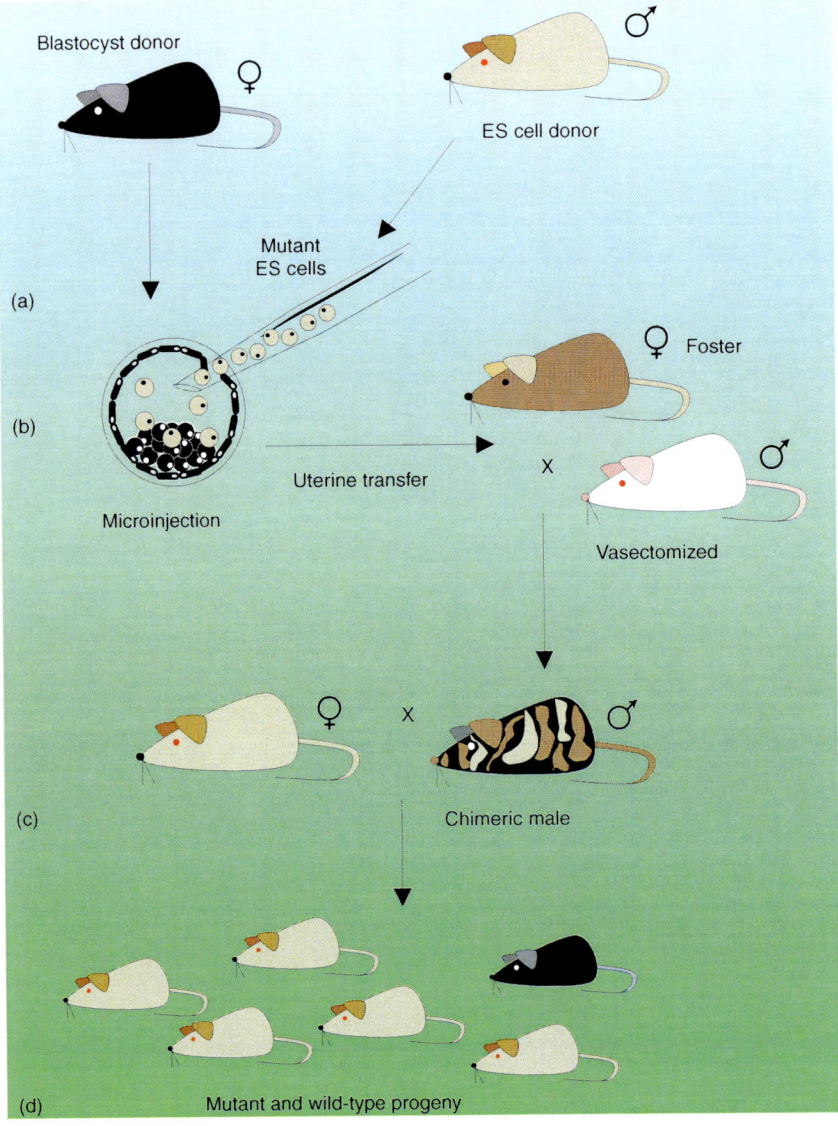

Fig. 10 (p. 884) (legend continued on next page) Biotechnological procedures in conventional gene targeting. (a) Inbred strains are used for superovulation and mating to generate sufficient blastocysts for injection. The procedures in use to obtain sufficient blastocysts for microinjection are natural matings or superovulation. Breeding properties of some inbred strains are relatively poor, hence, natural mating often require large numbers of mating pairs. Super ovulation is often not very efficient in inbred strains either, but will usually yield adequate embryo numbers for injection. (b) Upon microinjection, the embryos are transferred into the oviduct of a pseudopregnant female (see also legend to Fig. 7). (c) Since male ES cells are used, researchers look for a gender bias among chimeric offspring: a potent male (XY) ES cell line forces even female recipient blastocysts (XX) into a male sex phenotype. As a general rule is considered: the higher the coat color mosaicism, the more likely it is that the ES cells participate in spermatogenesis, the higher the chance for germ-line transmission of

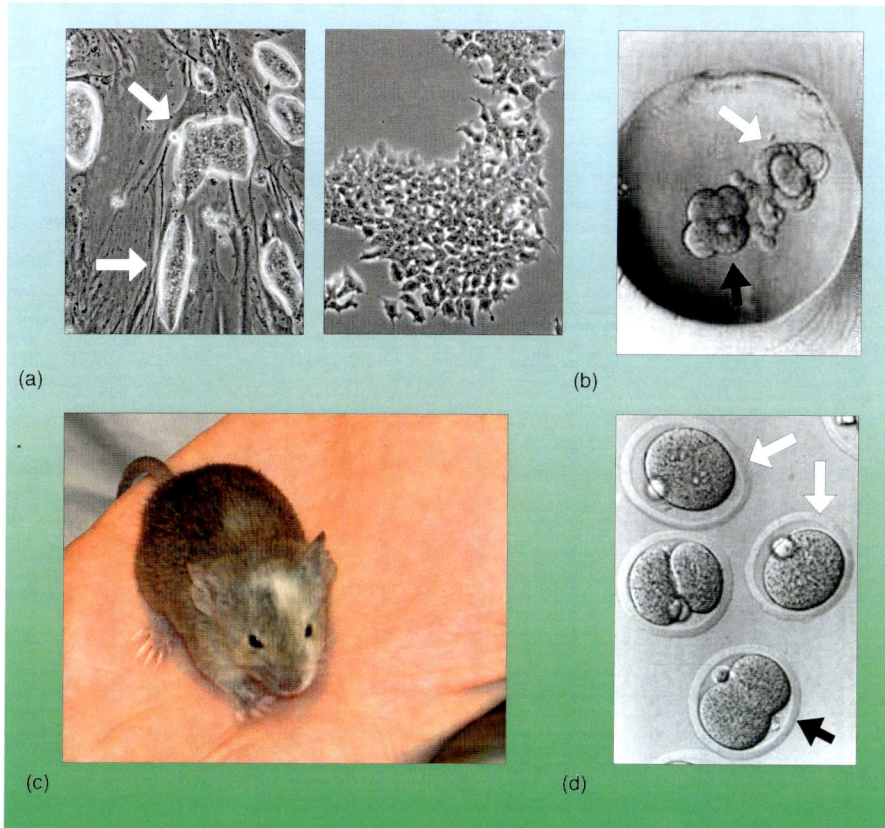

Fig. 9 (p. 883) Embryonic stem cells, embryos, and chimeric mice. (a) Left panel: ES cells grown on mouse embryonic fibroblasts (MEFs); white arrows point at colonies of tightly coherent ES cells; right panel: feeder-free ES cells. (b) Aggregation of 8-cell-stage morulas (black arrow) with genetically modified ES cells (white arrow). (c) Chimeric mice beautifully displaying the coat color mosaicism resulting from contribution of ES cells of two different genetic backgrounds to the developing embryo the mix of, in this case, C57Bl (recipient blastocyst – black) and 129Ola (donor ES cells – sand) becomes manifest as Agouti, patches of purely 129Ola-derived tissue is sand colored (e.g. over left eye). (d) Electrofusion of two-cell-stage embryos (middle left) results in fusion and formation of one-cell-stage tetraploid embryos (white arrows). Tetraploidy does not hamper placenta formation, but does not permit normal embryogenesis. Tetraploid blastocysts injected with diploid ES cells will support fully ES cell–derived embryonic development (source image: (http://www.mshri.on.ca/nagy/cre.htm)).

Fig. 10 (legend continued from previous page) the mutation; (d) – Germ-line transmission is revealed by the coat color of the donor (ES cell; sand-colored in the figure; several ES cell lines are in use) strain; 50% of the sand-colored offspring will be heterozygous for the germ-line mutation.

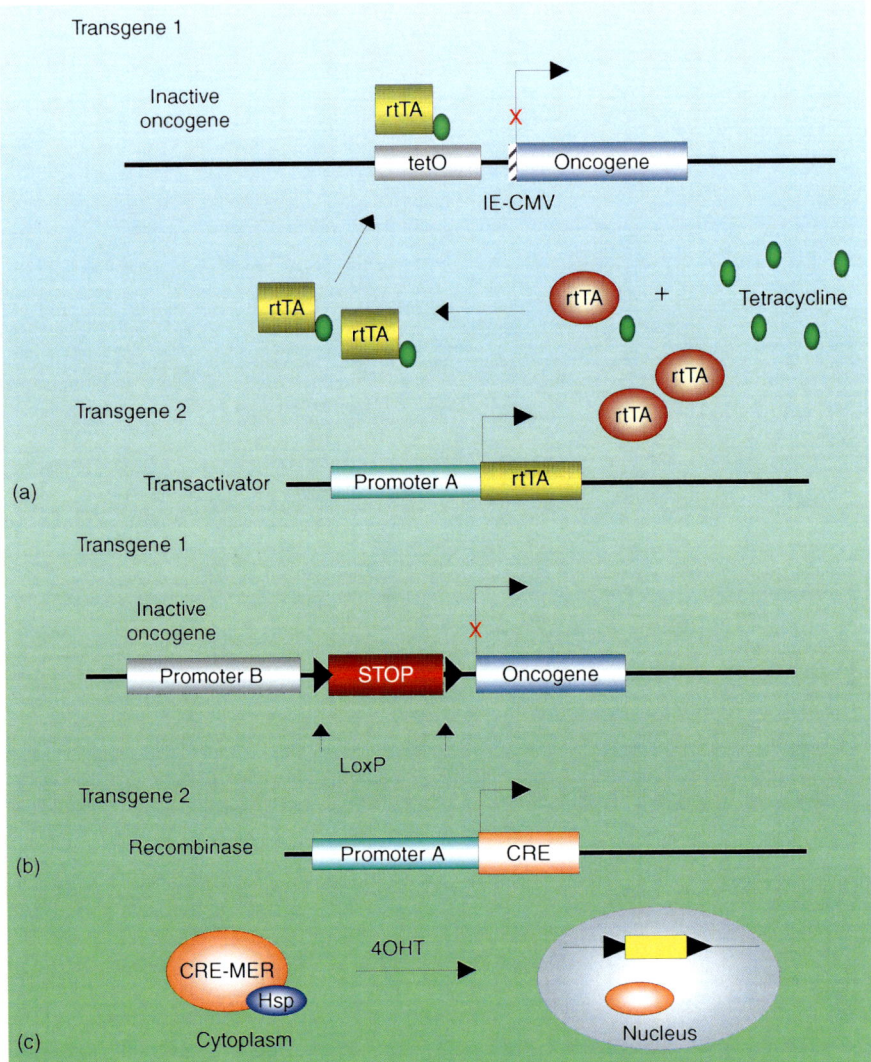

Fig. 11 (p. 886) (legend continued on next page) Binary models. (a) The Tet-system requires two independently generated transgenic lines, which are combined by interbreeding. Transgene 1 carries a minimal ("enhancer-less") Cytomegalovirus immediate early (IE–CMV) promoter fused to heptamerized prokaryotic tetO regulatory sequences. Transgene 1 is transcriptionally silent until the transactivator line (transgene 2) provides transactivator protein. In the original Tet-system, transgene 2 expressed a tetracycline (Tc)-controlled transcriptional activator, tTA. tTA is a fusion between the Tet-repressor of the *Tn10* Tc resistance operon of *E. coli* and a C-terminal part of the herpes simplex transactivator protein VP16. tTA will induce transcription from the tetO-controlled transgene up to five orders of magnitude, but is inhibited by Tc. The example depicts rtTA (red ovals), a reverse Tc-controlled transactivator, which is a mutant form of tTA and needs Tc-derivatives to bind tetO (yellow squares). So rather than an inhibitable model, rtTA is an inducible model. Refinements on the Tet-system include choice of regulatory sequences (promoter A) in transgene 2 (cell type–specific, general promoter). In addition, several derivatives of Tc exist, among which Doxocyclin (Dox) has

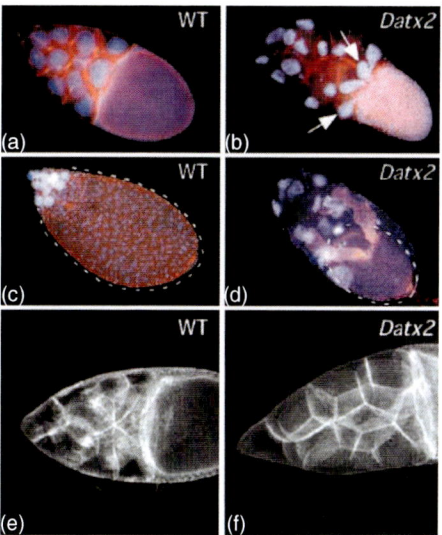

Fig. 4 (p. 1004) Studies of the ataxin-2 ortholog in *Drosophila melanogaster* reveals a role for Drosophila ataxin-2 (*Datx2*) in actin filament formation during oogenesis. Egg chambers from normal (Wild-Type (WT)) and mutant flies with reduced expression (*Datx2*) of Drosophila ataxin-2 were stained with DAPI (blue) and phalloidin (red) to indicate nuclei and filamentous actin respectively (a–d). The egg chambers of WT flies prior to cytoplasmic transport (a) display well demarcated, separated but interconnected cells (blue) as expected, while the egg chambers of *Datx2* flies (b) contain irregularly arranged cells. After cytoplasmic transport, the egg chambers of WT flies (c) show one greatly enlarged oocyte (dashed line) with only a small section of compressed cells, while the egg chambers of *Datx2* flies (d) have failed to yield an enlarged oocyte, instead retaining dispersed and large adjacent cells. Confocal images of egg chambers prior to cytoplasmic transport stage reveal a prominent actin filament network in WT flies (e), but a remarkably transparent actin filament network in *Datx2* flies (f). The decreased density of the actin filament network underlies the cytoplasmic "dumping" defect in the *Datx2* flies. (From Satterfield, T.F., Jackson, S.M., Pallanck, L.J. (2002) A Drosophila homolog of the polyglutamine disease gene *SCA2* is a dosage-sensitive regulator of actin filament formation, *Genetics* **162**, 1687–1702, used with permission of *Genetics*.)

Fig. 11 (legend continued from previous page) better tissue penetration and pharmacokinetic properties than Tc itself and very high and selective affinity for TetR. (b) The use of bacteriophage P1 CRE-LoxP recombination–dependent removal of a strong transcriptional block (STOP) in a transgene construct may be helpful in defining the molecular events at the onset of tumorigenesis and in addition circumvents possible embryonic lethal effects of the transgene. The choice of regulatory sequences (promoter A) in transgene 2 is determined by the model. Although the excision of a transcriptional block is, in principle, reversible, for practical reasons this system is rarely used as a binary switch, since it would entail obligatory screening for genetic modifications (insertions) in target cells: this obviously compromises the usefulness of the models in terms of precision and rapidity. (c) Fusion of CRE to the mutant forms of the ligand-binding domain of nuclear hormone receptors, such as the estrogen receptor (MER), has generated a recombinase CRE-MER that can be "induced" at the posttranslation level: CRE-MER is retained cytoplasmic through interaction with heatshock proteins (Hsp) such as HSsp90; upon addition of 4-hydroxytamoxifen (4OHT), a synthetic steroid, the Hsp-binding is released, and the recombinase shuttles to the nucleus to excise floxed sequences (also see Fig. 12).

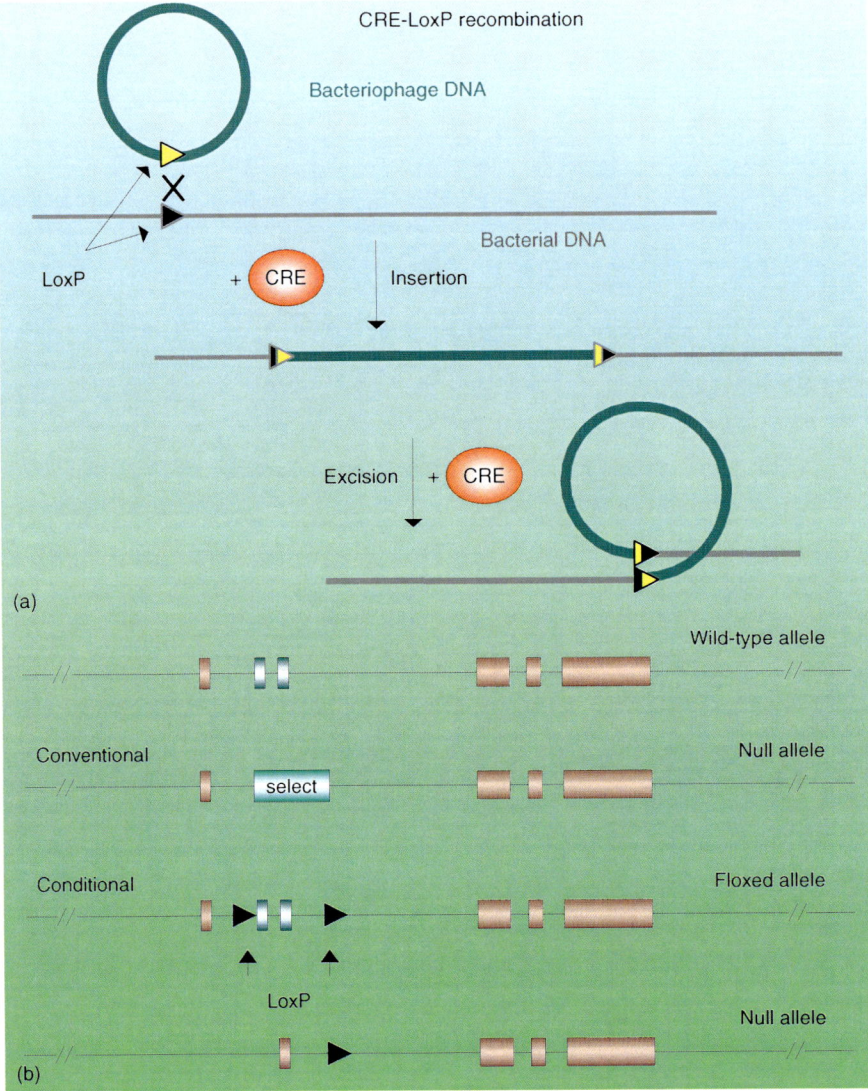

Fig. 12 (p. 889) (a) CRE-LoxP system: LoxP sites are short sequences harboring inverted 13 nucleotide repeats separated by an 8-nucleotide asymmetric spacer. The LoxP are recognition sites for the recombinase CRE. CRE drives site-specific recombination of two identical LoxP sites, for clarity depicted as a yellow and a black triangle. Two similarly oriented LoxP sites on one contiguous DNA strand will be recombined by CRE and the sequence in between the LoxP sites is excised. (b) This recombination principle is used in conditional gene-targeting vectors: a crucial coding region (two blue exons) is "floxed" (i.e. flanked by LoxP sites) in ES cells and excised *in vivo*. The LoxP sites in the "floxed" allele are assumed not to interfere with wild-type gene expression. The figure compares the structure of a conventional and a conditional knockout allele. A variation on this theme is the FLP-*frt* system, which is derived from yeast. The FLP gene product, like CRE, is a site-specific recombinase; *frt* (FLP recognition target) represents the small inverted repeats, analogous to LoxP, recognized by the recombinase.

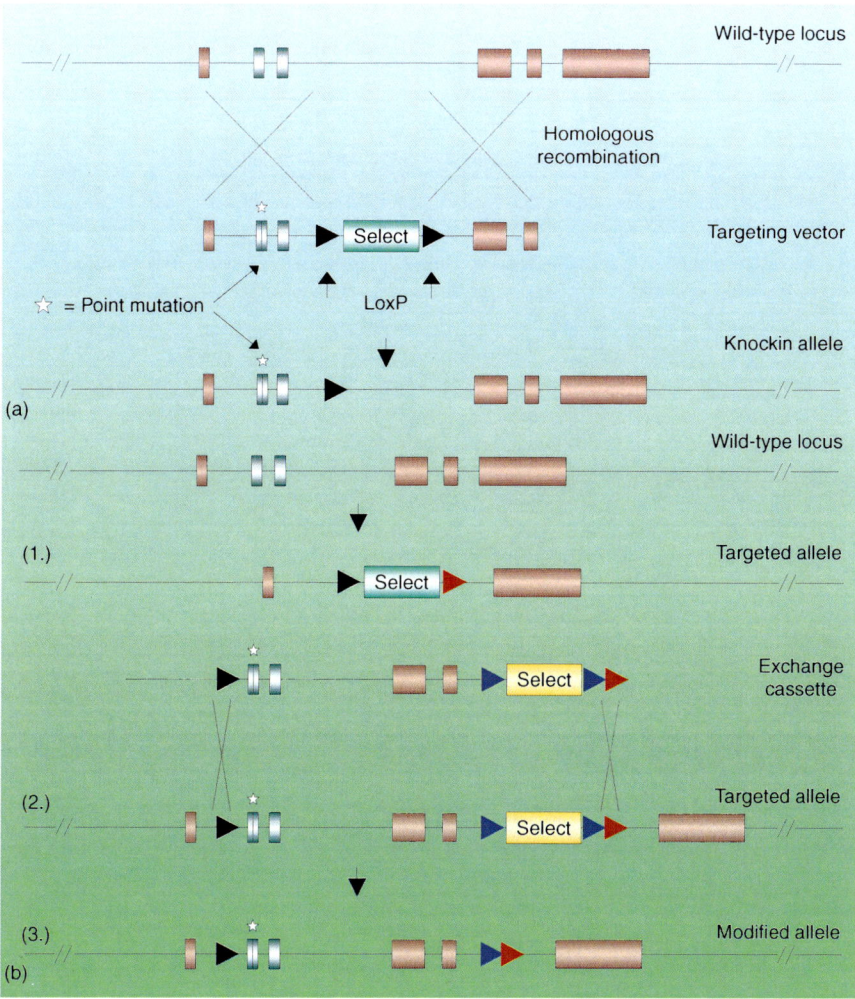

Fig. 13 (p. 892) Applications of site-directed recombination in molecular mouse genetics. (a) A conditional targeting construct usually harbors an antibiotic resistance gene (select) flanked by recombinase recognition sites (either LoxP or *frt*; see legend to Fig. 12). Prior to microinjection into blastocysts or aggregation with morulas, the selection marker is excised in ES cells by recombination (by transient expression of either CRE or FLP). The targeting construct depicted contains a specific point mutation (asterix) in one of its exons. Homologous recombination results in introduction of the mutation in the endogenous locus. (b) The use of multiple nonidentical LoxP and or *frt* sites (black, red, and blue triangles) creates the possibility to repeatedly exchange sequences between vectors and a locus via site-specific recombination. The procedure is referred to as *recombination-mediated cassette exchange (RCME)*. First, a locus is targeted and an asymmetrically "floxed" selection marker is introduced (1). Using the same LoxP variants, an exchange cassette is introduced (2) upon which the selection marker is removed by recombinase-mediated excision (3). The exchange procedure can be repeated multiple times, since the position of the most peripheral recombinase recognition sites (black and red triangles) is maintained.

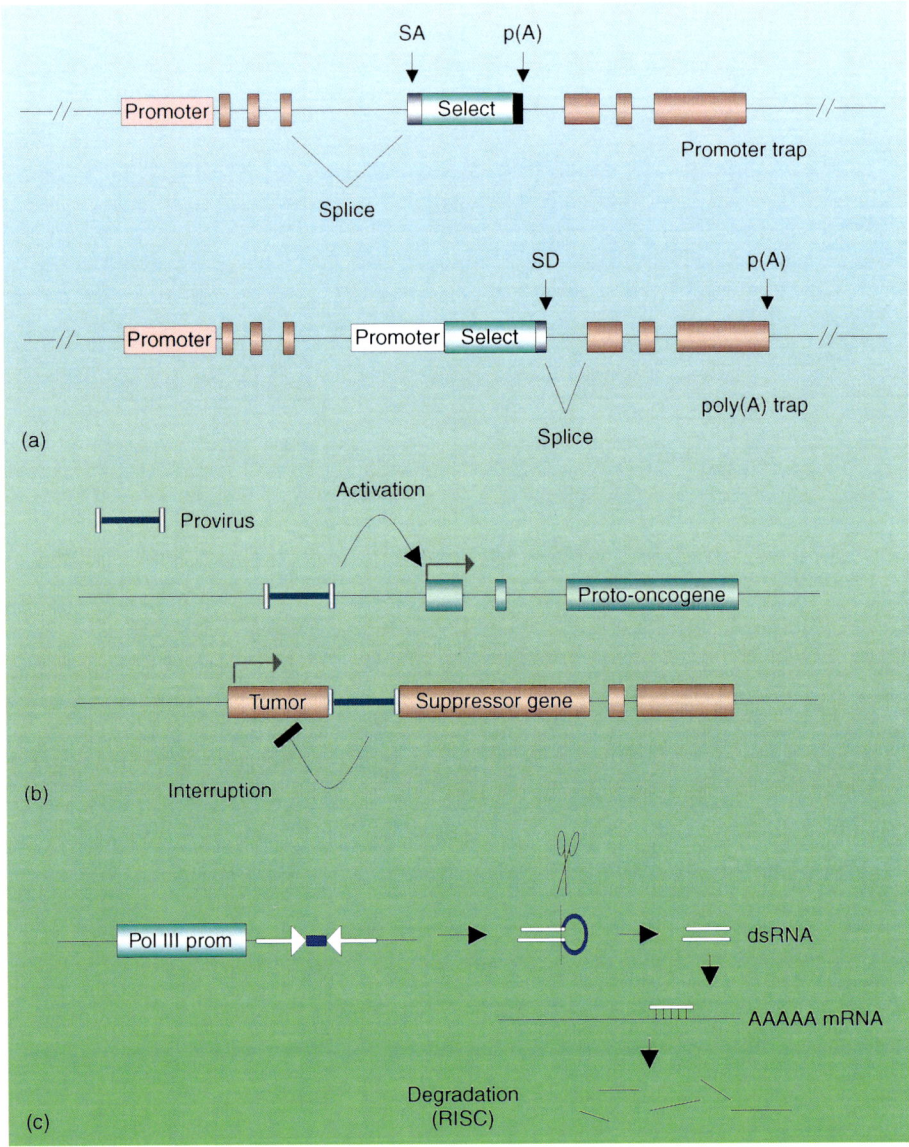

Fig. 15 (p. 895) (legend continued on next page) Random mutagenesis strategies. (a) Promoter-trap and poly(A)-trap vectors, respectively, catch or induce transcriptional activity and have the potential to identify most, if not all, of the active and inactive genes in the genome, including alternatively spliced forms and low-abundance transcripts, and is thus an important tool in genome annotation. (b) Gene mutation through random insertion of retroviral vectors (or transposable elements; see Sect. 5.2). Shown is an oncogene, which is activated by a nearby proviral insert, and a tumorsuppressor gene, which is inactivated proviral insertion into the gene. Inserted elements are engineered such that they simultaneously function as sequencing tags to identify the surrounding genomic DNA. (c) Stable or inducible RNA interference as an alternative to targeted mutagenesis to study gene function in mice: a short double-strand RNA (ds) stem-loop-stem structure

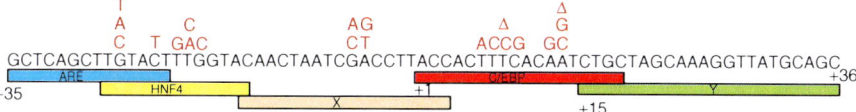

Fig. 9 (p. 1074) Hemophilia B mutations affecting the promoter of F9. Mutant residues are in red, Δ indicates single base deletion. Bars below promoter sequence indicate mutation-affected regions known to bind transactivating factors. ARE = androgen responsive element, HNF4 = liver-specific transcription factor binding site, C/EBP = binding site for members of the C/EBP family of transactivating factors, X = region binding HNF4 and members of the steroid hormone receptor super family (ARP1, Coop/Ear3) that appear to exercise repressor activity on F9 promoter.

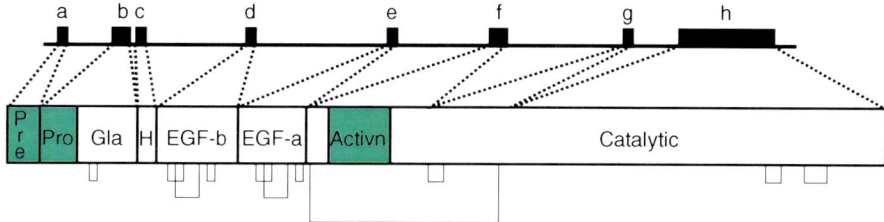

Fig. 3 (p. 1064) Factor IX gene (F9) and protein. Diagram of F9 gene (upper line) and FIX (lower bar). Exons are indicated by black boxes and labeled (a–h). Dotted lines show domains encoded by individual exons. Green areas are protein domains cleaved prior to secretion (pre, pro) or at activation: pre = prepeptide; pro = propeptide; gla = gla region; H = hydrophobic stack; EGF-b = first epidermal growth factor domain with Ca^{++} binding site; EGF-a = second epidermal growth factor domain; activn = activation peptide flanked by carboxyl end of light chain of activated factor IX and amino end of the catalytic region. Disulfide bridges in factor IX are shown below the protein domains.

Fig. 15 (legend continued from previous page) (white arrows, blue loop) is produced from a RNA polymerase III-promoted vector. The loop is enzymatically removed inside the cell; the resulting short interfering RNAs will bind their complementary mRNA, upon which the target mRNA is degraded by a complex called *RISC (RNA-induced silencing complex)*. RNA interference–mediated gene silencing does not require targeted mutagenesis of a gene, and may therefore represent a more efficient and rapid method to assess gene function.

xxvi | Color Plates

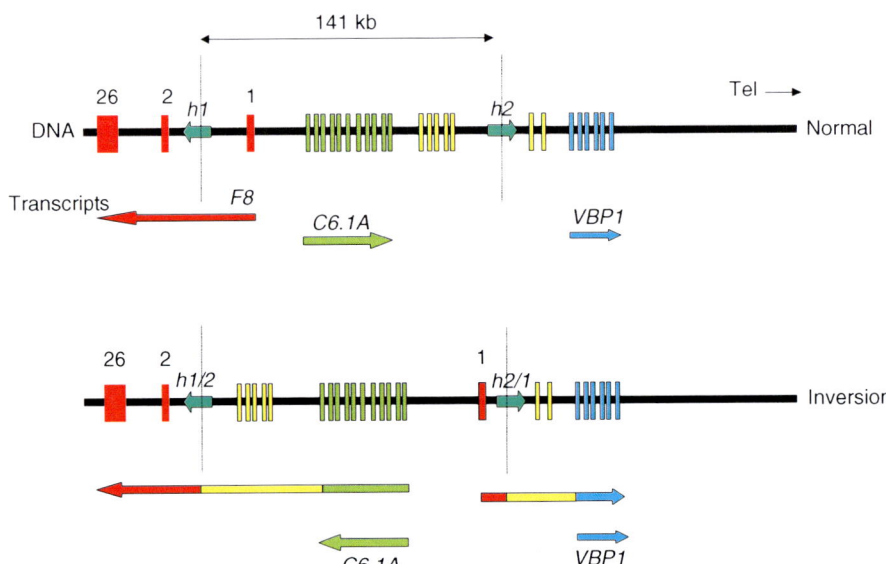

Fig. 6 (p. 1068) Diagram of *int1h*-related inversion, showing its transcriptional consequences. Top: normal structure of DNA region and relevant transcripts (not drawn to scale). Red bars indicate numbered exons of *F8*. Light green and blue bars are exons of the *C6.1A* and *VBP1* gene respectively; yellow bars are exons unrelated to the above three genes. Green arrows in DNA are *int1h*-1 (*h1*) and *int1h*-2 (*h2*). Vertical dotted lines indicate sites where intranemic homologous recombination occurs and leads to inversion. Arrow point in transcripts indicates 3′ end. Bottom: structure of DNA region with inversion and relevant transcripts. *H1/2* and *h2/1* are recombined *int1h* repeats. Transcripts show that *F8* contributes to two mRNAs: one containing all but the last exon of *C6.1A*, some novel exons and exons 2 to 26 of *F8* and one containing exon 1 of *F8* plus novel exons and all but the first exon of *VBP1*. The patients' cells also produce normal *C6.1A* and *VBP1* mRNA.

Part 4
Genomic Technologies

Genomics and Genetics. Edited by Robert A. Meyers.
Copyright © 2007 Wiley-VCH Verlag GmbH & Co. KGaA, Weinheim
ISBN: 978-3-527-31609-0

20
Genomic Sequencing

Todd Charles Wood[1] *and Jeffrey P. Tomkins*[2]
[1] *Bryan College, Dayton, TN, USA*
[2] *Clemson University, Clemson, SC, USA*

1	Introduction to Genome Sequencing	583
2	The Sequencing Technique	584
3	Sequencing Bacterial and Archaeal Genomes	592
4	Sequencing Eukaryotic Genomes	594
5	Interpreting the Sequence	600
6	Summary and Principles	601
	Bibliography	602
	Books and Reviews 602	
	Primary Literature 602	

Keywords

Annotation
Discovery or prediction of biologically important features (e.g. genes, repeat sequences, etc.) in DNA sequence.

Assembly
Recovery of a single consensus sequence from random shotgun reads.

Genomics and Genetics. Edited by Robert A. Meyers.
Copyright © 2007 Wiley-VCH Verlag GmbH & Co. KGaA, Weinheim
ISBN: 978-3-527-31609-0

Bacterial Artificial Chromosome (BAC)
A single-copy plasmid cloning system for large DNA molecules (100–300 kb).

Base-calling
Derivation of nucleotide assignments from sequencing chromatograms.

Fingerprinting
Enzymatic restriction of large-insert clones electrophoretically separated and viewed on an agarose gel.

Genetic Map
A set of genetic markers ordered by their recombination frequencies.

Genome
The complete complement of DNA for an organism consisting of all genes and organized on one or more chromosomes.

Physical Map
An ordered set of large-insert clones providing a physical coverage of a genome.

Production or Shotgun Sequencing
A generation of redundant, raw sequence data from random subclones of a longer target clone.

Beginning with the landmark work of Watson and Crick and continuing with the technological breakthroughs of the 1970s, 1980s, and the 1990s, molecular genetics and genomics has revolutionized all aspects of biology. Today, scientists are able to decipher the precise order of every nucleotide of an organism's complete genome, using a conceptually simple approach called *shotgun sequencing*. Briefly described, shotgun sequencing involves randomly breaking the target genome into smaller pieces, which are sequenced and reassembled into a complete sequence using computer software. Genome sequences provide a basis for further research into gene regulation, pathology, evolution, and metabolism. In order to appreciate a complete genome sequence for the experimental data that it is, it is necessary to review the steps taken to generate genomic sequence. This article will review both the technical aspects of genome sequencing as well as the various strategies employed for sequencing large, eukaryotic genomes.

1
Introduction to Genome Sequencing

The history of modern genomics developed from a fusion of genetics, cytogenetics, biochemistry, and molecular biology. Each of these fields has contributed to our understanding of the material basis of inheritance and how genes code for organismal phenotypes. We may briefly define a genome as the complete set of DNA for a given organism, organized into chromosomes and containing all the genes, which code for the organism's traits (Fig. 1). Thus, historically separate fields of study find union in genomics, the study of organismal genomes. Genome sequencing represents one of the most fundamental steps in genomic analysis. To sequence a genome is to decipher the precise order of nucleotides (the sequence) of that genome. Genome sequencing serves as a foundation for future analyses of gene regulation, chromosome structure, genetic pathologies, and evolution.

With the historic deciphering of the chemical structure of DNA by Watson and Crick in 1953, the modern era of molecular biology and genomics began in earnest. Efficient methods of DNA sequencing were introduced in the 1970s, and the complete 16 569 bp human mitochondrial genome sequence was published in 1980. By the middle of the 1980s, researchers were already beginning to discuss the feasibility of deciphering the sequence of the entire human genome, and the official Human Genome Project began in 1990. The successful application of shotgun sequencing to bacterial genomes by the Institute for Genomic Research in the mid-1990s ushered in a period of rapid expansion of genome sequencing. Today, the complete genomes of more than 100 bacteria and archaea have been sequenced, together with several eukaryotes. Two different drafts of the human genome sequence were published in 2001, and the human genome sequence was completed in 2003.

Despite the enormous changes wrought on modern biology by genomics, the basic techniques of DNA sequencing have not fundamentally changed in the 30 years since their development. This article will begin with an overview of Sanger's sequencing technique and the shotgun strategy utilized in modern genome-sequencing projects. We will then discuss strategies for obtaining the sequence of large genomes, including the hierarchical mapping/sequencing method and the whole-genome shotgun (WGS) that has recently become popular for sequencing eukaryotic genomes. The article will conclude with a brief summary of methods for identifying biological features in raw DNA sequence data.

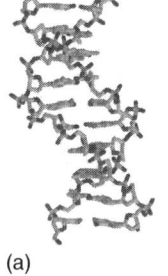 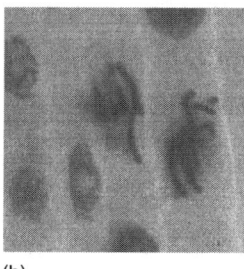

Fig. 1 An organism's genome is the complete complement of DNA, (a) consisting of all genes and (b) organized into chromosomes.

(a)　　　　　　　(b)

2
The Sequencing Technique

The basic sequencing technique developed in the 1970s begins with the polymerization of synthetic strands of DNA from a template strand. The template strand is a cloned piece of DNA that originates from the organism of interest (see below). The polymerization of the synthetic strand of DNA takes place in a mixture of normal deoxynucleoside triphosphates (dNTPs) and a dideoxynucleoside triphosphate (ddNTP). ddNTPs lack a 3' hydroxyl but in all other ways are identical to normal dNTPs. They can substitute for normal dNTPs in DNA polymerization reactions, but the missing 3' hydroxyl prevents elongation of any DNA strand when a ddNTP is incorporated (Fig. 2). As a result, the chain terminates at ddNTP incorporation, giving ddNTPs their colloquial name "chain terminators."

Chain terminators are used in sequencing by including a small amount of them in a synthetic DNA polymerization reaction. The result is random incorporation of chain terminators into growing DNA strands, yielding a complete set of fragments that begin at the same 5' end but terminate at 3' positions where a chain terminator has been incorporated (Fig. 2). For example, when ddATP is included in synthetic DNA polymerization, the resulting strands will vary in length but all will terminate at the 3' end with an adenine. Using the other three ddNTPs, fragments can be generated that terminate with a known base at the 3' end of the fragment (Fig. 3).

Chain-terminated fragments are then size-fractionated by electrophoresis on a polyacrylamide slab gel. Because the gel has a very fine pore size, fragments differing in length by a single nucleotide can be separated. This allows each chain-terminated fragment to be distinguished on the polyacrylamide gel. Traditionally, fragments in the gel were visualized by incorporating radioactively labeled dATP in the sequencing reactions and exposing the gel to X-ray film after electrophoresis.

One of the major innovations that rendered genome sequencing practical is the introduction of fluorescent labels, which alleviated the need for radioactive dATP and X-ray film. In fluorescent sequencing, the DNA fragments corresponding to the four different ddNTP chain terminators are labeled with a different fluorescent tag. These tags are excited by a laser at the end of the gel and the fluorescent light is detected at the different wavelengths of each of the four dyes. Detection of the fluorescent tags can take place as the electrophoresis is still occurring, eliminating the need to stop the electrophoresis and expose the gel to X-ray film. For fluorescent tags attached to the sequencing primer (dye-primer), the four sequencing reactions corresponding to the four bases must occur in isolation. Alternatively, fluorescent tags can be attached directly to the ddNTP (dye terminator), which allows for a single sequencing reaction in a single container for each template to be sequenced. Generally, dye-terminators reduce the frequency of anomalous mobility during electrophoresis but are poorly recognized by the DNA polymerase, presumably because of the added bulk of the attached fluorescent dye. Dye primers, while requiring isolated preps and an extra mixing stage prior to electrophoresis, provide a consistently readable signal.

A record of fluorescent tags as they pass by the detector during electrophoresis is called a *sequence trace*. A trace can be converted to a visual representation called a *chromatogram* (Fig. 4) for translation

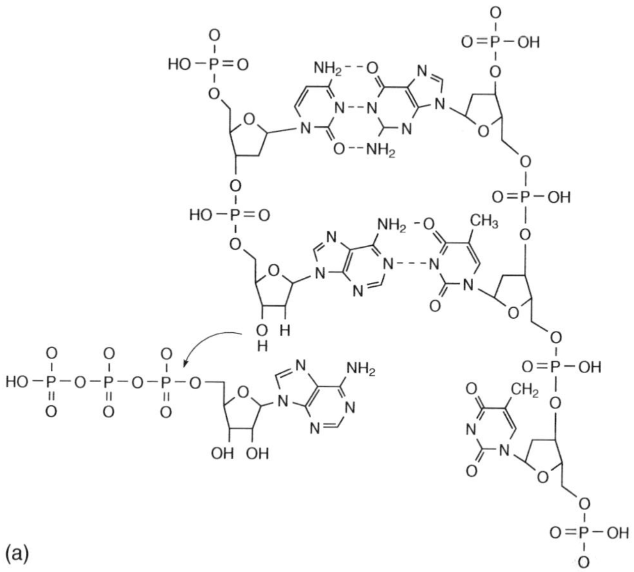

Fig. 2 (a) During normal DNA polymerization, the 3′ hydroxyl is necessary for formation of the phosphate bond that elongates the nascent DNA strand. (b) If the 3′ hydroxyl is missing, as when a ddNTP (chain terminator) is incorporated, the nascent strand cannot be elongated further.

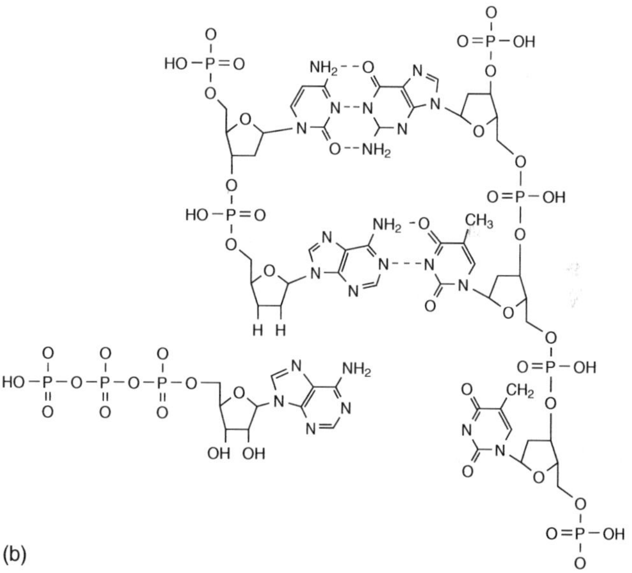

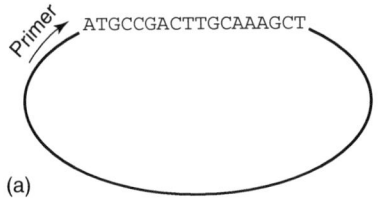

ddATP	ddCTP	ddGTP	ddTTP
ATGCCGACTTGCAAA	ATGCCGACTTGCAAAGC	ATGCCGACTTGCAAAG	ATGCCGACTTGCAAAGCT
ATGCCGACTTGCAA	ATGCCGACTTGC	ATGCCGACTTG	ATGCCGACTT
ATGCCGACTTGCA	ATGCCGAC	ATGCCG	ATGCCGT
ATGCCGA	ATGCC	ATG	AT
A	ATGC		

(b)

Sequence	A	C	G	T
ATGCCGACTTGCAAAGCT				—
ATGCCGACTTGCAAAGC		—		
ATGCCGACTTGCAAAG			—	
ATGCCGACTTGCAAA	—			
ATGCCGACTTGCAA	—			
ATGCCGACTTGCA	—			
ATGCCGACTTGC		—		
ATGCCGACTTG			—	
ATGCCGACTT				—
ATGCCGACT				—
ATGCCGAC		—		
ATGCCGA	—			
ATGCCG			—	
ATGCC		—		
ATGC		—		
ATG			—	
AT				—
A	—			

(c)

Fig. 3 (a) The sequencing reaction generates fragments from the template, which is originally cloned into a suitable sequencing vector. (b) For each of the four chain terminators, all possible fragments that terminate at the 3' end with that nucleotide are generated in the sequencing reaction. (c) When electrophoresed on a polyacrylamide gel, the fragments can be visualized, and the sequence can be read directly from the gel.

into a nucleotide sequence. Translation of a chromatogram into a sequence is called *base calling*. In the chromatogram, nucleotides correspond to peaks of fluorescence, which are displayed with different colors representing different nucleotides. Because of the obvious need for rapid base calling in large-scale genome projects, software has been developed to automate the process. There are several base-calling

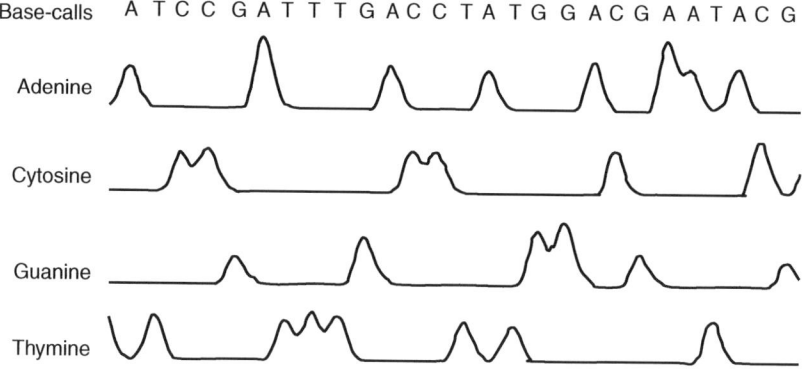

Fig. 4 A sequence chromatogram generated by an automatic sequencer and visualized as fluorescent intensity (y-axis) and electrophoretic mobility (x-axis).

programs available, including commercial software provided by manufacturers of automated DNA sequencers. The Phred program, developed by Phil Green at the University of Washington, is the most commonly used program today.

Phred base-calls in a four-step process. First, peak locations are predicted using Fourier methods. Prediction of peak location is necessary for regions where nucleotides are unreadable or ambiguous. In the next two steps, the program identifies observed peaks and matches them to the predicted peak locations. As observed peaks are matched to the predicted peak locations, the program is able to identify and resolve ambiguities in the sequence. Finally, any unambiguous observed peak that was not matched to a predicted peak is called and inserted into the sequence.

Owing to problems inherent in the sequencing reaction and in electrophoresis, sequence traces rarely yield a perfect sequence. Aberrant mobility of short sequences, unreacted sequencing reagents, DNA hairpin structures, and even simple diffusion can all produce regions of ambiguity in sequence traces. Because of these problems, Phred assigns an error probability to each base in the sequence.

In calculating the error probability, Phred considers peak spacing, the ratio of the largest uncalled peak to the called peaks, and the peak resolution for the short sequence surrounding each nucleotide. Phred converts the error probabilities into "quality values," which are 10 times the negative logarithm of the error probability. Thus, a quality value of 30 corresponds to an error probability of 1 in 1000. Higher Phred quality values are to be desired.

A second technical advance that facilitated the rate of genome sequencing is the advent of practical capillary electrophoresis. Originally, for each limited set of sequencing reactions, a separate polyacrylamide gel needed to be created for the electrophoresis step, and the slab gel had to be exposed to X-ray film to visualize the radioactively labeled DNA fragments. With fluorescent labels, the sequence can be read by a fluorescent detector during electrophoresis. In either case, the step of creating the slab gel was time consuming, and electrophoresis in a slab gel separates the sequencing fragments slowly. With capillary sequencers, the DNA fragments are fractionated by electrophoresis through a gel embedded in a capillary tube with an internal diameter of 100 μ. Capillary sequencers require fewer reagents for the gel and completely eliminate the time-consuming process of manually preparing, loading, and editing the electrophoretic gel.

The addition of robotics to DNA sequencing processes provided a means to automate the more repetitive processes. When combined with fluorescent detection and capillary electrophoresis, highly automated sequencing devices such as the ABI PRISM 3730 or 3100 DNA Analyzers and the Amersham Pharmacia MegaBACE 4000 automated sequencing system can sequence hundreds of thousands of nucleotides everyday (Fig. 5). Special plates containing the results of sequence reactions are merely loaded into the machines. Robotics within the machine load each sample into a capillary, and detectors record the fluorescence as the fractionated sequencing fragments approach the end of the capillary.

Owing to limitations of the electrophoresis technique, approximately 500 to 1000 nucleotides can be reliably read from each template, even when using a capillary sequencer. To sequence even a small genome like the 0.5 Mb genome of *Mycoplasma genitalium*, the DNA molecule must be broken into small pieces that are able to be sequenced. The general strategy employed for recovering the complete sequence of a large DNA molecule is called *shotgun sequencing*. First proposed in the early 1980s, shotgun sequencing requires random fragmentation of the longer DNA molecule, cloning the fragments into a sequencing vector, and sequencing the fragments (Fig. 6). The sequence of the original DNA molecule is then assembled from the fragments using computational algorithms.

The first step in shotgun sequencing is the subcloning of the DNA sequencing target, also called *library construction*. In this step, the target DNA is broken into smaller pieces, which are then cloned into vectors that can be used in sequencing reactions. Any collection of clones that represent complete coverage of a larger piece of DNA is called a *library*. To create a sequence-ready shotgun library, the DNA of the target molecule is a fragment enzymatically by a partial digestion with a nuclease or restriction enzyme, or the DNA can be physically sheared, for example, by sonication. Physical shearing generally produces a more randomized sample of the target sequence and is more suitable for high throughput sequencing centers.

Two types of cloning vectors are available for shotgun sequencing, each with its advantages and disadvantages. The

Fig. 5 A modern capillary sequencer, the ABI Prism 3100 Genetic Analyzer.

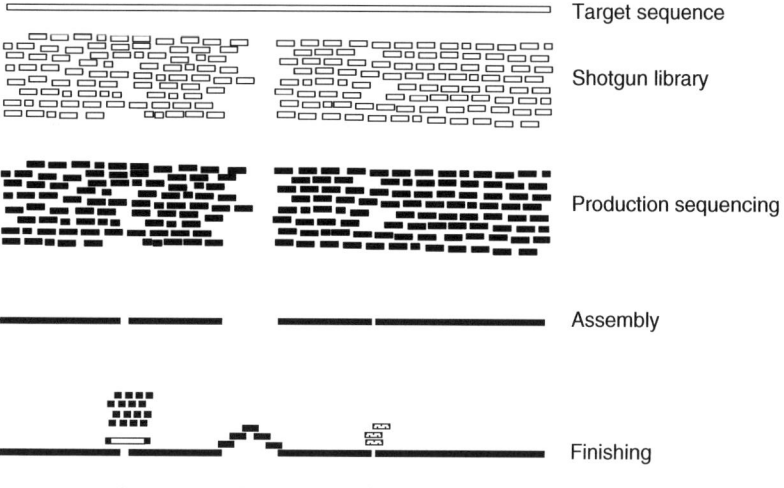

Fig. 6 In the shotgun-sequencing strategy, the target sequence is fragmented into random pieces, which are cloned in a sequencing vector to form the shotgun library. Production sequencing generates sequence data for the shotgun clones to at least a sixfold coverage. Computer programs assemble the shotgun reads into contiguous sequences. During finishing, various strategies are used to close the remaining gaps, including identification by end sequences and shotgun sequencing of a clone that spans the gap A, primer walking B, and alternative chemistry sequencing C.

first is bacteriophage M13, which naturally exists as single-stranded DNA. Since DNA sequencing templates must be single-stranded DNA, M13 naturally provides sequence-ready DNA templates. M13 clones exhibit a bias, in that repetitive sequences are poorly represented in M13 libraries. Alternatively, double-stranded plasmid clones can also serve as the vector for subcloning shotgun libraries. Plasmids do not exhibit the same cloning bias as M13 vectors, and thus represent the target DNA sequence more evenly. The disadvantage of plasmids is that they are double-stranded and therefore require additional steps to purify single-stranded DNA sequencing templates. A large genome-sequencing center will often use a combination of M13 and plasmid subclones for different sequencing needs.

Having created a suitable shotgun library, the next step is typically referred to as *production sequencing*, because it resembles the factory production typical of an assembly line. A random selection of shotgun clones is prepared and sequenced using the automated sequencing techniques discussed above. The number of clones necessary to reconstruct the original target sequence depends on the average length of sequence obtained from a single shotgun clone, the length of the target sequence, and the desired accuracy of the completed sequence. For production sequencing, genome-sequencing centers typically strive for a six- or eightfold coverage of the target sequence; that is, each nucleotide of the target is sequenced six (or eight) different times. This level of redundancy assures that most of the target sequence will be covered by the randomly positioned shotgun clones, and whatever gaps or ambiguities remain after the shotgun sequencing can be quickly resolved.

After generating a sixfold redundant shotgun sequence, the sequence traces are then processed computationally. First, the traces are base-called using Phred or a comparable program. Second, the sequences are "assembled" into contiguous sequences called *contigs*. A contig is composed of two or more sequence reads that originate from an overlapping region of the target sequence. Contig assembly begins with an automated step, and concludes with a manual editing stage. The automated step is carried out by an assembler program such as Phrap, the TIGR Assembler, or CAP3. The assembler examines the sequence reads for regions of near identity (allowing for errors in the sequence read) and attempts to reconstruct the original target sequence. If the sequence reads were a perfect and redundant random representation of the target sequence, the assembler could theoretically reconstruct the entire target sequence from the shotgun sequence. In reality, variations in the quality of sequence reads, regions of the target with low shotgun representation, and repetitive sequences can result in assembly of many different contigs (depending on the size of the target sequence and the redundancy of the shotgun clones).

All assembly programs work in very similar ways. First, regions of overlap are identified by doing standard sequence comparisons between the sequence reads. Unlike a normal sequence comparison program, however, assemblers also consider the quality values of the sequence reads in calculating similarity scores between reads. From the overlap information, the assembler then reconstructs sequence contigs and some also calculate quality values of each nucleotide in the contig, based on the number of reads available at that position. To be certain that the contig comes only from the target sequence, sequence reads are typically filtered before assembly in order to remove contaminating DNA, such as vector sequence, that often appears at the 5' end of a sequence read.

Having completed the automatic assembly, the sequence rarely resolves into a single contig of uniform quality. Reasons for obtaining multiple contigs include failure of the assembly program, cloning bias, regions that are difficult to sequence, and regions that were poorly represented in shotgun sequence reads by chance. To resolve these problems and to produce a single contig of uniform quality, manual inspection and editing of the contigs is necessary, together with additional sequencing reactions where appropriate. This manual phase is called either *gap closure* or more commonly *finishing*.

Finishing is aided by the use of a graphical contig editor such as Consed. For each assembly, Consed displays a list of the contigs and sequence reads. After selection of a contig, Consed displays the consensus sequence together with each sequence read that was assigned to that region by the assembler. The consensus sequence is displayed at the top of the assembly window, and the sequence reads appear in an alignment format just below the consensus (Fig. 7). Important features such as base quality are color-coded for easy identification of low-quality regions. Other windows in Consed display the actual chromatograms of any desired sequence reads.

The additional work necessary to close a gap in the sequence assembly depends on the cause of the gap. Regions of low shotgun representation can often be resolved by sequencing the opposite end

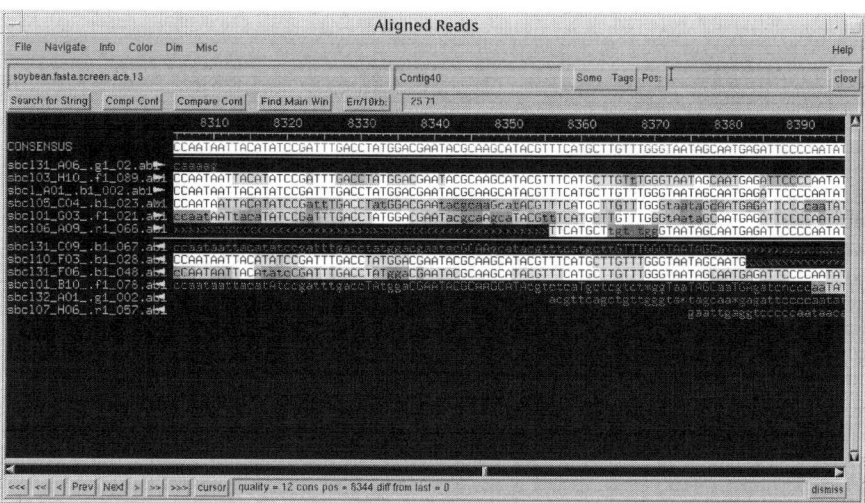

Fig. 7 A sequence assembly shown in a window of Consed. The consensus contig sequence is displayed at the top of the window, with individual sequence reads aligned and displayed beneath. The background color indicates the relative quality of base-calls. Uppercase letters with white backgrounds are of high quality; lowercase letters with black background are of poor quality. Nucleotides represented as x come from contaminating vector sequences, which have been removed prior to assembly.

of the clone insert for reads adjacent to the low-quality region. This will generate sequence reads for both ends of the clone insert and will usually provide sufficient sequence coverage to allow the gap to be closed. If the gap is caused by vector bias, it will be necessary to clone the missing region into a different vector. For example, as we discussed above, M13 does not effectively clone regions of repetitive DNA. Utilization of both M13 and plasmid clones can resolve problems caused by this cloning bias. In cases in which the assembly algorithm has failed to derive the correct consensus sequence, the statistical parameters of the assembly algorithm can be manipulated to generate the desired assembly. For example, assembly algorithms assemble regions with tandem repeats poorly. These errors are easily identified because of the abnormally large number of sequence reads assigned to a single region.

The most difficult class of problems encountered in finishing are regions that are intrinsically difficult to sequence, such as simple sequence repeats, homopolymeric regions, or regions with secondary structure. These types of gaps are easily identified because of the uniformity with which the quality drops off. In other types of gaps, the low quality may be observed in only a few of the shotgun reads, but in areas that are difficult to sequence, all shotgun reads will exhibit the same low quality at the same point in the sequence.

Resolution of sequence gaps requires a number of different strategies, each designed to deal with a particular sequencing problem. For example, specialized chemistry designed for simple sequence repeats can produce high-quality reads of repetitive regions. Larger repeats or secondary structures could require a targeted subcloning strategy designed to break up the difficult region prior to sequencing. In another finishing strategy, called *primer walking*, sequencing primers are designed from the sequences at the boundaries of gaps. The sequencing reaction is then repeated, generating new sequence. This process of primer development and sequencing is repeated until the gap is closed.

After *finishing* has produced a single contig, the consensus sequence is validated in a number of ways. First, the length of the sequence is compared to the expected length of the target sequence. The expected length is typically measured by a restriction digest of the target sequence. Second, the length of the restriction fragments observed for the target sequence are compared to the length of the restriction fragments predicted for the consensus sequence. Whereas matching the length to the expected length confirms that a sequence of the correct size has been generated, matching the predicted and observed restriction fragment lengths ensures that the general order of the sequence assembly is good. This procedure can be repeated with multiple restriction enzymes to verify the assembly even further.

3
Sequencing Bacterial and Archaeal Genomes

As outlined in the previous section, the basic technique of shotgun sequencing is suitable for contiguous sequences of 10 Mb or less with few repeat sequences. For the small genomes of many bacteria and archaea, a straightforward shotgun strategy works well. The size and repetitive nature of eukaryotic genomes prevent the ready application of a shotgun-sequencing technique, necessitating specialized strategies for sequencing these genomes. Depending on the level of completion required, eukaryotic genomes are sequenced by a mapping strategy (for high-quality sequences) or by a whole-genome shotgun (WGS) strategy (for rough draft sequences), or by a hybrid of both. Strategies currently in development attempt to target the genes of large eukaryotic genomes by specialized cloning techniques. In this section, we will discuss the shotgun strategy used to generate the first bacterial genome sequence, *Haemophilus influenzae*. The following section will focus on strategies used in eukaryotic genome sequencing.

The first organismal genome completely sequenced by the shotgun method was that of *H. influenzae*, sequenced by a team from the Institute for Genomic Research (TIGR) and published in 1995. As a bacterial genome, the repetitive DNA content was much lower than most eukaryotes, and the size of the complete chromosome was only 1.8 million nucleotides. Despite being the first organismal genome, the issues considered and strategies employed in sequencing are still representative of modern genome projects.

Because the success of the shotgun method depends on a random selection of sequence templates, the TIGR team expended much care on template library construction. Two types of libraries were made for the *H. influenzae* project: a short-insert (1.6–2 kb) plasmid library and two long-insert (15–20 kb) λ phage libraries. The plasmid library was used as the primary sequencing template, while the λ libraries were reserved for finishing and validation (we will discuss the utilization of the λ libraries below). For all libraries, the genomic DNA was mechanically sheared rather than digested by restriction enzymes. As mentioned previously, shearing the source DNA ensures a more random library sample than restriction digest.

On the basis of simple statistical calculations, we know that the probability that a nucleotide will be unsequenced in a shotgun-sequencing project is $P_o = e^{-m}$, where m is the genome coverage. Sequencing random clones sufficient to cover the genome five times (fivefold coverage) results in a probability of 0.0067 that a nucleotide will be unsequenced. For the 1.8 million-nucleotide *H. influenzae* genome, a fivefold shotgun coverage should result in approximately 12 000 unsequenced nucleotides, distributed randomly throughout the genome in sequence gaps between the contigs.

Given an average high-quality read-length of 460 nucleotides (the limits of the sequencing technology at that time), approximately 19 000 sequence reads would be necessary to cover the *H. influenzae* five times. The TIGR team sequenced 19 687 short-insert templates using dye-primer sequencing chemistry (*forward reads*). To supplement these reads and to provide contig assembly information, 9297 templates were re-sequenced at the opposite end of the insert (known as *reverse reads*), also using dye-primer chemistry. The result is a pair of 460-nucleotide sequence reads that are known to be 700–1100 nucleotides apart, a significant advantage during finishing.

Automatic assembly of forward and reverse reads yielded 210 contigs. Because the parameters used in assembly are optimized for general use, local variations can occur due to differences in repeat- or GC-content. As a result, manual inspection of potential overlaps resulted in a reduction to only 140 contigs, which could not be further combined without additional sequencing reactions. The remaining gaps could be categorized into two types depending on the orientation of paired sequence reads. If the forward reads at the end of one contig matched the corresponding reverse reads from the same template at the end of a second contig, the TIGR team called the gap a *sequence gap*. Because they are spanned by a single, small-insert plasmid, sequence gaps are known to be small (less than 1500 nucleotides), and known templates (the plasmid inserts spanning the gaps) are available for immediate, additional sequencing. For the *H. influenzae* genome, 98 gaps were sequence gaps.

The second type of gap occurred when forward/reverse reads of the same template did not span the gap. The TIGR team labeled these gaps *physical gaps*, because a clone insert that spanned the gap was not immediately available. Additional strategies were devised to identify and prepare suitable sequencing templates that spanned the physical gaps. For two gaps, protein sequences could be used to orient the adjacent contigs. This was possible because the gap occurred within a protein-coding gene, the 5′ end of which was sequenced on one contig and the 3′ end on another. By comparing the contig sequences to known protein sequences, these overlaps could be detected and PCR used to generate suitable sequencing templates.

The majority of the gaps were closed using one of two strategies. First, DNA fingerprinting and hybridization to oligonucleotides prepared from the ends of contigs revealed possible overlapping DNA segments, from which templates could be prepared. For example, if oligonucleotide probes prepared from the ends of two contigs hybridized to the same restriction fragment of the genomic DNA, it is likely that the hybridizing fragment contains the sequence spanning the gap between the two contigs. The other successful gap closure method was the paired forward/reverse sequences from the λ libraries. Small gaps are more likely to be spanned by the large inserts (15–20 kb) of the λ libraries than the small inserts (~2 kb) of the plasmid library. The remaining physical gaps were closed by simple combinatorial PCR, systematically using oligonucleotide primers from each possible contig pair.

These strategies resulted in a single contig with a consensus sequence of 1 830 137 nucleotides. The assembly of the consensus sequence was validated using additional λ library forward/reverse sequence reads and restriction fragments and restriction site locations. Since the λ library has a known insert size of 15–20 kb, paired reads that deviate significantly from that distance would indicate a misassembly. Sizes of restriction fragments generated by three different enzymes could also be matched to the predicted sizes of the consensus sequence. Restriction site locations from restriction mapping could also be matched to the restriction sites found on the consensus sequence. For all of these validations, the *H. influenzae* consensus sequence matched the observed characteristics of the genome.

Despite being the very first bacterial genome to be sequenced by the shotgun method, much of the techniques and strategies developed are the same as those used today for bacterial and archaeal genome projects. Library construction remains an important step to assure a random coverage of the genome. Sequencing both ends of the inserts and using different sequencing chemistries are both common strategies for gap closure and finishing. The major difference between a modern genome project and the *H. influenzae* project is the rate. Facilitated by capillary sequencers and more powerful computers for assembly and finishing, a bacterial genome of the same size as *H. influenzae* can be completed in a fraction of the time and at a fraction of the cost.

4
Sequencing Eukaryotic Genomes

Eukaryotic genomes differ from bacterial and archaeal genomes in several important ways that make genome-sequencing projects more complex. Although several bacteria and archaea are known to have multiple chromosomes or *megaplasmids*, a typical prokaryotic genome is a single, circular chromosome. In contrast, eukaryotic genomes are divided into several (sometimes many) linear chromosomes. Eukaryotic genomes are also much larger, on average: the largest bacterial genome sequenced to date is the 9 Mb chromosome from *Bradyrhizobium japonicum*. At 3200 Mb, the human genome is approximately 355 times larger (see also Fig. 8). The size and multiple chromosomes alone would challenge the assembly algorithms and finishers, but the sequence of eukaryotic genomes also contain a high repeat content, rendering cloning, sequencing, and assembly of particular regions extremely difficult.

Because of these limitations, extra pre-sequencing strategies must be utilized to make the genome amenable to sequencing, and the definition of "finished genome" is often project-specific. The most common strategy employed for eukaryotic genome sequencing is a "map-then-sequence" approach that seeks to

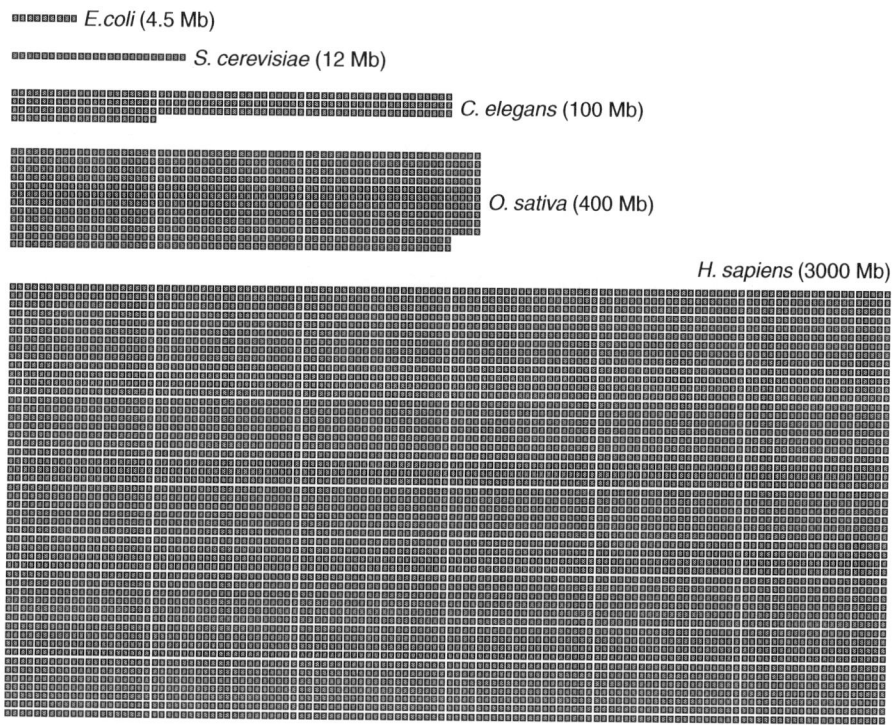

Fig. 8 Relative sizes of sequenced genomes. Each square represents 500 000 nucleotides, approximately the size of the complete M. genitalium genome. Genomes shown represent the K12 strain of E. coli, baker's yeast S. cerevisiae, the nematode C. elegans, rice O. sativa, and the human genome H. sapiens.

obtain a physical map of the genome prior to the shotgun-sequencing phase. The physical map is composed of ordered, overlapping large-insert clones anchored to markers in the genome, identified by genetic or cytogenetic mapping. On the basis of the order and orientation of the map, individual clones are selected and subjected to the standard shotgun sequencing and finishing, as described above (Fig. 9). There are several advantages to this strategy, including an overall reduction in the number of repeats that need to be resolved during any particular finishing phase and the identification of problem regions before sequencing even begins.

Ideally, the physical map and finished genome sequence should cover the entire genome, but this level of accuracy is rarely achieved. Regions of concentrated repeats, such as telomeres and centromeres, do not clone well and are extremely difficult to sequence. Regions of high repeat content are called *heterochromatin*. Euchromatin has a much lower repeat concentration and is believed to contain the majority of the genes. As a result, most eukaryotic sequencing projects aim to completely sequence the euchromatic regions of the genome. For example, approximately one-third of the fruitfly (*Drosophila melanogaster*) genome was heterochromatic and consequently not sequenced. Even when dealing with just

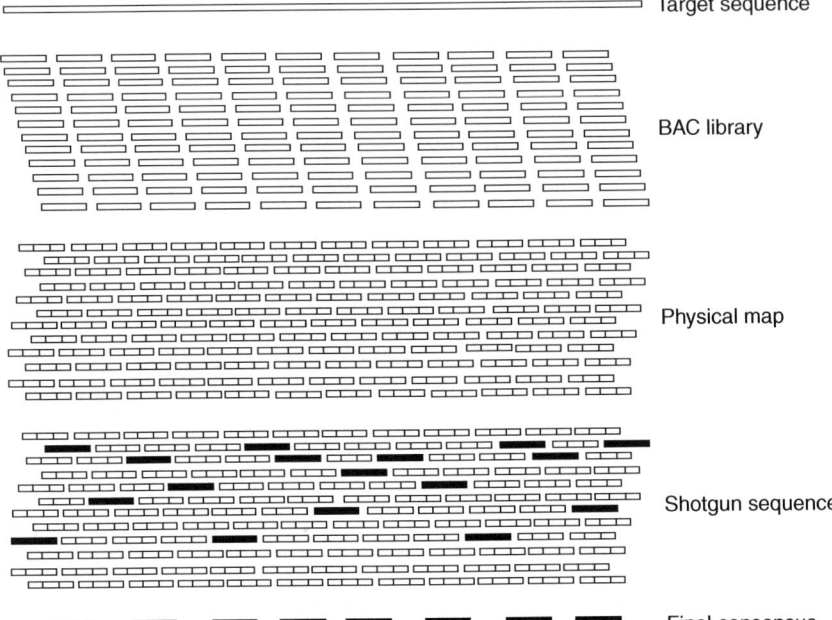

Fig. 9 The map-then-sequence strategy used for large, eukaryotic genomes. The target sequence is fragmented by partial restriction digest and cloned into a large-insert BAC library. The BAC inserts are fingerprinted and overlaps detected to generate a physical map. On the basis of the physical map, clones are selected for shotgun sequencing (see Fig. 6), and the final consensus consists of a series of overlapping clones.

euchromatin, small regions that are difficult to sequence can still occur, and most eukaryotic genome projects allow for a certain number of gaps in the "finished" product, provided the gap length is known. The public Human Genome Project allows an unspecified number of gaps, as long as their length is less than 150 kb.

Because physical mapping is beyond the scope of this article, we will here provide only a cursory overview of the map and sequence strategy. A physical map begins with the construction of large-insert libraries. Currently, the vector of choice for mapping is the bacterial artificial chromosome (BAC). BACs can hold approximately 100–300 kb of insert and are not prone to recombination. Unlike shotgun libraries, the source DNA for a BAC library is typically fragmented by partial restriction digest. Protocols for BAC cloning randomly sheared DNA are still in development. Because of the inherent biases introduced by partial restriction digests, multiple libraries constructed with different restriction enzymes are often necessary to get complete, random coverage of the genome.

After the construction of the genomic library, BAC inserts are ordered and aligned by "fingerprint" analysis. Each clone is completely digested with the same restriction enzyme, and the fragments are fractionated and viewed by agarose gel electrophoresis (Fig. 10). A new technology called *high information*

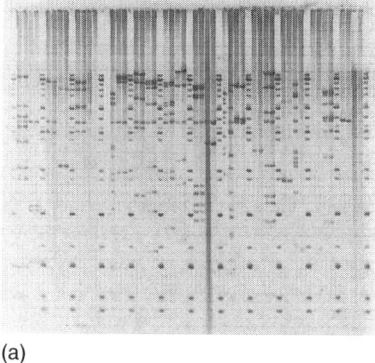
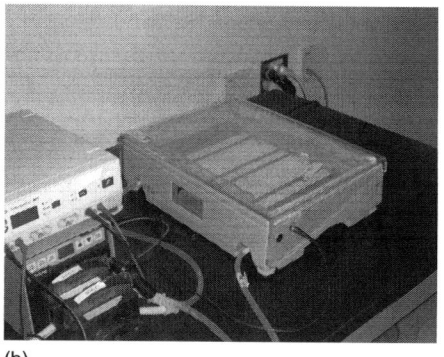

Fig. 10 (a) A fingerprint gel generated from a large-insert BAC library of rice (b) and the electrophoresis apparatus used for fingerprinting.

content fingerprinting (HICF) utilizes up to five different restriction enzymes followed by dye labeling and electrophoresis on a DNA capillary sequencer rather than a standard agarose gel. HICF reduces labor, increases automation and the number of fragments per BAC, and allows automatic measurement of fragment mobility from chromatograms. Owing to the increased fragment number, more robust contigs can be generated from the fingerprint data.

The relative mobilities of each clone's restriction fragments are loaded into a computer program (such as FPC) that determines overlapping regions between clones based on shared fragment mobilities. The result is a set of physical map contigs (Fig. 11). Because restriction fingerprints provide less data than actual sequencing, the assembly of physical map contigs is less precise than sequence contigs and consequently requires a deeper coverage of the genome. Many physical maps cover the genome 10 to 15 times. As with shotgun sequencing, the coverage is uneven, and many difficult-to-clone regions of the genome may be missing from the physical map contigs. Closure of gaps in physical maps takes place through both computational and additional laboratory experiments, using some strategies analogous to gap closure in sequence finishing.

The goal of the physical mapping effort is to produce a set of minimally overlapping clones that cover the entire euchromatic portion of the genome, also called a *sequence-ready* map. Because the coverage can be uneven, the physical map must be "anchored" by hybridization with marker sequences of a known, unique genomic position. These markers can be identified by a number of mapping methods, including traditional genetic mapping. Once the physical map is anchored, any remaining gaps in the euchromatin coverage can be identified, and strategies to fill those gaps can be devised and implemented.

As an aid to the creation of the sequence-ready map, the ends of the BAC clone inserts can be sequenced. Although some of these "BAC-end sequences" or "sequence-tagged connectors" (STCs) will undoubtedly come from repetitive sequence, many will be unique sequences that represent the ends of only one clone insert. STCs of clones bordering gaps in the physical map can be used to generate novel probes to identify clones and possibly contigs that span the gap. During the

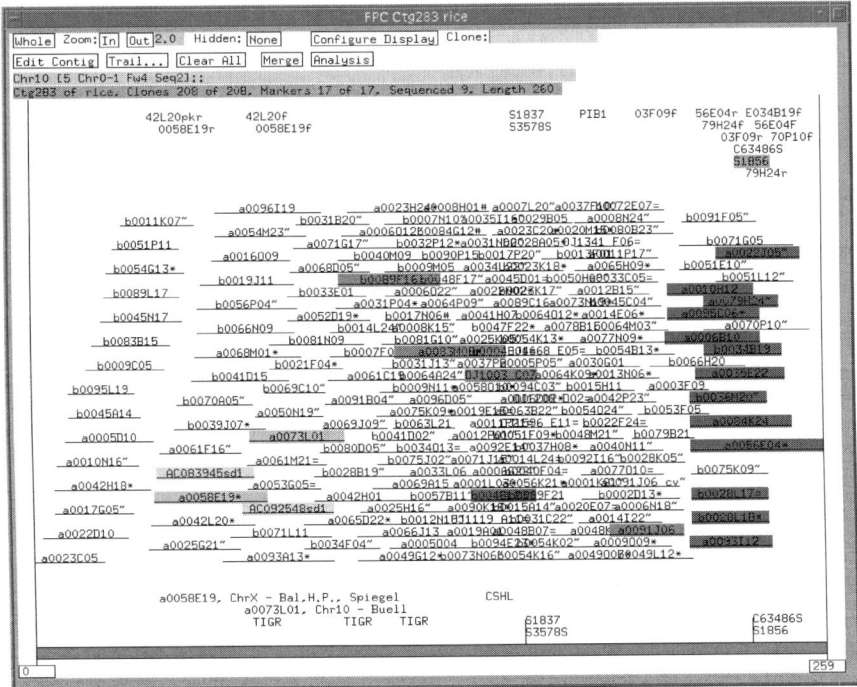

Fig. 11 A physical map contig from the rice genome, as displayed by the physical mapping software FPC.

sequencing phase, STCs can be used to choose the next minimally overlapping clone with a great deal of precision. STCs can also be a valuable dataset in their own right by providing a preliminary survey of the contents of a eukaryotic genome.

After (or even during) construction of the physical map, sequencing can commence using the standard shotgun method. A sequence-ready clone is chosen, shotgun libraries are prepared, and the clone is sequenced. Additional clones are selected and sequenced with minimal overlap with the existing sequence. Finishing is limited to the regions that do not overlap with an existing, finished sequence.

With a physical map, another advantage becomes immediately apparent. Unlike a pure shotgun approach for small genomes, the sequencing of a mapped genome can be divided between more than one sequencing center. The genome can be divided into regions or even whole chromosomes, and sequencing can take place in many different labs simultaneously. Additionally, even at a single sequencing center, sequencing can commence at different positions in the genome. As new sequence-ready clones are identified by physical mapping, shotgun sequencing and finishing can commence on other clones. Thus, the mapping and sequencing at a single lab can proceed simultaneously.

The map-then-sequence strategy has been successfully applied to several prominent eukaryotic genome projects. Genome sequences of the nematode *Caenorhabditis elegans* and the plant *Arabidopsis thaliana* were both completed in this way, and the public Human Genome Project also used this method. The alternative strategy is the whole-genome shotgun (WGS) or a mixture of mapping and WGS. WGS sequences are rarely finished to the same quality as a mapped-and-sequenced genome, but for certain genome projects, WGS provides a reasonable and affordable alternative to a completed sequence. Because of the unfinished quality of WGS sequences, they are often referred to as *draft sequences*.

Though the vast majority of genome sequence is obtained during the shotgun-sequencing phase, the majority of effort (and money) is invested in the finishing phase. With the complete human genome sequence, finished genome sequences from other vertebrates may be an unnecessary expense for comparatively little information. Because genomes of mammals and vertebrates have conserved gene content and order, the human genome sequence can serve as a reference template for other genome sequences. A WGS project could generate four- or fivefold coverage of the genome and still theoretically achieve 98% coverage of the euchromatic region of the genome. A WGS for a vertebrate would produce a very large number of sequence contigs that could be ordered by reference to the human genome.

As with a standard shotgun-sequencing project, a eukaryotic WGS begins with careful construction of genomic clone libraries. Celera's attempt to sequence the human genome by WGS showed that multiple clone libraries with differing insert sizes are essential to the assembly of sequence contigs. For each library, both ends of the clone inserts must be sequenced to provide pairs of sequences of a known distance apart on the chromosome. Paired-end sequences are necessary to order and orient sequence contigs into "*scaffolds*." Finally, the actual assembly of eukaryotic WGS sequence reads usually proceeds in a multistep process, even with advanced computational hardware. Typically, the repeat sequences will be masked in the early assembly steps to prevent misassembly due to highly conserved repetitive sequences. Once contigs and scaffolds of unique sequences have been assembled, the repeat sequences can be added back to the assembly with the unique sequences acting as an assembly framework.

A third application of WGS is to use a mixed approach, with elements of both map-then-sequence and WGS. For example, a low-coverage WGS can be generated rapidly at a large genome-sequencing center while finishing could be completed on a clone-by-clone basis at smaller sequencing centers. This hybrid approach was used for the *Drosophila* genome project. Celera provided a nearly 15-fold shotgun coverage of the *Drosophila* genome, and finishing was completed in a number of publicly funded laboratories.

Despite the advantages of WGS for some eukaryotes, the drawbacks are also significant. Experience with the rice genome has shown that contigs produced by WGS may be a poor substrate for *ab initio* gene prediction (see below). In a detailed analysis of the complete sequence of rice chromosome 10, researchers predicted twice as many genes as had been predicted by previous WGS sequencing efforts of the same chromosome. Additionally, genes predicted from the WGS contigs were on average one-third shorter than genes predicted from the finished sequence.

Several new techniques have been proposed for specifically targeting gene-rich sequences during the initial library construction, prior to shotgun sequencing. In one method, methylated DNA is eliminated prior to library construction. In eukaryotic genomes, transposons are often methylated at CpG dinucleotides as a means of downregulating their transcription. By excluding methylated DNA from library construction, only unmethylated genes should be included in the library. The primary drawback of this method is that most organisms use methylation as a means of transcriptional control of cellular genes as well as transposons. The second method of gene-targeted cloning, called *cot-based cloning and sequencing* (CBCS), relies on the fact that repetitive DNA reanneals more quickly than unique DNA after denaturation. By isolating the DNA that reanneals the slowest, nonrepetitive DNA can be cloned without eliminating methylated genes. Both these techniques are still in the development phase and have not yet been widely applied to genome sequencing.

5
Interpreting the Sequence

Once the genome sequence has been completed to a previously defined set of quality standards, the biologically important features of the sequence can be identified in a process called *annotation*. Biologically important features identified in a typical genome project include genes and transposable elements. For bacterial or archaeal genomes, gene identification is relatively straightforward, while gene identification in eukaryotic genomes is much more difficult due to the presence of introns. Three basic methods are used for protein-coding gene identification: *ab initio* gene prediction, sequence similarity searches, and expressed-sequence tag (EST) sequencing. All genome projects use both gene prediction and homology searching, while EST sequencing is primarily a technique for eukaryotic gene identification.

The goal of *ab initio* gene prediction is to correctly identify the start and stop codons (and intron/exon boundaries, if relevant) of a gene given only the DNA sequence and statistical parameters derived from known genes. For the simple structure of a bacterial or archaeal genome, genes can be identified from open reading frames that match a known set of codon preferences. Glimmer is a popular program for prokaryotic gene prediction. Statistical parameters considered for eukaryotic genes can include codon preferences, frequencies of longer combinations of codons, and attributes of intron/exon boundaries. These statistical parameters are derived from analyses of a large number of experimentally identified and sequenced genes. Because eukaryotic gene-prediction algorithms must correctly identify all possible exons, some of which can be very short, gene prediction in eukaryotic DNA is much less accurate than in prokaryotic DNA. Genscan and FGenesH are commonly used eukaryotic gene-prediction programs.

Genes can also be identified by their similarity to known sequences. Programs such as BLAST or FASTA compare a query sequence to a database of sequences. For each sequence in the database, a similarity score is calculated, and the statistical significance of each similarity score can be estimated with a high degree of accuracy. If the query sequence matches a sequence in the database with a statistically significant similarity score, we can infer with confidence that the sequences are related either through a common biochemical or physiological function. Because the statistical properties of sequence similarity scores are well understood, sequence similarity searching is the most reliable computational method of identifying protein-coding genes, RNA genes, pseudogenes, and transposable elements. Because sequence similarity programs do not identify start and stop codons or intron/exon boundaries, additional manual editing is necessary to derive a gene prediction from the significant similarity to a known sequence.

Because eukaryotic gene-prediction algorithms do not produce completely reliable results, experimental evidence is often sought to supplement computational gene identification efforts. The most common experimental method of gene identification is to isolate messenger RNAs, clone their cDNAs into a sequencing vector, and generate a single sequence read from each clone. Although the single read is insufficient to sequence the entire cDNA to high quality, it is generally enough to generate 500–700 nucleotides of the cDNA. This sequence information serves as a tag that can mark the location of a gene in genomic sequence, hence the name expressed-sequence tag. EST sequences are collected into databases, which are then compared to genomic sequence using sequence comparison programs (BLAST and FASTA). Gene identification by EST is limited by the isolation and preparation of pure mRNA sequences, because rRNA and other stable RNA can contaminate the cDNA library. Furthermore, any one tissue only transcribes a fraction of potential genes at any one time. cDNA libraries should be constructed from multiple tissues at multiple growth stages.

A typical eukaryotic genome project will utilize every gene identification method during annotation. Several different *ab initio* gene-prediction programs will be run on the genomic sequence. Additionally, sequence similarity searches will be performed on several different databases, including previously characterized protein sequences, known transposable elements, and ESTs from the target organism (and closely related species where available). The results of these computational analyses are then manually edited into a final annotation by a trained genome researcher. The annotation and the sequence can be deposited in one of the public DNA sequence databases, such as GenBank.

6
Summary and Principles

As consumers of genomic information, the majority of biologists are not directly involved in the generation of genome sequence data. Consequently, it is important for biologists to understand how genome sequence data is obtained in order to appreciate the advantages and limitations of using complete genomes. As we have explained, there is no single definition of "complete genome" to which all genome projects conform. Different levels of completion provide their own strengths and weaknesses. Sequences completed to the quality of most bacterial genomes and early eukaryotic genomes (yeast, *Caenorhabditis elegans*, *Arabidopsis*) give excellent insight into both the content and organization of the genome. For eukaryotes, such high-quality genome sequences require significant investment of time and funding. Draft sequences created by WGS are excellent for comparative genomics with close relatives (especially if the close relative has a high-quality completed sequence) and can provide insight into gene content. Draft sequences are probably unsuitable for comparative genomics with distantly related organisms and may not be adequate for *ab initio* gene prediction.

It is helpful to remember that the completed genome sequence is experimental data and subject to experimental error. Even high-quality, finished genome sequences will have errors. Most genomic sequence that is subject to finishing is completed to an accuracy of 1 error in 10 000 nucleotides. With three billion nucleotides, we should expect 300 000 single-nucleotide errors in the human genome sequence. For most researchers, this level of error will hardly ever be noticeable, but occasionally researchers may stumble across one of these errors.

Finally, researchers should also keep in mind that efforts to annotate genomic sequence, and in particular to identify protein-coding genes, are subject to much higher and less quantifiable error rates than actually obtaining the sequence itself. For important research projects on particular genes or gene families, predicted genes should be subjected to experimental verification, based on the evidence used to predict the gene. Genes predicted from a combination of EST similarity, *ab initio* predictions, and significant similarity to a closely related protein sequence will require less verification than those predicted from *ab initio* gene-prediction software alone.

Presently, genome technology and sequencing continues to grow at an amazing rate. Many of the strategies and limitations discussed in this article could be resolved at some time in the future. As genomics continues to grow, our understanding of the chemical basis of life will also grow, providing new platforms and methods for understanding disease, inheritance, and evolution. The benefits of investing in genome sequencing will continue to be discovered for years to come.

See also Molecular Basis of Genetics.

Bibliography

Books and Reviews

Gibson, G., Muse, S.V. (2002) *A Primer of Genome Science*, Sinauer, Sunderland, MA.

Green, E.D. (2001) Strategies for the systematic sequencing of complex genomes, *Nat. Rev. Genet.* **2**, 573–583.

Green, E.D., Birren, B., Klapholz, S., Myers, R.M., Hieter, P. (Eds.) (1999) *Genome Analysis: A Laboratory Manual*, 4 vols, CSHL Press, Cold Spring Harbor, New York.

Mount, D.W. (2001) *Bioinformatics: Sequence and Genome Analysis*, CSHL Press, Cold Spring Harbor, New York.

Primary Literature

Adams, M.D., et al. (2000) The genome sequence of *Drosophila melanogaster*, *Science* **287**, 2185–2195.

Altschul, S.F., Madden, T.L., Schäffer, A.A., Zhang, J., Zhang, Z., Miller, W., Lipman, D.J. (1997) Gapped BLAST and PSI-BLAST: a new generation of protein database search programs, *Nucleic Acids Res.* **25**, 3389–3402.

Anderson, S., Bankier, A.T., Barrell, B.G., de Bruijn, M.H., Coulson, A.R., Drouin, J., Eperon, I.C., Nierlich, D.P., Roe, B.A., Sanger, F., Schreier, P.H., Smith, A.J., Staden, R., Young, I.G. (1981) Sequence and organization of the human mitochondrial genome, *Nature* **290**, 457–465.

Balding, D.J., Bishop, M., Cannings, C. (Eds.) (2001) *Handbook of Statistical Genetics*, Wiley, New York.

Bult, C.J., White, O., Olsen, G.J., Zhou, L., Fleischmann, R.D., Sutton, G.G., Blake, J.A., FitzGerald, L.M., Clayton, R.A., Gocayne, J.D., Kerlavage, A.R., Dougherty, B.A., Tomb, J.F., Adams, M.D., Reich, C.I., Overbeek, R., Kirkness, E.F., Weinstock, K.G., Merrick, J.M., Glodek, A., Scott, J.L., Geoghagen, N.S. and Venter, J.C. (1996) Complete genome sequence of the methanogenic archaeon, *Methanococcus jannaschii*, *Science* **273**, 1058–1073.

Burge, C., Karlin, S. (1997) Prediction of complete gene structures in human genomic DNA, *J. Mol. Biol.* **268**, 78–94.

Chen, M., Presting, G., Barbazuk, W.B., Goicoechea, J.L., Blackmon, B., Fang, G., Kim, H., Frisch, D., Yu, Y., Sun, S., Higingbottom, S., Phimphilai, J., Phimphilai, D., Thurmond, S., Gaudette, B., Li, P., Liu, J., Hatfield, J., Main, D., Farrar, K., Henderson, C., Barnett, L., Costa, R., Williams, B., Walser, S., Atkins, M., Hall, C., Budiman, M.A., Tomkins, J.P., Luo, M., Bancroft, I., Salse, J., Regad, F., Mohapatra, T., Singh, N.K., Tyagi, A.K., Soderlund, C., Dean, R.A., Wing, R.A. (2002) An integrated physical and genetic map of the rice genome, *Plant Cell* **14**, 537–545.

Delcher, A.L., Harmon, D., Kasif, S., White, O., Salzberg, S.L. (1999) Improved microbial gene identification with GLIMMER, *Nucleic Acids Res.* **27**, 4636–4641.

Ewing, B., Green, P. (1998) Base-calling of automated sequencer traces using *Phred*. II. Error probabilities, *Genome Res.* **8**, 186–194.

Ewing, B., Hillier, L., Wendl, M.C., Green, P. (1998) Base-calling of automated sequencer traces using *Phred*. I. Accuracy assessment, *Genome Res.* **8**, 175–185.

Fleischmann, R.D., Adams, M.D., White, O., Clayton, R.A., Kirkness, E.F., Kerlavage, A.R., Bult, C.J., Tomb, J.F., Dougherty, B.A., Merrick, J.M., et al. (1995) Whole-genome random sequencing and assembly of *Haemophilus influenzae* Rd, *Science* **269**, 496–512.

Goff, S.A., Ricke, D., Lan, T., Presting, G., Wang, R., Dunn, M., Glazebrook, J., Sessions, A., Oeller, P., Varma, H., Hadley, D., Hutchison, D., Martin, C., Katagiri, F., Lange, B.M., Moughamer, T., Xia, Y., Budworth, P., Zhong, J., Miguel, T., Paszkowski, U., Zhang, S., Colbert, M., Sun, W., Chen, L., Cooper, B., Park, S., Wood, T.C., Mao, L., Quail, P., Wing, R., Dean, R., Yu, Y., Zharkikh, A., Shen, R., Sahasrabudhe, S., Thomas, A., Cannings, R., Gutin, A., Pruss, D., Reid, J., Tavtigian, S., Mitchell, J., Eldredge, G., Scholl, T., Miller, R.M., Bhatnagar, S., Adey, N., Rubano, T., Tusneem, N., Robinson, R., Feldhaus, J., Macalma, T., Oliphant, A., Briggs, S. (2002) A draft sequence of the rice genome (*Oryza sativa* L. ssp. *japonica*), *Science* **296**, 92–100.

Gordon, D., Abajian, C., Green, P. (1998) *Consed*: a graphical tool for sequence finishing, *Genome Res.* **8**, 195–202.

Hillier, L., Lennon, G., Becker, M., Bonaldo, M.F., Chiapelli, B., Chissoe, S., Dietrich, N., DuBuque, T., Favello, A., Gish, W., Hawkins, M., Hultman, M., Kucaba, T., Lacy, M., Le, M., Le, N., Mardis, E., Moore, B., Morris, M., Parsons, J., Prange, C., Rifkin, L., Rohlfing, T., Schellenberg, K., Soares, M.B., Tan, F., Thierry-Meg, J., Trevaskis, E., Underwood, K., Wohldman, P., Waterston, R., Wilson, R., Marra, M. (1996) Generation and analysis of 280 000 human expressed sequence tags, *Genome Res.* **6**, 807–828.

Huang, X., Madan, A. (1999) CAP3: a DNA sequence assembly program, *Genome Res.* **9**, 868–877.

International Human Genome Mapping Consortium. (2001) A physical map of the human genome, *Nature* **409**, 934–941.

International Human Genome Sequencing Consortium. (2001) Initial sequencing and analysis of the human genome, *Nature* **409**, 860–921.

Marra, M.A., Kucaba, T.A., Dietrich, N.L., Green, E.D., Brownstein, B., Wilson, R.K., McDonald, K.M., Hillier, L.W., McPherson, J.D., Waterston, R.H. (1997) High throughput fingerprint analysis of large-insert clones, *Genome Res.* **7**, 1072–1084.

McPherson, J.D. (1997) Sequence ready – or not? *Genome Res.* **7**, 1111–1113.

Myers, E.W., Sutton, G.G., Delcher, A.L., Dew, I.M., Fasulo, D.P., Flanigan, M.J., Kravitz, S.A., Mobarry, C.M., Reinert, K.H., Remington, K.A., Anson, E.L., Bolanos, R.A., Chou, H.H., Jordan, C.M., Halpern, A.L., Lonardi, S., Beasley, E.M., Brandon, R.C., Chen, L., Dunn, P.J., Lai, Z., Liang, Y., Nusskern, D.R., Zhan, M., Zhang, Q., Zheng, X., Rubin, G.M., Adams, M.D., Venter, J.C. (2000) A whole-genome assembly of *Drosophila*, *Science* **287**, 2196–2204.

Pearson, W.R. (2000) Flexible sequence similarity searching with the FASTA3 program package, *Methods Mol. Biol.* **132**, 185–219.

Rabinowicz, P.D., McCombie, W.R., Martienssen, R.A. (2003) Gene enrichment in plant genomic shotgun libraries, *Curr. Opin. Plant Biol.* **6**, 150–156.

Rabinowicz, P.D., Schutz, K., Dedhia, N., Yordan, C., Parnell, L.D., Stein, L., McCombie, W.R., Martienssen, R.A. (1999) Differential methylation of genes and retrotransposons facilitates shotgun sequencing of the maize genome, *Nat. Genet.* **23**, 305–308.

Rice Chromosome 10 Sequencing Consortium. (2003) In-depth view of structure, activity, and evolution of rice chromosome 10, *Science* **300**, 1566–1569.

Sanger, F., Air, G.M., Barrell, B.G., Brown, N.L., Coulson, A.R., Fiddes, J.C., Hutchison, C.A., Slocombe, P.M., Smith, M. (1977a) Nucleotide sequence of bacteriophage ΦX174 DNA, *Nature* **265**, 687–695.

Sanger, F., Nicklen, S., Coulson, A.R. (1977b) DNA sequencing with chain-terminating inhibitors, *Proc. Natl. Acad. Sci. U.S.A.* **74**, 5463–5467.

Venter, J.C., et al. (2001) The sequence of the human genome, *Science* **291**, 1304–1351.

Yu, J., et al. (2002) A draft sequence of the rice genome (*Oryza sativa* L. ssp. *indica*), *Science* **296**, 79–92.

21
Construction and Applications of Genomic DNA Libraries

Eugene R. Zabarovsky
Microbiology and Tumor Biology Center, Karolinska Institute, Stockholm, Sweden

1	**Principles** 607	
2	**Techniques** 609	
2.1	General Characteristics of λ-based Vectors Used for Construction of Genomic Libraries 609	
2.2	Construction of General Genomic Libraries 611	
2.3	Construction of Jumping and Linking Libraries. Use of Linking and Jumping Clones to Construct a Physical Chromosome Map 615	
3	**Applications and Perspectives** 619	
3.1	Cloning DNA Markers Specific for a Particular Chromosome 619	
3.2	CpG Islands as Powerful Markers for Genome Mapping; CpG Islands and Functional Genes 621	
3.3	Alu-PCR and Subtractive Procedures to Clone CpG Islands from Defined Regions of Chromosomes 622	
3.4	IBD (Identical-by-descent) Fragments for Identification of Disease Genes 624	
3.5	Strategies to Map and Sequence Genomes; Hierarchical, Whole-genome, and Slalom Sequencing Approaches 624	
3.6	Restriction-site-tagged Microarrays to Study CpG-Island Methylation 626	
3.7	Restriction-site-tagged Sequences to Study Biodiversity 628	
4	**Summary** 629	
	Bibliography 629	
	Books and Reviews 629	
	Primary Literature 629	

Genomics and Genetics. Edited by Robert A. Meyers.
Copyright © 2007 Wiley-VCH Verlag GmbH & Co. KGaA, Weinheim
ISBN: 978-3-527-31609-0

21 Construction and Applications of Genomic DNA Libraries

Keywords

Blue–white Selection
Not really selection but color identification. Vectors carrying the β-galactosidase (*lacZ*) gene (or part of it) produce blue plaques in the presence of 5-bromo-4-chloro-3-indolyl-β-D-galactopyranoside (X-gal). If this gene is located in a stuffer fragment, then all recombinants will form white plaques and parental vectors will produce blue plaques in the presence of X-gal.

Genetic Selection
Usually, in cloning, selection against parental, nonrecombinant molecules in favor of the recombinant. For λ-based vectors used for construction of genomic libraries, the two most commonly used types of selection are *Spi* and *supF*. *Spi*+ phages carrying *red* and *gam* genes cannot grow in *E. coli* lysogens carrying prophage P2; since, however, the majority of the vectors contain these genes in a stuffer fragment, only recombinant phages can grow in such *E. coli* strains. Selection for *supF* exploits λ vectors carrying amber mutations. These vectors cannot replicate without the *supF* gene, which must be present either in the host or in the cloned insert. If the insert carries the *supF* gene, only recombinant phages will be able to replicate in an *E. coli* host without the suppressor gene.

Polylinker
A short DNA fragment (in the vector) that contains recognition sites for many restriction enzymes, which can be used for cloning DNA fragments into this vector.

Restriction Enzyme
An enzyme that recognizes a specific sequence in DNA and can cut at or near this sequence. In cloning procedures, the most commonly used enzymes produce specific protruding (sticky) ends at the ends of the DNA molecule. Each enzyme produces unique sticky ends. The DNA molecules possessing the same sticky ends can be efficiently joined with the aid of DNA ligase (see ligation).

(STS) Sequence-tagged Site
A short (200–500 bp) sequenced fragment of genomic DNA that can be specifically amplified using PCR. STS represents or is linked to some kind of marker (i.e. it is mapped to a specific locus on a chromosome).

By virtue of the powerful technology developed in molecular biology, it is possible to isolate any DNA fragment in the genome of an organism and, after reverse transcription, any transcribed gene in the form of a complementary DNA. The isolation (cloning) procedure involves the insertion of the DNA fragment into a vector, capable of replication in a microorganism, which allows production of large quantities of the DNA fragment for physical or biological analysis. Upon determination of the location in the genome from which the particular DNA fragment was derived, that fragment acquires the property of a DNA marker. Such DNA markers are a prerequisite for physical and genetic mapping of the genome of the organism. DNA markers are also of importance for the diagnosis of genetic diseases. DNA markers can be divided into several different classes depending on the way in which the markers were selected among the fragments of genomic DNA. Examples of such classes are anonymous, micro- and minisatellites, restriction fragment length polymorphism (RFLP) markers, and *Not*I linking clones.

Vectors and clone libraries of different types can be used to clone markers. Lambda-based vectors and genomic libraries of different kinds are commonly used for this purpose. Many different variants of λ-based vectors that combine features of different cloning vehicles (plasmids, M13 and P1 phages) have been created for this purpose. The use of each vector is usually limited to a specific task: the construction of general genomic libraries (which contain all genomic DNA fragments) or special genomic libraries (which contain only a particular subset of genomic DNA fragments). Among these special libraries, *Not*I linking and jumping libraries have particular value for physical/genetic mapping and sequencing of the human genome. Shotgun and slalom libraries are usually used for sequencing purpose and comparative genomics.

1
Principles

In molecular biology, "cloning" is the insertion of DNA with interesting information into a specific vector that allows replication and transfer of the cloned DNA from one host to another. The vector containing the inserted DNA is called a *recombinant vector* to distinguish it from its parental vector, which does not contain any foreign DNA. Usually, "interesting information" is a piece of DNA obtained from any target organism; it can be a gene (or part of a gene) or simply anonymous DNA sequences for which no function is yet known. It can originate directly from DNA or can be obtained from reverse transcription of RNA molecules. The main idea of cloning is to obtain the interesting piece of DNA in a quantity large enough for analysis and further experiments. Now, the vectors and strategies used for cloning come in many different types. This chapter concentrates on the widely used λ-based vectors and the construction of genomic libraries, which played an important role in the Human Genome Project.

A genomic library is a collection of recombinant vectors; it contains DNA fragments representing the genome of a particular organism. Genomic libraries can be either general, containing DNA fragments covering the whole genome, or special, containing only specific genomic fragments that differ in certain parameters. Some are CG rich whereas others contain only particular size fragments of DNA obtained after digestion with a particular restriction enzyme, contain specific repeats, and so on. Important special genomic libraries are the jumping and linking types (see Sect. 2.3).

Cloned DNA fragments can be located to a specific site of a chromosome, after which they can serve as markers for physical and genetic mapping. Different types of markers are used. The so-called anonymous markers represent randomly cloned DNA fragments whose functions or specific features are not known. Other DNA markers can possess specific features. They can contain a known gene or expressed sequences with unknown function, CpG islands (also associated with genes, see Sect. 3.2), or recognition sites for rare cutting restriction enzymes convenient for long-range mapping. Such markers can be polymorphic, that is, they have different structures in different individuals (they are usually distinguished on the basis of different mobility in gel electrophoresis). Such markers are extremely important in mapping and cloning human disease genes and for construction of genetic maps. Three types of polymorphic markers are commonly used (Fig. 1).

Single-nucleotide polymorphism (SNP) markers are DNA fragments that have a point mutation in some individuals of a population. The advantages of SNPs are their abundant numbers ($>10^6$) and the fact that they can be detected by nonelectrophoretic methods, for example, using microarrays. However, usually SNP has only two alleles. A subtype of SNPs are RFLP markers that recognize genomic fragments containing polymorphic recognition sites for a particular restriction endonuclease (e.g. *TaqI*, *MspI*). The same chromosomal regions in different individuals contain or lack this recognition site.

A second form of DNA polymorphism results from variation in the number of tandemly repeated (VNTR) DNA sequences in a particular locus. Usually, they are divided in two types – mini- and microsatellites. Minisatellites are DNA fragments 0.1 to 20 kb long that contain many copies (from 3 to more than 40) of

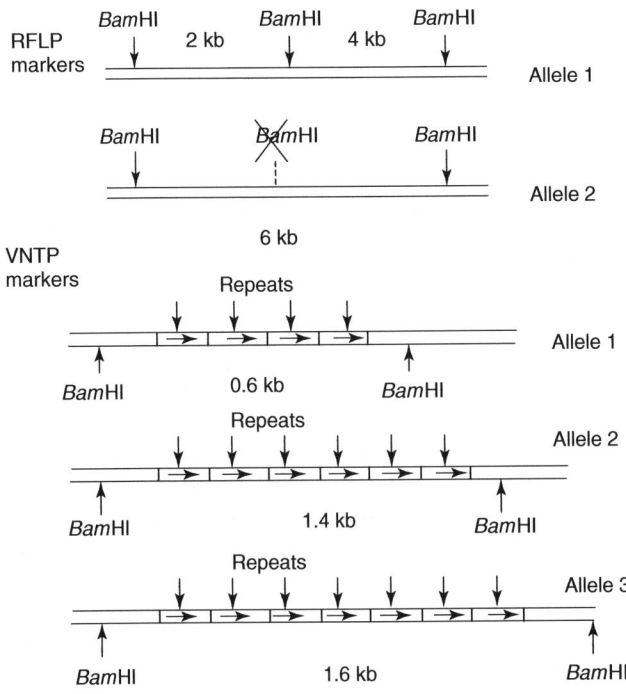

Fig. 1 The difference between RFLP and VNTR polymorphism.

6 to 60 bp repeats. All these repeats share a 10 to 15 bp core sequence similar to the generalized recombination signal (*chi*) of *E. coli*. When DNA from different individuals is digested with a restriction enzyme that does not cut inside these repeats, the length of the fragments produced will depend on the number of repeats at the locus. Since the minisatellites and repeats constitute a relatively large fragment, it is possible to discriminate between different alleles using ordinary nondenaturing gel electrophoresis and Southern blot analysis. Many different alleles (usually more than five) can be distinguished at a locus containing minisatellites. Minisatellites cluster around the distal ends of human chromosomes and, sometimes, are located near the genes.

Microsatellites are relatively short DNA fragments (usually <100 bp) with repeat units from 1 to 5 bp (such as A, AC, AG, AAC, AAAG, etc.). They are numerous (5×10^5) and uniformly distributed throughout the human genome with an estimated average spacing of about 6 kb.

Microsatellites can be very polymorphic (more than 10 alleles at the same locus), and polymorphism usually increases with the number of repeats. Microsatellites with fewer than 10 copies are usually not polymorphic. Since microsatellites are short, they can be analyzed quickly by using the polymerase chain reaction (PCR) with primers flanking each locus. Different alleles can be resolved using denaturing gel electrophoresis. Very frequently, microsatellites are associated with Alu repeats (see Sect. 3.1) and this creates problems for their use with PCR, since Alu repeats are conserved in the human genome. Thus, flanking primers for the microsatellite located in the Alu repeat are unlikely to be useful because they prime from many Alu repeats and, instead of discrete bands, give a smear after electrophoretic separation of the PCR products.

An optimal marker should have the features of all the different types of markers just discussed. The ideal marker should contain (1) a gene, (2) a CpG island that has been shown to be very conserved in the genome and can be used for comparative gene mapping in different species, (3) a rare cutting restriction site useful for physical mapping, and (4) polymorphic sequences (e.g. microsatellites). One of the best candidates for having all these features together is *Not*I linking clones, that is, recombinant clones containing the *Not*I restriction site.

2 Techniques

2.1 General Characteristics of λ-based Vectors Used for Construction of Genomic Libraries

Among other more modern vectors used for construction of genomic libraries (yeast, bacterial and P1 artificial chromosomes; YAC, BAC, and PAC respectively), phage λ-based vectors are still very popular. The reason is that the genetics and features of both λ phage and *E. coli* (the host for λ phage) are well known. The size of the phage DNA that can be packaged into viable phage particles is limited between 37.7 and 52.9 kb. This means that it is possible to biologically regulate the size range of the cloned DNA fragment. There exist extremely efficient *in vitro* systems for packaging such DNA into λ phage particles (10^9 plaque-forming units per μg of DNA) to produce viable phages. To combine the features of different vector systems, extensive modifications of λ phage vectors were developed (Fig. 2).

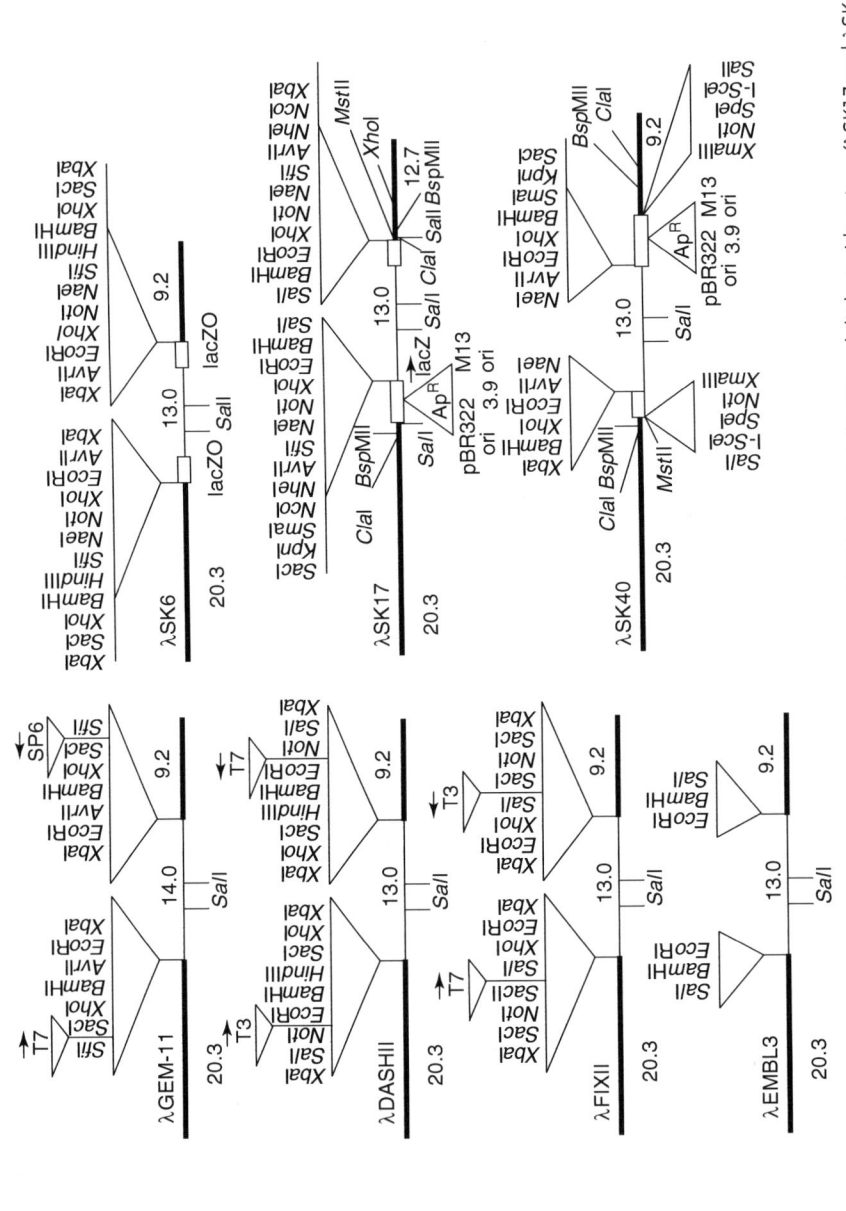

Fig. 2 Examples of λ-based vectors: standard λ vectors (λGEM11, λFIXII, λDASH, λEMBL3, λSK6) and diphasmid vectors (λSK17 and λSK40). Sizes are in kilobases. Not all restriction sites are shown. Heavy lines represent vector arms, thin lines denote the stuffer fragment; open boxes mark plasmid and M13 sequences; lacZO is the lac operator sequence. T3, T7, SP6 – promoters for T3, T7, and SP6 RNA polymerases.

Cosmids are essentially plasmids that contain the *cos* region of phage λ responsible for packaging of DNA into the phage particle. The advantages of cosmids are easy handling (as with plasmids) and large cloning capacity. Since the plasmid body is usually small (3–6 kb), large DNA molecules (46–49 kb) can be cloned in these vectors.

Phasmids are λ phages that have an inserted plasmid. They have the same basic features as λ phage vectors, but the inserted foreign DNA fragment can be separated from the body of the phage DNA and converted into the plasmid form. After the conversion, the cloned DNA fragment will exist as a recombinant plasmid.

Hyphages represent another type of λ-based vectors. They were constructed from M13 vectors with a built-in *cos* site of λ. Since these vectors have the main features of M13 vectors, they can be obtained in single-stranded form. Their distinctive feature is the capability to be packaged with high efficiency into λ-phage-like particles. This decreases the chance of recovering nonrecombinant vectors and opens the possibility of constructing a representative genomic library in single-stranded form.

Diphasmids are vectors, which offer the opportunity to combine the advantages of phages (λ and M13) and plasmids. Diphasmids can be divided into two classes: those that can replicate as phage λ (an improvement over phasmids) and those that are incapable of replicating as phage λ (i.e. a cosmid capable of being packaged into phage M13 particles).

In some cases, it is more convenient to work with a genomic library in plasmid than in λ phage form. The construction of a representative genomic library directly in a plasmid vector has several drawbacks and difficulties. However, all these problems can be easily solved with the help of phasmid and diphasmid vectors. Often a genomic library is constructed in a λ phage and then converted in its entirety to plasmid form.

2.2
Construction of General Genomic Libraries

Representativity is one of the most important features of a genomic library. In a "representative" genomic library, every genomic DNA fragment will be present in at least one of the recombinant phages of the library. In practice, however, this is difficult to achieve. Some genomic fragments are not clonable because of the strategy used for construction of the genomic library. For example, if the maximal cloning capacity of the vector is 18 kb and *Eco*RI digestion is used to construct the library, no genomic *Eco*RI fragments bigger than 18 kb will be present. In some cases, genomic DNA fragments can suppress the growth of the vector or the host cell, with the result that its cloning can be restricted to specific vector systems.

The important reason for decreased representativity is the different replication potential of different recombinant phages. Most researchers work with amplified libraries. The ligated DNA molecules are packaged into the λ phage particles and plated on a lawn of *E. coli* cells (usually many petri dishes are used for such plating). Then all λ phage particles are eluted from the plaques, and the liquid eluted from all the petri dishes is mixed. Glycerol or dimethyl sulfoxide is added to the eluate, and the aliquots are kept at $-76\,°C$. This procedure is called *amplification* of the library.

With amplification, a library can be kept for many years and can be used for many experiments by many researchers. On the other hand, each recombinant phage present in the library gives a single plaque at the first plating, and since different recombinant phages differ in their growth potential, there may be differences of 100 times in the abundance of clones after amplification. This means that in the amplified library, some of the phages are present 100 times more often than others. In this case, to recover all recombinant phages obtained after packaging into λ phage particles, one needs to plate 100 times more phages than that obtained after the original plating. Since such quantities are difficult to achieve in practice, some recombinant phages are virtually lost from the library after amplification.

How is the representativity of a library estimated? A library is considered to be representative if after the first plating (before amplification) it contains a number of recombinant clones together, containing genomic DNA fragments equal to 7 to 10 genome equivalents. For example, human genomic DNA contains approximately 3×10^9 bp. If the vector contains on an average 15 kb inserts, the representative library should contain 1.4 to 2.0×10^6 recombinant clones.

The way in which the genomic DNA fragments are produced for cloning is also important. The more randomly the genomic DNA is broken, the more representative a library can be obtained. Clearly, the *Eco*RI enzyme (6 bp recognition site) will cut genomic DNA less randomly than *Sau*3AI (4 bp recognition). Probably, the shearing of DNA molecules using physical methods (e.g. syringe, sonication) is the most reliable way to obtain randomly broken DNA molecules.

An important characteristic feature of a library is the percentage of recombinants. For most purposes, if a library contains more than 80% of recombinant phages, it is better to omit the genetic selection procedure because it usually decreases the representativity of the library. To calculate the percentage of recombinants, one can use genetic (*Spi*) selection as in the case of λEMBL vectors. Another approach relies on blue–white color identification (e.g. λ-Charon series). A third class of vectors has both genetic selection and blue–white color identification (λSK4, λSK6). There are three commonly used ways to construct genomic libraries (Table 1, Fig. 3).

The original ("classical") method includes generation of sheared genomic DNA fragments using physical or enzymatic manipulations followed by the physical separation of fragments of a particular size using, for example, ultracentrifugation or gel electrophoresis. The vector DNA is digested with two (or even three) restriction enzymes whose recognition sites are located in the polylinker and are separated by a few base pairs. The arms and the stuffer piece are purified from the small oligonucleotide molecules released after the digestion by, for example, precipitation with polyethylene glycol (PEG) 6000. The stuffer piece and both arms will now have different sticky ends, preventing re-creation of the original vector molecules during subsequent ligation with genomic DNA fragments.

In the second "dephosphorylation" approach, the phage arms are prepared by simultaneous digestion with two restriction enzymes as shown earlier. Genomic DNA is partially digested to the extent that DNA fragments with sizes in the range of 15 to 20 kb will represent the majority of the products. These DNA fragments are dephosphorylated to prevent their ligation to each other and are then ligated to the vector arms. If too big or too small genomic DNA fragments are ligated to the phage arms, size limitations will make it impossible for these recombinant molecules to yield viable phages. Compared to the preceding

Tab. 1 Comparison of three basic methods used in construction of representative genomic libraries.

Method	DNA [µg] needed to construct a representative genomic library	DNA fractionation	Self-ligation of vector DNA	Self-ligation of genomic DNA	Effectiveness of packaging Per microgram of vector DNA	Effectiveness of packaging Per microgram of genomic DNA	Number of packaging reactions to get representative library at maximal efficiency per microgram of genomic DNA	Genetic selection necessary to remove nonrecombinants
Classical	100–1000	Yes	Yes	Yes	$10^5–10^7$	$10^5–10^7$	1	Yes
Dephosphorylation	5–10	No	Yes	No	$10^4–10^5$	$10^5–10^6$	3–5	Yes
Partial filling in	5–10	No	No	No	$10^5–10^7$	$10^5–10^7$	1	No

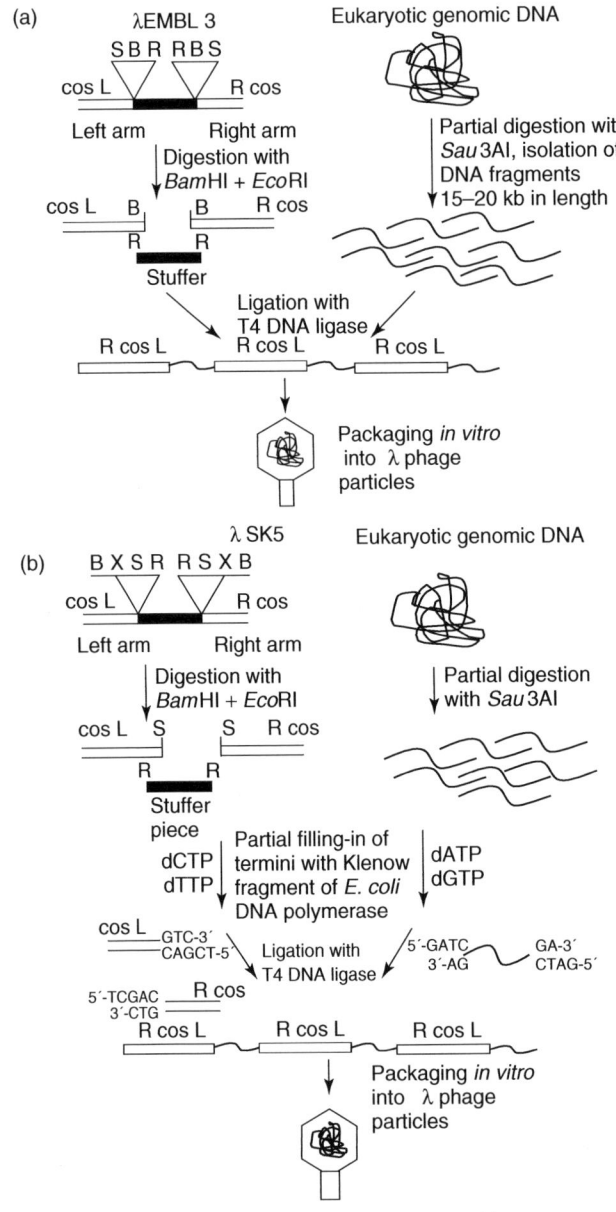

Fig. 3 Two approaches for constructing genomic libraries: (a) classical method and (b) partial filling-in method. Lcos and Rcos – left and right parts of the *cos* site correspondingly; B, BamHI; R, EcoRI; S, SalI; X, XmaIII.

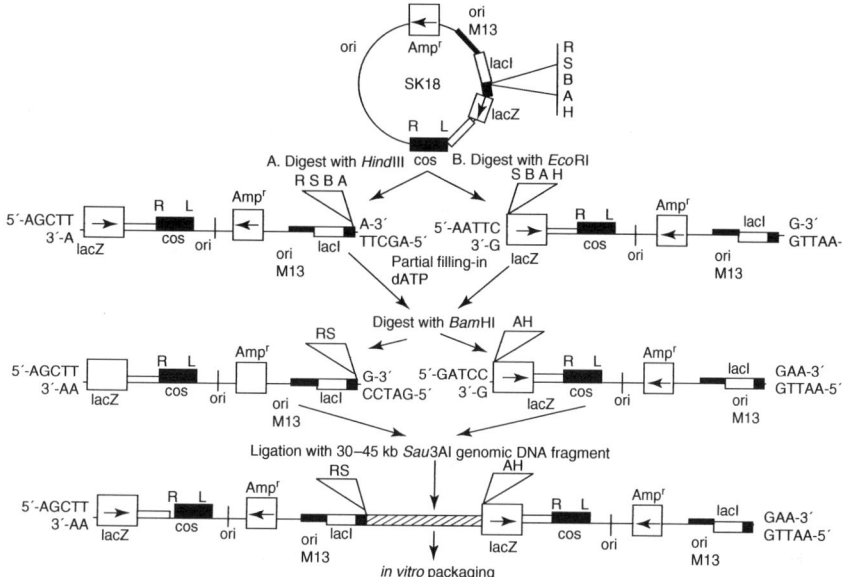

Fig. 4 One approach to the construction of genomic libraries in cosmid vectors (not all restriction sites are shown): A, *Acc*I; B, *Bam*HI; H, *Hind*III; R, *Eco*RI; S, *Sma*I.

method, this procedure is quick, and representative libraries can be obtained from a small quantity of genomic DNA.

The third "partial filling-in" method, also avoids fractionation steps (Fig. 3). Phage arms are prepared as described before (in this particular case *Sal*I and *Eco*RI are shown, but many other combinations can be used), and the sticky ends produced are partially filled in with the Klenow fragment of DNA polymerase I (or other DNA polymerase) in the presence of dTTP and dCTP. Genomic DNA partially digested with *Sau*3AI is also partially filled in, but in the presence of dATP and dGTP. Under such conditions, self-ligation of vector arms or genomic DNA is impossible.

Genomic libraries can be constructed in cosmids using the same approaches just described. The absence of selection against nonrecombinant vector and the possibility of packaging into phage particles concatemers composed solely of cosmid fragments make it even more important to prevent self-ligation of vector fragments. Many similar approaches have been suggested and one of them is shown in (Fig. 4) where the formation of vector-concatemers is prevented by partial filling in.

A similar effect can be achieved by dephosphorylation or by digestion at the first step with *Acc*I and *Sma*I instead of *Eco*RI and *Hind*III. *Sma*I produces blunt ends and *Acc*I gives sticky ends with only two protruding base pairs. The ligation of these ends will be far less effective than that for *Bam*HI and *Sau*3AI sticky ends (four protruding base pairs).

2.3
Construction of Jumping and Linking Libraries. Use of Linking and Jumping Clones to Construct a Physical Chromosome Map

For long-range mapping and cloning of large stretches of genomic DNA, the two most widely used methods are construction of overlapping DNA sequences (contigs) using chromosome walking (e.g. BAC cloning) and chromosome jumping. The technique of BAC cloning is now used by many laboratories. Still, this approach is not devoid of problems and drawbacks. These problems could be diminished, however, by using jumping/linking libraries. Moreover, jumping and linking libraries can be used independently for construction of a long-range restriction map using pulsed field gel electrophoresis (PFGE). Jumping clones contain DNA sequences adjacent to neighboring *Not*I sites, and linking clones contain DNA sequences surrounding the same restriction site.

The two best-known kinds of jumping libraries are the *Not*I jumping library and the "general" jumping (hopping) library. The basic principle of both methods is to clone only the ends of large DNA fragments rather than continuous DNA segments, as in BAC clones. Internal DNA is deleted by controlled biochemical techniques. The main difference is that in the first type of library, complete digestion with a rare cutting enzyme (*Not*I is the most popular) is used, and the second is based on a partial digestion with a frequently cutting enzyme, followed by isolation of DNA fragments of desired size. Using the first type of library, it is possible to jump over long distances (>1000 kb), but only from certain starting points (i.e. those containing the recognition site for the rare cutting enzyme). Using the hopping library, it is possible to start jumping from practically any point and to cover a defined but shorter distance (<150 kb). Only the first type of jumping libraries can be used in conjunction with linking libraries to create genomic maps, as described next.

There are two main approaches for the construction of *Not*I jumping and linking libraries. According to the first method (Fig. 5a), jumping libraries are constructed as follows: DNA of high molecular mass, isolated in low-melting agarose, is completely digested with *Not*I.

The DNA is ligated at very low concentration, in the presence of a dephosphorylated plasmid containing a marker (*supF* gene), and is then circularized, trapping the *supF* gene, which acts like a marker to select clones that contain the ends of a long fragment. Another enzyme, one that has no recognition site in the plasmid, is used to digest the large circular molecules into small fragments, each of which is cloned in a vector phage carrying amber mutations. Recombinant phages containing the plasmid with the two terminal fragments are selected in an *E. coli* strain lacking the suppressor gene.

The linking library can be constructed in different ways. In the original protocol, the genomic DNA is partially digested with *Sau*3A and size selected to obtain 10 to 20 kb fragments. The DNA is then diluted and circularized in the presence of *supF* marker plasmid. The circular products are digested with *Not*I, ligated into a *Not*I-digested suppressor-dependent vector (NotEMBL3A), and plated on a suppressor-negative host.

In another approach, DNA from a total genomic library in a circular form (e.g. cosmid) is digested with *Not*I, and a selectable marker (e.g. resistance to the antibiotic) is inserted into recombinants containing this site. Then these recombinants are selected by their resistance to the antibiotic.

The most important drawback is that all these methods used to construct linking libraries exploit strategies and vectors different from those used to construct jumping libraries. Thus, some fragments

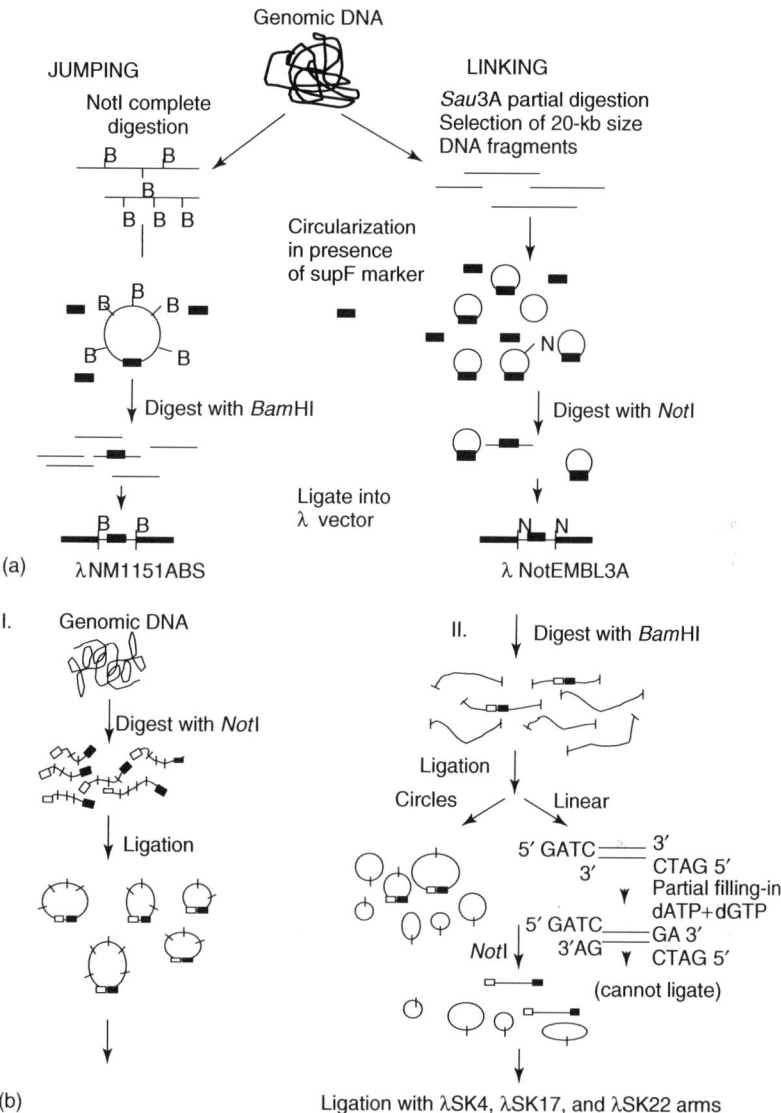

Fig. 5 Two approaches for the construction of jumping and linking libraries: (a) using supF marker and (b) using partial filling in. (a) Black boxes, supF marker; B, BamHI. (b) Black and white bars denote NotI sites and vertical slashes represent BamHI sites; (b I), (b II), construction of the jumping library; (b II), construction of the linking library. In this case, digestion of the genomic DNA with BamHI is the first step.

present in one library (e.g. jumping) will be absent in another (e.g. linking), which creates serious problems for the use of these libraries in mapping.

An integrated approach for construction of jumping and linking libraries is outlined in Fig. 5(b). The most important feature here is that the same vectors and protocol are used for construction of both libraries.

For the linking library, genomic DNA is completely digested with BamHI. Subsequently, the DNA is self-ligated at a very low concentration (without a supF marker) to yield circular molecules as the main product. To eliminate any linear molecules, the sticky ends are partly filled in with the Klenow fragment in the presence of dATP and dGTP. Since the Klenow fragment also has exonuclease activity, all the BamHI sticky ends are neutralized and nearly all ends generated upon random DNA breakage become unavailable for ligation. Subsequently, the DNA is cut with NotI and cloned in λSK4, λSK17, and λSK22 vectors.

The same strategy is applied in the construction of a NotI jumping library. One initial step is added: genomic DNA is fully digested with NotI and self-ligated at a very low concentration. The subsequent steps are the same as for the linking library. An obvious difference between this approach and the preceding ones is that the procedure combines a biochemical selection for NotI jumping fragments with improved ligation kinetics during the preparation of the libraries (see Table 2 for a comparison of two methods).

In theory, a linking library is sufficient to construct a physical chromosomal map. When linking clones are used to probe a PFGE genomic blot (NotI-digested genomic DNA), each clone should reveal two DNA fragments, which are adjacent in the genome. Thus, in principle, one should be able to order the rare cutting sites with just a single library and one digest, although it will generally not be possible to distinguish between two fragments of the same size. To resolve such ambiguities, it is important to use several different libraries, each for a particular enzyme, and to overlap the resulting patterns just as in ordinary restriction

Tab. 2 Comparison of two main methods for construction of Not I jumping libraries.

Method I (with supF marker)	Method II (with partial filling in)
Materials required for construction of jumping library	
2 µg genomic DNA (10 mL ligation volume)	1 µg genomic DNA (5 mL ligation volume)
40 µg vector arms	1 µg vector arms
500 µL-sonicated extract (SE) for *in vitro* packaging	15 µL SE
2000 µL freeze-thaw extract for *in vitro* packaging (FTL)	10 µL FTL
Cloning capacity	
0–12 kb	0.2–23 kb
Expected yield	
$1-5 \times 10^4$/µg genomic DNA	$1-5 \times 10^5$/µg genomic DNA
Percentage of recombinants before genetic selection	
<1%	45–60%
Maximum sizes of jumps	
450 kb	>1000 kb

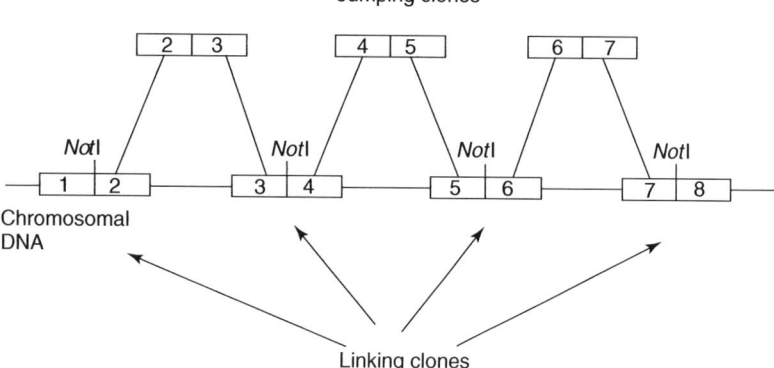

Fig. 6 Long-range mapping using jumping and linking libraries.

fragment analysis. To accomplish this for the whole human genome will be very laborious. However, for small stretches of the genome containing 5 to 10 *Not*I sites, this approach can be efficient. The use of jumping and linking libraries in a complementary fashion simplifies this approach (Fig. 6).

Moreover, by cross-screening the two libraries, it should be possible in principle to jump from clone to clone (–jumping–linking–jumping, etc.), to generate an ordered map without using PFGE techniques at all. One of the main problems with this approach is that the presence of very CG-rich regions and repeats in the human DNA may result in cross-hybridization between different clones.

In the shotgun sequencing approach for long-range genome mapping, the hybridization technique is replaced by sequencing. Jumping and linking clones are sequenced from the ends and, subsequently, the linear order of the *Not*I clones on a chromosome can be established using a computer program. Even a 20 bp sequence is likely to uniquely identify a sequence in the human genome. The sequence data provide a means to discriminate even between different instances of the same class of repeats.

3
Applications and Perspectives

3.1
Cloning DNA Markers Specific for a Particular Chromosome

It is important, for many purposes, to clone individual DNA markers for a specific chromosome. The most straightforward approach for such purposes is to prepare special libraries that contain recombinant clones from a particular chromosome. One approach is based on the use of the fluorescence-activated cell sorter (FACS) system. FACS operates on the principle of rapid analysis of suspended particles, single cells, or even chromosomes. A suspension of chromosomes stained with fluorescent dyes (usually Hoechst 33258 and chromomycin A3) passes through the focus of a laser beam that excites the DNA-bound fluorescent dyes. Equal-sized droplets are formed by ultrasonic dispersion, and the droplets containing the desired chromosome as indicated by the fluorescence measurement are deflected from the main stream by an electric field and collected in a tube. DNA isolated from sorted chromosomes can be used for construction of genomic libraries.

Another approach is based on the use of hybrid cell lines. To obtain such somatic cell hybrids, human cells are fused with rodent cells (e.g. mouse or hamster). When the resulting hybrid cells are grown in culture, there is a progressive loss of human chromosomes until only one or a few of them are left. At this step, some of the segregant hybrid cells can be quite stable. In a modification of this technique, a human cell line is transfected with a plasmid vector or infected with a retroviral vector that contains a selectable marker (e.g. *gpt* or *neo*). Transfected clones are screened for those that contain only a single integrated plasmid per cell (and thus containing a single marked chromosome). These transformants are micronucleated by prolonged colcemid treatment and the microcells, containing only one chromosome, are produced using special techniques. The microcells are fused to mouse cells, and the resulting human–mouse microcell hybrids, containing the single marked chromosome, are isolated by growth in selective medium. Hybrid cells containing only fragments of human chromosomes can be produced using human chromosomes with translocations and deletions, or the fragments can be produced experimentally by X-irradiation (radiation hybrids). Such hybrid cell lines are very useful not only for constructing genomic libraries but also for physical mapping of already isolated DNA markers and genes. Another advantage of somatic hybrid cells is that DNA is available in large amounts and can be used for different purposes. Human-specific clones can be isolated from the library by hybridization to total human DNA.

At least 40% of the human genome consists of repetitive sequences of different kinds. Among them, about 50% are repeats randomly distributed in the human genome – different kinds of short (SINE) and long (LINE) interspersed (repeating) elements. The most abundant among them is the Alu repeat family. Alu repeats are present at $>1.0 \times 10^6$ copies per genome with an estimated average spacing of about 3 kb. Alu repeats have a length of about 300 bp and consist of 2 homologous units. Related repeats also exist in other mammals. Different members of this family from the same species usually have homology of 80 to 90% but are only about 50% identical in different species. These repeats have conserved and variable regions. It is possible to find conserved sequences that are species-specific. These conserved sequences can be used as primers for PCR, to specifically amplify human sequences in the presence of nonhuman DNA. These features are the basis for using Alu repeats for isolation of human chromosome–specific sequences from hybrid cell lines containing human and nonhuman DNA sequences. Moreover, if a hybrid cell line contains only a short piece of human chromosome, the Alu-PCR approach can be used for isolation of markers specific for a defined region of the chromosome. The principles of the approach are shown in Fig. 7.

In the case of two hybrid cell lines, one containing a complete human chromosome (HCL1) and the other carrying the same chromosome but with a deletion (HCL2), this method offers the possibility of obtaining markers specific for the deletion. This variant can be called the differential Alu-PCR approach for obtaining

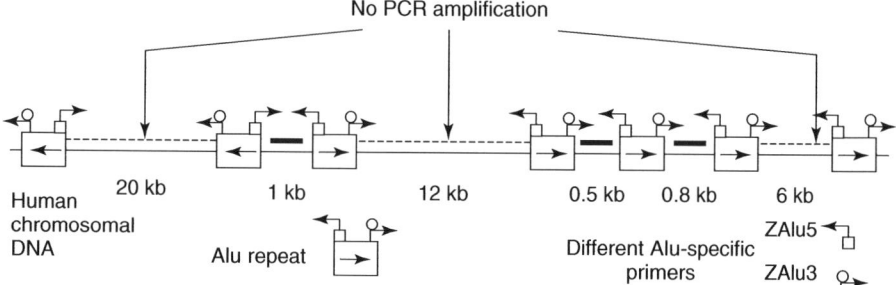

Fig. 7 General scheme for Alu-PCR.

DNA markers. The approach is mainly used in two variants.

In one variant, Alu-PCR is done using DNA from both cell lines, and the products of the reactions are separated by agarose gel electrophoresis. Some bands present in the products from HCL1 will be absent among the products from HCL2. These bands can be excised and cloned, giving markers localized in the deletion. The disadvantage of this approach is that usually such Alu-PCR results in a large number of products that have a very complex pattern and look like a smear on the gel. Among the solutions to this problem that have been suggested is the use of more specific primers (only for a subset of the Alu repeats), or genomic DNA digested with restriction enzyme. Another solution is to use hybrid cell lines that contain only small pieces of human chromosomes.

The second variant of the Alu-PCR approach is mainly used in connection with sources that contain only a limited amount of human material: YAC clones and radiation hybrid cell lines containing small pieces of the human chromosomes. The YACs can, for example, be used for Alu-PCR and the total products of the PCR reaction can be used as a probe to screen genomic libraries (e.g. in cosmids). The hybridization pattern reveals which cosmids are present in one YAC, which in other YACs, and which are present in one but absent in another. Such an approach is also useful for mapping.

Another approach to obtaining region-specific libraries is to use chromosome microdissection to physically remove the chromosomal region of interest; the minute quantities of microdissected DNA can be subjected to a microcloning procedure. Spreads of human chromosomes are made and stained using standard cytogenetic techniques. DNA from an individual band is then cut from the chromosome using ultrafine glass needles or is isolated with the help of laser equipment. In the latter case, all other chromosomes are destroyed by the laser, and intact DNA is present only in the chromosome of interest. DNA obtained from only a few (2–20) chromosomes is enough for constructing a region-specific library. This DNA is amplified using PCR and cloned in plasmid or λ vectors.

3.2
CpG Islands as Powerful Markers for Genome Mapping; CpG Islands and Functional Genes

Although human DNA is highly methylated, stably unmethylated sequences (about 1% of the genome) have been observed in human chromosomal DNA. Such sequences occur as discrete "islands," usually 1 to 2 kb long, that are dispersed in the genome. They are usually called *CpG (rich) islands* because they contain more than 50% of CG (human genome contains, on average, about 40% of CG contents). Their distinctive feature is the presence of CpG pairs at a predicted frequency, whereas elsewhere in the genome, it is present at a frequency less than 25%. Altogether, there are about 30 000 islands in the haploid genome (the average spacing is about 1 per 100 kb). It is now clear that the majority (if not all) of CpG islands are associated with genes. It has been shown that recognition sites for many of the rare cutting enzymes are closely associated with CpG islands. For example, at least 82% of all *Not*I and 76% of all *Xma*III sites are located in the CpG islands. More than 20% of CpG-island-containing genes have at least one *Not*I site in their sequence, while about 65% of these genes have *Xma*III site(s).

3.3
Alu-PCR and Subtractive Procedures to Clone CpG Islands from Defined Regions of Chromosomes

The Alu-PCR approach is used successfully for cloning DNA markers, but it does result in cloning small DNA fragments (500 bp) between Alu sequences. Alu sequences are distributed in a random fashion and are not linked with genes or other markers. An obvious suggestion for making Alu-PCR more useful for mapping is to use not simply genomic DNA from different sources but linking libraries constructed from these sources. This modification has certain advantages: using isolated probes, it is easy to clone a parental linking clone (e.g. *Not*I), which is a natural marker on the chromosome, convenient for linkage with other markers. Furthermore, linking clones are located in CpG-rich islands that are associated with genes. According to this scheme, linking libraries are constructed from different hybrid cell lines containing either whole or deleted human chromosomes. Then, total DNA isolated from these libraries can be used for Alu-PCR in the manner described in Sect. 3.1. However, in this case every PCR product (either discrete bands or total product) is used as a probe to isolate linking clones from the defined region of the chromosomes.

Genomic subtractive methods represent potentially powerful tools for identification of deleted sequences and cloning region–specific markers. This approach has given rewarding results in less-complex systems such as yeast or cDNA libraries, but the great complexity of the human genome has generated serious problems. These problems can be overcome by reducing the complexity of the human genomic sequences. Two approaches have been suggested to achieve this aim. In one (representational difference analysis), only a subset of genomic sequences (e.g. *Bam*HI fragments less then 1 kb) is used for subtractive procedures; this approach will result in cloning of random sequences. In the other, *Not*I linking libraries are used instead of whole genomic DNA. Intermediate products, that is, circles after the first ligation step can also be successfully utilized for subtraction (Fig. 8).

The *Not*I linking library is at least 100 times lower in complexity than the whole human genome. It is approximately equal in complexity to the yeast genome. Since this approach is not linked with Alu repeats, it offers the possibility of isolating *Not*I linking clones that are unavailable for cloning using Alu-PCR.

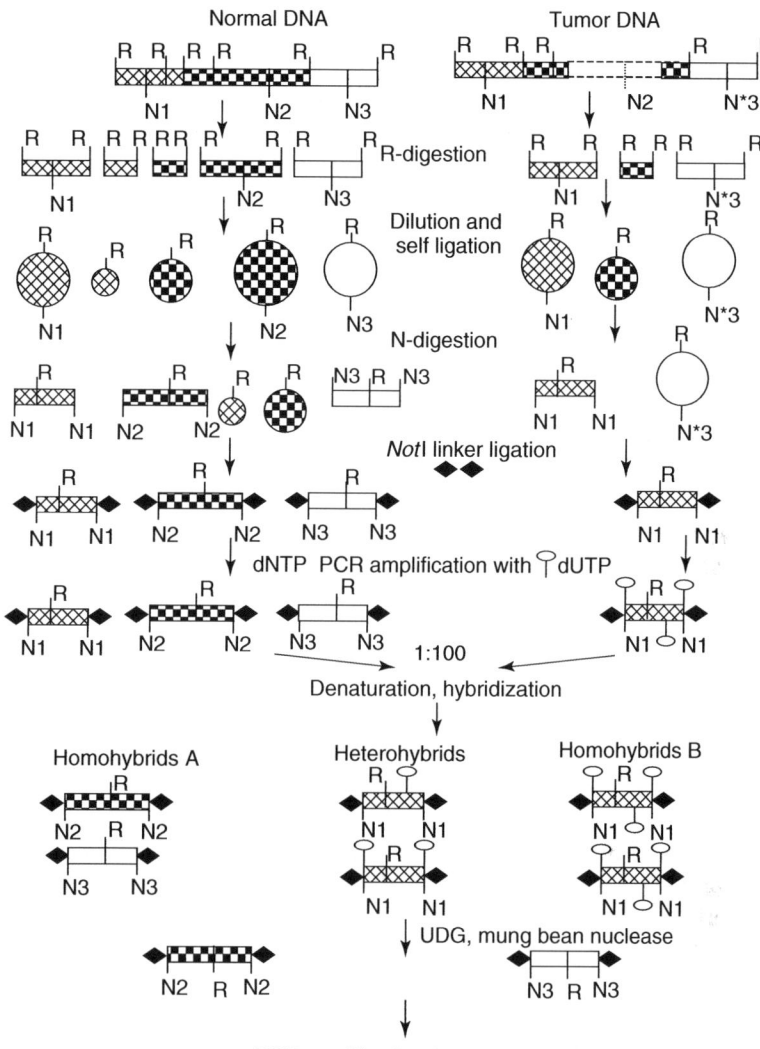

Fig. 8 General scheme for the *Not*I-CODE (Cloning Of DEleted_ sequences) procedure. N, *Not*I sites; R, restriction enzymes recognizing a 4 to 6 bp sequence. Methylated *Not*I site is indicated by an asterisk. UDG, uracil-DNA glycosylase destroys all DNA molecules containing dUTP. Mung bean nuclease digests double-stranded DNA with mismatches and all single-stranded molecules. Circles digested with *Not*I and PCR-amplified are called NR, *Not*I representation, because only sequences surrounding *Not*I sites are present in this amplification product. The N2 site is deleted and the N*3 site is methylated in tumor DNA.

3.4
IBD (Identical-by-descent) Fragments for Identification of Disease Genes

Development of the methods permitting cloning of identical sequences (CIS) between two sources of DNA can be very useful for many purposes, including isolation of disease genes. Identical-by-descent(IBD) sequences refer to segments of the human genome shared by two individuals because they are inherited from a common ancestor. Regions that are IBD between individuals affected with a disease conceivably can contain the disease gene(s). Two approaches were suggested to clone such IBD sequences.

In GMS (genomic mismatch scanning), each DNA preparation is digested with *Pst*I to yield fragments with protruding 3′ ends. The 3′ protruding ends are protected from digestion by exonuclease III (*Exo*III) in later steps. One of the DNA preparations is fully methylated at all GATC sites with *E. coli Dam* methylase (DAM+). The other DNA preparation remains unmethylated. The two DNA pools are then mixed in equal ratios, denatured, and allowed to reanneal. Digestion of the reannealed DNA with both *Dpn*I and *Mbo*I, which cut at fully methylated and unmethylated GATC sites respectively, results in digestion of the homohybrids. The heterohybrids are resistant to both *Dpn*I and *Mbo*I digestion and survive this treatment. Discrimination between perfect, mismatch-free heterohybrids and those with mismatches is done by the MutHLS enzyme. Only perfect duplexes will escape nicking during this step. All DNA molecules, except mismatch-free ones, are degraded further with *Exo*III. Thus, the full-length, unaltered heterohybrids are purified from the other DNA fragments.

Another method was called *CIS* (cloning of identical sequences). The scheme of the CIS-procedure is shown in Fig. 9.

DNA A and B is digested with *Bam*HI and ligated to special linkers containing two recognition sites for *Mvn*I. DNA A is PCR-amplified in the presence of dUTP and m5dCTP; thus, all cytosines will be methylated. DNA B is PCR-amplified in the presence of normal dCTP and biotinylated primers. The two DNA preparations are mixed in equal ratios, denatured, and hybridized. Subsequently, the DNA is digested with *Mvn*I. This enzyme can digest only dsDNA molecules without methylcytosine and will digest all homohybrids B (they contain at least four sites for *Mvn*I).

The DNA mixture is next treated with mung bean nuclease. This nuclease destroys all imperfect hybrids and ssDNAs. Thus, after this treatment we will have only perfect homohybrids A and perfect (without any mismatches) heterohybrids. The DNA mixture is then treated with UDG (uracil-DNA glycosylase). This enzyme removes the uracil base from the DNA and thus destroys all DNA from individual A. As a result, there will be only ssDNA from individual B, which is identical to the DNA in individual A.

3.5
Strategies to Map and Sequence Genomes; Hierarchical, Whole-genome, and Slalom Sequencing Approaches

During the last few years, impressive progress has been made in mapping and sequencing whole genomes of various organisms. Two basic strategies have so far been employed for genome sequencing.

According to one scheme (hierarchical approach), the whole genome is mapped using different types of markers, and a minimal set of large-insert clones, such

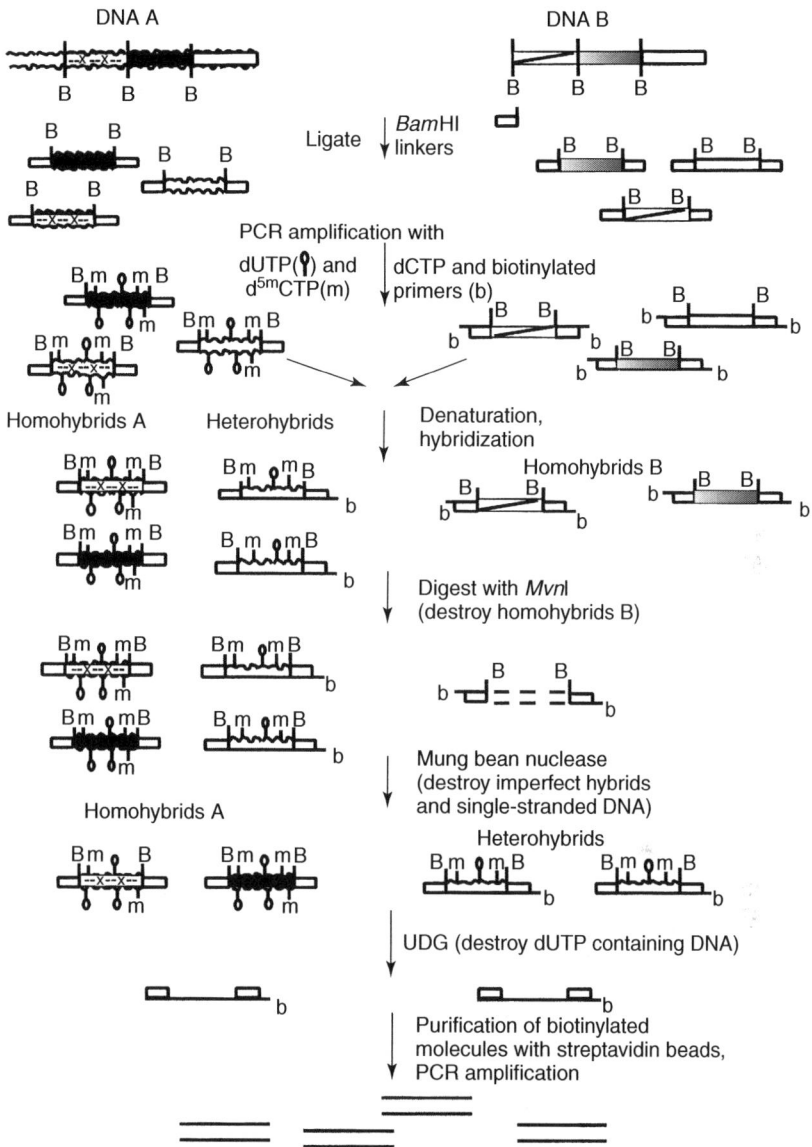

Fig. 9 General scheme of the CIS, cloning of identical sequences, procedure. The same enzymes are used as in the CODE method but in a different order, and the result is opposite.

as cosmids, PAC, or BAC clones, is established. Subsequently, these large-insert clones are sequenced using a shotgun sequencing strategy: small-insert libraries, containing randomly sheared fragments of the large-insert clones, are constructed and sequenced.

A variant of this strategy, the whole-genome shotgun (WGS) sequencing strategy, was developed later and has proved valuable. This method involves end sequencing of large- (PAC, BAC, cosmids) and small-insert (2 and 10 kb) clones. DNA fragments in the small-insert clones are generated by physical shearing of whole genomic DNA. All resulting reads are joined in one sequence with special computer programs. The WGS method requires the generation of sequences covering the whole genome 10 to 15 times. If sequence coverage is less, then the contig assembly process cannot be completed, and sequences and clones will just represent islands, without connection or order. Therefore, despite impressive technological progress, mapping and sequencing of even small bacterial genomes is expensive and laborious.

After completion of the genomic sequence from one organism, there will be a great demand, in many cases, for comparison with the genomes of other individuals, related species, pathogenic and nonpathogenic strains, and so on, in the growing field of comparative genomics. Such comparisons are highly relevant to our understanding of human and animal health, evolution, and ecology.

An efficient strategy for simultaneous genome mapping and sequencing was recently developed. The approach is based on slalom libraries, which combine features of general genomic, jumping, and linking libraries. First experiments demonstrated the feasibility of the approach, and showed that the efficiency (cost-effectiveness and speed) of existing mapping/sequencing methods can be improved at least 5- to 10-fold. The slalom allows the establishment of a physical map, with minimal sets of overlapping clones, which will pinpoint differences in genome organization between organisms. At the same time, considerable sequence coverage of the genome (about 50%) will be achieved. This will make it possible to locate virtually every gene in a genome, for more detailed (comparative) study. Furthermore, since the efficiency of contig assembly in the slalom approach is virtually independent of sequence read length, even short sequences, as produced by rapid high-throughput sequencing techniques suffice to complete a physical map and sequence scan of a small genome. A combination of these new sequencing techniques with the slalom approach increases the power of the method 10 to 50 times more and makes it an efficient tool for comparative genomics. The main principle of the slalom libraries is shown in Fig. 10.

Two standard genomic *Eco*RI- and *Bam*HI-digested libraries are constructed and they will completely cover the whole genome. The problem is how to put *Eco*RI and *Bam*HI fragments in the correct order. It can be solved using the connecting library. The connecting library can be constructed as follows: DNA isolated "en masse" from an *Eco*RI library is digested with *Bam*HI, circularized in the presence of the Kan^r gene, and plated on agar with kanamycin. The clones isolated in this manner will be identical in structure to the clones from an *Eco*RI jumping library prepared in the classical way. By comparing end sequences in these three slalom libraries, all clones can be positioned relative to each other and a minimal contig of overlapping clones may be established.

3.6
Restriction-site-tagged Microarrays to Study CpG-Island Methylation

Methylation, deletions, and amplifications of cancer genes constitute important mechanisms in carcinogenesis, and CGI (CpG-island-containing) microarrays were

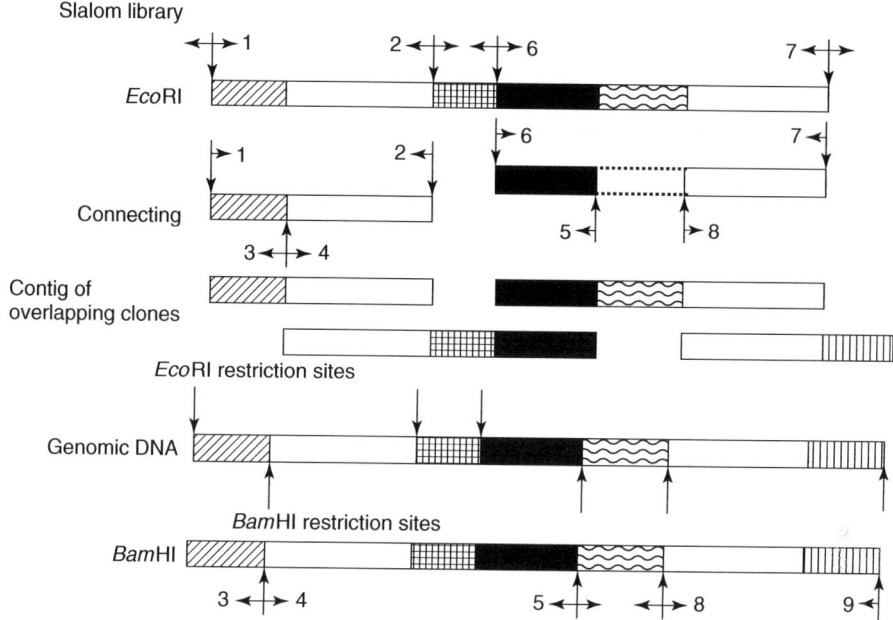

Fig. 10 Simplified slalom library approach scheme. Numbers designate identical end sequences in different libraries that can be joined by a computer program in a contig of overlapping clones. Dashed lines show genomic DNA sequences deleted in the connecting library.

suggested to study hypermethylation in cancer cells. These microarrays can detect methylation changes in tumor DNA. However, it is unclear whether these microarrays can be used to detect hemizygous methylation or copy-number changes. As the whole human genome DNA is used for labeling and the clones are small (0.2–2 kb), this creates a serious problem. Oligonucleotide-based microarrays also can be used to study methylation changes in cancer cells; however, only a limited number of genes can be tested in such experiments.

A rough estimation is that the human genome contains 15 000 to 20 000 *Not*I sites. Therefore, thousands of genes could be tested with *Not*I clone microarrays. The fundamental problems for genome-wide screening using *Not*I clones are (1) the size and complexity of the human genome, (2) the number of repeat sequences, and (3) the comparatively small sizes of the inserts in *Not*I clones (on average 6–8 kb). A special procedure was developed to amplify only regions surrounding *Not*I sites, the so-called *Not*I representations (NRs, see Fig. 8). Other DNA fragments were not amplified. Therefore, only 0.1 to 0.5% of the total DNA is labeled. Interestingly, sequences surrounding *Not*I sites contain 10-fold fewer repetitive sequences than the human genome on average, and therefore, these microarrays are not as sensitive as other methods to the background hybridization caused by repeats. The main idea of this application is clear from Fig. 8 (tumor DNA). If a particular *Not*I site is present in the DNA, then the circle will be opened with *Not*I and labeled. However, if this *Not*I site is deleted or methylated, then the NR will not contain the corresponding DNA sequences. The *Not*I microarrays can simultaneously detect copy-number changes and methylation and, therefore, they allow the simultaneous study of genetic and epigenetic factors.

The technique underlying the preparation and use of *Not*I microarrays is applicable to any restriction enzyme and represents a new type of microarray, referred to as *restriction-site-tagged* (RST) microarrays. Such RST microarrays can be used for different purposes, for example, to study species composition of complex microbial systems.

3.7
Restriction-site-tagged Sequences to Study Biodiversity

There is still much to learn about the human normal microflora. The human gut contains approximately 1 to 2 kg of bacterial cells. The number of these cells in the intestine is 10 to 100 times larger than the number of cells in the human body, but at best 10 to 15% of the microbial species are known. To be able to analyze complex microbial mixtures is of great importance for many applications. For instance, differences between individual compositions of the normal flora will be instrumental for future analysis of the effects on the normal flora composition of diet, foods, geographical location, and medication. Conversely, the effects of gut microflora on aging, autoimmunity, and colonic cancer risk can be studied.

Analysis of human *Not*I flanking sequences (see Sect. 2.3) have demonstrated that even short sequences surrounding *Not*I sites can yield information sufficient to isolate new genes or uniquely describe eukaryotic or prokaryotic genomes.

These results led to the realization that it would be possible to use short sequences surrounding *Not*I sites or, in general, restriction-site-tagged sequences (RSTS) for the analysis of complex microbial mixtures. The collection of *Not*I tags represents the *Not*I sequence passport or in short *Not*I passport. *Not*I passporting means the process of creating *Not*I tags/passports.

The general design of the experiment is as follows. Genomic DNA is digested with *Not*I and ligated to a linker with *Not*I sticky ends. This linker contains the *Bpm*I recognition sites. This restriction nuclease cuts 16/14 bp outside of the recognition site. The ligation mixture is PCR-amplified with special primers and, finally, 19 bp tags flanking *Not*I sites are generated. DNA from, for instance, fecal samples and surgical specimens is digested with *Not*I, and a *Not*I passport for the particular specimen is generated. A comparison of such passports from different individuals or from the same individual before and after drug treatment will reveal the differences between them.

Analysis of tags for *Not*I, *Pme*I, and *Sbf*I for 70 completely sequenced bacteria revealed that more than 95% of tags are species-specific and even different strains of the same species can be distinguished. None of these tags matched human or rodent sequences. Therefore, the approach allows analysis of complex microbial mixtures such as those in the human gut and identification with high sensitivity of a particular bacterial strain on a quantitative and qualitative basis.

A similar approach can be used for eukaryotic cells, for example, for analysis of cancer cells.

RSTS-passporting and RST-microarray approaches are mutually complementary. These two approaches are based on completely different biochemical techniques but aim to solve the same problems.

4
Summary

While several different strategies are available to obtain and use DNA markers for identifying and mapping DNA sequences in complex organisms, no single system is likely to suffice for obtaining a complete and accurate map and sequence of the human genome. Rather, a combination of different approaches and vector systems is needed to corroborate data from different sources.

See also Molecular Basis of Genetics.

Bibliography

Books and Reviews

Ausubel, F.M., Kingston, R.E., Brent, R., Moore, D.D., Sedman, J.G., Struhl, K., Smith, J.A. (1987–2003) *Current Topics in Molecular Biology*, Wiley, New York.

Bird, A. (2002) DNA methylation patterns and epigenetic memory, *Genes Dev.* **16**, 6–21.

Brown, T.A. (1999) *Genomes*, Wiley, New York co-published with BIOS Scientific Publishers, Oxford.

Collins, F.S. (1988) Chromosome Jumping, in: Davis, K.E. (Ed.) *Genome Analysis: A Practical Approach*, IRL Press, Oxford, pp. 73–94.

Mueller, R.F., Young, I.D. (2001) *Emery's Elements of Medical Genetics*, Churchill Livingstone, Edinburgh.

Poustka, A., Lehrach, H. (1988) *Chromosome Jumping: A Long Range Cloning Technique*, in: Setlow, J.K. (Ed.) Genetic Engineering Principles and Methods, Vol. 10, Brookhaven National Laboratory, Upton, New York and Plenum Press, New York, London, pp. 169–193.

Sambrook, J., Fritsch, E.F., Maniatis, T. (1989) *Molecular Cloning: A Laboratory Manual*, 2nd edition, Cold Spring Harbour Laboratory Press, Cold Spring Harbour, New York.

Strachan, T., Read, A. (1999) *Human Molecular Genetics*, Wiley, New York co-published with BIOS Scientific Publishers, Oxford.

Zabarovsky, E.R., Kashuba, V.I., Gizatullin, R.Z., Winberg, G., Zabarovska, V.I., Erlandsson, R., Domninsky, D.A., Bannikov, V.M., Pokrovskaya, E., Kholodnyuk, I., Petrov, N., Zakharyev, V.M., Kisselev, L.L., Klein, G. (1996) NotI jumping and linking clones as a tool for genome mapping and analysis of chromosome rearrangements in different tumors, *Cancer Detect. Prev.* **20**, 1–10.

Zabarovsky, E.R., Winberg, G., Klein, G. (1993) The SK-diphasmids – vectors for genomic, jumping and cDNA libraries, *Gene* **127**, 1–14.

Primary Literature

Adorjan, P., Distler, J., Lipscher, E., Model, F., Muller, J., Pelet, C., Braun, A., Florl, A.R., Gutig, D., Grabs, G., Howe, A., Kursar, M., Lesche, R., Leu, E., Lewin, A., Maier, S., Muller, V., Otto, T., Scholz, C., Schulz, W.A., Seifert, H.H., Schwope, I., Ziebarth, H., Berlin, K., Piepenbrock, C., Olek, A. (2002) Tumour class prediction and discovery by microarray-based DNA methylation analysis, *Nucleic Acids Res.* **30**, e21, 1–9.

Allikmets, R.L., Kashuba, V.I., Bannikov, V.M., et al. (1994) NotI linking clones as tools to join physical and genetic mapping of the human genome, *Genomics*, **19**, 303–309.

Bicknell, D.C., Markie, D., Spurr, N.K., Bodmer, W.F. (1991) The human chromosome content in human x rodent somatic cell hybrids analyzed by a screening technique using Alu PCR, *Genomics* **10**, 186–192.

Bird, A., Taggard, M., Frommer, M., Miller, O.J., Macleod, D. (1985) A fraction of the mouse genome that is derived from islands of nonmethylated, CpG-rich DNA, *Cell* **40**, 91–99.

Brenner, S., Johnson, M., Bridgham, J., et al. (2000) Gene expression analysis by massively parallel signature sequencing (MPSS) on microbead arrays, *Na. Biotechnol.* **18**, 630–634.

Brookes, A.J., Porteous, D.J. (1991) Coincident sequence cloning, *Nucleic Acids Res.* **19**, 2609–2613.

Brown, P.O., Botstein, D. (1999) Exploring the new world of the genome with DNA microarrays, *Nat. Genet.* **21**(Suppl. 1), 33–37.

Broder, S., Venter, J.C. (2000) Sequencing the entire genomes of free-living organisms: the foundation of pharmacology in the new millennium, *Annu. Rev. Pharmacol. Toxicol.* **40**, 97–132.

Carninci, P., Shibata, Y., Hayatsu, N., et al. (2001) Balanced-size and long-size cloning of full-length, cap-trapped cDNAs into vectors of the novel lambda-FLC family allows enhanced gene discovery rate and functional analysis, *Genomics* **77**, 79–90.

Cheung, V.G., Gregg, J.P., Gogolin-Ewens, K.J., et al. (1998) Linkage-disequilibrium mapping without genotyping, *Nat. Genet.* **18**, 225–230.

Collins, F.S., Weissman, S.M. (1984) Directional cloning of DNA fragments at a large distance from an initial probe: a circularization method, *Proc. Natl. Acad. Sci. U.S.A.* **81**, 6812–6816.

Collins, F.S., Drumm, M.L., Cole, J.L., et al. (1987) Construction of a general human chromosome jumping library, with application to cystic fibrosis, *Science* **235**, 1046–1049.

Costello, J.F., Fruhwald, M.C., Smiraglia, D.J., et al. (2000) Aberrant CpG-island methylation has non-random and tumour-type-specific patterns, *Nat. Genet.* **24**, 132–138.

Cross, S.H., Charlton, J.A., Nan, X., Bird, A.P. (1994) Purification of CpG islands using a methylated DNA binding column, *Nat. Genet.* **6**, 236–244.

Eads, C.A., Danenberg, K.D., Kawakami, K., et al. (2000) MethyLight: a high-throughput assay to measure DNA methylation, *Nucleic Acids Res.* **28**, e32, 1–8.

Galm, O., Rountree, M.R., Bachman, K.E., et al. (2002) Enzymatic regional methylation assay: a novel method to quantify regional CpG methylation density, *Genome Res.* **12**, 153–157.

Gonzalgo, M.L., Liang, G., Spruck, C.H., et al. (1997) Identification and characterization of differentially methylated regions of genomic DNA by methylation-sensitive arbitrarily primed PCR, *Cancer Res.* **57**, 594–599.

Hayashizaki, Y., Hirotsune, S., Okazaki, Y. (1993) Restriction landmark genomic scanning method and its various applications, *Electrophoresis* **14**, 251–258.

Kunkel, L.M., Monaco, A.P., Middlesworth, W., Ochs, H.D., Latt, S.A. (1985) Specific cloning of DNA fragments absent from the DNA of a male patient with an X chromosome deletion, *Proc. Natl. Acad. Sci. U.S.A.* **82**, 4778–4782.

Kutsenko, A., Gizatullin, R., Al-Amin, A.N., et al. (2002) NotI flanking sequences: a tool for gene discovery and verification of the human genome, *Nucleic Acids Res.* **30**, 3163–3170.

Lamar, E.E., Palmer, E. (1984) Y-encoded, species-specific DNA in mice: evidence that the Y chromosome exists in two polymorphic forms in inbred strains, *Cell* **37**, 171–177.

Lander, E.S., Linton, L.M., Birren, B., et al., International Human Genome Sequencing Consortium (2001) Initial sequencing and analysis of the human genome, *Nature* **409**, 860–921.

Larsen, F., Gundersen, G., Prydz, H. (1992) Choice of enzymes for mapping based on CpG islands in the human genome, *Genet. Anal. Tech. Appl.* **9**, 80–85.

Li, J., Protopopov, A., Wang, F., et al. (2002) NotI subtraction and NotI-specific microarrays to detect copy number and methylation changes in whole genomes, *Proc. Natl. Acad. Sci. U.S.A.* **99**, 10724–10729.

Li, J., Wang, F., Kashuba, V., et al. (2001) Cloning of deleted sequences (CODE): a genomic subtraction method for enriching and cloning deleted sequences, *Biotechniques* **31**, 788–793.

Li, J., Wang, F., Zabarovska, V.I., et al. (2000) COP – a new procedure for cloning single nucleotide polymorphisms, *Nucleic Acids Res.* **28**, e1, 1–5.

Lindblad-Toh, K., Tanenbaum, D.M., Daly, M.J., et al. (2000) Loss-of-heterozygosity analysis of small-cell lung carcinomas using single-nucleotide polymorphism arrays, *Nat. Biotechnol.* **18**, 1001–1005.

Lisitsyn, N., Lisitsyn, N., Wigler, M. (1993) Cloning the differences between two complex genomes, *Science* **259**, 946–951.

Lucito, R., West, J., Reiner, A. (2000) Detecting gene copy number fluctuations in tumor cells by microarray analysis of genomic representations, *Genome Res.* **10**, 1726–1736.

Mirzayans, F., Mears, A.J., Guo, S.W., Pearce, W.G., Walter, M.A. (1998) Identification of the human chromosomal region containing the iridogoniodysgenesis anomaly locus by genomic-mismatch scanning, *Am. J. Hum. Genet.* **61**, 111–119.

Myers, E.W., Sutton, G.G., Delcher, A.L., et al. (2000) A whole-genome assembly of Drosophila, *Science* **287**, 2196–2204.

Nelson, S.F. (1995) Genomic mismatch scanning: current progress and potential applications, *Electrophoresis* **16**, 279–285.

Nelson, S.F., McCusker, J.H., Sander, M.A. (1993) Genomic mismatch scanning: a new approach to genetic linkage mapping, *Nat. Genet.* **4**, 11–18.

Nussbaum, R.L., Lesko, J.G., Lewis, R.A., Ledbetter, S.A., Ledbetter, D.H. (1987) Isolation of anonymous DNA sequences from within a submicroscopic X chromosomal deletion in a patient with choroideremia, deafness, and mental retardation, *Proc. Natl. Acad. Sci. U.S.A.* **84**, 6521–6525.

Palmisano, W.A., Divine, K.K., Saccomanno, G., et al. (2000) Predicting lung cancer by detecting aberrant promoter methylation in sputum, *Cancer Res.* **60**, 5954–5958.

Pinkel, D., Segraves, R., Sudar, D., et al. (1998) High resolution analysis of DNA copy number variation using comparative genomic hybridization to microarrays, *Nat. Genet.* **20**, 207–211.

Poustka, A., Pohl, T.M., Barlow, D.P., Frischauf, A.M., Lehrach, H. (1987) Construction and use of human chromosome jumping libraries from NotI-digested DNA, *Nature* **325**, 353–355.

Protopopov, A., Kashuba, V., Zabarovska, V.I., et al. (2003) An integrated physical and gene map of the 3.5-Mb chromosome 3p21.3 (AP20) region implicated in major human epithelial malignancies, *Cancer Res.* **63**, 404–412.

Ronaghi, M., Pettersson, B., Uhlen, M., Nyren, P. (1998) A sequencing method based on real-time pyrophosphate, *Science* **281**, 363–365.

Rosenberg, M., Przybylska, M., Straus, D. (1994) 'RFLP subtraction': a method for making libraries of polymorphic markers, *Proc. Natl. Acad. Sci. U.S.A.* **91**, 6113–6117.

Shi, H., Maier, S., Nimmrich, I., et al. (2003) Oligonucleotide-based microarray for DNA methylation analysis: principles and applications, *J. Cell Biochem.* **88**, 138–143.

Smith, C.L., Lawrance, S.K., Gillespie, G.A., et al. (1987) Strategies for mapping and cloning macroregions of mammalian genomes, *Methods Enzymol.* **151**, 461–489.

Snijders, A.M., Nowak, N., Segraves, R., et al. (2001) Assembly of microarrays for genome-wide measurement of DNA copy number, *Nat. Genet.* **29**, 263, 264.

Sugimura, T., Ushijima, T. (2000) Genetic and epigenetic alterations in carcinogenesis, *Mutat. Res.* **462**, 235–246.

Ushijima, T., Morimura, K., Hosoya, Y., et al. (1997) Establishment of methylation-sensitive-representational difference analysis and isolation of hypo- and hypermethylated genomic fragments in mouse liver tumors, *Proc. Natl. Acad. Sci. U.S.A.* **94**, 2284–2289.

Velculescu, V.E., Zhang, L., Vogelstein, B., Kinzler, K.W. (1995) Serial analysis of gene expression, *Science* **270**, 484–487.

Venter, J.C., Adams, M.D., Myers, E.W., et al. (2001) The Sequence of the Human Genome, *Science* **291**, 1304–1351.

Waterston, R.H., Lindblad-Toh, K., Birney, E., et al. (2002) Initial sequencing and comparative analysis of the mouse genome, *Nature* **420**, 520–562.

Worm, J., Aggerholm, A., Guldberg, P. (2001) In-tube DNA methylation profiling by fluorescence melting curve analysis, *Clin. Chem.* **47**, 1183–1189.

Yan, P.S., Chen, C.M., Shi, H., et al. (2001) Dissecting complex epigenetic alterations in breast cancer using CpG island microarrays, *Cancer Res.* **61**, 8375–8380.

Zabarovska, V.I., Gizatullin, R.G., Al-Amin, A.N., et al. (2002) Slalom libraries: a new approach to genome mapping and sequencing, *Nucleic Acids Res.* **30**, e6, 1–8.

Zabarovska, V., Kutsenko, A., Petrenko, L., et al. (2003) NotI passporting to identify species composition of complex microbial systems, *Nucleic Acids Res.* **31**, e5, 1–10.

Zabarovska, V., Li, J., Fedorova, L., et al. (2000) CIS – cloning of identical sequences between two complex genomes, *Chromosome Res.* **8**, 77–84.

Zabarovsky, E.R., Allikmets, R.L. (1986) An improved technique for the efficient construction of gene library by partial filling-in of cohesive ends, *Gene* **42**, 119–123.

Zabarovsky, E.R., Boldog, F., Thompson, T., et al. (1990) Construction of a human chromosome

3 specific *Not*I linking library using a novel cloning procedure, *Nucleic Acids Res.* **18**, 6319–6324.

Zabarovsky, E.R., Boldog, F., Erlandsson, R., et al. (1991) A new strategy for mapping the human genome based on a novel procedure for constructing jumping libraries, *Genomics* **11**, 1030–1039.

Zabarovsky, E.R., Kashuba, V.I., Zakharyev, V.M., et al. (1994) Shot-gun sequencing strategy for long range genome mapping: first results, *Genomics* **21**, 495–500.

Zabarovsky, E.R., Winberg, G., Klein, G. (1993) The SK-diphasmids – vectors for genomic, jumping and cDNA libraries, *Gene* **127**, 1–14.

Zardo, G., Tiirikainen, M.I., Hong, C., et al. (2002) Integrated genomic and epigenomic analyses pinpoint biallelic gene inactivation in tumors, *Nat. Genet.* **32**, 453–458.

Zoubak, S., Clay, O., Bernardi, G. (1996) The gene distribution of the human genome, *Gene* **174**, 95–102.

22
Shotgun Sequencing (SGS)

Jun Yu, Gane Ka-Shu Wong, Jian Wang, and Huanming Yang
Beijing Genomics Institute, Beijing & James D. Watson Institute of
Genome Sciences, Hangzhou, China

1	**Introduction** 635	
1.1	A Brief History of DNA Sequencing 638	
1.2	Overview of SGS Methodology and Application 641	
2	**Essentials for SGS** 645	
2.1	Unbiased Clone Coverage 646	
2.2	Sequence Quality Control 648	
2.3	Repetitive Contents 649	
2.4	Physical and Sequence Gaps 650	
2.5	Validation of Sequence Assembly 650	
2.6	Integration of Genomic Information 651	
3	**Technical and Experimental Basics** 651	
3.1	Genomic DNA Libraries 652	
3.1.1	Source DNA 653	
3.1.2	Quality Assessment 653	
3.2	Sequencing Data Acquisition 654	
3.2.1	WG-SGS and CBC-SGC 654	
3.2.2	Directed Cloning and Random SGS 655	
3.2.3	Process Control 655	
3.3	Sequence Assembly 656	
3.3.1	An Overview of General Procedures 656	
3.3.2	Classification of Repetitive Sequences 656	
3.3.3	Software Packages for Sequence Assembly 657	
3.3.4	Building Contigs and Scaffolds 657	
3.4	Sequence Finishing 658	
3.4.1	Improving Sequence Quality 660	
3.4.2	Primer-walking and End-sequencing 660	

Genomics and Genetics. Edited by Robert A. Meyers.
Copyright © 2007 Wiley-VCH Verlag GmbH & Co. KGaA, Weinheim
ISBN: 978-3-527-31609-0

3.4.3   Closing Sequence Gaps   661
3.4.4   Physical Mapping and Clone Fingerprinting   662
3.5   Genome Annotation and Analysis   662
3.5.1   Compositional and Structural Dynamics   663
3.5.2   Genome Annotation   664
3.5.3   Sequence Variations   666
3.5.4   Genome Evolution   667

4   **Other Useful Resources for SGS**   **669**
4.1   ESTs and Full-length cDNA   669
4.2   Physical and Genetic Maps   670
4.3   Comparative Genome Analyses   671

5   **Perspectives**   **671**
5.1   Large Genomes   671
5.2   Genomes with High Repetitive Sequence Content   672
5.3   Polyploidy: Ancient and New   672
5.4   Mixed and Meta Genomes   672
           Acknowledgment   673

           Bibliography   **673**
           Books and Reviews   673
           Primary References   673

Keywords

Large-insert Clones
Clones made as cosmids, fosmids, BACs or YACs, with an insert size around or over 40 Kb.

Minimal-tiling-path Clones
Physically mapped clones that cover their source DNA molecule with a single path and least amount of overlaps.

Physical Gap
Gaps not covered by available clones in a clone assembly.

Physical Map
DNA sequences or chromosome segments delineated by a set of ordered and oriented markers or fingerprinted clones.

Scaffold
A group of ordered and oriented contigs with physical contiguity but without sequence contiguity.

Sequence Contig
A contiguous sequence assembly composed of overlapping sequences.

Sequence Gap
Gaps not covered by sequences in a sequence assembly.

Shotgun Sequencing
A DNA sequencing technique used for achieving sequence contiguity by acquiring random sequences from clone libraries made of the target DNA.

Shotgun sequencing (SGS) is primarily a large-scale sequencing (LSS) technique that does not rely on precise guiding information about the target DNA, which includes large-insert clones and single genomes, ranging from thousands of basepairs (Kb) to billions of basepairs. A mixture of smaller genomes can also be sequenced in a similar way when retrospective means are available to assemble and distinguish them. It provides a fast and cost-effective way of sequencing large genomes regardless of whether the project as a whole takes a "whole-genome" (WG) or "clone-by-clone" (CBC) approach. The success of SGS essentially depends on random sampling, high-quality data, sufficient sequence coverage, effective assembly, and gap closing procedures. SGS has already demonstrated its tremendous power in sequencing not only microbial genomes but also large eukaryotic genomes, such as those of the cultivated rice and the laboratory mouse. Together with improved sequencing technologies and computing tools, SGS is expected to play a central role in the field of genomics, especially when the latter has to constantly face some major challenges, scientific, managerial, and political. Some of the scientific challenges relate to the effective sequencing of large, polyploid, and mixed genomes, as well as those with highly repetitive sequence contents. A key managerial challenge is for the operators to consistently produce high-quality data while increasing throughput and reducing cost. It is always a tough decision for a steering committee organizing a genome project to choose between WG and CBC, but SGS is always the basic technique of choice.

1
Introduction

DNA sequencing has become one of the essential laboratory methods used in molecular biology, together with DNA cloning, DNA microarray, polymerase chain reaction (PCR), electrophoresis, chromatography, and centrifugation. It has become not only a fundamental method of determining genome sequences but also a basic tool for identifying DNA methylation sites and sequence polymorphisms. In both scale and scope, DNA sequencing outperforms all other molecular technologies in acquisition of basic genomic information, thereby paving the way for other genome-scale, technology-oriented disciplines and, to use contemporary terminology, the numerous "-omics" (transcriptomics and proteomics being well-accepted examples) that will be leading biologists into the future decades, providing new data, ample information, fresh knowledge, and innovative insights.

At the leading edge of genomics, the principal ideas of DNA sequencing are not fundamentally different from the technical innovations demonstrated in the early days of technology development in molecular biology. When a cloned DNA molecule is larger than one (from one end) or two (from both ends) sequencing readouts (often referred as *sequencing reads*; screenshots of a sequence trace and a sequence alignment are shown in Fig. 1) from an electrophoresis-based instrument, its full-length sequence has to be obtained in one of the following two ways: (1) a stepwise primer-directed "walking" strategy, based

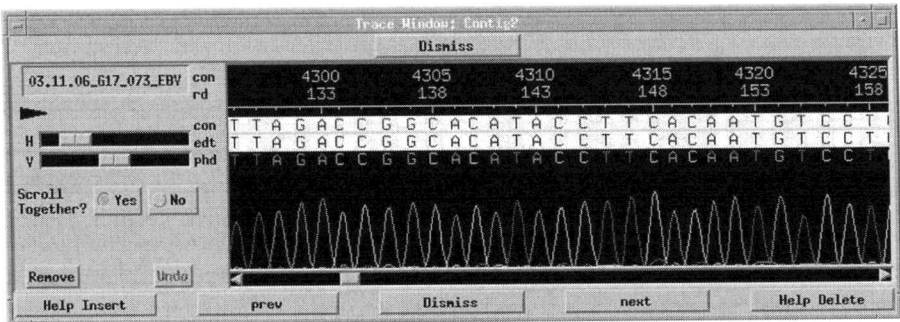

(a)

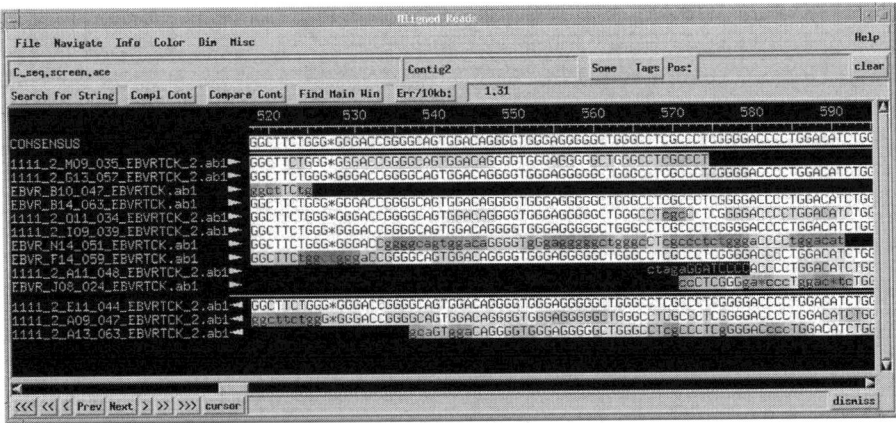

(b)

Fig. 1 Screenshots of a partial sequence read (a) and a sequence assembly with aligned sequences (b). Brightness in (b) indicates the quality of the sequence as estimated and displayed by a software tool kit (sequence assembler), Phrad-Phrep-Consed.

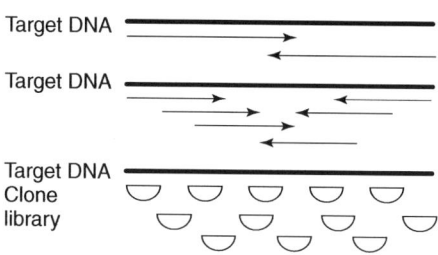

Fig. 2 Two basic ways of sequencing DNA beyond a size of two sequencing reads: (a) to "walk" in a stepwise way from both ends by designing primers based on newly acquired sequence information from a previous walk, and (b) to "shotgun" the target DNA into smaller pieces that are carried by a clone library (or subclones). Random sequencing of the subclones with high redundancy leads to an assembly whose consensus sequence is equivalent to that of the target. Arrows depict sequencing reads and their directions.

on newly acquired sequences; (2) a parallel subcloning strategy, based on directed or "shotgun" cloning techniques by breaking the target DNA into smaller pieces so that a collection of sequencing reads from overlapping "subclones" sufficiently covers the entire target length (Fig. 2). It is obvious that the choice of approaches depends on the capability and throughput of a sequencing operation. Whereas one is managed serially and is therefore limited in scale, the other involves paralleling and is suitable for high-throughput, large-scale operations. The current capability of a typical sequencing instrument limits the sequencing length in a single run to just 500 to 2000 basepairs (bp). Besides solutions involving instrumentation (such as parallelization and incremental increase of read length), the choice of approaches for any large-scale sequencing (LSS) project has to take advantage of SGS for large-insert clones over tens of kilobasepairs (Kb), or genomes in a size range from a few kilobases (such as large viral genomes) to a few megabasepairs (Mb; such as bacterial genomes), right up to a few hundred or even a few thousand Mb in length (such as higher plants and animals).

SGS has been successfully applied for sequencing many large genomes, essentially at two levels, targeting either clones or whole genomes. Large genome-sequencing projects are often initiated by large-insert clone-based physical mapping, especially for important model organisms from representative taxonomic groups; the Human Genome Project (HGP) is the most famous example. These physical mapping techniques include YAC (yeast artificial chromosome)-based or BAC (bacterium artificial chromosome)-based STS (sequence-tagged site) mapping, and restriction-digest mapping with different types of large-insert clones. These clones are subsequently sequenced by SGS, guided by a collection of clones that cover the targeted region with least overlaps, often termed as minimal tiling path (MTP). Physical maps providing anchored clones for sequencing are often referred to as *sequence-ready maps*. In a map-directed process, sequence data are captured at clone level and merged, after data from overlapping clones are completed individually. Although SGS is utilized at clone level, such an approach is often referred as a *clone-by-clone SGS* or *CBC-SGS*, as opposed to a whole-genome SGS, or WG-SGS. In the second scenario, when sufficient information (including estimated genome size, number of chromosomes, GC-content, repeat contents, and genetic homogeneity) about a genome has been acquired either from the organism itself or from a closely related one, the WG-SGS approach becomes feasible, and is often the method of choice. In most cases, SGS has its "sweet range" in sequencing genomes, from a few Mb to a few hundred Mb. However, the quality and completeness of a final assembly habitually depends on repetitive sequence content, clonability, and complexity of the genome, coupled with other factors such as funding, project management, and homogeneity of the source DNA (sequence variations among individuals that donate DNA sometimes are known to delay a quality sequence assembly for months or even years). Before long, SGS will have to face even greater challenges, when it begins to be applied to genomes in a size range around thousands of Mb, polyploid genomes, and even a mixture of genomes. Novel approaches, techniques, and computing tools will have to be gradually introduced in order to make SGS a more efficient and robust technique. Two major significant factors are driving this effort:: the ever-increasing desire for more and more sequence information, and the incremental cost reductions, as evidenced in the history of genomics over the past decade.

1.1
A Brief History of DNA Sequencing

The concept and technology of DNA sequencing have been evolving over the past thirty years. The initial phase of DNA sequencing was born with the development of the *basic technology*, fundamentally a series of technology advancements in DNA sequencing and molecular cloning around the early 1970s. The second phase began with the proposals to set up the Human Genome Project in 1983 and 1984. The decade following this, before the advent of LSS of HGP in 1995, was the most critical period for *special technology and concept* development, most notably physical mapping and DNA sequencing. The past 10 years have been the *Large-scale Sequencing Era*, which is undoubtedly going to last at least another decade or even longer. In this period, alongside additional technological and strategic improvements, there has been a collective effort on the part of those involved in sponsored academic research as well as those belonging to the profit-making biotech industry, to make improvements in the areas of policy development, organizational reforms, and management practices. This has resulted in a continual reduction in the cost of sequencing from the earlier standard level of "$1 per base" (calculated as $40 per sequence read on the assumption of 8x coverage of a target genome) when HGP first started, to the real cost of $15 per read when HGP transitioned into the Large-scale Sequencing Era, to a current cost estimate of $1 per read. These are the estimated direct costs for academic operations and not the retail price offered by private companies. Some academic institutions have claimed lower costs in a rather variable range from $0.3 to $0.8. The cost of sequencing is expected to go down continually over the next 5 to 10 years, dropping at a rate of 50 percent every two years (Fig. 3). Meanwhile, of course, some new ground-breaking technology could emerge and dramatically change the current practice.

The DNA polymerase-based sequencing method and a variety of molecular cloning techniques are the centerpieces of contemporary DNA sequencing technology. In the early days, primer-dependent tactics and shotgun principles were already being employed in tackling large pieces of DNA molecules and certain levels of

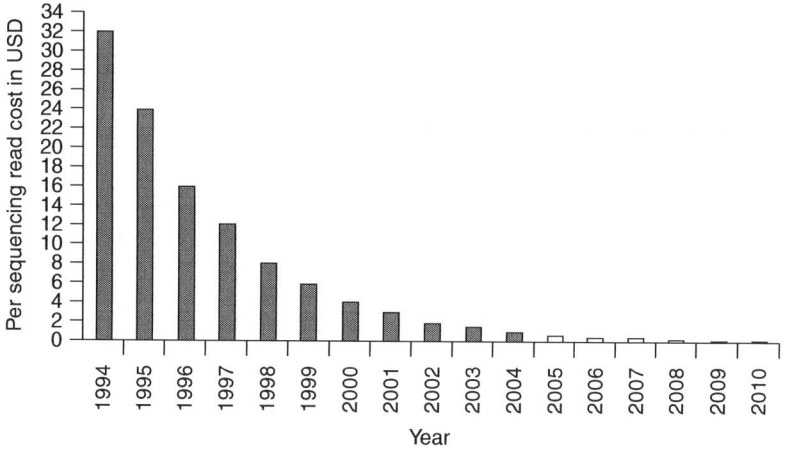

Fig. 3 The trend of estimated cost reductions since the beginning of large-scale sequencing (LSS). Although the legendary estimate, "one dollar per basepair," at the beginning of the HGP should be marked as a starting point, LSS was not launched on a full scale until the middle of 1990s when capillary sequencers and robotics-aided basic sequencing protocols became available. Real academic costs are calculated over a course of 10 years (1995–2004; solid bars). The costs beyond 2004 are estimated according to the trend (open bars). An empirical rule becomes clear, namely, that the sequencing cost is halved every two years.

complexity, such as overhang sequences that are common in bacteriophages. It should be remembered that it took the pioneers over a couple of decades to achieve an increase in sequencing read length from a few tens of basepairs to a couple of hundred basepairs. The first genome sequenced by a "shotgun" design was the bacteriophage phiX174 (5386 bp in length and containing 10 genes) published in 1977. Since then, bacterial cloning systems have leveraged the biology of both bacteriophages and plasmids. They have been designed to achieve higher yields for cloned DNA (such as high-copy plasmid), and equipped with polylinkers composed of multiple cutting sites of commonly used restriction enzymes, as well as several choices in the matter of antibiotic resistance.

The WG-SGS approach was first successfully applied to a microbial genome, *H. influenzae* (1 830 138 bp in size; a bacterium causing upper respiratory infection) in 1995. The success of the WG-SGS approach has opened the floodgates for microbial genome sequencing. As a result, collections of completed microbial genome sequences are updated almost on a daily basis, now number in the hundreds. The first metazoan genome to be sequenced using the WG-SGS method was the *Drosophila* genome with a euchromatic content of slightly over 122 Mb. The SGS data were subsequently combined with those from the public sequencing effort on the same organism, thereby unlocking another information gateway towards insect genomics. Insect genomes, such as mosquitoes and silkworms, were sequenced subsequently by the WG-SGS method in different institutions. The Human Genome Project was carefully planned and initiated with a genome-scale physical mapping effort, followed by SGS of large-insert clones, including YACs and cosmids, but mostly BACs. It was therefore labeled as an SGS approach, although the large-insert clones were sequenced one by one using SGS. The landmarks for CBC-SGS were *E. coli* K-12 (the first bacterium), *S. cerevisiae* (the first eukaryotic species), and *C. elegans* (the first metazoan species). Private teams later used the WG-SGS approach to interrogate the human and mouse genomes, but the wealth of public data were also carefully merged to arrive at the final assemblies. The mouse genome was the first trial of the WG-SGS strategy by the publicly supported consortium and the rat genome was the next. Following the *Arabidopsis* genome (150 Mb), the first plant genome deciphered with a CBC-SGS approach, the first crop plant genome – the rice genome (450 Mb) – has been sequenced four times. Two of the efforts were led by private companies with the WG-SGS approach. An international consortium took a CBC approach and has completed sequencing several chromosomes of the rice genome from the same cultivar as the one sequenced by the private companies, namely, a *japonica* rice, Nipponbare. The fourth rice genome, an *indica* variety, was sequenced by our own institute as part of its Hybrid Rice Genome Project using WG-SGS strategies. The WG-SGS wave has reached new heights. Large genomes of various species, even certain taxonomic groups (such as a suite of vertebrates, mammals, and grasses), are being placed in undergoing sequencing operations, and these are expected to be completed in the next 5 to 10 years. The selection of these targets for sequencing is justified because of their relevance to human biology and diseases, as well as to broader areas of economic interest such as domestic crops, farm animals, and their wild relatives.

From a technical perspective, two major breakthroughs have revolutionized large-scale sequencing. The first is the invention of large-insert clones, YAC being the first and BAC, the second. They offered unprecedented substrates for both LSS and physical mapping. The timely invention of PCR technology also stimulated the new physical mapping concept of STS. Since mapped YACs have to be subcloned into bacterial vectors for sequencing (because YACs are intermingled with host chromosomes, their DNAs are not readily separable by straightforward chemical methods), BACs have now become the preferred substrate. The second breakthrough was the invention of four-colorfluorescence-based methodology and instrumentation, allowing for a high degree of parallelization and miniaturization to come into play. Capillary-based sequencers are the current choice of high-throughput sequencing, although versions of slab-gel-based sequencers emerge from time to time. A central computational support for large-scale sequence acquisition and analysis is the concept of a quality score system for each sequenced base, pioneered by Phil Green and his colleagues. Other emerging technologies such as "single molecule" sequencing technology are being sponsored by certain funding agencies, but these are still in their infancy or at a proof-of-principle stage, and may take years of development and research in order to be introduced to a sequencing equipment assemblage.

On the whole, it is clear that SGS has becoming a mainstay of LSS because of collective experience, sharpened software tools, and increasing knowledge about the structure of target genomes. It is most likely that many large genomes will be sequenced by the SGS method rather than by the CBC one in the years to come.

1.2
Overview of SGS Methodology and Application

Other than biological significance, two categories of practical issues have to be considered for any large genome project beyond a few Mb in target size: managerial and technical. In general, whereas the total number of sequence reads required to cover a genome length is comparable in the two versions of SGS, namely, WG and CBC, WG-SGS is both money- and time-saving because it largely skips a labor-consuming step: clone library construction based on sequence-ready maps. For a CBC-SGS approach, each of the large-insert clones (often BACs) has to be shotgun cloned into small-insert clones (often plasmids). The larger the genome (or the more the MTP clones), the more the libraries that have to be constructed. For instance, to sequence a 500-Mb genome, over 3500 physically ordered BAC (assuming 150 Kb as an average length for BACs) clones have to be made into subclone libraries, whereas only a few carefully prepared small-insert clone libraries are adequate for WG-SGS. If carefully planned, a genome project could start with randomly picked BACs from a whole-genome library that sufficiently covers an entire length, followed by a low-pass sequence sampling (such as a few hundred sequence reads) and sequence overlap assessment, and finished with parallel procedures to complete each BAC of MTP. Although such a strategy appears feasible, it is as yet untried. It is composed of multiple tasks at different levels, for which the project has to be managed in a serial fashion. In contrast WG-SGS, in its unique path, provides a highly parallel process avoiding managerial pressures often created at intermediate levels. Table 1 lists some completed genome sequences in categories of sequence methods, WG-SGS and CBC-SGS.

SGS has never been a stand-alone technology. It includes many other cloning and computing techniques and strategic approaches along the path of its expanding applications. The first set of technologies has to do with molecular cloning. Currently, the preferred cloning systems are BACs, that are best for cloning large DNA fragments and for clone stability; and plasmids, best as regards high yield and readiness for preparing DNA templates for sequencing. BACs are also easier to fingerprint by six-base cutting enzymes (as opposed to YACs), and the system has a copy number control scheme that is believed to enhance clone stability in bacterial hosts, as compared to cosmids. Other merits of using BACs include their simple, robust DNA preparation procedures for parallel and scaled operations, and the fact that they are better reagents for physical mapping for long-range coverage and contiguity (BACs can be made in an insert size range of 100 to 200 Kb) than the lambda packaging-based cosmids and fosmids, which are limited by an insert size range of 30 to 40 Kb. However, one has to realize that cosmids and fosmids are both suitable for SGS on a limited scale and BAC libraries are not as easy to make in large quantity as the cosmid system. Regardless of which of the large-insert cloning systems is employed, plasmids provide an excellent direct substrate for SGS sequencing, whereas others can only be directly end-sequenced on a limited scale for achieving better contiguity over a variable range. On certain occasions, miniature BACs, or mini-BACs

Tab. 1 Representative sequenced genomes with WG-SGS and CBC-SGS strategies.

Species	Genome Size [Mb]	Strategy	# of Contigs	# of Scaffolds
B. mori (Silkworm)	530	WG-SGS	213 289	N/A
			N/A	N/A
P. chrysosporium (White rot fungus)	30	WG-SGS	1323	767
G. gallus (Chicken)	1200	WG-SGS	111 864	N/A
A. mellifera (Honey bee)	300	WG-SGS	30 074	N/A
C. merolae (Red algae)	16.5	WG-SGS	N/A	N/A
R. norvegicus (Rat)	2750	WG-SGS	137 910	N/A
N. crassa (Bread mold)	40	WG-SGS	821	N/A
P. troglodytes (Chimpanzee)	3000	WG-SGS	361 864	37 931
C. briggsae (Nematode)	104	WG-SGS	578	N/A
C. familiaris (Dog)	2500	WG-SGS	1 089 636	N/A
M. musculus (Mouse)	2500	WG-SGS	224 713	N/A
P. falciparum (Malaria parasite)	23	WG-SGS	N/A	N/A
O. sativa (Rice)	450	WG-SGS	103 044	N/A
			N/A	N/A
T. rubripes (Pufferfish)	365	WG-SGS	12 381	20 379
T. nigroviridis (Pufferfish)	350	WG-SGS	108 177	N/A
C. intestinalis (Sea squirt)	160	WG-SGS	2501	2501
A. gambiae (Mosquito)	278	WG-SGS	69 724	8987
H. Sapiens (Human)	3000	CBC-SGS	N/A	N/A
		WG-SGS	221 036	N/A
D. melanogaster (Fruit fly)	165	WG-SGS	2775	N/A
				N/A
A. thaliana (Mouse-ear Cress)	125	CBC-SGS	N/A	N/A
C. elegans (Nematode)	97	CBC-SGS	N/A	N/A
E. coli	4.6	CBC-SGS	N/A	N/A
S. cerevisiae (Yeast)	12.1	CBC-SGS	N/A	N/A

Note: Some of the genomes have not yet been completed, such as B. mori, R. norvegicus, C. briggsae and M. musculus.

(10 to 20 Kb) are used to compensate some of the systemic drawbacks when insert size and clone stability are of major considerations. Occasionally, mini-insert (100 to 200 bp) clones and YAC clones have to be utilized in coping with low GC-content DNA that is extremely unstable in bacterial hosts, such as is the case in the sequencing of the malaria parasite, P. falciparum.

The second set of technologies crucial for SGS comprises various DNA preparation methods for sequencing substrates. Two basic criteria for choosing a practical method are scalability and cost; both demand robust and automatable protocols. The M13 bacteriophage system used in the early days of LSS was soon given up, largely because of its requirement for repeated precipitation steps in isolating phage particles and DNA, which significantly hinders the automation process, although it is capable of producing high-quality data. Being part of the rate-limiting steps in large-scale sequencing, plasmid DNA preparation methods have evolved a great deal over the past few years. Besides alkaline lysis protocols, absorbent-based and isothermal amplification protocols are also widely used in plasmid DNA preparations. In the absorbent-based protocol, centrifugation steps are mostly reduced to simple pluming, and the isothermal amplification protocol is aimed at eliminating bacterial precipitation and plasmid DNA preparation all together. PCR-based protocols also seem to be convenient in meeting the need for preparation of sequencing templates, but we have yet to see a robust one that yields high-quality data. Despite the difficulty in applying PCR technology to template preparations for large-scale SGS, it is still used in all sequencing reactions, catalyzed by a thermal stable polymerase. It is also vital in primer-directed walking and PCR product sequencing for joining sequence contigs in the finishing phase of genome-sequencing projects.

The most demanding component of all SGS-related technologies is biocomputing, broadly termed as bioinformatics, especially for WG-SGS. Bioinformatic tools are often packaged when applied to almost all aspects of SGS, such as data management, sequence contig assembly, and genome annotation. While being less selective concerning databases and data management systems, SGS relies heavily on sequence assembly and gene-finding algorithms. A WG-SGS project can be initiated only when a qualified sequence assembler is put in place. There are several versions of sequence assembly software packages that are able to handle a target size range of a few tens of Kb to a few Mb (Table 2). By far, the most popular package is the Phred-Phrap-Consed developed by Phil Green and his colleagues. For large genomes in a size range of a few tens of Mb to hundreds of Mb, even thousands of Mb, additional algorithms are required to facilitate the complicated assembly process, including repeat identification and masking, contig and scaffold assembly, and customized annotation. Finally, in most of the cases, a specialized database has to be built to host the integrated information for a sequenced genome, in order to better serve the relevant research community in addition to timely submissions of the generated data to the public databases across continents.

Even though most of the genome projects are aimed at producing a "fully finished" genome sequence (the expected end-product for representative or important genomes), WG-SGS is also capable of producing three intermediate products: a genome survey sequence, targeted genome sequences, and a working draft

Tab. 2 Examples of sequence assembly software packages.

Software package	Author's name	Institute
PhredPhrap	Phil Green, et al.	University of Washington
CAP	Xiaoqiu Huang[1], Anup Madan[2]	[1]Michigan Technological University, [2]University of Washington
Atlas	Paul Havlak, et al.	Human Genome Sequencing Center
TIGR Assembler	G. Sutton, et al.	The Institute for Genomic Research
RePS	Jun Wang, et al.	Beijing Genomics Institute
ARACHNE	Serafim Batzoglou, et al.	Whitehead Institute/MIT
Celera Assembler	Eugene W. Myers, et al.	Celera Genomics Inc.
Phusion Assembler	James C. Mullikin, Zemin Ning	The Wellcome Trust Sanger Institute
PCAP	Xiaoqiu Huang, et al.	Iowa State University

of the whole genome. Each serves a special purpose and is vital for predicting the outcomes of a project in many circumstances. For instance, the genome survey sequence that covers less than 1x of the genome could provide information about the repeat contents and gene structures. It can also help in gathering polymorphisms when DNA sources from several closely related breeds are compared with a complete or nearly complete reference genome. A noteworthy example is the chicken polymorphism study where DNA sequences from three major domesticated breeds with nearly 1x coverage collectively are able to offer over three million SNPs (single nucleotide polymorphisms). One apparent application for targeted genome sequences is that the sequencing reads from an ongoing project can provide markers for identifying large-insert clones for the target regions before pictures of a genome or a locus become clear. A genome working draft (about 6x coverage of a genome) is the most popular product of many genome projects. It often covers 95 to 99% of the functional content of a genome and also anchors genes and their regulatory regions physically on the chromosomes when physical and genetic maps of the organism become available. Since finishing a genome sequence to a reasonable contiguity is difficult and expensive, sometimes impossible, not all genomes are expected to be finished. In fact, in many cases a finishing standard is debatable and occasionally highly controversial. Therefore, genome working draft sequences of many large eukaryotic organisms are expected to stay as such for a long time, though the community standard for finished prokaryotic genomes remains a gap-free sequence.

When zeroing in on the structural dynamics of large eukaryotic genomes, one will soon realize that there are numerous differences among genomes of taxonomically distant species to be considered before a genome-sequencing project is proposed. The most notable differences are between animal and plant genomes. Among animal genomes, genes spread over almost the entire length of a genome and transposable elements are inserted into introns, enlarging the genes over the evolutionary timescale. Most of the animal genes have more than a few alternatively spliced variants. The conjecture is that animal RNA splicing (spliceosomal) machinery is better established and complex enough to be capable of handling large introns ranging from a few tens of Kb to hundreds of Kb. Plant genomes are different from animal genomes in these same two areas. First, plant genes remain small in size owing to slower growth of the intron size, and transposable elements are mostly inserted into intergenic spaces. Transposable elements jump onto each other, forming gene-free clusters in high densities between gene "islands." As plant genomes evolve, only repetitive contents grow thus enlarging repeat clusters and not the gene islands. This is the case with wheat (1700 Mb) and corn (1200 Mb) as opposed to rice, which has a 500-Mb genome, nearly half of it as repeats of different kinds. Second, alternative splicing among plant genes is not as prevalent as it is among animal genes, perhaps as a result of less capable splicing machineries. Figure 4 illustrates the different genome structural models.

There are some minor issues relating to insect genomes wherein the heterochromatic contents are high. Insect genomes in sizes of hundreds of Mb seem not much different from other animal genomes of different taxonomic groups, but the basic organization of larger ones in the order of thousands of Mb are unknown so far.

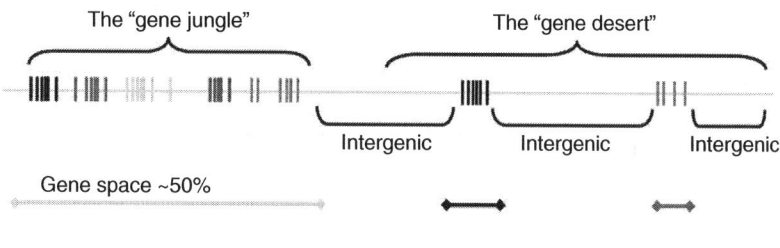

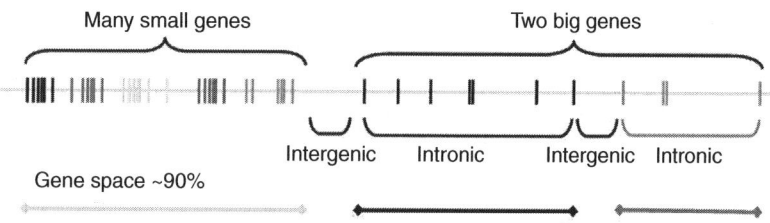

Fig. 4 A schematic illustration explaining major differences in basic genome structures between animals (represented by human) and plants (represented by rice). Genes are depicted as vertical bars in different shades. Gene spaces (diamond-headed lines) are estimates based on real experimental data. Spaces (horizontal brackets) between and within genes are defined as intergenic and intronic sequences, respectively. "Gene jungles" and "gene deserts" exist in plant genomes but are not known in animal genomes. In animal genomes, most of the chromosomal spaces are occupied by genes at different densities. Gene-poor regions most probably contain large genes or large introns.

Protists have the most diverged genomes studied to date and their genome sizes vary from a few tens of Mb to a few thousand Mb (Fig. 5). The GC-content of these organisms varies dramatically; the most extreme case was found in the malaria parasite, *P. falciparum*, which has a GC-content of 18%. As more genomes of these primitive but unusual organisms are sequenced, we will know better about their genome organizations.

2
Essentials for SGS

Regardless of whether it is associated with WG or CBC, SGS has five irreducible steps (Fig. 6): clone library construction, clone sequencing, sequence assembly, sequence annotation, and data analysis. The rest are either managerial (such as data handling, integration, and release) or political (such as cost accounting and choice of strategies) issues. The first two steps are apparently "wet-bench" related (other than data and database management) and the others are mostly bioinformatics, except a hidden "time bomb" in sequence assembly – sequence finishing, where close coordination between the two seemingly distinct disciplines is of the essence for completing a project successfully. The wet-bench operators must make sure that there are no obvious technical flaws in the raw data and that the assembled sequence is in agreement with preexisting data and knowledge, and the bioinformatics

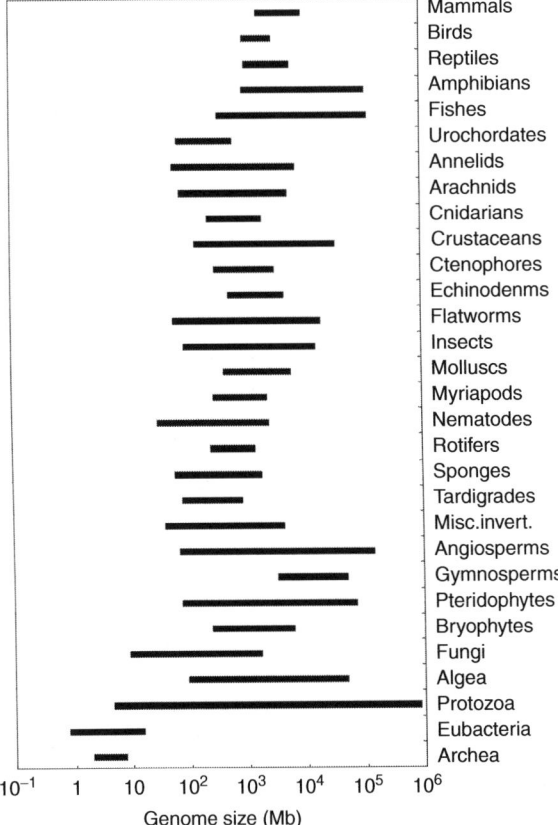

Fig. 5 Genome size distribution from organisms of selected taxonomic groups. The genome size is converted to nucleotide lengths from c-values. Horizontal bars mark the range of genome sizes for a given group. Polyploidy is not factored in to normalize the data, and so that the broader ranges in lower vertebrates, angiosperms, and protists are largely due to this effect. Data are limited to experimentally determined genomes, available from the *Database of Genome Sizes (DOGS)*.

operators must tell their counterparts if something seems wrong, or when certain problems are not solvable by computers alone. Technical details are useful but principles are extremely important.

2.1
Unbiased Clone Coverage

The essence of achieving unbiased clone coverage lies in making the clone libraries properly, avoiding possible negative effects. In a large SGS project (say for a 500-Mb genome and 6x genome coverage that is calculated as genome size divided by average read length or 6x 500 000 Kb/0.5 Kb = 6 million sequence reads), the first 4x sequence coverage is to cover the target sequence evenly and sufficiently. In this regard, the insert size of clone libraries should be small, usually approximately 1.5 Kb or twice the size

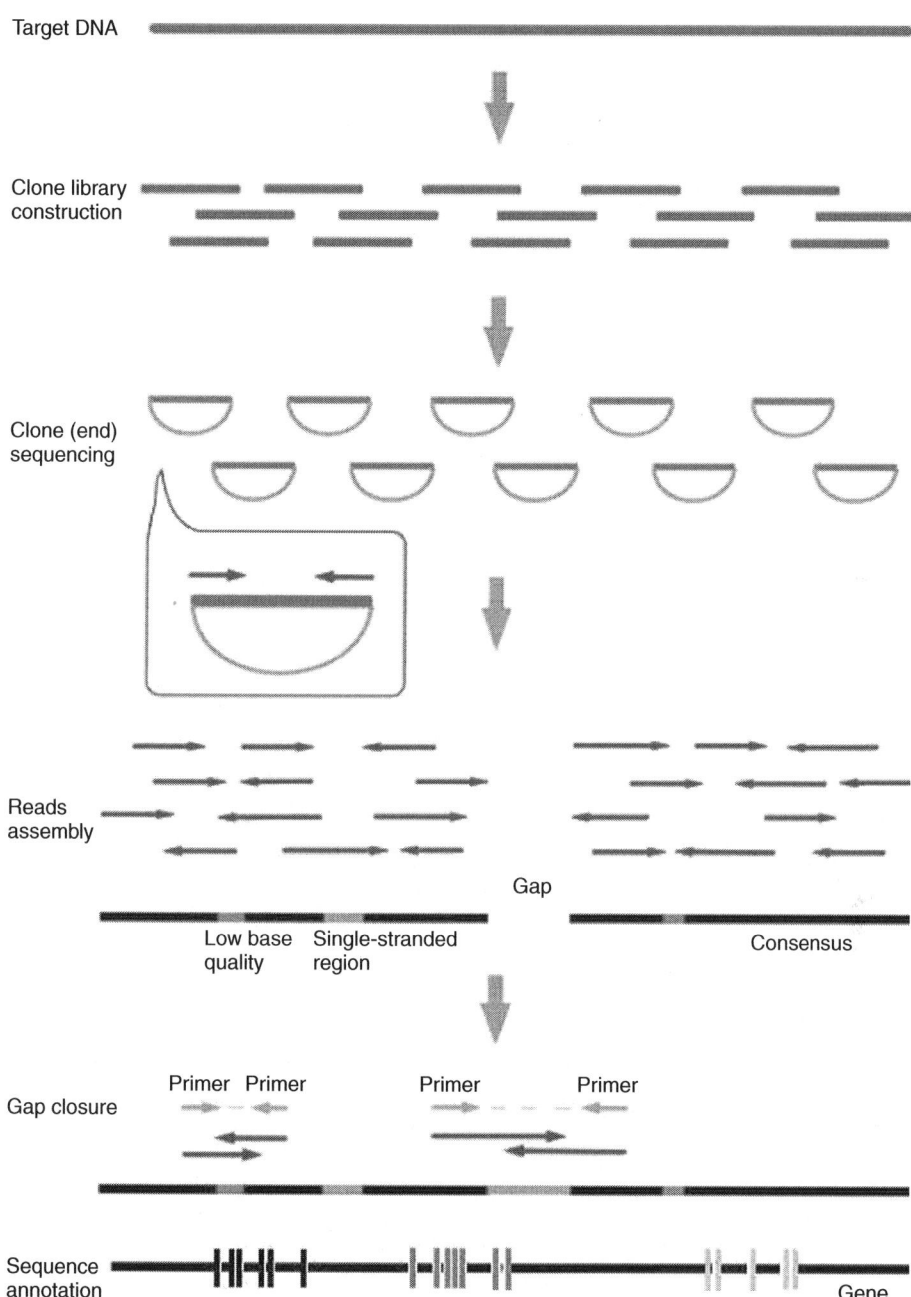

Fig. 6 A schematic illustration of the SGS strategy. Starting from a target DNA, the strategy contains five essential steps: clone library construction, clone (end) sequencing primed by universal primers from vector-borne sequences, reads assembly, gap closure, and sequence annotation. The inset illustrates that the sequencing is done from clone ends with universal primers. Detailed processes are described in the text.

of an average read length. The merits of smaller inserts are multifold. They pose fewer problems for cloning efficiency, randomness, DNA preparation, and sequence assembly. It is also easier to achieve uniformity in size fractionation since DNA fragments of such a size are readily separated in agarose gel electrophoresis. The next 2x is for accomplishing contiguity, and so the insert size should vary according to the repeat contents, including types, ages (in an evolution sense), cluster sizes, and whether there are active transposons that often litter nearly identical repetitive sequences in the genome. When numerous, they create real problems for sequence assembly. The choices of insert size and cloning systems are abundant, anywhere from 5 to 10 Kb for plasmids, from 20 to 30 Kb for mini-BACs, and from 35 to 40 Kb for cosmids or fosmids. BACs are always end-sequenced because in all cases BAC libraries are usually made for building physical maps and targeted gene sequencing, and often constructed before a sequencing project is launched. The only wet-bench concern is that sometimes DNA of large-insert clones is not easily prepared in the same way as small-insert clones and is a poorer substrate for standard sequencing reactions. A subtle computational concern is the "interleaving" problem, where repeat clusters are a few-fold smaller than the size of a large-insert clone and longer than a few sequence read lengths, such that sequence contigs between the two end-sequences could be lost (1.2 and 2.3).

Inadequate clone coverage always results in both sequence and clone gaps. Depending on the compositional dynamics of the genome and the data quality (mostly the average of sequence read lengths), sufficient coverage ranges from 6x to 8x. More often than not, a WG-SGS stops at 4x when contiguity is not a major concern. The Lander–Waterman curves are commonly used for estimating such coverage. The growth of number of contigs can be followed closely while the project is ongoing. The only variable for fitting the curve is the average read length when a genome size is known. The read length can be estimated in several ways. The most popular one is to use the Q values calculated by Phred. The value Q20 cut-off (a base error rate of 1:100) is adequate for this purpose although hard-nosed sequencing operators sometimes use Q30 (a base error rate of 1:1000) for assembly quality measurement. The reads in a length shorter than 200 bp should not be used for the measurement but they can be incorporated later into the final assembling process. Another way to estimate read length is to calculate the fraction that contributes to the consensus sequence. No matter what is used for the read length estimates, more sequence coverage (redundancy) is always a faithful friend during a project that is pressed for time. Sequence read coverage could be pushed up to 10x for smaller projects, such as a bacterial genome. After redundancy is maximized, anomalies should be carefully inspected in an assembly, not limited to assembled contigs and remaining gaps, physical or sequence, but also leftover singlets (stand-alone sequencing read) or singletons (short contigs that are single-read length and assembled by a coverage lower than average). Finishing strategies should be plotted with special attention to the distribution of physical gaps and the nature of sequence gaps.

2.2
Sequence Quality Control

Poor sequence quality that can be improved experimentally usually comes from poor DNA preparation, since reagents and equipments are all provided by a supply company and process-optimized. Average read length at Q20 is a superior estimator of daily assessment of data quality, and feedbacks from the analysis to the wet-bench operation are necessary. Read length can be improved in several ways. In addition to better DNA preparation protocols, these include prolonging the separation (run) time, adjusting template concentration, and optimizing sequencing reaction conditions. These should be done and managed separately in a small R&D group, rather than constantly revising pipeline protocols. Poor signal-to-noise ratio is also attributable to poor DNA preparation in most cases, although malfunction of the process prior to DNA preparation may also be to blame, such as poor viability or growth of bacterial cultures. System contamination and malfunction of sequencing reactions can also result in poor signal-to-noise ratios.

2.3
Repetitive Contents

All genomes seem to have repetitive sequences; some are simple and others complex. Repetitive sequences can be defined in two ways: mathematically (mathematically defined-repeat or MDR) and biologically (biologically defined-repeat or BDR). When a sequence of a given length is seen in a genome more than once, it is mathematically defined as a repeat. The only variable is length that depends on the complexity or the size of a genome (assuming a random model for four nucleotides). BDRs are functionally defined and usually refer to *sequence-recognizable transposable elements* that are prevalent in genomes of all life forms. BDRs are usually identified with a software tool (such as RepeatMasker) composed of a database and a search engine. Since BDRs have to be identified by sequence homology, there are several problems in classifying them. First, the discovery of BDRs is very much limited by prior knowledge of their existence although fragments of the repeats are easily defined as MDRs. Second, it is difficult to classify BDRs because they do diverge over time according to a neutral rate of sequence mutation and some of the transposable elements are known to be old, such as LINE (long-interspersed-nuclear-element) in vertebrate genomes. Third, BDRs are often fragmented because of the dynamics of DNA sequencing of a genome (such as mutation and recombination) and the biological nature of various transposons (such as active and nested insertion). Therefore, parameters of a search engine can be adjusted to maximize or to ignore certain diverged BDRs, but to find them all is an unattainable goal. The quantitative relationship among MDRs, BDRs and apparent BDRs (BDR's) is depicted in Fig. 7.

Two basic parameters are always a concern for repetitive sequence analysis: the length of repeats and their complexity in a given genome. There are three additional lengths to be aware of in the analysis: those of a read, a clone, and a targeted DNA fragment (such as a chromosome or a genome). When two sequence reads are nearly identical, their positions in the target sequence become ambiguous, especially when sequencing reads are highly redundant for covering the target. In this sense, the smaller the target size, the easier the assembly process. However, when complexity of a target DNA and density of repeats are both high, it poses a serious problem that often results in sequence gaps in the target sequence. In practice, the

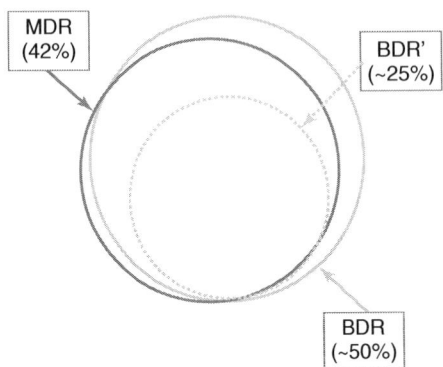

Fig. 7 Distributions of MDRs and BDRs in the rice genome. MDRs can be defined precisely (solid line) by counting exactly repeated elements in a genome. A portion of BDRs (broken line) is defined by degrees of homology (>50%) to known transposon families. The rest is too divergent over the evolutionary timescale to be readily identified but they are clearly present in the genome (gray) and can be estimated through homology analysis with low stringency.

longer the read length (assuming high quality), the less an assembly suffers with sequence gaps. Subclones of SGS project with slightly larger inserts will sometimes help in eliminating sequence gaps but cloning efficiency, size variation due to limited separation power of agarose-based electrophoresis, and other technical details will limit the increase of insert sizes.

2.4 Physical and Sequence Gaps

There are essentially two types of gaps encountered in sequencing a genome. When there is no clone available to bridge two contigs in any libraries for mapping and sequencing, a physical gap is inevitable; it is thus also called a *clone gap*. It is difficult to close such gaps because technology to carry out targeted cloning directly from isolated DNA of a large genome has yet to be developed. It is, however, possible to measure the lengths of gaps by using standard Southern techniques. The reasons for this type of gaps are often complex, but the gaps are usually the results of either inadequate coverage of the library or poor clonability of the missing DNA. Employing directed PCR amplification, adding coverage to the existing library and changing the cloning system, and varying the insert size are all useful remedies, coming from the wet-bench part of a sequence operation. The other type is sequence-related, in which a sequence assembly remains fragmented. In such cases, computer-aided analysis is required to provide clues for the wet-bench workers after they promise to deliver data quality, and have clones of various types in hands. If the nature of the gaps (oftentimes they are either nasty repeats in eukaryotic genomes or toxic genes in microbial genomes) is worked out, they can be left alone after some measurements on the gap sizes. In any case, a finishing plan should be drawn to handle the two types of gaps before they become a problem for sequence analysis, since the more gaps a project has, the more contigs have to be ordered. In a WG-SGS project, at least some effort has to be devoted to sequence finishing because genes or parts of genes may get lost in a massive number of contigs when sequence coverage is low (such as 4x).

2.5 Validation of Sequence Assembly

Multiple estimators should be used to validate a sequence assembly, because oftentimes estimators are biased because of differences in community interests, emphasizing either functional or targeting aspects preferentially. The available resources are (1) full-length cDNAs and EST clusters (Unigenes); (2) physical maps and their markers (STSs); (3) genetic maps and markers; and (4) close-related genome sequences.

There are three essential parameters to be measured in validating an assembly. The first is average coverage, which can be calculated in a straightforward way such as a percentage of hits of the markers and sequences aligned to the target assembly, although the definition of a hit is always subjective because of the subtle details involved, such as real sequence polymorphisms. The second parameter is to determine how complete those matches are from a study of the markers and sequences to the targeted assembly. The exact matching sequence has to be statistically presented precisely down to each basepair. None of the above parameters would tell much about the correctness of the assembly in terms of long-range contiguity and local misassemblies except in the case of using restriction-digest (fingerprinting)-based physical maps in which the digested fragments are precisely size-measured and even ordered. A retrospective comparison between the sequence assembly and its corresponding physical maps, from a BAC clone or genome-wide, is always an excellent estimator for sequence validation. It is, however, difficult to assess contiguity and accuracy of an assembly when resources for estimation are sparse.

2.6
Integration of Genomic Information

Together with the rapid increase in genome sequences, data integration at multiple levels from genomics-scale studies is of the essence for understanding the basic genomic information and functional annotation of sequenced genomes. In a vertical integration scheme, the top layer consists of genes (their structural and regulatory elements) and other basic genomic elements (repeats, telomeres, and centromeres). The intermediate layer includes gene expression information, from mRNA to protein. The bottom layer is the relationship of a gene product, RNA or protein, and its interacting partners in the context of location (defined by spatial connection and timing), function (defined chemically or physiologically), and mechanisms (defined by specific roles of each of the components and their cooperative functions). Genetic information, including markers and sequence polymorphisms, is usually gene-centric or marker-centric and must be integrated into the top layer as part of the physical map before correlations can be made precisely between genotypes and phenotypes. Database systems are usually built not only to manage and search the raw data. They are also integrated with software packages for visualizing, editing, annotating, and analyzing the data. An integration scheme for rice genomic data is illustrated in Fig. 8.

3
Technical and Experimental Basics

SGS as an operational entity, especially in LSS, is highly specialized, with more than 10 years of experience in process improvement and automation. It is relatively mature in the sense that its basic process cannot change overnight. For instance, sample preparations and all subsequent liquid handling procedures are based on microtiter plates with 96 and 384 wells. The

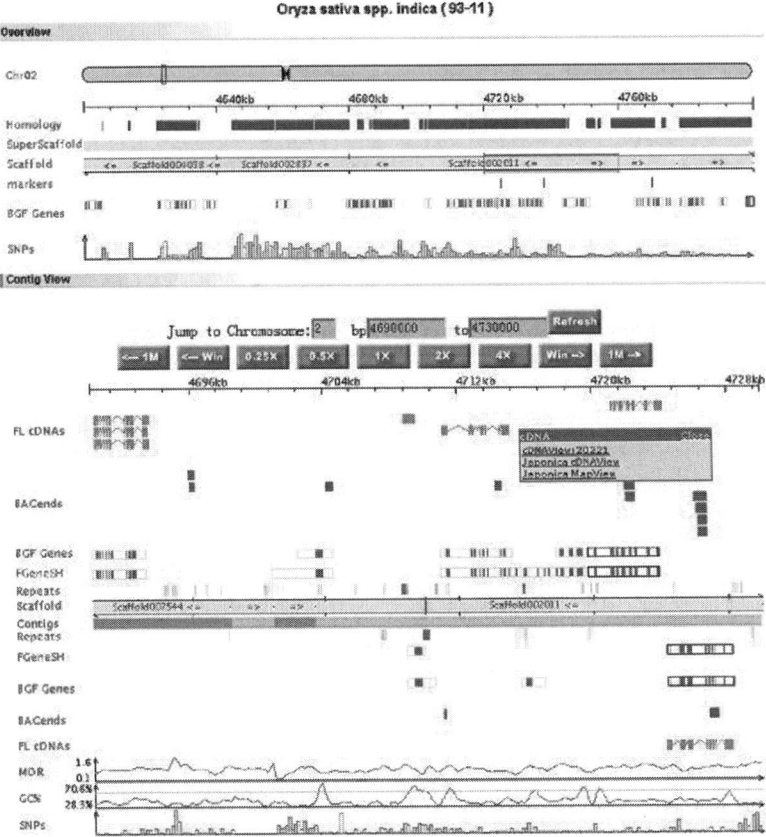

Fig. 8 An example of sequence annotation schemes from the rice genome. Gene-finding results from two gene-finding software packages, FgeneSH and BGF are shown in the annotation.

sequence technology may change some day in the future, but sequencing and cloning systems will live on.

3.1
Genomic DNA Libraries

The preparation of insert DNA is one of the crucial steps in making SGS libraries. For constructing small-insert libraries or subcloning of large-insert clones, physical power, rather than enzymes, is usually preferred for shearing DNA into fragments. There are two ways of doing this – sonication and nebulization. Commercial instruments are readily available and a skillful technician can easily master the technique even for large-scale operations. Partial restriction-digest is not commonly used for preparing insert DNA, especially in large batches. It is, however, very useful when clone libraries with insert sizes ranging from 5 to 20 Kb are constructed for achieving better contiguity.

Size-selection of insert DNA is another important step in securing high-quality libraries, but this is often neglected. Better performance in size-selection results in not only narrower insert size variation but also lower numbers of "empty" (a consequence of too small an insert) clones, thus improving the standards of library quality control. A more uniform insert size is very helpful in sequence assembly when information on clone ends is used to join contigs (3.3). To achieve a better size-selection, two key factors should be kept in mind: electrophoretic conditions, and appropriate loadings (micrograms of DNA per well in an agarose gel). Rapid electrophoresis, by increasing either currents or voltages of the electric field, reduces separation power, while overloading damages the filtration power of the matrix (agarose). For this simple but key step, an optimized system should be carefully established.

3.1.1 Source DNA

One can never over-emphasize the importance of selecting source DNA for a genome project, regardless of whether WG or CBC was used as the downstream strategy. The potential problem lies in the genetic heterogeneity of DNA samples. It is only a minor problem in large organisms because they have sufficient tissue or body mass, but in small organisms of which tissue samples from multiple individuals are used for extracting genomic DNA, it can be a major problem. In the case of sequencing small-sized genomes, such as those of bacteria or archaea, DNA samples should be prepared starting from single colonies. When different batches of DNA preparations are unavoidable, special care should be given to watch for transposon insertions and segmental duplications. Such a situation pleads for restriction fingerprinting analysis (using pulse-field gel electrophoresis if necessary) to make sure of sequence integrity among the samples. For small multi-cellular organisms such as worms, insects, and small plants, it is crucial to prepare genomic DNA for library construction from a homogenous (inbred) population. It is not unheard of that a WG-SGS project might have to be abandoned because of the impossibility of attaining assembling sequence reads at an identity of 95 to 99%. When a non-laboratory strain has to be used for a project, physical maps should be built with large-insert clones, from which a set of MTP clones is fed into sequence pipelines. It thus becomes a CBC-SGS project. Extreme caution should be exercised lest other invisibly small organisms contaminate the source DNA isolated from a wild species, especially those from open waters, such as large protists, small marine animals (mollusks, sponges, and corals), and miniature plants.

The final concern, though not the least, is about sex chromosomes. In a sexual organism, sex chromosomes are often unevenly distributed between males and females. Other than human males who have a miniature sex chromosome, Y, sex chromosomes in other sexual species are sometimes comparable in size. However, they may not have the same distribution as human sex chromosomes, where XY is male and XX is female. In birds, the male is homozygous, ZZ, and the female is heterozygous, ZW. Oddities are often observed in life forms, with even temperature and crowdedness being determinants for sex and genetic makeup. Whenever possible, a haploid genome provides much better source DNA for WG-SGS.

3.1.2 Quality Assessment

There are a few parameters to be measured experimentally to assess the quality of an SGS clone library: the average insert size, the rate of non-informative clones (small inserts and background vector self-ligation), and clone recovery rate. The first two parameters can be measured by colony-PCR (using a small number of bacterial cells as PCR templates after heat-denaturation) or by sampling isolated plasmids. To avoid false-negative results of PCR, tested samples should be in duplicates with appropriate controls. The last parameter is a result of several potential tribulations, including mixed or defective clones, and non-optimal growth conditions. It has to be tested by sequence sampling. For a large-scale operation, it might be more economical to just test all the parameters by sequencing random samples.

3.2
Sequencing Data Acquisition

The sequencing data acquisition process is the engine of SGS. It is calibrated in a very subjective and precise way as throughput and cost. The throughput has shaped the way LSS is organized globally, with only a dozen or so academic institutions being focused on such an operation. The cost structure of DNA sequencing is composed of capital equipment, reagent/supply, and labor. At least two-thirds of the total cost is on account of reagents and supplies, so there is still plenty of room for cost reduction. Another important factor is the management of the two major components of sequencing business: one, the experimental process, and the other, computation. Only a short feedback loop between these two components can ensure operational efficiency.

3.2.1 WG-SGS and CBC-SGC
Operationally, there are major differences between CBC-SGS and WG-SGS. First, CBC-SGS is a map-directed operation and it relies heavily on the completion of a physical map containing targeted regions. In most of the multi-institutional operations, work is allocated according to chromosomal regions (in the case of HGP) or simply individual chromosomes (in the case of the international rice genome consortium). A potential danger is that loosely affiliated members may all have their own priorities and agendas, and that their schedules for carrying the project to the finishing line may vary accordingly. As a result, the satellite projects have to be finished and published separately; such an operation not only reduces the scientific impact of a project but also faces the risk of being "scooped." WG-SGS is in a much better position to succeed, since data are to be acquired by the same group or groups with comparable operating styles. Second, CBC-SGS requires an intermediate management layer to handle maps and clones, to construct more shotgun libraries, and to assemble each clone, each of which requires a different sort of attention to complete. In other words, wherever genomes with high repetitive sequence contents are involved, they may never be finished to a similar standard, and therefore CBC-SGS could face a disastrous situation. Large plant genomes are illustrative of such cases, where the repeat contents exceed gene contents. For smaller animal genomes, however, the problem is less hazardous. Third, WG-SGS poses major challenges for sequence assembly when the target genomes are large and contain high repetitive content, but assembling sequences for large-insert clones in typical CBC-SGS are not. In CBC-SGS, sequence contigs are ordered according to clone orientation in physical maps, whereas in WG-SGS, contigs can be ordered within scaffolds or according to physical markers. In some cases of WG-SGS, contig and scaffolds may not be ordered until

both physical and genetic maps become available. WG-SGS data, however, provide a foundation for retrospective physical (gene) and genetic mapping efforts.

3.2.2 Directed Cloning and Random SGS

During the early phase of LSS, there were two essential strategies developed to evaluate the cost and efficiency for an operational process. The first system utilized a random shotgun protocol and M13 phage (single-stranded DNA) cloning system. DNA preparations for M13 cloning, when carried out manually, are usually easy and clean, giving rise to high-quality sequences. It is unfortunate that only one end of the single-stranded DNA can be primed for sequencing. The method was soon replaced because the success of a complex experimental protocol depends on the collective merits of all the procedures. The second system is the plasmid cloning system (double-stranded DNA). Two simple advantages of using plasmids are: sequences from both ends of the insert are obtainable from a single DNA preparation, and the length of DNA inserts is also a piece of powerful information for sequence assembly when their sizes are prepared relatively uniformly. Although other attempts were made in the early days of LSS, such as transposon-directed plasmid sequencing, the current practice is focused on a plasmid-based random cloning system, oftentimes coupled with controlled insert sizes.

3.2.3 Process Control

SGS remains merely a process rather than an approach or strategy when it is not evaluated and compared with other sequencing techniques. It is quite complex, involving integration of sequence acquisition, data processing, and analysis. As a process, it has to be versatile because relevant technologies are still underdeveloped and any part of it can be changed over a short period of time. For instance, capillary sequencers now dominate all large-scale sequencing operations, although the slab gel format may continue for special purposes such as for small-scale operations and for achieving longer read lengths. Of course, the most practical consideration in building a sequencing operation has to be capital investment. Any such investment on a process-unstable technology should be carefully calculated. With an estimated annual budget exceeding $1 billion US dollars globally for large-scale genome sequencing, more than half of which comes from public funding agencies worldwide, LSS is undoubtedly going to stay with us for some time.

SGS creates major managerial challenges for process control. On the wet-bench side, the managers are Ph.D. scientists and skillful technical staff, who are able to master cloning techniques and to have troubleshooting capabilities for all protocols used; each has precise quality-control parameters that are under close surveillance with fast feedback loops. For instance, whether or not a clone library is going into the pipeline depends on its "pass rate," in which contaminations from *E. coli* host and vectors, average lengths of the insert, and identities of the target sequences (such as signature repeats) are all factored in. However, things can go wrong in between too, such as failures in DNA preparation and sequencing reactions, equipment-related errors, and even data tracking mistakes. In most of the cases, a custom-designed software system has to be implemented for a process-controlled operation. The challenge in building a robust system is to decide on the extent of automation *versus* human intervention, where fully automated systems are impossible to build. On the computing side, things are even worse. Data of different kinds are voluminous and instantly accumulating, including raw trace files (calling for data management and database building) from different projects (calling for project management when target DNAs are from different sources), quality assessment (involving quality management to distinguish usable and unusable data), assembled intermediates (requiring process management dealing with reads, contigs, and scaffolds), and annotations (calling for biological research to interrelate the results by searching against all available data); and even these data are always incomplete. Because automation is being continually expanded, other groups of people have also entered the process, namely, engineers and manufacturers. Three indispensable groups with different cultures have to work together on the same process. When physical mapping is involved, another group of people would have to join this outsized team.

A process should be evaluated by its throughput and efficiency. There are many reasons for there being only a few large genome centers to do the basic jobs, but management and experience matter the most. In addition, scales, automation, and robust protocols are all major concerns for a sizable sequencing operation. As the size of the target genome gets bigger, the scale of an SGS operation goes up. Currently, projects of the size of mammalian genomes (3 billion basepairs) are usually handled by a joint effort of more than one genome center. There are only half a dozen genome centers in the world that are able to handle the sequencing of large genomes independently.

3.3
Sequence Assembly

Sequence assembly is the most time-consuming and computing-intensive part of SGS, especially for WG-SGS. Software tools for sequence assembly, and also for gene-finding and sequence annotation, have to be developed and implemented before a real project starts. In addition, the skills and experience of the operating teams are also important since the entire process is very complex and demanding, as is described below.

3.3.1 An Overview of General Procedures

There are distinct steps in the procedure for sequence assembly. First, repetitive sequences need to be defined, classified, and masked for each sequence read before being placed into an assembly pipeline. Second, low-quality reads (usually at the ends of a sequence read) and contaminated reads (from library hosts and other projects) are removed through different computational filters. Third, computer-assembled sequence contigs are evaluated based on mathematic models (such as the Lander-Waterman model) and examined manually after each finishing protocol is applied (3.4). Fourth, contigs are further linked into scaffolds based on additional information. Finally, contigs and scaffolds are ordered and oriented according to chromosomal organizations based on physical markers or other experimental data, such as those from fluorescent in-situ hybridization (FISH).

3.3.2 Classification of Repetitive Sequences

Almost all eukaryotic genomes contain significant portions of complex repetitive sequences. Despite the fact that these repeats are mostly definable by their repetitive natures, being created over evolutionary timescales, they have not only jumped in and out of genomes at different points of time (lineage-specific) but are also being mutated and recombined among themselves, making their precise identification an intricate task. Repetitive sequences are often classified into two basic categories, simple and complex. Simple repeats are easily definable and of less concern for sequence assembly, because they usually constitute only a small percentage of the genome sequences in vertebrates and seed plants. They do occasionally pose minor problems for sequencing (3.4.3) and may become a major nuisance for insect genomes, as they could account for a major part of heterochromatins. Complex repeats are mostly defined biologically by detailed evolutionary analyses and by building sequence libraries for featured annotations of discrete taxonomic groups (2.3). Another class of repetitive sequences is the so-called *low-complexity repeats* (LCR). The origin of LCRs is recent segmental duplications, and they may cause significant problems when concentrated in a proximal region of a chromosome. However, they are not very frequent, and experienced researchers can deal with them quite easily with the aid of denser physical markers and unique sequence signatures. The final class of repeats concerns special regions of chromosomes, such as centromeres, telomeres, and their nearby regions, such as subtelomeric regions. Different research groups usually handle these repeats according to their own specific interests.

3.3.3 Software Packages for Sequence Assembly

Sequence information is acquired through an electrophoresis-based instrument that is capable of resolving differences among DNA fragments at single base level. An important parameter is the quality of each resolved base, which is measured as error probability or error rate. In the most popular software package for sequence assembly, Phred-Phrap-Consed developed by Phil Green and his colleagues, Phred is a base caller that is capable of assigning an error rate to each sequenced base from a machine-processed trace file. Using stringent quality measurement, all eligible bases are used for assembly. Phrap carries these quality scores when it assembles consensus sequences from a set of sequencing reads. Consed is essentially a viewer that allows users to examine the results and edit the sequences. Nearly all sequence assemblers have capabilities similar to the Phred-Phrap-Consed package, but some have been outfitted with more user-friendly features. For instance, CAP, another sequence assembly program developed by Xiaoqiu Huang and Anup Madan, is able to clip 5′ and 3′ low-quality regions and uses forward–reverse constraints to correct assembly errors. Phusion, an assembler developed at the Sanger Institute, is competent in assembling mammalian genomes, and was used in the mouse genome project with a WG-SGS dataset. Phusion incorporates end-pairing information and has an ability to group large numbers of sequence reads before performing a parallel assembly protocol. Some of the major software packages for sequence assembly are listed in Table 2.

3.3.4 Building Contigs and Scaffolds

The fundamental nature of sequence assembly software is to define identity among sequences of different reads and to detect their overlaps. Several software packages have been developed for DNA sequence assembly but by far the most popular one is the Phred-Phrap-Consed package. For large WG-SGS, a special package has to be built in to handle additional information that is not dealt with by Phrap. For the purpose of illustrating the basic principles and methodology, RePS (an assembler, repeat-masked Phrap with scaffolding, developed at Beijing Genomics Institute) is taken as an example (Fig. 9).

The first step of a sequence assembly is to define "repeats," and these are the major obstacles whenever sequence comparison is the engine of an assembler. Repeats are usually defined mathematically as exact matching sequences of oligonucleotides (MDRs), as a series of "oligomers." The lengths are often defined empirically, such as 20mers for large eukaryotic genomes. Repeats are identified from each sequencing read and "masked" before being placed in an assembly process. Some of these reads may be completely masked so that they have little chance to contribute to intermediate assemblies, but they are factored in as anonymous coverage only in the final product. Phrap is often used as an assembly engine for most assemblers although other algorithms have been developed for general sequence comparison and assembly.

In nearly every LSS project, another powerful piece of information is also used, namely, paired sequence reads from the ends of the same clone (acquired from opposite directions), which allow users to order and orient additional contigs that are not obviously overlapping in sequences but definable by clone-length criteria. When the insert sizes of a clone library are not significantly deviated from a median (usually measured experimentally), a uniform size is assigned to all the clones sequenced from both directions. The insert sizes may vary among libraries used for a particular project but not within the libraries. As a result, the ends of a clone can be paired, though sequences of the middle portion of an insert may not be covered from both ends, and this information is utilized to "scaffold" neighboring contigs that are anchored by these end sequences. Figure 9 depicts the assembly process involving the two primary components of RePS: assembly of repeat-masked reads with Phrap, and scaffolding based on clone-end information.

The final process of an assembly is to close the gaps. At this point, two types of gaps are in evidence: artificially created ones due to repeat masking, and physical ones that occur statistically by chance. The former is referred as *repeat gaps* (seen within sequence reads) and the latter is called *Lander–Waterman gaps*. Repeat gaps can be filled after reads are assembled and Lander–Waterman gaps are to be closed experimentally because they often result from inadequate coverage or purely by chance. In practice, both gaps are usually smaller than the lengths of single reads and can be bridged by clone-end information, when paired ends of a clone happen to join their two neighboring contigs.

3.4
Sequence Finishing

Sequence finishing, a process for achieving better contiguity of a target sequence, is always a time-consuming process, and requires the collective experience in LSS built up over a decade. Efforts in automating such a process have had some successes, but have suffered because of the inadequacy of standards as well as funding. There is no question that certain important genomes have to be finished

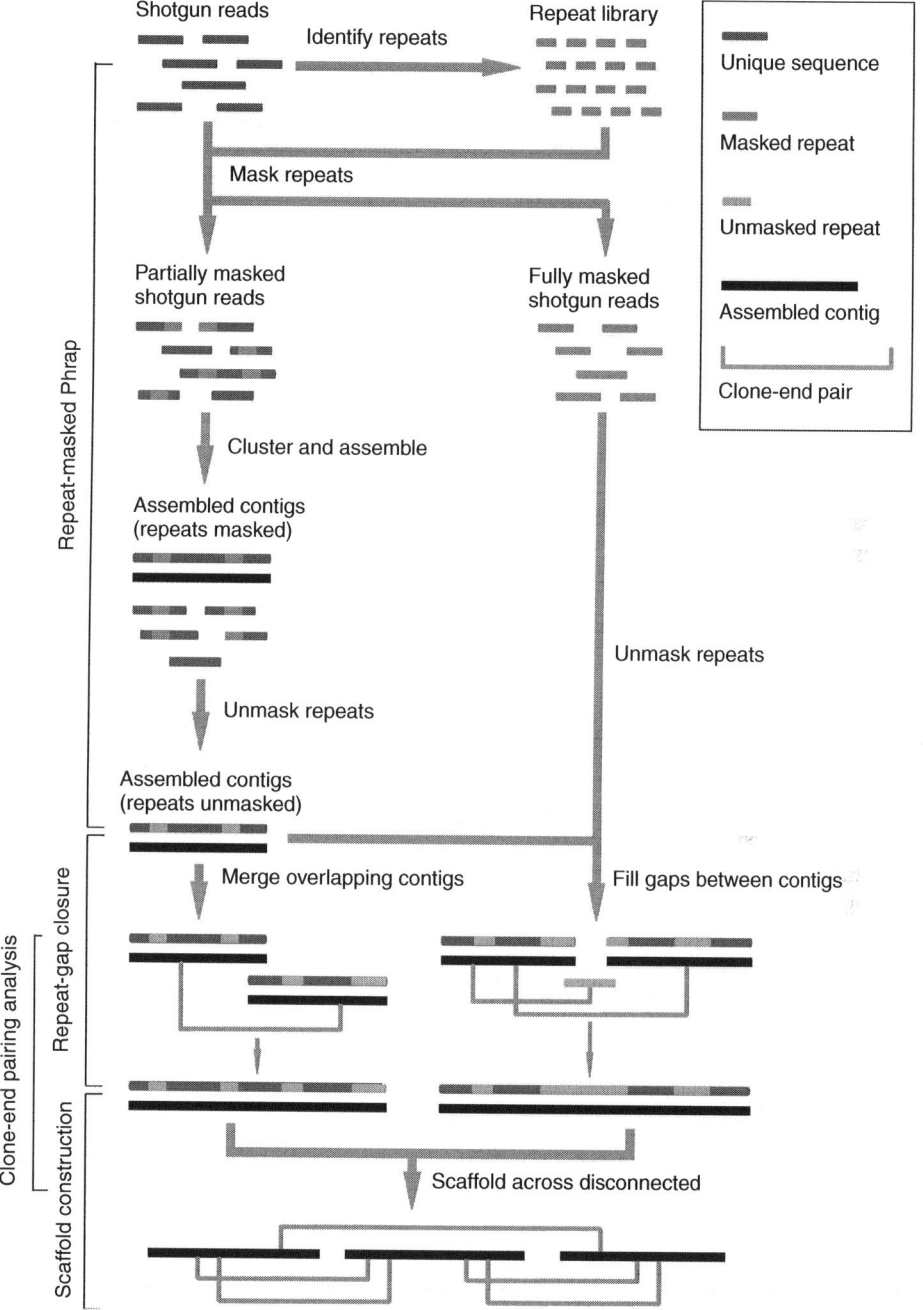

Fig. 9 Schematic flowchart shows the process of sequence assembly.

to a standard that is acceptable to the scientific community. Examples of such genomes are the human and laboratory mouse genomes, but for other genomes a working draft (6–8x coverage) as the final product of a genome project should be satisfactory. With either WG-SGS or CBC-SGS, the basic finishing process for sequence gaps is essentially the same. It includes: (1) improving quality in the weak regions (often defined as those covered by a single read generated from a single type of sequencing chemistry); (2) primer-walking on clones at contig ends; (3) PCR-sequencing to close gaps; and (4) checking consistency guided by physical maps. These activities are normally performed by a team of well-trained technicians and led by a highly experienced scientist with a Ph.D. degree.

3.4.1 Improving Sequence Quality

Before one looks into the relationship among sequence contigs, the first thing to do in the finishing stage is to analyze the weakly covered sequences (poor sequence quality and single-read coverage) that are often the reason for misassemblies and false joints between contigs. Poor reads assembled into contigs should be removed first and repeated experimentally with or without new DNA preparations. Plasmid clones (not M13 clones) can be sequenced from the opposite directions if the poor reads are due to the nature of DNA sequences (such as long simple repeat tracks or extreme GC-contents). This seemingly less impressive practice is in fact very powerful, and at times reduces the number of contigs by half in a project on bacterial genomes of a few Mb in size. The second major effort should be to figure out reasons why certain contigs stop growing; some may be obvious and even straightforward to determine, and others may not. For instance, when contigs repeatedly end with poor-quality reads, the most obvious solution is to add more reads to increase the total sequence coverage. When contigs end with large numbers of reads, higher than the estimated average coverage, and stop sharply with a pile of high-quality reads, the reasons are usually misassemblies due to repetitive sequences. In large genomes, low-copy repeats, which may be either large (such as recent segmental duplications) or small (such as recent transposon insertions and duplications), from time to time create serious headaches for sequence finishers and need to be spotted as early as possible.

3.4.2 Primer-walking and End-sequencing

All sequence contigs end for some particular reason. These reasons include, but are not limited to: (1) low-quality reads or failed sequence in one end of a clone that happens to be in a poorly covered region; (2) high-copy repeats that are too long to be linked by any clones in the assembly; (3) low-copy repeats that are assembled together; and (4) false joints that are made within the contigs. These reasons can be placed in one of two categories: repeat-related and quality related. Quality-related contig-breakers can be eliminated by primer-walking, in which specific primers are designed to improve the local sequence quality. When a PCR product is too long to be "walked" over (for example, if it is more than 3 Kb in size), a clone library can sometimes be made from the PCR product for a small SGS project. Repeat-related contig-breakers are, however, very hard to eliminate. Software tools are required to examine sequence contigs and their contents, and to relate repetitive sequences among contigs. In the assembly software assembly software package, Phred-Phrap-Consed, there is a useful little tool, called *PhrapView*, allowing users to see distributions of repetitive sequences among contigs within a project. A general remedy for the repeat problem is to add some large-insert and end-sequenced clones into the assembly, such as BACs or cosmids/fosmids. The hope is that these long-range sequences can solve, in part, this type of assembly problem. When repeat-related problems are explicitly understood, editing of software parameters and read assemblies becomes an easy solution.

3.4.3 Closing Sequence Gaps

There are always mixed feelings about closing gaps. On the one hand, it is satisfactory to close some simple gaps, such as by PCR-based methods. On the other hand, when thinking about the nature of these gaps, one may wonder why we have to take so much trouble over closing them. This is because most of the gaps are caused by short (a couple of Kb or less), low-complexity (a few unique copies in the genome), and simple repeats (such as {AT}n and {CCT}n). Time and again, gaps are found due to degenerate simple repeats (such as degenerate repeat tracks of pentamers or hexamers, often classified as MDRs, 2.3) that are poorly maintained by the bacterial hosts. In other words, they are mostly physical gaps caused by cloning failures, and are especially present in animal genomes, such as those of human, mouse, and insects. Short BDRs of high density can also result in similar gaps, especially in plant genomes. When large repeat clusters are encountered, gap closure can be difficult and sometimes impossible. Extreme GC-contents, often caused by simple AT-rich or GC-rich repeats, can also result in gaps because they are tricky to sequence with a standard sequence reaction protocol. Both annealing and extension temperatures of a sequencing reaction have to be adjusted to repeat the experiments. In the genomes of grasses and warm-blooded vertebrates, a negative GC-content gradient along the transcription direction exists in a significant portion of their genes. This is the result of a mutation drive that is believed to originate from the transcription-coupled DNA repair mechanism. When the GC-content of these genes reaches 70%, the 5'-end is occasionally missing in the assembly. Though this is a particular case seen in sequencing the rice genome, the phenomenon may extend to other grass genomes (such as those of wheat and corn). In summary, there are no universal rules about how to close gaps, and what the sequence characteristics of the gaps are. It is, however, worth making an effort to close a few gaps just to know what is out there.

It is still debatable whether all gaps should be closed in a genome sequence and it is up to the research communities and their funding agencies to take a final decision. It is our opinion that, in a WG-SGS project, certain gaps should be closed and their sizes measured where feasible; these should include gaps that separate a coding sequence and that are shorter than a few Kb, within a PCR-reachable distance. On the other hand, when gaps are known to be the result of running against long-range repeats, they should not and cannot be closed with currently available techniques. On an experimental note, when closing gaps that are poorly covered by bacterial clones, one can use isolated YAC or genomic DNAs as PCR templates for gap sizing and filling since they may have resulted from DNA sequences that have poor clonability in bacterial hosts.

3.4.4 Physical Mapping and Clone Fingerprinting

Because both physical mapping and clone fingerprinting are very important techniques for SGS, almost all large-scale sequencing projects are initiated along with a parallel project to make large-insert clone libraries. Physical maps can be constructed by using three basic substrates: chromosomal segments (such as radiation hybrid mapping, in which overlapping chromosomal segments are carried by a cell line originated from different species); large-insert clones; and restriction fragments (or even DNA fragments). Two basic markers are used for physical mapping, namely, STSs, and restriction-digest fingerprints. The essence of physical mapping is to order these markers, clones, and DNA fragments. Large-insert clones are the most useful reagents for sequence-ready physical maps, especially BCAs and cosmids/fosmids.

In CBC-SGS, physical mapping becomes a prerequisite and a sequence-ready map is the starting point for large-scale SGS. Since BACs (100 Kb to 200 Kb) and cosmids (30 Kb to 40 Kb) and also fosmids and phages are better sequencing substrates than YACs, both directly (such as for end-sequencing) and indirectly (their subclones are the direct substrates), they are preferred for constructing a sequence-ready map. YACs are generally useful for long-range physical mapping, such as YAC-based STS mapping, but they have to be subcloned into cosmids or fosmids again in order to make a sequence-ready map. For WG-SGS, it is foreseeable that physical maps will be needed for hanging sequence contigs or scaffolds on chromosomes when projects on large genomes are carried over their SGS phase. With unlimited STS candidates in hands, it is quite feasible to construct a contig-based STS map retrospectively, even if the number of scaffolds may exceed a few hundreds. Another way to order sequence contigs is to use the FISH technique, in which representative large-insert clones can be placed and ordered over chromosomes experimentally.

Clone fingerprints are also useful in validating an assembly, especially for CBC-SGS. When large-insert clones are fingerprinted with high redundancy, each clone and also the genome-wide sequence quality can be systematically checked, aided by software tools such as FPC, which is an integrated program for assembling sequence-ready clones, developed at the Sanger Center.

3.5 Genome Annotation and Analysis

Genome annotation has now become a specialty in genome-sequencing projects. It gets more and more complex because data of different types are being accumulated exponentially and most of them are cross-referenced both vertically and horizontally. In a vertical sense, data of a single species have to be integrated from genome to transcriptome and proteome, as well as information from other levels of functional studies (such as metabolisms, networks, and interactions). In the future, all data that can be cross-referenced will be cross-referenced without any delay. In the horizontal sense, data from closely related species (within a species, genus, and even within the same phylum) have to be compared and analyzed with regard to the details relating to the genes and their functions. The norm is going to be changed over time until species representing all life forms of the extant taxa have been somewhat sequenced. This is not a dream but a future reality.

3.5.1 Compositional and Structural Dynamics

Genome sequences are dynamic entities, deriving from common ancestral genomes, and evolving to become more divergent, complex, while remaining related at different evolutionary scales. Eukaryotic genomes are known to undergo compositional transitions driven by mutation mechanisms that are by and large governed by various DNA polymerases involved in DNA replication and repair. Although eukaryotic genomes with extreme average GC-contents are rare, heterogeneous distributions of DNA composition are regularly observed. Figure 10 shows several examples of such a dynamic nature. Major compositional transitions have been discovered among genomes, between warm- and cold-blooded vertebrates, and between grasses (one of the major monocotyledonous plant groups) and the majority of dicotyledonous plants. Moreover, not only is such an increase in GC-content directional (Fig. 11), from 5' end of the transcription unit toward its 3' end, but it also involves significant portions of the genomes and highly expressed genes. The exact molecular mechanism has not yet been demonstrated experimentally, but it is speculated that it could be largely attributable to transcription-coupled DNA repairs. The relevance of genome compositional dynamics to SGS is multifold. Extremes of GC-content (lower than 20% and higher than 60%) often pose problems

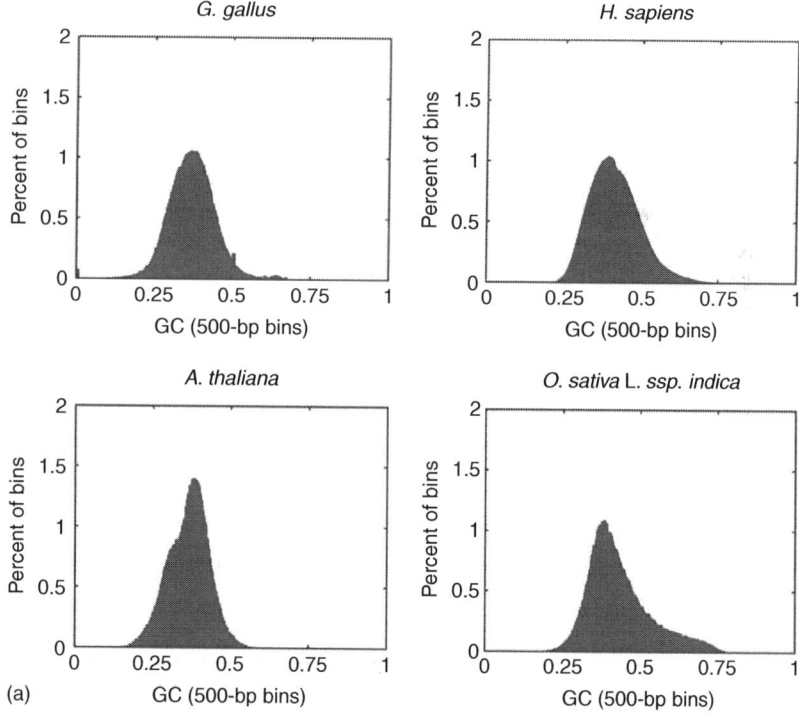

Fig. 10 Genomic GC-content distributions of selected genomes: G. gallus, H. sapiens, A. thaliana, O. sativa, D. rerio, X. laevis, C. intestinalis, and C. elegans. GC-contents are computed over a bin size of 500 bp.

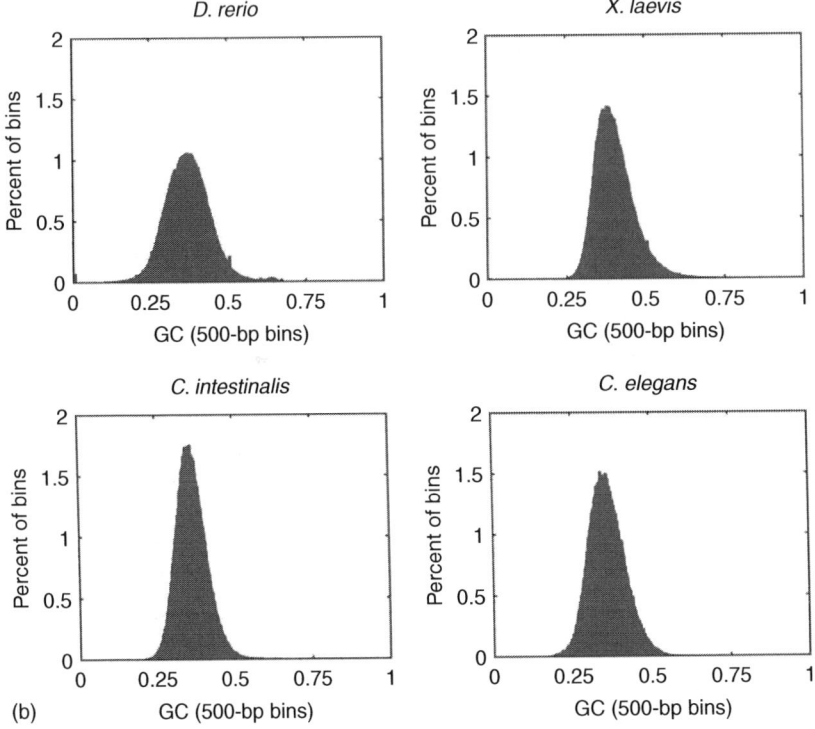

Fig. 10 (Continued)

in cloning and sequencing procedures. Sequence heterogeneity interferes with gene-finding algorithms, especially when training data sets are not readily available. It also biases sequence homology comparison, making genome annotation more difficult.

The structural dynamics of evolving genomes are studied at higher levels than single or a few dozen basepairs, generally involving large duplications and deletions, which could occur genome-wide and in the whole genome, in a chromosome segment, and at gene levels. Broadly speaking, structural dynamics within genes and gene families should also be dealt with in a similar way, such as alternative spliced variants and differential gene expression though transcription mechanisms.

In-depth understanding of genome and gene structural dynamics is of importance in genome assembly and annotation.

3.5.2 Genome Annotation

The keys in genome annotation are gene-finding and information integration of experimental evidence and, to a lesser extent, repeat identification. For LSS, these are always handled with automated software tools and organized in databases. Table 3 lists some of the major sequence, genome, and annotation databases as well as their relevant information.

Since sequence data are being accumulated much faster than experimental evidence, development of *ab initio* gene-finding software becomes a major research activity for bioinformatics. Table 4

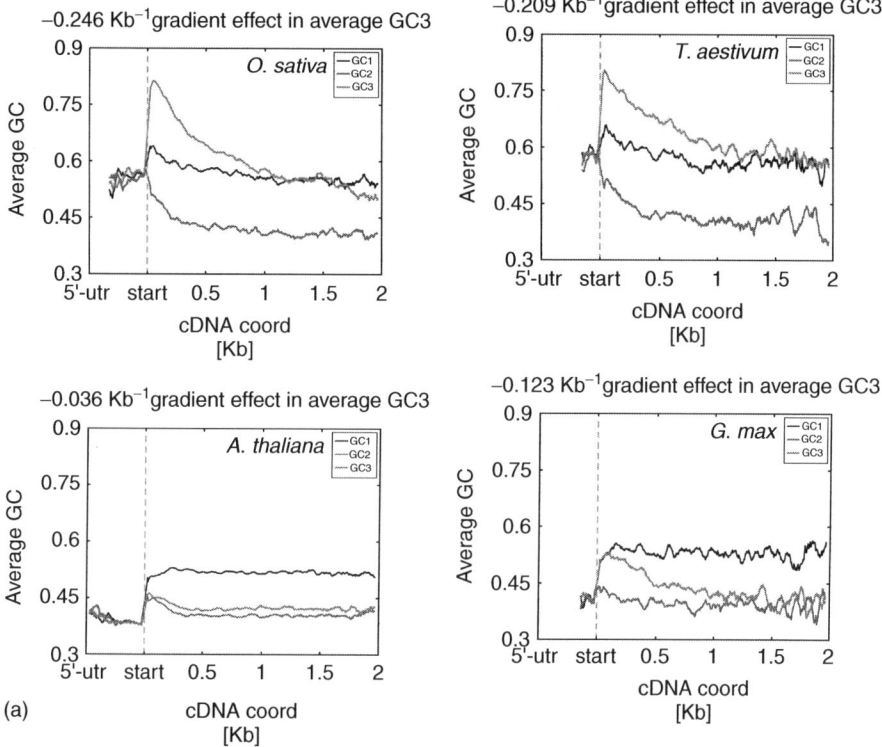

Fig. 11 GC-content increases in cDNAs of plants and animal genomes. GC1, GC2, and GC3 content are plotted as a function of cDNA position, relative to the start of the coding region, and averaged over all cDNAs with a 51-bp sliding window. Phase information is extended into the 5′-UTR. Eight examples are shown here: (a) from plants, O. sativa, T. aetivum, A. thaliana and G. max, and (b) from animals, H. sapiens, G. gallus, D. rerio, and X. tropicalis.

shows selected examples of such software packages, their methods, and immediate applications. Regardless of which mathematical model (Hidden Markov Models or neural network) is used and what parameters (sequence composition or signal to weight), a common misperception about gene finding is that a universal logarithm must be developed for all genomes. In addition to support from experimental evidence, two common factors that determine the performance of a gene-finding program are sensitivity (a measure of prediction that is equivalent to one minus the false-negative rate) and specificity (a measure of prediction that is equivalent to one minus the false-positive rate). These two parameters often obscure the possibility that there are classes of genes for which the performance is exceptionally better or worse, such as those in different size classes. It is necessary to consider how performance varies as a function of some continuous variables that describe the properties of a test gene set. Gene size as a variable is particularly important in predicting genes for animal genomes because most of them are accounted as gene

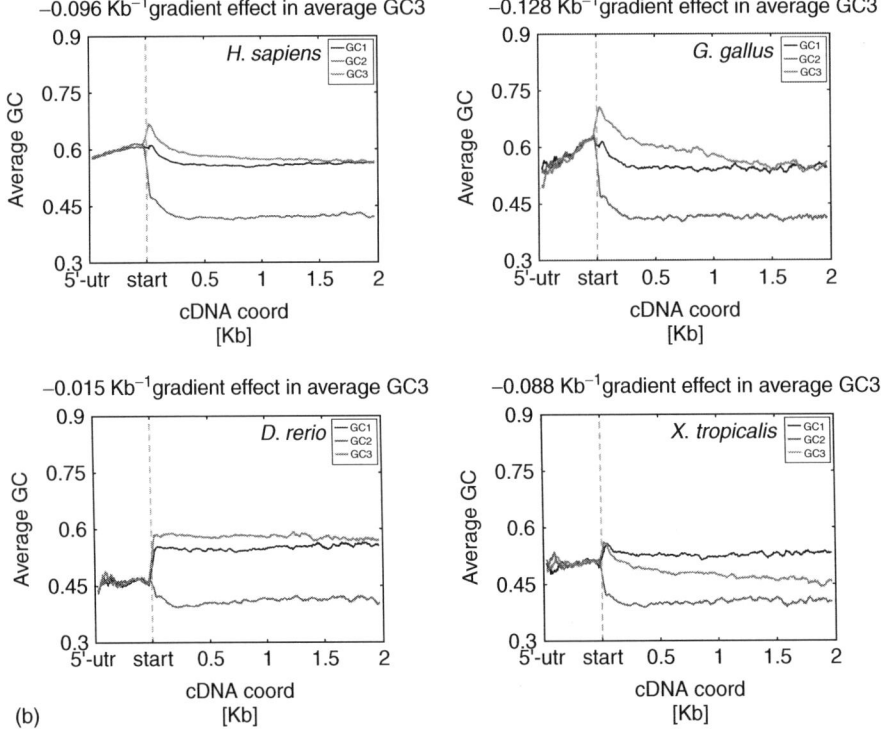

Fig. 11 (Continued)

spaces (including exons and introns, 1.2). Not only do mammalian genomes contain large genes that are megabases in size, but insect genomes too are known to have large genes across heterochromatin intervals. Therefore gene predictions should be carefully evaluated and interpreted based on the dynamic features of the target genome. Algorithms that succeed in analyzing one genome should not be blindly used in others, especially when they are by and large unrelated, such as (in an extreme case), between animal and plant genomes.

3.5.3 Sequence Variations

It is becoming increasingly common for a genome project to include an effort to generate sequence variation data through sequence sampling of wild counterparts (the chicken genome project did this in the opposite way, choosing the wild type, the red junglefowl, as the target), diverged breeds and closely related species. The study of sequence variation involves two basic steps: discovery in order to find candidate variations, and validation to acquire frequency information of verified variations in a background population. SGS provides a powerful procedure for the first step and could be used for the second in a rather limited way in consideration of cost differences between sequencing and genotyping. The cost of the latter is believed to be dropping at a much faster pace than that for sequencing.

Tab. 3 Major databases for sequence storage, genome information, and annotation.

Name	URL	Description
Sequence database		
GenBank	http://www.ncbi.nlm.nih.gov/	Nucleotide sequences
EMBL	http://www.ebi.ac.uk/embl/	Nucleotide sequences
DDBJ	http://www.ddbj.nig.ac.jp/	DNA Data Bank of Japan
SWISS-PROT	http://us.expasy.org/sprot/	Curated protein sequences
PIR	http://pir.georgetown.edu/	Protein sequences
Genome database		
ENSEMBL	http://www.ensembl.org/	A large collection of genomes
UCSC	http://www.genome.ucsc.edu/	A large collection of genomes
AceDB	http://www.acedb.org/	C. elegans
FlyBase	http://flybase.net/	D. melanogaster
SGD	http://www.yeastgenome.org/	S. cerevisiae
WormBase	http://www.wormbase.org/	Worm genome
GDB	http://www.gdb.org/	H. sapiens
RGD	http://rgd.mcw.edu/	R. norvegicus
MGD	http://www.informatics.jax.org/	M. musculus
TAIR	http://www.arabidopsis.org/	A. thaliana
BGI-RIS	http://rise.genomics.org.cn	O. sativa
Major databases for sequence annotation		
dbEST	http://www.ncbi.nlm.nih.gov/dbEST/	ESTs
RefSeq	http://www.ncbi.nlm.nih.gov/RefSeq/	Reference sequences
UniGene	http://www.ncbi.nlm.nih.gov/entrez/	Non-redundant gene clusters
dbSNP	http://www.ncbi.nlm.nih.gov/SNP/	SNPs
COG	http://www.ncbi.nlm.nih.gov/COG/	Cluster of Orthologous Groups of proteins
GO	http://www.geneontology.org/	Gene Ontology
InterPro	http://www.ebi.ac.uk/interpro/	Protein families and domains
Pfam	http://www.sanger.ac.uk/Software/Pfam/	Protein families
PRINTS	http://bioinf.man.ac.uk/dbbrowser/PRINTS/	Protein fingerprints
PROSITE	http://us.expasy.org/prosite/	Protein families and domains
ProDom	http://protein.toulouse.inra.fr/prodom/	Protein domain families
DIP	http://dip.doe-mbi.ucla.edu/	Database of Interacting Proteins
BIND	http://www.blueprint.org/bind/	Biomolecular interaction networks
KEGG	http://www.genome.jp/kegg/	Molecular interaction networks

3.5.4 Genome Evolution

The study of genome evolution in the Genomics Era has focused largely on lineage evolution (such as within vertebrates and fungi) in a vaguely defined context of comparative genomics, but nonetheless on the scale of genomes. Vertebrate genomes appear to be evolving to become more complex, as opposed to insect genomes that are evolving to become more diverged. Complexity has to build upon intricate networks of interacting genes or, more precisely, their products; these intricate networks are in turn defined functionally as molecular mechanisms and cellular processes. It would be of great interest to see how a new mechanism or a new process, created or modified in terms of new genes or

Tab. 4 A list of widely used *ab initio* gene-finding programs and the organisms applied.

Name	Method	Organism	Access
Glimmer	Interpolated Markov model	Microbes	http://www.tigr.org/software/glimmer/
Genie	General Hidden Markov model	Vertebrates	http://www.hgc.lbl.gov/projects/genie.html
GeneID	Perceptron, rules	Vertebrates	http://www.imim.es/GeneIdentification/Geneid/
FgeneSH	Hidden Markov model	Eukaryotes	http://www.softberry.com/
BGF	Hidden Semi Markov model	Eukaryotes	http://61.50.158.108:8080/GeneFinder/
GeneMark	5th-Markov	Eukaryotes	http://genemark.biology.gatech.edu/GeneMark/
GenScan	Semi Markov model	Vertebrates, plants, *C. elegans*	http://genes.mit.edu/GENSCAN.html
GenView	Linear combination	Vertebrates, diptera	http://www.itba.mi.cnr.it/webgene
GRAIL	Neural network	Vertebrates, Plants, *D. melanogaster*, and *E. coli*	http://compbio.ornl.gov/Grail-1.3

their definable variations over evolutionary timescales, results in building new organs, gaining new functions, and giving birth to new species.

As genomes of major domesticated crops and farm animals (oftentimes also their wild counterparts) are being sequenced, learning the rules that underlie many of the evolutionary changes in an artificial process (domestication) over a relatively short time period (10 000 years or so) should command the attention of many research teams, because of its scientific and economic importance. Ample collections of plant germplasms and animal breeds provide invaluable genomic resources, and domesticated animals can also serve as model organisms for human disease research.

Genome sequences evolve in both macro- and micro-scales. At the macro-scale, there are whole-genome duplications and subsequent chromosomal duplications, deletions, and rearrangements. At the micro-scale, there are gene duplications and deletions as well as enormous sequence variations within genes and gene families, and all are complicated by transposable element insertions over a variable evolutionary timescale. Genome evolution should be studied in a much broader context than what has been defined by the classic field of molecular evolution.

4
Other Useful Resources for SGS

The ultimate goal of a genome-sequencing project is to be able to present a whole picture of a genome with all genes positioned on its chromosomes. In the pursuit of this goal, one faces at least two major challenges: identifying all the genes, and ordering them. Algorithms for *ab initio* gene identification are not yet mature, as proved by the fact that the numerous versions of such software have had only moderate success thus far. Under current circumstances, independent data are oftentimes required for a better annotation of the WG-SGS data; these include ESTs and cDNAs for validating the presence of gene transcripts, physical and genetic maps for gene ordering and referencing to inheritable traits, and comparative analysis to exploit evolutionary inference.

4.1
ESTs and Full-length cDNA

When compared with a WG-SGS project, EST sequencing requires little effort. For instance, to cover a 500-Mb genome once, it takes one million reads. To acquire significant coverage for a transcriptome for a given metazoan cell, 10 000 reads ought to be enough. It is worthwhile to allot 1 to 10% of the budget of a genome project to this effort, either as part of the main project or independent of it. As an example, the Sino-Danish Swine Genome Project has initiated a "one-million EST" Project, allowing researchers to acquire powerful transcriptomic data from one hundred different tissues and cell types, nearly 10 000 ESTs from each. Table 5 summarizes the top 20 EST collections from the public databases.

Full-length cDNA data are superior, but require much expertise to generate. It involves making full-length cDNA libraries, sequencing cDNA clones that are longer than typical sequencing reads, strenuous management inputs to follow up each clone, and reiterating the libraries for novel clones. Yoshihide Hyashizaki's group at RIKEN and Shoshi Kicuchi's group at NIAS, Japan, have done excellent jobs on

Tab. 5 A list of top 20 EST collections from selected species.

Species	# of ESTs	# of UniGene entries
H. sapiens	5 657 997	106 934
M. musculus	4 235 142	76 298
R. norvegicus	661 663	39 050
O. sativa	284 007	33 563
T. aestivum	559 149	24 735
B. Taurus	467 943	24 195
X. laevis	434 889	23 754
S. scrofa	328 573	21 947
D. rerio	532 545	20 978
A. thaliana	322 641	20 783
G. gallus	495 089	20 155
C. elegans	298 805	15 942
C. familiaris	154 720	15 665
X. tropicalis	423 107	14 632
Z. mays	415 211	14 179
O. mykiss	157 996	14 178
C. intestinalis	684 280	14 098
A. gambiae	150 042	14 085
D. melanogaster	382 439	13 928

Note: This list is ordered by the total number of UniGene entries/ESTs by 2004-8-20.

this front to have curated adequate among of human, mouse, and rice full-length cDNAs. Full-length cDNAs do provide irreplaceable resources for gene identification and sequence quality validation, and thus demand independent efforts.

4.2
Physical and Genetic Maps

The mapping of information, whether physical (other than sequence maps that are, strictly speaking, also a form of physical map) or genetic, is always a plus point for WG-SGS. Physical maps allow the assembled sequence and genes therein to be referenced to the chromosomes in physical order, and genetic maps link genes associated with inheritable traits to a physical location on a chromosome through polymorphic sequence sites. For large genomes in the size range over a billion basepairs, such information becomes essential. Another yet-to-be tackled challenge is to map genes on small chromosomes that have no obvious chromosome banding patterns, such as the microchromosomes in the chicken genome and numerous minute chromosomes in the genomes of crabs and shrimps. High-quality WG-SGS is capable of producing an assembled genome sequence map within one to a few scaffolds, which can be mapped physically by several methods, such as different versions of FISH and large-insert clone-based physical maps. Since both mapping processes are expensive, time-consuming, and resource-demanding, they can be constructed retrospectively after a WG-SGS, over a period of time. Nevertheless, sequence data, nevertheless, provide ample candidate markers from genes, microsatellites, and even SNPs, for mapping activities. Physical mapping methods are essentially based on three laboratory techniques. The most common technique is restriction digests of large-insert clones, or clone fingerprinting. It relies on redundancy and precise fragment size measurement to determine overlapping clones, and thus achieve ordering. The second technique is large-insert clone-based STS mapping. It leverages PCR technology and calls overlap according to whether the large-insert clones share short unique sequences in a given genome. The third technique is hybridization-based, in which short oligonucleotide sequences (also termed as sequence-tagged connectors generated from end sequences of large-insert clones) serve as probes to tag clones that are gridded in a very high density on reusable nylon filters.

Genetic maps are always needed for genetic and functional studies, where the relationship of each locus is linked by the frequencies of a recombination event as measured in pedigrees or populations. For SGS, genetic maps provide guidance in validating physical maps and sequence assembly, and SGS provides the most powerful tool for sequence polymorphism discovery.

4.3 Comparative Genome Analyses

Finally, one must emphasize the importance of comparative analyses. The NIH initiative of Mammalian Gene Collection (MGC) provides a precedent to direct focus towards this aspect. Representative mammal species from distinct taxonomic groups are all lined up in various sequencing pipelines, if not already sequenced. This is also being extended to other lower vertebrates and species in the root of chordates. Such comparative and systematic studies offer profound and broad information for not only genome annotation but also evolutionary studies on functional aspects of genes, gene families, and gene networks, thereby guiding molecular biologists to explore basic molecular mechanisms among all life forms. The next wave of genome sequencing will unquestionably go deeper and broader into taxa of the extant life forms, prioritized by economic and scientific interests. Comparative genomics offers a guidepost for targets of choice in genome sequencing, deviating from human-centric to species-focused.

5 Perspectives

Looking into the future of SGS, it is not only genomicists who have reason to be excited but also molecular biologists, and even field biologists in all disciplines. Genome sequencing is now well past its infancy and out of its crib. It is exploring new territories, checking out its skills, and learning all the time as it grows. Genomics has already broken out of its original definition and morphed into genome biology or systems biology without losing scale or momentum. What are some of the immediate challenges it has to face?

5.1 Large Genomes

Given the increasing requirements for sequence information, the need to tackle larger genomes is inevitable. The current records for finished genome projects on the basis of genome sizes are the human genome (among animal genomes) and rice (among plant genomes). Can WG-SGS strategy be applied to animal genomes larger than the human genome, or to plant genomes larger than that of rice? The answer is obviously yes but not without caveats. Genomes of different taxonomic groups vary quite significantly (Fig. 5). For instance, the genomes of mammals deviate very minimally from the mean size, about 3000 Mb, and are therefore suitable for simple WG-SGS. The next large plant genome to be sequenced may perhaps be the maize genome that is 2500 Mb in size. It is debatable, however, whether the wheat or barley genomes should be sequenced by the WG-SGS approach, because they pose an unprecedented challenge for sequence assembly on account of their repetitive contents (over 90%), their large sizes, and polyploid nature (in the case of wheat alone). Large insect and crustacean genomes, such as those of grasshoppers and shrimps, and large protist genomes that have been largely untouched so far, should all be carefully evaluated before placed into a sequencing pipeline.

5.2
Genomes with High Repetitive Sequence Content

The repeat content, especially that of BDRs, varies widely among different taxonomic groups. Most notable are the plant genomes, some of which may have a repeat content as high as 95%. It is not a severe problem when the size of the target genome remains small, because the gene content can be put together easily and the cost is usually low. A balance between demands and costs is achievable within a threshold (say for genomes sized less than 3000 Mb). Large genomes in a size range of a few thousands of Mb have to be handled with great caution. Here, WG-SGC would not be suitable because of its high cost, rendering over 90% of the investment ineffective. Despite the fact that some of the gene content can be enriched experimentally by different methods (such as taking advantage of differential methylation between the gene and repeat contents among plant genomes), the goal of a dedicated genome project is to acquire a complete map of all its genes. In this regard, prioritization becomes a key for acquiring other resources (4) that are of great assistance in sequence assemblies and analyses over a relative prolonged period of time, in comparison with an ordinary genome project.

5.3
Polyploidy: Ancient and New

Polyploidy is very common among the extant species, some of it being ancient (formed naturally), and some recent (formed either naturally or artificially). A straightforward consideration for sequencing targets is to choose a diploid species (preferably the wild counterpart or the ancestor species) rather than a polyploid one, especially in the cases of crop species. However, in most cases, the preference is for the crop species to be sequenced directly first. In an ideal situation, the wild counterpart, the diploid species that gave rise to the modern polyploid crop, and the crop itself should all be sequenced in series. When there is no option, the genome of palepolyploid species should be evaluated first by physical mapping or sequence sampling for sequence divergence and feasibility of whole-genome assembly before a WG-SGS project is launched.

5.4
Mixed and Meta Genomes

In theory, when a mixture of multiple genomes is to be sequenced, it can be regarded as one organism with multiple chromosomes, perhaps in different ratios. For instance, genomes of a microbial community indigenous to a certain habitat are often referred to as *metagenomes*; examples of such habitats are the oral cavity, portions of a ruminant stomach, and even drinking water. However, microbes residing in the same environment are not guaranteed to propagate with the same speed and to respond to nutrient changes in similar ways. As a common characteristic under a defined circumstance, a microbial group is dominated by a limited number of species and contaminated by a large number of minor ones. Therefore, in an SGS project that utilizes DNA extracted from a meta-organism, genomes of the dominant species would be easily assembled, whereas those of the minor ones would have little chance of reaching full coverage. Unless novel strategies and techniques are developed, such an endeavor provides only a survey of the metagenome and complete sequences of its few dominant members.

Acknowledgment

We apologize to our readers and colleagues for having only a limited number of citations from a recent explosion of genomic literature due to space limits. We would like to thank Ms. Wei Gong and Jing Wang for their valuable assistance in preparing the figures and tables, and Mr. Xin Zhang for proofreading the manuscript. The authors are supported by grants from National Natural Science Foundation (No. 30270748), Ministry of Science and Technology (No. 2002AA229021, 863-306-ZD11-04-1) and Chinese Academy of Sciences (KSCX2-SW-223).

See also Molecular Basis of Genetics; Construction and Applications of Genomic DNA Libraries; Genomic Sequencing; Whole Genome Human Chromosome Physical Mapping.

Bibliography

Books and Reviews

Clayton, J., Dennis, C. (2003) *50 years of DNA*, Palgrave Macmillan, Hampshire, England.
Collins, F.S., Green, E.D., Guttmacher, A.E., Guyer, M.S. (2003) A vision for the future of genomics research-A blueprint for the genomic era, *Nature* **422**, 835–847.
Dulbecco, R. (1986) A turning point in cancer research: sequencing the human genome, *Science* **231**, 1055–1056.
Marshall, E. (1995) A strategy for sequencing the genome 5 years early, *Science* **267**, 783–784.
Olson, M.V. (1993) The human genome project, *Proc. Natl. Acad. Sci. U.S.A.* **90**, 4338–4344.
Olson, M.V. (1995) A time to sequence, *Science* **270**, 394–396.
Olson, M.V. (2001) The maps. Clone by clone by clone, *Nature* **409**, 816–818.
Sulston, J., Ferry, G. (2002) *The Common Thread: A Story of Science, Politics, Ethics and the Human Genome*, Joseph Henry Press, United States.
Waterston, R.H., Lander, E.S., Sulston, J.E. (2002) On the sequencing of the human genome, *Proc. Natl. Acad. Sci. U.S.A.* **99**, 3712–3716.
Watson, J.D. (1990) The human genome project: past, present, and future, *Science* **248**, 44–49.

Primary References

Altschul, S.F., Gish, W., Miller, W., Myers, E.W., Lipman, D.J. (1990) Basic local alignment search tool, *J. Mol. Biol.* **215**, 403–410.
Adams, M.D., Celniker, S.E., Holt, R.A., Evans, C.A., Gocayne, J.D., Amanatides, P.G., Scherer, S.E., Li, P.W., Hoskins, R.A., Galle, R.F., et al. (2000) The genome sequence of *Drosophila melanogaster*, *Science* **287**, 2185–2195.
Aparicio, S., Chapman, J., Stupka, E., Putnam, N., Chia, J.M., Dehal, P., Christoffels, A., Rash, S., Hoon, S., Smit, A., et al. (2002) Whole-genome shotgun assembly and analysis of the genome of *Fugu rubripes*, *Science* **297**, 1301–1310.
Arratia, R., Lander, E.S., Tavare, S., Waterman, M.S. (1991) Genomic mapping by anchoring random clones: a mathematical analysis, *Genomics* **11**, 806–827.
Blattner, F.R., Plunkett, G. III, Bloch, C.A., Perna, N.T., Burland, V., Riley, M., Collado-Vides, J., Glasner, J.D., Rode, C.K., Mayhew, G.F., et al. (1997) The complete genome sequence of *Escherichia coli* K-12, *Science* **277**, 1453–1474.
Burke, D.T., Carle, G.F., Olson, M.V. (1987) Cloning of large segments of exogenous DNA into yeast by means of artificial chromosome vectors, *Science* **236**, 806–812.
Celniker, S.E., Wheeler, D.A., Kronmiller, B., Carlson, J.W., Halpern, A., Patel, S., Adams, M., Champe, M., Dugan, S.P., Frise, E., et al. (2002) Finishing a whole-genome shotgun: release 3 of the *Drosophila*

melanogaster euchromatic genome sequence, *Genome Biol.* **3**, RESEARCH0079.

Cohen, S.N., Chang, A.C., Boyer, H.W., Helling, R.B. (1973) Construction of biologically functional bacterial plasmids in vitro, *Proc. Natl. Acad. Sci. U.S.A.* **70**, 3240–3244.

Coulson, A., Sulston, J., Brenner, S., Karn, J. (1986) Toward a physical map in the genome of the nematode *Caenorhabditis elegans*, *Proc. Natl. Acad. Sci. U.S.A.* **83**, 7821–7825.

Dehal, P., Satou, Y., Campbell, R.K., Chapman, J., Degnan, B., De Tomaso, A., Davidson, B., Di Gregorio, A., Gelpke, M., Goodstein, D.M., et al. (2002) The draft genome of *Ciona intestinalis*: insights into chordate and vertebrate origins, *Science* **298**, 2157–2167.

Ewing, B., Green, P. (1998) Base-calling of automated sequencer traces using phred. II. Error probabilities, *Genome Res.* **8**, 186–194.

Ewing, B., Hillier, L., Wendl, M.C., Green, P. (1998) Base-calling of automated sequencer traces using phred. I. Accuracy assessment, *Genome Res.* **8**, 175–185.

Galagan, J.E., Calvo, S.E., Borkovich, K.A., Selker, E.U., Read, N.D., Jaffe, D., Fitz-Hugh, W., Ma, L.J., Smirnov, S., Purcell, S., et al. (2003) The genome sequence of the filamentous fungus *Neurospora crassa*, *Nature* **422**, 859–868.

Gardner, M.J., Hall, N., Fung, E., White, O., Berriman, M., Hyman, R.W., Carlton, J.M., Pain, A., Nelson, K.E., Bowman, S., et al. (2002) Genome sequence of the human malaria parasite *Plasmodium falciparum*, *Nature* **419**, 498–511.

Gibbs, R.A., Weinstock, G.M., Metzker, M.L., Muzny, D.M., Sodergren, E.J., Scherer, S., Scott, G., Steffen, D., Worley, K.C., Burch, P.E., et al. (2004) Genome sequence of the Brown Norway rat yields insights into mammalian evolution, *Nature* **428**, 493–521.

Goff, S.A., Ricke, D., Lan, T.H., Presting, G., Wang, R., Dunn, M., Glazebrook, J., Sessions, A., Oeller, P., Varma, H., et al. (2002) A draft sequence of the rice genome (*Oryza sativa* L. ssp. *japonica*), *Science* **296**, 92–100.

Goffeau, A., Barrell, B.G., Bussey, H., Davis, R.W., Dujon, B., Feldmann, H., Galibert, F., Hoheisel, J.D., Jacq, C., Johnston, M., et al. (1996) Life with 6000 genes, *Science* **274**, 546–567.

Gordon, D., Abajian, C., Green, P. (1998) Consed: a graphical tool for sequence finishing, *Genome Res.* **8**, 195–202.

Green, E.D., Olson, M.V. (1990) Systematic screening of yeast artificial-chromosome libraries by use of the polymerase chain reaction, *Proc. Natl. Acad. Sci. U.S.A.* **87**, 1213–1217.

Gregory, S.G., Sekhon, M., Schein, J., Zhao, S., Osoegawa, K., Scott, C.E., Evans, R.S., Burridge, P.W., Cox, T.V., Fox, C.A., et al. (2002) A physical map of the mouse genome, *Nature* **418**, 743–750.

Holt, R.A., Subramanian, G.M., Halpern, A., Sutton, G.G., Charlab, R., Nusskern, D.R., Wincker, P., Clark, A.G., Ribeiro, J.M., Wides, R., et al. (2002) The genome sequence of the malaria mosquito *Anopheles gambiae*, *Science* **298**, 129–149.

Kirkness, E.F., Bafna, V., Halpern, A.L., Levy, S., Remington, K., Rusch, D.B., Delcher, A.L., Pop, M., Wang, W., Fraser, C.M., et al. (2003) The dog genome: survey sequencing and comparative analysis, *Science* **301**, 1898–1903.

Lander, E.S., Waterman, M.S. (1988) Genomic mapping by fingerprinting random clones: a mathematical analysis, *Genomics* **2**, 231–239.

Martinez, D., Larrondo, L.F., Putnam, N., Gelpke, M.D., Huang, K., Chapman, J., Helfenbein, K.G., Ramaiya, P., Detter, J.C., Larimer, F., et al. (2004) Genome sequence of the lignocellulose degrading fungus *Phanerochaete chrysosporium* strain RP78, *Nat. Biotechnol.* **22**, 695–700.

Matsuzaki, M., Misumi, O., Shin-I, T., Maruyama, S., Takahara, M., Miyagishima, S.Y., Mori, T., Nishida, K., Yagisawa, F., Nishida, K., et al. (2004) Genome sequence of the ultrasmall unicellular red alga *Cyanidioschyzon merolae* 10D, *Nature* **428**, 653–657.

Maxam, A.M., Gilbert, W. (1977) A new method for sequencing DNA, *Proc. Natl. Acad. Sci. U.S.A.* **74**, 560–564.

Messing, J., Crea, R., Seeburg, P.H. (1981) A system for shotgun DNA sequencing, *Nucleic Acids Res.* **9**, 309–321.

Mita, K., Kasahara, M., Sasaki, S., Nagayasu, Y., Yamada, T., Kanamori, H., Namiki, N., Kitagawa, M., Yamashita, H., Yasukochi, Y., et al. (2004) The genome sequence of silkworm, *Bombyx mori*, *DNA Res.* **11**, 27–35.

Myers, E.W., Sutton, G.G., Delcher, A.L., Dew, I.M., Fasulo, D.P., Flanigan, M.J., Kravitz, S.A., Mobarry, C.M., Reinert, K.H., Remington, K.A., et al. (2000) A whole-genome assembly of Drosophila, *Science* **287**, 2196–2204.

Olson, M., Hood, L., Cantor, C., Botstein, D. (1989) A common language for physical mapping of the human genome, *Science* **245**, 1434–1435.

Port, E., Sun, F., Martin, D., Waterman, M.S. (1995) Genomic mapping by end-characterized random clones: a mathematical analysis, *Genomics* **26**, 84–100.

Riethman, H.C., Moyzis, R.K., Meyne, J., Burke, D.T., Olson, M.V. (1989) Cloning human telomeric DNA fragments into *Saccharomyces cerevisiae* using a yeast-artificial-chromosome vector, *Proc. Natl. Acad. Sci. U.S.A.* **86**, 6240–6244.

Riles, L., Dutchik, J.E., Baktha, A., McCauley, B.K., Thayer, E.C., Leckie, M.P., Braden, V.V., Depke, J.E., Olson, M.V. (1993) Physical maps of the six smallest chromosomes of *Saccharomyces cerevisiae* at a resolution of 2.6 kilobase pairs, *Genetics* **134**, 81–150.

Sanger, F., Coulson, A.R. (1975) A rapid method for determining sequences in DNA by primed synthesis with DNA polymerase, *J. Mol. Biol.* **94**, 441–448.

Sanger, F., Nicklen, S., Coulson, A.R. (1977) DNA sequencing with chain-terminating inhibitors, *Proc. Natl. Acad. Sci. U.S.A.* **74**, 5463–5467.

Sanger, F., Air, G.M., Barrell, B.G., Brown, N.L., Coulson, A.R., Fiddes, C.A., Hutchison, C.A., Slocombe, P.M., Smith, M. (1977) Nucleotide sequence of bacteriophage phi X174 DNA, *Nature* **265**, 687–695.

Shizuya, H., Birren, B., Kim, U.J., Mancino, V., Slepak, T., Tachiiri, Y., Simon, M. (1992) Cloning and stable maintenance of 300-kilobase-pair fragments of human DNA in *Escherichia coli* using an F-factor-based vector, *Proc. Natl. Acad. Sci. U.S.A.* **89**, 8794–8797.

Smit, A.F.A., Green, P. RepeatMasker [http://ftp.genome.washington.edu/RM/RepeatMasker.html]

Smith, M., Brown, N.L., Air, G.M., Barrell, B.G., Coulson, A.R., Hutchison, C.A. III, Sanger, F. (1977) DNA sequence at the C termini of the overlapping genes A and B in bacteriophage phi X174, *Nature* **265**, 702–705.

Smith, L.M., Sanders, J.Z., Kaiser, R.J., Hughes, P., Dodd, C., Connell, C.R., Heiner, C., Kent, S.B.H., Hood, L.E. (1986) Fluorescence detection in automated DNA sequence analysis, *Nature* **321**, 674–679.

Stein, L.D., Bao, Z., Blasiar, D., Blumenthal, T., Brent, M.R., Chen, N., Chinwalla, A., Clarke, L., Clee, C., Coghlan, A., et al. (2003) The genome sequence of *Caenorhabditis briggsae*: a platform for comparative genomics, *PLoS Biol.* **1**, E45.

The Arabidopsis Genome Initiative. (2000) Analysis of the genome sequence of the flowering plant *Arabidopsis thaliana*, *Nature* **408**, 796–815.

The *C. elegans* Sequencing Consortium. (1998) Genome sequence of the nematode *C. elegans*: a platform for investigating biology. *Science* **282**, 2012–2018.

The International Human Genome Sequencing Consortium. (2001) Initial sequencing and analysis of the human genome, *Nature* **409**, 860–921.

The Mouse Genome Sequencing Consortium. (2002) Initial sequencing and comparative analysis of the mouse genome, *Nature* **420**, 520–562.

Vieira, J., Messing, J. (1982) The pUC plasmids, an M13MP7-derived system for insertion mutagenesis and sequencing with synthetic universal primers, *Gene* **19**, 259–268.

Venter, J.C., Adams, M.D., Myers, E.W., Li, P.W., Mural, R.J., Sutton, G.G., Smith, H.O., Yandell, M., Evans, C.A., Holt, R.A., et al. (2001) The sequence of the human genome, *Science* **291**, 1304–1351.

Whitfield, C.W., Band, M.R., Bonaldo, M.F., Kumar, C.G., Liu, L., Pardinas, J.R., Robertson, H.M., Soares, M.B., Robinson, G.E. (2002) Annotated expressed sequence tags and cDNA microarrays for studies of brain and behavior in the honey bee, *Genome Res.* **12**, 555–566.

Wong, G., Yu, J., Thayer, E.C., Olson, M.V. (1997) Multiple-complete-digest restriction fragment mapping: generating sequence-ready maps for large-scale DNA sequencing, *Proc. Natl. Acad. Sci. U.S.A.* **94**, 5225–5230.

Wong, G.K.-S., Passey, D.A., Huang, Y.Z., Yang, Z.Y., Yu, J. (2000) Is "junk" DNA mostly intron DNA? *Genome Res.* **10**, 1672–1678.

Wong, G.K.-S., Wang, J., Tao, L., Tan, J., Zhang, J.G., Passey, D.A., Yu, J. (2002) Compositional gradients in *Gramineae* genes, *Genome Res.* **12**, 851–856.

Wu, R. (1970) Nucleotide sequence analysis of DNA. I, Partial sequence of the cohesive ends of bacteriophage lamda and 186 DNA, *J. Mol. Biol.* **51**, 501–521.

Wu, R., Taylor, E. (1971) Nucleotide sequence analysis of DNA. II, Complete nucleotide sequence of the cohesive ends of bacteriophage lambda DNA, *J. Mol. Biol.* **57**, 491–511.

Yu, J., Hu, S., Wang, J., Wong, G.K., Li, S., Liu, B., Deng, Y., Dai, L., Zhou, Y., Zhang, X., et al. (2002) A draft sequence of the rice genome (*Oryza sativa* L. ssp. *indica*), *Science* **296**, 79–92.

23
Whole Genome Human Chromosome Physical Mapping

Cassandra L. Smith, Giang H. Nguyen, Denan Wang, and Nickolaev Bukanov
Boston University, MA, USA

1	Introduction	681
2	Genomic Mapping Terms and Concepts	682
3	Evolving Approaches for Physical Mapping	683
4	Genomic Restriction Maps	683
5	Genomic Clone Libraries	687
6	Cloneless Genomic Libraries	688
7	Gap Closure	690
8	Prospectus	690

Bibliography 691
Books and Reviews 691
Primary Literature 691

Keywords

Alu Repeat
A 300-bp DNA interspersed repeat sequence that occurs about once every 5000 bp. The most common repeat element in the human genome, it is CG rich and appears to occur preferentially in light Giemsa bands on condensed chromosomes.

Genomics and Genetics. Edited by Robert A. Meyers.
Copyright © 2007 Wiley-VCH Verlag GmbH & Co. KGaA, Weinheim
ISBN: 978-3-527-31609-0

Bacterial Artificial Chromosome (BAC)
An artificial chromosome cloning system in *E. coli* based on elements of the native circular chromosome.

Cloneless Library
A library of genomic fragments consisting of gel slices of electrophoretically fractionated DNA.

Comparative Genomic Hybridization (CGH)
A molecular cytogenetic method capable of detecting and locating relative genomic sequence copy number differences between pairs of DNA samples. Equal concentrations of differentially labeled DNAs are hybridized simultaneously to metaphase chromosome spreads. Regions with deleted or amplified DNA sequences are seen as changes in the ratio of the intensities of these two labels along the target chromosome.

Expressed Sequence Tag (EST)
A segment of a sequence from cDNA clone that corresponds to an mRNA. Mapping ESTs helps make maps and points directly to expressed genes.

Fluorescence *In Situ* Hybridization (FISH)
Hybridization of nucleic acid probes to chromosomes immobilized on microscope slides or filters.

Inter-*Alu* PCR
A polymerase chain reaction method that uses primers contained in *Alu* repeat element to amplify single-copy sequences between adjacent *Alu* elements.

Jumping Libraries
A library of clones containing DNA sequences from the ends of the same restriction fragment.

Kpn Repeat
The second most commonly occurring repeat sequence in the human genome, occurring on average once every 50 000 bp. It is a relatively AT-rich, LINE repeat and occurs preferentially in dark Giemsa bands on condensed chromosomes.

Linking Libraries
Libraries of clones containing DNA sequences that span a selected restriction enzyme recognition site.

Locus
A region on a chromosome linked to a functional unit such as a gene, telomere, centromere, and replication origin.

Long Interspersed Repeat Elements (LINEs)
Long interspersed repeating sequences that appear to be similar to retroposons.

Multiplex Analysis
An analytical approach to increase throughput by collecting multiple data simultaneously.

P1-derived Artificial Chromosomes (PAC)
An *E. coli* bacteriophage cloning system used for cloning of fragments up to 150 kb in size.

Polonies
Isolation of DNA sequences on beads rather than in clones.

Polymerase Chain Reaction (PCR)
A method for amplifying DNA by alternatively denaturing double-stranded DNA, annealing pairs of primers located near each other on complementary strands, and synthesizing the DNA between the primers using DNA polymerase.

Polymorphism Link-up
A mapping approach that establishes continuity between restriction fragments by taking advantage of the naturally occurring polymorphism in different DNAs. In some cell lines, hybridized probes appear to identify different fragments, whereas in others they appear to identify the same fragment. The pattern of occurrence of these fragments can be used to assess whether the probes identify the same or adjacent fragments.

Pulsed Field Gel (PFG) Electrophoresis
A method of electrophoresis that exposes nucleic acids to alternating electrical fields. Fractionation is based on the speed at which the molecules can change directions.

Radiation Hybrid Maps
Radiation hybrid maps are created by analyzing randomly broken DNA cloned into hybrid cell lines. The breakage is done using X rays and the centiRay distance is a function of how often two loci remain together in a library of radiation hybrid cell lines.

RARE (RecA-assisted Restriction Endonuclease) Cleavage
A site-specific DNA cleavage method based on the ability of RecA protein from *E. coli* to pair an oligonucleotide to its homologous sequence in duplex DNA. This three-stranded DNA complex is protected from methylase. After methylation and removal of RecA, restriction endonuclease cleavage is limited to the site previously protected from methylation. If pairs of oligonucleotides are used, a specific fragment can be cleaved out of genomes.

Retroposons
A group of DNA sequence elements that appear to transpose through an RNA intermediate. Retroposon elements do not code for reverse transcriptase, do not have terminally redundant sequences, and do have a 39-poly(A)$_n$ stretch. A variable-sized target duplication occurs at the site of integration.

Sequence-tagged Restriction Site (STAR)
A short DNA sequence used to identify the DNA surrounding a restriction nuclease cleavage site.

Sequence-tagged Site (STS)
A short DNA sequence used to identify a DNA segment.

Short Interspersed Repeat Elements (SINEs)
Repeated sequences less than 500-bp long and present in a high copy number (10^5 per human genome). The *Alu* repeat element consists of SINEs.

Yeast Artificial Chromosome (YAC)
An artificial yeast chromosome constructed by cloning genomic fragments into vectors that can replicate in yeast. YACs have the following characteristics: a yeast centromere, two telomeric sequences, and a selectable marker.

■ This entry reviews the types of physical map that can be constructed, the methods used to construct them and their most likely current use. In particular, the emphasis is on the construction of genomic restriction maps and ordering overlapping libraries using top-down mapping approaches to enable *de novo* genome sequencing and for gap filling in sequencing projects. Also explored is the use of these approaches in functional studies.

1 Introduction

In the past, most genome studies were limited to organisms having well-developed genetic systems. Thus, a few model systems were studied intensely (e.g. bacteriophage lambda, *Eschericia coli*, Drosophila, mice) in a large number of different ways. The accumulation of diverse knowledge on these model organisms further promoted research on these model systems. Now, a number of molecular approaches have been developed that allow the construction of physical maps for virtually any chromosome. This means that analysis of an uncharacterized organism can begin with physical dissection including sequencing. Knowledge gained from this initial foray provides the foundation for functional studies. Hence, a new synergism in biology is provided by bottom-up genome studies.

Bacterial genomes are small and range from ~1 to 15 Mb in size, reflecting in part, the relative ability of these organisms to be free living. The entire genomic sequence of an increasing number of bacterial genomes is becoming available. These genomes are small enough so that sequencing of random clones is used to obtain the entire genome sequence. Hence, physical genomic maps for these organisms are used to study global gene expression or other global genomic activities.

In the past, the study of some lower eukaryotic genomes such as protozoa was hampered by the lack of well-developed genetic systems. Also, these chromosomes did not condense during cell division and could not be visualized microscopically. The application of pulsed field gel electrophoresis (PFG) to examine the genetic make up of these organisms was quite useful. PFG separates DNA chromosomal DNAs or large fragments that range up to ~10 Mb in size. Protozoan genomes range in size from the length that overlaps large bacterial genomes, up to about 100 Mb. Surprisingly, one parasitic protozoan genome, the *Giardia lamblia* genome, was found to be only about 12 Mb in size (i.e. smaller than the *Saccharomyces* genome).

Until recently, large, complex genomes, like the human genome (~3000 Mb in size) have been particularly recalcitrant to molecular dissection. Human and other higher eukaryotic chromosomes are 50 to 300 Mb in size, condense and are visible microscopically. Differential staining of the condensed chromosomes allows a finer division into regions ("bands") estimated to be 5 to 30 Mb in size that appear to reflect regional differences in GC content. The banding patterns serve as anchors to a large amount of genomic data. Also, the division of the genome into chromosomes and chromosomal bands provides convenient pieces for top-down mapping approaches.

In top-down mapping approaches, the genome is divided into units to facilitate study. For instance, chromosomes naturally divide the genome, and chromosome-banding patterns represent another division. Further division will depend on the method of analysis. Conventional recombinant DNA approaches to genome analysis allowed the characterization of molecules up to only about 0.05 Mb. The "resolution gap," 0.05 to 10 Mb was exactly the size range that was most amenable to study by PFG techniques and the entire size range of DNA molecules could be analyzed. PFG also promoted the extension of recombinant DNA methods to large cloning system such as yeast artificial chromosomes (YACs), P1 artificial chromosomes (PACs), and bacterial artificial chromosomes (BACs).

2
Genomic Mapping Terms and Concepts

Many molecular techniques with varying degrees of resolution can be used to characterize chromosomes. Hence, there is no unifying concept of what constitutes a chromosomal physical map. Further, there are ambiguities in the literature with respect to the distinction of physical versus genetic maps.

Any map will consist of markers or objects. The order, and perhaps the distance, between pairs of objects will be known (some might argue that a map must consist of both order and distance). Object ordering along a chromosome should be maintained irrespective of the method used to construct the map, whereas map distances are method dependent. For instance, the amount of recombination along a chromosome is not constant. Hence, it is not surprising that a comparison of the physical and genetic distances along the long arm of chromosome 21 revealed at least a sixfold variation in the distances. It is quite clear that the ultimate map is the entire sequence of a chromosomal DNA. All maps and objects will be anchored to the DNA sequence once it is available. Some imprecision will always exist because map objects may be imprecisely defined, or have imprecisely defined locations by their very nature.

Classically, a genetic map was composed of chromosomal loci and the amount of recombination between the loci was the genetic distance. For eukaryotic organisms, genetic distance, expressed in centimorgans (cM), is a measure of the coinheritance of genetic markers. This type of analysis requires the examination of two loci in multiple generations. In bacteria like *E. coli*, recombination is measured as the time of transfer and integration of DNA from one cell into the chromosome of another cell.

In some instances, the placement of genes on various physical maps has been referred to as a *genetic map* (i.e. "a gene map"), although only physical distances and locations were known. Cytogenetic maps lead to further confusion in terminology. In these maps, the presence or absence of a gene or map object is correlated with an observable genomic location (i.e. a chromosome band).

Another concept whose use results in some confusion is "locus." Classically, this term defined the location genes (i.e. "genetic locus"). More generally, however, a chromosomal locus, representing a location on a chromosome, can consist of objects other than genes. Examples include DNA sequences, sequenced tagged sites (STS), expressed sequenced tags (ESTs), probe sites, restriction enzyme sites, clone sites, centromeres, telomeres, and chromosomal breakpoints (such as those that occur naturally or are induced by ultraviolet breakage). To add to the confusion, new genetic markers based on anonymous DNA sequences now define genetic loci because they are used in genetic mapping experiments. In this entry, a *chromosomal locus* is any chromosomal location that has been identified in a distinctive manner.

Here, a physical map is considered to be any map consisting of objects that have been located by physical rather than by genetic methods. Thus, a physical map can consist of objects located along the chromosome, such as a chromosomal band, a breakpoint, a genomic restriction fragment, and the location of a clone on a chromosome. Here, the ordering of restriction fragments and clones will be used. These maps are constructed using molecular methods.

3
Evolving Approaches for Physical Mapping

The term "physical maps" has been used to describe both genomic restriction maps and genomic clone libraries. Low-resolution whole-genome restriction maps are created by ordering fragments that have been fractionated by size electrophoretically. These maps have been created by hybridization experiments using cloned sequences as probes or by PCR (polymerase chain reaction) testing for the presence of specific STSs/ESTs.

Another physical type of map is a radiation hybrid map. In this case, randomly fragmented DNA is cloned to create a hybrid cell line library. Different fragments are present in each hybrid cell line that is tested for the presence of a library of loci. The distance loci are reflected in the frequency of co-occurrence of two loci in the same cell line.

Also, genome restriction maps may be created by analysis of restriction sites contained on overlapping clones. The resolution of such restriction maps depends on the restriction enzyme used, and the frequency at which its recognition site occurs within the genomic DNA sample. Furthermore, an overlapping library may in itself represent a map consisting of ordered objects whose size can be approximated but not stated with certainty.

More efficient "cloneless" library approaches use PCR to analyze gel slices of lanes containing specific fragments. Here, the gel slices represent the cloneless library units that can be used in place of cloned DNA. In the future, cloneless approaches will likely replace time-consuming clone library construction and ordering.

4
Genomic Restriction Maps

The first step in creating a genomic restriction map is choosing the DNA source. For many organisms or chromosomes, this is obvious, since there is a well-characterized isolate or a cell line that may be useful. For most eukaryotic genomes, if available, some complications associated with the analysis of polymorphic diploid DNA can be avoided by using DNA from hybrid cell lines containing a single chromosome.

Usually, the genomic DNA is extracted and purified intact, in agarose to prevent shear damage. Small, 10-Mb chromosomal DNAs may be sized directly by PFG. Other chromosomal DNAs must be cleaved with a restriction enzyme before they are subjected to PFG analysis. The largest size standard for PFG electrophoresis is 6 Mb. Thus, analytical results use this limit, although enhanced versions of the technique allow the fractionation of molecules greater than 6 Mb.

Usually, maps are constructed using restriction enzymes that have large recognition sequences or a recognition site that occurs infrequently in the genome of interest. In some cases, it has been useful to test a battery of enzymes. The usefulness of a particular enzyme may be estimated roughly from the size of the site or the GC content of the test organism. Even so, these predictions are somewhat inaccurate because the genomic DNA sequences do not occur at random, whereas calculated occurrences assume randomness. Furthermore, the frequency of occurrence in known sequences may not be representative of the entire genome because in the past many molecular studies have focused on gene sequencing.

Besides restriction enzymes, a number of other enzymes or methods have been adapted for cutting genomic DNA into specific, large pieces. These methods usually depend on the formation of an unusual structure (e.g. a triplex or D-loop structure) having an associated single-stranded region. D-loops are formed with the aid of RecA protein at targeted sites. In one strategy, termed the *Achilles' heel* strategy, or RARE (RecA-assisted restriction endonuclease) cleavage, genomic DNA containing a D-loop is treated with DNA methylase. The methylase recognizes and modifies specific sequence except when the sites occur within the D-loop. The methylase is removed, and the DNA is then subjected to digestion with the cognate restriction enzyme. Only the restriction enzyme recognition site in the previously formed D-loop is now susceptible to cleavage by the restriction enzyme.

This type of approach allows for site-directed cleavage of genomic DNA at specific locations and has been used to map the end of human chromosomes, as only one RARE cleavage is required. The combination of two RARE reactions from nearby sites would produce an interstitial fragment whose length would be equivalent to the distance between the two sites. Furthermore, fragments smaller than 10 Mb in size could be purified from the remaining genomic DNA using PFG electrophoresis.

Genomic map construction is similar to putting a puzzle together. It is much easier if all the pieces are identified in advance. For small genomes, this is accomplished with a restriction enzyme that produces reasonable number of fragments resolvable by PFG analysis. In these experiments, all genomic DNA is visualized by simple ethidium bromide staining.

A method was developed for visualizing all the megabase restriction fragments for individual chromosomes from large genomes. This approach analyzes PFG-fractionated restriction fragments from genomic DNA that is obtained from monosomic hybrid cell lines. Here, a species-specific interspersed repetitive hybridization probe is used to identify the megabase restriction fragments. For example, the megabase restriction fragments from chromosome 20 are visualized with a human-specific *Alu* probe in Fig. 1. This approach works as long as the restriction enzyme digestion goes to completion. Otherwise, partial digestion products can obscure the complete and especially larger digestion fragments. For the human genome, the restriction enzymes *Not* I, *Sgr*A I, *Fse* I, and *Asc* I have this desired characteristic and cut the genome into fragments that appear to average 1 Mb in size. In higher eukaryotic organisms, partial cleavage by a restriction enzyme is due to partial CpG methylation of their recognition sites inhibiting cleavage.

Most organisms will have several interspersed repeats that are of different lengths. The most commonly occurring interspersed repeat in the human genome is the *Alu* repeat. This repeat is a short interspersed repeating element (SINE) that is estimated to occur about every 5000 bp. The second most commonly occurring interspersed human repeat is the *Kpn* repeat. The *Kpn* element, a LINE (long interspersed) repeat, is estimated to occur at a frequency tenfold less than that of the *Alu* repeat. Theoretically, these two repeats used as hybridization probes to hybrid cell line DNA cleaved with the appropriate enzyme should reveal all the

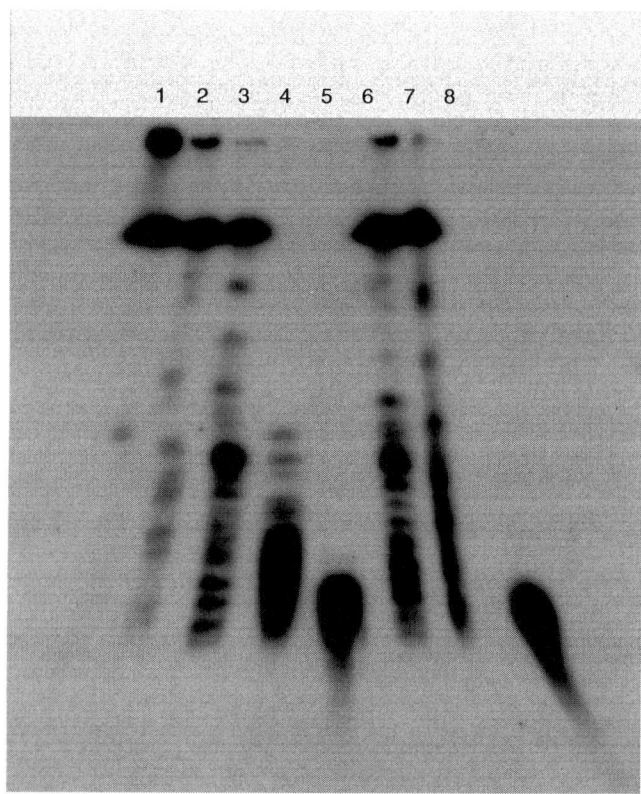

Fig. 1 Detection of PFG-fractionated restriction fragments of human chromosomes. DNA from a monosomic hybrid cell line containing chromosome 20 was digested to completion with different enzymes, PFG fractionated and hybridized to the human-specific *Alu* probe. The recognition sites for the enzymes used are: (1) *Sgf* I (GCGATCGC); (2) *Asc* I (GGCGCGCC); (3) *Fse* I (GGCCGG/CC); (4) *Pme* I (GTTTAAAC); (5) *Pac* I (TTAATTAA); (6) *Not* I (GCGGCCGC); (7) *Sgr*AI (CRCCGGYG); and (8) *Swa* I (ATTT/AAAT).

human megabase fragments. However, this approach may not distinguish two fragments of the same or very similar sizes even with high-resolution PFG fractionations.

The use of repeat sequences as hybridization probes can reveal information about the size and distribution of restriction fragments, but do not reveal order. Regional specific repetitive sequences, like those that occur at telomeric and centromeric regions may be used to identify the ends of physical map, as well as to provide an important anchor for regions of condensed chromosomes visualized microscopically. In human-hybrid cell lines, use of such sequences identifies ends of individual chromosomes.

In conventional mapping experiments, interstitial megabase fragments are linked to single-copy sequences by hybridization experiment or by the more efficient

PCR approach described in the following. Single-copy sequences, also called *STSs*, that have been located on genetic maps can be used for the regional assignment of restriction fragments and can serve as anchors between genetic and physical maps. The accuracy of fragment location will be a function of the resolution of the genetic map. When this approach is applied to a sample like that in Fig. 1, the cataloged megabase restriction fragments containing interspersed repeats are linked to specific single-copy sequences and, if known, genomic locations. ESTs are STS linked to mRNAs and their use would make a map of expressed genes. Note that any PCR primer pair can be used for mapping even if the genetic location of the specific sequence is unknown.

Complete genomic restriction maps do not identify neighboring fragments. At least one neighbor can be identified unambiguously using an STS to analyze partially digested DNA. The difficulty of interpreting the partial digestion data increases dramatically with the number of partial digestion products. For instance, it is important to consider that both the neighboring fragments may be of the same size, that is, the possibility that a single partial product band could represent two different products of the same size must always be borne in mind. This means that neighboring fragments must be confirmed in partial digestion experiments using STSs from the two complete digest fragments. Many times, the confusion associated with partial digest data can be sorted out by obtaining partial digest data on nearly adjacent fragments. In this approach, different sets of partial digest fragments are identified by sequences on different and complete digest fragments. The correct map interpretation is one that is consistent with all the data.

The analysis of partial products is simplified if the STS is from a telomere region or is situated on a small fragment located next to a very large fragment. In both cases, the partial digest information reflects the order of fragments in only one direction. Otherwise, the partial digest products results will represent bidirectional information. This approach also eliminates the problem of comigration of same-size products associated with bidirectional partial mapping data.

Megabase restriction fragments themselves may also be fingerprinted by cleavage with a second enzyme. The products are best analyzed using probes located at the end of the original megabase fragment. The partial digest product data will be different for the two probes located at different positions. The interpretation of the distribution of the cleavage sites of the second enzyme must be consistent with data obtained using the two interstitial (or terminal) probes. This approach is particularly useful for analyzing megabase fragments that are at the limit of the PFG technique, because the products of the second enzyme will be within the PFG limit.

There are several other approaches to proving that two restriction fragments are adjacent. By far the most powerful is the use of linking libraries – small-insert libraries that contain DNA segments from adjacent megabase restriction fragments. Hence, a *Not* I linking clone used as a hybridization probe to genomic DNA digested with *Not* I will identify adjacent *Not* I fragments. A complete linking library would suffice to construct, in the most efficient manner, a complete genomic restriction map. Without a complete linking library, partial digest strategies combined with polymorphism link-up and double restriction enzyme strategies can provide information on neighboring fragments. In both approaches, parallel analysis of a region provides a regional signature.

The same polymorphism link-up refers to the results obtained when one probe is used to analyze restriction enzyme digested DNA from different cells. For instance, a restriction site may be missing in one cell line, either because there is a mutation at the restriction enzyme site or because methylation is interfering with restriction enzyme cleavage. When a number of DNA sources are examined, it is often possible to fingerprint the polymorphism of a restriction site (i.e. assuming that the site is polymorphic). The probes that are usually located on separate, but adjacent, fragments can, in some DNAs, be found on the same fragment. Many times this latter fragment will be equal to the total size of the two smaller fragments seen in other cell lines. In such a case, the pattern of occurrence of the two distinct smaller fragments, detected by the different probes, is self-consistent. This is very similar to the more familiar approach of fingerprinting a DNA using many different restriction enzymes.

5
Genomic Clone Libraries

The construction and ordering of an overlapping library begins with the selection of the genomic DNA sample as discussed earlier. The next decision entails choice of the type of vector that will be used. The most distinguishing feature of vector possibilities is the size of the DNA that can be cloned into them. It is helpful to begin whole-genome mapping projects using the largest DNA fragments possible, minimizing the number of clones to order. In the past, the largest cloning vectors were cosmids that contained 40-kb cloned sequences. Today, several systems are available for the cloning of larger segments. For instance, megabase fragments can be cloned into YACs (yeast artificial chromosomes) in yeast cells. These libraries have been replaced by the easier to use and more stable PAC (P1 artificial chromosomes) and BAC (bacterial articificial chromosomes) libraries that clone fragments of about 100 kb in size. Today, the construction and ordering of PAC/BAC libraries still remains a major effort.

Usually, today's top-down sequencing approaches begin by creating large clone BAC libraries. Then, random, small overlapping fragments of each BAC are subcloned into sequencing vectors. The large clone libraries are ordered using a variety of techniques (see the following), whereas sequencing is done on an unordered library. Ultimately, the small-insert library is ordered when overlaps are identified while the sequence is put together.

Clone Ordering. In the past, chromosome walking (or "crawling") experiments, involved hybridizing individual clones to an entire genomic library to detect overlapping clones. Although this approach is guaranteed to detect overlapping clones, it is slow and laborious.

Today, many efficient genomic approaches are used to order all the clones in a library. The genomic approaches analyze many samples at the same time in a "multiplex" approach and generally use methods that are easy to automate. For instance, clones are analyzed in pools to minimize the number of experiments that are needed to order the entire library. In some cases, a tiered pooling strategy is used, where the first testing is done on a small group of superpools, each composed of different pools. Only the pools that make up a positive superpool needs to be further analyzed. Some designed pooling strategies have each clone present in multiple but distinct pools. The pattern of positive signals to such a set of pools will reveal the identity of the overlapping clones.

Most bottom-up, fingerprinting methods require that libraries be constructed using partial restriction enzyme digests to ensure that overlaps exist between different clones and that all or most of the genome is cloned. There are a number of ways to fingerprint clones to identify overlaps. Bottom-up strategies for library ordering usually involve testing individual clones to search for an overlapping restriction fragment or restriction site pattern. This type of fingerprinting is easy to automate. Clones may also be fingerprinted using interspersed repetitive sequences. Here, clones, or restriction fragments of the clones, are hybridized to a set of short oligonucleotides or longer interspersed repetitive sequences, and the patterns of hybridization are used to link clones.

A variation in this approach that has proven to be quite efficient is the use of genomic DNA restriction fragments as hybridization probes to order clone libraries. Megabase restriction fragment probes allow the assignment of clones to particular fragments. If the fragments are mapped, the clones can then be regionally assigned in the genome. Clones located in different regions are ordered simultaneously by further hybridization experiments using probes generated from a pool of restriction fragments. Here, the pools are created from gel slices containing small restriction fragments. Each pool consists of genomic restriction fragments of a particular size range from different regions of the genome. Overlapping clones will be located on the same megabase fragment and hybridize to the same pool of small fragments. This approach increases the efficiency of library ordering by eightfold.

Another ordering approach involves detecting overlaps by DNA sequencing. Here, the sequences may be chosen at random sites (e.g. STSs) or the sequences can be collected at specific locations originally called *sequence-tagged restriction sites* (STARs). The STAR approach may be used with both partial and complete digest libraries. DNA sequence information may be collected at specific restriction sites, including the ends of the cloned sequences. Clone linking is done using a linking clone library that spans the restriction enzyme site when the first library was created from completely digested DNA. Clones that generated from partial restriction enzyme digestions have ends that overlap. The later method eliminates the need for up-front clone ordering, but a highly representative library is required, and sequence read lengths must be long enough to span most small repeats.

6
Cloneless Genomic Libraries

Cloneless libraries have several advantages over clone libraries. These libraries can be made rapidly and inexpensively from any genomic source and all genomic DNA is present in the library. These libraries are made from agarose or acrylamide gels containing size fractionation DNA of low or high complexity. The gel slice containing the cloneless fraction is treated in the same manner as a cloned DNA sample.

In most of the discussion below, only complete digest cloneless libraries are discussed. However, partially digested or random broken DNAs could be used to generate a cloneless library. In essence, the analysis and use of a cloneless library is very similar to that of a clone library.

The gel lane is carefully sliced into thin pieces using a clean coverslip for each cut to insure minimum cross-contamination. The DNA in each slice is used as a source for testing, for instance, with PCR. For small and medium size genome, the number of unique fragments within a slice will be small. For large genomes, there are many fragments in a single slice. In the case of samples from the gels, like those shown in Fig. 1, the only visible fragment is from the human DNA but other rodent fragments are present. The cloneless approach is a hybrid between genomic mapping and clone ordering.

One consideration with the cloneless approach is that DNA in agarose is believed to be difficult to manipulate enzymatically. However, others and we have developed robust and reliable simple methods for using DNA in agarose that avoids purification. Most methods were developed some time ago for PFG applications. Additional methods were developed more recently for PCR that involve the addition of preservatives to the stored fraction and the PCR.

Another potential problem with cloneless fractions is the low amount of DNA in each slice. There are a number of PCR methods that solve this problem by randomly amplifying the DNA in a fraction (or clone) to generate a large amount of additional material. Furthermore, these methods have been developed to minimize the introduction of mutations. Hence, the amplified DNA appears to be representative of the starting template.

Cloneless fractions can be used for a variety of experiments in addition to that of genomic sequencing. For instance, a recently developed method termed *comparative genome hybridization* (CGH) uses differentially labeled cDNA, or genomic DNA, from two samples. The labeled samples are mixed together in equal amounts and used as a hybridization probe. The target samples are metaphase chromosomes or clones. Quantitative analysis of the hybridization signal provides information on the relative amounts of DNA from different chromosomal regions. In summary, CGH experiments provide positional information of differences between samples.

An alternative to conventional CGH would utilize genome DNA arrays that have been generated from size-fractionated genomic restriction fragments. An array consisting of 2-mm slices from a 15-cm gel lane would have ~40-Mb resolution. Each fraction could be cleaved with a second enzyme and subjected to a second electrophoretic size fractionation. The ~1000 slices generated from the entire second fractionation would provide DNA at ~3-Mb resolution.

The cloneless, like a clone library approach does not provide positional genomic information directly. For instance, clones are linked to genomic position by FISH (fluorescence *in situ* hybridization) experiments to metaphase chromosomes. Similarly, the DNA in each slice can be used in a FISH experiment to metaphase chromosome and locate it to a specific genomic region. Genomic restriction maps, or ordered clone, or cloneless libraries can be used in place of metaphase chromosomes especially in those organisms, like protozoan, whose chromosomes do not condense during cell division.

The approaches used to order clone libraries may be applied to cloneless libraries. For instance, fraction can be linked to STSs/ESTs. Alternatively, hybridization experiments or other methods that were described earlier can be used to detect overlaps with clone libraries or a second cloneless library generated with a second enzyme.

Cloneless libraries can be used in place of X-ray hybrid maps. X-ray hybrid maps are created by determining the frequency that two loci are separated by random breaks in DNA generated by X rays. In this approach, randomly broken DNA is cloned into a hybrid cell line and a library of different cell lines are created with different pieces of DNA. Then, each cell line is tested for the presence of a marker and a centiRay distance is calculated that reflects the frequency at which two loci occur in the same cell line. Like genetic mapping, the closer two loci are, the more likely they will be on the same randomly generate fragments. Here, instead of cell lines, randomly broken DNA fractionated by size could be used to create a cloneless library. Then each fraction could be tested for each loci and the distance between loci calculated as a function of the number of fractions that contain both loci.

7
Gap Closure

It is easy to start maps but difficult to finish them. Each gap in each map presents a unique problem. Gaps may be true gaps or pseudogaps. Pseudogaps contain repetitive sequences and require different strategies for closure than true gaps.

True gaps contain unique sequences and need to be treated individually. The best strategy will depend on the putative size of the gap, the amount of polymorphism in the region, the number of unassigned megabase fragments or clones. Gaps that arise from unclonable sequences may be filled using a cloneless library. For instance, STSs could be developed from cloneless fraction that overlap two regions as we have done for chromosome 21. For instance, single-copy human sequences located between *Alu* elements can be generated from inter-*Alu* PCR amplifications and used as probes to identify fragments in gap regions. Alternatively, the template contained within a fraction can be labeled and used as a hybridization probe to an array of samples at the borders of gaps. Ultimately, the DNA within a gap is identified in a cloneless library and directly sequenced using a direct genomic sequences approach.

8
Prospectus

What is the role of physical maps today and in the future? An increasing sequencing efficiency is coupled to a decreasing need to create and order expensive and time-consuming clone libraries. For small genomes, whole-genome sequencing projects use a bottom-up sequencing approach with unordered clone libraries. Soon a cloneless sequencing approach using "polonies" may replace the need for cloning of these small genomes. The polonies approach sequences pools of single DNA molecules that have been immobilized on beads. Sequencing is done on a library of immobilized DNAs simultaneously. Today, this method allows sequencing read of up to $\sim$14 bases per fragment end, or $\sim$28 bases per polony, and up to 1×10^9 beads/slide. Alternatively, imaging methods that directly view the sequence of individual DNA molecules may be used. Meanwhile, adaption of the cloneless approach for larger genome sequencing projects should considerably reduce *de novo* sequencing by eliminating the need for creating clone libraries, as well as enable gap closure.

See also Organization of Genes and Genome Domains; Construction and Applications of Genomic DNA Libraries; Genomic Sequencing; Shotgun Sequencing (SGS).

Bibliography

Books and Reviews

Cantor, C.R., Smith, C.L., Matthew, M.K. (1988) Pulsed field gel electrophoresis of very large DNA molecules, *Annu. Rev. Biophys. Biophys. Chem.* **17**, 287–304.

Ledbetter, S.A., Nelson, D.L. (1991) Genome Amplification Using Primers Directed to Interspersed Repetitive Sequences (Irs-Pcr), in: McPherson, M.J., Quirke, P., Taylor, G.R. (Eds.) *Polymerase Chain Reaction I: A Practical Approach*, IRL Press, Oxford.

Murphy, M.I., Hammond, H.A., Caskey, C.T. (1996) Solid-phase automated sequencing of Pcr-amplified genomic DNA, *Methods Mol. Biol.* **65**, 163–176.

Schalkwyk, L.C., Francis, F., Lehrach, H. (1995) Techniques in mammalian genome mapping, *Curr. Opin. Biotechnol.* **6**(1), 37–43.

Smith, C.L., Condemine, G. (1990) New approaches for physical mapping of small genomes, *J. Bacteriol.* **172**(3), 1167–1172.

Smith, C.L., Lawrance, S.K., Gillespie, G.A., Cantor, C.R., Weissman, S.M., Collins, F.S. (1987) Strategies for mapping and cloning macroregions of mammalian genomes, *Methods Enzymol.* **151**, 461–489.

Smith, C.L., Klco, S., Zhang, T.Y., Fang, H., Oliva, R., Fan, J.B., Bremer, M., Lawrence, S. (1993) Analysis of Megabase DNA Using Pulsed Field Gel Electrophoresis, in: Adolph, K.W. (Ed.) *Methods in Molecular Genetics*, Academic Press, San Diego, CA.

Van der ploeg, L.H.T., Gottesdiener, K.M., Korman, S.H., Weiden, M., Le Blancq, S. (1992) Protozoan Genomes, in: Ulanovsky, L., Burmeister, M. (Ed.) *Methods in Molecular Biology, Pulsed-Field Gel Electrophoresis. Protocols, Methods, and Theories*, Humana Press, Totowa, NJ.

Welsh, J., McClelland, M. (1994) Fingerprinting Using Arbitrarily Primed Pcr: Application to Genetic Mapping, Population Biology, Epidemiology, and Detection of Differentially Expressed Rnas, in: Mullis, K.B., Ferre, F., Gibbs, R.A. (Eds.) *The Polymerase Chain Reaction*, Birkhauser, Boston, MA..

Primary Literature

Albarghouthi, M.N., Barron, A.E. (2000) Polymeric matrices for DNA sequencing by capillary electrophoresis, *Electrophoresis* **21**(18), 4096–4111.

Barnes, W.M. (1994) Pcr amplification of up to 35-Kb DNA with high fidelity and high yield from lambda bacteriophage templates, *Proc. Natl. Acad. Sci. U.S.A.* **91**(6), 2216–2220.

Cai, W., Jing, J., Irvin, B., Ohler, L., Rose, E., Shizuya, H., Kim, U.J., Simon, M., Anantharaman, T., Mishra, B., Schwartz, D.C. (1998) High-resolution restriction maps of bacterial artificial chromosomes constructed by optical mapping, *Proc. Natl. Acad. Sci. U.S.A.* **95**(7), 3390–3395.

Dederich, D.A., Okwuonu, G., Garner, T., Denn, A., Sutton, A., Escotto, M., Martindale, A., Delgado, O., Muzny, D.M., Gibbs, R.A., Metzker, M.L. (2002) Glass bead purification of plasmid template DNA for high throughput sequencing of mammalian genomes, *Nucleic Acids Res.* **30**(7), e32.

Ferrin, L.J., Camerini-Otero, R.D. (1994) Long-range mapping of gaps and telomeres with Reca-assisted restriction endonuclease (Rare) cleavage, *Nat. Genet.* **6**(4), 379–383.

Fleischmann, R.D., Adams, M.D., White, O., Clayton, R.A., Kirkness, E.F., Kerlavage, A.R., Bult, C.J., Tomb, J.F., Dougherty, B.A., Merrick, J.M. et al. (1995) Whole-genome random sequencing and assembly of *Haemophilus influenzae* rd, *Science* **269**(5223), 496–512.

Florijn, R.J., Bonden, L.A., Vrolijk, H., Wiegant, J., Vaandrager, J.W., Baas, F., den Dunnen, J.T., Tanke, H.J., van Ommen, G.J., Raap, A.K. (1995) High-resolution DNA fiber-fish for genomic DNA mapping and colour barcoding of large genes, *Hum. Mol. Genet.* **4**(5), 831–836.

Fonstein, M., Haselkorn, R. (1995) Physical mapping of bacterial genomes, *J. Bacteriol.* **177**(12), 3361–3369.

Goldenberger, D., Perschil, I., Ritzler, M., Altwegg, M. (1995) A simple "Universal" DNA extraction procedure using Sds and proteinase K is compatible with direct Pcr amplification, *PCR Methods Appl.* **4**(6), 368–370.

Green, E.D. (2001) Strategies for the systematic sequencing of complex genomes, *Nat. Rev. Genet.* **2**(8), 573–583.

Hatada, I., Hayashizaki, Y., Hirotsune, S., Komatsubara, H., Mukai, T. (1991) A genomic scanning method for higher organisms using restriction sites as landmarks, *Proc. Natl. Acad. Sci. U.S.A.* **88**(21), 9523–9527.

Heiskanen, M., Kallioniemi, O., Palotie, A. (1996) Fiber-fish: experiences and a refined protocol, *Genet. Anal.* **12**(5–6), 179–184.

Hozier, J., Graham, R., Westfall, T., Siebert, P., Davis, L. (1994) Preparative in situ hybridization: selection of chromosome region-specific libraries on mitotic chromosomes, *Genomics* **19**(3), 441–447.

Kallioniemi, A., Kallioniemi, O.P., Sudar, D., Rutovitz, D., Gray, J.W., Waldman, F., Pinkel, D. (1992) Comparative genomic hybridization for molecular cytogenetic analysis of solid tumors, *Science* **258**(5083), 818–821.

Kim, U.J., Birren, B.W., Slepak, T., Mancino, V., Boysen, C., Kang, H.L., Simon, M.I., Shizuya, H. (1996) Construction and characterization of a human bacterial artificial chromosome library, *Genomics* **34**(2), 213–218.

Kovacic, R.T., Comai, L., Bendich, A.J. (1995) Protection of megabase DNA from shearing, *Nucleic Acids Res.* **23**(19), 3999–4000.

Lawrence, J.B., Carter, K.C., Gerdes, M.J. (1992) Extending the capabilities of interphase chromatin mapping, *Nat. Genet.* **2**(3), 171–172.

Marra, M.A., Kucaba, T.A., Dietrich, N.L., Green, E.D., Brownstein, B., Wilson, R.K., McDonald, K.M., Hillier, L.W., McPherson, J.D., Waterston, R.H. (1997) High throughput fingerprint analysis of large-insert clones, *Genome Res.* **7**(11), 1072–1084.

Michalet, X., Ekong, R., Fougerousse, F., Rousseaux, S., Schurra, C., Hornigold, N., van Slegtenhorst, M., Wolfe, J., Povey, S., Beckmann, J.S., Bensimon, A. (1997) Dynamic molecular combing: stretching the whole human genome for high-resolution studies, *Science* **277**(5331), 1518–1523.

Newell, W., Beck, S., Lehrach, H., Lyall, A. (1998) Estimation of distances and map construction using radiation hybrids, *Genome Res.* **8**(5), 493–508.

Paez, J.G., Lin, M., Beroukhim, R., Lee, J.C., Zhao, X., Richter, D.J., Gabriel, S., Herman, P., Sasaki, H., Altshuler, D., Li, C., Meyerson, M., Sellers, W.R. (2004) Genome coverage and sequence fidelity of Phi29 polymerase-based multiple strand displacement whole genome amplification, *Nucleic Acids Res.* **32**(9), e71.

Ricke, D.O., Liu, Q., Gostout, B., Sommer, S.S. (1995) Nonrandom patterns of simple and cryptic triplet repeats in coding and noncoding sequences, *Genomics* **26**(3), 510–520.

Smith, C.L., Warburton, P., Gaal, A., Cantor, C.R. (1986) Analysis of genome organization and rearrangements by pulsed gradient gel electrophoresis, *Genet. Eng.* **8**, 45–70.

Wu, C.C., Nimmakayala, P., Santos, F.A., Springman, R., Scheuring, C., Meksem, K., Lightfoot, D.A., Zhang, H.B. (2004) Construction and characterization of a soybean bacterial artificial chromosome library and use of multiple complementary libraries for genome physical mapping, *Theor. Appl. Genet.* **109**(5), 1041–1050.

Zabarovska, V.I., Gizatullin, R.Z., Al-Amin, A.N., Podowski, R., Protopopov, A.I., Lofdahl, S., Wahlestedt, C., Winberg, G., Kashuba, V.I., Ernberg, I., Zabarovsky, E.R. (2002) A new approach to genome mapping and sequencing: slalom libraries, *Nucleic Acids Res.* **30**(2), E6.

24
Serial Analysis of Gene Expression

Jacques Marti[1] *and Jean-Marc Elalouf*[2]
[1] *Institut de Génétique Humaine, Montpellier, France*
[2] *Commissariat à l'Energie Atomique/Saclay, France*

1	**Methodology** 695	
1.1	Rationale for SAGE 695	
1.2	How SAGE Libraries are Built: The Original Method and Further Improvements 695	
1.3	Processing Sequence Data 700	
1.4	Statistical Analysis of Digital Transcriptomes 700	
1.5	Tag-to-gene Mapping 701	
2	**Making Sense of SAGE Data** 703	
2.1	Unmatched Tags and Novel Transcripts: Using SAGE to Annotate the Genome and the Transcriptome 703	
2.2	From SAGE to Functional Genomics: Linking High-throughput Expression Analysis, Studies on Genome Organization and Complex Biological Networks 704	
2.3	SAGE, Molecular Classification of Cell Types and Cell Physiology 705	
2.4	SAGE and Disease-induced Alterations of the Transcriptome 705	
3	**Perspectives** 706	
	Bibliography 707	
	Books and Reviews 707	
	Primary Literature 708	

Genomics and Genetics. Edited by Robert A. Meyers.
Copyright © 2007 Wiley-VCH Verlag GmbH & Co. KGaA, Weinheim
ISBN: 978-3-527-31609-0

Keywords

Expressed Sequence Tag (EST)
A partial, usually 200–800 nucleotides long fragment of cDNA, obtained by single-pass sequencing of a randomly selected clone isolated from a cell or tissue cDNA library.

Microarray
A glass slide that contains hundreds to several thousands ordered cDNA probes to measure gene expression levels by hybridization analysis of labeled cDNAs.

Polymerase Chain Reaction
Referred to as PCR, it consists of an *in vitro* enzymatic amplification of specific DNA sequences.

Single Nucleotide Polymorphism
A DNA sequence variation that occurs when a single nucleotide in the genome sequence is changed to another one.

Transcriptome
The molecular phenotype of a cell, as defined by the diversity and abundance of mRNAs, differs between normal, developmental, and disease conditions.

Type IIS Restriction Endonuclease
A restriction enzyme that cleaves at a defined distance away from its asymmetric recognition site.

■ Serial Analysis of Gene Expression (SAGE) is a method designed to measure the abundance of a large number of mRNAs in tissues and cells to characterize their transcriptome. By the end of the twentieth century, when determining the entire genomic sequence of higher organisms began to be perceived as a realistic and attainable goal, the interest for genome-wide analysis of gene expression rapidly increased. One of the first strategies was to generate expressed sequence tags (ESTs), that is, partial sequences of cDNA clones randomly selected in libraries prepared from various tissue samples. The EST approach was highly valuable for gene discovery. Moreover, as long as the abundance of a given mRNA could be correlated with the frequency of its EST, it might also provide quantitative information. However, this approach evaluated only a limited number of genes per library. Obviously, getting larger quantitative expression profiles required to increase the rate of acquisition of experimental data. SAGE rests on the sequencing of short diagnostic tags, and as such increased by more than 1 order of magnitude the number of transcripts analyzed by current sequencing methods. Sets of 100 000 SAGE tags are now routinely sequenced, making possible the quantification of low copy number mRNAs.

1
Methodology

1.1
Rationale for SAGE

A major principle of serial analysis of gene expression (SAGE) was to reduce expressed sequence tags (ESTs) to their minimal size by using a type IIS restriction enzyme (Fig. 1). Type IIS endonucleases cleave DNA at a defined distance from their asymmetric recognition site. *Bsm* FI was chosen to generate the conventional SAGE 14-bp tags from a defined (anchoring) site created by a conventional restriction enzyme near the 3' end of each cDNA. Following the four common nucleotides of this anchoring site, SAGE tags differ by the next 10 variable nucleotides. Associated with the positional information, this size is usually sufficient for identifying each transcript, considering that 1 048 576 (4^{10}) different sequences can be distinguished. However, it may not be sufficient to map tags directly onto the complex genome of higher vertebrates: another endonuclease, *Mme* I, generating 21-bp tags, was preferred when using SAGE to annotate the genome by the so-called LongSAGE method.

One of the most appealing features of SAGE was the strategy used to amplify the whole population of tags while preserving the same quantitative distribution as in the initial mRNA population. Another characteristic of SAGE was to assemble tags into concatemers before DNA sequence analysis so that multiple tags (20–30 or more) could be read from each sequencing lane. Specific computer tools were developed for extracting the list of tags from sequence files, counting the number of times each one is observed and identifying the corresponding genes.

A SAGE analysis eventually results in a large sequence file in which tag sequences are punctuated by the four nucleotides of each anchoring site. Because experimental datasets are obtained *de novo* and do not rely on the calibration of any physical instrument, revocable external data or historical context, SAGE data can be considered as definitively acquired. Absolute transcript numbers are directly provided in a digital format, and data from multiple laboratories are easily assembled in comprehensive databases (Table 1). Tag-to-gene assignment can be updated at any time to keep in pace with progress in genome annotation, and an increasing body of functional annotations can be registered in the database. SAGE has now become one of the leading methods for gene expression profiling.

1.2
How SAGE Libraries are Built: The Original Method and Further Improvements

Poly(A) RNAs are preferably isolated using oligo(dT)$_{25}$ covalently linked to paramagnetic beads (Fig. 1). Other methods such as oligo(dT)-cellulose chromatography provide highly variable yields of mRNAs contaminated by substantial amounts of ribosomal RNAs (see Sect. 1.5). Thus, it is not surprising that the very first SAGE protocol, described in 1995, required 5 µg poly(A) RNA (i.e. several millions of mammalian cells). The need to work on small biological samples prompted several investigators to adapt the method to downsized extracts as in the SADE, microSAGE, and miniSAGE protocols. It then became possible to build SAGE libraries starting from 1 µg total RNA and to analyze biological samples containing 10^5 cells or even less. It is still possible to lower sample size and to start from 2500 cells or as little as

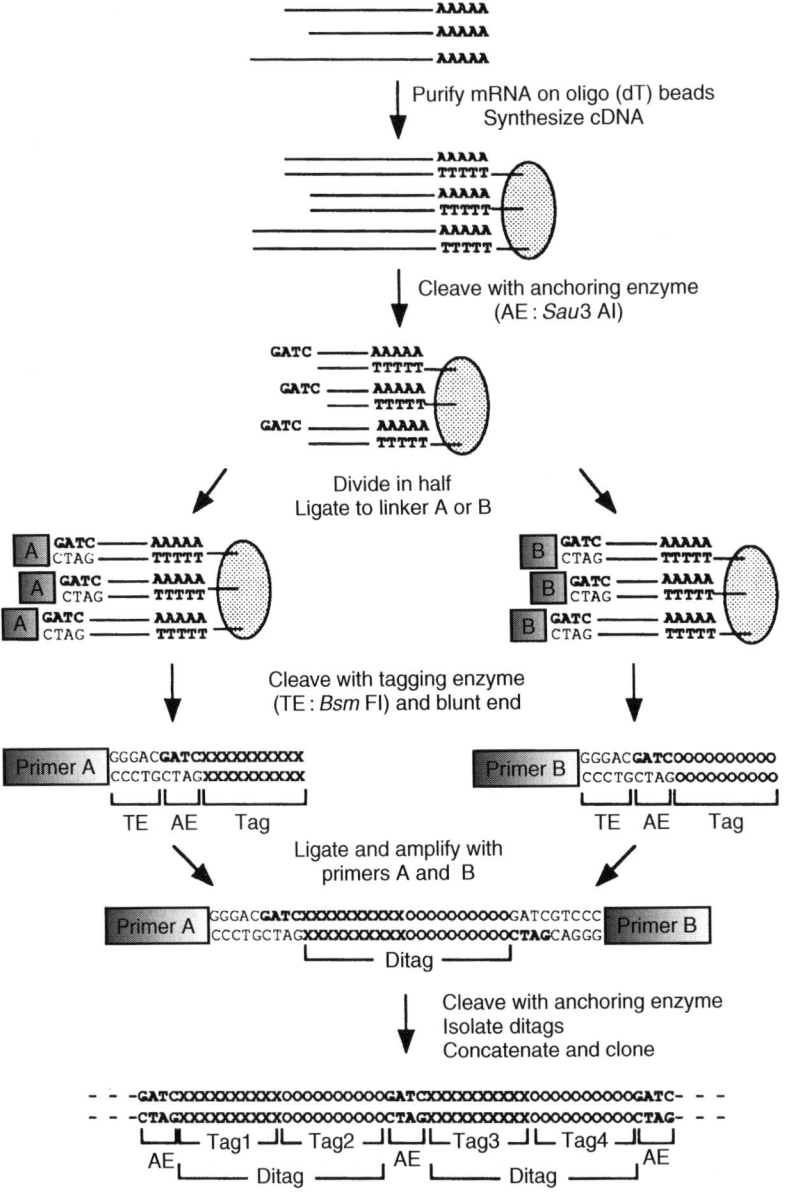

Fig. 1 Outline of procedures for constructing SAGE libraries. Poly(A) RNAs are isolated from tissue lysate using oligo(dT)$_{25}$ covalently linked to paramagnetic beads, and cDNA is synthesized under solid-phase condition. Bold face characters correspond to biologically relevant sequences, whereas light characters represent linker-derived sequences. The anchoring enzyme (AE) is *Sau* 3AI, whereas the tagging enzyme (TE) is *Bsm* FI. See text for details (Reprinted with permission from Cheval, L., Virlon, B., Elalouf, J.-M. (2000) SADE: A Microassay for Serial Analysis of Gene Expression, in: Hunt, S.P., Livesey, F.J. (Eds.) *Functional Genomics: A Practical Approach*, Oxford University Press, Oxford).

Tab. 1 Example of SAGE output (reproduced by courtesy of R. Quere and G. Bertrand, PhD theses, Montpellier University). Comparison between the expression profiles of **(A)** proliferating acute promyelocytic leukemia cells (NB4 human cell line), **(B)** cells incubated for 24 h in the presence of 1 µM retinoic (inducing cell growth arrest and granulocytic differentiation), and **(C)** normal blood granulocytes. SAGE was performed, using Sau3AI as anchoring enzyme. Comparison with **(D)** the sum of tag counts for the same genes in a compilation of 260 published SAGE libraries, built from various human tissues using NlaIII as anchoring enzyme was also performed. These 36 entries represent about 0.7% of the total number of genes (>5000) revealed in this experiment. Examples of ambiguities in tag assignment are provided by B2M and CAPN10, sharing the same Sau 3AI-anchored tag but distinguished with NlaIII, and by COX7C identified only with Sau 3AI since its transcript is devoid of NlaIII restriction site. The upper rows illustrate the detection of differentiation markers. The lowest rows show the case of components of the translation machinery, ribosomal proteins, and elongation factor transcripts, abundant in leukemia cells, but barely detectable in terminally differentiated normal cells.

Tag (Sau 3AI)	A	B	C	Tag (NlaIII)	D	Gene name/symbol/accession number
GATCATGCTCCATC	0	2	27	CATGCTCCATCCAG	359	Colony stimulating factor 3 receptor (granulocyte)/CSF3R/NM_156039
GATCATCTTTCCTG	0	6	55	CATGGTGCCGTGAG	3923	Major histocompatibility complex, class I, C/HLA-C/NM_002117
GATCACATAAAACA	0	2	131	CATGGCCTTAACAA	1040	Pre-B-cell colony-enhancing factor/PBEF/NM_005746
GATCACAGTGGCCA	0	7	551	CATGGTGGCCACGG	5094	S100 calcium binding protein A9 (calgranulin B)/S100A9/NM_002965
GATCCAGGTTTTTA	0	8	73	CATGACTCAGCCCG	750	Tumor necrosis factor, α-induced protein 2/TNFAIP2/NM_006291
GATCTTCTTTATAA	1	9	251	CATGGTTGTGGTTA	33 742	β-2-microglobulin/B2M/NM_004048
GATCTTCTTTATAA	1	9	251	CATGAGCGCTGATG	543	Calpain 10/CAPN10/NM_023087
GATCAAAGTTGGCG	1	5	27	CATGTTAGGGAGGA	233	Intercellular adhesion molecule3/ICAM3/NM_002162
GATCCACAAGTCCT	1	3	28	CATGTGAAGCACT	2464	Interleukin 8/IL8/NM_000584
GATCCATGGATGTA	1	7	117	CATGGATGTAGTGG	273	Myeloid cell nuclear differentiation antigen/MNDA/NM_002432
GATCACACTGACCT	2	20	38	CATGCTGACCTGTG	5927	Major histocompatibility complex, class I, B/HLA-B/NM_005514
GATCCTGTACAGCC	2	19	18	CATGGTGCTGAATG	14 044	Myosin, light polypeptide 6, alkali.../MYL6/NM_079425
GATCTTCATGCCCT	3	10	74	CATGCCCTGGTTC	16 750	Ferritin, light polypeptide/FTL/NM_000146
GATCACCTGGGTTT	3	3	28	CATGTCACCTTAGG	2424	Integral membrane protein 2B/ITM2B/NM_021999
GATCTCTGGGCTGG	6	4	32	CATGACGATGGCCG	319	Glia maturation factor, γ/GMFG/NM_004877

(continued overleaf)

Tab. 1 (Continued)

Tag (Sau 3AI)	A	B	C	Tag (Nla III)	D	Gene name/ symbol/ accession number
GATCAACAAGAAAT	8	17	62	CATGGGCTGGGGC	12 427	Profilin 1/ PFN1/NM_005022
GATCCAAGTGTTTA	9	10	49	CATGTTAACCCTCT	1877	H3 histone, family 3B (H3.3B)/H3F3B/NM_005324
GATCAGGAATCTTG	9	2	0	CATGTGAATCTGGG	1222	SET translocation (myeloid leukemia-associated)/SET/NM_003011
GATCGAGCTCGCCT	14	14	13	CATGGTACTGTGCC	1847	Chloride intracellular channel 1/CLIC1/NM_001288
GATCAAGTTTAAAT	16	37	279	CATGTTGGTGAAGG	24 592	Thymosin, $\beta 4$, X chromosome/TMSB4X/NM_021109
GATCACACTGACCT	20	38	2	CATGCTGACCTGTG	5927	Major histocompatibility complex, class I,B/HLA-B/NM_005514
GATCAGTTCAGCCA	20	21	3	CATGGCGGTTGTGG	2308	Lysosomal-associated multispanning membrane protein-5/LAPTM5/NM_006762
GATCTATCACCTGT	24	12	16	CATGTAGGTTGTCT	43 934	Tumor protein, translationally controlled 1/TPT1/NM_003295
GATCTCATTCCATG	28	26	20	CATGGGAAAAAAAA	9029	ATP synthase, H+ transporting, mitochondrial F1 complex/ ATP5E/NM_006886
GATCCCCTCAGAAG	28	3	33	CATGTGAGGAATA	9867	Triosephosphate isomerase 1/TPI1/NM_000365
GATCTGGGAGTGGC	31	85	29	CATGGCCATAAAAT	1740	Proteoglycan 1, secretory granule/ PRG1/NM_002727
GATCAAGAATTTGG	37	49	62	CATGTCTTTACTTG	1209	Actin related protein 2/3 complex, subunit 3, 21 kDa/ARPC3/NM_005719
GATCAAACTAGAAC	43	36	2	none	4434	Cytochrome c oxidase subunit VIIc/COX7C/NM_001867
GATCTGCGCTTCCA	50	57	144	CATGTTTTTAATGT	2107	H3F3A histone/H3F3A/NM_002107
GATCCACTACCGGA	100	73	1	CATGCTCCTCACCT	9461	Ribosomal protein L13a/RPL13A/NM_012423
GATCGGGTAGCTCA	142	95	1	CATGCTGCTATACG	7957	Ribosomal protein L5/RPL5/NM_000969
GATCTGTGTTTGCT	148	123	2	CATGTACAAGAGGA	6980	Ribosomal protein L6/RPL6/NM_000970
GATCCAGTTCTAAG	198	165	1	CATGGTTAACCTCC	4635	Ribosomal protein L36a/RPL36a/NM_021029
GATCCTACTCTCTT	257	238	1	CATGGGACTGGACA	4556	Ribosomal protein L18/ RPL18/BF677228
GATCAGACTCTGAA	448	263	3	CATGCCGTGTCCG	15 586	Ribosomal protein S6/RPS6/AK027187
GATCCGCCGTTCTGG	643	592	0	CATGTGTGTTGAGA	46 979	Eukaryotic translation elongation factor $1\alpha 1$/EEF1A1/NM_001402

50 ng of RNA, by adding a loop of RNA amplification in the SAGE protocol as in SAR-SAGE or aRNA-longSAGE. It must be kept in mind that modifications aimed at reducing the amount of starting material may introduce representation biases. When this risk is carefully evaluated, it is possible to build a SAGE library starting from minute amounts of material.

The synthesis of double-stranded cDNA is performed according to classical methods. In the original protocol, a soluble biotinylated oligo(dT) primer was used to initiate the reaction and cDNA fragments were purified at a later step by capture on streptavidin-coated beads. In more recent protocols, oligo(dT) covalently linked to paramagnetic beads were used to bind poly(A) RNA. Performing all subsequent reactions on this solid phase greatly improved the overall yield.

The next step consists of cleaving cDNA by a restriction enzyme with a 4-bp recognition site (SAGE-anchoring enzyme). Only the most 3′ fragment of each cDNA remains on magnetic beads and the other fragments are eliminated. Since the enzyme cleaves every 256 bp (4^4) on average, it is expected to generate a 3′ fragment from the vast majority of cDNAs. The choice of an anchoring enzyme (AE) is limited by the necessity to create a protruding 4-bp single strand, further used to bind a synthetic linker. The most commonly used one is *Nla* III recognizing CATG sites, but *Sau* 3AI (GATC) works equally well. The bead suspension, bearing the purified 3′ cDNA fragments, is then divided in half. Each aliquot is ligated to one of two linkers containing the cognate sequence for the type IIS restriction endonuclease (*Bsm* FI or *Mme* I), used as a tagging enzyme (TE). The enzyme releases DNA fragments, each containing a SAGE tag associated to a synthetic linker.

The two pools of released fragments are ligated to form ditags. For being properly coupled tail to tail, DNA fragments must be blunted on their "biological" side and chemically blocked on their "synthetic" side. Ditags then serve as templates for amplification by polymerase chain reaction (PCR) with primers specific to each linker. Since there is a small probability of any two tags being assembled twice in the same combination, amplifying randomly assembled ditags rather than separate tags eliminates most PCR biases. Following amplification, the 4-bp restriction enzyme initially used to generate anchoring sites is now used to release the synthetic linkers. Purified ditags, now bearing sticky ends, are concatenated by ligation and the concatemers inserted in a cloning vector.

The SAGE library now consists of a collection of bacterial clones, each one harboring a unique random collection of tags. Before submitting them to sequence analysis, it is advisable to evaluate the average size of concatemers. It is indeed rather difficult to control ligation kinetics and concatemer elongation. Although it has no incidence on individual tag detection and does not alter the overall quality of the analysis, it has a direct effect on the cost and rate of acquisition of data. In some cases, it may be wiser preparing a new reaction rather than analyzing very small concatemers.

Technically, building a SAGE library was perceived as a heavy task, involving multiple steps critical for the success of the overall project. Several laboratories have devoted specific studies to the solution of various problems: efficiency of SAGE adaptor ligation, efficiency of *Bsm*FI digestion, prevention of amplification biases, improved *Nla*III digestion of 102-bp ditags, synthesis of concatemers and cloning efficiency, easier clone selection. Eventually, the commercial availability of dedicated reagents (Invitrogen Corporation, Carlsbad, CA) greatly facilitated the preparation of SAGE libraries. With all these improvements, the SAGE technology is now being considered in many laboratories as a standard for gene expression profiling.

1.3
Processing Sequence Data

Raw SAGE data are provided as computer files containing the sequences of numerous clones. Merging these files into a unique one, usually in FASTA format, provides a permanent record of the analysis. A dedicated computer program is then needed for parsing this file and identifying individual tag boundaries. Since the first software created at the inception of SAGE in the laboratory of K. Kinzler (Johns Hopkins University, Baltimore, MD, USA), other parsers were developed with more or less extended functions. They usually include the detection of linker artifacts and sequence ambiguities, and must provide a measure of ditag average length. Indeed, type IIS restriction enzymes such as *Bsm* FI yield fragments of variable length. Since tags are ligated tail to tail, delineation of tag ends becomes so difficult in short ditags that they are preferably rejected. This problem has been discussed elsewhere. Ditags with twice the expected length can also be detected in sequence files. Their frequency may help estimate the sequencing error rate, assuming that the apparent loss of the 4-bp anchoring enzyme site defining ditag boundaries results from sequencing errors. If obtained early enough during the sequencing process, this indication may prompt the investigators to improve the quality of the sequencing machine or base calling software. When dealing with already obtained SAGE data, correcting the sequencing error noise requires more complex treatments. Identical ditags can also be observed. In principle, considering the complexity of the transcriptome in eukaryotic cells expressing thousands of genes, there is a low probability, which can be evaluated, to find the same tags associated several times in identical ditags. A low transcriptome complexity, increasing the number of identical ditag, may be observed in terminally differentiated, highly specialized cells. Alternatively, a high frequency of identical ditags may indicate that the cDNA input was not large enough to reproduce the true complexity of the natural mRNA population: this problem may be encountered when sampling minute amounts of biological material. The decision to discard or include repeatedly observed ditags in the calculation of tag frequencies eventually depends on a biologist's appraisal.

While in any case, the ultimate goal is to generate a reliable list of tags along with the number of times they occur, sequence data contain additional information, which, when extracted by appropriate software, provides elements of diagnosis for assessing the quality of SAGE libraries.

1.4
Statistical Analysis of Digital Transcriptomes

Methods for the quantitative analysis of gene expression can be broadly classified into two categories. "Analog" methods, using homemade microarrays or commercial DNA chips, are based on the measurement of analogical signals, correlated with the extent of hybridization of labeled transcripts to arrayed cDNAs or oligonucleotides. "Digital" methods are based on tag counting in large collections of ESTs, SAGE, or Massively Parallel Signature Sequencing libraries (MPSS). The quantitative value of both kinds of methods has been thoroughly investigated and it is possible to delineate the limits within which each experimental dataset can be used. Digital profiles can now be interpreted on a sound theoretical framework. In analysis of SAGE data, the first assumption is usually that the number of tags is proportional to the number of cognate transcripts in the original biological sample, implying that each mRNA copy has the same chance of ending up as a tag in the library. The second assumption is that, when prepared for sequence characterization, SAGE tags are picked at random in a highly redundant, unbiased library. The problem of distinguishing biological values from technical errors becomes to set the limits between genuine variation and sampling fluctuation. Focusing on this sampling variability, several tests have been published giving essentially the same values when applied to the comparison of two SAGE libraries. Assessing fluctuations between groups of libraries requires more complex treatments.

1.5
Tag-to-gene Mapping

Identification of the gene transcripts revealed by SAGE tags is independent of experimental data acquisition. It might thus be conceivable to initiate a SAGE project on a given species before having an extensive knowledge of its genome, and indeed, investigations on human cells began several years before completion of the human genome sequencing project. Even when the full-length genome sequence is available, mapping SAGE tags directly to chromosomes is possible only for species with a small compact genome, such as yeast and *Arabidopsis thaliana*. For mammalian species, the size of noncoding regions is such that only LongSAGE (21-bp) tags can be mapped on to the genome. When dealing with conventional 10-bp tags, it is better to start from collections of expressed sequences such as UniGene, maintained at the National Center for Biotechnology Information (NCBI, http://www.ncbi.nlm.nih.gov/). This experimental system organizes GenBank sequences into a nonredundant set of clusters, each one containing sequences that represent a unique gene. A sequence representative of each cluster can be retrieved from UniGene, as well as related information about the biological properties of the corresponding gene.

Using such a collection, an efficient strategy is to prepare a list of putative tags, predicted by performing virtual SAGE *in silico*. Lists of already predicted tags can be retrieved from SAGEmap, another public resource of NCBI dedicated to the dissemination of SAGE data. SAGEmap gives both tag-to-gene and gene-to-tag mapping, and calculates a score indicating the quality of the connection. Lists of predicted and experimental tags are easily recorded in the table format of current relational database management systems, either from commercial (Microsoft Access) or open source (MySQL), so that all tag-to-gene connections can be performed at once. The rate of success obviously depends on the number of well-annotated genes in the species under investigation. In human samples, more than 50% of tag sequences can now be connected immediately to known genes, representing about 70% of the total number of tag counts. The discrepancy between these two values simply reflects the fact that abundant transcripts correspond more often to already well-known genes.

A small number of genes, which can be predicted, will systematically escape SAGE analysis because their transcript lacks the 4-bp anchoring site required to adapt SAGE linkers. Obviously, they must be studied by other means. A solution would be to prepare two libraries in parallel using each of the two enzymes (*Nla* III and *Sau* 3AI) already selected for their efficiency as anchoring enzymes. Table 1, as well as computer simulation on UniGene sequences, show that identification would then be nearly complete, but with a rather limited benefit compared to the twofold increase of the cost for sequencing analysis.

Another problem is that different genes may share the same tag. Although this risk seems statistically low, real-life examples show that multiple matches are actually observed at a rather high rate. In human samples, this problem results essentially from the presence of retrotransposed repetitive elements (mostly Alu sequences) inserted in the 3' end of otherwise unrelated transcripts. These sequences can be retrieved from a dedicated database such as RepBase Update (http://www.girinst.org/Repbase_Update.html), and the corresponding tag and gene sequences registered in specific files.

Conversely, a gene may have more than one tag, either as the consequence of alternative splicing or because longer mRNA species encompassing the first polyadenylation site are actually present in cell extracts. The existence of alternative polyadenylation sites increases the number of tags and this phenomenon must be taken into account when analyzing SAGE data.

Upon completion of the tag-to-gene mapping procedure, it is tempting to consider that unmatched tags actually signal yet unknown transcript. However, before developing strategies for their identification, it is necessary to purge SAGE datasets from spurious sequences generated by technical artifacts. Tags matching the mitochondrial genome are sometimes considered as contaminants. However, it is worth mentioning that transcription of the mitochondrial genome produces two polycistronic RNAs (one from each strand) giving rise to a series of poly(A) RNAs. Tags matching the mitochondrial genome are therefore biologically relevant. By contrast, ribosomal RNA fragments may be considered as contaminants, because mRNAs represent a low percentage of the overall nucleic acid cell content. These fragments amounted to up to 1% of the total tag population in libraries generated using the initial SAGE protocol. With more stringent conditions being used for the purification of poly(A) RNAs, their abundance was reduced to less than 0.01% in most libraries. Finally, a significant number of incorrect tags are created by sequencing errors. Typically, SAGE tags are obtained by single-pass sequencing. With 1% of nucleotides determined incorrectly, about 1 of 10 tags will be erroneous. However, since the probability of the same real tag being erroneously sequenced several times is low, the major effect is an increasing number of tags observed at low to moderate frequency. With a small number of SAGE libraries at hand, authors usually discarded tags observed only once. Now that a large number of libraries are publicly available, it is possible to distinguish artifacts from genuine tags of rare transcripts by checking their occurrence in multiple samples. Taken together, technical artifacts essentially increase the apparent number of different tag sequences. However, because their individual frequencies are low, they alter only marginally the quantification of authentic mRNAs.

As a whole, tag-to-gene mapping can be perceived as a sieving process, in which the list of experimental tags is percolated through multiple filters. At the end of this process, investigators obtain a list of tags identifying authentic mRNAs and a list of tags that are susceptible to revealing the existence of novel transcripts.

2
Making Sense of SAGE Data

2.1
Unmatched Tags and Novel Transcripts: Using SAGE to Annotate the Genome and the Transcriptome

In species with a well-annotated genome, the vast majority of known mRNAs are identified without ambiguity. However, in spite of progress in genome annotation, SAGE libraries often contain unassigned tags susceptible to reveal rarely expressed genes, new alternate transcripts of known genes, genetic variants, or polyadenylated transcripts of noncoding genes.

As could be expected, the characterization of new transcripts proved to be more efficient when using the LongSAGE method instead of the classical one, as shown for the human and more recently for the mouse genome. When dealing with a conventional 10-bp short tag, its assignment to entirely new loci will be easier if it can be matched with a yet unassigned EST, providing an extended probe for experimental investigations and computer searches. When such information is lacking, it is still possible to use the tag itself as a probe, and to amplify an extended cDNA stretch starting from the 10-bp SAGE tag.

With respect to the size of the genome, evaluating the size of the transcriptome remains a controversial issue. Alternative splicing and alternative polyadenylation, each concerning perhaps half the human genes, contribute to increasing the number of transcripts. SAGE can detect alternative exons in the very 3' end of mRNAs, but does not make distinction between splice variants differing in more 5' sites. On the contrary, SAGE is highly sensitive to alternative polyadenylation. The existence of transcripts of variable length, seldom registered in current transcript database, has been clearly documented by using dedicated computer programs. These results suggest that nearly half the human genes might generate alternate transcripts, depending on tissue- or disease-specific regulatory mechanisms.

SAGE is sensitive to genetic polymorphism. Alternative tags can be generated by single nucleotide polymorphism (SNP), either because an SNP disrupts the anchoring site of the most commonly observed tag, creates a new anchoring site, or changes the sequence usually registered in databases. A database of human SNP-associated alternative SAGE tags has recently been integrated in SAGE Genie, improving the interpretation of SAGE data.

Detailed analyzes of chromosomes 21 and 22 by massively parallel hybridization on genomic DNA probes revealed a mammalian transcriptome larger than previously estimated. In fact, just as SAGE itself, these strategies made no distinction between conventional mRNAs and noncoding polyadenylated transcripts. The ability of SAGE to reveal such untranslated transcripts deserves interest, considering the regulatory function they may exert.

An initially intriguing observation was that of a subpopulation of tags mapping on to the opposite strand of known mRNAs. For a part, they might be assigned to repetitive elements inserted in opposite orientations in multiple mRNAs. More than one million copies of the approximate 300-bp Alu elements are interspersed in the human genome and a fraction of them are inserted in transcribed 3'-UTRs. Other pairs of tags were found associated with unique pairs of transcripts. Some of them are encoded in trans, the corresponding genes being found on remote loci, while other pairs of tags reveal cis-encoded genes. These observations are corroborated by recent computer searches at the genome level, identifying a large number of contiguous, oppositely oriented genes, potentially able to express complementary mRNAs. SAGE is well adapted for investigating whether such genes are expressed simultaneously, a condition obviously required for them for producing functional double-stranded RNAs.

The detection of genuine tags that were left unpredicted by the analysis of conventional transcript sequences contributes to improving the quality of genome annotation. SAGE can now be considered as more sensitive than EST sequencing for detecting low-abundantly and rarely expressed genes in mammalian genomes.

2.2
From SAGE to Functional Genomics: Linking High-throughput Expression Analysis, Studies on Genome Organization and Complex Biological Networks

The first comprehensive view of a transcriptome was obtained by representing the frequency of SAGE tags on the physical map of yeast chromosomes. More recently, a human transcriptome map, based on the computer-assisted representation of SAGE tags mapping on to chromosomes, has illustrated the power of SAGE for investigating the organization of chromatin and the distribution of genes in actively transcribed regions.

A contemporary challenge is to establish relationships between global variations of gene expression profiles and changes in complex metabolic or signaling networks. Until now, findings in molecular and cell biology were reported in textual form, using an uncontrolled vocabulary, and disseminated in a million publications. The heterogeneity of biological knowledge and, even more critically, the fuzziness of essential concepts including the concept of biological function itself, make the task highly difficult. Collective efforts are being made for calling genes unequivocally (HUGO Gene Nomenclature Committee, http://www.gene.ucl.ac.uk/nomenclature/) and to control the vocabulary used for defining the semantic links that describe functional relations among biological entities (Gene Ontology™ (GO) Consortium: http://www.geneontology.org/). Lists of descriptors provided by this consortium can be linked to SAGE data, so that tags may be automatically classified according to the functional properties of their corresponding genes. Most algorithms for large-scale gene expression analysis were developed to facilitate the interpretation of microarray data. In fact, methods such as association-rules discovery or clustering analysis can be used as well for analyzing SAGE data.

Several interactive Web sites provide synthetic views of an increasing number of networks. For instance, by connecting to the Web site of the Kyoto Encyclopedia of Genes and Genomes (KEGG, http://www.genome.jp/kegg/), it is possible to automatically stain a large number of metabolic charts according to the results of SAGE and to get in a few hours a synthetic view of events occurring in most metabolic and some signaling pathways.

Interpretation of the set of data generated by each individual laboratory is usually not extensive. Since SAGE data are cumulative, comparisons with remote data may enrich the interpretation of local results. For this purpose, the Gene Expression Omnibus collects expression datasets issued from SAGE and microarray technologies, making them accessible for further analysis and reinterpretation. More specifically, SAGE Genie, a bioinformatic project dedicated to the management of large amounts of experimental SAGE data provides, among other intuitive tools, visualization displayers relating gene expression and tissue localization.

2.3
SAGE, Molecular Classification of Cell Types and Cell Physiology

Conceptually, a given cell type might be defined by its transcriptome, considered as the whole set of genes responsible for its phenotype. However, cells are under constant evolution. They may express functions dictated by proteins whose mRNAs were expressed at a former stage of development, but already catabolized at the time of sampling. Conversely, the presence of a transcript does not always mean that it was being translated at the time of sampling. A gene expression profile must therefore be considered as a snapshot and interpreted as such. With this reservation, transcriptome analysis possesses an advantage over other high-throughput strategies, such as proteome analysis, which, because proteins cannot be amplified, is less sensitive and identifies a smaller number of cell components.

Evaluation of SAGE data from a variety of normal and diseased tissues soon provided an enlarged view of the human transcriptomes.

The possibility to work on tiny cell samples was decisive for defining the profile of discrete cell subtypes. Analysis of microdissected anatomical structures provided panoramic views of complex tissues such as kidney (Fig. 2), retina, or brain. Homogeneous samples of circulating cells are more easily purified than cells from solid tissues and numerous blood cell types have now been described by SAGE at the molecular level.

Most biological response modifiers do not act only on rapid cell kinetics through the activation of signaling pathways, but also exert long-lasting effects by changing gene expression patterns. Large-scale transcriptome analysis thus added a new dimension to studies involving the experimental induction of a signaling pathway and the effects of cytokines, hormones, or differentiation inducers.

2.4
SAGE and Disease-induced Alterations of the Transcriptome

It was readily perceived that SAGE offered new possibilities for characterizing global expression profile alterations in infectious diseases, for instance, upon exposition to bacterial toxins or viral infection. However, the major incentive for implementing the SAGE technology came from cancer research. Molecular classification of cancer cells, expected for many years, is now coming of age, opening new perspectives for cancer therapies. The power of SAGE for identifying molecular markers of cancer was rapidly demonstrated, and since the pioneering works, a large number of SAGE libraries have been obtained

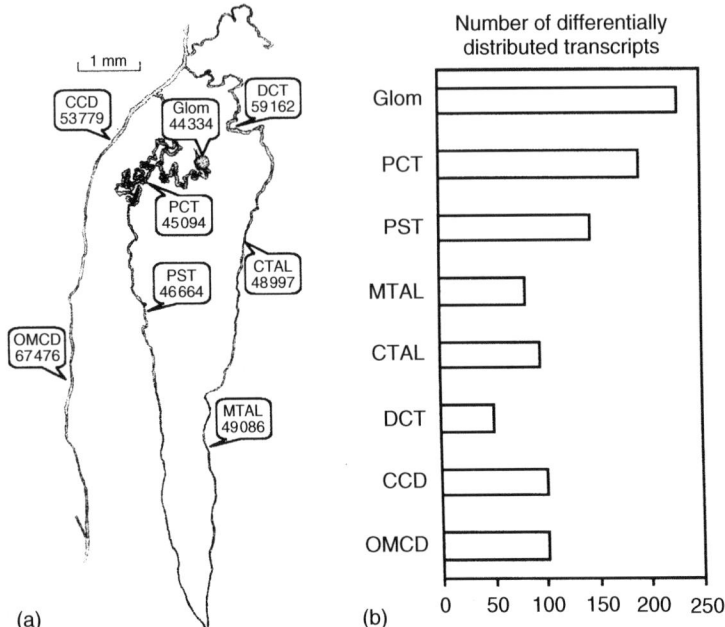

Fig. 2 Transcriptome characterization of individual nephron portions of the human kidney. (a) A microdissected human nephron. Structures analyzed for trancriptome characterization, together with the number of SAGE tags sequenced are indicated. Glom, glomerulus; PCT, proximal convoluted tubule; PST, proximal straight tubule; MTAL, medullary thick ascending limb of Henle's loop; CTAL, cortical thick ascending limb of Henle's loop; DCT, distal convoluted tubule; CCD, cortical collecting duct; OMCD, outer medullary collecting duct. (b) Overview of mRNA differential distribution. The number of mRNAs displaying a significant differential distribution ($p < 0.01$ as compared with three nephron portions) is indicated for each structure. The overall number of heterogeneously distributed transcripts is 998. Data from Chabardes-Garonne, D., Mejean, A., Aude, J.C., et al. (2003) A panoramic view of gene expression in the human kidney, *Proc. Natl. Acad. Sci. U.S.A.* **100**, 13710–13715.

with the purpose of comparing normal cells and their malignant counterparts. SAGE was also used to investigate tumor angiogenesis or drug sensitivity of malignant cells. It is not possible to summarize here all the projects developed since the inception of SAGE. A comprehensive view can be obtained from SAGEmap (http://www.ncbi.nlm.nih.gov/SAGE) and SAGE Genie (http://cgap.nci.nih.gov/SAGE), two components of the Cancer Genome Anatomy Project intended to provide a central database for comparing gene expression profiles in normal and malignant tissues.

3 Perspectives

SAGE was developed at the same time as DNA chip and microarrays technologies.

More recently, MPSS offered an alternative approach to SAGE, but the need for specific equipment limited its use to a small number of laboratories. Each technology has its specific advantages. Methods based on hybridization are well adapted for the comparison of multiple samples, but are limited to the study of the genes represented in the array. SAGE and MPSS are more laborious, but do not require *a priori* knowledge on gene sequences. Moreover, their results are cumulative and comparison between independent datasets are easily performed, while exchanging microarray data requires a complex standardization protocol. Rather than a competing methodology with other global gene expression approaches, it thus seems better considering SAGE as a complementary approach, and the decision to initiate a new SAGE project as depending on the biological problem is under investigation.

SAGE is still receptive to methodological improvements. It is obvious that it will benefit from any progress in DNA sequencing technology that increases the rate of acquisition and reduces the cost of analysis. Until now, the conventional method identified mRNAs by tagging their 3′ end. However, there was still a need for a global method of also identifying transcripts through their 5′ end. Protocols enabling the high-throughput extraction of tags from mRNA 5′ ends have now been published. The 5′-end SAGE is expected to improve the characterization of transcriptional start sites and the functional annotation of poorly known genes.

It is worth noticing that SAGE has already produced a huge amount of experimental results from which latent biological knowledge may be extracted by mining the existing dataset. With the progress of computational biology, the integration of SAGE data in larger information systems, seamlessly coupling data from various sources, opens the way to global analyses on the regulation of gene expression at a higher level of organization. Comparative analysis between man and animal models will be highly useful, and facilitated by the development of new databases such as the Mouse SAGE site (http://mouse.biomed.cas.cz/sage/).

SAGE has already contributed, in conjunction with other methodologies, in achieving a more comprehensive and quantitative picture of transcriptomes. It will probably continue to provide new data until their complete disclosure.

See also Molecular Basis of Genetics; Construction and Applications of Genomic DNA Libraries.

Bibliography

Books and Reviews

Blackshaw, S., Kim, J.B., St. Croix, B., Polyak, K. (2002) Serial Analysis of Gene Expression, in: Ausubel, F.M., Brent, R., Kingston, R.E., Moore, D.D., Seidman, J.G., Smith, J.A., Struhl, K. (Eds.) *Current Protocols in Molecular Biology*, Wiley, NY.

Cheval, L., Virlon, B., Elalouf, J.-M. (2000) SADE: A Microassay for Serial Analysis of Gene Expression, in: Hunt, S.P., Livesey, F.J. (Eds.) *Functional Genomics: A Practical Approach*, Oxford University Press, London.

Lorkowski, S., Cullen, P. (2003) *Analysing Gene Expression; A Handbook of Methods Possibilities and Pitfalls*, Wiley-VCH, Weinheim, Germany.

Shimkets, R.A. (2004) *Gene Expression Profiling: Methods and Protocols (Methods in Molecular Biology)*, Humana Press, Totowa, NJ.

Wang, S.M. (2005) *SAGE Technologies: Current Innovations and Future Trends*, Horizon Scientific Press, Norwich, CT.

Primary Literature

Adams, M.D., Kelley, J.M., Gocayne, J.D., et al. (1991) Complementary DNA sequencing: expressed sequence tags and human genome project, *Science* **252**, 1651–1656.

Akmaev, V.R., Wang, C.J. (2004) Correction of sequence-based artifacts in serial analysis of gene expression, *Bioinformatics* **20**, 1254–1263.

Allinen, M., Beroukhim, R., Cai, L., et al. (2004) Molecular characterization of the tumor microenvironment in breast cancer, *Cancer Cell* **6**, 17–32.

Angelastro, J.M., Klimaschewski, L.P., Vitolo, O.V. (2000) Improved NlaIII digestion of PAGE-purified 102-bp ditags by addition of a single purification step in both the SAGE and microSAGE protocols, *Nucleic Acids Res.* **28**, E62.

Angelastro, J.M., Ryu, E.J., Torocsik, B., Fiske, B.K., Greene, L.A. (2002) Blue-white selection step enhances the yield of SAGE concatemers, *Biotechniques* **32**, 484–486.

Angelastro, J.M., Klimaschewski, L., Tang, S., Vitolo, O.V., Weissman, T.A., Donlin, L.T., Shelanski, M.L., Greene, L.A. (2000) Identification of diverse nerve growth factor-regulated genes by serial analysis of gene expression (SAGE) profiling, *Proc. Natl. Acad. Sci. U.S.A.* **97**, 10424–10429.

Audic, S., Claverie, J.M. (1997) The significance of digital gene expression profiles, *Genome Res.* **7**, 986–995.

Baggerly, K.A., Deng, L., Morris, J.S., Aldaz, C.M. (2004) Overdispersed logistic regression for SAGE: modelling multiple groups and covariates, *BMC Bioinformatics* **5**, 144.

Beaudoing, E., Gautheret, D. (2001) Identification of alternate polyadenylation sites and analysis of their tissue distribution using EST data, *Genome Res.* **11**, 1520–1526.

Becquet, C., Blachon, S., Jeudy, B., Boulicaut, J.F., Gandrillon, O. (2002) Strong-association-rule mining for large-scale gene-expression data analysis: a case study on human SAGE data, *Genome Biol.* **3**, RESEARCH0067.

Beissbarth, T., Hyde, L., Smyth, G.K., Job, C., Boon, W.M., Tan, S.S., Scott, H.S., Speed, T.P. (2004) Statistical modeling of sequencing errors in SAGE libraries, *Bioinformatics* **20**(Suppl. 1), I31–I39.

Bertrand, G., Coste, J., Segarra, C., Schved, J.F., Commes, T., Marti, J. (2004) Use of serial analysis of gene expression (SAGE) technology reveals new granulocytic markers, *J. Immunol. Methods* **292**, 43–58.

Bonafoux, B., Lejeune, M., Piquemal, D., Quere, R., Baudet, A., Assaf, L., Marti, J., Aguilar-Martinez, P., Commes, T. (2004) Analysis of remnant reticulocyte mRNA reveals new genes and antisense transcripts expressed in the human erythroid lineage, *Haematologica* **89**, 1434–1438.

Boon, K., Caron, H.N., Van Asperen, R., et al. (2001) N-myc enhances the expression of a large set of genes functioning in ribosome biogenesis and protein synthesis, *EMBO J.* **20**, 1383–1393.

Boon, K., Osorio, E.C., Greenhut, S.F., et al. (2002) An anatomy of normal and malignant gene expression, *Proc. Natl. Acad. Sci. U.S.A.* **99**, 11287–11292.

Brazma, A., Hingamp, P., Quackenbush, J., et al. (2001) Minimum information about a microarray experiment (MIAME)-toward standards for microarray data, *Nat. Genet.* **29**, 365–371.

Brenner, S., Johnson, M., Bridgham, J., et al. (2000) Gene expression analysis by massively parallel signature sequencing (MPSS) on microbead arrays, *Nat. Biotechnol.* **18**, 630–634.

Buckhaults, P., Zhang, Z., Chen, Y.C., et al. (2003) Identifying tumor origin using a gene expression-based classification map, *Cancer Res.* **63**, 4144–4149.

Cai, L., Huang, H., Blackshaw, S., Liu, J.S., Cepko, C., Wong, W.H. (2004) Clustering analysis of SAGE data using a Poisson approach, *Genome Biol.* **5**, R51.

Caron, H., van Schaik, B., van der Mee, M., et al. (2001) The human transcriptome map: clustering of highly expressed genes in chromosomal domains, *Science* **291**, 1289–1292.

Chabardes-Garonne, D., Mejean, A., Aude, J.C., et al. (2003) A panoramic view of gene expression in the human kidney, *Proc. Natl. Acad. Sci. U.S.A.* **100**, 13710–13715.

Chen, J., Sun, M., Lee, S., Zhou, G., Rowley, J.D., Wang, S.M. (2002) Identifying novel transcripts and novel genes in the human genome by using novel SAGE tags, *Proc. Natl. Acad. Sci. U.S.A.* **99**, 12257–12262.

Chen, J., Sun, M., Kent, W.J., Huang, X., Xie, H., Wang, W., Zhou, G., Shi, R.Z., Rowley, J.D. (2004) Over 20% of human transcripts might form sense-antisense pairs, *Nucleic Acids Res.* **32**, 4812–4820.

Datson, N.A., van der Perk-de Jong, J., van den Berg, M.P., de Kloet, E.R., Vreugdenhil, E. (1999) MicroSAGE: a modified procedure for serial analysis of gene expression in limited amounts of tissue, *Nucleic Acids Res.* **27**, 1300–1307.

de Chaldee, M., Gaillard, M.C., Bizat, N., Buhler, J.M., Manzoni, O., Bockaert, J., Hantraye, P., Brouillet, E., Elalouf, J.M. (2003) Quantitative assessment of transcriptome differences between brain territories, *Genome Res.* **13**, 1646–1653.

de Jonge, R.R., Vreijling, J.P., Meintjes, A., Kwa, M.S., van Kampen, A.H., van Schaik, I.N., Baas, F. (2003) Transcriptional profile of the human peripheral nervous system by serial analysis of gene expression, *Genomics* **82**, 97–108.

DeRisi, J.L., Iyer, V.R., Brown, P.O. (1997) Exploring the metabolic and genetic control of gene expression on a genomic scale, *Science* **278**, 680–686.

Divina, P., Forejt, J. (2004) The ouse SAGE Site: database of public mouse SAGE libraries, *Nucleic Acids Res.* **32**, D482–D483.

Du, Z., Scott, A.D., May, G.D. (2003) Amplification of high-quantity serial analysis of gene expression ditags and improvement of concatemer cloning efficiency, *Biotechniques* **35**, 66–67.

El-Rifai, W., Moskaluk, C.A., Abdrabbo, M.K., Harper, J., Yoshida, C., Riggins, G.J., Frierson, H.F., Powell, S.M. (2002) Gastric cancers overexpress S100A calcium-binding proteins, *Cancer Res.* **62**, 6823–6826.

Fizames, C., Munos, S., Cazettes, C., et al. (2004) The arabidopsis root transcriptome by serial analysis of gene expression. Gene identification using the genome sequence, *Plant Physiol.* **134**, 67–80.

Gnatenko, D.V., Dunn, J.J., McCorkle, S.R., Weissmann, D., Perrotta, P.L., Bahou, W.F. (2003) Transcript profiling of human platelets using microarray and serial analysis of gene expression, *Blood* **101**, 2285–2293.

Hashimoto, S., Nagai, S., Sese, J., et al. (2003) Gene expression profile in human leukocytes, *Blood* **101**, 3509–3513.

Hashimoto, S., Suzuki, T., Dong, H.Y., Yamazaki, N., Matsushima, K. (1999) Serial analysis of gene expression in human monocytes and macrophages, *Blood* **94**, 837–844.

Hashimoto, S., Suzuki, Y., Kasai, Y., Morohoshi, K., Yamada, T., Sese, J., Morishita, S., Sugano, S., Matsushima, K. (2004) 5′-end SAGE for the analysis of transcriptional start sites, *Nat. Biotechnol.* **22**, 1146–1149.

Heidenblut, A.M., Luttges, J., Buchholz, M., et al. (2004) aRNA-longSAGE: a new approach to generate SAGE libraries from microdissected cells, *Nucleic Acids Res.* **32**, e131.

Hough, C.D., Sherman-Baust, C.A., Pizer, E.S., Montz, F.J., Im, D.D., Rosenshein, N.B., Cho, K.R., Riggins, G.J., Morin, P.J. (2000) Large-scale serial analysis of gene expression reveals genes differentially expressed in ovarian cancer, *Cancer Res.* **60**, 6281–6287.

Hu, Y., Sun, H., Drake, J., et al. (2004) From mice to humans: identification of commonly deregulated genes in mammary cancer via comparative SAGE studies, *Cancer Res.* **64**, 7748–7755.

Iseli, C., Stevenson, B.J., de Souza, S.J., Samaia, H.B., Camargo, A.A., Buetow, K.H., Strausberg, R.L., Simpson, A.J., Bucher, P., Jongeneel, C.V. (2002) Long-range heterogeneity at the 3′ ends of human mRNAs, *Genome Res.* **12**, 1068–1074.

Jurka, J. (2000) Repbase update: a database and an electronic journal of repetitive elements, *Trends Genet.* **16**, 418–420.

Kampa, D., Cheng, J., Kapranov, P., et al. (2004) Novel RNAs identified from an in-depth analysis of the transcriptome of human chromosomes 21 and 22, *Genome Res.* **14**, 331–342.

Kenzelmann, M., Muhlemann, K. (1999) Substantially enhanced cloning efficiency of SAGE (Serial Analysis of Gene Expression) by adding a heating step to the original protocol, *Nucleic Acids Res.* **27**, 917–918.

Kenzelmann, M., Muhlemann, K. (2000) Transcriptome analysis of fibroblast cells immediate-early after human cytomegalovirus infection, *J. Mol. Biol.* **304**, 741–751.

Lal, A., Lash, A.E., Altschul, S.F., et al. (1999) A public database for gene expression in human cancers, *Cancer Res.* **59**, 5403–5407.

Lash, A.E., Tolstoshev, C.M., Wagner, L., Schuler, G.D., Strausberg, R.L., Riggins, G.J., Altschul, S.F. (2000) SAGEmap: a public gene expression resource, *Genome Res.* **10**, 1051–1060.

Lee, S., Chen, J., Zhou, G., Wang, S.M. (2001) Generation of high-quantity and quality tag/ditag cDNAs for SAGE analysis, *Biotechniques* **31**, 348–350.

Lee, J.Y., Eom, E.M., Kim, D.S., Ha-Lee, Y.M., Lee, D.H. (2003) Analysis of gene expression profiles of gastric normal and cancer tissues by SAGE, *Genomics* **82**, 78–85.

Liang, P. (2002) SAGE genie: a suite with panoramic view of gene expression, *Proc. Natl. Acad. Sci. U.S.A.* **99**, 11547–11548.

Lockhart, D.J., Dong, H., Byrne, M.C., et al. (1996) Expression monitoring by hybridization to high-density oligonucleotide arrays, *Nat. Biotechnol.* **14**, 1675–1680.

Margulies, E.H., Kardia, S.L., Innis, J.W. (2001) Identification and prevention of a GC content bias in SAGE libraries, *Nucleic Acids Res.* **29**, E60.

Menssen, A., Hermeking, H. (2002) Characterization of the c-MYC-regulated transcriptome by SAGE: identification and analysis of c-MYC target genes, *Proc. Natl. Acad. Sci. U.S.A.* **99**, 6274–6279.

Muschen, M., Lee, S., Zhou, G., Feldhahn, N., Barath, V.S., Chen, J., Moers, C., Kronke, M., Rowley, J.D., Wang, S.M. (2002) Molecular portraits of B cell lineage commitment, *Proc. Natl. Acad. Sci. U.S.A.* **99**, 10014–10019.

Pauws, E., van Kampen, A.H., van de Graaf, S.A., de Vijlder, J.J., Ris-Stalpers, C. (2001) Heterogeneity in polyadenylation cleavage sites in mammalian mRNA sequences: implications for SAGE analysis, *Nucleic Acids Res.* **29**, 1690–1694.

Pauws, E., Veenboer, G.J., Smit, J.W., de Vijlder, J.J., Morreau, H., Ris-Stalpers, C. (2004) Genes differentially expressed in thyroid carcinoma identified by comparison of SAGE expression profiles, *FASEB J.* **18**, 560–561.

Piquemal, D., Commes, T., Manchon, L., Lejeune, M., Ferraz, C., Pugnere, D., Demaille, J., Elalouf, J.M., Marti, J. (2002) Transcriptome analysis of monocytic leukemia cell differentiation, *Genomics* **80**, 361–371.

Polyak, K., Riggins, G.J. (2001) Gene discovery using the serial analysis of gene expression technique: implications for cancer research, *J. Clin. Oncol.* **19**, 2948–2958.

Porter, D.A., Krop, I.E., Nasser, S., Sgroi, D., Kaelin, C.M., Marks, J.R., Riggins, G.,

Polyak, K. (2001) A SAGE (serial analysis of gene expression) view of breast tumor progression, *Cancer Res.* **61**, 5697–5702.

Powell, J. (1998) Enhanced concatemer cloning-a modification to the SAGE (Serial Analysis of Gene Expression) technique, *Nucleic Acids Res.* **26**, 3445–3446.

Quere, R., Manchon, L., Lejeune, M., Clement, O., Pierrat, F., Bonafoux, B., Commes, T., Piquemal, D., Marti, J. (2004) Mining SAGE data allows large-scale, sensitive screening of antisense transcript expression, *Nucleic Acids Res.* **32**, e163.

Rinn, J.L., Euskirchen, G., Bertone, P., et al. (2003) The transcriptional activity of human Chromosome 22, *Genes Dev.* **17**, 529–540.

Robert-Nicoud, M., Flahaut, M., Elalouf, J.M. et al. (2001) Transcriptome of a mouse kidney cortical collecting duct cell line: effects of aldosterone and vasopressin, *Proc. Natl. Acad. Sci. U.S.A.* **98**, 2712–2716.

Ruijter, J.M., Van Kampen, A.H., Baas, F. (2002) Statistical evaluation of SAGE libraries: consequences for experimental design, *Physiol. Genomics* **11**, 37–44.

Saha, S., Sparks, A.B., Rago, C., Akmaev, V., Wang, C.J., Vogelstein, B., Kinzler, K.W., Velculescu, V.E. (2002) Using the transcriptome to annotate the genome, *Nat. Biotechnol.* **20**, 508–512.

Seth, P., Krop, I., Porter, D., Polyak, K. (2002) Novel estrogen and tamoxifen induced genes identified by SAGE (Serial Analysis of Gene Expression), *Oncogene* **21**, 836–843.

Sharon, D., Blackshaw, S., Cepko, C.L., Dryja, T.P. (2002) Profile of the genes expressed in the human peripheral retina, macula, and retinal pigment epithelium determined through serial analysis of gene expression (SAGE), *Proc. Natl. Acad. Sci. U.S.A.* **99**, 315–320.

Silva, A.P., De Souza, J.E., Galante, P.A., Riggins, G.J., De Souza, S.J., Camargo, A.A. (2004) The impact of SNPs on the interpretation of SAGE and MPSS experimental data, *Nucleic Acids Res.* **32**, 6104–6110.

So, A.P., Turner, R.F., Haynes, C.A. (2004) Increasing the efficiency of SAGE adaptor ligation by directed ligation chemistry, *Nucleic Acids Res.* **32**, e96.

St Croix, B., Rago, C., Velculescu, V. et al. (2000) Genes expressed in human tumor endothelium, *Science* **289**, 1197–1202.

Stein, W.D., Litman, T., Fojo, T., Bates, S.E. (2004) A serial analysis of gene expression (SAGE) database analysis of chemosensitivity: comparing solid tumors with cell lines and comparing solid tumors from different tissue origins, *Cancer Res.* **64**, 2805–2816.

Strausberg, R.L., Simpson, A.J., Old, L.J., Riggins, G.J. (2004) Oncogenomics and the development of new cancer therapies, *Nature* **429**, 469–474.

Sun, M., Zhou, G., Lee, S., Chen, J., Shi, R.Z., Wang, S.M. (2004) SAGE is far more sensitive than EST for detecting low-abundance transcripts, *BMC Genomics* **5**, 1.

Suzuki, T., Hashimoto, S., Toyoda, N., Nagai, S., Yamazaki, N., Dong, H.Y., Sakai, J., Yamashita, T., Nukiwa, T., Matsushima, K. (2000) Comprehensive gene expression profile of LPS-stimulated human monocytes by SAGE, *Blood* **96**, 2584–2591.

Velculescu, V.E., Zhang, L., Vogelstein, B., Kinzler, K.W. (1995) Serial analysis of gene expression, *Science* **270**, 484–487.

Velculescu, V.E., Zhang, L., Zhou, W., Vogelstein, J., Basrai, M.A., Bassett, D.E., Hieter, P., Vogelstein, B., Kinzler, K.W. (1997) Characterization of the yeast transcriptome, *Cell* **88**, 243–251.

Velculescu, V.E., Madden, S.L., Zhang, L., et al. (1999) Analysis of human transcriptomes, *Nat. Genet.* **23**, 387–388.

Vencio, R.Z., Brentani, H., Patrao, D.F., Pereira, C.A. (2004) Bayesian model accounting for within-class biological variability in serial analysis of gene expression (SAGE), *BMC Bioinformatics* **5**, 119.

Versteeg, R., van Schaik, B.D., van Batenburg, M.F., Roos, M., Monajemi, R., Caron, H., Bussemaker, H.J., van Kampen, A.H. (2003) The human transcriptome map reveals extremes in gene density, intron length, GC content, and repeat pattern for domains of highly and weakly expressed genes, *Genome Res.* **13**, 1998–2004.

Vilain, C., Libert, F., Venet, D., Costagliola, S., Vassart, G. (2003) Small amplified RNA-SAGE: an alternative approach to study transcriptome from limiting amount of mRNA, *Nucleic Acids Res.* **31**, e24.

Virlon, B., Cheval, L., Buhler, J.M., Billon, E., Doucet, A., Elalouf, J.M. (1999) Serial microanalysis of renal transcriptomes, *Proc. Natl. Acad. Sci. U.S.A.* **96**, 15286–15291.

Wei, C.L., Ng, P., Chiu, K.P., Wong, C.H., Ang, C.C., Lipovich, L., Liu, E.T., Ruan, Y. (2004) 5' Long serial analysis of gene expression (LongSAGE) and 3' LongSAGE for transcriptome characterization and genome annotation, *Proc. Natl. Acad. Sci. U.S.A.* **101**, 11701–11706.

Yamamoto, M., Wakatsuki, T., Hada, A., Ryo, A. (2001) Use of serial analysis of gene expression (SAGE) technology, *J. Immunol. Methods* **250**, 45–66.

Ye, S.Q., Zhang, L.Q., Zheng, F., Virgil, D., Kwiterovich, P.O. (2000) miniSAGE: gene expression profiling using serial analysis of gene expression from 1 microg total RNA, *Anal. Biochem.* **287**, 144–152.

Yu, J., Zhang, L., Hwang, P.M., Rago, C., Kinzler, K.W., Vogelstein, B. (1999) Identification and classification of p53-regulated genes, *Proc. Natl. Acad. Sci. U.S.A.* **96**, 14517–14522.

Zhang, L., Zhou, W., Velculescu, V.E., Kern, S.E., Hruban, R.H., Hamilton, S.R., Vogelstein, B., Kinzler, K.W. (1997) Gene expression profiles in normal and cancer cells, *Science* **276**, 1268–1272.

Zhou, G., Chen, J., Lee, S., Clark, T., Rowley, J.D., Wang, S.M. (2001) The pattern of gene expression in human CD34(+) stem/progenitor cells, *Proc. Natl. Acad. Sci. U.S.A.* **98**, 13966–13971.

25
Gene Mapping by Fluorescence In Situ Hybridization

Barbara G. Beatty[1] *and Henry H.Q. Heng*[2]
[1] *Department of Pathology, University of Vermont College of Medicine, Burlington, VT, USA*
[2] *Center for Molecular Medicine and Genetics, Wayne State University School of Medicine, Detroit, MI, USA*

1	**Introduction** 716	
2	**Probes for Gene Mapping by FISH** 717	
2.1	Identification of Probes from Genomic Databases	717
2.2	Probes for FISH Mapping 718	
2.2.1	Genomic DNA Probes 718	
2.2.2	Unique Sequence cDNA and EST Probes 719	
2.2.3	Synthetic Oligonucleotides 719	
2.3	Probe Labeling 720	
2.3.1	Purification 720	
2.3.2	Labeling Methods 720	
2.3.3	Postlabeling Purification and Storage 722	
3	**DNA Targets for FISH Mapping** 722	
3.1	Metaphase Chromosomes 722	
3.1.1	Metaphase Chromosome Preparation 723	
3.2	Interphase Nuclei 724	
3.2.1	Interphase Nuclei Preparation 725	
3.3	Chromatin and DNA Fibers 725	
3.3.1	Fiber FISH Target Preparation 726	
4	**Denaturation, Hybridization, and Detection** 729	
4.1	Pretreatment and Denaturation 729	
4.2	Hybridization 730	
4.2.1	Stringency Washes 730	

Genomics and Genetics. Edited by Robert A. Meyers.
Copyright © 2007 Wiley-VCH Verlag GmbH & Co. KGaA, Weinheim
ISBN: 978-3-527-31609-0

4.3	Detection	730
4.3.1	Counterstaining and Banding	731
5	**Microscopy and Image Analysis**	**732**
6	**Applications**	**733**
6.1	Metaphase Mapping	736
6.1.1	Single-copy Genes	736
6.1.2	Segmental Duplications and FISH Mapping	737
6.1.3	Relational Mapping	737
6.1.4	FISH Mapping Combined with CGH and spectral karyotyping (SKY)	740
6.1.5	Comparative Mapping by ZOO-FISH	740
6.2	Interphase FISH Mapping	741
6.3	Chromatin and DNA Fiber Mapping	741
	Bibliography	**742**
	Books and Reviews	742
	Primary Literature	742

Keywords

BAC
Bacterial artificial chromosome vector that is used to clone large genomic DNA inserts of 100 to 300 kb.

Chromosomal Bands
Light and dark staining patterns along the length of the chromosome that are unique for each chromosome and are produced by differential binding of dyes.

Chromatin Fibers
Strands of complexed DNA, and proteins (histones) organized into compact nucleosomes that are disk-like structures 10 nm in diameter and comprise 140 to 146 base pairs of DNA wound around histone octomers.

Released Chromatin Fibers
Linearized 30-nm interphase chromatin fibers experimentally released from interphase nuclei.

Released DNA Fiber
Linearized DNA strands experimentally released from nuclei or chromosomes serving as target for high-resolution FISH mapping.

FISH
Fluorescence *in situ* hybridization is the technique of localizing fluorescent nucleic acid probes (DNA, RNA) to complementary sequences of target DNA on a glass slide and visualizing them in a fluorescence microscope.

Interphase Nuclei
Nuclei that are between one mitosis and the next and contain decondensed chromatin.

Metaphase Chromosome
Fully condensed chromosome present at the mitotic stage (metaphase) in which chromosomes are attached to the mitotic spindle but have not yet segregated to the opposite spindle poles.

PAC
P1-derived artificial chromosome vector used to clone large genomic DNA inserts of 130 to 150 kb.

YAC
Yeast artificial chromosome vector used to clone large genomic DNA inserts of a few hundred kb to 1 Mb.

■ Gene mapping by fluorescence *in situ* hybridization (FISH) is the localization of a unique DNA sequence (probe) to a specific position (chromosomal band) within the genome (target). The DNA probe localization on the target is visualized using a fluorescent reporter molecule directly or indirectly bound to the probe. Although gene mapping by FISH usually refers to the localization of a gene or DNA sequence to a specific metaphase chromosome band, it can also include the relative position of genes along a chromatin or DNA fiber, or the identification of specific genomic BAC/PAC clones present in an array. FISH combines the specificity of recombinant DNA technology with basic cytogenetic techniques for target DNA preparation.

Valuable information regarding the biological and/or clinical significance of a gene can be obtained by identifying its subchromosomal location in both normal and disease conditions. Mapping a gene to certain critical chromosomal regions that may be deleted, amplified, or involved in translocation breakpoints can provide important information about the role of that gene in disease processes and may have significant clinical applications. Important information regarding gene function can also be obtained from mapping genes from individual tissues or disease states, identifying transgene insertion sites, studying interspecies synteny, and determining the order of and distance between two or more genes within a chromosome band.

1 Introduction

Fluorescence *in situ* hybridization (FISH) combines the chromosome preparation and banding techniques of classical cytogenetics with recombinant DNA technology. Specific DNA sequences (probes) when denatured will form a heteroduplex with denatured DNA on a glass microscope slide (target), and the resulting hybridization is visualized by a fluorescent reporter molecule (Fig. 1). The use of fluorescence over radioactivity as a mode of detection became feasible in the early 1980s when methods for labeling nucleic acids with haptens such as biotin were developed. This indirect approach to visualizing hybridized probe requires further detection steps with fluorescently labeled reporter molecules such as avidin or specific antibodies. Direct incorporation of fluorochromes into nucleic acids has now greatly simplified detection.

The first gene was mapped to human metaphase chromosomes using FISH in the mid-1980s, and now over 5000 articles utilizing gene mapping by FISH have been published. FISH is one of the few mapping techniques that does not require clones to overlap, to be positioned and ordered, and thus has broad applications in positional cloning studies. FISH now plays an important role in confirming *in silico* mapping information obtained through the human genome project, which has produced genome sequence databases for both human and other species.

FISH can localize a single gene to a subchromosomal band, can allow ordering of several genes within a chromosome band, and can aid in determining locus copy number, probe chimerism (YACs), pseudogenes, and location of gene family members. FISH mapping resolution is dependent on the level of chromatin condensation present in the target DNA and covers a wide range from high-resolution DNA fiber mapping (1–5 kb) to metaphase chromosome mapping at a resolution of 1 to 2 Mb, making it an exceptionally versatile mapping tool.

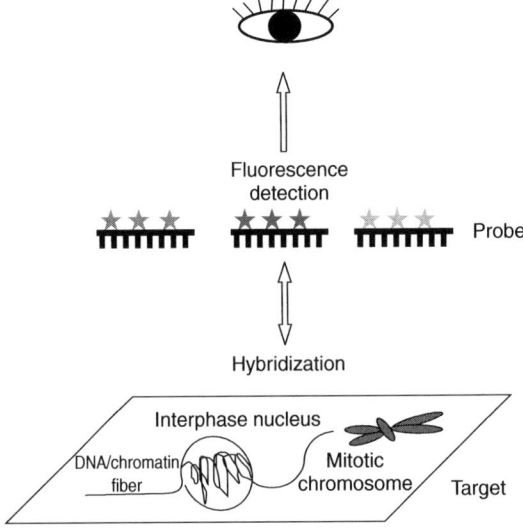

Fig. 1 Schematic representation of FISH. Basic FISH protocols consist of (a) preparation of target DNA in the form of metaphase chromosomes, interphase nuclei, or DNA/chromatin fibers on a slide; (b) labeling of probes directly with a fluor or indirectly with a hapten followed by a secondary fluor; (c) hybridization of the probe to the target; and (d) detection of the fluorescent hybridized probe under an epifluorescence microscope. (We acknowledge Dr Christine Ye and Miss Wei Lu for their help with the diagram.)

2
Probes for Gene Mapping by FISH

2.1
Identification of Probes from Genomic Databases

The human genome project has provided a wealth of data regarding DNA sequences suitable for gene mapping by FISH. Several databases containing information about expressed sequence tags (ESTs) and genes that have been sequenced or partially sequenced are available publicly on the World Wide Web. It may be necessary to search several websites in order to determine the relationship and ordering of these clones as no single database currently contains all this information. Although the quality of the clones is variable, use of these databases and *in silico* methods provide the best initial approach to identifying or localizing a particular DNA sequence to a chromosomal band or region. Validation of the localization can then be performed by FISH using a corresponding genomic DNA probe. Libraries containing bacterial artificial chromosome (BAC), P1 artificial chromosome (PAC), yeast artificial chromosome (YAC), and cosmid clones are now available for most of the human genome, and confirmation of the physical chromosomal location of these DNA sequences by FISH mapping is extremely useful.

There are several approaches to identifying clones (probes) suitable for human FISH mapping on the web. The first is to use the cytogenetic location. One can search for premapped FISH probes using the Genome Database http://www.gdb.org/hugo/ under "Chromosome Resources" and clones based on cytogenetic position can be obtained using one of the three following sites:

- Genome database site: http://www.gdb.org/gdb/regionSearch.html,
- GeneCards: http://bioinfo.weizmann.ac.il/cards/index.html, or
- Genetic Location Database: http://cedar.genetics.soton.ac.uk/public_html/gmap.html.

A second approach is to obtain BAC, PAC, and cosmid clones specific for a particular gene or DNA marker by searching either the Genome Database (http://www.gdb.org) or the Locus link (http://www.ncbi.nlm.nih.gov/Locus Link/). In each case, the gene or marker name can be used to link to a DNA sequence database that can be screened to identify the clone name. For YACs, a similar approach is used at the Whitehead/MIT website (http://carbon.wi.mit.edu.8000/cgi-bin/conyig/phys_map) to identify both the clone and the contig containing overlapping clones along the chromosome.

To search for a corresponding genomic clone by gene or marker sequence, the sequence can be retrieved from a text search of http://www.ncbi.nlm.nih.gov:80/entrez/query.fcgi?db=Nucleotide. The accession number or DNA sequence can then be copied and pasted into http://www.ncbi.nih.gov/blast/blast.cgi and databases NR (finished sequences), HTGS (rough draft sequence), or GSS (overlapping BAC ends) screened for those containing the corresponding genomic clones. If identification of nearby or overlapping clones is required, this can be achieved on the basis of knowing the BAC clone (1) name or sequence (http://www.TIGR.org/tdb/humgen/bac_end_search/bac_end_intro.html), (2) fingerprint patterns (http://genome.wustl.edu/gsc/human/human_database.shtml), or (3) DNA sequence alignment [(Ensembl (http://www.ensembl.org/), NCBI (http://

www.ncbi.nlm.nih.gov/cgi-bin/Entrez/hum_srch?chr=hum_chr.inf), or UCSF (http://genome.ucsc.edu/)].

Once the specific clones or sequences have been identified, there are several sources from which they can be obtained and the websites are listed as follows:

Cosmids, BACs, PACs, and YACS

- The Centre for Applied Genomics (http://tcag.bioinfo.sickkids.on.ca/) – Canada
- DHGP (http://www.rzpd.de/) – Germany
- HGMP Resource Centre (http://www.hgmp.mrc.ac.uk/) – UK

BACs, PACs, and YACs

- TIGEM (http://www.spr.it/iger/home.html) – Italy
- Research Genetics (http://www.resgen.com/) – USA
- Genome Systems (http://reagents.incyte.com/) – USA

BACs and PACs

- Sanger Center (http://www.sanger.ac.uk/HGP/) – UK
- Pieter de Jonge's lab (http://www.chori.org/bacpac/) – USA

Genome databases are also available on the web for other species such as: http://bos.cvm.tamu.edu/bovgbase.html (cattle), http://ratmap.gen.gu.se/DownloadGeneData.html (rat), and http://www.ncbi.nlm.nih.gov/entrez/query.fcgi?db=unigene (EST data for mouse, cow, zebra fish, clawed frog, fruit fly, and mosquito as well as wheat, barley, maize, and cress) as sources for genomic clones for FISH mapping.

2.2
Probes for FISH Mapping

The probes for FISH range from chemically synthesized oligonucleotides to total genomic DNA. Typically, DNA sequences used for FISH mapping should be of such a size and nature as to be readily visualized by a fluorescence reporter molecule under a 60 or 100× objective of an epifluorescence microscope. The two most common single-copy probes used in mapping experiments are cloned genomic sequences (YACs, BACs, PACs, cosmids, and plasmids) and enzymatically amplified unique sequences of cDNAs and expressed sequence tags. In all cases, the probe purity, type, and size are critical considerations for optimal mapping results. The degree of probe purity required is dependent on the labeling method to be used. The type and size of probe affects the labeling approach, prehybridization treatment, and detection method.

2.2.1 Genomic DNA Probes

Cloned single-copy genomic probes range in size from ~5 kb (plasmids) to ~1 Mb (YACs) and their signal strength is usually directly proportional to their size. Genomic probes can be labeled by nick translation or PCR and are visualized by direct or indirect labeling (see Sect. 2.3). Since genomic probes consist of contiguous genome sequences, they contain both unique sequences specific for the gene or marker, and commonly repeated sequences (SINEs/alu, LINEs) interspersed throughout the genome. In order to visualize the chromosomal location of the unique genomic sequences, with minimal nonspecific hybridization background, the common repeat sequences must be suppressed or blocked prior to hybridization.

This is achieved by preannealing the probe DNA with competitor total DNA or Cot-1 DNA of the same species. Cot-1 DNA is genomic DNA enriched in the commonly repeated genome sequences that is commercially available but only for human or mouse. This suppression hybridization approach is essential for mapping genomic probes to their unique location on target chromosomes.

Among the single-copy genomic probes, BACs and PACs are used most frequently for FISH mapping due to the availability of BAC and PAC libraries, their stability, and appropriate insert size (100 to 200 kb) for easy visualization. YACs must be checked for chimerism prior to use as they tend to be unstable and have a high rate of noncontiguous coligated insert sequences. Consequently, caution must be taken when using YACs for mapping purposes.

Other genomic probes important in FISH applications in molecular diagnostics, human cytogenetics, and genomics that also have mapping related functions include total genomic DNA used in comparative genomic hybridization (CGH) and chromosome painting probes. The application of these probes and their role in genome mapping will be discussed in Sect. 6. Satellite tandem repeat probes recognizing chromosome-specific tandem repeat sequences in the chromosomal centromeric and telomeric regions can be used in conventional and quantitative FISH mapping studies to identify specific chromosomes and measure telomeric repeats.

2.2.2 Unique Sequence cDNA and EST Probes

Unique sequence probes are smaller (<5 kb) than genomic probes and contain only expressed exonic sequences. These probes do not require blocking of common repeat sequences by Cot-1 DNA. As these probes contain noncontiguous genomic DNA sequences, their signal strength is related to the size and structure of the gene target and thus can be unpredictable. The transcribed (expressed) sequences within a gene may comprise a number of large contiguous exons or a number of short, widely separated exons. The former configuration would have a higher likelihood of producing a visible signal compared to that of the latter for the same sized cDNA probe. Thus, both the size and organization of the gene can determine the mapping results and should be tested on an individual basis. In general, cDNA or EST probes are best visualized by indirect labeling techniques and tyramide amplification detection methods (see Sects. 2.3.2 and 4.3)

2.2.3 Synthetic Oligonucleotides

Chemically synthesized oligonucleotides are usually too small to be visualized as unique sequence probes in mapping studies. However, they can be used for detection of high-frequency tandem repeat sequences such as those found in alpha satellite, telomeric, and Alu repeats present in centromeres, telomeres, and chromosome bands. These probes can be important in determining the chromosome-specific location of single-copy probes, and as an alternative to chromosome banding (see Sect. 4.3.1), especially in species other than human or mouse. Specialized artificial peptide nucleic acid (PNA) probes that have a polyaminoethyl glycine backbone replacing the pentose-phosphate backbone have the advantage that they bind DNA more stably at low salt concentrations. These probes have been useful in quantitative examination of telomeric repeat sequence lengths.

2.3
Probe Labeling

2.3.1 Purification

Prior to labeling, the starting probe DNA must be purified. The labeling method determines the extent of the required chemical purity of the probe. For preparations to be labeled by nick translation, DNA purification procedures are required for removal of RNA, which, if present, would result in significant nonspecific background. PCR labeling methods do not require the same purity of DNA. If cloned probes are being used, it is usually not necessary to remove probe DNA inserts from the host vector DNA prior to labeling since vector DNA does not reduce the labeling efficiency of insert DNA and will not hybridize to mammalian DNA targets. Cloned genomic probe preparations are most often isolated and purified using standard CsCl or alkaline lysis procedures, or commercially available kits (Qiagen, Roche Diagnostics).

2.3.2 Labeling Methods

The goal of labeling a probe for FISH is to both incorporate labeled nucleotides into the DNA and to produce short DNA fragments of 100 to 500 bp. This fragment size allows probe access to target chromosomal DNA while maintaining high specificity. Longer fragments can result in reduced target access, producing reduced specific signal and increased background. Shorter fragments usually require reduced stringency wash conditions and may result in significant nonspecific chromosomal background.

Nick translation, the preferred method of labeling for most genomic probes, both introduces a labeled nucleotide into the probe DNA, and cuts the DNA into appropriate fragments. Nick translation combines a DNase I enzyme, which randomly nicks single strands of the DNA template producing available 3'ends, with *Escherichia coli* DNA polymerase I, which both removes nucleotides by its 5'-3' exonuclease activity and incorporates nucleotides at the available 3' ends by its 5'-3' polymerase activity. The labeling mixture consists of probe DNA, the enzymes DNAse 1 and DNA polymerase 1, a dNTP mixture and a fluor, aminoallyl, or hapten-coupled nucleotide (Table 1), usually in final volume of 100 µL or less. The reaction mixture is placed in a 15 to 16 °C water bath, for 1.5 to 2 h or longer depending on the labeled nucleotide being incorporate. Following incubation, the reaction is placed on ice and the size of the labeled fragments checked by electrophoresis on a 1.5 to 2% agarose minigel containing ethidium bromide. A standard sizing ladder should be included in the run. Samples can be heat-denatured prior to loading on the gel, but this is not necessary. If the fragment size is too large, the reaction can be continued with further incubation; if the fragment size is too small, the labeling procedure must be repeated for a shorter time. If the fragment size is within the correct range, the reaction is stopped by addition of 0.5 M ethylenediaminetetraacetic acid (EDTA), pH 8.0. High prime or random primer labeling is another less popular approach that may provide a higher efficiency of labeled nucleotide incorporation, but fragment size cannot be well controlled, and further digestion by DNase I is often required. PCR labeling using specific primers can be performed when amplifying DNA from any source, but requires some knowledge of the probe sequence. Alternatively,

Tab. 1 Common fluors and haptens for FISH.

	Peak wavelengths [nm]	
	Excitation	Emission
Fluor		
Fluorescein (FITC)	494	518
Alexa fluor 488	495	519
Cyanine Cy2	492	510
Oregon green 488	496	524
Spectrum green	497	524
Tetramethylrhodamine (TRITC)	550	570
Indocarbocyanine Cy3	550	570
Alexa 546	556	573
Spectrum orange	559	588
Texas red	596	620
Alexa 647	650	668
Indodicarbocyanine Cy5	650	670
DNA counterstain[a,b]		
4'-6-diamidino-2-phnylindole (DAPI)	365	450
Propidium iodide (PI)	540	615
Hoechst 33258	360	465
Chromomycin	445	570
Quinacrine	455	495
Haptens		
Biotin		
Digoxigenin (DIG)		
Dinitrophenyl aminohexanoic acid (DNP)		
Bromodeoxyuridine (BrdU)		

[a] Dye/DNA complex.
[b] Spectral properties are dependent on dye/DNA ratio, and ionic strength and pH of the dye solution.

using degenerate oligonucleotide primer (DOP)-PCR, complex probe libraries such as chromosome-specific painting probes can be prepared and labeled by incorporation of fluor, hapten, or amine group-derivatized nucleotides.

The type of labeling usually depends on the size of the probe and/or the target.

1. *Direct labeling*: Single-copy genomic probes such as PACs, BACs, or YACs that are >100 kb, and probes for large targets such as alpha satellite repeats, can usually be detected by the direct labeling approach. Fluorochrome-coupled or aminoallyl-derivatized nucleotides that bind amine reactive fluorescent dyes can be obtained commercially or prepared in the laboratory and are incorporated into probe DNA by nick translation, sequence-specific PRC, or DOP-PCR labeling. Fluorochrome-labeled DNA probes usually do not require further detection/amplification systems, however,

antibodies directed against some fluors are available if further amplification is required.

2. *Indirect labeling*: Unique sequence cDNA, EST, and smaller genomic probes (cosmids and plasmids) are too small to be visualized directly and require some form of amplification. These probes are best labeled with a hapten such as biotin or digoxigenin (DIG), which must be further reacted with a fluorescent reagent to be visualized. DNA probes labeled with biotin or DIG-conjugated nucleotides require secondary detection systems involving fluorochrome-labeled avidin or anti-DIG antibodies. In general, although direct labeling is usually a longer process then indirect labeling, detection is significantly expedited. The most common fluors and haptens used for FISH mapping are shown in Table 1.

2.3.3 Postlabeling Purification and Storage

Following the labeling process, the labeled DNA is purified by ethanol precipitation or with spin columns. Addition of carrier DNA (salmon or herring sperm DNA) or glycogen increases the yield of DNA during ethanol precipitation and reduces nonspecific background binding to both the target DNA and the glass slide. Genomic probes also require suppression hybridization with competitor Cot-1 or genomic DNA to block the common interspersed repeat sequences present (see Sect. 2.2.1). Following precipitation, the probe is centrifuged, washed with cold 70% ethanol, recentrifuged, and air-dried. The DNA pellet can be left as a dry pellet or resuspended in distilled water, Tris EDTA (TE) buff, or hybridization buffer and stored at $-20\,°C$.

3 DNA Targets for FISH Mapping

3.1 Metaphase Chromosomes

Metaphase chromosomes have been widely used to assign a chromosomal map location to the many genes and DNA sequences (both human and other species) arising from the human genome project. Genomic DNA in the form of metaphase chromosomes is in its most condensed state and mapping resolution is limited to interprobe distances of >1 Mb. Localization of a single labeled probe on a chromosome can be determined by either the relative position of the signal along the length of the chromosome, or by its localization within a specific chromosomal band. Relative positioning is achieved by measuring the distance between the telomeric end of the p arm (pter) and the probe as a fraction of the total chromosome length (Flpter, or fractional length from the pter). Difficulties with this approach include nonuniform condensation along the chromosome and the presence of repeat sequence length polymorphisms in different individuals. This method has, for the most part, been replaced by band or subband localization.

Reproducible banding patterns unique for each chromosome have been developed using DNA staining procedures. Banding techniques that allow simultaneous visualization of the probe signal and the banding pattern provide an efficient and precise approach to metaphase gene mapping.

Multicolor probes can be used to visualize the order of several genes or markers along metaphase chromosomes. The labeled probes must be separated by at least 1 Mb to be distinguished, but since most chromosomal bands are several Mb in width, ordering probes within a single band can be accomplished by this multicolor FISH. Mapping to telomeric and centromeric regions should be performed with caution and verified by other approaches, as chromatin folding in these regions tends to produce artifacts and order changes.

3.1.1 Metaphase Chromosome Preparation

The most common source of human and large mammal metaphase chromosomes suitable for gene mapping is mitogen-stimulated short-term cultures of peripheral blood lymphocytes. Amniocyte cultures, fibroblasts, lymphoblasts, and cell cultures can also produce good results. For mouse and rat, bone marrow, spleen, lymph nodes, and thymus are commonly used for chromosome preparation. For precise mapping, high quality, well-spread chromosomes with distinct banding patterns are essential. Preparation of mapping quality metaphase chromosomes is based on minor modifications of standard cytogenetic techniques. Cell synchronization may be required for some types of banding (see Sect. 4.3.1) and usually enhances the number of cells in metaphase, however, it is generally not necessary for most mapping situations.

There are four basic steps involved in the preparation of metaphase chromosomes from cultured cells:

1. Stimulation of peripheral T lymphocytes by the mitogen phytohaemagglutinin (PHA).
2. Inhibition of cell cycle by addition of a mitotic spindle inhibitor (colcemid) to arrest the cells in metaphase, that is, when the genomic DNA is condensed into chromosomes.
3. Hypotonic swelling of the cells to reduce cytoplasm and enhance chromosome spreading.
4. Fixation to inhibit further swelling, arrest cell function, and preserve good cell morphology.

The following comments pertain to human cells, but can easily be adapted to other species. Heparanized peripheral blood should be cultured within three to four days of collection in a medium containing 1% PHA to stimulate the division of T lymphocytes. When the cultured cells reach maximum mitotic activity (within 60–70 h), addition of colcemid for 20 to 30 min prior to harvest disrupts spindle formation and arrests the cells in metaphase. In this phase of the cell cycle, the nuclear membrane disintegrates, allowing the chromosomes to disperse within the cell. Production of good metaphase chromosome spreads depends on both the concentration of colcemid and the length of exposure. Too long a treatment results in short, condensed chromosomes and tight metaphases unsuitable for mapping purposes. Some protocols for preparation of murine chromosomes omit colcemid altogether. For adherent cells, subculturing 48 h and feeding 24 h prior to harvest at 50 to 80% confluence can enhance the mitotic index, and addition of colcemid for a longer time (4–6 h to overnight) is suggested for optimal metaphase chromosome production.

If cell synchronization is required, methotrexate (amethopterin) can be added to growing cells. Methotrexate inhibits incorporation of thymidine into DNA, thus arresting the cells in early S-phase. Following the removal of the S-phase block and addition of thymidine or its derivative bromodeoxyuridine (BrdU), the cells resume DNA synthesis and proceed through the cell cycle in a synchronized manner. Substitution of BrdU is used when replication banding is required (see Sect. 4.3.1).

Harvested cells are swollen with a hypotonic solution of KCl or medium to increase fragility and lyse any remaining red blood cells. The loss of nuclear membrane results in the release of chromosomes throughout the enlarged cells facilitating their dispersion. The cells are slowly fixed in methanol:glacial acetic acid to enhance nucleoprotein cross-linking and gradually dehydrate the cells, preserving them in the swollen state. Proper fixation is critical for good chromosome morphology and attachment to the slides. Improperly fixed cells may result in cell contraction and poorly spread tight chromosomes. Overfixation can result in probe and reagent penetration problems, whereas underfixation may result in loss of chromatin and chromosome morphology. The fixed cells can be stored as pellets in tubes containing fixative at $-20\,°C$ for several months. Resuspension in fresh fixative prior to each use, and minimizing exposure to room temperature air prevents evaporation of methanol and fixative esterification due to uptake of moisture. The fixed cells are dropped onto clean, wet glass slides and allowed to run down the slide held at a $45°$ angle, and dried. Metaphase spreads should be aged at room temperature for at least two days before being used for FISH mapping. Slides can be stored at room temperature up to four weeks, or for several months at $-20°$ in a sealed dry container.

Preparation of high-quality metaphase spreads with clear banding resolution suitable for mapping studies is dependent on many factors. Colcemid treatment and fixation, humidity, temperature, and drying time are all critical steps during slide preparation and affect chromosome spreading, morphology, and the degree of residual cytoplasm present on the slides. Residual cytoplasmic proteins and RNA can increase background and reduce visualization of signal. Under ideal conditions of temperature, humidity, and drying time, when viewed under a phase-contrast microscope, cytoplasm should not be visible and chromosomes should be dark gray with sharp borders. The use of commercially available controlled environment chambers has significantly improved the standardization of slide making.

3.2
Interphase Nuclei

Interphase nuclei mapping provides improved resolution for ordering probes situated too close to each other to be distinguished on metaphase chromosomes. Since the folding of DNA in somatic interphase nuclei is about $20\times$ less than that found in metaphase DNA, the mapping resolution is increased to 50 to 100 kb, compared to 1 to 3 Mb in metaphase DNA. In certain regions of the genome, the distance between probe signals in interphase nuclei are proportional to their genomic distance, with a linear relationship existing between the mean square of the distance ($\leq 1.5\,\mu m$) and the genetic distance ($< 1-2\,Mb$). Since probes separated by > 1 to 2 Mb do not follow this relationship because of chromatin looping, mapping such probes may result in incorrect distance measurements.

Somatic interphase mapping does not provide information on chromosomal banding localization or location relative to the centromere or telomere region. Probe order, however, can be determined by two approaches: in the first approach, three or more probes labeled with a single fluor can be ordered on the basis of distance measurements (mean-square distance) of pairwise hybridization experiments provided they are in the 50 kb to 1–2 Mb range. Use of probes with a high hybridization efficiency and evaluation of ≥ 100 nuclei are necessary for statistically accurate mapping results using this method. In the second approach, three probes, each labeled with a different fluor, or combination of fluors, can be hybridized simultaneously and the visualized order of signals can be examined in at least 50 nuclei.

The use of pronuclei, produced by the *in vitro* fusion of hamster eggs with hamster or human sperm, can further increase interphase mapping resolution by ~ 3 times as they have a diameter of ~ 50 µm compared to 10 to 20 µm for somatic interphase nuclei. Thus, a resolution of 20 to 25 kb can be achieved.

3.2.1 Interphase Nuclei Preparation

Interphase mapping is best performed on G0/G1(2 N) fibroblast nuclei grown to confluence to reduce the number of G2 (4 N) nuclei. The presence of G2 nuclei in uncultured preparations (peripheral blood lymphocytes) makes results more difficult to interpret. Fibroblast nuclei are large and flatten well following hypotonic treatment allowing visualization of signals on the same plane of focus. Fibroblasts grown under standard culture conditions until they reach confluence should be left for a further 6 to 7 days at 37 °C until no mitotic cells are observed. The cells are washed in a citrate saline solution, trypsinized, and collected by centrifugation. The pelleted cells are then treated with a hypotonic solution of KCl or medium, fixed with careful addition of 3 : 1 methanol : glacial acetic acid, and stored as a pellet at $-20\,°C$ as described for the preparation of metaphase chromosomes. The fixed cells are dropped onto clean glass slides and as the fixative evaporates, the cells flatten. The cell concentration should be such that there are sufficient nonoverlapping nuclei for adequate analysis. The air-dried slides can be stored at room temperature for up to four weeks or longer at $-20\,°C$ in a sealed container containing desiccant.

3.3 Chromatin and DNA Fibers

High-resolution fiber FISH methodology was developed as a result of the demand arising from the Human Genome Project. In order to group the increasing number of genes being mapped during the early 1990s, and to provide order within the same chromosomal region (or band), a high-resolution mapping approach was developed. High-resolution mapping has allowed for the construction of a finely integrated genetic and physical map, forming the foundation for the sequencing phase of the Human Genome Project.

As mentioned earlier, signal resolution using metaphase chromosomes is significantly limited. Chromatin fibers are highly compacted, and a single, small chromosomal band can contain up to a few Mb pairs of DNA, and include many genes. To take advantage of the less compacted status of chromatin, methodologies have been developed to produce released chromatin or DNA fibers for direct visualization. Originally named "free chromatin FISH," fiber FISH has demonstrated the feasibility of FISH signal detection on extended chromatin fiber for chromosomal structure analysis and physical mapping. A variety of modifications to improve the original methodology, such as increasing the mapping resolution, simplifying the fiber preparation, reducing the length variation of released fiber, quantifying the measurements of FISH signals, performing multiple-color detection, and combining DNA-protein and codetection using released fibers quickly followed. Various names corresponding to the methods of fiber preparation include: extended chromatin or DNA fiber FISH, elongated chromatin FISH, DNA halo FISH, individual stretch DNA molecule FISH, direct visual hybridization (DIRVSH), quantitative DNA fiber mapping, and molecular combing. The original and alternative approaches are now collectively referred to as high-resolution fiber FISH.

3.3.1 Fiber FISH Target Preparation

The released fibers can be categorized as either chromatin fiber or DNA fiber according to the preparation protocols and the morphological features following DNA staining. However, in many cases, mixtures of both types of fibers are produced. Generally speaking, the harsher the releasing conditions, the more DNA fiber will be generated. Chromatin fiber can be further released into DNA fiber by changing the release conditions to strip chromatin proteins. Fibers that are generated from cloned DNA fragments or the molecular combing procedure will generate no chromatin fibers.

Each type of fiber has advantages and disadvantages in terms of resolution and coverage. The resolution of chromatin fiber FISH is around 20 kb with coverage up to a few megabases; the ideal range to study larger regions of the genome in detail. In contrast, resolution of DNA fiber FISH is a few kb with coverage around 200 to 500 kb, which is best suited to a small area requiring the highest resolution. Thus, it is necessary to choose the type of fibers according to the desired resolution and coverage. Often, a combination of both types of fiber is desirable. It should be noted that the use of chromatin fiber has a unique benefit based on the following two considerations. Chromatin fiber may be more useful than DNA fiber when a high-resolution approach is required to study DNA–protein complexes. In addition, when comparing pattern or size variations between homologous regions, it is easier to trace individual cells using partially released chromatin fiber than DNA fibers.

Chromatin fiber preparation Chromatin fiber preparation protocols are based on effective release of chromatin fibers without destroying the 30-nm chromatin fiber structure. Four typical approaches have been developed to achieve this purpose: (1) chemical treatment to interfere with chromosomal condensation; (2) alkaline buffer or extensive hypotonic treatments to lyse the nuclear envelope; (3) accumulation of late G2 or early G1 phase cells to generate free chromatin; and (4) physical stretching of the nuclei with a cell centrifuge to produce chromatin fibers.

The chemical treatment method was the original approach that was used to increase the frequency of free chromatin. Although the steps for this approach are tedious when compared to other methods, an advantage is that different drugs may be used to delineate the chromosome condensation process, an important strategy to study the high-order structure of the chromosome. Preparation of chromatin fibers from cultured lymphocytes by drug treatment is accomplished by culturing lymphocytes for 48 to 52 h followed by short-term treatment with ethidium bromide (EtBr), BrdU, or mAMSA. Harvesting chromatin fiber uses the same standard protocols as for harvesting chromosomes. After checking the chromatin fiber density under phase-contrast microscopy, if it is determined that the fibers are too crowded on the slide, the concentration of the suspension can be adjusted by adding more fixation solution. Once a high-quality batch of slides is completed, they should be air-dried at room temperature for a day, and sealed in slide containers with parafilm. Slides may then be stored for several weeks at $-20\,°C$. It should be noted that irreversible damage occurs to chromatin fiber slides that are overdehydrated. Fixed free chromatin suspensions can be stored at $-20\,°C$ for a much longer time.

A second approach to increase the amount of chromatin fiber uses protocols designed to open the nuclear envelope. The use of alkaline buffer to release chromatin fibers from cultured cells is based on the fact that nuclear lamins can be interrupted by high pH treatment, destabilizing the nuclear envelope, and rendering it susceptible to rupture by hypotonic treatment. Prolonged alkaline treatment can cause breakage of the nuclear envelope without hypotonic shock. Aliquots of a harvested fresh cell suspension are transferred into tubes containing alkaline buffer, and at various time intervals (3 to 5 min), the alkaline treatment is terminated by adding fixation solution followed by refixation. The fixed cell suspension is dropped onto slides and air-dried. Chromatin fiber density is checked using a phase-contrast microscope and the concentration of the suspension is adjusted accordingly. The optimal alkaline treatment for each particular cell line has to be obtained empirically using a brief screening test.

Generating chromatin fibers by this approach is affected by many factors including the length of treatment by the alkaline buffer and the pH (10–11.5). In general, a high pH over a longer period of time promotes the lysis of nuclei. The optimal combination for a given cell line can be found by systematically varying the conditions. Excessive treatment should be avoided since it reduces the number of useful chromatin fibers because of aggregation that follows release from the nuclei, and it can destroy the 300 Å structure leading to the production of the naked DNA fiber. Optimal results are best obtained by fixing the chromatin fibers quickly before they become aggregated following the release from the nuclei. The use of a small volume of alkaline buffer and a large volume of the prefixation solution is also a good way to avoid aggregation of chromatin fibers.

A cytospin also works well for generating chromatin fiber for many cell lines. Cells are resuspended in CSK buffer, incubated for a short time on ice, and then spun onto a microscope slide using the cytospin. The slides are then rinsed with fixation solution and examined under a phase-contrast microscope to determine the chromatin quality. Optimal cell concentrations vary for different types of cells and should be adjusted accordingly.

It should be noted that different protocols and even the same protocol will generate chromatin fibers with varied resolution, as it is difficult to control the releasing process that consists of multiple levels of DNA condensation. It is also common to obtain a mixture of 30-nm chromatin fibers and naked DNA fibers, or fibers between chromatin and naked DNA including 10-nm chromatin fibers. An important step for fiber FISH detection is to include an internal marker probe of a known molecular size. By cohybridizing the probes in question with such a marker (using contrasting differential color detection), the size of the marker can be used to compare and measure the probes or regions in question. As long as an internal marker probe of known length is used, quantitative measurements can be performed, allowing the comparison of different chromatin fibers. For other types of analysis using fiber FISH where precise mapping information is not necessary, the use of a known marker for measuring purposes is not needed.

DNA fiber preparation Protocols for preparation of released DNA fibers fall into the following four categories: (1) complete release of DNA using detergent/alkaline treatment then linearizing fibers by gravity or by pulling with a coverslip; (2) generation of DNA fiber from a "Halo" preparation by protein extraction; (3) nuclear lysis and DNA release in a gel block then linearizing fibers by a mechanical or an electronic pulling force; and (4) linearizing DNA molecules by molecular combing. Two protocols that are very popular due to their simplicity and the quality of fibers produced will be described. The release of DNA fiber using alkali and linearization of the fiber by mechanical "pulling" is one of the simplest protocols to perform. Preparation of DNA fiber by pretreating cells within gel blocks and then mechanically "pulling" the fiber reduces interference from other nuclei. These two protocols should be useful as they exemplify most cases and can serve as a technical gate for additional protocols.

The first simple and popular method for preparing DNA fiber is the method of alkaline release and linearization of the DNA using mechanical pulling. Cells quickly fixed after hypotonic treatment are dropped onto a slide and treated with alkaline solution. One edge of a coverslip is placed on the slide and the coverslip is then pulled along the slide from one end to the other. The slides are ready for FISH after dehydration. The duration of alkaline solution treatment can be adjusted according to the desired release, with a longer treatment resulting in a greater amount of release from the nuclei. By adjusting the release conditions, this protocol can also be used to generate chromatin fiber. The alkaline solution can also be replaced with SDS-releasing buffer as described by others.

The second popular way of fiber preparation utilizes pretreatment of cells within gel blocks and mechanically linearizing the fiber. Lymphocytes are embedded in low-melting-point agarose and the resulting blocks of cells are treated with proteinase K and RNase A. A small piece of agarose-embedded DNA is placed at the end of a poly-L-lysine-coated slide. The slide is placed in a microwave oven, heated until the agarose melts, and the DNA is extended on the slide using a second slide that is pulled across the DNA fiber slide from one end to the other. The slide is then air-dried.

The fiber FISH molecular combing approach, that permits the stretching and aligning of deproteinized DNA on a solid surface, has recently drawn attention by demonstrating its effectiveness to high-resolution mapping. The advantage of this DNA fiber preparation method is the consistency of the length of prepared DNA fibers, an essential element of acceptable statistical analysis. Because of the requirements of specific computer software and special instrumentation, this application has been limited to only a few research groups.

4 Denaturation, Hybridization, and Detection

4.1 Pretreatment and Denaturation

Prior to denaturation, pretreatment of slides with RNase and/or Proteinase K may be required to reduce signal background due to nonspecific binding, and to increase probe access to its target DNA. RNase pretreatment is usually required when using the counterstain propidium iodide, as it stains both DNA and RNA.

In order for probe DNA to bind to its homologous sequences on the target, both must be denatured, that is, made single-stranded. Denaturation of probe and target DNA can be performed separately or simultaneously. In either case, precipitated labeled probe DNA (see Sect. 2.3.3) is resuspended in a hybridization mixture containing formamide and dextran sulphate. Formamide destabilizes DNA such that it dissociates at a more convenient lower temperature than the normal "melting temperature." This is critical for maintaining target chromosome morphology. The concentration of formamide determines the stringency of the reaction, and a higher concentration should be used with repeat sequence probes to minimize cross-reaction. As oxidation of formamide can result in DNA depurination, it is critical that the formamide be of high quality, deionized, and have a pH of 7.0. Dextran sulphate changes the void volume increasing the "apparent" DNA concentration. If probe and target are to be denatured separately, the probe in hybridization buffer is placed in a 75 °C water bath for 5 min (genomic probes) or a 95 °C water bath for 10 min (unique and repeat sequence probes). Probes containing unique or tandem repeat sequences should be placed immediately on ice and added to the target as soon as possible. Probes requiring suppression hybridization must be denatured prior to the target DNA to allow for preannealing of the competitor DNA to take place. As discussed in Sect. 2.2.1, same species competitor (Cot-1 or genomic) DNA is added to any probe containing interspersed repeats to suppress the binding of these sequences to the homologous target repeat sequences. Slides containing the target DNA are denatured in 70% formamide for 2 min at 70 °C and immediately placed in ice cold 70% ethanol, dehydrated in 90 and 100% ethanol, and air dried. The denatured probe is then added to the slide, covered with a coverslip, sealed with rubber cement, and incubated at 37 °C overnight in a humidified chamber.

Alternatively, codenaturation of probe and target can be accomplished using a modified programmed slide warmer or PCR machine. The undenatured probe is added to the slide, covered with a coverslip, sealed with rubber cement, and placed in the slide warmer set to 70 to 95 °C. Both probe and target DNA are denatured simultaneously for 2 to 5 min, and the temperature is then ramped down to 37 °C for hybridization. This approach simplifies the denaturation procedure and minimizes handling of formamide, a toxic teratogen. Codenaturation is appropriate for probes not requiring the preannealing hybridization suppression step.

4.2
Hybridization

If probe and target DNA are denatured separately, the denatured probe is placed on the dried slide, covered with a coverslip, sealed with rubber cement, and placed in a humidified box in a 37 °C oven for hybridization. This temperature is critical for optimal hybridization. At temperatures below 37 °C, lower stringency hybridization can result, causing increased nonspecific chromosomal background signal. Conversely, higher temperatures may reduce signal intensity because of increased stringency conditions. If the slides dry out during this step, nonspecific background will be increased significantly. Although most probes are allowed to hybridize overnight, centromeric alpha satellite probes and some other larger genomic probes (BACs, YACs) may produce sufficient signal for visualization after 2 to 4 h. Hybridization timing should be determined empirically for individual probes.

4.2.1 Stringency Washes

Following hybridization, unbound DNA sequences are removed by washing with a solution of slightly higher stringency than that used for the hybridization. Stringency is determined by a combination of formamide concentration, temperature, and salt concentration. Varying any one of these will alter the stringency. Increasing the temperature and formamide concentration or lowering the salt concentration, all result in a higher stringency and require a more exact matching of probe and target sequence (lower % mismatched base pairing). Stringency conditions of hybridization and washing are key factors in optimizing the signal-to-noise ratio. If the stringency is too high, the background may be clean but may result in a weak signal; if the stringency is too low, high nonspecific background may obscure the true signal. Stringency conditions vary with the type of probe and target, and should be optimized for each mapping condition. The most common wash protocols use either a 50% formamide wash at a low temperature of 42 to 44 °C or a high temperature wash without formamide at 72 °C in 1 to 2x SSC (saline sodium citrate). Following the wash step, the slides can be counterstained and viewed immediately (direct labeled probes) or processed further for detection (indirect labeled probes).

4.3
Detection

Probes that have incorporated a fluorescently labeled nucleotide directly into the DNA sequence (direct labeling) can be detected immediately after the stringency wash and are thus the simplest to use. Biotin-labeled probes (indirect labeling) can be detected with reporter molecules such as avidin or streptavidin labeled with a fluorochrome. Signal amplification is achieved by addition of biotinylated anti-avidin/streptavidin antibody and another round of avidin/streptavidin labeled with fluorochrome. Similarly, DIG-labeled probes can be detected with an anti-DIG antibody followed by a DIG-labeled secondary antibody specific for the primary antibody and an anti-DIG antibody labeled with fluorochrome. Both systems work well for most probes. Although theoretically the number of rounds of amplification can be increased several times to enhance the signal, in reality, after one or two rounds, the nonspecific background increases substantially and lowers the signal-to-noise ratio. Instead, for weak signals or very small DNA probe sequences

(≤1 kb), another signal amplification approach, termed *tyramide signal amplification* (TSA) or *catalyzed reporter deposition* (CARD) amplification, has been developed on the basis of the reaction of horseradish peroxidase and tyramine. A peroxidase-labeled hapten or antibody specific to the labeled probe is added to the bound probe, followed by biotinylated tyramine. The reaction of tyramine and peroxidase results in precipitation of a large number of biotin molecules at the site of probe hybridization. Biotin is then readily detected by fluorochrome-labeled avidin. This has resulted in successful localization of probes <1 kb in size.

Following the detection steps, slides can be counterstained directly or kept in the dark for 1 to 2 days at 4 °C in a 0.1 M phosphate buffer, pH 8.0 containing a detergent (0.05% NP-40). Slides should not be allowed to dry out before addition of the counterstain.

4.3.1 Counterstaining and Banding

The aim of physical mapping by metaphase FISH is to identify the specific chromosomal band to which the sequence is localized. This requires a method of visualizing both the gene or DNA sequence and the chromosomal bands simultaneously. Chromosomal bands range from 3 to 10 Mb in width and the number of bands per chromosomal spread can vary from 500 (low resolution) to 1000 (high resolution). Bands are named according to ISCN nomenclature and are numbered from the centromere to the telomere on each chromosome arm.

Several DNA binding dyes are available for staining chromosomes, some of which produce banding patterns similar to those produced in classical cytogenetics by the absorption dye Geimsa (G-banding). The two most common dyes (counterstains) used for chromosome banding following *in situ* hybridization are 4′,6′-diamidino-2-phenylindole (DAPI) and propidium iodide (PI). Other fluorescent DNA stains such as Hoechst 33258(QFH banding), quinacrine (Q banding), or a combination of chromomycin A3 and distamycin A (R banding) are infrequently used. Banding can also be achieved by cohybridization of PCR labeled long or short interspersed repetitive sequences (IRS) LINEs and SINEs (G-bands) or Alu (R-bands) repeat sequences.

DAPI binds to AT minor groove double-stranded DNA sequences and produces a reverse G-banding pattern with bright blue bands corresponding to AT rich, gene-poor regions and dark bands corresponding to GC and gene-rich DNA regions. Further treatment of DAPI-stained chromosomes with actinomycin D, or cell synchronization and incorporation of the thymidine analog BrdU into the late replicating DNA may enhance banding patterns in cases where the bands are not sufficiently distinct. Using software such as Adobe Photoshop to electronically invert grayscale images of DAPI-banded chromosome to the conventional G-banding pattern is also a convenient approach to chromosome identification. DAPI stain requires excitation by UV light and thus must be used with microscope systems that have a UV light source.

PI is a nonspecific double-stranded DNA and RNA intercalating (binding) dye that can produce an R (reverse)-banding pattern on metaphase chromosomes when cells in culture are synchronized by a thymidine block and then released from the block in the presence of BrdU. BrdU is then incorporated into the late replicating DNA (G positive, AT rich bands). Staining with Hoechst 33258 followed by UV irradiation reduces the PI staining of these bands, resulting in bright red staining of the early replicating CG-rich R-bands and dark staining of late replicating G-bands. High-quality banding is dependent on appropriate synchronization, and the timing of the block release and cell harvest.

The choice of counterstain is in part determined by the fluorochromes used to label the probes being mapped. Probes labeled with red fluors such as TRITC, Texas red, spectrum Orange, Cy3 should not be used with target DNA counterstained with the red dye PI. DAPI can be used with both red and green fluorochromes and does not require cell synchronization for metaphase chromosome preparation. When using either counterstain, too strong a stain can obscure the banding pattern or result in a weakened probe signal. Both counterstains can be used on interphase nuclei and fiber FISH targets.

In order to preserve signal fluorescence, antifade chemicals such as p-phenylenediamine dihydrochloride have been developed to decrease fading caused by exposure to UV light. Antifade is usually combined with the counterstain and added to the slides after the detection step.

5
Microscopy and Image Analysis

Accurate mapping of genes by FISH is dependent on the spatial and spectral characteristics of the epifluorescence microscope, the sensitivity of the camera for capturing and recording images, and the digital-imaging resolution for analysis. High-quality epifluorescence microscopes equipped with a properly aligned high-intensity light source designed for high-light collection power (detection sensitivity) are essential for mapping analyses. The three commonly used light sources are mercury lamps, xenon lamps, and lasers. Mercury arc lamps are used most frequently as their emissions are similar to the excitation peaks of many commonly used fluors (254–646 nm). Xenon lamps tend to give a more even excitation across the 250- to 1000-nm range and are most useful with fluors excited above 700 nm. Both mercury and xenon lamps require filter sets that control the wavelength of light reaching the fluor. The most commonly used laser sources for FISH, argon ion lasers (emission 488 and 514 nm), and helium-neon lasers (emission 543, 594, and 633 nm), do not require excitation filters and only have a narrow range of emitted light, thus limiting their use to a restricted number of fluors.

The objective lenses must be of the maximum numerical aperture possible, of low self-fluorescing UV transmitting glass, and demonstrate minimum spherical and chromatic aberrations (apochromatic) to produce optimal sensitivity of fluorescent signal visualization. It is useful to have a range of objective magnifications – dry $10\times$ and $20\times$ for scanning metaphases and nuclei, and oil 60x and 100x for visualization and localization of the signal. A nonfluorescing oil must be used for the high-power objectives. Mixing oils from different sources may cause cloudiness, and use of oil stored for long periods of time can result in increased autofluorescence, both of which reduce signal visualization.

The interference filters should possess low autofluorescing properties and be properly aligned to reduce unwanted reflections. Filter sets for viewing fluorescence signals and counterstained chromosomes are determined by the fluorochromes that are being used and the type of mapping being performed. To map a single probe, narrowband pass filters, which have optimized peak excitation and emission wavelengths for a single fluorochrome, are used to provide maximum signal intensity. Wideband pass filters are used in relational mapping to view two (dual-band pass) or three (triple-band pass) colors simultaneously. These filters result in a reduction of individual signal intensity due to "bleed through" of emission signals. The mechanism of switching excitation and emission filters to visualize the different fluors sequentially has been a critical issue for mapping with respect to proper alignment of signal with exact chromosomal location. Careful matching and mechanical adjustment, and use of internal reference microspheres, multiband pass filters, or automatic filter wheels have all been employed to minimize image shifts. Currently, automatic computer-controlled filter wheels and software packages that enable the capture of real-time images are used to minimize this registration problem.

Direct visualization of signals from small single-copy or unique DNA sequences is often difficult, even at high magnification (100x oil). Cooled charged-couple device (CCD) cameras are used for fine mapping studies as they provide excellent sensitivity, spatial resolution, and linearity with high dynamic range (12–16 bit). Cooling CCD cameras to minus 38 °C reduces dark current noise and background signals caused by the camera itself. Most high-resolution CCD cameras are cooled monochrome cameras that capture grayscale images based on light intensity. Grayscale images must be captured separately for each fluorochrome and transferred into an image analysis software program such as Adobe Photoshop for addition of color and merging. Modified color CCD cameras are also available, however, the spatial resolution is limited in the single-chip color CCD cameras with the result that some signals may be missed. The use of these cameras with real-time image capture and dual or triple filters eliminates the need for merging images. Further, image software has developed rapidly over the past decade and now facilitates quantitative FISH, automatic image focusing, acquisition at multiple wavelengths, visualization of multicolor probes, multicolor pseudocolor simultaneous imaging, enhancement of DAPI-banded chromosomes, chromosome classification, slide location of cells, fluorescence intensity, morphometry, image storage, and many forms of data tabulation making FISH mapping and FISH-based analyses more feasible in research and clinical laboratories.

6
Applications

The main function of gene mapping by FISH is the precise physical localization of genes of known biological activity. Although many genes and DNA sequences can now be positioned using *in silico* methods and the vast information available in the genome databases discussed in Sect. 2.1, visual verification is often required to finalize and verify the precise chromosomal location. Thus, FISH has played an important role in positional cloning as it is the most direct way to map and order DNA clones within a chromosomal band. To date, thousands of genes and markers obtained by standard cloning approaches have been mapped by FISH. FISH has also provided an effective approach for anchoring YAC and BAC contigs within a chromosomal band.

The resolution of FISH mapping depends on the degree of target DNA

Tab. 2 FISH mapping on DNA targets of increasing resolution.

Visual mapping of genes and DNA clones on	Resolution	Application	Advantages (+) and disadvantages (−)
Metaphase chromosome	1–3 Mb	Gene mapping on chromosomal regions; detection of translocations, hemizygosity and regional amplification; detection of chimerism or homologous regions; whole genome mapping (SKY/M-FISH, CGH and ZOO-FISH)	+ Telomere–centromere orientation + Mapping to specific band + Whole genome approach + Highest coverage + Chromosome identity − Low resolution
Interphase Nucleus	50–100 kb	Detection of translocation; visualization of chromatin domain interaction; monitor replication status; study gene amplification; ordering of clones	+ Distance measurement + Detect the level of gene amplification − Telomere-centromere orientation not possible − No chromosome identity

Released chromatin fiber	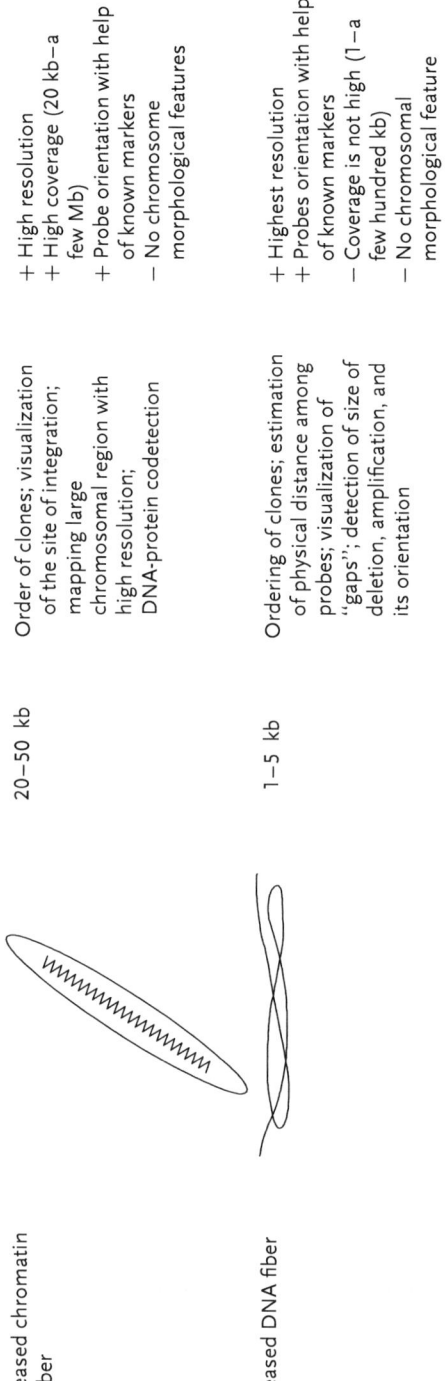	20–50 kb	Order of clones; visualization of the site of integration; mapping large chromosomal region with high resolution; DNA-protein codetection	+ High resolution + High coverage (20 kb–a few Mb) + Probe orientation with help of known markers − No chromosome morphological features
Released DNA fiber		1–5 kb	Ordering of clones; estimation of physical distance among probes; visualization of "gaps"; detection of size of deletion, amplification, and its orientation	+ Highest resolution + Probes orientation with help of known markers − Coverage is not high (1–a few hundred kb) − No chromosomal morphological feature

condensation, and spans several orders of magnitude ranging from 1 to 3 Mb on metaphase chromosomes, 50 to 100 kb in interphase nuclei, 20 to 50 kb on chromatin fiber, and 1 to 5 kb on DNA fibers. The type of target DNA used for FISH mapping will depend on the type of information required (see Table 2). Aspects of these three levels of gene mapping by FISH are presented in the following sections.

6.1
Metaphase Mapping

6.1.1 Single-copy Genes

Metaphase chromosomes present the most compacted form of DNA and thus provide the least-sensitive mapping resolution (1–3 Mb). Metaphase mapping has provided a subchromosomal banding localization for single-copy genes, viral insert sites, specific breakpoints, and regions of amplification or deletion. FISH mapping on metaphase chromosomes also orients genes relative to chromosome centromeres and telomeres. Key to metaphase FISH mapping is identification of true signal, the hallmark of which is a doublet signal on each chromosome (one per chromatid) of a homologous pair (see Fig. 2). The doublet signals must be visualized consistently at the same band location in each metaphase spread. The number of metaphase chromosomes needed to establish a chromosomal localization is determined by hybridization efficiency (% metaphases showing two pairs of consistent doublet signals). Hybridization efficiency is determined by the type and size of both probe and target DNA. Genomic BACs and PACs usually hybidize with >90% efficiency requiring analysis of 15 to 20 high-quality banded chromosomes. Smaller genomic probes (<10 kb) and cDNAs may show <20% hybridization efficiency and thus require analysis of more chromosomes (25–50).

Visualization of cDNAs can vary with the size of the target gene and its intron/exon

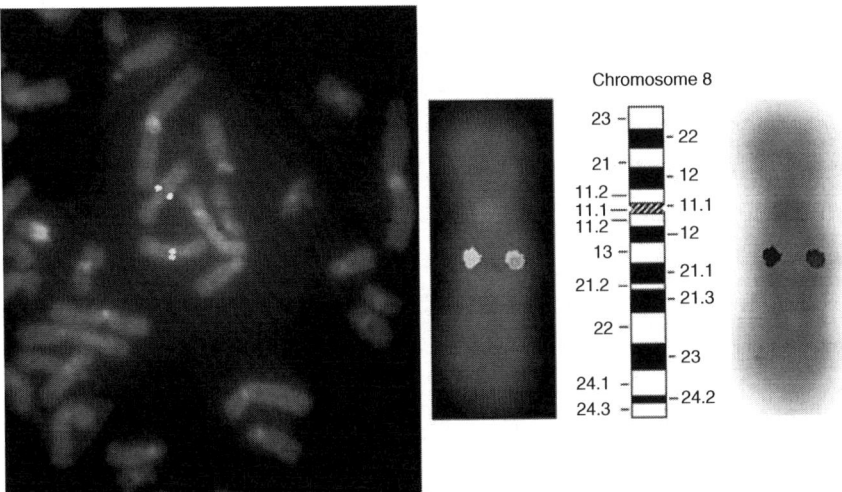

Fig. 2 Metaphase FISH mapping of a single-copy probe. Biotinylated DNA PAC probe (∼100 kb) hybridized to normal human metaphase chromosomes counterstained with DAPI and detected with FITC. Doublet probe signals map to 8q13 on both chromosome 8 homologs.

structure, and usually requires amplification by approaches such as tyramide-based detection. Alternatively, cDNA screening of a PAC or BAC library can provide homologous genomic clones that are then easily mapped by FISH. Since one or more clones may be identified from such a screen, care must be taken in the interpretation of the results. Even if one genomic clone is identified, FISH mapping may result in either localization to a true mapping site on a single chromosome, or localization to two different chromosomes, suggesting the presence of a double or chimeric clone. In situations when several PAC/BAC clones are identified for one cDNA, PCR verification of the insert sequence should be performed prior to proceeding with FISH mapping. If two or more "verified" clones still result, localization to the same chromosomal site would indicate the true mapping site and redundancy within the library, whereas localization to different chromosome sites would suggest the presence of a gene family or a pseudogene. Since PAC/BAC libraries are publicly available (see Sect. 2.1), this approach has the advantage of providing both genomic clones and localization data for expressed sequences. FISH mapping on metaphase chromosomes is also a straightforward approach to the identification of the presence of chimerism within YAC clones, which are noted for their high rate of chimerism. Doublet hybridization signals at more than one chromosomal location indicate the presence of a chimeric clone and further analysis by somatic cell hybrids or other molecular techniques is advised.

Although the resolution of metaphase FISH is usually not sufficient for the fine mapping requirements of positional cloning, it provides an initial banding assignment that can be followed by more precise FISH mapping obtained with released chromatin or fiber FISH (Sect. 6.3). It should be noted that mapping in the subtelomeric and pericentromeric chromosomal regions is less accurate and should be avoided or used with caution as chromatin folding and duplications in these regions can result in inverted and/or erroneous ordering of probes.

6.1.2 Segmental Duplications and FISH Mapping

The presence of segmental duplications ≥ 1 kb with 90 to 98% identity has been reported to comprise 3.6% of the genome and has been the cause of some physical mapping ambiguities. These duplications show a strong pericentromeric and subtelomeric bias of approximately 5- to 10-fold, however, this clustering of duplications in the pericentromeric and subtelomeric regions is not uniformly distributed among all the chromosomes. BAC clones containing these duplicated sequences tend to be found more frequently in unordered and misassigned contigs as their chromosomal localization is difficult to assign. FISH mapping of BAC clones containing interchromosomal duplications has been used to determine their coverage and duplication patterns. Less than half the chromosomes positive by FISH have a corresponding chromosomal localization of these clones by BLAST sequence identity. FISH mapping data can thus be used as a standard with which to identify chromosomal localization of these duplicated sequences, and to compare the assembly completeness and accuracy of interchromosomal duplications.

6.1.3 Relational Mapping

Relational mapping is used when the order of two or more genes localized to the same chromosomal band or region is important. A probe's relation to a cryptic deletion or amplification, a specific breakpoint (Fig. 3), or confirmation of probe order (Fig. 4) based on computational approaches, are situations in which FISH can provide valuable information. The condensed nature of metaphase DNA requires probe sequences to be separated by at least 1 Mb of DNA, and at least 25 metaphases should be scored for statistically accurate results. Multicolor probes are usually used for this type of analyses and the number of genes that can be mapped simultaneously will depend on fluorochrome availability and filter capabilities of the microscope. Mapping probes relative to a specific chromosomal breakpoint can provide information regarding their biological and/or diagnostic

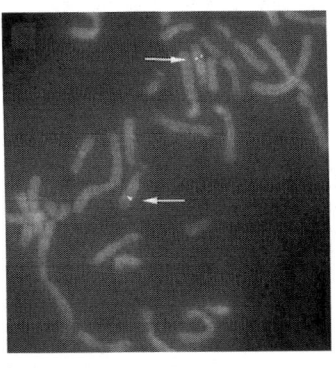
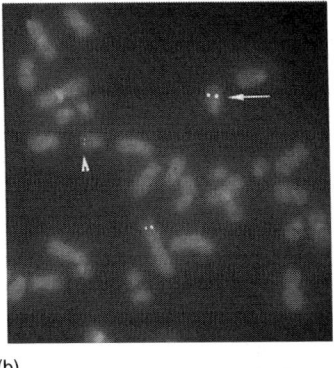

(a)   (b)

Fig. 3 Mapping across a chromosome breakpoint. (a) A cosmid probe (14 kb) hybridized to normal human metaphase chromosomes showed doublet signals at 14q24 on both chromosome homologs. (b) Hybridization of the same cosmid probe to metaphase chromosomes from a patient with a 2 : 14 (p25;q24) translocation. The probe demonstrates doublet signals on both normal (arrow) and derivative (arrowhead) chromosome 14 as well a doublet signal on the short arm of derivative chromosome 2 (no arrow). A consistently stronger signal was noted on der2, indicating that the breakpoint is located closer to the proximal (centromeric) end of the cosmid. (From Barbara G. Beatty, Stephen W. Scherer. (2002), FISH: A Practical Approach reprinted by permission of Oxford University Press).

Fig. 4 Relational mapping on metaphase chromosomes using two fluorochromes. (a) Three probes: Pairwise mapping was used to order two YACs and one cosmid probe within a YAC contig, which mapped to 14q24.2–24.3. The two YACs were labeled with DIG or biotin and the cosmid with biotin. DIG was detected with rhodamine (red) and biotin with FITC (yellow). Left chromosome: DIG-YAC 1 plus cosmid; middle chromosome: DIG-YAC 2 plus cosmid; right chromosome: Biotin-YAC 1 plus DIG-YAC 2. Thus, the order shown is YAC 1 – cosmid – YAC 2. Note: YAC 2 often demonstrated double-hybridization signals in a lateral orientation suggesting two domains of strong hybridization or clustering of Alu sequences within the YAC (see right chromosome). (b) Two probes: Two PAC probes localized to 14q24-31 by FISH were labeled with different haptens (DIG-PAC A and biotin-PAC B) and hybridized simultaneously to normal human metaphase chromosomes counterstained with DAPI. PAC A (detected with rhodamine, red) localized telomeric to PAC B (detected with FITC, yellow). [From Barbara G. Beatty, Stephen W. Scherer. (2002), Fish: A Practical Approach, reprinted by permission of Oxford University Press.] (See color plate p. vii.)

significance. Genes that span a breakpoint can be identified if the sequence length on either side of the breakpoint is sufficient to produce a visible signal as shown in Fig. 3. Mapping DNA probes relative to other chromosomal abnormalities, regions of amplification or deletion, or to tandem repeats can be performed using metaphase chromosomes, provided probes are available for the abnormal region and the distance between them is >1 Mb. It is worth mentioning that quantitative FISH (Q-FISH) has demonstrated its ability to study the length of telomeric repeats for individual chromosomes. Using FITC-PNA probes and quantitative

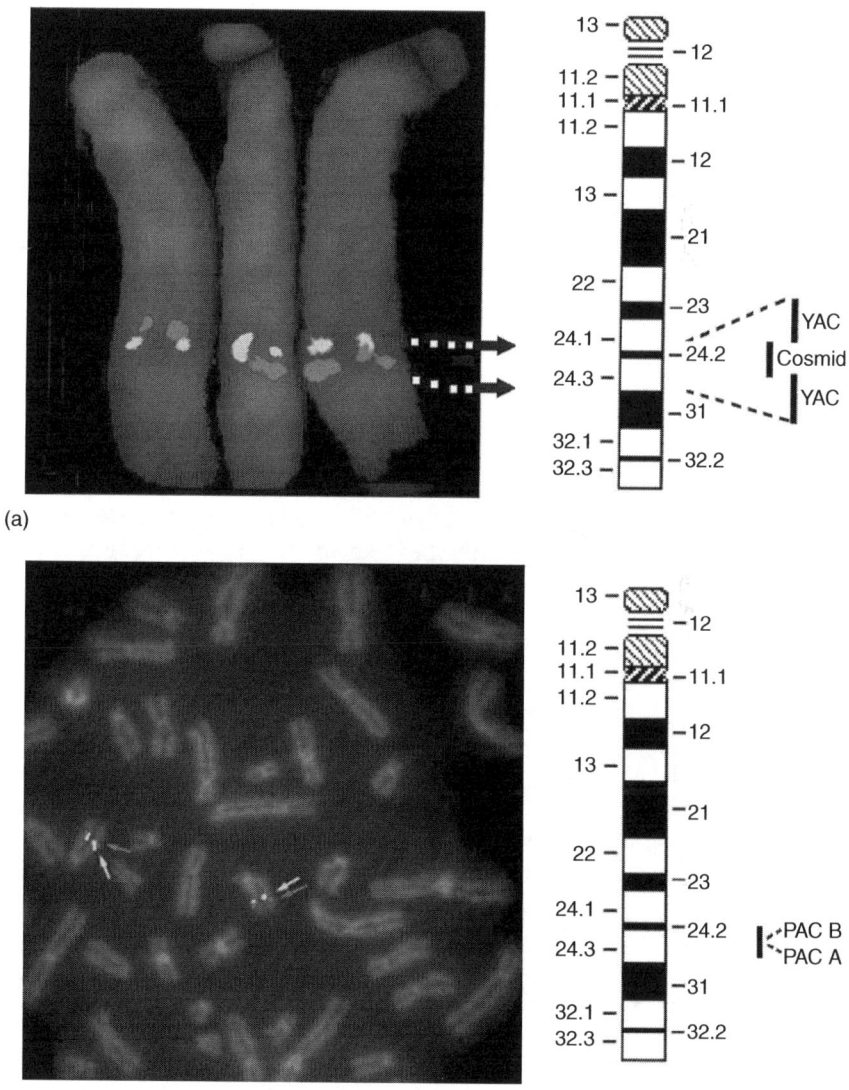

software, Q-FISH has been used to monitor changes in the telomeric repeats under different developmental, pathological, and disease conditions.

6.1.4 FISH Mapping Combined with CGH and spectral karyotyping (SKY)

With the development of the genome project, individual gene mapping has been expanded into genome mapping of chromosomal regions, whole chromosomes, and the entire genome. By combining individual gene mapping with entire genome or chromosomal FISH approaches such as CGH, which identifies regions of whole genome chromosome amplification and deletion, or SKY and M-FISH, which identify interchromosomal rearrangements, genes involved in these critical regions can be identified and localized.

CGH utilizes competitive hybridization of two probes (control and test) labeled with two different fluors onto normal metaphase chromosomes to reveal the presence and location of DNA copy number imbalances by comparing the relative signal intensities of each probe along the length of each chromosome. Special digital image analysis software programs developed to analyze and evaluate CGH data are available commercially. CGH has been particularly useful in identifying regions of DNA sequence copy number amplification and deletion in solid tumors. FISH mapping of single-copy or unique sequences within these regions can provide important information about the identification of candidate oncogenes and tumor suppressor genes respectively. Resolution of CGH on metaphase chromosomes is limited to several megabases. A significantly higher mapping resolution can be achieved using targets such as cloned DNA fragments (BACs, PACs, etc.) and cDNAs arrayed on glass slides. This high-resolution CGH, termed *matrix CGH*, can be used to map and study microdeletions and sequence overrepresentations.

Similarly, multicolor chromosome paint cocktails have now been developed for identifying interchromosomal rearrangements and mapping translocation breakpoints within the whole genome. Two commercially available approaches, SKY, and M-FISH utilize FISH probes derived from flow-sorted or microdissected chromosomes labeled by DOP-PCR (see Sect. 2.3.2). Each chromosome is identified by a unique spectral signature (SKY) or combination of fluors (M-FISH), permitting unambiguous identification of marker chromosomes and the different chromosome fragments involved in interchromosomal rearrangements. Sequential application of conventional G-banding and SKY on the same metaphase, as well as the combinatorial use of SKY and FISH, makes it possible to define the individual chromosomal bands involved in the structural rearrangements. Identification of putative genes or markers adjacent to or spanning specific breakpoints by *in silico* methods can then be verified and mapped by conventional FISH analysis.

6.1.5 Comparative Mapping by ZOO-FISH

Another application of gene mapping is the comparative mapping of map-rich species (human/mouse) with experimentally and economically important map-poor species (cattle, pig, horse, chicken). The location of conserved sequence segments between species can be visualized using composite chromosome-specific DNA probes (paints) from one species to hybridize to the chromosomes of another species (target). This cross-species painting approach is termed *ZOO-FISH* and has a resolution of approximately 7 Mb. The probes used are generally chromosome paints from chromosome-specific libraries, but can be cDNAs or genomic probes (PACs/BACs) obtained from corresponding species-specific libraries. ZOO-FISH provides a mechanism for identifying major regions of conserved synteny between the two species being examined. This application of FISH has provided valuable information on regions of homology between the human genome and other map-poor species such as pig, chicken, and cattle, and between different members of a species including mammals, insects, and plants. It is expected that a better understanding of cross-species homologies will provide new models for the study of disease as well as new hypotheses of genome evolution.

6.2
Interphase FISH Mapping

FISH mapping using interphase nuclei provides a higher order of resolution than metaphase mapping and is used for clone ordering and generating physical maps with a resolution of <1 Mb. It is also useful for verifying distance measurements obtained from genome database contig information in the 50 kb to 1 Mb range. FISH mapping with pronuclei provides more precise ordering and estimation of genomic and physical distances and has been used in preparing high-resolution physical maps of chromosome 19. The concerns regarding pronuclei interphase mapping are that they are time consuming, laborious, and not suited to large-scale mapping approaches. Further, although the relationship between genomic and physical distance is considered linear from ~50 to 800 kb, the extent of DNA condensation within these nuclei cannot be controlled. Interphase mapping provides relative mapping distance measurements and should contain an internal control of markers a known distance apart. No information on chromosome location is provided by this approach.

6.3
Chromatin and DNA Fiber Mapping

Fiber FISH has demonstrated its advantages in a variety of applications. The majority of publications have involved physical mapping and chromosome structure

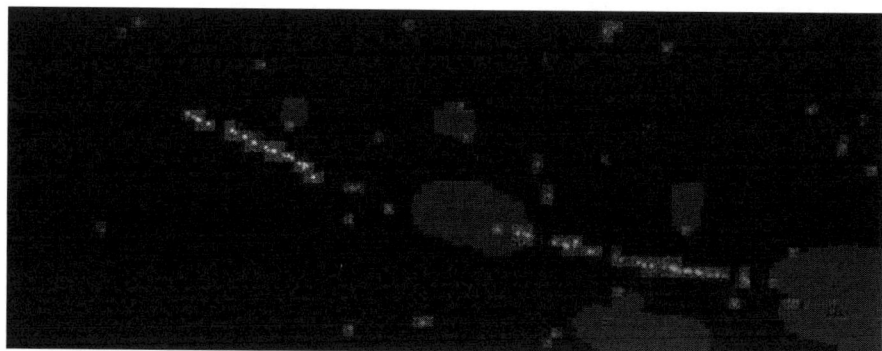

Fig. 5 Sizing the "gap" in current physical map using fiber FISH: Two BAC probes were detected by two-color fiber FISH and the distance between two BAC probes represents the size of "gap".

studies of both plant and animal genomes. These applications include: gap estimation for the current Human Genome Project as shown in Fig. 5; order and orientation for groups of genes/ESTs or DNA fragments; quantification of the size of amplified or duplicated fragments of special genes or chromosomal regions or integrated foreign inserts; identification or exclusion of genes or chromosomal regions defined by genetic markers; comparison of evolutionary conserved regions among various species; illustration of multiple repetitive sequences within chromosomal regions and direct visualization of genome organization; and length measurement of telomeric regions for the study of chromosomal packaging.

See also Organization of Genes and Genome Domains; Molecular Basis of Genetics.

Bibliography

Books and Reviews

Beatty, B., Mai, S., Squire, J. (Eds.) (2002) *FISH: A practical Approach*, Oxford University Press, London.

Heng, H.H.Q., Tsui, L.-C., Windle, B., Parra, I. (1995) High resolution FISH analysis in: Dracopoli, N., Haines, J., Korf, B., Moir, D., Morton, C., Seidman, C., Seidman, J., Smith, D. (Eds.) *Current Protocols in Human Genetics*, John Wiley and Sons, New York, pp. 4.5.1–4.5.25.

Heng, H.H.Q., Tsui, L.-C. (1994) FISH detection on DAPI-banded chromosomes, in: Choo, K.H.A. (Ed.) *Methods in Molecular Biology*, Vol. 33, Humana Press, Totowa, pp. 35–49.

Knuutila, S., Bjorkqvist, A.-M., Autio, K., Tarkkanen, M., Wolf, M., Monni, O., Szymanska, J., Larramendy, M.L., Tapper, J., Pere, H., El-Rifai, W.E., Hemmer, S., Wasenius, V.-M., Vidgren, V. Zhu, Y. (1998) DNA copy number amplifications in human neoplasms: Review of comparative genomic hybridization studies. *Am J Path* **152**, 1107–1123.

Lebo, R.V., Su, Y. (1994) Positional cloning and multicolor *in situ* hybridization, in: Choo, K.H.A. (Ed.) *Methods in Molecular Biology* Vol. 33, Humana Press, Totowa, pp. 409–438.

Leversha, M.A. (1997) Fluorescence *in situ* hybridization, in: Dear, P.H. (Ed.) *Genome mapping: A practical approach*, Oxford University Press, London, pp. 199–225.

Leversha, M.A. (2001) Mapping of genomic clones by fluorescence *in situ* hybridization. in: Starkey, M.P. and Elaswarapu, R. (Eds.) *Methods in Molecular Biology*, Vol. 175, Humana Press, Totowa, pp. 109–127.

Trask, B., (1999) Flourescence in situ hybridization, in: Birren, B., Green, E.D., Hieter, P., Klapholz, S., Myers, R.M., Riethman, H., Roskams, J. (Eds.) *Genome Analysis: A Laboratory Manual*, Volume 4, Cold Spring Harbor Laboratory Press, New York.

Primary Literature

Al-Bayati, H.K., Duscher, S., Kollers, S., Rettenberger, G., Fries, R., Brenig, B. (1999) Construction and characterization of a porcine P1-derived artificial chromosome (PAC) library covering 3.2 genome equivalents and cyntenic assignment of six type I and type II loci. *Mamm Genome* **10**(6), 569–572.

Bailey, J.A., Yavor, A.M., Massa, H.F., Trask, B.J., Eichler, E.E. (2001) Segmental duplications: Organization and impact within the current human genome project assembly. *Genome Research* **11**, 1005–1017.

Baldini, A., Ross, M., Nizetic, D., Vatcheva, R., Lindsay, E.A., Lehrach, H., Siniscalco, M. (1992) Chromosomal assignment of human YAC clones by fluorescence in situ hybridization: use of single-yeast-colony PCR and multiple labeling. *Genomics* **14**, 181–184.

Bensimon, A., Simon, A., Chiffaudel, A., Croquette, V., Heslot, F., Bensimon, D. (1994) Alignment and sensitive detection of DNA by a moving interface. *Science* **265**, 2096–2098.

Brandriff, B.F., Gordon, L.A., Trask, B.J. (1991) DNA sequence mapping by fluorescence in situ hybridization. *Environmental and Molecular Mutagenesis* **18**, 259–262.

Brandriff, B., Gordon, L., Trask, B. (1991) A new system for high-resolution DNA sequence mapping in interphase pronuclei. *Genomics* **10**, 75–82.

Burkholder, G.D. (1993) The basis of chromosome banding. *Applied Cytogenetics* **19**, 181–186.

Conti, C., Bensimon, A. (2002) A combinatorial approach for fast, high-resolution mapping, *Genomics* **80**, 135–137.

Du Manoir, S., Speicher, M.R., Joos, S., et al. (1993) Detection of complete and partial chromosome gains and losses by comparative genomic in situ hybridization. *Hum Genet* **90**, 590–610.

Dunham, I., Shimizu, N., Roe, B.A., Chissoe, S., Hunt, A.R., Collins, J.E., Bruskiewich, R., Beare, D.M., Clamp, M., Smink, L.J., Ainscough, R., Almeida, J.P., Babbage, A., Bagguley, C., Bailey, J., Barlow, K., Bates, K.N., Beasley, O., Bird, C.P., Blakey, S., Bridgeman, A.M., Buck, D., Burgess, J., Burrill, W.D., O.'Brien, K.P., et al. (1999) The DNA sequence of human chromosome 22, *Nature* **402**, 489–495.

Dutrillaux, B., Viegas-Pequignot, E. (1981) High resolution R- and G-banding on the same preparation. *Hum Genet* **57**, 93–95.

Eppig, J.T. (1996) Comparative maps:adding pieces to the mammalian jigsaw puzzle. *Curr Opin Genet Dev* **6**(6), 723–730.

Fan, Y.S., Davis, L., Shows, T. (1990) Mapping small DNA sequences by fluorescence in situ hybridization directly on banded metaphase chromosomes. *Proc Natl Acad Sci USA* **87**, 6223–6227.

Fauth, C., Speicher, M.R. (2001) Classifying by colors: FISH-based genome analysis. *Cytogenet. Cell Genet.* **93**, 1–10.

Fidlerova, H., Senger, G., Kost, M., Sanseau, P., Sheer, D. (1994) Two simple procedures for releasing chromatin from routinely fixed cells for fluorescence in situ hybridization. *Cytogenet Cell Genet* **65**, 203–205.

Florijn, R.J., Bonden, L.A.J., Vrolijk, H., et al. (1995) High resolution DNA fiber-FISH for genomic DNA mapping and color bar-cording of large genes. *Hum Mol Genet* **4**, 831–836.

Gervasini, C., Bentivegna, A., Venturin, M., Corrado, L., Larizza, L., Riva, P. (2002) Tandem duplication of the NF1 gene detected by high-resolution FISH in the 17q11.2 region, *Hum Genet.* **111**, 465–7.

Haaf, T., Ward, D.C. (1994) Structural analysis of α-satellite DNA and centromere proteins using extended chromatin and chromosomes, *Hum. Mol. Genet.* **3**, 697–709.

Haaf, T. (1996) High-resolution analysis of DNA replication in released chromatin fibers containing 5-bromodeoxyuridine, *Biotechniques* **21**, 1050–4.

Heiskanen, M., Hellsten, E., Kallioniemi, O.-P., et al. (1995) Visual mapping by fiber FISH. *Genomics* **30**, 31–36.

Heiskanen, M., Karhu, R., Hellsten, E., Peltonen, L., Kallioniemi, O.P., Palotie, A. (1994) High resolution mapping using fluorescence in situ hybridization to extended DNA fibers prepared from agarose-embedded cells, *BioTechniques* **17**, 928–934.

Heiskanen, M., Peltonen, L., Palotie, A. (1996) Visual mapping by high resolution FISH, *Trends Genet.* **12**, 379–382.

Heng, H.H.Q., Squire, J., Tsui, L.-C. (1992) High resolution mapping of mammalian genes by in situ hybridization to free chromatin, *Proc. Natl. Acad. Sci. USA* **89**, 9509–9513.

Heng, H.H.Q., Tsui, L.C., (1993) Modes of DAPI banding and simultaneous in situ hybridization. *Chromosoma* **102**, 325–332.

Heng, H.H.Q., Chamberlain, J.W., Shi, X.-M., Spyropoulos, B., Tsui, L.-C., Moens, P.B. (1996) Regulation of meiotic chromatin loop size by chromosomal position, *Proc. Natl. Acad. Sci. USA* **93**, 2795–2800.

Heng, H.H.Q., Spyropoulos, B., Moens, P.B. (1997) FISH technology in chromosome and genome research, *Bioessays* **19**, 75–84.

Heng, H.H.Q., Tsui, L.C. (1998) High resolution free chromatin/DNA fiber fluorescent in situ hybridization. *J of Chromatography* **806**, 219–229.

Heppell-Parton, A.C., Albertson, D.G., Fishpool, R., Rabbitts, P.H. (1994) Multicolour fluorescence in situ hybridization to order small, single-copy probes on metaphase chromosomes. *Cytogenet Cell Genet* **66**, 42–47.

Herrick, J., Xavier, M., Conti, C., Schurra, C., Bensimon, A. (2000) Quantifying single gene copy number by measuring fluorescent probe lengths on combed genomic DNA, *Proc. Natl. Acad. Sci. USA* **97**, 222–227.

Hirai, M., Suto, Y., Kanoh, M. (1994) A method for simultaneous detection of fluorescent G-bands and in situ hybridization signals. *Cytogenet Cell Genet.* **66**, 149–151.

Horelli-Kuitunen, N., Aaltonen, J., Yaspo, M.L., Eeva, M., Wessman, M., Peltonen, L., Palotie, A. (1999) Mapping ESTs by fiber-FISH. *Genome Res* **9**, 62–71.

Houseal, T.W., Dackowski, W.R., Landes, G.M., Klinger, K.W. (1994) High resolution mapping of overlapping cosmids by fluorescence in situ hybridization. *Cytometry* **15**, 193–198.

Inoue, K., Osaka, H., Imaizumi, K., Nezu, A., Takanashi. J., Arij, J., Murayama, K., Ono, J., Kikawa, Y., Mito, T., Shaffer, L., Lupski, J. (1999) Proteolipid protein gene duplications causing Pelizaeus-Merzbacher Disease: molecular mechanism and phenotypic manifestations, *Ann. Neurol.* **45**, 624–632.

Ioannou, P.A., Amemiya, C.T., Garnes, J., et al. (1994) A new bacteriophage P1-derived vector for the propagation of large human DNA fragments. *Nature Genet* **6**, 84–89.

Jiang, F., Lin, F., Price, R., Gu, J., Medeiros, L.J., Zhang, H.Z., Xie, S.S., Caraway, N.P., Katz, R.L. (2002) Rapid detection of IgH/BCL2 rearrangement in follicular lymphoma by interhase fluorescence in situ hybridization with bacterial artificial chromosome probes, *J. Mole. Diagn.* **4**, 144–149.

Joos, S., Fink, T.M., Ratsch, A., Lichter, P. (1994) Mapping and chromosome analysis: the potential of fluorescence in situ hybridization. *J Biotech* **35**, 135–153.

Kallioniemi, A., Kallioniemi, O.-P, Sudar, D., Rutovitz, D., Gray, J.W., Waldman, F., Pinkel, D. (1992) Comparative genome hybridization for molecular cytogenetic analysis of solid tumors. *Science* **258**, 818–821.

Kallioniemi, A., Kallioniemi, O.-P., Piper, J., et al. (1994) Detection and mapping of amplified DNA sequences in breast cancer by comparative genome hybridization. *Proc Natl Acad Sci USA* **91**, 2156–2160.

Karhu, R., Rummukainen, J, Lorch, T., Isola, J. (1999) Four-color CGH: A new method for quality control of comparative genomic hybridization. *Genes, Chromosomes & Cancer* **24**, 112–118.

Korenberg, J.R., Chen, X.-N. (1995) Human cDNA mapping using a high resolution R-banding technique and fluorescence in situ hybridization. *Cytogen Cell Genet.* **69**, 196–200.

Korenberg, J.R., Chen, X.N., Adams, M.D., Venter, J.C. (1995) Toward a cDNA map of the human genome. *Genomics* **29**, 364–370.

Korenberg, J.R., Yang-Feng, T., Schreck, R., Chen, X.N. (1992) Using fluorescence in situ hybridization (FISH) in genome mapping. *Trends Biotechnol* **10**, 27–32.

Langer, P.R., Waldrop, A.A., Ward, D.C. (1981) Enzymatic synthesis of biotin-labeled polynucleotides: novel nucleic acid affinity probes. *Proc Natl Acad Sci USA* **78**, 6633–6637.

Lawrence, J.B., Villnave, C.A., Singer, R.H. (1988) Interphase chromatin and chromosome gene mapping by fluorescence detection of in situ hybridization reveals the presence and orientation of two closely integrated copies of EBV in a human lymphoblastoid cell line. *Cell* **52**, 51–61.

Lawrence, J.B. (1990). A fluorescence in situ hybridization approach for gene mapping and the study of nuclear organization, *Genome Analysis* **1**, 1–38.

Lawrence, J.B., Singer, R., McNeil, J. (1990) Interphase and metaphase resolution of different distances within the human dystrophin gene. *Science* **249**, 928–932.

Lemieux, N., Malfoy, B., Fetni, R., Muleris, M., Vogt, N., Richer, C.L. (1994) In situ hybridization approach at infragenic level on metaphase chromosomes. *Cytogenet Cell Genet* **66**, 107–112.

Lestou, V.S., Strehl, S., Lion, T., Gadner, H., Ambros, P.F. (1996) High-resolution FISH of the entire integrated Epstein-Barr virus genome on extended human DNA, *Cytogenet. Cell Genet.* **74**, 211–217.

Lichter, P., Cremer, T., Borden, J., Manuelidis, L., Ward, D.C. (1988) Delineation of individual human chromosomes in metaphase and interphase cells by in situ suppression hybridization using recombinant DNA libraries. *Hum Genet* **80**, 224–234.

Lichter, P., Ledbetter, S.A., Ledbetter, D.H., Ward, D.C. (1990) Fluorescence in situ hybridization with Alu and L1 polymerase chain reaction probes for rapid characterization of human chromosomes in hybrid cell lines. *Proc Natl Acad Sci USA* **87**, 6634–6638.

Lichter, P., Tang, C.C., Call, K., Hermanson, G., Evans, G.A., Housman, D, Ward, D.C. (1990) High resolution mapping of human chromosome 11 by in situ hybridization with cosmid clones. *Science* **247**, 64–69.

Michalet, X., Ekong, R., Fougerousse, F., Rousseaux, S., Schurra, C., Hornigold, N., van Slegtenhorst, M., Wolfe, J., Povey, S., Bechmann, J.S., Bensimon, A. (1997) Dynamic molecular combing: stretching the whole human genome for high resolution studies. *Science* **277**, 1518–1523.

Nath, J., Johnson, K.L. (1999) A review of fluorescence in situ hybridization (FISH): current status and future prospects. *Biotech & Histochem* **75**, 54–78.

Nederlof, P.M., Robinson, D., Abuknesha, R., Wiegant, J., Hopman, A.H.N., Tanke, H.J., Raap, A.K. (1989) Three-color fluorescence in situ hybridization for the simultaneous detection of multiple nucleic acid sequences. *Cytometry* **10**, 20–27.

Nederlof, P.M., van der Flier, S., Wiegant, J., Raap, A.K., Tanke, H.J., Ploem, J.S., van der Ploeg, M. (1990) Multiple fluorescence in situ hybridization. *Cytometry* **11**, 126–131.

Palotie, A., Heiskanen, M., Laan, M., Horelli-Kuitunen. (1996) High-resolution fluorescence in situ hybridization: a new approach in genome mapping. *Finn Med Soc Ann Med* **28**, 101–106.

Pandita, A., Godbout, R., Zielenska, M., Thorner, P., Bayani, J., Squire, J.A. (1997) Relational mapping of MYCN and DDX1 in band 2p24 and analysis of amplicon arrays in double minute chromosome and homogeneously staining regions by use of free chromatin FISH, *Genes Chromosomes Cancer* **20**, 243–252.

Parra, I., Windle, B. (1993) High resolution visual mapping of stretched DNA by fluorescent hybridization. *Nature Genet* **5**, 17–21.

Pinkel, D., Straume, T., Gray, J.W. (1986) Cytogenic analysis using quantitative, high sensitivity, fluorescence hybridization. *Proc Natl Aad Sci USA* **83**, 2934–2938.

Pinkel, D., Landegent, J., Collins, C., Fuscoe, J., Segraves, R., Lucas, J., Gray, J.W. (1988) Fluorescence in situ hybridization with human chromosome-specific libraries: detection of trisomy 21 and translocations of chromosome 4. *Proc Natl Acad Sci USA* **85**, 9138–9142.

Popescu, N.C., Zimonjic, D.B. (1997) Molecular cytogenetic characterization of cancer cell alterations. *Cancer Genet. Cytogenet.* **93**, 10–21.

Poulsen, T.S., Silahtaroglu, A.N., Gisselo, C.G., Gaarsdal, E., Rasmussen, T., Tommerup, N., Johnsen, H.E. (2001) Detection of illegitimate rearrangement within the immunoglobulin locus on 14q32.3 in B-cell malignancies using end-sequenced probes, *Genes Chromosomes Cancer* **32**, 265–274.

Raap, A.K., van de Corput, M.P., Vervenne, R.A., van Gijlswijk, R.P., Tanke, H.J., Wiegant, J. (1995) Ultra-sensitive FISH using peroxidase-mediated deposition of biotin- or fluorochrome tyramides. *Hum Mol Genet* **4**, 529–534.

Raderschall, E., Golub, E.I., Haaf, T. (1999) Nuclear foci of mammalian recombination proteins are located at single-stranded DNA regions formed after DNA damage, *Proc. Natl. Acad. Sci. USA* **96**, 1921–1926.

Rapp, A. (1998) Advances in fluorescence in situ hybridization, *Mutation Research* **400**, 287–298.

Raudsepp, T., Mariat, D., Guerin, G., Chowdhary, B.P. (2001) Comparative FISH mapping of 32 loci reveals new homologous regions between donkey and horse karyotypes. *Cytogenet Cell Genet* **94**, 180–185.

Rettenberger, G., Klett, C., Zechner, U., Kunz, J., Vogel, W., Hamesiter, H. (1995) Visualization of the conservation of synteny between humans and pigs by heterologous chromosomal painting. *Genomics* **26**, 372–378.

Reid, T., Arnold, N., Ward, D.C., Wienberg, J. (1993) Comparative high-resolution mapping of human and primate chromosomes by fluorescence in situ hybridization. *Genomics* **18**, 381–386.

Ried, T., Baldini, A., Rand, T.C., Ward, D.C. (1992) Simultaneous visualization of seven different DNA probes by in situ hybridization using combinatorial fluorescence and digital imaging microscopy. *Proc Natl Acad Sci USA* **89**, 1388–1392.

Riemersma, S.A., Jordanova, E.S., Schop, R.F., Philippo, K., Looijenga, L.H., Schuuring, E., Kluin, P.M. (2000) Extensive genetic alterations of the HLA region, including homozygous deletions of HLA class II genes in B-cell lymphomas arising in immune-privileged sites, *Blood* **96**, 3569–3577.

Rottger, S., Yen, P.H., Schempp, W. (2002) A fiber-FISH contig spanning the non-recombining region of the human Y chromosome. *Chromosome Res.* **10**, 621–35.

Scherthan, H., Cremer, T., Amason, U., Weier, H.-U., Lima-de-Faria, A., Fronicke, L. (1994) Comparative chromosome painting discloses homologous segments in distantly related mammals. *Nat Genet* **6**, 342–347.

Schriml, L.M., Padilla-Nash, H.M., Coleman, A., Moen, P., Nash, W.G., Menninger, J., Jones, G., Tied, T., Dean, M. (1999) Tyramide signal amplification (TSA)-FISH applied to mapping PCR-labeled probes less than 1 kb in size. *BioTechniques* **27**, 608–613.

Schrock, E., duManoir, S., Veldman, T., et al. (1996) Multicolor spectral karyotyping of human chromosomes. *Science* **273**, 494–497.

Senger, G., Ragoussis, J., Trowsdale, J., Sheer, D. (1991) Fine mapping of the human MHC class II region within chromosome band 6p21 and evaluation of probe ordering using interphase fluorescence in situ hybridization. *Am J Hum Genet* **48**, 1–15.

Senger, G., Jones, T.A., Fidlerova, H., Sanseau, P., Trowsdale, J., Duff, M., Sheer, D. (1994) Released chromatin: linearized DNA for high resolution fluorescence in situ hybridization. *Hum Mol Genet* **3**, 1275–1280.

Shizuya, H., Birren, B., Kim, U.-U. (1992) Cloning and stable maintenance of 300-kilobase-pair fragments of human DNA in Escherichia coli using an F-factor-based vector. *Proc Natl Acad Sci USA* **89**, 8794–8797.

Slijepcevic P. (2001) Telomere length measurement by Q-FISH. *Methods Cell Sci* **23**(1–3), 17–22.

Speel, E.J.M., Ramaekers, F.C.S., Hopman, A.H.N. (1997) Sensitive multicolor fluorescence in situ hybridization using catalyzed reporter deposition (CARD) amplification. *J Histochem Cytochem* **45**, 1439–1446.

Speicher, M.R., Ballard, S.G., Ward, D.C. (1996) Karyotyping human chromosomes by combinatorial multi-fluor FISH. *Nature Genet* **12**, 368–375.

Speicher, M.R., duManoir, S., Schrock, E., et al. (1993) Molecular cytogenetic analysis of formalin-fixed, paraffin-embedded solid tumors by comparative genomic hybridization after universal DNA-amplification. *Hum Mol Genet* **2**, 1907–1914.

Takahashi, E., Koyama, K., Hirai, M., Itoh, H., Nakamura, Y. (1995) A high resolution cytogenetic map of human chromosome 2: localization of 434 cosmid markers by direct R-banding fluorescence in situ hybridization. *Cytogenet Cell Genet* **68**, 112–114.

Tanke, H.J., Florijn, R.J., Wiegant, J., Raap, A.K., Vrolijk, J. (1995) CCD microscopy and image analysis of cells and chromosomes stained by fluorescence in situ hybridization. *Histocheml J* **27**, 4–14.

Trask, B., Pinkel, D., van den Engh, G. (1989) The proximity of DNA sequences in interphase cell nuclei is correlated to genomic distance and permits ordering of cosmids spanning 250 kilobase pairs. *Genomics* **5**, 710–717.

Trask, B.J. (1991) Gene mapping by in situ hybridization. *Genet and Dev* **1**, 82–87.

Trask, B.J., Massa, H., Kenwrick, S., Gitschier, J. (1991) Mapping of human chromosome Xq28 two-color fluorescence in situ hybridization of DNA sequences to interphase cell nuclei. *Am J Hum Genet* **48**, 1–15.

Trask, B.J., Massa, H., Brand-Arpon, V., Chan, K., Friedman, C., Nguyen, O.T., Eichler, E., van den Engh, G., Rouquier, S., Shizuya, H., Giorgi, D. (1998) Large multi-chromosomal duplications encompass many members of the olfactory receptor gene family in the human genome. *Hum Mol Genet* **7**, 2007–2020.

Trower, M.K., Orton, S.M., Purvis, I.J., Sanseau, P., Riley, J., Christodoulou, C., Burt, D., See, C.G., Elgar, G., Sherrington, R., Rogaev, E.I., St George-Hyslop, P., Brenner, S., Dykes, C.W. (1996) Conservation of synteny between the genome of the pufferfish (Fugu rubripes) and the region on human chromosome 14 (14q24.3) associated with familial Alzheimer disease (AD3 locus). *Proc Natl Acad Sci USA* **93**, 1366–1369.

Vaandrager, J.W., Schuuring, E., Rapp, T., Phillippo, J., Klelverda, K., Kluin, P. (2000) Interphase FISH detection of BCL2 rearrangement in follicular lymphoma using breakpoint-flanking probes. *Genes, Chromo & Cancer* **27**, 85–94.

Valdes, J., Tagle, D.A. (1997) Gene mapping goes from FISH to surfing the net. *Methods in Mol Biol* **68**, 1–10.

van den Engh, G., Sachs, R. and Trask, B. (1992) Estimating genomic distances from DNA sequence location in cell nuclei by a random walk model. *Science* **257**, 1410–1412.

van Ommen, G.J., Breuning, M.H., Raap, A.K. (1995) FISH in genome research and molecular diagnostics, *Curr Opin Genet Dev.* **5**, 304–308.

Vesa, J., Hellsten, E., Verkruyse, L.A., Camp, L.A., Rapola, J., Santavuori, P., Hofmann, S.L., Peltonen, L. (1995) Mutations in the palmitoyl protein thioesterase gene causing infantile neuronal ceroid lipofuscinosis. *Nature.* **376**, 584–587.

Viegas-Pequignot, E., Dutrillaux, B., Magdelenat, H., Coppey-Moisan, M. (1989) Mapping of single-copy DNA sequences on human chromosomes by in situ hybridization with biotinylated probes: Enhancement of detection sensitivity by intensified-fluorescence digital-imaging microscopy. *Proc Natl Acad Sci USA* **86**, 582–586.

Volante, M., Pecchioni, C., Bussolati, G. (2000) Post-incubation heating significantly improves tyramide signal amplification. *The J of Histochem & Cytochem* **48**, 1583–1585.

Weier, H.U., Wang, M., Mullikin, J.C., Zhu, Y., Cheng, J.-F., Greulich, K.M., Bensimon, A., Gray, J.W. (1995) Quantitative DNA fiber mapping, *Hum. Mol. Genet.* **4**, 1903–1910.

Weier, H.U. DNA fiber mapping techniques for the assembly of high-resolution physical maps. (2001) *J Histochem & Cytochem* **49**, 939–948.

Wiegant, J., Ried, T., Nederlof, P., van der Ploeg, M., Tanke, H.J. and Raap, A.K. (1991) In situ hybridization with fluoresceinated DNA. *Nucleic Acids Res* **19**, 3237–3241.

Wiegant, J., Kalle, W., Mullenders, L., Brookes, S., Hoovers, J.M.N., Dauwerse, J.G., van Ommen, G.J.B. and Raap, A.K. (1992) High-resolution in situ hybridization using DNA halo preparations. *Hum Mol Genet* **1**, 587–592.

Ye, J., Lu, W., Moens, P., Liu, G., Bremer, S., Wang Y., Hughes, M., Krawetz, S., and Heng, H.H.Q. (2001) The combination of SKY and specific loci detection with FISH or Immunostaining. *Cytogenet Cell Genet* **93**, 195–202.

Yokota, H., Singer, M.J., van den Engh, G.J., Trask, B.J. (1997) Regional differences in the compaction of chromatin in human G0/G1 interphase nuclei. *Chromosome Res* **5**, 157–166.

26
Gene Mapping and Chromosome Evolution by Fluorescence–Activated Chromosome Sorting

Malcolm A. Ferguson-Smith
Cambridge University Centre for Veterinary Science, Cambridge, UK

1 Introduction 751

2 Fluorescence-activated Chromosome Sorting (FACS) 752

3 Chromosome Painting 754

4 Chromosome Microdissection 755

5 Gene Mapping 756

6 Chromosome Painting in the Diagnosis of Chromosome Aberrations 756

7 Karyotype Evolution 759

8 Conclusions 762

Acknowledgements 762

Bibliography 762
Books and Reviews 762
Primary Literature 763

Keywords

BAC (Bacterial Artificial Chromosome)
A vector used in DNA cloning that is capable of carrying DNA inserts measuring 100 to 200 kb.

Genomics and Genetics. Edited by Robert A. Meyers.
Copyright © 2007 Wiley-VCH Verlag GmbH & Co. KGaA, Weinheim
ISBN: 978-3-527-31609-0

Chromosome-specific Paint Probe
The product of PCR amplification of sorted chromosomes, using random PCR primers. When labeled with fluorochromes coupled to nucleotides in the amplification procedure, the paint probe can be hybridized by fluorescence *in situ* hybridization (FISH) techniques to chromosome preparations on microscope slides. This reveals an even distribution of FISH signals along the chromosomes from which the paint was derived.

Contig
A contiguous series of cloned DNA sequences with overlapping ends.

Cross-species Reciprocal Chromosome Painting
A technique used to identify blocks of chromosome homology shared between species. A chromosome-specific paint probe from species A is hybridized to the chromosomes of species B. Paint probes from species B are hybridized back to the chromosomes of species A. This reveals the chromosomal identity of each homologous region in both species.

Cosmid
A vector used in DNA cloning that is capable of carrying DNA inserts measuring 20 to 40 kb.

FISH (Fluorescence *in situ* Hybridization)
A method whereby fluorescence–labeled DNA probes are hybridized to their complementary sequences on denatured chromosomes fixed on microscope slides.

Flow Karyotype
Graphic representation of mitotic chromosomes separated by a dual laser flow cytometer according to size and base-pair ratio.

Interchromosomal Rearrangement
A structural chromosome change involving exchange between two or more nonhomologous chromosomes.

Intrachromosomal Rearrangement
A structural change involving inversion or insertion of material within a chromosome.

Plasmid Vector
A vector that can be used for cloning small DNA inserts of 3 to 8 kb.

PCR (Polymerase Chain Reaction)
A method for the primer-directed amplification of DNA sequences.

26 Gene Mapping and Chromosome Evolution by Fluorescence–Activated Chromosome Sorting

The ability to sort and collect chromosomes by flow cytometry provides studies on gene mapping and chromosome evolution with a unique source of chromosome-specific DNA. Chromosome-specific gene libraries made from this DNA have yielded genetic markers used in constructing ordered chromosome maps. This was one of the key steps in the project which led to sequencing the human genome. PCR amplification of sorted chromosomal DNA has been used to prepare fluorescence-labeled chromosome-specific probes. When hybridized *in situ* to chromosome preparations, these probes "paint" the entire chromosome. The judicious use of different combinations of fluorochromes allows each chromosome pair in the cell to be given a different color. This procedure is helpful in the diagnosis of chromosome rearrangements present in malignant cells or associated with handicapping syndromes. Paint probes from one species can be used to identify homologous chromosome regions in other species. Comparison of the patterns of homology produced by reciprocal cross-species painting adds information on the evolutionary history of a species by distinguishing ancient chromosome rearrangements from those that have occurred more recently. The study of comparative genomics and karyotype evolution owes much to the development of these molecular cytogenetic techniques.

1
Introduction

The fluorescence-activated cell sorter was first used successfully in 1975 to separate chromosomes from fluid suspensions. It was soon found that the method allowed the collection of substantial quantities of chromosomes of one type without cross-contamination with other chromosomes. The chromosome-specific DNA so produced was used to prepare chromosome-specific libraries of DNA in plasmid and other vectors. In the early days of the human genome project, these plasmid libraries were of great importance in cloning chromosome-specific markers, which were used to map individual chromosomes. Linkage of polymorphic markers in family studies allowed the order of markers along the chromosome to be determined, and the detection of linkage between markers and cloned sequences from known genes similarly allowed the assignment of disease and other genes to their location on chromosomes. In the late 1980s, mapping by genetic linkage was supplemented by fluorescence *in situ* hybridization (FISH) mapping in which suitably labeled DNA markers were hybridized *in situ* to their complementary sequences on human chromosomes. With the development of BAC, cosmid, and other vectors for DNA cloning, it became possible to build up series of contiguous sequences of DNA (contigs) along each chromosome on the basis of this preliminary marker map. This approach, coupled with large-scale sequencing of contigs, has been one of the key strategies used in sequencing the human genome. Now that a draft sequence, albeit with gaps, of the entire human chromosome complement has been achieved, chromosome sorting has less application in mapping human genes. However, it still has application in the building of preliminary maps of unmapped species.

Chromosome-specific DNA prepared from sorted chromosomes may also be used to prepare chromosome-specific probes using the polymerase chain reaction (PCR) with random DNA primers to amplify and label the probes with fluorescent dyes. The resulting labeled DNA is hybridized by FISH to air-dried metaphases on microscope slides in such a way that the respective chromosomes are "painted" with fluorescence along the entire length of the chromosome. The technology has advanced to the extent that each chromosome can be recognized by painting with different combinations of fluorochromes (multicolor FISH) in one hybridization experiment. The main application of M-FISH is in the diagnosis of interchromosomal aberrations that cause physical and mental childhood handicap and in the cytogenetic analysis of cancer (see chapter on FISH).

Chromosome painting has played an important part in characterizing regions of genetic homology between chromosomes of different species. The first studies in the early 1990s used chromosome-specific plasmid libraries, but more recent work uses chromosome-specific painting probes amplified from sorted chromosomes. The first comparisons were between humans and other primates. It was confirmed that the karyotype of humans differed from that of the great apes by a tandem fusion of two chromosomes, which formed human chromosome 2 and which occurred after the divergence of the human lineage from the common ancestor with the chimpanzee. Numerous inversions have also occurred during the evolution of the great apes, but the only other interchromosomal rearrangement found was a reciprocal chromosome translocation between human chromosome equivalents 5 and 17 in the divergence of the gorilla. Interchromosomal rearrangements between human and other species have been observed since, some involving whole chromosome arms and some involving smaller segments. The remarkable feature has been the high level of conservation between genomes. The range in the number of conserved (syntenic) autosomal blocks between human and the mammalian species studied to date varies from 23 to 200, the majority in the order of 40 to 50 and the highest in the house mouse. Some rearrangements are more ancient than others and so are shared by a wider range of species. The pattern of rearrangements in a species reflects the history of the genome evolution of that species, and the comparison of patterns between species provides information about their relationship and their place on the evolutionary tree.

2
Fluorescence-activated Chromosome Sorting (FACS)

Chromosomes are sorted using a dual laser flow cytometer equipped with two 5 W argon ion lasers. One laser is tuned to emit 300 mW of light in the UV (351–364 nm) to excite Hoechst fluorescence, and the second laser is tuned to emit 300 mW at 458 nm to excite chromomycin A3 fluorescence.

Chromosome preparations are made by standard methods from short-term blood cultures, lymphoblastoid cell lines, and a variety of tissue culture cells including fibroblasts from biopsy material. Colcemid ($0.1\ \mu g\ mL^{-1}$) is added to the actively dividing cell culture to arrest cells in metaphase over several hours. The mitotic cells are collected and resuspended in 75 mM KCl for 15 min and are transferred to a buffer containing polyamines and Triton X-100. Chromosomes are released into suspension by rapid vortexing for 10 s and are stained with chromomycin A3 (final concentration 40 µg mL^{-1}) and Hoechst 33258 (final concentration 2 µg mL^{-1}) followed by 2 h of incubation at 4 °C. Fifteen minutes prior to flow sorting, sodium citrate (final concentration 10 mM) and sodium sulfate (final concentration 25 mM) are added to the sample. Excessive disruption of the mitotic cells should be avoided as this leads to chromosome fragmentation indicated by an unacceptable level of debris at the lower end of the flow karyotype (Fig. 1).

Cultures that contain a high proportion of cells in the anaphase/telophase may also lead to reduced resolution owing to contamination from the inclusion of anaphase chromosomes.

The chromosomes suspended in polyamine buffer, and surrounded by sheath fluid, flow under pressure through the two laser beams sequentially to permit the fluorescence signal emitted from each chromosome to be collected separately and stored in the computer. After the chromosomes have passed through the two lasers, the fluid stream breaks into a series of droplets, some of which will contain a single chromosome. Sorting is achieved by applying an electrical charge to the droplets

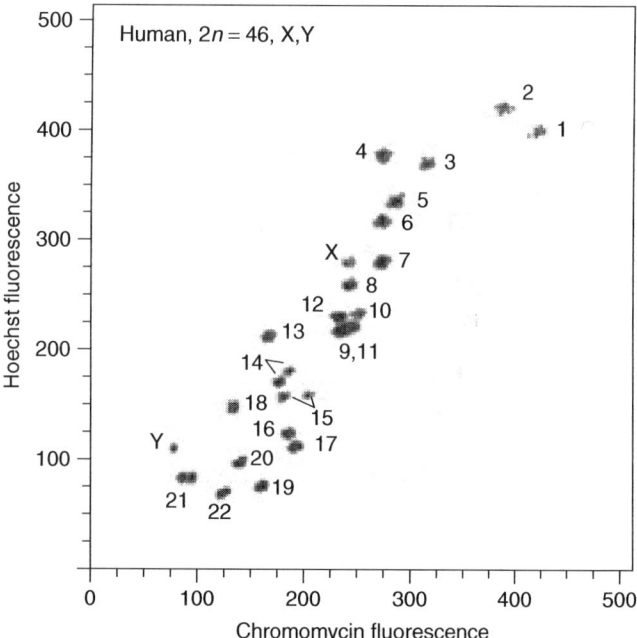

Fig. 1 Flow karyotype constructed from a fluid suspension of metaphase chromosomes from a normal male. The numbered peaks correspond to the numbers used in the classification of the human chromosome complement. Chromosomes are separated on the basis of DNA content and base-pair ratio. Note the separation of homologues of chromosomes 14, 15, and 21 due to heteromorphic blocks of repetitive DNA.

containing the chromosome of interest so that they can be deflected into a container as they pass between two high-voltage plates. Highly pure samples of two types of chromosome can be collected simultaneously. The fluorescence measurements from each chromosome are accumulated in large numbers in the computer and are used to construct a flow karyotype (Fig. 1), which reveals discrete clusters of signals, each cluster representing one or more chromosome types arranged in order of size and base-pair ratio. A–T-rich chromosomes sort above a diagonal line drawn through the middle of the chromosome clusters, while G–C-rich chromosomes sort below the line. The accumulation of signals can be observed on a video monitor and the chromosomes of interest selected for collection by a simple gating procedure. It takes less than 5 min to collect the 300 to 500 chromosomes required for PCR mapping and for the production of chromosome-specific paint probes. Larger samples of 2 to 3 million are required for preparing chromosome-specific DNA libraries, and this requires prolonged sorting.

3
Chromosome Painting

The DNA of 300 to 500 chromosomes sorted by FACS is amplified by degenerate oligonucleotide-primed PCR (DOP-PCR). After a primary round of amplification using the 6-MW primer (5′ CCGACT CGA GNN NNN NAT GTGG 3′) and unlabeled deoxynucleotide triphosphates, the DNA is subjected to a secondary round of amplification, this time incorporating a labeled nucleotide such as biotin-11-dUTP. The biotinylated probe is hybridized to denatured chromosomes in air-dried metaphase preparations using standard FISH methods and is detected using fluorochrome-conjugated streptavidin or anti-biotin antibodies. Many labeling techniques now use fluorescent dyes, such as Cy-dyes, coupled to the nucleotides used in the amplification procedure, and this avoids the need for an additional detection step.

The FISH signals are examined by fluorescence microscopy. An image processing system is helpful in analyzing preparations of low luminosity and in situations in which complex chromosome rearrangements are to be expected. Most digital fluorescence microscopy analysis systems employ a sensitive monochromatic CCD camera and excitation and emission filters appropriate to the fluorescence dye used as a label. When several paint probes, labeled with different combinations of dyes, are used, the gray-scale image of the fluorescence emitted by each fluorochrome is acquired sequentially and merged to produce a composite image in which each paint probe is assigned a separate false color on the computer monitor. Probe sets capable of assigning a different color combination to each pair of chromosomes in the complement are now available for human and mouse, using M-FISH and systems for classification and analysis of karyotypes with interchromosomal rearrangements (Fig. 2). The stored images acquired by these computerized systems are available for color adjustment, image enhancement, image reversal, and a number of other procedures, which assist in chromosome measurement and analysis. Spatial filters and contrast adjustment are particularly useful in enhancing the weak chromosome banding patterns obtained by DAPI counterstaining. Instead of the analysis of images of each individual

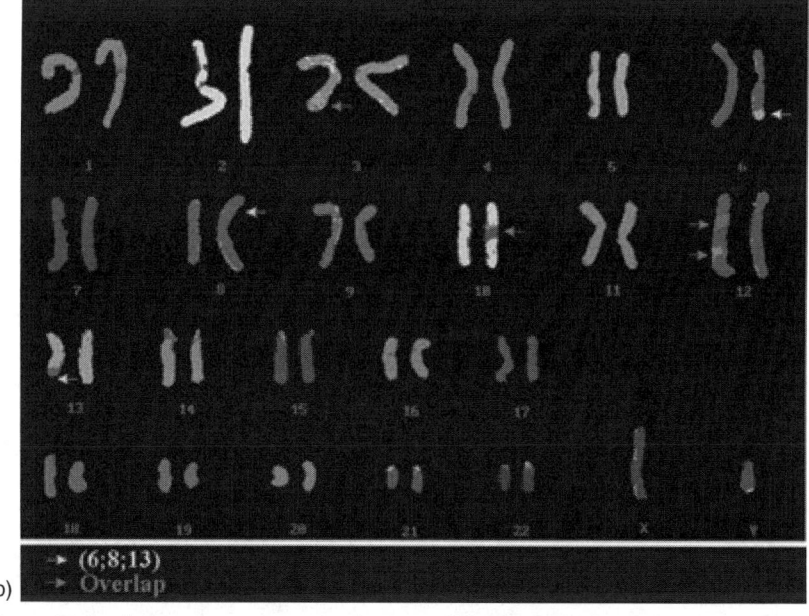

Fig. 2 Multicolor FISH. (a) Using a combination of five fluorochromes, each chromosome pair has a distinctive color. (b) This allows the identification of interchromosomal translocations between chromosomes 6, 8, and 13 in the karyotype. (See color plate p. ix.)

color, the alternative spectral karyotyping (SKY) system analyzes the spectral characteristics of each pixel in one image using an interferometer. Both M-FISH and SKY provide similar resolution, each capable of detecting rearrangements as small as 2 Mb.

4 Chromosome Microdissection

Microdissection is an alternative method for preparing labeled chromosome-specific DNA. Glass microneedles are used under direct vision with an inverted microscope and phase-contrast optics to scrape the chromosome of interest from air-dried chromosome preparations on a microscope slide. About 20 chromosomes are removed into a tube containing distilled water. DOP-PCR is used to make amplified primary products, which can then be used, in a further round of amplification with labeled nucleotides, to achieve the desired painting probe. The technique has been used with considerable success to generate single-chromosome-arm probes and region-specific probes. One application termed *multicolor banding* (MCB) is based on the microdissection of up to four to five regions along each chromosome. The resulting fragments are labeled with different combinations of fluorochromes so that hybridization to a metaphase produces a series of colored bands along the chromosome. This has been used to detect inversions and insertions in diagnostic cytogenetics. However, the technology is complex and is at present confined to the research laboratory.

5
Gene Mapping

Chromosome-specific DNA produced by flow sorting of chromosome suspensions can contribute to the assignment of genes to their respective chromosomes, provided that a DNA sequence from within the gene is known. In this method, PCR primers from both ends of the DNA sequence are used to amplify the gene in a panel made from primary amplification products of each chromosome, both in pools of five to six chromosomes and separately. It has been found that more than 85% of known genes can be mapped in this way. The method provides a rapid procedure for assigning linkage groups to the chromosomes of poorly mapped species, as has been shown recently in the construction of the genetic map of the dog.

Gene mapping in many species has been advanced by the construction of homology maps based on the results of reciprocal cross-species chromosome painting with chromosome-specific probes from the well-mapped human and mouse chromosomes (Fig. 3). While it is straightforward to assign linkage and radiation hybrid groups to their respective syntenic blocks in other species, their orientation and gene order has to be determined by FISH mapping on metaphase preparations with locus-specific probes (see chapter on FISH methods for the types of probe available). This is because intrachromosomal rearrangements, that is, inversions and insertions, are known to be relatively common events in species divergence. A higher level of resolution in obtaining gene order can be obtained from the longer chromosomes observed in interphase nuclei. However, the highest resolution by cytological methods is achieved using FISH on extended DNA fibers (Fig. 4). These fibers are produced by methods that release the DNA molecule from its associated proteins, especially histones, and allow spreading of the naked DNA molecules on microscope slides. Cloned DNA sequences of 1 kb or less can be hybridized successfully, and these can reveal such detail as the order and relative size of the exons and introns comprising a single gene. For further detail, DNA sequencing can be used, which is the ultimate level of resolution in gene maps.

6
Chromosome Painting in the Diagnosis of Chromosome Aberrations

Abnormal chromosomes can be sorted as well as normal ones provided they

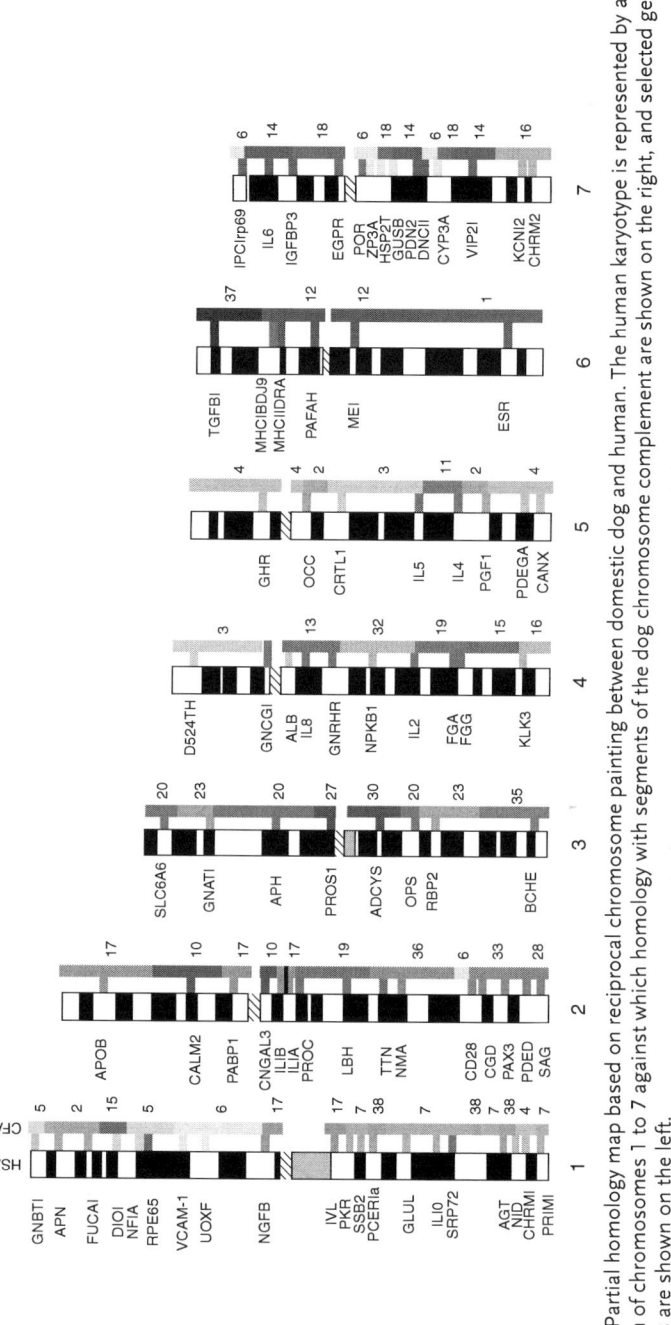

Fig. 3 Partial homology map based on reciprocal chromosome painting between domestic dog and human. The human karyotype is represented by a diagram of chromosomes 1 to 7 against which homology with segments of the dog chromosome complement are shown on the right, and selected gene markers are shown on the left.

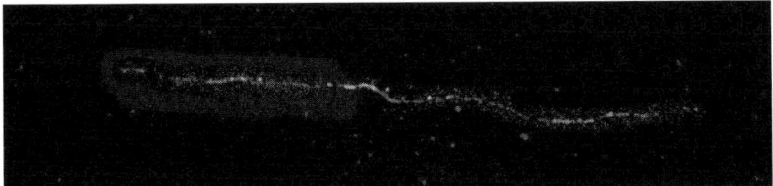

Fig. 4 DNA fiber FISH showing order of three 130-kb BAC clones from the DMTR1 locus on human chromosome 9.

are separated from others in the flow karyotype. When paints are made from such aberrations and hybridized to normal metaphases (reverse painting), their origin from normal chromosomes can be determined rather precisely. Translocations involving two or more chromosomes are evident from the number of chromosome regions highlighted by the paint probe. Intrachromosomal duplications, which are among the most frequent aberrations, are evident from the observation that only one pair of chromosomes is painted. The duplicated region may be revealed by an increased dosage of label over the duplication.

Reverse painting can also be applied to probes prepared from the microdissection of abnormal chromosomes, especially small supernumerary chromosomes. However, the more usual approach is to use M-FISH or SKY systems in the analysis of complex aberrations. Where these are not available, single whole-chromosome paints (WCP) can be used in sequential rounds of denaturation and hybridization until the nature of the aberration is revealed.

As neither SKY nor M-FISH, nor WCP methods, can resolve inversions, color-banding methods have been developed to solve this problem. The published methods include bar code approaches using differentially labeled fragments of chromosome derived from radiation hybrids of human and rodent cells, each clone containing only a small number of human chromosome fragments. Different labels are used for each clone, so that the resulting paint probe set produces a series of color bands along each chromosome. The Harlequin approach (Cambio) uses M-FISH-labeled gibbon chromosomes, which when painted onto human chromosomes produce a color-banded appearance reflecting the 80 or more interchromosomal rearrangements that have occurred during the divergence of human and gibbon species. The MCB approach achieves a banding appearance by using several fragments removed by microdissection from each chromosome and distinguishing each fragment by a different combination of fluorochromes. All these methods can be used in the analysis of intrachromosomal rearrangements in both constitutional and neoplastic cytogenetics.

One of the aims of using multicolor painting probes in human and cancer cytogenetics is to identify the genes involved at the breakpoints of rearrangements. It has been found that these breakpoints may be associated with deletion or disruption of DNA followed by loss of gene function relevant to the phenotype. The strategy is first to identify the sites of the rearrangement and then to FISH-map the region using a series of BAC or other types of cloned DNA known, from the gene map, to be within the region of interest. Once a clone that bridges the breakpoint has been found, subclones in plasmids can be used to narrow down the area to a point when DNA sequencing can be used to identify the pathological change. A more effective approach, based on comparative genome hybridization (CGH), has recently been introduced. Flow cytometry is used to isolate and collect the two derivative chromosomes involved in a rearrangement. Each is amplified, labelled with red and green respectively, and the mixture used to probe a microarray of genomic DNA reference BAC clones. Some 3000 reference clones (unlabeled), selected because they span the entire human genome at approximately 1-Mb intervals, are arrayed on glass slides for this type of analysis. The arrays are scanned to detect the ratio of red : green in the hybridization signals generated in each clone by the probe mixture. The information is examined in relation to the genetic map. This reveals the constitution of each derivative and the breakpoints of the rearrangement. A clone that gives an intermediate red : green ratio may be one that spans the breakpoint. The technique termed *array-CGH* has particular application in the identification of cancer genes and genes involved in dysmorphic and handicapping syndromes associated with constitutional chromosome aberrations. It is likely that the reference set of 3000 BAC clones will become an important part of the cytogeneticist's armamentarium in the future, both for array-CGH and for FISH mapping.

7
Karyotype Evolution

The extensive conservation of genetic information in the animal kingdom is evident from gene mapping studies, which show that the same groups of genes are often linked together in a wide range of species. In closely related species, this conservation often extends to the pattern of G-banding revealed by conventional karyotype analysis. The X-chromosome of placental mammals is a particularly striking example of conservation as it contains virtually all the same transcribed genes in every mammalian species. In marsupials, the homology with the human X is confined to the long arm of the human X. Human X short-arm genes are autosomal in marsupials. In birds, the Z-chromosome in all species studied has close homology with human chromosome 9, and the equivalent of the human X-linked genes are to be found mostly in two autosomes, for example, in chicken chromosomes 1 and 4.

The results from comparative G-banding studies are often difficult to interpret, and it has been found that cross-species chromosome painting gives much more accurate evidence of chromosome homology between species. As indicated in the Introduction, there are two main limitations of the painting method. The first is that intrachromosomal rearrangements cannot be detected unless interspersed with chromosomal exchanges (see Fig. 3). The second is that the efficiency of cross-species hybridization diminishes as the relationship between the two species becomes more distant. In general, reciprocal chromosome painting works well between placental mammals, but is difficult between mammals and marsupials and between mammals and birds. There are exceptions, and paints prepared from the tamar wallaby X hybridize, albeit weakly, with the long arm of the human X. In addition, paints prepared from human chromosome 4 have been reported to hybridize to chicken chromosome 4 in accordance with comparative mapping data; this result illustrates a remarkable degree of genomic conservation between species estimated to have diverged over 300 million years ago.

Chromosome rearrangements that have occurred during evolution can be used to study the degree of relationship between species and to speculate on the ancestral karyotype. When the same rearrangement is found in two species from separate families, but which is absent in all other species, it can be assumed that the shared rearrangement is a recent event in evolution. On the other hand, a rearrangement that is common to a wide range of divergent species (outgroup species) can be regarded as ancestral. It can usually be assumed that when large blocks of DNA homology identified by chromosome painting are retained intact in distantly related species, this arrangement represents the ancestral condition. When these blocks are divided between different chromosomes in a single species, or in a group of closely related species, the rearrangements responsible may be regarded as recent evolutionary events. Occasionally, an ancient rearrangement may be reconstituted by a more recent event, that is, there is a reversion to the ancestral state. Much can be learned about phylogenomic relationships and karyotype evolution from these chromosome-painting studies between closely related species and outgroup species. As a result, some conclusions can be drawn about the nature of the ancestral karyotype of all placental mammals.

A recent evolutionary event is the well-known fusion of ancestral chromosomes to form human chromosome 2. All extant great apes do not show this fusion, which must have arisen in the human lineage following divergence of the chimpanzee and human lineages from their common ancestor. In another example, which relates to the evolution of the primate chromosome destined to become the short arm of human chromosome 2, the gorilla and orangutan have a pericentric inversion, which is absent in both human and chimpanzee. At first it was thought that the arrangement in human and chimpanzee had occurred subsequent to the divergence of the gorilla and orangutan and therefore served to link chimpanzee and human. It was later found that the inversion was absent in Old World monkeys (outgroup species), indicating that the human arrangement in this case is the ancestral one. The inversion in the gorilla and orangutan must have occurred subsequent to the divergence of humans and chimpanzees from the lineages, leading to the other great apes.

Chromosome painting with human probes reveals a number of ancient syntenic blocks in outgroup species that are split between different human chromosomes (Table 1). For example, the homologues of human chromosomes 14 and 15 are fused in one or two blocks in most mammalian species with the exception of the apes, dog, and giant panda. In the latter two cases, independent fission events have occurred as comparatively recent evolutionary events. Other ancient syntenies are found to characterize orders and families and, sometimes, small groups of species. Their discovery can be helpful in assigning a species to its place in the phylogenetic tree. For example, it has been shown by comparative painting with dog paints that the common ancestor of the domestic dog, red fox, Arctic fox, and raccoon dog had a small chromosome number composed of 38 autosomal segments and that a number of specific

Tab. 1 Ancient syntenies revealed by the association of chromosome segments homologous to human chromosomes in the genomes of various mammals. Chromosome numbers refer to individual human chromosomes.

Species	Ancient syntenies								
	14/15	3/21	12/22	16/19	4/8	7/16	3/19	1/2	5/19
Great apes	−	−	−	−	−	−	−	−	−
Lesser apes	−	−	−	−	−	−	−	−	−
Old World monkeys	+	−	−	−	−	−	−	−	−
New World monkeys	±	±	−	−	−	−	−	−	−
Lemurs	+	+	+	−	−	−	−	−	−
Cattle	+	+	+	+	+	+	−	+	+
Sheep	+	+	+	+	+	+	−	+	+
Muntjac	+	+	+	+	−	+	−	+	+
Pig	+	+	+	+	−	−	−	−	+
Horse	+	+	+	+	+	+	−	−	−
Zebra	+	+	+	+	+	+	−	+	−
Dolphin	+	+	+	+	−	−	+	−	+
Seal	+	+	+	+	+	−	+	−	−
Cat	+	+	+	+	+	+	+	+	−
Mink	+	+	+	+	+	+	+	−	−
Dog	−	+	+	−	+	+	+	+	−
Panda	−	+	+	+	+	−	+	+	−
Rabbit	+	+	+	+	+	+	−	−	−
Shrew	+	+	−	+	−	−	−	−	−

interchromosomal rearrangements distinguish the karyotypes of each species. As the dog has a chromosome number of 78, the change from low numbers has been achieved by a series of fission events, with the formation of new centromeres. Chromosome fusion, on the other hand, has been the mechanism by which the three pairs of chromosomes of the Indian muntjac ($2n = 6, 7$) have evolved from an ancestral deer karyotype of $2n = 70$. This is illustrated by painting chromosome-specific probes of the Chinese muntjac ($2n = 46$) onto Indian muntjac chromosomes. Each Indian muntjac chromosome can be shown to be formed by the fusion of several apparently complete Chinese muntjac chromosomes (Fig. 5). Remnants of ancestral centromeres can be demonstrated at most of the fusion sites, indicating frequent centromere–telomere events, although this remains to be confirmed by molecular analysis.

Comparative studies of the genomes of mouse and human at the DNA sequence level now show that evolutionary breakpoints tend to occur within clustered gene families. It seems that the most likely mechanism is meiotic recombination between homologous sequences on nonhomologous chromosomes or chromosome segments, that is, a similar mechanism that occurs in the formation of pathological rearrangements both in somatic and in germ cells. Speciation can now be seen as resulting largely from the gain or loss of sequences within gene families, leading to the degeneration of some genes of known function and the generation of new genes of different function. It is believed that duplication events are the main source of evolutionary novelty.

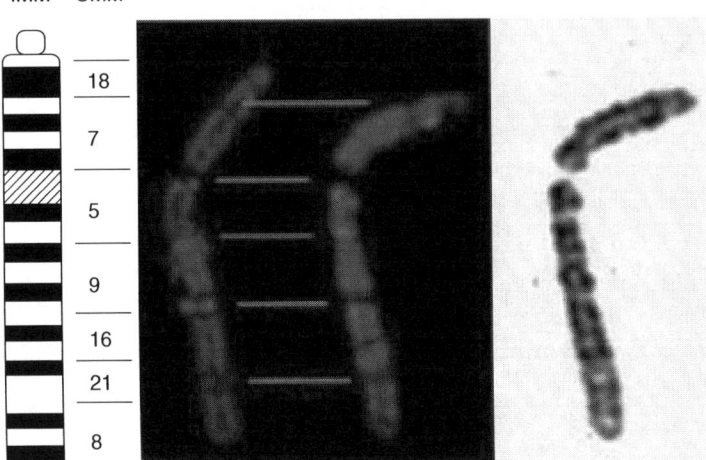

Fig. 5 Indian muntjac (IMM) chromosome 3 to which chromosome-specific paint probes of the Chinese muntjac (CMM) have been hybridized. This reveals that this chromosome has been formed by the tandem fusion of seven entire chromosomes of the Chinese muntjac. A centromere probe (red) indicates that remnants of ancestral centromeres are still present at the sites of fusion. (See color plate p. vi.)

8
Conclusions

The utility of chromosome painting extends from diagnostic cytogenetics to its application in comparative gene mapping and to studies of karyotype evolution. The development of highly effective techniques for sorting pure samples of chromosomes in fluid suspension has played a central role in theses advances. Future understanding of genome functions relating to gene regulation, chromosome structure and behavior, genome evolution, and the significance of noncoding conserved DNA will depend on the construction of comparative maps in a wide range of species. Only by such comparisons will it be possible to determine which parts of the genome are conserved in evolution because they retain vital functions. In these comparisons, chromosome sorting can be expected to continue to play a useful role.

Acknowledgements

The author is grateful to Dr Fengtang Yang, Dr Willem Rens, Dr Fumio Kasai, Mrs Patricia O'Brien, and other colleagues at the Cambridge Resource Centre for Comparative Genomics for the illustrations and for useful discussions in the preparation of this review.

See also Organization of Genes and Genome Domains; Molecular Basis of Genetics.

Bibliography

Books and Reviews

Dutrillaux, B. (1986) Evolution chromosomique chez les primates, les carnivores, et les rongeurs, *Mammalia* **50**, 1–203.

Ferguson-Smith, M.A. (1997) Genetic analysis by chromosome sorting and painting:

phylogenetic and diagnostic applications, *Eur. J. Hum. Genet.* **5**, 253–265.

Ferguson-Smith, M.A., Smith, K. (2002) Cytogenetic analysis. Chapter 25, in: Rimoin, D.L., Connor, J.M., Pyeritz, R.E., Korf, B.R. (Eds.) *Emery & Rimoin's Principles and Practice of Medical Genetics*, Fourth Edition, Churchill Livingstone, London, pp. 690–722.

Gray, J.W., Carrano, A.V., Steinmetz, L.L., Van Dilla, M.A., Moore, D.H., Mayall, B.H., Mendelsohn, M.L. (1975) Chromosome measurement and sorting by flow systems, *Proc. Natl. Acad. Sci. U.S.A.* **72**, 1231–1234.

Meltzer, P.S., Guan, X.Y., Burgess, A., Trent, J.M. (1992) Rapid generation of region specific probes by chromosome microdissection and their application, *Nat. Genet.* **1**, 24–28.

Murphy, W.J., Stanyon, R., O'Brien, S. (2001) Evolution of mammalian genome organisation inferred from comparative gene mapping, *Gen. Biol.* **2**, 1–8.

Scherthan, H., Cremer, T., Arnason, U., Weier, H.-U., Lima-de-Faria, A., Fronicke, L. (1994) Comparative chromosome painting discloses homologous segments in distantly related mammals, *Nature Genetics* **6**, 342–347.

Schröck, E., du Manoir, S., Veldman, T., Schoell, B., Wienberg, J., Ferguson-Smith, M.A., Ning, Y., Lebetter, D., Bar-Am, I., Soenksen, D., Garini, Y., Ried, T. (1996) Multicolor spectral karyotyping of human chromosomes, *Science* **273**, 494–497.

Speicher, M.R., Ballard, S.G., Ward, D.C. (1996) Karyotyping human chromosomes by combinatorial multi-fluor FISH, *Nat. Genet.* **12**, 368–375.

Trask, B.J. (2002) Human cytogenetics: 46 chromosomes, 46 years and counting, *Nature Reviews, Genetics* **3**, 769–778.

Van Dilla, M.A., Deaven, L.L., Albright, K.L., Allen, N.A., Aubuchon, M., Bartoldi, M., Brown, N., Campbell, A., Carrano, A., Clark, L., Cram, L., Crawford, B., Fuscoe, J., Gray, J., Hildebrand, E., Jackson, P., Jett, J., Longmire, J., Lozes, C., Luedemann, M., Martin, J., McNinch, J., Meincke, L., Mendelson, M., Meyne, J., Moyzis, R., Munk, A., Perlman, J., Peters, D., Silvam, A., Trask, B. (1986) Human chromosome-specific DNA libraries: construction and availability, *Biotechnology* **4**, 537–552.

Wienberg, J., Stanyon, R. (1998) Comparative chromosome painting of primate genomes, *ILAR J.* **39**, 77–91.

Primary Literature

Carter, N.P., Ferguson-Smith, M.A., Perryman, M.T., Telenius, H., Pelmear, A.H., Leversha, M.A., Glancy, M.T., Wood, S.L., Cook, K., Dyson, H.M., Ferguson-Smith, M.E., Willatt, L.R. (1992) Reverse chromosome painting: a method for the rapid analysis of aberrant chromosomes in clinical cytogenetics, *J. Med. Genet.* **29**, 299–307.

Chowdhary, B.P., Raudsepp, T. (2000) HSA 4 and GGA 4: remarkable conservations despite 300 Myr divergence, *Genomics* **64**, 102–105.

Chudoba, I., Plesch, A., Lorch, T., Lemke, J., Claussen, U., Senger, G. (1999) High resolution multicolour-banding: a new technique for refined FISH analysis of human chromosomes, *Cytogenet. Cell Genet.* **84**, 156–160.

Davies, K.E., Young, B.D., Elles, R.G., Hill, M.E., Williamson, R.W. (1981) Cloning of a representative genomic library of the human X chromosome after sorting by flow cytometry, *Nature* **293**, 374–376.

Dehal, P., Predki, P., Olsen, A.S., Kobayashi, A., Folta, P., Lucas, S., Land, M., Terry, A., Zhou, C.L.E., Rash, S., Zhang, Q., Gordon, L., Kim, J., Elkin, C., Pollard, M.J., Richardson, P., Rokhsar, D., Uberbacher, E., Hawkins, T., Branscomb, E., Stubbs, L. (2001) Human chromosome 19 and related regions in mouse: conservative and lineage-specific evolution, *Science* **293**, 104–111.

Dutrillaux, B. (1979) Chromosomal evolution in primates: tentative phylogeny from Microcebus murinus (Prosimian) to man, *Human Genetics* **48**, 251–314.

Fiegler, H., Carr, P., Douglas, E.J., Burford, D.C., Hunt, S., Smith, J., Vetrie, D., Gorman, P., Tomlinson, I.P.M., Carter, N.P. (2003) DNA microarrays for comparative genomic hybridisation based on DOP-PCR amplification of BAC and PAC clones, *Genes, Chromosomes & Cancer* **36**, 361–374.

Florijn, R.J., Blonden, L.A.J., Vrolijk, J., Wiegant, J., Vaandrager, J.W., Baas, F., Den Dunnen, J.T., Tanke, H.J., Van Ommen, G.J.B., Raap, A.K. (1995) High resolution DNA fibre-FISH genomic DNA mapping and colour bar coding of large genes, *Hum. Mol. Genet.* **4**, 831–836.

Glas, R., Graves, J.A.M., Toder, R., Ferguson-Smith, M.A., O'Brien, P.C.M. (1999) Cross-species chromosome painting between human and marsupial demonstrates the ancient region of the mammalian X, *Mamm. Genome* **10**, 1115, 1116.

Graphodatsky, A.S., Yang, F., O'Brien, P.C.M., Perelman, P., Milne, B.S., Serdukova, N., Kawada, S.I., Ferguson-Smith, M.A. (2001) Phylogenetic implications of the 38 putative ancestral chromosome segments for four canid species, *Cytogenet. Cell Genet.* **92**, 243–247.

Graves, J.A.M., Watson, J.M. (1991) Mammalian sex chromosomes: evolution of organisation and function, *Chromosoma* **101**, 63–68.

Griffin, D.K., Sanoudou, D., Adamski, E., Mcgiffert, C., O'Brien, P., Wienberg, J., Ferguson-Smith, M.A. (1998) Chromosome specific comparative genome hybridisation for determining the origin of intrachromosomal duplications, *J. Med. Genet.* **35**, 37–41.

Harris, P., Boyd, E., Ferguson-Smith, M.A. (1985) Optimising human chromosome separation for the production of chromosome-specific DNA libraries by flow sorting, *Human Genetics* **70**, 59–65.

Jauch, A., Wienberg, J., Stanyon, R., Arnold, N., Tofanelli, S., Ishida, T., Cremer, T. (1992) Reconstruction of genomic rearrangements in great apes and gibbons by chromosome painting, *Proc. Natl. Acad. Sci. U.S.A.* **89**, 8611–8615.

John, H., Birnstiel, M., Jones, K. (1969) RNA:DNA hybrids at the cytological level, *Nature* **223**, 582–587.

Lichter, P., Cremer, C., Borden, J., Manuelidis, L., Ward, D.C. (1988) Delineation of individual human chromosomes in metaphase and interphase cells by in situ suppression hybridisation using recombinant DNA libraries, *Human Genetics* **80**, 224–234.

Lichter, P., Tang, C.C., Call, K., Hermanson, G., Evans, G.A., Housman, D., Ward, D.C. (1990) High resolution mapping of human chromosome 11 by in situ hybridisation with cosmid clones, *Science* **247**, 64–69.

Mann, S.M., Burkin, D.J., Griffin, D.K., Ferguson-Smith, M.A. (1997) A fast, novel approach for DNA fibre fluorescence *in situ* hybridisation analysis, *Chromosome Res.* **5**, 145–147.

Muller, S., O'Brien, P.C.M., Ferguson-Smith, M.A., Wienberg, J. (1998) Cross species colour segmenting: a novel tool in human karyotype analysis, *Cytometry* **33**, 445–452.

Muller, S., Rocchi, M., Ferguson-Smith, M.A., Wienberg, J. (1997) Towards a multicolor chromosome bar code for the entire human karyotype by fluorescence *in situ* hybridisation, *Hum. Genet.* **100**, 271–278.

Pinkel, D., Straume, T., Gray, J.W. (1986) Cytogenetic analysis using quantitative, high-sensitivity, fluorescence hybridisation, *Proc. Natl. Acad. Sci. U.S.A.* **83**, 2934–2938.

Pinkel, D., Landegent, J., Collins, C., Fuscoe, J., Seagraves, R., Lucas, J., Gray, J. (1988) Fluorescence in situ hybridisation with human chromosome-specific libraries: detection of trisomy 21 and translocations of chromosome 4, *Proc. Natl. Acad. Sci.* **85**, 9138–9142.

Rens, W., Yang, F., O'Brien, P.C.M., Solanky, N., Ferguson-Smith, M.A. (2001) A classification efficiency test of SKY and MFISH: identification of chromosomal homologies between *Homo sapiens* and *Hylobates leucogenys*, *Genes, Chromosomes Cancer* **31**, 65–74.

Ried, T., Baldini, A., Rand, T.C., Ward, D.C. (1992) Simultaneous visualisation of seven different DNA probes by in situ hybridisation using combinatorial fluorescence and digital imaging microscopy, *Proc. Natl. Acad. Sci.* **89**, 1388–1392.

Sargan, D.R., Yang, F., Squire, M., Milne, B.S., O'Brien, P.C.M., Ferguson-Smith, M.A. (2000) Use of flow-sorted canine chromosomes in the assignment of canine linkage, radiation, hybrid, and syntenic groups to chromosomes: refinement and verification of the comparative chromosome map for dog and human, *Genomics* **69**, 182–195.

Schmid, M., Nanda, I., Guttenbach, M., Steinlein, C., Hoehn, H., Schartl, M., Haaf, T., et al. (2000) First report on chicken genes and chromosomes 2000, *Cytogenet. Cell Genet.* **90**, 169–218.

Sillar, R., Young, B.D. (1981) A new method for the preparation of metaphase chromosomes for flow analysis, *J. Histochem. Cytochem.* **29**, 74–78.

Telenius, H., Pelmear, A.H., Tunnacliffe, A., Carter, N.P., Behmel, A., Ferguson-Smith,

M.A., Nordenskjold, M., Peragner, R., Ponder, B.A.J. (1992) Cytogenetic analysis by chromosome painting using DOP-PCR amplified flow-sorted chromosomes, *Genes, Chromosomes Cancer* **4**, 257–263.

Yang, F., Alkalaeva, E.Z., Perelman, P.L., Pardini, A.T., Harrison, W.R., O'Brien, P.C.M., Fu, B., Graphodatsky, A.S., Ferguson-Smith, M.A., Robinson, J.J. (2003) Reciprocal chromosome painting among human, aardvark, and elephant (superorder Afrotheria) reveals the likely eutherian ancestral karyotype, *Proc. Natl. Acad. Sci. U.S.A.* **100**, 1062–1066.

Yang, F., O'Brien, P.C.M., Milne, B.S., Graphodatsky, A.S., Solanky, N., Trifonov, V., Rens, W., Sargan, D., Ferguson-Smith, M.A. (1999) A complete comparative chromosomal map of the dog, red fox and human and its integration with canine genetic maps, *Genomics* **62**, 189–202.

Yang, F., O'Brien, P.C.M., Wienberg, J., Ferguson-Smith, M.A. (1997) A reappraisal of the tandem fusion theory of karyotype evolution in the Indian muntjak using chromosome painting, *Chromosome Res.* **5**, 109–117.

Part 5
Genetic Engineering

Genomics and Genetics. Edited by Robert A. Meyers.
Copyright © 2007 Wiley-VCH Verlag GmbH & Co. KGaA, Weinheim
ISBN: 978-3-527-31609-0

27
Gene Targeting

Michael M. Seidman[1] and John H. Wilson[2]
[1] *National Institute on Aging, NIH, Baltimore, MD, USA*
[2] *Baylor College of Medicine, Houston, TX, USA*

1	**Gene Targeting** 771	
1.1	The Problem 771	
2	**Homologous Recombination** 771	
2.1	Introduction 771	
2.2	Repair of Broken Replication Forks 773	
2.3	Targeted Recombination 775	
2.4	Stimulation of Targeted Recombination 779	
3	**Polyamides** 781	
3.1	Netropsin and Distamycin 781	
3.2	The Development of Sequence Specific Minor Groove Ligands 782	
3.3	Affinity and Specificity 784	
3.4	Biological Applications 786	
4	**Triple Helix–Forming Oligonucleotides** 787	
4.1	Triple Helical DNA 787	
4.2	Triplex Formation under Physiological Conditions 788	
4.3	Oligonucleotide Modifications Improve TFO Activity 789	
4.4	Activity of TFOs *In Vivo* 790	
4.5	TFOs Linked to DNA Reactive Compounds 790	
4.6	Triplex Formation, by itself, can Promote Mutagenesis 791	
4.7	Triplex-induced Recombination and Gene Correction 792	
4.8	Expansion of the Triplex Binding Code 792	
5	**Mutation Repair by Small DNA Fragments** 793	
5.1	Small Fragment Homologous Replacement 793	
5.2	Chimeric Oligonucleotides 794	

Genomics and Genetics. Edited by Robert A. Meyers.
Copyright © 2007 Wiley-VCH Verlag GmbH & Co. KGaA, Weinheim
ISBN: 978-3-527-31609-0

6	**Peptide Nucleic Acids** 794
7	**Summary** 795
	Bibliography 796
	Books and Reviews 796
	Primary Literature 797

Keywords

Gene Targeting
The recognition and binding of a specific sequence in the chromosome of living cells by a reagent constructed for that application. The purpose may be sequence manipulation or modulation of transcription.

Homologous Recombination
A process through which disrupted replication forks are restored, double-strand breaks are repaired, and genetic exchange occurs.

Polyamide
A short term for a synthetic molecule consisting of pyrrole and imidazole units joined by amide linkages. These bind in a sequence-restricted manner in the minor groove of an intact duplex.

DNA Triple Helix
A three-stranded structure in which a third strand of nucleic acid lies in the major groove of an intact polypurine:polypyrimidine duplex. Binding is sequence-specific, and stabilized by hydrogen bonds between the bases in the third strand and the purine strand of the duplex.

Peptide Nucleic Acid (PNA)
DNA analog in which the sugar phosphate backbone has been replaced by peptide linkages.

The recognition and binding of a specific sequence in the chromosome of living cells by a reagent constructed for that application. The purpose may be sequence manipulation or modulation of transcription.

1
Gene Targeting

Gene targeting reagents bind specific chromosomal sequences in living cells. They can, or would, be used to manipulate genomic sequences for gene inactivation, modification, or restoration of function; suppress or activate gene transcription; deliver site-specific DNA damage; and probe chromatin and chromosome structure. Practical applications include the construction of transgenic cell lines and animals for research and commercial purposes. They are also under development as pharmaceuticals for the treatment of infectious diseases, gene therapy of genetic disorders and so on. Current strategies are based on either recombinant DNA technology (to produce DNA or protein constructs) or synthetic molecules that are based on oligonucleotides or natural product antibiotics. While many technologies are the subject of active research, only one, homologous recombination (HR), is widely employed at this time. In this review, we will discuss the current status of gene targeting strategies and challenges to the development of gene targeting reagents for use in mammalian, including human, cells. As will become apparent, most of the literature describes the work in progress.

1.1
The Problem

Gene targeting reagents must recognize and bind a specific DNA sequence in living cells, and be in residence long enough to provoke the desired outcome. Thus, a successful reagent must be stable in a physiological environment; be deliverable to the nucleus of the target cell; be able to bind a target in the context of mammalian chromatin structure; have orders of magnitude preference for the intended target sequence over any other sequence; and must find the target with sufficient frequency to be of practical use. Normal DNA metabolism requires sequence recognition for HR and the regulation of transcription. The molecules involved in these processes, DNA, and the proteins that regulate transcription serve as explicit models for some of the reagents under development. One fundamental consideration is the nature of the recognition process (Fig. 1). Reagents that recognize sequence in an intact duplex must do so from the vantage point of the major groove, which is rich in recognition sites, or the minor groove, which is relatively poor. Those designed to ultimately pair with the target sequence must engage a pathway through which the complementary strand becomes accessible and available for hybridization.

2
Homologous Recombination

2.1
Introduction

Targeted gene modification was first demonstrated in mammalian cells nearly 20 years ago. These initial studies targeted chromosomal genes in established cell lines and in mouse embryonic stem (ES) cells. Collectively, they laid the foundations for subsequent studies of gene function in cells and mice, and for the creation of mouse models of human disease. They also raised the enticing possibility that genetic defects might be precisely corrected as the ultimate approach to gene therapy. These considerations have provided the impetus to understand the

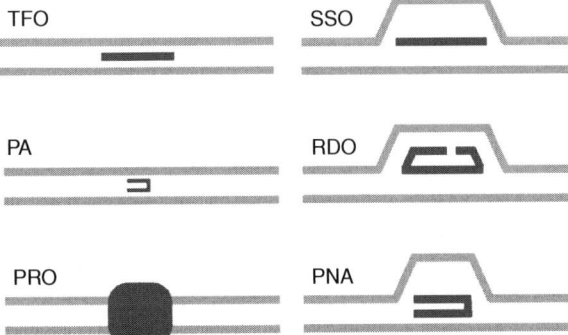

Fig. 1 Recognition of DNA sequences by various gene-targeting reagents. Targeting reagents either bind to duplex DNA or to the separated strands. Triplex forming oligonucleotides (TFOs) recognize their target DNA via interactions in the major groove. Polyamides (PAs) bind through the minor groove. Proteins (PROs) can make binding contacts in both grooves, although most commonly with moieties in the major groove. Single-stranded oligonucleotides (SSOs) pair with one of the separated strands and may mimic an intermediate in HR (see Fig. 3). It may be that small fragment homologous recombination (SFHR) proceeds through such an intermediate, but mechanistic studies have not been done. RNA:DNA oligonucleotides (RDOs) are envisioned to bind to both separated strands, but such binding has not been demonstrated. Peptide nucleic acids (PNAs) clamp onto one of the separated strands by Watson–Crick pairing and by forming Hoogsteen hydrogen bonds with moieties in the major groove of the PNA:DNA hybrid.

mechanism of gene targeting and to define the key experimental parameters for this technology.

Targeted gene modification can occur by HR and engages the cells own recombination machinery, which has been fine-tuned by evolution. For that reason, it possesses many of the attributes of an ideal system. HR occurs with very high specificity and selectivity, discriminating readily between genes that differ by only a few percent, and is extremely accurate, introducing very few errors. Two types of vectors – replacement (ends-out) and insertion (ends-in) – have been used for targeted recombination in mammalian cells (Fig. 2), with replacement vectors being far more common. The main drawback of targeted recombination in mammalian cells is low frequency; roughly 1 event per 10^5 to 10^7 treated cells under commonly used conditions. This presents a problem for therapeutic applications where targeted modification of a high fraction of a cell population is critical. Low frequency is less of a barrier for genome manipulation in proliferating cells in culture because the rare, properly modified cells can be identified and grown into a pure population. Nevertheless, a substantial increase in the frequency of targeted recombination would make genome modification less tedious and improve the chances for targeted gene

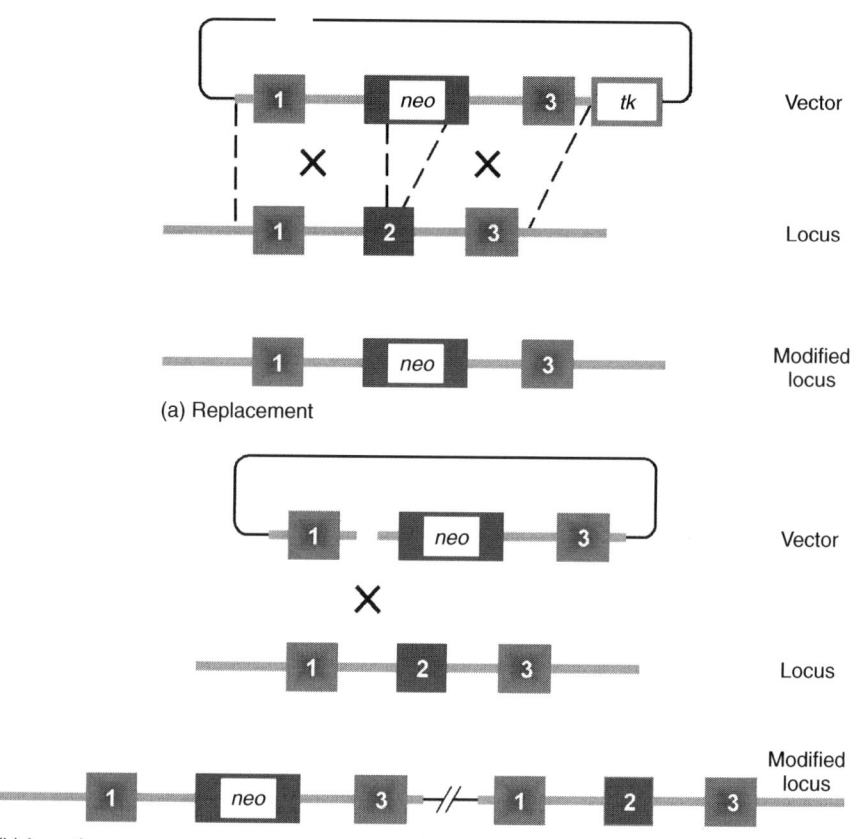

Fig. 2 Types of targeting vectors. (a) Replacement vector. Positive–negative selection is illustrated. Exon-2 is disrupted by incorporation of the gene for neomycin resistance (neo), which constitutes the positive selection marker. The herpesvirus *thymidine kinase* (*tk*) gene is included at one end of the homologous sequences and constitutes the negative selectable marker. The vector is linearized in the plasmid sequences (thin black line). HR in the homologous sequences on either side of the *neo* gene (indicated by Xs) replaces the chromosomal copy of exon-2 with the one from the plasmid vector. The *tk* gene is lost upon HR so that targeted cells survive selection against the *tk* gene. Random integrants (not shown) incorporate the *tk* gene and are killed by selection. (b) Insertion vector. The vector is linearized within the homologous DNA, in this instance between exons 1 and 2. HR at the double-strand ends inserts the vector into the locus, causing a duplication of the homologous sequences with the plasmid sequences located between the two copies. (After Muller, U. (1999) Ten years of gene targeting: targeted mouse mutants, from vector design to phenotype analysis, *Mech. Dev.* **82**, 3–21. Reprinted with permission of Elsevier Science Limited.)

therapy. Rational improvements depend on understanding the underlying process of HR. Clearly, genome manipulation cannot be the main function of HR in cells.

2.2 Repair of Broken Replication Forks

The initial demonstration that HR occurred in cultured mammalian cells was somewhat surprising, as recombination had been thought about and studied mainly in the context of meiotic recombination during gamete formation in the germ line. Over time, it has become clear that HR in somatic cells is one of a number of cellular systems for dealing with DNA damage. It is specifically called into play when both strands of the DNA duplex have been damaged, as occurs, for example, with a double-strand break or when the strands have been cross-linked. Quantitatively, the most important role for HR is the recombinational repair of stalled or collapsed replication forks. In *Escherichia coli*, in the absence of the principal recombinase, RecA, about half the cells in a culture are dead, suggesting that a majority of replication forks are compromised in each replication cycle. In mammalian cells, with their much larger genomes, it is thought that replication forks stall more often, which is the likely reason why cells defective for the RecA homolog, Rad51, are inviable.

It is useful to consider, in overview, one of the ways by which HR can function to restart a broken replication fork because it provides an analog for the main reactions of targeted recombination. As shown in Fig. 3, when a replication fork suffers a break, or encounters one, it collapses, leaving a partially replicated duplex with a protruding single-stranded tail. With the aid of several accessory proteins, the Rad51 recombinase is loaded onto the single strand to form a helical filament of Rad51 with a central single strand of DNA. With the aid of additional accessory proteins, this structure coordinates the search for the corresponding sequence in the intact duplex. When the matching sequence is located, Rad51 catalyzes strand invasion, which pairs the invading single strand with

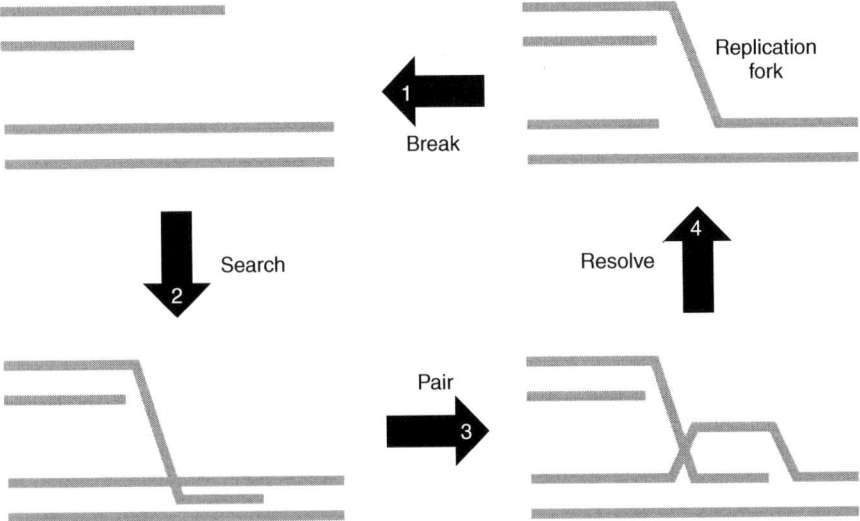

Fig. 3 Repair of a broken replication fork. The single-stranded tail at the broken end is coated with Rad51 recombinase, which then catalyzes the search for homology and the pairing of the single strand with its complement in the duplex. As shown in more detail in Fig. 4, this heteroduplex structure can be resolved by strand cleavage to regenerate a fork-like DNA structure. Several additional proteins are required to convert this branched structure into a functional replication fork.

its complement in the target duplex. This structure then serves as the substrate for a group of replication proteins that modify it to reconstitute the replication fork.

Two steps in this pathway – the search for homology and pairing with the complementary strand – provide structural analogies for the key steps in duplex recognition by all methods of gene targeting (Fig. 1). The search for homology and DNA pairing are thought to take place within the filament of Rad51 and single-stranded DNA, as originally proposed for RecA. In principle, the single-stranded DNA in the filament could approach the duplex DNA via the major groove or the minor groove. Studies using groove-specific DNA binding reagents, modified nucleotides, and cross-linkers, as well as structural analysis of the products, indicate that the initial attack most likely occurs through the minor groove. The search for homology is catalyzed by the recombinase, which binds to duplex DNA and transiently underwinds it, exposing the minor groove for interaction with the single strand in the Rad51 filament.

In the minor groove, there are insufficient hydrogen-bonding moieties to allow specific recognition of homology by base–base interactions. Instead, the recombinase apparently tests the single strand for homology with the duplex by catalyzing an exchange of pairing between bases in the duplex (mainly A and T) and those in the single strand. This trial pairing is accomplished by rotation of the bases through the minor groove in a plane perpendicular to the helix axis, breaking the Watson–Crick pairing in the duplex and reforming it, if possible, with the bases in the single strand. In regions of heterology, the imperfect pairing is rapidly reversed, allowing a new segment of duplex to be tested. When homology is encountered, the pairing between the single strand and its complement in the duplex is extended to form a heteroduplex (Fig. 3). The displaced single strand lies in the major groove of the newly formed heteroduplex. Although this is a triple-stranded structure, it is not the same as that in a triple helix, in which the duplex is intact (see below).

The search for homology and the formation of the heteroduplex intermediate are aided by Rad54 protein, which is a member of the Swi2/Snf2 family of DNA-dependent ATPases with helicase motifs. Rad51 and Rad54 interact *in vitro*. Rad54 acts upon the duplex target DNA to introduce negative and positive supercoiling. Domains of negative supercoiling around the target site would favor the unwinding of the duplex that occurs upon binding the Rad51 filament, which likely explains the ability of Rad54 to increase the efficiency of Rad51-mediated strand pairing. Both Rad51 and Rad54 are present in greater amounts in S-phase relative to the G1-phase, consistent with a role in the repair of replication forks.

2.3
Targeted Recombination

In its role in replication, HR must be nearly 100% effective in reconstituting broken or stalled replication forks. If the basic reactions of fork repair and targeted recombination are similar, then why are so few treated cells successfully modified? Examining some of the possibilities is illuminating; it records our step-by-step progress in understanding the nature of targeted recombination and points the way toward increasing efficiency.

One obvious difference between targeted recombination and fork repair is that the broken arm of a replication fork is already inside the nucleus, whereas the targeting vector must be delivered from outside the cell. Two general methods – nuclear injection and mass delivery – have been used to introduce targeting vectors into cells. Early experiments using microinjection yielded about 1 targeted recombinant per 1000 injected cells. Mass delivery methods, which are vastly simpler and cheaper, typically yield targeted recombinants at frequencies 1000-fold less than microinjection. The basis for this difference is unclear. It does not appear to be a problem with transport to the nucleus, since 10 to 100% of treated cells can express genes transfected by the common methods of mass delivery such as electroporation, calcium phosphate precipitation, Fugene-6, and LipofectAmine. Thus, entry into the nucleus is not the fundamental barrier to efficient gene targeting.

A second, more subtle difference between targeted recombination and fork repair is the number of DNA ends. The broken arm of a replication fork has only one end, whereas a targeting vector has two. It was shown earlier on that linear vectors give several fold higher targeting frequencies than circular ones, as expected from targeting experiments in yeast. The presence of multiple ends, however, makes the transfected targeting vector a substrate for nonhomologous end joining (NHEJ), which along with HR is a major pathway for repair of double-strand breaks in mammalian cells. NHEJ operates throughout the cell cycle, whereas HR is prominent in mid to late S-phase and G2. A reasonable expectation is that NHEJ might compete with HR, effectively eliminating targeting substrates by circularization and multimerization. In addition, NHEJ is thought to be responsible for random integration, which occurs at roughly 1000-fold higher frequency than targeted recombination. The high background of random integration events can be minimized by any of the several tricks such as positive–negative selection that render most random integrants inviable (Fig. 2). Consistent with the idea that HR and NHEJ compete for substrates, repair of chromosomal double-strand breaks by HR is increased 2- to 25-fold when components of the NHEJ pathway are defective or inhibited. Surprisingly, however, neither the joining of transfected DNA ends nor random integration appears to be affected by the absence of NHEJ, and targeted recombination is not substantially increased. These observations and others suggest that there may be additional mechanisms for dealing with DNA ends. Although it is unclear to what extent NHEJ might affect recombination, the absence of a substantial effect in NHEJ-defective cells suggests that competition between NHEJ and HR is not the key to the low frequency of targeted recombination.

Another obvious difference between targeted recombination and fork repair is the proximity of the DNA to the target sequence. The DNA end at a broken replication fork is tethered in the general vicinity of the target sequence by virtue of its attachment to the oppositely directed replication fork. By contrast, the exogenous DNA must find its target among all the sequences in the genome: a daunting task. Nevertheless, the search for homology does not limit the frequency of targeted recombinants in mammalian cells. Varying the number of exogenous DNA molecules in the nucleus over a 1000-fold range (delivered by nuclear injection), or the number of targets in the genome by up to 400-fold had no significant effect on the frequency of targeted recombination. These counterintuitive results have been confirmed in competition experiments. Thus, the frequency of targeted recombination is limited by something other than the search for homology. These results stand in stark contrast to those in yeast, where the process was found to be dependent on the number of target copies.

Targeted recombination and fork repair also differ significantly in the extent of homology, which, for the broken arm of a replication fork is as long as the replication bubble. For practical reasons, the length of homology in targeting vectors is limited to less than about 20 kb. Over this range, the frequency of targeting in mammalian cells shows a steep dependence on the length of homology, with a 2- to 3-fold change in length giving a 10-fold change in targeting frequency. Once again, these results are distinct from those in yeast, where gene targeting is linearly dependent on length of homology. They also stand in contrast to the linear length dependence of intrachromosomal HR in mammalian cells. The peculiar dependence of mammalian gene targeting on the length of homology suggests that the limiting step in the pathway is downstream of finding the target gene. In some undefined way, longer stretches of homology must exponentially improve the chances of a productive outcome.

Beyond the length of homology, there is also a potential difference in the quality of homology. A broken arm of a replication fork is identical to the intact arm (with the exception of very rare replication errors). The homologous segment of a recombination vector, however, is often isolated from nonisogenic sources and thus may carry nucleotide polymorphisms. Single-nucleotide differences between homologous sequences lead to mismatches in heteroduplex recombination intermediates. In bacteria, yeast, and mammalian cells, such mismatches reduce the frequency of HR in a way that is dependent on the mismatch-repair machinery and certain DNA helicases. In mammalian cells, use of nonisogenic DNA (up to about 1% different) can reduce the efficiency of targeting up to 20-fold. Although this can be a substantial effect, even isogenic DNA will function at very low absolute frequencies. In contrast to the effects of distributed mismatches, blocks of nonhomology at the ends or within the homologous regions (as shown in Fig. 2) have little or no effect.

Perhaps the most basic difference between fork repair and targeted recombination is the intended outcome. For reconstitution of the replication fork, the heteroduplex recombination intermediate must be resolved so that a fork is created (Fig. 4). For recombination, a further resolution is needed. Breakage of one strand at the fork can either release the exogenous DNA or integrate it into the chromosome to give a targeted replacement of the underlying DNA. In the parlance of recombination, release corresponds to a conversion event and integration corresponds to a crossover event. It is notable that repair of broken DNA in mammalian cells occurs with a very strong bias (up to 100-fold or more) against crossover events, perhaps as a protective mechanism against loss of heterozygosity. In mammalian cells, it seems that broken ends often invade a homologous duplex, prime DNA synthesis, and then disengage from the duplex. It is common to find randomly integrated targeting vectors that have been extended up to 10 kb or so at one or both ends by a prior homologous interaction with the target duplex. Thus, the inherent bias toward conversion-like resolution of targeting intermediates may constitute the principal barrier to gene targeting in mammalian cells.

The view that a preference for conversion limits the frequency of targeted recombination is supported by results in which

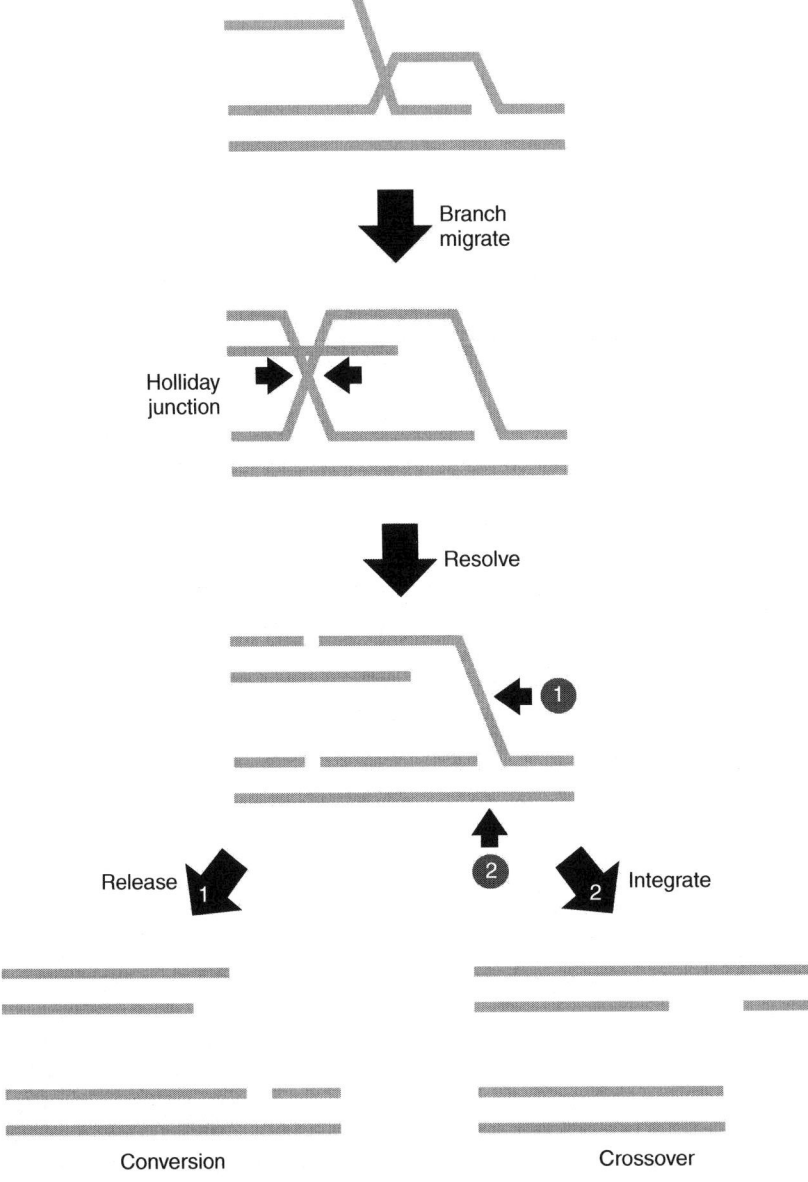

Fig. 4 Resolution of the initial heteroduplex intermediate. The initial paired structure can branch migrate to form a Holliday junction. Resolution of the Holliday junction by cleavage of the pair of strands indicated by the small arrows yields a branched structure equivalent to a replication fork. Cleavage of the branched structure at arrow 1 releases the targeting DNA, whereas cleavage at arrow 2 integrates the targeting DNA.

double-strand breaks are introduced into the target sequence. By embedding the recognition site for a rare-cutting endonuclease such as I-SceI in the chromosomal target, site-specific double-strand breaks can be introduced by transfection of an expression vector for I-SceI. When the targeting vector is included, targeted recombination can be stimulated up to 10 000-fold with the specific recombinants constituting several percent of treated cells. One or both of the chromosomal ends at the break are thought to invade the targeting duplex, extend themselves by DNA synthesis, and then disengage and pair with one another to reconstitute the duplex and eliminate the double-strand break. This mechanism, which is known as synthesis-dependent strand annealing (SDSA), provides a productive way of thinking about targeted recombination.

2.4
Stimulation of Targeted Recombination

An appreciation of the normal role HR plays in somatic cells, coupled with an understanding of the probable pathway for targeted recombination, allows one to devise rational strategies for increasing the frequency of the event. Three general approaches have been tried; altering levels of HR enzymes, introducing damage at the target site, and delivering DNA via viral infection. These approaches are not mutually exclusive, and significant improvements may require some combination of them. Below we discuss the rationale for each of those strategies and their current status.

In principle, it should be possible to increase frequencies by temporarily altering the cellular mix of HR enzymes in a way that favors targeted recombination. For example, one might try to overexpress enzymes that favor HR and inhibit those involved in competing pathways such as NHEJ. Thus far, such experiments have met with limited success. As mentioned above, inhibition of NHEJ by knocking out key components had no significant effect. Three HR enzymes – Rad51, Rad52, and Rad54 – have been tested by over expression. A fourfold overexpression of the Rad51 recombinase stimulated targeted recombination some two- to fourfold. By contrast, overexpression of Rad52 protein, which helps to load Rad51 onto single-stranded DNA, was inhibitory by about twofold. Overexpression of Rad51 and Rad52 together had little effect on targeting frequencies. Finally, overexpression of Rad54, which helps prepare the target duplex for invasion, stimulated recombination by about 10-fold (Z. Songyang, personal communication).

The positive results with Rad51 and Rad54 serve as proof of principle and encourage further experimentation along these lines, especially with enzymes in the more poorly defined, downstream portion of a pathway. The results with Rad52 raise two cautionary notes. First, it was shown that overexpression of Rad52 stimulated both intrachromosomal and extrachromosomal HR two- to fivefold. These assays measured effects on the single-strand annealing pathway of HR, which is independent of Rad51 and distinct from the likely pathway for targeted recombination. The effects of manipulating cellular enzyme levels should always be measured in targeted recombination assays. Second, stable overexpression of Rad52 was found to be toxic to cells, as was Rad51 in some cell lines. Consequently, careful adjustments in enzyme levels may be essential for optimal effects.

As mentioned above, the most dramatic effects on targeted recombination have been generated by I-SceI-mediated cleavage of the target sequence on the chromosome. The requirement for prior modification, however, means that rare-cutting endonucleases such as I-SceI cannot be used as general tools to stimulate targeted HR, nor can restriction enzymes be used because their sites are all too common. However, zinc finger proteins can be engineered to recognize a diverse set of DNA sequences. By combining a zinc finger recognition domain with a nonspecific DNA-cleavage domain, zinc finger nucleases have been created that offer a flexible strategy for delivering site-specific double-strand breaks to the genome. In current designs, the zinc finger nuclease consists of a DNA binding domain composed of three zinc fingers linked to the nonspecific DNA-cleavage domain of FokI, a type IIS restriction enzyme. Each zinc finger recognizes a specific 3-nucleotide stretch of DNA so that the three fingers in a domain recognize a 9-nucleotide DNA segment. Because efficient DNA cleavage requires dimerization through the FokI nuclease domain, the effective recognition site is 18 nucleotides in length, long enough to be unique in mammalian genomes.

Zinc finger nucleases have been shown to stimulate intramolecular HR in 50% of plasmid molecules injected into *Xenopus* oocytes, and to cause site-specific somatic mutations in 50% of males, and germ-line mutations in 5% of males, when expressed in *Drosophila* larvae. They also stimulate targeted recombination in *Drosophila* by 100-fold, and in human cells by several 1000-fold. It should be noted that observations in *Drosophila* and in human cells indicate that high-level expression of some zinc finger nucleases can be toxic. If a broad range of zinc finger nucleases with high affinity and single-site specificity can be developed, it may be possible in the future for an investigator to select from a bank of such nucleases the one that will cleave a specific gene to stimulate localized gene targeting.

In the section on triple helix forming oligonucleotides (TFOs) in this review, we discuss two methods for using TFOs to induce HR and gene correction. First, by linking a DNA damaging agent such as psoralen to a TFO, it is possible to introduce site-specific damage to a genomic target specified by TFO binding. Introduction of a TFO into cells has been shown to stimulate up to several thousandfold intrachromosomal HR at a site containing the TFO target sequence. Second, it is possible to link a correcting segment of DNA to the TFO itself, using the binding specificity in the TFO to deliver the DNA to an adjacent target. Directed toward a plasmid target in mammalian cells, this approach stimulated homologous correction 5- to 10-fold above background levels. Improvements in design and delivery of such TFO reagents, as discussed later, will likely be required to enhance the effectiveness of this method for stimulating HR in cells.

The unexplained difference in targeting frequencies between nuclear injection and mass delivery methods raises the distinct possibility that alternative ways to introduce DNA into cells might yield positive benefits. One promising approach is the delivery of DNA by viral vectors. Although retroviral vectors have been shown to target at low frequencies, adeno-associated virus (AAV), which carries a single-stranded DNA genome, has been investigated most extensively. AAV viral vectors have been shown to target chromosomal sequences at frequencies ranging from 0.01 to 1% depending on the multiplicity of infection, and on the type of alteration introduced, with insertions into the target giving generally higher frequencies than deletions. Chromosomal targets carrying an I-SceI site, which were corrected at low frequencies by the AAV vector alone, could be targeted at about 100-fold higher frequencies in the presence of I-SceI. The relatively high frequencies of targeted recombination by AAV may reflect the nuclear delivery of a high concentration of DNA by the viral preparations, or they may relate to the single-stranded nature of AAV DNA with its hairpin-protected ends. This appears to be a promising technology although it has had relatively little exposure.

3
Polyamides

3.1
Netropsin and Distamycin

Netropsin and distamycin are natural product antibiotics consisting of two or three pyrrole (Py) rings respectively connected by amide linkages. In addition, the molecules have a positively charged propylamidine at the end designated as the tail and either formyl (distamycin) or guanidine (netropsin) groups at the head (Fig. 5). The compounds have received attention for many years and it was established some time ago that they bound AT-rich sequences in the minor groove. Both drugs have a curvature that fits quite well with the floor of the minor groove. Crystallographic studies showed that a single netropsin molecule binds in the minor groove displacing the spine of hydration. There are nonspecific ionic interactions between the positively charged head and tail groups and the negatively charged phosphates of the duplex. Hydrogen bonds between the proton donors of netropsin (the amide NH) and the DNA bases position the drug, while the actual sequence recognition results from close van der Waals contacts between atoms on the drug and adenine residues. The poor binding to GC pairs is due to steric hindrance from the 2–amino group guanine, which interferes with occupancy of the groove. The structural studies detailed the interactions that determine the affinity and specificity

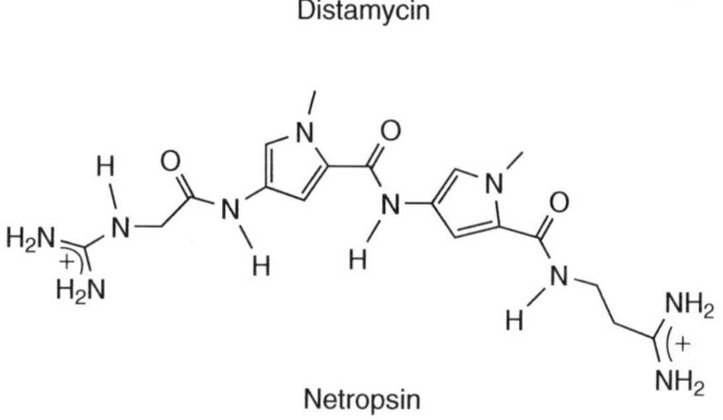

Fig. 5 The structures of distamycin and netropsin, two minor groove binding antibiotics. Each contains pyrrole residues joined by amide linkages. Note the positively charged tails and the sickle shape.

of the natural products and numerous synthetic derivatives. It was appreciated that these compounds might serve as leads for the development of sequence-specific DNA binders that could inhibit transcription of specific genes. If the suppressed genes were linked to diseases, such compounds could have pharmaceutical application. This has been a motivation for much of the work in the field.

3.2
The Development of Sequence Specific Minor Groove Ligands

A thorough discussion of the chemistry and structural properties of the numerous polyamide derivatives (sometimes called lexitropsins) is beyond the scope of this review. These features of the field have been reviewed in depth in several recent publications. We will summarize the important developments and then consider the current understanding of their biological activity and prospects for pharmacological utility.

In order to expand the binding options to GC pairs, Kopka et al. had suggested, and Lown had independently synthesized netropsin derivatives in which pyrrole was substituted with imidazole (Im). Replacement of the ring CH with N provided space for the exocyclic amine of guanine, and the N served as an acceptor of the H bond from the guanine donor. Although the new compounds recognized GC sites, they also continued to bind AT pairs, albeit with reduced affinity. These somewhat disappointing observations were rescued by one of the most important discoveries in the field. NMR studies of distamycin bound to AT sequences revealed that when the binding site contained five AT pairs, distamycin bound as an antiparallel, side-by-side, dimer, with the positively charged tails at either end (2:1 mode). Each drug molecule made contact with one of the duplex strands, with the N terminus at the 5' end of the strand. In contrast, netropsin bound only in a 1:1 mode, because of the presence of positive charges at both ends of the molecule. The Dervan and Lown groups synthesized distamycin analogs with the pyrroles replaced by imidazole. The new compounds bound target sequences containing GC pairs, as predicted. However, further analysis by both groups determined that the optimal side-by-side arrangement for G:C binding was an imidazole:pyrrole pairing, with the imidazole interacting with the G and the pyrrole interacting with the C. Specificity was extended when the Dervan group introduced 3-hydroxypyrrole (Hp) which showed a preference for interaction with T. The Hp/Py pair recognized T:A and Py/Hp recognized A:T with discrimination greater than 10-fold depending on the orientation and target. The preference by Hp/Py for T:A is the result of shape selection by an asymmetric cleft in the minor grove and also hydrogen bonding between Hp and T. These results established pairing rules for the 2:1 binding motif: the pyrrole/imidazole pair (Py/Im) for C:G, Hp/Py for T:A, Py/Hp for A:T, while Py/Py is degenerate and recognizes both A:T and T:A (Fig. 6). The tails of the molecules (dimethylaminopropyl, in the compounds from the Dervan group) are placed at A:T pairs.

Additional affinity was gained when the two chains were linked (Fig. 7). Two methods of linkage have been employed. In one format, the chains are coupled via bonds between internal rings. While this did provide enhanced affinity, the more effective

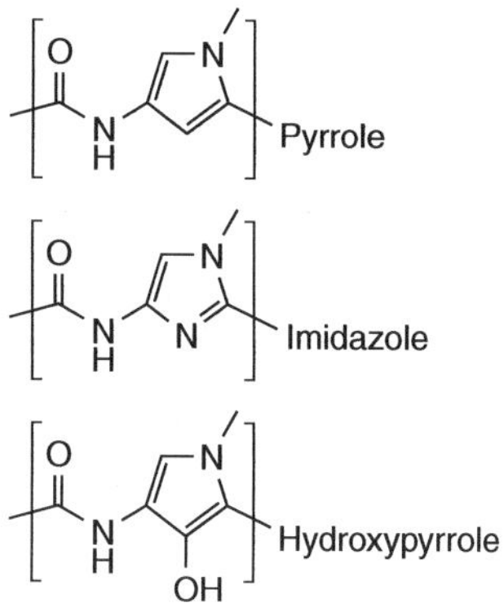

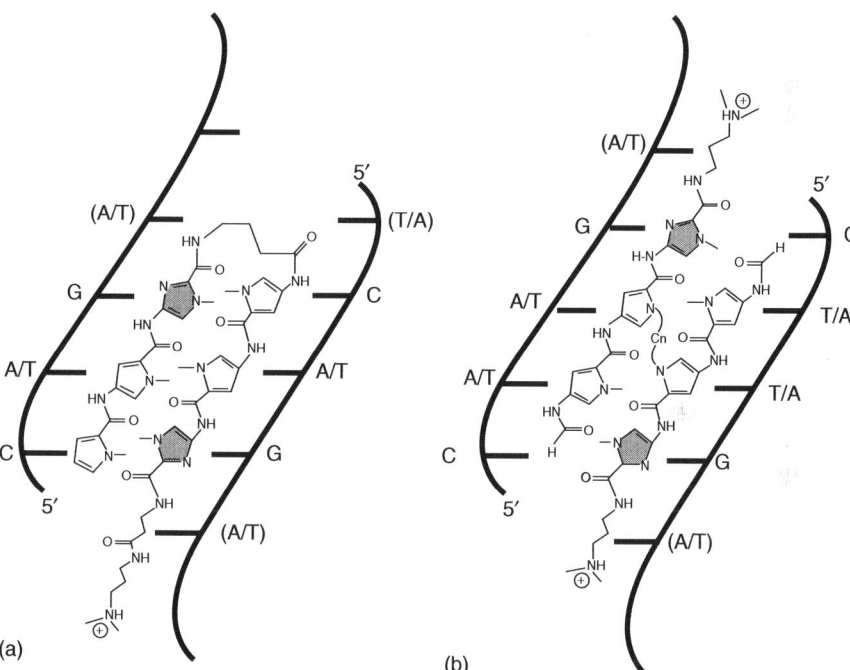

Fig. 6 The structures of the subunits of synthetic polyamide minor groove binders. When binding occurs in the 2 : 1 mode, pyrrole binds to A, T, or C, imidazole binds to G, and hydroxypyrrole binds to T. (From Goodsell, D. S. 2001, Goodsell, Sequence recognition of DNA by lexitropsins, *Curr. Med. Chem* **8**, 509–516. Reprinted with permission from Bentham Science Publishers Ltd.)

Fig. 7 (a) Hairpin and (b) "stapled" polyamides. Each contains 2 imidazole (stippled) residues and 4 pyrrole rings. The shifted phasing of the stapled version relative to the arrangement in the hairpin changes the target sequence. (A/T) or (T/A) base pairs at the ends of each sequence are recognized by the atoms in the linkers and tails. (Reprinted from Walker et al., 1998, PNAs **95**, 4315–4320 with permission from the National Academy of Sciences, USA.)

approach involved linkage via a head-to-tail format with the three-carbon linker, γ-aminobutyric acid. This has been employed in most subsequent syntheses. The hairpin locks the position of the subunits relative to one another, thus preventing slipped pairing schemes. This linker also binds at AT pairs, adding another component to the recognition scheme. Although hairpin polyamides have been studied more extensively, the cross-linked polyamides continue to receive attention.

Another important development was the recognition that there was a limitation on the length of the polyamide chains. Sequence specificity and affinity were not improved beyond five rings. This was attributed to an overcurvature of the longer polyamide chain (as compared to the 3- and 4-ring compounds that match the minor groove curvature), resulting in a loss of register with the binding sites. A solution to this problem was provided by the placement of β-alanine residues after three or four polyamides. These insertions appear to act as "springs" to reset the phasing. When paired with each other in hairpin polyamides, or with Py, they interact with T:A and A:T pairs. When paired with Im, they can interact with C in G:C or C:G pairs, and may offer a solution to the otherwise "difficult" sequences. At this time, it is fair to say that considerable progress has been made in the development of the polyamide minor-groove binders. This reflects the persistent effort and ingenuity of several laboratories, the development of solid-phase methods of synthesis, and the allure of high affinity, sequence-specific ligands for ultimate pharmaceutical application.

3.3
Affinity and Specificity

The introduction of the hairpin polyamides advanced the affinity for a five-base target sequence by two orders of magnitude relative to the free three-ring compound. This brought the association constants into the nanomolar range, competitive with transcription factors. Subnanomolar affinities were reported for hairpin polyamides with four side-by-side rings binding to a six base target. These results clearly demonstrated high affinity binding by the polyamides, and suggested that they might be effective as inhibitors of transcription factor binding, and thus transcription.

The issue of specificity is, however, another matter. This question has been addressed in a number of publications from the Dervan laboratory. In one study, the binding of two eight-ring hairpin polyamides, designed to recognize sequences that contained one or two G:C pairs, was analyzed by DNAse footprinting. The constructs contained Py/Py pairs which do not distinguish A:T from T:A. Binding by one of the constructs to the specific site was subnanomolar, with two log poorer binding to two other sites with replacement of G:C by A:T or T:A. However, the other construct had a tenfold lower affinity for the specific target and only an eightfold discrimination between the specific target and the single mismatch sequence. In addition, weaker, but measurable binding (1 nM range) was seen with a double-mismatch site. In this example, both constructs bound the mismatched target with similar affinity. They differed substantially, however, in their affinity to the specific target.

Although replacement of Py/Py pairs with Hp/Py, does afford some discrimination between A:T and T:A pairs (in the range of 10 fold), it does so at the expense of affinity. Thus, the more effective binders are the compounds with no discrimination between A:T and T:A, and perhaps because of this, the Hp subunit has received little use in the recent work from the Dervan laboratory.

Binding to a single site in the human genome would require specific recognition of about 16 contiguous base pairs. In the initial efforts to increase the target specificity of the polyamides the Dervan group constructed six- or eight-ring polyamides with internal β-alanine linkers. These bound via a "slipped pairing" mode in which there was a partial overlap of the individual polyamide chains in minor groove, with extension of the remaining unpaired chains along the target sequence. These did bind sequences as long as 16 bases. However, binding to mismatch sites was also observed with affinities relatively close (5–20-fold) to the perfect match targets. The Laemmli group has also prepared linked hairpin polyamides designed to bind insect telomeric repeat sequences. While the dimer bound with 10- to 20-fold higher affinity than the unlinked monomer, at higher concentrations it also bound almost all the sequences in the probe (termed "coating" by the Laemmli group). One unavoidable difficulty with all these studies is that analyses of specificity are limited by the sequences chosen for the binding experiments. Obviously, the array of potential binding sites in the genome cannot be represented with this strategy. However, even within this limitation it is apparent from visual inspection of the DNAse footprints from virtually all of the polyamide literature that binding occurs to many sites other than those of specific interest (Fig. 8).

A recent publication highlights some of these points. Colon cancer cells in culture were incubated with a polyamide designed to inhibit binding of the LEF-1 transcription factor to an eight-base

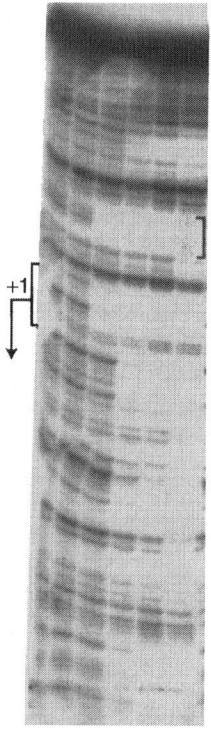

Fig. 8 DNAse footprint analysis of polyamide binding to the upstream promoter region of the *Xenopus TyrD* tRNA gene. A ^{32}P labeled restriction fragment was incubated with the polyamide at the indicated concentrations and then digested with DNAse. Protected regions are not digested and appear as blanks in the pattern. The brackets indicate the regions the polyamide was designed to protect. The arrow denotes the transcription start site. (From McBryant et al. (1999) *J. Mol. Biol.* **286**, 973–981. Reprinted with permission by Elsevier Science.)

0 1 3 10 30 100 nM

sequence in the promoter of the cyclin D1 gene (CD1). The authors were careful to demonstrate nuclear localization of the polyamide. This was an important issue as another study suggested limitations in nuclear uptake of these compounds. The effect on message expression was examined by hybridization array analysis. Although the levels of a number of messages were affected, it was noteworthy that the levels of the targeted CD1 gene were unchanged. A computer search of the promoters of the 11 000 genes represented on the hybridization array yielded 37 380 candidate binding sites for the anti-CD1 polyamide.

3.4
Biological Applications

Early in the development of polyamides they were seen as having the potential to become rationally designed transcriptional inhibitors. Important support for this concept was provided by the demonstration that a hairpin polyamide could suppress transcription of a *Xenopus* 5S RNA gene. The compound was designed to bind a specific sequence in the binding site of the 5S RNA gene transcription factor, TFIIIA. The polyamide reduced 5S transcription in both cell extracts and live cells. In an experiment with therapeutic implications, combinations of appropriately designed polyamides were shown to inhibit HIV transcription in a cell-free assay, and to block HIV replication in peripheral blood lymphocytes. Since then there have been several studies describing biological applications of polyamides. They can inhibit transcription factor binding, and transcription *in vitro*. Most, but not all, sites on nucleosomal DNA are bound by polyamides, which block the progress of RNA polymerase through nucleosomes, apparently by inhibiting repositioning of nucleosomes during transcription. In an interesting new direction, polyamides were linked to transcriptional activation domain proteins. The polyamides provided sequence-specific DNA binding and the conjugates functioned as artificial transcription factors in transcription assays *in vitro*.

Perhaps the most spectacular publications are from Laemmli and colleagues, who synthesized polyamides targeted at the AT-rich, satellite scaffolding–associated regions of Drosophila. The satellites are found in condensed heterochromatin, and genes near these regions are silenced because of the chromatin structure. The satellite sequences are bound by the MATH (for multi-AT hooks) protein that interacts with AT tracts through minor groove contacts. The experimental rationale was based on the expectation that the polyamides would interfere with MATH dependent heterochromatin condensation. Treatment of nuclei resulted in extensive chromatin unfolding and selective staining of polytene chromosomes in the expected satellite region. Even more remarkable was the result of feeding the compounds to flies. The flies normally showed a strong variegation in eye (mottled red/white) color due to a juxtaposition of the *white* gene adjacent to the heterochromatin satellite region (resulting in variable silencing of the gene for the red color). Depending on the polyamide and the intended target, the flies showed an increase in the red color phenotype. This was interpreted as due to an opening of the chromatin resulting in relief of the suppression of the red color gene. Additional experiments with a different polyamide targeted to a different satellite at another locus also had pronounced effects on gene expression.

The results of both series of experiments suggest that extensive chromatin remodeling could result from polyamide treatment. While the effects on transcription were indirect, these results argued that polyamides could reach nuclei of living cells in whole organisms, with remarkable consequences for chromosome structure and gene expression.

Although considerable effort and intelligence has been brought to the development of polyamides, there are serious questions on the likelihood of their becoming successful drugs. Concerns about target specificity and effects on chromatin structure will have to be addressed before pharmacological applications can be considered.

4
Triple Helix–Forming Oligonucleotides

4.1
Triple Helical DNA

The DNA triple helix was discovered more than forty years ago and has been studied persistently ever since. The most stable triplexes form on polypurine:polypyrimidine tracts, which are abundant in the human genome, particularly in promoters and introns. The third strand of nucleic acid, which may consist of purines or pyrimidines (depending on the target sequence), lies in the major groove of an intact duplex. The complex is stabilized by two "Hoogsteen" hydrogen bonds between third strand bases and the bases in the purine strand of the duplex. These are formed without interfering with the hydrogen bonds between the bases of the duplex strands. Pyrimidine motif (Y.R:Y) third strands bind in a direction parallel to the purine strand in the duplex, with canonical base triplets of T.A:T and C.G:C (by convention the first base of the indicated base pair is the purine). Purine motif (R.R:Y) third strands bind antiparallel to the purine strand, with base triplets of A.A:T and G.G:C. Typically, the purine motif is used for G:C rich targets, and the pyrimidine motif for A:T rich targets (Fig. 9).

For most of the three decades following the initial discovery, triplex research was largely the province of DNA structural and physical chemists until methods were developed for automated, solid phase–based, oligonucleotide synthesis. The demonstration that synthetic oligonucleotides could form stable triplexes led to the explicit suggestion that TFOs could be developed as sequence-specific targeting reagents in living cells. Although largely conducted in the absence of biological assays, the many years of biochemical and biophysical research did define a number of obstacles that had to be, and still must be, overcome in order for TFOs to become a legitimate option for gene targeting. Consideration of biological applications raises four major issues: the activity, under physiologically relevant conditions, of TFOs directed against the classical homopurine:homopyrimidine targets in purified DNA; the activity against more general sequences; the activity against chromosomal targets, in the context of mammalian chromatin structure; and, the molecular consequences of successful triplex formation, that is, what happens after binding. Solutions to the first two have been, and will continue to be addressed by developments in oligonucleotide chemistry, while the third and fourth are the subject of active research from more biochemical and biological perspectives.

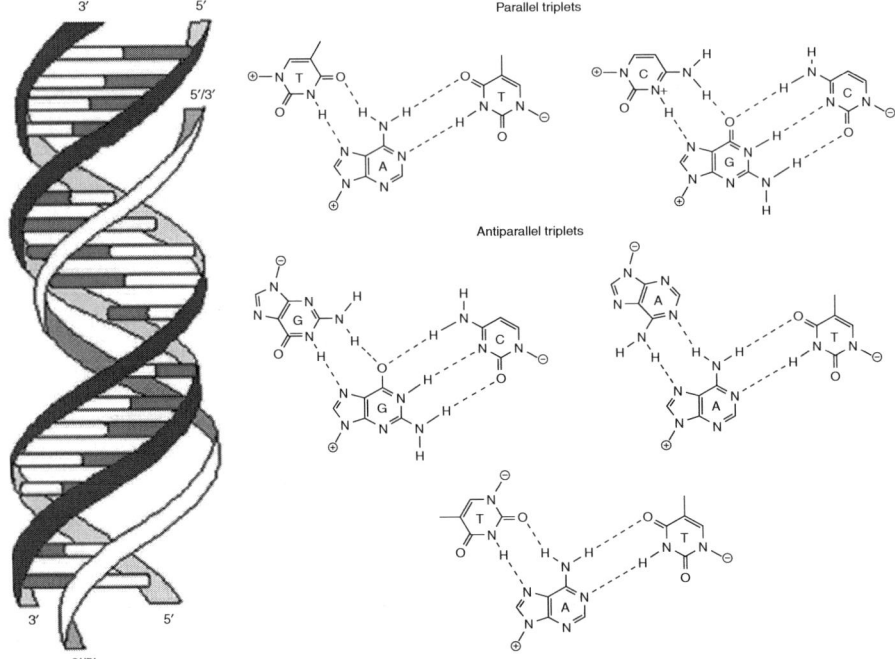

Fig. 9 The structure of the DNA triple helix. A third strand of nucleic acid lies in the major groove of an intact duplex. The target sequence consists of purines on one strand and pyrimidines on the other. Specificity and stability are provided by Hoogsteen hydrogen bonding between the bases in the third strand and the purine bases in the duplex. These hydrogen bonds do not perturb the Watson–Crick bonds of the duplex strands. Third strand orientation relative to the purine duplex strand is a function of the third strand composition – those in the pyrimidine motif bind with the same polarity as the duplex purine strand ("parallel"), while purine motif third strands bind in the opposite polarity (antiparallel). In pyrimidine, third strands T bind to A:T pairs, and C to G:C pairs. In purine, third strands G bind to G:C, A to A:T, and T can bind to A:T pairs. Note that the TA:T pairs in the two different motifs are not identical.

4.2 Triplex Formation under Physiological Conditions

Triplex formation is an inherent property of DNA without the requirement for enzymes or proteins. Nonetheless, the probability of complex formation is affected by fundamental biophysical considerations, which are further complicated by the constraints of a physiological environment. A negatively charged third strand must approach and bind a doubly negatively charged duplex, via a zipper type mechanism following an initial nucleation event involving three to five bases. In the test tube charge repulsion can be reduced by use of levels of Mg^{++} (5–10 mM) that are much higher than likely to be available as the free ion in cells. Pyrimidine motif triplexes do not form at physiological pH because of the requirement for cytosine protonation that occurs at relatively acidic pH ($pK_a = 4.5$). This is necessary for one of the hydrogen bonds (Fig. 9), and the resultant positive charge also makes an important contribution to triplex stability. Purine motif third strands (which are G rich) can form G-tetrad structures in physiological levels of K^+, inhibiting triplex formation. Finally, conformational changes are required by both the third strand and the duplex target. All these factors contribute to the much slower kinetics of triplex formation relative to duplex formation, and reduce triplex stability (most triplexes, even under optimal conditions *in vitro*, are less stable than the underlying duplex).

4.3
Oligonucleotide Modifications Improve TFO Activity

Certain base and sugar modifications improve TFO activity under physiologically relevant conditions. For example, 5-methylcytosine partially alleviates the pH restriction of TFOs in the pyrimidine motif, due to the contribution of the methyl group to base stacking and/or the exclusion of water molecules from the major groove.

RNA third strands form more stable pyrimidine motif triplexes than the corresponding DNA strands, and that observation led to the use of RNA analog sugar residues such as 2'-O-Methyl (2'-OMe) (Fig. 10). Other analogs, such as the 2', 4' bridged ribose (Locked Nucleic Acid, LNA, or Bridged Nucleic Acid, BNA) also improve triplex stability. These modifications stabilize the C-3' endo configuration of the sugar. This has the salutary effect of preorganizing the third strand in a conformation that is compatible with triplex formation, while also imposing minimal distortion on the underlying duplex. Many other modifications have been considered. These include intercalators linked to TFOs and the base analog propynyl-deoxyuridine, both of which stabilize triplexes. The Mg^{++} dependence of pyrimidine motif TFOs has also been reduced by modifications to the backbone, such as replacement of the phosphate linkage, or a bridging oxygen with a nitrogen. Substitution of a nonbridging oxygen in the backbone with a charged amine reduces G-tetrad formation by purine TFOs in physiological K^+, making more oligonucleotide available for triplex formation and reducing the charge repulsion.

A positive charge and an RNA-like sugar conformation have been combined in the 2'-O-(2-aminoethyl) (AE) ribose derivatives of Cuenoud and colleagues. TFOs carrying these substitutions can form triplexes at rates over 1000-fold faster than corresponding deoxy third strands. The stability of the resultant complex at physiological pH and low Mg^{++} concentration is also enhanced. NMR analysis indicates a specific interaction between the positively charged amines (at physiological pH) and phosphate groups in the purine strand of the duplex.

Although the biochemical data indicate that these modifications improve TFO binding, few have been tested in biological assays. In the following section, we will

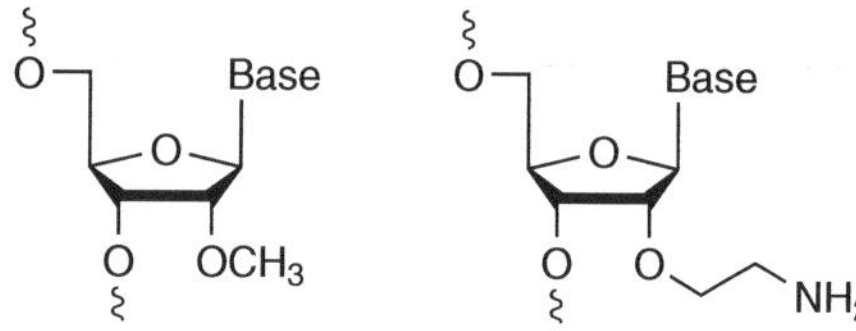

Fig. 10 The structure of modified sugars incorporated in triple helix forming oligonucleotides, 2'-O-methyl (2'-OMe) and 2'-O-aminoethyl (2'-AE).

discuss the biological activity of TFOs carrying some of these substitutions.

4.4
Activity of TFOs *In Vivo*

Any effort to develop TFO for gene targeting in living cells must be cognizant of the potential restriction to triplex formation by eukaryotic chromatin structure. It has been shown in biochemical experiments that target sequences in nucleosomes are poor substrates for triplex formation. However, chromatin structure is dynamic *in vivo*, and there are recent indications that manipulation of the biology of the cell can influence targeting efficiency (see below).

In the early biological studies, TFOs were envisioned as tools to inhibit gene expression by blocking transcription initiation or elongation (the antigene strategy). Triplex formation within promoter sites can block transcription factor access, and inhibit gene activation *in vitro*. Promoters of active genes are often nucleosome free or have nucleosomes whose interaction with the DNA has been relaxed by remodeling or histone modification. Results from a number of investigators suggest that TFOs designed to bind specific promoters can decrease expression of the targeted gene in mammalian cells. The results of these and similar studies suggest that TFOs can find chromosomal targets and produce a measurable biological endpoint. However, the evidence for target interaction is necessarily inferential and there is no direct demonstration of TFO binding at the target site. Furthermore, as is well documented in the antisense literature, oligonucleotides can have multiple effects on cells and transcript levels, via mechanisms unrelated to their original design.

In an alternative approach, TFOs have been used for genome modification, resulting in a change in target sequence. This has the advantage of introducing permanent changes in the target sequence, which simplifies interpretation of the experiments. Two strategies have been employed: in the first the TFO is linked to a DNA reactive compound, while in the second the binding of the TFO is sufficient to elicit a response.

4.5
TFOs Linked to DNA Reactive Compounds

TFOs linked to DNA reactive compounds have been used to introduce DNA adducts at specific target sites. TFOs conjugated to cleavage and alkylating reagents have been described, although these reagents have been largely restricted to biochemical experiments. The most useful, and most frequently used reagent is psoralen, a DNA cross-linker requiring photoactivation by long wave ultraviolet light. In the initial biological experiments, purine motif TFOs conjugated to psoralen were used to target cross-links to a specific site on mutation reporter plasmids or phage *in vitro*. Passage of the modified vectors through host cells resulted in mutagenesis of the target site. Psoralen-conjugated TFOs that transfected into cells were shown to induce base pair-specific mutations within a *supF* mutation reporter gene in a shuttle vector plasmid either cotransfected or previously transfected into cells. These results demonstrated intracellular targeting by a TFO, and also demonstrated that the mutagenesis assay could be used to relate the affinity of the TFO to the biological activity.

The shuttle vector experiments were followed by the demonstration of targeted mutagenesis of a chromosomal site by a psoralen-linked pyrimidine TFO containing 2'-OMe sugars and a pyrene intercalator. The target sequence was in the endogenous hamster *HPRT* (hypoxanthine phosphoribosyltransferase) gene. Cells with inactivating mutations in *HPRT* can be quantitatively identified in a simple and unambiguous selection procedure. TFOs with uniform 2'-OMe substitutions were inactive, but those with the pyrene intercalator showed *HPRT* knockout activity at frequencies in the range of 10^{-4} to 10^{-3}. Sequence analysis of the mutant clones confirmed the localization of mutations to the target region. More recently, psoralen-linked TFOs containing several 2'-AE-residues as well as 2'-OMe substitutions have been prepared. These mixed substitution TFOs form triplexes at lower levels of Mg^{++} than required for the TFOs with only the 2'-OMe residues. Thermal melting experiments showed that the AE triplexes were more stable than the 2'-OMe only triplexes, and notably, were more stable than the underlying duplex at physiological pH. The AE TFOs were much more active in the *HPRT* knockout assay than the 2'-OMe TFOs. As discussed above, the enhanced biochemical and biological properties are due to the positive charge of the amine group. Recently it has been shown that psoralen-TFOs, carrying optimal levels of AE substitutions, can introduce mutations at the HPRT target site at frequencies in the range of 5%. This is sufficiently high to allow identification of cell clones with targeted mutations in direct screens without a requirement for selection.

Access to chromatin targets has been considered in studies with nuclei and in permeabilized cells and has been shown for psoralen-linked TFOs, as well as for a TFO alkylator conjugate. In these experiments, the frequency of cross-linking served as a measure of TFO binding at the time of photoactivation. Targeting activity was greatest in cells in which the target region was undergoing active transcription prior to permeabilization. In recent work, we have found that the frequency of targeted cross-links in living cells (and the resultant mutagenesis) is greatest in S-phase cells.

These results argue that transcription and replication can influence target access. However, the relationship between specific chromatin structures and target access remains to be determined.

4.6
Triplex Formation, by itself, can Promote Mutagenesis

Triplex structures are recognized by DNA repair proteins such as XPA/RPA. The possibility that this recognition could lead to mutagenesis was raised in experiments in cultured mammalian cells cotransfected with a plasmid containing a *supF* reporter gene with an embedded triplex target site and a purine motif TFO, designed to bind the target. Plasmids with mutations in the *supF* gene were recovered. These included deletions and point mutations distributed throughout the gene, and resembled those recovered from experiments with nicked plasmids. Mutagenesis was independent of psoralen, but did require functional DNA repair and transcription coupled repair functions. It seems likely that some triplex structures are recognized as lesions and trigger DNA repair with an error-prone outcome, as well as other events associated with DNA metabolism (see below).

Chromosomal targets in both cultured cells and transgenic mice have been mutagenized by treatment of cells or mice with purine motif TFOs. Unconjugated purine motif TFOs were employed, and the mutations included frameshifts (in the homopolymer runs of the target sequence) and distributed point mutations. In the mouse experiment, all tissues tested showed TFO-induced mutagenesis except the brain, which had no mutagenesis over background, consistent with TFOs being unable to cross the blood–brain barrier. The mutations were consistent with error-prone filling of gaps or slipped pairing during gap filling. Thus TFO targeting is possible in whole animals.

4.7
Triplex-induced Recombination and Gene Correction

The suggestion that triplexes would be recognized as a substrate for DNA repair prompted the notion that they might also trigger recombination. Experiments with a TFO and shuttle vector plasmid carrying two (different) mutant copies of the *supF* gene with the TFO target site between them demonstrated that triplex formation, with or without psoralen cross-linking, could induce recombination to produce a functional *supF* gene to psoralen. Gene correction was dependent on a functional XPA protein, a DNA damage–recognition protein and a component of the nucleotide excision repair machinery.

This approach has been extended to a chromosomal target in mouse fibroblasts. This was prepared by integration into a single chromosomal location of a construct containing a triplex target sequence with different mutant *thymidine kinase (TK)* genes placed on either side. When the appropriate TFOs were microinjected into the nuclei of the cells, the yield of recombinants was 1 to 2%, more than 1000-fold over background. Analysis of the recombinant clones revealed that all the recombination events involved gene conversion. Similar studies using TFOs to stimulate recombination have also been described using a locus in CHO cells containing duplicated *APRT* genes as a target.

The observation that third strand binding could provoke DNA repair and stimulate recombination led to the development of a strategy to mediate targeted gene conversion. In this protocol, a TFO was linked to a short DNA fragment homologous to the target site (except for the base pair to be corrected). The conception of the approach is that the TFO domain mediates site-specific binding to target the molecule to the desired gene. This binding would also trigger repair to initiate recombination or gene conversion with the "tethered" donor sequence. This idea was tested with an oligomer containing a 40-mer donor element and a 30-mer TFO domain. Correction of a single base pair mutation in the *supF* reporter gene in a shuttle vector plasmid in mammalian cells was shown. *In vitro* studies in human cell-free extracts revealed a requirement for XPA and Rad51 in the recombination pathway.

4.8
Expansion of the Triplex Binding Code

Although research on the biological applications of TFOs must be considered to be in the early stages, it does appear that three of the important issues raised above have been addressed: it is possible to synthesize TFOs that are functional *in vivo*; chromosomal targets are accessible, and accessibility can be modulated by manipulating the cell biology; TFOs can be used to target mutagenesis and recombination. Thus, although there is a great deal to be done to extend these conclusions, expansion of target options to mixed purine/pyrimidine sequences would appear to be the most important challenge facing the field. It is likely that solutions will come from synthetic chemistry via novel base analogs. Although there have been some interesting directions identified in the effort to expand the triplex binding code, it is fair to say that at this time there are no analogs that permit binding to general sequences. This remains an area of opportunity.

5 Mutation Repair by Small DNA Fragments

A seemingly attractive approach to introducing defined sequence changes into chromosomal loci would be based on conventional DNA fragments, homologous to the target region except for the intended change. The strategy is based on the notion that heteroduplex formation between the target and the donor sequence would be resolved, at least some of the time, in the experimentalist's favor. The method would take advantage of the four base recognition code, and relative ease and low cost of preparation of the correcting reagents. The formation of the critical heteroduplex would result from the action of the recombination apparatus in the host cells, or, (perhaps in the case of single-strand fragments) as a result of transient exposure of single-strand regions during replication, transcription, or repair. That such an approach might work is supported by the early demonstration of "marker rescue" by Edgall and Hutchison, in *E. coli*. Measurable correcting activity of single-strand oligonucleotides against a chromosomal target in yeast, an extrachromosomal target in mammalian cells, and, recently, the chromosomal *HPRT* gene in mammalian cells have also been reported. The frequency of correction in all these reports was too low to have practical utility. Short double-strand fragments were also shown to have low correcting activity in mammalian cells. In a recent work, oligonucleotides have been used to manipulate chromosomal gene sequences in *E. coli* at frequencies high enough to have practical utility. The much higher frequencies in the bacterial host may reflect fundamental differences in the exposure of single-strand regions in *E. coli* versus mammalian cells.

5.1 Small Fragment Homologous Replacement

Gruenert and colleagues have developed an approach based on double-strand fragments approximately 500-bp long (termed *small fragment homologous replacement, SFHR*). They have been largely concerned with the correction of mutations in the cystic fibrosis transmembrane conductance regulator (CFTR). They have introduced fragments designed to correct the common ΔF508 mutation (a deletion of three bases) and presented PCR-based evidence for conversion at the level of the chromosome and mRNA. They have argued that the conversion frequency was in the range of 1 to 10% of treated cells. Recently, this approach has been extended to the correction of a mutation in the dystrophin gene in mouse cells. Although evidence for correction was presented, expression of wild-type dystrophin was not observed. While the SFHR method may prove useful, it has received attention from few laboratories and no mechanistic studies have been performed. One of the limitations in developing the methodology has been the focus on disease genes as targets. Although of obvious interest, they do not lend themselves to straightforward quantitative measurement of correction events.

5.2 Chimeric Oligonucleotides

A review of oligonucleotide and DNA fragment–based gene correction technologies would not be complete without mention of chimeric oligonucleotides and the claims for their activity by Kmiec and coworkers. In the original description, these were foldback molecules containing a DNA double-strand region that contained a "correction" base pair flanked by a stretch of DNA: 2′-OMe hybrid (RNA–DNA oligonucleotide, RDO). Treatment of cells carrying a β-globin sickle-cell mutation with a "wild-type" chimeric oligonucleotide was reported to result in very high levels of correction (30–50%) after 6 h of treatment. However, there has been widespread failure to reproduce this and other claims from this group, and it is not clear that the original report is credible.

6 Peptide Nucleic Acids

Peptide nucleic acids (PNAs) are DNA analogs consisting of nucleobases attached to a peptide backbone of N-(2-aminoethyl)glycine residues (Fig. 11). The spacing of the bases along the backbone maintains the register relative to canonical DNA-permitting base pairing between PNA strands and DNA or RNA strands. The replacement of the phosphate backbone by a neutral linkage greatly reduces the charge repulsion between PNA and DNA or RNA strands. Thus, very stable PNA:DNA or PNA:RNA hybrids are formed. They can also form four stranded complexes with polypurine:polypyrimidine DNA duplex targets. Binding to duplex DNA involves triplex formation with one strand of

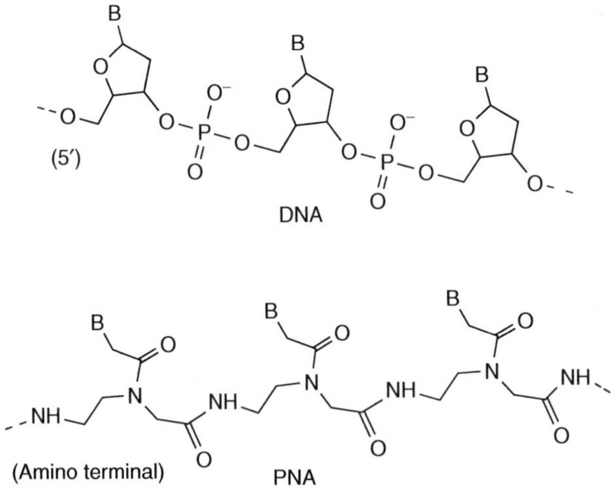

Fig. 11 The structure of a peptide nucleic acid backbone (PNA) as compared to DNA. (Reprinted from with permission from Elsevier Science).

PNA acting as a third strand paired via Hoogsteen hydrogen bonds to the purine strand of the duplex (as in a conventional triplex), while another PNA strand forms Watson–Crick pairs with the same purine strand (PNAPNA:DNA). The latter interaction occurs by way of what has been termed *strand invasion* in which the duplex DNA partner is eventually displaced by the PNA strand. This results in a very stable "clamp" structure with the pyrimidine strand of the duplex looped out. Linkage of the two PNA strands to form bis-PNAs improves binding by reducing entropic barriers. Although PNAs bind with high affinity under optimal conditions, it is important to note that binding is dependent on breathing of the duplex. At physiological salt concentrations, binding is almost completely suppressed, suggesting that strand capture, rather than strand invasion, may be a better description of the process of complex formation between PNAs and duplex targets. Basic peptides have been attached to bis-PNAs in order to accelerate binding to linear duplex DNA. Recently, this approach was shown to partially ameliorate the restriction on binding by salt concentrations of 100 mM KCl.

The limitation on binding activity by physiological salt concentrations would appear to mitigate against gene targeting by PNAs *in vivo*. However, reports from the Glazer laboratory suggest that at least some binding occurs at chromosomal targets *in vivo*. PNAs were designed to form clamp complexes with polypurine sites in a *supF* gene, which was integrated into the genome of murine cells. Treatment of the cells with the PNAs resulted in mutagenesis of the *supF* gene, at frequencies 10-fold above the background. The mutations were single-base insertions and deletions, as well as single-base substitutions, within a run of guanines. The mutations could be rationalized by a replication slippage model. In another experiment, PNAs were targeted to polypurine sites upstream of the transcription start site for the human $G\gamma$-globin gene. Treatment of K562 cells, which have low levels of γ-globin message, resulted in the enhanced expression of the gene from start sites at the PNA binding site as well as from the native promoter. The interpretation of both experiments is that the PNA clamp structure can form *in vivo* on a chromosomal target and the resultant "lesion" can stimulate error-prone DNA repair synthesis and/or transcription on the opened template strand. This structure also stimulated the recombination in cell-free extracts.

7
Summary

At the outset of this review, we noted that targeting reagents could be classified according to their approach to target recognition via the major or minor groove, or single strand. This is a useful way of thinking about the problem of sequence recognition and discrimination. At the end of the review, it is appropriate to consider the problem from another perspective and ask which strategies work in terms of selective sequence recognition. The simplest answer is that the successful approaches are those with the benefit of selection, during evolution, such as HR, or in the laboratory such as the zinc finger proteins reviewed elsewhere in this series. In contrast, those approaches based on chemistry supported by molecular modeling have been less effective. Thus, the considerable efforts to construct base analogs that expand the triplex binding code, or polyamides with high sequence selectivity, have had limited success. To a certain extent, this is the result of gaps in our understanding of the molecular architecture of target sequences under physiological conditions. Although molecular models are often compelling and promise success, the results often expose the limitations of the models rather than confirm the efficacy of the constructions. It may be that substantial improvements of reagents, based on chemistry, will require the application of combinatorial, or quasi-combinatorial, strategies in which effective solutions are derived by evolution in the laboratory, rather than by reliance on models.

See also Molecular Basis of Genetics.

Bibliography

Books and Reviews

Bailly, C., Chaires, J.B. (1998) Sequence specific DNA minor groove binders. Design and synthesis of netropsin and distamycin analogues, *Bioconjugate Chemistry* **9**, 513–538.

Braasch, D.A., Corey, D.R. (2002) Novel antisense and peptide nucleic acid strategies for controlling gene expression, *Biochemistry* **41**, 4503–4510.

Choo, Y., Klug, A. (1995) Designing DNA-binding proteins on the surface of filamentous phage, *Curr. Opin. Biotechnol.* **6**, 431–436.

Cox, M.M. (2001) Recombinational DNA repair of damaged replication forks in Escherichia coli: questions, *Annu. Rev. Genet.* **35**, 53–82.

Dean, D.A. (2000) Peptide nucleic acids: versatile tools for gene therapy strategies, *Adv. Drug Deliv. Rev.* **44**, 81–95.

Frank-Kamenetskii, M.D., Mirkin, S.M. (1995) Triplex DNA structures, *Annu. Rev. Biochem.* **64**, 65–95.

Fry, C.J., Peterson, C.L. (2002) Transcription. Unlocking the gates to gene expression, *Science* **295**, 1847–1848.

Gilbert, D.E., Feigon, J. (1999) Multistranded DNA structures, *Curr. Opin. Struct. Biol.* **9**, 305–314.

Goodsell, D.S. (2001) Sequence recognition of DNA by lexitropsins, *Curr. Med. Chem.* **8**, 509–516.

Gowers, D.M., Fox, K.R. (1999) Towards mixed sequence recognition by triple helix formation, *Nucleic Acids Res.* **27**, 1569–1577.

Haber, J.E. (1999) DNA recombination: the replication connection, *Trends Biochem. Sci.* **24**, 271–275.

Luyten, I.A., Herdewijn, P. (2002) Hybridization properties of base-modified oligonucleotides within the double and triple helix motif, *Eur. J. Med. Chem.* **33**, 515–576.

Muller, U. (1999) Ten years of gene targeting: targeted mouse mutants, from vector design to phenotype analysis, *Mech. Dev.* **82**, 3–21.

Lusetti, S.L., Cox, M.M. (2002) The bacterial RecA protein and the recombinational DNA repair of stalled replication forks, *Annu. Rev. Biochem.* **71**, 71–100.

Nielsen, P.E. (2001) Targeting double stranded DNA with peptide nucleic acid (PNA), *Curr. Med. Chem.* **8**, 545–550.

Roth, D.B., Wilson, J.H. (1988) in: Kucherlapati, R., Smith, G. (Eds.) *Genetic Recombination*, American Society of Microbiology, Washington, DC, pp. 621–651.

Thuong, N.T., Helene, C. (1993) Sequence specific recognition and modification of double helical DNA by oligonucleotides, *Angewandte Chemie, Intl. Ed.* **32**, 666–690.

van der Weyden, L., Adams, D.J., Bradley, A. (2002) Tools for targeted manipulation of the mouse genome, *Physiol. Genomics* **11**, 133–164.

Vasquez, K.M., Marburger, K., Intody, Z., Wilson, J.H. (2001b) Manipulating the mammalian genome by homologous recombination, *Proc. Natl. Acad. Sci. U.S.A.* **98**, 8403–8410.

Wemmer, D.E. (1999) Ligands recognizing the minor groove of DNA: development and applications, *Biopolymers* **52**, 197–211.

Wemmer, D.E. (2000) Designed sequence-specific minor groove ligands, *Annu. Rev. Biophys. Biomol. Struct.* **29**, 439–461.

Wemmer, D.E., Dervan, P.B. (1997) Targeting the minor groove of DNA, *Curr. Opin. Struct. Biol.* **7**, 355–361.

Wolfe, S.A., Nekludova, L., Pabo, C.O. (2000) DNA recognition by Cys2His2 zinc finger proteins, *Annu. Rev. Biophys. Biomol. Struct.* **29**, 183–212.

Primary Literature

Adair, G.M., Nairn, R.S., Wilson, J.H., Seidman, M.M., Brotherman, K.A., MacKinnon, C., Scheerer, J.B. (1989) Targeted homologous recombination at the endogenous adenine phosphoribosyltransferase locus in Chinese hamster cells, *Proc. Natl. Acad. Sci. U.S.A.* **86**, 4574–4578.

Alberti, P., Arimondo, P.B., Mergny, J.L., Garestier, T., Helene, C., Sun, J.S. (2002) A directional nucleation-zipping mechanism for triple helix formation, *Nucleic Acids Res.* **30**, 5407–5415.

Aratani, Y., Okazaki, R., Koyama, H. (1992) End extension repair of introduced targeting vectors mediated by homologous recombination in mammalian cells, *Nucleic Acids Res.* **20**, 4795–4801.

Arimondo, P.B., Garestier, T., Helene, C., Sun, J.S. (2001) Detection of competing DNA structures by thermal gradient gel electrophoresis: from self-association to triple helix formation by (G,A)-containing oligonucleotides, *Nucleic Acids Res.* **29**, E15.

Arimondo, P.B., Moreau, P., Boutorine, A., Bailly, C., Prudhomme, M., Sun, J.S., Garestier, T., Helene, C. (2000) Recognition and cleavage of DNA by rebeccamycin- or benzopyridoquinoxaline conjugated of triple helix-forming oligonucleotides, *Bioorg. Med. Chem.* **8**, 777–784.

Arya, D.P., Bruice, T.C. (1999) Triple-helix formation of DNA oligomers with methylthiourea-linked nucleosides (DNmt): a kinetic and thermodynamic analysis, *Proc. Natl. Acad. Sci. U.S.A.* **96**, 4384–4389.

Asensio, J.L., Carr, R., Brown, T., Lane, A.N. (1999) Conformational and thermodynamic properties of parallel intramolecular triple helixes containing a DNA, RNA, or 2′-OMeDNA third strand, *J. Am. Chem. Soc.* **121**, 11063–11070.

Asensio, J.L., Lane, A.N., Dhesi, J., Bergqvist, S., Brown, T. (1998) The contribution of cytosine protonation to the stability of parallel DNA triple helices, *J. Mol. Biol.* **275**, 811–822.

Bailis, A.M., Rothstein, R. (1990) A defect in mismatch repair in Saccharomyces cerevisiae stimulates ectopic recombination between homologous genes by an excision repair dependent process, *Genetics* **126**, 535–547.

Baird, E.E., Dervan, P.B. (1996) Solid phase synthesis of polyamides containing imidazole and pyrrole amino acids, *J. Am. Chem. Soc.* **118**, 6141–6146.

Baliga, R., Singleton, J.W., Dervan, P.B. (1995) RecA.oligonucleotide filaments bind in the minor groove of double-stranded DNA, *Proc. Natl. Acad. Sci. U.S.A.* **92**, 10393–10397.

Belitsky, J.M., Leslie, S.J., Arora, P.S., Beerman, T.A., Dervan, P.B. (2002) Cellular uptake of N-methylpyrrole/N-methylimidazole polyamide-dye conjugates, *Bioorg. Med. Chem.* **10**, 3313–3318.

Belmaaza, A., Wallenburg, J.C., Brouillette, S., Gusew, N., Chartrand, P. (1990) Genetic exchange between endogenous and exogenous LINE-1 repetitive elements in mouse cells, *Nucleic Acids Res.* **18**, 6385–6391.

Belousov, E.S., Afonina, I.A., Kutyavin, I.V., Gall, A.A., Reed, M.W., Gamper, H.B., Wydro, R.M., Meyer, R.B. (1998) Triplex targeting of a native gene in permeabilized intact cells: covalent modification of the gene for the chemokine receptor CCR5, *Nucleic Acids Res.* **26**, 1324–1328.

Bentin, T., Nielsen, P.E. (1996) Enhanced peptide nucleic acid binding to supercoiled DNA: possible implications for DNA "breathing" dynamics, *Biochemistry* **35**, 8863–8869.

Besch, R., Giovannangeli, C., Kammerbauer, C., Degitz, K. (2002) Specific inhibition of ICAM-1 expression mediated by gene targeting with Triplex-forming oligonucleotides, *J. Biol. Chem.* **277**, 32473–32479.

Bibikova, M., Breumer, K., Trautman, J.K., Carroll, D. (2003) Enhanced gene targeting by target cleavage with designed zinc-finger nucleases, *Science* **300**, 764.

Bibikova, M., Carroll, D., Segal, D.J., Trautman, J.K., Smith, J., Kim, Y.G., Chandrasegaran, S. (2001) Stimulation of homologous recombination through targeted cleavage by chimeric nucleases, *Mol. Cell Biol.* **21**, 289–297.

Bibikova, M., Golic, M., Golic, K.G., Carroll, D. (2002) Targeted chromosomal cleavage and mutagenesis in drosophila using zinc-finger nucleases, *Genetics* **161**, 1169–1175.

Blume, S.W., Lebowitz, J., Zacharias, W., Guarcello, V., Mayfield, C.A., Ebbinghaus, S.W.,

Bates, P., Jones D.E. Jr., Trent, J., Vigneswaran, N., Miller, D.M. (1999) The integral divalent cation within the intermolecular purine*purine. pyrimidine structure: a variable determinant of the potential for and characteristics of the triple helical association, *Nucleic Acids Res.* **27**, 695–702.

Bollag, R.J., Liskay, R.M. (1988) Conservative intrachromosomal recombination between inverted repeats in mouse cells: association between reciprocal exchange and gene conversion, *Genetics* **119**, 161–169.

Brown, P.M., Fox, K.R. (1996) Nucleosome core particles inhibit DNA triple helix formation, *Biochem. J.* **319**, 607–611.

Bruscia, E., Sangiuolo, F., Sinibaldi, P., Goncz, K.K., Novelli, G., Gruenert, D.C. (2002) Isolation of CF cell lines corrected at DeltaF508-CFTR locus by SFHR- mediated targeting, *Gene Ther.* **9**, 683–685.

Campbell, C.R., Keown, W., Lowe, L., Kirschling, D., Kucherlapati, R. (1989) Homologous recombination involving small single-stranded oligonucleotides in human cells, *New Biol.* **1**, 223–227.

Capecchi, M.R. (1989) Altering the genome by homologous recombination, *Science* **244**, 1288–1292.

Carlomagno, T., Blommers, M.J., Meiler, J., Cuenoud, B., Griesinger, C. (2001) Determination of aliphatic side-chain conformation using cross-correlated relaxation: application to an extraordinarily stable 2′-aminoethoxy-modified oligonucleotide triplex, *J. Am. Chem. Soc.* **123**, 7364–7370.

Chan, P.P., Lin, M., Faruqi, A.F., Powell, J., Seidman, M.M., Glazer, P.M. (1999) Targeted correction of an episomal gene in mammalian cells by a short DNA fragment tethered to a triplex-forming oligonucleotide, *J. Biol. Chem.* **274**, 11541–11548.

Chandrasegaran, S., Smith, J. (1999) Chimeric restriction enzymes: what is next? *Biol. Chem.* **380**, 841–848.

Chen, Y.H., Yang, Y., Lown, J.W. (1996) Optimization of cross-linked lexitropsins, *J. Biomol. Struct. Dyn.* **14**, 341–355.

Chiang, S.Y., Burli, R.W., Benz, C.C., Gawron, L., Scott, G.K., Dervan, P.B., Beerman, T.A. (2000) Targeting the ets binding site of the HER2/neu promoter with pyrrole–imidazole polyamides, *J. Biol. Chem.* **275**, 24246–24254.

Cole-Strauss, A., Yoon, K., Xiang, Y., Byrne, B.C., Rice, M.C., Gryn, J., Holloman, W.K., Kmiec, E.B. (1996) Correction of the mutation responsible for sickle cell anemia by an RNA–DNA oligonucleotide, *Science* **273**, 1386–1389.

Cooney, M., Czernuszewicz, G., Postel, E.H., Flint, S.J., Hogan, M.E. (1988) Site-specific oligonucleotide binding represses transcription of the human c-myc gene in vitro, *Science* **241**, 456–459.

Cox, M.M., Goodman, M.F., Kreuzer, K.N., Sherratt, D.J., Sandler, S.J., Marians, K.J. (2000) The importance of repairing stalled replication forks, *Nature* **404**, 37–41.

Cuenoud, B., Casset, F., Husken, D., Natt, F., Wolf, R.M., Altmann, K.H., Martin, P., Moser, H.E. (1998) Dual recognition of double stranded DNA by 2′-aminoethoxy-modified oligonucleotides, *Angewandte Chemie, Intl. Ed.* **37**, 1288–1291.

Dagle, J.M. Weeks, D.L. (1996) Positively charged oligonucleotides overcome potassium-mediated inhibition of triplex DNA formation, *Nucleic Acids Res.* **24**, 2143–2149.

Datta, H.J., Chan, P.P., Vasquez, K.M., Gupta, R.C., Glazer, P.M. (2001) Triplex-induced recombination in human cell-free extracts: Dependence on XPA and HsRad51, *J. Biol. Chem.* **276**, 18018–18023.

Delacote, F., Han, M., Stamato, T.D., Jasin, M., Lopez, B.S. (2002) An xrcc4 defect or Wortmannin stimulates homologous recombination specifically induced by double-strand breaks in mammalian cells, *Nucleic Acids Res.* **30**, 3454–3463.

Deng, C., Capecchi, M.R. (1992) Reexamination of gene targeting frequency as a function of the extent of homology between the targeting vector and the target locus, *Mol. Cell Biol.* **12**, 3365–3371.

Dickinson, L.A., Gulizia, R.J., Trauger, J.W., Baird, E.E., Mosier, D.E., Gottesfeld, J.M., Dervan, P.B. (1998) Inhibition of RNA polymerase II transcription in human cells by synthetic DNA-binding ligands [see comments], *Proc. Natl. Acad. Sci. U.S.A.* **95**, 12890–12895.

Doetschman, T., Gregg, R.G., Maeda, N., Hooper, M.L., Melton, D.W., Thompson, S., Smithies, O. (1987) Targeted correction of a mutant HPRT gene in mouse embryonic stem cells, *Nature* **330**, 576–578.

Dwyer, T.J., Geierstanger, B., Bathini, Y., Lown, J.W., Wemmer, D. (1992) Design and binding of a distamycin A analog to d(CGCAAGTTGGC):d(GCCAACTTGCG): synthesis, NMR studies, and implications for the design of minor groove binding oligopeptides, *J. Am. Chem. Soc.* **114**, 5911–5919.

Egholm, M., Buchardt, O., Christensen, L., Behrens, C., Freier, S.M., Driver, D.A., Berg, R.H., Kim, S.K., Norden, B., Nielsen, P.E. (1993) PNA hybridizes to complementary oligonucleotides obeying the Watson–Crick hydrogen-bonding rules, *Nature* **365**, 566–568.

Egholm, M., Christensen, L., Dueholm, K.L., Buchardt, O., Coull, J., Nielsen, P.E. (1995) Efficient pH-independent sequence-specific DNA binding by pseudoisocytosine-containing bis-PNA, *Nucleic Acids Res.* **23**, 217–222.

Ellis, J., Bernstein, A. (1989) Gene targeting with retroviral vectors: recombination by gene conversion into regions of nonhomology, *Mol. Cell Biol.* **9**, 1621–1627.

Escude, C., Giovannangeli, C., Sun, J.S., Lloyd, D.H., Chen, J.K., Gryaznov, S.M., Garestier, T., Helene, C. (1996) Stable triple helices formed by oligonucleotide N3′ →P5′ phosphoramidates inhibit transcription elongation, *Proc. Natl. Acad. Sci. U.S.A.* **93**, 4365–4369.

Escude, C., Sun, J.S., Rougee, M., Garestier, T., Helene, C. (1992) Stable triple helices are formed upon binding of RNA oligonucleotides and their 2′-O-methyl derivatives to double-helical DNA, *C. R. Acad. Sci. III* **315**, 521–525.

Espinas, M.L., Jimenez-Garcia, E., Martinez-Balbas, A., Azorin, F. (1996) Formation of triple-stranded DNA at d(GA.TC)n sequences prevents nucleosome assembly and is hindered by nucleosomes, *J. Biol. Chem.* **271**, 31807–31812.

Essers, J., Hendriks, R.W., Wesoly, J., Beerens, C.E., Smit, B., Hoeijmakers, J.H., Wyman, C., Dronkert, M.L., Kanaar, R. (2002) Analysis of mouse Rad54 expression and its implications for homologous recombination, *DNA Repair (Amst)* **1**, 779–793.

Evans, E., Alani, E. (2000) Roles for mismatch repair factors in regulating genetic recombination, *Mol. Cell Biol.* **20**, 7839–7844.

Faruqi, A.F., Datta, H.J., Carroll, D., Seidman, M.M., Glazer, P.M. (2000) Triple-helix formation induces recombination in mammalian cells via a nucleotide excision repair-dependent pathway, *Mol. Cell Biol.* **20**, 990–1000.

Faruqi, A.F., Egholm, M., Glazer, P.M. (1998) Peptide nucleic acid-targeted mutagenesis of a chromosomal gene in mouse cells, *Proc. Natl. Acad. Sci. U.S.A.* **95**, 1398–1403.

Faruqi, A.F., Seidman, M.M., Segal, D.J., Carroll, D., Glazer, P.M. (1996) Recombination induced by triple-helix-targeted DNA damage in mammalian cells, *Mol. Cell Biol.* **16**, 6820–6828.

Felsenfeld, G., Davies, D.R., Rich, A. (1957) Formation of a three stranded polynucleotide molecule, *J. Am. Chem. Soc.* **79**, 2023–2024.

Ferguson, D.O., Alt, F.W. (2001) DNA double strand break repair and chromosomal translocation: lessons from animal models, *Oncogene* **20**, 5572–5579.

Flygare, J., Falt, S., Ottervald, J., Castro, J., Dackland, A.L., Hellgren, D., Wennborg, A. (2001) Effects of HsRad51 overexpression on cell proliferation, cell cycle progression, and apoptosis, *Exp. Cell Res.* **268**, 61–69.

Folger, K.R., Wong, E.A., Wahl, G., Capecchi, M.R. (1982) Patterns of integration of DNA microinjected into cultured mammalian cells: evidence for homologous recombination between injected plasmid DNA molecules, *Mol. Cell Biol.* **2**, 1372–1387.

Giovannangeli, C., Diviacco, S., Labrousse, V., Gryaznov, S., Charneau, P., Helene, C. (1997) Accessibility of nuclear DNA to triplex-forming oligonucleotides: the integrated HIV-1 provirus as a target, *Proc. Natl. Acad. Sci. U.S.A.* **94**, 79–84.

Giovannangeli, C., Perrouault, L., Escude, C., Gryaznov, S., Helene, C. (1996) Efficient inhibition of transcription elongation in vitro by oligonucleotide phosphoramidates targeted to proviral HIV DNA, *J. Mol. Biol.* **261**, 386–398.

Golub, E.I., Kovalenko, O.V., Gupta, R.C., Ward, D.C., Radding, C.M. (1997) Interaction of human recombination proteins Rad51 and Rad54, *Nucleic Acids Res.* **25**, 4106–4110.

Gottesfeld, J.M., Belitsky, J.M., Melander, C., Dervan, P.B., Luger, K. (2002) Blocking transcription through a nucleosome with synthetic DNA ligands, *J. Mol. Biol.* **321**, 249–263.

Gottesfeld, J.M., Melander, C., Suto, R.K., Raviol, H., Luger, K., Dervan, P.B. (2001)

Sequence-specific recognition of DNA in the nucleosome by pyrrole–imidazole polyamides, *J. Mol. Biol.* **309**, 615–629.

Gottesfeld, J.M., Neely, L., Trauger, J.W., Baird, E.E., Dervan, P.B. (1997) Regulation of gene expression by small molecules, *Nature* **387**, 202–205.

Gupta, R.C., Folta-Stogniew, E., O'Malley, S., Takahashi, M., Radding, C.M. (1999) Rapid exchange of A:T base pairs is essential for recognition of DNA homology by human Rad51 recombination protein, *Mol. Cell* **4**, 705–714.

Haber, J.E., Heyer, W.D. (2001) The fuss about Mus81, *Cell* **107**, 551–554.

Havre, P.A., Gunther, E.J., Gasparro, F.P., Glazer, P.M. (1993) Targeted mutagenesis of DNA using triple helix-forming oligonucleotides linked to psoralen, *Proc. Natl. Acad. Sci. U.S.A.* **90**, 7879–7883.

Helene, C. (1991) The antigene strategy: control of gene expression by triplex-forming-oligonucleotides, *Anti-Cancer Drug Des.* **6**, 569–584.

Helene, C., Thuong, N.T., Harel-Bellan, A. (1992) Control of gene expression by triple helix-forming oligonucleotides. The antigene strategy, *Ann. N. Y. Acad. Sci.* **660**, 27–36.

Hirata, R., Chamberlain, J., Dong, R., Russell, D.W. (2002) Targeted transgene insertion into human chromosomes by adeno-associated virus vectors, *Nat. Biotechnol.* **20**, 735–738.

Howard-Flanders, P., West, S.C., Stasiak, A. (1984) Role of RecA protein spiral filaments in genetic recombination, *Nature* **309**, 215–219.

Hunger-Bertling, K., Harrer, P., Bertling, W. (1990) Short DNA fragments induce site specific recombination in mammalian cells, *Mol. Cell Biochem.* **92**, 107–116.

Hutchison, C.A. III, Edgell, M.H. (1971) Genetic assay for small fragments of bacteriophage ϕ X174 deoxyribonucleic acid, *J. Virol.* **8**, 181–189.

Inoue, N., Hirata, R.K., Russell, D.W. (1999) High-fidelity correction of mutations at multiple chromosomal positions by adeno-associated virus vectors, *J. Virol.* **73**, 7376–7380.

Janssen, S., Cuvier, O., Muller, M., Laemmli, U.K. (2000a) Specific gain- and loss-of-function phenotypes induced by satellite-specific DNA-binding drugs fed to drosophila melanogaster, *Mol. Cell* **6**, 1013–1024.

Janssen, S., Durussel, T., Laemmli, U.K. (2000b) Chromatin opening of DNA satellites by targeted sequence-specific drugs, *Mol. Cell* **6**, 999–1011.

Jasin, M., Elledge, S.J., Davis, R.W., Berg, P. (1990) Gene targeting at the human CD4 locus by epitope addition, *Genes Dev.* **4**, 157–166.

Johnson, R.D., Jasin, M. (2000) Sister chromatid gene conversion is a prominent double-strand break repair pathway in mammalian cells, *EMBO J.* **19**, 3398–3407.

Kabotyanski, E.B., Gomelsky, L., Han, J.O., Stamato, T.D., Roth, D.B. (1998) Double-strand break repair in Ku86- and XRCC4-deficient cells, *Nucleic Acids Res.* **26**, 5333–5342.

Kaihatsu, K., Braasch, D.A., Cansizoglu, A., Corey, D.R. (2002) Enhanced strand invasion by peptide nucleic acid-peptide conjugates, *Biochemistry* **41**, 11118–11125.

Kapsa, R., Quigley, A., Lynch, G.S., Steeper, K., Kornberg, A.J., Gregorevic, P., Austin, L., Byrne, E. (2001) In vivo and in vitro correction of the mdx dystrophin gene nonsense mutation by short-fragment homologous replacement, *Hum. Gene Ther.* **12**, 629–642.

Kelly, J.J., Baird, E.E., Dervan, P.B. (1996) Binding site size limit of the 2:1 pyrrole-imidazole polyamide-DNA motif, *Proc. Natl. Acad. Sci. U.S.A.* **93**, 6981–6985.

Kenner, O., Kneisel, A., Klingler, J., Bartelt, B., Speit, G., Vogel, W., Kaufmann, D. (2002) Targeted gene correction of hprt mutations by 45 base single-stranded oligonucleotides, *Biochem. Biophys. Res. Commun.* **299**, 787–792.

Khorlin, A.A., Krylov, A.S., Grokhovsky, S.L., Zhuze, A.L., Zasedatelev, A.S., Gursky, G.V., Gottikh, B.P. (1980) A new type of AT-specific ligand constructed of two netropsin-like molecules, *FEBS Lett.* **118**, 311–314.

Kim, H.G., Miller, D.M. (1998) A novel triplex-forming oligonucleotide targeted to human cyclin D1 (bcl-1, proto-oncogene) promoter inhibits transcription in HeLa cells, *Biochemistry* **37**, 2666–2672.

Kim, Y.G., Cha, J., Chandrasegaran, S. (1996) Hybrid restriction enzymes: zinc finger fusions to Fok I cleavage domain, *Proc. Natl. Acad. Sci. U.S.A.* **93**, 1156–1160.

Kochetkova, M., Iversen, P.O., Lopez, A.F., Shannon, M.F. (1997) Deoxyribonucleic acid triplex formation inhibits granulocyte macrophage colony-stimulating factor gene expression and suppresses growth in juvenile myelomonocytic leukemic cells, *J. Clin. Invest.* **99**, 3000–3008.

Kopka, M.L., Yoon, C., Goodsell, D., Pjura, P., Dickerson, R.E. (1985) The molecular origin of DNA-drug specificity in netropsin and distamycin, *Proc. Natl. Acad. Sci. U.S.A.* **82**, 1376–1380.

Kuhn, H., Demidov, V.V., Frank-Kamenetskii, M.D., Nielsen, P.E. (1998) Kinetic sequence discrimination of cationic bis-PNAs upon targeting of double-stranded DNA, *Nucleic Acids Res.* **26**, 582–587.

Kukreti, S., Sun, J.S., Garestier, T., Helene, C. (1997) Extension of the range of DNA sequences available for triple helix formation: stabilization of mismatched triplexes by acridine-containing oligonucleotides, *Nucleic Acids Res.* **25**, 4264–4270.

Kumar, K.A., Muniyappa, K. (1992) Use of structure-directed DNA ligands to probe the binding of recA protein to narrow and wide grooves of DNA and on its ability to promote homologous pairing, *J. Biol. Chem.* **267**, 24824–24832.

Lacroix, L., Lacoste, J., Reddoch, J.F., Mergny, J.L., Levy, D.D., Seidman, M.M., Matteucci, M.D., Glazer, P.M. (1999) Triplex formation by oligonucleotides containing 5-(1-propynyl)-2'-deoxyuridine: decreased magnesium dependence and improved intracellular gene targeting, *Biochemistry* **38**, 1893–1901.

Lee, J.S., Woodsworth, M.L., Latimer, L.J., Morgan, A.R. (1984) Poly(pyrimidine).poly(purine) synthetic DNAs containing 5-methylcytosine form stable triplexes at neutral pH, *Nucleic Acids Res.* **12**, 6603–6614.

Letai, A.G., Palladino, M.A., Fromm, E., Rizzo, V., Fresco, J.R. (1988) Specificity in formation of triple-stranded nucleic acid helical complexes: studies with agarose-linked polyribonucleotide affinity columns, *Biochemistry* **27**, 9108–9112.

Liang, F., Han, M., Romanienko, P.J., Jasin, M. (1998) Homology-directed repair is a major double-strand break repair pathway in mammalian cells, *Proc. Natl. Acad. Sci. U.S.A.* **95**, 5172–5177.

Liang, F., Jasin, M. (1996) Ku80-deficient cells exhibit excess degradation of extrachromosomal DNA, *J. Biol. Chem.* **271**, 14405–14411.

Liang, F., Romanienko, P.J., Weaver, D.T., Jeggo, P.A., Jasin, M. (1996) Chromosomal double-strand break repair in Ku80-deficient cells, *Proc. Natl. Acad. Sci. U.S.A.* **93**, 8929–8933.

Liskay, R.M., Letsou, A., Stachelek, J.L. (1987) Homology requirement for efficient gene conversion between duplicated chromosomal sequences in mammalian cells, *Genetics* **115**, 161–167.

Lown, J.W., Krowicki, K., Bhat, U.G., Skorobogaty, A., Ward, B., Dabrowiak, J.C. (1986) Molecular recognition between oligopeptides and nucleic acids: novel imidazole-containing oligopeptides related to netropsin that exhibit altered DNA sequence specificity, *Biochemistry* **25**, 7408–7416.

Lukacsovich, T., Waldman, B.C., Waldman, A.S. (2001) Efficient recruitment of transfected DNA to a homologous chromosomal target in mammalian cells, *Biochim. Biophys. Acta* **1521**, 89–96.

Luo, Z., Macris, M.A., Faruqi, A.F., Glazer, P.M. (2000) High-frequency intrachromosomal gene conversion induced by triplex-forming oligonucleotides microinjected into mouse cells, *Proc. Natl. Acad. Sci. U.S.A.* **97**, 9003–9008.

Macris, M.A., Glazer, P.M. (2002) Transcription dependence of chromosomal gene targeting by triplex-forming oligonucleotides, *J. Biol. Chem.* **278**, 3357–3362.

Maeshima, K., Janssen, S., Laemmli, U.K. (2001) Specific targeting of insect and vertebrate telomeres with pyrrole and imidazole polyamides, *EMBO J.* **20**, 3218–3228.

Majumdar, A., Khorlin, A., Dyatkina, N., Lin, F.L., Powell, J., Liu, J., Fei, Z., Khripine, Y., Watanabe, K.A., George, J., Glazer, P.M., Seidman, M.M. (1998) Targeted gene knockout mediated by triple helix forming oligonucleotides, *Nat. Genet.* **20**, 212–214.

Majumdar, A., Puri, N., Cuenoud, B., Natt, F., Martin, P., Khorlin, A., Dyatkina, N., George, A.J., Miller, P.S., Seidman, M.M. (2003) Cell cycle modulation of gene targeting by a triple helix forming oligonucleotide, *J. Biol. Chem.* **278**, 11072–11077.

Manor, H., Rao, B.S., Martin, R.G. (1988) Abundance and degree of dispersion of genomic d(GA)n.d(TC)n sequences, *J. Mol. Evol.* **27**, 96–101.

Mansour, S.L., Thomas, K.R., Deng, C.X., Capecchi, M.R. (1990) Introduction of a lacZ reporter gene into the mouse int-2 locus by homologous recombination, *Proc. Natl. Acad. Sci. U.S.A.* **87**, 7688–7692.

Mapp, A.K., Ansari, A.Z., Ptashne, M., Dervan, P.B. (2000) Activation of gene expression by small molecule transcription factors, *Proc. Natl. Acad. Sci. U.S.A.* **97**, 3930–3935.

Mayfield, C., Ebbinghaus, S., Gee, J., Jones, D., Rodu, B., Squibb, M., Miller, D. (1994) Triplex formation by the human Ha-ras promoter inhibits Sp1 binding and in vitro transcription, *J. Biol. Chem.* **269**, 18232–18238.

Mazin, A.V., Bornarth, C.J., Solinger, J.A., Heyer, W.D., Kowalczykowski, S.C. (2000) Rad54 protein is targeted to pairing loci by the Rad51 nucleoprotein filament, *Mol. Cell* **6**, 583–592.

McCulloch, R.D., Read, L.R., Baker, M.D. (2003) Strand invasion and DNA synthesis from the two 3′ ends of a DSB, *Genetics* **163**, 1439–1447.

Melander, C., Herman, D.M., Dervan, P.B. (2000) Discrimination of A/T sequences in the minor groove of DNA within a cyclic polyamide motif, *Chemistry* **6**, 4487–4497.

Merrihew, R.V., Marburger, K., Pennington, S.L., Roth, D.B., Wilson, J.H. (1996) High-frequency illegitimate integration of transfected DNA at preintegrated target sites in a mammalian genome, *Mol. Cell Biol.* **16**, 10–18.

Miller, D.G., Petek, L.M., Russell, D.W. (2003) Human gene targeting by adeno-associated virus vectors is enhanced by DNA double strand breaks, *Mol. Cell Biol.* **23**, 3550–3557.

Moerschell, R.P., Das, G., Sherman, F. (1991) Transformation of yeast directly with synthetic oligonucleotides, *Methods Enzymol.* **194**, 362–369.

Mollegaard, N.E., Buchardt, O., Egholm, M., Nielsen, P.E. (1994) Peptide nucleic acid.DNA strand displacement loops as artificial transcription promoters, *Proc. Natl. Acad. Sci. U.S.A.* **91**, 3892–3895.

Moser, H.E., Dervan, P.B. (1987) Sequence-specific cleavage of double helical DNA by triple helix formation, *Science* **238**, 645–650.

Mrksich, M., Parks, M.E., Dervan, P.B. (1994) Hairpin peptide motif: A new class of oligopeptides for sequence specific recognition in the minor groove of double helical DNA, *J. Am. Chem. Soc.* **116**, 7983–7988.

Mrksich, M., Wade, W.S., Dwyer, T.J., Geierstanger, B.H., Wemmer, D.E., Dervan, P.B. (1992) Antiparallel side-by-side dimeric motif for sequence-specific recognition in the minor groove of DNA by the designed peptide 1-methylimidazole-2-carboxamide netropsin, *Proc. Natl. Acad. Sci. U.S.A.* **89**, 7586–7590.

Myung, K., Datta, A., Chen, C., Kolodner, R.D. (2001) SGS1, the Saccharomyces cerevisiae homologue of BLM and WRN, suppresses genome instability and homeologous recombination, *Nat. Genet.* **27**, 113–116.

Nielsen, P.E., Egholm, M., Berg, R.H., Buchardt, O. (1991) Sequence-selective recognition of DNA by strand displacement with a thymine-substituted polyamide, *Science* **254**, 1497–1500.

Nielsen, P.E., Egholm, M., Buchardt, O. (1994) Evidence for (PNA)2/DNA triplex structure upon binding of PNA to dsDNA by strand displacement, *J. Mol. Recognit.* **7**, 165–170.

Nishinaka, T., Shinohara, A., Ito, Y., Yokoyama, S., Shibata, T. (1998) Base pair switching by interconversion of sugar puckers in DNA extended by proteins of RecA-family: a model for homology search in homologous genetic recombination, *Proc. Natl. Acad. Sci. U.S.A.* **95**, 11071–11076.

O'Hare, C.C., Mack, D., Tandon, M., Sharma, S.K., Lown, J.W., Kopka, M.L., Dickerson, R.E., Hartley, J.A. (2002) DNA sequence recognition in the minor groove by crosslinked polyamides: The effect of N-terminal head group and linker length on binding affinity and specificity, *Proc. Natl. Acad. Sci. U.S.A.* **99**, 72–77.

Orr-Weaver, T.L., Szostak, J.W., Rothstein, R.J. (1981) Yeast transformation: a model system for the study of recombination, *Proc. Natl. Acad. Sci. U.S.A.* **78**, 6354–6358.

Orson, F.M., Thomas, D.W., McShan, W.M., Kessler, D.J., Hogan, M.E. (1991) Oligonucleotide inhibition of IL2R alpha mRNA transcription by promoter region collinear triplex formation in lymphocytes, *Nucleic Acids Res.* **19**, 3435–3441.

Paques, F., Haber, J.E. (1999) Multiple pathways of recombination induced by double-strand breaks in saccharomyces cerevisiae, *Microbiol. Mol. Biol. Rev.* **63**, 349–404.

Park, M.S. (1995) Expression of human RAD52 confers resistance to ionizing radiation in mammalian cells, *J. Biol. Chem.* **270**, 15467–15470.

Pazin, M.J., Kadonaga, J.T. (1997) SWI2/SNF2 and related proteins: ATP-driven motors that disrupt protein–DNA interactions? *Cell* **88**, 737–740.

Peffer, N.J., Hanvey, J.C., Bisi, J.E., Thomson, S.A., Hassman, C.F., Noble, S.A., Babiss, L.E. (1993) Strand-invasion of duplex DNA by peptide nucleic acid oligomers, *Proc. Natl. Acad. Sci. U.S.A.* **90**, 10648–10652.

Pelton, J.G., Wemmer, D.E. (1989) Structural characterization of a 2:1 distamycin A.d(CGCAAATTGGC) complex by two-dimensional NMR, *Proc. Natl. Acad. Sci. U.S.A.* **86**, 5723–5727.

Pennington, S.L., Wilson, J.H. (1991) Gene targeting in Chinese hamster ovary cells is conservative, *Proc. Natl. Acad. Sci. U.S.A.* **88**, 9498–9502.

Perkins, B.D., Wilson, J.H., Wensel, T.G., Vasquez, K.M. (1998) Triplex targets in the human rhodopsin gene, *Biochemistry* **37**, 11315–11322.

Pesco, J., Salmon, J.M., Vigo, J., Viallet, P. (2001) Mag-indo1 affinity for Ca(2+), compartmentalization and binding to proteins: the challenge of measuring Mg(2+) concentrations in living cells, *Anal. Biochem.* **290**, 221–231.

Petukhova, G., Stratton, S., Sung, P. (1998) Catalysis of homologous DNA pairing by yeast Rad51 and Rad54 proteins, *Nature* **393**, 91–94.

Petukhova, G., Van Komen, S., Vergano, S., Klein, H., Sung, P. (1999) Yeast Rad54 promotes Rad51-dependent homologous DNA pairing via ATP hydrolysis-driven change in DNA double helix conformation, *J. Biol. Chem.* **274**, 29453–29462.

Pierce, A.J., Hu, P., Han, M., Ellis, N., Jasin, M. (2001) Ku DNA end-binding protein modulates homologous repair of double-strand breaks in mammalian cells, *Genes Dev.* **15**, 3237–3242.

Podyminogin, M.A., Meyer, R.B., Gamper, H.B. (1995) Sequence-specific covalent modification of DNA by cross-linking oligonucleotides. Catalysis by RecA and implication for the mechanism of synaptic joint formation, *Biochemistry* **34**, 13098–13108.

Porteus, M.H., Baltimore, D. (2003) Chimeric nucleases stimulate gene targeting in human cells, *Science* **300**, 763.

Porteus, M.H., Cathomen, T., Weitzman, M.D., Baltimore, D. (2003) Efficient gene targeting mediated by AAV and DNA double strand breaks, *Mol. Cell Biol.* **23**, 3558–3565.

Porumb, H., Gousset, H., Letellier, R., Salle, V., Briane, D., Vassy, J., Amor-Gueret, M., Israel, L., Taillandier, E. (1996) Temporary ex vivo inhibition of the expression of the human oncogene HER2 (NEU) by a triple helix-forming oligonucleotide, *Cancer Res.* **56**, 515–522.

Puri, N., Majumdar, A., Cuenoud, B., Natt, F., Martin, P., Boyd, A., Miller, P.S., Seidman, M.M. (2002) Minimum number of 2'-O-(2-aminoethyl) residues required for gene knockout activity by triple helix forming oligonucleotides, *Biochemistry* **41**, 7716–7724.

Rapozzi, V., Cogoi, S., Spessotto, P., Risso, A., Bonora, G.M., Quadrifoglio, F., Xodo, L.E. (2002) Antigene effect in k562 cells of a peg-conjugated triplex-forming oligonucleotide targeted to the bcr/abl oncogene, *Biochemistry* **41**, 502–510.

Rayssiguier, C., Thaler, D.S., Radman, M. (1989) The barrier to recombination between Escherichia coli and Salmonella typhimurium is disrupted in mismatch-repair mutants, *Nature* **342**, 396–401.

Rice, K.P., Chaput, J.C., Cox, M.M., Switzer, C. (2000) RecA protein promotes strand exchange with DNA substrates containing isoguanine and 5-methyl isocytosine, *Biochemistry* **39**, 10177–10188.

Richardson, C., Moynahan, M.E., Jasin, M. (1998) Double-strand break repair by interchromosomal recombination: suppression of chromosomal translocations, *Genes Dev.* **12**, 3831–3842.

Roberts, R.W., Crothers, D.M. (1992) Stability and properties of double and triple helices: dramatic effects of RNA or DNA backbone composition, *Science* **258**, 1463–1466.

Rogers, F.A., Vasquez, K.M., Egholm, M., Glazer, P.M. (2002) Site-directed recombination via bifunctional PNA-DNA conjugates, *Proc. Natl. Acad. Sci. U.S.A.* **99**, 16695–16700.

Rommerskirch, W., Graeber, I., Grassmann, M., Grassmann, A. (1988) Homologous recombination of SV40 DNA in COS7 cells occurs with high frequency in a gene dose independent fashion, *Nucleic Acids Res.* **16**, 941–952.

Russell, D.W., Hirata, R.K. (1998) Human gene targeting by viral vectors, *Nat. Genet.* **18**, 325–330.

Sangiuolo, F., Bruscia, E., Serafino, A., Nardone, A., Bonifazi, E., Lais, M., Gruenert, D.C., Novelli, G. (2002) In vitro correction of cystic fibrosis epithelial cell lines by small fragment homologous replacement (SFHR) technique, *BMC. Med. Genet.* **3**, 8–21.

Sargent, R.G., Brenneman, M.A., Wilson, J.H. (1997) Repair of site-specific double-strand breaks in a mammalian chromosome by homologous and illegitimate recombination, *Mol. Cell Biol.* **17**, 267–277.

Sargent, R.G., Wilson, J.H. (1998) Recombination and gene targeting in mammalian cells, *Curr. Res. Mol. Ther.* **1**, 584–692.

Scheerer, J.B., Adair, G.M. (1994) Homology dependence of targeted recombination at the Chinese hamster APRT locus, *Mol. Cell Biol.* **14**, 6663–6673.

Seeman, N.C., Rosenberg, J.M., Rich, A. (1976) Sequence-specific recognition of double helical nucleic acids by proteins, *Proc. Natl. Acad. Sci. U.S.A.* **73**, 804–808.

Seidman, M.M., Bredberg, A., Seetharam, S., Kraemer, K.H. (1987) Multiple point mutations in a shuttle vector propagated in human cells: evidence for an error-prone DNA polymerase activity, *Proc. Natl. Acad. Sci. U.S.A.* **84**, 4944–4948.

Selva, E.M., New, L., Crouse, G.F., Lahue, R.S. (1995) Mismatch correction acts as a barrier to homeologous recombination in Saccharomyces cerevisiae, *Genetics* **139**, 1175–1188.

Shimizu, M., Konishi, A., Shimada, Y., Inoue, H., Ohtsuka, E. (1992) Oligo(2'-O-methyl)ribonucleotides. Effective probes for duplex DNA, *FEBS Lett.* **302**, 155–158.

Shulman, M.J., Nissen, L., Collins, C. (1990) Homologous recombination in hybridoma cells: dependence on time and fragment length, *Mol. Cell Biol.* **10**, 4466–4472.

Smih, F., Rouet, P., Romanienko, P.J., Jasin, M. (1995) Double-strand breaks at the target locus stimulate gene targeting in embryonic stem cells, *Nucleic Acids Res.* **23**, 5012–5019.

Smith, J., Bibikova, M., Whitby, F.G., Reddy, A.R., Chandrasegaran, S., Carroll, D. (2000) Requirements for double-strand cleavage by chimeric restriction enzymes with zinc finger DNA-recognition domains, *Nucleic Acids Res.* **28**, 3361–3369.

Smithies, O., Gregg, R.G., Boggs, S.S., Koralewski, M.A., Kucherlapati, R.S. (1985) Insertion of DNA sequences into the human chromosomal beta-globin locus by homologous recombination, *Nature* **317**, 230–234.

Smolik-Utlaut, S., Petes, T.D. (1983) Recombination of plasmids into the Saccharomyces cerevisiae chromosome is reduced by small amounts of sequence heterogeneity, *Mol. Cell Biol.* **3**, 1204–1211.

Song, K.Y., Schwartz, F., Maeda, N., Smithies, O., Kucherlapati, R. (1987) Accurate modification of a chromosomal plasmid by homologous recombination in human cells, *Proc. Natl. Acad. Sci. U.S.A.* **84**, 6820–6824.

Sonoda, E., Sasaki, M.S., Buerstedde, J.M., Bezzubova, O., Shinohara, A., Ogawa, H., Takata, M., Yamaguchi-Iwai, Y., Takeda, S. (1998) Rad51-deficient vertebrate cells accumulate chromosomal breaks prior to cell death, *EMBO J.* **17**, 598–608.

Stasiak, A., Di Capua, E., Koller, T. (1981) Elongation of duplex DNA by recA protein, *J. Mol. Biol.* **151**, 557–564.

Stein, C.A. (1997) Controversies in the cellular pharmacology of oligodeoxynucleotides, *Antisense Nucleic Acid Drug Dev.* **7**, 207–209.

Supekova, L., Pezacki, J.P., Su, A.I., Loweth, C.J., Riedl, R., Geierstanger, B., Schultz, P.G., Wemmer, D.E. (2002) Genomic effects of polyamide/DNA interactions on mRNA expression, *Chem. Biol.* **9**, 821–827.

Szostak, J.W., Wu, R. (1979) Insertion of a genetic marker into the ribosomal DNA of yeast, *Plasmid* **2**, 536–554.

Takasugi, M., Guendouz, A., Chassignol, M., Decout, J.L., Lhomme, J., Thuong, N.T., Helene, C. (1991) Sequence-specific photo-induced cross-linking of the two strands of double-helical DNA by a psoralen covalently linked to a triple helix-forming oligonucleotide, *Proc. Natl. Acad. Sci. U.S.A.* **88**, 5602–5606.

Takata, M., Sasaki, M.S., Sonoda, E., Morrison, C., Hashimoto, M., Utsumi, H., Yamaguchi-Iwai, Y., Shinohara, A., Takeda, S. (1998) Homologous recombination and non-homologous end-joining pathways of DNA double-strand break repair have overlapping roles in the maintenance of chromosomal integrity in vertebrate cells, *EMBO J.* **17**, 5497–5508.

Taubes, G. (2002a) Gene therapy. Pioneering papers under the microscope, *Science* **298**, 2118–2119.

Taubes, G. (2002b) Gene therapy. The strange case of chimeraplasty, *Science* **298**, 2116–2120.

te Riele H., Maandag, E.R., Berns, A. (1992) Highly efficient gene targeting in embryonic stem cells through homologous recombination with isogenic DNA constructs, *Proc. Natl. Acad. Sci. U.S.A.* **89**, 5128–5132.

Thomas, K.R., Capecchi, M.R. (1987) Site-directed mutagenesis by gene targeting in mouse embryo-derived stem cells, *Cell* **51**, 503–512.

Thomas, K.R., Folger, K.R., Capecchi, M.R. (1986) High frequency targeting of genes to specific sites in the mammalian genome, *Cell* **44**, 419–428.

Torigoe, H., Hari, Y., Sekiguchi, M., Obika, S., Imanishi, T. (2001) 2′-O,4′-C-methylene bridged nucleic acid modification promotes pyrimidine motif triplex DNA formation at physiological pH: thermodynamic and kinetic studies, *J. Biol. Chem.* **276**, 2354–2360.

Trauger, J.W., Baird, E.E., Dervan, P.B. (1998) Recognition of 16 base pairs in the minor groove of DNA by a pyrrole–imidazole polyamide dimer. *J. Am. Chem. Soc.* **120**, 3534–3535.

Trauger, J.W., Baird, E.E., Mrksich, M., Dervan, P.B. (1996a) Extension of sequence specific recognition in the minor groove of DNA by pyrrole–imidazole polyamides to 9–13 base pairs, *J. Am. Chem. Soc.* **118**, 6160–6166.

Trauger, K.W., Baird, E.E., Dervan, P.B. (1996b) Recognition of DNA by designed ligands at subnanomolar concentrations, *Nature* **382**, 559–561.

Tzung, T.Y., Runger, T.M. (1998) Reduced joining of DNA double strand breaks with an abnormal mutation spectrum in rodent mutants of DNA-PKcs and Ku80, *Int. J. Radiat. Biol.* **73**, 469–474.

van der Steege, G., Schuilenga-Hut, P.H., Buys, C.H., Scheffer, H., Pas, H.H., Jonkman, M.F. (2001) Persistent failures in gene repair, *Nat. Biotechnol.* **19**, 305–306.

Vasquez, K.M., Christensen, J., Li, L., Finch, R.A., Glazer, P.M. (2002) Human XPA and RPA DNA repair proteins participate in specific recognition of triplex-induced helical distortions, *Proc. Natl. Acad. Sci. U.S.A.* **99**, 5848–5853.

Vasquez, K.M., Dagle, J.M., Weeks, D.L., Glazer, P.M. (2001a) Chromosome targeting at short polypurine sites by cationic triplex-forming oligonucleotides, *J. Biol. Chem.* **276**, 38536–38541.

Vasquez, K.M., Narayanan, L., Glazer, P.M. (2000) Specific mutations induced by triplex-forming oligonucleotides in mice, *Science* **290**, 530–533.

Vasquez, K.M., Wang, G., Havre, P.A., Glazer, P.M. (1999) Chromosomal mutations induced by triplex-forming oligonucleotides in mammalian cells, *Nucleic Acids Res.* **27**, 1176–1181.

Vasquez, K.M., Wensel, T.G., Hogan, M.E., Wilson, J.H. (1995) High-affinity triple helix formation by synthetic oligonucleotides at a site within a selectable mammalian gene, *Biochemistry* **34**, 7243–7251.

Vasquez, K.M., Wilson, J.H. (1998) Triplex-directed modification of genes and gene activity, *Trends Biochem. Sci.* **23**, 4–9.

Wake, C.T., Wilson, J.H. (1979) Simian virus 40 recombinants are produced at high frequency during infection with genetically mixed oligomeric DNA, *Proc. Natl. Acad. Sci. U.S.A.* **76**, 2876–2880.

Wake, C.T., Wilson, J.H. (1980) Defined oligomeric SV40 DNA: a sensitive probe of general recombination in somatic cells, *Cell* **21**, 141–148.

Waldman, A.S., Liskay, R.M. (1988) Dependence of intrachromosomal recombination in mammalian cells on uninterrupted homology, *Mol. Cell Biol.* **8**, 5350–5357.

Wang, G., Levy, D.D., Seidman, M.M., Glazer, P.M. (1995) Targeted mutagenesis in mammalian cells mediated by intracellular triple helix formation, *Mol. Cell Biol.* **15**, 1759–1768.

Wang, G., Seidman, M.M., Glazer, P.M. (1996) Mutagenesis in mammalian cells induced by triple helix formation and transcription-coupled repair, *Science* **271**, 802–805.

Wang, G., Xu, X., Pace, B., Dean, D.A., Glazer, P.M., Chan, P., Goodman, S.R., Shokolenko, I. (1999) Peptide nucleic acid (PNA) binding-mediated induction of human gamma-globin gene expression, *Nucleic Acids Res.* **27**, 2806–2813.

Wartell, R.M., Larson, J.E., Wells, R.D. (1974) Netropsin. A specific probe for A-T regions of duplex deoxyribonucleic acid, *J. Biol. Chem.* **249**, 6719–6731.

White, S., Szewczyk, J.W., Turner, J.M., Baird, E.E., Dervan, P.B. (1998) Recognition of the four Watson-Crick base pairs in the DNA minor groove by synthetic ligands [see comments], *Nature* **391**, 468–471.

Wilson, J.H. (1979) Nick-free formation of reciprocal heteroduplexes: a simple solution to the topological problem, *Proc. Natl. Acad. Sci. U.S.A.* **76**, 3641–3645.

Wilson, J.H., Leung, W.Y., Bosco, G., Dieu, D., Haber, J.E. (1994) The frequency of gene targeting in yeast depends on the number of target copies, *Proc. Natl. Acad. Sci. U.S.A.* **91**, 177–181.

Wolfe, S.A., Greisman, H.A., Ramm, E.I., Pabo, C.O. (1999) Analysis of zinc fingers optimized via phage display: evaluating the utility of a recognition code, *J. Mol. Biol.* **285**, 1917–1934.

Wurtz, N.R., Pomerantz, J.L., Baltimore, D., Dervan, P.B. (2002) Inhibition of DNA binding by NF-kappa B with pyrrole–imidazole polyamides, *Biochemistry* **41**, 7604–7609.

Xodo, L.E., Manzini, G., Quadrifoglio, F., van der Marel, G.A., van Boom, J.H. (1991) Effect of 5-methylcytosine on the stability of triple-stranded DNA – a thermodynamic study, *Nucleic Acids Res.* **19**, 5625–5631.

Yamamoto, A., Taki, T., Yagi, H., Habu, T., Yoshida, K., Yoshimura, Y., Yamamoto, K., Matsushiro, A., Nishimune, Y., Morita, T. (1996) Cell cycle-dependent expression of the mouse Rad51 gene in proliferating cells, *Mol. Gen. Genet.* **251**, 1–12.

Yanez, R.J., Porter, A.C. (1999) Gene targeting is enhanced in human cells overexpressing hRAD51, *Gene Ther.* **6**, 1282–1290.

Yanez, R.J., Porter, A.C. (2002) Differential effects of Rad52p overexpression on gene targeting and extrachromosomal homologous recombination in a human cell line, *Nucleic Acids Res.* **30**, 740–748.

Yu, D., Ellis, H.M., Lee, E.C., Jenkins, N.A., Copeland, N.G., Court D.L. (2000) An efficient recombination system for chromosome engineering in Escherichia coli, *Proc. Natl. Acad. Sci. U.S.A.* **97**, 5978–5983.

Zheng, H., Hasty, P., Brenneman, M.A., Grompe, M., Gibbs, R.A., Wilson, J.H., Bradley, A. (1991) Fidelity of targeted recombination in human fibroblasts and murine embryonic stem cells, *Proc. Natl. Acad. Sci. U.S.A.* **88**, 8067–8071.

Zheng, H., Wilson, J.H. (1990) Gene targeting in normal and amplified cell lines, *Nature* **344**, 170–173.

Zhou, X., Adzuma, K. (1997) DNA strand exchange mediated by the Escherichia coli RecA protein initiates in the minor groove of double-stranded DNA, *Biochemistry* **36**, 4650–4661.

Zimmer, A., Gruss, P. (1989) Production of chimaeric mice containing embryonic stem (ES) cells carrying a homeobox Hox 1.1 allele mutated by homologous recombination, *Nature* **338**, 150–153.

28
Genetic Engineering of Antibody Molecules

Manuel L. Penichet and Sherie L. Morrison
Department of Microbiology, Immunology, and Molecular Genetics,
University of California, Los Angeles, CA, USA

1	**Antibody Structure and Engineering**	**809**
1.1	The Basic Structure of Antibodies	809
1.2	Classes and Subclasses of Antibodies	810
2	**From Mouse to Human Antibodies**	**812**
2.1	Murine Monoclonal Antibodies	812
2.2	Chimeric Antibodies	814
2.3	Humanized Antibodies	815
2.4	Human Monoclonal Antibodies in Mice	815
3	***In Vitro* Antibody Production by Phage Libraries**	**816**
4	**Further Genetic Modifications of Antibodies**	**818**
4.1	Engineering Antibody Fragments: Monovalent, Bivalent, and Multivalent scFvs	818
4.2	Bispecific Antibodies	820
4.3	Polymers of Monomeric Antibodies	821
4.4	Antibody Fusion Proteins	822
5	**Expression Systems**	**824**
6	**Conclusion**	**824**
	Bibliography	**825**
	Books and Reviews	825
	Primary Literature	825

Genomics and Genetics. Edited by Robert A. Meyers.
Copyright © 2007 Wiley-VCH Verlag GmbH & Co. KGaA, Weinheim
ISBN: 978-3-527-31609-0

Keywords

Antibody/Antigen
Antibodies are proteins known as *immunoglobulins* (Igs), which are produced by the immune system in response to the presence of a foreign substance. Antigens are molecules (usually proteins) on the surface of cells, viruses, fungi, bacteria, and some nonliving substances such as toxins, chemicals, drugs, and foreign particles, which are recognized and specifically bound by antibodies. The immune system recognizes immunogenic antigens (also known as *immunogens*), and produces antibodies that destroy or neutralize substances or organisms containing antigens.

Antibody-dependent Cellular Cytotoxicity (ADCC)
Cell-killing reaction in which the Fc receptor bearing killer cells recognizes target cells via specific antibodies.

Bacteriophage
Bacteriophages or phages are viruses that infect bacteria. Bacteriophage λ is a temperate phage, which can grow lytically lysing the bacteria and forming a clear plaque on a lawn of bacteria. Infection with filamentous phages such as M13 is not lethal and the host bacteria do not lyse. Instead, their rate of growth slows and they form turbid plaques on the bacterial lawn.

Complement
Group of serum proteins participating in the lysis of foreign cells and pathogens [a process known as *complement-dependent cytotoxicity* (CDC)]; they also play an important role in phagocytosis.

Constant Region
Portion of the antibody molecule exhibiting little variation and determining the isotype (class or subclass) of the antibody.

Fab
The Fab fragment is a monovalent antigen-binding fragment of an antibody that consists of one light chain and part of one heavy chain (the variable region and the first constant region domain). It can be obtained by digestion of intact antibody with papain or by genetic engineering techniques.

Fc
The Fc is a nonantigen binding fragment of an antibody that consists of the carboxy-terminal portion of both heavy chains. It can be obtained by papain digestion of an intact antibody. Two Fabs and one Fc fragment comprise a complete IgG antibody.

Fv/scFv
The Fv is a monovalent antigen-binding fragment of an antibody composed of the variable regions from the heavy and light chains. scFv fragments are composed of the

variable domains of heavy and light chains (V_L and V_H) joined by a synthetic flexible linker peptide. Thus, the scFv provides a fully functional antigen binding domain that is encoded in a single gene and expressed as a single polypeptide.

Hybridoma
Cell derived by a fusion between a normal cell, usually a lymphocyte, and a tumor cell, usually a myeloma cell.

Variable Region
Variable portion of the antibody molecule that is responsible for antigen binding.

> Antibodies have long been appreciated for their exquisite specificity. With the development of the hybridoma technology, it was possible to produce rodent (mouse or rat) monoclonal antibodies that are the product of a single clone of antibody-producing cells and have only one antigen-binding specificity. Advances in genetic engineering and expression systems have been applied to overcome problems of immunogenicity of rodent-produced antibodies and to improve their ability to trigger human immune effector mechanisms. The production of chimeric, humanized, and totally human antibodies as well as antibodies with novel structures and functional properties has resulted in improved monoclonal antibodies. As a consequence, recombinant antibody-based therapies are now used to treat a variety of diverse conditions that include infectious diseases, inflammatory disorders, and cancer. This article summarizes and compares different strategies for developing recombinant antibodies and their derivatives.

1
Antibody Structure and Engineering

1.1
The Basic Structure of Antibodies

Antibodies are molecules with multiple properties that make them a critical component of the immune system. These properties include the ability to recognize a vast array of different molecules known as *antigens* and to interact with and activate the host effector systems.

The basic structure of all antibodies, also known as *immunoglobulins* (Igs), is a unit consisting of two identical light polypeptide chains and two identical heavy polypeptide chains linked together by disulfide bonds (Fig. 1). Heavy and light chains are encoded by separate genes and are organized into discrete globular domains separated by short peptide segments. The amino-terminus end of both heavy and light chains is the antigen binding site and consists of one domain characterized by sequence variability (variable region or V) in both the heavy and light chains, called the V_H and V_L regions respectively. The rest of the molecule has a relatively constant (C) structure. The

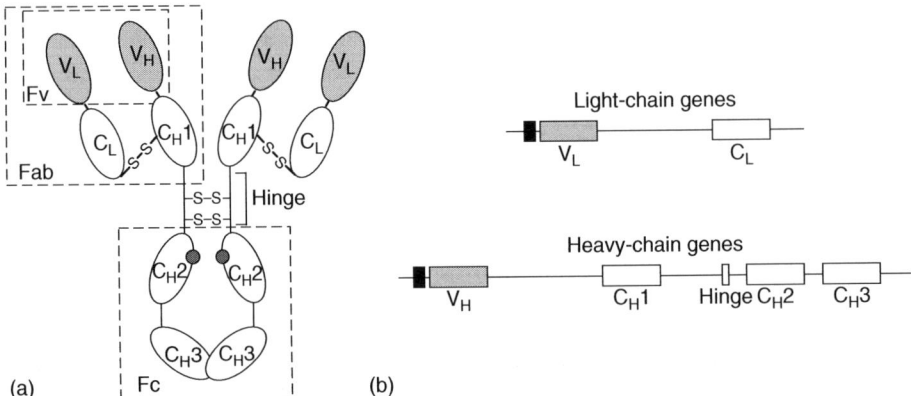

Fig. 1 (a) Diagram of an immunoglobulin G (IgG) molecule (the most abundant antibody in serum) and the active fragments that can be derived from it. The antibody molecule is divided into discrete functional domains: two domains constitute the light chain (V_L and C_L), while four domains make up the heavy chain (V_H, C_H1, C_H2, and C_H3). The variable region domains make the antibody binding site and are designated as the Fv region. The effector functions of the antibody are properties of the constant region domains. The carbohydrate units (black circles) present within the C_H2 domains contribute to the functional properties of the antibody. The hinge region provides flexibility to the antibody molecule, facilitating antigen binding and some effector functions. The enzyme papain cleaves the antibody into two Fab fragments containing the antigen binding sites and an Fc fragment responsible for the effector functions. (b) Genes that encode the heavy and light chains. In the genes, each domain is encoded by a discrete exon (indicated by boxes) separated by intervening sequences (introns) indicated by the line; the intervening sequences are present in the primary transcript but are removed from the mature mRNA by splicing. Both heavy and light chains contain hydrophobic leader sequences (indicated by the black exon) necessary for their secretion. This leader sequence is present in the newly synthesized heavy and light chains but is cleaved from them after they enter the endoplasmic reticulum and, therefore, is not present in the mature chains.

constant region of the light chain is termed the C_L region. The constant region of the heavy chain is further divided into three structural domains stabilized by intrachain disulfide bonds: C_H1, C_H2, C_H3 (Fig. 1). The C_H3 domain of the heavy chain represents its carboxy-terminus. The domain structure of the antibodies is very important for genetic engineering because it facilitates protein engineering, allowing the exchange between molecules of functional domains carrying antigen-binding activities (Fabs or Fvs) or effector functions (Fc).

The hinge region, a segment of heavy chain between the C_H1 and C_H2 domains, provides flexibility in the molecule. Papain digestion of IgG yields two Fab fragments and one Fc fragment. The Fab region binds antigen, while the Fc region mediates effector functions such as complement activation, antibody-dependent cellular cytotoxicity (ADCC), and placental transmission. All antibodies are glycoproteins, and the carbohydrate present in the constant region has been shown to be essential for many of its effector functions.

1.2
Classes and Subclasses of Antibodies

The constant region of the heavy chain determines the class or isotype of the

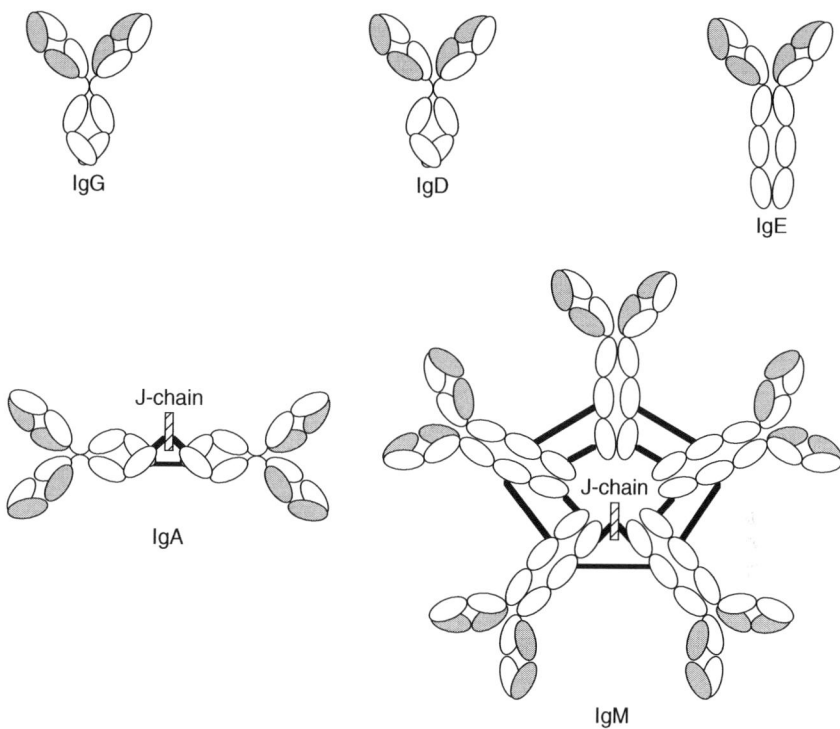

Fig. 2 The five classes of antibodies: IgG, IgD, and IgE are monomeric antibodies; IgA and IgM are polymeric antibodies.

antibody. Figure 2 shows the five different classes of antibodies, which differ in the constant region of their respective heavy chains. The different heavy-chain isotypes that define these classes of antibodies are designated by lower-case Greek letters: γ for IgG, δ for IgD, ε for IgE, α for IgA, and μ for IgM. IgM and IgE heavy chains contain an extra C_H domain (C_H4) and lack the hinge region found in IgG, IgD, and IgA. However, the absence of a hinge region does not imply that IgM and IgE lack flexibility; electron micrographs of IgM molecules binding to ligands have shown that the Fab arms can bend, relative to the Fc fragment. There are also two different light-chain isotypes, which are designated by the lower-case Greek letters κ and λ. Light chains of both isotypes are found associated with all the heavy-chain isotypes.

In humans, there is only one class of IgD, IgE, and IgM with monomeric structures having a molecular weight (MW) of around 190 kDa. However, human IgA antibodies can be further subdivided in two subclasses (IgA1 and IgA2) with a molecular weight of around 160 kDa for each monomer, and human IgG antibodies can be subdivided into four subclasses: IgG1, IgG2, IgG4 (all three with a molecular weight of around 150 kDa) and IgG3 with a molecular weight of around 165 kDa. The higher molecular weight exhibited by human IgG3 is due to the presence of an extended hinge region, which provides extraordinary flexibility. In fact, IgG3 is the most flexible human IgG.

Although all antibody molecules are constructed from the basic unit of two heavy and two light chains (H_2L_2), both IgA and IgM form polymers (Fig. 2). IgA forms a dimeric structure in which two H_2L_2 units are joined by a J-chain [$(H_2L_2)_2$], and IgM forms either a pentameric structure with five H_2L_2 units joined by a J-chain [$(H_2L_2)_5$J] or a hexameric structure [$(H_2L_2)_6$], which does not have the J-chain (not shown). IgM and IgA heavy-chain constant regions contain a "tailpiece" of 18 amino acids that contains a cysteine residue essential for polymerization. The J-chain is a 15 kDa polypeptide produced by B-lymphocytes and plasma cells (the same cells that produce the antibodies) that promotes polymerization by linking to the cysteine of the tailpiece (Fig. 2). Differences in the isotype of the heavy chain determine the number of carbohydrate units and the ability to engage in various effector functions such as complement activation, ADCC, and placental transmission. Differences in the isotype of the light chain do not appear to significantly influence the structure or the effector functions of the antibody molecule.

2
From Mouse to Human Antibodies

2.1
Murine Monoclonal Antibodies

During the normal immune response, a wide variety of antibodies is produced. These antibodies, known as *polyclonal antibodies* (i.e. they are the product of many different antibody-producing cells), include antibodies with different variable regions as well as antibodies with the same variable regions associated with different constant regions. Rarely do different individuals mount an identical immune response. This heterogeneity in the immune response plus ethical and safety concerns has made it difficult to use polyclonal antibodies for many applications.

A significant breakthrough was made when it became possible to produce stable cell lines that synthesize a single homogeneous antibody (Fig. 3). By fusing a normal B-cell from the splenocytes of an immunized animal (initially a mouse or a rat) with a myeloma cell, it is possible to generate a "hybridoma," which possesses the immortality of the myeloma cell and secretes the antibody characteristic of the normal B-cell. Antibodies produced by hybridomas are monoclonal (i.e. they are the product of a single antibody-producing cell) and therefore have a single variable region associated with only one constant region. The immortality of the hybridoma ensures the continued availability of a well-characterized antibody. Once a hybridoma cell line is developed, it can be grown *in vitro* or *in vivo* for large-scale production of the monoclonal antibody, or it can be used to clone the variable regions of the monoclonal antibody for genetic engineering purposes.

Owing to their high affinity and exquisite specificity, murine monoclonal antibodies seemed to be the ideal "magic bullets" for diagnosis or therapy of multiple diseases including cancer. However, the progression of "magic bullets" from dream to reality has been slow because mouse (murine) monoclonal antibodies are not the ideal agents to be administered into a human. Murine monoclonal antibodies compared to human antibodies require more frequent dosing to maintain a therapeutic level of monoclonal antibodies because of a shorter circulating half-life in humans than human antibodies. In addition, the

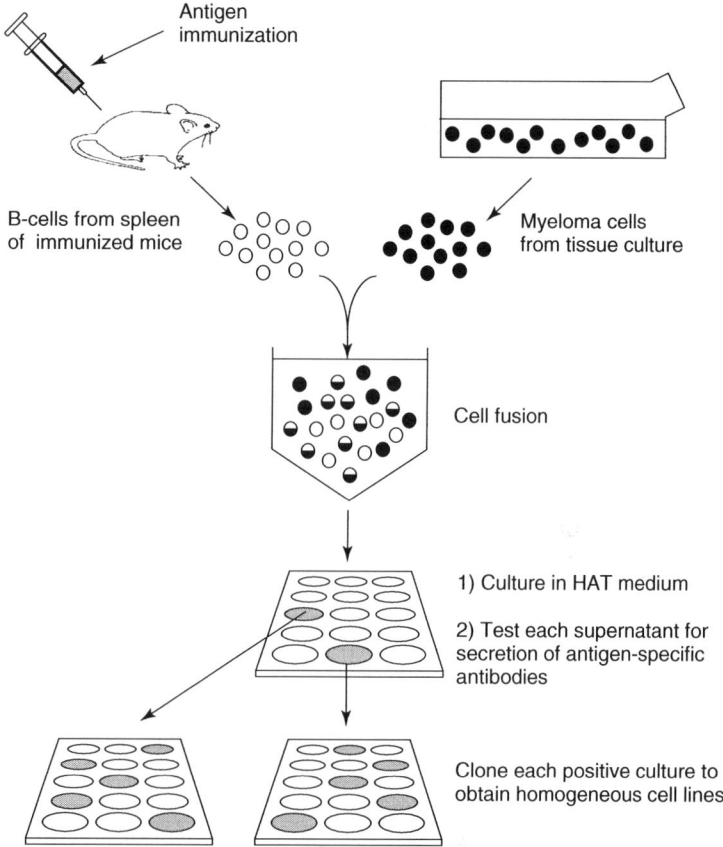

Fig. 3 Production of monoclonal antibodies. Mice are immunized with the antigen of interest and spleen cells are isolated and then fused with myeloma cells. HAT (hypoxanthine, aminopterin, and thymidine) medium is used to separate unfused myeloma cells from fused hybridoma cells. It takes advantage of the fact that normal mammalian cells can synthesize nucleotides by both a *de novo* and a salvage pathway, while the myeloma cells used have a defect in the salvage pathway. When the *de novo* pathway is blocked by aminopterin, cells must then utilize the enzymes of the salvage pathway. Thus, in the presence of aminopterin, unfused myeloma cells will die. Normal spleen cells do not grow. Only cells that have acquired the enzymes of the salvage pathway from the normal cells and the capacity for continuous growth from the myeloma cell will survive. Supernatants of hybridomas are screened for the presence of the antibody with the desired specificity, and positive cell cultures are subcloned to obtain a homogeneous cell line.

human immune system recognizes the mouse protein as foreign, generating a human antimouse antibody (HAMA) response, which results in an even more rapid clearance of the murine antibody (rendering the therapeutic useless) and, in some cases, in a severe allergic reaction. Moreover, murine constant regions can be ineffective in interacting with the human immune effector system. These problems would be overcome by producing human monoclonal antibodies using human B-cells from immunized human donors. However, human monoclonal antibodies are difficult to produce using the hybridoma technology originally designed for the production of murine antibodies: the cell lines are unstable and frequently produce antibodies of the IgM isotype, which have low affinity for the antigen and are difficult to purify and handle. In addition, there are ethical and safety problems in obtaining humans immunized with certain antigens.

2.2
Chimeric Antibodies

To overcome the problems associated with the administration of murine monoclonal antibodies to humans, protein engineering has been used to convert murine monoclonal antibodies to mouse/human chimeric antibodies by genetically fusing the mouse variable regions to the human constant regions (Fig. 4), a procedure that is facilitated by the structure of antibodies.

Initially, variable region genes were obtained from genomic or cDNA libraries produced using DNA or mRNA from antibody-producing cells. More recently, cloning of genes encoding for the antibody variable region has been greatly facilitated by polymerase chain reaction (PCR)-based procedures. The variable region consists of hypervariable complementarity determining regions CDRs that are responsible for antibody specificity supported by relatively conserved framework regions. A limited number of different hydrophobic leader sequences are also found associated with the different variable regions. Therefore, it is possible to design sets of oligonucleotide primers on the basis of either the framework or the leader regions that will bind to virtually all mouse variable regions. A limited number of primers are required for all constant regions. Using upstream consensus primers for framework or leader regions in light- and heavy-chain variable regions and downstream primers for the constant regions, PCR can be used to amplify cDNAs generated by reverse-transcription (RT-PCR) directly from hybridoma mRNA.

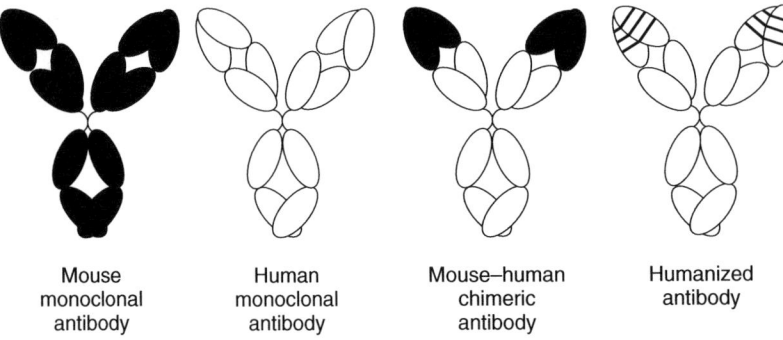

Mouse monoclonal antibody
Human monoclonal antibody
Mouse–human chimeric antibody
Humanized antibody

Fig. 4 Schematic representation of a murine monoclonal antibody, a human monoclonal antibody, a mouse–human chimeric antibody, and a CDR-grafted or humanized antibody. Chimeric antibodies have murine derived variable regions and binding specificities joined to human constant regions with their corresponding effector functions. Humanized antibodies are composed mostly of human sequences except for the areas in contact with the antigen (CDR regions), which are derived from mouse sequences.

This product is cloned and sequenced, and then used for the construction of chimeric antibodies or for further genetic modifications, as described below.

For the most part, chimeric antibodies retain their target specificity and show reduced HAMA responses. An example of a successful mouse–human chimeric antibody approved for clinical use is Rituximab (Rituxan/Mabthera), which targets the CD20 antigen and is now widely used to treat non-Hodgkin's lymphoma.

2.3
Humanized Antibodies

Although mouse–human chimeric antibodies are less immunogenic than mouse antibodies, in some cases they can still elicit a significant human antichimeric antibody (HACA) response. One approach to overcome this problem is to further manipulate the antibody variable region encoding for the antigen binding site resulting in humanized antibodies (Fig. 4). Each variable domain consists of a β-barrel with seven antiparallel β-strands connected by loops. Among the loops are the CDR regions. It is feasible to move the CDRs and their associated specificity from one scaffolding β-barrel to another, thereby creating "CDR-grafted" or "humanized" antibodies. However, it is rarely sufficient to move only the CDRs from a murine antibody onto a completely human framework because the resulting antibody frequently has reduced or no binding activity. In these cases, additional mouse residues near the CDRs are incorporated until binding is restored.

An example of a successful humanized monoclonal antibody that has been approved for clinical use is Trastuzumab (Herceptin), which has demonstrated significant antitumor activity in patients affected with breast cancer overexpressing the tumor-associated antigen HER2/neu.

2.4
Human Monoclonal Antibodies in Mice

Recently, mice have been produced that make antigen-specific antibodies that are totally human. To accomplish this goal, transgenic mice carrying portions of the human IgH and Igκ loci [in germ line configuration using megabase-sized YACs (yeast artificial chromosome)] were obtained that included the majority of the variable region repertoire, the genes for Cμ, Cδ and Cγ1, Cγ2, or Cγ4, as well as the *cis* elements required for their function. The IgH and Igκ transgenic was then bred into a genetic background deficient in the production of murine immunoglobulin. Therefore, the resulting mouse model, named *XenoMouse*, has elements of the human heavy- and light-chain loci in a murine context in which endogenous mouse heavy and light chains were disrupted.

The large and complex human variable region repertoire encoded on the immunoglobulin transgenes in XenoMouse strains support the development of large peripheral B-cell compartments and the generation of a diverse primary immune repertoire similar to that of adult humans. The human genes are compatible with mouse enzymes mediating class switching from IgM to IgG as well as somatic hypermutation and affinity maturation. Importantly, the immune system of the XenoMouse recognizes human antigens as foreign, with a concomitant strong human humoral immune response. The use of XenoMouse mice in conjunction with well-established hybridoma procedures (see Fig. 3) reproducibly results in human IgG monoclonal antibodies with high affinity for human antigens and suitable for repeated administration to humans. To date, these engineered mice can produce only some of the human isotypes, but further genetic modification of these mice promises to expand their potential.

3
In Vitro Antibody Production by Phage Libraries

The production of monoclonal antibodies frequently requires several injections of antigen, which can take weeks to months. Hybridoma construction and screening require additional time. In addition, the immune response is often biased toward certain "immunodominant" epitopes of the antigen, making it difficult or impossible to produce monoclonal antibodies with the specificity desired for a particular use. Moreover, the production of antibodies against antigens conserved among species may be difficult or impossible. Phage display, a technology that allows the expression of immunoglobulin genes in bacteriophages (viruses that infect bacteria) without the need for developing hybridomas, provides an alternative approach that overcomes many of these problems.

In phage display, the heavy and light V-gene obtained from the spleen of an immunized mouse or from the peripheral blood of a naive or immunized donor are expressed as Fab or single-chain Fv (scFv) on the surface of filamentous phages (f1, M13, and fd). Phage libraries can be generated using variable antibody gene repertoires from any species including humans, or even synthetic sequences. Figure 5 shows the development of scFv libraries using a human donor.

Both scFv and Fab fragments can be expressed on the surface of filamentous bacteriophages (f1, M13, and fd) as either single or multiple copies of the antibody of interest depending on the phage protein used for fusion. Expression of multiple copies facilitates the identification and isolation of low-affinity antibodies. When functional antibody V domains are displayed on the surface of filamentous phages, the resulting phages bind specifically to antigen, and rare phages can be isolated on the basis of their ability to bind antigen. Multiple rounds of enrichment consisting of binding to immobilized antigen, expanding the bound phage, and further enriching by again binding to immobilized antigen can yield specific phages even if the desired specificity was present on less than 1 in 10^6 of the original phages in the library. The selected variable regions will generally have affinities similar to monoclonal antibodies and can be expressed as antibody fragments in the bacterium *Escherichia coli* or they can be used to produce complete antibodies and can be expressed in mammalian hosts. In addition, specific variable regions can be mutagenized and phages that express variable regions with increased affinity can be selected. The bacteriophage expression systems are designed to allow the genes encoding for heavy and light chains to undergo random combinations, which are tested for their ability to bind the desired antigen.

The decision of whether to produce scFv or Fab libraries depends partly on the intended use. The single gene format of the scFv is an advantage for the construction of fusion proteins such as "immunotoxins" (antibodies or antibody fragments fused to a toxin) or for targeted gene therapy approaches where the *scFv* gene is fused to a viral envelope protein gene. The use of scFv also appears to be a better option for "intracellular

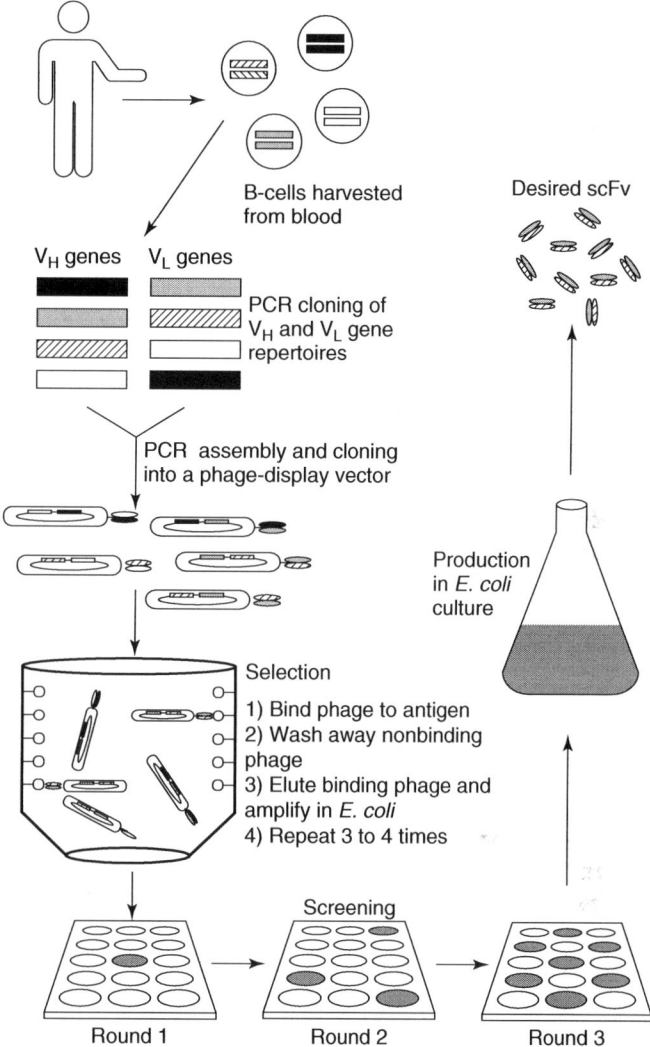

Fig. 5 *In vitro* human antibody production using phage libraries (phage-display technology). Peripheral blood mononuclear cells are harvested from a nonimmunized or immunized human donor, and the heavy- and light-chain V-gene regions (V_L and V_H) are amplified by PCR and assembled as a single-chain Fv region (scFv). Note that the original heavy- and light-chain pairings become scrambled during scFv assembly. The scFv genes are cloned into filamentous bacteriophages, where the encoded scFvs are displayed in a functional form on the phage surface. Multiple rounds of selection with a solid-phase antigen allow the isolation of even rare phages from the original library. The selected scFv can be expressed in *E. coli* (The illustration was adapted from Powers D.B. and Marks J.D. Monovalent Phage Display of Fab and scFv fusions. In *Antibody fusion proteins*. John Wiley & Son, Inc., New York, 1999).

immunization," a procedure in which the gene encoding the antibody binding site is delivered intracellularly to achieve phenotypic knockout. However, the use of Fab fragments offers the advantage that heavy- and light-chain libraries can be produced independently and reassorted. The use of two vectors also facilitates the construction of hierarchical libraries in which a fixed heavy or light chain is paired with a library of partners. In theory, Fab libraries may be preferred where the final product will be a complete antibody, since, in some instances, removal of the scFv linker might alter the antigen-binding properties. However, scFvs have been successfully converted into complete antibodies.

Phage display has proved to be very effective in generating a variety of antibodies that are difficult to obtain using the hybridoma technology. For preexisting mouse monoclonal antibodies, phage display can be used to obtain antibodies that are entirely human in sequence, but which bind to the same part of the antigen (epitope) as the mouse monoclonal antibodies. Antibodies to targets previously inaccessible using immunization approaches (for example, self-antigens, ubiquitous compounds, or toxic compounds) have been isolated by selection of phage on antigen *in vitro*. Building on the concept of phage display, alternative display strategies such as ribosome display and cell-surface display have been recently developed.

4
Further Genetic Modifications of Antibodies

4.1
Engineering Antibody Fragments: Monovalent, Bivalent, and Multivalent scFvs

As explained in Sect. 3, scFv fragments consist of the variable domains of heavy and light chains (V_L and V_H) genetically fused by a flexile linker peptide. scFvs are much smaller than intact antibodies (Fig. 6) (25–27 kDa vs around 150 kDa for intact IgG), yet can retain the specificity and the affinity of the parental molecule. The small size of scFvs is a tremendous

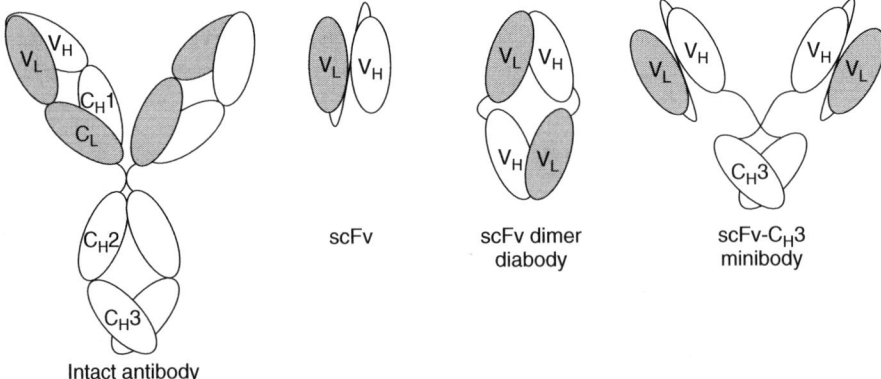

Fig. 6 Schematic representation showing the relative size and domain relationships between intact IgG (around 150 kDa) and engineered single-chain Fv (scFv, 25–27 kDa), scFv dimer or diabody (55 kDa), and minibody (scFv-C_H3, 80 kDa) fragments.

advantage in certain applications such as tumor targeting for detection and/or therapeutic purposes. In fact, scFv fragments display a combination of rapid, high-level tumor targeting with concomitant clearance from normal tissues and circulation that have made radiolabeled scFvs important tools for detection and treatment of cancer metastasis in both preclinical models and patients. An example of a successful scFv in the clinic is MFE-23. This scFv is specific for carcinoembryonic antigen (CEA), a glycoprotein that is highly expressed in colorectal adenocarcinomas. MFE-23, expressed in bacteria, has been used in two clinical trials: a gamma camera imaging trial using ^{123}I-MFE-23 and a radioimmunoguided surgery trial using ^{125}I-MFE-23, in which tumor deposits are detected by a handheld probe during surgery. Both trials showed that MFE-23 is safe and effective in localizing tumor deposits in patients with cancer.

Despite their successful use in some tumor-targeting studies, a significant limitation of using scFvs for targeting *in vivo* is their monovalent binding to antigen. Intact antibodies have significant avidity as a result of the presence of two antigen binding sites. To address this problem, antibody engineers have used the monomeric scFv as a building block for larger engineered fragments. One approach consists of producing dimers of scFv by incorporating a carboxy-terminal cysteine residue so that a disulfide bridge forms, yielding (scFv)$_2$ fragments. Alternately, the single-chain concept has been extended by using an additional linker peptide to join the two scFv molecules in tandem. Another approach for the production of scFv dimers results from the observation that the use of a very short linker peptide to connect the antibody variable regions caused the formation of "cross-paired" dimers, in which the V_L of one molecule associates with the V_H of a second and the V_L of the second molecule associates with the V_H of the first (Fig. 6). These noncovalent dimers, also known as *diabodies*, are capable of bivalent binding to antigen. scFv dimers and diabodies have a molecular weight similar to that of the antibody Fab fragments (55–60 kDa) but contain two antigen binding sites. Diabodies show significant improvement in tumor targeting compared to monovalent scFv. scFv fragments can also be fused to an immunoglobulin C_H3 domain, resulting in a self-assembling bivalent "minibody" (Fig. 6).

To further increase the avidity of antibody fragments, several laboratories have generated fragments with an increased valence. One strategy to develop fragments with increased valence has been to extend the diabody approach by decreasing the length of the interdomain linker peptide, which may result in the formation of tribodies and tetrabodies. It is also possible to obtain larger bivalent or multivalent fragments by fusing scFvs to protein domains normally involved in protein association such as helix bundles or leucine zippers. An alternative approach for producing multivalent fragments has been the fusion of scFvs to the bacterial protein streptavidin. Since streptavidin is a tetramer composed of four noncovalently linked monomers, four scFvs assemble to form a tetrameric structure with four antigen binding sites. A practical example of this technology was the development of Rh-specific scFv fused to streptavidin. This fusion protein named scFv::strep was able to directly agglutinate antigen-positive red blood cells (a reaction that is impossible to achieve by using antigen-specific monomeric scFv or IgG), suggesting the potential use of scFv::strep as a blood-typing reagent.

4.2
Bispecific Antibodies

In contrast to normal antibodies, which are monospecific, bispecific antibodies contain binding sites with two different specificities. Antibodies with two different specificities have been prepared by chemical modification to combine univalent fragments of different pepsin-treated antibodies, by fusing two hybridomas secreting antibodies of different specificities yielding a quadroma and by joining two different single-chain antibodies (scFvs). However, chemical modification of antibodies is inefficient and can lead to side reactions that damage the combining site. In a quadroma, it is difficult to separate the desired bispecific antibodies from the mixed population of heavy and light chains produced by the two hybridomas, and scFvs lacking constant regions also lack the antibody effector functions, which may be critical in certain applications. To address these problems, novel bispecific antibodies have been produced in which an scFv of one specificity has been genetically fused after the hinge (hinge-scFv) or at the carboxy-terminus (C_H3-scFv) of an antibody with a different specificity. Both fusion proteins were expressed by gene transfection in the context of a murine variable region. Transfectomas secreted a homogeneous population of the recombinant antibody with the two different specificities, one at the amino-terminus (anti-dextran) and the other at the carboxy-terminus (anti-dansyl). The C_H3-scFv antibody, which maintains the constant region of human IgG3, has some of the associated effector functions such as long half-life and Fc receptor binding. As expected, hinge-scFv antibody, which lacks the C_H2 and C_H3 domains, has no known effector functions.

Bispecific antibodies provide potential tools for use in immunotherapy. They take advantage of the great specificity of variable regions for their antigens and can be envisioned as transporters of therapeutic drugs or molecules or even immune effector cells to the specific targets identified by one of their binding sites. In fact, bispecific antibodies have been shown to be beneficial in the recruitment of immune cells for the treatment of cancer. Depending on the effector/target interaction, bispecific antibodies enhance cytotoxicity and phagocytosis. Several of these antibodies are currently undergoing evaluation in phase I and II clinical trials. Among the most extensively studied bispecific antibodies are 2B1 and MDX-210. 2B1 produced from a hybrid hybridoma is specific for the tumor-associated antigens HER2/*neu* and FcγRIII (CD16, the Fc gamma receptor expressed by key cytotoxic effector cells such as natural killer cells, neutrophils, and activated mononuclear phagocytes), while MDX-210 is a chemically conjugated hetero F(ab)'2 fragment specific for HER2/*neu* and FcγRI (CD64, the Fc gamma receptor expressed by key cytotoxic effector cells such as monocytes, macrophages, and IFN-γ–activated granulocytes). Both antibodies showed therapeutic promise in late-stage cancer patients whose tumor was refractory to conventional therapy. Since conventionally prepared bispecific antibodies have already shown promise in clinical trials and results from preclinical studies of recombinant bispecific antibodies are encouraging, in the future recombinant bispecific antibodies will undoubtedly be used in multiple clinical trials.

4.3 Polymers of Monomeric Antibodies

Although IgM is a naturally occurring polymeric antibody (see Fig. 2), for several applications it is desirable to have other classes of antibodies in an IgM-like (polymeric) structure. For example, IgM does not bind the Fc receptors on phagocyte cells, whereas IgG effectively triggers effector functions mediated through gamma-specific Fc receptors. In addition, since in IgG both heavy and light variable regions are usually somatically mutated (because of isotype switching to IgG), it is expected that polymeric IgG may be of higher intrinsic and functional affinity than the currently available IgM, resulting in more sensitivity and/or specificity. In addition, many methods such as protein-A binding are available that facilitate the isolation of IgG.

Vectors have been developed for the construction and expression of human polymeric IgG. It was observed that the 18–amino acid carboxyl-terminal tailpiece from human μ heavy chain is sufficient for polymer assembly. This finding was exploited to produce IgM-like polymers of IgG by fusing the 18–amino acid carboxyl-terminal tailpiece from human μ to the carboxyl-terminal of γ constant regions (Fig. 7). Using this technique, IgM-like polymers of IgG1, IgG2, IgG3, and IgG4 have been produced. IgGs obtained by this approach possess up to 6 Fcs and 12 antigen-combining sites, greatly increasing the avidity of their interactions with other molecules. These polymeric antibodies possess the Fcγ receptor–binding properties of IgG. Not surprisingly, the complement activity of normally active IgG1 and IgG3 and somewhat less-active IgG2 antibodies is dramatically enhanced upon polymerization. An unexpected result is that IgG4, normally devoid of complement activity, when polymerized in the same fashion directs complement-mediated lysis of target cells almost as effectively as the other polymers. These experiments demonstrate that polymerization of monomeric antibodies such as IgG is an effective approach to obtain antibodies with broader and more powerful effector functions than their wild-type counterparts.

An alternative strategy to make polymers of IgG is to genetically fuse chicken

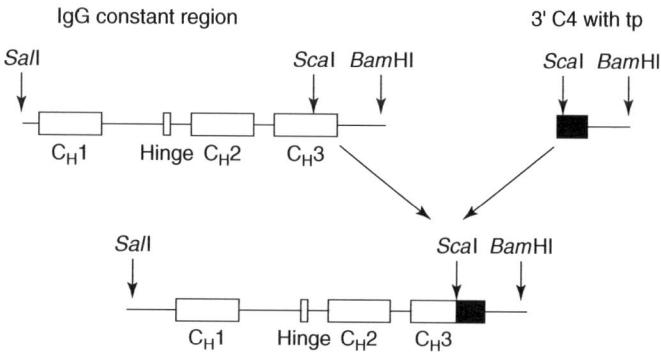

Fig. 7 Strategy for the construction of polymeric IgG. Using appropriate restriction sites, the μ tailpiece (μtp) of human IgM is genetically fused to the end of the heavy chain of human IgG.

avidin to the carboxy-terminus of the heavy chain of IgG. This approach is similar to that described in Sect. 4.1 (production of tetramers of scFv by fusing it with streptavidin) based on the fact that both streptavidin and avidin are tetramers of four noncovalently linked monomers. Since each antibody-avidin protein contains two molecules of avidin (one genetically fused at the carboxy-terminus of each heavy chain), two independent antibody fusion proteins bind to each other through their respective avidins forming a dimeric structure. One example of this approach is a recently developed human IgG3-avidin fusion protein specific for the transferrin receptor (TfR). The anti-TfR IgG3-avidin was able to function as a universal vector to deliver different biotinylated compounds into cancer cells overexpressing the TfR. Furthermore, it was unexpectedly discovered that anti-TfR IgG3-avidin, but not a recombinant anti-TfR IgG3 or a nonspecific IgG3-avidin, possesses a strong antiproliferative/proapoptotic activity against hematopoietic malignant cell lines. Studies confirmed that anti-TfR IgG3-avidin exists as a dimer, suggesting that cross-linking of the surface transferrin receptor may be responsible for the cytotoxic activity. These findings demonstrate that it is possible to transform an antibody specific for a growth factor receptor that does not exhibit inhibitory activity into a novel drug with significant intrinsic cytotoxic activity against selected cells by fusing it with avidin. The antitumor activity may be enhanced by delivering biotinylated therapeutics into cancer cells. Further development of this technology may lead to effective therapeutics for *in vivo* eradication of hematological malignancies and *ex vivo* purging of cancer cells in autologous transplantation.

4.4
Antibody Fusion Proteins

Fusion proteins with nonantibody molecules fused to antibodies can be produced using different approaches. Antibody fusion proteins that contain an intact antigen binding site should retain the ability to bind antigen, while the attached nonantibody partner should be able to exert its function. Such molecules, which have been called *immunoligands*, can be produced in several different ways (Fig. 8 a–f). When the nonantibody partner is fused to the end of the C_H3 domain (C_H3-ligand) (Fig. 8a), the antibody-combining specificity can be used to deliver an associated biological activity as well as antibody-related effector functions. An example is the anti-TfR IgG3-avidin fusion protein described in Sect. 4.3. Other examples are antibodies targeting cancer cells fused with interleukin-2 and GM-CSF. The goal of this approach to cancer therapy is to concentrate the cytokine in the tumor microenvironment and, by so doing, enhance the tumoricidal effect of the antibody and/or the host immune response against the tumor, while limiting severe toxic side effects associated with a high dose of cytokine administration. Such antibody–cytokine fusion proteins have shown significant antitumor activity in mice bearing tumors, leading to clinical trials. Immunoligands with the nonantibody partner fused immediately after the hinge (H-ligand) (Fig. 8b) or to the C_H1 domain (C_H1-ligand) (Fig. 8c) may be useful when the antibody-related effector functions are unnecessary or harmful. In addition, for many applications such as tumor targeting, the small size of the H-ligand and C_H1-ligand may be an advantage over the larger C_H3-ligand.

Fig. 8 Schematic representation of antibody fusion proteins. (a) to (c) represent different antibody fusion proteins in which the nonantibody partner was fused at the carboxy-terminus after the C_H3 domain (a), immediately after the hinge (b), or after the C_H1 domain (c). (d) to (f) represent antibody fusion proteins in which the nonantibody partner has been joined to the amino-terminus of the full-length heavy chain (d) or the truncated heavy chain (e and f). (g) and (h) represent two fusion proteins with the nonantibody partner fused to the amino-terminus of the C_H1 domain (g) or immediately before the hinge (h).

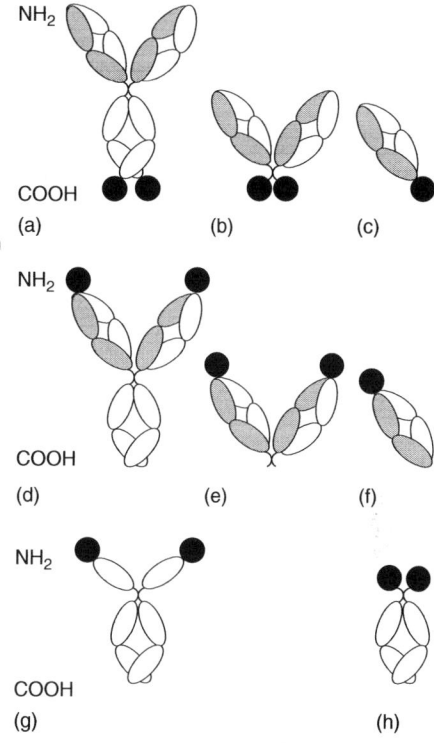

An alternative approach is to construct antibody fusion proteins with the ligand fused to the amino-terminus of the heavy chain (Fig. 8 d–f). This may be necessary for proteins that require N-terminal processing or folding for activity such as nerve growth factor (NGF), the costimulatory molecule B7.1, and interleukin-12 (IL-12). In fact, antibody-NGF, antibody-(B7.1), and antibody-(IL-12) fusion proteins containing the ligand fused to the amino-terminus of the antibody retain both the ability to bind antigen and the activity of the nonantibody partner.

Nonantibody sequences can also be used to replace the V_H domain or the V_H–C_H1 domains (Fig. 8 g and h). These molecules, which lack the ability to bind antigen, have been called *immunoadhesins* because they contain an adhesive molecule linked to the immunoglobulin Fc effector domains. In these proteins, the fused moiety acquires antibody-associated properties such as effector functions or improved pharmacokinetics. An example is the tumor necrosis factor (TNF) receptor IgG fusion protein, which binds to TNF (a mediator of inflammation) and neutralizes its activity. In fact, this molecule has been demonstrated to be efficacious for the treatment of rheumatoid arthritis.

Although Fig. 8 shows the nonantibody partner fused to the heavy chain, it should be appreciated that the nonantibody partner can also be fused to the light chain. It is also possible to construct antibody fusion proteins that combine more than one kind of nonantibody partner at the amino- or carboxy-termini of the heavy and/or light chains. In addition, a nonantibody molecule can be fused to an scFv molecule as described in Sect. 4.1. Moreover, two antibody Fc fragments can be genetically fused resulting in a molecule with novel biological properties. An example of this approach is the protein GE2, which is the product of the fusion between the Fc fragments of human IgG and human IgE (Fcγ–Fcε fusion protein). GE2 does not have an Fab and, as a consequence, is unable to target an antigen. However, GE2 was able to form complexes with both FcγRII and FcεRI, resulting in an inhibition of mast cell and basophil activation that results in the blocking of the anaphylaxis in transgenic mice expressing human FcεRIα. This approach has therapeutic potential in IgE- and FcεRI-mediated diseases such as allergic asthma, allergic rhinitis, chronic urticaria, angioedema, and anaphylaxis.

5
Expression Systems

A large variety of expression systems have been used for the production of genetically engineered antibodies and antibody fragments. These expression systems include bacteria, yeast, plants, baculovirus, and mammalian cells.

Antibody fragments are commonly expressed in bacteria and yeast. The bacterium *E. coli* is a frequently used expression system owing to its rapid growth and easy genetic manipulation. However, proteins expressed in *E. coli* are frequently insoluble and/or inactive, and refolding may be required to obtain functional fragments. Secretion of fragments into the bacterial periplasm or culture supernatant provides an alternative means to obtain the desired functional fragments without the need for refolding. Although the results are highly antibody-dependent, there are many examples in which the bacterial expression system is successful.

Complete functional antibodies have been most successfully expressed in mammalian cells, as these cells possess the mechanisms required for correct immunoglobulin assembly, posttranslational modification (glycosylation), and secretion. Posttranslational modifications can influence the biologic properties and effector functions, important considerations especially when the antibody is to be used for therapy. Examples of mammalian cells that have been successfully used to express properly assembled and glycosylated antibodies and antibody fusion proteins are the mouse myeloma cell lines P3X63Ag8.653, Sp2/0-Ag14, and NS0/1. These three myeloma cell lines have lost the ability to produce endogenous H and L chains and are derived from the parent myeloma. Antibodies produced in nonlymphoid cell lines such as Chinese hamster ovary (CHO), HeLa, C6, and PC12 are also properly assembled and glycosylated. Owing to the slower growth of mammalian cells, mammalian expression requires a longer time frame and higher costs than bacterial or yeast expression, but it is preferred when complete functional antibodies with proper glycosylation and disulfide bonds are required. Other expression systems that have been extensively used to produce complete functional antibodies include insect cells and plants. There is no "universal" expression system – each system has its advantages and disadvantages.

6
Conclusion

Rapid progress has been made in producing genetically engineered antibodies. The ability to express foreign DNA in a variety of host cells has made it possible to produce chimeric, humanized, and human antibodies as well as antibodies with novel structures and functional properties in quantities sufficiently large for many applications including clinical therapy. The available experience suggests that antibody-based therapies can be successfully developed for use in clinical situations in which no alternative effective therapy is available. However, continued progress in the development of antibody-based therapies will require extensive research to further define the mechanism of antibody action and on how to optimally use the novel proteins with unique functional properties.

See also Molecular Basis of Genetics.

Bibliography

Books and Reviews

Carter, P. (2001) Improving the efficacy of antibody-based cancer therapies, *Nat. Rev. Cancer* **1**, 118–29.

Chamow, S.M., Ashkenazi, A. (Eds.) (1999) *Antibody fusion proteins*, John Wiley & Son, New York.

Helguera, G.F., Morrison, S.L., Penichet, M.L. (2002) Antibody-cytokine fusion proteins: harnessing the combined power of cytokines and antibodies to cancer therapy, *Clin. Immunol.* **105**, 233–246.

Janeway, C.A., Travers, P., Walport, M., Shlomchik, M. (Eds.) (2001) The Generation of Lymphocyte Antigen Receptors, *Immunobiology: The Immune System in Health and Disease*, 5th edition, Garland Publisher, New York, pp. 123–154.

Kriangkum, J., Xu, B., Nagata, L.P., Fulton, R.E., Suresh, M.R. (2001) Bispecific and bifunctional single chain recombinant antibodies, *Biomol. Eng.* **18**, 31–40.

Rader, C., Barbas, C.F. (1997) Phage display of combinatorial antibody libraries, *Curr. Opin. Biotechnol.* **8**, 503–508.

Sensel, M.G., Coloma, M.J., Harvill, E.T., Shin, S.U., Smith, R.I., Morrison, S.L. (1997) Engineering novel antibody molecules, *Chem. Immunol.* **65**, 129–158.

Verma, R., Boleti, E., George, A.J. (1998) Antibody engineering: comparison of bacterial, yeast, insect and mammalian expression systems, *J. Immunol. Methods* **216**, 165–181.

Wu, A.M., Yazaki, P.J. (2000) Designer genes: recombinant antibody fragments for biological imaging, *Q. J. Nucl. Med.* **44**, 268–283.

Yoo, E.M., Chintalacharuvu, K.R., Penichet, M.L., Morrison, S.L. (2002) Myeloma expression system, *J. Immunol. Methods* **261**(1–2), 1–20.

Primary Literature

Abramowicz, D., Crusiaux, A., Goldman, M. (1992) Anaphylactic shock after retreatment with OKT3 monoclonal antibody, *N. Engl. J. Med.* **327**, 736.

Atwell, J.L., Breheney, K.A., Lawrence, L.J., McCoy, A.J., Kortt, A.A., Hudson, P.J. (1999) scFv multimers of the anti-neuraminidase antibody NC10: length of the linker between VH and VL domains dictates precisely the transition between diabodies and triabodies, *Protein Eng.* **12**, 597–604.

Bajorin, D.F., Chapman, P.B., Wong, G.Y., Cody, B.V., Cordon-Cardo, C., Dantes, L., Templeton, M.A., Scheinberg, D., Oettgen, H.F., Houghton, A.N. (1992) Treatment with high dose mouse monoclonal (anti-GD3) antibody R24 in patients with metastatic melanoma, *Melanoma Res.* **2**, 355–362.

Barbas, C.F., Bain, J.D., Hoekstra, D.M., Lerner, R.A. (1992) Semisynthetic combinatorial antibody libraries: a chemical solution to the diversity problem, *Proc. Natl. Acad. Sci. U.S.A.* **89**, 4457–4461.

Biocca, S., Pierandrei-Amaldi, P., Campioni, N., Cattaneo, A. (1994) Intracellular immunization with cytosolic recombinant antibodies, *Biotechnology (NY)* **12**, 396–399.

Boulianne, G.L., Hozumi, N., Shulman, M.J. (1984) Production of functional chimaeric mouse/human antibody, *Nature* **312**, 643–646.

Challita-Eid, P.M., Penichet, M.L., Shin, S.-U., Mosammaparast, N., Poles, T.M., Mahmood, K., Slamon, D.L., Morrison, S.L., Rosenblatt, J.D. (1997) A B7.1-Ab fusion protein retains antibody specificity and ability to activate via the T cell costimulatory pathway, *J. Immunol* **160**, 3419–3426.

Chester, K.A., Bhatia, J., Boxer, G., Cooke, S.P., Flynn, A.A., Huhalov, A., Mayer, A., Pedley, R.B., Robson, L., Sharma, S.K., Spencer, D.I., Regent, R.H. (2000) Clinical applications of phage-derived sFvs and sFv fusion proteins, *Dis. Markers* **16**, 53–62.

Chester, K.A., Mayer, A., Bhatia, J., Robson, L., Spencer, D.I., Cooke, S.P., Flynn, A.A., Sharma, S.K., Boxer, G., Pedley, R.B., Begent, R.H. (2000) Recombinant anti-carcinoembryonic antigen antibodies for targeting cancer, *Cancer Chemother. Pharmacol.* **46**, S8–S12.

Coloma, M.J., Hastings, A., Wims, L.A., Morrison, S.L. (1992) Novel vectors for the expression of antibody molecules using variable regions generated by polymerase chain reaction, *J. Immunol. Methods* **152**, 89–104.

Coloma, M.J., Morrison, S.L. (1997) Design and production of novel tetravalent bispecific antibodies [see comments], *Nat. Biotechnol.* **15**, 159–163.

Dangl, J.L., Wensel, T.G., Morrison, S.L., Stryer, L., Herzenberg, L.A., Oi, V.T. (1988) Segmental flexibility and complement fixation of genetically engineered chimeric human, rabbit and mouse antibodies, *EMBO J.* **7**, 1989–1994.

Davis, A.C., Roux, K.H., Pursey, J., Shulman, M.J. (1989) Intermolecular disulfide bonding in IgM: effects of replacing cysteine residues in the mu heavy chain, *EMBO J.* **8**, 2519–2526.

Davis, A.C., Roux, K.H., Shulman, M.J. (1988) On the structure of polymeric IgM., *Eur. J. Immunol.* **18**, 1001–1008.

Davis, A.C., Shulman, M.J. (1989) IgM–molecular requirements for its assembly and function, *Immunol. Today* **10**, 127, 128.

Dubel, S., Breitling, F., Fuchs, P., Zewe, M., Gotter, S., Welschof, M., Moldenhauer, G., Little, M. (1994) Isolation of IgG antibody Fv-DNA from various mouse and rat hybridoma cell lines using the polymerase chain reaction with a simple set of primers, *J. Immunol. Methods* **175**, 89–95.

Dubel, S., Breitling, F., Kontermann, R., Schmidt, T., Skerra, A., Little, M. (1995) Bifunctional and multimeric complexes of streptavidin fused to single chain antibodies (scFv), *J. Immunol. Methods* **178**, 201–209.

Ernst, M., Meier, D., Sonneborn, H.H. (1999) From IgG monoclonals to IgM-like molecules, *Hum. Antibodies* **9**, 165–170.

Figini, M., Marks, J.D., Winter, G., Griffiths, A.D. (1994) In vitro assembly of repertoires of antibody chains on the surface of phage by renaturation, *J. Mol. Biol.* **239**, 68–78.

George, A.J., Titus, J.A., Jost, C.R., Kurucz, I., Perez, P., Andrew, S.M., Nicholls, P.J., Huston, J.S., Segal, D.M. (1994) Redirection of T cell-mediated cytotoxicity by a recombinant single- chain Fv molecule, *J. Immunol.* **152**, 1802–1811.

Glennie, M.J., Brennand, D.M., Bryden, F., McBride, H.M., Stirpe, F., Worth, A.T., Stevenson, G.T. (1988) Bispecific F(ab' gamma)2 antibody for the delivery of saporin in the treatment of lymphoma, *J. Immunol.* **141**, 3662–3670.

Glennie, M.J., McBride, H.M., Worth, A.T., Stevenson, G.T. (1987) Preparation and performance of bispecific F(ab' gamma)2 antibody containing thioether-linked Fab' gamma fragments, *J. Immunol.* **139**, 2367–2375.

Goldenberg, M.M. (1999) Trastuzumab, a recombinant DNA-derived humanized monoclonal antibody, a novel agent for the treatment of metastatic breast cancer, *Clin. Ther.* **21**, 309–318.

Green, N.M. (1990) Avidin and streptavidin, *Methods Enzymol.* **184**, 51–67.

Green, L.L. (1999) Antibody engineering via genetic engineering of the mouse: XenoMouse strains are a vehicle for the facile generation of therapeutic human monoclonal antibodies, *J. Immunol. Methods* **231**, 11–23.

Griffiths, A.D., Malmqvist, M., Marks, J.D., Bye, J.M., Embleton, M.J., McCafferty, J., Baier, M., Holliger, K.P., Gorick, B.D., Hughes-Jones, N.C., Hoogboom, H., Winter, H. (1993) Human anti-self antibodies with high specificity from phage display libraries, *EMBO J.* **12**, 725–734.

Hanes, J., Jermutus, L., Weber-Bornhauser, S., Bosshard, H.R., Plückthun, A. (1998) Ribosome display efficiently selects and evolves high-affinity antibodies *in vitro* from immune libraries, *Proc. Natl. Acad. Sci. U.S.A.* **95**, 14130–14135.

He, M., Taussig, M.J. (1997) Antibody-ribosome-mRNA (ARM) complexes as efficient selection particles for *in vitro* display and evolution of antibody combining sites, *Nucleic Acids Res.* **25**, 5132–5134.

Higuchi, K., Araki, T., Matsuzaki, O., Sato, A., Kanno, K., Kitaguchi, N., Ito, H. (1997) Cell display library for gene cloning of variable regions of human antibodies to hepatitis B surface antigen, *J. Immunol. Methods* **202**, 193–204.

Holliger, P., Prospero, T., Winter, G. (1993) "Diabodies": small bivalent and bispecific antibody fragments, *Proc. Natl. Acad. Sci. U.S.A.* **90**, 6444–6448.

Hu, S., Shively, L., Raubitschek, A., Sherman, M., Williams, L.E., Wong, J.Y., Shively, J.E., Wu, A.M. (1996) Minibody: a novel engineered anti-carcinoembryonic antigen antibody fragment (single-chain Fv-CH3) which exhibits rapid, high-level targeting of xenografts, *Cancer Res.* **56**, 3055–3061.

Hudson, P.J., Kortt, A.A. (1999) High avidity scFv multimers; diabodies and triabodies, *J. Immunol. Methods* **231**, 177–189.

Iliades, P., Kortt, A.A., Hudson, P.J. (1997) Triabodies: single chain Fv fragments without a linker form trivalent trimers, *FEBS Lett.* **409**, 437–441.

Jaffers, G.J., Fuller, T.C., Cosimi, A.B., Russell, P.S., Winn, H.J., Colvin, R.B. (1986) Monoclonal antibody therapy. Anti-idiotypic and non-anti-idiotypic antibodies to OKT3 arising despite intense immunosuppression, *Transplantation* **41**, 572–578.

James, N.D., Atherton, P.J., Jones, J., Howie, A.J., Tchekmedyian, S., Curnow, R.T. (2001) A phase II study of the bispecific antibody MDX-H210 (anti-HER2 × CD64) with GM-CSF in HER2+ advanced prostate cancer, *Br. J. Cancer* **85**, 152–156.

Jespers, L.S., Roberts, A., Mahler, S.M., Winter, G., Hoogenboom, H.R. (1994) Guiding the selection of human antibodies from phage display repertoires to a single epitope of an antigen, *Biotechnology (NY)* **12**, 899–903.

Kohler, G., Milstein, C. (1975) Continuous cultures of fused cells secreting antibody of predefined specificity, *Nature* **256**, 495–497.

Kontsekova, E., Kolcunova, A., Kontsek, P. (1992) Quadroma-secreted bi(interferon alpha 2–peroxidase) specific antibody suitable for one-step immunoassay, *Hybridoma* **11**, 461–468.

Kortt, A.A., Lah, M., Oddie, G.W., Gruen, C.L., Burns, J.E., Pearce, L.A., Atwell, J.L., McCoy, A.J., Howlett, G.J., Metzger, D.W., Webster, R.G., Hudson, P.J. (1997) Single-chain Fv fragments of anti-neuraminidase antibody NC10 containing five- and ten-residue linkers form dimers and with zero- residue linker a trimer, *Protein Eng.* **10**, 423–433.

Kostelny, S.A., Cole, M.S., Tso, J.Y. (1992) Formation of a bispecific antibody by the use of leucine zippers, *J. Immunol.* **148**, 1547–1553.

Kuus-Reichel, K., Grauer, L.S., Karavodin, L.M., Knott, C., Krusemeier, M., Kay, N.E. (1994) Will immunogenicity limit the use, efficacy, and future development of therapeutic monoclonal antibodies? *Clin. Diagn. Lab. Immunol.* **1**, 365–372.

Leget, G.A., Czuczman, M.S. (1998) Use of rituximab, the new FDA-approved antibody, *Curr. Opin. Oncol.* **10**, 548–551.

Lewis, L.D., Cole, B.F., Wallace, P.K., Fisher, J.L., Waugh, M., Guyre, P.M., Fanger, M.W., Curnow, R.T., Kaufman, P.A., Ernstoff, M.S. (2001) Pharmacokinetic-pharmacodynamic relationships of the bispecific antibody MDX-H210 when administered in combination with interferon gamma: a multiple-dose phase-I study in patients with advanced cancer which overexpresses HER-2/neu, *J. Immunol. Methods* **248**, 149–165.

Liu, S.J., Sher, Y.P., Ting, C.C., Liao, K.W., Yu, C.P., Tao, M.H. (1998) Treatment of B-cell lymphoma with chimeric IgG and single-chain Fv antibody-interleukin-2 fusion proteins, *Blood* **92**, 2103–2112.

Lloyd, F. Jr., Goldrosen, M. (1991) The production of a bispecific anti-CEA, anti-hapten (4-amino-phthalate) hybrid-hybridoma, *J. Natl. Med. Assoc.* **83**, 901–904.

Mallender, W.D., Voss, E.W. Jr. (1994) Construction, expression, and activity of a bivalent bispecific single-chain antibody, *J. Biol. Chem.* **269**, 199–206.

Marasco, W.A. (1997) Intrabodies: turning the humoral immune system outside in for intracellular immunization, *Gene Ther.* **4**, 11–15.

Marks, J.D., Hoogenboom, H.R., Bonnert, T.P., McCafferty, J., Griffiths, A.D., Winter, G. (1991) By-passing immunization. Human antibodies from V-gene libraries displayed on phage, *J. Mol. Biol.* **222**, 581–597.

McGrath, J.P., Cao, X., Schutz, A., Lynch, P., Ebendal, T., Coloma, M.J., Morrison, S.L., Putney, S.D. (1997) Bifunctional fusion between nerve growth factor and a transferrin receptor antibody, *J. Neurosci. Res.* **47**, 123–133.

Milstein, C., Cuello, A.C. (1983) Hybrid hybridomas and their use in immunohistochemistry, *Nature* **305**, 537–540.

Moreland, L.W., Baumgartner, S.W., Schiff, M.H., Tindall, E.A., Fleischmann, R.M., Weaver, A.L., Ettlinger, R.E., Cohen, S., Koopman, W.J., Mohler, K., Widmer, M.B., Blosch, C.M. (1997) Treatment of rheumatoid arthritis with a recombinant human tumor necrosis factor receptor (p75)-Fc fusion protein [see comments], *N. Engl. J. Med.* **337**, 141–147.

Moreland, L.W., Schiff, M.H., Baumgartner, S.W., Tindall, E.A., Fleischmann, R.M., Bulpitt, K.J., Weaver, A.L., Keystone, E.C., Furst, D.E., Mease, P.J., Ruderman, E.M., Horwitz, D.A., Arkfeld, D.G., Garrison, L., Burge, D.J., Blosch, C.M., Lange, M.L., McDonnell, N.D., Weinblatt, M.E. (1999) Etanercept therapy in rheumatoid arthritis. A randomized, controlled trial, *Ann. Intern. Med.* **130**, 478–486.

Morrison, S.L., Johnson, M.J., Herzenberg, L.A., Oi, V.T. (1984) Chimeric human antibody molecules: mouse antigen-binding domains with human constant region domains, *Proc. Natl. Acad. Sci. U.S.A.* **81**, 6851–6855.

Ng, P.P., Dela Cruz, J.S., Sorour, D.N., Stinebaugh, J.M., Shin, S.U., Shin, D.S., Morrison, S.L., Penichet, M.L. (2002) An anti-transferrin receptor-avidin fusion protein exhibits both strong proapoptotic activity and the ability to deliver various molecules into cancer cells, *Proc. Natl. Acad. Sci. U.S.A.* **99**, 10706–10711.

Orlandi, R., Gussow, D.H., Jones, P.T., Winter, G. (1989) Cloning immunoglobulin variable domains for expression by the polymerase chain reaction, *Proc. Natl. Acad. Sci. U.S.A.* **86**, 3833–3837.

Pack, P., Plückthun, A. (1992) Miniantibodies: use of amphipathic helices to produce functional, flexibly linked dimeric FV fragments with high avidity in *Escherichia coli*, *Biochemistry* **31**, 1579–1584.

Parmley, S.F., Smith, G.P. (1988) Antibody-selectable filamentous fd phage vectors: affinity purification of target genes, *Gene* **73**, 305–318.

Peng, L.S., Penichet, M.L., Morrison, S.L. (1999) A single-chain IL-12 IgG3 antibody fusion protein retains antibody specificity and IL-12 bioactivity and demonstrates antitumor activity, *J. Immunol.* **163**, 250–258.

Penichet, M.L., Dela Cruz, J.S., Shin, S.U., Morrison, S.L. (2001) A recombinant IgG3-(IL-2) fusion protein for the treatment of human HER2/neu expressing tumors, *Hum. Antibodies* **10**, 43–49.

Penichet, M.L., Harvill, E.T., Morrison, S.L. (1998) An IgG3-IL-2 fusion protein recognizing a murine B cell lymphoma exhibits effective tumor imaging and antitumor activity, *J. Interferon Cytokine Res.* **18**, 597–607.

Plückthun, A., Pack, P. (1997) New protein engineering approaches to multivalent and bispecific antibody fragments, *Immunotechnology* **3**, 83–105.

Raso, V., Griffin, T. (1981) Hybrid antibodies with dual specificity for the delivery of ricin to immunoglobulin-bearing target cells, *Cancer Res.* **41**, 2073–2078.

Russell, S.J., Hawkins, R.E., Winter, G. (1993) Retroviral vectors displaying functional antibody fragments, *Nucleic Acids Res.* **21**, 1081–1085.

Shin, S.U., Wu, D., Ramanathan, R., Pardridge, W.M., Morrison, S.L. (1997) Functional and pharmacokinetic properties of antibody-avidin fusion proteins, *J. Immunol.* **158**, 4797–4804.

Shusta, E.V., Kieke, M.C., Parke, E., Kranz, D.M., Wittrup, K.D. (1999) Yeast polypeptide fusion surface display levels predict thermal stability and soluble secretion efficiency, *J. Mol. Biol.* **292**, 949–956.

Smith, G.P. (1985) Filamentous fusion phage: novel expression vectors that display cloned antigens on the virion surface, *Science* **228**, 1315–1317.

Smith, R.I., Coloma, M.J., Morrison, S.L. (1995) Addition of a mu-tailpiece to IgG results in polymeric antibodies with enhanced effector functions including complement-mediated cytolysis by IgG4, *J. Immunol.* **154**, 2226–2236.

Smith, R.I., Morrison, S.L. (1994) Recombinant polymeric IgG: an approach to engineering more potent antibodies, *Biotechnology (NY)* **12**, 683–688.

Somia, N.V., Zoppe, M., Verma, I.M. (1995) Generation of targeted retroviral vectors by using single-chain variable fragment: an approach to *in vivo* gene delivery, *Proc. Natl. Acad. Sci. U.S.A.* **92**, 7570–7574.

Tada, H., Kurokawa, T., Seita, T., Watanabe, T., Iwasa, S. (1994) Expression and characterization of a chimeric bispecific antibody against fibrin and against urokinase-type plasminogen activator, *J. Biotechnol.* **33**, 157–174.

Valone, F.H., Kaufman, P.A., Guyre, P.M., Lewis, L.D., Memoli, V., Deo, Y., Graziano, R., Fisher, J.L., Meyer, L., Mrozek-Orlowski, M., Wardwell, K., Guyre, V., Morley, T.L., Arvizu, C., Fanger, M.W. (1995) Phase Ia/Ib trial of bispecific antibody MDX-210 in patients with advanced breast or ovarian cancer that overexpresses the proto-oncogene HER-2/neu, *J. Clin. Oncol.* **13**, 2281–2292.

van Dijk, M.A., van de Winkel, J.G. (2001) Human antibodies as next generation therapeutics, *Curr. Opin. Chem. Biol.* **5**, 368–374.

Vaughan, T.J., Williams, A.J., Pritchard, K., Osbourn, J.K., Pope, A.R., Earnshaw, J.C., McCafferty, J., Hodits, R.A., Wilton, J., Johnson, K.S. (1996) Human antibodies with sub-nanomolar affinities isolated from a large non-immunized phage display library, *Nat. Biotechnol.* **14**, 309–314.

Verhoeyen, M., Milstein, C., Winter, G. (1988) Reshaping human antibodies: grafting an antilysozyme activity, *Science* **239**, 1534–1536.

Weiner, L.M., Clark, J.I., Ring, D.B., Alpaugh, R.K. (1995) Clinical development of 2B1, a bispecific murine monoclonal antibody targeting c-erbB-2 and Fc gamma RIII, *J. Hematother.* **4**, 453–456.

Weinstein, J.N., Eger, R.R., Covell, D.G., Black, C.D., Mulshine, J., Carrasquillo, J.A., Larson, S.M., Keenan, A.M. (1987) The pharmacology of monoclonal antibodies, *Ann. N.Y. Acad. Sci.* **507**, 199–210.

Zhu, D., Kepley, C.L., Zhang, M., Zhang, K., Saxon, A. (2002) A novel human immunoglobulin Fc gamma Fc epsilon bifunctional fusion protein inhibits Fc epsilon RI-mediated degranulation, *Nat. Med.* **8**, 518–521.

29
Transgenic Fish

Pinwen Peter Chiou, Jenny Khoo, Chung Zoon Chun, and Thomas T. Chen
University of Connecticut, Storrs, CT, USA

1	Introduction 833	
2	**Selection of Model Fish Species** 833	
2.1	Small-size Fish Model 834	
2.2	Medium-size Fish Model 834	
2.3	Large-size Fish Model 834	
3	**Types of Transgenes** 835	
3.1	Gain-of-function Transgenes 835	
3.2	Reporter-function Transgenes 837	
3.3	Loss-of-function Transgenes 838	
4	**Methods of Gene Transfer** 838	
4.1	Microinjection of DNA into Embryos or Unfertilized Eggs 838	
4.2	Electroporation of DNA into Embryos or Sperm 839	
4.3	Infection with Replication-defective Pantropic Retroviral Vectors 840	
4.4	Emerging New Methodology 841	
5	**Characterization of Transgenic Fish** 842	
5.1	Identification of Transgenic Fish 842	
5.2	Pattern of Transgene Integration 842	
5.3	Inheritance of Transgenes 843	
5.4	Expression of Transgenes 843	
6	**Application of Transgenic Fish** 843	
6.1	Transgenic Fish in Basic Research 844	
6.1.1	Vertebrate Developmental Biology 845	
6.1.2	Functional Analysis of Promoter/Enhancer Elements 846	

Genomics and Genetics. Edited by Robert A. Meyers.
Copyright © 2007 Wiley-VCH Verlag GmbH & Co. KGaA, Weinheim
ISBN: 978-3-527-31609-0

6.1.3	Signaling Pathways 847
6.1.4	Human Disease Modeling 848
6.2	Growth Hormone Transgenic Fish 850
6.3	Transgenic Fish with Enhanced Resistance to Pathogen Infection 850
6.4	Transgenic Fish with Different Body Color 851
6.5	Transgenic Fish as Environmental Biomonitors 852
6.6	Other Examples of Transgenic Fish 854
7	**Concerns and Future Perspectives** 855

Bibliography   856
Books and Reviews   856
Primary Literature   857

Keywords

Electroporation
A process utilizing a series of short electrical pulses to permeate cell membranes, thereby permitting the entry of DNA molecules into animal or plant cells.

Polymerase Chain Reaction (PCR)
An *in vitro* method for the enzymatic synthesis of a specific DNA sequence, using two oligonucleotide primers hybridizing to opposite strands and flanking the region of interest in the target DNA.

Transgene
A piece of nonself origin gene or noncoding DNA fragment that is introduced into plants or animals for the production of transgenic species.

Transgenic Fish
Fish into which a foreign gene or noncoding DNA fragment has been introduced and stably integrated into its genome.

F_1 Transgenic Fish
Transgenic progeny derived from a cross between a P_1 transgenic individual and a nontransgenic individual or between two P_1 transgenic individuals.

P_1 Transgenic Fish
Transgenic fish developed from embryos that received foreign gene transfer.

> Fish into which foreign DNA is artificially introduced and integrated in their genomes are called *transgenic fish*. Since the development of the first transgenic fish in 1985, techniques to produce transgenic fish have improved tremendously, resulting in the production of genetically modified (GM) fish of many species. In the past few years, the fast booming application of transgenic fish technology has led to many valuable discoveries in different disciplines of the biological sciences, many of which could potentially lead to the next breakthrough in new treatments for human diseases. Transgenic fish technology has also been applied to produce fish with beneficial traits, such as fast growth rate, enhanced disease resistance, and environmental biomonitors, in the hope of improving the quality of both human life and the earth environments.

1
Introduction

Fish into which a foreign gene or noncoding DNA fragment is artificially introduced and integrated in their genomes are called *transgenic fish*. Since 1985, a wide range of transgenic fish species have been produced mainly by microinjecting or electroporating homologous or heterologous transgenes into newly fertilized or unfertilized eggs and, sometimes, sperm. To produce a desired transgenic fish, several factors should be considered. First, an appropriate fish species must be chosen. This decision would depend on the nature of each study and the availability of the fish holding facility. Second, a specific gene construct must be designed on the basis of the special requirement of each study. For example, the gene construct may contain an open reading frame encoding a gene product of interest and regulatory elements that regulate the expression of the gene in a temporal, spatial, and/or developmental manner. Third, the gene construct has to be introduced into sperm or embryos in order for the transgene to be integrated stably into the genome of the embryonic cells. Fourth, since not all instances of gene transfer are efficient, a screening method must be adopted for identifying transgenic individuals.

Since the development of the first transgenic fish, techniques to produce transgenic fish have improved tremendously, resulting in the production of genetically modified (GM) fish of many species. In recent years, transgenic fish have been established as valuable models for different disciplines of biological research as well as for human disease modeling. The application of transgenic fish technology to produce fish with beneficial traits, such as environmental biomonitors and enhanced disease resistance, is also rising. In this chapter, we will review the progress of the transgenic fish technology and its application in basic research and biotechnological application.

2
Selection of Model Fish Species

Gene transfer studies have been conducted in several different fish species such as salmon, rainbow trout, common carp, goldfish, loach, channel catfish, tilapia, northern pike, walleye, zebrafish, and Japanese medaka. Depending on the purpose of the transgenic fish studies, the embryos of some fish species prove to be more suitable than others. In selecting a fish species for gene transfer studies, a series of parameters are considered. These parameters are (1) length of the life cycle; (2) year round supply of eggs and sperm; (3) culture conditions; (4) size of the adult at maturity; and (5) availability of background information on genetics, physiology, and endocrinology of the fish species.

2.1
Small-size Fish Model

Japanese medaka (*Oryzias latipes*) and zebrafish (*Danio rerio*) are regarded as ideal fish species for conducting gene transfer studies. Both the fish species have short life cycles (3 months from eggs to mature adults), produce hundreds of eggs on a regular basis without exhibiting a seasonal breeding cycle, and can be maintained easily in the laboratory for two to three years. Their eggs are relatively large, 1.0 to 1.5 mm diameter, and possess very thin and semitransparent chorions. These features allow easy microinjection of DNA into the eggs if appropriate glass needles are used. Furthermore, inbred lines and various morphological mutants of both the fish species are available. These fish species are suitable candidates for producing transgenic fish for (1) studying developmental regulation of gene expression and gene action; (2) identifying regulatory elements that regulate the expression of a gene; (3) measuring the activities of promoters; and (4) producing transgenic models for human diseases and environmental toxicology. However, a major drawback of these two fish species is that their small body size makes them unsuitable for endocrinological or biochemical analysis.

2.2
Medium-size Fish Model

The handicap of using small body–size fish species such as medaka or zebrafish in gene transfer studies can be easily overcome by the use of fish species such as loach, killifish (*Fundulus*), goldfish, and tilapia. The body sizes of these fish species are big enough so that sufficient amounts of tissues can be isolated from individual fish for biochemical and endocrinological studies. Another important attribute of these fish species is their shorter maturation time compared to rainbow trout or salmon. In particular, tilapia is a very appropriate medium-size fish species for gene transfer studies since it requires only about three months to complete its entire life cycle. It is possible to produce three generations of transgenic tilapia in one year. Unfortunately, the lack of a well-defined genetic background and asynchronous reproductive behavior of these fish species present some problems for conducting gene transfer studies.

2.3
Large-size Fish Model

Rainbow trout, salmon, channel catfish, and common carp are commonly used large body–size model fish species for transgenic fish studies. Since the knowledge of endocrinology, reproductive biology, and physiology of these fish species have been well worked out, they are well suited for studies on comparative endocrinology as well as on aquaculture applications. However, the maturation time for each of these fish species is relatively long. For example, rainbow trout and salmon require two to three years to reach reproductive maturity, and common carp or channel catfish about one to two years.

Thus, it will prolong inheritance studies. In addition, since these fish species have a single spawning cycle per year, it restricts the number of gene transfer attempts that can be conducted per year.

3
Types of Transgenes

A transgene is a piece of nonself origin gene or DNA fragment that is introduced into fish for the production of transgenic fish. Transgenes are usually constructed in plasmids that contain appropriate promoter/enhancer elements for proper expression of the transgenes in fish. A variety of promoters from nonfish species have been employed to control the expression of transgenes in fish (Table 1). Additionally, increasing number of fish promoters (Table 1) are available for the purpose of generating "all-fish" expression cassettes, among which the carp β-actin promoter has been shown to drive strong expression of transgenes in various fish cell types. Tissue-specific promoters and inducible promoters such as zebrafish *heat shock protein 70 (hsp70)* have also been used successfully to control the expression of transgene in the desired tissue(s) at the desired time(s). For example, activation of *hsp70*-controlled transgene can be achieved by heat shock and, intriguingly, by focusing a sublethal laser microbeam onto individual cells that carry the transgene. The targeted cells appear normal after the treatment.

Depending on the purpose of the gene transfer studies, transgenes can be grouped into three major types: (1) *gain-of-function*; (2) *reporter-function*; and (3) *loss-of-function*. The basic organization and the function of these transgenes will be discussed below.

3.1
Gain-of-function Transgenes

Gain-of-function transgenes are designed to add new functions to the transgenic

Tab. 1 Examples of promoters used in production of transgenic fish.

	Origin Species	Promoter
Nonfish origin:		
	Rat	GAP43[a]
	Chicken	β-actin
	Xenopus	elongation factor 1α
	Cytomegalovirus	CMV intermediate–early
	Moloney murine leukemia virus	MoMLV LTR
Fish origin:		
	Carp	β-actin
	Goldfish	α1 tubulin[a]
	Medaka	elongation factor 1α, β-actin
	Zebrafish	H2A.F/Z, α-actin[a], heat shock protein 70[b]

[a] tissue-specific promoters- GAP43: nervous system; α1 tubulin: nervous system; α-actin: muscle (see able 3 for more examples).
[b] Inducible promoters; LTR: long terminal repeat; H2A.F/Z: histone H2A.F/Z.

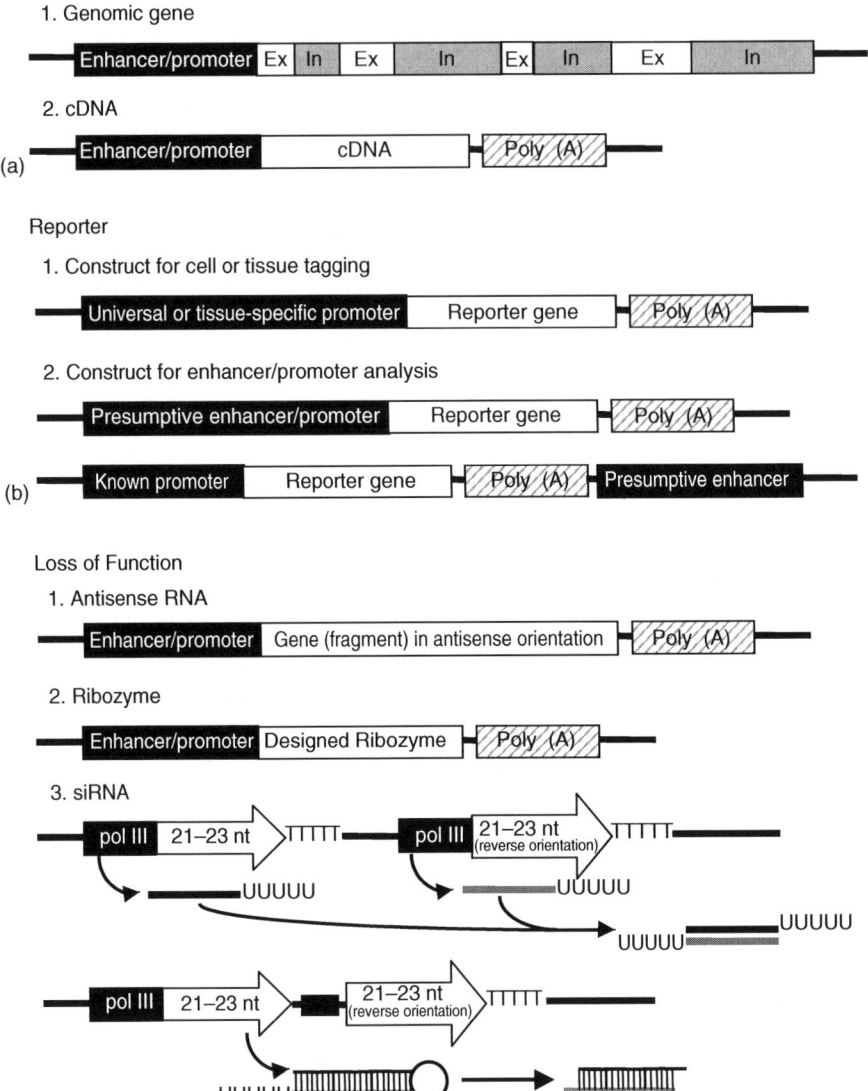

Fig. 1 Prototypes of transgenes for fish transgenesis. Depending on the experimental purpose, the promoter used in the transgene construct can be either constitutive or an inducible one. Ex: exon; in: intron; pol III: RNA polymerase III promoters.

individuals or to facilitate the identification of the transgenic individuals if the genes are expressed properly in the transgenic individuals. The coding regions of the transgenes are usually homologous or heterologous genomic or cDNA sequences, which encode polypeptide products (Fig. 1a). The expression of the transgenes is usually driven by promoter/enhancer sequences of homologous or heterologous sources. If the transgenes contain functional signal peptide sequences, the gene products will be secreted out of cells once they are synthesized. Transgenes containing the structural gene of mammalian or fish growth hormones (GH) or the respective cDNA fused to a functional promoter/enhancer of chicken or fish β-actin gene are examples of the gain-of-function transgene constructs. Expression of such transgenes in transgenic individuals has been shown to result in growth enhancement. Winter flounder antifreeze protein gene and chicken δ-crystalline gene with their own promoter/enhancer sequences are other commonly used *gain-of-function* transgenes used in various gene transfer studies.

Another type of commonly used *gain-of-function* transgenes are the "reporter-function transgenes," which are discussed in the next section.

3.2
Reporter-function Transgenes

Genes such as *bacterial chloramphenicol acetyltransferase* (CAT), β-*galactosidase* (β-gal), *neomycin phosphotransferase* II (neo^R), *luciferase*, and *green fluorescent protein* (GFP), are frequently used as reporters. These reporter genes serve as a convenient marker for indicating the success of gene transfer into the target cells since the translational products of these genes can be easily monitored. Furthermore, because these gene products can be measured quantitatively, they are also routinely adopted for identifying the sequence of an undefined promoter/enhancer or for measuring the relative activity of a promoter/enhancer element in question.

A less frequently used reporter type is the one with target-specificity. One such example is cameleon, a fluorescent calcium sensor, which was originally used to detect calcium transients in culture cells and pharyngeal muscle of *Caenorhabditis elegans*. Recently, *cameleon* and several other indicator genes have been used to establish transgenic zebrafish for monitoring neuronal activity. The prototypes of a *reporter-function* transgene are shown in Fig. 1b.

Among the commonly used reporters, the jellyfish (*Aequorea victoria*) GFP and its variants have become the choice markers for transgenic fish because the detection of GFP signal is direct and easy, requiring no exogenous substrate. There are numerous GFP variants available, including those producing green, blue, and yellow fluorescence light. Together with the red fluorescence protein (RFP), these reporters can be easily detected in live transgenic fish such as zebrafish and medaka whose embryos and body are transparent, without the need to sacrifice the fish. Additionally, these reporters can be used to study multiple transgenes simultaneously in the same fish. Since the first GFP transgenic fish in 1995, GFP-tagged transgenic embryos and fish have become the most convenient and widely used tools in basic biological research, particularly in the regulation of gene expression and morphogenetic movements during embryonic development (see Sect. 6.1). In many cases, similar applications can be achieved with the other reporter genes mentioned above. However, the detection of these reporters is indirect and involves additional steps such as enzymatic reaction with substrate and immunohistochemical staining with specific antibody to the reporter.

3.3 Loss-of-function Transgenes

The *loss-of-function* transgenes are constructed for interfering with the expression of host genes at the transcriptional or translational level. Mechanistically, there are at least three groups of "loss-of-function" transgenes. The transgenes might encode antisense RNAs or oligonucleotides that selectively hybridize to a target endogenous mRNA, resulting in degradation of the mRNA with RNase H or interference with translation of the mRNAs via preventing the binding of ribosomes. A variation of the antisense RNA approach is to introduce a group of chemically modified oligonucleotides known as morpholino phosphorodiamidate oligonucleotides (MOs) into fish embryos. In contrast to ordinary antisense molecules, MOs exhibit long-term stability with less toxicity to living cells and appear to function through translation blockage only. The second group of "loss-of-function" transgenes is catalytic RNA (ribozymes) that can cleave specific target mRNAs at a specific site and thereby "knock down" the normal gene expression. The most recent development of the *loss-of-function* transgenes is to introduce a group of small RNAs known as small (short) interfering RNAs (siRNAs) into cells. siRNA will guide a ribonuclease known as RISC (the RNA-induced silencing complex) to recognize mRNA containing a sequence homologous to the siRNA, resulting in degradation of the mRNA by the ribonuclease. The prototype of *loss-of-function* transgenes is shown in Fig. 1c. Although there is currently no report of stable lines of transgenic fish carrying this type of transgenes, the feasibility of this approach in silencing genes has been demonstrated in numerous fish cell lines *in vitro*. Moreover, many recent studies also prove that MOs and siRNA molecules can silence their targets specifically and effectively when introduced into fish embryos, such as zebrafish, medaka, and rainbow trout.

4 Methods of Gene Transfer

Techniques such as calcium phosphate precipitation, direct microinjection, retrovirus infection, electroporation, and particle gun bombardment have been widely used to introduce foreign DNA into animal cells, plant cells, and germ lines of mammals and other vertebrates. Among these methods, direct microinjection and electroporation of DNA into newly fertilized eggs have proven to be the most reliable methods of gene transfer in fish systems.

4.1 Microinjection of DNA into Embryos or Unfertilized Eggs

Microinjection of foreign DNA into newly fertilized eggs was first developed for the production of transgenic mice in the early 1980s. Since then, the technique of microinjection has been adopted for introducing transgenes into Atlantic salmon, common carp, catfish, goldfish, loach, medaka, rainbow trout, tilapia, and zebrafish. The gene constructs that were used in these studies include human or rat growth hormone gene, rainbow trout or salmon GH cDNA, chicken δ-crystalline protein gene, winter flounder antifreeze protein gene, *E. coli* β-galactosidase gene, and *E. coli* hygromycin resistance gene. In general, gene transfer in fish by direct microinjection is conducted as follows. The parameters for microinjection are summarized in Table 2. Eggs and sperm are collected in separate, dry containers.

Tab. 2 Parameters of fish transgenesis by microinjection and electroporation.

Parameters	Gene Transfer Method	
	Microinjection	*Electroporation*
Developmental stage	1 to 2 cells	1 to 2 cells
DNA size	<10 Kb	<10 Kb
DNA concentration	10^{6-7} molecules/embryo	100 µg ml^{-1}
DNA topology	Linear	Linear
Chorion barrier	Dechorionated/micropyle	Intact chorion
Electrical field strength	N/A[a]	500 to 3000 v
Pulse shape	N/A[a]	exponential/square
Pulse duration	N/A[a]	ms[c] to s
Temperature	RT[b]	RT[b]
Medium	PBS/saline[d]	PBS/saline

[a] N/A, not applicable.
[b] RT, room temperature (25 °C).
[c] ms, millisecond.
[d] PBS/saline: phosphate-buffered saline.

Fertilization is initiated by mixing sperm and eggs, then adding water, with gentle stirring to enhance fertilization. Eggs are microinjected within the first few hours after fertilization. The injection apparatus consists of a dissecting stereomicroscope and two micromanipulators, one with a micro-glass-needle for injection and the other with a micropipette for holding fish embryos in place. Routinely, about 10^6 to 10^8 molecules of a linearized transgene are injected into the egg cytoplasm. Following injection, the embryos are incubated in water until hatching. Since natural spawning in zebrafish or medaka can be induced by adjusting photoperiod and water temperature, precisely staged newly fertilized eggs can be collected from the aquaria for gene transfer. If the medaka eggs are maintained at 4 °C immediately after fertilization, the micropyle on the fertilized eggs will remain visible for two hours. The DNA solution can be delivered into the embryos by injecting through this opening during this time period.

Depending on the species, the survival rate of injected fish embryos ranges from 35 to 80% while the rate of DNA integration ranges from 10 to 70% in the survivors. The tough chorions of the fertilized eggs in some fish species, for example, rainbow trout and Atlantic salmon, can make insertion of glass needles difficult. This difficulty has been overcome by any one of the following methods: (1) inserting the injection needles through the micropyle; (2) making an opening on the egg chorions by microsurgery; (3) removing the chorion by mechanical or enzymatic means; (4) preventing chorion hardening by initiating fertilization in a solution containing 1 mM glutathione; or (5) injecting the unfertilized eggs directly.

4.2
Electroporation of DNA into Embryos or Sperm

Electroporation is a successful method for transferring foreign DNA into bacteria, yeast, and plant and animal cells in culture.

In the past years, this method has become popular for transferring foreign genes into fish embryos or sperm. Electroporation utilizes a series of short electrical pulses to permeate cell membranes, thereby permitting the entry of DNA molecules into embryos or sperm. The patterns of electrical pulses can be emitted in a single pulse of exponential decay form (i.e. exponential decay generator) or high-frequency multiple peaks of square waves (i.e. square-wave generator). The basic parameters are summarized in Table 2. Studies conducted in our laboratory and those of others have shown that the rate of DNA integration in electroporated embryos is on the order of 20% or higher in the survivors. Although the overall rate of DNA integration in transgenic fish produced by electroporation was equal to or slightly higher than that of microinjection, the actual amount of time required for handling a large number of embryos by electroporation is orders of magnitude less than that required for microinjection. Several reports have also appeared in the literature describing successful transfer of transgenes into fish by electroporating sperm instead of embryos. With the success seen in sperm electroporating, electroporation can therefore be considered as an efficient and versatile mass gene transfer technology.

4.3
Infection with Replication-defective Pantropic Retroviral Vectors

By microinjection or electroporation, plasmid-encoded transgenes can be introduced at a satisfactory efficiency into many fish species to produce transgenic offspring. However, the resulting P_1 transgenic individuals are almost always mosaics as a result of delayed transgene integration. Furthermore, both methods are not efficient or applicable in transferring foreign DNA into embryos of life-bearing fish, marine fish, and invertebrates because of the restrictions imposed by the unique spawning behavior or anatomical structure of these species. Modification of transgenes has been adopted to increase the efficiency of gene transfer by microinjection or electroporation. Hsiao et al. and Thermes et al. have taken the following approaches: (1) using the inverted terminal repeat DNA from adeno-associated virus to increase the stability of transgene integration; and (2) employing I-SceI meganuclease to mediate high transportation efficiency of the transgene into the nucleus. Despite the improvement, these modified approaches cannot be applied to the mentioned species, whose transgenic offspring can not be produced by either microinjection or electroporation. Instead, transformation of mature or immature germ cells by infection with transgene-encoded retroviral vectors may be an effective approach for transgenesis in these fish.

A replication-defective pantropic retroviral vector containing the long terminal repeat (LTR) sequence of Moloney murine leukemia virus (MoMLV) and transgenes in a viral envelop with the G-protein of vesicular stomatitis virus (VSV), was developed by Burns et al. Since entry of VSV into host cells is mediated by interaction of the VSV-G protein with a phospholipid component of the host cell membrane, this pseudotyped retroviral vector has a very broad host range and is able to transfer transgene into many different cell types. Using the pantropic pseudotyped defective retrovirus as a gene transfer vector, transgenes containing neo^R or β-galactosidase were transferred into zebrafish, medaka, and dwarf surf clams.

More recently, Sarmasik et al. and Chen et al. also used the pantropic retroviral vector to transfer genes into the immature gonads of crayfish, desert guppy, and shrimp, respectively. They found that, by using the pantropic retroviral vector as a gene transfer vector, the problem of transgene mosaicism in P_1 transgenic fish is eliminated as the transgenes are introduced into immature male gonads by the pantropic retroviral vector.

4.4
Emerging New Methodology

Nuclear transplantation is a key technique in animal cloning. This method involves the transfer of nuclei from donor cells to enucleated eggs, stimulating eggs into cleavage and early phase of embryonic development, and in the case of mammals, the developing embryos are transferred into pseudopregnant females to complete the final stages of development. The first successful cloning of animals by nuclear transplantation was established for Northern Leopard Frog (*Rana pipiens*) by Briggs and King in 1952. However, Dolly, the lamb, is the first mammal successfully cloned by the nuclear transplantation technique, and since then, many laboratories throughout the world have adopted this technique to clone other mammals and farm animals. To date, production of diploid and fertile fish by the nuclear transplantation technique have been achieved in two small laboratory fish species, medaka and zebrafish. Three types of donor cells have been used for nuclear transfer: sperm, long-term cultured somatic cells, and embryonic cells.

In the study reported by Wakamatsu et al., embryonic cells from two transgenic medaka lines were used as donor cells to generate transgenic medaka. Blastoderm cells from the donor embryos in the mid-blastula stage were collected and dissociated into single cells. These were subsequently transplanted into the cytoplasm of the recipient enucleated eggs at the animal pole through the micropyle. In this study, 1 out of 588 eggs transplanted with the first donor and 5 out of 298 eggs transplanted with the second donor developed successfully to adulthood. The transgene of the donor nuclei was also transmitted to the F_1 and F_2 offspring in a Mendelian fashion.

Two versions of nuclear transfer technique have been established for zebrafish recently. Jesuthasan and Subburaju reported the use of sperm nuclei as donor for nuclear transfer. Sperm collected from mature males were demembranated by treatment with lysolecithin, digitonin, or by freeze-thawing. The demembranated sperm were then incubated with transgene DNA at room temperature prior to microinjecting into the animal pole region of the unfertilized egg. This method could produce nonmosaic expression of transgene in all cells of the embryos if incubation of the demembranated sperm with DNA was prolonged to 20 min. Few of the treated eggs however developed into fertile adults and the transgene was shown to be transmitted through the germ line. In the report by Lee et al., embryonic fibroblast cells from disaggregated embryos were used as donor cells. These cells were first modified by retroviral insertions expressing GFP as a reporter and were cultured for at least 12 weeks before their nuclei were transplanted into enucleated, unfertilized eggs. The method resulted in 2% of the transplants developing into fertile, diploid offspring. This study clearly demonstrated that nuclei from slowly dividing cultured fish somatic cells could be reprogrammed to support rapid embryonic development and used as donor cells for nuclear transfer in fish. Moreover, the promising use of modified somatic cells provides an alternative choice of vehicle for gene targeting and fish cloning since fish ES cells are still unavailable. Overall, these studies have demonstrated the feasibility of using germinal and somatic cells from fish as donors for conducting nuclear transfer studies in different important fish species for both basic research and application purposes. In addition, the problem of transgene mosaicism in P_1 transgenic fish can be overcome by the nuclear transplantation method.

5
Characterization of Transgenic Fish

5.1
Identification of Transgenic Fish

Identification of transgenic individuals is the most time-consuming step in the production of transgenic fish. Traditionally, dot blot and Southern blot hybridization of genomic DNA were common methods used to determine the presence of transgenes in the presumptive transgenic individuals. These methods involve isolation of genomic DNA from tissues of presumptive transgenic individuals, digestion of DNA samples with restriction enzymes, and Southern blot hybridization of the digested DNA products. Although this method is expensive, laborious, and insensitive, it offers a definitive answer as to whether a transgene has been integrated into the host genome. Furthermore, it also reveals the pattern of transgene integration if appropriate restriction enzymes are employed in the Southern blot hybridization analysis. In order to handle a large number of samples efficiently and economically, a polymerase chain reaction (PCR) based assay has been adopted. The strategy of the assay involves the isolation of genomic DNA from a very small piece of fin tissue, PCR amplification of the transgene sequence, and Southern blot analysis of the amplified products. Although this method does not differentiate whether the transgene is integrated in the host genome or exists as an extrachromosomal unit, it serves as a rapid and sensitive screening method for identifying individuals that contain the transgene at the time of analysis. This method has been used in our laboratory as a routine preliminary screen for the presence of transgenes in thousands of presumptive transgenic fish.

5.2
Pattern of Transgene Integration

Studies conducted in many fish species have shown that following injection of linear or circular transgene constructs into fish embryos, the transgenes are maintained as an extrachromosomal unit through many rounds of DNA replication in the early phase of the embryonic development. At a later stage of embryonic development, some of the transgenes are randomly integrated into the host genome while others are degraded, resulting in the production of mosaic transgenic fish. To determine the pattern of transgene integration, genomic DNA from PCR positive fish is digested with a series of restriction endonucleases and the resulting products resolved in agarose gels for Southern blot analysis. In many fish species studied to date, it is found that multiple copies of transgenes are integrated in a head-to-head, head-to-tail or tail-to-tail form, except in transgenic common carp and channel catfish where a single copy of transgene was integrated at multiple sites on the host chromosomes.

5.3
Inheritance of Transgenes

Stable integration of transgenes is an absolute requirement for continuous vertical transmission to subsequent generations and establishment of a transgenic fish line. To determine whether the transgene is transmitted to subsequent generations, P_1 transgenic individuals are mated to nontransgenic individuals and the progeny are assayed for the presence of transgene by the PCR assay method described earlier. Although it has been shown that the transgene may persist in the F_1 generation of transgenic zebrafish as extrachromosomal DNA, detailed analysis of the rate of transmission of transgenes to F_1 and F_2 generations in many transgenic fish species indicates true and stable incorporation of gene constructs into the host genome. If the entire germ line of the P_1 transgenic fish is transformed with at least one copy of the transgene per haploid genome, at least 50% of the F_1 transgenic progeny will be expected in a backcross involving a P_1 transgenic with a nontransgenic control. In many such crosses, only about 20% of the progeny are transgenic. When the F_1 transgenic is backcrossed with a nontransgenic control, however, at least 50% of the F_2 progeny are transgenics. These results clearly suggest that the germ lines of the P_1 transgenic fish are mosaic as a result of delayed transgene integration during embryonic development.

5.4
Expression of Transgenes

An important aspect of gene transfer studies is the detection of transgene expression. Depending on the levels of transgene products in the transgenic individuals, the following listed methods are commonly used to detect transgene expression: (1) RNA northern or dot blot hybridization; (2) RNase protection assay; (3) reverse transcription-polymerase chain reaction (RT-PCR); (4) immunoblotting assay; and (5) other biochemical assays for determining the presence of the transgene protein products. Among these assays, RT-PCR is the most sensitive method and only requires a small amount of sample. The strategy of this assay is summarized in Fig. 2. Briefly, it involves the isolation of total RNA from a small piece of tissue, synthesis of single-stranded cDNA by reverse transcription, and PCR amplification of the transgene cDNA by employing a pair of oligonucleotide specific to the transgene product as amplification primers. The resulting products are resolved on agarose gels and analyzed by Southern blot hybridization using a radiolabeled transgene as a hybridization probe. Transgene expression can also be quantified by a quantitative RT-PCR method or a quantitative real-time RT-PCR.

6
Application of Transgenic Fish

As techniques of producing various species of transgenic fish have become available in the last two decades, there has been a rapid boom in applying transgenic fish strategy to different disciplines of basic research and biotechnological applications. In this section, we will discuss recent advances in applying the transgenic fish technology in basic research as well as in producing transgenic fish with beneficial traits such as enhanced growth and disease resistance for aquaculture.

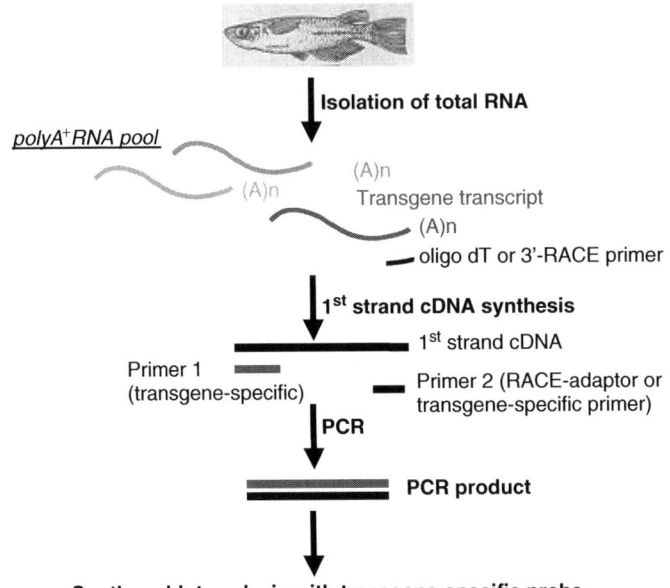

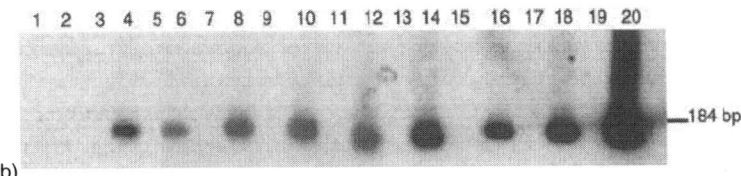

(a)

(b)

Fig. 2 Determination of transgene expression by RT-PCR or RACE-PCR. (a) Strategy of RT-PCR and RACE-PCR. (b) An example of detecting cecropin transgene in transgenic medaka. Total RNA was isolated from fin tissues of F_2 transgenic medaka. Expression of cecropin transgene was detected by reverse transcribing poly A^+ RNA into complementary cDNA and PCR amplification of the cecropin cDNA by using cecropin-specific primers. The amplified product was further confirmed by Southern hybridization using a cecropin-specific probe. RT-PCR was conducted with mRNA from transgenic fish without prior reverse transcription (lanes 5, 7, 9, 11, 13, 15, 17, 19) or with first cDNA transcribed from mRNA isolated from the transgenic fish (lanes 4, 6, 8, 10, 12, 14, 16, 18). Lane 1: RT-PCR with no mRNA input; lane 2: RT-PCR with mRNA from negative control fish (c-mRNA); lane 3: PCR with c-mRNA without prior reverse transcription; lane 20: PCR with control plasmid containing cecropin gene. From Sarmasik, A., Warr, G., Chen, T.T. (2002) Production of transgenic medaka with increased resistance to bacterial pathogens, Mar. Biotechnol. 4, 310–322 with permission.

6.1
Transgenic Fish in Basic Research

Teleost fish serve as extremely valuable models for basic research in vertebrate biology as they are functionally similar to mammalian species, and because they are sufficiently evolutionarily distant to mammals, they are well suited for comparative genomics. Transgenic fish can be easily and economically produced and reared in large quantities, granting them superiority over the transgenic mouse model, especially in studies requiring a large number of animals. In recent years, transgenic fish tagged with GFP have emerged as important tools in studies of vertebrate developmental biology, basic molecular and cell biology, and human disease modeling. Additionally, gene misexpression and knockdown are also widely used in these studies.

6.1.1 Vertebrate Developmental Biology

Transgenic zebrafish and medaka and their embryos are ideal models for studying the complex developmental mechanisms that transform a single-cell zygote into a multicellular organism with diverse functions. Both the fish species provide the following advantages: (1) short generation time (about 3 months); (2) constantly available and easily reared; and (3) most importantly, transparent embryos granting clear visualization of all stages of embryonic development. Additionally, there are mutant zebrafish and medaka whose transparent bodies consist of less interfering pigmentation. Although most of the developmental studies have so far been conducted in zebrafish, medaka embryos provide a larger window for observation at smaller developmental intervals because of their longer embryonic developmental time (2 days versus 10 days).

As shown in Table 3, researchers have established many lines of transgenic reporter zebrafish that express GFP in specific cells or tissues, including germ cells, blood, lymphoid cells, epithelia, muscle, heart (Fig. 4a), pancreas, liver, intestine (Fig. 4b), vasculature, and the nervous system. This is achieved by introducing a

Tab. 3 Examples of cell-type and tissue-type specific transgenic zebrafish lines with GFP or EGFP reporter.

Specificity	Promoter
Cell-specific	
Germ cells	Xenopus efl α (-vasa 3'UTR)
	Zebrafish versa
Blood	Zebrafish gata 1
Lymphoid cells	Zebrafish rag1
Epithelia	Zebrafish krt8
Tissue-specific	
Muscle	Zebrafish α-actin
Pancreas	Zebrafish pdx-1
	Zebrafish insulin
Liver	Zebrafish l-fabp
Intestine	Zebrafish i-fabp
Vasculature	Zebrafish fli 1
Heart	Zebrafish cmlc2
Nervous system	Zebrafish islet1
	Zebrafish huC
	Goldfish α1 tubulin

GFP, enhanced green fluorescence protein (EGFP), or RFP reporter that is placed under control of tissue- or cell-specific promoter/enhancer. Because the same fluorescence-tagged cells can be followed over time, these transgenic fish have been successfully used in the study of morphogenesis and gene regulation during embryonic development. For example, transgenic zebrafish with GFP-tagged blood vessels have been used to study the blood vessels assembly during development. Zebrafish fli1 promoter is able to drive expression of GFP in all blood vessels throughout embryogenesis. By coupling this feature with high-resolution dynamic imaging techniques, confocal microangiography has clearly illustrated the anatomical architecture of the vasculature of zebrafish and show that blood vessels undergoing angiogenesis display pathfinding behavior similar to that of neuronal growth cones.

Application of fish transgenesis on developmental biology can also be achieved by gene misexpression or knockdown. So far, such an approach has been carried out exclusively in fish embryos. For example, zebrafish embryos have been used to study the role of Sema3D protein in guiding retinal ganglion cell (RGC) axons during embryonic development. Sema3D belongs to the semaphorin family, many members of which are known to repel or attract specific growth cones. Sema3D may participate in guiding RGC axons along the dorsoventral axis of the tectum, as indicated by the differentially higher expression of the protein in the ventral versus dorsal tectum. To understand the function of Sema3D, homozygous zebrafish embryos have been established from stable transgenic zebrafish fish carrying the *sema3D* gene under control of the zebrafish *heat shock protein 70* (*hsp70*) promoter. Ubiquitous misexpression (overexpression) of Sema3D in the embryos can be achieved by heat shock and has been found to inhibit ventral but not dorsal RGC axon growth. Additionally, mosaic Sema3D misexpression can be induced by heat shock in the embryos microinjected with the *hsp70-*controlled *sema3D* transgene. In these embryos, ventral RGC axons have been found to avoid or stop at individual cells misexpressing Sema3D along their pathway. Furthermore, knockdown of Sema3D with morpholino antisense causes errors in retinotectal mapping along the dorsoventral axis. These analyses suggest that Sema3D in the ventral tectum directs the correct innervation of ventral RGCs to dorsal tectum by inhibiting them from extending into ventral tectum. A similar approach has been employed to study the function of Sema4E protein in embryonic development.

6.1.2 Functional Analysis of Promoter/Enhancer Elements

Transgenic zebrafish and medaka have become popular tools in the identification and functional analysis of promoter/enhancer elements of vertebrates. The fast-growing genomic information of zebrafish and medaka facilitates the identification of evolutionarily conserved promoter/enhancer elements via cross-species comparison of genome sequences among fish, mouse, and human. Transgenic GFP-tagged zebrafish and medaka have been applied to define the essential regions of promoter/enhancer elements as well as to unmask the regulatory functions of these elements. For example, transgenic medaka and zebrafish have been used to characterize promoters of carp rhodopsin gene (cRh) and zebrafish *myf-5* gene (myf-5), respectively. In these studies, transgenic fish were designed to carry EGFP gene fused with different lengths of the upstream fragments of the presumed promoter/enhancer regulatory sequences. The role of motifs, enhancers, and minimal cis-regulatory elements involved in tissue-specific expression or the translocation of progenitor muscle cells have been identified by monitoring the expression of the EGFP reporter.

Another example of this approach is the investigation of the regulatory function of the LTRs of human ERV-9 endogenous retrovirus. The LTRs of endogenous ERV-9 retrovirus are conserved during primate evolution, but their function in the primate genomes is unknown. In human, the solitary LTRs of ERV-9 are identified as middle repetitive DNAs associated with up to 4000 human gene loci including the β-globin gene locus where the ERV-9 LTR is juxtaposed to the locus control region (β-LCR) far upstream of the globin genes. The regulatory function of ERV-9 LTRs in the human genome has recently been demonstrated via using transgenic zebrafish carrying the β-globin ERV-9 LTR coupled to the GFP gene. The analysis suggests a regulatory role of ERV-9 LTR enhancers, during oogenesis, in the synthesis of maternal mRNAs that are required for early embryogenesis. Alternatively, the ERV-9 LTR enhancers could transcriptionally potentiate and preset chromatin structure of the cis-linked gene loci in oocytes and adult stem/progenitor cells.

6.1.3 Signaling Pathways

In addition to traditional cell-culture system, transgenic fish and their embryos have become an important tool for studying different signaling pathways in recent years. So far, most of the transgenic fish studies are conducted in zebrafish, thanks to the abundant genetic information of the species. Several achievements by the new approach are described here. Perz-Edwards et al. have identified specific regions within a whole developing zebrafish embryo where cellular transcription are responsive to retinoic acid signaling, which has been shown to be critical in normal vertebrate development. Lawson et al. have identified the downstream factors of receptor tyrosine kinase signaling that are necessary during *in vivo* blood vessel formation and arterial development. Additionally, Delta-Notch signaling in oligodendrocyte specification and the regulatory function of chemokine signaling in sensory cell migration have also been investigated by using transgenic zebrafish system. In a non-zebrafish system, Petrausch et al., by using transgenic goldfish, have identified the effects of several trophic factors converging on a purine-sensitive signaling mechanism that controls axonal outgrowth and the expression of multiple growth-associated proteins.

Despite the valuable discoveries achieved by using cell culture systems in studying signaling pathways in the last two decades, there are limitations to the use of cell culture, mainly being an artificial system that is isolated from the physiological environment of the animal. On the contrary, the most significant advantage of using transgenic fish for signaling pathway study is that transgenic fish provide an easy and representative system to dissect out signaling pathways and their biological implications with relevance to time and space in an animal. For example, while in many vertebrate species Wnt/b-catenin signaling is known to be involved in the development of neural crest cells and the specification of pigment cells derived from them at later stage, it is unclear which Wnts are involved, and when they are required. To address these issues, Lewis et al. established a transgenic zebrafish line that carried an inducible dominant-negative repressor of Wnt/β-catenin signaling, T-cell Factor 3 (Tcf-3a). The N-terminus of zebrafish Tcf-3a was conjugated with GFP as a reporter and was placed under the control of the zebrafish *hsp70* promoter. By activating the repressor to inhibit endogenous Wnt/b-catenin signaling at discrete times in development, a critical period for Wnt signaling in the initial induction of neural crest was defined. Wnt/b-catenin signaling is crucial for neural crest induction from the end of gastrulation through the 3-somite stage, and the cells lose sensitivity to transgene activation between the 3- and 6-somite stages despite the later requirement of the signaling for the pigment cell lineage. Researchers have also identified a specific Wnt, Wnt8, and demonstrated its crucial role in initial neural crest induction, by blocking Wnt8 with antisense morpholino oligonucleotides. As demonstrated by these examples, transgenic fish is an excellent system to study a common scenario for many signaling pathways, such as BMP and Notch, in which the same signaling pathways are used in different places and at different times throughout development.

6.1.4 Human Disease Modeling

By providing a general proof of principle or reproducing specific aspects of the human diseases, animal models may be used to identify disease-associated factors or genes and to screen for new treatments for these diseases. In addition to murine models, fish have been used to study human diseases for decades, particularly in cancer research. In recent years, transgenic zebrafish and other species are emerging as valuable tools for modeling human diseases, especially in the domains of neurodegenerative diseases and cancers.

Among the neurodegenerative diseases, transgenic fish have been used for studies of Alzheimer's and Parkinson's diseases, two devastating diseases that affect tens of millions of people worldwide. The formation of neurofibrillary tangles (NFT), along with the β-amyloid plaques, are two central events in the development of Alzheimer's disease (AD). A microtubule-associated protein, tau, is critical in the process of NFT formation, in which tau protein is compromised in its normal association with microtubules, leading to the formation of paired helical filaments. Hyperphosphorylation of tau protein and mutations in the genes encoding the protein have been implicated in the pathogenesis of AD and a variety of hereditary dementias, collectively termed *frontotemporal dementia* with parkinsonism linked to chromosome 17 (FTDP-17).

The cellular consequence of NFT *in vivo* was first illustrated in a fish model by Hall et al. In this model, the giant neurons (anterior bulbar cells [ABCs]) of the central nervous system in the sea lamprey (*Petromyzon marinus*) were used to study the cytopathological changes caused by chronical overexpression of human tau protein. The unique advantages of this system include the already established morphological and cellular characterization of ABCs in detail, and most importantly, the ready accessibility of the neurons for manipulation by microinjection. This system demonstrates the association of filamentous tau deposits with stereotyped cytopathological changes, including the loss of dendritic microtubules and synapses, plasma membrane degeneration, and eventually the formation of extracellular tau deposits and cell death. Furthermore, this sequence of change is spatiotemporally correlated with the AD-related hyperphosphorylation of tau protein. By using this system, a lipid-soluble, low-molecular-weight proprietary compound has been shown to significantly retard the progressive degeneration of ABCs that express human tau protein, suggesting that development of tau-inhibitors may be a promising approach for AD treatment.

A transgenic zebrafish embryo model has also been established to study the functional consequence of FTDP-related mutations of tau protein in neurodegeneration. Zebrafish embryos at the 1–2 cell stage were microinjected with *tau* mutants under control of the neural specific GATA-2 promoter. An FTDP-17 mutant form of human tau has been shown to produce a cytoskeletal disruption that closely resembled the NFT in human disease.

This model system is therefore useful in the study of other mutant taus in vertebrate neurons *in vivo* and in the study of other molecules that are involved in the pathogenesis of human neurodegenerative diseases.

Parkinson's disease (PD) is characterized by the degeneration of mesencephalic dopaminergic neurons (MDNs) in the substantia nigra (SN). One of the most important functions of MDNs is the control of voluntary movement; loss of MDNs will eventually lead to a loss of dopamines, its metabolites, and the dopamine transporter. One of the most promising approaches to develop treatments for PD is to study the normal development of MDNs, in the hope that an understanding of their development will lead to means that can replenish damaged MDNs. Zebrafish have been used to screen for mutations that affect MDN development, and fish transgenesis can further assist in the understanding of the roles of these mutated genes in MDN development. Currently, a transgenic zebrafish model has been established to study the *ubiquitin C-terminal hydrolase L1* (*uch-L1*), a gene associated with the inherited form of PD. Zebrafish embryos were microinjected with *in vitro* synthesized zebrafish *uch-L1* mRNA or a plasmid construct containing GFP reporter under control of the *uch-L1* promoter. Together, these assays demonstrate the preferential activity of uch-L1 promoter in neurons and the spatiotemporal expression of the *uch-L1* in the ventral diencephalons, a functionally homologous region to the SN in human. The transgenic zebrafish system has thus provided further evidence for the connection of *uch-L1* in the development and degeneration of neural cells and can be further used to characterize other genes involved in neurodegenerative diseases.

In cancer research, stable lines of transgenic tumor-bearing zebrafish have been established to study the conserved cancer pathways involved in oncogene-induced lymphoblastic leukemia. In this model, zebrafish embryos at one-cell stage were injected with mouse oncogene *c-myc* (*mMyc*) and chimerical EGFP-*mMyc* transgenes under control of the zebrafish *Rag2* promoter, a lymphocyte specific promoter. In the resultant P_1 founder fish, the *myc*-induced clonally derived T-cell acute lymphoblastic leukemia was observed in almost 100% of the fish expressing green fluorescence. By tracing migration of the EGFP-tagged cells and confirming by RNA *in situ* hybridization, the leukemic T cells were clearly shown to arise in the thymus. The cells then spread to the gills and eyes and finally disseminate to the skeletal muscle and abdominal organs. The homing of leukemic T cells back to thymus was also illustrated in irradiated fish transplanted with GFP-tagged leukemic lymphoblasts. To establish stable lines of *mMyc*-transgenic fish, an *in vitro* fertilization technique was adopted because of the prompt onset of leukemia in the F_1 transgenic fish after germ-line transmission. By fertilizing the eggs collected from normal females with the sperm from 90-day-old leukemic males, the resultant transgene-positive progeny were shown to exhibit expansion of GFP-positive leukemic T cells from the thymus by one month of age. This transgenic zebrafish model can serve as a very powerful tool in the "forward-genetic" screening for modifier genes that are associated with the *myc*-induced leukemia or other types of cancers. The model can also be applied to large-scale screen for new drugs to treat or prevent human leukemia in the future.

In addition to the examples mentioned above, transgenic fish model can be applied in the studies of many other types of human diseases. We expect to see a plethora of such studies in the near future.

6.2
Growth Hormone Transgenic Fish

The initial drive of transgenic fish research came from attempts to increase production of economically important fish for human consumption. The worldwide supply of fishery products has traditionally depended upon commercial harvest of finfish, shellfish, and crustaceans from freshwater and marine-water sources. In recent years, while the worldwide commercial catch of fish experienced a sharp reduction, the worldwide demand of fish products has risen steeply. In order to cope with the demand of fish products, many countries have turned to aquaculture. Although aquaculture has the potential to meet the world demand for fish products, innovative strategies are required to improve its efficiency. What can transgenic fish technology offer in this regard?

There are three aspects of fish growth characteristics that could be improved for aquaculture: (1) increasing the initial growth rate of fry for an early head start; (2) enhancing somatic growth rate in adults to provide larger market body size; and (3) improving feed conversion efficiency of fish to achieve effective utilization of feeds. Among these three, enhanced somatic growth rates via manipulation of the GH gene shows considerable promise. Studies conducted by Agellon et al. and Paynter et al. showed that treatment of yearling rainbow trout and oysters with recombinant rainbow trout growth hormone resulted in significant growth enhancement. Similar results of growth enhancement in fish treated with recombinant GH have been reported by many investigators. These results point to the possibility of improving the somatic growth rate fish by manipulating fish GH or its gene.

Zhu et al. reported the first successful transfer of a human GH gene fused to a mouse metallothionein (MT) gene promoter into goldfish and loach. The F_1 offspring of these transgenic fish grew to be twice as large as their nontransgenic siblings. Since then, similar enhanced growth effect has been observed in the GH-transgenic fish of many other species, such as tilapia, carp, catfish, sea bream, and salmon. These studies have shown that the expression of a foreign human or fish GH gene could result in significant growth enhancement in the P_1, F_1, and F_2 transgenic fish. Additionally, Dunham et al. have demonstrated that transgenic common carp carrying rainbow trout GH transgene display, consistently in two consecutive generations, a favorable body shape, better dress-out yield, and better quality of flesh compared with controls. These studies have demonstrated that the application of GH-transgenic fish could be beneficial to the worldwide aquaculture, and hence could help to alleviate the starvation in many economically poor countries by providing more efficient, cheaper and yet high-quality protein sources.

6.3
Transgenic Fish with Enhanced Resistance to Pathogen Infection

Disease is one of the most severe bottlenecks in aquaculture. For the past few decades, efforts to control infectious diseases in commercially important teleosts have primarily focused on the development of suitable vaccines for fish and selection of fish strains with robust resistance to infectious pathogens. Although effective vaccines have been developed for several important fish pathogens, the current vaccination practice is expensive, laborious, and time consuming. Genetic selection based on traditional crossbreeding techniques is time consuming and the outcome is frequently unpredictable, and sometimes unachievable because of the lack of desired genetic traits. More effective approaches for controlling fish disease in aquaculture are highly desirable.

Transgenic fish technology can facilitate the genetic selection process by directly modifying the undesirable genetic traits that confer vulnerability to pathogens or by introducing specific genes that are related to disease resistance into fish. The introduced transgenes can be fish-originated or characterized genes from other species. We have recently demonstrated the feasibility of introducing the antimicrobial peptide genes, cecropin B and cecropin P1, into fish to enhance disease resistance. Cecropins, first identified in *Cecropia* moth, are members of the antimicrobial peptide family that are evolutionarily conserved in species from insects to mammals. These peptides possess activities against a broad range of microorganisms, including bacteria, yeasts, and even viruses. *In vitro* studies conducted by Sarmasik et al. showed that recombinant cecropin B inhibited the propagation of common fish pathogens such as *Pseudomonas fluorescens, Aeromonas hydrophila,* and *Vibrio anguillarum*. Furthermore, Chiou et al. showed that cecropin and a designed analog, CF-17 peptide, effectively inhibited the replication of fish viruses such as infectious hematopoietic necrosis virus (IHNV), viral hemorrhagic septicemia virus (VHSV), snakehead rhabdovirus (SHRV), and infectious pancreatic necrosis virus (IPNV) (Fig. 3). More recently, our laboratory has shown that synthetic peptide CF-17 is also effective in inhibiting the propagation of insect baculovirus. Additionally, Jia et al. demonstrated the enhanced resistance to bacterial infection in fish which were continuously transfused with a cecropin–melittin hybrid peptide, CEME, and pleurocidin amide, a C-terminally amidated form of the natural flounder peptide. These results led to the hypothesis that production of disease-resistant fish strain may be achieved by introducing the known antimicrobial peptide genes into fish through the application of transgenic fish technology.

To test this hypothesis, we introduced gene constructs containing preprocecropin, procecropin, mature cecropin B, and cecropin P1 into medaka embryos by electroporation. The resulting F_2 transgenic medaka were subjected to bacterial challenges at a LD_{50} dose with *Pseudomonas fluorescens* and *Vibrio anguillarum*, respectively. The resulting relative percent survival (RPS) of the tested transgenic F_2 fish ranged from 72 to 100% against *P. fluorescens* and 25 to 75% against *V. anguillarum* (Table 4). These results clearly demonstrate the potential application of transgenic fish technology in producing fish with more robust resistance to infectious pathogens.

6.4
Transgenic Fish with Different Body Color

There are two main motives in producing fish with altered body color: (1) to generate novel varieties of ornamental fish with rare colors for the purpose of rearing as pets, and (2) to use the change of body color as indicators of environmental changes. GFP and several GFP variants are the most used genes for such purposes.

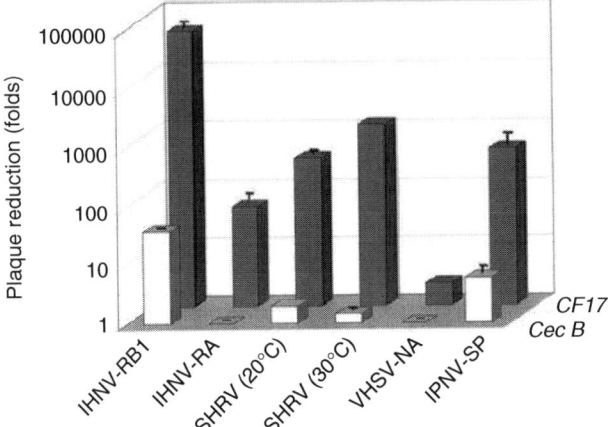

Fig. 3 Effect of peptides cecropin and CF-17 on inhibition of fish virus propagation *in vitro*. The antiviral activity of the peptides cecropin B (Cec B) and CF17 was evaluated against one nonenveloped birnavirus, the SP isolate of infectious pancreatic necrosis virus (IPNV), and three enveloped rhabdoviruses: the RB1 (type2) and RA (type 1) isolates of infectious hematopoietic necrosis virus (IHNV), snakehead rhabdovirus (SHRV), and an avirulent North America isolate (NA) of viral hemorrhagic septicemia virus (VHSV). Viral infection was carried out in EPC cells for SHRV, and CHSE-214 cells for the rest viruses. From Chiou, P.P., Lin, C.M., Perez, L., Chen, T.T. (2002) Effect of cecropin B and a synthetic analogue on propagation of fish viruses *in vitro*, Mar. Biotechnol. **4**, 294–302, with permission.

Additionally, melanin-concentrating hormone (MCH) isolated from chum salmon has also been exploited to generate transgenic medaka fish with altered body color. So far, researchers have successfully produced several colorful transgenic zebrafish displaying whole-body green, red, yellow, or orange fluorescent colors under daylight, dim light, or UV light. Some of these fish have been sterilized to avoid contaminating the wild population if they should accidentally be released from aquariums to the environments. For research purpose, mutant zebrafish and medaka with less interfering pigmentation for optical observation have also been developed. One such example is a line of "see-through" transgenic medaka which are transparent throughout their entire life, thus allowing clear visualization of GFP that is introduced into the fish as a reporter. The application of transgenic fish as biomonitors will be discussed in the next section.

6.5
Transgenic Fish as Environmental Biomonitors

Fish have long been used as models in environmental toxicology studies. Prompted by concerns of human health, tissues of wild-caught fish have been monitored as indicators for the presence of dangerous pollutants such as polycyclic hydrocarbons, oxidants, and heavy metals in the waters or in fish themselves. Different assays

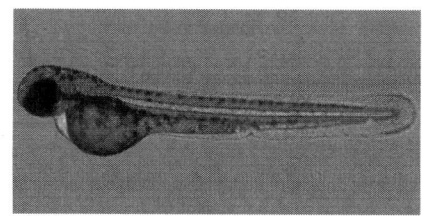
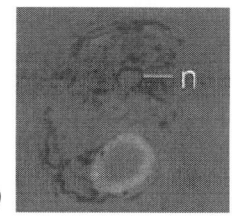

(a) (b)

Fig. 4 Examples of transgenic zebrafish that express GFP or RFP reporter in specific tissues. (a) A "green-heart" fish. Transgenic zebrafish carrying GFP reporter placed under control of the promoter of zebrafish cardiac myosin light chain 2 gene (*cmlc2*). Green fluorescence was intensively and specifically expressed in the myocardial cells located both around the heart chambers and the atrioventricular canal. The transgenic fish is kindly provided by Dr. Tsai (Huang, C.J., Tu, C.T., Hsiao, C.D., Hsieh, F.J., Tsai, H.J. (2003). Germ-line transmission of a myocardium-specific GFP transgene reveals critical regulatory elements in the cardiac myosin light chain 2 promoter of zebrafish, *Dev. Dyn.* **228**, 30–40). (b) A fish with "red guts." Transgenic zebrafish carrying RFP reporter placed under control of the promoter of zebrafish intestinal fatty acid binding protein gene (*i-fabp*). Homogenous red fluorescent signals are shown in the gut epithelial cells as demonstrated in this transverse section of a transgenic larva. n: notochord. From Her, G.M., Yeh, Y.H., Wu, J.L. (2003) 435-bp liver regulatory sequence in the liver fatty acid binding protein (L-FABP) gene is sufficient to modulate liver regional expression in transgenic zebrafish, *Dev. Dyn.* **227**, 347–356, with permission.

Tab. 4 Relative percent survival (RPS) of F_2 cecropin transgenic medaka challenged with *P. fluorescens* and *V. anguillarum*.

Transgenic lines	*P. fluorescene RPS*[a]	*V. anguillarum RPS*[a]
Preprocecropin B: CMV – pre – pro – cecropin B	100% (1)	50% (1)
Procecropin B: CMV – Ig – pro – cecropin B	72–100% (5)	25–50% (3)
Cecropin B: CMV – Ig – cecropin B	72–100% (2)	ND
Cecropin P1: CMV – Ig – cecropin P1	100% (2)	75% (1)

[a] RPS = [1 − %mortality of F_2 transgenic fish/%mortality of F_2 nontransgenic fish] × 100 CMV: cytomegalovirus promoter; Ig: catfish IgG signal peptide; ( ): number of F_2 families tested; ND: nondetermined.
Source: Adapted from Sarmasik, A., Warr, G., Chen, T.T. (2002) Production of transgenic medaka with increased resistance to bacterial pathogens, *Mar. Biotechnol.* **4**, 310–322 with permission.

have been employed to measure biological parameters that are impacted by toxic pollutants, including DNA damage, defense enzymes (e.g. glutathione peroxidase and superoxide dismutase), genes inducible by toxic chemicals (e.g. cytochrome P450 1A1 and 1A2), and factors that regulate redox potential (e.g. glutathione and ascorbic acid). Although these assays are very sensitive, they require specialized techniques and equipment, and thus cannot be performed in the field when the samples are collected. In addition, the obtained data from wild-caught fish samples give no indication of the exact time when those fish were exposed to the toxicants.

Transgenic fish can potentially serve as sensitive sentinels for aquatic pollution, with the benefit of avoiding the inconveniences mentioned above. The principle of such biomonitor fish is to introduce into the fish reporters that are easy to observe and are under the control of promoter/enhancer elements responsive to pollutants. GFP and its variants are thus the best reporters of choice for easy optical observation, whereas luciferase is for extreme sensitivity. So far, researchers have made significant progress in establishing transgenic zebrafish carrying luciferase or GFP reporter under control of the aromatic hydrocarbon response elements (AHREs), electrophile response elements (EPREs), or metal response elements (MREs). The specificity of each type of response elements is as follows: AHREs respond to numerous polycyclic hydrocarbons and halogenated coplanar molecules such as 2,3,7,8-tetrachlorodibenzo-p-dioxin (TCDD) and polychlorinated biphenyls; EPREs respond to quinones and numerous other potent electrophilic oxidants; MREs respond to heavy metal cations such as mercury, copper, nickel, cadmium, and zinc. In the zebrafish cell line ZEM2S, these three types of elements can drive the expression of luciferase reporter in a dose-dependent, chemical-class-specific manner in response to more than 20 environmental pollutants. The expression of GFP transgenes can be detected in the F_1 and, sometimes, in the F_2 generations; however, the transgenes are eventually lost following that. There are still technical obstacles in generating transgenic biomonitor fish, but the problems can be possibly overcome in the near future.

6.6
Other Examples of Transgenic Fish

Other applications include the use of transgenic fish as a model for studying mechanisms of seawater adaptation and generating cold-resistant fish carrying antifreeze proteins (AFPs). AFPs are found in many cold-water fish species, such as the winter flounder, ocean pout, and sea raven. Expression of *afp* genes permit these fish to survive in freezing seawater under as low as $-1.8\,^\circ$C. However, AFP transgenic Atlantic salmon does not demonstrate a measurable increase of cold tolerance.

Transgenic fish can potentially serve as bioreactors for producing therapeutic proteins beneficial to human health. The advantages of fish as bioreactors include (1) relatively low cost as compared to mammals such as cows; (2) easy maintenance; (3) producing large quantity of progeny; and (4) most importantly, no known viruses or prions have been reported to infect both human and fish. Successful expression of bioactive human proteins such as coagulation factor VII in fish embryos demonstrate this application of transgenic fish to produce therapeutic chemicals for human in the future.

7 Concerns and Future Perspectives

Transgenic fish serve as very valuable model systems for different disciplines of biological research as well as for human disease modeling. The application of transgenic fish technology to produce fish with beneficial traits, such as environmental biomonitors and GM food, is also on the rise. There has been groundbreaking discoveries made in the past few years based on the transgenic fish models, and such application is expected to continue to thrive in the future. The development of transgenic fish as environmental biomonitors is also encouraging and will continue to expand. The technique to produce transgenic fish has improved tremendously in the last two decades; nonetheless, there are still concerns about low efficiency, mosaic expression, nonfish components of the transgenes, and the long generation time for certain fish species. The following advances would thus be essential in developing the next generation transgenic fish: (1) developing more efficient mass gene transfer technologies; (2) developing targeted gene transfer technology such as embryonic stem cell gene transfer method; (3) identifying more suitable fish-origin promoters to direct the expression of transgenes at optimal levels and desired time; and (4) developing methods to shorten the time required to generate homozygous offspring.

Despite the promising future of transgenic fish as model systems in biological research, there are great concerns about the environmental impacts of such applications of transgenic fish technology to produce GM food or hobby pets. On the basis of mathematical modeling, a "Trojan Gene Effect" has been demonstrated in the male transgenic medaka carrying salmon growth hormone gene. Release of such fish into the environments could ultimately lead to the extinction of the wild type population owing to their advantage in mating with the wild type female. The study provides a sound analysis of the potential impact of the accidental release of transgenic fish into the environment, reminding us of the delicate interaction between a newly introduced species and the native population and environment. However, we would reason that the potential environmental impact by transgenic fish is controllable and even avoidable. First of all, the "Trojan Gene Effect" might not necessarily exist for other types of transgenes or in other fish species. In practice, transgenic fish can be maintained in confined areas to avoid the escape of fish into the nearby waters. Moreover, as the desirable genetic traits can be introduced into fish, the undesirable behavior traits could be possibly removed from the transgenic population by screening the behavior pattern of each transgenic individual. Ultimately, establishment of sterile transgenic population, such as a sterile triploid all-female strain, can be applied to avoid the spreading of transgenes into the wild population. Ideally, such sterility should be reversible; otherwise it would mean the loss of the goose that lays golden eggs. Finally, from a historical perspective, the concerns on the environmental impacts in the early days of the recombinant DNA technology have not been realized and many lessons can be learned from this experience.

Acceptance of GM fish as food by society is a key factor for the development of transgenic fish as a new source of cheaper and high-quality food. Largely, the concern is due to the public perception that GM food is unsafe to human health. Safety is the ultimate goal of GM fish development and relies on thorough characterization of the transgenes and safety assessment of the GM fish, resembling the development of new drugs. Unfortunately, it is a fact that there is always a subpopulation of people who are allergic, to different degrees, to many of the existing natural food sources, such as peanuts and shellfish. Therefore, the challenge is great in developing GM fish with absolutely no adverse effect to every single individual in the human population. A thorough evaluation process to fully analyze any health risk of GM fish to human should be established and the information should be clearly delivered to the public and well marked on the food product. Development of "gene-inactivated" transgenic fish may be more acceptable by people who are concerned of the safety of the "new" genes that are introduced into fish. One candidate of such GM fish is transgenic fish strains with their myostatin gene being inactivated. Inactivation of myostatin gene causes double muscling in animals, resulting in significant increase in mass of skeletal muscle. Natural mutation of myostatin gene has been found in the meaty Belgian Blue and Piedmontes cattles, and the double muscling effect has been observed in mice whose myostatin gene is artificially inactivated. Such "gene-inactivation" approach would be very valuable in producing transgenic fish with more flesh and higher food conversion rate, particularly for the economically important aquaculture species.

In summary, the application of transgenic fish has resulted in many valuable discoveries in basic biological science, many of which could lead to the next breakthrough for new cures for human diseases. On the other hand, the potential of GM fish as bioreactors or as new food sources should not be deterred by the environmental and safety concerns but instead it requires more research efforts in the future to develop new GM species with beneficial traits that are also safe to human health and carry no harm to the environment.

See also Molecular Basis of Genetics; Zebrafish (*Danio rerio*) Genome and Genetics.

Bibliography

Books and Reviews

Boman, H.G. (1995) Peptide antibiotics and their role in innate immunity, *Annu. Rev. Immunol.* **13**, 61–92.

Chen, T.T., Powers, D.A. (1990) Transgenic fish, *Trends Biotechnol.* **8**, 209–215.

Chen, T.T., Lin, C-M., Zhu, Z., Gonzalez-Villasenor, L.I., Dunham, R.A., Powers, D.A. (1990) Gene transfer, expression and inheritance of rainbow trout growth hormone genes in carp and loach, in: Church, R. (Ed.) *Transgenic Models in Medicine and Agriculture*, Wiley, New York, pp. 127–139.

Di Berardino, M.A., McKinnell, R.G., Wolf, D.P. (2003) The golden anniversary of cloning: a celebratory essay, *Differentiation* **71**, 398–401.

Duxbury, M.S., Whang, E.E. (2004) RNA interference: a practical approach, *J. Surg. Res.* **117**, 339–344.

Götz, J., Streffer, J.R., David, D., Schild, A., Hoerndli, F., Pennanen, L., Kurosinski, P., Chen, F. (2004) Transgenic animal models of Alzheimer's disease and related disorders: histopathology, behavior and therapy, *Mol. Psychiatry* **9**, 1–20.

Hackett, P.B. (1993) The molecular biology of transgenic fish, in: Hochachka, P.W., Mommsen, T.P. (Eds.) *Biochemistry and Molecular Biology of Fish*, 2, Elsevier, B.V., pp. 207–240.

Muller, F., Blader, P., Strahle, U. (2002) Search for enhancers: teleost models in comparative genomic and transgenic analysis of cis regulatory elements, *BioEssays* **24**, 564–572.

Stern, H.M., Zon, L.I. (2003) Cancer genetics and drug discovery in the zebrafish, *Nat. Rev. Cancer* **3**, 533–539.

Udvadia, A.J., Linney, E. (2003) Windows into development: historic, current, and future perspectives on transgenic zebrafish, *Dev. Biol.* **256**, 1–17.

Primary Literature

Agellon, L.B., Emery, C.J., Jones, J.M., Davies, S.L., Dingle, A.D., Chen, T.T. (1988) Growth hormone enhancement by genetically engineered rainbow trout growth hormone, *Can. J. Fish. Aquat. Sci.* **45**, 146–151.

Amsterdam, A., Lin, S., Hopkins, N. (1995) The *Aequorea victoria* green fluorescent protein can be used as a reporter in live zebrafish embryos, *Dev. Biol.* **171**, 123–129.

Ballagi-Pordany, A., Ballagi-Pordany, A., Funa, K. (1991) Quantative determination of mRNA phenotypes by polymerase chain reaction, *Anal. Biochem.* **196**, 88–94.

Boonanuntanasarn, S., Yoshizaki, G., Takeuchi, T. (2003) Specific gene silencing using small interfering RNAs in fish embryos, *Biochem. Biophys. Res. Commun.* **310**, 1089–1095.

Boonanuntanasarn, S., Yoshizaki, G., Takeuchi, Y., Morita, T., Takeuchi, T. (2002) Gene knock-down in rainbow trout embryos using antisense Morpholino phosphorodiamidate oligonucleotides, *Mar. Biotechnol.* **4**, 256–266.

Buono, R.J., Linser, P.J. (1992) Transient expression of RSVCAT in transgenic zebrafish made by electroporation, *Mol. Mar. Biol. Biotechnol.* **1**, 271–275.

Burns, J.C., Friedmann, T., Driever, W., Burrascano, M., Yee, J.K. (1993) VSV-G pseudotyped retroviral vector: concentration to very high titer and efficient gene transfer into mammalian and nonmammalian cells, *Proc. Natl. Acad. Sci. USA* **90**, 8033–8037.

Carvan, M.J. III, Dalton, T.P., Stuart, G.W., Nebert, D.W. (2000) Transgenic zebrafish as sentinels for aquatic pollution, *Ann. N.Y. Acad. Sci.* **919**, 133–147.

Chen, T.T., Powers, D.A., Lin, C.M., Kight, K., Hayat, M., Chatakondi, N., Ramboux, A.C., Duncan, P.L., Dunham, R.A. (1993) Expression and inheritance of RSVLTR-rtGH1 cDNA in common carp, *Cyprinus carpio*, *Mol. Mar. Biol. Biotechnol.* **2**, 88–95.

Chiou, P.P., Lin, C.M., Perez, L., Chen, T.T. (2002) Effect of cecropin B and a synthetic analogue on propagation of fish viruses *in vitro*, *Mar. Biotechnol.* **4**, 294–302.

Cotten, M., Jennings, P. (1989) Ribozyme mediated destruction of RNA *in vivo*, *EMBO J.* **8**, 3861–3866.

Du, S.J., Gong, G.L., Fletcher, G.L., Shears, M.A., King, M.J., Idler, D.R., Hew, C.L. (1992) Growth enhancement in transgenic Atlantic salmon by the use of an "all fish" chimeric growth hormone gene construct, *Biotechnology* **10**, 176–181.

Dunham, R.A., Chatakondi, N., Nichols, A.J., Kucuktas, H., Chen, T.T., Powers, D.A., Weete, J.D., Cummins, K., Lovell, R.T. (2002) Effect of rainbow trout growth hormone complementary DNA on body shape, carcass yield, and carcass composition of F1 and F2 transgenic common carp (*Cyprinus carpio*), *Mar. Biotechnol.* **4**, 604–611.

Dunham, R.A., Ramboux, A.C., Duncan, P.L., Hayat, M., Chen, T.T., Lin, C-M., Kight, K., Gonzalez-Villasenor, L.I., Powers, D.A. (1992) Transfer, expression and inheritance of salmonid growth hormone genes in channel catfish, *Ictalurus punctatus*, and effects on performance traits, *Mol. Mar. Biol. Botechnol.* **1**, 380–389.

Fletcher, G.L., Davis, P.L. (1991) Transgenic fish for aquaculture, in: Setlow, J.K. (Ed.) *Genetic Engineering* Vol. 13, Plenum Publishing, New York, pp. 331–370.

Fletcher, G.L., Shears, M.A., King, J.J., Davis, P.L., Hew, C.L. (1988) Evidence of antifreeze protein gene transfer in Atlantic salmon (*Salmo salar*), *Can. J. Fish. Aquat. Sci.* **45**, 352–357.

Fronhoffs, S., Totzke, G., Stier, S., Wernert, N., Rothe, M., Bruning, T., Koch, B., Sachinidis, A., Vetter, H., Ko, Y. (2002) A method for the rapid construction of cRNA standard curves in quantitative real-time reverse transcription polymerase chain reaction, *Mol. Cell. Probes* **16**, 99–110.

Gill, J.A., Stumper, J.P., Donaldson, E.M., Dye, H.M. (1985) Recombinant chicken and bovine growth hormone in cultured juvenile Pacific salmon, *Biotechnology* **3**, 4306–4310.

Goldman, D., Hankin, M., Li, Z., Dai, X., Ding, J. (2001) Transgenic zebrafish for studying nervous system development and regeneration, *Transgenic Res.* **10**, 21–33.

Gong, Z., Wan, H., Tay, T.L., Wang, H., Chen, M., Yan, T. (2003) Development of transgenic fish for ornamental and bioreactor by strong expression of fluorescent proteins in the skeletal muscle, *Biochem. Biophys. Res. Commun.* **308**, 58–63.

Gong, Z., Ju, B., Wang, X., He, J., Wan, H., Sudha, P.M., Yan, T. (2002) Green fluorescent protein expression in germ-line transmitted transgenic zebrafish under a stratified epithelial promoter from keratin8, *Dev. Dyn.* **223**, 204–215.

Hall, G.F., Lee, S., Yao, J. (2002) Neurofibrillary degeneration can be arrested in an in vivo cellular model of human tauopathy by application of a compound which inhibits tau filament formation in vitro, *J. Mol. Neurosci.* **19**, 253–260.

Hall, G.F., Lee, V.M., Lee, G., Yao, J. (2001) Staging of neurofibrillary degeneration caused by human tau overexpression in a unique cellular model of human tauopathy, *Am. J. Pathol.* **158**, 235–246.

Halloran, M.C., Sato-Maeda, M., Warren, J.T., Su, F., Lele, Z., Krone, P.H., Kuwada, J.Y., Shoji, W. (2000) Laser-induced gene expression in specific cells of transgenic zebrafish, *Development* **127**, 1953–1960.

Her, G.M., Chiang, C.C., Wu, J.L. (2004) Zebrafish intestinal fatty acid binding protein (I-FABP) gene promoter drives gut-specific expression in stable transgenic fish, *Genesis* **38**, 26–31.

Her, G.M., Yeh, Y.H., Wu, J.L. (2003) 435-bp liver regulatory sequence in the liver fatty acid binding protein (L-FABP) gene is sufficient to modulate liver regional expression in transgenic zebrafish, *Dev. Dyn.* **227**, 347–356.

Higashijima, S., Hotta, Y., Okamoto, H. (2000) Visualization of cranial motor neurons in live transgenic zebrafish expressing green fluorescent protein under the control of the islet-1 promoter/enhancer, *J. Neurosci.* **20**, 206–218.

Higashijima, S., Masino, M.A., Mandel, G., Fetcho, J.R. (2003) Imaging neuronal activity during zebrafish behavior with a genetically encoded calcium indicator, *J. Neurophysiol.* **90**, 3986–3997.

Higashijima, S., Okamoto, H., Ueno, N., Hotta, Y., Eguchi, G. (1997) High-frequency generation of transgenic zebrafish which reliably express GFP in whole muscles or the whole body by using promoters of zebrafish origin, *Dev. Biol.* **192**, 289–299.

Holden, C. (2003) "That special glow" (a news report in Random Samples). Holden, C. (Ed.), *Science* **300**, 1368.

Howard, R.D., DeWoody, J.A., Muir, W.M. (2004) Transgenic male mating advantage provides opportunity for Trojan gene effect in a fish, *Proc. Natl. Acad. Sci. USA* **101**, 2934–2938.

Hsiao, C.D., Hsieh, F.J., Tsai, H.J. (2001) Enhanced expression and stable transmission of transgenes flanked by inverted terminal repeats from adeno-associated virus in zebrafish, *Dev. Dyn.* **220**, 323–336.

Huang, C.J., Tu, C.T., Hsiao, C.D., Hsieh, F.J., Tsai, H.J. (2003) Germ-line transmission of a myocardium-specific GFP transgene reveals critical regulatory elements in the cardiac myosin light chain 2 promoter of zebrafish, *Dev. Dyn.* **228**, 30–40.

Huang, H., Vogel, S.S., Liu, N., Melton, D.A., Lin, S. (2001) Analysis of pancreatic development in living transgenic zebrafish embryos, *Mol. Cell. Endocrinol.* **177**, 117–124.

Hwang, G., Muller, F., Rahman, M.A., Williams, D.W., Murdock, P.J., Pasi, K.J., Goldspink, G., Farahmand, H., Maclean, N. (2004) Fish as Bioreactors: Transgene Expression of Human Coagulation Factor VII in Fish Embryos, *Mar. Biotechnol.* (http://www.springerlink.com/media/3G267U4FJG5JWHE03N2G/contributions/C/X/A/N/CXAN5MRVF4817U62_html/fulltext.html. Published online Apr. 29, 2004)

Inoue, K., Takei, Y. (2003) Asian medaka fishes offer new models for studying mechanisms of seawater adaptation, *Comp. Biochem. Physiol. B Biochem. Mol. Biol.* **136**, 635–645.

Jessen, J.R., Willett, C.E., Lin, S. (1999) Artificial chromosome transgenesis reveals long-distance negative regulation of rag1 in zebrafish, *Nat. Genet.* **23**, 15–16.

Jesuthasan, S., Subburaju, S. (2002) Gene transfer into zebrafish by sperm nuclear transplantation, *Dev. Biol.* **242**, 88–95.

Jia, X., Patrzykat, A., Devlin, R.H., Ackerman, P.A., Iwama, G.K., Hancock, R.E. (2000) Antimicrobial peptides protect coho salmon from Vibrio anguillarum infections, *Appl. Environ. Microbiol.* **66**, 1928–1932.

Kambadur, R., Sharma, M., Smith, T.P., Bass, J.J. (1997) Mutations in myostatin (GDF8) in double-muscled Belgian Blue and Piedmontese cattle, *Genome Res.* **7**, 910–916.

Kinoshita, M., Morita, T., Toyohara, H., Hirata, T., Sakaguchi, M., Ono, M., Inoue, K., Wakamatsu, Y., Ozato, K. (2001) Transgenic medaka overexpressing a melanin-concentrating hormone exhibit lightened body color but no remarkable abnormality, *Mar. Biotechnol.* **3**, 536–543.

Knaut, H., Steinbeisser, H., Schwarz, H., Nusslein-Volhard, C. (2002) An evolutionary conserved region in the vasa 3'UTR targets RNA translation to the germ cells in the zebrafish, *Curr. Biol.* **12**, 454–466.

Krøvel, A.V., Olsen, L.C. (2002) Expression of a vas::EGFP transgene in primordial germ cells of the zebrafish, *Mech. Dev.* **116**, 141–150.

Langenau, D.M., Traver, D., Ferrando, A.A., Kutok, J.L., Aster, J.C., Kanki, J.P., Lin, S., Prochownik, E., Trede, N.S., Zon, L.I., Look, A.T. (2003) Myc-induced T cell leukemia in transgenic zebrafish, *Science* **299**, 887–890.

Lawson, N.D., Mugford, J.W., Diamond, B.A., Weinstein, B.M. (2003) Phospholipase C gamma-1 is required downstream of vascular endothelial growth factor during arterial development, *Genes Dev.* **17**, 1346–1351.

Lawson, N.D., Weinstein, B.M. (2002) In vivo imaging of embryonic vascular development using transgenic zebrafish, *Dev. Biol.* **248**, 307–318.

Lee, K.Y., Huang, H., Ju, B., Yang, Z., Lin, S. (2002) Cloned zebrafish by nuclear transfer from long-term-cultured cells, *Nat. Biotechnol.* **20**, 795–799.

Lewis, J.L., Bonner, J., Modrell, M., Ragland, J.W., Moon, R.T., Dorsky, R.I., Raible, D.W. (2004) Reiterated Wnt signaling during zebrafish neural crest development, *Development* **131**, 1299–1308.

Li, Q., Shirabe, K., Kuwada, J.Y. (2004) Chemokine signaling regulates sensory cell migration in zebrafish, *Dev. Biol.* **269**, 123–136.

Lin, S., Gaiano, N., Culp, P., Burns, J.C., Friedmann, T., Yee, J-K., Hopkins, N. (1994) Integration and germ-line transmission of a pseudotyped retroviral vector in zebrafish, *Science* **265**, 666–668.

Liu, Y., Berndt, J., Su, F., Tawarayama, H., Shoji, W., Kuwada, J.Y., Halloran, M.C. (2004) Semaphorin3D guides retinal axons along the dorsoventral axis of the tectum, *J. Neurosci.* **24**, 310–318.

Long, Q., Meng, A., Wang, H., Jessen, J.R., Farrell, M.J., Lin, S. (1997) GATA-1 expression pattern can be recapitulated in living transgenic zebrafish using GFP reporter gene, *Development* **124**, 4105–4111.

Lu, J.K., Burns, J.C., Chen, T.T. (1994) Retrovirus-mediated transfer and expression of transgenes in medaka, in: *Proceedings of the Third International Marine Biotechnology Conference at Tromso, Norway*, pp. 72.

Lu, J.K., Fu, B.H., Wu, J.L., Chen, T.T. (2002) Production of transgenic silver sea bream (*Sparus sarba*) by different gene transfer methods, *Mar. Biotechnol.* **4**, 328–337.

Lu, J.K., Chen, T.T., Allen, S.K., Matsubara, T., Burns, J.C. (1996) Production of transgenic dwarf surfclams, *Mulinia lateralis*, with Pantropic retroviral vectors, *Proc. Natl. Acad. Sci. USA* **93**, 3482–3486.

Lu, J.K., Chrisman, C.L., Andrisani, O.M., Dixon, J.E., Chen, T.T. (1992) Integration expression and germ-line transmission of foreign growth hormone genes in medaka, *Oryzias latipes, Mol. Mar. Biol. Biotechnol.* **1**, 366–375.

Martinez, R., Estrada, M.P., Berlanga, J., Guillen, I., Hernandez, O., Cabrera, E., Pimentel, R., Morales, R., Herrera, F., Morales, A., Pina, J.C., Abad, Z., Sanchez, V., Melamed, P., Lleonart, R., de la Fuente, J. (1996) Growth enhancement in transgenic tilapia by ectopic expression of tilapia growth hormone, *Mol. Mar. Biol. Biotechnol.* **5**, 62–70.

McPherron, A.C., Lee, S.J. (1997) Double muscling in cattle due to mutations in the myostatin gene, *Proc. Natl. Acad. Sci. USA* **94**, 12457–12461.

Moav, Boaz, Liu, Z., Groll, Y., Hackett, P.R. (1992) Selection of promoters for gene transfer into fish, *Mol. Mar. Biol. Biotechnol.* **1**, 338–345.

Moriyama, S., Takahashi, A., Hirano, T., Kawauchi, H. (1990) Salmon growth hormone is transported into the circulation of rainbow trout (*Oncorhynchus mykiss*) after intestinal administration, *J. Comp. Physiol. B* **160**, 251–260.

Neumann, E., Schaefer-Ridder, M., Wang, Y., Hofschneider, P.H. (1982) Gene transfer into mouse lyoma cells by electroporation in high electric fields, *EMBO J.* **1**, 841–845.

Ozato, K., Kondoh, H., Inohara, H., Iwanatsu, T., Wakamatsu, Y., Okada, T.S. (1986) Production of transgenic fish: introduction and expression of chicken δ-crystallin gene in medaka embryos, *Cell Differ.* **19**, 237–244.

Palmiter, R.D., Brinster, R.D. (1986) Germ-line transformation in mice, *Ann. Rev. Genet.* **20**, 465–499.

Park, H.C., Appel, B. (2003) Delta-Notch signaling regulates oligodendrocyte specification, *Development* **130**, 3747–3755.

Park, H.C., Kim, C.H., Bae, Y.K., Yeo, S.Y., Kim, S.H., Hong, S.K., Shin, J., Yoo, K.W., Hibi, M., Hirano, T., Miki, N., Chitnis, A.B., Huh, T.L. (2000) Analysis of upstream elements in the HuC promoter leads to the establishment of transgenic zebrafish with fluorescent neurons, *Dev. Biol.* **227**, 279–293.

Pauls, S., Geldmacher-Voss, B., Campos-Ortega, J.A. (2001) A zebrafish histone variant H2A.F/Z and a transgenic H2A.F/Z:GFP fusion protein for in vivo studies of embryonic development, *Dev. Genes Evol.* **211**, 603–610.

Paynter, K., Chen, T.T. (1991) Biological activity of biosynthetic rainbow trout growth hormone in the eastern oyster, (*Crassostrea virginica*), *Biol. Bull.* **181**, 459–462.

Perz-Edwards, A., Hardison, N.L., Linney, E. (2001) Retinoic acid-mediated gene expression in transgenic reporter zebrafish, *Dev. Biol.* **229**, 89–101.

Petrausch, B., Tabibiazar, R., Roser, T., Jing, Y., Goldman, D., Stuermer, C.A., Irwin, N., Benowitz, L.I. (2000) A purine-sensitive pathway regulates multiple genes involved in axon regeneration in goldfish retinal ganglion cells, *J. Neurosci.* **20**, 8031–8041.

Pi, W., Yang, Z., Wang, J., Ruan, L., Yu, X., Ling, J., Krantz, S., Isales, C., Conway, S.J., Lin, S., Tuan, D. (2004) The LTR enhancer of ERV-9 human endogenous retrovirus is active in oocytes and progenitor cells in transgenic zebrafish and humans, *Proc. Natl. Acad. Sci. USA* **101**, 805–810.

Potter, H., Weir, L., Leder, P. (1984) Enhancer-dependent expression of human k immunoglobulin genes introduced into mouse pre-B lymphocytes by electroporation, *Proc. Natl. Acad. Sci. USA* **81**, 7161–7165.

Powers, D.A., Hereford, L., Cole, T., Creech, K., Chen, T.T., Lin, C.M., Kight, K., Dunham, R.A. (1992) Electroporation: a method for transferring genes into gametes of zebrafish (*Brachydanio rerio*), channel catfish (*Ictalurus punctatus*), and common carp (*Cyprinus carpio*), *Mol. Mar. Biol. Biotechnol.* **1**, 301–308.

Rahman, M.A., Ronyai, A., Engidaw, B.Z., Jauncey, K., Hwang, G-L., Smith, A., Roderick, E., Penman, D., Varadi, L., Maclean, N. (2001) Growth and nutritional trials on transgenic Nile tilapia containing an exogenous fish growth hormone gene, *J. Fish Biol.* **59**, 62–78.

Reinhard, E., Nedivi, E., Wegner, J., Skene, J.H., Westerfield, M. (1994) Neural selective activation and temporal regulation of a mammalian GAP-43 promoter in zebrafish, *Development* **120**, 1767–1775.

Riddle, R., Pollock, J.D. (2003) Making connections: the development of mesencephalic dopaminergic neurons, *Brain Res. Dev. Brain Res.* **147**, 3–21.

Sarmasik, A., Warr, G., Chen, T.T. (2002) Production of transgenic medaka with increased resistance to bacterial pathogens, *Mar. Biotechnol.* **4**, 310–322.

Schreurs, R.H., Legler, J., Artola-Garicano, E., Sinnige, T.L., Lanser, P.H., Seinen, W., Van der Burg, B. (2004) In vitro and in vivo antiestrogenic effects of polycyclic musks in zebrafish, *Environ. Sci. Technol.* **38**, 997–1002.

Sekine, S., Miizukzmi, T., Nishi, T., Kuwana, Y., Saito, A., Sato, M., Itoh, H., Kawauchi, H. (1985) Cloning and expression of cDNA for salmon growth hormone in E. coli, *Proc. Natl. Acad. Sci. USA* **82**, 4306–4310.

Shears, M.A., Fletcher, G.L., Hew, C.L., Gauthier, S., Davies, P.L. (1991) Transfer, expression, and stable inheritance of antifreeze protein genes in Atlantic salmon (*Salmo salar*), *Mol. Mar. Biol. Biotechnol.* **1**, 58–63.

Shigekawa, K., Dower, W.J. (1988) Electroporation of eukaryotes and prokaryotes: a general approach to introduction of macromolecules into cells, *Biotechniques* **6**, 742–751.

Sin, F.Y., Walker, S.P., Symonds, J.E., Mukherjee, U.K., Khoo, J.G., Sin, I.L. (2000) Electroporation of salmon sperm for gene transfer: efficiency, reliability, and fate of transgene, *Mol. Reprod. Dev.* **56**(Suppl. 2), 285–288.

Son, O.L., Kim, H.T., Ji, M.H., Yoo, K.W., Rhee, M., Kim, C.H. (2003) Cloning and expression analysis of a Parkinson's disease gene, uch-L1, and its promoter in zebrafish, *Biochem. Biophys. Res. Commun.* **312**, 601–607.

Stuart, G.W., McMurray, J.V., Westerfield, M. (1988) Replication, integration, and stable germ-line transmission of foreign sequence injected into early zebrafish embryos, *Development* **109**, 403–412.

Stuart, G.W., Vielkind, J.V., McMurray, J.V., Westerfield, M. (1990) Stable lines of transgenic zebrafish exhibit reproduction patterns of transgene expression, *Development* **109**, 293–296.

Symonds, J.E., Walker, S.P., Sin, F.Y.T. (1994) Development of mass gene transfer method in Chinook salmon: optimization of gene transfer by electroporated sperm, *Mol. Mar. Biol. Biotechnol.* **3**, 104–111.

Thermes, V., Grabher, C., Ristoratore, F., Bourrat, F., Choulika, A., Wittbrodt, J., Joly, J.S. (2002) I-SceI meganuclease mediates highly efficient transgenesis in fish, *Mech. Dev.* **118**, 91–98.

Tomasiewicz, H.G., Flaherty, D.B., Soria, J.P., Wood, J.G. (2002) Transgenic zebrafish model of neurodegeneration, *J. Neurosci. Res.* **70**, 734–745.

Tseng, F.S., Lio, I.C., Tsai, H.J. (1994). Introducing the exogenous growth hormone cDNA into loach (*Misgurnus anguillicaudatus*) eggs via electroporated sperms as carrier, in: *3rd International Marine Biotechnology Conference*, Tromso, Norway. abstract pp. 71.

Wakamatsu, Y., Pristyazhnyuk, S., Kinoshita, M., Tanaka, M., Ozato, K. (2001b) The see-through medaka: a fish model that is transparent throughout life, *Proc. Natl. Acad. Sci. USA* **98**, 10046–10050.

Wakamatsu, Y., Ju, B., Pristyaznhyuk, I., Niwa, K., Ladygina, T., Kinoshita, M., Araki, K., Ozato, K. (2001a) Fertile and diploid nuclear transplants derived from embryonic cells of a small laboratory fish, medaka (Oryzias latipes), *Proc. Natl. Acad. Sci. USA* **98**, 1071–1076.

Wan, H., He, J., Ju, B., Yan, T., Lam, T.J., Gong, Z. (2002) Generation of two-color transgenic zebrafish using the green and red fluorescent protein reporter genes gfp and rfp, *Mar. Biotechnol.* **4**, 146–154.

Wang, T.M., Chen, Y.H., Liu, C.F., Tsai, H.J. (2002) Functional analysis of the proximal promoter regions of fish rhodopsin and myf-5 genes using transgenesis, *Mar. Biotechnol.* **4**, 247–255.

Wilmut, I., Schnieke, A.E., McWhir, J., Kind, A.J., Campbell, K.H. (1997) Viable offspring derived from fetal and adult mammalian cells, *Nature* **385**, 810–813.

Xiao, T., Shoji, W., Zhou, W., Su, F., Kuwada, J.Y. (2003) Transmembrane sema4E guides branchiomotor axons to their targets in zebrafish, *J. Neurosci.* **23**, 4190–4198.

Zhang, P., Hayat, M., Joyce, C., Gonzalez-Villasenor, L.I., Lin, C-M., Dunham, R., Chen, T.T., Powers, D.A. (1990) Gene transfer, expression and inheritance of pRSV-Rainbow Trout-GH-cDNA in the carp, *Cyprinus carpio* (Linnaeus), *Mol. Reprod. Dev.* **25**, 3–13.

Zhu, Z., Li, G., He, L., Chen, S.Z. (1985) Novel gene transfer into the goldfish (*Carassius auratus* L 1758), *Angew Ichthyol.* **1**, 31–34.

30
Transgenic Mice in Biomedical Research

J. Willem Voncken and Marten Hofker
University of Maastricht, Maastricht, The Netherlands

1	**A Brief History of Mice** 866	
1.1	Mouse Genetics 866	
1.2	Molecular Mouse Genetics 868	
1.2.1	Brief Overview of Developments in Transgenesis 869	
1.2.2	Brief Overview of Developments in Gene Targeting 870	
2	**Transgenesis; Basics and Technology** 871	
2.1	Transgene Design 871	
2.1.1	Origin and Structure of the Transgene 871	
2.1.2	Regulatory Elements 873	
2.2	Pronuclear Injection 876	
2.3	Biotechnological Procedures 876	
3	**Gene Targeting; Basics and Technology** 879	
3.1	Targeting Vector Design 879	
3.2	Embryonic Stem (ES) Cells 882	
3.3	Genesis of Mosaic Embryos 882	
3.4	Biotechnological Procedures 884	
4	**Refinements to the Models** 885	
4.1	Fully ES Cell–derived Embryos 885	
4.2	Conditional Transgenesis 887	
4.3	Conditional Gene Targeting 888	
4.3.1	Controlled CRE Expression 890	
4.4	Knockins: Humanized Mice 891	
5	**Random Mutagenesis in ES Cells** 894	
5.1	Gene Trap Mutagenesis 895	

Genomics and Genetics. Edited by Robert A. Meyers.
Copyright © 2007 Wiley-VCH Verlag GmbH & Co. KGaA, Weinheim
ISBN: 978-3-527-31609-0

| 5.2 | Retroviruses and Transposable Elements in Random Mutagenesis 896 |
| 5.3 | Random Enu Mutation 897 |

6 Phenotype Analysis 897

7 Perspectives 898
7.1 Partial Transgenesis; Viral Shuttles 898
7.2 RNA Interference; Posttranscriptional Gene Silencing 899

Bibliography 900
Books and Reviews 900
Primary Literature 901
Websites (References Indicated as they Appear in the Original Text) 903

Keywords

Autologous
Derived from endogenous sequences.

Binary Model
Mouse model in which gene activity can be switched on and off.

Chimeric Mouse
Mouse derived from donor ES cells and recipient embryo, displaying coat color mosaicism.

Chromosome Substitution
A single full-length chromosome from one inbred strain has been transferred onto the genetic background of a second strain by repeated backcrossing.

Conditional Model
Mouse model in which a modification can be activated on command.

CRE-LoxP
Bacterio phage P1-derived recombination system: the recombinase CRE-mediated recombination between two identical recombinase recognition sites: LoxP.

CRE-MER
Fusion of a CRE recombinase with a mutated ligand-binding domain of the estrogen receptor.

Ectopic Expression
Expression out of place or out of time.

Embryonic Stem Cells
Pluripotent stem cells derived from preimplantation embryos.

Gene Targeting
Modification of endogenous genes through use of replacement vectors.

Gene Trapping
Process by which transcription of a marker gene is coupled to that of an endogenous locus in which the gene trap vector is inserted and transcripts are trapped.

Heterologous
Derived from exogenous sequences.

Inbred Strain
A mouse strain bred to homozygocity for all loci by 20 brother-to-sister back crosses.

Insertional Mutagenesis
Gene modification by insertion of foreign DNA vectors.

Isogenic DNA
DNA derived from the same genetic background.

Knockin
Targeted mutation of an existing locus, mostly replacing endogenous coding sequences with functional sequences

Pluripotent
Unlimited differentiation capacity.

Recombinant Inbred Strain
Recombinant inbred strains are derived from crosses of two inbred strains.

RNA interference
RNA-mediated posttranscriptional gene silencing.

Spatio-temporal Expression Pattern
Gene activity determined in time and location.

Tetraploid
Containing four haploid genome copies.

Transgene
A(n expression) construct that is inserted into the mouse genome.

Genetically modified mouse models have become a crucial tool in present day biomedical research. Modifying genes in mice provides an excellent approach to unravel gene function at the molecular and cellular level, as well as its role in the physiology and pathology of the intact organism. By doing so, numerous models have been developed that reliably replicate diseases in humans, thereby providing insight in disease mechanisms and paving the way toward prevention and therapy. Crucial for these advances has been the ability to modulate gene expression at will. It is possible to increase or decrease gene expression, or eliminate the expression of a gene completely. These alterations can be made cell type–specific and even inducible or reversible. Moreover, it is possible to replace mouse genes with the cognate human gene carrying a specific mutation. This chapter will deal with the essential approaches in transgenic mouse research and its impact on biomedical research.

1
A Brief History of Mice

1.1
Mouse Genetics

The resources directed toward studying the mouse have been exponentially growing. After the completion of the human genome sequence, the mouse was the second mammalian genome for which the complete sequence was determined. There is no mammalian system that comes close to rival the molecular and genetic possibilities that the mouse has to offer to researchers. Many physiological methods that worked fine in larger experimental animals have been miniaturized to make use of the genetic approaches uniquely available to mouse researchers. What makes the mouse so special?

The advantages that come to mind first are obvious: for biomedical researchers, mammals are highly preferred model systems, because the physiology of man and most mammals is very similar and the pathology of most diseases is reproduced well in animal models. The mouse stands out from the other mammals as being the smallest. Mouse husbandry is therefore easy and economical; its requirements for food are modest, and mice are excellent breeders with a short generation time, allowing around four successive generations each year. Upon completion of the human and mouse genome sequence, the usefulness of the mouse as a model animal for the human has been further reinforced. The number of genes in both species is around 30 000. With the exception of about 200 genes, which were shown unique to mouse or man, the vast majority of genes is conserved between mouse and man. These combined properties make the mouse an excellent animal model to study gene function and extrapolate the outcome to human disease.

Second, the mouse has advantages that may be less obvious, but which relate to the long tradition of mouse research starting in the early 1900s. Up until then, hobbyist mouse keepers had already been breeding mice for their coat color variation. By making use of these mice, it was possible to demonstrate that this variation followed the laws of Mendelian inheritance. Soon, it was also realized that there was a need for genetically homogeneous mouse strains and the most widely used *inbred strains* such as BALB/c and C57BL6 originate from that period. An excellent historical overview on mouse genetics is provided by Dr. Silver.

At present, hundreds of inbred strains have been generated (see Sect. 6). This resource is the key to the success of mouse research. Inbred strains ensure that experiments can be carried out in the absence of inter-individual genetic variation, because mice within an inbred strain are all *genetically identical* to each other and *homozygous* throughout the genome. Moreover, each strain has one or more unique and, often, well-established properties that make the strain particularly useful for a specific area of research. For instance, some mouse strains appeared susceptible to cancer or cardiovascular diseases, while others are better suitable for behavioral research. The genetics of such complex diseases and biological processes is very difficult to study in man. This is mainly due to the fact that multiple genes are involved and that the genetic basis for susceptibility to, for instance, cancer will vary from person to person. In the mouse, it is possible to crossbreed a susceptible and a resistant strain. As a result, the susceptibility to cancer will vary in the progeny. Genetic variation in these mice is confined to the differences between the two parental strains, and every given locus will only have two different alleles. Hence, the genetics is confined to a maximum of two alleles per locus, which greatly simplifies genetic analysis. These properties allowed generating specific resources to study complex diseases, including *recombinant inbred* (RI) strains (http://jaxmice.jax.org/info/recombin-bred.html). These RI strains combine a specific genetic contribution of one mouse strain with the genetic background of another strain. Along these lines, a very powerful resource has been developed recently, consisting of a panel of inbred mouse strains carrying only a single chromosome from the other strain (http://jaxmice.jax.org/library/notes/493c.html). These *chromosome substitution* strains are very similar to recombinant inbred strains, but offer a less costly strategy to move from a susceptibility locus toward the gene proper. The principles of mouse genetics are crucial when working with transgenic animals, because it is vital to control the genetic background in the experiments (see below): the interaction between the transgene and the genetic background will ultimately determine the phenotype; for example, changing the genetic background on a given genetic trait may prevent an early death of a particular mouse model.

While mouse genetics has a major impact on our understanding of the mechanisms underlying complex diseases, a constant influx of mice carrying single gene mutations greatly stimulates mouse research. These mutations have either occurred spontaneously in mouse breeding programs or originated from research on the gene toxicity of radiation and chemical compounds. The latter approach has been rejuvenated upon the availability of better technology to identify the causative mutations (see below). A good example of a spontaneous mutation is the leptin gene mutation that occurred in the C57BL6 mouse. The mutation was named *ob*, after obese, because the mice have no control over their feeding behavior and become overwhelmingly obese. Hence, the phenotype could be discovered easily. The subsequent discovery of the causative gene marks one of the most dramatic discoveries in this research field.

1.2 Molecular Mouse Genetics

Despite all the possibilities for the basic biomedical research offered by "traditional" mouse genetics and the discoveries of relevant mutations through random mutagenesis, the true breakthrough for the mouse as the animal of choice came with the accomplishment of (1) transgenesis through injecting DNA in fertilized oocytes and (2) gene targeting via homologous recombination in embryonic stem cells. The latter mice are commonly

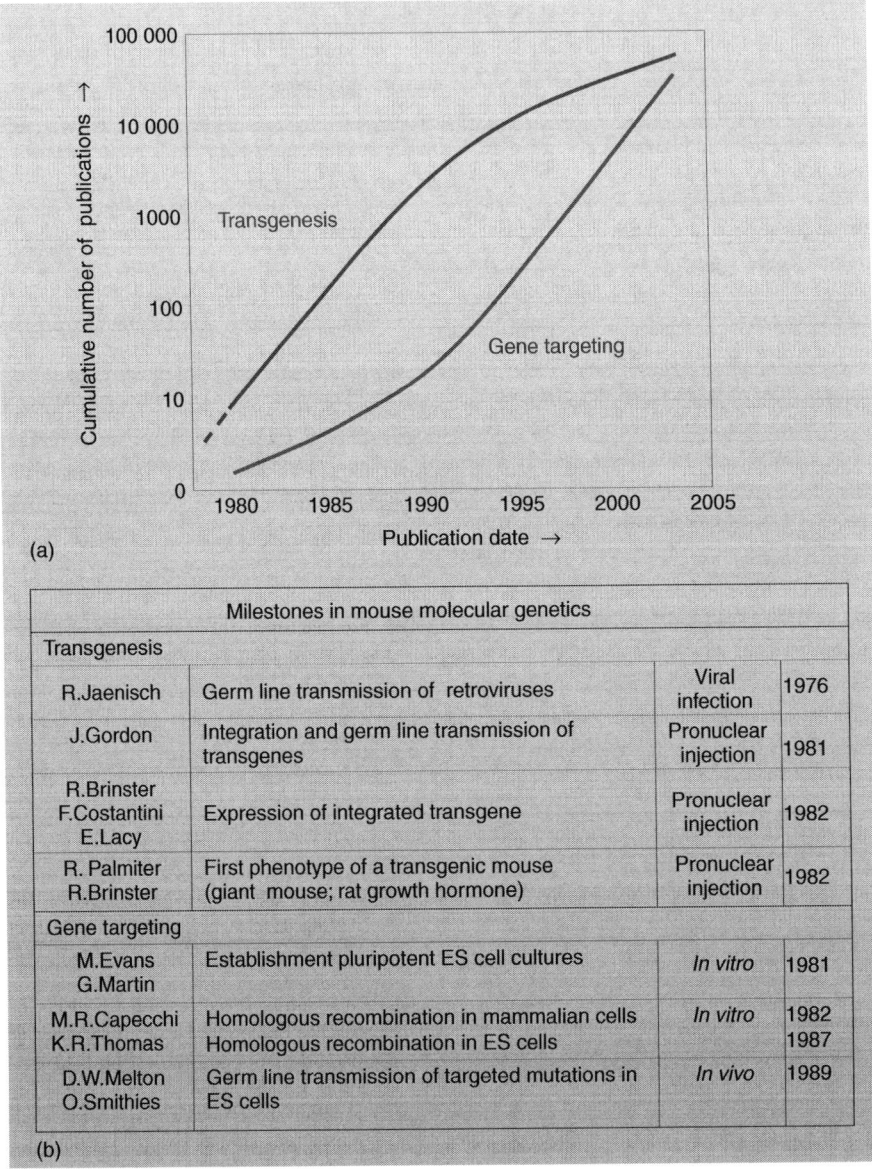

known as *knockout* mouse models. This combination of technologies offers the possibility to design models with "gain of function" and "loss of function," which is an ideal situation for a (mouse) geneticist (Fig. 1a).

Notably, these technological breakthroughs were facilitated by the availability of inbred mouse strains. Transgenesis works most efficient when the donor eggs are produced by first generation (F1) hybrids between C57Bl6 and BALB/c. In the case of the gene-targeting approach using embryonic stem cells, it is critical to make use of the appropriate strain combinations when injecting the cells into blastocysts and propagating the embryos in foster strains. This dependence on the genetic background provides the ability to choose the optimal strain and may be one of the reasons why, mice are, as of yet, the only mammalian species suitable for gene targeting (see Sect. 3). The goal of altering the gene function through transgenesis is to generate animal models in which the role(s) of the gene of interest either in normal (e.g. development) or abnormal biological processes (e.g. disease) is revealed. Other important applications in fundamental science include the study of mammalian gene expression regulation. Currently, genetic modification is carried out in a wide range of invertebrate and vertebrate species, including mammals such as mice, rats, rabbits, and livestock such as sheep, pigs, cattle, and recently even primates. The choice of the models is primarily determined by its objective (e.g. improvement of economical value or development of pharmaceutical proteins). Although fundamental biological processes can be well studied in relatively simple organisms such as the nematode *Caenorhabditis elegans*, or the fruitfly *Drosophila melanogaster*, for experimental biomedical research the laboratory mouse is by far the most used animal model. Below, we will briefly outline the emergence of the two technologies and sketch the initial experimental strategies that caused great excitement within the scientific community.

1.2.1 Brief Overview of Developments in Transgenesis

By classical definition, a *transgenic* organism carries extra, often foreign (i.e. from a different organism) DNA in its genome, called a *transgene*. Transgenesis literally crosses (*trans*) species barriers. One of the most obvious conceptual characteristics is that a transgene integrates at random in the recipient host genome, whereas gene targeting, by homologous recombination, selectively alters an existent locus. A number of milestone achievements,

---

Fig. 1 Genetically modified mouse models in biomedical science. (a) A rough estimation of published scientific output using genetically modified mouse models. The blue line represents reports on transgenic models and the red line on gene targeting. Note: with the advent of conditional gene targeting, which utilizes transgenesis in conjunction with targeted mutagenesis, a strict separation of transgenesis and gene targeting into the twentieth century is difficult. (b) Milestones in mouse molecular genetics. The first report on germ-line transmission of integrated retroviruses (via infection) dates back to the mid-1970s. Subsequently, pronuclear microinjection of fertilized oocytes and subsequent proof of transgene integration, expression, and germ-line transmission, were pioneered in different laboratories. Similarly, breakthroughs in the use of embryonic stem (ES) cell technology are listed. Many other excellent laboratories have since contributed to refinement of technologies and strategies; these are listed throughout this chapter (see color plate p. x).

which were pivotal for the development of modern-day mouse molecular genetics, are summarized in Fig. 1(b).

In the mid 1970, robust technology to clone genes and manipulate genes was still about five years away. Subsequently, as the structure of eukaryotic genes was revealed and recombinant DNA technology developed, gene constructs could be generated comprising an eukaryotic promoter fused to the full-length coding sequence of the gene of interest (consisting of a DNA copy (cDNA) of the mRNA molecule). Generating mice with a strong promoter driving an extra copy of the gene resulted in mouse models with high-level expression of the gene of interest. Interesting phenotypes did occur when the gene was markedly overexpressed. In addition, however, genes would become expressed in the wrong tissue or at the wrong place. This *ectopic* expression was also a powerful way to obtain phenotypic information. Initially, researchers were often confronted with a high failure rate: gene constructs would be integrated in high-copy head-to-tail arrays (see Sect. 2), but would not show a corresponding increase in expression level, or transgene expression would diminish in consecutive generations. A much-improved understanding of eukaryotic gene regulation in the mid-1980s provided the insight, leading to improved construct design. For instance, DNA elements located tens of kilobases outside of the β-globin gene were shown to control high-level tissue-specific gene expression through interaction with DNA sequences immediately 5′-prime of the gene. Also, sequence elements were discovered that isolate a specific gene from influences of flanking DNA. In addition, evidence was obtained for a role of splicing in mRNA stability. These advances led to the use of much larger constructs harboring the complete intron–exon structure of the genes, and including large up- and downstream DNA segments. Such fragments were obtained through cloning in cosmids, bacterial artificial chromosomes and yeast artificial chromosomes, allowing handling DNA fragments in the range from 40 up to 1000 kb.

1.2.2 Brief Overview of Developments in Gene Targeting

The development of gene targeting by homologous recombination in embryonic stem cells (ES) can be attributed to two crucial technological advances that ultimately compounded. One line of research involves ES research. Initially, ES cells were used to study cellular differentiation and cancer, because when injected subcutaneously into nude mice, teratomas would develop. Figure 1(b) itemizes a number of crucial advances in ES cell technology that helped develop ES cell technology to its current status. Improved culturing conditions ensured sustenance of pluripotence, a property of ES cells crucial for their application as vehicles to introduce germ-line mutations. Soon after the first mouse was made that showed germ-line transmission of genes introduced through genetic modification of ES cells *in vitro*, evidence was obtained that an endogenous locus could be exchanged for a mutated sequence by homologous recombination *in vitro* (see Sect. 3). These strategies targeted the murine *HPRT* gene and were based on the fact that HPRT is an excellent selectable marker expressed in the genome on the X chromosome of male cells. Spurred by success, the question arose whether gene targeting would also be feasible for nonselectable autosomal genes; this was proven possible shortly after.

Sections 2 and 3 will elaborate on the conventional transgenesis and gene-targeting technology. Noteworthy, with the introduction of novel concepts in molecular genetics, our ability to manipulate gene expression in the mouse currently knows few limitations. Sections 4, 5, and 7 of this chapter will elaborate on some of the more recent technological advances such as binary systems (see Sect. 4), random mutagenesis (see Sect. 5), and RNA interference (see Sect. 7).

2
Transgenesis; Basics and Technology

2.1
Transgene Design

Shortly after its introduction in the early 1980s, the main scientific applications of transgenesis in the mouse could be roughly divided in two areas: (1) studies of gene promoter activity (and/or other regulatory elements) and (2) studies of the effect(s) of gene overexpression (or a mutated form thereof; Fig. 2a). Gene activity is, simply put, controlled by a *promoter*. A promoter is a stretch of DNA sequence, which usually precedes the coding part of a gene. A promoters' task is to regulate gene expression: it controls the "when (at which time point) and where" (in what cells, tissues) of gene activity. Eukaryotic genes also carry a termination element. In order to function properly in a eukaryotic genome, a transgene minimally meets these requirements: it carries a promoter, a coding part, and a stop signal. The easiest way to read out promoter activity in cell or a developing animal is by placing a so-called *reporter gene* downstream of it. Reporter genes encode proteins, often found in lower organisms such as yeast, fireflies, or jellyfish, which reveal their presence by a microscopically or macroscopically clearly visible mark (Fig. 2a, b). This approach is standardly used to examine *spatio-temporal* gene expression patterns and/or tissue specificity of native promoter sequences *in vivo* (Fig. 2b).

By far, transgenesis is used to achieve and/or study the effects of overexpression (higher gene activity than normal) or *ectopic* expression (gene activity in different cell types than normal) of a gene of interest, or of a mutated form thereof. The common denominator of these studies is the analysis of the resulting altered phenotype. This can be abnormal development, altered physiology or behavior, development of disease, and so on. The choice of regulatory sequences (see below) determines whether the expression of a transgene follows the expression pattern of its endogenous counterpart or is limited to distinct cell types or particular developmental stages.

2.1.1 Origin and Structure of the Transgene

The origin of a transgene, or elements therein, may range from prokaryotes or from lower (e.g. reporter genes) to higher eukaryotes, like mice or men. In experimental biomedical research, human transgenes offer several advantages. Importantly, many known genetic disorders in humans have been extensively characterized at the molecular genetic level: mutations or deletions are charted and such "diseased alleles" are readily available. In addition, most genes and proteins are well conserved between mouse and man. This, of course, is important when the biological activity of a gene product depends on proper interaction with other cellular proteins. Finally, the sequence divergence between mouse (transgene recipient) and human (transgene donor) of a given gene can be used to determine whether a mouse is transgenic or not, what its copy number is (number of transgene integrations), and whether a transgene is actively expressed, since it allows for discrimination between transgene and the endogenous gene. Figure 3 gives an example of such an analysis.

Most eukaryotic genes display a typical intron/exon structure, with only exon sequences represented in mRNA. Transgenes, in general, are more reliably expressed if the intron/exon boundaries are preserved in a transgene construct. Solely cDNA-based expression vectors (i.e. exons sequences only) frequently show low expression levels and are often silenced. Since many integrating DNA or RNA viruses do not carry the eukaryotic intron/exon structures, an expressed genetic sequence without such boundaries may be recognized as foreign and potentially harmful by the eukaryotic host cell and will be inactivated (silenced) by cellular defense mechanisms. The native intron/exon structure of a gene need not be preserved in its entirety: indeed, inclusion of only one intron in a transgene has been shown

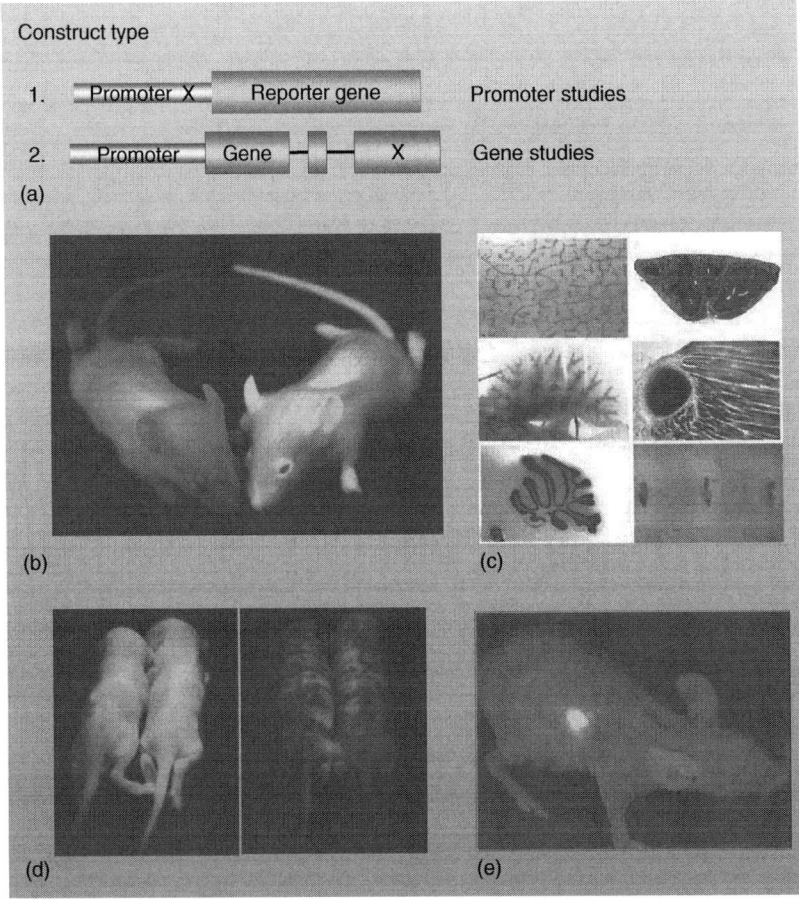

to augment transgene expression. Many successful transgene constructs include both genomic and cDNA sequences derived form the same gene (Fig. 4a).

2.1.2 Regulatory Elements

Any transgene comprises a number of essential elements that control gene expression (Fig. 4b). The most basic and essential elements are the promoter and termination signals for transcription (RNA synthesis). A promoter controls gene expression by providing binding sites for proteins (RNA polymerase transcriptional machinery, cell type–specific transcription factors) that regulate gene transcription (i.e. activation). The real-life situation is often much more complicated than this; many more regulatory elements exist that all play a role in establishing stable and faithful gene expression patterns. Among these are enhancers, which typically act in an orientation-independent manner, Matrix attachment regions (MARs), scaffold attachment regions (SARs), Locus control regions (LCR), chromosomal insulators, and antirepressor elements. Some of these elements are involved in subnuclear localization of genes to areas of active transcription and/or insulate gene expression from influences of surrounding chromatin. Some of these autologous (i.e. endogenous) regulatory elements may actually be many thousands of base pairs removed from the coding sequences of a gene. Indeed, if the purpose of a study is to examine faithful gene expression patterns, for instance throughout embryonic development, a transgene should include such endogenous elements. The exact location of such elements within a locus is often not known and hence there is an obvious advantage in using large DNA segments as transgenes. The development of molecular vectors, which can harbor large pieces of DNA, such as so-called artificial

---

Fig. 2 Transgenesis; concepts. (a) Original scientific applications of transgenesis in the mouse were roughly divided in two areas: (1) studies of gene promoter activity (and/or other regulatory elements) and (2) studies of the effect(s) of overexpression (or a mutated form) of gene X. Promoter studies typically make use of *reporter* genes (green box) that reveal their presence through specific catalytic activity. Often-used reporter genes include the green fluorescent protein jelly fish (b, d), bacterial beta-galactosidase (c), (also see Fig. 3), and fire fly luciferase (d). (b) Transgene expression is controlled by a keratinocyte-specific promoter (source image: http://www.mshri.on.ca/nagy/cre.htm; Maatman, R., Gertsenstein, M., de Meijer, E., Nagy, A. et al. (2003) Aggregation of embryos and embryonic stem cells, *Methods Mol. Biol.* **209**, 201–230). (c) Differential expression of beta-galactosidase (upper left to lower right): Lymphatic vessels (intestinal surface), marginal zones between red and white pulp (spleen), bronchial epithelium (lungs), Large skeletal muscles (limb), granule cell layer (cerebellum), Intervertebral discs (source images: Valenzuela, D.M., Murphy, A.J., Frendewey, D., Gale, N.W. et al. (2003) High-throughput engineering of the mouse genome coupled with high-resolution expression analysis, *Nat. Biotechnol.* **21**, 652–659). (d) Fluorescent images of an X-linked enhanced green fluorescent protein (EGFP) transgene; transgenic male offspring (left) of a mating between an X-linked EGFP transgenic female and a nontransgenic male have ubiquitous green fluorescence in the skin owing to the presence of a single active X chromosome. Transgenic female pups (right) have mosaic (tortoiseshell) green fluorescence in the skin owing to random inactivation of one X chromosome (source image: Hadjantonakis, A.K., Dickinson, M.E., Fraser, S.E., Papaioannou, V.E. (2003) Technicolour transgenics: imaging tools for functional genomics in the mouse, *Nat. Rev. Genet.* **4**, 613–625). (e) An external image of adenovirus-encoded GFP gene expression acquired from a nude mouse in the light box 72 h after portal vein injection (see color plate p. xii).

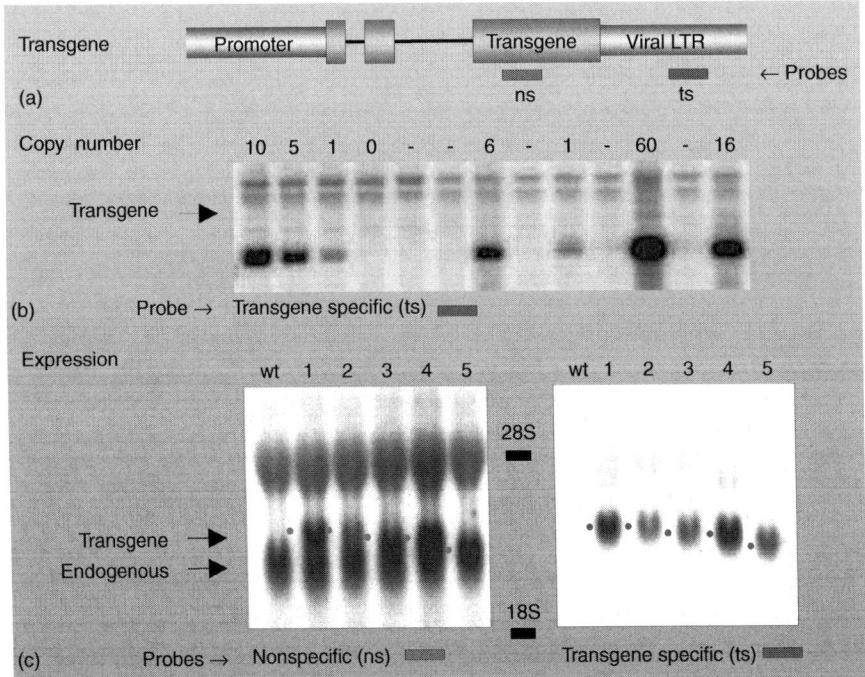

Fig. 3 Transgene construction and detection. (a) Schematic representation of a hypothetical transgene. The transgene is derived from mouse genomic sequences (brown). Transgene expression is under control of a general promoter and terminates in a retroviral long terminal repeat (LTR) element. Probes used for detection of transgene DNA or RNA are depicted as a brown box (nonspecific (ns) probe, which detects both the endogenous gene and the transgene) or a blue box (a transgene-specific (ts) probe, which only detects the transgene). (b) Copy number determination by Southern blot analysis with a ts probe. Bold print above image represent DNA samples with known copy numbers, resp. 10, 5, 1, and no transgene copies inserted; the remaining lanes show four founders with varying copy numbers (image adapted from:(http://www.healthsystem.virginia.edu/internet/transgenic-mouse/southerndesign.cfm)) (c) Detection of transgene expression by Northern blot hybridization clearly demonstrates the need for transgene specific probes (right panel). Left panel: expression signal of endogenous gene masks transgene expression (see color plate p. xiii).

chromosomes that can be propagated in yeast (YACs), phages P1 (bacterial viruses; PACs), or bacteria (BACs) has been imperative for the cloning and expression of such transgenic constructs. Exceedingly large constructs (>500 000 bp) are first transferred into embryonic stem cells (see Sect. 3), which are then used to generate transgenic mice.

Regulatory elements need not necessarily be autologous in nature, they may be heterologous. For instance, when overexpression or ectopic expression is required, general type heterologous promoters (e.g. viral promoters) and/or enhancers are widely used. The use of heterologous and autologous regulatory elements is often combined. The first published transgenic mouse model displaying a specific phenotype made use of the general-type metallothionein promoter (pMT). The MT promoter is inducible *in vitro* and *in*

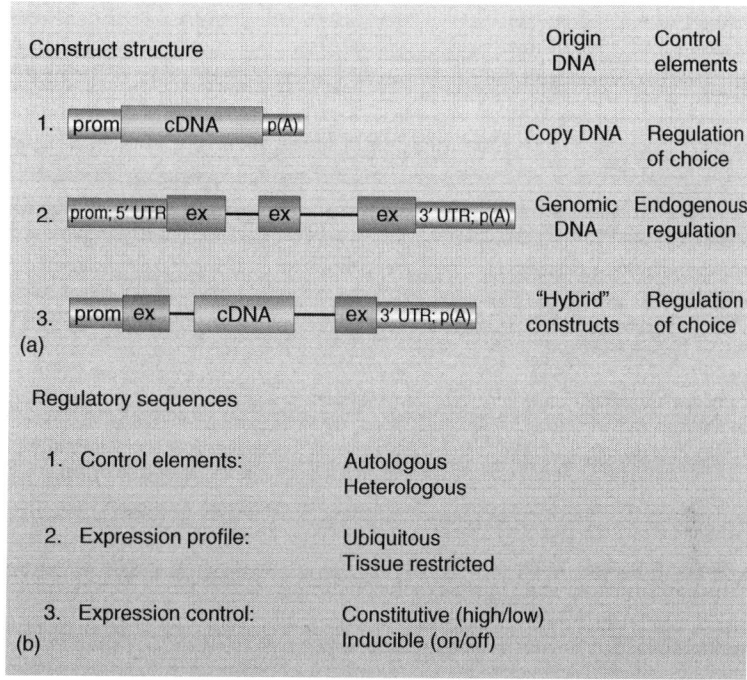

Fig. 4 Transgene structure and regulation. (a) Schematic overview of basic transgenic construct design. For all constructs (1,2,3), eukaryotic coding sequence is the starting material. Regulatory sequences are cloned separately into the construct, which is entirely cDNA based (1). Endogenous regulatory sequences may be included when genomic DNA is used (2). A transgene can be tailored to specific requirements. The hybrid genomic/cDNA construct depicted (3) combines intro-exon boundary inclusion with facile cloning options: the use of cDNA may reduce the size of a vector considerably. prom: promoter sequences, ex: exon, cDNA: copy DNA, 3' UTR: 3-prime untranslated region, p(A): polyadenylation signal. (b) Regulatory sequences in transgene design. Depending on the specific purpose of the model, many different systems are available to control transgene expression. As indicated, regulatory element can be autologous (i.e. derived from an endogenous locus) or heterologous in nature (1). The required transgene expression profile may need to be ubiquitous, or cell type specific. This determines the choice of promoter (2). Finally, the model may require controlled expression; instead of constitutive expression, inducible expression would be preferred (3). (Figures adapted from: Voncken, J.W. (2003b) Transgene design, *Methods Mol. Biol.* **209**, 51–67).

vivo with glucocorticoids, heavy metals, or bacterial endotoxin (LPS). Heterologous promoters and other regulatory elements are applied widely, and as indicated before, their incorporation into constructs is determined primarily by the aim of the animal model (see Fig. 2). Not only general-type promoters, like the ones driving housekeeping genes (e.g. *phosphoglycerate kinase PGK*) or other genes that are widely expressed throughout eukaryotic cell types (e.g. *histones, β-actin*), but also viral promoters are often used to ensure high and ubiquitous transgene expression. LCRs are included in transgenic constructs to both enhance transgene expression and to achieve position-independent and copy number–dependent expression of a transgene, often with cell lineage–specific enhancer activity. Finally, promoter choice is often dictated by whether expression of a transgene is deleterious to the mouse or not. For example, many oncogenes play an important role in embryonic development, interfering with their normal expression (pattern) results in embryonic lethality. To bypass this problem, either tissue-specific or inducible promoters (or combinations of such regulatory systems are used (see Sect. 4)).

Taken together, the origin of DNA, structure and choice of regulatory elements for a transgene are largely determined by the scientific aim of the experimental model itself. For biomedical studies employing the laboratory mouse as a model system, the use of human transgenes is often preferred.

2.2
Pronuclear Injection

Transgenes can be introduced into living cells in many different ways. Among the most standard procedures *in vitro* ("in the test tube") are transfection methods either using calcium phosphate/DNA precipitates or liposomes (small DNA-containing vesicles surrounded by a lipid membrane), which are taken up by cells, electroporation (electroshock), viral vectors (e.g. adenovirus, retrovirus, which carry the transgene), particle bombardment, and microinjection. Some of these methods, however, cannot be used in combination with mouse oocytes. The very first proof-of-principle report on germ-line transmission of a foreign DNA sequence in the mouse was achieved through retroviral transduction. As indicated above, viruses of this class are often inactivated in eukaryotic cells. Although recent technological developments in the field of transgenesis again make use of retroviral constructs (see Sect. 7), pronuclear injection has been the golden standard for generating transgenic animals over the last two decades.

The principle of pronuclear injection is simple: a sterile DNA solution is prepared, which contains a very low concentration of the transgene (nanograms ($10e{-}12$ kg)/microliter ($10e{-}15$ l)). A very fine glass needle containing the transgene DNA solution is inserted into one of the pronuclei and DNA solution is expelled into the nucleoplasma (Fig. 5a, 6a). The amount of DNA injected is extremely low (femtograms ($10e{-}18$ kg)/picoliters ($10e{-}12$ l)). Injected transgene DNA in the nucleus tends to integrate at random into the genome. Often, this occurs in head-to-tail tandem arrays called *concatamers*. (Fig. 5b). The repetitive nature of such concatamers may lead to transgene inactivation, as does the random character of the insertion: it often causes transgenes to land in inactive heterochromatin, which constitutes most of a eukaryote cells' DNA. The use of insulator elements (see Sect. 2.1) or targeted transgenesis (see Sect. 4) provides solutions to these problems.

2.3
Biotechnological Procedures

Figure 6 gives an overview of the biotechnological procedures involved in the production of transgenic mice. Because generating transgenic mice by pronuclear injection is a rather inefficient process, a relatively large number of fertilized oocytes

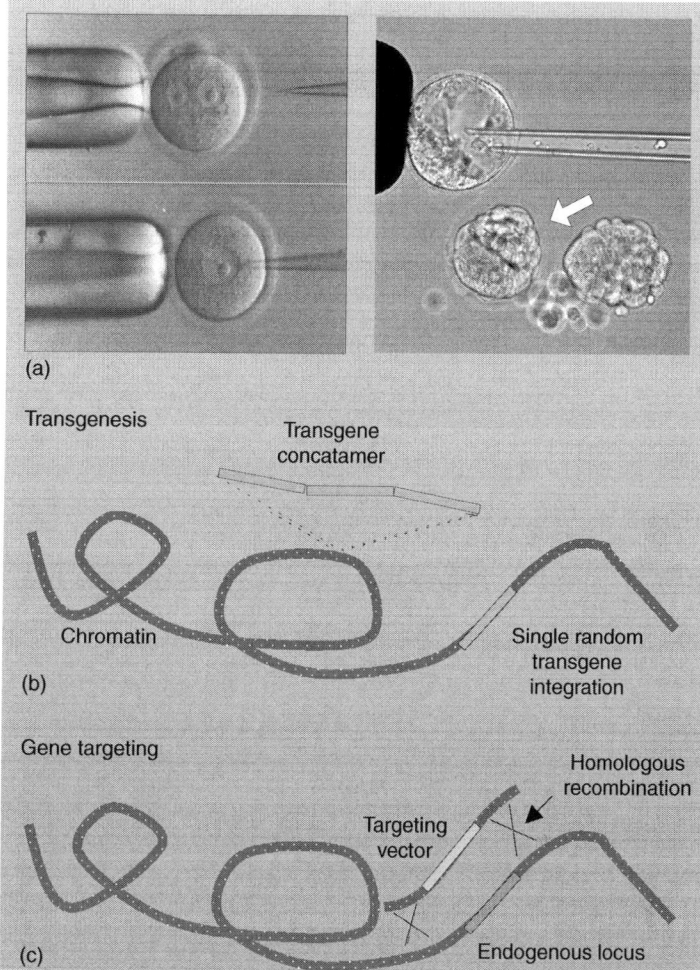

Fig. 5 Microinjection. (a) Pronuclear injection of DNA into a fertilized one-cell-stage embryo. In the early stages following fertilization, two pronuclei are visible in the zygote: one from the oocyte and one from the sperm cell; these will eventually fuse to form a diploid nucleus. Both prefusion pronuclei are clearly visible in the cytoplasm (the male pronucleus is mostly used for injection, since it is the largest) (left panel; source image: Hogan, B., Beddington, R., Costantini, F., Lacey, E. (1994) *Manipulating the Mouse Embryo. A Laboratory Manual*, 2nd edition, Cold Spring Harbor Laboratory Press, Cold Spring Harbor, New York). Microinjection of genetically modified ES cells into blastocysts; arrow indicates collapsed, injected blastocysts; ES cells are clearly visible in the blastocoel (right panel). (b) Schematic representation of random integration of transgenesis in the mouse genome: transgene either integrated single or as concatameric head-to-tail arrays. (c) Gene targeting is homology-driven: a targeting vector will exchange with an endogenous locus by homologous recombination.

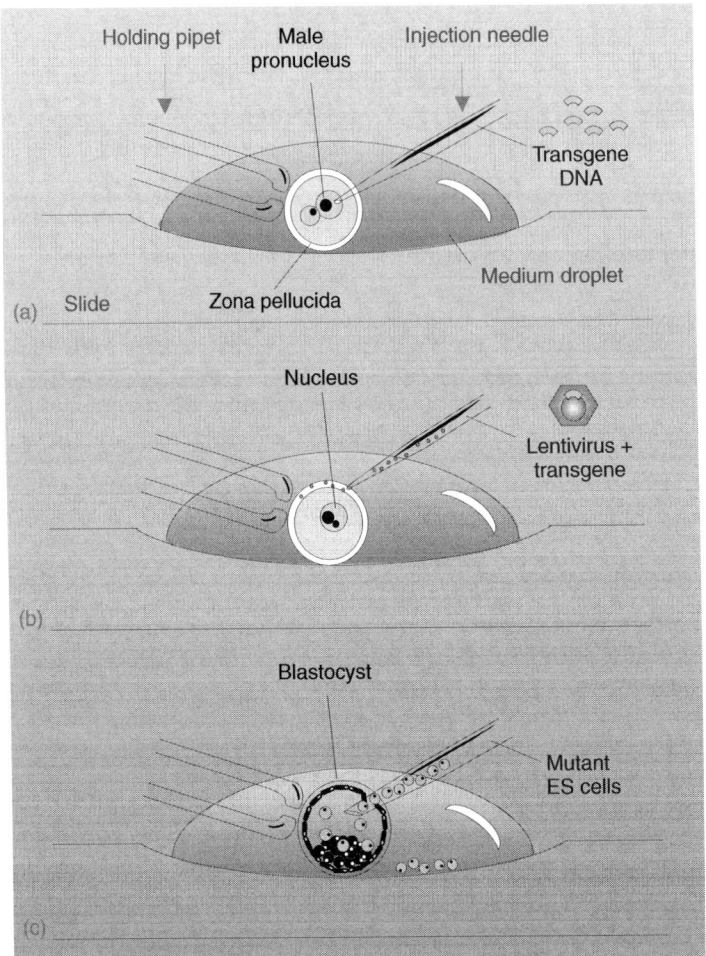

Fig. 6 Schematic representation of a number of techniques to generate genetically modified mouse models. (a) Pronuclear injection of transgene DNA. (b) Injection of infectious viral particle in the perivitellin space; the advantages of using lentiviral shuttle vector are significant: it obviates potentially harmful procedures otherwise used to introduce transgenes into a cell (pronuclear microinjection, electroporation) and offers some degree of control over transgene copy numbers. Lentiviral vectors are also used to generate transgenic lines through infection of ES cells. One of the few restrictions on recombinant lentiviral vector use is the size limitation on the transgene inserts. (c) Microinjection of pluripotent, genetically modified ES cells into blastocysts.

needs to be injected to obtain the so-called founder mice: transgenic founders developed *in utero* from successfully injected zygotes that carry an integrated transgene in their genome. A typical transgenic study uses at least three expressing transgenic founders to establish independent transgenic mouse lines from.

Convenient numbers of fertilized eggs are obtained by superovulation, much the same as way as in humans for *in vitro* fertilization purposes: exogenous hormones (follicle-stimulating hormone and luteinizing hormone analogs) are injected at a 48-h interval to achieve multiple ovulations per donor female, which are subsequently mated to fertile males. Simultaneously, recipient females, the so-called foster females, are endocrinologically primed for pregnancy by copulation to vasectomized males (see Fig. 7). Microsurgery is carried out under aseptic conditions and general anesthesia. A number of different mouse strains are useful for biotechnological procedures, many of which have their own advantages and specific traits (see Sect. 1). Detailed descriptions of the technology, useful strains, and husbandry are to be found in a number of media.

3
Gene Targeting; Basics and Technology

First established in the late 1980s and early 1990s of the last century, gene targeting is now a well-established technology for generating genetically modified mouse models. Gene targeting allows for germ-line manipulation at predetermined genomic loci by a process called *homologous recombination*: part of an endogenous allele is replaced or mutated by vector sequences that carry flanking sequences homologous to the endogenous gene (Sect. 3.1). It is in this aspect where gene targeting fundamentally differs from conventional transgenic technology: the possibility to change existing genes offers a very important and powerful extension of the molecular genetic tools to create experimental animal models in biomedical research. Whereas transgenesis asks for genes and their phenotypes to be dominant in nature (e.g. expression of (mutated) oncogenes), gene targeting now permits study of recessive gene function (e.g. tumor suppressor genes). The concept of gene targeting is simple: eradication of a gene may reveal its function by a resultant phenotype. Current applications range from conventional *null mutation* of a given gene, the so-called gene knockouts, to replacement of endogenous alleles with mutated alleles and even inducible (*conditional;* see Sect. 4.3) genetic modifications.

3.1
Targeting Vector Design

Generally speaking, two types of vectors exist by which genes can be mutated.

1. Replacement vectors
2. Random insertion vectors

All classical *replacement type vectors* have in common that they carry stretches of DNA, which are highly homologous or identical to the gene, which is to be targeted. The mechanism by which gene replacement is accomplished, *homologous recombination*, uses these identical or very similar sequences of DNA: homologous recombination is a sequence homology-initiated and driven process (Fig. 5c, 8a). It appears to be a highly selective process, which only takes place between relatively large (several kilobase pairs) DNA segments. In fact, it is important that the constructs used for homologous recombination are being derived from the exact same genetic background (*isogenic*) as the cell line used for gene targeting. This requirement implies that a sequence divergence of around 1 in 100 bp will cause a marked drop in

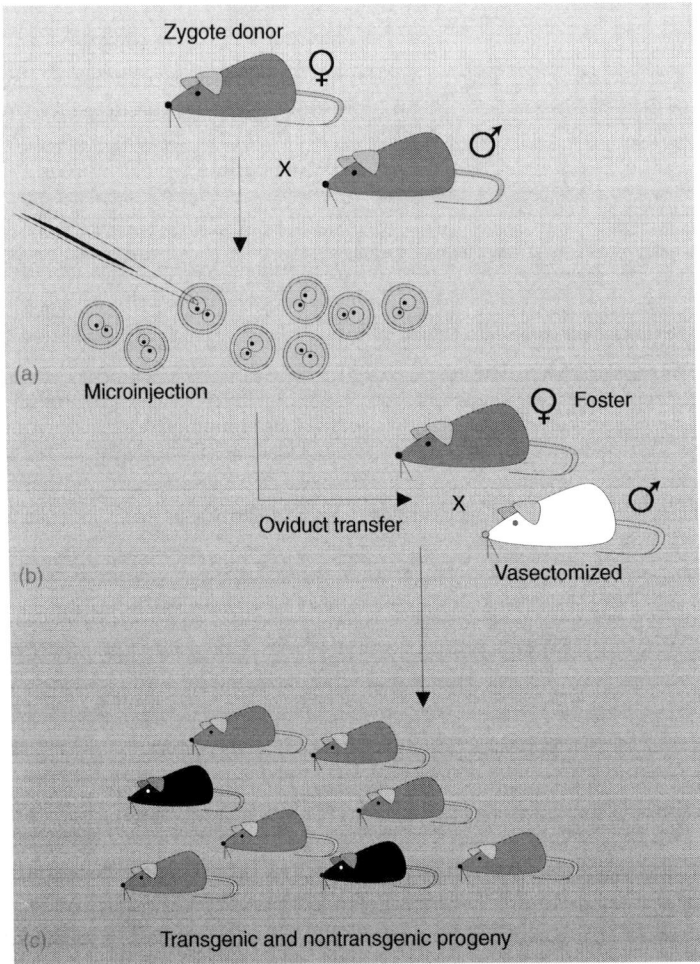

Fig. 7 Biotechnological procedures in conventional transgenesis. (a) Hybrid F1 animals are used for superovulation and mating to generate sufficient zygotes for injection. (b) Upon microinjection, the embryos are transferred into the oviduct of a pseudopregnant female; *pseudo-pregnancy*, is achieved by mating to a vasectomized male and exists for a maximum of 3 days; the surgical transfer of microinjected one- or two-cell-stage embryos directly into the fallopian tube at 0.5 day *post coitum* ensures sustenance of the pregnant state. (c) Offspring will manifest coat colors of both strains (e.g. C57Bl6 and CBA) used to generate the F1 hybrids (e.g. B6CBA/F1). Roughly, 10% of the offspring will carry an integrated transgene, and will be further examined for transgene expression and transmission. Such an animal is referred to as a *transgenic founder*. In addition, these media provide comprehensive information on historical and genetic backgrounds of in- and outbred strains on mouse embryology and dissection of specific developmental stages.

the frequency of homologous recombination. Conventional gene-targeting constructs mostly harbor a selection marker (neomycin, hygromycin, or puromycin resistance gene), which serves a number of purposes: it replaces endogenous gene sequences, thereby interrupting the gene and rendering it nonfunctional (Fig. 8a). In addition, the marker is used to positively select for embryonal stem cells that have integrated the targeting vector. A second feature often included in a replacement

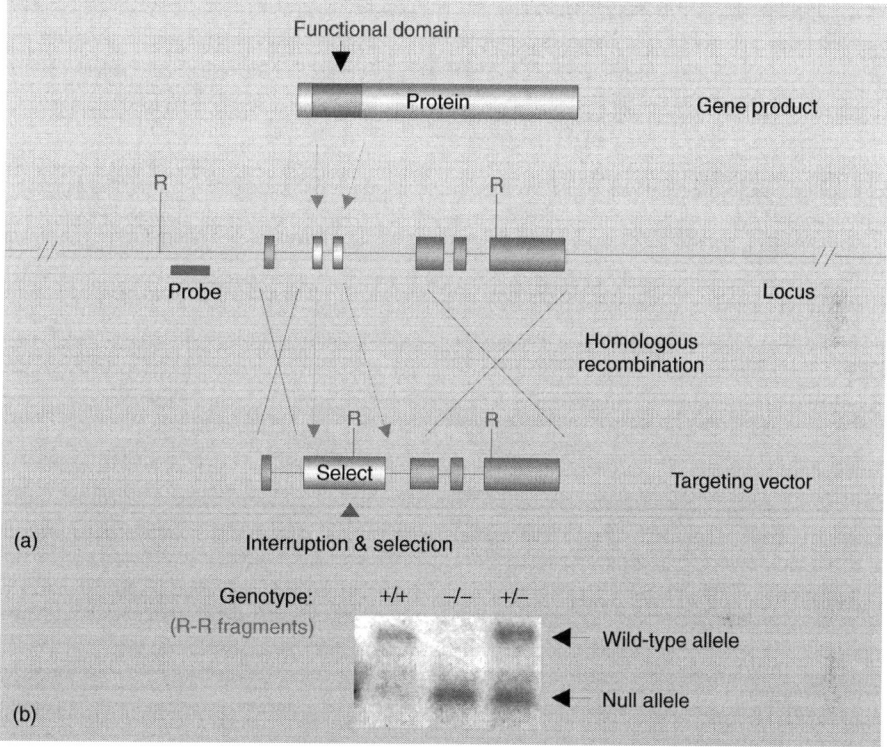

Fig. 8 Basic design of a conventional targeting vector. (a) The targeting vector duplicates part of the locus to be mutated (homologous sequences), but is interrupted by a positive selection marker (select), for instance, an antibiotic resistance gene. Many conventional targeting vectors also contain a negative selection marker. The herpes simplex virus Thymidine Kinase gene is often included: the TK gene product converts a harmless compound, for instance, Ganciclovir, into a cytotoxic substance. The positioning of the TK cassette at one of the ends of a targeting vector results in exclusion of TK from the genome upon homologous recombination, whereas TK is cointegrated in case of random integration of a targeting vector. (b) The vector is designed such that by means of an "external probe" (blue-grey box: probe; from within the to be targeted locus, but outside the homology regions used in the targeting construct) and strategically chosen restriction endonuclease sites (R) a shift is detected upon homologous recombination. The figure shows the three possible genotypes; two wild-type (+/+) alleles, two knockout (−/−) alleles (homozygous knockout), and a combination of both in the heterozygous (+/−) condition (see also Fig. 10). Mutagenesis vectors may be introduced into ES cells via several means; the most commonly used method is electroporation (electroshock) by which the cellular membrane is damaged, allowing uptake of DNA in cells (see color plate p. xi).

vector is a negative selection marker, for example, the *Herpes simplex virus* thymidine kinase (HSV-TK). Upon positive (negative) selection, a number of separately isolated ES cell clones are "genotyped" to ensure proper recombination on both ends of the targeting vector. Typically, these cells carry a monoallelic mutation, that is, only one of two alleles in the ES clone is replaced by the targeting vector (Fig. 8b).

Random insertion type vectors insert at will in the genome and, therefore, do not rely on sequence homology with target genes (see Sect. 5.1). Integration of a vector in a locus may, however, cause inactivation of a gene. The best-known application of insertion type vectors is *gene trapping* (see Sect. 4 and 5). Gene trap vectors are constructed such that they either trap promoter activity or trap a polyadenylation signal from an endogenous locus (Fig. 15). In this manner, not only actively transcribed but also inactive genes can be trapped (with promoter-trap or poly(A)-trap vector respectively). Variations on this concept are described in Sect. 5.

3.2
Embryonic Stem (ES) Cells

In contrast to classical transgenesis, where mutagenesis is done directly in the embryo, gene targeting is done in *embryonic stem cells* (ES cells) (Fig. 9a). These are very early stem cells isolated from preimplantation embryos (*blastocysts*). ES cells are *pluripotent*, that is, they are fully undifferentiated and, in essence, have the capacity to participate in mouse embryogenesis and become any cell type in the animal. The first steps in a gene targeting experiment are carried out *in vitro*. ES cell culturing conditions are aimed at maintaining their pluripotency. ES cells are either cultured on a layer of supporting feeder cells (mouse embryo fibroblasts (MEFs)) in the presence of factors that block differentiation of stem cells in general (like *Leukemia Inhibitory Factor,*) or are "feeder-independent." It is their pluripotency that is ultimately called upon: when generating targeted mutations in the mouse, male mutant ES cells are reintroduced into preimplantation embryos and made to participate in embryogenesis and spermatogenesis in particular. If the latter occurs, the targeted mutation can be stably transmitted via the germline to offspring and bred to homozygocity.

3.3
Genesis of Mosaic Embryos

ES cells may be used in a number of ways to generate genetically mosaic embryos. The most commonly used procedure is *blastocyst injection*: preimplantation blastocysts are used to introduce ES cell into (Fig. 5a). The donor mouse line (i.e. of which ES cells are derived) and the recipient line (i.e. of which blastocysts are procured) differ in coat color genetics. Typically, an inbred strain such as *C57blk* (black coat color) is used for blastocyst production to inject *129Sv* or a *129Ola* (sand-colored coat), some of the most commonly used ES cell lines, into. The procedure described above is schematically represented in Fig. 6c.

Alternatively, ES cells can be fused to morula stage embryos in a procedure termed *morula aggregation*. Embryos in the morula stage typically comprise 8 to 16 blastomeres (individual embryonic cells) surrounded by a protective layer called the *zona pellucida*. In order to achieve aggregation, this zone is first removed and the embryos together with a small number of ES cells are cocultured in a small depression, in which gravity forces them to

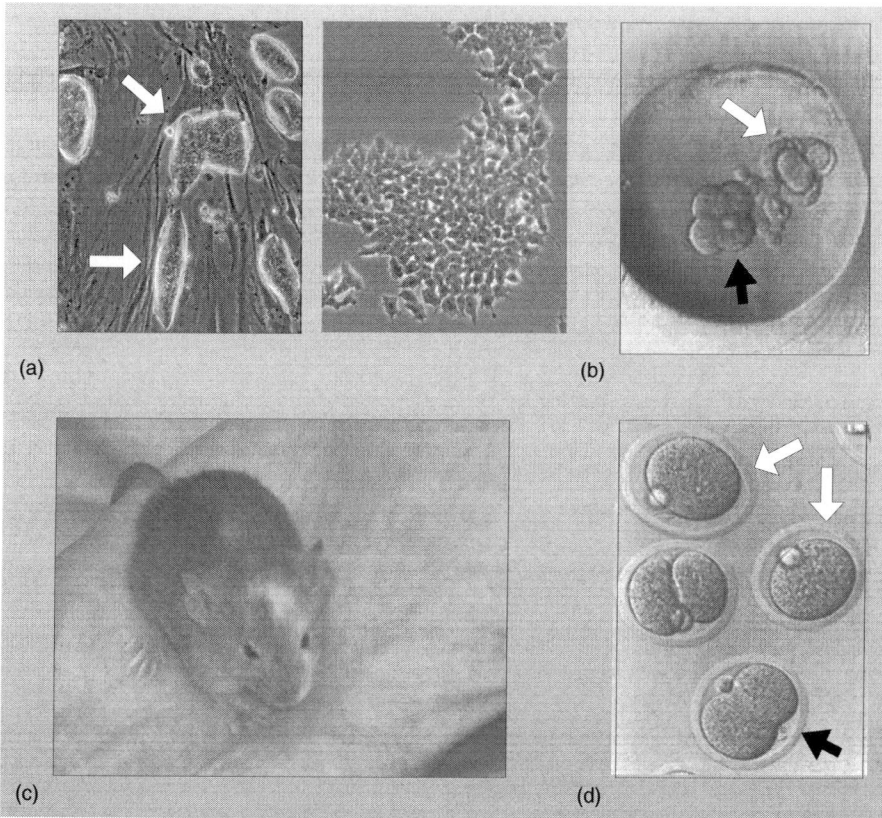

Fig. 9 Embryonic stem cells, embryos, and chimeric mice. (a) Left panel: ES cells grown on mouse embryonic fibroblasts (MEFs); white arrows point at colonies of tightly coherent ES cells; right panel: feeder-free ES cells. (b) Aggregation of 8-cell-stage morulas (black arrow) with genetically modified ES cells (white arrow). (c) Chimeric mice beautifully displaying the coat color mosaicism resulting from contribution of ES cells of two different genetic backgrounds to the developing embryo the mix of, in this case, C57Bl (recipient blastocyst – black) and 129Ola (donor ES cells – sand) becomes manifest as Agouti, patches of purely 129Ola-derived tissue is sand colored (e.g. over left eye). (d) Electrofusion of two-cell-stage embryos (middle left) results in fusion and formation of one-cell-stage tetraploid embryos (white arrows). Tetraploidy does not hamper placenta formation, but does not permit normal embryogenesis. Tetraploid blastocysts injected with diploid ES cells will support fully ES cell–derived embryonic development (source image: (http://www.mshri.on.ca/nagy/cre.htm)) (see color plate p. xv).

be in close contact with each other (Fig. 9b; http://www.mshri.on.ca/nagy/cre.htm).

Both methods yield the typical mosaic coat (e.g. a mixture of different coat colors, like black and sand). The mice that display the coat color mosaicism are referred to as *chimeric mice* (Fig. 9c). Chimeric males are bred to wild-type mice to generate a mouse that is heterozygous for the mutant allele in all its cells; the subsequent mating of two heterozygous animals allows the mutation to be bred into a homozygous state.

3.4
Biotechnological Procedures

In contrast to transgenic technology, preimplantation embryos are isolated at the blastocyst stage. Microinjected embryos are transferred directly into the uterus of a 2.5-day pseudopregnant female, where they develop into chimeric embryos (Fig. 9b, 10). A typical targeting experiment uses at least two independent mutant embryonic stem cell lines for microinjection to verify and confirm the phenotype

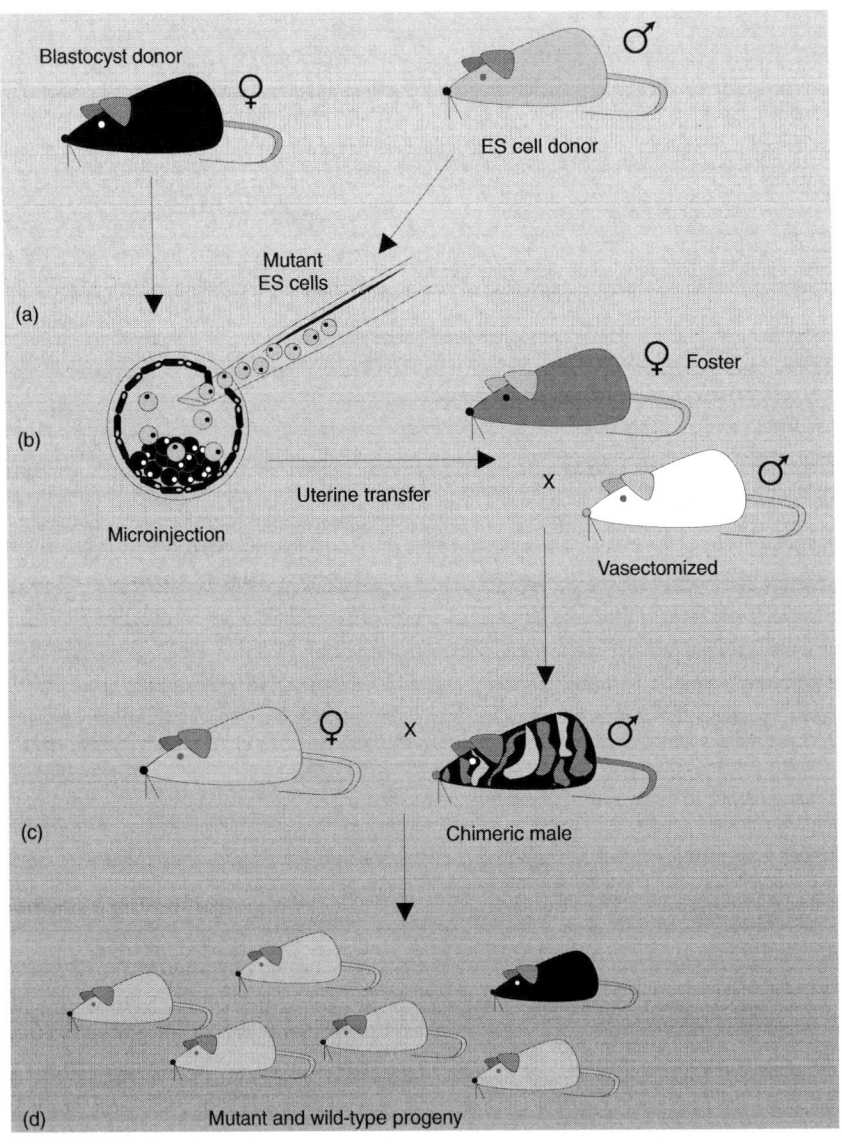

in two independently established homozygous knockout animals.

4
Refinements to the Models

Transgenes or gene knockouts may cause unexpected embryonic or neonatal lethality, either by interfering with embryo development or placentation. If the aim of an animal model is to, for example, study the effect of a transgene or a gene knockout on adult physiology or development of a disease, embryonic lethality limits the usefulness of such a model. To circumvent this problem, animal molecular geneticists have sought to refine animal models. The resulting mouse models often combine aspects of transgenesis and gene targeting and may include a molecular switch, by which null mutation of a gene now is rendered *conditional*. Such on/off models are referred to as *binary models*. The most pivotal advantages of such molecular switches are gain in accuracy and specificity of the model and an important concomitant reduction of disease burden for the animals. Some exciting models and applications that have been developed over the last decade are described below.

4.1
Fully ES Cell–derived Embryos

The placenta is one of the first organ systems to develop during early embryogenesis. It provides an essential means by which the developing embryo and the mother animal exchange essential factors such as nutrients and oxygen. A surprisingly large number of null mutations result in defective placentation. Defective placentation in a conventional genetic approach thwarts all possibilities to uncover a gene function during early embryogenesis. J. Rossant and coworkers developed an elegant solution to this problem during the early 1990s: tetraploid embryos were used to reintroduce diploid ES cells into. The concept is elegantly simple: tetraploid cells will contribute to placentation (polyploidy is a normal feature within the developing placenta); they will, however, not contribute to the development of the embryo itself. It was shown that in this approach, reintroduced normal diploid wild-type ES cells were fully capable of supporting normal embryonic development. The data demonstrated that with this technology

Fig. 10 Biotechnological procedures in conventional gene targeting. (a) Inbred strains are used for superovulation and mating to generate sufficient blastocysts for injection. The procedures in use to obtain sufficient blastocysts for microinjection are natural matings or superovulation. Breeding properties of some inbred strains are relatively poor, hence, natural mating often require large numbers of mating pairs. Super ovulation is often not very efficient in inbred strains either, but will usually yield adequate embryo numbers for injection. (b) Upon microinjection, the embryos are transferred into the oviduct of a pseudopregnant female (see also legend to Fig. 7). (c) Since male ES cells are used, researchers look for a gender bias among chimeric offspring: a potent male (XY) ES cell line forces even female recipient blastocysts (XX) into a male sex phenotype. As a general rule is considered: the higher the coat color mosaicism, the more likely it is that the ES cells participate in spermatogenesis, the higher the chance for germ-line transmission of the mutation; (d) –Germ-line transmission is revealed by the coat color of the donor (ES cell; sand-colored in the figure; several ES cell lines are in use) strain; 50% of the sand-colored offspring will be heterozygous for the germ-line mutation (see color plate p. xiv).

the study of genetically manipulated ES cells in all fetal lineages was possible, while, at the same time, potential negative effects of a null mutation on placental development were circumvented. Tetraploid recipient embryos were generated by electrofusion: normal diploid two-cell-stage embryos were subjected to a defined electroshock that results in fusion of the cells to form one tetraploid one-cell-stage embryo (Fig. 9d). This tetraploid early embryo is allowed to further develop *in vitro*, to be used at a later stage for aggregation or microinjection (see Sect. 3.3). It should be noted that although the role of a gene can now be studied independent of placentation, specific lethality as a result of gene mutation is not prevented. To bypass

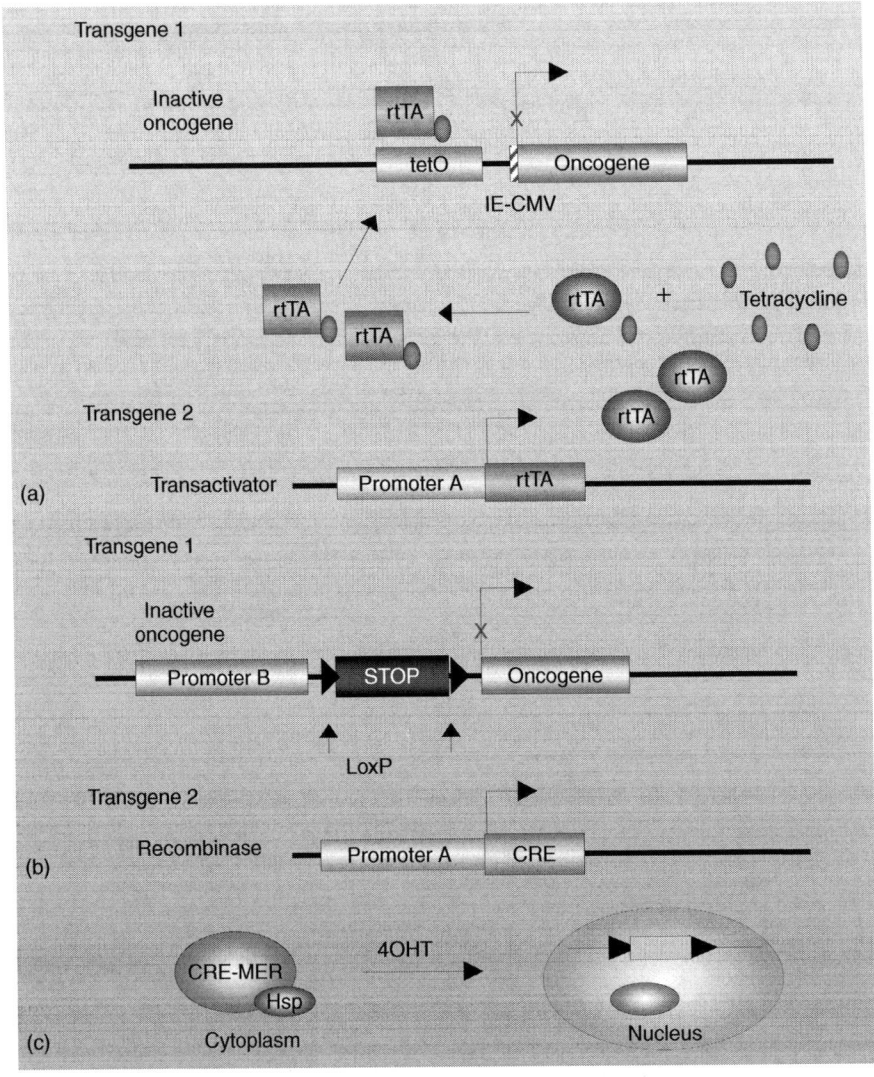

these problems, inducible or binary models are used.

4.2
Conditional Transgenesis

Conditional transgenesis is employed to have better control over (trans)gene expression in a given model. This may be advantageous for a number of reasons, for instance, to circumvent lethality during embryonic development caused by toxic or teratogenic properties of transgenes or to study the onset of disease from its earliest stages onward.

In its simplest form, conditionality is integrated in an animal model by means of tissue-specific promoters and/or control element usage (see Sect. 2.1). Transgene expression is confined to a defined target tissue or select cells within a tissue, for instance, epithelial lung cells (surfactant protein promoter) or anterior pituitary cells (POMC, pro-opiomelanocortin promoter). Several tens to hundreds of models have been created and used in this fashion to successfully study the specific effect(s) of transgene expression in development or disease.

Truly conditional transgenesis, in which the investigator not only determines where a transgene is turned on but also when and how strong, can be achieved with binary models. Binary models incorporate a switching mechanism. Several molecular switches exist to turn on or off transgenes. Figure 11 lists examples of these methods. Most binary switches are reversible. Two main concepts are

---

Fig. 11 Binary models. (a) The Tet-system requires two independently generated transgenic lines, which are combined by interbreeding. Transgene 1 carries a minimal ("enhancer-less") Cytomegalovirus immediate early (IE–CMV) promoter fused to heptamerized prokaryotic tetO regulatory sequences. Transgene 1 is transcriptionally silent until the transactivator line (transgene 2) provides transactivator protein. In the original Tet-system, transgene 2 expressed a tetracycline (Tc)-controlled transcriptional activator, tTA. tTA is a fusion between the Tet-repressor of the Tn10 Tc resistance operon of E. coli and a C-terminal part of the herpes simplex transactivator protein VP16. tTA will induce transcription from the tetO-controlled transgene up to five orders of magnitude, but is inhibited by Tc. The example depicts rtTA (red ovals), a reverse Tc-controlled transactivator, which is a mutant form of tTA and needs Tc-derivatives to bind tetO (yellow squares). So rather than an inhibitable model, rtTA is an inducible model. Refinements on the Tet-system include choice of regulatory sequences (promoter A) in transgene 2 (cell type–specific, general promoter). In addition, several derivatives of Tc exist, among which Doxocyclin (Dox) has better tissue penetration and pharmacokinetic properties than Tc itself and very high and selective affinity for TetR. (b) The use of bacteriophage P1 CRE-LoxP recombination–dependent removal of a strong transcriptional block (STOP) in a transgene construct may be helpful in defining the molecular events at the onset of tumorigenesis and in addition circumvents possible embryonic lethal effects of the transgene. The choice of regulatory sequences (promoter A) in transgene 2 is determined by the model. Although the excision of a transcriptional block is, in principle, reversible, for practical reasons this system is rarely used as a binary switch, since it would entail obligatory screening for genetic modifications (insertions) in target cells: this obviously compromises the usefulness of the models in terms of precision and rapidity. (c) Fusion of CRE to the mutant forms of the ligand-binding domain of nuclear hormone receptors, such as the estrogen receptor (MER), has generated a recombinase CRE-MER that can be "induced" at the posttranslation level: CRE-MER is retained cytoplasmic through interaction with heatshock proteins (Hsp) such as HSsp90; upon addition of 4-hydroxytamoxifen (4OHT), a synthetic steroid, the Hsp-binding is released, and the recombinase shuttles to the nucleus to excise floxed sequences (also see Fig. 12) (see color plate p. xvi).

used: (1) a combination of nonmammalian promoters and transcription factors and (2) integration of transcriptional interruptions that are removed by recombination. The use of inducible transcription by means of nonmammalian, for example, regulatory elements borrowed from bacteria or lower eukaryotes, like the fruit fly (*D. melanogaster*), allows transgenes to be activated and silenced "on command." The system that utilizes removal of a strong transcriptional block by means of recombination, although in principal reversible, is, for practical reasons, mostly used one-way.

Several binary switches are itemized below:

- *Tetracycline-controlled transgene expression: the Tet-system*. The transgene of interest is placed under the control of a Tetracycline resistance operon from *Escherichia coli*, the tet-operator (*tetO*). A second, independent transgenic line expressing a derivative of TetR, a transcriptional regulator of tetO, confers Tetracycline (Tc)-inducibility or repressiveness to the transgene of interest (see Fig. 11a). Transcription factors with regulatory activity toward the tetO are absent in eukaryotic cells, which means that theoretically the system cannot be leaky for transcription and pleiotropic effects are therefore not expected.
- *Insect hormone-controlled transgene expression*. The system uses the molting steroid hormone *ecdysone* for transgene induction. It requires a heterodimeric receptor and an ecdysone responsive element, which is incorporated as a regulatory element in the transgene of interest. Administration of *ecdysone* (or a derivative *muristerone A*) rapidly and selectively induces transgene expression. Upon withdrawal of the steroid hormone, expression is silenced. Again, since mammalian cells lack insect hormone receptors, the binary system theoretically works as an on/off system, much like the Tet-system. Ecdysone nor its derivatives are toxic or teratogenic. Owing to excellent pharmacokinetic properties, the method is ideally suited for very precise conditional control over gene expression.
- *Transgene induction by removal of a transcriptional block*. The system makes use of a recombination process derived from bacteriophage P1: phages (bacterial viruses) use a site-specific recombinase (CRE) to integrate their DNA into the genome of their bacterial host. An essential second component of this process is a recombinase recognition site, called an *LoxP site*: two LoxP sites are recognized and recombined by CRE (Fig. 12a). The two essential components, LoxP and CRE, also work in eukaryotic systems, and are commonly maintained on two separate transgenic lines. Flanking a sequence with two similarly oriented LoxP sites will lead to its excision in the presence of CRE. Hence, an on/off transgenic system employs a silenced transgene, in which a strong transcriptional block is bracketed by two similarly oriented LoxP sites; a second independent transgenic line, which expresses the CRE recombinase, may be driven off a general, tissue-specific or inducible promoter, depending on the objective of the study (Fig. 11b). More details on this recombination-based switching mechanism will be discussed in Sect. 4.3.

4.3 Conditional Gene Targeting

Gene targeting is widely used to study gene function in the mouse. Hundreds

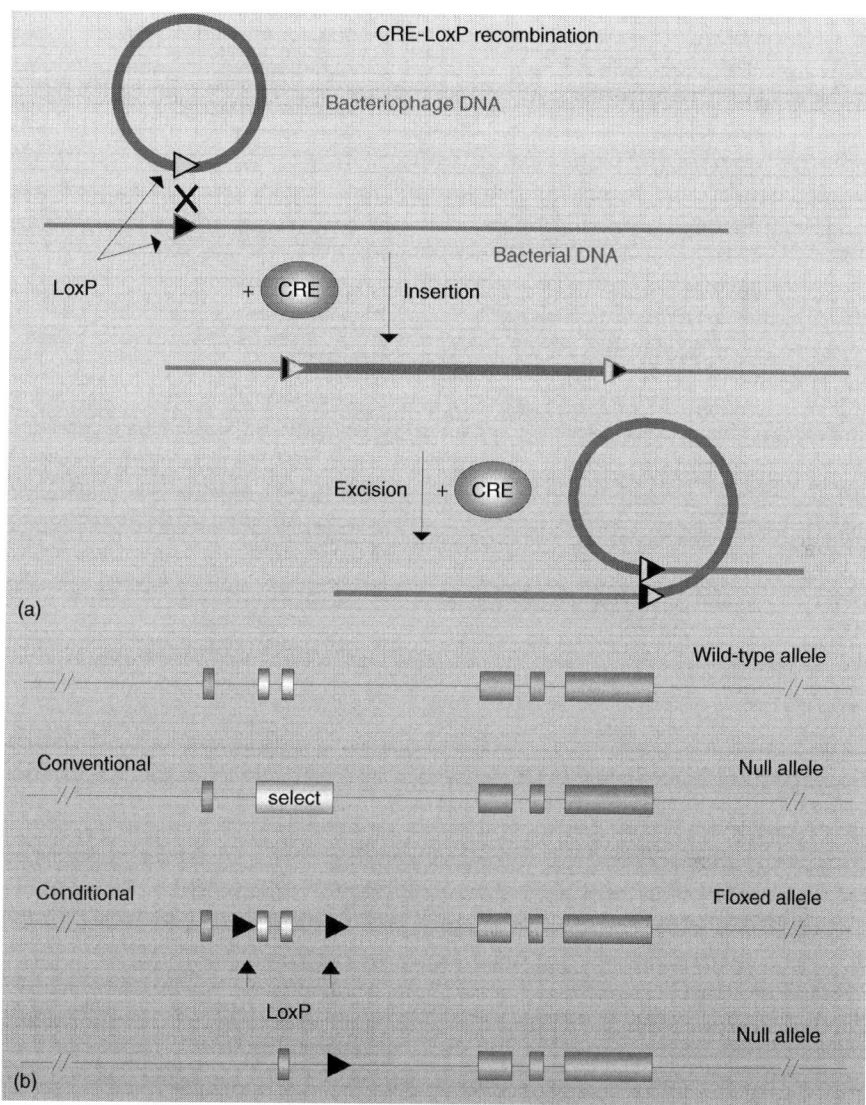

Fig. 12 (a) CRE-LoxP system: LoxP sites are short sequences harboring inverted 13 nucleotide repeats separated by an 8-nucleotide asymmetric spacer. The LoxP are recognition sites for the recombinase CRE. CRE drives site-specific recombination of two identical LoxP sites, for clarity depicted as a yellow and a black triangle. Two similarly oriented LoxP sites on one contiguous DNA strand will be recombined by CRE and the sequence in between the LoxP sites is excised. (b) This recombination principle is used in conditional gene-targeting vectors: a crucial coding region (two blue exons) is "floxed" (i.e. flanked by LoxP sites) in ES cells and excised *in vivo*. The LoxP sites in the "floxed" allele are assumed not to interfere with wild-type gene expression. The figure compares the structure of a conventional and a conditional knockout allele. A variation on this theme is the FLP-*frt* system, which is derived from yeast. The FLP gene product, like CRE, is a site-specific recombinase; *frt* (FLP recognition target) represents the small inverted repeats, analogous to LoxP, recognized by the recombinase (see color plate p. xviii).

of genes have been "knocked out" in the conventional approach, in which the null mutation is transmitted through the germline, as discussed in Sect. 3.1. An estimated one in three null mutations results in embryonic or early postpartum lethality. Therefore, germ-line mutations may be appropriate models to recapitulate human congenital genetic disorders. They may, however, not be the best tools to study gene function in adult mice, since this is hampered by often unexpected and unexplained lethality. K. Rajewsky and coworkers were among the first to develop a conditional gene-targeting strategy that allows inactivation of a gene in a defined *spatio-temporal* setting: the null mutation can, for instance, be confined to selected cell types or to a specific stage during development.

Conditional gene targeting makes use of the CRE-LoxP principle, as described in Sect. 4.2. Briefly, LoxP sites flank a genomic sequence in a targeting vector, which will be deleted *in vivo*; this allele is referred to as a *floxed* allele (Fig. 12b). This final "floxed" wild-type allele only carries recombinase recognition sites at positions within the locus where they are presumed inert, for example, introns: up until the moment of recombination-driven excision *in vivo*, the allele should function as a wild-type allele. Since the CRE-LoxP system is applied more widely than the FLP-*frt* system *in vivo*, further descriptions of conditional gene targeting will focus mainly on the CRE-LoxP system.

The current standard experimental conditional approach requires two separate mouse lines:

1. the "floxed" line (as described above)
2. the "deleter" line, expressing CRE.

The transcriptional control directing CRE expression determines the specificity and degree of regulation of the conditional knockout. CRE-transgene expression is controlled either by constitutively cell type–specific or inducible regulatory elements (also see Sect. 2.1 and 4.2). Deleter strains are most often generated by conventional transgenic technology. Alternatively, the CRE-cDNA can be inserted via targeted mutagenesis to a locus of interest, under control of the endogenous promoter. Although the latter method does not require a thorough knowledge of the regulatory elements controlling the locus at hand, the approach is not as straightforward as conventional transgenesis. Drawbacks of the former method comprise all known problems linked to conventional transgenesis: transgene expression is dependent on integration site and copy number. Consequently, a number of independent founders will have to be generated to select a useful strain. A substantial collection of tissue-specific CRE mouse lines is currently available.

Inducible CRE expression allows for a high level of control (see Sect. 4.2). Again, it should be noted that conditional gene targeting is mainly used unidirectionally, that is, once a gene of interest is inactivated through CRE-mediated recombination, reversion of the genotype is not easily accomplished.

4.3.1 Controlled CRE Expression

Inducibility can be conferred by rather simple measures such as CRE expression through infectious viral particles. Such viruses can be locally administered or systemically. In addition, their tropism (i.e. host-cell range) can be manipulated.

Inducible mammalian promoters have been used, as well as the Tet-system (see Sect. 4.2). Besides the induction of CRE transcription and translation, CRE induction has also been achieved at the posttranslational level. This was accomplished by fusing CRE-encoding sequences in frame to the ligand-binding domain (LBD) of nuclear steroid hormone receptors, such as the estrogen or progesterone receptor (ER, PR respectively). Ligand binding transfers the CRE-ER fusion product into the nucleus, where the recombinase exerts its catalytic activity toward LoxP sites. Mutation of the ER moiety, has generated *CRE-MER* (mutated estrogen receptors) fusions, which are only induced by synthetic hormones and not by their natural ligands (Fig. 11c). Ongoing improvements of these and alike inducible systems clearly generate a highly sensitive and specific means to control transgene expression and enable the study of gene ablation and overexpression that, without such tools, were impossible. An increasing number of published models employing CRE-MER variants attest to the feasibility of the approach *in vivo*.

4.4
Knockins: Humanized Mice

Many studies in human disease have uncovered mutations that are relevant to the etiology of the pathological condition. Examples of these are large congenital chromosomal duplications or deletions in inborn genetic disease, specific dominant mutations in oncogenes, recessive mutations in tumorsuppressor genes, or polymorphisms in genes involved in lipid metabolism linked to atherosclerosis. In order to understand the exact molecular mechanism underlying the development of disease, such mutations need to be exactly recreated in the mouse. Clearly, conventional and conditional genomic manipulations are not specific enough to address such issues. CRE-mediated recombination has provided possibilities and strategies to manipulate the genome beyond conditional gene knockouts or transgenesis. As illustrated in Fig. 13 and 14, the CRE-LoxP system can be used to manipulate the genome in various manners:

- *Gene knockin* or *gene exchange strategies* apply CRE-LoxP-mediated recombination to exchange endogenous coding regions for exogenous or mutated sequences in a locus (Fig. 13a). Especially, the latter application is of significant interest to the scientific community, since it permits the development of animal models, which reflect human genetic disorders more closely than models applying overexpression or ablation of a gene. Initial approaches inserted reporter genes into an endogenous locus; this simple variation on conventional gene targeting achieves the same purpose as a transgene harboring a reporter preceded by endogenous regulatory sequences (see Sect. 2.1). The main difference, of course, is that this strategy obviates the handling of large DNA fragments with potentially unidentified regulatory elements needed to achieve faithful expression patterns, since these elements are provided by the targeting into the endogenous locus. Recently developed strategies take knockin significantly further and allow for specific introduction of (point) mutations and repeated exchange of genetic sequences: recombination-mediated cassette exchange (RMCE) applies strategically positioned pairs of wt and mutated LoxP sites or LoxP and *frt* sites to achieve stepwise and directional replacement of genomic sequences with other of mutated sequences (Fig. 13b).

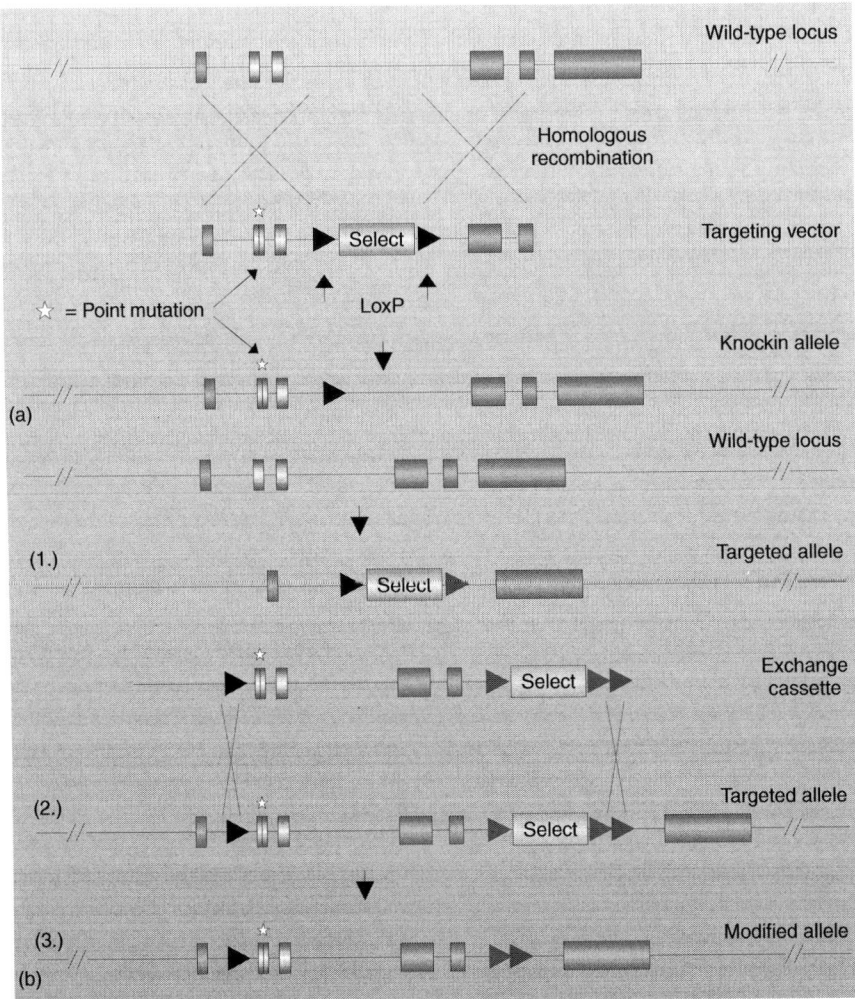

Fig. 13 Applications of site-directed recombination in molecular mouse genetics. (a) A conditional targeting construct usually harbors an antibiotic resistance gene (select) flanked by recombinase recognition sites (either LoxP or frt; see legend to Fig. 12). Prior to microinjection into blastocysts or aggregation with morulas, the selection marker is excised in ES cells by recombination (by transient expression of either CRE or FLP). The targeting construct depicted contains a specific point mutation (asterix) in one of its exons. Homologous recombination results in introduction of the mutation in the endogenous locus. (b) The use of multiple nonidentical LoxP and or frt sites (black, red, and blue triangles) creates the possibility to repeatedly exchange sequences between vectors and a locus via site-specific recombination. The procedure is referred to as *recombination-mediated cassette exchange (RCME)*. First, a locus is targeted and an asymmetrically "floxed" selection marker is introduced (1). Using the same LoxP variants, an exchange cassette is introduced (2) upon which the selection marker is removed by recombinase-mediated excision (3). The exchange procedure can be repeated multiple times, since the position of the most peripheral recombinase recognition sites (black and red triangles) is maintained (see color plate p. xix).

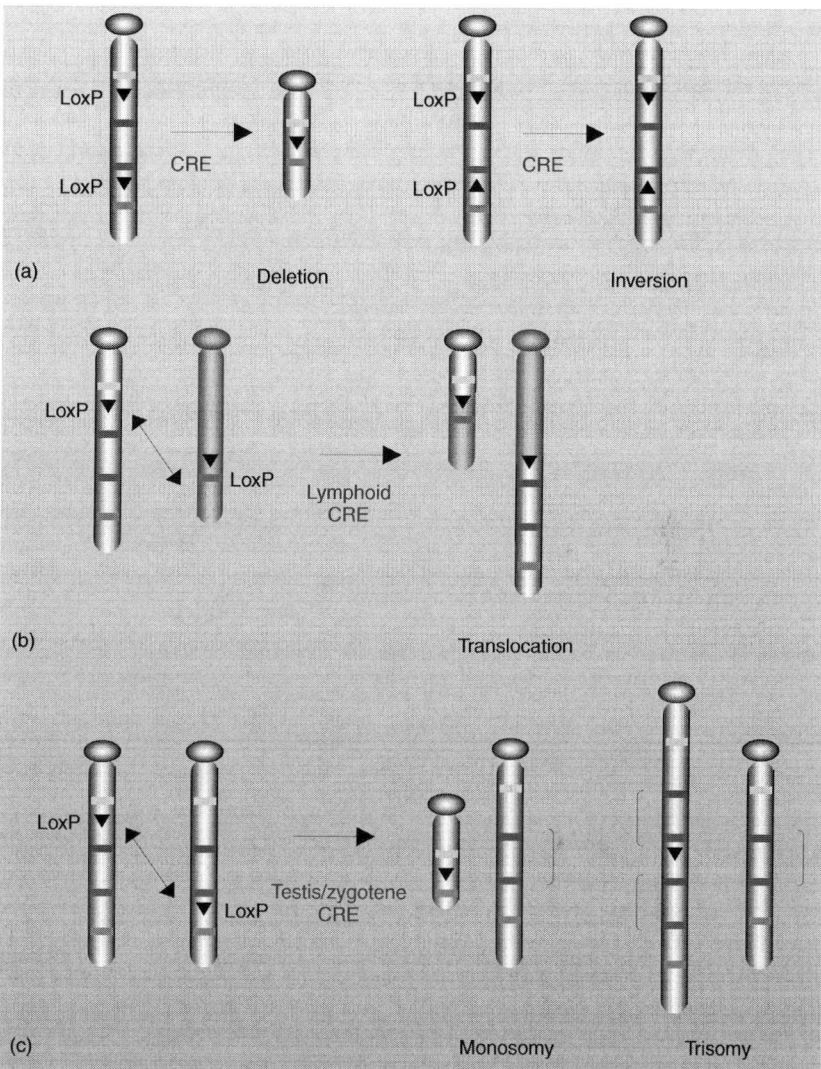

Fig. 14 CRE-LoxP applications in chromosome engineering. (a) Chromosome engineering is an important tool that facilitates functional analysis of large chromosomal regions: large segments of DNA can be either deleted or inversed, depending on the orientation of the LoxP sites in respect to each other. Series of overlapping deletions can be generated by combined use of one targeted LoxP site and retrovirus-mediated random insertion of additional LoxP sites. CRE-LoxP recombination may be used to generate balancer chromosomes to maintain lethal recessive mutations and has the potential to facilitate large-scale mutagenesis screens. Deletion or inversion carriers can be used to uncover and instantly map, for example, ENU-induced recessive phenotypes to deleted or inversed chromosomal regions. (b) Interchromosomal recombination by positioning LoxP sites on nonhomologous chromosomes will recreate translocations, which occur in human disease. (c) The use of CRE-mediated recombination to create defined autosomal trisomies in the mouse, for example, through meiosis-/zygotene-specific CRE-deleter strains that are solely active in testis.

- *Chromosomal deletions, inversions, translocations.* Simultaneous specific positioning of LoxP sites on one chromosome allows for gene inversions or complete deletions, depending on the orientation of the recombinase recognition sites relative to each other (Fig. 14). Positioning of such recombinase recognition sites on nonhomologous chromosomes has proven to be a feasible strategy to achieve chromosomal translocation, alike translocations that are implicated in some leukemias.
- *Targeted integration of transgenes* to loci, which are known to be "open," that is, not embedded in heterochromatin. The gain of this approach is a high probability of transgene expression. An often-used target locus is the *ROSA26* locus, an ubiquitously transcribed locus that does not encode a protein, but which is active in many different cell types throughout development and adult life. Transgenes are flanked by DNA sequences homologous to the *ROSA26* locus and targeted via conventional methods (see Sect. 3).
- A substantial number of other applications for site-directed recombination has been reported on in the scientific literature. Since the principal mechanism remains the same in all instances, a few of these applications are listed here, but not worked out in detail: *selection marker recycling, exchange of reporter genes, removal of reporter genes, or selection markers* (as in the generation of floxed alleles).

5
Random Mutagenesis in ES Cells

The completion of the mouse and human genome sequences has sparked renewed interest in developing tools for genome-wide mutagenesis in the mouse. Among the main molecular genetic technologies currently pursued are targeted mutagenesis (see Sect. 3 and 4), *insertional mutagenesis*, by insertion of gene trap vectors or viral or transposable sequences, and chemical mutagenesis. These technologies all represent random mutagenesis protocols, which, in principle, can be applied to mouse ES cells. The important advantage of random mutagenesis is that new mutations can be generated in mice at

---

Fig. 15 Random mutagenesis strategies. (a) Promoter-trap and poly(A)-trap vectors, respectively, catch or induce transcriptional activity and have the potential to identify most, if not all, of the active and inactive genes in the genome, including alternatively spliced forms and low-abundance transcripts, and is thus an important tool in genome annotation. (b) Gene mutation through random insertion of retroviral vectors (or transposable elements; see Sect. 5.2). Shown is an oncogene, which is activated by a nearby proviral insert, and a tumorsuppressor gene, which is inactivated proviral insertion into the gene. Inserted elements are engineered such that they simultaneously function as sequencing tags to identify the surrounding genomic DNA. (c) Stable or inducible RNA interference as an alternative to targeted mutagenesis to study gene function in mice: a short double-strand RNA (ds) stem-loop-stem structure (white arrows, blue loop) is produced from a RNA polymerase III-promoted vector. The loop is enzymatically removed inside the cell; the resulting short interfering RNAs will bind their complementary mRNA, upon which the target mRNA is degraded by a complex called *RISC (RNA-induced silencing complex)*. RNA interference–mediated gene silencing does not require targeted mutagenesis of a gene, and may therefore represent a more efficient and rapid method to assess gene function (see color plate p. xx).

a rate far exceeding conventional gene targeting. In addition, it is possible to identify and maintain recessive lethal phenotypes at any developmental stage, which, for instance, is not possible with chemical mutagenesis (see Sect. 5.3).

5.1
Gene Trap Mutagenesis

Gene trapping provides a method to mutate genes in the genome through random insertion throughout the genome.

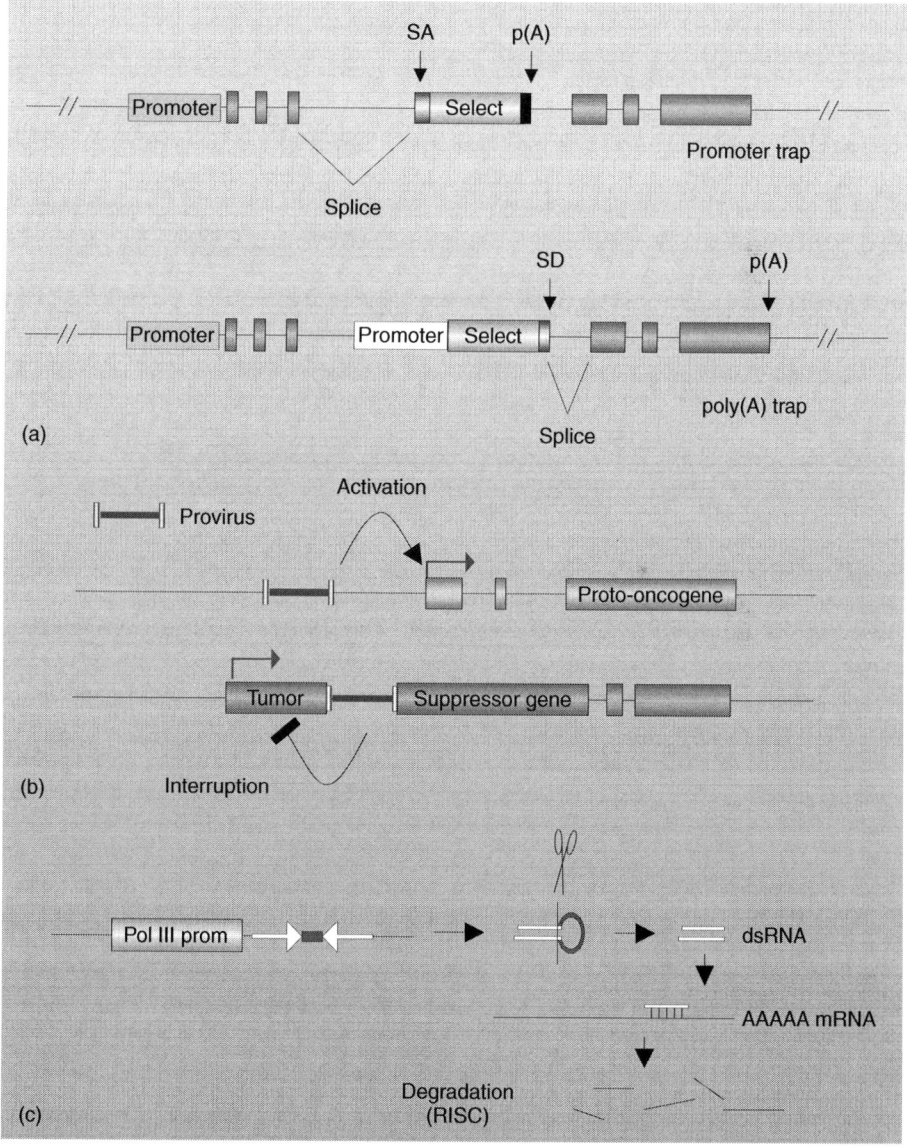

Essentially, all currently designed gene trap vectors insert a selectable marker or reporter gene lacking essential regulatory sequences into an endogenous gene. Through marker insertion the gene is interrupted, while marker expression is complemented by endogenous regulatory elements (Fig. 15a). In this manner, transcriptionally active as well as inactive loci can be trapped. Recent reports provided proof that gene trapping can also be used to selectively mutagenize genes that share certain properties, in this case, the fact that their gene products are all transmembrane proteins. A "secretory" trap vector was constructed such that it captured signal sequences in preceding exons by means of including a "capturing" transmembrane domain in the gene trap construct. These and other approaches collectively demonstrate the power and versatility of gene trapping in genome-wide functional analysis of molecular mechanisms important in development and physiology. A collective gene trap database has recently been initiated by a number of consortia (http://www.igtc.ca/index.html – http://www.escells.ca/ – http://www.mmrrc.org/ – http://tikus.gsf.de/project/web_new/index.html – http://www.sanger.ac.uk/PostGenomics/genetrap/ – http://baygenomics.ucsf.edu/overview/welcome.html).

5.2
Retroviruses and Transposable Elements in Random Mutagenesis

The use of *retroviruses* in random mutagenesis is adapted from applications in which proviruses are used to screen for collaborating oncogenes in leukemogenesis. Slow transforming retroviruses such as Moloney Murine Leukemia Virus (Mo-MuLV) cause transformation by integration into the genome. Proviral integration may activate oncogenes or interrupt tumor suppressor genes (Fig. 15b). This is an extremely useful asset in random mutagenesis, since the integrated proviruses now serve as a *molecular tag* for cancer genes: it can be used to identify the locus at which it is integrated by sequencing the flanking integration site. Retroviral vectors have been adapted for use in mouse embryos and ES cells as well, and now permits germ-line mutagenesis.

In analogy to random proviral insertion, also *transposable elements* from yeast, insects, fish, and mammals, among other organisms, can be used for functional genomic analyses. Transposable elements (TEs) are a heterogeneous class of genetic elements that have the capacity to move from one site on a chromosome to another: they "jump." TEs are very efficient at integrating into DNA, and some elements can carry extra DNA. TEs are therefore useful vectors for transferring new genetic material into genomes and for random insertional mutagenesis. By way of example, among the properties that make the Tc1/*mariner* superfamily of TEs useful for molecular genetic research are the facts that they are easy to work with, their efficient and stable integration, and persistent, long-term transgene expression following transposon-mediated gene transfer. Importantly, such TEs "jump" in species other than their original hosts, which underscores the need for further development and use of these and alike molecular tools for functional genomics in the mouse. Several reports on germline induced TE transposition and germline attest to the mutagenic potential of the approach.

5.3
Random Enu Mutation

For many years, mutant mouse models were obtained largely as a "spin-off" from research on the mutagenic properties of chemical compounds and radiation. With the recent advent of more genome resources, it has become possible to identify such induced mutations more rapidly. Initially, substantial efforts were undertaken, involving collaborations of many laboratories, each with their own phenotyping expertise. At present, robust technology is in place to mutagenize mice. One of the *caveats* remained however, because, although screens for recessive mutations are the most promising ones, such screens are exceedingly difficult and expensive. Recently, it was shown that a narrowly focused screening system for recessive mutations can be highly productive, while the expenses remained acceptable. Moreover, M. Justice, A. Bradley, and colleagues made use of a set of strains carrying engineered coat color markers and a chromosomal inversion (see Fig. 14a), enabling to further reduce costs by focusing the screen on a predefined chromosomal segment. Given the large number of loci controlling complex genetic disorders, this strategy permits an ordered approach to finding disease genes. The strategy is highly suitable to focus on some syntenic regions in the mouse genome, which are known to be associated with disease susceptibility in man.

6
Phenotype Analysis

Genetically engineered mouse models serve to study genotype–phenotype relationships and can show an impressive variety of phenotypes, ranging from embryonic lethality to increased susceptibility to chronic diseases such as cardiovascular disease and cancer. The models allow defining primary gene functions as well as the role of a particular gene mutation in disease.

Once mouse models have been generated, it is important to carry out an in-depth analysis of the pathology of the entire mouse and not restrict the analysis to the target organs that are the main subject of the intended study. A detailed strategy is outlined elsewhere. Often, one will find that the dramatic alterations in the gene expression pattern will show unexpected effects on the phenotype. For instance, in the case of a study on apolipoprotein C1, the expected perturbation of lipid metabolism was accompanied by a dramatic and progressive hair loss of the mice with the highest level of expression of the transgene. The most common problem encountered, however, in studies using genetically modified mouse models is a lack of phenotype, even when a unique and evolutionarily conserved gene has been altered for which there are no obvious functionally redundant homologs present in the mouse genome. The main cause of this is the large difference in environmental stress experienced by wild mice as opposed to inbred laboratory mice. Often, phenotypes need to be evoked by either increasing the environmental stress (i.e. high-fat diets, mutagens) or increasing the genetic stress using a second transgenic strain with a mutation in a related pathway. A good example is provided by the use of mouse strains in cardiovascular research. Wild-type laboratory mice are resistant to develop atherosclerosis because of its highly antiatherogenic lipoprotein profile. By knocking out genes controlling lipoprotein metabolism, such as the apolipoprotein E gene, the mice become sensitized. Upon feeding a moderated high-fat diet, such mice will develop complex lesions in the aorta within 2 months. Many gene alterations will have no effect on atherogenesis when studied separately. However, when these mice are crossed with apolipoprotein E–deficient mice, it is now possible to study the role of such genes. Thus, the apolipoprotein E–deficient mouse serves as a sensitized mouse model that is highly susceptible to the effects of additional genetic changes. Likewise, changing the genetic background of the mice will affect susceptibility to disease; this approach has been used to define novel loci in human disease.

Recent technological advances permit analysis of transgene expression and the biological consequences thereof by noninvasive methods. In vital imaging, very sensitive cameras measure bioluminescence from, for example, tumors or embryos *in utero*, which express luciferase of GFP reporter transgenes in living animals (Fig. 2d). Clearly, the impact of technological developments like these is substantial not only in terms of increased specificity and resolution of the model but also in terms of reduced animal usage.

7
Perspectives

7.1
Partial Transgenesis; Viral Shuttles

Recent technological advances allow more rapid approaches to study gene function without the need for germ-line modification. Remarkably, partial transgenesis can be achieved through local electroporation of transgenic vectors into tissues. Using this approach, local muscle tissue in rats has been specifically modified to study its function. Another example is the study of cell migratory and (trans)differentiation processes during neural tube closure; in these studies, one side of the tube is selectively electroporated with an expression construct, while the other side functions as a control.

The use of viral vectors to deliver transgenes to mammalian cells is another method to achieve partial transgenesis. The use of several classes and combinations of viral vectors is being explored in gene therapeutic applications. Similar vectors are very useful in accomplishing partial transgenesis, restricted to certain tissues such as bone marrow (which is easily harvested and cultured *ex vivo* and reintroduced into a recipient mouse). The latter approach can also be employed to, for instance, study isolated effects of gene knockout in the hematopoietic compartment in combination with bone marrow transplantation. Local and transient transgenesis is currently used to activate conditional floxed transgenes (see Sect. 5), for instance by adenovirus-encoded CRE. This approach was proven useful in, for example, somatic activation of transforming oncogenes in the lung.

Lentiviral vectors, unlike the classical slow replicating retroviruses, harbor the advantage that they are not silenced in embryonic stem cells, which makes expression characteristics of such vectors good. Recombinant lentiviruses were recently used to introduce transgenes into fertilized oocytes and ES cells (Fig. 6b). The advantages of using lentiviral shuttle vector are significant: it obviates potentially harmful procedures otherwise used to introduce transgenes into a cell (pronuclear microinjection, electroporation) and offers some degree of control over transgene copy numbers.

7.2
RNA Interference; Posttranscriptional Gene Silencing

The abundance of biological information that has become available upon completion of the genome sequence of both mouse and human is exciting and overwhelming at the same time. An estimated 10 000 genes have been identified currently, which represents an estimated one-third of the total gene complement in a human or mouse. Of these genes, only a few thousand have been functionally studied. Clearly, one of the remaining challenges in molecular genetics is to analyze the function of novel genes. Gene targeting is beyond any doubt one of the most successful and informative technologies when it comes to elucidating gene function. Despite improvements in cloning technologies that employ recombination in prokaryotes instead of "standard" recombinant DNA technology and that significantly simplify otherwise laborious cloning strategies, gene targeting remains time-consuming and expensive.

Recent scientific and technological advancements may, however, provide faster means to address novel gene function. One of the most exciting developments finds its origin in the realization that gene expression is controlled at a post-transcriptional level by double-strand (ds) RNA-mediated degradation of cytoplasmic RNAs. This process, called *RNA silencing* – or *posttranscriptional gene silencing* – was first discovered in plants as an unexpected gene silencing in transgenic plants: silencing of an endogenous gene often occurred in concert with silencing of the transgene. Transgene-mediated RNA silencing is now recognized as a form of RNA silencing that uses small interfering dsRNAs (siRNA), which serve as a guide for specific target mRNA recognition and its subsequent enzymatic degradation. Investigations by many groups showed that RNA interference (RNAi) is a regulatory process used by many organisms to control gene expression. Many of the protein factors involved in RNAi are conserved, from protozoans to humans. RNAi has recently been developed into a technology by which endogenous gene function can be assessed in many vertebrates, including mammals. One obvious advantage of RNAi-mediated gene silencing is that it does not rely on genetic modification. This makes the technology very rapid compared to (conditional) gene targeting. Proof of principle has been achieved in many vertebrate systems including several human cells lines. RNAi also works in early mouse embryos, and was recently demonstrated to be transmittable via the germline in transgenic mice. Whereas initial strategies in cell lines made use of synthetic dsRNA sequences, a relative costly approach, RNAi, is currently achieved via stable expression. In addition, recent reports show that random siRNA libraries can be generated from cDNA libraries via ingenious cloning strategies. The potential held by this approach is that such libraries are self-selecting, thereby enabling investigators to set-up high-throughput genetic screens. RNAi can be made tissue specific, inducible, and can be delivered to mouse embryos using viral vectors. Such improvements will clearly enhance the application of RNAi in functional genomic research.

Although genetic manipulation is possible in tissue culture, the study of genetic interaction of transgenes with other genes and local or systemic effects of transgene expression in the intact organism provides a much more complete and physiologically relevant picture of a gene's function than can ever be achieved *in vitro*. Genetic modification therefore represents an important and biologically relevant tool to complement *in vitro* gene expression studies aimed at, for example, delineation of signal transduction pathways or identification of tissue-specific regulatory elements. Refinements to genetically modified mouse models in biomedical research are of importance for a number of reasons. Conditional models recapitulate the pathophysiology as it occurs in humans and permit a high degree of selectivity and resolution, which not only renders studies more specific and focused but also significantly reduces the burden on the laboratory animal. The positive effects thereof are reduced stress on the system and, concurrently, improved specificity.

Bibliography

Books and Reviews

Bass, B.L. (2000) Double-stranded RNA as a template for gene silencing, *Cell* **101**, 235–238.

Beier, D.R. (2002) ENU mutagenesis: a work in progress, *Physiol. Genomics* **11**, 111–113.

Berns, A. (1988) Provirus tagging as an instrument to identify oncogenes and to establish synergism between oncogenes, *Arch. Virol.* **102**, 1–18.

Botteri, F.M., van der Putten, H., Miller, A.D., Fan, H. et al. (1986) Recombinant retroviruses in transgenic mice, *Ann. N. Y. Acad. Sci.* **478**, 255–268.

de Winther, M.P., Heeringa, P. (2003) Bone marrow transplantations to study gene function in hematopoietic cells, *Methods Mol. Biol.* **209**, 281–292.

Dudley, J.P. (2003) Tag, you're hit: retroviral insertions identify genes involved in cancer, *Trends Mol. Med.* **9**, 43–45.

Fazio, S., Linton, M.F. (2001) Mouse models of hyperlipidemia and atherosclerosis, *Front. Biosci.* **6**, D515–D525.

Gijbels, M.J., de Winther, M.P. (2003) Autopsy and histologic analysis of the transgenic mouse, *Methods Mol. Biol.* **209**, 267–279.

Hernandez, D., Fisher, E.M. (1999) Mouse autosomal trisomy: two's company, three's a crowd, *Trends Genet.* **15**, 241–247.

Hofker, M., van Deursen, J. (2002) *Transgenic Mouse: Methods and Protocols (Methods in molecular biology)*, Humana press, Totowa, NJ, U.S.A.

Hogan, B., Beddington, R., Costantini, F., Lacey, E. (1994) *Manipulating the Mouse Embryo. A Laboratory Manual*, 2nd edition, Cold Spring Harbor Laboratory Press, Cold Spring Harbor, New York.

Ivics, Z., Izsvak, Z. (2004) Transposable elements for transgenesis and insertional mutagenesis in vertebrates: a contemporary review of experimental strategies, *Methods Mol. Biol.* **260**, 255–276.

Jackson, I.J., Catherine, M., Abbott, C.M. (2000) *Mouse Genetics and Transgenics: A Practical Approach*, Oxford University press, New York, U.S.A.

Justice, M.J., Noveroske, J.K., Weber, J.S., Zheng, B. et al. (1999) Mouse ENU mutagenesis, *Hum. Mol. Genet.* **8**, 1955–1963.

Kootstra, N.A., Verma, I.M. (2003) Gene therapy with viral vectors, *Annu. Rev. Pharmacol. Toxicol.* **43**, 413–439.

Kuhn, R., Schwenk, F. (2003) Conditional knockout mice, *Methods Mol. Biol.* **209**, 159–185.

Li, Q., Harju, S., Peterson, K.R. (1999) Locus control regions: coming of age at a decade plus, *Trends Genet.* **15**, 403–408.

Maatman, R., Gertsenstein, M., de Meijer, E., Nagy, A., Vintersten, K. (2003) Aggregation of embryos and embryonic stem cells, *Methods Mol. Biol.* **209**, 201–230.

Massoud, T.F., Gambhir, S.S. (2003) Molecular imaging in living subjects: seeing fundamental biological processes in a new light, *Genes Dev.* **17**, 545–580.

McCormick, S.P., Liu, C.Y., Young, S.G., Nielsen, L.B. (2003) Manipulating large insert clones for transgenesis, *Methods Mol. Biol.* **209**, 105–123.

Mikkers, H., Berns, A. (2003) Retroviral insertional mutagenesis: tagging cancer pathways, *Adv. Cancer Res.* **88**, 53–99.

Nagy, A. (2000) Cre recombinase: the universal reagent for genome tailoring, *Genesis* **26**, 99–109.

Nagy, A., Gertsenstein, M., Vintersten, K., Behringer, R. (2002) *Manipulating the Mouse Embryo: A Laboratory Manual*, 3rd edition,

Cold Spring Harbor Laboratory Press, Cold Spring Harbor, New York.

Osumi, N., Inoue, T. (2001) Gene transfer into cultured mammalian embryos by electroporation, *Methods* **24**, 35–42.

Pedersen, Ra.R.J. (1989) *Transgenic Techniques in Mice: A Video Guide*, Cold Spring Harbor Laboratory, Cold Spring Harbor, New York.

Ramirez-Solis, R., Liu, P., Bradley, A. (1995) Chromosome engineering in mice, *Nature* **378**, 720–724.

Roebroek, A.J., Wu, X., Bram, R.J. (2003) Knockin approaches, *Methods Mol. Biol.* **209**, 187–200.

Rossant, J., Cross, J.C. (2001) Placental development: lessons from mouse mutants, *Nat. Rev. Genet.* **2**, 538–548.

Saunders, T.L. (2003) Reporter molecules in genetically engineered mice, *Methods Mol. Biol.* **209**, 125–143.

Sharp, P.A. (1999) RNAi and double-strand RNA, *Genes Dev.* **13**, 139–141.

Silver, L. (1995) *Mouse Genetics: Concepts and Applications*, Oxford University press, New York, U.S.A.

Soewarto, D., Blanquet, V., Hrabe de Angelis, M. (2003) Random ENU mutagenesis, *Methods Mol. Biol.* **209**, 249–266.

Somia, N., Verma, I.M. (2000) Gene therapy: trials and tribulations, *Nat. Rev. Genet.* **1**, 91–99.

Soriano, P. (1999) Generalized lacZ expression with the ROSA26 Cre reporter strain, *Nat. Genet.* **21**, 70–71.

Voncken, J.W. (2003a) Genetic modification of the mouse. General technology; pronuclear and blastocyst injection, *Methods Mol. Biol.* **209**, 9–34.

Voncken, J.W. (2003b) Transgene design, *Methods Mol. Biol.* **209**, 51–67.

Yang, M., Baranov, E., Moossa, A.R., Penman, S. et al. (2000) Visualizing gene expression by whole-body fluorescence imaging, *Proc. Natl. Acad. Sci. U. S. A.* **97**, 12278–12282.

Primary Literature

Baharvand, H., Matthaei, K.I. (2004) Culture condition difference for establishment of new embryonic stem cell lines from the C57BL/6 AND BALB/c mouse strains, *In Vitro Cell. Dev. Biol. Anim.* **40**, 76–81.

Brummelkamp, T.R., Bernards, R., Agami, R. (2002) A system for stable expression of short interfering RNAs in mammalian cells, *Science* **296**, 550–553.

Carmell, M.A., Zhang, L., Conklin, D.S., Hannon, G.J. et al. (2003) Germline transmission of RNAi in mice, *Nat. Struct. Biol.* **10**, 91–92.

Choi, T., Huang, M., Gorman, C., Jaenisch, R. (1991) A generic intron increases gene expression in transgenic mice, *Mol. Cell. Biol.* **11**, 3070–3074.

Christ, N., Droge, P. (2002) Genetic manipulation of mouse embryonic stem cells by mutant lambda integrase, *Genesis* **32**, 203–208.

Doerfler, W. (1991) Patterns of DNA methylation–evolutionary vestiges of foreign DNA inactivation as a host defense mechanism. A proposal, *Biol. Chem. Hoppe Seyler* **372**, 557–564.

Dupuy, A.J., Fritz, S., Largaespada, D.A. (2001) Transposition and gene disruption in the male germline of the mouse, *Genesis* **30**, 82–88.

Escary, J.L., Perreau, J., Dumenil, D., Ezine, S. et al. (1993) Leukaemia inhibitory factor is necessary for maintenance of haematopoietic stem cells and thymocyte stimulation, *Nature* **363**, 361–364.

Evans, M.J., Kaufman, M.H. (1981) Establishment in culture of pluripotential cells from mouse embryos, *Nature* **292**, 154–156.

Feil, R., Brocard, J., Mascrez, B., LeMeur, M. et al. (1996) Ligand-activated site-specific recombination in mice, *Proc. Natl. Acad. Sci. U. S. A.* **93**, 10887–10890.

Folger, K.R., Wong, E.A., Wahl, G., Capecchi, M.R. (1982) Patterns of integration of DNA microinjected into cultured mammalian cells: evidence for homologous recombination between injected plasmid DNA molecules, *Mol. Cell. Biol.* **2**, 1372–1387.

Forrester, L.M., Nagy, A., Sam, M., Watt, A. et al. (1996) An induction gene trap screen in embryonic stem cells: Identification of genes that respond to retinoic acid in vitro, *Proc. Natl. Acad. Sci. U. S. A.* **93**, 1677–1682.

Ghoshal, K., Jacob, S.T. (2000) Regulation of metallothionein gene expression, *Prog. Nucleic. Acid Res. Mol. Biol.* **66**, 357–384.

Gordon, J.W., Ruddle, F.H. (1981) Integration and stable germ line transmission of genes injected into mouse pronuclei, *Science* **214**, 1244–1246.

Gossen, M., Bujard, H. (1992) Tight control of gene expression in mammalian cells by tetracycline-responsive promoters, *Proc. Natl. Acad. Sci. U. S. A.* **89**, 5547–5551.

Gossler, A., Joyner, A.L., Rossant, J., Skarnes, W.C. (1989) Mouse embryonic stem cells and reporter constructs to detect developmentally regulated genes, *Science* **244**, 463–465.

Hadjantonakis, A.K., Dickinson, M.E., Fraser, S.E., Papaioannou, V.E. (2003) Technicolour transgenics: imaging tools for functional genomics in the mouse, *Nat. Rev. Genet.* **4**, 613–625.

Ikawa, M., Tanaka, N., Kao, W.W., Verma, I.M. (2003) Generation of transgenic mice using lentiviral vectors: a novel preclinical assessment of lentiviral vectors for gene therapy, *Mol. Ther.* **8**, 666–673.

Jaenisch, R. (1976) Germ line integration and Mendelian transmission of the exogenous Moloney leukemia virus, *Proc. Natl. Acad. Sci. U. S. A.* **73**, 1260–1264.

Jahner, D., Haase, K., Mulligan, R., Jaenisch, R. (1985) Insertion of the bacterial gpt gene into the germ line of mice by retroviral infection, *Proc. Natl. Acad. Sci. U. S. A.* **82**, 6927–6931.

Jong, M.C., Gijbels, M.J., Dahlmans, V.E., Gorp, P.J. et al. (1998) Hyperlipidemia and cutaneous abnormalities in transgenic mice overexpressing human apolipoprotein C1, *J. Clin. Invest.* **101**, 145–152.

Kisseberth, W.C., Brettingen, N.T., Lohse, J.K., Sandgren, E.P. (1999) Ubiquitous expression of marker transgenes in mice and rats, *Dev. Biol.* **214**, 128–138.

Koller, B.H., Hagemann, L.J., Doetschman, T., Hagaman, J.R. et al. (1989) Germ-line transmission of a planned alteration made in a hypoxanthine phosphoribosyltransferase gene by homologous recombination in embryonic stem cells, *Proc. Natl. Acad. Sci. U. S. A.* **86**, 8927–8931.

Lacy, E., Roberts, S., Evans, E.P., Burtenshaw, M.D. et al. (1983) A foreign beta-globin gene in transgenic mice: integration at abnormal chromosomal positions and expression in inappropriate tissues, *Cell* **34**, 343–358.

Lyons, S.K., Meuwissen, R., Krimpenfort, P., Berns, A. (2003) The generation of a conditional reporter that enables bioluminescence imaging of Cre/loxP-dependent tumorigenesis in mice, *Cancer Res.* **63**, 7042–7046.

Mansour, S.L., Thomas, K.R., Capecchi, M.R. (1988) Disruption of the proto-oncogene int-2 in mouse embryo-derived stem cells: a general strategy for targeting mutations to nonselectable genes, *Nature* **336**, 348–352.

Mao, X., Fujiwara, Y., Orkin, S.H. (1999) Improved reporter strain for monitoring Cre recombinase-mediated DNA excisions in mice, *Proc. Natl. Acad. Sci. U. S. A.* **96**, 5037–5042.

Martin, G.R. (1981) Isolation of a pluripotent cell line from early mouse embryos cultured in medium conditioned by teratocarcinoma stem cells, *Proc. Natl. Acad. Sci. U. S. A.* **78**, 7634–7638.

Metzger, D., Clifford, J., Chiba, H., Chambon, P. (1995) Conditional site-specific recombination in mammalian cells using a ligand-dependent chimeric Cre recombinase, *Proc. Natl. Acad. Sci. U. S. A.* **92**, 6991–6995.

Meuwissen, R., Linn, S.C., Linnoila, R.I., Zevenhoven, J. et al. (2003) Induction of small cell lung cancer by somatic inactivation of both Trp53 and Rb1 in a conditional mouse model, *Cancer Cell* **4**, 181–189.

Mural, R.J., Adams, M.D., Myers, E.W., Smith, H.O. et al. (2002) A comparison of whole-genome shotgun-derived mouse chromosome 16 and the human genome, *Science* **296**, 1661–1671.

Nagy, A., Gocza, E., Diaz, E.M., Prideaux, V.R. et al. (1990) Embryonic stem cells alone are able to support fetal development in the mouse, *Development* **110**, 815–821.

Nagy, A., Rossant, J., Nagy, R., Abramow-Newerly, W. et al. (1993) Derivation of completely cell culture-derived mice from early-passage embryonic stem cells, *Proc. Natl. Acad. Sci. U. S. A.* **90**, 8424–8428.

No, D., Yao, T.P., Evans, R.M. (1996) Ecdysone-inducible gene expression in mammalian cells and transgenic mice, *Proc. Natl. Acad. Sci. U. S. A.* **93**, 3346–3351.

Ostertag, E.M., DeBerardinis, R.J., Goodier, J.L., Zhang, Y. et al. (2002) A mouse model of human L1 retrotransposition, *Nat. Genet.* **32**, 655–660.

Paddison, P.J., Caudy, A.A., Hannon, G.J. (2002) Stable suppression of gene expression by RNAi in mammalian cells, *Proc. Natl. Acad. Sci. U. S. A.* **99**, 1443–1448.

Palmiter, R.D., Chen, H.Y., Brinster, R.L. (1982) Differential regulation of metallothionein-thymidine kinase fusion genes in transgenic mice and their offspring, *Cell* **29**, 701–710.

Palmiter, R.D., Brinster, R.L., Hammer, R.E., Trumbauer, M.E. et al. (1982) Dramatic growth of mice that develop from eggs microinjected with metallothionein-growth hormone fusion genes, *Nature* **300**, 611–615.

Rajewsky, K., Gu, H., Kuhn, R., Betz, U.A. et al. (1996) Conditional gene targeting, *J. Clin. Invest.* **98**, 600–603.

Robertson, E., Bradley, A., Kuehn, M., Evans, M. (1986) Germ-line transmission of genes introduced into cultured pluripotential cells by retroviral vector, *Nature* **323**, 445–448.

Schonig, K., Bujard, H. (2003) Generating conditional mouse mutants via tetracycline-controlled gene expression, *Methods Mol. Biol.* **209**, 69–104.

Sen, G., Wehrman, T.S., Myers, J.W., Blau, H.M. (2004) Restriction enzyme-generated siRNA (REGS) vectors and libraries, *Nat. Genet.* **36**, 183–189.

Shirane, D., Sugao, K., Namiki, S., Tanabe, M. et al. (2004) Enzymatic production of RNAi libraries from cDNAs, *Nat. Genet.* **36**, 190–196.

Skarnes, W.C., Auerbach, B.A., Joyner, A.L. (1992) A gene trap approach in mouse embryonic stem cells: the lacZ reported is activated by splicing, reflects endogenous gene expression, and is mutagenic in mice, *Genes Dev.* **6**, 903–918.

Su, H., Wang, X., Bradley, A. (2000) Nested chromosomal deletions induced with retroviral vectors in mice, *Nat. Genet.* **24**, 92–95.

te Riele, H., Maandag, E.R., Berns, A. (1992) Highly efficient gene targeting in embryonic stem cells through homologous recombination with isogenic DNA constructs, *Proc. Natl. Acad. Sci. U. S. A.* **89**, 5128–5132.

Thomas, K.R., Capecchi, M.R. (1987) Site-directed mutagenesis by gene targeting in mouse embryo-derived stem cells, *Cell* **51**, 503–512.

Thompson, S., Clarke, A.R., Pow, A.M., Hooper, M.L. et al. (1989) Germ line transmission and expression of a corrected HPRT gene produced by gene targeting in embryonic stem cells, *Cell* **56**, 313–321.

Valenzuela, D.M., Murphy, A.J., Frendewey, D., Gale, N.W. et al. (2003) High-throughput engineering of the mouse genome coupled with high-resolution expression analysis, *Nat. Biotechnol.* **21**, 652–659.

van Deursen, J., Wieringa, B. (1992) Targeting of the creatine kinase M gene in embryonic stem cells using isogenic and nonisogenic vectors, *Nucleic Acids Res.* **20**, 3815–3820.

Van Deursen, J., Fornerod, M., Van Rees, B., Grosveld, G. (1995) Cre-mediated site-specific translocation between nonhomologous mouse chromosomes, *Proc. Natl. Acad. Sci. U. S. A.* **92**, 7376–7380.

van de Wetering, M., Oving, I., Muncan, V., Pon Fong, M.T. et al. (2003) Specific inhibition of gene expression using a stably integrated, inducible small-interfering-RNA vector, *EMBO Rep.* **4**, 609–615.

van der Putten, H., Botteri, F.M., Miller, A.D., Rosenfeld, M.G. et al. (1985) Efficient insertion of genes into the mouse germ line via retroviral vectors, *Proc. Natl. Acad. Sci. U. S. A.* **82**, 6148–6152.

Venter, J.C., Adams, M.D., Myers, E.W., Li, P.W. et al. (2001) The sequence of the human genome, *Science* **291**, 1304–1351.

Vicat, J.M., Boisseau, S., Jourdes, P., Laine, M. et al. (2000) Muscle transfection by electroporation with high-voltage and short-pulse currents provides high-level and long-lasting gene expression, *Hum. Gene Ther.* **11**, 909–916.

Waterston, R.H., Lindblad-Toh, K., Birney, E., Rogers, J. et al. (2002) Initial sequencing and comparative analysis of the mouse genome, *Nature* **420**, 520–562.

Zhang, Y., Buchholz, F., Muyrers, J.P., Stewart, A.F. (1998) A new logic for DNA engineering using recombination in Escherichia coli, *Nat. Genet.* **20**, 123–128.

Zhang, Y., Proenca, R., Maffei, M., Barone, M. et al. (1994) Positional cloning of the mouse obese gene and its human homolog, *Nature* **372**, 425–432.

Zheng, B., Sage, M., Cai, W.W., Thompson, D.M. et al. (1999) Engineering a mouse balancer chromosome, *Nat. Genet.* **22**, 375–378.

Websites (References Indicated as they Appear in the Original Text)

http://jaxmice.jax.org/info/recombinbred.html.

http://jaxmice.jax.org/library/notes/493c.html.

http://www.mshri.on.ca/nagy/cre.htm.

http://www.healthsystem.virginia.edu/internet/transgenic-mouse/southerndesign.cfm.

http://www.igtc.ca/index.html
http://www.escells.ca/
http://www.mmrrc.org/
http://tikus.gsf.de/project/web_new/index.html
http://www.sanger.ac.uk/PostGenomics/genetrap/
http://baygenomics.ucsf.edu/overview/welcome.html

31
Transgenic Plants for Food Use

Susanne Stirn and Horst Lörz
University of Hamburg, Hamburg, Germany

1	**Methods to Establish Transgenic Plants** 907	
1.1	Transformation Methods 907	
1.1.1	Agrobacterium Transformation 907	
1.1.2	Direct Gene Transfer 908	
1.2	Tissue Requirements 908	
1.3	Molecular Requirements 909	
1.3.1	Promoter 909	
1.3.2	Codon Usage 909	
1.3.3	Selectable Marker and Reporter Genes 909	
2	**GM Plants Already on the Market** 912	
2.1	Herbicide Resistance in Soybean, Maize, Oilseed Rape, Sugar Beet, Rice, and Cotton 912	
2.2	Insect Resistance in Maize, Potatoes, Tomatoes, and Cotton 912	
2.3	Virus Resistance, Male Sterility, Delayed Fruit Ripening, and Fatty Acid Contents in GMPs 912	
3	**GM Plants "In the Pipeline"** 920	
3.1	Input Traits 920	
3.1.1	Insect Resistance 920	
3.1.2	Virus, Fungal, Bacterial, and Nematode Resistance 921	
3.1.3	Tolerance Against Abiotic Stress 924	
3.1.4	Improved Agronomic Properties 925	
3.2	Traits Affecting Food Quality 926	
3.2.1	Increased Vitamin Content 926	
3.2.2	Production of Very-long-chain Polyunsaturated Fatty Acids 927	
3.2.3	Increased Iron Level 927	
3.2.4	Improved Amino Acid Composition 927	
3.2.5	Reduction in the Content of Antinutritive Factors 928	

Genomics and Genetics. Edited by Robert A. Meyers.
Copyright © 2007 Wiley-VCH Verlag GmbH & Co. KGaA, Weinheim
ISBN: 978-3-527-31609-0

3.2.6	Production of "Low-calorie Sugar" 928
3.2.7	Seedless Fruits and Vegetables 928
3.3	Traits that Affect Processing 929
3.3.1	Altered Gluten Level in Wheat to Change Baking Quality 929
3.3.2	Altered Grain Composition in Barley to Improve Malting Quality 929
3.4	Traits of Pharmaceutical Interest 929
3.4.1	Production of Vaccines 929
3.4.2	Production of Pharmaceuticals 931

4   Outlook 932

Acknowledgment 932

Bibliography 932
Books and Reviews 932
Primary Literature 932

Keywords

B.t. Toxins
B.t. toxins are proteins produced by the soil bacterium *Bacillus thuringiensis*. The toxin recognizes certain receptors on the surface of insect midgut epithelial cells. A pore complex is formed through the membrane resulting in the loss of potassium ions (necessary for the osmotic balance). The insect will die due to massive water uptake.

Herbicide Resistance
By transferring the respective genes, plants can become tolerant to certain herbicides. Herbicide-resistant plants already on the market are mostly tolerant to the broad-spectrum herbicides glyphosate and glufosinate, which do not discriminate between certain plant species (as do selective herbicides).

Hypersensitive Reaction
Plants possess a natural defense mechanism against pathogens like fungi, bacteria, or viruses. During early infection, infected cells will be triggered to die (programmed cell death). Together with the plant cells, the pathogens die and their spread will be prevented. With the help of gene technology, the hypersensitive reaction shall be accelerated to make plants more resistant to pathogens.

Marker Gene
Marker genes are necessary to select early transformation events since plant transformation is of low efficiency (1–10% transformation success). Mostly, antibiotic resistance genes have been used as marker genes, since bacterial genes are readily available and antibiotics can be easily used as selective agents in culture media.

Transgenic or Genetically Modified Organism
A transgenic or genetically modified organism is an organism in which the genetic material has been modified in a way that does not occur naturally (e.g. by conventional breeding or natural recombination). Genetic modification involves moving, inserting, or deleting genes within or between species.

> Here we describe the state of the art of genetically modified plants intended for food use. First, an introduction to the different methods of plant transformation as well as the molecular requirements for the stable introduction and expression of the gene(s) of interest is given. Then, we present an overview on the transgenic plants already on the market or approved for placing on the market worldwide. Finally, examples of new transgenic plants in different stages of development are presented. These comprise of well-established traits like herbicide or insect resistance, being transferred to additional crop species as well as new approaches to enhance the nutritional quality of food crops or to produce pharmaceuticals in transgenic plants.

1
Methods to Establish Transgenic Plants

As an introduction to transgenic or genetically modified plants (GMP), we would first like to describe the transformation methods that are currently used, as well as the molecular requirements for the stable introduction and expression of the gene(s) of interest.

1.1
Transformation Methods

In general, two different approaches to transferring foreign DNA into plant cells can be distinguished: (1) a vector-mediated transformation method (via *Agrobacterium*); and (2) direct gene transfer methods.

1.1.1 Agrobacterium Transformation
In the first case, the natural ability of the bacterial phytopathogen *Agrobacterium* to transfer DNA into plant cells is exploited. *Agrobacterium tumefaciens* and *Agrobacterium rhizogenes* are soil microorganisms that incite crown gall tumors and hairy root disease, respectively, in a wide range of dicotyledonous as well as in some monocotyledonous plants. For plant transformation, mainly *A. tumefaciens* is used, and this bacterium contains a large tumor-inducing plasmid (Ti). During infection, a specific segment of the plasmid DNA referred to as T-DNA (transferred DNA) is inserted into the plant genome. The T-DNA contains genes for phytohormones, which are responsible for cell proliferation and the formation of the crown gall, as well as genes for the formation of special nutrients (opines) in the plant cell. In order to use *Agrobacterium* as a tool in genetic engineering, these genes have been deleted and replaced by the genes of interest. This was possible since only the 25-bp T-DNA border sequences on the right and left border are needed for DNA transfer. A wide

variety of such "disarmed" (nontumor-inducing) vectors have been developed. *Agrobacterium*-mediated transformation is the most commonly used method for most dicotyledonous plants. Advantages include the typical insertion of only one or a few copies of the transgene, and the transfer of relatively large segments of DNA with only minimal rearrangements.

1.1.2 Direct Gene Transfer

For many years, *Agrobacterium* could not be used to transform monocotyledonous and other recalcitrant species, so direct gene transfer methods have been developed as an alternative. These include DNA uptake into protoplasts (protoplast transformation) and the shooting of DNA-coated particles into tissues (particle bombardment).

1.1.2.1 Transformation of protoplasts.
In contrast to animal cells, plant cells possess a solid cell wall, which is the first barrier to overcome when foreign genes are to be transferred into them. One way to circumvent this barrier is to use plant cells in which the cell wall has been digested enzymatically, resulting in the so-called *protoplasts*. DNA uptake into protoplasts can then be stimulated by the use of either polyethylene glycol (PEG-transformation) or electric pulses (electroporation). Where an appropriate protoplast-to-plant regeneration system is available, large numbers of transformed clones can be regenerated to fertile transgenic plants.

In cases where a barrier to gene transfer has not been detected, virtually every protoplast system has proven transformable, though with considerable differences in the efficiency. However, problems can arise from the regeneration of fertile plants from protoplasts as the regeneration is strongly species- and genotype-dependent, and undesired somaclonal variation can occur due to the relatively long tissue culture phase. One advantage of the protoplast transformation method is its great independence from other patented techniques.

1.1.2.2 Particle bombardment.
The new method of using high-velocity microprojectiles to deliver DNA into plant tissue was developed by Sanford and colleagues in 1987. A particle gun is used to accelerate DNA-coated microprojectiles into cells, past the cell wall and the cell membrane. The microprojectile is small enough (0.5 – 5 µm) to enter the plant cell without causing too much damage, yet it is of large enough mass to penetrate the cell wall and carry an appropriate amount of DNA on its surface.

The main advantage of particle bombardment is the absence of biological incompatibilities that are found when using biological vectors. Organelles such as chloroplasts have also been transformed using particle bombardment. Unfortunately, particle bombardment, as well as other direct DNA uptake methods often result in complex insertion loci, which can cause gene silencing.

1.2 Tissue Requirements

The ability to cultivate plant tissue *in vitro* is a prerequisite in almost all current transformation protocols. Transformation requires competent (i.e. transformable) cultured cells that are embryogenic or organogenic. Plant cells suitable for regeneration are either cocultivated with *Agrobacterium* or bombarded. For *Agrobacterium* transformation, mostly leaf discs or immature embryos are used for particle bombardment in monocotyledonous plant scutellar tissues.

1.3
Molecular Requirements

A typical plant gene consists of a promoter, a coding sequence, a transcription terminator, and a polyadenylation signal. The expression level of the gene is mainly determined by these components, but can also be affected by the surrounding sequences.

1.3.1 Promoter

The promoter is the main determinant of the expression pattern in the transgenic plant. Constitutive promoters direct expression in all or almost all tissues, independently of developmental or environmental signals. The promoter directing the synthesis of the cauliflower mosaic virus 35S RNA is the most frequently used constitutive promoter; other constitutive promoters are derived from agrobacterial T-DNA, such as the nopaline synthase (nos) promoter or the octopine synthase (ocs) promoter. Both are mainly used in dicotyledonous plants.

In monocotyledonous plants, it is mostly promoters from the rice actin 1 gene (*act1*) or the maize ubiquitin 1 gene (*ubi1*) that are used. To enhance expression levels, the first intron of the respective genes has been added to the expression cassettes. In dicotyledonous plants, the addition of introns seems to have less pronounced effects on the expression level. Nevertheless, the insertion of an intron is always necessary when the expression of the gene product in bacteria must be completely avoided (e.g. for genes conferring bacterial resistance; see Sect. 3.1.2).

In cases where transgene expression should be directed to certain tissues or developmental stages, regulated promoters are required. Promoter elements responsible for expression in seeds, tubers, vegetative organs, and leaves have been isolated.

Environmental influences can induce gene expression after wounding, heat- or cold stress, or anaerobiosis. The use of natural inducible promoters has the disadvantage of causing pleiotropic effects, since endogenous genes will also be turned on. Chemically induced promoters are favored for the production of pharmaceuticals in transgenic plants, as production of the desired compound can be restricted to a certain time point.

1.3.2 Codon Usage

Expression of the gene of interest can also be influenced at the level of translation: the codon usage differs between plants and bacteria. Therefore, the removal of rare codons, e.g. codons not frequently used in plant cells, from genes of bacterial origin can enhance expression considerably. This was first described for a modified *B. thuringiensis* toxin gene where expression was enhanced up to 100-fold.

1.3.3 Selectable Marker and Reporter Genes

In plant transformation systems, the efficacy of stable gene transfer is low; hence, systems that allow the selection of the transformed cells are required. A selection system consists of a selective agent (which interferes with the plant metabolism) and a selectable marker gene (which codes for a protein, enabling inactivation or evasion of the selective agent).

The most commonly used selectable marker genes in transgenic plants code for antibiotic or herbicide resistance including the following:

1.3.3.1 Antibiotic resistance genes.

- *nptII* gene: The *neo* or *nptII* gene, isolated from transposon Tn5 from *Escherichia coli* K12 codes for the neomycin phosphotransferase (NPTII). This enzyme detoxifies a range of aminoglycoside antibiotics such as neomycin, kanamycin, and geneticin. These antibiotics are added to the culture medium after the transformation procedure. Only transformed cells or tissues will survive and will be regenerated to plants.

- *hpt* gene: The *hpt* gene has been isolated from *E. coli* and codes for hygromycin phosphotransferase, which detoxifies the antibiotic hygromycin B. Most plant tissues show a higher sensitivity to hygromycin B than to kanamycin or geneticin. In particular, cereals that are resistant to kanamycin and geneticin can be selected with hygromycin.

- *bla* gene: The bla_{TEM-1} gene codes for the TEM-1 β-lactamase, the most encountered ampicillin resistance marker in molecular biology. TEM-1 β-lactamase attacks narrow-spectrum cephalosporins and all the anti-gram-negative-bacterium penicillins, except temocillin. Mostly, the ampicillin resistance in transgenic plants is not used as a selectable marker for plant transformation, but is a reminiscence of the transformation method. With direct gene transfer methods, it is used as a selectable marker during the cloning procedure prior to plant transformation. When present on the plasmid used for transformation, the ampicillin resistance gene is transferred and integrated together with the genes of interest, and remains under the control of its prokaryotic promoters.

- *aadA*: The *aadA* gene confers resistance to streptomycin and spectinomycin. The gene has been found in association with several transposons (Tn7, Tn21, ...) and is ubiquitous among gram-negative bacteria.

1.3.3.2 Herbicide resistance genes.

- *bar*/pat gene: the *bar* gene from *Streptomyces hygroscopicus* and the *pat* gene from *Streptomyces viridichromogenes* code for phosphinotricin acetyltransferase (PAT), which confers resistance to the herbicide compound, phosphinotricin. It is an inhibitor of glutamine synthetase, a plant enzyme involved in ammonia assimilation. Besides phosphinotricin, the commercially available nonselective herbicides Basta and bialaphos can be used as selective agents. Selection can be applied in the culture medium or by spraying of regenerated plantlets.

- *EPSPS* gene: The 5-enolpyruvylshikimate-5-phosphate synthase (EPSPS) gene codes for the enzyme 5-enolpyruvylshikimate-5-phosphate synthase (EPSPS) that is involved in the skimatic acid pathway of the aromatic amino acid biosynthesis. EPSPS is present in all plants, bacteria, and fungi, but not in animals. The nonselective herbicide glyphosate binds to and inactivates EPSPS. By transferring a glyphosate-resistant EPSPS gene from the soil bacterium *Agrobacterium* strain CP4 to plants, glyphosate-resistant crop plants have been obtained. Glyphosate is the active ingredient of the commercially available herbicide Roundup.

1.3.3.3 Reporter genes.
In contrast to selectable marker genes, reporter genes do not confer resistance to selective agents inhibitory to plant development. Reporter genes code for products that can be detected directly or that catalyze reactions whose products are detectable.

- *gusA* gene: The *gusA* (uidA) gene encodes β-glucuronidase (GUS), which hydrolyzes a wide range of β-glucuronides. Substrates for GUS are available for spectrometric, fluorometric, and histochemical assays. The GUS reporter gene system allows easy quantification with high sensitivity. One disadvantage of this system is its nonviability: examined cells and tissues cannot be regenerated into transgenic plants. Therefore, this system is mainly used to optimize a specific transformation protocol.
- *luc* gene: The *luciferase* reporter gene assay uses bioluminescence reactions with substrate/enzyme combinations, which lead to detectable light emission. The *luc* gene originates from the firefly *Photinus pyralis*, and codes for a luciferase, which decarboxylates beetle luciferin. In contrast to the GUS reporter system, the luciferase reporter gene is nonlethal to plant cells.

1.3.3.4 Alternative marker and excision systems.

Owing to discussions on the possible risk of a gene transfer of antibiotic resistance genes from transgenic plants to clinically important microorganisms, alternative marker systems have been developed. Additionally, through the use of excision systems the presence of antibiotic or herbicide resistance genes in the final product can be avoided.

Alternative markers Several alternative marker systems have recently been developed. For example, Chua and colleagues successfully selected transgenic plants using an inducible *Agrobacterium* gene encoding for isopentenyltransferase (IPT). This enzyme catalyzes the first step in biosynthesis of cytokinins, a class of phytohormones. Only cells containing this gene are able to form shoots and differentiate into mature plants.

As another alternative selection system, phosphomannose isomerase (PMI) was successfully used in transgenic sugar beet and maize. PMI catalyzes the interconversion of mannose-6-phosphate and fructose-6-phosphate. Except for leguminous plants, PMI is absent from most plants, and therefore, only transgenic plants expressing the *E. coli manA* gene can survive on media containing mannose as a carbon source.

Excision systems Another approach is the use of recombination systems that enable the excision of marker genes in the successfully transformed plant. Four different site-specific recombination systems have been shown to function in plant cells. The best-characterized system in plants is the CRE/lox system of bacteriophage P1: the CRE recombinase enzyme recognizes specifically and catalyzes recombination between two lox sequences. Marker genes flanked by these lox sites can be precisely excised in the presence of the CRE recombinase. One disadvantage of the system was that the CRE protein had to be introduced in the transgenic plant via secondary transformation or sexual crossing with a CRE transformant. When inducible developmental stage promoters or tissue-specific promoters are used, both components of the CRE/lox system can be transferred to one transformant. After selection of the transformants, the promoter will be intentionally induced, resulting in excision of the marker gene. This system is also applicable to vegetatively propagated plants.

A recent publication describes the efficient excision of selectable marker genes that are framed by *attP*-sites without the induction of a recombinase. The molecular basis of the reaction is still unknown, but presumably it is due to illegitimate recombination. Further experiments are necessary to reveal the underlying mechanism and to verify its general applicability.

2
GM Plants Already on the Market

In Europe, 10 genetically modified (crop) plants have been approved for placing on the market according to 90/220 EEC. These approvals covered different uses as applied for by the applicant: in two cases, the approvals included use as food and feed, namely, Roundup Ready soybean (Monsanto) and insect-resistant maize (Novartis, formerly Ciba-Geigy). Nevertheless, only few commercial plantings are carried out in Europe, and in Germany GMPs are grown only on an experimental scale.

Since 1997, GMPs intended for food or feed use in the EU must be notified or approved according to the "Novel Food Regulation." Since then, refined oil from herbicide-tolerant oil seed rape, refined oil from herbicide-tolerant and insect-resistant cotton, and processed products from herbicide-resistant and insect-resistant maize have been notified (Table 1).

Worldwide, GMPs have been grown on 67.7 million hectares in 2003 according to ISAAA. Since 1996, the global area of transgenic crops increased 40-fold, with an increasing proportion grown by developing countries. The main producers of GMPs are the US (42.8 million hectares), Argentina (13.9 million hectares), Canada (4.4 million hectares), Brazil (3 million hectares), China (2.8 million hectares), and South Africa (0.4 million hectares).

Genetically modified soybeans occupied 41.4 million hectares; GM maize was planted on 15.5 million hectares, transgenic cotton on 7.2 million hectares, and GM oilseed rape on 3.6 million hectares.

Some 73% of the GMPs contain a herbicide resistance gene, followed by 18% carrying insect resistance genes, and 8% both herbicide- and insect-resistance genes (stacked genes).

2.1
Herbicide Resistance in Soybean, Maize, Oilseed Rape, Sugar Beet, Rice, and Cotton

As mentioned above, herbicide resistance is the leading trait in commercialized GMP, with 25 lines having been approved for cultivation and/or food and feed use worldwide (Table 2).

2.2
Insect Resistance in Maize, Potatoes, Tomatoes, and Cotton

Insect resistance is the trait found second most often in approved GMPs. Insect-resistant lines, as well as lines that acquire an insect resistance gene in combination with a herbicide resistance or virus-resistance gene, are listed in Table 3.

2.3
Virus Resistance, Male Sterility, Delayed Fruit Ripening, and Fatty Acid Contents in GMPs

GMPs with virus resistance (potato, squash, and papaya; Table 4), male sterility (oilseed rape, maize, and chicory; Table 5), delayed fruit ripening (tomato; Table 6) as well as modified fatty acid content (soybean, oilseed rape; Table 7) have also been approved in some countries.

Tab. 1 Genetically modified crop plants and products approved in the EU.

Plant (line)	Trait (gene)	Applicant	Approved use	Year
Soybean (GTS-40-3-2)	HR (CP4 EPSPS)	Monsanto	Import and processing	1996
Maize (Bt 176)	IR (cryIAb), HR (bar), ABR (bla)	Ciba-Geigy	Cultivation; import and processing	1997
Maize (Bt 11)	IR (cryIAb), HR (pat)	Northrup King	Import and processing; food use	1998; 2004
Maize (MON 810)	IR (cryIAb)	Monsanto	Cultivation; flour, semolina, starch and starch products, oil	1998; 1997
Maize (MON 809)	IR (cryIAb), HR (CP4 EPSPS, gox)	Pioneer	Food additives	1998
Maize (T 25)	HR (pat), ABR (partial bla)	AgrEvo	Cultivation; refined oil, starch and starch products, fermented or heat-treated products from maize flour	1998
Oil seed rape (MS1 × RF1; MS1 × RF2)	MS/RF (barnase/barstar), HR (bar), ABR (nptII)	PGS	Cultivation; refined oil	1997
Oil seed rape (MS8 × RF3)	MS/RF (barnase/barstar), HR (bar)	PGS	Refined oil	1999
Oil seed rape (Topas 19/2)	HR (pat), ABR (nptII)	AgrEvo	Import and processing (feed); refined oil	1995; 1997
Oil seed rape (Falcon GS 40/90)	HR (pat)	AgrEvo	Refined oil	1999
Oil seed rape (Liberator L62)	HR (pat)	AgrEvo	Refined oil	1999
Oil seed rape (GT 73)	HR (CP4 EPSPS, gox)	Monsanto	Refined oil	1997
Cotton (MON 1445/1698)	HR (CP4 EPSPS), ABR (nptII, aad)	Monsanto	Refined oil	2002
Cotton (MON 531/757/1076)	IR (cryIAc), ABR (nptII, aad)	Monsanto	Refined oil	2002

Notes: HR: herbicide resistance (glyphosate resistance CP4 EPSPS gene, glyphosate resistance oxidoreductase gene (gox), glufosinate resistance pat or bar gene);
ABR: antibiotic resistance (ampicillin resistance (bla), kanamycin resistance (nptII)), streptomycin resistance gene (aad);
IR: insect resistance (Bacillus thuringiensis cryIA(b));
MS: male sterility (barnase);
RF: restorer of fertility (barstar).

Tab. 2 Herbicide-resistant crop plants and products worldwide.

Plant (line)	Trait (gene)	Applicant	Country
Soybean (GTS-40-3-2)	HR (CP4 EPSPS)	Monsanto	Argentina, Australia, Brazil, Canada, China, Czech Republic, EU, Japan, Korea, Mexico, Philippines, Russia, Switzerland, South Africa, UK, Uruguay, USA
Soybean (A 2704-12, A 2704-21, A 5547-35)	HR (pat)	AgrEvo	Canada, Japan, USA
Soybean (A 5547–127)	HR (pat), ABR (partial bla)	AgrEvo	USA
Soybean (GU 262)	HR (pat), ABR (partial bla)	AgrEvo	USA
Soybean (W 62, W 98)	HR (bar), RP (gus)	AgrEvo	USA
Maize (GA 21)	HR (maize EPSPS)	Monsanto	Argentina, Australia, Canada, China, Japan, Korea, Philippines, USA
Maize (NK 603)	HR (CP4 EPSPS)	Monsanto	Argentina, Australia, Canada, Japan, Philippines, South Africa, USA
Maize (MON 832)	HR (CP4 EPSPS, gox), ABR (nptII)	Monsanto	Canada
Maize (B 16 = DLL 25)	HR (bar), ABR (bla)	DeKalb	Canada, Japan, Philippines, USA
Maize (T 14, T 25)	HR (pat), ABR (partial bla)	AgrEvo	Argentina, Australia, Canada, EU, Japan, Philippines, USA
Oilseed rape (GT 73)	HR (CP4 EPSPS, gox)	Monsanto	Australia, Canada, China, EU, Japan, Philippines, USA

Oilseed rape (GT 200)	HR (CP4 EPSPS, gox)	Monsanto	Canada, USA
Oilseed rape (HCN 92 = Topas 19/2)	HR (pat), ABR (nptII)	AgrEvo	Canada, EU, Japan, USA
Oilseed rape (HCN 10)	HR (pat)	AgrEvo	Canada, Japan, USA
Oilseed rape (HCN 28 = T 45)	HR (pat)	AgrEvo	Australia, Canada, Japan, USA
Oilseed rape (Oxy 235)	HR (bxn)	Rhone Poulenc	Australia, Canada, Japan, USA
Brassica rapa (ZSR 500, 501, 502)	HR (CP4 EPSPS, gox)	Monsanto	Canada
Cotton (MON 1445/1698)	HR (CP4 EPSPS), ABR (nptII, aad)	Monsanto	Argentina, Australia, Canada, China, Japan, Philippines, South Africa, USA
Cotton (BXN)	HR (bxn), ABR (nptII)	Calgene	Australia, Canada, Japan, USA
Cotton (LL Cotton 25)	HR (bar)	Bayer Crop Science	Canada, USA
Cotton (19-51A)	HR (als)	DuPont	USA
Sugar beet (T120-7)	HR (pat), ABR (nptII)	AgrEvo	Canada, Japan, USA
Sugar beet (GTSB 77)	HR (CP4 EPSPS, gox)	Novartis	Australia, USA
Wheat (MON 71800)	HR (CP4 EPSPS)	Monsanto	USA
Rice (LLRICE 06, 62)	HR (bar)	AgrEvo	USA
Flax (FP967)	HR (als) ABR (nptII)	University of Saskatoon	Canada, USA

Notes: HR: herbicide resistance (glyphosate resistance CP4/maize EPSPS gene, glyphosate resistance oxidoreductase gene (gox), glufosinate resistance pat or bar gene, bromoxynil herbicide resistance nitrilase gene (bxn), sulfonylurea resistance als gene);
ABR: antibiotic resistance (ampicillin resistance gene bla, kanamycin resistance gene nptII, streptomycin resistance gene (aad));
RP: reporter gene (gus).

Tab. 3 Insect-resistant crop plants and products worldwide.

Plant (line)	Trait (gene)	Applicant	Country
Maize (MON 810)	IR (cryIAb)	Monsanto	Argentina, Australia, Canada, EU, Japan, South Africa, Switzerland, USA
Maize (MON 802)	IR (cryIAb), HR (CP4 EPSPS, gox), ABR (nptII)	Monsanto	Canada, Japan, USA
Maize (MON 80100)	IR (cryIAb), HR (CP4 EPSPS, gox), ABR (nptII)	Monsanto	USA
Maize (MON 863)	IR (cryIIIBb2), ABR (nptII)	Monsanto	Australia, Canada, Japan, Philippines, USA
Maize (MON 809)	IR (cryIA(b)), HR (CP4 EPSPS, gox)	Pioneer	Canada, Japan, USA
Maize (Bt 176)	IR (cryIAb), HR (bar), ABR (bla)	Ciba-Geigy	Argentina, Australia, Canada, EU, Japan, Switzerland, USA
Maize (Bt 11)	IR (cryIAb), HR (pat)	Northrup King (Syngenta)	Argentina, Australia, Canada, EU, Japan, Philippines, South Africa, Switzerland, UK, USA
Maize (DBT 418)	IR (cryIAc, protease inhibitor II (pinII)); HR (bar), ABR (bla)	DeKalb	Argentina, Australia, Canada, Japan, Philippines, USA
Maize (TC1507)	IR (cryIFa2), HR (pat)	Mycogen/Pioneer	Canada, Japan, Philippines, South Africa, USA

Potato (Bt 6, Russet Burbank New Leaf)	IR (crylIIA), ABR (nptII)	Monsanto	Canada, Japan, Philippines, USA
Potato (ATBT04-6 etc. 4 lines Atlantic and Superior New Leaf)	IR (crylIIA), ABR (nptII)	Monsanto	Australia, Canada, Japan, Philippines, USA
Potato (SEMT 15–15, New Leaf Y)	IR (crylIIA), VR (coat protein PVY) ABR (nptII, aad)	Monsanto	Australia, Canada, Philippines, USA
Potato (RBTM 21–350, Russet Burbank New Leaf Plus)	IR (crylIIA), VR (replicase PLRV, helicase PLRV) ABR (nptII)	Monsanto	Australia, Canada, Japan, USA
Cotton (MON 531, 757, 1076, Bollgard)	IR (crylAc), ABR (nptII, aad)	Monsanto	Argentina, Australia, Canada, China, India, Japan, Mexico, Philippines, South Africa, USA
Cotton (MON 15985, Bollgard II)	IR (crylAc, crylAb), ABR (nptII, aad), RP (uidA)	Monsanto	Australia, Canada, Japan, Philippines, USA
Cotton (31807/31808)	IR (crylAc), HR (bxn) ABR (nptII)	Calgene	Japan, USA
Tomato (5345)	IR (crylAc), ABR (nptII, aad)	Monsanto	Canada, USA

Notes: IR: insect resistance (delta endotoxin genes of *Bacillus thuringiensis* (crylAb, crylAc, crylIIA, crylFa2; protease inhibitor II (pinII));
HR: herbicide resistance (glyphosate resistance CP4 EPSPS gene, glyphosate resistance oxidoreductase gene (gox), glufosinate resistance pat or bar gene, bromoxynil herbicide resistance nitrilase gene (bxn));
VR: virus resistance (potato virus Y (PVY) coat protein gene, replicase, and helicase gene of potato leafroll virus (PLRV), respectively);
ABR: antibiotic resistance (ampicillin resistance (bla), kanamycin resistance (nptII), streptomycin resistance (aad));
RP: reporter gene (β-D-glucuronidase (gus)).

Tab. 4 Virus-resistant crop plants and products worldwide.

Plant (line)	Trait (gene)	Applicant	Country
Potato (SEMT 15–15, New Leaf Y)	VR (coat protein PVY), IR (cryIIIA), ABR (nptII, aad)	Monsanto	Australia, Canada, Philippines, USA
Potato (RBTM 21–350, Russet Burbank New Leaf Plus)	VR (replicase PLRV, helicase PLRV), IR (cryIIIA), ABR (nptII)	Monsanto	Australia, Canada, Japan, USA
Squash (CZW-3)	VR (coat protein CMV, coat protein ZYMV, coat protein WMV 2), ABR (nptII)	Asgrow	Canada, USA
Squash (ZW20)	VR (coat protein ZYMV, coat protein WMV 2)	Upjohn	Canada, USA
Papaya (55-1/63-1)	VR (coat protein PRSV), RP (gus), ABR (nptII)	Cornell University	Canada, USA

Notes: VR: virus resistance (coat protein genes of potato virus Y (PVY), cucumber mosaic virus (CMV), zucchini yellows mosaic virus (ZYMV) and watermelon mosaic virus (WMV) 2, respectively; replicase and helicase genes of potato leafroll virus (PLRV));
IR: insect resistance (delta endotoxin gene of *Bacillus thuringiensis* (cryIIIA));
ABR: antibiotic resistance (kanamycin resistance (nptII), streptomycin resistance (aad));
RP: reporter gene (gus).

Tab. 5 Male sterility in crop plants worldwide.

Plant (line)	Trait (gene)	Applicant	Country
Oilseed rape (MS1, RF1 (PGS1))	MS (barnase), FR (barstar), HR (bar), ABR (nptII)	PGS	Australia, Canada, EU, Japan, USA
Oilseed rape (MS1, RF2 (PGS2))	MS (barnase), FR (barstar), HR (bar), ABR (nptII)	PGS	Australia, Canada, EU, Japan, USA
Oilseed rape (MS8× RF3)	MS (barnase), FR (barstar), HR (bar)	PGS	Australia, Canada, Japan, USA
Oilseed rape (PHY14, PHY35)	MS (barnase), FR (barstar), HR (bar)	PGS	Japan
Oilseed rape (PHY36)	MS (barnase), FR (barstar), HR (bar)	PGS	Japan
Maize (MS3)	MS (barnase), HR (bar), ABR (bla)	PGS	Canada, USA
Maize (MS6)	MS (barnase), HR (bar), ABR (bla)	PGS	USA
Maize (676, 678, 680)	MS (dam), HR (pat)	Pioneer	USA
Chicory (RM3-3, RM3-4, RM3-6)	MS (barnase), HR (bar), ABR (nptII)	Bejo Zaden	EU, USA

Notes: MS: male sterility (barnase from Bacillus amyloliquefaciens, dam (DNA adenine methylase from E. coli);
FR: fertility restoration (barstar from Bacillus amyloliquefaciens);
HR: herbicide resistance (glufosinate resistance bar gene);
ABR: antibiotic resistance (ampicillin resistance (bla), kanamycin resistance (nptII)).

Tab. 6 Delayed fruit ripening in tomatoes and tomato products worldwide.

Plant (line)	Trait (gene)	Applicant	Country
Tomato (FLAVR SAVR)	DR (*PG*, antisense) ABR (*nptII*)	Calgene	Canada, Japan, Mexico, USA
Tomato (B, Da, F)	DR (*PG*, sense or antisense) ABR (*nptII*)	Zeneca	Canada, USA
Tomato (8338)	DR (*ACCd*) ABR (*nptII*)	Monsanto	USA
Tomato (35 1 N)	DR (*sam-K*) ABR (*nptII*)	Agritope	USA
Tomato (1345-4)	DR (*ACC*, sense) ABR (*nptII*)	DNA Plant Technology Corporation	Canada, USA

Notes: DR: delayed ripening (polygalacturonase gene in sense or antisense direction (*PG*); 1-amino-cyclopropane-1-carboxylic acid deaminase gene (*ACCd*); S-adenosylmethionine hydrolase gene with Kozak consensus sequence (*sam-K*); aminocyclopropane cyclase synthase gene in sense orientation (*ACC*));
ABR: antibiotic resistance (kanamycin resistance (*nptII*)).

Tab. 7 Modified fatty acid content in crop plants and products worldwide.

Plant (line)	Trait (gene)	Applicant	Country
Soybean (G94-1, G94-19, G168, high-oleic soybean)	MFA (*GmFad2-1*), ABR (*bla*), RG (*gus*)	DuPont	Australia, Canada, Japan, USA
Oilseed rape (23-18-17, 23–198, high-laurate canola)	MFA (*BayTE*), ABR (*nptII*)	Calgene	Canada, USA

Notes: MFA: modified fatty acid content (delta 12 desaturase gene (*GmFad2-1*); thioesterase gene from *Umbellaria californica* (Bay TE);
ABR: antibiotic resistance (ampicillin resistance (*bla*), kanamycin resistance (*nptII*));
RP: reporter gene (*gus*).

3
GM Plants "In the Pipeline"

3.1
Input Traits

In the following chapters, selected examples will be presented of studies being conducted in the genetic engineering of crop plants. First, attempts to reduce the production costs of crop plants are summarized. These studies refer to crop plants with genetically engineered resistance against insects, diseases, abiotic stresses, and improved agronomic properties.

3.1.1 Insect Resistance

The best-characterized insecticidal proteins are the delta-endotoxins of *B. thuringiensis* (*B.t.*-toxins), which have been used as biopesticides in agriculture (also organic farming), forestry, and as a mosquito vector control for many years.

B.t. insecticidal activity is highly specific in that the endotoxins are nontoxic to nontarget insects, birds, and mammals.

Besides the commercially available *B.t.*-maize, -potato, -cotton and -tomato lines (see Table 3), insect-resistant rice (*cryIAb* and *cryIAc*), soybeans (*cryIAc*), oilseed rape (*cryIAc*), and eggplants (*cryIIIB*) have been developed. Field trials have also been performed in the United States with insect-resistant sunflower, lettuce, grapefruit, sugarcane, apple, walnut, grape and peanut.

In order to control corn rootworm infections, Mycogen, and Pioneer Hi-Bred are developing transgenic maize, which contains two novel proteins from *B. thuringiensis* strain PS149B1. These proteins belong to another class of insecticidal proteins with no homology to the delta-endotoxins (Bt PS149B1 toxins).

Another approach to achieve insect-resistant plants is to use plant defense proteins such as proteinase inhibitors, lectins or α-amylase inhibitor. Lectins are thought to confer resistance toward sap-sucking insects (e.g. the rice brown planthopper), which act as vector for virus transmission.

Morton and colleagues transferred the α-amylase inhibitor 1 gene from beans to peas and conducted field trials with the insect-resistant pea plants, and reported high insect mortality in transgenic plants. One advantage of using α-amylase inhibitor genes is the long history of human consumption of beans, and the fact that bean α-amylase has no effect on starch digestion in humans.

In an attempt to confer broad-spectrum resistance to storage pests, maize was transformed with the chicken avidin gene. The mode of action of avidin is to cause a deficiency of the vitamin biotin in insects; hence, a thorough safety testing for human consumption must be carried out before the commercial use of biotin maize.

3.1.2 Virus, Fungal, Bacterial, and Nematode Resistance

3.1.2.1 Virus resistance.
One strategy to obtain virus-resistant plants is to transfer genes from the pathogen itself into the plant (pathogen-derived resistance). The most widely used approach is to express the virus coat protein in transgenic plants. In theory, the expression of viral genes disrupts viral infection or symptom development. All but one of the commercially available virus-resistant plants contain viral coat proteins (see Table 4), and this technique is extended to other plants such as rice, plum tree, tomato, pea, and peanut. Additionally, field trials have been performed in the United States with coat protein–mediated virus-resistant wheat, soybean, sugarcane, sugar beet, cucumber, sweet potato, grapefruit, pineapple, and papaya.

Another form of pathogen-derived resistance is the use of viral replicase genes (or RNA-dependent RNA polymerase genes), which presumably act via posttranscriptional gene silencing. This technique has been used to confer resistance to potato leafroll virus in potato (see Table 4), to barley yellow dwarf virus in oats, cucumber mosaic virus in tomato, rice tungro spherical virus in rice, and wheat streak mosaic virus in wheat.

Since varying degrees of virus resistance have been obtained with coat protein–mediated resistance, approaches have been formulated to ameliorate resistance against cucumber mosaic virus via satellite RNA, especially in tomato. However, this approach has raised controversy as a single point mutation in the satellite RNA can transform it into a harmful necrogenic form.

In order to protect plants against more than one virus, ribosome-inactivating proteins (RIPs) have been expressed in transgenic plants. RIPs are strong inhibitors of protein synthesis and, depending on the plant species from which they originate, they show different levels of toxicity against different hosts. Pokeweed antiviral protein (PAP) confers resistance to PVX and PVY in transgenic potatoes, while PAPII confers resistance to TMV, PVX, and fungal infections in tobacco, respectively.

On a more experimental scale are approaches to achieve virus resistance using antibodies against the virus coat protein. Such antibodies are able to neutralize virus infection, presumably by interacting with newly synthesized coat protein and disrupting viral particle formation. Like RIPs, broad-spectrum antibodies might be used to protect plants against a wider range of viruses, as has been demonstrated in the case of potyviruses.

3.1.2.2 **Fungal resistance.** Fungal resistance can be conferred via the activation of specific self-defense mechanisms in the plant. One of the mechanisms is the so-called *hypersensitive response* (HR), which enables plants to enclose the pathogen in the infected area by the formation of necrotic lesions. HR induces many defense-related signal molecules such as salicylic acid, ethylene, and phytoalexin. HR is also characterized by an accumulation of pathogenesis-related (PR) proteins that include fungal cell wall–degrading enzymes, antimicrobial peptides, thionins, lipid transfer proteins, and proteinase inhibitors.

In rice, the introduction of chitinase and thaumatin-like protein, respectively, led to increased resistance to sheath blight (*Rhizoctonia solani*). Enhanced resistance to the rice blast fungus *Magnaporthe grisea* was observed upon constitutive expression of chitinase and defense-related protein genes in transgenic rice.

Pathogenesis-related proteins from plants have been used to confer fungal resistance in alfalfa, cucumber, oilseed rape, tomatoes, wheat, grape vine, and oranges.

Other antifungal genes of plant origin comprise genes for RIPs, genes for phytoalexins, and anthocyanin genes. An example of an antifungal gene from nonplant sources that has been transferred to plants is the human lysozyme gene.

In the United States, field trials have been performed with fungal-resistant wheat, barley, cotton, rice, and bananas using different antifungal proteins.

Individual PR-proteins, however, have a narrow spectrum of antifungal activity, and need to function collectively in order to provide a modest but long-term resistance. Therefore, research is currently focusing on genes from mycoparasitic fungi as a source for improving resistance to fungal pathogens. An endochitinase of the mycoparasitic fungus *Trichoderma harzianum* has been transferred to tobacco and potato, and has been reported to confer a high level and a broad spectrum of resistance.

A similar approach has been taken by transferring an antifungal protein from a virus that persistently infects *Ustilago maydis*, to wheat. Transgenic wheat plants exhibit increased resistance against stinking mut (*Tilletia tritici*).

3.1.2.3 **Bacterial resistance.** Resistance to bacterial infections is not yet as developed as virus- and fungal resistance, due partly to the fact that bacterial diseases are a major problem in only some crop plants such as potato, tomato, rice, and certain fruit trees.

The most efficient form of protection is genetic resistance, which is based on single dominant or semidominant genes. These R genes usually confer race-specific resistance, and their effectiveness is based on their interaction with complementary pathogen avirulence (AV) genes in the pathogen, the so-called *gene-for-gene interaction*. Resistance to bacterial blight caused by *Xanthomonas oryzae* pv. *oryzae* was achieved by transferring the resistance gene *Xa21* from a wild rice species to the elite indica rice (IR) variety 'IR72'. Accordingly, the resistance gene *Bs2* from pepper was transferred to tomato, which then exhibited resistance to bacterial spot disease.

The tomato disease resistance gene *Pto* specifies race-specific resistance to *Pseudomonas syringae* pv. *tomato* carrying the *avrPto* gene. By overexpressing *Pto*, a race-nonspecific resistance was observed in transgenic tomatoes.

The resistance based on single dominant gene expression always bears the danger of early evolution of counterresistance in the pathogen due to the emergence of strains that no longer express the specific avirulence gene product. Therefore, new resistance genes are being investigated for use in pyramiding strategies (combination of resistance genes against the same pathogen, but with different targets). One example is the *AP1* gene from sweet pepper, which delays the hypersensitive response when expressed in transgenic rice plants, and which can be used in combination with *Xa21* or other resistance genes.

In several plant species, bifunctional enzymes with lysozyme activity have been detected, which are thought to be involved in defense against bacteria. After transfer of the bacteriophage T4 lysozyme gene, transgenic potatoes showed a decreased susceptibility toward *Erwinia carotovora atroseptica* infections.

Insects produce antimicrobial peptides as major defense response to pathogen attack, and these include sarcotoxins, cecropins, and attacins. The latter has been transferred to apples and pears, which have improved resistance to *Erwinia amylovora*, causing fire blight. Similarly, the expression of synthetic antimicrobial peptide chimeras in transgenic tobacco led to broad-spectrum resistance against both bacterial and fungal pathogens. Bacterial resistant grapes, transformed with the antibiotic protein mangainin from toads are being field tested in the United States.

3.1.2.4 **Nematode resistance.** In the United States, field tests are being conducted with nematode-resistant carrots, tomatoes, potatoes, and pineapples. In the case of carrot, tomato, and potato, cysteine proteinase inhibitors from cowpea and rice, respectively, have been expressed in the transgenic plants. Proteinase inhibitors are an important element of natural plant defense strategy. The cowpea trypsin inhibitor (CpTI), a serine proteinase inhibitor, and oryzacystatin (Oc-I), an inhibitor of cysteine proteinase, have been shown to be effective against proteinases of a cyst nematode. In field trials, no detrimental effect on plant growth was observed in transgenic potatoes. When two partial nematode-resistant potato varieties were transformed with the cystatin, full resistance to potato cyst nematode *Globodera* was achieved. Cysteine proteinase inhibitors are of particular interest because they are the only class of proteinase that is not expressed in the digestive system of mammals.

3.1.3 Tolerance Against Abiotic Stress

The effects of weather, erosion, and depleted soils expose plants to a variety of stresses. Genetic engineering has been used to provide plants with additional stress response genes in order to counteract these environmental stresses. These genes can be grouped into two categories: (1) genes that respond directly against a particular stress; and (2) genes that regulate stress gene expression and signal transduction. Transfer or overexpression of both types of genes has been used in transgenic approaches to enhance tolerance against abiotic stresses. In the following section, salt, drought, and cold stress are considered together, as, on a cellular basis, all act as dehydration stress.

3.1.3.1 Genes conferring dehydration-stress tolerance in transgenic plants.

Plants react to these stresses by displaying complex, quantitative traits that involve the function of many genes. Expression of these genes leads to the accumulation of low molecular weight components such as osmolytes, synthesis of late-embryogenesis-abundant (LEA) proteins, and activation of detoxifying enzymes. In transgenic approaches, genes for the enzymes that are responsible for the production of these compounds have been transferred to nontolerant plants. Most of these studies have been performed in model plants such as *Arabidopsis* or tobacco, but some investigations have been extended to food crops such as rice.

Bajaj and colleagues have summarized the results from these studies in a recent review. After transfer of genes encoding enzymes for osmoprotectants such as glycinebetaine, proline or putrescine, transgenic rice plants exhibited increased tolerance to salt and drought stress. Constitutive overexpression of the oat arginine decarboxylase gene in rice led to severely affected developmental patterns. When the arginine decarboxylase gene was under the control of an ABA-inducible promoter, the transgenic rice plants showed an increase in biomass under salinity-stress conditions.

Transgenic rice plants expressing the arginine decarboxylase gene (*adc*) from *Datura* produced high levels of the polyamine putrescine under stress, protecting the plants from drought.

The expression of a LEA protein in the transgenic rice plants led to increased tolerance to water deficit and salt stress in the transgenic plants. As it is believed that abiotic stress primarily affects plants through oxidative damage, genes for detoxifying enzymes have been transferred to sensitive plants. McKersie et al. showed that transgenic alfalfa overexpressing Mn superoxide dismutase to reduce free radicals was more tolerant of water deficit and freezing, and had better winter survival rates. It is envisaged that the use of stress-inducible promoters and the introduction of multiple genes will improve dehydration-stress tolerance.

3.1.3.2 Regulatory genes encoding transcription factors.

The other promising approach to confer tolerance to dehydration stress is to transfer regulatory genes. As the products of these genes regulate gene expression and signal transduction under stress conditions, their overexpression can activate the expression of many stress-tolerance genes simultaneously. The overexpression of the transcription factor DREB1A in transgenic *Arabidopsis* led to increased tolerance to drought, salt, and freezing stresses. However, the constitutive overexpression of DREB1A also resulted in severe growth retardation under normal growth conditions.

In contrast, the stress-inducible expression of this gene had minimal effects on plant growth, and provided greater tolerance to stress conditions than genes driven by the 35S promoter. Drought tolerant wheat lines transformed with DREB1A are currently being field tested in Mexico.

Lee and colleagues cloned the functional homolog of the yeast Dbf2 kinase that enhances salt, drought, cold, and heat tolerance upon overexpression in yeast, as well as transgenic plant cells. However, the utilization of this gene to engineer transgenic crops with enhanced stress tolerance has still to be shown.

3.1.3.3 Additional genes that confer tolerance to salt stress.

The detrimental effects of salt on plants are a consequence of both a water deficit resulting in osmotic stress, and the effects of excess sodium ions on key biochemical processes. To tolerate high levels of salts, plants should be able to use ions for osmotic adjustment and internally to distribute these ions to keep sodium away from the cytosol. The first transgenic approaches introduced genes that modulated cation transport systems. Hence, transgenic tomato plants overexpressing a vacuolar Na^+/H^+ antiport were able to grow, flower, and produce fruits in the presence of 200 mM sodium chloride.

3.1.3.4 Genes conferring tolerance to low iron.

In arid and semiarid regions of the world, the soils are alkaline in nature and therefore, crop yields are limited by a lack of available iron. Under iron stress, some plants release specific Fe(III)-binding compounds, known as siderophores, which bind the otherwise insoluble Fe(III) and transport it to the root surface. To increase the quantity of siderophores released under conditions of low iron availability, Takahashi et al. transferred two barley genes coding for the enzyme nicotianamine aminotransferase (*naat-A* and *naat-B*) together with the endogenous promoters into rice plants. Transgenic rice plants excreted about 1.8 times more siderophore under iron stress than wild-type rice plants. This relatively small increase allowed transgenic rice plants to withstand iron deprivation remarkably better, resulting in a fourfold increase in grain yield as compared with wild-type plants.

Alternately, increasing the rate-limiting step of Fe(III) chelate reduction, which reduces iron to the more soluble Fe(II) form, might enhance iron uptake in alkaline soils.

3.1.3.5 Aluminum tolerance.

Aluminum toxicity is one of the major factors limiting crop productivity in acidic soils, which comprise about 40% of the world's arable land. One of the most visible symptoms of aluminum toxicity is the inhibition of root elongation. One possible mechanism of aluminum tolerance is the chelation of aluminum by organic ions like citrate or malate in the rhizosphere or within root cells. Overproduction of citrate was shown to result in aluminum tolerance in transgenic tobacco, papaya, and oilseed rape. Expression of the *ALMT1* gene (aluminum-activated malate transporter (ALMT)) from wheat in transgenic barley plants resulted in high-level aluminum tolerance.

3.1.4 Improved Agronomic Properties

3.1.4.1 Acceleration of sprouting time in potato.

The ability to control sprouting time in potato tubers is of considerable economic importance to the potato industry. At present, the potato industry uses a range of chemical treatments in order to obtain the desired control. Using a biotechnological approach, the *pyrophosphatase* gene from *E. coli* under the control of the tuber-specific patatin promoter was transferred into potatoes. The *pyrophosphatase* gene was chosen because of the central role of inorganic phosphate in starch degradation and sucrose biosynthesis. It is believed that starch breakdown and subsequent formation of various metabolites is needed for growth of the sprout.

Transgenic potatoes displayed a significantly accelerated sprouting: the transgenic tubers sprouted on average six to seven weeks faster than did control tubers. In addition, after cold storage the majority of transgenic tubers sprouted within one week, whereas the wild-type tubers needed eight weeks or more.

3.1.4.2 Reduction of generation time in citrus.
Citrus trees have a long juvenile phase that delays their reproductive development by between 6 and 20 years. With the aim of accelerating flower time, juvenile citrus seedlings were transformed with the *Arabidopsis* LEAFY (*LFY*) and APETALA (*AP1*) genes, which promote flower initiation in *Arabidopsis*. Both types of transgenic citrus produced fertile flowers as early as the first year. These traits are submitted to the offspring as dominant traits, generating trees with a generation time of one year from seed to seed. Constitutive expression of *LFY* also promoted flower initiation in transgenic rice.

3.2 Traits Affecting Food Quality

The following sections provide examples of how the nutritional value of food crops can be improved through genetic engineering, though none of these genetically engineered food plants has yet reached the marketplace. In addition to the safety of the GMPs for health and environment, the effectiveness of this approach in comparison to alternative methods must be shown for these traits.

3.2.1 Increased Vitamin Content
One of the most advanced projects to fortify food crops with vitamins is the so-called *golden rice*. To enable provitamin A biosynthesis in the rice endosperm, four additional enzymes are required. Immature rice embryos were cotransformed with two *Agrobacterium* constructs containing the *psy* and the *lcy* gene from daffodil, coding for the phytoene synthase and the lycopene β-cyclase, respectively, as well as the *crtI* gene from *Erwinia uredovora*, coding for a bacterial phytoene desaturase. Ten plants harboring all transferred genes were recovered, and all showed a normal vegetative phenotype and were fertile. In most cases, the transformed endosperms were yellow, indicating carotenoid formation. After transfer of the new trait into locally best-adapted varieties, either by traditional breeding or *de novo* transformation, it is hoped that vitamin-A deficiency in the developing countries can be prevented.

In contrast to rice, tomato plants already produce carotenoids, and tomato products are viewed as the principal dietary source of lycopene and one of the major sources of β-carotene. Since lycopene and β-carotene are considered beneficial to health (reducing chronic conditions such as coronary heart disease and certain cancers), an attempt was made to enhance the carotenoid content and profile of tomato fruits via genetic engineering. Therefore, the bacterial carotenoid gene (*crtI*), which encodes for the phytoene desaturase, has been introduced into tomato plants. Transgenic plants showed an increased β-carotene content (up to 45% of the total carotenoid content), though the total carotenoid content was not elevated.

The regular uptake of vitamin C (ascorbic acid) is essential for humans since they lack the ability to synthesize it. In plants, ascorbic acid can be regenerated from its oxidized form in a reaction catalyzed by dehydroascorbate reductase (DHAR). The overexpression of a DHAR cDNA from wheat in maize and tobacco led to a two- to fourfold increase in ascorbic acid levels in foliar and kernels.

3.2.2 Production of Very-long-chain Polyunsaturated Fatty Acids

Very-long-chain polyunsaturated fatty acids (VLCPUFAs) like arachidonic, eicosapentaenoic, or docosahexaenoic acid have important roles in human health and nutrition. They are components of membrane phospholipids, precursors of prostaglandins, required for the development of the fetal neuronal system and reported to reduce the incidence of cardiovascular diseases. The diet of most modern societies is relatively low in ω 3-PUFAs with a concomitant increased level of ω 6-PUFA intake. An important source of VLCPUFAs is oily fish, whereas all plant oils do not contain VLCPUFAs. Now first results have been obtained to genetically engineer the capacity to synthesize these fatty acids in oilseed species. The seed-specific expression of cDNAs encoding fatty acid acyl-desaturases and elongases resulted in the accumulation of up to 5% of VLCPUFAs in linseed.

3.2.3 Increased Iron Level

As cereal grains are deficient in certain essential mineral nutrients, including iron, several approaches have been used to increase iron accumulation and alter iron metabolism. Since ferritin is a general iron storage protein in all living organisms, ferritin genes have been introduced into rice and wheat plants. Goto et al. generated transgenic rice plants expressing soybean ferritin under the control of the seed-specific rice Glu-B1 promoter. Transgenic rice seeds accumulated up to three times more iron as wild-type seeds. Likewise, increased iron levels were found in transgenic rice seeds expressing bean ferritin under the control of the related Gt-1 promoter.

In order to not only increase iron accumulation, but also improve its absorption in the human intestine, two approaches have been adopted. First, the level of the main inhibitor of iron absorption, phytic acid, was decreased by the introduction of a heat-tolerant phytase from *Aspergillus fumigatus*. Transgenic rice plants exhibited at least double phytase activity, whereas in one individual transgenic line, the phytase activity of the grains increased about 130-fold. Second, as cysteine peptides are considered a major enhancer of iron absorption, the endogenous cysteine-rich metallothionein-like protein gene (*rgMT*) was overexpressed, and the cysteic acid content increased significantly in transgenic seeds. The authors suggested that high-phytase rice, with increased iron content and rich in cysteine-peptide, has the potential to greatly improve iron supply in rice-eating populations.

3.2.4 Improved Amino Acid Composition

Potatoes contain limited amounts of the essential amino acids lysine, tryptophan, methionine, and cysteine. In order to improve the nutritional value of potatoes, a nonallergenic seed albumin gene (*AmA1*) from *Amaranthus hypochondriacus* was transferred to potato plants. The seed-specific albumin was under the control of a tuber-specific and a constitutive promoter, respectively. In both transgenic lines, a 35 to 45% increase in total protein content was reported in transgenic tubers, which corresponded to an increase in most essential amino acids. Additionally, a twofold increase in tuber number and a 3.0- to 3.5-fold increase in tuber yield were observed.

3.2.5 Reduction in the Content of Antinutritive Factors

Cassava is one of the few plants in nature that contains toxic cyanogenic glycosides in the leaves and roots. Sufficient processing of the harvested roots normally renders the cassava safe, although the processing that renders toxic cassava varieties safe also removes certain nutritional value. Therefore, Moller et al. have isolated the genes responsible for cyanogenic glycoside production and are now transforming cassava with antisense constructs of the respective genes *CYP79D1* and *CYP79D2*.

Another strategy for reducing the cyanide toxicity of cassava roots is to introduce a gene that codes for the enzyme hydroxynitrile lyase (HNL). This enzyme breaks down the major cyanogen acetone cyanohydrin, and is expressed only in leaves. After transformation of the cDNA encoding HNL, the HNL activities in cassava roots were comparable with those in leaves. Field trials will determine whether the expression of HNL in roots in fact reduces the cyanide toxicity of cassava food products.

3.2.6 Production of "Low-calorie Sugar"

Koops and colleagues have developed a new sugar beet that produces fructan, a low-calorie sweetener, by inserting a single gene from Jerusalem artichoke that encodes an enzyme for converting sucrose to fructan (1-sucrose:sucrose fructosyl transferase [1-sst]). Short-chain fructans have the same sweetness as sucrose, but provide no calories as humans lack the fructan-degrading enzymes necessary to digest them. Longer-chain fructans form emulsions having a fatlike texture, while fructans also promote the growth of beneficial bacteria present in the gut. Transgenic sugar beet roots produce the same amount of total sugar, but expression of the *1-sst* gene resulted in conversion of more than 90% of the stored sucrose into fructans. Under greenhouse conditions, the "fructan beets" developed normally and had almost the same amount of root dry weight as normal sugar beets. The yield of 110 µmol g^{-1} fresh weight of fructan makes the extraction of these compounds economically interesting.

3.2.7 Seedless Fruits and Vegetables

In plants that are able to develop fruits without fertilization (parthenocarpic fruits), the seeds are absent – a feature that can increase fruit acceptance by consumers. To achieve parthenocarpic development, it is common practice to treat flower buds with synthetic auxinic hormones. To mimic the hormonal effects by genetic engineering, the expression of a gene able to increase auxin content and activity should be induced in the ovule. Rotino and colleagues used the *iaaM* gene from *P. syringae* pv. *savastanoi* under the control of an ovule-specific promoter (*DefH9*) from *Antirrhinum majus* to induce parthenocarpic development in transgenic tobaccos and eggplants. In transgenic eggplants, fruit setting took place even under environmental conditions prohibitive for the untransformed line. Fruit size and weight were similar to that obtained by pollination in transgenic and control plants. It was envisaged that this approach might also be valuable in other horticultural species such as pepper and tomatoes, in which fruit quality is susceptible to uneven pollination and seed set during lower temperature fluctuations associated with "winter season" production. A similar approach might also be used to produce seedless watermelons rather than the cumbersome triploid seed production system now employed.

3.3 Traits that Affect Processing

3.3.1 Altered Gluten Level in Wheat to Change Baking Quality

The bread-making quality of wheat flour depends primarily on the presence of high molecular weight (HMW) glutenins. Wheat is unique among cereals in having this property. The glutenin proteins are encoded by six genes, and the total glutenin content of the grain is proportional to the expression of these genes. It has been shown that the quality of dough can be influenced both by the quantity and quality of the expressed genes. When specific glutenin genes were added back to wheat lines missing some of the glutenin genes, the dough-mixing characteristics were improved significantly. This same technique could be applied to wheat lines that have already been optimized for baking, and could result in flour with glutenin levels higher than the current maximum of 10% of total protein. Concordantly, the quality of durum wheat (*Triticum turgidum* L. var. *durum*) for bread and pasta making has been modified by insertion of HMW glutenin subunit genes.

3.3.2 Altered Grain Composition in Barley to Improve Malting Quality

β-Glucan is the major constituent of the cell wall of the starchy endosperm of barley. These cell wall molecules are very large, water-soluble, and produce viscous worts if not sufficiently reduced in size by hydrolysis. This causes slow filtration of wort and beer, as well as glucan precipitate in the finished product. β-Glucan hydrolysis is likely to be a function of the level of (1–3, 1–4)-β-glucanase produced by the aleurone, and how much of the enzyme survives high-temperature kilning and mashing. Doubling the amount of β-glucanase activity is likely to ensure sufficient β-glucan hydrolysis, and this could be achieved either by increasing the amount of enzyme synthesized or by changing the heat stability of β-glucanase. The latter has been achieved by transferring a fungal thermotolerant endo-1,4-β-glucanase to two barley cultivars. The amount of heterologous enzyme has been shown to be sufficient to reduce wort viscosity by decreasing the soluble β-glucan content.

In another approach, biochemically active wheat thioredoxin *h* has been overexpressed in the endosperm of transgenic barley grain. Such overexpression in germinated grain effected an up to fourfold increase in the activity of the debranching enzyme pullulanase, a rate-limiting enzyme in the breakdown of starch. The breaking down of starch is a key step in the malting process, and tests with transgenic varieties showed that the time required could be reduced by up to a day.

3.4 Traits of Pharmaceutical Interest

3.4.1 Production of Vaccines

Vaccines are designed to elicit an immune response without causing disease. While typical vaccines are composed of killed or attenuated disease-causing organisms, recombinant vaccines are desirable as an alternative, as they generally cause fewer side effects than those that occur when the whole organism is delivered. Many candidate proteins have been identified that may function as effective subunit vaccines. Currently, the most common large-scale production of recombinant proteins are genetically engineered bacteria and yeast, due to the ease of manipulation and their rapid growth. However, recombinant proteins overexpressed in microorganisms must be extensively purified to remove host proteins and compounds. Transgenic plants provide an alternative system, with the great practical advantage of directly producing an edible plant tissue for oral immunization. In addition, edible plant-based recombinant vaccines are safe, provide nutrition, and are easy to administer.

At present, a great deal of research is focused on the understanding of transgene expression, stability, and processing in plants. Typical experiments investigating in-planta protein expression employ plant model systems such as potato, tomato, and maize. The ideal plant for human vaccination would be bananas, because they are readily eaten by babies, are consumed uncooked, and are indigenous to many developing countries. Since regeneration and growth to maturity can take up to three years in bananas, it is nevertheless necessary to optimize high-level expression of a vaccine antigen in model plants.

For viral infections, virus-like particles (VLPs), which form by self-assembly of viral surface proteins, are effective as vaccines. The hepatitis B surface antigen (HbsAg) has successfully been used as a vaccine against hepatitis B virus. The vaccine was first produced in yeast by Merck and SmithKline Beecham, and was the first recombinant subunit vaccine. Hence, it served as a model for a first attempt to produce a plant-based vaccine. HbsAg was expressed in tobacco and potato plants and shown to assemble into VLPs that are similar to the yeast-derived commercial vaccine.

In the United States, preliminary clinical trials have been performed, and two out of three volunteers who ate transgenic lettuce carrying a hepatitis B antigen displayed a good systemic response. Likewise, 19 of 20 people who ate a potato vaccine aimed at the Norwalk virus showed an immune response.

Currently, clinical trials with a plant-based vaccine against a bacterial pathogen are underway in the United States. Arntzen and colleagues inserted a gene for a part of an *E. coli* enterotoxin (LT-B subunit gene) that caused diarrhea in humans, into potato plants, and all but one of the 11 volunteers who ate the transgenic raw potatoes produced antibodies to the toxin.

A common problem of vaccine antigens expressed in plants has been the low level of expression. In LT-B-expressing plants, this has been overcome by using a "plant codon usage optimized" synthetic gene encoding LT-B and the use of the tuber-specific patatin promoter.

Future considerations must include containment of the transgenes as well as quality control for antigen content. These considerations are particularly important since inappropriate dosing of vaccines can impair their effectiveness, and constant dosing can lead (potentially) to immunological tolerance. In a more recent discussion, the feasibility of direct oral immunization by eating transgenic plants was doubted because of the problem of lot-to-lot consistency and administering pills prescribed by doctors was favored.

3.4.2 Production of Pharmaceuticals

Some recombinant proteins are already available commercially, including human erythropoietin or glucocerebrosidase. Until now, commercial production has used fermentation in *E. coli* and yeast or mammalian cell systems. However, these expression systems have significant limitations: bacteria cannot perform the complex posttranslational modifications required for bioactivity of many human proteins and production of proteins such as growth regulators or cell cycle inhibitors would negatively impact the transgenic animal cell culture.

Plants have several advantages compared with traditional systems for the molecular farming of pharmaceutical proteins. These include: (1) low production costs; (2) reduced time to market; (3) unlimited supply; (4) eukaryotic protein processing; and (5) safety from blood- or animal tissue-borne human pathogens. The production of a range of therapeutic proteins in transgenic plants has proven their capability for production of bioactive human proteins of pharmaceutical value: the proteins appear fully functional and structurally comparable with the analogous proteins produced in animal cell culture.

Examples of human therapeutic proteins include serum proteins such as hemoglobin as blood substitute, interferons as viral protection agents, lysosomal enzymes lacking in patients with Gaucher or Fabry disease, and other proteins such as hirudin, which is effective as an anticoagulant.

The majority of these pharmaceuticals are produced in model plants that are easy to transform, such as tobacco and potato, by using the constitutive 35S promoter from the cauliflower mosaic virus. However, as high levels of protein accumulation may negatively impact on yield and/or growth of the transgenic plants, inducible or tissue-specific promoters are preferable. Tobacco is an excellent biomass producer, but does produce toxic compounds. Crop-based production systems (e.g. wheat, rice, corn) lack these toxic substances and provide an existing infrastructure for their cultivation, harvest, distribution, and processing. This makes food crops highly attractive but leads to the potential for the unintended presence of therapeutic proteins in human food. Recently publicized incidents in which genetically modified crops have inadvertently mixed with those destined for human consumption have highlighted the need for mechanisms to ensure the segregation of plants that express pharmaceuticals.

Cereal crops (rice and wheat) were first used as production and storage system for a single-chain antibody against carcinoembryogenic antigen (CEA), a well-characterized tumor-associated marker antigen. In fact, in dried seeds, the antibodies could be stored for at least five months without significant loss of activity of the antibody. An alternate technique employs targeting of the recombinant protein to the oil bodies of *Brassica napus* seeds. A synthetic gene coding for a hirudin variant was fused to an *Arabidopsis* oleosin gene and transferred to oilseed rape. The recombinant hirudin was correctly targeted and accumulated on the oilbodies of transgenic seeds, which, because of their lower density, could easily be separated by flotation centrifugation. The functional biopharmaceutical was then released by protease treatment.

A recent publication showed the potential for producing recombinant proteins in the guttation fluid of tobacco. Guttation, which is the loss of water and dissolved materials from uninjured plant organs, leads to the production of a guttation fluid. This can be collected throughout the plant's life, thereby providing a continuous and nondestructive system for recombinant protein production.

4
Outlook

It is likely that, in the future, a rapid development of molecular and technological methods will take place, and this will comprise the initial development of new transgenic approaches in model plants such as *Arabidopsis*, followed by the transfer of ready-established methods to most crop plants. The combination of genetic engineering and molecular marker technology with conventional crossings and subsequent selection will enable a rapid progress in plant breeding.

Despite these possibilities, limitations of widespread application can be foreseen in the patent situation, and this will lead to a partial unpredictability in production costs. A hindrance to the worldwide trade – and therefore also to the marketability of genetically modified foods – is founded in the different legal regulations, for example, in the United States and Europe. Additionally, the lack of acceptance by the public of genetically modified food, especially in central Europe, might delay the commercial potential of transgenic plants, even when there might be clear advantages for the consumer.

Acknowledgment

We thank Nina Mitra for her valuable technical assistance in preparing this manuscript.

See also Molecular Basis of Genetics.

Bibliography

Books and Reviews

AGBIOS (Agriculture & Biotechnology Strategies (Canada) Inc.) (2002) GMO Database. http://www.agbios.com/default.asp (data from December 2004).

Bajaj, S., Targolli, J., Liu, L.-F., Ho, T.-H., Wu,, R. (1999) Transgenic approaches to increase dehydration-stress tolerance in plants, *Mol Breeding* **5**, 493–503.

Daniell, H., Khan, M.S., Allision, L. (2002) Milestones on chloroplast genetic engineering: an environmentally friendly era in biotechnology, *Trends in Plant Sci.* **7**, 84–91.

Dixon, R.A., Lamb, C.J., Masoud, S., Sewalt, V.J.H., Paiva, N.L. (1996) Metabolic engineering: prospects for crop improvement through the genetic manipulation of the phenylpropanoid biosynthesis and defense responses – a review, *Gene* **179**(1), 61–71.

Hammond, J., McGarvey, P., Yushibov, V. (Eds.) (1999) *Plant Biotechnology: New Products and Applications*, Springer-Verlag, Berlin, New York.

ISAAA (International Service for the Acquisition of Agri-biotech Applications) (2004) ISAAA Briefs No 30-2003. Global Status of Commercialized Transgenic Crops: 2003. http://www.isaaa.org/ (accessible December 2004).

Ma, J.K.-C., Drake, P.M.W., Christou, P. (2003) The production of recombinant pharmaceutical proteins in plants, *Nat. Rev. Genet.* **4**, 794–805.

Pew Initiative on Food and Biotechnology (2001) Harvest on the horizon: future uses of agricultural biotechnology. http://pewagbiotech.org/research/harvest/harvest.pdf. (accessible December 2004).

Potrykus, I., Spangenberg, G. (Eds.) (1993) *Gene Transfer to Plants*, Springer Lab Manual, Berlin.

Primary Literature

Abbadi, A., Domergue, F., Bauer, J., Napier, J.A., Welti, R., Zähringer, U., Cirpus, P., Heinz, E.

(2004) Biosynthesis of Very-Long-Chain polyunsaturated fatty acids in transgenic oilseeds: Constraints on their accumulation, *Plant Cell* **16**, 2734–2748.

Arakawa, T., Chong, D.K.X., Langridge, W.H.R. (1998) Efficacy of a food plant-based oral cholera toxin B subunit vaccine, *Na. Biotechnol.* **16**, 292–297.

Arencibia, A.D., Carmona, E.R., Cornide, M.T., Castiglione, S., O'Reilly, J., Chinea, P., Oramas, P., Sala, F. (1999) Somaclonal variation in insect-resistant sugarcane (Saccharum hybrid) plants produced by electroporation, *Transgenic Res.* **8**, 349–360.

Atkinson, H.J., Urwin, P.E., Hansen, E., McPherson, M.J. (1995) Designs for engineered resistance to root-parasitic nematodes, *Trends Biotechnol..* **13**, 369–374.

Barro, F., Rooke, L., Békés, F., Gras, P., Tatham, A.S., Fido, R., Lazzeri, P., Shewry, P.R., Barcelo, P. (1997) Transformation of wheat with high molecular weight subunit genes results in improved functional properties, *Nat. Biotechnol.* **15**, 1295–1299.

Capell, T., Bassie, L., Christou, P. (2004) Modulation of the polyamine biosynthetic pathway in transgenic rice confers tolerance to drought stress, *PNAS* **101**, 9909–9914.

Celis, C., Scurrah, M., Cowgill, S., Chumbiauca, S., Green, J., Franco, J., Main, G., Kiezebrink, D., Visser, R.G.F., Atkinson, H.J. (2004) Environmental biosafety and transgenic potato in a centre of diversity for this crop, *Nature* **432**, 222–225.

Chakraborty, S., Chakraborty, N., Datta, A. (2000) Increased nutritive value of transgenic potato by expressing a nonallergenic seed albumin gene from *Amaranthus hypochondriacus*, *Proc. Natl. Acad. Sci. U.S.A.* **97**(7), 3724–3729.

Cheng, X., Sardana, R., Kaplan, H., Altosaar, I. (1998) *Agrobacterium*-transformed rice plants expressing synthetic *cryIA(b)* and *cryIA(c)* genes are highly toxic to striped stem borer and yellow stem borer, *Proc. Natl. Acad. Sci. U.S.A.* **95**, 2767–2772.

Chen, Z., Young, T.E., Ling, J., Chang, S.-C., Gallie, D.R. (2003) Increasing vitamin C content of plants through enhanced ascorbate recycling, *PNAS* **100**, 3525–3530.

Cho, M.-J., Wong, J.H., Marx, C., Jiang, W., Lemaux, P.G., Buchanan, B.B. (1999) Overexpression of thioredoxin *h* leads to enhanced activity of starch debranching enzyme (pullulanase) in barley grain, *Proc. Natl. Acad. Sci. U.S.A.* **96**(25), 14641–14646.

Clausen, M., Kräuter, R., Schachermayr, G., Potrykus, I., Sautter, C. (2000) Antifungal activity of a virally encoded gene in transgenic wheat, *Nat. Biotechnol.* **18**, 446–449.

Corneille, S., Lutz, K., Svab, Z., Maliga, P. (2001) Efficient elimination of selectable marker genes from the plastid genome by the cre-lox site-specific recombination system, *Plant J.* **27**, 171–178.

De Block, M., Botterman, J., Vandewiele, M., Dockx, J., Thoen, C., Gossele, V., Movva, N.R., Thompson, C., Van Montague, M., Leemans, J. (1987) Engineering herbicide resistance in plants by expression of a detoxifying enzyme, *EMBO J.* **6**, 2513–2518.

De la Fuente, J.M., Ramirez-Rodriguez, V., Cabrera-Ponce, J.L., Herrera-Estrella, L. (1997) Aluminum tolerance intransgenic plants by alteration of citrate synthesis, *Science* **276**, 1566–1568.

Delhaize, E., Ryan, P.R., Hebb, D.M., Yamamoto, Y., Sasaki, T., Matsumoto, H. (2004) Engineering high-level aluminium tolerance in barley with the ALMT1 gene, *PNAS* **101**, 15249–15254.

Düring, K., Porsch, P., Fladung, M., Lörz, H. (1993) Transgenic potato plants resistant to the phytopathogenic bacterium *Erwinia carotovora*, *Plant J.* **3**(4), 587–598.

Farré, E.M., Bachmann, A., Willmitzer, L., Trethewey, R.N. (2001) Acceleration of potato tuber sprouting by the expression of a bacterial pyrophosphatase, *Nat. Biotechnol.* **19**, 268–272.

Fujimoto, H., Yamamoto, M., Kyozuka, J., Shimamoto, K. (1993) Insect-resistant rice generated by introduction of a modified δ-endotoxin gene from *Bacillus thuringiensis*, *Biotechnology* **11**, 1151–1155.14.

Gal-On, A., Wang, Y., Faure, J.E., Pilowsky, M., Zelcer, A. (1998) Transgenic resistance to cucumber mosaic virus in tomato: blocking of long-distance movement of the virus in lines harbouring a defective viral replicase, *Phytopathology* **88**(10), 1101–1107.

Gandikota, M., de Kocho, A., Chen, L., Ithal, N., Fauquet, C., Reddy, A.R. (2001) Development of transgenic rice plants expressing maize anthocyanin genes and increased blast resistance, *Mol. Breeding* **7**(1), 73–83.

Gatehouse, A.M.R., Hilder, V.A., Powell, K.S., Wang, M., Davison, G.M., Gatehouse, L.N., Down, R.E., Edmonds, H.S., Boulter, D., Newell, C.A., Merryweather, A., Hamilton, W.D.O., Gatehouse, J.A. (1994) Insect-resistant transgenic plants: choosing the gene to do the 'job', *Biochem. Soc. Trans.* **22**, 944–949.

Goto, F., Yoshihara, T., Shigemoto, N., Toki, S., Takaiwa, F. (1999) Iron fortification of rice seeds by the soybean ferritin gene, *Nat. Biotechnol.* **17**, 282–286.

Hain, R., Reif, H.J., Krause, E., Langbartels, R., Kindl, H., Vornam, B., Wiese, W., Schmelzer, E., Schreier, P.H., Stocker, R.H., Stenzel, K. (1993) Disease resistance results from foreign phytoalexin expression in a novel plant, *Nature* **361**, 153–156.

Hayakawa, T., Zhu, Y., Itoh, K., Izawa, T., Shimamoto, K., Toriyama, S. (1992) Genetically engineered rice resistant to rice stripe virus, an insect-transmitted virus, *Proc. Natl. Acad. Sci. U.S.A.* **89**(20), 9865–9869.

He, X.Z., Dixon, R.A. (2000) Genetic manipulation of isoflavone 7-O-methyltransferase enhances biosynthesis of 4'-O-methylated isoflavonoid phytoalexins and disease resistance in alfalfa, *Plant Cell* **12**(9), 1689–1702.

He, Z., Fu, Y., Hu, G., Zhang, S., Yu, Y., Sun, Z. (2004) Phosphomannose-isomerase (*pmi*) gene as selectable marker for rice transformation via *Agrobacterium*, *Plant Sci.* **166**, 17–22.

He, Z., Zhu, Q., Dabi, T., Li, D., Weigel, D., Lamb, C. (2000) Transformation of rice with the *Arabidopsis* floral regulator *LEAFY* causes early heading, *Transgenic Res.* **9**, 223–227.

Jefferson, R.A., Kavanagh, T.A., Bevan, M.W. (1987) GUS fusions: β-glucuronidase as a sensitive and versatile gene fusion marker in higher plants, *EMBO J.* **6**, 39001–33908.

Joersbo, M., Donaldson, I., Petersen, S.G., Brunstedt, J., Okkels, F.T. (1998) Analysis of mannose selection used for transformation of sugar beet, *Mol. Breeding* **4**, 111–117.

Kasuga, M., Liu, Q., Yamaguchi-Shinozaki, K., Shinozaki, K. (1999) Improving plant drought, salt, and freezing tolerance by gene transfer of a single stress-inducible transcription factor, *Nat. Biotechnol.* **17**, 287–291.

Komarnytsky, S., Borisjuk, N.V., Borisjuk, L.G., Alam, M.Z., Raskin, I. (2000) Production of recombinant proteins in tobacco guttation fluid, *Plant Physiol.* **124**, 927–933.

Kramer, K.J., Morgan, T.D., Throne, J.E., Dowell, F.E., Bailey, M., Howard, J.A. (2000) Transgenic avidin maize is resistant to storage insect pests, *Nat. Biotechnol.* **18**, 670–674.

Kunkel, T., Niu, Q.-W., Chan, Y.-S., Chua, N.-H. (1999) Inducible isopentenyl transferase as a high-efficiency marker for plant transformation, *Nat. Biotechnol.* **17**, 916–919.

Langridge, W.H.R. (2000) Edible vaccines, *Sci. Am.* **283**(3), 66–71.

Lin, W., Anuratha, C.S., Datta, K., Potrykus, K., Muthukrishnan, S., Datta, S.K. (1995) Genetic engineering of rice for resistance to sheath blight, *Biotechnology* **13**, 686–691.

Lodge, J.K., Kaniewski, W.K., Tumer, N.E. (1993) Broad-spectrum resistance in transgenic plants expressing pokeweed antiviral protein, *Proc. Natl. Acad. Sci. U.S.A.* **90**, 7089–7093.

Lorito, M., Woo, S.L., Fernandez, I.G., Colucci, G., Harman, G.E., Pintor-Toro, J.A., Filippone, E., Muccifora, S., Lawrence, C.B., Zoina, A., Tuzun, S., Scala, F. (1998) Genes from mycoparasitic fungi as source for improving plant resistance to fungal pathogens, *Proc. Natl. Acad. Sci. U.S.A.* **95**, 7860–7865.

Mason, H.S., Lam, D.M.K., Arntzen, C.J. (1992) Expression of hepatitis B surface antigen in transgenic plants, *Proc. Natl. Acad. Sci. U.S.A.* **89**, 11745–11749.

McElroy, D., Brettell, R.I.S. (1994) Foreign gene expression in transgenic cereals, *Trends Biotechnol.* **12**, 62–68.

Moellenbeck, D.J., Peters, M.L., Bing, J.W., et al. (2001) Insecticidal proteins from *Bacillus thuringiensis* protect corn from corn root worms, *Nat. Biotechnol.* **19**, 668–672.

Morton, R.L., Schroeder, H.E., Bateman, K.S., Chrispeels, M.J., Armstrong, E., Higgins, T.J. (2000) Bean alpha-amylase inhibitor 1 in transgenic peas (*Pisum sativum*) provides complete protection from pea weevil (*Bruchus pisorum*) under field conditions, *Proc. Natl. Acad. Sci. U.S.A.* **97**(8), 3820–3835.

Nuutila, A.M., Ritala, A., Skadsen, R.W., Mannonen, L., Kaupinnen, V. (1999) Expression of fungal thermotolerant endo-1,4-beta-glucanase in transgenic barley seeds during germination, *Plant Mol. Biol.* **41**(6), 777–783.

Oldach, K., Becker, D., Lörz, H. (2001) Heterologous expression of genes mediating enhanced fungal resistance in transgenic wheat, *Mol. Plant-Microbe Interact.* **14**(7), 832–838.

Osusky, M., Zhou, G., Osuska, L., Hancock, R.E., Kay, W.W., Misra, S. (2000) Transgenic plants expressing cationic peptide chimeras exhibit broad-spectrum resistance to phytopathogens, *Nat. Biotechnol.* **18**, 1162–1166.

Ow, D.W., Wood, K.V., DeLuca, M., de Wet, J.R., Helinski, D.R., Howell, S.H. (1986) Transient and stable expression of the firefly luciferase gene in plant cells and transgenic plants, *Science* **234**, 856–859.

Parmenter, D.L., Boothe, J.G., Van Rooijen, G.J.H., Yeung, E.C., Moloney, M.M. (1995) Production of biological active hirudin in plant seeds using oleosin partitioning, *Plant Mol. Biol.* **29**, 1167–1180.

Pena, L., Martin-Trillo, M., Juarez, J., Pina, J.A., Navarro, L., Martinez-Zapater, J.M. (2001) Constitutive expression of *Arabidopsis LEAFY* or *APETALA1* genes in citrus reduces their generation time, *Nat. Biotechnol.* **19**, 263–267.

Rao, K.V., Rathore, K.S., Hodges, T.K., Fu, X., Stoger, E., Sudhakar, D., Williams, S., Christou, P., Bharathi, M., Bown, D.P., Powell, K.S., Spence, J., Gatehouse, A.M.R., Gatehouse, J.A. (1998) Expression of snowdrop lectin (GNA) in transgenic rice plants confers resistance to rice brown planthopper, *Plant J.* **15**(4), 469–477.

Römer, S., Fraser, P.D., Kiano, J.W., Shipton, C.A., Misawa, N., Schuch, W., Bramley, P.M. (2000) Elevation of the provitamin A content of transgenic tomato plants, *Nat. Biotechnol.* **18**, 666–669.

Rotino, G.L., Perri, E., Zottini, M., Sommer, H., Spena, A. (1997) Genetic engineering of parthenocarpic plants, *Nat. Biotechnol.* **15**, 1398–1401.

Roy, M., Wu, R. (2001) Arginine decarboxylase transgene expression and analysis of environmental stress tolerance in transgenic rice, *Plant Sci.* **160**, 869–875.

Sanford, J.C., Klein, T.M., Wolf, E.D., Allen, N. (1987) Delivery of substances into cells and tissues using a particle bombardment process, *J. Part. Sci. Tech.* **5**, 27–37.

Sayre, R.T. (2000) Cyanogen reduction in transgenic cassava. Generation of safer food product for subsistence farmers, ISB News Report, August 2000. http//www.isb.vt.edu/articles/aug0003.htm (accessible December 2004).

Schaffrath, U., Mauch, F., Freydell, E., Schweizer, P., Dudler, R. (2000) Constitutive expression of the defense-related *Rir1b* gene in transgenic rice plants confers enhanced resistance to the rice blast fungus *Magnaporthe grisea*, *Plant Mol. Biol.* **43**, 59–66.

Sevenier, R., Hall, R.D., van der Meer, I., Hakkert, H.J.C., van Tunen, A.J., Koops, A.J. (1998) High-level fructan accumulation in a transgenic sugar beet, *Nat. Biotechnol.* **16**, 843–846.

Stöger, E., Vaquero, C., Torres, E., Sack, M., Nicholson, L., Drossard, J., Williams, S., Keen, D., Perrin, Y., Christou, P., Fischer, R. (2000) Cereal crops as viable production and storage systems for pharmaceutical scFv antibodies, *Plant Mol. Biol.* **42**, 583–590.

Stommel, J.R., Tousignant, M.E., Wai, T., Pasini, R., Kaper, J.M. (1998) Viral satellite RNA expression in transgenic tomato confers field tolerance to cucumber mosaic virus, *Plant Dis.* **82**(4), 391–396.

Tacket, C.O., Mason, H.S., Losonsk, G., Clements, J.D., Levine, M.M., Arntzen, C.J. (1998) Immunogenicity in humans of a recombinant bacterial antigen delivered in a transgenic potato, *Nat. Med.* **4**, 607–609.

Tai, T.H., Dahlbeck, D., Clark, E.T., Gajiwala, P., Pasion, R., Whalen, M.C., Stall, R.E., Staskawicz, B.J. (1999) Expression of the *Bs2* pepper gene confers resistance to bacterial spot disease in tomato, *Proc. Natl. Acad. Sci. U.S.A.* **96**(24), 14153–14158.

Takahashi, M., Nakanishi, H., Kawasaki, S., Nishizawa, N.K., Mori, S. (2001) Enhanced tolerance of rice to low iron availability in alkaline soils using barley nicotianamine aminotransferase genes, *Nat. Biotechnol.* **19**, 466–469.

Takaichi, M., Oeda, K. (2000) Transgenic carrots with enhanced resistance against two major pathogens, *Erysiphe heraclei* and *Alternaria dauci*, *Plant Sci.* **153**, 135–144.

Tang, X., Xie, M., Kim, Y.J., Zhou, J., Klessig, D.F., Martin, G.B. (1999) Overexpression of *Pto* activates responses and confers broad resistance, *Plant Cell* **11**(1), 15–29.

Tavladoraki, P., Benvenuto, E., Trinca, S., De Martinis, D., Cattaneo, A., Galeffi, P. (1993) Transgenic plants expressing a functional single-chain Fv antibody are specifically protected from virus attack, *Nature* **366**, 469–472.

Tepfer, M. (1993) Viral genes and transgenic plants, *Biotechnology* **11**, 1125–1132.

Tu, J., Ona, I., Zhang, Q., Mew, T.W., Khush, G.S., Datta, S.K. (1998) Transgenic rice variety 'IR 72' with Xa21 is resistant to bacterial blight, *Theoret. Appl. Genet.* **97**(1/2), 31–36.

Tu, J., Zhang, G., Datta, K., Xu, C., He, Y., Zhang, Q., Khush, G.S., Datta, S.K. (2000) Field performance of transgenic elite commercial hybrid rice expressing *Bacillus thuringiensis* δ-endotoxin, *Nat. Biotechnol.* **18**, 1101–1104.

Urwin, P.E., Green, J., Atkinson, H.J. (2003) Expression of a plant cystatin confers partial resistance to Globodera, full resistance is achieved by pyramiding a cystatin with natural resistance, *Mol Breeding.* **12**, 263–269.

Urwin, P.E., Lilley, C.J., McPherson, M.J., Atkinson, H.J. (1997) Resistance to both cyst and root-knot nematodes conferred by transgenic *Arabidopsis* expressing a modified plant cystatin, *Plant J.* **12**(2), 455–461.

Valanzuolo, S., Catello, S., Colombo, M., Dani, M., Monti, M.M., Uncini, L., Petrone, P., Spingo, P. (1994) Cucumber mosaic virus resistance in transgenic San marzano tomatoes, *Acta Hortic.* **376**, 377–386.

Ye, X., Al-Babili, S., Klöti, A., Zhang, J., Lucca, P., Beyer, P., Potrykus, I. (2000) Engineering the provitamin A (β-carotene) biosynthetic pathway into (carotenoid-free) rice endosperm, *Science* **287**, 303–305.

Zhang, H.-X., Blumwald, E. (2001) Transgenic salt-tolerant tomato plants accumulate salt in the foliage but not in fruit, *Nat. Biotechnol.* **19**, 765–768.

Zubko, E., Scutt, C., Meyer, P. (2000) Intrachromosomal recombination between attP regions as a tool to remove selectable marker genes from tobacco transgenes, *Nat. Biotechnol.* **18**, 442–445.

Zuo, J., Niu, Q.-W., Møller, S.G., Chua, N.-H. (2001) Chemical-regulated, site-specific DNA excision in transgenic plants, *Nat. Biotechnol.* **19**, 157–161.

**Part 6
Gene Medicine and Disease**

Genomics and Genetics. Edited by Robert A. Meyers.
Copyright © 2007 Wiley-VCH Verlag GmbH & Co. KGaA, Weinheim
ISBN: 978-3-527-31609-0

32
Human Genetic Variation and Disease

Lynn B. Jorde
University of Utah Health Sciences Center, Salt Lake City, Utah, USA

1	Introduction 941	
2	**Mutation** 942	
2.1	Gene Size and Duchenne Muscular Dystrophy 943	
2.2	Mutation Hot Spots 943	
3	**Natural Selection** 944	
3.1	Heterozygote Advantage and Sickle Cell Disease 944	
3.2	Natural Selection and HIV 945	
4	**Genetic Drift and Founder Effect** 945	
4.1	Haplotype Analysis 946	
4.2	Mutation Analysis 946	
5	**Drift or Selection?** 947	
5.1	Hemochromatosis in Europe 947	
5.2	G6PD Deficiency in Africa and the Mediterranean 948	
6	**Gene Flow** 948	
6.1	Cystic Fibrosis in African-Americans 948	
6.2	Tay–Sachs Disease in French Canadians 948	
7	**Genetic Variation and Common Diseases** 949	

Genomics and Genetics. Edited by Robert A. Meyers.
Copyright © 2007 Wiley-VCH Verlag GmbH & Co. KGaA, Weinheim
ISBN: 978-3-527-31609-0

8   Perspectives 949

     Bibliography 950
     Books and Reviews 950
     Primary Literature 950

Keywords

Fitness
An individual's relative success in transmitting his or her genes to the next generation.

Genetic Drift
Random intergenerational change in gene frequencies due to finite population size.

Haplotype
The allelic constitution of several loci on a single chromosome ("haploid genotype").

Heterozygote Advantage
A fitness differential in which the heterozygote has higher fitness than any of the homozygotes.

Natural Selection
An evolutionary process in which individuals with a specific genetic variant have higher fitness than those lacking the variant.

Polymorphism
The occurrence of two or more alleles at a locus in a population, where at least two alleles occur with frequencies greater than 1%.

> Variation in the prevalence of genetic diseases results from the action of the evolutionary processes of mutation, natural selection, genetic drift, and gene flow. Genetic markers can be studied in populations to help determine the ways in which these processes interact to influence disease prevalence. This enhances our understanding of the origins and evolution of genetic diseases. This article will review the ways in which genetic markers are used to study disease variation in populations, using specific genetic diseases as examples.

1
Introduction

The prevalence of many genetic diseases varies widely among human populations (see Fig. 1 for examples). Cystic fibrosis (CF) is the most common lethal single-gene disorder among Caucasians, affecting approximately 1 in 2500 individuals. Yet, it is quite uncommon among Asians, with a prevalence of only 1 in 90 000 births. Sickle cell disease affects 1 in 400 to 600 African Americans, but it is rare among individuals of northern European descent. Tay–Sachs disease, a lethal recessive disorder seen in 1 in 3600 Ashkenazi Jewish births, is uncommon in most other populations.

Variation in the prevalence of genetic disease is but one aspect of human genetic variation. The evolutionary processes affecting "normal" human variation (e.g. traits such as height, skin color, blood groups, and DNA polymorphisms) also affect genetic disease variation. Four major evolutionary processes can be identified: mutation, natural selection, genetic drift, and gene flow. The relative contribution of each of these processes to genetic disease variation has been the subject of considerable controversy.

Much of this controversy arises from difficulties in specifying the precise nature of genetic variation. Until fairly recently, population variation in genetic disease was assessed primarily in terms of

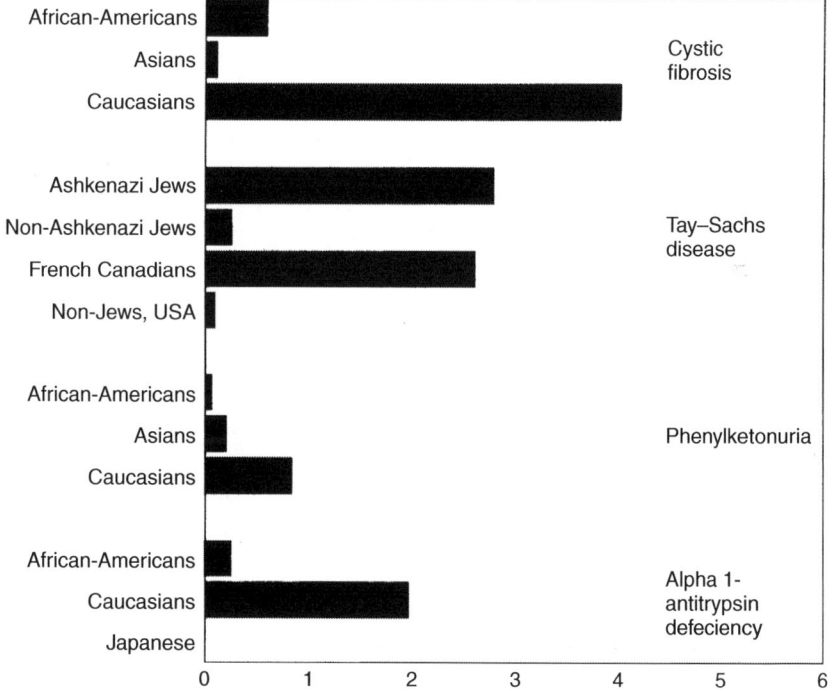

Fig. 1 Prevalence of selected genetic diseases (per 10 000 births) for a series of populations. Note that the French Canadian population refers only to those living in eastern Quebec.

protein variants. Protein electrophoresis could sometimes establish that different populations had different amino acid substitutions, leading to a disease (e.g. for sickle cell disease), but variation at the DNA level was largely unmeasurable. Furthermore, for most disease genes, the gene product, and thus the amino acid sequence, was unknown. Consequently, tracing the origins and evolution of genetic diseases in populations was often a highly speculative endeavor.

The molecular genetic techniques developed during the past two decades have improved this situation substantially. With millions of polymorphisms now identified in the human genome, it is feasible to examine polymorphisms near and within any disease locus. These polymorphisms, or markers, can take the form of restriction fragment length polymorphisms (RFLPs), short tandem repeat polymorphisms (STRPs), or single nucleotide polymorphisms (SNPs). It is becoming increasingly common to sequence entire genes or regions containing many genes in each study subject so that all variation in a region can be assessed. If several polymorphisms are measurable in the vicinity of the disease locus, haplotypes can be constructed for chromosomes carrying the disease mutation and for normal chromosomes. For example, a series of four two-allele SNPs, in which alleles are labeled 1 or 2, could form a haplotype composed of alleles 1, 2, 1, and 1 on the chromosomes bearing a particular disease mutation. Other haplotypes may be observed on normal chromosomes. Because recombination between closely linked polymorphisms is rare, these haplotypes can often be used to trace the evolutionary history of a disease mutation (i.e. under appropriate conditions, all affected individuals with the same chromosome haplotype should be descendants of the same individual). As will be shown in this article, comparisons of haplotypes in different populations can reveal whether a genetic disease had independent origins in each population.

As more and more disease genes are cloned, it is becoming possible to analyze not only marker haplotypes but also disease mutations themselves. This can be accomplished through techniques such as single-strand conformation polymorphism analysis (SSCP) or through direct DNA sequencing. Analysis of mutations sometimes reveals a tremendous diversity of different mutations at a single disease locus. Since the cloning of the CF gene (cystic fibrosis transmembrane conductance regulator, *CFTR*) in 1989, more than 1000 different mutations have been identified at this locus. As will be discussed below, these mutations vary in frequency from one population to another.

As a result of advances in molecular technology, our understanding of the role of evolutionary forces in creating and maintaining genetic variation among populations (including variation in genetic diseases) has increased considerably. This article will review the evolutionary mechanisms responsible for such variation, illustrating each mechanism with specific disease examples. It will also discuss the ways in which molecular genetic markers have contributed to our understanding of these mechanisms.

2
Mutation

Mutation, which is a heritable alteration in DNA sequence, is the ultimate source of genetic variation. While many mutations have no biological consequence for the organism, some may produce changes in amino acid sequence that result in genetic disease. Mutations can occur as single-base substitutions (missense mutations that alter a single amino acid or nonsense mutations that produce a stop codon), deletions and duplications (some of which may produce frameshifts), alterations of splice sites (resulting in inappropriate inclusion of introns or exclusion of exons in the mature messenger RNA), promoter alterations (resulting in decreased transcription of messenger RNA), or insertions of mobile elements such as *Alu* or LINE1 repeats.

The average mutation rate in humans is approximately 10^{-9} per nucleotide per year. There is little evidence that mutation rates vary substantially among human populations. Thus, mutation itself does not account for much interpopulation variation in genetic disease. However, mutation rates do vary substantially from one disease gene to another and thus help explain why some diseases are relatively common and others are relatively rare.

2.1
Gene Size and Duchenne Muscular Dystrophy

Duchenne muscular dystrophy (DMD) is one of the more common lethal genetic diseases, affecting approximately 1 in 3500 males in most surveyed populations. Because it is an X-linked recessive disorder, it seldom affects females. Affected males usually die of cardiac or respiratory failure before age 20 and seldom reproduce. It is estimated that the DMD gene has one of the highest known mutation rates, approximately 10^{-4} (mutations per gene per generation). This high mutation rate helps to account for the fact that DMD is quite common in spite of its lethality in males.

Cloning the DMD gene has helped to explain why the mutation rate is so high. The DMD gene, which spans 2.3 million DNA bases and includes 14 000 bp (14 kilobases or kb) of coding DNA, is the largest gene known in the human. Its large size, quite simply, presents a large "target" for mutation. A similar example is given by neurofibromatosis type 1 (NF1), an autosomal dominant disorder that produces benign tumors associated with peripheral nerves. This disorder also has a high mutation rate, estimated as approximately 10^{-4}. The NF1 gene is again very large, spanning over 350 kb. The size of the coding DNA, 13 kb, is similar to that of the DMD gene. In general, mutation rate appears to correlate positively with gene size.

2.2
Mutation Hot Spots

A second factor influencing mutation rates involves DNA base-pair composition. It is known that the CG dinucleotide is highly susceptible to mutation. The cytosine base is usually methylated when it occurs next to a guanine, and methylated cytosine is likely to spontaneously lose an amino group. This produces a transition from cytosine to thymine. Surveys have shown that an inordinately large proportion of disease-causing mutations occur at CG dinucleotides. Thus, loci rich in such mutational "hot spots" are likely to have higher mutation rates. For example, achondroplasia, a reduced-stature condition, affects up to 1 in 15 000 live births, but the responsible gene (*FGFR3*, fibroblast growth factor receptor 3) spans 16.5 kb and is thus much smaller than the *NF1* or *DMD* genes. Remarkably, 99% of all mutations that cause achondroplasia occur at a single nucleotide (a methylated CG hot spot) of *FGFR3*, where the mutation rate is several orders of magnitude higher than that of other nucleotides. Examples of other genetic diseases that have such mutational hot spots include phenylketonuria (PKU) and hemophilia A and B.

3 Natural Selection

Natural selection can be thought of as a screening process to which genetic variation is subjected. Advantageous variants are positively selected: individuals carrying the variants are more likely to survive and reproduce. This increases the frequency of the gene in the population. Disadvantageous variants will be selected against, reducing the frequency of the variant. For most genetic diseases, an equilibrium state will eventually be reached in which mutation introduces new copies of a disease-causing gene into the population while selection removes them at the same rate.

3.1 Heterozygote Advantage and Sickle Cell Disease

Certain recessive diseases present a slightly more complicated picture. In some cases, the homozygous recessive genotype may undergo negative selection, while the heterozygote may have a selective advantage. In this situation, the disease allele will be maintained at a higher frequency than if the heterozygote had no advantage. A good example of this is found in sickle cell disease. This autosomal recessive disorder, in which the erythrocytes assume a characteristic sickle shape and thus cannot move easily through the capillaries, is often fatal in homozygotes. The heterozygote has no health problems and, in a malarial environment, has a decided survival advantage. This is because sickle cell heterozygotes are resistant to infection by the *Plasmodium falciparum* malaria parasite. Consequently, heterozygotes have a survival advantage over both types of homozygotes. The frequency of the sickle cell gene is quite high in those parts of the world in which malaria has been endemic. It reaches its highest frequency in west-central Africa, where up to one in four individuals is a heterozygous carrier, and one in 45 is affected with the disorder. Sickle cell disease is also seen in parts of the Middle East, India, and the Mediterranean region. It is interesting that the frequency of the sickle cell mutation among African-Americans is substantially lower than in western Africa. This partly reflects admixture with Caucasians (see below), but it is likely also a result of a lack of natural selection in the malaria-free North American environment.

The best-studied of the sickle cell mutations is a single-base substitution (thymine instead of adenine) at the DNA triplet that specifies the sixth amino acid of the β-globin chain. This mutation results in a substitution of valine for glutamic acid known as the hemoglobin S (HbS) variant. At one time, it was thought that the presence of HbS in India and the Mideast was due to migration of Africans who carried the *HbS* gene. However, more recent investigations using RFLPs show that this is probably not the case. Marker haplotypes have been examined, using RFLPs in and around the β-globin locus. They show that the disease haplotypes occurring in Africa differ substantially from those occurring in the Mideast and India. Some haplotype variation can be expected as a result of gene conversion or recombination subsequent to the origin of a mutation, but it is often possible to distinguish these events from multiple mutations. The RFLP haplotypes show that the HbS mutation arose at least twice, once in Africa and once in Saudi Arabia or India. Natural selection then operated in each environment independently to raise the frequency of the mutant gene. Molecular studies have shown that $\Psi\beta$, a pseudogene (i.e. a nontranscribed gene highly similar to β-globin in DNA sequence), does *not* increase in frequency among populations in malarial environments.

3.2
Natural Selection and HIV

Another example of natural selection came to light because of the human immunodeficiency virus (HIV). Some strains of this virus gain entry to macrophages and helper T cells via a cell-surface receptor that normally binds to a cytokine. The resulting viral destruction of helper T cells, a critical component of the body's immune system, leads to acquired immune deficiency syndrome (AIDS). Individuals who are homozygous for a 32-bp deletion of the gene that encodes one of the cytokine receptors (*CCR5*) lack the receptor on their cell surfaces and are thus remarkably resistant to HIV infection (because other cell-surface receptors can perform the same role as CCR5, the mutation has no ill effects). This deletion is especially common in northeastern European populations, where its frequency reaches 0.20. It is virtually absent in Asian and African populations. Analysis of haplotype variation in the chromosomal region that contains *CCR5* indicates that the deletion arose in European populations about 700 to 2000 years ago. Since HIV appeared in humans only a few decades ago, the high deletion frequency in northeastern Europe must be due to a selective force other than HIV or perhaps to genetic drift. Considering the age of the deletion, it was first suggested that it had conferred resistance against the bacterium, *Yersinia pestis*, that causes bubonic plague. However, recent research with mouse models demonstrated that Y. *pestis*' entry into cells is not affected by the absence of the CCR5 molecule. Other evidence suggests that the selective agent may instead have been the smallpox virus.

4
Genetic Drift and Founder Effect

Often, small, isolated populations exhibit high frequencies of genetic diseases that are rare in other populations. For example, Ellis van Creveld syndrome, a very rare disorder that involves reduced stature, is seen with increased frequency among the genetically isolated Old Order Amish population of Lancaster County, Pennsylvania. In fact, nearly as many cases have been observed in a single Amish kindred as have been observed in the rest of the world. In the Finnish population, which has also been quite isolated until recently, over two dozen otherwise rare autosomal recessive diseases occur with substantially elevated frequencies. Such variation in disease prevalence is usually the result of the related processes of genetic drift and founder effect.

Genetic drift refers to the random fluctuation in gene frequencies that occurs from one generation to the next as a result of sampling a limited number of gametes. In small populations, the degree of fluctuation in gene frequency increases (just as a coin-tossing experiment using only 10 coins is quite likely to result in a large deviation from the expected proportion of

50% heads and 50% tails). Genetic drift can thus lead to high frequencies of genetic diseases in small populations. Founder effect occurs when a small number of individuals form a new population. Because these individuals form a small sample of an original population, their genes are unlikely to form a representative sample of the original population from which they came. This can again result in a highly skewed distribution of disease genes among their descendants. The Old Order Amish population, for example, was founded by only about 50 couples. This effect is further exaggerated when certain founders have a disproportionately large number of descendants.

Traditionally, demographic history and genealogies have been used to demonstrate drift and founder effect. Genealogical analysis demonstrated, for example, that all 30 000 cases of porphyria variegata in the South African Afrikaner population could be traced to a single couple who emigrated from Holland in the 1680s. This approach is obviously limited by the availability of extensive genealogies.

4.1
Haplotype Analysis

More recently, molecular genetic markers have made it possible to assess founder effect at the DNA level. Haplotype analysis is particularly helpful in this regard. It is reasonable to assume that each different haplotype background associated with a disease represents a different founder. Thus, the number of distinct disease haplotypes can be counted to infer the number of founders who introduced a disease into a population. In northern Finland, it was shown that all individuals affected with X-linked choroideremia, a disease that causes blindness, had the same RFLP haplotype (using markers closely linked to the choroideremia gene). This provided evidence that each of these individuals was descended from the same common ancestor. Genealogical analysis proved this to be the case: all affected individuals shared a common ancestor that lived 12 generations ago. A similar exercise showed that the high frequency of PKU among Yemeni Jews can be attributed to the descendants of a single founding couple who lived in the seventeenth century. Haplotype analysis of markers closely linked to the *CFTR* locus demonstrated that all studied cases of CF in the Hutterite population could be traced to three founders.

The disease consequences of founder effect and genetic drift have been especially well characterized in the Ashkenazi Jewish population. For example, torsion dystonia, a rare recessive disorder, has an elevated frequency in this population. Analysis of haplotype variation has shown that the disease-causing mutation is only about 350 years old and has likely been amplified in frequency because of population bottlenecks and subsequent expansion. Similarly, three mutations in the *BRCA1* and *BRCA2* genes have reached a combined frequency of 2.5% in Ashkenazi Jews and account for a substantial proportion of inherited breast cancer cases. A mutation in the *APC* gene is found in about 1/20 members of this population and confers a twofold increased risk of developing colon cancer. This mutation alone accounts for about 10% of colon cancer cases in the Ashkenazi population.

4.2
Mutation Analysis

The cloning of disease genes often permits direct analysis of population variation in mutations. One of the best examples is the ΔF508 (deletion of a phenylalanine residue at position 508) mutation in the *CFTR* locus, a three-base deletion that produces a loss of function and accounts for approximately 70% of the *CFTR* mutations found in Caucasians. Extensive population analysis has shown that the proportion of *CFTR* mutations accounted for by ΔF508 has a north–south cline, with a maximum of about 70 to 90% in northern Europe (where the disease itself is most common) and a minimum of about 50% in southern Europe. The proportion of ΔF508 mutations is even lower among other surveyed populations, such as African-Americans (37%) and Ashkenazi Jews (30%). The elevated frequency of CF among Europeans can be traced primarily to the ΔF508 mutation, since the frequencies of non-ΔF508 mutations are quite similar among Europeans, Asians, and Africans. Among Europeans, the great majority (>90%) of ΔF508 mutations occur on a single haplotype background, indicating that most current copies of the mutation may have descended from the same common ancestor.

Cloning of the *CFTR* gene has helped explain why CF has a high frequency in European populations. *CFTR* encodes a chloride channel found on the surfaces of epithelial cells that line the gut and the airway. The number of chloride channels in these cells is substantially reduced in transgenic mice that carry one copy of the ΔF508 mutation. The bacterium that causes typhoid fever, *Salmonella typhi*, gains access to intestinal cells via these chloride channels, and, as predicted, heterozygous mice are quite resistant to typhoid fever infection. It is reasonable to conclude that humans who were heterozygous for this mutation also benefited from resistance to typhoid fever.

5
Drift or Selection?

When a genetic disease is seen with high frequency in a population, it is often unclear whether it is caused by a selective advantage or genetic drift/founder effect. In a few cases (such as sickle cell disease or CF), a demonstrable selective agent is known. But in most cases, the evidence is far more equivocal. Here again, molecular markers have been helpful.

5.1
Hemochromatosis in Europe

Idiopathic hemochromatosis is an autosomal recessive disorder in which excess iron is absorbed in the gut and deposited in many parts of the body, including the skin, joints, liver, pancreas, and heart. Consequently, some homozygotes develop liver disease, heart failure, arthropathy, diabetes, and other disease-related conditions late in life. Approximately 1 in 300 Europeans is homozygous for mutations in the responsible gene, *HFE*. Although several different *HFE* mutations have been reported, 85% of them consist of a cysteine→tyrosine substitution that results in a loss of function of the *HFE* product. Haplotype analysis has shown that this mutation has occurred only once and is confined almost exclusively to the European population. Analysis of variation in markers near *HFE* has shown that the primary disease-causing mutation took place only about 60 to 70 generations ago. The mutation could have achieved a high frequency in this population either through drift or natural selection, but extensive analysis of marker variation near the *HFE* gene has shown that it is extremely improbable that drift alone could have elevated the mutation frequency so quickly. In addition, population surveys show that heterozygotes have a significant reduction in the frequency of iron-deficiency anemia. Thus, a heterozygote advantage is likely to account for a rapid increase in the frequency of this mutation.

5.2
G6PD Deficiency in Africa and the Mediterranean

G6PD (glucose-6-phosphate-dehydrogenase) deficiency is in X-linked disorder that affects 400 million individuals worldwide and can result in hemopathology when mutation carriers are exposed to specific infections, drugs (e.g. primaquine), or foods (e.g. fava beans). G6PD deficiency, like sickle cell disease, confers resistance to infection by the malaria parasite, *P. falciparum*. Molecular analysis has shown that a major mutation, A-, is common in Africa, while another major mutation, Med, is common in Mideastern and Mediterranean populations. The A-mutation occurred approximately 6000 years ago, and the Med mutation occurred approximately 3000 years ago. The recent ages of these mutations reflect limited haplotype diversity and indicate that genetic drift alone could not have elevated these mutations to high frequency in just a few thousand years. Natural selection, as a result of malarial resistance, must have played an important role in raising the frequencies of these mutations. It is interesting that the form of *P. falciparum* responsible for severe malaria probably did not arise until about 10 000 years ago, after the introduction of agriculture in these areas.

6
Gene Flow

Gene flow reduces genetic differences among populations. When two populations both manifest a genetic disease, the question often arises whether the disease mutation arose independently in the two populations or was communicated from one to another via gene flow. When only a disease phenotype can be observed, it is difficult to answer this question. However, more precise molecular characterization of disease loci often illuminates the issue.

6.1
Cystic Fibrosis in African-Americans

An example is given by CF in the African-American population. This disease affects approximately 1 in 17 000 African-Americans, a prevalence much lower than that of Europeans. One hypothesis is that the disorder was introduced into the African-American population through gene flow from the European-American population. However, analysis of CF mutations shows that $\Delta F508$ is quite rare among African-Americans and that two-thirds of the observed CF mutations are unique to this population. Thus, CF almost certainly had independent origins in the African-American population and is not principally derived from European-Americans.

6.2
Tay–Sachs Disease in French Canadians

Studies of Tay–Sachs disease in Quebec provide another example. As one would expect, the Ashkenazi Jewish population of this region has an elevated prevalence of this disorder, and the frequencies of the two major mutations closely match those of other Ashkenazi Jewish populations. Tay–Sachs disease is also seen among the French Canadian population of Quebec. In fact, in the eastern portion of the province, the disease gene frequency is as high as among Ashkenazi Jews. Does this reflect gene flow from the Jewish population, or was there an independent origin? Direct examination of the Tay–Sachs gene shows that the French Canadians usually carry a 7.6-kb deletion that is not found among Jews. Thus, an independent origin is established. Further investigations may ultimately establish whether the high frequency of the disorder among French Canadians is the result of founder effect or natural selection.

7
Genetic Variation and Common Diseases

The examples used thus far have all involved relatively rare single-gene diseases. In each case, mutations have increased in frequency as a result of factors such as drift or selection. Increasing attention is now being devoted to mutations responsible for common diseases such as diabetes, heart disease, and common cancers. The most significant of these mutations are those that have reached high frequency (e.g. >10%). Typically, such mutations are relatively old and are thus shared among most worldwide populations. Their frequencies vary among populations, however. For example, the $\epsilon 4$ allele of the apolipoprotein E (APOE) gene, which confers an increased risk of developing Alzheimer's disease, exhibits frequencies above 30% among some African populations (e.g. Pygmies, Khoi San, Sudanese, and Nigerians) but substantially lower (<15%) among southern Europeans and Asians (there is also evidence, however, that the association between $\epsilon 4$ and Alzheimer's disease is weaker among individuals of African descent than among those of European descent). A gain-of-function mutation in the SCN5A gene, which encodes a sodium channel present in cardiac myocytes, confers an eightfold increased risk of cardiac arrhythmia and was found in approximately 13% of African-Americans but was absent in samples of 511 Europeans and 578 Asians. A promoter variant of the angiotensinogen (AGT) gene is associated with an increased risk of hypertension and reaches high frequencies in African populations. Haplotype analysis has demonstrated that natural selection has acted to decrease the frequency of this variant in non-African populations, possibly as a result of their adaptation to nontropical, high-sodium environments after they left the African continent. Differences in the frequencies of mutations such as these may help account for differences in the prevalence of common diseases in populations, although environmental factors such as diet must also have important effects. It is critical to point out that, because of the recent origin of modern humans from a small founder population and continued gene flow among populations through time, most of these variants tend to be shared among major populations and typically vary only in their frequencies.

8
Perspectives

This review has highlighted some ways in which the study of disease variation in populations can be related to evolutionary processes. Such studies are significant in understanding evolutionary biology. They have other implications as well. An awareness of population variation in disease can aid in clinical diagnosis. For example, a physician evaluating a child who presents with recurrent respiratory infections and failure to thrive should include ethnic background in considering a differential diagnosis. If the patient is of African descent, sickle cell disease is a distinct possibility. If the patient is of northern European descent, sickle cell disease would be unlikely (but not impossible), whereas cystic fibrosis should be considered. In some cases, knowledge of the potential adaptive significance of a genetic disease (e.g. the heterozygote advantage for sickle cell disease, cystic fibrosis, or hemochromatosis) may lead to increased understanding of gene function and pathophysiology.

When considering genetic variants that contribute to common diseases, as well as those that influence response to therapeutic drugs, it should be emphasized that most such variants are shared among most major human populations, varying only in their frequencies. Consequently, population affiliation (or "race") is, at best, a very approximate indicator of critical factors such as drug response.

Advances in molecular genetics have greatly improved our understanding of the evolutionary processes leading to population variation in genetic disease. Yet many important questions remain unanswered, particularly for common, complex diseases. As more polymorphic markers are analyzed, more DNA sequenced, and more disease genes cloned, new and more profound insights are certain to be gained.

Bibliography

Books and Reviews

Bamshad, M., Wooding, S.P. (2003) Signatures of natural selection in the human genome, *Nat. Rev. Genet.* 4(2), 99–111.

Crow, J.F. (2000) The origins, patterns and implications of human spontaneous mutation, *Nat. Rev. Genet.* 1(1), 40–47.

Goldstein, D.B., Chikhi, L. (2002) Human migrations and population structure: what we know and why it matters, *Annu. Rev. Genomics Hum. Genet.* 3, 129–152.

Jobling, M.A., Hurles, M.E., Tyler-Smith, C. (2003) *Human Evolutionary Genetics: Origins, Peoples, and Disease*, Garland Science, New York.

Jorde, L.B., Watkins, W.S., Bamshad, M.J. (2001) Population genomics: a bridge from evolutionary history to genetic medicine, *Hum. Mol. Genet.* 10, 2199–2207.

Jorde, L.B., Carey, J.C., Bamshad, M.J., White, R.L. (2003) *Medical Genetics*, 3rd edition, Mosby, St. Louis, MO.

Kwiatkowski, D. (2000) Genetic susceptibility to malaria getting complex, *Curr. Opin. Genet. Dev.* 10(3), 320–324.

McKusick, V.A. (Ed.) (1978) *Medical Genetic Studies of the Amish: Selected Papers*, Johns Hopkins University Press, Baltimore.

Ostrer, H. (2001) A genetic profile of contemporary Jewish populations, *Nat. Rev. Genet.* 2(11), 891–898.

Tishkoff, S.A., Verrelli, B.C. (2003) Patterns of human genetic diversity: implications for human evolutionary history and disease, *Annu. Rev. Genomics Hum. Genet.* 4(1), 293–340.

Primary Literature

Abadie, V., Lyonnet, S., Maurin, N., Berthelon, M., Caillaud, C., Giraud, F., Mattei, J.F., Rey, J., Rey, F., Munnich, A. (1989) CpG dinucleotides are mutation hot spots in phenylketonuria, *Genomics* 5(4), 936–939.

Ajioka, R.S., Jorde, L.B., Gruen, J.R., Yu, P., Dimitrova, D., Barrow, J., Radisky, E., Edwards,

C.Q., Griffen, L.M., Kushner, J.P. (1997) Haplotype analysis of hemochromatosis: evaluation of different linkage-disequilibrium measures and evolution of disease chromosomes, *Am. J. Hum. Genet.* **60**, 1439–1447.

Allison, A.C. (1961) Genetic factors in resistance to malaria, *Ann. N.Y. Acad. Sci.* **91**, 710–729.

Avigad, S., Cohen, B.E., Bauer, S., Schwartz, G., Frydman, M., Woo, S.L., Niny, Y., Shiloh, Y. (1990) A single origin of phenylketonuria in Yemenite Jews, *Nature* **344**(6262), 168–170.

Bamshad, M., Wooding, S., Salisbury, B.A., Stephens, J.C. (2004) Deconstructing the relationship between genetics and race, *Nat. Rev. Genet.* **5**(8), 598–609.

Bamshad, M.J., Mummidi, S., Gonzalez, E., Ahuja, S.S., Dunn, D.M., Watkins, W.S., Stone, A.C., Jorde, L.B., Weiss, R.B., Ahuja, S.K. (2002) A strong signature of balancing selection in the promoter of CCR5, *Proc. Natl. Acad. Sci. U. S. A.* **99**, 10539–10544.

Bobadilla, J.L., Macek, M. Jr, Fine, J.P., Farrell, P.M. (2002) Cystic fibrosis: a worldwide analysis of CFTR mutations-correlation with incidence data and application to screening, *Hum. Mutat.* **19**(6), 575–606.

Cawthon, R.M., Weiss, R., Xu, G.F., Viskochil, D., Culver, M., Stevens, J., Robertson, M., Dunn, D., Gesteland, R., O'Connell, P. (1990) A major segment of the neurofibromatosis type 1 gene: cDNA sequence, genomic structure, and point mutations, *Cell* **62**(1), 193–201.

Chen, S.-H., Zhang, M., Lovrien, E.W., Scott, C.R., Thompson, A.R. (1991) CG dinucleotide transitions in the factor IX gene account for about half of the point mutations in hemophilia B patients: a Seattle series, *Hum. Genet.* **87**, 177–182.

Cutting, G.R., Curristin, S.M., Nash, E., Rosenstein, B.J., Lerer, I., Abeliovich, D., Hill, A., Graham, C. (1992) Analysis of four diverse population groups indicates that a subset of cystic fibrosis mutations occur in common among Caucasians, *Am. J. Hum. Genet.* **50**, 1185–1194.

Dean, M., Carrington, M., Winkler, C., Huttley, G.A., Smith, M.W., Allikmets, R., Goedert, J.J., Buchbinder, S.P., Vittinghoff, E., Gomperts, E., Donfield, S., Vlahov, D., Kaslow, R., Saah, A., Rinaldo, C., Detels, R., O'Brien, S.J. (1996) Genetic restriction of HIV-1 infection and progression to AIDS by a deletion allele of the CKR5 structural gene. Hemophilia Growth and Development Study, Multicenter AIDS Cohort Study, Multicenter Hemophilia Cohort Study, San Francisco City Cohort, A, *Science* **273**(5283), 1856–1862.

Diamond, J.M., Rotter, J.I. (1987) Observing the founder effect in human evolution, *Nature* **329**, 105–106.

Eisensmith, R.C., Okano, Y., Dasovich, M., Wang, T., Guttler, F., Lou, H., Guldberg, P., Lichter-Konecki, U., Konecki, D.S., Svensson, E. (1992) Multiple origins for phenylketonuria in Europe, *Am. J. Hum. Genet.* **51**, 1355–1365.

European Working Group on CF Genetics. (1990) Gradient of distribution in Europe of the major CF mutation and of its associated haplotype, *Hum. Genet.* **85**, 436–445.

Farrer, L.A., Cupples, L.A., Haines, J.L., Human, B., Kukull, W.A., Mayeux, R., Myers, R.H., Pericak-Vance, M.A., Risch, N., van Duijn, C.M. (1997) Effects of age, sex, and ethnicity on the association between apolipoprotein E genotype and Alzheimer disease: a meta-analysis, *JAMA* **278**, 1349–1356.

Feder, J.N., Gnirke, A., Thomas, W., Tsuchihashi, Z., Ruddy, D.A., Basava, A., Dormishian, F., Domingo, R. Jr., Ellis, M.C., Fullan, A., Hinton, L.M., Jones, N.L., Kimmel, B.E., Kronmal, G.S., Lauer, P., Lee, V.K., Loeb, D.B., Mapa, F.A., McClelland, E., Meyer, N.C., Mintier, G.A., Moeller, N., Moore, T., Morikang, E., Wolff, R.K. (1996) A novel MHC class I-like gene is mutated in patients with hereditary haemochromatosis, *Nat. Genet.* **13**(4), 399–408.

Frisch, A., Colombo, R., Michaelovsky, E., Karpati, M., Goldman, B., Peleg, L. (2004) Origin and spread of the 1278insTATC mutation causing Tay-Sachs disease in Ashkenazi Jews: genetic drift as a robust and parsimonious hypothesis, *Hum. Genet.* **114**(4), 366–376.

Galvani, A.P., Slatkin, M. (2003) Evaluating plague and smallpox as historical selective pressures for the CCR5-Delta 32 HIV-resistance allele, *Proc. Natl. Acad. Sci. U.S.A.* **100**(25), 15276–15279.

Hill, A.V.S. (1989) Molecular Markers of Ethnic Groups, in: Cruickshank, J.K., Beevers, D.G. (Eds.) *Ethnic Factors in Health and Disease*, Wright, London, pp. 25–31.

Hoyer, L.W. (1994) Hemophilia A, *N. Engl. J. Med.* **330**, 38–47.

Jeunemaitre, X., Soubrier, F., Kotelevstev, Y.V., Lifton, R.P., Williams, C.S., Charru, A., Hunt, S.C., Hopkins, P.N., Williams, R.R., Lalouel, J.-M., Corvol, P. (1992) Molecular basis of human hypertension: role of angiotensinogen, *Cell* **71**, 169–180.

Kere, J. (2001) Human population genetics: lessons from Finland, *Annu. Rev. Genomics Hum. Genet.* **2**, 103–128.

Kerem, B., Rommens, J.M., Buchanan, J.A., Markiewicz, D., Cox, T.K., Chakravarti, A., Buchwald, M., Tsui, L. (1989) Identification of the cystic fibrosis gene: genetic analysis, *Science* **245**, 1073–1080.

Kirchhoff, T., Satagopan, J.M., Kauff, N.D., Huang, H., Kolachana, P., Palmer, C., Rapaport, H., Nafa, K., Ellis, N.A., Offit, K. (2004) Frequency of BRCA1 and BRCA2 mutations in unselected Ashkenazi Jewish patients with colorectal cancer, *J. Natl. Cancer Inst.* **96**(1), 68–70.

Klinger, K., Horn, G.T., Stanislovitis, P., Fujiwara, T.M., Morgan, K. (1990) Cystic fibrosis mutations in the Hutterite brethren, *Am. J. Hum. Genet.* **46**, 983–987.

Koenig, M., Hoffman, E.P., Bertelson, C.J., Monaco, A.P., Feener, C., Kunkel, L.M. (1987) Complete cloning of the Duchenne muscular dystrophy (DMD) cDNA and preliminary genomic organization of the DMD gene in normal and affected individuals, *Cell* **50**(3), 509–517.

Kondrashov, A.S. (2003) Direct estimates of human per nucleotide mutation rates at 20 loci causing Mendelian diseases, *Hum. Mutat.* **21**(1), 12–27.

Kwiatkowski, D. (2000) Genetic susceptibility to malaria getting complex, *Curr. Opin. Genet. Dev.* **10**(3), 320–324.

Myerowitz, R., Hogikyan, N.D. (1986) Different mutations in Ashkenazi Jewish and non-Jewish French Canadians with Tay-Sachs disease, *Science* **232**, 1646–1648.

Nakajima, T., Jorde, L.B., Ishigami, T., Umemura, S., Emi, M., Lalouel, J.M., Inoue, I. (2002) Nucleotide diversity and haplotype structure of the human angiotensinogen gene in two populations, *Am. J. Hum. Genet.* **70**(1), 108–123.

Nakajima, T., Wooding, S., Sakagami, T., Emi, M., Tokunaga, K., Tamiya, G., Ishigami, T., Umemura, S., Munkhbat, B., Jin, F., Guan-Jun, J., Hayasaka, I., Ishida, T., Saitou, N., Pavelka, K., Lalouel, J.M., Jorde, L.B., Inoue, I. (2004) Natural selection and population history in the Human Angiotensinogen Gene (AGT): 736 complete AGT sequences in chromosomes from around the world, *Am. J. Hum. Genet.* **74**(5), 898–916.

Niell, B.L., Long, J.C., Rennert, G., Gruber, S.B. (2003) Genetic anthropology of the colorectal cancer-susceptibility allele APC I1307K: evidence of genetic drift within the Ashkenazim, *Am. J. Hum. Genet.* **73**(6), 1250–1260.

O'Brien, S.J. (1998) AIDS: a role for host genes, *Hosp. Pract. (Off. Ed.)* **33**(7), 53–6, 59–60, 66–7. passim.

Pier, G.B., Grout, M., Zaidi, T., Meluleni, G., Mueschenborn, S.S., Banting, G., Ratcliff, R., Evans, M.J., Colledge, W.H. (1998) Salmonella typhi uses CFTR to enter intestinal epithelial cells, *Nature* **393**, 79–82.

Risch, N., de Leon, D., Ozelius, L., Kramer, P., Almasy, L., Singer, B., Fahn, S., Breakfield, X., Bressman, S. (1995) Genetic analysis of idiopathic torsion dystonia in Ashkenazi Jews and their recent descent from a small founder population, *Nat. Genet.* **9**, 152–159.

Rousseau, F., Bonaventure, J., Legeal-Mallet, L., Pelet, A., Rozet, J., Maroteaux, P., Le Merrer, M., Munnich, A. (1994) Mutations in the gene encoding fibroblast growth factor receptor-3 in achondroplasia, *Nature* **371**, 252–254.

Ruwende, C., Khoo, S.C., Snow, R.W., Yates, S.N.R., Kwiatkowski, D., Gupta, S., Warn, P., Allsopp, C.E., Gilbert, S.C., Peschu, N. (1995) Natural selection of hemi- and heterozygotes for G6PD deficiency in Africa by resistance to malaria, *Nature* **376**, 246–249.

Sabeti, P.C., Reich, D.E., Higgins, J.M., Levine, H.Z., Richter, D.J., Schaffner, S.F., Gabriel, S.B., Platko, J.V., Patterson, N.J., McDonald, G.J., Ackerman, H.C., Campbell, S.J., Altshuler, D., Cooper, R., Kwiatkowski, D., Ward, R., Lander, E.S. (2002) Detecting recent positive selection in the human genome from haplotype structure, *Nature* **419**(6909), 832–837.

Sankila, E.-M., de la Chapelle, A., Kärnä, J., Forsius, H., Frants, R., Eriksson, A. (1987) Choroideremia: close linkage to DXYS1 and DXYS12 demonstrated by segregation analysis and historical-genealogical evidence, *Clin. Genet.* **31**, 315–322.

Scriver, C.R. (2001) Human genetics: lessons from Quebec populations, *Annu. Rev. Genomics Hum. Genet.* **2**, 69–101.

Splawski, I., Timothy, K.W., Tateyama, M., Clancy, C.E., Malhotra, A., Beggs, A.H., Cappuccio, F.P., Sagnella, G.A., Kass, R.S., Keating, M.T. (2002) Variant of SCN5A sodium channel implicated in risk of cardiac arrhythmia, *Science* **297**(5585), 1333–1336.

Stephens, J.C., Reich, D.E., Goldstein, D.B., Shin, H.D., Smith, M.W., Carrington, M., Winkler, C., Huttley, G.A., Allikmets, R., Schriml, L., Gerrard, B., Malasky, M., Ramos, M.D., Morlot, S., Tzetis, M., Oddoux, C., Giovine, F.S., Nasioulas, G., Chandler, D., Aseev, M., Hanson, M., Kalaydjieva, L., Glavac, D., Gasparini, P., Dean, M. (1998) Dating the origin of the CCR5-Δ32AIDS-resistance allele by the coalescence of haplotypes, *Am. J. Hum. Genet.* **62**(6), 1507–1515. di.

Struewing, J.P., Hartge, P., Wacholder, S., Baker, S.M., Berlin, M., McAdams, M., Timmerman, M.M., Brody, L.C., Tucker, M.A. (1997) The risk of cancer associated with specific mutations of BRCA1 and BRCA2 among Ashkenazi Jews, *N. Engl. J. Med.* **336**, 1401–1408.

Tishkoff, S.A., Varkonyi, R., Cahinhinan, N., Abbes, S., Argyropoulos, G., Destro-Bisol, G., Drousiotou, A., Dangerfield, B., Lefranc, G., Loiselet, J., Piro, A., Stoneking, M., Tagarelli, A., Tagarelli, G., Touma, E.H., Williams, S.M., Clark, A.G. (2001) Haplotype diversity and linkage disequilibrium at human G6PD: recent origin of alleles that confer malarial resistance, *Science* **293**(5529), 455–462.

Toomajian, C., Ajioka, R.S., Jorde, L.B., Kushner, J.P., Kreitman, M. (2003) A method for detecting recent selection in the human genome from allele age estimates, *Genetics* **165**(1), 287–297.

Volkman, S.K., Barry, A.E., Lyons, E.J., Nielsen, K.M., Thomas, S.M., Choi, M., Thakore, S.S., Day, K.P., Wirth, D.F., Hartl, D.L. (2001) Recent origin of Plasmodium falciparum from a single progenitor, *Science* **293**(5529), 482–484.

33
Alzheimer's Disease

Jun Wang, Silva Hecimovic and Alison Goate
Department of Psychiatry, Washington University School of Medicine,
St. Louis, MO, USA

1	**Pathology and Epidemiology of Alzheimer's Disease**	958
2	**The Genetics of Alzheimer's Disease** 960	
2.1	The Genetics of Familial Alzheimer's Disease	960
2.1.1	β-amyloid Precursor Protein Mutations	960
2.1.2	Presenilin Mutations 961	
2.2	The Genetics of Sporadic Alzheimer's Disease	963
2.2.1	Apolipoprotein E 963	
2.2.2	The Search for New Alzheimer's Disease Susceptibility Genes	964
3	**Molecular and Cellular Mechanisms Underlying Alzheimer's Disease**	965
3.1	β-amyloid Precursor Protein 965	
3.2	Production of Aβ – Proteolytic Processing of APP	965
3.3	β-secretase – BACE1 966	
3.4	γ-secretase – Presenilins and Cofactors	967
3.4.1	Presenilins 967	
3.4.2	Cofactors 968	
3.5	Neurofibrillary Tangles and Tau Protein	969
3.6	Apolipoprotein E 970	
4	**Transgenic Mouse Models of Alzheimer's Disease** 971	
4.1	APP Transgenic Mice 971	
4.2	Presenilin 1/2 Transgenic Mice 972	
4.3	Tau Transgenic Mice 973	
4.4	Apolipoprotein E Transgenic Mice 974	
4.5	Modulating and Enhancing the Phenotype: Transgenic Crosses	974

Genomics and Genetics. Edited by Robert A. Meyers.
Copyright © 2007 Wiley-VCH Verlag GmbH & Co. KGaA, Weinheim
ISBN: 978-3-527-31609-0

5	**Potential Treatments** 975
5.1	Secretase Inhibitors 976
5.2	Vaccine Approaches 977
5.3	Other Therapeutic Approaches 977
6	**Summary** 978

Bibliography 978
Books and Reviews 978
Primary Literature 978

Keywords

Association Study
Tests for differences in allele or genotype frequencies between unrelated, affected, and unaffected individuals.

Autosomal Dominant Trait
A phenotype that is expressed in a heterozygote, which maps to a chromosome other than the sex chromosomes.

β-secretase
Also known as BACE1; a transmembrane aspartyl protease involved in processing of APP-generating N-terminal end of Aβ peptides.

Candidate Gene Study
An association study performed on a particular gene(s) selected according to the biology of the disease and/or their location within chromosome regions found to be implicated in disease (by linkage or association study).

Full Genome Scan
A linkage or association analysis performed between the disease and polymorphic markers that are spaced throughout the genome.

γ-secretase
A multiprotein complex involved in intramembranous cleavage of APP-generating C-terminal end of Aβ peptides. It contains four proteins: the catalytic component (presenilins) and three cofactors (nicastrin, APH-1, and PEN-2).

Neurofibrillary Tangles (NFTs)
Abnormal intraneuronal deposits abundant in the brains of patients with AD and some other neurodegenerative diseases. The main constituent of NFTs is hyperphosphorylated tau protein.

Senile Plaques
Complex extracellular deposits abundant in the brains of AD patients. The main proteinaceous component of plaques is amyloid-β peptide.

Susceptibility Locus/Gene
A region of a chromosome or a specific gene that may influence the risk to a certain condition/disease.

Familial Alzheimer's Disease
Transmitted as an autosomal dominant trait; usually has an age of onset <60 years.

Sporadic Alzheimer's Disease
Shows modest familial clustering and probably results from the synergistic action of genetic and environmental factors. Usually has an age of onset >60 years.

Abbreviations

Aβ	β-amyloid peptide
AD	Alzheimer's disease
AICD	APP Intracellular domain
APLP	amyloid precursor-like protein
APOE	apolipoprotein E
APP	β-amyloid precursor protein
BBB	blood brain barrier
CAA	cerebral amyloid angiopathy
CNS	central nervous system
CTF	C-terminal fragment
FAD	familial Alzheimer's disease
FTDP-17	frontotemporal dementia with parkinsonism linked to chromosome 17
KPI	Kunitz-type protease inhibitor
MT	microtubule
NFT	neurofibrillary tangles
NSAID	nonsteroidal anti-inflammatory drugs
NTF	N-terminal fragment
PDGF	platelet-derived growth factor
PHF	paired helical filaments
PS	presenilin
TM	transmembrane

Alzheimer's disease (AD) is the most common cause of dementia in the elderly. It is characterized by a progressive loss of memory, reasoning, judgment, and orientation, by the presence of large numbers of extracellular senile plaques and intracellular neurofibrillary tangles, as well as substantial cell loss in the brain. Genetic studies in early onset families have identified mutations in three genes that cause AD, while genetic studies in sporadic AD have identified the apolipoprotein E4 allele as a risk factor for disease. *In vitro* cell biology and transgenic mouse studies implicate all of these genes in the biology of Aβ metabolism. A consistent effect of these genetic factors is an increase in Aβ deposition. Furthermore, cell biology studies have identified targets for possible treatment of AD. Transgenic models are now being used to test the validity of these targets and therapeutic approaches to the treatment of AD.

1
Pathology and Epidemiology of Alzheimer's Disease

The German psychiatrist Alois Alzheimer first described the clinical and pathological features of Alzheimer's disease (AD) in a 55-year-old female patient; he published the discovery as a case report in 1907. It was thought that the "presenile dementia" described by Alzheimer was distinct from "senile dementia" seen in older patients, but it is now widely accepted that the two conditions are the same disease, differing primarily in the age of onset.

Alzheimer's disease is the most common cause of dementia in the western world, and the fourth leading cause of deaths in the United States. The disease is confined mainly to aged individuals and is progressive. It affects 1% of people in the developed nations and is likely to become a major problem in developing countries as the proportion of elderly persons increases in these populations. The incidence of AD rises with increasing age: 1 to 5% of people aged 65 years are affected, while 10 to 20% of those over the age of 80 have AD.

The first clinical manifestation is usually a loss of short-term memory; this is followed, over the next 6 to 20 years, by a progressive loss of memory, reasoning, judgment, and orientation (dementia). Death occurs, on average, about 10 years after the onset of symptoms. Many epidemiological surveys have been carried out to identify risk factors. There is a clear correlation between AD and a positive family history in about one-third of cases; the risk of AD is also elevated in individuals with Down syndrome. Severe head trauma with loss of consciousness increases risk for AD, particularly in those individuals who also have an apolipoprotein E4 allele. In contrast, the use of nonsteroidal anti-inflammatory agents or cholesterol-lowering agents, such as the statins, may reduce risk for AD. Another factor that may potentially decrease risk for AD is the years of education; the more the years of education, the lower the risk of disease.

Alzheimer's disease is characterized neuropathologically by the presence of large numbers of neuritic plaques and neurofibrillary tangles (NFT) within the brain cortex (Fig. 1). In addition, there is massive

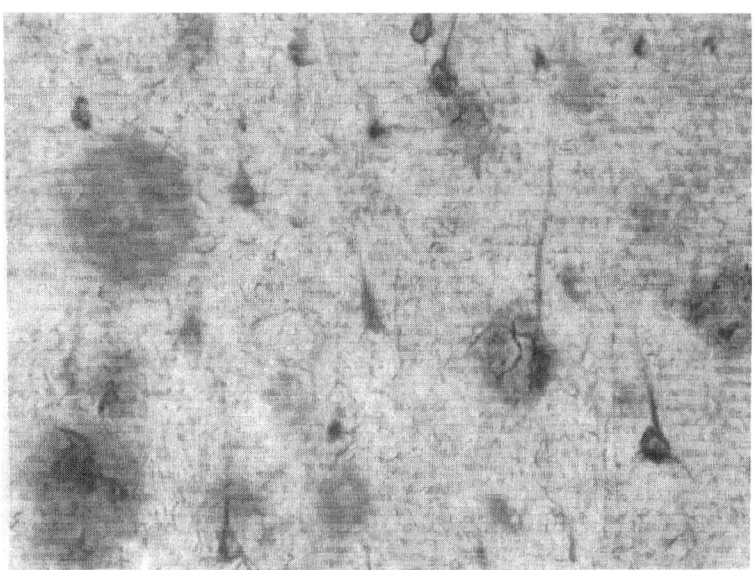

Fig. 1 Plaques and tangles in an AD brain. Shown is a brain section from an individual with mild AD. The section was double-stained with antibody PHF-1 specific to tangles and 10D5 specific to β-amyloid. Courtesy of Dr. Joel Price.

neuronal cell loss, and amyloid deposits that occur in the walls of the blood vessels of the meninges and cerebral cortex. Neither plaques nor tangles are specific to AD – they occur in intellectually normal, elderly people and in certain other diseases. However, they are more numerous and more widely distributed throughout the brains of Alzheimer's patients. Senile plaques are extracellular deposits, while tangles initially appear within the degenerating neurons. The major proteinaceous constituent of plaques is a 39 to 43 amino acid peptide called β-amyloid, or $A\beta$. The microtubule-associated protein, tau, is a major constituent of tangles. In the AD brain, there is extensive activation of astrocytes and microglia, the two primary mediators of inflammation in the central nervous system (CNS). A large number of inflammatory mediators and complement components are elevated in the AD brain, and colocalized with both plaques and tangles. The inflammatory response might be induced by increased $A\beta$ levels or by plaque formation.

The deposition of $A\beta$ has also been observed in association with several other pathological conditions. Individuals with Down syndrome, who live into their late thirties and beyond, develop a neuropathology indistinguishable from that of AD. Temporal studies in Down syndrome suggest that $A\beta$ deposition is the earliest of the known pathological markers to appear, and that neurofibrillary tangles and neuronal cell loss occur decades after the first $A\beta$ deposition. However, dementia correlates more closely with neurofibrillary tangles and neuron loss than with plaques. According to one hypothesis based on these observations, amyloid deposition precedes and induces tangle formation; tangles (and perhaps plaques) cause neuronal death, and the resultant neuronal damage destroys important neurotransmission pathways in the brain, producing dementia.

2
The Genetics of Alzheimer's Disease

There are two forms of AD: familial Alzheimer's disease (FAD), in which the disease is transmitted as an autosomal dominant trait; and sporadic AD, which shows modest familial clustering and probably results from the synergistic action of genetic and environmental factors. Sporadic AD can have an early (<60 yrs) or late (>60 yrs) age of onset. FAD does not appear to be clinically or neuropathologically different from the more common sporadic form of AD, except that it generally has an earlier age of onset.

2.1
The Genetics of Familial Alzheimer's Disease

The existence of families in which AD segregates as a fully penetrant, autosomal, dominant trait presents the most striking evidence for the involvement of genetic factors in the etiology of AD. Mutations in three distinct genes, β-amyloid precursor protein (APP), *PS1* (presenilin 1) and *PS2* (presenilin 2) have been identified to cause FAD.

2.1.1 β-amyloid Precursor Protein Mutations

The main component of the AD-associated senile plaques, Aβ, is derived by the proteolytic processing of APP. The gene coding for APP is located on the long arm of chromosome 21. It contains 18 exons, three of which can be alternatively spliced to produce a variety of mRNA species.

The first mutation shown to cause FAD was a valine to isoleucine substitution at residue 717 of APP (Fig. 2). To date, 22 APP mutations have been reported, of which 17 are linked to early-onset AD (www.alzforum.org). Most mutations lie within the transmembrane (TM) domain of APP in the region that is involved in the generation of the C-terminal end of Aβ. All TM domain FAD-APP mutations result in an increased production of the longer, more amyloidogenic form of Aβ (Aβ42/43). The most severe mutation (T714I) causes a 11-fold increase in the Aβ42/Aβ40 ratio; it causes both a decrease in Aβ40 and an increase in Aβ42 production. Patients carrying the T714I FAD-APP mutation die in their 40s. FAD mutations in APP can differentially affect Aβ-peptide formation by either increasing Aβ42 or decreasing Aβ40, or both. The Swedish mutation (K670N, M671L) lies in the region outside of the TM domain in which cleavage by β-secretase generates the N-terminal end of the Aβ, resulting in an increase in both Aβ40 and Aβ42 levels. Five mutations are located within the Aβ sequence but outside the TM domain: Flemish (A692G), Dutch (E693Q), Arctic (E693G), Italian (E693K), and Iowa (D694N). Patients carrying these mutations deposit Aβ in senile plaques and/or in cerebral vascular walls (cerebral amyloid angiopathy – CAA) and present clinically with either AD or hemorrhagic strokes, or both. For example, Flemish A692G patients present with both AD and strokes, while most Dutch E693Q patients suffer from strokes followed by a progressive multi-infarct dementia (HCHWA-D, hereditary cerebral hemorrhage with amyloidosis Dutch type). Pathologically, Flemish A692G patients demonstrate a classical AD pathology although with a strong CAA component, while Dutch APP patients

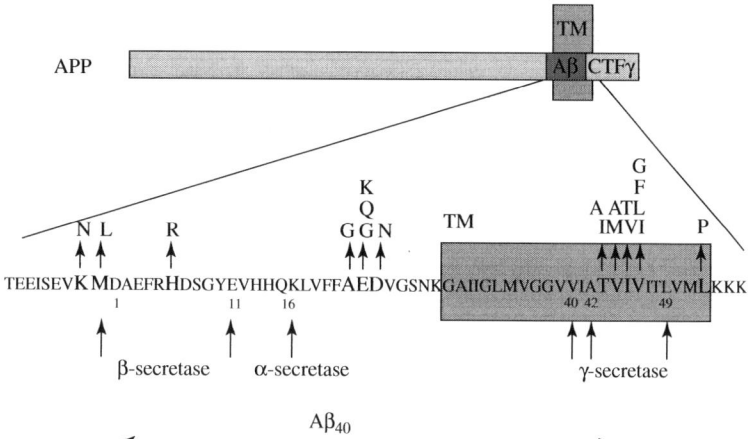

Fig. 2 FAD mutations in APP. Shown is a schematic representation of the APP protein showing the regions coding for Aβ peptide and C-terminal intracellular domain-CTFγ/AICD. The location of each FAD mutation is shown. Letters in bold are FAD-mutation sites, letters in gray are FAD substitutions. Major sites of APP processing, α-, β-, and γ-secretase sites, are also labeled. The amino acid numbering below the letters is according to the Aβ sequence.

have mainly vascular Aβ in the absence of neurofibrillary pathology. Recent analysis of pathogenic effect on Aβ levels of Arctic, Italian, Dutch, and Flemish APP mutations revealed that all, except the Flemish mutation, cause a decrease in both Aβ40 and Aβ42. In contrast, the Flemish mutation increases both Aβ40 and Aβ42 levels.

2.1.2 Presenilin Mutations

The discovery that mutations within the presenilin 1 *(PS1)* gene located on chromosome 14 (14q24.3) were responsible for most FAD cases opened up a whole new area of research into the molecular mechanisms of the pathogenesis of AD. Shortly after the discovery of *PS1*, DNA sequence database searches led to the identification of a homologous gene named presenilin 2 *(PS2)*, located on chromosome 1 (1q41.2). Mutations in *PS2* also result in FAD.

Pathogenic mutations in both of these genes have been identified; overall, mutations in *PS1*, *PS2*, and *APP* genes account for the majority of FAD cases. Although the proteins encoded by the two presenilin genes are 63% homologous at the amino acid level, intensive mutation and epidemiological analysis have revealed some interesting differences between the two proteins. The number of mutations identified in *PS1* is far greater than those in *PS2*. To date, more than 107 individual point mutations in *PS1* have been identified (Fig. 3, http://molgen-www.uia.ac.be/ADMutations) and found to be responsible for the majority of the early onset FAD. Most of these alterations are missense mutations that result in single amino acid changes in residues that are conserved between the PS1 and PS2 proteins. An intronic mutation (referred as ΔE9) results in the disruption of a splice acceptor site leading to the in-frame deletion of exon 9/10. This is the only *PS1*-FAD mutation that results in pathologically active protein that does not undergo proteolysis. Endoproteolytic

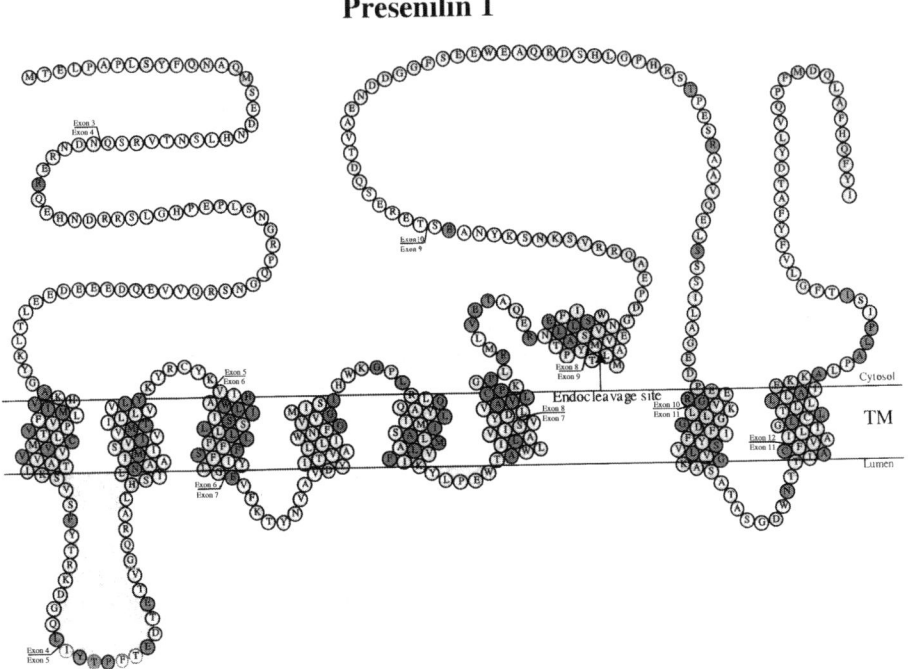

Fig. 3 FAD mutations in presenilin-1. Shown is a schematic representation of human presenilin 1 protein. The protein is predicted to have eight transmembrane domains. Arrow indicates the endocleavage site. Shaded residues are those associated with FAD mutations.

processing of the presenilins is an important aspect of the biochemistry of both PS1 and PS2 proteins, since it has been implicated in the formation of the functionally active complex (see Sect. 3). The most consistent effect mediated by the presenilin FAD mutations is increased production of Aβ42. This has been confirmed both in transfected cells and in transgenic mice. Although the mechanism by which FAD mutations in *PS1* and *PS2* cause AD is not fully understood, it is suggested that FAD mutations may alter γ-secretase activity toward cleavage of the more amyloidogenic peptide of Aβ, Aβ42/43. Although point mutations are distributed throughout the *PS1* gene, the majority of them are clustered within the TM domains and large hydrophilic loop between TM6 and TM7 (Fig. 3). The mean age of onset in all *PS1* families is about 45 years with a range of 28 to 62 years, suggesting that mutations in *PS1* result in a more aggressive form of FAD than APP mutations.

In contrast to *PS1*, only eight point mutations have been identified in *PS2* (http://molgen-www.uia.ac.be/ADMutations). The first *PS2* mutation to be identified was the N141I mutation that is responsible for disease in a large Volga German kindred. Whether the preponderance of *PS1* versus *PS2* mutations is due to the differing genomic environment of the two genes or the factors related to differing biological functions of the two molecules, remains unclear. Mutation analysis of the two proteins has also led to the identification of a phenotypic difference. Overall, the mutations in *PS1* result in relatively early and constant age of onset while those in *PS2* lead to a somewhat later and more variable age of onset. The identification of a *PS1* N135I mutation, a position and substitution that corresponds to the *PS2* Volga German mutation, and the finding that the affected members of this pedigree have a relatively constant age of onset suggests that the observed variations in age of onset are dependent on the molecule itself as opposed to the specific position or nature of the amino acid substitution.

2.2 The Genetics of Sporadic Alzheimer's Disease

While the genetics of FAD is fairly well understood, our understanding of the more common sporadic AD remains much less complete. The majority of cases have onset >60 years. Some families have stronger clustering of AD cases than others, suggesting that the genetic component of the disease can be quite variable. The estimated cumulative risk to first-degree relatives of AD-affected probands approaches 50% by age 90 compared to a disease risk of 10 to 15% in the general population. To date, only the Apolipoprotein E4 (*APOE4*) allele has been linked to increased risk for sporadic AD. While it is clear that *APOE* is a major risk factor for AD, epidemiological studies estimate that 42 to 68% of AD cases do not have an *APOE4* allele, indicating that additional genetic and environmental factors are involved in this form of the disease. The balance of these risk factors most likely determines both the age of onset of the disease and the number of affected first-degree relatives.

2.2.1 Apolipoprotein E

Linkage analysis in late-onset AD pedigrees led to the observation that these families inherit a genetic risk factor located on chromosome 19. The defect involved in these particular pedigrees is a susceptibility defect as opposed to a causative defect such as those present in the gene coding for APP and the presenilins. The inheritance of the susceptibility gene defect alone is not sufficient to cause the disease, and one or more secondary events, either environmental or genetic, must accompany the inheritance of the primary defect to cause the disease. Association studies have shown that the chromosome 19 susceptibility gene (localized to 19q13.2) is within the *APOE* gene locus.

APOE protein is a lipid-transport molecule and the primary apolipoprotein observed in the CNS. It is encoded by a polymorphic gene that exists as three alleles designated *APOE2*, *APOE3*, and *APOE4*. These genetic variations result in amino acid substitutions (arginine or cysteine) at positions 112 and 158 of the protein. Thus, whereas the *APOE2* polypeptide has a cysteine at both of these positions, *APOE3* has a cysteine at 112 and an arginine at 158, and *APOE4* has arginine at both positions. In most Caucasian populations, the *APOE3* is the most common allele (frequency 0.78); *APOE4* (frequency 0.14) and *APOE2* (frequency 0.08) are considered variants.

When the association of the three common *APOE* alleles with AD was examined in late-onset AD pedigrees, it was found that the *APOE4* allele frequency was 52%, compared to 16% for age-matched controls. In addition, in sporadic AD cases, the *APOE4* allele frequency was 40%. The *APOE4* association with AD was rapidly and widely confirmed in patients with familial and sporadic AD. These multiple confirmations established the *APOE4* allele as the most important genetic marker of risk for the disease identified so far, accounting for approximately 50% of the genetic component of AD. The frequency of *APOE4* allele among AD and control cases can vary in different populations, however, the association with AD is consistent.

Individuals inheriting two copies of the *APOE4* allele have an increased risk and earlier age of onset of AD than individuals with one *APOE4* allele. Likewise, individuals with one *APOE4* allele have a greater risk and earlier onset of disease than those with no *APOE4* alleles. These data demonstrate that the effect of *APOE4* on risk and age of onset is dose-related. In contrast to the effect of *APOE4* allele, the *APOE2* allele has a protective effect on AD, as evidenced by a decreased frequency of this allele in AD cases. The effect of the most commonly inherited allele, *APOE3*, on risk and age of onset falls between *APOE4* and *APOE2*. There also appears to be an interaction of *APOE* genotypes with the age of onset in FAD. When survival analysis was used in a large Colombian kindred carrying a presenilin mutation, the presence of an *APOE4* allele was observed to decrease age of onset, while presence of an *APOE2* allele increased age of onset. Although it is still not clear how an *APOE4* allele modifies risk for AD, several studies suggest that APOE may act via an $A\beta$-dependent mechanism.

2.2.2 The Search for New Alzheimer's Disease Susceptibility Genes

There are two genetic approaches to finding susceptibility loci: full genome scans and candidate gene studies. In candidate gene studies, genes are selected on the basis of the known biology of the disease or their location within chromosomal regions implicated in risk for late-onset AD, and are assessed individually in order to determine whether variants in each candidate are associated with disease. An alternative approach, a genome scan, does not select genes *a priori*, but instead tests for linkage or association between the disease and polymorphic markers that are spaced evenly through the genome. Association methods use unrelated case-control samples or siblings who are discordant for disease, and test for differences in allele frequencies between affected and unaffected individuals. In contrast, linkage studies use extended families or affected sibling pairs and look for regions of the genome in which there is an increased allele sharing between affected relatives.

To date, only five late-onset AD genome-wide screens have been reported. The most consistent findings among these studies are evidences for AD susceptibility loci on chromosomes 19 (*APOE*), 9, 10, and 12. Since FAD is closely associated with elevated $A\beta42$ levels, some investigators have hypothesized that loci, which modify plasma $A\beta42$, might also modify risk for AD. Using this approach, evidence for linkage was observed in the same region of chromosome 10 as was observed in the AD-linkage studies. Over the past few years, a large number of biologically interesting candidate genes as well as candidate genes that map within chromosome regions implicated in risk for late-onset AD have been evaluated in case-control populations. The genes that have received most attention are those encoding *APOE* on chromosome 19, $\alpha2$-macroglobulin and the LDL-receptor related protein on chromosome 12 and insulin degrading enzyme and urokinase on chromosome 10. All of these genes have been closely linked to $A\beta$ metabolism. However, with the exception of the *APOE* gene, whose association with

AD has been universally replicated, the relative significance of these associations is still in question.

3
Molecular and Cellular Mechanisms Underlying Alzheimer's Disease

3.1
β-amyloid Precursor Protein

β-Amyloid precursor protein (APP) is a large, type I transmembrane protein that is expressed in almost all cell types except erythrocytes. It consists of a large extracellular N-terminal region, a transmembrane domain, and a short intracellular C-terminal tail. The Aβ-peptide sequence begins close to the membrane on the extracellular part of APP and ends within the membrane-spanning domain. At least five different mRNAs arise as a result of alternative splicing of the primary transcript of the APP gene: APP770, 751, 714, 695, and 563. Transcript APP695 lacks the Kunitz-type protease inhibitor (KPI) domain and is the major form expressed in neurons, whereas APP751, which contains the KPI domain, is the major transcript in non-neuronal cell types. The APP transcripts appear to be developmentally regulated, and all cell types so far tested contain at least one of the transcripts.

In addition to differential RNA splicing, APP undergoes posttranslational modification, including phosphorylation, N- and O-linked glycosylation, sulfation and proteolytic processing. It is not clear how these different events modify APP metabolism or function. A detailed knowledge of posttranslational events is undoubtedly crucial to understanding amyloid plaque formation and to elucidating the biological functions of APP and Aβ.

The normal function of APP is still unclear, but it has been implicated in neuroprotective action. APP is a member of a multigene family that contains at least two other homologs known as amyloid precursor-like protein 1 and 2 (APLP1 and APLP2). Both $APP^{-/-}$ and $APLP2^{-/-}$ mice are viable while APP/APLP2 double knockout mice are early postnatal lethal suggesting that APP and APLPs may share important physiological functions and the APLPs may compensate for the function of APP.

3.2
Production of Aβ – Proteolytic Processing of APP

It is clear that Aβ production does occur in normal cells, suggesting that excessive Aβ production in AD is due to enhanced or altered processing of APP by normal pathways that generate Aβ, rather than by an abnormal route specific to Alzheimer's cells. APP undergoes a series of proteolytic processing events (Fig. 4) by several different proteases called secretases. It is first cleaved by either α- or β-secretase. Two disintegrin metalloproteases, ADAM 10 and TACE (TNF-α converting enzyme), are involved in α-cleavage of APP. This cleavage produces a large soluble N-terminal fragment (NTF), APPsα and an 83-amino acid membrane-bound C-terminal fragment, C83 (CTFα). Alternatively, cleavage by the β-secretase results in an N-terminal fragment APPsβ and a 99-amino acid membrane-bound fragment, C99 (CTFβ). The β-secretase has recently been identified as a transmembrane aspartyl protease called BACE1, for β-site APP-cleaving enzyme. Both C83 and C99 are substrates for γ-secretase, a mysterious enzyme that carries out an unusual proteolysis in the middle of the transmembrane domain of

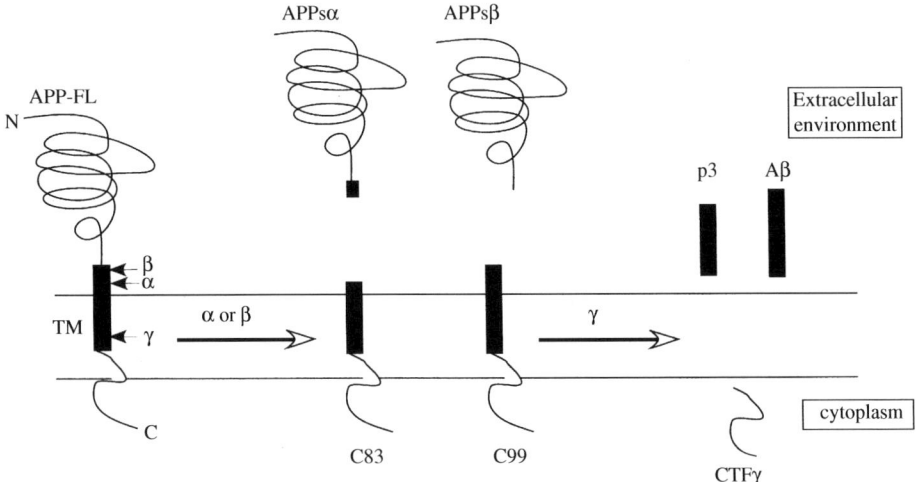

Fig. 4 Proteolytic processing of APP. APP is initially cleaved by either α- or β-secretase to generate membrane-bound C-terminal fragments, C83 or C99, which are subsequently cleaved by the γ-secretase (in the transmembrane domain) to produce p3 or Aβ.

APP, resulting in the release and secretion of the 4-kD Aβ from C99 and the 3-kD p3, a nonpathogenic peptide, from C83. γ-secretase cleavage is promiscuous and produces a spectrum of Aβ peptides varying from 39 to 43 residues in length. Over 90% of the Aβ in the brain is Aβ40 and the remaining 5 to 10% is the longer species Aβ42/43.

The C-terminal product of the γ-secretase cleavage is called CTFγ or AICD (from APP Intracellular Domain). Interestingly, CTFγ, purified from brain tissue, is not C59/57, which would be the predicted length of the C-terminal product of Aβ cleavage. Instead, sequence analysis reveals that it is C50/49, several amino acids shorter. The mechanism that accounts for such differential cleavages is currently under investigation. Several reports have suggested a role of CTFγ in APP signaling to the nucleus. Indeed, it was shown that CTFγ, when together with Fe65 protein, translocates to the nucleus and that this complex can activate transcription of a reporter gene *in vitro*.

3.3
β-secretase – BACE1

In late 1999, four groups almost simultaneously reported the discovery of BACE1, a novel aspartyl protease that exhibited all the known functional characteristics of the β-secretase. BACE1 is a type I transmembrane protein with an open reading frame of 501 amino acids. It is most closely related to the pepsin family of aspartyl proteases. BACE1 is expressed in most tissues and cell types at modest levels, but shows much higher expression levels in neurons. Structurally, it contains an N-terminal 21 amino acid signal peptide, a propeptide domain, a large lumenal domain followed by a transmembrane domain, and a short cytosolic C-terminal tail. BACE1 contains two aspartyl protease active site motifs in the lumenal domain: DTGS (residues 93–96) and DSGT (residues 289–292). Both aspartic acid residues have been shown to be essential for enzymatic activity. BACE1 contains intrinsic proteolytic activity; as a purified recombinant, BACE1 can directly cleave APP *in vitro*. Overexpression of BACE1 in cell lines dramatically increases β-secretase activity and β-cleavage products: APPsβ and C99. In addition, mice deficient in BACE1 show a normal phenotype, but Aβ generation is abolished, further supporting that BACE1 is indeed a β-secretase.

BACE1 is initially synthesized in the ER as an immature precursor protein, proBACE1, which undergoes rapid maturation involving N-glycosylation as well as removal of the propeptide domain by a furin-like convertase in the Golgi apparatus. Unlike most zymogens, proBACE1 possesses β-secretase activity. The propeptide domain does not appear to significantly inhibit protease activity, but rather acts as a chaperone to assist proper folding of the protease domain. Phosphorylation of BACE1 at Ser498 appears to be important for the intracellular trafficking of the protein between early endosomes and late endosomes/TGN compartments.

A homologous aspartyl protease, BACE2 was shortly identified by searching the EST database with the BACE1 sequence. BACE1 and BACE2 share ~64% amino acid similarity. Together they define a novel family of transmembrane aspartyl proteases. Brain expression of BACE2 is very low and most evidence suggests that it does not play a major role in β-cleavage of APP.

3.4
γ-secretase – Presenilins and Cofactors

The enzyme that is clearly central to AD pathogenesis is the γ-secretase, which cleaves C99 and produces Aβ. Evidence suggests that it is an aspartyl protease with the unusual property of cleaving substrates within the transmembrane domain. Independent from its role in processing APP, γ-secretase also mediates the intramembranous cleavage of several other type I transmembrane proteins including Notch and ErbB-4. γ-secretase cleavage of Notch releases the Notch intracellular domain, which translocates to the nucleus and acts as a cofactor to modify transcription of target genes.

The identity of the γ-secretase is rather enigmatic. It appears to be a multiprotein, high molecular weight complex that consists of presenilins (PS) and three other cofactors: nicastrin, APH-1 and PEN-2.

3.4.1 Presenilins

Presenilins are a family of polytopic transmembrane proteins that are implicated in AD pathogenesis. PS homologs have been found in species as diverse as *Drosophila melanogaster*, *Caenorhabditis elegans* and *Arabidopsis*, but not in yeast. The function of PS appears to be conserved. Deficiency in sel-12, one of the nematode PS homolog, causes egg-laying defects. This phenotype can be rescued by expressing human PS1 in the mutant worms. In mammals, there are two PS genes, *PS1* and *PS2*, that share 65% identity. *PS1* is believed to account for the majority of PS functions, while *PS2* seems to play a complementary role.

Presenilins are found primarily within intracellular membranes including the ER and Golgi as well as the plasma membrane. Structurally, they are predicted to have eight transmembrane domains with both the amino and carboxyl termini oriented toward the cytoplasm (Fig. 3). A large cytoplasmic loop is postulated between TM6 and TM7. PS are synthesized as single polypeptides that rapidly undergo endoproteolysis within the cytoplasmic loop, generating a 30-kD N-terminal fragment (NTF) and a 20-kD C-terminal fragment (CTF). The NTF and CTF form a stable complex, which appears to be the functional form of presenilins. PS endoproteolysis is tightly regulated such that overexpression of PS leads to the accumulation of a full-length protein, but does not result in an increase in the fragments. The NTF/CTF complex, but not the holoprotein, are components of a high molecular weight complex, estimated at ~250 kD or even larger. This high molecular weight complex is believed to be the γ-secretase complex.

Many lines of evidence indicate that PS1 is the catalytic component of the γ-secretase. The two membrane-embedded aspartyl residues in PS1 (D257 in TM6 and D385 in TM7) are essential for γ-secretase activity. The mutation of either residue abolishes γ-secretase processing of APP and Notch. The knockout of the PS genes blocks the Aβ production and Notch function. While in PS1 single K/O cells, there is residual Aβ production, in PS1/2 double knockout cells, there is no detectable Aβ, demonstrating that there is no PS-independent γ-secretase activity. Inhibitors designed to be transition-state analogs of the γ-secretase bind to presenilin fragments. Recently, a family of signal peptide peptidases was discovered, which are apparently *bona fide* intramembranous aspartyl proteases, with similar active site motifs and similar topology to the presenilins. This finding further supports the hypothesis that presenilin is an aspartyl protease.

3.4.2 Cofactors

Nicastrin is a large type I transmembrane glycoprotein that was immunopurified from PS1 immunoprecipitates. It has no significant sequence homology to other functionally characterized proteins. Nicastrin is a 709 amino acid protein containing a putative signal peptide, a long N-terminal hydrophilic domain with numerous glycosylation sites, a TM domain and a short hydrophilic C-terminal tail of 20 residues. The mature protein migrates at ~150 kD owing to heavy glycosylation. There is evidence suggesting that the association of nicastrin with γ-secretase is tightly regulated via glycosylation. Nicastrin binds to presenilins and the active γ-secretase complex. Nicastrin deficiency in *Drosophila* abolishes Notch signaling and γ-secretase cleavage of APP. In addition, presenilin stability was compromised in the absence of nicastrin. All these data support that nicastrin is a component of the γ-secretase complex and an important regulator of the γ-secretase activity.

Genetic screens in *C. elegans* identified two proteins, APH-1 and PEN-2, which interact with worm homolog of PS1 and nicastrin. Loss-of-function mutations of either protein produces phenotypes that resemble that of null mutants in nicastrin and presenilins. APH-1 is an integral protein containing seven predicted TM domains. There are two mammalian APH-1 genes (mAPH-1a and mAPH-1b). mAPH-1a has at least two splice variants: mAPH-$1a^L$ has seven exons and encodes a longer protein of 265 residues; mAPH-$1a^S$ has six exons and encodes a shorter protein of 247 residues. All three variants are widely expressed, although their abundance seems to be coordinately regulated. To date, no functional difference among these variants has been documented. PEN-2 is a small protein of 101 amino acids containing two TM domains. Topology study suggests that both the N- and C-terminals of the protein face the lumen of the ER.

Human APH-1 and PEN-2 can partially rescue the *C. elegans* mutants, suggesting functional conservation among eukaryotes. The human genes must be present together to rescue the phenotypes and adding PS1 improves the rescue, indicating that these proteins act in concert. Consistent with the genetic studies, biochemical studies showed that reduction of APH-1 and PEN-2 expression in mammalian cells by RNA inhibition reduced γ-secretase activity.

While the exact functions of these cofactors remain to be elucidated, the data obtained so far suggest that all four proteins, presenilins, nicastrin, APH-1, and PEN-2, are necessary components of the γ-secretase. They appear to regulate one another for protein stability, proper maturation, and functional γ-secretase complex formation. It is also very likely that the least abundant component becomes the cellular-limiting factor that controls overall γ-secretase activity.

3.5
Neurofibrillary Tangles and Tau Protein

Neurofibrillary tangles are intracellular deposits that consist of aggregates of a fibrous substance, the paired helical filaments (PHF). These filaments are highly insoluble, even resistant to strong detergents such as SDS and guanidine hydrochloride. A major component of PHF is tau, a microtubule (MT)-associated phosphoprotein. Tau is abundantly expressed in the brain. At least six isoforms of tau are present in the brain, resulting from alternative mRNA splicing of a single gene on chromosome 17q.

Unlike APP, whose functions are not clear, tau has a biological role that is reasonably well understood. One function of tau is to promote microtubule assembly, a process controlled by phosphorylation of tau. In addition, it has been proposed that tau has a specific role in the generation of axon morphology. Increased phosphorylation of tau decreases its ability to promote tubulin polymerization to microtubules.

Tau in PHF is hyperphosphorylated, and is insoluble under conditions in which normal tau is soluble. It is believed that tangles are formed when tau becomes hyperphosphorylated and dissociates from microtubules and spontaneously forms insoluble PHF. Over 25 phosphorylation sites have been identified in PHF tau purified from AD brains. Most of the phosphorylation sites are located in the flanking region of the MT-binding domain. A variety of protein kinases, including glycogen synthase kinase-3β (GSK3β), cyclin-dependent kinase 5 (cdk5), mitogen-activated protein kinase (MAPK) and protein kinase A, have been found to be involved in hyperphosphorylation of tau. This suggests that multiple phosphorylation cascades may be activated in NFT-bearing neurons. However, it remains unclear whether these cascades are activated independently or whether one or more kinases are primarily responsible for initiating the cascade *in vivo*.

In addition to their presence in AD, filamentous tau inclusions accompanied by extensive gliosis and loss of neurons are the neuropathological hallmarks of other neurodegenerative diseases that are sometimes designated as tauopathies. There has been considerable debate in the AD literature concerning the primacy of Aβ versus NFT in the pathogenesis of AD. This was largely resolved by the identification of mutations in the tau gene in families with frontotemporal dementia with parkinsonism (FTDP-17). FTDP-17 is characterized clinically by the presence of prominent behavioral changes early in the disease. The neuropathology associated with FTDP-17 is characterized by large numbers of NFT but no senile plaques. Mutations in the tau gene in FTDP-17 can alter splicing and produce a shift from the short tau isoform with three repeats to the longer isoform, which contains four repeats. Several other point mutations have been discovered in the coding region of tau that impact the functional integrity of the tau protein itself, P301L being the most common. So far, genetic analysis has not linked tau mutation directly to AD, but tau dysfunction may still have a significant role to play in the disease, if not in its initiation, then in its progression.

3.6
Apolipoprotein E

Apolipoprotein E is a 299 amino acid glycoprotein with a molecular weight of ~34 kD. Its N-terminal domain consists of four amphipathic α-helices and the C-terminal domain contains most of the lipid-binding activity. APOE is a major apolipoprotein that regulates cholesterol uptake and release. The CNS contains high levels of APOE with abundance only second to the liver. In the CNS, APOE is synthesized predominantly by astrocytes and is secreted into the CSF, where it is a major lipid carrier.

APOE appears to have multiple functions in the brain – any or all of which could potentially influence AD pathogenesis. The molecular mechanisms by which APOE protein is involved in AD pathogenesis are unclear because reports on the allele-specific effects of APOE are controversial. Nevertheless, numerous studies have demonstrated that APOE can affect AD pathogenesis either by influencing $A\beta$ deposition or by a direct effect on neuronal survival. APOE is a major component of senile plaques and specifically binds to soluble $A\beta$s. Allele-specific differences in $A\beta$ binding, aggregation, and fibrillogenesis have been reported. Most histopathologic studies demonstrated a correlation between senile plaque density and *APOE4* allele dose in AD. Animal studies also show that APOE has isoform-specific effects on $A\beta$ deposition and plaque formation (see Sect. 4.4).

There is also evidence that different alleles of *APOE* have different effects on neurite extension, neuronal survival, and the cell's response to oxidative stress. *In vitro* studies with primary neurons and neuronal cell cultures show that lipid-bound E3 promotes neurite extension, while E4 is neutral or inhibitory. These *in vitro* findings are consistent with reported defects in neuronal remodeling in AD brains carrying an E4 allele. APOE can have allele-specific differences on both neurotoxic or neuroprotective activities. High levels of APOE or some APOE synthetic peptides were shown to cause neurite degeneration or neuronal death in cultured cells, with the E4 having the most neurotoxic effects. On the other hand, APOE appears to play a role in protection against oxidative stress, a prominent phenomenon in AD pathogenesis. Apoe -/- mice exhibit increased markers of lipid peroxidation within the brain and appear to have increased susceptibility to neuronal damage. Different *APOE* alleles are shown to have isoform-specific neuroprotective effects (E2 > E3 > E4) on oxidative cytotoxicity caused by hydrogen peroxide, or $A\beta$ peptides. A possible mechanism for the different antioxidant activities of the *APOE* alleles is their ability to detoxify 4-hydroxynonenal, a neurotoxic lipid-peroxidation product that is believed to play a key role in neuronal death in

AD. 4-hydroxynonenal covalently binds to APOE, with more binding to E2 and E3 than E4, probably because of the cysteine residues that are present in E2 and E3. E2 and E3 may therefore protect neurons by binding to, and thereby detoxifying, 4-hydroxynonenal.

4
Transgenic Mouse Models of Alzheimer's Disease

A major hindrance to basic research into the molecular pathogenesis of AD has been the absence of a mouse or rat model. As a result of the identification of mutations that cause AD in humans, this has begun to change. Although initial efforts to make transgenic models of AD failed, more recent efforts have recapitulated at least some of the key features of the disease. Many genetically altered mice have been designed to reproduce the neuropathology of AD, however, the majority of them share only some of the neuropathological and/or cognitive impairment characteristics of AD (http://www.alzforum.org). These include mice expressing human forms of the APP, the PS, APOE and, more recently, tau. Transgenic mice expressing human APP develop amyloid plaques, but neurodegeneration and neurofibrillary tangles are not observed. The Presenilin transgenic mice produced only increased $A\beta 42/43$ levels and did not develop signs of AD pathology, even though the same mutations caused some of the earliest forms of inherited AD in humans. Strategies emphasizing tau resulted in increased phosphorylation of tau and tangle formation, although amyloid plaques were absent. Nevertheless, crossing transgenic animals expressing mutated tau and APP has produced a mouse that closely recapitulates the neuropathology of AD. This section provides a review of the various murine models and their role in understanding the pathogenesis of AD.

4.1
APP Transgenic Mice

The most extensively studied APP transgenic mouse lines are known as PDAPP and Tg2576. The PDAPP transgenic mouse expresses a human APP770 mini gene containing the V717F FAD-APP mutation, while Tg2576 expresses human APP695 that contains the Swedish FAD-APP mutation. In PDAPP, expression of the *APP* gene is under the control of the human platelet–derived growth factor (PDGF)-β chain neuronal promoter that targets expression preferentially to neurons in the cortex, hippocampus, hypothalamus, and cerebellum of the transgenic animals. This mouse was made on a mixed-strain background (C57BL/6, DBA/2, and Swiss-Webster), while Tg2576 was made on a single C57B6/SJL background. Expression of the *APP* gene in Tg2576 mice is under the control of the prion protein promoter resulting in a less-limited expression of APP both in the CNS and the periphery. The success of both models was due to the high level of APP expression achieved (>10-fold overexpression of human APP in PDAPP mouse and >6-fold higher expression of APP in Tg2576 compared to endogenous murine APP levels). Cortical and limbic amyloid deposits begin slightly earlier in the PDAPP mouse: by 3 months of age in homozygotes and at 6 to 9 months in heterozygotes, while in Tg2576 mouse they occur between 9 to 12 months of age, elevated $A\beta$ production is observed as early as 3 months of age. Both models develop β-amyloid deposits (in hippocampus and neocortex but limited in striatum and cerebellum), and the deposits are associated with dystrophic neurites, punctate immunoreactivity to hyperphosphorylated tau, astrocytosis, microgliosis, and vascular amyloidosis. The elevation of Aβ in the Tg2576 mouse correlates with the appearance of memory and learning deficits in the oldest group of transgenic mice as well. However, both APP transgenic models fail to show all the pathological features of human AD; no significant neuronal loss and no formation of NFTs were observed in either line.

Several other APP transgenic models were generated in order to obtain a complete or more comprehensive picture of the hallmarks of AD. The TgAPP22 transgenic mice overexpress a double mutant APP with Swedish (K670N, M671L) and London (V717I) FAD mutations in cis, while TgAPP23 only carries the Swedish mutation. In both models, the human APP751 isoform is used and its expression is under the control of the murine Thy-1 promoter. The TgAPP22 mice show twofold overexpression of the transgene over endogenous APP, and Aβ deposits were detected in the neocortex and hippocampus at 18 months of age, while the TgAPP23 has higher, sevenfold overexpression of human APP and typical Aβ plaques appear at the age of six months. Interestingly, these two murine AD models showed differences in plaque type; the majority of amyloid deposits in TgAPP22 mice brains are of the "diffuse" type, while in the TgAPP23 mice almost all extracellular amyloid deposits were fibrillar. Both substantial neurodegeneration and a reduction of neuron numbers were apparent in TgAPP23. In 14–18 months old TgAPP23 neuronal loss was 14% and reached 25% in mice with high Aβ plaque load. Additional pathological features included dystrophic neurites surrounding the plaques, hyperphosphorylated tau, but no NFTs developed.

Each of the models described above is remarkable in that the anatomical pattern of plaque formation parallels that seen in human AD. Furthermore, the morphology of amyloid plaques in aged APP-transgenic mice recapitulates amyloid pathology in human AD: the plaques span a continuum from diffuse Aβ deposits to compact core plaques with inflammation and neuritic dystrophy. However, none of these models reflects a complete picture of the neuropathology of AD.

4.2
Presenilin 1/2 Transgenic Mice

Several PS1 transgenic mice have been created to study presenilin biology *in vivo*. These mice differ in the promoter and the strains of mice used, but are similar in achieving high levels of protein production (1–3 fold over endogenous) in neuronal regions of the brain. Knockin transgenes were created expressing either the wild-type human PS1 or PS1 containing the FAD mutation A246E under the transcriptional control of the human Thy-1 promoter in PS1 null background. Both mice rescued the PS1 knockout mouse from embryonic lethality, confirming that PS1 mutation does not cause a loss of function of PS1 protein but has similar physiological properties as wild-type PS1. A second PS1 transgenic model expressing human mutant (M146L or M146V) and wild-type PS1 was generated under the control of the PDGF promoter. All three PS1 transgenics showed similar results: overexpressing mutant PS1 in the brains of transgenic mice lead to the elevation of

Aβ42/43, but not Aβ40. This effect was a direct result of the mutation and not overexpression of the human protein as the overexpression of the wild-type PS1 did not have any significant effect on Aβ levels. The physiological significance of the specific elevation of Aβ42/43 appears to be the observation that AD patients with PS1 mutations show plaques composed primarily of Aβ42/43. However, all three PS1 transgenic mice, in contrast to APP transgenics, revealed no Aβ deposition or other AD-associated pathology.

Recently, a PS1 mouse model carrying a PS1-P264L FAD mutation was developed using gene-targeting approach. The gene-targeted models are distinct from transgenic models because the mutant gene is expressed at normal levels, in the absence of the wild-type protein. $PS1^{P264L/P264L}$ mice had normal expression of PS1 mRNA, but levels of the N- and C-terminal protein fragments of PS1 were reduced while levels of the holoprotein were increased. $APP^{sw/sw}/PS1^{P264L/P264L}$ double gene-targeted mice had elevated levels of Aβ42, sufficient to cause Aβ deposition beginning at six months of age. This was the first animal model that exhibited Aβ deposition without overexpression of APP.

Three transgenic mice that express PS2-FAD mutation N141I under the control of PDGFβ chain promoter, chicken β-actin promoter or the neuron-specific enolase promoter were designed. The first two PS2 transgenic mice showed increased levels of Aβ42 in the brain of 12-month old mice. The third model utilizing neuron-specific enolase promoter did not show any difference in Aβ42 levels compared to the PS2 wild-type transgenic mice and both had no obvious differences in AD phenotypes (both had amyloid deposits).

4.3
Tau Transgenic Mice

One of the deficits of most AD transgenic models is the lack of tau pathology. APP and APP/PS transgenic mice, which exhibit extensive amyloid deposition also develop hyperphosphorylated tau around the amyloid deposits but it does not progress to resemble human NFTs. This suggests that some disruption of tau does occur in response to amyloid accumulation, but it does not progress to resemble human tauopathy in the mouse model. The idea that a species barrier prevents this progression has been proposed suggesting that primates and rodents differ in their response to injected amyloid not only in terms of neurodegeneration, but also in the accumulation of abnormal tau around the injected amyloid.

The most exciting development in the tau field is the creation of a transgenic mouse model with robust and reproducible tangle pathology. This transgenic mouse line, designated JNPL3, express human four repeat tau containing the most common FTDP-17 mutation, P301L, under the control of the mouse prion promoter. The mice develop motor and behavioral deficits that are associated with age- and gene-dosage-dependent development of congophilic NFT. Tau-immunoreactive pretangles are found in cortex, hippocampus, and basal ganglia, but at lower levels than commonly found in human disease. The tangles are associated with neurodegeneration, especially in the spinal cord where motor neurons were reduced by approximately 48%. This mouse has recently been used in breeding double and triple transgenic mouse models to produce a model system that more accurately recapitulates the hallmark pathology of AD (see further in the text). Recently, another

P301L tau transgenic model has been described that develops fivefold increase in the numbers of NFT soon after 18 days after the intracerebral injection of Aβ42 into the CA1 region of the hippocampus. These experiments indicate that an interaction of β-amyloid with the P301L tau mutation accelerates NFT formation.

4.4
Apolipoprotein E Transgenic Mice

A consistent consequence of carrying the *APOE4* allele is an increased number of amyloid plaques in the brain and more abundant amyloid deposition in the cerebral vasculature. The mechanism by which *APOE4* contributes to the development of neurodegeneration remains unknown, although the evidence suggests that this may be linked to the ability of APOE to interact with the amyloid β peptide and influence its concentration and structure. Recent studies with transgenic mice that overexpress the human APOE isoforms APOE2, APOE3, and APOE4 in a PDAPP hemizygous mouse (Apoe-/- background) show that isoforms that are known to increase the risk of AD enhance amyloid load and increase the neuritic pathology associated with fibrillar plaque development in aged animals (by 15 months of age APOE4 expressing mice having a 10-fold greater amyloid burden than APOE3 mice). The time course of Aβ deposition, Aβ levels, structure, and anatomic and subcellular distribution showed profound age-, species-, and isoform-dependent effect of APOE. It was also found that APOE not only impacts on the nature of aggregated Aβ, but also on the clearance of soluble Aβ from the brain across the blood–brain barrier (BBB). Aβ40 was rapidly cleared from the brain, while Aβ42 was cleared much less effectively, supporting the idea that Aβ42 production may favor amyloid deposition due to a reduced clearance across the BBB. However, Aβ clearance was not APOE isoform specific.

These recent findings are consistent with earlier work on Apoe null mice. When Tg2576 and PDAPP mice were crossed with Apoe null mice it was revealed that the absence of Apoe altered the quantity, character, and distribution of Aβ deposits in the transgenic animals, confirming that APOE affects the neuropathological phenotype in AD brains. The Aβ deposition is significantly reduced and is not thioflavine-S-positive.

4.5
Modulating and Enhancing the Phenotype: Transgenic Crosses

One of the great advantages of transgenic animals is that different mice can be mated together so that the effect of multiple transgenes on the disease phenotype can be observed. In particular, since most transgenic AD models do not exhibit all the pathological features of AD, different crosses have been made in order to design a model that will more closely describe the pathology, neurodegeneration, and memory loss seen in AD patients.

Double transgenic mice that carry *PS1-A246E* and *APPsw* FAD mutations show an elevated Aβ42/Aβ40 ratio in brain homogenates compared to the ratio observed in transgenic mice expressing *APPsw* alone or transgenic mice coexpressing wild-type human *PS1* and *APPsw*. In addition, double transgenic mice *APPswxPS1-A246E* developed numerous amyloid deposits much earlier (9 months of age) than age-matched mice expressing *APPsw* (18 months of age) and wild-type *huPS1* and *APPsw* alone. Interestingly, the majority of Aβ deposits in the double transgenic mice were not immunoreactive to $A\beta42$ but instead were stained with antisera to $A\beta40$. Similarly, crossing the Tg2576 transgenic mice, which express mutant APPsw, with mice transgenic for *PS1-M146L* (PDPS1) revealed 41% increase in $A\beta42/43$ levels and formation of $A\beta$ deposits in the cortex and hippocampus as early as 12 weeks of age, compared to development of $A\beta$ deposits in Tg2576 mice of 9–12 months of age and approximately 1.5-fold elevation of $A\beta42/43$ from birth in PDPS1 line. The early $A\beta$ deposits were found to be primarily composed of fibrillar $A\beta$ and resembled compact amyloid plaques. As the mice aged, these fibrillar deposits did not increase substantially beyond the 12 months of age. Interestingly, the diffuse deposits appeared only until later, this is opposite to the general perception that in AD the compact deposits are formed by condensation of the diffuse material.

Studies on APPxPS1 double transgenics, therefore, revealed that marginal increases in $A\beta42/43$ levels, evoked by PS1 mutant transgene, can accelerate the deposition process by several months. As these mice have a severe $A\beta$ pathology for an extended time relative to singly transgenic APP mice, other features of the disease such as tau abnormalities or major cell loss may become apparent as the animals age.

Double mutants produced from crossing JNPL3 transgenic mice expressing mutant P301L tau with Tg2576 mice expressing the APPsw mutation offered proof that $A\beta$ influences the development of NFTs. The double mutant exhibited NFT pathology that was substantially enhanced in the limbic system and the olfactory cortex. These results suggest that APP or $A\beta$ augments the formation of NFTs in the regions of the brain vulnerable to the formation of these lesions. Recently, a triple transgenic mouse was generated expressing APPsw, PS1-M146L, and P301L tau mutant protein that exhibited both plaques and NFT. However, other characteristics of AD such as neurodegeneration and memory loss have yet to be examined.

Although most of the transgenic models for AD do not completely recreate the disease phenotype, they have shown considerable utility, both for studying the disease mechanisms and for the preliminary testing of therapeutic agents, particularly those that are designed to modulate $A\beta$ deposition.

5
Potential Treatments

The ultimate goal of AD research is to prevent disease and/or develop treatments for this devastating disease. Unfortunately, there is still no effective treatment available. Currently, the only FDA-approved therapy for AD is cholinesterase inhibitors, which can enhance cholinergic activity and temporarily improve cognitive function in some individuals in the early stages of disease. However, these drugs only treat the symptoms but have no impact on progression of the disease.

Recent advances in dissecting the molecular and cellular mechanisms that are involved in AD pathogenesis have provided substantial knowledge for determining novel targets for drug development. Several new therapeutic approaches targeted at distinct aspects of the disease are currently being pursued. Potential therapies include: anti-inflammatory agents, cholesterol-lowering drugs, antioxidants, hormonal therapy, and approaches that inhibit $A\beta$ production or increase $A\beta$ degradation and clearance. Strategies targeted directly at $A\beta$ may be most effective given the fact that Aβ deposition is the central and fundamental event that causes AD. Other approaches affecting downstream events in the neurodegenerative process are likely to be used as supplementary treatments.

5.1
Secretase Inhibitors

Two prime targets for drug development are β- and γ-secretases, the two enzymes that process APP to Aβ. Inhibitors to these two proteases are expected to have the greatest Aβ-lowering effect since inhibiting either enzyme can completely inhibit Aβ generation.

γ-secretase is considered to be central to AD pathogenesis because altering its activity invariantly favors Aβ42 production and causes AD. The major concern in developing a drug that targets the γ-secretase is its potential for side effects. It is clear that γ-secretase is important for several physiological functions through its role as the protease that mediates the intramembranous cleavage of a number of TM proteins, including Notch. One possible solution is to develop an inhibitor that selectively lowers Aβ (Aβ42 in particular) production without significantly affecting the cleavage of other γ-secretase substrates. The development of such inhibitors has been reported. Another possible approach is partial inhibition of γ-secretase activity, at inhibitor concentrations that reduce Aβ production without affecting Notch signaling. Indeed, it has been shown that compounds with no reported selectivity allow significant Aβ reduction without changing the Notch-related function. Some inhibitors, however, are reported to preferentially increase Aβ42 production when used at low concentrations.

Most evidence suggests that β-secretase (BACE1) may be a better therapeutic target than γ-secretase. BACE1 knockout mice do not generate Aβ, but are otherwise healthy and fertile without obvious deficits in the basal, neurological, and physiological functions. In addition, the X-ray structure of the BACE1 protease domain has been solved and should provide valuable knowledge for inhibitor design.

Although the knockout data are reassuring that the detrimental effects of inhibiting β-secretase may be minimal, concerns regarding the possible side effects remain. First, there are likely other β-secretase substrates since it is counterintuitive to assume that BACE1 has evolved just to generate Aβ. Indeed, one recent study suggests that β-secretase may be responsible for the cleavage and secretion of a Golgi resident sialyltransferase. It is very important to identify these substrates because they are valuable for predicting the possible side effects of BACE1 inhibitors. It will also be important to know the phenotype of BACE1/2 double knockout mice. Although BACE2 is not primarily involved in Aβ generation, it may have an important physiological role and inhibitors of BACE1 may also inhibit BACE2 because of the high degree of homology between these two proteins.

While inhibitors to both secretases may potentially be effective in AD treatment, the challenges facing drug development are significant. The inhibitors need to be highly selective for the target enzyme and target substrate to reduce possible side effects to a minimum. The inhibitors need to be able to efficiently cross the BBB. Potent peptide-based inhibitors to BACE1 have been developed. However, such large compounds are not viable drug candidates because they will not penetrate the BBB to a sufficient extent. Instead, the enzyme inhibitors with therapeutic potential are preferably small organic molecules with high specificity. The X-ray structure of BACE1 suggests that its active site is more open and less hydrophobic than that of other aspartyl proteases. It may pose challenges for the development of small-molecule inhibitors.

5.2
Vaccine Approaches

Another novel approach for Alzheimer therapy is $A\beta$ immunization. Many studies have shown that direct immunization with $A\beta$ peptide or anti-$A\beta$ antibodies can greatly reduce plaque formation, neuritic dystrophy, and other AD-like pathology in APP transgenic mice. One study also shows that such an immunization strategy is also effective in reversing behavioral deficits in the transgenic mouse model. It appears that antibodies to $A\beta$ can enhance $A\beta$ clearance, prevent $A\beta$ fibril formation, disrupt $A\beta$ fibrils, as well as block $A\beta$ toxicity.

Despite the great effects in mice, there are serious concerns about the safety of this approach when used in humans. One concern is that autoimmunity may occur in humans, although it has not been reported in mice. Another major concern is the toxicity of $A\beta 42$, which can cross the BBB and seed fibril formation in the brain, and therefore may actually promote plaque formation. In addition, $A\beta 42$ may cause inflammation and neurotoxicity. A phase II clinical trial using $A\beta 42$ vaccination was eventually terminated because of cerebral inflammation observed in several patients. The first autopsy result from this trial was just published. It shows exactly the two sides of this approach. Vaccination has such a powerful effect that the patient's cerebral cortex was almost cleared of $A\beta$ deposition, but the side effect of cerebral inflammation was so severe that the patient developed meningoencephalitis, which led to death.

Safer therapeutic approaches including using nonamyloidogenic/nontoxic $A\beta$ derivatives as an immunogen or passive immunization with $A\beta$ antibodies are being developed. One study shows that passive immunization with (Fab')$_2$ fragments of $A\beta$-specific antibody can clear $A\beta$ in a mouse model. This is promising because (Fab')$_2$ fragments do not interact with Fc receptors and so will not activate the cellular-immune response which has been the major cause for the severe side effects.

5.3
Other Therapeutic Approaches

Several FAD-approved nonsteroidal anti-inflammatory drugs (NSAIDs), including ibuprofen, indomethacin, and sulindac have been shown to selectively reduce $A\beta 42$ levels in cultured cells. In addition, treatment of APP transgenic mice with ibuprofen reduces $A\beta 42$ levels in the brain and suppresses plaque pathology. It appears that NSAIDs can subtly alter γ-secretase activity so that cleavage is shifted from $A\beta 42$ to $A\beta 38$.

Many studies also indicate that cholesterol influences $A\beta$ metabolism and cholesterol-reducing drugs may have a beneficial effect on AD. Individuals taking cholesterol-lowering drugs such as statins show greatly reduced risk for developing AD, whereas individuals with elevated cholesterol are at a higher risk. Statins and other cholesterol-lowering drugs have been shown to reduce $A\beta$ levels and $A\beta$ deposition in both cell culture systems and animal models. The role of cholesterol in $A\beta$ metabolism appears to be quite complex and the mechanisms are not clear.

6
Summary

We have seen a dramatic increase in our knowledge of the underlying molecular mechanisms of AD pathology during the last ten years. Understanding the effects of causative mutations in *APP* and the presenilins and the genetic risk factor, *APOE4* has led to the conclusion that changes in Aβ metabolism and deposition are central to the disease process. These studies suggest that similar mechanisms underlie both FAD and sporadic AD. Transgenic models of AD now provide a new hope for the rapid development in rational treatments for AD.

Bibliography

Books and Reviews

Duff, K. (2001) Transgenic mouse models of Alzheimer's disease: phenotype and mechanisms of pathogenesis, *Biochem. Soc. Symp.* **67**, 195–202.

Golde, T.E. (2003) Alzheimer disease therapy: Can the amyloid cascade be halted? *J. Clin. Invest.* **11**, 11–18.

Hardy, J., Selkoe, D.J. (2002) The amyloid hypothesis of Alzheimer's disease: progress and problems on the road to therapeutics, *Science* **297**, 353–356.

Myers, A., Goate, A.M. (2001) The genetics of late-onset Alzheimer's disease, *Curr. Opin. Neurol.* **14**, 433–440.

Richardson, J.A., Burns, D.K. (2002) Mouse models of Alzheimer's disease: a quest for plaques and tangles, *ILAR J.* **43**, 89–99.

Selkoe, D.J., Podlisny, M.B. (2002) Deciphering the genetic basis of Alzheimer's disease, *Annu. Rev. Genomics Hum. Genet.* **3**, 67–99.

Tandon, A., Fraser, P. (2002) The presenilins, *Genome Biol.* **3**, 3014.1–3014.9; reviews.

Vassar, R. (2002) β-Secretase (BACE) as a drug target for alzheimer's disease, *Adv. Drug Delivery Rev.* **54**, 1589–1602.

Primary Literature

Alzheimer, A. (1907) Über eine eigenartige Erkrankung der Hirnrinde, *Allg. Zeitsch. Psychiatrie Psychisch-gerichtliche Med.* **64**, 146–148.

Biernat, J., Gustke, N., Drewes, G., Mandelkow, E.M., Mandelkow, E. Phosphorylation of Ser262 strongly reduces binding of tau to microtubules: distinction between PHF-like immunoreactivity and microtubule binding, *Neuron* **11**, 153–163.

Blacker, D., Bertram, L., Saunders, A.J., Moscarillo, T.J., Albert, M.S., Wiener, H., Perry, R.T., Collins, J.S., Harrell, L.E., Go, R.C., Mahoney, A., Beaty, T., Fallin, M.D., Avramopoulos, D., Chase, G.A., Folstein, M.F., McInnis, M.G., Bassett, S.S., Doheny, K.J., Pugh, E.W., Tanzi, R.E. (2003) Results of a high-resolution genome screen of 437 Alzheimer's disease families, *Hum. Mol. Genet.* **12**, 23–32.

Borchelt, D.R., Ratovitski, T., van Lare, J., Lee, M.K., Gonzales, V., Jenkins, N.A., Copeland, N.G., Price, D.L., Sisodia, S.S. (1997) Accelerated amyloid deposition in the brains of transgenic mice coexpressing mutant presenilin 1 and amyloid precursor proteins, *Neuron* **19**, 939–945.

Borchelt, D.R., Thinakaran, G., Eckman, C.B., Lee, M.K., Davenport, F., Ratovitsky, T., Prada, C.M., Kim, G., Seekins, S., Yager, D., Slunt, H.H., Wang, R., Seeger, M., Levey, A.I., Gandy, S.E., Copeland, N.G., Jenkins, N.A., Price, D.L., Younkin, S.G., Sisodia, S.S. (1996) Familial Alzheimer's disease-linked presenilin 1 variants elevate Aβ1-42/1-40 ratio *in vitro* and *in vivo*, *Neuron* **17**, 1005–1013.

Braak, H., Braak, E. (1991) Neuropathological stageing of Alzheimer-related changes, *Acta Neuropathol. (Berl)* **82**, 239–259.

Brendza, R.P., Bales, K.R., Paul, S.M., Holtzman, D.M. (2002) Role of apoE/Aβ interactions in Alzheimer's disease: insights from transgenic mouse models, *Mol. Psychiatry* **7**, 132–135.

Cai, X.D., Golde, T.E., Younkin, S.G. (1993) Release of excess amyloid β protein from a mutant amyloid β protein precursor, *Science* 259, 514–516.

Chung, H.M., Struhl, G. (2001) Nicastrin is required for Presenilin-mediated transmembrane cleavage in Drosophila, *Nat. Cell. Biol.* 3, 1129–1132.

Citron, M., Oltersdorf, T., Haass, C., McConlogue, L., Hung, A.Y., Seubert, P., Vigo-Pelfrey, C., Lieberburg, I., Selkoe, D.J. (1992) Mutation of the β-amyloid precursor protein in familial Alzheimer's disease increases β-protein production, *Nature* 360, 672–674.

Citron, M., Westaway, D., Xia, W., Carlson, G., Diehl, T., Levesque, G., Johnson-Wood, K., Lee, M., Seubert, P., Davis, A., Kholodenko, D., Motter, R., Sherrington, R., Perry, B., Yao, H., Strome, R., Lieberburg, I., Rommens, J., Kim, S., Schenk, D., Fraser, P., St. George Hyslop, P., Selkoe, D.J. (1997) Mutant presenilins of Alzheimer's disease increase production of 42-residue amyloid β-protein in both transfected cells and transgenic mice, *Nat. Med.* 3, 67–72.

Corder, E.H., Saunders, A.M., Risch, N.J., Strittmatter, W.J., Schmechel, D.E., Gaskell, P.C. Jr., Rimmler, J.B., Locke, P.A., Conneally, P.M., Schmader, K.E., et al. (1994) Protective effect of apolipoprotein E type 2 allele for late onset Alzheimer disease, *Nat. Genet.* 7, 180–184.

Corder, E.H., Saunders, A.M., Strittmatter, W.J., Schmechel, D.E., Gaskell, P.C., Small, G.W., Roses, A.D., Haines, J.L., Pericak-Vance, M.A. (1993) Gene dose of apolipoprotein E type 4 allele and the risk of Alzheimer's disease in late onset families, *Science* 261, 921–923.

DeMattos, R.B., Bales, K.R., Cummins, D.J., Dodart, J.C., Paul, S.M., Holtzman, D.M. (2001) Peripheral anti-Aβ antibody alters CNS and plasma Aβ clearance and decreases brain Aβ burden in a mouse model of Alzheimer's disease, *Proc. Natl. Acad. Sci. U.S.A.* 98, 8850–8855.

DeMattos, R.B., Bales, K.R., Cummins, D.J., Paul, S.M., Holtzman, D.M. (2002) Brain to plasma amyloid-β efflux: a measure of brain amyloid burden in a mouse model of Alzheimer's disease, *Science* 295, 2264–2267.

De Strooper, B., Annaert, W., Cupers, P., Saftig, P., Craessaerts, K., Mumm, J.S., Schroeter, E.H., Schrijvers, V., Wolfe, M.S., Ray, W.J., Goate, A., Kopan, R.A. (1999) presenilin-1-dependent γ-secretase-like protease mediates release of Notch intracellular domain, *Nature* 398, 518–522.

De Strooper, B., Saftig, P., Craessaerts, K., Vanderstichele, H., Guhde, G., Annaert, W., von Figura, K., Van Leuven, F. (1998) Deficiency of presenilin-1 inhibits the normal cleavage of amyloid precursor protein, *Nature* 391, 387–390.

Duff, K., Eckman, C., Zehr, C., Yu, X., Prada, C.M., Perez-Tur, J., Hutton, M., Buee, L., Harigaya, Y., Yager, D., Morgan, D., Gordon, M.N., Holcomb, L., Refolo, L., Zenk, B., Hardy, J., Younkin, S. (1996) Increased amyloid-β42(43) in brains of mice expressing mutant presenilin 1, *Nature* 383, 710–713.

Edbauer, D., Winkler, E., Haass, C., Steiner, H. (2002) Presenilin and nicastrin regulate each other and determine amyloid β-peptide production via complex formation, *Proc. Natl. Acad. Sci. U.S.A.* 99, 8666–8671.

Ertekin-Taner, N., Graff-Radford, N., Younkin, L.H., Eckman, C., Baker, M., Adamson, J., Ronald, J., Blangero, J., Hutton, M., Younkin, S.G. (2000) Linkage of plasma Aβ42 to a quantitative locus on chromosome 10 in late-onset Alzheimer's disease pedigrees, *Science* 290, 2303–2304.

Esch, F.S., Keim, P.S., Beattie, E.C., Blacher, R.W., Culwell, A.R., Oltersdorf, T., McClure, D., Ward, P.J. (1990) Cleavage of amyloid β peptide during constitutive processing of its precursor, *Science* 248, 1122–1124.

Estus, S., Golde, T.E., Younkin, S.G. (1992) Normal processing of the Alzheimer's disease amyloid β protein precursor generates potentially amyloidogenic carboxyl-terminal derivatives, *Ann. N. Y. Acad. Sci.* 674, 138–148.

Fagan, A.M., Watson, M., Parsadanian, M., Bales, K.R., Paul, S.M., Holtzman, D.M. (2002) Human and murine ApoE markedly alters Aβ metabolism before and after plaque formation in a mouse model of Alzheimer's disease, *Neurobiol. Dis.* 9, 305–318.

Farrer, L.A., Bowirrat, A., Friedland, R.P., Waraska, K., Korczyn, A.D., Baldwin, C.T. (2003) Identification of multiple loci for Alzheimer disease in a consanguineous Israeli-Arab community, *Hum. Mol. Genet.* 12, 415–422.

Flood, D.G., Reaume, A.G., Dorfman, K.S., Lin, Y.G., Lang, D.M., Trusko, S.P., Savage, M.J., Annaert, W.G., De Strooper, B., Siman, R.,

Scott, R.W. (2002) FAD mutant PS-1 gene-targeted mice: increased Aβ42 and Aβ deposition without APP overproduction, *Neurobiol. Aging* **23**, 335–348.

Francis, R., McGrath, G., Zhang, J., Ruddy, D.A., Sym, M., Apfeld, J., Nicoll, M., Maxwell, M., Hai, B., Ellis, M.C., Parks, A.L., Xu, W., Li, J., Gurney, M., Myers, R.L., Himes, C.S., Hiebsch, R., Ruble, C., Nye, J.S., Curtis, D. (2002) aph-1 and pen-2 are required for Notch pathway signaling, γ-secretase cleavage of βAPP, and presenilin protein accumulation, *Dev. Cell.* **3**, 85–97.

Games, D., Adams, D., Alessandrini, R., Barbour, R., Berthelette, P., Blackwell, C., Carr, T., Clemens, J., Donaldson, T., Gillespie, F. (1995) Alzheimer-type neuropathology in transgenic mice overexpressing V717F β-amyloid precursor protein, *Nature* **373**, 523–527.

Glenner, G.G., Wong, C.W. (1984) Alzheimer's disease: initial report of the purification and characterization of a novel cerebrovascular amyloid protein, *Biochem. Biophys. Res. Commun.* **120**, 885–890.

Goate, A., Chartier-Harlin, M.C., Mullan, M., Brown, J., Crawford, F., Fidani, L., Giuffra, L., Haynes, A., Irving, N., James, L. (1991) Segregation of a missense mutation in the amyloid precursor protein gene with familial Alzheimer's disease, *Nature* **349**, 704–706.

Goedert, M., Wischik, C.M., Crowther, R.A., Walker, J.E., Klug, A. (1988) Cloning and sequencing of the cDNA encoding a core protein of the paired helical filament of Alzheimer disease: identification as the microtubule-associated protein tau, *Proc. Natl. Acad. Sci. U.S.A.* **85**, 4051–4055.

Goldgaber, D., Lerman, M.I., McBride, O.W., Saffiotti, U., Gajdusek, D.C. (1987) Characterization and chromosomal localization of a cDNA encoding brain amyloid of Alzheimer's disease, *Science* **235**, 877–880.

Gómez-Isla, T., Price, J.L., McKeel, D.W., Morris, J.C., Growdon, J.H., Hyman, B.T. (1996) Profound loss of layer II entorhinal cortex neurons occurs in very mild Alzheimer's disease, *J. Neurosci.* **16**, 4491–4500.

Gotz, J., Chen, F., van Dorpe, J., Nitsch, R.M. (2001) Formation of neurofibrillary tangles in P301l tau transgenic mice induced by Aβ42 fibrils, *Science* **293**, 1491–1495.

Goutte, C., Tsunozaki, M., Hale, V.A., Priess, J.R. (2002) APH-1 is a multipass membrane protein essential for the Notch signaling pathway in *Caenorhabditis elegans* embryos, *Proc. Natl. Acad. Sci. U.S.A.* **99**, 775–779.

Grundke-Iqbal, I., Iqbal, K., Quinlan, M., Tung, Y.C., Zaidi, M.S., Wisniewski, H.M. (1986) Microtubule-associated protein tau. A component of Alzheimer paired helical filaments, *J. Biol. Chem.* **261**, 6084–6089.

Haass, C., Koo, E.H., Mellon, A., Hung, A.Y., Selkoe, D.J. (1992) Targeting of cell-surface β-amyloid precursor protein to lysosomes: alternative processing into amyloid-bearing fragments, *Nature* **357**, 500–503.

Haass, C., Schlossmacher, M.G., Hung, A.Y., Vigo-Pelfrey, C., Mellon, A., Ostaszewski, B.L., Lieberburg, I., Koo, E.H., Schenk, D., Teplow, D.B. (1992) Amyloid β-peptide is produced by cultured cells during normal metabolism, *Nature* **359**, 322–325.

Holcomb, L., Gordon, M.N., McGowan, E., Yu, X., Benkovic, S., Jantzen, P., Wright, K., Saad, I., Mueller, R., Morgan, D., Sanders, S., Zehr, C., O'Campo, K., Hardy, J., Prada, C.M., Eckman, C., Younkin, S., Hsiao, K., Duff, K. (1998) Accelerated Alzheimer-type phenotype in transgenic mice carrying both mutant amyloid precursor protein and presenilin 1 transgenes, *Nat. Med.* **4**, 97–100.

Holtzman, D.M. (2002) Aβ conformational change is central to Alzheimer's disease, *Neurobiol. Aging* **23**, 1085–1088.

Hong, C.S., Caromile, L., Nomata, Y., Mori, H., Bredesen, D.E., Koo, E.H. (1999) Contrasting role of presenilin-1 and presenilin-2 in neuronal differentiation *in vitro*, *J. Neurosci.* **19**, 637–643.

Hsiao, K., Chapman, P., Nilsen, S., Eckman, C., Harigaya, Y., Younkin, S., Yang, F., Cole, G. (1996) Correlative memory deficits, Aβ elevation, and amyloid plaques in transgenic mice, *Science* **274**, 99–102.

Hu, Y., Ye, Y., Fortini, M.E. (2002) Nicastrin is required for γ-secretase cleavage of the *Drosophila* Notch receptor, *Dev. Cell.* **2**, 69–78.

Hutton, M., Lendon, C.L., Rizzu, P., Baker, M., Froelich, S., Houlden, H., Pickering-Brown, S., Chakraverty, S., Isaacs, A., Grover, A., Hackett, J., Adamson, J., Lincoln, S., Dickson, D., Davies, P., Petersen, R.C., Stevens, M., de Graaff, E., Wauters, E., van Baren, J., Hillebrand, M., Joosse, M., Kwon, J.M., Nowotny, P., Heutink, P. (1998) Association of missense and 5'-splice-site mutations in tau with the inherited dementia FTDP-17, *Nature* **393**, 702–705.

Iqbal, K., Zaidi, T., Thompson, C.H., Merz, P.A., Wisniewski, H.M. (1984) Alzheimer paired helical filaments: bulk isolation, solubility, and protein composition, *Acta Neuropathol. (Berl)* **62**, 167–177.

Kang, J., Lemaire, H.G., Unterbeck, A., Salbaum, J.M., Masters, C.L., Grzeschik, K.H., Multhaup, G., Beyreuther, K., Müller-Hill, B. (1987) The precursor of Alzheimer's disease amyloid A4 protein resembles a cell-surface receptor, *Nature* **325**, 733–736.

Leissring, M.A., Murphy, M.P., Mead, T.R., Akbari, Y., Sugarman, M.C., Jannatipour, M., Anliker, B., Muller, U., Saftig, P., De Strooper, B., Wolfe, M.S., Golde, T.E., LaFerla, F.M. (2002) A physiologic signaling role for the γ-secretase-derived intracellular fragment of APP, *Proc. Natl. Acad. Sci. U.S.A.* **99**, 4697–4702.

Levy, E., Carman, M.D., Fernandez-Madrid, I.J., Power, M.D., Lieberburg, I., van Duinen, S.G., Bots, G.T., Luyendijk, W., Frangione, B. (1990) Mutation of the Alzheimer's disease amyloid gene in hereditary cerebral hemorrhage, Dutch type, *Science* **248**, 1124–1126.

Levy-Lahad, E., Wasco, W., Poorkaj, P., Romano, D.M., Oshima, J., Pettingell, W.H., Yu, C.E., Jondro, P.D., Schmidt, S.D., Wang, K. (1995) Candidate gene for the chromosome 1 familial Alzheimer's disease locus, *Science* **269**, 973–977.

Lewis, J., Dickson, D.W., Lin, W.L., Chisholm, L., Corral, A., Jones, G., Yen, S.H., Sahara, N., Skipper, L., Yager, D., Eckman, C., Hardy, J., Hutton, M., McGowan, E. (2001) Enhanced neurofibrillary degeneration in transgenic mice expressing mutant tau and APP, *Science* **293**, 1487–1491.

Lewis, J., McGowan, E., Rockwood, J., Melrose, H., Nacharaju, P., Van Slegtenhorst, M., Gwinn-Hardy, K., Paul Murphy, M., Baker, M., Yu, X., Duff, K., Hardy, J., Corral, A., Lin, W.L., Yen, S.H., Dickson, D.W., Davies, P., Hutton, M. (2000) Neurofibrillary tangles, amyotrophy and progressive motor disturbance in mice expressing mutant (P301L) tau protein, *Nat. Genet.* **25**, 402–405.

Lopez-Schier, H., St. Johnston, D. (2002) *Drosophila* nicastrin is essential for the intramembranous cleavage of notch, *Dev. Cell.* **2**, 79–89.

Masters, C.L., Simms, G., Weinman, N.A., Multhaup, G., McDonald, B.L., Beyreuther, K. (1985) Amyloid plaque core protein in Alzheimer disease and Down syndrome, *Proc. Natl. Acad. Sci. U.S.A.* **82**, 4245–4249.

Mayeux, R., Lee, J.H., Romas, S.N., Mayo, D., Santana, V., Williamson, J., Ciappa, A., Rondon, H.Z., Estevez, P., Lantigua, R., Medrano, M., Torres, M., Stern, Y., Tycko, B., Knowles, J.A. (2002) Chromosome-12 mapping of late-onset Alzheimer disease among Caribbean Hispanics, *Am. J. Hum. Genet.* **70**, 237–243.

Murrell, J., Farlow, M., Ghetti, B., Benson, M.D. (1991) A mutation in the amyloid precursor protein associated with hereditary Alzheimer's disease, *Science* **254**, 97–99.

Nicoll, J.A., Wilkinson, D., Holmes, C., Steart, P., Markham, H., Weller, R.O. (2003) Neuropathology of human Alzheimer disease after immunization with amyloid-β peptide: a case report, *Nat. Med.* **9**, 448–452.

Nilsberth, C., Westlind-Danielsson, A., Eckman, C.B., Condron, M.M., Axelman, K., Forsell, C., Stenh, C., Luthman, J., Teplow, D.B., Younkin, S.G., Naslund, J., Lannfelt, L. (2001) The 'Arctic' APP mutation (E693G) causes Alzheimer's disease by enhanced Aβ protofibril formation, *Nat. Neurosci.* **4**, 887–893.

Price, J.L., Davis, P.B., Morris, J.C., White, D.L. The distribution of tangles, plaques and related immunohistochemical markers in healthy aging and Alzheimer's disease, *Neurobiol. Aging* **12**, 295–312.

Rogaev, E.I., Sherrington, R., Rogaeva, E.A., Levesque, G., Ikeda, M., Liang, Y., Chi, H., Lin, C., Holman, K., Tsuda, T. (1995) Familial Alzheimer's disease in kindreds with missense mutations in a gene on chromosome 1 related to the Alzheimer's disease type 3 gene, *Nature* **376**, 775–778.

St. George-Hyslop, P.H., Tanzi, R.E., Polinsky, R.J., Haines, J.L., Nee, L., Watkins, P.C., Myers, R.H., Feldman, R.G., Pollen, D., Drachman, D. (1987) The genetic defect causing familial Alzheimer's disease maps on chromosome 21, *Science* **235**, 885–890.

Schenk, D., Barbour, R., Dunn, W., Gordon, G., Grajeda, H., Guido, T., Hu, K., Huang, J., Johnson-Wood, K., Khan, K., Kholodenko, D., Lee, M., Liao, Z., Lieberburg, I., Motter, R., Mutter, L., Soriano, F., Shopp, G., Vasquez, N., Vandevert, C., Walker, S., Wogulis, M., Yednock, T., Games, D., Seubert, P. (1999) Immunization with amyloid-β attenuates

Alzheimer-disease-like pathology in the PDAPP mouse, *Nature* **400**, 173–177.

Scheuner, D., Eckman, C., Jensen, M., Song, X., Citron, M., Suzuki, N., Bird, T.D., Hardy, J., Hutton, M., Kukull, W., Larson, E., Levy-Lahad, E., Viitanen, M., Peskind, E., Poorkaj, P., Schellenberg, G., Tanzi, R., Wasco, W., Lannfelt, L., Selkoe, D., Younkin, S. (1996) Secreted amyloid β-protein similar to that in the senile plaques of Alzheimer's disease is increased *in vitro* by the presenilin 1 and 2 and APP mutations linked to familial Alzheimer's disease, *Nat. Med.* **2**, 864–870.

Seubert, P., Oltersdorf, T., Lee, M.G., Barbour, R., Blomquist, C., Davis, D.L., Bryant, K., Fritz, L.C., Galasko, D., Thal, L.J. (1993) Secretion of β-amyloid precursor protein cleaved at the amino terminus of the β-amyloid peptide, *Nature* **361**, 260–263.

Seubert, P., Vigo-Pelfrey, C., Esch, F., Lee, M., Dovey, H., Davis, D., Sinha, S., Schlossmacher, M., Whaley, J., Swindlehurst, C. (1992) Isolation and quantification of soluble Alzheimer's β-peptide from biological fluids, *Nature* **359**, 325–327.

Sherrington, R., Rogaev, E.I., Liang, Y., Rogaeva, E.A., Levesque, G., Ikeda, M., Chi, H., Lin, C., Li, G., Holman, K. (1995) Cloning of a gene bearing missense mutations in early-onset familial Alzheimer's disease, *Nature* **375**, 754–760.

Shoji, M., Golde, T.E., Ghiso, J., Cheung, T.T., Estus, S., Shaffer, L.M., Cai, X.D., McKay, D.M., Tintner, R., Frangione, B. (1992) Production of the Alzheimer amyloid β protein by normal proteolytic processing, *Science* **258**, 126–129.

Sinha, S., Anderson, J.P., Barbour, R., Basi, G.S., Caccavello, R., Davis, D., Doan, M., Dovey, H.F., Frigon, N., Hong, J., Jacobson-Croak, K., Jewett, N., Keim, P., Knops, J., Lieberburg, I., Power, M., Tan, H., Tatsuno, G., Tung, J., Schenk, D., Seubert, P., Suomensaari, S.M., Wang, S., Walker, D., John, V. (1999) Purification and cloning of amyloid precursor protein β-secretase from human brain, *Nature* **402**, 537–540.

Skovronsky, D.M., Zhang, B., Kung, M.-P., Kung, H.F., Trojanowski, J.Q., Lee, V.M. (2000) *In vitro* detection of amyloid plaques in a mouse model of Alzheimer's disease, *Proc. Natl. Acad. Sci. U.S.A.* **97**, 7609–7614.

Stenh, C., Nilsberth, C., Hammarback, J., Engvall, B., Naslund, J., Lannfelt, L. (2002) The Arctic mutation interferes with processing of the amyloid precursor protein, *Neuro Report* **13**, 1857–1860.

Strittmatter, W.J., Saunders, A.M., Schmechel, D., Pericak-Vance, M., Enghild, J., Salvesen, G.S., Roses, A.D. (1993) Apolipoprotein E: high-avidity binding to β-amyloid and increased frequency of type 4 allele in late-onset familial Alzheimer disease, *Proc. Natl. Acad. Sci. U.S.A.* **90**, 1977–1981.

Sturchler-Pierrat, C., Abramowski, D., Duke, M., Wiederhold, K.H., Mistl, C., Rothacher, S., Ledermann, B., Burki, K., Frey, P., Paganetti, P.A., Waridel, C., Calhoun, M.E., Jucker, M., Probst, A., Staufenbiel, M., Sommer, B. (1997) Two amyloid precursor protein transgenic mouse models with Alzheimer disease-like pathology, *Proc. Natl. Acad. Sci. U.S.A.* **94**, 13287–13292.

Sturchler-Pierrat, C., Staufenbiel, M. (2000) Pathogenic mechanisms of Alzheimer's disease analyzed in the APP23 transgenic mouse model, *Ann. N. Y. Acad. Sci.* **920**, 134–139.

Suzuki, N., Cheung, T.T., Cai, X.D., Odaka, A., Otvos, L., Eckman, C., Golde, T.E., Younkin, S.G. (1994) An increased percentage of long amyloid β protein secreted by familial amyloid β protein precursor (β APP717) mutants, *Science* **264**, 1336–1340.

Tanzi, R.E., Gusella, J.F., Watkins, P.C., Bruns, G.A., St. George-Hyslop, P., Van Keuren, M.L., Patterson, D., Pagan, S., Kurnit, D.M., Neve, R.L. (1987) Amyloid β protein gene: cDNA, mRNA distribution, and genetic linkage near the Alzheimer locus, *Science* **235**, 880–884.

Terry, R.D., Masliah, E., Salmon, D.P., Butters, N., DeTeresa, R., Hill, R., Hansen, L.A., Katzman, R. (1991) Physical basis of cognitive alterations in Alzheimer's disease: synapse loss is the major correlate of cognitive impairment, *Ann. Neurol.* **30**, 572–580.

Thinakaran, G., Borchelt, D.R., Lee, M.K., Slunt, H.H., Spitzer, L., Kim, G., Ratovitsky, T., Davenport, F., Nordstedt, C., Seeger, M., Hardy, J., Levey, A.I., Gandy, S., Jenkins, N.A., Copeland, N.G., Price, D.L., Sisodia, S.S. (1996) Endoproteolysis of presenilin 1 and accumulation of processed derivatives *in vitro*, *Neuron* **17**, 181–190.

Thinakaran, G., Harris, C.L., Ratovitski, T., Davenport, F., Slunt, H.H., Price, D.L., Borchelt, D.R., Sisodia, S.S. (1997) Evidence that levels of presenilins (PS1 and PS2) are coordinately regulated by competition for limiting cellular factors, *J. Biol. Chem.* **272**, 28415–28422.

Van Nostrand, W.E., Schmaier, A.H., Farrow, J.S., Cunningham, D.D. (1990) Protease nexin-II (amyloid β-protein precursor): a platelet α-granule protein, *Science* **248**, 745–748.

Vassar, R., Bennett, B.D., Babu-Khan, S., Kahn, S., Mendiaz, E.A., Denis, P., Teplow, D.B., Ross, S., Amarante, P., Loeloff, R., Luo, Y., Fisher, S., Fuller, J., Edenson, S., Lile, J., Jarosinski, M.A., Biere, A.L., Curran, E., Burgess, T., Louis, J.C., Collins, F., Treanor, J., Rogers, G., Citron, M. (1999) B-secretase cleavage of Alzheimer's amyloid precursor protein by the transmembrane aspartic protease BAC, *Science* **286**, 735–741.

Weggen, S., Eriksen, J.L., Das, P., Sagi, S.A., Wang, R., Pietrzik, C.U., Findlay, K.A., Smith, T.E., Murphy, M.P., Bulter, T., Kang, D.E., Marquez-Sterling, N., Golde, T.E., Koo, E.H. (2001) A subset of NSAIDs lower amyloidogenic Aβ42 independently of cyclooxygenase activity, *Nature* **414**, 212–216.

Weihofen, A., Binns, K., Lemberg, M.K., Ashman, K., Martoglio, B. (2002) Identification of signal peptide peptidase, a presenilin-type aspartic protease, *Science* **296**, 2215–2218.

Wolfe, M.S., Xia, W., Ostaszewski, B.L., Diehl, T.S., Kimberly, W.T., Selkoe, D.J. (1999) Two transmembrane aspartates in presenilin-1 required for presenilin endoproteolysis and γ-secretase activity, *Nature* **398**, 513–517.

Yu, G., Nishimura, M., Arawaka, S., Levitan, D., Zhang, L., Tandon, A., Song, Y.Q., Rogaeva, E., Chen, F., Kawarai, T., Supala, A., Levesque, L., Yu, H., Yang, D.S., Holmes, E., Milman, P., Liang, Y., Zhang, D.M., Xu, D.H., Sato, C., Rogaev, E., Smith, M., Janus, C., Zhang, Y., Aebersold, R., Farrer, L.S., Sorbi, S., Bruni, A., Fraser, P., St. George-Hyslop, P. (2000) Nicastrin modulates presenilin-mediated notch/glp-1 signal transduction and βAPP processing, *Nature* **407**, 48–54.

34
Triplet Repeat Diseases

Stephan J. Guyenet and Albert R. La Spada
University of Washington, Seattle, WA 98195

1	**A Novel Mechanism of Genetic Mutation Emerges** 987	
1.1	Repeat Sequences of All Types and Sizes 987	
1.2	Trinucleotide Repeat Expansion as a Cause of Disease: Unique Features Explain Unusual Genetics 988	
2	**Repeat Diseases and Their Classification** 989	
2.1	Summary of Repeat Diseases 989	
2.2	Differences in Repeat Sequence Composition and Location within Gene 990	
2.3	Classification Based upon Mechanism of Pathogenesis and Nature of Mutation 990	
3	**Type 1: The CAG/Polyglutamine Repeat Diseases** 993	
3.1	Spinal and Bulbar Muscular Atrophy 993	
3.2	Huntington's Disease 996	
3.3	Dentatorubral Pallidoluysian Atrophy 999	
3.4	Spinocerebellar Ataxia Type 1 1000	
3.5	Spinocerebellar Ataxia Type 2 1002	
3.6	Spinocerebellar Ataxia Type 3/Machado–Joseph Disease 1003	
3.7	Spinocerebellar Ataxia Type 6 1005	
3.8	Spinocerebellar Ataxia Type 7 1006	
3.9	Spinocerebellar Ataxia Type 17 1009	
3.10	Role of Aggregation in Polyglutamine Disease Pathogenesis 1009	
3.11	Protein Context 1013	
3.12	Transcriptional Dysregulation 1014	
3.13	Proteolytic Cleavage 1014	
4	**Type 2: the Loss-of-function Repeat Diseases** 1017	
4.1	Fragile X Syndrome 1017	

Genomics and Genetics. Edited by Robert A. Meyers.
Copyright © 2007 Wiley-VCH Verlag GmbH & Co. KGaA, Weinheim
ISBN: 978-3-527-31609-0

4.2	Fragile XE Mental Retardation 1018
4.3	Friedreich's Ataxia 1020
4.4	Progressive Myoclonus Epilepsy Type 1 1023

5	**Type 3: the RNA Gain-of-function Repeat Diseases 1024**
5.1	Myotonic Dystrophy Type 1 1024
5.2	Myotonic Dystrophy Type 2 1026
5.3	Spinocerebellar Ataxia Type 8 1027
5.4	The Fragile X Tremor – Ataxia Syndrome (FXTAS) 1028

6	**Type 4: The Polyalanine Diseases 1030**
6.1	Overview 1030
6.2	Oculopharyngeal Muscular dystrophy 1030
6.3	Synpolydactyly (Syndactyly Type II) 1032
6.4	Cleidocranial Dysplasia 1032
6.5	Holoprosencephaly 1032
6.6	Hand-foot-genital Syndrome 1033
6.7	Blepharophimosis-ptosis-epicanthus Inversus Syndrome 1033
6.8	Syndromic and Nonsyndromic X-linked Mental Retardation 1033
6.9	Congenital Central Hypoventilation Syndrome 1034

7	**Unclassified Repeat Diseases Lacking Mechanistic Explanations 1034**
7.1	Spinocerebellar Ataxia Type 10 1034
7.2	Spinocerebellar Ataxia Type 12 1035
7.3	Huntington's Disease Like 2 (HDL2) 1036

Bibliography   1038
Books and Reviews   1038
Primary Literature   1038

Keywords

Aggregate
Accumulation of proteinaceous and/or ribonucleic acid material into a structure that is visible in a cell at the light microscope level.

Anticipation
Worsening severity of a disease phenotype as the causal (typically dominant) genetic mutation is transmitted from one generation to the next in a family segregating the disease of interest.

Gain of function
Refers to a type of mutation that imparts a novel activity or action to the gene product containing the mutation.

Loss of function
Refers to a type of mutation that eliminates the action of the gene product encoded by the gene within which the mutation resides.

Repeat Expansion
An elongation of a repeat to a larger size that no longer falls within the size distribution range typically seen in the normal population; this process is now recognized as a mechanism of human genetic mutation.

Trinucleotide
Three DNA base pairs of specific sequence composition (e.g. cytosine-adenine-guanine).

■ The repeat expansion disorders are a group of human diseases that are caused by the elongation of a DNA repeat sequence. In this chapter, we provide an overview of the discovery of repeat expansion as an important cause of human disease, and we summarize the molecular genetics and mechanistic basis of 27 microsatellite repeat disorders. Comparison of the many repeat expansion disorders reveals distinct categories of repeat diseases, allowing us to propose a classification of the repeat expansion disorders based upon mutation sequence and pathogenic mechanism. The four types of repeat expansion disorders defined by this approach are the CAG/polyglutamine repeat diseases; the loss-of-function repeat diseases; the RNA gain-of-function repeat diseases; and the polyalanine diseases. Although the genetic basis for most of these diseases was determined less than a decade or so ago, considerable advances have been made in our understanding of how "dynamic mutations" produce molecular pathology and human disease.

1
A Novel Mechanism of Genetic Mutation Emerges

1.1
Repeat Sequences of All Types and Sizes

Long before the sequencing of the human genome was undertaken, the discovery of bacterial enzymes that recognize specific DNA sequences ("recognition sites") and cleaved them, yielded a new methodology for differentiating individuals ("molecular fingerprinting"). This methodology also found application in the mapping of inherited genetic diseases, taking advantage of human variation in the form of so-called restriction fragment length polymorphisms (RFLP's). In the search for even more informative genetic markers, investigators uncovered a variety of very short repeat sequences ("short tandem repeats" – STR's; "simple sequence length polymorphisms" – SSLP's) that were highly dispersed throughout the human genome. These latter repeat sequences came to be known as "microsatellites" to differentiate them from the "minisatellites" that were being used for molecular fingerprinting. Minisatellites were originally defined as tandem arrays of 14–100 bp repeating sequences. In the case of minisatellites, the repeat sequence was in essence a "consensus" as deviations from the exact repeat sequence were common. Microsatellites, however, were typically pure perfect repeats of less than 13 bp, the most common microsatellites being dinucleotide repeats, trinucleotide repeats, or tetranucleotide repeats. The discovery of minisatellites and microsatellites yielded a virtual bonanza of reagents for molecular fingerprinting and genetic linkage mapping, while also providing evolutionary biologists and population geneticists with intriguing material to attempt to reconstruct evolutionary relationships and inter- or intraspecies relationships. Although human geneticists and evolutionary biologists were applying microsatellites in different ways for their own studies, they shared the commonly held belief that such repeats were neutral and therefore unlikely to be of much functional consequence, let alone play a role in causing human disease. Of course, all that would soon change in the last decade of the twentieth century.

1.2
Trinucleotide Repeat Expansion as a Cause of Disease: Unique Features Explain Unusual Genetics

In 1991, two groups working on seemingly unrelated inherited genetic diseases independently made paradigm-shifting discoveries. In one case, a CAG trinucleotide repeat expansion within the first coding exon of the androgen receptor (*AR*) gene was found to be the cause of an X-linked neuromuscular disorder known as spinal and bulbar muscular atrophy (SBMA or Kennedy's disease). CAG encodes the amino acid glutamine; thus, elongation of a polyglutamine tract within the AR protein was hypothesized to be the molecular basis for the motor neuron degeneration in SBMA. In the other case, a disorder known as the fragile X syndrome of mental retardation (FRAXA), also X-linked but much more common, was reported to result from expansion of a CGG repeat. In the latter case, although originally envisioned to encode a polyarginine tract, the CGG repeat turned out to be in the 5′ untranslated region of a novel gene, the so-called *FMR-1* gene (for "<u>f</u>ragile X <u>m</u>ental <u>r</u>etardation-1").

The identification of triplet repeat expansions as the cause of two inherited diseases was an exciting turning point in the field of human molecular genetics not only because of the novel nature of these findings, but also because of the unusual genetic characteristics of this new type of mutation. Analysis of families segregating FRAXA revealed the existence of three distinct allele categories defined by the length of the pure CGG repeat: a normal size range; a disease size range; and an intermediate, "premutation" size range. As discussed below, individuals carrying premutation-sized CGG repeats never develop the mental retardation phenotype, but instead are at risk for passing on even larger CGG repeats to their children or grandchildren who then display the mental retardation phenotype. An important tenet that emerged from these early studies was that expanded CGG repeats (larger than the normal size range) displayed an exceptionally high rate of further mutation. This observation reversed a commonly held view of the genetic material – the notion that any single nucleotide in the human genome displayed a mutation rate of $\sim 1 \times 10^{-5}$. In the case of expanded CGG repeats in premutation carriers, the rate approached unity (10^0)! Besides displaying this high mutation rate, there were other unusual features: (1) the CGG repeats showed a marked tendency to further expansion, suggesting that repeat mutation was a polar process; and (2) the sex of the individual transmitting the premutation-sized CGG allele determined whether a large expansion into the disease range would be possible. For FRAXA, expansion into the disease range could only occur if the premutation allele was transmitted by a female carrier. All of these unusual aspects of FRAXA CGG repeat genetics thoroughly accounted for the bizarre non-Mendelian inheritance patterns described in FRAXA families as the "Sherman paradox."

The recognition of repeat expansion as the cause of the neurodegenerative disorder SBMA and as an explanation for the puzzling genetics of FRAXA set the stage for further discoveries in the field of neurogenetics (the study of inherited neurological disorders). One disorder known as *myotonic dystrophy* (DM), the most common of all muscular dystrophies, had been the center of a genetic controversy that had gone on for nearly a century. The controversy involved the debate over whether the clinical phenomenon of anticipation truly existed or was simply an artifact of clinical study (Anticipation may be defined as a progressively earlier age of disease onset with increasing disease severity in successive generations of a family segregating an inherited disorder). Although anticipation was initially proposed as a defining feature of DM in 1918, a number of leading geneticists, among them Penrose, L.S., dismissed its authenticity, claiming that it was a product of ascertainment bias due to better clinically defining the profound variable expressivity in this disorder. This view persisted from its promulgation in the 1950s, reinforced by the concept that the genetic material is seldom subject to alteration or modification that could be heritable. However, with the discovery that expanded trinucleotide repeats could further expand, and that indeed the expansion process was a prerequisite for the FRAXA disease phenotype – accounting for maternal inheritance and greater percentages of affected individuals in more recent generations – a role for repeats in diseases displaying anticipation was entertained. Consequently, the third repeat mutation disorder to be identified – just one year after the SBMA and FRAXA discoveries – was DM. Studies of the causal CTG repeat expansion in DM demonstrated a strong correlation between disease severity (i.e. age of onset and rate of progression) and the length of the CTG repeat. In this way, it became clear that anticipation was a genuine phenomenon and that expanded repeat mutational instability was its long awaited molecular explanation. The mutational instability characteristics of expanded repeats have led to their designation as so-called "dynamic mutations."

2
Repeat Diseases and Their Classification

2.1
Summary of Repeat Diseases

The list of diseases caused by microsatellite repeat expansions (involving repeats of

3–12 bp) now includes more than 25 disorders (Table 1). Certain aspects of the list of repeat diseases deserve emphasis. Twenty of the repeat diseases either principally or exclusively affect the neuraxis – that is, anywhere from the brain to the cerebellum/brainstem, to the spinal cord, to the peripheral nerve or muscle – and are degenerative disorders. Almost all of these inherited neurological disorders are caused by large expanded repeats that display the property of pronounced genetic instability (dynamic mutation). On the other hand, the developmental malformation syndromes for which repeat expansion mutations have been implicated all involve modestly sized disease repeats by comparison, and these repeats do not exhibit such dynamic mutation genetic instability.

2.2
Differences in Repeat Sequence Composition and Location within Gene

When faced with the task of categorizing the various repeat expansion diseases into different classes, a number of approaches are possible. We have found that consideration of the sequence of the repeat and its location within the gene are the most useful characteristics to apply for grouping the different repeat diseases. As shown in Table 1, there are many different types of repeats varying in length and sequence composition. However, among the recurrent sequence types are CAG trinucleotide repeats, CTG trinucleotide repeats, and GCG trinucleotide repeats. The rest of the repeat sequences are unique in composition and differ widely in size, ranging from 3 to 12 bp as noted above. Comparison of the location of a repeat within the gene it is affecting also yields different types of repeats. The largest single-repeat location category is within the coding region of a gene, which applies to both CAG (glutamine) and GCG (alanine) repeats, and has been proposed but not yet demonstrated for CTG (leucine) repeats. The rest of the locations defined for repeats vary widely, ranging from the gene's promoter to its 5′ untranslated region to an intron to the 3′ untranslated region, and no more than two to three repeat expansions can be placed in each of these categories at this time.

2.3
Classification Based upon Mechanism of Pathogenesis and Nature of Mutation

To allow us to reconstruct how the different repeat expansion mutations cause molecular pathology in the various disorders that they cause, we have chosen to categorize the 25 repeat disorders into four classes (Table 2). The first class of disorders, the Type 1 repeat diseases, are the "CAG-polyglutamine disorders." This class of repeat diseases includes nine inherited neurodegenerative disorders (SBMA, Huntington's disease, dentatorubral pallidoluysian atrophy, and six forms of spinocerebellar ataxia) that all share the common feature of being caused by a CAG repeat located within the coding region of a gene. Upon CAG repeat expansion, a mutant protein with an extended polyglutamine tract is produced, making the protein then adopt an abnormal conformation and misfold to initiate the pathogenic cascade. The resultant pathology is believed to primarily stem from a gain of function of the mutant protein imparted by the expanded polyglutamine tract. As discussed below, much effort has gone into trying to define what the gain-of-function effect is for each polyglutamine disease protein and into determining if

Tab. 1 Compilation of repeat expansion diseases in humans.

Disease	Gene	Repeat	Normal	Premutation	Disease
Blepharophimosis-ptosis-epicanthus inversus	FOXL2	GCG	14	–	24
Cleidocranial dysplasia	RUNX2	GCG	11–17	–	27
Congenital central hypoventilation	PHOX-2B	GCG	20	–	25–29
Dentatorubralpallidoluysian atrophy (Haw River syndrome)	Atrophin-1	CAG	7–35	–	49–88
Fragile X syndrome of mental retardation	Fmr-1	CGG	6–53	45–200[a]	200–>2000
FRAXE mental retardation	Fmr-2	GCC	6–40	61–200	200–1000
Friedreich's ataxia	Frataxin	GAA	7–38	33–60	66–1700
Hand-foot-genital	HOX-A13	GCG	18	–	24–27
Huntington's disease	Htt/IT15	CAG	6–35	27–36	39–250
Huntington's disease like 2	JPH3	CAG	7–27	–	50–60
Holoprosencephaly	ZIC2	GCG	15	–	25
Myoclonus epilepsy type 1	Cystatin B	CCCCGCCCCGCG	2–3	12–17	30–150
Myotonic dystrophy type 1	DMPK	CTG	5–37	50–80	50–4000
Myotonic dystrophy type 2	ZNF9	CCTG	104–176	?	175–11 000
Oculopharyngeal muscular dystrophy	PABP2	GCG	6–7	–	8–17
Spinocerebellar ataxia type 1	Ataxin-1	CAG	6–44	–	39–>10
Spinocerebellar ataxia type 2	Ataxin-2	CAG	13–33	–	31–>10
Spinocerebellar ataxia type 3 (Machado-Joseph disease)	Ataxin-3	CAG	12–40	–	55–84
Spinocerebellar ataxia type 6	CACNA1A	CAG	4–18	–	19–33
Spinocerebellar ataxia type 7	Ataxin-7	CAG	4–35	–	37–306
Spinocerebellar ataxia type 8	?	CTG	16–>800	?	71–>1000
Spinocerebellar ataxia type 10	Ataxin-10	ATTCT	10–22	?	800–4500
Spinocerebellar ataxia type 12	PPP2R2B	CAG	7–32	–	55–78
Spinocerebellar ataxia type 17	TBP	CAG	25–48	–	43–66
Syndromic and nonsyndromic X-linked mental retardation	ARX	GCG	10–16	–	17–23
Synpolydactyly	HOX-D13	GCG	9–15	–	16–29
X-linked spinal & bulbar muscular atrophy	AR	CAG	5–34	–	37–70

[a] Male premutation carriers of the FMR-1 CGG repeat develop the Fragile X tremor-ataxia syndrome.
–, no premutation alleles identified, so they may not exist.
?, existence of premutation alleles is unknown.

Tab. 2 Classification of the repeat expansion diseases.

CAG/polyglutamine disorders	Loss-of-function disorders	RNA gain-of-function disorders	Polyalanine disorders
Dentatorubral pallidoluysian atrophy	Fragile X syndrome of MR	Myotonic dystrophy type 1	Blepharophimosis-ptosis-epicanthus inversus
Huntington's disease	FRAXE MR	Myotonic dystrophy type 2	Cleidocranial dysplasia
Spinal and bulbar muscular atrophy	Friedreich's ataxia	Spinocerebellar ataxia type 8	Congenital central hypoventilation
Spinocerebellar ataxia type 1	Myoclonus epilepsy type 1	Fragile X tremor-ataxia syndrome	Hand-foot-genital syndrome
Spinocerebellar ataxia type 2			Holoprosencephaly
Spinocerebellar ataxia type 3			Oculopharyngeal muscular dystrophy
Spinocerebellar ataxia type 6			Synpolydactyly
Spinocerebellar ataxia type 7			X-linked MR (syndromic & nonsyndromic)
Spinocerebellar ataxia type 17			

MR = mental retardation.

shared pathways of toxicity are initiated in the different diseases. At least one, and perhaps a greater number of these disorders may principally involve a simultaneous dominant-negative partial loss of the normal function of the disease protein. The next class of disorders, the Type 2 repeat diseases, are a much more disparate group of repeat disorders. The Type 2 repeat diseases are the "Loss-of-function repeat disorders." These disorders include different repeats that vary in sequence composition and gene location, but share a final common pathway of disease pathogenesis – a loss of function of the disease gene within which they occur. This group includes various classic trinucleotide repeat disorders such as the two fragile X syndromes of mental retardation (FRAXA and FRAXE) and Friedreich's ataxia – but also encompasses the dodecamer repeat expansion in progressive myoclonic epilepsy type 1, and possibly the CAG repeat expansion in Huntington's disease like-2 (*HDL2*) gene. Strong evidence for a loss-of-function pathway in the form of nonrepeat loss-of-function mutations supports many of these classifications. The third group of repeat diseases, the Type 3 disorders, comprise a shared class because all of them have been proposed to involve the production of a toxic RNA species. This category of repeat diseases is thus called the *RNA gain-of-function disorders*. Included among these disorders are two closely related forms of DM, the common and classic myotonic dystrophy type 1 (DM1) and its uncommon phenocopy, myotonic dystrophy type 2 (DM2). Another member of this group is the recently described fragile X tremor-ataxia syndrome (FXTAS) in male premutation carriers – a fascinating example of two different disease pathways operating upon the same expanded repeat mutation based upon size range differences. One form of spinocerebellar ataxia (SCA8) with an unclear mechanism of pathogenesis has also been provisionally placed into this category, based upon current working models of how its repeat causes disease. The last class of repeat disease, the Type 4 disorders, are the "GCG-polyalanine disorders" that are grouped together because all involve short GCG repeat tracts falling within the coding regions of unrelated genes that become expanded to moderately sized GCG repeats. With the exception of oculopharyngeal muscular dystrophy, all are developmental malformation syndromes, and while gain-of-function polyalanine toxicity has been proposed for a number of these disorders, loss of function due to the polyalanine expansion seems more likely for others. Finally, a number of repeat disorders, spinocerebellar ataxia type 10 (SCA10), spinocerebellar ataxia type 12 (SCA12), and Huntington's disease like 2 (HDL2), currently defy classification because very little is known about their molecular basis. These diseases will be considered in the final section of this chapter.

3
Type 1: The CAG/Polyglutamine Repeat Diseases

3.1
Spinal and Bulbar Muscular Atrophy

Spinal and bulbar muscular atrophy (SBMA; Kennedy's disease) is a late-onset neurodegenerative disease with an inheritance pattern resembling X-linked recessive. It has a prevalence of about 1 in 50 000 males. Patients suffer a late-onset, progressive degeneration of primarily lower motor neurons in the anterior horn of the spinal cord and in the bulbar region of the brainstem; however, variable involvement of sensory neurons in the dorsal root ganglia also occurs. SBMA typically presents with cramps, followed by proximal muscle weakness and atrophy. Patients often exhibit dysarthria, dysphagia, and fasciculations of the tongue and lips. Affected individuals have symptoms of mild androgen insensitivity, such as gynecomastia, reduced fertility, and testicular atrophy.

SBMA is caused by a polymorphic (CAG)n repeat in the first exon of the *AR* gene, which is expressed as a glutamine tract. Unaffected individuals carry 5 to 34 triplet repeats, while affected individuals carry 37 to 70 repeats. SBMA exhibits a paternal expansion bias. The disease does not appear to involve a simple loss-of-function mechanism, as complete loss of AR does not result in motor neuron degeneration.

AR is widely expressed in males and females, and is a member of the steroid receptor–thyroid receptor superfamily with a highly conserved DNA binding domain, ligand binding domain and transactivation domain (Fig. 1). In its inactive state, it forms an apo-receptor complex with heat-shock proteins (HSPs) 70 and 90, and resides in the cytoplasm. Upon binding androgen, it dissociates from these HSP chaperone proteins and translocates to the nucleus. Once in the nucleus, AR dimers transactivate certain genes, many of which are responsible for generating and maintaining male characteristics. Although the glutamine expansion does not affect the binding of its ligand, androgen (testosterone), the glutamine tract is in the major transactivation domain, and may affect transactivation competence. However, the effect of the polyglutamine expansion upon AR transactivation competence remains controversial.

Many lines of evidence suggest that AR must translocate to the nucleus to exert its toxicity. Nuclear inclusions (NIs) are present in motor neurons of the spinal cord and brainstem in SBMA patients. In a Drosophila model of SBMA, retinal expression of mutant AR only yielded a degenerative phenotype in the presence of ligand. As in humans, androgen binding causes the nuclear translocation of AR in mice. Transgenic male SBMA mice produce testosterone and will only

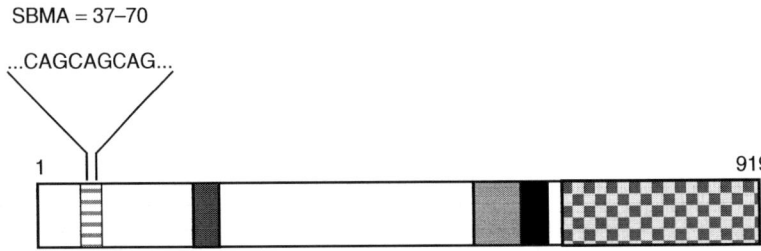

Fig. 1 Diagram of the androgen receptor. The androgen receptor is a member of the steroid receptor-thyroid hormone receptor superfamily, and consequently displays a stereotypical architecture. The CAG repeat – polyglutamine tract (striped box) resides within the amino-terminal domain, which mediates transcription activation through an "activation function" domain (charcoal gray box). Additional conserved domains include the DNA binding domain (light gray box), nuclear localization signal (black box), and the ligand binding domain (checkered box). Expansion of the CAG repeat to alleles of ≥37 triplets is the cause of spinal and bulbar muscular atrophy (SBMA).

develop motor neuron disease if expressing polyglutamine-expanded full-length AR protein, yet when castrated, such SBMA transgenic mice do not develop a phenotype. In the same study, female transgenic mice hemizygous for expanded AR develop motor neuron degeneration when exposed to exogenous androgen. Importantly, even rare human females who are homozygous for an AR CAG repeat expansion mutation do not develop SBMA, despite widespread expression of mutant AR throughout the CNS (central nervous system). Thus, SBMA is not a true X-linked recessive disorder, but rather is classified as a sex-limited disorder, since expression of the disease phenotype is dependent upon male levels of androgen.

A number of studies on SBMA patients and mouse models have characterized the NIs seen in this disease. While NIs are widely distributed in lower motor neurons of the spinal cord and brainstem, NIs also occur in a variety of nonneuronal tissues that appear to function normally. NIs colocalize with components of the proteasome and with molecular chaperones. How this contributes to the phenotype is unknown, although HSP70 overexpression attenuates toxicity in cell culture and in transgenic mouse. The presence of proteasome components and HSPs in NIs may thus be revealing a protective intervention by the cell, or alternatively may be deleterious due to depletion of these important cellular proteins. Interestingly, only antibodies raised to amino-terminal fragments of AR detect NIs, indicating that proteolysis may play a role in the disorder. Caspase-3 has been shown to cleave AR in a polyglutamine tract length-dependent manner *in vitro*, and this may have pathogenic significance, as truncated AR is more toxic than full-length protein in cell culture studies and transgenic mouse models. The phosphorylation of AR is modulated by androgen, and this posttranslational modification appears to enhance caspase-3 cleavage.

One possible mechanism of expanded AR toxicity is through transcription interference. Polyglutamine-expanded AR interacts with the transcription coactivator CREB-binding protein (CBP) in a polyglutamine tract length-dependent manner, colocalizes with CBP in spinal cord NIs from patients, and can interfere with CBP-dependent transcription. CBP is a transcription cofactor that regulates the expression of vascular endothelial growth factor (VEGF), among other genes. VEGF is important in motor neuron health, as deletion of a portion of its promoter called the *hypoxia response element* (HRE) causes motor neuron degeneration even in normoxic mice. Pathologically expanded AR reduces VEGF transcript expression in males, with VEGF165 isoform expression reduced at both the RNA and protein levels. Adding VEGF165 to a motor neuron-like cell line (MN-1) expressing polyglutamine-expanded AR significantly rescues its cell death, again supporting the role of transcription interference in SBMA and suggesting that VEGF165 may serve as a neurotrophic factor for motor neurons.

While no effective treatment for SBMA has yet been validated in human patients, the role of testosterone in SBMA disease progression has received considerable attention. Interestingly, testosterone supplementation was initially used as a treatment for SBMA. Rather than aggravating the disease as one might fear, it was reported to attenuate the phenotype slightly, but did not significantly retard disease progression. Such a beneficial effect of testosterone may be due to anabolic effects on muscle strength or to a downregulation of AR in response to elevated androgen levels. Still another possibility is that AR-mediated transcription in motor neurons somehow performs a trophic function in the face of damage or injury. Very recent work, however, strongly indicates that elimination of ligand by surgical or pharmacological castration is a very effective treatment in SBMA transgenic mouse models, even showing efficacy when SBMA mice display an advanced phenotype. As ligand binding is associated with nuclear translocation of mutant AR and aberrant effects upon nuclear transcription appear crucial for SBMA disease progression, abrogation of nuclear localization may account for the success of castration. Consistent with this hypothesis are results of studies with flutamide, a drug that appears to block AR-dependent transactivation without preventing its nuclear translocation. While pharmacological castration with leuprolide is highly effective therapeutically, flutamide does not ameliorate symptoms or disease progression in an SBMA mouse model. Attempted translation of these preclinical trial results to human SBMA patients is currently underway.

3.2
Huntington's Disease

Huntington's disease (HD) is an autosomal dominant disorder with a prevalence of 1 per 15 000 persons worldwide. It is a debilitating disease that often presents clinically in the fourth or fifth decade of life with chorea (i.e. spontaneous, involuntary dancelike movements). Personality change and cognitive impairment may precede the clinical onset by years, and ultimately culminate in dementia after onset of the movement disorder. Chorea gives way to bradykinesia and rigidity late in the disease. CNS atrophy occurs most prominently in the striatum, which is reduced to a fraction of its original size (Fig. 2). However, significant neurodegeneration and neuron loss in the cortex is also typical, while cerebellum, brainstem, and spinal cord are relatively spared – except in juvenile-onset cases.

HD is caused by a (CAG)n repeat expansion in the *huntingtin* gene (*htt*), which encodes a 350-kDa protein containing 67 exons. Unaffected individuals carry 6 to 35 repeats, while affected individuals carry 39 to 250 repeats. The largest repeats cause juvenile-onset HD and display a paternal transmission bias. The htt protein is ubiquitously expressed in brain tissue,

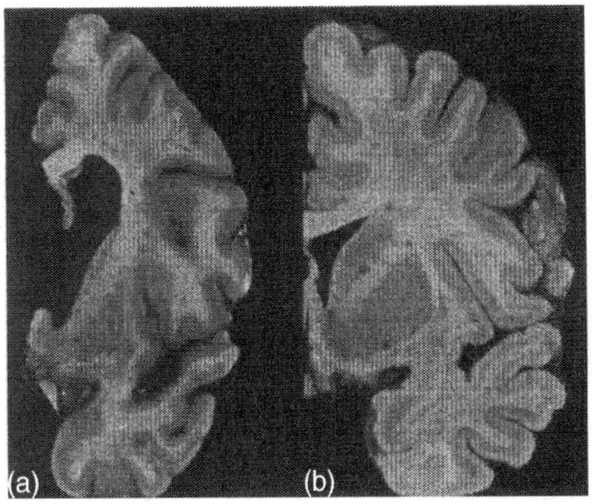

Fig. 2 Huntington's disease (HD) neuropathology. Hemi-coronal sections of postmortem brains from (a) a classic, adult-onset HD patient and (b) a normal control reveal marked degeneration of the striatum (midmedial region) and considerable atrophy of cortical regions. (From Robataille et al. (1997) *Brain pathol.* **7**, 901. Used with permission).

with highest levels in striatal interneurons and cortical pyramidal cells. Two htt mRNA transcripts have been detected: one 10 kb and the other 13 kb in length, expressed most highly in CNS neurons. Ultrastructurally, the wild-type full-length protein is predominantly cytoplasmic and is located in the pre- and postsynaptic regions of dendrites and axons. Htt protein associates with microtubules, vesicles, and organelles.

HD appears to have a predominantly dominant mechanism due to its inheritance pattern, evidence that homozygotes are no more severely affected than heterozygotes, and the observation that heterozygous deletion of huntingtin does not cause HD. The fact that no HD-causing loss-of-function mutations have been documented in the huge *htt* gene further supports a gain-of-function mechanism. However, postnatal elimination of htt expression in regions of the cortex can cause striatal degeneration in mice, so the notion that gain of function fully accounts for the HD phenotype is being reexamined. As htt is a regulator of transcription activation and/or mediator of vesicular brain-derived neurotrophic factor (BDNF) transport up and down axons, many investigators now envision the effects of expanded htt as twofold: simultaneously causing protein misfolding leading to gain-of-function toxicity, and inducing partial loss of an ill-defined normal function.

In 1997, it was first reported that mutant htt forms dense amyloid-type aggregates in the nucleus, perikarya, and neuropil of neurons from a transgenic mouse model and in human patients. The role of aggregation in the polyglutamine diseases and in a wide range of neurodegenerative disorders including Alzheimer's disease, Parkinson's disease, amyotrophic lateral sclerosis, and the prion diseases, thus emerged as an important theme at the end of the last decade. While initially it was thought that the formation of large aggregates is the basis of polyglutamine neurotoxicity, the weight of evidence now suggests little correlation between visible aggregate formation and disease pathology. However, the occurrence of aggregates along with neuropathology suggests that aggregation is inextricably linked to the pathogenicity of polyglutamine disease proteins. We will address the role of polyglutamine aggregation in neurotoxicity in detail in a separate section.

Many theories have been proposed to account for how polyglutamine-expanded htt causes neurotoxicity. A thorough discussion of this literature goes beyond the scope of this chapter, so the reader is referred to the relevant books and reviews in the Bibliography for a more intensive treatment of this topic. Major theories of htt neurotoxicity that will be considered herein include transcription dysregulation, proteasome inhibition, mitochondrial dysfunction/excitotoxicity, and proteolytic cleavage. As multiple independent toxicity events may be occurring concomitantly, these disease pathways are not mutually exclusive.

Expanded htt protein may interfere with transcriptional processes. Many transcription factors, such as CBP, contain glutamine tracts that mediate important protein–protein interactions. CBP interacts directly with htt, and mutant htt toxicity is ameliorated in striatal neurons *in vitro* when CBP lacking the htt interaction domain is overexpressed. CBP mediates the transcription of a number of neuronal survival factors, such as BDNF, and functions by acetylating histones, one aspect of chromatin remodeling that permits transcription to occur. Studies of HD fruit fly and mouse models have highlighted the potential importance of histone acetylation by demonstrating that histone deacetylase inhibitors (HDAC Is) can successfully rescue degenerative phenotypes in these model organisms. Htt can affect transcription mediated by p53, Sp1, and REST interaction, thereby potentially altering the expression of a large number of genes. Differential expression of survival factors, neurotransmitter receptors, and a number of other genes may thus contribute to HD pathogenesis.

The ubiquitin-proteasome protein degradation pathway has also been implicated in HD. Nuclear inclusions of htt colocalize with ubiquitin, which indicates that the expanded protein has been identified as misfolded and thus targeted for proteasomal degradation. However, polyQ sequences longer than 9 glutamines are impossible for eukaryotic proteasomes to cleave. The proteasomes of htt-transfected cells are consequently less capable of degrading proteins other than htt, as demonstrated by the reduced degradation of GFP-tagged proteins in culture. If proteasome components are clogged with mutant protein and/or sequestered into inclusions, they may be unable to degrade other misfolded or damaged proteins that carry out important functions. Alternatively, accumulation of improperly degraded proteins may interfere with other normal cellular processes, such as autophagy or mitochondrial oxidative phosphorylation.

Perhaps the longest and most thoroughly studied htt toxicity pathway involves metabolic disturbances and excitotoxicity. Glucose metabolism and oxygen consumption are reduced in HD brains as measured by PET. Severe deficits in the activity of complexes II/III and IV of the mitochondrial electron transport chain (ETC) are evident in HD brains. Inhibitors of complex II of the mitochondrial electron transport chain, such as 3-nitropropionic acid (3-NPA), can cause a selective degeneration of the striatum in rat and primate models when injected systemically, since the striatum has among the highest energy demands of all neuronal regions. 3-NPA reduces levels of ATP, resulting in mitochondrial and cellular depolarization with activation of voltage-dependent NMDA receptors. Excitotoxic damage may act in concert with increased production of free radicals to cause selective striatal degeneration in HD. Some of the earliest animal models of HD were thus generated by exposing the striatum or entire brain of rats or primates to metabolic or excitotoxic insults. In 1976, the first such model of HD was created by injection of the glutamate analog and excitotoxin kainic acid into the striatum of rats. Such lesioned rats exhibited a selective degeneration of neurons in the striatum reminiscent of HD. Recent studies have suggested that impaired Ca^{++} flux due to aberrant interaction of htt with the inositol phosphate-3 receptor (IP3-R) may underlie excitotoxic pathology. Studies of mitochondria from HD patients reveal abnormal mitochondrial Ca^{++} handling and decreased mitochondrial depolarization thresholds, in support of this hypothesis. One very recent study has found that deletion of peroxisome proliferator-activated receptor (PPAR) gamma coactivator 1 alpha (PGC-1α), a key mediator of mitochondrial biogenesis, yields HD-like striatal degeneration in mice. Thus, metabolic insults and excitotoxicity may be key steps in HD disease pathogenesis.

Another important theory of htt neurotoxicity posits that proteolytic cleavage of htt is a required step in neuronal dysfunction and the degeneration process. As noted above, the htt protein is enormous, with full-length product consisting of >3100 amino acids. Analysis of human

HD material has indicated an absence of midprotein and C-terminal epitopes in htt aggregates. Careful biochemical studies of htt have characterized a variety of putative caspase and calpain cleavage sites. Various studies suggest that the more truncated the polyglutamine-expanded htt protein, the more toxic it is in cell culture and in animal models, and the more likely it is to enter the nucleus and produce toxicity there. In the case of htt, a series of proteolytic cleavage steps culminating with cleavage to an ~100 amino acid peptide fragment by an aspartyl protease has been proposed to yield a final "toxic fragment." As it turns out, the "toxic fragment hypothesis" (as it has also been called) may be applicable to a number of polyQ diseases, which will be reviewed later in this section.

3.3
Dentatorubral Pallidoluysian Atrophy

Dentatorubral pallidoluysian atrophy (DRPLA) is a rare, autosomal dominant neurodegenerative disorder most prevalent in Japan. Adult-onset DRPLA typically involves progressive cerebellar ataxia, choreoatheosis, epilepsy, and dementia, while juvenile-onset cases also display myoclonus, epilepsy, and mental retardation. Neuropathological abnormalities include degeneration of the dentate nucleus of the cerebellum, rubral nucleus, and globus pallidus, as well as a more generalized degeneration and gliosis involving the brainstem, cerebellum, cortex, and pons. DRPLA is caused by a polymorphic $(CAG)n$ repeat in the carboxy-terminal coding region of the *atrophin-1* gene on chromosome 12. Normal individuals carry 6–35 repeats, while affected individuals carry 49–88 repeats. Anticipation is prominent in DRPLA, as very large repeat expansions can occur in a single generation, typically via paternal transmission.

The DRPLA disease protein, atrophin-1, is widely expressed and appears to be predominantly cytoplasmic. Atrophin-1 contains both a putative nuclear localization signal (NLS) and nuclear export signal (NES), and nuclear localization of normal atrophin-1 is observed. Nuclear localization of polyglutamine-expanded atrophin-1 has been linked to increased toxicity in cell culture models. Ubiquitinated NIs are present in neurons and glia from patient brains, and are also immunoreactive for small ubiquitinlike modifier (SUMO) protein. Atrophin-1 is a substrate for c-Jun N-terminal kinase (JNK), with JNK's affinity for atrophin-1 inversely proportional to the size of the polyglutamine tract expansion.

Although the relevance of JNK phosphorylation to atrophin-1 function is unknown, other insights into the function of atrophin-1 have been reported. Using *Drosophila melanogaster* as a model system, Zhang et al. took advantage of the existence of a fly ortholog of atrophin-1 (*Atro*) and created lines of flies carrying mutations in the *Atro* gene. These flies demonstrated severe developmental abnormalities due to complex patterning defects, and subsequent experiments revealed that *Atro* is a transcription corepressor whose activity diminishes with increasing polyglutamine tract length. Independent studies of human atrophin-1 in cell culture studies found evidence for transcription interference of polyglutamine-expanded atrophin-1 with CREB-mediated gene activation. Transcription dysregulation may thus play a prominent role in DRPLA. Another study suggests that disturbed carbohydrate metabolism may contribute to the DRPLA neurodegenerative phenotype.

3.4 Spinocerebellar Ataxia Type 1

Spinocerebellar ataxia type 1 (SCA1) is an autosomal dominant disorder characterized primarily by a progressive cerebellar ataxia. It accounts for about 6% of all autosomal dominant cerebellar ataxias (ADCAs) worldwide. SCA1 patients suffer from coordination difficulties including dysarthria, dysphagia, and ophthalmoplegia. The cerebellum undergoes atrophy, gliosis, and severe loss of Purkinje cells, accompanied by degeneration of the dentate nucleus, inferior olive, and some brainstem nuclei. Disease onset typically occurs in the third or fourth decade of life, but presentation in childhood or adolescence to late life may be seen. SCA1 is caused by the expansion of a coding $(CAG)n$ repeat in the amino-terminal coding region of the *ataxin-1* gene. The ataxin-1 CAG repeat is highly polymorphic, ranging in size from 6 to 44 triplets in unaffected individuals. The repeat length associated with disease ranges from 39 to more than 100 CAGs. Unaffected individuals with more than 20 repeats have CAT triplet interruptions within their $(CAG)n$ repeat tracts. Such interruptions stabilize the repeat expansion, while absence of the CAT repeat interspersion is noted in SCA1 patient alleles. As in many other polyglutamine disorders, there is strong evidence supporting a gain-of-function mechanism in SCA1.

Ataxin-1 is widely expressed in the CNS and throughout the periphery, although expression levels are several-fold higher in nervous system tissues. The protein is predominantly nuclear in the CNS, however, some cytoplasmic staining is apparent in Purkinje cells of the cerebellum and in brainstem nuclei. Ataxin-1 knockout mice do not develop SCA1, although they do exhibit impairments in motor and spatial learning. These mice also have decreased paired-pulse facilitation in the CA1 region of the hippocampus, suggesting that ataxin-1 may normally function in synaptic plasticity and learning.

Large ataxin-1 containing NI's occur in the brainstem of affected individuals, and are immunoreactive for ubiquitin, the 20 S proteasome subunit, and HSPs HDJ2 and Hsc70. Work done on SCA1 in transgenic mice by the Orr and Zoghbi laboratories has been crucial in formulating models of not only SCA1 disease pathogenesis but also for influencing views of the molecular basis of all polyglutamine diseases. Indeed, the first mouse model for a polyglutamine disease was generated by transgenic overexpression of polyglutamine-expanded ataxin-1 in Purkinje cells (Fig. 3). This SCA1 transgenic mouse model has laid the foundation for

Fig. 3 The original spinocerebellar ataxia type 1 (SCA1) mouse model. (a) Diagram of the *Pcp2-SCA1* transgene construct. A Purkinje cell-specific expression cassette based upon inclusion of the promoter (straight line), first two noncoding exons (black boxes), and first intron (bent line) of the Purkinje cell protein 2 (*Pcp2*) gene and a SV40 polyadenylation sequence (open box) was the basis for this landmark work. Ataxin-1 cDNAs (gray box) containing either 30 CAGs (control) or 82 CAGs (expanded) were inserted into the *Pcp2* expression cassette. Sites of various PCR primer sets are also shown. (b) *Pcp2-SCA1* CAG-82 mice display ataxia. Still photographs of a 30-week-old *Pcp2-SCA1* CAG-82 mouse from a transgenic line that greatly overexpresses the mutant ataxin-1 transgene illustrates the inability of this mouse to maintain its balance when ambulating. Loss of balance when walking is consistent with the gait ataxia seen in human SCA1 patients. (From Burright et al. (1995) *Cell* **82**, 937, used with permission).

numerous follow-up studies of polyglutamine disease pathogenesis. For example, expression of mutant ataxin-1 lacking an intact self-association domain precluded aggregate formation, but permitted neurotoxicity, demonstrating that NIs are not required for SCA1 in mice. In a later study, crossing of SCA1 transgenic mice with mice lacking a ubiquitin ligase enzyme yielded SCA1 mice incapable of aggregate formation. These mice were more severely affected than their transgenic counterparts

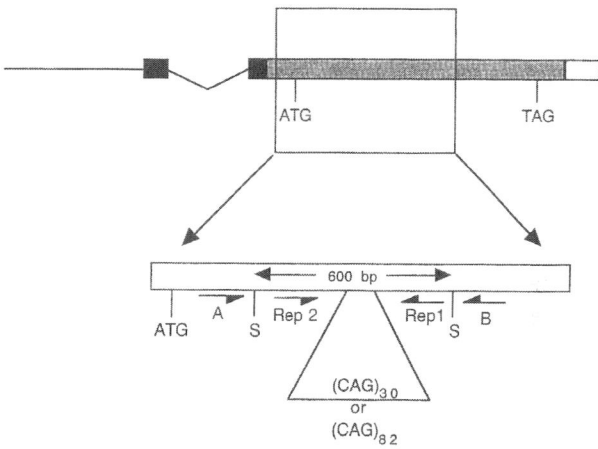

(a)

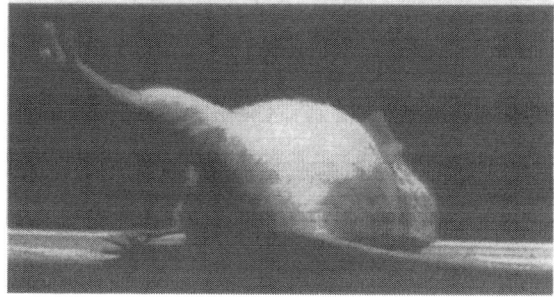

(b)

due to absence of the ubiquitin ligase. This work supported the view that visible aggregate formation may represent a protective cellular response for neutralizing misfolded polyglutamine-containing peptides. In another study, mice expressing ataxin-1 with a mutated NLS were found not to develop SCA1, suggesting that nuclear localization is absolutely required for SCA1 molecular pathology. Although targeting ataxin-1 to Purkinje cells appears sufficient to recapitulate a dramatic SCA1-like disease in mice, a subsequent knockin mouse model of SCA1 indicated that expression in other regions of the CNS yields a more representative disease phenotype.

Over the past few years, numerous leads have emerged in the search for the pathogenic basis of SCA1. In one line of investigation, phosphorylation of serine 776 of the ataxin-1 protein was shown to affect pathogenesis, highlighting the importance of this posttranslational modification. SCA1 transgenic mice in which serine 776 had been mutated to an alanine, exhibited a dramatically attenuated phenotype and a complete lack of NI's. In an accompanying study, interaction of polyglutamine-expanded ataxin-1, but not wild-type ataxin-1, with several isoforms of the phosphoserine/threonine binding protein 14-3-3 was reported. This extremely abundant peptide is thought to serve a regulatory function by binding proteins and determining their subcellular localization, among other things. The interaction between ataxin-1 and 14-3-3 was shown to be dependent upon the Akt-mediated phosphorylation of serine 776. A novel mechanism for ataxin-1 toxicity was thus proposed: upon Akt phosphorylation of polyglutamine-expanded ataxin-1, ataxin-1 binds 14-3-3, is stabilized, and ultimately accumulates in the nucleus. The downstream effects of the nuclear accumulation of mutant ataxin-1 on neuronal function, however, remain undefined.

Transcriptional dysregulation is also a reasonable hypothesis for SCA1 pathogenesis. One study has demonstrated that ataxin-1 interacts with polyglutamine binding protein 1 (PQBP-1), and that this interaction results in interference with RNA polymerase-dependent transcription. Independent studies have supported a role for ataxin-1 as a transcription corepressor. In pull-down assays and in Drosophila, ataxin-1 interacts with the proteins SMRT (silencing mediator of retinoid and thyroid hormone receptors) and HDAC3 (histone deacetylase 3), both transcriptional repressors. Aggregates of polyglutamine-expanded ataxin-1 sequester SMRTER, the Drosophila ortholog of SMRT. Transcription repressors and histone deacetylases also appeared in an earlier screen for modulators of the Ataxin-1 phenotype in the fly. In addition, two separate studies of gene expression alterations in presymptomatic SCA1 transgenic mice have uncovered changes in the levels of transcripts encoding proteins involved in Ca^{++} flux and metabolism. It thus appears that transcription dysregulation may be a key step in SCA1 disease pathogenesis.

3.5
Spinocerebellar Ataxia Type 2

Spinocerebellar ataxia type 2 (SCA2) is an autosomal dominant, progressive cerebellar ataxia that accounts for about 13% of all ADCA cases. Although patients display problems with voluntary coordinated movements as in the rest of the SCAs, its main distinguishing clinical feature is extremely slow saccadic eye movements. Other symptoms may include hyporeflexia, myoclonus, and action tremor. Patients suffer a gradual degeneration of the cerebellum, inferior olive, pons, and spinal cord. Both Purkinje and granule cells degenerate in the cerebellum, and CNS atrophy in some patients can be widespread. Interestingly, in certain cases, there is involvement of the substantia nigra, with such patients displaying a prominent degree of parkinsonism. The disorder is caused by a coding $(CAG)n$ expansion in a novel gene of unknown function, named *ataxin-2*. The SCA2 CAG repeat displays little polymorphism in the normal population, with 95% of the population possessing 22 or 23 repeats in each allele. The remaining 5% of alleles in unaffected individuals do nonetheless range from 15–31 CAG repeats. Such unaffected individuals typically have two CAA interruptions in their CAG repeat tract. Affected SCA2 patients typically display CAG repeats numbering from 32–63 triplets, and such disease alleles are always uninterrupted CAG tracts. There is a nonlinear inverse correlation between expansion size and age of onset in SCA2, with some studies documenting a paternal transmission bias for larger expansions.

Both wild-type and expanded ataxin-2 mRNA is widely expressed, with highest levels in the substantia nigra and Purkinje cells of the cerebellum. The 140-kDa protein is cytoplasmic and its function remains uncertain. Since it contains Sm1 and Sm2 motifs common in proteins involved in RNA splicing and protein–protein interactions, involvement in RNA processing has been proposed. A yeast two-hybrid screen indicated that ataxin-2 has a binding partner, ataxin-2 binding protein-1 (A2BP1). A2BP1 is highly conserved throughout the animal kingdom and its expression pattern corresponds well with that of ataxin-2. A2BP1 contains a domain that is also conserved in RNA-binding proteins, the RNP motif. Thus, a complex including ataxin-2 and A2BP1 may be involved in RNA processing or metabolism.

The *ataxin-2* gene is evolutionarily conserved. The murine ortholog of ataxin-2 does not contain a polyglutamine repeat tract, however, but instead possesses a single glutamine residue at the analogous location. (Absence of a substantial glutamine repeat tract is typical for mouse orthologs of polyglutamine disease proteins.) In the case of ataxin-2, study of the Drosophila ortholog (*Datx2*), which does show two regions of marked amino acid similarity and does contain polyglutamine repeat regions, has yielded some potentially important insights into ataxin-2s normal function. Modulation of *Datx2* dosage resulted in mutant phenotypes whose cause could be traced to aberrant actin filament formation (Fig. 4). This work suggests that alteration of ataxin-2-mediated regulation of cytoskeletal structure could affect dendrite formation or other aspects of neuronal function in SCA2. As SCA2 is one of the few polyglutamine disorders to display prominent cytosolic aggregates, such a model of SCA2 pathogenesis seems plausible. In other experiments, eliminating the *C. elegans* ortholog of ataxin-2 (ATX-2) yielded a lethal phenotype, indicating that ATX-2 is required for early embryonic development of this nematode worm. It remains to be seen how these observations in worms and flies will apply to ataxin-2 function in mammals, especially since no simple or conditional knockout of the mouse *ataxin-2* gene has been performed.

3.6
Spinocerebellar Ataxia Type 3/Machado–Joseph Disease

Spinocerebellar ataxia type 3 (SCA3) is the most common inherited dominant

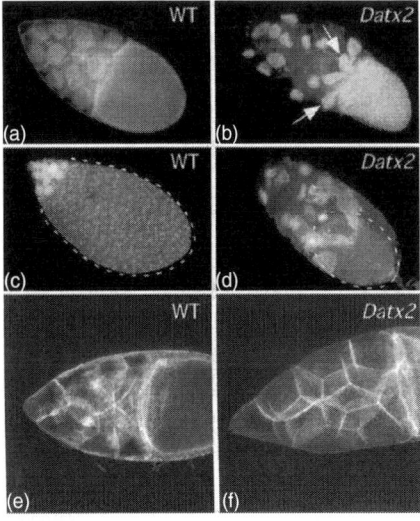

Fig. 4 Studies of the ataxin-2 ortholog in *Drosophila melanogaster* reveals a role for Drosophila ataxin-2 (*Datx2*) in actin filament formation during oogenesis. Egg chambers from normal (Wild-Type (WT)) and mutant flies with reduced expression (*Datx2*) of Drosophila ataxin-2 were stained with DAPI (blue) and phalloidin (red) to indicate nuclei and filamentous actin respectively (a–d). The egg chambers of WT flies prior to cytoplasmic transport (a) display well demarcated, separated but interconnected cells (blue) as expected, while the egg chambers of *Datx2* flies (b) contain irregularly arranged cells. After cytoplasmic transport, the egg chambers of WT flies (c) show one greatly enlarged oocyte (dashed line) with only a small section of compressed cells, while the egg chambers of *Datx2* flies (d) have failed to yield an enlarged oocyte, instead retaining dispersed and large adjacent cells. Confocal images of egg chambers prior to cytoplasmic transport stage reveal a prominent actin filament network in WT flies (e), but a remarkably transparent actin filament network in *Datx2* flies (f). The decreased density of the actin filament network underlies the cytoplasmic "dumping" defect in the *Datx2* flies. (From Satterfield, T.F., Jackson, S.M., Pallanck, L.J. (2002) A Drosophila homolog of the polyglutamine disease gene *SCA2* is a dosage-sensitive regulator of actin filament formation, *Genetics* **162**, 1687–1702, used with permission of *Genetics*.) (See color plate p. xvii).

spinocerebellar ataxia worldwide. SCA3 is also known as Machado–Joseph disease (MJD) because of its initial description in a group of Portuguese residents of the Azores islands. It is a progressive, autosomal dominant cerebellar ataxia whose clinical features include ophthalmoplegia, dystonia, dysarthria, and signs of lower motor neuron disease, such as tongue and facial fasciculations. Degeneration occurs in the spinocerebellar tracts, dentate nuclei, red nuclei, substantia nigra, and spinal cord. This disease is unique among the SCAs because the cerebellar cortex and inferior olive are largely spared. SCA3 is caused by a $(CAG)n$ repeat near the $3'$ end of the coding region of a novel gene (*ataxin-3*). Normal alleles range from 12–40 CAG repeats while affected individuals carry 55–84 repeats. Unlike many other triplet repeat disorders, there is a substantial gap between the largest normal repeat allele and the smallest disease-causing repeat allele. Some researchers have proposed that this could be due to a SCA3 founder effect. The presence of SCA3 in every major racial population worldwide, however, would require the founder to be truly ancient. There is the typical inverse correlation between repeat number and age of onset in SCA3, and in this disease, paternal expansion bias is characteristic, as documented by repeat sizing of sperm.

Ataxin-3 is a ubiquitously expressed 42-kDa protein, making it the smallest of the polyglutamine proteins. Ataxin-3 is highly conserved in eukaryotes, with homology to ENTH and VHS domain proteins involved in regulatory adaptor functions and membrane trafficking. Ataxin-3 has several splice variants which reside in the nucleus and the cytoplasm, and it appears developmentally regulated. The existence of NIs was first documented in the brains of patients with SCA3, and this feature of polyglutamine-expanded ataxin-3 remains evident in SCA3 cell culture and mouse models. Ataxin-3 NIs are ubiquitinated and contain numerous transcription factors.

A number of recent discoveries regarding the domain structure and normal function of ataxin-3 suggest pathways by which polyglutamine repeat expansion could result in disease pathogenesis. Comparison of ataxin-3 amino acid sequences across a wide range of eukaryotic species revealed the presence of an extremely highly conserved amino-terminal sequence that was named the *josephin domain*. Study of this region indicates that it may play a role in aggregate formation in concert with the expanded polyglutamine tract. A rather intriguing feature of ataxin-3 that was discovered independently is the presence of multiple ubiquitin interaction motifs (UIMs), and the demonstration that ataxin-3 is a polyubiquitin binding protein. This suggests a role for ataxin-3 in mediating protein refolding and degradation.

In addition to a possible normal role in protein surveillance, other studies have found ataxin-3 directly interacts with the histone acetyltransferases CBP and p300, and can block histone acetyltransferase activity by inhibiting access of such coactivators to their histone substrates. This appears to be mediated by the interaction of ataxin-3 with histones. *In vitro* and *in vivo* studies of ataxin-3 further revealed an interaction with histones and the chromatin remodeling machinery that led to a repression of transcription activation. Thus, polyglutamine-expanded ataxin-3 may also affect transcriptional processes once it begins to accumulate in the nuclear compartments of the cell types where it is expressed.

3.7
Spinocerebellar Ataxia Type 6

Spinocerebellar ataxia type 6 (SCA6) is an autosomal dominant, slowly progressing cerebellar ataxia that accounts for ~10–20% of ADCA worldwide. It is characterized predominantly by cerebellar dysfunction that may have an episodic component. Other common features may include dysarthria, nystagmus, loss of vibration sense and proprioception, and imbalance. Histopathological changes include loss of Purkinje cells, cerebellar granule neurons, and neurons in the dentate nucleus and inferior olive. SCA6 is caused by a coding $(CAG)n$ expansion in exon 47 of the gene *CACNA1A*, which encodes the α_{1A} subunit of the P/Q-type voltage-gated calcium channel. This gene is located on chromosome 19 at band p13. The SCA6 CAG repeat is small compared to other polyglutamine disorders, with a pathogenic range of only 19–33 triplets. Unaffected individuals carry alleles of 4–18 repeats. There is an inverse correlation between repeat length and age of disease onset, and minimal intergenerational and somatic instability has been reported for the SCA6 repeat expansion in affected patients.

The 9.8 kb *CACNA1A* transcript implicated in SCA6 is expressed throughout the CNS, and most highly in cerebellar Purkinje cells and granule neurons. The α_{1A} subunit is the pore-forming component of the channel, which is important in neurotransmitter release at the synapse. Polyglutamine expansions in this subunit cause variable changes in calcium transmission rates, depending on the system and the β subunit coexpressed. Expansions do not cause a reduction in membrane channel density in HEK293 cells, suggesting that aggregation does not occur at pathogenic repeat lengths. Ubiquitin-negative neuronal inclusions are visible in the cytoplasm of SCA6 patient Purkinje cells, however.

In addition to displaying a disease allele range that does not overlap with the other polyQ diseases, there are several other reasons to suspect that SCA6 is different from the rest of the polyglutamine repeat group. The first difference is the existence of two disorders, episodic ataxia type 2 (EA2) and familial hemiplegic migraine (FMH), that are both allelic to SCA6, as both are caused by point mutations in the *CACNA1A* gene. EA2 resembles SCA6 as affected EA2 individuals suffer from a slowly progressive form of episodic ataxia, accompanied by nystagmus, dysarthria, loss of balance and sometimes cerebellar atrophy. EA2 is usually caused by truncation mutations in *CACNA1A*, resulting in loss-of-function of the calcium channel. CACNA1A knockout mice similarly develop ataxia and late-onset cerebellar atrophy, characteristic of SCA6 and EA2 patient phenotypes. Another reason SCA6 is different from other polyQ disorders is that the CACNA1A protein probably does not undergo a structural change that converts it into an aggregate-prone, beta-sheet adopting, amyloid-like conformer. Indeed, even the largest SCA6 polyglutamine tract is below the threshold required for stable β-pleated sheet formation in other polyglutamine disorders. Consistent with this prediction, channel localization of mutant polyglutamine-expanded CACNA1A protein is not affected in cell culture models, and its channel function is not completely abolished. Cytoplasmic aggregates in patient material are ubiquitin-negative, suggesting that the protein may not be grossly misfolded. This evidence supports a model whereby a dominant-negative loss of function of the P/Q-type voltage-gated calcium channel due to association of the mutant α_{1A} subunit with other subunits causes SCA6. The coincidence that this disorder is caused by a polyQ tract expansion and causes a progressive cerebellar ataxia cannot be ignored, however, more data will be required to soundly refute a possible concomitant toxic gain-of-function effect.

3.8
Spinocerebellar Ataxia Type 7

Spinocerebellar ataxia type 7 (SCA7) is an autosomal dominant, progressive cerebellar ataxia. It is unique among autosomal dominant SCAs, as patients typically develop visual impairment in addition to their cerebellar ataxia. The visual impairment is due to a cone-rod dystrophy that results in retinal degeneration. Patients first develop problems distinguishing colors, but ultimately go blind in this type of retinal degeneration. SCA7 patients may present either with cerebellar ataxia or visual impairment. The likelihood of their presentation is dictated by the size of their CAG repeat disease allele, with larger repeats typically favoring presentation with visual impairment. Affected individuals display prominent dysarthria, and can develop increased reflexes, decreased vibration sense, and oculomotor disturbances. Neuronal degeneration and reactive gliosis occur in the cerebellar cortex, dentate nucleus, inferior olive, pontine nuclei, and occasionally in the basal ganglia. NIs are widespread. Infantile-onset SCA7 has been documented, and in this fatal form of the disease, nonneuronal tissues such as the heart and the kidney are severely affected. SCA7 is caused by a highly polymorphic (CAG)n repeat in the 5′ coding region of the *ataxin-7* gene. Unaffected individuals carry 4–35 repeats, while affected individuals carry 37–306 repeats. The SCA7 trinucleotide repeat is one of the most unstable of all polyglutamine disease genes, sometimes expanding by as much as 250 repeats in a single generation. There is a pronounced paternal expansion bias, with large expansions occurring in male germ cells, frequently causing embryonic lethality that results in reduced transmission. Marked repeat instability can occur in the brain, and produce large somatic expansions on occasion.

Ataxin-7 is a ubiquitously expressed protein of 892 amino acids. It is expressed most highly in heart, skeletal muscle, and pancreas. Expression levels within the CNS are highest in the cerebellum and brainstem. One splice variant, ataxin-7b, is expressed predominantly in the CNS (32). Ataxin-7 contains a functional arrestin domain, a protein interaction domain that is highly selective for phosphorylated forms of its interacting protein(s). This suggests that it interacts with specific phospho-proteins, although none have been discovered to date. It also contains two SH3 domains, which are protein interaction domains that bind proline-rich sequences and mediate a number of cell signaling processes. Ataxin-7 also contains three putative NLSs and one NES.

Ataxin-7 is conserved throughout eukaryotes, and its yeast ortholog SGF73 is part of a multisubunit histone acetyltransferase complex called *SAGA (Spt/Ada/Gcn5 acetyltransferase)*. The human orthologs of SAGA comprise the so-called STAGA (SPT3/TAF9/ADA2/GCN5 acetyltransferase) complex, and are essential transcription coactivators required for the transcription of certain genes. STAGA components immunoprecipitate with ataxin-7. Although pathogenic expansion of ataxin-7 does not alter its ability to be integrated into the STAGA complex, the presence of the polyglutamine-expanded ataxin-7 has a dominant-negative effect upon the GCN5 histone acetyltransferase activity of the STAGA complex, resulting in transcription dysregulation. The transcription dysregulation caused by polyglutamine-expanded ataxin-7 likely causes a disease phenotype by altering the ability of certain transcription factors to activate expression of their target genes.

The best characterized example of transcription dysregulation by polyglutamine-expanded ataxin-7 is its interference with the cone-rod homeobox protein (CRX), a glutamine domain containing transcription factor expressed only in the retina and the pineal gland. Ataxin-7 interacts directly and functionally with CRX, according to studies performed *in vitro* and in a mouse model of SCA7. Importantly, the interaction between ataxin-7 and CRX appears to involve the glutamine tract regions found in both proteins. Autosomal dominant mutations in CRX can cause a cone-rod dystrophy in humans, further supporting a model in which CRX's diminished transactivation competence is central to the SCA7 retinal degeneration phenotype. Several transcription factors, including CBP, can be found in SCA7. CRX may be but one of a number of transcription factors whose function is diminished by polyQ-expanded ataxin-7 interaction and dysregulation of STAGA complex-mediated gene expression.

One intriguing feature of SCA7 was discovered upon generation of transgenic mice expressing ataxin-7 with the mouse prion protein promoter. This promoter drives expression in every tissue, with the occasional notable exception of the Purkinje cells of the cerebellum. Despite lack of Purkinje cell expression of ataxin-7, mice

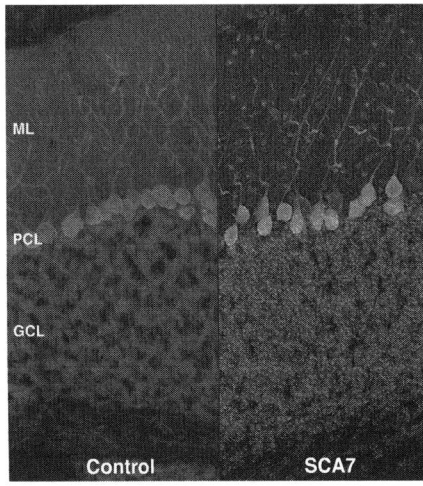

Fig. 5 Noncell autonomous Purkinje cell degeneration in a mouse model of spinocerebellar ataxia type 7 (SCA7). Confocal microscopy analysis of cerebellar sections from a SCA7 transgenic mouse (SCA7) created with an ataxin-7 CAG-92 containing murine prion protein expression vector and from an age- and sex-matched nontransgenic littermate (Control). Staining with an anti-ataxin-7 antibody (magenta), a calbindin antibody (green), and DAPI (blue) reveals a healthy, normal-appearing cerebellum characterized by properly oriented Purkinje cells with extensive dendritic arborization in the "Control" mice. However, SCA7 transgenic mice display pronounced Purkinje cell degeneration as evidenced by decreased dendritic arborization and displacement of Purkinje cell bodies. Interestingly, although numerous neurons in the granule cell layer (GCL) and the molecular layer (ML) display aggregates of ataxin-7, there is no accumulation of mutant ataxin-7 in the degenerating Purkinje cells due to lack of appreciable expression there. As the Purkinje cells degenerate without expressing the mutant protein, the degeneration is described as noncell autonomous. (Adapted from Garden, G.A., Libby, R.T., Fu, Y.H., Kinoshita, Y., Huang, J., Possin, D.E., Smith, A.C., Martinez, R.A., Fine, G.C., Grote, S.K., et al. (2002) Polyglutamine-expanded ataxin-7 promotes noncell-autonomous Purkinje cell degeneration and displays proteolytic cleavage in ataxic transgenic mice, *J. Neurosci.* **22**, 4897–4905, used with permission of the *Journal of Neuroscience*.) (See color plate p. viii).

developed a cerebellar ataxia accompanied by degeneration of the Purkinje cells (Fig. 5). This noncell-autonomous degeneration may point to a disease mechanism involving withdrawal of trophic support by communicating neurons (olivary, deep cerebellar, brainstem, or granule neurons) or dysfunction of glutamate transporters expressed by surrounding glia. Damage to inferior olivary neurons or Bergmann glia can indeed cause the degeneration of Purkinje cells, lending credence to this theory.

Another interesting feature of SCA7 that is common to other neurodegenerative disorders is the prominence of morphological and functional degeneration without pronounced apoptosis. Some neurons in SCA7 humans and mouse models exhibit indentations in the nuclear envelope, reduced arborization, ectopy, and increased autophagy. (Autophagy is the bulk degradation of cellular components by a membrane-bound autophagosome and is accelerated by a number of cellular insults.) Caspase-3, a proteolytic enzyme classically associated with apoptosis, is activated at abnormally high levels in SCA7 patient brains. This may suggest that caspase activation, rather than causing apoptosis, is contributing to a nonapoptotic degenerative process.

In normal human neurons, ataxin-7 is variably located in the nucleus or the cytoplasm. In SCA7 patients, ataxin-7 gradually undergoes a shift in localization into NIs. This process has been replicated in SCA7 transgenic mice, in which ataxin-7 immunoreactivity shifts from the cytoplasm to the nucleus, and ultimately forms foci there. In patient's brains, these NIs colocalize with promyelocytic leukemia (PML) protein, which is an integral part of nuclear bodies (NBs). NBs are associated with transcription regulation and the ubiquitin-proteasome pathway; thus, accumulation of ataxin-7 in NBs may represent an attempt by the cell to degrade misfolded protein.

3.9
Spinocerebellar Ataxia Type 17

Spinocerebellar ataxia type 17 (SCA17) is the most recently discovered polyglutamine repeat disease. It is an autosomal dominant, progressive disorder characterized by dementia as well as cerebellar ataxia and involuntary movement abnormalities. Its symptoms are diverse and heterogeneous, but typically begin with behavioral disturbances and cognitive impairment. This is followed by ataxia, rigidity, dystonia, hyperreflexia, and rarely parkinsonism. Neuropathologically, patients may suffer degeneration of the cortex, cerebellum (including Purkinje cells), inferior olive, caudate nucleus, and medial thalamic nuclei. SCA17 is caused by the expansion of a coding $(CAG)n$ repeat in the TATA-binding protein (TBP) gene on chromosome 6q27. Unaffected individuals carry 25–48 repeats, while affected individuals carry 43–66 repeats. The area of overlap between affected and unaffected alleles indicates incomplete penetrance at intermediate repeat lengths. The pathogenic threshold is higher than for most other polyglutamine disorders, yet there is an inverse correlation between repeat length and age of onset.

TBP is a ubiquitously expressed transcription initiation factor that is a core component of the RNA polymerase II transcription factor D (TFIID) complex. TBP possesses DNA binding activity in the TFIID complex, and is therefore required for the transcription of numerous genes. SCA17 patients have NIs immunoreactive for TBP, polyglutamine, and ubiquitin. Given TBP's central role in transcription regulation, the basis of SCA17 disease pathogenesis is proposed to involve transcription dysregulation, but this is yet to be proven.

3.10
Role of Aggregation in Polyglutamine Disease Pathogenesis

In most polyglutamine disorders and in many neurodegenerative diseases in general, protein aggregation is a prominent feature. It occurs in almost all polyglutamine diseases, despite the lack of domain or structural similarity between the different disease proteins. Aggregates have long been considered reliable indicators of disease, although their pattern and onset often do not correspond well with the cell-type specificity of disease pathology. As the role of aggregation in pathogenesis has been one of the most hotly debated issues in the field of neurodegeneration, it may have implications for Alzheimer's disease, Parkinson's disease, and amyotrophic lateral sclerosis. Why? There are several characteristics of polyglutamine-mediated aggregation that are presumably common to all of the disorders that show aggregation. First of all, aggregates are rich in β-pleated sheets and have many of the properties of amyloid. This is supported by Congo red birefringence and immunoreactivity with antiamyloid antibodies. Various studies have shown that the conformational change is associated with the production of visible aggregates rather than in the soluble nonaggregated phase, where polyglutamine tracts are thought to remain in a random coil conformation.

In vitro, the kinetics of aggregation are consistent with nucleated-growth polymerization, in which the rate-limiting step is the formation of misfolded peptide assemblies, often referred to as *oligomers*. A mechanism involving nucleated-growth polymerization would predict that only one (or a few) visible aggregates would appear in each affected cell, and this appears to be the case for most polyglutamine disorders. In the cellular milieu however, where membranes and intermolecular interactions compartmentalize proteins, this prediction may not be valid. The discovery of polyglutamine microaggregates confirms this suspicion. The threshold for aggregation of polyglutamine tracts is similar to the disease threshold for most polyglutamine diseases. *In vitro*, longer polyglutamine tracts have a lower concentration threshold for aggregation and nucleate more quickly than shorter tracts. This is one possible explanation for the correlation between tract length and disease onset and progression.

Some data support a role for aggregation in polyglutamine disease pathogenesis. Some proponents of this theory invoke the fact that aggregates sequester many proteins besides the disease protein. Intermolecular interactions between transcription factors often involve glutamine tracts or glutamine-rich regions. Some studies have shown that normal proteins with glutamine tracts or glutamine-rich regions are enriched in polyglutamine aggregates in cell culture and animal models. Several transcription factors colocalize with such aggregates, including CBP, TBP, and numerous TBP-associated factors (TAFs). CBP is responsible for the prosurvival effects of BDNF, and its soluble concentration is lowered in HD patient's brains. Postnatal mice lacking CREB and its homolog CREM develop a progressive degeneration of the hippocampus and striatum. Titration of enzymes and factors required for protein refolding and degradation away from the soluble phase and into aggregates has been proposed as a potential cause of cell toxicity in neurons with aggregates. Caspase activation is another way in which aggregation could be linked to pathology. Caspase recruitment into aggregates can lead to their activation, which could result in dysfunction or cell death in neurons. Further supporting the role of aggregates in polyglutamine protein toxicity, injection of preformed polyglutamine aggregates into the nuclei of cultured cells causes cell death, while the injection of nonpolyglutamine aggregates does not. At the same time, there is strong evidence that soluble polyglutamine protein, rather than aggregates, is the primary source of toxicity. Importantly, the pattern of aggregates observed in human patients often does not correlate with the pattern of neuronal dysfunction. For example, in the striatum of HD patients, large interneurons contain aggregates more frequently than medium spiny neurons, yet the former neurons are largely spared, while the latter are most vulnerable. HD transgenic mice expressing full-length mutant htt will develop inclusions in many brain regions many of which are neuropathologically unaffected, while few inclusions are detected in the striatum, the region of the brain displaying the most prominent pathology.

Aggregation and toxicity have been directly dissociated in other polyglutamine disease model systems. Studies of the SCA1 knockin mouse model revealed that neurons lacking aggregates were more susceptible to dysfunction and demise, while those neurons displaying prominent aggregates were protected. Similarly, SCA7 transgenic mice exhibit retinal pathology before the occurrence of visible aggregates. Finally, in a very provocative study, SCA1 transgenic mice lacking the ubiquitin ligase E6-AP were significantly less capable of forming large and numerous aggregates in neurons in comparison to SCA1 transgenic mice on a wild-type background, but displayed earlier onset of an ataxic phenotype and accelerated neurodegeneration. This suggests that the aggregates may be protective, although the toxicity of compromising the proteasome may have contributed to the accelerated phenotype. Thus ensued a contentious debate over the role of aggregate formation in polyglutamine disease – with some workers espousing the view that aggregates were responsible for disease pathology, others suggesting that the aggregation process was a protective coping mechanism of the cell and thereby beneficial, and still others insisting that aggregates were incidental and irrelevant. This debate was complicated by the existence of "microaggregates," small clumps of aggregated protein visible only at the electron microscope level.

Over the last few years, studies deconstructing polyglutamine tract aggregation into a multistep process suggest a parsimonious explanation for the divergent views. Using a variety of biophysical approaches, including transmission electron microscopy (TEM), Fourier transform infrared spectroscopy (FTIR), and atomic force microscopy (AFM), one study dissected the process of huntingtin (htt) exon 1 peptide aggregation and found evidence for a number of sequential morphological and structural intermediates (Fig. 6). By showing that misfolded htt exon 1–44Q adopted intermediate structures, such work opened up the possibility that intermediates (not visible at the light microscope level) are the toxic species and that the ultimate visible aggregated forms of htt exon 1–44Q are neutralized versions of mutant protein. To examine the role of aggregate formation in polyglutamine toxicity, another group then tracked survival time versus diffuse, soluble htt protein expression levels in htt-transfected striatal neurons undergoing aggregate formation. Comparison of neurons that developed aggregates over time versus those that did not confirmed that the level of soluble nonaggregated mutant polyglutamine protein was the more reliable predictor of cellular toxicity. Such studies have shifted our attention to the role of the intermediates (oligomers, protofibrils, etc.) as the toxic species instead of the final, visible aggregates.

A reasonable model for polyglutamine toxicity predicts that the process starts with a protein that misfolds because of the presence of an expanded polyglutamine tract. The misfolded protein is initially detectable in the soluble phase due to the cell's ability to maintain the protein in a properly folded state and direct it to the degradation machinery. However, the refolding capacity of the cell is ultimately exceeded, and since the degradation machinery cannot turnover the misfolded mutant polyglutamine protein, accumulation occurs. Misfolded polyglutamine proteins can spontaneously change their structural properties and adopt abnormal conformations. These abnormally folded proteins can then nucleate, forming oligomers. Oligomers then form protofibrils that grow into fibrils. The transition to the fibril stage is characterized by the attainment of a β-sheet structure, so at this point the structures are amyloid-like. Fibrils then grow into fibers (also known as microaggregates), which then assemble into aggregates visible under a light microscope. According to such a model, blocking an intermediate step could be therapeutically effective. Consistent with this, one

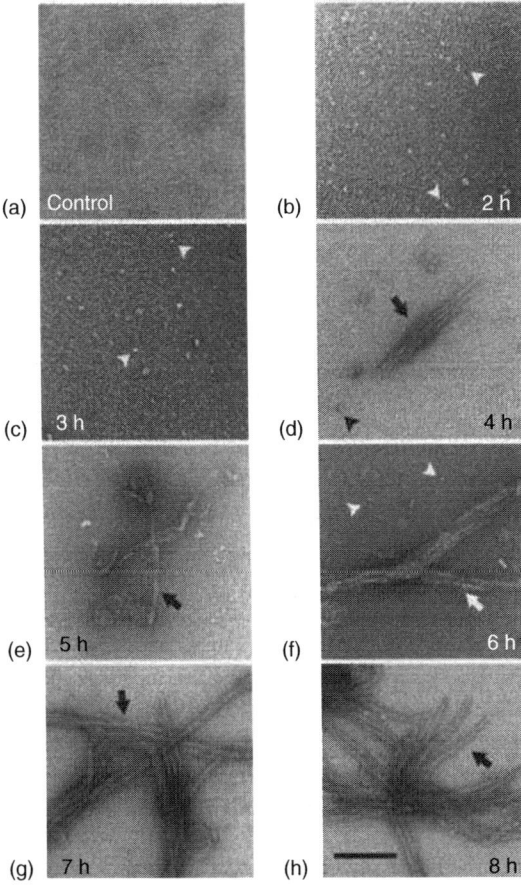

Fig. 6 Polyglutamine-expanded huntingtin protein undergoes a variety of structural alterations on its way to becoming a visible aggregate. In this experiment, transmission electron microscopy tracked the structure of huntingtin exon 1 peptide with 44 glutamines after release from a linked affinity tag. (a) Prior to affinity tag release, nothing is observed. (b-c) After affinity tag release, globular structures become apparent. (d) This is followed by formation of short fibers (4–5 nm in diameter). (e-f) Fiber formation becomes more prominent. However, during this time, large globular assemblies remain (arrowheads). (g-h) Eventually, only fibers are present, and the fibers begin to adopt a uniform appearance, suggesting consolidation into protofibrils. (From Poirier, M.A., Li, H., Macosko, J., Cai, S., Amzel, M., Ross, C.A. (2002) Huntingtin spheroids and protofibrils as precursors in polyglutamine fibrilization, *J. Biol. Chem.* **277**, 41032–41037, used with permission of the *Journal of Biological Chemistry*.).

group has nicely shown that Congo red can bind polyglutamine peptides once they are amyloid-like and prevent their conversion into fibers, while an independent group has shown that Congo red delivery to an HD mouse model is a highly effective treatment intervention.

One satisfying aspect of this model is that it allows us to simultaneously view aggregates as harmful, protective, and innocuous. How? First, aggregates clearly must be toxic as their creation is predicated upon the production of earlier intermediate toxic forms. At the same time, the final visible aggregates are less toxic than the earlier intermediates, so anything that enhances their sequestration into visible aggregates is beneficial. Finally, since oligomers are difficult to detect, it is often not possible to know whether cells are successfully sequestering the toxic precursors into aggregates or if high levels of toxic intermediates are building up. So, aggregates are thus also incidental, since their presence does not provide us with insight into the steady state levels of the toxic precursors. In conclusion, advances in our understanding of the role of aggregate formation in polyglutamine disease processes suggest that aggregates may be simultaneously viewed as harmful, beneficial, and incidental.

3.11
Protein Context

The importance of protein context in the pathogenesis of polyglutamine disorders is a subject of great interest to many researchers, because it is directly related to the mechanisms of toxicity in each disease. It is evident in the polyglutamine disorders that protein context plays a role in cell-type specificity. For example, a polyglutamine expansion in the Htt protein causes a severe degeneration of the striatum and cortex, while the same size glutamine tract in ataxin-1 causes a degeneration of structures in the cerebellum and brainstem, while sparing the striatum and cortex. This cannot be attributed to gross differences in expression patterns, since both proteins are pan-neural (Fig. 7).

One effect that protein context may have on pathology is to affect subcellular localization. For example, the presence of a functional NLS or NES dictates preferential subcellular localization in the nucleus or cytoplasm. Mice expressing expanded ataxin-1 with a mutated NLS do not develop the SCA1 phenotype, while those without mutated NLSs do. Interaction domains also affect polyglutamine protein toxicity. The ability of expanded polyglutamine proteins to interact with other molecules is a promising avenue in the search for mechanisms of cell-type specificity. Yeast two-hybrid screens and other techniques have yielded interaction partners for a number of polyglutamine proteins. For example, Htt's interactions with transcription factors may prove to be central to its pathological effects. At the same time, it is also evident that polyglutamine tracts are innately toxic. Pure polyglutamine tracts cause toxicity in cell culture. Mice that express a glutamine tract of 150 amino acids in the Hprt protein, which does not naturally contain any such tracts, exhibit progressive neurological deficits and NIs. Thus, it appears that pathology in each polyglutamine disease is due to a gain of function of the glutamine tract that is then modulated by the protein context in which it resides.

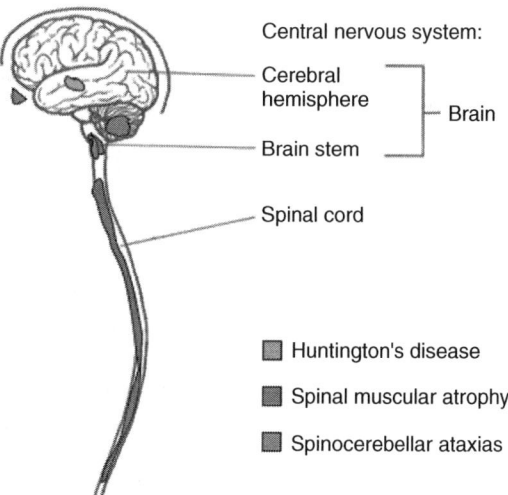

Fig. 7 The conundrum of cell-type specificity in the polyglutamine diseases. Although the different polyglutamine disease proteins are expressed throughout the central nervous system, only select populations of neurons degenerate in the different disorders. The principal regions of selective vulnerability are shown for certain of the polyglutamine diseases.

3.12
Transcriptional Dysregulation

Transcriptional dysregulation is one theory of polyglutamine toxicity that has gained support from many lines of evidence in several disorders, and it appears increasingly likely that it is one of the fundamental causes of pathogenesis. Microarray technology and other genomic techniques are facilitating rapid advances in this area of the polyglutamine field. The majority of polyglutamine disorders are caused by proteins that are either transcription factors/cofactors or interact closely with transcription factors. TBP and AR are transcription factors, atrophin-1 is a transcriptional corepressor, ataxin-7 is part of a transcriptional coactivator complex, and ataxin-1, htt, and ataxin-3 all interact with various transcription factors. It is therefore reasonable to suspect that an abnormally long polyglutamine tract could interfere with the transcriptional activity of these proteins and/or their interactors.

CBP is one of the most studied transcription factors in the polyglutamine field. It is a transcription activator that mediates part of the cellular response to cAMP, and it can interact with numerous polyglutamine proteins in the soluble or insoluble phase, including huntingtin, ataxin-3, ataxin-7, and AR. Expanded htt represses CBP-regulated genes, and overexpression of CBP causes a considerable rescue of the cell death phenotype in HD and SBMA cell culture models. Postnatal mice lacking CBP's upstream activator CREB and its homolog CREM display a profound degeneration of the striatum and hippocampus. Thus, interference with CBP appears to be a common theme in polyglutamine pathology.

Another common theme in the transcription dysregulation equation is the tendency for the polyglutamine disease proteins to alter transcription factor/coactivator-mediated histone acetyltransferase (HAT) activity. The ability of a gene to be transcribed depends upon its chromatin structure, and thus upon the degree of histone acetylation in its vicinity. This is because acetylated histones cause chromatin to be in an "open," transcription-friendly conformation. Acetylation status depends upon the balance between the activity of HATs and their countervailing counterparts, the HDACs. CBP, p300, and p300/CBP-associated factor (PCAF) are all HATs that are inhibited by polyglutamine-expanded htt exon 1 peptide and ataxin-3. In cell culture, expression of a mutant htt fragment reduces global histone acetylation.

Finally, according to a very recent study, ataxin-7 directly interacts with GCN5 as part of the STAGA complex, and upon polyglutamine expansion, mutant ataxin-7 causes a dominant-negative effect upon GCN5 HAT activity. This results in CRX transcription interference that may contribute to the SCA7 retinal degeneration phenotype. Inhibiting HDACs with drugs known as *HDAC inhibitors* (HDAC Is) attenuates the phenotype of HD and SBMA in mouse and fly models, presumably by shifting the cells' acetylation status. HDAC Is such as sodium butyrate and especially suberoylanilide hydroxamic acid (SAHA) have thus emerged as possible candidates for therapeutic trials in human polyglutamine disease patients.

3.13
Proteolytic Cleavage

The occurrence of proteolytic cleavage in polyglutamine diseases first became apparent when it was shown in HD that htt can be cleaved by extracts derived

Fig. 8 Proteolytic cleavage in the polyglutamine diseases. Production of amino-terminal truncated proteins is observed in many of the polyglutamine diseases, as is shown here for SCA7. In this experiment, nuclear fractions of retinal lysates from age- and sex-matched nontransgenic (nt), ataxin-7 CAG-24 (24Q), and ataxin-7 CAG-92 (92Q) mice were prepared and immunoblotted with an antibody directed against the amino-terminal region of the protein ataxin-7. As shown here, in addition to soluble full-length ataxin-7 (which is typically reduced in cells where it is aggregating into insoluble inclusions), an ~50–60 kD

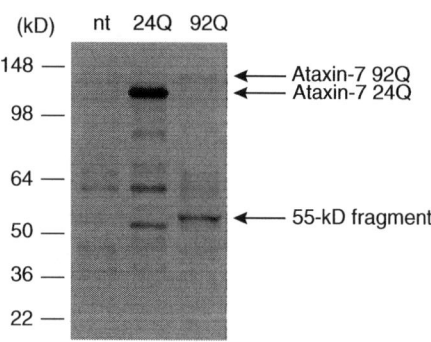

fragment is detected. (Adapted from Garden, G.A., Libby, R.T., Fu, Y.H., Kinoshita, Y., Huang, J., Possin, D.E., Smith, A.C., Martinez, R.A., Fine, G.C., Grote, S.K., et al. (2002) Polyglutamine-expanded ataxin-7 promotes noncell-autonomous Purkinje cell degeneration and displays proteolytic cleavage in ataxic transgenic mice, *J. Neurosci.* **22**, 4897–4905. Used with permission of the *Journal of Neuroscience*.).

from apoptotic cells. Subsequent publications showed that only amino-terminal epitopes of htt protein are detectable in aggregates. SBMA, DRPLA, and SCA7 were then added to the list of diseases in which aggregates are only immunoreactive for a glutamine-containing fragment of the disease protein. Since then, evidence of proteolytic cleavage has been published for nearly all known polyglutamine disorders. Experimental studies revealed that inhibition of caspase cleavage reduces aggregate formation and toxicity in a cell culture model of HD, underscoring the relevance of proteolytic cleavage in polyglutamine disease pathogenesis. Cleavage promotes aggregation and/or toxicity in SCA3 and SCA7 (Fig. 8). Polyglutamine tract containing htt fragments are more toxic in cell culture than full-length htt, and the most widely used HD mouse model expresses only exon 1 of the *htt* gene.

There are several mechanisms by which cleavage may modulate polyglutamine neurotoxicity. Abnormal proteolysis of polyglutamine proteins may lead to a toxic species that can cause damage in the soluble or insoluble phase. Alternatively, proteolysis may be a normal event in protein turnover and the inability to clear cleaved protein may be the problem. Another possibility is that soluble polyglutamine proteins may cause the activation of proteases such as caspases, which then cleave the disease protein and send the cell on a path to degeneration and ultimately apoptosis. As we will see, each theory has support and not all are mutually exclusive.

Abnormal proteolysis of polyglutamine disease proteins is one possible mechanism of toxicity. Some studies have shown that polyglutamine repeat length modulates susceptibility to proteolytic cleavage, but results have been inconclusive. Few studies have been performed using human brain tissue, however. Considerable evidence exists to show that polyglutamine tracts themselves are resistant to degradation by mammalian proteasomes. The 20S and 26S eukaryotic proteasomes are incapable of cutting within polyglutamine tracts longer than 9 amino acids *in vitro*. The inability of the proteasome to digest glutamine tracts may contribute to the toxicity and aggregation of expanded polyglutamine peptide fragments.

Proteolysis may be a normal part of the turnover of wild-type and mutant htt. One group published evidence that caspase-3 cleaves both mutant and wild-type htt in the cytoplasm of HD brains, suggesting that this particular cleavage is a normal event in its turnover. Since this fragment was located exclusively in the cytoplasm, it also implies that further cleavage occurs before the nuclear translocation of htt. Another study suggested that pepstatin-sensitive aspartyl endopeptidases such as the presenilins and cathepsins D and E may be involved in the normal turnover of htt. This process, which may normally aid in maintaining the balance between htt synthesis and degradation, could generate a toxic, highly aggregation-prone (and hence intermediate oligomer/protofibril-prone) species when it cleaves mutant htt.

Another possible source of toxic cleavage products is the ubiquitin-proteasome system, which is responsible for labeling and digesting misfolded proteins, among other things. Polyglutamine tracts beyond a threshold of about 35 glutamines adopt an abnormal β-pleated sheet conformation, which may cause the host protein to be ubiquitinated and thus targeted for degradation. In support of this theory, a common feature of polyglutamine diseases is the presence of ubiquitin-positive inclusions. Arguing against this idea is an experiment showing that in a cell culture model of SCA1, increased proteasome degradation due to the introduction of a degradation signal reduces aggregates and toxicity. Another potential source of toxic fragments is the autophagy pathway, which is responsible for the bulk degradation of cellular components. According to one study, degradation of htt by autophagy and the autophagosome-associated cathepsin D creates toxic htt fragments.

There are many proteases that are suspected to play a role in the proteolytic cleavage of polyglutamine proteins, the most studied of which are the caspases. Caspases are cysteine-dependent aspartyl proteases that play a crucial role in apoptosis, but are increasingly being investigated for their nonapoptotic functions. In healthy cells, caspases are present predominantly as inactive proenzymes, which must be cleaved for full activity. Caspase-3 is the only protease that has been shown to cleave htt in HD patient's material, although caspases 1 and 6 have also been implicated in cell culture and in vivo. Inhibiting the cleavage of htt by any of these three proteases is protective, and an HD mouse model expressing dominant-negative caspase-1 showed significantly delayed disease onset. Caspase-3 has also been implicated in the proteolytic cleavage of atrophin-1, ataxin-3 and AR. Caspases are activated in response to a number of cellular insults, and in postmitotic cells such as neurons where inhibitors of apoptosis proteins (IAPs) are highly expressed, this may cause cellular damage that does not result in classical apoptosis.

Calpains are another family of proteases implicated in the cleavage of polyglutamine proteins. They are calcium-activated cysteine proteases that exist predominantly as proenzymes. As mouse models of HD and material from HD patients exhibit abnormal mitochondrial calcium regulation, some investigators have proposed that altered calcium flux is the mechanism of calpain activation. Indeed, several cell culture studies support the role of calpain proteases I, II, and "m" in htt cleavage. As mentioned previously, other aspartyl proteases such as the Alzheimer's disease-associated presenilins and autophagy-associated cathepsins have also been implicated in htt cleavage.

4
Type 2: the Loss-of-function Repeat Diseases

4.1
Fragile X Syndrome

Fragile X syndrome (FRAXA) is an X-linked disorder with a prevalence of 1 in 4000 in males and 1 in 8000 in females, making it the most common form of inherited mental retardation. Neurological presentation frequently includes mild to severe mental retardation, hyperactivity, poor eye contact, high-pitched speech, and flapping or biting hand movements. Physical signs in males include long, prominent ears, and jaws, macrocephaly, postpubescent macroorchidism, and occasionally, connective tissue abnormalities. The mutation responsible for FRAXA is typically an expansion of a polymorphic $(CGG)n$ repeat found in the $5'$-untranslated region of the fragile X mental retardation-1 (FMR1) gene. Expanded chromosomes have a folate-sensitive fragile site at Xq27.3 that can be viewed under a light microscope under special cell culture conditions. Normal alleles contain 6–53 triplets punctuated by one or more AGGs, which are considered to have a stabilizing influence on the repeat. Disease alleles contain expansions beyond 200 and up to 2000 repeats, with no AGG interruptions. Pathogenic expansions result exclusively from maternal transmission. FRAXA appears to be a loss-of-function disorder, since deletion of FMR1 and loss-of-function point mutations can also cause FRAXA. This view is further supported by the FMR1 knockout mouse, which reproduces certain aspects of the human disorder.

FRAXA patients have reduced levels of the *FMR1* gene product, FMRP, and there is a linear correlation between reduced protein levels and IQ test scores. Expansion of the disease allele results in hypermethylation of the $(CGG)n$ tract, which spreads to a nearby CpG island in the FMR1 promoter region. Some of the proposed secondary structures formed by the FRAXA repeat contain C–C mispairs, which are good targets for human DNA methyltransferase. Another theory invokes the RNAi protein Dicer's ability to cleave CGG repeat RNA, postulating that the resulting siRNAs recruit DNA methyltransferases to the FMR1 locus. Affected alleles also display condensed chromatin, loss of histone acetylation and increased histone methylation. These data support a process whereby DNA methylation recruits transcription silencing machinery, which subsequently suppresses FMR1 transcription in the nearby promoter region. Interestingly, while premutation carriers had been viewed as perfectly normal for decades, recent work indicates that some premutation carriers (60–200 uninterrupted CGG repeats) display a phenotype distinct from FRAXA. Females have a predisposition to premature ovarian failure (POF), and males may develop late-onset ataxia and tremor with the presence of neuronal intranuclear inclusions (NIs) that consist of RNA. (This new FXTAS is discussed in a separate section.) Expression studies have shown that premutation carriers may express up to seven times more FMR1 mRNA than normal individuals. This upregulation is probably not due to compensation for loss of function, because individuals with point mutations resulting in loss of function of FMRP do not have higher levels of the transcript. Abnormal transcript levels due to premutations or full mutations are thought to be as a result of the expansion's effect on transcription initiation rather than on mRNA stability.

The FMR1 promoter lacks a functional TATA box and initiator sequence. Four functional transcription factor binding sites have been identified in normal individuals: two sites binding USF1/USF2 and Nrf-1, and two GC boxes that bind members of the "Sp" family of transcription factors. Sp1 and Nrf-1 binding is disrupted in cells from FRAXA patients. The *FMR1* gene encodes the widely expressed FMRP, which is an RNA-binding protein most highly expressed in the brain and testes. It is 60% homologous to two other proteins, FXR1P and FXR2P, with which it interacts. The protein is thought to bind approximately 4% of all brain mRNA transcripts, through its RGG box domain. It selectively recognizes RNA-containing hairpin or tetraplex secondary structures ("G-quartet"), and shuttles into and out of the nucleus and associates with polyribosomes in messenger ribonucleoprotein complexes (mRNPs). The FMRP-containing mRNP complexes also contain Pur α and mStaufen, proteins involved in the transport of neuronal granules. These granules, which contain RNA and associated proteins, are transported to dendritic spines in a metabotropic glutamate receptor 5 (mGluR5)-dependent manner. FMRP can suppress the translation of certain transcripts *in vitro* and *in vivo*. FMRP may thus regulate the transport, localization, and translation of certain mRNAs in an activity-dependent manner (Fig. 9). In Drosophila, FMRP interacts with components of the RNAi machinery, which is involved in gene silencing and thus translational control. Although FMRP does not seem to affect the siRNA pathway, it may regulate microRNAs (miRNAs), which are noncoding RNAs thought to control the translation of mRNAs by binding to their 3'-untranslated region. FMRP associates with miRNAs, and proteins in miRNA-containing complexes in mammals, interactions that may be relevant to FMRPs regulation of translation.

FMRP is important for the development of dendritic spines and synaptic plasticity. It regulates the expression of MAP1B, an important regulator of microtubule stability. FMRP knockout mice have abnormally high levels of MAP1B, resulting in increased microtubule stability. This may affect the development of dendrites and/or dendritic spines. The absence of FMRP in hippocampal neurons results in immature dendritic spine morphology and delayed synaptic connections, perhaps contributing to the neurological phenotype observed in FRAXA. FMRP may also affect mGluR5-dependent long term depression (LTD). LTD is a process by which neuronal activity can cause a lasting desensitization of neurons to depolarization. LTD-associated protein synthesis at synapses may be enhanced by mGluR5 activation, while it appears to be suppressed by FMRP. This is supported by evidence of enhanced LTD in the FMR-1 knockout mouse. FMRP suppression and mGluR5 activation of LTD-dependent local protein synthesis may be opposing forces that are out of balance in FRAXA, perhaps accounting for part of the cognitive abnormalities.

4.2
Fragile XE Mental Retardation

Fragile XE mental retardation (FRAXE) is an X-linked disorder with a prevalence of about 1–4% that of FRAXA. It accounts for approximately two-thirds of families with nonspecific X-linked mental retardation (MRX), a classification in which mental retardation is the only consistent

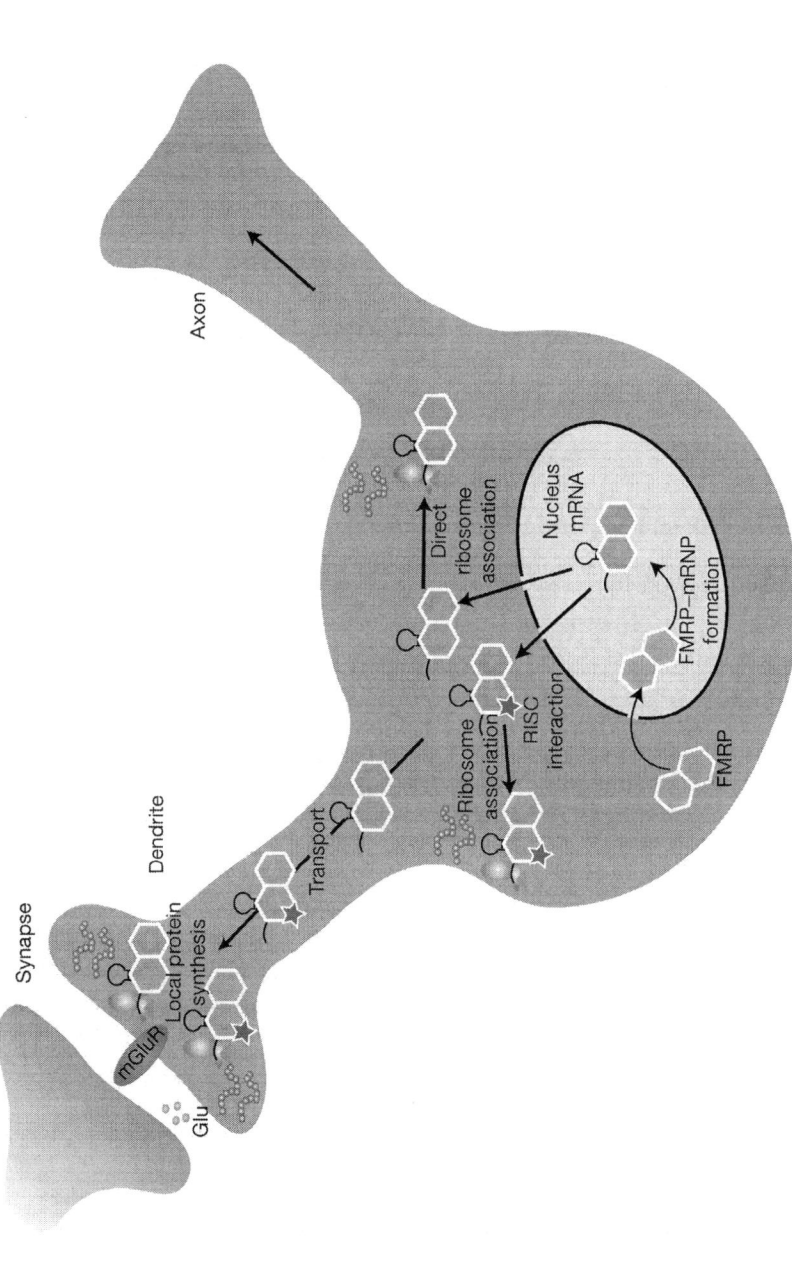

Fig. 9 Advances in the understanding of FMRP's normal function. FMRP binds to recently transcribed messenger RNAs in the nucleus to form a ribonucleoprotein complex (mRNP). The FMRP–mRNP complex moves out of the nucleus and then either directly associates with ribosomes or interacts with the RNA-induced silencing complex (RISC). The FMRP–mRNP complex is also transported to dendrites. Whether in the perinuclear cytosol or in dendrites, FMRP is believed to regulate protein translation. (From Jin, P., Alisch, R.S., Warren, S.T. (2004) RNA and microRNAs in fragile X mental retardation, *Nat. Cell Biol.* **6**, 1048–1053. Used with permission of *Nature Cell Biology*.).

clinical feature. Other characteristics are variable and may include hyperactivity, mild facial hypoplasia and nasal abnormalities. FRAXE is caused by the expansion of a (GCC)n repeat in the promoter region of *FMR2*, a gene 600 kb downstream of *FMR-1*. It also overlaps with the putative promoter of a gene called *FMR3*, transcribed in the opposite direction. Both maternal and paternal transmission can result in expansion. Normal individuals bear 6–35 GCC repeats, premutation carriers bear 61–200 repeats, and full mutation carriers bear >200 repeats. Pathogenic expansion results in a folate-sensitive fragile site at the disease locus.

In mice, FMR2 mRNA is expressed in adult and fetal brain, kidney, lung, and placenta, with highest brain levels in the hippocampus and amygdala. Transcripts of both FMR2 and FMR3 are reduced beyond detection in FRAXE patients. This may be due to a methylation-dependent silencing process similar to FRAXA. The presence of the repeat within the preinitiation region of the FMR2 promoter may also suggest a more direct disruption of transcription. FMR2 mRNA is highly expressed in the hippocampus, an area critical for learning and memory. Its paralogs AF4 and LAF4 are both transcription transactivator proteins, and the FMR2 protein seems to be a potent transcription activator itself. Furthermore, the FMR2 protein is nuclear, consistent with its proposed role in transcription regulation. The Drosophila ortholog of FMR2/AR4, Lilliputian, is essential for proper organ development, and its loss of function is lethal. FMR2 knockout mice display impairment in the conditioned fear test and enhanced LTP (long-term potentiation) in the hippocampus. The role of FMR3 in FRAXE, if any, is unknown. The mechanism of FRAXE pathogenesis may therefore involve silencing of the human *FMR2* gene resulting in altered transcription in many parts of the developing nervous system and mature brain. The neurological deficits characteristic of FRAXE may involve effects of altered transcription on the hippocampus, amygdala, and possibly other brain structures.

4.3
Friedreich's Ataxia

With a prevalence of about one per 50 000 individuals in the Caucasian population, Friedreich's ataxia (FRDA) is the most common inherited ataxia in this ethnic group. It is a multisystem degenerative disease that is unusual among the triplet repeat disorders due to its autosomal recessive inheritance. Neurological symptoms include gait, limb and truncal ataxia, loss of position and vibration senses, diminished tendon reflexes, and dysarthria. Neuropathology changes include degeneration of the posterior columns of the spinal cord, loss of large primary sensory neurons in the dorsal root ganglia (DRG), and mild, late-onset degeneration of the cerebellar cortex. Other common clinical features are cardiomyopathy, diabetes mellitus, scoliosis, and other skeletal abnormalities. Patients often present with symptoms in childhood, become wheelchair-bound by their late teens or early twenties, and have reduced lifespans due to cardiac failure. Adult presenting patients, however, can have nearly normal lifespans with more protracted progression and less severe nonneuronal involvement. Indeed, until the identification of the causal mutation, many of these adult-onset cases went undiagnosed.

FRDA results from the expansion of a polymorphic (GAA)n repeat in the first intron of the gene *X25*, now known as frataxin. FRDA patients have at least one expanded allele. For the disease to occur, the second allele must contain either an expanded repeat, or rarely, a copy of frataxin containing a loss-of-function mutation. No patients have been described with loss-of-function allele mutations in both *FRDA* genes. This is probably due to prenatal lethality, as frataxin-null mice die in utero. Disease severity and onset age are determined by the size of the (GAA)n repeat of the smaller expansion allele. Unaffected individuals carry at least one allele between 7–38 repeats, while affected individuals carry two alleles of 66–1700 repeats. Interruptions of the (GAA)n repeat result in later onset and attenuated presentations. GAA expansion at the FRDA locus can result in enormous repeats in a single generation. Maternal transmission can result in expansion or contraction, while paternal transmission results primarily in contraction.

FRDA patients exhibit reduced frataxin RNA and protein levels, and evidence suggests a defect in transcription or RNA maturation. Current models propose that frataxin RNA elongation may be disturbed by triplex structure formation between expanded DNA strands during transcription. DNA triplexes are formed when one strand of a double-stranded DNA molecule folds back upon itself, and interacts with two previously annealed strands. This creates a local structure comprising three DNA strands held together by hydrogen bonds. Certain sequences favor this process, among them extended (GAA)n repeats. Triplex formation in the first intron would presumably affect transcript elongation but not initiation. In support of this model are *in vitro* transcription experiments demonstrating no effect of repeat length on transcript initiation.

Frataxin is a 210 amino acid protein that is well conserved from prokaryotes to mammals. It contains mitochondrial targeting signal sequences and localizes to the mitochondrial matrix. Frataxin expression occurs at the primary sites of pathology: dorsal root ganglia, spinal cord, sensory nerves, heart, and pancreas. These are tissues that rely upon high levels of oxidative metabolism and consequently are rich in mitochondria. In addition, such cell types are often postmitotic, meaning that most dividing cell types are spared in FRDA patients.

Studies with FRDA patient material have demonstrated an increased heart iron content and a deficiency in iron–sulfur cluster-containing proteins, including aconitase, a protein involved in iron homeostasis. In addition, fibroblasts derived from FRDA patients are abnormally sensitive to iron and hydrogen peroxide induced stress. Ablation of the yeast frataxin homolog results in respiratory dysfunction, abnormal accumulation of mitochondrial iron, impaired biogenesis of iron–sulfur proteins, and increased sensitivity to oxidative stress. Conditional knockout of frataxin in striated muscle results in a heart-specific phenotype resembling the cardiac abnormalities in human FRDA (Fig. 10). This supports a genetic mechanism involving loss of function of the disease protein. Frataxin may be part of a complex that delivers iron to Iron–sulfur clusters (ISCs), which are cofactors essential to the activity of many important cellular proteins. The accumulation of iron outside the mitochondria, although probably not central to FRDA pathogenesis, supports a deficiency in iron delivery to ISCs. Among ISC-dependent cofactors are proteins involved in mitochondrial electron transport and thus respiration. Disturbances in oxidative metabolism are often associated with increases in the production of

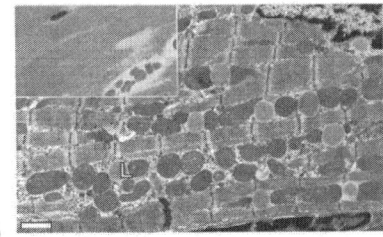
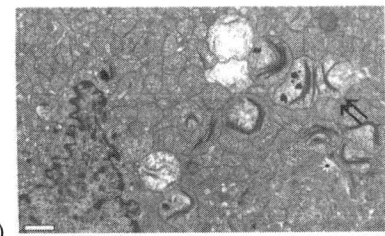
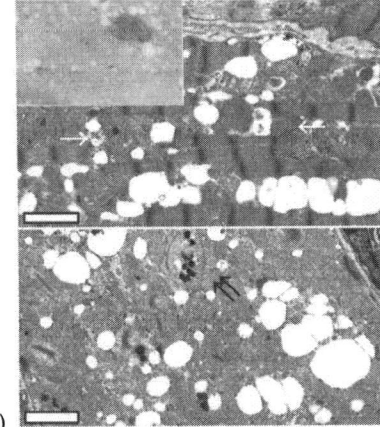

Fig. 10 Friedreich's ataxia mice display cardiac muscle pathology.
(a) Transmission electron microscopy of mice lacking expression of frataxin in only their muscle reveals the abnormal accumulation of lipid droplets (L) in cardiac muscle at 4 weeks of age. (b) By 7 weeks of age, mitochondria appear abnormal and iron deposits are visible (arrows). (c) With further progression, large vacuoles emerge (inset-top) as mitochondria undergo continued prominent degeneration. Ultimately, the mitochondria become engorged with electron-dense material consistent with iron deposits. (From Seznec, H., Simon, D., Monassier, L., Criqui-Filipe, P., Gansmuller, A., Rustin, P., Koenig, M., Puccio, H. (2004) Idebenone delays the onset of cardiac functional alteration without correction of Fe-S enzymes deficit in a mouse model for Friedreich ataxia, *Hum. Mol. Genet.* **13**, 1017–1024, used with permission of *Human Molecular Genetics*).

toxic reactive oxygen species (ROS). FRDA therefore is thought to result from reduced frataxin levels, leading to abnormal iron–sulfur metabolism, mitochondrial dysfunction, oxidative stress, and tissue degeneration. FRDA is thus caused by the dynamic mutation of a nuclear-encoded mitochondrial protein. The disease shares features with classic mitochondrial disorders such as MELAS (mitochondrial myopathy, encephalopathy, lactic acidosis, and strokelike episodes syndrome) and MERRF (myoclonus epilepsy with ragged red fibers), which are caused by stable mutations to mitochondrial-encoded proteins. FRDA appears to be a classic mitochondrial disorder with an unusual genetic basis – dynamic mutation.

Given the data suggesting a role for oxidative metabolism and iron transport in FRDA pathogenesis, antioxidants and iron transport molecules are being evaluated as treatments for this disease. Several clinical trials have been conducted in FRDA patients using idebenone, a synthetic analog of coenzyme Q10 and a potent antioxidant. Some studies have shown substantial improvements in the heart function of patients, although the lack of randomized placebo controls and the variability of the disorder render the results controversial. A recent conditional knockout mouse model of FRDA also supports the use of idebenone as a treatment for the disorder, making treatment of FRDA patients with this antioxidant a distinct possibility.

4.4
Progressive Myoclonus Epilepsy Type 1

Progressive myoclonus epilepsy type 1 (EPM1) is a rare, autosomal recessive neurological disorder most prevalent in Finland and parts of North Africa. Its most prominent symptoms are progressive photosensitive myoclonus and tonic-clonic epilepsy. Some patients also experience a progressive cerebellar ataxia and cognitive decline. Neurodegeneration occurs in the thalamus, spinal cord, and the Purkinje and granular neurons of the cerebellum. Typical age of onset is 6 to 16 years of age. EPM1 is sometimes caused by missense mutations in the *cystatin B* gene (CSTB); however, analysis of affected patients lacking such mutations revealed a dodecamer repeat expansion upstream of CSTB. The dodecamer repeat is located between 66 and 77 bp 5′ of two putative transcription start sites, and has the sequence CCCCGC-CCCGCG. Unaffected individuals have a repeat number of 2–3, premutation carriers have 12–17 repeats, and affected individuals carry 30–150 repeat alleles.

CSTB is a highly conserved gene in the cystatin family of cysteine protease inhibitors. CSTB mRNA is ubiquitous, with high transcript levels in the hippocampus. The protein binds to and inhibits lysosomal proteases such as cathepsins B, H, L, and S. Pathogenic expansion of the repeat causes a reduction in CSTB transcription in some cell types. Cstb knockout mice develop symptoms similar to those seen in EPM1 patients, and missense mutations causing the disease disrupt the ability of CSTB to bind cysteine proteases. Together, these data suggest a loss-of-function mechanism for EPM1. Lowered inhibition of cysteine proteases may result in neuronal damage, causing the phenotype observed in EPM1.

There are several ways in which the repeat might reduce CSTB expression. The first is that reduced gene expression may be due to an increase in the distance between promoter elements and the transcription start site. AP-1 binding sites have been located upstream of the repeat. One study demonstrated a two- to fourfold reduction in promoter activity when the repeat is expanded. An equivalent reduction also occurred when similarly sized heterologous DNA was used in place of the expanded repeat. Independent studies have confirmed a large reduction in promoter activity due to repeat expansion. Another way the repeat could affect CSTB expression is through alterations in chromatin structure. The repeat region is G-C rich and might exclude nucleosomes or promote the formation of secondary structures. Abnormal DNA secondary structures have been observed in the EPM1 repeat region. These alterations could affect the expression of *CSTB* or other genes in the vicinity. A third possible mechanism for expression changes lies in the sequence of the repeat region. Each dodecamer repeat contains a GC box, which could act as an Sp1 binding site *in vivo*. In cases of abnormally large repeat numbers, increased Sp1 binding could contribute to the suppression of CSTB transcription.

5
Type 3: the RNA Gain-of-function Repeat Diseases

5.1
Myotonic Dystrophy Type 1

Myotonic dystrophy or DM is an autosomal dominant, multisystem disorder with a prevalence of 1 in 100 000 worldwide, and 1 in 8000 in European and North American Caucasian populations. Patients typically present with proximal or distal muscle dysfunction including weakness, pain, and myotonia (failure of muscle relaxation). DM exhibits a combination of characteristic symptoms: cardiac conduction abnormalities, subcapsular iridescent cataracts, and unusual endocrine changes. Other features include testicular atrophy, type II diabetes, and late-onset cognitive impairment. In its most severe form, congenital DM, mental retardation, craniofacial deformities, and other developmental abnormalities are present. It exhibits both paternal and maternal transmission, although there is an almost exclusive maternal transmission in congenital DM. Myotonic dystrophy type 1 (DM1) is caused by a $(CTG)n$ expansion in the 3′-untranslated region of the gene, dystrophica myotonica protein kinase (DMPK), on chromosome 19. Unaffected individuals carry 5–37 repeats, while affected individuals carry 50–4000 repeats. DM exhibits pronounced anticipation and dramatic somatic instability. Interestingly, the clinical phenomenon of anticipation (worsening severity as a disease gene is transmitted from one generation to the next) was first described nearly a century ago in a family segregating DM1. Although the anticipation phenomenon was dismissed as an artifact of ascertainment by geneticists of the mid twentieth century, anticipation is now known to be a genuine feature of dynamic mutation diseases. Anticipation results from the tendency of disease repeats to expand and the inverse correlation between repeat length and age of disease onset.

Several theories have been advanced to explain the molecular basis of DM1, including haploinsufficiency of DMPK, local chromatin effects on neighboring genes, and gain of function exerted by expanded RNAs. DMPK is proposed to have many functions, some of which could relate to the disease, such as modulation of skeletal muscle sodium channels, RNA metabolism, calcium homeostasis, and the cell stress response. Initial expression studies reported a decrease in DMPK RNA and protein levels, but a DMPK knockout mouse designed to test the haploinsufficiency theory exhibited only mild myopathy that was inconsistent with the DM phenotype. The $(CTG)n$ tract is a strong nucleosome assembly site, and it was therefore hypothesized to have trans-effects on the expression of neighboring genes. According to this model, the myriad symptoms of DM1 are caused by disturbances in the expression of multiple nearby genes due to the expansion – making DM1 a "contiguous gene syndrome." There are several genes in close proximity to the DMPK gene (i.e. <5 kb) whose roles in DM1 were considered; the most studied of these genes has been SIX5. The SIX5 homolog in Drosophila is required for normal eye development and its mouse homolog is involved in regulating distal limb muscle development. Expression data on SIX5 in

DM1 patients has been inconsistent, however. A Six5 knockout mouse developed cataracts, although they were not of the type observed in DM1. Another gene implicated by this model is *DMWD*, which is expressed in the testis and suspected to be involved in male infertility. Studies indicated that the expression of DMWD is not altered.

The third theory for the molecular basis of DM proposes that an RNA gain-of-function mechanism is responsible. RNA foci, which are accumulations of expanded transcripts, accumulate in the nuclei of patient cells (Fig. 11). The RNA gain-of-function theory was buttressed by the discovery that DM2, a disorder with symptoms almost identical to DM1, is caused by a (CCTG)n expansion in intron 1 of the zinc finger 9 protein (ZNF9). The two genes responsible for DM1 and DM2 are unrelated and reside on different chromosomes. Genes near the two loci bear no obvious resemblances. The only striking parallels between the two expansions are: (1) they both contain CTG triplets; and (2) they both occur in transcribed but noncoding regions of the genome. More support for the RNA gain-of-function model came from a mouse model generated by Mankodi et al. in 2000, which contained a (CTG)n expansion in the 3′ untranslated region of the skeletal actin (*HSA*) gene. This mouse exhibited myopathy typical of DM1, although the skeletal muscle-restricted expression pattern of HSA precludes broader conclusions. Thus, there are several lines of evidence suggesting that CUG expansion-containing RNAs are capable of causing DM.

RNA-binding proteins and transcription factors colocalize with the RNA foci found in DM, potentially altering nuclear processes. Elevated levels of CUG-containing RNA has been shown to alter gene splicing in specific transcripts which could be relevant to DM: cardiac troponin T (cTNT), involved in cardiomyopathy; Insulin Receptor (IR), involved in diabetes; and Clc-1, the main chloride channel in muscle. Abnormal splicing of cTNT, IR, and ClC-1 are proposed to account for the cardiac abnormalities, insulin insensitivity and myotonia observed in DM.

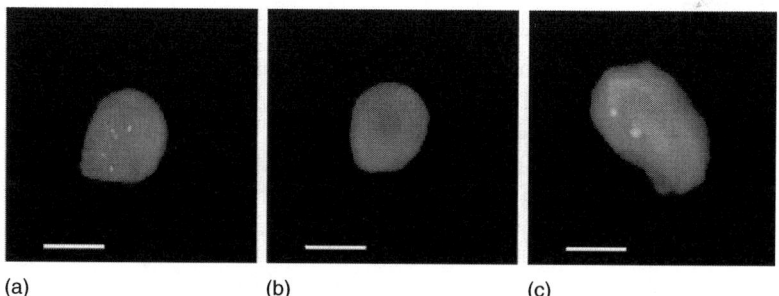

(a) (b) (c)

Fig. 11 *In situ* hybridization of muscle sections with fluorescently labeled antisense oligonucleotide probes reveals accumulation of mutant RNA in DM1 and DM2.
(a) Probing of DM2 muscle with a CAGG probe indicates that multiple RNA foci are present. (b) Probing of normal muscle with a CAGG probe demonstrates absence of RNA foci. (c) Probing of DM1 muscle with a CAG probe yields prominent RNA foci. (From Liquori, C.L., Ricker, K., Moseley, M.L., Jacobsen, J.F., Kress, W., Naylor, S.L., Day, J.W., Ranum, L.P. (2001) Myotonic dystrophy type 2 caused by a CCTG expansion in intron 1 of ZNF9, *Science* **293**, 864–867, used with permission of *Science*).

Splicing abnormalities cause a reduction in the membrane concentration of ClC-1 and reduce chloride conductance to levels consistent with myotonia. These spicing alterations are thought to be due to the repeat-expanded RNA's effects on two families of RNA-binding proteins: "CUG-BP1 and ETR-3-like factors" (CELF) and "muscleblind-like proteins" (MBNL). CELF proteins regulate pre-mRNA splicing in cTNT, IR, and ClC-1. In patient tissue and cell culture, CUG-BP1 levels and activity are increased in response to elevated levels of CUG-containing RNA. This may be due to a lengthening of the protein's half-life. MBNL proteins were named after their Drosophila ortholog *muscleblind*, which is required for photoreceptor and muscle differentiation in flies. The three known proteins in this family (MBNL1 (MBNL), MBNL2 (MBLL), and MBNL3 (MBXL)) are splicing regulators that are thought to act antagonistically to CELF family proteins. They colocalize with RNA foci *in vivo*, and a mouse model lacking specific isoforms of MBNL1 recapitulates the myotonia, cataracts, and splicing dysregulation observed in DM.

The splicing alterations seen in DM patients are consistent with loss of function of MBNL proteins or an increase in CELF protein activity in muscle and brain. Because of the colocalization of MBNL proteins with RNA foci, it has been proposed that sequestration and subsequent depletion of these proteins from the cellular milieu is responsible for the symptoms of DM. Although this process probably plays a critical role in the disease, recent evidence suggests it is not solely responsible. The most plausible current theory explaining DM pathogenesis proposes an imbalance between the antagonistic MBNL and CELF proteins, resulting in specific splicing abnormalities that cause DM's diverse range of symptoms.

5.2
Myotonic Dystrophy Type 2

Myotonic dystrophy type 2 (DM2) is an autosomal dominant, multisystem disorder very similar to DM1. The majority of its symptoms resemble DM1: progressive weakness, myotonia, cardiac disturbances, iridescent cataracts, and insulin insensitivity. There are some notable differences, however. DM2, unlike DM1, predominantly affects proximal muscles at its onset, which is why many cases were originally classified as proximal myotonic myopathy (PROMM). Other interesting differences are that mental retardation is not observed in DM2, DM2 patients show increased sweating, and DM2 congenital forms have not been observed. DM2 is caused by a $(CCTG)n$ tetranucleotide expansion in the first intron of the zinc finger protein 9 (*ZNF9*) gene on chromosome 3q21.3. The tetranucleotide repeat can expand to stunning lengths, with the longest cases comprising 44 kb of DNA, making them the longest tracts observed in the repeat expansion disorders. The DM2 locus exhibits marked somatic instability. Over the course of a patient's lifetime, the average repeat length increases substantially, as judged by blood drawings.

The similarities between DM1 and DM2 are not restricted to the clinical presentation. The expansions are both large, CTG-containing tracts that are transcribed, but not translated. Both disorders cause nuclear RNA foci that sequester specific RNA-binding proteins (Fig. 11), including CELF and MBNL family members. Similar splicing abnormalities are also observed in DM2. Since the genes associated with the repeat expansions in DM1 and DM2 are unrelated, it is likely that the cause of both disorders is a toxic gain of function of long tracts of CUG-containing RNA. A mouse model generated by Mankodi et al. in 2000, which contained a (CTG)n expansion in the 3' untranslated region of the skeletal actin (*HSA*) gene, exhibited myopathy typical of DM. Thus, even outside the context of the DM1 and DM2 loci, CTG-containing expansions can cause DM-like symptoms. This supports the RNA gain-of-function theory, as transcribed CTG-containing expansions are inherently capable of causing disease. Differences between DM1 and DM2 may be due to regional or temporal expression patterns or differences in the affinity of RNA-binding proteins for CTG or CCTG tracts.

5.3
Spinocerebellar Ataxia Type 8

Spinocerebellar ataxia type 8 (SCA8) is a dominantly inherited cerebellar ataxia. Affected individuals suffer from late-onset, slowly progressing gait ataxia as well as dysarthria, oculomotor incoordination, spasticity, and decreased vibration sense. The cerebellar cortices and vermis undergo a slowly progressing but dramatic atrophy, while the brainstem exhibits little evidence of degeneration. Patients may become wheelchair-bound as early as their fourth decade of life. Using a direct method for cloning expanded triplet repeats (i.e. RAPID cloning), a gene containing an expanded CTG tract was identified in a large family (MN-A) segregating the SCA8 phenotype. Interestingly, this CTG is contained within a gene on the long arm of chromosome 13 that is transcribed, but apparently not translated into a protein product. Thus, it was proposed that the production of an RNA transcript containing this expanded CTG repeat tract is the cause of SCA8. Numerous subsequent studies, however, have indicated that possession of the expanded CTG repeat tract appears necessary, but not sufficient, for the production of the SCA8 phenotype. Thus, reduced penetrance is viewed as a key feature of SCA8 at this time. On the basis of the available genetic data, normal individuals always carry fewer than 70 CTG repeats, while affected SCA8 patients can carry anywhere from 71 to >1000 CTG repeats. As extremely large CTG repeat alleles (>800 triplets) were shown to not cause disease in early reports, this was initially attributed to lack of stability of the mutant RNA product. However, further work has revealed considerable overlap between disease-causing CTG repeat expansions and nonpenetrant CTG repeat alleles, indicating that a secondary factor – either a trans-acting genetic factor or an environmental factor – must be present to yield the SCA8 disease phenotype. In the case of the original MN-A family, evidence for a *cis* modifier appears responsible for the extremely high penetrance in this large pedigree. Thus, while the causality of the CTG repeat expansion in SCA8 has been somewhat controversial, review of the current literature suggests that the CTG repeat expansion is directly involved in SCA8, but may not alone be sufficient to produce the disorder.

Another unique feature of the SCA8 CTG repeat is its extreme and unusual genetic instability. Paternal transmissions generally result in contractions, while maternal transmissions generally result in expansions. Expansions of up to 600 repeats have occurred in one generation through maternal transmission. Large deletions in expanded alleles often occur in sperm cells, offering an explanation for the paternal contraction bias. Pathogenic expansions often have 5′ triplet interruptions, the role of which is unclear.

The molecular basis of how the expanded SCA8 RNA causes disease remains unclear. The transcript has been detected exclusively in the brain, and appears to be transcribed in the CTG orientation. Translation of a polyglutamine tract from the CAG-containing transcript in the opposite direction is precluded by the existence of stop codons flanking the repeat. The longest transcript identified to date contains 6 exons, and many alternatively spliced forms of the gene have been detected, although all these isoforms are expressed at very low levels. None contain significant open reading frames. In 1999, Koob et al. reported that a gene partially overlaps the 5′ end of the *SCA8* gene locus on the antisense strand. This gene, called *Kelch-like 1 (KLHL1)*, for its homology to the Drosophila *KELCH* gene, is highly conserved and predicted to encode an actin-binding protein of 748 amino acids. Its expression overlaps with that of the SCA8 transcript, suggesting that the *SCA8* gene may produce an antisense RNA whose function is to regulate KLHL1 expression. Whether or how this is occurring remains uncertain at this time.

With the discovery of the DM RNA toxic gain-of-function pathway, an emerging theory for SCA8 pathology holds that the production of a CUG-expanded transcript results in RNA gain-of-function toxicity within the restricted neuronal populations where the *SCA8* gene is expressed. In support of this hypothesis, transgenic mice derived with a human bacterial artificial chromosome containing the entire *SCA8* gene with a 118 CTG repeat expansion develop neurological disease. The severity of their phenotype depends on the expression level of the transgene. In the more moderate expressing lines, the SCA8 mice display a slowly progressive gait ataxia reminiscent of the human disease. While no protein product has been detected, 1C2 and ubiquitin antibody staining reveal intranuclear inclusions in cerebellar neurons, consistent with the RNA-containing inclusions observed in FXTAS brains. An alternative interpretation is that translation of CTG into a polyleucine tract is occurring. Further work will be needed to distinguish these two possibilities.

Studies of the SCA8 CTG repeat expansion in *D. melanogaster* have also been informative. Whether directing expression of the *SCA8* gene to fly retina with a normal CTG repeat tract or an expanded CTG repeat tract, a neurodegenerative eye phenotype results. Modifier screens using fly stocks carrying mutations in genes encoding RNA-binding protein yielded a number of genes that could either enhance or suppress this retinal degeneration phenotype. Interestingly, Drosophila *muscleblind*, whose mammalian counterpart has been implicated in the DM RNA toxicity pathway, modified the retinal degeneration caused by expanded SCA8 CTG repeat expression more so than the retinal degeneration caused by normal SCA8 CTG repeat expression. Such data supports the hypothesis that the SCA8 CTG repeat expansion is producing neurotoxicity by altering the function of RNA-binding proteins as in DM. Although it would be premature to definitively conclude that SCA8 is an RNA gain-of-function repeat disease, provisional classification in this category is appropriate.

5.4
The Fragile X Tremor – Ataxia Syndrome (FXTAS)

Fragile X tremor-ataxia syndrome (FXTAS) is a rare and unusual disorder that is associated with premutation of the FRAXA locus in males. Patients develop a progressive ataxia and intention tremor. This is sometimes accompanied by dementia, parkinsonism, and autonomic dysfunction. Neuropathological changes include degeneration of the cerebellum and ubiquitin-positive intranuclear inclusions in neurons and glia. FXTAS is associated with the expansion of a $(CGG)n$ tract in the 5′-untranslated region of the *FMR1* gene, found on chromosome Xq27.3. Expansion of this tract beyond 200 repeats causes FRAXA, a disorder resulting from a reduction in the expression of FMR1. Males with tracts between 55 and 200 CGG repeats are considered to be in the "premutation range," and are now considered to be at risk for FXTAS beyond middle age.

In contrast to FRAXA, FXTAS patients show increased expression of the FMR1 transcript, with longer repeats corresponding to higher transcript levels. Premutation carriers with more than 100 repeats have an average of five times more FMR1 mRNA than individuals in the normal repeat range. Despite the increase in transcript level, the protein product of the *FMR1* gene, FMRP, is slightly reduced in FXTAS patients. Since elevated transcript

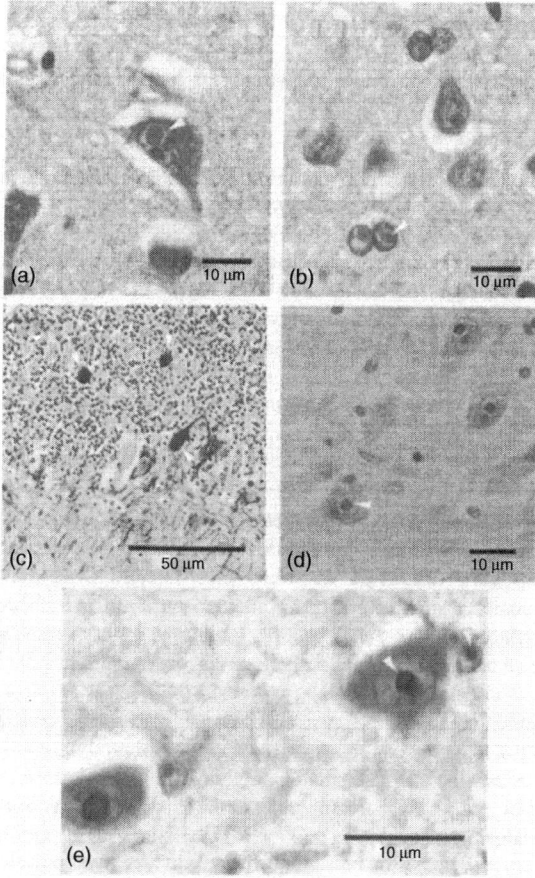

Fig. 12 Intranuclear inclusions are present in the brains of FXTAS patients. (a) Hematoxylin & eosin staining of cerebral neurons reveals a refractile, eosiniphilic nuclear inclusion of ~5 µM in diameter (white arrowhead). (b) Hematoxylin & eosin staining of cerebral astrocytes reveals refractile, eosiniphilic nuclear inclusions of ~2 µM in diameter (white arrowhead). (c) Antineurofilament antibody staining of cerebellum demonstrates presence of dystrophic neurites, consistent with ongoing Purkinje cell degeneration. (d) Antiubiquitin antibody staining of cerebral neurons labels intranuclear inclusions. (e) Antiubiquitin positive intranuclear inclusions form in both neurons (white arrowhead; larger cell) and astrocytes (white arrowhead; smaller cell). (From Greco et al. (2002) *Brain* **125**, 1760, used with permission).

levels are present without great alteration in protein levels, this disorder may be due to an RNA gain of function of the FMR1 transcript. Indeed, immunostaining of FXTAS patients' brain sections reveal intranuclear inclusions in neurons and glia, believed to be comprised of accumulated Fmr1 RNA transcripts and various proteins (Fig. 12). To determine the molecular basis of FXTAS, a knockin mouse model of FXTAS was generated using a human 98 CGG repeat, and was noted to produce elevated levels of the Fmr1 transcript and display intranuclear ribonucleoprotein inclusions, containing ubiquitin, Hsp40 and the 20S proteasome subunit. The inclusions may be a response to RNA toxicity or a result of the aggregation of CGG-binding proteins. While the mechanism of FXTAS may be due to a dominant gain of function of triplet-expanded RNA on RNA-binding proteins as in DM, there is currently no direct evidence supporting this theory.

6
Type 4: The Polyalanine Diseases

6.1
Overview

Polyalanine disorders are characterized by small expansions in trinucleotide repeats encoding alanine tracts. Most are rare, autosomal dominant developmental malformations. Although the normal function of the alanine tract is unknown, they tend to be found in transcription factors. The majority of alanine tract expansions associated with disease loci are also located in transcription factors, most of which play a role in development. In contrast to polyglutamine expansions, polyalanine expansions are small and stable, not exceeding 30 triplets. Most are composed of "imperfect" alanine tracts, including combinations of GCG, GCA, GCC, and GCT codons. This composition argues against an expansion mechanism involving single-strand, hairpin-like secondary structures as in other repeat disorders. A more plausible mechanism for polyalanine expansion is unequal crossing-over, in which alleles mispair during meiotic crossing-over, resulting in one expanded and one contracted tract.

6.2
Oculopharyngeal Muscular dystrophy

Oculopharyngeal muscular dystrophy (OPMD) is a predominantly autosomal dominant, late-onset disorder characterized by progressive drooping eyelids, dysphagia, and proximal limb weakness. Certain skeletal muscles in affected patients degenerate and contain nuclear inclusions and rimmed vacuoles. This effect is particularly striking in the levator palpebra and pharyngeal muscles, responsible for lifting the eyelids and swallowing. OPMD is caused by an alanine expansion in the gene polyadenine-binding protein 2 (PABP2). PABP2 normally contains a 10-alanine tract encoded by $(GCG)_6(GCA)_3GCG$. Affected individuals carry 12–17 alanines, probably resulting from unequal crossing-over of the two alleles. Repeat length appears to correlate with disease severity.

PABP2 is an abundant, ubiquitously expressed pre-mRNA-binding protein that plays a role in controlling the formation and length of mRNA polyA tails. NIs immunoreactive for PABP2 are present in the skeletal muscle cells of OPMD patients and sequester polyA-containing transcripts. The aggregates are filamentous and contain PABP2, ubiquitin and proteasome subunits, leading investigators to conclude that expansion of the alanine tract probably causes PABP2 to misfold and aggregate. Aggregation in OPMD has been linked to toxicity in several experimental systems. Polyalanine-expanded PABP2 aggregates and causes cell death in cultured cells and transgenic mice. Indeed, widespread expression of the human *PABP2* gene in transgenic mice yielded polyalanine length-dependent muscle pathology, including rimmed vacuoles, central nuclei, and numerous dystrophic changes (Fig. 13). Deleting the C-terminal oligomerization domain, overexpressing chaperones, or exposing cells to aggregation inhibitors such as Congo red and doxycycline, reduces aggregate formation and toxicity in cell culture. A polyalanine-expanded peptide, similar to the amino-terminal region of PABP2 was shown to adopt a β-sheet conformation, whereas the same peptide with 7 alanines adopted an α-helical conformation. This is reminiscent of polyglutamine proteins, which also adopt a β-sheet conformation when the polyglutamine tract exceeds a threshold of about 35 glutamines. Expanded polyalanine proteins also activate caspase-3 and -8 in cultured cells, another feature reminiscent of polyglutamine toxicity.

In 2004, Wirtschafter et al. proposed a model for the selective vulnerability of

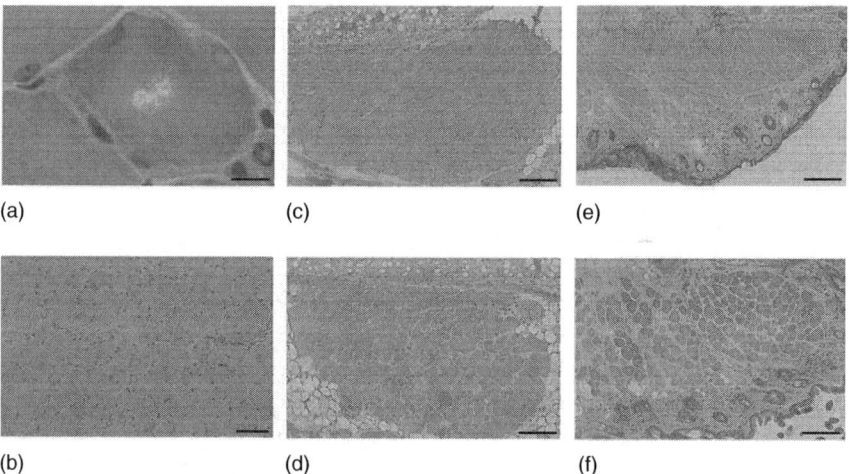

Fig. 13 Mice expressing polyalanine-expanded PABP2 protein display prominent muscle pathology consistent with the OPMD phenotype. Transgenic mice were generated using the pCAGGS expression vector that consists of the chicken beta-actin promoter and a CMV enhancer, and thus ubiquitously express either human PABP2 protein with six alanines (c, e) or nine alanines (a, b, d, f) at roughly comparable levels. Hematoxylin & eosin staining of sections from the soleus muscle (a, b), the pharynx muscle (c, d), and the eyelid muscle (e, f) reveal normal muscle histology in the six-alanine PABP2 expressing mice (c, e). However, mice expressing PABP2 with nine alanines display nonuniform muscle fiber size and prominent connective tissue in pharynx and eyelid musculature (d, f). At high power, cytoplasmic vacuoles, reminiscent of OPMD "rimmed vacuoles" are apparent in soleus muscle from the nine-alanine PABP2 expressing mice (a). Low power examination further indicates frequent central nuclei in the soleus muscle from such mice (b). (From Hino et al. (2004) *Hum. Mol. Genet.* **13**, 181, used with permission).

extraocular muscles in OPMD. Unlike other skeletal muscles, extraocular muscles are not postmitotic and they continually undergo remodeling. This requires the frequent upregulation of genes involved in cell cycling and protein synthesis. Failure of correct mRNA polyadenylation or transport in these cells may have a cumulative toxic effect resulting in the progressive degeneration of these muscles. The mechanism of OPMD could be due to an interference with polyadenylation, disturbances in intracellular trafficking of mRNA, or toxicity due to the aggregation of misfolded and/or aggregated PABP2 species. Further research will be required to distinguish between these possibilities.

6.3
Synpolydactyly (Syndactyly Type II)

Synpolydactyly (SPD) is a rare, autosomal dominant developmental disorder characterized by fused and extra digits (syndactyly and polydactyly, respectively). It is caused by the expansion of a polyalanine tract in the amino-terminal region of the transcription factor HOX-D13. *HOX* genes act in concert to coordinate axial patterning in animals. Tracts of 7–14 alanines in HOX-D13 have been linked to SPD, and disease severity is proportional to repeat length. Multiple studies support a "dominant-negative" role for expanded HOX-D13 protein in SPD. Mice null for Hox-d13 have a phenotype less severe than SPD, while mice with alanine tract expansions have a form that more closely resembles the disorder. Mice lacking Hox11, Hox12, and Hox13 have a phenotype similar to SPD, suggesting that the alanine expansion in SPD antagonizes the function of other *HOX* genes. This is supported by a genetic complementation study. In further support of this theory, humans with suspected loss-of-function mutations in the *HOX-D13* gene do not have a phenotype consistent with SPD.

6.4
Cleidocranial Dysplasia

Cleidocranial dysplasia (CCD) is a rare, autosomal dominant developmental disorder characterized by holes in the skull, dental malformations, absent or hypoplastic clavicles and maxillae, and other skeletal malformations. The primary cause of CCD is thought to be loss-of-function mutations in the gene *RUNX2*. This gene affects osteoblast differentiation and is a member of the Runt family of transcription factors. In one family, phenotypically distinct from classic CCD, an expansion from 17 to 27 alanines in RUNX2 has been detected. This family exhibited brachydactyly and a mild CCD phenotype. The difference in phenotype between the polyalanine expansion mutation family and typical presumed haploinsufficient, loss-of-function CCD patients supports a gain-of-function effect of the expanded alanine tract in this atypical family. Contraction of the tract in RUNX2 is common and does not cause a detectable phenotype.

6.5
Holoprosencephaly

Holoprosencephaly (HPE) is a common developmental malformation resulting in partial or full cyclopia, failure to develop midline structures in the ventral forebrain, and prenatal lethality. In rare cases, HPE is caused by the heterozygous expansion of a 15-amino acid alanine tract in the protein ZIC2. This protein is one member of a family of zinc finger proteins believed to regulate neurulation, left–right axis formation and other developmental processes. Other individuals with heterozygous loss-of-function mutations have a phenotype indistinguishable from those with alanine tract expansions. Partial loss of function of Zic2 in mice similarly causes developmental abnormalities similar to HPE. Therefore, expansion of the alanine tract likely causes loss of function of ZIC2 in the case of HPE.

6.6
Hand-foot-genital Syndrome

Hand-foot-genital syndrome (HFGS) is a rare, dominantly inherited developmental abnormality characterized by malformation of the distal limbs and lower urogenital tract. Short thumbs, short great toes, and abnormal carpals and tarsals are some of its salient features. Some HFGS patients have alanine tract expansions in the protein HOX-A13, which is in the same family as the SPD-associated protein HOX-D13. HOX-A13 contains three alanine tracts, the most C-terminal of which is the most commonly mutated. Expansions enlarge the second or third tract by 6 to 9 alanines. In humans, deletion of HOX-A13 causes a phenotype that is mild in comparison to HFGS caused by alanine tract mutations. Hoxa13-null mice also have a milder phenotype than mice carrying a frameshift deletion suspected to confer gain of function. This evidence suggests a dominant-negative mechanism for HFGS caused by alanine tract expansions.

6.7
Blepharophimosis-ptosis-epicanthus Inversus Syndrome

Blepharophimosis-ptosis-epicanthus inversus syndrome (BPES) is a rare, autosomal dominant developmental disorder resulting in malformation of the upper eyelids and forehead, and occasionally premature ovarian failure in women. It is caused by a number of different mutations in the gene *Forkhead L2* (*Foxl2*), the most common of which is expansion of its carboxy-terminal 14-amino acid alanine tract. FOXL2 is a highly conserved transcription factor whose role in the ovary has been studied most thoroughly. It is expressed in the ovaries and eyelids during development and adulthood. Studies in which Foxl2 have been ablated in mice show that it is important for the differentiation of ovarian granulosa cells, and its absence causes accelerated follicle cell depletion leading to POF. Knockout mice also display craniofacial abnormalities consistent with BPES, suggesting that a dominant-negative mechanism may be responsible in humans.

6.8
Syndromic and Nonsyndromic X-linked Mental Retardation

Several loosely related disorders are associated with alanine tract expansions in the Aristaless related homeobox (ARX) protein on chromosome Xp22.13. One is nonsyndromic X-linked mental retardation (XLMR), a heterogeneous condition in which mental retardation is the main consistent feature. Several syndromic XLMR disorders linked to alanine expansion in ARX include West syndrome (WS) and Partington syndrome (PRTS). WS causes progressive mental retardation with abnormal EEG and infantile seizures, while PRTS causes mental retardation, dysarthria, and dystonic movements of the hands. Brain anatomy appears normal in these disorders.

ARX is a paired-class homeodomain protein that is expressed in the ventricular and marginal zones of the developing mouse brain and continues to be expressed in adult cortex. It is suspected to play a role in neuroepithelial cell differentiation and maintenance of neuronal subtypes in the adult cortex. ARX-null mice have small brains and neuronal migration deficits. Expansions of two different alanine tracts in ARX cause XLMR: a 12-alanine tract at amino acids 144–155, and a second tract at amino acids 100–115. Both sites cause highly variable forms of XLMR, with expansions in the former alanine tract resulting in nonsyndromic XLMR, WS or PRTS. Alanine tract expansions in ARX probably result in partial loss of function of the protein. Heterozygous female carriers of expanded ARX do not exhibit XLMR, suggesting that one normal copy of the protein is sufficient for normal cognitive ability. Also, humans with null mutations of ARX suffer from X-linked lissencephaly with abnormal genitalia (XLAG), a much more severe condition causing major developmental abnormalities in the brain and genitalia. Thus, XLMR due to alanine expansions in ARX may be because of partial loss of function leading to subtle developmental defects and/or failure to maintain specific neuronal populations.

6.9
Congenital Central Hypoventilation Syndrome

Congenital central hypoventilation syndrome (CCHS) is a rare, autosomal dominant disorder causing a failure of autonomic control of breathing. It attenuates or abolishes responses to hypercarbia and hypoxemia. In the majority of cases, it is caused by the expansion of one of two alanine tracts in the protein PHOX-2B. Mutations expand the 20-residue tract to 25–29 alanines. PHOX-2B is a paired-class transcription factor containing a homeodomain. A loss-of-function mutation in murine Phox2b is homozygous lethal, and specifically prevents the development of parasympathetic ganglia. Heterozygous mice show chronic pupil dilation but no parasympathetic or respiratory disturbances. Also, a patient hemizygous for a 5-Mb deletion including Phox2b does not have CCHS. No cases of CCHS have been reported in which PHOX2B is truncated before the homeobox domain. This evidence suggests that the expansion of alanine tracts in PHOX2B may cause a subtle, dominant-negative effect on the development of respiratory control pathways.

7
Unclassified Repeat Diseases Lacking Mechanistic Explanations

7.1
Spinocerebellar Ataxia Type 10

Spinocerebellar ataxia type 10 (SCA10) is an autosomal dominant, progressive ataxia that exhibits nearly pure cerebellar signs. It appears restricted to individuals of Mexican ethnicity. SCA10 patients typically present with gait ataxia, followed by dysarthria, dysphagia, and ocular dysmetria, and most patients also experience recurrent motor seizures. Cerebellar atrophy is the most prominent neuropathological change. SCA10 is caused by the expansion of a highly polymorphic pentanucleotide repeat, $(ATTCT)n$, on chromosome 22. Unaffected individuals carry 10–22 repeats. While expanded alleles have not been successfully PCR amplified, transcript sizes on Northern blots and fragment sizes on Southern blots indicate that disease alleles can range from ~800 to 4500 repeats. Anticipation occurs in SCA10, independently supporting the causal nature of the repeat expansion. The ATTCT repeat is located in intron 9 of the *SCA10* gene, a previously unrecognized 66-kb gene that encodes a novel putative 475 amino acid protein of unknown function with few recognizable motifs or domains. The 2-kb SCA10 transcript is ubiquitously expressed though the highest levels of expression occur in the brain, testis, and adrenal glands. Within the brain, it is most highly expressed in the cerebellum and associated structures. The carboxy-terminal portion of the protein appears to contain "armadillo repeats," which are responsible for the membrane association of β-catenins. Ataxin-10 does not seem to associate with membranes, however. The entire protein is highly conserved between humans and rodents, and potential orthologs exist in Arabidopsis and Drosophila.

There are several proposed mechanisms for SCA10 pathogenesis. The most obvious hypothesis is that the large intronic expansion affects ataxin-10 expression, perhaps by altering local chromatin structure. siRNA knockdown of ataxin-10 in cell culture experiments yields higher rates of cell death in cerebellar neurons than in cortical neurons. However, the expression levels of the SCA10 transcript are not reduced in patient's lymphoblast cells, arguing against simple haploinsufficiency. Consequently, RNA gain-of-function toxicity has been proposed as the potential mechanism; however, there are no further data at this time to support such a hypothesis. Additional work will need to be done to distinguish between these and other possibilities, and will need to account for the cell-type specific pattern of neurodegeneration that occurs in the face of apparently widespread expression of the *SCA10* gene mutation.

7.2
Spinocerebellar Ataxia Type 12

Among the more recent additions to the unstable repeat disease group is spinocerebellar ataxia type 12 (SCA12), a rare, autosomal dominant disorder that may be most prevalent in Indian populations. Its symptoms are distinct from the other SCAs, typically beginning with an action tremor of the upper extremities and progressing to include hyperreflexia, mild cerebellar dysfunction, bradykinesia, increased muscle tone, psychiatric symptoms, and dementia. The brains of SCA12 patients likely undergo a slow, generalized atrophy that is most prominent in the cortex, but also results in loss of Purkinje cells in the cerebellum. Disease onset typically occurs in the third or fourth decade, and a gradually progressive disease course is typical.

SCA12 is caused by a (CAG)n expansion in chromosome 5q31 – q33. The expansion is 5' to the *PPP2R2B* gene, encoding a regulatory subunit of the protein phosphatase 2A enzyme (PP2A). This gene has many transcription start sites, some of which include the repeat (but many of which do not). GENSCAN predicts an exon including the expansion that would encode a polyserine tract, but this prediction is of low probability and considered unlikely. Other evidence suggests that the expansion is located in the promoter region, and this is currently the most widely accepted view. Unaffected individuals carry 7–32 CAG repeats while affected individuals carry 55–78 CAG triplets. The most common repeat size in unaffected individuals is 10 CAGs. The repeat is fairly stable, with only modest expansions and contractions resulting equally from maternal and paternal transmission. A significant correlation between repeat length and age of onset has not been documented.

The protein PP2A is an essential serine/threonine phosphatase expressed in all known eukaryotic cells. It is involved in diverse cellular functions, including cell growth, differentiation, DNA replication, neurotransmitter release, and apoptosis. PPP2R2B is a brain-specific regulatory subunit of PP2A. The class of regulatory subunits including PPP2R2B may affect PP2A's phosphatase activity for certain substrates, including histone-1, vimentin, and tau. It may also affect PP2A's subcellular localization.

There are several possible explanations for SCA12 pathogenesis. The first is that expanded PPP2R2B may generate a polyamino acid tract-containing protein, resulting in toxicity. Northern blots probing for sequence flanking the CAG repeat did not detect a PPP2R2B transcript, indicating that if a repeat-containing exon exists and is transcribed, it is not present at appreciable levels. Nevertheless, the possibility of polyglutamine, polyserine, or polyalanine toxicity, though unlikely, cannot be completely ruled out at this time. A second possible cause of SCA12 pathogenesis is RNA gain of function, as occurs in DM and FXTAS. The repeat expansions are smaller in SCA12 than in DM, however, and more importantly, CAG tract-containing transcripts appear to be rather scarce. A third possibility is altered splicing of the PPP2R2B transcript. Several different amino-termini are possible, the ratio of which may affect PP2A's subcellular localization. Yet another theory of SCA12 pathogenesis is that the expansion affects PPP2R2B transcript levels. Repeat expansion causes a substantial increase in PPP2R2B expression as measured in reporter assays using a neuroblastoma cell line. Altered levels of the protein could affect PP2A's specificity or subcellular localization. This has the potential to disturb a multitude of processes in the CNS. At this time, all of the above theories of SCA12 CAG repeat expansion neurotoxicity remain plausible.

7.3
Huntington's Disease Like 2 (HDL2)

Perhaps the most exciting and enigmatic recent discovery in the repeat expansion field is that of the mutational basis of a disorder known as Huntington's disease like 2 or HDL2. HDL2 is so named because it is in essence a genocopy of classical HD, as the original HDL2 pedigree was labeled with a diagnosis of HD until HD CAG repeat testing indicated that this family's HD-like disease did not result from a CAG repeat expansion in the *htt* gene. HDL2 patients present with weight loss and diminished coordination, and then develop tremors, dysarthria, hyperreflexia, and rigidity. Patients display psychiatric involvement, chorea, and dystonia, and ultimately become demented. Death occurs 15–25 years after onset, when patients become bedridden as in typical HD cases. MRI findings reveal marked atrophy of the caudate and of the cerebral cortex, making the neuropathology indistinguishable from classic HD. HDL2 patients do not show cerebellar signs or neuropathology.

As soon as it was found that HDL2 patients do not have the HD CAG repeat expansion, direct cloning methods for triplet repeat expansions of the CAG/CTG type were applied to patient samples and an expanded CAG/CTG repeat ($n = 55$) was isolated. Sequence flanking this repeat indicated that HDL2 is caused by CTG repeat expansions in the *junctophilin-3* (*JPH3*) gene, one of a family of structural proteins whose function is to link

Ca++ channels on the ER with voltage sensors on the plasma membrane. Analysis of JPH3 CTG repeat indicates that normal individuals typically carry 7–27 CTG repeats while affected HDL2 patients usually have expansions of 50–60 CTG repeats. Widespread screening of HD-like patients from around the globe suggests that the HDL2 JPH3 CTG repeat expansions are most common in individuals of African ethnicity.

The question of how the JPH3 CTG repeat expansion causes HDL2 remains unknown, but there are at least four possible explanations. These alternative (but not mutually exclusive) theories stem from the documented alternative processing of the *JPH3* gene, which permits the prediction of the CTG repeat tract as: (1) part of intron 1; (2) as part of the 3′ untranslated region; or (3) as encoding either a polyleucine or polyalanine tract. One theory is that haploinsufficiency of the Ca++ regulating brain- and testes-specific junctophilin-3 protein is responsible for HDL2. While JPH3 knockout mice display motor incoordination, no histological abnormalities are found in their brains. Further work with these mice is ongoing to evaluate this hypothesis. Evidence against simple haploinsufficiency has come from study of HDL2 patient's brain material, however. 1C2 and ubiquitin antibody immunostaining reveal intranuclear inclusions in neurons throughout the brains of these patients, with dramatic similarity in distribution to patients with classic HD (Fig. 14). At this time, the molecular basis of HDL2 is unknown; however, the incredible overlap between HD and HDL2 in terms of clinical phenotype and neuropathology strongly suggests that solving HDL2 should have profound implications for our mechanistic understanding of HD.

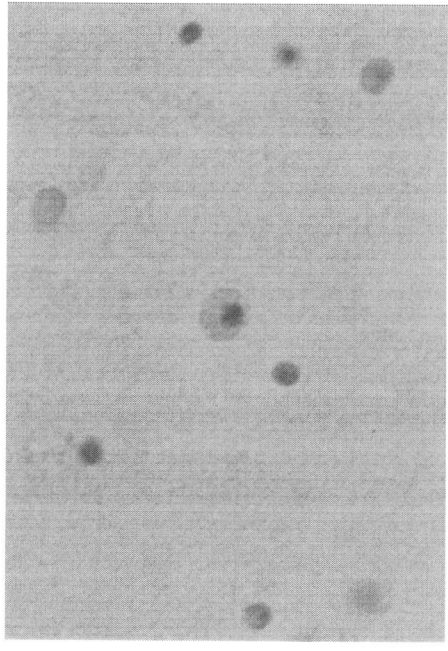

Fig. 14 HDL2 patients have 1C2-positive intranuclear inclusions. 1C2 staining of cerebral cortex (frontal lobe) from an HDL2 patient reveals a prominent intranuclear inclusion that resembles the intranuclear inclusions seen in classic HD patients. As the 1C2 antibody is directed against expanded, misfolded polyglutamine tracts, which are not predicted to be expressed from the causal *HDL2* gene, the explanation for the presence of such nuclear inclusions in HDL2 patients remains unknown. (From Margolis, R.L., O'Hearn, E., Rosenblatt, A., Willour, V., Holmes, S.E., Franz, M.L., Callahan, C., Hwang, H.S., Troncoso, J.C., Ross, C.A. (2001) A disorder similar to Huntington's disease is associated with a novel CAG repeat expansion, *Ann. Neurol.* **50**, 373–380, used with permission of *Annals of Neurology*, and John Wiley & Sons, publisher).

See also Molecular Basis of Genetics.

Bibliography

Books and Reviews

Brown, L.Y., Brown, S.A. (2004) Alanine tracts: the expanding story of human illness and trinucleotide repeats, *Trends Genet.* **20**, 51–58.

Harper, P.S. (2001) *Myotonic Dystrophy*, Saunders, London, UK.

Harper, P.S., Perutz, M. (2001) *Glutamine Repeats and Neurodegenerative Diseases: Molecular Aspects*, Oxford University Press, New York.

Jin, P., Alisch, R.S., Warren, S.T. (2004) RNA and microRNAs in fragile X mental retardation, *Nat. Cell Biol.* **6**, 1048–1053.

Ranum, L.P., Day, J.W. (2004) Pathogenic RNA repeats: an expanding role in genetic disease, *Trends Genet.* **20**, 506–512.

Wells, R.D., Warren, S.T. (1998) *Genetic Instabilities and Hereditary Neurological Diseases*, Academic Press, San Diego, CA.

Primary Literature

Albrecht, M., Golatta, M., Wullner, U., Lengauer, T. (2004) Structural and functional analysis of ataxin-2 and ataxin-3. *Eur. J. Biochem.* **271**, 3155–3170.

Amiel, J., Laudier, B., Attie-Bitach, T., Trang, H., de Pontual, L., Gener, B., Trochet, D., Etchevers, H., Ray, P., Simonneau, M. et al. (2003) Polyalanine expansion and frameshift mutations of the paired-like homeobox gene *PHOX2B* in congenital central hypoventilation syndrome, *Nat. Genet.* **33**, 459–461.

Arrasate, M., Mitra, S., Schweitzer, E.S., Segal, M.R., Finkbeiner, S. (2004) Inclusion body formation reduces levels of mutant huntingtin and the risk of neuronal death, *Nature.* **431**, 805–810.

Banfi, S., Servadio, A., Chung, M., Capozzoli, F., Duvick, L.A., Elde, R., Zoghbi, H.Y., Orr, H.T. (1996) Cloning and developmental expression analysis of the murine homolog of the spinocerebellar ataxia type 1 gene (*Sca1*), *Hum. Mol. Genet.* **5**, 33–40.

Brais, B., Bouchard, J.P., Xie, Y.G., Rochefort, D.L., Chretien, N., Tome, F.M., Lafreniere, R.G., Rommens, J.M., Uyama, E., Nohira, O. et al. (1998) Short GCG expansions in the *PABP2* gene cause oculopharyngeal muscular dystrophy, *Nat. Genet.* **18**, 164–167.

Brook, J.D., McCurrach, M.E., Harley, H.G., Buckler, A.J., Church, D., Aburatani, H., Hunter, K., Stanton, V.P., Thirion, J.P., Hudson, T. et al. (1992) Molecular basis of myotonic dystrophy: expansion of a trinucleotide (CTG) repeat at the 3' end of a transcript encoding a protein kinase family member, *Cell.* **69**, 385.

Brown, V., Jin, P., Ceman, S., Darnell, J.C., O'Donnell, W.T., Tenenbaum, S.A., Jin, X., Feng, Y., Wilkinson, K.D., Keene, J.D. et al. (2001) Microarray identification of FMRP-associated brain mRNAs and altered mRNA translational profiles in fragile X syndrome, *Cell.* **107**, 477–487.

Brown, L.Y., Odent, S., David, V., Blayau, M., Dubourg, C., Apacik, C., Delgado, M.A., Hall, B.D., Reynolds, J.F., Sommer, A. et al. (2001) Holoprosencephaly due to mutations in ZIC2: alanine tract expansion mutations may be caused by parental somatic recombination, *Hum. Mol. Genet.* **10**, 791–796.

Burnett, B., Li, F., Pittman, R.N. (2003) The polyglutamine neurodegenerative protein ataxin-3 binds polyubiquitynated proteins and has ubiquitin protease activity, *Hum. Mol. Genet.* **12**, 3195–3205.

Calado, A., Tome, F.M., Brais, B., Rouleau, G.A., Kuhn, U., Wahle, E., Carmo-Fonseca, M. (2000) Nuclear inclusions in oculopharyngeal muscular dystrophy consist of poly(A) binding protein 2 aggregates which sequester poly(A) RNA, *Hum. Mol. Genet.* **9**, 2321–2328.

Chen, S., Peng, G.H., Wang, X., Smith, A.C., Grote, S.K., Sopher, B.L., La Spada, A.R. (2004) Interference of Crx-dependent transcription by ataxin-7 involves interaction between the glutamine regions and requires the ataxin-7 carboxy-terminal region for nuclear localization, *Hum. Mol. Genet.* **13**, 53–67.

Chen, H.K., Fernandez-Funez, P., Acevedo, S.F., Lam, Y.C., Kaytor, M.D., Fernandez, M.H., Aitken, A., Skoulakis, E.M., Orr, H.T., Botas, J. et al. (2003) Interaction of Akt-phosphorylated ataxin-1 with 14-3-3 mediates neurodegeneration in spinocerebellar ataxia type 1, *Cell.* **113**, 457–468.

Cummings, C.J., Reinstein, E., Sun, Y., Antalffy, B., Jiang, Y., Ciechanover, A., Orr, H.T., Beaudet, A.L., Zoghbi, H.Y. (1999) Mutation of the E6-AP ubiquitin ligase reduces nuclear inclusion frequency while accelerating polyglutamine-induced pathology in SCA1 mice, *Neuron.* **24**, 879–892.

David, G., Abbas, N., Stevanin, G., Durr, A., Yvert, G., Cancel, G., Weber, C., Imbert, G., Saudou, F., Antoniou, E. et al. (1997) Cloning of the *SCA7* gene reveals a highly unstable CAG repeat expansion, *Nat. Genet.* **17**, 65–70.

Davies, S.W., Turmaine, M., Cozens, B.A., DiFiglia, M., Sharp, A.H., Ross, C.A., Scherzinger, E., Wanker, E.E., Mangiarini, L., Bates, G.P. (1997) Formation of neuronal intranuclear inclusions underlies the neurological dysfunction in mice transgenic for the HD mutation, *Cell.* **90**, 537–548.

Day, J.W., Ricker, K., Jacobsen, J.F., Rasmussen, L.J., Dick, K.A., Kress, W., Schneider, C., Koch, M.C., Beilman, G.J., Harrison, A.R. et al. (2003) Myotonic dystrophy type 2: molecular, diagnostic and clinical spectrum, *Neurology.* **60**, 657–664.

De Baere, E., Beysen, D., Oley, C., Lorenz, B., Cocquet, J., De Sutter, P., Devriendt, K., Dixon, M., Fellous, M., Fryns, J.P. et al. (2003) FOXL2 and BPES: mutational hotspots, phenotypic variability, and revision of the genotype-phenotype correlation, *Am. J. Hum. Genet.* **72**, 478–487.

DiFiglia, M., Sapp, E., Chase, K.O., Davies, S.W., Bates, G.P., Vonsattel, J.P., Aronin, N. (1997) Aggregation of huntingtin in neuronal intranuclear inclusions and dystrophic neurites in brain, *Science.* **277**, 1990–1993.

Dunah, A.W., Jeong, H., Griffin, A., Kim, Y.M., Standaert, D.G., Hersch, S.M., Mouradian, M.M., Young, A.B., Tanese, N., Krainc, D. (2002) Sp1 and TAFII130 transcriptional activity disrupted in early Huntington's disease, *Science.* **296**, 2238–2243.

Ellerby, L.M., Hackam, A.S., Propp, S.S., Ellerby, H.M., Rabizadeh, S., Cashman, N.R., Trifiro, M.A., Pinsky, L., Wellington, C.L., Salvesen, G.S. et al. (1999) Kennedy's disease: caspase cleavage of the androgen receptor is a crucial event in cytotoxicity, *J. Neurochem.* **72**, 185–195.

Fu, Y.H., Kuhl, D.P., Pizzuti, A., Pieretti, M., Sutcliffe, J.S., Richards, S., Verkerk, A.J., Holden, J.J., Fenwick, R.G., Jr., Warren, S.T., et al. (1991) Variation of the CGG repeat at the fragile X site results in genetic instability: resolution of the Sherman paradox, *Cell.* **67**, 1047–1058.

Garden, G.A., Libby, R.T., Fu, Y.H., Kinoshita, Y., Huang, J., Possin, D.E., Smith, A.C., Martinez, R.A., Fine, G.C., Grote, S.K., et al. (2002) Polyglutamine-expanded ataxin-7 promotes noncell-autonomous Purkinje cell degeneration and displays proteolytic cleavage in ataxic transgenic mice, *J. Neurosci.* **22**, 4897–4905.

Gecz, J. (2000) The *FMR2* gene, FRAXE and non-specific X-linked mental retardation: clinical and molecular aspects, *Ann. Hum. Genet.* **64**, 95–106.

Goodman, F.R., Scambler, P.J. (2001) Human *HOX* gene mutations, *Clin. Genet.* **59**, 1–11.

Grabczyk, E., Usdin, K. (2000) The GAA*TTC triplet repeat expanded in Friedreich's ataxia impedes transcription elongation by T7 RNA polymerase in a length and supercoil dependent manner, *Nucleic Acids Res.* **28**, 2815–2822.

Greene, E., Handa, V., Kumari, D., Usdin, K. (2003) Transcription defects induced by repeat expansion: fragile X syndrome, FRAXE mental retardation, progressive myoclonus epilepsy type 1, and Friedreich ataxia, *Cytogenet. Genome Res.* **100**, 65–76.

Hagerman, R.J., Leehey, M., Heinrichs, W., Tassone, F., Wilson, R., Hills, J., Grigsby, J., Gage, B., Hagerman, P.J. (2001) Intention tremor, parkinsonism, and generalized brain atrophy in male carriers of fragile X, *Neurology.* **57**, 127–130.

Helmlinger, D., Hardy, S., Sasorith, S., Klein, F., Robert, F., Weber, C., Miguet, L., Potier, N., Van-Dorsselaer, A., Wurtz, J.M., et al. (2004) Ataxin-7 is a subunit of GCN5 histone acetyltransferase-containing complexes, *Hum. Mol. Genet.* **13**, 1257–1265.

Hillman, M.A., Gecz, J. (2001) Fragile XE-associated familial mental retardation protein 2 (FMR2) acts as a potent transcription activator, *J. Hum. Genet.* **46**, 251–259.

Hodgson, J.G., Agopyan, N., Gutekunst, C.A., Leavitt, B.R., LePiane, F., Singaraja, R., Smith, D.J., Bissada, N., McCutcheon, K., Nasir, J., et al. (1999) A YAC mouse model for Huntington's disease with full-length mutant huntingtin, cytoplasmic toxicity, and selective striatal neurodegeneration, *Neuron.* **23**, 181–192.

Holmes, S.E., O'Hearn, E., McInnis, M.G., Gorelick-Feldman, D.A., Kleiderlein, J.J., Callahan, C.A., Ingersoll-Ashworth, R.G., Sherr, M., Sumner, A.J. (1999) Expansion of a novel CAG repeat in the 5′ region of a gene encoding a subunit of protein phosphatase 2A is associated with spinocerebellar ataxia type 12, *Am. J. Hum. Genet.* **65**, A4.

Holmes, S.E., O'Hearn, E., Rosenblatt, A., Callahan, C., Hwang, H.S., Ingersoll-Ashworth, R.G., Fleisher, A., Stevanin, G., Brice, A., Potter, N.T., et al. (2001) A repeat expansion in the gene encoding junctophilin-3 is associated with Huntington disease-like 2, *Nat. Genet.* **29**, 377–378.

Huntington's Disease, Collaborative, Research and Group (1993) A novel gene containing a trinucleotide repeat that is expanded and unstable on Huntington's disease chromosomes, *Cell* **72**, 971–983.

Huynh, D.P., Figueroa, K., Hoang, N., Pulst, S.M. (2000) Nuclear localization or inclusion body formation of ataxin-2 are not necessary for SCA2 pathogenesis in mouse or human, *Nat. Genet.* **26**, 44–50.

Ishikawa, K., Fujigasaki, H., Saegusa, H., Ohwada, K., Fujita, T., Iwamoto, H., Komatsuzaki, Y., Toru, S., Toriyama, H., Watanabe, M., et al. (1999) Abundant expression and cytoplasmic aggregations of [alpha]1A voltage-dependent calcium channel protein associated with neurodegeneration in spinocerebellar ataxia type 6, *Hum. Mol. Genet.* **8**, 1185–1193.

Jiang, H., Mankodi, A., Swanson, M.S., Moxley, R.T., Thornton, C.A. (2004) Myotonic dystrophy type 1 is associated with nuclear foci of mutant RNA, sequestration of muscleblind proteins and deregulated alternative splicing in neurons, *Hum. Mol. Genet.* **13**, 3079–3088.

Katsuno, M., Adachi, H., Kume, A., Li, M., Nakagomi, Y., Niwa, H., Sang, C., Kobayashi, Y., Doyu, M., Sobue, G. (2002) Testosterone reduction prevents phenotypic expression in a transgenic mouse model of spinal and bulbar muscular atrophy, *Neuron.* **35**, 843–854.

Kitamura, K., Yanazawa, M., Sugiyama, N., Miura, H., Iizuka-Kogo, A., Kusaka, M., Omichi, K., Suzuki, R., Kato-Fukui, Y., Kamiirisa, K., et al. (2002) Mutation of ARX causes abnormal development of forebrain and testes in mice and X-linked lissencephaly with abnormal genitalia in humans, *Nat. Genet.* **32**, 359–369.

Klement, I.A., Skinner, P.J., Kaytor, M.D., Yi, H., Hersch, S.M., Clark, H.B., Zoghbi, H.Y., Orr, H.T. (1998) Ataxin-1 nuclear localization and aggregation: role in polyglutamine-induced disease in SCA1 transgenic mice, *Cell.* **95**, 41–53.

Koob, M.D., Moseley, M.L., Schut, L.J., Benzow, K.A., Bird, T.D., Day, J.W., Ranum, L.P. (1999) An untranslated CTG expansion causes a novel form of spinocerebellar ataxia (SCA8), *Nat. Genet.* **21**, 379–384.

Kremer, E.J., Pritchard, M., Lynch, M., Yu, S., Holman, K., Baker, E., Warren, S.T., Schlessinger, D., Sutherland, G.R., Richards, R.I. (1991) Mapping of DNA instability at the fragile X to a trinucleotide repeat sequence p(CCG)n, *Science.* **252**, 1711–1714.

La Spada, A.R., Wilson, E.M., Lubahn, D.B., Harding, A.E., Fischbeck, K.H. (1991) Androgen receptor gene mutations in X-linked spinal and bulbar muscular atrophy, *Nature.* **352**, 77–79.

La Spada, A.R., Roling, D.B., Harding, A.E., Warner, C.L., Spiegel, R., Hausmanowa Petrusewicz, I., Yee, W.C., Fischbeck, K.H. (1992) Meiotic stability and genotype-phenotype correlation of the trinucleotide repeat in X-linked spinal and bulbar muscular atrophy, *Nat. Genet.* **2**, 301–304.

La Spada, A.R., Fu, Y., Sopher, B.L., Libby, R.T., Wang, X., Li, L.Y., Einum, D.D., Huang, J., Possin, D.E., Smith, A.C., et al. (2001) Polyglutamine-expanded ataxin-7 antagonizes CRX function and induces cone-rod dystrophy in a mouse model of SCA7, *Neuron.* **31**, 913–927.

Lafreniere, R.G., Rochefort, D.L., Chretien, N., Rommens, J.M., Cochius, J.I., Kalviainen, R., Nousiainen, U., Patry, G., Farrell, K., Soderfeldt, B. et al. (1997) Unstable insertion in the 5′ flanking region of the *cystatin B* gene is the most common mutation in progressive myoclonus epilepsy type 1, EPM1, *Nat. Genet.* **15**, 298–302.

Lalioti, M.D., Antonarakis, S.E., Scott, H.S. (2003) The epilepsy, the protease inhibitor and the dodecamer: progressive myoclonus epilepsy, cystatin b and a 12-mer repeat expansion, *Cytogenet. Genome Res.* **100**, 213–223.

Lalioti, M.D., Scott, H.S., Antonarakis, S.E. (1999) Altered spacing of promoter elements due to the dodecamer repeat expansion contributes to reduced expression of the cystatin B gene in EPM1, *Hum. Mol. Genet.* **8**, 1791–1798.

Lin, X., Antalffy, B., Kang, D., Orr, H.T., Zoghbi, H.Y. (2000) Polyglutamine expansion down-regulates specific neuronal genes before pathologic changes in SCA1, *Nat. Neurosci.* **3**, 157–163.

Liquori, C.L., Ricker, K., Moseley, M.L., Jacobsen, J.F., Kress, W., Naylor, S.L., Day, J.W., Ranum, L.P. (2001) Myotonic dystrophy type 2 caused by a CCTG expansion in intron 1 of ZNF9, *Science.* **293**, 864–867.

Loesch, D.Z., Huggins, R.M., Bui, Q.M., Epstein, J.L., Taylor, A.K., Hagerman, R.J. (2002) Effect of the deficits of fragile X mental retardation protein on cognitive status of fragile x males and females assessed by robust pedigree analysis, *J. Dev. Behav. Pediatr.* **23**, 416–423.

Lunkes, A., Lindenberg, K.S., Ben-Haiem, L., Weber, C., Devys, D., Landwehrmeyer, G.B., Mandel, J.L., Trottier, Y. (2002) Proteases acting on mutant huntingtin generate cleaved products that differentially build up cytoplasmic and nuclear inclusions, *Mol. Cell.* **10**, 259–269.

Mangiarini, L., Sathasivam, K., Seller, M., Cozens, B., Harper, A., Hetherington, C., Lawton, M., Trottier, Y., Lehrach, H., Davies, S.W., et al. (1996) Exon 1 of the *HD* gene with an expanded CAG repeat is sufficient to cause a progressive neurological phenotype in transgenic mice, *Cell.* **87**, 493–506.

Mankodi, A., Logigian, E., Callahan, L., McClain, C., White, R., Henderson, D., Krym, M., Thornton, C.A. (2000) Myotonic dystrophy in transgenic mice expressing an expanded CUG repeat, *Science.* **289**, 1769–1773.

Margolis, R.L., O'Hearn, E., Rosenblatt, A., Willour, V., Holmes, S.E., Franz, M.L., Callahan, C., Hwang, H.S., Troncoso, J.C., Ross, C.A. (2001) A disorder similar to Huntington's disease is associated with a novel CAG repeat expansion, *Ann. Neurol.* **50**, 373–380.

Matsuura, T., Fang, P., Lin, X., Khajavi, M., Tsuji, K., Rasmussen, A., Grewal, R.P., Achari, M., Alonso, M.E., Pulst, S.M., et al. (2004) Somatic and germline instability of the ATTCT repeat in spinocerebellar ataxia type 10, *Am. J. Hum. Genet.* **74**, 1216–1224.

Matsuura, T., Yamagata, T., Burgess, D.L., Rasmussen, A., Grewal, R.P., Watase, K., Khajavi, M., McCall, A.E., Davis, C.F., Zu, L., et al. (2000) Large expansion of the ATTCT pentanucleotide repeat in spinocerebellar ataxia type 10, *Nat. Genet.* **26**, 191–194.

McCampbell, A., Taylor, J.P., Taye, A.A., Robitschek, J., Li, M., Walcott, J., Merry, D., Chai, Y., Paulson, H., Sobue, G., et al. (2000) CREB-binding protein sequestration by expanded polyglutamine, *Hum. Mol. Genet.* **9**, 2197–2202.

Monckton, D.G., Cayuela, M.L., Gould, F.K., Brock, G.J., Silva, R., Ashizawa, T. (1999) Very large (CAG)(n) DNA repeat expansions in the sperm of two spinocerebellar ataxia type 7 males, *Hum. Mol. Genet.* **8**, 2473–2478.

Mortlock, D.P., Innis, J.W. (1997) Mutation of HOXA13 in hand-foot-genital syndrome, *Nat. Genet.* **15**, 179–180.

Mosemiller, A.K., Dalton, J.C., Day, J.W., Ranum, L.P. (2003) Molecular genetics of spinocerebellar ataxia type 8 (SCA8), *Cytogenet. Genome Res.* **100**, 175–183.

Mundlos, S., Otto, F., Mundlos, C., Mulliken, J.B., Aylsworth, A.S., Albright, S., Lindhout, D., Cole, W.G., Henn, W., Knoll, J.H., et al. (1997) Mutations involving the transcription factor CBFA1 cause cleidocranial dysplasia, *Cell.* **89**, 773–779.

Muragaki, Y., Mundlos, S., Upton, J., Olsen, B.R. (1996) Altered growth and branching patterns in synpolydactyly caused by mutations in HOXD13, *Science.* **272**, 548–551.

Mutsuddi, M., Marshall, C.M., Benzow, K.A., Koob, M.D., Rebay, I. (2004) The spinocerebellar ataxia 8 noncoding RNA causes neurodegeneration and associates with staufen in Drosophila, *Curr. Biol.* **14**, 302–308.

Nagafuchi, S., Yanagisawa, H., Sato, K., Shirayama, T., Ohsaki, E., Bundo, M., Takeda, T., Tadokoro, K., Kondo, I., Murayama, N., et al. (1994) Dentatorubral and pallidoluysian atrophy expansion of an unstable CAG trinucleotide on chromosome 12p, *Nat. Genet.* **6**, 14–18.

Nakamura, K., Jeong, S.Y., Uchihara, T., Anno, M., Nagashima, K., Nagashima, T., Ikeda, S., Tsuji, S., Kanazawa, I. (2001) SCA17, a novel autosomal dominant cerebellar ataxia caused by an expanded polyglutamine in TATA-binding protein, *Hum. Mol. Genet.* **10**, 1441–1448.

Okazawa, H., Rich, T., Chang, A., Lin, X., Waragai, M., Kajikawa, M., Enokido, Y., Komuro, A., Kato, S., Shibata, M., et al. (2002) Interaction between mutant ataxin-1 and

PQBP-1 affects transcription and cell death, *Neuron.* **34**, 701–713.

Oostra, B.A., Willemsen, R. (2003) A fragile balance: FMR1 expression levels, *Hum. Mol. Genet.* **12**, **Spec No 2**, R249–R257.

Orr, H.T., Chung, M.Y., Banfi, S., Kwiatkowski, Jr., T.J., Servadio, A., Beaudet, A.L., McCall, A.E., Duvick, L.A., Ranum, L.P., Zoghbi, H.Y. (1993) Expansion of an unstable trinucleotide CAG repeat in spinocerebellar ataxia type 1, *Nat. Genet.* **4**, 221–226.

Paulson, H.L., Perez, M.K., Trottier, Y., Trojanowski, J.Q., Subramony, S.H., Das, S.S., Vig, P., Mandel, J.L., Fischbeck, K.H., Pittman, R.N. (1997) Intranuclear inclusions of expanded polyglutamine protein in spinocerebellar ataxia type 3, *Neuron.* **19**, 333–344.

Penrose, L.S. (1948) The problem of anticipation in pedigrees of dystrophica myotonica, *Ann. Eugen.* **14**, 125–132.

Perutz, M.F., Johnson, T., Suzuki, M., Finch, J.T. (1994) Glutamine repeats as polar zippers: their possible role in inherited neurodegenerative diseases, *Proc. Natl. Acad. Sci. U. S. A.* **91**, 5355–5358.

Poirier, M.A., Li, H., Macosko, J., Cai, S., Amzel, M., Ross, C.A. (2002) Huntingtin spheroids and protofibrils as precursors in polyglutamine fibrilization, *J. Biol. Chem.* **277**, 41032–41037.

Pulst, S.M., Nechiporuk, A., Nechiporuk, T., Gispert, S., Chen, X.N., Lopes-Cendes, I., Pearlman, S., Starkman, S., Orozco-Diaz, G., Lunkes, A., et al. (1996) Moderate expansion of a normally biallelic trinucleotide repeat in spinocerebellar ataxia type 2, *Nat. Genet.* **14**, 269–276.

Rasmussen, A., Matsuura, T., Ruano, L., Yescas, P., Ochoa, A., Ashizawa, T., Alonso, E. (2001) Clinical and genetic analysis of four Mexican families with spinocerebellar ataxia type 10, *Ann. Neurol.* **50**, 234–239.

Sanchez, I., Mahlke, C., Yuan, J. (2003) Pivotal role of oligomerization in expanded polyglutamine neurodegenerative disorders, *Nature.* **421**, 373–379.

Sanpei, K., Takano, H., Igarashi, S., Sato, T., Oyake, M., Sasaki, H., Wakisaka, A., Tashiro, K., Ishida, Y., Ikeuchi, T., et al. (1996) Identification of the *spinocerebellar ataxia type 2* gene using a direct identification of repeat expansion and cloning technique, DIRECT, *Nat. Genet.* **14**, 277–284.

Satterfield, T.F., Jackson, S.M., Pallanck, L.J. (2002) A Drosophila homolog of the polyglutamine disease gene *SCA2* is a dosage-sensitive regulator of actin filament formation, *Genetics.* **162**, 1687–1702.

Scherzinger, E., Lurz, R., Turmaine, M., Mangiarini, L., Hollenbach, B., Hasenbank, R., Bates, G.P., Davies, S.W., Lehrach, H., Wanker, E.E. (1997) Huntingtin-encoded polyglutamine expansions form amyloid-like protein aggregates in vitro and in vivo, *Cell.* **90**, 549–558.

Schilling, G., Wood, J.D., Duan, K., Slunt, H.H., Gonzales, V., Yamada, M., Cooper, J.K., Margolis, R.L., Jenkins, N.A., Copeland, N.G., et al. (1999) Nuclear accumulation of truncated atrophin-1 fragments in a transgenic mouse model of DRPLA, *Neuron.* **24**, 275–286.

Seznec, H., Simon, D., Monassier, L., Criqui-Filipe, P., Gansmuller, A., Rustin, P., Koenig, M., Puccio, H. (2004) Idebenone delays the onset of cardiac functional alteration without correction of Fe-S enzymes deficit in a mouse model for Friedreich ataxia, *Hum. Mol. Genet.* **13**, 1017–1024.

Sopher, B.L., Thomas, P.S., Jr., LaFevre-Bernt, M.A., Holm, I.E., Wilke, S.A., Ware, C.B., Jin, L.W., Libby, R.T., Ellerby, L.M., La Spada, A.R. (2004) Androgen receptor YAC transgenic mice recapitulate SBMA motor neuronopathy and implicate VEGF164 in the motor neuron degeneration, *Neuron.* **41**, 687–699.

Stevanin, G., Camuzat, A., Holmes, S.E., Julien, C., Sahloul, R., Dode, C., Hahn-Barma, V., Ross, C.A., Margolis, R.L., Durr, A., et al. (2002) CAG/CTG repeat expansions at the Huntington's disease-like 2 locus are rare in Huntington's disease patients, *Neurology.* **58**, 965–967.

Stromme, P., Mangelsdorf, M.E., Shaw, M.A., Lower, K.M., Lewis, S.M., Bruyere, H., Lutcherath, V., Gedeon, A.K., Wallace, R.H., Scheffer, I.E., et al. (2002) Mutations in the human ortholog of Aristaless cause X-linked mental retardation and epilepsy, *Nat. Genet.* **30**, 441–445.

Sutherland, G.R., Baker, E. (1992) Characterisation of a new rare fragile site easily confused with the fragile X, *Hum. Mol. Genet.* **1**, 111–113.

Uda, M., Ottolenghi, C., Crisponi, L., Garcia, J.E., Deiana, M., Kimber, W., Forabosco, A., Cao, A., Schlessinger, D., Pilia, G. (2004) Foxl2 disruption causes mouse ovarian failure by pervasive blockage of follicle development, *Hum. Mol. Genet.* **13**, 1171–1181.

Ueno, S., Kondoh, K., Kotani, Y., Komure, O., Kuno, S., Kawai, J., Hazama, F., Sano, A. (1995) Somatic mosaicism of CAG repeat in dentatorubral-pallidoluysian atrophy (DRPLA), *Hum. Mol. Genet.* **4**, 663–666.

Wallis, J., Williamson, R., Chamberlain, S. (1990) Identification of a hypervariable microsatellite polymorphism within D9S15 tightly linked to Friedreich's ataxia, *Hum. Genet.* **85**, 98–100.

Warren, S.T. (1997) Polyalanine expansion in synpolydactyly might result from unequal crossing-over of HOXD13, *Science.* **275**, 408–409.

Watase, K., Weeber, E.J., Xu, B., Antalffy, B., Yuva-Paylor, L., Hashimoto, K., Kano, M., Atkinson, R., Sun, Y., Armstrong, D.L., et al. (2002) A long CAG repeat in the mouse Sca1 locus replicates SCA1 features and reveals the impact of protein solubility on selective neurodegeneration, *Neuron.* **34**, 905–919.

Wellington, C.L., Ellerby, L.M., Gutekunst, C.A., Rogers, D., Warby, S., Graham, R.K., Loubser, O., van Raamsdonk, J., Singaraja, R., Yang, Y.Z., et al. (2002) Caspase cleavage of mutant huntingtin precedes neurodegeneration in Huntington's disease, *J. Neurosci.* **22**, 7862–7872.

Wellington, C.L., Ellerby, L.M., Hackam, A.S., Margolis, R.L., Trifiro, M.A., Singaraja, R., McCutcheon, K., Salvesen, G.S., Propp, S.S., Bromm, M., et al. (1998) Caspase cleavage of gene products associated with triplet expansion disorders generates truncated fragments containing the polyglutamine tract, *J. Biol. Chem.* **273**, 9158–9167.

Willemsen, R., Hoogeveen-Westerveld, M., Reis, S., Holstege, J., Severijnen, L.A., Nieuwenhuizen, I.M., Schrier, M., van Unen, L., Tassone, F., Hoogeveen, A.T., et al. (2003) The FMR1 CGG repeat mouse displays ubiquitin-positive intranuclear neuronal inclusions; implications for the cerebellar tremor/ataxia syndrome, *Hum. Mol. Genet.* **12**, 949–959.

Wirtschafter, J.D., Ferrington, D.A., McLoon, L.K. (2004) Continuous remodeling of adult extraocular muscles as an explanation for selective craniofacial vulnerability in oculopharyngeal muscular dystrophy, *J. Neuroophthalmol.* **24**, 62–67.

Yoo, S.Y., Pennesi, M.E., Weeber, E.J., Xu, B., Atkinson, R., Chen, S., Armstrong, D.L., Wu, S.M., Sweatt, J.D., Zoghbi, H.Y. (2003) SCA7 knockin mice model human SCA7 and reveal gradual accumulation of mutant ataxin-7 in neurons and abnormalities in short-term plasticity, *Neuron.* **37**, 383–401.

Yvert, G., Lindenberg, K.S., Picaud, S., Landwehrmeyer, G.B., Sahel, J.A., Mandel, J.L. (2000) Expanded polyglutamines induce neurodegeneration and trans-neuronal alterations in cerebellum and retina of SCA7 transgenic mice, *Hum. Mol. Genet.* **9**, 2491–2506.

Zakany, J., Duboule, D. (1996) Synpolydactyly in mice with a targeted deficiency in the HoxD complex, *Nature.* **384**, 69–71.

Zhang, S., Xu, L., Lee, J., Xu, T. (2002) Drosophila atrophin homolog functions as a transcriptional corepressor in multiple developmental processes, *Cell* **108**, 45–46.

Zhuchenko, O., Bailey, J., Bonnen, P., Ashizawa, T., Stockton, D.W., Amos, C., Dobyns, W.B., Subramony, S.H., Zoghbi, H.Y., Lee, C.C. (1997) Autosomal dominant cerebellar ataxia (SCA6) associated with small polyglutamine expansions in the alpha 1A-voltage-dependent calcium channel, *Nat. Genet.* **15**, 62–69.

Zuccato, C., Tartari, M., Crotti, A., Goffredo, D., Valenza, M., Conti, L., Cataudella, T., Leavitt, B.R., Hayden, M.R., Timmusk, T., et al. (2003) Huntingtin interacts with REST/NRSF to modulate the transcription of NRSE-controlled neuronal genes, *Nat. Genet.* **35**, 76–83.

35
Molecular Genetics of Down Syndrome

Charles J. Epstein
University of California, San Francisco, CA, USA

1	**Phenotype of Down Syndrome**	1046
2	**Cytogenetics**	1047
3	**Structure of Chromosome 21**	1047
4	**Phenotypic Mapping**	1048
5	**Pathogenesis**	1049
6	**Animal Models**	1049
6.1	Trisomic Mice	1050
6.2	Transgenic Mice	1052
6.3	Gene Expression Analyses	1053
	Bibliography	1053
	Books and Reviews	1053
	Primary Literature	1054

Keywords

Acrocentric
A type of chromosome in which the centromere is very close to one end.

Autosomes
All of the chromosomes in the genome except for the sex chromosomes (X and Y).

Genomics and Genetics. Edited by Robert A. Meyers.
Copyright © 2007 Wiley-VCH Verlag GmbH & Co. KGaA, Weinheim
ISBN: 978-3-527-31609-0

Centromere
The structure within each chromosome at which the fibers required to move the chromosomes during meiosis or cell division (mitosis) attach.

Meiosis
The process within the germ cells during which genetic recombination occurs, and in which the number of chromosomes within the egg or sperm is reduced from the 46 found in somatic cells to 23.

Trisomy
The presence in the genome of three rather than two copies of a specific chromosome.

Down syndrome (DS) is the commonest of the genetically caused forms of mental retardation. It occurs with a frequency of approximately 1 per 800 to 1 per 1000 live births, and its incidence increases with increasing maternal age. DS is caused by the presence of an extra chromosome 21 within the genome, which, in turn, results in a 50% increase in the expression of the genes contained on the chromosome. By mechanisms currently undefined, the increased expression of several genes on human chromosome 21 results in a syndrome characterized by mental and growth retardation, a distinctive set of major and minor congenital malformations, a variety of cellular abnormalities, and, later in life, by the development of Alzheimer disease.

1
Phenotype of Down Syndrome

The most immediately apparent, if not the most serious, manifestations of Down syndrome (DS) are the many minor dysmorphic features that collectively constitute its distinctive physical phenotype. Salient among these are upslanting palpebral fissures, epicanthic folds, flat nasal bridge, brachycephaly, short broad hands, incurved fifth fingers, loose skin of the nape of the neck, open mouth with protruding tongue, and transverse palmar creases. Although any individual with DS will have many of the characteristic features and can be easily recognized as having the disorder, none of these features is present in all persons with DS. DS affects the nervous system in three principal ways: hypotonia, which occurs in virtually all newborns and infants; delayed psychomotor development in infancy and mental retardation throughout life; and neuronal degeneration during the adult years. The latter process, which is pathologically identical with Alzheimer disease (presenile and/or senile dementia), results in significant pathologic changes in the brain and may further compromise the already impaired mental functioning. DS is associated with two types of major congenital malformations. Most frequent (about 40%) is congenital heart disease, usually of the endocardial cushion type or one of its variants. Gastrointestinal tract abnormalities occur in about 4.5% of individuals, more than half of whom have duodenal stenosis or atresia. Structural abnormalities of the thymus and functional defects in T-cell function leading to an increased susceptibility to infection are present, and there is a 10- to 18-times normal incidence of childhood leukemia with a frequent occurrence of acute megakaryoblastic leukemia.

2
Cytogenetics

Down syndrome is the phenotypic manifestation of trisomy 21. As such, it occurs when a third copy of chromosome 21 is present in the genome, either as a free chromosome or as part of a Robertsonian fusion chromosome (in which the long arms of two acrocentric chromosomes are joined at the centromeres). Except in 2 to 4% of cases in which mosaicism, with two populations of cells, one diploid and one trisomic, exists, all cells of the body are trisomic. Although these cytogenetic abnormalities involve most or all of chromosome 21, there are rare cases involving translocations in which only part of the long arm of the chromosome is triplicated. Depending on the region of the chromosome that is involved, such cases may or may not express the classical DS phenotype.

It has long been recognized that the risk of having a child with DS increases with maternal age and that the distribution of maternal age in the population of women having children is the primary determinant of the overall incidence of DS. A variety of estimates of the incidence of DS in the newborn population have been made, and most recent figures are in the vicinity of 1 per 1000. The figures can be broken down to provide maternal age-specific rates per 1000 live births: 0.6 at 20 years, 1.0 at 30 years, 2.6 at 35 years, 9.1 at 40 years, 24.9 at 44 years, and 41.2 at 46 years. Analyses using DNA markers have shown that maternal nondisjunction, the failure of paired chromosomes to separate properly, accounts for 92% of all cases of trisomy 21, 65% occurring at meiosis I, 23% at meiosis II, and 3% postzygotically (at mitosis). Eight percent of cases are paternal in origin, 3% at meiosis I and 5% at meiosis II. A reduced rate of recombination is associated with meiosis I errors and an increased rate with meiosis II errors.

3
Structure of Chromosome 21

Chromosome 21 is an acrocentric chromosome with a genetic length estimated to be 46 cM. It is the smallest of the human autosomes constituting approximately 1.7% of the length of the haploid genome. In physical terms, chromosome 21 is an acrocentric chromosome with the centromere very close to one end and with a very small short arm. The short arm (21p) terminates in a satellite region that may vary in size. Proximal to the satellite is the stalk (secondary constriction) that, as the nucleolar organizer region (NOR), contains multiple copies of the ribosomal RNA genes (RNR4) and stains characteristically with silver. The degree of silver staining appears to be a representation of the molecular activity of the ribosomal RNA genes that the chromosome contains. The region of 21p adjacent to the centromere contains highly repeated DNA sequences that consist of the satellite (including alphoid) and the "724" families of sequences. None of these gene families is unique to chromosome 21. It is believed that these families of repeated gene sequences may be involved in the juxtaposition or association of the satellite regions (satellite association) of the acrocentric chromosomes during mitotic metaphase and with the formation of the nucleolus during interphase.

The major part of chromosome 21 is the long arm (21q) that has a characteristic banding pattern consisting of 3 or 4 bands at low resolution and as many as 11 dark and light bands resolvable by prometaphase banding. With one possible exception, all genes of known function (other than those for ribosomal RNA) are located on this arm of chromosome 21, and only this arm is essential for normal development and function. The presence of a Robertsonian fusion chromosome in which the short arms of two acrocentric chromosomes (sometimes both chromosomes 21) are deleted does not cause detectable abnormalities if the genome is otherwise balanced.

The sequence of the long arm of chromosome 21 was completed in 2000 with an estimated 99.7% coverage. This arm has a length of 33.7 Mbp and is relatively gene-poor, with less than half the number of identified genes found on the similarly sized chromosome 22. In the original annotation, it was judged to contain approximately 127 known genes, 98 predicted genes (of which 69% have no similarity to known proteins), and 59 pseudogenes. These numbers have been repeatedly revised as annotation of the sequence has proceeded, and the most recent figures are 178 confirmed and 36 predicted genes. There is evidence that not all expressed genes have been identified and that some predicted genes do not exist. Among the known genes are at least 10 kinases, 5 genes in the ubiquitination pathway, 5 cell adhesion molecules, 7 ion channels, 5 members of the interferon receptor family, and several transcription factors. About 22.4% of the chromosome consists of interspersed Alu sequences and LINE1 elements.

4
Phenotypic Mapping

Cases in which only part of chromosome 21 is triplicated have been intensively studied to arrive at a phenotypic map that will permit a correlation between particular phenotypic features of DS and specific regions or loci on the chromosome. The consensus of studies carried out before molecular markers became available is that the full DS phenotype, as manifested by mental retardation, congenital heart disease, characteristic facial appearance, hand anomalies, and dermatoglyphic changes, appears when band 21q22 is duplicated, and of this, subbands 21q22.1 and probably 21q22.2 are required. Molecular analysis has been used to define the extent of the triplication of regions of chromosome 21, and the results have been used to generate phenotypic maps. Although interpretations vary, it seems clear that many of the phenotypic features appear to be associated with imbalance in the region surrounding and distal to *D21S55*. However, there is also evidence for contributions of genes outside the *D21S55* region, especially in the proximal part of 21q, although the region responsible for congenital heart disease seems to be confined to the distal part of 21q. The genes that, when present in extra copies, contribute to impaired cognition appear to be located in several regions of the chromosome.

5
Pathogenesis

The immediate consequence of an aneuploid state is a gene dosage effect for each of the loci present on the unbalanced chromosome or chromosome segment. Such gene dosage effects have been reported for several chromosome 21 loci. With the exception of the amyloid precursor protein (APP) in the brain from fetuses with DS, and possibly a few other chromosome 21–encoded genes, the measured increase in activities or concentrations in trisomic cells is close to the theoretically expected value of 1.5. Taken in the aggregate, these results confirm the existence of quite precise dosage effects in cells aneuploid for chromosome 21. However, changes in the synthesis of proteins coded for by genes elsewhere in the genome have been observed, presumably as the result of genomic dysregulation resulting from the increased expression of chromosome w31 genes. However, this dysregulation is limited and certainly does not affect the vast majority of expressed genes.

On the basis of studies in animal models (see the following section), decreased platelet serotonin uptake and prostaglandin synthesis, and abnormal myoneural junctions, all of which occur in DS, have been ascribed to increased activity of CuZn-superoxide dismutase (SOD1). With these exceptions, it has not been possible to relate any component of the DS phenotype to overexpression of specific loci. However, a great amount of attention has been devoted to the pathogenesis of the cognitive deficits and Alzheimer disease in DS. The effects of increased expression of several of the loci on chromosome 21 have been considered, and it has been possible to construct plausible explanations for how overexpression of each locus might affect the development, function, and integrity of the central nervous system. A recent case in point is the gene for cystathionine synthase, overexpression of which in DS results in depressed levels of plasma homocysteine. It has been suggested that the altered homocysteine metabolism could compromise the folate-dependent resynthesis of methionine and thereby create a functional folate deficiency, the consequences of which on the central nervous system function could be many. However, in considering such possibilities, it is important to keep in mind that whereas imbalance of an individual locus could well have significant effects on the central nervous system, it is unlikely that it will provide a unitary explanation for the entire range of deficits found in DS. Nevertheless, it is possible that the development of Alzheimer disease in DS is related to the overproduction of APP, perhaps superimposed on an intrinsically defective nervous system.

6
Animal Models

The fact that many of the consequences of aneuploidy in humans arise during the period of morphogenesis and affect the development and function of the central nervous system places special stumbling blocks in the way of their investigation. Research on events occurring during gestation, especially early gestation, is both technically impractical and, currently and for the foreseeable future, ethically and legally impossible. Similarly, there are severe limitations to the study of the central nervous system function in living individuals, and experimental genetic alterations are impossible. It is for this reason that interest has turned to the development of animal models for DS.

Since ultimate concern is with humans and with human disease, we require models that will duplicate the human condition – in developmental and functional terms – as closely as possible. No model can be an exact one, since no other organism duplicates the human with respect to all of his or her biologic and genetic attributes. Nevertheless, models based on other mammals, which share numerous biologic similarities with humans, seem most appropriate. Primates with the genetic and clinical equivalent of trisomy 21 and DS have been observed but do not offer significant advantages over humans for investigational purposes. Therefore, the mouse has been the model animal of choice for several reasons, including ease of manipulation and genetic control and the similarities to humans in the processes of morphogenesis and probably of central nervous system function, in neurobiological if not psychological terms. Furthermore, despite considerable rearrangement of the mammalian genome, sizable chromosomal regions carrying many genes remain intact and structurally similar in both humans and mice.

Models based on the mouse are not without their problems. It may not always be easy or even possible to obtain postnatally viable animals when sizable regions of the genome are unbalanced. Furthermore, even if it were possible, these animals will never be able to duplicate the higher central nervous system functions, such as cognition, which are so vulnerable to the effects of aneuploidy in humans. As difficult as it is to define mental retardation in an aneuploid human, it is even more difficult to specify the proper functional homology in an aneuploid mouse. Nevertheless, elementary forms of learning are common to all animals with an evolved nervous system, and there appears to be a conservation of the relevant biochemical mechanisms. Therefore, the mouse can be considered to be a legitimate model for learning and memory in humans.

6.1
Trisomic Mice

The modeling of human trisomy 21 in the mouse started at the two ends of the possible spectrum of such models. On one end is the trisomic mouse with an extra copy of mouse chromosome 16, the mouse chromosome that most closely resembles human chromosome 21. At the other extreme is the transgenic mouse with an extra copy or copies of a single human chromosome 21 gene. More recently, however, several intermediate possibilities, with duplications of only parts of mouse chromosome 16 or the insertion of mouse or human yeast artificial chromosome (YAC) transgenes, have been developed.

With perhaps one exception, all the loci in human 21q from the centromere to the *MX1* locus in the distal part of band 21q22.3, approximately 130 to 150 in number, map to the distal part of mouse chromosome 16 (see http://www.informatics.jax.org). The remaining loci, not on mouse chromosome 16, have been mapped to mouse chromosomes 10 and 17. These loci are outside the region shown by phenotypic mapping to be responsible for the facial features of DS. In addition to coding regions, there are also a large number of sequences conserved between the human and mouse genomes that appear to have a functional significance but cannot be identified as protein-coding or RNA genes.

Because of the genetic discrepancies between mouse chromosome 16 and human chromosome 21, trisomy for all of mouse chromosome 16, while reproducing much of the genetic imbalance associated with human trisomy 21, results in an overall degree of imbalance that is more extensive than in human trisomy and in an overall phenotype that is more severe. For this reason, the results of studies with mouse trisomy 16 (Ts16) must be viewed with concern insofar as modeling DS is concerned. To correct this situation, segmentally trisomic mice (designated Ts65Dn) in which only the human chromosome 21 orthologous region is triplicated were produced. In these animals, approximately 115 of the original number of 225 genes on human chromosome 21q are represented. The trisomic animals do not have the major anomalies present in Ts16. Rather, their salient phenotypic abnormalities are confined to the craniofacial structures and the central nervous system, although they do exhibit male sterility. As in persons with DS, there is a reduction in the size and granule cell density of the cerebellum, a reduced sensitivity to pain, elevation of brain *myo*-inositol, and abnormalities of the dendritic spines. Although motor learning is normal, Ts65Dn mice have major deficits in learning and behavior, and show abnormalities in long-term potentiation (LTP) and depression and in hippocampal cAMP generation. The pattern of behavioral and learning deficits suggests impairment of both hippocampal spatial learning and of prefrontal functions, and the LTP and cAMP abnormalities are also indicative of hippocampal dysfunction. Further evidence for abnormality of the hippocampus of the Ts65Dn mouse is provided by the presence of giant dendritic spines with aberrant connections in hippocampal neurons and by suppression of LTP by inhibitory inputs. The finding in Ts65Dn mice of an age-dependent loss of function and atrophy of cholinergic neurons in the medial septum of the basal forebrain, which correlates with the loss of cognitive function and appears to be related to an abnormality of nerve growth factor transport, is also consistent with hippocampal dysfunction. It is of note, in this regard, that a reduction in Trk-A immunoreactive cholinergic neurons has been observed in the nucleus basalis of persons with DS.

A second segmentally trisomic mouse, Ts1Cje, is functionally trisomic for the region of mouse chromosome 16 distal to *Sod1*. This trisomy involves about 87 chromosome 21–orthologous genes and still includes the region responsible for the major phenotypic abnormalities in DS. Their learning deficits are similar to but somewhat less severe than those of the Ts65Dn animals. Similarly, there is a reduction in the volume of the cerebellum, but granule cell density is much less affected. However, Ts1Cje animals do not appear to exhibit degeneration of the cholinergic neurons, suggesting that the gene(s) responsible for this phenotype are located in the region of difference between the two segmental trisomies, a region containing about 30 chromosome 21–orthologous genes. A third segmentally trisomic mouse, designated Ms1Ts65, carries a triplication of this region of difference. These animals have only minimal learning deficits, but they are still not normal.

Taken together with the comparison of Ts65Dn and Ts1Cje, the results with Ms1Ts65 indicate that the major genes responsible for the behavioral and learning abnormalities of Ts65Dn mice are located distal to *Sod1*. These studies constitute the first step in phenotypic mapping of the segmental-trisomy DS models by a subtractive approach in which a known phenotype (as in Ts65Dn) is compared with the phenotype obtained after reduction of the region of trisomy (as in Ts1Cje). The same approach has been used to compare the alterations in craniofacial morphogenesis produced by Ts65Dn and Ts1Cje. Both trisomics demonstrate alterations in the formation of the craniofacial skeleton that are similar to what is observed in DS, with the changes in Ts1Cje being largely the same as those in Ts65Dn. This again demonstrates that the major genes responsible for these skeletal abnormalities in the Ts65Dn mice are located distal to *Sod1*. The ability to generate these various segmental trisomics and to compare their phenotypes directly provides a precedent for the systematic creation of smaller and smaller segmental trisomics and the dissection of the trisomic phenotype.

6.2
Transgenic Mice

In contrast with the formation of trisomic animals, an approach that permits the analysis of the effects of increased dosage of individual human chromosome 21 genes is the construction of transgenic mice. Several strains of such transgenic mice carrying *SOD1*, the gene for human CuZnSOD, have been made and shown to have from 1.6- to 6-fold increased activities of CuZnSOD in the brain and other tissues. Several effects of this increased activity of CuZnSOD that resemble changes found in persons with DS have been found. Transgenic mouse platelets have diminished levels of serotonin and decreased rates of serotonin accumulation that are attributed to a diminished pH gradient across the granule membrane. There is a reduction in prostaglandin E_2 and prostaglandin D_2 biosynthesis in transgenic primary fetal cells and in the cerebellum and hippocampus. The neuromuscular junctions of the transgenic tongue muscles are abnormal, with atrophy, degeneration, withdrawal and destruction of terminal axons, development of multiple small terminals, decreased ratio of terminal axon area to postsynaptic membrane, and hyperplastic secondary folds. Transgenic mice also have anatomical abnormalities of the thymus and premature thymic involution. Impairment of LTP has also been reported.

Many of the abnormalities in transgenic tissues just discussed are presumed to be mediated by alterations in the metabolism of oxygen free radicals resulting from the increased CuZnSOD activity (this enzyme mediates the first step in the ultimate conversion of superoxide anions to water).

Limited work has been done with other types of transgenic mice carrying individual human chromosome 21 genes or their murine homologues. Although they do not develop the full pathology of AD, mice expressing the human APP-751 isoform gene do have diffuse and amorphous extracellular deposits of β-amyloid that resemble those of early AD and manifest deficits in spatial learning. Transgenic mice overexpressing the Ets2 transcription factor develop craniofacial and other skeletal anomalies, with brachycephaly and short necks, thought to resemble anomalies found in DS and mouse Ts16. However, similar anomalies are not found in any of the segmental trisomics of chromosome 16, even though *Ets2* is present in three copies. Overexpression of *HMG14* results in abnormalities of the thymus and epithelial cysts, and mice transgenic for the mouse single-minded gene, *mSim2*, have impaired spatial learning.

The transgenes just described involve loci present on mouse chromosome 16 as well as human chromosome 21. Therefore, the absence of similar findings in trisomic mice involving the same regions is of concern. The following two loci, however, are represented on mouse chromosome 10, and the trisomy 16 models are, therefore, not relevant. Transgenic mice overexpressing the gene for the protein subunit $S100\beta$ were reported to be hyperactive, to have learning impairments, and to manifest astrocytosis, axonal proliferation, and a significant loss of dendrites in the hippocampus. Mice overexpressing liver-type phosphofructokinase had elevated activity during fetal life but not during adult life.

Another approach to the generation of transgenic models for DS has been the insertion of genomic segments larger than individual genes in the form of yeast artificial chromosomes. A mouse produced in this manner had abnormalities of learning and behavior too, but less severe than that observed with the segmental trisomies, Ts1Cje and Ts65Dn. Since the major and possibly only gene in the integrated segment of human DNA was the human minibrain gene (*DYRK*), it was suggested that imbalance of this locus is responsible for the observed abnormalities. Support for this inference is provided by mice transgenic for the *Dyrk1A* gene. These animals display delayed neuromotor development and impairment in spatial learning with defective reference memory, findings considered indicative of hippocampal and prefrontal cortex dysfunction.

6.3
Gene Expression Analyses

The cloning and sequencing of both the mouse and human genomes has made it possible to undertake *in situ* hybridization studies of the expression of chromosome 21–orthologous genes in diploid mouse embryos, fetuses, and postnatal animals. Broad expression of the human chromosome 21 orthologues is present in early embryos and becomes more restricted as development proceeds. The number of genes expressed and their patterns of expression differ among tissues, but it is possible to identify several clusters of coexpressing genes. By correlating expression patterns with specific features of the DS phenotype, candidate genes have been proposed for congenital heart and intestinal defects.

See also Molecular Basis of Genetics.

Bibliography

Books and Reviews

Antonarakis, S.E., Lyle, R., Chrast, R., Scott, H.S. (2001) Differential gene expression studies to explore the molecular pathophysiology of Down syndrome, *Brain Res. Rev.* **36**, 265–274.

Capone, G.T. (2001) Down syndrome: advances in molecular biology and the neurosciences, *Dev. Behav. Pediatr.* **22**, 40–59.

Dierssen, M., Fillat, C., Crnic, L., Arbonés, M., Flórez, J., Estivill, X. (2001) Murine models for Down syndrome, *Physiol. Behav.* **73**, 859–871.

Epstein, C.J. (1986) *The Consequences of Chromosome Imbalance. Principles, Mechanisms, and Models*, Cambridge University Press, New York.

Epstein, C.J. (Ed.) (1993) *The Phenotypic Mapping of Down Syndrome and Other Aneuploid Conditions*, Wiley-Liss, New York.

Epstein, C.J. (2001) Down Syndrome (trisomy 21), in: Scriver, C.R., Beaudet, A.L., Sly, W.S., Valle, D. (Eds.) *The Metabolic and Molecular Bases of Inherited Disease*, 8th edition, McGraw-Hill, New York, pp. 1123–1256.

Epstein, C.J., Hassold, T., Lott, I.T., Nadel, L., Patterson, D. (Eds.) (1995) *Etiology and*

Pathogenesis of Down Syndrome, Wiley-Liss, New York.

Galdzicki, Z., Siarey, R., Pearce, R., Stoll, J., Rapoport, S.I. (2001) On the cause of mental retardation in Down syndrome: extrapolation from full and segmental trisomy 16 models, *Brain Res. Rev.* **35**, 115–145.

Hassold, T., Hunt, P. (2001) To err (meiotically) is human: the genesis of human aneuploidy, *Nat. Rev. Genet.* **2**, 280–291.

Pueschel, S.M., Rynders, J.E. (Eds.) (1982) *Down Syndrome. Advances in Biomedicine and the Behavioral Sciences*, Ware, Cambridge, MA.

Reeves, R.H., Baxter L.L., Richtsmeier, J.T. (2001) Too much of a good thing: mechanisms of gene action in Down syndrome, *Trends Genet.* **17**, 83–88.

Primary Literature

Altafaj, X., Dierssen, M., Baamonde, C., Martí, E., Visa, J., Guimerà, J., Oset, M., González, J.R., Flórez, J., Fillat, C., Estivill, X. (2001) Neurodevelopmental delay, motor abnormalities and cognitive deficits in transgenic mice overexpressing *Dyrk1A* (*minibrain*), a murine model of Down's syndrome, *Hum. Mol. Genet.* **10**, 1915–1923.

Chrast, R., Scott, H.S., Papasavvas, M.P., Rossier, C., Antonarakis, E.S., Barras, C., Davisson, M.T., Schmidt, C., Estivill, X., Dierssen, M., Pritchard, M., Antonarakis, S.E. (2000) The mouse brain transcriptome by SAGE: differences in gene expression between P30 brains of the partial trisomy 16 mouse model of Down syndrome (Ts65Dn) and normals, *Genome Res.* **10**, 2006–2021.

Cooper, J., Salehi, A., Delcroix, J.-D., Howe, C.L., Belichenko, P.V., Chua-Couzens, J., Kilbridge, J.K., Carlson, E.J., Epstein, C.J., Mobley, W.C. (2001) Failed retrograde transport of NGF in a mouse model of Down's syndrome: reversal of cholinergic neurodegenerative phenotypes following NGF infusion, *Proc. Natl. Acad. Sci. U.S.A.* **98**, 10439–10444.

Dermitzakis, E.T., Reymond, A., Lyle, R., Scamuffa, N., Ucla, C., Deutsch, S., Stevenson, B.J., Flegel, V., Bucher, P., Jongeneel, C.V., Antonarakis, S.E. (2002) Human chromosome 21 gene expression atlas in the mouse, *Nature* **420**, 578–582.

Ema, M., Ikegami, S., Hosoya, T., Mimura, J., Ohtani, H., Nakao, K., Inokuchi, K., Katsuki. M., Fujii-Kuriyama, Y. (1999) Mild impairment of learning and memory in mice overexpressing the mSim2 gene located on chromosome 16: an animal model of Down's syndrome, *Hum. Mol. Genet.* **8**, 1409–1415.

Engidawork, E., Lubec, G. (2001) Protein expression in Down syndrome brain, *Amino Acids* **21**, 331–361.

Gitton, Y., Dahmanane, N., Balk, S., Ruiz, I., Altaba, A., Neidhardt, L., Scholtze, M., Herrmann, B.G., Kahlem, P., Benkahla, A., Schrinner, S., Yildirimman, R., Herwig, R., Lehrach, H., Yaspo, M.-L. (2002) A gene expression map of human chromosome 21 orthologues in the mouse, *Nature* **420**, 586–600.

Hattori, M., Fujiyama, A., Taylor, T.D., Watanabe, H., Yada, T., Park, H.-S., Toyoda, A. et al. (2000) The DNA sequence of human chromosome 21, *Nature* **405**, 311–319.

Holtzman, D.M., Bayney, R.M., Li, Y.W., Khosrovi, H., Berger, C.N., Epstein, C.J., Mobley, W.C. (1992) Dysregulation of gene expression in mouse trisomy 16, an animal model of Down syndrome, *EMBO J.* **11**, 619–627.

Hyde, L.A., Frisone, D.F., Crnic, L.S. (2001) Ts65Dn mice, a model for Down syndrome, have deficits in context discrimination learning suggesting impaired hippocampal function, *Behav. Brain Res.* **118**, 53–60.

Korenberg, J.R., Chen, X., Schipper, R., Sun, Z., Gonsky, R., Gerwehr, S., Carpenter, N., Daumer, C., Dignan, P., Disteche, C., Graha, J.M. Jr., Hudgins, L., McGillivray, B., Miyazaki, K., Ogasawara, N., Park, J.P., Pagon, R., Pueschel, S., Sack, G., Say, B., Schuffenhauer, S., Soukup, S., Yamanaka, T. (1994) Down syndrome phenotypes: the consequences of chromosome imbalance, *Proc. Natl. Acad. Sci. U.S.A.* **91**, 4997–5001.

Kapranov, P., Cawley, S.E., Drenkow, J., Bekiranov, S., Strausberg, R.L., Fodor, S.P.A., Gingeras, T.R. (2002) Large-scale transcriptional activity in chromosomes 21 and 22, *Science* **296**, 916–919.

Lamb, N.E., Feingold, E., Savage, A., Avramopoulos, D., Freeman, S., Gu, Y., Hallberg, A., Hersey, J., Karadima, G., Pettay, D., Saker, D., Shen, J., Taft, L., Mikkelsen, M., Petersen, M.B., Hassold, T., Sherman, S.L. (1997) Characterization of susceptible chiasma configurations that increase the risk for maternal nondisjunction of chromosome 21, *Hum. Mol. Genet.* **6**, 1391–1399.

Mouse Genome Sequencing Consortium (2002) Initial sequencing and comparative analysis of the mouse genome, *Nature* **420**, 520–562.

Mural, R.J., Adams, M.D., Myers, E.W., Smith, H.O., Miklos, G.L.G., Wides, R. et al. (2002) A comparison of whole-genome shotgun-derived mouse chromosome 116 and the human genome, *Science* **296**, 1661–1671.

Pinter, J.D., Eliez, S., Schmitt, J.E., Capone, G.T., Reiss, A.L. (2001) Neuroanatomy of Down's syndrome: a high resolution MRI study, *Am. J. Psychiatry* **158**, 1659–1665.

Pletcher, M.T., Wiltshire, T., Cabin, D.E., Villanueva, M., Reeves, R.H. (2001) Use of comparative physical and sequence mapping to annotate mouse chromosome 16 and human chromosome 21, *Genomics* **74**, 45–54.

Pogribna, M., Melnyk, S., Pogribny, I., Chango, A., Yi, P., James, S.J. (2001) Homocysteine metabolism in children with Down syndrome: in vitro modulation, *Am. J. Hum. Genet.* **69**, 88–95.

Reeves, R.H., Irving, N.G., Moran, T.H., Wohn, A., Kitt, C., Sisodia, S.S., Schmidt, C., Bronson, R.T., Davisson, M.T. (1995) A mouse model for Down syndrome exhibits learning and behaviour deficits, *Nat. Genet.* **11**, 177–184.

Reymond, A., Camargo, A.A., Deutsch, S., Stevenson, B.J., Parmigiani, R.B., Ucla, C., Bettoni, F., Rossier, C., Lyle, R., Guipponi, M., de Souza, S., Iseli, C., Jongeneel, C.V., Bucher, P., Simpson, A.J., Antonarakis, S.E. (2002) Nineteen additional unpredicted transcripts from human chromosome 21, *Genomics* **79**, 824–832.

Reymond, A., Friedli, M., Henrichsen, C.N., Chapot, F., Deutsch, S., Ucla, C., Rossier, C., Lyle, R., Guipponi, M., Antonarakis, S.E. (2001) From PREDs and open reading frames to cDNA isolation: revisiting the human chromosome 21 transcription map, *Genomics* **78**, 46–54.

Reymond, A., Marigo, V., Yaylaoglu, M.B., Leoni, S., Ucla, C., Scamuffa, N., Caccioppoli, C., Dermitzakis, E.T., Lyle, R., Banfi, S., Eichele, G., Antonarakis, S.E., Ballabio, A. (2002) Human chromosome 21 gene expression atlas in the mouse, *Nature* **420**, 582–586.

Richtsmeier, J.T., Zumwalt, A., Carlson, E.J., Epstein, C.J., Reeves, R.H. (2002) Craniofacial anomalies in segmentally trisomic mouse models for Down syndrome, *Am. J. Med. Genet.* **107**, 317–324.

Sago, H., Carlson, E.J., Smith, D.J., Rubin, E.M., Crnic, L.S., Huang, T.-T., Epstein, C.J. (2000) Genetic dissection of the region associated with behavioral abnormalities in mouse models of Down syndrome, *Pediatr. Res.* **48**, 606–613.

Sinet, P.M., Théophile, D., Rahmani, Z., Chettouh, Z., Blouin, J.L., Prieur, M., Noel, B., Delabar, J.M. (1994) Mapping of the Down syndrome phenotype on chromosome 21 at the molecular level, *Biomed. Pharmacother.* **48**, 247–252.

Weitzdoerfer, R., Fountoulakis, M. (2002) Reduction of actin-related protein complex 2/3 in fetal Down syndrome brain, *Biochem. Biophys. Res. Commun.* **293**, 836–841.

36
Molecular Genetics of Hemophilia

Francesco Giannelli and Peter M. Green
Department of Medical and Molecular Genetics, Division of Genetics and Development, Guy's, King's and St. Thomas' School of Medicine, London, UK

1	**Genetics of the Hemophilias** 1059	
1.1	The Factor VIII Gene and its Product 1061	
1.2	The Factor IX Gene and its Product 1063	
2	**Hemophilia A Mutations** 1065	
2.1	Gross Sequence Changes 1066	
2.2	Small Sequence Changes 1067	
2.2.1	Changes Affecting RNA Processing 1067	
2.2.2	Changes Affecting mRNA Translation 1068	
2.2.3	Mutations Causing Alterations of the Fine Structure of Factor VIII 1069	
2.3	Functional Interpretation of Observed Sequence Changes and Correlations between Genotype and Phenotype 1071	
3	**Hemophilia B Mutations** 1072	
3.1	Gross Sequence Changes 1073	
3.2	Small Sequence Changes 1073	
3.2.1	Mutations Affecting Transcription, RNA Processing, and Translation 1074	
3.2.2	Missense Mutations and Single Amino Acid Deletions 1075	
3.3	Functional Interpretation of Observed Sequence Changes and Correlations between Genotype and Phenotype 1076	
4	**Considerations Arising from the Study of Mutations in Hemophilia A and B** 1077	
5	**Mutational Heterogeneity and Genetic Counseling in the Hemophilias** 1078	
6	**Progress in the Treatment of Hemophilia** 1079	

Genomics and Genetics. Edited by Robert A. Meyers.
Copyright © 2007 Wiley-VCH Verlag GmbH & Co. KGaA, Weinheim
ISBN: 978-3-527-31609-0

36 Molecular Genetics of Hemophilia

Bibliography 1080
Books and Reviews 1080
Primary Literature 1080

Keywords

CpG and CpG island
CpG is a dinucleotide formed by C followed by G, and CpG islands are genomic regions unusually rich in CpGs and found before genes and especially genes with ubiquitous expression patterns.

Founder Mutation
A mutation traceable to a common ancestor and found relatively frequently in some populations.

Frameshift
Mutation of a coding sequence due to loss or addition of one or two nucleotides or multiples thereof. This alters the amino acid sequence from the point of mutation onward and frequently leads to premature termination of the encoded polypeptide.

Male-driven Evolution
A situation where the male, by having a higher mutation rate than the female, contributes a greater proportion of the variation acted on by natural selection.

Missense
Mutation resulting in the substitution of one amino acid for another.

Mutational Load
The sum of the deleterious mutations present in a zygote.

Nonsense
Mutation converting a codon specifying an amino acid into one determining the end of a polypeptide sequence.

Uninemic
Of events affecting a single molecule of double-stranded DNA as that present in a chromatid or a prereplication chromosome.

Zygote
The cell formed by the fusion of a male and female gamete and representing the origin of a new potential individual.

■ The two X-linked hemorrhagic diseases hemophilia A and hemophilia B have made an important contribution to the understanding of human genetics. The first historical record of the inheritance of a disease and of "genetics"-motivated advice relates to these diseases, as do the first description of the X-linked pattern of inheritance, the appreciation of equilibria between selection and mutation, and the first attempt to estimate mutation rates. In modern times, the hemophilias have clearly illustrated the successes and problems of replacement therapy and have provided some insight into the causes of immunological complication associated with the therapeutic administration of the missing gene products. It is therefore not surprising that nowadays they attract the attention of scientists and clinicians involved in the development of gene therapy. The hemophilias, which affect 1/5000 males in the population, are members of a large group of diseases characterized by very high mutational heterogeneity. This group is of particular importance because, on the one hand, the wealth of natural mutants that each member possesses allows accumulation of detailed information on the structural features that are important to the functions of the gene and gene product involved in the disease; on the other, this creates the least favorable situation for the development of fully efficient strategies for the provision of the carrier and prenatal diagnoses that are necessary for genetic counseling. Surprisingly, mutation analysis in hemophilia A has revealed that homologous intranemic recombination between duplicated sequences can represent an important cause of recurring gene inactivation.

1
Genetics of the Hemophilias

A clear reference to the X-linked diseases, now called *hemophilias*, is found in the Babylonian Talmud of the fifth century A.D., wherein the Rabbi advises mothers who had lost two sons as a result of uncontrolled bleeding after circumcision not to expose other sons to such a risk. Furthermore, the X-linked recessive pattern of inheritance of these diseases, characterized by affected males and unaffected females who transmit the disorder, was lucidly described in 1803, more than a century before the discovery of sex-linked inheritance and 50 years ahead of Mendel's work. In 1935, it was realized that the loss of hemophilia genes due to the affected males' low probability of reproduction would have led to the elimination of the disease if an equilibrium had not existed between such loss and the gain of hemophilia genes by new mutations. Under such an equilibrium, the rate of renewal of hemophilia genes in the population is $1/3\,(1-f)$ per generation, where f is the effective fertility or genetic fitness of the affected male and 1/3 expresses the well-known fact that males account for one-third of the X-linked genes in the population, as they have only one X chromosome, whereas females have two. In 1947, investigation of the linkage between hemophilia and color blindness (another X-linked trait) offered the first hint that hemophilia was not a single disease, because while most families showed close linkage between hemophilia and color blindness, one did not. Five years later, however, the existence of two forms of hemophilia was clearly demonstrated by complementation of the coagulation defect in a mixture of plasma from different hemophiliacs. Now, we know that deficiency of coagulation factor VIII (FVIII) causes hemophilia A, while deficiency of factor IX (FIX) causes hemophilia B and that the identical clinical features of these diseases are entirely to be expected, since FIX and FVIII are respectively the enzyme and cofactor of a functional unit that contributes to the proteolytic cascade, which converts fibrinogen into fibrin by activating coagulation factor X (Fig. 1).

Hemophilia A is 5 to 6 times more frequent than hemophilia B and, currently, in the United Kingdom, it has an incidence of approximately 1/5000 males. Until the introduction of modern therapy, the effective fertility of patients with hemophilia A or B was half that of normal individuals and, therefore, the rate of renewal of the pool of hemophilia A and B genes in the population was 1/6 per generation. This is in keeping with the clinical heterogeneity of the diseases. Since hemophilia A is 5 to 6 times more frequent than hemophilia B, and the rate of renewal of hemophilia A and B genes is similar, it follows that the rate of mutation for hemophilia A is about 5 to 6 times greater than that for hemophilia B. Indirect estimates of mutation rates in hemophilia have been based on the equilibrium postulated above, but recently, a direct estimate independent of the postulate was made in hemophilia B, through the study of the UK population, and this gave a value of 7.7×10^{-6} mutations per gene per generation, with a 95% confidence interval of 6.2 to 9.1. However, the mutation rate is higher in the male than in the female germline as the estimate of the former is 18.8×10^{-6} and that of the latter is 2.18×10^{-6} so that the ratio

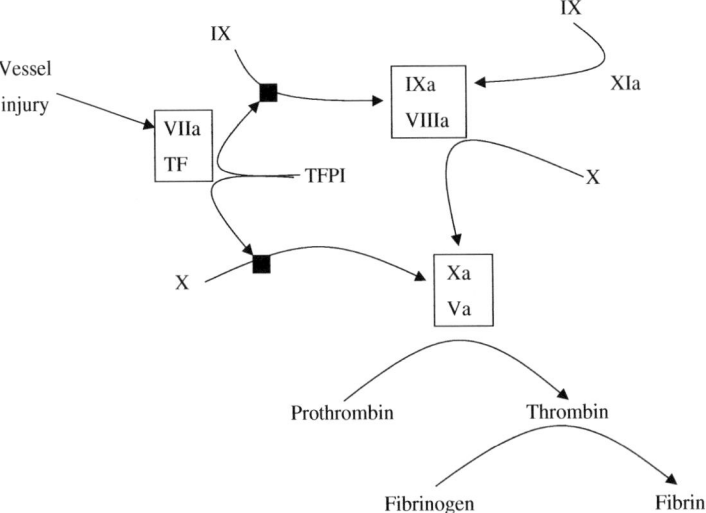

Fig. 1 The role of FVIII and FIX in blood coagulation. After tissue damage, activated FVII (FVIIa) in complex with tissue factor (TF) activates both FIX and FX, and, in turn, activated FX (FXa) and thrombin activate FVIII (not shown). Inhibition of the FVIIa–TF complex by TFPI makes further conversion of FIX to its active form dependent on FXIa and activation of FX on the FVIIIa–FIXa complex.

of male to female mutation rates is 8.64, with a 95% confidence interval of 5.46 to 14.5. This high ratio is in keeping with the theory of male-driven evolution and may be explained by the different development and physiology of the male and female gonads. Extrapolation from these data led to the suggestion that the deleterious mutation rate in the human genome is greater than 1 and possibly as great as 4 per zygote per generation. This is important because such a high mutational load in turn suggests that the long-term survival of humans depends on genetic recombination and, hence, sexual reproduction, which allows the elimination of chromosome regions carrying deleterious mutations by a process of truncating selection. This is a process that allows the elimination of mutations in bunches by preferentially eliminating zygotes with multiple deleterious mutations. In this way, it can prevent the decline in genetic fitness from generation to generation that could follow deleterious mutation rates as high as suggested above.

1.1
The Factor VIII Gene and its Product

The FVIII gene (*F8*) is 186-kb long (Fig. 2) and contains 26 exons ranging in size from 69 to 3106 bp. The latter is an internal exon (number 14) and as such is of exceptional size. The introns also vary considerably in size, from 200 bp (intron 17) to 32.4 kb (intron 22). The latter intron is of particular interest because it contains a CpG island associated with two additional RNA transcripts, one of opposite polarity to the FVIII transcript, entirely contained within intron 22 and defining an intronless gene, *F8A*, of about 2 kb and one of the same polarity as FVIII, *F8B*. This contains a first exon derived from intron 22 that contributes eight

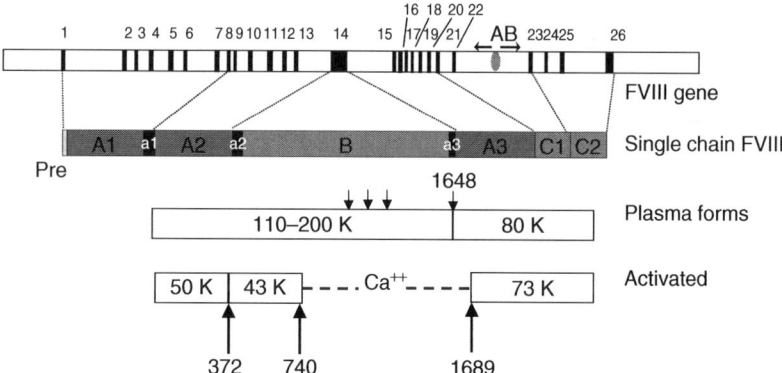

Fig. 2 Factor VIII gene (*F8*) and protein. Diagram of *F8* gene (top bar) and FVIII (lower bars). Exons are indicated by black boxes and numbered (1–26). Arrows in opposite directions over intron 22 indicate the presence of CpG island (green box) associated with the start of two transcripts of opposite orientation (A, B). Dotted lines indicate exons coding for the different protein domains (see text). Arrows in plasma and activated forms indicate site of proteolytic cleavages (see text) and numbers associated with arrows indicate position of amino acid at the amino terminal end of cleavage sites. Numbers in bars are weight in Daltons of individual protein segments. (See color plate p. vi.)

codons to a message comprising exons 23 to 26 of the *F8* gene. The functional significance of these two transcripts is still uncertain. A protein encoded by *F8B* has not been found yet, while the protein encoded by *F8A* has been found associated to Huntingtin. The above CpG island in intron 22 and the *F8A* gene are part of a 9.5-kb sequence called *int22h* that is repeated 500 and 600 kb more telomerically.

The FVIII mRNA is thought to be 9028-nt long with a 170 leader sequence followed by 2351 codons and an 1805-nt untranslated 3' tail. The primary translation product starts with a prepeptide of 19 residues needed for intracellular transport and secretion. The remainder of the protein consists of six domains and three small acidic peptides (a_1 to a_3) arranged as follows: $A_1a_1A_2a_2Ba_3A_3C_1C_2$. The A domains consist of approximately 330 amino acids (aa) and are homologous to ceruloplasmin. The B domain (encoded by exon 14) contains 983 aa and is unique. The C domains consist of approximately 150 aa and are homologous to a recently described milk fat globule binding protein and the slime mold's discoidin. Factor V, the cofactor of factor X that is responsible for the activation of prothrombin, is a homolog of FVIII and has a similar protein domain structure; however, it lacks the three acidic peptides (a_1, a_2, and a_3) and has a B domain with no homology to that of FVIII.

The FVIII polypeptide undergoes extensive intracellular modification including glysosylation, sulfation, and proteolytic cleavage. Glycosylation is both O and N linked, and FVIII contains 26 glycosylatable asparagines of which 19 are in the B domain. Sulfation is known to occur at the following tyrosines: Tyr_{346}, Tyr_{718}, Tyr_{719}, Tyr_{723}, Tyr_{1664}, and Tyr_{1680}. Secretion of FVIII requires not only post translational glycosylation of the asparagines in the B domain but also processing of the glycosidic chains to fully glucose-trimmed mannose 9 structures. The transfer of FVIII from the endoplasmic reticulum to the Golgi apparatus is achieved with the intervention of trafficking factors, as revealed by the fact that absence of endoplasmic reticulum Golgi intermediate compartment -53, or ERGIC-53 for short, causes combined deficiency of both FVIII and FV.

Prior to secretion, FVIII is cleaved not only to release the prepeptide but also at the B/a_3 boundary (aa 1648–1649) to form a molecule with a light chain consisting of $a_3A_3C_1C_2$ (aa 1649–2232) and a heavy chain consisting of $A_1a_1A_2a_2B$. This chain, however, is of variable length since the B domain is also cleaved at variable internal positions.

Circulating FVIII is bound to and protected by the von Willebrand factor (vWF). Two FVIII regions are important for this binding: the a_3 peptide and in particular the sulfated tyrosine at position 1680 within this peptide and a region at the carboxyl end of the C_2 domain comprising residues 2303–2332. Conversion of FVIII to its active form entails cleavage at the a_1/A_2 and the a_3/A_3 boundaries (aa Arg_{372}-Ser_{373} and Arg_{1689}-Ser_{1690}). The former cleavage transforms factor VIII into a three-chain molecule and the latter releases factor VIII from von Willebrand factor. Activation is also accompanied by cleavage at the a_2/B boundary. The activated FVIII is relatively unstable and, in particular, the dissociation of the A_2 domain from the active heterotrimer results in loss of activity. Activated protein C, an important regulator of coagulation, cleaves FVIII at the A_1/a_1 boundary (Arg_{336}-Met_{337}) and causes its inactivation, but this protein also cleaves at Arg_{562}

and this may represent an important target for inactivation by protein C. In addition, work on the clearance of FVIII from the circulation has suggested that in the absence of vWF binding FVIII is rapidly cleared by a process that involves a member of the family of endocytic receptors: the low-density lipoprotein receptor–related protein also called the α_2-*macroglobulin receptor*.

The large size and multiple forms of the FVIII protein in circulation have hindered crystallographic investigation, but a model of the tridimensional structure of the A domains of FVIII has been derived, by analogy, from the crystal structure of ceruloplasmin. Crystals of the C_2 domain of FVIII have provided data on the tridimensional structure of this module and, finally, two-dimensional analysis of crystals formed by activated FVIII in association with phospholipidic membranes has yielded a picture of the arrangement of the activated FVIII domains relative to each other, the phospholipidic membrane and activated FIX.

The C_2 domain essentially consists of an eight-stranded antiparallel β-barrel formed by two β-sheets tightly packed against each other. Loops projecting from one end of the barrel have been shown to be important for the binding to phospholipidic membranes. Thus, it appears that three hydrophobic feet formed by the side chains of Met_{2199} and Phe_{2220}, Val_{2223} and Leu_{2251} plus Leu_{2252} embed into the membrane, while the four basic residues Arg_{2215}, Arg_{2220}, Lys_{2227}, and Lys_{2249} interact with the negative charges of the phospholipids.

The C_1 domain of FVIII can be modeled on the C_2 domain, especially in the core regions that are more conserved. The model of the three A domains of the activated FVIII suggests that these domains are arranged in a triangular structure, while the two-dimensional crystals of activated FVIII prepared on monolayers of negatively charged phospholipids indicate that the C_2 domain penetrates the membrane with four loops. The C_1 domain is almost at a right angle to C_2 and has its major axis parallel to the membrane. The A_3 domain is in close association with the C_1 and C_2 domains near the membrane, and the heterotrimer of A domains is tilted approximately 65° relative to the membrane plane, with the A_2 domain protruding partially between the A_1 and A_3 domains.

1.2
The Factor IX Gene and its Product

The FIX gene (*F9*) is approximately 33-kb long (Fig. 3) and contains eight exons varying in size from 25 to 1935 bp. The introns also vary in size from 188 to 9473 bp. The gene promoter has been partly characterized and binding sites for a number of transcription factors (NF1-L, androgen receptor, LF-A1/HNF4, DBP, and members of the C/EBP family) have been found close to the transcription start site. The FIX mRNA is 2802-nt long and includes a leader sequence 29- to 50-nt long, depending on the true position of the translation start, and 1390 nt of 3′ untranslated tail. There is a close correspondence between *F9* exons and protein domains. These are modules shared by large families of proteins, and the structure of the *F9* gene is consistent with the exon-shuffling hypothesis of gene evolution. Three other proteins of blood coagulation are coded by genes with the same structure as *F9*, that is, factors VII, X, and protein C. These genes are probably derived from the same ancestral gene as FIX. More remotely related to *F9* gene is the prothrombin gene, as these two genes

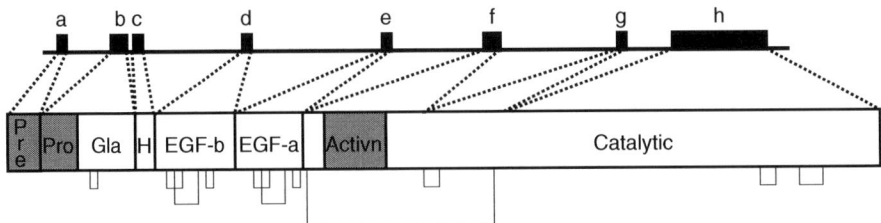

Fig. 3 Factor IX gene (*F9*) and protein. Diagram of *F9* gene (upper line) and FIX (lower bar). Exons are indicated by black boxes and labeled (a–h). Dotted lines show domains encoded by individual exons. Green areas are protein domains cleaved prior to secretion (pre, pro) or at activation: pre = prepeptide; pro = propeptide; gla = gla region; H = hydrophobic stack; EGF-b = first epidermal growth factor domain with Ca^{++} binding site; EGF-a = second epidermal growth factor domain; activn = activation peptide flanked by carboxyl end of light chain of activated factor IX and amino end of the catalytic region. Disulfide bridges in factor IX are shown below the protein domains. (See color plate p. xxi.)

show identical arrangements only for the first three exons and are homologous only in this region and in that encoding the catalytic domain.

The first exon of *F9* codes for the predomain of the leader peptide that is involved in the transport and secretion of FIX. The second exon codes for the prodomain of the leader peptide and the gla region. These two domains are functionally related since the propeptide binds the enzyme responsible for the γ-carboxylation of 11 glutamic acid residues that give the name to the gla region (γ-carboxyglutamic acid = gla). The third exon codes for a small hydrophobic region that contains the 12th gla residue of FIX. The gla region confers on FIX affinity for phospholipidic membranes. The fourth and fifth exons code for domains homologous to epidermal growth factor (EGF domains). The first of these domains is called *type B* and contains a high-affinity Ca^{++} binding site contributed by Asp_{47}, Asp_{49}, Gln_{50}, and Asp_{64}. The latter residue also undergoes β-hydroxylation in a proportion of the FIX molecules while Ser_{53} and Ser_{61} are glycosylated. The second EGF domain that lacks the Ca^{++} binding site and β-hydroxylated residue is called *type A*. The EGF domains are involved in protein–protein interaction. Thus, the gla plus the EGF1 domain of FIX seems to interact with the activated FVII and tissue factor complex that activates FIX and the interface between the two EGF domains with residues 1811–1818 of FVIII. The sixth exon codes for the activation domain that comprises a peptide of 35 aa (residues 146 to 180) that is cleaved off during activation. This contains two glycosylated asparagines (Asn_{157} and Asn_{167}). Finally, the last two exons of the *F9* gene code for a catalytic domain that has all the features of serine proteases (e.g. trypsin). Thus, His_{221}, Asp_{269}, and Ser_{365} are considered the cardinal residues of the active center and Asp_{359} the residue conferring affinity for basic residues. Residues 301–303 and 333–339 of FIX appear to interact with residues 558–565 and 698–712 of the A_2 domain of activated FVIII.

FIX is synthesized in the liver as a precursor molecule with a leader peptide of 46, 41, or 39 aa; it is not known which of these is correct, but 39 is probable. This peptide is cleaved prior to secretion in two independent steps for the pre- and prodomains. Circulating FIX is a monomer of 415 aa that upon activation by FXI or FVII plus tissue factor is converted into a two-chain molecule.

Crystallographic and other structural studies have shown that the gla domain that contains the phospholipids binding site has at the N terminus a tightly hydrogen-bonded complex involving two calcium ions and a pseudoplanar network of six to seven calcium ions that are tightly associated with gla residues. A helix then connects the calcium-associated fragment to the EGF1 domain. The calcium ion bound to the EGF1 domain is presumed to fit in the interface between the gla and EGF1 domain. The principal axes of the oblong EGF domains form an angle of 100° stabilized by an interdomain salt bridge. The catalytic region is localized above the EGF2 domain so that the activated FIX molecule shows a curved configuration with the catalytic center about 80 Å above the phospholipidic membrane surface. The concave surface of activated FIX contains interacting sites for FVIII and sites affected by hemophilia B–causing mutations.

2
Hemophilia A Mutations

The large size and complexity of the *F8* gene has meant that adequate screening of this gene for hemophilia A mutations has had to wait for the development of novel, efficient, and fast methods. Of particular value in this respect was the use of traces of FVIII mRNA in the RNA prepared from peripheral lymphocytes. This, in particular, has revealed that 20% of all cases and 50% of those with severe hemophilia are due to gross unexpected disruptions of the *F8* gene. The other hemophilia A mutations are mostly small changes that involve single base substitutions or the insertion and deletion of a few nucleotides, while gross deletions or duplications are relatively rare (Fig. 4).

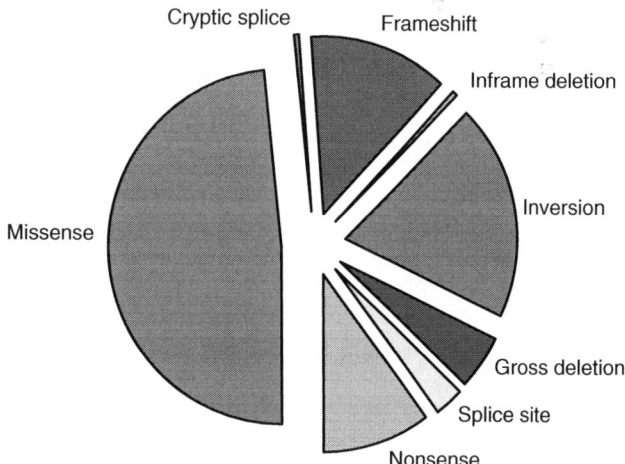

Fig. 4 Pie chart showing the different types of mutations reported in hemophilia A. Cryptic splice refers to mutations generating novel splice signals. In-frame deletion refers to mutations causing loss of one or more codons.

2.1
Gross Sequence Changes

In the United Kingdom, the study of 25% of the hemophilia A population has shown that 1.2% of patients have gross gene deletions. The world database of hemophilia A mutations (HAMSTeRS at http://europium.csc.mrc.ac.uk) records 116 gross deletions. These may involve the whole or a part of the gene (e.g. a single exon), and are usually associated with severe disease. In this series, the four deletions that are accompanied by moderate hemophilia affect exons 5 and 6; exon 22; exons 23 and 24; and exon 25. The mRNA in these cases is expected to maintain the normal reading frame and, presumably, the deletion simply leads to the loss of the aa encoded by the missing exons.

Large insertions are quite rare, but two interesting patients have been reported with insertion of the repetitive LINE sequence in AT-rich regions of exon 14. These have been shown to be recent transposition events from an active LINE sequence in chromosome 22.

The major cause of gross F8 gene rearrangements is inversions. These, as mentioned above, cause 50% of the instances of severe hemophilia A or about 20% of all cases of the disease. Two main types of inversion exist: one that breaks the F8 gene in intron 22 and one that breaks intron 1. The former accounts for 45% and the latter for 5% of patients with severe hemophilia A.

The inversions breaking intron 22 (Fig. 5) were the first to be discovered, and formal proof was obtained that the inversions result from intranemic homologous recombination between repeats of a 9.5-kb sequence called int22h. One copy of this sequence (int22h-1) is part of intron 22 of the F8 gene, while a second (int22h-2) and third copy (int22h-3) are respectively located 500 and 600 kb more telomerically and are both in inverted orientation relative to int22h-1. These repeats are more than

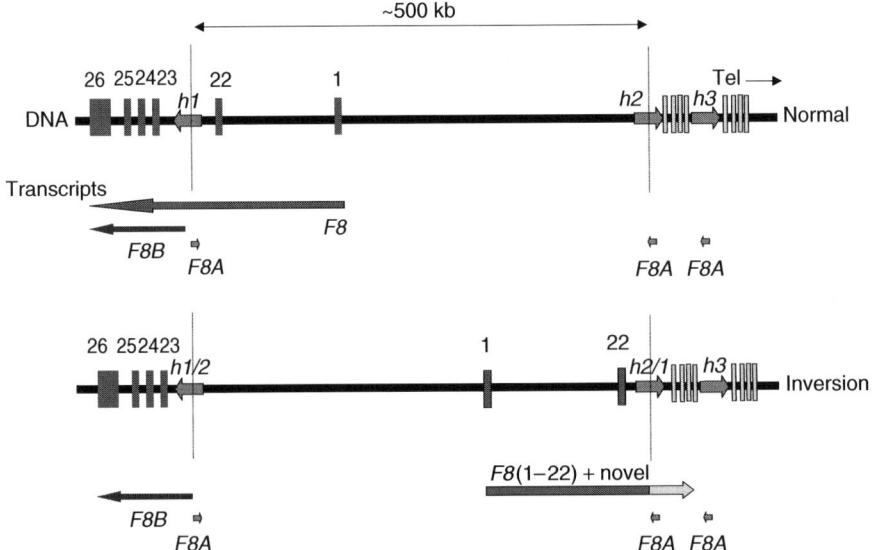

99.9% identical. Inversions due to homologous recombination between *int22h-1* and *int22h-3* are 5 to 6 times more common than those involving *int22h-1* and *int22h-2*, but the reason for this bias is unknown. It also appears that these inversions arise predominantly during male gametogenesis. Since the above inversions cause severe disease and hence are under strong negative or purifying selection, their high frequency is explained by a high rate of mutation. Estimates of 4 and 7×10^{-6} *int22h*-related inversions per gamete per generation have been offered. The *int22h*-related inversions are the first mutations sought in patients with severe hemophilia A, and this is done either by Southern blotting or by a more rapid PCR-based procedure.

The inversions breaking intron 1 of *F8* (Fig. 6) arise by intranemic homologous recombination like the *int22h*-related inversions. The sequence involved in this recombination is a 1041-bp segment of intron 1 called *int1h-1* that is duplicated in inverted orientation 140 kb more telomerically to yield *int1h-2*. These two repeats are more than 99.9% identical.

The *int1h*-related inversions are about ninefold rarer than the *int22h*-related inversions, and this reflects the size difference between the *int1h* and *int22h* repeats, in keeping with the idea that the frequency of homologous recombination is related to the length and degree of identity of the sequences involved.

2.2 Small Sequence Changes

A variety of small sequence changes has been reported, but none so far affecting the putative *F8* promoter. The known mutations, therefore, may primarily affect RNA processing, translation, or the fine structure of FVIII.

2.2.1 Changes Affecting RNA Processing

The presence of traces of FVIII mRNA in peripheral lymphocytes offers the opportunity to detect mutations altering the structure of mRNA irrespective of where they are located within the *F8* gene or, indeed, to confirm that mutations postulated to alter mRNA processing actually do so. Many of these mutations can be expected to alter the donor or acceptor splice signal with consequent skipping of the affected exon. In some patients, however, alteration of the normal splice signal is accompanied by the activation of cryptic splice signals in the affected exon or in the adjacent intron, thus respectively causing partial exon exclusion and exon extension. Finally, more rarely, patients are found to produce mRNA with an additional exon

Fig. 5 Diagram of *int22h*-related inversion, showing its transcriptional consequences. Top: normal structure of DNA region and *F8* gene transcripts (not drawn to scale). F8 = transcript encoding coagulant factor, F8A and F8B = transcripts arising from CpG island in *int22h* repeat. Green arrows in genomic DNA represent *int22h* repeats (h1 = *int22h*-1, h2 = *int22h*-2, h = *int22h*-3). Red bars indicate some numbered exons of the *F8* gene, yellow bars are exons unrelated to the *F8* gene. Vertical dotted lines are sites where intranemic homologous recombination occurs and leads to the inversion. Tel → shows direction to telomere. Arrow point in transcripts indicates 3' end. Bottom: structure of DNA region with inversion resulting from recombination of *int22h*-1 with *int22h*-2. H1/2 and h2/1 are the recombined *int22h* repeats. The transcripts show that the *F8* transcript contains only exons 1 to 22 of *F8* attached to the normally unrelated yellow exons and that exons 23 to 26 are present only in the *F8B* transcript. NB inversions involving *int22h*-3 yield analogous transcripts. (See color plate p. v.)

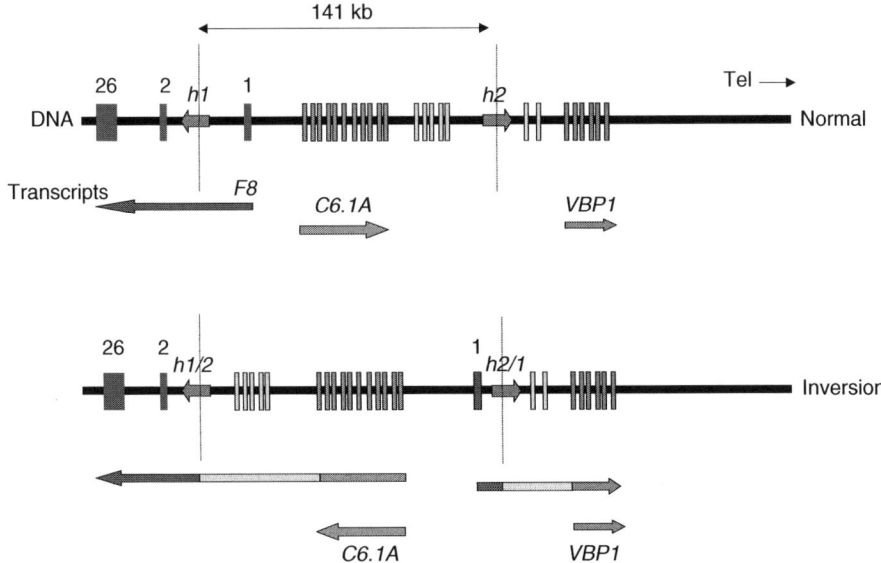

Fig. 6 Diagram of *int1h*-related inversion, showing its transcriptional consequences. Top: normal structure of DNA region and relevant transcripts (not drawn to scale). Red bars indicate numbered exons of *F8*. Light green and blue bars are exons of the *C6.1A* and *VBP1* gene respectively; yellow bars are exons unrelated to the above three genes. Green arrows in DNA are *int1h*-1 (*h1*) and *int1h*-2 (*h2*). Vertical dotted lines indicate sites where intranemic homologous recombination occurs and leads to inversion. Arrow point in transcripts indicates 3' end. Bottom: structure of DNA region with inversion and relevant transcripts. H1/2 and h2/1 are recombined *int1h* repeats. Transcripts show that *F8* contributes to two mRNAs: one containing all but the last exon of *C6.1A*, some novel exons and exons 2 to 26 of *F8* and one containing exon 1 of *F8* plus novel exons and all but the first exon of *VBP1*. The patients' cells also produce normal *C6.1A* and *VBP1* mRNA. (See color plate p. xxii.)

because of single base substitutions generating a novel splice signal in a region of an intron close enough to a cryptic splice signal appropriate to the definition of a novel exon.

The severity of the disease in patients with the above mutations appears to be related to whether a fraction of the pre-mRNA can be spliced into wild-type molecules or whether the abnormal mRNA has retained a normal reading frame. When the normal reading frame has been lost and no normally spliced mRNA is produced, the patient tends to have severe hemophilia A.

2.2.2 Changes Affecting mRNA Translation

HAMSTeRS lists nonsense mutations at 95 different codons, and at 12 of these sites, the mutations were transitions at a CpG dinucleotide. These are usually observed in multiple patients because of the relatively high rate of transition mutations at this type of dinucleotide.

Nonsense mutations should result in premature termination of translation and hence in truncated FVIII, which may or may not be stable enough to be secreted. However, a phenomenon called *nonsense mediated mRNA degradation* may lead to mRNA instability and hence prevent or grossly reduce FVIII synthesis. Conversely, some nonsense mutations may also impair exon recognition and splicing. In this case, exon skipping may restore the reading frame in at least some mRNA molecules, which can then be expected to produce FVIII lacking the amino acids encoded by the exon that has been skipped.

Nonsense mutations are fairly evenly distributed along the FVIII coding sequence and are generally associated with severe hemophilia A.

HAMSTeRS lists small deletions causing frameshifts at 138 sites and small insertions also causing frameshifts at 48 sites of the FVIII coding sequence. At most of these sites, the mutations appear to be unique events but there are some sites that are repeatedly affected and thus represent hotspots of deletion or insertion. The most spectacular hotspots are represented by regions of single base repeats such as strings of A that are present at various locations in exon 14. These appear to undergo slippage during DNA replication, thus acquiring or losing a base. High mutation rates at mononucleotide repeats could lead to somatic mosaicism in the patient with reversion of the mutation, and this would have beneficial effects. Alternatively, errors occurring during transcription could result in some mRNA molecules with a restored reading frame. It is interesting, therefore, to note that while patients with frameshift mutations usually have severe disease, a few with mild or moderate disease have been listed and these bear frameshifts affecting polyA strings. In the absence of the above corrections, occurring either in some cells at DNA replication or in some transcripts during gene expression, frameshifts tend to cause functional defects similar, by and large, to gross deletions and nonsense mutations.

2.2.3 Mutations Causing Alterations of the Fine Structure of Factor VIII

HAMSTeRS contains two mutations resulting in amino acid addition: Ile at codon 613 in one case and a string of six amino acids after residue 1888 in the other. Conversely, the database contains eight amino acid deletions, of which five affect a single residue, two affect two residues, and one affects four residues. However, the mutations that affect the fine structure of FVIII are mostly amino acid substitutions. The above database, after exclusion of the UK patients studied in the course of establishing a confidential national database of hemophilia A mutations and pedigrees, contains 437 different missense mutations, 324 of which are unique events while 113 have been observed two or more times. The above UK study has examined 767 unrelated patients and can be contrasted with the data summarized in HAMSTeRS in order to estimate the total number of FVIII missense mutations that can cause hemophilia A. The 437 missense mutations in HAMSTeRS represent the group of missense mutations causing hemophilia A that is already known. The size of the group formed by hemophilia A missense mutations that are still unknown can be estimated if the missense mutations found in the UK sample are divided into those that are listed in HAMSTeRS and those that are not. The ratio of the latter to the former is proportional to the number of different missense mutations that can cause hemophilia A and are not found in HAMSTeRS. In fact, 102 of 202 missense mutations found in the UK study are novel relative to HAMSTeRS and, hence, the missense mutations causing hemophilia A and yet not in this collection should be 221 (i.e. $102/202 \times 437$), so that the total number of missense mutations capable of causing hemophilia A can be assumed to be about 221 + 437 = 658. This is not a particularly high figure if one considers that the *F8* gene encodes 2351 amino acids and that 15,588 missense mutations can occur in this gene.

The distribution of missense mutations causing hemophilia A is far from being homogeneous (Fig. 7). Thus, the B domain is bereft of such mutations while it has a proper share of nonsense mutations and even an excess of frameshift mutations. This is explained by the fact that the B domain does not contribute to the coagulant activity and hence to the antihemophilic function of FVIII.

The three A domains of the patients' FVIII show similar numbers of missense mutations and these preferentially appear to occur at residues that are identical in the three A domains, as the known mutations affect 44% of these residues while they involve only 28% and 24% of those that are, respectively, identical in only two domains or different in all three domains. Similarly, the known missense mutations in the C domains affect 42% of the residues that are identical in all the C domains of both factor VIII and factor V, while they affect only 22% of the residues that are identical in the C domains of factor VIII and 11% of the residues that are not identical in the C domain of factor VIII.

These data indicate that substitution of residues important for the basic structure of the A and C domains are much more likely to cause hemophilia A than substitution of other amino acids.

A number of missense mutations affect residues of known functional importance. Thus, Arg has been found to be substituted for Ser_1 at the cleavage site of the prepeptide. Pro, His, or Cys, and Pro or Leu, respectively, have been found substituted for Arg_{372} and Ser_{373}, which form an activation cleavage site. Similarly, Cys or His, and His or Arg, respectively, have been found substituted for Arg_{1689} and Ser_{1690}, which also form an activation cleavage site.

Conversely, some missense mutations have indicated residues important to the stability of the FVIIIa heterotrimer because they affect the interfaces between the A_1/A_2, A_1/A_3, and A_2/A_3 domains. This group of mutations includes the following: R284E, R284P, S289L, R527W, R531H, R531G, Q694I, R698L, R698W, L1932F, M1947I, G1948D, and H1954L. Interestingly, coagulant assays on patients with these mutations show higher values when they include FVIII activation in

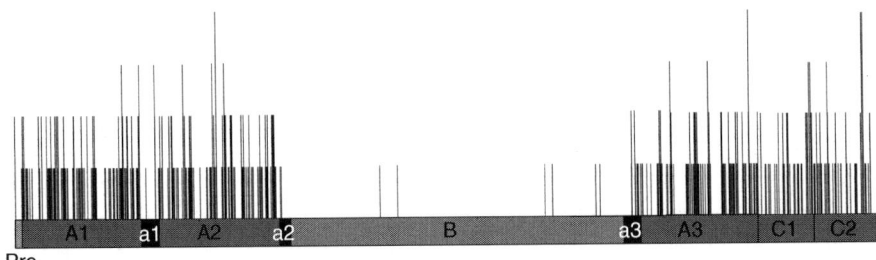

Fig. 7 Missense mutations of *F8* associated with hemophilia A. The horizontal bar shows the domains of the FVIII polypeptide. The vertical lines show the positions of missense mutations, and their heights are proportional to the number of different substitutions (from 1 to 4) found at each residue. Any line thicker than the minimum thickness indicates mutations at two or more residues close to one another but not necessarily adjacent in FVIII (data from HAMSTeRS).

the measuring phase (1-stage assays) than when FVIII activation is allowed to precede the measuring phase (2-stage assays) as this is likely to allow greater FVIII inactivation, prior to measurement, if the FVIIIa heterotrimer is unstable. *Ad hoc* experiments for some of these mutations have demonstrated abnormally rapid dissociation of the A_2 domain from the heterotrimer.

Missense mutations in hemophilia A may also offer some information on the importance of sites of protein interaction. Three regions of FVIII are considered binding sites for FIX: residues 558–565, 698–712, and 1811–1818. The first two interact respectively with the region of FIX residues 333–339 and 301–303; the third appears to interact with the interface between the two EGF domains of FIX. In HAMSTeRS, eight missense mutations are listed for the first site, nine for the second, and none for the third. With regard to the interaction of FVIII with vWF, the Y1680C and Y1680F mutations found in hemophilia A patients bear witness to the importance of this sulfated tyrosine for the binding to vWF. Fifteen different missense mutations affecting the binding of FVIII to vWF have been found in the C domains.

2.3
Functional Interpretation of Observed Sequence Changes and Correlations between Genotype and Phenotype

A proportion of the observed FVIII mutations can undoubtedly be considered detrimental, that is, gross deletions or other gross rearrangements involving the essential sequences of the gene, such as the *int22h-* and *int1h-*related inversions, alteration of absolutely conserved elements of the splice site consensuses, frameshifts, and nonsense mutations. However, the accurate prediction of the functional consequences of mutations detected in genomic DNA may demand mRNA analysis.

Many mutations are not overtly detrimental and do not lead to gross mRNA aberrations. In order to determine the detrimental nature of such mutations, it is important to have accurate diagnoses and to perform a very thorough mutation screen on the *F8* gene. In particular, one should exclude diseases that can mimic hemophilia A. At present, we know of two such conditions: the combined deficiency of FV and FVIII associated with defects in secretion pathways and the Normandy-type von Willebrand disease due to mutations that impair the ability of vWF to bind FVIII. Mutation screening of the *F8* gene should ensure levels of detection as close as possible to 100%. Currently, this can be done best by procedures that include the use of mRNA. Under these conditions, mutations that represent the only change detected in the *F8* gene are likely to be the cause of the disease. In some cases, family studies can show that the mutation found in a patient is of recent origin and this strongly supports its deleterious nature. Also, the independent occurrence of the same mutation in two or more patients strongly supports the idea that the mutation is the cause of the disease. Less useful indicators of the disease-causing role of observed sequence changes are considerations on the degree of conservation of the mutated residues of FVIII or the chemical difference between the normal and the mutant FVIII residues. The accumulation of thorough hemophilia A mutation data should also help identify neutral polymorphisms, as these may appear as second changes common to patients with different disease-causing mutations.

With regard to the correlation between phenotype and genotype, missense mutations are frequently associated with mild or moderate disease.

Variability in the severity of the disease has been reported among hemophilia A patients with the same mutation, and this may be due to polymorphic variation in genes that encode factors important for the maturation, secretion, transport, clearance, and modulation of the activities of FVIII.

The most worrying phenotypic feature of patients with hemophilia A is the propensity to develop antibodies against therapeutic FVIII. The risk of developing such antibodies or inhibitors is strongly correlated with the type of mutation causing the disease. Thus, a risk of about 30% is associated with the *int22h*-related inversions, gross deletions, and nonsense mutations, while frameshifts show a lower risk of about 20% and missense mutations an overall risk of 8%. HAMSTeRS shows 39 missense mutations associated with the presence of inhibitors and for 20 of these mutations, there are two or more patients per mutations with information on the inhibitor status, forming a total of 161 patients, 48 of which (29.8%) had inhibitors. This suggests that most of the risk of inhibitors associated with missense mutations is due to a small group of residues and their substituents that possibly create novel antigenic epitopes. Some evidence that this might be the case has been obtained for inhibitors associated with the mutations R593C and R2150H. With regard to the first mutation, a patient has been reported with inhibitors directed exclusively against the wild-type FVIII, while, with regard to the second mutation, two patients with high titer inhibitors were reported who still displayed significant levels of FVIII activity, thus suggesting that the patients' mutant FVIII was not recognized and hence not blocked by the patients' inhibitors. The high risk of inhibitors associated with gross rearrangements and nonsense and frameshift mutations is best explained by the inability of patients with these mutations to produce FVIII. This could result in lack of immunological tolerance to FVIII, which would not be recognized as "self." As nonsense and frameshift mutations are expected to cause similar effects on FVIII synthesis, the reasons why frameshifts are associated with a somewhat lower risk of inhibitors should be considered. The insertion or deletion mutants in polyA regions ($\geq$6A residues) appear to be mainly responsible as they show a very low risk of inhibitors. Thus, only five (6.5%) of 76 patients with these frameshifts are reported as having inhibitors in HAMSTeRS. Possibly, this is due to errors of replication or transcription of the mutated polyA regions, which, as mentioned above, may allow production of traces of FVIII that may serve to develop immune tolerance to FVIII.

3
Hemophilia B Mutations

The vast majority of hemophilia B mutations are due to single base-pair substitution, or deletion and insertion of a few base pairs, while gross deletions and other rearrangements are rare (Fig. 8). So far, 2847 mutations involving small sequence changes and 90 gross deletions have been reported in the world database (http://www.kcl.ac.uk/ip/petergreen/haemBdatabase.html). However, large population samples provide more precise information on the relative frequency of gross and small sequence changes.

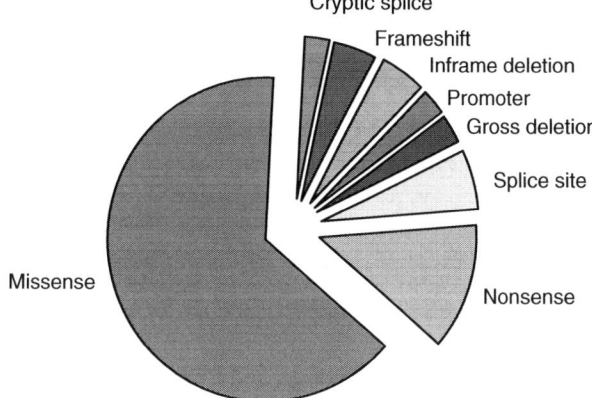

Fig. 8 Pie chart showing the different types of mutations reported in hemophilia B. Promoter refers to mutations affecting the *F9* promoter.

3.1 Gross Sequence Changes

A fairly unbiased sample representing >70% of the UK hemophilia B population has shown that gross sequence changes account for about 1.5% of the patients. Most gross mutations are deletions that may involve the whole or part of the *F9* gene. They usually cause severe disease with no measurable FIX protein in circulation. However, in one patient, with deletion of the fourth exon (exon d), significant amounts of circulating FIX have been reported. Such a deletion should not alter the reading frame of the mRNA. Large insertions are very rare and among those of particular interest is the insertion of an Alu repeat, as it raises the issue of ongoing retrotranspositions in the human genome.

3.2 Small Sequence Changes

The 2847 hemophilia B mutations due to small sequence changes (<20 bp) consist of 2646 single base substitutions, 160 deletions, 38 insertions, and 3 combined deletions and insertions. Furthermore, 28 patients showed two, and one patient three *F9* gene mutations, probably, and in some cases certainly, representing the association of a detrimental mutation with additional "private" neutral sequence changes. Excluding such additional changes, 916 unique molecular events have been identified, while the remainder are repeats of previously observed changes. Many of these are CG > TG or CA changes that define mutational hotspots but, in some instances, frequent repeated observation may be due to founder mutations. This probably applies to some of the −6G > A promoter mutations, to the three missense mutations G60S, T296M, and I397T, and to the 17810A > G base substitution in intron 5 that creates a novel donor splice signal. This fits the consensus better than the normal site, is located 13-nt downstream of exon e and causes a mild hemophilia, accounting for 5.5% of hemophilia B patients in the United Kingdom.

3.2.1 Mutations Affecting Transcription, RNA Processing, and Translation

Detrimental promoter mutations should, by definition, impair transcription. Twenty-one different promoter mutations have been reported (Fig. 9). They affect binding sites for different transcription factors (LF.A1/HFN4, DBP, C/EBP), and usually cause a disease characterized by a clinical and hematological age-related improvement (especially postpubertally) that may lead to complete loss of symptoms. However, no age-related improvement has been reported in patients with substitution of residue −26 (G > C, G > A or G > T), which is part of a weak androgen response element (nt −36 to −22) that in its 3′ region overlaps with an LF.A1/HNF4 binding site. It seems probable that the improvement of patients with promoter mutations is due to the greater dependence of the promoter on developmentally related stimulatory events such as increases in androgen levels.

Very high amplification of nucleic acid sequences has allowed examination of tissue-specific mRNAs in tissues that do not express these mRNAs as traces of the relevant mRNA are nevertheless found. However, the F9 gene, mainly expressed in the liver, is unusual because while trace levels of FIX mRNA are present in lung, skeletal muscle, and heart, no traces of full-length FIX mRNA are found in peripheral lymphocytes, lymphocyte-derived cell lines, or pancreas. Instead, a novel mRNA is found that contains only the last two exons of the F9 gene. This has prevented thorough analysis of the F9 mutations that affect mRNA processing. Nevertheless, several mutations that can be expected to affect mRNA splicing have been reported as probable causes of hemophilia B. These include base substitution or deletion of the very highly conserved dinucleotides at the start and end of each intron, alteration of less conserved elements of the splice signal and also the generation of novel splice signals such as that described in the previous section as causing 5.5% of all UK hemophilia B and a new donor splice in the 3′ untranslated tail of the message. The latter has been observed repeatedly in patients with severe disease and appears likely to disrupt the processing of the 3′ end of the message.

Mutations affecting translation, that is, frameshifts and nonsense codons, are scattered throughout the FIX coding sequence. Interestingly, these mutations, irrespective of their location, are usually associated with absence of FIX protein in circulation. Thus, even translation stops very close to the end of the protein (e.g. codon 411) have been found in patients

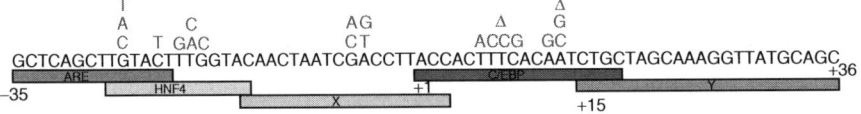

Fig. 9 Hemophilia B mutations affecting the promoter of F9. Mutant residues are in red, Δ indicates single base deletion. Bars below promoter sequence indicate mutation-affected regions known to bind transactivating factors. ARE = androgen responsive element, HNF4 = liver-specific transcription factor binding site, C/EBP = binding site for members of the C/EBP family of transactivating factors, X = region binding HNF4 and members of the steroid hormone receptor super family (ARP1, Coop/Ear3) that appear to exercise repressor activity on F9 promoter. (See color plate p. xxi.)

with no circulating FIX. This suggests that even moderately truncated FIX and/or FIX mRNA containing premature stop codons are unstable.

3.2.2 Missense Mutations and Single Amino Acid Deletions

Twelve different single aa deletions and 549 substitutions have been reported so far plus five aa substitutions that are probably or certainly neutral R-44H, I-40F, A147T, F178L, and V328I. The above 549 missense mutations affect 225 residues of FIX (Fig. 10) plus seven codons that contribute to splice consensuses. It is difficult to know whether these seven mutations mainly act by changing an amino acid in FIX or by impairing mRNA formation.

The leader peptide is relatively lightly affected and its prodomain shows mutations essentially in a region important for its cleavage from the mature protein. The Val-$_{10}$, which is important for the binding of the γ-carboxylase that modifies the gla residues, has shown two substitutions (A-10T and A-10V) that do not cause hemophilia B but are associated with hypersensitivity to warfarin, an antagonist of vitamin K, which is the cofactor of γ-carboxylase. Several residues of the gla domain are affected by mutation, including gla residues (except those at positions 15, 36, and 40) and the Cys$_{18}$ and Cys$_{23}$, which form a disulfide bridge. All together 55 missense mutations have been found in this domain, which contains 38 amino acids. The EGF type B domain shows 66 different missense mutations affecting, among others, residues involved in the high affinity Ca^{++} binding site and the cysteines of the three disulfide bridges characteristic of these domains. The EGF type A domain shows a lower number of missense mutations (43), many of which affect the six cysteines involved in disulfide bridges. The activation domain shows relatively few mutations and these are concentrated in regions of obvious functional importance such as Cys$_{132}$, which is part of a disulfide bridge that holds together the light and heavy chain of activated FIX and its immediate neighbors; the residues of the cleavage sites necessary for the excision of the activation peptide; and the first residues of the heavy chain of activated FIX that are part of the catalytic domain. Three hundred and nine different amino acid substitutions are distributed throughout the catalytic domain, and bear witness to the fairly highly constrained nature of

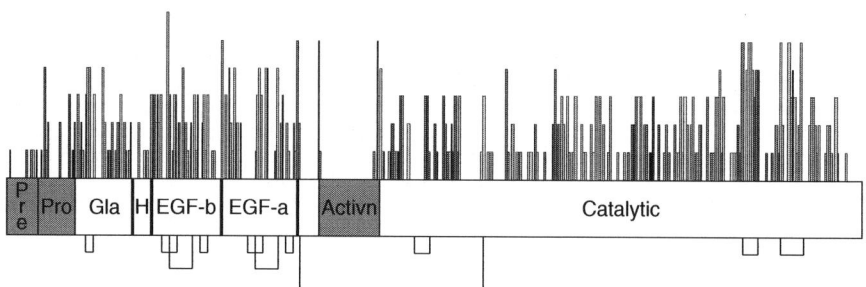

Fig. 10 Missense mutations of *F9* associated with hemophilia B. The horizontal bar shows the domains of the FIX polypeptide with its disulfide bridges indicated below. The vertical lines mark the positions of missense mutations and their heights are proportional to the number of different substitutions (from 1 to 6) found at each residue (data from http://www/kcl.ac.uk/ip/petergreen/haemBdatabase.html).

this domain. The highest concentration of missense mutations is found in the region comprising Ser_{365} of the active center and also in a region comprising residues that form the entrance to the activated FIX specificity pocket. The α-helix comprising residues 330 to 338 is also rich in missense mutations. A number of loops border the substrate-binding groove. Of these, two: (residues 199–204 and 223–229) seem mutation-poor, while the others (235–245, 256–268, 312–322, and 340–347) show average to high numbers of mutations.

In general, the missense mutations found in hemophilia B show a strong preference for residues that are conserved in the homologs of FIX: FVII, FX, and protein C. Thus, 291 different substitutions have been found at the 107 residues that are absolutely conserved, 113 at the 73 amino acids that show only one or two conservative changes, 58 at the 45 residues that show one nonconservative change, and 90 at the 229 less-conserved amino acids.

The current list of hemophilia B mutations also identifies residues that are mandatory for the structural and/or functional integrity of FIX because different types of substitutions are associated with hemophilia B. Thus, 53 residues show two different substitutions, 48 show three, 32 show four, 8 show five, and 5 show six. The last two groups comprise Glu_{33}, Pro_{55}, Cys_{88}, Cys_{95}, Cys_{132}, Arg_{145}, Arg_{180}, Cys_{222}, Cys_{361}, Asp_{364}, Ser_{365}, Thr_{380}, and Ser_{384}. The abundance of FIX residues that are known to have suffered multiple substitutions also suggests that the data available on hemophilia B mutations already form a fairly good representation of the mutational spectrum of hemophilia B. We have also calculated the target for single base substitutions causing hemophilia B. This comprises 825 substitutions of which 189 are expected to cause nonsense and 636 missense mutations. The current list of observed hemophilia B mutations in fact contains 73 different nonsense and 554 different missense mutations that represent 38.6% and 87%, respectively, of the total number expected.

3.3
Functional Interpretation of Observed Sequence Changes and Correlations between Genotype and Phenotype

The criteria that can be used to decide whether an *F9* sequence change observed in a hemophilia B patient is the cause of the disease are essentially those already mentioned for mutations of the *F8* gene and hemophilia A. The lack of a convenient source of FIX mRNA of course prevents the analysis of most mRNA processing defects and the clear identification of intronic mutations capable of causing hemophilia. Fortunately, the systematic analysis of hemophilia B populations indicates that no more than 1–2% of the *F9* mutations are likely to be in regions other than the proximal promoter, exons, and normal RNA processing signals. Thus, the lack of mRNA precludes the rapid detection of only a very small proportion of the hemophilia B mutations.

As in hemophilia A, gross deletions, frameshifts, and nonsense mutations are usually associated with a severe phenotype and predisposition to the inhibitor complication. The latter was recognized as early as 1983 in hemophilia B. Patients with hemophilia B bearing identical mutations appear to show less phenotypic variation than analogous groups of patients with hemophilia A. Comparison of FIX coagulant activity and protein concentration in blood may indicate whether mutations merely affect the function or also the secretion and/or stability of Factor IX. Thus, the prepeptide mutation I-30N, which alters one of the essential features of the prepeptide (its hydrophobic core) and may thus affect secretion, and also substitutions at cysteines involved in disulfide bridges, which may affect stability, are associated with low protein levels. Conversely, mutations affecting key functional residues such as Asp_{369} and Ser_{365} show normal protein levels.

4
Considerations Arising from the Study of Mutations in Hemophilia A and B

It is instructive to compare some features of the molecular genetics of hemophilia A and B. These two diseases are clinically indistinguishable and their differential diagnosis requires specific coagulant assays. Nevertheless, once hemophilic patients are separated into the A and B groups, some differences emerge. In hemophilia A, a much larger proportion of patients is severely affected. This is explained by the F8 gene proclivity to disruptive inversions involving the *int1h* and *int22h* sequences. This difference in the hemophilias' mutational spectra (see Figs. 4 and 8) also largely explains why 20% of hemophilia A and only 3% of hemophilia B patients develop inhibitors during the course of replacement therapy.

The amino acid sequence of FIX appears much more constrained than that of FVIII because while 4.2% of all *F8* missense mutations are thought to be able to cause hemophilia A, that is so for 21.2% of the *F9* missense mutations. Since the B domain of FVIII represents 39.3% of the FVIII polypeptide and does not participate in coagulation, the above difference is largely but by no means entirely due to this domain. Thus, the apparent correspondence between the ratio of the sizes of *F8* and *F9* and the ratio of the incidence of hemophilia A and B appears to result from three main factors: the predisposition of the *F8* gene to disruption by inversion, the presence in FVIII of the large B domain that tolerates amino acid substitutions, and the more constrained structure of FIX.

The study of an unbiased sample of hemophilia B patients representing 72% of the UK population has allowed a direct estimate of the hemophilia B mutation rate and of the rates of different types of mutations. Thus, transitions at CpG sites appear to occur at the rate of 97×10^{-9} per nucleotide per generation, transitions at other sites at the rate of 7.3×10^{-9}, transversion at CpG sites at 5.4×10^{-9}, and transversions at other sites at 7.0×10^{-9}. Hence, the rate of transitions at CpG sites is 13-fold greater than other base substitutions. While this explains the deficit of CpG doublets observed in the human genome, it is still insufficient to explain the proportion of hemophilia B (34.8%) or hemophilia A (32.3%) mutations due to transitions at CpG sites. A full explanation must consider natural selection at essential sites of the coding regions of *F8* and *F9*. At these sites, selection will oppose the tendency to lose CpG doublets. Consequently, CpG sites are relatively more frequent at critical sites of gene coding regions, and their mutations are, therefore, particularly likely to cause disease. In keeping with this idea, we observe that of the 96 base substitutions that can occur at CpG sites in the coding sequence of the *F9* gene, 58, or 60%, are expected to cause hemophilia B, while only 767, or 24% of the 3179 base substitutions at other sites are expected to cause the disease.

5
Mutational Heterogeneity and Genetic Counseling in the Hemophilias

The great mutational heterogeneity of hemophilia A and B is a problem with regard to the carrier and prenatal diagnoses that are required for precise genetic counseling.

In the hemophilias, coagulant assays often do not provide definite carrier diagnoses because the vagaries of X chromosome inactivation result in broad variation of FVIII or FIX values in carriers and considerable overlap between carrier and normal females.

The analysis of the intrafamilial segregation of intragenic polymorphic markers may provide definite carrier diagnoses as well as first trimester prenatal diagnoses but it inevitably fails in a large proportion of cases. The diagnostic failures are due to the limited informativity of available polymorphic markers in the population of interest, to the logistic problem of obtaining samples from all required individuals, and especially to the large proportion of families that have a single affected individual. For example, in the United Kingdom and Sweden, approximately 50% of the families with hemophilia A or B have a single affected individual and the polymorphic marker approach is of very modest value in these families.

Only the direct detection of the gene defect can offer maximum diagnostic success. Therefore, we have argued that ideal strategies to optimize genetic counseling in the field of hemophilia should be based on the construction of national confidential databases of mutation and pedigree information. This can be achieved by characterizing the mutation of an index individual in each family and collecting his/her pedigree. Such national databases are now being constructed in the United Kingdom and other developed countries. They, of course, provide information on the molecular biology of the disease and allow very rapid carrier and prenatal diagnoses on the blood relatives of the index individuals because such diagnoses can be based simply on the analysis of the region of the gene that is defective in the index person.

A further concern in genetic counseling is the issue of somatic or gonadal mosaicism for the hemophilia A or B mutations. Since mutations in the *F8* or *F9* gene may arise postzygotically, individuals who have suffered such a mutational event may have a mixture of cells that either have or have not a hemophilia mutation. If this mixture is present in somatic tissues, the individual is a somatic mosaic, while if the mixture is confined to germline tissue he/she is a gonadal mosaic. Anecdotal evidence for somatic mosaicism is available for both hemophilia A and B, and evidence for gonadal mosaicism has been presented for Duchenne muscular dystrophy. In general, mosaic status will be found in an individual who is the first in the family to carry the relevant mutation, but exceptionally a mosaic may result from a zygote with two or more X chromosomes such that the X chromosome carrying a preexisting hemophilia mutation is lost from some of the cells during postzygotic development.

Obviously, it is impossible to predict the risk of a mosaic mother transmitting the mutant gene to her offspring as this will depend on the proportion of eggs carrying the mutant gene and their chance of maturation. Genetic counseling will therefore generally have to rely on estimation of average empiric risks. This can be obtained from careful, unbiased population studies. Thus, in the UK population, we investigated how many hemophilia B patients or carriers could be attributed to gonadal mosaicism in mothers designated as noncarriers when their leucocyte DNA was tested. Since we observed no patients or carriers from 47 informative births in the UK hemophilia B population, we concluded that the empiric risk of a "noncarrier" mother of a hemophilia B patient manifesting as a gonadal mosaic by transmitting the mutation to a second child was less than 0.062.

In the case of detectable somatic mosaicism, the upper limit of risk given should be that appropriate to carrier mothers, and, if requested, prenatal diagnosis should be undertaken.

Of course, the chance of detecting somatic mosaics depends on how efficient the laboratory test is in identifying the mutant DNA mixed with that of wild type. For this purpose, mutation detection methods based on chemical mismatch cleavage or denaturing gradient high-performance liquid chromatography are preferable to sequencing.

6
Progress in the Treatment of Hemophilia

In the second half of the last century, therapies based on concentrates of FVIII and FIX prepared from human plasma increased the patients' life expectancy to near-normal values. However, in the 1980s, HIV-infected blood products created a cohort of patients suffering from this lethal disease. Similarly, high rates of hepatitis C infection have marred the progress achieved with the introduction of blood-derived coagulant concentrates. Moreover, recently, concern has arisen about the transmission through blood products of the Creutzfeldt–Jakob prion disease.

Fortunately, the early cloning of the *F9* and *F8* genes has allowed the development of industrial procedures for the production of both FIX and FVIII from mammalian cell cultures created to express these factors *in vitro* so as to minimize risks from the above-mentioned viruses and prions.

In the last 18–20 years, gene therapy has been an important treatment aim. The hemophilias appear well suited to this type of approach because there is no need to ensure tight regulation of the expression of the corrective gene, and any site of expression allowing the coagulant factor to reach the blood circulation should be suitable. Moreover, the enhancement of coagulant level required for the success of this approach is relatively modest because even an increase of 1–2% of the normal level can produce important clinical effects such as transforming severe disease into moderate disease and reducing the number of spontaneous bleeding episodes experienced by the patient. A substantial body of work has been done to develop suitable expression constructs, using retroviral vectors or vectors derived from DNA viruses such as adenovirus or adenovirus-associated virus, and systems of delivery for both *in vivo* or *ex vivo* approaches to treatment (i.e. the injection of expression constructs *in vivo* or the modification with the expression constructs of the patient's cells *ex vivo* followed by culture and expansion *in vitro* and then reinjection into the patient).

Model animals were then used to test various therapy schemes and some experiments were positive enough to allow initiation of clinical trials, mainly in the United States. Thus, 13 hemophilia A patients, known to have HIV and hepatitis C, and expected to be at very low risk of developing inhibitors, were infused with amphotrophic retroviral vectors for FVIII expression at doses between 3×10^7 and 9×10^8 viral particles (vp) per kilogram; no significant benefit ensued.

Eight hemophilia B patients were injected intramuscularly with an associated adenovirus vector of serological type 2 (AAV2) for FIX expression and one patient receiving the lowest dose was reported to show 1–2% FIX levels for 40 months, but at higher viral doses, no significant FIX level was observed and in all patients the development of antiviral antibodies precluded readministration of the vector.

Administration of factor IX constructs in AAV2 vectors via the hepatic artery in two patients at a dose of 10^{12} vp/kg resulted in transient FIX levels of 10–12%.

In a further trial, autologous skin fibroblasts from six patients were transformed by electroporation with an FVIII expressing plasmid and expanded *in vitro* so as to inject $1-4 \times 10^8$ cells in the patients' greater omentum. Four patients showed FVIII levels above pretreatment values but these returned to baseline values in all patients after 12 months.

Thus, clinical trials have not yet led to satisfactory treatment schedules but since they have not revealed significant toxicity in the patients treated so far, they leave the door open for future advances. These, of course, will always have to consider patients' safety as paramount because established replacement therapy is reasonably effective in both hemophilias and any new treatment must achieve better results with no increase in risk.

See also Heterochromatin and Eurochromatin – Organization, Packaging, and Gene Regulation.

Bibliography

Books and Reviews

Bloom, A.L., Forbes, C.D., Thomas, D.P., Tuddenham, E.G.D. (1994) *Haemostasis and Thrombosis*, 3rd edition, Churchill Livingstone, Edinburgh, Scotland.

Lee, C.A. (1996) *Haemophilia*, Vol. 2, Baillière's Clinical Haematology 9, Baillière Tindall, London, UK.

Mannucci, P.M., Tuddenham, E.G. (2001) The haemophilias – from royal genes to gene therapy, *N. Engl. J. Med.* **344**, 1773–1779.

Oldenburg, J., Brackmann, H.H., Hanfland, P., Schwaab, R. (2000) Molecular genetics of haemophilia A, *Vox Sanguinis* **78**(Suppl. 2), 33–38.

Tuddenham, E.G.D., Cooper, D.N. (1994) *The Molecular Genetics of Haemostasis and its Inherited Disorders*, Oxford Monographs on Medical Genetics 25, Oxford University Press, Oxford, UK.

Primary Literature

Bagnall, R.D., Waseem, N., Green, P.M., Giannelli, F. (2002) Recurrent inversion breaking intron 1 of the factor VIII gene is a frequent cause of severe hemophilia A, *Blood* **99**, 168–174.

Bagnall, R.D., Waseem, N., Green, P.M., Colvin, B., Lee, C., Giannelli, F. (1999) Creation of a novel donor splice site in intron 1 of the factor VIII gene leads to activation of a 191 bp cryptic exon in two haemophilia A patients, *Br. J. Haematol.* **107**, 766–771.

Brandstetter, H., Bauer, M., Huber, R., Lollar, P., Bode, W. (1995) X-ray structure of clotting factor IXa: active site and module structure related to Xase activity and hemophilia B, *Proc. Natl. Acad. Sci. U.S.A.* **92**, 9796–9800.

Chen, S.-W.W., Pellequer, J.-L., Schved, J.-F., Giansily-Blaizot, M. (2002) Model of a ternary complex between activated factor VII, tissue factor and factor IX, *Thromb. Haemost.* **88**, 74–82.

Crow, J.F. (1999) The odds of losing at genetic roulette, *Nature* **397**, 293–294.

Dombroski, B.A., Mathias, S.L., Nanthakumar, E., Scott, A.F., Kazazian, H.H. Jr. (1991) Isolation of an active human transposable element, *Science* **254**, 1805–1808.

Evatt, B. (2000) Creutzfeldt-Jakob disease and haemophilia: assessment of risk, *Haemophilia* **6**(Suppl. 1), 94–99.

Fijnvandraat, K., Turenhout, E.A., van den Brink, E.N., ten Cate, J.W., van Mourik, J.A., Peters, M., Voorberg, J. (1997) The missense mutation Arg593 > Cys is related to antibody formation in a patient with mild hemophilia A, *Blood* **89**, 4371–4377.

Fressinaud, E., Mazurier, C., Meyer, D. (2002) Molecular genetics of type 2 von Willebrand disease, *Int. J. Hematol.* **75**, 9–18.

Giannelli, F., Green, P.M. (2000) The X chromosome and the rate of deleterious mutations in humans, *Am. J. Hum. Genet.* **67**, 515–517.

Giannelli, F., Anagnostopoulos, T., Green, P.M. (1999) Mutation rates in humans. II. Sporadic mutation-specific rates and rate of detrimental human mutations inferred from haemophilia B, *Am. J. Hum. Genet.* **65**, 1580–1587.

Giannelli, F., Choo, K.H., Rees, D.J., Boyd, Y., Rizza, C.R., Brownlee, G.G. (1983) Gene deletions in patients with haemophilia B and anti-factor IX antibodies, *Nature* **303**, 181–182.

Gitschier, J., Wood, W.I., Goralka, T.M., Wion, K.L., Chen, E.Y., Eaton, D.H., Vehar, G.A., Capon, D.J. (1984) Characterisation of the human factor VIII gene, *Nature* **312**, 326–330.

Green, P.M., Rowley, G., Giannelli, F. (2003) Unusual expression of the *F9* gene in peripheral lymphocytes hinders investigation of *F9* mRNA in hemophilia B patients, *J. Thromb. Haemost.* **1**, 2675–2676.

Green, P.M., Saad, S., Lewis, C.M., Giannelli, F. (1999) Mutation rates in humans. I. Overall and sex-specific rates obtained from a population study of haemophilia B, *Am. J. Hum. Genet.* **65**, 1572–1579.

Green, P.M., Waseem, N.H., Bagnall, R.D., Giannelli, F. (1998) Mutation analysis and genetic service: the construction and use of national confidential databases of mutations and pedigrees, *Genet. Test.* **1**, 181–188.

Hakeos, W.H., Miao, H., Sirachainan, N., Kemball-Cook, G., Saenko, E.L., Kaufman, R.J., Pipe, S.W. (2002) Hemophilia A mutations in the factor VIII A2-A3 subunit interphase destabilize factor VIIIa and cause one-stage/two-stage activity discrepancy, *Thromb. Haemost.* **88**, 781–787.

High, K.A. (2001) Gene transfer as an approach to treating haemophilia, *Circ. Res.* **88**, 137–144.

Lakich, D., Kazazian, H.H., Antonarakis, S.E., Gitschier, J. (1993) Inversions disrupting the factor VIII gene are a common cause of severe haemophilia A, *Nat. Genet.* **5**, 226–241.

Lenting, P.J., Neels, J.G., van den Berg, B.M.M., Clijsters, P.P.F.M., Meijerman, D.W.E., Pannekock, H., van Mourik, J.A., Mertens, K., van Zonneveld, A.J. (1999) The light chain of factor VIII comprises a binding site for low density lipoprotein receptor-related protein, *J. Biol. Chem.* **274**, 23734–23739.

Levinson, B., Kenwrick, S., Gamel, P., Fisher, K., Gitschier, J. (1992) Evidence for a third transcript from the human factor VIII gene, *Genomics* **14**, 585–589.

Levinson, B., Kenwrick, S., Lakich, D., Hammonds, G. Jr., Gitschier, J. (1990) A transcribed gene in an intron of the human factor VIII gene, *Genomics* **7**, 1–11.

Li, X., Scaringe, W.A., Hill, K.A., Roberts, S., Mengos, A., Careri, D., Pinto, M.T., Kasper, C.K., Sommer, S.S. (2001) Frequency of recent retrotransposition events in the human factor IX gene, *Hum. Mutat.* **17**, 511–519.

Miyata, T., Hayashida, H., Kuma, K., Mitsuyasu, K., Yasunaga, T. (1987) Male-driven molecular evolution: a model and nucleotide sequence analysis, *Cold Spring Harbor Symp. Quant. Biol.* **58**, 863–867.

Moussalli, M., Pipe, S.W., Hauri, H.P., Nichols, W.C., Ginsburg, D., Kaufman, R.J. (1999) Mannose-dependent endoplasmic reticulum (ER)-Golgi intermediate compartment-53-mediated ER to Golgi trafficking of coagulation factors V and VIII, *J. Biol. Chem.* **274**, 32539–32542.

Naka, H., Brownlee, G.G. (1996) Transcriptional regulation of the human factor IX promoter by the orphan receptor superfamily factors, HNF4, ARP1 and COUP/Ear3, *Br. J. Haematol.* **92**, 231–240.

Naylor, J.A., Green, P.M., Rizza, C.R., Giannelli, F. (1992) Factor VIII gene explains all cases of haemophilia A, *Lancet* **340**, 1066–1067.

Naylor, J.A., Brinke, A., Hassock, S., Green, P.M., Giannelli, F. (1993) Characteristic mRNA abnormality found in half the patients with severe haemophilia A is due to large DNA inversions, *Hum. Mol. Genet.* **2**, 1773–1778.

Naylor, J.A., Green, P.M., Montandon, J.A., Rizza, C.R., Giannelli, F. (1991) Detection of three novel mutations in two haemophilia A patients by rapidly screening whole essential regions of the factor VIII gene, *Lancet* **337**, 635–639.

Naylor, J.A., Buck, D., Green, P.M., Williamson, H., Bentley, D.R., Giannelli, F. (1995) Investigation of the factor VIII intron 22 repeated region (*int22h*) and the associated inversion functions, *Hum. Mol. Genet.* **4**, 329–333.

Nichols, W.C., Seligsohn, U., Zivelin, A., Terry, V.H., Colette, H.E., Wheatley, M.A., Moussalli, M.J., Hauri, H.P., Ciavarella, N., Kaufman, R.J., Ginsburg, D. (1998) Mutations in the ER-Golgi intermediate compartment protein ERGIC-53 cause combined deficiency of coagulation factors V and VIII, *Cell* **93**, 61–70.

O'Donovan, M.C., Oefner, P.J., Roberts, S.C., Austin, J., Hoogendoorn, B., Guy, C., Speight, G., Upadhyaya, M., Sommer, S.S., McGuffin, P. (1998) Blind analysis of denaturing high-performance liquid chromatography as a tool for mutation detection, *Genomics* **52**, 44–49.

Oldenburg, J., Quenzel, E.M., Harbrecht, U., Fregin, A., Kress, W., Muller, C.R., Hertfelder, H.J., Schwaab, R., Brackmann, H.H., Hanfland, P. (1997) Missense mutations at ALA-10 in the factor IX propeptide: an insignificant variant in normal life but a decisive cause of bleeding during oral anticoagulant therapy, *Br. J. Haematol.* **98**, 240–244.

Peerlinck, K., Jacquemin, M.G., Arnout, J., Hoylaerts, M.F., Gilles, J.G., Lavend'homme, R., Johnson, K.M., Freson, K., Scandella, D., Saint-Remy, J.M., Vermylen, J. (1999) Antifactor VIII antibody inhibiting allogeneic but not autologous factor VIII in patients with mild hemophilia A, *Blood* **93**, 2267–2273.

Perera, L., Darden, T.A., Pedersen, L.G. (2001) Modelling human zymogen factor IX, *Thromb. Haemost.* **85**, 596–603.

Peters, M.F., Ross, C.A. (2001) Isolation of a 40-kDa Huntingtin-associated protein, *J. Biol. Chem.* **276**, 3188–3194.

Pittman, D.D., Wang, J.H., Kaufman, R.J. (1992) Identification and functional importance of tyrosine sulfate residues within recombinant factor VIII, *Biochemistry* **31**, 3315–3325.

Pratt, K.P., Shen, B.W., Takeshima, K., Davie, E.W., Fujikawa, K., Stoddard, B.L. (1999) Structure of the C2 domain of human factor VIII at 1.5 Å resolution, *Nature* **402**, 439–442.

Ragni, M.V. (2002) Safe passage: a plea for safety in hemophilia gene therapy, *Mol. Therapy* **6**, 436–440.

Rizza, C.R., Spooner, R.J.D. (1983) Treatment of haemophilia and related disorders in Britain and Northern Ireland during 1976–1980: report on behalf of the directors of haemophilia centres in the United Kingdom, *Br. J. Haematol.* **286**, 929–933.

Rowley, G., Saad, S., Giannelli, F., Green, P.M. (1995) Ultrarapid mutation detection by multiple, solid-phase chemical cleavage, *Genomics* **30**, 574–582.

Schwaab, R., Brackmann, H.H., Meyer, C., Seehafer, J., Kirchgesser, M., Haack, A., Olek, K., Tuddenham, E.G.D., Oldenburg, J. (1995) Haemophilia A: mutation type determines risk of inhibitor formation, *Thromb. Haemost.* **74**, 1402–1406.

Stoilova-McPhie, S., Villoutreix, B.O., Mertens, K., Kemball-Cook, G., Holzenburg, A. (2002) 3-Dimensional structure of membrane-bound coagulation factor VIII: modeling of the factor VIII heterodimer within a 3-dimensional density map derived by electron crystallography, *Blood* **99**, 1215–1223.

Stoylova, S.S., Lenting, P.J., Kemball-Cook, G., Holzenburg, A. (1999) Electron crystallography of human blood coagulation factor VIII bound to phospholipids monolayers, *J. Biol. Chem.* **274**, 36573–36578.

Toole, J.J., Knopf, J.L., Wozney, J.M., Sultzman, L.A., Brecker, J.L., Pittman, D.D., Kaufman, R.J. (1984) Molecular cloning of a cDNA encoding human antihaemophilic factor, *Nature* **312**, 342–347.

Vielhaber, E., Jacobson, D.P., Ketterling, R.P., Liu, J.Z., Sommer, S.S. (1993) A mutation in the 3' untranslated region of the factor IX gene in four families with hemophilia B, *Hum. Mol. Genet.* **2**, 1309–1310.

Walsh, C.E. (2003) Gene therapy progress and prospects: gene therapy for the haemophilias, *Gene Ther.* **10**, 999–1003.

Yoshitake, S., Schach, B.G., Foster, D.C., Davie, E.W., Kurachi, K. (1985) Nucleotide sequence of the gene for human factor IX (antihemophilic factor B), *Biochemistry* **24**, 3736–3750.

37
Gene Therapy and Cardiovascular Diseases

Michael E. Rosenfeld[1] and Alan D. Attie[2]
[1] *Department of Pathobiology, University of Washington, Seattle, WA, USA*
[2] *Department of Biochemistry, University of Wisconsin-Madison, Madison, WI, USA*

1	Overview of Gene Therapy	1091
2	Atherosclerosis: A Chronic Inflammatory Disease	1092
3	Gene Therapy Targets for Cardiovascular Disease in the Lipoprotein Pathways	1095
3.1	The Lipoprotein Pathways Provide Many Potential Gene Therapy Targets	1095
3.2	Specific Genes are Critical for Dietary Lipid Absorption in the Intestine	1096
3.3	The Intestine Exports Lipids on Chylomicron Particles	1097
3.4	The Liver is a Major Producer of Plasma Lipoproteins	1097
3.5	Lipoprotein Lipase as a Target for Gene Therapy	1098
3.6	The LDL and VLDL Receptors are Attractive Targets for Gene Therapy	1099
3.7	Increased Cholesterol and Phospholipid Efflux via ABCA1 Can Prevent Atherosclerosis	1100
3.8	Some Apolipoproteins Can Prevent Atherosclerosis	1101
3.9	The Liver SR-B1 Receptor Protects against Atherosclerosis	1102
4	Vascular and Inflammatory Targets for Gene Therapy	1103
4.1	Adhesion Molecules and Leukocyte Counter Receptors	1103
4.2	Cytokines	1104
4.3	Chemokines and Chemokine Receptors	1104
4.4	Interleukins, the TNF-α family, and Interferon-γ	1105
4.5	Growth Factors and Receptors	1106
5	Pro- and Antioxidant Enzymes	1107

Genomics and Genetics. Edited by Robert A. Meyers.
Copyright © 2007 Wiley-VCH Verlag GmbH & Co. KGaA, Weinheim
ISBN: 978-3-527-31609-0

6	Scavenger Receptors 1108
7	Transcription Factors 1109
8	Noninflammatory Gene Targets 1110
8.1	Cell Death 1110
8.2	Proteases 1110
8.3	Regulators of Vascular Tone and Blood Pressure 1111
8.4	Coagulation Related Genes 1111
8.5	Osteogenic Proteins 1112

Bibliography 1112
Books and Reviews 1112
Primary Literature 1113

Keywords

ABCA1
A membrane lipid–transport protein involved in cholesterol and/or phospholipid efflux from cells. Its action is necessary for the extracellular assembly of lipoprotein particles.

Adhesion Molecules
These are proteins on the luminal surface of endothelial cells, bind to counter receptors on the membrane of leukocytes, and enable the leukocytes to attach tightly to the surface of the endothelial cells.

Angina Pectoris
Chest pain caused by reduced blood flow to the heart due to the blockage of coronary arteries by blood clots.

Angiogenesis
The process of formation of new blood vessels.

Angioplasty
An intervention to increase blood flow in a clogged artery. It involves insertion of a balloon catheter into the blocked artery and subsequent inflation of the balloon to break open or squash the atherosclerotic lesion and/or blood clot that is reducing blood flow.

Apical Membrane
In polarized cells (e.g. hepatocytes in the liver, enterocytes in the intestine), the membrane exposed to the outside world; in the hepatocytes, the bile canalicular membrane; in enterocytes, the membrane exposed to the intestinal lumen.

ApoA1
An apolipoprotein principally associated with HDL, an activator of lecithin cholesterol: acyltransferase. It interacts with the cells to mediate delivery of cholesterol ester from HDL particles.

ApoB100
An apolipoprotein associated with VLDL and LDL particles, synthesized in the liver, a ligand for the LDL receptor.

ApoB48
An apolipoprotein associated with chylomicrons, synthesized in the intestine; it is a truncated form of apoB100 and does not bind to the LDL receptor.

ApoE
An apolipoprotein principally associated with VLDL and chylomicrons, is responsible for the receptor-mediated clearance of IDL and chylomicron remnants, is a ligand for most members of the LDL-receptor superfamily. The apoE4 isoform is associated with increased risk of Alzheimer's Disease.

Apolipoproteins
The protein components of plasma lipoproteins.

Apoptosis
The programmed cell death of a cell or cellular suicide. It is a defined molecular pathway where proteolytic enzymes are activated to destroy key cellular proteins.

Atherosclerosis
An inflammatory disease in the arterial wall leading to the accumulation of cellular outgrowths that can become unstable, break off, and cause the formation of blood clots.

Basolateral Membrane
In polarized cells, the membrane exposed to the inside world; in hepatocytes, the sinusoidal membrane, which is exposed to venous circulation, in enterocytes, the basolateral membrane, which is exposed to the lymphatics and venous circulation.

Bile Acids
Detergent-like molecules formed from cholesterol. They are secreted by the liver and together with cholesterol and phospholipids, form bile. Bile forms micelles that emulsify lipids in the intestinal lumen aiding in their absorption.

Chemokines
A category of cytokines that function specifically to chemically attract leukocytes to sites of inflammation.

Chylomicron Remnants
Chylomicron particles that have been depleted of triacylglycerol after the lipoprotein lipase–mediated hydrolysis of their triacylglycerols.

Chylomicrons
Lipoprotein particles produced in the intestine to package and secrete dietary lipids. Chylomicrons are secreted into the mesenteric lymph.

Cytokines
Cell regulatory proteins secreted by many cell types that are key factors in the initiation and control of inflammation.

Familial Hypercholesterolemia
Elevation in LDL cholesterol due to mutations at the LDL-receptor locus.

Familial Hypoalphalipoproteinemia
Deficiency in HDL cholesterol due to mutations at the ABCA1 locus.

Fatty Streak
Earliest stage in the development of an atherosclerotic lesion. It consists of aggregates of macrophage-derived foam cells and lymphocytes that form underneath the endothelial lining of an artery.

Fibrous Cap
A thin layer of smooth muscle cells and connective tissue that encapsulates areas of dead cells and lipid in a developing atherosclerotic lesion. If the fibrous cap is too thin, it can break and enable formation of blood clots that can clog arteries.

Foam Cells
Primarily macrophages that have taken up large numbers of modified lipoprotein particles and have stored the excess lipid in non membrane bound cholesteryl-ester droplets and in increased numbers of secondary lysosomes.

Gallstones
Large crystals, usually composed of cholesterol, that form in the biliary tract and/or gall bladder when cholesterol levels in bile are too high relative to phospholipid and bile acids.

Growth Factors
Secreted by many cell types and by binding to specific receptors on the plasma membrane of adjacent cells, are potent regulators of cellular functions, including proliferation, migration, differentiation, and survival/apoptosis.

HDL
High-density lipoprotein, carries about 20% of plasma cholesterol. Its levels negatively correlate with risk of coronary heart disease. Is thought to mediate "reverse cholesterol transport."

Hemorrhage
Bleeding into a tissue. In atherosclerotic lesions, it occurs when the lesion forms fissures or ruptures and is an indication that the lesion is unstable.

IDL
An intermediate density lipoprotein, also called "VLDL remnant." It is the VLDL particle that has been depleted of triacylglycerol through the action of lipoprotein lipase.

Intima
The inner layer of a muscular artery. It is the space immediately behind the endothelium where atherosclerotic lesions develop. In normal, nondiseased human arteries, there is a diffuse thickening of the intima that contains connective tissue and sparse numbers of macrophages and smooth muscle cells.

Ischemia
The inadequate perfusion of a tissue with blood leading to the reduced availability of oxygen and nutrients in the tissue.

LDL
Low-density lipoprotein carries approximately two-thirds of plasma cholesterol. Its levels are positively correlated with the risk of coronary heart disease and is formed in the circulation from the catabolism of VLDL.

LDL Receptor
Expressed in most tissues and is mainly necessary for normal clearance of LDL from the bloodstream.

Lipoprotein Lipase
Catalyzes the hydrolysis of VLDL and LDL triacylglycerols to free fatty acids and glycerol, and is present on the luminal surface of the capillary endothelium.

Lipoproteins
Particles that transport lipids in the bloodstream.

Media
The middle layer of a muscular artery that contains concentric rings of smooth muscle cells oriented perpendicular to blood flow. The orientation allows the contraction of the smooth muscle cells to constrict the artery. Atherosclerotic lesions do not initially form in the media but can expand and replace the media in very advanced stages of the disease.

Myocardial Infarction
The necrotic death of heart tissue due to a sudden blockage of blood flow to the affected tissue. It most frequently occurs following the formation of a blood clot on top of an atherosclerotic lesion that has ruptured in a main coronary artery.

Necrotic Zones
Areas within an atherosclerotic lesion that contain dead cells and debris, and lipid droplets formed from dead foam cells and aggregated lipoproteins. The expansion of necrotic zones into a *necrotic core* can make the lesion unstable and at risk of rupture.

Neovascularization
The formation of new, small blood vessels within an advanced atherosclerotic lesion. Neovessels are often formed deep in the lesion where there is limited diffusion of oxygen and nutrients, and can be a conduit for bringing more leukocytes into the advanced lesions.

Restenosis
The process of reformation of an arterial lesion that reduces blood flow following an intervention procedure such as angioplasty. A restenotic lesion is not the same as an atherosclerotic lesion as it forms primarily because of the migration and proliferation of smooth muscle cells following injury to the artery caused by the intervention procedure. It occurs in up to 40% of people who have angioplasty to increase blood flow to the heart.

Scavenger Receptors
Bind to a wide range of molecules, including modified forms of LDL. Are involved in the accumulation of cholesterol in macrophages and smooth muscle cells in the arterial wall.

Statins
Drugs that inhibit the synthesis of cholesterol. They increase the abundance of the LDL receptor and thus reduce plasma LDL levels.

Tangier Disease
A severe HDL deficiency syndrome caused by homozygous mutations at the ABCA1 locus.

Thrombosis
The formation of blood clots. A thrombus consists of variable numbers of aggregated platelets, red blood cells, and fibrin. In the heart, brain, or peripheral blood vessels, it can be an *occlusive thrombus* that blocks blood flow, or a smaller *mural thrombus* that is attached to an atherosclerotic lesion and does not block blood flow.

Vasospasm
The transient constriction of a blood vessel. Depending on the extent of constriction, blood flow can be reduced leading to tissue ischemia.

VLDL
Very low-density lipoprotein, a triacylglycerol-rich lipoprotein assembled in the endoplasmic reticulum and Golgi of hepatocytes and then secreted into the bloodstream. While in the circulation, its triacylglycerol core is depleted through the action of lipoprotein lipase and it gives rise to low-density lipoprotein.

1
Overview of Gene Therapy

Gene therapy refers to the expression of a recombinant gene in a patient. The recombinant gene can be delivered to the patient directly or it can be incorporated into cells *in vitro*, followed by delivery of the genetically modified cells to the patient. DNA delivery can occur in the patient as naked DNA or within a biological or artificial delivery vehicle. The biological delivery vehicle is usually a virus. A wide variety of artificial delivery vehicles have been developed. They usually consist of lipid vesicles (liposomes) or positively charged polymers that complex with DNA and have a high affinity for cell surfaces that are negatively charged.

Genetic therapy for the treatment of cardiovascular disease (CVD) is in its infant stage despite the large number of potential target genes associated with the development of atherosclerosis. Cardiovascular disease lends itself to genetic therapy because of the focal nature of atherosclerotic lesions and the availability of techniques for direct delivery of the genes to the affected areas. The desire to conduct gene therapy for atherosclerosis is in large part motivated by the large number of transgenic mouse studies in which expression or deletion of a gene has a profound therapeutic effect on the animal. For example, overexpression of the apolipoprotein-A1 gene protects animals from atherosclerosis. A logical extension is to mimic these studies with gene therapy.

Although the gene therapy field is now more than 20 years old, it has not fulfilled its initial promise. From the beginning, it has been impossible to achieve sustained expression of foreign genes at high levels. There is still an incomplete understanding of the fundamental processes by which DNA enters the cells, is transported to the nucleus, integrates into chromosomal DNA, and interacts with transcriptional machinery.

Regardless of the vehicle, DNA must be administered into the circulatory system in order to gain access to a high proportion of cells in a solid tissue. However, most tissues are not in direct contact with blood. They are shielded from blood by a tight layer of endothelial cells that comprise the luminal surface of arteries and veins. Particles as large as viruses or DNA/polymer aggregates cannot cross the small gaps (fenestrae) between endothelial cells. An important exception is the vasculature that perfuses the liver. These vessels have large fenestrae, allowing large particles to gain access to the hepatocytes in the liver. For this reason, it is relatively easy to target the liver and quite difficult to target extrahepatic tissues.

There are three predominant types of viral vectors used in gene therapy; adenoviruses, adeno-associated viruses, and lentiviruses. *Adenoviruses* belong to a family of human DNA tumor viruses that can cause noncancerous respiratory tract infections in humans. Upon infection, the viral DNA remains episomal (outside the host genome), thus making expression transient. Adenoviruses can infect both dividing and nondividing cells. Adenoviral infection causes an inflammatory response, which curtails the infection and can produce severe reactions in some individuals.

Adeno-associated virus is a small nonpathogenic single-stranded DNA virus. On its own, it cannot replicate. It depends on simultaneous expression of genes provided by adenovirus. The virus infects a broad range of dividing and nondividing cells. It can maintain its DNA as an episome or can, at low frequency, integrate into the host cell's chromosomal DNA. Adeno-associated virus is potentially attractive because its expression tends to be long-lived and it does not elicit an inflammatory reaction the way adenoviruses do.

Lentiviruses are retroviruses, RNA viruses that replicate through a DNA intermediate. HIV is a lentivirus. Lentiviruses have been modified to infect a broad range of cells. They can infect nondividing cells. As is the case with adeno and adeno-associated virus, lentiviruses are produced that are replication-defective for gene therapy applications.

Injection of *naked DNA* into muscle results in transfection of the cells *in vivo*. Intravascular injection of naked DNA leads to uptake by hepatocytes in the liver. Some of the cells stably incorporate the foreign DNA and express its genes. Since animals are continually exposed to foreign DNA coming from the diet and from symbiotic and infectious microorganisms, they have evolved numerous mechanisms to protect their genomes from invasion by foreign DNA. These include degradation by nucleases, inefficient transport to the nucleus, cell death upon DNA uptake, and inactivation of promoters. Naked DNA is still very attractive because it avoids the complications of viral vectors. This justifies a continued effort to overcome the obstacles to efficient and prolonged expression of genes from the naked DNA.

Liposomes are vesicles formed from lipids, usually, phospholipids. Cationic lipids form stable complexes with DNA and can therefore be incorporated into liposomes. These complexes are routinely used by researchers to introduce DNA into cell lines *in vitro*. Their use *in vivo* is attractive because it does not raise the safety issues associated with viral vectors. However, they are far less efficient than viruses so their utility is still limited.

An alternative to direct injection of DNA or a vehicle into a patient is to obtain cells from a patient and incorporate a foreign gene into the cells *in vitro*. Cells that have been selected for stably incorporating the foreign gene can then be given back to the patient. The benefits of this step cannot be overstated; instability of transfected DNA is a major unsolved problem with *in vivo* gene therapy. This approach has been used in atherosclerosis research with cells derived from bone marrow. Since the bone marrow contains precursors of red blood cells and cells of the immune system, this is a practical way to specifically target such cells. After reintroduction to the bone marrow, cells containing a foreign gene enter into the circulation and repopulate tissues. For example, the bone marrow is a source of circulating phagocytic cells ("monocytes" in the circulation, "macrophages" in most tissues, and "Kupffer cells" in the liver). Animal experiments have shown dramatic alterations in atherosclerosis susceptibility with the replacement of macrophages in the bone marrow. Partial hepatectomy stimulates a robust regenerative response in the liver. This makes possible the introduction of hepatocytes that have been genetically modified into a partially hepatectomized liver. The new cells respond to the proliferative signals and contribute to the new cell mass of the liver.

2
Atherosclerosis: A Chronic Inflammatory Disease

Atherosclerosis is thought to be the result of a chronic fibro-proliferative inflammatory response. Animal studies suggest that this inflammatory disease is initiated following the deposition, modification, and retention of lipoproteins within the artery wall. The initial inflammatory response involves the upregulation of expression of *adhesion molecules* and *chemokines* by endothelial cells and smooth muscle cells, leading to the recruitment of monocytes and lymphocytes into the arterial *intima*. The monocytes rapidly convert into tissue macrophages that have the capacity to scavenge modified lipoproteins, secrete proinflammatory *cytokines* and *growth factors*, and further regulate immune functions via presentation of antigens. The *scavenger receptor* mediated uptake of modified lipoproteins transforms the macrophages into lipid-loaded *foam cells*, the accumulation of which is the hallmark of the fatty streak, the initial stage of atherosclerosis (Fig. 1).

The transition of the *fatty streak* into more advanced stages of the disease appears to involve the coalescence of deposited lipids coupled with the death of

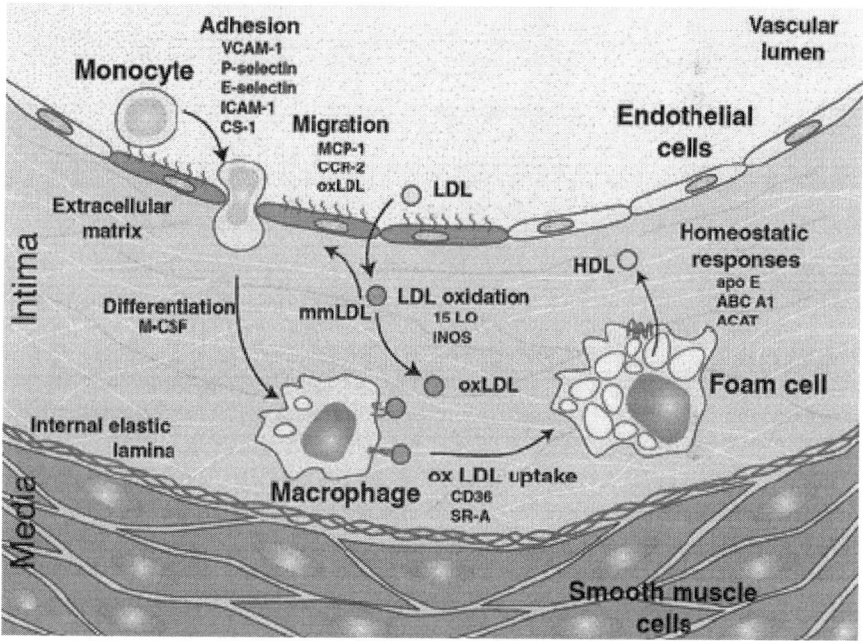

Fig. 1 Initiation of a fatty streak. Lipoproteins such as LDL become trapped within the extracellular matrix of the artery. The trapped lipoproteins are modified by processes such as oxidation or glycosylation (mmLDL, oxLDL) and become proinflammatory leading to the activation of endothelial cells to express adhesion molecules such as VCAM-1 and chemokines such as MCP-1. Monocytes are then recruited into the arterial intima and differentiate into macrophages most likely in response to increased expression of M-CSF. The modified lipoproteins are then taken up by scavenger receptors such as CD36 and the SRA1 expressed by the macrophages and this leads to foam cell formation. The excess cholesterol taken up by the foam cells is esterified by ACAT and stored in lipid droplets. It can be converted back to the more soluble free cholesterol and exported to extracellular HDL acceptors via cholesterol transporters, such as *ABC-A1*. (From Glass, C. K., Witztum, J. L. (2001) Atherosclerosis. The road ahead, *Cell* **104**(4), 503–516, included with permission of the Publisher.)

foam cells to form acellular *necrotic zones*. As in any classical inflammatory response, formation of the necrotic zones is accompanied by a wound-healing response where smooth muscle cells are recruited to encapsulate the necrotic zones by forming a *fibrous cap* around the lipid pools and necrotic debris (Fig. 2). With ongoing hyperlipidemia, however, there is a continuous influx, trapping, and modification of lipoproteins in the developing lesion that recruits additional inflammatory cells and leads to formation of new fatty streaks adjacent to or on top of the initial lesions. The combination of layering of new fatty streaks and erosion of the fibrous cap destabilizes the plaques allowing fissures, ruptures, and intraplaque *hemorrhage* to occur. While the exact mechanisms that cause plaques to rupture are currently unknown, the activation and/or death of macrophages with release of proteolytic enzymes such as *matrix metalloproteinases*, likely play a role. The resulting erosion and rupture exposes procoagulant proteins such as *tissue factor* to the blood and facilitates the formation of mural and *occlusive thrombi* (Fig. 3). The occlusion of the main coronary arteries causes the clinical outcomes of myocardial ischemia, *angina*, and ultimately *myocardial infarction*. Occlusion of cerebral vessels can lead to ischemic stroke while occlusion of peripheral blood vessels

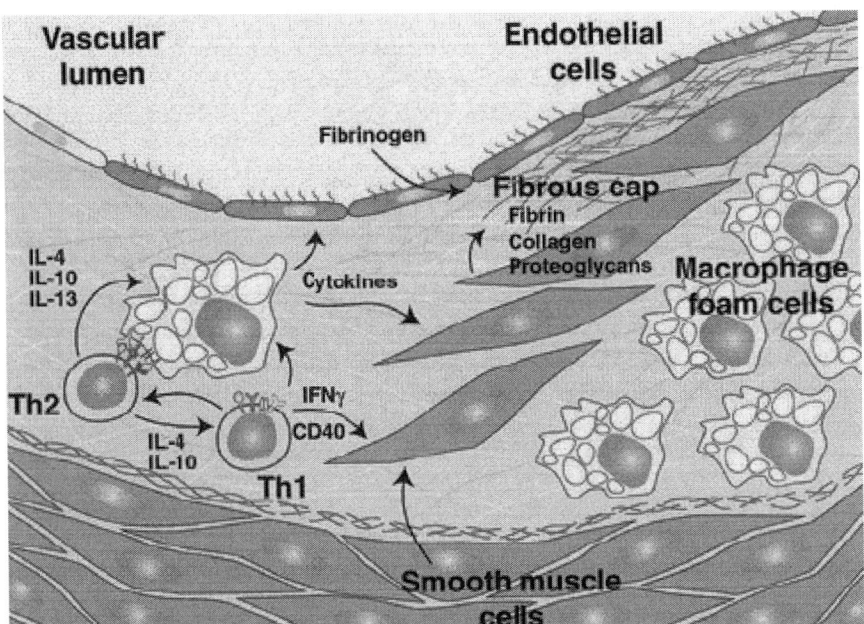

Fig. 2 Atherosclerotic lesion progression. Continued deposition of lipoproteins coupled with foam cell formation and interactions between foam cells, Th1 and Th2 T Helper cells and smooth muscle cells contribute to a chronic inflammatory response. Cytokines secreted by lymphocytes and macrophages exert both pro- and antiatherogenic effects. As part of a wound-healing response smooth muscle cells migrate from the medial portion of the arterial wall, proliferate, and secrete extracellular matrix proteins that form a fibrous cap. (From Glass, C.K., Witztum, J.L. (2001) Atherosclerosis. The road ahead, *Cell* **104**(4), 503–516, included with permission of the Publisher.)

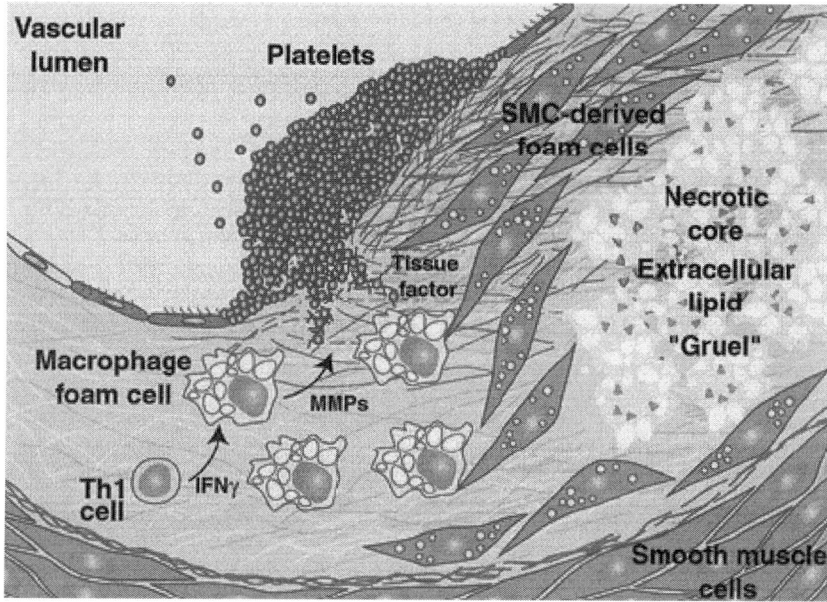

Fig. 3 Plaque rupture and thrombosis. The death of macrophage-derived foam cells and coalescence of trapped lipoproteins leads to the formation of an acellular necrotic zone and accumulation of extracellular cholesterol. Macrophage secretion of matrix metalloproteinases and neovascularization contribute to weakening of the fibrous cap. Plaque rupture exposes blood components to tissue factor, initiating coagulation, the recruitment of platelets, and the formation of a thrombus. (From Glass, CK and Witztum JL, Cell 2001; **104**(4): 503–516, included with permission of the Publisher.)

is associated with critical limb ischemia and gangrene.

3
Gene Therapy Targets for Cardiovascular Disease in the Lipoprotein Pathways

3.1
The Lipoprotein Pathways Provide Many Potential Gene Therapy Targets

A major risk factor for premature atherosclerosis is abnormalities in lipoprotein metabolism. In addition, people with diabetes mellitus frequently have lipoprotein abnormalities that elevate the risk of premature atherosclerosis that is already associated with diabetes.

Lipoprotein metabolism involves the interplay of the intestine, the liver, adipose tissue, and muscle. Endocrine organs also play a role, chiefly pancreatic islets, adrenals, the hypothalamus, and adipose tissue. However, the following discussion will emphasize the role of the liver because it is the most readily accessible target for gene therapy that is based upon the intravenous injection of DNA or DNA within a vector or carrier.

Animals transport large quantities of lipids through the bloodstream on carrier particles called lipoproteins. Lipoproteins consist of a shell of proteins called *apolipoproteins*, amphipathic lipids, primarily phospholipid and unesterified cholesterol, surrounding a core of

Tab. 1 Density, diameter, and composition of plasma lipoproteins.

Class	Density [g mL^{-1}]	Diameter [nm]	Chol	PL	Protein	TG	CE
Chylomicrons	0.93	75–1200	2	7	2	86	3
VLDL	0.93–1.006	30–80	7	18	8	55	12
IDL	1.006–1.019	25–35	9	19	19	23	29
LDL	1.019–1.063	18–25	8	22	22	6	42
HDL2	1.063–1.125	9–12	5	33	40	5	17
HDL3	1.125–1.210	5–9	4	35	55	3	13

Notes: Values represent percentage of dry mass. Chol = cholesterol; PL = phospholipids; TG = triacylglycerols; CE = cholesterol esters.

nonpolar lipid. The core composition consists of variable proportions of triacylglycerol and cholesterol ester.

When plasma is subjected to ultracentrifugation, most proteins sediment at the bottom of the centrifuge tube. Because they are associated with lipid and have a low buoyant density, lipoproteins float in the ultracentrifuge. The density at which they float is the basis for their classification (Table 1). In fasting plasma, most of the triacylglycerol is carried by very low-density lipoprotein (VLDL) particles. Cholesterol and cholesterol ester are carried by low-density lipoprotein (LDL) and high-density lipoprotein (HDL) particles, with LDL carrying about two-thirds of all cholesterol in human plasma.

3.2
Specific Genes are Critical for Dietary Lipid Absorption in the Intestine

It is important to remember that ordinarily, the major component of dietary fat is always triacylglycerol, not cholesterol. For example, milk and butter have very little cholesterol but are very high in triacylglycerol. Triacylglycerol (and other glycerolipids, such as phospholipids) is hydrolyzed in the intestinal lumen to yield monoglycerides and free fatty acids. These lipolysis products are then absorbed by the intestinal epithelial cells and resynthesized as triacylglycerols, phospholipids, and cholesterol esters. Thus, the intestine mediates both lipolysis (in the lumen) and reesterification (within the epithelial cells) of dietary lipids. The molecular mechanisms underlying intestinal absorption of fatty acids are still not understood. There is no consensus about whether a specific transporter is involved in fatty acid transport into the enterocytes.

Cholesterol absorption into the enterocyte is also not well understood. A specific inhibitor of cholesterol absorption, Ezetimibe, is in clinical use and is one line of evidence that there is a specific transporter for cholesterol at the membrane of the enterocyte that faces the intestinal lumen (the apical membrane).

The free fatty acids coming from intestinal lipolysis are solubilized into micelles within the intestinal lumen. These micelles consist of bile acids, cholesterol, and phosphatidylcholine. Bile acids are detergent-like molecules produced from cholesterol by the liver. Together with phospholipid and cholesterol, bile acids are secreted from the apical membrane of hepatocytes into the bile canaliculi, which drain into the bile duct and finally into the duodenum. As discussed below, conversion of cholesterol to bile acids and secretion of biliary cholesterol are the principal routes of cholesterol elimination.

Two membrane cholesterol transporters have been identified in the intestine. One consists of a heterodimer of two proteins, ABCG5, and ABCG8. This pair is located at the apical membrane and transports cholesterol out of the enterocyte into the lumen. In addition, the transporter transports plant sterols and thus protects animals from accumulation of these nonphysiological sterol molecules. Mutations in ABCG5 or ABCG8 cause a rare disease called sitosterolemia, which involves elevated levels of cholesterol and plant sterols in the bloodstream. ABCA1 also transports cholesterol out of enterocytes. In contrast to ABCG5/ABCG8, this transporter appears to transport cholesterol across the basolateral membrane of the enterocyte, to the bloodstream and/or lymphatics. Overexpression of ABCG5/ABCG8 in transgenic mice reduces cholesterol absorption, making this an attractive target for therapeutic intervention.

3.3
The Intestine Exports Lipids on Chylomicron Particles

Since the intestine is primarily an absorptive organ, it must have the means of exporting newly absorbed lipids. The enterocyte reesterifies fatty acids and monoglycerides to form triacylglycerols and phospholipids. Absorbed cholesterol is esterified to form cholesterol esters. Triacylglycerols and cholesterol esters are then packaged into the core of lipoprotein particles unique to the intestine – *chylomicrons*. Rather than being secreted into the bloodstream, chylomicrons are secreted into the lymphatics. By secreting chylomicrons into the lymphatics, they gain entrance into the general circulation via the thoracic duct. This guarantees that extrahepatic tissues, principally adipose tissue and muscle, are the first to be exposed to the newly secreted chylomicrons – if chylomicrons were secreted directly into the bloodstream, they would first be delivered to the liver, via the portal vein.

After entry into the bloodstream, chylomicrons interact with the luminal surface of the capillary beds of adipose tissue and muscle. Here resides the enzyme *lipoprotein lipase*, which hydrolyzes the triacylglycerols all the way to free fatty acids and glycerol. (Cholesterol esters are not substrates for lipoprotein lipase.) The free fatty acids are then absorbed where they are reesterified and stored in triacylglycerol droplets (in adipocytes) or oxidized for energy (in muscle). The chylomicron depleted of its triacylglycerol (a chylomicron remnant) is rapidly cleared from the circulation by the liver. There is significant variation in the abundance and activation state of lipoprotein lipase. This affects the efficiency of clearance of triacylglycerol-rich lipoproteins and is therefore an attractive target for gene therapy.

3.4
The Liver is a Major Producer of Plasma Lipoproteins

Adipose tissue and liver are quite active in the conversion of carbohydrate into fat. Adipose tissue stores the fat in the form of triacylglycerol droplets. However, these droplets are quite dynamic; they are continuously being formed and turned over. The turnover of adipocyte droplets occurs through the action of hormone-sensitive lipase. The free fatty acids that are released from adipocytes can go to the liver where they can be used to make ketone bodies and/or are reesterified to form triacylglycerol again. Whether derived from adipose tissue or from *de novo* synthesis, the liver does not generally store triacylglycerol. Triacylglycerol accumulation in the liver is generally regarded as a pathological state, termed *fatty liver* or *hepatic steatosis*.

3.5
Lipoprotein Lipase as a Target for Gene Therapy

In a healthy liver, triacylglycerol is packaged for secretion within the secretory pathway. This involves transfer of triacylglycerol to the endoplasmic reticulum lumen and interaction with apoB100 (Fig. 4). Within the secretory pathway, there are several steps in which triacylglycerol is incorporated into a growing VLDL particle. VLDL is then secreted into the bloodstream. Like chylomicrons, VLDL interacts with *lipoprotein lipase* while in the circulation. However, unlike chylomicrons, the depletion of triacylglycerol from VLDL yields a VLDL remnant, also termed *intermediate density lipoprotein* (IDL), which goes on to produce a stable lipoprotein,

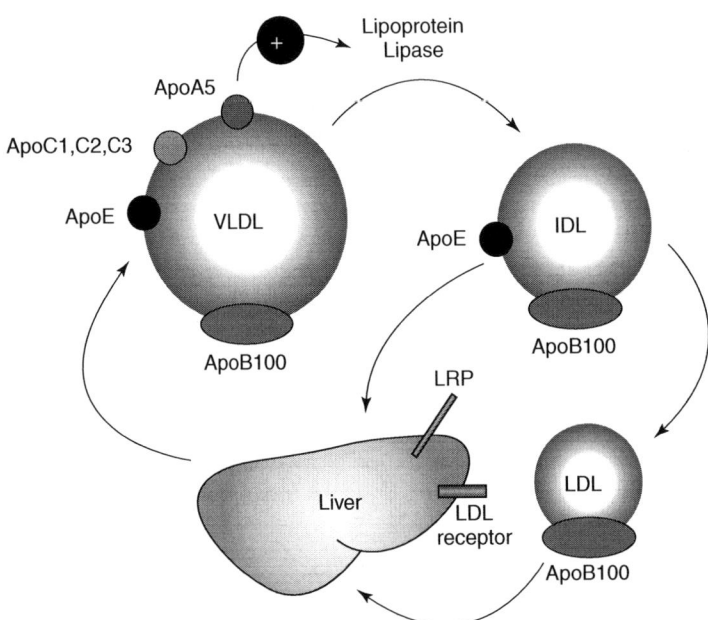

Fig. 4 Metabolic interrelationships of liver-derived plasma lipoproteins. The liver secretes VLDL, a particle enriched in triacylglycerol. It carries various apolipoproteins. ApoB100 is the only nonexchangeable protein on VLDL particles. ApoC2 and possibly, apoA5, are activators of lipoprotein lipase. This enzyme resides on the luminal surface of the capillary endothelium of adipose tissue and muscle and hydrolyzes the fatty acids from VLDL triacylglycerols. The resulting particle, IDL, can be cleared from the circulation by the liver or can become LDL. Clearance of IDL is mediated by apoE. Clearance of LDL is mediated by apoB100. The LDL receptor plays a critical role in the clearance of IDL and LDL; both apoE and apoB100 can bind to the LDL receptor. The LRP acts as a backup for IDL clearance; in the absence of the LDL receptor, it can clear IDL by binding to apoE.

LDL. Owing to the loss of most of the triacylglycerol, the predominant neutral lipids of LDL are cholesterol and cholesterol ester.

The catabolism of VLDL is dependent upon efficient hydrolysis of its triacylglycerol core by lipoprotein lipase. Lipoprotein lipase deficiency results in dramatic hypertriglyceridemia, liver enlargement (hepatomegaly), and inflammation of the pancreas (pancreatitis). However, elevated plasma VLDL can result from many other as-yet unidentified causes. It is very common in the human population and is associated with obesity, diabetes, and in some individuals, high-carbohydrate diets. Although these individuals have normal levels of lipoprotein lipase, increasing the level of this enzyme might still be a useful therapeutic strategy to lower plasma triacylglycerols. Since it is presently not feasible to target genes to the normal sites of expression of lipoprotein lipase (muscle and adipose tissue), expression in a nonphysiological site, the liver, has been tested in mice and in a naturally occurring feline model of lipoprotein lipase deficiency.

In addition to the expected role of lipoprotein lipase in catalyzing the hydrolysis of lipoprotein triacylglycerol, there is also a nonenzymatic property of potential therapeutic value. Lipoprotein lipase binds to lipoproteins and also binds to lipoprotein receptors in the LDL receptor family, most notably, the LDL receptor-related protein (LRP). In addition, lipoprotein lipase can bind to cell surface proteins that are enriched in acidic carbohydrate residues, the proteoglycans. The best evidence that these nonenzymatic functions are therapeutically relevant is the demonstration that a mutant form of lipoprotein lipase lacking enzymatic activity is still able to reduce plasma triacylglycerol by about 20 to 30%.

Lipoprotein lipase-mediated hydrolysis of triacylglycerol causes fatty acids to be taken up at the site of hydrolysis. In the liver, this leads to reesterification of the fatty acids and can result in an increase in cellular triacylglycerols. Through mechanisms not clearly understood, excessive triacylglycerol accumulation dampens the cell's responsiveness to insulin. In the liver, insulin normally suppresses glucose production. Thus, one potential side effect of expression of lipoprotein lipase in the liver is to reduce the ability of insulin to suppress glucose production, a potential problem for people who are already insulin-resistant or diabetic.

3.6
The LDL and VLDL Receptors are Attractive Targets for Gene Therapy

LDL is cleared from the circulation in large part through its interaction with the *LDL receptor*. Mutations affecting the expression or function of the LDL receptor are responsible for a common inherited disorder, *familial hypercholesterolemia*. Increased expression of the LDL receptor reduces LDL levels; thus, the LDL receptor has been used in gene therapy for hypercholesterolemia.

The VLDL receptor is a member of the LDL receptor family. It is expressed in a wide range of tissues, but not in the liver. However, adeno-associated virus was used to express this receptor in the livers of mice lacking the LDL receptor. These mice experienced a 40% reduction of cholesterol, due to enhanced clearance of both VLDL and LDL. Since patients who do not express the LDL receptor are immunologically naïve to this antigen, gene therapy with the VLDL receptor might be an effective treatment that would avoid the danger of immune rejection.

3.7 Increased Cholesterol and Phospholipid Efflux via ABCA1 Can Prevent Atherosclerosis

Epidemiological studies show an inverse relationship between HDL levels and risk of premature atherosclerosis. Unlike VLDL and chylomicrons, HDL is formed from its protein and lipid components in the bloodstream and interstitial fluids rather than within the secretory pathway of cells. The major apolipoproteins of HDL are apoA1 and apoA2. These proteins are secreted from hepatocytes and intestinal epithelial cells independently and also as minor components of VLDL and chylomicrons. ApoA1 and apoA2 bind to phospholipids. Phospholipids are available from the surface of VLDL after lipolysis. In addition, cells are able to efflux phospholipids and cholesterol through the action of ABCA1 (Fig. 5). Its crucial role in this process was established by the discovery that two types of severe inherited HDL deficiency syndromes are caused by mutations in ABCA1, Tangier Disease, and familial hypoalphalipoproteinemia (FHA).

Tangier Disease is a rare recessive disorder in which patients have almost no HDL. Cholesterol ester accumulates in macrophages and macrophage-rich tissues like spleen and liver. FHA is a very common dominant disorder in which people have low HDL (typically <30 mg dL^{-1}) and suffer from premature heart disease even without an elevation in LDL. Approximately 40% of patients with premature coronary heart disease have low HDL, making it the most common lipid disorder of heart disease patients.

Tangier Disease and FHA are caused by mutations in ABCA1. Tangier Disease patients are homozygous (or inherit two different mutant alleles). FHA patients are heterozygous for mutations at the ABCA1 locus. ABCA1 mutations lower HDL because they prevent phospholipid and/or cholesterol from effluxing and becoming associated with apoA1 and apoA2. In the absence of sufficient lipid, apoA1 is rapidly cleared by the kidneys.

Why do mutations in ABCA1 lead to premature heart disease? ABCA1 fulfills a rate-limiting step in the pathway by which cells get rid of cholesterol, and thus might be a critical protector against cholesterol overload. With the exception of hepatocytes, cells are unable to catabolize large quantities of cholesterol and must therefore protect themselves from cholesterol overload by expelling cholesterol to an appropriate extracellular carrier. It is interesting that ABCA1 is especially abundant in macrophages; macrophages can become engorged with cholesterol esters and form foam cells in the arterial wall. A mutation in ABCA1 might therefore predispose an individual to atherosclerosis by impeding cholesterol efflux (Fig. 1).

Proof-of-principle studies in mice nominate ABCA1 for the gene therapist's arsenal. Moderate overexpression of ABCA1 decreases atherosclerosis in mice, despite having a minimal effect on HDL levels. The most plausible explanation is that macrophage ABCA1 function is more important than HDL levels in protecting an animal from atherosclerosis. Indeed, if animals are transplanted with macrophages derived from mice lacking ABCA1, they show increased atherosclerosis. These animal studies suggest that increased expression of ABCA1 in macrophages might be a sensible gene therapy strategy for protection against or treatment of atherosclerosis.

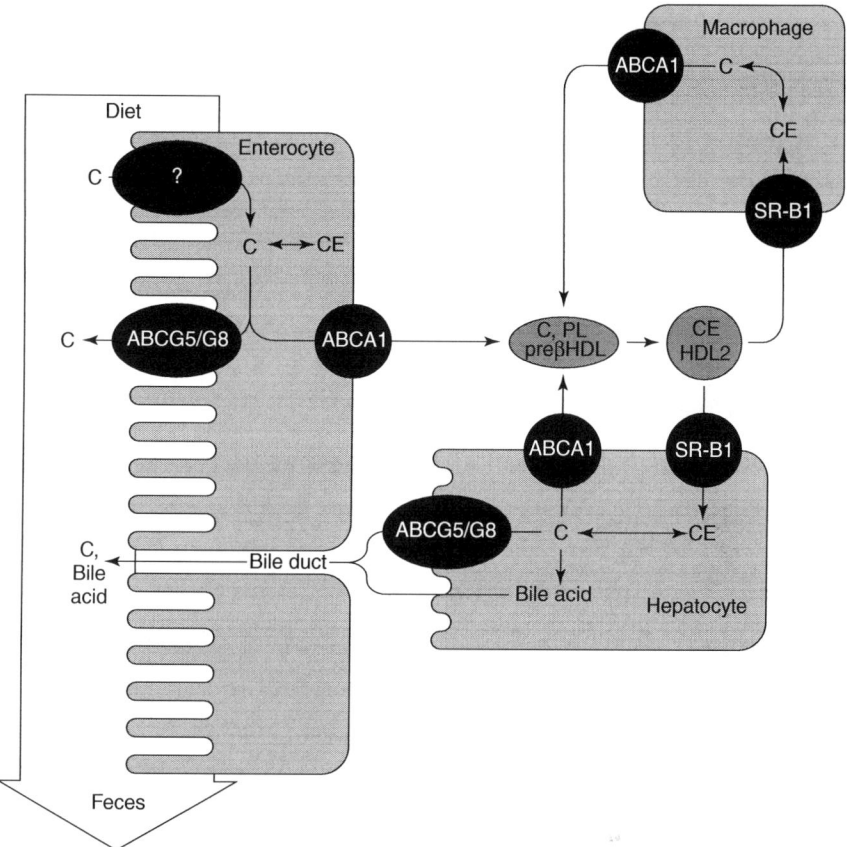

Fig. 5 ABC transporters involved in cholesterol transport. ABCG5 and ABCG8 function together as a heterodimer and mediate the transport of cholesterol from hepatocytes across the canalicular (apical) membrane into the bile. The ABCG5/G8 complex is also involved in cholesterol and plant sterol efflux from the enterocytes of the intestine to the intestinal lumen. This reduces the net absorption of cholesterol and prevents absorption of plant sterols. ABCA1 transports cholesterol and phospholipids on the sinusoidal (basolateral) surface of hepatocytes and enterocytes. In addition, ABCA1 is abundant in macrophages where it mediates cholesterol and phospholipid efflux. ApoA1 interacts with ABCA1 and mediates the formation of precursors to HDL (preβHDL). The esterification of the cholesterol in the preβHDL by the blood-borne enzyme, lecithin cholesterol acyl transferase (LCAT), produces mature HDL. Mature HDL can bind to the SR-B1 receptor and deliver cholesterol esters to hepatocytes. Much of the cholesterol ester is thus delivered and hydrolyzed by a cellular cholesterol esterase and secreted into bile as unesterified cholesterol.

3.8
Some Apolipoproteins Can Prevent Atherosclerosis

ApoA1 is a major protein constituent of HDL. It is an activator of an enzyme in the bloodstream, LCAT. This enzyme transfers a fatty acid from the sn-2 position of phosphatidylcholine to cholesterol to form a cholesterol ester. It enables HDL to "accept" cholesterol from cells or other lipoproteins and then to fill its core with the cholesterol by esterifying it. ApoA1 also interacts with ABCA1 to mediate cholesterol and phospholipid efflux from cells.

Overexpression of apoA1 in transgenic mice or injection of large quantities of the apoA1 protein produces dramatic effects on atherosclerosis – both prevention and regression of atherosclerotic lesions. ApoA1 overexpression is a very attractive prospect for gene therapy. The biggest difficulty is that apoA1 levels in human plasma are already quite high, ~ 1 mg mL^{-1}. Thus, with present technology, it would be quite difficult to achieve a significant increase in this steady state level.

ApoA5 is a newly discovered apolipoprotein that is associated with VLDL. Its deletion in mice results in elevated triacylglycerols and its overexpression in transgenic mice drives down triacylglycerol levels. It is likely that it plays a role in the ability of lipoprotein lipase to hydrolyze VLDL triacylglycerol. ApoA5 might be an attractive candidate for gene therapy of hypertriglyceridemia.

ApoE is a ligand for the LDL receptor and virtually all other members of the LDL receptor family. It is associated with VLDL and chylomicrons and to a varying extent, with HDL. Chylomicrons depend on apoE for their clearance from the circulation. In addition, although VLDL does have another competent LDL receptor ligand, apoB100, VLDL remnants are inefficiently cleared in the absence of apoE. There are three common allelic variants of apoE – apoE2, apoE3, and apoE4. ApoE2 has a greatly reduced affinity for its receptors and is therefore associated with impaired clearance of VLDL and chylomicron remnants. From these properties, one might predict that increasing apoE levels might be therapeutically desirable by promoting efficient removal from the circulation of remnant lipoproteins. However, studies in transgenic mice and rabbits revealed two unexpected consequences of apoE overexpression. First, increased apoE synthesis in the liver leads to increased VLDL production. Second, increased apoE in the VLDL particle slows the hydrolysis of triacylglycerol by lipoprotein lipase.

3.9
The Liver SR-B1 Receptor Protects against Atherosclerosis

The scavenger receptor-B1 (SR-B1) is highly expressed in tissues that actively convert cholesterol into steroid hormones; for example, adrenals, ovaries, and testis. It is also highly expressed in hepatocytes. SR-B1 binds to HDL and mediates the selective uptake of cholesterol ester from the HDL particle into the cell. Unlike the uptake of LDL particles, the particle itself is spared and can then recycle back to pick up another cargo of cholesterol. The overexpression of SR-B1 in transgenic mice causes a drastic drop in HDL levels and protects the animals from atherosclerosis. Conversely, disruption of the gene raises HDL and increases the susceptibility of animals to atherosclerosis. Interestingly, increased SR-B1 expression increases the secretion of cholesterol into the bile, indicating that uptake of cholesterol from HDL effectively targets the HDL toward biliary secretion.

The foregoing might seem to contradict the known inverse relationship between HDL levels and atherosclerosis. However, increased HDL can be due to increased HDL production or decreased HDL turnover. Increased HDL production is thought to be beneficial because it promotes cholesterol flow back to the liver. However, increased HDL, if it is due to an impairment in its ability to deliver cholesterol to the liver, is not beneficial. The transgenic experiments in mice suggest that increased SR-B1 expression in the liver might be a viable approach to gene therapy for atherosclerosis.

4 Vascular and Inflammatory Targets for Gene Therapy

Small clinical trials of gene therapy for cardiovascular disease have been limited to induction of *angiogenesis* in the heart and legs for the treatment of *angina* and critical limb *ischemia* or to blocking of the proliferative response following injury induced by *angioplasty* and vascular grafting as a means of preventing *restenosis*. These trials have involved use of adenoviruses and DNA-containing liposomes for localized overexpression of angiogenic growth factors such as vascular endothelial growth factor (VEGF), fibroblast growth factor (FGF), hepatocyte growth factor (HGF) and antisense oligonucleotides or decoy oligonucleotides for inhibiting expression of cell cycle genes such as *c-Myc* or for preventing binding of transcription factors such as E2F to cis-acting response elements respectively. The localized overexpression of VEGF and other growth factors in the heart and limbs have had mixed results with some recorded successes in formation of collateral vessels, increased blood flow, and reduced pain. Large-scale clinical trials in the United States have been slowed because of the problems with virus-related inflammation and with VEGF-induced edema.

As discussed in the first section of this chapter, genes involved in lipid and lipoprotein metabolism are potential targets for reducing hyperlipidemia and the consequent deposition and retention of lipids and lipoproteins in the artery wall. However, there are still a considerable number of individuals who develop cardiovascular disease who do not have appreciably elevated lipids or are resistant to current therapies. Thus, other targets need to be considered. Genes involved in regulating leukocyte function and the inflammatory response within the artery wall are potential candidates.

4.1 Adhesion Molecules and Leukocyte Counter Receptors

The *adhesion molecules*, vascular cell adhesion molecule (VCAM-1) and intercellular adhesion molecule (ICAM-1), are members of the immunoglobulin superfamily of proteins. Together with their leukocyte counter receptors, the integrins VLA4 and MAC-1/CD11b:CD18, are potential targets for gene therapy. VCAM-1 and ICAM-1 are expressed by endothelial cells and support the tight binding of monocytes and lymphocytes to the endothelial cell plasma membrane. The selectins (E-, P-, and L-selectins) are also adhesion molecules that play a role in the initial tethering and rolling of leukocytes along the surface of the blood vessel. They are lectin-like molecules that recognize carbohydrate containing counterreceptors.

Knockouts of ICAM-1, E-, and P-selectins in mice all partially inhibit development of atherosclerosis. Thus, localized inhibition of expression of these genes might prevent initiation of lesions and limit progression of established lesions due to layering and the continuous influx of leukocytes. Adenovirus-mediated gene transfer to endothelial cells has been successful in both *in vitro* and *in vivo* studies and suggests that this approach could be used for inhibiting expression of the adhesion molecules.

The transient nature of adenoviral-mediated gene transfer is a cause for concern. Given that the atherosclerotic lesions develop over several decades, effective adenoviral therapy would require repeated injections. The use of retroviruses or adeno-associated viruses to incorporate the gene into the cellular genome might be a better alternative. However, this approach is also limited by the small percentage of endothelial cells that proliferate under normal circumstances. Blocking expression of the leukocyte integrins is even more problematic. Currently, this would require transfection of bone marrow stem cells and would likely compromise the capacity to mount an inflammatory response at other sites. Overexpression of secreted forms of the adhesion molecules may be an effective alternate approach.

4.2
Cytokines

Cytokines are cell regulatory proteins that are key factors in the initiation and control of inflammation. They are classified as members of several different families of proteins that include: the tumor necrosis factor (TNF) family, interleukins, interferons, colony-stimulating factors, growth factors, and chemokines. There are significant overlaps in the functions of many of the cytokines. This overlap is in part due to the convergence of signal transduction pathways from the different cytokine specific receptors leading to activation of the same transcription factors. However, in other cases cytokines can have opposite effects and are frequently designated as being either proinflammatory or anti-inflammatory. Serum cytokine levels are elevated in people with established cardiovascular disease. Virtually, every major pro- and anti-inflammatory cytokine is expressed by cells within atherosclerotic lesions and all vascular cell types are capable of responding to these proteins. Thus, there is significant potential for the inhibition and treatment of CVD by manipulating the production of or response to cytokines using gene transfer approaches.

4.3
Chemokines and Chemokine Receptors

Monocyte chemotactic protein (MCP-1) and IL-8 are two potent leukocyte chemoattractants (*chemokines*) that play a role in recruiting leukocytes into the sites of inflammation. MCP-1 is expressed by all of the cell types within atherosclerotic lesions. Recent data from studies on atherosclerosis-prone mice with targeted deletion of MCP-1 or the chemokine receptors CCR-2 or CXCR-2, and studies of bone marrow transplantation of cells overexpressing MCP-1 or devoid of IL-8, strongly suggest that these molecules play a critical role in the initiation of fatty streaks.

Recent gene therapy studies with an N-terminal deletion mutant of the human MCP-1 gene transfected into the skeletal muscle of the atherosclerosis-prone apoE knockout mice have demonstrated the potential for inhibiting atherosclerosis by blocking MCP-1 activity. However, as with the adhesion molecules, there are caveats with regard to targeting the MCP-1, IL-8, and/or CCR-2 and CXCR-2 genes for CVD gene therapy. Because MCP-1 is expressed by all of the cells within the lesions and IL-8 primarily by macrophages, it is unclear whether adenoviral or retroviral approaches would be successful in inhibiting MCP-1 or IL-8 expression by macrophages.

Primary macrophages are extremely difficult to transfect, therefore, blocking their capacity for expressing MCP-1, IL-8, or the receptors likely would require repeatedly knocking out the genes in bone marrow cells. This again raises the question, to what extent would this compromise the inflammatory response at other sites. Thus, localized overexpression of mutated forms of these proteins such as the N-terminal deletion mutant sited above is the most logical approach to targeting chemokines for prevention and treatment of cardiovascular disease.

4.4
Interleukins, the TNF-α family, and Interferon-γ

Currently, *in vitro* and limited *in vivo* data suggest that IL-1, IL-2, TNF-α, IL-3, IL-6, IL-12, IL-15, and IL-18 appear to have proatherogenic properties, while the IL-1 receptor antagonist (IL-1ra), IL-9, IL-10, IL-11 and possibly IL-4, and IL-13 are likely to be antiatherogenic. Thus, localized inhibition of proatherogenic cytokines and overexpression of antiatherogenic cytokines with gene therapy has significant potential for prevention and treatment of cardiovascular disease.

TNF-α and IL-1-β are the proinflammatory cytokines that have received the most attention. *In vitro* studies have clearly demonstrated that these two cytokines potently stimulate the expression of adhesion molecules and chemokines and suggest that they play an important role in the atherogenic process. Surprisingly, targeted deletion of TNF-α or the p55 TNF-α receptor is not protective of atherosclerosis in mouse models. Furthermore, the chronic administration of a TNF-α binding protein reduces fatty streak formation only in female apoE knockout mice. The effects of deletion of IL-1 on atherosclerosis have not been reported. However, in knockouts of the IL-1ra, there was a trend toward increased foam-cell lesion area compared with the wild-type littermate controls. Successful gene therapy studies in animal models of cerebral, pancreatic, and articular inflammation with IL-1ra further support the potential for this approach to cardiovascular disease.

CD40 and CD40 ligand are members of the TNF receptor and protein families respectively. They are expressed by monocytes, macrophages, dendritic cells, lymphocytes, endothelial cells, and smooth muscle cells and play a role in many inflammatory processes including the expression of adhesion molecules, cytokines, chemokines, and matrix metalloproteinases. Recent data suggests that CD40/CD40 ligand play fundamental roles in the atherogenic process as CD40 ligand–apoE double knockout mice have significantly reduced atherosclerosis and lesions that contain reduced numbers of inflammatory cells, are connective tissue rich, and are likely resistant to plaque rupture. Thus, CD40 or CD40 ligand are both additional potential targets for localized CVD gene therapy.

Interferon gamma (IFN-γ) is an immuno-stimulatory cytokine that increases antigen presentation by macrophages and the activation of T lymphocytes. As macrophages and T lymphocytes are key cellular components of the atherosclerotic lesions at all stages of lesion development, gene therapy that targets IFN-γ expression could have a significant impact on the disease process. This is supported by several studies in mice that demonstrate a critical role for this cytokine. For example, atherosclerotic lesions are significantly reduced in IFN-γ-deficient mice crossed with either LDL-receptor knockout or apoE knockout mice and are increased with chronic administration of recombinant IFN-γ. Interleukin-18 (IL-18) in part promotes inflammatory responses through the release of IFN-γ. Administration of recombinant IL-18 to apoE knockout mice, like the administration of IFN-γ, significantly increases lesion size. Not surprisingly, in IFN-γ/apoE double knockout mice, there are no effects of IL-18 on lesion development.

IL-2 is another interleukin with probable proatherogenic properties. It also functions to alter T-lymphocytes by affecting a shift to a Th1 T helper cell phenotype. It is expressed within atherosclerotic lesions and serum levels are increased in people with ischemic heart disease and unstable angina pectoris. However, to date, IL-2 has not been shown to play a direct role in the development of lesions in animal models. In contrast to IL-2, IL-4 and IL-10 in part function to cause a shift toward an antiatherogenic Th2, T helper cell phenotype. There is reduced atherosclerosis in LDL-receptor knockout mice transplanted with bone marrow from IL-4 deficient mice but no increase in atherosclerosis in cholesterol-fed, IL-4 deficient mice. On the other hand, fatty streak formation in IL-4-deficient mice immunized with HSP65 or Mycobacterium tuberculosis is significantly reduced when compared with lesions in wild-type C57BL/6 mice.

In addition to causing a Th2 shift, IL-10 inhibits differentiation of monocytes to macrophages, macrophage presentation of antigens, generation of reactive oxygen and nitrogen species, and endothelial cell expression of ICAM-1. There is accelerated formation of atherosclerotic lesions in IL-10 knockout mice and reduced formation of lesions in IL-10 transgenic mice. Thus, localized overexpression of IL-4 and IL-10 could provide some protection against cardiovascular disease. There is also recent evidence that IL-6 may play a role in the development of atherosclerosis. Treatment of fat-fed C57BL/6 or apoE knockout mice with recombinant IL-6 significantly increases fatty streak formation.

4.5
Growth Factors and Receptors

Growth factors are potent regulators of cellular functions including proliferation, migration, differentiation, and survival/apoptosis. Many growth factors and growth factor receptors are expressed or deposited in atherosclerotic lesions. These include all forms of platelet-derived growth factor (PDGF), basic fibroblast growth factor (bFGF), vascular endothelial growth factor (VEGF), insulin-like growth factors (IGF-1 and IGF-2), thrombin, endothelin-1, angiotensin-II, heparin binding-epidermal growth factor (Hb-EGF), several forms of transforming growth factor-beta (TGF-β), and the hematopoietic growth factors: macrophage colony-stimulating factor (M-CSF), granulocyte colony–stimulating factor (G-CSF), and granulocyte-macrophage colony-stimulating factor (GM-CSF).

The original "Response to Injury" hypothesis of Glomset and Ross posited that denuding injury to the endothelium led to the activation and attachment of platelets with release of PDGF. This was thought to lead to increased smooth muscle cell migration and proliferation and development of atherosclerotic lesions. Over the past three decades, the paradigm has shifted to account for the crucial role of inflammatory cells and the potential protective role of smooth muscle migration and proliferation in the formation of the fibrous cap (Figs. 1–3). Thus, for gene therapy approaches to CVD, localized overexpression of growth factor genes could help stabilize lesions at risk of rupture and thrombosis. As proof of principle, transplant of fetal liver cells from mice deficient in PDGF-BB or chronic treatment with a PDGF-receptor antagonist or anti-PDGF receptor antibodies in apoE knockout mice leads to the formation of lesions that contain mostly macrophages, appear less mature, and have a reduced frequency of fibrous cap formation as compared with control mice. In contrast to primary atherosclerotic lesions, smooth muscle migration and proliferation contributes to *restenosis* following angioplasty and other interventions. Gene therapy approaches to inhibiting growth factor expression postinjury have been in development for several years.

Stimulating or inhibiting the expression of individual growth factors alone may not be sufficient to control vascular smooth muscle cell migration and proliferation. This is because both smooth muscle cell migration and proliferation are exceedingly complex processes and involve dissociation from the extracellular matrix through the activation and secretion of proteases, engagement of multiple growth factor receptors with activation of a variety of receptor associated and cytoplasmic kinases, activation of transcription factors, and production and activation of proteins involved in cell cycle traverse and DNA synthesis such as cyclins and cyclin-dependent kinases. Thus, there are potentially many targets other than growth factors or their specific receptors for gene therapy. This has been demonstrated by the ongoing clinical trials previously cited where antisense oligonucleotides or decoy oligonucleotides for inhibiting expression of cell cycle genes such as *c-Myc* or for preventing binding of transcription factors such as E2F are being tested as the means of preventing restenosis following angioplasty.

The hematopoietic growth factors M-CSF and GM-CSF play a role in the differentiation of monocytes and possibly in regulating macrophage proliferation and death within atherosclerotic lesions. Thus, these factors may also be good targets for gene therapy for both early and unstable lesions. This has been supported by data showing that M-CSF deficient mice crossed with apoE knockout mice have significantly reduced atherosclerosis. Furthermore, chronic administration of an antibody that blocks the M-CSF receptor (c-fms) also reduces early lesions in apoE knockout mice.

Angiogenic growth factors such as the VEGFs, HGF, and the angiopoietins have already been mentioned with regard to ongoing clinical trials for the induction of collateral circulation in people with occluded coronary and peripheral arteries. However, small *neovessels* also form in advanced atherosclerotic lesions and are additional conduits for recruitment of leukocytes into the established lesions. Blocking *neovascularization* could help reduce inflammation and stabilize the lesions. Thus, localized inhibition of expression of these factors or their respective receptors (FLT and FLK for VEGFs and Tie receptors for angiopoietins) are potential additional CVD targets.

5
Pro- and Antioxidant Enzymes

There is now ample evidence that polyunsaturated fatty acids, phospholipids, and cholesteryl-esters within lipoproteins that become trapped within the extracellular matrix of the artery wall can become oxidized. *In vitro* studies have demonstrated that macrophage scavenger receptors

(MSR) recognize oxidized lipoproteins and that foam cell formation is due in part to the accumulation of oxidized lipoproteins. The uptake of oxidized lipids can lead to intense oxidative stress within the foam cells and likely contributes to cell death.

It is currently unclear exactly how lipoproteins are oxidized within the artery wall. Several lines of evidence indicate that proinflammatory enzymes such as myeloperoxidase and the lipoxygenases can oxidize lipoproteins. Macrophages also express the bacteriocidal enzymes NADPH oxidase and the inducible form of nitric oxide synthase that generate the superoxide anion and nitric oxide respectively. These reactive oxygen and nitrogen species can also contribute to lipoprotein modification and oxidative stress within foam cells. Thus, it is possible that gene therapy could be used to inhibit expression of these enzymes within atherosclerotic lesions and thus reduce formation of the necrotic zones and help stabilize lesions that are susceptible to plaque rupture and thrombosis. This is supported by studies showing that mice deficient in 5-lipoxygenase or 12/15 lipoxygenase are protected from atherosclerosis and that mice transgenic for 12/15 lipoxygenase have accelerated formation of lesions.

In contrast to the lipoxygenases, knockout of myeloperoxidase and the NADPH oxidase subunits do not protect mice from atherosclerosis. As with other essential proinflammatory factors, a deficiency of these enzymes can have a significant effect on the capacity to mount an inflammatory response as is seen in people with a deficiency of NADPH oxidase and the resulting chronic granulomatous disease. Thus, future gene therapy to inhibit the expression or activity of these enzymes will require viral targeting to leukocytes only within the artery wall.

All cell types have protective mechanisms against the buildup of pro-oxidants that could disrupt cellular functions by oxidizing lipids, proteins, and DNA. These protective mechanisms include formation of the primary endogenous antioxidant glutathione by the enzyme complex glutamate cysteine ligase. Glutathione acts by enzymatically and chemically converting electrophilic centers to thioether bonds and as a substrate in the glutathione peroxidase mediated destruction of hydroperoxides. Complete glutathione redox systems are located in both the mitochondria and cytoplasm. Reduction of oxidized glutathione by glutathione reductase utilizes reducing equivalents supplied by NADPH and consumes NADPH at a higher rate than most other NADPH-dependent enzymes. Under conditions of oxidative stress with active glutathione redox cycling (as would occur in foam cells containing increased lipid peroxides), the cell's supply of glutathione and NADPH can be depleted, thus limiting many important redox sensitive and biosynthetic reactions and can lead to cell injury and death. Cells also contain superoxide dismutases that convert the superoxide anion to hydrogen peroxide and the enzyme catalase that converts hydrogen peroxide to water. Thus, targeted overexpression of the enzymes involved in glutathione metabolism – the superoxide dismutases and/or catalase could help reduce oxidative stress and foam cell death.

6
Scavenger Receptors

There are now known to be at least three families of broad specificity *scavenger receptors* that are expressed primarily by macrophages within human atherosclerotic lesions. These include the type A family (subtypes I–III and the macrophage receptor with collagenous structure or MARCO), type B receptors that include CD36, SR-B1/CLA-1 and the splice variant SR-B2, and an additional group of receptors such as the mucin-like receptor macrosalin/CD68 and LOX-1. All of these receptors bind oxidized LDL, but CD36 now appears to be the major receptor on macrophages responsible for accumulation of oxidized LDL.

Recent studies of hyperlipidemic mice with targeted deletions of SRA I/II or CD36 have clearly established that both types of receptors play a fundamental role in foam cell formation and lesion development in vivo. Macrophages within human lesions also express a specific receptor that recognizes advanced glycosylation end-product (AGE) proteins, the LDL receptor, VLDL receptor, and the LDL-receptor related protein/α_2-macroglobulin receptor (LRP). Because all of these receptors are expressed primarily by macrophages, this again brings up the problem that primary macrophages are extremely difficult to transfect and blocking their capacity for expressing scavenger receptors would likely require repeatedly knocking out the genes in bone marrow cells. Like adhesion molecules, an alternate approach that should be effective is the overexpression of soluble decoy scavenger receptors that would bind and clear modified lipoproteins and reduce lipoprotein trapping in the artery wall and the subsequent formation of foam cells. This is supported by a recent study showing that there is reduced atherosclerosis in LDL-receptor deficient mice treated with an adenoviral construct containing the human MSR AI extracellular domains.

7
Transcription Factors

Decoy oligonucleotides have been used to block binding of activated transcription factors to their consensus DNA response elements in the promoters of many genes. This may be a viable approach to reducing expression of many cytokines and other proinflammatory factors in vivo because activation of transcription factors such NF-κB, AP-1, and the STAT proteins are end points common to signal transduction pathways activated by various proinflammatory stimuli. Blocking the activation or expression of NF-κB could be a particularly effective therapy as there is activated NF-κB in all of the cell types within atherosclerotic lesions. In addition to blocking activation of NF-κB with decoy oligonucleotides, stimulating the activity of the natural inhibitors of NF-κB, the I-κB proteins, or inhibiting the expression or activity of the I-κB kinases that phosphorylate the I-κB proteins leading to the proteosomal degradation of the NF-κB inhibitors, may also be viable approaches. Targeting the expression of the NF-κB subunits p65 and p50 is another possible approach. Other transcription factors such as Egr-1, Nrf-1, and Nrf-2 may also be potential targets for cardiovascular disease. Egr-1 is an essential factor for regulating the expression of tissue factor, a key procoagulant protein expressed by cells in atherosclerotic lesions. Nrf-1 and Nrf-2 are important factors for regulating the expression of pro- and antioxidant enzymes and bind to consensus antioxidant response elements of these genes.

8
Noninflammatory Gene Targets

There are a large number of proteins that are not directly part of the inflammatory response that are potential targets for CVD gene therapy. These include proteins that regulate cell death, cellular proteases that may play a role in breaking down the arterial extracellular matrix, proteins that play a role in regulating blood pressure and vascular tone, pro- and anticoagulant proteins, and osteogenic proteins that may contribute to vascular calcification.

8.1
Cell Death

As noted, the death of macrophages likely plays a fundamental role in the formation of the necrotic zones of advanced lesions. There is ample evidence showing the presence of macrophages with fragmented DNA (a marker of cell death) located within or adjacent to these necrotic zones. The death of smooth muscle cells also plays a role in the thinning of the fibrous cap, and like the macrophages, smooth muscle cells with fragmented DNA have been documented within atherosclerotic lesions from both humans and experimental animals. Thus, strategies designed to reduce cell death within atherosclerotic lesions could help stabilize the plaques and prevent plaque rupture and formation of occlusive thrombi.

Although it is currently unclear what induces the death of either cell type *in vivo*, there is *in vitro* evidence that apoptosis (programmed cell death or cellular suicide) can be induced in both macrophages and smooth muscle cells by oxidized lipids and other proinflammatory factors. Thus, proteins known to play a role in apoptotic death are potential targets for gene therapy. These include death-promoting proteins such as Fas (a death receptor that is a member of the TNF-α receptor family) and the Fas ligand, the cysteine proteases (caspases 3, 8, 9), and other death pathway effector proteins such as Fas-associated death domain (FADD) (Fas-associated protein with death domain), TNF receptor associated death domain (TRADD) (TNF receptor associated death domain protein), TRAF (TNF receptor associated factor), and receptor interacting protein (RID). There are also a number of antiapoptotic proteins such as c-FLIP (Fas-associated death domain (FADD) protein (FADD)-like interleukin-1 beta-converting enzyme [FLICE (caspase-8)]-inhibitory protein), BCL-2 and BCL-X that could provide protection against cell death if locally overexpressed.

8.2
Proteases

Secretion of proteolytic enzymes such as the matrix metalloproteinases (MMPs) by cells within the blood vessel plays an essential role in enabling the cells to migrate and proliferate. However, excess secretion of these enzymes by activated leukocytes or release of lysosomal enzymes such as cathepsins following the death of cells likely contributes to the breakdown of the extracellular matrix and to the fissure and rupture of the plaques. This is supported by data showing the presence of activated MMPs in unstable human atherosclerotic plaques. Thus, strategies designed to inhibit the expression of these proteolytic enzymes or the activation of the zymogen forms by membrane-associated enzymes such as the ADAM family of proteases could help prevent plaque destabilization and formation of occlusive thrombi. Cells within atherosclerotic lesions also express specific tissue inhibitors of metalloproteinases (TIMPs) and thus localized overexpression of these inhibitors could also be an effective gene therapy approach for cardiovascular disease.

8.3
Regulators of Vascular Tone and Blood Pressure

Elevated blood pressure is a known risk factor for cardiovascular disease. It is well established that as atherosclerotic lesions progress, the affected blood vessel loses its capacity to adequately vasodilate. Controlling acute changes in vascular tone that cause *vasospasm* may be even more fundamental to CVD as evidence suggests that vasospasm may precipitate plaque rupture and the formation of occlusive thrombi. There are currently many drugs available for chronically controlling high blood pressure but very few for managing vasospasm. Thus, localized gene therapy that targets the production of vasoactive substances by arterial wall cells or the response of smooth muscle cells to these vasoactive substances leading to increased vasodilation and reduced vasospasm may be effective approaches for the treatment of cardiovascular disease.

Nitric oxide is an extremely potent vasodilator and most endothelial cells express an endothelium specific form of nitric oxide synthase (e-NOS). However, e-NOS expression and NO production are reduced in atherosclerotic arteries. Thus, a localized increase in the expression of e-NOS could help alleviate the impaired vasodilatory properties of atherosclerotic arteries. Endothelial cells also produce the vasodilatory prostaglandin, prostacyclin (PGI_2). PGI_2 is produced from arachidonic acid by the action of the cyclooxygenases followed by PGI_2 synthase. Thus, increased expression of PGI_2 synthase could be an effective approach for increasing PGI_2 synthesis. Endothelial cells also produce a variety of vasoconstrictors such as endothelin-1 and endothelial cell hyperpolarizing factors. Thus, inhibiting expression of the proform of endothelin or the endothelin-converting enzyme that activates endothelin could help control vasoconstriction. There are also endogenous inhibitors of endothelin activation such as the endothelin-converting enzyme inhibitor that could also be locally overexpressed and play a beneficial role in regulating blood pressure. Similarly, endothelial cells express a form of the angiotensin-converting enzyme (ACE). Angiotensin-II is a potent vasoconstrictor. It is produced following the conversion of angiotensinogen to angiotensin-I by renin and from angiotensin-I to angiotensin-II by ACE. Thus, inhibition of endothelial expression of ACE could also be an effective therapy. Finally, blocking the response of smooth muscle cells to these various vasoconstrictors or increasing the response to the vasodilators by targeting expression of the specific smooth muscle cell receptors or the down stream signal transduction pathways may also be effective gene therapy targets.

8.4
Coagulation Related Genes

There are a variety of pharmacological agents that effectively inhibit *thrombosis*, and chronic administration of these drugs is associated with reduced frequencies of angina and myocardial infarction in people with established cardiovascular disease. However, because we still do not know what causes atherosclerotic plaques to rupture and because plaque rupture is unpredictable, administration of anticoagulants throughout the three to four decades that lesions develop may not be practical for most people due to the accompanying bleeding disorders. Thus, specifically blocking the expression of procoagulant proteins such as tissue factor by cells within atherosclerotic lesions, may be an effective approach to preventing localized formation of occlusive thrombi following plaque rupture. Tissue factor participates in the extrinsic pathway of coagulation and binds to and activates factor VIIa, which in turn activates factor X enabling it to convert prothrombin to thrombin and thrombin to convert fibrinogen to fibrin, the primary protein component of blood clots. Cells within atherosclerotic lesions also express a specific tissue factor inhibitor; thus an additional possibility would be to stimulate expression of this inhibitor. Another potential approach for reducing formation of occlusive thrombi is to increase the production of thrombolytic factors such as plasmin by increasing the localized expression of tissue plasminogen activator or reducing the expression of the plasminogen activator inhibitors.

8.5
Osteogenic Proteins

Bone mineral (calcium-phosphate) is deposited in most advanced atherosclerotic plaques and in heart valves. However, it is still controversial as to whether calcification of the plaques is a good prognostic indicator of subsequent CVD events. Nevertheless, preventing vascular calcification is beneficial for interventions such as angioplasty and for reducing the need for heart valve replacement. Vascular calcification is now known to be an active cellular mediated process that is analogous to the process by which cartilage is converted to bone. It thus involves a balance between bone forming osteoblast type cells and bone removing osteoclast type cells and the expression of proteins that are both pro-osteogenic and antiosteogenic. These include the matrix gla proteins, osteopontin, osteoprotegrin, osteonectin, and the bone morphogenic proteins. Thus, localized gene therapy designed to either increase or inhibit expression of these proteins could have dramatic effects on reducing plaque calcification and complications resulting from "stiffening" of the blood vessels.

Bibliography

Books and Reviews

Freedman, S.B. (2002) Clinical trials of gene therapy for atherosclerotic cardiovascular disease, *Curr. Opin. Lipidol.* **13**, 653–661.

Glass, C.K., Witztum, J.L. (2001) Atherosclerosis. the road ahead, *Cell* **104**, 503–516.

Herweijer, H., Wolff, J.A. (2003) Progress and prospects: naked DNA gene transfer and therapy, *Gene Ther.* **10**, 453–458.

Khurana, R., Martin, J.F., Zachary, I. (2001) Gene therapy for cardiovascular disease: a case for cautious optimism, *Hypertension* **38**, 1210–1216.

Morishita, R. (2002) Recent progress in gene therapy for cardiovascular disease, *Circ. J.* **66**, 1077–1086.

Newby, A.C., Zaltsman, A.B. (2000) Molecular mechanisms in intimal hyperplasia, *J. Pathol.* **190**, 300–309.

Ross, R. (1999) Atherosclerosis – an inflammatory disease, *N. Engl. J. Med.* **340**, 115–126.

Springer, T.A., Cybulsky, M.I. (1996) Traffic signals on endothelium for leukocytes in health, inflammation, and atherosclerosis, in: Fuster, V., Ross, R., Topol, E.J. (Eds.) *Atherosclerosis and Coronary Artery Disease*, Lippincott-Raven, Philadelphia.

Tangirala, R.K., Tsukamoto, K., Chun, S.H., Usher, D., Pure, E., Rader, D.J. (1999) Regression of atherosclerosis induced by liver-directed gene transfer of apolipoprotein A-I in mice, *Circulation* **100**, 1816–1822.

Thomas, C.E., Ehrhardt, A., Kay, M.A. (2003) Progress and problems with the use of viral vectors for gene therapy, *Nat. Rev. Genet.* **4**, 346–358.

Von Der Thusen, J.H., Kuiper, J., Van Berkel, T.J., Biessen, E.A. (2003) Interleukins in atherosclerosis: molecular pathways and therapeutic potential, *Pharmacol. Rev.* **55**, 133–166.

Primary Literature

Aiello, R.J., Bourassa, P.A., Lindsey, S., Weng, W., Natoli, E., Rollins, B.J., Milos, P.M. (1999) Monocyte chemoattractant protein-1 accelerates atherosclerosis in apolipoprotein E-deficient mice, *Arterioscler. Thromb. Vasc. Biol.* **19**, 1518–1525.

Aiello, R.J., Brees, D., Bourassa, P.A., Royer, L., Lindsey, S., Coskran, T., Haghpassand, M., Francone, O.L. (2002) Increased atherosclerosis in hyperlipidemic mice with inactivation of ABCA1 in macrophages, *Arterioscler. Thromb. Vasc. Biol.* **22**, 630–637.

Babaev, V.R., Gleaves, L.A., Carter, K.J., Suzuki, H., Kodama, T., Fazio, S., Linton, M.F. (2000) Reduced atherosclerotic lesions in mice deficient for total or macrophage-specific expression of scavenger receptor-A, *Arterioscler. Thromb. Vasc. Biol.* **20**, 2593–2599.

Boisvert, W.A., Black, A.S., Curtiss, L.K. (1999) ApoA1 reduces free cholesterol accumulation in atherosclerotic lesions of ApoE-deficient mice transplanted with ApoE-expressing macrophages, *Arterioscler. Thromb. Vasc. Biol.* **19**, 525–530.

Boisvert, W.A., Curtiss, L.K. (1999) Elimination of macrophage-specific apolipoprotein E reduces diet-induced atherosclerosis in C57BL/6J male mice, *J. Lipid Res.* **40**, 806–813.

Boisvert, W.A., Santiago, R., Curtiss, L.K., Terkeltaub, R.A. (1998) A leukocyte homologue of the IL-8 receptor CXCR2 mediates the accumulation of macrophages in atherosclerotic lesions of LDL receptor-deficient mice, *J. Clin. Invest.* **101**, 353–363.

Boring, L., Gosling, J., Cleary, M., Charo, I. (1998) Decreased lesion formation in CCR2-/- mice reveals a role for chemokines in the initiation of atherosclerosis, *Nature* **394**, 894–897.

Brennan, M.L., Anderson, M.M., Shih, D.M., Qu, X.D., Wang, X., Mehta, A.C., Lim, L.L., Shi, W., Hazen, S.L., Jacob, J.S., Crowley, J.R., Heinecke, J.W., Lusis, A.J. (2001) Increased atherosclerosis in myeloperoxidase-deficient mice, *J. Clin. Invest.* **107**, 419–430.

Brown, D.L., Hibbs, M.S., Kearney, M., Isner, J.M. (1997) Differential expression of 92-kDa gelatinase in primary atherosclerotic versus restenotic coronary lesions, *Am. J. Cardiol.* **79**, 878–882.

Buono, C., Come, C.E., Stavrakis, G., Maguire, G.F., Connelly, P.W., Lichtman, A.H. (2003) Influence of interferon-gamma on the extent and phenotype of diet-induced atherosclerosis in the LDLR-deficient mouse, *Arterioscler. Thromb. Vasc. Biol.* **23**, 454–460.

Caligiuri, G., Rudling, M., Ollivier, V., Jacob, M.P., Michel, J.B., Hansson, G.K., Nicoletti, A. (2003) Interleukin-10 deficiency increases atherosclerosis, thrombosis, and low-density lipoproteins in apolipoprotein E knock-out mice, *Mol. Med.* **9**, 10–17.

Chen, S.J., Rader, D.J., Tazelaar, J., Kawashiri, M., Gao, G., and Wilson, J.M. (2000) Prolonged correction of hyperlipidemia in mice with familial hypercholesterolemia using an adeno-associated viral vector expressing very-low-density lipoprotein receptor, *Mol. Ther.* **2**, 256–261.

Collins, R.G., Velji, R., Guevara, N.V., Hicks, M.J., Chan, L., Beaudet, A.L. (2000) P-Selectin or intercellular adhesion molecule (ICAM)-1 deficiency substantially protects against atherosclerosis in apolipoprotein E-deficient mice, *J. Exp. Med.* **191**, 189–194.

Cyrus, T., Pratico, D., Zhao, L., Witztum, J.L., Rader, D.J., Rokach, J., FitzGerald, G.A., Funk, C.D. (2001) Absence of 12/15-lipoxygenase expression decreases lipid peroxidation and atherogenesis in apolipoprotein e-deficient mice, *Circulation* **103**, 2277–2282.

Dawson, T.C., Kuziel, W.C., Osahar, T.A., Madea, N. (1999) Absence of CC chemokine receptor-2 reduces atherosclerosis in apolipoprotein E-deficient mice, *Atherosclerosis* **143**, 205–211.

Devlin, C.M., Kuriakose, G., Hirsch, E., Tabas, I. (2002) Genetic alterations of IL-1 receptor antagonist in mice affect plasma cholesterol level and foam cell lesion size, *Proc. Natl. Acad. Sci. U.S.A.* **99**, 6280–6285.

Elhage, R., Maret, A., Pieraggi, M.T., Thiers, J.C., Arnal, J.F., Bayard, F. (1998) Differential effects of interleukin-1 receptor antagonist and tumor necrosis factor binding protein on fatty-streak formation in apolipoprotein E-deficient mice, *Circulation* **97**, 242–244.

Fabunmi, R.P., Sukhova, G.K., Sugiyama, S., Libby, P. (1998) Expression of tissue inhibitor of metalloproteinases-3 in human atheroma and regulation in lesion-associated cells: a potential protective mechanism in plaque stability, *Circ. Res.* **83**, 270–278.

Fazio, S., Babaev, V.R., Burleigh, M.E., Major, A.S., Hasty, A.H., Linton, M.F. (2002) Physiological expression of macrophage apoE in the artery wall reduces atherosclerosis in severely hyperlipidemic mice, *J. Lipid Res.* **43**, 1602–1609.

Febbraio, M., Podrez, E.A. (2000) Targeted disruption of the class B scavenger receptor CD36 protects against atherosclerotic lesion development in mice, *J. Clin. Invest.* **105**, 1049–1056.

Forlow, S.B., Ley, K. (2001) Selectin-independent leukocyte rolling and adhesion in mice deficient in E-, P-, and L-selectin and ICAM-1, *Am. J. Physiol. Heart Circ. Physiol.* **280**, H634–H641.

Frenette, P.S., Wagner, D.D. (1997) Insights into selectin function from knockout mice, *Thromb. Haemost.* **78**, 60–64.

Galis, Z.S., Johnson, C., Godin, D., Magid, R., Shipley, J.M., Senior, R.M., Ivan, E. (2002) Targeted disruption of the matrix metalloproteinase-9 gene impairs smooth muscle cell migration and geometrical arterial remodeling, *Circ. Res.* **91**, 852–859.

George, J., Afek, A., Shaish, A., Levkovitz, H., Bloom, N., Cyrus, T., Zhao, L., Funk, C.D., Sigal, E., Harats, D. (2001) 12/15-Lipoxygenase gene disruption attenuates atherogenesis in LDL receptor-deficient mice, *Circulation* **104**, 1646–1650.

George, J., Mulkins, M., Shaish, A., Casey, S., Schatzman, R., Sigal, E., Harats, D. (2000) Interleukin (IL)-4 deficiency does not influence fatty streak formation in C57BL/6 mice, *Atherosclerosis* **153**, 403–411.

George, J., Shoenfeld, Y., Gilburd, B., Afek, A., Shaish, A., Harats, D. (2000) Requisite role for interleukin-4 in the acceleration of fatty streaks induced by heat shock protein 65 or mycobacterium tuberculosis, *Circ. Res.* **86**, 1203–1210.

Gupta, S., Pablo, A.M., Jiang, X., Wang, N., Tall, A.R., Schindler, C. (1997) IFN-gamma potentiates atherosclerosis in ApoE knock-out mice, *J. Clin. Invest.* **99**, 2752–2761.

Harats, D., Shaish, A., George, J., Mulkins, M., Kurihara, H., Levkovitz, H., Sigal, E. (2000) Overexpression of 15-lipoxygenase in vascular endothelium accelerates early atherosclerosis in LDL receptor-deficient mice, *Arterioscler. Thromb. Vasc. Biol.* **20**, 2100–2105.

Huber, S.A., Sakkinen, P., Conze, D., Hardin, N., Tracy, R. (1999) Interleukin-6 exacerbates early atherosclerosis in mice, *Arterioscler. Thromb. Vasc. Biol.* **19**, 2364–2367.

Inoue, S., Egashira, K., Ni, W., Kitamoto, S., Usui, M., Otani, K., Ishibashi, M., Hiasa, K., Nishida, K., Takeshita, A. (2002) Anti-monocyte chemoattractant protein-1 gene therapy limits progression and destabilization of established atherosclerosis in apolipoprotein E-knockout mice, *Circulation* **106**, 2700–2706.

Jalkanen, J., Leppanen, P., Narvanen, O., Greaves, D.R., Yla-Herttuala, S. (2003) Adenovirus-mediated gene transfer of a secreted decoy human macrophage scavenger receptor (SR-AI) in LDL receptor knock-out mice, *Atherosclerosis* **169**, 95–103.

Jong, M.C., van Dijk, K.W., Dahlmans, V.E., Van der Boom, H., Kobayashi, K., Oka, K., Siest, G., Chan, L., Hofker, M.H., Havekes, L.M. (1999) Reversal of hyperlipidaemia in apolipoprotein C1 transgenic mice by adenovirus-mediated gene delivery of the low-density-lipoprotein receptor, but not by the very-low-density-lipoprotein receptor, *Biochem. J.* **338**(Pt 2), 281–287.

Joyce, C.W., Amar, M.J., Lambert, G., Vaisman, B.L., Paigen, B., Najib-Fruchart, J., Hoyt, R.F., Jr., Neufeld, E.D., Remaley, A.T., Fredrickson, D.S., Brewer, H.B., Jr., Santamarina-Fojo, S. (2002) The ATP binding cassette transporter A1 (ABCA1) modulates the development of aortic atherosclerosis in C57BL/6 and apoE-knockout mice, *Proc. Natl. Acad. Sci. U.S.A.* **99**, 407–412.

Kawashiri, M., Zhang, Y., Usher, D., Reilly, M., Pure, E., Rader, D.J. (2001) Effects of coexpression of the LDL receptor and apoE on cholesterol metabolism and atherosclerosis in LDL receptor-deficient mice, *J. Lipid Res.* **42**, 943–950.

Kim, I.H., Jozkowicz, A., Piedra, P.A., Oka, K., Chan, L. (2001) Lifetime correction of genetic deficiency in mice with a single injection of helper-dependent adenoviral vector, *Proc. Natl. Acad. Sci. U.S.A.* **98**, 13282–13287.

King, V.L., Szilvassy, S.J., Daugherty, A. (2002) Interleukin-4 deficiency decreases atherosclerotic lesion formation in a site-specific manner in female LDL receptor−/− mice, *Arterioscler. Thromb. Vasc. Biol.* **22**, 456–461.

Kirk, E.A., Dinauer, M.C., Rosen, H., Chait, A., Heinecke, J.W., LeBoeuf, R.C. (2000) Impaired superoxide production due to a deficiency in phagocyte NADPH oxidase fails to inhibit atherosclerosis in mice, *Arterioscler. Thromb. Vasc. Biol.* **20**, 1529–1535.

Kozaki, K., Kaminski, W.E., Tang, J., Hollenbach, S., Lindahl, P., Sullivan, C., Yu, J.C., Abe, K., Martin, P.J., Ross, R., Betsholtz, C., Giese, N.A., Raines, E.W. (2002) Blockade of platelet-derived growth factor or its receptors transiently delays but does not prevent fibrous cap formation in ApoE null mice, *Am. J. Pathol.* **161**, 1395–1407.

Kozarsky, K.F., Donahee, M.H., Glick, J.M., Krieger, M., Rader, D.J. (2000) Gene transfer and hepatic overexpression of the HDL receptor SR-BI reduces atherosclerosis in the cholesterol-fed LDL receptor-deficient mouse, *Arterioscler. Thromb. Vasc. Biol.* **20**, 721–727.

Lemaitre, V., O'Byrne, T.K., Borczuk, A.C., Okada, Y., Tall, A.R., D'Armiento, J. (2001) ApoE knockout mice expressing human matrix metalloproteinase-1 in macrophages have less advanced atherosclerosis, *J. Clin. Invest.* **107**, 1227–1234.

Linton, M.F., Atkinson, J.B., Fazio, S. (1995) Prevention of atherosclerosis in apolipoprotein E-deficient mice by bone marrow transplantation, *Science* **267**, 1034–1037.

Luoma, J., Hiltunen, T. (1994) Expression of alpha 2-macroglobulin receptor/low density lipoprotein receptor-related protein and scavenger receptor in human atherosclerotic lesions, *J. Clin. Invest.* **93**, 2014–2021.

Lutgens, E., Daemen, M.J. (2002) CD40-CD40L interactions in atherosclerosis, *Trends Cardiovasc. Med.* **12**, 27–32.

Mallat, Z., Gojova, A., Marchiol-Fournigault, C., Esposito, B., Kamate, C., Merval, R., Fradelizi, D., Tedgui, A. (2001) Inhibition of transforming growth factor-beta signaling accelerates atherosclerosis and induces an unstable plaque phenotype in mice, *Circ. Res.* **89**, 930–934.

Mehrabian, M., Allayee, H., Wong, J., Shi, W., Wang, X.P., Shaposhnik, Z., Funk, C.D., Lusis, A.J., Shih, W. (2002) Identification of 5-lipoxygenase as a major gene contributing to atherosclerosis susceptibility in mice, *Circ. Res.* **91**, 120–126.

Murayama, T., Yokode, M., Kataoka, H., Imabayashi, T., Yoshida, H., Sano, H., Nishikawa, S., Nishikawa, S., Kita, T. (1999) Intraperitoneal administration of anti-c-fms monoclonal antibody prevents initial events of atherogenesis but does not reduce the size of advanced lesions in apolipoprotein E-deficient mice, *Circulation* **99**, 1740–1746.

Ni, W., Egashira, K., Kitamoto, S., Kataoka, C., Koyanagi, M., Inoue, S., Imaizumi, K., Akiyama, C., Nishida, K.I., Takeshita, A. (2001) New anti-monocyte chemoattractant protein-1 gene therapy attenuates atherosclerosis in apolipoprotein E-knockout mice, *Circulation* **103**, 2096–2101.

Pennacchio, L.A., Olivier, M., Hubacek, J.A., Cohen, J.C., Cox, D.R., Fruchart, J.C., Krauss, R.M., Rubin, E.M. (2001) An apolipoprotein influencing triglycerides in humans and mice revealed by comparative sequencing, *Science* **294**, 169–173.

Pinderski, L.J., Fischbein, M.P., Subbanagounder, G., Fishbein, M.C., Kubo, N., Cheroutre, H., Curtiss, L.K., Berliner, J.A., Boisvert, W.A. (2002) Overexpression of interleukin-10 by activated T lymphocytes inhibits atherosclerosis in LDL receptor-deficient mice by altering lymphocyte and macrophage phenotypes, *Circ. Res.* **90**, 1064–1071.

Ross, R., Glomset, J.A. (1973) Atherosclerosis and the arterial smooth muscle cell: proliferation of smooth muscle is a key event in the genesis of the lesions of atherosclerosis, *Science* **180**, 1332–1339.

Sano, H., Sudo, T., Yokode, M., Murayama, T., Kataoka, H., Takakura, N., Nishikawa, S., Nishikawa, S.I., Kita, T. (2001) Functional blockade of platelet-derived growth factor receptor-beta but not of receptor-alpha prevents vascular smooth muscle cell accumulation in fibrous cap lesions in apolipoprotein E-deficient mice, *Circulation* **103**, 2955–2960.

Savontaus, M.J., Sauter, B.V., Huang, T.G., Woo, S.L. (2002) Transcriptional targeting of conditionally replicating adenovirus to dividing endothelial cells, *Gene Ther.* **9**, 972–979.

Schreyer, S.A., Peschon, J.J., LeBoeuf, R.C. (1996) Accelerated atherosclerosis in mice lacking tumor necrosis factor receptor p55, *J. Biol. Chem.* **271**, 26174–26178.

Shichiri, M., Tanaka, A., Hirata, Y. (2003) Intravenous gene therapy for familial hypercholesterolemia using ligand-facilitated transfer of a liposome: LDL receptor gene complex, *Gene Ther.* **10**, 827–831.

Simons, M., Edelman, E.R. (1997) Antisense oligonucleotide inhibition of PDGFR-ß receptor subunit expression directs suppression of intimal thickening, *Circulation* **95**, 669–676.

Singaraja, R.R., Bocher, V., James, E.R., Clee, S.M., Zhang, L.H., Leavitt, B.R., Tan, B., Brooks-Wilson, A., Kwok, A., Bissada, N., Yang, Y.Z., Liu, G., Tafuri, S.R., Fievet, C., Wellington, C.L., Staels, B., Hayden, M.R. (2001) Human ABCA1 BAC transgenic mice show increased high density lipoprotein cholesterol and ApoAI-dependent efflux stimulated by an internal promoter containing liver X receptor response elements in intron 1, *J. Biol. Chem.* **276**, 33969–33979.

Smith, J.D., Trogan, E., Ginsberg, M., Grigaux, C., Tian, J., Miyata, M. (1995) Decreased atherosclerosis in mice deficient in both macrophage colony-stimulating factor (op) and apolipoprotein E, *Proc. Natl. Acad. Sci. U.S.A.* **92**, 8264–8268.

Sukhova, G.K., Schonbeck, U., Rabkin, E., Schoen, F.J., Poole, A.R., Billinghurst, R.C., Libby, P. (1999) Evidence for increased collagenolysis by interstitial collagenases-1 and -3 in vulnerable human atheromatous plaques, *Circulation* **99**, 2503–2509.

Suzuki, H., Kurihara, Y., Takeya, M., Kamada, N., Kataoka, M., Jishage, K., Sakaguchi, H., Kruijt, J.K., Higashi, T., Suzuki, T., van Berkel, T.J., Horiuchi, S., Takahashi, K., Yazaki, Y., Kodama, T. (1997) The multiple roles of macrophage scavenger receptors (MSR) in vivo: resistance to atherosclerosis and susceptibility to infection in MSR knockout mice, *J. Atheroscler. Thromb.* **4**, 1–11.

Thorngate, F.E., Rudel, L.L., Walzem, R.L., Williams, D.L. (2000) Low levels of extrahepatic nonmacrophage ApoE inhibit atherosclerosis without correcting hypercholesterolemia in ApoE-deficient mice, *Arterioscler. Thromb. Vasc. Biol.* **20**, 1939–1945.

Tsukamoto, K., Tangirala, R.K., Chun, S., Usher, D., Pure, E., Rader, D.J. (2000) Hepatic expression of apolipoprotein E inhibits progression of atherosclerosis without reducing cholesterol levels in LDL receptor-deficient mice, *Mol. Ther.* **1**, 189–194.

Whitman, S.C., Ravisankar, P., Daugherty, A. (2002) IFN-gamma deficiency exerts gender-specific effects on atherogenesis in apolipoprotein E−/− mice, *J. Interferon Cytokine Res.* **22**, 661–670.

Whitman, S.C., Ravisankar, P., Daugherty, A. (2002) Interleukin-18 enhances atherosclerosis in apolipoprotein E(−/−) mice through release of interferon-gamma, *Circ. Res.* **90**, E34–E38.

Whitman, S.C., Ravisankar, P., Elam, H., Daugherty, A. (2000) Exogenous interferon-gamma enhances atherosclerosis in apolipoprotein E−/− mice, *Am. J. Pathol.* **157**, 1819–1824.

Yoshimura, S., Morishita, R., Hayashi, K., Yamamoto, K., Nakagami, H., Kaneda, Y., Sakai, N., Ogihara, T. (2001) Inhibition of intimal hyperplasia after balloon injury in rat carotid artery model using cis-element 'decoy' of nuclear factor-kappaB binding site as a novel molecular strategy, *Gene Ther.* **8**, 1635–1642.

Yu, L., Li-Hawkins, J., Hammer, R.E., Berge, K.E., Horton, J.D., Cohen, J.C., Hobbs, H.H. (2002) Overexpression of ABCG5 and ABCG8 promotes biliary cholesterol secretion and reduces fractional absorption of dietary cholesterol, *J. Clin. Invest.* **110**, 671–680.

Zhao, L., Cuff, C.A., Moss, E., Wille, U., Cyrus, T., Klein, E.A., Pratico, D., Rader, D.J., Hunter, C.A., Pure, E., Funk, C.D. (2002) Selective interleukin-12 synthesis defect in 12/15-lipoxygenase-deficient macrophages associated with reduced atherosclerosis in a mouse model of familial hypercholesterolemia, *J. Biol. Chem.* **277**, 35350–35356.

38
Somatic Gene Therapy

M. Schweizer, E. Flory, C. Münk, Uwe Gottschalk, and K. Cichutek
Paul-Ehrlich-Institut, Langen, Germany

1	Introduction 1118	
2	**Gene Transfer Methods** 1120	
2.1	Nonviral Vectors and Naked Nucleic Acid 1122	
2.2	Viral Vectors 1123	
2.2.1	Retroviral Vectors 1123	
2.2.2	Lentiviral Vectors 1124	
2.2.3	Adenoviral Vectors 1124	
2.2.4	AAV (Adeno-associated Viral) Vectors 1125	
2.2.5	Poxvirus Vectors 1125	
3	**Clinical Use** 1125	
3.1	Overview on Clinical Gene Therapy Trials 1125	
3.2	Gene Therapy of Monogeneic Congenital Diseases 1125	
3.3	Tumor Gene Therapy 1127	
3.4	Gene Therapy of Cardiovascular Diseases 1128	
3.5	Preventive Vaccination and Gene Therapy of Infectious Diseases 1128	
3.6	Clinical Gene Therapy for the Treatment of Other Diseases 1129	
4	**Manufacture and Regulatory Aspects** 1129	
5	**First Experience with the Clinical Use of Gene Transfer Medicinal Products** 1133	
	Bibliography 1133	
	Books and Reviews 1133	
	Primary Literature 1134	

Genomics and Genetics. Edited by Robert A. Meyers.
Copyright © 2007 Wiley-VCH Verlag GmbH & Co. KGaA, Weinheim
ISBN: 978-3-527-31609-0

1
Introduction

Innovative biotechnologicals of the future include gene transfer medicinal products. It can be assumed that by mid-2003, ~4000 patients or healthy individuals would have been treated within a clinical gene therapy trial, ~600 of those in Europe, and ~260 in Germany (The data originate from some statistics published in the Internet by Wiley Genetic Medicine Clinical Trials Online, http:///www.wiley.co.uk.genetherapy, and are reproduced by courtesy of the publisher.). Most of the clinical trials are currently in phase I or II due to a great diversity of ongoing developments, clinical experience must first be gained, before target-orientated product development and phase III clinical trials can be initiated. In this regard, investigator-driven gene therapy strategies developed by biomedical laboratories together with special clinic teams are very distinct from those developed by pharmaceutical industry. Investigator-driven gene therapy strategies are being invented by teams of biomedical researchers and physicians while developing a new approach for the treatment of a special disease in a defined stage. This is used, for the first time, on a selected group of patients in first clinical trials of phase I/II and aimed at proving the safety of the medicinal product. In clinical trials sponsored by pharmaceutical industry, this phase of orientation has often already been completed and further development in phase II or III is aimed at dose finding or proving efficacy. Concerning product development, there are no standard approaches because, at this stage of development, little experience has been gained and the types of gene transfer medicinal products are very diverse. Therefore, in the following sections, the main current clinical developments will be described while a brief outline of a single example of a manufacturing process is given, also due to manifold diversity.

Gene transfer medicinal products for human use are medicinal products used for *in vivo* diagnosis, prophylaxis, or therapy (Fig. 1). They contain or consist of:

1. genetically modified cells,
2. viral vectors, nonviral vectors or so-called naked nucleic acids, or
3. recombinant replication-competent microorganisms used for purposes other than the prevention or therapy of the infectious diseases that they cause.

The aim of the nucleic acid or gene transfer is the genetic modification of human somatic cells, either in the human body, that is, *in vivo*, or outside the human body, that is, *ex vivo*, in the latter case followed by transfer of the modified cells to the human body. The simplest case of genetic modification of a cell results from addition of a therapeutic gene encompassed by an expression vector. At least in theory, nucleic acid transfer may also be aimed at exchange of individual point mutations or other minimal genetic aberrations. Scientifically, this process is termed *homologous recombination* with the aim of repairing a defective endogenous gene at its locus. In principle, this can be achieved by so-called homologous recombination achieved by transferring oligonucleotides, where – thanks to 5' and 3' flanking homology regions – the new correct DNA sequence is replacing the existing defective one. In practice, homologous recombination is technically not yet achievable with the efficiency that will be required for clinical use.

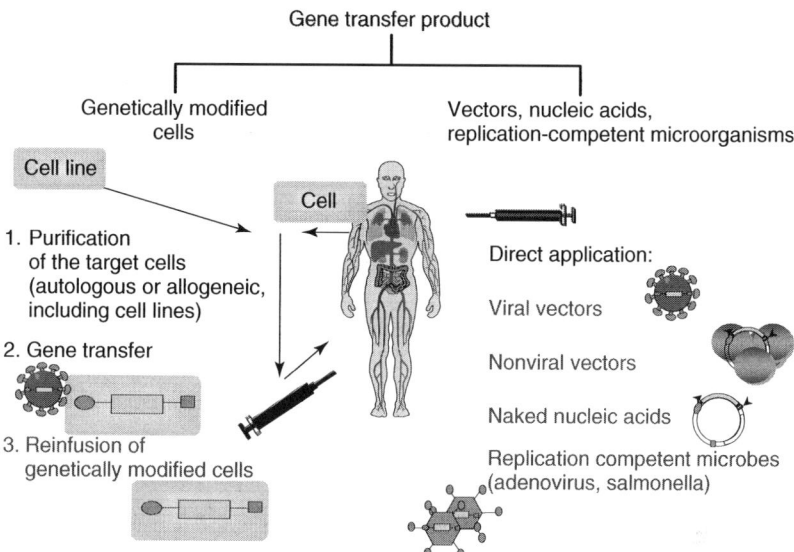

Fig. 1 Gene transfer medicinal products. The gene transfer medicinal products mentioned here are identical with those described in Table 1 of the European "Note for guidance on the quality, preclinical, and clinical aspects of gene transfer medicinal products (CPMP/BWP/3088/99)." The definition given is in compliance with the legally binding definition of gene therapeutics in Part IV, Annex I of Directive 2003/63/EC amending Directive 2001/83/EC.

Normally, genetic modification of cells is nowadays achieved by the transfer of an expression vector on which the therapeutic gene is located. The vector is transferred to cells via a delivery system (Fig. 2) such as a viral vector particle, a nonviral vector complex or a plasmid. In the latter case, the expression vector is inserted into and therefore part of a bacterial plasmid, which allows its manufacture and amplification in bacteria. Viral expression vectors contain the sequence signals (nucleic acid sequences) required for transfer by a particular viral vector particle. For retroviral vectors, for example, such signals are encompassed by the flanking "long terminal repeat" (LTR) sequences, the packaging signal psi (Ψ) required for incorporation of the expression vector by the retroviral vector particle, and other sequence signals. For nonviral vector complexes and naked nucleic acid, the expression vector is part of a bacterial carrier, the so-called plasmid DNA. Nonviral vectors are, for example, plasmid DNA mixed with a transfection reagent, whereas naked DNA does not contain a transfection reagent.

Another example of a gene transfer medicinal product is a recombinant microorganism such as conditionally replication-competent adenoviruses for tumor therapy. Here, neither is an endogenous cellular gene repaired by homologous recombination nor is a nonadenoviral therapeutic gene transferred. The transfer of conditional replicating adenoviruses to the malignant tumor cells induces cell lysis and local tumor ablation. The entire genome of the adenovirus is transferred without an additional therapeutic gene. The adenoviral genome may therefore be considered as the therapeutic gene.

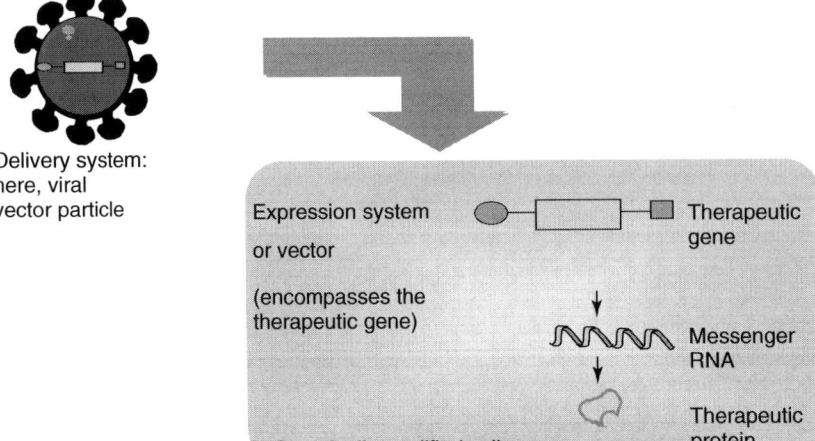

Fig. 2 Delivery system and expression vector used as gene transfer medicinal products. The terminology complies with the definition of gene therapeutics in Part IV, Annex I of Directive 2003/63/EC amending Directive 2001/83/EC.

Gene transfer efficiency plays a central role in gene transfer. It depends on a number of factors, for example, target cell, type of application (*ex vivo* or *in vivo* strategy), the tissue or organ containing the target cells, the physiological situation, and the disease and disease stage. Table 1 shows the most common viral vectors currently in clinical use. The vectors shown are replication incompetent and only transfer the expression vector void of any viral genes as much as possible. So-called integrating vectors mediate chromosomal integration of the expression vector (e.g. retroviral vectors), whereas nonintegrating vectors lead to an episomal status of the expression vector in the cell (e.g. adenoviral vectors), or to its cytoplasmatic replication (e.g. α-virus-derived vectors, vaccinia). Vectors derived from vaccinia, for example, used for tumor vaccination, may be replication incompetent such as modified vaccinia ancara (MVA) or ALVAC or replication competent, but attenuated like vaccinia.

After uptake by human somatic cells, the expression vector is transcribed like a normal cell gene. The resulting messenger RNA (mRNA) is translated and the therapeutic protein is synthesized by the cellular machinery. When so-called ribozyme genes are used, the mRNA acts like a catalytic enzyme and is itself the therapeutic gene product. As already mentioned, when a recombinant microorganism such as a conditionally replication-competent adenovirus (RCA) is used, the genome of the microorganism may be seen as the therapeutic gene.

2
Gene Transfer Methods

The objective of clinical gene transfer is the transfer of nucleic acids for the purpose of genetically modifying human cells (Fig. 3).

Whether a viral, a nonviral vector, or a naked plasmid DNA is used depends on

Tab. 1 Gene transfer methods (vectors/delivery systems).

Delivery system	Description	Chromosomal integration
Naked nucleic acid	Plasmid DNA, in absence of transfection reagents	No (after im inoculation)
Nonviral vector	Plasmid DNA/transfection reagent mixture	No (application dependent)
Viral vector		
Retroviral vector	Derived from murine leukemia virus (MLV)	Yes
Lentiviral vector	Derived from HIV-1	Yes
Adenoviral vector	Deletions in the virus genes E1, E3 or E4, E2ts, combinations thereof or "gutted" (gene-depleted)	No
Conditionally replication-competent adenovirus	No therapeutic gene except for the virus genome	No
Adeno-associated virus (AAV) vector	Wild-type AAV-derived	Yes /no (application dependent)
Smallpox virus vector	MVA ("Modified Vaccinia Ancara")	No
	ALVAC ("Avian Vaccinia")	No
	Vaccinia	No
Alphavirus vector	SFV	No
Herpes-viral vector	Herpes simplex virus	No

Note: SFV: Semliki Forest virus; MLV: murine leukemia virus.

the target cell of the genetic modification and whether an *in vivo* modification of the cell is at all possible. For a monogeneic disease affecting immune cells, it is for example, possible to purify CD34-positive cells or lymphocytes from the peripheral blood (e.g. by leukapharesis), to genetically modify the cells in culture, and to return the treated cells. Before reapplication, the treated cells may or may not be enriched. Currently, long-term correction of cells is only possible when integrating vectors such as retroviral or lentiviral vectors are used. Owing to chromosomal integration of the expression vector, the genetic modification is passed on to the daughter cells during cell division and persists. Only long-term expression may still be a problem. For therapy of a monogeneic disease such as cystic fibrosis, the target cells are primarily the endothelial cells of the bronchopulmonary tract, which can only be subjected to *in vivo* modification attempts. Although long-term correction would be desirable, *in vivo* modification using adenoviral vectors appeared to be more promising, because the target cells are largely in a resting state of the cell cycle amenable to adenoviral gene transfer due to expression of the cell surface receptors used by adenoviruses for cell entry. In addition, the amount and titers of adenoviral vectors seemed suitable. These examples illustrate that a number of factors contribute to the choice of the treatment strategy, the vector, and the route of administration. No single "ideal" vector is therefore suitable

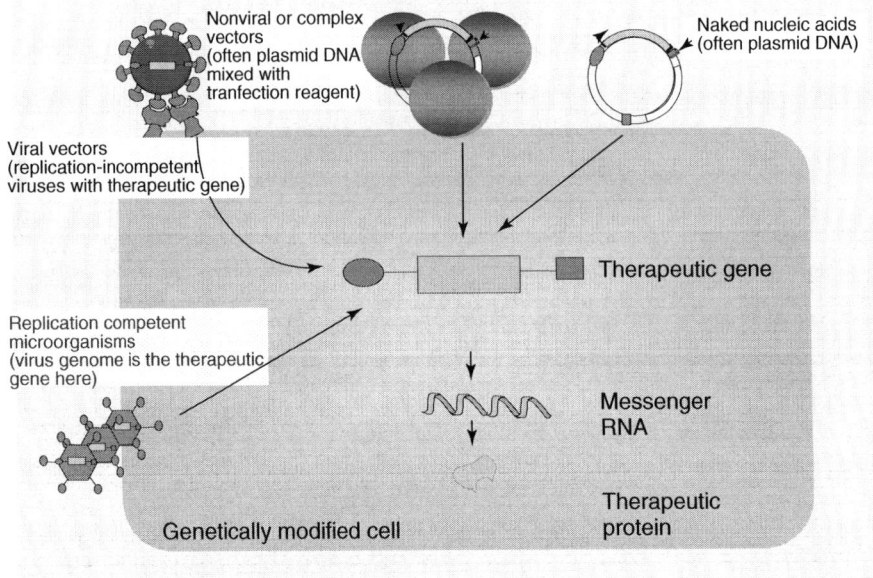

Fig. 3 Delivery systems used in clinical gene transfer. During gene therapy, an expression vector (therapeutic genes) is transferred to somatic cells via a delivery system, for example, a viral or nonviral vector (replication incompetent), a naked nucleic acid, or a recombinant, mostly conditionally replication-competent microorganism. The gene transfer, termed *transfection* when a viral vector is used or termed *transfection* when naked DNA or a nonviral vector is used, leads to genetic modification of the cell. The gene transfer can be carried out *in vivo*, that is, directly in or on the human body, or *ex vivo*, that is, in cell culture followed by the transfer of the modified cells to the human body.

for a large variety of gene therapies. In the past 15 years, many novel gene transfer techniques have been developed and used in clinical studies. In the following section and in Table 1, specific characteristics of the vectors most commonly used in the clinic are summarized.

2.1
Nonviral Vectors and Naked Nucleic Acid

The advantage of nonviral gene transfer systems compared with viral gene transfer systems is the smaller size limitations for the genes to be transferred. The expression vectors are nowadays usually part of a bacterial plasmid that can easily be amplified and grown in bacterial cultures. Plasmid DNA of up to 20 kb pairs encompassing an expression vector of up to 17 kb pairs can easily be manufactured. Promising methods for the *in vivo* administration of plasmid DNA include intradermal or intramuscular injection for the so-called naked nucleic acid transfer. Needle injection or application by medical devices such as gene guns can be used for this purpose. For so-called nonviral vectors, for example, synthetic liposomes or other transfection reagents mixed with plasmid DNA, the DNA-binding liposomes mediate contact with the cellular plasma membrane thus releasing the DNA into the cytoplasma of the cell where uptake by the nucleus has to occur subsequently. During receptor-mediated uptake of nonviral vectors, cell surface proteins (receptors), for example, asialoglycoprotein or the transferrin receptor, mediate cellular uptake of the DNA complex containing a specific receptor ligand.

2.2
Viral Vectors

During evolution, viruses have been optimized to efficiently enter mammalian cells and replicate. Infected mammalian cells transcribe the viral genes and synthesize the viral gene products with high efficiency, sometimes to the disadvantage of endogenous protein production. Viral vectors are replication-incompetent particles derived from viruses by genetic engineering, which no longer transfer to cells the complete set or any viral genes. Instead, an expression vector with one or more therapeutic genes is transferred to cells. Since no complete viral genome is transferred, virus replication is impossible or, in some cases, it is impaired like with first- or second-generation adenoviral vectors. The following section briefly describes the properties of the current frequently used viral vectors.

2.2.1 Retroviral Vectors

The retroviral vectors in clinical use have mainly been derived from murine leukemia virus (MLV). MLV causes leukemia in mice and replication-competent retrovirus (RCR) in a contaminated vector preparation was shown to cause leukemia in severely immunosuppressed monkeys. RCR absence has therefore to be verified before human use of retrovirally modified cells. MLV vector use *in vivo* has been very rare. The genome of the retroviral vectors consists of two copies of single-stranded RNA, which contains one or more coding regions flanked by the viral control elements, the so-called "long-terminal repeat" regions. In the infected cells, the RNA is translated into double-stranded viral DNA and integrated into the cell. The integrated vector DNA is the expression vector. MLV vectors allow efficient genetic modification of proliferating cells by chromosomally integrating the expression vector.

Advantages of retroviral vectors include high gene transfer (transduction) efficiency, and long-term modification of cells due to stable integration of the expression vector into the chromosome of the cells. In addition, the MLV envelope proteins can be exchanged against those from other viruses (which is termed *vector pseudotyping*). This allows preparation of MLV vectors with improved transduction efficiency for certain cell types. Disadvantages of retroviral vectors include the small size of the coding region (~9 kb pairs or less), the restriction of transduction to proliferating cells only, insertional mutagenesis due to integration and the low titer of usually not more than 10^8 transducing units per milliliter of vector preparation. Although chromosomal integration occurs generally at random, it may lead to activation of cellular cancer genes, so-called proto-oncogenes, or, theoretically, to inactivation of tumor suppressor cells. In conjunction with additional genetic mutations, this may result in very low frequency in malignant cell transformation. Hundreds of patients that have been treated with retrovirally modified hematopoietic cells years ago have not shown any signs of cancer related to the gene transfer except for two patients treated during an SCID-X1 (severe combined immunodeficiency disease) gene therapy trial in France. In the latter two leukemia cases, the vector-mediated overexpression of the proto-oncogene LMO2, possibly in conjunction with the therapeutic γc chain gene (which may influence cell proliferation and signal transduction) and the SCID-X1 disease, are the probable cause of the leukemia (see the following sections).

2.2.2 Lentiviral Vectors

Lentiviral vectors have been derived from human immunodeficiency virus type 1 (HIV-1), simian immunodeficiency virus (SIV) isolated from various old-world monkeys, feline immunodeficiency virus (FIV) and equine infectious anemia virus (EIAV) isolated from horses. Lentiviruses cause an acquired immunodeficiency syndrome and a replication-competent virus has therefore to be excluded before human use by batch-to-batch analysis and verification of the absence of replication-competent lentivirus (RCL). Lentiviral vectors may transfer coding regions of up to 9 kb pairs and allow pseudotyping just like MLV vectors. Their advantage is the dual capacity to transfer therapeutic genes into nonproliferating cells in conjunction with persistent genetic modification due to chromosomal integration. This could be useful for *ex vivo* modification of stem cells and *in vivo* modification of neuronal cells. Most lentiviral vectors have been pseudotyped with the G-protein of vesicular stomatitis virus (VSV-G) or the envelope proteins of Gibbon ape leukemia virus. The first clinical study using lentiviral vectors was started in 2003 and involves the *ex vivo* modification of autologous lymphocytes of HIV-infected patients with a therapeutic ribozyme gene shown *in vitro* to inhibit HIV-1 replication.

2.2.3 Adenoviral Vectors

The adenoviral genome consists of double-stranded DNA that persists episomally, that is, inside the nucleus, but not integrated into the chromosome of the cell. Therefore, the genetic modification may be lost during cell proliferation. Adenoviral vectors are the currently preferred vectors for the *in vivo* transduction of a variety of human somatic cells including nonproliferating cells. In contrast to lentiviral vectors, they allow insertion of larger coding regions of therapeutic genes above 10 kb pairs and are not associated with a detectable risk of insertional oncogenesis. In addition, vector titers above 10^{11} transducing units per milliliter can usually be achieved. The lack of long-term expression is in part due to the fact that certain adenovirus genes have been kept on first or second-generation adenoviral expression vectors, and because of the frequent generation of RCA during production. So-called gutless vectors are void of any adenoviral genes, but have to be purified from RCA after production.

Some wild-type (replicating) adenovirus strains cause inflammations of the airways and the conjunctivae. Adenoviral vectors may therefore also be transferred by inhalation of aerosols and inflammations observed following vector applications are mainly local, transient, and associated with very high titer applications. High-titer adenoviral vectors are no longer systemically administrated, because one patient died during systemic administration of a maximum dose of $\sim 10^{13}$ vector particles during gene therapy of the monogeneic disease OTCOTC ("ornithine transcarbamylase") deficiency, a life-threatening metabolic disorder.

2.2.4 AAV (Adeno-associated Viral) Vectors

Adeno-associated viruses (AAVs) belong to the family of parvoviruses. Their genome consists of single-stranded DNA. Wild-type AVV can only replicate in the presence of helper viruses like adenovirus or herpesvirus and has not been associated with disease. AAV can infect hematopoietic cells including nonproliferating cells. Integration in infected human somatic cells is often confined to a distinct locus on human chromosome 19. AAV-derived vectors are usually classified as integrating vectors, although vector integration is unfortunately no longer confined to chromosome 19, but absence of integration may be observed, for example, following intramuscular administration The size of the coding region is very limited ($\sim$4 kb pairs).

2.2.5 Poxvirus Vectors

Poxvirus vectors encompass vaccinia derived from the smallpox vaccine and more attenuated variants like ALVAC or MVA. Their genome consists of single-stranded DNA of 130 to 300 kb pairs. Replication is restricted to the cytoplasm of cells and high amounts of protein are synthesized by the cell following transduction. Most applications, therefore, involve intramuscular vaccination.

3 Clinical Use

3.1 Overview on Clinical Gene Therapy Trials

A number of clinical trials show promising results (see Table 2). In the past few years, it has become increasingly clear that for each disease, the development of a particular and specific gene transfer method in connection with a particular treatment approach will probably be necessary. The first standard use of an approved gene transfer medicinal product is to be expected within the next seven years since $\sim$1% of the clinical gene therapy studies are in an advanced stage of phase II or phase III clinical trial.

Clinical gene therapy studies have been performed initially in North America and Europe. About 50 clinical gene transfer studies have been registered in Germany, with slightly more than 250 patients that have been treated (http://www.pei.de) (http://www.zks.uni-freiburg.de/dereg.html). A general overview on registered studies is listed on the following Web sites: (http://www.wiley.co.uk/genetherapy, or www.pei.de). In Germany, a public registry has been available since 2004.

Target diseases in most clinical gene therapy trials have been cancer, cardiovascular diseases, infectious diseases such as AIDS, or monogeneic congenital disorders. The vectors most frequently used *ex vivo* are MLV vectors derived from murine leukaemia virus (MLV), whereas vectors derived from adenovirus, pox viruses, and AAV are usually used *in vivo*. A growing number of studies involves the use of nonviral vectors or naked DNA.

3.2 Gene Therapy of Monogeneic Congenital Diseases

The idea underlying gene therapy is the replacement of a defective gene by its normal, functional counterpart; for example, a mutation of the gene encoding the γc chain of the interleukin-2 and other receptors is the cause of the congenital immune disorder SCID-X1.

Tab. 2 Promising clinical gene therapy trials.

Disease	Therapeutic gene	Pharmaceutical form/vector	Target cell	Remarks
Human severe combined immunodeficiency SCID-X1	γc Chain gene (e.g. interleukin-2-receptor part)	MLV vector	Bone marrow stem cells *ex vivo*	4 of 1 patient cured, 2 leukemias
PAOD	VEGF gene	Plasmid DNA	Muscle/endothelial cells *in vivo* (im)	Improved blood flow
Head and neck tumor	Adenovirus genome (cell lysis/apoptosis)	Tumor cell specific replicating adenovirus	p53-Negative tumor cells *in vivo*	Local tumor remission in combination with chemotherapy
Graft versus host disease in donor lymphocyte transfer for leukemia treatment	Thymidin kinase gene of the herpes simplex virus, followed by treatment with Ganciclovir	T cells, MLV vectors	T-lymphocytes *ex vivo*	Successful treatment of host-versus-graft disease
Hemophilia B	Coagulation factor IX gene	AAV vector	Muscle cells *in vivo* (im)	Improved coagulation factor concentration

Note: PAOD: peripheral artery occlusive disease; VEGF: vascular endothelial growth factor; AAV: Adeno-associated virus; MLV: murine leukemia virus.

Owing to this defect, immunologically relevant receptors are unable to mediate the normal differentiation and immune function of lymphoid cells such as T cells and natural killer lymphocytes (NK). Therefore, newborn babies suffering from SCID-X1 have a very limited immune system and must live in a germ-free environment. Their life expectancy is strongly reduced. Conventional treatment, that is, bone marrow transplantation, can provide a cure to a certain extent, but involves a high risk if no HLA haploidentical donor is available. For the latter situation, gene therapy within the framework of a clinical study was considered in France.

In this study, autologous CD34-positive bone marrow stem cells were retrovirally modified to express the functional γc chain gene. T cells and other hemaotpoietic cells derived from corrected stem cells were shown to repopulate the hematopoietic cell compartment, and over a period of up to three years, 11 treated patients, mostly newborns, displayed a functional and nearly normal immune system. This represents the first reproducible cure of a disease by gene therapy.

A leukemia-like lymphoproliferative disease was diagnosed roughly three years after treatment of two obviously cured patients. Treatment had been started at the age of a few months. Subsequent analysis revealed that the leukemia-like disease was indeed caused by the MLV vector; the disease mechanism is termed *insertional oncogenesis* resulting from insertional mutagenesis of the proto-oncogene LMO2 (mentioned earlier). According to current knowledge, up to 50 cells with an integration in LMO2 may have been administered together with the $\sim 10^8$ genetically modified CD34-positive bone marrow cells. Owing to the expression vector integration, the transcription of the LMO2 gene was deregulated and activated. Under normal circumstances, the body can cope with individual cells presenting preneoplastic changes like the one described. In the two treated children, however, further genetic changes must have accumulated to finally result in leukemia. Contributing factors discussed include the effect of the therapeutic γc chain gene, the product of which influences cell proliferation and differentiation, and other so far unknown genetic changes that may have occurred during the massive *in vivo* cell replication. In SCID-X1, the T-cell compartment is completely depleted, and is replenished after gene therapy by differentiation and replication of a few genetically corrected blood stem cells. During this process, genetic aberrations may occur with substantial frequency. However, further analysis will be required to understand the exact cause of leukemia development in SCID-X1 gene therapy. Since hundreds of patients treated with retrovirally modified cells in the past 10 years have not developed leukemia up to now, it is currently assumed that a practical risk of leukemia only exists in SCID-X1 gene therapy.

Gene therapy of hemophilia B also seems promising. Here, AAV vectors encoding a smaller, but functional version of the human coagulation factor IX gene were administered by intramuscular injection. A detectable increase in factor IX plasma concentration was observed. Even repeated AAV injections were well tolerated.

3.3
Tumor Gene Therapy

There are various gene therapy approaches that are being developed for the treatment of cancer. They are aimed at inhibiting molecular pathways underlying malignant cell transformation. In other cases, tumor cell ablation by directly applying cell-killing mechanisms or, more indirectly, by improving immunological defense mechanisms directed against tumor cells are attempted.

A number of gene therapy studies involving the adenoviral transfer of tumor suppressor genes like p53 have already been performed. This is aimed at reverting malignant cell transformation or at inducing apoptosis. However, transduction following, for example, needle inoculation into tumors has been shown to be limited to a few cells close to the needle tracks. Direct tumor cell ablation by local injection of conditional RCAs in head and neck tumors led to detectable local tumor regression by direct virus-mediated cell lysis, especially when chemotherapy was used in parallel. Here, virus replication improved transduction efficiency *in vivo*. For the treatment of malignant brain tumors, variant herpesviruses have been inoculated into the tumor in order to lyse the tumor cells *in vivo*, especially if prodrugs have been administered that are converted by the viral thymidin kinase gene to a toxic drug.

In addition, a number of clinical approaches have already been tested that led to an improvement of immune recognition of tumors. They involved intratumoral injection of vectors, which transfer foreign MHC genes, such as B7.1 or B7.2, or cytokine genes, for example, interleukin-2 or granulocyte-macrophage colony-stimulating factor (GM-CSF). Here, vaccinia derived vectors such as MVA or ALVAC have often been used. Autologous or allogeneic tumor cells were also modified *ex vivo* by transfer of immunostimulating genes. Promising results have been reported from a phase I-study in which autologous tumor cells were adenovirally modified with the GM-CSF gene and rapidly reinoculated to stimulate antitumor immunity.

3.4
Gene Therapy of Cardiovascular Diseases

Local intramuscular injection of plasmid DNA or adenoviral vectors encoding vascular epithelial growth factor or fibroblast growth factor, both able to induce the formation of new blood vessels, has been used to improve microcirculation in ischemic tissue. Needle injection of plasmid DNA has been used in leg muscle, catheter application, or needle injection was also tried in ischemic heart muscle. The formation of new blood vessels and an improvement in the microcirculation has been observed.

A narrowing of the blood vessels (restenosis) often occurs after coronary blood vessel dilatation by stent implantation. This is probably caused by the proliferation of smooth muscle cells following injuring of the blood vessel endothelium by the stent. Here, the role of adenoviral or plasmid DNA mediated transfer of the gene encoding inducible nitroxide synthase (iNOS) is thought to result in reduced cell proliferation.

3.5
Preventive Vaccination and Gene Therapy of Infectious Diseases

During the past five to ten years, effective medicines have been developed for the treatment of AIDS. Combinations of effective chemotherapeutics are able to inhibit various steps of the replication cycle of HIV-1. This often results in reduction of the viral load in the peripheral blood, sometimes down to a level barely detectable with modern techniques. Because of the requirement for long-term treatment and the massive adverse effects related to conventional treatment by chemotherapy, gene therapy of HIV infection could offer additional therapy options. *Ex vivo* retroviral transfer of HIV-inhibiting genes into peripheral blood lymphocytes or CD34-positive human cells has been attempted, so far with little success. The therapeutic molecules used include (1) decoy-RNA specifying multiple copies of the Rev- or the Tat-responsive element, so-called poly-TAR or poly-RRE sequences, (2) mini antibodies (single chain Fv; scFv), able to capture viral gene products within the cell, (3) trans-dominant negative mutants of viral proteins, for example, RevM10, or (4) ribozyme RNA, which enzymatically cleaves RNA. Other genes still under development are designed to prevent entry or chromosomal integration of HIV. It remains to be shown whether such gene therapy approaches present a suitable therapeutic option compared with existing chemotherapy.

The best prevention of infectious diseases is achieved by prophylactic vaccines. Clinical trials using vectored vaccines based on ALVAC or MVA have been initiated. Other clinical trials pursue the goal of developing vaccines against HIV-1, malaria, hepatitis B, tuberculosis, and influenza A virus infections. Vaccination regiments using poxvirus vectors such as ALVAC or MVA in combination with naked DNA as a prime vaccine, sometimes followed by further booster injections of recombinant viral antigens, are being tested in humans. Such regiments have been shown to prevent disease progression after lentivirus infection of monkeys. This illustrates the complexity of vaccination strategies that are currently pursued in vaccine research.

3.6
Clinical Gene Therapy for the Treatment of Other Diseases

Clinical gene therapy can also be used for the treatment of diseases not necessarily caused by single known gene defects, if promising therapeutic genes can be reasonably applied. Patients with chronic rheumatoid arthritis, for instance, should benefit from a reduction of the inflammations in joints. Such inflammations are caused or at least maintained by a cascade of events including the overexpression and increased release of a number of inflammatory cytokines. Monoclonal antibodies able to reduce the local concentration of the tumor necrosis factor (TNF) have already been successfully used to treat disease. Here, clinical gene transfer approaches involve the transfer of autologous synovial cells modified *ex vivo* by a therapeutic gene encoding interleukin-1 receptor antagonist. Alternatively, adenoviral vectors with the same gene have been directly injected into the affected joint.

4
Manufacture and Regulatory Aspects

The regulation of gene therapy is very complex and differs considerably in the European Union and the United States. In Part IV, Annex I of Directive 2003/63/EC (which replaces Annex I of Directive 2001/83/EC), a definition of so-called gene therapeutics is given. As gene therapy not only includes therapeutic but also preventive and diagnostic use of vectors, nucleic acids, certain microorganisms and genetically modified cells, the term "gene transfer medicinal products" as used in the relevant European guideline "Note for guidance on the quality, preclinical, and clinical aspects of gene transfer medicinal products (CPMP/BWP/3088/99)" seems more exact. An accurate listing of the medicinal products that belong to the group of gene transfer medicinal products can be found in the table contained in the guideline. The definition given in Sect. 1 of this article is in accordance with this guideline and is in agreement with the definition of gene therapeutic products of Directive 2003/63/EC. The annex of the latter Directive contains legally binding requirements for quality and safety specifications of gene transfer products. Although targeted at product licensing, these requirements may have a bearing on their characterization before clinical use. Active ingredients of gene transfer as defined medicinal products may include, for example, vectors, naked plasmid DNA, or certain microorganisms such as conditionally replicated adenovirus. For the *ex vivo* strategy, the active ingredients are the genetically modified cells.

Written approval by a competent authority in conjunction with positive appraisal by an ethics committee will in the future be necessary for the initiation of clinical gene therapy trials. Respective regulatory processes are currently established in all EU member states during transformation of Directive 2001/20/EC. The manufacture of clinical samples in compliance with good manufacturing practice (GMP) will become compulsory. Germline therapy is illegal in the EU. The law relevant for clinical gene therapy trials and manufacture of gene transfer medicinal products in Germany is the German Drug Law (AMG) and respective decrees and operation ordinances. The law governing the physicians' profession stipulates in the "Guidelines on gene transfer into human somatic cells" ("Richtlinien zum Gentransfer in menschlichen Körperzellen") that the competent ethics committee may seek advice from the central "Commission of Somatic Gene Therapy" of the Scientific Council of the German Medical Association before coming to its vote. The Paul-Ehrlich-Institut is the competent authority in Germany and offers information on current clinical trial regulations.

Gene transfer medicinal products will be licensed via the centralized procedure by the European Commission. The licensing process is coordinated by the EMEA ("European Agency for the Evaluation of Medicinal Products") following submission of a licensing application. The marketing authorization is governed by Council Regulation (EC) No. 2309/93. The recommendation in favor or against marketing authorization is made on the basis of Directives 75/319/EEC and 91/507/EEC by experts of the national competent authorities which are members of the "Committee for Proprietary Medicinal Products" (CPMP).

In the United States, the Center for Biologics Evaluation and Research "(CBER) of the "Food and Drug Administration" (FDA) is responsible for clinical trial approval and marketing authorization.

The assessment of the licensing application focuses on the quality, safety, efficacy, and environmental risk of a gene transfer medicinal product. The manufacturing process has to be designed and performed according to GMP regulations. Like other biologicals, gene therapy products have considerably larger size and complexity compared to chemicals, and analysis of the finished product is not sufficient to control their quality and safety. A suitable process management, in-process control of all critical parameters identified within process validation are decisive factors. Gene transfer medicinal products containing or consisting of genetically modified organisms are also subject to contained use regulations before licensing and until these organisms are applied to humans.

From the economic point of view, procedures for the manufacture of therapeutic DNA must be scalable and efficient, and at the same time simple and robust. Manufacturing processes are as manifold as the gene transfer methods used in gene therapy. As an example, manufacture of plasmid DNA for naked nucleic acid transfer can be briefly described as follows. The methods available for plasmid production today largely originate from lab procedures for the production of DNA for analytical purposes (mini preparations) and have been adapted to fit process scale. Toxic substances and those that present a hazard to the environment, expensive ingredients, and nonscalable methods must be avoided. In this context, the experience gained from industrial manufacture of raw materials with the aid of bacterial cultures and virus production for the purpose of vaccine production are useful for fermentation. Suitable methods for downstream processing above all include chromatographic methods with high

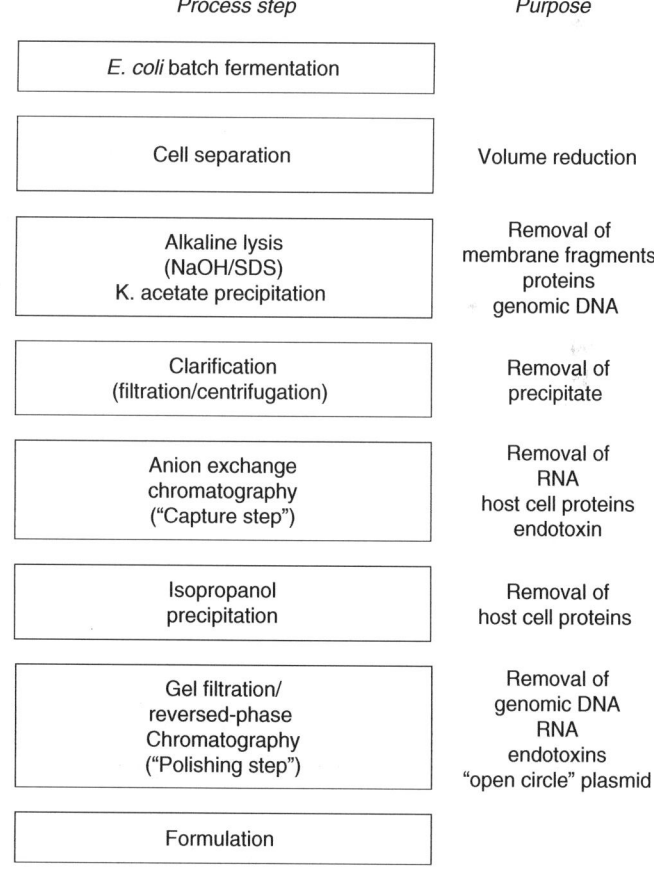

Fig. 4 Therapeutic plasmid DNA: typical manufacturing process.

Test	Specification (method)
Appearance	Clear colorless solution (visual inspection)
Size, restriction interfaces (identity)	Agreement with plasmid card (agarose gel electrophoresis, restriction enzyme assay)
Circular plasmid DNA (ccc)	>95% (Agarose gelelektrophorese, HPLC)
E. coli DNA	<0.02 µg/µg plasmid DNA (southern blot)
Protein	Not detectable (BCA protein assay)
RNA	Not (Agarose gel electrophoresis)
Endotoxin	<0.1 EU/µg plasmid DNA (LAL assay)
Sterility	No growth after 14 days (USP)
Specific activity	Conforms to reference standard (in vitro transfection)

Fig. 5 Therapeutic plasmid DNA: typical release specifications.

dynamic capacity and selectivity as well as high throughput.

In a typical procedure for the manufacture of therapeutic plasmid DNA (cf. Figs. 4 and 5), the first step is batch fermentation of Escherichia coli cells from a comprehensively characterized "Master Working Cell Bank" (MWCB). For this purpose, modern methods use high-density fermentation with optimized and safe E. coli K12 strains bearing a high number of copies of the required plasmid. The bacterial cells are harvested for further processing, resuspended in a small buffer volume, and lysed in an alkaline lysis procedure. By neutralization, the plasmid DNA is renatured while a large quantity of proteins, membrane components, and genomic DNA remain denatured. After separation of the precipitate by filtration, a chromatographic step can be performed as "capture step." Because of the anionic character of the nucleic acid, anion exchange chromatography (AEX) is the method of choice. In fractionated gradient elutions, differences in the charge enable the separation from contaminated RNA. Gel filtration (GF) or reversed phase (RP) steps can be used for fine purification. For final product analysis, evidence must be provided batch by batch that besides the correct identity and homogeneity, critical impurities like microorganisms, host cell proteins, genomic DNA, RNA, or endotoxins have been reduced below the specified limits. Removal of endotoxins is critical for in vivo gene transfer efficiency achieved with naked DNA.

Some established methods from protein chemistry can be used for processing therapeutic DNA. In parallel with the processing of proteins, however, the fact cannot be concealed that nucleic acids have some very specific properties. These include the extremely high viscosity of DNA solutions, the high sensitivity of nucleic acids to gravity, the low static and dynamic capacity of their chromatographic adsorption, and the ability to penetrate filtration media with porosities well below their molecular weight ("spaghetti effect").

After first experience, plasmid concentrations of ~200 mg L^{-1} fermentation broth can be obtained in high-density fermentation (optical density >50), corresponding to an yield of ~800 mg plasmid DNA per kg of dry biomass. Thus, from a fermenter of 1000 L usable volume, ~100 g plasmid DNA can be isolated per run in a batch fermentation at a purification yield of ~50%. Consequently, capacities for production of kilogram amounts can be built up with existing technologies.

5
First Experience with the Clinical Use of Gene Transfer Medicinal Products

The development of somatic gene therapy is still in its infancy. A number of theoretical risks of gene therapy have been listed, and numerous approaches and gene transfer methods are being developed in the clinic, even more in preclinical experiments.

In spite of this, a few SCID patients have been apparently cured by gene therapy using retrovirally modified bone marrow stem cells. At the same time, the occurrence of leukemia in two of the ~10 successfully treated children showed that theoretical risks cannot be clearly distinguished from clinically relevant risks due to the so far insufficient clinical experience. Trends, however, show that each pathological situation will require the development of a certain adapted gene therapy approach. Thus, in the long run, gene therapy will present real therapy or prevention options, especially for a number of up to now insufficiently treatable or untreatable diseases.

See also Gene Therapy and Cardiovascular Diseases.

Bibliography

Books and Reviews

Anderson, F. (1995) Gene therapy, *Sci. Am.* **273**, 96B–98B.

Armentato, D., Sookdeo, C., White, G. (1994) Second generation adenovirus vectors for cystic fibrosis gene therapy, *J. Cell Biochem.* **18A**, 102–107.

Birnboim, H.C. (1983) A rapid alkaline extraction method for the isolation of plasmid DNA, *Methods Enzymol.* **100**, 243–249.

Buchholz, C.J., Stitz, J., Cichutek, K. (1999) Retroviral cell targeting vectors, *Curr. Opin. Mol. Ther.* **5**, 613–621.

Cichutek, K. (2000) DNA vaccines: development, standardization and regulation, *Intervirology* **43**, 331–338.

Cohen-Haguenauer, O., Rosenthal, F., Gansbacher, B., Cichutek, K., et al. (2002) Euregenethy network. Opinion paper on the current status of the regulation of gene therapy in Europe, *Hum. Gene Ther.* **13**, 2085–2110.

Culver, K.W. (1994). *Gene Therapy*, Mary Ann Liebert, New York

Green, A.P., Prior, G.M., Helveston, N.M., Taittinger, B.E., et al. (1997) Preparative purification of supercoiled plasmid DNA for therapeutic applications, *Biopharmacology* **10**, 52–62.

Gottschalk, U.. (1997) The Industrial Perspective of Somatic Gene Therapy, in: Müller, S., Simon, J.W., Vesting J.W. (Eds) *Interdisciplinary Approaches to Gene Therapy*, Springer.

Gottschalk, U., Chan, S. (1998) Somatic gene therapy. Present situation and future perspective, *Arzneimittelforschung/Drug Res.* **48**, 1111–1120.

Hodgson, C.P. (1995) The vector void in gene therapy, *Biotechnology* **13**, 222–229.

Jain, K.K. (1996) *Vectors for Gene Therapy. Scrip Reports*, PJB Publications, Surrey.

Kotin, R.M. (1994) Prospects for the use of adeno-associated virus as a vector for human gene therapy, *Hum. Gene Ther.* **5**, 793–797.

Marquet, M., Horn, N.A., Meek, J.A. (1995) Process development for the manufacture of plasmid DNA vectors for the use in gene therapy, *Biopharmacology* **10**, 26–37.

Marquet, M., Horn, N.A., Meek, J.A. (1997) Characterization of plasmid DNA vectors for use in human gene therapy. Part 1. *Biopharmacology* **10**, 42–50; Part 2. (1997) *Biopharmacology* **10**, 40–45.

Morgan, R.A., Anderson, W.F. (1993) Human gene therapy, *Annu. Rev. Biochem.* **62**, 191–217.

Müller, M. (2003) Considerations for the Scale-up of Plasmid DNA Purification, in: Bowlen, B., Dürre, P. (Eds) *Nucleic Acid Isol. Meth.*, American Scientific Publishers, New York.

Mulligan, R.C. (1993) The basic science of somatic gene therapy, *Science* **260**, 926–932.

Nussenzweig, R.S., Long, C.A. (1994) Malaria vaccines: multiple targets, *Science* **265**, 1381–1384.

Prior, C., Bay, P., Ebert, B. (1995) Process development for the manufacture of inactivated HIV-1, *Pharm. Technol.* **19**, 30–52.

Randrianarison-Jewtoukoff, V., Perricaudet, M. (1995) Recombinant adenoviruses as vaccines, *Biologicals* **23**, 145–147.

Spooner, R.A., Deonarian, M.P., Epenetos, A.A. (1995) DNA vaccination for cancer treatment, *Gene Ther.* **2**, 1–11.

Tolstoshev, P., Anderson, W.F. (1993) Gene transfer techniques for use in human gene therapy, *Genome Res. Mol. Med. Virol.* **7**, 35–47.

Vile, R.G., Russell, S.J. (1995) Retroviruses as vectors, *Br. Med. Bull.* **51**, 12–15.

Primary Literature

Chandra, G., Patel, P., Kost, T.A., Gray, J.G. (1992) Large scale purification of plasmid DNA by fast protein liquid chromatography using a Hi-load Q sepharose column, *Anal. Biochem.* **203**, 169–177.

Gansbacher, B., Zier, K., Daniels, B. (1990) Interleukin-2 gene transfer into tumor cells abrogates tumorigenicity and induces protective immunity, *J. Exp. Med.* **172**, 1217–1222.

Horn, N.A., Meek, J.A., Budahazi, G., et al. (1995) Cancer gene therapy using plasmid DNA: purification of DNA for human clinical trials, *Hum. Gene Ther.* **6**, 565–573.

Mannino, R.J., Gould-Fogerite, S. (1988) Liposome-mediated gene transfer, *Biotechniques* **6**, 682–688.

Michael, S.I., Curiel, D.T. (1994) Strategies to achieve targeted gene delivery via the receptor-mediated endocytosis pathway, *Gene Ther.* **1**, 223–232.

Stitz, J., Buchholz, C.J., Cichutek, K., et al. (2000) Lentiviral vectors pseudotyped with envelope glycoproteins derived from gibbon ape leukemia virus and murine leukemia virus 10A1, *Virology* **273**, 16–20.

Stitz, J., Muhlebach, M.D., Cichutek, K., et al. (2001) A novel lentivirus vector derived from apathogenic simian immunodeficiency virus, *Virology* **291**, 191–197.

Wilson, J.W., Grossmann, M., Cabrera, J.A., et al. (1992) A novel mechanism for achieving transgene persistence in vivo after somatic gene transfer into hepatocytes, *J. Biol. Chem.* **267**, 11483–11489.

39
Mutagenesis, Malignancy and Genome Instability

Garth R. Anderson, Daniel L. Stoler, and Jeremy D. Bartos
Roswell Park Cancer Institute, Buffalo, NY, USA

1	**The Need for Genomic Instability** 1137	
1.1	The Conceptual Framework 1137	
1.2	Multiple Forms of Genomic Damage 1138	
2	**When Does Genomic Instability Begin?** 1141	
2.1	Onset Early in Tumor Progression 1141	
2.2	The Leukemia Exception 1142	
3	**The Maintenance of Genomic Integrity** 1143	
3.1	DNA Damage Repair 1143	
3.2	High-fidelity DNA Replication 1144	
3.3	Chromosomal Segregation during Mitosis 1144	
3.4	Destabilized Chromosomes: Telomeres 1145	
3.5	When All Else Fails: Apoptosis. And without it, Trouble 1145	
4	**In the Clinic** 1146	
4.1	Heterogeneous Tumor Cell Populations and the Therapeutic Target Problem 1146	
4.2	Genomic Instability as a Clinical Prognostic Tool 1147	
5	**A Difficult Future** 1148	
	Bibliography 1148	
	Books and Reviews 1148	
	Primary Literature 1149	

Genomics and Genetics. Edited by Robert A. Meyers.
Copyright © 2007 Wiley-VCH Verlag GmbH & Co. KGaA, Weinheim
ISBN: 978-3-527-31609-0

Keywords

Aneuploidy
A cell containing other than the normal diploid number of chromosomes.

Arbitrarily primed PCR and Inter-(Simple Sequence Repeat) PCR
Sampling methods to monitor genomic damage, in which polymerase chain reaction products are compared for tumor and corresponding normal tissue DNA. The former method utilizes arbitrary PCR primer pairs to copy the sampled set, while the latter method utilizes a single PCR primer to copy sequences (200–2000 bp) between relatively closely spaced repeat sites.

Intrachromosomal Instability
A form of genomic instability in which alterations occur within individual chromosomes, but not particularly to repeat sequences.

Microsatellite Instability
A form of genomic instability arising from defects in the DNA mismatch repair system, leading particularly to alterations of small numbers of bases within DNA-repeat sequences.

Spectral Karyotype (SKY)
A technique by which individual chromosomes are labeled with distinctive fluorescent probes, enabling the direct viewing of both the number of chromosomes in a cell and the translocations between chromosomes.

■ Genomic instability is a widespread and essential feature of the common adult-onset cancers. For solid tumors to arise, several key genes must become mutated (to generate growth, facilitate invasion, enlist nutrient supply, evade the immune response, avoid apoptosis, and so on) and this degree of mutation requires abnormally high mutation rates. To achieve such rates, genes whose normal role is to maintain genomic integrity become mutated, leading to ongoing genomic instability. Exogenous mutagens (carcinogens) additionally contribute to overall genomic damage, particularly in tobacco smokers. Genomic destabilization slowly leads to cancer as cells with mutations in genes advantageous for tumor development become selected, expand their populations, and evolve further. A considerable majority of human genes are involved in multicellular coordination and are not essential for maintaining the viability and reproductive potential of individual cells; thus, a window exists wherein an appropriate degree of genomic instability leads to tumor progression, as somatic evolution selects cells driven by self-interest. As with Darwinian evolution, such somatic evolution inherently leads to genomic diversity, seen clinically as tumor heterogeneity. Although genomic instability and the ensuing genomic heterogeneity render therapeutic targeting difficult, measurement of the degree and form of genomic instability has considerable potential as a clinical prognostic tool.

1
The Need for Genomic Instability

1.1
The Conceptual Framework

Cancer. What a strange disease! For in biology we are so used to seeing order, stepwise differentiation and development, and precise control. Yet, with the common adult-onset solid tumors, we see imprecision, unpredictable behavior, evasion of targeted therapies, disorganized growth, and chaos. And death. But why is this so? What is the nature of this disease, and why does it not behave like other life? Or does it, and we just need to recognize that cancer has more in common with the random nature of evolutionary biology than it does with the fine coordination of organismic biology governed by precise processes of molecular biology?

To comprehend these cancers, we must first recognize the basic nature of multicellular organisms such as ourselves. Going from a unicellular organism to a multicellular organism was a huge leap in evolutionary history. Single-celled organisms have to cope with the formidable tasks of creating and replicating genetic information, controlling its expression, metabolically capturing energy to synthesize new components, and then coordinating the whole process to replicate their entire selves. These cells have the essential genes and systems to allow self-replication, but their only concern about their neighbors is to outcompete them. In contrast, for complex multicellular organisms like ourselves, with a wide variety of highly differentiated specialized tissues, spatial and temporal control and coordination issues come to predominate. Over the course of our evolutionary history, genes evolved, which enable cells to communicate with one another, responding to signals from other cells in a highly specific manner, while also enabling those same cells to respond to other programs locked in during differentiation. Our cells must cooperate, proliferating only when and where appropriate, and willingly dying when and where appropriate; the necessary genes to effect these behaviors evolved. Further gene-based regulatory mechanisms arose, which cause specific cells to adhere only to appropriate partners, and move only in appropriate manners.

The intricate, evolved coordination process generating this complex cellular behavior is contained within the information stored in each duplicated copy of our genome, present in each of our hundred trillion or so cells, and this coordination process continues, as most of these cells are themselves turning over, at widely varied rates. The complexities of this coordination process are far more intricate than those fundamentally required for unicellular life, and this is manifested within our genomes. For the unicellular eukaryote yeast, only about 4000 genes are essential for its cellular growth and replication

(Saccharomyces Genome Database). In contrast, humans contain approximately 10 times as many genes; some of these additional genes generate more complex cellular structures and functions, but most are utilized to coordinate multicellular growth and organization, giving rise to differentiated cells and tissues signaling and responding to one another.

What happens when the quality of a cell's genetic information deteriorates? When damage is severe and extensive, essential gene functions will be lost and the cell dies. But when genetic damage is less severe, and is ongoing over a long period of time, cancer can slowly arise. For if genetic damage occurs at a rate insufficient to kill a cell, a population of mutant cells has time to develop, expand, and acquire additional mutations. With most human genes involved in multicellular coordination, it becomes substantially more likely that a coordination gene will be lost than a gene essential for cellular survival. If both alleles of a gene eventually become lost or mutated, less coordinated growth then occurs. And these daughter cells in turn evolve further, and expand their populations. Selection will occur again and again for cells with greater abilities to proliferate, spread, and evade attack in the form of host defenses or therapeutic treatments. Different daughter cells are free to evolve in different directions; unlike classical Darwinian evolution where natural selection is stringent and competition fierce, the human body is a hospitable environment for many of the successful diverging lineages. This somatically evolved, genomically heterogeneous cell population becomes what we see clinically presenting in the patient as a cancer.

Human evolution occurred in spite of this same basic problem inherent in all multicellular organisms, namely, where loss of genomic stability will lead to cells evolving toward unicellular behavior patterns, with disastrous consequences. We therefore contain within each cell a series of redundant safeguards necessary to maintain the quality of our genetic information for many decades, enabling us to survive as individuals to a point where our progeny are fully capable of surviving and reproducing on their own. If a single gene were sufficient to start the process of somatic evolution eventually leading to malignancy, then somewhere within our large population of cells that individual gene loss would inevitably occur, and we would never be able to survive to adulthood. Redundant safeguards to ensure replication fidelity and efficient damage repair therefore exist to preserve our genomic integrity. Additional apoptotic safeguards destroy cells with badly damaged genomes, reducing the probability of developing cancer to a tolerable level. In spite of this, the disease manages to eventually develop clinically in about one-fourth of us, and is present in nearly all the elderly at a microscopic level. Low mutation rates acting over a long period of time provide a powerful means for even the most stringent safeguards to be overcome by natural selection and cell lineage expansion.

1.2 Multiple Forms of Genomic Damage

Unlike leukemias where a single translocation can complete the two basic steps needed to produce this malignancy, solid tumors require several mutational events before a normal cell becomes an uncontrollably proliferating invasive cancer cell; as described above, this slowly develops through an ongoing process of cell population expansion and natural selection of advantageous mutations. To a large degree, this arises from the mutation of genes whose role is for the maintenance of genomic integrity, yielding the genomic instability that underlies tumor progression; the continuing genomic instability of cultured tumor cells conclusively establishes a genetic component. Extrinsic carcinogens such as tobacco smoke can initiate the process by mutating genes for the preservation of genomic integrity, and expedite the process by directly generating numerous additional mutational events. Genomic instability and ensuing tumor progression thus become the combined result of endogenous (genetic) and exogenous factors.

Genome instability in cancer is seen in multiple forms. Roughly 10% of adult-onset solid tumors exhibit abundant single or oligobase events seen as microsatellite instability. As discussed below, this arises from defects in DNA mismatch repair during DNA replication. Since repeat sequences are inherently more likely to generate small replication errors because of slipped mispairing during replication, the phenomenon is most readily seen in repetitive microsatellite sequences. A small number of genes such as *TGFβ-RII* contain coding repeat sequences, and in tumors arising by microsatellite instability, mutations in such genes become selected for. These tumors have very few other mutations within nonrepeat coding sequences.

Genome instability in cancer is more often seen on a larger scale, in the form of chromosomal instability or aneuploidy. Entire chromosomes are missing or present in extra copies, arising from inappropriate chromosomal segregation during mitosis generating aneuploidy. Translocations generating recombinant and/or truncated chromosomes additionally occur, and multiple recombination events can arise within a single chromosome. Spectral karyotypic analysis of multiple cells from a single tumor reveals differences between cells, as chromosomal instability continues during tumor growth and genomic heterogeneity develops (Fig. 1).

Intrachromosomal instability adds a third fundamental element to the genomic complexity of solid tumors, in the form of numerous insertions, deletions, inversions, amplifications, and other smaller events. The estimated number of such events is increasing as our ability to detect and resolve them improves, with current rough estimates of around ten thousand events per tumor cell. The failure to detect genomic damage in the forms of microsatellite instability or chromosomal instability should not be taken to mean there is no genomic instability at all in a tumor.

Inherent advantages exist for aneuploidy to cooperate with intrachromosomal instability during tumor progression. In order to remove both alleles of a key regulatory gene through aneuploidy alone, both copies of the chromosome will have to be lost. But along with the gene of interest, other genes essential for cellular viability inevitably become lost, bringing the somatic evolutionary process of tumor progression to an abrupt end. On the other hand, mechanisms, which alone generate small events, also have problems; the probability of hitting both alleles of a given gene become very low as the event size shrinks. But when a process generating small events cooperates with a process generating chromosomal instability, critical genes can readily be mutated and the normal allele lost, still without killing the cell (Fig. 2). The evolutionary process of tumor progression is then free to slowly unfold.

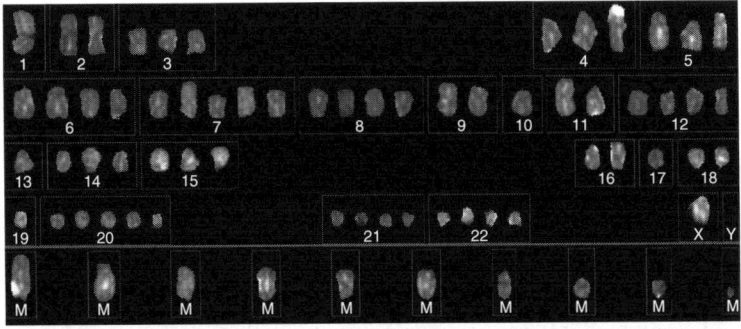

Colon 5

Cell carrying 92 chromosomes with various translocations and derivatives

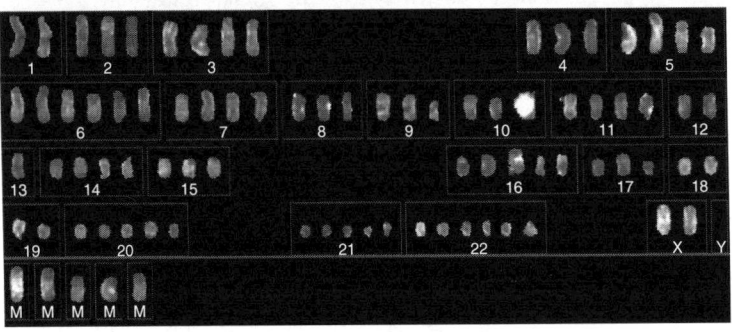

Cell carrying 84 chromosomes with various translocations and derivatives

Spectral karyotypes of two cells from a single tumor

Fig. 1 Genomic instability gives rise to genomic heterogeneity within each tumor. Two cells from the same tumor, colon 5, differ extensively at the genomic level when analyzed by spectral karyotyping. Chromosomes designated M represent a variety of different translocations. Image is from Bartos et al. (2004).

Multiple forms of instability occur sequentially

A. Early events: small

- Detected by inter-(simple sequence repeat) PCR, but not by spectral karyotyping (SKY) or Bac-arrays
- Low probability of hitting both copies of a gene
- Tumors are relatively benign

B. Later events: large

- Detected by SKY, ordered Bac-arrays, inter-SSR PCR
- Early silent mutations revealed
- Tumors are more aggressive

Fig. 2 Multiple forms of genomic instability arise during tumor progression. Inherent advantages exist for small-event instability to cooperate with large-event instability.

This article focuses on genomic instability at the genetic level, where cells acquire permanent heritable changes. An additional level of genomic instability occurs at the epigenetic level, where changes in DNA methylation patterns, particularly of CpG islands, can block the expression of otherwise normal genes. Clinically, this is a significant source of microsatellite instability, when the mismatch repair gene *MLH1* is silenced epigenetically. Complicating the study of epigenetic events in cancer is the finding by Smiraglia and colleagues that DNA methylation patterns of cell lines grown in culture exhibit major differences from the patterns of tumors themselves, presumably reflecting the silencing of genes superfluous to the artificial culture environment.

2
When Does Genomic Instability Begin?

2.1
Onset Early in Tumor Progression

Lawrence Loeb was first to clearly point out the need for genomic destabilization to permit the several requisite steps to convert a normal cell into a malignant tumor cell. This became more evident upon the seminal demonstration by Land et al. that mutation of two cooperating oncogenes alone was insufficient to produce malignancy in mice, and the observations by White, Vogelstein, and others that several events in addition to suppressor gene loss occurred during human colorectal tumor progression. On the basis of known normal mutation rates and the need for more than two mutational events, somatic evolution to malignancy had to somehow be facilitated by decreasing genomic stability.

However, that viewpoint is still not universally accepted. Rubin, Duesberg, and others postulated that normal mutation rates applied to a large, rapidly proliferating cell population, such as the stem cells in colonic crypts, would be likely to generate a rare cell, with exceptional bad luck, containing all the necessary mutations to produce malignancy. This cell would then expand its population inexorably, generating the malignant tumor. By this reasoning, tumors should be essentially homogeneous at the genomic level, but recent spectral karyotyping analyses of multiple tumor cells clearly show this not to be the case.

Verification of the Loeb model has come from multiple directions, centering on demonstration of extensive genomic damage within tumor cells, the finding of extensive genomic heterogeneity within tumors, and tissue culture assays of genomic instability using tumor cell lines showing substantially elevated gene amplification rates. Since genomic instability describes an ongoing process, it is important that these separate forms of evidence exist. Extensive genomic damage occurring in a single burst, as can be seen clinically when induced by radiation therapy or chemotherapy, does not establish that instability exists; in contrast, true genomic instability generates progressively accumulating genomic alterations and genomic diversity within a tumor.

Measuring genomic damage is a complex task, reflecting the diversity of forms of damage that may be present. Flow cytometry measurement of DNA content per cell provides a rough measurement of overall aneuploidy, but karyotyping is essential for identification of specific chromosomal gains and losses. The advent of spectral karyotyping has greatly simplified such analyses, and while also revealing more complex chromosomal translocation events. Comparative genomic hybridization to entire metaphase chromosomes reveals amplifications and deletions of chromosomes and chromosomal fragments, ranging in size down to a few megabases. These may be integrated within other chromosomes, or may be existing freely in the form of double minutes. Substantially higher resolution is now achievable with the development of ordered Bac-arrays, in which chromosomal regions as small as 150 kb may be analyzed. Finer analyses down to single base size may be achieved through genomic sampling, in which several smaller regions distributed throughout the genome are studied in detail. Arbitrarily primed PCR (polymerase chain reaction) utilizes multiple primers with arbitrary sequences to amplify a small set of broadly distributed products. Inter-(simple sequence repeat) PCR, inter-SSR PCR, utilizes a single primer, anchored within repeat sequences at their 3′ ends facing outward, to amplify those non-repeat sequences between two repetitive elements and generates roughly forty different genomic products per reaction. Both of these techniques provide means of readily sampling the genome to estimate overall damage levels. Direct DNA sequencing of the entire tumor genome is not currently practical, and inevitably becomes complicated by genomic heterogeneity within the tumor cell population. But eventual comparisons of sequence data for individual tumor cell genomes can be expected to yield fascinating data.

Examination of premalignant colorectal adenomas demonstrated that extensive genomic damage begins early in tumor progression. Inter-(simple sequence repeat) PCR analysis shows several thousand genomic events have occurred in early adenomas, a level similar to that seen in frank carcinomas. Such results do not mean that further genomic damage does not accumulate during progression, however. In the inter-SSR PCR analyses, DNA from million cell specimens were used, making later events not present in a majority of cells of the specimen undetectable by this method. Karyotypic analyses of early adenomas, including spectral karyotypes, confirm early extensive genomic damage in the forms of aneuploidy and translocations. Recent work by Pretlow and colleagues, utilizing arbitrarily primed PCR indicates that certain forms of genomic instability begin even prior to the adenoma stage, and can be detected within approximately one-fourth of aberrant crypt foci. These observations are highly important in establishing that genomic instability begins early, putting it in the proper position to facilitate the somatic evolutionary process of tumor progression. Early reports proposing that genomic instability might simply represent one of the numerous phenotypic differences between malignant and normal cells, being a consequence and not a cause of malignancy, are no longer tenable.

2.2
The Leukemia Exception

Acute leukemias, some chronic leukemias, and a few other uncommon childhood/adolescent cancers present exceptions to this model, affecting cell types when they are particularly vulnerable to tumor development. If a normally proliferating cell type in a nonlimiting environment uncontrollably increases its proliferation rate, or loses the ability to differentiate, mature, and ultimately die, the population of this cell type will expand dramatically. To generate these tumors, the number of mutated genes can be as few as two,

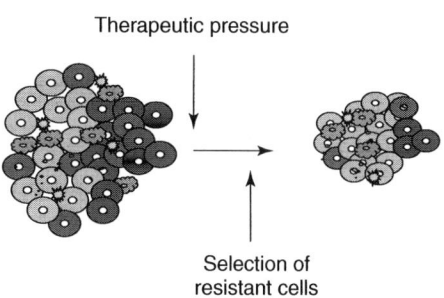

Fig. 3 Extensive genomic heterogeneity within a tumor underlies a considerable part of the eventual failure of pharmacologic therapies for the common, genomically unstable cancers. When therapeutic agents are first applied, susceptible cells are destroyed. But eventually, resistant cells are able to repopulate and further expand the tumor cell population.

which can be generated in a single step by a translocation event that simultaneously rearranges two genes. Such tumors arise rapidly, but due to their genomic homogeneity they also generally respond well to treatment; chronic myelogenous leukemia and its response to Glivec is the classical example of a drug targeted to the translocation event, but childhood acute leukemias also respond well to more conventional therapies. This picture can abruptly change, as such a tumor progresses through blast crisis, giving rise to a genomically unstable heterogeneous tumor cell population capable of evading therapeutic interventions, and in doing so, leukemias can come to resemble the adult-onset solid tumors (Fig. 3).

3
The Maintenance of Genomic Integrity

3.1
DNA Damage Repair

DNA repair is a versatile process evolved to cope with diverse types of DNA damage, which naturally occur. Single base damage elicits base excision repair, while dinucleotide damage (as occurs with UV-induced thymidine dimmers) utilizes nucleotide excision repair. Both of these processes exploit the genetic information present in the undamaged DNA strand. Larger events where breakage occurs can exploit the genetic information present in a sister chromatid or homologous chromosome, through homologous recombinational repair, and still maintain full integrity of the genetic information. While this is feasible in a diploid cell or in a haploid cell in cell cycle phase G2 using the sister chromatid, when allelic losses have occurred or the sister chromatid has not yet been synthesized, the process can fail because of the absence of an intact template. When homologous sequences have been lost or when breakage is extensive, repairs of multiple double-stranded DNA breakage events can still utilize nonhomologous end joining, but with the consequential loss of genomic integrity at the translocated repair sites. Further problems arise if chromosomal fragments are not reintegrated into whole chromosomes; either the genetic information is lost or the small fragment circularizes, with the risk of overreplication and amplification.

In cancer, examples are known for defects in each of these systems, as well as germ-line defects giving rise to hereditary cancer syndromes. The first such case identified was the causal relationship between mismatch repair defects and hereditary nonpolyposis colorectal cancer. Although germ-line defects are found in the family cancer syndrome, with mutations predominantly in hMLH1 and hMSH2, microsatellite unstable sporadic colon cancers (about 10% of sporadic colon cancers) also exhibit defects in this same system, either arising from somatic mutations or epigenetic silencing. Identification of BRCA1 and BRCA2 as the major genes behind familial breast cancer, and the subsequent elucidation of the roles of these genes in homologous recombinational repair, reinforced the concept of defective repair genes underlying both genomic instability and the subsequent cancer arising after decades of somatic evolution.

3.2
High-fidelity DNA Replication

DNA replication is the point where genomic integrity inherently becomes most vulnerable, and most malignancies correspondingly arise out of the relatively rapidly proliferating cell populations such as those lining the gut or in the basal layer of the skin. During replication, the three billion base pairs of the human genome must be copied accurately, completely, and only once, during the few hours that constitute S-phase. This may be occurring in an environment contaminated with carcinogens, reactive oxygen species, or other agents that may damage DNA. And once the DNA is replicated, it must be accurately segregated to daughter cells during mitosis. In spite of these difficulties, and the fact that billions of cells endure these constraints over seven decades, most of us complete our lives never having developed a malignancy.

How is such precision achieved? DNA replication involves over two hundred separate gene products, in a most intricate orchestration. Aside from the highly functional gene products themselves, two keys are the ability to monitor the quality of the replication product, repairing damage immediately, and the ability to temporarily halt replication through checkpoint activation, waiting to resume until the more major damage is repaired. And as a final safeguard, the cellular suicide process of apoptosis is ready to eliminate any cell with irreparable damage to its genomic integrity.

While DNA repair draws most attention as a foundation for genomic instability and tumor progression, uncorrected errors introduced during DNA replication are equally serious and are known to underlie some familial cancers. Mismatched repair errors, in particular, give rise to hereditary nonpolyposis colorectal cancers and endometrial cancers. Errors arising during normal DNA replication are in general within the capability of repair systems to process, but high-level error rates have the potential to exceed normal repair capacities.

3.3
Chromosomal Segregation during Mitosis

The first known form of genomic instability in cancer was aneuploidy, ranging from extra or missing copies of single chromosomes up through tetraploid or higher states, where entire additional copies of the complete genome are present. These phenomena arise out of inaccurate segregation of daughter chromosomes during mitosis. One common diagnostic hallmark of solid tumors is the presence of multipolar mitoses leading to inappropriate segregation.

Centrosomal anomalies typically underlie such inaccurate segregation; these in turn can be a consequence of oncogene activation as shown by Duensing et al. Thus, a stimulus for overproliferation can promote further tumor progression by giving rise to this particular form of chromosomal instability in turn leading to the loss of suppressor or other growth inhibitory genes, extra copies of growth-promoting genes, and the like. Qualitative changes in these genes, however, do not arise out of aneuploidy; for this reason there is a selective advantage during tumor progression to bring intrachromosomal instability into play. Apparent centrosomal abnormalities can also arise from improper separation of sister chromatids during mitosis, as is seen with separate defects.

3.4
Destabilized Chromosomes: Telomeres

Chromosomal ends are capped with unique structures, telomeres, which are recognized as end structures, and thus are not subjected to ligation-mediated repair. Human telomeres contain eight-base DNA-repeat sequences, totaling around 10 kb in length in young individuals. Telomeres shorten in somatic cells as they undergo repeated rounds of replication as part of the normal maturation process, since the telomerase that generates these cap structures is only active during embryonic development and in germ cell generation.

When telomeres are short or absent, chromosomes become very sticky at their ends as repair is attempted through ligation. This generates a dicentric chromosome, with two centromeres. Alternatively, exonuclease processing may play a significant role in generating dicentric chromosomes. During mitosis, dicentric chromosomes are broken randomly as the daughter cells separate. In turn, these breakpoints are highly unstable, fusing into other chromosomes or generating small circular chromosomal structures. This entire process is known as *bridge-breakage-fusion*, and represents one means of producing translocations and deletions. Amplifications can also arise as circular structures overreplicate and randomly reintegrate.

Telomere deficiency is able to promote tumor development in mice, and human cells with severely shortened telomeres are associated with tumor formation. Since telomerase is reactivated in many tumor cells as part of the acquisition of cellular immortality, telomere-mediated instability is likely to play a larger role relatively early in tumor progression. Attempts at treating cancer by interfering with this reactivated telomerase may be counterproductive as genomic instability is driven higher, although this would relate to the relative rate of increase in genomic instability compared to the relative rate at which the cells then progress to senescence.

3.5
When All Else Fails: Apoptosis. And without it, Trouble

Maintenance of genomic integrity includes the fail-safe mechanism of apoptosis to cause the self-destruction of cells, which in spite of their several DNA damage repair systems and replication checkpoints, still manage to contain significant unrepaired or incompletely repaired damage. Apoptosis utilizes sensors to detect DNA damage; these particularly rely on the ataxia telangiectasia mutated (ATM) protein family to sense broken DNA and trigger a kinase cascade leading through capase activation to cell destruction. Antiapoptotic mechanisms exist to suppress uncalled-for suicide, utilizing in particular the Bcl2 system. Overexpression of Bcl2 can promote hematologic malignancies simply by preventing proper numerical control of specific cell populations.

For solid tumors accumulating extensive DNA damage during their years of progression, defects in the apoptotic pathway are widely seen and play their own crucial role. As genomic damage accumulates, these cells survive and progress further, instead of altruistically eliminating themselves.

4
In the Clinic

4.1
Heterogeneous Tumor Cell Populations and the Therapeutic Target Problem

Medical oncology has achieved notable curative success in the treatment of acute childhood leukemias responding to agents targeting proliferating cells, and excellent if not curative responses have been achieved in a few adult tumors such as chronic myelogenous leukemia, where the etiologic Philadelphia Chromosome translocation provides a specific target in the recombined Abl kinase.

However, medical oncologists have achieved far less success in the treatment of adult-onset solid tumors, where death may be delayed but cure is rarely achieved. The same genome instability necessary for tumor progression to occur has an additional negative consequence as genomic diversity permeates the tumor cell population. Genetic events frequently seen in a tumor type are not present in all cells of the tumor, unless the event occurred very, very early in progression and all tumor cells have descended from a cell with that event. Apc evidently represents such a target for colorectal tumor progression, but even here alternative mutations occurring later in progression may create alternatives to the initiating event. And targeting a totally lost gene, such as with adenovirus cytocidal to p53-negative cells, is proving to be even more difficult than targeting a specific alteration.

This problem can be further illustrated with two antibody-based therapies. Herceptin is used in the treatment of breast cancer, directed against the frequently amplified Her2-neu homolog of the epidermal growth factor receptor, which confers a proliferative advantage. Therapeutic failure occurs after several months, largely as a subpopulation of existing resistant tumor cells become selected for and repopulates the tumor (Fig. 3). Antiangiogenic therapies have had considerable theoretical appeal, in that they are directed against the genomically stable endothelial cells of the tumor vasculature. Experimental studies of mouse models with two peptides (angiostatin and endostatin) were highly successful, but have so far yielded disappointing results in human clinical trials. This has been followed by new approaches using monoclonal antibodies targeted at the major angiogenic factor, vascular endothelial growth factor and its receptor. But even here therapeutic failure occurs within a few months. Heterogeneous tumor cells produce several different angiogenic factors, and the elimination of one system soon selects for cells utilizing other factors.

It now appears reasonable to conclude that therapeutic rehabilitation of adult-onset solid tumor cells is likely to be an exercise in futility. And based on results seen up to the present with even the newest classes of tumor cell targeted agents, the same could relate to therapies, even combination therapies, targeted at specifically killing the tumor cells. For within each such tumor there is a genomically diverse cell population, exploiting a variety of pathways, and with a broad spectrum of genomic defects in individual cells. Common events will be present in various lineages, akin to the branches of a tree, and various recurring means of effecting proliferation and evading host defenses will be selected, but diversity and heterogeneity will almost always be present to some extent.

So how can cancer ever be cured? We must never forget that currently almost half of all cancers are indeed cured, and those successes are principally the result of surgical resection. If the patient is physically separated from the heterogeneous tumor cell population, the heterogeneity arising out of tumor cell genome instability no longer matters. Surgery, although relatively costly and associated with its own morbidities and mortalities, is the most effective approach to curing cancer, and efforts directed at improving surgical capabilities have genuine potential. Improvements in diagnosis and detection through imaging, molecular diagnostics, and patient-friendly screening approaches, can significantly improve the ability of the surgeon to cure cancer patients. This, combined with proven prevention strategies directed at tobacco and other environmental factors, should not be neglected in favor of glamorous but marginally effective new therapies.

4.2
Genomic Instability as a Clinical Prognostic Tool

Clinical application of genomic instability for purposes of diagnosis and prognosis is constrained by the limitations of assaying a biopsy specimen *in vitro*, and generally at a single point in time. Thus, what is being measured is more accurately genomic damage, but even from this, meaningful clues often can be obtained. Tumors with more extensive damage typically will reflect more rapidly evolving populations, when comparisons are made on a stage-for-stage basis and patients with prior DNA damaging therapies are excluded. Multiple biopsies can reveal genomic heterogeneity within the tumor cell population, which in turn further reflect the likelihood that cells resistant to any given therapy are already present.

The form of genomic instability present also has major clinical implications, in terms of genomic diversity. Microsatellite instability generates approximately 600 000 events, but since few genes contain coding repetitive sequences, relatively little damage to genetic information occurs. And such tumors generally have a good prognosis. With hematologic malignancies, acute childhood diseases driven by a single translocation provide a homogeneous target, and also respond to therapy. But once aneuploidy and intrachromosomal damage becomes extensive, as in blast crisis, therapeutic outcome becomes poor.

The selection of a clinically practicable assay methodology is not a trivial matter. A few approaches have demonstrated their practicability for providing genome-wide information in a clinical setting. Spectral karyotyping, in which chromosomes are painted with specific probes, requires short-term cell culture but reveals considerable information in terms of aneuploidy and chromosomal rearrangements. The cost and expertise level required for this technique presently make it of rather limited general clinical utility.

Array-based approaches comparing tumor DNA with normal DNA effectively reveal amplifications and deletions, in a higher resolution form of comparative genomic hybridization. Such approaches can exploit conventional expression arrays, but with nuclear DNA copies substituting for cDNAs copied from mRNA. Alternatively, specialized arrays utilizing ordered bacterial artificial chromosomes (Bacs) can representatively scan the entire genome, at a resolving power of about 150 kb. Future improvements in resolution, exploiting single nucleotide polymorphisms, can be anticipated. Current analyses of array data are not yet practical for widespread clinical application due to cost and limitations in data evaluation, although this should be surmounted in the near future.

An alternative approach is the application of genomic sampling methodologies, arbitrarily primed PCR and its variant, inter-(simple sequence repeat) PCR. In these approaches, small pieces of the genome are amplified by PCR, and these genomic samples prepared from tumor DNA are compared with genomic samples prepared from the patient's normal tissue DNA. An electrophoretic comparison of the PCR products reveals what fractions of the bands are altered, providing a quantitative estimate of the degree of genomic damage.

These methodologies have been applied as a prognostic tool for breast cancer, ulcerative colitis, and colorectal cancer, along with Barrett's esophagus. Genomic instability quantitated by inter- (simple sequence repeat) PCR has further been shown to be capable of distinguishing between thyroid tumors and benign lesions, which should be of substantial utility in dealing with relatively common thyroid nodules of indeterminate cytology, which heretofore have been surgically removed as a precautionary measure.

5
A Difficult Future

Genomic destabilization is inherent in the process of development of the common adult-onset cancers. This same instability gives rise to genomic and cellular heterogeneity within the tumor as it presents in the patient; such heterogeneity generates serious problems for nonsurgical therapeutic approaches.

Yet, all is not hopeless. Most cancer cures are currently achieved with surgery, and there remains considerable room for improvement in this area. Promising routes for the future include improving early detection, particularly through the use of proteomic technologies, in turn enabling more timely surgical intervention before metastasis has occurred. Identification of at-risk individuals with familial cancer genes, and definition of tumor specific markers and tests, both fall in the area of early detection. Targeting the genomically stable endothelial cells of the tumor vasculature has its own inherent advantages over targeting the genomically unstable and genomically heterogeneous tumor cell population. And basic prevention measures such as eliminating tobacco use must not be taken lightly.

The cancer problem will be solved neither quickly nor easily. But as the nature of the disease becomes better appreciated, research efforts will hopefully come to focus more on truly productive areas.

Bibliography

Books and Reviews

Anderson, G.R., Stoler, D.L., Brenner, B.M. (2001) Cancer: the evolved consequence of a destabilized genome, *BioEssays* **23**, 1037–1046.

Lengauer, C., Kinzler, K.W., Vogelstein, B. (1998) Genetic instabilities in human cancers, *Nature* **396**, 643–649.

Lindahl, T. (1996) *Genetic Instability in Cancer*, Cold Spring Harbor Press, New York.

Loeb, L.A. (2001) A mutator phenotype in cancer, *Cancer Res.* **61**, 3230–3239.

Moses, R.E. (2001) DNA damage processing defects and disease, *Annu. Rev. Genomics Hum. Genet.* **2**, 41–68.

Nyberg, K.A., Michelson, R.J., Putnam, C.W., Weinert, T.A. (2002) Toward maintaining the genome: DNA damage and replication checkpoints, *Annu. Rev. Genet.* **36**, 617–656.

Shiloh, Y. (2003) ATM and related protein kinases: safeguarding genomic integrity, *Nat. Rev. Cancer* **3**, 155–178.

Van Brabant, A.J., Stan, R., Ellis, N.A. (2000) DNA helicases, genomic instability and human genetic disease, *Annu. Rev. Genomics Hum. Genet.* **1**, 409–459.

Primary Literature

Almasan, A., Linke, S.P., Paulson, T.G., Huang, L.C., Wahl, G.M. (1995) Genetic instability as a consequence of inappropriate entry into and progression through S-phase, *Cancer Metast. Rev.* **14**, 59–73.

Anderson, G.R., Brenner, B.M., Swede, H., Chen, N., Henry, W.M., Conroy, J.M., Karpenko, M.J., Issa, J.P., Bartos, J.D., Brunelle, J.K., Jahreis, G.P., Kahlenberg, M.S., Basik, M., Sait, S., Rodriguez-Bigas, M.A., Nowak, N.J., Petrelli, N.J., Shows, T.B., Stoler, D.L. (2001) Intrachromosomal genomic instability in human sporadic colorectal cancer measured by genome wide allelotyping and inter-(simple sequence repeat) PCR, *Cancer Res.* **61**, 8274–8283.

Basik, M., Stoler, D.L., Kontzoglou, K.C., Rodriguez-Bigas, M.A., Petrelli, N.J., Anderson, G.R. (1997) Genomic instability in sporadic colorectal cancer quantitated by inter-simple sequence repeat PCR analysis, *Genes Chromosomes Cancer* **18**, 19–29.

Ball, C.A., Jin, H., Sherlock, G., Weng, S., Matese, J.C., Andrada, R., Binkley, G., Dolinski, K., Dwight, S.S., Harris, M.A., Issel-Tarver, L., Schroeder, M., Botstein, D., Cherry, J.M. (2001) Saccharomyces genome database provides tools to survey gene expression and functional analysis data, *Nucleic Acids Res.* **29**, 80–81.

Bartos, J.D., Stoler, D.L., Matsui, S.-I., Swede, H., Willmott, L.J., Sait, S.N., Petrelli, N.J., Anderson, G.R. (2004) Genomic heterogeneity and instability in colorectal cancer: spectral karyotyping, glutathione transferase-M1 and Ras, *Mutation Res.* in press.

Bertram, J.S. (2000) The molecular biology of cancer, *Mol. Aspects Med.* **21**, 167–223.

Cahill, D.P., Lengauer, C., Yu, J., Riggins, G.J., Willson, J.K., Markowitz, S.D., Kinzler, K.W., Vogelstein, B. (1998) Mutations of mitotic checkpoint genes in human cancers, *Nature* **392**, 300–303.

Chen, R., Rabinovitch, P.S., Crispin, D.A., Emond, M.J., Koprowicz, K.M., Bronner, M.P., Brentnall, T.A. (2003) DNA fingerprinting abnormalities can distinguish ulcerative colitis patients with dysplasia and cancer from those who are dysplasia/cancer-free, *Am. J. Pathol.* **162**, 665–672.

Cheng, K.C., Loeb, L.A. (1993) Genomic instability and tumor progression: mechanistic considerations, *Adv. Cancer Res.* **60**, 121–156.

Cheung, A.M., Hande, M.P., Jalali, F., Tsao, M.S., Skinnider, B., Hirao, A., McPherson, J.P., Karaskova, J., Suzuki, A., Wakeham, A., You-Ten, A., Elia, A., Squire, J., Bristow, R., Hakem, R., Mak, T.W. (2002) Loss of Brca2 and p53 synergistically promotes genomic instability and deregulation of t-cell apoptosis, *Cancer Res.* **62**, 6194–6204.

Coleman, W.B., Tsongalis, G.J. (1999) The role of genomic instability in human carcinogenesis, *Anticancer Res.* **19**, 4645–4664.

Da Costa, L.T., Lengauer, C. (2002) Exploring and exploiting instability, *Cancer Biol. Ther.* **1**, 212–225.

Davidson, J.F., Guo, H.H., Loeb, L.A. (2002) Endogenous mutagenesis and cancer, *Mutat. Res.* **509**, 17–21.

Duensing, S., Munger, K. (2002) Human papillomaviruses and centrosome duplication errors: modeling the origins of genomic instability, *Oncogene* **21**, 6241–6248.

Duker, N.J. (2002) Chromosome breakage syndromes and cancer, *Am. J. Med. Genet.* **115**, 125–129.

Feldser, D.M., Hackett, J.A., Greider, C.W. (2003) Telomere dysfunction and the initiation of genome instability, *Nat. Rev. Cancer* **3**, 623–627.

Fishel, R., Lescoe, M.K., Rao, M.R., Copeland, N.G., Jenkins, N.A., Garber, J., Kane, M., Kolodner, R. The human mutator gene homolog MSH2 and its association with hereditary nonpolyposis colon cancer, *Cell* **75**, 1027–1038.

Garkavtsev, I.V., Kley, N., Grigorian, I.A., Gudkov, A.V. (2001) The Bloom syndrome protein interacts and cooperates with p53 in regulation of transcription and cell growth control, *Oncogene* **20**, 8276–8280.

Gatenby, R.A., Frieden, B.R. (2002) Application of information theory and extreme physical information to carcinogenesis, *Cancer Res.* **62**, 3675–3684.

Gaudet, F., Hodgson, J.G., Eden, A., Jackson-Grusby, L., Dausman, J., Gray, J.W., Leonhardt, H., Jaenisch, R. (2003) Induction of tumors in mice by genomic hypomethylation, *Science* **300**, 489–492.

Haigis, K.M., Caya, J.G., Reichelderfer, M., Dove, W.F. (2002) Intestinal adenomas can develop with a stable karyotype and stable microsatellites, *Proc. Natl. Acad. Sci. U.S.A.* **99**, 8927–8931.

Heinen, C.D., Schmutte, C., Fishel, R. (2002) DNA repair and tumorigenesis: lessons from hereditary cancer syndromes, *Cancer Biol. Ther.* **1**, 477–485.

Hermsen, M., Postma, C., Baak, J., Weiss, M., Rapallo, A., Sciutto, A., Roemen, G., Arends, J.W., Williams, R., Giaretti, W., De Goeij, A., Meijer, G. (2002) Colorectal adenoma to carcinoma progression follows multiple pathways of chromosomal instability, *Gastroenterology* **123**, 1109–1119.

Hoglund, M., Gisselsson, D., Hansen, G.B., Sall, T., Mitelman, F. (2003) Ovarian carcinoma develops through multiple modes of chromosomal evolution, *Cancer Res.* **63**, 3378–3385.

Ionov, Y., Peinado, M.A., Malkhosyan, S., Shibata, D., Perucho, M. (1993) Ubiquitous somatic mutations in simple repeated sequences reveal a new mechanism for colonic carcinogenesis, *Nature* **363**, 558–561.

Jackson, A.L., Loeb, L.A. (1998a) On the origin of multiple mutations in human cancers, *Semin. Cancer Biol.* **8**, 421–429.

Jackson, A.L., Loeb, L.A. (1998b) The mutation rate and cancer, *Genetics* **148**, 1483–1490.

Jackson, A.L., Loeb, L.A. (2001) The contribution of endogenous sources of DNA damage to the multiple mutations in cancer, *Mutat. Res.* **477**, 7–21.

Kahlenberg, M.S., Stoler, D.L., Basik, M., Petrelli, N.J., Rodriguez-Bigas, M., Anderson, G.R. (1996) p53 tumor suppressor gene status and the degree of genomic instability in sporadic colorectal cancers, *J. Natl. Cancer Inst.* **88**, 1665–1670.

Khong, H.T., Restifo, N.P. (2002) Natural selection of tumor variants in the generation of "tumor escape" phenotypes, *Nat. Immunol.* **3**, 999–1005.

Kolodner, R.D., Putnam, C.D., Myung, K. (2002) Maintenance of genome stability in Saccharomyces cerevisiae, *Science* **297**, 552–557.

Kucherlapati, M., Yang, K., Kuraguchi, M., Zhao, J., Lia, M., Heyer, J., Kane, M.F., Fan, K., Russell, R., Brown, A.M., Kneitz, B., Edelmann, W., Kolodner, R.D., Lipkin, M., Kucherlapati, R. (2002) Haploinsufficiency of Flap endonuclease (Fen1) leads to rapid tumor progression, *Proc. Natl. Acad. Sci. U.S.A.* **99**, 9924–9929.

Kunkel, T.A. (2003) Considering the cancer consequences of altered DNA polymerase function, *Cancer Cell* **3**, 105–110.

Kuraguchi, M., Yang, K., Wong, E., Avdievich, E., Fan, K., Kolodner, R.D., Lipkin, M., Brown, A.M., Kucherlapati, R., Edelmann, W. (2001) The distinct spectra of tumor-associated Apc mutations in mismatch repair-deficient Apc1638N mice define the roles of MSH3 and MSHK6 in DNA repair and intestinal tumorigenesis, *Cancer Res.* **61**, 7934–7942.

Lengauer, C., Kinzler, K.W., Vogelstein, B. (1997) Genetic instability in colorectal cancers, *Nature* **386**, 623–627.

Levitt, N.C., Hickson, I.D. (2002) Caretaker tumour suppressor genes that defend genome integrity, *Trends Mol. Med.* **8**, 179–186.

Loeb, L.A. (2001) A mutator phenotype in cancer, *Cancer Res.* **61**, 3230–3239.

Loeb, L.A., Christians, F.C. (1996) Multiple mutations in human cancers, *Mutat. Res.* **350**, 279–286.

Loeb, K.R., Loeb, L.A. (1999) Genetic instability and the mutator phenotype. Studies in ulcerative colitis, *Am. J. Pathol.* **154**, 1621–1626.

Loeb, L.A., Loeb, K.R., Anderson, J.P. (2003) Multiple mutations and cancer, *Proc. Natl. Acad. Sci. U.S.A.* **100**, 776–781.

Luo, L., Li, B., Pretlow, T.P. (2003) DNA alterations in human aberrant crypt foci and colon cancers by random primed polymerase chain reaction, *Cancer Res.* **63**, 6166–6169.

Marchetti, M.A., Kumar, S., Hartsuiker, E., Maftahi, M., Carr, A.M., Freyer, G.A., Burhans, W.C., Huberman, J.A. (2002) A single unbranched S-phase DNA damage and replication fork blockage checkpoint pathway, *Proc. Natl. Acad. Sci. U.S.A.* **99**, 7472–7477.

Myung, K., Kolodner, R.D. (2003) Induction of genome instability by DNA damage in *Saccharomyces cerevisiae*, *DNA Repair (Amst.)* **2**, 243–258.

Nowak, M.A., Komarova, N.L., Sengupta, A., Jallepalli, P.V., Shih, I.e.M., Vogelstein, B., Lengauer, C. (2002) The role of chromosomal instability in tumor initiation, *Proc. Natl. Acad. Sci. U.S.A.* **99**, 16226–16231.

O'Sullivan, J.N., Bronner, M.P., Brentnall, T.A., Finley, J.C., Shen, W.T., Emerson, S., Emond, M.J., Gollahon, K.A., Moskovitz, A.H., Crispin, D.A., Potter, J.D., Rabinovitch, P.S. (2002) Chromosomal instability in ulcerative colitis is related to telomere shortening, *Nat. Genet.* **32**, 280–284.

Perucho, M., Welsh, J., Peinado, M.A., Ionov, Y., McClelland, M. (1995) Fingerprinting of DNA and RNA by arbitrarily printed polymerase chain reaction: applications in cancer research, *Methods Enzymol.* **254**, 275–290.

Pihan, G.A., Wallace, J., Zhou, Y., Doxsey, S.J. (2003) Centrosome abnormalities and chromosome instability occur together in pre-invasive carcinomas, *Cancer Res.* **63**, 1398–1404.

Powell, S.N., Willers, H., Xia, F. (2002) BRCA2 keeps Rad51 in line. High-fidelity homologous recombination prevents breast and ovarian cancer? *Mol. Cell* **10**, 1262–1263.

Przybytkowski, E., Girouard, S., Allard, B., Lamarre, L., Basik, M. (2003) Widespread bimodal intrachromosomal genomic instability in sporadic breast cancers associated with 13q allelic imbalance, *Cancer Res.* **63**, 4588–4593.

Reid, T.M., Fry, M., Loeb, L.A. (1991) Endogenous mutations and cancer, *Princess Takamatsu Symp.* **22**, 221–229.

Rouse, J., Jackson, S.P. (2002) Interfaces between the detection, signaling, and repair of DNA damage, *Science* **297**, 547–551.

Rowley, J.D. (2001) Chromosome translocations: dangerous liaisons revisited, *Nat. Rev. Cancer* **1**, 245–250.

Schmutte, C., Fishel, R. (1999) Genomic instability: first step to carcinogenesis, *Anticancer Res.* **19**, 4665–4696.

Shih, I.M., Zhou, W., Goodman, S.N., Lengauer, C., Kinzler, K.W., Vogelstein, B. (2001) Evidence that genetic instability occurs at an early stage of colorectal tumorigenesis, *Cancer Res.* **61**, 818–822.

Sieber, O.M., Heinimann, K., Gorman, P., Lamlum, H., Crabtree, M., Simpson, C.A., Davies, D., Neale, K., Hodgson, S.V., Roylance, R.R., Phillips, R.K., Bodmer, W.F., Tomlinson, I.P. (2002) Analysis of chromosomal instability in human colorectal adenomas with two mutational hits at APC, *Proc. Natl. Acad. Sci. U.S.A.* **99**, 16910–16915.

Snijders, A.M., Fridlyand, J., Mans, D.A., Segraves, R., Jain, A.N., Pinkel, D., Albertson, D.G. (2003) Shaping of tumor and drug-resistant genomes by instability and selection, *Oncogene* **22**, 4370–4379.

Stewart, G., Elledge, S.J. (2002) The two faces of BRCA2, a FANCtastic discovery, *Mol. Cell* **10**, 2–4.

Stoler, D.L., Chen, N., Basik, M., Kahlenberg, M.S., Rodriguez-Bigas, M.A., Petrelli, N.J., Anderson, G.R. (1999) The onset and extent of genomic instability in sporadic colorectal tumor progression, *Proc. Natl. Acad. Sci. U.S.A.* **96**, 15121–15126.

Tauchi, H., Matsuura, S., Kobayashi, J., Sakamoto, S., Komatsu, K. (2002) Nijmegen breakage syndrome gene, NBS1, and molecular links to factors for genome stability, *Oncogene* **21**, 8967–8980.

Tlsty, T.D., Margolin, B.H., Lum, K. (1989) Differences in the rates of gene amplification in nontumorigenic and tumorigenic cell-lines as measured by Luria-Delbruck fluctuation analysis, *Proc. Natl. Acad. Sci. U.S.A.* **86**, 9441–9445.

Wang, X., Zou, L., Zheng, H., Wei, Q., Elledge, S.J., Li, L. (2003) Genomic instability and endoreduplication triggered by RAD17 deletion, *Genes Dev.* **17**, 965–970.

Wei, K., Clark, A.B., Wong, E., Kane, M.F., Mazur, D.J., Parris, T., Kolas, N.K., Russell, R., Hou, H. Jr., Kneitz, B., Yang, G., Kunkel, T.A., Kolodner, R.D., Cohen, P.E., Edelmann, W. (2003) Inactivation of exonuclease 1 in mice results in DNA mismatch repair defects, increased cancer susceptibility, and male and female sterility, *Genes Dev.* **17**, 603–614.

Windle, B., Draper, B.W., Yin, Y.X., Ogorman, S., Wahl, G.M. (1991) A central role for chromosome breakage in gene amplification, deletion formation, and amplicon integration, *Genes Dev.* **5**, 160–174.

Zou, L., Cortez, D., Elledge, S.J. (2002) Regulation of ATR substrate selection by Rad17-dependent loading of Rad9 complexes onto chromatin, *Genes Dev.* **16**, 198–208.

Cumulative List of Contributors

Garth R. Anderson
Roswell Park Cancer Institute,
Buffalo, NY,
USA

Werner Arber
Department of Molecular Microbiology,
Biozentrum, University of Basel,
Basel,
Switzerland

Henry Rudolph Victor Arnstein
National Institute for
Medical Research,
London, England,
UK

Alan D. Attie
Department of Biochemistry,
University of Wisconsin,
Madison, WI,
USA

Jeremy D. Bartos
Roswell Park Cancer Institute,
Buffalo, NY,
USA

Barbara G. Beatty
Department of Pathology,
University of Vermont,
Burlington, VT,
USA

Nickolaev Bukanov
Boston University,
Boston, MA,
USA

Ralph A. Bungard
University of Canterbury,
Christchurch,
New Zealand

Thomas T. Chen
University of Connecticut,
Storrs, CT,
USA

Pinwen Peter Chiou
University of Connecticut,
Storrs, CT,
USA

Chung Zoon Chun
University of Connecticut,
Storrs, CT,
USA

K. Cichutek
Paul-Ehrlich-Institut,
Langen,
Germany

Frank H. Collins
University of Notre Dame,
Notre Dame, IN,
USA

Genomics and Genetics. Edited by Robert A. Meyers.
Copyright © 2007 Wiley-VCH Verlag GmbH & Co. KGaA, Weinheim
ISBN: 978-3-527-31609-0

Josep M. Comeron
The University of Iowa,
Iowa City, IA,
USA

Robert Ashley Cox
National Institute for Medical Research,
London, England,
UK

Ralf Dahm
Medical University of Vienna,
Vienna Austria

Raymond Devoret
Institute Curie,
Orsay,
France

Ingo Ebersberger
Institute for Bioinformatics,
University of Düsseldorf,
Germany

Jean-Marc Elalouf
Commissariat à l'Energie
Atomique/Saclay,
France

Sarah C. R. Elgin
Washington University,
St. Louis, MO,
USA

Charles J. Epstein
Department of Pediatrics,
University of California,
San Francisco, CA,
USA

Robert Feil
Institute of Molecular Genetics,
CNRS,
Montpellier,
France

Malcolm A. Ferguson-Smith
Department of Clinical Veterinary
Medicine,
Cambridge University,
Cambridge,
UK

E. Flory
Paul-Ehrlich-Institut,
Langen,
Germany

Javier Garcia-Frias
Department of Electrical and Computer
Engineering,
University of Delaware,
Newark, DE,
USA

Robert Geisler
Max Planck Institute for
Developmental Biology,
Tübingen,
Germany

Francesco Giannelli
Department of Medical and
Molecular Genetics,
Division of Genetics and Development,
Guy's, King's and St. Thomas'
School of Medicine,
London,
UK

Alison Goate
Department of Psychiatry,
Washington University School of
Medicine,
St. Louis, MO,
USA

Yuji Goto
Institute of Molecular Genetics,
CNRS,
Montpellier,
France

Uwe Gottschalk
Paul-Ehrlich-Institut,
Langen,
Germany

Peter M. Green
Department of Medical and
Molecular Genetics,
Division of Genetics and Development,
Guy's, King's and St. Thomas'
School of Medicine,
London,
UK

Preethi Gunaratne
Human Genome Sequencing Center,
Baylor College of Medicine,
Houston, TX,
USA

Stephan J. Guyenet
University of Washington,
Seattle, WA,
USA

Necat Havlioglu
Department of Pediatrics and
Department of Molecular Biology
and Pharmacology,
Washington University School of
Medicine,
St. Louis, MO,
USA

Silva Hecimovic
Department of Psychiatry,
Washington University School of
Medicine,
St. Louis, MO,
USA

Jack A. Heinemann
University of Canterbury,
Christchurch,
New Zealand

Henry H. Q. Heng
Center for Molecular Medicine and
Genetics,
Wayne State University,
Detroit, MI,
USA

Marten Hofker
University of Maastricht,
Maastricht,
The Netherlands

Robert A. Holt
Canada's Michael Smith
Genome Science Centre,
Vancouver, BC,
Canada

Takashi Horiuchi
National Institute for Basic Biology,
Okazaki,
Japan

Lynn B. Jorde
University of Utah Health Sciences Center,
Salt Lake City, Utah,
USA

Jerzy Jurka
Genetic Information Research Institute,
Mountain View, CA,
USA

Vladimir V. Kapitonov
Genetic Information Research Institute,
Mountain View, CA,
USA

Jenny Khoo
University of Connecticut,
Storrs, CT,
USA

Albert R. La Spada
University of Washington,
Seattle, WA,
USA

Boris A. Leibovitch
Washington University,
St. Louis, MO,
USA

Horst Lörz
University of Hamburg,
Hamburg,
Germany

Hisaji Maki
Graduate School of Biological Sciences,
Nara Institute of Science and Technology,
Ikoma,
Japan

Adam G. Marsh
Department of Marine Studies,
University of Delaware,
Lewes, DE,
USA

Jacques Marti
Institut de Génétique Humaine,
Montpellier,
France

Hirotada Mori
Research and Education Center for Genetic Information,
Nara Institute of Sciences and Technology,
Ikoma,
Japan

Sherie L. Morrison
Department of Microbiology, Immunology, and Molecular Genetics,
University of California,
Los Angeles, CA,
USA

C. Münk
Paul-Ehrlich-Institut,
Langen,
Germany

Yusaku Nakabeppu
Medical Institute of Bioregulation,
Kyushu University,
Fukuoka,
Japan

Giang H. Nguyen
Boston University,
Boston, MA,
USA

Christiane Nüsslein-Volhard
Max Planck Institute for Developmental Biology,
Tübingen,
Germany

Adam Pavlicek
Genetic Information Research Institute,
Mountain View, CA,
USA

Manuel L. Penichet
Department of Microbiology, Immunology, and Molecular Genetics,
University of California,
Los Angeles, CA,
USA

Michael E. Rosenfeld
Department of Pathobiology,
University of Washington,
Seattle, WA,
USA

Robert D. C. Saunders
Department of Biological Sciences,
The Open University,
Milton Keynes,
UK

Jeffrey H. Schwartz
University of Pittsburgh,
Pittsburgh, PA,
USA

Matthias Schweizer
Paul-Ehrlich-Institut,
Langen,
Germany

Michael M. Seidman
Laboratory of Molecular Gerontology,
National Institute on Aging,
National Institute of Health,
Baltimore, MD,
USA

Mutsuo Sekiguchi
Biomolecular Engineering Research Institute,
Osaka,
Japan

Cassandra L. Smith
Boston University,
Boston, MA,
USA

D. Peter Snustad
Department of Plant Biology,
University of Minnesota,
St. Paul, MN,
USA

Susanne Stirn
University of Hamburg,
Hamburg,
Germany

Daniel L. Stoler
Roswell Park Cancer Institute,
Buffalo, NY,
USA

Jeffrey P. Tomkins
Department of Genetics and Biochemistry,
Clemson University,
Clemson, SC,
USA

David Umlauf
Institute of Molecular Genetics,
CNRS,
Montpellier,
France

J. Willem Voncken
University of Maastricht,
Maastricht,
The Netherlands

Denan Wang
Boston University,
Boston, MA,
USA

Jian Wang
Beijing Genomics Institute,
Beijing & James D. Watson
Institute of Genome Sciences,
Hangzhou,
China

Jun Wang
Department of Psychiatry,
Washington University School
of Medicine,
St. Louis, MO,
USA

John H. Wilson
Department of Biochemistry and Molecular Biology,
Baylor College of Medicine,
Houston, TX,
USA

Gane Ka-Shu Wong
Beijing Genomics Institute,
Beijing & James D. Watson
Institute of Genome Sciences,
Hangzhou,
China

Todd Charles Wood
Division of Natural Sciences,
Bryan College,
Dayton, TN,
USA

Kim C. Worley
Human Genome Sequencing Center,
Baylor College of Medicine,
Houston, TX,
USA

Jane Y. Wu
Northwestern University Feinberg,
School of Medicine
Center for Genetic Medicine,
Chicago, IL,
USA

Huanming Yang
Beijing Genomics Institute,
Beijing & James D. Watson
Institute of Genome Sciences,
Hangzhou,
China

Liya Yuan
Department of Pediatrics and
Department of Molecular Biology
and Pharmacology,
Washington University School of Medicine,
St. Louis, MO,
USA

Jun Yu
Beijing Genomics Institute,
Beijing & James D. Watson
Institute of Genome Sciences,
Hangzhou,
China

Eugene R. Zabarovsky
Microbiology and Tumor Biology Center,
Karolinska Institute,
Stockholm,
Sweden

Yujing Zeng
Department of Electrical and Computer Engineering,
University of Delaware,
Newark, DE,
USA

Subject Index

a

14-3-3 1002
Aβ 966
Aβ immunization 977
Aβ production 965
ab initio gene-finding software 664
acetylation 1014ff
acetylcholinesterase 458
Achilles' heel strategy 684
actin 1003ff, 1004
"activation function" domain 994
adaptation 401
Adeno-associated viruses (AAVs) 1125
age of onset 958
Agrobacterium tumefaciens 343
AIDS 571
Akt 1002
Alan Templeton 369
albright hereditary osteodystrophy 202
albumin 364
alignment 371
alkylation of bases
 mispairing by 183
Allan Wilson 363, 375ff
allosteric transitions 31
alternate transcripts *see* alternative splicing 703ff
alternative splicing 702
Alu repeat 677, 684
 Alu-PCR 620ff
Alzheimer's disease 960, 997ff
amyloid 959, 1009ff
amyotrophic lateral sclerosis 997ff
ancestral core 518
ancestral polymorphism 559
anchor markers 40
androgenetic 192, 193
 development 489
aneuploid 213
aneuploidy 4

Angelman syndrome 198, 201
anion exchange chromatography (AEX) 1132
annotation 452, 600
Anopheles 445
antibiotic resistance 342
antibodies 977
anticipation 989ff
anticodon 24, 76, 84, 85
 codon interaction 85
 definition 76
 tRNA loop 84
antifreeze proteins (AFPs) 854
anti-oncogene 184
APH-1 968ff
apolipoprotein E 963ff
 allele frequency 963
 alleles 963, 970
 functions 970
 gene 963
 genetic risk factor 963
 molecular mechanisms 970
 protein 963
apoptosis 1145
 ATM 1145
 Bcl2 1145
 DNA damage 1145
APP 965ff
 alternative splicing 965
 FAD mutation 961
 gene 960, 965
 homologs 965
 mRNAs 965
 mutations 960
 normal function 965
 proteolytic processing 965ff
Aristaless related homeobox (ARX) protein 1033
armadillo repeats 1035
arrestin domain 1007
assembly 590

Genomics and Genetics. Edited by Robert A. Meyers.
Copyright © 2007 Wiley-VCH Verlag GmbH & Co. KGaA, Weinheim
ISBN: 978-3-527-31609-0

Subject Index

assembly software package 661
  PhrapView 661
  Phred-Phrap-Consed 661
associative expression networks 254
AT-rich or GC-rich repeats 661
ataxin-1 1000
ataxin-2 1003
ataxin-2 binding protein-1 (A2BP1) 1003
ataxin-3 1004
ataxin-7 1006ff
atrophin-1 999
attenuation 4, 31
autologous 864, 873
autophagy 1008
autosomal 11
  inheritance 4
autosomes 174
average linkage 369

b

BAC (bacterial articificial chromosomes) 496, 504, 596, 641, 678, 682, 687
  clone stability 641
  cloning large DNA fragments 641
  plasmids 641
BAC-end sequences 597
BACE1 965ff
Bacillus subtilis 341
bacteria
  mutagenesis in 183, 184
bacterial artificial chromosomes *see* BAC
bacterial cloning systems 639
  bacteriophages and plasmids 639
  cloned DNA 639
  polylinkers 639
bacterial conjugation 391
bacterial genomes 681
bacteriophage-mediated transduction 391
balanced lethal stocks 420
Barr body 213
λ-based vectors 609ff
  cosmids 611ff
  diphasmids 611
  hyphages 611
  phasmids 611
base analog
  base replacement by 183
base calling 586
base excision repair (BER) 174
batched segregant analysis 494
Battista Grassi 444
Bcl2 rearrangement
  effect of 178

Beckwith–Wiedemann syndrome 194, 196, 202
binary model 864, 885
biodiversity 404
BLAST 600
blepharophimosis-ptosis-epicanthus inversus syndrome (BPES) 1033
blood meal 455
blue–white selection 606ff
  β-galactosidase (lacZ) gene 606
bottom-to-top strategy 567
brain-derived neurotrophic factor (BDNF) 997
branching order 553

c

C83 965ff
C99 965ff
C-value paradox 326, 327
CACNA1A 1005ff
Caenorhabditis elegans 869
calcium channel 1005ff
calpains 1016
β-catenins 1035
cancer 190, 203, 706ff
candidate gene studies 964
CAP3 590
capillary 587
capillary-based sequencers 640
capture step 1132
carboxylesterases 457
cardiac troponin T (cTNT) 1025
cardiovascular diseases 1128
  gene therapy of 1128
"carriers" 13
caspase 995ff
cathepsins D and E 1016
caveats 897
CBC-SGC 654
CBC-SGS 654
  map-directed operation 654
  shotgun libraries 654
CBC-SGS approach 640, 641
  crop plant genome – the rice genome 640
  large-insert clones 641
  shotgun cloned 641
  small-insert clones 641
Celera effort 235
Center for Biologics Evaluation and Research (CBER) 1130
centimorgan (cM) 4, 39
centromeres 11, 211ff, 511, 513
CFTR mutations 174, 176
chaperone binding site 106
Charles Laveran 444

Charles Sibley 368
chemical mutagenesis 478ff
  breeding scheme 480ff
chemicals 393
chemosensation 537, 538
chimeric mouse 864, 883
chimpanzee 569–570, 572
chimpanzee genome project
  finishing phase 574
  sequencing errors 575
  sequencing strategy 573
  shotgun phase 574
chloroquine 444
chorea 996
chromatin 200, 210ff, 211ff
chromatogram 584
chromosomal banding
  C-bands 238
  G-bands 238, 243
  H3⁻ R-bands 243
  R-bands 238
  T-bands 238, 243
chromosomal locus 682
chromosomal segregation 1144
  aneuploidy 1144
  centrosomal anomalies 1144
chromosome aberrations 35
  deletion or deficiency 35
  duplication 35
  inversion 35
  translocation 35
chromosomes 9, 173, 211ff
  banding 561
  forms 447
  inversions 447
  microdissection 621
  substitution 864, 867
  subtelomeric region 562ff
cladistics 373, 374ff
class switch recombination (CSR) 184
cleidocranial dysplasia (CCD) 1032
clinical gene therapy 1129
  treatment of diseases 1129
  trials 1125
clone coverage 646
  clone libraries proper 646
  Unbiased 646
clone Fingerprinting 662
clone fingerprints 662
clone gap 650
  standard Southern techniques 650
clone linking 688
clone ordering 687, 688

clone recovery rate 654
  measured by colony-PCR 654
clone-by-clone (CBC) 635
clone-by-clone SGS or CBC-SGS 637
cloneless genomic libraries 688
cloneless library 678, 683, 689
cloning 606ff, 607, 841
closing sequence gaps 661
  genomic DNAs as PCR templates 661
  isolated YAC 661
  physical gaps caused by cloning failures 661
CMAH 563
codon 5, 24, 76, 82, 85, 97, 111
  anticodon interaction 85
  start 82, 97
  termination 82, 97
  usage 111
coenzyme Q10 1023
coevolution, virus–host 571
cofactors 968ff
colinearity 20
Committee for Proprietary Medicinal Products (CPMP) 1130
comparative genome analyses 671
  gene families 671
  gene networks 671
  genome annotation 671
  guiding molecular biologists 671
  Mammalian Gene Collection (MGC) 671
comparative genome hybridization (CGH) 678, 689
comparative genomics 671
  genome sequencing 671
  human-centric to species-focused 671
competition model 350
complementation test 5, 9
complexity hypothesis 350
compositional constraints 402
concatemer 695, 699, 700ff
conditional gene targeting 888, 891
conditional model 864
conditional transgenesis 887, 888
cone-rod dystrophy 1006ff
cone-rod homeobox protein (CRX) 1007
congenital central hypoventilation syndrome (CCHS) 1034
Congo red birefringence 1009
conjugation 342, 397
  bacterial conjugation 342
  plasmid 342
  T-DNA 342
  Ti plasmid 342
conserved synteny 496

Subject Index

construction of genomic library
  partial filling-in 615
contigs 41, 504, 590, 657
contigs and scaffolds 658
  paired sequence reads 658
  "scaffold" neighboring contigs 658
coordinately transcribed genes 255
CoT analysis 449
CpG Islands 519, 621ff, 1017ff
  density of CpG islands 250
  DNA methylation 248
  GC suppression 248
  gene silencing 249
  tissue-specific expression patterns 250
CRE expression 890, 891
CRE-LoxP 864, 890
CRE-MER 864
CREB 1010
CREB-binding protein (CBP) 995ff
CREM 1010
cross-reactivity 555
crossing-over 5, 38
  intrachromosomal 563
CUG-BP1 and ETR-3-like factors (CELF) 1026
cystatin B gene (CSTB) 1023ff
cystic fibrosis 174
cytochrome P450s 457
cytoplasmic "dumping" 1004
cytosine deaminase
  induced activation of 178, 184

d

2D electrophoresis 556
data integration 651
  basic genomic information 651
  functional annotation, of sequenced genomes 651
  gene-centric or marker-centric 651
  Genetic information 651
  markers 651
  sequence polymorphisms 651
DDT 444
D. E. Kohne 367
deamination 182
"dead-on-arrival" (DOA) elements 328
  Helena 328
  nonallelic 328
deletion bias (DB) 326, 327, 329
  based on small indels 328, 329
  drosophila 328
dentatorubral pallidoluysian atrophy 990ff
derived features 374
detoxification 539
developmental genetics 361

diabetes 202
  mellitus 202
  transient neonatal 202
diploids 9
directed cloning 655
  M13 phage 655
  plasmid cloning system 655
  plasmid-based random cloning system 655
  random shotgun protocol 655
disease, human 569
  environmental influences 571
  genetic predispositions 571
diseased alleles 871
disomy 191, 194
  uniparental 191, 194
distamycin 781
diuretic hormones 459
DM2 1025
DNA 212, 215–217, 219, 221, 223, 583
  carcinogens leave no definite mutational signatures on 184
  electroporation into embryos 839, 840
  electroporation into sperm 839, 840
  gains of 515, 518
  losses of 515, 518
  methylation 217, 222, 223
  methyltransferases 217
  microinjection into embryos 838, 839
  microinjection into unfertilized eggs 838, 839
  mutagenesis of 171, 185
  repair of 171, 185
  repetitive 212, 215, 216
  satellite 216, 219, 221
  structure of 173, 176
DNA acquisition 396
DNA chip 700, 706ff
DNA cloning 635
DNA gain 330
DNA heteroduplex 557
DNA hybridization 366, 367ff
DNA lesion
  vs mutation 176
DNA loss
  due to small indels and genome size 329
DNA marker 607ff
DNA methylation 190, 191, 199, 200
DNA methyltransferases 200
DNA microarray 635
DNA polymerase-based sequencing 638
DNA polymerization errors 181
DNA rearrangements 177, 393, 398
DNA repair 172
  systems 393

DNA replication   1144
  checkpoint activation   1144
DNA restriction–modification systems   391
DNA sequence   371, 556ff
  deletion   564ff
  diversity   560
  evolutionary rate   564
  insertion   564ff
  nonfunctional   558
  repetitive   563
  substitution   563ff
    human-specific   568
    multiple/parallel   566
  transposable elements   563
  variation   560
DNA sequence assembly   657, 658
  close the gaps   658
  Lander–Waterman gaps   658
  Phred-Phrap-Consed package   658
  repeat gaps   658
  RePS (an assembler, repeat-masked Phrap with scaffolding   658
DNA sequence comparison   556ff
  DNA–DNA hybridization   557
  mitochondrial DNA   558
  nuclear DNA   558ff
DNA sequence homologies   390
DNA sequencing   635, 638
  basic technology   638
  concept and technology of   638
  DNA methylation sites   635
  primer-dependent tactics   638
  sequence polymorphisms   635
  shotgun principles   638
DNA transformation   391
DNA triplex   1021
DNA-damaging agents   172, 176
Dobzhansky   366, 378ff
dominance   5, 32
  codominant   33
  partial dominance   33
  recessive   32
dominant-negative   993
Drosophila melanogaster   213, 869
duality of genomic information   403
duplex   173
duplication, segmental   563
dynamic expression linkages   259
dystrophica myotonica protein kinase (DMPK)   1024ff

e
ecosystem   388
ectopic expression   864, 871, 874
effective population size (Ne)   326, 327
electron transport chain (ETC)   998
electrophoresis   584
electroporation   832, 838
elongation factors *see* mRNA translation, protein factors   87
embryonic stem (ES) cells   865, 870, 882, 894
embryos   885
  ES cell–derived   885, 887
EMEA (European Agency for the Evaluation of Medicinal Products)   1130
Emile Zuckerkandl   363
end-sequencing   660
endogenous retroviruses
  derivatives of   536
endosymbiosis   400
enhancer   31, 218
enhancer mutations   34
Ensembl system   234
environmental conditions   388
enzymatic repair systems   393
epidemiology   958ff
epigenesis   376
epigenetic   190–192
  modifications   191
  inheritance   211, 221
episodic ataxia type 2 (EA2)   1006
epistasis   5, 28
epitope   555
equine infectious anemia virus (EIAV)   1124
Erika Hagelberg   370
ES cells *see* embryonic stem cells
  random mutagenesis in   894, 897
EST see expressed sequence tag   694
EST (expressed sequence tag)   694ff, 695, 703, 704ff
euchromatin   210ff, 420, 423, 427, 430, 433, 595
eukaryotes
  noncoding DNA evolution in   327, 333
  noncoding DNA, size of   327, 328
eukaryotic genomes   330, 644
  influence of recombination   330
evo-devo   361, 376
evolution   1061
  biological   565
  DNA sequence   565ff
  male-driven   1061
  molecular   565
  of phenotypes   565
  species-specific   563
evolution and development   376
evolution genes   391, 399
evolutionary change   379ff

evolutionary fitness   399, 400
evolutionary function of viruses   402
excitotoxicity   997ff
exon   5, 21
expanding trinucleotide repeats   36
expressed sequence tag (EST)   40, 600, 678, 682, 694ff
expression controls   261
expressivity   28

f
F_1 transgenic fish   832
F_2 transgenic fish   850
F8A   1061
F8B   1061
factor V   1062
factor VII   1063
factor VIII   1060, 1061
    gene   1061
factor FVIII   1062
    mRNA   1062
    protein domains   1062
factor IX   1060, 1063, 1065, 1076
    gene   1063
    inhibitor   1076
    mRNA   1063
    protein domains   1063
    tridimensional structure   1065
factor X   1062, 1063
familial hemiplegic migraine (FMH)   1006
FASTA   600
feline immunodeficiency virus (FIV)   1124
FGenesH   600
fibrinogen domains   454
fingerprint   596
fingerprint contig (FPC)   508
fingerprinting   496
finished sequence   504
FISH see fluorescence in situ hybridization   689
fish model
    large-size   834, 835
    medium-size   834
    small-size   834
fish species
    selection of   833, 835
FISH technique   662
fish transgenesis   846
flip-flop system   394
fluorescence in situ hybridization (FISH)   678
fluorescence-activated cell sorter (FACS)   619
FMR2   1020
FMR3   1020

Food and Drug Administration (FDA)   1130
Forkhead L2 (Foxl2)   1033
forms of genomic damage   1138, 1139
    chromosomal instability   1139
    intrachromosomal instability   1139
    microsatellite instability   1139
forward genetics   478ff
founder effect   1004ff
FOXP2   567
FPC   597, 598, 662
    assembling of, sequence-ready clones   662
fragile site   1017
fragile X mental retardation-1 (FMR1) gene   1017ff
fragile X syndrome (FRAXA)   1017ff
fragile X tremor-ataxia syndrome (FXTAS)   993ff, 1028ff
fragile XE mental retardation (FRAXE)   988ff, 1018ff
frameshift mutation   5, 34
frameshifts   177
frataxin   1020ff
Friedreich's ataxia   993ff
Friedreich's ataxia (FRDA)   1020ff
FTDP-17   969ff
full genome scans   964
functional annotation see genome annotation   707ff
functional coordination   255
functional genomics   704
fusion, telomeric   561
FVIII   1063, 1071, 1072
    binding sites for FIX   1071
    inactivation by protein C   1063
    inhibitors   1072
    interaction with vWF   1071
    normandy-type von willebrand disease   1071
    tridimensional structure   1063
FVIII and FV   1062
    combined deficiency   1062
FVIIIa   1070
    heterotrimer stability   1070

g
α_{2u} globulin pheromones   538, 539
G + C content   519
G-protein of vesicular stomatitis virus (VSV-G)   1124
G-quartet   1018
gain-of-function mutation   5, 34, 990ff
gametogenesis   375
gap closure   690
    pseudogaps   690
    true gaps   690

GC content 239, 557, 663
 problems in cloning, and sequencing
  procedures 664
GC_3 values 241
Gel filtration (GF) 1132
gene 8, 172, 376
 evolution of 522, 532
 fusion 394
 homeotic 376
 loss 529, 530
 map 236
 model 245
 number 232
 prediction 233, 600
 regulatory 376
 set
  construction of 522, 523
 size 665
 spaces 666
  empty space 241
  genome core 241
 structural 376
 structure 241
  exon length 242
  intron sizes 242
  number of exons 242
 targeting 771, 865, 870, 879, 884
  basics of 879, 885
  biotechnological procedures of 884, 885
  developments in 870, 871
  technology of 879, 885
  vector design of 879, 882
 therapy 1128
  infectious diseases of 1128, 1129
 transfer 1120
  methodology of 841, 842
  methods of 838, 842, 1120, 1122, 1123, 1125
   naked nucleic acid 1122, 1123
   Nonviral Vectors 1122, 1123
   viral Vectors 1123
   viral vectors 1125
 trap 865, 895
  mutagenesis of 895, 896
 tree 559
 vector 397
gene density 238
gene duplications 526, 529
gene expression *see* gene expression profiling
gene expression profile *see* gene expression profiling 705ff
gene expression profiling 377, 568, 695, 700ff, 704ff
gene-finding algorithms 664

gene-finding program 665
gene-regulatory regions
 conservation of 529
generators of genetic variations 401
Genesis 403
genetic code 77, 82
 deviations from 82
 reading frame 82
 standard 82
genetic distance 554ff
genetic diversity 340, 559ff
genetic drift 560
genetic engineering 404
genetic instability 990ff
genetic locus 682
genetic map 5, 493ff
genetic maps 38, 670f, 682
 assembled, genome sequence map 670
 genetic and functional studies 671
 sequence polymorphism discovery 671
genetic polymorphism 390
genetic screens 478
genetic selection 606ff
 Spi 606
 supF 606
genetic sequence 173
genetic variation 388
genetically identical 867
genetically modified organisms 339, 349, 404
 lateral gene transfer 339
 transgenic crops 349
genetics 960ff
 familial Alzheimer's disease 960
 sporadic Alzheimer's disease 963
genome 172, 504
genome annotation 662–664, 666, 695, 703, 704ff
 and analysis 662
 compositional and structural dynamics 663
 genome and gene structural dynamics, importance of 664
 heterogeneous distributions of DNA composition 663
 sequence variations 666
 SGS 666
genome duplication 475
genome evolution 667, 669
 gene duplications and deletions 669
 insect genomes 667
 Vertebrate genomes 667
genome human chromosome 677
 physical mapping of 677, 691
Genome organization 390

Genome sequencing   671
  Large Genome, maize   671
  Large Genomes   671
  WG-SGS strategy for, animal genomes   671
genome sequencing   497
  BAC clones   497
  ESTs   497
  whole-genome shotgun   497
genome size   511
genome working draft   644
genome-sequencing project   662, 669
  EST sequencing   669
  ESTs and cDNAs   669
  Full-length cDNA   669
  NIAS   669
  RIKEN   669
  sequencing cDNA clones   669
genomic clone libraries   687, 688
genomic DNA Libraries   652
genomic features
  covariation of   522
genomic imprinting   190, 191
genomic instability as a clinical prognostic tool   1147, 1148
  array-based approaches   1147
  clinically practicable assay methodology   1147
  genomic sampling methodologies   1148
genomic instability begin   1141, 1142
  colorectal adenomas   1142
  early in tumor progression   1141
  measuring genomic damage   1141
  the Loeb model   1141
genomic library   607ff
  amplified libraries   611
  classical
    dephosphorylation   612
    construction   612
      classical   612
      dephosphorylation   612
    genomic libraries   612
    general   607
    jumping   615ff
    linking   615ff
    percentage of recombinants   612
    representativity   611ff
    special   607
genomic mapping   682
  concepts of   682, 683
genomic restriction maps   683, 687, 689
genomics   15, 573 583
genotype   174
genotype–phenotype correlation   567
Genscan   600

George H. F. Nuttall   362
George Todaro   367
GGG trinucleotide   251
gibbon   553ff
Glimmer   600
glutathione reductase   459
glutathione-S-transferases   457
glycosylation   1062
good manufacturing practice (GMP)   1130
Goodman   365ff
granulocyte-macrophage colony-stimulating factor (GM-CSF)   1128
great apes   553ff
green fluorescent protein   492ff
gynogenetic development   489

h
Haemophilus influenzae   341
hand-foot-genital syndrome (HFGS)   1033
haploinsufficiency   1024ff
haplotypes   451
HDAC3 (histone deacetylase 3)   1002
HDL2   1036ff
α-helical   1031
heat shock proteins (HSP)   375, 994ff
heat shock response   379
hemizygous   13
hemochorial placentation   365
hemoglobin   363, 367ff
hemophilia   1059, 1060, 1077–1080
  genetic counseling   1078
  genetics   1059
  gonadal mosaicism   1078
  incidence   1060
  missense mutation   1077
  mutation at CpG sites   1077
  treatment   1079
    gene therapy   1079
  treatment, gene therapy   1080
    clinical trial   1080
hemophilia A   1059–1060, 1065–1069, 1077
  gross gene deletions   1066
  inhibitors   1077
  inversions   1066
  large insertions   1066
  mutation   1067–1069
    affecting mRNA translation   1068
    affecting RNA processing   1067
    amino acid addition   1069
    amino acid deletions   1069
    deletion or insertion hotspots   1069
    missense   1069
    nonsense   1068
    small sequence changes   1067

mutation rate   1060
mutations   1065
hemophilia B   1059–1060, 1072, 1073, 1075, 1077
  inhibitors   1077
  mutation   1073, 1075
    founder mutations   1073
    large insertions   1073
    missense   1075
    small sequence changes   1073
  mutation affecting   1073
    mRNA processing   1074
    RNA Processing   1074
    transcription   1074
    translation   1074
  mutation rate   1060
  mutations   1072
    gross deletions   1072
hepatitis   571
heritability   5, 14
heterochromatin   210ff, 420, 422, 423, 430, 433, 595
  satellite   786
heterologous   865, 874
HGT   340, 348, 351
  barriers   340, 348, 351
  measurements of   340
hidden Markov models (neural network)   665
  sequence composition   665
  signal to weight   665
high repetitive sequence content   672
histone acetylation   200, 998
histone deacetylase inhibitors (HDAC Is)   998
histone methylation   200
histones   997ff
holoprosencephaly (HPE)   1032ff
homeodomain   1033ff
homeostasis   375, 380ff
hominoids   553ff
homologous recombination   771, 879
homology searching   600
homozygous   867
horizontal gene transfer   339, 391, 397, 398
host preference   457
HOX   1032
HSP   379ff
Human Genome Project (HGP)   637, 639f
  BAC (bacterium artificial chromosome)-based
    STS (sequence-tagged site)   637
  cosmids   640
  genome-scale physical mapping effort   640
  large-insert clones   640
  minimal tiling path (MTP)   637
  physical mapping   637
  YAC (yeast artificial chromosome)-based   637, 640
human immunodeficiency virus type 1 (HIV-1)   1124
human transcriptome map (HTM)   255
human–chimpanzee comparison   561
  biomedical differences   570ff
  cytogenetic differences   561
  DNA sequence differences   575
  ethical considerations   572
  human rights   573ff
  infectious diseases   571
  qualitative difference   572
huntingtin   996
Huntington's disease   990ff, 1036ff
hybrid cell lines   620
  radiation hybrids   620
hybridization   340, 345, 348
  hybrid-bridge   345
  hybrid-bridges   348
β-hydroxylation   1064
hypoxia response element   995

i

IBD (identical-by-descent) fragments   624
  CIS (cloning of identical sequences)   624
  GMS (genomic mismatch scanning)   624
    CIS (cloning of identical sequences)   624
idebenone   1023
IGF/INS pathway   194, 196
immune-response genes   460
immunological distance   364
imprint   191
imprinted genes   196
imprinting-control region   190, 198, 199–201
imprinting-control regions   198
imprints   192
improvement of a biological function   398
in situ hybridization   420, 425, 450
inbred strain   865, 882
inclusions   998ff
indels   327, 329
  polymorphic   329
independent assortment   6, 37
induced mutations   182, 183
induction   30
  inducible enzymes   30
  inducible genes   30
infidelities of DNA replication   392
initiation factors *see* mRNA translation, protein factors   91
innate immune system   460
inositol phosphate-3 receptor (IP3-R)   998

insecticide resistance   457
insertional mutagenesis   480ff, 865, 894
  plasmid DNA   480ff
  pseudotyped retroviral vectors   480ff
  rate of mutagenesis   482
  retroviral vector   480ff
  transposon   480ff
insulin receptor (IR)   1025
inteins   25
inter-Alu PCR   678, 690
intergenic noncomplementation   29
interphase   211ff
interspersed simple sequence repeats (SSRs)   536
intertwined complementary strands   173
intragenic complementation   29
introgression   340, 352
  introgress   340
intronic splice signals
  conservation of   526
introns   6, 21, 330
inversion   561ff
iron–sulfur cluster-containing proteins   1021
Iron–sulfur clusters (ISCs)   1021
IS elements   395
isochores   240
isogenic DNA   865
isolation   389
  reproductive   562

j
Jon Ahlquist   368
Jonathan Marks   371, 380ff
josephin domain   1005
jumping library   678
  construction of Not I jumping   616ff
  "general" jumping (hopping)   616
  jumping clones   616
  Not I jumping   616
Jun N-terminal kinase (JNK)   999
junctophilin-3 (JPH3)   1036ff

k
KELCH   1028
Kelch-like 1 (KLHL1)   1028
kidney   705, 706ff
kinetochore   216
King   381ff
knockdown   484ff
knockin   865
  mutant   569
knockout   486ff
  mutant   569
Kpn repeat   678, 684

l
laboratory strains   475ff
lamina   216
  nuclear   216
Lander–Waterman curves   648
  clone coverage   648
large genome-sequencing projects   637
  large-insert clone-based physical mapping   637
large-insert clones   634, 640, 662
  BAC   640, 662
  BACs or YACs   634
  YAC   640
large-scale sequencing (LSS)   635, 640
  LSS and physical mapping   640
Large-scale Sequencing Era   638
law of large numbers   361, 369, 380, 381ff
lentiviral vectors   898
leptospirosis   506
lethal mutations   396
Lewis Wolpert   376
library   588
ligand-binding domain (LBD)   891
ligand-binding proteins   525
LINE (long interspersed) repeat   684
lineage sorting   559
linkage   38
linking libraries   616, 678, 686
Linus Pauling   363
local DNA sequence change   392, 398
locus   679, 690
long interspersed repeat elements (LINEs)   679
long term depression (LTD)   1018
long terminal repeat (LTR)   1119
loss-of-function mutation   6, 34, 993ff
low-complexity repeats (LCR)   657
LSS   655

m
M13   589
Machado–Joseph disease   1003ff
maintenance of genomic integrity   1143
  DNA damage repair   1143
  homologous recombinational repair   1143
  mismatch repair   1143
  nonhomologous end joining   1143
major histocompatibility complex (MHC)   571

malaria 444
malignant *see* cancer 706ff
mammals
 carcinogenesis in 183, 184
marker rescue 793
markers 494, 607
 anonymous markers 608
 polymorphic 608
  microsatellites 608ff
  minisatellites 608
  single-nucleotide polymorphism (SNP) 608
Mary-Claire King 375ff
Maryellen Ruvolo 373
massive parallel signature sequencing 701
massively parallel hybridization *see* DNA chip 703ff
master gene 568
Master Working Cell Bank (MWCB) 1132
maximum likelihood 373
mechanisms of genetic variation 389
meiosis 6
melting temperature 557
membrane 212, 216
 nuclear 212, 216
mendelian 14
mesothelioma 184
messenger ribonucleoprotein complexes (mRNPs) 1018
messenger RNA (mRNA) 76–78, 1120
meta genomes 672
 portions of ruminant stomach 672
metabolic pathway 705
metabolic resistance 457
metabotropic glutamate receptors 459, 1018
metaphase 211ff
methylation 379, 1017ff
microaggregates 1010
microarray 694, 700, 704–707ff
 hybridization 493
microbial genome sequencing 639
microcomplement fixation 363
microdeletion 519
microinjection 492
microRNAs 1018
microsatellites 41, 493, 988
microtubule 216
minimal tiling path (MTP) 634, 637
mismatch-repair (MMR) 342
mitochondrial genome 702
mitosis 6, 213
mixed genomes 672
MMR 348
 mutators 348

mobile genetic elements 391, 395
mobility pattern 556
model organism 569ff
modified vaccinia ancara (MVA) 1120
molecular and cellular mechanisms 965
molecular anthropology 361
molecular assumption 363, 369, 370, 373–375ff
molecular change 361, 364, 365, 367, 373–375ff
molecular clock 361, 363–365ff, 368, 375ff
molecular cloning techniques 638
molecular evolution 388, 403
molecular systematics 361, 362, 367, 370ff
moloney murine leukemia virus (MoMLV) 840
monogeneic congenital diseases 1125
 gene therapy of 1125
morphogenetic fields 377
morpholino 484ff, 838
 structure of morpholino 484ff
Morris Goodman 364
morula aggregation 882
mosaic embryos 882
 genesis of 882, 884
mosquitoes 443
mouse 866, 868, 891
 colocalization of SINEs in 535, 536
 genetics of 866, 867
 genome 519
 history of 866, 871
 knockins of 891, 894
 lineage
  different activity of SINEs in 534, 535
 molecular genetics of 868, 871
 shift in substitution spectra 519, 520
MPSS see massive parallel signature sequencing 701, 707ff
mRNA *see* messenger RNA 76
mRNA structure 78–081, 105, 107–111
 5′ cap 81
 eukaryote 81
 ferritin 107, 108
 frameshift 78
 histone 79
 internal RNA entry site 110, 111
 polarity 78
 polyadenylation 81
 polycistronic 108, 109
 prokaryote 81
 pseudoknot 108
 Shine–Dalgarno sequence 81, 111
 stability 105
 translational control 108

mRNA translation control 76, 104, 105, 107, 108, 111, 113
  amino acid abundance 111
  antisense polynucleotide 111
  codon usage 111
  elongation factor activity 108
  frameshift 108
  initiation factor activity 108
  internal RNA entry sites 110, 111
  mRNA/protein interactions 107
  mRNA stability 105
  mRNA structure 108
  ribosome activity, modulation 111
  ribosome inactivation 112
mRNA translation mechanism 76, 80, 90, 95, 97, 99, 100, 103, 104, 113
  initiation of translation 90, 95
  peptide bond formation 80, 100, 103
  peptide chain elongation 95
  peptide chain elongation/translocation cycle 97
  termination 97, 99
  translocation 103, 104ff
mRNA translation, protein factors 90–93, 95, 97, 99
  elongation factors, eukaryotic 97, 99
  elongation factors, prokaryotic 95, 97
  initiation factors, eukaryotic 91, 93
  initiation factors, prokaryotic 90–092
  termination factors 95, 99
mt DNA 370
  hypervariable zone 370
multiplex analysis 679, 687
murine leukemia virus (MLV) 1123
muscleblind 1026ff
muscleblind-like proteins (MBNL) 1026
mutagenesis 172, 391, 393, 478ff
  to produce immunity repertoire 184, 185
mutagenic conditions 393
mutagens 172, 176, 393, 478ff
mutation 6, 32, 172, 378–380, 390, 478ff, 1061
  at regulatory sequences 178
  chromosomal rearrangement 378
  gene duplication 378
  load 1061
  macromutation 378, 379, 380ff
  micromutation 378
  point 378
  types of 176, 179
mutation hotspots 178
mutation rate, germ line-specific 566
mutation reversions 178
Mycobacterium leprae 347
Mycobacterium tuberculosis 347

myoclonic epilepsy type 1 993ff
myoglobin 365
myotonic dystrophy 989ff
myotonic dystrophy type 1 (DM1) 1024ff
myotonic dystrophy type 2 (DM2) 1026ff

n

natural habitats 475
natural killer lymphocytes (NK) 1127
natural selection 365, 388
natural strategies of genetic variation 398
N-ethyl-N-nitrosourea 478ff
  mechanism of mutagenesis 479
  rate of mutagenesis 479
nearest-neighbor joining 373
Neisseria gonorrhoeae 341
nephron 706
netropsin 781
neurofibrillary tangles 969
neurotransmitter 1005ff
neutral substitution rate 527
new world monkeys 554
nicastrin 968
non-Mendelian inheritance 989
noncoding DNA 327, 331
noncoding lesions 183
noncoding RNA genes 532
nondisjunction 6, 35
  monosomy 35
  trisomy 35
nonsynonymous mutation 180, 505
nuclear bodies (NBs) 1008ff
nuclear DNA 371
nuclear inclusions 994ff
nuclear transplantation 841
nucleic acids 8
nucleic base
  loss of 182
nucleoid 9
nucleolus 212, 216
nucleosome 210ff
nucleotide excision repair (NER) 174
nucleotide substitution 392, 557
nucleus 9, 211ff
nutrient transfer 190, 203, 204

o

oculopharyngeal muscular dystrophy 993ff, 1030ff
odorant binding proteins 457
odorant receptors 454
Okazaki fragments 17
old world monkeys 554

oligomers 1010
operon fusion 394
operons 31
1 : 1 orthologs 453
ORF 505
ORs 457
orthologous chromosomal segments 513
   large-scale rearrangements 513
orthologous genes
   properties of 523, 524
orthology
   determination of 522, 523
OTC (ornithine transcarbamylase) 1124
Otto 235
outgroup 575
ovary ecdysteroids 459
overall similarity 363
oxidized base
   incorporation of 182

p

P_1 transgenic fish 832, 843
P1 artificial chromosomes (PACs) 682, 687
P1-derived artificial chromosomes (PAC) 679
p53 998
p300/CBP-associated factor (PCAF) 1014
parasites 343, 344
   parasitic plants 344
Parkinson's disease 849, 997ff, 1003
parthenogenesis 192, 193, 203, 488
partial transgenesis 898, 900
Partington syndrome (PRTS) 1033
pathogenicity islands 344
pathogens 343
pathology 958ff
PAUP 373
PCR *see* polymerase chain reaction
PCR-based protocols 643
   preparation of sequencing templates 643
PEN-2 968ff
penetrance 28
peptide bond, formation 80, 100, 103
peptide nucleic acids 794
pericentric inversion 562
"perturbed expression" profiling 254
peroxidases 454
PEST (Pink Eye STandard) strain 447
phenotype 174, 388
   analysis 897, 898
   emergence of new 567
phosphoglycerate kinase (PGK) 875
PHOX-2B 1034
Phrap 590
Phred 587

Phred-Phrap-Consed 643
phylogenetic analysis using parsimony 373
phylogeny 475ff
   vertebrate lineages 477
phylogeny, human and great ape 553ff
physical colocation and dislocation 257
physical gaps 634, 648, 650
physical mapping 40, 496, 597, 634, 662, 670, 683
   approaches for 683
   radiation hybrid mapping 662
   restriction-digest fingerprints 662
   STSs 662
physical mapping methods 670
   clone fingerprinting 670
   hybridization-based 670
   large-insert clones 670
   large-insert, clone-based STS mapping 670
   sequence-tagged connectors 670
plant genomes 644, 672
   diploid species 672
   polyploid crop 672
plaques 959
plasmid 589
plasmid DNA 1119
   preparation methods 642, 643
   absorbent-based 642
   alkaline lysis protocols 642
   centrifugation steps 643
   isothermal amplification protocols 642
β-pleated sheets 1009
pleiotropy 28
ploidy 35
   aneuploidy 35
   euploidy 35
   polyploid 35
pluripotent 865
point mutations 176, 374
polonies 679, 690
polyadenine-binding protein 2 (PABP2) 1030ff
polyadenylated transcripts *see* polyadenylation 703ff
polyadenylation 702
polyalanine 993ff
polyamide 781
   imidazole 782
   pyrrole 781
polyglutamine 988ff
   binding protein 1 (PQBP-1) 1002
polylinker 606
polymerase chain reaction (PCR) 533, 635, 640, 679, 683, 694, 699ff, 832, 842

polymorphic
  microsatellites   608ff
  minisatellites   608
  single-nucleotide polymorphism (SNP)   608
polymorphism   694, 703ff
  link-up   679, 686
polypeptide chain, polarity   80
polypeptide synthesis *see* mRNA translation   77
polyploidy   6
polypurine (polypyrimidine)   537, 787
polytene chromosomes   420, 421, 423–427, 430, 434, 435, 447
polyubiquitin   1005
POMC (pro-opiomelanocortin promoter)   887
population genetic theory   327, 328
population size, effective   561
position effect variegation (PEV)   213ff
positional cloning   41
positional information   377
postsegregational killing genes   351
  suicide genes   351
posttranscriptional gene silencing   31, 899, 900
posttranslational modification   1002
PPP2R2B   1035ff
Prader–Willi syndrome   198, 201
premature ovarian failure (POF)   1017
premutation   988ff
presenilin   967ff
  CTF   968
  endocleavage   962
  endoproteolysis   961, 968
  FAD mutations   962
  gene   961, 967
  homologs   967
  mutations   961
  NTF   968
  structure   967
preventive vaccination   1128
  infectious diseases of   1128, 1129
primate phylogeny   555
primate-rodent ancestor   518
primer walking   592, 660
primitive features   374
prion diseases   997
process control   655–656
  data tracking mistakes   655
  equipment-related errors   655
  failures in DNA preparation   655
  managerial challenges for   655
  scales, automation, and robust protocols   656
  throughput and efficiency   656

progressive myoclonus epilepsy type 1 (EPM1)   1023ff
promoter   219, 221, 222
promoter/enhancer elements
  functional analysis of   846, 847
promyelocytic leukemia (PML) protein   1008
proofreading   16
prosimians   554
proteasome   1016ff
protein C   1063
protein comparison
  amino acid sequence comparison   556
  electrophoretic comparison   555
  immunological comparison   555
protein domains   454
protein family expansions   454
protein phosphatase 2A enzyme (PP2A)   1035ff
protein synthesis *see* mRNA translation   78
protein-coding sequences
  indels in   524, 525
  repeats in   524, 525
proteolysis   539, 995ff
proteome   453
prothrombin   1062
protozoan genomes   681
PS1   960
PS2   960
pseudogenes   252, 329, 529, 530
psoralen   790
public consortium   234
pulsed field gel (PFG) electrophoresis   679, 681
Purkinje cells   1000ff

q
quality assessment   653
  quality of SGS clone library   653
quantitative traits   14

r
R. J. Britten   367
radiation hybrid mapping   495ff
radiation hybrid maps   679, 683
radiation hybrids   620
radiation mutagenesis   483
  γ-radiation   483
  ionising radiation   483
  mechanism of mutagenesis   483
  rate of mutagenesis   483
random Enu mutation   897
random insertion type   882

random mutagenesis 896
  retroviruses elements of 896
  transposable elements of 896
random SGS 655
Raoul Benveniste 367
rapid electrophoresis 653
RARE (RecA-assisted restriction endonuclease) cleavage 680, 684
rat 506, 507
  colocalization of SINEs in 535, 536
  different activity of SINEs in 534, 535
  genes 530, 532
  Shift in Substitution Spectra 519, 520
rat genome 503, 506, 507, 545
  assembly strategy 507, 510
  evolutionary hotspots of 520, 522
  features of 511, 522
  human disease gene orthologs 539, 544
  substitution rates 518, 519
  transcription-associated substitution 525, 526
rat lineage 519
  LINE-1 activity in 532, 534
rat-specific biology 537, 539
reactive oxygen species (ROS) 1022
rearrangement 213
  chromosomal 213, 562
  subchromosomal 562ff
recombinant inbred strain 865, 867
recombination 6, 36
  rate 566
reduced penetrance 1027
regions of increased gene expression (RIDGE) domains 255
regulation of development 376
regulator genes 30
  corepressors 31
  inducers 31
regulatory peptides 459
repeated DNA 368
RepeatMasker 649
Repetitive Sequences 656–657
  Classification of 656
  eukaryotic genomes 656
  simple and complex 657
replacement type vectors 879
replication 211, 217, 218, 222
replication forks 773
replication-competent adenovirus (RCA) 1120
replication-competent lentivirus (RCL) 1124
replication-competent retrovirus (RCR) 1123
replisome 19
reporter-function transgenes 837

repression 30
repressible genes 30
research, biomedical 569
reshuffling of DNA segments 398
restriction endonucleases 397
restriction enzyme 606
  cleavage 687
restriction fragment length polymorphisms (RFLPs) 40, 395
retinal degeneration 1006ff
retinoblastoma protein (Rb) 221
retroposons 680
retroviruses 348
reverse genetics 483
reversed phase (RP) 1132
RGG box 1018
ribosomal protein 697, 698ff
ribosomal RNA 695, 702ff
ribosome 86
ribosome function 76, 78, 84, 86, 95, 97, 100–0104, 106, 110–112
  aminoacyl tRNA binding 97, 102
  chaperone binding 106
  decoding mRNA 76, 104
  elongation factor-G binding 104
  elongation factor-Tu binding 103
  fidelity 76, 78, 84, 86
  GTP hydrolysis 103, 104
  internal RNA entry site 110, 111
  peptide bond formation 100, 103
  peptide chain elongation 95, 100
  peptidyl-tRNA binding 97, 102
  proof reading 101
  ribosome inactivating proteins 112
  translocation 103, 104ff
  tRNA binding 97, 102
ribosome structure 76, 102–104, 106
  aminoacyl-tRNA binding site 102
  conformational change 103ff
  elongation factor-G binding site 104
  elongation factor-Tu binding site 104
  peptidyl-tRNA binding site 102
  tunnel 106
rice genomic data 651
Richard Goldschmidt 378
risk factors 958
RNA 223
  interference 223
RNA analog sugar 789
  2'-O-(2-aminoethyl) (AE) 789
  2'-O-Methyl (2'-OMe) 789
  Bridged Nucleic Acid 789
  Locked Nucleic Acid, LNA 789
RNA editing 22

RNA interference (RNAi)   31, 488, 865, 899, 900
RNA polymerase   219
RNA polymerase II transcription factor D (TFIID) complex   1009
RNA splicing   1003
RNA-induced silencing complex (RISC)   1019
RNP motif   1003
rodents
   medium-length duplications in   537
Ronald Ross   444
rRNA   216
RST
   microarrays
      Not I clone microarrays   627
RST microarrays   626ff
   CpG-Island   626
      CGI (CpG-island-containing) microarrays   626
RUNX2   1032

S

SAGA   1007
Sarich   364, 365ff
scaffold   505, 634, 657
SCID-X1 (severe combined immunodeficiency disease)   1124
α-secretase   961, 965
β-secretase   961, 965ff
   active site motifs   966
   structure   966
γ-secretase   961, 965ff
second-order selection   399, 400
secondary crossover site   394
secretase inhibitors   976
segmental duplications   513, 515
segments, chromosomal   561
segregation   7, 36
selection   1061
   truncating   1061
self-organization   402
semiconservative replication   15
sequence
   consensus   592
sequence assembly   650, 656, 657, 658
   assembly process   658
   base caller   657
   computational filters   656
   contaminated reads   656
   electrophoresis-based instrument   657
   Lander-Waterman model   656
   low-quality reads   656
   Multiple estimators   650
   operating teams   656
   Phred-Phrap-Consed   657
   Repeats   658
   repetitive sequences   656
   Software Packages for   657
   to define "repeats"   658
   Validation of   650
sequence assembly software   657
   identifying different reads, and detect overlaps   657
sequence contigs   635, 654
sequence finishing   658, 660, 662
   better contiguity of a target sequence   658
   Clone Fingerprinting   662
   End-sequencing   660
   high-copy repeats   660
   low-copy repeats   660
   PCR-sequencing   660
   Physical Mapping   662
   Primer-walking   660
   Quality-related contig-breakers   660
   sequence contigs   660
   single-read coverage   660
sequence gap   635, 648, 650
sequence heterogeneity   664
sequence homology comparison   664
sequence quality control   648
   poor DNA preparation   649
   poor signal-to-noise ratio   649
   system contamination and malfunction   649
   wet-bench operation   649
sequence slippage   524
sequence-tagged connectors   597
sequence-tagged restriction site (STAR)   680, 688
sequence-tagged site (STS)   40, 606ff 680, 682
sequencing
   DNA   584
   genome   583
   shotgun   583, 588
sequencing data acquisition   654
   engine of SGS   654
sequencing reads   636, 637
   a parallel subcloning strategy   637
   stepwise primer-directed "walking" strategy   636
serine proteases   454, 1064
sex-limited   995
sex-linked inheritance   7, 11
SGS   641, 645, 655
   clone library construction   645
   clone sequencing   645
   cloning   641
   computing techniques   641
   data analysis   645

libraries 652
Methodology 641
related technologies 643
sequence annotation 645
sequence assembly 645
sequencing 641
wet-bench operators 645
SH3 domains 1007
Sherman paradox 989
short interspersed repeating element (SINE) 680, 684
shotgun sequencing see SGS 635
signal transduction pathways 376, 380ff, 705
silencers 31
silent base substitutions 179, 181
simian immunodeficiency virus (SIV) 1124
single nucleotide polymorphism (SNPs) 450, 494
  RFLP markers 608
single-copy DNA 367, 368
site-specific DNA inversion 394
site-specific recombination 394
SIX5 1024ff
slab-gel-based sequencers 640
slipped-strand mispairing 182
small fragment homologous replacement, SFHR 793
small indels 328
small ubiquitinlike modifier (SUMO) 999
SMRT (silencing mediator of retinoid and thyroid hormone receptors) 1002
solar radiations 176
somatic gene therapy 1117, 1118, 1125, 1129, 1133
  clinical use of 1125, 1129
  homologous recombination 1118
  manufacture of 1129, 1132
  regulatory aspects of 1129, 1132
SOS mutagenic repair 172
source DNA 653
  genetic heterogeneity of DNA samples 653
  using pulse-field gel electrophoresis 653
southern blot hybridization 842
Sp1 998
spaghetti effect 1132
spatio-temporal expression pattern 865
species tree 559
speech impairment 567
spinal and bulbar muscular atrophy 988ff
spinocerebellar ataxia 990ff
  type 1 (SCA1) 1000ff
  type 2 (SCA2) 1002ff
  type 3 (SCA3) 1003ff
  type 6 (SCA6) 1005ff
  type 7 (SCA7) 1006ff
  type 8 (SCA8) 1027ff
  type 10 (SCA10) 1034ff
  type 12 (SCA12) 1035ff
  type 17 (SCA17) 1009ff
splicing control elements (SCEs) 251
spontaneous mutations 181, 390
  generation mechanisms 181, 182
sporogenic developmental cycle 445
STAGA 1007
standard genetic sequence 174
strategies of genetic variation 389
strategies to map and sequence genomes 624ff
  hierarchical 624
  slalom 624
  whole-genome 624
Streptococcus pneumoniae 341
STS see sequence tagged site
subtraction
  Not I-CODE 623
subtraction procedures 622ff
  Not I linking libraries 622
  Not I subtraction see *Not* I-CODE Fig. 8 622
  representational difference analysis 622
sulfation 1062
supF mutation reporter 790
suppressor mutation 7, 34
susceptibility genes 964
symbionts 343
symbiosis islands 344
symbiotic associations 400
synapsis 38
synonymous 505
synonymous mutations 180, 181
synpolydactyly (SPD) 1032
synteny 453

t

tangles 959
TATA-binding protein (TBP) 1009
tau 959
  biological role 969
  hyperphosphorylation 969
  isoforms 969
tautomeric forms of 392
TBP-associated factors (TAFs) 1010
telomeres 11, 211ff, 511, 513, 1145
  bridge-breakage-fusion 1145
  dicentric chromosome 1145
  telomerase 1145
test gene set 665
tetraploid 865, 885
theodicy 403

theory of molecular evolution 389, 400, 401
therapeutic approaches 977
TILLING 486ff
top-to-bottom strategy 567
training data sets 664
trans heterozygote 9
trans-lesion DNA-synthesis 176
transcript densities 236
transcription 20, 214, 219, 224
  dysregulation 997ff
  factors 31
  interference 995ff
transcriptome 694, 700, 703–705, 706ff, 707ff
transduction 343
  bacteriophage 343
  generalized 343
  specialized 343
transfer RNA see tRNA
transformation 341, 397
  competence 341
  transformants 341
transgene 832, 835, 865, 869, 871, 894
  design of 871, 876
  expression of 843
  gain-of-function 835, 837
  inheritance of 843
  loss-of-function 838
  origin of 871, 873
  reporter-function 837
  structure of 871, 873
  types of 835, 838
transgene donor 872
transgene expression 843
transgene integration
  pattern of 842, 843
transgene recipient 872
transgenesis 492, 869, 871, 873, 876, 885
  basics of 871, 879
  biotechnological procedures in 876, 879
  developments in 869, 870
  pronuclear injection 876
  rates of integration 492
  refinements of 885, 894
  regulatory elements of 873, 876
  stable expression 492
  technology of 871, 879
  transient expression 492
transgenic fish 831–833, 856
  application of 843, 854
  as environmental biomonitors 852, 854
  characterization of 842, 843
  examples of 854
  growth hormone 850
  human disease modeling 848, 850

identification of 842
in basic research 844, 850
signaling pathways 847, 848
vertebrate developmental biology 845, 846
with different body color 851, 852
with resistance to pathogen infection 850, 851
transgenic mice 863, 971ff
  apolipoprotein E transgenic mice 974
  APP transgenic mice 971
  in biomedical research 863, 900
  presenilin 1/2 transgenic mice 972ff
  tau transgenic mice 973ff
  transgenic crosses 974ff
transgenic mosquitoes 461
transition 7, 33
translation 23
translocation 194
transmembrane regions 525
transposable element (TE) 7, 39, 215–217, 223, 224, 253, 328, 330
  evolution of 532, 537
  insertion of 178
transposition 213, 394, 395
transposons 39
transversion 7, 33, 177
treatments 975ff
tree
  genealogical 560
  phylogenetic 560
tricarboxylic acid pathways (TCA) 257
trinucleotide repeat expansion 988ff
triple helical DNA 787
triple helix–forming oligonucleotides 787
tRNA (transfer RNA) 76, 83, 84, 86, 101, 111
  abundance 111
  aminoacylation 84, 86
  anticodon loop 83ff
  elongation factor Tu•GTP complex 86, 101
  nomenclature 83
  structure 83ff
trojan gene effect 855
tumor 1127
  gene therapy of 1127
tumor necrosis factor (TNF) 1129
tumor therapy 1119
  tumor therapy replication-competent adenoviruses 1119
tumor vaccination 1120

u
UAR 368, 369, 375ff
ubiquitin interaction motifs (UIMs) 1005
ultraviolet radiation 375

uniform average rate of genomic change   368
universality of the genetic code   397

v

vaccine   977
variable number tandem repeats   41
variation generator   396f
vascular endothelial growth factor (VEGF)   995
vector
   BAC   343, 607, 609, 625
   PAC   609, 625
   parental vector   607
   recombinant vector   607
   YAC   609
vertical gene transfer   340
vesicular stomatitis virus (VSV)   840
Vincent Sarich   363
viral shuttles   898, 900
viral vectors   1123–1125
   AAV (adeno-associated viral) vectors   1125
   adenoviral vectors   1124
   lentiviral vectors   1124
   poxvirus vectors   1125
   retroviral vectors   1123, 1124
virus-mediated transduction   397
vomeronasal receptors   537

w

West syndrome (WS)   1033
WG-SGS   639–640, 643, 654
   contigs   654
   Drosophila genome   639
   genome survey sequence   643
   Insect genomes   639
   microbial genome, H. influenzae   639
   mouse genome   640
   ordered within scaffolds   654
   rat genome   640
   targeted genome sequences   643
   working draft   643
WGS (whole-genome shotgun)   448, 505, 507, 635, 637
whole-genome shotgun sequencing   574ff
Wilson   364, 381ff
World Health Organization-directed worldwide Malaria Eradication Campaign   444
worldview   403

w

X chromosome   11
X-linked mental retardation (XLMR)   1033ff

y

Y chromosome   11
YACs (yeast artificial chromosomes)   680f, 687
yeast   213, 217, 222, 223, 704

z

zebrafish development   472ff
   external development   473ff
   generation time   473
   numbers of offspring   474
   speed of development   472
   transparency   473ff
ZIC2   1032ff
zinc finger 9 protein (ZNF9)   1025
zona pellucida   882

Printed and bound in the UK by
CPI Antony Rowe, Eastbourne